Earlier

HANDBOOK OF
Biological Control

HANDBOOK OF
Biological

Control

Principles and Applications of Biological Control

Editors:

Thomas S. Bellows
University of California
Riverside, California

T. W. Fisher
University of California
Riverside, California

Associate Editors:

L. E. Caltagirone
University of California
Berkeley, California

D. L. Dahlsten
University of California
Berkeley, California

G. Gordh
Agricultural Research Service
Kika de la Garza Subtropical
Agricultural Research Center

C. B. Huffaker
University of California
Berkeley, California

ACADEMIC PRESS

A Harcourt Science and Technology Company

San Diego   San Francisco   New York   Boston
London   Sydney   Tokyo

This book is printed on acid-free paper.

Copyright © 1999 by ACADEMIC PRESS

All Rights Reserved.
No part of this publication may be reproduced or transmitted in any form or by any
means, electronic or mechanical, including photocopy, recording, or any information
storage and retrieval system, without permission in writing from the publisher.

Academic Press
a division of Harcourt Brace & Company
525 B Street, Suite 1900, San Diego, California 92101-4495, USA
http://www.apnet.com

Academic Press
24-28 Oval Road, London NW1 7DX, UK
http://www.hbuk.co.uk/ap/

Library of Congress Catalog Card Number: 98-89454

International Standard Book Number: 0-12-257305-6

PRINTED IN THE UNITED STATES OF AMERICA
99 00 01 02 03 04 EB 9 8 7 6 5 4 3 2 1

To S. C. C. J.

May your light never fail.

Contents

PART

I

INTRODUCTION

1. Scope and Significance of Biological Control

C. B. HUFFACKER AND D. L. DAHLSTEN

2. Theories and Mechanisms of Natural Population Regulation

T. S. BELLOWS, Jr. AND M. P. HASSELL

8. Life Table Construction and Analysis for Evaluating Biological Control Agents

T. S. BELLOWS AND R. G. VAN DRIESCHE

9. Evaluation of Biological Control with Experimental Methods

ROBERT F. LUCK, B. MERLE SHEPARD, AND PETER E. KENMORE

10. Evaluation of Results

A. P. GUTIERREZ, L. E. CALTAGIRONE, AND W. MEIKLE

11. Periodic Release and Manipulation of Natural Enemies

GARY W. ELZEN AND EDGAR G. KING

PART

III

AGENTS, BIOLOGY, AND METHODS

22. Nutrition of Entomophagous Insects and Other Arthropods

S. N. THOMPSON AND K. S. HAGEN

23. Sex Ratio and Quality in the Culturing of Parasitic Hymenoptera: A Genetic and Evolutionary Perspective

R. F. LUCK, L. NUNNEY, AND R. STOUTHAMER

PART

V

RESEARCH AND THE FUTURE

41. Whither Hence, Prometheus? The Future of Biological Control

T. S. BELLOWS

Contributors

Numbers in parentheses indicate the pages on which the authors' contributions begin.

M. T. AliNiazee (743) Department of Entomology, Oregon State University, Corvallis, OR 97331

Miguel A. Altieri (319, 975) Division of Insect Biology, University of California, Berkeley, CA 94720

L. A. Andrés (103, 871) USDA-ARS, Western Research Center, 800 Buchanan Street, Albany CA 94710

J. W. Beardsley (45, 713) Department of Entomology, University of Hawaii at Manoa, Honolulu, HI 96822

T. S. Bellows, Jr. (17, 87, 199, 505, 699, 841, 1011) Department of Entomology, University of California, Riverside, CA 92521

L. E. Caltagirone (243, 355) Division of Biological Control, University of California, Berkeley, CA 94720

R. Charudattan (891) Plant Pathology Department, University of Florida, Gainesville, FL 32611

Bret Cooper (557) Division of Plant Biology, Department of Cell Biology, The Scripps Research Institute, La Jolla, CA 92037

B. A. Croft (743) Department of Entomology, Oregon State University, Corvallis, OR 97331

D. L. Dahlsten (1, 761, 919) Division of Biological Control, University of California, Berkeley, CA 94720

J. Allan Dodds (549) Department of Plant Pathology, University of California, Riverside, CA 92521

Gary W. Elzen (253) Subtropical Agricultural Research Center, U.S. Department of Agriculture, Agricultural Research Sevice, Weslaco, TX 78596

L. K. Etzel (125) Center for Biological Control, University of California, Berkeley, CA 94720

B. A. Federici (517, 575) Department of Entomology, University of California, Riverside, CA 92521

T. W. Fisher (103) Department of Entomology, University of California, Riverside, CA 92521

D. L. Flaherty (853) Cooperative Extension, Agricultural Guidling, County Civic Center, Visalia, CA 93291

Dennis W. Fulbright (691) Department of Botany and Plant Pathology, Michigan State University, East Lansing, MI 48824

R. Garcia (935, 993) Division of Biological Control, University of California, Albany, CA 94706

Dan Gerling (789) Department of Zoology, Tel Aviv University, Ramat Aviv, Israel

R. D. Goeden (871) Department of Entomology, University of California, Riverside, CA 92521

G. Gordh (45, 355, 383) USDA, Agricultural Research Service, Kika de la Garza Subtropical Agricultural Research Center, Weslaco, TX 78596

A. P. Gutierrez (243) Center for Biological Control, University of California, Berkeley, CA 94720

K. S. Hagen (383, 593) Division of Biological Control, University of California, Berkeley, Albany, CA 94706

R. W. Hall (919) Department of Entomology, Ohio State University, Columbus, OH 43210

Lise Stengård Hansen (819) Danish Pest Infestation Laboratory, Ministry of Foods, Agriculture, and Fisheries, Lyngby, Denmark

M. P. Hassell (17) Department of Biology, Imperial College, Silwood Park, Ascot, Berkshire, SL5 7PY, United Kingdom

D. H. Headrick (505) Department of Crop Science, California Polytechnic State University, San Luis Obispo, CA 93407

Mark S. Hoddle (955) Department of Entomology, University of California, Riverside, CA 92521

Majorie A. Hoy (271) Department of Entomology and Nematology, University of Florida, Gainesville, FL 32611

C. B. Huffaker (1) Center for Biological Control, University of California, Berkeley, CA 94720

Marshall W. Johnson (673, 297) Department of Entomology, University of Hawaii, Honolulu, HI 96822

J. B. Kadir (891) Department of Plant Protection, Faculty of Agriculture, Universiti Putra Malaysia, Selangor, Malaysia

Peter E. Kenmore (225) Inter-Country IPC (Rice) Programme, Manila, The Philippines

C. E. Kennett (713) Division of Biological Control, University of California, Albany, CA 94706

Edgar G. King (253) Subtropical Agricultural Research Center, U.S. Department of Agriculture, Agricultural Research Sevice, Weslaco, TX 78596

Marcos Kogan (789) Integrated Plant Protection Center, Oregon State University, Corvallis, OR 97331

E. F. Legner (87, 125, 355, 935) Department of Entomology, University of California, Riverside, CA 92521

Joop Van Lenteren (819) Department of Entomology, Agricultural University, Wageningen, The Netherlands

D. K. Letourneau (319) Department of Environmental Studies, University of California, Santa Cruz, CA 95064

R. F. Luck (225, 653) Department of Entomology, University of California, Riverside, CA 92521

Joseph V. Maddox (789) Center of Agricultural Entomology, Illinois Natural History Survey, Champaign, IL 61820

W. Meikle (243) Center for Biological Control, University of California, Berkeley, CA 94720

N. J. Mills (383, 761) Division of Biological Control, University of California, Berkeley, CA 94720

J. A. McMurtry (383, 713) Department of Entomology, University of California, Riverside, CA 92521

Clara I. Nicholls (975) Department of Entomology, University of California, Davis, CA

L. Nunney (653) Department of Biology, University of California, Riverside, CA 92521

Michael P. Parrella (819) Department of Entomology, University of California, Davis, CA 95616

John H. Perkins (993) The Evergreen State College, Olympia, WA 98505

E. N. Rosskopf (891) USDA-ARS, Fort Pierce, FL 34945

B. Merle Shepard (225) International Rice Research Institute, Manila, The Philippines

R. Stouthamer (653) Department of Entomology, Agricultural University, Wageningen, The Netherlands

Bruce E. Tabashnik (673, 297) Department of Entomology, University of Arizona, Tucson, AZ 85721

S. N. Thompson (593) Division of Biological Control, University of California, Riverside, CA 92521

Thomas R. Unruh (57) Yakima Agricultural Rsearch Laboratory, USDA Agricultural Research Service, Wapato, WA 98951

R. G. Van Driesche (199) Department of Entomology, University of Massachusetts, Amherst, MA 01003

M. J. Whitten (271) FAO Intercountry Programme on IPM in South and Southeast Asia, Makati City, The Philippines

L. T. Wilson (853) Department of Entomology, Texas A&M University, College Station, TX 77843

James B. Woolley (57) Department of Entomology, Texas A&M University, College Station, TX 77843

Preface and Acknowledgments

At a meeting of 22 biological control workers from the Berkeley and Riverside campuses of the University of California in January 1986, a new biological control-oriented reference text was conceived, and collectively there was agreement to contribute to the task of preparing such a volume. The end product was not perceived as a revision of *Biological Control of Insect Pests and Weeds,* edited by DeBach (1964), which was "California School" classical biological control oriented and had a strong technique, or application, bias. We initially hoped that the present volume would be made available for the Vedalia Centennial Symposium on Biological Control held at Riverside in April, 1989. However, further consideration of the project, and of the greatly expanded scope of biological control in the last thirty years, indicated that a much larger work than was at first envisaged would be necessary to cover the field. The project was refined and reshaped, additional authors were recruited to add chapters suited to the scope of the field, and the volume grew in both breadth and depth. The intent was to cover, as much as possible, the great breadth of biological control science and application as is practiced in so many fields today. The result is the volume before you.

Reported in this volume are certain aspects of the science derived from biological control research (Principles and Processes), detailed discussion of many of the agents and practices that form the operational characteristics of biological control (Agents, Biology, and Methods), and detailed reports on its use in many field programs (Applications). Of the 62 authors involved in the present volume, 21 received advanced degrees from U.C. departments of entomology or biological control. Three—Fisher, Hagen and Huffaker—also participated in the writing of *Biological Control of Insect Pests and Weeds* which, because of the mass of information it contains, continues to be viewed as an important classic general biological control reference. We hope that a reprinting will be forthcoming as a companion to the present book.

The intentional use of natural enemies to control pests of agriculture and forestry was practiced well before biological control, per se, became part of entomological and agricultural curricula. Formal course presentation in the U.S. was first offered in 1900 at Massachusetts Agricultural College (Sweetman 1958). In 1947 a course titled "Biological Control" which emphasized population dynamics and applied biological control entered the curriculum at the University of California, Berkeley (then the teaching campus), under the guidance and participation of Professor Harry S. Smith, who founded the biological control research unit at the University of California Citrus Experiment Station, Riverside, in 1923. Lecture and laboratory courses were taught at UC Berkeley by H. S. Smith, S. E. Flanders, J. K. Holloway (USDA) and R. L. Doutt. "Insect Pathology" under E. A. Steinhaus also became a formal and continuing component of the entomological curriculum at UCB. In 1962 similar courses were added to the entomological curriculum at UCR (P. DeBach and I. M. Hall, instructors). Between the two campuses (UCB 1940–1997, UCR 1964–1997) approximately 90 M.S. and 210 Ph.D. biological control-oriented degrees have been awarded. Approximately 54% were to foreign students.

Many of the students and postdoctoral associates who worked at the two campuses subsequently were employed by universities or industry in the U.S. or abroad. Others were employed by their respective country's or state's agricultural agencies. Those who became involved in their own teaching and/or research programs, and their students, further extended an ecological, if not strictly classical, biological control perception. From this cadre and their students, the more sophisticated academic approaches (population and systems modeling, genetics, biotechnology, etc.) aimed at better understanding of ecological mechanisms and their interrelatedness, have greatly broadened the understanding of many crop ecosystems.

Beginning in 1964, a regional research program, titled

Project W-84, was funded by the USDA for the purpose of developing effective biological control procedures and incorporating them into pest management programs. Scientists, mostly entomologists, from agricultural experiment stations in nine western states and the USDA, participated in that project. The practical applied results of this project for the years 1964–1969 were published by Davis *et al.* [eds.] (1979), and the results of the years 1970–1990 are presented in another text, Nechols *et al.* [eds.] 1992. From 1971–1979, funding by the U.S. National Science Foundation and the Environmental Protection Agency (NSF/EPA) permitted expansion, refinement and deepening of this type of research, and included personnel from 19 universities across the U.S. Popularly known as "the Huffaker Project" (after its director, C. B. Huffaker), the formal title was "Integrated Pest Management: The principles, strategies and tactics of pest population regulation and control in major crop ecosystems"; the results of the research conducted in this project were published in many papers and books, and were broadly treated by Huffaker (1980). The W-84 program (now continued with a new number, W-185) and the "Huffaker Project" served to fix irrevocably ecological awareness in the thinking of pest control management practices in the U.S. and have served as models and catalysts for similar studies on other crop systems in the U.S. and other countries.

The term "biological control" connotes a broad variety of meanings and concepts depending on the age, perspective (or bias) and training of those who use the term. In the literal, applied sense, classical biological control, as well-stated by Waterhouse & Norris (1987), means "the importation and permanent establishment of exotic natural enemies" (and the consequences therefrom—[eds.]), and they list over 50 species of introduced weed and arthropod pests as being worthy of further study in the south and western Pacific regions. Recent publications which further emphasize the importance of continuing study, importation, and use of naturally occurring biological control organisms are those by Rosen (1990) and Andrews and Quezada (1989). DeBach and Rosen (1991) put forth several compelling reasons in support of strengthening research in classical biological control. Baker and Dunn (1990) and Maramorosch and Starkey (1992) include information on a wide miscellany of entomophagous microorganisms and present a broader view of "biological control" as perceived by many current workers who hold a strong bio-technological perspective of the subject. Largely from that perspective, two journals were inaugurated in 1991, "Biological control: theory and application in pest management," Academic Press, San Diego, CA, U.S.A., and "Biocontrol science and technology," Carfax Publ. Co., Abingdon, Oxfordshire, U.K. Other current articles on biological control are published by Intercept, Andover, Hampshire, U.K. In 1992 the Entomological Society of America began publishing a spe-

cialty journal titled "Biological Control." The text "Biological Control" by Van Driesche and Bellows (1996) focuses on both underlying principles and their practical application in employing biological control to solve real-world problems.

Improved biological control of pests is widely considered the cornerstone of "sustainable agriculture," today's buzzword which implies incorporation of all known components in food and fiber production programs—plant breeding, soil fertility, pest control, socio-economics, and so on—with safer human, animal, and environmental health being the main objective. Prospects presented by genetic engineering toward enhancement of pest control along several avenues (including beneficial organisms) offer much to ponder and conjecture about, but their application, much less their effectiveness over time in the field, has a very short track record. Biological control explorers continually search for presumed biotypes, which genetically may be more effective as natural enemies than those previously introduced. Possibly, this activity could eclipse the need for genetic engineering of natural enemies.

Ironically, increased interest in biological control in the U.S. as currently expressed by federal (USDA, EPA) and state departments of agriculture parallels a deemphasis, *i.e.,* reduced support, of classical biological control programs, especially in universities where most of the pioneering research and development in this field has occurred. This is a trend we hope to see reversed!

Royalties from the previous *Biological Control of Insect Pests and Weeds* volume and from this present one accrue to the Harry Scott Smith Memorial Fund, a UC Regents endowment fund which is used at the University of California, Riverside to bring guest lecturers to UC and to assist students in their biological control training program through travel grants or modest stipends, and at UC Berkeley, to provide stipends for outstanding high school students to conduct biological control research in a laboratory setting.

The editorial committee has tried to accomplish a meaningful presentation of the subject matter as submitted by the authors who wrote from their individual perspectives and subspecialties in the broader field of biological control. As an unavoidable consequence, certain of the biological control programs and the natural enemies or hosts involved are treated in more than one chapter.

Our original intent was to list only currently accepted scientific names (from a U.S. perspective) in the index, but to include recently used synonyms in the text when appropriate. Names shown only as binomens are the responsibility of the chapter author(s) who used them.

We wish to acknowledge with gratitude the assistance of our associate editors. As the chapters came in, a division of editorial labor evolved. Editors at Berkeley (L. E. Caltagirone, D. L. Dahlsten and C. B. Huffaker) reviewed and commented on the manuscripts and returned them to Riverside, where the heavily time consuming task of technical editing was performed, mainly by G. Gordh and T. S. Bellows. T. W. Fisher and T. S. Bellows managed the collative technical editing and management of the development of the project. The critique of outside reviewers has been carefully considered and is gratefully acknowledged. Further appreciation is extended to the secretarial staff at the authors' respective institutions for their skill and patience in producing initial drafts for most of the chapters. Special thanks are due to Pam Hoatson (Administrative Assistant, UCR), who prepared numerous revisions of chapters as they developed. Her professional expertise, dedication, and good natured handling of the task are indeed greatly appreciated. C. Meisenbacher and S. Kokel assisted greatly during the proof stage, especially in tracking down missing references.

During the gestation of this project, a few of our colleagues have retired from institutions were they long advanced the field of biological control. Among these are L. Andres, J. Beardsley, L. Caltagirone, T. Fisher, D. Flaherty, R. Garcia, K. Hagen, C. Huffaker, C. Kennett, E. F. Legner, J. Maddox, J. McMurtry, and M. Whitten; C. Huffaker and K. Hagen are now deceased. We have enjoyed the privilege of working with these individuals during this project. When we consider the years of experience in the field of biological control represented by the authors who contributed to this Handbook, we recognize more than 1500 scientist-years of expertise, gathered here between two covers. Our thanks to each of them for their patience, endurance, and contribution.

And to the reader, our best wishes as you delve into a most rewarding and fascinating field.

T. W. Fisher
T. S. Bellows, Jr.
UC Riverside,
March, 1999

References Cited

Andrews, K. L. & Quezada, J. R. (1989). Manejo entegrado de plagas insectides en la agricultura. Library School of Agriculture Pan-Americana, Apt. Post. 93, Tegucigalpa, Honduras. 623 pp.

Baker, R. R. & P. E. Dunn. 1990 (1989). New directions in biological control: alternatives for suppressing agricultural pests and diseases. Proceedings of a UCLA colloquium, Frisco, Colorado, Jan. 20–27, 1989. Alan R. Liss, Inc., New York. 815+ pp.

Davis, D. W., S. C. Hoyt, J. A. McMurtry & M. T. AliNiazee [eds.]. 1979. Biological control and insect pest management. University of California, Division of Agricultural Sciences, Bulletin 4096 (reprinted as Bull. 1911). 102 pp.

DeBach, P. [ed.]. 1964. Biological control of insect pests and weeds. Chapman & Hall. 844 pp.

DeBach, P. & D. Rosen [eds.]. 1991. Biological control by natural enemies, 2nd ed. Cambridge University Press. 440 pp.

Huffaker, C. B. [ed.]. 1980. New technology of pest control. John Wiley & Sons. 500 pp.

Maramorosch, K. & R. L. Starkey [eds.]. 1992. Biotechnology for biological control of pests and vectors. CRC Press, Inc., Boca Raton, Fla.

Nechols, J., Jr., L. A. Andres, J. W. Beardsley, R. D. Goeden & C. G. Jackson [eds.]. 1992. Biological control in the U.S. western region: Accomplishments and benefits of regional research project W-84, 1964–1989. UC Press (D.A.N.R.) Oakland, CA.

Rosen, D. [ed.]. 1990. World crop pests: armored scale insects— their biology, natural enemies, and control. Elsevier Scientific Publishers, Amsterdam. 2 vols., 800+ pp.

Sweetman, H. L. 1958. The principles of biological control. Wm. C. Brown, Co., Dubuque, IA. 560 pp.

Van Driesche, R. G. & Bellows, T. S. Jr. (1996). Biological Control. New York: Chapman & Hall.

Waterhouse, D. F. & K. R. Norris. 1987. Biological control: Pacific prospects. Australian Centre for Int. Agricultural Research. Inkata Press, Melbourne. 443+ pp.

1

Scope and Significance of Biological Control

C. B. HUFFAKER and D. L. DAHLSTEN

Center for Biological Control
201 Wellman Hall
University of California
Berkeley, California

The amount of food for each species of course gives the extreme limit to which each can increase, but very frequently it is not the obtaining of food, but the serving as prey to other animals, which determines the average number of a species.   Darwin on the Origin of Species (1859)

Biological control, when considered from the ecological viewpoint as a phase of natural control, can be defined as the action of parasites, predators or pathogens in maintaining another organism's population density at a lower average than would occur in their absence

P. DeBach (1964, p. 6)

INTRODUCTION

Biological control has commonly had both pure and applied definitions and connotations. In this book the authors adhere mostly to DeBach's quotation, with the proviso that parasites, predators, and pathogens include "antagonists" of plant pathogens. The significance of biological control is clear in the quotations from both Darwin and DeBach, residing in the roles such natural enemies have in determining the densities of other organisms in nature and thus their impacts on their environments.

Huffaker and Rabb (1984) considered, "Nature as a complex of interacting systems is widely accepted, but many questions remain as to how these systems are arranged and to what degree they are interdependent. To what extent are structure and function deterministic or probabilistic? To what degree should we treat elements of the system

as relatively fixed or subject to [and altered by] evolutionary change?" These are questions various authors in this book have considered and have offered some insights. DeBach's usage of biological control in the antecedent volume to this book (DeBach, 1964a) was "demographic and ecological in context but does not explain the mechanisms of control or regulation."

Bellows and Fisher's preface to this book discusses the various concepts of biological control and its implications in practice. This section also details the beginning of curricular biological control at the University of California and discusses the antecedent volume on this subject edited by DeBach (1964b) for its historical relevance. Readers of this chapter should be sure to read that preface as a prelude to this introductory chapter and the rest of the book.

There are some rather basic differences among some chapter authors relative to what is meant by biological control, and in other cases the value of different approaches to evaluating the role of natural enemies. For example, Federici in Chapter 21 includes as biological control certain biotechniques not accepted as such by the authors of many of the chapter. Also, the methods of evaluation of natural enemies utilized by certain authors in the book *Critical Issues in Biological Control,* edited by Mackauer, Ehler, and Roland (1990) differ distinctly from those described in Chapter 2 by Bellows and Hassell, in which they consider that theories of interactions "accord well with observed outcomes of experimental populations in both laboratory and natural settings," especially so for simpler systems.

DeBach and Rosen (1991) give data on some 164 species of insect pests being permanently controlled by introduced natural enemies (classical biological control): for 75

1

species "complete," for 74 species "substantial," and for 15 species "partial" control. Because many of these species were also controlled in more than one country, they noted that up to 1988 there had been some "384 successful projects worldwide." They also state: "This is by far the greatest number of successes achieved by any one non-chemical method, with the possible exception of cultural control, which to our knowledge has not been tabulated." All other nonchemical methods combined would add only a few pest species up to now. It was noted that nearly one-third of the species first controlled in a country were again controlled in other countries. Information is drawn from works of DeBach (1964a), Clausen (1978), Laing and Hamai (1976), and Luck (1984); and from BIOCAT information furnished by D. J. Greathead in 1988 and "some additions from their own records."

With the preceding records of "4 successes in 10," directed against only 415 pest species, it is an indictment of environmentally interested sponsorship worldwide that none of the others among some 10,000 species of insect pests (95%) have been investigated for possible importations. Moreover, DeBach and Rosen (1991) note that of some "4226 natural enemy species introduced worldwide, 1251 have become established, 2038 failed, and the fate of 932 is still unknown." This gives an establishment rate of "38% of those with known outcome." Yet, clearly it is not the percentage, or number, of species established, but the success of the *project* that is important. As often happens, a highly effective species may displace or prevent establishment of less effective competing species, and better biological control commonly results (see "competitive displacement," Huffaker & Messenger 1976; also DeBach & Sundby, 1963; Huffaker & Kennett, 1969). Also, Huffaker *et al.* (1971) present specific cases where competitive displacements have improved biological control, often with reductions in the complex of enemy species in the system.

Successes worldwide have been directly related to the effort expended; for this reason only, such areas as Hawaii (31), California (24), the rest of the United States (20), Chile (16), Australia (13), New Zealand (13), Canada (11), Israel (11), France (11), and Mauritius (10) have had corresponding degrees of success. Moreover, DeBach and Rosen (1991) conclude that the costs of biological control importations are very low compared to the costs of developing pesticides, which are offering declining rewards, and that if all countries expended the effort on biological control that "the leading twelve [do now] the number of problems solved would jump enormously." As seen in this volume, classical importations include only a part of modern biological control. Nechols (1995) discusses the accomplishments and benefits of a number of W-84 control projects in the western United States from 1964 to 1989. Rosen and DeBach (1991) discuss the importance foreign exploration plays in classical biological control.

DeBach's definition of biological control (cited earlier) does not describe the discipline of biological control, which is broader and includes what the profession must do (see Chapters 40 and 41). In that same antecedent volume the density-regulating roles of natural enemies and the mechanisms were discussed by Huffaker and Messenger (1964), including interdependencies and deterministic or probabilistic features, for different types of habitats and weather. These ideas are used and modified by various authors in this book.

Biological control of other organisms is exemplified far beyond entomology. Its research and practice have been greatly expanded relative to plant and animal diseases and weed control. It was also utilized very early in attempts to control rats, for example, by use of the mongoose. Davis *et al.* (1976) also dealt with the use of other vertebrates to control vertebrates, and with the use of disease for control of European rabbit hordes in Australia. Since then, biological control has become an increasingly strong component in a broader vital area, that of conserving and promoting a viable, sustainable agriculture, forestry, and biosphere (see Chapters 26, 29, 34, and 36 to 39).

To regulate a population, factors (singly or in combination) must act either in a direct density-dependent way (Huffaker *et al.,* 1971; Huffaker *et al.,* 1984) or in an inverse density-dependent way associated with prey patches within a host generation (Hassell, 1986). Huffaker *et al.* (1984) dealt at length with the functioning of natural control in general, and with the roles of the specific density-dependent regulating performance of various natural enemies or other negative feedback processes. They noted that even while some such natural control factors may be responsible for density regulation in a specific limited time and place frame, evolutionary processes may be operating to alter the case in time (see Chapters 12, 13, 15, and 23). They presented a chart illustrating that intraspecific competition of predators for prey and the linked intraspecific competition of the prey for hiding places or predator-free space can explain the various manifestations of density regulation for specific biotic and abiotic conditions. They also cited several authors who refer to competition for enemy-free space as a means of reducing the species diversity of herbivorous insects on a particular plant species. Yet, most of the authors of chapters in this book dealing with theory of natural-enemy control of prey (or host) species look at it as linked predator–prey population interaction and do not consider it one of prey competing for predator-free places. We also present here in Fig. 1, Huffaker *et al.*'s illustrations of various density-related performance functions for insects and other relevant factors.

In this volume, considerable general attention is given to the question of evolutionary processes, for example, in Chapters 15 and 23, and to the question of pesticide resistance in Chapters 13 and 24.

It is important now to place the forms of biological control in the context of animal and plant protection

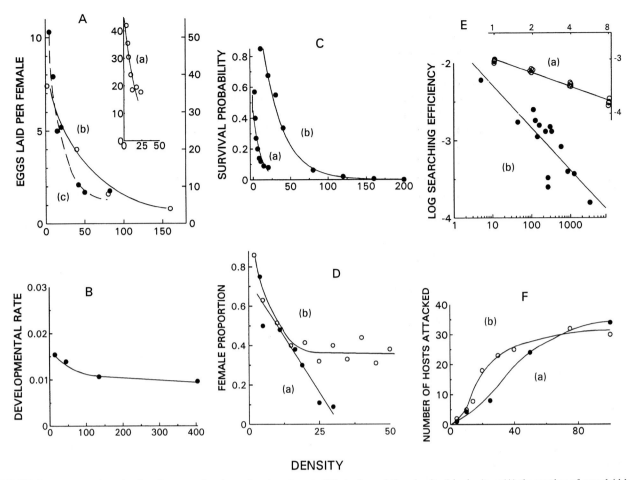

FIGURE 1   Some density-related performance functions of various insects. Effect of population density (abscissa) on (A) the number of eggs laid by a bark beetle, (a) *Dendroctonus pseudotsugae* (McMullen & Atkins, 1961), a flour beetle, (b) *Cryptolestes* (Varley *et al.,* 1973), and a psyliid, (c) *Cardiaspina albitextura* (Clark, 1963); (B) the developmental rate of *Endrosis sarcitrella* (Anderson, 1956); (C) the probability of survival of a grain beetle, (a) *Rhizopertha dominica* (Crombie, 1944) and the blowfly, (b) *Lucilia cuprina* (Nicholson, 1954); (D) the proportion of females of (a) *Bracon hebetor* (Benson, 1973) and (b) *Nasonia vitripennis* (Walker, 1967); (E) the host-searching efficiency (log10a) of (a) *Pseudeucoila bocheri* parasitizing *Drosophila* larvae (Bakker *et al.,* 1967), and (b) *Nemeritis canescens* parasitizing *Anagasta kuhniella* (Hassell & Huffaker, 1969); and (F) the sigmoid relation in number of hosts attacked by (a) *Aphidius uzbeckistanicus* parasitizing the aphid *Hyalopteroides humilis* (Dransfield, 1975), and (b) mosquito larvae attacked by the waterboatman, *Plea atomaria* (A. Reeve, unpubl., from Hassell *et al.,* 1977). *Note:* These latter show the effect of increase in prey density in attracting a greater intensity of predation. (After Huffaker, C. B., Berryman, A. & Laing, J. E. [1984]. In C. B. Huffaker, & R. L. Rabb (Eds.), *Ecological entomology,* (p. 375). New York: John Wiley & Sons.

schemes often discussed today. Figure 2 from Garcia *et al.* (1988) presents this in terms of biological, cultural, and chemical control methods. It is seen that we view true biological control as that which is self-sustaining or partially so; and that which is achieved by parasites, predators, and pathogens, including antagonists of plant pathogens. There are three forms recognized—that which is naturally self-sustained, that in which augmentation is utilized, and that which uses conservation of natural enemies through management practices. While "autocidal" methods and breeding for host-plant resistance are biotechniques, they do not utilize parasites, predators, or pathogens and they are not self-sustaining. Note, too, that breeding, selection, and genetic engineering may result in improvements that could become useful in several of the plant protection methods, including that of true biological control.

We also reproduce in Fig. 3 a diagram from Krebs (1972) of the general areas of biology that interact and overlap in the ecology of any organism. The four areas of genetics, physiology, behavior, and evolution impinge strongly as shown. Behavior is often directly classed as ecology, and physiology and ecology interact in ecophysiology. Genetics lies at the roots of each of the other expressions and evolution is the product of change in any of these over time. We define ecology as the total relationships of organisms with their environment—with their own kind, with other organisms, and with the abiotic factors. Huffaker *et al.* (1984) noted: "Ecology's special domain has become the analysis of a great variety of *ecosystems:* subdivisions of the global environment containing a restricted number of living species adapted to survive within well defined abiotic and biotic conditions." Often, the populations inhabiting an ecosystem can only be estimated from samples and mathematical estimates of densities, and spatial distribution tends to be abstract and probabilistic. Yet these methods sometimes offer enough prediction to be useful in making pest control decisions.

The question of risks and conflicts of interest in introducing biological control agents presents a variety of concerns, greater in the case of biological control of weeds than of insect pests, but of concern in any case (e.g., Caltagirone & Huffaker, 1980; Price *et al.*, 1980; Batra, 1982; Julien, 1982; Howarth, 1983; Clarke *et al.*, 1984; Maws, 1984; Pimentel *et al.*, 1984; Turner, 1985; U.S. Congress, 1995; and others). Because of a variety of concerns a seminar was organized in 1996 by the United States Department of Agriculture (USDA) to draft a plan for the coordination, regulation, and accountability of biological control within the USDA (Carruthers & Petroff, 1997). Choice of target plants and choice of specific biological control organisms to introduce against plants are the key decisions for resolving any conflict of interest. Three major mechanisms suggested by Turner (1985) are (1) authoritative decision, (2) mediation, and (3) litigation. Each method has been used; for example, litigation over the *Echium* project in Australia went to the High Court of Australia before it was resolved in favor of introducing biological control agents, which led to the passage of an enabling act for biological control. Gutierrez *et al.* (see Chapter 10) also note the difficulty of including in the costs and benefits of biological control introductions, not only the poorly assessed direct costs and benefits (in our projects) but also those to which the external costs, risks, and benefits can only be surmised. Various chapters in this volume go considerably further in establishing real costs, benefits, and risks of conflicts than was possible in DeBach (1964b).

PRINCIPLES AND PROCESSES

Ecology and biological control has had such a long history and has been handled in such a diversity of ways that it is not possible to present an exhaustive discussion in one chapter. Luck (1984) reviewed the use of three general types of mathematical models, "those that treat continually growing predator–prey populations (overlapping generations), those that treat predator–prey populations with nonoverlapping generations, and those that view the predation process as one of resource exploitation (i.e., foraging theory)." Luck concluded that all are "to one degree or another inadequate." Yet he pointed to the value of such models in focusing attention on certain assumptions and processes associated with predation (or parasitism). He noted that several assumptions appear capable of stabilizing the predator–prey process: "predator aggregation on dense prey patches, spatial and temporal habitat heterogeneity, presence of a sigmoid functional response to a surge of prey

species occurring in patches that differ in density, or to switching in multiple prey species (complexes) as a frequency-dependent response, and competition in the prey species." It was only through the exploitation of prey patches that such stabilization was envisioned. Luck, Shepard, and Kenmore see little of value in any of the mathematical approaches, but see their Chapter 9 in this book.

Huffaker (1988) also reviewed in a general way the question of ecology of insect pest control. Theoretical predator–prey models were not dealt with, but the work of Teng (1985) was considered a supplemental and different approach. Huffaker and Rosen (1990) dealt with the attributes of effective natural enemies, and Huffaker (1990) reviewed the role of weather factors as they may influence the host–parasite regulation process. Van Driesche and Bellows (1996) covered the principles behind biological control techniques and their implementation, and incorporated practical examples from the biological control of a variety of pests.

Chapter 2

The authors Bellows and Hassell deal with theories and mechanisms of natural population regulation, starting with single-species populations; and leading on to interspecific competition, host–parasitoid systems, host–pathogen systems, and multispecies systems. They state that the principal aim of biological control is the reduction of negative impact of pest species, including both pest suppression and pest regulation. The dynamic behavior involved can be of a density-dependent nature for the pest and the natural enemy; and can involve specific search behavior, more than a single natural enemy, the patchiness of the environment, and both behavioral and stochastic interactions (particularly related to stabilizing populations). The authors develop their chapter according to an analytical framework of difference equations, but in cases consider also continuous-time systems. They deal primarily with insect parasitoids, predators, and pathogens, for which there is a large body of data; and do not deal with insect herbivory effects on plant populations because of the scarcity of suitable data (Crawley, 1989).

Chapters 3 through 14 deal rather more specifically with the processes in biological control, although certain other chapters throughout the book bear on such principles and processes and are used for illustrations. Topics dealt with in this section are taxonomic and evolutionary closeness (Chapter 3), molecular methods (Chapter 4), foreign exploration (Chapter 5), quarantine (Chapter 6), culture and colonization (Chapter 7), evaluation (Chapters 8, 9, and 10), manipulation (Chapter 11), genetic improvement (Chapter 12), and enhancement of biological controls (Chapters 13 and 14).

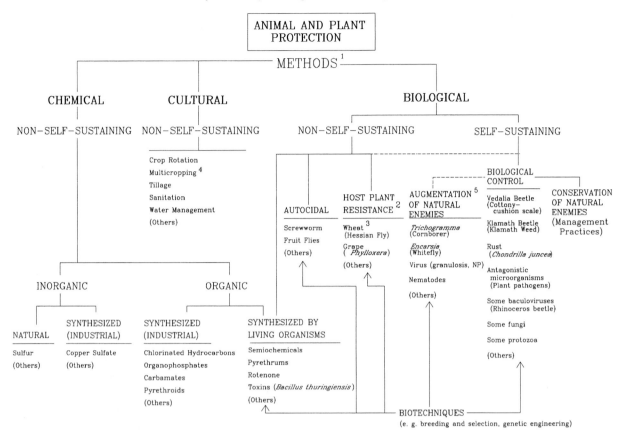

FIGURE 2 Methods for animal and plant protection, with some specific examples. (After Garcia, R., Caltagirone, L. E., & Gutierrez, A. P. [1988]. *BioScience, 35*(10), 692–694.)

Chapter 3

Gordh and Beardsley point to the confusion and misleading inferences about an organism if it is not properly classified. A correct name, alone, furnishes much information. Seemingly minor structural differences "can mean the difference between pest and non-pest, or establishment or failure for a natural enemy." For example, of the species of *Pectinophora,* only *P. gossypiella* is of major importance and its ability to diapause within cotton seeds is a major reason for its widespread distribution.

Confusion of *Aonidiella aurantii* and *A. citrina* in early work led to many introductions of *A. citrina* parasites that were unsuitable for *A. aurantii*. The incorrect identification of the coffee mealybug, *Phenococcus kenyae,* led to years of fruitless effort to achieve biological control from importing ineffectual parasites of two Asiatic *Planococcus,* whereas effective parasites of *P. kenyae* were already in Africa and when established in Kenya quickly effected complete biological control. Confusion concerning natural enemies is also of great importance. For example, the genera *Aphytis* and *Marietta* are very similar morphologically, but *Aphytis* species are primary parasites of armored scales while *Marietta* species are hyperparasites, and generally viewed as harmful to biological control. The highly significant problems of the identity of various species of *Trichogramma,* crucial to their use, are also discussed.

Chapter 4

Unruh and Woolley consider molecular methods for biological control, with particular reference to their utility in biosystematics and the study of field populations of insect biotypes. This chapter reviews isozyme electrophoresis; restriction fragment analysis; and sequencing of ribosomal RNA, mitochondrial DNA, and genomic DNA. DNA–DNA hybridization and immunological distances are also discussed.

Very little has been done, for example, on enzyme electrophoresis of natural enemies of insects, but possibilities are illustrated by work on the systematics of insects. Insect biological control agents have not been studied using mo-

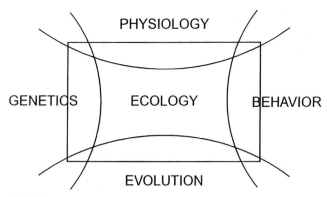

FIGURE 3 Relation of ecology to other biological sciences. (After Krebs, C. J. (1972). *Ecology: The Experimental Analysis of Distribution and Abundance.* New York: Harper & Row.)

lecular techniques for nucleic acids but a few have been studied using the isozyme method. Considerable study has been done on insect viruses and entomogenous bacteria. Primarily, this chapter illustrates a methodology for studying natural-enemy species and biotypes with reference to their ecological capabilities and roles in managed and natural settings. These genetic approaches are needed in biological control to more fully understand and utilize the adaptive diversity in populations of insects as natural enemies.

This section deals also with the essentials of classical biological control: exploration, introduction, culture, colonization, and evaluation. The material on this "backbone" of biological control work is covered in Chapters 5 to 7 and 10, and is outlined briefly here.

Chapter 5

Legner and Bellows describe criteria that will help delineate search areas of choice before exploration for entomophagous organisms begins. Ascertaining region or origin of the target pest or its plant hosts, if possible, will prioritize areas to be searched. Key considerations are taxonomy, distribution, hosts of record, biologies of closely related species, and input from foreign colleagues in the proposed areas of search. The authors also cite several successful biological control programs by organisms secured in areas other than the native home of the target pest.

A generalized categorization of environmental risk factors associated with introduction of exotic natural enemies is presented. Terrestrial vertebrates present the greatest potential risk, followed by phytophagous arthropods, phytopathogens, and terrestrial scavengers. Because of their restrictive habits, uses of parasitic and predaceous arthropods and pathogens of arthropods are considered the least risky introductions, but even they require careful evaluation from several perspectives before being released. Criteria to aid in

assessment of their capabilities prior to colonization are presented.

The mechanics of organizing and executing foreign exploration for beneficials are described. Bureaucratic requirements at home and abroad for permits, shipping methods, etc., must be anticipated and followed to expedite safe arrival of shipments at home quarantine receiving facility.

Personal and technical problems faced by explorers in the field are briefly discussed.

Chapter 6

This chapter by Fisher and Andrés on quarantining or isolation work contains concepts, including principles, facilities needed, and procedures to be used. They note that Kahn in 1988 presented a solid and up-to-date review of the quarantine exclusion process. The authors cover these areas fully, citing both the important older and newer works.

The principal function of the biological control quarantine operation is to furnish a safe area and competent staff to screen incoming material to exclude unwanted organisms and determine the identities and biologies of those desired to be propagated and colonized. Proving the environmental safety of herbivorous organisms for the biological control of weeds can be very demanding and crucial. It is desirable to conduct these studies in the country of origin; otherwise, the entry country must conduct the necessary testing under quarantine.

Other references cited in Chapter 6 include Coulson and Soper, who in 1989 presented an excellent review of the role of quarantine in biological control. Fisher and Andrés deal with (1) national, state, and county regulations; (2) quarantine laboratory design and equipment; (3) personnel; and (4) operating procedures. The authors note that the mechanics have not changed appreciably for entomophagous species since Fisher's account in 1964 (see DeBach, 1964b); thus, the emphasis in this chapter is on weed-feeding candidates, the area undergoing much expansion and ever-more-demanding concerns for environmental safety and involving close consideration of pathogens.

Chapter 7

This chapter by Etzel and Legner is primarily, but not exclusively, on insect parasites and predators and relates to three major purposes: (1) permanent establishment, (2) periodic colonization and augmentation, and (3) inundative releases. The programs needed vary with each of these purposes and with the types of organisms. The first goal is that associated with successful classical biological control through introductions. The second is illustrated by the phenomenal success with yearly (periodic) spring releases of the parasitoid *Pediobius foveolatus* for control of the Mexican bean beetle in southeastern United States, and by releases of the fish *Tilapia zillii* in irrigation canals for aquatic weed and mosquito-habitat control in southeastern California. The third is where very large releases are made to achieve quick reduction of the pest, as with releases of a tachinid parasite of the sugarcane borer, and of hydra against mosquito larvae.

Host foods used in production are living plants, harvested plant parts, vegetables or fruits, and prepared diets. The choice to be made depends on the species of the host and its intended use. Prepared diets have perhaps been the most researched new approach, and 22 multiple species rearing diets have been used for dozens of insect species.

Etzel and Legner also discuss both production and colonization of host and natural-enemy production in detail. Contamination of cultures by insects or disease-producing organisms is a common problem, and methods of handling such a problem are dealt with. The authors also discuss advantages and disadvantages of high or low genetic composition in a natural enemy, with the former desired for field testing of effectiveness and the latter desired for insects used for assay purposes. The questions of diapause forms, rearing conditions (temperature, light, and humidity), mating, and other behaviors are covered. Most of their extensive references are recent. Changes in laboratory culture are also discussed, and the problems (including inbreeding, sex ratios, founder effects, laboratory environment, genetic drift, mating type, mutation, etc., with suggested solutions) are considered. Finally, the authors include an addendum that provides an invaluable list of important reviews covering in detail the necessary information on entomophage biologies, production problems, and techniques and facilities so important in all culture and colonization programs. (Extensive references are cited.)

Chapter 8

The authors Bellows and Van Driesche discuss the construction and analysis of field life tables in the evaluation of biological control agents. They describe the types of life tables and use them to evaluate natural enemy actions to answer two basic questions; the first deals with their impact on their hosts (prey), and the second considers their ecological roles, particularly in stabilizing populations.

The construction of life tables for these purposes requires good estimates of numbers entering the different stages and numbers dying within them due to the various specific causes. Methods they describe to obtain such estimates include stage–frequency or age–frequency analysis, recruitment, growth-rate analysis, and death rate analysis; the measure of recruitment of both hosts and parasites is the most direct means. For any of these methods, sampling programs "must avoid potential biases caused by behavioral changes of parasitized hosts and by host patchiness" (in distribution). Mortalities from natural enemies and other causes are typically measured as rates and their relative influences in population growth can be expressed as impacts on the net reproductive rate (R_o) of the target population; this rate must be reduced to below unity for the population to decrease. They note that the *marginal attack rate* is the only measure whose calculation permits correct interpretation of the impact of factors, even contemporaneous mortality factors, between different systems.

The authors also discuss various means of using methods for analyzing joint host–parasitoid systems, for dealing with both discretely breeding populations and continuously breeding ones, and for analyzing sequentially acting and contemporaneously acting factors. Indispensable mortality, a most important concept, is unfortunately rather neglected, perhaps because of problems in its establishment. Indispensable mortality is an important aspect that bears on the question of multiple versus single introductions. They deal with *horizontal* life table construction [based on real mortalities of a given group (cohort)] over a period of time, and with *vertical* life table construction by examining age structures at given times. The notion of vertical life tables has its origins in the work of R. D. Hughes in 1963 (see Southwood, 1966).

The authors detail extensive studies on tests for density dependence (employing, e.g., Monte Carlo simulation), and cite methods of Pollard *et al.* (1987) and Reddingius and van den Boer (1989). Many problems of solid evaluation of density-dependent performance for many situations remain.

Chapter 9

Luck, Shepard, and Kenmore update and suggest methods to achieve greater precision in the evaluation of natural-enemy effectiveness (components of Chapter 14 in DeBach, 1964b, and especially of Chapter 11 in DeBach *et al.,* 1976). In the former DeBach book, the superior place of using experimental "check methods" (exclusion, inclusion, additions, and interference), including removal methods, to evaluate the efficacy of natural enemies was closely considered; the same subject was covered in a long series of papers by DeBach and associates (some cited in Chapter 9). DeBach *et al.* (1976) noted that although key-factor and K-factor analyses may offer some possibilities of assessing the regulatory power of a natural enemy, "In the field there are so many other contemporaneously interacting factors . . . that use of these life tables and regression techniques gives little assurance that the true role of a natural enemy will necessarily be discovered by their use. We reiterate that the best, the closest, and only reliable means of evaluation is through experimental or comparison methods." Chapter 8 in this book goes a great deal farther, especially in delineating appropriate sampling schemes and experimental designs to achieve such improved evaluations.

Chapter 10

Gutierrez, Caltagirone, and Meikle present an economic evaluation of biological control efforts. The authors start with an enlightening preamble from DeBach, who in 1974 stated that farmers have become convinced they "must use pesticides regularly or perish" and notes that this conviction has had biased origins and little solid economic evaluation. The work relates to classical and naturally occurring biological control by natural enemies. They follow the scheme in Fig. 2, considering *Bacillus thuringiensis* (B.t.) toxins, ryania, pyrethrum, and other pesticides derived from organisms; and other biotechniques, such as sterile male releases, as being outside of the traditional biological control concept. Yet they go beyond the traditional use of the term and include not only parasites, predators, and pathogens but also competitors, possibly to include antagonists in the control of plant pathogens.

The authors note that most of the benefits of naturally occurring biological control have come to light from use of pesticides that have interfered with natural enemies. Many examples (see review of DeBach & Rosen, 1991) are discussed and referenced.

Costs and benefits are discussed fully, including both the direct ones (assessable) and certain indirect, external ones. The latter are assignable to the cost–benefit equations, both negatively and positively (e.g., benefits society derives from not using objectionable pesticides as a result of biological control). The authors also discuss various case histories, some with measurable economic gains and others with vast benefits that are still not adequately definable (e.g., cottony-cushion scale on citrus worldwide).

The authors also present mathematical formulations for assessing costs and benefits. They discuss in dollar values, for example, the biological control of ice plant scales along California highways as reported by Tassan *et al.* in 1982, and note the biological control of wheat aphids in South America and of cassava pests in Africa, where economic analyses have not been conducted but the net benefits are nevertheless impressive. Cassava mealybug was solved at a cost of 8 cents per affected person over a vast region of Africa. (See references in Chapter 10.)

Applications of "Enhancements of Biological Control" is reviewed in Chapter 11 by Elzen and King on manipulation of natural enemies, in Chapter 12 by Whitten and Hoy on genetic improvements; in Chapter 13 by Johnson and Tabashnik on improving the use of chemicals; and in Chapter 14 by Letourneau and Altieri on environmental management.

Chapter 11

Elzen and King consider the use of natural enemies in a "rational insect management program that considers all available methods, such as selective use of pesticides, use of semiochemicals and/or supplements of natural enemies." Earlier background manipulations to enhance parasitoid activity as reviewed by Powell in 1986 were cited. The chapter includes an up-to-date review of augmentative releases of *Trichogramma* spp., the most extensively employed organism.

The wind tunnel flight chamber is considered a useful tool for examining flight behavior and foraging. Elzen and King also discuss methods of assessing augmentative releases and other management procedures, for example, those in Chapters 8 and 9. They discuss selective use of chemicals and the increasing problem with chemical pesticides. For example, even low doses may reduce parasitoid fecundity and egg viability. They emphasize the need to integrate biological control with needed chemicals to conserve natural enemies by using selective pesticides and minimal dosages integrated with cultural practices and host-plant resistance, and improved methods of assessing abundance and effectiveness of the enemies. They point to the need to use the best biotypes, genetically improved stocks, use of behavior-modifying substances, and *in vitro* techniques for culturing.

Chapter 12

Whitten and Hoy deal with breeding and selection for genetic improvements of natural enemies—mites are stressed because more work has been done on mite predators of mites than on other groups. They review the "overwhelmingly favorable record" for biological control in the classical sense, with reference to cost effectiveness; environmental acceptability; and safety record for control of arthropods, nematodes, weeds, and plant pathogens. They note, too, that biological control has been attempted on less than 5% of the 5000 or so arthropod pests, and that a substantial proportion of efforts have had limited or no success, as various others have noted. They cite reviews on genetic improvements of pathogens and nematodes but do not deal with them.

Whitten and Hoy consider the rather low success rate for introductions, and consider the question of genetic improvement. They state, however, that "the track record of achievement of genetic improvement of arthropod natural enemies by 1971 was sufficiently unimpressive to elicit in 1971 the pithy but fair assessment of Messenger and van den Bosch: 'artificial selection of natural enemies has been considered by many, attempted by few, and, unfortunately, proven practicable by no one.'" They question whether any species had been "actually improved" and proved of value in the field. Whitten and Hoy then consider 14 cases since 1971 of parasites and predators being genetically improved in the laboratory. They consider three categories: (1) cases where the pest status is due to lack of effective enemies;

(2) cases involving biological control of secondary pests triggered to pest status by pesticides; and (3) cases in ill-defined, novel, or disturbed situations, where over time stocks have become better adapted to the conditions and presumably could be hastened to that end by genetic improvements.

Chapter 13

Johnson and Tabashnik discuss how pesticides can disrupt biological control through both direct and indirect effects. Sublethal effects may lead to poor searching or habitat orientation, for example. Short-term direct mortality and reduction of hosts have the greatest impact. Due to cumulative uptake of toxins, biological control agents that are better searchers (more mobile) may be at a disadvantage. The authors propose that pesticide companies furnish information on the detrimental effects of pesticides on natural enemies relative to the specific crop and locality. They note also that there are relatively few documented pest population thresholds for initiating pesticide treatments, even though such thresholds are vital to this issue.

Chapter 14

Letourneau and Altieri discuss various aspects of environmental management to enhance biological control. This naturally overlaps with other chapters, especially Chapters 12, 28, and 33.

They note especially that trends in agriculture toward decreasing environmental heterogeneity, increasing use of chemicals and machinery, and decreasing genetic diversity in the crops have increased problems for effective action by natural enemies. They note, too, the increasing need for integrated pest management (IPM) programs in view of the "cumulative restrictions on various pesticides and of public concerns about their use." Effects of reduced tillage and vegetational diversification have received much attention, while other approaches have been minimal, except for the "novel advances in the use of food sprays and kairomones to enhance natural enemy action" (see Chapters 16 and 22).

The authors list Grigg's (1974) main types of agricultural systems in the world: (1) shifting cultivation systems; (2) semipermanent, rain-fed cropping systems; (3) permanent rain-fed cropping systems; (4) arable irrigation systems; (5) perennial crop systems; (6) grazing systems; and (7) systems alternating arable cropping and sown pasture.

AGENTS, BIOLOGY, AND METHODS

In this section the biology of parasitoids, predators, and pathogens used for the biological control of insects are discussed (Chapters 15, 16, 18, and 21). Other factors that

may influence natural-enemy efficacy such as nutrition (Chapter 22), sex ratios (Chapter 23), and pesticide resistance of natural enemies (Chapter 24) are also dealt with. Phytophagous insects used in the biological control of weeds are discussed in Chapter 17, and the various agents and strategies used in the rapidly expanding field of the biological control of plant pathogens are dealt with in Chapters 19, 20, and 25.

Chapter 15

Gordh, Caltagirone, and Legner deal specifically with the biology and mechanisms of parasitoids and parasitism.

A parasitoid lives in or on a single host organism, unlike true predators. Seven basic criteria that distinguish parasitoids from other parasitic animals are listed (but there are exceptions to each criterion). They develop internally or externally, or initially as one and then the other. The classical forms of parasitism are defined and differentiated. Hyperparasitism is highly developed in Hymenoptera compared to Diptera and Coleoptera. Anatomical features related to their modes of life are specified.

The Parasitica (Hymenoptera) is the largest group and most important to biological control. The various superfamilies and families are categorized briefly as to their biology and in some cases their biological control potential. Development from egg to adult is outlined and differences between various developmental forms are noted, as are the general physiologies and behaviors of adults. It is noted that "parasitoids have ovipositional strategies which are essential in optimizing lifetime performance." Interestingly, gregarious parasitoids may estimate host size and adjust the number of eggs accordingly.

Sex regulation, host feeding, habitat and host location, and host acceptance and suitability are also considered.

Chapter 16

Hagen, Mills, Gordh, and McMurtry discuss the biology of predators and mechanisms of predation.

The authors note that predaceous arthropods have important roles in the three main tactics employed in biological control of pests: through importation, manipulation, or conservation. Each tactic has a relevance in our overall integrated approach to pest management. Brief biologies of the families represented are given, including eggs and immature stages. The prey range of immature and adult forms, and the cues that are employed to find prey are also presented.

The roles of behavioral chemicals and some of the semiochemicals that often involve tritrophic interrelationships are discussed, as are the basic nutritional requirements of these predators. The roles of generalist and specialist predators in biological control are compared, and the controversy over releasing polyphagous species in new regions is discussed. The authors consider that in the future an increase in periodic releases of natural enemies is likely as the curtailments in use, or the effectiveness, of chemicals develop further. They also see an increased need to mass culture natural enemies on natural or artificial diets, along with use of food supplements and shelters in field situations.

The authors review the Chelicerata and the Uniramia, excluding the Crustacea of the former phylum Arthropoda. In the Chelicerata, the most important to applied biological control is the Arachnida; only the spiders (Araneae) and the mites (Acariformes) are considered here. The spiders are considered to be more regulators of complexes in the ecosystem than of individual species. Of the 331 known families of Acariformes, at least 20 include predaceous species but only 11 are dealt with. The Phytoseiidae are the principal predators of phytophagous mites. Their roles in biological control are discussed, as well as examples of attempts to improve their qualities through genetic manipulation, selection, and periodic releases. There also has been some utilization of Tydeidae, Bdellidae, Chyletidae, and Stigmaeidae.

In the phylum Uniramia, the insects (Hexapoda) are of crucial importance both as predators and as parasitoids. The different foraging strategies of predators, such as random search in hunting, ambush, and trapping, are all represented in the insects. Some are generalists and others highly prey (and host) specific. The authors review the main orders of importance, including the historical significance of predatory Coccinellidae in establishing the significance of predatory insects in controlling pest species.

The authors also discuss the sequence patterns of prey finding and generalist versus specific predators. They also compare the nutritional requirements of predators and herbivores and suggest the most likely avenues for improving biological control.

Chapter 17 (and Chapter 34)

These chapters deal with biological control of weeds. In the former, Bellows and Headrick discuss the attributes of insects used to control alien weeds.

There has been a great increase in the number of biological control projects against weeds (Chapters 17 and 34). The number of species established has been proportional to the number of agents introduced, but not necessarily successful as controllers.

Chapter 18 and 21

Federici considers the use of insect pathogens in biological control. Emphasis is not on biological control per se, but on its use in IPM. He takes a pessimistic view of introduction of pathogens to achieve lasting (classical) bio-

logical control, particularly if control of weeds (aquatic and terrestrial) is included along with insect pests. Federici considers *Bacillus thuringiensis* (B.t.) to have had the greatest success, pointing out that it kills via a toxin, which places this tactic in a category different from biological control in the sense used in Fig. 2 and most other chapters. Also, he includes as biological control other biotechniques not accepted as such by the other authors of this book.

Federici notes the paradox that the greater specificity that gives these agents (pathogens) their biological utility also makes them less marketable than chemicals, even others derived from organisms. He considers that recombinant DNA technology to improve the use and efficacy of microbial insecticides (or to produce crops resistant to the pests) will have an increasing role in the future. He then discusses the development, current usage, limitations, and likely modes for improvement for each major pathogen group (bacteria, viruses, fungi, and protozoa).

Several B.t. strains are promising, including *B.t. israelensis* for mosquito and blackfly larvae, and *B.t. tenebrionis* for Coleoptera, the latter now used for Colorado potato beetle in the United States and Russia and also being developed for use against the elm leaf beetle. The cloning of a gene encoding the B.t. toxin has resulted in transfers to tobacco, potato, and tomato, with others to follow. This gives protection to the plant from specific insects, and the possibilities seem to be mushrooming. Studies suggested by the new genetic technologies are also discussed for viruses, where registration problems exist. While there are no fungi currently registered for use in the United States, some possibilities are suggested. Protozoans are discussed, but the prospects are not considered very promising.

Chapter 19

Dodds provides a review of two vital plant processes that limit the disease process. Systemic acquired resistance occurs when a plant is inoculated with a virulent pathogen against which it is able to mount a resistance reaction. Later, should the plant be inoculated by a pathogen to which it is normally susceptible, infection is greatly limited by the resistance in the plant induced by the first pathogen. The second pathogen may or may not be related to the first. Cross protection is a more specific protection, when prior challenge by a microbe limits the disease caused by later infection with a related microbe.

Chapter 20

A variety of mechanisms are currently being investigated to genetically engineer plants to be resistant to pathogenic viruses. Cooper shares insights into several of these mechanisms, including coat-protein mediated protection, RNA mediated protection, and several mechanisms that relate to the pathogen genome or to changes in plant enzymes relied on by the pathogen for reproduction. A simple, widely effective resistance mechanism that can be engineered into plants in general has thus far eluded researchers, and a combination of several mechanisms may turn out to be important in achieving resistance to a broad spectrum of heterologous viruses.

Chapter 22

Thompson and Hagen deal with the principles of nutrition of entomophagous insects. They note many nutritional advances since the reviews by Doutt (1964) and Hagen (1964) cited in DeBach (1964b). Rearing some of these insects artificially has revealed that their nutritional needs embrace a "complex of behavioral, physiological, and nutritional factors as cited by Slansky (1982, 1986)." The host may influence development, sex ratio, fecundity, longevity, and vigor of the subsequent adult parasitoids. This is often the result of the nutrition afforded by the host. Comparatively few similar studies have been conducted with predators—limited work has been reported on 10 coccinellid species. Thompson and Hagen also report that numerous studies show that entomophagous insects have "no unusual qualitative nutritional requirements," but that dietary suitability involves a "quantitative balance" beyond mere qualitative essentials. They also note that "nutritional requirements of adult entomophagous insects remain obscure."

The addition of semiochemicals to supplemental foods attracts *Chrysoperla carnea* adults and causes them to "settle down." Flight patterns and attraction to prey are affected by tryptophan breakdown products from honeydew. Coccinellids and other predators aggregated around artificial honeydew. Spraying of corn with sugar or molasses increased predation on, and lowered populations of, aphids and corn borers. Currently, practical applications are "restricted to use of food and food supplements to enhance entomophagous insects in the fields where synchrony is lacking or the enemies are isolated from environments having nectars and honeydew they commonly use." Feeding of *Trichogramma* prior to release markedly increased fecundity and longevity, and feeding *Bracon brevicornis* honey solution sprayed on sorghum stalks in winter increased parasitoid cocoons, while hosts declined. (See extensive citations by Thompson and Hagen.)

Chapter 23

Luck, Nunney, and Stouthamer deal more specifically with the evolutionary ecology of parasitoids. They discuss the impact of culturing techniques on sex ratios and quality of the insectary product, the natural enemy, both from a theoretical perspective of first principles and from a practical perspective of the insectary workers.

The genetic (hence evolutionary) significance of factors such as male diploidy and son-killer bacteria, which may eradicate or substantially reduce the culture or may produce a thelytokous result, is discussed at length. When cultures of parasitoids attenuate or start yielding proportionately more males than females, it is important that the insectary supervisor understands the possible explanations for it to take corrective action.

The most important factors controllable within a confined rearing program that contribute to quality (i.e., highest parasitization in the field) are the number of females and their size and individual heterozygosity, especially in outbreeding species such as those of Ichneumonoidea, in which the effect of inbreeding on female function may be severe.

Chapter 24

Tabashnik and Johnson note that natural enemies account for only 3% of insect and mite species reported as resistant to pesticides. They review three main hypotheses proposed to explain this pattern: (1) documentation bias, (2) differential preadaptation, and (3) differences in population ecology.

Analysis of studies of 14 species of natural enemies published from 1979 to 1987 showed that with the exception of phytoseiid mites, substantial resistance development in field populations of natural enemies was rare. High levels of resistance in Hymenoptera were virtually absent. The authors conclude that pesticide resistance is more likely to be documented in pests than in natural enemies, yet the lack of resistance in several carefully studied natural enemy species suggests that resistance does evolve faster in pests than in natural enemies.

The authors report that intrinsic levels of detoxification enzymes were not consistently lower in natural enemies than in pests. They suggest that various preadaptation hypotheses based on differences in detoxification enzymes have limited explanatory power. Differences in intrinsic tolerance to pesticides and differences in genetic systems were also found to be unlikely general explanations for why pests become resistant more readily than their arthropod natural enemies.

Tabashnik and Johnson suggest that food limitation due to reduction in host or prey densities caused by pesticides is a major factor that retards resistance development in natural enemies (see also Huffaker, 1971). To maintain biological control, they recommend sparing and judicious use of selective pesticides. (Extensive references are cited.)

Chapter 25

Fulbright outlines the history and findings of the research that has followed the fate of chestnut blight, a devastating disease of chestnuts in both North America and Europe. Following widespread destruction of chestnut after the introduction of the pathogenic fungus, a few stands of chestnut were discovered that were recovering from infection. Milder, nonlethal strains of the fungus appeared to outcompete, or at least to moderate the impact of, the virulent strain. The hypovirulent trait appears transmittable to virulent strains through transfer of genetic material inside an infected tree. This work holds promise that chestnut may someday be recovered as an important forest tree in deciduous forests, and indicates that hypovirulence may be a promising area for future developments against other fungal diseases.

APPLICATIONS

It was intended that the principles would largely have been handled in the preceding sections; yet, inevitably there are in this section various possibilities and consequences from applications that bear on, and contribute examples of, principles from classical or manipulative biological controls or enhancements. Some examples relate to broader ecological aspects than to just biological control, for example, in general resource, agroecosystem, crop, forest, or range *management.* These areas embrace an enormous spread of environments and human enterprise—for the simplest row-crop situation and multicrop interplantings in single-objective agriculture, multiobjective small farm units, greenhouse cultures; and perhaps the most complex of all, for that of urban multiplantings of great complexity, objectives, and vague potentials for estimating cost-benefit functions. Also, there are medical and veterinary problems that include solutions involving management of disease vectors in ways that utilize biological control, but are not classifiable specifically to the environmental categories. Biological control of soil-borne and foliar plant pathogens are discussed (Chapters 26 and 32) as well as the use of plant pathogens to control weeds (Chapter 35). Biological control of vertebrate pests is considered in Chapter 38. Chapter 40 deals with social and economic factors of human orientation, and omits the basic ecological underpinnings.

Chapters 26 and 32

Controlling plant pathogens through biological control or manipulation of microbial populations is the subject of Chapters 26 and 32. As Bellows points out, this area of biological control is essentially ecological management at the microbial level. Manipulations of microbial populations to favor nonpathogenic types and limit pathogens can take place in a variety of ways, through conservation of populations, manipulating the environment to favor saprotrophic organisms over pathogens, or by adding organisms directly to a cropping environment. The groups of organisms and the ways in which they are manipulated vary with the system that they are targeted for, with principal differences

occurring between those used on the aerial parts of plants (Chapter 32), and those found in the soil or rhizosphere (Chapter 26).

Chapter 27

Kennett, Beardsley, and McMurtry discuss tropical and subtropical crops and provide by far the most examples of solid classical biological control. They also illustrate a wide diversity of the principles in ecology and natural enemy behavior, particularly in relation to results from introduced natural enemies. This is not unexpected because the vast majority of efforts have been centered in the tropical and subtropical areas against the armored and unarmored scale insects, mealybugs, whiteflies, psyllids, and mites. Included are work on citrus (the earliest and most researched), coconut, banana, coffee, tea, and olive. These illustrate some of the more striking successful examples and include the more recent efforts. However, sugarcane, pineapple, macadamia nut, and passion fruit are not covered (but see Bennett *et al.,* Chapter 15, in Huffaker & Messenger, 1976).

Examples include those where competitive displacement has improved biological control, and the culture and release of mycelia of a pathogenic fungus was used to control the citrus red mite.

Chapter 28

AliNiazee and Croft review biological control work on pests of deciduous tree fruits and the use of natural enemies and IPM for these crops. These crops are characterized by large pest complexes; since biological control agents are inadequate against some of these, integrated programs are generally needed, and in some cases strictly chemical control programs are necessary. The authors discuss their programs in terms of key, sporadic, and secondary pests. Successful introductions of exotic natural enemies have been few, but striking results from introducing pesticide-resistant phytoseiids into Australia for control of mites (a pesticide-triggered problem) are reported. Otherwise, classical biological control introductions for these crops have been limited mainly to efforts on woolly apple aphid, winter moth, and codling moth.

Chapter 29

In this chapter on biological control of forest insects, Dahlsten and Mills discuss the great differences between forest management and agricultural crop management. Forests are a multiple-use resource and differ widely from single-crop agriculture. Time to harvest is 20 to 30 years at the shortest and 50 to 100 years in other cases. The cost of treating such a long-term crop annually would be impossible economically. Some 78% of biological control importations for forest use have involved parasitoids (Hymenoptera and tachinids); 40 species of enemies have been used in unsuccessful efforts against the balsam woolly aphid, *Adelges piceae*. Augmentation and conservation of natural enemies are promising routes.

The effort on the larch casebearer has been a long one—from 1928—and is continuing. Cited are Ryan *et al.* (1987), who report a steady decline of the casebearer since releases of *Chrysocharis laricinellae* and *Agathis pumila.*

Chapter 30

Kogan, Gerling, and Maddox review the methods of enhancing biological control on several transient crops. They note that disturbances in such crops are difficult to correct, but nevertheless many less disturbed crops benefit from a high level of resident natural enemies. Such crops are also susceptible to secondary pest outbreaks triggered by pesticide use.

These circumstances make transient crops "particularly suited for augmentative releases of natural enemies and use of microbial pesticides or methods to trigger natural epizootics of pathogens."

Much of the treatment is based on soybean as a model. The instability of the tritrophic interactions is a major problem for classical biological control. Hence, colonization (addition) of natural enemies is stressed, including "(1) native host-specific enemies that overwinter in or near the fields, (2) polyphagous species that develop in various nearby habitats and move in as opportunists, and (3) migrant colonizers from more subtropical areas." The authors also state, "Despite the inherent ecological instability—most [of the] herbivore populations are effectively regulated by a complement of natural enemies." They use examples for predators, parasites, and disease pathogens (e.g., *Nomurea*) and discuss cotton, soybean, corn, melon, alfalfa, and cassava.

The microsporidian *Nosema pyraustae* (inadvertently introduced) is the major enemy of corn borer in many areas and may have displaced a previously significant introduced tachinid, *Lydella thompsoni. Trichogramma* spp. are used in inundative and augmentative releases for various crops: rice, millet, crucifer, beet, corn, cassava, and others. Many successes have been documented, but the potential for even greater successes exists.

Chapter 31

Parrella, Hansen and van Lanteren deal with biological control in greenhouse environments, some 100,000 to 150,000 ha worldwide. Manipulations in greenhouses are easier than those outdoors—for the pests, the natural enemies, the environment, and IPM in general. Greenhouses offer food, warmth, protection from weather, and physical barriers to prevent dispersal by drift. Work has been mostly on vegetables and flowers, but there has been a dramatic

increase since 1968. Yet it is estimated that biological control is used regularly on only about 3000 ha, excluding some undocumented use in Russia.

Cucumbers and tomatoes are by far the most important. It is interesting that "the intensive harvesting (2 to 3 times a week) is difficult to coordinate with statutory re-entry and pesticide residue restrictions for pesticide usage. Phytotoxicity from pesticides, especially young plants in winter, also favors biological control."

The high tolerance of cucumbers to spider mite damage and the very high and quick efficiency of *Phytoseiulus persimilis* account for its widespread usage for cucumbers. It is a welcome "solution to increasing problems" with resistance of *Trichogramma urticae* to acaricides. The initial density of pest mites and the rate of introduction are critical. Cucumbers are deliberately infested with spider mites immediately after planting, or both spider mites and *P. persimilis* are introduced at the first natural appearance of spider mites.

Greenhouse whitefly was controlled on cucumbers and tomatoes by *Encarsia formosa* in the United Kingdom from 1927 until the 1940s when synthetic organic pesticides began to be used instead. The development of resistance in *T. urticae* to pesticides ushered in biological control on these crops and this necessitated a return to biological control for greenhouse whitefly. *Encarsia* has many desirable traits for achieving good control on tomatoes in greenhouses, but less so for cucumbers. Leaf miners on cucumbers, tomatoes, and melons have also been controlled by three species of parasitoids in the Netherlands, United Kingdom, and Sweden. Related problems in other countries and on other greenhouse crops are also receiving attention. (See references in Chapter 31.)

Chapter 33

In this chapter on insects and mites of grapes, Flaherty and Wilson review the culture and pest problems of cultivated grapes worldwide, noting their origin in Asia Minor. Included is consideration of the various groups of pests; they note that for so valuable a crop, very little work on biological control has been done. In California (where most work has been done), the egg parasitoid *Anagrus epos* is a key factor in biological control and IPM. In favorable places this parasitoid parasitizes about 80% of the native grape leafhopper, but only 20% of the variegated grape leafhopper, which now poses additional problems. Proper integration of chemicals for control of primary pests is crucial to avoid major problems with pesticide-generated pests, such as spider mites.

Chapter 34

Goeden and Andrés cover the biological control of weeds, both terrestrial and aquatic. They detail the methodology in choosing and carrying out a project (reviews cited are those by Goeden, 1978 in Clausen, 1978, Zwölfer & Harris, 1971, and Turner, 1985). For importations, (1) project selection, (2) search for natural enemies, (3) tests and biological studies of host range, (4) evaluation of host–range studies, (5) importation and release, and (6) evaluation form the components of a program. They also deal with conserving and augmenting existing biological control agents. They noted more than 80 projects in biological weed control in the United States and Canada alone, and cite Julien (1982), who states that 101 species of weeds have been targeted worldwide for 174 projects. They note that while insects have been used mostly, phytopathogens are showing promise and some success. Their chapter and Chapter 17 cover details with the Coleoptera, Lepidoptera, Diptera, and Hemiptera–Homoptera being the main agents introduced; and 19 to 44% of these providing *effective control* (not just being *established*).

Goeden and Andrés also detail the histories of early projects. They detail work on alligator weed in the United States and Australia; puncture vine in the southwestern United States, Hawaii, and St. Kitts; water hyacinth in the southeastern United States; and three other successes (i.e., for *Cuscuta* in China, *Ambrosia* (ragweed) in Russia, and *Salvinia* in Australia and New Guinea). In the latter case the significance of solid taxonomy and host relationship was found to be crucial. Two other points are worth noting here: (1) the authors downplay the promise of genetic improvements in natural enemies, but in Chapter 12 on this subject, considerable promise is suggested; (2) the authors state that before undertaking an introduction effort, "there must be assurances that the weed has few, if any, redeeming virtues and that there is little or no public opposition to the project," citing Turner (1985). Yet Turner's paper is a closely considered, comprehensive analysis, not just the cautions to be considered. It also gives suggestions on how apparent conflicts of interest may be resolved, even where there is not complete public acceptance and there may be some redeeming virtue. There is now a closer concern about making unwise releases; more kinds of agents are being used, adverse environmental consequences are being posed, and many more personnel who have different views and qualifications concerning these complex risks are involved. So it is proper that protocols are being tightened and ecologically disruptive possibilities are being more closely considered (see Turner, 1985).

Chapter 35

Pathogens in weed control is the subject of Chapter 35 by Rosskopf, Charudattan, and Kadir. In this chapter they discuss the opportunities for biological control of weeds using pathogens. Pathogens, with the possibility of high selectivity, may be viewed with favor both in programs that seek to permanently introduce new pathogens into an environment, and also by programs that seek a pathogen that can be mass-reared and then added to a particular system in an inundative way. Some opportunity for commercial development of pathogens may find fruition in this field.

Chapter 36

In this chapter on urban environments, Dahlsten and Hall note that these environments are the most complex, diverse, and discontinuous due to the many introduced ornamental plants in fragmented patches. These environments offer special opportunities for IPM and biological control. There are problems in assessing aesthetic and nuisance pests. There is a higher and more diversified pesticide usage per acre than in any crop or forest, especially prior to the 1970s, with no regard for IPM or biological controls. Overlapping problems include medical, psychological, architectural, agricultural, silvicultural, floricultural, and horticultural concerns. Improved incentive for biological control comes from public concerns for clean air and water and fear of pesticide exposure.

For example, in Berkeley, California, 10 tree species constitute the native tree flora; now there are 123 species, with nearly 300 on the University of California at Berkeley campus alone. They have come from most moderate climate regions of the world and have a wide variety of introduced pests. Most of the successes with introduced natural enemies have been with homopterous insects on perennials, as in agriculture, for example, woolly whitefly on citrus, acacia psyllids on acacia, and aphids on elm.

That urban fly and mosquito control with chemicals can interfere with biological controls on urban trees is illustrated by outbreaks of *Lecanium corni* on fruit trees on Mackinac Island, Michigan; and of pine needle scale in South Lake Tahoe, California. There is currently a Berkeley campus program on control of cockroaches using two egg parasitoids.

Chapter 37

In this chapter on biological control of medical and veterinary pests, Garcia and Legner deal with earlier reviews and bring in newer developments. A few references to snail vectors of digenetic trematodes are made, but most of the work has been against pest Diptera.

The agents used for mosquito control have been mostly fish, insects, and other arthropods; and the microbial insecticide *B. thuringiensis* var. *israelensis* (H-14). Several nematodes and fungi are being tested. Interest at the turn of the century on dragonflies quickly waned. Soon thereafter the mosquito fish, *Gambusia affinis,* was intensively used for some 40 years. Synthetic organic insecticides curtailed all this work shortly after World War II, and it was revived only when the problems of resistance to the chemicals loomed ever greater. *Gambusia* is not now widely accepted but is still used in some situations; it is better against non-anophelines. There is a real problem with fixing mosquito tolerance levels, but more important is the temporary, unstable habitat of the pest species, which is not conducive to sustained biological control (Legner & Sjogren, 1984 cited).

Of the synanthropic Diptera, muscoid species, including the housefly, are the most important. In Australia, dung-burrowing beetles introduced for bush fly control have aided agriculture through their manure mixing in the soil, but appear not to have much affected densities of the bush fly.

For snails, biological control is very limited and restricted to special situations. Sciomyzid flies are highly specific and have shown some success in Hawaii (Bay *et al.,* 1976; Garcia & Huffaker, 1979) to control the intermediate host snail of the giant liver fluke of cattle. Suppression of *Biomphalaria glabrata,* the intermediate host of human schistosomiasis, through competitive displacement by *Marisa cornuarietis* and also *Helisoma duryi,* shows promise.

Chapter 38

Biological control of vertebrates is a challenging and important area worldwide. Hoddle provides in this chapter a review of the types of damage exotic vertebrates can cause to both natural and agricultural ecosystems. Historical approaches, employing vertebrate predators, are reviewed, and the risks and benefits of such programs are discussed. Other techniques that include biological sterilization, and use of specific pathogens, are reviewed as perhaps more likely to be implemented in the future than would be additional movement of vertebrate predators.

Chapter 39

Altieri and Nicholls utilize programs in Latin America as a model for dealing with these types of problems in countries where assessment technology is limited; or pesticide laws are lax or even favor pesticide abuses, or allow the "dumping" of disallowed pesticides from more environmentally conscious countries (e.g., the United States). The authors deal with these questions and with the complex factors associated with using biological and cultural controls. Of the biological control work considered, Chile has had the most introductions, with 66 introductions being made from 1903 to 1984 against pests of citrus, grape, peach, apple, tomato, and other crops. They report that some 42 species of natural enemies were established; and 60% of the targeted host species are under complete or substantial biological control, 38% by introduced predators and 24% by introduced parasites.

Chapter 40

Perkins and Garcia discuss the general social, political, economic, and philosophical factors that affect research and implementation of biological control. They define biological control as "the identification of indigenous and exotic natural enemies, the importation and release of exotic natural enemies, and the evaluation of the abilities of natural enemies to suppress a pest." This is an excellent disciplinary expression in that it specifies the identification, thus the taxonomic relevance, and the evaluation—what workers must do. They also summarize the various economic and social factors discussed and give recommendations. These relate, not to the basic ecological factors, for example, weather, allies, and competition, that affect natural enemy success, but to those social, political, and economic factors so important to biological control and other resource management efforts.

Chapter 41

In this final chapter, Bellows provides a view toward the future of where biological control can take us, or more precisely, where it will likely fit in our world's future. He covers possible developments in combating adventive and native pests, advances in use of microbial agents, possibilities of protecting natural environments from invading species, the roles of experts and institutions, and the importance of training future biological control scientists.

Acknowledgments

The authors wish to express appreciation of the cooperation of all subsequent chapter authors in preparing this introductory chapter, especially to A. P. Gutierrez for insightful suggestions; and to David L. Rowney and the secretarial staff for much patience and help.

References

Batra, S. W. T. (1982). Biological control in agroecosystems. Science, 215, 134–139.

Bay, E. C., Berg, C. O., Chapman, H. C., & Legner, E. F. (1976). Biological control of medical and veterinary pests. In C. B. Huffaker & P. S. Messenger (Eds.), Theory and practice of biological control (pp. 457–479). New York: Academic Press.

Caltagirone, L. E., & Huffaker, C. B. (1980). Benefits and risks of using natural enemies for controlling pests. In B. Lundholm & M. Stuckerud (Eds.), Environmental protection and biological forms of control of pest organisms. Stockholm: Ecol. Bull.

Carruthers, R. I., & Petroff, J. K. (Eds.). (1997). Proceedings of the Invitational Workshop on USDA Activities in Biological Control. Oct. 8–11, 1996, Riverdale, Maryland and Washington, D. C. USDA, Agr. Res. Service 1997–01, 109 pp.

Clarke, B., Murray, J., & Johnson, M. S. (1984). The extinction of endemic species by a program of biological control. Pac. Sci., 38, 97.

Clausen, C. P. (ed.). (1978). Introduced parasites and predators of arthropod pests and weeds: A world review. Agriculture Handbook No. 480. Washington, DC: USDA.

Crawley, M. J. (1989). Insect herbivores and plant population dynamics. Annual Review of Entomology, 34, 531–564.

Davis, D. E., Myers, K., & Hoy, J. B. (1976). Biological control among vertebrates. In C. B. Huffaker & P. S. Messenger (Eds.), Theory and practice of biological control (pp. 501–519). New York: Academic Press.

DeBach, P. (1964a). The scope of biological control. In P. DeBach (Ed.), Biological control of insect pests and weeds (pp. 3–20). London: Chapman & Hall.

DeBach, P. (Ed.). (1964b). Biological control of insect pests and weeds. London: Chapman & Hall.

DeBach, P., & Rosen, D. (1991). Biological control by natural enemies (2nd ed.). Cambridge, England: Cambridge Univ. Press.

DeBach, P., & Sundby, R. A. (1963). Competitive displacement between ecological homologues. Hilgardia, 34, 105–166.

DeBach, P., Huffaker, C. B., & MacPhee, A. W. (1976). Evaluation of the impact of natural enemies. In C. B. Huffaker & P. S. Messenger (Eds.), Theory and practice of biological control (pp. 255–285). New York: Academic Press.

Garcia, R., & Huffaker, C. B. (1979). Ecosystem management for suppression of vectors of human malaria and schistosomiasis. Agro-Ecosystems, 5, 295–315.

Garcia, R., Caltagirone, L. E., & Gutierrez, A. P. (1988). Comments on a redefinition of biological control. BioScience, 38, 692–694.

Gilbert, N. E., Gutierrez, A. P., Fraser, B. D., & Jones, R. E. (1976). Ecological relationships. London: Freeman.

Hassell, M. P. (1986). Parasitoids and population regulation. In J. Waage & D. Greathead (eds.), Insect parasitoids (pp. 201–224). London: Academic Press.

Howarth, F. G. (1983). Classical biocontrol: Panacea or Pandora's box? Proceedings of the Hawaiian Entomological Society, 24, 239–244.

Huffaker, C. B. (1971). The ecology of pesticide interference with insect populations (upsets and resurgences). In J. E. Swift (Ed.), Agricultural chemicals—harmony or discord for food, people and the environment (pp. 92–104). Proceedings of Symposium, Univ. of Calif., Div. of Agric. Sci.

Huffaker, C. B. (1988). Ecology of insect pest control (Vol. 3). In R. Goren & K. Mendel (Eds.), Proceedings, Sixth International Citrus Congress, Tel-Aviv, Israel, March 6–11, 1988 (pp. 1047–1066). Philadelphia, Rehovot: Balaban Publ.

Huffaker, C. B. (1990). Effects of environmental factors on natural enemies of armored scale insects (Vol. 4B). In David Rosen (Ed.), The armored scale insects, their biology, natural enemies and control (pp. 205–220). Amsterdam: Elsevier.

Huffaker, C. B., & Gutierrez, A. P. (1990). Evaluation of efficiency of natural enemies in biological control (Vol. 4B). In David Rosen (ed.), The armored scale insects, their biology, natural enemies and control (pp. 443–495). Amsterdam: Elsevier.

Huffaker, C. B., & Kennett, C. E. (1969). Some aspects of assessing efficiency of natural enemies. Canadian Entomology, 101, 425–447.

Huffaker, C. B., & Messenger, P. S. (1964). The concept and significance of biological control. In P. DeBach (Ed.), Biological control of insect pests and weeds (pp. 45–117). London: Chapman & Hall.

Huffaker, C. B., & Messenger, P. S. (Eds.). (1976). Theory and practice of biological control. New York: Academic Press.

Huffaker, C. B., & Rabb, R. L. (Eds.). (1984). Ecological entomology. New York: Wiley-Interscience, John Wiley & Sons.

Huffaker, C. B., & Rosen, D. (1990). The attributes of effective natural enemies (Vol. 4B). In D. Rosen (Ed.), The armored scale insects, their biology, natural enemies and control (pp. 197–204). Amsterdam: Elsevier.

Huffaker, C. B., Berryman, A., & Laing, J. E. (1984). Natural control of insect populations. In C. B. Huffaker & R. L. Rabb (Eds.), Ecological entomology (pp. 359–398). New York: Wiley-Interscience, John Wiley & Sons.

Huffaker, C. B., Gordon, H. T., & Rabb, R. L. (1984). Meaning of ecological entomology—The ecosystem. In C. B. Huffaker & R. L. Rabb (Eds.), Ecological entomology (pp. 3–17). New York: Wiley-Interscience, John Wiley & Sons.

Huffaker, C. B., Messenger, P. S., & DeBach, P. (1971). The natural enemy component in natural control and the theory of biological control. In C. B. Huffaker (Ed.), Biological control (pp. 16–67). New York, London: Plenum Press.

Julien, M. H. (Ed.). (1982). Biological control of weeds. A world catalogue of agents and their target weeds. Slough, England: Commonwealth Agricultural Bureau, Commonwealth Institute of Biological Control.

Krebs, C. J. (1972). Ecology: The experimental analysis of distribution and abundance. New York: Harper & Row.

Laing, J. E., & Hamai, J. (1976). Appendix. In C. B. Huffaker & P. S. Messenger (Eds.), Theory and practice of biological control (pp. 685–743). New York: Academic Press.

Luck, R. F. (1984). Principles of arthropod predation. In C. B. Huffaker & R. L. Rabb (Eds.), Ecological entomology (pp. 497–527). New York: Wiley-Interscience, John Wiley & Sons.

Mackauer, M., Ehler, L. E., & Roland, J. (Eds.). (1990). Critical issues in biological control. Andover, Herts: Intercept.

Maws, M. G. (1984). Ambrosia artemisiifolia L., common ragweed (Compositae). In J. S. Kelleher & M. A. Hulme (Eds.), Biological control programmes against insects and weeds in Canada 1969–1990 (pp. 111–112). Commonwealth Agric. Bureau.

Nechols, J. R. [Exec. Ed.]. (1995). Biological control in the western United States. Univ. of Calif., Div. of Agric. and Nat. Res. Pub. 3361.

Pimentel, D., Glenister, C., Fast, S., & Callahan, D. (1984). Environmental risks of biological pest controls. Oikos, 42, 283–290.

Price, P. W., Bouton, C. G., Gross, P., McPheron, B. A., Thompson, J. N., & Weiss, A. E. (1980). Interactions among three trophic levels: influence of plants on interactions between herbivores and natural enemies. Annual Review of Ecology & Systems, 11, 41–65.

Reynolds, H. T. (1971). A world review of insect population upsets and resurgences caused by pesticide chemicals. In J. E. Swift (Ed.), Agricultural chemicals—harmony and discord for food, people and environment (pp. 108–112). Proceedings of Symposium, Univ. of Calif., Div. of Agric. Sci.

Rosen, D., & DeBach, P. (1991). Foreign exploration: The key to classical biological control. Florida Entomology, 75, 409–413.

Ryan, R. B. (in press). Evaluation of biological control: introduced parasites of larch casebearer (Lepidoptera: Coleophoridae) in Oregon. Environmental Entomology,

Southwood, T. R. E. (1966). Ecological methods. London: Methuen & Co., Ltd.

Teng, P. S. (1985). Integrating crop and pest management: the need for comprehensive management of yield constraints in cropping systems. Journal of Plant Prot. Tropics, 2(1), 15–26.

Turner, C. E. (1985). Conflicting interests and biological control of weeds. In E. S. Delfosse (Ed.), Proceedings, VI, International Symposium on Biological Control of Weeds, Vancouver, Canada (pp. 203–225).

U.S. Congress, Office of Technology Assessment (1995). Biologically based technologies for pest control (OTA-ENV-636). Washington, D.C.: U.S. Government Printing Office.

Van Driesche, R., & Bellows, T. S., Jr. (1996). Biological control. London: Chapman and Hall.

Zwölfer, H., & Harris, P. (1971). Host specificity determination of insects for biological control of weeds. Annual Review of Entomology, 16, 159–178.

Theories and Mechanisms of Natural Population Regulation

T. S. BELLOWS

Department of Entomology
University of California
Riverside, California

M. P. HASSELL

Department of Biology
Imperial College
Silwood Park, Ascot
Berkshire, SL5 7PY, United Kingdom

INTRODUCTION

The concepts of population equilibria and population regulation lie at the heart of biological control. Characteristic of "successful" biological control programs are the reduction of pest populations and their maintenance, in a stable interaction with their natural enemies, about some low, nonpest level. Such outcomes have frequently been recorded (DeBach, 1964; Debach & Rosen, 1991; Van Driesche & Bellows, 1996), but documented evidence is less common (Beddington *et al.,* 1978; Bellows *et al.,* 1992; Bellows, 1993). Four instances tracking the decline of a population following the introduction of natural enemies are shown in Figs. 1 and 2. Biological control programs seek to enhance such natural control of populations, and an understanding of the principals involved requires a fundamental understanding of many aspects of population ecology (see also Mills & Getz, 1996).

In this chapter, we consider mechanisms contributing to the reduction of pest population levels and to population regulation. We begin with single-species systems, followed by competition between species, before turning to interactions between a single host or prey and a natural enemy. Finally, we turn to systems with more than one species of either prey or natural enemy and to systems of populations, or metapopulations. Throughout the chapter, we develop these topics generally using simple model frameworks that are, for the most part, supported by a large body of experimental evidence. Much of these data come from studies on insect predator–prey and host–parasitoid systems (Hassell, 1978). Although there is abundant information on the effects of herbivory on the performance of *individual* plants,

there are little data on the effects of insect herbivory on the dynamics of plant *populations* (Crawley, 1983, 1989). For this reason, our discussion focuses primarily on insect predator–prey, host–parasitoid, and host–pathogen interactions.

SINGLE-SPECIES POPULATIONS

Homogeneous Environments

Single Age–Class Systems

The study of single-species population dynamics has a long history of both theoretical and empirical development both on systems in continuous time and with discrete generations. In this section, we examine the dynamics of populations with discrete generations, which is appropriate for many insect populations. The algebraic framework is straightforward:

$$N_{t+1} = Fg(N_t)N_t. \qquad (1)$$

Here N is the host population denoted in successive generations t and $t + 1$, and $Fg(N_t)$ is the per capita net rate of increase of the population dependent on the per capita fecundity F and the relationship between density and survival g (which is density dependent for $dg/dN < 0$).

By assuming that $g(N_t)$ is density dependent, the fundamental concept embodied in Eq. (1) is that some resource, critical to population reproduction or survival, occurs at a finite and limiting level (when $g = 1$, there is no resource limitation and the population grows without limit). Individuals in the population compete for the limiting resource, for

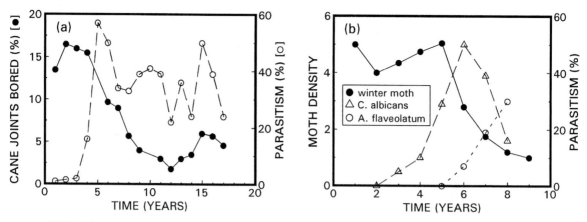

FIGURE 1 Biological control of two insect pests. (a) Sugarcane borers (*Diatraea* spp.) following introductions of the tachinid *Lyxophaga diatraeae* (Townsend) and the braconid *Apanteles flavipes* (Cameron) (after Anonymous, 1980). (b) Winter moth [*Operophtera brumata* (Linnaeus)] in Nova Scotia following introductions of the tachinid *Cyzenis albicans* (Gravenhorst) and the ichneumonid *Agrypon flaveolatum* (Gravenhorst) (after Embree, 1966).

example, as adults for oviposition sites (Utida, 1941; Bellows, 1982a) or food (Nicholson, 1954), or as larvae for food (Nicholson, 1954; Bellows, 1981).

The dynamics of such single-species systems span the range of behavior from geometric (or unconstrained) growth (when competition does not occur), monotonic approach to a stable equilibrium, damped oscillations approaching a stable equilibrium, stable cyclic behavior, and, finally, systems characterized by aperiodic oscillations or

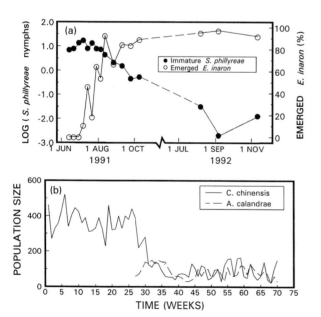

FIGURE 2 (a) Population dynamics in the field of the whitefly *Siphoninus phillyreae* (Haliday) following the introduction of the parasitoid *Encarsia inaron* (Walker) (after Bellows *et al.*, 1992). (b) Dynamics of the weevil *Callosobruchus chinensis* (Linnaeus) and the parasitoid *Anisopteromalus calandrae* (Howard) populations in the laboratory (after May & Hassell, 1988).

"chaos" (Fig. 3) (May, 1975; May & Oster, 1976). Which type of behavior occurs in a particular case depends both on the reproductive rate and on the strength (nonlinearity) of density-dependent competition. Thus, species showing contest competition have more stable dynamics than species showing scramble competition, which tend to show more cyclic behavior as long as the reproductive rates are sufficiently high (Hassell, 1975; May, 1975; Bellows, 1981).

The exact form of the function used to describe *g* is not particularly critical to these general conclusions and many forms have been proposed (May, 1975; May *et al.*, 1978; Bellows, 1981), although different forms may have specific attributes more applicable to particular cases (Bellows, 1981). The most flexible function is that proposed by Maynard Smith and Slatkin (1973), where $g(N)$ takes the form

$$g(N) = [1 + (N/a)^b]^{-1}. \tag{2}$$

Here the relationship between proportionate survival and density is defined by the two parameters *a*, the density at which density-dependent survival is 0.5, and *b*, which determines the severity of the competition. As *b* approaches 0, competition becomes less severe until it no longer occurs ($b = 0$). When $b = 1$, density dependence results in *contest* competition, with the number of survivors reaching a plateau as density increases. Finally, for $b > 1$, *scramble* competition occurs, with the number of survivors declining as the density exceeds $N = a$. Examples of different types of dynamical behavior represented by Eq. (1) with $g(N)$ defined by Eq. (2) are shown in Fig. 4.

Some reviews of insect populations showing density dependence in natural and laboratory settings have suggested that most populations exhibit monotonic or oscillatory damping toward a stable equilibrium (Hassell *et al.*, 1976; Bellows, 1981). Of natural populations, only the Colorado potato beetle, *Leptinotarsa decimlineata* (Say) (Harcourt,

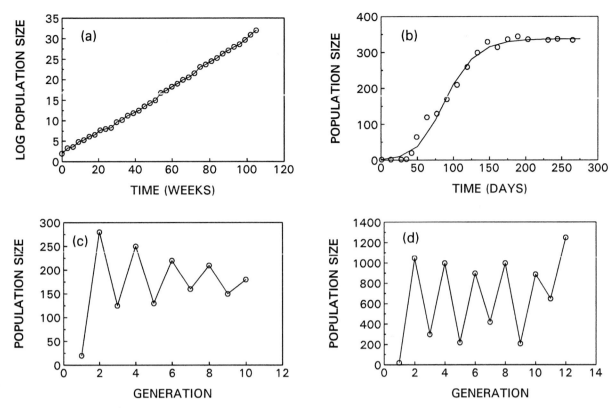

FIGURE 3 Dynamical behavior of single-species populations. (a) Unrestricted growth of *Lasioderma serricorne* (Fabricius) (from Lefkovitch, 1963). (b) Logistic growth of *Rhizopertha dominica* (Fabricius) (after Crombie, 1945). (c) Oscillatory growth of *Callosobruchus chinensis* (Linnaeus) (after Utida, 1967). (d) Cyclic dynamics of *Callosobruchus maculatus* (Fabricius) (after Fujii, 1968).

1971), was predicted to show limit cycles, and only Nicholson's (1954) blowflies fell within the region of chaos (Hassell, 1975; Bellows, 1981). By collapsing the dynamics into a single-species framework, however, there is the distinct possibility that cycles and higher order behavior arising from interactions with other species have been overlooked (Hassell *et al.,* 1976; Godfray & Blythe, 1990).

Multiple Age–Class Systems

Most populations are separable into distinct age or stage classes, which can bear importantly on the outcome of competition. In most insects the preimaginal stages must compete for resources for growth and survival, while adults may additionally compete for resources for egg maturation and oviposition sites. In such cases, competition within populations divides naturally into sequential stages. Equation (1) may now be extended to the case of two age classes (May *et al.,* 1974; Bellows, 1982a) and, assuming competition occurs primarily within stages (larvae compete with larvae and adults with adults), gives

$$A_{t+1} = g_l(L_t)L_t, \qquad (3a)$$

$$L_{t+1} = Fg_a(A_t)A_t, \qquad (3b)$$

where A and L denote the adult and larval populations. Here larval competition, governed by $g_l(L_t)$ in Eq. (3a), leads to A_t adults that compete for resources ($g_a(A_t)$) and produce the next generation of larvae. In such multiple age–class systems, the dynamical behavior of the population is dominated by the outcome of competition in the stage in which b is nearest to one. For example, a population in which adults contest for oviposition sites while larvae scramble for food will show a monotonic approach to a stable equilibrium. More complex approaches to constructing models of age- or stage-structured single–species populations have been developed (Bellows, 1982b; Nisbet & Gurney, 1982).

Patchy Environments

For many populations, resources are not distributed either continuously or uniformly over the environment, but instead occur in discrete units or patches. Consider an environment divided into j discrete patches (such as individual plants) that are utilized by an insect species. Adults (N) disperse and distribute their progeny among the patches.

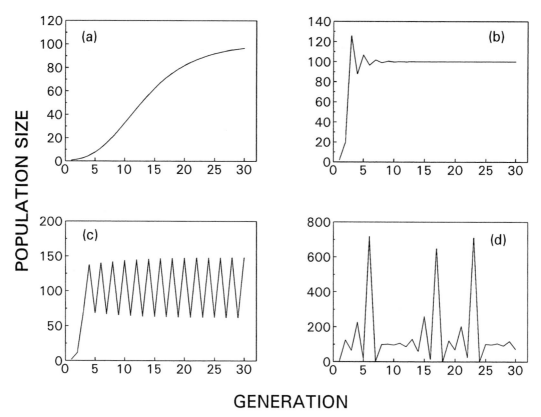

FIGURE 4 Examples of the population dynamics exhibited by the single-species model of Eqs. (1) and (2). In all cases, the equilibrium density N^* = 100. (a) Monotonic population growth to stable equilibrium (F = 5, b = 0.2); (b) damped oscillations to a stable equilibrium (F = 10, b = 1.7); (c) stable limit cycles (F = 6, b = 2.5); (d) chaotic behavior (F = 50, b = 3).

Progeny compete for resources only within their patch. The population dynamics is now dependent partly on the distribution of adults reproducing in patches, Γ, and partly on the density-dependent relationship g that characterizes pre-imaginal competition. Population reproduction from time t to t + 1 over the entire environment (all patches) is characterized by the relationship (de Jong, 1979)

$$N_{t+1} = jF\sum_i \{\Gamma(n_{ti})n_{ti}g[Fn_{ti}]\}, \qquad (4)$$

where n_{ti} is the number of adults in a particular patch at time t, $\Gamma(n_i)$ is the proportion of patches colonized by n_i adults, F is the per capita reproductive rate, and g is the density-dependent survival rate.

De Jong (1979) considers four distinct distributions of adults among patches. In the case of uniform dispersion, Eq. (4) is equivalent to Eq. (1) for homogeneous environments. For three random cases, positive binomial, independent (Poisson), and negative binomial, the outcome depends somewhat on the form taken for the function g. For most reasonable forms of g, the general outcomes are a lower equilibrium population level and enhanced stability compared to a homogeneous environment with the same F and g. In addition, for a fixed amount of resource, stability tends

to increase as the number of patches increases (the more finely divided the resource the more stable the interaction). Finally, one additional feature arises in systems with overcompensatory or scramble competition: there is an optimal fecundity for maximal population density, with fecundities both above and below the optimum resulting in fewer surviving progeny.

INTERSPECIFIC COMPETITION

Although Strong *et al.* (1984), Lawton and Strong (1981), and Lawton and Hassell (1984) suggest that competition is not commonly a dominant force in shaping many herbivorous insect communities, it certainly is important in some communities such as the social insects and, to a lesser extent, insects feeding on ephemeral resources (*Drosophila* spp., Atkinson & Shorrocks, 1977), insect parasitoids (Luck & Podoler, 1985), and predatory beetles. The processes and outcomes of interspecific competition in insects have been studied widely in the laboratory (Crombie, 1945; Fujii, 1968, 1970; Bellows & Hassell, 1984) (Fig. 5) as well as in the field (see review by Lawton & Hassell, 1984).

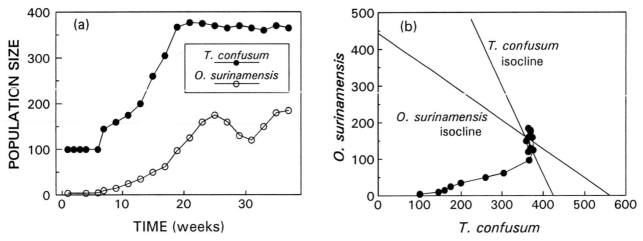

FIGURE 5 Stable equilibrium in a competitive interaction between *Tribolium confusum* Duval and *Oryzaephilus surinamensis* (Linnaeus). (a) Population trajectories over time. (b) Population trajectories in phase space. [After Crombie, A. D. (1946). *Proceedings of the Royal Society of London, Series B, 133,* 76–109.]

Homogeneous Environments

Single Age–Class Systems

Many of the same mechanisms implicated in intraspecific competition for resources (competition for food and oviposition sites) also occur between species (Combie, 1945, 1946; Park, 1948; Fujii, 1968, 1970). The dynamics of these interspecific systems can be considered in a framework very similar to that for single-species populations.

Equation (1) may be extended to the case for two (or more) species by allowing the function g to depend on the density of both competing species (Hassell & Comins, 1976). The reproduction of species X now depends not only on the density of X but also on the density of Y (and similarly for species Y)

$$X_{t+1} = F_x g_x (X_t + \alpha Y_t) X_t, \qquad (5a)$$

$$Y_{t+1} = F_y g_y (Y_t + \beta X_t) Y_t. \qquad (5b)$$

Here the parameters α and β are the competition coefficients reflecting the severity of interspecific competition with respect to intraspecific competition.

Population interactions characterized by Eq. (5) may have one of four possible outcomes: the two species may coexist, species X may always exclude species Y, species Y may always exclude species X, or either species may exclude the other depending on their initial relative abundances. Coexistence is possible only when the product of the interspecific competition parameters $\alpha\beta < 1$. As for single-species systems, Eq. (5) shows a range of dynamic behavior from stable points (Fig. 5), to cycles and higher order behavior determined by the severity of the intraspecific competition described by g (Hassell & Comins, 1976).

Multiple Age–Class Systems

Many insect populations compete in both preimaginal and adult stages, perhaps by competing as adults for oviposition sites and subsequently as larvae for food (Fujii, 1968), and in some cases the superior adult competitor may be inferior in larval competition (Fujii, 1970). The properties of such multiple age–class systems may be considered by extending Eq. (3) as follows (Hassell & Comins, 1976):

$$X_{t+1} = x_t g_{xl}(x_t + \alpha_1 y_t), \qquad (6a)$$

$$Y_{t+1} = y_t g_{yl}(y_t + \beta_1 x_t), \qquad (6b)$$

$$x_{t+1} = X_t F_x g_{xa}(X_t + \alpha_a Y_t), \qquad (6c)$$

$$y_{t+1} = Y_t F_y g_{ya}(Y_t + \beta_a X_t), \qquad (6d)$$

where x and y are the preimaginal or larval stages and X and Y are the adults. Larval survival of each species is dependent on the combined larval densities, and, similarly, adult reproduction of each species is dependent on the combined adult densities. Interspecific larval competition is characterized by the larval competition coefficients α_l and β_l while adult competition is characterized by α_a and β_a.

The simple extension of competition to more than one age class has important effects on dynamics. The isoclines of zero population growth are now no longer linear (as in the case of single age–class systems, compare Figs. 5 and 6); in addition, multiple equilibria may now occur involving in some cases more than one stable equilibrium (Hassell & Comins, 1976). Such curvilinear isoclines are in accord with those found for competing populations of *Drosophila* spp. (Ayala *et al.,* 1974).

More complex systems can be envisaged with additional age classes and with competition between age classes (Bel-

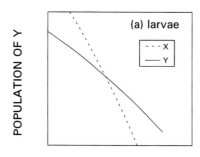 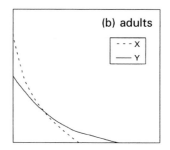

POPULATION OF X

FIGURE 6   Zero-growth isoclines for populations with two age classes may not be linear, but curved to greater or lesser degree. The isoclines are differently shaped for the different age classes. In this example a single stable equilibrium is indicated at the place where the isoclines intersect. In other cases multiple intersections may occur, indicating multiple equilibria. [After Hassell, M. P., & Comins, H. N. (1976). *Theoretical Population Biology, 9,* 202–221.]

lows & Hassell, 1984). The general conclusions from studies of these more complex systems are similar to those for the two age–class system, viz that more complex systems have nonlinear isoclines and consequently may have more complicated dynamical properties. More subtle interactions may also affect the competitive outcome, such as differences in developmental time between two competitors. In the case of *Callosobruchus chinensis* (Linnaeus) and *C. maculatus* (Fabricius) (Coleoptera: Bruchidae), the intrinsically superior competitor *(C. maculatus)* can be outcompeted by *C. chinensis* because of the latter's faster devel-

opment that allows it earlier access to resources in succeeding generations. This confers sufficient competitive advantage to *C. chinensis* that it eventually excludes *C. maculatus* from mixed species systems (Bellows & Hassell, 1984) (Fig. 7).

Patchy Environments

Many insect populations are dependent on resources that occur in patches (fruit, fungi, dung, flowers, dead wood). Dividing the resource into discrete patches can have impor-

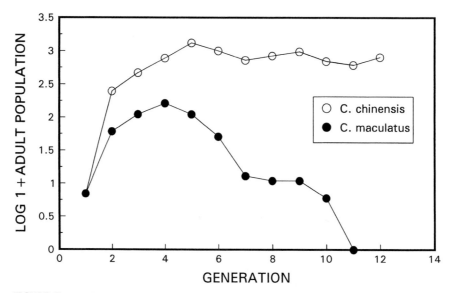

FIGURE 7   In mixed populations of *Callosobruchus chinensis* (Linnaeus) and *C. maculatus* (Fabricius), *C. chinensis* excludes the competitively superior *C. maculatus*. This is a result of the faster developmental time of *C. chinensis,* which provides increasingly earlier access to resources in succeeding generations. This earlier access allows *C. chinensis* to successfully colonize resources (in this case, dried beans) and reproduce before *C. maculatus* can fully occupy the beans, leading eventually to complete exclusion of *C. maculatus*.

tant effects on the conditions for coexistence of the competing species. This is well illustrated by the models of Atkinson and Shorrocks (1981), de Jong (1981), Hanski (1981), Ives and May (1985), and Comins and Hassell (1987). This body of work emphasizes the importance of the spatial distribution of the competing species for coexistence. Thus, Atkinson and Shorrocks (1981) concluded that coexistence becomes more likely if the patches are more finely divided and if the competitors are aggregated in their distribution between patches independently of one another. Particularly important for coexistence is the marked aggregation of the superior competitor, thus providing more patches in which it is absent and in which the inferior competitor can survive.

HOST–PARASITOID SYSTEMS

Equation (1) for a single-species population in an homogeneous environment can be extended to include the additional effect of mortality caused by a natural enemy. Following previous work (Nicholson & Bailey, 1935; Hassell & May, 1973, 1974; Beddington *et al.,* 1978; May *et al.,* 1981), we will consider principally insect protolean parasites, or parasitoids. Such systems have been the focus of considerable work, both theoretical and experimental [see Hassell (1978) for a review]. By continuing with the discrete-generation framework of the preceding sections and by assuming a coupled, synchronized parasitoid population, we can write the following generalized model:

$$N_{t+1} = Fg[f(N_t, P_t)]N_t f(N_t, P_t), \tag{7a}$$

$$P_{t+1} = cN_t[1 - f(N_t, P_t)]. \tag{7b}$$

Here N and P are the host and parasitoid populations. $Fg[f(N_t, P_t)]$ is the per capita net rate of increase of the host population, with intraspecific competition defined as before by the function g with density dependence for $dg/dN < 0$. The function f defines the proportion of hosts that survive parasitism and embodies the functional and numerical responses of the parasitoid, and c is the average number of adult female parasitoids that emerge from each attacked host (c therefore includes the average number of eggs laid per host parasitized, the survival of these progeny, and their sex ratio). In such discrete-generation frameworks with both host density dependence (g) and parasitism (f), different dynamics can arise depending on the sequence in the host's life cycle in which these occur (Wang & Guttierrez, 1980; May *et al.,* 1981; Hassell & May, 1986). In effect, therefore, the model represents a minimally complicated age-structured host population with pre- and postparasitism stages. Equation (7) describes the particular case of parasitism acting first followed by the density dependence defined

by g [May *et al.* (1981) provide a discussion of alternatives].

Within the framework of Eqs. (7a) and (7b) the degree to which the parasitoid population can reduce the average host population level (leaving aside whether or not this is a stable equilibrium) can be defined by the ratio, q, of average host abundances with and without the parasitoid (i.e., $q = N^*/K$, where K is the carrying capacity of the host population in the absence of the parasitoid and N^* is the parasitoid-maintained equilibrium). The magnitude of this depression depends on the balance between (1) the host's net rate of increase ($Fg[f(N_t, P_t)]$) in Eq. (7a) and (2) the various factors affecting overall parasitoid performance contained within the function $f(N_t, P_t)$ and the term c. These include the per capita searching efficiency and maximal attack rate of adult females, the spatial distribution of parasitism in relation to that of the host (see later), and the sex ratio and survival of parasitoid progeny.

The general framework of Eq. (7) has been explored with many variants for the functions f and g. The original, and most familiar, version is that of Nicholson (1933) and Nicholson and Bailey (1935), where $c = g(N_t) = 1$ (i.e., no host density dependence and a single female parasitoid progeny from each host attacked):

$$N_{t+1} = FN_t\exp(-aP_t), \tag{8a}$$

$$P_{t+1} = N_t\{1 - \exp(-aP_t)\}. \tag{8b}$$

Here $f(N, P)$ is represented by the zeroth term of the Poisson distribution implying that each host is equally susceptible to parasitism by the P_t adult parasitoids, and a is the per capita searching efficiency of the parasitoid (the area of discovery) that sets the proportion of hosts encountered per parasitoid per unit time. The model thus implicitly assumes a type I functional response (Holling, 1959b) with no upper limit to the number of hosts that a parasitoid can successfully parasitize. Handling time is thus assumed to be negligible and parasitoid egg supply to be unlimited. The model assumes that: (1) each host is equally subject to attack (random search); (2) the parasitoids have a linear functional response; (3) each host parasitized produces one female progeny for the next generation (i.e., $c = 1$); and (4) the host population suffers no additional density dependence due, for example, to resource limitation ($g = 1$). The model predicts expanding oscillations of host and parasitoid populations around an unstable equilibrium. The inclusion of a finite handling time, and thus a type II functional response, makes this instability more acute (Hassell & May, 1973).

Because such unstable interactions have only been observed from a few simple laboratory experiments (Burnett, 1958) (Fig. 8), there has been much interest in factors that could be important in promoting the persistence or stability of host–parasitoid interactions. One possibility is that host–parasitoid systems persist by grace of additional den-

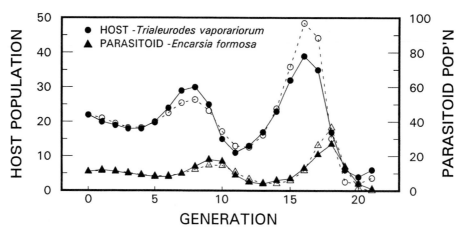

FIGURE 8 Dynamics of a simple host–parasitoid laboratory system consisting of *Trialeurodes vaporariorum* (Westwood) and *Encarsia formosa* Gahan. Solid symbols are observed population numbers; hollow symbols are those predicted from Eq. (7) with $F = 2$, $a = 0.067$, $g = 1$. [After Burnett, T., (1958). *Proceedings of the Tenth International Congress of Entomology, 2, 679–686.*]

sity-dependent factors ($g < 1$) affecting the host population, although mechanisms for this become hard to invoke at low levels of host abundance. Alternatively, the description of parasitism in Eq. (8) may be inadequate and parasitism itself may be a regulatory process. A number of such regulatory mechanisms have been proposed, including sigmoid functional responses (Murdoch & Oaten, 1975; Nunney, 1980), mutual interference between searching predators (Hassell & Varley, 1969; Hassell & May, 1973), density-dependent sex ratios (Hassell *et al.*, 1983; Comins & Wellings, 1985), and heterogeneity in the distribution of parasitism or predation among the prey population (see later).

An example of how readily functional forms of $f(N_t, P_t)$ can be found that stabilize host–parasitoid interactions is given by May (1978), in which the distribution of parasitoid attacks among hosts is described by a clumped distribution, the negative binomial, instead of the independently random Poisson. Host survival, f, in Eq. (7) is thus given by the zeroth term of this distribution (May, 1978)

$$f(N, P) = \left[1 + \frac{aP}{k} \right]^{-k} \qquad (9)$$

Here k describes the contagion in the distribution of parasitoid attacks among host individuals. Contagion increases as $k \to 0$, whereas in the opposite limit of $k \to \infty$ attacks become distributed independently and the Poisson distribution [Eq. (8)] is recovered. As May and Hassell (1988) have discussed, the outcome of a parasitoid's searching behavior cannot usually be fully characterized as simply as Eq. (9) (cf. Hassell & May, 1974; Chesson & Murdoch, 1986; Perry & Taylor, 1986; Kareiva & Odell, 1987). Nonetheless, the use of Eq. (9) with a constant k permits the dynamic effects of nonrandom or aggregated parasitoid

searching behavior to be explored without introducing a large list of behavioral parameters. More complex cases, such as the value of k varying with host density, can be considered (Hassell, 1980), but have little effect on the dynamic properties of the host–parasitoid interaction.

The simple change from independently random search in Eq. (8) to the more general case of Eq. (9) that includes some heterogeneity in the risk of parasitism between individual hosts (see later) can have profound effects on the dynamics of the interaction. The populations are stable for $k < 1$ and show increasing oscillations for $k > 1$ (Fig. 9). A modification of this model in which k is a function of average host density per generation is described by Hassell (1980) in relation to the winter moth, *Operophtera brumata* Linnaeus, in Nova Scotia parasitized by the tachinid, *Cyzenis albicans* (Gravenhorst) (cf. Embree, 1966).

Host Density Dependence

The preceding discussion has assumed no additional host density dependence ($g = 1$). This is likely to be appropriate for many situations, particularly where biological control agents are established and populations are maintained substantially below their environmentally determined carrying capacity. In other cases, however, we must address the relative contributions to regulation of both host density dependence and parasitism.

The framework presented in Eq. (7) can be used to explore the joint effects of density dependence in the host and the action of parasitism (Beddington *et al.*, 1975; May *et al.*, 1981). One important feature of such discrete systems incorporating both host and parasitoid density dependence is that the outcomes of the interactions will depend on whether the parasitism acts before or after the density de-

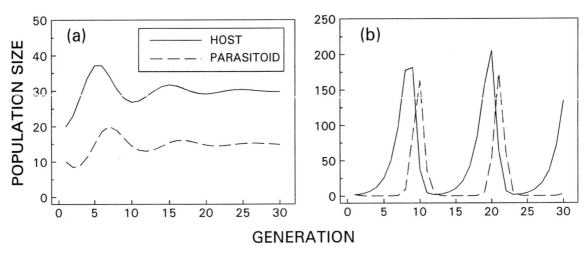

FIGURE 9   Examples of (a) stable ($F = 2$, $a = 0.1$, $k = 0.5$) and (b) unstable ($F = 2$, $a = 0.1$, $k = 1$) dynamics of the parasitoid–host model with negative binomial distribution of attacks [Eqs. (7) and (9)].

pendence in the host population. May *et al.* (1981) envisaged two general cases, the first where host density dependence acts first and the second where parasitism acts first (their models 2 and 3). They employed Eq. (9) for function f, with host density dependence described by $g = \exp(-xN_t)$. With host density dependence acting before parasitism, we have

$$N_{t+1} = Fg(N_t)N_t f(P_t), \tag{10a}$$

$$P_{t+1} = N_t g(N_t)\{1 - f(P_t)\}; \tag{10b}$$

and with parasitism acting first:

$$N_{t+1} = Fg[f(P_t)]N_t f(P_t), \tag{11a}$$

$$P_{t+1} = N_t[1 - f(P_t)]. \tag{11b}$$

Beddington *et al.* (1975) and May *et al.* (1981) have explored the stability properties of such interactions in terms of two biological features of the system—the host's intrinsic rate of increase (log F) and the level of the host equilibrium in the presence of the parasitoid (N^*) relative to the carrying capacity of the environment (K, the host equilibrium due only to host density dependence in the absence of parasitism). This ratio between the parasitoid-induced equilibrium N^* and K is termed q, $q = N^*/K$.

When Eq. (9) is employed for function f, the behavior of these systems with both density dependence and parasitism depends on three parameters: the host rate of increase (log F), the degree of suppression of the host equilibrium (q), and the degree of contagion in the distribution of parasitoid attacks (k of Eq. 9). The specific relationships between these parameters that determine the dynamics of the system depend on whether parasitism occurs before or after density dependence in the life cycle of the host (Fig. 10). In both cases the degree of host suppression possible increases with increased contagion of attacks. The new par-

asitoid-caused equilibrium may be stable or unstable, and for unstable equilibria the populations may exhibit geometric increase or oscillatory or chaotic behavior (Fig. 11). For density dependence acting after parasitism (Fig. 10 c and d) and for $k < 1$ any population reduction is stable. May *et al.* (1981) provide additional detail on the possible dynamical outcomes of such systems, but in general much of the parameter space for both cases implies a stable reduced population whenever $k < 1$, that is, whenever parasitism is more clumped than in the independently random Poisson distribution.

Spatial Patterns of Parasitism

In the same way that we have treated single species and competing species in heterogeneous or patchy environments, we now turn to host–parasitoid systems where the host populations are distributed among discrete patches. In this way we consider in more detail the consequences of clumped or aggregated distributions of natural enemy attacks in the environment. The early work by Hassell and May (1973, 1974) led the way for much of this work (see also Holt & Hassell, 1993; Jones *et al.,* 1993).

The consequences of heterogeneous and patchy host distributions on the dynamics of the host–parasitoid system depend importantly on the variation in the distribution of parasitism among patches. Several mechanisms exist that tend to lead to variable parasitism rates per patch. For example, aggregations of natural enemies in patches of relatively high prey densities may result from such patches being more easily discovered by natural enemies (Sabelis & Laane, 1986), or from changes in search behavior on discovery of a host (e.g., Hassell & May, 1974; Waage, 1979).

The considerable interest in the effects of heterogeneity

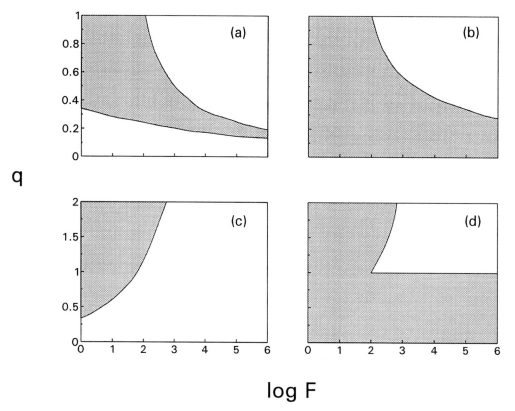

q

log F

FIGURE 10 Regions of stable and unstable population dynamics for Eqs. (10a), (10b), (11a) and (11b), where both parasitism and density dependence act on the host population. In (a) and (b) density dependence acts before parasitism, while in (c) and (d) density dependence acts after parasitism in the host life cycle. In (a) and (c), parasitoid attacks are distributed independently in the host population (Poisson distribution, with negative binomial parameter $k = \infty$). In (b) and (d), attacks are aggregated ($k = 0.5$). Stable two-species equilibria are possible only in the shaded areas. In general, stability is enhanced by aggregated attacks.

on host–parasitoid dynamics has led many workers to record the distribution of parasitism in the field in relation to the local density of hosts per patch. Of 194 different examples listed in the reviews of Lessells (1985), Stiling (1987), and Walde and Murdoch (1988), 58 show variation in attack rates among patches depending directly on host density (Fig. 12 a and c), 50 show inversely density-dependent relationships (Fig. 12 b) and 86 show variation uncorrelated with host density (density independent) (Fig. 12 d and e).

A popular interpretation of these data has been that only the direct density-dependent patterns promote the stability of the interacting populations. This, however, is an incomplete picture. Both inverse density-dependent patterns (Hassell, 1984; Walde & Murdoch, 1988), and variation in parasitism that is *independent* of host density (Chesson & Murdoch, 1986; May & Hassell, 1988; Pacala *et al.*, 1990; Hassell *et al.*, 1991), can also be just as important to population regulation. The reason for this is that *any* variation in levels of parasitism from patch to patch has the net effect

of reducing the per capita parasitoid searching efficiency (measured over all hosts) as average parasitoid density increases [the so-called "pseudo-interference" effect of Free *et al.*, (1977)]. This has obvious implications for the design of field studies on host–parasitoid systems, because no longer can the effects of such heterogeneity be inferred simply from the shape of the relationships between percentage parasitism and local host density.

To explore the dynamic effects of such heterogeneous parasitism, let us consider a habitat that is divided into discrete patches (e.g., food plants for an herbivorous insect) among which adult insects with discrete generations distribute their eggs. The immature stages of these insects are hosts for a specialist parasitoid species whose adult females forage across the patches according to some as-yet-unspecified foraging rule. We also assume that parasitism dominates host mortality such that the hosts are on average kept well below their carrying capacity. With this scenario, Eq. (7) applies with $c = g = 1$ and $f(N_t, P_t)$ representing across all patches the *average* of the fraction of hosts escaping

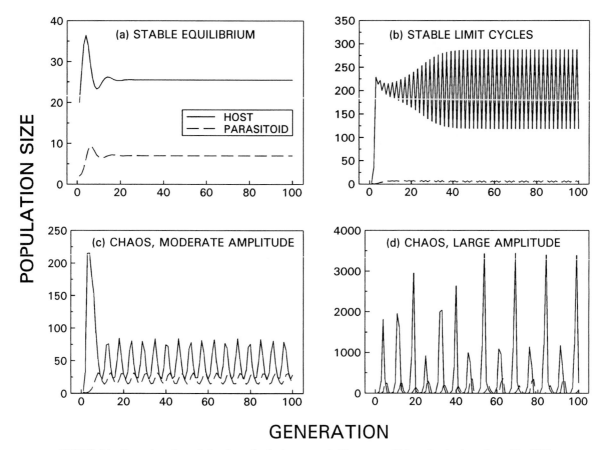

FIGURE 11   Examples of population dynamics for host–parasitoid systems with host density dependence [Eq. (10)]. In all cases $a = 0.1$. (a) Stable equilibrium ($F = 2$, $k = 0.5$, $b = 0.01$); (b) stable limit cycles ($F = 10$, $k = 0.2$, $b = 0.01$); (c) chaotic cycles of moderate amplitude ($F = 10$, $k = 5$, $b = 0.01$); (d) chaotic cycles of larger amplitude ($F = 10$, $k = 2$, $b = 0.001$).

parasitism. The distribution of hosts in such a patchy setting either can be random or can vary in some other prescribed way. Similarly, the density of searching parasitoids in each patch can be either a random variable independent of local host density or a deterministic function of local host density. Pacala *et al.* (1990), Hassell *et al.* (1991), and Pacala and Hassell (1991) call these patterns of heterogeneity in parasitoid distribution *host density-dependent (HDD) heterogeneity* and *host density-independent (HDI) heterogeneity,* respectively. Comparable terms have been coined by Chesson and Murdoch (1986) who labeled models with randomly distributed parasitoids as "pure error models" and those with parasitoids responding to host density in a deterministic way as "pure regression models."

May and Hassell (1988) suggested a very simple and approximate stability condition for model 7 in a patchy environment, namely, that the populations will be stable if the distribution of parasitoids from patch to patch (measured as the square of the coefficient of variation, CV^2) is greater than one. Since then, Pacala *et al.* (1990), Hassell

and Pacala (1990), and Hassell *et al.* (1991) have extended this work and have showed that a very similar criterion applies across a broad range of models in discrete time. Their criterion differs in that the density of searching parasitoids per patch is now weighted by the number of hosts in that patch.

The $CV^2 > 1$ rule is in terms of the distribution of searching parasitoids. Such data, however, are rarely available from natural populations; most of the information is in the form of relationships between percentage parasitism and host density per patch (Fig. 12). Pacala and Hassell (1991), however, show that it is possible to estimate values of all the parameters necessary to calculate the CV^2 from such data. In particular, they show that $CV^2 > 1$ from this general model may be approximated as:

$$CV^2 = C_I C_D - 1 \qquad (12)$$

Here $C_I = 1 + \sigma^2$ where σ^2 is the variance of a gamma-distributed random variable describing the density-independent component of parasitoid distribution, and $C_D =$

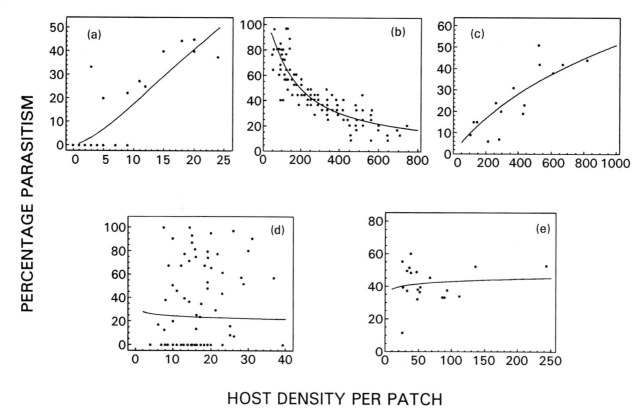

FIGURE 12 Examples of field studies showing percentage of parasitism as a function of host density. (a) The eucoilid *Trybliographa rapae* (Westwood) parasitizing the anthomyiid *Delia radicum* (Linnaeus) (Jones & Hassell, 1988). (b) The encyrtid *Ooencyrtus kuwanai* (Howard) parasitizing the lymantriid *Lymantria dispar* (Linnaeus) (Brown & Cameron, 1979). (c) The aphelinid *Aspidiotiphagus citrinus* (Craw) parasitizing the diaspidid *Fiorinia externa* Ferris (McClure, 1977). (d) The eulophid *Tetrastichus* sp. parasitizing the cecidomyiid *Rhopalomyia californica* Felt (Ehler, 1986). (e) The aphelinid *Coccophagoides utilis* Doutt parasitizing the diaspidid *Parlatoria oleae* (Colvée) (Murdoch *et al.*, 1984). Curves are predicted values given by Hassell & Pacala (1990) from a host density-independent probabilistic model relating percentage parasitism to host density per patch. [After Hassell, M. P., & Pacala, S. (1990). *Philosophical Transactions of the Royal Society of London, Series B, 330,* 203–220.]

$1 + \gamma^2\mu^2$ where γ^2 describes the degree of contagion in the host's distribution and μ describes the strength of any density dependence (positive or negative) in the parasitoid response to host distribution. Full details are given in Pacala and Hassell (1991).

Hassell and Pacala (1990) analyzed 65 examples from field studies reporting percentage of parasitism versus local host density per patch. For each they obtained estimates of σ^2, μ, and γ^2, and then calculated C_I, C_D, and CV^2. It was also possible in each case to predict the mean percentage of parasitism in relation to host density per patch from the expression that provides the fitted lines in Fig. 12. Interestingly, of the five examples in Fig. 12, heterogeneity in parasitism is sufficient (if typical of the interactions) to stabilize the dynamics only in Fig. 12 a and d. The inverse pattern in Fig. 12 b would have been sufficient had the host population been more clumped in its distribution.

On surveying the results of the 65 examples, Pacala and Hassell (1991) found that in 18 of the 65 cases $CV^2 > 1$,

indicating that heterogeneity at this level ought to be sufficient to stabilize the populations. Interestingly, in 14 of these 18 cases heterogeneity in C_I alone was sufficient to make $CV^2 > 1$. Thus, contrary to the popular view, this analysis suggests that density-independent spatial patterns of parasitism (e.g., Fig. 12 d) may be more important in promoting population regulation than density-dependent patterns.

The work of Pacala and Hassell shows how relatively simple models of host–parasitoid interactions can be applied to field data on levels of parasitism in a patchy environment. Such heterogeneity has often been regarded as a complicating factor in population studies, and one that rapidly leads to analytical intractability of models. Clearly, this need not necessarily be so. The $CV^2 > 1$ rule explains the consequences of heterogeneity for population dynamics in terms of a simple description of the heterogeneity itself. The rule gives a rough prediction of the effects of heterogeneity and also identifies the kinds of heterogeneity that contribute to population regulation. Taylor (1993) reviews the application of the CV^2 rule as an expression of "aggregation of risk" of attack by a parasitoid and discusses further applications to data on parasitism from natural systems.

Age-Structured Systems

Insects grow through distinct developmental stages, and hence the concepts of age and stage structure are linked (more closely in some systems than in others). Many of the general models described in the previous sections take some account of developmental stages [Eqs. (4), (10), and (11)]. A more detailed treatment of age or stage structure, however, requires different approaches to developing population models and theories of population regulation.

One such approach has been to construct more complex models, often called system or simulation models, which incorporate more biological detail at the expense of analytical tractability. Such models can be used to explore not only population dynamics but also other features such as population developmental rate, biomass and nutrient allocation, community structure, and management of ecosystems. In this chapter, we will be concerned only with those features of such systems that bear on population regulation in ways not directly addressable in the simpler analytical frameworks presented earlier.

Synchrony of Parasitoid and Host Development

Insect populations in continuously favorable environments (laboratory populations, some tropical environments) may develop continuously overlapping generations. Interestingly, in the presence of parasitism as a major cause of mortality such populations often exhibit more-or-less discrete generations (Tothill, 1930; Taylor, 1937; Van der Vecht, 1954; Utida, 1957; Wood, 1968; Hassell & Huffaker, 1969; White & Huffaker, 1969; Metcalf, 1971; Bigger, 1976; Banerjee, 1979; Perera, 1987). Godfray & Hassell (1987, 1989) and Godfray *et al.* (1994) have discussed these findings and have developed both a simulation model and a delayed differential equation model in which they consider an insect host population growing in a continuously favorable environment (with no intraspecific density dependence) that passes through both an adult (reproductive) and preimaginal stages.

The dynamical behavior of these systems was characterized by either (1) stable populations in which all stages were continuously present in overlapping generations, (2) populations that were stable but occurred in discrete cycles of approximately the generation period of the host, and (3) unstable populations. These dynamics were dependent principally on two parameters, the degree of contagion in parasitoid attacks, k, and the relative lengths of preimaginal

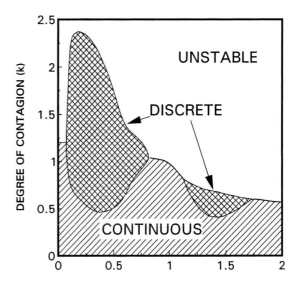

FIGURE 13   Regions of continuous reproduction, discrete generations, and unstable behavior for a system with developmental asynchrony between host and parasitoid. [After Godfray, H. C. J., & Hassell, M. P. (1987). *Nature (London), 327,* 144–147.]

host and parasitoid developmental time (Fig. 13). Low values of k (strong contagion) promoted continuous, stable generations. Moderate values of k (less strong contagion) were accompanied by continuous generations when the parasitoid had developmental times approximately the same length as the host, approximately twice as long, or very short. When developmental times of the parasitoid were approximately half or 1.5 times that of the host, discrete generations arose. Unstable behavior arose for larger values of k. Asynchrony between host and parasitoid can thus be an important factor affecting the dynamic behavior of continuously breeding populations, particularly for parasitoids that develop faster than their hosts.

Parasitism and Competition in Asynchronous Systems

Utida (1950, 1957) described the dynamics of a host–parasitoid system that had unusual dynamic behavior characterized by bounded, but aperiodic, cyclic oscillations that may well prove to be chaotic (Fig. 14 a). The laboratory system consisted of a regularly renewed food source, a phytophagous weevil, and a hymenopteran parasitoid. Salient characteristics of the system were host–parasitoid asynchrony (the parasitoid developed in two-thirds of the weevil developmental time), host density dependence (the weevil adults competed for oviposition sites and larvae for food resources), and age specificity in the parasitoid–host relationship (parasitoids could attack and kill three larval weevil stages and pupae, but could only produce female progeny on the last larval stage and pupae).

Bellows and Hassell (1988) describe a model of a similar system that incorporated detailed age-structured host and parasitoid populations, intraspecific competition among host larvae and adults, and age-specific interactions between host and parasitoid. The dynamics of the model (Fig. 14 b) had characteristics similar to those shown by the experimental populations and distinct from those of any simpler model. In particular, asynchrony of host and parasitoid, the attack by the parasitoid of young hosts (on which reproduction was limited to male offspring), and intraspecific competition by the host were all important features contributing to the observed dynamics. The interaction of these three factors caused continual changes in both host density and age–class structure. In generations where parasitoid emergence was synchronized with the presence of late larval hosts, there was substantial host mortality and parasitoid reproduction. This produced a large parasitoid population in the succeeding generation that, when emerging coincident with young host larvae, killed many host larvae but produced few female parasitoids. The reduced host larval population suffered little competition (because of reduced density). This continual change in intensity of competition and parasitism contributed significantly to the cyclic behavior of the system.

Invulnerable Age Classes

The two previous models both incorporated susceptible and unsusceptible stages, ideas that are inherent to any stage-specific modeling construction for insects where the parasitoid attacks a specific stage such as egg, larvae, or pupa. The consequences of invulnerable stages in a population have been considered by Murdoch *et al.* (1987) in relation to the interaction between California red scale [*Aonidiella aurantii* (Maskell)] and its external parasitoid, *Aphytis melinus* DeBach. Their models include both vulnerable and invulnerable host stages, juvenile parasitoids, and adult parasitoids; they also contain no explicit density dependence in any of the vital rates or attack parameters, but did contain time delays in the form of developmental times from juvenile to adult stages of both populations.

Murdoch *et al.* (1987) developed two models, one in which the adult hosts are invulnerable and one in which the juvenile hosts are invulnerable. They found that either model can have stable equilibria (approached either monotonically or via damped oscillations), stable cyclic behavior, or chaotic behavior. The realm of parameter space that permitted stable populations was substantially larger for the model in which the adult was invulnerable than for the model when the juvenile was invulnerable. Whether or not the stabilizing effect on an invulnerable age class was sufficient to overcome the destabilizing influence of parasitoid

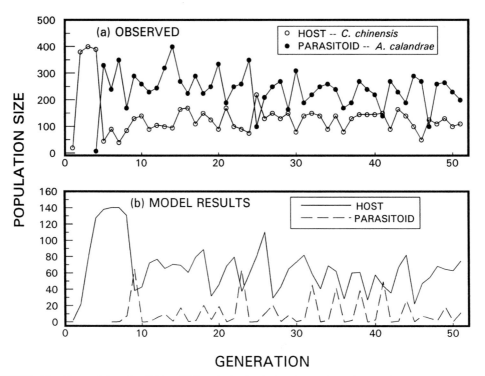

FIGURE 14   Observed (a, after Utida, 1950) and simulated (b, after Bellows & Hassell, 1988) dynamics for laboratory host–parasitoid systems with both developmental asynchrony and host competition.

developmental delay depended on the relative values of several parameters, but short adult parasitoid life span, low host fecundity, and a long adult invulnerable age class all promoted stability.

Many insect parasitoids attack only one or few stages of a host population (although predators may be more general), and hence many populations contain potentially unattacked stages. In addition, however, many insect populations are attacked by more than one natural enemy, and general statements concerning the aggregate effect of a complex of natural enemies attacking different stages of a continuously developing host population are not yet possible. Nonetheless, it appears that, at least for the California red scale–*A. melinus* system, the combination of an invulnerable adult stage and overlapping generations is likely a factor contributing to the observed stability of the system (Reeve & Murdoch, 1985; Murdoch *et al.*, 1987).

Spatial Complexity and Asynchrony in Metapopulations

In predator–prey or parasitoid–host systems that occur in a patchy or heterogeneous environment, we may distinguish between dynamics that occur between the species within a patch and dynamics of the regional or global system. Here we distinguish between "local" dynamics (those within a patch) and "global" dynamics (the characteristics of the system as a whole). Such systems where the entire population is distributed in smaller units, or patches, each with its own intrinsic dynamics, are termed "metapopulations." The properties and dynamics of such metapopulations have received considerable attention, and much research is still addressing questions of biological parameters suited to the investigation of such populations and of suitable mathematical frameworks for use in describing them. Considerable attention has been directed at questions of movement between subpopulations (patches) and at how such movement, combined with unequal distributions of individuals among patches, can affect population dynamics.

While still interested in such dynamical behavior as stability of an equilibrium, we also seek to understand what features of the system might lead to global persistence (the maintenance of the interacting populations) in the face of unstable dynamical behavior at the local level (see review by Taylor, 1988). One set of theories concerned with the global persistence of predator–prey systems emphasizes the importance of asynchrony of local, within-patch, predator–prey cycles (Den Boer, 1968; Reddingius & Den Boer, 1970; Levin, 1974, 1976; Maynard Smith, 1974; Crowley, 1977, 1978, 1981). In this context, asynchrony between patches implies that, on a regional basis, unstable predator–prey cycles may be occurring in each patch but out of phase with one another (prey populations may be increasing in

some fraction of the environment while they are being driven to extinction by predators in another). Such asynchrony may reduce the likelihood of global extinction and thus promote the persistence of the populations. (This asynchrony in populations among patches is distinct from the developmental asynchrony between host and parasitoid of the three preceding systems.)

An example of one such system comes from the model of interacting populations of the spider mite, *Tetranychus urticae* Koch, and the predatory mite, *Phytoseiulus persimilis* Athias-Henriot, described by Sabelis and Laane (1986). This is a regional model of a plant–phytophage–predator system that incorporates patches of plant resource that may be colonized by dispersing spider mites; colonies of spider mites may in turn be discovered by dispersing predators. The dynamics of the populations within the patch are unstable (Sabelis, 1981; Sabelis *et al.*, 1983; Sabelis & van der Meer, 1986), with overexploitation of the plant by the spider mite leading to decline of the spider mite population in the absence of predators. When predators are present in a patch, they consume prey at a rate sufficient to cause local extinction of the prey and subsequent extinction of the predator.

In contrast to the local dynamics of the system, the regional or global dynamics of the system was characterized by two states. In one, the plant and spider mite coexisted but exhibited stable cycles (driven by the intraspecific depletion of plant resource in each patch and the time delay of plant regeneration). The other state was one in which all three species coexisted. This latter case was also characterized by stable cycles, but these were primarily the result of predator–prey dynamics; the average number of plant patches occupied by mites in the three-species system was less than 0.01 of the average number occupied by spider mites in the absence of predators. The overall system thus persisted despite the unstable dynamics at the patch level (Fig. 15). Principal among the models features that contributed to global persistence was asynchrony of local cycles, making it very unlikely that prey would be eliminated in all patches at the same time. This asynchrony could be disturbed when the predators became so numerous that all prey patches were likely to be simultaneously discovered, thereby leading to global extinction of both prey and predator.

These results are broadly in accord with the classic experiments of Huffaker (1958) and Huffaker *et al.* (1963) (Fig. 16), who found that increasing spatial heterogeneity enhanced population persistence. Three features of these experiments are in accord with the behavior of the model of Sabelis and Laane (1986): (1) overall population numbers did not converge to an equilibrium value but oscillated with a more or less constant period and amplitude; (2) facilitation of prey dispersal relative to predator dispersal enhanced the persistence of the populations (Huffaker,

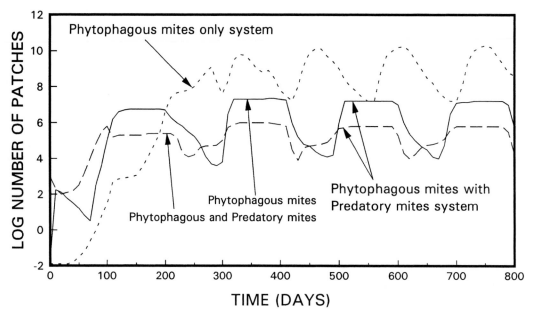

FIGURE 15 Regional dynamics of a system, or simulation, model for a spatially discrete predator–prey system (metapopulation). Dotted line is number of plant patches occupied by mites for system with only plants and phytophagous mites. Solid line (patches with only phytophagous mites) and dashed line (patches with either both phytophagous mites and predatory mites, or only predatory mites) are for system with plants, prey, and predator. Note the reduction in the number of patches occupied by phytophagous mites by more than two orders of magnitude when the predator is included in the system. [After Sabelis, M. W., & Laane, W. E. M. (1986). In J. A. Metz & O. Diekmann (Eds.), *The Dynamics of Physiologically Structured Populations.* New York: Springer-Verlag.]

1958); and (3) increased food availability per prey patch resulted in increased predator production at times of high prey density, which in turn led to synchronization of the local cycles and then to regional extinction (Huffaker *et al.,* 1963).

Results reported in larger scale systems, particularly greenhouses, include elimination of prey and subsequently of predator (e.g., Chant, 1961; Bravenboer & Dosse, 1962; Laing & Huffaker, 1969; Takafuji, 1977; Takafuji *et al.,* 1981a, 1981b), fluctuations of varying amplitude (Hamai & Huffaker, 1978), and wide fluctuations of increasing amplitude (Burnett, 1979; Nachman, 1981). Interpreting these

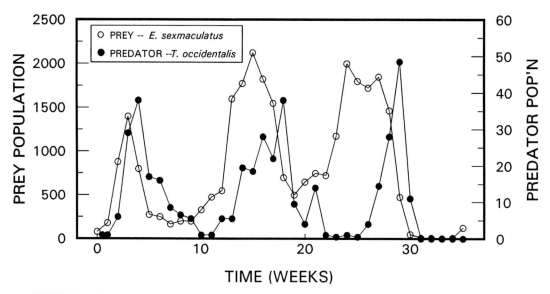

FIGURE 16 Observed dynamics of a mite predator–prey system in the laboratory. [After Huffaker, C. B. (1958). *Hilgardia, 27,* 343–383.]

results requires caution because of differences in scale, relation of the experimental period to the period of the local cycles, and relative differences in ease of prey and predator redistribution in different systems. Nonetheless, it is clear that asynchrony among local patches can play an important role in conferring global stability or persistence to a system composed of locally unstable population interactions. These general findings are supported in additional theoretical and experimental work (e.g., Walde, 1994; Ruxton, 1996; Stelter *et al.*, 1997; Ruxton *et al.*, 1997) both for single-species metapopulations and for parasitoid–host systems.

In addition to the issues of asynchrony among local populations, much theoretical attention has been given to the effects of nonuniform distribution of individuals among patches, the effects of density dependence, and the movement among patches for both single-species and for host–parasitoid systems (e.g., Murdoch & Stewart-Oaten, 1989; Murdoch *et al.*, 1992; Godfray & Pacala, 1992; Holt & Hassell, 1993; Jones *et al.*, 1993; Hassell *et al.*, 1995; Hanski *et al.*, 1996; Holyoak & Lawler, 1996; Murdoch *et al.*, 1996). While the specific findings often depend on details in the formulation of the models used (cf. Murdoch & Stewart-Oaten, 1989; Godfray & Pacala, 1992), many of the findings support the general concept that the dynamics of interpatch movement and the intrinsic population dynamics within each patch each have an effect on the overall system behavior. Generally, systems are more stable with more patches, and the dynamics of the system as a whole move closer to the dynamic behavior within patches as movement between patches becomes more and more instantaneous. Mortality during movement between patches (Ruxton *et al.*, 1997), aggregation among patches, density dependence within patches or in the metapopulation as a whole, and refuges (explicit or implicit in the model) can all contribute to stability (Murdoch *et al.*, 1992; Godfray & Pacala, 1992; Holt & Hassell, 1993; Jones *et al.*, 1993; Hassell *et al.*, 1995, Hanski *et al.*, 1996; Holyoak & Lawler, 1996; Murdoch *et al.*, 1996).

Generalist Natural Enemies

The preceding discussions have focused on natural enemies that are specific and synchronized with their hosts. Many species of natural enemies, however, are generalists and feed or reproduce on a variety of different hosts, making their population dynamics more or less independent of a particular host population.

Equation (11a) may be modified to represent a host population subject to a generalist natural enemy,

$$N_{t+1} = FN_t[\{1 + aG_t/[k(1 + aT_hN_t)]\}^{-k}], \qquad (13)$$

where G_t is the number of generalist natural enemies attacking N_t hosts, and the other parameters have the same meaning as before. This equation embodies a type II functional response (with handling time T_h) for a generalist whose interactions with the host population may be aggregated or independently distributed (depending on the value of k).

Central to such a model are the details of the numerical response of the generalist that determine the values of G_t in Eq. (13). Insofar as they have been reported in the literature, the data tend to show a tendency for G_t to rise with increasing N_t to an upper asymptote (Holling, 1959a; Mook, 1963; Kowalski, 1976) (Fig. 17). This simple relationship may be described by (Southwood & Comins, 1976; Hassell & May, 1986):

$$G_t = m[1 - \exp(-N_t/b)]. \qquad (14)$$

Here m is the saturation number of predators and b determines the prey density at which the number of predators reaches a maximum. Such a numerical response implies that the generalist population responds to changes in host density quickly relative to the generation time of the host, as might occur from rapid reproduction relative to the timescale of the host or by switching from feeding on other prey to feeding more prominently on the host in question (Murdoch, 1969; Royama, 1979). The complete model for this host–generalist interaction [incorporating Eq. (14) into Eq. (13)] now becomes

$$N_{t+1} = FN_t\left[1 + \frac{am[1 - \exp(-N_t/b)]}{k(1 + aT_hN_t)} \right]^{-k}. \qquad (15)$$

This equation represents a reproduction curve with implicit density dependence. Hassell and May (1986) present an analysis of this interaction together with the following conclusions. First, the action of the generalist reduces the growth rate of the host population (which in the absence of the natural enemy grows without limit in this analysis). Whether the growth rate has been reduced sufficiently to produce a new equilibrium depends on the attack rate a and maximum number of generalists m being sufficiently large relative to the host fecundity F. The host equilibrium decreases as predation by the generalist becomes less clumped, as the combined effect of search efficiency and maximum number of generalists (the overall measure of natural enemy efficiency am) increases, and as the host fecundity (F) decreases. A new equilibrium may be stable or unstable (in which case populations will show limit cycle or chaotic dynamics). These latter persistent but nonsteady state interactions can arise when the generalists cause sufficiently severe density-dependent mortality, promoted by low degrees of aggregation (high values for k), large am, and intermediate values of host fecundity F.

HOST–PATHOGEN SYSTEMS

Insect populations can be subject to infection by viruses, bacteria, protozoa, and fungi, the effects of which may vary from reduced fertility to death. In many cases these have

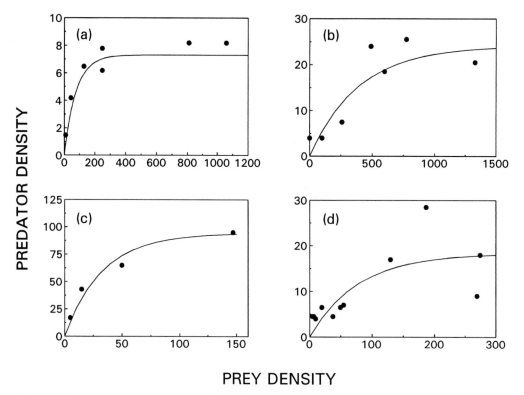

FIGURE 17 Numerical responses by generalist predators. (a) *Peromyscus maniculatus* How & Kennicott, and (b) *Sorex cinereus* Kerr (both as numbers per acre) in relation to the density of larch sawfly [*Neodiprion sertifer* (Geoffroy)] cocoons (thousands per acre) (after Holling, 1959a). (c) The bay-breasted warbler *(Dendroica fusca)* (nesting pairs per 100 acres) in relation to third instar larvae of the spruce budworm (*Choristoneura fumiferana* (Clemens)) (numbers per 10 ft² of foliage) (after Mook, 1963). (d) The staphylinid *Philonthus decorus* (Gr.) (pitfall trap index) in relation to winter moth (*Operophtera brumata* (Linnaeus)) larvae per square meter (after Kowalski, 1976).

been intentionally manipulated against insect populations; reviews of case studies have been presented by Tinsley and Entwistle (1974), Tinsley (1979), Falcon (1982), and Entwistle (1983).

Much of this early work was largely empirical, and a theoretical framework for interactions between insects populations and their pathogens has been developed by Anderson and May (1981) (see also Regniere, 1984; May, 1985; May & Hassell, 1988; Hochberg, 1989; Hochberg *et al.*, 1990).

Let us consider first a host population with discrete, nonoverlapping generations [a univoltine temperate Lepidoptera such as the gypsy moth, *Lymantria dispar* (Linnaeus), and its nuclear polyhedrosis virus disease], affected by a lethal pathogen that is spread in an epidemic fashion via contact between infected and healthy individuals of each generation prior to reproduction. We may apply a variant of Eq. (7) to describe the dynamics of such a population (where $g = 1$ so that there is no other density-dependent mortality):

$$N_{t+1} = FN_t S(N_t). \tag{16}$$

Here $S(N_t)$ represents the fraction escaping infection as an epidemic spreads through a population of density N_t. It is given implicitly by the Kermack–McKendrick expression, $S = \exp\{-(1 - S)N_t/N_T\}$ (Kermack & McKendrick, 1927), where N_T is the threshold host density (which depends on the virulence and transmissibility of the pathogen) below which the pathogen cannot maintain itself in the population. For populations of size N_t less than N_T, the epidemic cannot spread and the population consequently grows geometrically while the infected fraction $1-S$ decreases to ever smaller values. As the population continues to grow, it eventually exceeds N_T and the epidemic can again spread. This very simple system has very complicated dynamical behavior; although completely deterministic, it has neither stable equilibrium nor stable cycles, but exhibits chaotic behavior (where the population fluctuates between relatively high and low densities) in an apparently random sequence (Fig. 18) (May, 1985).

Hochberg *et al.* (1990) have explored an extension of Eq. (16) where transmission is via free-living stages of the pathogen (rather than direct contact between diseased and healthy individuals). They found that the ability of a path-

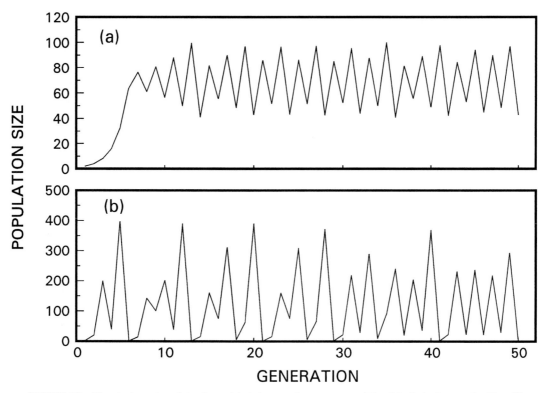

FIGURE 18   Chaotic dynamics of the deterministic host–pathogen system of Eq. (16). In both examples $N_T = 50$ and the starting population size was $N_1 = 2$. In (a) $F = 2$; in (b) $F = 10$.

ogen to produce long-lived external stages contributed to the persistence of the pathogen and tended to dampen the chaos described by Eq. (16). However, if stages were too long-lived, then the pathogen could build up a reservoir resulting in population fluctuations of long period.

Additionally, many such populations may have generations that overlap to a sufficient degree that differential, rather than difference, equations are a more appropriate framework for their analysis. The study of many insect host–pathogen systems have thus been framed in differential equations. Following Anderson and May (1980, 1981), we first assume that the host population has constant per capita birth rate a and death rate (from sources other than the pathogen) b. We divide the host population $N(t)$ into uninfected $[X(t)]$ and infected $[Y(t)]$ individuals, so that $N = X + Y$. For consideration of insect systems, the model does not require the separate class of individuals that have recovered from infection and are immune, as may be required in vertebrate systems, because current evidence does not clearly indicate that insects are able to acquire immunity to infective agents. This basic model further assumes that infection is transmitted directly from infected to uninfected hosts at a rate characterized by the parameter β, so that the rate at which new infections arise is βXY (Anderson & May, 1981). Infected hosts either recover at rate g or die from the disease at rate γ. Both infected and healthy hosts

continue to reproduce at rate a and to be subject to other causes of death at rate b.

The dynamics of the infected and healthy portions of the population are now characterized by

$$dX/dt = a(X + Y) - bX - \beta XY + gY, \qquad (17a)$$

$$dY/dt = \beta XY - (\gamma + b + g)Y. \qquad (17b)$$

The healthy host population increases from both births and recovery of infected individuals. Infected individuals appear at rate βXY and remain infectious for average time $1/(\gamma + b + g)$ before they die from disease or other causes or recover. The dynamics of the entire population are characterized by

$$dN/dt = rN - \gamma Y, \qquad (18)$$

where $r = a - b$ is the per capita growth rate of the population in the absence of the pathogen. There is no intraspecific density dependence or self-limiting feature in the host population, so that in the absence of the pathogen the population will grow exponentially at rate r.

We may now consider the consequences of introducing a few infectious individuals into a population previously free from disease. The disease will spread and establish itself provided the right-hand side of Eq. (17b) is positive, or more generally if the total population of the host exceeds a threshold density N_T,

$$N_T = (\gamma + b + g)/\beta. \tag{19}$$

Because the population in this simple analysis increases exponentially in the absence of the disease, the population will eventually increase beyond the threshold. In a more general situation where other density-dependent factors may regulate the population around some long-term equilibrium level K (in the absence of disease), the pathogen can only establish in the population if $K > N_T$ (Anderson & May, 1981).

Once established in the host population, the disease can (in the absence of other density-dependent factors) regulate the population so long as it is sufficiently pathogenic, with $\gamma > r$. In such cases, the population of Eq. (17) will be regulated at a constant equilibrium level $N^* = [\gamma/(\gamma - r)]N_T$. The proportion of the host population infected is $Y^*/N^* = r/\gamma$. Hence, the equilibrium fraction infected is inversely proportional to disease virulence, thereby decreasing with increasing virulence of the pathogen. If the disease is insufficiently pathogenic to regulate the host ($\gamma < r$), then the host population will increase exponentially at the reduced per capita rate $r' = r - \gamma$ (until other limiting factors affect the population).

The relatively simple system envisaged by Eq. (17) permits some additional analysis. First, pathogens cannot in general drive their hosts to extinction, because the declining host populations eventually fall below the threshold for maintenance of the pathogen. Additionally, we may consider what features of a pathogen might be implicated in maximal reduction of pest density to an equilibrium regulated by the disease, most particularly, what degree of pathogenicity produces optimal host population suppression. Pathogens with low or high virulence lead to high equilibrium host populations, while pathogens with intermediate virulence lead to optimal suppression (Anderson & May, 1981). This is an important point, because many control programs (and indeed many genetic engineering programs) often begin with an assumption that high degrees of virulence are desirable qualities. While this may be true in some special cases of inundation, it is not true for systems that rely on any degree of perpetual host–pathogen interaction (Anderson, 1982; May & Hassell, 1988; Hochberg et al., 1990).

A number of potentially important biological features are not considered explicitly in Eq. (17) (Anderson & May, 1981). Several of these have fairly simple consequences for the general conclusions presented earlier. Pathogens may reduce the reproductive output of infected hosts prior to their death (which renders the conditions for regulation of the host population by the pathogen less restrictive). Pathogens may be transmitted between generations (vertically) from parent to unborn offspring (which reduces N_T and thus

permits maintenance of the pathogen in a lower density host population). The pathogen may have a latency period where infected individuals are not yet infectious (which increases N_T and also makes population regulation by the pathogen less likely). The pathogenicity of the infection may depend on the nutritional state of the host, and hence indirectly on host density. Under these conditions the host population may alternate discontinuously between two stable equilibria. Anderson and May (1981) give further attention to these cases.

A more significant complication arises when the free-living transmission stage of the pathogen is long-lived relative to the host species. Such is the case with the spores of many bacteria, protozoa, and fungi and the encapsulated forms of many viruses (Tinsley, 1979). Most of the conclusions from Eq. (17) still hold, but the regulated state of the system now may be either a stable point or a stable cycle with period of greater than two generations. Anderson and May (1981) show that the cyclic solution is more likely for organisms that have high pathogenicity [and many insect pathogens are highly pathogenic, see Anderson & May (1981) and Ewald (1987)] and that produce large numbers of long-lived infective stages. The cyclic behavior results from the time delay introduced into the system by the pool of long-lived infectious stages. Such cyclic behavior appears characteristic of populations of several forest Lepidoptera and their associated diseases (Anderson & May, 1981). In one particular case where sufficient data are available to estimate the necessary parameters, there is substantial agreement between the expected and observed period of population oscillation (Anderson & May, 1981; McNamee et al., 1981).

Hochberg (1989) has shown that heterogeneity in the structure of the pathogen population can have important effects on pathogen persistence and the ability of pathogens to regulate their hosts. In particular, the formation of a "reservoir" of long-lived stages can dampen the tendency to cycle that was identified in the models of Anderson and May (1980, 1981). This buffering effect gives a plausible explanation for the known successes of biological control programs using pathogens (Hochberg, 1989).

MULTISPECIES SYSTEMS

The preceding emphasis has centered on the dynamics of single- and two-species systems, divorced from the more complex webs of multispecies interactions of which they will often be an integral part. If we are to understand how population dynamics can influence the structure of simple communities, it is clearly important in the first place to determine how the dynamics of two interacting species are influenced by the additional linkages typically found with other species in the food web. With this in mind, the dynamics of a wide range of different three-species systems have been examined. These range from a natural enemy species (predator or pathogen) attacking competing prey or host species (e.g., Roughgarden & Feldman, 1975; Comins & Hassell, 1976; Anderson & May, 1986), competing natural enemy species sharing a common prey or host species (e.g., May & Hassell, 1981; Kakehashi *et al.,* 1984; Anderson & May, 1986; Hochberg *et al.,* 1990), a prey species attacked by both generalist and specialist natural enemies (Hassell & May, 1986), and various three trophic level interactions (Beddington & Hammond, 1977, May & Hassell, 1981; Anderson & May, 1988). Interestingly, in some of these systems the dynamics are not just the expected blend of the component two-species interactions. Instead, the additional nonlinearities introduced by the third species can lead to quite unexpected dynamical properties.

We now turn to four examples illustrating how population dynamics are affected when moving from two- to three-species interactions.

Competing Natural Enemies

In many natural systems phytophagous species are attacked by a suite of natural enemies, and plants are attended by a complex of herbivores. In biological control programs, attempts to reconstruct such multiple-species systems have often met with some debate in spite of their ubiquitous occurrence. Some workers have suggested that interspecific competition among multiple natural enemies will tend to reduce the overall level of host suppression (e.g., Turnbull & Chant, 1961; Watt, 1965; Kakehashi *et al.,* 1984). Others view multiple introductions as a potential means to increase host suppression with no risk of diminished control (e.g., Huffaker *et al.,* 1971; van den Bosch & Messenger, 1973; May & Hassell, 1981; Waage & Hassell, 1982). The significance of this issue probably varies in different systems, but the basic principles may be addressed analytically.

The dynamics of a system with a single host and two parasitoids may be addressed by extending the single-host–single-parasitoid model of Eq. (7) (with no host density dependence, $g = 1$) to include an additional parasitoid. One possibility is the case described by the following (May & Hassell, 1981):

$$N_{t+1} = FN_t h(Q_t)f(P_t), \qquad (20a)$$

$$Q_{t+1} = N_t\{1 - g(Q_t)\}, \qquad (20b)$$

$$P_{t+1} = N_t h(Q_t)\{1 - f(P_t)\}. \qquad (20c)$$

Here the host is attacked sequentially by parasitoids Q and P. The functions h and f represent the fractions of the host population surviving attack from Q and P, respectively, and are described by Eq. (9); the distribution of attacks by one species is independent of attacks by the other. Variations

on this theme have also been considered, such as when P and Q attack the same stage simultaneously (May & Hassell, 1981). The general qualitative conclusions, however, are the same.

Three general conclusions arise from an examination of this system. First, the coexistence of the two species of parasitoids is more likely if both contribute in some measure to the stability of the interaction [both species have values of $k < 1$ in Eq. (9)]. Second, if we consider the system with the host and parasitoid P already coexisting and attempt to introduce parasitoid Q, then coexistence is more likely if Q has a searching efficiency higher than P. If Q has too low a searching efficiency, it will fail to become established, precluding coexistence. If the search efficiency of Q is sufficiently high, it may suppress the host population below the point at which P can continue to persist, thus leading to a new single-host–single-parasitoid system. Apparent examples of such competitive displacement include the introductions of parasitoids against *Dacus dorsalis* Hendel in Hawaii (Bess *et al.,* 1961; Fig. 19) and the displacement of *Aphytis lingnanensis* Compere by *A. melinus* in interior southern California (Luck & Podoler, 1985). Finally, the successful establishment of a second parasitoid species (Q) will in almost every case further reduce the equilibrium host population. For certain parameter values, it can be shown that the equilibrium might have been lower still if only the host and parasitoid Q were present, but this additional depression is slight. In general, the analysis points to multiple introductions as a sound biological strategy.

Kakehashi *et al.* (1984) have considered a case similar to Eq. (20), but where the distributions of attacks by the two parasitoid species are identical instead of independent. This assumes, therefore, that the two species of parasitoids respond in the same way to environmental cues involved in locating hosts. This modification does not change appreciably the stability properties of Eq. (20), but does change the equilibrium properties. In particular, a system with the superior parasitoid alone now can cause a greater host population depression than does the three-species system. In natural systems, however, complete covariance of parasitism between species may be less likely than more independent distributions (Hassell & Waage, 1984). Nonetheless, this is an example where general, tactical predictions can be affected by changes in detailed model assumptions.

Generalist and Specialist Natural Enemies

We now turn to interactions between populations of specialist and generalist natural enemies sharing a common host species. As noted earlier, discrete systems with more than one mortality factor may have different dynamics depending on the sequence of mortalities in the host life cycle. We will consider here a situation where the specialist natu-

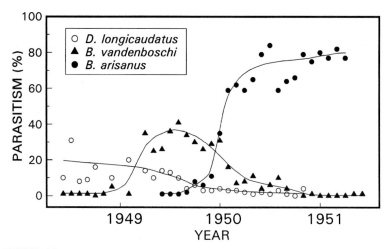

FIGURE 19 The sequential introduction of three fruit fly parasitoids (*Diachasmimorpha longicaudatus* [Ashmead], *Biosteres vandenboschi* [Fullaway], and *Biosteres arisanus* [Sonan]) into Hawaii shows the sequential competitive replacement of one species by another together with increased overall parasitism rates. [After Bess, H. A., *et al.* (1961). *Proceedings of the Hawaiian Entomological Society, 17,* 367–378.]

ral enemy acts first, followed by the generalist, both preceding reproduction of the adult host (Hassell & May, 1986):

$$N_{t+1} = FN_t f(P_t)g[N_t f(P_t)], \tag{21a}$$

$$P_{t+1} = N_t\{1 - f(P_t)\}. \tag{21b}$$

Here $g(N_t)$ is the host survival from the generalist that incorporates a numerical response together with the negative binomial distribution of attacks (see earlier discussion). By assuming handling time to be small relative to the total searching time available, we have

$$g(N) = \left[1 + \frac{am[1 - \exp(-N/b)]}{k}\right]^{-k}. \tag{22}$$

The function $f(P)$ is the proportion surviving parasitism and, again assuming a negative binomial distribution of parasitoid attacks, is given by

$$f(P) = [1 + a'P/k]^{-k'}. \tag{23}$$

We now consider the conditions for generalist and specialist to coexist and their combined effects on the host population. In particular, a specialist natural enemy can coexist with the host and generalist most easily if the effect of the generalist is small (k and am are small, indicating low levels of highly aggregated attacks), if the efficiency of the specialist is high, and if there is low-density dependence in the numerical response of the generalist (Hassell & May, 1986). In simple terms, if the effect of the generalist is small, there is greater potential that the host population can support an additional natural enemy (the specialist). Conversely, if the host rate of increase F is low or the efficiency of the generalist population (am) is too high, then a specialist is unlikely to be able to coexist in the host–generalist

system. In general, the parameter values allowing coexistence of the specialist and generalist are somewhat more relaxed for the case of the specialist acting before the generalist in the host life history, because there are then more hosts present for the specialist to attack. In each case the equilibrium population of the host is further reduced by adding the third species. Further details are given in Hassell and May (1986).

Parasitoid–Pathogen–Host Systems

We now consider interactions in which a host is attacked by both a parasitoid (or predator) and a pathogen (Carpenter, 1981; Anderson & May, 1986; May & Hassell, 1988; Hochberg *et al.*, 1990). These may be considered as cases of two-species competition, where the natural enemies compete for the resource represented by the host population. As for interspecific competition discussed earlier, the systems are characterized by these possible outcomes: the parasitoid and pathogen may coexist with the host, either parasitoid or pathogen may regulate the population at a density below the threshold for maintenance of the other agent, or there may be two alternative stable (or unstable) states (one with host and parasitoid and one with host and pathogen). The outcome of any particular situation may depend on the initial condition of the system.

Consider a population that is first attacked by a lethal pathogen (spread by direct contact) and then the survivors are attacked by parasitoids, represented by combining the models of Eq. (7) (with $g = 1$) and Eq. (16):

$$N_{t+1} = FN_t S(N_t)f(P_t), \tag{24a}$$

$$P_{t+1} = cN_t S(N_t)[1 - f(P_t)]. \tag{24b}$$

Here $S(N)$ is the fraction surviving the epidemic given earlier [Eq. (16)] by the implicit relation $S = \exp[-(1 - S)N_t/N_T]$, f has the Nicholson–Bailey form $f(P) = \exp(-aP)$ representing independent, random search by parasitoids, F is the per capita host reproductive rate, and c is the number of parasitoids produced by a single parasitized host.

The dynamic character of this system has been summarized by May and Hassell (1988) and has been considered in some detail by Hochberg *et al.* (1990). For $acN_T(\ln F)/(F - 1) < 1$, the pathogen excludes the parasitoid by maintaining the host population at levels too low to sustain the parasitoid. For parasitoids with greater searching efficiency or with greater degrees of gregariousness, or for systems with higher thresholds (N_T), so that $acN_T(\ln F)/(F - 1) > 1$, a linear analysis suggests that the parasitoid would exclude the pathogen in a similar manner. However, the diverging oscillations of the Nicholson–Bailey system eventually lead to densities higher than N_T, and the pathogen can repeatedly invade the system as the host population cycles to high densities. The resulting dynamics (Fig. 20) can be quite complex, even from the simple and purely deterministic interactions of Eq. (24). Here the basic period of the oscillation is driven by the Nicholson–Bailey model, with the additional effects of the (chaotic) pathogen–host interaction leading to approximately stable (instead of diverging) oscillations. As May and Hassell (1988) discuss, in such complex interactions it can be relatively meaningless to ask whether the dynamics of the system are determined mainly by the parasitoid or by the pathogen. Both contribute significantly to the dynamical behavior, the parasitoid by setting the average host abundance and the period of the oscillations, and the pathogen by providing long-term "stability" in the sense of limiting the amplitude of the fluctuations and thereby preventing catastrophic overcompensation and population extinction. Extensions to the model of May and Hassell where the pathogen is capable of producing long-lived external stages, and where the parasitoid and pathogen compete within host individuals have been explored by Hochberg *et al.* (1990).

Competing Herbivores and Natural Enemies

The presence of polyphagous predators in communities of interspecific competitors can have profound effects on the number of species in the community and the relative roles that predation and competition play in population dynamics. Paine (1966, 1974) demonstrated that intertidal communities contain more species when subject to predation by the predatory starfish, *Pisaster ochraceus* (Brandt), than when it is absent. Since then, there has been much theoretical attention to the relative roles of predation and competition in multispecies communities (e.g., Parrish & Saila, 1970; Cramer & May, 1972; Steele, 1974; van Valen, 1974; Murdoch & Stewart-Oaten, 1975; Roughgarden & Feldman, 1975; Comins & Hassell, 1976; Fujii, 1977; Hassell, 1978, 1979; Hanski, 1981). One general conclusion of this work is that natural enemies, under certain conditions can enable competing species to coexist where they otherwise could not. This effect is enhanced if the natural enemy shows some preference for the dominant competitors or switches between prey species as one becomes more abundant than the other.

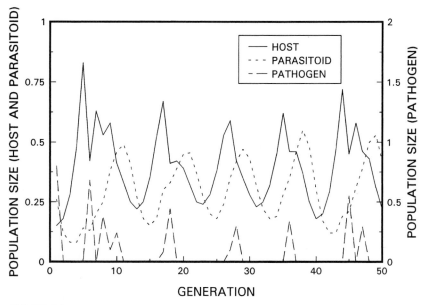

FIGURE 20 Dynamical behavior of host–parasitoid–pathogen system of Eq. (24). In this example $F = 2$, $a = 2$, $c = 1$. [After May, R. M., & Hassell, M. P. (1988). *Philosophical Transactions of the Royal Society of London, Series B, 318*, 129–169.]

This work also has been extended to the case of competing prey and natural enemies existing in a patchy environment. Comins and Hassell (1987) considered the cases of patchily distributed, competing prey and either a generalist natural enemy (whose dynamics were unrelated to the dynamics of the prey community) or a more specialized natural enemy coupled to the competing prey in the community. In both cases, the natural enemy population could, under certain conditions, add stability to an otherwise unstable competition community. This occurred more readily with the generalist than the specialist. In all cases, aggregation by the natural enemy in patches of high prey density (which leads to a "switching" effect) was an important attribute contributing to stability. Predation that was independently random across patches was destabilizing for both the generalist and the specialist cases. Coexistence of competing prey species was possible in this spatially heterogeneous model even when the distributions of the prey species in the environment were correlated and when interspecific competition was extreme.

Acknowledgments

This work was supported in part (T. S. B.) by U. S. National Science Foundation Grant BSR-8604541 and in part (M. P. H.) by the Natural Environment Research Council (United Kingdom). We are grateful to M. Hochberg for his comments on the manuscript.

References

Anderson, R. M. (1982). Theoretical basis for the use of pathogens as biological control of pest species. Parasitology, 84, 3–33.

Anderson, R. M., & May, R. M. (1980). Infectious diseases and population cycles of forest insects. Science, 210, 658–661.

Anderson, R. M., & May, R. M. (1981). The population dynamics of microparasites and their invertebrate hosts. Philosophical Transactions of the Royal Society of London, Series B, 291, 451–524.

Anderson, R. M., & May, R. M. (1986). The invasion, persistence and spread of infectious diseases within animal and plant communities. Philosophical Transactions of the Royal Society of London, Series B, 314, 533–570.

Anderson, R. M., & May, R. M. (1988). The invasion, persistence and spread of infectious diseases within animal and plant communities. Proceedings of the Royal Society of London, Series B, 314, 533–570.

Anonymous. (1980). Biological control Service. 25 years of achievement. Slough, U. K.: Commonwealth Agricultural Bureau.

Atkinson, W. D., & Shorrocks, B. (1977). Breeding site specificity in the domestic species of Drosophila. Oecologia (Berlin), 29, 223–232.

Atkinson, W. D., & Shorrocks, B. (1981). Competition on a divided and ephemeral resource: A simulation model. Journal of Animal Ecology, 50, 461–471.

Ayala, F. J., Gilpin, M. E., & Ehrenfield, J. G. (1974). Competition between species: Theoretical models and experimental tests. Theoretical Population Biology, 4, 311–356.

Banerjee, B. (1979). A key-factor analysis of population fluctuation in Andraca bipunctata (Walker) (Lepidoptera: Bombycidae). Bulletin of Entomological Research, 69, 195–201.

Beddington, J. R., & Hammond, P. S. (1977). On the dynamics of host-parasite-hyperparasite interactions. Journal of Animal Ecology, 46, 811–821.

Beddington, J. R., Free, C. A., & Lawton, J. H. (1975). Dynamic complexity in predator-prey models framed in difference equations. Nature (London), 255, 58–60.

Beddington, J. R., Free, C. A., & Lawton, J. H. (1978). Characteristics of successful natural enemies in models of biological control of insect pests. Nature (London), 273, 513–519.

Bellows, T. S., Jr. (1981). The descriptive properties of some models for density dependence. Journal of Animal Ecology, 50, 139–156.

Bellows, T. S., Jr. (1982a). Analytical models for laboratory populations of Callosobruchus chinensis and C. maculatus (Coleoptera, Bruchidae). Journal of Animal Ecology, 51, 263–287.

Bellows, T. S., Jr. (1982b). Simulation models for laboratory populations of Callosobruchus chinensis and C. maculatus (Coleoptera, Bruchidae). Journal of Animal Ecology, 51, 597–623.

Bellows, T. S., Jr. (1993). Introduction of natural enemies for suppression of arthropod pests. In R. D. Lumsden and J. L. Vaughn (Eds.), Pest management: Biologically based technologies (pp. 82–89). Washington, DC: American Chemical Society.

Bellows, T. S., Jr., & Hassell, M. P. (1984). Models for interspecific competition in laboratory populations of Callosobruchus spp. Journal of Animal Ecology, 53, 831–848.

Bellows, T. S., Jr., & Hassell, M. P. (1988). The dynamics of age-structured host-parasitoid interactions. Journal of Animal Ecology, 57, 259–268.

Bellows, T. S., Paine, T. D., Gould, J. R., Bezark, L. G., Ball, J. C., & Bentley, W. (1992). Biological control of ash whitefly: A success in progress. California Agriculture, (Jan.–Feb.), 46(1), 24, 27–28.

Bess, H. A., van den Bosch, R., & Haramoto, F. H. (1961). Fruit fly parasites and their activities in Hawaii. Proceedings of the Hawaiian Entomological Society, 17, 367–378.

Bigger, M. (1976). Oscillation of tropical insect populations. Nature (London), 259, 207–209.

Bravenboer, L., & Dosse, G. (1962). Phytoseiulus riegeli Dosse als prädator einiger schadmilben aus der Tetranychus urticae-gruppe. Entomologia Experimentalis et Applicata, 5, 291–304.

Brown, M. W., & Cameron, E. A. (1979). Effects of disparlure and egg mass size on parasitism by the gypsy moth egg parasite, Ooencyrtus kuwanai. Environmental Entomology, 8, 77–80.

Burnett, T. (1958). A model of host-parasite interaction. Proceedings of the Tenth International Congress of Entomology, 2, 679–686.

Burnett, T. (1979). An acarine predator-prey population infesting roses. Researches on Population Ecology (Kyoto), 20, 227–234.

Carpenter, S. R. (1981). Effect of control measures on pest populations subject to regulation by parasites and pathogens. Journal of Theoretical Biology, 92, 181–184.

Chant, D. A. (1961). An experiment in biological control of Tetranychus telarius (L.) (Acarina: Tetranychidae) in a greenhouse using the predaceous mite Phytoseiulus persimilis Athias-Henriot (Phytoseiidae). Canadian Entomologist 93, 437–443.

Chesson, P. L. & Murdoch, W. W. (1986). Relationships among host-parasitoid models. American Naturalist, 127, 696–715.

Comins, H. N., & Hassell, M. P. (1976). Predation in multi-prey communities. Journal of Theoretical Biology, 62, 93–114.

Comins, H. N., & Hassell, M. P. (1987). The dynamics of predation and competition in patchy environments. Theoretical Population Biology, 31, 393–421.

Comins, H. N., & Wellings, P. W. (1985). Density-related parasitoid sex ratios: Influence on host-parasitoid dynamics. Journal of Animal Ecology, 54, 583–594.

Cramer, N. F., & May, R. M. (1972). Interspecific competition, predation and species diversity: A comment. Journal of Theoretical Biology, 34, 289–290.

Crawley, M. J. (1983). Herbivory, the dynamics of animal-plant interactions. Berkeley: University of California Press.

Crawley, M. J. (1989). Insect herbivores and plant population dynamics. Annual Review of Entomology, 34, 531–564.

Crombie, A. D. (1945). On competition between different species of gramnivorous insects. Proceedings of the Royal Society of London, Series B, 132, 362–395.

Crombie, A. D. (1946). Further experiments on insect competition. Proceedings of the Royal Society of London, Series B, 133, 76–109.

Crowley, P. H. (1977). Spatially distributed stochasticity and the constancy of ecosystems. Bulletin of Mathematical Biology, 39, 157–166.

Crowley, P. H. (1978). Effective size and the persistence of ecosystems. Oecologia, 35, 185–195.

Crowley, P. H. (1981). Dispersal and the stability of predator-prey interactions. American Naturalist, 118, 673–701.

de Jong, G. (1979). The influence of the distribution of juveniles over patches of food on the dynamics of a population. Netherlands Journal of Zoology, 29, 33–51.

de Jong, G. (1981). The influence of dispersal pattern on the evolution of fecundity. Nethlands Journal of Zoology, 32, 1–30.

DeBach, P. (1964). Successes, trends, and future possibilities. (In P. DeBach (Ed.), Biological control of insect pests and weeds (pp. 673–713). New York: Reinhold.

DeBach, P., & Rosen, D. (1991). Biological control by natural enemies, 2nd ed. Cambridge: Cambridge University Press.

Den Boer, P. J. (1968). Spreading of risk and stabilization of animal numbers. Acta Biotheoretica, 18, 165–194.

Ehler, L. E. (1986). Distribution of progeny in two ineffective parasites of a gall midge (Diptera: Cecidomyiidae). Environmental Entomology, 15, 1268–1271.

Embree, D. G. (1966). The role of introduced parasites in the control of the winter moth in Nova Scotia. Canadian Entomologist, 98, 1159–1168.

Entwhistle, P. F. (1983). Control of insects by virus diseases. Biocontrol News, 4, 202–225.

Enwald, P. W. (1987). Pathogen induced cycling of outbred insect populations. In P. Barbosa & J. C. Schultz (Eds.) Insect outbreaks (pp. 269–286). San Diego: Academic Press.

Falcon, L. A. (1982). Use of pathogenic viruses as agents for the biological control of insect pests. In R. M. Anderson & R. M. May (Eds.), Population biology of infectious diseases (pp. 191–210). New York: Springer-Verlag.

Free, C. A., Beddington, J. R., & Lawton, J. H. (1977). On the inadequacy of simple models of mutual interference for parasitism and predation. Journal of Animal Ecology, 46, 543–554.

Fjukii, K. (1968). Studies on interspecies competition between the azuki bean weevil and the southern cowpea weevil. III. Some characteristics of strains of two species. Research on Population Biology, 10, 87–98.

Fujii, K. (1969). Studies on interspecies competition between the azuki bean weevil and the southern cowpea weevil. IV. Competition between strains. Researches on Population Ecology, (Kyoto), 11, 84–91.

Fujii, K. (1970). Studies on interspecies competition between the azuki bean weevil and the southern cowpea weevil. V. The role of adult behavior in competition. Researches on Population Ecology, (Kyoto), 12, 233–242.

Fujii, K. (1977). Complexity-stability relationships of two-prey–one-predator species systems model: Local and global stability. Journal of Theoretical Biology, 69, 613–623.

Godfray, H. C. J., & Blythe, S. P. (1990). Complex dynamics in multispecies communities. Philosophical Transactions of the Royal Society of London, Series B, 330, 2221–2233.

Godfray, H. C. J., & Hassell, M. P. (1987). Natural enemies may be a cause of discrete generation in tropical insects. Nature (London), 327, 144–147.

Godfray, H. C. J., & Hassell, M. P. (1989). Discrete and continuous insect populations in tropical environments. Journal of Animal Ecology, 58, 153–174.

Godfray, H. C. J., & Pacala, S. W. (1992). Aggregation and the population dynamics of parasitoids and predators. American Naturalist, 140, 30–40.

Godfray, H. C. J., Hassell, M. P., & Holt, R. D. (1994). The population dynamic consequences of phenological asynchrony between parasitoids and their hosts. Journal of Animal Ecology, 63, 1–10.

Goss-Custard, J. D. (1970). The response of redshank [Tringa totanus (Linnaeus)] to spatial variations in the density of their prey. Journal of Animal Ecology, 39, 91–113.

Hamai, J., & Huffaker, C. B. (1978). Potential of predation by Metaseiulus occidentalis in compensating for increased nutritionally induced, power of increase of Tetranychus urticae. Entomophaga, 23, 225–237.

Hanski, I. (1981). Coexistence of competitors in patchy environment with and without predation. Oikos, 37, 306–312.

Hanski, I., Foley, P., & Hassell, M. (1996). Random walks in a metapopulation: How much density dependence is necessary for long-term persistence? Journal of Animal Ecology, 65, 274–282.

Harcourt, D. G. (1971). Population dynamics of Leptinotarsa decemlineata (Say) in eastern Ontario. III. Major population processes. Canadian Entomologist, 103, 1049–1061.

Hassell, M. P. (1968). The behavioural response of a tachinid fly [Cyzenis albicans (Fallen)] to its host, the winter moth [Operophtera brumata (Linnaeus)]. Journal of Animal Ecology, 37, 627–639.

Hassell, M. P. (1971). Mutual interference between searching parasites. Journal of Animal Ecology, 40, 473–486.

Hassell, M. P. (1975). Density dependence in single-species populations. Journal of Animal Ecology, 42, 693–726.

Hassell, M. P. (1978). The dynamics of arthropod predator-prey systems. Princeton, NJ: Princeton University Press.

Hassell, M. P. (1979). The dynamics of predator-prey interactions: Polyphagous predators, competing predators and hyperparasitoids. In R. Anderson, B. Turner, & L. R. Taylor (Eds.), Population dynamics. Oxford: Blackwell Scientific Publishers.

Hassell, M. P. (1980). Foraging strategies, population models and biological control: A case study. Journal of Animal Ecology, 48, 603–628.

Hassell, M. P. (1984). Insecticides in host-parasitoid interactions Theoretical Population Biology, 25, 378–386.

Hassell, M. P. & Comins, H. N. (1976). Discrete time models for two-species competition. Theoretical Population Biology, 9, 202–221.

Hassell, M. P., & Huffaker, C. B. (1969). Regulatory processes and population cyclicity in laboratory populations of Anagasta kuhniella (Zeller) (Lepidoptera: Phycitidae). III. The development of population models. Researches on Population Ecology, (Kyoto), 11, 186–210.

Hassell, M. P., & May, R. M. (1973). Stability in insect host-parasite models. Journal of Animal Ecology, 42, 693–726.

Hassell, M. P., & May, R. M. (1974). Aggregation in predators and insect parasites and its effect on stability. Journal of Animal Ecology (Kyoto), 43, 567–594.

Hassell, M. P., & May, R. M. (1986). Generalist and specialist natural enemies in insect predator-prey interactions. Journal of Animal Ecology, 55, 923–940.

Hassell, M. P., & Pacala, S. (1990). Heterogeneity and the dynamics of host-parasitoid interactions. Philosophical Transactions of the Royal Society of London, Series B, 330, 203–220.

Hassell, M. P., & Varley, G. C. (1969). New inductive model for insect parasites and its bearing on biological control. Nature (London), 223, 1133–1137.

Hassell, M. P., & Waage, J. K. (1984). Host-parasitoid population interactions. Annual Review of Entomology, 29, 89–114.

Hassell, M. P., Lawton, J. H., & May, R. M. (1976). Patterns of dynamical behavior in single species populations. Journal of Animal Ecology, 45, 471–486.

Hassell, M. P., Miramontes, O., Rohani, P., & May, R. M. (1995). Appropriate formulations for dispersal in spatially structured models: Comments on Bascompte & Solé. Journal of Animal Ecology, 64, 662–664.

Hassell, M. P., Pacala, S., May, R. M., & Chesson, P. L. (1991). The persistence of host-parasitoid associations in patchy environments. I. A general criterion. American Naturalist, 138, 584–605.

Hassell, M. P., Waage, J. K., & May, R. M. (1983). Variable parasitoid sex ratios and their effect on host-parasitoid dynamics. Journal of Animal Ecology, 52, 889–904.

Hochberg, M. E. (1989). The potential role of pathogens in biological control. Nature (London), 337, 262–264.

Hochberg, M., Hassell, M. P. & May, R. M. (1990). The dynamics of host-parasitoid-pathogen interactions. American Naturalists, 94, 74–94.

Holling, C. S. (1959a). The components of predation as revealed by a study of small mammal predation of the European pine sawfly. Canadian Entomologist, 91, 293–320.

Holling, C. S. (1959b). Some characteristics of simple types of predation and parasitism. Canadian Entomologist, 91, 385–398.

Holling, C. S. (1966). The functional response of invertebrate predators to prey density. Memoirs of the Entomological Society of Canada, 48, 1–86.

Holt, R. D., & Hassell, M. P. (1993). Environmental heterogeneity and the stability of host-parisitoid interactions. Journal of Animal Ecology, 62, 89–100.

Holyoak, M., & Lawler, S. P. (1996). The role of dispersal in predator-prey metapopulation dynamics. Journal of Animal Ecology, 65, 640–652.

Hufaker, C. B. (1958). Experimental studies on predation: Dispersion factors and predator–prey oscillations. Hilgardia, 27, 343–383.

Huffaker, C. B., Messenger, P. S., & DeBach, P. (1971). the natural enemy component in natural control and the theory of biological control. In C. B. Huffaker (Ed.), Biological control (pp. 16–67). Proceedings of an AAAS Symposium on Biological Control, Boston, MA, December 30–31, 1969, New York, London: Plenum.

Huffaker, C. B., Shea, K. P., & Herman, S. G. (1963). Experimental studies on predation: Complex dispersion and levels of food in an acarine predator-prey interaction. Hilgardia, 34, 305–330.

Ives, A. R., & May, R. M. (1985). Competition within and between species in a patchy environment: Relations between microscopic and macroscopic models. Journal of Theoretical Biology, 115, 65–92.

Jones, T. H. & Hassell, M. P. (1988). Patterns of parasitism by Trybliographa rapae, a cynipid parasitoid of the cabbage root fly, under laboratory and field conditions. Ecological Entomology, 13, 309–317.

Jones, T. H., Hassell, M. P., & Pacala, S. W. (1993). Spatial heterogeneity and the population dynamics of a host-parasitoid system. Journal of Animal Ecology, 62, 251–262.

Kakehashi, N., Suzuki, Y., & Iwasa, Y. (1984). Niche overlap in parasitoids in host-parasitoid systems: Its consquence to single versus multiple introduction controversy in biological control. Journal of Applied Ecology, 21, 115–131.

Kareiva, P., & Odell, G. M. (1987). Swarms of predators exhibit "prey-taxis" if individual predators use area restricted search. American Naturalist, 130, 233–270.

Kermack, W. O., & McKendrick, A. G. (1927). A contribution to the mathematical theory of epidemics. Proceedings of the Royal Society of London, Series A, 115, 700–721.

Kowalski, R. (1976). Philonthus decorus (Gr.) (Coleoptera: Staphylinidae): Its biology in relation to its action as a predator of winter moth pupae (Operophtera brumata) (Lepidoptera: Geometridae). Pediobiologia, 16, 233–242.

Laing, J. E., & Huffaker, C. B. (1969). Comparative studies of predation by Phytoseiulus persimilis Athias-Henriot and Metaseiulus occidentalis (Acarina: Phytoseiidae) on populations of Tetranychus urticae Koch (Acarina: Tetranychidae). Researches on Population Ecology (Kyoto), 11, 105–126.

Lawton, J. H., & Hassell, M. P. (1984). Interspecific competition in insects. In C. B. Huffaker & R. L. Rabb (Eds.), Ecological entomology (pp. 451–495). New York: John Wiley & Sons.

Lawton, J. H., & Strong, D. R. (1981). Community patterns and competition in folivorous insects. American Naturalist, 118, 317–338.

Lefkovitch, L. P. (1963). Census studies on unrestricted populations of Lasioderma serricorne (F.) (Coleoptera: Anobiidae). Journal of Animal Ecology, 32, 221–231.

Lessells, C. M. (1985). Parasitoid foraging: Should parasitism by density dependent? Journal of Animal Ecology, 54, 27–41.

Levin, S. A. (1974). Dispersion and population interactions. American Naturalist, 108, 207–228.

Levin, S. A. (1976). Population dynamics in heterogeneous environments. Annual Review of Ecology and Systematics, 7, 287–310.

Luck, R. F., & Podoler, H. (1985). Competitive exclusion of Aphytis lingnanensis by A. melinus: Potential role of host size. Ecology, 66, 904–913.

May, R. M. (1975). Biological population obeying difference equations: Stable points, stable cycles and chaos. Journal of Theoretical Biology, 51, 511–524.

May, R. M. (1978). Host-parasitoid systems in patchy environments: A phenomenological model. Journal of Animal Ecology, 47, 833–843.

May, R. M. (1985). Regulation of population with nonoverlapping generations by microparasites: A purely chaotic system. American Naturalist, 125, 573–584.

May, R. M., & Hassell, M. P. (1981). The dynamics of multiparasitoid-host interactions. American Naturalist, 117, 234–261.

May, R. M., & Hassell, M. P. (1988). Population dynamics and biological control. Philosophical Transactions of the Royal Society of London, Series B, 318, 129–169.

May, R. M., & Oster, G. F. (1976). Bifurcations and dynamic complexity in simple ecological models. American Naturalist, 110, 573–599.

May, R. M., Conway, G. R., Hassell, M. P., & Southwood, T. R. E. (1974). Time delays, density dependence, and single-species oscillations. Journal of Animal Ecology, 43, 747–770.

May, R. M., Hassell, M. P., Anderson, R. M., & Tonkyn, D. W. (1981). Density dependence in host-parasitoid models. Journal of Animal Ecology, 50, 855–865.

Maynard Smith, J. (1974). Models in ecology. New York: Cambridge University Press.

Maynard Smith, J., Slatkin, M. (1973). The stability of predator-prey systems. Ecology, 54, 384–391.

McClure, M. S. (1977). Parasitism of the scale insect Fiorinia externa (Homoptera: Diaspididae) by Aspidiotiphagus citrinus (Hymenoptera: Eulophidae) in a Hemlock forest: Density dependence. Environmental Entomology, 6, 551–555.

McNamee, P. J., McLeod, J. M., & Holling, C. S. (1981). The structure and behavior of defoliating insect/forest systems. (Publication R-25, pp. 1–89). Vancouver: University of British Columbia Institute of Research Ecology.

Metcalf, J. R. (1971). Observations on the ecology of Saccharosydne saccharivora (Westw.) (Hem., Hom., Delphacidae) in Jamaican sugarcane fields. Bulletin of Entomological Research, 60, 565–596.

Mills, N. J., & Getz, W. M. (1996). Modelling the biological control of insect pests—A review of host-parasitoid models. Ecological Modelling, 92, 121–143.

Mook, L. J. (1963). Birds and the spruce budworm. In R. F. Morris (Ed.), The dynamics of epidemic spruce budworm populations. Memoirs of the Entomological Society of Canada, 31, 268–271.

Murdie, G., & Hassell, M. P. (1973). Food distribution, searching success and predator-prey models. In M. S. Bartlett & R. W. Hiorns (Eds.), The mathematical theory of the dynamics of biological populations (pp. 87–101). London: Academic Press.

Murdoch, W. W. (1969). Switching in general predators: Experiments on predator and stability of prey populations. Ecological Monographs, 39, 335–354.

Murdoch, W. W., & Stewart-Oaten, A. (1975). Predation and population stability. Advances in Ecological Research, 9, 2–131.

Murdoch, W. W., & Stewart-Oaten, A. (1989). Aggregation by parasioids and predators: Effects on equilibrium and stability. American Naturalist, 134, 288–310.

Murdoch, W. W., Reeve, J. D., Huffaker, C. B., & Kennett, C. E. (1984). Biological control of scale insects and ecological theory. American Naturalist, 123, 371–392.

Murdoch, W. W., Nisbet, R. M., Blythe, S. P., Gurney, W. S. C., & Reeve, J. D. (1987). An invulnerable age class and stability in delay-differential parasitoid-host models. American Naturalist, 129, 263–282.

Murdoch, W. W., Briggs, C. J., Nisbet, R. M., Gurney, W. S. C., & Stewart-Oaten, A. (1992). Aggregation and stability in metapopulation models. American Naturalist, 140, 41–58.

Murdoch, W. M., Swarbrick, S. L., Luck, R. F., Walde, S., & Yu, D. S. (1996). Refuge dynamics and metapopulation dynamics: An experimental test. American Naturalist, 147, 424–444.

Nachman, G. (1981). Temporal and spatial dynamics of an acarine predatory-prey system. Journal of Animal Ecology, 50, 435–451.

Nicholson, A. J. (1933). The balance of animal populations. Journal of Animal Ecology, 2, 131–178.

Nicholson, A. J. (1954). An outline of the dynamics of animal populations. Australian Journal of Zoology, 2, 9–65.

Nicholson, A. J., & Bailey, V. A. (1935). The balance of animal populations. Part I. Proceedings of the Zoological Society of London, 3, 551–598.

Nisbet, R. M., & Gurney, W. S. C. (1982). Modeling fluctuating populations. New York: John Wiley & Sons.

Nunney, L. (1980). The influence of the type 3 (sigmoid) functional response upon the stability of predator-prey difference models. Theoretical Population Biology, 18, 257–278.

Pacala, S., & Hassell, M. P. (1991). The persistence of host-parasitoid associations in patchy environments. II. Evaluation of field data. American Naturalist 138(3), 584–605.

Pacala, S., Hassell, M. P., & May, R. M. (1990). Host-parasitoid associations in patchy environments. Nature (London), 344, 150–153.

Paine, R. T. (1966). Food web complexity and species diversity. American Naturalist, 100, 65–75.

Paine, R. T. (1974). Intertidal community structure. Oecologia, 15, 93–120.

Park, T. (1948). Experimental studies of interspecies competition. I. Competition between populations of the flour beetles, Tribolium confusum Duval and Tribolium castaneum Herbst. Ecological Monographs, 18, 265–308.

Parrish, J. D., & Saila, S. B. (1970). Interspecific competition, predation and species diversity. Journal of Theoretical Biology, 27, 207–220.

Perry, J. N., & Taylor, L. R. (1986). Stability of real interacting population in space and time: Implications, alternatives and the negative binomial kc. Journal of Animal Ecology, 55, 1053–1068.

Reddingius, J., & Den Boer, P. J. (1970). Simulation experiments illustrating stabilization of animal numbers by spreading of risk. Oecologia, 5, 240–248.

Reeve, J. D., & Murdoch, W. W. (1985). Aggregation by parasitoids in the successful control of the California red scale: A test of theory. Journal of Animal Ecology, 54, 858–873.

Regniere, J. (1984). Vertical transmission of disease and population dynamics of insects with discrete generations. Journal of Theoretical Biology, 107, 287–301.

Roughgarden, J., & Feldman, M. (1975). Species packing and predator pressure. Ecology, 56, 489–492.

Royama, T. (1970). Factors governing the hunting behaviour and selection of food by the great tit (Parus major L.). Journal of Animal Ecology, 39, 619–669.

Ruxton, G. D. (1996). Dispersal and chaos in spatially structured models: An individual-level approach. Journal of Animal Ecology, 65, 161–169.

Ruxton, G., Gonzalez-Andujar, J. L., & Perry, J. N. (1997). Mortality during dispersal stabilizes local population fluctuations. Journal of Animal Ecology, 66, 289–292.

Sabelis, M. W. (1981). Biological control of two-spotted spider mites using phytoseiid predators. Modelling the predator-prey interaction at the individual level (Agricultural Research Report 910). Waginengin, Netherlands: Pudoc.

Sabelis, M. W., & Laane, W. E. M. (1986). Regional dynamics of spider-mite populations that become extinct locally because of food source depletion and predation by phytoseiid mites (Acarina: Tetranychidae, Phytoseiidae). In J. A. J. Metz & O. Diekmann (Eds.), The dynamics of physiologically structured populations (pp. 345–376). New York: Springer-Verlag.

Sabelis, M. W., & van der Meer, J. (1986). Local dynamics of the interaction between predatory mites and two-spotted spider mites. In J. A. J. Metz & O. Diekmann (Eds.), The dynamics of physiologically structured populations (pp. 322–344). New York: Springer-Verlag.

Sabelis, M. W., Alebeek, F., van Bal, A., Bilsen, J., van Heijningen, T., van kaizer, P., Kramer, G., Snellen, H., Veenebos, R., & Vogelezang, J. (1983). Experimental validation of a simulation model of the interaction between Phytoseiulus persimilis and Tetranychus urticae on cucumber. OILB-Bulletin SROP/WPRS 6(3), 207–229.

Southwood, T. R. E., & Comins, H. N. (1976). A synoptic population model. Journal of Animal Ecology, 45, 949–965.

Steele, J. H. (1974). The structure of marine ecosystems. Cambridge, MA: Harvard University Press.

Stelter, C., Reich, M., Grimm, V., & Wissel, C. (1997). Modelling persistence in dynamic landscapes. Lessons from a metapopulation of the grasshopper Bryodema tuberculata. Journal of Animal Ecology 66, 508–518.

Stiling, P. D. (1987). The frequency of density dependence in insect host-parasitoid systems. Ecology, 68, 844–856.

Strong, D. R., Lawton, J. H., & Southwood, T. R. E. (1984). Insects on plants. Oxford: Blackwell Scientific Publishers.

Takafuji, A. (1977). The effect of the rate of successful dispersal of a phytoseiid mite, Phytoseiulus persimilis Athias-Henriot (Acarina: Phytoseiidae) on the persistence in the interactive system between the predator and its prey. Researches on Population Ecology (Kyoto), 18, 210–222.

Takafuji, A., Inoue, T., & Fujita, K. (1981a, September). Analysis of an acarine predator-prey system in glasshouse, (pp. 144–153). First Japan/USA Symposium on Integrated Pest Management, Tsukuba, Japan.

Takafuji, A., Tsuda, Y., & Miki, T. (1981b). System behaviour in predator-prey interaction, with special reference to acarine predator-prey systems. Researches on Population Ecology (Suppl. 3), 75–92.

Taylor, A. D. (1988). Parasitoid competition and the dynamics of host-parasitoid models. American Naturalist 132, 417–436.

Taylor, A. D. (1993). Heterogeneity in host parasitoid interactions—aggregation of risk and the cv(2) greater-than 1 rule. Trends in Ecology & Evolution, 8, 400–405.

Taylor, T. H. C. (1937). The biological control of an insect in Fiji. London: Imperial Institute of Entomology.

Tinsley, T. W. (1979). The potential of insect pathogenic viruses as pesticide agents. Annual Review of Entomology, 24, 63–87.

Tinsley, T. W., & Entwhistle, P. F. (1974). The use of pathogens in the control of insect pests. In D. Price-Jones & M. E. Solomon (Eds.), Biology in pests and disease control (pp. 115–129). Oxford: Blackwell Science Publishers.

Tothill, J. D., Taylor, T. H. C., & Paine, R. W. (1930). The coconut moth in Fiji. London: Imperial Institute of Entomology.

Turnbull, A. L., and Chant, P. A. (1961). The practice and theory of biological control of insects in Canada. Canadian Journal of Zoology, 39, 697–753.

Utida, S. (1941). Studies on experimental population of the azuki bean weevil, Callosobruchus chinensis (L.). IV. Analysis of density effect with respect to fecundity and fertility of eggs. Memoirs of the College of Agriculture, Kyoto Imperial University, 49, 21–42.

Utida, S. (1950). On the equilibrium state of the interacting population of an insect and its parasite. Ecology, 31, 165–175.

Utida, S. (1957). Cyclic fluctuations of population density intrinsic to the host-parasite system. Ecology, 38, 442–449.

Utida, S. (1967). Damped oscillation of population density at equilibrium. Researches on Population Ecology (Kyoto), 9, 1–9.

van den Bosch, R., & Messenger, P. S. (1973). Biological control. New York: Intext Press.

Van der Vecht, J. (1954). Parasitism in an outbreak of the coconut moth (Artona cataxantha (Hamps.)) in Java (Lep.). Entomologische Berichten Amsterdam, 15, 122–132.

van Valen, L. (1974). Predation and species diversity. Journal of Theoretical Biology, 44, 19–21.

van Driesche, R. G., & Bellows, T. S. Jr. (1996). Biological control. New York: Chapman & Hall.

Waage, J. K. (1979). Foraging for patchily distributed hosts by the parasitoid, Nemeritis canescens. Journal of Animal Ecology, 48, 353–371.

Waage, J. K., & Hassell, M. P. (1982). Parasitoids as biological control agents—a fundamental approach. Parasitology, 84, 241–268.

Walde, S. J. (1994). Immigration and the dynamics of a predator-prey interaction in biological control. Journal of Animal Ecology, 63, 337–346.

Walde, S. J., & Murdoch, W. W. (1988). Spatial density dependence in parasitoids. Annual Review of Entomology, 33, 441–466.

Wang, Y. H., & Guttierrez, A. P. (1980). An assessment of the use of stability analyses in population ecology. Journal of Animal Ecology, 49, 435–452.

Watt, K. E. F. (1965). Community stability and the strategy of biological control. Canadian Entomologist, 97, 887–895.

White, E. G. & Huffaker, C. B. (1969). Regulatory processes and population cyclicity in laboratory populations of Anagasta kuhniella (Zeller) (Lepidoptera: Phycitidae). II. Parasitism, predation, competition and protective cover. Researches on Population Ecology 11 (Kyoto), 150–185.

Wood, B. J. (1968). Pests of oil palms in Malaysia and their control. Kuala Lumpur: Incorporated Society of Planters.

3

Taxonomy and Biological Control

GORDON GORDH

USDA, Agricultural Research Service
Kika de la Garza Subtropical Agricultural Research Center
2301 S. International Blvd.
Weslaco, Texas 78596

JOHN W. BEARDSLEY

Department of Entomology
University of Hawaii
Honolulu, Hawaii

INTRODUCTION

Taxonomy is the branch of biology that involves the naming, identifying, and classifying of organisms. The importance of various aspects of taxonomy relative to biological control has been discussed earlier (Clausen, 1942; Sabrosky, 1955; Schlinger & Doutt, 1964; Delucchi, 1966; Compere, 1969; Gordh, 1977, 1982). Taxonomy is important to biological control from theoretical and applied viewpoints. The applied biological control worker needs names for zoological entities involved in pest problems. The names provide an important mechanism for the dissemination of information about the involved organisms. Taxonomists provide scientific names through identification of taxa that are known to science or description of taxa that are new to science. The accuracy or correctness of identifications and descriptions is vitally important to biological control workers.

From a theoretical viewpoint, taxonomy is important to biological control workers because taxonomists develop classifications that are intended to reflect evolutionary relationships. Natural (evolutionary) classifications are helpful to biological control workers because these classifications are intended to predict details of biology and distribution.

In the following chapter, we develop lines of reasoning intended to show the importance of taxonomy to biological control and illustrate instances where taxonomic work has been vital in the success or occasionally responsible for the failure, of biological control programs. We delineate areas of responsibility in collaboration between biological control workers and taxonomists. Finally, we note the importance of voucher specimens to the future success of biological control programs; and the collation of biological information about hosts, prey, and their natural enemies.

TAXONOMY: THE HISTORICAL PERSPECTIVE

Taxonomy has a pervasive influence on all human activities. The urge to arrange, organize, describe, name, and classify seems fundamental. This urge operates at all levels of social organization, from the alphabetical arrangement of family names in the telephone directory to the development of legends for street maps. In a very real sense, a world without organization cannot be imagined. These efforts at organization, protracted over time, have been called taxonomy or systematics. In biology, taxonomy is the branch that deals with the description of new taxa and the identification of taxa known to science. Classification involves the arrangement of taxa based on morphological and biological characteristics.

Beginning with the ancient civilizations, names have been applied to organisms. Insects were apparently among the objects of naming because Beelzebub, of biblical reference, was (among other things) "lord of the flies." These so-called "common" names of many organisms are in widespread usage today, thus testifying to the utility of such a practice. Despite the usefulness of common names to the layman, several problems are inherent in common names. Most serious of these problems is synonymy. Frequently, more than one common name is applied to a single organism (synonyms), or the same common name is used for different organisms (homonyms). Synonymy has created confusion and misunderstanding because the biological characteristics and habits of similar organisms can be profoundly different.

In its earliest form, the scientific name given to an organism was often impractical and unwieldy. Scientific names during the lifetime of John Ray (= Wray) (1628 to 1705) consisted of a series of Latin adjectives catenated in such a way as to describe the animal. The system was less ambiguous than the common name system, but it had the problem of being cumbersome because the name of an animal frequently was several lines or a paragraph long. Thus, to mention the scientific name of an animal in conversation required an excellent memory and plenty of time.

A major contribution in the naming of organisms was provided by the Swedish natural historian and physician Carl Linnaeus (1707 to 1778). Linnaeus is credited with developing the binomial system of naming organisms that is in use today. The start of zoological nomenclature is taken as the tenth edition of Linnaeus' monumental work, *Systema Naturae*. The notable exception is the nomenclature of spiders (*Aranei svecici*), which originates with the work of Karl Alexander Clerck (1710 to 1765). The accepted date of publication of these works is January 1, 1758, the official starting point of zoological nomenclature.

During the following century taxonomic zoologists followed the lead of Linnaeus and prepared descriptions of species for publication but named the animal with a binomen. The binomen consists of two parts, the generic name and the specific epithet. With the accumulation of taxonomic descriptions, problems developed with synonymy, homonymy, inconsistent application of binomens, and related nomenclatural difficulties. The first attempt at addressing these problems was the so-called Strickland code, prepared in 1846. The code was developed by a panel of taxonomists, including Charles Darwin, who was a noted taxonomist of barnacles. Subsequently, an International Code of Zoological Nomenclature was developed (1906). This code has been altered slightly from time to time, but continues to represent the basic guidelines for the formation and validation of zoological names for taxa. The most recent revision of the code was published in 1985. More complicated problems of nomenclature or matters requiring the fixation of names in the interest of stability are referred to the International Commission of Zoological Nomenclature, which serves as a taxonomic supreme court.

The Importance of Taxonomy to Biological Control

The importance of taxonomy to entomology has been reviewed by Danks (1988). Concerning the importance of taxonomy and biological control, the earliest comments were prepared by Clausen (1942). Subsequent important reviews include Sabrosky (1955), Schlinger and Doutt

(1964), Delucchi (1966), Compere (1969), Gordh (1977, 1982), and Knutson (1981). These discussions provide detailed documentation of earlier treatment in the literature.

The Scientific Name

From the perspective of applied biological control, the scientific name of an organism is of utmost importance (Gordh, 1977). The scientific name provides a key to the published literature concerning any zoological taxon. Without the correct name the investigator has no access to knowledge published about an animal of interest.

In a sense, the scientific name is a shorthand method for conveying an enormous amount of information about an organism that is available in published literature. Thus, when we say *"Musca domestica,"* we are providing a reference key to all the volumes of information that have been published about the housefly. Similarly, all the information that has been developed about any organism important in biological control work is stored beneath the scientific name for that organism. In this regard, we cannot overemphasize that the correctness of the name is vitally important because information under an erroneous name is misfiled and thus is unavailable to biological control workers.

Need and Importance of Accurate Identification

In biological control the need for identification is great, but the importance of accurate identification is greater. Two species that are very similar morphologically are not necessarily similar biologically. Subtle differences in morphology or biology of closely related species frequently can be profound. Distinguishing between variation in taxonomic characters within a species and difference in character states between species (individual variation versus interspecific differences) is frequently difficult. Understanding the functional significance of the observed anatomical features that serve to distinguish between species is an area of research that has lagged far behind orthodox taxonomic studies. Seemingly slight anatomical differences may reflect significant differences in the biology of two organisms. Thus, so-called minor structural differences can mean the difference between pest and non-pest status for species that are potential threats to agriculture, or between establishment and failure to establish in the case of natural enemies of pests. A few examples follow that should serve to illustrate this point.

The Pink Bollworm, Pectinophora gossypiella *(Saunders)*

The gelechiid genus *Pectinophora* contains three described species: *P. scutigera, P. endema,* and *P. gossypiella. Pectinophora scutigera* is found in Australia, Papua New Guinea, Micronesia, and Hawaii; *P. endema* is restricted to eastern Australia (Common, 1958); *P. gossy-*

piella occurs in Western Australia, India, Southeast Asia, Africa, various Pacific islands, and the Americas. All species consume the flowers, seeds and seed capsules of Malvaceae: *P. endema* consumes only native *Hibiscus* in Australia and is not regarded as an agricultural pest. The remaining species consume other Malvaceae, including *Gossypium* spp. (cotton). *Pectinophora gossypiella* is one of the most serious pests of cotton worldwide. Larvae of this species can diapause within the seeds of its host plant. This apparently accounts for its widespread distribution. In contrast, *P. scutigera* apparently does not diapause within seeds, is limited in distribution, and is not considered a major pest of cotton.

Holdaway (1926) provided a name for *P. scutigera* based on the larval differences. Subsequently, Holdaway (1929) described the structural characters of the adult genitalia to separate the species. The validity of *P. scutigera* as a species was questioned by some taxonomists, but is now accepted (Zimmerman, 1979).

The importance of correct identification of the bollworms focuses on the pest status of these insects and quarantine enforcement. In Australia, *P. scutigera* is not a significant pest of cotton and its distribution is limited by intrinsic biological characteristics. It does not play a significant role in quarantine efforts. In contrast, *P. gossypiella* is a very significant pest of cotton worldwide. In Australia, the moth occurs in the Northern Territory and in Western Australia, but not in Queensland. Quarantine serves as an important barrier restricting movement of *P. gossypiella*. Quarantine is expensive to the state and the commercial enterprise. Unnecessarily restricting movement of material infested with *P. scutigera* is a questionable practice of limited value. The practical implication of correct identification should be apparent.

The California Red Scale, Aonidiella aurantii *(Maskell)*

The history of the California red scale provides an excellent example of the potential costs of incomplete taxonomic and biogeographic knowledge of a pest species. This scale is a member of a complex of species native to the tropics and subtropics of the Old World (Africa through Southeast Asia and the Orient) (McKenzie, 1937). It became a pest of citrus when introduced into the New World without its associated natural enemies (Compere, 1961). Many parasites associated with closely related *Aonidiella* species will not attack or are not effective against *A. aurantii*. The failure of early attempts at biological control of this scale were due, at least in part, to the inability to differentiate it from such closely related species as *A. citrina* (a less important pest species also introduced into the New World). Some parasite species in the Orient appeared promising to exploratory entomologists. However, these species failed when introduced into California because their preferred hosts were other species of *Aonidiella*. This became apparent after scale taxonomist Howard McKenzie made a careful revision of the genus *Aonidiella* and showed that the species could be separated on the basis of microscopic morphological differences.

The Coffee Mealybug, Planococcus kenyae *(LePelley)*

The coffee mealybug of Kenya presents another interesting example of early failure and delayed success in biological control caused by misidentification of the pest species. The pest first appeared during the 1930s and caused serious losses to the coffee crop of Kenya. First, it was identified as the common, widespread, citrus mealybug [*Planococcus citri* (Risso)]. Later, it was determined as a related Philippine species, *P. lilacinus* (Cockerell). Finally, both of these identifications were shown to be incorrect. Unfortunately, on the basis of these names, a great amount of effort and expense was devoted to searching for and shipping natural enemies of *Planococcus* in the Asiatic tropics. Parasites that appeared promising when collected could not be established in Kenya. The problem was clarified when the mealybug taxonomist LePelley examined specimens of the pest. He found relatively inconspicuous but consistent morphological differences that indicated that the coffee mealybug was an undescribed species, which he named (LePelley, 1935, 1943). It was then discovered that this mealybug also occurred in what are now Uganda and Tanzania, where it was under natural biological control. Parasites imported into Kenya from those areas produced complete biological control in a relatively short time. However, by then the incorrect identification of the pest had resulted in several years of wasted time and effort. Had the identity of the pest not been confused with previously named species, biological control might have been achieved much sooner at much less expense.

Accurate Identification of Natural Enemies

Of equal importance to accurate determination of pest species in biological control work is the correct identification of the entomophagous organisms that are found in association with target pest species and that may be considered for utilization in biological control programs. In many cases, such natural enemies belong to groups of small-to-minute insects, the species of which often resemble one another very closely. Taxonomic knowledge needed to differentiate species-level taxa in such groups has accumulated slowly with great effort. In many taxonomic groups, knowledge remains woefully incomplete. Some examples of the sorts of taxonomic problems involving natural enemy taxa important to biological control work are discussed later.

Among the Aphelinidae, an important family of entomophagous Chalcidoidea, the genera *Aphytis* and *Marietta* appear closely related on the basis of morphology. Indeed, superficially, it is difficult for nonspecialists to place some species in the correct genus. Biologically, the differences between the genera are profound. *Aphytis* species are primary parasites of armored scale insects; *Marietta* species are hyperparasites, usually associated with armored scale insects or other Coccoidea. Because hyperparasites are generally viewed as deleterious to biological control programs, importation or deliberate movement of *Marietta* could impact negatively on control efforts. Hence, the ability to differentiate between these two genera is of major importance in the applied biological control of scale insects.

The family Encyrtidae, another large group within the Chalcidoidea, contains a vast array of genera whose species are primary parasites of phytophagous insects. Yet the same family also contains genera whose species are mostly, or largely, secondary parasites (*Cheiloneurus, Quaylea*). Recognition of these hyperparasites and their elimination requires taxonomic knowledge of the family Encyrtidae. Failure to do so could result in the introduction and establishment of such unwanted species, and, in a few instances, this occurred. A few genera of encyrtids (e.g., *Psyllaephagus*) contain both primary and secondary parasite species, calling for careful biological and taxonomic study to separate the beneficial primary and undesirable hyperparasites prior to making releases for biological control.

Returning to the case of the California red scale, not only did difficulty in distinguishing the pest from related species retard biological control, but also this endeavor was hindered by a lack of knowledge about a very important group of armored scale parasites, the genus *Aphytis*. DeBach *et al.* (1971) showed that this lack of taxonomic knowledge delayed achievement of biological control of California red scale by 50 years. Early exploratory work for natural enemies of red scale revealed the presence of *Aphytis* parasites associated with red scale at several localities in the Orient. Specimens from these collections were determined as *A. chrysomphali* Mercet, a species already present in California that was not particularly effective against red scale. As a consequence, no effort was made to propagate and release new oriental *Aphytis* for California red scale until after World War II (Compere, 1961). The two most effective natural enemies of red scale presently known, *A. lingnanensis* Compere and *A. melinus* DeBach, were not recognized as distinct species until 1948 and 1956, respectively. These species might have been introduced into California many years earlier had a proper understanding of the taxonomy of *Aphytis* existed. A similar occurrence involves *A. holoxanthus* DeBach, the most effective parasite presently known for the Florida red scale, *Chrysomphalus aonidum* (Linnaeus). *Aphytis holoxanthus* apparently was collected first around 1900, but was ignored because it was confused with another species. *Aphytis holoxanthus* became available for biological control in 1960 when DeBach recognized that it was a distinct species (DeBach *et al.,* 1971).

Trichogramma is a cosmopolitan genus of minute parasites that presently contains more than 120 species. All species for which the biology is known develop as primary internal parasites of eggs. *Trichogramma* has been used extensively against lepidopterous pests in classical biological control introductions and inundative release programs. Some programs have produced contradictory results, with some workers claiming success and others admitting failure. Poor taxonomic knowledge has contributed to conflicting assessment of biological control programs involving *Trichogramma.*

Until recently, *Trichogramma* was taxonomically chaotic. Early workers rarely deposited voucher specimens for their research, and without bona fide exemplars it is difficult or in some instances impossible to determine what species of *Trichogramma* was used in a release program. For instance, most references to *T. minutum* Riley, *T. evanescens* Westwood, and *T. semifumatum* (Perkins) made prior to 1980 probably are erroneous. We now know that *Trichogramma* contains many anatomically similar species that can be distinguished only by microscopic differences in antennae and genitalia. Traditional reliance on body coloration is of limited utility and has been shown to depend on environmentally induced variation. Many species display dark coloration at the base of the forewings, and the name *T. semifumatum* was often applied to such forms by early workers. We now recognize *T. semifumatum* as endemic to the Hawaiian Islands and based on one collection (Pinto *et al.,* 1978). The classification and systematics of *Trichogramma,* and related genera, must involve careful anatomical and biological studies to identify taxa.

CONTRIBUTIONS OF BIOLOGICAL CONTROL TO TAXONOMY

An element of reciprocity exists between the biological control worker and taxonomist that must be fully developed to maximize the usefulness of taxonomy as an adjunct to biological control. From the standpoint of supplemental information, biological control workers can offer taxonomists important data necessary to complete taxonomic projects. Kinds of information that are important include zoogeographical, biological, behavioral, ecological, and hybridizational data.

Zoogeographical Data

Biological control workers frequently engage in foreign exploration. The work is time consuming and expensive. Often, the results of this work go unpublished and the imported material remains unstudied. Yet such material can provide potentially important data for taxonomic studies in terms of understanding geographic variation and expanding known limits of distribution. Biological control workers should be encouraged to publish the results of their work or make collections available for study.

Biological Data

The kinds of information a biological control worker can provide a taxonomist are frequently important in elucidating biological principles concerning the interrelationships between entomophagous species and their hosts. Such data may also be of value to taxonomists attempting to analyze taxonomic relationships within and between groups of entomophagous species. For instance, many specialists believe there are trends toward habitat specialization and host specificity in many groups of parasitic Hymenoptera. Data on host range and host preference is obtained in the field and in the insectary by biological control specialists. This information can be used by taxonomists to refine their taxonomic analyses of groups. Biological control workers must be encouraged to publish or otherwise make available biological data that are obtained during the course of applied control projects. Similarly, information on pest species, such as host-plant preferences, should be shared with appropriate taxonomic specialists.

Behavioral Data

Subtle differences in behavior between populations of what appears to be one species may point to taxonomic differences between two or more closely related species. Behavioral differences between populations are the sort of data that cannot be easily obtained by the museum taxonomist who must rely primarily on preserved specimens. Yet taxonomists must be made aware of such differences. Once the existence of behavioral differences is known, the taxonomist may be encouraged to search more diligently for minor anatomical differences that can be used to distinguish between closely related taxa.

The kinds of potentially important behavioral differences are diverse. For instance, courtship behavior in *Aphytis* appears to be controlled largely by species-specific sex pheromones released by virgin females. Males are strongly attracted to the pheromone released by conspecific females. Also, males produce a pheromone that appears to calm the virgin female and render her sexually receptive. Males and females do not normally respond to members of the opposite sex belonging to other, even closely related species (Rosen & DeBach, 1979). In addition to courtship and mating behavior, other kinds of behavior, such as host finding, may also be indicative of taxonomic difference between populations that show no readily apparent anatomical differences.

Ecological Data

Closely related species often differ materially in their ecological requirements. Important data must be kept on the ecological associations of entomophagous arthropods collected for biological control purposes. Factors such as elevation and season are important, but less apparent ecological data, such as the type of plant community in which the species occurs, can also provide valuable clues to the taxonomist who is attempting to differentiate anatomically similar forms. Host specificity among related species of parasitic Hymenoptera is often reflected in their association with specific plants that harbor their insect hosts. Thus, information on the plant hosts on which parasites are collected may prove useful to taxonomists.

Hybridization Studies

Most classical taxonomists are confined to museums and have limited access to insect rearing facilities. As a consequence, these taxonomists are in a situation that restricts their ability to test reproductive compatibility and therefore to make judgments involving the biological species concepts. While most museum taxonomists would acknowledge reproductive compatibility as a viable approach to the study of species limits, in reality they usually are limited to conceptual acknowledgment only. Biological control workers with access to laboratory and insectary facilities are in a position to provide detailed information concerning reproductive compatibility and reproductive isolation. This kind of information is important because if taxonomists have information concerning hybridization studies, they can then proceed to examine material more critically for morphological characters. Such studies are essential for groups such as *Trichogramma* (Pinto *et al.,* 1986).

SOURCES OF TAXONOMIC EXPERTISE

As was the case when biological control research began, it is often difficult to find specialists sufficiently expert in the taxonomy of pests and natural enemies who are willing and able to provide biological control workers with the unequivocal identifications that they require. This has been particularly true for groups of minute parasites that are of major importance to biological control. Dwindling public support for natural history museums and for taxonomic research in general has intensified this problem during the past several decades. Many biological control specialists have been required to undertake systematic research in an effort to solve taxonomic problems associated with their own biological control research. Thus, we find that a considerable amount of basic research, particularly with entomophagous forms, has been carried out by scientists whose taxonomic interests originated with their involvement in applied biological control projects. An example is the detailed study of the aphelinid genus *Aphytis* by Rosen and DeBach (1979). As a result of this research, *Aphytis* now is recognized as among the best understood genera of Hymenoptera used in biological control. Professor Earl Oatman and his colleagues have done similar work with *Trichogramma*. Their careful biosystematic studies on this important group of egg parasites has clarified taxonomic problems in *Trichogramma* and related genera. Nevertheless, much more careful research will be required before an adequate understanding of the systematics of parasitic Hymenoptera as a whole is achieved.

Finding appropriate taxonomic specialists can frequently be troublesome to a biological control worker. Directories of taxonomic specialists are published periodically (Blackwelder & Blackwelder, 1961). These documents are useful but frequently become dated and require major modification. Word of mouth has been used frequently for discovering taxonomic talent, but this method is flawed. Workers can sometimes obtain the names of taxonomists from curators at large institutional collections. An effective method of locating taxonomic expertise is by consulting the most recent volumes of the Zoological Record. Examining the record will reveal nearly all publications on a taxonomic group and provide the names and addresses of taxonomists involved in the work.

Many factors influence the availability of taxonomic expertise. Some taxonomic groups receive little or no attention for long periods of time. Taxonomists may be interested in a problem, but necessary nomenclatural work or work in assembling material for study may represent a substantial investment in time and effort. Taxonomists may change the direction of emphasis in their research programs. Taxonomists change jobs. Taxonomists retire and they are not replaced. However, apparently insurmountable obstacles are generally overcome by dedicated effort on the part of workers interested in the identity of their specimens.

Establishing a good working relationship with any taxonomist consulted on a problem involving identification is important. A letter indicating the nature of the problem should precede any specimens sent through the mail. A letter should also inquire as to the best method of preservation of the specimens. Taxonomic specialists are in the best position to know how material should be preserved and mounted for study. Frequently, a taxonomist will prefer personally to prepare the specimens for study. At least the taxonomists should be given the option of indicating how the material should be prepared. No matter how carefully prepared, specimens may be inappropriately curated. This is particularly true of slide-mounted specimens, as proper slide mounting technique is often critical for accurate identification.

Taxonomists frequently invest considerable time and effort in providing identifications. Often, the names are provided *au gratis,* but to partially recover salary costs, museums are charging for this service. Biological control workers should provide taxonomic specialists with the opportunity of retaining specimens for a collection. In this sense the specimens are voucher specimens, and we discuss this topic elsewhere in this chapter.

Acknowledgment of taxonomic assistance in publications is very important for several reasons. Recognition of personal effort is important to many taxonomic workers, and this recognition may represent the only form of payment or satisfaction for completion of the task. Acknowledgment in print also represents tangible evidence of a taxonomist's efforts. This may serve to document time investment to an employer when used to justify continued employment or promotion. Finally, acknowledgment in publication alerts the scientific community to taxonomic authority behind scientific names used in publications.

Specimens submitted for identification often involve taxa that are new to science. Particularly, biological control programs involving parasitic wasps frequently reveal undescribed species and uncharacterized genera. This necessitates providing names for the organisms involved. Manuscript names are scientific names intended for validation as new to science. Taxonomists frequently are pressured by biological control workers for scientific names of taxa so they may be used in research reports, dissertations, and scientific publications. Knowledgeable taxonomists guard these names until they are validated in published descriptions. Careless or nonvigilant taxonomists may inadvertently reveal these names to biological control workers interested in publishing their own work. It is important for biological control workers to understand that scientific names should not be used in print before the names are validated.

LITERATURE, REFERENCE COLLECTIONS, AND VOUCHER SPECIMENS

Specimens in Biological Control Research

Every biological scientist is aware of his or her need to have readily available a comprehensive library that can provide basic and current information in his or her specialty. The taxonomist must have appropriate taxonomic keys and other reference publications. Important among these are revisionary studies, monographs, and catalogs.

Revisionary studies are the results of taxonomic effort including analysis of previously unstudied material. Revisionary studies are essentially descriptive in nature. Poorly known species are redescribed, new species are described, keys may be presented, new synonymies are proposed, existing synonymies are scrutinized or summarized, and geographic distributions are reviewed. Revisions should include an analysis of previous taxonomic work on the group and discuss biological information that has been developed. Revisions can provide a wealth of information concerning the taxa they include, and biological control workers should carefully study such publications.

Monographs are essentially revisions undertaken on a larger scale. Both kinds of research publication are descriptive in nature. Taxonomic works that carry the title of monograph suggest an exhaustive effort, implicitly justifying belabored discussion of minutiae. Biological control workers may glean potentially useful information embedded within comments that might otherwise go unnoticed or omitted from more telegraphic forms of taxonomic publication.

Taxonomic catalogs are invaluable to biological control specialists and taxonomists because such documents provide a compendium of information arranged by taxonomic categories. Taxonomic catalogs include citation of the original publication for each species, with authors' names and date of publication, synonyms of genera and species, and important taxonomic references; geographic distributions are given; hosts or host plants are listed; and biological references are generally provided.

Taxonomists also need collections of specimens of the organisms with which they work. Frequently, published literature that deals with a particular group of organisms is inadequate, contradictory, or outdated. Often, the final recourse to solving questions of identity is to consult authoritatively identified specimens preserved in museum reference collections. Therefore, the creation and maintenance of reference collections should be an important concern to all individuals involved in biological control, especially quarantine workers.

Collections are the basic tool of taxonomic research and they must be properly maintained and available for study by taxonomists. Financial support for private, government, and university natural history museums and insect reference collections has not kept pace with the development of biological science in general. Often facilities are old and obsolete, curatorial care is inadequate, and staff taxonomists and managers are few in number. This is a trend that all those who are involved in biological control work should seek to reverse, because the need for authoritative identifications and properly maintained reference collections to serve biological control projects is certain to continue.

The value and utility of biological reference specimens is strongly correlated with the curatorial care they receive and their prior use in research. Thus, the most valuable specimens are those that are authoritatively identified and properly labeled with relevant collection and determination information. The specimens on which the published descriptions of biological taxa are based are of the utmost value. These primary types (holotypes, lectotypes, and neo-

types) are the ultimate resource required to resolve questions involving the identity of taxa. Type specimens must receive extra care because they are essentially irreplaceable and should be regarded as the property of science in general. All primary types should be placed permanently in well-curated museum facilities where they can be made available whenever necessary for examination by qualified taxonomists.

Any specimen that has been studied and identified by a competent taxonomist becomes more valuable for reference purposes than similar unstudied specimens. The determination labels affixed to specimens by taxonomic specialists bear the scientific name of the taxon, the name of the identifier, and usually the date or year when the identification was made. Such labels should never be removed, even if the determination is later shown to be incorrect, which sometimes happens. The label, even if not currently correct, indicates the identifier's concept of the taxon at the time the determination was made, and this knowledge may be of value to present or future workers.

Collections of properly labeled specimens, whether identified or unidentified, are a major source of raw material for taxonomic research. The taxonomist who desires to carry out a comprehensive revisionary study of a taxonomic group must see material from as many sources as possible. Specimens obtained in the course of biological control research often are particularly valuable because they are frequently associated with host and other ecological data of taxonomic significance. Biological control workers often collect or rear reasonably large numbers of specimens of organisms with which they are working. These may prove useful to taxonomists by indicating the range of intraspecific variation in taxonomic characters. Even though it may be impractical to individually mount and label long series of specimens of every species, these should be preserved for future study.

Specimens deposited in permanent collections that serve to document research projects are known as voucher specimens. Voucher specimens are valuable in biological control programs because they document introductions, release programs, and recoveries. The need for voucher specimens has not been fully appreciated until relatively recently, and in the past such specimens often were not preserved. As a result, it is difficult or sometimes impossible to properly document or evaluate many of the biological control projects that were conducted before the deposit of vouchers became an accepted practice.

The importance of voucher specimens goes beyond active biological control efforts. Large amounts of information are generated in the form of published research. This research is based on material that has been referred to a known taxon (identified) or has been judged to belong to an undescribed taxon (new genus, new species, etc.). Frequently, however, material is misidentified or misdetermined. All the biological information published about a particular animal is potentially useless if voucher specimens are not deposited in museum collections for subsequent study. This is painfully evident in the case of *Trichogramma*. During the past 50 years, thousands of scientific reports have been published under various names, such as *T. pretiosum* and *T. minutum*. Current research now indicates that many determinations were incorrect (Pinto *et al.,* 1978, 1986). Unfortunately, it is difficult (and sometimes impossible) to determine the zoological species to which published information refers in the absence of voucher specimens.

The importance of attaching accurate information to specimens cannot be over-emphasized. Basic collection information that should be provided with specimens includes precise geographic location, date, host (in the case of parasites), host plant (in the case of phytophages), and important ecological data. In a sense, there is never too much information. Some workers have replaced collection information on labels with cryptic numbers of codes that refer to more extensive data in a field or laboratory notebook. This practice should be strongly discouraged. Over time, specimens become separated from notebooks and thus the material becomes scientifically worthless. Some workers replace named localities with data on longitude and latitude. This procedure has merit, but most entomologists still rely on geographic names. Spelling of place names should be carefully checked for accuracy.

Local and vernacular names for collecting sites should be avoided, and abbreviations should be clearly understood. Dates of collection should be clear, and numerical shorthand should be avoided. Collection dates such as "6–8–1986" could mean June 8 or August 6 to different workers. Roman numerals should be used for the month of collection and should follow the day of collection. Thus, "6–VIII–1986" would mean 6 August 1986 and could not be confused with "6–8–1986." Some workers tend to abbreviate the year of collection by using the last two numerals. By such convention, "86" could mean 1886, 1986, or another century. Admittedly, this is less likely a source of confusion.

MODERN SYSTEMATIC TOOLS AND TECHNIQUES

Preparation of Specimens

Each taxonomic group of organisms is unique in that it possesses suites of features that are used to differentiate this group from similar groups. Traditionally these differences have been morphological. To view these features to the best advantage of the taxonomist, specimens must be prepared for study in a manner appropriate to the group.

Microscope slide-mounted specimens are frequently employed to view morphological features that require higher magnification and better resolution than are available with reflected light and the traditional dissecting microscope. Small-bodied, weakly sclerotized organisms frequently are placed on slides for study.

Mounting media used in slide preparation may be characterized as temporary and permanent. Hoyer's medium is a popular temporary mounting among many entomologists because of the ease with which specimens can be prepared and because of the favorable refractive index of the medium. Unfortunately, temporary mounts usually are short-lived and break down. The widespread usage of this mounting among taxonomists is regrettable. Specimens so preserved are often rendered worthless for study within a comparatively short period of time. Future students of taxonomy and biological control should avoid temporary mountants. Convenience is no justification to jeopardizing scientifically valuable material. Collections of temporary mounts must be periodically examined for mountant breakdown and damage to the material. Important specimens must be periodically remounted. Thus, collections of specimens mounted in temporary mountants potentially require more time for conservation, therefore negating the advantage of time when compared to specimens preserved in media of a permanent nature. Particularly disturbing is the potential threat to type-material that must be remounted in a permanent medium if it is to be properly preserved.

Permanent microscope slide mounts are time consuming to prepare and require an element of acquired skill. Canada balsam is the permanent mountant most commonly used by entomologists. Permanent mounts should not be attempted for the first time when valuable material is involved. Despite the seeming inconvenience of this method, the amount of time invested is repaid by the permanency of the preparation. Thus, generations of taxonomic specialists and biological control specialists can study and restudy material in the search for new or unstudied characters. Type-material that must be slide mounted is preserved indefinitely in this manner.

A hybrid technique involving Hoyers medium and Canada balsam has been developed by acarologists and has been adapted by some biological control workers. The process involves mounting specimens in a temporary mountant, placing a coverslip over the material, drying the slide for an extended period of time (up to several months), and then sealing the mount with a permanent mountant and second coverslip over the first coverslip. Elimination of all water from the Hoyers medium during the drying process seems critical for suitable preservation (Les Ehler, personal communication). Preliminary results suggest this system is acceptable, but critical, long-term observation (ca. 10 to 20 years) should be made before such a procedure is universally adopted.

Point Mounts

Point mounts consist of small, acutely shaped triangular pieces of celluloid or paper. A conventional insect pin is thrust through the base of the mount, and the insect is attached to the acute-angled point with glue. Point punches are commercially available that make standardized points. Glues should be water soluble so specimens can be removed from the point if this becomes necessary. Paper used to make points should be high-quality, acid-free, 100% rag content. Two-ply bristol board makes points that are enduring; construction paper or note card paper should be avoided because they disintegrate over time.

In many parts of the world, particularly in North America, point mounts have been an accepted practice for preserving small-to-minute insect specimens. In this category fit most of the parasitic Hymenoptera that are too small to accept an insect pin without damage. Most taxonomists prefer material point mounted; however, before considerable effort is expended, the taxonomist in question should be consulted.

Card Mounts

Mounting specimens on rectangular cards has been a widespread practice in Europe for more than a century. In other areas it is becoming more commonly adopted. The technique has advantages and disadvantages, and workers should be aware of both. Well-prepared card mounts are superior to point mounts as a form of protecting the specimen from damage. Examining old collections containing card mounts and point mounts illustrates this point well. A potential disadvantage to card mounts is that they can obscure important diagnostic characters. This renders the specimen potentially vulnerable if it must be remounted because older specimens are more fragile than recently killed specimens.

Minuten Nadelns

Minuten nadelns should be avoided because they corrode with time and destroy or obscure important characters on small-bodied specimens.

Alcohol as a Preservative

Long-term storage of specimens preserved in 70% ethanol is not recommended for natural enemies, particularly parasitic Hymenoptera. Specimens frequently lose their color with time. Specimens become brittle or easily damaged if stored in alcohol for long periods of time. Perhaps more dangerous is the evaporation of alcohol with time. Collections are rendered scientifically worthless if the alcohol evaporates completely.

Specimens preserved in alcohol should be critically point dried before mounting on cards or points (Gordh & Hall, 1979).

Transmittal of Specimens

The transmittal of specimens between biological control specialist and taxonomist deserves some consideration. Specimens must reach the taxonomic specialist in a condition suitable for study. Labeling of material should be accurate and unambiguous.

Material sent for identification should be carefully curated. Pinned specimens should be firmly embedded in pinning bottoms and supplementary pins should be used such that specimens do not "cartwheel" on pins during transit. This action frequently has disastrous results. Also, a cardboard or styrofoam top should be placed between the lid of the box and top of the insect pins. This practice helps ensure that pins will not free themselves from the pinning bottom during transit in the mails.

Wrapping parcels is exceedingly important, and care should be exercised to ensure that packages conform to all national and international postal regulations. Parcels should have addresses legibly printed in indelible ink. Addresses should also be included within the package in the event of damage by the postal service. All customs regulations should be understood. Parcels should be sent airmail whenever possible.

Additional Technologies

Taxonomy should represent a dynamic area of investigation. As such, taxonomic research should respond to new concepts in classification and embrace new techniques in the search for new characters to distinguish among zoological entities. This notion was developed by Huxley (1940) and discussed in relation to biological control by Gordh (1982).

Scanning electron microscopy (SEM) has been used as a taxonomic tool for about 25 years and its use is becoming more widespread. As a method for illustration, micrographs provide a three-dimensional representation of form. While SEM does not replace traditional techniques of pen and ink for illustrating characters and character states, it does provide an unambiguous representation of structure and ultrastructure that cannot be obtained with optical microscopy. The employment of SEM can only increase the objectivity of taxonomic research. SEM micrographs are particularly useful for illustrating details of cuticular sculpture that are often among the best taxonomic characters available in many groups of parasitic Hymenoptera.

Biochemical analysis using the techniques of electrophoresis provides another as yet underutilized tool for discriminating between closely related taxa. Electrophoretic isozyme analyses have been used to demonstrate differences within closely related groups of *Trichogramma* species (Voegelé & Berge, 1976; Jardak *et. al.,* 1979; Pintureau & Babault, 1981; Hung, 1982). Nur (1977) demonstrated this technique can be used to differentiate between sibling species of mealybugs that are difficult or impossible to separate on strictly morphological grounds.

Distinguishing Similar Species

Most modern taxonomists subscribe to the biological species concept in which total reproductive isolation between organisms is taken as an indication of species status. Although the concept has considerable merit, several problems are associated with implementation. For example, most taxonomists work primarily with preserved museum specimens, and reproductive isolation cannot be tested in museum preserved material. Biological control workers, with laboratory and insectary facilities available, are better equipped than most museum taxonomists to carry on reproductive isolation studies.

The confirmation of reproductive isolation through hybridization studies in some cases has led to reevaluation of the comparative morphology of sibling species, which, in turn, has elicited minor but consistent anatomical differences. Examples of hybridization studies with closely related taxa pertinent to biological control include *Muscidifurax* (Kogan & Legner, 1970), *Aphytis* (Rao & DeBach, 1969a, 1969b, 1969c) and *Trichogramma* (Nagarkatti and Nagaraja, 1977). It is important to emphasize that the taxonomy of the cultures and species involved must be carefully researched before taxonomic decisions are made based on hybridization work.

The extent of reproductive isolation has been shown to vary among organisms. Hybridization experiments with natural enemies for use in biological control projects often yield living samples of closely related natural enemies from geographically and ecologically diverse localities. These can provide the raw material for the basic hybridization studies needed to clarify the taxonomic status of similar entomophagous forms.

Not all organisms reproduce sexually. In so-called uniparental organisms, the biological species concept cannot be used to test reproductive isolation because males do not exist or exist at very low percentages of the offspring and may not be functional. The phenomenon of female-only species is called thelytoky by workers in biological control. Unfortunately for biological control workers, thelytoky is common among natural enemies of agricultural pests and presents an obstacle to accurate identification. In the absence of tests for reproductive isolation, morphometric analysis may provide clues to identity of closely related or morphologically nearly identical forms.

Parlatoria pergandii Comstock (chaff scale) represents a problem on citrus in Texas. *Aphytis hispanicus* (Mercet) and *A. comperei* (DeBach & Rosen) are among the natural enemies found on chaff scale. Both species are thelytokous and similar (cryptic species). Key anatomical characters used to distinguish the species overlap. However, Woolley and Browning (1987) have used principal component analysis and canonical variate analysis to distinguish between the species. These and other statistical techniques may be used by museum taxonomists when electrophoretic analysis is not possible.

References

Blackwelder, R. E., & Blackwelder, R. M. (1961). Directory of zoological taxonomists of the world. Carbondale, IL: Southern Illinois University Press.

Clausen, C. P. (1942). The relation of taxonomy to biological control. Economic Entomology, 35, 744–748.

Common, I. F. B. (1958). A revision of the pink bollworms on cotton (*Pectinophora* [Lepidoptera: Gelechiidae]) and related genera in Australia. Australian Journal of Zoology 6, 268–306.

Compere, H. (1961). The red scale and its natural enemies. Hilgardia, 31, 1–271.

Compere, H. (1969). The role of systematics in biological control: A backward look. Israel Journal of Entomology 4, 5–10.

Danks, H. V. (1988). Systematics in support of entomology. Annual Review of Entomology, 33, 271–296.

DeBach, P., Rosen, D., & Kennett, C. E. (1971). Biological control of coccids by introduced natural enemies. In C. B. Huffaker (Ed.), Biological control (pp. 165–194). New York: Plenum Press.

Delucchi, V. L. (1966). The significance of biotaxonomy to biological control. Mushi, 39, 119–125.

Gordh, G. (1977). Biosystematics of natural enemies. In R. L. Ridgway & S. B. Vinson (Eds.), Biological control by augmentation of natural enemies (pp. 125–148). New York: Plenum Press.

Gordh, G. (1982). Taxonomic recommendations concerning new species important to biological control. International Journal of Entomology, 1(1), 15–19.

Gordh, G., & Hall, J. (1979). A critical point drier used as a method of mounting insects from alcohol. Entomological News, 90(1), 57–59.

Holdaway, F. G. (1926). The pink bollworm of Queensland. Bulletin of Entomological Research, 17, 67–83.

Holdaway, F. G. (1929). Confirmatory evidence of the validity of the species *Pectinophora scutigera* Holdaway (Queensland pink bollworm, from a study of the genitalia). Bulletin of Entomological Research, 20, 179–185.

Hung, A. C. F. (1982). Chromosome and isozyme studies in *Trichogramma* (Hymenoptera: Trichogrammatidae). Proceedings of the Entomological Society of Washington, 84(4), 791–796.

Huxley, J. (1940). Toward the new systematics. In J. Huxley (Ed.), The new systematics (pp. 1–46). London: Oxford University Press.

Jardak, T., Pintureau, B., & Voegelé, J. (1979). Mise en evidence d'une nouvelle espece de *Trichogramma* (Hym., Trichogrammatidae). Phenomene d'intersexualite; etude enzymatique. Annales de la Societe Entomologique France (N.S.) 15, 635–642.

Knutson, L. (1981). Symbiosis of biosystematics and biological control. In G. C. Papavizas (Ed.), Biological control in crop production (pp. 61–78). Beltsville Symposium in Agricultural Research, Symposium 5. Totowa: Allanheld, Osmun.

Kogan, M., & Legner, E. F. (1970). A biosystematic revision of the genus *Muscidifurax* (Hymenoptera: Pteromalidae) with description of four new species. Canadian Entomologist, 102, 1268–1290.

LePelley, R. H. (1935). The common coffee mealybug of Kenya (Hem., Coccidae). Stylops, 4, 185–188.

LePelley, R. H. (1943). The biological control of a mealy bug on coffee and other crops in Kenya. Empire Journal of Experimental Agriculture, 11(42), 78–88.

McKenzie, H. L. (1937). Morphological differences distinguishing California red scale, yellow scale, and related species (Homoptera: Diaspididae). University of California, Berkeley, Publications in Entomology 6, 323–335.

Nagarkatti, S. V., & Nagaraja, H. (1977). Biosystematics of Trichogramma and Trichogrammatoidea species. Annual Review of Entomology, 22, 157–176.

Nur, U. (1977). Electrophoretic comparison of enzymes of sexual and parthenogenetic mealybugs (Homoptera: Coccoidea: Pseudococcidae). Virginia Polytechnical Institute State University Research Division Bulletin, 127, 69–84.

Pinto, J. D., Oatman, E. R., & Platner, G. R. (1986). Trichogramma pretiosum and a new cryptic species occurring sympatrically in southwestern North America (Hymenoptera: Trichogrammatidae). Annals of the Entomological Society of America, 80, 1019–1028.

Pinto, J. D., Platner, G. R., & Oatman, E. R. (1978). Clarification of the identity of several common species of North American Trichogramma (Hymenoptera: Trichogrammatidae). Annals of the Entomological Society of America, 71, 169–181.

Pintureau, B., & Babault, R. (1981). Caracterisation enzymatique de Trichogramma evanescens et de T. maidis (Hym.: Trichogrammatidae): Etude des hybrides. Entomophaga, 26, 11–22.

Rao, S. V., & DeBach, P. (1969a). Experimental studies on hybridization and sexual isolation between some Aphytis species (Hymenoptera: Aphelinidae). I. Experimental hybridization and an interpretation of evolutionary relationships among the species. Hilgardia, 39, 515–553.

Rao, S. V., & DeBach, P. (1969b). Experimental studies on hybridization and sexual isolation between some Aphytis species (Hymenoptera: Aphelinidae). II. Experiments on sexual isolation. Hilgardia, 39, 555–567.

Rao, S. V., & DeBach, P. (1969c). Experimental studies on hybridization and sexual isolation between some Aphytis species (Hymenoptera: Aphelinidae). III. The significance of reproductive isolation between interspecific hybrids and parental species. Evolution, 23, 525–533.

Rosen, D., & DeBach, P. (1979). Species of Aphytis of the World (Hymenoptera: Aphelinidae). The Hague: W. Junk.

Sabrosky, C. W. (1950). Taxonomy and ecology. Ecology, 31, 151–152.

Sabrosky, C. W. (1955). The interrelations of biological control and taxonomy. Economic Entomology, 48, 710–714.

Schlinger, E. I., & Doutt, R. L. (1964). Systematics in relation to biological control. In P. DeBach et al. (eds.), Biological control of insect pests and weeds (pp. 247–280). London: Chapman & Hall.

Voegelé, J., & Bergé, J. B. (1976). Les Trichogrammes (Insectes, Hymenop., Chalcidiens, Trichogrammatidae), caracteristiques isoesterasiques de deux especes: Trichogramma evanescens Westw. et T. achaeae Nagaraja et Nagarkatti. Comtes Rendus del' Academie des Sciences Paris, 283, 1501–1503.

Woolley, J. B., & Browning, H. W. (1987). Morphometric analysis of uniparental Aphytis reared from chaff scale, Parlatoria pergandii Comstock, on Texas citrus (Hymenoptera: Aphelinidae; Homoptera: Diaspididae). Proceedings of the Entomological Society of Washington, 89(1), 77–94.

Zimmerman, E. C. (1979). Insects of Hawaii: Vol. 9. Microlepidoptera. Honolulu: University of Hawaii Press.

Molecular Methods in Classical Biological Control

THOMAS R. UNRUH

Yakima Agricultural Research Laboratory
USDA. Agricultural Research Service
5230 Konnowac Pass Road
Wapato, Washington

JAMES B. WOOLLEY

Department of Entomology
Texas A&M University
College Station, Texas

INTRODUCTION AND RATIONALE

The goal of classical biological control is to import and establish natural enemies that will control pest populations in selected habitats. Although dramatic successes have resulted from biological control introductions, the outcome is far from predictable. Only about 30% of introductions have resulted in establishments, and successful pest suppression has occurred less frequently (Hall & Ehler, 1979; Greathead & Greathead, 1992). We have written this chapter because we believe that increased understanding of natural enemy and pest groups will lead to increased introduction and success rates in biological control. Molecular markers provide new characters for study of phylogenetic relatedness, for identification of cryptic species and biotypes, and for assessment of heritable variation for population genetics and ecological investigations.

With some pest insects, decades of effort and many score of introductions have provided little or no control (e.g., Bermuda scale and gypsy moth, Clausen, 1978) suggesting an empirical practice of introducing new natural enemies as they are found may not be justified [but see van Lenteren (1980) for a lucid counter argument]. Heightened concern over introduction of exotic species including concern over past introductions of biological control agents (Howarth, 1991; Simberloff, 1992; Strong, 1997; Louda *et al.,* 1997) will mandate cautious and clearly defined selection criteria. One critical criterion is that the natural enemy with which we are dealing is taxonomically well characterized. In other words, we should know whether we are dealing with a single homogeneous population or one that includes descendents from several populations, encompass-ing broad geographic ranges or ecological capacities. Molecular characterization of the genetic structure of candidate natural enemy populations can provide this. To be explicit, the key to successful biological control may not be in another species but instead in different geographic races or biotypes (Diehl & Bush, 1984) of a species already introduced. A classic example was the success of the Persian strain of the walnut aphid parasite *Trioxys pallidus* (Haliday) in the harsh climate of central California where a previously introduced French strain had failed (van den Bosch *et al.,* 1970). Similarly, the perpetrator of nontarget effects may be a specific race of a polytypic species (Unruh & Goeden, 1987). Can molecular characterization of candidate natural enemy populations allow us to predict such adaptive differences? Probably not in all cases. However, to the extent that phylogeny can recapitulate capacities such as host range, climatic tolerances, and other traits of concern, molecular investigations can yield valuable insights and (hopefully) organizing principles for selecting natural enemies. We are just beginning to explore the diversity of nature in this way.

This chapter is organized into two sections. The first represents a primer on the nature of genetic variation and the major techniques employed to measure that variation for systematic and population studies. These include isozyme electrophoresis, mitochondrial and nuclear DNA restriction analysis and sequencing, DNA fingerprinting, randomly amplified DNA methods, and techniques to discover variable DNA sequence suitable for study. Our goal is to provide access to the concepts these methods embody and to direct the reader to useful references for learning the details. In the second half we describe how these methods may be used to address key systematic and population genetic questions in biological control.

MOLECULAR MARKERS

The Genetic Material

Nuclear DNA

The nuclear genome of insects consists of 200 million to 3 billion base pairs (bp) organized as a mosaic of 20 to 50 thousand genes and noncoding DNA (Spradling & Rubin, 1981; Lewin, 1995). Genes (or *cistrons*) occur in single copies, others may be moderately repeated, and still others—those coding for products that occur abundantly in the cell—are repeated hundreds or even thousands of times. Many cistrons are interrupted, divided into coding regions (or *exons*) and intervening, noncoding regions (or *introns*). Genes are variously interspersed and organized in a background of noncoding DNA. This includes variously sized stretches and types of repeating sequences, some in repeating patterns of only two to six nucleotides (microsatellites) and others in repeating motifs 7 to 200 bp long (minisatellites), which occur in many to many thousands of copies. Between 10 and 70% of arthropod nuclear DNA consists of moderately to highly repetitive sequences (Palmer *et al.,* 1994). Much of the longer minisatellite DNA is found in the telomeres and centromeres of the chromosomes. Much of the smaller repeats or microsatellites are dispersed throughout the genome (Lewin, 1995). Also the genome is littered with silent remnants of genes that have lost their active role, with transposons or mobile genetic elements and viruses (many of which have become stranded), and with other forms of variation including those yet to be explained (Lewin, 1995).

Ribosomal RNA

Ribosomal RNA (rRNA) is an important structural and functional component of the ribosome. DNA coding for rRNA (rDNA) occurs in multiple, tandemly repeated copies, usually localized in a part of the nuclear genome on one or several chromosomes. Each copy of the eukaryotic nuclear rDNA is organized as three genes, 18s, 5.8s, and 28s, separated from one another by two internal, transcribed spacer regions, ITS-1 and ITS-2. Separating the tandem copies are two flanking regions, an external transcribed spacer (ETS) and a nontranscribed spacer (NTS). The duplicated copies of the nuclear rDNA are arranged in tandem with all three coding regions and intergenic and intragenic noncoding regions in each repeat. Although there are multiple copies of this gene family, heterogeneity among copies is typically low or absent, homogeneity being maintained by concerted evolution (Lewin, 1995). The organization of nuclear rDNA in insects was reviewed by Beckingham (1982), and evolution and phylogenetic utility of eukaryotic rDNA was reviewed by Hillis and Dixon (1991).

Mitochondrial DNA

In contrast to the magnificently complex nuclear genome is the simple and elegant mitochondrial genome. It is particularly amenable to genetic and population studies because of several attributes of this extrachromosomal DNA (Brown, 1983; Moritz *et al.,* 1987; Harrison, 1989): (1) mitochondrial DNA (mtDNA) is a relatively small molecule of 15 to 20 thousand base pairs that are arranged in a covalently closed circle; its small size allows the molecule or a few fragments from it to be analyzed directly by electrophoresis. (2) mtDNA is readily isolated because mitochondria are discretely sized organelles easily segregated by centrifugation and occur abundantly in almost all cells. Also, mtDNA often has a buoyant density significantly different from nuclear DNA, further facilitating its purification from nuclear DNA by centrifugation. (3) The molecule is simple compared to nuclear DNA. It is about four orders of magnitude smaller; it has the same 36 to 37 genes in all eukaryotes, each gene being represented by only a single copy; and there are virtually no intergenic spacers, introns, or other noncoding sequence in the mtDNA molecule with the exception of the short control or AT-rich region (Zhang & Hewitt, 1996). Two of the 37 genes code for the 12s and 16s subunits of the mitochondrial ribosomal RNA. (4) mtDNA is haploid, which allows straightforward analysis of the resulting fragments on the gels. (5) It is maternally inherited, which simplifies genetic analysis. (6) Usually, all mitochondria within an individual have identical DNA sequences [homoplasmy; see Moritz *et al.* (1987) and Harrison (1989) for instances of heteroplasmy]. By 1995, the complete sequence of the mitochondrial DNA was known for *Drosophila yakuba* (Clary & Wolstenholme, 1985), the honeybee (Crozier & Crozier, 1993), two *Anopheles* species, and *D. melanogaster* (Lewis *et al.,* 1995). In some exceptional instances parts of the mitochondrial genome have been found duplicated in the nuclear genome, which may present difficulties in phylogenetics (Zhang & Hewitt, 1996).

Molecular Variation

Three nonexclusive forms of nucleic acid variation occur: single base pair substitutions, loss or gain of blocks of DNA, (e.g., of a repeat in satellite DNA), or inversion or translocation of a block of DNA. Each of these processes can be combined by genetic recombination and crossover during the sexual cycle in diploid organisms, with the latter being facilitated by recombination during meiosis. Differences in DNA sequence arising from these phenomena are expected among all organisms, except in clonal lineages, identical twins, and the like. These differences can be measured for a subset of the genome by DNA sequencing, providing exact and abundant information. Differences can also be measured indirectly, although often quite accurately, by methods that include isozyme electrophoresis, restriction fragment analysis, randomly amplified polymorphic DNA, DNA fingerprinting, and related techniques. We survey these methods later. For greater detail on the acquisition, interpretation, and analysis of nucleic acid data for systematics and population studies, we recommend Hillis *et al.* (1996c) and Avise (1994).

Two qualitatively distinct methods to measure sequence divergence are immunological distance and DNA–DNA hybridization. Both techniques yield pairwise measures of overall sequence differences between samples. Such distances can be used in generating phylogenetic hypotheses, but there are conceptual problems on this front (Swoford *et al.* 1996). Because no discrete characters are elaborated from distance techniques, they are ill suited to population biology studies. We do not present these distance approaches because of these weaknesses. Good reviews of immunological distance techniques (Beverley & Wilson, 1985; Maxson & Maxson, 1990) and DNA–DNA hybridization methods (Werman *et al.,* 1996) are available. We restrict our presentation to methods that both estimate DNA sequence variability and can provide discrete characters.

Enzyme Polymorphism

Gel electrophoresis and histochemical localization of enzymes within the supporting gel matrix (Hunter & Markert, 1957) comprise a powerful method for measuring genetic variation within and between animal populations (Hubby & Lewontin, 1966; Lewontin, 1974) including pest species and their natural enemies (Loxdale & den Hollander, 1989). This allozyme method consists of separating enzyme molecules in a crude or purified homogenate by placing the homogenate in a supporting matrix (e.g., starch, acrylamide, cellulose acetate) saturated with an appropriate buffer. An electric field is applied to the support matrix; and the proteins are separated by their differences in size, conformation, and molecular charge. The enzymes are visualized by applying a stain that develops a color when substrates in the stain are modified by a specific enzymatic activity. Over 100 different enzymes can be visualized with specific stain recipes (Harris & Hopkinson, 1976; Richardson *et al.,* 1986; Manchenko, 1994; Murphy *et al.,* 1996).

Allozyme data consist of the frequency of mobility variants or alleles at one or more enzyme loci (genes) for one to several populations. Phenotypes of individuals as seen on the stained gels are usually interpretable as genotypes, given knowledge of the insect's ploidy and the subunit structure of the enzyme. Because more than one locus may exist for a specific enzyme function, it is possible to have several bands (isozymes) on a gel, arising from an unknown numbers of loci and alleles, and occasionally from nongenetic phenomena. These situations are usually obvious but require breeding and genetic analysis to identify alleles segregating at each locus (Richardson *et al.,* 1986).

The allozyme method is based on the assumption that each enzyme is coded by one gene and substitutions in its nucleotide sequence will result in amino acid substitutions in the coded protein that can be visualized by electrophoresis (Lewontin, 1974). Due to redundancy of the genetic code, about one-third of the possible nucleotide substitutions in a gene cause no amino acid substitutions and only about 10 to 20% of the amino acid substitutions that occur are detectable by electrophoresis (Nei & Chakraborty, 1973; Nei, 1987). Nei (1987) estimates for the average enzyme locus that about 8%, or about 100 nucleotides will, if substituted, result in detectable variation in mobility in the enzyme or protein product. Hence, protein electrophoresis, although highly correlated with underlying DNA sequence variation, is an inexact measure of sequence differences. Given that a sample of 10 enzyme loci is roughly equivalent to seeking sequence differences in 1000 nucleotides, enzyme electrophoresis is the most cost effective of the methods described here for estimating genetic differentiation.

Isoelectric focusing (electrophoresis in a pH gradient) is not significantly superior to zone electrophoresis in detecting this variation [Aquadro & Avise, 1982, but see Coyne (1982)] but is superior in its ability to concentrate proteins while it separates them (Righetti, 1983). Hence, isoelectric focusing is valuable for studies of very small organisms (Kazmer, 1991), where enzyme concentrations in homogenates are dilute. Cellulose acetate electrophoresis also has proved valuable for very small organisms (Easteal & Boussey, 1987).

Techniques for Studying DNA

Many books and thousands of primary references exist on methods to characterize DNA, and exhaustive review of these is unwarranted. However, there are key techniques for population and systematic studies that we review briefly. We do not review methods to clone DNA, the procedure of inserting DNA in bacteria or their parasites, to magnify and isolate DNA sequence. A general introduction to cloning and analysis of cloned sequences is provided by Brown (1990) and by Warner *et al.* (1992). Cloning strategies relevant to systematics and population studies together with protocols are provided by Hillis *et al.* (1996b).

The Polymerase Chain Reaction

The polymerase chain reaction (PCR) is a method of amplifying a specific 50- to >5000-bp region of DNA by a factor of about 10^9 with high accuracy (Saiki *et al.*, 1988; Stoflet *et al.*, 1988; Innis *et al.*, 1990). Standard PCR requires knowledge of the sequences flanking both ends of the target DNA sequence. Short oligonucleotide primers (20 to 40 bp each) that are complementary to the flanking regions are added to a solution along with DNA containing the target sequence and the enzyme DNA polymerase. PCR consists of multiple cycles of heating, which denatures the double-stranded target DNA into single strands, followed by cooling, which allows the primers and the target DNA to anneal. In the cool phase, the enzyme DNA polymerase makes a copy of the region flanked by the primers. Because both the original sequence and those synthesized in previous cycles are copied, the reaction doubles the amount of the sequence with each cycle. Variations on the general theme of PCR include long-PCR, allowing amplifications up to 35 kb (Rohrdanz, 1995); inverse PCR that copies the region upstream and downstream from a single primer position by circularizing the area of interest (Ochman *et al.*, 1988); and anchored PCR, which uses a primer for only one side of the target sequence (Loh *et al.*, 1989). PCR can also be used to quantify the number of targets in a solution (Wang & Mark, 1990). PCR products can be taken directly into or combined with the dideoxy sequencing techniques (Gyllenstein & Erlich, 1988) or other methods of interpreting nucleotide variation as discussed later.

Randomly Amplified Polymorphic DNA

Randomly amplified polymorphic DNA (RAPD, pronounced "rapid") is an extension of the PCR that can uncover extensive molecular variation useful for genomic mapping and for providing markers that vary between populations and species (Williams *et al.*, 1990). The RAPD technique consists of using a single 10-nucleotide primer of arbitrary sequence in the PCR. Amplification occurs when the sequence complementary to the primer exists in the appropriate orientation on both strands of the sample DNA within 50 to 5000 bp of each other. The random occurrence of complementary sites throughout the genome produces several different sizes of oligonucleotides with most arbitrary primers, and many of these will be polymorphic among individuals and populations (Williams *et al.*, 1990).

RAPDs have proved extremely valuable in genomic mapping, particularly in plants (Michelmore *et al.*, 1991), and can be used for parentage analyses (Hedrick, 1992; Scott *et al.*, 1992). The technique has also been used to fingerprint insect biotypes (Puterka *et al.*, 1993), identify species (Landry *et al.*, 1993), and estimate genetic differentiation (Kambhampati *et al.*, 1992). RAPD polymor-

phisms have been used in studies of aphids (Black *et al.*, 1992; Puterka *et al.*, 1993; Nicol *et al.*, 1997), fruit flies (Haymer & Mcinnis, 1994), dragonflies (Hadrys *et al.*, 1993), weevils (Williams *et al.*, 1994), pyralid moths (Dowdy & Mcgaughey, 1996), honeybees (Hunt & Page, 1992), leaf-cutting bees (Lu & Rank, 1996), paper wasps (O'Donnell, 1996), ants (Hasegawa, 1995), parasitoids (Edwards & Hoy, 1993, 1995; Kazmer *et al.*, 1995, 1996; Landry *et al.*, 1993; Roehrdanz *et al.*, 1993; Vanlerberghe-Masutti, 1994a, 1994b), and insect pathogens (Hajek *et al.*, 1996), among others.

RAPDs are very inexpensive compared with most other DNA techniques, but unfortunately RAPDs also have serious problems. While careful studies of the transmission genetics and linkage relationships of RAPD bands have shown that they are usually inherited as Mendelian dominants, seen as the presence and absence of bands on a gel (Black, 1993; Hunt & Page, 1992), many cases are known in which RAPD bands have a more complex inheritance. Nonparental bands may appear in progeny of hybrid crosses (Kazmer *et al.*, 1995), polymorphic bands may represent fragment length polymorphisms (FLP) or FLP-like loci (Hunt & Page, 1992; Kazmer *et al.*, 1995), and sometimes bands may be found in diploids only [apparently resulting from heteroduplex formations from alternate alleles in heterozygotes (Hunt & Page, 1992)]. Many studies have demonstrated that lack of repeatability of RAPD banding patterns may occur through time, between labs, or even between PCR machines [Ellsworth *et al.* (1993), Penner *et al.* (1993), and Grosberg *et al.* (1996) provide long lists of references]. RAPDs reveal much more variation than seen with allozymes which may be critically important for insects groups such as parasitoids where enzyme polymorphism can be low (Unruh *et al.*, 1986). However, the heritable nature of RAPD variation must be tested in each case and strict standardization of procedures is imperative (e.g., Kazmer *et al.*, 1995). Because the genetic basis for RAPD bands is usually unknown, this means that homology of bands between taxa cannot be determined with any certainty (Grosberg *et al.*, 1996) without sequencing or methods such as heteroduplex analysis (see later). RAPDs are very inexpensive compared with most other DNA techniques, but this crippling problem dramatically reduces the utility of RAPDs for phylogenetic studies (Black, 1993; Rieseberg, 1996; Smith *et al.*, 1994) and can obviously influence their accuracy in population genetics studies (Black, 1993). We wish to emphasize that uncertainty about homology of features is not an esoteric, abstract problem, but represents profound difficulties at the very core of comparative or evolutionary studies.

Also unfortunate is the apparent abandonment of common sense that has often accompanied the use of RAPDs. Because these markers are typically dominant, they have significantly less power in resolving genetic structure than do allozymes or other codominant traits (Lynch & Milligan, 1994). Nevertheless, many studies have appeared with much smaller sample sizes than would be used in an allozyme study. Those considering a molecular genetic study of populations should consider costs of proving RAPD polymorphisms and their lower power against the lower variation found in allozymes or the higher costs of uncovering and measuring microsatellite variability (see later).

The resolving power of the RAPD technique can be enhanced by analytical thin-layer acrylamide electrophoresis and single-strand conformation polymorphism (see later; Hiss *et al.,* 1994; Antolin *et al.,* 1996a), but the degree to which these methods eliminate homology problems with anonymous DNA remains to be quantified. Lynch and Milligan (1994) and Milligan and McMurry (1993) refine the statistical underpinnings and discuss the logic of using dominant variation such as found in RAPDs and amplified fragment length polymorphisms (AFLPs, pronounced "aflips").

DNA Restriction Fragment Polymorphisms

Restriction endonucleases are enzymes that cleave DNA at characteristic 4 to 6 base sequences (Smith, 1979). For example, the restriction enzyme known as EcoRI cleaves 5′-GAATT-3′ between the G and A. Differences in the DNA of individuals or populations in the presence and position of these specific recognition sequences can be measured through a multistep process as follows. DNA is purified from tissue or amplified by PCR, restriction enzymes are used to cleave a sample of the DNA, and resulting molecules are separated by electrophoresis and visualized by staining. The resulting different numbers and sizes of DNA molecules between the samples are called restriction fragment length polymorphisms (RFLPs) or restriction polymorphisms. Over 2000 restriction enzymes are now known and information on their specificity and commercial availability can be accessed on the world wide web (Roberts & Macelis, 1997). Restriction enzyme patterns have been used to characterize polymorphisms in natural populations since 1979 (Avise *et al.,* 1979). Dowling *et al.* (1996) and Aquadro *et al.* (1992) present detailed methods.

Loss or gain of a restriction site implies a change in one nucleotide in the recognition sequence. Thus, if a six-base restriction enzyme cuts one sequence at five places and another at only four of the five, then a one-base difference at one of the recognition sites is assumed. Because of the unique properties of mtDNA described earlier and especially because of it being haploid, most of the research using RFLPs in systematics and evolutionary biology has utilized this molecule. Many studies have capitalized on the fact that individuals from different geographic populations of a species often differ in RFLP banding patterns. Moreover, differences are few in number with the choice of the

appropriate enzymes, and the investigator is not overwhelmed with differences that are difficult to interpret. Establishing the homology of fragments and presumed cut sites is usually straightforward (Avise & Lansman, 1983, Dowling *et al.,* 1996).

These and other properties of mtDNA RFLP data also make it quite suitable for phylogenetic analysis. [Avise (1994) and Swofford *et al.* (1996) discuss the unique properties of these data, some of which are advantageous.] Studies using RFLP of mtDNA have consistently produced clear models of the evolutionary history of related geographic populations, or "intraspecific phylogeographies" (Avise *et al.,* 1987; Avise, 1992, 1994), the level of resolution depending on the rate of evolution of the mtDNA in the targeted group (Chang *et al.,* 1989). In some vertebrate species, mtDNA restriction sites are so variable that restriction patterns are virtually individual specific, allowing their use as a "DNA fingerprint" (Avise *et al.,* 1989). For comparisons among higher taxa, the utility of mtDNA RFLPs is limited by the rapid accumulation of transitional nucleotide substitutions (Brown, 1983), which decreases the signal to noise ratio to unacceptably low levels. Interpretation of RFLP from a diploid genome is more problematic. For example, multiple cuts in different places in a sequence amplified from chromosome pairs can produce confusing results. Thus, RFLP analysis of genomic sequence is typically preceded by cloning. Methods of statistical inference from restriction fragment data are reviewed by Engels (1981), Nei (1987), and Avise (1994).

DNA Fingerprinting: Minisatellites

Multilocus minisatellite DNA fingerprinting is a specialized technique with limited but powerful applications in population biology (Lewin, 1989). Minisatellites are tandemly arranged repeats of sequences seven to several hundred base pairs in length. Multilocus DNA fingerprinting consists of fragmenting total genomic DNA with restriction enzyme(s), separating the cut DNA by electrophoresis, and identifying a subset of the thousands of bands with genomic probes (a complementary sequence previously isolated and labeled by fluorescence or isotopes to allow visualization [Warner *et al.,* 1992]). Many probes have been derived from the human genome and consist of several copies of characteristic core repeat motifs (Jeffreys *et al.,* 1987). Synthetic oligonucleotides with simpler repeat motifs can also be used, as was done by DeBarro *et al.* (1994) for an aphid.

The fragments visualized are so numerous and variable in occurrence among individuals that they display banding patterns often individually unique. This has led to the use of DNA fingerprinting in forensic sciences, parentage analyses, pathogen identification, and various other medical uses (Kirby, 1992). Because the regions in the genome giving rise to the visualized fragments are often unknown, because the presence of a band is usually a dominant trait, and because several bouts of genetic recombination can produce a very different banding pattern, the method has limited use as a multigenerational marker for population studies [see Lynch (1988, 1990)]. However, DNA finger-printing is extremely powerful in identifying the relation-ship of close relatives (parent, sibling) that can be used to assess the fitness of individuals in nature by identifying their progeny (Lewin, 1989). Multilocus fingerprints are also valuable in selective breeding programs (Hillel *et al.,* 1989) and recent theory suggests fingerprints can be used to estimate population heterozygosity (Stephans *et al.,* 1992). Danforth and Freeman-Gallant (1996) present a more conservative view along with corrections for esti-mates of relatedness between individuals. Haymer *et al.* (1992) show that DNA fingerprints can be used to identify closely related Mediterranean fruit fly strains.

Amplified Fragment Length Polymorphisms

Use of AFLPs comprises a PCR-based method to visu-alize selected products of a total genomic DNA restriction digest (Vos *et al.,* 1995; Zabeau & Vos, 1993). This method consists of digesting purified DNA with a pair of restriction enzymes that create fragments with enzyme-specific sticky ends (i.e., overhanging short segments at the cut sites). These are ligated to adapters that complement the cut ends of the digested DNA and contain the additional known sequence. All fragments with ligated adapters are amplified in a first round of PCR using primers matching the known sequence of the adapters and also containing one to five extra nucleotides extending into the anonymous sequence (i.e., beyond the cut point). These extra nucleotides are arbitrarily chosen and cause the primers to work for only a subset of the molecules—those that are homologous for these arbitrarily chosen nucleotides (Vos *et al.,* 1995). A second round of PCR is often employed with one to four more extra arbitrary nucleotides attached to the same prim-ers. These selective amplification procedures act to amplify only a subset of the many thousands of cut bands. The resulting products are separated by high-resolution electro-phoresis and can be visualized either by silver staining, or by using fluorescent or radiolabeled primers in the last round of PCR (Vos *et al.,* 1995). The final results are dozens of amplified bands for each restriction enzyme and arbitrary linker combination, many displaying presence–absence polymorphism. The method has been applied pre-dominantly to plants (e.g., Travis *et al.,* 1996; Beismann *et al.,* 1997) but should prove extremely valuable for genome mapping and population genetic studies of insects. AFLPs represent a different source of variation that may have fewer reliability and homology problems; however, the sta-tistical caveat we described for dominant RAPD bands also applies to multilocus fingerprinting and AFLPs.

DNA Sequencing

Two major DNA sequencing techniques exist (Brown, 1990; Hillis *et al.,* 1996a). The most used one consists of synthesizing complementary strands from a single-stranded DNA template (the target DNA) and a short primer in the presence of low concentrations of a chain-terminating di-deoxynucleotide. This produces a family of bands that vary in length, running from the starting primer sequence and ending where the chain-terminating base is integrated. These are then separated by analytical electrophoresis and are visualized by autoradiography or fluorometry. Prior to sequencing, the target DNA must be isolated, usually by restriction enzymes, and produced in abundance, usually by cloning or PCR. However, cycle sequencing, a combination of the dideoxy method and PCR, allows sequencing from very small samples of starting DNA (Gyllenstein & Erlich, 1988). Sequencing strategies and methods are provided by Hillis *et al.* (1996b). Sequencing is still generally too costly to use for the many samples required for population studies but is the method of choice for molecular phylogenetics.

Heterologous Sequence and Primer Design

A major challenge in acquiring DNA sequence data is isolating a part of the immense insect genome to sequence. One approach, suitable for both nuclear and mitochondrial DNA, is to use the DNA sequence already known from studies of related organisms to design PCR primers for the region of interest in a different species. Simon *et al.* (1994) reviewed studies of insectan mtDNA and provided a com-pilation of primers suitable for amplifying most segments of the molecule. Brower and DeSalle (1994) provide a similar compilation for several nuclear genes, with special references to rDNA and intron-crossing primers. The Uni-versity of British Columbia currently sells primer kits for insectan mitochondrial and nuclear genes, partly based on these publications. An exhaustive source of information on DNA sequence is the world gene banks, accessible through the world wide web (e.g., GenBank; Benson *et al.,* 1997). From this source one can locate the primary references for the sequences of organisms most related to the taxa of interest.

The use of these heterologous sequences for primer de-sign is facilitated by the mosaic nature of DNA variability throughout the genome. In general, protein-coding regions are much more conserved than noncoding regions (Lewin, 1995). Hence, primers can be designed from the coding regions (exons) of a gene so that they amplify noncoding DNA in an intervening intron. This method, termed exon primed intron crossing polymerase chain reaction (EPIC-PCR) is described by Lessa and Applebaum (1993). The organization of rDNA (described earlier) is analogous and has extremely useful properties for analysis of closely related entities. Because the genic regions tend to be considerably more conserved than either the internal or the external spacer regions, it is possible to design PCR primers to span these. Variation useful at the level of conspecific populations and closely related species is attainable. This strategy was used by Campbell *et al.* (1993) to find a useful sequence in a study of cryptic species of *Nasonia* (Hymenoptera: Pteromalidae), previously recognized by Darling and Werren (1990) on the basis of postzygotic reproductive isolation, host preferences, differences in allozymes, and subtle morphological differences. Even within coding regions of genes, including rDNA, significant differences in the amount of nucleotide variability may exist among portions of the gene. Similarly in mtDNA, some genes and domains within genes are likely to be more variable than others. Thus, the investigator can choose that region of the genome most appropriate for the degree of genetic differentiation likely to be found among taxa or populations to be studied.

Seeking Variation for Population Studies

Given the availability of primers from many parts of the genome, it is now possible to amplify by PCR a sequence that is likely to show high sequence variation among individuals, populations, or subspecies. However, in many cases, direct sequences of many such amplified regions will be cost prohibitive. Several new methods have been developed to determine if sequence differences exist in similarly sized segments of DNA, indicating if sequencing would be fruitful. These include the venerable restriction analysis (see earlier) and newer techniques including single- and double-strand conformation polymorphism, denaturing and temperature gradient gel electrophoresis, and various forms of DNA and RNA heteroduplex analysis. These approaches can be used to both reveal new variation and also to group samples into classes displaying characteristic patterns and then sequencing only a subset of these classes. Slade *et al.* (1993) used EPIC primers to amplify introns from several genes to discover a useful variation for population studies. They sequenced amplification products from geographically disparate individuals and chose restriction enzymes that would cut at the variable sites they observed. Subsequently, restriction analysis was used to assess population-level variation. They suggested this approach was more efficient than using a panel of restriction enzymes in an exhaustive empirical screening (Slade *et al.,* 1993).

Lessa and Applebaum (1993) review three additional methods to screen for variation. Single-strand conformation polymorphism (SSCP) separates DNA based on the conformation that each single-stranded DNA segment will take when it is quickly chilled and folded onto itself (Orita *et al.,* 1989). The method is most sensitive with shorter sequences (~200 bp), but conspecific variation in DNA over

700 bp has been identified by this approach, indicating sensitivity is both size and sequence dependent (Orti *et al.,* 1997). Hiss *et al.* (1994) described the suitability of SSCP and silver staining for mtDNA products in a variety of insects. SSCP has proved to be an important adjunct to RAPD PCR, providing greater sensitivity in identifying RAPD polymorphisms and increasing the utility of RAPDs in insect genome mapping by exposing more codominant variability (Antolin *et al.,* 1996a). Double-strand (ds) conformation polymorphism is a method to uncover variation in DNA that has significant tracts of adenine, such as the control region of insect mtDNA. Adenine imparts curvature to dsDNA and differences in their quantity and position will cause molecules to change shape enough to be separated in analytical electrophoresis conditions (Atkinson & Adams, 1997).

Denaturing gradient gel electrophoresis (DGGE), and temperature gradient gel electrophoresis (TGGE) are often combined with heteroduplex analysis (HA) (Lessa & Applebaum, 1993). In HA one utilizes a standard DNA source and a test source, denatures each by heating, mixes them, and allows the strands to reanneal. These are then separated by electrophoresis. Homoduplex and heteroduplex molecules show different mobilities because those with base pair mismatches partially denature and migrate more slowly. These mismatch influences are capitalized on when heteroduplex molecules are electrophoresed through denaturing conditions, or more specifically through gradients of increasing denaturing conditions. Bands with greater mismatch disassociate and their mobility in gels decline at less extreme denaturing conditions than do bands with lower mismatch. Of the two methods, TGGE is more expensive because it requires specialized equipment to impose a temperature gradient on the gel. Both methods are improved for longer sequence (>250 bp) if a ~40 bp gc clamp repeat is added [Lessa & Applebaum (1993) and their cited references]. Tek Tay *et al.* (1997) utilized DGGE to characterize mtDNA haplotypes in a study of the population structure of a *Rhytidoponera* ant, sequencing DNA from 18 individuals to validate DGGE-determined haplotypes of over 1000 ants.

A new and powerful modification of the heteroduplex approach is the RNA–RNA heteroduplex (Lenk & Wink, 1997), in which RNA is transcribed from one strand of the PCR product using a special primer and the alternate strand in another sample; the resulting RNA bands are mixed and allowed to anneal. These are digested with RNAase A, which cuts double-stranded RNA at the mismatches. The potential strength of this method is it can detect all mismatches, even in long sequences. In contrast, SSCP and DDGE only work reliably with segments less than 500 bp, and optimally with those having 200 bp or less. However, RNA duplex studies require rigorous electrophoretic conditions because mismatches that result in cleavage pieces that are very small (i.e., are close together or near the ends of the sequence) may be missed by most electrophoretic techniques.

Microsatellites: The Cool New Tool for Population Genetics

Microsatellites consist of moderately repeated tandem copies of simple sequence repeats (SSR; 2 to 6 bp in length) flanked by unique sequences. Primer sequences are determined from sequencing candidate microsatellites (identified by probing a genomic library with synthetic SSRs or by screening RAPD bands (Ender *et al.*, 1996). Zietkiewicz *et al.* (1994) offer an alternative approach to finding microsatellites using degenerate primers and anchored PCR. Once primers are found and proven, unique microsatellites can be amplified by PCR from total genomic DNA. Microsatellites are often highly polymorphic, with as many as 20 alleles segregating in natural populations. Banding is visualized on analytical polyacrylamide gels (sequencing gels) with ethidium bromide, silver staining, or autoradiography; and then it is interpretable as a simple codominant Mendelian variation, like allozymes.

Microsatellite variation is thought to arise from mismatch errors leading to unequal crossover among the repeats. In other words, when an unequal crossover occurs, one allele gains one or more repeat unit(s) and the other allele loses one or more. Alleles are not independent of each other—those different by one repeat unit are more likely to change into similar sized alleles than to change into a strongly different size class. Similarly, when the repeat number of an allele is high, it is more likely that a crossover event will change the repeat number. Slatkin (1995) corrected statistics for the mutation model underpinning high microsatellite variation. The high variation in microsatellites is generated by high mutation frequencies making the method inappropriate for phylogenetic analysis at all but the lowest taxonomic levels. Goldstein and Pollack (1997) describe the pitfalls and prospects of using microsatellites for phylogenetics.

For looking at closely related individuals or populations, microsatellites are unequaled once the effort to design primers has been completed. However, for comparative work with populations or higher levels, allozymes should be tried prior to development of microsatellites because they are much more cost effective if sufficient variation occurs. Many studies have appeared in the last 4 years describing the breeding systems or nest mate kinship patterns in eusocial to colonial insects [e.g., *Megachile rotundata* in Blanchetot (1992); numerous examples in Hedrick and Parker (1997)]. Microsatellites may allow significantly more studies of population structure of parasitic Hymenoptera that show little or no allozyme variation (Unruh *et al.*, 1986; Hughes & Quellar, 1993). One microsatellite was found in the parasitoid *Aphelinus abdominalis* (Vanlerberghe-Masutti & Chavigny, 1997). Goodman (1997) provides corrected *F* statistics for estimating genetic structure of natural populations with microsatellites and reviews previous conceptual contributions.

APPLICATIONS IN BIOLOGICAL CONTROL

Systematics

Because biological control projects typically utilize or study many closely related entities in the range between geographic races and cryptic species, it is not surprising that we typically (perhaps inevitably) encounter taxonomic problems. The decision as to whether such entities are to be treated as subspecific entities (subspecies, races, populations, etc.) or species rests initially on one's view or concept of the nature of a species. Unfortunately, this is one of the most complex and refractory areas of evolutionary biology. Almost 150 years after Darwin, we are far from a consensus as to the nature of animal species (Otte & Endler, 1989), and therefore far from a consensus as to what sorts of evidence are regarded as conclusive for recognizing species. Nevertheless, we will make some recommendations.

First, biological control workers need to be aware of the complexities underlying such decisions, and they should adopt a realistic but pragmatic view of species. No one criterion is likely to be conclusive (or even satisfactory) in all cases. For example, many would suggest that reproductive compatibility should be the basis for deciding if different entities represent different species (i.e., the biological species concept). However, *Wolbachia*-induced reproductive incompatibility may be common in insects (Werren, 1997), and complex patterns of reproductive isolation may therefore be caused by factors that have little to do with evolutionary divergence and that may be quite unstable in evolutionary time. On the other hand, information about reproductive compatibility between entities is obviously useful if multiple collections of natural enemies are made from different geographic areas.

Many have suggested that evolutionarily distinct lineages with independent evolutionary fates should be treated as species, regardless of reproductive status. [See Cracraft (1989) and Templeton (1989) for two very different points of view.] Such a phylogenetic species concept may well be useful for biological control, because it represents an excellent way to integrate various kinds of data, particularly different kinds of molecular data. Explicit hypotheses of phylogenetic relationship between populations or species of interest to biological control will provide a powerful framework for studying any number of issues important to biological control (Coddington, 1988). For example, the ability to place host relationships in a phylogenetic context will allow researchers to pinpoint the exact points at which host shifts have occurred and to study particular adaptations or genetic changes associated with host shifts.

Phylogenetically sound classifications enhance our ability to predict many characteristics of species new to science. For example, because all known species in the genus *Marietta* (Aphelinidae, Hymenoptera) are hyperparasites (Clausen, 1940), one would predict that a newly encountered *Marietta* would also have this habit. Likewise, because all known species of *Aphytis* are primary, external parasitoids of armored scale (Rosen & DeBach, 1979), it is extremely unlikely that a new species of *Aphytis* will have a different habit. These examples work because species of *Marietta* and *Aphytis* almost certainly share common ancestors with the same respective biologies, and precisely this common ancestry underlying natural classifications provides their predictive power. Unfortunately, classifications of many important parasitoid groups are not natural in this sense, because it has often been quite difficult to finds reliable morphological characters on which to base hypotheses of common ancestry.

Analysis of molecular data in a phylogenetic context requires that one be aware of the particular properties of different types of data and of different analytical approaches to data analysis. As far as the data itself, the issue of homology between character states (particular markers) is paramount. Two similar features in two different organisms are said to be homologous if the presence of the feature in both is due to inheritance from a common ancestor. In other words, shared homologous features provide evidence for taxonomically natural groups. Complex or evolutionarily conservative characters provide the best source of evidence for common ancestry. For example, because all vertebrates (including extinct species) share a common type of internal skeleton, we assume that the common ancestor of vertebrates had such a skeleton. Such arguments can also be framed in terms of the rate of evolutionary change; for example, it appears that major changes to the general form of the vertebrate skeletons are most uncommon, and a person with an understanding of the detailed structure of extant vertebrates would probably have no problem recognizing the corresponding bones in a *Tyrannosaurus rex* skeleton in a museum. This example also works because of the complexity of the character system: the complex shapes and positional relationships of vertebrate bones provide a wealth of information on which to base hypotheses of homology.

Exactly the same criteria apply to molecular data. Because the biochemical procedures to stain for allozymes are complex and quite specific, it is commonly assumed that alleles with similar mobility patterns on gels are homologous. Such hypotheses are less strong in systems with lots of allelic polymorphism such as esterases because of the increased probability of confounding two different alleles with similar mobility patterns. A major message to emerge from the molecular revolution of the last two decades is that all genes, or DNA sequences, do not evolve at the same rates. This became apparent in the earliest interspecific comparisons with allozymes (Johnson, 1974) and has been a recurrent theme in evolutionary studies of the eukaryotic genome up to now. This variation in evolutionary rates has implications for the analytical procedures used for phylogeny estimation (Swofford *et al.,* 1996) and has practical implications for choosing an appropriate molecular technique for addressing specific phylogeny at different levels of evolutionary divergence (Hillis *et. al.,* 1996b). Thus, the rate at which particular molecular markers are evolving will influence to a large degree their usefulness or reliability for phylogenetic models. Markers that are evolving too slowly will not provide sufficient information. Markers that are evolving too rapidly provide abundant information, but because the probability of independent evolution of particular characters is high (or the gain of a marker followed by its loss), they are unreliable for determining patterns of common ancestry because the ratio of phylogenetic signal to noise is too low.

Therefore, there are three questions that must be asked if one is contemplating phylogenetic study of molecular markers. First, is the marker evolving at a rate appropriate for the biological system under study, so that we can be reasonably confident that individuals that share markers do so because of common ancestry and not because of coincidence? One of the reasons that mtDNA has been used so successfully in many studies of infraspecific phylogeography (see earlier), is that six-base-cutter restriction endonucleases typically generate a useful and informative level of variation at the level of local populations of animals. Conversely, many investigators are reluctant to use microsatellites for phylogenetic analysis of local populations because the large number of different alleles typically encountered makes individual hypotheses of homology questionable. The second question that must be asked is whether we understand enough about the source of variation in the marker to be reasonably sure we are comparing the "same" (homologous) features in different taxa. This is precisely the problem with RAPD bands. Typically, one has no idea of the genetic basis for RAPD bands; they are simply fragments of DNA that may or may not display similar density and mobility patterns under particular conditions. A third question that must be asked is whether our analytical methods are robust to variation in rates of evolution among different marker systems or between different taxa for a single marker system.

There are three quite different approaches to the comparative analysis or synthesis of molecular information for systematics purposes: (1) phylogenetic methods, (2) distance methods, and (3) maximum likelihood methods.

Swofford *et al.* (1996) provide a detailed comparison of the three approaches. Phylogenetic methods are based on character by character assumptions of homology that are used to assemble nested statements of common ancestry, commonly depicted in tree form as cladograms. Of the many introductions to the logic and jargon of phylogenetics, we have found Kavanaugh (1978) to be among the most readable. Most phylogenetic methods are based on the assumption that if we can be reasonably certain of the homology of similar character states, the hypothesis that requires fewer assumptions of convergent acquisition or gains followed by losses of character states, the most "parsimonious" model, provides the best estimate of phylogenetic history (Sober, 1989).

Phylogenetic methods have several distinct advantages. They provide a simple and rigorous context to directly compare information from various sources, for example, different sequences of DNA, allozymes, and morphology, Another advantage is that the evidential support for particular "clades" (groups or lineages) can usually be directly determined, along with the degree to which this evidence conflicts with other data. As discussed earlier, many problems of long-standing and current interest in biological control can be usefully approached using phylogenetic methods. Phylogenetic analysis of large data sets is computationally intensive and can be quite tricky, although for up to 20 terminal taxa (populations, species, whatever) it is usually straightforward. This size of problem will certainly encompass many projects of interest to biological control. In addition, some forms of data are not suited to phylogenetic analysis, because parsimony methods typically require data with discrete character states. For example, to analyze allozyme data in a parsimony analysis requires that many fixed differences in alleles be present between samples. If difference between samples are mainly in the form of allele frequency differences, other methods such as distance methods must be used. Phylogenetic methods are generally robust to rate differences among markers or taxa, and in fact will reveal such differences clearly (which may in itself be of interest) although extreme cases may be problematic (Felsenstein, 1979).

Distance methods typically operate in a two-stage process. First, differences between all pairs of taxa of interest are computed using an appropriate distance measure. A great many distance measures are available, many specifically designed for particular types of molecular markers. Nei (1987) and Swofford *et al.* (1996) review distance measures for molecular data. Some measures attempt to correct for known properties of particular kinds of molecular data. For example, some measures attempt to correct for the underestimation of genetic distance caused by saturation of substitution sites in DNA sequences. Users should be aware of the particular assumptions underlying particular distance measures. For example, Nei's genetic distance, a very widely used measure for allozymes and other molecular data, assumes that all marker systems (e.g., loci) are evolving at the same rate (Hillis, 1984), an assumption that is rarely (perhaps never) met in the typical allozyme survey in which several different loci are used. Such loci would be expected to evolve at quite different rates. Other distance measures, including a corrected Nei's distance, do not assume homogeneity of rates in different markers (Hillis, 1984; Swofford *et al.*, 1996).

In the second step of a distance analysis, the pairwise distances between taxa are summarized into a single diagram, commonly using clustering algorithms that produce treelike branching diagrams (dendrograms). The extent to which such diagrams approximate phylogenetic relationships is determined by how well the data fit the assumptions of the particular distance measure used and the properties of the clustering algorithm. For example, the very widely used UPGMA (Unweighted pair-group method using arithmetic averages) clustering method belongs to a class of methods known as ultrametrics (Swofford *et al.*, 1996). These methods produce diagrams in which the distance from any tip in the diagram to the base is always equal. In effect, this assumes that all taxa are evolving at the same rate, and if the data do not fit this assumption, serious distortion will result. Other commonly used clustering methods such as neighbor joining (Saitou & Nei, 1987) and Distance Wagner (Farris, 1972) are not ultrametrics and they allow internal branch lengths in the diagrams to be individually fitted to the data. However, empirical work demonstrates that the shape of neighbor-joining trees may be sensitive to the order in which species are input to the clustering algorithm (Farris *et al.*, 1996), a computational artifact common in methods that produce a single tree topology from a matrix of distances or similarities (Sneath & Sokal, 1973). Distance methods can be applied to a great range of data types and they are computationally fast and efficient, even for very large data sets (Felsenstein, 1984).

Maximum likelihood methods (Felsenstein, 1981) attempt to produce a model for the evolution of a set of taxa that is most consistent with what is known about the behavior of the data that is used. As a simple example, assume the probability of transversions and transitions are defined, or estimated from the data, such that transitions (substitution of a purine for a purine or a pyrimidine for a pyrimidine) are twice as likely as transversions (substitution of a purine for a pyrimidine or vice versa). Given the sequence data for two taxa, a series of changes from one taxa to the next can be postulated and its likelihood calculated by taking the product of all the postulated steps. The overall likelihood would be maximized by having the appropriate combination of the fewest possible steps and the highest proportion of transitions. As the statistical properties and behavior of various kinds of DNA data are becoming better understood, maximum likelihood methods are finding increasing application. Maximum likelihood methods depend on detailed understanding of the statistical probabilities of particular kinds of state changes, and this is possible only with certain kinds of data. However, programs will now empirically estimate such probabilities directly from the sequences under study (Swofford *et al.,* 1996). Maximum likelihood methods are computationally intensive, particularly as the complexity of the underlying model increases, which limits their use to relatively small problems or relatively simple models of character evolution. Maximum likelihood methods appear to be robust to rate heterogeneity problems that may be serious with distance methods or even parsimony methods in extreme cases (Swofford *et al.,* 1996).

Biotypes: Or "A Taxon by Any Other Name . . ."

In classical biological control of arthropod pests, multiple geographic races of natural enemies (biotypes of Gonzalez *et al.,* 1979) are commonly introduced (Ehler, 1990). Such an approach is justified by the successes achieved when new biological races have been introduced [e.g., *Trioxys pallidus* against walnut and filbert aphids; reviewed in Unruh and Messing (1993)]. However, it makes biotype determination a common problem in biological control programs. For arthropods introduced for weed biological control, biotypic differences in host-plant affinities are now well known and require careful elaboration of geographic and host-plant-related variation in candidate species (McEvoy, 1996). Diehl and Bush (1984) classify insect biotypes using a combination of genetic and geographic relationships. Given that some biological trait is variable (identifying the biotype), they suggest five categories into which biotypic variation falls: nongenetic polyphenisms, genetic polymorphisms, geographic variation, host races, and species. Each case must be examined in terms of how the insect's life system may reinforce isolation among subpopulations (host-mediated mate finding or allochrony) or which promotes divergent selection.

In contrast to Diehl and Bush (1984), Gonzalez *et al.* (1979) defined biotypes as reproductively compatible populations that display differences in some biological attributes. Claridge and Den Hollander (1983), on the other hand, suggest that the term "biotype" should be used only for cases in which a particular gene for virulence in a pest is known to be associated with a gene for resistance in a host, a very limited usage indeed. These authors argue that widespread usage of a term such as "biotype" for all cases in which forms are biologically different obscures the very different historical and genetic mechanisms that may be involved. We concur and note that the categories outlined by Diehl and Bush (1984) provide a richer, more meaningful vocabulary to refer to the forms used in biological control. For example, many of the classic cases involving biotypes of natural enemies (e.g., walnut aphid, olive scale) probably exploited different *geographic races* of natural enemies, which differed in critical traits such as temperature tolerances and diapause timing. To refer to these entities simply as biotypes obscures the evolutionary basis for these differences.

Thus, a major use of molecular markers for biological control is to assist in assigning biotypes to one of the evolutionary stages outlined by Diehl and Bush (1984). In the following discussion we derive some general rules for interpretation of molecular variation to disentangle biotypes from Diehl and Bush (1984), Berlocher (1989), and Menken (1989), among others. *Nongenetic polyphenisms* will show no molecular variation associated with the phenotype unless that is confounded by some other factor such as geographic differentiation.

Genetic polymorphism at single and polygenic traits is to be expected among all sexually breeding forms, but it usually will not be consistently associated with molecular markers unless they are linked to the polymorphism. Such linkages can be discovered by extensive screening of molecular markers such as RAPDs Such efforts often show that so-called biotypes have a high diversity of molecular variants with none closely associated with the biological trait(s) of interest [e.g., grape phylloxera (Fong *et al.,* 1995) and greenbug (Shufran *et al.,* 1997)]. Also, when trait variation is evident among clones of parthenogenetic or cyclically parthenogenetic forms, it is possible for multiple individuals sampled to show the same molecular patterns and these may be associated with a particular phenotype because the whole genome is linked by absence of meiotic recombination. Such observed linkages will break down with occasional bouts of sexual reproduction, such as that observed among thelytokous and arrhenotokous lineages of some parasitoid species (Stouthamer & Kazmer, 1994), or in the sexual generation of many aphids (Shufran *et al.,* 1997).

Biotypes associated with different hosts (or habitats, etc.) in sympatry are probably species if each is also characterized by diagnostic molecular markers, because this indicates that they are reproductively isolated from one another in nature. Such observations should be validated by careful and conservatively interpreted laboratory mating studies where feasible. Of particular difficulty, from the perspective of classification, is the presence of sympatric populations that are reproductively isolated because of cytoplasmic incompatibility inducing microbes (Perrot-Minnot *et al.,* 1995). None of the nomenclature we have introduced for disentangling biotypes may fit this intriguing natural phenomenon. It is notable that such infections may cause asymmetrical reproductive incompatibility (Perrot-Minnot *et al.,* 1995), which is commonly encountered crossing closely related entities of parasitoids. A related set of problems are posed by races of parasitoids that are reproductively isolated by virtue of *Wolbachia*-induced thely-toky. This appears to be common in many species of chalcidoids (Stouthamer, 1993), notably *Trichogramma* (Stouthamer & Werren, 1993; Stouthamer *et al.,* 1996). One approach to classifying such entities is to remove *Wolbachia* from test strains, thereby obtaining males, and to perform crossing studies (Stouthamer *et al.,* 1990). Another approach is to ignore reproductive relationships, to assess the degree of differentiation among uniparental entities using molecular or biochemical markers, and to apply a phylogenetic (Cracraft 1989) or "ethological" (DeBach, 1969) species concept.

Many cases of *geographic races, semispecies,* or *subspecies* have been encountered in biological control (Rosen, 1986; DeBach, 1969; Wharton *et al.,* 1990). Historically, similar looking forms from different geographic origins were allowed to die in quarantine because they appeared identical to organisms already imported and released (Sawyer, 1996; DeBach & Rosen, 1991). Diagnostic molecular differences between allopatric populations cannot alone indicate species differences, but they may provide important clues. For example, Pashley *et al.* (1983) discovered nearly diagnostic allozyme differences among allopatric populations of the fall armyworm. Later these differences were observed in recently sympatric, host-associated populations (Pashley, 1986). Subsequently, reproductive incompatibility (Pashley & Martin, 1987) and diagnostic differences in mtDNA (Yang-Jiang *et al.,* 1992) were demonstrated among the populations supporting the hypothesis that they are indeed species.

Allozymes that differ in frequency among geographic populations are more problematic measures of specific status. For example, host-adapted races of the apple maggot show very subtle differences in gene frequencies that reoccur in several distinct regions of sympatry (McPheron *et al.,* 1988). This pattern can occur from genetic isolation of two forms or from similar selective forces in each region. Smith (1988) shows that host-induced shifts in apple maggot phenology occur and may produce allochronic isolation, but the similarity in gene frequencies over regions for each race also implicates selection. Experimental studies by Feder *et al.* (1997) have demonstrated how selection on fly diapause duration (associated with periods of host suitability) can shift allozyme frequencies. Other species show host races with patterns similar to *Rhagoletis pomonella* (Walsh) [e.g., the larch bud moth (Emelianov *et al.,* 1995), *Larinus* weevils (Briese *et al.,* 1996)].

Different allopatric or even sympatric populations of weed-feeding species may show striking differences in their host-plant spectrum, influencing their potential nontarget host range and their efficacy against the target weed. An example of notoriety for its shift to nontarget hosts is the weevil *Rhinocyllus conicus* Froelich (Louda *et al.,* 1997;

Strong, 1997). Three biotypes of this species specialize on the flower heads of thistles in the Palearctic and were introduced into North America to control milk (*Silybum marianum* [Linnaeus]), musk (*Carduus nutans* Linnaeus), and Italian (*C. pycnocephalus* Linnaeus) thistles (Goeden & Kok, 1986). In 1985, about a decade after the milk and Italian thistle races were established in southern California, *R. conicus* was discovered feeding and reproducing on two species of native North American thistles, *Cirsium californicum* Gray and *C. proteanum* Howell. Allozymes studies showed that the Italian thistle race had shifted to the native thistles (Unruh & Goeden, 1987), as one might expect from earlier observations that the *Carduus*-attacking biotype uses European *Cirsium* spp. in their native range (Zwölfer & Preiss, 1983). The insights provided by allozyme analysis and other postrelease studies of this system are difficult to evaluate. It is not clear that introductions would have been altered given the importance of the pest weeds and the differences in perceptions of risk to nontarget species at the time of the introduction. However, it is clear that *R. conicus* would not be released now given the information at hand. Allozymes and RAPDs are now being routinely used to develop a detailed genetic picture of host-associated populations of weed natural enemies prior to their introduction (Laroche *et al.,* 1996; Manguin *et al.,* 1993; Nowierski *et al.,* 1996; Briese *et al.,* 1996).

Intraspecific Phylogeography

Intraspecific phylogeography is the study of the geographic distribution of a genealogical lineage (Avise *et al.,* 1987). This type of historical biogeography was born at the end of the 1970s when studies of variation in mtDNA began to accumulate and researchers realized that this molecule was ideal for generating phylogenetic hypothesis for populations of a species (Avise, 1991). Losses and gains of mtDNA restriction sites can often be unambiguously ordered (by parsimony) into simple but explicit phylogenetic models, whereas variation in allozymes, or other measures of nuclear sequence variation, often cannot (Avise, 1994, but see Swofford *et al.,* 1996). The earliest applications of this approach produced clear models for the evolutionary history of local populations of organisms (Avise *et al.,* 1979) that were consistent with, but far more informative than, electrophoretic or morphological studies. The power of the method was quickly appreciated and it has been very widely used (Avise *et al.,* 1987, Avise 1992). In many insect species, males are the most dispersive sex, and gene flow by males followed by recombination can obscure the footprints of founder events leading to range expansions or spread of a new mutation throughout the range of a species. This footprint may remain in the mtDNA because it is maternally inherited. One notable insect study applied mtDNA RFLP to the intriguing and well-studied case of *Drosophila silvestris* and *D. heteroneura* in Hawaii, where it provided new clues to the evolutionary history at the population level (DeSalle & Giddings, 1986). Even in this case that had been well studied using polytene chromosomes (which themselves are amenable to phylogenetic interpretation), allozymes, morphology, behavior, and so on, mtDNA offered important new insights. Another approach, arguably less powerful but perhaps more practical in some cases, is to construct phylogenetic hypotheses that are not explicitly genealogical but are based on genetic distances from allozymes or other measures of nuclear DNA variation (Roderick, 1996). Statistical techniques suitable for allozyme, microsatellite, and RFLP variation may allow stringent classification of immigrant individuals, based on knowledge of gene frequencies from source and colonized areas (Rannala & Mountain, 1997).

Hence, phylogeography represents a way, perhaps the only way in many instances, to identify the origin(s) of exotic pest populations. For example, Latorre *et al.* (1986) used an mtDNA restriction map to show that *D. subobscura* Collin populations that recently colonized the Nearctic were derived from European and not African progenitors. Additional examples of phylogeography based on mtDNA in insects include *Eurostis solidaginis* (Brown *et al.,* 1996), several *Papilio* species (Sperling & Harrison, 1994), *Greya politella* (Brown *et al.,* 1997), honeybee (Arias & Sheppard, 1996), Mediterranean fruit fly (Gasparich *et al.,* 1995), Colorado potato beetle (Azeredo-Espin *et al.,* 1996; Zehnder *et al.,* 1992), screwworm fly (Roehrdanz, 1989), and southwestern corn borer (McCauley *et al.,* 1995).

Puterka *et al.* (1993) used both allozymes and RAPDs to suggest that the Russian wheat aphid populations in North America probably came from Turkey, France, or South Africa but not other areas of the Old World. mtDNA (Harrison & Odell, 1989), microsatellites (Bogdanowicz *et al.* 1997), and RAPDs (Schreiber *et al.,* 1997) have been used to identify Asian versus European gypsy moth races in North America.

The origins of various races of established natural enemies can also further our understanding of biological control, for example, to demonstrate that climatic matching is as important as believed. An *a posteriori* allozyme analysis indicated that *Aphidius ervi* populations released for pea aphid control in southern California were derived from European or North African progenitors and not from other races from throughout its wide Palearctic distribution (Unruh *et al.,* 1986, 1989). Kazmer *et al.* (1996) used RAPDs to examine the phylogeographic relationship of races or incipient species of *Aphelinus asychis* imported for biological control of Russian wheat aphid in North America. This work is marked by careful mating compatibility studies among the populations, which together with RAPDs indi-

cated that Central Asian populations were distinct. Antolin *et al.* (1996b) used RAPD variation to identify the origin of a gregarious parasitoid of flies that appears to have colonized North America unaided, or as a contaminant in releases of other *Muscidifurax* species.

Interpretation of Genetic Distances

The notion that various taxonomic ranks, such as local populations, races, subspecies, and cryptic species, could be diagnosed by characteristic levels of genetic distance has seduced many biologists. Unfortunately, although one would expect generally increasing levels of genetic distance as one ascends such a scale, numerous exceptions were noted even in the earliest reviews (e.g., Lewontin, 1974). Although some general trends do emerge from such studies, there are typically as many exceptions as cases that fit a general pattern. This is not surprising because different modes of speciation and evolution are likely to have very different genetic consequences (Templeton, 1981), some of which involve broad-based genetic differentiation as measured by protein electrophoresis and some of which do not.

Table 1 summarizes some of the relevant literature that provides genetic distance estimates at different taxonomic levels. The association of genetic distance to nominal taxonomic rank is clearly inconsistent (Table 1), particularly for infraspecific forms. Interspecific comparisons are more consistent, but range from near 0 (for one cryptic species) to 1.4. Comparisons among species groups or genera show *D* values of one or greater.

In some interpopulation comparisons, morphology, behavior, or cytology support one hypothesis of specific status while allozyme data support another. Examples are seen in the 10 *Speyeria* (Nymphalidae) species in Table 1. These included one cryptic species pair: *S. mormonia* (Boisduval) and *S. egleis* (Behr), which were among the most genetically distinct of the 10 species examined (*D* = 0.224). Another example is the cryptic species pair, *Gryllus veletis* Alexander and Bigelow and *G. pennsylvanicus* (Burmeister) (Grylliae; Table 1), which were found to be distinct at several enzyme loci (*D* = 0.165) (Harrison, 1979). In contrast, three morphologically distinct species differed very little in allozymic traits (*G. pennsylvanicus, G. firmus, G. ovisopis* (Walker) (*D* < 0.03), but the first two at least are clearly distinguished by mtDNA RFLP markers (Harrison and Rand, 1989).

Examples of extremely low electrophoretic differentiation between morphologically or biologically distinct species include *Magicicada* spp. and *D. silvestri* (Perkins) and *D. heteroneura* (Perkins) (Table 1). Interestingly, *D. silvestri* and *D. heteroneura* are strikingly different in morphology (Spieth, 1981). Each of these groups display aspects of biology and evolutionary history that are consistent with

TABLE 1   Genetic Distances (Nei, 1972) among Taxa of Various Ranks for 29 Groups of Insects

Group	No. Loci	Local populations	Races (R) incipient (I) sub (s) or Semi (S) species	Cryptic species	Morphological species	Species group (s) genus (g)	Source
Diptera							
Drosophila obscura group	68			0.292	0.637		Cabrera *et al.*, 1983[a]
D. willonstoni group	36	0.030	0.23 (s)	0.574	1.044		Ayala *et al.*, 1974b[a]
D. equinoxialis	27	0.021	0.255 (s)				Ayala *et al.*, 1974a[a]
D. mulleri group	16	0.001	0.130 (s)	0.203	0.828		Zouros, 1973[a]
D. melanogaster group	55			0.377			Gonzalez *et al.*, 1982
D. silvestri and *D. heteroneura*	24				0.079		Sene & Carson, 1977
Rhagoletis	15	0.008	0.012 (I)	0.066 (s)[b]	0.307	0.818 (s) 2.170 (g)	Morgante *et al.*, 1980; Berlocher & Bush, 1982
Aedes aegypti	18	0.020	0.042 (I)				Tabachnick *et al.*, 1979
A. atropalpus and *A. epactius*	17		0.163 (R)		0.547		Munstermann, 1980
A. scutellaris group	24	0.06			0.593		Pashley *et al.*, 1985
Hymenoptera							
Iridomyremex purpareus group	15	0.020		0.059			Halliday, 1981[a]
Rhytidoponera impessa group	22	0.015		0.136			Ward, 1980[a]
Polistes fuscatus group	18–20			0.084	0.853		Metcalf *et al.*, 1984
Aphidius spp.	16	0.018	0.203 (I)	0.190	0.718	2.656 (s)	Unruh *et al.*, 1989
Solenopsis spp.		0.003		0.176[c]	0.534		Ross *et al.*, 1987
Orthoptera							
Allonemobius spp.	20			0.247	0.519		Howard, 1983
Caledia captiva group	20	0.078		0.270			Daly *et al.*, 1981[a]
Trogliphilus spp.	17	0.060			0.373		Sbordoni *et al.*, 1981[a]
Gryllus spp.	17	0.008		0.174	0.145		Harrison, 1979[a]
Hadenoecus and *Euhadenoecus*	41	0.099			0.526	1.433	Caccone & Sbordoni, 1987
Ephemeroptera							
Eurylophella and *Ephemerella*	17–28	0.006			1.392	1.551 (g)	Sweeney *et al.*, 1987
Lepidoptera							
Speyeria	16	0.013	0.023		0.182		Brittnacher *et al.*, 1978[a]
Pieridae	20		0.008		0.317	0.851[d] 1.406[d]	Geiger, 1980[a]
Euphydryas	28	0.126	0.036 (s) 0.153 (s)	0.178	0.394	1.23	Brussard *et al.*, 1985
Yponomeutidae	28–51	0.006		0.203	1.331		Menken, 1981
Choristoneura	18	(0.017)	(.110)	0.076			Stock & Castrovillo, 1981

(continues)

TABLE 1 *(continued)*

Group	No. Loci	Local populations	Races (R) incipient (I) sub (s) or Semi (S) species	Cryptic species	Morphological species	Species group (s) genus (g)	Source
Homoptera							
Magicicada	20	0.01	0.01	0.02	0.13		Simon, 1979[e]
Coleoptera							
Dendroctonus	18		0.004		0.568	1.68 (g)	Bentz & Stock, 1986
Bathyplectes curculionis	22	0.008[f]	0.056[g]				Hsiao & Stutz, 1985

[a] Modified from Brussard *et al.,* 1985.
[b] *Pomonella* group excluding *carnivora.*
[c] Comparison between *S. richter* and *S. invicta.*
[d] Genera within and between subfamilies, respectively.
[e] Only Rogers' distance presented; Nei's distances all smaller.
[f] Within Egyptian and eastern U.S. populations and within western U.S. populations.
[g] Between western and (Egyptian) eastern.

rapid development of reproductive isolation. In *Magicicada,* reproductive isolation is imposed by 13- and 17-year life cycles and isolated populations. This isolation may have been reinforced by a history of glaciations that fragmented the species ranges (Simon, 1979). In *D. silvestri* and *D. heteroneura,* sexual selection on highly ritualized courtship behavior (Spieth, 1981) may have mediated rapid reproductive isolation. In this case, founder events between islands triggered strong shifts in particular genetic systems such as those underlying courtship and mate recognition, but speciation apparently did not involve broad-based genetic differentiation, as measured by allozymes (Templeton, 1980). mtDNA RFLP patterns (DeSalle & Giddings, 1986) and mtDNA sequences (DeSalle, 1992) have been used to clarify phylogenies previously addressed by chromosomes, morphology, behavior, allozymes, and DNA–DNA hybridization for the Hawaiian *Drosophila.* Similarly, mtDNA restriction and sequence data have helped resolve *Magicicada* (Martin & Simon, 1988; Simon *et al.,* 1990).

Low genetic distances among some species contrast with high distances among geographic populations of others. In cave crickets, genetic distance between species that inhabit forests where populations are free to interbreed was 10-fold lower than that for comparisons between isolated populations of a cave-inhabiting species where interbreeding is precluded (Caccone & Sbordoni, 1987). Genetic differences above 0.2 have been associated with postreproductive isolating barriers in cave-inhabiting invertebrates (Caccone & Sbordoni, 1987). High distance was also seen for intraspecific populations of butterflies in the genus *Euphydryas* (Brussard *et al.,* 1985).

In summary, if genetic distances are higher than expected, research should focus on whether cryptic species have been overlooked. Low genetic distances among taxa that are distinct in other traits cause us to seek understanding of the forces that cause rapid reproductive isolation. However, we hope it is clear from the foregoing discussion that basing taxonomic rank on measures of genomic divergence is a misadventure, either with allozymes or with other molecular markers (see Thorpe, 1983).

Applications

Allozymes have been used for making systematic inferences in insects for more than 20 years (Bush & Kitto, 1978; Buth, 1984). Berlocher (1984b) identified three major tasks of such investigations: species discrimination, species identification, and hierarchical classification. The use of allozymes to discriminate species is now common. Examples include polyacrylamide electrophoresis to characterize genetic divergence among cryptic species of the *Chrysoperla carnea* complex in Europe (Thierry *et al.,* 1997) and cellulase acetate electrophoresis to provide diagnostic characters for identification of larval leaf miners (*Liriomyza* spp.) (Collins, 1996). At the lower formal levels of classification (intrageneric), allozymes have proved useful in many insect groups (see examples in Table 1). At higher taxonomic ranks, allozymes have been less valuable because of the high level of divergence and the likelihood of homoplasy at the intergeneric range and above [e.g., in tortricoid moths, Pashley (1983), but see Brussard *et al.* (1985)].

Powell and Walton (1989) and Unruh *et al.* (1986) review the applications of electrophoresis in parasitic Hymenoptera. Most work on parasitoids has concentrated on the genus *Trichogramma,* well known for its taxonomic difficulties. Development of isoelectric focusing electrophoresis procedures for *Trichogramma* (Kazmer, 1991) led to research integrating morphological study, crossing experiments, and protein electrophoretic characterization of both natural populations and laboratory cultures. This work has clarified the taxonomy and evolutionary relationships in the difficult *T. minutum* group (Pinto *et al.*, 1992), in which varying levels of reproductive isolation are found between presumably conspecific geographic populations, illustrating the difficulty in using crossing tests alone as a criterion for species limits. Pinto's group is continuing to use a multilateral approach to *Trichogramma* taxonomy (Pinto *et al.*, 1997).

Starch gel electrophoresis has been widely used in studies of European *Trichogramma* (Pintureau, 1993a, 1993b; Neto & Pintureau, 1995; Ram *et al.*, 1995a, 1995b), particularly in the identification of cryptic species. Other electrophoretic work in Chalcidoidea has included species of Torymidae (Izawa *et. al.*, 1995, 1996), Eurytomidae (Roux & Roques, 1996), and Aphelinidae (Strong, 1993). Hunter *et al.* (1996) used isoelectric focusing electrophoresis in combination with morphology and study of courtship behavior and reproductive relationships to demonstrate that *Eretmocerus* spp. (Aphelinidae) reared from *Bemisia argentifolii* Bellows and Perring (Aleyrodidae) in Texas are not conspecific with those from Arizona and California; in fact, these populations represented two cryptic species.

Several comprehensive studies have used electrophoretic evidence to assist in identification of cryptic species of Braconidae, particularly Aphidiinae (Walton *et al.*, 1990), and two studies have used explicitly phylogenetic procedures to infer relationships among geographic populations and closely related species of *Aphidius* (Unruh *et al.*, 1989) and *Cotesia* (Kimani *et al.*, in press).

As with other DNA sequence data, rDNA sequences can be analyzed using rigorous phylogenetic methods, greatly facilitating its direct comparison with other data. For example, Vogler and DeSalle (1994) constructed an intraspecies phylogeny of populations of *Cicindela dorsalis* (Coleoptera: Cicindelidae) based on ITS-1 sequences that was largely congruent with patterns inferred using RFLP analysis of mtDNA. Schlotterer *et al.* (1994) found a phylogeny of the *D. melanogaster* group based on ITS sequences to be consistent with relationships inferred from other data. Tang *et al.* (1996) used ITS to study populations in the *Simulium damnosum* complex and Williams *et al.* (1985) studied variation in the NTS region in populations of *D. mercatorum*. The ITS-2 region apparently evolves somewhat more slowly than ITS-1, and apparently contains fewer insertions and deletions (Schlotterer *et al.*, 1994) that complicate phylogenetic analysis. The ITS-2 region has proved useful for phylogenetic studies of species and infraspecific populations in both Hymenoptera (Campbell *et al.*, 1993) and Diptera (Porter & Collins, 1991; Wesson *et al.*, 1993, Fritz *et al.*, 1994).

Phylogenetic analyses of Hymenoptera using sequence data are presently underway in many laboratories, but several studies have already appeared. Sequence data have led to new phylogenetic hypotheses for species within *Apis* (Cornuet & Garnery, 1991; Smith, 1991) and tribes of the Apidae (Cameron, 1991) at higher taxonomic levels. Dowton and Austin (1994) and Derr *et al.* (1992a, 1992b) used 16s rDNA to examine relationships among the major lineages of Hymenoptera. In studies perhaps more directly relevant to biological control, Gimeno *et al.* (1997) used cytochrome b, and both 16s and 28s rDNA to examine relationships among Alysiinae and Opiinae (Braconidae); and Belshaw and Quicke (1997) used elongation factor 1a, cytochrome b, and a portion of the 28s gene to produce a phylogeny of Aphidiinae (Braconidae).

Population Biology

Although it may be overly pessimistic to characterize the first two decades of allozyme studies in population genetics as the "dark age of electrophoresis" (Scharloo, 1989), many studies have produced very few answers about how genetic variation is maintained in natural populations (Lewontin, 1974). Tighter linkage between the function of molecular variants (allozymes) in the organism and their distribution within and among populations seems the best way to understand the selective nature of genetic variation (Carter & Watt, 1988). Through much of the following discussion, neutrality (Kimura, 1983) of the genetic markers is tacitly assumed. However, evidence is accumulating in various model systems that patterns of allozyme differences among populations are often selectively maintained by either selection on the allozyme function itself or by virtue of linkage to traits under strong selection (e.g., apple maggot; Feder *et al.*, 1997); or by reflection of a complicated mosaic of selective, historical, and gene-flow-mediated phenomena (Begun & Aquadro, 1992). Differences in other forms of variation such as RAPDs, RFLPs, microsatellites, and sequence variation at anonymous loci or introns are currently thought to be much less likely to reflect selective phenomena (Avise, 1994).

Applications of genetic markers to various questions in the population biology of pests and natural enemies can be a truly exciting endeavor. We address four applications here: population structure and gene flow, inbreeding and postcolonization adaptation, genetic mapping, and miscellaneous ecological applications. A detailed review of other applications of molecular markers is presented for insects (in general) by Roderick (1996) and for life (in general) by Avise (1994).

Population Structure, Gene Flow, and Mating Systems

Population genetic structure is the partitioning of genetic variation among individuals, local populations, and regional populations or other similar hierarchical organizations (Wright, 1978). Typically gene flow is more important than mutation, selection, or genetic drift in shaping the distribution of the variation among groups (Slatkin, 1987; but see later). Currently there are three major approaches to estimating population structure and subsequently gene flow among populations. The original partitions total genetic variance using the F statistic such that $(1 - F_{it}) = (1 - F_{is})(1 - F_{st})$, where the subscripts denote individual, subpopulation, and total population, and F is a measure of genetic variation (correlation among alleles) that varies from 0 to 1 [see Avise (1994) or Wright (1978) for straightforward explanations and Weir & Cockerham (1984) and Excoffier et al. (1992) for improved analytical methods]. From this, gene flow rates can be estimated from the approximate relationships $F_{st} = 1/4(N_e m + 1)$ (Wright, 1951). $N_e m$, the number of alleles moving among populations, is the product of a typically large number, N_e, the effective population size, and a small number, m, the migration rate per generation. The F statistic approach assumes that populations are in genetic equilibrium and that there have been no significant deviations from equilibrium in recent evolutionary time.

A second approach is based on the prediction that if populations seldom exchange genes, unique or rare alleles may accumulate in some subset of the populations, whereas with high gene exchange these "rare alleles" will more likely be shared among populations at frequencies closer to the population-wide average (Slatkin, 1987). This rare allele approach also assumes genetic equilibrium but may be more robust to deviations from it than are F statistics in some cases.

A third approach, based on coalescence theory, is more explicitly phylogenetic (Slatkin & Madison, 1989) and must be applied to haplotypes that have no history of recombination, such as mtDNA (see Hudson et al., 1992). First, a phylogeny, or gene tree (Avise, 1994), is developed, and then the minimum number of migration events required to explain haplotype distributions is estimated. As stated by Avise (1994), none of these approaches provide precise measures of gene flow; instead they provide estimates like high, low, or intermediate. Before undertaking a study of the genetic population structure, biological control workers should decide what hypothesis will be tested and determine whether the generally vague result of very high, intermediate, or very low gene flow provides an adequate test of the hypothesis.

Selection, gene flow, mutation rates, and stochastic changes in small populations or local demes shape gene frequencies within and among populations (Wright, 1978). All these forces can leave historical footprints in current populations. The potential of past genetic bottlenecks, events of hybridization, or other historical phenomena on the existing pattern of genetic variation within and among populations remains a challenge for theoreticians. Methods to estimate gene flow are based on the assumptions of no severe, recent colonization events or other gross perturbations of a presumed genetic equilibrium (Slatkin & Barton, 1989). Such perturbations may be commonplace in natural enemies imported for biological control, and may be characteristic of many organisms in agricultural systems. The traditional F statistic approach seems generally the most robust of the three methods to deviations from genetic equilibrium (Slatkin & Barton, 1989). The cladistic approach is more robust when recombination rate is reduced and thus is most appropriate when gene flow is estimated from mtDNA (Neigel, 1997).

Many agricultural pests have been the subject of population genetic structure studies, often for the purpose of understanding the potential for spread of insecticide resistance [e.g., Heliothis (Daly, 1989), pear psylla (Unruh, 1990), diamondback moth (Caprio & Tabashnick, 1992)]. Another important application of genetic structure studies is to seek evidence of host associations as discussed in the section on biotypes. Sheppard et al. (1991) used both allozymes and mtDNA to document gene flow among African and European honeybees (1991). Studies of gene flow within and among populations of insect predators include Coleomegilla (Krafsur et al., 1995), Harmonia (Coll et al., 1994; Krafsur et al., 1997), and Hippodamia (Krafsur et al., 1996).

Mating behavior and the level of inbreeding it may engender have important evolutionary implications for Hymenopteran parasitoids. An inbreeding lifestyle in haplodiploids may influence the optimal sex ratio of a population (Hamilton, 1967; Charnov, 1982; Waage, 1986). The optimal sex ratio always exceeds 50% females, and may dramatically do so when only a single female utilizes a batch of hosts. One can define discrete patches of hosts as used by the founding females and her progeny and compare the level of inbreeding estimated from molecular variation such as allozymes or microsatellites and from observed foundress numbers. Both measures of inbreeding can be used to predict sex ratio from local mate competition (LMC) theory (Luck et al., 1993), and these predictions can be compared to observed sex ratios. Deviations from LMC predictions may arise from fitness differences between females and males developing in different size hosts (Charnov, 1982) and from various other deviations from the assumptions of LMC (Luck et al., 1993).

The parasitic Hymenoptera, like other Hymenoptera, show lower levels of genetic variation as assessed by allozymes than do diploid insects (Hughes & Quellar, 1993; Unruh et al., 1986). Thus, allozymes may not provide enough variation to characterize mating systems in some species. Microsatellites have proved to be highly useful for Hymenoptera because of the much higher level of heterozygosity evident in this type of variation (Hedrick & Parker, 1997). Several statistical methods for estimating relatedness among individuals using variation such as allozymes have been developed (Pamilo, 1984; Weir & Cockerham, 1984; Queller & Goodnight, 1989).

Colonization, Inbreeding, and Postcolonization Adaptation

Maintaining fitness and adaptability of the natural enemy is crucial for the success of biological control introductions. Unfortunately, the practices used in collection, importation, quarantine, rearing (Unruh et al., 1983), and release (Hopper & Roush, 1993) phases of an introduction may lead to reduced fitness or, at least, reduced genetic variation (Roush, 1990; Hopper et al., 1993; Unruh & Messing, 1993). Parasitoids and predators, like other insects, are likely to show reduced genetic variation when cultured in the laboratory. Electrophoretic evidence has been used to evaluate genetic changes in laboratory cultures of parasitoids [Aphidius ervi (Unruh et al., 1983), Cotesia flavipes (Omwega & Overholt, 1996)] but the degree to which such predominately neutral variation indicates the fate of variation for more adaptive traits is unclear (Roush, 1990; Hopper et al., 1993). Several other studies monitored changes in allozyme variation in insects (Berlocher & Friedman, 1981; Pashley & Proverbs, 1981; Stock & Robertson, 1982), and loss of variation in these provides estimates of inbreeding and drift but not of selective changes (Roush, 1990). These contrast with the now classic work on mass cultures of screwworm where allele frequency changes were related to a loss in flight ability and competitiveness compared with field population (Bush & Neck, 1976).

Roush and Hopper (1995) demonstrated that maintaining many isofemale lines is theoretically superior to a few large populations for preventing allele loss by drift, while Unruh et al. (1983) favored a few large populations as most practical. If selection is significant in the laboratory, it would be valuable to produce many isogenic lines to prevent selective loss of alleles as maintained by Roush and Hopper (1995). However, in most cases each original pair initiating an isoline will contain much of the genetic variation housed in the population, including most alleles segregating for quantitative traits [see discussion by Lande & Barrowclough (1987)]. Each sibmating will reduce that variation in the isoline by roughly half and several bouts of sibmating are required before quantitative traits would be fixed at all loci. During this inbreeding, selection could occur in parallel in all lines making fixation of alleles nonrandom. Thus, the advice of Hopper et al. (1993) to minimize time in the laboratory and to have the laboratory mimic the field as much as possible seems the best solution.

Inbreeding, including that proposed for preservation of specific alleles in the previous discussion, is known to often produce negative effects (inbreeding depression; Charlesworth & Charlesworth, 1987). However, the haplodiploid Hymenoptera may be less prone to inbreeding depression than are diploids [reviewed in Unruh & Messing (1993)]. In haplodiploids a significant portion of the genetic load is "filtered out" of the population through expression (= genetic death) in haploid males (White, 1948; Werren, 1993). However, predator species and diploid parasitoids such as the Tachinidae should be similar to other diploids in their susceptibility to inbreeding.

In contrast, several Hymenopteran groups, notably Ichnuemonoidea, may be extremely sensitive to inbreeding because of a heterozygosity-dependent mechanism of sex determination (Whiting, 1943; Stouthamer et al., 1992; Unruh & Messing, 1993). A single gene controls sex, and haploidy at the gene—arising from unfertilized eggs—yields normal haploid males. Heterozygosity at the sex gene yields females, but the chance occurrence of homozygous individuals at the sex gene results in diploid, nonfunctional males (Whiting, 1943; Crozier, 1977). In many instances, these diploid males are viable to adulthood and both cytological methods and allozymes can be used to identify them (e.g., Hedderwick et al., 1988). The frequency of diploid males in the population can then be used to estimate the number of sex alleles segregating in the population (Owen, 1993) and the level of the genetic load (Stouthamer et al., 1992). Microsatellites should be even more sensitive for detecting diploid males because of their higher levels of heterozygosity (Hughes & Quellar, 1993; Hedrick & Parker, 1997). The importance of this form of genetic load to biological control was elaborated by Stouthamer et al. (1992). Unruh and Messing (1993) suggest that the production of diploid males by ecologically dominant Hymenoptera, such as ants, may be useful markers of habitat fragmentation in conservation studies.

Changes in the genetic makeup of a population are expected to follow a population bottleneck, such as might occur during introductions of biological control agents (Roderick, 1992). First, most rare alleles from the endemic source population will be lost (Nei et al., 1975). Second, inbreeding and gametic phase and linkage disequilibrium should increase in inverse proportion to the number of founders (Templeton, 1980). Third, if multiple populations are founded independently, they should display more interpopulation variation in gene frequencies (greater F_{ST}) than the source population(s). Each of these phenomena can be estimated using polymorphic enzyme loci, microsatellites, or other nuclear variants. However, only one mtDNA haplotype is passed by a mother to her descendants, while four alleles of each nuclear gene are passed. Hence, effective population size of mtDNA is one-fourth that of chromosomal genes, making mtDNA a more sensitive indicator of genetic bottlenecks (DeSalle & Giddings, 1986), assuming adequate mtDNA variation exists to monitor these effects. Note, however, that large founding populations, rapid population growth, and extensive gene flow after colonization can significantly reduce the effect of a genetic bottleneck on heterozygosity (Nei, *et al.,* 1975; Kambahmpati *et al.,* 1990) of nuclear genes.

Molecular studies of colonized populations are consistent with the predicted effects. Berlocher (1984a) noted higher F_{ST} and linkage disequilibrium for eight colonized populations of the walnut husk fly. The cynipid gall wasp, *Andricus quercuscalicis,* apparently expanded its range into North and Western Europe following human assisted range expansion of its host plant, *Quercus cerris.* Populations showed lower allelic diversity and average heterozygosity in the colonized range, with the intensity of the effect correlated with distance from its original range. Furthermore, F_{ST} increased and a striking isolation by distance pattern was evident in the new range but not in the original (Stone & Sunnucks, 1993). The pattern of genetic variation seen in the new range, if characterized with F statistics (but without an historical perspective), would indicate quite limited gene flow compared with the same species further south in Europe. Although populations are probably more isolated in the North, the correlative evidence of reduced genetic variation and the known history of range expansion strongly implicate founder events in elevating F_{ST}. Caveats about estimating gene flow by F_{ST} or rare alleles in non-equilibrium populations are reinforced by this study (see earlier). In other examples, fewer rare alleles were observed in colonized populations of myna birds (Baker & Moeed, 1987) and the Eurasian tree sparrow (St. Louis & Barlow, 1988). The adaptive nature of these changes was not addressed.

Quantitative genetic changes associated with colonization of biological control agents have been studied but molecular traits were not used. Myers and Sabath (1980) detected postcolonization changes in polygenic traits of the cinnabar moth [*Tyria jacobaeae* (Linnaeus)], an enemy of Tansy ragwort (*Senecio jacobaea* Linnaeus). Stearns (1983) observed changes in life history traits of mosquito fish (*Gambusia*) introduced to water reservoirs in Hawaii that corresponded with the stability of water levels in the reservoirs through time. Genetic bottlenecks were not implicated in either study. Adaptive changes in diapause duration and intensity are commonly observed when insects invade new ranges (Danks, 1987). Such quantitative genetic traits may be far less susceptible to allele and heterozygosity loss that may accompany genetic bottlenecks (Lande & Barrowclough, 1987).

Many past natural enemy introductions and pest colonizations would be fruitful ground for molecular and quantitative genetic studies of the effects of colonization. A world review of parasite and predator introductions (Clausen, 1978) documents many cases where genetic bottlenecks were likely. Examples include the Chinese race of *Comperiella bifasciata* Howard established in California from a stock of 5 gravid females; 24 *Gambusia affinis* Baird and

Girard females formed the genetic base for establishment in the Philippines; several severe bottlenecks are indicated for *Ophimyia lantanae* (Froggatt) establishments on Pacific islands to control Lantana; *Carabus nemoralis* Mueller, a predator of gypsy moth in the United States, stems from a release of 136 individuals.

Genetic Mapping

The mapping of genes of an organism can be considered a fairly specialized endeavor appropriate for fundamental investigations of model organisms. However, newer techniques to find many variable markers in the genome of most organisms makes mapping feasible for more poorly known organisms. The overall approach to linkage mapping and its application to insects has been reviewed (Heckel, 1993). Heckel's list of insect linkage maps includes two parasitoids, both model species with maps based on morphological traits (*Habrobracon juglandis,* and *Nasonia vitripennis*). Most of the remaining species are *Drosophila,* mosquitoes, or other pest Diptera. Subsequent to Heckel's review, additional linkage maps have been developed on several Hymenopteran taxa using RAPDs including *Aphelinus asychis* (Kazmer *et al.,* 1995), *Bracon hebetor* (Antolin *et al.,* 1996a), and honeybee (Hunt & Page, 1995). Eventually AFLP polymorphisms (see earlier) should provide even more abundant variation for genetic mapping, making it possible to develop a fairly saturated linkage map quickly and providing the ability to localize genes of interest for further molecular characterization (Heckel, 1993). These include genes involved in quantitative traits [Lander & Botstein, 1989; e.g., foraging behaviors in honeybees (Hunt *et al.,* 1995)] and may eventually allow us to describe the genetic basis of traits such as host range. Other markers such as RFLPs, microsatellites, and allozymes, as well as visible traits, remain useful for mapping. For an example of the use of these for mapping and for localizing a region of the genome subject to selection see Roethele *et al.* (1997).

Ecological Markers

Isozymes have proved useful for detecting past predation by detecting prey enzymes in predator guts (Luck *et al.,* 1988; Murray *et al.,* 1989). Isozymes have also been used to detect parasitoids of Lepidoptera (Menken, 1982), aphids (Wool *et al.,* 1978; Castañera *et al.,* 1983), and whiteflies (Wood *et al.,* 1984). Isoelectric focusing (Kazmer, 1991) of individual predatory mites, *Euseius tularensis* Congdon, was used to show when they had fed on citrus thrips, *Scirtothrips citri* (Moulton), for up to 40 h after feeding occurred (Jones & Morse, 1995).

Nucleic acid based techniques should be much more sensitive for detection of parasitism. cDNA blotting methods (using probes complementary to the target DNA), especially PCR, have proved to be very sensitive in detection of disease in humans, plants, and their disease vectors (e.g., Lopez-Moya *et al.,* 1992; Suzuki *et al.,* 1992). These methods have recently been applied to detection of internal parasitoids used for biological control (Greenstone & Edwards 1998, Zhu & Greenstone, in press). Disease detection may be particularly important prior to the introduction of biological weed control agents.

Molecular markers are also valuable for studies of insect dispersal and for direct measurements of mating success, parasitism, or other measures of fitness. Kazmer and Luck (1995) compared the parasitism of large and small *Trichogramma pretiosum* females in replicated field trials testing theories on size-related fitness and local mate competition. Edwards and Hoy (1995) used RAPDs to monitor changes in mixed cage populations of insecticide resistant and susceptible *Trioxys pallidus* to determine their relative fitness.

CONCLUSIONS

G. Evelyn Hutchinson likened evolution to a play and the ecological environment to a theater (Hutchinson, 1965). Biological control introductions represent a disruption of the evolutionary context; using Hutchinson's analogy, the play is moved to a new theater in which both the supporting cast and the stage may be different. We disassociate organisms from their ecological and geographic context, giving rise to our preoccupation with the concepts of biotype, race, geographic, sibling, semispecies, and cryptic species (DeBach, 1969; Rosen & DeBach, 1973; Rosen, 1978; Gonzalez *et al.,* 1979). Molecular characterization of genetic variation among populations along this evolutionary continuum may shed considerable light on these concerns. We must question how previous evolutionary history affects adaptation to a novel ecological milieu and we must consider the ecological and genetic dynamics that may arise when two or more genetically distinct populations are released into the same area against the same host. The notion that biological control programs represent evolutionary experiments on a grand scale (Myers, 1978) is certainly valid, but few studies have examined explicitly postrelease adaptations of natural enemies.

Molecular biology has provided highly informative characters for natural enemy and pest systematics providing for a more stable classification system and facilitating rapid identification and placement of newly discovered species. Molecular variation among geographic populations of exotic species introduced into new environments allows us to monitor interbreeding and life history changes following colonization. The use of molecular markers to study mating systems (especially pertaining to sex ratio), sex determination, diapause, temperature tolerance, and other fitness-

related traits of natural enemies should yield permanent rewards.

References

Antolin, M. F., Bosio, C. F., Cotton, J., Sweeney, W., Strand, M. R., & Black, W. C., IV. (1996a). Intensive linkage mapping in a wasp (*Bracon hebetor*) and a mosquito *Aedes aegypti* with single-strand conformation polymorphism analysis of random amplified polymorphic DNA markers. Genetics, 143, 1727–1738.

Antolin, M. F., Guertin, D. S., & Petersen, J. J. (1996b). The origin of gregarious *Muscidifurax* (Hymenoptera: Pteromalidae) in North America: An analysis using molecular markers. Biological Control, 6, 76–82.

Aquadro, C. F., & Avise, J. C. (1982). Evolutionary genetics of birds. VI. A reexamination of protein divergence using varied electrophoretic conditions. Evolution, 36, 1003–1019.

Aquadro, C. F., Noon, W. A., & Begun, D. J. (1992). RFLP analysis using heterologous probes. In A. R. Hoelzel (Ed.), Molecular genetic analysis of populations: A practical approach (pp. 115–158). Oxford: (IRL) Oxford University Press.

Arias, M. C., and Sheppard, W. S. (1996). Molecular phylogenetics of honey bee subspecies (*Apis mellifera* L.) Inferred from mitochondrial DNA sequence. Molecular Phylogenetics and Evolution, 5, 557–566.

Atkinson, L., & Adams, E. S. (1997). Double-strand conformation polymorphism (DSCP) analysis of the mitochondrial control region generates highly variable markers for population studies in a social insect. Insect Molecular Biology, 6(4), 369–376.

Avise, J. C. (1991). Ten unorthodox perspectives on evolution prompted by comparative population genetic findings on mitochondrial DNA. Annual Review of Genetics, 25, 45–69.

Avise, J. C. (1992). Molecular population structure and the biogeographic history of a regional fauna: A case history with lessons for conservation biology. Oikos, 63, 62–76.

Avise, J. C. 1994. Molecular markers, natural history and evolution. New York: Chapman & Hall.

Avise, J. C., & Lansman, R. A. (1983). Polymorphism of mitochondrial DNA in populations of higher animals. In N. Nei (Ed), Evolution of genes and proteins (pp. 147–164). Sunderland ME: Sinauer Associates.

Avise, J. C., Bowen, B. W., & Lamb, T. (1989). DNA fingerprints from hypervariable mitochondrial genotypes. Molecular Biology and Evolution, 6(3), 258–269.

Avise, J. C., Giblin-Davidson, C., Laerm, J., Patton, J. C., & Lansman, R. A. (1979). Mitochondrial DNA clones and matriarchal phylogeny within and among geographic populations of the pocket gopher, *Geomys pinetis*. Proceedings of the National Academy of Sciences, USA, 76, 6694–6698.

Avise, J. C., Arnold, J., Ball, R. M., Bermingham, E., Lamb, T., & Neigel, J. E. (1987). Intraspecific phylogeography: The mitochondrial bridge between population genetics and systematics. Annual Review of Ecology and Systematics, 18, 489–522.

Ayala, F. J., Tracey, M. L., Barr, L. G., & Ehrenfeld, J. G. (1974a). Genetic and reproductive differentiation of the subspecies, *Drosophila equinoxialis caribbensis*. Evolution, 28, 24–41.

Ayala, F. J., Tracey, M. L., Hedgecock, D., & Richmond, R. C. (1974b). Genetic differentiation during the speciation process in *Drosophila*. Evolution, 28, 576–592.

Azeredo-Espin, A. M. L., Schroder, R. F. W., Roderick, G. K., & Sheppard, W. S. (1996). Intraspecific mitochondrial DNA variation in the Colorado potato beetle, *Leptinotarsa decemlineata* (Coleoptera: Chrysomelidae). Biochemical Genetics, 34, 7–8.

Baker, A. J., & Moeed, A. (1987). Rapid genetic differentiation and founder effect in colonizing populations of common Mynas (*Acridotheres tristis*). Evolution, 42, 525–538.

Beckingham, K. (1982). Insect rDNA. Cell Nuclei, 10, 205–269.

Begun, D. J., & Aquadro, C. F. (1992). Evolutionary inferences from DNA variation at the 6-phosphogluconate dehydrogenase locus in natural populations of Drosophila: Selection and geographic differentiation. Genetics, 136, 155–171.

Beismann, H. J., Barker, H. A., Karp, A., & Speck, T. (1997). AFLP analysis sheds light on distribution of two *Salix* species and their hybrid along a natural gradient. Molecular Ecology, 6, 989–993.

Belshaw, R., & Quicke, D. L. J. (1997). A molecular phylogeny of the Aphidiinae (Hymenoptera: Braconidae). Molecular Phylogenetics and Evolution, 7(3), 281–293.

Benson, D. A., Boguski, M. S., Lipman, D. J., & Ostell, J. (1997). GenBank. Nucleic Acids Research, 25(1), 1–6.

Bentz, B. J., & Stock, M. W. (1986). Phenetic and phylogenetic relationships among ten species of *Dendroctonus* bark beetles (Coleoptera: Sclytidae). Annals of the Entomological Society of America, 69, 527–534.

Berlocher, S. H. (1984a). Genetic changes coinciding with the colonization of California by the walnut husk fly, *Rhagoletis completa*. Evolution, 38, 906–918.

Berlocher, S. H. (1984b). Insect molecular systematics. Annual Review of Entomology, 29, 403–433.

Berlocher, S. H. (1989). The complexities of host races and some suggestions for their identification by enzyme electrophoresis. In H. D. Loxdale and J. D. Hollander (Eds.), Electrophoretic studies on agricultural pests (pp. 51–68). Oxford: Clarendon Press.

Berlocher, S. H., & Bush, G. L. (1982). An electrophoretic analysis of *Rhagoletis* (Diptera: Tephritidae) phylogeny. Systematic Zoology, 31, 136–155.

Berlocher, S. H., & Friedman, S. (1981). Loss of genetic variation in laboratory colonies of *Phormia regina*. Entomologia experimentalis et Applicata, 30, 205–208.

Beverley, S. M., & Wilson, A. C. (1985). Ancient origin for Hawaiian Drosophilinae inferred from protein comparisons. Proceedings of the National Academy of Sciences, USA, 82, 4753–4757.

Black, W. C., IV (1993). PCR with arbitrary primers: Approach with care. Insect Molecular Biology, 2(1), 1–6.

Black, W. C., IV, DuTeau, N. M., Puterka, G. J., Nechols, J. R., & Pettorini, J. M. (1992). Use of random amplified polymorphic DNA polymerase chain reaction (RAPD-PCR) to detect DNA polymorphisms in aphids. Bulletin of Entomological Research, 82, 151–159.

Blanchetot, A. (1992). DNA fingerprinting analysis in the solitary bee *Megachile rotundfata:* Variability and nest mate genetic relationships. Genome, 35, 681–688.

Bogdanowicz, S. M., Mastro, V. C., Prasher, D. C., & Harrison, R. G. (1997). Microsatellite DNA variation among Asian and North American gypsy moths (Lepidoptera: Lymantriidae). Annals of the Entomological Society of America, 90, 768–775.

Briese, D. T., Espiau, C., & Pouchot-Lermans, A. (1996). Micro-evolution in the weevil genus *Larinus:* The formation of host biotypes and speciation. Molecular Ecology, 5, 531–545.

Brittnacher, J. G., Sims, S. R., & Ayala, F. J. (1978). Genetic differentiation between species of the genus *Speyeria* (Lepidoptera: Nymphalidae). Evolution, 32, 199–210.

Brower, A. V. Z., & DeSalle, R. (1994). Practical and theoretical considerations for choice of a DNA sequence region in insect molecular systematics, with a short review of published studies using nuclear gene regions. Annals of the Entomological Society of America, 87, 702–716.

Brown, J. M., Leebens-Mack, J. H., Thompson, J. N., Pellmyr, O., & Harrison, R. G. (1997). Phylogeography and host association in a pollinating seed parasite *Greya politella* (Lepidoptera: Prodoxidae). Molecular Ecology, 6, 215–224.

Brown, J. M., Abrahamson, W. G., & Way, P. A. (1996). Mitochondrial DNA phylogeography of host races of the goldenrod ball gallmaker, *Eurosta solidaginis* (Diptera: Tephritidae). Evolution, 50, 777–786.

Brown, M. A. D. (1990). Sequencing with Taq DNA polymerase. In M. A. Innis, D. H. Gelfand, J. J. Sninsky, & T. J. White (Eds.), PCR protocols, a guide to methods and applications (pp. 189–196). San Diego: Academic Press.

Brown, T. A. (1990). Gene cloning, an introduction (2nd ed). London: Chapman & Hall.

Brown, W. M. (1983). Evolution of animal mitochondrial DNA. In M. Nei (Ed.), Evolution of genes and proteins (pp. 62–88). Sunderland, M.: Sinauer Associates.

Brussard, P. F., Ehrlich, P. R., Murphy, D. D., Wilcox, B. A., & Wright, J. (1985). Genetic distances and the taxonomy of checkerspot butterflies (Nymphalidae: Nymphalinae). Journal at Kansas Entomological Society, 58, 403–412.

Bush, G. L., & Kitto, G. B. (1978). Application of genetics to insect systematics and analysis of species differences (Vol. 2). In J. A. Rhomberger et al. (Eds.), Biosystematics in agriculture (pp. 89–118). Beltsville Symposia in Agricultural Research, Rouberger.

Bush, G. L., & Neck, R. W., Jr. (1976). Ecological genetics of the screwworm fly, *Cochliomyia hominivorax* (Diptera: Calliphoridae) and its bearing on the quality control of mass-reared insects. Environmental Entomology, 5, 821–826.

Buth, D. G. (1984). The application of electrophoretic data in systematic studies. Annual Review of Ecology and Systematics, 15, 501–522.

Cabrera, V. M., Gonzalez, A. M., Larruga, J. M., & Gullon, A. (1983). Genetic distance and evolutionary relationships in the *Drosophila obscura* group. Evolution, 37, 675–689.

Caccone, A., & Sbordoni, V. (1987). Molecular evolutionary divergence among North American cave crickets. I. Allozyme variation. Evolution, 41, 1198–1214.

Cameron, S. A. (1991). A new tribal phylogeny of the Apidae inferred from mitochondrial DNA sequences. In D. R. Smith (Ed.), Diversity in the genus Apis (pp. 51–70). Boulder, CO: Westview Press.

Cameron, S. A., Derr, J. N., Austin, A. D., Woolley, J. B., & Wharton, R. A. (1992). The application of nucleotide sequence data to phylogeny of the Hymenoptera: A review. Journal of Hymenoptera Research, 1, 63–79.

Campbell, B. C., Steffen-Campbell, J. D., & Werren, J. H. (1993). Phylogeny of the *Nasonia* species complex (Hymenoptera: Pteromalidae) inferred from an internal transcribed spacer (ITS2) and 28s rDNA sequences. Insect Molecular Biology 2(4), 225–237.

Caprio, M. A., Tabashnik, B. E. (1992). Allozymes used to estimate gene flow among populations of diamondback moth (Lepidoptera: Putellidae) in Hawaii. Environmental Entomology, 21, 808–16.

Carter, P. A., & Watt, W. B. (1988). Adaptations at specific loci. V. Metabolically adjacent enzyme loci may have very distinct experiences of selective pressures. Genetics, 119, 913–924.

Castañera, P., Loxdale, H. D., & Nowak, K. (1983). Electrophoretic study of enzymes from cereal aphid populations. II. Use of electrophoresis for identifying aphidiid parasitoids (Hymenoptera) of *Sitobion avenae* (Fabricius) (Hemiptera: Aphididae). Bulletin of Entomological Research, 73, 659–665.

Chang, H. Y., Wang, D., & Ayala, F. J. (1989). Mitochondrial DNA evolution in the *Drosophila nasuta* subgroup of species. Journal of Molecular Evolution, 28, 337–348.

Charlesworth, D., & Charlesworth, B. (1987). Inbreeding depression and its evolutionary consequences. Annual Review of Ecology and Systematics, 18, 237–268.

Charnov, E. L. (1982). The theory of sex allocation. Princeton, NJ: Princeton University Press.

Claridge, M. F., & Den Hollander, J. (1983). The biotype concept and its application to insect pests of agriculture. Crop Protection, 2, 85–95.

Clary, D. O., & Wolstenholme, D. R. (1985). The mitochondrial DNA molecule of *Drosophila yakuba:* Nucleotide sequence, gene organization, and genetic code. Journal of Molecular Evolution, 22, 252–271.

Clausen, C. P. (1940) Entomophagous insects. New York: McGraw Hill. (Reprinted 1972 by New York: Hafner Publishers.)

Clausen, C. P. (ed.). (1978). Introduced parasites and predators of arthropod pests and weeds: A world review. USDA Handbook No. 480. Washington, DC: U.S. Government Printing Office

Coddington, J. A. (1988). Cladistic tests of adaptational hypotheses. Cladistics, 4, 3–22.

Coen, E. S., Thoday, J. M., & Dover, G. (1982). Rate of turnover of structural variants in the rDNA gene family of *Drosophila melanogaster.* Nature (London), 295, 564–568.

Coll, M., Garcia de Mendoza, L., & Roderick, G. K. (1994). Population structure of a predatory beetle: The importance of gene flow for intertrophic level interactions. Heredity, 72, 228–236.

Collins, D. W. (1996). The separation of *Liriomyza huidobrensis* (Diptera: Agromyzidae) from related indigenous and non-indigenous species encountered in the United Kingdom using cellulose acetate electrophoresis. Annals of Applied Biology, 128(3), 387–398.

Cornuet, J., & Garnery, L. (1991). Genetic diversity in *Apis mellifera.* In D. R. Smith (Ed.), Diversity in the genus Apis (pp. 103–116). Boulder, CO: Westview Press.

Coyne, J. A. (1982). Gel electrophoresis and cryptic protein variation. Isozymes Current Topics in Biological and Medical Research, 6, 1–32.

Coyne, J. A., & Kreitman, M. (1986). Evolutionary genetics of two sibling species, *Drosophila simulans* and *D. sechellia.* Evolution, 404, 673–691.

Cracraft, J. (1989). Speciation and its ontology: The empirical consequences of alternative species concepts for understanding patterns and processes of evolution. In D. Otte and J. A. Endler (Eds.), Speciation and Its Consequences, (pp. 28–59). Sunderland, MA: Sinauer Associates.

Crozier, R. H. (1977). Evolutionary genetics of the Hymenoptera. Annual Review of Entomology, 22, 263–288.

Crozier, R. H., & Crozier, Y. C. (1993). The mitochondrial genome of the honeybee *Apis mellifera:* Complete sequence and genome organization. Genetics, 133, 97–117.

Daly, J. C. (1989). The use of electrophoretic data in a study of gene flow in the pest species *Heliothis armigera* (Hhbner) and *H. Punctigera* Wallengren (Lepidoptera: Noctuidae). In H. D. Loxdale and J. D. Hollander (Eds.), Electrophoretic studies on agricultural pests (pp. 115–141). New York: Oxford University Press.

Daly, J. C., Wilkinson, P., & Shaw, D. D. (1981). Reproductive isolation in relation to allozymic and chromosomal differentiation in the grasshopper *Caledia captiva.* Evolution, 35, 1164–1179.

Danforth, B. N., & Freeman-Gallant, C. R. (1996). DNA fingerprinting data and the problem of non-independence among pairwise comparisons. Molecular Ecology, 5, 221–227.

Danks, H. V. (1987). Insect dormancy: A biological perspective: Biological survey of Canada Monograph (Series No. 1) Ottowa.

De Barro, P., Sherratt, T., Wratten, S., & Maclean, N. (1994). DNA fingerprinting of cereal aphids using (GATA)$_4$. European Journal of Entomology, 91, 109–114.

DeBach, P. (1969). Uniparental, sibling and semi-species in relation to taxonomy and biological control. Israel Journal of Entomology, 4, 11–28.

Debach, P., & Rosen, D. (1991). Biological control by natural enemies (2nd ed.). Cambridge, NY: Cambridge University Press.

Derr, J. N., Davis, S. K., Woolley, J. B., & R. A. Wharton, (1992a). Variation and the phylogenetic utility of the large ribosomal subunit of mitochondrial DNA from the insect order Hymenoptera. Molecular Phylogenetics and Evolution, 1(2), 136–147.

Derr, J. N., Davis, S. K., Woolley, J. B., & Wharton, R. A. (1992b).

Reassessment of the 16S rRNA nucleotide sequence from members of the parasitic Hymenoptera. Molecular Phylogenetics and Evolution, 1(4), 338–341.

DeSalle, R. (1992). The origin and possible time of divergence of the Hawaiian Drosophilidae: Evidence from DNA sequences. Molecular Biology and Evolution, 9, 905–916.

DeSalle, R., & Giddings, L. V. (1986). Discordance of nuclear and mitochondrial DNA phylogenies in Hawaiian *Drosophila.* Proceedings of the National Academy of Sciences, USA, 83, 6902–6906.

Diehl, S. R., & Bush, G. L. (1984). An evolutionary and applied perspective of insect biotypes. Annual Review of Entomology, 29, 471–504.

Dowdy, A. K., & Mcgaughey, W. H. (1996). Using random amplified polymorphic DNA to differentiate strains of the Indianmeal moth (Lepidoptera: Pyralidae). Environmental Entomology, 25(2), 396–400.

Dowling, T. E., Moritz, C., Palmer, J. D., & Rieseberg, L. H. (1996). Nucleic acids III: Analysis of fragments and restriction sites. In D. M. Hillis, C. Moritz, and B. K. Mable (Eds.), Molecular systematics (pp. 249–320). Sunderland, MA: Sinauer Associates.

Dowton, M., & Austin, A. D. (1994). Molecular phylogeny of the insect order Hymenoptera: Apocritan relationships. Proceedings of the National Academy of Sciences, USA, 91, 9911–9915.

Easteal, S., & Boussy, I. A. (1987). A sensitive and efficient isoenzyme technique for small arthropods and other invertebrates. Bulletin of Entomological Research, 77, 407–415.

Edwards, O. R., & Hoy, M. A. (1993). Polymorphism in two parasitoids detected using random amplified polymorphic DNA polymerase chain reaction. Biological Control, 3, 243–257.

Edwards, O. R., & Hoy, M. A. (1995). Monitoring laboratory and field biotypes of the walnut aphid parasite, *Trioxys pallidus,* in population cages using RAPD-PCR. Biocontrol science and technology, 5, 313–327.

Ehler, L. E. (1990). Introduction strategies in biological control of insects. In M. Mackaue, L. E. Ehler, & J. Roland (Eds.), Critical issues in biological control (pp. 111–134). Andover, UK: Intercept Limited.

Ellsworth, D. L., Rittenhouse, K. D., & Honeycutt, R. L. (1993). Artificial variation in randomly amplified polymorphic DNA banding patterns. Biotechniques, 14, 214–217.

Emelianov, I., Mallet, J., & Baltensweiler, W. (1995). Genetic differentiation in *Zeiraphera diniana* (Lepidoptera: Tortricidae, the larch budmoth): polymorphism, host races or sibling species? Heredity, 75, 416–424.

Ender, A., Schwenk, K., Stadler, T., Streit, B., & Schierwater, B. (1996). RAPD identification of microsatellites in *Daphnia.* Molecular Ecology, 5, 437–441.

Engels, W. R. (1981). Estimating genetic divergence and genetic variability with restriction endonucleases. Proceedings of the National Academy of Sciences, USA, 78, 6329–6333.

Excoffier, L., Smouse, P. E., & Quattro, J. M. (1992). Analysis of molecular variance inferred from metric distances among DNA haplotypes: Application to human mitochondrial DNA restriction data. Genetics, 131, 479–491.

Farris, J. S. (1972). Estimating phylogenetic trees from distance matrices. American Naturalist, 106, 645–668.

Farris, J. S., Albert, V. A., Kallersjo, M., Lipscomb, D., & Kluge, A. G. (1996). Parsimony jackknifing outperforms neighbor-joining. Cladistics, 12(2), 99–124.

Feder, J. L., Roethele, J. B., Wlazlo, B., & Berlocher, S. H. (1997). Selective maintenance of allozyme differences among sympatric host races of the apple maggot fly. Proceedings of the National Academy of Sciences, USA, 94, 11417–11421.

Felsenstein, J. (1979). Cases in which parsimony and compatibility methods will be positively misleading. Systematic Zoology, 27, 401–410.

Felsenstein, J. (1981). Evolutionary trees from DNA sequences: A maximum likelihood approach. Journal of Molecular Evolution, 17, 368–376.

Felsenstein, J. (1984). Distance methods for inferring phylogenies: A justification. Evolution, 38, 16–24.

Felsenstein, J. (1988). Phylogenies from molecular sequences: Inference and reliability. Annual Review of Genetics, 22, 521–565.

Fong, G., Walker, M. A., Granett, J. (1995). RAPD assessment of California phylloxera diversity. Molecular Ecology, 4, 459–464.

Fritz, G. N., Conn, J., Cockburn, A., and Seawright, J. (1994). Sequence analysis of the ribosomal DNA internal transcibed spacer 2 from populations of Anopheles nuneztovari (Diptera: Culicidae). Molecular Biology and Evolution, 11, 406–416.

Gasparich, G. E., Sheppard, W. S., Han, H.-Y., McPheron, B. A., & Steck, G. J. (1995). Analysis of mitochondrial DNA and development of PCR-based diagnostic molecular markers for Mediterranean fruit fly (Ceratitis capitata) populations. Insect Molecular Biology, 4, 61–67.

Geiger, H. J. (1980). Enzyme electrophoretic studies on the genetic relationships of pierid butterflies (Lepidoptera: Pieridae). I. European taxa. Journal of Research on the Lepidoptera, 19, 181–195.

Gimeno, C., Belshaw, G., & Quicke, D. L. J. (1997). Phylogenetic relationships of the Alysiinae/Opiinae (Hymenoptera: Braconidae) and the utility of cytochrome b, 16s and 28s D2 rRNA. Insect Molecular Biology, 6(3), 273–284.

Goldstein, D. B., and Pollock, D. D. (1997). Launching microsatellites: A review of mutation processes and methods of phylogenetic inference. Journal of Heredity, 88, 335–342.

Gonzalez, A. M., Cabrera, V. M., Larruga, J. M., & Gullon, A. (1982). Genetic distances in the sibling species Drosophila melanogaster, Drosophila similans and Drosophila mauritiana. Evolution, 36, 517–522.

Gonzalez, D., Gordh, G., Thompson, S. N., & Adler, J. (1979). Biotype discrimination and its importance to biological control. In M. A. Hoy & J. J. McKelvey, Jr. (Eds.), Genetics in relation to insect management (pp. 129–136). New York: Rockefeller Foundation.

Goodman, S. J. (1997). R_{ST} calc: A collection of computer programs for calculating estimates of genetic differentiation from microsatellite data and determining their significance. Molecular Ecology, 6, 881–885.

Goeden, R. D., & Kok, L. T. (1986). Comments on a proposed new approach for selecting agents for the biological control of weeds. Canadian Entomology, 118, 51–58.

Greathead, D. J., & Greathead, A. H. (1992). Biological control of insect pests by insect parasitoids and predators: the BIOCAT database. Biocontrol News and Information, 13, 61N–68N.

Greenstone, M. H., & Edwards, M. J. (1998). DNA hybridization probe for endoparasitism by Microplitis croceipes (Hymenoptera: Braconidae). Annals of the Entomological Society of America 91, 415–421.

Grosberg, R. K., Levitan, D. R., & Cameron, B. B. (1996). Characterization of genetic structure and genealogies using RAPD-PCR markers: A random primer for the novice and nervous. In J. D. Ferraris & S. R. Palumbi (Eds.). Molecular Zoology: Advances, Strategies and Protocols, (pp. 67–100). New York: Wiley–Liss.

Gyllensten, U. B., & Erlich, H. A. (1988). Generation of single-stranded DNA by the polymerase chain reaction and its application to direct sequencing of the HLA-DQA locus. Proceedings of the National Academy of Sciences, USA, 85, 7652–7656.

Hadrys, H., Schierwater, B., Dellaporta, S. L., Desalle, R., & Buss, L. W. (1993). Determination of paternity in dragonflies by random amplified polymorphic DNA fingerprinting. Molecular Ecology, 2, 79–87.

Hajek, A. E., Hodge, K. T., Liebherr, J. K., Day, W. H., & Vandenberg, J. D. (1996). Use of RAPD analysis to trace the origin of the weevil pathogen Zoophthora phytonomi in North America. Mycological Research, 100(3), 349–355.

Hall, R. W., & Ehler, L. E. (1979). Rate of establishment of natural enemies in classical biological control. Bulletin of the Entomology Society of America, 25(4), 280–282.

Halliday, R. B. (1981). Heterozygosity and genetic distance in sibling species of meat ants (Iridomyrmex purpureus group). Evolution, 35, 234–242.

Hamilton, W. D. (1967). Extraordinary sex ratios. Science, 156, 477–488.

Harris, H., & Hopkinson, D. A. (1976). Handbook of enzyme electrophoresis in human genetics. Oxford: North Holland.

Harrison, R. G. (1979). Speciation in North American field crickets: Evidence from electrophoretic comparisons. Evolution, 33, 1009–1023.

Harrison, R. G. (1989). Animal mitochondrial DNA as a genetic marker in population and evolutionary biology. Trends in Ecological Evolution, 4, 6–11.

Harrison, R. G., & Rand, D. M. (1989). Mosaic hybrid zones and the nature of species boundaries. In D. Otte and J. A. Endler (Eds.), Speciation and Its Consequences. Sunderland, MA: Sinauer Associates.

Harrison, R. G., & Odell, T. M. (1989). Mitochondrial DNA as a tracer of gypsy moth origins. In W. E. Wallner and K. A. McManus (Eds.), Proceedings, Lymantriidae: A comparison of features of new and old world tussock moths, (pp. 265–273). Broomall, PA: Northeastern Forest Experiment Station.

Hasegawa, E. (1995). Parental analysis using RAPD markers in the ant Colobopsis nipponicus: A test of RAPD markers for estimating reproductive structure within social insect colonies. Insectes Sociaux-Social Insects, 42(4), 337–346.

Haymer, D. S., McInnis, D. O., & Arcangeli, L. (1992). Genetic variation between strains of Mediterranean fruit fly, Ceratitis capitata, detected by DNA fingerprinting. Genome, 35, 528–533.

Haymer, D. S., & Mcinnis, D. O. (1994). Resolution of populations of the Mediterranean fruit fly at the DNA level using random primers for the polymerase chain reaction. Genome, 37(2), 244–248.

Heckel, D. G. (1993). Comparative genetic linkage mapping in insects. Annual Review of Entomology, 38, 381–408.

Hedderwick, M. P., El Agoze, M., & Periquet, G. (1988). Enzymatic polymorphism, inbreeding and occurrence of diploid males in Diadromus pulchellus. Les Colloques de l'INRA, 48, 109–110.

Hedrick, P. (1992). Shooting the RAPDs. Nature (London), 335, 679–680.

Hedrick, P. W., and Parker, J. D. (1997). Evolutionary genetics and genetic variation of Haplodiploids and X-linked genes. Annual review of ecology and systematics, 28, 55–83.

Hillel, J., Plotzy, Y., Haberfeld, A., Lavi, U., Cahaner, A., & Jeffreys, A. J. (1989). DNA fingerprints of poultry. Animal Genetics, 20, 145–155.

Hillis, D. M. (1984). Misuse and modification of Nei's genetic distance. Systematic Zoology, 33, 238–240.

Hillis, D. M. (1987). Molecular versus morphological approaches to systematics. Annual Review of Ecology and Systematics, 18, 23–42.

Hillis, D. M., & Dixon, M. T. (1991). Ribosomal DNA: Molecular evolution and phylogenetic interference. Quarterly Review of Biology, 66(4), 411–453.

Hillis, D. M., Mable, M. K., & Moritz, C. (1996a). Applications of molecular systematics. In D. M. Hillis, C. Moritz, and B. K. Mable (Eds.), Molecular systematics. Sunderland, MA: Sinauer Associates.

Hillis, D. M., Mable, B. K., Larson, A., Davis, S. K., & Zimmer, E. A. (1996b). Nucleic acids IV: Sequencing and cloning. In D. M. Hillis, C. Moritz, & B. K. Mable (Eds.), Molecular systematics (pp. 321–381). Sunderland, MA: Sinauer Associates.

Hillis, D. M., Moritz, C., & Mable, B. K. (Eds.) (1996c). Molecular systematics. Sunderland, MA: Sinauer Associates.

Hiss, R. H., Norris, D. E., Dietrich, C. H., Whitcomb, R. F., West, D. F., & Bosio, C. F. (1994). Molecular taxonomy using single-strand conformation polymorphism (SSCP) analysis of mitochondrial ribosomal DNA genes. Insect Molecular Biology, 3(3), 171–182.

Hopper, K. R., & Roush, R. T. (1993). Mate finding, dispersal, number released, and the success of biological control introductions. Ecological Entomology, 18, 321–331.

Hopper, K. R., Roush, R. T., & Powell, W. (1993). Management of genetics of biological-control introductions. Annual Review of Entomology 38, 27–51.

Howard, D. J. (1983). Electrophoretic survey of eastern North American *Allonemobius* (Orthoptera: Gryllidae): Evolutionary relationships and the discovery of three new species. Annals of the Entomological Society of America, 76, 1014–1021.

Howarth, F. G. (1991). Environmental impacts of classical biological control. Annual Review of Entomology, 36, 485–509.

Hsiao, T. H., & Stutz, J. M. (1985). Discrimination of alfalfa weevil strains by allozyme analysis. Entomologia Experimentalis et Applicata, 37, 113–121.

Hubby, J. L., & Lewontin, R. C. (1996). A molecular approach to the study of genetic heterozygosity in natural populations. I. The number of alleles at different loci in *Drosophila pseudoobscura*. Genetics, 54, 477–594.

Hudson, R. R., Slatkin, M., & Madison, W. P. (1992). Estimation of levels of gene flow from DNA sequence data. Genetics, 132, 583–589.

Hughes, C. R., & Queller, D. C. (1993). Detection of highly polymorphic microsatellite loci in a species with little allozyme polymorphism. Molecular Ecology, 2, 131–137.

Hunt, G. J., & Page, R. E. (in press). Patterns of inheritance with RAPD molecular markers reveal novel types of polymorphism in the honey bee. Theoretical and Applied Genetics.

Hunt, G. J., & Page, R. E. (1995). Linkage map of the honey bee, *Apis mellifera*, based on RAPD markers. Genetics, 139, 1371–1382.

Hunt, G. J., Page, R. E., Jr., Fondrk, M. K., & Dullum, C. J. (1995). Major quantitative trait loci affecting honey bee foraging behavior. Genetics, 141, 1537–1545.

Hunter, M. S., Antolin, M. F., & Rose, M. (1996). Courtship behavior, reproductive relationships, and allozyme patterns of three North American populations of *Eretmocerus* nr. *californicus* (Hymenoptera: Aphelinidae) parasitizing the whitefly *Bemisia* sp., *tabaci* Complex (Homoptera: Aleyrodidae). Proceedings of the Entomological Society of Washington, 98(1), 126–137.

Hunter, R. L., & Markert, C. L. (1957). Histochemical demonstration of enzymes separated by zone electrophoresis in starch gels. Science, 125, 1294–1295.

Hutchinson, G. E. (1965). The ecological theater and the evolutionary play. New Haven, CT: Yale University Press.

Innis, M. A., Gelfand, D. H., Sninsky, J. J., & White, T. J. (Eds.) (1990). PCR protocols, a guide to methods and applications. San Diego: Academic Press.

Izawa, H., Osakabe, M., & Moriya, S. (1995). Relation between banding patterns of malic enzyme by electrophoresis and a morphological character in exotic and native *Torymus* species. Applied Entomology and Zoology, 30(1), 37–41.

Izawa, H., Osakabe, M., Moriya, S., & Toda, S. (1996). Use of malic enzyme to detect hybrids between *Torymus sinensis* and *T. beneficus* (Hymenoptera: Torymidae) attacking *Dryocosmus kuriphilus* (Hymenoptera: Cynipidae) and possibility of natural hybridization. [Japanese]. Japanese Journal of Applied Entomology and Zoology, 40(3), 205–208.

Jeffreys, A. J., Hillel, J., Hartley, N., Bulfield, G., Morton, D., & Wilson, V. (1987). Hypervariable DNA and genetic fingerprints. Animal Genetics, 18, 141–142.

Johnson, G. B. (1974). Enzyme polymorphism and metabolism. Science, 184, 28–37.

Johnson, G. B. (1977). Assessing electrophoretic similarity: The problem of hidden heterogeneity. Annual Review of Ecology and Systematics, 8, 309–328.

Jones, S. A., & Morse, J. G. (1995). Use of isoelectric focusing electrophoresis to evaluate citrus thrips (Thysanoptera: Thripidae) predation by *Euseius tularensis* (Acari: Phytoseiidae). Environmental Entomology, 24(5), 1040–1051.

Kambhampati S., Black, W. C., Rai, K. S., & Sprenger, D. (1990). (Temporal variation in genetic structure of a colonising species—*Aedes albopictus* in the United States. Journal of Heredity, 64, 281–287.

Kambhampati, S., Black, W. C., & Rai, K. S. (1992). Randomly amplified polymorphic DNA of mosquito species and populations (Diptera; Culicidae): Techniques, statistical analysis, and applications. Journal of Medical Entomology, 29, 939–945.

Kavanaugh, D. H. (1978). Hennigian phylogenetics in contemporary systematics: Principles, methods and uses. In L. Knutson (Ed.) Biosystematics in Agriculture (pp. 139–150). Montclair, NJ: Allanheld, Osmun.

Kazmer, D. J. (1991). Isoelectric focusing procedures for the analysis of allozymic variation in minute arthropods. Annals of the Entomological Society of America, 84, 332–339.

Kazmer, D. J., & Luck, R. F. (1995). Field tests of the size-fitness hypothesis in the egg parasitoid *Trichogramma pretiosum*. Ecology, 76, 412–425.

Kazmer, D. J., Hopper, K. R., Coutinot, D. M., & Heckel, D. G. (1995). Suitability of random amplified polymorphic DNA for genetic markers in the aphid parasitoid, *Aphelinus asychis* Walker. Biological Control, 5, 503–512.

Kazmer, D, J., Maiden, K., Ramualde, N., Coutinot, D., & Hopper, K. R. (1996). Reproductive compatibility, mating behavior, and random amplified polymorphic DNA variability in some *Aphelinus asychis* (Hymenoptera: Aphelinidae) derived from the old world. Annals of the Entomological Society of America, 89, 212–220.

Kimani Njogu, S. W., Overholt, W. A., Woolley, J. B., & Omwega, C. O. (1998). Electrophoretic and phylogenetic analyses of selected allopatric populations of the *Cotesia flavipes* complex (Hymenoptera: Braconidae) parasitoids of cereal stemborers. Biochemical Systematics and Ecology 26, 285–296.

Kimura, M. (1983). The neutral theory of molecular evolution. Cambridge, MA: Cambridge University Press.

Kirby, L. T. (1992). DNA fingerprinting: An introduction. New York: W. H. Freeman.

Krafsur, E. S., Obrycki, J. J., & Nariboli, P. (1996). Gene flow in colonizing *Hippodamia variegata* ladybird beetle populations. Journal of Heredity, 87, 41–47.

Krafsur, E. S., Obrycki, J. J., & Schaeffer, P. W. (1995). Genetic heterozygosity and gene flow in *Coleomegilla maculata* De Geer (Coleoptera: Coccinellidae). Biological Control, 5 (1), 104–111.

Krafsur, E. S., Kring, T. J., Miller, J. C., Naiiboli, P., Obrycki, J. J., & Ruberson, J. (1997). Gene flow in the exotic colonizing ladybeetle *Harmonia axyridis* in North America. Biological Control, 8, 207–214.

Lande, R., and Barrowclough, F. (1987). Effective population size, genetic variation, and their use in population management. In Soule, M. E. (Ed.), Viable populations for conservation (pp. 87–123). Cambridge, New York: Cambridge University Press.

Lander, E. S., & Botstein, D. (1989). Mapping mendelian factors underlying quantitative traits using RFLP linkage maps. Genetics, 121, 185–199.

Landry, B. S., Dextraze, L., & Boivin, G. (1993). Random amplified polymorphic DNA markers for DNA fingerprinting and genetic variability assessment of minute parasitic wasp species (Hymenoptera: Mymaridae and Trichogrammatidae) used in biological control programs of phytophagous insects. Genome, 36, 580–587.

Laroche, A., DeClerck-Floate, R. A., LeSage, L., Floate, K. D., & Demeke, T. (1996). Are *Altica carduorum* and *Altica cirsicola* (Coleoptera: Chrysomelidae) different species? Implications for the release of *A. Cirsicola* for the biocontrol of Canada thistle in Canada. Biological Control, 6, 306–314.

Latorre, A., Moya, A., & Ayala, F. J. (1986). Evolution of mitochondrial DNA in *Drosophila subobscura*. Proceedings of the National Academy of Sciences, USA 83, 8649–8653.

Lenk, P., & Wink, M. (1997). A RAN/RNA heteroduplex cleavage analysis to detect rare mutations in populations. Molecular Ecology, 6, 687–690.

Lessa, E. (1992). Rapid surveying of DNA sequence variation in natural populations. Molecular Biological Evolution, 9, 323–330.

Lessa, E. P., & Applebaum, G. (1993). Screening techniques for detecting allelic variation in DNA sequences. Molecular Ecology, 2, 119–129.

Lewin, B. M. (1995). Genes, V. New York: John Wiley & Sons.

Lewin, R. (1989). Limits to DNA fingerprinting. Science, 243, 1549–1551.

Lewis, D. L., Farr, C. L., & Kaguni, L. S. (1995). Drosophila melanogaster mitochondrial DNA: Completion of the nucleotide sequence and evolutionary comparisons. Insect Molecular Biology, 4, 263–278.

Lewontin, R. C. (1974). The genetic basis of evolutionary change. New York: Columbia University Press.

Loh, E. Y., Elliott, J. F., Cwirla, S., Lanier, L. L., & Davis, M. M. (1989). Polymerase chain reaction with single-sided specificity: Analysis of T cell receptor chain. Science, 243, 217–220.

Lopez-Moya, J. J., Cubero, J., Lopez-Abella, D., & Diaz-Ruiz, J. R. (1992). Detection of cauliflower mosaic virus (CaMV) in single aphids by the polymerase chain reaction (PCR). Journal of Virology Methods, 37, 129–138.

Louda, S. M., Kendall, D., Connor, J., Simberloff, D. (1997). Ecological effects of an insect introduced for the biological control of weeds. Science, 277, 1088–1090.

Loxdale, H. D., & Brookes, C. P. (1989). Use of genetic markers (allozymes) to study the structure, overwintering and dynamics of pest aphid populations. In H. D. Loxdale and J. D. Hollander (Eds.), Electrophoretic studies on Agricultural pests (pp. 231–270). Oxford: Clarendon Press.

Loxdale, H. D., & Den Hollander J. (Eds.) (1989). Electrophoretic studies on agricultural pests. New York: Oxford University Press.

Lu, R., & Rank, G. H. (1996). Use of RAPD analyses to estimate population genetic parameters in the alfalfa leaf-cutting bee, Megachile rotundata. Genome, 39(4), 655–663.

Luck, R. F., Shepard, B. M., & Kenmore, P. E. (1988). Experimental methods for evaluating arthropod natural enemies. Annual Review of Entomology, 33, 367–392.

Luck, R. F., Stouthamer, R., & Nunney, L. P. (1993). Sex determination and sex ratio patterns in parasitic Hymenoptera. In D. L. Wrensch & M. A. Ebbert (Eds.), Evolution and diversity of sex ratio in haplodiploid insects and mites (pp. 442–476). Routledge, New York: Chapman & Hall.

Lynch, M. (1988). Estimation of relatedness by DNA fingerprinting. Molecular and Biology Evolution, 5, 584–599.

Lynch, M. (1990). The similarity index and DNA fingerprinting. Molecular Biology and Evolution, 7, 478–484.

Lynch, M., & Milligan, B. G. (1994). Analysis of population genetic structure with RAPD markers. Molecular Ecology, 3, 91–99.

Manchenko, G. P. (1994). Handbook of detection of enzymes on electrophoretic gels. Boca Raton, FL: CRC Press.

Manguin, S., White, R., Blossey, B., & Hight, S. D. (1993). Genetics, taxonomy, and ecology of certain species of Galerucella (Coleoptera: Chrysomelidae). Annals of the Entomological Society of America, 86, 397–410.

Martin, A. P., & Simon, C. (1988). Anomalous distribution of nuclear and mitochondrial DNA markers in periodical cicadas. Nature (London), 336, 237–239.

Maxson, L. R., & Maxson, R. D. (1990). Proteins II: Immunological techniques. In D. M. Hillis & C. Moritz (Eds.), Molecular systematics (pp. 127–155). Sunderland, MA: Sinauer Associates.

McCauley, D. E., Schife, N., Breden, F. J., & Chippendale, G. M. (1995). Genetic differentiation accompanying range expansion by the southwestern corn borer (Lepidoptera: Pyralidae). Annals of the Entomological Society of America, 88, 357–361.

McEvoy, P. (1996). Host specificity and biological pest control. BioScience, 46, 401–405.

McPheron, G. A., Smith, D. C., & Berlocher, S. H. (1988). Genetic differences between host races of Rhagoletis pomonella Nature (London), 336(6194), 64–66.

Menken, S. B. J. (1982). Enzymatic characterization of nine endoparasite species of small ermine moths (Yponomeutidae). Experientia, 38, 1461–1462.

Menken, S. B. J. (1981). Host races and sympatric speciation in small ermine moths, Yponomeutidae. Entomologia Experimentalis et Applicata, 30, 280–292.

Menken, S. B. J. (1989). Electrophoretic studies on geographic populations, host races, and sibling species of insect pests. In Electrophoretic studies on agricultural pests (pp. 181–202). In H. D. Loxdale & J. D. Hollander (Eds.), New York: Oxford University Press.

Metcalf, R. A., Marlin, J. C., & Whitt, G. S. (1984). Genetics of speciation within the Polistes fuscatus species complex. Journal of Heredity, 75, 117–120.

Michelmore, R. W., Paran, I., & Kesseli, R. V. (1991). Identification of markers linked to disease-resistance genes by bulked segregant analysis. A rapid method to detect markers in specific genomic regions by using segregating populations. Proceedings of the National Academy of Sciences, USA, 88, 9829–9832.

Milligan, B. G., & McMurry, C. K. (1993). Dominant vs. codominant genetic markers in the estimation of male mating success. Molecular Ecology, 2, 275–283.

Miyamoto, M. M., & Cracraft, J. (Eds.). (1991). Phylogenetic analysis of DNA sequences. New York: Oxford University Press.

Morgante, J. S., Malavasi, A., & Bush, G. L. (1980). Biochemical systematics and evolutionary relationships of neotropical Anastrepha. Annals of the Entomological Society of America, 73, 622–630.

Moritz, C., Dowling, T. E., & Brown, W. M. (1987). Evolution of animal mitochondrial DNA: Relevance for populations biology and systematics. Annual Review of Ecology Systematics, 18, 269–292.

Munstermann, L. E. (1980). Distinguishing geographic strains of the Aedes atropalpus group (Diptera: Culicidae) by analysis of enzyme variation. Annals of the Entomological Society of America 73, 699–704.

Murphy, R. W., Sites, J. W., Jr., Buth, D. G., & Haufler, C. H. (1996). Proteins: Isozyme electrophoresis. In D. M. Hillis, C. Moritz, and B. K. Mable (Eds.), Molecular systematics (pp. 51–120). Sunderland, MS: Sinauer Associates.

Murray, R. A., Solomon, M. G., & Fitzgerald, J. D. (1989). The use of electrophoresis for determining patterns of predation in arthropods. In H. D. Loxdale and J. D. Hollander (Eds.). Electrophoretic studies on agricultural pests (pp. 467–483). New York: Oxford University Press.

Myers, J. (1978). Biological control introductions as grandiose field experiments: Adaptations of the cinnabar moth to new surroundings (pp. 181–188). Proceedings of the Fourth International Symposium of Biological Control of Weeds, Gainesville, FL, 1976.

Myers, J. H., & Sabath, M. D. (1980). Genetic and phenotypic variability, genetic variance, and the success of establishment of insect introductions for the biological control of weeds (pp. 91–102). Proceedings of the Fifth International Symposium on Biological Control of Weeds, Brisbane, Australia.

Nechols, J. R., Kauffman, W. C., & Schaefer, P. W. (1992). Significance of host specificity in classical biological control. In Proceedings, 1990 Annual Meeting of the Entomological Society of America (pp. 41–52). New Orleans, LA: Thomas Say Publications in Entomology.

Nei, M. (1972). Genetic distance between populations. American Naturalist, 106, 283–292.

Nei, M. (1987). Molecular evolutionary genetics. New York: Columbia University Press.

Nei, M., & Chakraborty, R. (1973). Genetic distance and electrophoretic identity of proteins between taxa. Journal of Molecular Evolution, 2, 323–328.

Nei, M., Maruyama, T., & Chakraborty, R. (1975). The bottleneck effect and genetic variability in populations. Evolution, 19, 1–10.

Neigel, J. E. (1997). A comparison of alternative strategies for estimating gene flow from genetic markers. Annual review of ecology and systematics, 28, 105–128.

Neto, L., & Pintureau, B. (1995). Taxonomic study of a population of Trichogramma turkestanica discovered in southern Portugal (Hymenoptera: Trichogrammatidae). Annales de la Societe Entomologique de France, 31,(1), 21–30.

Nicol, D., Armstrong, K. F., Wratten, S. D., Cameron, C. M., Frampton, C., & Fenton, B. (1997). Genetic variation in an introduced aphid pest (Metopolophium dirhodum) in New Zealand and relation to individuals from Europe. Molecular Ecology, 6, 255–265.

Nowierski, R. M., McDermott, G. J., Bunnell, J. E., Fitzgerald, B. C., & Zeng, Z. (1996). Isozyme analysis of Aphthona species (Coleoptera: Chrysomelidae) associated with different Euphorbia species (Euphorbiaceae) and environmental types in Europe. Annals of the Entomological Society of America, 89, 858–868.

Ochman, H. A., Gerber, S., Hartl, D. L. (1988). Genetic applications of an inverse polymerase chain reaction. Genetics, 120, 621–625.

O'Donnell, S. (1996). RAPD markers suggest genotypic effects on forager specialization in a eusocial wasp. Behavioral Ecology and Sociobiology, 38(2), 83–88.

Omwega, C. O., & Overholt, W. A. (1996). Genetic changes occurring during laboratory rearing of Cotesia flavipes Cameron (Hymenoptera: Braconidae), an imported parasitoid for the control of gramineous stem borers in Africa. African Entomology, 4(2), 231–237.

Orita, M., Iwahana, H., Kanazawa, H., Hayashi, K., and Sekiya, T. (1989). Detection of polymorphisms of human DNA by gel electrophoresis as single-strand conformation polymorphisms. Proceedings of the National Academy of Sciences, USA, 86, 2766–2770.

Orti, G., Hare, M. P., & Avise, J. C. (1997). Detection and isolation of nuclear haplotypes by PCR-SSCP. Molecular Ecology, 6, 575–580.

Otte, D. and Endler, J. A. (Eds.). (1989). Speciation and Its Consequences. Sunderland, MA: Sinauer Associates.

Owen, R. E. (1993). Genetics of parasitic Hymenoptera. In S. K., Narang, A. C. Bartlett, and R. M. Faust (Eds.). Applications of genetics to arthropods of biological control significance (pp. 69–89). Boca Raton, FL: CRC Press.

Packer, L., & Owen, R. E. (1990). Allozyme variation, linkage disequilibrium and diploid male production in a primitively social bee Augochlorella striata (Hymenoptera; Halictidae). Heredity, 65, 241–248.

Palmer, M. J., Bantle, J. A., Guo, X., & Fargo, W. S. (1994). Genome size and organization in the ixodid tick Amblyomma americanum (L.). Insect Molecular Biology, 3, 57–62.

Pamilo, P. (1984). Genotypic correlation and regression in social groups: Multiple alleles, multiple loci and subdivided populations. Genetics, 107, 307–320.

Pashley, D. P. (1983). Biosystematic study in Tortricidae, (Lepidoptera), with a note on evolutionary rates of allozymes. Annals of the Entomological Society of America, 76, 139–148.

Pashley, D. P. (1986). Host-associated genetic differentiation in fall armyworm (Lepidoptera: Noctuidae): A sibling species complex? Annals of the Entomological Society of America, 79(6), 898–904.

Pashley, D. P. (1989). Host-associated differentiation in armyworms (Lepidoptera: Noctuidae): An allozymic and mitochondrial DNA perspective. In H. D. Loxdale and J. D. Hollander (Eds.), Electrophoretic studies on agricultural pests (pp. 103–114). New York: Oxford University Press.

Pashley, D. P., & Martin, J. A. (1987). Reproductive incompatibility between host strains of the fall armyworm (Lepidoptera: Noctuidae). Annals of the Entomological Society of America, 80(6), 731–733.

Pashley, D. P., & Ke, L. D. (1992). Sequence evolution in mitochondrial ribosomal and ND-1 genes: Implications for phylogenetic analysis. Molecular Biology and Evolution, 9, 1061–1075.

Pashley, D. P., & Proverbs, M. D. (1981). Quality control by electrophoretic monitoring in a laboratory colony of codling moths. Annals of the Entomological Society of America, 74, 20–23.

Pashley, D. P., Rai, K. S., & Pashley, D. N. (1985). Patterns of allozyme relationships compared with morphology, hybridization, and geologic history in allopatric island-dwelling mosquitoes. Evolution, 39, 985–997.

Pashley, D. P., Johnson, S. J., & Sparks, A. N. (1983). Genetic population structure of migratory moths: The fall armyworm (Lepidoptera: Noctuidae). Annals of the Entomological Society of America, 78(6), 756–761.

Penner, G. A., Bush, A., Wise, R., Kim, W., Domier, L., & Kasha, K. (1993). Reproducability of random amplified polymorphic DNA (RAPD) analysis among laboratories. PCR Methods Applications 2, 341–345.

Perrot-Minnot, M.-J., Guo, L. R., & Werren, J. H. (1995). Single and double infections with Wolbachia in the parasitic wasp Nasonia vitripennis: Effects on compatibility. Genetics, 143, 961–972.

Pinto, J. D., Kazmer, D. J., Platner, G. R., & Sassaman, C. A. (1992). Taxonomy of the Trichogramma minutum complex (Hymenoptera: Trichogrammatidae): Allozymic variation and its relationship to reproductive and geographic data. Annals of the Entomological Society of America, 85, 413–422.

Pinto, J. D., Stouthamer, R., & Platner, G. R. (1997). A new cryptic species of Trichogramma (Hymenoptera: Trichogrammatidae) from the Mojave Desert of California as determined by morphological, reproductive and molecular data. Proceedings of the Entomological Society of Washington, 99(2), 238–247.

Pinto, J. D., Stouthamer, R., Platner, G. R., & Oatman, E. R. (1991). Variation in reproductive compatibility in Trichogramma and its taxonomic significance. Annals of the Entomological Society of America, 84, 37–46.

Pintureau, B. (1993a). Enzymatic analysis of the genus Trichogramma (Hym.: Trichogrammatidae) in Europe. Entomophaga, 38(3), 411–431.

Pintureau, B. (1993b). Enzyme polymorphism in some African, American and Asiatic Trichogramma and Trichogrammatoidea species Hymenoptera Trichogrammatidae. Biochemical Systematics and Ecology, 21(5), 557–573.

Porter, C. H., & Collins, F. H. (1991). Species-diagnostic differences in ribosomal DNA internal transcribed spacer from sibling species Anopheles freeborni and Anopheles hermsi (Diptera: Culicidae). American Journal of Tropical Medicine and Hygiene, 45, 271–279.

Powell, W., & Walton, M. P. (1989). The use of electrophoresis in the study of hymenopteran parasitoids of agricultural pests. In H. D. Loxdale and J. D. Hollander (Eds.), Electrophoresis studies on agricultural pests, (pp. 443–465). Oxford: Clarendon Press.

Puterka, G. J., Black, IV, W. C., Steiner, W. M., & Burton, R. L. (1993). Genetic variation and phylogenetic relationships among worldwide collections of the Russian wheat aphid, Diuraphis noxia (Mordvilko), inferred from allozyme and RAPD-PCR markers. Heredity, 70, 604–618.

Queller, D. C., & Goodnight, K. F. (1989). Estimating relatedness using genetic markers. Evolution, 43(2), 258–275.

Ram, P., Tshernyshev, W. B., Afonina, V. M., & Greenberg, S. M. (1995a). Studies on the strains of Trichogramma evanescens Westwood (Hym., Trichogrammatidae) collected from different hosts in Northern Maldova. Journal of Applied Entomology, 119(1), 79–82.

Ram, P., Tshernyshev, W. B., Afonina, V. M., Greenberg, S. M., & Ageeva, L. I. (1995b). Studies on strains of Trichogramma evanescens Westwood from different regions of Eurasia. Biocontrol Science and Technology, 5(3), 329–338.

Rannala, B., & Mountain, J. L. (1997). Detecting immigration by using multilocus genotypes. Proceedings of the National Academy of Sciences, USA, 94, 9197–9201.

Richardson, B. J., Baverstock, P. R., & Adams, M. (1986). Allozyme electrophoresis, a handbook for animal systematics and population studies. New York: Academic Press.

Rieseberg, L. H. (1996). Homology among RAPD fragments in interspecific comparisons. Molecular Ecology, 5, 99–105.

Righetti, P. G. (1983). Isoelectric focusing: Theory, methodology and applications. Amsterdam: Elsevier.

Roberts, R. J. (1992). Restriction enzymes. In A. R. Hoelzel (Ed.), Molecular genetic analysis of populations: A practical approach (pp. 281–298). Oxford: (IRL) Oxford University Press.

Roberts, R. J., & Macelis, D. (1997). REBASE-restriction enzymes and methylases. Nucleic Acids Research, 25(1), 248–262.

Roderick, G. K. (1992). Postcolonization evolution of natural enemies. In Proceedings, 1990 Annual Meeting of the Entomological Society of America (pp. 71–86). New Orleans, LA: Thomas Say Publications in Entomology.

Roderick, G. K. (1996). Geographic structure of insect populations: Gene flow, phylogeography, and their uses. Annual Review Entomology, 41, 325–352.

Roehrdanz, R. L. (1989). Intraspecific genetic variability in mitochondrial DNA of the screwworm fly (Cochliomyia hominivorax). Biochemical Genetics, 27, 9–10.

Roehrdanz, R. L. (1995). Amplification of complete insect mitochondrial genome in two easy pieces. Insect Molecular Biology, 4(3), 169–172.

Roehrdanz, R. L., Reed, D. K., & Burton, R. L. (1993). Use of polymerase chain reaction and arbitrary primers to distinguish laboratory-raised colonies of parasitic hymenoptera. Biological Control, 3, 199–206.

Roethele, J. B., Feder, J. L., Berlocher, S. H., Kreitman, M. E., & Lashkari, D. A. (1997). Toward a molecular genetic linkage map for the apple maggot fly (Diptera: Tephritidae): Comparison of alternative strategies. Annals of the Entomological Society of America, 90, 470–479.

Rossler, Y., & DeBach, P. (1973). Genetic variability in a thelytokous form of Aphytis mytilaspidis (Le Baron) (Hymenoptera: Aphelinidae). Hilgardia, 42, 149–176.

Rosen, D. (1978). The importance of cryptic species and specific identifications as related biological control: Vol. 2. In J. A. Rhomberger et al. (Eds.), Biosystematics in agriculture (pp. 23–35). Beltsville Symposium on Agricultural Research.

Rosen, D. (1986). The role of taxonomy in effective biological control programs. Agriculture, Ecosystems and Environment, 15, 121–129.

Rosen, D., & DeBach, P. (1973). Systematics, morphology and biological control. Entomophaga, 18, 215–222.

Rosen, D., & DeBach, P. (1979). Species of Aphytis of the world (Hymenoptera: Aphelinidae). The Hague: W. Junk.

Ross, K. G., Vargo, E. L., & Fletcher, D. J. C. (1987). Comparative biochemical genetics of three fire ant species in North America, with special reference to the two social forms of Solenopsis invicta (Hymenoptera: Formicidae). Evolution, 41(5), 979–990.

Roush, R. T. (1990). Genetic variation in natural enemies: Critical issues for colonization in biological control. In M. Mackauer, L. E. Ehler, & J. Roland (Eds.), Critical issues in biological control (pp. 263–288). Andover: Intercept.

Roush, R. T., & Hopper, K. R. (1995). Use of single family lines to preserve genetic variation in laboratory colonies. Annals of the Entomological Society of America, 88, 713–717.

Roux, G., and Roques, A. (1996). Biochemical genetic differentiation among seed chalcid species of genus Megastigmus (Hymenoptera: Torymidae). Experientia, 52(6), 522–530.

Saiki, R. K., Gelfand, D. H., Stoffel, S., Scharf, S. J., Higuchi, R., & Horn, G. T. (1988). Primer-directed enzymatic amplification of DNA with a thermostable DNA polymerase. Science, 239, 487–491.

Saitou, N., & Nei, M. (1987). The neighbor-joining method: A new method for reconstructing phylogenetic trees. Molecular Biology and Evolution, 4, 406–425.

Sawyer, R. C. (1996). To make a spotless orange: Biological control in California. Ames, IA: Iowa State University Press.

Sbordoni, V., Allegrucci, G., Caccone, A., Cesaroni, D., Cobolli Sbordoni, M., & deMatthaeis, E. (1981). Genetic variability and divergence in cave populations in Troglophilus cavicola and T. andreinii (Orthoptera: Ryaphidophoridae). Evolution, 35, 226–233.

Scharloo, W. (1989). Developmental and physiological aspects of reaction norms. BioScience, 39(7), 465–471.

Schlotterer, C., Hauser, M. T., Von Haeseler, A., & Tautz, D. (1994). Comparative evolutionary analysis of rDNA ITS regions in Drosophila. Molecular Biology and Evolution, 11, 513–522.

Schreiber, D. E., Garner, K. J., & Slavicek, J. M. (1997). Identification of three randomly amplified polymorphic DNA-polymerase chain reaction markers for distinguishing Asian and North American gypsy moths (Lepidoptera: Lymantriidae). Annals of the Entomological Society of America, 90, 667–674.

Scott, M. P., Haymes, K. M., & Williams, S. M. (1992). Parentage analysis using RAPD PCR. Nucleic Acids Research, 20, 5493.

Sene, F. M., & Carson, H. L. (1977). Genetic variation in Hawaiian Drosophila. IV. Allozymic similarity between D. silvestri and D. heteroneura from the island of Hawaii. Genetics, 86, 187–198.

Sheppard, W. S., Rinderer, T. E., Mazzoli, J. A., Steizer, J. A., & Shimanuki, H. (1991). Gene flow between African- and European-derived honey bee populations in Argentina. Nature (London), 349, 782–784.

Shufran, K. A., Peters, D. C., & Webster, J. A. (1997). Generation of clonal diversity by sexual reproduction in the greenbug, Schizaphis graminum. Insect Molecular Biology, 6, 203–209.

Simberloff, D. (1992). Conservation of pristine habitats and unintended effects of biological control. In W. C. Kauffman & J. E. Nechols (Eds.), Selection criteria and ecological consequences of importing natural enemies. (pp. 103–116). New Orleans, LA: Thomas Say Publications in Entomology.

Simon, C. M. (1979). Evolution of periodical cicadas: Phylogenetic inferences based on allozymic data. Systematic Zoology, 28, 22–39.

Simon, C., Fratl, F., Beckenbach, A., Crespi, B., Liu, H., & Flook, P. (1994). Evolution, weighting, and phylogenetic utility of mitochondrial gene sequences and a compilation of conserved polymerase chain reaction primers. Annals of the Entomological Society of America, 87, 651–701.

Simon, C., Paabo, S., Kocher, T. D., & Wilson, A. C. (1990). Evolution of mitochondrial ribosomal RNA in insects as shown by the polymerase chain reaction. In M. T. Clegg & S. J. O'Brien (Eds.), UCLA Symposium on molecular and cellular biology: Vol. 122 (pp. 235–244). New York: Alan R. Liss.

Slade, R. W., Moritz, C., Heideman, A., & Hale, P. T. (1993). Rapid assessment of single-copy nuclear DNA variation in diverse species. Molecular Ecology, 2, 359–373.

Slatkin, M. (1987). Gene flow and the geographic structure of natural populations. Science, 236, 787–792.

Slatkin, M. (1995). A measure of population subdivision based on microsatellite allele frequencies. Genetics, 139, 457–62.

Slatkin, M., & Barton, N. H. (1989). A comparison of three indirect methods for estimating average levels of gene flow. Evolution, 43, 1349–68.

Slatkin, M., & W. P. Maddison, (1989). A cladistic measure of gene flow inferred from the phylogenies of alleles. Genetics, 123, 603–613.

Smith, D. C. (1988). Heritable divergence of Rhagoletis pomonella host races by seasonal asynchrony. Nature (London), 336, 66–67.

Smith, D. R. (1991). Mitochondrial DNA and honey bee biogeography. In D. R. Smith (Ed.), Diversity in the genus Apis (pp. 131–176). Boulder, CO: Westview Press.

Smith, H. O. (1979). Nucleotide sequence specificity of restriction endonucleases. Science, 205, 455–462.

Smith, J. J., Scott-Craig, J. S., Leadbetter, J. R., Bush, G. L., Roberts, D.

L., & Fulbright, D. W. (1994). Characterization of randomly amplified polymorphic DNA (RAPD) products from *Xanthomonas campestris* and some comments on the use of RAPD products in phylogenetic analysis. Molecular Phylogenetics and Evolution, 3(2): 135–145.

Sneath, P. H. A., & Sokal, R. R. (1973). Numerical taxonomy. San Francisco; W. H. Freeman.

Sober, E. (1989). Reconstructing the past: Parsimony, evolution, and inference. Cambridge, MA: MIT Press.

Sperling, F. A. H., & Harrison, R. G. (1994). Mitochondrial DNA variation within and between species of the *Papilio machaon* group of swallowtail butterflies. Evolution, 48, 408–422.

Spieth, H. T. (1981). *Drosophila heteroneura* and *Drosophila silvestri:* Head shapes, behaviors and evolution. Evolution 35, 921–930.

Spradling, A. C., & Rubin, G. M. (1981). *Drosophila* genome organization: Conserved and dynamic aspects. Annual Review of Genetics, 15, 219–264.

St. Louis, V. L., & Barlow, J. C. (1988). Genetic differentiation among ancestral and introduced populations of the Eurasian tree sparrow (*Passer montanus*). Evolution, 42, 266–276.

Stearns, S. C. (1983). A natural experiment in life-history evolution: Field data on the introduction of mosquitofish (*Gambusia affinis*) to Hawaii. Evolution, 37, 601– 617.

Stephans, J. C., Gilbert, D. A., Yuhki, N., & O'Brien, S. J. (1992). Estimation of heterozygosity for single-probe multilocus DNA fingerprints. Molecular Biology and Evolution, 9, 729–743.

Stock, M. W., & Robertson, J. L. (1982). Quality assessment and control in a western spruce budworm laboratory colony. Entomologia Experimentalis et Applicata 32, 28–32.

Stock, M. W., & Castrovillo, P. J. (1981). Genetic relationships among representative populations of five *Choristoneura* species: *C. occidentalis, C. retiniana, C. biennis, C. lambertiana,* and *C. fumiferana* (Lepidoptera: Tortricidae). Canadian Entomologist, 113, 857–865.

Stoflet, E. S., Koeberl, D. D., Sarkar, G., & Sommer, S. S. (1988). Genomic amplification with transcript sequencing. Science, 239, 491–494.

Stone, G. N., & Sunnucks, P. (1993). Genetic consequences of an invasion through a patchy environment–the cynipid gallwasp *Andricus quercuscalicis* (Hymenoptera: Cynipidae). Molecular Ecology, 2, 251–268.

Stouthamer, R. (1993). The use of sexual versus asexual wasps in biological control. Entomophaga, 38, 3–6.

Stouthamer, R., & Werren, J. H. (1993). Microbes associated with parthenogenesis in wasps of the genus *Trichogramma*. Journal of Invertebrate Pathology, 61, 6–9.

Stouthamer, R., & Kazmer, D. J. (1994). Cytogenetics of microbe-associated parthenogenesis and its consequences for gene flow in *Trichogramma* wasps. Heredity, 73, 317–327.

Stouthamer, R., Pinto, J. D., Platner, G. R., & Luck, R. F. (1990). Taxonomic status of thelytokous forms of *Trichogramma* Hymenoptera: Trichogrammatidae. Annals of the Entomological Society of America, 83, 475–481.

Stouthamer, R., Luck, R. F., & Werren, J. H. (1992). Genetics of sex determination and the improvement of biological control using parasitoids. Environmental Entomology, 21, 427–435.

Stouthamer, R., Luck, R. F., Pinto, J. D., Platner, G. R., & Stephens, B. (1996). Non-reciprocal cross-incompatibility in *Trichogramma deion*. Entomologia Experimentalis et Applicata, 80, 481–489.

Strong, D. R. (1997). Fear no weevil? Science, 277, 1058–1059.

Strong, K. L. (1993). Electrophoretic analysis of two strains of *Aphelinus-varipes* Foerster Hymenoptera Aphelinidae. Journal of the Australian Entomological Society, 32(1), 21–22.

Suomalainen, E., Saura, A., Lokki, J., & Teeri, T. (1980). Genetic polymorphism and evolution in parthenogenetic animals. Theoretical and Applied Genetics, 57, 129–132.

Suzuki, K., Okamoto, N., Watanabe, S., & Kano, T. (1992). Chemoluminescent microtiter method for detecting PCR amplified HIV-1 DNA. Journal of Virological Methods, 38, 113–122.

Sweeney, B. W., Funk, D. H., & Vannote, R. L. (1987). Genetic variation in stream mayfly (Insecta: Ephemeroptera) populations of Eastern North America. Annals of the Entomological Society of America, 80, 600–612.

Swofford, D. L., Olsen, G. J., Waddell, P. J., & Hillis, D. M. (1996). Phylogenetic inference (pp. 407–514). In D. M. Hillis, C. Moritz, & B. K. Mable (Eds.), Molecular systematics Sunderland, MA: Sinauer Associates.

Tabachnick, W. J., Munstermann, L. E., & Powell, J. R. (1979). Genetic distinctness of sympatric forms of *Aedes aegypti* in East Africa. Evolution, 33(1), 287–295.

Tang, J., Toe, L., Back, C., & Unnasch, T. R. (1996). Intra-specific heterogeneity of the rDNA internal transcribed spacer in *Simulium damnosum* (Diptera: Simuliidae) complex. Molecular Biology and Evolution, 13, 244–252.

Tek Tay, W., Cook, J. M., Rowe, D. J., & Crozier, R. H. (1997). Migration between nests in the Australian arid-zone ant *Rhytidoponera* sp. 12 revealed by DGGE analyses of mitochondrial DNA. Molecular Ecology, 6, 403–411.

Templeton, A. R. (1980). The theory of speciation via the founder principle. Genetics, 94, 1011–1038.

Templeton, A. R. (1981). Mechanisms of speciation—a population genetic approach. Annual Review Ecology Systematics, 12, 23–48.

Templeton, A. R. (1989). The meaning of species and speciation: A genetic perspective. In D. Otte & J. A. Endler (Eds.), Speciation and its consequences (pp. 3–27). Sunderland, MA: Sinauer Associates.

Thierry, D., Ribodeau, M., Foussard, F., & Jarry, M. (1997). Allozyme polymorphism in a natural population of *Chrysoperla carnea sensu lato* (Neuroptera: Chrysopidae): A contribution to the status of the constitutive taxons in western Europe. European Journal of Entomology, 94(2), 311–316.

Thorpe, J. P. (1983). Enzyme variation, genetic distance and evolutionary divergence in relation to levels of taxonomic separation. In G. S. Oxford & Rollinson, D. (Eds.), Protein polymorphism, adaptive and taxonomic significance (pp. 130–152). London: Academic Press, Systematics Associates (Special Vol. No K24).

Travis, S. E., Maschinski, J., & Keim, P. (1996). An analysis of genetic variation in *Astragalus cremnophylax* var. *Cremnophylax,* a critically endangered plant, using AFLP markers. Molecular Ecology, 5, 735–745.

Unruh, T. R. (1990). Genetic structure among 18 West Coast pear *Psylla* populations: Implications for the evolution of resistance. American Entomologist, 36, 37–43.

Unruh, T. R., & Goeden, R. D. (1987). Electrophoresis helps to identify which race of the introduced weevil, *Rhinocyllus conicus* (Coleoptera: Curculionidae), has transferred to two native southern California thistles. Environmental Entomology, 16, 979–983.

Unruh, T. R., & Messing, R. H. (1993). Intraspecific biodiversity in Hymenoptera: Implications for conservation and biological control. In J. LaSalle & I. D. Gauld (Eds.), Hymenoptera and biodiversity (pp. 27–52). Wallingford, UK: C A B International.

Unruh, T. R., White, W., Gonzalez, D., & Woolley, J. B., (1989). Genetic relationships among seventeen *Aphidius* (Hymenoptera: Aphidiidae) populations, including six species. Annals of the Entomological Society of America, 82(6), 754–768.

Unruh, T. R., White, W., Gonzalez, D., & Luck, R. F. (1986). Electrophoretic studies of parasitic Hymenoptera and implications for biological control. Miscellaneous Publications of the Entomological Society of America, 61, 150–163.

Unruh, T. R., White, W., Gonzalez, D., Gordh, G., & Luck, R. F. (1983). Heterozygosity and effective size in laboratory populations of *Aphidius ervi* (Hym.: Aphidiidae). Entomophaga, 28, 245–258.

van den Bosch, R., Fraser, R. D., Davis, C. S., Messenger, P. S., & Hom, R. (1970). *Trioxys pallidus*—an effective new walnut aphid parasite from Iran. California Agriculture, 24, 8–10.

van Lenteren, J. C. (1980). Evaluation of control capabilities of natural enemies: Does art have to become science? Netherlands Journal of Zoology, 30, 369–381.

Vanlerberghe-Masutti, F. (1994a). Detection of genetic variability in *Trichogramma* populations using molecular markers. Norwegian Journal of Agricultural Sciences (Suppl.) 16, 171–176.

Vanlerberghe-Masutti, F. (1994b). Molecular identification and phylogeny of parasitic wasp species (Hymenoptera: Trichogrammatidae) by mitochondrial DNA, RFLP and RAPD markers. Insect Molecular Biology, 3(4), 229–237.

Vanlerghe-Masutti, F., & Chavigny, P. (1997). Characterization of a microsatellite locus in the parasitoid wasp *Aphelinus abdominalis* (Hymenoptera: Aphelinidae). Bulletin of Entomological Research, 87(3), 313–318.

Vogler, A. P., & DeSalle, R. (1994). Evolution and phylogenetic information content of the ITS-1 region in the tiger beetle *Cicindella dorsalis*. Molecular Biology and Evolution, 11(3), 393–405.

Vos, P., Hogers, R., Blecker, M., Reijans, M., Van de Lee, T., & Hornes, M. (1995). AFLP: A new technique for DNA fingerprinting. Nucleic Acids Research, 23, 4407–4414.

Waage, J. K. (1986). Family planning in parasitoids: Adaptive patterns of progeny and sex allocation. In J. Waage & D. Greathead (Eds.), Insect parasitoids. 13th Symposium of the Royal Entomological Society of London (pp. 63–95). New York: Academic Press.

Walton, M. P., Loxdale, H. D., & Williams, L. A. (1990). Electrophoretic keys for the identification of parasitoids (Hymenoptera: Braconidae: Aphelinidae) attacking *Sitobion avenae* (Fabricius) (Hemiptera: Aphididae). Biological Journal of the Linnean Society, 40, 333–346.

Wang, A. M., & Mark, D. F. (1990). Quantitative PCR. In M. A. Innis, D. H. Gelfand, J. J. Sninsky, & T. J. White (Eds.), PCR protocols, a guide to methods and applications (pp. 70–75). San Diego: Academic Press.

Ward, P. S. (1980). Genetic variation and population differentiation in the *Rhytidoponera impresso* group, a species complex of ponerine ants (Hymenoptera: Formicidae). Evolution, 34, 1060–1076.

Warner, P. J., Yuille, M. A. R., & Affara, N. A. (1992). Genomic libraries and the development of species-specific probes. In A. R. Hoelzel (Ed.), Molecular genetic analysis of populations: A practical approach (pp. 189–224). Oxford: (IRL) Oxford University Press.

Weir, B. S., & Cockerham, C. C. (1984). Estimating F-statistics for the analysis of population structure. Evolution, 38(6), 1358–1370.

Weller, S. J., Friedlander, T. P., Martin, J. A., & Pashley, D. P. (1992). Phylogenetic studies of ribosomal RNA variation in higher moths and butterflies (Lepidoptera: Ditrysia). Molecular Phylogenetics of Evolution, 1, 312–337.

Werman, S. D., Springer, M. S., & Britten, R. J. (1996). Nucleic acids I: DNA-DNA hybridization. In D. M. Hillis, C. Moritz, & B. K. Mable (Eds.), Molecular Systematics (pp. 169–203). Sunderland, MA: Sinauer Associates.

Werren, J. H. (1993). The evolution of inbreeding in haplodiploid organisms. In N. Thornhill (Ed.), The natural history of inbreeding and outbreeding. Chicago: University of Chicago Press.

Werren, J. H. (1997). Biology of *Wolbachia*. Annual Review of Entomology, 42, 587–609.

Wesson, D. M., & Porter, C. H. (1993). Sequence and secondary structure comparisons of ITS rDNA in mosquitoes (Diptera: Culicidae). Molecular Phylogenetics and Evolution, 1, 253–269.

Wharton, R. A., Woolley, J. B. & Rose, (1990). Relationship and importance of Taxonomy to classical biological control (pp. 11–16). In D. H. Habeck, F. D. Bennett & J. H. Frank (Eds.), Classical biological control in the southern United States. Southern Cooperative

Series Bulletin No. 355. IFAS Editorial, University of Florida, Gainesville.

White, M. J. D. (1948). Animal cytology and evolution. Cambridge, U.K.: Cambridge University Press.

Whiting, P. W. (1943). Multiple alleles in complementary sex determination of *Habrobracon*. Genetics, 28, 365–382.

Williams, C. J., & Evarts, S. (1989). The estimation of concurrent multiple paternity probabilities in natural populations. Theoretical Population Biology, 35, 90–112.

Williams, C. L., Goldson, S. L., Baird, D. B., & Bullock, D. W. (1994). Geographical origin of an introduced insect pest, *Listronotus bonariensis* (Kuschel), determined by RAPD analysis. Heredity, 72(4), 412–419.

Williams, G. K., Kubelik, A. R., Livak, K. J., & Rafalski, J. A., & Tingey, S. V. (1990). DNA polymorphisms amplified by arbitrary primers are useful as genetic markers. Nucleic Acids Research, 18, 6531–6535.

Williams, S. M., DeSalle, R., & Strobeck C. (1985). Homogenization of geographical variants at the nontranscribed spacer of rDNA in *Drosophila mercatorum*. Molecular Biology and Evolution, 2(4), 338–346.

Wood, D., Gerling, D., & Cohen, I. (1984). Electrophoretic detection of two endoparasite species, *Encarsia lutea* and *Eretmocerus mundus* in the whitefly, *Bemisia tabaci* (Genn.) (Hom., Aleurodidae). Zeitschrift fuer Angewandte Entomologie, 98, 276–279.

Wool, D., Van Emden, H. F., & Bunting, S. W. (1978). Electrophoretic detection of the internal parasite, *Aphidius matricariae* in *Myzus persicae*. Annals of Applied Biology, 90, 21–26.

Wright, S. (1951). The genetical structure of populations. Annals of Eugenics, 15, 323–53.

Wright, S. (1978). Evolution and the genetics of populations (Vol 4). Variability within and among natural populations, Chicago, IL: University of Chicago Press.

Yang-Jiang, L., Adang, M. J., Isenhour, D. J., & Kochert, G. D. (1992). RFLP analysis of genetic variation in North American populations of the fall armyworm moth *Spodoptera frugiperda* (Lepidoptera: Noctuidae). Molecular Ecology, 1, 199–208.

Zabeau, M., & Vos, P. (1993). Selective restriction fragment amplification: A general method for DNA fingerprinting. European Patent Application, Publication #0534858-A1, Office europeen des brevets, Paris.

Zchori-Fein, E., Faktor, O., Zeidan, M., Gottlieb, Y., Czosnek, H., & Rosen, D. (1995). Parthenogenesis-inducing microorganisms in *Aphytis* (Hymenoptera: Aphelinidae). Insect Molecular Biology, 4(3), 173–178.

Zehnder, G. W., Sandall, L., Tisler, A. M., & Powers, T. O. (1992). Mitochondrial DNA diversity among seventeen geographic populations of *Leptinotarsa decemlineata* (Coleoptera: Chrysomelidae). Annals of the Entomological Society of America, 85, 234.

Zhang, D.-X. & Hewitt, G. M. (1996). Highly conserved nuclear copies of the mitochondrial control region in the desert locust *Schistocerca gregaria:* Some implications for population studies. Molecular Ecology, 5, 295–300.

Zhu, Y-C. & Greenstone, M. H. (in press). PCR techniques for distinguishing three species and two strains of *Aphelinus* (Hymenoptera: Aphelinidae) from *Diuraphis noxia* and other cereal aphids (Homoptera: Aphidiidae) Annuals of the Entomological Society of America.

Zietkiewicz, E., A. Rafalski & Labuda. D., (1994). Genome fingerprinting by simple sequence repeat (SSR)-anchored polymerase chain reaction amplification. Genomics, 20, 176–183.

Zouros, E. (1973). Genetic differentiation associated with the early stages of speciation in the *Mulleri* subgroup of *Drosophila*. Evolution, 27, 601–621.

Zwolfer, H., & Preiss, M. (1983). Host selection and oviposition behavior in West European ecotypes of *Rhinocyllus conicus* Froel. (Col., Curculionidae). Zeitschrift fuer Angewandte Entomologie, 95, 113–122.

5

Exploration for Natural Enemies

E. F. LEGNER and T. S. BELLOWS

Department of Entomology
University of California
Riverside, California

INTRODUCTION

The procurement of exotic natural enemies to suppress pest populations has long been an integral part of biological control, which has repeatedly proved extremely valuable in eliminating pest problems (Van Driesche & Bellows, 1996). This tactic has been applied to pests in a wide variety of natural, agricultural, and urban settings. Introduced natural enemies include vertebrates, invertebrates, and microbes, and have been deployed against pest plants, insects, vertebrates, and plant diseases. The potential for control of many pest organisms in diverse environments using introduced natural enemies is substantial, and consequently the practice of exploration for new natural enemies continues to be an indispensable part of many research programs.

Several steps are included in the development of an exploration program, the duration of which may span several years or decades. A natural enemy exploration and introduction program is best viewed as a continuing process of natural enemy introductions over a span of time rather than as a single foreign exploration event. Many of the technical and biological considerations relative to acquiring and shipping biological agents remain much the same as those described for entomophagous and phytophagous arthropods by Bartlett and van den Bosch (1964), Boldt and Drea (1980), Klingman and Coulson (1982), Schroeder and Goeden (1986), Coulson and Soper (1989), and Bellows and Legner (1993) and Van Driesche and Bellows (1996). Nonetheless, there are particular aspects of the process that must be incorporated into every program. These include selecting search locations, the candidate natural enemy species, and planning and conducting the exploration and collection. We consider each of these in turn.

AREAS FOR SEARCH

Considerations on Geographic Origin of Pest

An initial realistic appraisal of the pest problem and the chances for success characterize a well-conducted natural enemy introduction program. Natural enemies may be sought in the native home of the pest and in areas that includes a climate similar to that of destination. The kinds of natural enemies sought may include those that attack the target species, taxonomically related species, or taxonomically unrelated species that share a common ecological niche with the target species. Natural enemies that restrict their attack to the pest or its close relatives are sometimes preferred (particularly in programs against plants). Those natural enemies that possess a high degree of preference for the pest are often chosen for final field release, although guidelines vary for such different taxa as phytophages, insect parasitoids and predators, and pathogens. During the search, natural enemies are optimally collected from all possible habitats and locations, especially those with climates similar to the planned release areas, to ensure the inclusion of cryptic forms, races, biotypes, or varieties. Special attention is generally given to searching on the species of host plant on which the target arthropod pest is problematic, or on the target pest weed in the case of plant pests. Biologies, host associations, and species attributes are ascertained automatically during the rearing and transfer process.

There are several challenges intrinsic to the determination of a native home for a species. These may be illustrated by reference to Nearctic insect faunas. The number of these faunas is large, including 30 orders, 500 families, and about 12,000 genera and 150,000 species (Ross, 1953). Knowledge about the origin and dispersal patterns of insects is in reality very spotty; therefore, the origin and evolution of the North American insect faunas, for example, is largely a subject for speculation. It is generally held that the Palearctic was the center or origin for many of the ancestors of new North American insects. The number of species in a genus can be used as a clue to the center of distribution of a particular group, and this may prove valuable in identifying the potential native home for members of the genus or species group. Exceptions may arise, however. The concentration of species in a particular area may, in fact, reflect more their common environmental needs than the center of dispersal after relatively recent speciation in that area. A large complex of natural enemies is often believed to indicate the site of longest residence (native home) of a species, especially if one or more natural enemies are host specific. However, much of our native insect faunas are so old that we have little basis for discussing their places of origin and we may consider such species as native to the Nearctic. If such species invade other areas and become pests, natural enemies could be sought in America.

Although much can be inferred about the past history of species from studies of their present range and recent changes in distribution, the fossil record may provide an important objective source for inference, as illustrated by a few examples. The staphylinid beetle *Oxytelus gibblus* (Eppelsheim) is presently restricted to the western Caucasus Mountains, even though the fossil records show that this species was extremely abundant in Britain during the last glaciation. The western Caucasus, then, probably represents the *last remaining stand* or relic population of a once widespread species rather than its place of origin. The carabid subgenus *Cryobius* has its center of distribution in northeastern Siberia and northwestern America. However, it was represented by a greater number of species in western Europe during the Wisconsin glaciation, although none are found there at the present time. The evolution of the mosquito genus *Culex* has been studied through observations of progressive changes in structures of the male genitalia. The genus apparently spread through Africa where it gave rise to a leading line, *C. guardi* Coquillett. Seven different lines were formed, and all but one remains confined to Africa. The exception, *C. pipiens* Linnaeus, which apparently spread to India, produced new lines. Some of these eventually reached North America via South America, giving rise among others to the species *C. tarsalis* Coquillett. Had it not been for the survival of connecting links in Africa and Asia, the current distribution of the species group might be taken as evidence that it originated in South America.

For biological control it may be sufficient to identify a place of *recent* origin of a species to locate the best natural enemies. *Recent* is a vague term. It could be a few hundreds or thousands of years. However, natural enemy complexes efficient in regulation of a targeted pest may evolve quicker

than generally imagined, so that paleoentomology may not be too important in biological control.

The region of origin of certain pest species is known with very little doubt. Included are cottony-cushion scale (*Icerya purchasi* Maskell), Klamath weed (*Hypericum perforatum* Linnaeus), *Opuntia* spp. cacti in India and Australia, European rabbit (*Oryctolagus cuniculus* [Linnaeus]) in Australia, eucalyptus snout beetle (*Gonipterus scutellatus* Gyllenhal), the fungi *Ceratocystis ulmi* (Boisman) C. Moreau (Dutch elm disease) and *Cryphonectria parasitica* (Murriu) Anderson & Anderson (chestnut blight), olive scale (*Parlatoria oleae* [Colvée]), walnut aphid (*Chromaphis juglandicola* [Kaltenbach]), navel orangeworm (*Amyelois transitella* [Walker]), grape leaf skeletonizer (*Harrisinia brillans* Barnes & McDunnough), and many more. The origin of other pests is open to speculation, with the following examples illustrating some of the difficulties involved.

Saissetia oleae (Bernard), the lecanine black scale on citrus, originally was believed to have originated in East Africa. Harold Compere (University of California, Riverside) imported 30 species of parasitoids from eastern and southern Africa during 1936–1937, with five becoming established in the United States. No appreciable reduction in host density was caused by most species. However, *Metaphycus helvolus* (Compere) from Capetown, South Africa proved to be very successful in reducing black scale densities. The host scale is now believed to have possibly originated in northern Africa and not within the range of its most effective known parasitoid.

Circulifer tennellus (Baker) (Cicadellidae), the beet leafhopper, is a serious pest of sugar beet, tomato and other crops because it vectors curly top virus. This leafhopper possesses strong migratory habits, moving hundreds of miles each spring and early summer to cultivated crops. Systematics has played an important role in determining the origin of this insect. It was originally thought to be from Australia under the genus name *Eutettix*. However, parasitoids shipped from Australia failed to establish in North America. Oman (1948) reclassified the leafhopper as *Circulifer* and thought it originated in arid portions of the Mediterranean and Central Asian regions. A total of 36 shipments of parasitic material was received during 1951 through 1954 (Huffaker, 1954). Twelve parasitoid species were involved, but the identity of all species is still not positive. Only two released species were ever recovered as *Lymaenon* "A" and *Polynema* "A." Further search in Asia might be advisable.

Helicoverpa zea (Boddie) (Noctuidae) has been described as several species due to a great variability in color and markings. It is widespread within 40° north to 49° south latitude. Originally it was thought to have originated in the West Indies because of the fact that it feeds there particularly on American plant genera, such as tomato and corn. It is of rare occurrence in Europe and this would indicate to some that it originated there with a better natural enemy complex. Various parasitoids were imported against it with some Braconidae and Tachinidae showing some success after their discovery in India. Native predators are presently thought to be most useful against *H. zea,* but not usually to a satisfactory control level. Further search in Asia and possibly Africa should be considered.

Carpocapsa pomonella (Linnaeus) (Olethreutidae), the codling moth, damages pome fruit and some stone fruits. This species was known in Europe in ancient times (around the fourth century B.C.). The codling moth is widely distributed throughout the world, but it is still apparently of lesser or no consequence in parts of China and Japan. Its northern limit is determined by minimum heat in summer. The presence of the moth in America, Australia, South America, South Africa, and eastern parts of the former USSR is thought to be of recent occurrence and its probable point of origin is in the central Palearctic where wild apples, *Malus silvestris,* occur (Sheldeshova, 1967). A long list of natural enemies is known, which undoubtedly reflects the large number of researchers who have studied this pest. Some natural enemies were introduced from Spain, eastern North America, and the Middle East to various countries, but only a few became established without showing any control (Clausen, 1978). In considering the great importance of this insect, it is surprising that a greater effort has not been made to control it biologically. In Europe, for example, several potentially very effective parasitoids are kept from expressing themselves fully due to hyperparasitism. Efforts to import the primary parasitoids to the United States have begun (see Chapter 28). Their capabilities could be fully expressed here without interference from hyperparasitoids and cleptoparasitoids (N. Mills, personal communications). Also, an intensive search in northern China, southern Siberia, and the Japanese islands might turn up some effective natural enemies, although the host insect is officially absent in the latter region.

Aonidiella aurantii (Maskell) (Diaspididae), the California red scale, is found between 40° north and 40° south latitude, but it is at pest status only in the subtropics and warmer temperate zones. This scale has a wide host range but prefers citrus and roses. By using the last of three leads for tracing the area of origin [(1) the area where the preferred host originated—citrus, (2) the area where the pest is present but kept at low densities by natural enemies—the Far East (China), and (3) the area where an abundance of related species exists—Neotropics], a search was begun in South America from 1934 to 1936, where 17 species in the scale genus *Chrysomphalus* were reported. No natural enemy from these was effective in biological control. China and Southeast Asia were searched subsequently, but it took from 1905 to 1941 to establish a parasitoid, *Comperiella bifasciata* Howard. Problems that interfered with finding the right species included the sago palm trap–host plant

approach for locating parasitoids, which failed because this particular plant species made host scales immune to attack. There were also numerous host misidentifications. The parasites *Aphytis lingnanensis* Compere and *A. melinus* DeBach were later successfully imported from the Far East. The resulting success was attributed to the fact that parasitoids were sought from the whole geographic range of the host in the East.

Tetranychus spp. of tetranychid mites are pests of worldwide distribution that are found primarily in the subtropics and warmer temperate zones. There are 200 to 250 species of economic importance, but there is only limited information on the native home of many species. Some inferences have been made on the native homes of the European red mite (*Panonychus ulmi* [Koch]) and citrus red mite (*Panonychus citri* [McGregor]), however (McMurtry *et al.,* 1978, Chapters 16 and 27). Introductions of predatory mites have been rather scanty and should be continued before any conclusions are drawn. The phytoseiid mite *Phytoseiulus persimilis* Athias-Henriot was released in periodic mass liberations and has been used successfully to prevent phytophagous mite outbreaks in greenhouses (see Chapter 31). In Israel 2 phytoseiid species were established on citrus mite pests, and in southern California 12 phytoseiids were released on avocado and 9 were released against citrus red mite. Results are too recent in the California introductions to determine permanency. Insecticide resistant strains of *Typhlodromus occidentalis* Nesbitt have been produced and released in various agroecosystems in California and Oregon (Croft, 1970; Hoy *et al.,* 1982; Croft & AliNiazee, 1983) (see also Chapter 28).

Phthorimaea operculella (Zeller), the potato tuberworm, is presently cosmopolitan. Some workers consider America to be the native area based on the origin of its principal food plants, most of the wild progenitors of which originated in South America. However, the first successful introduction against this presumably American species was to France of *Bracon gelechiae* Ashmead, where infestations were reduced. This parasitoid was later successfully introduced in Australia, Asia, Africa, and other portions of North America. Other species are currently being tried and it is probable that some finds in South America (Oatman and Platner, 1974) will provide even greater control.

Nezara viridula (Linnaeus) (Pentatomidae), the southern green stinkbug, is probably of African origin. This pest is especially important in tropical and subtropical areas of the world, attacking 60 or more unrelated species of plants. The Egyptian scelionid egg parasitoid, *Microphanurus basalis* (Wollaston), was successfully introduced into Australia, Fiji, and Hawaii, where it was effective only in the more tropical areas (Davis, 1967). Further attempts should be made to secure parasitoids from the cooler portions of the pest's range in Africa.

Pectinophora gossypiella (Saunders), the pink bollworm, is now cosmopolitan. However, advances in the tax-onomy of the genus have placed its origin in the vast Australasian region. Importations of natural enemies from the northwestern part of Australia and southeastern Indonesia and Malaysia have just begun. However, this was preceded by 60 years of work with parasitoids that were secured from Europe, Africa, and India, yielding poor results (Common, 1958; Wilson, 1972; Legner & Medved, 1979). Unfortunately, this breakthrough in locating the probable area of origin has been overlooked, and the previous 60 years of failure are held as proof for being unable to control pink bollworm biologically. Renewed effort at securing natural enemies from northwestern Australia has begun in California, Arizona, New Mexico, Texas, and Egypt.

Diptera of medical and veterinary importance include species in the genera *Musca, Fannia, Culex, Aedes,* and *Anopheles,* which are becoming more difficult to control with insecticides and by cultural means. Emphasis is now being placed on biological control. Unlike agricultural pests, control of medical and veterinary pests is often more difficult to evaluate because economic loss data are not readily available. Therefore, although significant achievements have been made with biological control of several species, permanent population reductions have not been reported (see Chapter 37).

Securing Natural Enemies in the Native Range

Some of the most dramatic successes in biological control, where the target pest's population density is permanently reduced to below the economic threshold, involved the introduction of one or two species of natural enemy (Franz, 1961a, 1961b; DeBach, 1964, 1974; van den Bosch, 1971; Hagen & Franz, 1973; Clausen, 1978; Franz and Krieg, 1982; Luck, 1982; DeBach & Rosen, 1991; Van Driesche & Bellows, 1996), in both stable and unstable habitats (Ehler & Miller, 1978; Hall & Ehler, 1979; Hall *et al.,* 1980; Greathead & Greathead, 1992). As previously mentioned, the usual procedure when a pest species invades a new area is to seek natural enemies in its native home. The first and most widely known biological control success, the cottony-cushion scale controlled by *Cryptochaetum iceryae* (Williston) (Cryptochetidae) and *Rodolia cardinalis* (Mulsant) (Coccinellidae), followed that pattern (Quezada & DeBach, 1973). The scale invaded California and the natural enemies were found in its native range, southern Australia. It seemed logical to follow the same format with subsequent biological control efforts, and indeed this is still considered one of the first approaches for a newly invaded pest. However, using this approach solely restricts the number and kinds of successes that can be realized.

Searching Outside the Native Range

In many parts of the world, especially Europe, Africa, and much of Asia, there are numerous native pests whose natural enemies are incapable of maintaining density levels below the economic threshold under prevailing agricultural management. What may be done other than costly and hazardous cultural and chemical control? Pimentel (1963) and Hokkanen and Pimentel (1984) suggested a potential approach for biological control. In many instances, the natural enemies that caused significant drops in population densities of organisms had never experienced evolutionary contact with their hosts. When we consider other cases than those exemplified by the host–parasitoid–predator relationship in cottony-cushion scale, we find a few examples of great reductions of a population density by organisms that originated in places other than the native home of the host. Such cases may include the devastation of desirable native species by accidentally invaded organisms. Well-known examples to illustrate this phenomenon are the American elm, *Ulmus americana* Linnaeus, destroyed by the fungus, *Ceratocystis ulmi*, of eastern hemispheric origin and vectored primarily by the European beetle, *Scolytus multistriatus* (Marsham); the American chestnut, *Castanea dentata* (Marsham) Borkhauser, practically eliminated by a fungus, *Cryphonectria parasitica* of Asian origin; and Asiatic citrus destroyed by *Icerya purchasi* from Australia before biological control efforts reduced the scale's density.

There is a history of successful biological control against invaded pests by organisms actively secured in areas other than the native home. Famous examples are the European rabbit regulated by a myxomatosis virus of South American origin and the black scale, *Saisettia oleae* (Bernard), of probable northern African origin (D. P. Annecke, personal communication), regulated by the aphelinid *Metaphycus helvolus* (Compere) from extreme southern Africa. Also, the sugarcane borer, *Diatraea saccharalis* (Fabricius), of the Neotropics, regulated by the braconid *Apanteles flavipes* (Cameron) from northern India; the coconut moth, *Levuana iridescens* Bethune-Baker, native to Fiji, regulated by a parasitoid, *Bessa remota* (Aldrich), secured from Malaya; and *Oxydia trychiata* (Guenée) in Colombia regulated by *Telenomus alsophilae* Viereck from eastern North America (Bustillo & Drooz, 1977; Drooz *et al.,* 1977). Other examples exist (Pimentel, 1963; Pimentel *et al.,* 1984), which has led to some speculation that the best natural enemies for biological control might be those that have not experienced close evolutionary contact with the target organism. The theory considers that the host and its natural enemies coevolve to a balanced point where the host may exist at a relatively higher density than where no coevolutionary balance has had a chance to evolve. Although the examples of drastic impact on a host by natural enemies without pre-evolutionary contact are numerous and impressive, there are equally impressive and thorough examples of host reduction in which natural enemies were obtained from the native home where coevolution has occurred (Bellows, 1993; Bellows & Legner, 1993). The cottony-cushion scale, Comstock mealybug, *Pseudococcus comstocki* (Kuwana) (Ervin *et al.*, 1983), and the ash whitefly, *Siphoninus phillyreae* (Haliday) (Bellows *et al.*, 1992), successes illustrate this clearly. Thus, the native home ought not to be deemphasized as suggested by Hokkanen and Pimentel (1984). In fact, an analysis by Hokkanen (1985) concluded that there are no differences of biocontrol success according to the origin of the pest, and that native hosts can be completely controlled by introduced natural enemies exactly as exotic ones. However, as time goes on, one would expect examples of the latter to drop proportionally because the former include accidental invasions and random acquisitions, as well as planned biological control attempts.

There are some antagonists to the approach endorsed by Hokkanen and Pimentel, especially among the biological control of weeds researchers (Goeden & Kok, 1986; Schroeder & Goeden, 1986). They argue that the record in the case of weeds supports the native home approach most strongly. However, the record may need careful historical review because many biological weed control efforts emphasized the native home in the search for natural enemies.

Summary Concerning Search Locations

Productive searches for natural enemies can take place both within and outside the presumed native range of the target pest. It is important to search with both a taxonomic and an ecological view of the target organism, so that searches may include natural enemies of closely related organisms or unrelated organisms that occupy similar ecological niches. Searches should include different seasons, elevations, and climates as the natural enemy faunas in each may vary significantly. Searches for natural enemies should include candidates attacking any life stage of the target species. Finally, exploration is a continuing process that may require considerable long-term effort to secure suitable natural enemies.

RISK ASSESSMENT AND EVALUATION OF NATURAL ENEMY POTENTIAL

One potential for maximizing biological control successes and limiting risks is the placement of various natural enemy groups into different risk categories before proceeding with decisions for introduction. Although the full impact of natural enemies that have never been studied is impossible to accurately predict before establishment

(DeBach, 1964, 1974; Coppel & Mertins, 1977; Ehler, 1979; Miller, 1983), and therefore involves empirical judgment, there is nevertheless a compelling desire that the process proceeds with an educated empiricism (Coppel & Mertins, 1977; Ehler & Hall, 1982; Legner, 1986b; Ehler, 1991). Two schools of thought are (1) narrowly targeted natural enemies based on available information; and (2) a broader approach, with the simultaneous importation of several potentially effective natural enemies, if the information base is weak. The debate among proponents of the two schools has been lengthy, but both ideas have merit when applied judiciously. Some taxonomic groups may be imported with little risk of potential problems, while others require more caution (e.g., phytophages of crop–plant relatives).

Whatever the theory behind biological control by natural enemies, it is still one of the most effective and astonishing weapons in our arsenal of pest management techniques. Imported organisms, once established, are not always easily extirpated; and in some instances their elimination is impossible altogether, regardless of the amount of effort and funding. There is, therefore, some risk involved in any biological control approach. However, risks are a companion to life itself, and any pest management tactic involves some degree of risk, with alternatives to importation of natural enemies undoubtedly being more formidable (Pimentel *et al.*, 1984; Legner, 1986b). Trying to eliminate too much risk through government regulation is not advisable because it can paradoxically make life more dangerous as well as more expensive and less convenient (Huber, 1983).

Because of the broad nature of biological control, including manipulation of vertebrates, arthropods, and pathogens, we may consider risks in two categories: (1) risks of affecting the environment and health of humans and domestic animals, and (2) risks of making choices that may preclude or adversely affect biological control at a later date. The first category is often the object of government regulations that provide obligate guidelines for the evaluation of candidate natural enemies.

Ways in which the potential capabilities of a biological control agent are judged, as well as the environmental risks, vary for different groups of organisms (Ehler, 1989; 1990a, 1990b, 1990c). If we consider in descending order of environmental risk—the terrestrial vertebrates first, followed by zoopathogens, phytopathogens, phytophagous arthropods, terrestrial scavengers (e.g., scarab beetles), aquatic vertebrates, and invertebrates, and finally parasitic–predatory arthropods and arthropod pathogens—it would be logical to screen the first group more thoroughly than the last. However, judgments of potential effectiveness may require more effort for the last group and least effort for the first, because of problems in measuring dispersal and other behavioral traits. Hence, it is often easier to examine and calculate potential risk, and thus provide a basis for excluding candidate species, than it is to demonstrate potential value and thus provide an *a priori* reason for importing a species.

Terrestrial Vertebrates

The use of imported vertebrates for biological control of terrestrial pests, in part because of their capacity for learned behavior, has high potential risk and is rarely practiced. An apparent desirable example is the common myna bird, *Acritotheres tristis* Linnaeus, importation from India to tropical areas for insect control. However, mongoose, *Herpestes auropunctatus birmanicus* (Hodgson), importation from India to tropical islands for rat (*Rattus* spp.) control and giant toad, *Bufo marinus* (Linnaeus), importation from America to tropical islands and Australia for insect control have produced undesirable side effects, either through numerical abundance or in the latter example by predation of beneficial dung beetles (Macqueen, 1975).

Such terrestrial vertebrates as the mongoose, myna bird, and giant toad are more readily observed than are invertebrates in their places of origin because of their size, so behavioral attributes may be more easily studied prior to introduction. However, they are also capable of becoming conspicuous additions to the general landscape, and without natural predators of their own, may have the capacity to soar in numbers in the areas of their introduction. Thus, they may pose the threat of becoming pests themselves because of their numerical abundance and often nonspecific feeding behavior, as well as the side effects this can have on native and other desirable faunas and floras.

Phytophagous Arthropods and Phytopathogens

Environmental risks are of especial concern in the biological control of weeds, where both arthropods and pathogens are candidates for importation. These groups have traditionally evoked the most thorough regulations and preintroduction studies to safeguard desirable plant species in the areas of introduction (Huffaker, 1957; Harris & Zwölfer, 1968; Harris, 1973; Klingman & Coulson, 1982; Ehler & Andrés, 1983; Goeden, 1983; 1993; Van Driesche & Bellows, 1996; see also Chapters 34 and 35). Screening has been so successful that among the numerous importations of beneficial phytophagous arthropods and pathogens around the world (Julien, 1992), there has never been any widespread occurrence of harmful behavior shown by the organisms imported. Rare reports of beneficial phytophagous arthropods feeding on desirable plant species after importation (Greathead, 1973) usually involved temporary, geographically restricted alterations of behavior during the establishment phase when the colony contracted in size due to its inability to survive and to reproduce on alternate host plants. Such reports have been, nevertheless, exaggerated and cited out of context as, for example, by Pimentel *et al.* (1984). A cultivated variety or relative of a targeted weed may come under attack by a phytophagous arthropod introduced earlier to combat the related weed, as has occurred after >40 years in northern California with imported natural enemies of Klamath weed, *Hypericum perforatum,* and a species of *Hypericum* used for roadside plantings (Andrés, 1981).

The benefits derived from importation of phytophagous arthropods and phytopathogens are so vast when levied against weeds that cannot be controlled effectively, economically, or safely with any other strategy, that biological control will continue to be a major effort. The risk involved has been and should continue to be minimized by the extensive studies required prior to importation.

Terrestrial Scavengers

Terrestrial scavengers include scarab beetles that remove cattle dung accumulated in grazing areas to both improve pastures and control symbovine flies, particularly *Haematobia irritans* (Linnaeus), *Musca autumnalis* DeGeer, and *M. vetustissima* Walker. Although dramatic successes have been achieved in the removal of dung by the importation of several species of exotic scarabs in Australia, Hawaii, California, and Texas (Waterhouse, 1974; Macqueen, 1975; Legner, 1978; Legner & Warkentin, 1983; M. M. Wallace, personal communication), there apparently have been no widespread concurrent reductions of fly densities (Legner, 1986a, 1986b; M. M. Wallace, personal communication). In some instances fly densities may have actually increased temporarily in the presence of established dung-burying scarabs. Although laboratory and field experiments predicted practical fly reductions by the dung-scattering activities of the scarabs, in pastures several forces interplay to thwart experimentally based predictions. Elimination of predatory arthropods and increase of available fly larval breeding habitat could be two of the principal causative factors (Legner & Warkentin, 1983; Legner, 1986a, 1986b).

Terrestrial organisms that alter large habitats, such as scarab beetles, are especially risky biological control candidates because their activity may overlap portions of the niche of other beneficial species (e.g., predators) so that potential disruptive side effects among organisms in different guilds exist. The outcome for future symbovine fly control may be limited in that some potentially regulative natural enemies, such as certain predatory arthropods, may be more difficult to establish in the disrupted habitat (Legner, 1986a). In California and Texas, the predatory staphylinid genus *Philonthus* is severely restrained from colonizing the drier dung habitat created by the activity of the dung beetle *Onthophagus gazella* Fabricius (Coleoptera:

Scarabaeidae) (Legner & Warkentin, 1983; Roth *et al.,* 1983; Legner, 1986b). Furthermore, various nongraminaceous weed species often invade California irrigated pastures that sustain large populations of exotic scarab beetles, so that mechanical pasture renovation again is required (Legner, 1986a).

Aquatic Vertebrates and Invertebrates

Aquatic vertebrates and invertebrates used for biological control include such groups as fish for biological aquatic weed and arthropod control and Turbellaria and Coelenterata for arthropod control. The minnows *Gambusia* and *Poecilia* are used worldwide in the biological control of mosquitoes (Legner *et al.,* 1974; Legner & Sjogren, 1984). However, the threat to endemic fish has caused widespread concern so that alternatives in the use of native fish are under consideration (Legner *et al.,* 1975; Walters & Legner, 1980). Because fish can be manipulated readily, the potential for resident species to increase their effectiveness as natural enemies is greater than that with terrestrial organisms where widespread natural dispersion may have already covered most possibilities.

Environmental risks may be minimized when vertebrates such as fish are imported to restricted aquatic ecosystems. Such fish can be studied under natural conditions, but in isolation ponds or in restricted waterways, for adverse effects before being widely disseminated. Yet, there are still possibilities that undesirable unforeseen behavioral and adaptive traits, such as spawn-feeding on other desirable fish species, or an extension of subtropical species into temperate climates (e.g., *Gambusia* spp.) may be expressed once populations are allowed to establish broadly (Legner and Sjogren, 1984).

Careful evaluation of ecological traits of closely related organisms may indicate important differences among candidate natural enemies within these groups. One example is a group of cichlid fish imported from Africa to the southwestern United States for the biological control of mosquitoes, mosquito habitats, and chironomid midges. *Tilapia zillii* (Gervais), *Sarotherodon* (= *Tilapia*) *mossambica* (Peters), and *S.* (= *Tilapia*) *hornorum* Trewazas were imported to California for the biological control of emergent aquatic vegetation (*Scirpus* spp., *Typha* spp., *Eleocharus* spp., etc.) that provides a habitat for such encephalitis vectors as the mosquito *Culex tarsalis* Coquillett and, additionally, as predators of mosquitoes and chironomid midges.

Studies under natural, but quarantine, areas in California showed that the different fish species each possessed certain attributes for combating the respective target pests (Legner & Medved, 1973). *Tilapia zillii* was best able to perform both as a habitat reducer (aquatic weed feeder) and as an insect predator. It also had a slightly greater tolerance to low water temperatures, which guaranteed its survival

through the winter months in southern California, while at the same time it did not pose a threat to salmon and other game fisheries in the colder waters of central California. It was the superior game species and was the most desirable for eating.

However, the agencies supporting the research (mosquito abatement and county irrigation districts) acquired and prematurely distributed all three species simultaneously throughout thousands of kilometers of irrigation system, storm drainage channels, and recreational lakes. The outcome was the permanent and semipermanent establishment of the two less desirable species, *S. mossambica* and *S. hornorum,* over a broader portion of the distribution range. This was achieved apparently by a competitive superiority rendered by an ability to mouth-brood their fry, while *T. zillii* did not have this attribute strongly developed. It serves as an example of competitive exclusion such as conjectured by Ehler and Hall (1982). In some clearer waters of lakes in coastal and southwestern California, the intense predatory behavior of *S. mossambica* males on fry of *T. zillii* could easily be observed, even though adults of the latter species gave a strong effort to fend off these attacks.

The outcome was not too serious for chironomid control because the *Sarotherodon* species were quite capable of permanently suppressing chironomid densities to below annoyance levels (Legner *et al.,* 1980). However, for control of submerged higher aquatic weeds—namely, *Potamogeton pectinatus* Linnaeus, *Myriophyllum spicatum* var. *exalbescens* (Fernald) Jepson, and *Hydrilla verticillata* Royle— and emergent aquatic weeds—such as *Typha* spp.—the *Sarotherodon* species showed no capability whatsoever (Legner & Medved, 1973). Thus, competition excluded *Tilapia zillii* from expressing its maximum potential in the irrigation channels of the lower Sonoran Desert of California and in recreational lakes of southwestern California. Furthermore, because the *Sarotherodon* species were of a more tropical nature, they died out annually in the colder waters of the irrigation canals and recreational lakes without thermal effluent. Although *T. zillii* populations could have been restocked, attention was later focused on a potentially more environmentally dangerous species, the white amur *Ctenopharyngodon idella* (Valenciennes), and other carps. The substitution of *T. zillii* in storm drainage channels of southwestern California is presently impossible because the *Sarotherodon* species are permanently established over a broad geographic area.

Parasitic and Predaceous Arthropods and Arthropod Pathogens

The risk of making wrong choices of parasitic and predatory arthropods and arthropod pathogens, especially host-specific ones, does not pose obvious environmental threats, because the outcome of the establishment of an innocuous natural enemy is the pest density remaining at status quo; although there is some theoretical debate on the issue (Turnbull, 1967; May & Hassell, 1981, 1988; Ehler 1982, 1990d). However, arguments have been made that wrong choices could possibly preclude the achievement of maximum biological control (Force, 1970, 1974; Legner, 1986b)—although theoretical considerations indicate that this would be unlikely (Hassell & May, 1973; Hassell & Comins, 1978)—and add to the list of failures, so that careful decisions are desirable (Franz, 1973a, 1973b, Hughes, 1973). The need for choosing the best biological control candidates has generated considerable discussion and controversy over the past 2 1/2 decades, and continues to stir controversy (Turnbull & Chant, 1961; Watt, 1965; van den Bosch, 1968; Huffaker et al., 1971; Zwölfer, 1971; Zwölfer & Harris, 1971; Ehler, 1976, 1982; Zwölfer et al., 1976; van Lenteren, 1980; Pimentel et al., 1984; Legner, 1986b). However, the manner in which the best biological control candidates can be chosen is not clearly delineated for most groups of organisms (DeBach, 1974; Coppel & Mertins, 1977; Pimentel et al., 1984); even though there is a common desire for prejudgment, if for no other reason than to expedite a biological control success.

Parasitic and predaceous arthropods (Insecta and Acarina) and arthropod pathogens (Nematoda, viruses, bacteria, fungi, etc.) are in a distinct category that usually defies accurate prejudgment of biological control potential. Theoretical guidelines based on laboratory studies and mathematical models are not always useful to judge performance in nature. The extremely small size of parasitic and predaceous arthropods, their high dispersal capacity, unique sex determination mechanisms, differential response to varying host densities and climate, distribution patterns and size, unreliable sample techniques, dependence on alternate hosts, and possibilities of rapid genetic change at the introduction site (Messenger, 1971; Eikenbary & Rogers, 1973; McMurtry et al., 1978; Attique et al., 1980; Mohyuddin et al., 1981; Legner, 1986b) all contribute to the impossibility of making universally applicable judgments about performance under natural conditions. Further, release from attacks by specific hyperparasitoids, cleptoparasitoids, and predators make predictions of their performance highly uncertain. Even sophisticated population models such as those developed for the winter moth, Operophtera brumata (Linnaeus), could not predict the exceptional performance of the tachinid parasitoid Cyzenis albicans Fallén against this host in Canada (Hassell, 1969a, 1969b, 1978, 1980; Embree, 1971; Waage & Hassell, 1982).

Because risks to the environment posed by parasitic and predaceous arthropods are very low, the inability to predict their impact has not been a major obstacle to their deployment in successful biological control. At the same time, selection of these biological control agents for importation has not been unsophisticated and lacking in scientific judgment as was suggested by van Lenteren (1980). There are valid scientific criteria for determining probable good candidates that are especially useful for elimination of those with little likelihood for success or that possess certain undesirable characteristics such as hyperparasitism (Luck et al., 1981) and cleptoparasitism.

Assessing Natural Enemy Potential

Coppel and Mertins (1977) proposed a list of 10 desirable attributes of beneficial organisms to aid in assessment of their capabilities prior to widespread dissemination that are closely interrelated and difficult to separate. However, they do categorize organisms and weigh their potential according to a scientific plan. The list, obviously developed from efforts on prior biological control projects, considers ecological capability, temporal synchronization, density responsiveness, reproductive potential, searching capacity, dispersal capacity, host specificity and compatibility, food requirements and habits, hyperparasitism, and culturability. Additional categories of importance to a broad understanding of the niche of a potential biological control agent are systematic relationships and anatomical attributes (fossorial structures, heavy sclerotization, type of spermatheca), physiological attributes (synovigenic or proovigenic), cleptoparasitism, and genetic data (number of founders, collection locality, strain characteristics). Consideration of these attributes in a more holistic approach to natural enemy acquisition is desirable because it incorporates interrelations among qualifications that cannot be detected from an accumulation of single, even well-quantified sets of data.

Biological control researchers traditionally have acquired data under these various categories whenever possible, which has greatly aided in the interpretations of the dynamics in biological control successes. However, there has not been a systematized plan for data acquisition and storage. The collection of facts in the different groups is meant not only to form a database for more accurate predictions of success, but also to raise biological control pursuits to a more intercommunicative realm. The absence of a data collection system in the past has unquestionably resulted in the loss of valuable information about biological control organisms. Critics of a more systematized data acquisition scheme point to the weaknesses of data secured in experimental fashion, but overlook the fact that even incomplete data can elevate one's understanding to the "educated empiricism" of Coppel and Mertins (1977). A step toward improving data acquisition (the tracking of field-released beneficial species) has begun in the United States at the U.S. Department of Agriculture Agricultural Research Service (USDA/ARS) Biological Control Documentation Center, Beltsville, Maryland, as described in Chapter 6. A different database, with slightly different emphases, has been developed for biological control programs by the International Institute of Biological Control (Greathead & Greathead, 1992).

Although there is even more room for additional information about the natural enemies of a given host to help the foreign explorer, it is becoming increasingly possible to appraise performance in nature. The knowledge comes from a combination of laboratory and field studies both in the places of origin of the respective natural enemies and at their introduction sites. For example, among the parasitoids attacking endophilous synanthropic flies, we find the different natural enemy species naturally distributed within certain ecological zones. Cooler, more humid environments harbor different species and strains than those found in hot and drier areas. Marked seasonal abundances, and host and microhabitat preferences are demonstrated by the different species (Legner, 1986b; Legner *et al.,* 1990; Legner & Gordh, 1992). Thus, it is possible to estimate which species are better suited for biological control in a given area based on knowledge of ecological requirements. Additionally, accumulated information on temporal synchrony with the target hosts allows a high degree of certainty in the prediction of which species is most capable of exerting a regulative effect on a single host species in a multihost habitat (Legner, 1986b). Data on density responsiveness, reproductive potential under different ambient temperatures and humidities, searching capacity in different environments and strata of the host habitat, dispersal rates, host specificity and parasitoid food requirements, systematics, synovigenic characteristics, mass production, and genetic variability provide a base from which to judge the likely performance of particular species in different areas. The data also provide a means to properly sample for host destruction (Legner, 1986b). All parameters probably will never be quantifiable, however, as demonstrated by the discovery of a group of polygenes in parasitic Hymenoptera that begin phenotypic expression in the mother prior to the appearance of her progeny (Legner, 1989, 1991).

Such ecological information can also provide better understanding about difficulties faced in prior biological control attempts. For example, in Australia the opinion of some researchers that predatory and parasitic natural enemies of symbovine flies have been duly tested in biological control is based on previous attempts at introduction. However, basic information now available on species that were tried, such as *Aphaereta pallipes* (Say), *Aleochara tristis* Gravenhorst, and *Heterotylenchus* spp., suggests that these candidates were poorly suited to their targeted hosts in the introduced environment. They were incapable of exerting much pressure against them because of different climatic preferences (Legner *et al.,* 1974; Legner, 1986b).

Research performed on the pink bollworm, *Pectinophora gossypiella*; navel orangeworm, *Amyelois transitella*; and carob moth, *Ectomyelois ceratoniae* (Zeller) has al-

ready laid the foundation for accurate decisions on which species of natural enemies have potential for reducing their respective host densities (Legner & Medved, 1979; Sands & Hill, 1982; Legner & Silveira-Guido, 1983; Naumann & Sands, 1984; Legner, 1986b). Clues were found to which regions of the world might be searched for additional candidates.

A common criticism of a systematized approach to data collection and evaluation is a lack of adequate and sustained funding, which suggests that a certain degree of adequacy in financial support ought to be secured before new projects are embraced. More biological control researchers are faced with the necessity for holistic studies with the outcome that basic data are obtained more frequently. An example is the biological control of chestnut gall wasp, *Dryocosmos kuriphilus* Yasumatsu, in Japan, which was guided largely by data acquired about the natural enemies in their place of origin, China (Murakami *et al.,* 1977, Murakami & Ao 1980). Another example is the Comstock mealybug biological control success in California (Meyerdirk & Newell, 1979; Meyerdirk *et al.,* 1981), which was guided by basic research on natural enemies in Japan (Murakami, 1966). Also, the success of the North American egg parasitoid, *Telenomus alsophilae*, against the South American geometrid *Oxydia trychiata* (Bustillo and Drooz, 1977) was based on results of studies on the parasitoid in North America (Fedde *et al.,* 1979).

PLANNING, PREPARATION, AND EXECUTION OF A FOREIGN COLLECTING TRIP

Planning and Preparation

Permits, Regulations, and Quarantine Facilities

The first steps in planning and organizing an exploration trip are to obtain the permits necessary to import the collected material and to coordinate with an authorized importation quarantine facility (see Chapter 6) for handling the shipped material. Permits may be necessary from both national agencies and local (state or provincial) agencies. Separate permits may be required for importation into quarantine and for release from quarantine (as is the case in the United States). Coordination with a quarantine facility should include expected dates of shipments; arrangements for clearance of shipped material through customs and agricultural inspection; and expectations for material handling, sorting, and subsequent shipping of natural enemies to the researcher.

Funding

There is a need for a long-term stable commitment so that chronic pest problems can be pursued. Ongoing search missions usually require a minimum of 12 to 18 months of preparation, and the explorer must have assurance that funds will be available when the trip finally is conducted. Funds should include planning for support staff, space, and facilities in the home institution to handle shipped material while the explorer is overseas. Funds will also usually be necessary for host colony development and maintenance for a period of several months before and after the exploration, as well as for maintenance of any imported natural enemies.

Planning funds for the actual exploration trip should include anticipated travel and subsistence costs as well as support for local guides, unanticipated travel, air transport for shipped material, and local purchases of supplies. In some countries local travel to and from field sites must be provided by official experiment station drivers, and funds for this service must also be incorporated into the budget planning process. Additional funding is often desirable to establish cooperative projects of extended collection and shipment with colleagues in foreign locations.

Traditional sources of funding include national and state or provincial governments, commodity groups, experiment stations, and universities. Experiment station staff might use sabbatical leaves for extended studies abroad, academic responsibilities permitting. A full-time person might be hired to carry on broad biological field studies abroad, as is currently practiced by the USDA/ARS in biological weed control.

The Explorer/Collector

The explorer must have a broad knowledge of the target pest and its natural enemies, including their geographic and host range, host plants, biology, and taxonomy. These duties have often been assigned to full-time foreign explorers, but more recently have been undertaken by academic or professional staff from state or federal agriculture experiment stations or other agencies. These staff are usually scientists already working on the target pest or its close relatives.

The explorer must be able to travel, often under adverse circumstances. A knowledge of the languages and customs of the areas where explorations are to be conducted is highly desirable. However, the cooperation of colleagues or contacts (such as former students) in the area being contemplated for search is highly important for maximizing the often compressed time frame within which an explorer may be obligated to perform his or her assignment.

Because of personal promotion constraints that require research, publication, and teaching, university academic appointees often cannot afford the time required to thoroughly search for and study targeted natural enemies in the field, either foreign or domestic. On the other hand, because of their own time commitments and responsibilities, it may not be appropriate to ask or expect foreign colleagues to

search and ship to the home quarantine facility the desired natural enemies.

Collections for a particular program may also be obtained through the services of intermediary organizations (USDA/ARS will usually respond to requests for natural enemy collection and shipment from U.S. collaborators, and other international organizations, such as the International Institute for Biological Control, are also able to provide such services), but experience has shown that biological control workers who know what they need and who physically participate in the collecting process tend to make a better showing in terms of successful introductions.

Planning the Trip

Before departing on the trip, a detailed itinerary should be developed for use in coordination with the quarantine facility and other staff involved in the program. Passports, visas, immunizations, etc. must be obtained, often significantly in advance; and for some countries this may require letters of cooperation or requests from national or provincial officials.

Contact with foreign collaborators well in advance of the proposed trip can provide valuable information on suitable times or seasons and locations to be included in the exploration. Where guides will be valuable or necessary, foreign colleagues can often advise on the availability and costs of these services.

The purpose of the itinerary in planning with other staff of the program does not preclude alterations in the itinerary once the trip is underway. However, the itinerary should include days and times to expect shipments; and, if possible, dates and times for planned contact by telephone, FAX, or telegram on a frequent basis to apprise the staff of the quarantine facility and home institution on developments and any changes in the expected dates of shipments and other itinerary changes.

Conducting the Exploration

The previously mentioned realities concerning the explorer notwithstanding, the procedure for setting up a pre-planned foreign exploration trip typically involves a literature search, taxonomy studies, and studies of museum material; or possibly voucher specimens and explorers' notes from earlier trips dealing with the same pest, and correspondence with collaborators abroad to determine the best season to search. The latter may include referrals from colleagues, local agricultural extension people, botanists, botanic gardens, and nature preserves (especially in the search area); and a letter of introduction from the host country's consulate to institutions requesting their cooperation and ideally temporary use of their facilities.

Permits are then obtained (refer to Chapter 6 for procedural details) that involve technical and bureaucratic requirements. In the United States the agency is USDA/APHIS (Animal and Plant Health Inspection Service). Permits are required not only for importation into the home country, but also often by the host country where regulations may govern the export of living as well as dead (museum) materials. Individual states or provinces may impose further restrictions or conditions on either import or export.

Advance arrangements must be made with the receiving quarantine laboratory to assure timely availability of host material. In addition to a valid passport, the necessary visas, and the immunization inoculations, the explorer may wish to obtain and carry letters of authorization from host institutions, USDA, and proper officials in the country of search, showing names of cooperating institutions and/or individual collaborators, as well as a supply of personal business cards. In the host country shipping materials may be difficult to find, so it is prudent to carry essential items, such as shipping boxes, tape, and labels.

The field journal recorded by the collector should include names of contacts, villages, farms, host plants, other pests noted, and possible sources of beneficial organisms for other pests. Ideally, photographs should supplement the written record, but even a small camera may pose an unacceptable complication to an already "overloaded" explorer.

Travel

The fundamentals of travel for foreign exploration are not different in principle to those of any foreign travel. Nonetheless, some special applications require some discussion.

Foreign explorers must travel "self-contained." Necessary supplies include drinking water, water purification tablets, medicine, and high-energy foods. Provision should especially be made for care of the feet (including antifungal medications) because long hours of field work in hot climates will subject them to risk of infection, and for intestinal disorders, that (although unlikely to be life-threatening) can limit time planned for travel and searching. Provisions in the expedition equipment must include all the necessary items for collections and shipment.

The explorer should have a thoroughly planned itinerary and tickets before leaving the home country, because tickets can be extremely difficult to obtain in some locations. Nonetheless, the explorer can be flexible and prepared to alter the itinerary should circumstances (such as particularly productive or unproductive collecting in a particular location) warrant.

Collections

The assistance of a local scientist can be invaluable in locating habitats or agricultural plantings suitable for collecting. Local graduate students often are excellent guides.

The explorer should collect as much material as possible. Searches are usually for native populations of natural enemies, and these may be at extremely low densities (substantially lower than more familiar pest populations). Accurate field notes should be kept, including notes on habitat types, plant communities, plant species, and host species located, as well as natural enemies found. Maps and photographs can be very helpful in defining search locations precisely. These records are especially valuable in planning future trips.

The explorer should be aware that constant field work may be more taxing than customary employment. Suitable precautions (carrying water in the field, wearing hats in warm climates, and not exceeding one's physical limits) should be taken, because these will be important in permitting continued, productive exploration.

Provisions should be made to maintain collected material in optimal condition while in the field. This usually will require a small, insulated ice chest or other similar container to limit changes in temperature, together with suitable collecting containers or media.

Processing Collected Material

Adequate time should be allowed daily for sorting and processing collected material for storage and shipment. Depending on the amount of material found in the field, time for processing collections may exceed the time spent collecting. Access to a local laboratory facility can be invaluable in providing working space, lights, microscopes, and so on, but in general the explorer should take all the materials and equipment necessary to process collections in a hotel room or similar accommodation.

Collected material should be shipped as frequently as necessary to ensure optimal condition of the material on arrival at the quarantine facility. Where possible, natural enemy stages least susceptible to the rigors of transport should be selected for shipment. Such stages would include pupae of holometabolous insects, larvae in diapause (and therefore not requiring food), or eggs; and cadavers, spores, fungal hyphal colonies in agar, etc. might be included for pathogens. Special provisions (such as ice packs and humidity-limiting materials) should be made in packaging the material where variations in temperature or humidity must be restricted during shipment. Additional details concerning packaging and shipping natural enemies are found in Chapter 6.

Shipping

Shipment of collected material should always be by the most effective way possible, which is usually air freight, or, where this is impossible, air mail. Shipping via air freight is the preferred method, but the collector may have to hold material for several days pending return to a city with air freight service. Once in the city with such service, travel to an airport facility and making arrangements for shipping can require several hours, and substantial time should be planned for this in the itinerary. Arrangements at the airport can be complicated by language barriers, so the assistance of a local colleague may be helpful. If air mail is the transport method, it is essential that the shipper while at the foreign post office witness application of stamps to the package *and* (most important) their *hand cancellation.*

If at all possible, air freight and air mail shipments should be dispatched at a time when the material will arrive in the home country from Monday through Thursday. This will facilitate rapid handling of the package and minimize delays, especially avoiding the delay characteristic of weekend arrivals at ports of entry. Personnel at the quarantine facility should be notified by phone, telegram, or FAX concerning the shipment, including the expected carrier and routing, time of arrival, and the air waybill number; and should also be notified of changes in travel itinerary and expected shipment dates.

The external packaging material should bear the necessary permits and address labels to facilitate recognition and handling by customs and inspection personnel. Chapter 6 provides details concerning labeling of contents, escape-proof wrapping, and affixation of proper shipping labels. These procedures *must* be meticulously performed to avoid delay at the port-of-entry customs and agriculture inspection.

Assistance

The assistance of a local colleague (perhaps a former student or associate of the collector) can be of exceptional value for maximizing the collection effort, especially if a relatively short stay in the area is mandated by prearranged travel itineraries, unexpected delays from local holidays or strikes, inclement weather, etc. Such a person can serve as guide, as driver of a rental vehicle, and as mentor concerning local customs; can expedite entry to private or public properties; and can assist in organizing shipment of collections. By knowing the local language, a host will be able to allay the suspicions or perhaps outright hostility of the local inhabitants toward the presence of a "strange acting" foreigner. Obviously the collector must be able to pay promptly for expenses incurred. Fluctuating exchange rates can be a problem difficult to anticipate when making long-range plans. Surges in the inflation rate can also create unexpected expenditures. Because of such contingencies, the collector is well advised to carry perhaps 20% more than estimated or anticipated costs in the form of $20 or $50 travelers' checks. An internationally acceptable credit card should also be carried.

U.S. explorers who are federal employees on overseas duty, such as USDA staff, may obtain official assistance at U.S. embassies. State and university employees are typically viewed by embassies as traveling citizens (because they are not federal employees) and typically cannot receive special privileges or assistance unless a USDA appointment (as "collaborator," etc.) has been authorized before the trip. Inquiries at USDA and other international laboratories abroad can often yield useful information and assistance, however.

Acknowledgments

We are grateful to T. W. Fisher, R. D. Goeden, and G. Gordh for their suggestions concerning this chapter.

References

Andres, L. A. (1981). Conflicting interests and the biological control of weeds. In E. S. Del Fosse (ed.), Proceedings of the Fifth International Symposium on Biological Control of Weeds, 1980 (pp. 11–120). Brisbane, Queensland, Melbourne, Victoria, Australia: CSIRO.

Attique, M. R., Mohyuddin, A. I., Inayatullah, C., Goraya, A. A., & Mustaque, M. (1980). The present status of biological control of *Chilo partellus* (Swinh.) (Lep.: Pyralidae) by *Apanteles flavipes* (Cam.) (Hym.: Braconidae) in Pakistan. Proceedings of the First Pakistan Congress of Zoology, Series B, 301–305.

Bartlett, B. R., & R. van den Bosch. (1964). Foreign exploration for beneficial organisms. In P. DeBach & E. I. Schlinger (Eds.), Biological control of insect pests and weeds. London: Chapman & Hall.

Bellows, T. S. Jr. (1993). Introduction of natural enemies for suppression of arthropod pests. pp. 82–89. In R. Lumsden, & J. Vaughn (eds.), Pest management: Biologically based technologies. Washington, D.C.: American Chemical Society.

Bellows, T. S. Jr., Paine, T. D., Gould, J. R., Bezark, L. G., Ball, J. C., Bentley, W., Coviello, R., Downer, J., Elam, P., Flaherty, D., Gouviea, P., Koehler, K., Molinar, R., O'Connell, N., Perry, E., & Vogel, G. (1992). Biological control of ash whitefly: A success in progress. California Agriculture, 46(1), 24, 27–28.

Bellows, T. S., & Legner, E. F. (1993). Foreign exploration. In R. O. Van Driesche & T. S. Bellows, Jr. (Eds.), Steps in classical biological control (pp. 25–41). Lanham, MD: Thomas Say Publications in Entomology, Entomological Society of America.

Boldt, P. E., & J. J. Drea. (1980). Packaging and shipping beneficial insects for biological control. FAO Plant Protection Bulletin, 28(2), 64–71.

Bustillo, A. E., & A. T. Drooz. (1977). Cooperative establishment of a Virginia (USA) strain of *Telenomus alsophilae* on *Oxydia trychiata* in Colombia. Journal of Economic Entomology, 70, 767–770.

Clausen, C. P. (ed.). (1978). Introduced parasites and predators of arthropod pests and weeds: A world review. Agricultural Handbook No. 48, Washington, DC: USDA.

Common, I. F. B. (1958). A revision of the pink bollworms of cotton [*Pectinophora* Busck (Lepidoptera: Gelechiidae)] and related genera in Australia. Australian Journal of Zoology, 6(3), 268–306.

Coppel, H. C., & J. W. Mertins. (1977). Biological insect pest suppression. New York: Springer-Verlag.

Coulson, J. R., & R. S. Soper. (1989). Protocols for the introduction of biological control agents in the United States. In R. Kahn (Ed.), Plant quarantine (pp. 1–35). Boca Raton, FL: CRC Press.

Croft, B. A. (1970). Comparative studies on four strains of *Typhlodromus occidentalis* Nesbitt (Acarina: Phytoseiidae). Unpublished doctoral thesis, University of California, Riverside.

Croft, B. A., & M. T. AliNiazee. (1983). Differential tolerance or resistance to insecticides in *Typhlodromus arboreus* Chant and associated phytoseiid mites from apple in the Willamette Valley, Oregon. Journal of Economic Entomology, 12, 1420–1423.

Davis, C. J. (1967). Progress in the biological control of the southern green stink bug, *Nezara viridula smaragdula* in Hawaii. Muschi, (Suppl.), 9–16.

DeBach, P. (1964). Successes, trends, and future possibilities. In P. De-Bach (Ed.), Biological control of insect pests and weeds (pp. 673–713). New York: Reinhold.

DeBach, P. (1974). Biological control by natural enemies. London: Cambridge University Press.

DeBach, P., & Rosen, D. (1991). Biological control by natural enemies (2nd ed.). Cambridge, MA Cambridge University Press.

Drooz, A. T., Bustillo, A. E., Fedde, G. F., & Fedde, V. H. (1977). North American egg parasite successfully controls a different host in South America. Science, 197, 390–391.

Ehler, L. E. (1976). The relationship between theory and practice in biological control. Bulletin of the Entomological Society of America, 22, 319–321.

Ehler, L. E. (1979). Assessing competitive interactions in parasite guilds prior to introduction. Environmental Entomology, 8, 558–560.

Ehler, L. E. (1982). Foreign exploration in California. Environmental Entomology, 11, 525–530.

Ehler, L. E. (1989). Environmental impact of introduced biological-control agents: Implications for agricultural biotechnology. In J. J. Marois & G. Bruening (Eds.), Risk assessment in agricultural biotechnology. Proceedings, International Conference, (University of California, Division Agric. Nat. Res. Publ. No. 1928), Oakland, CA.

Ehler, L. E. (1990a). Introduction strategies in biological control of insects. pp. 111–134. In MacKauer, M., Ehler, L. E., Roland, J. (eds.) Critical issues in biological control. New York: VCH Publishers.

Ehler, L. E. (1990b). Revitalizing biological control. Issues in Science and Technology, 7, 91–96.

Ehler, L. E. (1990c). Environmental impact of introduced biological control agents: Implications for agricultural biotechnology. In J. J. Marois & G. Bruening (Eds.), Risk assessment in agricultural biotechnology (pp. 85–96). Proceedings, International Conference, (University of California, Division of Agriculture and Natural Resources Publication No. 1928) Oakland, CA.

Ehler, L. E. (1990d). Some contemporary issues in biological control of insects and their relevance to the use of entomopathogenic nematodes. In R. Gaugler & H. K. Kaya (eds.), Entomopathogenic nematodes in biological control (pp. 1–9). Boca Raton, FL: CRC Press.

Ehler, L. E. (1991). Planned introductions in biological control. In L. R. Ginzburg (ed.), Assessing ecological risks of biotechnology (pp. 21–39). Boston: Butterworth–Heinemann.

Ehler, L. E., & J. C. Miller. (1978). Biological control in temporary agroecosystems. Entomophaga, 23, 207–212.

Ehler, L. E., & R. W. Hall. (1982). Evidence for competitive exclusion of introduced natural enemies in biological control. Environmental Entomology, 11, 1–4.

Ehler, L. E., & L. A. Andres. (1983). Biological control: Exotic natural enemies to control exotic pests. In C. L. Wilson & C. L. Graham (Eds.), Exotic plant pests and North American agriculture (pp. 395–418). New York: Academic Press.

Eikenbary, R. D., & C. E. Rogers. (1973). Importance of alternate hosts in establishment of introduced parasites. Proceedings of the Tall Timbers Conference on Ecological Animal Control Habitat Management, 5, 119–133.

Embree, D. G. (1971). The biological control of the winter moth in eastern Canada by introduced parasites. In C. B. Huffaker (Ed.), Biological control, (pp. 217–226). New York: Plenum Press.

Ervin, R. T., Moffitt, L. J., & Meyerdirk, D. E. (1983). Comstock mealybug (Homoptera: Pseudococcidae): Cost analysis of a biological control program in California. Journal of Economic Entomology, 76, 605–609.

Fedde, G. F., Fedde, V. H., & Drooz, A. T. (1979). Biological control prospects of an egg parasite, *Telenomus alsophilae* Viereck. In Current topics in forest entomology (USDA Forest Service General Technical Report WO-8) (pp. 123–127). Selected papers from the Fifteenth International Congress on Entomology,

Force, D. C. (1970). Competition among four hymenopterous parasites of an endemic insect host. Annals of the Entomological Society of America. 63, 1675–1688.

Force, D. C. (1974). Ecology of insect host-parasitoid communities. Science, 184, 625–632.

Franz, J. M. (1961a). Biologische Schädlingsbekämpfung. In P. Sorauer (Ed.), Handbuch der Pflanzenkrankheiten (Band VI, pp. 1–302). Berlin–Hamburg: Paul Parey Verlag.

Franz, J. M. (1961b). Biological control of pest insects in Europe. Annual Review of Entomology, 6, 183–200.

Franz, J. M. (1973a). Quantitative evaluation of natural enemy effectiveness. Introductory review of the need for evaluation studies in relation to integrated control. Journal of Applied Ecology, 10, 321–323.

Franz, J. M. (1973b). The role of biological control in pest management. Bollettino del Laboratorio di entomologia agraria "Filippo Silvestri", Portici, 30, 235–243.

Franz, J. M., & Krieg, A. (1982). Biologische Schädlingsbekämpfung (3 Auflage). Berlin–Hamburg: Verlag Paul Parey.

Goeden, R. D. (1983). Critique and revision of Harris' scoring system for selection of insect agents in biological control of weeds. Protection Ecology, 5, 287–301.

Goeden, R. D. (1993). Arthropods for suppression of terrestrial weeds. In R. D. Lumsden and J. L. Vaughn (Eds.), Pest management: Biologically based technologies (pp. 231–237). Washington, DC: American Chemical Society.

Goeden, R. D., & L. T. Kok. (1986). Comments on a proposed "new" approach for selecting agents for the biological control of weeds. Canadian Entomologist, 118, 51–58.

Greathead, D. J. (1973). Progress in the biological control of *Lantana camara* in East Africa and discussion of problems raised by the unexpected reaction of some of the more promising insects to *Seasamum indicum*. In P. H. Dunn (Ed.), Proceedings of the 2nd International Symposium on the Biological Control of Weeds (Commonwealth Institute of Biological Control Miscellaneous Publication 6, pp. 89–92).

Greathead, D. J., & A. H. Greathead. (1992). Biological control of insect pests by insect parasitoids and predators: The BIOCAT database. Bicontrol News and Information 13(4), 61N–68N.

Hagen, K. S., & J. M. Franz. (1973). A history of biological control. Annual Review of Entomology, 18, 433–476.

Hall, R. W., & L. E. Ehler. (1979). Rate of establishment of natural enemies in classical biological control. Bulletin of the Entomological Society of America, 25, 280–282.

Hall, R. W., Ehler, L. E., & Bisabri-Ershadi, B. (1980). Rate of success in classical biological control of arthropods. Bulletin of the Entomological Society of America, 26, 111–114.

Harris, P. (1973). The selection of effective agents of the biological control of weeds. Canadian Entomologist, 105, 1495–1503.

Harris, P., & H. Zwölfer. (1968). Screening of phytophagous insects for biological control of weeds. Canadian Entomologist, 100, 295–303.

Hassell, M. P. (1969a). A study of the mortality factors acting upon *Cyzenis albicans* (Fall.), a tachinid parasite of the winter moth, *Operophtera brumata* (Linnaeus). Journal of Animal Ecology, 38, 329–339.

Hassell, M. P. (1969b). A population model for the interaction between *Cyzenis albicans* (Fall.) (Tachinidae) and *Operophtera brumata* (Linnaeus) (Geometridae) at Wytham, Berkshire. Journal of Animal Ecology, 38, 567–576.

Hassell, M. P. (1978). The dynamics of arthropod predator-prey systems. Princeton, NJ: Princeton University Press.

Hassel, M. P. (1980). Foraging strategies, population models and biological control: A case study. Journal of Animal Ecology, 49, 603–628.

Hassell, M. P., & H. N. Comins. (1978). Sigmoid functional response and population stability. Theoretical Population Biology, 14, 62–67.

Hassell, M. P., & R. M. May. (1973). Stability in insect host-parasite models. Journal of Animal Ecology, 42, 693–726.

Hokkanen, H. M. T. (1985). Success in classical biological controls. CRC Critical Reviews in Plant Science, 3(1), 35–72.

Hokkanen, H., & Pimentel, D. (1984). New approach for selecting biological control agents. Canadian Entomologist, 116, 1109–1121.

Hoy, M. A., Castro, D., & Cahn, D. (1982). Two methods for large-scale production of pesticide-resistant strains of the spider mite predator Metaseiulus occidentalis. Zeitschrift fuer Angewandte Entomologie, 94, 1–9.

Huber, P. (1983). Exorcists vs. gatekeepers in risk regulation. Regulation, 7(6), 23–32.

Huffaker, C. B. (1954). Introduction of egg parasites of the beet leafhopper. Journal of Economic Entomology, 47, 785–789.

Huffaker, C. B. (1957). Fundamentals of biological control of weeds. Hilgardia, 27, 101–157.

Huffaker, C. B., Messenger, P. S., & DeBach, P. (1971). The natural enemy component in natural control and the theory of biological control. In C. B. Huffaker (Ed.), Biological control (pp. 16–17). New York: Plenum Press.

Hughes, R. D. (1973). Quantitative evaluation of natural enemy effectiveness. Journal of Applied Ecology, 10, 321–351.

Julien, M. M. (Ed.). (1992). Biological control of weeds: A world catalogue of agents and their target weeds (3rd Ed.). U. K.: Commonwealth Agricultural Bureau, Wallingford.

Klingman, D. L., & J. R. Coulson. (1982). Guidelines for introducing foreign organisms into the United States for biological control of weeds. Weed Science, 30, 661–667.

Legner, E. F. (1978). Natural enemies imported in California for the biological control of face fly, Musca autumnalis DeGeer, and horn fly, Haematobia irritans (Linnaeus). Proceedings of the California Mosquito and Vector Control Association, 46, 77–79.

Legner, E. F. (1986a). The requirement for reassessment of interactions among dung beetles, symbovine flies and natural enemies. Entomological Society of America Miscellaneous Publications, 61, 120–131.

Legner, E. F. (1986b). Importation of exotic natural enemies. In J. M. Franz (Ed.); Biological control of plant pests and of vectors of human and animal diseases. (Band 32, pp. 19–30). Forschrifte der Zoologie Stuttgart; New York: G. Fischer.

Legner, E. F. (1989). Wary genes and accretive inheritance in Hymenoptera. Annals of Entomological Society of America, 82, 245–249.

Legner, E. F. (1991). Estimations of number of active loci, dominance and heritability in polygenic inheritance of gregarious behavior in Muscidifurax raptorellus [Hymenoptera: Pteromalidae]. Entomophaga, 36, 1–18.

Legner, E. F., & G. Gordh. (1992). Lower navel orangeworm (Lepidoptera: Phycitidae) population densities associated with establishment of Goniozus legneri (Hymenoptera: Bethylidae) in California. Journal of Economic Entomology, 85,(6), 2153–2160.

Legner, E. F., & Medved, R. A. (1973). Influence of Tilapia mossambica (Peters), T. zillii (Gervais) (Cichlidae) and Molienesia latipinna LeSueur (Poeciliidae) on pond populations of Culex mosquitoes and chironomid midges. Journal of American Mosquito Control Association, 33, 354–364.

Legner, E. F., & Medved, R. A. (1979). Influence of parasitic Hymenoptera on the regulation of pink bollworm, Pectinophora gossypiella on cotton in the Lower Colorado Desert. Environmental Entomology, 8, 922–930.

Legner, E. F., & Silveira-Guido, A. (1983). Establishment of Goniozus emigratus and Goniozus legneri [Hym.: Bethylidae] on navel orangeworm, Amyelois transitella [Lep: Phycitidae] in California and biological control potential. Entomophaga, 28, 97–106.

Legner, E. F., Sjogren, & R. D. (1984). Biological mosquito control furthered by advances in technology and research. Journal of American Mosquito Control Association, 44, 449–456.

Legner, E. F., & Warkentin, R. W. (1983). Questions concerning the dynamics of Onthophagus gazella (Coleoptera: Scarabaeidae) with symbovine flies in the Lower Colorado Desert of California. Proceedings of the California Mosquito Vector Control Association, 51, 91–101.

Legner, E. F., McKeen, W. D., & Warkentin, R. W. (1990). Inoculation of three pteromalid was species (Hymenoptera: Pteromalidae) increases parasitism and mortality of Musca domestica Linnaeus pupae in poultry manure. Bulletin of the Society of Vector Ecology, 15(2), 149–155.

Legner, E. F., Medved, R. A., & Hauser, W. J. (1975). Predation by the desert pupfish, Cyprinodon macularius on Culex mosquitoes and benthic chironomid midges. Entomophaga, 20, 23–30.

Legner, E. F., Medved, R. A., & Pelsue, F. (1980). Changes in chironomid breeding patterns in a paved river channel following adaptation of cichlids of Tilapia mossambica-hornorum complex. Annals of the Entomological Society of America, 73, 293–299.

Legner, E. F., Sjogren, R. D., & Hall, I. M. (1974). The biological control of medically important arthropods. Critical Reviews of Environmental Control, 4, 85–113.

Luck, R. F. (1982). Parasitic insects introduced as biological control agents for arthropod pests. In. D. Pimentel (Ed.), Pest management in agriculture, CRC Handbook (Vol 2, pp. 125–284). Boca Raton, FL: CRC Press.

Luck, R. F., Messenger, P. S., & Barbieri, J. (1981). The influence of hyperparasitism on the performance of biological control agents. In D. Rosen (Ed.), The role of hyperparasitism in biological control: A symposium (pp. 33–42). University of California Division of Agricultural Science.

Macqueen, A. (1975). Dung as an insect food source: Dung beetles as competitors of other coprophagous fauna and as targets for predators. Journal of Applied Ecology, 12, 821–827.

May, R. M., & M. P. Hassell. (1981). The dynamics of multiparasitoid-host interactions. American Naturalist, 117, 234–261.

May, R. M. & Hassell, M. P. (1988). Population dynamics and biological control. Philosophical Transactions of the Royal Society of London, Series B, 318, 129–169.

McMurtry, J. R., Oatman, E. R., Phillips, P. A., & Wood, C. W. (1978). Establishment of Phytoseiulus persimilis (Acari: Phytoseiidae) in southern California. Entomophaga, 23, 175–179.

Messenger, P. S. (1971). Climatic limitations to biological controls. In Proceedings of the Tall Timbers Conference on Ecological Animal Control Habitat Management 3 (pp. 97–114). Tallahassee, FL.

Meyerdirk, D. E., & I. M. Newell. (1979). Importation, colonization and establishment of natural enemies on the Comstock mealybug in California. Journal of Economic Entomology, 72, 70–73.

Meyerdirk, D. E., Newell, I. M., & Warkentin, R. W. (1981). Biological control of Comstock mealybug. Journal of Economic Entomology, 74, 79–84.

Miller, J. C. (1983). Ecological relationships among parasites and the practice of biological control. Environmental Entomology, 74, 79–84.

Mohyuddin, A. I., Inayatullah, C., & King, E. G. (1981). Host selection and strain occurrence in Apanteles flavipes (Cameron) (Hymenoptera: Braconidae) and its bearing on biological control of graminaceous stem-borers (Lepidoptera: Pyralidae). Bulletin of Entomological Research, 71, 575–581.

Murakami, Y. (1966). Studies on the natural enemies of the Comstock mealybug. II. Comparative biology on two types of internal parasites, *Clausenia purpurea* and *Pseudaphycus melinus* (Hymenoptera, Encyrtidae). Bulletin of the Horticultural Research Station, Series A, 5, 139–163.

Murakami, Y., & Ao, H.-B. (1980). Natural enemies of the chestnut gall wasp in Hopei Province, China (Hymenoptera: Chalcidoidea). Applied Entomology and Zoology, 15, 184–186.

Murakami, Y., Umeya, K., & Oho, N. (1977). A preliminary introduction and release of a parasitoid (Chalcidoidea, Torymidae) of the chestnut gall wasp, *Dryocosmos kuriphilus* Yasumatsu (Cynipidae) from China. Japan Journal of Applied Entomology, 21, 197–203.

Naumann, I. D., & Sands, D. P. A. (1984). Two Australian *Elasmus* spp. (Hymenoptera: Elasmidae), parasitoids of *Pectinophora gossypiella* (Saunders) (Lepidoptera: Gelechiidae): Their taxonomy and biology. Journal of the Australian Entomological Society, 23, 25–32.

Oatman, E. R., & Platner, G. R. (1974). Parasitization of the potato tuberworm in southern California. Environmental Entomology, 3, 262–264.

Oman, P. W. (1948). Notes on the beet leafhopper *Circulifer tenellus* (Baker) and its relatives. Journal of Kansas Entomological Society, 21, 10–14.

Pimentel, D. (1963). Introducing parasites and predators to control native pests. Canadian Entomologist, 92, 785–792.

Pimentel, D., Glenister, C., Fast, S., & Gallahan, D. (1984). Environmental risks of biological controls. Oikos, 42, 283–290.

Quezada, J. R., & DeBach, P. (1973). Bioecological and population studies of the cottony-cushion scale, *Icerya purchasi* Mask., and its natural enemies, *Rodolia cardinalis* Muls., and *Cryptochaetum iceryae* Will., in southern California. Hilgardia, 41, 631–688.

Ross, H. H. (1953). On the origin and composition of the Nearctic insect fauna. Evolution, 7, 145–158.

Roth, J. P., Fincher, G. T., & Summerlin, J. W. (1983). Competition and predation as mortality factors of the horn fly, *Haematobia irritans* (Linnaeus) (Diptera: Muscidae) in a central Texas pasture habitat. Environmental Entomology, 12, 106–109.

Sands, D. P. A., & Hill, A. R. (1982). Surveys for parasitoids of Pectinophora gossypiella (Saunders) (Lepidoptera: Gelechiidae) in Australia (Rep. No. 29: 1–18). U.K: Commonwealth Scientific Industrial Research Organization, Division of Entomology.

Schroeder, D., & Goeden, R. D. (1986). The search for arthropod natural enemies of introduced weeds for biological control—in theory and practice. Biocontrol News and Information, 7(3), 147–155.

Sheldeshova, G. G. (1967). Ecological factors determining distribution of the codling moth, *Laspeyresia pomonella* Linnaeus (Lepidoptera: Tortricidae) in the northern and southern hemispheres. Entomological Review, 46, 349–361.

Turnbull, A. L. (1967). Population dynamics of exotic insects. Bulletin of the Entomological Society of America, 13, 333–337.

Turnbull, A. L., & Chant, D. A. (1961). The practice and theory of biological control of insects in Canada. Canadian Journal of Zoology, 39, 697–753.

van den Bosch, R. (1968). Comments on population dynamics of exotic insects. Bulletin of the Entomological Society of America, 14, 112–115.

van den Bosch, R. (1971). Biological control of insects. Annual Review of Ecological Systems, 2, 45–66.

Van Driesche, R. G., & Bellows, T. S. (1996). Biological control. New York: Chapman & Hall.

van Lenteren, J. C. (1980). Evaluation of control capabilities of natural enemies: Does art have to become science? Netherlands Journal of Zoology, 30, 369–381.

Waage, J. K., & Hassell, M. P. (1982). Parasitoids as biological control agents—a fundamental approach. Parasitology, 82, 241–268.

Walters, L. L., & Legner, E. F. (1980). Impact of the desert pupfish, *Cyprinodon macularius* and *Gambusia affinis* on fauna in pond ecosystems. Hilgardia, 48(3), 1–18.

Waterhouse, D. F. (1974). The biological control of dung. Scientific American, 230, 100–109.

Watt, K. E. F. (1965). Community stability and the strategy of biological control. Canadian Entomologist, 97, 887–895.

Wilson, A. G. L. (1972). Distribution of pink bollworm, *Pectinophora gossypiella* (Saund.), in Australia and its status as a pest in the Ord irrigation area. Journal of the Australian Institute of Agricultural Science, 38, 95–99.

Zwölfer, H. (1971). The structure and effect of parasite complexes attacking phytophagous host insects. In P. J. den Boer & G. R. Gradwell (Eds.), Dynamics of populations (pp. 405–418). Wageningen: Centre for Agricultural Publishing and Documentation.

Zwölfer, H., & P. Harris. (1971). Host specificity determination of insects for biological control of weeds. Annual Review of Entomology, 16, 159–178.

Zwölfer, H., Ghani, M. A., & Rao, V. P. (1976). Foreign exploration and importation of natural enemies. In C. B. Huffaker & P. S. Messenger (Eds.), Theory and practice of biological control (pp. 189–207). New York: Academic Press.

6

Quarantine

Concepts, Facilities, and Procedures

T. W. FISHER

Department of Entomology
University of California
Riverside, California

L. A. ANDRÉS

USDA, ARS
Albany, California

INTRODUCTION

The concept of quarantining materials and organisms to keep unwanted elements from entering new areas originated almost simultaneously with the distinction between "good and bad." Much of the information on excluding unwanted plants and plant pests from world commerce through the use of quarantine has been reviewed by Kahn (1989). Regulations governing the arrival of questionable materials to a country, state, province, or county arose with the establishment of quarantine facilities where transported items could be examined before passing on to their destination and assimilation into the general economy and environment. Ooi (1986) gives a summary overview of biological control quarantine from the perspective of developing countries. In this chapter we focus on movement of beneficial or biological control agents, facilities, regulations, and procedures involved.

The primary function of a biological control quarantine facility is to provide a secure area where the identity of all incoming biological control candidates can be confirmed, and unwanted organisms, especially hyperparasites, parasitoids of predators, extraneous host or host-plant material, and so on, can be eliminated. In fact, the quarantine laboratory often represents the last chance to study and evaluate potential biological control agents in the sequence of collection, importation, and release so that only desired beneficial arthropods and/or pathogens will be released. The less known about an incoming organism, the more important the role of the quarantine laboratory becomes.

The number of quarantine facilities in the United States certified to handle incoming shipments of biological control agents has increased from 4 to 26 over the past 50 years. In addition to the 23 listed by Coulson and Hagan (1986), new quarantine facilities have been constructed at Bozeman, MT (weeds); MD Department of Agriculture at Annapolis (weeds); and University of California at Riverside (UCR), CA (nematodes) (see Table 1 later for explanation of postal and other abbreviations). A new entomology insectary and quarantine facility at UCR was funded in 1998. Worldwide, new, or expanded quarantine facilities have been constructed in many countries (Great Britain, Canada, Australia, Thailand, Mexico, and Germany), and others are planned in the United States.

This steady increase in quarantine need and capacity is due in part to (1) increased interest in biological and nonpollutive methods of pest control, and the desire to expand on the many successes already achieved through the importation of exotic natural enemies; (2) the number of new pests that are transported throughout the world each year and that are amenable to biological control; and (3) stricter prerelease information requirements about the behavior and safety of biological control candidates to assure that they will not adversely impact nontarget crops or indigenous flora or fauna. This latter concern often means prolonged studies that tie up quarantine areas for long periods thereby increasing the need for greater quarantine capacity to avoid limiting the amount of materials that can be handled. For example, in the United States, proving the environmental safety of plant-feeding arthropods for biological control of weeds can include studies with as many as 10 to 20 North American native plant species related to the (introduced) target weed. When such studies are not permitted or feasible in the country of origin of the biological control agent, these tests must be conducted in a domestic (U.S.) quarantine facility. Similarly, testing parasites against indigenous insect species that have been declared legally threatened or endangered and that may be present in areas near or contiguous to insect pest infested agricultural croplands targeted for parasite release may not only require more quarantine space but may also delay or prevent the colonization of newly imported biological control agents. The longer imported biological control agents remain in quarantine to carry out these tests, the greater the risk that subtle genetic changes may occur, altering the potential fitness of the biological control agent.

Increasing concern over the quality of the environment is also causing a proliferation of the regulations governing the import and release of biological control agents; and explorer/collectors, quarantine officers, and project scientists are spending more time studying and complying with these domestic and foreign regulations. Air travel has reduced the amount of time required to move shipments of biological control agents from one continent to another, but the proliferation of international airports has spawned a logistical nightmare of unpredictable package routing, delayed agricultural and customs inspection, and unscheduled reloading and shipment to final destination. Frequently, material arrives dead or in weakened condition and on occasion may not arrive.

Despite the increasing quarantine work load, current methods of shipping and handling quarantined biological control agents continue to prove highly workable and represent the coevolution of practical experience and regulatory concerns. Worldwide, there have been remarkably few escapes, considering the hundreds of species and millions of specimens that have been processed. This safety record is the result of a surprisingly uniform set of international protocols and procedures that have arisen out of the common problems and concerns shared by quarantine personnel and regulatory officials in each country. An excellent overview of the history and continuing role of quarantine within the context of classical biological control is presented by Coulson and Soper (1989), Coulson et al. (1991), and Ertle (1993).

In this chapter, mainly from a U.S. perspective, we discuss the several statutory and technological elements of which the explorer/collector/shipper should be aware and which shape the operation of the quarantine laboratory, beginning with the collection, selection, and packaging of exotic biological control candidates in their native habitat or country of origin until their release from the quarantine laboratory or termination of the study. These include (1) national and state regulations (including required permits) as they pertain to the certification of quarantine facilities and the importation, handling, and release of natural enemies; (2) quarantine laboratory design and equipment; (3) personnel working in quarantine; and (4) quarantine operating procedures. In many respects this information supplements that presented by Fisher (1964), who dealt broadly with quarantine facilities and handling techniques. Although we emphasize the handling of entomophagous and phytophagous arthropods, the area of our greatest expertise, the principles and general procedures discussed also should be applicable to the quarantine handling of any restricted organism as well as other biological control agents and plant material.

Because the main emphasis in the 1964 quarantine article (Fisher, 1964) was on the processing of entomophagous beneficials (the mechanics of which have not changed appreciably) rather than on weed-feeding biological control agents, the latter will receive greater emphasis here.

ESTABLISHING QUARANTINE FACILITIES

Minimally, a quarantine facility provides a tight security room for opening and examining incoming shipments of beneficials prior to release to other laboratories for further study, or to cooperators for field release. U.S. Department of Agriculture (USDA) certified quarantine facilities may range from a one- or two-room unit in an existing building (University of California, Davis, CA), to multi-room complex structures designed to meet specific quarantine needs, such as screening (1) weed feeding biological control agents (USDA/ARS, Albany, CA), (2) plant pathogens (USDA/ARS, Frederick, MD), or (3) entomophagous parasites and predators (UCR, CA and USDA/ARS, Newark, DE). A listing of primary biological control facilities in the United States is given in Table 1.

A primary quarantine facility is one certified by the USDA (Animal and Plant Health Inspection Service, APHIS) to receive direct shipments from foreign sources that may contain live pest host material as well as the candidate natural enemies. Rigorous screening out of hyperparasites and nontarget arthropods and entomopathogens, along with studies to ascertain the host specificity and safety of releasing the candidate biological control agents, is the primary mission of a primary quarantine facility. If space or the supply of suitable host material is inadequate at the primary quarantine laboratory, the cleared material can be consigned to the secondary quarantine lab for additional prerelease safety and other tests.

Location and Utilities

Preferably, primary quarantine laboratories are located near (<2 h) a major port of entry (POE) (i.e., an international airport). This proximity becomes increasingly important in proportion to the number of shipments received throughout the year. The quarantine facility should be physically sited where water, electricity, natural gas or propane, road access, and so on, are available. A standby electrical generator powered by natural gas or propane is considered

TABLE 1   Partial List of Primary Quarantine Facilities Certified for Handling Exotic Beneficial Organisms in the United States[a,b]

Location (state, city)[c]	Bioagents handled[d]
U.S. Department of Agriculture, Agricultural Research Service Facilities	
CA, Albany	Phyto.
CT, Ansonia (Forest Service)	Entom., Entpath.
DE, Newark	Entom., Phyto., Poll., Vect., Compet., Entpath. (Ertle & Day, 1978)
MA, Otis AFB (APHIS)	Entom.
MD, Frederick	Phyto., Planpath. (Melching et al., 1983)
MS, Stoneville	Entom., Phyto. (Bailey & Kreasky, 1978; Jones et al., 1985)
MT, Bozeman	Phyto.
NY, Ithaca	Entpath.; Entom. (funded)
TX, Temple	Phyto. (Boldt, 1982)
TX, College Station	Compet.
TX, Mission	Entom., Phyto.
State University and State Department of Agriculture Research Facilities	
CA, Albany, University of California	Entom., Compet., Entpath. (Etzel, 1978)
CA, Davis, University of California	Entom.
CA, Riverside, University of California	Entom., Phyto., Compet., Entpath. (Fisher, 1978)
CA, Riverside, University of California	Nema. (J. Baldwin, personal communication)
FL, Gainesville, University of Florida	Planpath., Phyto.[e]
FL, Gainesville, Department of Agriculture	Entom., Phyto. (Denmark, 1978)
GU, Mangilao, University of Guam	Entom., Phyto.
HI, Oahu, HI Department of Agriculture	Entom., Phyto.
HI, Hilo, HI Volcanoes National Park	Phyto., Planpath.
MD, Annapolis, Department of Agriculture	Phyto.[e]
MT, Bozeman, Montana State University	Entom., Phyto.[e]
NC, Raleigh, NC Department of Agriculture	Entom.
OH, Columbus, Ohio State University	Entpath.
TX, College Station, Texas A&M University	Entom.
VA, Blacksburg, VA Polytechnic Institute	Entom., Phyto
WA, Pullman, Washington State University	Entom.

[a] Modified from Coulson & Hagan, 1986; and J. Coulson, personal communication

[b] An updated listing of all U.S. federal and state quarantine facilities is maintained by the USDA-ARS Biological Control Documentation Center, Insect Biocontrol Laboratory, Plant Sciences Institute, National Agricultural Library, Beltsville, MD 20705-2350 (J. Coulson, personal communication).

[c] State postal abbreviations: CA (California), CT (Connecticut), DE (Delaware), FL (Florida), GU (Guam), HI (Hawaii), MA (Massachusetts), MD (Maryland), MS (Mississippi), MT (Montana), NC (North Carolina), NY (New York), OH (Ohio), TX (Texas), UT (Utah), VA (Virginia), and WA (Washington).

[d] Categories of beneficials: Antag.—pathogens antagonistic to plant pathogens; Compet.—competitors, parasites, and predators of synanthropic flies; Entom.—entomophagous arthropods; Entpath.—entomopathogens; Nema.—nematodes; Phyto.—phytophagous arthropods; Planpath.—plant pathogens; Poll.—pollinators; and Vect.—vectors man/animals.

[e] USDA/ARS and state personnel.

essential for supplying power to selected circuits during power outages. Less obvious factors to be considered in the siting of a facility are freedom from windborne pollutants such as industrial smoke, dust, and pesticide drift. A telephone for communicating with collectors worldwide, and between quarantine personnel and federal or state regulatory personnel is essential. Use of computers to catalog incoming and outgoing material is now standard procedure, as is use of electronic mail (e-mail) and facsimile (FAX) for rapid communication.

Structural Design

Structural details of the quarantine building itself will be dictated in large part by government and local construction codes. Key features of all quarantine laboratories are (1) the sealed nature of the rooms or buildings, (2) a vestibule system for entry and exit with positive closure doors, and (3) a network of filters through which air enters and leaves the facility. Leppla and Ashley (1978) include diagrams of the floor plans of five biological control quarantine facilities in the United States.

To offset long-term heating and cooling costs, walls and ceilings (or roof) should be well insulated. To minimize transfer of heat in or out of the building and to deter window breakage, double glazing of windows is advised. The outer pane(s) should be of tempered or wire-reinforced glass. If vandalism is a problem, the quarantine facility could be encircled with a sturdy fence at least 2 m in height. Added precautions include alarms that signal unlawful entry and fire.

In the United States, the USDA/APHIS, Plant Protection and Quarantine (PPQ), Biological Assessment and Taxonomic Support Staff (BATS) approves the design of new and/or modification of established quarantine facilities. Final certification includes an on-site verification by an APHIS official to confirm compliance with structural and operational criteria. A document titled "Quarantine Facility Guidelines for the Receipt and Containment of Nonindigenous Arthropod Hervibores, Parasitoids and Predators" is currently available from USDA/APHIS, which outlines the most important structural design criteria (Robert V. Flanders, USDA BATS, personal communication).

Selected Design Criteria

The structural criteria will vary according to the beneficial organisms and their hosts to be handled and the risks posed to the environment. Introduction of organisms of interest to biotechnological research probably will require specialized containment facilities as well as tighter strictures on handling procedures.

Arthropods

Rooms for handling beneficial parasitoids and predators or beneficial phytophagous arthropods should have temperature, relative humidity, light, and air exchange control systems to meet the environmental needs of the contained species. When the biological requirements of organisms under study cover a relatively narrow range of environmental parameters, heating, ventilation, air conditioning (HVAC) delivery system(s) can be relatively simple. If, on the other hand, several species of beneficial arthropods having widely divergent environmental requirements must be handled simultaneously, each room or suite of rooms will require independent state-of-the art environmental controls to provide the variety of rearing conditions needed. The diversity of conditions can be greatly increased by the use of individual temperature or environmental chambers. However, these units should be viewed as temporary at best because of their limited size and because of the restricted numbers of beneficials that can be produced in them; of perhaps more serious concern is the typically undependable long-term performance of such cabinets.

Quarantine facilities for handling phytophagous arthropods should have one or more greenhouse containment rooms included in the quarantine complex and should be directly accessed from the quarantine laboratory.

Pathogens of Pest Arthropods

Basically, physical safeguards, equipment, and so on, are similar to modern laboratories that deal with microorganisms. A backup insectary facility for rearing and maintaining a dependable supply of host material is an essential component. For information on quarantine construction and operation criteria for the handling of pathogens, USDA, APHIS, PPQ, BATS (Hyattsville, MD 20782) may be contacted for latest information on facility design criteria.

Pathogens of Weeds

The criteria for a quarantine laboratory certified to handle plant pathogens are discussed by Melching et al. (1983) in their description of the USDA/ARS Plant Disease Research Laboratory (PDRL), Frederick, MD, of which a portion is allotted to weed pathogen studies. The buildings are sealed and air-conditioned via a tandem set of filters designed to remove particles larger than 0.5 μm. Exhaust air is also passed through a third, deep-bed filter unit, before being discharged to the outside. Each filter is capable of removing airborne bacteria or fungal spores. The air pressure within the unit is negative to the outside atmosphere, so in the event of any leakage, air flows inward. Wastewater is sterilized before discharge from the area. Workers must shower before leaving the laboratory, leaving their laboratory garments inside the quarantine. To minimize contamination among study areas, the laboratory and greenhouse are divided into a series of cubicles of varying size. Some of the work in progress at that facility is described by Bruckart and Dowler (1986). Another description of a facility designed to contain weed plant pathogens is given by Watson and Sackston (1985). A simplified pathology quarantine that incorporates all the essential features of these units is that described by Inman (1970), who converted a room in an older building into a quarantine facility.

Nematodes

Certain species of nematodes attack a narrow range of introduced weeds. Others are narrowly host specific, or pathogenic, on pest arthropods. Because beneficial as well as phytophagous (pest) species of nematodes are closely tied to the soil environment, the safe handling of imported species requires a quarantine facility capable of handling and sterilizing plants and soil. In the unique Isolation and Nematode Quarantine Facility recently constructed and USDA-certified at UCR, soil containment is the primary concern. Security measures include restricted entry, use of disposable overboots, arthropod control, and stringent disposal methods. Details can be obtained from the Department of Nematology, UCR, Riverside, CA 92521.

Equipment and Amenities

The equipment and amenities needed in quarantine will vary depending on the class of organism and the studies to be conducted. Most of the items are standard in biological (entomological) laboratories and include various dissecting tools, holding cages, microhabitat monitoring equipment, and illuminators. An olfactometer and video recording equipment for studying the behavioral biologies and host relationships of organisms would also be useful. We have categorized equipment into three main categories: (1) hardware (cages, microscopes, temperature cabinets, etc.), (2) reference items (literature files, identified voucher specimens, records), and (3) cleanup and disposal.

A quarantine laboratory to process incoming shipments solely for identification requires only a handling cage, a microscope, identified reference specimens and other identification aids, and containers for reshipment. When necessary to hold entomophagous or weed-feeding arthropods throughout their life cycles, hosts and host plants in various stages of development; several sizes of cages; and special lighting, temperature, and humidity controls will also be needed.

Because quarantine space is often limited, there is a need to avoid overstocking with equipment. Ample, enclosed storage within the facility should be provided to keep work surfaces clean and free of clutter. Normally, equipment used in quarantine should remain inside the facility. When a special piece of equipment is required for a one-time project or used only infrequently (e.g., olfactometer, camera), it may be removed from quarantine only after careful examination and cleaning. At the USDA plant pathology quarantine facility at Frederick, MD, equipment and notes are removed from the security area through a cold gas fumigation chamber, utilizing 10% ethylene oxide and 90% CO_2 (to reduce flammability) for a minimum of 2 h (Melching et al., 1983). With most arthropods, wiping down the equipment with 95% ethyl alcohol will be sufficient.

Handling Cages

For maximum procedural safety, shipments should be first opened while within a larger cage. A cage design

FIGURE 1   Basic sleeved cage, University of California.

originally conceived by S. E. Flanders (UCR) that has proved highly satisfactory in handling arthropods for over 40 years measures approximately 55 cm high, 44 cm deep, and 46 to 60 cm wide, as well as being constructed of wood with a sloping glass top and fine meshed cloth or screen on the back. A pair of cloth sleeves allows easy, yet effective escape-proof access to the cage's contents (Fig. 1). A USDA variation of this cage is constructed of lucite plastic (Fig. 2), the architecture of which eases the handling and recollection of large numbers of organisms by making them more accessible. The lucite cage's limited ventilation may allow moisture to condense on interior surfaces if large volumes of fresh plant material are held, but this presents no problem in quickly processed samples. Another UC variation (Fig. 3) can accommodate larger shipment containers or potted plants.

When processing an incoming shipment, the handling cage should be equipped inside with a knife or scissors to

FIGURE 2   Sleeved cage, USDA, ARS.

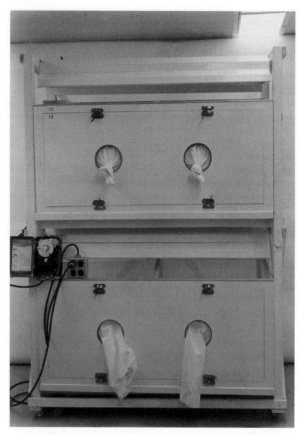

FIGURE 3 Sleeved cages, University of California.

open packages, tweezers, an aspirator and camel's hair brush to use in the transfer or capture of the organisms, vials and cartons to hold the biological control agents, and paper and pencil for recording the number of organisms and other observations. A CO_2 unit for anesthetizing organisms while in the handling cage or even refrigeration of the package prior to placement in the cage will reduce their activity and facilitate identification, sorting, and processing.

Microscope

A binocular dissecting microscope ($10-60\times$) and high-quality illuminator (fiber optic) are usually adequate for assessing the general conditions of quarantined arthropod material, including the identification and determining the sex of specimens. A microscope mounted on a pedestal with an adjustable arm is versatile and can also be used to view organisms on plants and in cages. In addition, a second microscope, such as a phase contrast microscope capable of detecting entomopathogens, will be needed to maintain healthy cultures (Poinar & Thomas, 1978).

Identified Voucher Specimens

The availability of identified reference specimens to compare with incoming material can greatly increase the speed and accuracy of workers as they select specimens for release or further study. Often a single box or, at most, a small cabinet with several trays of specimens, plus a file containing taxonomic keys and other aids will suffice.

Reference File

The following references and information have proved useful in the operation of the quarantine facility. These need not be kept in quarantine, but should be readily available to workers.

Borror *et al.* (1989). General; keys to insect families
DeBach (Ed.) (1964). Good general theory and techniques of biological control
Clausen (1940). Good general work on biologies of a wide range of arthropod parasites and predators
Clausen (Ed.). (1978). History of introductions to 1968 including biologies; 2600 references
Fuxa & Tanada (Eds.). (1987). Insect diseases
King & Leppla (Eds.). (1984). Broad treatment of arthropod rearing problems
Poinar (1977). Nematode identification
Poinar & Thomas. (1978). Entomopathogen preparation and tentative identification
Stehr (Ed.). (1987). Arthropod larval identification
Steinhaus (1963). Broad treatment of pathogens
Van Driesche & Bellows (Eds.). (1993). Steps in classical arthropod biological control
Van Driesche & Bellows (1996). Biological control
Waage & Greathead (Eds.). (1986). Biology, ecology of entomophagous parasitoids

Other information that should be readily available is annually updated listings of (1) quarantine officials at federal and state levels and at major ports of entry, and (2) sources of biological control agents for pest control in greenhouses. Taxonomic references such as catalogs and keys that treat groups of organisms to be imported, and a listing of systematic specialists who can help with plant and arthropod identification should also be available. See personnel sections.

Lights

Daylight–fluorescent and halide-type plant growth lamps are satisfactory for indoor plant culture and greenhouse containment areas that require supplemental light. Heat from halide lamps can speed plant growth, but is harmful if excessive heat buildup is a problem. These lamps also require special wiring and circuitry. Time switches are needed to simulate day length. The entry vestibules of the quarantine laboratory should be equipped with blacklight traps to attract insects that may have gained entry to the vestibule (anteroom) from the outside or from within the quarantine handling area.

Temperature Cabinets

Temperature cabinets are essential for experiments requiring closely regulated temperature cycles. Units with good records of reliability are preferred to minimize repairs and the frequent presence of repair people in quarantine. Ideally the quarantine will be designed so that the mechanical components of such equipment are located outside the quarantine handling and processing area (as on the opposite side of a wall). Temperature and humidity recording devices will be needed.

Refrigerator

Cooling incoming shipments of arthropods to approximately 5°C not only extends the longevity of the biological control agents but also facilitates handling during transfer and identification. Refrigerators may range from small, under-the-counter units, up to the large double door, restaurant models when space permits. The latter can accommodate large packages, and will be especially useful when large amounts of material are being handled. Household refrigerators with thermostatic controls adapted to operate at predetermined minimum temperatures also may be used. Care should be taken when selecting a temperature for long-term holding periods to avoid excessive mortality.

Cold Room

A built-in cold room should be considered if large amounts of plant material or hibernating immature arthropods are held under simulated winter conditions. For temporary storage of securely constrained dormant material, portable refrigerated walk-in units can be rented and placed near the quarantine facility.

Carbon Dioxide

Judiciously used, CO_2 anesthetization (1 min maximum) can facilitate the handling of entomophagous and phytophagous arthropods. However, Nicolas and Sillens (1989) point out that CO_2 narcosis may have adverse short- and long-term effects. CO_2 anesthetization time can be extended by passing the gas over ether. Carbon dioxide is best supplied with portable cylinders provided with pressure regulators that are under the direct control of the personnel using them. Central CO_2 installations that serve several stations, on the other hand, often develop leaks, or gas is wasted by someone inadvertently leaving it on.

Tools

A selection of hand tools such as hammers, screwdrivers, pliers, and small ladders are all useful for cage and equipment maintenance. Flashlights and fire extinguishers are essential for emergency purposes.

Vacuum and Pressure Pumps

Vacuum devices are useful when collecting large numbers of living specimens and when conducting olfactometer experiments. Air pressure is handy for cleaning cages, aerating hydroponic tanks of aquatic plants, and so on. Positive and negative (vacuum) air supplies can be from a central source in the building, or can be provided by portable units moved about the quarantine facility as needed.

Pest Control

Ants, whiteflies, aphids, spider mites, and other pests frequently pose problems to plant and insect cultures in quarantine. Control by nonpesticide methods are preferred (e.g., light traps, sticky boards, soaps, biological control agents, and handpicking of infested leaves). Commercial insectaries and some farm and garden supply stores are sources of biological control agents (Anonymous, 1989; Bezark, 1989). If chemical sprays are to be used, a unit to confine the treatments to the plants and to exhaust odors and drift outside the quarantine will be needed. Insecticidal dusts should not be used. Boric acid powder can be used for cockroach control (Ebeling, 1978). Rarely should it be necessary to tent and thoroughly fumigate the entire facility. Pest control without the use of traditional pesticides will pose a challenge and demand considerable attention.

Cleanup and Disposal Equipment

Vacuum cleaners (preferably the disposable bag type), brooms, sponges, mops, and other janitorial equipment will be needed. Incinerators were at one time commonly used to destroy waste materials but may now be banned where air pollution is a problem. Most containment facilities are equipped with pass-through steam autoclaves (doors on both ends) that allow direct removal of treated materials from quarantine without the possibility of recontamination. Use requirements will dictate the autoclave size to be installed. The juncture where the units pass through the walls must be well sealed so that organisms cannot escape from the quarantine room.

For treatment of small amounts of material, regular household ovens or specially constructed electrically heated boxes may supplement or substitute for the autoclave in purely arthropod handling facilities. The steam autoclave is preferred if soil and other compacted materials are used in containment. In any event, pathogen infected waste material should be sterilized at 121°C at 15 lb pressure for 25 min, and double treatment may be required.

There is some question concerning the usefulness of microwave ovens in treating material to be disposed. Hertelandy and Pinter (1986a, 1986b) report on the use of microwave to control stored product pests, but its effectiveness in killing spores of certain pathogens remains questionable.

Records

A file cabinet will be useful for keeping equipment operating instructions, quarantine handling records, taxonomic keys, pertinent literature, correspondence, and appropriate phone numbers. The cabinet may be kept in the quarantine anteroom, but should be readily accessible to workers in the quarantine area. Computer equipment will facilitate record keeping and exchange of information among quarantine laboratories and regulatory agencies.

Communication Units

To reduce traffic in and out of the quarantine facility a telephone communication capability between personnel working in quarantine and elsewhere in the building or entirely outside of the facility should be included. An intercom system permitting nonmanual operated response will minimize worker disruption.

QUARANTINE PROCEDURES

Introduction

The intent and design of quarantine procedures are to speed the safe importation and release clearance of candidate biological control agents, beginning with the federal and state permitting process through the actual shipping, receipt, processing, release, and finally documentation of the work. As noted earlier, interest and participation in biological control programs by an increasing number of government and university scientists and industry have resulted in the demand for more quarantine facilities. To assure that rigorous standards for handling imported materials are met, there has been a concurrent tightening of domestic and foreign regulations governing the collection and shipment permitting process; the certification of receiving and handling facilities; and a move toward the standard-

ization of procedures and the documentation of all shipping, handling, and release efforts. Pertinent treatises for certain "classes" of biological control agents include those by Charudattan (1982) (pathogens of weeds), Nickle *et al.* (1988) (insect-parasitic nematodes), Klingman and Coulson (1982) (arthropods and pathogens that attack weeds), and Coulson and Soper (1989) (general). Certification of new facilities and recertification of existing facilities require each quarantine lab (primary and secondary) to develop a written statement concerning policy and procedures (guidelines) for use by their collectors, shippers, and quarantine workers. Such a document should be required reading for all personnel, especially temporary or part-time assistants, to teach them correct methodologies and to impress on them the importance and responsibility of the quarantine function before they begin their duties. Copies of the guidelines document are filed with federal and state agricultural quarantine officials as evidence that laboratory personnel have been instructed in the proper handling of imported material. The procedures vary depending on the circumstances involved. For example, USDA shipments originating from federal research facilities abroad are well screened before shipment and often can be passed quickly through quarantine for direct field release. On the other hand, shipments originating from explorers in remote areas who do not have access to laboratory facilities often contain a mix of beneficial species, as well as pest host arthropods and host plants, requiring careful quarantine processing.

Federal Regulations*

As noted by Coulson and Soper (1989), "no single Federal statute specifically addresses regulation of the importation, movement and release of biological control agents per se." The authors then list six federal regulations that impact on biological control activities: (1) the Plant Quarantine Act, 1912 (initial legislation to restrict movement of potential pests into the United States); (2) the Federal Plant Pest Act of 1957 (regulates the importation and movement of plant pests and plant parts that may harbor pests); (3) the Public Health Services Act (regulates movement of insects and vectors of human disease agents); (4) the Federal Insecticide Fungicide and Rodenticide Act (FIFRA) [authorizes the Environmental Protection Agency (EPA) to regulate pesticides, which by broad definition includes biological controls] (5) the National Environmental Policy Act

---

*We refer mainly to USDA regulations. In the United States, other agencies such as Environmental Protection Agency (EPA), U.S. Public Health Service (PHS), U.S. Fish & Wildlife (F&W), etc., also have certain regulations that may directly or indirectly impact biological control importation and implementation programs. Many foreign governments have similar regulations and enforcement agencies. International agreement concerning standardization of regulations pertaining to collection and transport of biological control agents conceivably could facilitate biological control programs worldwide.

(NEPA) (requires an assessment of actions that may affect the quality of the environment); and (6) the Endangered Species Act (attempts to avoid impact on indigenous rare and endangered species).

Since 1983, in the United States regulation of movement of biological control agents, package inspection at ports of entry, and quarantine certification inspection have rested with the (former) Biological Assessment Support Staff (BASS), now BATS, of the PPQ section, of APHIS, USDA, Hyattsville, MD 20782 (Lima, 1983).

APHIS PPQ regulatory actions set the standards and guidelines for all federal and state biological control quarantine activities. However, individual states may attach additional regulations concerning treatment accorded specific pests within their geographic jurisdictions.

To facilitate its responsibility, PPQ has placed the organisms that must be quarantined into three categories:

Category A: Foreign plant pests not present or limited in distribution throughout the United States; domestic plant pests of limited U.S. distribution, including program pests; state regulated pests; and exotic strains of domestic pests

Category B: Biological control agents and pollinators

B1: High risk: weed antagonists; shipments accompanied by prohibited plant material or category A pests

B2: Low risk: pure cultures of known beneficial organisms

Category C: Domestic pests that have attained their ecological ranges, non-pest organisms and other organisms for which courtesy permits may be issued

Specifically, all exotic biological control organisms (category B) entering the United States accompanied by category A pests (hosts of the biological control agents) or weed antagonists must be received in a primary PPQ certified quarantine facility (Lima, 1983).

Permits

Federal and state permits are required for almost all movement of beneficial organisms into and throughout the United States.

Importation into the United States

Permits are required for all importations of live beneficial arthropods and microorganisms into the United States. Application for these permits currently must be made on PPQ Form 526 (Application and Permit to Move Live Plant Pests) obtained from APHIS/PPQ, Hyattsville, MD, or from state agricultural department quarantine officials. The completed forms should be routed to those state agricultural officials in whose jurisdiction the receiving quarantine facility is located. Following approval by state officials, the

forms are then forwarded to APHIS/PPQ for concurrence or instruction of further conditions required in handling the material. At this point PPQ Form 526 becomes the permit and the duly signed form, and appropriate PPQ shipping permit labels are forwarded to the applicant. These labels are to be affixed to the outside of the packages by the explorer/collectors or foreign shipper and addressed to the quarantine officer. Packages hand carried into the United States also should bear the proper shipping labels and should be accompanied by a copy of the approved PPQ Form 526 to avoid delays at the port of entry. At least 6 months should be allowed for routine processing of the 526 application through USDA (APHIS) and state departments of agriculture. However, in emergency situations the process may be speeded considerably through judicious use of the telephone, e-mail, and FAX.

With biological control of weeds work, it is often necessary to import exotic plants for host–range studies. Applications to import these plants can be obtained from the Permit Unit, Plant Protection and Quarantine Programs, APHIS, USDA, Federal Building, Hyattsville, MD 20782. Plants to be imported under the quarantine permit also may be subject to the provisions of the Convention on International Trade in Endangered Species of Wild Fauna and Flora (CITES). For information on which plants come under these provisions, how to obtain permits for their importation, and which foreign agencies should be contacted to obtain the proper export permits, one should contact the Wildlife Permit Office, U.S. Fish and Wildlife Service, U.S. Department of Interior, Washington, DC 20240. In addition, permits to move restricted plants from a foreign source (Anonymous, 1985) or from one state to another should be obtained from quarantine officials in the receiving state.

Biological control explorer/collectors are often unaware that in some countries (Australia, China, and Mexico) permits are required not only to collect but also to export living or dead specimens (museum material) of indigenous species. It is reasoned that because biological control explorers sometimes work with relatively obscure elements of a country's flora and fauna, it is not uncommon for them to discover new species and compile new biological information. By requiring export permits, officials of foreign governments are better able to monitor the biota of their countries and to assure that type specimens of newly described species remain in their national repositories. Arrangements for necessary travel, collection, and export permits needed by the explorer/collector can usually be handled through the Ministry of Agriculture, Plant Quarantine or Plant Protection Service officials of the host country. These arrangements must be initiated at least 12 months in advance of travel and should be completed prior to arrival in the country. To avoid bureaucratic delay, all U.S. and foreign permits (plus several copies of each) should be in hand before packaging materials for shipment to the United States.

Biological control materials exported from the United States to other countries will require an import permit from the requesting country to be furnished by the receiving scientist or agency.

Interstate Shipment

Some states in the United States have regulations governing the importation, movement, and release of arthropods within their boundaries (California, Florida, North Carolina, Oregon, and Texas). In these instances, APHIS-PPQ issues PPQ shipping permit labels at a state's request. This label covers the transport of biological control agents only and does not include living host material. (However, certain PPQ permits will allow interstate shipment of host material if it is of domestic origin and/or if the receiving state approves.) Packages of biological control agents shipped or mailed interstate should bear this label. Adherence to these regulations is the responsibility of the quarantine officer or project scientist. In any case, state regulatory officials should be routinely informed of releases within their areas, whether permits are required or not. Technically, shipments containing only living beneficial species do not require a permit, but their movement should be made a matter of record at the Biological Control Documentation Center, Beltsville, MD 20705.

In the case of weed-feeding arthropods and pathogens of weeds, the import application and labeling process has two added steps to that described earlier (Klingman & Coulson, 1982). Prior to completing and filing PPQ Form 526, the applicant must prepare a proposal justifying the planned importation of weed arthropods whether for study or for release. The proposal should (1) address the importance of the weed problem and whether the target plant has any redeeming features that may lead to objections against its control; (2) state the organism to be introduced for study and/or release; and (3) summarize the information known about the host range and biology of the organism, noting if studies are still needed and how it is to be handled in quarantine. The proposal is forwarded to the APHIS Technical Advisory Group (APHIS-TAG), which will consider the potential hazards and benefits of the proposed importation. APHIS-TAG will then either authorize importation or release, recommend further study, or reject the proposal. If it is concluded that the proposed action is safe, a completed PPQ Form 526 and the proposal are submitted to the agricultural officials in whose state the quarantine laboratory and/or the release site is situated. When approved by the state, APHIS-PPQ issues the import labels. If release is contingent on further study, a final summary of the required data is prepared and routed in the form of a release proposal to APHIS-TAG and to state agricultural officials for concurrence. To meet the requirements of NEPA, BATS currently requires data from the researchers to allow APHIS to prepare an environmental assessment prior to issuance of a permit for the initial release in the United States of an exotic weed biological control agent. In all instances, the project scientist and quarantine officer are responsible for the proper clearance and identification of the material released.

Procedural Problems

Permits

As currently designed, PPQ 526 (Application and Permit to Move Live Plant Pests) pertains to pest species only and in effect lumps biological control agents with plant pests. Packages so labeled are handled the same as plant pests. If, on the other hand, a new form plus a biological control shipping label were designed that accurately reflected the biological control package contents, several benefits would result. Not only would the package be identified clearly as containing live biological control material—thereby mandating prompt inspection—but also information would be more readily transferrable to the USDA 941, 942, and 943 biological control agent shipment forms for forwarding to the documentation center (see the section on records and reports later).

Importation Clearance at Port of Entry

Packages that the collector/shipper improperly wraps, labels, and/or fails to attach the proper permits often cause time-consuming dialogue and interaction between customs and agricultural inspectors, or airline freight personnel and the authorized quarantine pickup persons. When everything is prepared properly, and the package is secure, it does not need to be held by POE U.S. customs or APHIS inspectors, but can be shipped directly to the receiving quarantine facility. Delay of 1 day or even several hours, especially during the summer, can greatly increase mortality in shipments. Under no circumstances is a properly wrapped and labeled package to be opened outside a quarantine area, because to do so would invite release of category B1 (high-risk) material, thereby violating the quarantine concept.

Customs brokers may be of help in clearing packages, but they are often costly and their services can add as much as 3 days to package transit. When speed is not a problem, as during cool weather when the material is at less risk or when the contained material is dormant, customs brokers can assist in clearing customs and agricultural inspection and send the package by overnight mail to the quarantine laboratory.

Literal interpretation by inspectors of the permits affixed to incoming shipments (i.e., precise POE, country of origin, and contents) often causes delays. Broadly worded permits, on the other hand, allow inspectors to follow the spirit and intent of the permit. For example, permits that allow entry at any major U.S. POE give the shipper and carrier flexibility to ship by the best routing available. Likewise, permits that provide a general description of package contents (the name of the species or genera of natural enemies that are the primary targets of the search) are adequate. In the case of target pests with a broad host range that may accompany the beneficials, such as diaspid scales or aphids, listing the pest as polyphagous instead of all the potential hosts should be sufficient. Permits that state a wide geographic area of shipment origin (e.g., Europe instead of specific countries or localities) allow collectors to move to those areas that can be more profitably searched depending on the weather and advice from local colleagues.

The foregoing and other problems are typical of all biological control importation efforts and their resolution requires close cooperation between research workers and regulatory officials. They also demonstrate the value of knowing, or knowing how to contact experienced senior grade APHIS inspectors who can expedite the clearance process. Until such problems can be resolved, the quarantine pickup person must carry copies and amendments of permits to the POE in addition to a letter under institutional letterhead authorizing him or her to retrieve a shipment from the airfreight agent.

Packaging and Shipping of Biological Control Agents

Although the successful transport of living organisms demands ingenuity and good judgment on the part of the collector/shipper for each species and situation, several generalities can be made about shipping entomophagous or phytophagous arthropods to or from the quarantine facility:

1. The least active stages of an organism (eggs, pupae, diapausing larvae, and adults) often survive the rigors of packaging and travel better than active stages.

2. Extreme temperatures and length of time the organisms must remain in the package are the chief causes of shipment deterioration. If possible, all packing materials should be cooled before shipping and kept cool en route. The shipments should be directed by the swiftest and most secure routing—usually airfreight, or other express services, such as international couriers.

3. Minimize the amount of fresh plant material included in the shipment. Fresh foliage breaks down rapidly when packages are inadvertently placed in the sun or become warmed, leading to arthropod death. Because the permit under which entomophagous candidate agents are shipped usually precludes the inclusion of plant material except for nonpropagable fragments, excluding excess plant material should not be a problem.

4. Control the amount of free moisture in packages, especially if fresh plant material must be included. Plant transpiration and respiration can raise humidity to excessive levels in closed packages, which may lead to increased mortality of the natural enemies.

5. Organisms to be shipped should be taken from expanding, healthy populations to minimize the inclusion of diseased or genetically impoverished material (Myers & Sabath, 1981). Early season generations of multivoltine species generally contain fewer parasites, including hyperparasites, than do late season generations.

6. Delay sealing the package before shipment as long as possible. Cooling packages (perhaps to 5°C) before opening in quarantine can help to reduce the activity of mobile stages.

7. Collectors should notify receivers well in advance of shipping schedules and size of shipments whenever possible, and advise when certain actions should be taken. It is highly important to include notes and observations that may help in processing the material.

8. Avoid overloading packages. Objects included in the package should be fastened or braced to avoid excess movement. Do not leave sticky surfaces exposed. Some of the methods and techniques most commonly used in packaging beneficial entomophagous and phytophagous arthropods were described by Bartlett and van den Bosch (1964) and Boldt and Drea (1980). Each life stage may require particular attention and handling. In essence the "double or triple wrapped" principle should be observed: following segregation to species or stages, biological control agents should be packaged in one container, which is placed in another container (or a closely woven cloth bag) to be enclosed within the shipping box.

A method of packaging shipments of beneficial arthropods to or from a quarantine facility is illustrated in Figure 4. Fabricated to fit snugly in a heavy cardboard carton, styrofoam provides insulation. The unit will accommodate 2-pint cartons and a reusable artificial ice pack. Excelsior provides a resting substrate for adults. This unit was developed in 1979 by R. J. Dysart and K. S. Swan, USDA, ARS, Beneficial Insects Research Laboratory, Newark, DE 19713.

Eggs

Keep eggs cool to minimize hatching enroute. If the eggs are attached to plant substrate, clip away the excess host material. Fasten the substrate firmly in the package or layer the eggs on soft absorbent tissue to avoid movement that may break eggs loose. Package eggs in several small

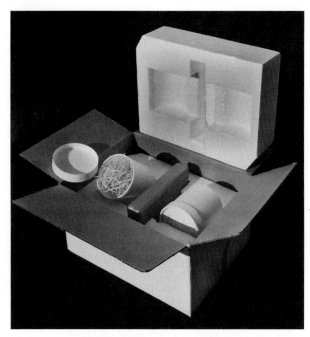

FIGURE 4   Shipping unit, USDA, ARS.

cartons (e.g., plastic petri dishes) instead of a single large unit.

Immatures

Actively feeding lepidopterous larvae or other naked immature stages should not be shipped if it can be avoided. Unless individually isolated, they tend toward cannibalism when the food supply is exhausted. Otherwise, avoid crowding, and provide excelsior (shredded wood) or other materials on which the organisms can rest, or try to package individually. Larvae and pupae in their insect hosts or in roots and stems should be left and shipped in the host *in situ*. After removing excess soil, roots should be packaged with lightly moistened sphagnum moss, wrapped in burlap and a cloth bag, and placed in an unwaxed cardboard or wooden container. To prevent excessive buildup of moisture, do not enclose materials in plastic. Lepidopterous pupae collected from the soil can be packed in lightly moistened sphagnum moss. If large numbers are shipped, the pupae should be divided into a number of small lots, and placed in a series of cartons instead of all in one carton. Dipterous pupae extracted from soil can be repackaged in moist sterilized soil in a series of small containers. Pupae on twigs or foliage can best be left attached to the host substrate and lightly mixed with shredded newsprint or sphagnum moss to hold the material firm. Dried flower heads containing diapausing larvae can be shipped in bulk in cloth bags enclosed in a cardboard or wooden container. Plant galls infested with actively developing stages (e.g.,

eriophyid mites in leaf and stem galls, cecidomyid larvae, and pupae in stems and foliar galls) can be packed with sphagnum moss or excelsior. Leave ample air space to avoid overheating and decomposition of materials.

Adults

Active adults should be cooled and packaged with access to a moist absorbent surface so they may obtain water. Avoid excess plant material and overcrowding. If coleopterous adults are well fed and in good condition, they can endure several days without plant food. If disease is a concern, adult beetles can be isolated in separate vials to minimize contamination of the entire shipment. A small amount of moistened shredded wood, pipe cleaners, or other materials should be provided as resting substrate for the adults. Adult Lepidoptera, Diptera, and Hymenoptera should have access to sugar water or honey, either as droplets on the side of glass vials in which they are shipped or on wicks of cotton wool, sponge, and so on.

The USDA/ARS is currently using a molded, high-density styrofoam unit for many of its shipments. The styrofoam is shaped to hold two cylindrical cartons ((9 × 9 cm)and a cold pack (see Fig. 4). The overall unit measures 22 1/2 × 15 × 29 1/2 cm, and is placed inside a cardboard box and wrapped with paper for added protection.

Coordination with Collectors

The role and responsibilities of the explorer–collector are discussed in Chapter 5. Certain problems can be alleviated or eliminated by the observance of procedural details on the part of foreign collaborators or explorer–collectors commissioned to send material.

From a quarantine worker's perspective, explorer–collectors should understand the importance of providing complete data concerning locality and host plant or host insect for each shipment and component thereof, especially if packaged material is from more than one locality or host. The University of California Senders and Receivers form (Fisher, 1964) or the USDA AD941 form (see section on records and reports later) provides key headings and spaces for entering this essential information, most of which can be supplied only by the collector.

Adherence to, or notification of, changes in a travel itinerary will greatly facilitate communication to and from collectors by telegram, telephone (or cellular telephone), or FAX. The collectors should notify the quarantine laboratory concerning anticipated time of arrival of a shipment, its routing (including carrier and flight number), and its airline freight waybill number. If possible, shipments should be dispatched in time to arrive at the port of entry and the home quarantine facility during the regular work week to avoid customs and agricultural clearance problems that characteristically surface on weekends.

Removal of Unwanted Parasites, Hyperparasites, and Pathogens

Parasites and pathogens contaminating incoming shipments of natural enemies must be eliminated if vigorous colonies are to be maintained or field releases are planned. Several publications have proved of value in recognizing entomopathogens (Steinhaus, 1963; Poinar & Thomas, 1978; Goodwin, 1984). To prevent the spread of disease, particular attention should be directed to the culture handling methods and to humidity, temperature, light, and substrate condition. Etzel et al. (1981) demonstrated that by simply using new, clean cartons that were changed every 2 to 3 days and by autoclaving and sterilizing vials and other equipment, a *Nosema* sp. infecting a culture of the chrysomelid *Galeruca rufa* Germar could be eliminated. In this case it appeared that the infection was transmitted through contaminated feces. The beetle had been imported for host-specificity studies with *Convolvulus arvensis* Linnaeus. Briese and Milner (1986) used egg surface sterilization to eliminate the microsporidian *Pleistophora schubergi* Kucera & Weiser from a laboratory colony of *Anaitis efformata* Gn., a candidate for the control of *Hypericum perforatum* Linnaeus. It was determined earlier that the pathogen was restricted to the host gut tissue and transmitted by contaminated feces (Milner & Briese, 1986).

Entomopathogens are difficult and costly to eliminate, especially from univoltine arthropods. The only remedies may be to destroy existing cultures; to sterilize rearing areas and equipment by scrubbing with appropriate chemicals, antibiotics, or sterilizing agents; and to start again with clean material. P. Dunn (retired USDA/ARS, Rome, Italy, personal communication) attempted to eliminate a *Nosema* sp. from the univoltine weevil *Trichosirocalus horridus* (Panzer) by isolating eggs from individual females, releasing only those eggs that had been produced by females found to be free of the pathogen in a postoviposition examination. The beetle was successfully colonized at several sites but a follow-up check for *Nosema* infection has not been made.

Identification

Any attempt to study or manipulate biological systems requires taxonomic or systematic assistance. Authoritative identification is the first step to unlocking information on host range, ecological relationships, biology (life history), and even previous uses as a biological control agent; and is especially helpful during pre-release evaluation periods in quarantine. Ultimately, information from the USDA/ARS

Biological Control Documentation Center at Beltsville, MD, should greatly facilitate acquisition of such information during the planning stages of a proposed collecting trip.

Identity of all live material passing out of quarantine should be reconfirmed. Although the explorer/shipper will attempt to include only beneficial species in a shipment, quarantine workers should have available, and be familiar with, appropriate aids and voucher material to verify the identification of the desired species. Most importantly, workers should know how to prepare specimens for identification and forwarding to appropriate specialists as discussed by Edwards *et al.* (1985) and Steyskal *et al.* (1986). A listing should be maintained of taxonomists at the USDA Systematic Entomology Laboratory and other scientists who might help with prompt identifications of arthropods and plants. The Association of Systematic Collections (ASC) Newsletter, c/o Museum of Natural History, University of Kansas, Lawrence, KS 66045, provides periodic updating of the systematic collection community internationally.

The majority of natural enemies handled in quarantine are from little-studied ecosystems in foreign areas, and often prove difficult to identify. Knutson (1981) cites R. J. Dysart (USDA/ARS, Newark, Delaware, Beneficial Insects Research Laboratory, a biological control quarantine facility, personal communication), that out of the 318 species of parasites and predators released through that quarantine laboratory from 1965 to 1979, 16% could be identified only to the generic level and 2% could be identified only to the family or higher level. Of the 42 species of arthropods released to control weeds in the United States at that time, all were identified to species, but 7 of them were new to science during the period when they first came under consideration as biological control agents (Knutson, 1981).

Often, incoming shipments will contain dead or dying specimens. These and others that may die during quarantine processing should be carefully mounted and labeled as voucher specimens and submitted for identification. The quarantine worker's first concern is establishing cultures of the new biological control agent, the precursor to specificity tests. A portion of the F_1 offspring should be retained as voucher specimens with the conserved parental material.

It is also important for the collector/shipper to supply a sample of the host material from which the natural enemy was collected. The collection of natural enemies from host plants or arthropods mistakenly believed to be the same as the target pest has resulted in failed establishment of the organisms and, at best, delayed biological control. For example, plant misidentification resulted in the collection of weed-feeding agents from the aquatic weed *Salvinia auriculata* Aublet instead of the targeted *S. molesta* D. S. Mitchell. Once the misidentification was realized, subsequent collections from *S. molesta* led to the virtual eradication of this weed in Australia, New Guinea, and Africa (Room *et al.,* 1981, 1984; also Chapter 34).

The identification process should be quick to alert the shipper about the need of any additional collections, and from which localities. A growing concern is that reduced numbers of identifiers at the USDA Systematic Entomology Laboratory, state and university facilities in the United States (and abroad), and insufficient support of taxonomy by public and private museums in general ultimately can have significant negative impact on biological control programs by causing delay of releases from quarantine. Misidentifications also can lead to duplicative reintroduction of species already established. Unfortunately, this erosion of support for classical systematics has been worldwide and has led to a shortage of specialists in a number of areas. It is a problem that is not quickly corrected because it may take at least 4 years for a taxonomist to gain credible proficiency in a group. This is especially true in the parasitic micro-Hymenoptera. The importance of alpha taxonomy in theoretical and applied biology, with special reference to biological control introductions, is clearly delineated by Knutson (1990).

The benefits to biological control from proper identification are obvious. Perhaps not so obvious are the reciprocal benefits to the biosystematist provided by well-documented biological control studies. The ability of the foreign explorer to collect specimens from populations of differing geographic range, habitat, and climate (biotypes), when combined with the quarantine workers' array of tests and studies, provides the information that permits biological features and identifications to be compared and correlated (Knutson, 1981). Chapter 3 provides additional information relative to the importance of taxonomy to biological control.

Culture of Hosts and Natural Enemies

The culture of entomophagous and weed-feeding arthropods requires host substrate materials free of pesticides or other harmful residues. Because space limitations are always a concern in quarantine, whenever possible, living plants needed for quarantine work are best cultured outside the containment area, (preferably out-of-doors) or in a screen or lath house structure or greenhouse; and the plants or plant parts are moved into quarantine as needed. Plants cultured in this way minimize the need for quarantine greenhouse space and are accessible to natural populations of predators and parasites that hold many of the common horticultural pests in check (whiteflies, leaf miners, aphids, and spider mites). Otherwise, the emended use and culture of plants totally within quarantine increases the need for quarantine equipment for storing and handling soil, pots, and fertilizer, but will also require various pest control measures [e.g., the use of nonpesticidal soaps, direct hand removal (of pests), nonresidual pesticides, and augmentation of natural enemies to maintain proper plant health]. In addition, living plants are often unwieldy and difficult to maintain in cages, as would be required in quarantine. Obviously, the use of artificial hosts is impractical for quarantine studies involving host-specific phytophagous arthropods and plant pathogens. In this case appropriate live hosts and test plant species must be available along with quarantine greenhouse facilities.

Phytophagous Arthropods

Cultures of host-specific phytophagous arthropods or plant pathogens can be maintained best only on live, healthy plants, and in some instances only on limited varieties of the plant. For example, before American workers were able to establish a viable colony of the highly host-specific rush skeletonweed eriophyid, *Eriophyes chondrillae* (Canestrini), in quarantine, they needed to return to Europe to find a population of the mite that would develop on the California strain of rush skeletonweed, *Chondrilla juncea* Linnaeus (Sobhian & Andres, 1978).

Many weed-feeding arthropods are univoltine in development, making their culture in quarantine difficult. To avoid problems, quarantine workers often request that appropriate stages of the candidate natural enemies be shipped from their native ranges abroad for direct use in specificity tests. For example, actively feeding *Longitarsus jacobaeae* (Waterhouse) and *Aphthona* spp. adults were imported directly from Europe for plant feeding, and ovipositional tests and neonate developmental studies on *Senecio jacobaea* Linnaeus and *Euphorbia esula* Linnaeus, respectively. As in all importations, these biological control agents are screened for unwanted parasites or pathogens that could alter their feeding or other behavior.

It is important to follow natural enemy development through the full life cycle in host-specificity tests. This may be difficult for univoltine species because it is not uncommon for 90 to 95% of a culture to die primarily due to deterioration of the host plant. In quarantine studies with the univoltine chrysomelid. *L. jacobaeae,* K. E. Frick (personal communication) was able to overcome this problem by carefully monitoring the host plant *S. jacobaea* and manually removing any pest insects. Similarly, larvae of the ststem and root mining *Oberea erythrocephala* (Schrank) (Cerambycidae) and *Chamaesphecia tenthrediniformis* (Schiffermuller) (Sesiidae) on *E. esula* Linnaeus, which feed and remain in the living plant 10 months of the year, are difficult to culture in quarantine. On the other hand, univoltine foliage feeding lepidopterans, such as *Tyria jacobaeae* Linnaeus on *S. jacobaea* and *Hyles euphorbiae* Linnaeus on *E. esula,* are relatively easy to culture in confinement because the larvae require fresh plant material only 1 or 2 months of the year, passing the remaining 10 to 11 months as dormant pupae away from their host.

Although multivoltine natural enemies are often more easily reared in confinement, their host plants must be carefully maintained in good health and frequently renewed. An adequately staffed greenhouse facility is required for quarantine culture of plants.

Entomophagous Arthropods

Host acceptability, space, light, humidity, supplemental food, mating, breaking of diapause (especially with shipments between hemispheres) and so on, can be serious constraints to rearing success. Many examples are presented in Chapter 7. Because of difficulties in maintaining live plants in close containment, it is often simpler to use artificial hosts, such as potato tubers and sprouts, squash, citrus fruits, or other plant parts to culture insects for use in quarantine studies (Finney & Fisher, 1964). For example, potatoes infested with scale insects that in many instances are readily attacked by parasites are easily handled, caged, observed, and disposed of. In addition, potatoes are relatively inexpensive and available throughout the year. "Seed" potatoes (certified disease free) should be sought. Culture principles, techniques, and problems are elaborated by King and Leppla (1984) and in Chapter 7.

Pathogens of Weeds

Because many species of weeds have become tolerant of chemical herbicides routinely used to control them, a highly specialized area of biological control research that utilizes plant pathogenic microorganisms is receiving increasing emphasis. The reader is referred to Charudattan (1982), Churchill (1982), and Melching *et al.* (1983) as introduction to this challenging field of research.

Artificial Diets

Artificial diets may be of value in quarantine and postquarantine studies, but the high degree of host specificity exhibited by most weed-feeding arthropods [quite likely in response to host-plant-generated synomones, Norlund & Lewis (1976)] coupled with the time required to develop acceptable diets will tend to limit their usefulness during the initial quarantine phase of the study. Also, when working with small numbers of natural enemies, quarantine workers usually prefer using live host-plant material to establish initial laboratory colonies. Artificial diets likewise should not be used during host-specificity testing of entomophagous natural enemies because behavioral characteristics, triggered in response to kairomones or other semiochemicals produced by the host insect or host plant, may be altered (Lewis & Norlund, 1984). Despite these constraints, a diet was employed to rear the univoltine cerambycid beetle, *Plagiohammus spinipennis* (Thompson), a candidate used against the weed *Lantana camara* Linnaeus in Australia (Harley & Wilson, 1968). Likewise, a diet also was used in quarantine to aid the root-infesting lepidopterous larvae, *Agapeta zoegana* Linnaeus and *Pterolonche inspersa* Stgr. incompleting their development once they had been removed from their knapweed host roots (S. S. Rosenthal, retired, USDA, Bozeman, MT, personal communication).

Diets may have their greatest value in increasing the numbers of organisms for release or for the handling of subsequent shipments to quarantine facilities in other areas. Singh (1972) and Singh and Moore (1985) listed a number of diets that have been developed for the culture and mass production of arthropods (see Chapter 22 for details). A continuing publication, "The Frass Newsletter" (Pershing, 1988 and 1989), contains much information on rearing techniques, problems, and diets of mainly phytophagous species.

Dilute honey or sugar water is commonly used to maintain stocks of lepidopteran, dipteran, and hymenopteran adults in the laboratory and during shipment for release.

Pathogens of Arthropods

Incoming entomogenous microorganisms are best handled by microbiologists following rigid, self-imposed quarantine procedures similar to those encountered in a bacteriology laboratory. Unfortunately, arthropod-oriented quarantine facilities usually lack the equipment and expertise required for the extensive handling, identification, and culture of imported pathogens. Ideally, quarantine handlers should be able to recognize the general symptoms caused by the various classes of pathogens (virus, bacteria, fungi, protozoa, etc.). To this end, Steinhaus (1963) or Fuxa and Tanada (1987) should be consulted. When expertise is lacking, the containers with the incoming cultures can be inspected for security (not opened), repackaged, and forwarded to a laboratory certified to handle such materials. Another option would be for the explorer/shipper, by prior arrangement through the APHIS permitting process, to bypass the biological control quarantine and to ship the diseased arthropod material directly to a certified pathogen laboratory (see Table 1). However, if neither option is available and the receipt of pathogens is anticipated, the names and addresses of persons who can help examine the material (e.g., Diagnostic Services, 992 Santa Barbara Road, Berkeley, CA 94707), taxonomic specialists, and resource personnel as well as procedural instructions should be available.

If, while searching for entomophagous insects or mites, an explorer chances on an active epizootic in a population of the target pest, care should be taken to send moribund as well as healthy-appearing material to the receiving quarantine laboratory or pathologist. By the time obviously diseased material collected in the field reaches an insect pathologist it probably will be difficult, if not impossible, to distinguish primary from invasive secondary organisms.

To be better prepared to cope with such an eventuality, collectors should consult with an insect pathologist concerning recommended techniques and procedures prior to departure.

Initial laboratory screening of pathogens is directed mainly toward accurate identification of isolates of the imported material. This procedure typically involves inoculating several recipes of nutrient media with the pathogen and/or inoculating them onto host arthropods from existing laboratory cultures to determine their host range. Goodwin (1984), Shapiro (1984), Granados and Federici (1986), Poinar and Thomas (1978), and Chapter 18 present broad treatment of the subject.

Containment of microorganisms to prevent contamination of insect colonies is of prime importance. Initial handling of microorganisms and culture procedures typically are conducted in a negative air pressure chamber or similar device to minimize the risk of unintentional spread.

Testing and Screening of Biological Control Agents

Often concurrent with the need to maintain or increase cultures of the candidate biological control agents will be a series of studies and tests to confirm their primary habit and host range.

Phytophagous Bioagents

Frick (1974) details the steps traditionally followed in studying and clearing weed biological control agents for release. Of key importance are the host-specificity tests delimiting the host range of each candidate (Zwölfer & Harris, 1971; Wapshere, 1974), plus biological and behavioral studies. Preferably, this research is carried out in the candidate's country of origin, but in almost all cases, some portion of the work must be done in a U.S. domestic quarantine facility. In this case, the candidates to be tested are field collected in their native area and are forwarded to the quarantine laboratory. Prior to the tests each specimen is identified, its sex is determined, and it is checked for the presence of parasites or entomopathogens that could affect host selection and development.

The cages used for these tests will depend on the stage of the biological agent and may range from vials or small cartons for leaf-feeding larvae or adults, up to whole plant cages for studying flower and seed parasites. In the case of univoltine insects, the plants and cages may have to remain in quarantine up to a year to allow for completion of the developing stages. If the results are to be statistically significant, 20 to 30 replications may be needed for each test plant (though this may prove cumbersome and space-monopolizing). When completed, the test information compiled must be reviewed and approved by both federal and state regulatory personnel prior to release of the organism from quarantine.

Entomophagous Bioagents

During initial handling of entomophages, especially those whose identity or host association is unclear and which also may be few in number, the main recourse is to offer them free choice of candidate host species in several stages of development. Direct observation then will reveal which hosts and which of their stages are, or are not, acceptable. Results of such efforts may vary depending on the propensity of the female(s) to oviposit amid the constraints imposed during the test period such as space, host substrate, light, and temperature.

Following initial observations, a larger supply of the observed acceptable host is made available to the candidate parasitoid or predator. Excessive host mutilation and host feeding without oviposition should be controlled by removal of females manually or by CO_2 anesthetization after a short oviposition period that may vary from minutes to days for different taxa. The obvious purpose of this early, time-consuming procedure is to obtain a colony of size suitable for further study in quarantine or for release from quarantine as soon as possible.

Predaceous mites can be propagated following the techniques developed by McMurtry and Scriven (1975) utilizing an artificial habitat consisting of a thin metal tile surrounded by a water barrier. Food consists of mass-produced eggs of *Tetranychus pacificus* McGregor (Scriven & McMurtry, 1971). Pollen has since been added to the diet.

The reader is referred to the procedures and techniques for the handling of entomophagous insects as elaborated by Fisher (1964) and Finney and Fisher (1964). They note various methods of culturing incoming entomophagous species that may salvage otherwise lost shipments if very low numbers are received. This is one area where the worker's knowledge, training, ingenuity, and patience will be challenged.

Disposal

When an experiment or task is completed, the area must be cleared of all debris and dead specimens. All disposable trash that may contain live arthropods, including sweeping and general litter, should be placed in paper bags and stapled securely before transfer to the disposal treatment area. Unwanted biological control agents can be collected in vials or in other small containers and either killed with cyanide or placed with the other rearing debris for heat treatment. Plastic bags may be used but they will disintegrate during the heat treatment and make subsequent handling of the trash difficult.

When using dry heat, as in the case of a small oven, certain hard plastic containers and material will be undamaged at 250°F (2-h exposure) and can be salvaged for reuse.

Whenever handling soil or other compactable materials, the container should have holes at the bottom to facilitate steam and heat penetration throughout.

Residues from all pathogen studies must be autoclaved (121°C at 15 lb pressure for 25 min). Once material for disposal has been treated, it should be carefully bagged and immediately removed from the quarantine area to avoid recontamination. Removal is simplified with the pass-through type of autoclave with one door in quarantine and the other leading to the outside.

In the Nematode Isolation and Quarantine Facility at UCR, a combination of heat and chemical treatment is used to sterilize water, soil, and plant residues on completion of testing procedures (see also Melching et al., 1983).

Voucher Specimens

An important step in the quarantine procedure is to see that the release of a new biological control agent is properly documented and that voucher specimens of natural enemies and hosts, along with research notes, are preserved. Voucher specimens aid workers in tracking the spread and success (or failure) of new agents and are of critical importance if reintroductions of the presumed same species occur at later dates. Knutson (1984) summed up the importance of retaining voucher materials, noting that they serve both to document the identity of the organisms released thus permitting a "retrogressive tracking of changes in names" if later needed, and to provide specimens and information for future studies that may not have been envisioned at the time of release. For example, voucher specimens helped unravel a 20-year-old taxonomy problem surrounding two trypetid seed head flies known to attack the weed *Centaurea solstitialis* Linnaeus in areas of the Mediterranean basin. An early importation of the fly, then identified as *Urophora sirunaseva* (Hering) (Zwölfer, 1969), failed to establish on this weed in California apparently due to an antibiosis reaction on the part of the California plant (D. M. Maddox, retired, USDA/ARS, Albany, CA, personal communication). Fifteen years later, a similar fly was observed attacking *C. solstitialis* plants of California origin in an experimental garden in Greece (Sobhian & Zwölfer, 1985). This latter fly was identified as the true *U. sirunaseva*

(White & Clement, 1987), while the earlier introduced Italian fly was confirmed to be *U. jacaluta* Rondani, as had been speculated by Steyskal (1979).

Voucher material should be prepared in the manner described by Steyskal et al. (1986) (also see Chapter 5) and should be saved from each source locality. In the case of weed-feeding arthropods, samples of the plants used for host-specificity tests should also be preserved (Klingman & Coulson, 1982). Similarly, host materials for entomophagous arthropods should also be preserved. When possible, the principal voucher material should be restricted to the P or F_1 generation of natural enemies.

Knutson (1984) proposed the establishment of the "U.S. National Voucher Collection for Introduced Beneficial Arthropods" in the USDA, Agricultural Research Service, Beneficial Insect Introduction Laboratory, Beltsville, MD 20705. "The collection will provide permanent reference material to substantiate the identity of the **first release** of any imported beneficial arthropod species in the United States." Until this important resource can be funded, quarantine workers are obliged to retain and curate their voucher material at their respective laboratories. Ideally 45 to 50 specimens or even more should be preserved, because a portion of the material should rest with the appropriate taxonomist at the USDA Systematic Entomology Laboratory, and other portions should rest with the voucher collection and the Canadian National Collection. The preservation of this many specimens may not always be practical, especially in the case of first release from limited stock. Similar central repositories may prove valuable in other countries. A minimal effort would require that voucher specimens of field-released material should be forwarded to a national museum.

Unpublished notes and observations also should be preserved, although as Knutson (1984) comments, "they may prove awkward to handle and guidelines are needed." Notes on behavior, host–range studies, environmental preferences, sources of material, and condition of material at the time of release are all extremely important in a postrelease assessment of project success or failure. Much of this information will not find its way into print, but should be kept at the quarantine laboratory and eventually consigned to a central repository, where it would be accessible to future workers. Since 1982, the USDA ARS Biological Control Documentation Center, Beltsville, MD, has had such a program in place (Coulson, 1988; Coulson et al., 1988). The importance of utilizing information systems in biological control is elaborated by Knutson et al. (1987).

Records and Reports

Shipping and Release

Each candidate biological control agent is tracked from the time of receipt until its final clearance and release into the environment. These records are of interest not only to biological control workers and taxonomists but also to USDA/APHIS-PPQ, which must monitor the importation and final disposition of biotic agents and associated plant pests. In the United States, originally each quarantine station developed its own system of record keeping. However, as the number of quarantine laboratories and the number of importations increased, the necessity of standardizing the shipping and receiving records became apparent and forms outlining the basic information to be recorded for each organism actually released in the field were developed by J. R. Coulson, USDA/ARS. Quarantine laboratories are making an effort to integrate these new forms into their work. Sets of these forms can be obtained from the Biological Control Documentation Center, Beneficial Insects Laboratory, USDA/ARS, Beltsville, MD 20705. The new forms include

Form AD-941: Biological Shipment Record—Foreign/ Overseas Source. This form covers the movement of exotic living biological control agents—including pollinators, beneficial competitors, antagonists, and so on, from all locations outside the contiguous United States, as well as from nonquarantine and quarantine sources in the United States—that are destined for study in a domestic quarantine facility.

Form AD-942: Biological Shipment Record—Quarantine Facility. This form records the movement of live beneficial organisms from quarantine to researchers working outside of the containment unit, whether at the same location as the quarantine facility or at other locations throughout the United States.

Form AD-943: Biological Shipment Record—Non Quarantine. This form covers the movement (consignment) of cultures from one nonquarantine laboratory or source to another as well as the collection and redistribution from field sites. Therefore, it usually is not used by primary quarantine receiving labs.

Instructions for completing and filing are contained on each form. The use of these or similar forms by quarantine laboratories in other countries could greatly enhance international exchange of information.

Quarantine Activity Reports

It has been customary for quarantine facilities to periodically document (usually every 6 or 12 months) a summary of material received and/or released. Although federal and state regulatory agencies may require such summaries, this has not been a general operational requirement and no standardized format has been followed for this reporting. The initial report on any biological control organism should detail whether it is a primary or a secondary parasite and

should assess the likelihood that it may attack other nontarget species. Klingman and Coulson (1982) enumerate the points that should be covered in reporting the biological and host-specificity information required to substantiate the importation and release of weed-controlling biological control agents. Published reports as well as unpublished reports should be kept on file and cataloged at the quarantine facility. The first attempt at a U.S. national documentation system, Release of Beneficial Organisms in the United States and Territories (ROBO), has been developed by the ARS Documentation Center (Coulson et al., 1988).

Written test reports and documentation of quarantine studies and observations can be time consuming and costly, but are needed to document decisions that may be taken with a natural enemy, especially if it is to be released into the environment.

Quarantine Procedural Problems

From time to time problems will arise in the importation and quarantine handling of beneficial organisms and the regulation of their movement. Some of the problems described earlier were broadly discussed and summarized at a quarantine technology workshop (Ridings and Fisher, 1983). Although the workshop participants recommended a number of specific actions, only those in the communication/information category have been expedited. Action was also recommended on the following problems:

1. Regulatory: APHIS should be legally authorized to regulate the importation of beneficial organisms, as well as pest species. This would entail the issuance of special labels to identify shipments containing biocontrol organisms and the training of APHIS inspectors on how to process rapidly these materials for consignment to a quarantine laboratory. Sufficient flexibility should be introduced into the wording and conditions stated on permits to allow the discretion needed by collectors, inspectors, and quarantine handlers to facilitate the movement of the natural enemies. The federal and state regulations governing the movement of natural enemies should be standardized.

2. Handling: The host–range screening of parasites and predators, especially among weed feeders and weed pathogens, should be preferably carried out in their country of origin. (This recommendation may be ideal from the safety standpoint, but may be impractical in areas where research facilities are lacking, or political or social unrest prevails.)

3. Quarantine facilities: Locate new facilities in areas where the premature escape of biological control agents would pose little threat (e.g., urbanized areas with few crops).

4. Taxonomy: Increase support for this vitally needed service.

5. Training: Create workshops to train quarantine handlers and collectors and develop improved, safe, and rapid shipping techniques. Such USDA quarantine workshops occurred in 1991.

PERSONNEL

To maximize security and assure that each organism is properly handled and tested, each containment facility should have a designated quarantine officer or supervisor who is responsible for its operation. An assistant should also be designated who can take charge in the absence of the supervisor. Only if the containment facility is large and has a variety of ongoing programs should persons other than the quarantine officer have free access to work in the security area. In this case, it is the quarantine officer's responsibility to see that these workers are familiar with and follow quarantine operating procedures, as stated in the policy and procedures document mentioned earlier, and meet work neatness standards. The quarantine officer and all persons who work regularly in the security area should be certified as to their familiarity with federal, state, and local regulations pertaining to importation and release of biological control agents, important quarantine techniques, and procedures. Periodic workshops (perhaps every 4 years) should be scheduled to familiarize "old and new" workers, as well as collectors, with changes in regulations and methodologies. Quarantine personnel should have knowledge and skill in the following five areas.

Biological Control and Biological Relationships of Natural Enemies

Workers should be familiar with the concepts, theory, and practice of classical biological control and how it pertains to the specific projects and organisms with which they are involved. The quarantine officer should have a broad knowledge of all projects conducted in the containment area.

Familiarity with the taxonomy and life cycles of targeted pests and possible candidate natural enemies is very important. Information of this type is valuable in planning host-specificity studies and in verifying the primary relationship between the parasite and its host. When information on specific candidate biological control agents is lacking, reported studies on related species or genera may be of help. Clausen (1940) remains one of the best sources of general information on the biologies of entomophagous biological control agents. The project leader, when planning a collecting trip, should provide quarantine personnel with appropriate background information on specific natural enemies.

Computer Skills

The increasing number of certified containment facilities, foreign explorers, shippers, and natural enemy importations and releases has highlighted the importance of improved record keeping on the part of quarantine workers. Workers with computer skills can quickly summarize collectors' reports and shipment records, revise the numbers of natural enemies released, match up foreign sources with domestic release sites, etc. Computer skills will also greatly facilitate exchanging information (i.e., via FAX and e-mail) with central documentation facilities such as the USDA/ARS Biological Control Documentation Center, Beltsville, MD.

Communication Skills

Maintaining contact with the network of explorers, foreign scientists, international and domestic shippers, customs and quarantine officers, agricultural officials (federal, state, and county) and domestic cooperators is an important task often shared by the quarantine officer and project investigators. Regulatory officials must be kept abreast of proposed importations and releases, and procedures should be developed for handling last minute shipment problems (alternate ports of entry, weekend and holiday arrivals, improperly or incompletely marked packages). Personal contact with local quarantine officials and inspectors is often helpful if not essential to speed the resolution of such problems. Correspondence by FAX, when available, greatly facilitates problem resolution.

Physical Plant

The quarantine supervisor and workers should be familiar with the physical operation of the containment facility and its equipment, especially air control, utilities, and function of the autoclave. The names and phone numbers of maintenance and repair personnel to be contacted to prevent breakdowns and hasten repair should be readily accessible for normal workdays as well as holidays or weekends.

Sanitation, Visitors, and Emergencies

Janitorial needs within the quarantine facility should be handled by quarantine staff. Facility and equipment service personnel must be appraised of the importance of maintaining quarantine security and should be accompanied by quarantine personnel when working in the containment area. Visiting scientists and regulatory officials should also be accompanied by trained personnel. Casual visitors, public groups, etc., are not permitted in the quarantine laboratory. Fire, earthquakes, vandalism, and illness may interrupt quarantine operations at any time. Emergency personnel should be advised by posted instructions on entry and exit procedures least likely to violate quarantine security. The off-duty (home) phone numbers of the quarantine officer and deputy personnel should be filed with local security personnel and prominently posted at the entrance to the quarantine facility. During periods when problems may be expected to surface, as with shipments on weekends or holidays, it would be appropriate for the supervisor or an assistant to carry a pager keyed to the quarantine phone.

References

Anonymous. (1985). Recommended measures for regulating the importation and movement of plants. Information Letter, FAO, Asia and Pacific Plant Protection Commission (1985, No. 132).

Anonymous. (1989). Directory of producers of natural enemies of common pests. IPM Practitioner, 11(4), 15–18 (BIRC, P.O. Box 7414, Berkeley, CA 94707).

Bailey, J. C., & Kreasky, J. B. (1978). Quarantine laboratory for plant-feeding insects. In N. C. Leppla & T. R. Ashley (Eds.), Facilities for insect research and production. (USDA, SEA. Tech. Bulletin 1576, pp. 53–56). Washington, DC: U.S. Department of Agriculture.

Bartlett, B. R., & van den Bosch, R. (1964). Foreign exploration for beneficial organisms, In P. De Bach (Ed.), Biological control of insect pests and weeds (pp. 283–304). London: Chapman & Hall.

Bezark, L. G. (1989). Suppliers of beneficial organisms in North America. (State of California Department of Food and Agriculture Biological Control Service Progress, BC 89-1).

Boldt, P. E. (1982). Quarantine facility for exotic phytophagous insects. FAO Plant Protection Bulletin, 30(2), 73–77.

Boldt, P. E., & Drea, J. J. (1980). Packaging and shipping beneficial insects for biological control. FAO Plant Protection Bulletin, 28(2), 64–71.

Borror, D. J., Triplehorn, C. A., & Johnson, N. F. (1989). An introduction to the study of insects (6th ed.). New York: Holt, Rinehart & Winston.

Briese, D. T., & Milner, R. J. (1986). Effect of the microsporidian Pleistophora schubergi in Anaitis efformata (Lepidoptera: Geometridae) and its elimination from a laboratory colony. Journal of Invertebrate Pathology, 48(1), 107–116.

Bruckart, W. L., & Dowler, W. M. (1986). Evaluation of exotic rust fungi in the United States for classical biological control of weeds. Weed Science, 34 (Suppl. 1), 11–14.

Charudattan, R. (1982). Regulation of microbial weed control agents. In R. Charudattan & H. L. Walker (Eds.), Biological control of weeds with plant pathogens (pp. 175–188). New York: John Wiley & Sons.

Churchill, B. W. (1982). Mass production of microorganisms for biological control. In R. Charudattan & H. L. Walker (Eds.), Biological control of weeds with plant pathogens (pp. 139–156). New York: John Wiley & Sons.

Clausen, C. P. (1940). Entomophagous insects. New York: McGraw-Hill.

Clausen, C. P. (Ed.). (1978). Introduced parasites and predators of arthropod pests and weeds: A world review Agriculture Handbook 480. Washington, DC: U.S. Department of Agriculture.

Coulson, J. R. (1987). Biological control data bases. In L. Knutson, F. C. Thompson, & R. W. Carlson (Eds.), Biosystematic and biological control information systems in entomology (pp. 378–388). Agricultural and Zoological Reviews, 2, 361–412.

Coulson, J. R. (1988). Biological control documentation center, Agricultural Research Service. In E. G. King, J. R. Coulson, & R. J. Coleman (Eds.), ARS National Biological Control Program, Proceedings of

Workshop on Research Priorities (pp. 23–26). U.S. Department of Agriculture, ARS, Washington, DC: U.S. Government Printing Office.

Coulson, J. R., & J. H. Hagan. (1986). (Biological control information document no. 00061). ARS Biological Control Document Center, Beltsville, MD: U.S. Department of Agriculture, ARS, BARC-East.

Coulson, J. R., & Soper, R. S. (1989). Protocols for the introduction of biological agents in the United States. In R. P. Kahn (Ed.), Plant protection and quarantine: Vol. 3. Special topics (pp. 1–35). Boca Raton, FL: CRC Press.

Coulson, J. R., Carrell, A., & Vincent, D. L. (1988). Releases of beneficial organisms in the United States and Territories—1981. (Miscellaneous Publication 1464). Washington, DC: U.S. Department of Agriculture.

Coulson, J. R., Soper, R. S., & Williams, D. W. (1991). Biological control quarantine: Needs and procedures. Workshop Proceedings (USDA-ARS, ARS-99). Washington, DC: U.S. Department of Agriculture.

DeBach, P. (Ed.). (1964). Biological control of insects pests and weeds. London: Chapman & Hall.

Denmark, H. A. (1978). Quarantine and biological-control laboratory. In N. C. Leppla & T. R. Ashley (Eds.), Facilities for insect research and production (USDA, SEA, Tech. Bulletin 1576, pp. 47–49). Washington, DC: U.S. Department of Agriculture.

Ebeling, W. (1978). Urban entomology. Berkeley: Division of Agricultural Science, University of California.

Edwards, S. R., Davis, G. M., & Nevling, L. I., Jr. (Eds.). (1985). The systematics community. Lawrence, KS: Association of Systems Collections, Museum of National History, University of Kansas.

Ertle, L. R. (1993). What quarantine does and what the collector needs to know. In R. G. Van Driesche & T. S. Bellows, Jr. (Eds.), Steps in classical arthropod biological control (pp. 53–65). Lanham, MD: Thomas Say Publications in Entomology: Proceedings. Entomological Society of American Publications.

Ertle, L. R., & Day, W. H. (1978). USDA quarantine facility, Newark, Delaware. In N. C. Leppla & T. R. Ashley (Eds.), Facilities for insect research and production (USDA, SEA, Tech. Bulletin 1576, pp. 49–52). Washington, DC: U.S. Department of Agriculture.

Etzel, L. K. (1978). University of California quarantine, Albany. In N. C. Leppla & T. R. Ashley (Eds.), Facilities for insect research and production (USDA SEA, Tech. Bulletin 1576, pp. 45–46). Washington, DC: U.S. Department of Agriculture.

Etzel, L. K., Levinson, S. O., & Andres, L. A. (1981). Elimination of Nosema in Galeruca rufa, a potential biological control agent for field bindweed. Environmental Entomology, 10(2), 143–146.

Finney, G. L., & Fisher, T. W. (1964). Culture of entomophagous insects and their hosts. In P. DeBach (Ed.), Biological control of insect pests and weeds, (pp. 328–355). London: Chapman & Hall. (Reprinted by Reinhold.)

Fisher, T. W. (1964). Quarantine handling of entomophagous insects. In P. DeBach (Ed.), Biological control of insect pests and weeds (pp. 305–326). London: Chapman & Hall.

Fisher, T. W. (1978). University of California quarantine facility, Riverside. In N. C. Leppla & T. R. Ashley (Eds.), Facilities for insect research and production (USDA, SEA, Tech. Bulletin 1576, pp. 56–60). Washington, DC: U.S. Department of Agriculture.

Frick, K. E. (1974). Biological control of weeds: Introduction, history, theoretical and practical applications. In F. G. Maxwell & F. A. Harris (Eds.), Proceedings of Summer Institute on Biological Control of Plants, Insects and Diseases (pp. 204–223). Jackson: University of Mississippi Press.

Fuxa, J. R., & Tanada, Y. (Eds.). (1987). Epizootiology of insect diseases. New York: John Wiley & Sons.

Goodwin, R. H. (1984). Recognition and diagnosis of diseases in insectaries and the effects of disease agents on insect biology. In E. G. King & N. C. Leppla (Eds.), Advances and challenges in insect rearing (pp. 96–130). USDA ARS Proceedings Conference, Atlanta, GA, March 1980.

Granados, R., & Federici, B. A. (Eds.). (1986). The biology of baculoviruses (Vols. 1, and Vol. 2). Boca Raton: CRC Press.

Harley, K. L. S., & Wilson, B. W. (1968). Propagation of cerambycid borer on a meridic diet. Canadian Journal of Zoology, 46(6), 1265–1266.

Hertelendy, L., & Pinter, A. (1986). A. Microwave method to control pests of stored products. Novenyvedelem, 1985, 21(5), 215. [From A. Szenessy et al. (Eds.). Abstracts and bibliography of Hungarian plant protection, 11, 38–39.]

Hertelendy, L., & Pinter, A. (1986). B. Application of microwave method to control pests of stored products. Novenyvedelem, 1985, 21(9), 423–425. [From A. Szenessy et al. (eds.), Abstracts and bibliography of Hungarian plant protection, 11, 39.]

Inman, R. E. (1970). A temporary quarantine laboratory for research with exotic plant pathogens. Plant Disease Reporter, 54(1), 3–7.

Jones, W. A., Powell, J. E., & King, E. G., Jr. (1985). Stoneville research quarantine facility: A national center for support of research on biological control of arthropod and weed pests. Bulletin of the Entomological Society of America, 31(2), 20–26.

Kahn, R. P. (1989). Plant quarantine: Principles, concepts, and problems (Vols. 1–3). Boca Raton, FL: CRC Press.

King, E. G., & Leppla, N. C. (Eds.). (1984). Advances and challenges in insect rearing. USDA ARS (southern region). Proceedings Conference in Atlanta, GA, March 4–6, 1980.

Klingman, D. L., & Coulson, J. R. (1982). Guidelines for introducing foreign organisms into the United States for biological control of weeds. Weed Science, 30, 661–667. (Also see Plant Disease, 66, 1205–1209, 1982; or Bulletin of the Entomological Society of America, 29, 55–61, 1983.)

Knutson, L. (1981). Symbiosis of biosystematics and biological control. In G. Papavizas (Ed.), Biological control in crop production. BARC Symposium No. 5 (pp. 61–78). Ottowa: Allanheld, Osmun.

Knutson, L. (1984). Voucher material in entomology: A status report. Entomological Society of America Bulletin, 30(4), 8–11.

Knutson, L. (1990). Alpha taxonomy: Seguy's metier and a modern need. Annals of the Entomological Society of France (N.S.), 26, 325–334.

Knutson, L., Thompson, F. C., & Carlson, R. W. (1987). Biosystematic and biological control information systems in entomology. Agricultural and Zoological Reviews, 2, 361–412.

Leppla, N. C., & Ashley, T. R. (Eds.). (1978). Facilities for insect research and production (USDA, SEA, Tech. Bulletin 1576). Washington, DC: U.S. Department of Agriculture.

Lewis, W. J., & Norlund D. A. (1984). Semiochemicals influencing fall armyworm parasitoid behavior: Implications for behavioral manipulation. Florida Entomologist, 67, 343–349.

Lima, P. J. (1983). Safeguard guidelines for containment of plant pests under permit (USDA, APHIS, PPQ: APHIS 81-61). (Revision of October 29, 1979 memo.)

McMurtry, J. A., & Scriven, G. T. (1975). Population increase of Phytoseiulus persimilis on different insectary feeding programs. Journal of Economic Entomology, 68(3), 319–320.

Melching, J. S., Bromfield, K. R., & Kingsolver, C. H. (1983). The plant pathogen containment facility at Frederick, MD. Plant Disease, 67(7), 717–722.

Milner, R. J., & Briese, D. T. (1986). Identification of the microsporidian Pleistophora schubergi infecting Anaitis efformata (Lepidoptera: Geometridae). Journal of Invertebrate Pathology, 48(1), 100–106.

Myers, J. H., & Sabath, M. D. (1981). Genetic and phenotypic variability, genetic variance, and the success of establishment of insect introductions for the biological control of weeds (pp. 91–102). In E. S. Del Fosse (Ed.), Proceedings of the Fifth International Symposium on Biological Control of Weeds, Brisbane, Australia, July, 22–29, 1980.

Nickle, W. R., Drea, J. J., & Coulson, J. R. (1988). Guidelines for introducing beneficial insect-parasitic nematodes into the United States. Annals of Applied Nematology, 2, 50–56.

Nicolas, G., & Sillens, D. (1989). Immediate and latent effects of carbon dioxide on insects. In T. E. Mittler, F. J. Radovsky, & V. H. Resh (Eds.), Annual Review in Entomology, 34, 97–116. (Annual Reviews, Inc., Palo Alto, CA.)

Norlund, A., & Lewis, W. J. (1976). Terminology of chemical releasing stimuli in interspecific and intraspecific interactions. Journal of Chemical Ecology, 2, 211–220.

Ooi, A. C. P. (1986). Quarantine and biological control. Biocontrol News Information, 7(4), 227–231 (Dec. Corp Protect. Branch, Department of Agriculture, Kuala Lumpur, Malaysia).

Pershing, J. C. (Ed.). (1988, 1989). Frass Newsletter, Vols. 11 and 12. Insect Rearing Group. c/o R. Wheeler, Chevron Chemical Corp., P.O. Box 4010, Richmond, CA 94804 (since 1975, various editors).

Poinar, G. O., Jr. (1977). CIH key to the groups and genera of nematode parasites of invertebrates. Slough, UK: Commonwealth Agricultural Bureau.

Poinar, G. O., Jr., & Thomas, G. M. (1978). Diagnostic manual for the identification of insect pathogens. New York: Plenum Press.

Ridings, W. H., & Fisher, T. W. (1983 February). Quarantine technology (pp. 51–52). In S. L. Battlefield (Ed.), USDA Proceedings of National Interdisciplinary Biological Control Conference, Las Vegas, NV.

Room, P. M., Forno, I. W., & Taylor, M. F. J. (1984). Establishment in Australia of two insects for biological control of the floating weed Salvinia molesta. Bulletin of Entomological Research, 74, 505–516.

Room, P. M., Harley, K. L. S., Forno, I. W., & Sands, D. P. A. (1981). Successful biological control of the floating weed Salvinia. Nature (London), 294, 78–80.

Scriven, G. T., & McMurtry, J. A. (1971). Quantitative production and processing of tetranychid mites for large-scale testing or predator production. Journal of Economic Entomology, 64(5), 1255–1257.

Shapiro, M. (1984). Micro-organisms as contaminants and pathogens in insect rearing. In E. G. King & N. C. Leppla (Eds.), Advances and challenges in insect rearing (pp. 130–143). USDA ARS Proceedings Conference, Atlanta, GA, March 1980.

Singh, P. (1972). Bibliography of artificial diets for insects and mites (Bull. 209). Wellington, New Zealand: New Zealand Department of Science and Industrial Research.

Singh, P., & Moore, R. F. (Eds.). (1985). Handbook of insects rearing (Vol. 2). Amsterdam: Elsevier.

Sobhian, R., & Andres, L. A. (1978). The response of the skeleton weed gall midge, Cystiphora schmidti (Diptera: Cecedomyiidae), and the gall mite, Aceria chondrillae (Eriophyidae) to North American strains of rush skeletonweed (Chondrilla juncea). Environmental Entomology, 7, 506–508.

Sobhian, R., & Zwölfer, H. (1985). Phytophagous insect species associated with flower heads of yellow starthistle (Centaurea solstitialis L.) in California. Angewardte Entomologia, 99, 301–321.

Stehr, F. W. (Ed.). (1987). Immature insects. Dubuque, IA: Kendall/Hunt.

Steinhaus, E. A. (1963). Insect pathology. An advanced treatise (Vols. 1 and 2). New York: Academic Press.

Steyskal, G. C. (1979). Taxonomic studies on fruitflies of the genus Urophora (Diptera: tephritidae). Miscellaneous Publications of the Entomological Society of Washington, 1–61.

Steyskal, G. C., Murphy, W. L. & Hoover, E. M. (Eds.). (1986). Insects and mites: Techniques for collection and preservation (Miscellaneous Publication No. 1443). Washington, DC: U.S. Department of Agriculture.

Van Driesche, R. G. & Bellows, T. S. (1996). Biological control. New York: Chapman & Hall. 539 pp.

Van Driesche, R. G., & T. S. Bellows Jr. (Eds.) (1993). Steps in classical arthropod biological control. Thomas Say Publications in Entomology, Lanham, MD: Entomological Society of America. 88 pp.

Waage, J., & Greathead, D. (Eds.). (1986). Insect parasitoids. New York: Academic Press.

Wapshere, A. J. (1974). Host specificity of phytophagous organisms and the evolutionary centers of plant genera and subgenera. Entomophaga, 19, 301–309.

Watson, A. K., & Sackston, W. E. (1985). Plant pathogen containment (quarantine) facility at Macdonald College. Canadian Journal of Plant Pathology, 7(2), 177–180.

White, I. M., & Clement, S. L. (1987). Systematic notes on *Urophora* (Diptera, Tephritidae) species associated with *Centaurea solstitialis* (Asteraceae, Cardueae) and other Palaearctic weeds adventive in North America. Proceedings of the Entomological Society of Washington, 89(3), 571–580.

Zwölfer, H. (1969). *Urophora siruna-seva* (HG.) (Dipt.: Trypetidae), a potential insect for the biological control of *Centaurea solstitialis* L. in California Technical Bulletin Commonwealth Institute of Biology Control, 11, 105–154.

Zwölfer, H., & Harris, P. (1971). Host specificity determination of insects for biological control of weeds. Annual Review of Entomology, 16, 159–178.

7

Culture and Colonization

L. K. ETZEL

Center for Biological Control
University of California
Berkeley, CA

E. F. LEGNER

Department of Entomology
University of California
Riverside, CA

CULTURE OF HOSTS FOR ENTOMOPHAGOUS ARTHROPODS

The culture and colonization of natural enemies are fundamental to biological control work. The three principle reasons for culturing parasitoids, predators, and pathogens are to provide field-release organisms for (1) permanent establishment, (2) periodic colonization, or (3) inundative releases.

For permanent establishment, insectaries typically propagate only relatively small numbers of a beneficial organism for release at several dispersed sites. Successful organisms will persist in the new environment, spread, and reduce the pest organism to a level that is below the economic injury threshold in what practitioners call "classical" biological control. Control of the pest then requires no further releases.

With periodic colonization and augmentation a beneficial organism is able to perform well when the pest is seasonally present in damaging numbers, even though it is unable to persist in sizable numbers year round. Sailer (1976) gave the example of 3000 *Pediobius foveolatus* Crawford, an eulophid parasitoid of the Mexican bean beetle, being released in midspring. The parasitoid spread 595 km by the end of October, and nearly eliminated the host populations at locations in north-central Florida. However, the parasitoid could not overwinter in the area and had to be recolonized annually, in a procedure called "inoculative periodic colonization." Similarly, control of aquatic weeds and mosquito habitats in irrigation canals in southeastern California usually requires periodic releases of the tropical fish *Tilapia zillii* (Gervais) (Legner & Murray,

1981). This fish cannot always overwinter in canals when water temperatures drop below 10°C (Legner, 1986), or when competition with predatory large-mouth bass devastates its population.

Inundative releases can sometimes effect short-term control of a pest. Such large liberations simulate a pesticide, and the agent simply reduces, instead of regulates, the pest population. Examples are releases of *Lixophaga diatraeae* (Townsend), a tachinid parasitoid on the sugarcane borer, *Diatraea saccharalis* (Fabricius) (King *et al.*, 1981); of the common green lacewing, *Chrysoperla carnea* (Stephens), which is predaceous on soft-bodied insects; of *Spalangia* spp. and *Muscidifurax* spp. (Peterson & Currey, 1996), pteromalid parasitoids on muscoid flies; and of hydra against mosquitoes (Yu *et al.*, 1974). However, the parasitoids most commonly released inundatively on a worldwide basis are egg parasitoids in the genus *Trichogramma* (Smith, 1996). Microbial insecticides and herbicides similarly belong in this category of inundatively released biological control agents. The bacterium *Bacillus thuringiensis* is sold commercially as an insecticide, and the fungus *Colletotrichum gloeosporiodes* f. sp. *Aeschynomene* is more than 90% effective against northern jointvetch, *Aeschynomene virginica,* a weed in U.S. rice fields (van den Bosch *et al.*, 1982). For more on inundative releases, see Chapter 11.

Host Food

Foods employed in rearing the hosts of entomophagous organisms are, in decreasing order of difficulty of provision: (1) living plants; (2) harvested plant parts; (3) tubers, fruit, and other produce; and (4) prepared diets.

Living Plants

It is often necessary to buy or grow natural host plants, and to maintain them at a considerable cost of space and labor to raise phytophagous insects. Losses from plant diseases or pest arthropods are not unusual. The required holding time is important and related to host and entomophage life cycles. For example, the life cycle of the black scale, *Saissetia oleae* (Olivier), is about 3 months at 21°C on potted oleanders. Because the scale must be nearly mature to be acceptable to some parasitoids, which themselves may have a life cycle of 3 to 6 weeks, the oleanders have to be alive and healthy for several months after infestation with scale crawlers. Diseases such as oleander knot or root rot, and contaminating pests such as mealybugs often complicate maintenance of oleanders.

Some plants used for insect production need only short durability, so that plant diseases are not usually a limiting factor. For instance, in raising certain parasitoids on the pea aphid, *Acyrthosiphon pisum* (Harris), which in turn grows on fava bean plants, the whole system is discarded quickly because of the rapid growth of the plants, and because of the short life cycles of the host insects and their parasitoids. Aphid feeding and root rot cause plants to collapse within 2 weeks, but parasitoid production is unaffected.

Any practice to alleviate a production problem first needs thorough testing for indirect effects. As an example, certain plant fungal diseases are treated with the fungicide Benlate®. Because Benlate® has a slight systemic action, aphids feeding on treated plants may consume sufficient quantities to kill their beneficial internal symbiotic microorganisms, the loss of which can cause their death. However, it is noteworthy that Benlate® can suppress certain protozoans that infect insectary-reared insects.

Another factor to consider when using living plants to rear hosts of entomophages is that plant quality can affect entomophage production by affecting production of the host insect. Forbes *et al.* (1985) indicated that use of young, vigorously growing plants for rearing aphids enabled rapid growth and reproduction. They noted that rates of development, body size, and fecundity can often be very different in reared, versus wild, aphids, and that these differences are partly due to variations between host plants in the field and in the laboratory. Furthermore, stressed laboratory plants can stimulate alate production, which may continue for several generations even after plant conditions improve.

Harvested Plant Parts

Food supplied to insects, especially those that are voracious feeders on perennials, often consists of plant parts. Potted perennials requiring lengthy developmental time might be destroyed in a day or two by a pest, such as occurs with alfalfa consumption by the Egyptian alfalfa weevil, *Hypera brunneipennis* (Boheman). Because the weevils consume so much food, it is necessary to feed them daily with cuttings taken from an alfalfa field and made into "bouquets" to retain foliage freshness.

Determining the type and condition of plant parts that are optimal for rearing pest insects often requires considerable experimentation. Willey (1985) found that the range grasshopper, *Arphia conspersa* Scudder, preferred dried dandelion greens to dried Romaine or head lettuce, or to assorted native grasses and alfalfa. However, the grasshopper did not favor fresh dandelion leaves. He noted that frozen storage of unprocessed dried leaves and buds of the dandelions in polyethylene bags for later use was satisfactory.

Tubers, Fruit, and Other Produce

The rearing of certain scale insects on potatoes, citron melons, and squashes is a common practice. Papacek and Smith (1985) reported that butternut pumpkins, *Cucurbita moschata,* provided the best substrate for oleander scale, *Aspidiotus nerii* Bouché. The scale in turn was the host for mass producing *Aphytis lingnanensis* Compere, a parasitoid of the California red scale, *Aonidiella aurantii* (Maskell). The annual production of 15 to 20 million parasitoids used a total of 1.5 to 2 tons of pumpkins per week. Likewise, the oriental scale, *A. orientalis* (Newstead), was successfully reared in the laboratory on butternut gramma for at least 30 generations, again for parasitoid production (Elder & Smith, 1995).

Rutabagas were the food used for growing cabbage maggot, *Delia radicum* (Linnaeus), a host for the predaceous and parasitic beetle *Aleochara bilineata* (Gyllenhal). Whistlecraft *et al.* (1985b) provided at least 1 g of rutabaga per cabbage maggot egg to ensure a uniform pupal size. Etzel (1985), in rearing the potato tuberworm, *Phthorimaea operculella* (Zeller), also found that 1 g of substrate was sufficient for one individual. Processed tuberworms were suitable as food for certain coccinellids and larvae of the common green lacewing.

Wight (1985) noted that insecticide residues can be troublesome with commercial produce. Stripping the outer leaves from heads of lettuce bought to feed the southern armyworm, *Spodoptera eridania* (Cramer), removed such residues.

The variety of produce is also important. The Russet potato is a "mealy" variety superior for tuberworm rearing, whereas White Rose, which has a smooth skin, is best for raising California red scale and oleander scale (Finney & Fisher, 1964).

Other significant problems associated with the use of produce are availability, durability, and consistency. Citron melons are useful for rearing the brown soft scale, *Coccus hesperidum* Linnaeus, but are not commercially available and must be specially grown. Much commercial produce varies greatly in consistency and durability, and sometimes rots rapidly when removed from cold storage. Control of relative humidity (RH) during storage and usage is important for reducing substrate deterioration. Decomposition not only ruins the food source, but also may generate toxic gases. Gases emitted by ripening grapefruit, for example, are lethal to some parasitoid and host species in a confined space (Finney & Fisher, 1964).

Chemical treatments might be useful to reduce deterioration of produce. In mass rearing the citrus mealybug, *Planococcus citri* (Risso), Krishnamoorthy and Singh (1987) treated ripe pumpkins, *Cucurbita moschata,* with 1% Benlate® and 5% formaldehyde solution.

Prepared Diets

Singh (1985) reviewed 22 multiple-species rearing diets that together have been widely useful to rear Lepidoptera and Diptera, in particular. Provided that they are nutritionally and physically adequate, diets provide the easiest and most consistent food source, and eliminate most problems involved with host plants, plant parts, or produce. However, adequate diets are more likely to be available for the least fastidious insects. With respect to mites, Hare and Bethke (1988) successfully developed an artificial feeding system for the citrus red mite, *Panonychus citri* McGregor.

Omnivorous or polyphagous insects are obviously much easier to rear than are monophagous ones. Moore (1985) presented a systematic procedure and guidelines for choosing and modifying an artificial diet for a phytophagous arthropod. He discussed stimulants, repellents, nutrient requirements, and microbial inhibitors, as well as physical and chemical adequacy, concentrations, and proportions. Grisdale (1984) emphasized that consistently good artificial diets require high-quality, fresh, and adequately mixed ingredients. For example, Keena *et al.* (1995) found that the Wesson salt mixture from one of two vendors used was found to cause slow, asynchronous development of gypsy moth larvae, *Lymantria dispar* (Linnaeus). The particular batch of wheat germ, casein, and vitamin mixture utilized, together with interaction of the method of diet production with the diet ingredients, also influenced the rate of larval development.

Moreover, both physical and chemical characteristics are important. Rearing success can sometimes hinge on a critical step or technique in the physical presentation of a diet. For example, Boller (1985) noted that cotton pads must have liquid larval diet only on one side to provide a moisture gradient suitable for optimal development of certain fruit flies, and Bay and Legner (1963) had to feed chloropid eye gnats with blood mixture diets on dried prunes or filter paper.

Weed-feeding arthropods used for biological control work are either monophagous or have a highly restricted host range. Therefore, their culture nearly always requires the use of plants because their high host specificity makes the formulation of prepared diets very difficult. A notable exception was the development of an artificial diet for the tephritid fly *Chaetorellia australis* Hering, which infests flower heads of the yellow star thistle *Centaurea solstitialis* Linnaeus (Pittara & Katsoyannos, 1992). Another exception was a diet adapted to rear mid- to late instar larvae of *Carmenta mimosa* Eichlin and Passoa, a stem-boring sesiid moth imported into Australia as a biological control agent for giant sensitive plant, *Mimosa pigra.* This diet enabled synchronous emergence of adult males and females, provided a useful larval transport medium, and afforded significant savings in handling time and space utilized (Smith & Wilson, 1995).

Provision of food for adults of holometabolous insects is generally not as complicated as is provision for larvae. However, for adult tephritid fruit flies, much basic research was required to understand their dietary requirements before relatively simple diets could be formulated (Tsiropoulos, 1992). Consequently, for example, adults of the Queensland fruit fly, *Bactrocera* (= *Dacus) tryoni* (Froggatt), were fed with sugar cubes, autolyzed brewer's yeast fraction, and water (Heather & Corcoran, 1985). Hydrolysis of the yeast made the protein available for egg production. Tolman *et al.* (1985) fed adult onion maggots, *Delia antiqua* (Meigen), with a dry diet consisting of 50% brewer's yeast, 33% yeast hydrolysate, and 17% soybean flour. Bartlett and Wolf (1985) fed pink bollworm moths, *Pectinophora gossypiella* (Saunders), with 10% sugar water plus 0.2% methyl parasept (to retard microbial growth). Sometimes feeding adults is not necessary. Etzel (1985) held potato tuberworm moths without food or water and still obtained adequate egg production.

Host Production

Many of the considerations applicable to host production also apply to entomophage rearing, but separate treatment simplifies an understanding of the often interacting factors.

Contamination

The most prevalent and often most serious problem in the production of organisms is contamination by unwanted ones, with resulting competition, interference, parasitism, predation, and/or disease. Efforts to control undesired elements require costly labor, supplies, equipment, and facilities. Some examples will indicate the range of contamination difficulties.

Phytophagous insects and mites frequently create problems in the production of host arthropods by competing for the substrate and by interfering with a host–parasitoid system. Mealybugs, mites, and aphids are frequent problems in rearing the black scale, *Saissetia oleae* (L. Etzel, unpublished). Likewise, aphid infestations were troublesome on fava bean plants used to rear larvae of the red-banded leaf roller, *Argyrotaenia velutinana* (Walker) (Glass & Roelofs, 1985).

Mites have caused difficulties in laboratory cultures of *Trogoderma* beetles (Speirs, 1985); *Drosophila* flies (Yoon, 1985); the lesser peach tree borer, *Synanthedon pictipes* (Grote & Robinson) (Reed & Tromley, 1985b); the plum curculio, *Conotrachelus nenuphar* (Herbst) (Amis & Snow, 1985); and the housefly, *Musca domestica* Linnaeus (Morgan, 1985). Phytoseiid mites have interfered with the mass rearing of spider mites (Ho & Chen, 1992).

Papacek and Smith (1985) reported that ants, citrus mealybugs, and the scale-eating coccinellids *Lindorus lophanthae* (Blaisdell) were contaminants of insectary diaspid scale cultures used to rear an aphelinid parasitoid, *Aphytis lingnanensis* Compere. Heather and Corcoran (1985) also had to cope with ants in a culture of the Queensland fruit fly. Horwood (1988) controlled the coastal brown ant, *Pheidole megacephala* (Fabricius), in insectaries with a bait of 0.5% methoprene in peanut butter.

Wight (1985) found that phorid fly maggots were occasional problems in rearing the southern armyworm, and rapidly destroyed prepupae and pupae in open pupation pans. Gardiner (1985c) reported that the parasitoids *Cotesia glomerata* (Linnaeus) and *Pteromalis puparum* Linnaeus are sometimes contaminants in laboratory cultures of the large white butterfly, *Pieris brassicae* Linnaeus.

While it is common for parasitic insects to be impediments in insectary cultures, it is unusual for other kinds of parasites to be troublesome. However, Gardiner (1985a) found that nematodes of the genus *Mermis* were occasionally parasitic in an insectary culture of the desert locust, *Schistocerca gregaria* Forskal.

The degree of arthropod contamination depends on the generation time of the desired organism. Friese *et al.* (1987) found that using spider mites from a clean source culture to infest initially clean host plants minimized contamination by unwanted organisms. Because a spider mite generation was a short 2 weeks, host plants were consequently rapidly cycled. However, they also noted that a low-dosage insecticide treatment could eliminate greenhouse contamination by indigenous phytoseiid predators for up to 3 weeks without interfering with spider mites. Ho and Chen (1992) found that contaminating phytoseiid mites could be differentially killed by dipping mite-infested soybean seedlings in water at 50°C, whereas the spider mite culture rapidly recovered.

Microorganisms can cause severe contamination problems by being plant pathogens, saprophytic contaminants, saprophytic facultative insect pathogens, saprophytic true insect pathogens, or obligatory true insect pathogens. Path-

ogens can readily destroy plants used to raise host insects, and saprophytic microorganisms compete with host insects for the same food and destroy it. Sikorowski (1984) noted that contaminating microbes growing on insect diets can biochemically change the nutritive value thereof, and may also produce harmful toxins. He later discussed other ramifications resulting from microbial contamination in insect rearing (Sikorowski & Lawrence, 1994). Shapiro (1984) concluded that fungi of the genus *Aspergillus* are the most common contaminants in insect cultures. These and other saprophytic fungi and bacteria are ubiquitous in nature and promptly appear in unsanitary conditions.

Although microorganisms are usually problems in insect diets, they occasionally can be beneficial. Tamashiro *et al.* (1990) showed that axenic rearing of the oriental fruit fly, *Bactrocera* (= *Dacus*) *dorsalis* Hendel, could be accomplished continuously, but that there was a highly significant reduction of fecundity, in comparison to xenically reared flies.

Saprophytic facultative pathogens include the bacterium *Serratia marcescens*, which can invade insects only through open wounds. After invasion, however, the bacterium causes acute disease.

Saprophytic true insect pathogens, which are capable of direct invasion, are not common problems in insectaries. However, the bacterium *Bacillus thuringiensis*, also commercially produced for insect control, is occasionally troublesome. Stewart (1984) reported that it had interfered with mass production of the pink bollworm.

Obligatory true insect pathogens among the fungi, protozoa, and viruses cause the most pervasive and difficult problems in host-insect production. The fungus *Nomuraea rileyi* (Farlow) was found in a culture of the velvetbean caterpillar, *Anticarsia gemmatalis* Hübner (Leppla, 1985), and the *Entomophthora* spp. attacked cultures of houseflies (Morgan, 1985) and onion maggot adults (Meigen) (Tolman *et al.*, 1985).

According to Goodwin (1984), protozoans (including Microsporidia) are the most important pathogens in insectaries, and many are not as host specific as originally thought. They can infect several closely related species and some may even infect insects in different orders or families. Protozoans are particularly troublesome because they typically cause chronic and debilitating diseases that are more difficult to detect and eliminate than are acute diseases. Protozoans of the microsporidian genus *Nosema* are prevalent. They cause problems in mass production of the spruce budworm, *Choristoneura fumiferana* (Clemens) (Grisdale 1984); the western spruce budworm, *C. occidentalis* Freeman (Robertson, 1985a); and the pink bollworm (Stewart, 1984). Guthrie *et al.* (1985) in fact noted that it is very difficult to start a clean culture of the European corn borer, *Ostrinia nubilalis* (Hübner), because most field-collected larvae contain *N. pyrausta*.

Mattesia is another bothersome protozoan genus. McLaughlin (1966) reported on efforts to eliminate *M. grandis* from a culture of the boll weevil, *Anthonomus grandis grandis* Boheman. In the entomophage insectary at the University of California at Berkeley, *M. dispora* causes a chronic disease in the Mediterranean flour moth, *Anagasta kuehniella* (Zeller). However, in the navel orangeworm, *Amyelois transitella* (Walker), it produces an acute disease that can destroy the culture, and thereby prevents its use for rearing the encyrtid *Copidosomopsis plethoricus* (Caltagirone) and the bethylid *Goniozus legneri* Gordh for field release. *Mattesia* is the only major problem interfering with rearing these parasitoids. Necessary measures to control the disease greatly restrict the level and ease of parasitoid production.

Three major groups of insect viruses can contaminate host-insect cultures, making rearing very difficult. The diseases caused are typically acute, however, and consequently are rather easily detected. Nuclear polyhedrosis viruses are the most prevalent. For example, such viruses were in cultures of the Douglas fir tussock moth, *Orgyia pseudotsugata* (McDunnough) (Robertson, 1985b); the forest tent caterpillar, *Malacosoma disstria* Hübner (Grisdale, 1985b); the Egyptian cotton leafworm, *Spodoptera littoralis* (Boisduval) (Navon, 1985); the beet armyworm, *S. exigua* (Hübner) (Patana, 1985b); and the cabbage looper, *Trichoplusia ni* (Hübner) (Guy *et al.*, 1985). A cytoplasmic polyhedrosis virus caused severe effects in mass production of the pink bollworm (Stewart, 1984); and Reed and Tromley (1985a) reported that a granulosis virus could interfere with rearing the codling moth, *Cydia pomonella* (Linnaeus).

Although one microorganism may severely disrupt a rearing program, a group of them is intolerable. Stewart (1984) reported that the fungus *Aspergillus niger*, the protozoan *Nosema* sp., a cytoplasmic polyhedrosis virus, and the bacterium *Bacillus thuringiensis* caused the greatest difficulties in mass producing the pink bollworm. Henry (1985), who reported another example of a complex of troublesome pathogens, noted that viruses, protozoa, bacteria, and fungi contaminated cultures of *Melanoplus* spp. grasshoppers.

Prevention and elimination of diseases and contamination problems are top priorities. However, the practical mechanics of achieving these goals can be very difficult and costly. Consideration of source provides clues to control. Saprophytic contaminants cause disease indirectly by depriving insects of proper nutrition or environment. Such microorganisms are ubiquitous, and can increase rapidly in insectaries with poor sanitation or design. The sources of obligate pathogens in an insectary have to be in or on insects introduced to initiate laboratory cultures, or on natural food used in rearing.

Shapiro (1984) recommended collecting insects from low-density population areas to decrease pathogen introduction in starting or adding to a culture, and Grisdale (1984, 1985b) suggested field-collecting insects only from new infestation areas where disease is still at a low level. This advice is particularly useful for insects with widespread, high-incidence pathogens, such as the nuclear polyhedrosis virus of the forest tent caterpillar (Grisdale, 1985b), and the microsporidian *Nosema fumiferanae*, which attacks the spruce budworm (Grisdale, 1984).

Field-collected larval stages are generally the most seriously infected by pathogens. If possible, it is best to collect another stage. Singh and Ashby (1985) noted that ". . . the egg is usually the best stage with which to start a culture since it is least likely to carry disease micro-organisms." However, virus or protozoan transmission sometimes occurs on the surface of the egg, and some probably occurs within the egg as well. For example, when establishing a new laboratory culture of the forest tent caterpillar, Grisdale (1985b) surveyed field sites for the presence of the protozoan *N. disstriae* by microscopic examination of fully formed larvae removed from field-collected eggs.

Obtaining eggs from field-collected adults is often easier than finding eggs in the field. By visual examination of field-collected adults, removal of dead ones, and surface sterilization of eggs laid in the laboratory, Leppla (1985) prevented fungus infection by *Nomuraea rileyi* in a culture of the velvetbean caterpillar.

Accidental introduction of pathogens into an insectary culture also occurs on natural food. Patana (1985b) attributed the loss of beet armyworm cultures from virus primarily to the use of natural food (cotton leaves in summer and chard in winter). After introducing prepared diet in 1965, Patana (1985b) reared the insect continuously without virus disease. Similarly, Gardiner (1985a) used *Brassica* instead of field grass for rearing the desert locust because of the danger of introducing diseases and nematode parasites of local grasshoppers.

Contaminating microorganisms can likewise enter insectaries on ingredients for prepared diets. Shapiro (1984) found that more than 95% of the total bacteria recovered from various ingredients of diet for the gypsy moth occurred on the raw wheat germ. The pathogenic protozoan *Mattesia dispora* and the bacterium *Bacillus thuringiensis* may contaminate stored grain products used for insect diets, inasmuch as these microbes were originally isolated from stored grain insects.

Ease of controlling contaminating microorganisms depends on the characteristics of the rearing programs, procedures, and facilities. Fisher (1984) listed sources of contamination in an insectary and possible measures to control it. Grisdale (1984) found that rearing several species of insects in the same facility could result in serious microbial contamination, particularly with the necessity of using both foliage and artificial diet. Even though Grisdale reared the spruce budworm on artificial diet, balsam fir foliage was still a required oviposition site, and was a principal source of fungal contamination. Stewart (1984) reported that cytoplasmic polyhedrosis virus caused severe disease problems in a pink bollworm culture until he discovered that moth scales carried virus polyhedra on air currents from oviposition areas to larval rearing areas. The introduction of major changes in procedures and facilities then virtually eliminated disease and highly increased insect production.

Rigorous sanitation, proper rearing procedures, and suitably designed insectaries can greatly reduce or eliminate microorganisms. Although there are usually specialists in large mass production facilities to recognize, monitor, and control microbes, all personnel should have some familiarity with microorganisms and with sanitation procedures. Poinar and Thomas (1978) presented a useful manual on the diagnosis of insect pathogens, and Goodwin (1984) reviewed the recognition and diagnosis of diseases in insectaries and the effects of disease agents on insect biology. Shapiro (1984) discussed microorganismal contaminants and pathogens in insect rearing; Sikorowski and Goodwin (1985), contaminant control and disease recognition in laboratory cultures; and Sikorowski (1984), occurrence, monitoring, prevention, and control of microbial contamination in insectaries. Rivers (1991) and Soares (1992) likewise presented information on pathogens and disease control in insect colonies. Later, Sikorowski and Lawrence (1994) further reviewed microbial contamination problems.

The first line of defense against contagious diseases in an insectary is exclusion by procedural, physical, and chemical techniques, both initially and continually. If possible, it is advisable to quarantine and rear insects individually for a few generations after laboratory introduction, and to monitor them for disease (Goodwin, 1984; Shapiro, 1984). The use of steam sterilization destroys any diseased insects. Although initial individual rearing is highly laborious, it may guarantee a pathogen-free culture. When Grisdale (1984) added field-collected spruce budworms to an existing culture, he reared the newly collected stock in laboratory isolation for two generations, and only cultured progeny from protozoan-free adults. Forbes *et al.* (1985) likewise recommended that only the progeny from field-collected aphids should be used to initiate laboratory cultures to reduce fungal disease.

In addition to quarantine for the elimination of pathogens, chemical surface disinfection of insect stages is often routine. This is particularly true with lepidopterous eggs, not only because of frequent transmission of obligate viruses and protozoans on these eggs but also because of bacterial and fungal contaminants creating problems on prepared diets typically used to rear lepidopterans.

Vail *et al.* (1968) and Sikorowski and Goodwin (1985) have recommended procedures for surface-disinfecting insect eggs. Various techniques using sodium hypochlorite are most popular. Usage of formalin occurs too, because it is a good viricide, but it is also carcinogenic.

Adjustments of sodium hypochlorite concentrations and exposure times are necessary for each insect species, depending on the chemical susceptibility of its eggs. Guy *et al.* (1985) used a very weak solution of 0.02% active ingredient (AI) for only 5 min to disinfect eggs of the cabbage looper. A solution commonly used contains 0.1% AI. Reed and Tromley (1985b) used this concentration for 5 min to disinfect eggs of the lesser peach tree borer, whereas Robertson (1985b) employed it for 15 min twice with strong mechanical stirring to treat eggs of the Douglas fir tussock moth; and Greenberg and George (1985) used it for 15 min with swirling to disinfect eggs of calliphorid flies.

Willey (1985) cautioned that a solution of 0.25% AI sodium hypochlorite used for 10 min to disinfect eggs of the range grasshopper was used infrequently because treated eggs had a much lower hatching success than those incubated *in situ*. Similarly, Etzel (unpublished) found that treatment of Mediterranean flour moth eggs for 5 min with 0.15% AI reduced hatchability by at least 50%, but was necessary to control disease caused by *Mattesia dispora*. Hatchability was best when treatments occurred after eggs were nearly completely embryonated. Even then, eggs were extensively dechorionated, but holding them on sponge-moistened filter paper in a petri dish prevented desiccation.

In culturing Egyptian alfalfa weevil parasitoids, L. Etzel (unpublished) found that weevil eggs collected from alfalfa stems required treatment with 1% AI sodium hypochlorite for 1 min to retard saprophytic fungal growth if storage at 4°C followed. As a final example, Grisdale (1985b) used full-strength household bleach (6% AI sodium hypochlorite) for 1.5 min to disinfect egg masses of the forest tent caterpillar, as well as to remove the spumilin coating.

Removing eggs from substrate can also make use of sodium hypochlorite. Shore (1990) utilized a 0.1% AI solution to extract eggs of the western hemlock looper, *Lambdina fiscellaria lugubrosa* (Hulst), from lichens to assess parasitism. Likewise, in mass-producing the hemerobiid *Micromus tasmaniae* Walker, Hussein (1986) used 0.75 to 1.00% AI sodium hypochlorite to remove eggs from a cloth substrate.

Parrott and Jenkins (1992) incorporated a sodium hypochlorite wash treatment in the design of an egg harvester for collecting eggs of mass-produced *Heliothis virescens* Fabricius. The machine was capable of collecting 700,000 eggs in 20 min. Nordlund and Correa (1995a) also developed a sodium hypochlorite based egg-harvesting system for obtaining stalk-free eggs in the mass production of *Chrysoperla carnea*.

Although not commonly used, formalin also disinfects egg surfaces. Bartlett and Wolf (1985) used 9.5% AI formaldehyde for 30 min to disinfect pink bollworm eggs.

Singh *et al.* (1985) noted that eggs of the light brown apple moth, *Epiphyas postvittana* (Walker), have to be 4 to 5 days old before they can withstand surface disinfection with 5% formalin solution for 20 min to prevent viral disease. Ashby *et al.* (1985) also cautioned that codling moth eggs should not be surface sterilized with 5% formalin until they are 4 to 6 days old. However, a satisfactory treatment for codling moth eggs is 0.15% sodium hypochlorite for 10 min. On the other hand, Toba and Howell (1991) developed an apparatus to surface-sterilize codling moth eggs with formaldehyde fumes.

Treatments of insect eggs with other chemicals occasionally occur. Speirs (1985) used 0.1% mercurous chloride in 70% ethanol plus 0.1 ml Triton X-100® per liter for 3 min to disinfect eggs of *Trogoderma* spp., and Moore and Whisnant (1985) utilized 18% cupric sulfate (a fungicide) and a 0.3% solution of Mikro-Quat® (alkyl dimethylbenzylammonium chloride) to surface disinfect boll weevil eggs. Grzegorczyk and Walker (1997) compared the use of 2.6% AI sodium hypochlorite solution with 50% ethanol for surface sterilization of grape phylloxera (*Daktulosphaira vitifoliae* Fitch) eggs, and chose the ethanol treatment for 7 min because it had the advantage of not needing a sterile water rinse after treatment, thus simplifying the procedure and reducing further risk of contamination.

Chemical treatment of insect larvae is also useful for disease prevention. Treatment of the European corn borer with a 0.01% phenylmercuric nitrate solution prior to diapause repressed *Nosema pyrausta* (Guthrie *et al.*, 1985).

It is not unusual for pupae to be surface disinfected to control contaminating microorganisms, where again sodium hypochlorite is the chemical of choice. Patana (1985b) treated pupae of the beet armyworm with a 0.03% AI solution for 5 min and Guy *et al.* (1985) used 0.1% AI solution for 10 min for cabbage looper pupae.

Sodium hypochlorite also dissolves cocoon silk, as well as disinfects the harvested larvae or pupae. Etzel (1985) used 1.3% AI sodium hypochlorite solution to dissolve cocoon silk of the potato tuberworm, and harvested larvae or pupae from the sand in which pupation had occurred. Likewise, Grisdale (1985b) collected pupae of the forest tent caterpillar after exposing cocoons to 3% AI sodium hypochlorite, and Bartlett and Wolf (1985) utilized the same concentration for 30 min to dissolve cocoons of the pink bollworm.

Other solutions used to surface-disinfect pupae include 5% phenol for calliphorids (Greenberg & George, 1985), and 0.2% mercuric chloride for the wood-boring scolytid *Xyleborus ferrugineus* (Fabricius) (Norris & Chu, 1985).

In addition to the use of chemicals to disinfect insect eggs, larvae, and pupae, disinfection for normal sanitation should be routine. Sikorowski (1984) reviewed different antimicrobials available for cleaning and disinfection, and

noted in particular that wet-mopping floors after flooding with disinfectant is preferable to sweeping and dry-mopping. Stewart (1984) reported that disinfection and cleaning of equipment and facilities with bleach, quaternary ammonium compounds, phenolic compounds, and stabilized chlorine dioxide solutions contributed significantly to controlling microbial disease in mass-produced pink bollworms.

As with surface disinfection of insects, sodium hypochlorite is the most common chemical for general sanitation. Concentrations range from about 0.026% to full-strength household bleach (5.25%), but 1% is usual. The lower concentrations are common for disinfecting rearing containers. Baumhover (1985) employed a 0.026% solution to soak clean rearing containers for a minimum of 4 h in culturing the tobacco hornworm, *Manduca sexta* (Linnaeus), and he mopped floors weekly with the same solution. Palmer (1985) used 0.05% sodium hypochlorite to soak water dishes and cheesecloth for 4 to 8 h in rearing the chalcidid *Brachymeria intermedia* (Nees). Moore and Whisnant (1985) prevented microsporidian infection of the boll weevil by washing adult cages and emergence boxes with soap and 0.5% sodium hypochlorite. A 1% concentration was useful for washing equipment and wiping down surfaces in the production of houseflies (Morgan, 1985), and *Melanoplus* spp. grasshoppers (Henry, 1985).

Some workers have used solutions of formaldehyde to spray walls, ceilings, cabinets, and counters; or to fumigate rearing rooms or containers. These practices are ill-advised because formaldehyde is a carcinogen.

Navon (1985) reported that treatment of rearing boxes overnight in 0.4% potassium hydroxide helped to prevent viral disease in rearing *Spodoptera littoralis*.

Insectary sanitation procedures have also included the use of commercial germicides, such as Roccal® (Guthrie *et al.*, 1985; Reed & Tromley, 1985a), Vesphene® (Riddiford, 1985), and Zephiran® (Morgan, 1985; ODell *et al.*, 1985). Morgan (1985) employed 0.13% Zephiran® as a surface disinfectant to kill the pathogenic fungus *Entomophthora* sp. Soares (1992) recommended the use of Exspore® as a safer, less corrosive, and better sterilant than solutions of sodium hypochlorite for insectary sanitation.

Physical means are likewise effective in insectaries for sterilization or disinfection. Steam sterilization (autoclaving) is most common for destroying unwanted laboratory organisms. However, steam deteriorates wooden cages. Legner (unpublished) found that steam sterilization of pink bollworm cages was unnecessary after the banning of smoking tobacco from rearing rooms, whereupon host production increased by several times.

Direct heat treatments on insects for disease control have had occasional usefulness. Etzel (1985) noted that hot water treatment of potato tuberworm eggs at 48.3°C for 20 min, as described by Allen and Brunson (1947), was advantageous for controlling the protozoan *Nosema*. However, Etzel *et al.* (1981) reported that the same treatment performed on eggs of the weed-feeding chrysomelid *Galeruca rufa* Germar destroyed them within 10 min. Shapiro (1984) reviewed other examples of heat treatment being helpful in disease control.

The physical design, structure, and equipment of an insectary, especially with respect to environmental control, are critical for the efficient production of healthy insects. In rearing gypsy moth larvae for parasitoid production, ODell *et al.* (1984) noted that in spite of egg disinfection and routine cleaning of work surfaces and equipment, there were still periodic severe problems with bacterial and fungal diseases, attributed to inadequate environmental control, other facility peculiarities, and stress of parasitization. Sikorowski and Goodwin (1985) remarked that proper facility design and traffic control aid significantly in controlling microbes. Dividing rearing facilities into a "clean" area for critical work and a "conventional" area for less critical work is advisable. Of particular benefit is the use of high-efficiency particulate air (HEPA) filters for clean rooms and laminar-air-flow workstations. Sikorowski (1984) believed that one of the best methods for controlling microorganisms when working with insect diet preparation or infestation, or when performing other procedures where contamination was a threat, was to use such a workstation. He also recommended HEPA-type exhaust filters for vacuum cleaners. Stewart (1984) virtually eliminated severe disease in mass-producing the pink bollworm by making major procedural and facility changes, including centralization of egg disinfection and larval transfer, positive air pressurization of rooms used for diet preparation and egg disinfection, and installation of HEPA filters for cleaning air in critical areas.

Careful regulation of temperature, humidity, moisture, and light is also important for disease control. Finney *et al.* (1947) reported that bacterial diseases caused by facultative pathogens in potato tuberworm cultures were suppressed by preventing high humidities and by rearing temperatures of <30.6°C. Thus, environmental stress is a contributing factor in disease causation. Greany *et al.* (1977) documented another case of temperature-caused stress, and subsequent insect disease. Rearing the Caribbean fruit fly, *Anastrepha suspensa* (Loew), and its braconid parasitoid *Diachasmimorpha longicaudata* (Ashmead) (= *Biosteres longicaudatus* [Ashmead]) above 30°C created stress that permitted the bacteria *Serratia marcescens* and *Pseudomonas aeruginosa* to become pathogenic, causing high mortality of both insects. Lowering the rearing temperature controlled the diseases.

Gardiner (1985c) found that gross overcrowding of large white butterfly larvae and excessive humidity contributed to occasional outbreaks of bacterial disease. He also noted that low humidities and avoidance of overcrowding were critical to preventing bacterial diseases in rearing the desert

locust (Gardiner, 1985a). Henry (1985) likewise recommended controlling various grasshopper diseases by maintaining RH at 30 to 35%.

Moisture and stagnant air particularly favor fungal development. Ankersmit (1985) found that holding rearing containers of the summer fruit tortrix, *Adoxophyes orana* Fischer von Röslerstamm, at a constant temperature reduced chances for moisture condensation, which correspondingly reduced microbial contamination. In rearing the beet armyworm, Patana (1985b) discovered that containers that allowed slight drying of the artificial diet controlled mold contamination. Likewise, Roberson and Wright (1984) used porous polyethylene to seal polystyrene trays in mass-producing the boll weevil, thus allowing air and moisture exchange in the rearing cavities. This measure, plus placing a sterile sand and corncob mixture on the diet to absorb moisture and to force hatching larvae to feed, greatly reduced microbial contaminants. Grisdale (1984) also recommended proper ventilation to control fungal contamination. Even under conditions of very high humidity that may be necessary for rearing some stages of certain insects, providing constant clean air movement can greatly reduce or control fungal growth.

Other environmental factors can also have an impact on microbial contamination. Insect activity by itself can be significant. Whistlecraft *et al.* (1985a) remarked that a seed-corn maggot population, *Hylemya platura* (Meigen), large enough to actively feed on the available artificial diet would prevent mold development. Even light can be a factor. Heather and Corcoran (1985) found that a contaminant yeast would grow on a carrot-based larval diet for the Queensland fruit fly in light but not in darkness. The handling of insect stages is likewise important. Henry (1985) recommended leaving grasshopper eggs *in situ* in the oviposition substrate to protect hatching nymphs from lethal bacterial and fungal diseases.

The preceding procedural, physical, and chemical means of controlling microbial contamination and insect diseases provide the best defenses. However, contamination and disease can still occur. Therefore, a further control is the frequent use of antimicrobial chemicals in insect food. Hartley (1990), for example, used a phosphoric–proprionic acid mix to control contaminants in the larval diet used for mass production of several noctuid species. However, Dunkel and Read (1991) discussed the insecticidal and insect growth regulatory effects of certain antimicrobials such as sorbic acid that are used in insect rearing diets, and they listed sensitive species. Shapiro (1984) provided an excellent review of chemical antimicrobials as ingredients in prepared diets. Sikorowski (1984) also reviewed different antimicrobial chemicals for diets. As did Goodwin (1984), he recommended against using antibiotics unless absolutely necessary because of the danger of selecting for resistant microbes.

Reinecke and Sikorowski (1989) experimented with combinations of heat treatment and antibiotics for control of microorganisms on artificial diet for the boll weevil. Similarly, Jang and Chan (1993) combined pasteurization with antibiotics to control acetic acid-producing bacteria in diet used for the mass production of the Mediterranean fruit fly, *Ceratitis capitata* (Wiedemann). The treated fruit fly diet not only improved pupal yields by 36 to 58% but also reduced atmospheric acetic acid levels in the rearing facility below the permissible level for humans set by the Occupational Safety and Health Administration (OSHA).

Once a disease caused by an obligate pathogen appears in a culture, it is usually best to destroy the insects, completely clean and sanitize the insectary, and start a new culture. However, a culture too valuable to discard warrants attempts to recover healthy specimens by use of isolation, quarantine, and labor-intensive sanitation procedures.

Contamination in production of beneficial organisms does not occur only from parasitoids, predators, pathogens, and interspecific competitors. The desired organism can also contaminate if it appears spatially or temporally where unwanted. Premature destruction of plants being grown for host-insect production might occur if contamination by that species precedes its purposeful infestation. Similarly, a contaminating entomophage can devastate a source culture of host insects used to rear it. In mass-producing pteromalids for filth fly control, one species may contaminate the culture of another. In such cases continuous manual elimination of contaminants is required if spatial separation of cultures is impractical (E. F. Legner, unpublished).

Intraspecific Competition

Intraspecific competition or cannibalism also can be troublesome, especially with host-insect production. In detailing the history of rearing *Heliothis* spp., Raulston and King (1984) noted that a major problem was cannibalism. Consequently, separating the reared larvae was necessary. One method was to use compartmentalized disposable plastic trays covered with Mylar film, as pioneered by Ignoffo and Boening (1970), and later automated (Sparks & Harrell, 1976). Hartley *et al.* (1982) described another type of compartmentalization. However, Patana (1985a) developed a different technique for separating larvae of these species. He placed 75 *Heliothis* larvae in a plastic box with a layer of diet covered by a layer of dried diet flakes. The dried flakes separated the larvae and greatly reduced cannibalism. Such rearing units yielded 65% pupae for the corn earworm, *H. zea* (Boddie), or 85% for the tobacco budworm, *H. virescens.*

Obviously, in mass production it is highly desirable to develop a system for rearing cannibalistic insects together, even though a major advantage of individual rearing is facilitation of disease control. Brinton *et al.* (1969) reared

another cannibalistic species "gregariously" by using a sawdust-based diet for codling moth larvae. Not only did the sawdust tend to separate the larvae but also the diet was more economical than if agar-based.

It is sometimes possible to avert cannibalism by seeking a naturally noncannibalistic race. This occurred with the planarian mosquito predator *Dugesia dorotocephala,* which is normally cannibalistic (Legner & Tsai, 1978).

There is no necessity to keep all cannibalistic insects physically separated. Grisdale (1985a) found that although the hemlock looper, *Lambdina fiscellaria fiscellaria* (Guenée), was cannibalistic, provision of an acceptable artificial diet allowed gregarious development. Rearing 10 to 20 larvae on diet in small, 22-ml cups until the third instar was satisfactory, whereupon placing four larvae in each new cup allowed completed development. Moreover, host cannibalism may be reduced by laboratory rearing of a parasitoid. Slovak (1987) reported significantly higher cannibalism among nonparasitized cabbage moth caterpillars, *Mamestra brassicae* Linnaeus, than those parasitized by the ichneumonid *Exetastes cinctipes* Retzlus.

Some insects are gregarious in nature, making rearing relatively easy. Grisdale (1985b) found that the first three instars of the forest tent caterpillar seemed to develop better when crowded on artificial diet. Mass production of *Nasonia vitripennis* Walker and *Muscidifurax raptorellus* Kogan & Legner, pteromalids for filth fly control, also occurs gregariously. The latter species exists in nature as several races that demonstrate development ranging from solitary to gregarious (Legner, 1987c, 1988c), suggesting that similar racial types might exist for other species.

Genetics

The genetic composition desired in a laboratory culture depends on its purpose. Either genetic uniformity or genetic variability is the preferable condition. A high homozygosity or genetic uniformity is desirable in a culture used for insecticide testing to provide a relatively stable standard for treatment comparisons (Wheeler, 1984). The same is true for insect cultures used to assay pathogens for microbial control. However, a high genetic variability is desirable in entomophages produced for biological control, as discussed later.

Concerning host provision for entomophage rearing, primary production goals are ease, rapidity, and quality maintenance. However, host-strain effects on parasitoid production are also important. For example, ODell *et al.* (1984) reported significantly different puparial weights between two groups of the tachinid *Blepharipa pratensis* Meigen reared on two different gypsy moth strains. The host-strain differences corresponded with their field densities and geographic sources.

Geographic-strain or biotype differences can also be important for ease of rearing. Diapause in the life cycle is a particularly aggravating production problem, and thus it is advantageous to obtain nondiapausing field strains. With the plum curculio, which has a northern biotype with diapause and a southern one without, Amis and Snow (1985) chose the southern one for culture. Bartlett and Wolf (1985) noted that the pink bollworm probably has a facultative diapause, with no diapause known for the insect in latitudes between 10° north and 10° south, such as in southern India. In California, pink bollworm diapausing strains intersperse with nondiapausing ones in different seasons (Legner, 1979c), whereas diapausing navel orangeworms occur at such a low frequency as to go largely undetected (Legner, 1983). Henry (1985) reported that the migratory grasshopper, *Melanoplus sanguinipes* (Fabricius), widely distributed in North America, has diapausing biotypes. Throughout most of the range it is univoltine, with an obligatory egg diapause. In southern areas, however, there may be two or three generations a year, and the eggs may simply become quiescent during the winter. Grasshoppers collected from a southern area thus would be best for initiating a laboratory culture.

Even if a nondiapausing field biotype does not exist, it may be possible to develop a nondiapausing laboratory biotype by selection over a number of generations. For example, Jackson (1985) noted that although the wild biotype of the western corn rootworm, *Diabrotica virgifera virgifera* LeConte, has a diapause in the egg stage, a laboratory nondiapausing biotype also exists.

Development of a nondiapause insect biotype illustrates planned genetic adaptation of a species to the laboratory. Whether planned or unplanned, some degree of such adaptation typically occurs before a species becomes easily reared. The problem is to balance the need for laboratory adaptation against the possible need to retain genetic variability or heterozygosity, and certainly to prevent genetic deterioration of the stock. Gardiner (1985c) noted that the large white butterfly was relatively easy to rear, but only after it had become adapted to the laboratory. In this case, the basic problem of adaptation was that adults required hand feeding for several generations until they would feed at artificial flowers. Heather and Corcoran (1985) used ripe whole fruit for rearing the Queensland fruit fly for the first couple of generations in the laboratory until the population increased and adaptation to a prepared diet began. In starting a culture of the Mediterranean fruit fly, Boller (1985) recommended rearing field-collected specimens at low densities during the early culture establishment period, because high adult fly mortality due to irritation and unnatural densities in laboratory cages can result in unwanted selection of laboratory ecotypes. Onyango and Ochieng'-Odero (1994) adapted the maize stem borer, *Busseola fusca* (Fuller), a noctuid, to laboratory culture by rearing larvae individually. Developing a nutritionally adequate diet containing 4- to 8-week-old sorghum powder, and exposing the insects to ambient laboratory conditions [25 to 30°C, 50 to 80% RH, and photoperiod of L12:D12], made it possible to rear 15 successive nondiapausing generations. By the 15th generation, the larval period was reduced to 32.3 days (from 70 days in the 1st generation), and the average fecundity increased from 158.0 to 394.6 eggs per female, with a concomitant increase of egg hatch from 44.8 to 79.6% in the 1st and 15th generation, respectively.

Once a species adapts to laboratory culturing, maintenance of genetic vigor in the culture depends on its genetic plasticity (environmental adaptability), the number of deleterious alleles in the population, the number of parent individuals, and the degree to which their mating mixes alleles in each generation. Maintenance of some insect cultures continues satisfactorily for years, whereas some require annual replenishment from field stock. Wight (1985) reported that continuously rearing the southern armyworm since 1938 resulted in remarkably consistent responses in pesticide testing, with this consistency being the consequence of genetic uniformity developed during long-term culturing. Guthrie *et al.* (1985) recounted the rearing of the European corn borer on artificial diet for 200 generations over 19 years with no genetic deterioration of fecundity, fertility, or pupal weight. However, a loss of adaptiveness to corn plants occurred after about 14 generations. Similarly, Baumhover (1985) continuously reared a laboratory culture of the tobacco hornworm for 170 generations (18 years) with no apparent genetic deterioration. Field tests of sterilized laboratory-reared male moths showed nearly complete competitiveness with native males.

In contrast, indefinite maintenance of most laboratory cultures cannot occur without replacement or replenishment with newly collected stock. Reed and Tromley (1985a) recommended renewing a codling moth laboratory culture after 20 to 30 generations on artificial diet. Leppla (1985) maintained genetic variability of a laboratory culture of the velvetbean caterpillar by annually mixing the eggs from about 50 wild-type and 50 laboratory females.

Many species genetically deteriorate rapidly in culture. Belloncik *et al.* (1985) found that the white cutworm, *Euxoa scandens* (Riley), and the darksided cutworm, *E. messoria* (Harris), genetically deteriorated after only four laboratory generations (about 1 year), witnessed by a loss of vigor and fertility, and the appearance of adult malformations. Jones (1985) discovered that annual recolonization with wild stock was necessary to maintain vigorous laboratory cultures of the southern green stink bug, *Nezara viridula* (Linnaeus). Starting five laboratory families from each of five field-collected females and then mating progeny to those from different families in a planned pattern minimized inbreeding depression.

Various workers have recommended planned mixing in a culture to reduce inbreeding depression. ODell *et al.*

(1985) advised mating males from one gypsy moth egg mass with females from another egg mass. In maintaining a culture of the beet armyworm for over 18 years, Patana (1985b) believed that continual mixing of larvae from different groups of parents provided a limited random mixing of genetic material that prevented the effects of absolute inbreeding. Young *et al.* (1976) improved mating, fecundity, and fertility in a corn earworm culture by studying genetic changes and developing a crossing procedure to reduce inbreeding. Hoffman *et al.* (1984) described a system using genetic selection to improve the characteristics of an already existing culture of the cabbage looper. To maintain genetic variability, they collected eggs from the parental culture on consecutive days, and set up 26 subcultures, each from a separate egg lot. Subcultures were discarded if they did not reach expected performance levels for hatch, larval development, pupation, and emergence in two consecutive generations. Hoffman *et al.* (1984) increased mean culture fecundity by 30% within three generations with subculture selection. This type of fractional colonization scheme also enables better control of insect diseases by immediate removal of contaminated subcultures.

Delpuech *et al.* (1993) provided further experimental support for the value of maintaining a laboratory culture as a series of subcultures. In their work with *Drosophila melanogaster* Meigen, they showed that although 32 isofemale lines held in culture for 23 generations might change individually, the global frequency of observed alleles remained relatively unchanged, and in fact the only rare allele observed was conserved. Tests of the biological capacity of the larvae to encapsulate a parasitoid revealed a change of genetic variation over the culture period, yet mass hybridization of individuals of the culture lines restored the variation close to that of the starting population.

Standard quality control tests such as size, fecundity, fertility, and longevity reveal the genetic vigor of laboratory cultures (Legner, 1988b). Sophisticated technical tests have also been used (Bush *et al.*, 1976; Goodenough *et al.*, 1978; Brown, 1984). Robertson (1985a) recommended using starch gel electrophoresis to monitor genetic quality of laboratory cultures of spruce budworms. As a result of her tests, she suggested introducing wild stock collected from the same area as the founder group into the culture at 2- to 3-year intervals to prevent excessive homozygosity.

Physical Environment

The actual laboratory production of insects is obviously dependent on environmental conditions. Combinations of light, temperature, and humidity, as well as sequences of these combinations, are particularly critical in managing development of insects that undergo facultative or obligatory diapause. Obligatory diapauses, in particular, cause severe production problems, but both facultative and obligatory diapauses are useful for long-term insect storage. For example, the darksided cutworm overwinters in the egg stage, which can be kept in storage at least 1 year at 4°C (Belloncik *et al.*, 1985).

Generally, light and temperature are the most important physical factors in initiating and ending diapause. To illustrate, the environmental regime for diapause prevention in cultures of the cabbage moth consists of 20°C, 60% RH, and a photophase of 18 h for larval rearing, after which the pupae are nondiapausing (Gardiner, 1985b). Rearing the larvae with a 9-h photophase initiates diapause. Gardiner (1985b) also noted that prevention of diapause in laboratory cultures of the cabbage moth had been difficult for many early workers, and that larval-food quality and insect strain had been two factors involved. Along with photophase, another factor that may affect diapause rates in the laboratory is the type of lighting used, as discussed by Goussard and Geri (1989) and Geri and Goussard (1989) concerning *Diprion pini* Linnaeus.

Moisture can also be a factor regulating diapause. According to Henry (1985), a subspecies (*nigricans*) of *Melanoplus differentialis* (Thomas), the differential grasshopper, occurs in the Central Valley of California and apparently undergoes a winter obligatory diapause possibly more conditioned by moisture than by temperature. Density is an occasional diapause factor as well. Speirs (1985) noted that overcrowding in *Trogoderma* cultures might increase the rate of diapause.

Long photophases with temperatures >20°C usually prevent facultative hibernal diapause in host insects, depending on the species. Such an environment mimics the natural summer when insects with a facultative hibernal diapause usually continue to reproduce. Daily photophases used to prevent diapause typically range from 16 h for the onion maggot (Tolman *et al.*, 1985) and spruce budworms (Robertson 1985a), to 18 h for the codling moth (Ashby *et al.*, 1985) and the large white butterfly (Gardiner, 1985c), to continuous light for the tobacco hornworm (Baumhover, 1985) and the European corn borer (Guthrie *et al.*, 1985).

Some insects, such as the Egyptian alfalfa weevil, have an estival diapause and are active in nature in the spring. New generation adults estivate until fall. Under laboratory conditions of 21°C and a daily photophase of 8 h, at least some individuals of each generation will forego estivation and produce eggs.

Chilling insects for several weeks to several months will typically break diapause, whether facultative or obligatory. Egg diapause has been broken in the grasshopper genus *Melanoplus* by exposure to 10°C for 3 to 12 months (Henry, 1985), in the Douglas-fir tussock moth by conditioning at 5 to 10°C for 4 to 6 months (Robertson, 1985b), and in the hemlock looper by storage at 2°C for 3 to 9 months (Grisdale, 1985a). Examples of chilling requirements to end diapause in larvae include 1°C for 18 to 35 weeks for spruce budworms (Grisdale, 1984); and 5 ± 2°C for 2 to 6 months for the red oak borer, *Enaphalodes rufulus* (Haldeman) (Galford, 1985).

Similar measures will terminate pupal hibernal diapauses. Tolman *et al.* (1985) were able to break diapause in the onion maggot by chilling at 1 ± 0.5°C for 2 to 12 months. The same procedure worked well for the cabbage maggot, except chilling had to be for a minimum of 4 months (Whistlecraft *et al.,* 1985b). Boller (1985) noted that pupae of the European cherry fruit fly, *Rhagoletis cerasi* (Linnaeus), required refrigeration at 4°C for 3 to 5 months to break an obligatory diapause. There is, however, a time limit for safely refrigerating insects.

The length of diapause conditioning of the egg stage can affect the sex ratio of emerging gypsy moth adults. After a short chilling period of 120 days, the sex ratio of the first 25% of hatching larvae was male biased; after a long chilling period of 180 days, it was female biased (ODell *et al.,* 1985). The obligatory diapause in the univoltine gypsy moth has always been a problem with respect to the mass production of gypsy moth parasitoids and viruses. However, Bell (1996) discovered that KK42, a novel insect growth regulator with antihormonal properties, effectively averted diapause in the gypsy moth. His results showed that KK42 treatment of 14-day-old eggs prevented diapause in 85% of the test populations. Eggs 14 to 15 days old could alternatively be placed into cold storage and retain complete sensitivity to KK42 whenever removed. The egg age at KK42 treatment, or when chilling was initiated, was critical. Eggs had to be held at 25°C for 14 to 15 days postoviposition to withstand prolonged chilling at 5°C and still hatch successfully when once again placed at the higher temperature. However, eggs that were 21 to 22 or 29 to 30 days old when chilling was started were not responsive to KK42 whenever removed, but had to be chilled for 4 to 6 weeks first. These findings will enable development of methods to manipulate diapause in stockpiled gypsy moth eggs for use in biological control programs.

Different host-insect stages and different species vary in developmental environmental requirements. Some examples indicate the range of variations and similarities. Phytophagous insect eggs frequently require moisture or high humidity to prevent desiccation, and providing just the right amount of moisture to maintain the eggs is critical. Singh *et al.* (1985) held eggs of the light brown apple moth in airtight containers to maintain egg turgidity. However, prevention of fungus contamination required frequent checking of the container to remove condensed moisture.

Clair *et al.* (1987) developed another way to control fungus contamination while providing moisture to eggs. They cut elm leaf beetle egg clusters, *Xanthogaleruca luteola* (Müller), from elm leaves and placed them on cloth and filter paper in a plastic petri dish. A wick of dental cotton extending through a basal hole in the petri dish to a water reservoir beneath kept the cloth and filter paper combination moist. Exposure of the eggs to air circulation then prevented stagnant air conducive to fungal growth. This type of system could be useful for maintaining a diversity of eggs.

Eggs treated with sodium hypochlorite will desiccate if not held on moist cloth and filter paper. However, closed containers are usually satisfactory for this purpose because the egg treatment also reduces fungal contamination.

In producing insects, aperiodic (constant) conditions for temperature and RH are commonly used, with periodic conditions only utilized for lighting. For example, Navon (1985) reared *Spodoptera littoralis* with a photophase of 16 h at 24°C and 50 to 70% RH. However, sometimes workers use completely aperiodic environmental conditions (*i.e.,* constant temperature, RH, and light) for insect rearing. Insects reared in this manner include the southern armyworm, (Wight, 1985), the lesser peachtree borer (Reed & Tromley, 1985b), the European corn borer (Guthrie *et al.,* 1985), and the tobacco hornworm (Baumhover, 1985).

Fluctuating environmental rearing conditions retain and promote insect vigor. For example, Greenberg and George (1985) cited Kamal (1958), who reported that fluctuating temperature and humidity conditions increased the longevity of several laboratory-reared calliphorid and sarcophagid species, as did a larger cage size.

Experimental determination of the optimal rearing temperature for each insect and strain is necessary. Orthopterans frequently require high rearing temperatures, although some need cool conditions. McFarlane (1985) found that crickets do best at temperatures between 28 and 35°C. When reared at 20°C, the mean weight of the emergent adults was greater than at higher temperatures, but they would not reproduce. However, the range grasshopper requires much lower laboratory rearing temperatures than some other species. Willey (1985) reared the various stages at 22°C and variable RH, with a photoperiod of 12 h. With these conditions, completion of a generation occurred in an average of 6 months. Temperatures above 30°C resulted in lower hatch and weak grasshoppers.

A few insects change forms (morphotypes) depending on rearing conditions. Forbes *et al.* (1985) reported that aphids reproduced parthenogenetically at 20 ± 1°C with a photophase of 16 h. The production of sexual forms required a maximum photophase of 8 to 12 h with a temperature of 15°C or less. Medrano and Heinrichs (1985) indicated, however, that nymphal density and food availability governed production of the two distinct morphotypes of the brown plant hopper, *Nilaparvata lugens* (Stål). They noted that a short-winged form developed with low nymphal density and plentiful food, whereas a long-winged form developed under opposite conditions.

Humidity, moisture, and substrate are often critical for

insect pupation. Baumhover (1985) noted that pupation requirements of the tobacco hornworm are precise, and RH close to 85% is necessary for adult ecdysis. A dehumidifier may be necessary to remove air moisture, because each prepupa loses 4 ml of water by the time of ecdysis. Further, prepupae require complete darkness to make them inactive, and must be held individually in flat cells to allow proper pupation. Pupal hardening must precede harvesting, because injury of teneral pupae occurs easily. Similar considerations of substrate moisture and pupal handling are important in rearing fruit flies for the mass production of parasitoids (Wong & Ramadan, 1992).

Pupation substrates for various insects include sand for the potato tuberworm (Etzel, 1985) and *Hippelates* eye gnats (Legner & Bay, 1964, 1965), sawdust for the Queensland fruit fly (Heather & Corcoran, 1985), a sawdust and ground corn cob mixture for the lesser peach tree borer (Reed & Tromley, 1985b), vermiculite for *Spodoptera littoralis* (Navon, 1985), and plastic soda straws for the greater wax moth, *Galleria mellonella* (Linnaeus) (Eischen & Dietz, 1990). The pupation medium can be quite critical, as it is in rearing the southern armyworm. Wight (1985) noted that vermiculite used for this insect must be no larger than 6-mm mesh, and must have the proper moisture content (400 ml of water in 1200 ml of vermiculite). If the medium is too wet, there is high pupal mortality, and if too dry, dead pupae or defective moths result.

Lighting conditions seem to be of particular importance to adult insects. The photoperiod for rearing the immature insects can even affect the subsequent adults. For example, McFarlane (1985) found a dramatic photoperiod effect on the house cricket, *Acheta domesticus* (Linnaeus), with adults surviving up to twice as long with a 14-h nymphal rearing photophase as they do with a 12-h one.

Adults of many insects mate and oviposit best when provided with natural light through laboratory windows. Such insects include the Queensland fruit fly (Heather & Corcoran, 1985); the saltmarsh caterpillar, *Estigmene acrea* (Drury) (Vail & Cowan, 1985); and the light brown apple moth (Singh *et al.*, 1985).

Lighting conditions required for different species vary greatly. Robertson (1985a) reported that spruce budworm adults mated most successfully in darkness within 24 h after emergence, and optimal oviposition also occurred in the dark at 23 to 26°C. Robertson also noted that the best laboratory conditions for oviposition by the Douglas-fir tussock moth were complete darkness and 23 to 26°C (Robertson, 1985b). However, hemlock loopers required a light–dark cycle (Grisdale, 1985a).

Sometimes adults mate and oviposit best if provided with a weak light during the scotophase. Guy *et al.* (1985) held cabbage looper moths under a photophase of 14 h, with a 0.25-W night light during the scotophase. Leppla and Turner (1975) earlier had shown that maximum fecund-

ity of the cabbage looper was possible with low-intensity night illumination. Gardiner (1985b) likewise provided low-intensity light during scotophase by placing a 7.5-W bulb at a distance of 3 to 6 ft from *Mamestra brassicae* to stimulate mating and oviposition. He used a 60- to 100-W bulb at the same location during the 12-h photophase.

Guthrie *et al.* (1985) employed a slight asynchrony in light and temperature phases to provide for mating and oviposition by European corn borer moths, with two more daily hours of higher temperature than of light. They used a room temperature of 27°C for 16 h, and of 18 to 20°C for 8 h. The lights were on for 14 h, starting 1 h after initiation of 27°C.

Temperature can obviously be critical by itself, however, without interacting with the photoperiod. Tolman *et al.* (1985) showed that survival and fecundity of the onion maggot were substantially greater at 20°C than at 15, 25, or 30°C.

Humidity must also be considered in providing optimal mating and oviposition conditions. Leppla (1985) reported that the velvet bean caterpillar mated and oviposited best with a RH higher than 80%, and with a source of liquid food. Wight (1985) held the oviposition cage for southern armyworm moths over a pan of water, and covered the cage with black cloth to encourage oviposition. The humidity in the cage had to be more than 50% to obtain good mating, oviposition, and egg hatch.

Mating can sometimes be quite difficult to achieve in the laboratory, and may involve a variety of factors. Although Reed and Tromley (1985b) reared the lesser peach tree borer under aperiodic conditions, they held the adults under a 16-h photophase for mating and oviposition, and noted that proper environment was important to achieve mating, with outdoor conditions used when possible. Indoor conditions required adequate lighting and ventilation (to avoid pheromone accumulation). Moths observed in copula were removed, and the females were used for oviposition.

Density may affect optimal mating and oviposition. Laboratory mating and oviposition of the large white butterfly required a relatively large cage (100 × 90 × 75 cm) that contained 200 adults (Gardiner, 1985c). Tobacco hornworm moths likewise required a large cage (137 × 121 × 125 cm), but for just 50 pairs (Baumhover, 1985). Low-light conditions were also necessary (25-W light for 12 h and rheostat-reduced 7.5-W light for 12 h). On the other hand, a high density was not detrimental to mating and oviposition of the spruce budworm. Grisdale (1984) reported that placing up to 300 pairs of these moths into a screened cage (35 × 35 × 25 cm) did not impair oviposition.

Even members of the same insect family can vary dramatically in ease of laboratory mating and oviposition. This is certainly true of the mosquito family Culicidae. Friend and Tanner (1985) reported that *Culiseta inornata* (Williston) males often initiate mating before females have completely emerged, without special flight cages. Munstermann and Wasmuth (1985a) noted that *Aedes aegypti* (Linnaeus) also mates easily in confined spaces. However, these workers had to use beheaded, impaled males of the eastern tree hole mosquito, *Aedes triseriatus* (Say), in a forced copulation technique (Munstermann & Wasmuth, 1985b). They observed, however, that the Walton strain of *A. triseriatus* mated satisfactorily in a cubical cage of at least 60 cm³. Bailey and Seawright (1984) reviewed a system useful to achieve rapid laboratory colonization of *Anopheles albimanus* Wiedemann, a vector of malaria, and discovered that field-collected females individually placed in 5-dram vials would produce more than 100 times the number of eggs of an equal number (500) of females placed together in a single cubical cage of 61 cm³. The degree of clustering of ovipositing adults and the amount of space provided can affect fertility as well as oviposition, and is a production program factor.

Although not directly related to the physical environment, the age of adult insects is a further consideration for mating and oviposition. ODell *et al.* (1985) noted that gypsy moth females would not mate once they began to lay eggs. Codling moth adults held for more than 5 days before mating had considerably reduced fecundity (Singh & Ashby, 1985). Similarly, Grisdale (1985b) recommended mating female moths of the forest tent caterpillar as soon after eclosion as possible for optimal results.

Frequently, insects will have a preoviposition period between emergence and oviposition, during which they feed and develop eggs. For example, adult cabbage maggots have a preoviposition period of 6 to 7 days at 19 ± 1°C (Whistlecraft *et al.*, 1985b), and adult onion maggots have a similar period but at 22 ± 1°C (Tolman *et al.*, 1985).

After insects have appropriate mating conditions, they need ovipositional stimulants. Insects that lay eggs in crevices will often oviposit on crinkled wax paper or on cloth. A good oviposition substrate for the hemlock looper is a six-layer-thick cheesecloth (Grisdale, 1985a), and Reed and Tromley (1985b) provided moist cotton balls to motivate the lesser peachtree borer to oviposit. Other insects can be far more fastidious in their ovipositional requirements, with ingenious systems necessary. Heather and Corcoran (1985) used hollowed-out half apples as an ovipositional substrate for the Queensland fruit fly to enable easy egg collection, and Boller (1985) devised a clever dome made of ceresin wax that served as an oviposition substrate for the European cherry fruit fly. In the automated mass production of the melon fly, *Bactrocera* (= *Dacus*) *cucurbitae* Coquillett, Nakamori *et al.* (1992) used cylindrical plastic oviposition receptacles perforated with numerous holes and containing a cellulose sponge saturated with diluted tomato juice. Baumhover (1985) described artificial leaves composed of

outdoor carpeting sandwiched between layers of polypropylene, which served as substitutes for tobacco leaves, and were sprayed daily with a tobacco leaf extract to stimulate oviposition by tobacco hornworm moths.

Behavior

Knowledge of insect behavior is obviously crucial to designing a successful rearing program. This is evident in devising an ovipositional substrate and stimulus. Blenk *et al.* (1985) discovered that six of seven reared noctuid species would oviposit on the underside of a paper towel on top of an oviposition cage, whereas the black cutworm, *Agrotis ipsilon* (Hufnagel), would only oviposit on paper toweling in the bottom of the unit. On the other hand, Guthrie *et al.* (1985) observed that European corn borer moths would oviposit only on smooth surfaces; and Hartley (1990) reported that polyester–cotton cloth used for oviposition by *Heliothis* spp. had to be replaced by a polyester–cotton cloth with a dimpled surface for oviposition by the beet armyworm, the velvetbean caterpillar, and the soybean looper, *Pseudoplusia includens* (Walker).

Chemical ovipositional stimuli may also be necessary. Adult seed corn maggots oviposit in response to moist soil, decaying vegetation, germinating seeds, and metabolites produced by seed-borne microorganisms (Whistlecraft *et al.*, 1985a). Likewise, rearing of the biological control agent *Weiseana barkeri* Jacoby, a chrysomelid that feeds on *Acacia nilotica,* requires the use of *A. nilotica* foliage to stimulate oviposition into strips of corrugated cardboard (Marohasy, 1994). On the other hand, insect-produced chemicals may deter oviposition. Boller (1985) found that the European cherry fruit fly and the Mediterranean fruit fly produced oviposition-deterring pheromones that lowered egg deposition in artificial devices. Consequently, the devices needed frequent washing.

Insect responses to stimuli can simplify rearing. For example, the positive phototaxis of scale crawlers makes them easily collected. Papacek and Smith (1985) lighted a rearing room for oleander scale 2 h before work began so scale crawlers would accumulate on top of butternut pumpkins. Such positive phototaxis is a common attribute of many insects and is similarly useful in their collection.

In some cases the combining of behavioral characteristics may be disadvantageous. Grisdale (1985a) reported that first-instar hemlock loopers are strongly photopositive and active, but also cannibalistic. Because the larvae drink readily, a light spray of distilled water and retention in the dark at 18°C reduce cannibalism.

Light can also affect insect emergence. Willey (1985) noted that range grasshopper nymphs hatch from eggs daily about 5 to 8 h after the start of the light cycle, so daily collection is timed accordingly. Similarly, Boller (1985) observed that the European cherry fruit fly and the Mediterranean fruit fly both emerge mostly during the morning and oviposit in the afternoon.

Techniques

There are many and varied techniques for arthropod rearing. Some were previously mentioned, and other examples follow.

Stockpiling refrigerated hosts is very helpful. Determination of the effects of refrigerated host material on parasitoid and predator production must be on a case-by-case basis. Legner (1979b) found that successful refrigeration of housefly pupae used to rear three pteromalid parasitoids could only occur at 10°C for <21 days. Parasitoids given nonrefrigerated pupae produced significantly more female progeny, however. Because the progeny did not differ significantly in biomass, the decreased reproductive potential with the use of refrigerated hosts was not readily apparent.

Precise storage temperatures can often be very critical in rearing insects. Ankersmit (1985) reported that newly laid eggs of the summer fruit tortrix perished when held at 5°C. However, refrigeration of embryonated eggs at 5°C was satisfactory if the holding period was not more than 4 days. There was also a critical temperature above which the tortrix eggs hatched. Eggs held at 13°C did not hatch, but at 15°C about 70% hatched.

The insect stage put in cold storage is also important. Adults of *Arphia* spp. grasshoppers lose vigor if stored in the cold or without food for more than 1 day. However, storage of diapausing eggs left *in situ* in soil was satisfactory at 2°C for 1 to 2 years if the soil remained moist, but storage of nondiapausing eggs was suitable at 10 to 17°C for several months (Willey, 1985).

The length of storage of newly hatched red-banded leaf rollers can be at least 7 days at 5 to 7°C with 100% RH (Glass & Roelofs, 1985). However, inducing diapause through larval exposure to an 11-h photophase enabled leaf roller pupal storage for 6 months at 5°C.

Differential cold storage of the sexes is useful for synchronizing emergence, because males of most insects develop more rapidly than females, and emerge first. To achieve synchronized emergence of male and female hemlock loopers, Grisdale (1985a) sexed freshly formed pupae and stored the males initially at a temperature 4°C lower than that for the females.

To maximize and quantify insect production, suitable methods of determining insect numbers are necessary. One way is to estimate by weight. Baumhover (1985) weighed tobacco hornworm eggs to ascertain their numbers (400 eggs weigh 0.534 g). He cautioned to weigh only fresh eggs because they lose 20% of their initial weight by hatch time. Similarly, Moore and Whisnant (1985) estimated numbers

of reared boll weevil adults by first weighing a sample of 10 before weighing the total collection.

Egg number determination is useful for adjusting available food per individual to maximize production and food usage. For example, Guy *et al.* (1985) reared cabbage looper larvae gregariously on artificial diet, with diet amount per individual adjusted by placing an appropriate number of eggs in each container. A medicine dropper facilitated applications of eggs to squares of paper toweling, with 50 to 60 eggs per spot. Casein glue affixed squares with dried egg spots to each container lid.

There are various techniques to separate insects from a substrate or from each other. Greenberg and George (1985) separated blowfly eggs by using 1% sodium sulfite to dissolve the adhesive holding them together. As previously discussed, the use of sodium hypochlorite is a similar method.

Rahalkar *et al.* (1985) described a technique to separate eggs of the red palm weevil, *Rhyncophorus ferrugineus* Oliver, from shredded sugarcane by placing this ovipositional substrate into a 30% aqueous solution of glycerol. After the sugarcane shreds sank, the floating eggs were collected with a strainer. Similarly, Martel *et al.* (1975) developed a method of extracting eggs of the carrot weevil, *Listronotus oregonensis* (LeConte), from carrot pieces.

Morgan (1985) separated viable from nonviable housefly eggs by placing them in water, where the viable eggs sank. The technique also worked to separate pupae from the larval medium, because the pupae floated. Likewise, Tolman *et al.* (1985) designed a simple flotation device to separate pupae of the onion maggot from the cut onion and sand larval substrate. However, the pupae must be at least 48 h old before they will float on water.

A separator and high-volume blower harvested pupae of the beet armyworm, the velvetbean caterpillar, and of *Heliothis* spp., but a special harvesting method was necessary for pupae of the soybean looper because of webbing (Hartley, 1990).

Anesthetizing insects makes handling easier. Carbon dioxide is useful when the brevity of its effect is not a hindrance. A combination of ethyl ether and carbon dioxide provides longer activity, as described by Etzel (1985). Munstermann and Wasmuth (1985b) made a device that used nitrogen gas saturated with water vapor to anesthetize adult eastern tree hole mosquitoes, and Nettles (1987) anesthetized the adult tachinid *Eucelatoria bryani* Sabrosky with nitrogen to enable sexing.

A general problem in producing lepidopterans is the accumulation of moth scales, which can be highly allergenic to humans. The usual method of removing the scales from the rearing environment is air filtration, as was utilized by Kumar and Jalali (1993) in the vacuum collection of moths and eggs of the rice meal moth, *Corcyra cephalonica*

(Stainton). However, Baumhover (1985) used another method in rearing the tobacco hornworm. He noted that 85% RH in mating and oviposition cages prevented moth–scale pollution if the moths remained inactive. Davis and Jenkins (1995) also addressed the issue of health hazards in lepidopteran-rearing facilities. They efficiently minimized this serious occupational hazard by facility and cage design, by installing an improved air filtration system, and by using appropriate sanitation procedures.

There are various methods of containing arthropods in rearing units. Sleeve cages of assorted sizes with wooden frames, organdy cloth sides, glass tops, and cloth sleeves to enable manipulation of cage contents are prevalent. Some arthropods, particularly mites, are often reared in open units. Margolies (1987) used a mixture of 4 ml of clove oil in 100 g of lanolin applied to the edge of a petri dish to confine tetranychid mites. Other arthropods that are only motile by walking can be contained by applying Fluon® to the edge of the unit in which they are contained (Morrison *et al.,* 1997).

Physical handling of insects can be critical to their rearing. Some insects are particularly fragile, at least in one or more stages. Boller (1985) observed that gentle handling prevented muscle ruptures of young fruit fly pupae.

The foregoing is a sampling of techniques and considerations in rearing insects generally. Each rearing system is different and may require development of unique methods. However, even after a seemingly satisfactory rearing system has been developed, difficulties whose causes are difficult to determine can develop. Such was the case with mass production of the gypsy moth, when a problem known as straggling appeared (ODell, 1992). Straggling, an apparent growth abnormality characterized by delayed and unpredictable larval growth, severely interfered with insect production for research and control purposes. ODell (1992) reported that the process initiated to systematically examine and evaluate gypsy moth rearing procedures could be generally used as a model for solving production problems.

Quality

The required quality of produced insects depends on their intended use. Waage *et al.* (1985) noted that species, size, and stage are factors affecting the quality of a host for parasitoids and predators.

Three aspects of quality are standards, assessment, and control. A survey of several papers revealed that the most commonly used quality criterion was fertility. The fertility test was used in producing Heteroptera (Jones, 1985), Coleoptera (Jackson, 1985; Rahalkar *et al.,* 1985), and Lepidoptera (Navon, 1985; Robertson, 1985a, 1985b; Bartlett & Wolf, 1985; Reed & Tromley, 1985a, 1985b; Guy *et al.,* 1985; ODell *et al.,* 1985). King and Hartley (1985a) used

this criterion in rearing the sugarcane borer as a host for the tachinid *Lixophaga diatraeae*, but also considered fecundity, larval and adult survival, and percentage of adult emergence. These quality criteria are in common usage, although workers often substitute mortality for survival. Another typical criterion is size, either in dimensions or weight. Vargas *et al.* (1994) used duration of larval development, pupal recovery, pupal weight, adult fecundity, and egg fertility to evaluate the quality of Mediterranean fruit flies reared on different diets, but emphasized fertility. An unusual criterion is "fluctuating asymmetry of bilateral characters," which Clarke and McKenzie (1992) found to be the best quality indicator in producing a sheep blowfly, *Lucilia cuprina* (Wiedemann).

Patana (1985b) believed that continued reproduction and survival are the most significant indicators of insect quality in long-term laboratory cultures of at least 1 to 5 years. Leppla *et al.* (1984) commented that even the best programs for mass-producing cabbage loopers can depend only on quantity produced, because quality control is difficult.

CULTURE OF ENTOMOPHAGOUS ARTHROPODS

Contamination

The culture of entomophagous arthropods for biological control involves many of the same factors as those for producing host organisms. However, there are a number of special considerations necessary. In biological control rearing systems the host-food and host-production levels are those that are primarily prone to contamination. Rarely do parasitoids of predaceous insects or hyperparasitoids of primary parasitoids cause impediments. Such difficulties usually occur only when a parasitoid or predator culture is first started from field-collected material. Chianese (1985) recommended that a culture of *Cotesia melanoscela* (Ratzeburg), a braconid parasitoid of gypsy moth larvae, be established from the first seasonal field generation to minimize hyperparasitism problems. Individual isolation of field-collected cocoons will prevent any emerging hyperparasitoids from attacking other cocoons. His advice is equally applicable to aphid parasitoids that frequently suffer from increasing hyperparasitism as the growing season progresses.

Morgan (1992) reported that crosscontamination between species of primary parasitoids of filth flies occurred during their mass production. Rearing *Spalangia* spp. at 28 to 29°C resulted in pure cultures, because the parasitoids *Muscidifurax raptor* Girault & Sanders and *Pachycrepoideus vindemiae* (Rondani) could not mature at that temperature. However, obtaining pure cultures when there was contamination between the latter two species, or between *Spalangia* spp., required the laborious examination of individual parasitoids with a cold table and a vacuum pencil. Purity of the parasitoid colonies was tested by electrophoretic analysis.

Microorganisms can affect entomophages as well as their hosts. Goodwin (1984) remarked that parasitoids developing in microbially diseased hosts may or may not contract the disease, but nonetheless would suffer physiologically and be less fit from host debilitation. On the other hand, Zchori-Fein *et al.* (1992a) discovered a situation where housefly hosts of *Muscidifurax raptor* in culture were not infected by a maternally transmitted microsporidian that severely debilitated the parasitoid. Geden *et al.* (1995) later found that higher (84%) levels of infection by the microsporidian *Nosema muscidifuracis* were observed on New York dairy farms where infected *M. raptor* from a commercial insectary had been released. They reduced disease incidence 35 to 93% in the laboratory by immersing infected parasitoid eggs within fly puparia in a 47°C water bath for 30 to 60 min. A 10-fold increase in parasitoid fecundity was obtained by eliminating the disease from an established *M. raptor* laboratory colony.

Brooks (1974) also reported the finding of protozoans in parasitic Hymenoptera, and Own and Brooks (1986) showed that *Pediobius foveolatus* — an egg parasitoid of the Mexican bean beetle — is highly susceptible to two protozoans. Because *P. foveolatus* is commonly mass-produced in the eastern United States for annual inoculative releases, the protozoans can be a serious limiting factor in parasitoid production. Sajap and Lewis (1988) similarly showed that *Nosema pyrausta* in eggs of the European corn borer significantly reduced emergence and fecundity of *Trichogramma nubilale* (Ertle & Davis). On the other hand, the same microsporidian had no effect on larvae of the common green lacewing that fed on infected corn borer eggs (Sajap & Lewis, 1989).

The bacterium *Serratia marcescens* caused mortality in cultures of the tachinid *Lixophaga diatraeae* on the sugarcane borer (King & Hartley, 1985a, 1985c). The technique to control the disease was to soak the maggots in 0.7% formalin solution for 5 min before they parasitized sugarcane borers, followed after larval development by disinfection of the puparia with 1% sodium hypochlorite solution for 3 min (King & Hartley, 1985c).

Serratia marcescens was also a problem in the laboratory production of the tachinid *Lydella jalisco* Woodley, a parasitoid of the Mexican rice borer, *Eoreuma loftini* (Dyar). *Serratia marcescens* contaminated first-instar tachinid maggots during harvesting, and then entered new hosts with the penetrating maggots, which resulted in high host mortality from septicemia (Rodriguez-Del-Bosque & Smith, 1996). The disease was controlled by surface sterilization of the harvested maggots with formalin.

Similarly, *S. liquefaciens* caused heavy mortality in rear-

ing of the tachinid *Ernestia consobrina* Meigen in larvae of *Mamestra brassicae* (Krieg & Huger, 1987). Along with hygienic measures, nalidixinic acid and chloramphenicol were suggested therapeutics for controlling this disease.

Goodwin (1984) listed some rickettsiae and closely related forms found to be pathogenic for entomophages. Chlamydiae causing stunting were in the nerve tissue of the sevenspotted lady beetle, *Coccinella septempunctata* Linnaeus, and in the ichneumonid alfalfa weevil parasitoid *Bathyplectes* sp. Also, *Enterella stethorae* (a rickettsia) caused an acute disease in the predaceous coccinellid *Stethorus* sp. by destruction of the midgut epithelium. A cytoplasmically inherited microorganism associated with male killing in the twospotted ladybird beetle, *Adalia bipunctata* (Linnaeus), was shown to be closely related to the genus *Rickettsia* (Werren *et al.,* 1994). Likewise, a maternally inherited "son-killer" bacterium causes the death of male embryos of infected females of the parasitoid *Nasonia vitripennis* (Balas *et al.,* 1996).

Production of entomophages on artificial diet is an expanding area of study, one in which microbial contaminants will be a constant potential problem. Bratti (1990b) discussed chemical–physical methods for controlling microorganism contamination of the environment and diets for *in vitro* rearing of parasitoids. Grenier and Liu (1990) tested fungicides in artificial host egg capsules for rearing *Trichogramma,* and found that most were ineffective or toxic. However, geneticin, amphotericin B, and nystatin at proper levels efficiently controlled molds without modifying *Trichogramma* development. These fungicides were also tested for *Trichogramma* mass production in China.

For *in vitro* rearing of the tachinid *Exorista larvarum* (Linnaeus), Mellini and Campadelli (1995) showed that transmission of bacteria and fungi to the diet by surface contamination of the eggs was prevented by egg disinfection with 60% alcohol, 0.26% AI sodium hypochlorite, or 1.3% formaldehyde.

Although some microorganisms are detrimental to entomophages, Goodwin (1984) noted that entomophagous parasitoids may have symbiotic microorganisms enabling them to successfully attack hosts. For example, Stoltz and Vinson (1979) showed that viruses present in the oviducts of braconids and ichneumonids suppressed the defensive hemocoelic encapsulation process in their hosts. With another system, Girin and Bouletreau (1995) hypothesized that the high infestation efficiency (60 eggs per 5 days per female) of a strain of *Trichogramma bourarachae* Pintureau and Babault was due to the presence of cytoplasmic–symbiotic microorganisms, which were not present in a low efficiency strain (25 eggs per 5 days per female). They supported their hypothesis with microscopic observations, and with antibiotic and heat treatments that reduced infestation efficiency only in the "high" strain. Another example of a microorganism advantageous to rearing a beneficial arthropod was the approximately 10-fold increase in production of the predaceous mite *Typhlodromus doreenae* Schicha when reared in the presence of mold mite *Tyrophagus putrescentiae* (Schrank) and the fungus *Rhizopus stolonifer* (James, 1993).

Microorganisms also can have direct involvement in reproductive processes of parasitoids, as first suggested for pteromalids (Legner, 1987b, 1989a, 1989b), trichogrammatids (Stouthamer *et al.,* 1990), and an aphelinid (Zchori-Fein *et al.,* 1992b). To what extent the acquisition of thelytoky (asexual production of female offspring) in Hymenopteran species involves chromosomal inheritance is uncertain. Evidence began to accumulate a decade ago that the process can involve extrachromosomal phenomena such as infection by microorganisms in the reproductive tract (Legner 1987a, 1987b, 1987d). Subsequent experimental data further indicated the probability of microorganisms being involved in thelytoky. That work, with *Trichogramma,* employed three kinds of antibiotics and high temperature to "cure" populations of their thelytoky (Stouthamer *et al.,* 1990). Likewise, Zchori-Fein *et al.* (1992b) obtained large numbers of males from antibiotic-fed females of thelytokous *Encarsia formosa* Gahan. Following further research, Stouthamer *et al.* (1993) concluded that the "parthenogenesis" microorganisms, together with the very closely related cytoplasmic incompatibility bacteria, belong to the Proteobacteria and "specialize" in manipulating chromosome behavior and reproduction of insects.

There is currently very active research in this field of relationships of microorganisms and reproductive processes in arthropods. Work has shown that bacteria in the genus *Wolbachia* (in the alpha Proteobacteria) are associated with reproductive alterations in their hosts, including parthenogenesis in insect parasitoids, cytoplasmic (reproductive) incompatibility in several insect and mite species, and feminization in various isopods (Werren *et al.* 1995; Breeuwer & Jacobs, 1996; Rigaud & Rousset, 1996; Min & Benzer, 1997). *Wolbachia* causes thelytoky in various parasitoids, including the pteromalid *Muscidifurax uniraptor* Kogan & Legner (Horjus & Stouthamer, 1995; Zchori-Fein *et al.,* 1995), species of *Trichogramma* (Louis *et al.,* 1993; Horjus & Stouthamer, 1995; Schilthuizen & Stouthamer, 1997), the encyrtid *Apoanagyrus diversicornis* Howard (Pijls *et al.,* 1996), and the aphelinids *Aphytis lingnanensis* Compere and *A. melinus* DeBach (Zchori-Fein *et al.,* 1995). On the other hand, *Wolbachia* can also cause cytoplasmic (reproductive) incompatibility in various insect and mite species, resulting in a severe reduction of the progeny produced, which consists of all males or mostly males (Lassy & Karr, 1996; Breeuwer, 1997; Frank, 1997; Clancy & Hoffmann, 1997). Cytoplasmic incompatibility caused by the maternally inherited *Wolbachia* in insects is characterized by crosses between infected males and uninfected fe-

males being sterile, whereas those between infected males and infected females are fertile (Frank, 1997). An insect parasitoid that exhibits cytoplasmic incompatibility from *Wolbachia* infections is *Nasonia vitripennis* (Breeuwer & Werren, 1993).

Several workers have indicated that *Wolbachia* can be transmitted not only vertically, but also horizontally, even between diverse insect or arthropod taxa (Werren *et al.,* 1995; Johanowicz & Hoy, 1996; Rigaud & Rousset; 1996; Schilthuizen & Stouthamer, 1997). Further information on *Wolbachia* can be found in a discourse on a phylogenetic analysis by Werren *et al.* (1995) and in a review by Werren (1997).

The implicated involvement of microorganisms in reproductive processes is obviously a further complication in the culturing of entomophages. For example, Stouthamer and Luck (1993) found that with unlimited host availability, parthenogenetic females produced fewer daughters than did bisexual females.

Pathogens are more a problem with weed-feeding insects used for biological control than with parasitoids and predators. Etzel *et al.* (1981) eliminated a *Nosema* disease interfering with production of the weed-feeding chrysomelid *Galeruca rufa,* and Bucher and Harris (1961) noted other cases.

It is well to remember that some microbial contaminants can be hazardous to rearing personnel because of pathogenicity or allergenicity (Sikorowski, 1984). For example, yeasts present in the cocoons of the parasitoids *Nasonia vitripennis* and *Muscidifurax* spp. apparently cause respiratory and intestinal allergic reactions (H. G. Wylie, personal communication; E. F. Legner, unpublished). Because of the various problems caused by microorganisms in insect production, Sikorowski (1984) and Sikorowski and Lawrence (1994) emphasized that basic sanitation must be an essential feature of insect mass-rearing programs. However, sanitation is critical not only because of microorganisms but also because of the human health hazards of any biological debris associated with arthropod rearing. Lugo *et al.* (1994) reported seven cases of allergic asthma and rhinoconjunctivitis in 13 workers who were producing beneficial arthropods in European "biofactories."

As a final note, Hopper *et al.* (1993) strongly stressed that unrecognized disease (particularly when transmitted maternally) is a serious problem when comparing performance of genetic strains of entomophages, and can also confuse the results of selection experiments.

Genetic Changes

Introduction of the Problem

Waage *et al.* (1985) noted that laboratory genetic changes resulting from long-term culturing of entomophages could unfavorably affect environmental fitness and be-

143

havioral characteristics such as host finding, host acceptance, and host preference. The potential for successful
introduction of parasitoids and predators into new areas is
linked to their genetic constitution and changes therein, as
considered by Legner and Warkentin (1985), Cilliers
(1989), Roush (1990), and others. Considerable controversy
exists as to the influences of homozygosity and heterozygosity on the fitness and capacity of biological agents to be
effective. There are theoretical considerations that do not
entirely support the requirement of heterozygosity (Remington, 1968; Legner, 1979a; Legner & Warkentin, 1985).

It is possible for even small numbers of insect parasitoids or predators introduced into a new area to become
successfully established and effective in controlling a pest.
There are examples of small numbers of pests, even single
fertilized females, invading new areas and establishing successful populations (Remington, 1968; Bartlett, 1985). Concerning biological control agents, Clausen (1977) noted the
Cuban experience with the predaceous coccinellid *Scymnus
smithianus* Clausen & Berry, imported from Sumatra in
1930 to control the citrus blackfly, *Aleurocanthus woglumi*
Ashby. Insectary stock of the coccinellid declined to a single female, but then increased to the point that releases at a
number of locations in 1931 resulted in establishment.

Bartlett (1985) indicated that fitness and genetic plasticity in insects reared for field release are essential for them
to exist in the environment, and stated that in every case
genetic plasticity decreases as genetic variability decreases.
He and Joslyn (1984) both noted that insects in laboratory
colonization will unavoidably become domesticated and
lose genotypes (variability), depending on a variety of factors. According to Bartlett (1984), decreased fitness often
occurs when homozygosity increases in a culture (i.e., genetic variability decreases). Such domesticated insects will
be less likely to adapt to a natural environment than those
with high fitness and high heterozygosity. It is important to
observe that the total number of alleles in a laboratory
culture may not change greatly over time. Instead, the important changes that can alter fitness are in distribution of
the alleles and in their arrangement (Bartlett, 1985). Myers
and Sabath (1980), Remington (1968), and Hopper *et al.*
(1993) offer a number of interesting theoretical viewpoints
on colonizing insects and on hetero- and homozygosity.

Types of Changes in Culture

Inbreeding depression, which produces a reduction of
physiological vigor and reproductive capacity, can occur in
small laboratory populations over time (Collins, 1984). Inbreeding and genetic deterioration of insectary stocks of
parasitoids and predators are always of concern, particularly
with cultures initiated from very few individuals or maintained as small cultures for extended periods.

Nakamura (1996) initiated laboratory culture lines from

10 pairs of field-collected specimens of the tachinid fly,
Exorista japonica Townsend, a natural enemy of many
lepidopterous larvae, and showed that the rates of adult
emergence and host mortality decreased even at the F2
generation, with lines derived from all pairs becoming extinct by F6 under inbreeding conditions. Rotational breeding of the lines prevented inbreeding depression up to the
F9 generation, when the study was completed.

Both males and females of tachinids such as *E. japonica*
are diploid. However, theoretically, significant inbreeding
depression should not develop in natural enemies with haplodiploid sex determination (Roush, 1990). Sorati *et al.*
(1996) inbred the egg parasitoid *Trichogramma* sp. nr.
brassicae for four generations and found no effects of inbreeding depression for any of the four female traits examined (fecundity, body length, head width, and hind tibia
length). They also found that measures of fitness were uncorrelated with wasp size. Kazmer and Luck (1995) likewise found that wasp size is not a reliable predictor of
individual or average cohort fitness in *T. pretiosum* (Riley).
Legner (1979a) studied the influences of inbreeding and
extended culturing on the reproductive potential of the pteromalids *Muscidifurax raptor* and *M. zaraptor* Kogan &
Legner. He found no reduced reproductive capacity of two
inbred lines established from an old laboratory culture of
M. raptor. With *M. zaraptor,* two of seven inbred lines
derived from a standard laboratory culture demonstrated
significantly greater intrinsic rates of increase (R_m) in three
comparisons with the standard culture. The other inbred
lines sometimes exhibited R_m values significantly lower
than that of the standard culture. However, the R_m value of
the standard culture was very close to the average of the
values of the seven inbred lines. This indicated that the
inbred clones represented a sample of genotypes from the
standard culture, with considerable genetic variability maintained overall.

Geden *et al.* (1992) cultured *M. raptor* for extended
periods as well, but found a rapid deterioration of searching
behavior, host destruction, and fecundity of the parasitoid.
After performing crossing experiments, however, they concluded that the deterioration was primarily caused by a
maternally transmitted pathogen, and only to a lesser extent
by inbreeding depression.

Legner (1979a) also compared a Danish strain of *M.
raptor* with an American wild strain, an old domestic laboratory culture, and two inbred lines started from the domestic culture. The Danish strain had markedly lower reproductive rates than did any of the other cultures. This finding
emphasizes the importance of the initial genetic composition of a laboratory culture of parasitoids or predators. Certain inbred long-cultured parasitoids or predators may also
be better candidates for field releases than are recently collected wild strains, or than are specimens from the original
parental culture for the inbred lines (Legner, 1979a).

Grinberg and Wallner (1991) likewise found no deleterious effects after culturing the braconid *Rogas lymantriae* Watanabe for 125 generations. Furthermore, culture of the predator *Geocoris punctipes* (Say) on artificial diet for 6 years did not cause degradation in its prey-selection characteristics (Hagler & Cohen, 1991). However, it is best to consider laboratory inbreeding to be deleterious to field establishment, and to plan culturing techniques accordingly. For example, Hoeller and Haardt (1993) were unable to achieve field establishment of two inbred lines of the aphid parasitoid *Aphelinus abdominalis* (Dalman) in winter wheat fields, even though in the laboratory both lines had high fecundities, female-biased sex ratios, and good longevity. Also, the parasitoid would not attack aphids in field cages.

Waage *et al.* (1985) noted that inbreeding can greatly affect the sex ratios of parasitic wasps and predaceous mites. This consequence of inbreeding seems related to the haplodiploid system of reproduction, where unfertilized haploid individuals are male and fertilized diploid ones are female. In at least some genera of ichneumonids and braconids, the sex of a parasitoid is determined by multiple alleles at a single chromosomal locus, with heterozygotes being female while homozygous or hemizygous individuals are male (Stouthamer *et al.*, 1992). A haploid individual is hemizygous (has only one of every pair of chromosomes found in a diploid individual) and therefore is effectively "homozygous" (the single allele at a genetic locus has the same effect as two identical alleles at the same locus in a diploid individual). With inbreeding, homozygosity will increase and may result in formation of diploid males (in addition to the normal haploid males) in at least some Ichneumonoidea, with a consequential male bias in the sex ratio (Cook, 1993a, 1993b; Periquet *et al.*, 1993; El Agoze *et al.*, 1994). It is generally believed that this is one of the key reasons why it is difficult to maintain cultures of ichneumonoid parasitoids (N. J. Mills, personal communication). Nevertheless, the effect of inbreeding on sex determination also probably depends on the degree of natural inbreeding (Waage *et al.*, 1985), so that gregarious wasps with a natural sex ratio biased toward females will likely suffer less from laboratory inbreeding than will solitary wasps with a 1:1 natural sex ratio (Waage, 1982). The extensive inbreeding performed with solitary species of the genus *Muscidifurax* (Legner, 1979a) did not skew the sex ratio toward males, however.

In predaceous Heteroptera, Neuroptera, Coleoptera, and Diptera, as well as in parasitic Diptera—where both males and females are of diploid origin—the sex ratio remains at about 1:1 (Waage *et al.*, 1985).

Causes of Genetic Change

Founder Effect (Field Sampling)

The loss of genotypes in laboratory culture is especially dependent on the founder effect (Bartlett, 1984), a random

event where there is initially a very restricted gene pool resulting from the selection of few founder individuals (Joslyn, 1984). The initial variation in the new laboratory culture will depend on the number of individuals collected, and the number of collection localities. As Bartlett (1985) stated, "The larger the original sample, the smaller the deviations of the sample from the original gene frequencies; the smaller the sample, the greater the observed deviation." However, for polygenically inherited traits, the loss of allelic diversity from having a small number of founder individuals will not necessarily be associated with a great reduction in the total genetic variance available for adaptation (Roush, 1990).

Natural Selection in the Laboratory

Bartlett (1984) noted that the variability and quality of the laboratory environment are other important influences on genetic variability in culture. Provision of a constant favorable laboratory environment changes the criteria that determine fitness and might eventually drastically change the capability of laboratory-reared insects to persist in nature when field released. He further observed that space restrictions in laboratory-rearing units could affect insect behavior—including mating, oviposition, and dispersal—and might result in a genetically selective impact on that behavior. Once started, there will be a natural selection for individuals in the culture that are most fit for the artificial environmental conditions of the laboratory, with a resulting decrease of genotypes not favored in the laboratory, but possibly favored in nature (Bartlett, 1985). Joslyn (1984) also believed that directional selection in the laboratory culture, whether planned or unplanned, can cause a loss of alleles that contribute to fitness in nature, particularly if environmental conditions in the insectary are constant. On the other hand, Hoy (1993) reviewed planned genetic selection for pesticide resistance in predators and parasitoids, and noted success with the phytoseiid mite *Metaseiulus* (*Typhlodromus*) *occidentalis* (Nesbitt), the common green lacewing, and the parasitoids *Aphytis melinus* DeBach and *Trioxys pallidus* Haliday. Hoy (1993) observed that often only a single major gene will determine pesticide resistance, thus more readily enabling the development of pesticide-resistant biotypes of natural enemies. Similarly, in the mass production of *Trichogramma maidis* Pintureau & Voegele, Qiu *et al.* (1992) achieved improved quality through selective laboratory rearing. With *T. brassicae* Bezdenko, Pintureau (1991) was able to select for improved fecundity, but not for improved host-finding capacity. Nonetheless, the improved fecundity had no clear influence on effectiveness of the parasitoid in greenhouse tests. In attempts to select for beneficial characteristics, Havron *et al.* (1987) discussed advantages of being able to do so on the basis of the hemizygous males in arrhenotokously reproducing Hymenoptera and Acarina.

Genetic Drift

Part of the loss of genetic variability in the laboratory is the result of genetic drift, a random event that occurs as the laboratory population fluctuates in size (Joslyn, 1984). The loss of alleles increases as population size decreases. However, Hopper *et al.* (1993) argued that, theoretically, drift-caused loss of genetic variation is not great, even with very low laboratory populations, and may be an unnecessary concern.

Mating Type

According to Bartlett (1985), the type of mating system will not change the allele frequencies in a laboratory population, but will change the genotypic frequencies (i.e., the relative degrees of homozygosity and heterozygosity). The three mating types are random mating, outbreeding, and inbreeding. Random mating in a large laboratory population should maintain existing genotypic frequencies. Outbreeding (hybridization) can cause increased heterozygosity and the accompanying hybrid vigor or degeneration, depending on the degree of integration or disparity between the gene pools of the two populations (Bartlett, 1985). Joslyn (1984) described inbreeding as a directional event that increases homozygosity across all loci. This decrease in heterozygosity may cause lower fitness through loss of hybrid vigor and through inbreeding depression produced by any harmful homozygous recessive alleles.

Experimental data with parasitic Hymenoptera, however, are grossly lacking, with most of the knowledge of genetics and probable outcomes in culture being derived from other animals. The haplodiploid system of sex determination in hymenopterous parasitoids complicates the process of hybridization. Because the haploid males develop from unfertilized eggs, the genome they carry is strictly maternal. However, their sisters develop from fertilized eggs and carry both maternal and paternal alleles. Therefore, a mother that mates with a male of a different strain or population will produce hybrid daughters, but her haploid sons will only have maternal alleles. Matings of hybrid sisters to their brothers would then effectively be backcrosses to the maternal line, and the grandsons of their mated mother would be the first male generation to exhibit hybrid characteristics (Legner, 1988b, 1988c). Because of this natural backcrossing effect in arrhenotokous parasitoids, Legner (1988b) suggested that there is a high potential for the production of a greater proportion of parental genotypes in succeeding generations, which in turn would set up conditions for additional hybrids and their accompanying heterosis. Hybrids demonstrating a greater reproductive capacity also should guarantee the persistence of both parental genotypes in the culture, with the possibility of the continual formation of superior hybrids (Legner 1988b, 1988c, 1991b).

In outbreeding certain parasitic Hymenoptera, Legner (1988c, 1989a) discovered a unique process involving partial gene expression ("wary genes") in mated females, and a stepwise pathway to inheritance termed "accretive inheritance." Polygenes can control parasitic Hymenoptera behavior, but quantitative inheritance might also involve extranuclear changes (Legner, 1991a, 1993). In outbreeding gregarious and solitary forms of *Muscidifurax raptorellus,* a South American species that parasitizes synanthropic Diptera, Legner (1988c, 1989a, 1993) found that partial expression of male polygenes controlling fecundity, gregarious oviposition, and other reproductive behavior occurred within hours in the inseminated female due to substances transferred at mating. The magnitude of this partial expression depended on the genome of the female with whom the male mated, but the behavioral changes were permanent, with no switchback following a second mating with a different male. Full expression occurred in the F1 virgin hybrid female.

In hybridization, wary genes may serve to quicken evolution by allowing natural selection to begin action in the mated mothers, where they are phenotypically expressed prior to their incorporation into the progeny genomes. Wary-gene alleles detrimental to the hybrid population might thus be more prone to elimination, and beneficial ones more likely expressed in the mother before the appearance of her active progeny (Legner, 1988a, 1988c, 1989a).

Thelytokous parthenogenetic entomophages are a special case with respect to genetic selection in the laboratory, in that virgin females produce female progeny. Selection can happen very quickly as with aphids. Forbes *et al.* (1985) cautioned that laboratory selection of aphids occurs rapidly as a result of parthenogenesis, because no sexual genetic recombination occurs under the usual rearing conditions.

However, Rossler and DeBach (1972) showed that mating and genetic mixing may occur in thelytokous Hymenoptera under certain conditions. They worked with arrhenotokous and thelytokous forms (sibling "species") of the aphelinid species complex *Aphytis mytilaspidis* (LeBaron) collected from different scale hosts in different geographic areas. By using genetic eye color markers found in the arrhenotokous form, mode of reproduction (arrhenotoky or thelytoky), and host preference as testable traits, Rossler and DeBach (1972) demonstrated that males of the arrhenotokous form could mate normally with thelytokous females. The greatest barrier to interbreeding seemed to be the precopulation period, where arrhenotokous males spent a greater time in courtship with thelytokous females than with arrhenotokous females (Rossler & DeBach, 1972). There was a tendency for the thelytokous form to be replaced eventually by the arrhenotokous form, and the persistence of thelytoky seemed to be dependent on the hybrids finding suitable hosts (Rossler & DeBach, 1972). However, as previously discussed, the extent to which chromosomal inheritance governs thelytoky is uncertain, because the bacterium *Wolbachia* has been shown to cause thelytoky in at least some hymenopterous parasitoids.

Recombination

If sampling bias or selection has reduced the number of alleles present at any locus, the role of genetic recombination in increasing the number of genotypes will be correspondingly less because of the fewer alleles available (Bartlett, 1985).

In the future, however, recombinant DNA techniques developed by molecular biologists might be used to genetically manipulate arthropod natural enemies (Hoy, 1993). Presnail and Hoy (1992) developed a "maternal microinjection" technique that was used to insert exogenous DNA into the predaceous mite *Metaseiulus occidentalis*. Further experimentation with the technique has occurred subsequently (Presnail & Hoy, 1994; Hoy *et al.*, 1995).

Mutation

The effect of mutation on increasing the genetic complement and the variability of a laboratory culture is very low because the natural mutation rate for beneficial mutants is also very low (Bartlett, 1985).

Retaining Genetic Diversity

Bartlett (1984, 1985) and Joslyn (1984) essentially suggested the same three most important methods for retaining genetic variability in laboratory cultures of insects. Bartlett (1984) recommended (1) beginning the culture with as many founding individuals as possible, (2) using an environment set to maintain the fittest genotype by using appropriate fluctuating temperatures and photoperiods throughout the life cycle, and (3) maintaining separate culture strains under unique conditions and crossing these systematically to increase F1 variability. Periodic culture infusions and monitoring of culture quality are further techniques used for genetic variability retention.

Culture Initiation

The collection of large numbers of pest insects to initiate a laboratory culture is not difficult, the main restriction being seasonal. However, the collection of large numbers of entomophages for classical biological control introduction work is the exception, rather than the rule. An effective entomophage that has reduced its host population to low subeconomic levels will be difficult to find. Thus, foreign exploration trips frequently yield very few specimens, and it is not unusual to initiate laboratory cultures from less than 10 individuals. From the practical point of view the best procedure is to process the agents through quarantine promptly and to begin field releases quickly to minimize the genetic loss of alleles. Naturally, insectary cultures are necessary for further releases. However, Hopper *et al.* (1993) recommended that culture time should not exceed five generations, if possible, prior to field release of entomophages.

One method to collect larger numbers of parasitoids is to expose laboratory-cultured host material to parasitoid attack in the field. Harris and Okamoto (1991) exposed papaya fruits infested with oriental fruit fly eggs to field populations of the braconid *Biosteres arisanus* (Sonan) (= *Biosteres oophilus* [Fullaway]), and then held the fruits in the laboratory where they used the emerging parasitoids as founders for a new culture. Similarly, Mills and Nealis (1992) collected numbers of the tachinid *Ceranthia samarensis* (Villeneuve) in Europe by placing gypsy moth larvae in areas with low density gypsy moth populations, and then by recovering the emerging parasitoids in the laboratory.

Obviously, field collecting the largest number of individuals that the laboratory can accommodate will maximize the number of feral population alleles present in the founder laboratory culture and will consequently minimize the magnitude of genetic drift and inbreeding (Bartlett, 1985). Roush (1990) noted that sampling from as many different sites as possible is the objective and suggested that the sample size should be between 20 and 100 individuals. However, individuals from the different areas must obviously have reproductive compatibility (Joslyn, 1984). Bartlett (1985) illustrated the mathematical procedure for estimating the number of field specimens to be collected to obtain a rare allele in the founder laboratory culture.

Wheeler (1984) suggested starting a laboratory insect culture with 300 to 500 individuals and collecting only a single developmental stage to increase the chances of obtaining maximum homozygosity, if that is the goal. Conversely, collecting a variety of stages at the same time should increase the chances of establishing a high heterozygosity in a new culture.

Culture Maintenance and Size

Unruh *et al.* (1983) believed that the best way to retain heterozygosity and prevent genetic drift in laboratory cultures is to maintain relatively large population sizes (>100) in the laboratory. After studying heterozygosity and effective size in laboratory populations of the aphidiid *Aphidius ervi* Haliday, Unruh *et al.* (1983) warned that genetic drift and loss of heterozygosity are more severe than would be expected from the number of individuals used to maintain cultures. Discussed were factors that make effective population sizes much smaller than would be apparent, including fluctuations between generations, haplodiploidy, sex linkage, highly variable progeny production by individual females, and highly skewed sex ratios.

Omwega and Overholt (1996) found that the effective population sizes for two geographic laboratory colonies of the braconid *Cotesia flavipes* Cameron from Pakistan were estimated at 3.4% and 9.0% of the number of breeding females used to continue the colonies. Approximately 1000 breeding females were used to perpetuate each colony during the first 12 laboratory generations, and during this time there was actually a slight increase in heterozygosity at the MDH (malate dehydrogenase) genetic locus in each colony. However, during the next 10 generations, only 500 females were used per generation, and there was a rapid decrease in genetic variation, as measured by electrophoresis. Therefore, using 1000 mated females to perpetuate the *C. flavipes* colonies maintained greater genetic variation than using 500 females.

Culture Maintenance and Inbred Lines

Another procedure for maintaining genetic variability in cultures, as recommended by Bartlett (1985), Joslyn (1984), Roush (1990), and Roush and Hopper (1995), is to develop and maintain a number of inbred subcultures. Joslyn (1984) suggested that the subcultures be exposed to different variable rearing environments on a rotating schedule. Individuals from these subcultures are systematically outbred to achieve hybrid vigor in progeny that are to be field released. Outcrossing and reisolating the strains are advisable after 4 to 6 generations to prevent development of isolating mechanisms that could result in hybrid degeneration. Roush and Hopper (1995) calculated that 25 single family lines for arrhenotokous species, or 50 lines for diploid species, are needed to preserve the common alleles that seem most likely to be important for field performance. Joslyn (1984) and Roush and Hopper (1995) also indicated the importance of maintaining a high effective population size (e.g., 500) in each subculture to reduce the possible decrease of variability caused by random drift and inbreeding. Cook (1993a), in an analysis of prevention of diploid male production in parasitic Hymenoptera, suggested that maintenance of both a single large culture as well as multiple small subcultures might be best because of the biological attributes of a given host–parasitoid system.

Legner and Warkentin (1985) have also suggested that one way to keep a broader range of genetic variability is to culture several separate lines from an explorer's initial acquisitions, especially because severe consequences of the founder effect that would reduce genetic variability occur in the first few generations of culture (Legner, 1979a; Unruh *et al.,* 1983). Although each culture might assume great homozygosity in time, different cultures would be homozygous for different characteristics through random founder effects. Specimens from the lines could then, if desired, be intermated prior to field release to restore genotypic variation via hybridization (Hopper *et al.,* 1993). However, there are insufficient data to decide whether homo- or heterozygosity is preferable in establishing beneficial species (Legner & Warkentin, 1985).

Unruh *et al.* (1983) (also see Chapter 4) considered the procedure of maintaining many inbred lines with periodical mixing to attempt restoration of variability to be an uncertain technique, and instead preferred to maintain large cultures. Unfortunately, because exotic collections of entomophages might yield very few founder individuals, sustaining genetic variability might require the maintenance of subcultures.

A method of reducing labor and material resources for maintaining genetic lines is to use cold storage. Gilkeson (1990) cold-stored cocooned last instars of the predatory cecidomyiid *Aphidoletes aphidimyza* (Rondani) for 8 months to preserve genetically important lines. Similarly, Tauber *et al.* (1993) stored diapausing adults of the common green lacewing for 31 weeks at 5°C with very good results; and recommended this approach for maintaining standardized stock for long-term ecological, physiological, or genetic research.

Culture Maintenance and Periodical Infusions

One approach to reduce loss of genetic variability in laboratory cultures from drift, selection, and inbreeding is to periodically introduce native individuals. Joslyn (1984) commented that simulating "migration" by occasionally adding wild specimens to the laboratory culture would also add new alleles. This introduction must be proportionately large enough to fix the new alleles into the population. One danger of the technique is the possible introduction of an unwanted insect disease, although Stone and Sims (1992), as well as Davis and Guthrie (1992), noted that using wild males to mate with laboratory females reduces this hazard. According to Bartlett (1984), the effectiveness of periodical infusions depends on regular introductions of relatively large numbers of individuals. These individuals are preferably obtained from the geographic area of collection because of possible incompatibility in intraspecific crosses between spatially and/or temporally separated populations. Data on *Muscidifurax* parasitoids support this idea (Legner, 1988c). Bartlett (1984) also remarked that without regular introduction of native alleles, selection would reestablish the previous laboratory allele frequencies. Further, the combined processes of selection and inbreeding would have definite rapid effects on changing these frequencies.

Chandra and Avasthy (1988) found that new stock of the tachinid *Sturmiopsis inferens* Townsend had to be introduced into the laboratory culture after three generations to restore the loss of parasitoid vigor due to inbreeding. King and Morrison (1984) felt that periodic replacement of a culture with field-collected material is costly; instead they recommended the routine monitoring of behavioral traits coupled with techniques to maintain essential characteristics, and even suggested selecting for desirable traits in mass-produced entomophages. However, in the mass production of *Trichogramma* spp., they stressed the need for annual culture replacement. They further recommended that culture replacement should only be from the target host on the affected crop, and in large enough numbers (>2000) to ensure a broad genetic base.

Culture Maintenance and Environmental Conditioning

Bartlett (1985) recommended rearing founder individuals as nearly as possible in natural environmental regimes and densities in the laboratory. Joslyn (1984) indicated that, "A static environment leads to a static genotype and ultimately to less fit insects." The laboratory population exposed to variable "natural" environmental regimes must be large enough to allow random mating to preserve genotypic variability. When few entomophages can be field collected, varying the laboratory environment can at least maintain a pressure for adaptive fitness. In addition to maintaining a field-type laboratory rearing environment, Hopper *et al.* (1993) advocated forcing parasitoids to search for mates or hosts as they would have to do in nature.

Culture Maintenance and Monitoring

In obtaining and maintaining genetic variability in the laboratory, it is first important to study the biology and behavior of a species very well so that information is available to compare attributes of the wild and domesticated populations, which will thereby enable detection of genetic differences (Bartlett, 1985). Hopper *et al.* (1993) repeatedly stressed that such genetic contributions to trait differences must be determined by reciprocal crosses or half-sib analyses, to eliminate poor nutrition and disease as contributing factors.

Singh and Ashby (1985) observed that in establishing a new laboratory culture, the first genetic selection to occur would be of developmental traits such as a shortened life cycle. Behavioral, physiological, and biochemical selection follows. Therefore, it is necessary to establish standards and tolerances for insect quality testing when newly colonized insects are still genetically close to the wild population. Galazzi and Nicoli (1996) did that when they determined intrinsic rate of natural increase values for the predatory mite *Phytoseiulus persimilis* Athias-Henriot at the time of field collection and after a year of mass production. They concluded that 1 year of mass rearing did not affect the potential of population increase of *P. persimilis* and that the substitution of the mass-reared strain therefore did not appear to be necessary if a large stock of predators is maintained. They also felt that field collection of wild predators to replace the mass-reared strain would involve the risk of inducing genetic bottlenecks during the initial laboratory

population increase. Hsin and Getz (1988) suggested that monitoring developmental variation in insectary cultures with an appropriate life table might be very useful in maintaining genetic viability.

Culture Maintenance Following Field Establishment

As soon as a released beneficial organism is field established, it is desirable to frequently collect individuals for further laboratory culturing and/or for translocation to other localities. A successfully field-established organism is often more vigorous and fit after confronting the hazards of nature during reproduction and development than is one raised in insectary conditions.

Waage *et al.* (1985) also advocated the use of this procedure, believing that the preservation of genetic variability and insect quality required culturing entomophages only briefly before field release. Field collections of established entomophages could then serve as sources for initiating new laboratory cultures. They further recommended climatic matching of release areas to the source areas to facilitate greater ease of entomophage establishment.

In concluding this section on the genetics of parasitoids and predators in culture, it is well to remember that not all laboratory culture deterioration is the result of genetic factors. Nutritional or environmental factors can also cause deterioration and might be correctable with no long-term effects, as noted by Bartlett (1985). In fact, Hopper *et al.* (1993), after reviewing genetic aspects of biological control, concluded that laboratory colony deterioration is more probably caused by poor nutrition and particularly by disease than by genetic changes.

Culture and Synthetic Diet

Singh (1984) reviewed work on rearing entomophagous parasitoids on synthetic diet; and noted that success to that date had been obtained with two ichneumonids, *Itoplectis conquistor* (Say) and *Exeristes roborator* (Fabricius); one trichogrammatid, *T. pretiosum;* and a pteromalid, *Pteromalus puparum*. Bratti (1990b) gave a detailed review of *in vitro* rearing techniques for entomophagous parasitoids, and observed that the extensive literature revealed that most work was still experimental, although *Trichogramma* spp. could be reared continuously on artificial diet. In a notable article, Dai *et al.* (1988) gave an account of rearing *Trichogramma* spp. on artificial eggs and then releasing them by the hundreds of thousands in cotton fields and pine forests in China, where the rate of parasitism was said to be similar to that of parasitoids reared on natural host eggs. Li (1992) also reported on the *in vitro* mass production and field release of *Trichogramma* spp., as well as that for another egg parasitoid, the eupelmid *Anastatus japonicus* Ashmead. The parasitoids were said to show good effectiveness in controlling cotton bollworms, pine caterpillars, sugarcane borers, and litchi stinkbugs in China.

Other work on the *in vitro* culturing of parasitic hymenopterans has included that done on the following egg parasitoids: various species of *Trichogramma* (Parra & Consoli, 1992; Grenier *et al.* 1993, 1995; J. F. Liu *et al.,* 1995; Z. C. Liu *et al.,* 1995; Consoli & Parra, 1996, 1997; Nordlund *et al.,* 1997; Xie *et al.,* 1997); the eupelmids *Anastatus* sp. (Li *et al.,* 1988; Liu *et al.,* 1988a) and *A. japonicus* (Xing & Li, 1990); the scelionids *Telenomus dendrolimusi* Chu (Li, 1992), *T. heliothidis* Ashmead (Strand *et al.,* 1988), and *Trissolcus basalis* (Wollaston) (Volkoff *et al.,* 1992); and the encyrtids *Ooencyrtus kuvanae* Howard (Lee & Lee, 1994), *O. nezarae* Ishii (Takasu & Yagi, 1992), and *O. pityocampae* (Mercet) (Battisti *et al.,* 1990; Masutti *et al.,* 1992, 1993). *In vitro* culturing of larval ectoparasitoids has occurred for the pteromalids *Dibrachys cavus* (Walker) (Li, 1992) and *Catolaccus grandis* (Burks) (Guerra *et al.,* 1993), and for the braconids *Bracon greeni* Ashmead (Xie *et al.,* 1989b; Li 1992) and *B. mellitor* Say (Guerra *et al.,* 1993). Guerra *et al.* (1993), in particular, were notably successful in culturing *C. grandis* and *B. mellitor,* both ectoparasitoids of the boll weevil, on an artificial diet devoid of insect components. Guerra and Martinez (1994) and Rojas *et al.* (1996) made later improvements to the artificial culture of *C. grandis.* Attempted *in vitro* culturing of hymenopterous larval endoparasitoids has also occurred, including the following: the braconids *Cardiochiles nigriceps* Viereck (Pennacchio *et al.,* 1992), *Habrobracon hebetor* (Say) (Xie *et al.,* 1989a; Li, 1992), and *Microplitis croceipes* (Cresson) (Greany *et al.,* 1989; Ferkovich *et al.,* 1991); and the chalcidid *Brachymeria intermedia* (Dindo, 1990, 1995; Dindo & Campadelli, 1993; Dindo *et al.,* 1995). In addition to the preceding hymenopterous egg and larval parasitoids, an effort was made to culture an aphidiid, *Lysiphlebus fabarum* (Marshall) (Rotundo *et al.,* 1988).

Considerable effort has been devoted to the *in vitro* culturing of tachinids as well. Nettles (1986) was able to obtain relatively good yields of *Eucelatoria bryani,* a specific parasitoid of *Heliothis* spp., by raising it first in *H. virescens* for 20 to 28 h and then on an artificial diet supplemented with defatted, toasted soy flour. Bratti and D'Amelio (1994) and Bratti and Nettles (1995) further experimented with artificial diets for *E. bryani.* Others have worked with unnatural diets for culturing *Pseudogonia rufifrons* Wiedemann (Bratti, 1990a; Fanti, 1991; Bratti & Benini, 1992; Dindo & Campadelli, 1993). Fanti (1991) and Bratti and Benini (1992) indicated that host ecdysone in the diet was important to the development of *P. rufifrons.* Attempts at the *in vitro* rearing of *Lydella thompsoni* Hertig and *Archytas marmoratus* (Town.) yielded very limited results (Bratti, 1994). However, Mellini *et al.* (1993) reported on the successful *in vitro* culturing of *Exorista larvarum.* This success was followed by considerable work on refin-

ing the *in vitro* diet with less expensive ingredients, and on developing techniques aimed at eventual mass production of the tachinid for augmentative control of lepidopterans feeding on tree foliage (Bratti & Campadelli, 1994; Mellini *et al.,* 1994; Bratti & Coulibaly, 1995; Bratti *et al.,* 1995; Mellini & Campadelli, 1996). Mellini *et al.* (1996) indicated that *E. larvarum* was relatively easy to rear *in vitro* because it is a gregarious larval idiobiontic parasitoid that can attack many Lepidoptera, because it has a simple relationship with its host, and because the larvae can produce primary integumental respiratory funnels either in the host or in the gelled diet so that they can display similar behavior in either. The major problem remaining in developing a mass production system is that *E. larvarum* oviposition on artificial substrates has not yet been obtained.

Researchers have also developed artificial diets for larvae or nymphs of predaceous arthropods, including the neuropterans *Chrysoperla carnea* (Hasegawa *et al.,* 1989; Letardi & Caffarelli, 1989, 1990), *Chrysopa septempunctata* Wesmael (Kaplan *et al.,* 1989; Niijima, 1995), and *C. scelestes* Banks (Gautam & Paul, 1987). Work on diets for coccinellids includes that of Hussein and Hagen (1991), Henderson and Albrecht (1988), Chen *et al.* (1989b), and Hattingh and Samways (1993). There has been considerable effort in culturing the predaceous lygaeid *Geocoris punctipes* on prepared diets (Cohen, 1981, 1985, 1992; Hagler & Cohen, 1991), and Grenier *et al.* (1989) have attempted to rear the mirid *Macrolophus caliginosus* Wagner on artificial media. Saavedra *et al.* (1992) experimented with an animal tissue based diet for the pentatomid *Podisus connexivus* Bergroth, and Saavedra *et al.* (1995, 1996) obtained development of *P. nigrispinus* (Dallas) on meat-based diets. Zanuncio *et al.* (1996) similarly used a meat-based diet to rear the pentatomid *Supputius cincticeps* (Stal) for two generations, and De Clercq and Degheele (1992) reported on a meat-based diet for *P. maculiventris* and *P. sagitta* (Fabricius). Sumenkova and Yazlovetskii (1992) also worked with *P. maculiventris* (Say), and developed an adequate artificial diet based on pupal homogenate. The predaceous phytoseiids *Amblyseius gossipi* El-Badry and *A. swirskii* Athias-Henriot were reared successfully on artificial diet by Abou-Awad *et al.* (1992), but Shih *et al.* (1993) found two artificial diets to be unsatisfactory for rearing *A. ovalis* (Evans).

Waage *et al.* (1985) noted that the greatest success in the development of artificial diets for entomophagous parasitoids had been with polyphagous parasitic wasps. Although there has been progress on rearing predaceous entomophages on artificial diets, production for field release still involves living or dead hosts in almost all cases. For future success in rearing entomophages on artificial diet, Yazlovetsky (1992) emphasized the importance of accumulating further knowledge of their physiology, nutritional and digestive biochemistry, and ecology.

Interactions between Entomophages, Hosts, and Host Food

Entomophage quality can be dependent on host quality, and in turn on host-food quality. Singh (1984) referred to two examples of interactions between prepared diets, herbivores, and natural enemies. Rearing the Asiatic rice borer, *Chilo suppressalis* (Walker), on artificial diet adversely affected its braconid parasitoid, *Apanteles chilonus* Munakata. Similarly, the tachinid *Lixophaga diatraeae* declined in quality when reared on larvae of the greater wax moth, unless vitamin E or wheat germ supplemented the larval beeswax pollen diet. Bloem and Duffy (1990) concluded that qualitative and quantitative protein differences in the diet of *Heliothis zea* affected growth of the endoparasitoid *Hyposoter exiguae* (Viereck), and Magrini and Botelho (1991) studied the influence of dietary proteins in the diet of *Diatraea saccharalis* on the egg parasitoid *Trichogramma galloi* Zucchi. Similarly, Bratti (1992) showed that development of the tachinid *Archytas marmoratus* was influenced by the diet of its factitious host, the galleriid *Galleria mellonella* Linnaeus.

ODell *et al.* (1984) reported that only one of three artificial diets was suitable for culturing *Rogas lymantriae* on gypsy moth larvae. On the other two diets the host larvae died shortly after parasitoid oviposition. The same parasitoid had significantly higher female weights when reared on gypsy moth larvae grown on a high wheat germ diet than when grown on a commercial diet (Moore *et al.*, 1985).

Artificial diet for the host insect can sometimes result in greater parasitoid production than does natural food. Rogers and Sullivan (1991) studied development of the pupal parasitoid *Brachymeria ovata* (Say) on three noctuid hosts reared on soybean varieties or artificial diet; and showed that soybeans as the host food adversely affected emergence rates, size, and sometimes development period. Beach and Todd (1986) found that parasitized soybean looper larvae fed a susceptible soybean variety yielded two and a half times more parasitoids than when fed a resistant variety. Parasitoid production increased another twofold when an artificial diet was the host food.

Bourchier (1991) introduced tannic acid into gypsy moth diet to study the effect on growth and development of the tachinid *Compsilura concinnata* (Meigan); and concluded that interactions between the host plant, the gypsy moth, and the parasitoid may provide a possible explanation for the irregular impact of the parasitoid in the field. Roth *et al.* (1997) similarly experimented with the effects of dietary phenolic glycosides and tannic acid on performance of the gypsy moth and its suitability as a host for the parasitoid *Cotesia melanoscela*. Gypsy moth growth was reduced to a greater extent on diets with phenolic glycosides than on those with tannic acid, with resultant prolonged development, reduced cocoon weight, and higher incidences of mortality for *C. melanoscela*. Additionally, *C. melanoscela* performance was reduced slightly more by phenolic glycosides than by tannic acid. The results showed that plant allelochemicals can have profound effects on plant–herbivore–natural enemy interactions.

The nutritive chemical composition of host plants also affects a host-entomophage system. Kaneshiro and Johnson (1996) found that the eulophid parasitoid *Chrysocharis oscinidis* (Ashmead) produced up to 40% more progeny when its agromyzid leaf miner host *Liriomyza trifolii* (Burgess) was reared on bean plants containing 4.88% leaf nitrogen (compared to other N levels in the 2.67 to 6.52% range). They noted that this highly significant fecundity difference could greatly affect pest/parasitoid populations in both agroecosystems and mass rearing programs.

Thus, it is important to recognize that some parasitoids vary in their rates of parasitism or other quantitative attributes, depending on the type and quality of host food plant (Waage *et al.*, 1985). With the chalcidid *Brachymeria intermedia*, Greenblatt and Barbosa (1981) discovered that gypsy moth larvae fed on red oak foliage, rather than on that of three other tree species, produced the largest and heaviest parasitoids. Likewise, Karowe and Schoonhoven (1992) found that the larval developmental rate and adult longevity of the braconid *Cotesia glomerata* (Linnaeus) were greatest when its host, *Pieris brassicae,* was reared on nasturtium instead of on Brussels sprouts, Swedish turnip, or rape. Similarly, Kumar *et al.*, (1988) showed that the aphidiid *Trioxys* (*Binodoxys*) *indicus* Subba Rao & Sharma produced more progeny when its host aphid *Aphis craccivora* Koch was reared on the plant *Cajanus cajan,* than when reared on *Dolichos lablab* or *Solanum melongena.* Moreover, a higher proportion of female parasitoid progeny was achieved on *C. cajan*-reared aphids than on *S. melongena*-reared aphids (Kumar & Tripathi, 1988). Bhatt and Singh (1989) found that fecundity and sex ratio of the same parasitoid (*T. indicus*) were significantly more reduced when it was reared on *A. gossypii* Glover on *Cucurbita maxima* than when reared on two other cucurbit plants, *Lagenaria vulgaris* and *Luffa cylindrica*. Further, the population growth of *T. indicus* was fastest when the aphid host was reared on *L. vulgaris* (Bhatt & Singh, 1991). Other work was done by Hare and Luck (1990), who studied the influence of four species of citrus host plants on the aphelinid *Aphytis melinus* reared on California red scale; by Ugur *et al.* (1987), who showed that rearing the noctuid *Agrotis segetum* (Denis & Schiff) on different plants maximized different developmental parameters of its ichneumonid parasitoid, *Pimpla turionellae* (Linnaeus); and by English-Loeb *et al.* (1993), who investigated the consequences on the tachinid parasitoid *Thelairia bryanti* Curran of feeding poison hemlock or lupine to its arctiid host *Platyprepia virginalis* (Bvd.).

Numerous studies have involved the effects of resistant plant varieties on parasitoids. These have included those of Dover *et al.* (1987), Farrar *et al.* (1994), Fuentes-Contreras *et al.* (1996), Grant and Shepard (1985), Kumar (1997), Mannion *et al.* (1994), Obrycki and Tauber (1984), Obrycki *et al.* (1985), Orr and Boethel (1985), Orr *et al.* (1985), Powell and Lambert (1984), Ruberson *et al.* (1989b), Tillman and Lambert (1995), Vogt and Nechols (1993), and Yanes and Boethel (1983).

Obviously, then, plants required for host production in entomophage rearing must be screened for detrimental effects on the entomophages. Some of these deleterious effects can come from plant chemicals consumed by the host, as noted previously. As a further example, Barbosa *et al.* (1982, 1990) and Bentz and Barbosa (1992) observed that the nicotine level in tobacco hornworms fed tobacco leaves affected the survival of its parasitoid, *Cotesia congregata* (Say). However, Farrar and Kennedy (1993) discovered an instance in which plant resistance had a direct effect on a parasitoid. Parasitism of *Heliothis zea* by the tachinid *Archytas marmoratus* (Townsend), which larviposits on the food plant of its host, was significantly reduced by methyl–ketone-mediated insect resistance of tomato plant lines. *Eucelatoria bryani*, another tachinid, was not as affected because it larviposited directly into its host, without its larvae touching the resistant foliage. Somewhat similarly, the scelionid parasitoid *Telenomus sphingis* (Ashmead) was inhibited from parasitizing eggs of *Manduca sexta* (Linnaeus) on foliage of a resistant wild tomato having high densities of glandular trichomes containing 2-tridecanone (Farrar & Kennedy, 1991). Tomato leaf trichomes also interfered with the encyrtid *Copidosoma koehleri* Blanchard, a parasitoid of the potato tuberworm, *Phthorimaea operculella* (Baggen & Gurr, 1995).

Characteristics other than resistance can likewise be important in choosing a host plant for a parasitoid production system. Flanders (1984) evaluated five varieties of bean plants for rearing the Mexican bean beetle, the required host for producing the parasitoid *Pediobius foveolatus*. Of the two varieties that were essentially equivalent in providing the highest net reproductive rate for the bean beetle, the preferred one had superior growth characteristics and was consequently the most economical to produce. Preference of an entomophage for a host on a particular plant is another important characteristic, as reported by Jalali *et al.* (1988), who demonstrated the preference of the braconid *Cotesia kazak* Telenga to attack *Heliothis armigera* (Hübner) larvae on cotton, tomato, or okra to that on dolichos, pigeon pea, cowpea, or chickpea. Also, the fecundity of the parasitoid was statistically greater on the first group of plants than on the second group.

Host plants can affect the suitability of phytophages not only for parasitoids, but also for predators such as coccinellids (Hodek, 1973; Nishida & Fukami, 1989; Rice & Wilde, 1989; Emrich, 1991; Reyd *et al.*, 1991), *Geocoris punctipes* (Rogers & Sullivan, 1986), and *Scolothrips longicornis* Priesner [a thrip predaceous on tetranychid mites (Sengonca & Gerlach, 1984)]. Legaspi *et al.* (1996) reported that if larvae of the predaceous lacewing *Chrysoperla rufilabris* (Burmeister) were fed on silverleaf whiteflies (*Bemisia argentifolii* Bellows & Perring) reared on cucumbers and cantaloupes, they developed satisfactorily, in comparison to those that fed on whiteflies reared on poinsettias and lima beans, which did not survive to the pupal stage. They concluded that whiteflies reared on poinsettias or lima beans either may have been nutritionally inadequate for *C. rufilabris* development, or may have accumulated plant compounds detrimental to the development of the lacewings. Host plants for phytophagous mites can also affect predatory mites, as shown by de Moraes and McMurtry (1987), who found an indication that adult female *Phytoseiulus persimilis* gained more weight when fed two-spotted spider mites, *Tetranychus urticae* Koch, reared on lima bean, *Phaseolus vulgaris,* than when fed on nightshade, *Solanum douglasii.* Grafton-Cardwell and Ouyang (1996) showed a direct effect of the nutritional status of host plants (citrus trees) on the reproductive potential of the predaceous phytoseiid *Euseius tularensis* (Congdon), indicating that the mite fed on leaf sap, as well as on phytophagous mites.

The type of produce used in a host–parasitoid production system can likewise affect the parasitoid yield. Simmonds (1944) reported that the encyrtid *Comperiella bifasciata* Howard effected 2 to 3 times more parasitism when oranges instead of lemons were the diet of its host, the California red scale.

The food source used to produce phytophages for weed control projects is similarly an important consideration. Frick and Wilson (1982) mass reared the weed-feeding tortricid *Bactra verutana* Zeller on a prepared diet and obtained adults that were 60% larger and twice as fecund as those reared on nutsedge plants. However, there were indications that the field flight capabilities were not as great with the diet-reared insects.

Price *et al.* (1980) reviewed earlier studies of interactions between plants, insect herbivores, and natural enemies; and Berlinger (1992) presented a fine review of host-plant effects on herbivore rearing, with reference also to effects on entomophagous predators and parasitoids. Van Emden (1995) summarized well the types of influences that plants have on parasitoids and predators that feed on their herbivores. These influences include host-plant allelochemicals that can affect natural enemies adversely through the passage of plant toxins up the trophic pyramid; the size of the host prey; the production of plant volatiles that attract natural enemies, even in the absence of prey; and the effects of different plants on predator/parasitoid functional responses, as mediated through prey size, differences in defensive, and other behavior of the prey, and through effects on searching time of natural enemies.

Food for Adult Entomophages

Although some adult entomophagous insects require special food, most parasitic Hymenoptera that do not host feed can naturally produce mature eggs with a source of carbohydrate such as honey (Waage *et al.*, 1985). Even with a parasitoid (*Aphytis melinus*) that does host feed, Heimpel *et al.* (1997) showed that the availability of honey is very important and interacts strongly with host feeding in influencing longevity and fecundity, and also has a strong direct effect on egg resorption. For *Trichogramma*, Morrison (1985a) used plump raisins or honey to feed adults. Likewise, Coombs (1997) increased the longevity and fecundity of adults of the tachinid *Trichopoda giacomellii* (Blanchard), an agent for biological control of *Nezara viridula*, by feeding them raisins and water. Munstermann and Leiser (1985) fed adult predatory *Toxorhynchites* mosquitoes with diluted honey absorbed onto strips of cellucotton; or with raisins, apple slices, or 15% sucrose solution. Mendel (1988) showed that the longevities of one pteromalid and two braconid parasitoids of scolytid bark beetles were directly related to provision of water and honey, and inversely related to temperature; and that longevities of parasitoids given only water were directly related to body size. In the mass production of the mymarid *Anaphes iole* Girault for augmentation against *Lygus hesperus* Knight, Jones and Jackson (1990) concluded that *not* providing food or water to adult parasitoids achieved the greatest efficiency.

Adults of some predators may require more complex diets. For adults of the common green lacewing, Morrison (1985b) used a diet composed of equal parts of sucrose and yeast flakes moistened with enough water to make a thick paste. Because this was a yeast commercially cultured on whey, it therefore contained about 65% animal protein, a necessity for high fecundity. Other ingredients may occasionally be important in adult entomophage food. For example, Moore (1985) reported that Nettles *et al.* (1982) showed a synergistic effect of potassium chloride and magnesium sulfate on oviposition by an insect parasitoid.

Factitious Hosts for Entomophages

Factitious (unnatural) hosts can sometimes simplify rearing. Similarly, a beneficial insect may have more than one type of natural host, and one of these may be easiest to rear in the laboratory. The quantity and quality of parasitoid or predator progeny on different hosts seem to vary in the insectary according to evolutionary contact, as mentioned by Legner and Thompson (1977). Their study compared the suitability of the potato tuberworm and the pink bollworm, the original source host, as hosts for a braconid, *Chelonus* sp. nr. *curvimaculatus* Cameron. The parasitoid increased its destruction of, and its fecundity on, the factitious host, after being reared for many generations on the potato tuberworm and then for one generation on the pink bollworm. The *Chelonus* spp. to which this parasitoid belongs respond to kairomones in the body scales of several lepidopterans (Chiri & Legner, 1982, 1986), and might most aptly be characterized as comprising a generalist group.

Fedde *et al.* (1982) reviewed guidelines for choosing factitious or unnatural hosts to be used for rearing hymenopterous parasitoids. They listed 43 examples of such hosts and emphasized that ease of rearing was the most important consideration, and that potential factitious hosts should be tested and not prejudged as to possible utility.

For example, Rojas *et al.* (1995) found that the easily reared *H. virescens* could be used as a somewhat unsuitable factitious host for the braconid *Bracon thurberiphagae* Muesebeck, a parasitoid of the boll weevil, but only if several amino acids were added to the adult parasitoid diet. The supplemental adult feeding provided a 100% improvement in the fecundity of *B. thurberiphagae*, restored a 1:1 sex ratio, and permitted the use of *H. virescens* as a factitious host.

A factitious host is especially advantageous if it is useful for rearing more than one beneficial insect. As an example, the greenbug, *Schizaphis graminum* (Rondani), was satisfactory not only for mass rearing *Aphidius matricariae* Haliday, an effective parasitoid of the green peach aphid, *Myzus persicae* (Sulzer) (Popov & Shijko, 1986; Shijko, 1989), but also for producing the predaceous cecidomyiid *Aphidoletes aphidimyza* (Belousov & Popov, 1989).

Egg parasitoids in the genus *Trichogramma* are generally reared on factitious hosts, but other egg parasitoids have been as well. Halperin (1990) produced *Ooencyrtus pityocampae* in eggs of the silkworm, *Bombyx mori* (Linnaeus). Hosts for *Trichogramma* spp. include the Angoumois grain moth, *Sitotroga cerealella* (Olivier); the Chinese oak silkworm, *Antheraea pernyi* Guérin-Méneville; another type of silkworm, *Samia cynthia ricini* (Boisduval); the Mediterranean flour moth; and the rice moth, *Corcyra cephalonica* (King & Morrison, 1984). Additionally, the eggs of common noctuids and lepidopteran stored grain insects are often used. However, a few *Trichogramma* spp. may be host specific (Morrison, 1985a).

There must be careful choosing of factitious hosts for *Trichogramma*. Pham Binh *et al.* (1995) addressed this topic. For example, in testing the suitability of pyralid species for *T. evanescens* Westwood, Brower (1983) found that the Mediterranean flour moth was by far a less suitable host than the tobacco moth, *Ephestia elutella* (Hübner); the almond moth, *Cadra cautella* (Walker); the raisin moth, *C. figulilella* (Gregson); or the Indian meal moth, *Plodia interpunctella* (Hübner). Only 59% of exposed flour moth eggs yielded emerged parasitoids, and 10% of these were runts. On the other hand, 88% of exposed almond moth eggs produced parasitoids, with only 0.4% being runts.

Trichogramma reared for long periods on a factitious host can still maintain a natural-host preference. Yu *et al.* (1984) collected a strain of the egg parasitoid *T. minutum* Riley from the codling moth, and then reared it for about 22 generations on eggs of the Mediterranean flour moth. Even after that time, the parasitoid still preferred eggs of the codling moth to eggs of the Mediterranean flour moth. Pavlik (1993) studied a total of nine strains of four species of *Trichogramma* and also found that some strains did not lose their ability to accept and develop in their natural hosts in spite of having been reared on Mediterranean flour moth eggs for several years.

However, rearing a *Trichogramma* sp. on a factitious host requires careful testing for retention of its host acceptance characteristics. In contrast to the preceding example, host acceptance of European corn borer eggs by *T. maidis* decreased with increasing numbers of generations of parasitoid rearing on the Mediterranean flour moth, and did not then improve after five generations of passaging on corn borer eggs (van Bergeijk *et al.*, 1989). In another case, Sukhova (1989) noted the widespread use of passaging *T. evanescens* through natural hosts (*Mamestra brassicae*) to maintain quality.

For rearing *T. evanescens* on factitious hosts, Mirzalieva and Mirzaliev (1988) recommended using eggs of the greater wax moth instead of those of *Sitotroga* because of the superior biological effectiveness of the produced parasitoids. In fact, Campadelli (1988) reported the greater wax moth to be an excellent substitute for rearing 47 species of Hymenoptera and 22 species of Tachinidae. For example, there are reports of its use in the mass production of *Lixophaga diatraeae*, a tachinid fly that parasitizes the sugarcane borer (Hartley *et al.*, 1977; King *et al.*, 1979; King & Hartley, 1985b).

Likewise, Stowell and Coppel (1990) used the greater wax moth to mass-produce *Brachymeria* spp., chalcidid parasitoids of the gypsy moth. Palmer (1985) previously showed *B. intermedia* to be more easily reared in the insectary on the greater wax moth. On the other hand, Rotheray *et al.* (1984) determined that the gypsy moth was a better host for *B. intermedia* because parasitoids produced from the wax moth were smaller and less able to oviposit in gypsy moth pupae. Palmer (1985) found that even rearing *Brachymeria* for 119 generations on the wax moth did not shift the host preference of the parasitoid, because it still readily attacked gypsy moth pupae. However, *B. intermedia* is apparently a generalist, and such a host preference shift is not expected. King and Morrison (1984) noted that rearing a parasitoid on a factitious host can eventually increase its acceptance of that host, but a parasitoid reared for only a few generations on an unnatural host can still react strongly to its natural host.

Predicting potential field effectiveness of generalist parasitoids and predators against specific target pests is a corollary benefit of testing host insects for production programs. Drummond *et al.* (1984) reported that the best host for rearing the spined soldier bug *Podisus maculiventris,* was the greater wax moth, in contrast to the Mexican bean beetle; the eastern tent caterpillar, *Malacosoma americanum* (Fabricius); or the Colorado potato beetle, *Leptinotarsa decemlineata* (Say). The Colorado potato beetle in fact was a suboptimal host in comparison to the other three prey species. Therefore, the spined soldier bug apparently has little potential as a field predator of this pest.

As with predaceous insects, factitious hosts are sometimes advantageous for rearing predaceous phytoseiid mites. Gilkeson (1992) noted in a review of the mass rearing of phytoseiid mites that they can be grouped into those that must be reared on their natural hosts and those that can be reared on alternate food sources. For example, *Amblyseius cucumeris* (Oudemans) was cultured on the American house dust mite *Dermatophagoides farinae* Hughes, a factitious host (Castagnoli, 1989, Castagnoli & Sauro, 1990). Rearing phytoseiids on pollen is also frequently successful, although the species of pollen can be critical (Affifi *et al.*, 1988; Englert & Maixner, 1988; James, 1989; Zhang & Li, 1989a, 1989b; Grout & Richards, 1992; Ouyang *et al.*, 1992; James & Whitney, 1993; Yue *et al.*, 1994; Fouly *et al.*, 1995; Yue & Tsai, 1996; Bruce-Oliver *et al.*, 1996; Reis & Alves, 1997; Abou-Setta *et al.*, 1997). Regardless of the food used to raise phytoseiids or polyphagous predators, Dicke *et al.* (1989) advocated caution in using novel diets or in using only one species as food because performance characteristics can be adversely affected. Even the stage of a host species fed to a phytoseiid can produce substantial differences in life table parameters, as shown by Bruce-Oliver and Hoy (1990), who compared *Metaseiulus occidentalis* fed eggs or mixed actives of the host *Tetranychus pacificus* McGregor. Consequently, they noted that laboratory diet is a factor to consider in evaluating potential efficacy of natural enemies.

Dead Hosts for Entomophages

The capability of rearing many predators and a few parasitoids on dead hosts simplifies production (Waage *et al.*, 1985). Etzel (1985) described the heat processing of potato tuberworms to be used as food for producing certain coccinellids and neuropterans. In successfully using heat-killed pupae of the greater wax moth to rear the parasitoid *Brachymeria intermedia,* Dindo (1990) observed that the one limiting factor seemed to be acceptance, instead of suitability of the killed hosts.

Sometimes, the extended refrigeration or freezing of insect eggs can facilitate their use for rearing egg parasitoids or predators. Such storage prevents the hatching of larvae that might interfere with entomophage production, as well as enables the stockpiling of host material. The technique is useful with eggs of the Mediterranean flour moth for rearing *Trichogramma* spp. (L. Etzel, unpublished). Nagarkatti *et al.* (1991) found prefrozen eggs of the tobacco hornworm to be superior for mass production of *T. nubilale,* with an average of about 10 large, robust parasitoids emerging from each egg. Similarly, Drooz and Weems (1982) used freeze-killed eggs of *Eutrapela clemataria* (J. E. Smith) to rear the encyrtid egg parasitoid *Ooencyrtus ennomophagus* Yoshimoto; prefrozen eggs of the southern green stinkbug facilitated propagation of the scelionid egg parasitoid *Trissolcus basalis* (Powell *et al.,* 1981; Powell & Shepard, 1982); and prefrozen eggs of the Colorado potato beetle were suitable for rearing the eulophid *Edovum puttleri* Gressell (Maini & Nicoli, 1990).

Refrigeration of eggs can typically occur for only relatively short storage periods before their suitability for parasitization decreases rapidly (Gautam, 1987; Hugar *et al.,* 1990). However, frozen eggs are suitable for much longer periods. In propagating *Trichogramma* spp., Ma (1988) found no significant differences in parasitism rate or number of emerged adult parasitoids when the use of fresh, untreated eggs of the Chinese oak silkworm was compared with that of eggs stored for 1 to 9 months in liquid nitrogen and then defrosted in water at 30°C before parasitization. Wang *et al.* (1988) obtained similar results in rearing *Trichogramma* spp. on *Antheraea pernyi* eggs stored in liquid nitrogen for over 1000 days. For *Trichogramma* production in the former USSR, Gennadiev *et al.* (1987) emphasized that adherence to precise freezing, storage, and thawing regimes was essential in handling eggs of the Angoumois grain moth.

As an alternative to prefreezing, Gross (1988) irradiated eggs of the corn earworm with 25 krad of ^{60}CO to prevent hatching and enable their use for producing *T. pretiosum.* Thorpe and Dively (1985) likewise irradiated tobacco budworm eggs for *Trichogramma* production. Harwalkar *et al.* (1987) treated female potato tuberworms with gamma irradiation, and used the sterile eggs to rear *T. brasiliensis* (Ashmead) with no ill effects. They found no significant differences between the *Trichogramma* reared on irradiated or prefrozen eggs. However, when Zhang and Cossentine (1995) irradiated codling moth adults with ^{60}CO to obtain nonviable eggs for rearing *T. platneri* Nagarkatti, they

found that given a choice, the *Trichogramma* parasitized significantly more viable than nonviable codling moth eggs, and took longer to develop in the nonviable eggs. Kfir and Van Hamburg (1988) used ultraviolet light to irradiate *Heliothis armigera* eggs for 2 h before parasitization by the egg parasitoids *Telenomus ullyetti* Nixon and *Trichogrammatoidea lutea* Girault. Others have also utilized ultraviolet irradiation to kill host eggs before *Trichogramma* production (Hirashima *et al.,* 1990a, 1990c; Sengonca & Shade, 1991; Nordlund *et al.,* 1997), and in one instance the workers first killed the eggs with ultraviolet irradiation and then froze them in liquid nitrogen for several months before their use for rearing *Trichogramma* (Hu & Xu, 1988).

Morrison (1985b) froze eggs at −10°C for >24 h in airtight containers before using them as food for rearing the predaceous common green lacewing. Likewise, Baumhover (1985) used frozen tobacco hornworm eggs to rear the predaceous stilt bug, *Jalysus wickhami* (Van Duzee), and found that these eggs would still be suitable predator food after storage for 2 years at −23°C. Similarly, Tommassini and Nicoli (1995) found that frozen Mediterranean flour moth eggs provided adequate nutrition for rearing the predaceous anthocorid *Orius laevigatus* (Fieber). Finally, Schanderl *et al.* (1988) utilized ultraviolet irradiation to kill host eggs fed to coccinellids, as did Chen *et al.* (1994) for raising the lacewing, *Mallada basalis* (Walker).

Even prefrozen insect pupae are occasionally useful for parasitoid production. Grant and Shepard (1987) raised the chalcidid parasitoid *B. ovata* on prefrozen pupae of seven species of Noctuidae. The velvet bean caterpillar was the best host because of the extended period of suitability of the frozen pupae (up to 256 days versus 30 to 90 days for the other species). Milward-De-Azevedo and Cardoso (1996) used prefrozen pupae of the calliphorid *Chrysomya megacephaia* (Fabricius) to successfully rear *Nasonia vitripennis.* The freezing and thawing process did not affect the time of preimaginal development of the parasite, but there was a marked increase in the female sex ratio. Another example was the use of prefrozen housefly pupae for producing two pteromalid parasitoids, *Muscidifurax zaraptor* (Petersen & Matthews, 1984; Petersen *et al.,* 1992, 1995) and *Pachycrepoideus vindemiae* (Pickens & Miller, 1978). Pawson *et al.* (1993) later compared the use of freeze-killed housefly pupae with freeze-killed stable fly (*Stomoxys calcitrans* Linnaues) pupae for rearing *M. zaraptor; P. vindemiae;* and a third pteromalid, *Spalangia nigroaenea* Curtis.

However, Morgan (1992) was able to produce *Muscidifurax, Pachycrepoideus,* and *Spalangia* on radiation-killed fly pupae. This technique enabled the shipping of killed pupae to France for parasitization and then their return to the United States with developing parasitoids. The irradiated pupae were suitable as hosts for up to 8 weeks.

There are examples of prefrozen arthropod stages other

than eggs being utilized for predator production. Sipayung *et al.* (1992) mass-produced predatory Pentatomidae and Reduviidae on prefrozen caterpillars, and Parajulee and Phillips (1993) reared the predaceous anthocorid *Lyctocoris campestris* (Fabricius) on previously frozen late instars of eight species of insects.

Spider mites also have been prefrozen before use in phytoseiid mite production. In a biological control program against the cassava green mite, *Mononychellus tanajoa* (Bondar) *sensu lato,* a primary reason for prefreezing mass-produced spider mites for 18 h was to eliminate contaminating predators, including phytoseiids, from food used to produce desired phytoseiids (Yaninek & Aderoba, 1986; Friese *et al.,* 1987).

Diapause

Diapause in the life cycle often interferes with parasitoid and predator production. For example, Eskafi and Legner (1974) showed that certain temperature and photoperiod combinations would induce diapause in adults and progeny of the eye gnat parasitoid *Hexacola* sp. nr. *websteri* (Crawford). Exposing larval parasitoids within their larval hosts to a long photophase of 16 h combined with a high temperature of 32°C caused the parasitoid prepupae to enter a diapause state, termination of which required contact of the host puparia with moisture for a few hours. However, this type of easily terminated diapause only occurred following a parental generation reared at 27°C with 14-h light. With the parasitoid parental generation reared at 32°C with 16-h light and the progeny held at 27°C with 14-h light, then >90% of the prepupal progeny entered diapause and could not be induced to terminate it by exposure to moisture. With another set of progeny from the same parents reared at 32°C with 16-h light, only 35% entered diapause. This example illustrates the great complexities involved in determining which combinations of environmental regimes in the insectary will prevent, induce, or terminate diapause.

In the case of *Trichogramma* spp., Zaslavski and Umarova (1990) showed that the interactive effects of environmental conditions in the maternal and filial generations governed larval diapause in the filial generation. Lowered temperature during development of the filial generation was the predominant diapause factor, but the photoperiod and temperature conditions of the maternal generation influenced the "norm" of this thermal reaction. Superimposed on these diapause reactions was an endogenous process running through the generations that changed diapause tendency and underlying reaction norms even under constant rearing conditions (Zaslavski & Umarova, 1990). Other examples of complex diapauses include the alfalfa weevil parasitoid system; *Chelonus* spp. parasitoids of the pink bollworm that terminate diapause at different intervals (Legner 1979c); and navel-orangeworm parasitoids, where diapause seems triggered by hormonal changes in the host at different latitudes (Legner, 1983).

Not only physical environmental conditions but also host conditions and host food can influence diapause in parasitoids. The physiological state of an alfalfa plant affects the yellow clover aphid (formerly called the spotted alfalfa aphid), *Therioaphis trifolii* (Monell), which then induces diapause in its aphidiid parasitoid *Praon exsoletum* Nees (Clausen, 1977). Similarly, the type of pollen used to rear predaceous mites can affect initiation of diapause. *Amblyseius potentillae* (Garman) and *A. cucumeris* entered diapause in a short-day photoperiodic regime when reared on ice plant pollen, but not when reared on pollen of the broad bean (Overmeer *et al.,* 1989). After studying various combinations of parasitoid species, aphid species, plants, and environmental conditions, Polgar *et al.* (1995) concluded that internal factors via the host aphid, such as host aphid life cycle (holocyclic versus anholocyclic), aphid morph (oviparae, etc.), and host-plant quality, as well as environmental cues (temperature and photoperiod), can all be interconnected in the induction of dormancy or diapause in aphid parasitoids.

Laing and Corrigan (1995) showed the importance of host species in initiating diapause in *Trichogramma minutum.* Diapause occurred at 15°C with a photophase of 12L:12D in eggs of *Lambdina fiscellaria fiscellaria,* but did not occur in eggs of *Anagasta kuehniella, Sitotroga cerealella,* or *Choristoneura fumiferana* held under these same conditions.

Appropriate environmental conditions or their combinations, particularly relating to light and temperature, are often useful for manipulating diapause (Singh & Ashby, 1985). For example, Tauber *et al.* (1997b) showed that a biotype of *Chrysoperla carnea* from San Pedro, Mexico, could be reared continuously without diapause with an intragenerational increase in photoperiod or could be reared with the regular intervention of a storage period. Diapause induction in individuals destined for storage was accomplished by rearing larvae under a long-day photoperiod and transferring the prepupae to a short-day photoperiod.

Waage *et al.* (1985) noted that one factor to consider in rearing programs is that entomophagous insects and their hosts may have different optimal developmental temperatures. As a corollary, natural enemies may enter diapause under conditions whereby their hosts remain active. The western flower thrips, *Frankliniella occidentalis* (Pergande), is a year-round greenhouse pest in northern climates, whereas commercially available strains of two predaceous mites, *Amblyseius cucumeris* and *A. barkeri* (Hughes), are ineffective in the winter because of reproductive diapause induced by short-light conditions. Van Houten *et al.* (1995), however, were able to genetically select for effective nondiapausing strains of these predators within 10 laboratory generations.

Environmental Conditions

Rearing conditions vary with the entomophages. For example, King and Hartley (1985c) cultured maggots of the parasitic tachinid *Lixophaga diatraeae* inside their larval host, the sugarcane borer, in complete darkness, whereas Morrison (1985a) reared *Trichogramma* spp. under constant light.

Optimal conditions for entomophages can also vary in the same genus. Morrison (1985b) reared the common green lacewing under constant light. However, *Chrysoperla rufilabris* required a 14-h photophase for high fecundity (Nguyen *et al.*, 1976). Singh and Ashby (1985) observed that light quality and photoperiod are important factors in insect mating and oviposition. In fact, sometimes darkness is necessary for a parasitoid to perform properly. Ortegon-E. *et al.* (1988) reported that the braconid *Meteorus laphygmae* (Viereck) parasitized *Spodoptera* at night.

Optimal rearing conditions even vary for different characteristics within a species. Miller (1995) studied rearing conditions for the coccinellid *Eriopis connexa* Mulsant, a predator of the Russian wheat aphid *Diuraphis noxia* (Mordvilko), and found that a different set of rearing conditions existed for optimal beetle size (11.8 mg), highest larval survival (91.7%), and most rapid larval growth (6.3 days). Miller (1995) developed an index based on larval developmental time, larval survival, and adult weight to determine that the composite optimal rearing conditions for beetle production were to use the Russian wheat aphid as prey at 34°C.

Morales-Ramos *et al.* (1996a) developed polynomial multiple regression models to estimate optimal levels of temperature, RH, and photoperiod to maximize production of the boll weevil parasitoid *Catolaccus grandis*. The maximum value of the intrinsic rate of increase (R_m) of *C. grandis* occurred at 29°C and 66% RH with a 16:8 (L:D) photoperiod.

Of course, optimal physical conditions for development of an entomophage might not also be optimal for its host and the host plant. Fabre (1989) found that maintenance of the aphidiid *Pauesia cedrobii* Stary & Leclant on its host *Cedrobium laportei* Remaudiere on potted Atlas cedar plants required holding the system at 12°C, which was satisfactory but not optimal for the parasitoid.

Mating

Waage *et al.* (1985) discussed mating problems in general and possible solutions concerning entomophages. Chianese (1985) noted that mating conditions were critical in the laboratory production of the gypsy moth parasitoid *Co-*

tesia melanoscela. Isolation of the cocooned parasitoids in gelatin capsules before adult emergence ensured that they remained virgins until needed. Then both sexes were combined in a screen cage under natural light. It was better to feed males before placement in the mating cage, while females were fed in the cage itself to reduce activity while males mated with them. Such females mated only once, and refrigeration of adults had to occur after mating (24 to 48 h). Nealis and Fraser (1988), who reared the braconid *Apanteles fumiferanae* Viereck on the spruce budworm, likewise reported on the importance of natural light for parasitoid mating.

Rappaport and Page (1985) were successful in maintaining a year-round culture of the ichneumonid *Glypta fumiferanae* (Viereck), a parasitoid of the western spruce budworm, and attributed part of their success to the mating procedure, which was to introduce a freshly emerged female and three males 2 to 4 days old into a 0.25-liter carton with mesh-screen ends. Greenberg *et al.* (1995) also found that an acceptable sex ratio (65% females) was obtained in the propagation system of *Catolaccus grandis,* a parasitoid of the boll weevil, if a ratio of one female per three males was used during the first hour of mating, immediately after emergence.

Galichet *et al.* (1985) also noted rather fastidious requirements for laboratory mating of the tachinid *Lydella thompsoni.* These included high humidity, high light intensity (at least 8,000 to 10,000 lux), and food of casein hydrolysate and honey. Mason *et al.* (1991) likewise reported that *L. thompsoni* required a strong light source to facilitate mating behavior and larviposition.

Godfray (1985) discovered extremely precise mating requirements for *Argyrophylax basifulva* (Bezzi), a tachinid parasitoid of the greater coconut spike moth, *Tirathaba complexa* Butler. Just to obtain 50% mating, he had to place the flies in a large outdoor cage in bright morning sunshine at 28°C and 90% RH, with a strong breeze provided by an electric fan. Further complications were a 3-day premating period and an 8-day preoviposition period.

Because the mating requirements of the predaceous mosquito *Toxorhynchites minimus* (Theobald) were unknown, Horio *et al.* (1989) found it necessary to use artificial mating for laboratory culturing, although it was not essential to remove the head of the male to achieve efficient mating, as is the case with some fastidious mosquitoes.

In contrast, King and Hartley (1985c) found mating by the tachinid *Lixophaga diatraeae* to be easily obtained without any exacting requirements. Although some light was necessary, the type and intensity were not critical. Mating occurred readily by placing about 200 adult *L. diatraeae* in a small screened cage under conditions of 26°C and 80% RH with a 14-h photophase.

The proportion of males placed together with females for mating can be quite low, which may be preferable to avoid problems caused by over-mating. Palmer (1985) found that adding 25 males and 300 females of the parasitoid *B. intermedia* to a 4-liter jar ensured complete mating. Females mated only once and unmated females produced males. Mating occurred at 24 to 27°C after emergence in bright artificial light, with resulting progeny being 60 to 85% females.

Some females require more than one mating. Bonde (1989) reported that the phytoseiid mite *Amblyseius barkeri* (Hughes) required multiple matings for optimal fertility.

In one instance, mating problems with parasitoids occurred because of insufficient taxonomic knowledge. Faulds (1990) surmounted difficulties in rearing an imported *Bracon* sp. in quarantine after discovering that the imported material included two closely related species with different mating requirements.

Sex-Ratio Changes

In prolonged culture of parasitoids, sex-ratio changes can be a complication, as noted before. However, modified rearing conditions can sometimes ameliorate this problem. For example, a 16-year-old culture of the thelytokous pteromalid *Muscidifurax uniraptor* gradually began producing predominantly male progeny despite no apparent changes in culturing techniques or the host insect (Legner, 1985). Improved female production occurred by allowing oviposition only by young mothers, and in moderate temperature conditions on alternate days. Nonetheless, the proportion of >95% females was no longer attainable. It was interesting that insectary production of two freshly collected wild cultures of *M. uniraptor* resembled that of the changed long-term culture. The involvement of microorganisms in thelytoky (Horjus & Stouthamer, 1995; Zchori-Fein *et al.*, 1995), as noted earlier, complicates interpretation of sex ratios in such species.

Waage *et al.* (1985) reviewed the factors related to rearing conditions that can significantly affect the sex ratio of parasitic wasps. These factors include degree of mating (including the effects of over-mating), host size, crowding, and high temperatures. At least some parasitic female wasps can respond to external cues by regulating the sex of eggs oviposited. These females tend to lay male eggs in small hosts, in already parasitized hosts, and in response to physical and chemical effects of crowding. Also, male larvae tend to be competitively superior, so there is a differential survival of males in superparasitized hosts. An example of female sterility correlated with crowding of the parental females during oviposition was reported by Gerling and Fried (1997), who found the phenomenon in the aphelinid whitefly parasitoid, *Eretmocerus mundus* Mercet. They did not determine whether the sterility was caused by crowding alone or possibly by the activity of microorganisms acting under crowded conditions.

Provision of abundant food and space can partially alleviate male-biased sex ratios in the laboratory (Waage *et al.*, 1985). A consequence of food and space limitations is that female progeny production can decline with increasing parasitoid to host ratios (Legner, 1967). Moreover, Fox *et al.* (1990) reported one instance in which the quality of host food affected the sex ratio of a parasitoid. A greater percentage of female parasitoids was achieved from diamondback moth larvae (*Plutella xylostella* [Linnaeus]) reared on collard plants with high leaf nitrogen than was achieved on unfertilized plants. Similarly, Shukla and Tripathi (1993) showed the offspring sex ratio of the aphidiid *Diaeretiella rapae* (McIntosh) to be significantly affected by three species of food plants upon which its host, *Lipaphis erysimi* Kalt, was reared. As noted previously, Bhatt and Singh (1989) likewise found a host-plant effect on the sex ratio of the aphidiid *Trioxys pallidus*.

Parasitoids vary in the combined factors that govern their sex ratios. For example, the sex ratio of the encyrtid *Epidinocarsis lopezi* (DeSantis) did not change in response to parasitoid density or to situations when the parasitoid oviposited in already parasitized hosts, but could vary due to host-size distribution and differential mortality in mass rearing (Van Dijken *et al.*, 1989).

Tillman *et al.* (1993) discovered that host species could also be a determinant in the sex ratio of parasitoids. *Microplitis croceipes* had a significantly greater female to male ratio than either *Cotesia kazak* or *M. demolitor* Wilkinson had when reared on *Heliothis zea*. However, the sex ratios were not significantly different when the three parasitoid species were reared on *H. armigera*. Likewise, Rojas *et al.* (1995) showed that rearing the boll weevil parasitoid *Bracon thurberiphagae* on the relatively unsuitable factitious host *H. virescens* resulted in a male-biased sex ratio, which, nonetheless, could be restored to a 50:50 ratio with the addition of several amino acids to the diet of the adult parasitoid. This supplemental nutrition also improved the fecundity 100%.

Besides the factors mentioned earlier, age of the parasitoid female can affect sex of the progeny. Hirashima *et al.* (1990b) reported that the trichogrammatids *Trichogramma chilonis* Ishii and *T. ostriniae* Pang & Chen produced the highest percentage of female progeny on the first day of adult life. Likewise, Leatemia *et al.* (1995) found that with *T. minutum* the offspring sex ratios of long-lived females were male-biased (50 to 62% males), while those of short-lived females were female-biased (74 to 82% females). Greenberg *et al.* (1995) showed that as female adults of the boll weevil parasitoid *Catolaccus grandis* aged from 5 to 10 days, the percentage of female progeny increased from 64.0 to 77.2, respectively. A further increase in female age, from 20 to 25 days, resulted in a declining percentage of female progeny from 54.0 to 48.5. Contrarily, the sex ratio of the progeny of the predaceous mite *Phytoseiulus persi-*

milis also changed with age, but in this case the older females produced more females than males (Chang *et al.,* 1990). In producing three species of fruit fly parasitoids, Wong and Ramadan (1992) obtained the highest numbers of female progeny when they allowed mating for the first 4 days of adult life, prior to host exposure. Thereafter, production of female progeny remained significantly high during adult female ages of 5 to 30 days for *Diachasmimorpha tryoni* (Cameron) (= *Biosteres tryoni* [Cameron]), 5 to 15 days for *D. longicaudata,* and 6 to 20 days for *Psyttalia fletcheri* (Silvestri). Wong and Ramadan (1992) also showed that oviposition in younger fruit fly larvae (second instar) by *D. tryoni* and *D. longicaudata* resulted in significantly fewer female progeny, whereas a similar effect did not occur with *P. fletcheri.* Hailemichael *et al.* (1994) found that in rearing the ichneumonid *Xanthopimpla stemmator* (Thunberg), a solitary endoparasitoid of pupae of Old World lepidopteran stalk borers, the sex ratio (female to male) of the parasitoid progeny increased with host age. Females comprised 47% of total parasitoid progeny emerging from 1-day-old laboratory-reared pupae of the sugarcane borer, and 84% of those from 6-day-old pupae.

Cold temperatures can also affect parasitoid sex ratios. Hoffman and Kennett (1985) demonstrated that prolonged exposure of *Aphytis melinus* to winter temperatures caused a male bias in F1 sex ratios, and they briefly reviewed published reports of similar effects of low temperatures on parasitoids. Even just cool temperatures can alter sex ratios, as indicated by Yadav and Chaudhary (1988), who showed that while the female offspring of the encyrtid egg parasitoid *Cheiloneurus pyrillae* Mani outnumbered males at all rearing temperatures, the proportion of females declined with decreasing constant temperatures from 30° to 17.5°C. Mani and Krishnamoorthy (1992) found that 30°C was also optimum for maximizing female production of the encyrtid *Anagyrus dactylopii* (Howard).

Autoparasitism introduces another complication in sex-ratio determination. As shown with the aphelinid *Coccophagus atratus* Compere, brood sex ratios are dependent on the relative availability of hosts for the male progeny and hosts for the female progeny (Donaldson & Walter, 1991). However, a further complication reported by Hunter *et al.* (1993) is the formation of males in an autoparasitoid species (the aphelinid *Encarsia pergandiella* Howard) by loss of one set of chromosomes from fertilized eggs. Such males then develop as primary parasitoids, and not in the normal fashion as hyperparasitoids on sibling females. Once formed, the "primary" male trait is apparently heritable, although at a low and variable rate.

Chromosome loss resulting in male formation has also been noted in a non-autoparasitoid species, *Nasonia vitripennis.* A supernumerary chromosome labeled paternal sex ratio (PSR) transmitted via the sperm causes supercondensation and destruction of paternal chromosomes in fertilized eggs, thus effectively converting diploid (female) eggs into haploid (male) eggs (Beukeboom & Werren, 1993a, 1993b; Werren & Beukeboom, 1993). However, an additional complication with this insect is the presence, as reported by Breeuwer and Werren (1993), of maternally inherited bacteria (later identified as *Wolbachia,* as noted previously) that cause reproductive incompatibility, also resulting in destruction of paternal chromosomes and conversion of diploid eggs into male ones. Dobson and Tanouye (1996) showed that the transmission and ability of PSR to induce chromosome loss is not dependent on *Wolbachia,* with comparisons suggesting an absence of interactions between PSR and *Wolbachia* when they occur together. McAllister and Werren (1997) studied the origin and evolution of the PSR chromosome in *Nasonia,* and Reed and Werren (1995) examined the early cytological events associated with paternal genome loss induced by the PSR chromosome and by *Wolbachia.* Contrary to paternal genome loss, the sex ratio in *N. vitripennis* can be distorted toward nearly all female families by a cytoplasmic element termed maternal sex ratio (MSR), as reported by Beukeboom and Werren (1992). Obviously, sex determination in *N. vitripennis,* and perhaps in other Chalcidoidea, is very complex, with additional research necessary to sort out genetic, cytoplasmic, microorganism-caused, and possibly other effects.

Further information on sex ratios can be found in King (1993), Luck *et al.* (1993), and Wrensch and Ebbert (1993). Also see Chapter 23.

Rearing Density and Entomophage to Host Ratios

Provision of adequate food and space is not only important for maintaining a suitable sex ratio but also for optimizing progeny production. In rearing the red scale parasitoid, *Aphytis lingnanensis,* Papacek and Smith (1985) recommended a uniform rearing density of 30 to 50 oleander scales per square centimeter on the surface of butternut pumpkins. Propp and Morgan (1983) stressed the importance of using suitable parasitoid to host ratios to prevent superparasitism in mass-producing the pteromalid *Spalangia endius* Walker on the housefly, and Raupp and Thorpe (1985) noted that increasing parasitoid to host ratios may result in multiple stinging and increased larval mortality. Wong and Ramadan (1992) minimized superparasitism in producing fruit fly parasitoids by exposing them for a limited period of time to excess numbers of host larvae. Chianese (1985) used the ratio of eight mated females of the parasitoid *Cotesia melanoscela* to 40 gypsy moth larvae for 3 h for oviposition, and achieved 70 to 80% parasitism. With a Korean strain of *C. melanoscela,* Kolodny-Hirsch (1988) achieved maximum parasitoid production with a ratio of five females (8 days old) to 50 first- and second-instar larvae for 24 h at 28°C with a light intensity of 1452 lux. For *Trichogramma* spp., Morrison (1985a) obtained maximum fecundity with a ratio of one female parasitoid per 100 host eggs. However, in culturing *T. perkinsi* Girault and *T. exiguum* Pinto & Platner on eggs of the pyralid *Corcyra cephalonica,* a ratio of one parasitoid per 16 eggs resulted in parasitism of over 95% (Balasubramanian & Pawar, 1988). With *T. maidis,* Wajnberg *et al.* (1989) reported that superparasitism seems to be genetically controlled, apparently for number of eggs deposited in each host as well as for frequency distribution of the eggs among hosts. Such evidence would have a bearing on mass-rearing conditions.

Obviously, then, experimentation must determine the proper entomophage to host ratio for maximizing progeny yield per unit of host material. For example, Harris and Bautista (1996) tested the host suitability of *Bactrocera dorsalis* for development of the braconid parasitoid *Biosteres arisanus* and showed that increases in host egg to female parasitoid ratios of 5:1, 10:1, 20:1, 25:1, and 30:1 corresponded with increases in parasitoid progeny yield up to a plateau at the 20:1 ratio. In another system, Mishra and Singh (1990) found that increasing the number of female parasitoids of *Lysiphlebus delhiensis* (Subba Rao & Sharma) from 1 to 8 in a cage of 200 *Rhopalosiphum maidis* (Fitch) for 24 h increased the percentage of parasitization as well as the number of parasitoids emerged, but decreased the rate of multiplication, the search rate, and the proportion of female offspring. Also, the presence of males can decrease progeny production, as Kumar *et al.* (1988) showed with the parasitoid *Trioxys indicus.* Whether or not a parasitoid is gregarious can be a further complication. In mass-producing the gregarious braconid ectoparasitoid *Allorhogas pyalophagus* Marsh on the pyralid host *Chilo partellus* (Swinhoe), Ballal and Kumar (1993) reported that the optimal ratio of one host per parasitoid provided a high percentage of parasitism with the maximal number of cocoons and healthy adults.

Munstermann and Leiser (1985) cautioned that in rearing the predaceous mosquito *Toxorhynchites amboinensis* (Doleschall), the ratio of predaceous larvae to prey larvae is critical. Too little prey can result in cannibalism among the predaceous larvae, and too much prey can result in adult prey emergence in the *Toxorhynchites* pans. For the common green lacewing, Morrison (1985b) used a modified ice cream carton as an oviposition unit, in which 500 adults produced approximately 79,000 eggs in 21 days. Morrison (1985b) emphasized that each adult needed 2.5 cm² of "resting" space to prevent a reduction of longevity and fecundity caused by overcrowding. For larval green lacewings, Bichao and Araujo (1989) reported that the best initial density was 2000 lacewing eggs per 32 × 22 × 7 cm container, and specified host egg quantities required for larval development. In propagating the predaceous mite *Phytoseiulus persimilis,* the greatest efficiency resulted

from a ratio of one predator to 20 prey, and a density of 250 to 300 females of the spider mite prey on kidney bean leaves (Osakabe *et al.,* 1988). For rearing the phytoseiids *Amblyseius victoriensis* (Womersley) and *Typhlodromus doreenae* Schicha, James and Taylor (1993) cautioned that high densities (7.7 prey per 4 cm² of *A. victoriensis* or 9.0 prey per 4 cm² of *T. doreenae,* versus one predator per 4 cm²) resulted in a reduced daily oviposition of about 70 and 50%, respectively.

Adequate space and abundant, well-fed hosts for cultures of entomophagous parasitoids and predators will usually prevent cannibalism, mortality, lowered longevity and fecundity, and reduced fitness (Waage *et al.,* 1985). An exception reported by Wajnberg *et al.* (1985) was that *Drosophila melanogaster* Meigen suitability for the eucoilid endoparasitoid *Leptopilina baulardi* (Barbotin, Carton & Kelner-Pillault) increased from 50% for *Drosophila* reared in optimal laboratory conditions to 90% for those reared in crowded conditions. In a somewhat similar situation, Kazimirova and Vallo (1992) found that significantly higher percentages of parasitization and parasitoid emergence occurred when smaller Mediterranean fruit fly pupae originating from a high larval density were exposed to the diapriid *Coptera occidentalis* Muesebeck.

Temperature is also a factor interacting with hosts and space in entomophage production. Isenhour (1985) found that at high host densities, the ichneumonid *Campoletis sonorensis* (Cameron) parasitized significantly more third-instar larvae of the fall armyworm, *Spodoptera frugiperda* (Smith), at 25°C than at 30°C.

Behavior

An understanding of behavior can be an important component of entomophage culture. For example, the positive phototaxis and negative geotaxis of *Trichogramma* spp. provide greater ease of manipulation (Morrison, 1985a). In fact, most entomophagous parasitoids exhibit positive phototaxis, which facilitates their collection. The anemophototactic response of predaceous mites likewise makes collection easy (Chalkov, 1989), although Zhurba *et al.* (1986) first starved *P. persimilis* to further stimulate migration into a collection container.

Waage *et al.* (1985) recommended rearing entomophagous insects on their natural hosts on natural food sources to provide all necessary behavioral stimuli. While this is desirable, it may not be possible in mass production. King and Hartley (1985c) noted that volatile substances produced by feeding sugarcane borers attract *Lixophaga diatraeae,* which then larviposits when it contacts borer frass. However, these researchers developed a mass production scheme for this tachinid by rearing it on the greater wax moth, a factitious host whose use did not simulate field conditions but did greatly facilitate production. Semiochemicals produced by plant-fed *H. zea* were found to be similarly important in stimulating responsiveness of the parasitoid *M. demolitor* Wilkinson (Herard *et al.,* 1988a, 1988b). However, such responsiveness was generally lacking in parasitoid females reared from hosts fed artificial diet, so Herard *et al.* (1988b) recommended "imprinting" the emerging females by exposing them to frass from plant-fed hosts. For *in vitro* rearing of the egg parasitoid *Ooencyrtus kuvanae,* Lee and Lee (1994) had to treat the artificial eggshell of polypropylene film with an ethanol extract from gypsy moth egg masses to stimulate ovipositing by the wasp. Thus, the ovipositional behavior of the parasitoid would not be expected to change in culture. In the case of the *in vitro* rearing of the ectoparasitoid *Catolaccus grandis,* however, Guerra *et al.* (1994) found that short-chain saturated hydrocarbons (alkanes) could elicit oviposition behavior, as could volatiles emanating from an artificial diet devoid of insect components that was specifically developed for the *in vitro* rearing of ectoparasitoids. They suggested that a synergistic combination of *n*-hexane and diet might be used to optimize the mechanized production of noncontaminated eggs. After using *n*-hexane for two consecutive generations to stimulate oviposition for *in vitro* rearing, Guerra and Martinez (1994) found no apparent adverse effects on the general behavior and reproduction of the parasitoid.

Feeding behavior is also important to entomophage production potential. Although a short life cycle of 15 days positively influences production of large numbers of the red scale parasitoid, *Aphytis lingnanensis* (Papacek & Smith, 1985), host feeding is a negative influence, with each female destroying an average of 46 scales, while laying an average of 57 eggs (Rosen & DeBach, 1979). Pteromalid parasitoids of Diptera that reproduce by arrhenotoky require host feeding early in their adult life for maximum fecundity, whereas early host feeding reduces fecundity of thelytokous populations (Legner & Gerling, 1967).

Host-feeding behavior of insect parasitoids complicates their propagation so that special attention to the parasitoid to host ratio and to parasitization exposure time is necessary. Lasota and Kok (1986a) determined that for optimal production of *Pteromalus puparum,* a gregarious endoparasitoid of the imported cabbageworm, *Pieris rapae* (Linnaeus), exposure of 10 freshly formed host pupae to each parasitoid pair for 6 days provided sufficient hosts for host feeding, and sufficient time for egg formation and oviposition before the pupae became too old. Lasota and Kok (1986b) concluded that balanced parasitoid to host ratios are important in the mass production of gregarious parasitoids to optimize host utilization while maintaining parasitoid quality.

Host feeding can likewise occur in adult tachinids. Nettles (1987) found that exposing adults of *Eucelatoria bryani*

to their host, the corn earworm, or to host hemolymph, significantly increased their fecundity.

Also, feeding by the host itself sometimes can be important. Chianese (1985) noted that poorly fed gypsy moth larvae would cannibalize other larvae and eat cocoons of their own parasitoid, *Cotesia melanoscela.*

Special Techniques and Considerations

Following are a few examples of the many special techniques developed for producing parasitoids and predators, and other considerations that are important. Some of these pertain to the handling of eggs. With eggs laid on a natural substrate, the deposition sites can be cut out and grouped for exposure to egg parasitoids (Morrison, 1985a). Clair *et al.* (1987) used a cork borer to cut out clusters of elm leaf beetle eggs from elm leaves so that they could be grouped together in a small petri dish for efficient exposure to attack by the eulophid egg parasitoid *Oomyzus* (= *Tetrastichus*) *gallerucae* (Fonscolombe).

In research that could prove useful in the future when *in vitro* culturing of parasitoids may become routine, Eller *et al.* (1990) studied factors affecting oviposition by the braconid *M. croceipes* into artificial substrates. Among other things, they discovered that hemispherical substrates with a volume of 35 to 40 μl were preferred; that more eggs were laid in colored substrates, with the highest number oviposited in blue substrate; and that the number of eggs laid increased with the color concentration. Kainoh and Brown (1994) found that *Chelonus* sp. nr. *curvimaculatus* females would actively oviposit into artificial eggs constructed with various amino acids or insect culture media sandwiched between Parafilm® sheets. The amino acids arginine, histidine, and lysine stimulated intense ovipositional activity from *Chelonus* adults, but more parasitoid eggs were oviposited into artificial eggs containing *Trichoplusia ni* larval hemolymph than were those containing amino acids or artificial insect culture media. In anticipation of the artificial rearing of the predaceous anthocorid *Orius insidiosus* (Say), Castane and Zalom (1994) successfully developed a gelatinous ovipositional medium made from carrageenan salt of potassium chloride and covered with paraffin wax. They noted that an artificial substrate is useful in mass rearing of insects because the medium is available for use at any time of the year, and the likelihood of colony contamination by pathogens or pesticide residues is reduced.

Age of adult entomophagous females is an important factor in determining the most efficient ovipositional period for a production program. Ramadan *et al.* (1989) reported that 94.4% of the total progeny of adult females of the braconid *Diachasmimorpha tryoni* were produced during the age interval of 5 to 20 days, and that progeny production and the ratio of female progeny declined significantly thereafter. Likewise, as mentioned previously, Greenberg *et al.* (1995) showed that in the mass production of the boll weevil parasitoid *Catolaccus grandis,* the maximum percentage of female progeny (between 54.0 and 77.2%) was attained by ovipositing females that were 5 to 20 days old.

Two factors to be considered in rearing egg parasitoids are the age of the host eggs and their fertilization status. Reznik and Umarova (1990) indicated that *Trichogramma embryophagum* (Hartig) and *T. semblidis* (Aurivillius) preferred to oviposit into young eggs of the Angoumois grain moth and suggested that old eggs inhibited oviposition by some *Trichogramma* spp. At least in some cases, successful parasitoid development may only occur in fertilized host eggs, and is possibly related to amino acid content, as indicated for the scelionid *Telenomus remus* Nixon, which parasitizes the noctuid *Spodoptera litura* (Fabricius) (Gautam, 1986).

In China, mass production of *Trichogramma* spp. involves the grinding of freshly emerged oak silkworm female moths to extract infertile eggs (King & Morrison, 1984). Cleaning and drying the eggs follows, after which storage at low temperatures occurs for several weeks before use.

Similarly, King and Hartley (1985c) extracted parasitoid maggots from adult females of the tachinid *Lixophaga diatraeae* with a small blender. Dispensing the maggots with a special machine followed their chemical treatment, collection, rinsing, and suspension in 0.15% agar solution (Gantt *et al.* 1976; King & Hartley, 1985c). Mechanical extraction of maggots from fecund females of *Archytas marmoratus* was performed by Gross (1994), who then applied them in aqueous suspension into corrugated cardboard disks (13-cm in diameter by 2-cm in depth) (19.5 maggots per cm^2) against approximately 600 mature larvae of *Galleria mellonella*. David *et al.* (1989) tried to increase parasitism by the tachinid *Sturmiopsis inferens* by removing the parasitoid chorion before inoculation, but found this to be unnecessary. They also discarded the technique of hand inoculating individual host larvae when they discovered that tachinid maggots suspended in 0.15% agar were just as effective in host finding.

It is not unusual for other tachinids to also require special rearing efforts. For instance, Mellini and Campadelli (1989) experimented with rearing *Pseudogonia rufifrons* on artificial diet, and with using chemical and physical techniques for *in vitro* hatching of the tachinid eggs. They found centrifugation of uterus-extracted eggs in saline solution to be the most effective procedure for obtaining high egg hatch without affecting larval viability.

Cannibalism among predaceous entomophages, and predation by hosts require special attention. To prevent larval cannibalism in green lacewings, Morrison (1985b) sealed larvae in separate chambers and fed them through an organdy cloth with prefrozen lepidopteran eggs. To reduce

similar cannibalism among the larvae of the carabid *Calosoma sycophanta* Linnaeus, a predator of the gypsy moth, Weseloh (1996) reared them in groups in plastic containers having 3 cm of moist peat moss in the bottom. By feeding the larvae with gypsy moth pupae placed on top of the peat moss in the ratio of at least 0.5 pupae per day per larva, the larval mortality from cannibalism was reduced to 50%. In a case of host predation of its parasitoid, Chianese (1985) reduced gypsy moth larval predation of parasitoid cocoons of *Cotesia melanoscela* by regularly collecting cocoons that were sufficiently hardened to prevent handling damage.

Kairomones and pheromones may concentrate in insect-rearing cages where there is little air movement, and hence no odor gradient to stimulate normal insect response. Chiri and Legner (1982) showed that the parasitoid *Chelonus* sp. nr. *curvimaculatus* responded to kairomones emitted by body scales not only from its natural host, the pink bollworm, but also from unnatural hosts such as the beet armyworm. Chiri and Legner (1982) speculated that high populations of beet armyworms in cotton fields would reduce pink bollworm parasitization because of kairomonal distractions. Similarly, a uniform accumulation of kairomones or pheromones in enclosed insectary spaces could interfere with insect production and would make necessary the adequate provision of fresh air.

In working with the parasitic tachinid *Bonnetia comta* (Fallén), Cossentine and Lewis (1986) used host odor to induce larviposition by moistening filter paper with an aqueous suspension of fecal material of the black cutworm. Rubink and Clement (1982) found that fecal pellets from late instar larvae of the black cutworm provided the greatest intensity of larviposition by this tachinid. Likewise, Grenier *et al.* (1993) reported that surface treating artificial host eggs with *Ostrinia nubilalis* scale extract, or a mixture of compounds found in *O. nubilalis* egg masses, stimulated oviposition by the egg parasitoid *Trichogramma brassicae.*

Waage *et al.* (1985) remarked that host defense reactions are important in rearing some parasitoids. It is inadvisable to produce parasitoids on hosts exhibiting defense reactions such as encapsulation of parasitoid eggs or larvae, or to field release them on such hosts. For example, even after rearing the braconid *Leiophron uniformis* (Gahan) for 11 generations on an alternate host, *Lygus lineolaris* (Palisot de Beauvois), DeBolt (1989) still found no increased parasitoid resistance to encapsulation by that host. However, Strand and Vinson (1982) noted that eggs lack such defense mechanisms and may be useful factitious hosts. They demonstrated that a recognition hormone from a normal host stimulated oviposition by a scelionid parasitoid, *Telenomus heliothidis,* in eggs of nonhosts. Such recognition hormones could thus be useful for producing specific egg parasitoids in nutritionally acceptable nonhosts.

As with herbivorous insect production, extended storage of entomophagous insects is very useful. Morrison and King (1977) reviewed various techniques and concluded that in almost all, low temperatures were used to reduce developmental rates. The potential duration of a low-temperature storage period varies with the entomophage species. Of parasitoids stored the longest, prepupae of the pteromalid *Pteromalus puparum* held at 10°C for up to 15 months still eclosed and produced viable progeny, although the mean number of adults emerging per host chrysalis declined progressively after 4 months (McDonald & Kok, 1990). Storage of adult females of the bethylid *Scleroderma guani* Xiao & Wu, also at 10°C, had no significant influence on their ability to parasitize after 6 months (Chen *et al.*, 1989a). Similarly, Palmer (1985) reported that storing adults of the gypsy moth parasitoid *Brachymeria intermedia* for at least 5 months at 10°C and 50% RH was possible, although there was a mortality of 30 to 50%, depending on sex ratio (males were unable to survive prolonged storage). However, it was best not to store *Brachymeria* intended for field release for longer than 48 h at 16°C. On the other hand, mummies of *Trioxys indicus* could be held for up to 1 month at 4°C prior to field release of the adults on aphid hosts (Singh & Agarwala, 1992).

Storage of other parasitoids is often only suitable for days instead of months. Papacek and Smith (1985) noted that storage of the California red scale parasitoid, *Aphytis lingnanensis,* could not exceed 3 days at 16°C. Maggots of the tachinid *Archytas marmoratus* contained within corrugated cardboard disks could also only be effectively stored for 3 days at 10 and 18°C before *Galleria mellonella* larvae were introduced for parasitism, without significantly affecting the percentage recovery of *A. marmoratus* adults (Gross, 1994). Pupae of *Campoletis chlorideae* Uchida, an ichneumonid parasitoid of the *Heliothis armigera,* could be stored at 8°C for just 10 to 15 days without reduced adult life span, and adult emergence fell to 42% after 35 days of storage (Patel *et al.,* 1988). Holding freshly formed cocoons or 1-day-old mated adults of the braconid *Cotesia marginiventris* (Cresson) at either 5 or 10°C for longer than 20 days caused a sharply rising mortality, significantly reduced adult longevity and fecundity, and produced a preponderance of male progeny (Jalali *et al.,* 1990). Similarly, cocoons of the braconid *Allorhogas pyralophagus* Marsh held at 5 or 10°C for 35 days yielded parasitoids that produced only male progeny (Ballal *et al.,* 1989). However, Adiyaman and Aktumsek (1996) were able to satisfactorily store the ichneumonid *Pimpla turionellae* as either pupae or adults at 4°C for 30 days. Jalali and Singh (1992) found that cold storage differentially affected four species of *Trichogramma,* with the maximum storage period being 49 days at 10°C.

Some workers have experimented with the use of conditioning periods before cold storage. Singh and Srivastava (1988) reported that a conditioning period of 72 h at 12 to 15°C for newly formed aphid mummies containing *Trioxys indicus* before a 10-day storage period at 6°C allowed maximum adult emergence (78%). However, they still obtained good emergence (72%) after a month of storage at 4°C following a 72-h prestorage period at 12 to 15°C.

Concerning entomophagous predators, Morrison (1985b) found that measurable viability reduction occurred with eggs of the common green lacewing held at 13 to 14°C and 70 to 80% RH for more than 10 days. However, Tauber *et al.* (1993) were able to store adults of the green lacewing at 5°C and short-day length for 31 weeks, with about 97% survival, greater than 90% of mating pairs producing fertile eggs, and fecundity 50 to 70% that of unstored females. Chang *et al.* (1995) duplicated these results of storing green lacewing adults for 31 weeks, and noted also that high survival and excellent poststorage reproduction occurred with a high carbohydrate diet before storage and a carbohydrate and protein diet during and after storage. Tauber *et al.* (1997a) found that adults of a Honduran population of a congeneric lacewing, *Chrysoperla externa* (Hagen), could be held at 10°C for up to 4 months with a high survival and a high poststorage oviposition rate. They reported that the long-term storage trait was just one of the characteristics that gave the species great potential for mass rearing and use in augmentative biological control.

Other predators can also be stored for extended periods. Freshly emerged adults of the clerid *Thanasimus formicarius* (Linnaeus) could withstand cold storage (4°C) for more than 8 months with only 4.2% mortality and no adverse fecundity effects, provided that they could feed at 3-month intervals on their bark beetle hosts (Faulds, 1988). Long-term storage of the predatory midge *Aphidoletes aphidimyza* was accomplished in its diapause state by Tiitanen (1988), who reported that 66% of diapausing larvae emerged after 7 months at 10°C; and by Gilkeson (1990), who indicated that less than 9% mortality occurred after larval storage at 5°C for 8 months. However, the best commercial storage regime to avoid protracted adult emergence was at 1°C for up to 2 months after conditioning for 10 days at 5°C (Gilkeson, 1990).

Various factors should be considered and tested before cold storing entomophages. As noted before, cold storage can affect progeny sex ratios. As further examples, Clausen (1977) reported that prolonged cold storage of adult *Aphidius smithi* Sharma & Subba Rao—a parasitoid of the pea aphid—inactivated sperm in males or mated females, and Chianese (1985) found that refrigeration of *Cotesia melanoscela* adults must not occur until after mating. Chilling can also affect entomophage behavior, as noted by Herard *et al.* (1988a), who observed that chilling pupae of *Microplitis demolitor* caused most emerging females to be unresponsive to semiochemicals produced by their plant-fed host insect.

In concluding this discussion on storing parasitoids and

predators, we must remember that diapause can be useful in this regard. Although diapause can be an aggravating annoyance in arthropod production, it can also be a beneficial tool in manipulating entomophage rearing. The key is to acquire a complete knowledge of all factors that govern inducement, maintenance, and termination of the diapause state in the particular species being reared.

Life cycles of entomophages in the laboratory usually vary between 8 and 42 days, depending on temperature and excluding the effect of diapause. However, it is common for most species to develop in 14 to 30 days. At 26.7°C and constant light, the egg parasitoid *Trichogramma pretiosum* averages 9.5 days, and *T. minutum* averages 8 days (Morrison, 1985a). *Brachymeria intermedia* develops in 15 to 30 days at 24°C (Palmer, 1985), and the common green lacewing requires about 30 days at 26.7°C and ca. 75% RH, with constant light (Morrison, 1985b).

Obviously, a thorough study and knowledge of the biological attributes and interactions involved in a rearing system will enable optimal production of a natural enemy. For example, after a detailed rearing study of the interaction of the tachinid *Archytas marmoratus* with a factitious host, the greater wax moth, Coulibaly *et al.* (1993) discovered rather precise information that resulted in more efficient production. The adult female tachinids were found to larviposit about 75% of their planidia between the 6th and 11th hours from the beginning of the photophase (14 h photophase, 70% RH, 27°C) on days 13 and 14 following adult eclosion. High adult yields of the parasitoid (ca. 57%) were then obtained by artificially parasitizing full-grown host larvae with two to eight planidia (Coulibaly *et al.*, 1993).

A variety of problems can arise in rearing any entomophage, with solutions requiring adjustments of rearing techniques. A good example of such problem solving was that of Palmer (1985), who developed a production system for rearing the parasitoid *B. intermedia* on the greater wax moth. Regular removal of moths and scales from oviposition units prevented the scales from interfering with parasitoids ovipositing in pupae. Inactive *Brachymeria* in the oviposition jars required stimulation by increasing light intensity and/or temperature. The situation where parasitoids were active but the parasitism rate was low involved one or more factors: the wax moth pupae were too old when presented to the parasitoids, the ratio of parasitoids to hosts was too low (25 females to 300 cocoons was necessary), or the photophase was too short (had to be more than 10 h). Furthermore, overheated oviposition jars caused a high mortality of ovipositing *Brachymeria* (the required temperature was 24 to 26°C).

Karelin *et al.* (1989) reported about a number of rearing system modifications that together resulted in a dramatic increase in the commercial production of the common green lacewing. Important changes involved cage shape and volume, oviposition substrate, feeding and egg collection

methods, and partial mechanization. The modified system yielded 5.6 to 8.7 million eggs per month from just one production line that consisted of 60 cages for adults, 60 cell cages for larvae, and 120 cages for group rearing. Nordlund and Morrison (1992) reviewed other techniques used for rearing *Chrysoperla* spp., and mentioned a hot wire vacuum removal device with potential for resolving the problem of egg collection. Later, Nordlund and Correa (1995b) described an adult feeding and oviposition unit (AFOU) and a hot wire egg harvesting system that could provide the basis for an automated adult handling and egg harvesting system for the mass production of green lacewings. With respect to oviposition by two species of the brown lacewing genus *Sympherobius,* the frayed fringe of a coarsely woven cloth stimulated copious egg laying in culture (T. W. Fisher, personal correspondence).

Another commercially produced entomophage is the aphelinid *Encarsia formosa* Gahan, a parasitoid of the greenhouse whitefly, *Trialeurodes vaporariorum* (Westwood). Popov *et al.* (1987) described a rearing system that evolved to produce 17 million parasitoids from a 150 m² greenhouse area in the same time in which only 1 million had been formerly produced with previous techniques.

Massive rearing programs for *Trichogramma* spp. exist in China, in particular. Tseng (1990) reported on a machine that dramatically improved *Trichogramma* production efficiency by preparing host eggs on cards, and Liu *et al.* (1988b) described a computer-controlled machine that successfully made artificial egg cards at the rate of 1200 per hour. It was possible to rear over 6 million *Trichogramma* individuals from each set of 1200 cards, and these *Trichogramma* reportedly parasitized sugarcane borer eggs at the rate of 75 to 86% when released inundatively in the field.

In mass production technology, notable advanced systems have been or are being developed for the Africa-wide Biological Control Program of the International Institute of Tropical Agriculture (IITA) (Herren, 1987, 1990; Haug *et al.,* 1987). Hydroponic culture techniques were devised for producing cassava, *Manihot esculenta* (Herren, 1987), and semiautomated systems were in use for producing organisms at three trophic levels (cassava; the cassava mealybug, *Phenacoccus manihoti* Matile-Ferrero; and its encyrtid parasitoid, *Epidinocarsis lopezi*) (Haug *et al.,* 1987; Neuenschwander *et al.,* 1989; Neuenschwander & Haug, 1992).

For planning the mass production of natural enemies of the cassava green mite, Klay (1990) proposed and described a population model to optimize use of prey material and to minimize labor. Similarly, Carey and Krainacker (1988) discussed extending basic principles of stable population theory in mite populations to include mite mass-rearing. For parasitoid rearing, Carey *et al.* (1988) used the example of rearing the braconid *Diachasmimorpha tryoni* on the

Mediterranean fruit fly to present demographic methods for determining the optimal balance in host–parasitoid rearing systems and for determining optimal harvest rates.

Quality

The quality of produced entomophages is dependent on their genotype, nutrition, and rearing environment; and is obviously critical to a biological control program. It is futile to expend resources for entomophage production if insufficient quality of the product results in ineffective field establishment or pest control.

Moore *et al.* (1985) discussed quality of laboratory-produced insects, and noted that rearing objectives must first be established. Selection and quantification of species traits pertinent to the objectives then enables establishment of quality standards, and tolerances. Routine assessment measurements of the variables made during insect production are compared to the standards, and quality control regulatory actions for correcting differences between actual and intended quality are then made.

It would be impractical to use all the possible quality assessment tests listed by Moore *et al.* (1985). However, even for very small production programs, it is advisable to use appropriate tests to achieve desired results. Pragmatically, Singh and Ashby (1985) felt that the standard life history measurements of fecundity, fertility, and weights of adults and pupae are adequate as quality indicators. However, entomophage release programs can also include percentage of parasitism or predation rate (Moore *et al.*, 1985).

Chambers and Ashley (1984) defined industrial quality control ideas and evaluated their applicability to insect rearing. Bigler (1989) discussed quality control requirements in three successful biological control programs and concluded that careful analysis of each particular situation was necessary for determining significant field performance parameters.

A review of quality assessment and control procedures used in mass production of several insect parasitoids revealed the following. King and Morrison (1984) noted that quality assessment of *Trichogramma* production typically consists of keeping records on numbers reared, parasitization rate, and sex ratio. Quality assessment tests used by Morrison (1985a) for *Trichogramma* production were percentage of parasitized eggs, percentage of emergence from parasitized eggs, and sex ratio. For performing the tests, it was necessary to take at least three samples of about 200 eggs from each oviposition unit. Accepted quality assessment standards were 80 ± 10% of parasitism of 48-h-old eggs, 90 ± 5% of adult emergence, and a sex ratio of 1.00 female to 1.25 males. Nordlund *et al.* (1997) used a number of quality control parameters, including development time,

adult longevity, sex ratio, pupation rate, percentage of pupae to emerge as adults, adult female body length, number of *H. zea* eggs parasitized by a female, and percentage of deformed females, to assess and to compare *T. minutum* reared *in vitro* for 10 generations to those reared *in vivo* on irradiated *H. zea* eggs. Cerutti and Bigler (1995) also used most of these criteria to assess the quality of *T. brassicae* but noted, in particular, the evaluation of fecundity on eggs of the natural host, an indirect measure of the acceptance and suitability of natural-host eggs. Dutton *et al.* (1996) likewise evaluated the fecundity of *T. brassicae* on its natural host (the European corn borer) but incorporated this measurement together with walking speed and life span to calculate a quality index. This index, plus a measure of the fecundity on the factitious host (*Anagasta kuehniella*), showed a correlation to *T. brassicae* field parasitism, for different *T. brassicae* populations. For quality control of *Trichogramma* spp. in China, vigor is maintained in laboratory cultures by forcing adult females to fly several feet in search of host eggs to eliminate weak individuals (King & Morrison, 1984). In the former USSR, combining quality control and quality assessment of *Trichogramma* was possible by use of a device through which the parasitoids had to pass a distance of 3 m to reach host eggs, with quality assessment then determined from the numbers of eggs parasitized (Torgovetskij *et al.*, 1988). Flight quality assessment tests have been used for *Encarsia formosa* (Doodeman *et al.*, 1994; Roskam *et al.*, 1996), *T. brassicae* (Dutton & Bigler, 1995), and the rhizophagid *Rhizophagus grandis* Gyllenhal—a predator of *Dendroctonus micans* Kugelann (Couillien & Gregoire, 1994).

ODell *et al.* (1985) used parasitoid size, longevity, and fecundity parameters to check quality of gypsy moth parasitoids. A variety of factors affected these parameters, including host diet, host density, microbial infection of the host, and environmental changes. Quality assessment in production of *Cotesia melanoscela*, a parasitoid of gypsy moth larvae, consisted of determining fecundity, defined as the number of progeny produced (ca. 100), instead of the number of eggs laid (ca. 1000) (Chianese, 1985). Palmer (1985) used the following quality assessment parameters in producing the gypsy moth parasitoid *B. intermedia:* (1) production totals per cage, per week, and per month; (2) percentage of successful parasitism; (3) female to male ratio; and (4) percentage of recovery from storage. The standard and minimal acceptable values for percentage of parasitism were 70 to 75 and 60%, respectively.

King and Hartley (1985c) used the following standards in assessing the quality of mass-produced *Lixophaga diatraeae:* (1) puparial weight (male = 14 mg, female = 20 mg), (2) percentage of parasitism (90%), (3) number of maggots per female (70), and (4) maximum of adult longevity (male = 29 days, female = 24 days).

COLONIZATION OF ENTOMOPHAGOUS ARTHROPODS

Liberation Procedures

Entomophages propagated in the insectary are field released to effect establishment. Establishment may occur with ease and rapidity or with difficulty over a long time, or may not occur at all.

The ease of insectary rearing does not correlate with ease of field establishment. In analyzing the successful biological control of the alfalfa blotch leaf miner, *Agromyza frontella* (Rondani), in the northeastern United States, Drea and Hendrickson (1986) noted that none of the most abundant and most easily reared European parasitoids became established. The most successful species were obtained by laboriously collecting 30,000 to 40,000 host puparia in Europe and subjecting them to specially developed laboratory recovery techniques to obtain healthy parasitoid individuals for field release.

Working with the same leaf miner, Harcourt *et al.* (1988) directly field released a genetically diverse group of 586 adults of the braconid *Dacnusa dryas* (Nixon) in eastern Ontario. The release site served as a "field nursery" for parasitoid reproduction, with specimens collected and released at various other sites. Within 3 years, the parasitoid had reduced leaf miner populations 50-fold. They subsequently collapsed to subeconomic levels. Desirable characteristics of this successful parasitoid were high dispersal capacity, host specificity, synchronization with the host life cycle, and high adaptability to diverse environmental conditions.

Another technique in field establishment is to make many releases over a short time. Krishnamoorthy and Singh (1987) obtained complete biological control of the citrus mealybug in southern India by introducing the encyrtid parasitoid *Leptomastix dactylopii* Howard. Repetition of field releases occurred 9 to 24 times over a short period of 2 to 4 months. Two orchards received 11,394 and 26,380 adults of the parasitoid, respectively.

It is obviously best to rapidly biologically control a new pest. Discovery of the citrus blackfly in Barbados in 1964 prompted quick biological control importations in the same year before the blackfly reached problematic levels. The aphelinid parasitoids *Eretomocerus serius* Silvestri and *Encarsia clypealis* increased rapidly and controlled the blackfly within 9 months (Bennett, 1966; Bennett & van Wherlin, 1966; Clausen, 1977).

Rapid biological control also occurred when the southern green stinkbug first appeared in Hawaii in late 1961. Importations of the parasitic scelionid *Trissolcus basalis* and the tachinid *Trichopoda pennipes pilipes* (Fabricius) in 1962 resulted in control of the pest by 1965 (Clausen, 1977).

An example of swift biological control was that of the ash whitefly, *Siphoninus phillyreae* (Haliday), an aleyrodid that invaded and rapidly spread throughout California beginning before mid-1988 when it was first detected (Pickett *et al.*, 1996). In 1989, outbreak populations occurred on ornamental trees in several major urban centers. Within 3 years of the whitefly's invasion of California, the aphelinid *Encarsia inaron* (Walker) was imported, mass reared, and released in 43 of 46 affected counties. Within 2 years of initial releases, *E. inaron* rapidly spread and established, and the infestation density of the ash whitefly decreased by 90 to 95%. The preserved economic value of the saved ornamental trees was estimated to be nearly $300 million in retail replacement values (Pickett *et al.*, 1996).

Field releases may consist of immature instead of mature entomophages. Katsoyannos and Argyriou (1985) released the aphelinid *Encarsia perniciosi* (Tower) against the San Jose scale, *Quadraspidiotus perniciosus* Comstock, by suspending squash fruit infested with parasitized scales in almond orchards. For field colonization of the aphidiid *Pauesia* sp., Kfir *et al.* (1985) suspended small logs heavily infested with parasitized black pine aphids 1.5 to 2.0 m high in trees. They reported rapid spread and establishment of the parasitoid, due to its high dispersal rate and searching ability.

For control of the Colorado potato beetle, Hough-Goldstein *et. al.* (1996) obtained excellent results by releasing nymphs of the pentatomid *Perillus bioculatus* (F.) in small plots. Control was nearly as good when the predator was released in the egg stage if the eggs were protected in screened cups.

Takahashi (1997) found that second-, third-, and fourth-instar larvae were the optimal stages of the coccinellid, *Coccinella septempunctata brucki* Mulsant, to release in alfalfa fields to control the blue alfalfa aphid, *Acyrthosiphon kondoi* Shinji. The optimal release time was early March when the maximum temperature was around 10°C.

With the predaceous coccinellid *Chilocorus nigritus* (Fabricius), Hattingh and Samways (1991) found that the adult was the most suitable stage for field release. This is because first-instar larvae cannot attack adults of the host scale, *Aonidiella aurantii,* and because adult beetles mass-reared on oleander scale will increase feeding rates when released on *A. aurantii,* whereas beetle larvae will not readily accept the dietary change (Samways & Wilson, 1988).

Shepherd (1996) also addressed the issue of which stadium of biological control insects should be field released, as well as the issue of timing the releases. Particular reference was made to the leaf-mining moth, *Dialectica scalariella* (Zeller), a gracillariid that attacks the weed Paterson's curse (*Echium plantagineum*).

Besides its utility in classical biological control, a "field insectary" is particularly useful for inoculative augmentation, where early season releases of small numbers of entomophages at key locations can achieve effective biological control. This is advisable when the entomophage has a short life cycle, high fecundity, and great vagility, yet cannot persist around the year. An example was the release of *Pediobius foveolatus* on Mexican bean beetles in small areas of snap beans, *Phaseolus vulgaris,* from which they spread to adjacent soybean fields (Stevens *et al.,* 1975; King & Morrison, 1984). Similar inoculative releases of this parasitoid protected urban gardens from damage by the Mexican bean beetle (Barrows & Hooker, 1981). Likewise, Halfhill and Featherston (1973) made inoculative releases of the parasitoid *Aphidius smithi,* in field cages on the pea aphid, and then allowed the millions of progeny to spread to adjacent alfalfa.

Inoculative and inundative releases of biological control agents are now rather common in greenhouses. Ravensberg (1992) provided an excellent historical and technical review of the use of natural enemies in European greenhouse crops. Hansen (1988) effectively protected cucumbers grown in greenhouses from the onion thrips, *Thrips tabaci* Lindeman, by making three to four releases of the predatory phytoseiid mite *Amblyseius barkeri* (Hughes) at rates of 300 to 600 per square meter. Establishing the predator before the thrips appeared enhanced success.

Periodic colonization of *Encarsia formosa* was successful against the greenhouse whitefly in Canada (Clausen, 1977). In certain areas of Australia the parasitoid has become permanently established and effective both in greenhouses and outdoors. In this regard, Gerling (1966) determined that temperatures above 24°C were necessary for the parasitoid to control the whitefly. King and Morrison (1984) also noted the extensive use of *E. formosa* in Europe to augmentatively control the greenhouse whitefly.

Stinner (1977) generally reviewed the efficacy of inundative field releases; and Goodenough (1984) reviewed improved packaging and distribution equipment, materials, and procedures for releasing the egg parasitoid, *Trichogramma pretiosum* (also see Reeves, 1975; Jones *et al.,* 1977; Jones *et al.,* 1979; Bouse *et al.,* 1980, 1981). Likewise, Smith and Wallace (1990) described ground systems for releasing *T. minutum* in plantation forests. Gardner and Giles (1996) obtained good results from the simulated mechanical distribution and release of commercially produced eggs of *Chrysoperla rufilabris* in an egg/vermiculite mixture, as well as from spraying an aqueous suspension of *C. rufilabris* eggs or of *Anagasta kuehniella* eggs parasitized with *T. pretiosum* (Gardner & Giles, 1997). Similarly, Sengonca and Loechte (1997) developed spray and atomizer techniques that allowed a targeted deposition of intact eggs of the common green lacewing on sugar beet plants in the field. Aircraft releases of entomophagous parasitoids have occurred with *Trichogramma* spp. (Ridgway *et al.,* 1977; Bouse *et al.,* 1981; Smith *et al.,* 1990), with *Lixophaga*

diatraeae (Ridgway *et al.,* 1977), and with *Chelonus* spp. in cotton fields (E. F. Legner, unpublished). A classic example of this type of release was the distribution by air of *Neodusmetia sangwani* (Rao) to control the Rhodes grass mealybug *Antonina graminis* (Maskell) in Texas in a very large and successful program (Dean *et al.* 1979).

Aerial-release technology was crucial for liberating the cassava mealybug parasitoid, *Epidinocarsis lopezi,* and cassava green mite predators in Africa, because ground releases would have been a major obstacle to achieving rapid control of the pests in the huge cassava belt (Herren, 1987). Notable features of the systems developed were an automatic acceleration of the parasitoids in the release device before ejection to reduce deceleration effects outside the aircraft, and a streamered container for predaceous mites that allowed retention in the cassava plant canopy for effective mite dispersal (Herren *et al.,* 1987).

Collection and Dissemination

Once entomophages establish at various loci in the range of a pest, a program of field collection and distribution to effect complete dissemination is advisable. For example, van den Bosch *et al.* (1959) used alfalfa cuttings and mechanical sweeper collections to increase the distribution of *Praon exsoletum* and *Trioxys utilis* Muesebeck on their host, the yellow clover aphid, a serious pest that invaded California in the 1950s.

Collection and distribution also followed the establishment of the braconid *Microctonus aethiopoides* (Nees) on the Egyptian alfalfa weevil in California. Parasitized adult weevils collected from estivation sites on trees produced parasitoids used for further field distribution (L. Etzel, unpublished). Two primary benefits of this technique were the high vigor and fitness of the recovered parasitoids, and the reduced costs resulting from using nature as an "insectary."

Similarly, the Animal and Plant Health Inspection Service (APHIS) of the United States Department of Agriculture (USDA) distributed the ichneumonid *Bathyplectes anurus* (Thomson) for biological control of the alfalfa weevil *Hypera postica* (Gyllenhal) and the Egyptian alfalfa weevil *H. brunneipennis* (Kingsley *et. al.,* 1993). This parasitoid has only one generation per year in nature, with two diapauses in its life cycle, and is thus very difficult to culture. Field establishment in the eastern United States required direct field releases of annually imported wild stock from Europe. After several years, *B. anurus* finally increased to sufficient abundance at some sites to enable the collection of large numbers that could be redistributed to other areas. A large program subsequently developed by APHIS effected a more general dispersal of this and other species (Kingsley *et al.,* 1993).

One of the largest collection and distribution programs

in the history of biological control occurred in Mexico in 1950 through 1953, following importation of several Indian and Pakistani species of parasitoids to be used against the citrus blackfly. A special gasoline tax levied to support this program funded a peak employment of 1600 workers (Clausen, 1977).

As mentioned, difficulties with mass production make collection and distribution programs particularly desirable. Harris and Okamoto (1983) reported that sex-ratio problems in culture prevented rearing the braconid fruit fly parasitoid *Biosteres arisanus* in large numbers. A method subsequently developed for parasitoid distribution first involved exposing papaya fruits preinfested in the laboratory with oriental fruit flies to field populations of the parasitoid for 24 h. After the recovered fruit fly larvae completed development on a diet in the laboratory, parasitoids emerged from the resulting puparia. This method allowed one technician to process over 11,000 parasitoids per day.

Small-scale collection and distribution can also be effective. Campbell (1975) developed a simple technique for citrus growers to distribute *Aphytis melinus* for control of California red scale. It involved placing a basket with scale-infested oranges in an orchard where the parasitoid was active, replacing half the oranges two weeks later, and taking the exposed oranges to new orchards for colonization.

Successful redistribution of native beneficial arthropods is also possible. The predaceous phytoseiid *Euseius hibisci* (Chant) was easily colonized in citrus orchards against the citrus thrips, *Scirtothrips citri* (Moulton), by transferring orange-branch terminals infested with the predator to six centrally located trees per 4 ha (Tanigoshi & Griffiths, 1982; Tanigoshi *et al.*, 1983). The mite readily dispersed aerially among groves within one season, resulting in a dramatic reduction of insecticide treatments. Another method of field colonizing this predator was to place bundles of lima bean seedlings containing the laboratory-reared mite into crotches of citrus trees.

An important consideration in relying on the field redistribution procedure is that a successful entomophage might reduce pest populations relatively rapidly, and thereby prevent easy collections. Furthermore, although field collection and distribution of established entomophages can be an expedient method of enlarging their establishment areas, caution is necessary to avoid the simultaneous dispersal of pest insects, hyperparasitoids, and other unwanted organisms. The Australian biological control program against black scale resulted in transferring many indigenous species around the country, including predaceous coccinellids and lepidopterans. Unfortunately, the distributed species also included the native hyperparasitoids *Coccidoctonus whittieri* (Girault) and *Myiocnema comperei* Ashmead (Clausen, 1977).

Principal Factors Influencing Establishment or Effectiveness

Species and Biotypes

It is not unusual for an entomophage species to have biotypes that vary in characteristics such as climatic or host population suitability. For example, King *et al.* (1978) demonstrated differences between populations of the tachinid *Lixophaga diatraeae*. Also, Jalali and Singh (1993) discovered that populations of *Trichogramma chilonis* Ishii collected from two of six cotton ecosystems in different agroclimatic zones of India were superior for mean fecundity, longevity, net reproductive rate, and rate of increase per female per generation. Likewise, Haardt and Hoeller (1992) showed pronounced differences among six isofemale lines of the aphid parasitoid *Aphelinus abdominalis* (Dalman) with respect to life history traits. In fact, crossing experiments revealed reproductive barriers that indicated the possible existence of sibling species. Work with species biotypes has included that for *T. evanescens* (Ram *et al.*, 1995), *T. exiguum* (Gomez *et al.*, 1995), and *Microctonus* spp. (Phillips & Baird, 1996; Goldson *et al.*, 1997; Phillips *et al.*, 1997).

Obrycki *et al.* (1987) reported that there are two observed biotypes of *Edovum puttleri*, an egg parasitoid of the Colorado potato beetle. They believed that matching biotypes to the agronomic and climatic conditions of the release areas would be important in achieving maximum control. Ruberson *et al.* (1989a) also compared the two biotypes and suggested using the intraspecific variability for genetic improvement.

Because of species biotypes, workers may seek the same entomophage species from many different areas and then rear the different collections separately to maximize field effectiveness of the released species. However, Clarke and Walter (1995) reviewed the literature on the classical biological control tactic of multiple introductions of conspecific populations of a natural enemy, and concluded that out of 178 identified projects, only 11 achieved eventual success with a second or later introduction of a conspecific population; and the success of at least 5 of these was thought to be due to sibling species. They suggested from their analysis that importation of different species, instead of additional biotypes, would be a more productive use of resources.

Sujii *et al.* (1996) performed a multivariate analysis on 21 potential biological control insect agents found feeding on the weed sicklepod (*Senna obtusifolia*) in Brazil, based on frequency of occurrence of each species, degree of insect/host association, and damage level. They determined on the basis of feeding guilds that introductions of a number of these species could be considered, and concluded that multivariate analysis can be a potentially useful tool in selecting candidates to be biological control agents. At the pest level, Barbosa and Segarra-Carmona (1993) reviewed criteria for the selection of pest arthropod species as candidates for biological control.

Especially in inundative release programs, the suitability of the released entomophage for its intended host must first be determined. For example, in South Africa, the egg parasitoid *T. pretiosum* was mass-produced and liberated against *Heliothis armigera,* but with poor results, caused at least partly by *H. armigera* being a generally unsuitable host (Kfir, 1981). Harrison *et al.,* (1985) likewise stressed the importance of precise taxonomic identification and biological testing of *Trichogramma* spp. before mass production for inundative releases. They found that *T. pretiosum* was preferable to *T. exiguum* to augmentatively control *Heliothis* spp. on Mississippi central delta cotton, because *T. pretiosum* could develop at the 35°C temperatures common in that area. To find suitable egg parasitoids for field release, Wuehrer and Hassan (1993) biologically tested 47 strains of *Trichogramma* and two strains of *Trichogrammatoidea* for their ability to attack the diamondback moth, *Plutella xylostella.* Likewise, Li *et al.* (1994) found that *Trichogramma* sp. nr. *sibericum* Sorokina, an indigenous species collected from eggs of the blackheaded fireworm, *Rhopobota naevana* (Hübner), a tortricid pest in cranberry fields, showed a high laboratory parasitism of fireworm eggs, whereas commercial *T. minutum* and *T. evanescens* displayed a low parasitism rate. Although the affinity to its natural host decreased after 17 generations reared on eggs of the noctuid *Peridroma saucia* (Hübner), percentages of acceptance and of parasitism by *T.* sp. nr. *sibericum* were still higher than those of the commercial *Trichogramma.* Another example of seeking a parasitoid species that is effective against its intended host was the selection of *T. (Trichogrammanza) carverae* (Oatman & Pinto), one of four candidate indigenous species of *Trichogramma* collected from vineyard egg masses of *Epiphyas postvittana,* as a potentially commercially viable biological control agent (Glenn & Hoffmann, 1997; Glenn *et al.,* 1997). Inundative release trials using small blocks within commercial vineyards showed that >75% of the egg masses in the vineyard could be parasitized by *T. carverae* at release rates of 70,000 *Trichogramma* per hectare with as few as 30 release sites per hectare (Glenn & Hoffmann, 1997).

Landry *et al.* (1993) developed a rapid DNA analysis system that enabled them to characterize individuals from species of the egg parasitoid genera *Anaphes* and *Trichogramma.* They stressed the importance to commercial biological control of being able to distinguish genetic variations that would be undetectable by traditional taxonomy. Similarly, Edwards and Hoy (1995a) used a randomly amplified polymorphic DNA polymerase chain reaction (RAPD-PCR) technique, together with discriminate analysis, to distinguish individuals of two biotypes of the walnut aphid parasitoid, *Trioxys pallidus,* in the laboratory. They

then used the technique to determine the fate of an insecticide-resistant biotype of the parasitoid after field release (Edwards & Hoy, 1995b). Likewise, Messing and Croft (1991) proposed that electrophoretic markers could be used to track the degree of field establishment and dispersal of imported foreign biotypes of phytoseiid mites.

Climate and Weather

As noted earlier, researchers generally make every effort to obtain entomophages from areas with climates similar to those at the release sites. In an examination of the establishment of introduced parasitoids, Stiling (1990) concluded that the climatic origin of a parasitoid had a large influence on its establishment rate.

Messenger and van den Bosch (1971) reported the classic example of the importation of two climatic biotypes of the parasitic braconid *T. pallidus* to control the walnut aphid, *Chromaphis juglandicola* (Kaltenbach), in California. A biotype of the parasitoid from France controlled the pest in the cooler coastal walnut-growing areas of California, whereas a biotype from Iran brought the aphid under control in the warmer Central Valley (Messenger & van den Bosch, 1971).

Current weather is likewise important in parasitoid releases. Laboratory experiments by Gross (1988) determined that unfavorable temperatures and RHs, and levels of free water at eclosion could have pronounced adverse effects on emergence of the egg parasitoid *Trichogramma pretiosum.* He noted the importance of identifying these effects on *Trichogramma* emergence at field liberation sites. The commonly erratic results of *Trichogramma* releases might be due to inattention to such factors (Gross, 1988). Bourchier and Smith (1996) confirmed this by analyzing the effects of weather conditions on the field efficacy of mass-released *T. minutum.* Temperature, the most important single variable, caused up to 75% of the variation in field parasitism, but there was also a significant negative relationship between the mean RH and the odds of parasitism in the field. Poor weather conditions following a release made laboratory quality parameters ineffective in predicting field efficacy. Chihrane and Lauge (1994, 1996, 1997) verified the importance of temperature on the effectiveness of inundative field releases of *T. brassicae* against the European corn borer. They showed that heat shocks of 44°C for 6 h resulted in lower fecundity and in drastically reduced proportion of female progeny. Wang *et al.* (1997) found that a higher mean temperature adversely affected the level of parasitism by *T. ostriniae* on European corn borer during hotter times of the season and conversely that lower temperatures (<17°C) reduced the egg parasitism during cooler periods.

Other investigators have also considered weather effects on entomophage releases. Releases of the coccinellid *Chilocorus bipustulatus* Linnaeus against the white date scale,

Parlatoria blanchardi (Targioni-Tozzetti), in date palm oases at 700- to 1600-m elevation in northern Niger were most successful during the rainy season (Stansly, 1984); and Smith (1988) considered the effect of wind and other factors on the fate of *T. minutum* released inundatively against the spruce budworm.

Weather, because of fluctuating temperatures, governs the seasonal appearance of pests. Consequently, Yu and Luck (1988) referred to the use of temperature-dependent, stage-specific developmental rates for timing parasitoid releases.

Habitats

If a pest insect attacks a variety of both economic and noneconomic plants, it is advisable to attempt to establish natural enemies of the insect on the greatest possible number of alternative plant hosts. Such establishment on noneconomic plants will increase reservoir populations of the beneficial organisms where they will be unaffected by pesticides (Argyriou, 1981). In South Africa, Hattingh and Samways (1991) established the coccinellid *Chilocorus nigritus* on the nontarget scale *Asterolecanium* sp. on giant bamboo. The coccinellid then easily moved to adjacent citrus orchards to attack the target prey, the California red scale.

Noncrop vegetation can be beneficial for introduced natural enemies even if their hosts are not found there. Dyer and Landis (1996, 1997) studied the habitat relationships of the ichneumonid *Eriborus terebrans* (Gravenhorst), the most abundant imported parasitoid of the European corn borer in Michigan. Their studies indicated that *E. terebrans* needs woodlot habitats adjacent to cornfields for sources of sugar and moderate microclimate unavailable in early season corn. They suggested that perennial habitats adjacent to annual agricultural crops may be necessary to provide resources needed for the successful maintenance of natural enemies and effective biological control. Dix *et al.* (1995) generalized that field edges where crops and edge vegetation meet provide a necessary habitat for maintaining natural enemies; and thus should be managed as to plant species and arrangement, consideration of optimal field sizes, and number of edges. Tietze (1994) also emphasized the importance of field edge grass, weed, and shrub communities that contain a high diversity of species due to their resource spectrum, thus having great regulation potential. Wissinger (1997) suggested that many indigenous and introduced natural enemies that attack pests of annual crops might be "cyclic colonizers" that, following disturbance in the crops, disperse to permanent refugia where they delay reproduction, overwinter, and then recolonize the following year. Therefore, effective biological control strategies in annual crop systems must provide permanent habitats that can act as reservoirs for natural enemies. Such strategies will depend on an increased information base about seasonal cycles, dispersal behavior, and overwintering ecology of natural enemies (Wissinger, 1997).

Pesticides in habitats always present a problem for entomophage releases. For example, Maier (1993) was only able to establish the encyrtid *Holcothorax testaceipes* (Ratzeburg) on the gracillariid leaf miner *Phyllonorycter blancardella* in unsprayed apple orchards. In another situation, Mansour *et al.* (1993) reported unusual effects of pesticides used in apple orchards where augmentative releases of the phytoseiid mite *Typhlodromus athiasae* Porath & Swirski were made. Laboratory tests revealed that the insect growth regulators triflumuron and fenoxycarb and the fungicide triadimenol caused only a slight mortality of the phytoseiid but a highly significant reduction in fecundity, whereas the same pesticides increased the fecundity of the phytophagous mite *Tetranychus cinnabarinus* Boisduval.

Adaptation

One possibility in the successful introduction of a biological control agent into a new environment is that the agent may have genetically adapted to the new pest population, habitat, and/or climate. Hopper *et al.* (1993) noted that instances where an introduced species is rare for a long time and then suddenly increases dramatically in population size may indicate genetic adaptation, but may also be explained by demographic models. Nonetheless, lack of postrelease adaptation of an entomophage to a new environment or specific host races can cause either failure in field colonization or reduction in effectiveness. For example, the encyrtid *Metaphycus luteolus* (Timberlake) from Californian brown soft scale would not adapt to the brown soft scale found in Texas (Clausen, 1977).

Dispersal

Entomophage dispersal varies greatly between species. Nevertheless, even entomophages that disperse slowly can be effective biological control agents, as shown with the Rhodes grass mealybug parasitoid, *Neodusmetia*. Also, the red wax scale, *Ceroplastes rubens* Maskell, which is serious on citrus in Japan, was controlled successfully by the encyrtid *Anicetus beneficus* Ishii & Yasumatsu. However, inasmuch as the natural spread of the parasitoid was at the rate of only 1 mi in 2 years, successful control of the scale required parasitoid distribution from the area of initial establishment (Clausen, 1977).

Hosts can often considerably assist parasitoid dispersal. As Clausen (1977) noted, the occurrence of alate females in many species of aphids can greatly facilitate dispersal of early parasitoid stages carried in their bodies. Rapid dispersal of *Praon exsoletum* occurs because it frequently parasitizes the winged adult of its host, the yellow clover aphid, which then carries the immature parasitoid for long distances during aphid migratory flights. *Trioxys utilis,* on the other hand, depends mainly on its own locomotion for dispersal, because it usually kills its host before the aphid can reach the winged stage (Schlinger & Hall, 1959).

Some entomophages can disperse phenomenally by their own locomotion. The encyrtid *Anagyrus indicus* Shafee *et al.* dispersed as much as 61 km in 1 year after being released in Jordan against *Nipaecoccus viridis* (Newstead), a citrus pest (Meyerdirk *et al.,* 1988). Releases of 41,054 had been made over an 18-month period, some directly into the trees and some into organdy sleeves tied around infested branches.

In Turkey, the aphelinid *Eretmocerus debachi* Rose and Rosen was imported from California in 1986 for release against the whitefly *Parabemisia myricae* (Kuwana), an aleyrodid that was one of the most serious citrus pests. *Eretmocerus debachi* was also capable of spreading 60 km in a year, and reduced the pest population from 17 immature stages per leaf to less than 0.1 stages per leaf within 3 years (Sengonca *et al.,* 1993).

Another instance of remarkable dispersal was reported by Agricola *et al.* (1989), who introduced an encyrtid parasitoid, *Gyranusoidea tebygi* Noyes, into Togo against the pseudococcid *Rastrococcus invadens* Williams. The parasitoid significantly controlled the pest in a range of ecological conditions at up to 100 km from each release point within a year. Contrarily, although Omwega *et al.* (1997) concluded that *Cotesia flavipes,* an exotic endoparasitoid of *Chilo partellus* in northern Tanzania, had dispersed about 60 km/year, and Mason *et al.* (1994) suggested that *Lydella thompsoni,* a parasitoid of the European corn borer, had dispersed from Delaware at about 50 km/year, there was no indication that either of these parasitoids was exerting effective biological control of its host.

Too much dispersal or uneven dispersion can be troublesome with augmentation programs. Augmentative releases of the coccinellid *Hippodamia convergens* Guérin in California field crops are useless because the beetles immediately leave the release sites (DeBach & Hagen, 1964). In commercial greenhouses, problems of obtaining even dispersion of coccinellids and chrysopids make them unsuitable for augmentative biological control (Chambers, 1986).

Augmentative releases of the tachinid *Lixophaga diatraeae* against the sugarcane borer resulted in rapid dispersal from the release sites, which negated the effects expected from releasing mated females at different rates in three adjacent fields (King *et al.,* 1981). However, there was some indication that the parasitoids remained more confined to sugarcane fields surrounded by woodlands.

In experiments on the problem of retaining augmentatively released parasitoids in a target area, *Trichogramma pretiosum* produced significantly higher parasitization rates on corn earworm eggs on field peas and cotton when it had prerelease exposure to corn earworm eggs in the laboratory (Gross *et al.,* 1981). There is also the possibility of using kairomones to improve the efficiency of augmentatively released parasitoids (Gross *et al.,* 1975, 1984). Hare *et al.* (1997) showed this in field experiments in which the parasitoid *Aphytis melinus,* which had been insectary reared on oleander scale on squash, was released to attack California red scale on citrus trees. Specimens of *A. melinus* that had been preconditioned by exposure to the red scale kairomone *O*-caffeoyltyrosine produced a greater than expected number of progeny compared to the control strain in all comparisons. They concluded that prior exposure of commercially reared *A. melinus* to *O*-caffeoyltyrosine may improve its effectiveness in augmentative release programs to control California red scale.

The augmentative use of parasitoids and predators with relatively low dispersal capabilities requires their introduction into a crop at multiple points of release, as for *Neoseiulus fallacis* (Garman), a phytoseiid predator of the two-spotted spider mite in strawberries (Coop & Croft, 1995); for *T. brassicae* released against the European corn borer (Greatti & Zandigiacomo, 1995); and for *T. platneri* released against the codling moth in apples (McDougall & Mills, 1997). An extrapolation of the work by Justo *et al.* (1997) indicates that 90% parasitism of eggs of the southern green stinkbug can be achieved by releasing 2000 *Trissolcus basalis* adults at 50-m intervals in a tomato field. To manage the dispersal of the curculionid *Apion fuscirostre* Fabricius for classical biological control of the weed Scotch broom (*Cytisus scoparius*), Isaacson *et al.* (1996) used Geographic Information Systems (GIS) distance measures.

A useful tool for quantifying dispersal parameters of parasitoids is the fluorescent dust marking and recapture method used by Corbett and Rosenheim (1996), who found that the tiny mymarid *Anagrus epos* (Girault), an egg parasitoid of the western grape leafhopper, *Erythroneura elegantula* Osborn, not only could disperse up to 13 m/day but also showed strong upwind displacement in one trial. Follett and Roderick (1996) recommended that gene-flow estimates be used to characterize dispersal capabilities in biological control organisms. They studied populations of the coccinellid *Curinus coeruleus* Mulsant, a predator of the leucaena psyllid, *Heteropsylla cubana* Crawford, to determine population structure and estimate levels of gene flow, and concluded that the inferior dispersal ability of *C. coeruleus* likely limits its rapid widespread establishment in release programs.

Each parasitoid or predator used in arthropod biological control work must be studied for its dispersal characteristics. However, in a study of four species of egg parasitoids in two families that attacked four genera of hemipteran pests in three families, Hirose *et al.* (1996) proposed the generalization that polyphagy and high dispersal rate are the main characteristics of effective natural enemies of naturally occurring mobile pests.

Numbers and Generation Time

It is preferable to release as many entomophages at a site as possible. Beirne (1975) declared that biological control projects in Canada were much more successful with releases of >800 individuals per liberation. Similarly, Hopper and Roush (1993), after analyzing previous introductions and using a mathematical model, suggested that a threshold of about 1000 insects per release would provide the best prospect for establishment of introduced parasitoids. Hopper *et al.* (1993) noted that the relationship of successful establishment of introduced biological control agents with the numbers released and the types of release habitats could be explained by a consideration of the Allee effect. Namely, the degree of dispersal of introduced species depends on the type of habitat into which they are released. With habitats that are less delimited, larger numbers of a species must generally be released to prevent population extinction by failure of the dispersing males and females to find each other (Allee effect). However, large numbers of some entomophages are difficult to obtain, which typically makes establishment much more troublesome. *Laricobius erichsonii* (Rosenhauer), a derodontid predator of the balsam woolly adelgid, *Adelges piceae* (Ratzeburg), not only has a single generation a year but also has a very slow annual dispersal rate. Considerably more effort was therefore necessary for its establishment than for that of other species (Clausen, 1977).

Some entomophages released in small numbers have become rapidly established. Such rapid establishment occurred with the encyrtid parasitoids *Metaphycus stanleyi* (Compere), *M. helvolus* (Compere), and *M. lounsburyi* (Howard), and with the pteromalid *Scutellista cyanea* Motschulsky, all released against the black scale in southern California. However, large releases of the encyrtid *Diversinervus elegans* Silvestri were necessary to achieve recoveries, but rapid spread then followed (Clausen, 1977).

Establishment of the braconid *Apanteles pedias* Nixon on the spotted tentiform leaf miner, *Phyllonorycter blancardella* (Fabricius), occurred in Ontario following release of only two females in May of 1978. For parasitization, Laing and Heraty (1981) had placed the females in a fine mesh sleeve cage over susceptible hosts on apple branches. By autumn of 1979, they found 25.7% parasitization at the original site, and recovered the parasitoid at a distance of 43 km. High rates of reproduction and dispersal were two factors that enabled establishment from such a small release.

Nealis and Quednau (1996) were only able to release between 200 and 600 gravid female adults of the European tachinid fly, *Ceranthia samarensis,* an endoparasitoid of the gypsy moth, in each of 5 years in southeastern Ontario. In two of the years they did free releases of the adults, with trees in the immediate vicinity of the releases seeded with susceptible age classes of gypsy moth larvae. In the other 3 years, they released into large cages enclosing entire oak trees onto which young gypsy moth larvae had been released. Both release techniques were successful, with parasitism by field-released females observed in every year.

Drea and Hendrickson (1986) attributed successful control of the alfalfa blotch leaf miner in the northeastern United States partly to a colonization procedure that emphasized timing, environmental conditions, and parasitoid numbers. Proper scheduling of diapause termination in groups of parasitized puparia enabled timing of periodic releases throughout the growing season. The release of parasitoids in an area of sequential alfalfa harvesting ensured the continual presence of susceptible hosts.

As Drea and Hendrickson (1986) noted, use of the term "adequate numbers" concerning parasitoid releases is vague. With the alfalfa blotch leaf miner, releases of very small numbers of the two parasitoid species that became the most important resulted in successful control. During 1977 and 1978, liberations in the original release fields totaled only 3307 *Chrysocharis punctifacies* Delucchi and 5207 *Dacnusa dryas.* Drea and Hendrickson (1986) used a "dribble release" technique whereby they liberated only a few dozen parasitoids weekly. They felt that repeated releases were more important than the liberation of large numbers at any one time.

Others have also achieved successful establishment with releases of small numbers of parasitoids. Fabre and Rabasse (1987) obtained establishment of the aphidiid *Pauesia cedrobii* by inserting 225 adults per sleeve cage placed on cedar branches with colonies of the cedar aphid *Cedrobium laportei.* In combating the arrowhead scale, *Unaspis yanonensis* Kuwana, Furuhashi and Nishino (1983) twice released 100 adults of the aphelinid *Aphytis yanonensis* DeBach & Rosen on each of three trees in citrus groves. Parasitism reached 80% and there was a marked decline of the scale within 6 months.

Seasonal characteristics can affect the necessary numbers of released parasitoids required for establishment. Campbell (1976) reported that successful establishment of the California red scale parasitoid *A. melinus* in the Riverland district of South Australia in summer and early autumn required colonizing 100 adult wasps into every third tree in every third row of a citrus orchard, but in late autumn establishment required the release of 1000 adults per tree. Widespread establishment of the same parasitoid in the Sunraysia district of New South Wales followed either about 50 small releases, each consisting of 100 to 300 parasitoids per tree, or placement of pumpkins covered with parasitized hosts in citrus trees.

Seasonal characteristics were also important for maintaining numbers of field-released *Cotesia melanoscela.* Wieber *et al.* (1995) reported that November/December field placements of *C. melanoscela* cocoons were most effective, with 74 to 92% emergence of adults during peak periods of susceptible stages of the host, the gypsy moth. Spring-placed cocoons were far more susceptible to natural enemies, and did not have the proper environmental diapause cues for synchrony of adult emergence with host stage susceptibility. On the other hand, Nealis and Bourchier (1995) offered the opposite view. In southern Ontario, they compared the vulnerability to predation and to hyperparasitism of an introduced European strain of *C. melanoscela* and an Asian strain of the parasitoid with no diapause stage. The nondiapause characteristic of the Asian strain decreased its exposure time and therefore reduced its vulnerability to predation and to hyperparasitism. They concluded that inundative release programs involving *Cotesia* spp. would be more effective by using a nondiapause strain and by making releases as early as possible in the season to minimize exposure to hyperparasitoids.

Fielding *et al.* (1991) studied the impact of release numbers of the predatory rhizophagid beetle *Rhizophagus grandis* Gyllenhal, the most important specific entomophage of the great European spruce bark beetle *Dendroctonus micans* (Kugelann). They concluded that colonization levels by the predator 4 weeks after release at >1000 predators per hectare were equal to those resulting from low-density introductions (50 to 100 predators per hectare) 2 years after release. Evans and Fielding (1994) later indicated that populations of *D. micans* were at low levels, with strong evidence that *R. grandis* was well established, spreading, and an important factor in regulating the bark beetle populations. In comparison with the low-density *R. grandis* releases, Kinawy (1991) used total releases of only 683 adults of the coccinellid *Chilocorus nigritus* to successfully control the coconut scale insect *Aspidiotus destructor* Sign over the Dhofar plain in southern Oman in only 24 months after the last release.

Augmentative inoculative releases of small numbers of parasitoids or predators can also be successful. Releases of the phytoseiid mite *Metaseiulus occidentalis* at the rate of only 64 per tree early in the season resulted in effective control of the McDaniel spider mite *Tetranychus mcdanieli* McGregor in California (Croft & McMurtry, 1972; McMurtry *et al.,* 1984). However, releases of nine phytoseiid species at 1200 mites per tree over 4 weeks to control the avocado brown mite, *Oligonychus punicae* (Hirst), were unsuccessful (McMurtry *et al.,* 1984). Pickett and Gilstrap (1986) controlled the Banks grass mite, *O. pratensis* (Banks), and the two-spotted spider mite on corn in Texas by making early season inoculative releases of the phyto-

seiid mites *Phytoseiulus persimilis* and *Amblyseius californicus* (McGregor). They noted, though, that a reduced cost of production and application of the predaceous mites would be necessary to make the procedure commercially feasible. Grafton-Cardwell *et al.* (1997) evaluated the effectiveness of six species of phytoseiid mites released at the rate of one per leaf on young potted citrus trees to control *T. urticae* and *Panonychus citri* in single- and mixed-species infestations. Infestations averaging 10 *T. urticae* per leaf were >85% controlled within a week, but infestations averaging 22 *P. citri* per leaf required 3 weeks for 74 to 80% control. The four more specialist phytoseiids—*Phytoseiulus longipes* Evans, *Galendromus occidentalis* (Nesbitt), *G. helveolus* (Chant), and *Neosieulus californicus* (McGregor)—were all effective predators. Of the two generalist phytoseiids, *Euseius stipulatus* Congdon and *Amblyseius limonicus* Garman & McGregor, *A. limonicus* Garman & McGregor was not effective in controlling *P. citri* at any time during the 6-week sampling period. *Galendromus occidentalis* was the most effective species in controlling both pest species in a mixed infestation. In a greenhouse environment, Rasmy and Ellaithy (1988) effectively controlled the two-spotted spider mite on cucumbers by releasing 10 predatory *P. persimilis* mites per plant at the first sign of spider mite damage.

However, large release numbers are essential for most entomophages used in augmentative biological control programs. While field testing the effectiveness of mass-produced *Trichogramma* strains, Hassan *et al.* (1988) released 400 to 9000 parasitoids per apple tree in four to six treatments to control the codling moth and the summer fruit tortrix, *Adoxophyes orana.* When Johnson (1985) used *T. pretiosum* against *Heliothis* spp. on cotton, he was unable to increase field parasitism by three low-level releases, two at 12,500 per hectare, followed by one at 37,500 per hectare made at 7-day intervals. Daane *et al.* (1996) had mixed results when they inundatively released the common green lacewing to suppress two vineyard pests, *Erythroneura variabilis* Beamer and *E. elegantula,* and they had to release at least 20,000 larvae per hectare to achieve any significant pest reductions.

Morales-Ramos *et al.* (1995) made inundative releases of the boll weevil parasitoid *Catolaccus grandis* over 2 years, and successfully reduced boll weevil survival (from egg to adult) to only 0.5 to 11.8%, in comparison to a survival of 72.8 to 78.2% in the control cotton fields. They used four methods of calculating the effectiveness of the parasitoid: percentage of parasitism, graphic Southwood and Jepson method, densities of emergence holes, and life table analysis. They later developed an age structured computer simulation model (ARCASIM) to understand the complexities of the interactions between cotton fruiting and population dynamics of the boll weevil and *C. grandis* (Morales-Ramos *et al.,* 1996b). Based on the simulations, the best strategy for augmentative releases of *C. grandis* against the boll weevil was to release the parasitoid at weekly intervals, instead of using a single release at an optimal time. The simulations revealed that success of the augmentative releases depended on the numbers of adult boll weevils immigrating into the release area, and also showed that cultivation practices can effectively disrupt the effect of the augmentative releases (Morales-Ramos *et al.*, 1996b).

Branquart *et al.* (1996) used a "Bombosh" model to determine the number of coccinellid larvae to field release for the control of potato aphids. The lower rate of increase of predators requires their introduction in large numbers early in the development of aphid populations. The parameters of this model were estimated by laboratory experiments on the observed rate of increase in aphid numbers and on the predatory efficiency of ladybird larvae.

Meadow *et al.* (1985) noted that augmentative releases of the predaceous cecidomyiid *Aphidoletes aphidimyza* had only been done on a large scale in greenhouses in Finland and in the former USSR. They experimented with using it to control the green peach aphid. In small plots of tomatoes and peppers in greenhouses and the field, they achieved effective control at varying rates, for example, two to three midge pupae per plant, with two sequential releases. A later monograph on this predator by Kulp *et al.* (1989) presented information on rearing methods and applications.

Stenseth and Aase (1983) investigated the numbers of *Encarsia formosa* required to control the greenhouse whitefly on cucumbers in Norwegian greenhouses. Three introductions of five parasitoids per plant at fortnightly intervals resulted in adequate control when there was an initial number of 10 to 30 adult whiteflies per 100 plants, whereas at lesser host densities only three parasitoids per plant were necessary. They noted that parasitoid introduction before the first of March in Norwegian greenhouses was not successful, probably because of the deleterious effect of low-light intensity on parasitoid reproduction.

Van de Veire and Vacante (1984) released the same parasitoid on greenhouse tomatoes by hanging 40 paper disks, each with ca. 110 parasitized whitefly pupae, at intervals over an area of 1500 m². This suggested rate was according to the recommendation of Woets (1978) for greenhouse whitefly control (also see Woets, 1973; and Woets & Van Lenteren, 1976). Rumei (1991) used an improved plant–pest–parasitoid (PPP) model to evaluate different methods (the Dutch method, the "pest in first" method and the "banker plant system" method) of releasing *E. formosa*. Subsequently, Van Roermund *et al.* (1997) developed a simulation model that could be used to evaluate a number of release strategies for *E. formosa* on the greenhouse whitefly on several crops and under various greenhouse climate conditions. The model was based on biological characteristics of both insect species, which re-

lated to development and to the searching and parasitization behavior of *E. formosa* in relationship to host-plant characteristics and greenhouse climate.

Hoddle *et al.* (1997) used the Beltsville strain of *E. formosa* to control the silverleaf whitefly, *Bemisia argentifolii,* on greenhouse-grown poinsettias. At a high release rate of three parasitoids per plant per week over 14 weeks, egg to adult survivorship for *B. argentifolii* was 1% and parasitism was 12%, versus the control, which had egg to adult survivorship of 71%. The net reproductive rate (R_o) for the control population of *B. argentifolii* was 17.1, indicating a rapidly increasing population. The net reproductive rate for the whitefly population subject to the high release rate was 0.32, indicating declining *B. argentifolii* population growth, which was attributed to high levels of in-house wasp reproduction. At harvest, the number of whiteflies per leaf at the high release rate was not significantly different from that found on commercially sold plants.

Del Bene (1990) successfully controlled three leaf miner species infesting chrysanthemum and gerbera plants in commercial greenhouses by releasing the eulophid ectoparasitoid *Diglyphus isaea* (Walker) at the rate of 5 to 30 adults per 10 m². However, because of a zero tolerance for mines on chrysanthemum, the parasitoid could only be used for the first weeks of plant development, after which time the early formed lower leaves with mines were removed at flower harvest.

Rutz and Axtell (1981) reported that weekly releases of a native strain of *Muscidifurax raptor* caused a significant reduction in the housefly population at a poultry farm. They released 5 parasitoids per bird per week (150,000 parasitoids per week) by placing parasitized housefly pupae at 10 to 15 spots on the manure in each poultry house. Petersen *et al.* (1992) developed a program to field-rear a similar parasitoid, *M. zaraptor,* to control houseflies in beef cattle pens. They exposed about 50,000 freeze-killed housefly pupae to a single field release of *M. zaraptor*. Subsequently, at 2-week intervals, an additional six lots of 50,000 freeze-killed pupae were set out for the parasitoids to attack. The result was a mean parasitoid emergence of 56.4% over the study period.

Kfir (1981) warned about published entomophage release figures. He stressed that citing the total number of *Trichogramma* released per unit of crop is meaningless without specifying the sex ratio and quality. An important principle in entomophage releases, particularly for inoculative releases of a classical biological control agent, is that quality should take precedence over quantity (Neuenschwander, 1989). An example of quality problems was given by Andress and Campbell (1994) who reported that commercially purchased fly pupal parasitoids, primarily *M. raptor* and *Spalangia nigroaenea*, were ineffective against stable flies despite weekly releases of high numbers. Parasitoid shipments contained neither the number requested nor the expected quality and species purity. Likewise, Lawson *et al.* (1997) found that variation in performance of *Trichogramma* spp. from different insectaries and among shipments from the same insectary was common, as did Vasquez *et al.* (1997), who noted inconsistent responses between shipments of commercially available trichogrammatid products, indicating potential problems with quality control.

Obrycki *et al.* (1997) also focused on quality control in their consideration that periodic augmentative releases of natural enemies may be the biological control method of choice for annually disturbed habitats. However, they also emphasized the importance of choosing appropriate natural enemies to match the characteristics of annual cropping systems; of enhancing the effectiveness of released individuals; and of studying biological, biosystematic, ecological, and economic aspects associated with the releases.

Biotic Interactions

Release methods have been developed to enhance the interaction between a pathogen, parasitoid, or predator, and the organism it attacks. In classical biological control, it is advantageous to release a beneficial organism when the susceptible stage of its host occurs in greatest numbers. For example, Nechols and Kikuchi (1985) recommended releasing the encyrtid *Anagyrus indicus* when the third nymphal stage of its host, *Nipaecoccus vastator* (Maskell), is the most numerous, to provide the longest exposure period for the most suitable host stages. For timing inundative releases (196,000 parasitoids per hectare) of *T. brassicae* to control the European corn borer on sweet corn, Yu and Byers (1994) found that using pheromone traps resulted in the greatest reduction in damage. Because European corn borer phenology varied among fields, general-area degree-day accumulation for postdiapause development could not adequately be used to determine the timing of releases for individual fields (Yu & Byers, 1994).

In augmentative biological control efforts, release of host material along with the beneficial organism may be desirable to increase the beneficial population. In a laboratory experiment, Nickle and Hagstrum (1981) successfully increased numbers of the braconid *Habrobracon hebetor* in a simulated peanut warehouse by releasing the parasitoid together with preparalyzed host individuals of the almond moth. In a greenhouse system, Parr (1972) placed spider mites on cucurbits to allow the predator *Phytoseiulus persimilis* to increase its population in time to control the increase of the endemic spider mite population. For the control of filth flies in dairies, Petersen (1986) made early-season releases of nonparasitized freeze-killed housefly pupae, as well as of housefly pupae parasitized with the pteromalid *M. zaraptor*. The freeze-killed pupae, which remained suitable as hosts for 4 weeks in the spring, appar-

ently provided substrate for sufficient parasitoid population increase to effectively control houseflies and stable flies in the dairies. As another example of this technique, releases of a field-crop insect, the imported cabbageworm, together with two parasitoids early in the growing season, successfully reduced pest damage (Parker & Pinnell, 1972).

A corollary technique in weed biological control projects is to fertilize the target weed to provide a better substrate for establishment of a biological control agent. Kuniata (1994) applied nitrogen fertilizer to the weed *Mimosa invisa*, a serious pest in rangeland, plantations, subsistence gardens, and nonproductive areas in Papua New Guinea, thereby significantly increasing numbers of a psyllid, *Heteropsylla spinulosa* Muddiman, Hodkinson & Hollis, an imported biological control agent.

Perring *et al.* (1988) discussed a different approach to augment natural biological control of the potato aphid, *Macrosiphum euphorbiae* (Thomas). In one season when tomatoes were planted earlier than normal, a population of green peach aphids developed that supported parasitoids. The presence of the early season parasitoid populations provided high levels of potato aphid parasitism when that pest later migrated to the tomatoes.

Cloutier and Johnson (1993) suggested that manipulation of the polyphagous phytoseiid *Amblyseius cucumeris* was important to promote development and survival of its protonymphs on the western flower thrips, *Frankliniella occidentalis*. The phytoseiid protonymphs are small relative to the thrips, and have difficulty attacking living thrips when they are the only available prey. Therefore, providing adult phytoseiids ensures the presence of dead thrips upon which the protonymphs can feed until they become independent predators.

Other biotic interactions that a released beneficial organism must face are not desirable. These include competition, disease, parasitism, and predation. For example, competition with native parasitoids may complicate field colonization of exotic parasitoids, as could have been the case with parasitoid releases against the beet leafhopper, *Circulifer tenellus* (Baker), in the Imperial Valley of California (Clausen, 1977). Competition between introduced parasitoids may also interfere with field colonization or effectiveness. Krause *et al.* (1990) assayed multiparasitism effects of two gypsy moth parasitoids, *Cotesia melanoscela* and *Glyptapanteles flavicoxis* Marsh, a recent import. Because of competitive interactions, they concluded that *G. flavicoxis* should be field released only after the oviposition period of the first generation of *C. melanoscela*.

Furthermore, McMurtry *et al.* (1984) suggested that competition with, or interference by, the native predator *Euseius hibisci* may have limited the abundance of nine species of phytoseiids augmentatively released at the rate of 1200 mites per tree to control the avocado brown mite, because the releases neither affected the average densities of the brown mite nor those of the total phytoseiids. However, in an orchard with few phytoseiids, Penman and Chapman (1980) were able to control the European red mite, *Panonychus ulmi* (Koch), with releases of the phytoseiid *Amblyseius fallacis* (Garman) at 300 per tree.

Heinz and Nelson (1996) investigated the interspecific relationships of two parasitoids, *Encarsia formosa* and *E. pergandiella,* and a coccinellid predator, *Delphastus pusillus* (LeConte), which were tested in all possible combinations for the inundative control of the silverleaf whitefly in greenhouse cage tests. While all combinations provided significant whitefly control, combinations of the predator with one or both of the parasitoids provided the greatest levels of suppression. The results suggested that the types of interspecific interactions, instead of the numbers, among natural enemies may be important to the outcome of inundative biological control programs (Heinz & Nelson, 1996).

Other interactions such as predation and cannibalism can also pose problems. Dreistadt *et al.* (1986) reported that the efficacy of inundatively releasing eggs of the common green lacewing to suppress the tulip tree aphid, *M. liriodendri* (Monell), was prevented by predation, cannibalism, highly erratic viability of the commercially produced green lacewing eggs, and lacewing larval entrapment on the sticky release tapes used.

Effects of host plants on natural enemies constitute another factor in the success of a project. Duso (1992) found that different grape varieties, characterized by glabrous or hairy leaf undersurfaces, caused dramatically different effects on the colonization of the phytoseiid mites *Amblyseius aberrans* (Oudemans) and *Typhlodromus pyri* Scheuten. Population densities of the two predators were markedly higher on the varieties with hairy leaf undersurfaces, and were largely independent of prey availability. The critical factors associated with the hairy leaves appeared to be a favorable microclimate, an improved protection from macropredators or from being dislodged, a large availability of shelter and oviposition sites, and an increased capacity for retaining wind pollen (an alternate food source).

In a greenhouse environment, Nihoul (1993) showed that high temperature and light intensity resulted in a higher density of glandular trichomes on tomato plants, which trapped a higher percentage of the predatory mite *Phytoseiulus persimilis,* allowing increased damage by spider mites. Also in greenhouses, Ekbom (1977) noted that control of the greenhouse whitefly with *E. formosa* was more effective on tomatoes than on cucumbers. Van Lenteren *et al.* (1995) showed that selection of less-hairy cucumber plants improved biological control of the greenhouse whitefly by *E. formosa,* with the finding and killing of more hosts per unit of time by the parasitoid. They concluded that plants can be selected that promote the effectiveness of natural enemies.

Heinz and Zalom (1996) studied the effect of tomato trichomes on a coccinellid, *Delphastus pusillus,* that attacks the silverleaf whitefly. They compared two tomato varieties, one of which had trichome-based resistance to the whitefly, and three times the number of trichomes as the other. They found that with this system, the plant resistance and the biological control agent did not act additively to suppress the whitefly. The predator had significantly slower walking speeds and reduced lifetime fecundities on the resistant tomato variety; however, this was compensated for by longer residency times, and so there was a neutral effect of the trichome-based resistance on silverleaf whitefly biological control.

Additional information on interactions between host-plant resistance and biological control agents can be found in a review by McAuslane (1994).

Another effect of host plants is that they can affect entomophages indirectly by controlling the phenology of the host. Schaefer *et al.* (1983) colonized *Pediobius foveolatus* against the Mexican bean beetle by placing parasitized larval mummies in nurse plots near soybean fields. They had planted the nurse plots with locally adapted bean varieties before normal soybean planting dates to provide reservoirs for early increases in bean beetle and parasitoid populations. Likewise, Correa-Ferreira and Moscardi (1996) planted a trap crop of early maturing soybeans in which they inoculatively released 15,000 per hectare of the egg parasitoid *Trissolcus basalis* when the first stinkbugs were detected. The inoculative releases resulted in the stinkbug population being held below the economic threshold level during the most critical stage of stinkbug attack on the main crop, which indicated that this technique could be an important component of the soybean integrated pest management (IPM) program in Brazil.

Autoparasitism

Autoparasitism can complicate field-release methods. Colonization of autoparasitic aphelinids to control armored scales requires special procedures, such as successive releases of mated and unmated females (Clausen, 1977).

ADDENDUM

This chapter has focused on factors involved at three trophic levels for producing arthropods for biological control, and on some aspects of field colonization. Space limitations precluded discussions of facilities or of other types of biological control agents, including microorganisms, invertebrates such as planarians or nematodes, and vertebrates such as fish. Moreover, examples of specific references for entomophage production could not be grouped and discussed by family. In general, Waage *et al.* (1985) noted that most parasitoids reared for biological control belong to the Ichneumonoidea, Chalcidoidea, and Tachinidae, whereas reared predators are primarily found in the Coleoptera, Diptera, Heteroptera, Neuroptera, and Phytoseiidae.

A knowledge of biologies is necessary for production of organisms. Good starting references for entomophage biologies are books by Clausen (1940, reprinted 1972), Clausen (1977), DeBach (1964), and Waage and Greathead (1986). A book by Van Driesche and Bellows (1996) also includes information on natural enemy biology, as well as on biological control methods. A review of entomophage mass production, with emphasis on specific systems, was given by Morrison and King (1977). Shaffer (1983) listed references that give mean laboratory developmental periods for 113 different insect and mite species at constant temperatures, including those for several predators and parasitoids. Thompson (1986) reviewed nutrition and *in vitro* culturing of insect parasitoids, and Bratti (1990b) presented a review of *in vitro* culturing. King and Leppla (1984), Singh and Moore (1985a, 1985b), and Anderson and Leppla (1992) edited useful books on general insect rearing, including information on entomophage production; and Leppla and Ashley (1978) edited a valuable book on insect-rearing facilities. King and Hartley (1992) described a program (for multiple-species insect rearing and the necessary facilities, and Singh and Clare (1992) introduced the concept of insect rearing management (IRM). Edwards *et al.* (1987) published a reference on arthropod species in culture, primarily in the United States, and Hunter (1997) compiled a list of North American suppliers of beneficial organisms.

Concerning field colonization, Ehler (1990) and Roush (1990) presented theoretical considerations of introduction strategies. Hopper (1996) reviewed techniques for making biological control introductions more effective, and Van Driesche (1993) discussed methods for the field colonization of new biological control agents.

For documentation of liberations, the Agricultural Research Service Biological Control Documentation Center of the USDA has established a computerized database called "releases of beneficial organisms" (ROBO), and has published annual reports that detail beneficial organism releases in the United States and its territories for the years 1981 to 1983 (Coulson *et al.,* 1988; Coulson, 1992, 1994). ROBO currently contains all importation and release records only for the years 1981 to 1985; these will be published on a CD-ROM, and skeletal importation records will be available on the world wide web for those years (current Internet address: http://www.ars-grin.gov/nigrp/robo.html) (J. R. Coulson, personal communication). Because of the impossibility of entering all historical and current data into the original and now archaic ROBO computer system, personnel in the Biological Control Documentation Center are in the process of developing a revised ROBO with a modern computer system in the Germplasm Resources Information Network (GRIN). Associated with this system will be the ability of all biological control workers with Internet access to directly enter importation/release data into ROBO, so that eventually the data can be current and usable for all researchers (J. R. Coulson, personal communication).

References

Abou-Awad, B. A., Reda, A. S., & Elsawi, S. A. (1992). Effects of artificial and natural diets on the development and reproduction of two phytoseiid mites *Amblyseius gossipi* and *Amblyseius swirskii* (Acari: Phytoseiidae). Insect Science Applications, 13, 441–445.

Abou-Setta, M. M., Fouly, A. H., & Childers, C. C. (1997). Biology of *Proprioseiopsis rotendus.* (Acari: Phytoseiidae) reared on *Tetranychus urticae* (Acari: Tetranychidae) or pollen. Florida Entomologist, 80, 27–34.

Adiyaman, N., & Aktumsek, A., (1996). The effects of low temperature on fecundity of pupae and adults of female *Pimpla turionellae* L. (Hymenoptera: Ichneumonidae). Turkish Journal of Zoology, 20, 1–5.

Afifi, A. M., Potts, M. F., Patterson, C. G., & Rodriguez J. G., (1988). Pollen diet of some predator mites. Transactions of the Kentucky Academy of Science, 49, 96–100.

Agricola, U., Agounke, D., Fischer, H. U., & Moore, D. (1989). The control of *Rastrococcus invadens* Williams (Hemiptera: Pseudococcidae) in Togo by the introduction of *Gyranusoidea tebygi* Noyes (Hymenoptera: Encyrtidae). Bulletin of Entomological Research 79, 671–678.

Allen, H. W., & Brunson, M. H., (1947). Control of *Nosema* disease of potato tuberworm, a host used in the mass production of *Macrocentrus ancylivorus.* Science, 105, 394.

Amis, A. A., & Snow, J. W. (1985). *Conotrachelus nenuphar.* In P. Singh & R. F. Moore (Eds.), Handbook of insect rearing (Vol. 1, pp. 227–235). New York: Elsevier.

Anderson, T. E., & Leppla, N. C. (Eds.). (1992). Advances in insect rearing for research and pest management. Boulder, CO: Westview Press.

Andress, E. R., & Campbell, J. B. (1994). Inundative releases of Pteromalid parasitoids (Hymenoptera: Pteromalidae) for the control of stable flies, *Stomoxys calcitrans* (L.) (Diptera: Muscidae) at confined cattle installations in west Central Nebraska. Journal of Economic Entomology, 87, 714–722.

Ankersmit, G. W. (1985). *Adoxophyes orana.* In E. G. King & N. C. Leppla (Eds.), Handbook of insect rearing (Vol. 2, pp. 165–175). New York: Elsevier.

Argyriou, L. C. (1981). Establishment of the imported parasite *Prospaltella perniciosi* (Hym.: Aphelinidae) on *Quadraspidiotus perniciosus* (Hom.: Diaspididae) in Greece. Entomophaga, 26, 125–130.

Ashby, M. D., Singh, P., & Clare. G. K., (1985). *Cydia pomonella.* In P. Singh & R. F. Moore (Eds.), Handbook of insect rearing (Vol. 2, pp. 237–248). New York: Elsevier.

Baggen, L. R., & Gurr, G. M. (1995). Lethal effects of foliar pubescence of solanaceous plants on the biological control agent *Copidosoma koehleri* Blanchard (Hymenoptera: Encyrtidae). Plant Protection Quarterly, 10, 116–118.

Bailey, D. L., & Seawright, J. A. (1984). Improved techniques for mass rearing *Anopheles albimanus*. In E. G. King & N. C. Leppla [Eds.], Advances and challenges in insect rearing (pp. 200–205) Washington, DC: U.S. Government Printing Office.

Balas, M. T., Lee, M. H., & Werren, J. H. (1996). Distribution and fitness effects of the son-killer bacterium in *Nasonia.* Evolutionary Ecology, 10, 593–607.

Balasubramanian, S., & Pawar, A. D. (1988). Improved technique for mass rearing of *Trichogramma* spp. (Hymenoptera: Trichogrammatidae) in the laboratory. Plant Protection Bulletin, India, 40, 12–13.

Ballal, C. R., & Kumar, P. (1993). Host-parasitoid interaction between *Chilo partellus* (Swinhoe) and *Allorhogas pyralophagus* Marsh. Biological Control, 7, 72–74.

Ballal, C. R., Singh, S. P., Jalali, S. K., & Kumar, P. (1989). Cold tolerance of cocoons of *Allorhogas pyralophagus* (Hym.: Braconidae). Entomophaga, 34, 463–468.

Barbosa, P., & Segarra-Carmona, A. (1993). Criteria for the selection of pest arthropod species as candidates for biological control. In R. G. Van Driesche & T. S. Bellows, Jr. (Eds.), Proceedings: Steps in classical arthropod biological control (pp. 5–23). Symposium, Baltimore, MD, September 30, 1990, Lanham, MD: Thomas Say Publications in Entomology: Entomological Society of America.

Barbosa, P., Saunders, J. A., & Waldvogel, M. (1982, March). Plant-mediated variation in herbivore suitability and parasitoid fitness (pp. 63–71). In Proceedings, 5th International Symposium on Insect-Plant Relationships, Wageningen, The Netherlands.

Barbosa, P., Kemper, J., Gross, P., & Martinat, P. (1990). Influence of dietary nicotine and colony source of *Manduca sexta* (Lepidoptera: Sphingidae) on its suitability as a host of *Cotesia congregata* (Hymenoptera: Braconidae). Entomophaga, 35, 223–232.

Barrows, E. M., & Hooker, M. E. (1981). Parasitization of the Mexican bean beetle by *Pediobius foveolatus* in urban vegetable gardens. Environmental Entomology, 10, 782–786.

Bartlett, A. C. (1984). Genetic changes during insect domestication. In E. G. King & N. C. Leppla (Eds.), Advances and challenges in insect rearing (pp. 2–8). Washington, DC: U.S. Government Printing Office.

Bartlett, A. C. (1985). Guidelines for genetic diversity in laboratory colony establishment and maintenance. In P. Singh & R. F. Moore (Eds.), Handbook of insect rearing (Vol. 1, pp. 7–17). New York: Elsevier.

Bartlett, A. C., & Wolf, W. W. (1985). *Pectinophora gossypiella*. In P. Singh & R. F. Moore (Eds.), Handbook of insect rearing (Vol. 2, pp. 415–430). New York: Elsevier.

Battisti, A. P. I., Milani, N., & Zanata, M. (1990). Preliminary accounts on the rearing of *Ooencyrtus pityocampae* (Mercet) (Hymenoptera, Encyrtidae). Journal of Applied Entomology 110, 121–127.

Baumhover, A. H. (1985). *Manduca sexta*. In P. Singh & R. F. Moore (Eds.), Handbook of insect rearing (Vol. 2, pp. 387–400). New York: Elsevier.

Bay, E. C., & Legner, E. F. (1963). Quality control in the production of *Hippelates collusor* (Tsnd.) for use in the search and rearing of their natural enemies. Proceedings of the New Jersey Mosquito Exterminating Commission 1963, 403–410.

Beach, R. M., & Todd, J. W. (1986). Foliage consumption and larval development of parasitized and unparasitized soybean looper, *Pseudoplusia includens* (Lep.: Noctuidae), reared on a resistant soybean genotype and effects on an associated parasitoid, *Copidosoma truncatellum* (Hym.: Encyrtidae). Entomophaga, 31, 237–242.

Beirne, B. P. (1975). Biological control attempts by introductions against pest insects in the field in Canada. Canadian Entomologist, 107, 225–236.

Bell, R. A. (1996). Manipulation of diapause in the gypsy moth, *Lymantria dispar* L., by application of KK-42 and precocious chilling of eggs. Journal of Insect Physiology, 42, 557–563.

Belloncik, S., Lavallée, C., & Quevillon, I. (1985) *Euxoa scandens* and *Euxoa messoria*. In P. Singh & R. F. Moore (Eds.), Handbook of insect rearing (Vol 2, pp. 293–299). New York: Elsevier.

Belousov, Y. V., & Popov, N. A. (1989). Rearing of *Aphidoletes aphidimyza* (Diptera, Cecidomyiidae) on greenbugs. Acta Entomologica Fennica, 53, 3–5.

Bennett, F. D. (1966). Current status of citrus blackfly and introduced parasites in Barbados. Agricultural Society of Trinidad, Tobago Journal, 66, 35–36.

Bennett, F. D., & van Wherlin, L. W. (1966). Occurrence of the citrus black fly in Barbados. Agricultural Society of Trinidad, Tobago Journal, 66, 31–34.

Bentz, J.-A., & Barbosa, P. (1992). Effects of dietary nicotine and partial starvation of tobacco hornworm, *Manduca sexta*, on the survival and development of the parasitoid *Cotesia congregata*. Entomologia Experimentalis et Applicata, 65, 241–245.

Berlinger, M. J. (1992). Importance of host plant or diet on the rearing of insects and mites. In T. E. Anderson & N. C. Leppla (Eds.), Advances in insect rearing for research and pest management (pp. 237–251). Boulder, CO: Westview Press.

Beukeboom, L. W., & Werren, J. H. (1992). Population genetics of a parasitic chromosome: Experimental analysis of PSR in subdivided populations. Evolution, 46, 1257–1268.

Beukeboom, L. W. (1993a). Deletion analysis of the selfish B chromosome, paternal sex ratio (PSR), in the parasitic wasp *Nasonia vitripennis*. Genetics, 13, 637–648.

Beukeboom, L. W. (1993b). Transmission and expression of the parasitic paternal sex ratio psr chromosome. Heredity, 70, 437–443.

Bhatt, N., & Singh, R. (1989). Bionomics of an aphidiid parasitoid *Trioxys indicus*: 30. Effect of host plants on reproductive and developmental factors. Biology of Agricultural Horticulture, 6, 149–158.

Bhatt, N., & Singh, R. (1991). Bionomics of an aphidiid parasitoid, *Trioxys indicus* Subba Rao & Sharma. 35. Influence of food plants on the life table statistics of the parasitoid through its host *Aphis gossypii* Glover. Insect Science Applications, 12, 385–390.

Bichao, M. H., & Araujo, J. (1989). Mass-rearing of *Chrysoperla carnea* (Stephens) (Neuroptera, Chrysopidae) larvae: Optimization of rearing unit yield. Boletim da Sociedade Portuguesa de Entomologia, 0(113), 117–124.

Bigler, F. (1989). Quality assessment and control in entomophagous insects used for biological control. Journal of Applied Entomology, 108, 390–400.

Blenk, R. G., Gouger, R. J., Gallo, T. S., Jordan, L. K., & Howell, E. (1985). *Agrotis ipsilon*. In P. Singh & R. F. Moore (Eds.), Handbook of insect rearing (Vol 2, pp. 177–187). New York: Elsevier.

Bloem, K. A., & Duffey, S. S. (1990). Effect of protein type and quantity on growth and development of larval *Heliothis zea* and *Spodoptera exigua* and the endoparasitoid *Hyposoter exiguae*. Entomologia Experimentalis et Applicata, 24, 141–148.

Boller, E. F. (1985). *Rhagoletis cerasi* and *Ceratitis capitata*. In P. Singh & R. F. Moore (Eds.), Handbook of insect rearing (Vol 2, pp. 135–144). New York: Elsevier.

Bonde, J. (1989). Biological studies including population growth parameters of the predatory mite *Amblyseius barkeri* [Acari: Phytoseiidae] at 25°C in the laboratory. Entomophaga, 34, 275–287.

Bourchier, R. S. (1991). Growth and development of *Compsilura concinnata* (Meigan) (Diptera: Tachinidae) parasitizing gypsy moth larvae feeding on tannin diets. Canadian Entomologist, 123, 1047–1056.

Bourchier, R. S., & Smith, S. M. (1996). Influence of environmental conditions and parasitoid quality on field performance of *Trichogramma minutum*. Entomologia Experimentalis et Applicata 80, 461–468.

Bouse, L. F., Carlton, J. B., & Morrison, R. K. (1981). Aerial application of insect egg parasites. Transactions of the ASAE, 14, 1093–1098.

Bouse, L. F., Carlton, J. B., Jones, S. L., Morrison, R. K., & Ables, J. R. (1980). Broadcast aerial release of an egg parasite for lepidopterous insect control. Transactions of the ASAE 23, 1359–1363, 1368.

Branquart, E., Hemptinne, J. L., Bruyere, M., Adam, B., & Gaspar, C. (1996). Biological control of potato aphids by ladybird beetles. Mededelingen Faculteit Landbouwkundige en Toegepaste Biologische Wetenschappen Universiteit Gent, 61, 905–909.

Bratti, A. (1990). *In vitro* rearing of *Pseudogonia rufifrons* Wied. (Diptera: Tachinidae) on pupal hemolymph of *Galleria mellonella* L. and on meridic diets. Bollettino dell'Istituto di Entomologia "Guido Grandi" della Università degli Studi di Bologna, 44, 11–22 (in Italian, with Italian and English summaries).

Bratti, A. (1990b). *In vitro* rearing techniques for entomophagous parasitoids. Bollettino dell'Istituto di Entomologia "Guido Grandi" della Università degli Studi di Bologna, 44, 169–220 (in Italian with Italian and English summaries).

Bratti, A. (1994). *In vitro* rearing of *Lydella thompsoni* Herting and *Archytas marmoratus* (Town.) (Dipt. Tachinidae) larval stages: Preliminary results. Bollettino dell'Istituto di Entomologia "Guido Grandi" della Università degli Studi di Bologna 48, 93–100.

Bratti, A., & Benini, S. (1992). *In vivo* rearing of *Pseudogonia rufifrons* Wied: Trials on meridic and host component diets. Bollettino dell'Istituto di Entomologia "Guido Grandi" della Università degli Studi di Bologna, 46, 71–85.

Bratti, A., & Campadelli, G. (1994). Comparison of insect-material in a meridic diet for *Exorista larvarum* L. (Dipt. Tachinidae) *in vitro* rearing. Bollettino dell'Istituto di Entomologia "Guido Grandi" della Università degli Studi di Bologna, 48, 59–65.

Bratti, A., & Constantini, W. (1992). Effects of new host artificial diets on the host-parasitoid couple *Galleria mellonella* L. (Lep. Galleridae) *Archytas marmoratus* (Town.) (Dipt. Tachinidae). Bollettino dell'Istituto di Entomologia "Guido Grandi" della Università degli Studi di Bologna, 46, 49–62.

Bratti, A., & Coulibaly, A. K. (1995). In vitro rearing of *Exorista larvarum* on tissue culture-based diets. Entomologia Experimentalis et Applicata, 74, 47–53.

Bratti, A., & D'Amelio, L. (1994). *In vitro* rearing of *Eucelatoria bryani* Sab. (Diptera Tachinidae) on tissue culture-based diets. Bollettino dell'Istituto di Entomologia "Guido Grandi" della Università degli Studi di Bologna, 48, 109–114.

Bratti, A., & Nettles, W. C., Jr. (1995). Comparative growth and development *in vitro* of *Eucelatoria bryani* Sab. and *Palexorista laxa* (Curran) (Diptera Tachinidae) fed a meridic diet and a diet of *Helicoverpa zea* (Boddie) (Lepidoptera Noctuidae) pupae. Bollettino dell'Istituto di Entomologia "Guido Grandi" della Università degli Studi di Bologna, 49, 119–129.

Bratti, A., Campadelli, G., & Mariani, M. (1995). *In vitro* rearing of *Exorista larvarum* (L.) on diet without insect components. Bollettino dell'Istituto di Entomologia "Guido Grandi" della Università degli Studi di Bologna, 49, 225–236.

Breeuwer, J. A. J. (1997). *Wolbachia* and cytoplasmic incompatibility in the spider mites *Tetranychus urticae* and *T. turkestani.* Heredity, 79, 41–47.

Breeuwer, J. A. J., & Jacobs, G. (1996). *Wolbachia:* Intracellular manipulators of mite reproduction. Experimental and Applied Acarology (Northwood), 20, 421–434.

Breeuwer, J. A. J., & Werren, J. H. (1993). Cytoplasmic incompatibility and bacterial density in *Nasonia vitripennis.* Genetics, 135, 565–574.

Brinton, F. E., Proverbs, M. D., & Carty, B. E. (1969). Artificial diets for mass production of the codling moth, *Carpocapsa pomonella* (L.) (Lepidoptera: Olethreutidae). Canadian Entomologist, 101, 577–584.

Brooks, W. M. (1974). Protozoan infections. In G. E. Cantwell (Eds.), Insect diseases (Vol. 1, pp. 237–300). New York: Marcel Dekker.

Brower, J. H. (1983). Eggs of stored-product Lepidoptera as hosts for *Trichogramma evanescens* (Hym.: Trichogrammatidae). Entomophaga, 28, 335–362.

Brown, H. E. (1984). Mass production of screwworm flies, *Cochliomyia hominivorax.* In E. G. King & N. C. Leppla (Eds.), Advances and challenges in insect rearing (pp. 193–199). Washington, DC: U.S. Government Printing Office.

Bruce-Oliver, S. J., & Hoy, M. A. (1990). Effect of prey stage on lifetable attributes of a genetically manipulated strain of *Metaseiulus occidentalis* (Acari: Phytoseiidae). Experimental and Applied Acarology (Northwood), 9, 201–218.

Bruce-Oliver, S. J., Hoy, M. A., & Yaninek, J. S. (1996). Effect of some food sources associated with cassava in Africa on the development, fecundity and longevity of *Euseius fustis* (Pritchard and Baker) (Acari:

Phytoseiidae). Experimental and Applied Acarology (Northwood), 20, 73–85.

Bucher, G. E., & Harris, P. (1961). Food-plant spectrum and elimination of disease of cinnabar moth larvae, *Hypocrita jacobaeae* (L.) (Lepidoptera: Arctiidae). Canadian Entomologist, 93, 931–936.

Bush, G. L., Neck, R. W., & Kitto, G. B. (1976). Screwworm eradication: Inadvertent selection of non-competitive ecotypes during mass rearing. Science, 193, 491–493.

Campadelli, G. (1988). *Galleria mellonella* L. as a substitute host for insect parasitoids. Bollettino dell'Istituto di Entomologia della Università degli Studi di Bologna, 42, 47–65 (in Italian with Italian and English summaries).

Campbell, M. M. (1975). Establishing *Aphytis melinus* in citrus orchards by a new simple method. Journal of the Australian Institute of Agricultural Science, 41, 62–53.

Campbell, M. M. (1976). Colonization of *Aphytis melinus* DeBach (Hymenoptera: Aphelinidae) in *Aonidiella aurantii* (Mask.) (Hemiptera: Coccidae) on citrus in South Australia. Bulletin of Entomological Research, 65, 659–668.

Carey, J. R., & Krainacker, D. A. (1988). Demographic analysis of mite populations: Extensions of stable theory. Experimental and Applied Acarology (Northwood), 4, 191–210.

Carey, J. R., Wong, T. T. Y., & Ramadan, M. M. (1988). Demographic framework for parasitoid mass rearing: Case study of *Biosteres tryoni,* a larval parasitoid of tephritid fruit flies. Theoretical Population Biology, 34, 279–296.

Castagnoli, M. (1989). Biology and prospects for mass rearing of *Amblyseius cucumeris* (Oud.) (Acarina: Phytoseiidae) using *Dermatophagoides farinae* Hughes (Acarina: Pyroglyphidae) as prey. Redia, 72, 389–402 (in Italian with Italian and English summaries).

Castagnoli, M., & Sauro, S. (1990). Biological observations and life table parameters of *Amblyseius cucumeris* (Oud.) (Acarina: Phytoseiidae) reared on different diets. Redia, 73, 569–584.

Castane, C., & Zalom, F. G. (1994). Artificial oviposition substrate for rearing *Orius insidiosus* (Hemiptera: Anthocoridae). Biological Control, 4, 88–91.

Cerutti, F., & Bigler, F. (1995). Quality assessment of *Trichogramma brassicae* in the laboratory. Entomologia Experimentalis et Applicata, 75, 19–26.

Chalkov, A. A. (1989). An appliance for separating *Phytoseiulus* from the plant mass and counting it in propagating greenhouses. Zashchita Rastenii, (Moscow), 4·89, 23 (in Russian).

Chambers, D. L., & Ashley, T. R. (1984). Putting the control in quality control in insect rearing. In E. G. King & N. C. Leppla (Eds.), Advances and challenges in insect rearing (pp. 256–260). Washington, DC: U.S. Government Printing Office.

Chambers, R. J. (1986). Preliminary experiments on the potential of hoverflies [Dipt.: Syrphidae] for the control of aphids under glass. Entomophaga, 31, 197–204.

Chandra, J., & Avasthy, P. N. (1988). Biological behaviour of *Sturmiopsis inferens* Towns., an indigenous parasite of moth borers of sugarcane. Indian Journal of Agricultural Research, 22, 85–91.

Chang, H.-Y., Fang, S., Hong, S.-J., & Chu, Y.-I. (1990). Some aspects on the reproduction of *Phytoseiulus persimilis.* Chinese Journal of Entomology, 10, 401–408.

Chang, Y. F., Tauber, M. J., & Tauber, C. A. (1995). Storage of the mass-produced predator *Chrysoperla carnea* (Neuroptera: Chrysopidae): Influence of photoperiod, temperature, and diet. Environmental Entomology, 24, 1365–1374.

Chen, S. M., Cheng, W. Y., & Wang, Z. T. (1994). Nonpartition mass rearing of a green lacewing, *Mallada basalis* (Walker) (Neuroptera: Chrysopidae). Report of the Taiwan Sugar Research Institute, 144, 25–32.

Chen, Z. H., Lu, M. J., & Lin, Q. Z., (1989a). Use of Asian cornborer,

Ostrinia furnacalis (Lep.:Pyralidae) as the factitious host for mass rearing *Scleroderma guani* (Hym.: Bethylidae). Chinese Journal of Biological Control, 5, 145–148 (in Chinese with Chinese and English summaries).

Chen, Z. H., Qin, J. D., & Shen, C. L. (1989b). Effects of altering composition of artificial diets on the larval growth and development of *Coccinella septempunctata*. Acta Entomologica Sinica, 32, 385–392 (in Chinese with Chinese and English summaries).

Chianese, R. (1985). *Cotesia melanoscelus*. In P. Singh & R. F. Moore (Eds.), Handbook of insect rearing (Vol. 1, pp. 395–400). New York: Elsevier.

Chihrane, J., & Lauge, G. (1994). Effects of high temperature shocks on male germinal cells of *Trichogramma brassicae* (Hymenoptera: Trichogrammatidae). Entomophaga, 39, 11–20.

Chihrane, J., & Lauge, G. (1996). Loss of parasitization efficiency of *Trichogramma brassicae* (Hym.: Trichogrammatidae) under High-temperature conditions. Biological Control, 7, 95–99.

Chihrane, J., & Lauge, G. (1997). Thermosensitivity of germ lines in *Trichogramma brassicae* Bezdenko (Hymenoptera): Impact on the parasitoid efficacy. Canadian Journal of Zoology, 75, 484–489.

Chiri, A. A., & Legner, E. F. (1982). Host-searching kairomones alter behavior of *Chelonus* sp. nr. *curvimaculatus,* a hymenopterous parasite of the pink bollworm. Environmental Entomology, 11, 452–455.

Chiri, A. A., & Legner, E. F. (1986). Response of three *Chelonus* (Hymenoptera: Braconidae) species to kairomones in scales of six Lepidoptera. Canadian Entomologist, 118, 329–333.

Cilliers, B. (1989). A model of balanced polymorphism in a host-parasite relationship. Phytophylactica, 21, 279–280.

Clair, D. J., Dahlsten, D. L., & Hart, E. R. (1987). Rearing *Tetrastichus gallerucae* (Hymenoptera: Eulophidae) for biological control of the elm leaf beetle, *Xanthogaleruca luteola*. Entomophaga, 32, 457–461.

Clancy, D. J., & Hoffmann, A. A. (1997). Behavior of *Wolbachia* endosymbionts from *Drosophila simulans* in *Drosophila serrata,* a novel host. American Naturalist, 149, 975–988.

Clarke, A. R., & Walter, G. H. (1995). "Strains" and the classical biological control of insect pests. Canadian Journal of Zoology, 73, 1777–1790.

Clarke, G. M., & Mckenzie, L. J. (1992). Fluctuating asymmetry as a quality control indicator for insect mass rearing processes. Journal of Economic Entomology, 85, 2045–2050.

Clausen, C. P. (1940) Entomophagous insects. New York: Hafner. (Reprinted 1972.)

Clausen, C. P. (Ed.). (1977, issued 1978). Introduced parasites and predators of arthropod pests and weeds: A world review. Agricultural Handbook No. 480. Washington, DC: U. S. Department of Agriculture.

Cloutier, C., & Johnson, S. G. (1993). Interaction between life stages in a phytoseiid predator: Western flower thrips prey killed by adults as food for protonymphs of *Amblyseius cucumeris*. Experimental and Applied Acarology (Northwood), 17, 441–449.

Cohen, A. C. (1981). An artificial diet for *Geocoris punctipes* (Say). Southwestern Entomology, 6, 109–113.

Cohen, A. C. (1985). Simple method for rearing the insect predator *Geocoris punctipes* (Heteroptera: Lygaeidae) on a meat diet. Journal of Economic Entomology, 78, 1173–1175.

Cohen, A. C. (1992). Using a systematic approach to develop artificial diets for predators. In T. E. Anderson & N. C. Leppla (Eds.), Advances in insect rearing for research and pest management pp. 77–91. Boulder, CO: Westview Press.

Collins, A. M. (1984). Artificial selection of desired characteristics in insects. *In* E. G. King & N. C. Leppla (Eds.), Advances and challenges in insect rearing pp. 9–19. Washington DC: U. S. Government Printing Office.

Consoli, F. L., & Parra, J. R. P. (1996). Comparison of hemolymph and holotissues of different species of insects as diet components for *in*

vitro rearing of *Trichogramma galloi* Zucchi and *T. pretiosum* Riley. Biological Control, 6, 401–406.

Consoli, F. L. & Parra, J. R. P. (1997). Development of an oligidic diet for *in vitro* rearing of *Trichogramma galloi* Zucchi and *Trichogramma pretiosum* Riley. Biological Control, 8, 172–176.

Cook, J. M. (1993a). Inbred lines as reservoirs of sex alleles in parasitoid rearing programs. Environmental Entomology, 22, 1213–1216.

Cook, J. M. (1993b). Sex determination in the hymenoptera: A reviews of models and evidence. Heredity, 71, 421–435.

Coombs, M. T. (1997). Influence of adult food deprivation and body size on fecundity and longevity of *Trichopida giacomellii:* A South American parasitoid of *Nezara viridula*. Biological Control, 8, 119–123.

Coop, L. B., & Croft, B. A. (1995). *Neoseiulus fallacis:* dispersal and biological control of *Tetranychus urticae* following minimal inoculations into a strawberry field. Experimental and Applied Acarology (Northwood), 19, 31–43.

Corbett, A., & Rosenheim J. A. (1996). Quantifying movement of a minute parasitoid, *Anagrus epos* (Hymenoptera: Mymaridae), using fluorescent dust marking and recapture. Biological Control, 6, 35–44.

Correa-Ferreira, B. S., & Moscardi, F. (1996). Biological control of soybean stink bugs by inoculative releases of *Trissolcus basalis*. Entomologia Experimentalis et Applicata, 79, 1–7.

Cossentine, J. E., & Lewis, L. C. (1986). Studies on *Bonnetia comta* (Diptera: Tachinidae) parasitizing *Agrotis ipsilon* (Lepidoptera: Noctuidae) larvae. Entomophaga, 31, 323–330.

Couillien, D., & Gregoire, J. C. (1994). Take-off capacity as a criterion for quality control in mass-produced predators, *Rhizophagus grandis* (Col.: Rhizophagidae) for the biocontrol of bark beetles, *Dendroctonus Micans* (Col.: Scolytidae). Entomophaga, 39, 385–395.

Coulibaly, A. K., Bratti, A., & Fanti, P. (1993). Rearing of *Archytas marmoratus* (Town.) (Diptera: Tachinidae) on *Galleria mellonella* L. (Lepidoptera: Galleriidae): parasitoid larviposition rhythm and optimum planidia number for parasitization. Bollettino dell'Instituto di Entomologia "Guido Grandi" della Università degli Studi di Bologna, 47, 13–25.

Coulson, J. R. (1992). Releases of beneficial organisms in the United States and territories—1982 (Miscellaneous Publication No. 1505). Washington, DC: U.S. Department of Agriculture.

Coulson, J. R. (1994). Releases of beneficial organisms in the United States and territories—1983 (Agricultural Research Service, ARS-131). Washington, DC: U.S. Department of Agriculture.

Coulson, J. R., Carrell, A., & Vincent, D. L. (1988). Releases of beneficial organisms in the United States and territories—1981 (Miscellaneous Publication No. 1464). Washington, DC: U.S. Department of Agriculture.

Croft, B. A., & McMurtry, J. A. (1972). Minimum releases of *Typhlodromus occidentalis* to control *Tetranychus mcdanieli* on apple. Journal of Economic Entomology, 65, 188–191.

Daane, K. M., Yokota, G. Y., Zheng, Y., & Hagen, K. S. (1996). Inundative release of common green lacewings, (Neuroptera: Chrysopidae) to suppress *Erythroneura variabilis* and *E. elegantula* (Homoptera: Cicadellidae) in vineyards. Environmental Entomology, 25, 1224–1234.

Dai, K. J., Zhang, L. W., Ma, Z. J., Zhong, L. S., Zhang, Q. X., & Cao, A. H. (1988). Research and utilization of artificial host egg for propagation of parasitoid *Trichogramma*. Colloques de l'INRA, No 43, 311–318.

David, H., Easwaramoorthy, S., Kurup, N. K., Shanmugasundaram, M., & Santhalakshmi, G. (1989). A simplified mass culturing technique for *Sturmiopsis inferens* Tns. Journal of Biological Control, 3, 1–3.

Davis, F. M., & Guthrie, W. D., (1992). Rearing Lepidoptera for plant resistance research. In T. E. Anderson & N. C. Leppla (Eds.), Advances in insect rearing for research and pest management (pp. 211–228). Boulder, CO: Westview Press.

Davis, F. M., & Jenkins, J. N. (1995). Management of scales and other insect debris: Occupational health hazard in a lepidopterous rearing facility. Journal of Economic Entomology, 88, 185–191.

De Clercq, P., & Degheele, D. (1992). A meat-based diet for rearing the predatory stinkbugs Podisus maculiventris and Podisus sagitta [Het.: Pentatomidae]. Entomophaga, 37, 149–157.

de Moraes, G. J., & McMurtry, J. A. (1987). Physiological effect of the host plant on the suitability of Tetranychus urticae as prey for Phytoseiulus persimilis (Acari: Tetranychidae, Phytoseiidae). Entomophaga, 32, 35–38.

Dean, H. A., Schuster, M. F., Boling, J. C., & Riherd, P. T. (1979). Complete biological control of Antonina graminis in Texas with Neodusmetia sangwani (a classic example). Bulletin of the Entomological Society of America, 25, 262–267.

DeBach, P. (Ed.). (1964). Biological control of insect pests and weeds. London: Chapman & Hall.

DeBach, P., & Hagen, K. S. (1964). Manipulation of entomophagous species. In P. DeBach (Ed.), Biological control of insect pests and weeds (pp. 429–458). London: Chapman & Hall.

DeBolt, J. W. (1989). Host preference and acceptance by Leiophron uniformis (Hymenoptera: Braconidae): Effects of rearing on alternate Lygus spp. (Heteroptera: Miridae). Annals of the Entomological Society of America, 82, 399–402.

Del Bene, G. (1990). Diglyphus isaea (Wlk.) in commercial greenhouses for the biological control of the leafminers Liriomyza trifolii (Burgess), Chromatomyia horticola (Goureau) and Chromatomyia syngenesiae Hardy on Chrysanthemum and Gerbera. Redia, 73, 63–78.

Delpuech, J.-M., Carton, Y., & Roush, R. T. (1993). Conserving genetic variability of a wild insect population under laboratory conditions. Entomologia Experimentalis et Applicata, 67, 233–239.

Dicke, M., DeJong, M., Alers, M. P. T., Stelder, F. C. T., Wunderink, R., & Post, J. (1989). Quality control of mass-reared arthropods: Nutritional effects on performance of predatory mites. Journal of Applied Entomology, 108, 462–475.

Dindo, M. L. (1990). Some observations on the biology of Brachymeria intermedia (Nees) (Hymenoptera Chalcididae) in vivo and in vitro. Bollettino dell'Istituto di Entomologia "Guido Grandi" della Università degli Studi di Bologna, 44, 221–232.

Dindo, M. L., & Campadelli, G. (1993). In vitro rearing of Pseudogonia rufifrons Wied. (Dipt. Tachinidae) and Brachymeria intermedia (Nees) (Hym. Chalcididae) on oligidic diets. Bollettino dell'Istituto di Entomologia "Guido Grandi" della Università degli Studi di Bologna 47, 151–154.

Dindo, M. L., Sama, C., & Farneti, R. (1995). Comparison of different commercial veal homogenates in artificial diets for Brachymeria intermedia (Nees) (Hymenoptera: Chalcididae). Bollettino dell'Istituto di Entomologia "Guido Grandi" della Università degli Studi di Bologna, 49, 15–19.

Dix, M. E., Johnson, R. J., Harrell, M. O., Case, R. M., Wright, & Hodges, L. (1995). Influences of trees on abundance of natural enemies of insect pests: A review. Agroforestry Systems, 29, 303–311.

Dobson, S., & Tanouye, M. (1996). The paternal sex ratio chromosome induces chromosome loss independently of Wolbachia in the wasp Nasonia vitripennis. Development Genes and Evolution, 206, 207–217.

Donaldson, J. S., & Walter, G. H. (1991). Brood sex ratios of the solitary parasitoid wasp, Coccophagus atratus. Ecology and Entomology, 16, 25–34.

Doodeman, C. J. A. M., Sebestyen, I., & Van Lenteren, J. C. (1994). Short-range flight test for quality control of Encarsia formosa. Mededelingen Faculteit Landbouwkundige en Toegepaste Biologische Wetenschappen Universiteit Gent, 59, 315–323.

Dover, B. A., Noblet, R., Moore, R. F., & Shepard, B. M. (1987). Development and emergence of Pediobius foveolatus from Mexican bean beetle larvae fed foliage from Phaseolus lunatus and resistant and susceptible soybeans. Journal of Agriculture and Entomology, 4, 271–279.

Drea, J. J., Jr., & Hendrickson, R. M., Jr., (1986). Analysis of a successful classical biological control project: The alfalfa blotch leafminer (Diptera: Agromyzidae) in the northeastern United States. Environmental Entomology, 15, 448–455.

Dreistadt, S., Hagen, K. S., & Dahlsten, D. L., (1986). Predation by Iridomyrmex humilis (Hym.: Formicidae) on eggs of Chrysoperla carnea (Neu.: Chrysopidae) released for inundative control of Illinoia liriodendri (Hom.: Aphididae) infesting Liriodendron tulipifera. Entomophaga, 31, 397–400.

Drooz, A. T., & Weems, M. L. (1982). Cooling eggs of Eutrapela clemataria (Lepidoptera: Geometridae) to −10°C forestalls decline in parasite production with Ooencyrtus ennomophagus (Hymenoptera: Encyrtidae). Canadian Entomologist, 114, 1195–1196.

Drummond, F. A., James, R. L., Casagrande, R. A., & Faubert, H. (1984). Development and survival of Podisus maculiventris (Say) (Hemiptera: Pentatomidae), a predator of the Colorado potato beetle (Coleoptera: Chrysomelidae). Environmental Entomology, 13, 1283–1286.

Dunkel, F. V., & Read, N. R. (1991). Review of the effect of sorbic acid on insect survival in rearing diets with reference to other antimicrobials. American Entomologist, 37, 172–178.

Duso, C. (1992). Role of Amblyseius aberrans (Oud.), Typhlodromus pyri Scheuten and Amblyseius andersoni (Chant) (Acari, Phytoseiidae) in vineyards. III. Influence of variety characteristics on the success of Amblyseius aberrans and Typhlodromus pyri releases. Journal of Applied Entomology, 114, 455–462.

Dutton, A., & Bigler, F. (1995). Flight activity assessment of the egg parasitoid Trichogramma brassicae (Hym.: Trichogrammatidae) in laboratory and field conditions. Entomophaga, 40, 223–233.

Dutton, A., Cerutti, F., & Bigler, F. (1996). Quality and environmental factors affecting Trichogramma brassicae efficiency under field conditions. Entomologia Experimentalis et Applicata, 81, 71–79.

Dyer, L. E., & Landis, D. A. (1996). Effects of habitat, temperature, and sugar availability on longevity of Eriborus terrebrans (Hymenoptera: Ichneumonidae). Environmental Entomology, 25, 1192–1201.

Dyer, L. E., & Landis, D. A. (1997). Influence of noncrop habitats on the distribution of Eriborus terebrans (Hymenoptera: Ichneumonidae) in cornfields. Environmental Entomology, 26, 924–932.

Edwards, D. R., Leppla, N. C., & Dickerson, W. A. (1987). Arthropod species in culture (Catalog), College Park, MD: Entomological Society of America.

Edwards, O. R., & Hoy, M. A. (1995a). Monitoring laboratory and field biotypes of the walnut aphid parasite, Trioxys pallidus, in population cages using RAPD-PCR. Biocontrol Science and Technology, 5, 313–327.

Edwards, O. R., & Hoy, M. A. (1995b). Random amplified polymorphic DNA markers to monitor laboratory-selected, pesticide-resistant Trioxys pallidus (Hymenoptera: Aphidiidae) after release into three California walnut orchards. Environmental Entomology, 24, 487–496.

Ehler, L. E. (1990). Introduction strategies in biological control of insects. In M. Mackauer, L. E. Ehler, & J. Roland (Eds.), Critical issues in biological control (pp. 111–134). XVIII International Congress of Entomology, Vancouver, British Columbia, Canada, July 3–9, 1988, New York: VCH Publishers.

Eischen, F. A., & Dietz, A. (1990). Improved culture techniques for mass rearing Galleria mellonella (Lepidoptera: Pyralidae). Entomological News, 101, 123–128.

Ekbom, B. S. (1977). Development of a biological control program for greenhouse whiteflies (Trialeurodes vaporariorum Westwood) using its parasite Encarsia formosa (Gahan) in Sweden. Zeitschrift fuer Angewandte Entomologie, 84, 145–154.

El Agoze, M., Drezen, J. M., Renault, S., & Periquet, G. (1994). Analysis of the reproductive potential of diploid males in the wasp Diadromus pulchellus (Hymenoptera: Ichneumonidae). Bulletin of Entomological Research, 84, 213–218.

Elder, R. J., & Smith, D. (1995). Mass rearing of Aonidiella orientalis (Newstead) (Hemiptera: Diaspididae) on butternut gramma. Journal of the Australian Entomolical Society, 34, 253–254.

Eller, F. J., Heath, R. R., & Ferkovich, S. M. (1990). Factors affecting oviposition by the parasitoid *Microplitis croceipes* (Hymenoptera: Braconidae) in an artificial substrate. Journal of Economic Entomology, 83, 398–404.

Emrich, B. H. (1991). Acquired toxicity of the lupine aphid, *Macrosiphum albifrons,* and its influence on the aphidophagous predators *Coccinella septempunctata, Episyrphus balteatus* and *Chrysoperla carnea.* Zeitschrift fuer Pflanzenkrankheiten und Pflanzenschutz, 98, 398–404.

Englert, W. D., & Maixner, M. (1988). Rearing of *Typhlodromus pyri* Scheuten in the laboratory and effects of pesticides on mortality and fecundity of this mite. Nachrichtenblatt des Deutschen Pflanzenschutzdienstes, 40, 121–124 (in German, with German and English summaries).

English-Loeb, G. M., Brody, A. K., & Karban, R. (1993). Host-plant-mediated interactions between a generalist folivore and its tachinid parasitoid. Journal of Animal Ecology, 62, 465–471.

Eskafi, F. M., & Legner, E. F. (1974). Fecundity, development, and diapause in *Hexacola* sp. nr. *websteri,* a parasite of *Hippelates* eye gnats. Annals of the Entomological Society of America, 67, 769–771.

Etzel, L. K. (1985). *Phthorimaea operculella.* In P. Singh & R. F. Moore (Eds.), Handbook of insect rearing (Vol. 2, pp. 431–442). New York: Elsevier.

Etzel, L. K., Levinson, S. O., & Andres, L. A. (1981). Elimination of *Nosema* in *Galeruca rufa,* a potential biological control agent for field bindweed. Environmental Entomology, 10, 143–146.

Evans, H. F., & Fielding, N. J. (1994). Integrated management of *Dendroctonus micans* in the UK. Forest Ecology and Management, 65, 17–30.

Fabre, J. P. (1989). Rearing of *Pauesia cedrobii* (Hymenoptera: Aphidiidae), a parasitoid of the Atlas cedar aphid: *Cedrobium laportei* (Homoptera: Lachnidae). Entomophaga, 34, 381–389 (in French, with French and English summaries).

Fabre, J. P., & Rabasse, J. M. (1987). Introduction dans le sud-est de la France d'un parasite: *Pauesia cedrobii* (Hym.: Aphidiidae) du puceron: *Cedrobium laportei* (Hom.: Lachnidae) du cèdre de l'atlas: *Cedrus atlantica.* Entomophaga, 32, 127–141.

Fanti, P. (1991). Hormonal factors triggering the first larval molt of the parasitoid *Pseudogonia rufifrons* Wied. (Diptera Tachinidae) reared *in vivo* and *in vitro.* Bollettino dell'Istituto di Entomolgia "Guido Grandi" della Università degli Studi di Bologna, 45, 47–59.

Farrar, R. R. J., & Kennedy, G. (1991). Inhibition of *Telenomus sphingis,* an egg parasitoid of *Manduca* spp. by trichome-2-tridecanone-based host plant resistance in tomato. Entomologia Experimentalis et Applicata, 60, 157–166.

Farrar, R. R. J., & Kennedy, G. (1993). Field cage performance of two tachinid parasitoids of the tomato fruitworm on insect resistant and susceptible tomato lines. Entomologia Experimentalis et Applicata, 67, 73–78.

Farrar, R. R., Jr., Barbour, J. D., & Kennedy, G. G. (1994). Field evaluation of insect resistance in a wild tomato and its effects on insects parasitoids. Entomologia Experimentalis et Applicata, 71, 211–226.

Faulds, W. (1988). Improved techniques for the laboratory rearing of *Thanasimus formicarius.* New Zealand Journal of Forestry Science, 18, 187–190.

Faulds, W. (1990). Introduction into New Zealand of *Bracon phylacteophagus,* a biocontrol agent of *Phylacteophaga froggati,* eucalyptus leaf-mining sawfly. New Zealand Journal of Forestry Science, 20, 54–64.

Fedde, V. H., Fedde, G. F., & Drooz, A. T. (1982). Factitious hosts in insect parasitoid rearings. Entomophaga, 27, 379–386.

Ferkovich, S. M., Dillard, C., & Oberlander, H. (1991). Stimulation of embryonic development in *Microplitis croceipes* (Braconidae) in cell culture media preconditioned with a fat body cell line derived from a nonpermissive host, gypsy moth, *Lymantria dispar.* Archives of Insect Biochemistry and Physiology, 18, 169–176.

Fielding, N. J., O'keefe, T., & King, C. J. (1991). Dispersal and host-finding capability of the predatory beetle *Rhizophagus grandis* Gyll. (Coleoptera, Rhizophagidae). Journal of Applied Entomology, 112, 89–98.

Finney, G. L., & Fisher, T. W. (1964). Culture of entomophagous insects and their hosts. In P. DeBach (Ed.), Biological control of insect pests and weeds (pp. 328–355). London: Chapman & Hall.

Finney, G. L., Flanders, S. E., & Smith, H. S. (1947). Mass culture of *Macrocentrus ancylivorus* and its host, the potato tuber moth. Hilgardia, 17, 437–483.

Fisher, W. R. (1984). The insectary manager. In E. G. King & N. C. Leppla (Eds.), Advances and challenges in insect rearing (pp. 295–299). Washington, DC: U.S. Government Printing Office.

Flanders, R. V. (1984). Comparison of bean varieties currently being used to culture the Mexican bean beetle (Coleoptera: Coccinellidae). Environmental Entomology, 13, 995–999.

Follett, P. A., & Roderick, G. K. (1996). Genetic estimates of dispersal ability in the leucaena psyllid predator *Curinus coeruleus* (Coleoptera: Coccinellidae): Implications for biological control. Bulletin of Entomological Research, 86, 355–361.

Forbes, A. R., Frazer, B. D., & Chan, C. K. (1985). Aphids. In P. Singh & R. F. Moore (Eds.), Handbook of insect rearing (Vol. 1, pp. 353–359). New York: Elsevier.

Fouly, A. H., Abou-Setta, M. M., & Childers, C. C. (1995). Effects of diet on the biology and life tables of *Typhlodromalus peregrinus* (Acari: Phytoseiidae). Environmental Entomology, 24, 870–874.

Fox, L. R., Letourneau, D. K., Eisenbach, J., & Van Nouhuys, S. (1990). Parasitism rates and sex ratios of a parasitoid wasp: Effects of herbivore and plant quality. Oecologia (Heidelberg), 83, 414–419.

Frank, S. A. (1997). Cytoplasmic incompatibility and population structure. Journal of Theoretical Biology, 184, 327–330.

Frick, K. E., & Wilson, R. F. (1982). Some factors influencing the fecundity and flight potential of *Bactra verutana.* Environmental Entomology, 11, 181–186.

Friend, W. G., & Tanner, R. J. (1985). *Culiseta inornata.* In P. Singh & R. F. Moore (Eds.), Handbook of insect rearing (Vol. 2, pp. 35–40). New York: Elsevier.

Friese, D. D., Megevand, B., & Yaninek, J. S. (1987). Culture maintenance and mass production of exotic phytoseiids. Symposium XI of the International Conference on Tropical Entomology: Africa-Wide Biological Control Programme of Cassava Pests. Insect Science Applications, 8, 875–878.

Fuentes-Contreras, J. E., Powell, W., Wadhams, L. J., Pickett, J. A., & Niemeyer, H. M. (1996). Influence of wheat and oat cultivars on the development of the cereal aphid and parasitoid *Aphidius rhopalosiphi* and the generalist aphid parasitoid *Ephedrus plagiator.* Annals of Applied Biology, 129, 181–187.

Furuhashi, K., & Nishino, M. (1983). Biological control of arrowhead scale, *Unaspis yanonensis,* by parasitic wasps introduced from the People's Republic of China. Entomophaga, 28, 277–286.

Galazzi, D., & Nicoli, G. (1996). Comparative study of strains of *Phytoseiulus persimilis* Athias-Henriot (Acarina Phytoseiidae). II. Influence of mass-rearing on population growth. Bollettino dell'Istituto di Entomologia "Guido Grandi" della Università degli Studi di Bologna, 50, 243–252.

Galford, J. R. (1985). *Enaphalodes rufulus.* In P. Singh & R. F. Moore (Eds.), Handbook of insect rearing (Vol. 1, pp. 255–264). New York: Elsevier.

Galichet, P. F., Riany, M., & Agounke, D. (1985). Bioecology of *Lydella thompsoni* Herting, (Dip.: Tachinidae) within the Rhone delta in southern France. Entomophaga, 30, 315–328.

Gantt, C. W., King, E. G., & Martin, D. F. (1976). New machines for use in a biological insect-control program. Transactions of the ASAE, 19, 242–243.

Gardiner, B. O. C. (1985a). *Schistocerca gregaria.* In P. Singh & R. F. Moore (Eds.), Handbook of insect rearing (Vol. 1, pp. 465–468). New York: Elsevier.

Gardiner, B.O.C. (1985b). *Mamestra brassicae.* In P. Singh & R. F. Moore (Eds.), Handbook of insect rearing (Vol. 2, pp. 381–385). New York: Elsevier.

Gardiner, B.O.C. (1985c). *Pieris brassicae.* In P. Singh & R. F. Moore (Eds.), Handbook of insect rearing (Vol. 2, pp. 453–457). New York: Elsevier.

Gardner, J., & Giles, K. (1996). Handling and environmental effects on viability of mechanically dispensed green lacewing eggs. Biological Control, 7, 245–250.

Gardner, J., & Giles, K. (1997). Mechanical distribution of *Chrysoperla rufilabris* and *Trichogramma pretiosum:* Survival and uniformity of discharge after spray dispersal in an aqueous suspension. Biological Control, 8, 138–142.

Gautam, R. D. (1986). Variations in amino acids in fertile and unfertile eggs of *Spodoptera litura* (Fabr.) contribute towards parasitism by *Telenomus remus* Nixon (Scelionidae: Hymenoptera). Journal of Entomological Research, 10, 161–165.

Gautam, R. D. (1987). Cold storage of eggs of host, *Spodoptera litura* (Fabr.) and its effects on parasitism by *Telenomus remus* Nixon (Scelionidae: Hymenoptera). Journal of Entomological Research, 11, 161–165.

Gautam, R. D., & Paul, A. V. N. (1987). An artificial diet for the larvae of green lacewing, *Chrysopa scelestes* Banks (Neuroptera: Chrysopidae). Journal of Entomological Research (New Delhi), 11, 69–72.

Geden, C. J., Long S. J., Rutz, D. A., & Becnel, J. J. (1995). *Nosema* disease of the parasitoid *Muscidifurax raptor* (Hymenoptera: Pteromalidae): Prevalence, patterns of transmission, management, and impact. Biological Control, 5, 607–614.

Geden, C. J., Smith, L., Long, S. J. & Rutz, D. A. (1992). Rapid deterioration of searching behavior, host destruction, and fecundity of the parasitoid *Muscidifurax raptor* (Hymenoptera: Pteromalidae) in culture. Annals of the Entomological Society of America, 85, 179–187.

Gennadiev, V. G., Khlistovskij, D. E., & Popov, L. A. (1987). Cryogenic storage of host eggs. Zashchita Rastenii, (Moscow), 5·87, 36–37 (in Russian).

Geri, C., & Goussard, F. (1989). Effect of light quality on the development and diapause of *Diprion pini* L. (Hymenoptera, Diprionidae). Journal of Applied Entomology, 108, 89–101.

Gerling, D. (1966). Biological studies on *Encarsia formosa* (Hymenoptera: Aphelinidae). Annals of the Entomological Society of America, 59, 142–143.

Gerling, D., & Fried, R. (1997). Density-related sterility in *Eretmocerus mundus.* Entomologia Experimentalis et Applicata, 84, 33–39.

Gilkeson, L. A. (1990). Cold storage of the predatory midge *Aphidoletes aphidimyza* (Diptera: Cecidomyiidae). Journal of Economic Entomology, 83, 965–970.

Gilkeson, L. A. (1992). Mass rearing of phytoseiid mites for testing and commercial application, In T. E. Anderson & N. C. Leppla (Eds.), Advances in insect rearing for research and pest management (pp. 489–506). Boulder, CO: Westview Press.

Girin, C., & Bouletreau, M. (1995). Microorganism-associated variation in host infestation efficiency in a parasitoid wasp, *Trichogramma bourarachae* (Hymenoptera: Trichogrammatidae). Experientia (Basel), 51, 398–401.

Glass, E. H., & Roelofs, W. L. (1985). *Argyrotaenia velutinana.* In P. Singh & R. F. Moore (Eds.), Handbook of insect rearing (Vol. 2, pp. 197–205). New York: Elsevier.

Glenn, D. C., & Hoffmann, A. A. (1997). Developing a commercially viable system for biological control of light brown apple moth (Lepidoptera: Tortricidae) in grapes using endemic *Trichogramma* (Hymenoptera: Trichogrammatidae). Journal of Economic Entomology, 90, 370–382.

Glenn, D. C., Hercus, M. J., & Hoffmann, A. A. (1997). Characterizing *Trichogramma* (Hymenoptera: Trichogrammatidae) species for biocontrol of light brown apple moth (Lepidoptera: Tortricidae) in grape-

vines in Australia. Annals of the Entomological Society of America, 90, 128–137.

Godfray, H. C. J. (1985). Mass rearing the tachinid fly *Argyrophylax basifulva,* a parasitoid of the greater coconut spike moth (*Tirathaba* spp.) (Lep.: Pyralidae). Entomophaga, 30, 211–215.

Goldson, S. L., Phillips, C. B., McNeill, M. R., & Barlow, N. D. (1997). The potential of parasitoid strains in biological control: Observations to date on *Microctonus* spp. intraspecific variation in New Zealand. Agriculture Ecosystems and Environment, 64, 115–124.

Gomez, L. L. A., Diaz, A. E., & Lastra, L. A. (1995). Selection of strains of *Trichogramma exiguum* for controlling sugarcane borers (*Diatraea* spp.) in the Cauca valley, Colombia. Colloques de l'INRA, 73, 75–78.

Goodenough, J. L. (1984). Materials handling in insect rearing. In E. G. King & N. C. Leppla (Eds.), Advances and challenges in insect rearing (pp. 77–86). Washington, DC: U.S. Government Printing Office.

Goodenough, J. L., Wilson, D. D., & Whitten, C. J. (1978). Visual sensitivity of four strains of screwworm flies. Annals of the Entomological Society of America, 71, 9–12.

Goodwin, R. H. (1984). Recognition and diagnosis of diseases in insectaries and the effects of disease agents on insect biology. In E. G. King & N. C. Leppla (Eds.), Advances and challenges in insect rearing (pp. 96–129). Washington, DC: U.S. Government Printing Office.

Goussard, F., & Geri, C. (1989). A continuous rearing of *Diprion pini* L. in laboratory. Agronomie (Paris), 9, 911–918 (in French with French and English summaries).

Grafton-Cardwell, E. E., & Ouyang, Y. (1996). Influence of citrus leaf nutrition on survivorship, sex ratio, and reproduction of *Euseius tularensis* (Acari: Phytoseiidae). Environmental Entomology, 25, 1020–1025.

Grafton-Cardwell, E. E., Ouyang, Y., & Striggow, R. A. (1997). Predaceous mites (Acari: Phytoseiidae) for control of spider mites (Acari: Tetranychidae) in nursery citrus. Environmental Entomology, 26, 121–130.

Grant, J. F., & Shepard, M. (1985). Influence of three soybean genotypes on development of *Voria ruralis* (Diptera: Tachinidae) and on foliage consumption by its host, the soybean looper (Lepidoptera: Noctuidae). Florida Entomologist, 68, 672–677.

Grant, J. F., & Shepard, M. (1987). Development of *Brachymeria ovata* (Say) (Hymenoptera: Chalcididae) in freezer-stored pupae of lepidopteran species. Environmental Entomology, 16, 1207–1210.

Greany, P. D., Ferkovich, S. M., & Clark, W. R. (1989): Progress towards development of an artificial diet and an *in vitro* rearing system for *Microplitis croceipes.* In J. E. Powell, D. L. Bull, & E. G. King (Eds.), Biological control of *Heliothis* spp. by *Microplitis croceipes.* Southwestern Entomology, (Suppl. 12), 89–94.

Greany, P. D., Allen, G. E., Webb, J. C., Sharp, J. L., & Chambers, D. L. (1977). Stress-induced septicemia as an impediment to laboratory rearing of the fruit fly parasitoid *Biosteres (Opius) longicaudatus* (Hymenoptera: Braconidae) and the Caribbean fruit fly *Anastrepha suspensa* (Diptera: Tephritidae). Journal of Invertebrate Pathology, 29, 153–161.

Greatti, M., & Zandigiacomo, P. (1995). Postrelease dispersal of *Trichogramma brassicae* Bezdenko in corn fields. Journal of Applied Entomology, 119, 671–675.

Greenberg, B., & George J. (1985). *Calliphora vicina, Phormia regina,* and *Phaenicia cuprina.* In P. Singh & R. F. Moore (Eds.), Handbook of insect rearing (Vol. 2, pp. 25–33). New York: Elsevier.

Greenberg, S. M., Morales-Ramos, J. A., King, E. G., Summy, K. R., & Rojas, M. G. (1995). Biological parameters for propagation of *Catolaccus grandis* (Hymenoptera: Pteromalidae). Environmental Entomology, 24, 1322–1327.

Greenblatt, J. A., & Barbosa, P. (1981). Effect of host's diet on two pupal parasitoids of the gypsy moth *Brachymeria intermedia* (Nees) and

Coccygomimus turionellae (L.). Journal of Applied Ecology, 18, 1–10.

Grenier, S., & Liu, W. H. (1990). Antifungals: Mold control and safe levels in artificial media for *Trichogramma* [Hymenoptera: Trichogrammatidae]. Entomophaga, 35, 283–292.

Grenier, S., Veith, V., & Renou, M. (1993). Some factors stimulating oviposition by the oophagous parasitoid *Trichogramma brassicae* Bezd. (Hym., Trichogrammatidae) in artificial host eggs. Journal of Applied Entomology, 115, 66–76.

Grenier, S., Guillaud, J., Delobe, B., & Bonnot, G. (1989). Nutrition and rearing of the polyphagous predator *Macrolophus caliginosus* (Heteroptera, Miridae) with artificial media. Entomophaga, 34, 77–86 (in French with French and English summaries).

Grenier, S., Yang, H., Guillaud, J., & Chapelle, L. (1995). Comparative development and biochemical analyses of *Trichogramma* (Hymenoptera: Trichogrammatidae) grown in artificial media with hemolymph or devoid of insect components. Comparative Biochemistry and Physiology [Part] B: Comparative Biochemistry and Molecular Biology, 111, 83–90.

Grinberg, P. S., & Wallner, W. E. (1991). Long-term laboratory evaluation of *Rogas lymantriae:* A braconid endoparasite of the gypsy moth, *Lymantria dispar.* Entomophaga, 36, 205–212.

Grisdale, D. G. (1984). A laboratory method for mass rearing the eastern spruce budworm, *Choristoneura fumiferana.* In E. G. King & N. C. Leppla (Eds.), Advances and challenges in insect rearing (pp. 223–231). Washington, DC: U.S. Government Printing Office.

Grisdale, D. G. (1985a). *Lambdina fiscellaria.* In P. Singh & R. F. Moore (Eds.), Handbook of insect rearing (Vol. 2, pp. 345–353). New York: Elsevier.

Grisdale, D. G. (1985b). *Malacosoma disstria.* In P. Singh & R. F. Moore (Eds.), Handbook of insect rearing (Vol. 2, pp. 369–379). New York: Elsevier.

Gross, H. R., Jr. (1988). Effect of temperature, relative humidity, and free water on the number and normalcy of *Trichogramma pretiosum* Riley (Hymenoptera: Trichogrammatidae) emerging from eggs of *Heliothis zea* (Boddie) (Lepidoptera: Noctuidae). Environmental Entomology, 17, 470–475.

Gross, H. R., Jr. (1994). Mass propagation of *Archytas marmoratus* (Diptera: Tachinidae). Environmental Entomology, 23, 183–189.

Gross, H. R., Jr., Lewis, W. J., & Nordlund, D. A. (1975) Kairomones and their use for management of entomophagous insects. III. Stimulation of *Trichogramma achaeae, T. pretiosum,* and *Microplitis croceipes* with host-seeking stimuli at time of release to improve their efficacy. Journal of Chemical Ecology, 1, 431–438.

Gross, H. R., Jr., Lewis, W. J., & Nordlund, D. A. (1981). *Trichogramma pretiosum:* Effect of prerelease parasitization experience on retention in release areas and efficiency. Environmental Entomology, 10, 554–556.

Gross, H. R., Jr., Lewis, W. J., Beevers, M., & Nordlund, D. A. (1984). *Trichogramma pretiosum* (Hymenoptera: Trichogrammatidae): Effects of augmented densities and distributions of *Heliothis zea* (Lepidoptera: Noctuidae) host eggs and kairomones on field performance. Environmental Entomology, 13, 981–985.

Grout, T. G., & Richards, R. I. (1992). The dietary effect of windbreak pollens on longevity and fecundity of a predacious mite *Euseius addoensis addoensis* (Acari: Phytoseiidae) found in citrus orchards in South Africa. Bulletin of Entomological Research, 82, 317–320.

Grzegorczyk, W., & Walker, M. A. (1997). Surface sterilization of grape phylloxera eggs in preparation for *in vitro* culture with *Vitis* species. American Journal of Enology and Viticulture, 48, 157–159.

Guerra, A. A., & Martinez, S. (1994). An *in vitro* rearing system for the propagation of the ectoparasitoid *Catolaccus grandis.* Entomologia Experimentalis et Applicata, 72, 11–16.

Guerra, A. A., Martinez, S., & Sonia Del Rio, H. (1994). Natural and synthetic oviposition stimulants for *Catolaccus grandis* (Burks) females. Journal of Chemical Ecology, 20, 1583–1594.

Guerra, A. A., Robacker, K. M., & Martinez, S. (1993). *In vitro* rearing of *Bracon mellitor* and *Catolaccus grandis* with artificial diets devoid of insect components. Entomologia Experimentalis et Applicata, 68, 303–307.

Guthrie, W. D., Robbins, J. C., & Jarvis, J. L. (1985). *Ostrinia nubilalis.* In P. Singh & R. F. Moore (Eds.), Handbook of insect rearing (Vol. 2, pp. 407–413). New York: Elsevier.

Guy, R. H., Leppla, N. C., Rye, J. R., Green, C. W., Barrette, S. L., & Hollien, K. A. (1985). *Trichoplusia ni.* In P. Singh & R. F. Moore (Eds.), Handbook of insect rearing (Vol. 2, pp. 487–494). New York: Elsevier.

Haardt, H., & Hoeller, C. (1992). Differences in life history traits between isofemale lines of the aphid parasitoid *Aphelinus abdominalis* (Hymenoptera: Aphelinidae). Bulletin of Entomological Research, 82, 479–484.

Hagler, J. R., & Cohen, A. C. (1991). Prey selection by *in vitro*-reared and field-reared *Geocoris punctipes.* Entomologia Experimentalis et Applicata, 59, 201–206.

Hailemichael, Y., Smith, J. W., Jr., & Wiedenmann, R. N. (1994). Host-finding behavior, host acceptance, and host suitability of the parasite *Xanthopimpla stemmator.* Entomologia Experimentalis et Applicata, 71, 155–166.

Halfhill, J. E., & Featherston, P. E. (1973). Inundative releases of *Aphidius smithi* against *Acyrthosiphon pisum.* Environmental Entomology, 2, 469–472.

Halperin, J. (1990). Mass breeding of egg parasitoids (Hymenoptera, Chalcidoidea) of *Thaumetopoea wilkinsoni* Tams (Lepidoptera, Thaumetopoeidae). Journal of Applied Entomology, 109, 336–340.

Hansen, L. S. (1988). Control of *Thrips tabaci* (Thysanoptera: Thripidae) on glasshouse cucumber using large introductions of predatory mites *Amblyseius barkeri* (Acarina: Phytoseiidae). Entomophaga, 33, 33–42.

Harcourt, D. G., Guppy, J. C., & Meloche, F. (1988). Population dynamics of the alfalfa blotch leafminer, *Agromyza frontella* (Diptera: Agromyzidae), in Eastern Ontario: Impact of the exotic parasite *Dacnusa dryas* (Hymenoptera: Braconidae). Environmental Entomology, 17, 337–343.

Hare, J. D., & Bethke, J. A. (1988). Egg production and survival of the citrus red mite on an artificial feeding system. Entomologia Experimentalis et Applicata, 47, 137–145.

Hare, J. D., & Luck, R. F. (1990). Influence of the host plant on life history parameters of *Aphytis melinus* when parasitizing California red scale reared on four *Citrus* species. 75th Annual Meeting of the Ecological Society of America on Perspectives in Ecology: Past, Present, and Future, Snowbird, UT, July 29–August 2, 1990. Bulletin of the Ecological Society of America, 71 (Suppl. 2), 181.

Hare, J. D., Morgan, D. J. W., & Nguyun, T. (1997). Increased parasitization of California red scale in the field after exposing its parasitoid, *Aphytis melinus,* to a synthetic kairomone. Entomologia Experimentalis et Applicata, 82, 73–81.

Harris, E. J., & Bautista, R. C. (1996). Effects of fruit fly host, fruit species, and host egg to female parasitoid ratio on the laboratory rearing of *Biosteres arisanus.* Entomologia Experimentalis et Applicata, 79, 187–194.

Harris, E. J., & Okamoto, R. Y. (1983). Description and evaluation of a simple method for the collection of the parasite *Biosteres oophilus* (Hym.: Braconidae). Entomophaga, 28, 241–243.

Harris, E. J., & Okamoto, R. Y. (1991). A method for rearing *Biosteres arisanus* (Hymenoptera: Braconidae) in the laboratory. Journal of Economic Entomology, 84, 417–422.

Harrison, W. W., King, E. G., & Ouzts, J. D. (1985). Development of *Trichogramma exiguum* and *T. pretiosum* at five temperature regimes. Environmental Entomology, 14, 118–121.

Hartley, G. G. (1990). Multicellular rearing methods for the beet armyworm, soybean looper, and velvetbean caterpillar (Lepidoptera: Noctuidae). Journal of Entomological Science, 25, 336–340.

Hartley, G. G., Gantt, C. W., King, E. G., & Martin, D. F. (1977). Equipment for mass rearing of the greater wax moth and the parasite Lixophaga diatraeae. (Rep. ARS-S-164). Washington, DC: U.S. Agricultural Research Service.

Hartley, G. G., King, E. G., Brewer, F. D., & Gantt, C. W. (1982). Rearing of the Heliothis sterile hybrid with a multicellular larval rearing container and pupal harvesting. Journal of Economic Entomology, 75, 7–10.

Harwalkar, M. R., Rananavare, H. D., & Rahaikar, G. W. (1987). Development of Trichogramma brasiliensis (Hym.: Trichogrammatidae) on eggs of radiation sterilized females of potato tuberworm, Phthorimaea operculella (Lep.: Gelechiidae). Entomophaga, 32, 159–162.

Hasegawa, M., Niijima, K., & Matsuka, M. (1989). Rearing Chrysoperla carnea (Neuroptera: Chrysopidae) on chemically defined diets. Applied Entomology and Zoology, 24, 96–102.

Hassan, S. A., Kohler, E., & Rost, W. M. (1988). Mass production and utilization of Trichogramma. X. Control of the codling moth Cydia pomonella and the summer fruit tortrix moth Adoxophyes orana (Lep.: Tortricidae). Entomophaga, 33 (4), 413–420.

Hattingh, V., & Samways, M. J. (1991). Determination of the most effective method for field establishment of biocontrol agents of the genus Chilocorus (Coleoptera: Coccinellidae). Bulletin of Entomological Research, 81, 169–174.

Hattingh, V., & Samways, M. J. (1993). Evaluation of artificial diets and two species of natural prey as laboratory food for Chilocorus spp. Entomologia Experimentalis et Applicata, 69, 13–20.

Haug, T., Herren, H. R., Nadel, D. J., & Akinwumi, J. B. (1987). Technologies for the mass-rearing of cassava mealybugs, cassava green mites and their natural enemies. Symposium XI of the International Conference on Tropical Entomology: Africa-Wide Biological Control Programme of Cassava Pests. Insect Science Applications, 8, 879–881.

Havron, A., Rosen, D., Rossler, Y., & Hillel, J. (1987). Selection on the male hemizygous genotype in arrhenotokous insects and mites. Entomophaga, 32, 261–268.

Heather, N. W., & Corcoran, R. J. (1985). Dacus tryoni. In P. Singh & R. F. Moore (Eds.), Handbook of insect rearing (Vol. 2, pp. 41–48). New York: Elsevier.

Heimpel, G. E., Rosenheim, J. A., & Kattari, D. (1997). Adult feeding and lifetime reproductive success in the parasitoid Aphytis melinus. Entomologia Experimentalis et Applicata, 83, 305–315.

Heinz, K. M., & Nelson, J. M. (1996). Interspecific interactions among natural enemies of Bemisia in an inundative biological control program. Biological Control, 6, 384–393.

Heinz, K. M., & Zalom, F. G. (1996). Performance of the predator Delphastus pusillus on Bemisia resistant and susceptible tomato lines. Entomologia Experimentalis et Applicata, 81, 345–352.

Henderson, S. A., & Albrecht, J. S. M. (1988). An artificial diet for maintaining ladybirds. Entomologists' Record and Journal of Variation, 100, 261–264.

Henry, J. E. (1985). Melanoplus spp. In P. Singh & R. F. Moore (Eds.), Handbook of insect rearing (Vol. 1, pp. 451–464). New York: Elsevier.

Herard, F., Keller, M. A., Lewis, W. J., & Tumlinson, J. H. (1988a). Beneficial arthropod behavior mediated by airborne semiochemicals. III. Influence of age and experience on flight chamber responses of Microplitis demolitor Wilkinson. Journal of Chemical Ecology, 14, 1583–1596.

Herard, F., Keller, M. A., Lewis, W. J., & Tumlinson, J. H. (1988b). Beneficial arthropod behavior mediated by airborne semiochemicals. IV. Influence of host diet on host-oriented flight chamber responses of Microplitis demolitor Wilkinson. Journal of Chemical Ecology, 14, 1597–1606.

Herren, H. R. (1987). A review of objectives and achievements. Symposium XI of the International Conference on Tropical Entomology: Africa-Wide Biological Control Programme of Cassava Pests. Insect Science Applications, 8, 837–840.

Herren, H. R. (1990). Biological control as the primary option in sustainable pest management: The cassava pest project. Mitteilungen der Schweizerischen Entomologischen Gesellschaft, 63, 405–414.

Herren, H. R., Bird, T. J., & Nadel, D. J. (1987). Technology for automated aerial release of natural enemies of the cassava mealybug and cassava green mite. Symposium XI of the International Conference on Tropical Entomology: Africa-Wide Biological Control Programme of Cassava Pests. Insect Science Applications, 8, 883–885.

Hirashima, J., Miura, K., & Miura, T. (1990a). Studies on the biological control of the diamondback moth, Plutella xylostella (Linnaeus). III. On the growth obstruction of host eggs utilized for mass culture of Trichogramma chilonis and Trichogramma ostriniae. Science Bulletin of the Faculty of Agriculture, Kyushu University, 44, 77–80 (in Japanese with Japanese and English summaries).

Hirashima, J., Miura, K., & Miura, T. (1990b). Studies on the biological control of the diamondback moth, Plutella xylostella (Linnaeus). IV. Effect of temperature on the development of the egg parasitoids Trichogramma chilonis and Trichogramma ostriniae. Science Bulletin of the Faculty of Agriculture, Kyushu University, 44, 81–87 (in Japanese, with Japanese and English summaries).

Hirashima, J., Miura, K., & Miura, T. (1990c). Studies on the biological control of the diamondback moth, Plutella xylostella (Linnaeus). VI. New technique for mass culture of the egg parasitoids Trichogramma chilonis and Trichogramma ostriniae. Science Bulletin of the Faculty of Agriculture, Kyushu University, 44, 95–100 (in Japanese with Japanese and English summaries).

Hirose, Y., Takasu, K., & Takagi, M. (1996). Egg parasitoids of phytophagous bugs in soybean: Mobile natural enemies as naturally occurring biological control agents of mobile pests. Biological Control, 7, 84–94.

Ho, C.-C., & Chen, W.-H. (1992). Control of phytoseiids in a spider mite mass-rearing system (Acari: Phytoseiidae, Tetranychidae). Experimental and Applied Acarology (Northwood), 13, 287–293.

Hoddle, M. S., Van Driesche, R. G., & Sanderson, J. P. (1997). Biological control of Bemisia argentifolii (Homoptera, Aleyrodidae) on Poinsettia with inundative releases of Encarsia formosa beltsville strain (Hymenoptera: Aphelinidae): Can parasitoid reproduction augment inundative releases? Journal of Economic Entomology, 90, 910–924.

Hodek, I. (1973). Biology of Coccinellidae. Prague, Czechoslovakia: Academia.

Hoeller, C., & Haardt, H. (1993). Low field performance of an aphid parasitoid, Aphelinus abdominalis efficient in the laboratory [Hym., Aphelinidae]. Entomophaga, 38, 115–124.

Hoffman, J. D., Ignoffo, C. M., Peters, P., & Dickerson, W. A. (1984). Fractional colony propagation—a new insect-rearing system. In E. G. King & N. C. Leppla (Eds.), Advances and challenges in insect rearing (pp. 232–233). Washington, DC: U.S. Government Printing Office.

Hoffmann, R. W., & Kennett, C. E. (1985). Effects of winter temperatures on the sex ratios of Aphytis melinus (Hym.: Aphelinidae) in the San Joaquin Valley of California. Entomophaga, 30, 125–132.

Hopper, K. R. (1996). Making biological control introductions more effective. In J. K. Waage (Ed.), BCPC Symposium Proceedings, No. 67; Biological control introductions: Opportunities for improved crop production (pp. 61–76). International Symposium, Brighton, UK, November 18, 1996. Farnham, UK: British Crop Protection Council.

Hopper, K. R., & Roush, R. T. (1993). Mate finding, dispersal, number released, and the success of biological control introductions. Ecology and Entomology, 18, 321–331.

Hopper, K. R., Roush, R. T., & Powell, W. (1993). Management of genetics of biological-control introductions. Annual Review of Entomology, 38, 27–51.

Horio, M., Tsukamoto, M., Jayasekera, N., & Kamimura, K. (1989). Laboratory colonization and bionomics of Toxorhynchites minimus (Diptera: Culicidae) from Sri Lanka. Japanese Journal of Sanitation Zoology, 40 (Suppl.), 11–24.

Horjus, M., & Stouthamer, R. (1995). Does infection with thelytoky-causing Wolbachia in the pre-adult and adult life stages influence the adult fecundity of Trichogramma deion and Muscidifurax uniraptor? In M. J. Sommeijer & P. J. Francke (Eds.), Proceedings of the Section Experimental and Applied Entomology of the Netherlands Entomological Society (N.E.V.) (Vol. 6, pp. 35–40). Sixth Meeting, Amsterdam, Netherlands, December 16, 1994. Amsterdam. Netherlands: Netherlands Entomological Society.

Horwood, M. A. (1988). Control of Pheidole megacephala (F.) (Hymenoptera: Formicidae) using methoprene baits. Journal of Australian Entomological Society, 27, 257–258.

Hough-Goldstein, J., Janis, J. A., & Ellers, C. D. (1996). Release methods for Perillus bioculatus (F.), a predator of the Colorado potato beetle. Biological Control, 6, 114–122.

Hoy, M. A. (1993). Biological control in US agriculture: Back to the future. American Entomology, 39, 140–150.

Hoy, M. A., Presnail, J. K., & Jeyaprakash, A. (1995). Transformation of beneficial arthropods by maternal microinjection. Journal of Cellular Biochemistry, (Suppl.) 194.

Hsin, C., & Getz, W. M. (1988). Mass rearing and harvesting based on an age-stage, two-sex life table: A potato tuberworm (Lepidoptera: Gelechiidae) case study. Environmental Entomology, 17, 18–25.

Hu, Z. W., & Xu, Q. Y. (1988). Studies on frozen storage of eggs of rice moth and oak silkworm. Colloques de l'INRA, 43, 327–338.

Hugar, P., Rao, K. J., & Lingappa, S. (1990). Effect of chilling on hatching and parasitism of eggs of Corcyra cephalonica (Stainton) by Trichogramma chilonis (Ishii). Entomon, 15, 49–52.

Hunter, C. D. (1997). Suppliers of beneficial organisms in North America. California Department Food Agriculture, Biological Control Services Program.

Hunter, M. S., Nur, U., & Werren, J. H. (1993). Origin of males by genome loss in an autoparasitoid wasp. Heredity, 70, 162–171.

Hussein, M. Y. (1986). A method of collecting eggs of Micromus tasmaniae Walker (Neuroptera: Hemerobiidae). Pertanika, 9, 449–454.

Hussein, M. Y., & Hagen, K. S. (1991). Rearing of Hippodamia convergens on artificial diet of chicken liver, yeast and sucrose. Entomologia Experimentalis et Applicata, 59, 197–199.

Ignoffo, C. M., & Boening, O. P. (1970). Compartmented disposable plastic trays for rearing insects. Journal of Economic Entomology, 63, 1696–1697.

Isaacson, D. L., Miller, G. A., & Coombs, E. M. (1996). Use of Geographic Information Systems (GIS) distance measures in managed dispersal of Apion fuscirostre for control of Scotch broom (Cytisus scoparius). In E. S. Delfosse & R. R. Scott (Eds.), Biological control of weeds (pp. 695–699). Eighth International Symposium, Canterbury, New Zealand, February 2–7, 1992, East Melbourne, Victoria, Australia: CSIRO Publications.

Isenhour, D. J. (1985). Campoletis sonorensis (Hym.: Ichneumonidae) as a parasitoid of Spodoptera frugiperda (Lep.: Noctuidae): Host stage preference and functional response. Entomophaga, 30, 31–36.

Jackson, J. J. (1985). Diabrotica spp. In P. Singh & R. F. Moore (Eds.), Handbook of insect rearing (Vol. 1, pp. 237–254). New York: Elsevier.

Jalali, S. K., & Singh, S. P. (1992). Differential response of four Trichogramma spp. to low temperatures for short term storage. Entomophaga, 37, 159–165.

Jalali, S. K., & Singh, S. P. (1993). Superior strain selection of the egg parasitoid Trichogramma chilonis Ishii: Biological parameters. Biological Control, 7, 57–60.

Jalali, S. K., Singh, S. P., Ballal, C. R., & Kumar, P. (1990). Response of Cotesia marginiventris (Cresson) (Hymenoptera: Braconidae) to low temperature in relation to its biotic potential. Entomon, 15, 217–220.

Jalali, S. K., Singh, S. P., Kumar, P., & Ballal, C. R. (1988). Influence of the food plants on the degree of parasitism of larvae of Heliothis armigera by Cotesia kazak. Entomophaga, 33, 65–71.

James, D. G. (1989). Influence of diet on development, survival and oviposition in an Australian phytoseiid, Amblyseius victoriensis (Acari: Phytoseiidae). Experimental Applications of Acarology, 6, 1–10.

James, D. G. (1993). Pollen mould mites and fungi improvements to mass rearing of Typhlodromus doreenae and Amblyseius victoriensis. Experimental and Applied Acarology (Northwood), 17, 271–276.

James, D. G., & Taylor, A. (1993). Predator population density influences oviposition rate in Amblyseius victoriensis Womersley and Typhlodromus doreenae Schicha. International Journal of Acarology, 19, 189–191.

James, D. G., & Whitney, J. (1993). Cumbungi pollen as a laboratory diet for Amblyseius victoriensis (Womersley) and Typhlodromus doreenae Schicha (Acari: Phytoseiidae). Journal of the Australian Entomological Society, 32, 5–6.

Jang, E. B., & Chan, H. T. J. (1993). Alleviation of acetic acid production during mass rearing of the Mediterranean fruit fly (Diptera: Tephritidae). Journal Economic Entomology, 86, 301–309.

Johanowicz, D. L., & Hoy, M. A. (1996). Wolbachia in a predator-prey system: 16S ribosomal DNA analysis of two phytoseiids (Acari: Phytoseiidae) and their prey (Acari: Tetranychidae). Annals of the Entomological Society of America, 89, 435–441.

Johnson, S. J. (1985). Low-level augmentation of Trichogramma pretiosum and naturally occurring Trichogramma spp. parasitism of Heliothis spp. in cotton in Louisiana. Environmental Entomology, 14, 28–31.

Jones, S. L., Morrison, R. K., Ables, J. R., & Bull, D. L. (1977). A new and improved technique for the field release of Trichogramma pretiosum. Southwestern Entomology, 2, 210–215.

Jones, S. L., Morrison, R. K., Ables, J. R., Bouse, L. F., Carlton, J. B., & Bull, D. L. (1979). New techniques for the aerial release of Trichogramma pretiosum. Southwestern Entomology, 4, 14–19.

Jones, W. A., Jr. (1985). Nezara viridula. In P. Singh & R. F. Moore (Eds.), Handbook of insect rearing (Vol. 1, pp. 339–343). New York: Elsevier.

Jones, W. A., Jr., & Jackson, C. G. (1990). Mass production of Anaphes iole for augmentation against Lygus hesperus: Effects of food on fecundity and longevity. Southwestern Entomology, 15, 463–468.

Joslyn, D. J. (1984). Maintenance of genetic variability in reared insects. In E. G. King & N. C. Leppla (Eds.), Advances and challenges in insect rearing (pp. 20–29). Washington, DC: U.S. Government Printing Office.

Justo, H. D., Jr., Shepard, B. M., & Elsey, K. D. (1997). Dispersal of the egg parasitoid Trissolcus basalis (Hymenoptera: Scelionidae) in tomato. Journal of Agricultural Entomology, 14, 139–149.

Kainoh, Y., & Brown, J. J. (1994). Amino acids as oviposition stimulants for the egg-larval parasitoid Chelonus sp. near curvimaculatus (Hymenoptera: Braconidae). Biological Control, 4, 22–25.

Kamal, A. S. (1958). Comparative study of thirteen species of sarcosaprophagous Calliphoridae and Sarcophagidae (Diptera). I. Bionom. Annals of the Entomological Society of America, 51, 261–271.

Kaneshiro, L. N., & Johnson, M. W. (1996). Tritrophic effects of leaf nitrogen on Liriomyza trifolii (Burgess) and an associated parasitoid Chrysocharis oscinidis (Ashmead) on bean. Biological Control, 6, 186–192.

Kaplan, P. B., Keiser, L. S., & Yazlovetskii, I. G. (1989). The artificial diet for larvae of Chrysopa septempunctata. Zoologicheskii Zhurnal, 68, 118–122 (in Russian with Russian and English summaries).

Karelin, V. D., Yakovchuk, T. N., & Danu, V. P. (1989). Development of techniques for commercial production of the common green lacewing, *Chrysopa carnea* (Neuroptera, Chrysopidae). Acta Entomologica Fennica, 53, 31–35.

Karowe, D. N., & Schoonhoven, L. M. (1992). Interactions among three trophic levels: The influence of host plant on performance of *Pieris brassicae* and its parasitoid, *Cotesia glomerata*. Entomologia Experimentalis et Applicata, 62, 241–251.

Katsoyannos, P. I., & Argyriou, L. (1985). The phenology of the San Jose scale *Quadraspidiotus perniciosus* (Hom.: Diaspididae) and its association with its natural enemies on almond trees in northern Greece. Entomophaga, 30, 3–11.

Kazimirova, M., & Vallo, V. (1992). Influence of larval density of Mediterranean fruit fly (*Ceratitis capitata*, Diptera, Tephritidae) on parasitization by a pupal parasitoid, *Coptera occidentalis* (Hymenoptera, Proctotrupoidea, Diapriidae). Acta Entomologica Bohemoslovaca, 89, 179–185.

Kazmer, D. J., & Luck, R. F. (1995). Field tests of the size-fitness hypothesis in the egg parasitoid *Trichogramma pretiosum*. Ecology, 76, 412–425.

Keena, M. A., Odell, T. M., & Tanner, J. A. (1995). Effects of diet ingredient source and preparation method on larval development of laboratory-reared gypsy moth (Lepidoptera: Lymantriidae). Annals of the Entomological Society of America, 88, 672–679.

Kfir, R. (1981). Effect of hosts and parasite density on the egg parasite *Trichogramma pretiosum* (Hym.: Trichogrammatidae). Entomophaga, 26, 445–451.

Kfir, R., & Van Hamburg, H. (1988). Interspecific competition between *Telenomus ullyetti* (Hymenoptera: Scelionidae) and *Trichogrammatoidea lutea* (Hymenoptera: Trichogrammatidae) parasitizing eggs of the cotton bollworm *Heliothis armiger* in the laboratory. Environmental Entomology, 17, 664–670.

Kfir, R., Kirsten, F., & Van Rensburg, N. J., (1985). *Pauesia* sp. (Hymenoptera: Aphidiidae): A parasite introduced into South Africa for biological control of the black pine aphid, *Cinara cronartii* (Homoptera: Aphididae). Environmental Entomology, 14, 597–601.

Kinawy, M. M. (1991). Biological control of the coconut scale insect (*Aspidiotus destructor* Sign, Homoptera: Diaspididae) in the southern region of Oman (Dhofar). Tropical Pest Management, 37, 387–389.

King, B. H. (1993). Sex ratio manipulation by parasitoid wasps. In D. L. Wrensch & M. A. Ebbert (Eds.), Evolution and diversity of sex ratio in insects and mites (pp. 418–441). New York: Chapman & Hall.

King, E. G., & Hartley, G. G. (1985a). *Diatraea saccharalis*. In P. Singh & R. F. Moore (Eds.), Handbook of insect rearing (Vol. 2, pp. 265–270) New York: Elsevier.

King, E. G., & Hartley, G. G. (1985b). *Galleria mellonella*. In P. Singh & R. F. Moore (Eds.), Handbook of insect rearing (Vol. 2, pp. 301–305). New York: Elsevier.

King, E. G., & Hartley, G. G. (1985c). *Lixophaga diatraeae*. In P. Singh & R. F. Moore (Eds.), Handbook of insect rearing (Vol. 2, pp. 119–123). New York: Elsevier.

King, E. G., & Hartley, G. G. (1992). Multiple-species insect rearing in support of research. In T. E. Anderson & N. C. Leppla (Eds.), Advances in insect rearing for research and pest management pp. 159–172. Boulder, CO: Westview Press.

King, E. G., & Leppla, N. C. (Eds.), (1984). Advances and challenges in insect rearing. Washington, DC: U.S. Government Printing Office.

King, E. G., & Morrison, R. K. (1984). Some systems for production of eight entomophagous arthropods. In E. G. King & N. C. Leppla (Eds.), Advances and challenges in insect rearing (pp. 206–222). Washington, DC: U.S. Government Printing Office.

King, E. G., Hatchett, J. H., & Martin, D. F. (1978). Biological characteristics of two populations of *Lixophaga diatraeae* (Tachinidae: Diptera) and their reciprocal crosses at three different temperatures in the labo-ratory. Proceedings of the International Society of Sugar-Cane Technology (pp. 509–516). 16th Congress of Entomology Section.

King, E. G., Sanford, J., Smith, J. W., & Martin, D. F. (1981). Augmentative releases of *Lixophaga diatraeae* (Dip.:Tachinidae) for suppression of early season sugarcane borer populations in Louisiana. Entomophaga, 26, 59–69.

King, E. G., Hartley, G. G., Martin, D. F., Smith, J. W., Summers, T. E., & Jackson, R. D. (1979). Production of the tachinid Lixophaga diatraeae on its natural host, the sugarcane borer, and on an unnatural host, the greater wax moth. (SEA, AAT-S-3/April 1979). Washington, DC: U.S. Department of Agriculture.

Kingsley, P. C., Bryan, M. D., Day, W. H., Burger, T. L., Dysart, R. J., & Schwalbe, C. P. (1993). Alfalfa weevil (Coleoptera: Curculionidae) biological control: Spreading the benefits. Environmental Entomology, 22, 1234–1250.

Klay, A. (1990). The use of a population model for the mass production of natural enemies of the cassava green mite. Mitteilungen der Schweizerischen Entomologischen Gesellschaft, 63, 339–404.

Kolodny-Hirsch, D. M. (1988). Influence of some environmental factors on the laboratory production of *Cotesia melanoscela* (Braconidae: Hymenoptera): A larval parasitoid of *Lymantria dispar*. Environmental Entomology, 17, 127–131.

Krause, S. C., Fuester, R. W., & Burbutis, P. P. (1990). Competitive interactions between *Cotesia melanoscelus* and *Glyptapanteles flavicoxis* (Hymenoptera: Braconidae): Implications for biological control of gypsy moth (Lepidoptera: Lymantriidae). Environmental Entomology, 19, 1543–1546.

Krieg, A., & Huger, A. M. (1987). Über den Erreger einer Bakteriose im Parasit-Wirts-System *Ernestia consobrina/Mamestra brassicae*: *Serratia liquefaciens*. Nachrichtenblatt des Deutschen Pflanzenschutzdienstes, 39, 132–134.

Krishnamoorthy, A., & Singh, S. P. (1987). Biological control of citrus mealybug, *Planococcus citri* with an introduced parasite, *Leptomastix dactylopii* in India. Entomophaga, 32, 143–148.

Kulp, D., Fortmann, M., Hommes, M., & Plate, H.-P. (1989). The predatory gall midge *Aphidoletes aphidimyza* (Rondani) (Diptera: Cecidomyiidae). An important aphid predator: A reference book to systematics, distribution, biology, rearing methods and application. (Communications from the Federal Biological Institute for Agriculture and Forestry, Berlin-Dahlem, No. 250): Berlin: Kommissionsverlag Paul Parey.

Kumar, A., & Tripathi, C. P. M. (1988). Parasitoid-host relationship between *Trioxys indicus* Subba Rao & Sharma (Hymenoptera: Aphidiidae) and *Aphis craccivora* Koch (Hemiptera: Aphididae): Effect of host plants on the sex ratio of the parasitoid. Entomon, 12, 95–100.

Kumar, A., Shanker, S., Pandey, K. P., Sinha, T. B., & Tripathi, C. P. M. (1988). Parasitoid-host relationship between *Trioxys (Binodoxys) indicus* (Hymenoptera: Aphidiidae) and *Aphis craccivora* (Hemiptera: Aphidiidae). VI. Impact of males on the number of progeny of the parasitoid reared on certain host plants. Entomophaga, 33, 17–23.

Kumar, H. (1997). Resistance in maize to *Chilo partellus* (Swinhoe) (Lepidoptera: Pyralidae): Role of stalk damage parameters and biological control. Crop Protection, 16, 375–381.

Kumar, P., & Jalali, S. K. (1993) A hygienic and efficient method of *Corcyra* moth and egg collection. International Journal of Pest Management, 39, 103–105.

Kuniata, L. S. (1994). Importation and establishment of *Heteropsylla spinulosa* (Homoptera; Psyllidae) for the biological control of *Mimosa invisa* in Papua New Guinea. International Journal of Pest Management, 40, 64–65.

Laing, J. E., & Corrigan, J. E. (1995). Diapause induction and postdiapause emergence in *Trichogramma minutum* Riley (Hymenoptera: Trichogrammatidae): The role of host species, temperature, and photoperiod. Canadian Entomologist, 127, 103–110.

Laing, J. E., & Heraty, J. M. (1981). Establishment in Canada of the parasite *Apanteles pedias* Nixon on the spotted tentiform leafminer, *Phyllonorycter blancardella* (Fabr.). Environmental Entomology, 10, 933–935.

Landry, B. S., Dextraze, L., & Boivin, G. (1993). Random amplified polymorphic DNA markers for DNA fingerprinting and genetic variability assessment of minute parasitic wasp species (Hymenoptera: Mymaridae and Trichogrammatidae) used in biological control programs of phytophagous insects. Genome, 36, 580–587.

Lasota, J. A., & Kok, L. T. (1986a). Parasitism and utilization of imported cabbageworm pupae by *Pteromalus puparum* (Hymenoptera: Pteromalidae). Environmental Entomology, 15, 994–998.

Lasota, J. A., & Kok, L. T. (1986b). Larval density effects on pupal size and weight of the imported cabbageworm (Lepidoptera: Pieridae), and their suitability as hosts for *Pteromalus puparum* (Hymenoptera: Pteromalidae). Environmental Entomology, 15, 1234–1236.

Lassy, C. W., Karr, T. L. (1996). Cytological analysis of fertilization and early embryonic development in incompatible crosses of *Drosophila simulans*. Mechanisms of Development, 57, 47–58.

Lawson, D. S., Nyrop, J. P., & Reissig, W. H. (1997). Assays with commercially produced *Trichogramma* (Hymenoptera: Trichogrammatidae) to determine suitability for obliquebanded leafroller (Lepidoptera: Tortricidae) control. Environmental Entomology, 26, 684–693.

Leatemia, J. A., Laing, J. E., & Corrigan, J. E. (1995). Effects of adult nutrition of longevity, fecundity, and offspring sex ratio of *Trichogramma minutum* Riley (Hymenoptera: Trichogrammatidae). Canadian Entomologist, 127, 245–254.

Lee, H. P., & Lee, K. S. (1994). Artificial rearing *in vitro* of *Ooencyrtus kuvanae* Howard (Hymenoptera: Encyrtidae): Artificial media, oviposition and development. Korean Journal of Entomology, 24, 311–316.

Legaspi, J. C., Nordlund, D. A., & Legaspi, B. C., Jr., (1996). Tri-trophic interactions and predation rates in *Chrysoperla* spp. attacking the silverleaf whitefly. Southwestern Entomology, 21, 33–42.

Legner, E. F. (1967). Behavior changes the reproduction of *Spalangia cameroni, S. endius, Muscidifurax raptor,* and *Nasonia vitripennis* (Hymenoptera: Pteromalidae) at increasing fly host densities. Annals of the Entomological Society of America, 60, 819–826.

Legner, E. F. (1979a). Prolonged culture and inbreeding effects on reproductive rates of two pteromalid parasites of muscoid flies. Annals of the Entomological Society of America, 72, 114–118.

Legner, E. F. (1979b). Reproduction of *Spalangia endius, Muscidifurax raptor* and *M. zaraptor* on fresh vs. refrigerated fly hosts. Annals of the Entomological Society of America, 72, 155–157.

Legner, E. F. (1979c). Emergence patterns and dispersal in *Chelonus* spp. near *curvimaculatus* and *Pristomerus hawaiiensis,* parasitic on *Pectinophora gossypiella.* Annals of the Entomological Society of America, 72, 681–686.

Legner, E. F. (1983). Patterns of field diapause in the navel orangeworm (Lepidoptera: Phycitidae) and three imported parasites. Annals of the Entomological Society of America, 76, 503–506.

Legner, E. F. (1985). Natural and induced sex ratio changes in populations of thelyolokous *Muscidifurax uniraptor* (Hymenoptera: Pteromalidae). Annals of the Entomological Society of America, 78, 398–402.

Legner, E. F. (1986). Importation of exotic natural enemies. In J. M. Franz (Ed.), Biological control of plant pests and vectors of human and animal diseases. New York: G. Fisher. (Series title: Fortschritte der Zoologie; Bd. 32.)

Legner, E. F. (1987a). Transfer of thelytoky to arrhenotokous *Muscidifurax raptor* Girault & Sanders (Hymenoptera: Pteromalidae). Canadian Entomology, 119, 265–271.

Legner, E. F. (1987b). Pattern of thelytoky acquisition in *Muscidifurax raptor* Girault & Sanders (Hymenoptera: Pteromalidae) Bulletin of the Society of Vector Ecology, 12, 1–11.

Legner, E. F. (1987c). Inheritance of gregarious and solitary oviposition in *Muscidifurax raptorellus* Kogan & Legner (Hymenoptera: Pteromalidae). Canadian Entomologist, 119, 791–808.

Legner, E. F. (1987d). Further insights into extranuclear influences on behavior elicited by males in the genus *Muscidifurax* (Hymenoptera: Pteromalidae). Proceedings of the California Mosquito Vector Control Association, 55, 127–130.

Legner, E. F. (1988a). *Muscidifurax raptorellus* (Hymenoptera: Pteromalidae) females exhibit postmating oviposition behavior typical of the male genome. Annals of the Entomological Society of America, 81, 522–527.

Legner, E. F. (1988b). Quantitation of heterotic behavior in parasitic Hymenoptera. Annals of the Entomological Society of America, 81, 657–681.

Legner, E. F. (1988c). Hybridization in principal parasitoids of synanthropic Diptera: The genus *Muscidifurax* (Hymenoptera: Pteromalidae). Hilgardia, 56(4), 36.

Legner, E. F. (1989a). Wary genes and accretive inheritance in Hymenoptera. Annals of the Entomological Society of America, 82, 245–249.

Legner, E. F. (1989b). Paternal influences in males of the parasitic wasp, *Muscidifurax raptorellus* Kogan & Legner (Hymenoptera: Pteromalidae). Entomophaga, 34, 307–320.

Legner, E. F. (1991a). Estimations of number of active loci, dominance and heritability in polygenic inheritance of gregarious behavior in *Muscidifurax raptorellus* (Hymenoptera: Pteromalidae). Entomophaga, 36, 1–18.

Legner, E. F. (1991b). Recombinant males in the parasitic wasp *Muscidifurax raptorellus* (Hymenoptera: Pteromalidae). Entomophaga, 36, 173–181.

Legner, E. F. (1993). Theory for quantitative inheritance of behavior in a protelean parasitoid, *Muscidifurax raptorellus* (Hymenoptera: Pteromalidae). European Journal of Entomology, 90, 11–21.

Legner, E. F., & Bay, E. C. (1964). Natural exposure of *Hippelates* eye gnats to field parasitization and the discovery of one pupal and two larval parasites. Annals of the Entomological Society of America, 57, 767–769.

Legner, E. F., & Bay, E. C. (1965). Culture of *Hippelates pusio* (Diptera: Chloropidae) in the West Indies for natural enemy exploration and some notes on behavior and distribution. Annals of the Entomological Society of America, 58, 436–440.

Legner, E. F., & Gerling, D. (1967). Host-feeding and oviposition on *Musca domestica* by *Spalangia cameroni, Nasonia vitripennis,* and *Muscidifurax raptor* (Hymenoptera: Pteromalidae) influences their longevity and fecundity. Annals of the Entomological Society of America, 60, 678–691.

Legner, E. F., & Murray, C. A. (1981). Feeding rates and growth of the fish *Tilapia zillii* [Cichlidae] on *Hydrilla verticillata, Potamogeton pectinatus* and *Myriophyllum spicatum* var. *exalbescens* and interactions in irrigation canals of southeastern California. Mosquito News, 41, 241–250.

Legner, E. F., & Thompson, S. N. (1977). Effects of the parental host on host selection, reproductive potential, survival and fecundity of the egg-larval parasitoid *Chelonus* sp. nr. *curvimaculatus,* reared on *Pectinophora gossypiella* and *Phthorimaea operculella.* Entomophaga, 22, 75–84.

Legner, E. F., & Tsai, S. C. (1978). Increasing fission rate of the planarian mosquito predator, *Dugesia dorotocephala,* through biological filtration. Entomophaga, 23, 293–298.

Legner, E. F., & Warkentin, R. W. (1985). Genetic improvement and inbreeding effects in culture of beneficial arthropods (pp. 156–161). Proceedings and Papers of the Fifty-Second Annual Conference of the California Mosquito and Vector Control Association, January 29–February 1, 1984.

Leppla, N. C. (1985). *Anticarsia gemmatalis.* In P. Singh & R. F. Moore

(Eds.), Handbook of insect rearing (Vol. 2, pp. 189–196). New York: Elsevier.

Leppla, N. C., & Ashley, T. R. (Eds.). (1978). Facilities for insect research and production (U.S. Department of Agriculture Tech. Bulletin 1576). Washington, DC: U.S. Government Printing Office.

Leppla, N. C., & Turner, W. K. (1975). Carbon dioxide output and mating in adult cabbage loopers exposed to discrete light regimes. Journal of Insect Physiology, 27, 1233–1236.

Leppla, N. C., Vail, P. V., & Rye, J. R. (1984). Mass rearing the cabbage looper, Trichoplusia ni. In E. G. King & N. C. Leppla (Eds.), Advances and challenges in insect rearing (pp. 248–253). Washington, DC: U.S. Government Printing Office.

Letardi, A., & Caffarelli, V. (1989). Use of a semi-artificial liquid diet for rearing larvae of Chrysoperla carnea (Stephens) (Planipennia, Chrysopidae). Redia, 72, 195–203 (in Italian with Italian and English summaries).

Letardi, A., & Caffarelli, V. (1990). Influence of the utilization of liquid artificial larval diet for rearing Chrysoperla carnea (Steph.) (Planipennia, Chrysopidae). Redia, 73, 79–88 (in Italian with Italian and English summaries).

Li, L. Y. (1992). In vitro rearing of parasitoids of insect pests in China. Korean Journal of Applied Entomology, 31, 241–246.

Li, L. Y., Liu, W. H., Chen, C. S., Han, S. T., Shin, J. C., & Du, H. S. (1988). In vitro rearing of Trichogramma spp. and Anastatus sp. in artificial eggs and the methods of mass production. Colloques de l'INRA, 43, 339–352.

Li, S. Y., Henderson, D. E., & Myers, J. H. (1994). Selection of suitable Trichogramma species for potential control of the blackheaded fireworm infesting cranberries. Biological Control, 4, 244–248.

Liu, J. F., Liu, Z. C., Wang, C. X., Yang, W. H., & Li, D. S. (1995). Study on rearing of Trichogramma minutum on artificial host eggs. Colloques de l'INRA, 73, 161–162.

Liu, Z. C., Liu, J. F., Wang, C. X., Yang, W. H., & Li, D. S. (1995). Mechanized production of artificial host egg for the mass-rearing of parasitic wasps. Colloques de l'INRA, 73, 163–164.

Liu, Z. C., Wang, Z. Y., Sun, Y. R., Liu, J. F., & Yang, W. H. (1988a). Studies on culturing Anastatus sp. a parasitoid of litchi stink bug, with artificial host eggs. Colloques de l'INRA, 43, 353–360.

Liu, Z. C., Sun, Y. R., Yang, W. H., Liu, J. F., Wang, Z. Y., & Wang, C. X. (1988b). Some new improvements in the mass rearing of egg parasitoids on artificial host eggs. Chinese Journal of Biological Control, 4, 145–148 [in Chinese with Chinese and English summaries.]

Louis, C., Pintureau, B., & Chapelle, L. (1993). Research on the origin of unisexuality—Thermotherapy cures both Rickettsia and thelytokous parthenogenesis in a Trichogramma species (Hymenoptera: Trichogrammatidae). Comptes Rendus de l'Academie des Sciences Serie III Sciences de la Vie, 316, 27–33.

Luck, R. F., Stouthamer, R., & Nunney, L. P. (1993). Sex determination and sex ratio patterns in parasitic Hymenoptera. In D. L. Wrensch and M. A. Ebbert (Eds.), Evolution and diversity of sex ratio in insects and mites (pp. 442–476). New York: Chapman & Hall.

Lugo, G., Cipolla, C., Bonfiglioli, R., Sassi, C., Maini, S., & Cancellieri M. P. (1994). A new risk of occupational disease: Allergic asthma and rhinoconjunctivitis in persons working with beneficial arthropods. International Archives of Occupational and Environmental Health, 65, 291–294.

Ma, H. Y. (1988). Studies on long-term storage of hosts for propagating Trichogramma. Colloques de l'INRA, No 43, 369–371.

Magrini, E. A., & M. Botelho, P. S. (1991). Influence of the food of the host Diatraea saccharalis (Lepidoptera, Pyralidae) on the egg parasite Trichogramma galloi (Hymenoptera, Trichogrammatidae). Anais da Sociedade Entomologica do Brazil, 20, 99–108.

Maier, C. T. (1993). Inoculative release and establishment of Holcothorax testaceipes (Hymenoptera: Encyrtidae), a Palaearctic agent for the biological control of Phyllonorycter spp. (Lepidoptera: Gracillariidae)

in Connecticut apple orchards. Journal of Economic Entomology, 86, 1069–1077.

Maini, S., & Nicoli, G. (1990): Edovum puttleri (Hym.: Eulophidae): biological activity and responses to normal and frozen eggs of Leptinotarsa decemlineata (Col.: Chrysomelidae). Entomophaga, 35, 185–193.

Mani, M., & Krishnamoorthy, A. (1992). Influence of constant temperatures on the developmental rate, progeny production, sex ratio and adult longevity of the grape mealybug parasitoid, Anagyrus dactylopii (How.) (Hymenoptera: Encyrtidae). Insect Science Applications, 13, 697–703.

Mannion, C. M., Carpenter, J. E., Wiseman, B. R., & Gross, H. R. (1994). Host corn earworm (Lepidoptera: Noctuidae) reared on meridic diet containing silks from a resistant corn genotype on Archytas marmoratus (Diptera: Tachinidae) and Ichneumon promissorius (Hymenoptera: Ichneumonidae). Environmental Entomology, 23, 837–845.

Mansour, F., Cohen, H., & Shain, Z. (1993). Integrated mite management in apples in Israel: Augmentation of a beneficial mite and sensitivity of tetranychid and phytoseiid mites to pesticides. Phytoparasitica, 21, 39–51.

Margolies, D. C. (1987). Conditions eliciting aerial dispersal behavior in banks grass mite, Oligonychus pratensis (Acari: Tetranychidae). Environmental Entomology, 16, 928–932.

Marohasy, J. (1994). Biology and host specificity of Weiseana barkeri (Col.: Chrysomelidae): A biological control agent for Acacia nilotica (Mimosaceae). Entomophaga, 39, 335–340.

Martel, P., Svec, H. J., & Harris, C. R. (1975). Mass rearing of the carrot weevil, Listronotus oregonensis (Coleoptera: Curculionidae), under controlled environmental conditions. Canadian Entomologist, 107, 95–98.

Mason, C. E., Jones, R. L., & Thompson, M. M. (1991). Rearing Lydella thompsoni (Diptera: Tachinidae), a parasite of the European corn borer (Lepidoptera: Pyralidae). Annals of the Entomological Society of America, 84, 179–181.

Mason, C. E., Romig, R. F., Wendell, L. E., & Wood, L. A. (1994). Distribution and abundance of larval parasitoids of European corn borer (Lepidoptera: Pyralidae) in the East Central United States. Environmental Entomology, 23, 521–531.

Masutti, L., Battisti, A., Milani, N., & Zanata, M. (1992). First success in the in vitro rearing of Ooencyrtus pityocampae (Mercet) (Hym. Encyrtidae): Preliminary note. Redia, 75, 227–232.

Masutti, L., Battisti, A., Milani, N., Zanata, M., & Zanazzo, G. (1993). In vitro rearing of Ooencyrtus pityocampae (Hym., Encyrtidae), an egg parasitoid of Thaumetopoea pityocampa (Lep., Thaumetopoeidae). Entomophaga, 38, 327–333.

McAllister, B. F., & Werren, J. H. (1997). Hybrid origin of a B chromosome (PSR) in the parasitic wasp Nasonia vitripennis. Chromosoma (Berlin), 106, 243–253.

McAuslane, H. J. (1994). Interactions between host-plant resistance and biological control agents. In D. Rosen, F. D. Bennett, & J. L. Capinera (Eds.), Pest management in the subtropics: Biological control: A Florida perspective (pp. 681–708). Andover, UK: Intercept Ltd.

McDonald, R. C., & Kok, L. T. (1990). Post refrigeration viability of Pteromalus puparum (Hymenoptera: Pteromalidae) prepupae within host chrysalids. Journal of Entomological Science, 25, 409–413.

McDougall, S. J., & Mills, N. J. (1997). Dispersal of Trichogramma platneri Nagarkatti (Hym., Trichogrammatidae) from point-source releases in an apple orchard in California. Journal of Applied Entomology, 121, 205–209.

McFarlane, J. E. (1985). Acheta domesticus. In P. Singh & R. F. Moore (Eds.). Handbook of insect rearing (Vol. 1, pp. 427–434). New York: Elsevier.

McLaughlin, R. E. (1966). Laboratory techniques for rearing disease-free insect colonies: Elimination of Mattesia grandis McLaughlin, and Nosema sp. from colonies of boll weevils. Journal of Economic Entomology, 59, 401–404.

McMurtry, J. A., Johnson, H. G., & Badii, M. H. (1984). Experiments to determine effects of predator releases on populations of *Oligonychus punicae* (Acarina: Tetranychidae) on avocado in California. Entomophaga, 29, 11–19.

Meadow, R. J., Kelly, W. C., & Shelton, A. M. (1985). Evaluation of *Aphidoletes aphidimyza* (Dip.: Cedidomyiidae) for control of *Myzus persicae* (Hom.: Aphididae) in greenhouse and field experiments in the United States. Entomophaga, 30, 385–392.

Medrano, F. G., & Heinrichs, E. A. (1985). *Nilaparvata lugens.* In P. Singh & R. F. Moore (Eds.), Handbook of insect rearing (Vol. 1, pp. 361–372). New York: Elsevier.

Mellini, E., & Campadelli, G. (1989). Tests in the extrauterine incubation and hatching of microtype eggs of *Pseudogonia rufifrons* Wied. Bollettino dell'Istituto di Entomologia "Guido Grandi" della Università degli Studi di Bologna, 43, 105–113 (in Italian with Italian and English summaries).

Mellini, E., & Campadelli, G. (1995). Qualitative improvements in the composition of oligidic diets for the parasitoid *Exorista larvarum* (L.). Bollettino dell'Istituto di Entomologia "Guido Grandi" della Università degli Studi di Bologna, 49, 187–196.

Mellini, E., & Campadelli, G. (1996). Formulas for "inexpensive" artificial diets for the parasitoid *Exorista larvarum* (L.). Bollettino dell'Istituto di Entomologia "Guido Grandi" della Università degli Studi di Bologna, 50, 95–106.

Mellini, E., Campadelli, G., & Dindo, M. L. (1993). Artificial culture of the parasitoid *Exorista larvarum* L. (Dipt. Tachinidae) on bovine serum-based diets. Bollettino dell'Istituto di Entomologia "Guido Grandi" della Università degli Studi di Bologna, 47, 223–231.

Mellini, E., Campadelli, G., & Dindo, M. L. (1994). Artificial culture of the parasitoid *Exorista larvarum* L. (Dipt. Tachinidae) on oligidic media: Improvements of techniques. Bollettino dell'Istituto di Entomologia "Guido Grandi" della Università degli Studi di Bologna, 48, 1–10.

Mellini, E., Campadelli, G., & Dindo, M. L. (1996). Actual possibilities of mass production of the parasitoid *Exorista larvarum* (L.) (Diptera: Tachinidae) on oligidic diets. Bollettino dell'Istituto di Entomologia "Guido Grandi" della Università degli Studi di Bologna, 50, 233–241.

Mendel, Z. (1988). Effects of food, temperature, and breeding conditions on the life span of adults of three cohabitating bark beetle (Scolytidae) parasitoids (Hymenoptera). Environmental Entomology, 17, 293–298.

Messenger, P. S., & van den Bosch, R. (1971). The adaptability of introduced biological control agents. In C. B. Huffaker (Ed.), Biological control (pp. 68–92). New York: Plenum Press.

Messing, R. H., & Croft, B. A. (1991). Biosystematics of *Amblyseius andersoni* and *Amblyseius potentillae* (Acarina: Phytoseiidae): Implications for biological control. Experimental and Applied Acarology (Northwood), 10, 267–278.

Meyerdirk, D. E., Khasimuddin, S., & Bashir, M. (1988). Importation, colonization and establishment of *Anagyrus indicus* (Hym.: Encyrtidae) on *Nipaecoccus viridis* (Hom.: Pseudococcidae) in Jordan. Entomophaga, 33, 229–237.

Miller, J. C. (1995). A comparison of techniques for laboratory propagation of a South American ladybeetle, *Eriopis connexa* (Coleoptera: Coccinellidae). Biological Control, 5, 462–465.

Mills, N. J., & Nealis, V. G. (1992). European field collections and Canadian releases of *Ceranthia samarensis* (Dipt.: Tachinidae), a parasitoid of the gypsy moth. Entomophaga, 37, 181–191.

Milward-De-Azevedo, E. M. V., & Cardoso, D. (1996). Frozen pupae of *Chrysomya megacephala* (Diptera: Calliphoridae): Preliminary study as breeding substrate for *Nasonia vitripennis* (Walker) (Hymenoptera: Pteromalidae). Arquivos de Biologia e Tecnologia (Curitiba), 39, 89–98.

Min, K. T., & Benzer, S. (1997). *Wolbachia,* normally a symbiont of *Drosophila,* can be virulent, causing degeneration and early death. Proceedings of the National Academy of Sciences, USA, 94, 10792–10796.

Mirzalieva, K. R., & Mirzaliev, B. T. (1988). The use *Trichogramma* on tomatoes. Zashchita Rastenii, (Mosow), 11, 30–31 (in Russian).

Mishra, S., & Singh, R. (1990). Effects of parasitization by the cereal aphid parasitoid *Lysiphlebus delhiensis* (Subba Rao & Sharma) (Hymenoptera, Aphidiidae) at its varying densities. Journal of Applied Entomology, 109, 251–261.

Moore, R. F. (1985). Artificial diets: Development and improvement. In P. Singh & R. F. Moore (Eds.), Handbook of insect rearing (Vol. 1, pp. 67–83). New York: Elsevier.

Moore, R. F., & Whisnant, F. F. (1985). *Anthonomus grandis.* In P. Singh & R. F. Moore (Eds.), Handbook of insect rearing (Vol. 1, pp. 217–225). New York: Elsevier.

Moore, R. F., ODell, T. M., & Calkins, C. O. (1985). Quality assessment in laboratory-reared insects. In P. Singh & R. F. Moore (Eds.), Handbook of insect rearing (Vol. 1, pp. 107–135). New York: Elsevier.

Morales-Ramos, J. A., Greenberg, S. M., & King, E. G. (1996a). Selection of optimal physical conditions for mass propagation of *Catolaccus grandis* (Hymenoptera: Pteromalidae) aided by regression. Environmental Entomology, 25, 165–173.

Morales-Ramos, J. A., Summy, K. R., & King, E. G. (1995). Estimating parasitism by *Catolaccus grandis* (Hymenoptera: Pteromalidae) after inundative releases against the boll weevil (Coleoptera: Curculionidae). Environmental Entomology, 24, 1718–1725.

Morales-Ramos, J., Summy, K. R., & King, E. G. (1996b). ARCASIM, a model to evaluate augmentation strategies of the parasitoid *Catolaccus grandis* against boll weevil populations. Ecological Modelling, 93, 221–235.

Morgan, P. B. (1985). *Musca domestica.* In P. Singh & R. F. Moore (Eds.), Handbook of insect rearing (Vol. 2, pp. 129–134). New York: Elsevier.

Morgan, P. B. (1992). Microhymenopterous pupal parasite production for controlling muscoid flies of medical and veterinary importance. In T. E. Anderson & N. C. Leppla (Eds.), Advances in insect rearing for research and pest management (pp. 379–392). Boulder, CO: Westview Press.

Morrison, L. W., Dall'aglio-holvorcem, C. G. & Gilbert, L. E. (1997). Oviposition behavior and development of *Pseudacteon* flies (Diptera: Phoridae), parasitoids of *Solenopsis* fire ants (Hymenoptera: Formicidae). Environmental Entomology, 3, 716–724.

Morrison, R. K. (1985a). *Trichogramma* spp. In P. Singh & R. F. Moore (Eds.), Handbook of insect rearing (Vol. 1, pp. 413–417). New York: Elsevier.

Morrison, R. K. (1985b). *Chrysopa carnea.* In P. Singh & R. F. Moore (Eds.), Handbook of insect rearing (Vol. 1, pp. 419–426). New York: Elsevier.

Morrison, R. K., & King, E. G. (1977). Mass production of natural enemies. In R. L. Ridgway & S. B. Vinson (Eds.), Biological control by augmentation of natural enemies (pp. 183–217). New York: Plenum Press.

Munstermann, L. E., & Leiser, L. B. (1985). *Toxorhynchites amboinensis.* In P. Singh & R. F. Moore (Eds.), Handbook of insect rearing (Vol. 2, pp. 157–164). New York: Elsevier.

Munstermann, L. E., & Wasmuth, L. M. (1985a). *Aedes aegypti.* In P. Singh & R. F. Moore (Eds.), Handbook of insect rearing (Vol. 2, pp. 7–14). New York: Elsevier.

Munstermann, L. E., & Wasmuth, L. M. (1985b). *Aedes triseriatus.* In P. Singh & R. F. Moore (Eds.), Handbook of insect rearing (Vol. 2, pp. 15–24). New York: Elsevier.

Myers, J. H., & Sabath, M. D. (1980). Genetic and phenotypic variability, genetic variance, and the success of establishment of insect introductions for biological control of weeds. Proceedings of the Fifth International Symposium of Biological Control of Weeds (pp. 91–102). Brisbane, Australia: Commonwealth Scientific and Industrial Research Organization, 1980.

Nagarkatti, S., Giroux, K. J., & Keeley, T. P. (1991). Rearing *Trichogramma nubilale* [Hymenoptera: Trichogrammatidae] on eggs of the tobacco hornworm, *Manduca sexta* [Lepidoptera: Sphingidae]. *Entomophaga, 36,* 443–446.

Nakamori, H., Kakinohana, H., & Yamagishi, M. (1992). Automated mass production system for fruit flies based on the melon fly *Dacus-cucurbitae* Coquillett (Diptera Tephritidae). In T. E. Anderson & N. C. Leppla (Eds.), <u>Advances in insect rearing for research and pest management</u> (pp. 441–454). Boulder, CO: Westview Press.

Nakamura, S. (1996). Inbreeding and rotational breeding of the parasitoid fly, *Exorista japonica* (Diptera: Tachinidae), for successive rearing. <u>Applied Entomology and Zoology, 31,</u> 433–441.

Navon, A. (1985). *Spodoptera littoralis.* In P. Singh & R. F. Moore (Eds.), <u>Handbook of insect rearing</u> (Vol. 2, pp. 469–475). New York: Elsevier.

Nealis, V. G., & Fraser, S. (1988). Rate of development, reproduction, and mass-rearing of *Apanteles fumiferanae* Vier. (Hymenoptera: Braconidae) under controlled conditions. <u>Canadian Entomologist, 120,</u> 197–204.

Nealis, V. G., & Bourchier, R. S. (1995). Reduced vulnerability to hyperparasitism in nondiapause strains of *Cotesia melanoscela* (Ratzeburg) (Hymenoptera: Braconidae). <u>Proceedings of the Entomological Society of Ontario, 126,</u> 29–35.

Nealis, V. G., & Quednau, F. W. (1996). Canadian field releases and overwinter survival of *Ceranthia samarensis* (Villeneuve) (Diptera: Tachinidae) for biological control of the gypsy moth, *Lymantria dispar* (L.) (Lepidoptera: Lymantriidae). <u>Proceedings of the Entomological Society of Ontario, 127,</u> 11–20.

Nechols, J. R., & Kikuchi, R. S. (1985). Host selection of the spherical mealybug (Homoptera: Pseudococcidae) by *Anagyrus indicus* (Hymenoptera: Encyrtidae): Influence of host stage on parasitoid oviposition, development, sex ratio, and survival. <u>Environmental Entomology, 14,</u> 32–37.

Nettles, W. C., Jr. (1986). Effects of soy flour, bovine serum albumin, and three amino acid mixtures on growth and development of *Eucelatoria bryani* (Diptera: Tachinidae) reared on artificial diets. <u>Environmental Entomology, 15,</u> 1111–1115.

Nettles, W. C., Jr. (1987). *Eucelatoria bryani* (Diptera: Tachinidae): Effect on fecundity of feeding on hosts. <u>Environmental Entomology, 16,</u> 437–440.

Nettles, W. C., Jr., Morrison, R. K., Xie, Z.-N., Ball, D., Shenkir, C. A., & Vinson, S. B. (1982). Synergistic action of potassium chloride and magnesium sulfate on parasitoid wasp oviposition. <u>Science, 218,</u> 164–166.

Neuenschwander, P., & Haug, T. (1992). New technologies for rearing *Epidinocarsis lopezi* (Hym.: Encyrtidae), a biological control agent against the cassava mealybug *Phenacoccus manihoti* (Hom.: Pseudococcidae). In T. E. Anderson & N. C. Leppla (Eds.), <u>Advances in insect rearing for research and pest management</u> (pp. 353–377). Boulder, CO: Westview Press.

Neuenschwander, P., Haug, T., Ajounu, O., Davis, H., Akinwumi, B., & Madojemu, E. (1989). Quality requirements in natural enemies used for inoculative release: Practical experience from a successful biological control program. <u>Journal of Applied Entomology, 108,</u> 409–420.

Nguyen, R. W., Whitcomb, W. H., & Murphy, M. (1976). Culturing of *Chrysopa rufilabris* (Neuroptera: Chrysopidae). <u>Florida Entomologist, 59,</u> 21–26.

Nickle, D. A., & Hagstrum, D. W. (1981). Provisioning with preparalyzed hosts to improve parasite effectiveness: A pest management strategy for stored commodities. <u>Environmental Entomology, 10,</u> 560–564.

Nihoul, P. (1993). Controlling glasshouse climate influences the interaction between tomato glandular trichome, spider mite and predatory mite. <u>Crop Protection, 12,</u> 443–447.

Niijima, K. (1995). Nutritional studies and artificial diet of *Chrysopa septempunctata.* <u>Bulletin of the Faculty of Agriculture Tamagawa University, 0(35),</u> 129–157.

Nishida, R., & Fukami, H. (1989). Host plant iridoid-based chemical defense of an aphid, *Acyrthosiphon nipponicus,* against ladybird beetles. <u>Journal of Chemical Ecology, 15,</u> 1837–1846.

Nordlund, D. A., & Correa, J. A. (1995a). Description of green lacewing adult feeding and oviposition units and a sodium hypochlorite-based egg harvesting system. <u>Southwestern Entomology, 20,</u> 293–301.

Nordlund, D. A., & Correa, J. A. (1995b). Improvements in the production system for green lacewings: An adult feeding and oviposition unit and hot wire egg harvesting system. <u>Biological Control, 5,</u> 179–188.

Nordlund, D. A., & Morrison, R. K. (1992). Mass rearing of *Chrysoperla* species. In T. E. Anderson & N. C. Leppla (Eds.), <u>Advances in insect rearing for research and pest management</u> (pp. 427–439). Boulder, CO: Westview Press.

Nordlund, D. A., Wu, Z. X., & Greenberg, S. M. (1997). *In vitro* rearing of *Trichogramma minutum* Riley (Hymenoptera: Trichogrammatidae) for ten generations, with quality assessment comparisons of *in vitro* and *in vivo* reared adults. <u>Biological Control, 9,</u> 201–207.

Norris, D. M., & Chu, H. M. (1985). *Xyleborus ferrugineus.* In P. Singh & R. F. Moore (Eds.), <u>Handbook of insect rearing</u> (Vol. 1, pp. 303–315). New York: Elsevier.

Obrycki, J. J., & Tauber, M. J. (1984). Natural enemy activity on glandular pubescent potato plants in the greenhouse: An unreliable predictor of effects in the field. <u>Environmental Entomology, 13,</u> 679–683.

Obrycki, J. J., Lewis, L. C., & Orr, D. B. (1997). Augmentative releases of entomophagous species in annual cropping systems. <u>Biological Control, 10,</u> 30–36.

Obrycki, J. J., Tauber, M. J., Tauber, C. A., & Gollands, B. (1985). *Edovum puttleri* (Hymenoptera: Eulophidae), an exotic egg parasitoid of the Colorado potato beetle (Coleoptera: Chrysomelidae): Responses to temperate zone conditions and resistant potato plants. <u>Environmental Entomology, 14,</u> 48–54.

Obrycki, J. J., Tauber, M. J., Tauber, C. A., & Gollands, B. (1987). Developmental responses of the Mexican biotype of *Edovum puttleri* (Hymenoptera: Eulophidae) to temperature and photoperiod. <u>Environmental Entomology, 16,</u> 1319–1323.

ODell, T. M. (1992). Straggling in gypsy moth production strains a problem analysis for developing research priorities. In T. E. Anderson & N. C. Leppla (Eds.), <u>Advances in insect rearing for research and pest management</u> (pp. 325–350). Boulder, CO: Westview Press.

ODell, T. M., Butt, C. A., & Bridgeforth, A. W. (1985). *Lymantria dispar.* In P. Singh & R. F. Moore (Eds.), <u>Handbook of insect rearing</u> (Vol. 2, pp. 355–367). New York: Elsevier.

ODell, T. M., Bell, R. A., Mastro, V. C., Tanner, J. A., & Kennedy, L. F. (1984). Production of the gypsy moth, *Lymantria dispar,* for research and biological control. In E. G. King & N. C. Leppla (Eds.), <u>Advances and challenges in insect rearing</u> (pp. 156–166). Washington, DC: U.S. Government Printing Office.

Omwega, C. O., & Overholt, W. A. (1996). Genetic changes occurring during laboratory rearing of *Cotesia flavipes* Cameron (Hymenoptera: Braconidae), an imported parasitoid for the control of gramineous stem borers in Africa. <u>African Entomology, 4,</u> 231–237.

Omwega, C. O., Overholt, W. A., Mbapila, J. C., & Kimani-Njogu, S. W. (1997). Establishment and dispersal of *Cotesia flavipes* (Cameron) (Hymenoptera: Braconidae), an exotic endoparasitoid of *Chilo partellus* (Swinhoe) (Lepidoptera: Pyralidae) in northern Tanzania. <u>African Entomology, 5,</u> 71–75.

Onyango, F. O., & Ochieng'-Odero, J. P. R. (1994). Continuous rearing of the maize stem borer *Busseola fusca* on an artificial diet. <u>Entomologia Experimentalis et Applicata, 73,</u> 139–144.

Orr, D. B., & Boethel, D. J. (1985). Comparative development of *Copidosoma truncatellum* (Hymenoptera: Encyrtidae) and its host, *Pseudoplusia includens* (Lepidoptera: Noctuidae), on resistant and susceptible soybean genotypes. <u>Environmental Entomology, 14,</u> 612–616.

Orr, D. B., Boethel, D. J., & Walker, W. A. (1985). Biology of *Telenomus chloropus* from eggs of *Nezara viridula* reared on resistant and susceptible soybean genotypes. <u>Canadian Entomologist, 117,</u> 1137–1142.

Ortegon-E., J., Torres-N., C., Luque, E., & Siabatto, A. (1988). Study on the longevity, habits, progeny and preliminary evaluation of *Meteorus laphygmae* (Viereck), parasite of *Spodoptera* sp. <u>Revista Colombiana</u>

Entomologia, 14, 7–12 (in Spanish with Spanish and English summaries).

Osakabe, M., Inoue, K., & Ashihara, W. (1988). A mass rearing method for *Phytoseiulus persimilis* Athias-Henriot (Acarina: Phytoseiidae). II. Prey densities and prey-predator ratios for efficient propagation of the phytoseiid mite. Bulletin of the Fruit Tree Research Station, E (Akitsu), Japan, 7, 59–70 (in Japanese with Japanese and English summaries).

Ouyang, Y., Grafton-Cardwell, E. E., & Bugg, R. L. (1992). Effects of various pollens on development, survivorship, and reproduction of *Euseius tularensis* (Acari: Phytoseiidae). Environmental Entomology, 21, 1371–1376.

Overmeer, W. P. J., Nelis, H. J. C. F., DeLeenheer, A. P., Calis, J. N. M., & Veerman, A. (1989). Effect of diet on the photoperiodic induction of diapause in three species of predatory mite, *Amblyseius potentillae, Amblyseius cucumeris* and *Typhlodromus pyri.* Experimental and Applied Acarology (Northwood), 7, 281–288.

Own, O. S., & Brooks, W. M. (1986). Interactions of the parasite *Pediobius foveolatus* (Hymenoptera: Eulophidae) with two *Nosema* spp. (Microsporida: Nosematidae) of the Mexican bean beetle (Coleoptera: Coccinellidae). Environmental Entomology, 15, 32–39.

Palmer, D. J. (1985). *Brachymeria intermedia.* In P. Singh & R. F. Moore (Eds.), Handbook of insect rearing (Vol. 1, pp. 383–393). New York: Elsevier.

Papacek, D. F., & Smith, D. (1985). *Aphytis lingnanensis.* In P. Singh & R. F. Moore (Eds.), Handbook of insect rearing (Vol. 1, pp. 373–381). New York: Elsevier.

Parajulee, M. N., & Phillips, T. W. (1993). Effects of prey species on development and reproduction of the predator *Lyctocoris campestris* (Heteroptera: Anthocoridae). Environmental Entomology, 22, 1034–1042.

Parker, F. D., & Pinnell, R. E. (1972). Further studies of the biological control of *Pieris rapae* using supplemental host and parasite releases. Environmental Entomology, 1, 150–157.

Parr, W. J. (1972). Biological control of glasshouse pests. Journal of the Royal Agricultural Society of England, 133, 48.

Parra, J. R. P., & Consoli, F. L. (1992). *In vitro* rearing of *Trichogramma pretiosum* Riley, 1879. Ciencia. Cultura (Sao Paulo), 44, 407–409.

Parrott, W. L., & Jenkins, J. N. (1992). Equipment for mechanically harvesting eggs of *Heliothis virescens* (Lepidoptera: Noctuidae). Journal of Economic Entomology, 85, 2496–2499.

Patana, R. (1985a). *Heliothis zea/Heliothis virescens.* In P. Singh & R. F. Moore (Eds.), Handbook of insect rearing (Vol. 2, pp. 329–334). New York: Elsevier.

Patana, R. (1985b). *Spodoptera exigua.* In P. Singh & R. F. Moore (Eds.). Handbook of insect rearing (Vol. 2, pp. 465–468). New York: Elsevier.

Patel, A. G., Yadav, D. N., & Patel, R. C. (1988). Effect of low temperature storage on *Campoletis chlorideae* Uchida (Hymenoptera: Ichneumonidae), an important endo-larval parasite of *Heliothis armigera* Hübner (Lepidoptera: Noctuidae). Gujarat Agricultural University Research Journal, 14, 79–80.

Pavlik, J. (1993). Variability in the host acceptance of European corn borer, *Ostrinia nubilalis* Hbn. (Lep., Pyralidae) in strains of the egg parasitoid *Trichogramma* spp. (Hym., Trichogrammatidae). Journal of Applied Entomology, 115, 77–84.

Pawson, B. M., Petersen, J. J., & Gold, R. E. (1993). Utilization of freeze-killed house fly and stable fly (Diptera: Muscidae) pupae by three pteromalid wasps. Journal of Entomological Science, 28, 113–119.

Penman, D. R., & Chapman, R. B. (1980). Integrated control of apple pests in New Zealand. 17. Relationships of *Amblyseius fallacis* to phytophagous mites in an apple orchard. New Zealand Journal of Zoology, 7, 281–287.

Pennacchio, F., Vinson, S. B., & Tremblay, E. (1992). Preliminary results on *in vitro* rearing of the endoparasitoid *Cardiochiles nigriceps* from egg to second instar. Entomologia Experimentalis et Applicata, 64, 209–216.

Periquet, G., Hedderwick, M. P., El Agoze, M., & Poirie, M. (1993). Sex determination in the hymenopteran *Diadromus pulchellus* (Ichneumonidae) validation of the one-locus multi-allele model. Heredity, 70, 420–427.

Perring, T. M., Farrar, C. A., & Toscano, N. C. (1988). Relationships among tomato planting date, potato aphids (Homoptera: Aphididae), and natural enemies. Journal of Economic Entomology, 81, 1107–1112.

Petersen, J. J. (1986). Augmentation of early season releases of filth fly (Diptera: Muscidae) parasites (Hymenoptera: Pteromalidae) with freeze-killed hosts. Environmental Entomology, 15, 590–593.

Peterson, J. J., & Currey, D. M. (1996). Timing of releases of gregarious *Muscidifurax raptorellus* (Hymenoptera: Pteromalidae) to control flies associated with confined beef cattle. Journal of Agricultural Entomology, 13, 55–63.

Petersen, J. J., & Matthews, J. R. (1984). Effects of freezing of host pupae on the production of progeny by the filth fly parasite, *Muscidifurax zaraptor* (Hymenoptera: Pteromalidae). Journal of Kansas Entomological Society, 56, 397–403.

Petersen, J. J., Watson, D. W., & Cawthra, J. K. (1995). Comparative effectiveness of three release rates for a pteromalid parasitoid (Hymenoptera) of house flies (diptera) in beef cattle feedlots. Biological Control, 5, 561–565.

Petersen, J. J., Watson, D. W., & Pawson, B. M. (1992). Evaluation of field propagation of *Muscidifurax zaraptor* (Hymenoptera: Pteromalidae) for control of flies associated with confined beef cattle. Journal of Economic Entomology, 85, 451–455.

Pham Binh, Q., Nguyen Van, S., & Nguyen Thi, D. (1995). Natural hosts of *Trichogramma* spp. and selection of factitious hosts for mass production. Colloques de l'INRA, 73, 165–167.

Phillips, C. B., & Baird, D. B. (1996). A morphometric method to assist in defining the South American origins of *Microctonus hyperodae* Loan (Hymenoptera: Braconidae) established in New Zealand. Biocontrol Science and Technology, 6, 189–205.

Phillips, C. B., Baird, D. B., & Goldson, S. L. (1997). South American origins of *Microctonus hyperodae* Loan (Hymenoptera: Braconidae) established in New Zealand as defined by morphometric analysis. Biocontrol Science and Technology, 7, 247–258.

Pickens, L. G., & Miller, R. W. (1978). Using frozen host pupae to increase the efficiency of a parasite-release program. Florida Entomologist, 61, 153–158.

Pickett, C. H., & Gilstrap, F. E. (1986). Inoculative releases of phytoseiids (Acari) for the biological control of spider mites (Acari: Tetranychidae) in corn. Environmental Entomology, 15, 790–794.

Pickett, C. H., Ball, J. C., Casanave, K. C., Klonsky, K. M., Jetter, K. M., & Bezark, L. G. (1996). Establishment of the ash whitefly parasitoid *Encarsia inaron* (Walker) and its economic benefit to ornamental street trees in California. Biological Control, 6, 260–272.

Pijls, J. W. A. M., Van Steenbergen, H. J., & Van Alphen, J. J. M. (1996). Asexuality cured: The relations and differences between sexual and asexual *Apoanagyrus diversicornis.* Heredity, 76, 506–513.

Pintureau, B. (1991). Selection of two characteristics in a *Trichogramma* sp.: Parasitic efficacy of the obtained strains (Hymenoptera: Trichogrammatidae). Agronomie (Paris), 11, 593–602.

Pittara, I. S., & Katsoyannos, B. I. (1992). Rearing of *Chaetorellia australis* (Dipt., Tephritidae) on an artificial diet. Entomophaga, 37, 253–257.

Poinar, G. O., Jr., & Thomas, G. M. (1978). Diagnostic manual for the identification of insect pathogens. New York: Plenum Press.

Polgar, L. A., Darvas, B., & Voelkl, W. (1995). Induction of dormancy in aphid parasitoids: Implications for enhancing their field effectiveness. Agriculture Ecosystems and Environment, 52, 19–23.

Popov, N. A., & Shijko, E. S. (1986). A parasitoid of the green peach aphid. Zashchita Rastenii, (Moscow), 10·86, 29–30 (in Russian).

Popov, N. A., Zabudskaya, I. A., & Burikson, I. G. (1987). The rearing of

Encarsia in biolaboratories in greenhouse combines. Zashchita Rastenii, (Moscow), 6·87, 33 (in Russian).

Powell, J. A., Shepard, M., & Sullivan, M. J. (1981). Use of heating degree day and physiological day equations for predicting development of the parasitoid *Trissolcus basalis.* Environmental Entomology, 10, 1008–1011.

Powell, J. E., & Lambert, L. (1984). Effects of three resistant soybean genotypes on development of *Microplitis croceipes* and leaf consumption by its *Heliothis* spp. hosts. Journal of Agricultural Entomology, 1, 169–176.

Powell, J. E., & Shepard, M. (1982). Biology of Australian and United States strains of *Trissolcus basalis,* a parasitoid of the green vegetable bug, *Nezara viridula.* Australian Journal of Ecology, 7, 181–186.

Presnail, J. K., & Hoy, M. A. (1992). Stable genetic transformation of a beneficial arthropod, *Metaseiulus occidentalis* (Acari: Phytoseiidae), by a microinjection technique. Proceedings of the National Academy of Sciences, USA, 89, 7732–7736.

Presnail, J. K., & Hoy, M. A. (1994). Transmission of injected DNA sequences to multiple eggs of *Metaseiulus occidentalis* and *Amblyseius finlandicus* (Acari: Phytoseiidae) following maternal microinjection. Experimental and Applied Acarology (Northwood), 18, 319–330.

Price, P. W., Bouton, C. E., Gross, P., McPheron, B. A., Thompson, J. N., & Weis, A. E. (1980). Interactions among three trophic levels: Influence of plants on interactions between insect herbivores and natural enemies. Annual Review of Ecology and Systematics, 11, 41–65.

Propp, G. D., & Morgan, P. B. (1983). Superparasitism of house fly, *Musca domestica* L., pupae by *Spalangia endius* Walker (Hymenoptera: Pteromalidae). Environmental Entomology, 12, 561–566.

Qiu, H.-G., Bigler, F., Van Bergeijk, K. E., Bosshart, S., & Waldburger, M. (1992). Quality comparison of *Trichogramma maidis* emerged on subsequent days. Acta Entomologica Sinica, 35, 449–455.

Rahalkar, G. W., Harwalkar, M. R., Rananavare, H. D., Tamhankar, A. J., & Shantram, K. (1985). *Rhynchophorus ferrugineus.* In P. Singh & R. F. Moore (Eds.), Handbook of insect rearing (Vol. 1, pp. 279–286). New York: Elsevier.

Ram, P., Tshernyshev, W. B., Afonina, V. M., Greenberg, S. M., & Ageeva, L. I. (1995). Studies on strains of *Trichogramma evanescens* Westwood from different regions of Eurasia. Biocontrol Science and Technology, 5, 329–338.

Ramadan, M. M., Wong, T. T. Y., & Beardsley, J. W., Jr., (1990). Insectary production of *Biosteres tryoni* (Cameron) (Hymenoptera: Braconidae), a larval parasitoid of *Ceratitis capitata* (Wiedemann) (Diptera: Tephritidae). Proceedings of the Hawaii Entomological Society (1989), 29, 41–48.

Rappaport, N., & Page, M. (1985). Rearing *Glypta fumiferanae* (Hym.: Ichneumonidae) on a multivoltine laboratory colony of the western spruce budworm (*Choristoneura occidentalis*) (Lep.: Tortricidae). Entomophaga, 30, 347–352.

Rasmy, A. H., & Ellaithy, A. Y. M., (1988). Introduction of *Phytoseiulus persimilis* for twospotted spider mite control in greenhouses in Egypt (Acari: Phytoseiidae, Tetranychidae). Entomophaga, 33, 435–438.

Raulston, J. R., & King, E. G. (1984). Rearing the tobacco budworm, *Heliothis virescens,* and the corn earworm, *Heliothis zea.* In E. G. King & N. C. Leppla (Eds.), Advances and challenges in insect rearing (pp. 167–175). Washington, DC: U.S. Government Printing Office.

Raupp, J. J., & Thorpe, K. W. (1985). Estimating total mortality caused by the parasitoid *Cotesia melanoscelus.* Gypsy Moth News, 9, 5; USDA Forest Service.

Ravensberg, W. J. (1992). Production and utilization of natural enemies in western European glasshouse crops. In T. E. Anderson & N. C. Leppla (Eds.), Advances in insect rearing for research and pest management (pp. 465–487). Boulder, CO: Westview Press.

Reed, D. K., & Tromley, N. J. (1985a). *Cydia pomonella.* In P. Singh & R. F. Moore (Eds.), Handbook of insect rearing (Vol. 2, pp. 249–256). New York: Elsevier.

Reed, D. K., & Tromley, N. J. (1985b). *Synanthedon pictipes.* In P. Singh & R. F. Moore (Eds.), Handbook of insect rearing (Vol. 2, pp. 477–482). New York: Elsevier.

Reed, K. M., & Werren, J. H. (1995). Induction of paternal genome loss by the paternal-sex-ratio chromosome and cytoplasmic incompatibility bacteria (*Wolbachia*): A comparative study of early embryonic events. Molecular Reproduction and Development, 40, 408–418.

Reeves, B. G. (1975). Design and evaluation of facilities and equipment for mass production and field release of an insect parasite and an insect predator. Unpublished doctoral dissertation, Texas A & M University, College Station.

Reinecke, J. P., & Sikorowski, P. P. (1989). Residual activity of four antibiotics in an artificial insect diet after three heat sterilization treatments. Southwestern Entomology, 14, 339–344.

Reis, P. R., & Alves, E. B. (1997). Biology of the predaceous mite *Euseius alatus* DeLeon (Acari: Phytoseiidae). Anais da Sociedade Entomologica do Brasil, 26, 359–363.

Remington, C. L. (1968). The population genetics of insect introduction. Annual Review of Entomology, 13, 415–427.

Reyd, G., Gery, R., Ferran, A., Iperti, G., & Brun, J. (1991). Voracity of *Hyperaspis raynevali* (Coleoptera: Coccinellidae): A predator of cassava mealybug, *Phenacoccus manihoti* (Homoptera: Pseudococcidae). Entomophaga, 36, 161–172 (in French with French and English summaries).

Reznik, S. Y., & Umarova, T. Y. (1990). The influence of host's age on the selectivity of parasitism and fecundity of *Trichogramma.* Entomophaga, 35, 31–38.

Rice, M. E., & Wilde, G. E. (1989). Antibiosis effect of sorghum on the convergent lady beetle (Coleoptera: Coccinellidae), a third-trophic level predator of the greenbug (Homoptera: Aphididae). Journal of Economic Entomology, 82, 570–573.

Riddiford, L. M. (1985). *Hyalophora cecropia.* In P. Singh & R. F. Moore (Eds.), Handbook of insect rearing (Vol. 2, pp. 335–343). New York: Elsevier.

Ridgway, R. L., King, E. G., & Carrillo, J. L. (1977). Augmentation of natural enemies for control of plant pests in the Western Hemisphere. In R. L. Ridgway & S. B. Vinson (Eds.), Biological control by augmentation of natural enemies (pp. 379–416). New York: Plenum Press.

Rigaud, T., & Rousset, F. (1996). What generates the diversity of *Wolbachia*—arthropod interactions? Biodiversity and Conservation, 5, 999–1013.

Rivers, C. F. (1991). The control of diseases in insect cultures. International Zoo Yearbook, 30, 131–137.

Roberson, J. L., & Wright, J. E. (1984). Production of boll weevils, *Anthonomus grandis grandis.* In E. G. King & N. C. Leppla (Eds.), Advances and challenges in insect rearing (pp. 188–192). Washington, DC: U.S. Government Printing Office.

Robertson, J. L. (1985a). *Choristoneura occidentalis* and *Choristoneura fumiferana.* In P. Singh & R. F. Moore (Eds.), Handbook of insect rearing (Vol. 2, pp. 227–236). New York: Elsevier.

Robertson, J. L. (1985b). *Orgyia pseudotsugata.* In P. Singh & R. F. Moore (Eds.), Handbook of insect rearing. (Vol. 2, pp. 401–406). New York: Elsevier.

Rodriguez-Del-Bosque, L. A., & Smith, J. W., Jr. (1996). Rearing and biology of *Lydella jalisco* (Diptera: Tachinidae), a parasite of *Eoreuma loftini* (Lepidoptera: Pyralidae) from Mexico. Annals of the Entomological Society of America, 89, 88–95.

Rogers, D. J., & Sullivan, M. J. (1986). Nymphal performance of *Geocoris punctipes* (Hemiptera: Lygaeidae) on pest-resistant soybeans. Environmental Entomology, 15, 1032–1036.

Rogers, D. J., & Sullivan, M. J. (1991). Development of *Brachymeria ovata* [Hymenoptera: Chalcididae] in three noctuid hosts reared on artificial diet and insect-resistant and susceptible soybean genotypes. Entomophaga, 36, 19–28.

Rojas, M. G., Morales-Ramos, J. A., & King, E. G. (1996). *In vitro* rearing of the boll weevil (Coleoptera: Curculionidae) ectoparasitoid *Catolaccus grandis* (Hymenoptera: Pteromalidae) on meridic diets. Journal Economic Entomology, 89, 1095–1104.

Rojas, M. G., Vinson, S. B., & Williams, H. J. (1995). Supplemental feeding increases the utilization of a factitious host for rearing *Bracon thurberiphagae* Muesebeck (Hymenoptera: Braconidae) a parasitoid of *Anthonomus grandis* Boheman (Coleoptera: Curculionidae). Biological Control, 5, 591–597.

Rosen, D., & DeBach, P. (1979). Species of Aphytis of the world (pp. 17–48). The Hague: W. Junk.

Roskam, M. M., Wessels, G. H., & Van Lenteren, J. C. (1996). Flight capacity and body size as indicators of quality for the whitefly parasitoid *Encarsia formosa*. In M. J. Sommeijer and P. J. Francke (Eds.), Proceedings of the Section Experimental and Applied Entomology of the Netherlands Entomological Society (N.E.V.) (Vol. 7, pp. 159–164). Seventh Meeting, Utrecht, Netherlands, December 15, 1995. Amsterdam, Netherlands: Netherlands Entomological Society.

Rossler, Y., & DeBach, P. (1972). The biosystematic relations between a thelytokous and an arrhenotokous form of *Aphytis mytilaspidis* (LeBaron) (Hymenoptera: Aphelinidae). I. The reproductive relations. Entomophaga, 17, 391–423.

Roth, S., Knorr, C., & Lindroth, R. L. (1997). Dietary phenolics affects performance of the gypsy moth (Lepidoptera: Lymantriidae) and its parasitoid *Cotesia melanoscela* (Hymenoptera: Braconidae). Environmental Entomology, 26, 668–671.

Rotheray, G. E., Barbosa, P., & Martinat, P. (1984). Host influences on life history traits and oviposition behavior of *Brachymeria intermedia* (Nees) (Hymenoptera: Chalcididae). Environmental Entomology, 13, 243–247.

Rotundo, G., Cavalloro, R., & Tremblay, E. (1988). *In vitro* rearing of *Lysiphlebus fabarum* [Hymenoptera: Braconidae]. Entomophaga, 33, 261–268.

Roush, R. T. (1990) Genetic variation in natural enemies: Critical issues for colonization in biological control. In M. Mackauer, L. E. Ehler, & J. Roland (Eds), Critical issues in biological control (pp. 263–288). XVIII International Congress of Entomology, Vancouver, British Columbia, Canada, July 3–9, 1988, Publishers. New York: VCH

Roush, R. T., & Hopper, K. R. (1995). Use of single family lines to preserve genetic variation in laboratory colonies. Annals of the Entomological Society of America, 88, 713–717.

Ruberson, J. R., Tauber, M. J., & Tauber, C. A. (1989a). Intraspecific variability in hymenopteran parasitoids: Comparative studies of two biotypes of the egg parasitoid *Edovum puttleri* (Hymenoptera: Eulophidae). Journal of the Kansas, Entomological Society, 62, 189–202.

Ruberson, J. R., Tauber, M. J., Tauber, C. A., & Tingey, W. M. (1989b). Interactions at three trophic levels: *Edovum puttleri* Grissell (Hymenoptera: Eulophidae), the Colorado potato beetle, and insect-resistant potatoes. Canadian Entomologist, 121, 841–851.

Rubink, W. L., & Clement, S. L. (1982). Reproductive biology of *Bonnetia comta* (Fallen) (Diptera: Tachinidae), a parasitoid of the black cutworm, *Agrotis ipsilon* Hufnagel (Lepidoptera: Noctuidae). Environmental Entomology, 11, 981–985.

Rumei, X. (1991). Improvements of the plant-pest-parasitoid (PPP) model and its application on whitefly *Encarsia* population dynamics under different release methods. Journal of Applied Entomology, 112, 274–287.

Rutz, D. A., & Axtell, R. C. (1981). House fly (*Musca domestica*) control in broiler-breeder poultry houses by pupal parasites (Hymenoptera: Pteromalidae): Indigenous parasite species and releases of *Muscidifurax raptor*. Environmental Entomology, 10, 343–345.

Saavedra, J. L. D., Zanuncio, J. C., Della Lucia, T. M. C., & Reis, F. P. (1992). Effect of artificial diet on fecundity and fertility of the predator *Podisus connexivus* Bergroth, 1891 (Hemiptera: Pentatomidae). Anais da Sociedade Entomologica do Brasil, 21, 69–76.

Saavedra, J. L. D., Zanuncio, J. C., Guedes, R. N. C., & De Clercq, P. (1996). Continuous rearing of *Podisus nigrispinus* (Dallas) (Heteroptera: Pentatomidae) on an artificial diet. Mededelingen Faculteit Landbouwkundige en Toegepaste Biologische Wetenschappen Universiteit Gent, 61, 767–772.

Saavedra, J. L. D., Zanuncio, J. C., Vilela, E. F., Sediyama, C. S., & De Clercq, P. (1995). Development of *Podisus nigrispinus* (Dallas) (Heteroptera: Pentatomidae) on meat-based artificial diets. Mededelingen Faculteit Landbouwkundige en Toegepaste Biologische Wetenschappen Universiteit Gent, 60, 683–688.

Sailer, R. I. (1976). Organizational responsibility for biocontrol research and development inputs. In Organized programs to utilize natural enemies of pests in Canada, Mexico, United States (APHIS 81-28), 154, pp. 10–19. Papers presented at the Joint Symposium, National Meeting of the Entomological Society of America, November 30–December 4, 1975, New Orleans, LA, Washington, DC: PPQ-APHIS-U.S. Department of Agriculture.

Sajap, A. S., & Lewis, L. C. (1988). Effects of the microsporidium *Nosema pyrausta* (Microsporida: Nosematidae) on the egg parasitoid, *Trichogramma nubilale* (Hymenoptera: Trichogrammatidae). Journal Invertebrate Pathology, 52, 294–300.

Sajap, A. S., & Lewis, L. C. (1989). Impact of *Nosema pyrausta* (Microsporida: Nosematidae) on a predator, *Chrysoperla carnea* (Neuroptera: Chrysopidae). Environmental Entomology 18, 172–176.

Samways, M. J., & Wilson, S. J. (1988). Aspects of the feeding behavior of *Chilocorus nigritus* (F.) (Coleoptera, Coccinellidae) relative to its effectiveness as a biocontrol agent. Journal of Applied Entomology, 106, 177–182.

Schaefer, P. W., Dysart, R. J., Flanders, R. V., Burger, T. L., & Ikebe, K. (1983). Mexican bean beetle (Coleoptera: Coccinellidae) larval parasite *Pediobius foveolatus* (Hymenoptera: Eulophidae) from Japan: Field release in the United States. Environmental Entomology, 12, 852–854.

Schanderl, H., Ferran, A. & Garcia, V. (1988). Rearing two coccinellids, *Harmonia axyridis* and *Semiadalia undecimnotata* on eggs of *Anagasta kuehniella* killed by exposure to UV radiation. Entomologia Experimentalis et Applicata, 49, 235–244.

Schilthuizen, M., & Stouthamer, R. (1997). Horizontal transmission of parthenogenesis-inducing microbes in *Trichogramma* wasps. Proceedings of the Royal Society of London, Series B, 264, 361–366.

Schlinger, E. I., & Hall, J. C. (1959). A synopsis of the biologies of three imported parasites of the spotted alfalfa aphid. Journal of Economic Entomology, 52, 154–157.

Sengonca, C., & Gerlach, S. (1984). Einfluss der Blattoberfläche auf die Wirksamkeit des Räuberischen Thrips, *Scolothrips longicornis* (Thysan.: Thripidae). Entomophaga, 29, 55–61.

Sengonca, C., & Loechte, C. (1997). Development of a spray and atomizer technique for applying eggs of *Chrysoperla carnea* (Stephens) in the field for biological control of aphids. Zeitschrift für Pflanzenkrankheiten und Pflanzenschutz, 104, 214–221.

Sengonca, C., & Schade, M. (1991). Sterilization of grapevine moth eggs by UV rays and their parasitization suitability for *Trichogramma semblidis* (Auriv.) (Hymenoptera, Trichogrammatidae). Journal of Applied Entomology, 111, 321–326 (in German with German and English summaries).

Sengonca, C., Uygun, N., Kersting, U., & Ulusoy, M. R. (1993). Successful colonization of *Eretmocerus debachi* (Hym.: Aphelinidae) in the Eastern Mediterranean citrus region of Turkey. Entomophaga, 38, 383–390.

Shaffer, P. L. (1983). Prediction of variation in development period of insects and mites reared at constant temperatures. Environmental Entomology, 12, 1012–1019.

Shapiro, M. (1984). Micro-organisms as contaminants and pathogens in insect rearing. In E. G. King & N. C. Leppla (Eds.), Advances and challenges in insect rearing (pp. 130–142). Washington, DC: U.S. Government Printing Office.

Shepherd, R. C. H. (1996). Releases of insects as biological control agents: Their timing and stadium for release, with reference to the Paterson's curse leaf-mining moth, *Dialectica scalariella* (Lepidoptera: Gracillariidae). In E. S. Delfosse and R. R. Scott (Eds.), Biological control of weeds (pp. 665–673). VIII International Symposium, Canterbury, New Zealand, February 2–7, 1992, East Melbourne, Victoria, Australia: CSIRO Publications.

Shih, C.-I. T., Chang, H. Y., Hsu, P. H., & Hwang, Y. F. (1993). Responses of *Amblyseius ovalis* (Evans) (Acarina: Phytoseiidae) to natural food resources and two artificial diets. Experimental and Applied Acarology (Northwood), 17, 503–519.

Shijko, E. S. (1989). Rearing and application of the peach aphid parasite, *Aphidius matricariae* (Hymenoptera, Aphidiidae). Acta Entomologica Fennica, 53, 53–56.

Shore, T. L. (1990). Recommendations for sampling and extracting the eggs of the western hemlock looper, *Lambdina fiscellaria lugubrosa* (Lepidoptera: Geometridae). Journal of the Entomological Society of British Columbia, 87, 30–35.

Shukla, A. N., & Tripathi, C. P. M. (1993). Effect of food plants on the offspring sex ratio of *Diaeretiella rapae* (Hymenoptera: Aphidiidae), a parasitoid of *Lipaphis erysimi* Kalt (Hemiptera: Aphididae). Biological Agriculture and Horticulture: An International Journal, 9, 137–146.

Sikorowski, P. P. (1984). Microbial contamination in insectaries. Occurrence, prevention, and control. In E.G. King & N. C. Leppla (Eds.), Advances and challenges in insect rearing (pp. 143–153). Washington, DC: U.S. Government Printing Office.

Sikorowski, P. P., & R. H. Goodwin, (1985). Contaminant control and disease recognition in laboratory colonies. In P. Singh & R. F. Moore (Eds.), Handbook of insect rearing (Vol. 1, pp. 85–105). New York: Elsevier.

Sikorowski, P. P., & Lawrence, A. M. (1994). Microbial contamination and insect rearing. American Entomologist, 40, 240–253.

Simmonds, H. W. (1944). The effect of the host fruit upon the scale *Aonidiella aurantii* Mask. in relation to its parasite *Comperiella bifasciata* How. Journal of Australian Institute of Agricultural Science, 10, 38–39.

Singh, P. (1984). Insect diets—Historical developments, recent advances, and future prospects. In E. G. King & N. C. Leppla (Eds.), Advances and challenges in insect rearing (pp. 32–44). Washington, DC: U.S. Government Printing Office.

Singh, P. (1985). Multiple-species rearing diets. In P. Singh & R. F. Moore (Eds.), Handbook of insect rearing (Vol. 1, pp. 19–44). New York: Elsevier.

Singh, P., & Ashby, M. D. (1985). Insect rearing management. In P. Singh & R. F. Moore (Eds.), Handbook of insect rearing (Vol. 1, pp. 185–215). New York: Elsevier.

Singh, P., & Clare, G. K. (1992). Insect rearing management IRM: An operating system for multiple-species rearing laboratories. In T. E. Anderson & N. C. Leppla (Eds.), Advances in insect rearing for research and pest management (pp. 135–157). Boulder, CO: Westview Press.

Singh, P., & Moore, R. F. (Eds.). (1985a). Handbook of insect rearing (Vol. 1). New York: Elsevier.

Singh, P., & Moore, R. F. (Eds.). (1985b). Handbook of insect rearing (Vol. 2) New York: Elsevier.

Singh, P., Clare, G. K., & Ashby, M. D. (1985). *Epiphyas postvittana*. In P. Singh & R. F. Moore (Eds.), Handbook of insect rearing (Vol. 2, pp. 271–282). New York: Elsevier.

Singh, R., & Agarwala, B. K. (1992). Biology, ecology and control efficiency of the aphid parasitoid *Trioxys indicus*: A review and bibliography. Biological Agriculture and Horticulture: An International Journal, 8, 271–298.

Singh, R., & Srivastava, M. (1988). Effect of cold storage of mummies of *Aphis craccivora* Koch subjected to different pre-storage temperature on per cent emergence of *Trioxys indicus* Subba Rao & Sharma. Insect Science Applications, 9, 647–657.

Sipayung, A., Desmier De Chenon, R., & Sudharto, P. (1992). Study of the Eocanthecona-Cantheconidea (Hemiptera: Pentatomidae, Asopinae) predator complex in Indonesia. Journal of Plant Protection of the Tropics, 9(6), 85–103.

Slovak, M. (1987). Cannibalism of cabbage moth caterpillars in relation to parasitization by ichneumonid *Exetastes cinctipes* Retz. Biologia (Bratislava), 42, 955–958.

Smith, C. S., & Wilson, C. G. (1995). Effect of an artificial diet on *Carmenta mimosa* Eichlin and Passoa (Lepidoptera: Sesiidae), a biological control agent for *Mimosa pigra* L. in Australia. Journal of the Australian Entomological Society, 34, 219–220.

Smith, S. M. (1988). Pattern of attack on spruce budworm egg masses by *Trichogramma minutum* (Hymenoptera: Trichogrammatidae) released in forest stands. Environmental Entomology, 17, 1009–1015.

Smith, S. M. (1996). Biological control with *Trichogramma*: Advances, successes, and potential of their use. Annual Review of Entomology, 41, 375–406.

Smith, S. M., & Wallace, D. R. (1990). Inundative release of the egg parasitoid, *Trichogramma minutum* (Hymenoptera: Trichogrammatidae), against forest insect pests such as the spruce budworm, *Choristoneura fumiferana* (Lepidoptera: Tortricidae): The Ontario Project 1882–1986: 3.2. Ground systems for releasing *Trichogramma minutum* Riley in plantation forests. Memoirs of the Entomological Society of Canada, no. 153, 31–37.

Smith, S. M., Wallace, D. R., Laing, J. E., Eden, G. M., & Nicholson, S. A. (1990). Inundative release of the egg parasitoid, *Trichogramma minutum* (Hymenoptera: Trichogrammatidae), against forest insect pests such as the spruce budworm, *Choristoneura fumiferana* (Lepidoptera: Tortricidae): The Ontario Project 1982–1986: 3.4. Deposit and distribution of *Trichogramma minutum* Riley following aerial release. Memoirs of the Entomological Society of Canada, no. 153, 45–55.

Soares, G. G. J. (1992). Problems with entomopathogens in insect rearing. In T. E. Anderson & N. C. Leppla (Eds.), Advances in insect rearing for research and pest management (pp. 289–322). Boulder, CO: Westview Press.

Sorati, M., Newman, M., & Hoffmann, A. A. (1996). Inbreeding and incompatibility in *Trichogramma* nr. *brassicae*: Evidence and implications for quality control. Entomologia Experimentalis et Applicata, 78, 283–290.

Sparks, A. N., & Harrell, E. A. (1976). Corn earworm rearing mechanization (USDA Tech. Bulletin 1554). Washington, DC: U.S. Department of Agriculture.

Speirs, R. D. (1985). *Trogoderma* spp. In P. Singh & R. F. Moore (Eds.), Handbook of insect rearing (Vol. 1, pp. 295–301). New York: Elsevier.

Stansly, P. A. (1984). Introduction and evaluation of *Chilocorus bipustulatus* (Col.: Coccinellidae) for control of *Parlatoria blanchardi* (Hom.: Diaspididae) in date groves of Niger. Entomophaga, 29, 29–39.

Stenseth, C., & Aase, I. (1983). Use of the parasite *Encarsia formosa* [Hym.: Aphelinidae] as a part of pest management on cucumbers. Entomophaga, 28, 17–26.

Stevens, L. M., Steinhauer, A. L., & Coulson, J. R. (1975). Suppression of Mexican bean beetle on soybeans with annual inoculative releases of *Pediobius foveolatus*. Environmental Entomology, 4, 947–952.

Stewart, F. D. (1984). Mass rearing the pink bollworm, *Pectinophora gossypiella*. In E. G. King & N. C. Leppla (Eds.), Advances and challenges in insect rearing (pp. 176–187). Washington, DC: U.S. Government Printing Office.

Stiling, P. (1990). Calculating the establishment rates of parasitoids in classical biological control. American Entomologist, 36, 225–230.

Stinner, R. E. (1977). Efficacy of inundative releases. Annual Review of Entomology, 22, 515–531.

Stoltz, D. B., & Vinson, S. B. (1979). Viruses and parasitism in insects. Advances in Virus Research, 24, 125–171.

Stone, T. B., & Sims, S. R. (1992). Insect rearing and the development of bioengineered crops. In T. E. Anderson & N. C. Leppla (Eds.), Advances in insect rearing for research and pest management (pp. 33–40). Boulder, CO: Westview Press.

Stouthamer, R., & Luck, R. F. (1993). Influence of microbe-associated parthenogenesis on the fecundity of Trichogramma deion and Trichogramma pretiosum. Entomologia Experimentalis et Applicata, 67, 183–192.

Stouthamer, R., Luck, R. F., & Hamilton, W. D. (1990). Antibiotics cause parthenogenetic Trichogramma (Hymenoptera/Trichogrammatidae) to revert to sex. Proceedings of the National Academy of Sciences, USA, 87, 2424–2427.

Stouthamer, R., Luck, R. F., & Werren, J. H. (1992). Genetics of sex determination and the improvement of biological control using parasitoids. Environmental Entomology, 21, 427–435.

Stouthamer, R., Breeuwer, J. A. J., Luck, R. F., & Werren, J. H. (1993). Molecular identification of microorganisms associated with parthenogenesis. Nature (London), 361, 66–68.

Stowell, S. D., & Coppel, H. C. (1990). Mass rearing the gypsy moth pupal parasitoids Brachymeria lasus and Brachymeria intermedia (Hymenoptera: Chalcididae) for small-scale laboratory studies. Great Lakes Entomologist, 23, 5–8.

Strand, M. R., & Vinson, S. B. (1982). Stimulation of oviposition and successful rearing of Telenomus heliothidis (Hym.: Scelionidae) on nonhosts by use of a host-recognition kairomone. Entomophaga, 27, 365–370.

Strand, M. R., Vinson, S. B., Nettles, W. C. J., & Xie, Z. N. (1988). In vitro culture of the egg parasitoid Telenomus heliothidis: The role of teratocytes and medium consumption in development. Entomologia Experimentalis et Applicata, 46, 71–78.

Sujii, E. R., Fontes, E. G., Pires, C. S. S., & Teixeira, C. A. D. (1996). Application of multivariate analysis for the selection of candidates for biological control agents. Biological Control, 7, 288–292.

Sukhova, V. I. (1989). Laboratory rearing of the cabbage moth. Zashchita Rastenii, (Moscow), 4·89, 29 (in Russian).

Sumenkova, V. V., & Yazlovetskii, I. G. (1992). A simple artificial nutritional diet for the predacious bug Podisus maculiventris (Hemiptera, Pentatomidae). Zoologicheskii Zhurnal, 71, 52–57.

Takahashi, K. (1997). Use of Coccinella septempunctata brucki Mulsant as a biological agent for controlling alfalfa aphids. JARQ, 31, 101–108.

Takasu, K., & Yagi, S. (1992). In-vitro rearing of the egg parasitoid Ooencyrtus nezarae Ishii (Hymenoptera: Encyrtidae). Applied Entomology and Zoology, 27, 171–173.

Tamashiro, M., Westcot, D. M., Mitchell, W. C., & Jones, W. E. (1991). Axenic rearing of the oriental fruit fly, Dacus dorsalis Hendel (Diptera: Tephritidae). Proceedings of the Hawaiian Entomological Society, (1990) 30, 113–120.

Tanigoshi, L. K., & Griffiths, H. J. (1982). A new look at biological control of citrus thrips. Citrograph, 67, 157–158.

Tanigoshi, L. K., Nishio-Wong, J. Y., & Fargerlund, J. (1983). Greenhouse and laboratory rearing studies of Euseius hibisci (Chant) (Acarina: Phytoseiidae), a natural enemy of the citrus thrips, Scirtothrips citri (Moulton) (Thysanoptera: Thripidae). Environmental Entomology, 12, 1298–1302.

Tauber, M. J., Tauber, C. A., & Gardescu, S. (1993). Prolonged storage of Chrysoperla carnea (Neuroptera: Chrysopidae). Environmental Entomology, 22, 843–848.

Tauber, M. J., Albuquerque, G. S., & Tauber, C. A. (1997a). Storage of nondiapausing Chrysoperla externa adults: Influence on survival and reproduction. Biological Control, 10, 69–72.

Tauber, M. J., Tauber, C. A., & Lopez-Arroyo, J. I. (1997b). Life-history variation in Chrysoperla carnea: Implications for rearing and storing a Mexican population. Biological Control, 8, 185–190.

Thompson, S. N. (1986). Nutrition and in vivo culture of insect parasitoids. Annual Review of Entomology, 31, 197–219.

Thorpe, K. W., & Dively, G. P. (1985). Effects of arena size on laboratory evaluations of the egg parasitoids Trichogramma minutum, T. pretiosum, and T. exiguum (Hymenoptera: Trichogrammatidae). Environmental Entomology, 14, 762–767.

Tietze, F. (1994). Interaction between the entomofauna of agro-ecosystems and neighboring, close to nature ecosystem structures. Berichte ueber Landwirtschaft Sonderheft, 136–150.

Tiitanen, K. (1988). Utilization of diapause in mass production of Aphidoletes aphidimyza (Rond.) (Dipt., Cecidomyiidae). Annales Agriculturae Fenniae, 27, 339–343.

Tillman, P. G., & Lambert, L. (1995). Influence of soybean pubescence on incidence of the corn earworm and the parasitoid, Microplitis croceipes. Southwestern Entomology, 20, 181–185.

Tillman, P. G., Laster, M. L., & Powell, J. E. (1993). Development of the endoparasitoids Microplitis croceipes, Microplitis demolitor, and Cotesia kazak (Hymenoptera: Braconidae) on Heliocoverpa zea and Heliocoverpa armigera (Lepidoptera: Noctuidae). Journal of Economic Entomology, 86, 360–362.

Toba, H. H., & Howell, J. F. (1991). An improved system for mass-rearing codling moths: Journal of the Entomological Society of British Columbia, 88, 22–27.

Tolman, J. H., Whistlecraft, J. W., & Harris, C. R. (1985). Delia antiqua. In P. Singh & R. F. Moore (Eds.), Handbook of insect rearing (Vol. 2, pp. 49–57). New York: Elsevier.

Tommassini, M. G., & Nicoli, G. (1995). Evaluation of Orius spp. as biological control agents of thrips pests: Initial experiments on the existence of diapause in Orius laevigatus. Mededelingen Faculteit Landbouwkundige en Toegepaste Biologische Wetenschappen Universiteit Gent, 60, 901–908.

Torgovetskij, A. V., Grinberg, S. M., & Chernyshev, V. B. (1988). A device for determining the quality of Trichogramma. Zashchita Rastenii, (Moscow), 7·88, 38 (in Russian).

Tseng, C.-T. (1990). The improved techniques for mass production of Trichogramma ostriniae. I. Egg card machine and preservation of egg cards. Chinese Journal of Entomology, 10, 101–108 (in Chinese with Chinese and English summaries).

Tsiropoulos, G. J. (1992). Feeding and dietary requirements of the tephritid fruit flies. In T. E. Anderson & N. C. Leppla (Eds.), Advances in insect rearing for research and pest management (pp. 93–118). Boulder, CO: Westview Press.

Ugur, A., Kansu, I. A., & Kedici, R. (1987). Investigations of the development of Pimpla turionellae (L.) (Hymenoptera: Ichneumonidae) in the pupae of Agrotis segetum (Denis and Schiff.) (Lepidoptera: Noctuidae) reared on various food plants. In Türkiye I. Entomoloji Kongresi Bildirileri, 13–16 Ekim 1987, Ege Üniversitesi, Bornova, Izmir. Bornova/Izmir, Turkey; Ege Üniversitesi, Atatürk Kültür Merkeze 481–490 (in Turkish with Turkish and English summaries).

Unruh, T. R., White W., Gonzalez, D., Gordh, G., & Luck, R. F. (1983). Heterozygosity and effective size in laboratory populations of Aphidius ervi (Hym.: Aphidiidae). Entomophaga, 28, 245–258.

Vail, P. V., & Cowan, D. K. (1985). Estigmene acrea. In P. Singh & R. F. Moore (Eds.), Handbook of insect rearing (Vol. 2, pp. 283–292). New York: Elsevier.

Vail, P. V., Henneberry, T. J., Kishaba, A. N., & Arakawa, K. Y. (1968). Sodium hypochlorite and formalin as antiviral agents against nuclear-polyhedrosis virus in larvae of the cabbage looper. Journal of Invertebrate Pathology, 10, 89–93.

van Bergeijk, K. E., Bigler, F., Kaashoek, N. K., & Pak, G. A. (1989). Changes in host acceptance and host suitability as an effect of rearing

Trichogramma maidis on a factitious host. Entomologia Experimentalis et Applicata, 52, 229–238.

van de Veire, M., & Vacante, V. (1984). Greenhouse whitefly control through the combined use of the colour attraction system with the parasite wasp *Encarsia formosa* (Hym.: Aphelinidae). Entomophaga, 29, 303–310.

van den Bosch, R., Messenger, P. S., & Gutierrez, A. P. (1982). An introduction to biological control. New York: Plenum Press.

van den Bosch, R., Schlinger, E. I., Dietrick, E. J., & Hall, I. M. (1959). The role of imported parasites in the biological control of the spotted alfalfa aphid in southern California. Journal of Economic Entomology, 52, 142–154.

Van Dijken, M. J., Van Alphen, J. J. M., & Van Stratum, P. (1989). Sex allocation in *Epidinocarsis lopezi:* Local mate competition. Entomologia Experimentalis et Applicata, 52, 249–256.

Van Driesche, R. G. (1993). Methods for the field colonization of new biological control agents. In R. G. Van Driesche & T. S. Bellows, Jr. (Eds.), Proceedings: Steps in classical arthropod biological control (pp. 67–86). Symposium, Baltimore, Maryland, September 30, 1990. Lanham, MD: Thomas Say Publications in Entomology: Entomological Society of America.

Van Driesche, R. G., & Bellows, T. S. (1996). Biological control. New York: Chapman & Hall.

Van Emden, H. F. (1995). Host plant-Aphidophaga interactions. Agriculture Ecosystems and Environment, 52, 3–11.

Van Houten, Y. M., Van Stratum, P., Bruin, J., & Veerman, A. (1995). Selection of non-diapause in *Amblyseius cucumeris* and *Amblyseius barkeri* and exploration of the effectiveness of selected strains for thrips control. Entomologia Experimentalis et Applicata, 77, 289–295.

Van Lenteren, J. C., Hua, L. Z., Kamerman, J. W., & Rumei, X. (1995). The parasite-host relationship between *Encarsia formosa* (Hym., Aphelinidae) and *Trialeurodes vaporariorum* (Hom., Aleyrodidae) XXVI. Leaf hairs reduce the capacity of *Encarsia* to control greenhouse whitefly on cucumber. Journal of Applied Entomology, 119, 553–559.

Van Roermund, H. J. W., Van Lenteren, J. C., & Rabbinge, R. (1997). Biological control of greenhouse whitefly with the parasitoid *Encarsia formosa* on tomato: An individual-based simulation approach. Biological Control, 9, 25–47.

Vargas, R. I., Mitchell, S., Hsu, C. L., & Walsh, W. A. (1994). Laboratory evaluation of diets of processed corncob, torula yeast, and wheat germ on four developmental stages of Mediterranean fruit fly (Diptera: Tephritidae). Journal of Economic Entomology, 87, 91–95.

Vasquez, L. A., Shelton, A. M., Hoffmann, M. P., & Roush, R. T. (1997). Laboratory evaluation of commercial trichogrammatid products for potential use against *Plutella xylostella* (L.) (Lepidoptera: Plutellidae). Biological Control, 9, 143–148.

Vogt, E. A., & Nechols, J. R. (1993). Responses of the squash bug (Hemiptera: Coreidae) and its egg parasitoid, *Gryon pennsylvanicum* (Hymenoptera: Scelionidae) to three Cucurbita cultivars. Environmental Entomology, 22, 238–245.

Volkoff, N., Vinson, S. B., Wu, Z. X., Nettles, W. C. J. (1992). *In vitro* rearing of *Trissolcus basalis* [Hym., Scelionidae], an egg parasitoid of *Nezara viridula* [Hem., Pentatomidae]. Entomophaga, 37, 141–148.

Waage, J. K. (1982). Sex ratio and population dynamics of natural enemies—some possible interactions. Annals of Applied Biology, 101, 159–167.

Waage, J. K., & Greathead, D. (Eds.). (1986). Insect parasitoids. London: Academic Press.

Waage, J. K., Carl, K. P., Mills, N. J., & Greathead, D. J. (1985). Rearing entomophagous insects. In P. Singh & R. F. Moore (Eds.), Handbook of insect rearing (Vol. 1, pp. 45–66). New York: Elsevier.

Wajnberg, E., Pizzol, J., & Babault, M. (1989). Genetic variation in progeny allocation in *Trichogramma maidis*. Entomologia Experimentalis et Applicata, 53, 177–188.

Wajnberg, E., Prévost, G., & Boulétreau, M. (1985). Genetic and epigenetic variation in *Drosophila* larvae suitability to a hymenopterous endoparasitoid. Entomophaga, 30, 187–191.

Wang, B., Ferro, D. N., & Hosmer, D. W. (1997). Importance of plant size, distribution of egg masses, and weather conditions on egg parasitism of the European corn borer, *Ostrinia nubilalis* by *Trichogramma ostriniae* in sweet corn. Entomologia Experimentalis et Applicata, 83, 337–345.

Wang, C. L., Wand, H. X., Lu, H., Gui, C. M., Jin, Z. T., Cheng, G. M. (1988). Studies on technique of long term keeping host eggs of *Trichogramma* under ultralow temperature. Colloques de l'INRA, 43, 399–401.

Werren, J. H. (1997). Biology of *Wolbachia*. Annual Review of Entomology, 42, 587–609.

Werren, J. H., & Beukeboom, L. W. (1993). Population genetics of a parasitic chromosome: Theoretical analysis of PSR in subdivided populations. American Naturalist, 142, 224–241.

Werren, J. H., Zhang, W., & Guo, L. R. (1995). Evolution and phylogeny of *Wolbachia:* Reproductive parasites of arthropods. Proceedings of the Royal Society of London, Series B, 261, 55–63.

Werren, J. H., Hurst, G. D. D., Zhang, W., Breeuwer, J. A. J., Stouthamer, R., & Majerus, M. E. N. (1994). Rickettsial relative associated with male killing in the ladybird beetle (Adalia bipunctata). Journal of Bacteriology, 176, 388–394.

Weseloh, R. M. (1996). Rearing the cannibalistic larvae of *Calosoma sycophanta* (Coleoptera: Carabidae) in groups. Journal of Entomological Science, 31, 33–38.

Wheeler, R. E. (1984). Industrial insect production for insecticide screening. In E. G. King & N. C. Leppla (Eds.), Advances and challenges in insect rearing (pp. 240–247). Washington, DC: U.S. Government Printing Office.

Whistlecraft, J. W., Tolman, J. H., & Harris, C. R. (1985a). *Delia platura*. In P. Singh & R. F. Moore (Eds.), Handbook of insect rearing (Vol. 2, pp. 59–65) New York: Elsevier.

Whistlecraft, J. W., Tolman, J. H., & Harris, C. R. (1985b). *Delia radicum*. In P. Singh & R. F. Moore (Eds.), Handbook of insect rearing (Vol. 2, pp 67–73). New York: Elsevier.

Wieber, A. M., Webb, R. E., Ridgway, R. L., Thorpe, K. W., Reardon, R. C., & Kolodny-Hirsch, D. M. (1995). Effect of seasonal placement of *Cotesia melanoscela* (Hym.: Braconidae) on its potential for effective augmentative release against *Lymantria dispar* (Lep.: Lymantriidae). Entomophaga, 40, 281–292.

Wight, D. P., Jr. (1985). *Spodoptera eridania*. In P. Singh & R. F. Moore (Eds.), Handbook of insect rearing (Vol. 2, pp.459–464). New York: Elsevier.

Willey, R. B. (1985). *Arphia conspersa*. In P. Singh & R. F. Moore (Eds.), Handbook of insect rearing (Vol. 1, pp. 435–450). New York: Elsevier.

Wissinger, S. A. (1997). Cyclic colonization in predictably ephemeral habitats: A template for biological control in annual crop systems. Biological Control 10, 4–15.

Woets, J. (1973). Integrated control in vegetables under glass in the Netherlands. Bulletin O.I.L.B./S.R.O.P., 1973/4, 26–31.

Woets, J. (1978). Development of an introduction scheme for *Encarsia formosa* Gahan (Hymenoptera: Aphelinidae) in greenhouse tomatoes to control the greenhouse whitefly, *Trialeurodes vaporariorum* (Westwood) (Homoptera: Aleyrodidae). Medelingen Faculteit Landbouwkundige, Rijksuniversiteit Gent, 43, 379–385.

Woets, J., & Van Lenteren, J. C. (1976). The parasite-host relationship between *Encarsia formosa* (Hymenoptera: Aphelinidae) and *Trialeurodes vaporariorum* (Homoptera: Aleyrodidae). Bulletin O.I.L.B./S.R.O.P., 1976/4, 151–164.

Wong, T. T. Y., & Ramadan, M. M. (1992). Mass rearing biology of larval parasitoids (Hymenoptera, Braconidae, Opiinae) of tephritid flies (Diptera, Tephritidae) in Hawaii. In T. E. Anderson & N. C. Leppla

(Eds.), Advances in insect rearing for research and pest management (pp. 405–426). Boulder, CO: Westview Press.

Wrensch, D. L., & Ebbert, M. A. (Eds.). (1993). Evolution and diversity of sex ratio in insects and mites. New York: Chapman & Hall.

Wuehrer, B. G., & Hassan, S. A. (1993). Selection of effective species/strains of Trichogramma (Hym., Trichogrammatidae) to control the diamondback moth Plutella xylostella L. (Lep., Plutellidae). Journal of Applied Entomology, 116, 80–89.

Xie, Z. N., Li, L., & Xie, Y. Q. (1989a). In vitro culture of Habrobracon hebetor (Say) (Hym: Braconidae). Chinese Journal of Biological Control, 5, 49–51 (in Chinese with Chinese and English summaries).

Xie, Z. N., Li, L. & Xie, Y. Q. (1989b). In vitro culture of the ectoparasitoid Bracon greeni Ashmead. Acta Entomologica Sinica, 32, 433–437 (in Chinese with Chinese and English summaries).

Xie, Z. N., Wu, Z. X., Nettles, W. C., Jr., Saldana, G., & Nordlund, D. A. (1997). In vitro culture of Trichogramma spp. on artificial diets containing yeast extract and ultracentrifuged chicken egg yolk but devoid of insect components. Biological Control, 8, 107–110.

Xing, J. Q., & Li, L. Y. (1990). In vitro rearing of an egg parasite, Anastatus japonicus Ashmead. Acta Entomologica Sinica, 33, 166–173 (in Chinese with Chinese and English summaries).

Yadav, R. P., & Chaudhary, J. P. (1988). Studies on combined effects of constant temperature and relative humidity on progeny production and sex-ratio of Cheiloneurus pyrillae Mani, an egg parasitoid of Pyrilla perpusilla Walker. Beiträge zur Entomologie, 38, 233–238.

Yanes, J., & Boethel, D. J. (1983). Effect of a resistant soybean genotype on the development of the soybean looper (Lepidoptera: Noctuidae) and an introduced parasitoid Microplitis demolitor Wilkinson (Hymenoptera: Braconidae). Environmental Entomology, 12, 1270–1274.

Yaninek, J. S., & Aderoba, O. T., (1986). Culture maintenance. In (IITA annual report for 1985, (pp. 70–71). Ibadan, Nigeria: International Institute of Tropical Agriculture.

Yazlovetsky, I. G. (1992). Development of artificial diets for entomophagous insects by understanding their nutrition and digestion. In T. E. Anderson & N. C. Leppla (Eds.), Advances in insect rearing for research and pest management (pp. 41–62). Boulder, Co: Westview Press.

Yoon, J. S. (1985). Drosophilidae. I. Drosophila melanogaster. In P. Singh & R. F. Moore (Eds.), Handbook of insect rearing (Vol. 2, pp. 75–84). New York: Elsevier.

Young, J. R., Hamm, J. J., Jones, R. L., Perkins, W. D., & Burton, R. L. (1976). Development and maintenance of an improved laboratory colony of corn earworm (U.S. Agricultural Research Service [Rep.] ARS-S-110) Washington DC: U.S. Department of Agriculture.

Yu, D. S., & Byers, J. R. (1994). Inundative release of Trichogramma brassicae Bezdenko (Hymenoptera: Trichogrammatidae) for control of European corn borer in sweet corn. Canadian Entomologist, 126, 291–301.

Yu, D. S., & Luck, R. F. (1988). Temperature-dependent size and development of California red scale (Homoptera: Diaspididae) and its effect on host availability for the ectoparasitoid, Aphytis melinus DeBach (Hymenoptera: Aphelinidae). Environmental Entomology, 17, 154–161.

Yu, D. S. K., Hagley, E. A. C., & Laing, J. E. (1984). Biology of Trichogramma minutum Riley collected from apples in southern Ontario. Environmental Entomology, 13, 1324–1329.

Yu, H.-S., Legner, E. F., & Sjogren, R. D. (1974). Mass release effects of Chlorohydra viridissima (Coelenterata) on field populations of Aedes nigromaculis and Culex tarsalis in Kern County, California. Entomophaga, 19, 409–420.

Yue, B., & Tsai, J. H. (1996). Development, survivorship, and reproduction of Amblyseius largoensis (Acari: Phytoseiidae) on selected plant pollens and temperatures. Environmental Entomology, 25, 488–494.

Yue, B., Childers, C. C., & Fouly, A. H. (1994). A comparison of selected plant pollens for rearing Euseius mesembrinus (Acari: Phytoseiidae). International Journal of Acarology, 20, 103–108.

Zanuncio, J. C., Saavedra Diaz, J. L., Zanuncio, T. V., & Santos, G. P. (1996). Development and reproduction of Supputius cincticeps (Heteroptera: Pentatomidae) under artificial diet for two generations. Revista de Biologia Tropical, 44, 247–251.

Zaslavski, V. A., & Umarova, T. Y. (1990). Environmental and endogenous control of diapause in Trichogramma species. Entomophaga, 35, 23–30.

Zchori-Fein, E., Geden, C. J., & Rutz, D. A. (1992a). Microsporidioses of Muscidifurax raptor (Hymenoptera: Pteromalidae) and other pteromalid parasitoids of muscoid flies. Journal of Invertebrate Pathology, 60, 292–298.

Zchori-Fein, E., Roush, R. T., & Hunter, M. S. (1992b). Male production induced by antibiotic treatment in Encarsia formosa (Hymenoptera: Aphelinidae), an asexual species. Experientia (Basel), 48, 102–105.

Zchori-Fein, E., Faktor, O., Zeidan, M., Gottlieb, Y., Czosnek, H., & Rosen, D. (1995). Parthenogenesis-inducing microorganisms in Aphytis (Hymenoptera: Aphelinidae). Insect Molecular Biology, 4, 173–178.

Zhang, N. X., Li, Y. X. (1989a). Rearing of the predacious mite, Amblyseius fallacis (German) (Acari: Phytoseiidae) with plant pollen. Chinese Journal of Biological Control, 5, 60–63 (in Chinese with Chinese and English summaries).

Zhang, N. X., & Li, Y. X. (1989b). An improved method of rearing Amblyseius fallacis (Acari: Phytoseiidae) with plant pollen. Chinese Journal of Biological Control, 5, 149–152 (in Chinese with Chinese and English summaries).

Zhang, Y., & Cossentine, J. E. (1995). Trichogramma platneri (Hym.: Trichogrammatidae): Host choices between viable and nonviable codling moth, Cydia pomonella, and three-lined leafroller, Pandemis limitata (Lep.: Tortricidae) eggs. Entomophaga, 40, 457–466.

Zhurba, G. T., Pamukchi, G. V., Popov, N. A., & Khudyakova, O. A. (1986). An improved method of rearing and using Phytoseiulus. Zashchita Rastenii, (Moscow), 10·86, 26–27 (in Russian).

Life Table Construction and Analysis for Evaluating Biological Control Agents

T. S. BELLOWS

Department of Entomology
University of California
Riverside, California

R. G. VAN DRIESCHE

Department of Entomology
University of Massachusetts
Amherst, Massachussetts

INTRODUCTION

Many approaches exist for evaluating the impact of natural enemies in biological systems. One method, presented in this chapter, is the construction and analysis of life tables. Other approaches include manipulative experiments (Chapter 9). A thorough examination of a particular system may require more than one approach to fully address questions concerning interactions among the species.

The topics covered in this chapter apply to natural enemies of all types, but much of the detail is presented with reference to insect parasitoids. The chapter is divided into five subsequent sections. We first review the types of life tables and data necessary for their construction. We next address measuring the quantitative impact of natural enemies on their target populations—these are the questions of "how much" mortality is caused by natural enemies. Next we consider how life tables may be employed to assess the ecological role of natural enemies—the questions of "what type" of impact is the natural enemy having on the dynamics of the system (e.g., stabilizing, destabilizing, neutral). We then suggest a general framework for the experimental use of life tables in the study of host–natural-enemy systems. We finish with a discussion on how the topics developed in this chapter should be applied to pathogens, predators, and beneficial herbivores.

DEFINITIONS AND DATA COLLECTION

Types of Life Tables

Life tables, first applied to the study of animal populations by Deevey (1947), are organized presentations of numbers of individuals surviving to fixed points in the life cycle, together with their reproductive output at those points. Mortality usually is assigned to specific causes. Such information can be organized by either age or stage, but age of individual insects rarely is known with precision in field populations, whereas developmental stages can usually be determined. Hence this information, for insects and other arthropods, most often is organized by stage, producing stage-specific instead of age-specific life tables. Inspection of such tables allows determination of stage survival rates and comparisons of the degree of mortality contributed by agents acting at differing points in the life cycle or in different populations (Bellows *et al.,* 1992a).

Life tables may be constructed in two ways. In the first, data are collected that present the fate of a real group or cohort, typically a generation of individuals, whose numbers and mortalities are determined over the course of time for each of a series of stages; this method has been referred to as a horizontal life table. An alternative approach, more applicable to continuously breeding populations than those breeding in discrete generations, is to examine the age structure of a population at one point in time and to infer from it the mortalities occurring in each stage. Such an approach requires assumptions that the population has reached a stable age distribution, and mortality factors acting on the population are constant. Theoretically, age distribution may be stable if the population is either expanding or declining exponentially or is remaining at an unchanging density. In practical terms, life tables of this type reflect only the type and magnitude of mortality acting in a short time period immediately preceding the sampling date. Consequently, one such life table will present an incomplete picture of the total pattern of mortality across the whole season, which may undergo major changes if specific factors act more strongly at some times than they do at others. Life tables developed in this manner are referred to as vertical life tables. Southwood (1978) provides a description of the terminology and conventional organization of both types of life tables. Both types of construction have application in the evaluation of natural enemies in insect life tables. Horizontal construction is most typical for insects breeding in discrete generations, and their construction and analysis is essentially a direct representation of the cohort for which the life table is constructed. Both horizontal and vertical construction are applicable for continuously breeding populations, but their relationships to the populations studied differ from one another. A horizontal life table may be constructed for a cohort defined somewhat arbitrarily by the researcher, and analyses then apply to that cohort. A horizontal life table cannot be constructed for the population as a whole, because the identity of the individuals studied relative to any particular generation is, by definition, not known. Multiple horizontal life tables, studied throughout a season, give a fuller picture of the dynamics of continuously breeding populations. Vertical life tables may be employed for continuously breeding populations in a similar way, recognizing that analysis of data collected in vertical life tables requires some specific assumptions when applying the results to the population as a whole.

The goal of constructing life tables for evaluating the impact of natural enemies is to obtain quantitative estimates of the mortality caused by each natural enemy. These estimates are typically measured as rates, the per capita number of individuals dying from a particular cause. Caution must be employed to distinguish between sequentially acting and contemporaneously acting factors. When data are collected, the sampling program also must permit factors that act contemporaneously to be distinguished from one another. Subsequently, suitable analytical procedures may be employed to correctly calculate the mortality caused by each factor. These matters are discussed more fully later.

Types of Data Used in Life Table Construction

The initial construction of a life table requires estimates of numbers entering successive stages in a life history. These may be obtained in two fundamentally different ways. The first of these methods is to measure the density of each stage several times during the generation or study, providing stage–frequency data. These data may then be analyzed by a variety of techniques to provide estimates of numbers entering successive stages (see Southwood, 1978; and McDonald et al., 1989 for reviews). These data do not, however, provide information on the causes of death in the separate stages. Assignment of causes of death must come from additional information collected during the study, such as dissections to determine parasitism or disease incidence, or by exclusion experiments as discussed later.

An important alternative method for obtaining estimates of the numbers entering successive stages is to measure the recruitment to each stage of interest (Van Driesche & Bellows, 1988; Bellows et al., 1989a; Van Driesche et al., 1991). This approach provides direct assessment of the processes that contribute to stage densities (Fig. 1), and thus permits immediate construction of the life table without recourse to stage–frequency analysis. The recruitment approach is particularly important because very few methods of stage–frequency analysis for two-species coupled systems (e.g., host–parasitoid systems, Bellows et al., 1989a, 1989b) have been developed.

The objective of life table construction usually is to assess the mortality rate assignable to a particular agent. The way in which the data are collected concerning the action of natural enemies can affect the accuracy of the estimates. Losses from parasitism must be assessed at the time of attack, in the host life stage in which the attack occurs. Attempts to score parasitism in a subsequent stage that is not the stage attacked but is the stage from which the parasitoid emerges will lead to incorrect estimates because losses may be obscured by subsequent mortality from other factors. Additionally, scoring parasitism at emergence may be further flawed if mortality levels are incorrectly associated with the host density of the more mature stage, rather than with the density of the earlier stage that was actually attacked.

Mortality rates can, in some circumstances, be estimated in the absence of stage density information without the formal construction of a life table. Gould et al. (1990) and Elkinton et al. (1990) have described an approach where groups of individuals are collected at frequent intervals (but without density information); these individuals are then held at field conditions, and their death rates during specific intervals are recorded. The cause of death of each individual dying during the interval is noted, and by a mathematical process the original mortality rates assignable to each cause are calculated. The process is repeated for samples collected throughout the season, and the interval-specific mortality rates may then be used to calculate the total mortality assignable to each cause during the study. When density information also is available, this approach is applicable to most mortality factors. In cases where density information is not available, the method is applicable to many, but not all, factors (Elkinton et al., 1990, 1992). This approach requires that recruitment to the host population be zero once mortality sampling has begun; if host recruitment is an ongoing process, then the analysis cannot correctly quantify mortality rates.

Finally, some mortality due to natural enemies (e.g., host feeding) is not readily quantifiable using the approaches discussed earlier. For these factors, experimental methods may be employed to estimate mortality rates. This is usually accomplished by measuring (either in the laboratory or in the field) the frequency of occurrence of these factors (such as host feeding) relative to some other, more readily quan-

(a) Single-Species System   (b) Joint Host-Parasitoid System

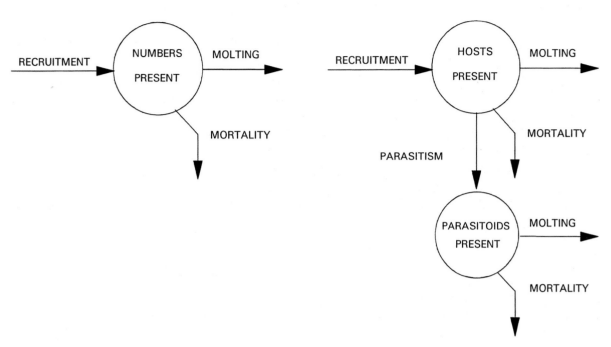

FIGURE 1   Processes affecting the number of individuals in a particular stage of a population at any particular instant in time. (a) In a single species system, the principal processes are recuitment to the population, molting to the next stage, and losses from mortality. (b) In a coupled host-parasitoid system, additional processes affect the numbers present: parasitism contributes losses to the host but recruitment to the parasitoid poulation, and both molting and mortality affect the parasitoid population.

tifiable, event such as parasitism. Once this relative frequency is known, extrapolation from the frequency of the observed event (e.g., parasitism) to the frequency of the unobserved event is possible (Van Driesche *et al.,* 1987).

Applications to Biological Control Systems

When life tables are constructed to assess the magnitude or role of mortality from natural enemies, three issues require careful attention: (1) accurate determination of total numbers entering successive stages and those dying from natural enemies *and* from all other sources of mortality; (2) assessment of all additional natural-enemy-caused mortality other than parasitism or predation per se, as, for example, host feeding by adult parasitoids; and (3) correct focusing of the sampling regimen in relation to the spatial and temporal scale of host distribution and natural-enemy attack.

Determining Total Numbers Entering Stages

Life table construction requires that estimates be obtained for numbers entering successive stages. More detail is required, however, to provide an evaluation for specific natural enemies. Estimates must be obtained for the numbers dying due to specific causes in each stage. These

causes might be specific natural enemies (the most explicit and generally desirable case), or general action of groups of natural enemies (e.g., "parasitism", cf. Carey, 1989, 1993). Several approaches to obtaining these estimates are available.

Stage–Frequency Analysis

Traditionally, methods for quantifying numbers entering a stage have made use of stage–frequency data, and a variety of techniques have been developed for treating such data to extract estimates of numbers entering stages [see Southwood (1978) and McDonald *et al.* (1989) for reviews]. These methods are not immediately applicable for use in quantifying processes in joint host–parasitoid or other natural-enemy systems (Bellows *et al.,* 1989a, 1989b) but must be modified to permit analysis of multispecies systems.

An exception to this case is where the natural histories of the species under study cause all members of the generation to be present in a single stage at a single moment of time (e.g., due to diapause at the end of the stage), and in these cases a single sample at that time may be an accurate estimate of total losses to parasitism provided significant losses have not occurred due to mortality from other factors. However, the more usual case is for recruitment, molting, and mortality to overlap broadly. In such cases no single sample provides an accurate estimate of total generational losses to parasitism (Simmonds, 1948; Miller, 1954; Van Driesche, 1983; Van Driesche et al., 1991). Several approaches have attempted to rectify the biases inherent in sample percentage of parasitism; and recommendations have included scoring parasitism after parasitoid oviposition in the host population is complete (Miller, 1954), using mathematical formulae for adding up successive levels of parasitism (Smith, 1964), and estimating parasitism from pooled samples of larvae in instars too old for parasitoid attack and too young for parasitoid emergence (Hill, 1988). None of these approaches provides an accurate estimate for the numbers dying due to a specific natural enemy for populations where recruitment, molting, and mortality overlap (Van Driesche, 1983).

However, methods developed for determining numbers entering a stage of one species (see Southwood, 1978; McDonald et al., 1989 for reviews) may be adapted to the problem of estimating total entries simultaneously for two species, the host and the parasitoid (Bellows et al., 1989a, 1989b). The graphic technique of Southwood and Jepson (1962), for example, may be used with certain modifications. Because the accuracy of this technique is strongly affected by mortality and because parasitism is a significant source of mortality, the application of the technique is limited. Bellows et al. (1989b) develop seven variants of the method applicable to different life histories and sampling requirements. The method appears to be suitable primarily for cases where independent estimates of host recruitment are available or where total mortality in the system is less than 20%, although specific cases discussed by Bellows et al. (1989b) permit its application in other situations. A modification of Richards and Waloff's (1954) second method may be used to estimate mortality for a stage where parasitism is the sole source of mortality (Van Driesche et al., 1989). Further work in extending single-species analytical techniques to the case of two interacting species will probably add to the methods available for analyzing systems in this manner. These modifications appear to be applicable to both discretely breeding and continuously breeding populations.

Recruitment

One important alternative to the stage–frequency approach is to directly measure the numbers recruited into each stage (Birley, 1977; Van Driesche & Bellows, 1988; Van Driesche, 1988a, 1988b; Lopez & Van Driesche, 1989). If this approach is taken, total numbers entering the stage are found by adding together recruitment for all time periods during the study or generation. Total numbers dying in each stage from parasitism also must be estimated in some manner. For parasitism this may be achieved by direct

measurement of recruited individuals into the "parasitized host" category (Van Driesche & Bellows, 1988; Van Driesche, 1988a, 1988b, Lopez & Van Driesche, 1989). Total parasitoid recruitment divided by total host recruitment then gives the proportion of hosts in the generation killed by the parasitoid. When applied to systems with discrete generations, this approach provides estimates of mortality per generation. When applied to systems with overlapping generations, this approach provides estimates of total mortality during the course of the study.

If recruitment cannot be directly measured for the stages of interest, it may be estimated from data on recruitment to a previous stage together with density estimates for the stage of interest (e.g., Bellows & Birley, 1981; Bellows et al., 1982). A review of techniques for estimating recruitment in host–parasitoid systems is presented by Van Driesche et al. (1991).

Population Growth Rates

When constructing life tables for continuously breeding populations, methods additional to those discussed earlier may be applied. These have as a unifying theme the use of population growth rates as predictors of population increase between samples, with the difference between observed and expected population sizes being an estimate of the numbers dying between sampling times. These methods differ in the approach used for calculating growth rates.

The approach was first developed by Hughes (1962, 1963) who proposed its use for such continuously breeding insect species as the cabbage aphid, Brevicoryne brassicae (Linnaeus). Hughes estimated the growth rate from the age–class distribution of a population in the field. An assumption of the method is that a stable age distribution, required for the estimation of the growth rate parameter R_m, has been attained when the population is studied. Carter et al. (1978) criticized the validity of this assumption and stated that instar distribution in the field should not be used to calculate R_m.

Hutchinson and Hogg (1984, 1985) used caged cohorts of the pea aphid, Acyrthosiphon pisum (Harris), to determine survival and fertility schedules and from these estimated the population growth rate R_m. This modified approach employs the assumption that small deviations from the stable age distribution would have little effect on the population growth rate. The difference between observed densities and those projected from the estimated population growth rates represent the aggregate effects of all causes of reduced reproduction, including mortality and reduced fertility of diseased or parasitized individuals. Quantifying the effects of separate factors requires additional information on the individual sources of mortality (Hutchinson & Hogg, 1985). Poswal et al. (1990) similarly constructed vertical life tables for the aphid Acyrthosiphon kondoi Shinji.

An alternative approach that avoids the general limitations of the methods discussed earlier is to measure directly in the field the per capita reproduction (e.g., recruitment) of adult females chosen randomly from the population over a short interval (Lopez & Van Driesche, 1989) and derive population rates of increase from these data. Such estimates of recruitment, together with density estimates of adult females, allow projections of population growth for comparison to actual population levels on subsequent sample dates. This approach has the advantage of not making any assumptions concerning age structure and does not compound the effects of mortality and reduced fertility of parasitized and diseased individuals.

Death Rates

One technique exists that permits the quantification of mortality rates (the goal of life table analysis) without first constructing a life table (Gould *et al.*, 1990). The method consists of scoring the death rates of individuals in the population at intervals throughout the study and analyzing the observed rates to provide estimates of the independent, or marginal, mortality rates assignable to each cause (cf. Royama, 1981a). This is accomplished by collecting samples of the stages of interest at frequent intervals and rearing the collected individuals under field conditions. These individuals are reared only until the next sampling date, and during the intervening period the numbers of individuals in the sample dying from specific causes are recorded. The proportions of individuals dying are used to calculate the marginal mortality rates for each cause or each factor using the equations given by Elkinton *et al.* (1992) [see also Gould *et al.* (1990)]. The aggregate losses in the population to a specific factor are calculated from the losses in each sampling interval during the study.

This method may be applied to a population provided that all hosts have entered the susceptible stage before the first sample (i.e., there is no recruitment to the population during the study). It has the particular advantage that population density data are not required to obtain estimates of mortality rates. The method is capable of providing estimated rates for factors that act contemporaneously. The method does not, however, provide the traditional stage-specific estimates of loss due to a particular factor if a factor can affect more than one developmental stage, because all stages are treated together during the study. The method does provide "interval-specific" loss rates, and calculates aggregate loss rates from these rather than from stage-specific loss rates. The method is applicable to many, but not all, types of natural-enemy and host interactions (see Elkinton *et al.*, 1992).

Comparison of Methods

Of the methods discussed earlier, directly measuring recruitment to both the host and parasitoid populations is preferable for most situations (Van Driesche & Bellows, 1988; Lopez & Van Driesche, 1989; Van Driesche *et al.*, 1991). It has the advantages of directly measuring the events of interest (the occurrence of parasitism), avoids compounding sequential and contemporaneous factors, and does not require complicated analytical techniques to construct the life table. It is applicable to both discretely breeding and continuously breeding populations.

When recruitment measurement is not possible, stage–frequency analysis provides a potential solution for obtaining estimates of numbers entering stages. A suitable stage–frequency analysis must be selected to extract estimates of numbers entering stages from the stage–frequency data. Although several techniques are available for use with single-species populations, few have been extended to incorporate the special considerations necessary for application to multispecies, host–natural-enemy systems (Bellows *et al.*, 1989a, 1989b; Van Driesche *et al.*, 1989).

The two other approaches, growth rate and death rate analysis, do not estimate numbers entering stages but instead estimate numbers or proportions dying. Growth rate analysis may be applied specifically to continuously breeding populations and provides a measure of total mortality during specific time periods. Separating this aggregate measure into component rates for specific factors requires additional information. Death rate analysis provides a method for estimating mortality rates for specific time periods without the need for data on stage density and allows the contributions of contemporaneous factors to be quantified separately.

Additional Sources of Mortality Caused by Parasitoids

The presence of a natural enemy in a system may result in host deaths not obviously attributable to the natural enemy. This applies particularly in the case of insect parasitoids (Table 1). Such losses may be difficult to quantify directly in field populations. They may resemble predation

TABLE 1   Types of Host Mortality Caused by Parasitoids

Factor	Impact in host population
Parasitism	Hosts in which parasitoid progeny develop
Host feeding	Hosts killed by adult parasitoid feeding
Ovipositor piercing and envenomization	Hosts killed by mechanical trauma or venom
Enhanced susceptibility to other factors	Unparasitized or parasitized hosts dying at increased rates following disturbance by foraging parasitoids
Lost natality	Reduced fertility of parasitized adult hosts prior to death due to parasitism

in that mortalities of these types often result in missing individuals that leave no traces or artifact (such as empty leaf mines, which can stand in the stead of the dead organisms). Such mortality is typically assigned to "predation" or "other" categories by default. Levels of these mortalities may not be trivial. They may equal or exceed losses attributed to demonstrable parasitism (DeBach, 1943; Alexandrakis & Neuenschwander, 1980; Jervis et al., 1996) and may be critical in explaining biological control successes in which observed levels of parasitism are low (Neuenschwander et al., 1986; Briggs et al., 1995).

Host Feeding

Host feeding has been recorded in over 20 families of Hymenoptera (Jervis & Kidd, 1986) and is nearly ubiquitous in some genera (e.g., Tetrastichus and Aphytis, Bartlett, 1964). Hosts killed in this manner may or may not have previously received an oviposition. The role of host feeding in field populations has received little study because the process usually does not leave identifiable remains. Field levels of host feeding of Sympiesis marylandensis Girault could be noted in life tables of Phyllonorycter crataegella (Clemens) as a distinct mortality factor because leaf mines preserved recognizable cadavers (Van Driesche & Taub, 1983). DeBach (1943) used field exclusion techniques to infer the level of mortality due to host feeding on the black scale, Saissetia oleae (Bern), by the parasitoid Metaphycus helvolus (Compere), and concluded that of the 70 to 97% mortality typically caused by this parasitoid, 45 to 77% was due to host feeding instead of parasitism. In a field study of the armored scale Aspidiotus nerii Bouché, host feeding by Aphytis chilensis Howard was found to contribute half of all host mortality based on field counts of dead and parasitized scales (Alexandrakis & Neuenschwander, 1980). For mobile hosts where cadavers neither adhere to plant surfaces nor are retained in galls or leaf mines, individuals killed by host feeding disappear and cannot be scored directly. In such cases laboratory data may be used to estimate parasitism to host-feeding ratios and, together with levels of field parasitism, to estimate host-feeding losses (Legner, 1979; Chua & Dyck, 1982; Van Driesche et al., 1987). Use of laboratory data must take into account such complexities as selective host feeding on hosts of ages different from those usually parasitized (often younger stages, Chua & Dyck, 1982), host feeding in habitat zones not suitable for oviposition (Legner, 1977), or changing host-feeding to parasitism ratios at varying host densities (Collins et al., 1981).

Deaths from Ovipositor Piercing and Envenomization

Hosts also may die from mechanical trauma of ovipositor piercing. This process is distinct from host feeding, and younger hosts may suffer this mortality more than older hosts (Rahman, 1970; Neuenschwander & Madojemu, 1986; Hammond et al., 1987; Neuenschwander & Sullivan, 1987; Van Driesche et al., 1987). Deaths unrelated to parasitism also occur in species that paralyze their hosts when host death occurs in paralyzed hosts in which no oviposition takes place [e.g., Sympiesis marylandensis, Van Driesche & Taub, 1983; the eulophid Pnigalio flavipes (Ashmead), Barrett & Brunner, 1990].

Enhanced Susceptibility to Other Factors

Parasitism may make hosts more susceptible to predation [Godwin & ODell, 1981 (p. 383); Jones, 1987] or disease (Godwin & Shields, 1984). Such events, occurring after parasitoid attack, do not change actual parasitoid-caused losses. Such factors may, however, obscure the actual rate of parasitoid attack, with deaths of parasitized hosts later eaten by predators being assigned in life tables to secondary agents of mortality instead of to parasitism. These deaths can be assigned correctly to the original cause (parasitism) by careful design of the sampling scheme, particularly by measuring recruitment, as discussed earlier. A more complicated situation arises in evaluating natural enemies of plants, because death may result from several factors acting together. In some cases, the presence of one factor can enhance the detrimental effect of another (Huffaker, 1953; Andres & Goeden, 1971; Harris, 1974). One approach to quantifying the relative contributions and interactions of these multiple factors is to use field experimental plots with different combinations of natural enemies (McEvoy et al., 1990, 1993).

The presence of parasitoids in systems can also lead to healthy individuals experiencing greater mortality from other factors. For example, Ruth et al. (1975) noted that when greenbugs, Schizaphis graminum (Rondani), were exposed to the braconid Lysiphlebus testaceipes (Cresson), 41.0 to 62.0% of the aphids left their feeding sites, often falling to the soil. Such aphids were more likely to die due to high soil temperature before reestablishing themselves on plants than were undisturbed aphids. Pea aphids also leave their host plants when disturbed by parasitoids (Tamaki et al., 1970).

In addition to effects on individual hosts, the presence of parasitoids may cause changes at population levels in other mortality factors. For example, introduction of exotic parasitoids suppressed winter moth, Operophtera brumata (Linnaeus), in British Columbia (Embree & Otvos, 1984), but apparently did so by making ground-inhabiting pupal predators more effective (Roland, 1988).

While the preceding types of losses are properly assignable in a life table to the actual cause of death, it is important to be aware of any enhancement in levels of mortality caused by the presence of a natural enemy. This enhancement may be significant and must be considered when evaluating the overall impact of a natural enemy in a system.

Lost Natality

Parasitoids may limit host population growth by suppressing natality through several mechanisms, including sterilization, reduced daily fertility, or reduced longevity. Some euphorine braconids sterilize host adults shortly after parasitoid attack (Smith, 1952; Loan & Holdaway, 1961; Loan & Lloyd, 1974). For example, *Microctonus aethiopoides* Loan attacks and sterilizes reproductively mature female alfalfa weevils, *Hypera postica* (Gyllenhal) (Loan & Holdaway, 1961; Drea, 1968), causing a rapid degeneration of already developed eggs. This results in a 50% loss in total population natality (Van Driesche & Gyrisco, 1979). Parasitism of *Nezara viridula* (Linnaeus) by the tachinid *Trichopoda pennipes* (Fabricius) reduces lifetime (but not daily) fecundity by 74% (Harris & Todd, 1982) by reducing adult life span. Dipteran parasitism [mainly the sarcophagid *Blaesoxipha hunteri* (Hough)] of the grasshopper *Melanoplus sanguinipes* (Fabricius) reduced both the proportion of females producing egg pods and the number of pods per laying, producing an overall reduction in natality of 76% (Rees, 1986). The myrmecolacid strepsipteran *Stichotrema dallatorreanum* Hofeneder reduced numbers of mature eggs in field-collected adults of the tettigoniid *Segestes decoratus* Redtenbacher in Papua, New Guinea by 67% (Young, 1987). Parasitism by the pteromalid *Tomicobia tibialis* Ashmead reduced progeny production of the bark beetle *Ips pini* (Say) by 50% (Senger & Roitberg, 1992). Parasitism of the sow thistle aphid *Hyperomyzus lacticae* (Linnaeus) by the aphidiid *Aphidius sonchi* Marshall reduced total fertility by a variable amount depending on the age of the host when parasitized. Aphids parasitized in the third, fourth, or adult stages suffered 92.4, 85.5, and 77.8%, respectively, loss of lifetime reproductive capacity (Liu & Hughes, 1984). Similar relationships have been reported for the pea aphid when parasitized by *Aphidius smithi* Sharma & Subba Rao (Campbell & Mackauer, 1975) and for the green peach aphid, *Myzus persicae* (Sulzer), when parasitized by *Ephedrus cerasicola* Stary (Hågvar & Hofsvang, 1986). Such effects appear to arise mainly from reduced adult longevity, but may also involve a reduced daily rate of progeny production prior to adult death. Polaszek (1986) showed that parasitized aphids experienced reductions in embryo number and length within 3 days after parasitoid attack. When life tables are constructed for such continuously breeding species as aphids, lost fecundity may be listed as a type of "mortality" (Hutchinson & Hogg, 1985).

Impact of Sample Design

The sampling regimen used to score mortality caused by a natural enemy must ensure adequate and unbiased sampling of both parasitized and unparasitized individuals. Sampling schemes also must use spatial and temporal scales appropriate to the species studied.

Behavioral Biases

Parasitized hosts may behave differently than unparasitized hosts in ways that render them more or less vulnerable to detection. They also may occupy different habitats than do healthy individuals. Many of these behaviors result from differences in mobility between parasitized and healthy individuals, and consequently these differences may be more likely to affect relative, instead of absolute, sampling regimes.

Traps may have differing efficiencies for parasitized and healthy individuals. Yano *et al.* (1985) reported that levels of parasitism in the leafhopper *Nephotettix cincticeps* Uhler were distinctly higher (13% versus 3%) in individuals taken in sweep nets than in those collected at the same date and location in light traps because parasitism damaged thoracic muscles and weakened the insect's flight ability. Wylie (1981) reported that levels of parasitism of flea beetles [*Phyllotreta striolata* (Fabricius) and *P. cruciferae* (Goege)] by the euphorine braconid *Microctonus vittatae* Muesebeck were lower in beetles collected in traps baited with allyl isothiocyanate than in beetles collected with a vacuum suction device, but only when beetles were reproductively active. Parasitized beetles are sterilized and hence, according to the author, reacted like nonreproducing beetles, which are less attracted to host-plant odors.

Parasitism also may influence movement of hosts between habitats. The potato aphid *Macrosiphum euphorbiae* (Thomas) when parasitized by diapause-bound *Aphidius nigripes* Ashmead leaves its habitat (Brodeur & McNeil, 1989), while those bearing parasites not bound for diapause do not. Wylie (1982) reports that flea beetles (*Phyllotreta cruciferae* and *P. striolata*) parasitized by *Microctonus vittatae* emerged from overwintering sites earlier than unparasitized beetles; consequently, samples of beetles in the crop exhibited a steady decline in percentage of parasitism over a 10-day emergence period, unrelated to changes in parasitism in the entire population. Ryan (1985) attributed decrease in percentage of parasitism of larvae of the larch casebearer, *Coleophora laricella* (Hübner), on larch foliage to the selective drop of parasitized larvae to the undergrowth, an unsampled habitat zone.

Parasitism can also affect rates of host movement within the habitat, making hosts more likely to be seen and collected (if done by hand) or caught in traps. The isopod *Armadillidium vulgare* Latreille moved farther and rested less often when parasitized by the acanthocephalan parasite *Plagiorhynchus cylindraceus* (Schmidt & Kuntz), making parasitized individuals more easily detectable in its habitat (Moore, 1983).

Most of the difficulties posed by these behaviors can be avoided by using absolute, instead of relative, measures of population density during sampling. Care must be taken to sample all occupied habitats and, where necessary, subsample different portions of the population to provide relative rates of parasitism in each. These partial rates may be weighted by the densities in each habitat to provide an overall estimate of numbers dying from parasitism in the population as a whole. Studies evaluating predation instead of parasitism may need to take into account similar effects.

Biases Affecting Detection of Density Dependency

Problems in detecting existing density dependency can occur if either the spatial scale or the timing of the sampling regime is inappropriate. If hosts are strongly clumped and clumps are distributed on a spatial scale that is meaningful to parasitoids, their activity may be concentrated on dense clumps, either from aggregation of foragers or from greater progeny production and retention in locally host-rich areas. In such cases, the sampling program must provide samples from patches of different densities, and each sample must consist of individuals from a given density instead of a mixture of hosts from high- and low-density patches (Heads & Lawton, 1983). If samples are based on mixtures of individuals from patches of strongly differing densities, any density dependency can be obscured (Hassell, 1985a, 1987; Hassell *et al.,* 1987). Pooling of samples from high- and low-density periods in a time series may have the same effect as pooling high- and low-density samples collected at one time from several locations, obscuring temporal density dependence.

Finally, it should be emphasized that parasitoid-caused mortality acts on hosts selected for oviposition, not hosts from which adult parasitoids emerge. Nevertheless, estimates of parasitism often are based on rearing parasitoids from host instars or stages subsequent to the one attacked. Mortality levels are then associated incorrectly with the density of the host at the time the samples were collected rather than with the density of the host when it was actually attacked. Density dependence of a mortality factor will only be detectable if its level is measured accurately and associated correctly with the host density upon which it acts.

ASSESSING QUANTITATIVE IMPACT OF NATURAL ENEMIES

With one or several well-constructed life tables for a host population affected by a natural enemy, questions concerning the amount of mortality (both in absolute terms and in relation to other sources) in the host population can be examined.

Parameters in the Life Table

The objective of life table analysis for natural-enemy evaluation is to estimate the attack rate of specific natural enemies to permit comparisons between agents or populations. Some of the methods discussed earlier under life table construction (such as measurement of recruitment) yield these rates directly and do not require further calculations from a life table. Where these methods have been used, construction of a life table and further analysis to determine the quantitative impact of the natural enemy may not be necessary. Construction of a life table in these cases may be useful if additional analyses, such as those relating attack rates to population densities, are desired. Other methods described previously will require that density and mortality information be subjected to further calculations to arrive at attack rates for the different factors in the life table.

The components of a life table typically include the numbers entering each of several life stages (l_x) in an insect's life cycle, numbers dying within each stage (d_x) due to specific factors, together with estimates of rates of loss in each stage (Table 2a) (Southwood, 1978). Mortality rates are typically expressed in proportions. Several different types of mortality rates have been included in life tables. These include real mortality, apparent mortality, indispensable mortality, marginal attack rates, and *k*-values.

It is likely that more than one mortality factor will act contemporaneously at some point in the life table. It is appropriate, therefore, when seeking an index for assessing the impact of natural enemies, to select one that will have the same meaning when describing both contemporaneous factors and those that act alone within a stage. Real mortality, apparent mortality, and indispensable mortality are primarily of value when considering factors that act alone in a stage. Marginal rates are applicable both to factors acting alone and contemporaneously acting factors.

Real mortality is the ratio of the number dying in a stage (d_x) to the number initially entering the first stage in a life table (l_o): real mortality = d_x/l_o. This column sums to the total mortality in the life table. This column may be useful for comparing the role of factors within generations (if calculated for sequential, noncontemporaneous factors) (Southwood, 1978).

Apparent mortality (denoted by q_x) is the ratio of the number dying in a stage to the number entering the stage, or the number dying from a factor to the number subject to that factor: $q_x = d_x/l_x$. When only one mortality factor occurs in a stage (or where more than one occurs and they act sequentially), then the apparent mortality (the proportion of animals dying from a factor) is the same as the proportion initially attacked by the factor (the marginal attack rate). Southwood (1978) suggests that this measure may be used for comparisons among independent (noncontemporaneous) factors or to compare the same factor in different life tables.

TABLE 2   Impact of Sequential versus Contemporaneous Factors in a Life Table

Stage	Factor	l_x	Stage d_x	Factor d_x	Marginal death rate	Apparent mortality Stage q_x	Apparent mortality Factor q_x	Real mortality Stage d_x/l_o	Real mortality Factor d_x/l_o
					(a)				
Stage 1		100	50			0.500		0.500	
	1.1			50	0.500		0.5		0.5
	1.2			0	0.000		0		0
Stage 2		50	25			0.500		0.250	
	2.1			25	0.500		0.5		0.25
	2.2			0	0.000		0		0
Adults		25							
Sex ratio		0.5							
Fertility		200							
F_1 progeny		2500							
R_o		25							
					(b)				
Stage 1		100	75			0.750		0.750	
	1.1			37.5	0.500		0.375		0.375
	1.2			37.5	0.500		0.375		0.375
Stage 2		25	0			0.000		0.000	
	2.1			0	0.000		0		0
	2.2			0	0.000		0		0
Adults		25							
Sex ratio		0.5							
Fertility		200							
F_1 progeny		2500							
R_o		25							

Apparent mortalities, because they are calculated on a stage- or factor-specific basis, are not additive, but the product of their associated stage survival rates (1 − stage apparent mortality) yields the total survival in the life table.

Indispensable mortality has been little used. It is described as "that part of the generation mortality that would not occur, should the mortality factor in question be removed from the life system, after allowance is made for the action of subsequent mortality factors" by Southwood (1978), who also describes its calculation. This type of calculation entails an assumption that subsequent mortality factors in the life history act in a density-independent manner. Huffaker and Kennett (1966) suggested that indispensable mortality may be used to assess the value of a factor in a biological control program, but this applies primarily to comparisons within a life table, rather than among several life tables, because its value depends on the quantitative level of other mortalities in the life table, which may vary in different systems.

The proportion of individuals entering a stage that are attacked by an agent is termed the marginal rate of attack or marginal attack rate (Royama, 1981b; Buonaccorsi & Elkinton, 1990; Elkinton et al., 1992). This measure of mortality has the most consistent interpretation among life tables or among factors within a life table; it is the only

measure whose calculation permits correct interpretation of the impact of contemporaneous mortality factors. The details of its calculation depend somewhat on the nature of a specific factor (Elkinton et al., 1992). For factors that act alone in a stage, the apparent mortality is the marginal attack rate. When two or more factors act contemporaneously, the apparent mortality will be different from (and smaller than) the marginal attack rate. For such contemporaneous factors, determining the number attacked by a factor must account for those that receive attacks from more than one type of agent. Two general approaches are available in these cases, either (1) assessing the attack rate as it occurs (e.g., measuring recruitment by dissection for parasitism), which directly estimates the marginal attack rates; or (2) calculating the attack rate from the observed death rates of individuals succumbing to the various factors (Buonaccorsi & Elkinton, 1990; Elkinton et al., 1992; Gould et al., 1992c). The equations employed in calculating marginal attack rates from observed numbers dying vary for different categories of natural enemies. Equations for contemporaneous parasitism differ slightly from those used when predation and parasitism occur together (Elkinton et al., 1992). The product of (1 − marginal rates) for all factors is equal to the overall survival rate for the life table.

In addition to these measures of mortality, k-values may

also appear in life tables. *k*-values are marginal survival rates on a logarithmic scale, and are the negative logarithm of (1 − the marginal attack rate) for a factor. Although equivalent in principle to the marginal rate, their calculation has been a source of difficulty in cases of contemporaneous factors. The explicit calculation of a *k*-value requires the number of attacked individuals and the number of individuals initially exposed or subject to the factor (Varley & Gradwell, 1960, 1968; Varley *et al.*, 1973), the same information necessary for calculating marginal rates. Use of the numbers observed dying due to a factor can only lead to correct calculation of a *k*-value if factors act strictly sequentially in a stage or in successive stages. *k*-values for contemporaneous factors cannot be calculated from the number observed dying because the action of each factor is obscured by the action of others. A lack of appreciation of this crucial distinction has led to the incorrect calculation of *k*-values in many studies. Because *k*-values are logarithms of survival rates, their sum (when each has been properly calculated) is equal to the logarithm of total survival, in the same way that the product of survival rates for separate factors yields the overall survival in the life table.

Evaluating the effects of the natural enemies in a system must be made with respect to some standard of host population growth potential. An appropriate standard for populations with discrete generations is the population net rate of increase R_o, which is the ratio of population sizes in two successive generations. Similarly, the intrinsic rate of increase R_m could be used for continuously breeding populations. Calculation of R_o from a life table requires data on fertility of the population, which often can be measured or estimated. The product of overall proportion survival and fertility yields an estimate of R_o. When $R_o = 1$ (or $R_m = 0$), the population is neither increasing nor decreasing. Values for R_o greater than unity (or $R_m > 0$) imply an increasing population, while values of R_o less than unity (or $R_m < 0$) imply that the population is decreasing in density. In the context of biological control programs, a value of R_o greater than unity implies a need for greater natural-enemy action to reduce R_o and contain the population.

Comparisons among factors and life tables are most easily accomplished with reference to marginal rates, the values of which are independent of the presence of additional, contemporaneous factors in the system (this is not true for either apparent or real mortality). Marginal rates assigned to a particular factor are directly comparable among different life tables, even when those life tables contain differing numbers or quantitative levels of other factors. When correctly quantified, *k*-values may be used equivalently.

Interpreting Life Tables

A series of examples will serve to illustrate the relationships among life table parameters together with their interpretation.

The simplest case for a life table is when each factor acts independently and sequentially, so that no overlap occurs among stages subject to individual factors (see Table 2a). In this case, the marginal death rate and the apparent mortality for each factor are the same. In this example, where 50% of the individuals die in each of two successive stages, real mortality declines from stage 1 to stage 2, because only 25 individuals die in stage 2.

When two factors act contemporaneously (see Table 2b) (in Tables 2, 3, and 4 factors occurring within a stage act contemporaneously), marginal rates and apparent rates differ. The proportion actually attacked is higher than the proportion ultimately dying from the factor. This is because some animals attacked by factor 1.1 are also attacked by factor 1.2. Because some animals may be attacked by both factors contemporaneously, but can die from only one, the total number of attacks exceeds the total number of animals dying. The total number dying in the population is the same as in Table 2a because the marginal rates of two factors are the same in both tables, but their sequential action in Table 2a leads to different numbers dying from each. This underscores an important feature of marginal rates that renders them so particularly valuable for comparison — the marginal rate is the proportion that would die due to that factor in the absence of other independent factors or when that factor is acting alone (Elkinton *et al.*, 1992). This feature is constant for marginal rates in any combination with other factors. No other measure of mortality has this uniformity of representation or meaning across different life tables.

It may be observed that factors with large "apparent" mortalities add only a small amount of additional "real" mortality to systems in which there is already substantial mortality. Table 3a is a hypothetical table with only one mortality factor, causing 0.90 loss in stage 1. This reduces numbers entering the final stage by 90%. In Table 3b, an additional source of mortality is added, attacking 0.80 of insects in stage 1 and acting contemporaneously with factor 1. The addition of this second factor, which attacked 80% of the insects in the stage (i.e., had a marginal rate of 0.80), raises the stage apparent mortality rate from 90%, (see Table 3a) to 98% (see Table 3b), thereby adding 8% "real" mortality. The effect of this second source of mortality, however, on the numbers entering the final stage is not an 8% but an 80% drop in density (from 10 adults in Table 3a to 2 in Table 3b). Again, the marginal rate reflects the contribution of the factor.

In each of the preceding life tables, values were given for fertility and sex ratio. These values may be multiplied by the total number surviving to the adult stage to provide an expected number of progeny in the first stage in the next generation. The ratio of this number of F_1 progeny to the original number entering the life table is an estimate of R_o. In the first three life tables, R_o is greater than unity, implying an increasing population. In Table 3b, $R_o = 1$, imply-

TABLE 3   Impact of Single versus Two Contemporaneous Factors in a Life Table

Stage	Factor	l_x	Stage d_x	Factor d_x	Marginal death rate	Apparent mortality Stage q_x	Apparent mortality Factor q_x	Real mortality Stage d_x/l_o	Real mortality Factor d_x/l_o
					(a)				
Stage 1		100	90			0.900		0.900	
	1.1			90	0.900		0.9		0.9
	1.2			0	0.000		0		0
Adults		10							
Sex ratio		0.5							
Fertility		100							
F_1 progeny		500							
R_o		5							
					(b)				
Stage 1		100	98			0.980		0.980	
	1.1			54	0.900		0.54		0.54
	1.2			44	0.800		0.44		0.44
Adults		2							
Sex ratio		0.5							
Fertility		100							
F_1 progeny		100							
R_o		1							

ing that this population is stationary. The difference between the R_o estimates for Table 3a and b is 80% and reflects the addition of the second mortality factor.

Life tables for actual systems typically contain more than one or two factors, but interpretations of the various rates are made in the same way. Thus for a population of *Pieris rapae* Linnaeus larvae (Tables 4a and 4b), high levels of both parasitism by *Cotesia glomerata* (Linnaeus) and

predation occurred, as expressed by the large marginal rates for these factors. These contributions are somewhat less clear in the apparent and real mortality rates. The contemporaneous action of these mortalities causes a much greater number of individuals to be attacked by each than to actually die from each, because a number of parasitized individuals are lost to predation prior to parasitoid emergence.

The contributions of a specific mortality agent may ad-

TABLE 4a   Life Table for a *Pieris rapae–Cotesia glomerata* System, Parasitoids Present

Stage	Factor l_x	Stage d_x	Factor d_x	Marginal death rate	Apparent mortality Stage q_x	Apparent mortality Factor q_x	Real mortality Stage d_x/l_o	Real mortality Factor d_x/l_o
Egg	10.669	0.128			0.012		0.012	
Infertility			0.128	0.012		0.012		0.012
Larvae	10.540	10.51			0.997		0.985	
C. glomerata			4.660	0.867		0.442		0.436
Predation			5.853	0.981		0.555		0.548
Pupae	0.0271	0.008			0.310		0.000	
Predation			0.008	0.310		0.31		0.001
Adults	0.0186							
Sex ratio	0.50							
Fertility	356[a]							
F_1 progeny	3.328							
R_o	0.311							

After Van Driesche, R. G., & Bellows, T. S., Jr. (1988). *Ecological Entomology, 13,* 215–222.
[a] Value from Norris (1935).

TABLE 4b　Life Table for *Pieris rapae–Cotesia glomerata* System, Parasitoids Removed

Stage	Factor l_x	Stage d_x	Factor d_x	Marginal death rate	Apparent mortality Stage q_x	Apparent mortality Factor q_x	Real mortality Stage d_x/l_o	Real mortality Factor d_x/l_o
Egg	10.669	0.128			0.012		0.012	
Infertility			0.128	0.012		0.012		0.012
Larvae	10.540	10.33			0.981		0.969	
C. glomerata			0	0.000		0		0
Predation			10.33	0.981		0.980		0.968
Pupae	0.2044	0.063			0.310		0.006	
Predation			0.063	0.310		0.31		0.005
Adults	0.1411							
Sex ratio	0.50							
Fertility	356[a]							
F_1 progeny	25.11							
R_o	2.354							

After Van Driesche, R. G., & Bellows, T. S., Jr., (1988). *Ecological Entomology, 13,* 215–222.
[a] Value from Norris (1935).

ditionally be evaluated by removing it from the life table and recalculating the survival and reproduction parameters. Comparisons between tables with and without the action of the natural enemy provide an index of its contributions to the system. This has been done by removing parasitism from Table 4a and recalculating the remaining parameters (Table 4b). Comparison of the two tables reemphasizes that "real mortality" is not very sensitive to the addition of another factor. For the cases in Tables 4a and 4b, the predation marginal rate was 98.1%, and the addition of an 86.7% parasitism marginal rate for parasitism added only 1.6% to the stage real mortality (Table 4a versus 4b). The presence of this factor, however, resulted in an 87% (equal to the marginal rate for this factor) reduction in numbers entering the last stage, reducing it from 0.141 to 0.0186. Thus parasitism was crucial in reducing the net rate of reproduction of the population from 2.4 to 0.31. The use of this technique assumes that the marginal rates of the other factors do not change with the removal of the factor of interest.

Evaluating the specific contribution of any particular factor in a life table requires the careful selection of an appropriate index. Because apparent mortality in a stage can rise only to 1.0, the value of additions of further mortality agents for a stage is not well reflected by rises in apparent mortality. In general, the higher the level of mortality within a stage from a pre-existing factor, the smaller will be the rise in apparent mortality to that stage from the addition of another factor. Thus, increases in apparent or real mortality in a stage due to the addition of a new mortality agent do not adequately reflect the contribution of the new mortality agent. In contrast, the marginal death rate of any factor in a system is a direct reflection of its impact on reducing the numbers entering the final stage in the table, and therefore its contribution in reducing host densities. Of the available measures for expressing mortality in life tables, marginal rates most accurately express the individual contributions of particular factors, particularly when two or more factors act contemporaneously. *k*-Values, when properly calculated to reflect the marginal rate, are similarly useful, because they are functionally interchangeable with marginal rates (on logarithmic scales) as expressions of mortality.

The overall contribution of specific mortality agents in life tables can be examined by addition or subtraction of such factors, manipulating numbers in the life table to reflect their absence or presence. Such manipulations allow hypotheses to be formulated concerning the impact of specific agents. Such hypotheses can be formulated in terms of changes in the net reproduction rate (R_o) of the population. R_o is a particularly suitable index because it expresses the ability of the population to reproduce itself given the state of all sources of mortality in the system.

It cannot be overemphasized that the percentage of mortality due to parasitism (or other agents) observed in populations is relatively meaningless in the absence of quantitative values for all mortalities acting in the parasitized stage. These additional mortalities are nearly always essential for estimating the marginal death rate due to any one specific factor (Royama, 1981b; Elkinton *et al.,* 1992). (One exception is when parasitism is being measured as a comparative parameter between treatments in experiments in which only a relative contrast is being measured, and other factors are known to be similar or constant between treatments.)

The relative importance of a mortality factor is most effectively expressed with respect to the reproductive dynamics of the insect it attacks, that is, the fertility of the host and a full quantitative description of all mortalities.

Even if any given natural enemy does not cause the population of the host immediately to decline, it may be valuable if it increases the overall mortality, because R_o may become less than unity after the addition of some additional factor or natural enemy (e.g., *Aphytis paramaculicornis* DeBach and *Coccophagoides utilis* Doutt on olive scale, Huffaker & Kennett, 1966).

DETERMINING ECOLOGICAL ROLES OF NATURAL ENEMIES

Role Questions and Density Relatedness

A basic precept of biological control is that effective natural enemies will contribute to a reduced and stable pest density. Both of these features are relative terms—the new pest density would be lower than the previous density, and would exhibit fewer fluctuations than the population without the natural enemies. In this context, natural enemies may play one or more of a variety of roles in the ecology of a natural-enemy and pest system. Most of the features desired in natural enemies fall into one of two categories: (1) the natural enemy will reduce the pest density and (2) the natural enemy will aid in stabilizing the pest density. This section deals with how the analysis of life table data can contribute to testing hypotheses concerning these and related roles for natural enemies.

Testing such hypotheses from life table data usually requires examining several life tables for trends or contrasts in the impact that natural enemies have on pest populations. Consequently, where in the previous sections we were concerned with the proper construction of, and quantifying factors in, life tables, here we will deal with the analysis of such features where several life tables are available for the same system. These might arise from sequential sampling of the same population over several generations, from contemporaneous sampling of several populations in different areas, or from both. The types of questions that can be addressed depend somewhat on the types of data that are available.

Natural enemies may play either or both of the previously mentioned roles in an ecological system, which leads to several possibilities in the structure of natural-enemy and pest interactions. The classical interaction envisaged by many authors is the situation where both roles are embodied in the same natural-enemy species, so that the natural enemy contributes quantitatively to the suppression of survival or reproduction (so that $R_o < 1$ or $R_m < 0$ at high densities) and also contributes to stabilizing the system at the new, reduced density. Such an outcome would indeed be optimal and desirable, because no further additions to the system are needed for success in the context of either reducing population density or maintaining stability. Two additional situations also are possible. The natural enemy

may contribute to reductions in survival or fertility (thus contributing mortality in the life table so that R_o will be reduced) without contributing to stability per se. In such a situation, the system may be stabilized by some other factor in the life table (e.g., Harcourt *et al.*, 1984), or may be relatively unregulated. Finally, the natural enemy may contribute to stability or regulation without increasing the total level of mortality in the life table, perhaps by replacing an existing factor with a new one that causes an equivalent level of mortality but acts with an increased level of density dependence.

Identifying the role of particular natural enemies in a particular system may not provide a comprehensive answer to the question of what features are significant in shaping the dynamics of pest and natural enemy populations. Addressing that question may require an evaluation of the role of several or all of the factors operating in the system.

Many of the available theories concerning host and parasitoid dynamics (Beddington *et al.*, 1978; May, 1978; Hassell, 1985b; Murdoch *et al.*, 1992, 1995, 1996; Bonsall & Hassel, 1995; Mills & Getz, 1996; Murdoch & Briggs, 1996) employ some density-related property as a stabilizing mechanism. These appear in various forms and can all be considered under the general heading of density dependence. These theories generally provide testable hypotheses concerning the role of natural enemies, although conducting the tests in a statistical sense can be difficult, as discussed later. Four cases concerning density dependence in a life table may be distinguished: (1) there may be no density dependence in the system, (2) density dependence may be attributable to a natural enemy under investigation, (3) density dependence may be due to some other factor in the life table, or (4) density-related factors may exist but may be masked by stochastic factors. In addition, more than one factor may be density dependent, which requires careful consideration in constructing tests of hypotheses. Hypotheses about density dependence are usually tested against the null hypothesis that no density dependence is present in the system.

Other theories have proposed dynamics of pest and natural-enemy systems that are not characterized by density-dependent stabilizing mechanisms (see Chapter 2). The hypotheses provided by such theories are not as readily testable by analyzing life table data, because they are characterized by dynamics that do not have deterministic relationships between measured variables (such as density and mortality). These theories may provide more readily testable hypotheses following further development.

Ecological Roles and Hypotheses

Before considering in detail some specific role questions and techniques for testing their related hypotheses, it is helpful to review some terms and their meanings. We will imply here by the term *regulation* the tendency of a population to move toward some mean value. This does not imply a reduction in density, which we will term *suppression*.

Regulation is often considered to be due to the action of some density-related factor. In general, density relatedness may be viewed as falling into one of three categories: *density dependence* (where proportional mortality increases as density increases), *inverse density dependence* (where proportional mortality decreases with increasing density), and *density independence* (where proportional mortality neither increases nor decreases with mortality). Density dependence may further be defined as direct density dependence, where the factor is related to the density of the generation in which it acts, or delayed density dependence, where the factor is related to the density of the generation prior to the one in which it acts. Density relatedness may be expressed among portions of a population in different locations (over space) or between successive generations of the same population (over time), or both. Finally, a *key factor* is the mortality factor most closely related to, or responsible for, change in total generational mortality among several generations in a population; in general, it is simply the most variable mortality factor in the system. The term key factor does not imply either that the factor is regulatory or that it is the factor most responsible for determining the mean density of the population. *Natural enemies may be important either as sources of mortality or as regulating factors without being the key factor in a system.*

The role questions best addressed by the examination of several life tables is whether or not the natural enemies function as regulating factors. Such regulation usually is reflected in hypotheses as density-dependent mortality, and consequently life tables are often examined to determine whether the mortality imposed by a natural enemy acts in a demonstrably regulating or a density-dependent manner. Several mechanisms have been proposed that fall into this category (see Chapter 2). In most cases the proportion of pests dying due to the natural enemy increases with pest density. Inverse density dependence also can act, in some cases, as a stabilizing factor (Hassell, 1984), as can simple spatial heterogeneity (May, 1978; May & Hassell, 1988; Pacala & Hassell, 1991).

It is crucial, when considering relationships between density and mortality, both to quantify correctly the proportional losses assigned to a factor and to associate this mortality with the density and stage on which the factor acts. For example, parasitoids attacking only young larvae are acting on a population whose density may be very different from the late larval population from which the parasitoids emerge (Van Driesche & Bellows, 1988). Similarly, when not all individuals in the population are susceptible to natural-enemy attack, the proportional mortality must be related to the density of susceptible individuals. A less rigorous approach will confound the underlying relationships by associating mortality rates with unrelated densities from inappropriate stages in the life table.

The possible alternative hypotheses related to natural enemies acting as a regulating factor are twofold: (1) they may act in a destabilizing manner (i.e., they are acting either in a destabilizing inverse density-dependent manner or in a delayed density-dependent manner), or (2) they may not contribute to regulation but serve solely as an additional density-independent mortality in the life table. In this second case, the density-independent mortality may have a small variance, or have a larger variance and be catastrophic in nature.

Analytical Techniques

The analysis of population and life table data for the purpose of detecting stability and regulation has followed two rather distinct approaches. The first of these addresses general questions of population stability with reference solely to density counts in successive generations. The second is concerned with density relatedness of specific factors in life tables. Although the overall objective of the two approaches is similar, they employ somewhat different analytical techniques and we will treat them here separately.

Tests for Population Stability

Tests for population stability focus on the general question of dynamical behavior of a population over several generations, without reference to causal mechanisms. The general framework for this question arises from Morris's (1959) proposal for the detection of stability in a population. In this context, stability is the tendency of a population to grow or decline in a manner that moves it toward an equilibrium when it is displaced from it. Such a population would tend to decrease when denser than the equilibrium value, and to increase when below the equilibrium value. Such populations are in contrast to those that either grow or decline exponentially and those that exhibit a random, but undirected, trajectory through time. In this light, if a population is characterized by the logarithm of its density in generation t, X_t, then the dynamics of an unstable population may be expressed by

$$X_{t+1} = r + X_t + e_t \qquad (1)$$

where r is the growth rate between generations and e_t is a stochastic error term representing random deviations in r. Stable populations may be represented, in contrast, by

$$X_{t+1} = r + bX_t + e_t$$

where b takes values between -1 and 1 and represents density-dependent restrictions on population growth.

Several analytical tests for detection of stability by examining series of population censuses have been developed. Much of this work has followed Morris's (1959) original proposal, offering criticisms and improvements. The original proposal involved regressing X_{t+1} against X_t and testing the slope of the regression for significant difference from 1, the null hypothesis value for no regulation. The general concept has been accepted by most workers, but its application to hypothesis testing has proved problematic. The first-order autocorrelation in the time series of Eq. (1), together with the presence of sampling errors in the abscissal values X_t, creates such significant biases in the regression slope that the test is generally considered inadequate (Varley & Gradwell, 1968; Bulmer, 1975; Pollard *et al.*, 1987) because it rejects the null hypothesis in a large proportion of cases when the null hypothesis is true (i.e., it has a large likelihood of a type I error). A number of parametric (regression) as well as simulation tests have been proposed to overcome this difficulty.

The first parametric test presented was that of Varley and Gradwell (1968), who proposed a modification of the criteria for rejecting the null hypothesis by suggesting that double regressions be performed, and that the slope estimates b (for the regression of X_{t+1} on X_t) and slope estimate b_{xy} (for the regression of X_t on X_{t+1}) be performed. The null hypothesis would be rejected only when both regression slopes differed significantly from unity and both b and $1/b_{xy}$ are less than unity. Simulation studies have indicated that, while it has a low likelihood of a type I error, this test also has relatively poor power (that is, it fails to reject the null hypothesis in a large proportion of cases when the population is stable); as a statistical test it is overly conservative.

Other parametric tests have been proposed. Bulmer (1975) introduced a test statistic based on the reciprocal of Von Neuman's ratio for time series analysis, and a modification of this statistic for cases when there are errors in sample estimates of population counts (the usual case). Slade (1977) suggested using two other statistics developed previously for estimating slopes of relationships where error occurs on both axes, the major axis test (Deming, 1943) and the standard major axis test (Ricker, 1973, 1975). Dennis and Taper (1994) proposed a similar regression model and outlined bootstrap and jackknife techniques to allow for hypothesis testing. A number of simulation studies have been conducted to assess the error rate (for type I errors) and the power of these various tests (Slade, 1977; Vickery & Nudds, 1984; Gaston & Lawton 1987; Holyoak, 1993; Holyoak & Lawton, 1993; Fox & Ridsdill-Smith, 1995). The general conclusions of these workers is that these tests are not all robust, and that many have acceptable error rates and power only in exceptional circumstances. In addition, the power of these types of tests is clearly related to the length of the census available for analysis, and censuses in excess of 15 to 20 generations appear to be necessary to ensure an acceptable degree of power (Hassell *et al.*, 1989; Bonsall & Hassell, 1995; Fox & Ridsdill-Smith, 1995). Fox and Ridsdill-Smith (1995) proposed that, as an initial or

general standard for testing for stability or regulation in a population census over several generations, Bulmer's (1975) (first) test be employed, because it seems to have the greatest degree of applicability and the fewest drawbacks in use. The variation on Bulmer's (1975) statistic proposed by Reddingius and den Boer (1989) may also be of value, although this test has not received the extensive attention of earlier proposals and has not yet been subject to testing by Monte-Carlo simulation, as have the earlier tests.

Alternatives to parametric, or regression, tests have been developed by several workers using Monte-Carlo techniques. These generally take the form of proposing population models for the two hypotheses under consideration (the null hypothesis of no stability and the alternative hypothesis of stability). The models, which incorporate various components of stochastic variation, are then used to simulate a long series of synthetic populations with parameter values taken from the natural population under study. The dynamics of these synthetic populations are then summarized in one or more statistics, and the same statistic is calculated for the natural population. The distribution of the statistic from synthetic populations is compared with observed value of the statistic from the natural population; and if the observed value lies near the extreme end of the synthetic distribution (usually beyond the 5% most extreme cases), the null hypothesis is rejected. This procedure has provided some very helpful insight into the behavior of the parametric tests discussed earlier, and has provided "simulation" tests that appear able to distinguish stable from unstable populations. Slade (1977) proposed one such test based on simulated distributions of the *t*-value associated with the usual regression slope *b*. Pollard *et al.* (1987) found this test insufficient, and developed a test based on likelihood ratios that appears both to have an acceptable type I error rate and a sufficient power to identify stable populations, although the matter of errors in density estimation was not addressed by this technique. Reddingius and den Boer (1989) developed a similar test that does provide for errors in density estimation, and provided a fuller examination of its power than have other researchers, although they did not provide any information on the error rate of their proposed index under the null hypothesis. Either one of the last two proposals would be suitable Monte-Carlo techniques for examining censuses for evidence of stability, but both require much more extensive use of computing than do the parametric tests of the previous paragraph.

Tests for Density Relatedness in a Specific Factor

Populations are affected by both biotic factors, which may show some form of density relatedness, and abiotic factors, which with few exceptions are density independent in their action. When the variation from year to year in the amount of mortality inflicted by density-independent factors is greater than the mortality caused by density-dependent factors, the population's dynamical behavior is dominated by these density-independent processes and, consequently, may not show stability in the sense discussed earlier. This does not preclude the presence of potentially regulatory mechanisms, but may make difficult their detection by examination of population census data (Hassell, 1985b). Consequently, tests have been developed to examine specific factors for attributes that could contribute to stability, even if they are acting in concert with other factors that obscure their effects.

The original approach was predicated on the view that temporal density dependence (*sensu* Nicholson, 1954) was the primary, or perhaps only, mechanism that could contribute to regulation of a system. Following this line of reasoning, Varley and Gradwell (1968) suggested plotting survival against density on log scales (the familiar plot of k-value versus log density). Regression analysis was used to determine if the slope of the relationship was significantly greater than 0, implying density dependence (because mortality rate was increasing with density). They recognized, however, that the estimate of density was employed on both axes (on the ordinal scale as a component of the k-value), and that the errors of estimation occur on both axes. These conditions preclude the application of usual tests for significance of the regression slope, complicating the issue of rejecting the null hypothesis of no density dependence. This issue received considerable attention subsequently, but no completely satisfactory solution has been proposed. Thus the technique of k-value analysis continues to be employed to provide initial assessments of density relatedness—density dependence, delayed density dependence, or inverse density dependence—in long-term studies of populations. Royama (1981a) suggested that an alternative approach might be to attempt to determine *a priori* what factors in a life table were density independent, identify them and quantify their impact on mortality, and subsequently examine the remaining factors for density relatedness. This proposal appears promising, but Royama did not address issues relating to statistical testing of hypotheses in this framework.

Examining factors for temporal density dependence in the host population is but one step in the search for regulating features in a life table. Other forms of density relatedness were soon appreciated as potential contributors to population stability, particularly density dependence occurring within a generation but over some spatial scale. These include interference among parasitoids (Hassell & Varley, 1969; Hassell, 1970), aggregation (Hassell & May, 1973, 1974; Beddington *et al.*, 1978), inverse density dependence (Hassell, 1984; Hassell *et al.*, 1985), host refuges (Reeve & Murdoch, 1986), specific types of natural enemy search behavior such as sigmoid functional response (Hassell *et al.*, 1977; Hassell & Comins, 1978), invulnerable life stages

or invulnerable fractions of populations (Murdoch *et al.*, 1987), and even simple spatial patchiness or heterogeneity (May, 1978). Not all these features have been found in natural field systems, although many are well known from laboratory systems. Some are known from some field systems and not from others [e.g., lack of aggregation of *Aphytis melinus* DeBach against *Aonidiella aurantii* (Maskell) (Reeve & Murdoch, 1985), but presence of aggregation of parasites attacking bivalves (Blower & Roughgarden, 1989)]. Occasionally a particular behavior usually considered to contribute to regulation via density dependence is found to be present, but stabilizing density dependence cannot be demonstrated (Smith & Maelzer, 1986). In some cases, the ability to detect certain mechanisms is dependent on the scale of measurement, for example, in cases of aggregation (Hassell *et al.*, 1987) or the assessment of patch sizes as perceived by the natural enemy (Heads & Lawton, 1983).

Our understanding of what types of behaviors and features of natural enemies and their host populations can enhance stability has advanced rapidly, faster than have statistical developments for handling these special testing needs. The intricate correlations and interdependencies among such life table variables as mortality and the density on which it acts are not completely understood for most of these types of factors. This makes the development of statistical tests that have acceptable error rates and have sufficient power a difficult task. In many of the articles cited in the preceding paragraph, researchers have employed various statistical techniques in an effort to demonstrate the presence (or absence) of a particular behavior. Most of these appear rational, but normal statistical assumptions are often violated. In addition, linear models relating behaviors to density have been employed when *a priori* considerations indicate that such models cannot apply and curvilinear models would be more appropriate. This is not to suggest that such studies have failed in their objectives, but only to point out that the suitability of most statistical techniques for use in detecting regulating behaviors has not been evaluated. Hence no standard statistical analytical technique has emerged for the evaluation of these behaviors.

In light of the plurality of features of biological systems that can affect their dynamics, and the potentially masking effects of random (density-independent) factors (Hassell, 1985b, 1987), no simple analysis will likely serve to provide definitive answers to questions of density relatedness or the presence of other stabilizing mechanisms in life tables. Carefully planned studies contrasting the behavior of systems both with and without natural enemies may permit simpler comparisons of system dynamical behaviors and testing of hypotheses. These ideas are outlined later.

Complementary to the use of statistical analytical techniques, computer models will continue to be a useful tool in the exploration of the role of any given factor in a life table. Such models can be used to generate hypotheses about the behavior of systems in the presence and absence of particular factors, such as natural enemies. Quantitative, explicit life tables may well be the key element in obtaining the necessary information with which to construct models from which hypotheses can be generated. These hypotheses subsequently can be tested only with reference to data from natural populations. This emphasizes the significance of accurate, well-constructed life tables for both the generation and the testing of hypotheses about system behavior and dynamics.

EXPERIMENTAL DESIGNS FOR LIFE TABLE STUDIES

Construction and analysis of life tables for separate host populations having and lacking a natural enemy is a powerful approach to natural-enemy assessment. Researchers can maximize the power of life table data to reveal both the total mortality contributed by an agent and the ecological role of the agent through careful use of such planned contrasts.

The required contrast of populations having and lacking a natural enemy can be achieved by organizing comparisons in one of three general ways:

1. Time may be used to organize the with and without contrast for cases of introduction of new agents where studies of the host population's dynamics can be initiated prior to the introduction (the "without" treatment) and then continued after the agent's establishment (the "with" treatment) (Quezada, 1974; Dowell *et al.,* 1979; Bellows *et al.,* 1992a).

2. Geography may be used, where plots in one location with the agent present are contrasted to plots in similar but separate locations lacking the agent (e.g. Gould *et al.,* 1992a,b). This is feasible chiefly with new agents that have not yet occupied their full potential range. This approach is less applicable to native or previously introduced agents, because sites having and lacking the agent are likely to differ in some factor of ecological importance to the agent. Contrasting life tables between the native home and the area of introduction (after establishment of a species) can be particularly helpful [e.g., winter moth in England (Varley & Gradwell, 1968; Hassell 1980) versus Nova Scotia (Embree, 1966)].

3. Exclusion may be used in which some type of barrier is erected to deny the agent access to a portion of the pest population. Methods to create such barriers have been reviewed by Luck *et al.* (1988) and also in Chapter 9. Generally, natural enemies may be excluded from plots by the use of cages, mechanical barriers at plot edges, selective insecticides, or in some cases by hand picking, by the presence of dust, or by ants.

Each method (time, geography, exclusion) of creating the desired with and without natural-enemy condition has certain limitations that may potentially confound the interpretation of results. Contrasts structured on time (before and after studies) are frequently criticized, for example, on the basis that no two years are ever identical in terms of weather, etc., and hence results may differ because of features other than the presence or absence of the natural enemy. Contrasts based on geography may be criticized because sites that appear similar to the researcher may in fact differ in nonapparent yet important ways. This may be compensated for by utilizing a set of three or more sites for both the with and without treatments. This, however, may be beyond the resources of many research projects, especially those attempting to construct life tables at each study site. Exclusion-based contrasts are criticized because the means used as barriers often change the physical or chemical environment of the pest population in one treatment group (the without) but do not do so in the other treatment. Cages, for example, may increase insect development due to within-cage greenhouse effects and also may prevent emigration of the pest under study. Selective pesticides may alter reproductive rates of pests in treated plots, either directly or through changes in plant chemistry. For reviews of such biases in exclusion studies see Luck *et al.* (1988) and Van Driesche and Bellows (1996).

In general terms, biases such as these are best controlled by the simultaneous use of two methods of establishing the desired with and without contrast. In such cases, each method (time, geography, exclusion) provides the researcher the opportunity to assess the degree of bias of the other method. For example, if, in terms of Fig. 2, the consequences of before and after contrasts are determined at

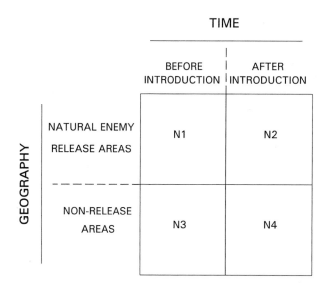

FIGURE 2 Determining consequences of before and after contrasts at a release (N1 versus N2) and a nonrelease (N3 versus N4) site.

both a release (N1 versus N2) and a nonrelease site (N3 versus N4), then the criticism that years may differ can be rebutted if there are no differences between years at the nonrelease site (N3 versus N4). Similarly, contrasts of sites having and lacking a new agent, after it is established (N2 versus N4), are stronger if these sites are studied prior to release and found not to differ (N1 versus N3). Van Driesche and Gyrisco (1979) illustrate the application of this "two-factor" method of constructing with and without contrasts, using geography and cage exclusion to measure the effect of an introduced braconid parasitoid *Microctonus aethiopoides* on oviposition by the alfalfa weevil, *Hypera postica.*

Conventionally, with and without contrasts, arrived at by whichever of these methods, have been evaluated by scoring the pest's density and the rate of mortality inflicted by the agent of interest. These may be determined either once at the termination of the experiment or several times during its progress. The additional construction of life tables for each of the two populations in the contrast provides an improved quantification of the agent's value by allowing marginal rates of mortality from each mortality agent in the system to be calculated, both in the presence and in the absence of the agent of interest. This contrast, in combination with a comparison of R_0 for the pest populations both attacked by and not attacked by the agent, provides a clear assessment of the value of the agent in suppressing the pest. Life tables for *Phyllonorycter crataegella,* modified from Van Driesche and Taub (1983), illustrate such a contrast. The value of R_0 for a population lacking parasites (9.45 Table 5b) was much higher than that of a nearby population

with parasitoids (1.79, Table 5a). This difference quantifies much of the value added to the system by the parasitoids. Had this study not been designed as a two plot, with and without contrast, but instead as a study solely of the plot where parasites were present, such an evaluation could not have been made. From a life table constructed from a single-plot study, the importance of the parasitoids to the host's R_0 could not be directly determined. A hypothesis about their likely impact could have been formulated by modifying the observed table through the deletion of the mortality due to parasitism. These calculations for values in Table 5a suggest that R_0 in the absence of parasitism, would have been 7.63. Actual observation of a similar value (9.45) in the adjacent plot that lacked parasitism confirms the hypothesis. Having tables for both sides of the with and without contrast thus allows a researcher to confirm what would otherwise remain a hypothesis about the impact of parasitism on R_0. Other examples of such comparisons include life tables for citrus blackfly (*Aleurocanthus woglumi* Ashby), in which populations attacked by the parasitoids *Prospaltella opulenta* Silvestri and *Amitus hesperidum* Silvestri had R_0's of 0.122 to 0.92 (collapsing populations) in contrast to R_0's of 2.4 to 4.5 for populations not affected by parasitism (Dowell *et al.,* 1979); and the whitefly *Siphoninus phillyreae* (Haliday), in which populations lacking parasitoids had values for R_0 of 2.9 to 6.0, while populations parasitized by *Encarsia inaron* (Walker) had values of R_0 of 0.4 to 1.1 (Gould *et al.,* 1992b).

In addition, contrasts of levels of mortality for factors in paired life tables can be helpful in understanding origins of such mortality. For example, in the *Phyllonorycter cratae-*

TABLE 5a   Life Tables for First Generation *Phyllonorycter crataegella* Populations at Buckland, MA, 1981, Unsprayed Orchard

Stage	Factor l_x	Stage d_x	Factor d_x	Marginal death rate	Apparent mortality Stage q_x	Apparent mortality Factor q_x	Real mortality Stage d_x/l_o	Real mortality Factor d_x/l_o
Egg	283	6			0.021		0.021	
Infertility[a]			6	0.021		0.021		0.021
Sap larvae	277	63			0.227		0.223	
Parasitism			35	0.134		0.126		0.123
Residual			28	0.108		0.101		0.098
Tissue larva	214	168			0.785		0.594	
Parasitism			140	0.729		0.654		0.494
Residual			28	0.206		0.130		0.098
Pupae[a]	46	23			0.500		0.081	
Adults	46							
Sex ratio[a]	0.499							
Fertility[a]	22							
F_1 progeny	506							
R_0	1.787							

After Van Driesche, R. G., & Taub, G. (1983). *Protection Ecology, 5,* 303–317.

[a] Values for *P. blancardella* (Fabricius) from Pottinger and LaRoux (1971).

TABLE 5b   Life Tables for First Generation *Phyllonorycter crataegella* Populations at Buckland, MA, 1981, Sprayed Orchard

Stage	Factor l_x	Stage d_x	Factor d_x	Marginal death rate	Apparent mortality Stage q_x	Apparent mortality Factor q_x	Real mortality Stage d_x/l_0	Real mortality Factor d_x/l_0
Egg	433	9			0.021		0.021	
Infertility[a]			9	0.021		0.020		0.020
Sap larvae	424	19			0.045		0.044	
Parasitism			1	0.002		0.002		0.002
Residual			18	0.043		0.042		0.041
Tissue larva	405	34			0.084		0.079	
Parasitism			17	0.043		0.041		0.039
Residual			17	0.043		0.041		0.039
Pupae[a]	371	0			0		0	
Adults	371							
Sex ratio[a]	0.499							
Fertility[a]	22							
F_1 progeny	4092							
R_0	9.450							

After Van Driesche, R. G., & Taub, G. (1983). *Protection Ecology, 5,* 303–317.
[a] Values for *P. blancardella* (Fabricius) from Pottinger and LaRoux (1971).

gella case, reductions in levels of mortality attributed to predation in pesticide-treated plots lacking parasitoids showed either that pesticide use suppressed predators or that some portion of the mortality listed as predation was in fact unrecognized deaths due to host feeding that could not be diagnosed as such due to decomposition of the host larva after death.

To examine the ecological role of a biotic agent, the action of the factor must be assessed over a range of host population densities. The spectrum of densities examined may be derived either from populations at a single point in time that exist as spatially distinct populations of varying density, or from assessment of a population at a single location over a series of generations through time. Density relatedness is more easily detected spatially when data are drawn from contemporaneous populations, because these data are not confounded by temporal autocorrelation in the host densities. In a single generation, however, only density dependence derived from intragenerational mechanisms, such as the functional or aggregation response of the agent, can be detected. To detect density relatedness from a numerical response of the agent through reproduction, a series of generations must be examined.

In both cases (spatial studies and temporal studies) the detection of density dependency on the part of the agent requires (1) accurate assessment of the pest's density for the stage on which the agent acts; and (2) accurate assessment of the true marginal rate of the agent's attack. Selection of apparent mortality rates (which are often significantly lower than marginal rates if rates for other contemporaneous factors are high) instead of marginal rates, greatly increases the likelihood that any conclusions about the density relatedness of the factor will be wrong. Because it is difficult to accurately assess marginal rates for mortality agents outside of the construction of full life tables (or at the very least with sufficient knowledge of the kinds and intensities of other mortalities), the construction of life tables is an essential step in the assessment of the density relatedness of a factor.

Where feasible, inclusion of with and without contrasts in the design of studies intended to cover a range of pest populations of varying densities (arranged either spatially or temporally) is desirable. Studies structured in this way allow the impact of the agent's presence on the pest population's R_0 value to be assessed, not just at a single pest density (as in the example of Table 5, discussed previously), but over a full range of pest densities. Further, such a study design allows simpler tests of hypotheses concerning the overall effects of density-related mortalities than does a study of a single population.

APPLICATIONS TO CATEGORIES OF NATURAL ENEMIES OTHER THAN PARASITOIDS

Many of the concepts developed in the chapter were conceived originally and developed for parasitoids as the source of mortality. Nevertheless, these techniques can be successfully applied directly to other classes of mortality agents or some can be applied with adaptations.

Pathogens of Arthropods

Nearly all the ideas in the chapter apply directly to pathogens. If marginal rates are to be assessed via direct observation of recruitment, two questions arise: (1) are all levels of pathogen titer lethal or will some be sublethal infections not ultimately killing the host, and (2) can diseased individuals be detected very early after infection? Early detection may be achieved by use of antigen-antibody technique (the monoclonal antibody process, McGuire & Henry, 1989; Hegedus & Khachatourans, 1993). If marginal rates for pathogens cannot be assessed via recruitment, analysis of death rates (Gould *et al.*, 1990; Elkinton *et al.*, 1992) can be used to calculate marginal rates from numbers dying due to disease and to other causes in reared samples.

Predators

In the construction of many life tables, some individuals disappear from the population and their disappearance cannot be reliably assigned to a particular factor. Hence life tables often use a category for such individuals such as "residual mortality" or "missing." The fraction of a population denoted as missing is the marginal rate for this category (Elkinton *et al.*, 1992). Caution must be employed to ensure that the disappearance of individuals is assigned to the correct stage when constructing the life table (Campbell *et al.*, 1982), particularly if intervals between samples are long.

Individuals eaten by predators disappear from the population. Consequently, mortality due to predation is often combined with other, unspecified sources of disappearance from the population. Caution should be exercised in assigning all disappearance to predation unless abiotic factors can be eliminated. In some cases, predation leaves artifacts (such as exoskeletal remains) that can be used to specifically assign deaths to this category (Gould *et al.*, 1992c). When this is possible, it permits marginal rates for predation to be separated from the general category of missing individuals. Other techniques have been suggested for quantifying predation rates (Sunderland, 1988). Generally, these do not allow marginal rates for predation to be divided into separate components for specific predators, but in some cases this may be accomplished or estimated by collateral evidence on the composition and relative significance of the members of the predator complex (Bellows *et al.*, 1982b).

Beneficial Herbivores and Plant Pathogens

Plants are rarely treated as a population of individuals whose births (recruitment) and deaths can be counted and assigned rates, although this very natural extension of life tables or actuarial tables would provide excellent quantitative information on effectiveness of natural enemies. The techniques presented here may be applied directly, considering plants as hosts and herbivores as "predators" or natural enemies whose impact does not directly eliminate entire plants but instead affects their reproduction through effects on their vital rates (such as fertility and death rates) (McEvoy *et al.*, 1993).

Other significant differences between plant and arthropod systems must be considered when evaluating effectiveness of natural agents via life table analysis. Life tables do not offer any direct way to measure herbivore impact on vigor or biomass except as these features are reflected in plant longevity and fertility (i.e., seed set, etc.). In some systems a useful approach might be to construct $l_x - m_x$ life tables for these systems, both with and without natural enemies (cf. Julien & Bourne, 1988), and to calculate estimates of population growth parameters from these tables. This would be particularly appropriate for biennial or perennial systems, were differences in fertility might be the major impact of some herbivores, for example, flower- or seed-feeding herbivores. Comparative life tables for populations with and without natural enemies are as valuable here as they are for any natural-enemy assessment (McEvoy *et al.*, 1990, 1993).

For pathogens of weeds, comparative $l_x - m_x$ or stage-specific life tables are equally applicable, but quantifying the dynamics of the upper trophic level population (the plant pathogen) may require very different sampling techniques. In some studies, the dynamics of the pathogen may be ignored (e.g., in the case of augmentation), but to document the natural effect of an introduced and established pathogen, some understanding of the dynamics of the pathogen population will be essential. Constructing life tables for the pathogen is a natural, if not often applied, approach for quantifying the relevant reproductive, recruitment, and survival rates. Finally, the seed population in the soil may have a temporal dynamic over a much longer timescale than the plants themselves, an issue that must be considered in the assessment of recruitment rates for these populations.

Augmentation

Mass rearing and release of natural enemies against pests is common in cropping systems (Legner & Medved, 1981; Frick *et al.*, 1983; van Lenteren & Woets, 1988; Bellows & Van Driesche, 1996). Hoddle *et al.* (1998), for example, used paired life tables in greenhouses to evaluate the efficacy of augmentative releases of the aphelinid whitefly parasitoid *Eretmocerus eremicus* Rose and Zolnerowich. The use of life tables in these settings can be particularly effective in evaluating the contribution of the released natural enemies. The effects of augmented populations of natural enemies can be treated identically to natural populations using the methods discussed throughout this chapter. The construction of a complete life table can provide unambiguous marginal rates for each factor acting on the host population. This permits the impact of released natural enemies to be quantified in relation to the mortality occurring naturally in the system, providing immediate and quantitative evaluation of the effectiveness of the augmentation.

The use of life tables to make planned comparisons is a simple and effective method for natural-enemy evaluation in augmentation studies. Augmentation studies imply the presence of a population with the natural enemy (the release location). The addition of a study in a nonrelease location permits the construction of life tables for a population without an augmented population of the natural enemy, and comparative analysis for the with and without situations then may be conducted. Such comparative studies may be especially valuable for systems where the natural enemy is also present continuously at low levels, and may be the only way in which any increased host mortality due to the augmentation program may be reliably quantified.

CONCLUSIONS

Life table analysis may be used in evaluating natural enemies to provide answers to two basic types of questions. The first of these deals with the quantitative impact of natural enemies. Net reproductive rates of pest populations (R_o) must be reduced to below unity for a population to decrease. Life table analysis permits assessment of the degree to which particular natural enemies contribute toward reaching this goal. The second type of question considers the ecological role of natural enemies, and life table analysis in this context is used to determine the degree to which natural enemies help stabilize pest populations.

Construction of life tables for the evaluation of natural enemies requires accurate estimates of numbers entering stages and numbers dying within stages due to specific causes. Methods for obtaining such estimates include stage–frequency analysis, sampling for recruitment, growth rate analysis, and death rate analysis. These approaches vary both in the types of data required for their use and in the types of information they can provide. Measurement of recruitment is the most direct method for obtaining the data required for life table construction and natural-enemy assessment. Regardless of the data collection procedures utilized, sampling programs must avoid potential biases caused

by behavioral changes of parasitized hosts and by host patchiness.

Life tables may contain several measures for expressing mortality caused by natural enemies. Principal among these are apparent mortality, real mortality, marginal attack rate, and *k*-values. Of these, marginal attack rates (or their associated *k*-values) provide the most suitable measure for comparison among factors within or between life tables, particularly when two or more factors act contemporaneously on a stage. The relative contributions of different natural enemies in reducing population growth may be evaluated by considering their impact on the reproductive rate (either R_o or R_m) of the host population.

Analyses of life tables for evaluating the ecological role of natural enemies have focused on the issue of natural-enemy contributions to population stability. Current methods are capable of detecting spatial density dependence. Detecting temporal density dependence is more difficult, because of the statistical difficulties in testing hypotheses from the temporally correlated types of data usually available. Some tests for detecting regulation in population censuses are available; but require relatively long censuses (15 to 20 generations or more) to be reliable.

A powerful approach for the evaluation of natural enemies is the combination of life table analysis and manipulation of host and natural-enemy populations. Studies that construct life tables for populations both with and without the natural enemy can provide exceptional opportunities for defining the quantitative level of natural-enemy impact in a system. In addition, such studies allow questions concerning ecological roles to be addressed in a comparative way, avoiding many of the statistical difficulties that frustrate the detection of density dependence and regulation in studies of single populations.

Broadly, the use of life tables in the evaluation of natural enemies is part of the iterative process of hypothesis development, data collection, analysis, and use of analytical results to pose further, more detailed hypotheses. While the focus of this chapter has been primarily the analytical processes that quantify interactions and test hypotheses, careful initial framing of questions and data collection procedures are essential if analytical methods are to be capable of providing decisive answers. Viewed in the larger context of the scientific method, life table analysis can be used— either alone or in combination with such other forms of natural-enemy evaluation as experimental manipulation— to address fundamental questions of population dynamics and regulation as well as practical problems of natural-enemy utilization.

Acknowledgment

Thanks to J. Elkinton for helpful discussion.

References

Alexandrakis, V., & Neuenschwander, P. (1980). Le role d'*Aphytis chilensis* [Hym.: Aphelinidae], parasite *D'Aspidiotus merii* [Hom.: Diaspididae] sur oliver en Crete. Entomophaga, 25, 61–71.

Andres, L. A., & Goeden, R. D. (1971). The biological control of weeds by introduced natural enemies. In C. B. Huffaker (Ed.), Biological control (pp. 143–164). New York: Plenum.

Barrett, B. A., & Brunner, J. F. (1990). Types of parasitoid-induced mortality, host stage preferences, and sex ratios exhibited by *Pnigalio flavipes* (Hymenoptera, Eulophidae) using *Phyllonorycter elmaella* (Lepidoptera: Gracillariidae) as a host. Environmental Entomology, 19, 803–807.

Bartlett, B. R. (1964). Patterns in the host-feeding habit of adult parasitic Hymenoptera. Annals of the Entomological Society of America, 57, 344–350.

Beddington, J. R., Free, C. A., & Lawton, J. H. (1978). Modeling biological control: On the characteristics of successful natural enemies. Nature (London), 273, 513–519.

Bellows, T. S., & Birley, M. H. (1981). Estimating developmental and mortality rates and stage recruitment from insect stage frequency data. Researches on Population Ecology (Kyoto), 23, 232–244.

Bellows, T. S., Jr., Ortiz, M., Owens, J. C., & Huddleston, E. W. (1982a). A model for analyzing insect stage-frequency data when mortality varies with time. Researches on Population Ecology (Kyoto), 24, 142–156.

Bellows, T. S. Jr., Owens, J. C., & Huddleston, E. W. (1982b). Predation of range caterpillar, *Hemileuca oliviae*, at various stages of development by different species of rodents in New Mexico during 1980. Environmental Entomology, 11, 1211–1215.

Bellows, T. S., Jr., Van Driesche, R. G., & Elkinton, J. S. (1989a). Life tables and parasitism: Estimating parameters in joint host-parasitoid systems. In L. McDonald, B. Manly, J. Lockwood, & J. Logan (Eds.), Estimation and analysis of insect populations (pp. 70–80). New York: Springer-Verlag.

Bellows, T. S., Jr., Van Driesche, R. G., & Elkinton, J. S. (1989b). Extensions to Southwood and Jepson's graphical method of estimating numbers entering a stage for calculating mortality due to parasitism. Researches on Population Ecology (Kyoto), 31, 169–184.

Bellows, T. S., Jr., Van Driesche, R. G., & Elkinton, J. S. (1992b). Life-table construction and analysis in the evaluation of natural enemies. Annual Review of Entomology, 37, 587–614.

Birley, M. (1977). The estimation of insect density and instar survivorship functions from census data. Journal of Animal Ecology, 46, 497–510.

Blower, S. M., & Roughgarden, J. (1989). Parasites detect host spatial pattern and density: A field experimental analysis. Oecologia, 78, 138–141.

Bonsall, M. B., & Hassell, M. P. (1995). Identifying density-dependent processes: A comment on the regulation of winter moth. Journal of Animal Ecology, 64, 781–784.

Briggs, C. J., Nisbet, R. M., Murdoch, W. W., Collier, T. R., & Metz, J. A. J. (1995). Dynamical effects of host feeding in parasitoids. Journal of Animal Ecology, 64, 403–416.

Brodeur, J., & McNeil, J. N. (1989). Seasonal microhabitat selection by an endoparasitoid through adaptive modification of host behavior. Science, 244, 226–228.

Bulmer, M. G. (1975). The statistical analysis of density dependence. Biometrics, 31, 901–911.

Buonaccorsi, J. P., & Elkinton, J. S. (1990). Estimation of contemporaneous mortality factors. Researches on Population Ecology (Kyoto), 32, 1–21.

Campbell, A., & Mackauer, M. (1975). The effect of parasitism by *Aphidius smithi* (Hymenoptera: Aphidiidae) on reproduction and population growth of the pea aphid (Homoptera: aphididae). Canadian Entomologist, 107, 919–926.

Campbell, R. W., Torgensen, T. R., & Beckwith, R. C. (1982). Disappearing pupae: A source of possible error in estimating population density. Environmental Entomology, 11, 105–106.

Carey, J. R. (1989). The multiple decrement life table: A unifying framework for cause-of-death analysis in ecology. Oecologia, 78, 131–137.

Carey, J. R. (1993). Applied demography for biologists, with special emphasis on insects. New York: Oxford University Press.

Carter, N., Aikman, D. P., & Dixon, A. F. G. (1978). An appraisal of Hughes' time-specific life table analysis for determining aphid reproductive and mortality rates. Journal Animal Ecology, 47, 677–687.

Chua, T. H., & Dyck, V. A. (1982). Assessment of *Pseudogonatopus flavifemur* E. & H. (Dyrinidae: Hymenoptera) as a biocontrol agent of the rice brown planthopper. Proceedings, International Conference on Plant Protection in the Tropics (pp. 253–265).

Collins, M. D., Ward, S. A., & Dixon, A. F. G. (1981). Handling time and the functional response of *Aphelinus thompsoni*, a predator and parasite of the aphid *Drepanosyphum plantanoides*. Journal of Animal Ecology, 50, 479–487.

DeBach, P. (1943). The importance of host-feeding by adult parasites in the reduction of host populations. Journal of Economic Entomology, 36, 647–658.

Deevey, E. S. (1947). Life tables for natural populations of animals. Quarterly Review of Biology, 22, 283–314.

Deming, W. E. (1943). Statistical adjustment of data. New York: John Wiley & Sons.

Dennis, B., & Taper, M. L. (1994). Density dependence in time series observation of natural populations: Estimation and testing. Ecological Monographs, 64, 205–224.

Dowell, R. B., Fitzpatrick, G. E., & Reinert, J. A. (1979). Biological control of citrus blackfly in southern Florida. Environmental Entomology, 8, 595–597.

Drea, J. J. (1968). Castration of male alfalfa weevils by *Microctonus* spp. Journal of Economic Entomology, 61, 1291–1295.

Elkinton, J. S., Gould, J. R., Ferguson, C. S., Liebhold, A. M., and Wallner, W. E. (1990). Experimental manipulation of gypsy moth density to assess impact of natural enemies. In Watt, A. D., Leather, S. R., Hunter, M. D., and Kidd, N. A. C., (Eds.), Population Dynamics of Forest Insects (pp. 275–287). Andover, U.K.: Inttercept.

Elkinton, J. S., Buonaccorsi, J. P., Bellows, T. S., Jr., & Van Driesche, R. G. (1992). Marginal attack rate, *k*-values and density dependence in the analysis of contemporaneous mortality factors. Researches on Population Ecology (Kyoto), 34, 29–44.

Embree, D. G. (1966). The role of introduced parasites in the control of the winter moth in Nova Scotia. Canadian Entomologist, 98, 159–1168.

Embree, D. G., & Otvos, I. S. (1984). *Operophtera brumata* (Linnaeus) winter moth (Lepidoptera: Geometridae). In J. S. Kelleher & M. A. Hulme (Eds.), Biological control programmes against insects and weeds in Canada 1969–1980 (pp. 353–357). Slough, United Kingdom: Commonwealth Agricultural Bureau.

Frick, K. E., Hartley, G. G., & King, E. G. (1983). Large scale production of *Bactra verutana* (Lep.: Tortricidae) for the biological control of nutsedge. Entomophaga, 28, 107–116.

Fox, D. R., & Ridsdill-Smith, J. (1995). Tests for density dependence revisited. Oecologia, 103, 435–443.

Gaston, K. J., & Lawton, J. H. (1987). A test of statistical techniques for detecting density dependence in sequential censuses of animal populations. Oecologia, 74, 404–410.

Godwin, P. A., & ODell, T. M. (1981). Intensive laboratory and field evaluations of individual species: *Blepharipa pratensis* (Meigin) (Diptera: Tachinidae). In C. C. Doane & M. L. McManus (Eds.), The gypsy moth: Research toward integrated pest management (U.S. Foreign Service Tech. Bulletin 1584, pp. 375–394).

Godwin, P. A., & Shields, K. S. (1984). Effects of *Blepharipa pratensis* [Dip.: Tachinidae] on the pathogenicity of nucleopolyhedrosis virus in stage V of *Lymantria dispar* (Lep.: Lymantriidae). Entomophaga, 29, 381–386.

Gould, J. R., Elkinton, J. S., & Wallner, W. E. (1990). Density-dependent suppression of experimentally created gypsy moth, *Lymantria dispar* (Lepidoptera: Lymantriidae), populations by natural enemies. Journal of Animal Ecology, 59, 213–233.

Gould, J. R., Bellows, T. S., & Paine, T. D. (1992a). Population dynamics of *Siphoninus phillyreae* in California in the presence and absence of a parasitoid. *Encarsia partenopea.* Ecological Entomology, 17, 127–134.

Gould, J. R., Bellows, T. S., & Paine, T. D. (1992b). Evaluation of biological control of *Siphoninus phillyreae* (Haliday) by the parasitoid *Encarsia partenopea* (Walker), using life-table analysis. Biological Control, 2, 257–265.

Gould, J. R., Elkinton, J. S., & Van Driesche, R. G. (1992c). Assessment of potential methods of measuring parasitism by *Brachymeria intermedia* (Hymenoptera: Chalcididae) of pupae of the gypsy moth, *Lymantria dispar* (Lepidoptera: Lymantriidae). Environmental Entomology, 21, 394–400.

Hågvar, E. B., & Hofsvang, T. (1986). Parasitism by *Ephedrus cerasicola* (Hym.: Aphidiidae) developing in different stages of *Myzus persicae* (Hom.: Aphididae). Entomophaga, 31, 337–346.

Hammond, W. N. O., Neuenschwander, P., & Herren, H. R. (1987). Impact of the exotic parasitoid *Epidinocarsis lopezi* on cassava mealybug *(Phenacoccus manihoti)* populations. Insect Science Applications, 8, 887–891.

Harcourt, D. G., Guppy, J. C., & Binns, M. R. (1984). Analysis of numerical change in subeconomic populations of the alfalfa weevil, *Hypera postica* (Coleoptera, Curculionidae), in eastern Ontario. Environmental Entomology, 13, 1627–1633.

Harris, P. (1974). The impact of the cinnabar moth on ragwort in eastern and western Canada and its implication for biological control strategy. Commwealth Institute Biological Control Trinidad, Miscellaneous Publications, 8, 119–123.

Harris, V. E., & Todd, J. W. (1982). Longevity and reproduction of the southern green stink bug, *Nezara viridula,* as affected by parasitization by *Trichopoda pennipes.* Entomologia Experimentalis et Applicata, 31, 409–412.

Hassell, M. P. (1970). Mutual interference between searching insect parasites. Journal of Animal Ecology, 39, 473–486.

Hassell, M. P. (1980). Foraging strategies, population models and biological control: A case study. Journal of Animal Ecology, 49, 603–628.

Hassell, M. P. (1984). Parasitism in patchy environments: Inverse density dependence can be stabilizing. IMA Journal of Mathematics Applied in Medicine and Biology, 1, 123–133.

Hassell, M. P. (1985a). Patterns of parasitism by insect parasitoids in patchy environments. Ecological Entomology, 7, 365–377.

Hassell, M. P. (1985b). Insect natural enemies as regulating factors. Journal of Animal Ecology, 54, 323–334.

Hassell, M. P. (1987). Detecting regulation in patchily distributed animal populations. Journal of Animal Ecology, 56, 705–713.

Hassell, M. P., & Comins, H. N. (1978). Sigmoid functional response and population stability. Theoretical Population Biology, 14, 62–67.

Hassell, M. P., & May, R. M. (1973). Stability in insect host-parasite models. Journal of Animal Ecology, 42, 693–726.

Hassell, M. P., & May, R. M. (1974). Aggregation of predators and insect parasites and its affects on stability. Journal of Animal Ecology, 43, 567–594.

Hassell, M. P., & Varley, G. C. (1969). New inductive population model and its bearing on biological control. Nature (London), 223, 1133–1137.

Hassell, M. P., Latto, J., & May, R. M. (1989). Seeing the wood for the trees: Detecting density dependence from existing life-table studies. Journal of Animal Ecology, 58, 883–892.

Hassell, M. P., Lawton, J. H., & Beddington, J. R. (1977). Sigmoid functional responses by invertebrate predators and parasitoids. Journal of Animal Ecology, 46, 249–262.

Hassell, M. P., Lessells, C. M., & McGavin, G. C. (1985). Inverse density dependent parasitism in a patchy environment: A laboratory system. Ecological Entomology, 10, 393–402.

Hassell, M. P., Southwood, T. R. E., & Reader, P. M. (1987). The dynamics of the Viburnim whitefly *(Aleurotrachelus jelinekii):* A case study of population regulation. Journal of Animal Ecology, 56, 283–300.

Heads, P. A., & Lawton, J. H. (1983). Studies on the natural enemy complex of the holly leaf-miner: The effects of scale on the detection of aggregative responses and the implication for biological control. Oikos, 40, 267–276.

Hegedus, D. D., & Khachatourians, G. G. (1993). Construction of cloned DNA probes for the specific detection of the entomopathogenic fungus *Beauveria bassiana* in grasshoppers. Journal of Invertebrate Pathology, 62, 233–240.

Hill, M. G. (1988). Analysis of the biological control of *Mythimna separata* (Lepidoptera: Noctuidae) by *Apanteles ruficrus* (Braconidae: Hymenoptera) in New Zealand. Journal of Applied Ecology, 25, 197–208.

Hoddle, M. S., Van Driesche, R. G., Sanderson, J. P., & Minkenberg, O. P. J. (1998). Biological control of *Bemisia argentifolii* with inundative release of *Eretmocerus eremicus* (Hymenoptera: Aphelinidae): Do release rates affect parasitism? Bulletin of Entomological Research, 88, 47–58.

Holyoak, M. (1993). The frequency of detection of density dependence in insect orders. Ecological Entomology, 18, 339–347.

Holyoak, M., & Lawton, J. H. (1993). Comments arising from a paper by Wolda and Dennis: using and interpreting the results of tests for density dependence. Oecologia, 95, 435–444.

Huffaker, C. B. (1953). Quantitative studies on the biological control of St. John's wort (Klamath weed) in California. Proceedings of the Seventh Pacific Science Congress, 4, 303–313.

Huffaker, C. B., & Kennett, C. E. (1966). Studies of two parasites of olive scale, *Parlatoria oleae* (Colvée). IV. Biological control of *Parlatoria oleae* (Colvée) through the compensatory action of two introduced parasites. Hilgardia, 37, 283–335.

Hughes, R. D. (1962). A method for estimating the effects of mortality on aphid populations. Journal of Animal Ecology, 31, 389–396.

Hughes, R. D. (1963). Population dynamics of the cabbage aphid, *Brevicoryne brassicae* (Linnaeus). Journal of Animal Ecology, 32, 393–424.

Hutchinson, W. D., & Hogg, D. B. (1984). Demographic statistics for the pea aphid (Homoptera: Aphididae) in Wisconsin and an comparison with other populations. Environmental Entomology, 13, 1173–1181.

Hutchinson, W. D., & Hogg, D. B. (1985). Time-specific life tables for the pea aphid, *Acyrthosiphon pisum* (Harris), on alfalfa. Researches on Population Ecology. (Kyoto), 27, 231–253.

Jervis, M. A., & Kidd, N. A. C. (1986). Host-feeding in hymenopteran parasitoids. Biological Review, 61, 395–434.

Jervis, M. A., Hawkins, B. A., & Kidd, N. A. C. (1996). The usefulness of destructive host feeding parasitoids in classical biological control: Theory and observation conflict. Ecological Entomology, 21, 41–46.

Jones, R. E. (1987). Ants, parasitoids, and the cabbage butterfly *Pieris rapae.* Journal of Animal Ecology, 56, 739–749.

Julien, M. H., & Bourne, A. S. (1988). Effects of leaf-feeding by larvae of the moth *Samea multiplicalis* Guen. (Lep., Pyralidae) on the floating weed *Salvinia molesta.* Journal of Applied Entomology, 106, 518–526.

Legner, E. F. (1977). Temperature, humidity and depth of habitat influencing host destruction and fecundity of muscoid fly parasites. Entomophaga, 22, 199–206.

Legner, E. F. (1979). The relationship between host destruction and parasite reproductive potential in *Muscidifurax raptor, M. zaraptor,* and *Spalangia endius* (Chalcidoidea: Pteromalidae). Entomophaga, 24, 145–152.

Legner, E. F., & Medved, R. A. (1981). Pink bollworm, *Pectinophora gossypiella* (Diptera [sic]: Gelechiidae) suppression with gossyplure, a pyrethroid, and parasite releases. Canadian Entomologist, 113, 355–357.

Liu, S.-S. & Hughes, R. D. (1984). Effects of host age at parasitization by *Aphidius sonchi* on the development, survival, and reproduction of the sowthistle aphid, *Hyperomyzus lactucae*. Entomologia Experimentalis et Applicata, 36, 239–246.

Loan, C. C., & Holdaway, F. G. (1961). *Microctonus aethiops* (Nees) and *Perilitis rutilus* (Nees) (Hymenoptera: Braconidae), European parasites of *Sitona* weevils (Coleop.: Curculionidae). Entomophaga, 19, 7–12.

Loan, C. C., & Lloyd, D. C. (1974). Description and field biology of *Microtonus hyperodae* Loan, n. sp. (Hymenoptera: Braconidae, Euphorinae), a parasite of *Hyperodes bonariensis* in South America (Coleoptera: Curculionidae). Entomophaga, 19, 7–12.

Lopez, E. R., & Van Driesche, R. G. (1989). Direct measurement of host and parasitoid recruitment for assessment of total losses due to parasitism in a continuously breeding species, the cabbage aphid *Brevicoryne brassicae* (L.) (Hemiptera: Aphididae). Bulletin Entomological Research, 79, 47–59.

Luck, R. F., Shepard, M., & Kenmore, P. E. (1988). Experimental methods for evaluating arthropod natural enemies. Annual Review of Entomology, 33, 367–391.

May, R. M. (1978). Host-parasitoid systems in patchy environments: A phenomenological model. Journal of Animal Ecology, 47, 833–844.

McDonald, L., Manly, B., Lockwood, J., & Logan, J. (1989). Estimation and analysis of insect populations. New York: Springer-Verlag.

McEvoy, P. B., Cox, C. S., James, R. R., & Rudd, N. T. (1990). Ecological mechanisms underlying successful biological weed control: Field experiments with ragwort *Senecio jacobaea*. Proceedings, Seventh International Symposium on Biological Control of Weeds (1st. Sper. Patol. Veg. (MPAS-P) 1988). Rome, Italy, March 6–11, 1988.

McEvoy, P. B., Rudd, N. T., Cox, C. S., & Huss, M. (1993). Disturbance, competition, and herbivory effects on ragwort *Senecio jacobaea* populations. Ecological Monographs, 63, 55–75.

McQuire, M. R., & Henry, J. E. (1989). Production and partial characterization of monoclonal antibodies for detection of entomopoxvirus from *Melanoplus sanguinipes*. Entomologia Experimentalis et Applicata, 51, 21–28.

Miller, C. A. (1954). A technique for assessing spruce budworm larval mortality caused by parasites. Canadian Journal of Zoology, 33, 5–17.

Mills, N. J., & Getz, W. M. (1996). Modelling the biological control of insect pests—A review of host-parasitoid models. Ecological Modelling, 92, 121–143.

Moore, J. (1983). Responses of an avian predator and its isopod prey to an acanthocephalan parasite. Ecology, 64, 1000–1015.

Morris, R. F. (1959). Single-factor analysis in population dynamics. Ecology, 40, 580–588.

Murdoch, W. W., Nisbet, R. M., Blythe, S. P., Gurney, W. S. C., & Reeve, J. D. (1987). An invulnerable age class and stability in delay-differential parasitoid-host models. American Naturalist, 129, 263–282.

Murdoch, W. W., Briggs, C. J., Nisbet, R. M., Gurney, W. S. C., & Stewart-Oaten, A. (1992). Aggregation and stability in metapopulation models. American Naturalist, 140, 41–58.

Murdoch, W. W., & Briggs, C. J. (1996). Theory for biological control—Recent developments. Ecology, 77, 2001–2013.

Murdoch, W. W., Swarbrick, S. L., Luck, R. F., Walde, S., & Yu, D. S. (1996). Refuge dynamics and metapopulation dynamics—An experimental test. American Naturalist, 147, 424–444.

Murdoch, W. W., Luck, R. F., Swarbrick, S. L., Walde, S., Yu, D. S., & Reeve, J. D. (1995). Regulation of an insect population under biological control. Ecology, 76, 206–217.

Neuenschwander, P., & Madojemu, E. (1986). Mortality of the cassava mealybug, *Phenacoccus manihoti* Mat. Ferr. (Hom. Pseudococcidae), associated with an attack by *Epidinocarsis lopezi* (Hym., Encyrtidae). Mitteilungen der Schweizerischen Entomologischen Gesellschaft, 59, 57–62.

Neuenschwander, P., Schulthess, F., and Madojemu, E. (1986). Experimental evaluation of the efficiency of *Epidinocarsis lopezi* a parasitoid introduced into Africa against the cassava mealybug, *Phenacoccus maniboti*. Entomologia Experimentalis et Applicata, 42, 133–138.

Neuenschwander, P., & Sullivan, D. (1987). Interactions between the endophagous parasitoid *Epidinocarsis lopezi* and its host, *Phenacoccus manihoti*. Insect Science Applications, 8, 857–859.

Nicholson, A. J. (1954). An outline of the dynamics of animal populations. Australian Journal of Zoology, 2, 9–65.

Norris, M. J. (1935). A feeding experiment on the adults of *Pieris rapae* Linnaeus (Lepid.: Rhop.). Entomologist, 68, 125–127.

Polaszek, A. (1986). The effects of two species of hymenopterous parasitoids on the reproductive system of the pea aphid, *Acyrthosiphon pisum*. Entomologia Experimentalis et Applicata, 40, 285–292.

Pollard, E., Lakhani, K. H., & Rothery, P. (1987). The detection of density-dependence from a series of annual censuses. Ecology, 68, 2046–2055.

Poswal, M. A., Berberet, R. C., & Young, J. L. (1990). Time-specific life tables for *Acyrthosiphon kondoi* (Homoptera: Aphididae) in first crop alfalfa in Oklahoma. Environmental Entomology, 19, 1001–1009.

Pottinger, R. P., & LeRoux, E. J. (1971). The biology and dynamics of *Lithocolletis blancardella* (Lepidoptera: Gracillariidae) on apple in Quebec. Memiors of the Entomological Society of Canada, 77, 437.

Quezada, J. R. (1974). Biological control of *Aleurocanthus woglumi* (Homoptera: Aleyrodidae) in El Salvador. Entomophaga, 19, 243–254.

Rahman, M. (1970). Mutilation of the imported cabbageworm by the parasite *Apanteles rubecula*. Journal of Economic Entomology, 63, 1114–1116.

Reddinguis, J., & Den Boer, P. J. (1989). On the stabilization of animal numbers. Problems of testing. I. Power estimates and estimation errors. Oecologia, 78, 1–8.

Rees, N. E. (1986). Effects of dipterous parasites on production and viability of *Melanoplus sanguinipes* eggs (Orthoptera: Acrididae). Environmental Entomology, 15, 205–206.

Reeve, J. D., & Murdoch, W. W. (1985). Aggregation by parasitoids in the successful control of the California red scale: A test of theory. Journal of Animal Ecology, 54, 797–816.

Reeve, J. D., & Murdoch, W. W. (1986). Biological control by the parasitoid *Aphytis melinus*, and population stability of the California red scale. Journal of Animal Ecology, 55, 1069–1082.

Richards, O. W., & N. Waloff (1954). Studies on the biology and population dynamics of British grasshoppers. Antilocust Bulletin, 17, 182 pp.

Ricker, W. E. (1973). Linear regressions in fishery research. Journal of Fisheries Research Board of Canada, 30, 409–434.

Ricker, W. E. (1975). A note concerning Professor Jolicoeur's comments. Journal of Fisheries Research Board of Canada, 32, 1494–1498.

Roland, J. (1988). Decline in winter moth population in North America: Direct versus indirect effect of introduced parasites. Journal of Animal Ecology, 57, 523–531.

Royama, T. (1981a). Fundamental concepts and methodology for the analysis of animal population dynamics, with particular reference to univoltine species. Ecological Monographs, 5, 473–493.

Royama, T. (1981b). Evaluation of mortality factors in insect life table analysis. Ecological Monographs, 5, 495–505.

Ruth, W. E., McNew, R. W., Caves, D. W., & Eckenbary, R. D. (1975). Greenbugs (Hom.: Aphididae) forced from host plants by *Lysiphlebus testaceipes* Hym.: Braconidae). Entomophaga, 20, 65–71.

Ryan, R. B. (1985). A hypothesis for decreasing parasitization of larch casebearer (Lepidoptera: Coleophoridae) on larch foliage by *Agathis pumila*. Canadian Entomologist, 117, 1573–1574.

Senger, S. E., & Roitberg, B. D. (1992). Effects of parasitism by *Tomicobia tibialis* Ashmead (Hymenoptera: Pteromalidae) on reproductive parameters of female pine engravers, *Ips pini* (Say). Canadian Entomologist, 124, 509–513.

Simmonds, F. J. (1948). Some difficulties in determining by means of field samples the true value of parasitic control. Bulletin of Entomological Research, 39, 435–440.

Slade, N. A. (1977). Statistical detection of density dependence from a series of sequential censuses. Ecology, 58, 1094–1102.

Smith, A. D., & Maelzer, O. A, (1986). Aggregation of parasitoids and density-independence of parasitism in field populations of the wasp *Aphytis melinus* and its host, the red scale *Aonidiella aurantii*. Ecological Entomology, 11, 425–434.

Smith, O. W. (1952). Biology and behavior of *Microctonus vittatae* Muesebeck (Braconidae). University of California, Publications in Entomology, 9, 315–344.

Smith, R. W. (1964). A field population of *Melanoplus sanguinipes* (Fabricius) (Orthoptera: Acrididae) and its parasites. Canadian Journal of Zoology, 43, 179–201.

Southwood, T. R. E. (1978). Ecological methods with particular reference to the study of insect populations (2nd ed.). London: Chapman & Hall.

Southwood, T. R. E., & Jepson, W. F. (1962). Studies on the populations of *Oscinella frit* L. (Diptera: Chloropidae) in the oat crop. Journal of Animal Ecology, 31, 481–495.

Sunderland, K. D. (1988). Quantitative methods for detecting invertebrate predation occurring in the field. Annals of Applied Biology, 112, 201–224.

Tamaki, G., Halfhill, J. E., & Hathaway, D. O. (1970). Dispersal and reduction of colonies of pea aphids by *Aphidius smithi* (Hymenoptera: Aphidiidae). Annals of the Entomological Society of America, 63, 973–980.

Van Driesche, R. G. (1983). Meaning of "percent parasitism" in studies of insect parasitoids. Environmental Entomology, 12, 1611–1622.

Van Driesche, R. G. (1988a). Survivorship patterns of *Pieris rapae* (Lep.: Pieridae) larvae in Massachusetts kale with special reference to mortality due to *Cotesia glomerata* (Hymen.: Braconidae). Bulletin of Entomological Research, 78, 397–405.

Van Driesche, R. G. (1988b). Field measurement of population recruitment of *Cotesia glomerata* (Hymenoptera: Braconidae), a parasitoid of *Pieris rapae* (Lepidoptera: Pieridae), and factors influencing adult parasitoid foraging success in collards. Bulletin of Entomological Research, 78, 199–208.

Van Driesche, R. G., & Bellows, T. S., Jr. (1988). Use of host and parasitoid recruitment in quantifying losses from parasitism in insect populations. Ecological Entomology, 13, 215–222.

Van Driesche, R. G., & Bellows, T. S. (1996). Biological control. New York: Chapman & Hall.

Van Driesche, R. G., & Gyrisco, G. G. (1979). Field studies of *Microctonus aethiopoides,* a parasite of the adult alfalfa weevil *Hypera postica,* in New York. Environmental Entomology, 8, 238–244.

Van Driesche, R. G., & Taub, G. (1983). Impact of parasitoids on *Phyllonorycter* leafminers infesting apple in Massachusetts, U.S.A. Protection Ecology, 5, 303–317.

Van Driesche, R. G., Bellotti, A., Herrera, C. J., & Castillo, J. A. (1987). Host feeding and oviposition trauma as additional sources of mortality in *Phenacoccus herreni* Cox & Williams (Homoptera: Pseudococcidae) caused by two encyrtid parasitoids, *Epidinocarsis diversicornis* (Howard) and *Acerophagus coccois* Smith (Hymenoptera: Encyrtidae). Entomologia Experimentalis et Applicata, 44, 97–100.

Van Driesche, R. G., Bellows, T. S., Jr., Ferro, D. N., Hazzard, R., & Maher, M. (1989). Estimating stage survival from recruitment and density data, with reference to egg mortality in the Colorado potato beetle, *Leptinotarsa decemlineata* (Say) (Coleoptera: Chrysomelidae). Canadian Entomologist, 121, 291–300.

Van Driesche, R. G., Bellows, T. S., Jr., Elkinton, J. S., Gould, J., & Ferro, D. N. (1991). The meaning of percentage parasitism revisited: Solutions to the problem of accurately estimating total losses from parasitism. Environmental Entomology, 20, 1–7.

van Lenteren, J. C., & Woets, J. (1988). Biological and integrated pest control in greenhouses. Annual Review of Entomology, 33, 239–269.

Varley, G. C., & Gradwell, G. R. (1968). Population models for the winter moth. In T. R. E. Southwood (Ed.), Insect abundance. Symposium of Research of the Entomological Society of London, No 4, 132–142.

Varley, G. C. & Gradwell, G. R. (1960). Key factors in insect population studies. Journal of Animal Ecology, 29, 399–401.

Varley, G. C., Gradwell, G. R., & Hassell, M. P. (1973). Insect Population Ecology: an analytical approach. Berkeley and Los Angeles: University of California Press. 212 pp.

Vickery, W. L., & Nudds, T. D. (1984). Detection of density-dependent effects in annual duck censuses. Ecology, 65, 96–104.

Wylie, H. G. (1981). Effects of collection method on estimates of parasitism and sex ratio of flea beetles (Coleoptera: Chrysomelidae) that infest rape crops in Manitoba. Canadian Entomologist, 113, 665–671.

Wylie, H. G. (1982). An effect of parasitism by *Microctonus vittatae* (Hymenoptera: Braconidae) on emergence of *Phyllotreta cruciferae* and *Phyllotreta striolata* (Coleoptera: Chrysomelidae) from overwintering sites. Canadian Entomologist, 114, 727–732.

Yano, K., Morakote, R., Satoh, M., & Asai, I. (1985). An evidence for behavioral change in *Nephotettix cincticeps* Uhler (Hemiptera: Delphacidae) parasitized by pipunculid flies (Diptera: Pipunculidae). Applied Entomology and Zoology, 20, 94–96.

Young, G. R. (1987). Some parasites of *Segestes decoratus* Redtenbacher (Orthoptera: Tettigoniidae) and their possible use in the biological control of tettigoniid pests of coconuts in Papua New Guinea. Bulletin of Entomological Research, 77, 515–524.

C H A P T E R

9

Evaluation of Biological Control with Experimental Methods

ROBERT F. LUCK

Department of Entomology
University of California
Riverside, California

B. MERLE SHEPARD

Clemson University
Coastal Research and Education Center
2865 Savannah Highway
Charleston, South Carolina

PETER E. KENMORE

Food and Agriculture Organization of the
United Nations
Room B-754
Viale delle Terme di Caracalla
00100 Rome, Italy

INTRODUCTION

The role of natural enemies in the regulation of prey or host populations is of interest to both population and applied ecologists. It is also of interest to biological control practitioners because natural enemies are the foundation for integrated pest management (IPM) (Stern et al., 1959; DeBach & Rosen, 1991) and the core of sustainable agriculture (DeBach & Rosen, 1991). In both management approaches, entomophagous arthropods are used to suppress arthropod populations of economic concern through: (1) their importation and permanent establishment, usually for the control of exotic pests (*classical biological control*), (2) their mass release as insectary reared natural enemies to augment indigenous populations (*augmentative biological control*), and (3) their conservation as resident natural enemies (*indigenous biological control*) for the control of potential or intermittent pests. The use of classical or augmentative biological control agents and their selection has been based on observation, trial and error, experience, and general theory (van Lenteren, 1980; Waage & Hassell, 1982; DeBach & Rosen, 1991; Kidd & Jervis, 1996). Indigenous natural enemies early were recognized as important sources of mortality in an agroecosystem because of the successes in classical biological control and later, because these sources of mortality were frequently disrupted by pesticide use (e.g., DeBach & Rosen, 1991).

These biological control approaches have been effective, but often the reasons for their successes or failures have not been well understood. Such understanding, however, is fundamental to improving biological control and its success rate and to the recognition of effective predator–prey (parasitoid–host) interactions in the field. Evaluating natural enemies and biological control projects, both successful and unsuccessful, is one means of gaining such understanding. These projects also can be used to test hypotheses about the desirable attributes of a natural enemy and to develop or test criteria that identify effective natural enemies in the field (e.g., Smith & DeBach, 1942; Huffaker & Kennett, 1966; Luck & Podoler, 1985; Kidd & Jervis, 1996; Pijls *et al.*, 1995; Pijls & van Alphen, 1996).

In general, two approaches have been used, alone or together, to evaluate natural enemies and their interaction with hosts or prey. These include (1) life table analysis (see Chapter 8); and (2) experimental manipulations under field conditions, including techniques that permit the identification of predator species and the estimation of their impact. These techniques aim to determine whether regulation of the herbivorous population exists and, if so, what agents are responsible for this regulation. This information is essential if pest management seeks to exploit the control provided by these natural enemies. We define regulation as the biological processes involving natural enemies that suppress prey or host densities below levels that would prevail in their absence. In an economic context, this regulation must exist below a pest's economic threshold. In this chapter we are concerned with the experimental evaluation of biological control. That is, whether regulation exists and which natural enemies are responsible for it.

Natural-enemy evaluations require the development of an appropriate sampling scheme. This scheme's design depends on the objective(s) of the experiment, the biology of the organisms, and the cost of acquiring the information. Without a clear objective and an understanding of an organism's biology, developing an appropriate experimental design and the associated sampling scheme becomes impossible. Furthermore, the sampling scheme's sample units must be taken randomly if they are to be representative of the population and the processes of interest and if traditional statistical methods are to be applied (Morris, 1955, 1960; Green, 1979; Crawley, 1993; Pedigo & Buntin, 1994). Randomness includes the location of field plots and the selection of sample units. It does not mean that samples are taken haphazardly. Each sample unit (within a stratum) must have an equal chance of being selected. Texts and articles on sampling and experimental design as well as statisticians experienced with population studies and sampling should be consulted *before* an evaluation of natural enemies or of biological control is begun (Morris, 1955, 1960; Cochran, 1963; Stuart, 1976; Elliot, 1977; Jessen, 1978; Southwood, 1978; Green, 1979; Crawley, 1993; Pedigo & Buntin, 1994). Ultimately the study's requirements must be balanced against the resources available. It is seldom the case that sufficient resources are available for a comprehensive study; thus a carefully conducted study of limited scope is more valuable than a larger one that suffers from incomplete, inadequate, and unreliable data.

Furthermore, an appropriate experimental design or sampling scheme often requires a preliminary study to identify sources of variation. Such a preliminary study can identify important sources of variation (Green, 1979). For example, Van Driesche (1983) describes some of the problems associated with estimating and interpreting percentage of parasitism from field samples while Van Driesche and Bellows (1988) discuss an analytical procedure for dealing with some of these problems. Sunderland (1996) describes the problems associated with estimating predation rates from consumption rates obtained with visual, serological, or electrophoretic approaches.

In evaluating indigenous populations, we need to determine whether biological control of the phytophagous pests exists and whether it is consistent. One means of doing so compares pest densities in an untreated area with those in an area treated with a pesticide. However, ceasing pesticide use in part of a field does not constitute the establishment of a previously unsprayed area. Prolonged pesticide use over several seasons reduces natural enemies and alternate prey or hosts on which the natural enemies depend. Thus, time is required to reestablish these interactions. Also, the untreated area must be large enough to buffer the plots from pesticide drift and to insulate arthropod populations within the plot from the dynamics and interactions of those in the adjacent areas. Estimating the degree of regulation exerted by the natural enemies residing in plots subjected to disruptive effects almost certainly underestimates the amount of potential biological control (DeBach & Rosen, 1991). Traditional pesticide trials in which a small, untreated block within a sprayed area is used to estimate the amount of control from factors other than the pesticide treatment is a case in point. Populations in the unsprayed area are overwhelmed by the dynamics of those in the surrounding treated blocks.

EVALUATION TECHNIQUES

Introduction and Augmentation of Natural Enemies

A host or prey population is expected to be self-sustaining in the case of classical and indigenous biological control; control arising from augmentative releases is temporary (lasting a season or less). The evaluation of each method poses different problems. For example, in classical biological control, a natural enemy's effectiveness can be demonstrated by comparing the change in a pest's abundance in the initial release site with a check site of similar characteristics but lacking the natural enemy (Huffaker *et al.,* 1962; Ryan, 1997). A decline in the pest's abundance in the release site compared with that in the check site suggests that the natural enemy is responsible for the pest's decline. This conclusion is further supported if the pest's abundance in the check site also declines following the subsequent introduction or immigration of the entomophage to that site. Replication of release and check sites can add confidence to the evaluation if the pattern of decrease is consistent across the plots.

A similar design can evaluate augmentative releases, but the results may be confounded if resident populations of the same or a closely related species attacks the same pest (e.g., *Trichogramma* species). How do you distinguish between the effects of released versus resident populations? Moreover, in such evaluations, a researcher or practitioner seldom has the luxury of sufficient replication in commercial agriculture, especially the no-release treatment. Few growers will tolerate the economic loss such replication may involve. Moreover, the limited number of plots also makes it unlikely that the plots are ecologically identical. Thus, the ability to discriminate among treatments is difficult unless they are dramatically different. These circumstances also raise issues about the generality of inferences made from such experiments. However, a release versus check plot experiment can be insightful (Oatman & Platner, 1971, 1978), especially when coupled with additional experimental approaches such as exclusion, inclusion, or interference methods. If they are consistent with the results from the release and no-release experiments, then increased confidence in the results is obtained. Also, releasing genetically marked individuals that differ from the resident population only in the genetic marker is another method that can provide an additional test of a hypothesis by distinguishing between resident and released populations (Kazmer & Luck, 1995).

Successful control of arthropod pests with introduced natural enemies demonstrates the potential role that indigenous entomophages play in population regulation (Wilson, 1960; Dowden, 1962; McGugan & Coppel, 1962; McLeod, 1962; DeBach, 1964; CIBC, 1971; Greathead, 1971; Rao *et al.*, 1971; Laing & Hamai, 1976; Clausen, 1978; Luck, 1981; Kelleher & Hulme, 1984; Cock, 1985). This interpretation is further substantiated when introductions are repeated at several locations with similar results (DeBach, 1964; Laing & Hamai, 1976).

EXCLUSION OR INCLUSION TECHNIQUES

Cages and Other Barriers

Exclusion or inclusion techniques utilizing cages have been frequently employed to evaluate natural enemies (Smith & DeBach, 1942; DeBach *et al.*, 1949; DeBach, 1955; Sparks *et al.*, 1966; Lingren *et al.*, 1968; Way & Banks, 1968; van den Bosch *et al.*, 1969; DeBach & Huffaker, 1971; Ashby, 1974; Campbell, 1978; Richman *et al.*, 1980; Aveling, 1981; Faeth & Simberloff, 1981; Frazer *et al.*, 1981b; Jones, 1982; Elvin *et al.*, 1983; Chambers *et al.*, 1983; Linit & Stephen, 1983; Barry *et al.*, 1984; Kring *et al.*, 1985; Parrella *et al.*, 1991; Rosenheim *et al.*, 1993; Heinz & Parella, 1994; Simmon & Minkenberg, 1994; Boavida *et al.*, 1995; Hopper *et al.*, 1995; Nechols *et al.*, 1996). The use of cages to exclude entomophages was pioneered by Smith and DeBach (1942), who used paired sleeve cages to test whether introduced parasitoid *Metaphycus helvolus* (Compere), controlled the black scale *Saissetia oleae* (Bern.). Comparison of the black scale in the open and closed cages showed that less black scale survived in the open cages. This technique was modified by using netting impregnated with insecticide to kill natural enemies that emerged in the closed cages when the method was adopted to evaluate other classical biological control projects (DeBach *et al.*, 1949; DeBach, 1955; DeBach & Huffaker, 1971). This modification eliminated the need to fumigate and reinfest the caged branches.

Small cages of different mesh size have been used to exclude classes of natural enemies based on their size (Campbell, 1978; Kring *et al.*, 1985). Usually three types of cages are employed: (1) a complete exclusion cage with small-mesh netting and sealed at both ends, (2) a control cage with similar netting and open at both ends, and (3) a partial exclusion cage with large-mesh netting and closed at both ends. The latter seeks to exclude large predators (ladybird beetles and earwigs) but allows access of small predators (anthocorids) and parasitoids.

Boavida *et al.* (1995) also used three treatments—a closed sleeve cage, open sleeve cage, and no sleeve cage—to evaluate the efficacy of *Gyranusoidea tebygi* Noyes, a classical biological control agent released in west Africa to control the mango mealybug *Rastrococcus invadens* Williams. The average density of mealybugs surviving differed significantly among treatments and each treatment differed significantly. The mealybug density per leaf in the closed sleeve cages was 2.7 times higher than that in the open sleeve cages and 6.2 times higher than that in the no sleeve cage treatments.

Rosenheim *et al.* (1993) used inclusion/exclusion cages to assess the effects of intraguild interactions among predators on the densities of an herbivore. They enclosed one of three predatory hemipterans or no hemipterans in replicated cages with green lacewing (*Chrysoperla* species). *Chrysoperla* is an important predator of the herbivore of interest. By using cages in which either lacewing larvae or eggs were introduced into cages alone or with the hemipteran predators in various combinations, they assessed the effect of the different hemipterans predators on lacewing numbers in the cage, and, in turn, indirectly on the herbivore densities in the cages. They sought to test the hypothesis that one or more of the species of hemipteran predators preyed on lacewing larvae and caused their rarity in San Joaquin Valley, California, cotton fields even though lacewing eggs in these fields were common. The predators most responsible for the mortality of lacewing larvae in the cages were *Zelus* spp. and *Geocoris* spp. These results were consistent with the field observations (Rosenheim, *et al.*, 1993).

A more complex design involving sleeve and field cages (large cages enclosing whole plants), coupled with samples of the prey and natural-enemy populations, showed that the spring increase of predators eliminated black bean aphid (*Aphis fabae* Scop.) colonies on its overwintering host (*Euonymus europaenus* Linnaeus) after June (Way & Banks, 1968). If the spring aphid populations on the tree had been dense, the predators that remained after the aphids had emigrated to their summer hosts prevented recolonization of spindle tree by late fundatrices during the summer, most years, even though the spindle tree was capable of supporting an increasing aphid population. Closed field cages covered with dieldrin-treated netting coupled with hand removal excluded natural enemies from some spindle trees, whereas open field cages constructed with slatted walls allowed access of the natural enemies to the aphids on the "uncaged" trees but provided the same degree of shading as the closed cage (Way & Banks, 1958, 1968). These sleeve cage experiments coupled with the population censusing on the sample twigs documented the importance of predators in excluding aphids from the overwintering host plant during the summers.

Large field cages and population samples were also used to evaluate indigenous natural enemies of cereal aphids. The experimental design combined field cages erected at several intervals after the aphids immigrated into a winter wheat field. The growth rates and peak densities of the aphid populations within the cages were compared with those in several open plots of similar size (Chambers et al., 1983). Samples showed that the abundance of *Coccinella 7-punctata* Linnaeus was negatively correlated with aphid abundance in the open plots, but the incidence of parasitism and disease was not negatively correlated with aphid abundance. These latter two factors were more common in the caged plots. If the difference between the aphid densities in the cage and open plots was converted to *per capita* aphid consumption, based on the sampled coccinellid densities, the calculated values were within the range of published values. Coccinellids appeared to be the key agents limiting the growth rate and peak abundance of cereal aphids during midseason, but they were unable to do so early in the season (Rabbinge et al., 1979; Carter et al., 1980).

Large field cages with open field controls were used to determine whether the predator complex aggregated at dense patches of the pea aphid *Acrythosiphon pisum* (Harris) (Frazer et al., 1981b). The cages excluded the predators and allowed the aphid population to increase to about five times that of the control (open) plots. When the cages were removed, the aphid populations declined to the densities that prevailed in the control plots and the decline was correlated with increased predator numbers aggregating at the denser aphid patches.

Large field cages have also been used to evaluate the potential of cotton predators to reduce egg and larval populations of the tobacco budworm *Helicoverpa* (= *Heliothis*) *virescens* (Fabricius) on cotton (Lingren et al., 1968). Evening releases of budworm moths initiated the prey populations within the cages. Less prey survived in the cages with predators than in cages excluding predators. Similar studies were conducted in California cotton to evaluate predation on the survival of larval populations of the cotton bollworm *Helicoverpa* (= *Heliothis*) *zea* (Boddie) (van den Bosch et al., 1969). The cotton plants within the "predator-free" cages were treated with an insecticide to eliminate resident predators before bollworm larvae were introduced. Significantly fewer prey survived in the untreated cages and significantly more predators were collected from the untreated cages.

Caged and uncaged plots and plots of similar size but enclosed with a cage within a cage were used to determine whether indigenous natural enemies or microclimatic changes within the cage explained the increased survival of the caged European corn borer *Ostrinia nubialis* (Hübner) larvae (Sparks et al., 1966). The double cage was designed such that the screened panels on the inside cage were opposite the unscreened panels on the outside cage and vice versa. This design allowed the predators access to the plants inside while maintaining the same level of shading and air flow in both the complete cage and the cage within a cage plots.

Cages were also used to evaluate the efficacy of the entomopathogenic fungi (Deuteromycotina) for the control of the black bug (*Scotinophara coarctata* Fabricius) in rice (Rombach et al., 1986a). Adult bugs were introduced into screen cages; and applications of the fungi *Beauveria bassiana* (Bals.) Vuill, *Metarhizium anisopliae* (Metsch.), and *Paecilomyces lilacinus* Thom. were made using a backpack sprayer. The black bugs were significantly less abundant in all treatments when compared with the untreated controls. The effects lasted up to 9 weeks. Similarly, caged brown plant hoppers (*Nilaparvata lugens* Stål) were treated with a group of entomopathogenic hyphomycetes (Rombach et al., 1986b). Mortality from the fungus infections ranged from 63 to 98% 3 weeks after application.

Trenches with insecticide-soaked straw were used to exclude ground predators, mainly carabids, from populations of the cabbage root fly [*Erioischia brassicae* (Bouché)] (Wright et al., 1960, Coaker, 1965). Polythene barriers were used to exclude predators from two of three treatments, in which the predator density was manipulated to determine their effect on the density of aphid populations (Winder, 1990). Sticky bands around selected branches of a spindle tree were used to exclude the walking predators of *Aphis fabae* (Way & Banks, 1968) and around the plant base to exclude walking predators of *Trichoplusia ni* (Hübner) (Jones, 1982). Sticky circles around *T. ni* eggs were used to exclude predators and parasitoids from attacking the eggs (Jones, 1982).

Exclusion or inclusion experiments with cages provide insight into the predation process and the predator complexes that attack aphids and Lepidoptera. Useful studies relate cage densities to the densities of resident field populations outside the cage (Frazer & Gilbert, 1976; Campbell, 1978; Aveling, 1981; Frazer et al., 1981b; Chambers et al., 1983) or provide specific tests of hypotheses (Way & Banks, 1968; van den Bosch et al., 1969; Campbell, 1978; Carter et al., 1980; Aveling, 1981; Faeth & Simberloff, 1981; Frazer et al., 1981b; Chambers et al., 1983; Rosenheim et al., 1993).

Cage studies can estimate predation rates (cf Elvin et al., 1983) but such estimates must be taken cautiously. They have limitations. Small sleeve cages inhibit predator or prey movement and are suited to experiments with sessile species or species with low vagility (Smith & DeBach, 1942). For example, the abundance of citrus red mite [*Panonychus* (= *Metatetranychus*) *citri* (McGregor)] within sleeve cages was up to 12 times greater than outside sleeve cages (Fleschner, 1958) even though the mite population outside the cages was kept predator-free by continuous hand removal of predators. The cage likely prevented the reproductive females from emigrating, the microclimate within the cages favored rapid growth of the mite population, or both factors influenced population growth (Fleschner et al., 1955; Fleschner, 1958). Moreover, open cages can inhibit predator and or parasitoid activity compared with that occurring in an uncaged treatment (Neuenschwander *et al.*, 1986; Boavida *et al.*, 1995). For example, Boavida *et al.* (1995) found that the population of mealybugs was 2.3 times denser per leaf in the open sleeve cages than in the nonsleeve cage treatment.

Exclusion cages cannot identify which members of a predator and/or parasitoid complex are controlling a host population unless the complex consists of one or a few species (Jones, 1982). Partial exclusion cages (i.e., small versus large meshed screening) may show whether small predators, disease, or parasitoids limit prey or host populations in the absence of large predators but they cannot show whether large predators limit the prey populations in the absence of parasitoids or small predators. We assume that small predator or parasitoid behavior is unaffected when these animals move through the mesh of partial exclusion cages.

Cages may also inhibit predator or prey movement or interfere with oviposition behavior. Two leaf-mining species on oak failed to reproduce within whole-tree cages and a third species failed to reproduce in one cage (Faeth & Simberloff, 1981). Also, aphid alates cannot emigrate from a cage. Thus caged versus uncaged aphid populations may show differences in density because alate immigration reduces the uncaged aphid population. Some predator species, perhaps most, aggregate at patches of high prey density (= numerical response) (Readshaw, 1973; Frazer *et al.*, 1981a; Kareiva, 1985; Hopper, 1991). This behavior may be inhibited by cage size because the spatial pattern in nature to which the predator species responds is larger than that present within the cage. Also, confining predators to a cage may cause them to search the caged areas repeatedly and thereby increase unrealistically the likelihood that they will encounter prey, especially when the prey densities are low. Under these conditions the predator may reduce prey densities to levels below those that can be expected under field conditions. In this sense, inclusion studies resemble laboratory experiments in which predators are confined with prey (van Lenteren & Bakker, 1976; Luck *et al.*, 1979).

Misleading results can occur when prey are put into a cage without understanding their preferences for oviposition sites, their density and distribution patterns, or their preferred feeding site(s) under field conditions. Some predators and parasitoids use kairomones to find their prey and hosts (Nordlund *et al.*, 1981; Vet & Dicke, 1992). Some kairomones are associated with feeding activity. Placing prey or hosts in new sites influences their risk of detection. Food quality may affect a phytophage's feeding time and increase its risk (exposure) to predation because of the kairomones released while feeding (Vet & Dicke, 1992). Unfortunately, abnormal behaviors usually cannot be de-

tected without study. Detailed studies of a predator's searching behavior and capture rates and of a prey's oviposition and feeding behavior are important (Fleschner, 1950; Dixon, 1959; Frazer & Gilbert, 1976; Gilbert *et al.*, 1976; Rabbinge *et al.*, 1979; Carter *et al.*, 1980; Baumgaetner *et al.*, 1981; Frazer & Gill, 1981; Sabelis, 1981; Hopper, 1991).

Cage studies employing predator-free controls are seldom predator free, even when they have been treated with insecticides (van den Bosch *et al.*, 1969; Irwin *et al.*, 1974; Elvin *et al.*, 1983; Rosenheim, *et al.*, 1993). Some predator stages pass through the screen (Sailer, 1966; Way & Banks, 1968), are very difficult to exclude because they bury themselves in the soil (Frazer *et al.*, 1981a; Elvin *et al.*, 1983), or their eggs are laid within the plant substrate and are difficult to detect prior to caging (Rosenheim *et al.*, 1993).

Cages also alter the microclimate through shading and inhibition of airflow. For example, exclusion and partial exclusion cages using Terylene (Dacron) netting reduced the light intensity inside the cages by 37 and 24%, respectively (Campbell, 1978), and saran screen reduced solar radiation by 19% (Hand & Keaster, 1967). Such shading required the use of a more shade-tolerant cotton cultivar than was normally planted in the region (van den Bosch *et al.*, 1969). Shading also affects plant physiology and thus may affect the plant's quality as a substrate for the host or prey population (Scriber & Slansky, 1981). Temperatures within cages used in a corn borer study were 8 to 10°F lower than the temperature outside. The humidity fluctuated more moderately within and was 5 to 10% higher than that outside the exclusion cages (Sparks *et al.*, 1966).

Changes in solar radiation cause changes in leaf temperature by as much as 13°C (Hand & Keaster, 1967). Leaf temperatures and moisture availability influence photosynthetic rates and evapotranspiration (Gates, 1980). Leaf temperatures probably affect the behavior and feeding rates of phytophagous hosts or prey. Temperature-related interactions between the growth rates of aphids and the searching rates of their predators are important (Frazer & Gilbert, 1976; Frazer *et al.*, 1981a). Screening also reduced wind speed within a cage by as much as 48% (Hand & Keaster, 1967), which—depending on humidity and wind velocity outside and inside a cager—will significantly influence the leaf's boundary layer within the cage (Gates, 1980; Ferro & Southwick, 1984). Instrumentation allows the monitoring of many of these effects but their influence on predator–prey interactions must be assessed.

The caveat from all these potential effects is, as with all experimental manipulation, to interpret the results with care and to test the results with one or more manipulations that use different methods with different assumptions. This provides a second independent test of the hypothesis. In spite of the limitations, cages remain a powerful technique with which to evaluate the efficacy of natural enemies.

Insecticide Removal

The application of insecticides is an excellent way to assess the impact of the natural-enemy complex. The technique was first used to kill natural enemies of the long-tailed mealybug *Pseudococcus longispinus* (Targioni-Tazzetti) without affecting the mealybugs (DeBach, 1946). Insecticide treatment has been used to determine whether indigenous predator populations in cotton suppress populations of the beet armyworm *Spodoptera exigua* (Hübner) and cabbage looper *Trichoplusia ni* (Ehler *et al.*, 1973; Eveleens *et al.*, 1973). Early-season insecticides applied to cotton were thought to interfere with natural controls (Ehler *et al.*, 1973; Eveleens *et al.*, 1973). Blocks 3 to 4 mi^2 were treated with an insecticide scheduled during (1) early season; (2) early and midseason; and (3) early, mid-, and late season. A fourth plot served as an unsprayed check. Samples and observations showed that the absence of predators in the treated plots was correlated with the increased survival of beet armyworm eggs and first generation small larvae of the cabbage looper. The hemipteran predators *Geocorus pallens* Stål, *Orius tristicolor* (White), and *Nabis americoferus* Carayon were implicated as the important predators because they were the most affected by the treatments whereas *Chrysoperla carnea* Stephen was not as drastically affected. Insecticide application showed that the suppression of cabbage looper densities in celery arising from egg parasitism by *Trichogramma* species, and predation of eggs and young larvae by *Hippodamia convergens* (Guérin-Méneville) and *O. tristicolor* were sufficient to prevent economic damage before the production of the first marketable petiole in celery (Jones, 1982).

Insecticide application was also used to test whether the coccinellid *Stethorus* species limited the density of the two-spotted spider mite *Tetranychus urticae* Koch in a previously unsprayed apple orchard in Australia (Readshaw, 1973). Two applications of malathion increased the density of the mite populations. *Tetranychus urticae,* unlike its predator faunas, was resistant to malathion. *Stethorus* limited the mite population by numerically responding both aggregatively and reproductively to the denser mite patches. Even with insecticide disruption and, perhaps, insecticidal stimulation of the mite reproduction (Charboussou, 1965; Bartlett, 1968; van de Vrie *et al.*, 1972; Dittrich *et al.*, 1974), *Stethorus* was able to prevent the mite population from attaining its economic density of 100 mites per leaf on most trees.

Neuenschwander *et al.* (1986) used both open, closed, and no sleeve cages along with a pesticide treatment to evaluate the efficacy of *Epidincarsis lopezi* against the cassava mealybug (*Planacoccus manihoti*). The use of the pesticide showed that *E. lopezi* suppressed cassava mealybug densities and was a second experimental test of the results that came from the sleeve cage experiments.

In a classic experiment, the compensatory action of two parasitoids of the olive scale [*Parlatoria oleae* (Colvée)] was also evaluated using insecticide applications (Huffaker & Kennett, 1966). The scale is bivoltine on olive in California's San Joaquin Valley: one generation occurs during the autumn and spring and the second generation occurs during the summer. *Aphytis paramaculicornis* DeBach & Rosen and *Coccophagoides utilis* Doutt had been introduced to control the scale (Rosen & DeBach, 1978). *Aphytis paramaculicornis* dominated during the autumn–spring scale generation; *C. utilis* dominated during the summer generation. Three DDT treatments were used to exclude the parasitoids: (1) a spring treatment to exclude *A. paramaculicornis*, (2) a summer treatment to exclude *C. utilis,* and (3) a spring and summer treatment to exclude both parasitoids. Untreated trees served as checks. Presumably, DDT residues on foliage and twigs inhibited the parasitoid's effectiveness but did not affect the scale's reproduction and survival. Treatments that excluded *C. utilis* had higher scale densities than did the untreated check but lower densities than treatments that excluded *A. paramaculicornis.* Treatments designed to exclude only one of the wasps had lower scale densities than treatments that excluded both wasps. The treatments also indicated that together the wasps provided better suppression than either did alone even though the mortality contributed by *C. utilis* was only about 5%. These experiments corroborated the results from the field samples that showed a higher fraction of scales were parasitized by *A. paramaculicornis.*

Interpreting the results from an insecticide treatment experiment poses several problems, including (1) pesticide-stimulated reproduction of the prey population, (2) pesticide-induced sex ratio bias, and (3) pesticide-induced physiological effects on the plant. Mites exposed to sublethal doses of pesticides are stimulated reproductively and on occasion evince increased female-biased sex ratios (Charboussou, 1965; Bartlett, 1968; van de Vrie *et al.*, 1972; Dittrich *et al.*, 1974; Maggi & Leigh, 1983; Jones & Parrella, 1984). Such effects may also extend to aphids (Bartlett, 1968; Mueke *et al.*, 1978) and delphacids (Chelliah *et al.*, 1980; Ressig *et al.*, 1982). Differential mortality resulting from pesticide treatments has also been reported. Male black pine leaf scale [*Nuculaspis californica* (Coleman)] (Edmunds & Alstad, 1985), and California red scale [*Aonidiella aurantii* (Maskell)] (Shaw *et al.*, 1973), are more susceptible to pesticides than females. Plant physiology is also affected by insecticide applications (Kinzer *et al.*, 1977; Jones *et al.*, 1983). Row crops treated with certain insecticides have become attractive oviposition sites for Lepidoptera (Kinzer *et al.*, 1977). Also, interactions between aphid reproduction, insecticides, and cultivars have been reported on alfalfa (Mueke *et al.,* 1978). As with other exclusion methods, knowledge of the biology and interactions is required to properly time the insecticide applications to disrupt the natural-enemy populations while minimizing their effects on the prey or host population. Because insecticides potentially stimulate arthropod reproduction and effect plant physiology, estimation of predation rates with this exclusion method should be conducted with caution. Although the insecticide treatments reproductively stimulated the brown plant hopper (*Nilaparvata lugens*) the amount of stimulation could not account for the high levels of resurgence. Only the reduction of the natural enemies by the chemicals could. Insecticides can be used to determine the relative importance of natural-enemy species when an entomophage complex is composed of a few species showing temporal separation of their affects, in seasonal occurrence or in the generations they attack.

Hand Removal

Hand removal is a method of assessing the effectiveness of natural enemies that has not been used often because the method is laborious. It has been used to evaluate the predators of tetranychid mites on citrus and avocado and to check results obtained with other exclusion methods (Fleschner *et al.,* 1955; Fleschner, 1958). It has also been used to evaluate the mirid *Crytorhinus fulvus* Knight introduced to control the taro leafhopper *Tarophagus proserpina* (Kirkaldy) (Matsumoto & Nishida, 1966). Predation of *Aphis fabae* was also assessed, in part, by hand removing adult predators that flew onto "predator-free" branches (Way & Banks, 1968). A sticky band at the base excluded walking predators from feeding on *A. fabae* individuals placed on the branch.

Huffaker and Kennett (1956) also used hand removal to determine whether excluding *Typhlodromus reticulatus* Oudemans or *T. cucumeris* Oudemans from several plots with an acaracide stimulated the cyclamen mite (*Tarsonemus pallidus* Banks) reproductively to increase its densities in these plots. They had used the acaricide to establish and maintain several predator-free check plots. When they removed the predacious mites by hand in several of the check plots, they found that cycalmen mite increased in the absence of predators acaricides. They therefore concluded that the acaracide treatments had not stimulated the cyclamen mite reproductively.

The hand removal method deserves more attention, especially as a method of checking for bias in other exclusion methods. It use is limited, however, to studies of predator–prey interactions involving species that are of low vagility, occur at reasonable densities, and are diurnally active or are undisturbed by night lights.

Outplants or Prey Enrichment

Insight into natural-enemy efficacy can be gained by placing prey on plants in the field. This method involves tethering prey to a substrate (Weseloh, 1974, 1982) or placing them on leaves or other plant parts where they normally occur (Ryan & Medley, 1970; Elvin *et al.,* 1983, van Sickle & Weseloh, 1974; Weseloh, 1974, 1978, 1982; Torgensen & Ryan, 1981). Some studies mark the prey with dyes before placing them in the field (Hawkes, 1972; Elvin *et al.,* 1983). The prey are visited at frequent intervals to measure predation. If predation is observed, the predator's identity is noted. Predators such as spiders can be observed in the field with their prey (Kiritani *et al.,* 1972). Web spinning spiders leave cadavers in or beneath their webs. The cadavers can be counted and identified to estimate predation (Turnbull, 1964).

Often, it is more practical to use greenhouse-grown plants of the same age, size, and variety as plants used in field studies. Plants can be caged in the greenhouse or field for oviposition by pests. These egg-laden plants are then transferred to the field and monitored for parasitism and predation. van der Berg *et al.* (1988) used eggs of several foliage-feeding rice pests to determine predation. The egg chorion showed that eggs were attacked by predators with chewing or sucking mouthparts.

The seasonal incidence of predation and parasitism of eggs of the yellow stemborer of rice [*Scirpophaga incertulas* (Walker)] was determined in rice using the preceding technique (Shepard & Arida, 1986). Their results suggested that predation and parasitism may alternate as major mortality factors during the year.

Parker (1971) used a prey enrichment method to test the hypotheses that the density of the imported cabbage butterfly [*Pieris rapae* (L.)] was too scarce during its first two generations to allow an early increase of sufficient natural enemies to suppress the third and subsequent butterfly generations. The butterfly's egg density during the first generation was one egg per 27 cabbage plants; in the second generation it was one egg per 4 cabbage plants. Moreover, these egg peaks were separated by 25 days with even lower egg densities between the peaks.

The prey enrichment technique is ideal with an evaluation program using cages and/or insecticides. Substantially more information can be gained from these techniques in combination, than from the use of either one alone. The major drawback is that outplants are somewhat limited to sessile forms (eggs, pupae, or some scale insects) but offer possibilities with tethered pests (Weseloh, 1974). Also, kairomones and other cues may be important in establishing the appropriate interaction (Nordlund *et al.,* 1981). Outplant techniques should be used more often in natural-enemy evaluation programs.

ASSAY TECHNIQUES

Serological Methods

Indigenous biological control arises from a combination of predation and parasitism. Most of the effort to characterize this natural-enemy complex has focused on the role of parasitoids (but see Baumgaetner *et al.*, 1981; Frazer & Gilbert, 1976; Frazer & Gill, 1981; Frazer *et al.*, 1981a, 1981b). However, in many agroecosystems predation is an important, if not the dominant source of mortality to its herbivorous and predaceous constitutents (*e.g.*, Ehler *et al.*, 1973; Eveleens *et al.*, 1973; Rosenheim *et al.*, 1993; Hagler & Naranjo, 1996). Techniques have been developed only within the last several decades to characterize the food-web relationships in predator-dominated systems, to estimate the relative contributions that each predator species makes to prey mortality in the system, and to estimate predation (i.e., consumption) rates (Sunderland, 1996). These methods mostly involve serological and isozyme analysis of the gut contents of predators [reviewed by Sunderland (1988) and Symondson and Liddell, (1996)].

Serological analyses have been the most widely used methods to investigate predator–prey interactions in a field context. They are used to associate predators with their prey (Dempster *et al.*, 1959; Dempster, 1960, 1964, 1967; Rothchild, 1966, 1970, 1971; Frank, 1967; Ashby, 1974; Vickermann & Sunderland, 1975; Boreham & Ohiagu, 1978; Sunderland & Sutton, 1980; Gardner *et al.*, 1981; Ragsdale *et al.*, 1981; Greenstone, 1983; Crook & Sunderland, 1984; Greenstone & Morgan, 1989; Hagler *et al.*, 1992; Sopp *et al.*, 1992; Hagler & Naranjo, 1994a, 1994b, 1996) and—in a more limited context with caveats—to estimate predation rates (Dempster *et al.*, 1959; Dempster, 1960, 1964, 1967; Rothchild, 1966; Sopp *et al.*, 1992).

Precipitin assays were the first serological methods used (Boreham & Ohiagu, 1978; Ohiagu & Boreham, 1978; Southwood, 1978; Greenstone, 1996). They involve the extraction of material from the prey species, which is then injected into a vertebrate host. This material typically involves large molecules, usually proteins or a carbohydrate, which serve as antigens. When they are injected into the vertebrate host (with several subsequent booster injections), they cause the vertebrate's immune system to produce antibodies against them. The host is then bled, the blood is allowed to clot, and the serum (which contains the antibodies) is collected. The specificity of the antibodies within the serum is then evaluated by immunoassays against antigens from the prey species of interest, to related prey species, and to other suspected prey species within the ecological system of interest. Typically, some cross-reaction to nontarget prey species occurs, especially if the species are related, that is, if they share a common ancestor evolutionarily. The antibodies within the serum are used to detect a predator has fed on the prey species of interest. This is accomplished by preparing a soluble fraction from a predator and combining it with a portion of the antibody solution. If the predator has fed on a prey item recently, the prey's antigens in the predator's gut combine with the antibodies to form a visible precipitate. This assay initially used serological tubes and large amounts of antibody and antigen, which were necessary to ensure sufficient precipitate for visibility. Techniques subsequently have been developed to improve the precipitin test's specificity and enhance its sensitivity [see Greenstone (1996) for detailed discussions of these techniques with references].

The precipitin test was first used to document arthropod predation of mosquito larvae (Bull & King, 1923; Hall *et al.*, 1953; Downe & West, 1954) and later applied to terrestrial predator–prey interactions (Downe & West, 1954). This technique's promise lay in its potential to quantify predation rates. The first prey for which such estimates were attempted from field samples was *Gonioctena* (= *Phytodecta*) *olivacea* (Forster), a chrysomelid beetle that feeds on broom (Dempster, 1960). The test identified six mirids, two anthocorids, a nabid, an earwig, and red mites as feeding on the beetle in the field. Laboratory studies showed that only the older mirid and anthocorid stages fed on *G. olivacea* and then only on the younger stages. One laboratory feeding by the mirids and anthocorids could be detected 24 h after they had ingested a meal.

The number of beetles preyed on depended on the degree of overlap between the older stages of the predator and the younger stages of the beetle. The densities of these prey and predators were estimated from field samples. The fraction of positive responses in predator samples estimated the fraction of the predator population that had fed on *G. olivacea*. Because in the field *G. olivacea* were scarce while alternative prey were abundant, encounters between *G. olivacea* and the predators were infrequent. Thus, if a predator tested positive to *G. olivacea* antibody, it was interpreted as a single predation event. Given this assumption the number of beetles preyed on by each predator could then be estimated using the following equation:

$$P_a = (N_{pi}F_{pi}T_{pi})/R_{pi} \qquad (1)$$

where P_a is the number of prey killed; N_{pi}, the density of the predator (or predator stage)i; F_{pi}, the fraction of positive tests of the ith predator species in a sample of predators; T_{pi}, the duration in days that the appropriate prey and predator stages coincident in the field; and R_{pi}, the length of time that a single prey feeding by i^{th} predator (or predator stage) could be detected based on laboratory tests. The estimates obtained from the precipitin test for egg and larval mortality due to predation during two beetle generations agreed quite closely with the independent estimates of "unknown" losses of eggs and young larvae during the same two beetle generations (Richards & Waloff, 1961).

The precipitin test was also used to identify the predator species and to determine the fraction of *Pieris rapae* (Linnaeus) eggs and young larvae that died due to predation (Dempster, 1967). Again, because of the relative scarcity of *P. rapae,* a positive precipitin test was interpreted as one predation event. Precipitin tests have also been used to identify the carabid predators of the black cutworm [*Agrotis ipsilon* (Hufnagel)], a periodic pest of corn (Lund & Turpin, 1977), and the predators of the tarnished plant bug [*Lygus lineolaris* (Palisot de Beauvois)]] (Whalon & Parker, 1978).

A study of the delphacid *Conomelus anceps* (Germar) also employed precipitin tests to identify 10 of 91 potential predator species as feeding on the delphacid (Rothchild, 1966). However, in this case the precipitin test could not be used to estimate predation rates because a positive reaction could not be assumed to represent a single predation event. Multiple predation events were likely because *C. anceps* was abundant and the most available prey.

Estimating consumption rates of the prey of interest with the precipitin test requires information about (1) predator and prey densities, (2) densities of alternate prey, (3) the period during which a meal can be detected in each predator, and (4) prey and predator stages involved. Precipitin tests can only estimate consumption rates of prey species that form a small fraction of the available prey or represent infrequent predation events. Biases in the estimates of predation rates may arise if predators have fed on other predators that have fed on the prey, if a suspected predator is phytophagous but ingests sessile prey stages while feeding on the plant, or if a suspected predator feeds on prey carrion (Boreham & Ohiagu, 1978; Sunderland, 1996). The precipitin tests may also yield biased estimates of predation rates from cross-reactions between the antibodies of closely related species. Thus, knowledge of the local fauna, which might serve as prey, and the predator's propensity for scavenging and its foraging behavior are essential for successful application of this technique (Boreham & Ohiagu, 1978; Sunderland, 1996).

The sensitivity of the percipitin test can be increased if the reaction between the antigen and antibody involves reagents that are bound to a solid surface and linked to a color-producing enzyme [i.e., the enzyme-linked immunosorbent assay (ELISA)] (Greenstone, 1996). The solid surface in this case is usually a molded polystyrene plate with 96 wells. The antibody is bound to the wall of the wells, which is then subjected to a period of incubation and washing, after which the extract from a predator is added to a well (one predator per well). If the extract contains the prey's antigen, it will bind to the antibody, which is attached to the wall of each well. After a second period of incubation and washing, additional antibody is added, but in this case the antibody is attached to an enzyme. Again this mixture is incubated and washed for a third time, after which a final reagent is added. In this case, the reagent is the substrate for the enzyme that is hydrolyzed by the enzyme producing a colored product that denotes an antibody–antigen reaction; that is, the predator is positive for the prey antigen. The amount of the reaction in each of the 96 wells can be scored with a multichanneled spectrophotometer that allows the reaction to be quantified. ELISA has become the assay of choice and is now widely used (Vickermann & Sunderland, 1975; Fichter & Stephen, 1979, 1981, 1984; Ragsdale *et al.,* 1981; Crook & Sunderland, 1984; Sunderland *et al.,* 1987; Sopp & Sunderland, 1989; Greenstone & Morgan, 1989; Hagler *et al.,* 1992; Sopp *et al.,* 1992; Hagler & Naranjo, 1994a, 1994b, 1996).

In the initial precipitin assays, the antibodies obtained in the serum extract from the vertebrate host consisted of a mixture of antibodies. This mixture is referred to as a polyclonal antibody [see Greenstone (1996) for a detailed discussion along with references]. The polyclonal nature of this initial extract makes the antibody nonspecific and insensitive. However, the sensitivity and specificity of an antibody antigen reaction can be improved substantially if, by screening the serum containing the polyclonal antibodies, a single antibody clone can be obtained that only reacts with the antigen of the prey species or stage of interest. This is referred to as a monoclonal antibody (MAb). Although this method is technically difficult, time consuming, and requires a substantial commitment of resources and expertise [see Greenstone (1996) and the cited references], it has several substantial advantages. First, it provides the most specific antibody attainable. Second, it provides the antibody in an unlimited, pure, and uniform supply. Third, the MAb can also be linked to an enzyme to create an ELISA assay, which allows the detection of very small amounts of prey antigen. This assay can detect hemolymph dilutions of more than 260,000 (Fichter & Stephen, 1981). Thus, it can differentiate among prey stages (Ragsdale *et al.,* 1981; Hagler & Naranjo, 1994a, 1994b, 1996). Moreover, it has been used by several groups for predation studies in systems in which the predator complex is the dominate natural-enemy component and the suspected source of substantial mortality to the herbivore species or stage of interest (Sopp & Sunderland, 1989; Greenstone & Morgan, 1989; Hagler *et al.,* 1992; Sopp *et al.,* 1992; Hagler & Naranjo, 1994a, 1994b, 1996). With this technique, as with the earlier precipitin assay, detection of prey in the gut of a predator depends on (1) the size of the meal, (2) the elapsed time since the predator ate, and (3) the rate of digestion (which includes the temperature at which digestion is occurring). This assay when coupled with ELISA is practical and rapid, and can be used to identify specific prey species or stages in a large sample of predators (e.g., Hagler & Naranjo, 1994b, 1996). In a few situations it can also be used to estimate predation (consumption) rates (Sopp *et al.,* 1992; Sunderland, 1988, 1996).

However, the assays cannot be used to estimate predation rates under all circumstances. Nonetheless, this method is valuable for identifying predator species or stages that feed on a prey species or stage. This method deserves more attention, especially now that more sensitive assays, such as ELISA, are available (Vickermann & Sunderland, 1975; Fichter & Stephen, 1981, 1984; Ragsdale *et al.,* 1981; Crook & Sunderland, 1984; Lövel *et al.,* 1985; Hance & Renier, 1987; Sunderland *et al.,* 1987; Sopp & Sunderland, 1989). Moreover, the development of highly specific monoclonal antibodies is likely to become much easier and less expensive with rapid advances in molecular techniques. In addition, it is also likely that the production of MAbs can be contracted as commercial facilities become available (Greenstone, 1996).

To estimate predator consumption rates more accurately, Sopp *et al.* (1992) modified Dempster's (1960) Eq. 1 to incorporate a more reasonable description of both the predator's digestion rate and the meal size initially ingested. No longer is it necessary to assume that a predator has fed on a single prey item. Instead, it is assumed that (1) the predator has fed on either a single prey item or the multiple prey items in a short interval of time; and (2) the capture of a predator with a meal is a random event (i.e., the meal contained in a predator's gut is, on average, half digested when the predator is captured). If the period during which the meal is detectable is known (i.e., R_{pi}), then f—the mean amount of food remaining in the gut—depends on the time since the meal was taken. By knowing the shape of the function describing the rate of digestion, f can then be estimated. Most of the published digestion curves decrease exponentially with time (Sopp *et al.,* 1992; Greenstone, 1996). Other modifications can be adopted that may increase the accuracy of this equation (Greenstone, 1996).

The equation also requires additional information: (1) the size distribution of the prey in the field so that the prey population can be converted to biomass, and (2) the size of meal a particular predatory stage is likely to take. The prey biomass ingested by the predator population can then be estimated from the following equation:

$$P_a = (N_{pi} \, Q_{pi})/f_{pi} \, R_{pi} \qquad (2)$$

where P_a is the biomass of prey killed; N_{pi}, the density of the predator (or predator stage) i; Q_{pi}, the quantity of prey recovered from the ith predator species in a sample; and R_{pi}, the length of time that feeding by i^{th} predator (or predator stage) can be detected based on laboratory tests. Sopp *et al.* (1992) evaluated prey consumption by predators estimated with this equation and compared it with laboratory data in which predators were allowed to feed on a population of aphids of known size for a fixed period of time at known laboratory temperatures. They also knew how many prey were taken. The predators were then assayed using MAbs and ELISA, and the estimates of consumed prey were com-

pared with the laboratory data. The equation and the laboratory data manifested substantial concordance.

Although this equation is more accurate than those previously used to estimate consumption rates, it still requires substantial information and makes several important assumptions. Much of the required information and the assumptions are the same as those required or used by previous equations [see Sopp *et al.* (1992) for a summary of these equations]. First, as with the other methods, the detection period during which a meal can be detected must be measured in laboratory experiments to calibrate the equation. This must be done for each predator species and stage. Digestive rates are known to differ substantially among species and even among some congeners (Sunderland, 1996). Second, it requires an accurate estimate of the density and age structure of each predator species of interest. Third, it requires an estimate of the prey density and age structure so that the prey population available to the predator can be converted into prey biomass. Finally, it requires knowledge of prey selection behavior for each predator species and stage. Thus, a study of this sort is, in essence, a commitment to a demographic study of a predator–prey system.

Moreover, the equation makes several important assumptions. First, the value for $f_{pi} \, R_{pi}$ assumes random feeding with respect to time of day. This will clearly be unrealistic for some predator species because some feed nocturnally while others feed diurnally. The equation also assumes that only a single meal has been taken in those predators that are assessed. The meal can consist of a single prey item or multiple prey items taken during a brief period. Finally, the equation assumes that a predator's gut contents are the result of predation and not scavenging. Sunderland (1996) points out that some predator species are known to obtain a substantial portion of their diet by scavenging as well by preying. He also points out that predators can also prey on other predators who have preyed on the prey species of interest (Fichter & Stephen, 1981, 1984; Sunderland, 1988, 1996; Rosenheim, 1993).

Electrophoretic and Isoelectric Focusing Methods

Electrophoresis is another technique that can be used to associate predators with their prey. Electrophoresis separates proteins based on charge and size differences in an electrical field. Differences in charge and size commonly occur among isoenzymes (proteins catalyzing the same reaction) from different taxa. If the prey and predator have isoenzymes with different electrophoretic mobilities, then analysis of homogenates prepared from predators fed on prey should exhibit protein bands corresponding to the predator and the prey. Similarly, if there are several potential prey of a predator, and if the prey have electrophoretically distinct isoenzymes, then analysis of predator homogenates should reveal the prey species inside the predator.

This approach will be successful if the prey isoenzymes are detectable after predator feeding, and electrophoretic variation occurs among the prey and predator isoenzymes. Isoenzyme detection depends on prey size, *in vitro* activity of the isoenzyme, presence and volume of the predator foregut, and type of electrophoresis employed (Murray & Solomon, 1978; Giller, 1984; Lister *et al.,* 1987; Sopp & Sunderland, 1989). Electrophoretic variation depends on the suite of isoenzymes available for comparison and the type of electrophoresis. Standard electrophoretic procedures (starch gel and polyacrylamide gel electrophoresis) can detect prey isoenzyme activity for several isoenzyme types involving relatively large prey (> 2 to 3 mm in body length). Below this size limit, the number of detectable prey isoenzymes is diminished and hence the likelihood of distinguishing closely related prey is decreased.

Electrophoretic techniques with enhanced sensitivity include conventional electrophoresis in cellulose acetate membranes (Easteal & Boussy, 1987; Höller & Braune, 1988) and isoelectric focusing (IEF) (Kazmer, 1991). IEF has advantages over other techniques involving small and large prey. In IEF, proteins are "focused" into narrow bands along relatively broad pH gradients. Focusing enhances the detectability of enzymes compared with other techniques that gradually spread the proteins into diffuse bands. In addition, because relatively broad pH gradients are used in IEF, enzymes with different charges, such as those that may occur between unrelated prey taxa, will remain sharply focused on the gel. The fine resolution of IEF does not affect the ability to distinguish enzymes with very similar charges. With standard techniques, these contrasting problems are difficult to solve simultaneously because one set of conditions (buffer type and pH, gel type) may be optimal for one prey type but not for others.

Electrophoretic techniques have been used to identify the prey of several arthropod predators. Polyacrylamide gradient gel electrophoresis was used to detect prey protein (esterases) in the gut of predators after they had fed on known prey (Murray & Solomon, 1978). The technique detected esterases of *Panonychus ulmi* (Koch) in predacious mite *Typhlodromus pyri* (Scheuten) and in anthocorids *Anthocoris nemoralis* (Fabricius) and *Orius minutus* (Linnaeus) that had fed on the mite in the laboratory. Dicke and De Jong (1986) used methods to determine whether *T. pyri* and *Amblyseius finlandicus* (Oudemans) also fed on apple rust mite *Aculus schlechtendali* (Nalepa), as an alternate host in the field. Electrophoresis was also used to identify the prey species exploited by *A. nemoralis* on alders in the field [an aphid, *Pterocallis alni* (De Geer)] (Murray & Solomon, 1978). Also, electrophoresis with polyacrylamide disk-gels detected esterases of several prey species in the gut of the water boatman (*Notonecta glauca* Linnaeus) (Giller, 1982, 1984, 1986). A meal was detectable from 17 to 48 h, depending on temperature and meal size, and was strongly correlated with the length of time the meal spent in the foregut (Giller, 1984). Giller (1986) used electrophoresis to identify the prey of *N. glauca* and *N. viridis* Delcourt in the field. Lister *et al.* (1987) used polyacrylamide gel electrophoresis and electrophoresis and esterase allozymes to determine the diet of several microarthropods and the predation rate by acarine predator *Gamasellus racovitzai* (Trousessart).

Estimating predation rates with serological methods and electrophoretic techniques requires a substantial commitment of resources. The techniques entail (1) the development of antibodies or techniques for identifying the isozymes of the prey species or stage; (2) the development of methods to estimate the predator and prey densities, including those needed to estimate the densities of alternate prey; and (3) the identification of the predator and prey stages involved. Initially the use of these techniques to estimate predation rates appears limited to prey populations that form a small fraction of the available prey or to those in which predation events by a predator are infrequent. Frequent predation events confound interpretation of a positive test because a single, large meal cannot be distinguished from several small meals (i.e., several small prey consumed).

Electrophoresis also has been used to detect immature parasitoids within aphids (Wool *et al.,* 1978; Castanera *et al.,* 1983) and whiteflies (Wool *et al.,* 1984). The presence of the parasitoid *Aphidius matricariae* Hal. in the green peach aphid *Myzus persicae* (Sulzer) was identified; and parasitism of whitefly *Bemisia tabaci* (Gennadius) by endoparasitoids *Encarsia lutea* (Masi) or *Eretmocerus mundus* (Merct) were detected with electrophoresis and histochemical staining for esterases. However, the whitefly parasitoids could not be identified to species (Wool *et al.,* 1984). Electrophoresis allows the processing of large numbers of hosts to estimate the fraction that are parasitized and, in some cases, the parasitoid species involved. This contrasts with the traditional methods in which field samples are reared or they are dissected while fresh. Electrophoresis requires an investment of time and resources to work out the methods and to identify useful isozymes for the parasitoid species attacking the host. However, this technique can detect immature parasitoids within a host without dissection. Moreover, the parasitoid's enzyme activity within a prey cannot be confused with that of the host.

Prey Marking

Marking techniques have been used to identify the predator species or to estimate predation rates. Markers include radioactive isotopes — [151]europium (Ito *et al.,* 1972),[32]phosphorus (Jenkins & Hassett, 1950; Pendleton & Grundmann, 1954; Jenkins, 1963; McDaniel & Sterling, 1979; McCarty *et al.,* 1980; Elvin *et al.,* 1983), ^{137}cesium (Moulder & Reichle, 1972), ^{14}carbon (Frank, 1967)—and rare elements (Stimann, 1974; Shepard & Waddill, 1976) and dyes (Hawks, 1972; Elvin *et al.,* 1983). Prey are fed (Frank, 1967; Room, 1977; Elvin *et al.,* 1983) or injected (McDaniel & Sterling, 1979; McCarty *et al.,* 1980) with the radioactive isotope; and the radioactivity is detected in a predator with use of scintillation, a Geiger counter, or autoradiography. For autoradiography, suspected predators are collected after exposure to labeled prey and are glued to paper, which is placed against X-ray film (McDaniel & Sterling, 1979). The film is developed, and dark spots on the film produced by the rays from ^{32}phosphorus indicate labeled predators.

Methods involving isotopes require training and necessary equipment to perform the assays. Safety regulations and environmental considerations eliminate the use of these methods. Other disadvantages, as with electrophoresis techniques and serological methods, include the inability to detect whether a predator had fed on other predators that had consumed labeled prey or a prey was scavenged. Experiments using isotopes, especially those using autoradiography, are simpler to conduct than serological and related techniques. Methods using labeled elements require several manipulations, but they provide more information per unit effort than do other kinds of marker tests.

Rare elements such as rubidium and strontium also have potential as labels. They can be sprayed on foliage or placed in the diet of the prey, incorporated into the prey's tissues, and then transferred to the predators or parasitoids who feed on labeled hosts (Stimmann, 1974; Shepard & Waddill, 1976). The mark—in theory—is retained for life, and self-marking is possible via a labeled plant. However, the technique requires an atomic absorption spectrophotometer, which is expensive. Moreover, placement of the labeled prey on plants may expose them to abnormal predation rates. Herbivores seldom choose feeding or oviposition sites on their plant hosts at random (Ives, 1978; Wolfson, 1980; Denno & McClure, 1983; Guerin & Städler, 1984; Whitham *et al.,* 1984; Myers, 1985; Papaj & Rausher, 1987). Parasitoids (Weseloh, 1974, 1982) and predators (Fleschner, 1950) do not search for their habitat uniformly. Thus, without behavior studies, the degree of bias in determining the natural-enemy complex or in estimating predation rates is unknown.

Rare elements have been used to mark parasitoids (Hopper, 1991; Hopper & Woolson, 1991; Corbett & Rosenheim, 1996). Corbett and Rosenheim (1996) documented the colonization of vineyards adjacent to the overwintering site of *Anagrus* spp., French prune trees planted adjacent to the vineyards. They also documented the relative contribution of movement by *Anagrus* spp. from the French prune trees versus that from the surrounding landscape. The prune tree refuges contributed 1% and 34% of the *Anagrus* spp. colonizing two of the experimental vineyards.

Visual Counts

The visual-count method had several advantages over many of the exclusion techniques: (1) no manipulation of the environment is required, (2) prey can be added or predators can be removed to determine the response of the predator to changes in prey density, and (3) it reveals the predators' diets in the field. The serological method and perhaps the electrophoretic techinique also share these three advantages. In contrast to the serological and electrophoretic methods, however, visual counts require little technology. Visual counts do require a substantial commitment of time to observe the predation and to determine the feeding rates for different combinations of predator and prey stages. The visual-count method cannot be used if the predator is cryptic, is easily disturbed, or flees from the observer. Also, the time a predator spends consuming a prey may vary depending on the range of prey stages attacked; the hunger level of the predator; the interference or stimulation by other predators or prey that are active in the vicinity; and the differences among individual predators due to genera, reproductive state, or molting. These in turn will determine the probability that a predator will be observed in the field with a prey item. Laboratory data on the time spent by four predators consuming prey was highly variable, leaving the investigators pessimistic about the visual-count method's utility for estimating predation rates (Kiritani *et al.,* 1972). However, the technique may still hold promise for some predator species. It has been used to determine the fraction of diurnal predation for each predator species in a complex (Elvin *et al.,* 1983).

Heimpel *et al.* (1997) also used visual observations to determine the life expectancy of adult *Aphytis* spp. in an almond grove. They observed searching *Aphytis* spp. for 89.6 h. These observations yielded 18 encounters with predators, 6 of which resulted in captures of an *Aphytis* spp. The predators involved two spiders, a salticid and a thomicid, the Argentine ant, *Linepthema humilae* Mayer, and nymphs of the reduviid *Zelus renardii*. The encounters occurred mostly in the autumn, September through November, which amounted to an encounter every 3.2 h. By using an exponential decay rate as the expectation of mortality for a searching *Aphytis* spp., they estimated the female's life expectancy to be 3 days.

SUMMARY

No single method will provide conclusive evidence that natural enemies are regulating a population. Natural enemies are not the only factor involved in many interactions. The plant can significantly affect the natural enemies' ability to regulate a population (Flanders, 1942; Starks *et al.,* 1972; Price *et al.,* 1980).

No research method is free of technical problems. We must make management decisions with insufficient knowledge. Thus, research aimed at developing an IPM program is an ongoing process in which hypotheses are continually being refined and tested (Way, 1973). Natural-enemy importations (classical biological control) and augmentative biological control remain important IPM tactics. They must be pursued and expanded aggressively to include situations for which they have not been emphasized (DeBach, 1964; DeBach & Rosen, 1991; Ridgway & Vinson, 1976; Carl, 1982; Hare, 1992). However, indigenous biological control forms the foundation for pest management and sustainable agriculture, and therefore it must be utilized if these strategies are to become more effective. Its presence in an agroecosystem can be demonstrated by disrupting it with insecticides (e.g., Folsom & Brondy, 1930; Woglum *et al.,* 1947; Pickett & Patterson, 1953; Ripper, 1956; Bartlett, 1968; Smith & van den Bosch, 1967; Wood, 1971; Ehler *et al.,* 1973; Eveleens *et al.,* 1973; Croft & Brown, 1975; Luck & Dahlsten, 1975; Luck *et al.,* 1977; Ressig *et al.,* 1982; Kenmore *et al.,* 1984; DeBach & Rosen, 1991) or by comparing unsprayed, abandoned orchards with treated orchards. Insecticidal disruption [the insecticidal check method of DeBach (1946)] provides one of the best experimental techniques for evaluating natural enemies. It can reveal the amount of control provided by indigenous entomophages (Stern *et al.,* 1959; Smith & van den Bosch, 1967; Falcon *et al.,* 1968; MacPhee & MacLellan, 1971; Wood, 1971; Flint & van den Bosch, 1981; Jones, 1982; Metcalf & Luckmann, 1982; Kenmore *et al.,* 1984; DeBach & Rosen, 1991). Indigenous natural enemies are the dominant form of biological control available to the IPM practitioner in most agroecosystems. Unfortunately, there is no shortcut to evaluating their effectiveness on an herbivorous population. A substantial research effort encompassing field and laboratory experiments, demographic studies, and models in conjunction with the testing of theory is required.

Acknowledgments

We thank T. Unruh, L. Nunney, A. Gutierrez, R. Stouthamer, and D. Kazmer for comments and discussion on several aspects of this chapter. A portion of the development of this chapter was supported by NSF grant No. BSR 8921100 and BARD Grant No. US-2359-93C to R.F.L.

References

Ashby, J. W. (1974). A study of arthropod predation of *Pieris rapae* L. using serological and exclusion techniques. Journal of Applied Ecology, 11, 419–425.

Aveling, C. (1981). The role of *Anthocoris* species (Hemiptera: Anthocoridae) in the integrated control of the Damson-hop aphid (*Phorodon humuli*). Annals of Applied Biology, 97, 143–153.

Barry, R. M., Hatchett, J. H., & Jackson, R. D. (1984). Cage studies with predators of the cabbage looper, *Trichoplusia ni,* and corn ear worm, *Heliothis zea* in soybeans. Journal of the Georgia Entomological Society, 9, 71–78.

Bartlett, B. R. (1968). Outbreaks of two-spotted spider mites and cotton aphids following pesticide treatment. I. Pest stimulation vs. natural enemy destruction as the cause of outbreaks. Journal of Economic Entomology, 61, 297–303.

Baumgaetner, J. U., Gutierrez, A. P., & Summers, C. G. (1981). The influence of aphid prey consumption on searching behavior, weight increase, developmental time and mortality of *Chrysopa carnea* (Neuroptera: Chrysopidae) and *Hippodamia convergens* (Coleoptera: Coccinellidae) larvae. Canadian Entomologist, 113, 1007–1014.

Boreham, P. F. L., & Ohiagu, C. E. (1978). The use of serology in evaluating invertebrate predator-prey relationships: A review. Bulletin of Entomological Research, 68, 171–194.

Boavida, D., Neuenschwander, P., & Herren, H. R. (1995). Experimental assessment of the inpact of the introduced parasitoid, *Gyranusoidea tebygi* Noyes, on the mango mealybug *Rastrococcus invadens* Willaims, by physical exdlusion. Biological Control, 5, 99–103.

Bull, C. G., & King, M. V. (1923). The identification of the blood meal of mosquitoes by means of the precipitin test. American Society of Hygiene, 3, 491–496.

Campbell, C. A. M. (1978). Regulation of the Damson-hop aphid (*Phorodon humuli* [Schrank]) on hops (*Humulus lupulus* Linnaeus) by predators. Journal of Horticultural Science, 53, 235–242.

Carl, K. P. (1982). Biological control of native pests by introduced natural enemies. Biocontrol News and Information, 3, 190–200.

Carter, N., McLean, I. F. G., Watt, A. E. & Dixon, A. G. G. (1980). Cereal aphids: A case study and review. In T. H. Cooker (Ed.), Applied biology (pp. 272–349). London: Academic Press.

Castanera, P., Loxdale, H. D., & Novak, K. (1983). Electrophoretic study of enzymes from cereal aphid populations. II. Use of electrophoresis for identifying aphidiid parasitoids (Hymenoptera) of *Sitobion avenae* (Fabricius) (Hemiptera: Aphididae). Bulletin of Entomological Research, 73, 659–665.

Chambers, R. J., Sunderland, K. D., Wyatt, I. J., & Vickerman, G. P. (1983). The effects of predator exclusion and caging on cereal aphids in winter wheat. Journal of Applied Ecology, 20, 209–224.

Charboussou, F. (1965). La multiplication par vole trophique des tetrayques a la suite des traitements presticides. Relations avec les phenomenes de resistance acquise. Bollettino di Zoologia Agraria Bachicoltura, 7, 143–184.

Chelliah, S., Fabellar, L., & Heinrichs, A. E. (1980). Effects of sublethal doses of three insecticides on the reproductive rate of the brown planthopper, *Nalaparvata lugens,* on rice. Environmental Entomology, 9, 778–780.

CIBC. (1971). Biological control programs against insects and weeds in Canada, 1959–1968. Vol. 4, Slough, United Kingdom: Commonwealth Agricultural Bureau Tech. Communication.

Clausen, C. P. (Ed.). (1978). Introduced parasites and predators of arthropod pests and weeds: A world review. Agriculture Handbook No. 480. Washington, DC: U.S. Department of Agriculture.

Coaker, T. H. (1965). Further experiments on the effects of beetle predation on the numbers of cabbage root fly, *Eriocschia brassicae* (Bouch) attacking brassica crops. Annals of Applied Biology, 56, 7–20.

Cochran, W. G. (1963). Sampling techniques (2nd ed.). New York: John Wiley & Sons.

Cock, M. J. W. (1985). A review of biological control of pests in the Commonwealth Caribbean and Bermuda up to 1982, Vol. 9. Slough, United Kingdom: Commonwealth Agricultural Bureau Tech. Communications.

Corbett, A., & Rosenheim, J. A. (1996). Impact of a natural enemy overwintering refuge and its interaction with the surrounding landscape. Ecological Entomology, 21, 155–164.

Crawley, M. J. (1993). GLIM for ecologists. Oxford: Blackwell Sciences.

Croft, B. A., & Brown, A. W. A. (1975). Responses of arthropod natural enemies to insecticides. Annual Review of Entomology, 20, 285–335.

Crook, N. E., & Sunderland, K. D. (1984). Detection of aphid remains in predators by ELISA. Annals Applied Biology, 105, 413–422.

DeBach, P. (1946). An insecticidal check method for measuring the efficacy of entomophagous insects. Journal Economic Entomology, 39, 695–697.

DeBach, P. (1955). Validity of insecticidal check method as a measure of the effectiveness of natural enemies of diaspine scale insects. Journal of Economic Entomology, 48, 584–588.

DeBach, P. (Ed.). (1964). Biological control of insect pests and weeds. New York: Reinhold.

DeBach, P., & Huffaker, C. B. (1971). Experimental techniques for evaluation of the effectiveness of natural enemies. In C. B. Huffaker (Ed.), Biological control (pp. 113–140) New York: Plenum Press.

DeBach, P., & Rosen, D. (1991). Biological control by natural enemies. New York: Cambridge University Press.

DeBach, P., Dietrick, E. J., & Fleschner, C. A. (1949): A new technique for evaluating the efficiency of entomophagous insects in the field. Journal of Economic Entomology, 42, 546.

Dempster, J. P. (1960). A quantitative study of the predators on the eggs and larvae of the broom beetle, *Phytodecta olivacea* (Forster), using the precipitin test. Journal of Animal Ecology, 29, 149–167.

Dempster, J. P. (1964). The control of *Pieris rapae* with DDT 1. The natural mortality of the young stages of *Pieris*. Journal of Applied Ecology, 4, 485–500.

Dempster, J. P. (1967). The feeding habits of the Miridae (Heteroptera) living on broom (*Sarothomnus scoparius* L.). Entomologia Experimentalis et Applicata, 7, 149–154.

Dempster, J. P., Richards, O. W., & Waloff, N. (1959). Carabids as predators on the pupal stage of the chrysomelid beetle *Phytodecta olivacea* (Forster). Oikos, 10, 65–70.

Denno, R. F., & McClure, M. S. (Eds.). (1983). Variable plants and herbivores in natural and managed systems. New York: Academic Press.

Dicke, M., & De Jong, M. (1986). Prey preference of predatory mites: Electrophoretic analysis of the diet of *Typhlodromus pyri* Scheuten and *Amislyseius finlandicus* (Oudemans) collected in Dutch orchards. Bulletin IOBC/WPRS, 9, 62–67.

Dittrich, V. P., Streibert, P., & Bathe, P. (1974). An old case reopened: Mite stimulation by insecticide residues. Environmental Entomology, 3, 534–539.

Dixon, A. F. G. (1959). An experimental study of the search behavior of the predatory coccinellid beetle, *Adalia compunctata* (Linnaeus). Journal Animal Ecology, 28, 259–281.

Dowden, P. B. (1962). Parasites and predators of forest insects liberated in the United States through 1960. Agriculture Handbook No. 226. Washington, DC: U.S. Department of Agriculture.

Downe, S. E. R., & West, A. S. (1954). Progress in the use of precipitin test in entomological studies. The Canadian Entomologist, 86, 181–184.

Easteal, S., & Boussy, I. A. (1987). A sensitive and efficient isozyme technique for small arthropods and other invertebrates. Bulletin of Entomological Research, 77, 407–415.

Edmunds, G. F., & Alstad, D. N. (1985). Malathion-induced sex ratio changes in black pine leaf scale (Hemiptera: Diaspididae). Annals of the Entomological Society of America, 78, 403–405.

Ehler, L. E., Eveleens, K. G., & van den Bosch, R. (1973). An evaluation of some natural enemies of cabbage looper on cotton in California. Environmental Entomology, 2, 1009–1015.

Elliott, J. M. (1977). Some methods for the statistical analysis of samples of benthic invertebrates (Scientific Publ. No. 25). Ferry House, United Kingdom: Freshwater Biological Association.

Elvin, H. K., Stimac, J. L., & Whitcomb, W. H. (1983). Estimating rates of arthropod predation on velvetbean caterpillar larvae in soybeans. Florida Entomologist, 66, 319–330.

Eveleens, K. G., van den Bosch, R., & Ehler, L. E. (1973). Secondary outbreak induction of beet armyworm by experimental insecticide applications in cotton in California. Environmental Entomology, 2, 497–503.

Faeth, S. H., & Simberloff, D. (1981). Population regulation of a leafmining insect, *Cameraria* sp. nov. at increased field densities. Ecology, 62, 620–624.

Falcon, L. A., van den Bosch, R., Ferris, C. A., Stromberg, L. K., Etzel, L. K., & Stinner, R. E. (1968). A comparison of season long cotton-pest-control programs in California during 1966. Journal of Economic Entomology, 61, 633–642.

Ferro, D. N., & Southwick, E. E. (1984). Microclimates of small arthropods: Estimating humidity within leaf boundary layer. Environmental Entomology, 13, 926–929.

Fichter, F. L., & Stephen, W. P. (1979). Selection and use of host-specific antigens. Miscellaneous Publications of the Entomological Society of America, 11, 25–33.

Fichter, F. L., & Stephen, W. P. (1981). Time related decay in prey antigens ingested by the predator *Podisus maculiventris* (Hemiptera: Pentatomidiae) as detected by ELISA. Oecologia, 51, 404–407.

Fichter, F. L., & Stephen, W. P. (1984). Time related decay in prey antigens ingested by arboreal spiders as detected by ELISA. Environmental Entomology, 13, 1583–1587.

Flanders, F. E. (1942). Absorptive development in parasitic Hymenoptera, Abortive induced by food plant of the insect host. Journal Economic Entomology, 35, 834–835.

Fleschner, C. A. (1950). Studies on searching capacity of the larvae of three predators of the citrus red mite. Hilgardia, 20, 233–265.

Fleschner, C. A. (1958). Field approach to population studies of tetranychid mites in citrus and avocado in California. Proceedings of the 10th International Congress of Entomology, 2, 669–674.

Fleschner, C. A., Hall, J. C., & Ricker, D. W. (1955). Natural balance of mite populations in an avocado grove. California Avocado Society Yearbook, 39, 155–162.

Flint, M. L., & van den Bosch, R. (1981). Introduction to integrated pest management. New York: Plenum.

Folsom, J. W., & Bondy, F. F. (1930). Calcium arsenate dusting as a cause of aphid infestations. (USDA Agriculture Circular 116). Washington, DC: U.S. Department of Agriculture.

Frank, J. H. (1967). The insect predators of the pupal stage of the winter moth *Operophtera boumata* (Linnaeus) (Lepidoptera: Hydriomenidae). Journal of Animal Ecology, 36, 375–389.

Frazer, B. D., & Gilbert, N. (1976). Coccinellids and aphids: A quantitative study of the impact of adult ladybirds (Coleoptera: Coccinellidae) preying on field populations of pea aphids (Homoptera: Aphididae). Journal of the Entomological Society of British Columbia, 73, 33–56.

Frazer, B. D., & Gill, B. (1981). Hunger, movement, and predation of *Coccinella californica* on pea aphids in the laboratory and in the field. Canadian Entomologist, 113, 1025–1033.

Frazer, B. D., Gilbert, N., Ives, N., & Raworth, D. A. (1981a). Predator reproduction and the overall predator-prey relationship. The Canadian Entomologist, 113, 1015–1024.

Frazer, B. D., Gilbert N., Nealis, V., & Raworth, D. A. (1981b). Control of aphid density by a complex of predators. The Canadian Entomologist, 113, 1035–1041.

Futuyma, D. J., & Peterson, S. C. (1985). Genetic variation in the use of resources by insects. Annual Review of Entomology, 30, 217–238.

Gardner, W. A., Shepard, M., & Noblet, R. (1981). Precipitin test for examining predator-prey interactions in soybean fields. The Canadian Entomologist, 113, 365–369.

Gates, D. M. (1980). Biophysical ecology. New York: Springer-Verlag.

Gilbert, N., Gutierrez, A. P., Frazer, B., & Jones, R. E. (1976). Ecological relationships. San Francisco: Freeman.

Giller, P. S. (1982). The natural diets of waterbugs (Hemiptera: Heteroptera): Electrophoresis is a potential method of analysis. Ecological Entomology, 7, 233–237.

Giller, P. S. (1984). Predator gut state and prey detectability using electrophoretic analysis of gut content. Ecological Entomology, 9, 157–162.

Giller, P. S. (1986). The natural diet of the Notonectidae: Field trials using electrophoresis. Ecological Entomology, 11, 163–172.

Greathead, D. S. (1971). A review of biological control in the Ethiopian region: Vol. 5. Slough, United Kingdom: Commonwealth Agricultural Bureau Tech. Communication.

Greathead, D. S. (Ed.). (1976). A review of biological control in western and southern Europe: Vol. 7. Slough, United Kingdom: Commonwealth Agricultural Bureau Tech. Communication.

Green, R. H. (1979). Sampling design and statistical methods for environmental biologists. New York: John Wiley & Sons.

Greenstone, M. H. (1983). Amblyospora site-specificity and site-tenacity in a wolf spider: A serological dietary analysis. Oecologia, 56, 79–83.

Greenstone, M. H. (1996). Serological analysis of arthropod predation: Past present and future. In W. O. C. Symondson & J. E. Liddell (Eds.), The ecology of agricultural pests (pp. 266–300). New York: Chapman & Hall.

Greenstone, M. H., & Morgan, C. E. (1989). Predation on Heliothis zea (Lepidoptera: Noctuidae). Annals of the Entomological Society of America, 82, 45–49.

Guerin, P. M., & Städler, E. (1984). Carrot fly cultivar preferences: Some influencing factors. Ecological Entomology, 9, 413–420.

Hagler, J. R., & Naranjo, S. E. (1994a). Qualitative survey of two coleopteran predators of Bemisia tabaci (Homoptera: Aleyrodidae) and Pectinophora gossypiella (Lepidoptera: Gelechiidae) using a multiple prey gut content ELISA. Environmental Entomology, 23, 193–197.

Hagler, J. R., & Naranjo, S. E. (1994b). Determining the frequency of heteropteran predation on seet potatio whitefly and pink bollworm using multiple ELISAs. Entomologia Experimentalis et Applicata, 72, 59–66.

Hagler, J. R., & Naranjo, S. E. (1996). Using gut content immunoassays to evaluate predaceous biological control agents: A case study. In W. O. C. Symondson & J. E. Liddell (Eds.), The ecology of agricultural pests (p. 383–399). New York: Chapman & Hall.

Hagler, J. R., Cohon, A. C., Bradley-Dunlop, D., & Enriquez, F. J. (1992). Field evaluation of predation on Lygus hesperus using a species and stage specific monoclonal antibody. Environmental Entomology, 21, 896–900.

Hall, R. R., Downe, A. E. R., McClellan, C. R., & Wiest, A. S. (1953). Evaluation of insect predator-prey relationships by precipitin test studies. Mosquito News, 13, 199–204.

Hance, T., & Renier, L. (1987). An ELISA technique for the study of the food of carabids. Acta Phytopathologica Entomologica Hungary, 22, 363–368.

Hand, L. F., & Keaster, A. J. (1967). The environment of an insect field cage. Journal of Economic Entomology, 60, 910–915.

Hare, J. D. (1992). Effects of plant variation on herbivore-natural enemy interactions. In R. S. Fritz & E. L. Simms (Eds.), Plant resistance to herbivores and pathogens: Ecology, evolution and genetics (pp. 278–298). Chicago: University of Chicago Press.

Hawks, R. B. (1972). A fluorescent dye technique for marking insect eggs in predation studies. Journal of Economic Entomology, 65, 1477–1478.

Heimpel, G. E., Rosenheim, J. A., & Mangel, M. (1997). Predation on adult Aphytis parasitoids in the field. Oecologia, 110, 346–342.

Heinz, K. M., & Parrella, M. P. (1994). Biological control of Bermesia argentifolii (Homoptera: Aleyrodidae) infesting Euphorbia pulcherrima: evaluations of releases of Encarsia luteola (Hymenoptera: Aphelinidae) and Delphastus pusillus (Coleoptera: Coccinellidae). Environmental Entomology, 23, 1346–1353.

Hopper, K. R. (1991). Ecological applications of elemental labeling, analysis of dispersal, density, mortality, and feeding. The Southwestern Entomology, (Suppl. 14), 71–83.

Hopper, K. E., & Woolson, E. A. (1991). Labeling a parasitic wasp, Microplitis croceipes (Hymenoptera: Braconidae), with trace elements for mark-recapture studies. Annals of the Entomological Society of America, 84, 255–262.

Hopper, K. E., Aidara, S., Agret, S., Cabal, J., Coutinot, D., & DaBire, R. (1995). Natural enemy impact on the abundance of Diauraphis noxia (Homoptera: Aphididae) in wheat is southern France. Environmental Entomology, 24, 402–408.

Höller, C., & Braune, J. J. (1988). The use of isoelectric focusing to assess percentage hymenopterous parasitism in aphid populations. Entomologia Experimentalis et Applicata, 47, 105–114.

Huffaker, C. B., & Kennett, C. E. (1956). Experimental studies on predation: Predation and cyclamen-mite populations on strawberries in California. Hilgardia, 26, 191–222.

Huffaker, C. B., & Kennett, C. E. (1966). Studies of two parasites of the olive scale, Parlatoria oleae (Colv e). IV. Biological control of Parlatoria oleae (Colv e) through the compensatory action of two introduced parasites. Hilgardia, 37, 283–335.

Huffaker, C. B., Kennett, C. E., & Finney, G. L. (1962). Biological control of olive scale, Parlatoria oleae (Colv e) in California by imported Aphytis maculicornis (Masi) (Hymenoptera: Aphelinidae). Hilgardia, 32, 541–636.

Irwin, M. E., Gill, R. W., & Gonzalez, D. (1974). Field cage studies of native egg predators of the pink bollworm in southern California cotton. Journal of Economic Entomology, 67, 193–196.

Ito, Y., Yamanaka, H., Nakasuji, F., & Kiritani, K. (1972). Determination of predator-prey relationship with an activable tracer, Europium-151. Konchu, 40, 278–283.

Ives, P. M. (1978). How discriminating are cabbage butterflies? Australian Journal of Entomology, 3, 261–276.

Jenkins, D. W. (1963). Use of radionuclides in ecological studies of insects. In V. Schult & A. W. Klement (Eds.), Radioecology (pp. 431–440). New York: Rheinhold.

Jenkins, D. W., & Hassett, C. C. (1950). Radioisotopes in entomology. Nucleonics, 6, 5–14.

Jessen, R. J. (1978). Statistical survey techniques. New York: John Wiley & Sons.

Jones, D. (1982). Predators and parasites of temporary row crop pests: Agents of irreplaceable mortality or scavengers acting prior to other mortality factors? Entomophaga, 27, 245–265.

Jones, V. P., & Parrella, M. P. (1984). The sublethal effects of selected pesticides on life table parameters of Panonychus citri (Acri: Tetranychidae). The Canadian Entomologist, 116, 1033–1040.

Jones, V. P., Youngman, R. R., & Parrella, M. P. (1983). The sublethal effects of selected insecticides on life table on photosynthetic rates of lemon and orange leaves in California. Journal of Economic Entomology, 76, 1178–1180.

Karieva, P. (1985). Patchiness, dispersal, and species interactions: Consequences for communities of herbivore insects. In T. Case & J. Diamond (Eds.), Community ecology (pp. 192–206). New York: Harper & Row.

Kazmer, D. J. (1991). Isoelectric focusing procedures for the analysis of allozymic variation in minute arthropods. Annals of the Entomological Society of America, 84, 332–339.

Kazmer, D. J., & Luck, R. F. (1995). Field tests of the size-fitness hypothesis in the egg parasitoid Trichogramma pretiosum. Ecology, 76, 412–425.

Kelleher, J. S., & Hulme, M. A. (1984). Biological control programmes against insects and weeds in Canada, 1969–1980: Vol. 8. Slough, United Kingdom: Commonwealth Agricultural Bureau Tech. Communication.

Kenmore, P. E., Carino, F. D., Perez, C. A., Dyck, V. A., & Gutierrez, A. P. (1984). Population regulation of the rice brown planthopper (Nila-

parvata lugens StŒl) within rice fields in the Philippines. Journal of Plant Protection of Tropics, 1, 19–37.

Kidd, N. A. C., & Jervis, M. A. (1996). Population dynamics. In M. A. Jervis & N. A. C. Kidd. (Eds), Insect natural enemies (pp. 293–374). New York: Chapman & Hall.

Kinzer, R. E., Cowan, C. B., Ridgway, R. L., Davis, F. J., Jr., & Copperage, R. J. (1977). Populations of arthropod predators and *Heliothis* spp. after applications of aldicarb and moncrotophos. Environmental Entomology, 6, 13–16.

Kiritani, K., Kawahera, S., Sasaba, T., & Nakasuji, F. (1972). Quantitative evaluation of predation by spiders on the green rice leafhopper, *Nephotettix cincticeps* Uhler, by a sight count method. Researches on Population Ecology (Kyoto), 13, 187–200.

Kring, T. J., Gilstrap, F. E., & Michels, F. J., Jr. (1985). Role of indigenous coccinellids in regulating greenbugs (Homoptera: Aphididae) on Texas grain sorghum. Journal of Economic Entomology, 78, 269–273.

Laing, J. E., & Hamai, J. (1976). Biological control of insect pests and weeds by imported parasites and predators. In C. B. Huffaker & P. S. Messenger (Eds.), Theory and practice of biological control (pp. 685–743). New York: Academic Press.

Lingren, P. D., Ridgway, R. L., & Jones, S. L. (1968). Consumption by several common arthropod predators of eggs and larvae of two *Heliothis* species that attack cotton. Annals of the Entomological Society of America, 61, 613–618.

Linit, M. J., & Stephen, F. M. (1983). Parasite and predator component of within-tree southern pine beetle (Coleoptera: Scolytidae) mortality. The Canadian Entomologist, 115, 679–688.

Lister, A., Usher, M. B., & Block, W. (1987). Description and quantification of field attack rates by predator mites: An example using an electrophoresis method with a species of Antarctic mite. Oecologia, 72, 185–191.

Lövel, G. L., Monostori, E., & Ando, I. (1985). Digestion rates in relation to starvation in the larva of a carabid predator, *Poecilus cupreus.* Entomologia Experimentalis et Applicata, 37, 123–127.

Luck, R. F. (1981). Parasitic insects introduced as biological control agents for arthropod pests. (pp. 125–284). In D. Pimentel (Ed.), CRC handbook of pest management in agriculture. Boca Raton, FL: CRC Press.

Luck, R. F., & Dahlsten, D. L. (1975). Natural decline of a pine needle scale (*Chionaspis pinifoliae* [Fitch]) outbreak at south Lake Tahoe, California, following cessation of adult mosquito control with malathion. Ecology, 56, 893–904.

Luck, R. F., & Podoler, H. (1985). Competitive exclusion of *Aphytis lingnanensis* by *A. melinus*: Potential role of host size. Ecology, 66, 904–913.

Luck, R. F., van den Bosch, R., & Garcia, R. (1977). Chemical insect control, a troubled pest management strategy. BioScience, 27, 606–611.

Luck, R. F., van Lenteren, J. C., Twine, P. H., Kuenen, L., & Unruh, T. (1979). Prey or host searching behavior that leads to a sigmoid functional response in invertebrate predators and parasitoids. Researches on Population Ecology (Kyoto), 20, 257–264.

Lund, R. D., & Turpin, F. T. (1977). Serological investigation of black cutworm larval consumption by ground beetles. Annals of the Entomological Society of America, 70, 322–324.

MacPhee, A. W., & MacLellan, C. R. (1971). Ecology of apple orchard fauna and development of integrated pest control in Nova Scotia. Proceedings of the Tall Timbers Conference on Ecological Animal Control Habitat Management, 3, 197–208 (Tallahassee, FL, Tall Timbers Research Station).

Maggi, V. L., & Leigh, T. F. (1983). Fecundity response of the two-spotted spider mite to cotton treated with methyl parathion of phosphoric acid. Journal of Economic Entomology, 76, 20–25.

Matsumoto, B. M., & Nishida, T. (1966). Predator-prey investigations of the taro leafhopper and its egg predator (Hawaiian Agricultural Experimental Station Tech. Bulletin 64). Honolulu: University of Hawaii.

McCarty, M. T., Shepard, M., & Turnipseed, S. G. (1980). Identification of predaceous arthropods in soybeans by using autoradiography. Environmental Entomology, 9, 199–203.

McDaniel, S. G., & Sterling, W. L. (1979). Predator determination and efficiency of *Heliothis virescens* eggs in cotton using ^{32}P. Environmental Entomology, 8, 1083–1087.

McGugan, B. M., & Coppel, H. C. (1962). Biological control of forest insects 1910–1958, In A review of the biological control attempts against insects and weeds in Canada Vol. 2, (pp. 35–127). Slough, United Kingdom: Commonwealth Agricultural Bureau Tech. Communication.

McLeod, J. H. (1962). Biological control of pests of crops, fruit trees, ornamentals and weeds in Canada up to 1959. In A review of the biological control attempts against insects and weeds in Canada Vol. 2, pp. 1–33. Slough, United Kingdom: Commonwealth Agricultural Bureau Tech. Communication.

Metcalf, R. L., & Luckmann, W. (1982). Introduction to insect pest management. New York: Wiley & Sons.

Miller, M. C. (Ed.). (1979). Serology in insect predator-prey studies. Miscellaneous Publications of the Entomological Society of America, 11(4), 1–84.

Morris, R. F. (1955). The development of sampling techniques for forest defoliators with particular reference to the spruce budworm. Canadian Journal of Zoology, 33, 225–294.

Morris, R. F. (1960). Sampling insect populations. Annual Review of Entomology, 5, 243–264.

Moulder, B. C., & Reichle, D. E. (1972). Significance of spider predation in energy dynamics of forest-floor arthropod communities. Ecological Monographs, 42, 473–498.

Mueke, J. M., Manglitz, G. R., & Kerr, W. R. (1978). Pea aphid: Interaction of insecticides and alfalfa varieties. Journal of Economic Entomology, 71, 61–65.

Murray, R. A., & Solomon, M. G. (1978). A rapid technique for analyzing diets of invertebrate predators by electrophoresis. Annals of Applied Biology, 90, 7–10.

Myers, J. H. (1985). Effect of physiological condition of the host plant on the ovipositional choice of the cabbage white butterfly, *Pieris rapae.* Journal of Animal Ecology, 54, 193–204.

Nechols, J. R., Obrycki, J. J., Tauber, C. A., & Tauber, M. J. (1996). Potential impact of native natural enemies on *Galerucella* spp. (Coleoptera: Chrysomelidae) imported for biological control of purple loosestrife: A field evaluation. Biological Control, 7, 60–66.

Neusenschwander, P., Schulthess, F., & Madojemu, E. (1986). Experimental evaluation of the efficiency of *Epidinocarsis lopezi*, a parasitoid introduced into Africa against the cassava mealybug, *Phenacoccus manihoti.* Entomologia Experimentalis et Applicata, 42, 133–138.

Nordlund, D. A., Jones, R. L., & Lewis, W. J. (Eds.). (1981). Semiochemicals: Their role in pest control. New York: John Wiley & Sons.

Oatman, E. R., & Platner, G. R. (1971). Biological control of the tomato fruitworm, cabbage looper, and hornworms on processing tomatoes in southern California, using mass releases of *Trichogramma pretiosum.* Journal of Economic Entomology, 64, 501–506.

Oatman, E. R., & Platner, G. R. (1978). Effects of mass releases of *Trichogramma pretiosum* against lepidopterous pests on processing tomatoes in southern California, with notes on host egg population trends. Journal of Economic Entomology, 71, 896–900.

Ohiagu, C. E., & Boreham, P. F. L. (1978). A simple field test for evaluating insect prey-predator relationships. Entomologia Experimentalis et Applicata, 23, 40–47.

Papaj, D. R., & Rausher, M. D. (1987): Components of conspecific host discrimination behavior in the butterfly, *Battus philenor.* Ecology, 68, 245–253.

Parker, F. D. (1971). Management of pest populations by maniplating densities of both hosts and parasites through periodic releases. In C. B. Huffaker. Biological Control, Proceedings of an AAAS Symposium on Biological Control (pp. 365–376). Boston, MA, December 30–31, 1969. New York: Plenum Press.

Parrella, M. P., Paine, T. D., Bethke, J. A., Robb, K. L., & Hall, J. (1991). Evaluation of *Encarsia formosa* (Hymenoptera: Aphelinidae) for biological control of sweet potato whitefly (Homoptera: Aleyrodidae) on poinsettia. Environmental Entomology, 20, 713–719.

Pedigo, L. P., & Buntin, G. D. (1994). Handbook of sampling methods for arthropods in agriculture. Boca Raton; FL: CRC Press.

Pendleton, R. C., & Grundmann, A. W. (1954). Use of ^{32}P in tracing some insect-plant relationships of the thistle, *Cirsium undulatum*. Ecology, 35, 187–191.

Pickett, A. D., & Patterson, N. A. (1953). The influence of spray programs on the fauna of apple orchards in Nova Scotia. IV. A review. The Canadian Entomologist, 85, 472–478.

Pijls, J. W. A. M., & van Alphen, J. J. M. (1996). On the coexistence of the cassava mealybug parasitoids *Apoanagyrus diversicornis* and *A. lopezi* (Hymenoptera: Encyrtidae) in their native South America. Bulletin of Entomological Research, 86, 51–59.

Pijls, J. W. A. M., Hofker, K. D., van Staalduinen, M. J., & van Alphen, J. J. M. (1995). Interspecific host discrimination and competition in *Apoanagyrus (Epidinocarsis) lopezi* and *A. (E.) diversicornis*, parasitoids of the cassava mealybug, *Phenacoccus manihoti*. Ecological Entomology, 20, 326–332.

Price, P. W., Bouton, C. E., Gross, P., McPheron, B. A., Thompson, J. N., & Weis, A. E. (1980). Interactions among three trophic levels: Influence of plants on interactions between insect herbivores and natural enemies. Annual Review of Ecology and Systematics, 11, 41–65.

Rabbinge, R., Ankersmit, G. W., & Pak, G. A. (1979). Epidemiology and simulation of population development of *Sitobion avenae* in winter wheat. Netherlands Journal of Plant Pathology, 85, 197–220.

Ragsdale, D. W., Larson, A. D., & Newson, L. D. (1981). Quantitative assessment of the predators of *Nezara viridula* eggs and nymphs within a soybean agroecosystem using an ELISA. Environmental Entomology, 10, 402–405.

Rao, V. P., Ghani, M. A., Sankaran, T., & Mather, K. C. (1971). A review of the biological control of insects and other pests in Southeast Asia and the Pacific region: Vol. 6. Slough, United Kingdom: Commonwealth Agricultural Bureau Tech. Communication.

Readshaw, J. L. (1973). The numerical response of predators to prey density. Journal of Applied Ecology, 10, 342–351.

Ressig, W. H., Heinrichs, E. A., & Valencia, S. L. (1982). Insecticide induced resurgence of the brown planthopper, *Nilaparvata lugens,* on rice varieties with different levels of resistance. Environmental Entomology, 11, 165–168.

Richards, O. W., & Waloff, N. (1961). A study of a natural population of *Phytodecta olivacea* (Forester) (Coleoptera, Chrysomelidae). Philosophical Transactions of the Royal Society of London, Series B, 244, 205–257.

Richman, D. B., Hemenway, R. C., Jr, & Whitcomb, W. H. (1980). Field-cage evaluation of predators of the soybean looper, *Pseudoplusia includens* (Lepidoptera: Noctuidae). Environmental Entomology, 9, 315–317.

Ridgway, R. L., & Vinson, S. B. (1976). Biological control by augmentation of natural enemies: Insect and mite control with parasites and predators. New York: Plenum Press.

Ripper, W. E. (1956). Effect of pesticides on balance of arthropod populations. Annual Review of Entomology, 1, 403–438.

Rombach, M. C., Aguda, R. M., Shepard, B. M., & Roberts, D. W. (1986a). Entomopathogenic fungi (Deuteromycotina) in the control of the black bug of rice, *Scotinophara coarctata* (Hemiptera: Pentatomidae). Journal of Invertebrate Pathology, 48, 174–179.

Rombach, M. C., Aguda, R. M., Shepard, B. M., & Roberts, D. W. (1986b). Infection of rice brown planthoppers, *Nilaparvata lugens*

(Homoptera: Delphacidae) by field application of entomopathogenic hyphomycetes (Deuteromycotina). Environmental Entomology, 15, 1070–1073.

Room, P. M. (1977). ^{32}P Labelling of immature stages of *Heliothis armigera* (Hübner) and *H. punctigera* Wallengren (Lepidoptera: Noctuidae): relationships of doses to radioactivity, mortality and label half-life. Journal of the Australian Entomological Society, 16, 245–251.

Rosen, D., & DeBach, P. (1978). Diaspididae. In C. P. Clausen (Ed.), Introduced parasites and predators of arthropod pests and weeds: A world review. Agriculture Handbook 480 (pp. 78–128). Washington, DC: U.S. Department of Agriculture.

Rosenheim, J. A., Wilhoit, L. R., & Armer, C. A. (1993). Influence of intraguild predation among generalist predators on the suppression of an herbivore population. Oecologia, 96, 439–449.

Rothchild, G. H. L. (1966). A study of natural populations of *Conomelus anceps* (Germar) (Homoptera: Delphacidae) including observations on predation using the precipitin test. Journal of Animal Ecology, 35, 413–434.

Rothchild, G. H. L. (1970). Observations on the ecology of the rice ear bug, *Leptocoris oratorius* (Hemiptera: Alydidae) in Sarawak (Malaysian Borneo). Journal of Applied Ecology, 7, 147–167.

Rothchild, G. H. L. (1971). The biology and ecology of the rice stem borer in Sarawak (Malaysian Borneo). Journal of Applied Ecology, 8, 287–332.

Ryan, R. B. (1997). Before and after evaluation of biological control of the larch casebearer (Lepidoptera: Coleophoridae) in the Blue Mountains of Oregon and Washington, 1972–1995. Environmental Entomology, 26, 703–715.

Ryan, R. B., & Medley, R. D. (1970). Test release of *Itoplectis quadricingulatus* against European pine shoot moth in an isolated infestation. Journal of Economic Entomology, 63, 1390–1392.

Sabelis, M. W. (1981). Biological control of two-spotted spider mites using phytoseiid predators. I. Modelling the predator-prey interaction at the individual level. Unpublished doctoral thesis, Agricultural University of Wageningen, The Netherlands.

Sailer, R. I. (1966). An aphid predator exclusion test. In J. Hodek (Ed.), Ecology of aphidiophagous insects (p. 263). The Hague: Junk.

Scriber, J. M., & Slansky, F., Jr. (1981). The nutritional ecology of immature insects. Annual Review of Entomology, 26, 183–211.

Shaw, J. G., Sunderman, R. P., Moreno, D. S., & Fargerlund, D. S. (1973). California red scale: Females isolated by treatments with dichlorvos. Environmental Entomology, 2, 1062.

Shepard, B. M., & Waddill, V. H. (1976). Rubidium as a marker for Mexican bean beetles, *Epilachna varivestis* (Coleoptera: Coccinellidae). The Canadian Entomologist, 108, 337–339.

Shepard, B. M., & Arida, G. S. (1986). Parasitism and predation of yellow stemborer *Scirpophaga incertulas* (Walker) (Lepidoptera: Pyralidae) eggs in transplanted and direct seeded rice. Journal of Entomological Science, 21, 26–32.

Simmon, G. S., & Minkenberg, O. P. J. M. (1994). Field-cage evaluation of augmentative biological control of *Bemisia argentifolii* (Homoptera: Aleyrodidae) in southern California cotton with the parasitoid, *Eretmocerus* nr *californicus* (Hymenoptera: Aphelinidae). Environmental Entomology, 23, 1552–1557.

Smith, H. S., & DeBach, P. (1942). The measurement of the effect of entomophagous insects on population densities of their hosts. Journal of Economic Entomology, 4, 231–234.

Smith, R. F., & van den Bosch, R. (1967). Integrated pest management. (pp. 295–340). In W. W. Kilgore & R. L. Doutt (Eds.), Pest control. New York: Academic Press.

Sopp, P. I., & Sunderland, K. D. (1989). Some factors affecting the detection of aphid remains in predators using ELISA. Entomologia Experimentalis et Applicata, 51, 11–20.

Sopp, P. I., Sunderland, K. D., Fenlon, J. S., & Wratten, S. D. (1992). An improved quantitative method for estimating invertebrated predation in the field using and enzyme-linked immunosorbent assay (ELSIA). *Journal of Applied Ecology, 79,* 295–302.

Southwood, T. R. E. (1978). *Ecological methods with particular reference to the study of insect populations.* New York: John Wiley & Sons.

Sparks, A. N., Chaing, H. C., Burkhardt, G. T., Fairchild, M. L. & Weekman, G. T. (1966). Evaluation of the influence of predation on corn borer populations. *Journal of Economic Entomology, 59,* 104–107.

Starks, K. J., Muniappan, R., & Eikenbary, R. D. (1972). Interaction between plant resistance and parasitism against the greenbug on barley and sorghum. *Annals of the Entomological Society of America, 65,* 650–655.

Stern, V. M., Smith, R. F., van den Bosch, R., & Hagen, K. S. (1959). The integration of chemical and biological control of the spotted alfalfa aphid. The integrated control concept. *Hilgardia, 29,* 81–101.

Stimmann, M. W. (1974). Marking insects with rubidium: Imported cabbage worm marked in the field. *Environmental Entomology, 3,* 327–328.

Stuart, A. (1976). *Basic ideas of scientific sampling. Griffin's statistical monograph no. 4.* New York: Hafner Press.

Sunderland, K. D. (1988). Quantitative methods for detecting invertebrate predation occurring in the field. *Annals of Applied Biology, 112,* 201–224.

Sunderland, K. D. (1996). Progress in quantifying predation using antibody techniques. In W. O. C. Symondson & J. E. Liddell (Eds.), *The ecology of agricultural pests* (pp. 419–455). New York: Chapman & Hall.

Sunderland, K. D., & Sutton, S. L. (1980). A serological study of arthropod predation on wood lice in a dune grassland ecosystem. *Journal of Animal Ecology, 49,* 987–1004.

Sunderland, K. D., Crook, N. W., Stacy, D. L., & Fuller, B. J. (1987). A study of feeding by polyphagous predators or cereal aphids using ELISA and gut dissections. *Journal of Applied Ecology, 24,* 907–933.

Symindson, W. O. C., & Liddell, J. E. (Eds). (1996). *The ecology of agricultural pests.* New York: Chapman & Hall.

Torgensen, T. R., & Ryan, R. B. (1981). Field biology of *Telenomus californicus* Ashmead, an important egg parasitoid of Douglas fir tussock moth. *Annals of the Entomological Society of America, 74,* 185–186.

Turnbull, A. L. (1964). The search for prey by a web-building spider *Achaearanea tepidariorum* (C. L. Koch) (Araneae: Theridiidae). *Canadian Entomologist, 96,* 568–579.

van den Bosch, R., Leigh, T. F., Gonzalez, D., & Stinner, R. E. (1969). Cage studies on predators of the bollworm in cotton. *Journal of Economic Entomology, 62,* 1486–1489.

van der Berg, H., Shepard, B. M., Litsinger, J. A., & Pantua, P. C. (1988). Impact of predators and parasitoids on the eggs of *Rivulz atimeta, Naranga venescens* (Lepidoptera: Noctuidae) and *Hydrellia philippina* (Diptera: Ephydridae) in rice. *Journal of Plant Protection of Tropics, 5,* 103–108.

van de Vrie, M. J., McMurtry, J. A., & Huffaker, C. B. (1972). Ecology of tetranychid mites and their natural enemies. III. Biology, ecology and pest status and host-plant relations of tetranychids. *Hilgardia, 41,* 343–432.

Van Driesche, R. G. (1983). Meaning of "percent parasitism" in studies of insect parasitoids. *Environmental Entomology, 12,* 1611–1622.

Van Driesche, R. G., & Bellows, T. S., Jr. (1988). Host and parasitoid recruitment for quantifying losses from parasitism, with reference to *Pieris rapae* and *Cotesia glomerata. Ecological Entomology, 13,* 215–222.

van Lenteren J. C. (1980). Evaluation of control capabilities of natural enemies: Does art have to be science? *Netherlands Journal of Zoology, 30,* 369–381.

van Lenteren, J. C., & Bakker, K. (1976). Functional responses in invertebrates. *Netherlands Journal of Zoology, 26,* 567–572.

van Sickle, D., & Weseloh, R. M. (1974). Habitat variables that influence the attack by hyperparasites of *Apanteles melanoscelus* cocoons. *Journal of New York Entomological Society, 82,* 2–5.

Vet, L. E. M., & Dicke, M. (1992). Ecology of infochemical use by natural enemies in a tritophic context. *Annual Review of Entomology, 37,* 141–172.

Vickermann, G. P., & Sunderland, K. D. (1975). Arthropods in cereal crops: Nocturnal activity, vertical distribution and aphid predation. *Journal of Applied Ecology, 12,* 755–766.

Waage, J. K., & Hassell, M. P. (1982). Parasitoids as biological agents—fundamental approach. *Parasitology, 84,* 241–268.

Way, M. J. (1973). Objectives, methods, and scope of integrated control. In P. W. Geier, L. R. Clark, D. J. Anderson, & H. A. Nix (Eds.), *Insects: Studies in population management* (Vol. 1, pp. 137–152). Memiors of the Ecological Society of Australia.

Way, M. J., & Banks, C. J. (1958). The control of *Aphis fabae* Scop. with special reference to biological control of insects which attack annual crops. *Proceedings of the 10th International Congress on Entomology, 4,* 907–909.

Way, M. J., & Banks, C. J. (1968). Population studies on the active stages of the black bean aphid *Aphis fabae* Scop., on its winter host *Euonymus europaeus* Linnaeus. *Annals of Applied Biology, 62,* 177–197.

Weseloh, R. M. (1974). Host-related microhabitat preference of gypsy moth larval parasite, *Parasetigena agilis. Environmental Entomology, 3,* 363–364.

Weseloh, R. M. (1978). Seasonal and spatial mortality patterns of *Apanteles melanoscelus* due to predators and gypsy moth hyperparasites. *Environmental Entomology, 7,* 662–665.

Weseloh, R. M. (1982). Implications of tree microhabitat preferences of *Compsilura concinnata* (Diptera: Tachinidae) for its effectiveness as a gypsy moth parasitoid. *The Canadian Entomologist, 114,* 617–622.

Whalon, M. E., & Parker, B. L. (1978). Immunological identification of tarnished plant bug predators. *Annals of the Entomological Society of America, 71,* 453–456.

Whitham, T. G., Williams, A. G., & Robinson, A. M. (1984). The variation principal, individual plants as temporal and spatial mosaics of resistance to rapidly evolving pests. In P. W. Price, C. N. Slobodchikoff, & W. S. Gaud (Eds.), *A new ecology—novel approaches to interactive systems* (pp. 15–51). New York: John Wiley & Sons.

Wilson, F. (1960). *A review of biological control of insects and weeds in Australia and Australian New Guinea: Vol. 1.* Slough, United Kingdom: Commonwealth Agriculture Bureau Tech. Communication.

Winder, L. (1990). Predation of cereal aphid *Sitobion avenae* by polyphagous predators on the ground. *Ecological Entomology, 15,* 105–110.

Woglum, R. S., LeFollette, J. R., Landon, W. E., & Lewis, H. C. (1947). The effect of field-applied insecticides on beneficial insects of citrus in California. *Journal of Economic Entomology, 40,* 818–820.

Wolfson, J. (1980). Oviposition responses of *Pieris rapae* to environmentally induced variation in *Brassica nigra. Entomologia Experimentalis et Applicata, 27,* 223–232.

Wood, B. J. (1971). Development of integrated control program for tropical perennial crops in Malaysia. In C. B. Huffaker (Ed.), *Biological control* (pp. 113–140). New York: Plenum Press.

Wool, D., Gerling, D., & Cohen, I. (1984). Electrophoretic detection of two endoparasitic species, *Encarsia lutea* and *Eretmocerus mundus* in the whitefly, *Bemisia tabaci* (Genn.) (Hom., Aleurodidae). *Zeitschrift fuer Angewandte Entomologie, 98,* 276–279.

Wool, D., van Emden, H. F., & Bunting, S. W. (1978). Electrophoretic detection of the internal parasite, *Aphidius matricariae* in *Myzus persicae. Annals of Applied Biology, 90,* 21–26.

Wright, D. W., Hughes, R. D., & Worell, J. (1960). The effect of certain predators on the numbers of cabbage root fly (*Erioschia brassicae* [Bouch]) and on the subsequent damage caused by the pest. *Annals of Applied Biology, 48,* 756–763.

10

Evaluation of Results
Economics of Biological Control

A. P. GUTIERREZ, L. E. CALTAGIRONE, and W. MEIKLE*

Center for Biological Control
University of California
Berkeley, California

The vast majority of objections to the application of chemical pesticides does not come from the people who use them most—the farmers. By and large, the man who depends for his livelihood on crop production is convinced that he must use pesticides regularly or perish. Such conviction has been actively fostered by the pesticide industry. All sorts of data and discussion in scientific and lay journals have emphasized the apparent need for, and the great good derived by farmers from, the faithful use of pesticides. DeBach (1974)

INTRODUCTION

The term "biological control" has been given various meanings (Caltagirone & Huffaker, 1980; NAS, 1987; Garcia *et al.,* 1988), which makes it difficult to discuss general aspects such as the analysis of costs and benefits without defining the area to which the discussion applies. In this chapter we distinguish classical and naturally occurring biological control of organisms—including pests—from other methods such as the use of pesticides derived from biological organisms (e.g., *Bacillus thuringiensis* toxins, ryania, and pyrethrum), and the use of sterile males obtained either through conventional breeding and selection or through genetic engineering [see Garcia *et al.,* (1988)]. We consider periodic colonization of natural enemies (inundative and inoculative) as an extension of biological control. Calling biological control any procedure of pest control that involves the use or manipulation of a biological organism or its products is confusing, as seen, for example, in the economic analyses of Reichelderfer (1979, 1981, 1985). Rei-

*Currently at the International Institute for Tropical Agriculture, van den Bosch Centre for Biological Control, Cotonu, Benin.

chelderfer's contribution has been to show how economic theory applies to an analysis of the economic benefits of augmentative releases of biological control agents, and in this sense the arguments are akin to those for estimating the benefits of using pesticides or any other control method.

As used here, the discipline of biological control as an applied activity concerns itself with the introduction and conservation of natural enemies that become, or are essential components of self-generating systems in which the interacting populations (principally prey/predator or host/parasite) are regulated. In biological control of pests, the manipulated organisms include predators, parasitoids, pathogens, and competitors.

Biological control of pests has been implemented worldwide in environments that are climatically, economically, and technologically diverse (Clausen, 1978). The net benefits derived from this tactic are difficult to quantify with any degree of accuracy. However, the considerable number of successful cases and the fact that no environmental damage has been detected in most of them make this tactic very desirable. Nevertheless, the classical biological control approach (the introduction of exotic natural enemies) has been challenged on the basis of possible negative effects on native organisms. For example, in Hawaii, Howarth (1983) proposed that the introduction of some natural enemies has adversely affected native faunas, and restoring the ecological situation by removal of these organisms is nearly impossible. This points to the possible environmental risk in using exotic biological control agents, especially those that may have polyphagous feeding habits. However, it has been accepted that these organisms, when introduced according to restrictions established by regulatory agencies [Animal and Plant Health Inspection Service (APHIS) in the United

States], are considered to pose no environmental hazard. Routinely, risk is recognized and must be evaluated when considering candidate natural enemies to control weeds. One case concerns the introduction of the Eurasian weevil *Rhinocyllus conicus* Froeh. for control of *musk thistle* in Canada and the United States. After its introduction, the weevil is thought to have expanded its host range, thus reducing the abundance if not causing the extinction of rare native thistles of the genus *Cirsium* and lowering the density of native tephritid flies (Louda *et al.,* 1997). A comprehensive discussion on this aspect of biological control is presented by Turner (1985) and Cullen and Whitten (1995).

Although rarely cited, the threat of extinction of native species might be greater from competition by unregulated accidental invaders. Extinctions most certainly occurred when Mediterranean annual grasses and other species were introduced into the United States and became pests [see Pimentel, (1995)]. The seriousness of this problem is illustrated in California by aggressive invaders such as Mediterranean annual grasses; Klamath weed (*Hypericum perforatum* L.), which choked Northern Californian pastures; or yellow star thistle *Centaurea solstitialis* H & A, which currently rages out of control [see Huffaker, (1957; Pimentel, (1995)]. The important point is that there may be trade-offs of not controling a pest using the biological control (or other) option, and these must be included in the equation.

The impact of exotic biological control agents on target pests is difficult to assess (Chapters 8 and 9). "Few cases have been rigorously documented" (Luck *et al.,* 1988; Neuenschwander *et al.,* 1989; Gutierrez *et al.,* 1993), making economic analysis correspondingly taxing. Even more difficult would be to include in the analysis the monetary value of the externalities, both the negative ones [referred to by Howarth (1983)] and the positive ones (e.g., the benefit that society derives from the reduction in, or the elimination of, the use of objectionable pesticides) as a result of the introduction of an effective natural enemy [see Cullen and Whitten (1995)].

NATURALLY OCCURRING BIOLOGICAL CONTROL

The economic benefits of naturally occurring biological control have been repeatedly demonstrated in cases where secondary pests became unmanageable as a result of overuse of chemical pesticides to control primary pests. DeBach (1974) showed clearly the effect of DDT in the disruptions of pests in many crops. The rice brown plant hopper, *Nilaparvata lugens* Stål, in Southeast Asia continued to be a pest as a result of it overcoming the new varieties' resistance and the use of pesticides to control it. Host-plant resistance may be overcome by natural selection of new biotypes of phytophages in the field in less than 7 years

(Gould, 1986). Kenmore (1980) and Kenmore *et al.* (1986) showed that the rice brown plant hopper is a product of the green revolution where the increased prophylactic use of pesticides destroyed its natural enemies and caused the secondary outbreak of this pest. Recognition of this problem led to the banning of many pesticides in rice by President Suharto of Indonesia. This prohibition has resulted in no losses in rice yields. Most of the pests in cotton in the San Joaquin Valley of California (Falcon *et al.,* 1971; Ehler *et al.,* 1973; Eveleens *et al.,* 1973; Ehler & van den Bosch, 1974; Burrows *et al.,* 1982), the Canete and other valleys in Peru (Lamas, 1980), Australia (Room *et al.,* 1981), Mexico (Adkisson, 1972), Sudan (von Arx *et al.,* 1983), and other areas are pesticide-induced, often causing these pests to become more important than the original target pest. These examples substantiate some of the benefits of naturally occurring biological control in controlling pests. Furthermore, these benefits are largely free of cost, unless special procedures are required to augment or reintroduce populations of the natural enemies.

ESTIMATING THE BENEFITS AND COSTS OF CLASSICAL BIOLOGICAL CONTROL

Calculating the costs (*C*) of a classical biological control project is relatively easy: sum the cost of the baseline research, the cost of foreign exploration, shipping, quarantine processing, mass rearing, field releases, and postrelease evaluation. The last cost must be evaluated judiciously because pursuing academic interests may push these costs beyond those required by the practical problem at hand. Harris (1979) proposed that costs be measured in scientist years (*SY*), with one *SY* being the administrative and technical support costs for one scientist for 1 year. For example, the U.S. Department of Agriculture (USDA) estimated that one *SY* in biological control cost $80,000 in 1976 (Andrés, 1977).

At the University of California, DeBach (1974) gives a rough estimate of the cost of importing natural enemies. He says, "I have imported several [natural enemies] into various countries with resulting impressive practical successes where the cost has been less than $100 per species. In other cases the cost may run much higher, but I believe not more than a few thousand dollars per entomophagous species at most." These tentative costs suggest that some classical biological control projects may be very inexpensive, but others may cost more because of the biological and other complexities encountered. In addition, the efficiency of the organization involved may also cause costs to vary considerably. Further, the cost of the biological control efforts on a per organization, per country, or worldwide basis must include the cost of fruitless efforts. Like any other endeavor, biological control must record not only its successes but also its failures (Ehler & Andrés, 1983).

In classical biological control, once establishment (and sometimes dispersal) in the new environment is obtained, no further costs for this natural enemy are incurred. Other uses of natural enemies may involve repeated (once a year or more often) releases of natural enemies in the field or greenhouse; these costs are analogous to the cost of pesticide applications. The release of *Aphytis* in California orange orchards (DeBach *et al.*, 1950); *Pediobius foveolatus* (Crawford) against Mexican bean beetle on soybean (Reichelderfer, 1979); *Trichogramma* spp. in many crops worldwide (Hassan, 1982; Li, 1982; Pak, 1988); *Encarsia formosa* Gahan against whiteflies in greenhouses (Hussey, 1970, 1985; Stenseth, 1985a); and phytoseiid mite predators in strawberries (Huffaker & Kennett, 1953), almonds (Hoy *et al.*, 1982, 1984), and greenhouses (Stenseth, 1985b) are examples in which costs of manipulation of natural enemies are incurred periodically. The use of sterile males in campaigns against screwworm [*Cochliomyia hominivorax* (Coqueral)], Mediterranean fruit fly (*Ceratitis capitata* Weld.), and pink bollworm (*Pectinophora gossypiella* Saunders) was aimed at eradication instead of regulation of the pest; the tactic is not part of biological control. Under these circumstances it is assumed that much higher costs can be tolerated.

The environmental costs of biological control derived from the possible suppression or (unlikely) eradication of native species by introduced exotic natural enemies (Howarth, 1983; Turner, 1985; Louda *et al.*, 1997) could be included in a benefit/cost analysis if monetary value for them was available. In most cases the costs of such factors cannot be accurately assessed in much the same way that increased cancer risks due to use of some pesticide cannot be priced except in some ad hoc way.

Computing the benefits of biological control is a difficult task. One of the most successful and historically the first case of biological control in California was the cottony-cushion scale, *Icerya purchasi* (Maskell), by the imported natural enemies *Rodolia cardinalis* (Mulsant) and *Cryptochetum iceryae* (Will.). In 1888 and 1989, when these natural enemies were imported to California at the cost of a few hundred dollars, the citrus industry was at the verge of collapse because of the scale; 1 year later shipments of oranges from Los Angeles County had increased threefold (Doutt, 1964). The current value of the citrus crop in California is approximately $500 million per year. What figures should we use to determine the benefits of such a program? Obviously the benefits continue to accrue to this day. In 1889, there was no other effective way to control the scale. So, is the yearly benefit the full net value of the citrus crop (assuming the uncontrolled pest would destroy all of the crop), or the total cost of using an effective pesticide?

Should we include the benefits of introducing these natural enemies from California to 26 other countries, in 23 of which the scale was completely controlled? Whichever method is chosen, the benefits of this project are vast but undocumented.

Still more difficult are cases in which partial noneconomic control occurs: the natural enemy is established, and regulates the population of the target species to a lower level, but not low enough as to prevent economic damage. Conceivably, in cases like these the natural enemies may make it easier to implement a more effective, complementary control tactic.

The effects of biological interactions (target pest/natural enemy) are complex. They are often influenced by other factors including weather, and the beneficial effects of the natural enemy may not be obvious. When the results of biological control are clear, increased production and increased land values may be only part of the benefit, because enhanced environmental and health effects may also occur but remain undocumented. The baseline for a comparison between the situation prior and after establishment of biological control must further consider the changing real value of money over time, the changing markets for the commodity involved, and the dynamics of land use. Further, enhanced yield may be due to reduced pest injury, but also to reduction in diseases the pest may vector.

The easiest benefits to estimate, albeit crudely, are those to the agricultural sector. Because of the permanent nature of biological control, the *net present value of benefits* (π) [i.e., benefit (B) − costs (C)] corrected for the present value of money using the discount rate $(1 + d)^{-i}$ accrue over t years ($i = 1, \ldots, t$) (Reichelderfer, 1981). Note that $d = 0.05$ is the assumed interest rate or price of money.

$$\pi = \sum_{i=1}^{t}(B_i - C_i)(1 + d)^i$$

Gross revenue (B) to the farmer equals $P(Y - DN(1 - E))$ with P being price; Y, the maximum possible yield; D, the damage rate per pest N; and E, the efficacy of the biological control. In reality, D may be a function of N (i.e., $D(N(1 - E))$), but for simplicity we assume that D is constant. In fact, the benefit of biological control for the ith year is $B_i = PDN_iE$, and in the extreme may equal PY when control is complete.

DeBach (1971, 1974), van den Bosch *et al.* (1982), Clausen (1978), and Greathead (1995) summarized several classical biological control projects worldwide. A few of them are reviewed here (Table 1), noting the ratio of total net benefits computed as indicated earlier to total costs over the first 10 years of the project (i.e., the benefit to cost ratios, *B:C*). In many cases the costs after establishment are zero. We realize that this ratio is dimensionless and tells

TABLE 1 Economic Analysis of Some Classical Biological Control Projects, Three Cases of Augmentative Releases of Natural Enemies, and the Screwworm Sterile Male Eradication Program

Area	Pest	Costs × $1000[a]	Benefits per annum × $1000[b]	Start date	Net benefits to 1993 × $1000[c]	Benefit to cost ratio for the first 10 years[d]	References
Caribbean							
Barbados	Sugarcane borer	150	1,000	1967	45,618	88	Huffaker *et al.*, 1976
Antigua	*Diatraea* spp.	21.25	1,280	1931	7,700	796	Huffaker *et al.*, 1976
St. Kitts	*Diatraea* spp.	0.5	125	1934	43,896	3,302	Huffaker *et al.*, 1976
St. Lucia	*Diatraea* spp.	2.5	30	1933	10,209	159	Huffaker *et al.*, 1976
Mauritius	*Cordia macrostachya*	25	250	1953	28,538	132	Huffaker *et al.*, 1979
St. Kitts	Cottony-cushion scale *Icerya purchasi*	1.88	5.6	1966	101	39.3	Simmonds, 1967
Nevis	*Opuntia triacantha*	1.36	25	1957	2,514	241	Simmonds, 1967
Australia	Skeletonweed *Chondrilla juncea*	—	14,674	1978	332,476	—	Cullen, 1985
	Opuntia spp.	600	3,900	1925	1,843,218	86	Huffaker & Caltagirone, 1986
	Salvinia molesta	—	(Very large)	—		—	Room *et al.*, 1981
Africa							
South Africa	*Opuntia megacantha*	42.5	237.5	1950	29,277	74	Huffaker *et al.*, 1976
Kenya	Coffee mealybug *Planococcus kenyae*	75	1,250	1939	319,262	202	Huffaker *et al.*, 1976
Principe	Coconut scale *Aspidiotus destructor*	10	180	1955	19,226	238	Huffaker *et al.*, 1976
North America							
Canada	Winter moth *Operophtera brumata*	150	175	1954	2,994	15	Huffaker *et al.*, 1976
	European spruce sawfly *Dyprion hercyniae*	250	375	1932	44,072	20	Huffaker *et al.*, 1976
United States	Rhodes grass mealybug *Antonina graminis*	200	19,468	1959	1,721,023	1,285	Huffaker & Caltagirone 1986; Dean *et al.*, 1979
	Black scale on citrus *Saissetia oleae*	—	2,100	1959	?	—	Huffaker *et al.*, 1976
	Citrophilus mealybug *Pseudococcus fragilis*	2.6	2,500	1959	223,068	12,698	Huffaker *et al.*, 1976
	Grape leaf skeletonizer *Harrisina brillans*	925?	1,000	1956	8,003?	14(?)	Huffaker *et al.*, 1976
	Klamath weed *Hypericum perforatum*	25	21,700	1953	2,749,251	11,464	Huffaker *et al.*, 1976; DeBach, 1964
	Olive parlatoria scale *Parlatoria oleae*	228	725	1962	36,936	42	Huffaker *et al.*, 1976; DeBach, 1964
	Spotted alfalfa aphid *Therioaphis maculata*	900	3,000	1958	199,156	44	Huffaker *et al.*, 1976; DeBach, 1964
	Walnut aphid *Chromaphis juglandicola*	?	250	1970	10,875	—	Huffaker *et al.*, 1976; DeBach, 1964
	Sugarcane leafhopper *Perkinsiella saccharicida*	?	5,000	1905	98,257	—	Huffaker & Caltagirone, 1986
Augmentative releases							
	Mexican bean beetle *Epilachna varivestis*	627	1,812	1977	26,849	3	Reichelderfer, 1979
	Alligator weed *Alternanthera philoxeroide*	50	400	1976	9.496	8	Andres, 1977
	Whiteflies *Trialeurodes vaporarium* and spider mites *Tetranychus* spp.	67	2,094	1978	20,323	31	Hussey & Scopes, 1985
Sterile male releases	Screwworms *Cochliomyia hominivorax*	50,000	90,000	1950	6,000,000	1.8	Hussey, 1970

[a] Costs of biological control are not recurring as is the case with augmentative releases and some sterile male programs.

[b] Benefits the first year.

[c] Benefits corrected for the current value of money (1993 dollars).

[d] The ratio or recurring or total benefit to total costs for a 10-year period.

nothing about the total gain; instead it is a useful measure of the rate of return per dollar invested. Some projects, such as Klamath weed and citrophilus mealybug have $B:C$ ratios in the thousands, while the ratios for most of the others are in the hundreds. These estimates are at best very rough approximations—for the reasons discussed earlier—but even if they overestimated the benefit by 50%, the B:C ratios still overwhelmingly favor the use of classical biological control. In fact, the estimates of benefits are conservative, and the errors are in the other direction.

There are many other examples of the benefits of biological control. Tassan *et al.* (1982) showed that the introduced natural enemies of two scale pests of ice plant, an ornamental used in California to landscape freeways, potentially saved the California Department of Transportation approximately $20 million in replanting costs (on 2428 ha, or 6000 acres). Chemical control at a cost of $185/ha ($75/acre), or $450,000 annually, did not prove satisfactory; hence if suitable biological control agents did not exist, the minimal long run cost would appear to be the replacement cost. The total cost of the project was $190,000 for 1 year, yielding a B:C ratio of 105. Clearly, this biological control project was cost effective.

The control of cassava mealybug by the introduced parasitoid *Apoanagyrus* (= *Epidinocarsis*) *lopezi* DeSantis over the vast cassava belt in Africa was a monumental task. Successful control of the mealybug enabled the continued cultivation of this basic staple by subsistence farmers, thus potentially helping to reduce hunger for 200 million people in an area of Africa larger than that of the United States and Europe. What monetary value would we assign to this biological control success? How does one price the reduction or prevention of human misery? This project has been characterized as the most expensive biological control project ever ($16 million to date) by some of its critics, but all things being relative, the costs of this program since its inception in 1982 are less than those of the failed attempt to eradicate pink bollworm from the southwestern United States, or roughly about the cost of a single fighter plane bought by many of these countries. The per capita cost of the project amounts to 8 cents per person affected in the region, which contrasted to average yield increases in the Savannah zones of west Africa of 2.5 metric tons per cultivated hectare, would appear a good return on the investment (Neuenschwander *et al.,* 1989). The project has done an excellent job of documenting nearly all phases of the project (Herren *et al.,* 1987; Gutierrez *et al.,* 1988a, 1988b, 1988c, 1993; Neuenschwander *et al.,* 1989) while satisfying as much as is humanly possible the concerns of Howarth (1983). Biological control of the cassava green mite in Africa is occurring, futher documenting the efficacy of the biological control tactic (S.J. Yaninek, personal communication).

There are also cases of successful biological control where the observable benefits are just as impressive but an economic analysis has not been conducted, such as the control of three Palearctic cereal aphids over the wheat-growing regions of South America during the early 1980s by the University of California (Berkeley)/INIA, Chile/EMBRAPA, Brazil/FAO cooperative effort (R. van den Bosch, A. P. Gutierrez, E. Zuniga, & L. E. Caltagirone, unpublished). This successful project reduced the pesticide load on the environment, causing direct enhancement in yields. New wheat varieties were being developed at the time, but their yield potential had not been stabilized; hence it is not possible to assess the maximum contribution of the biological control effort. However, if as a result of the establishment of natural enemies there was a saving of one application of pesticide per annum, the total savings in Argentina, Brazil, and Uruguay on 8,996,000 ha of wheat alone (FAO, 1987) would be substantial, especially if it is contrasted with the cost of the biological control component, which was probably less than $300,000.

The economic benefits of several successful classical biological control projects can be compared in Table 1 with the use of inundative releases of natural enemies in soybean for control of Mexican bean beetle and for greenhouse pests, and with the well-known sterile male eradication program. The release of resistant predatory mites in almond gave a *B:C* ratio of 3 (Headley & Hoy, 1987), and the screwworm eradication project is estimated to have given a ratio of 1.8. These *B:C* ratios are impressive, but on average they are still not as high as those achieved using classical biological control that is self-sustaining.

In augmentative release and especially eradication programs, the cost of starting and maintaining them may be very high. In some cases, a particular pest may be perceived to be of such damaging nature and effective natural controls may not be available that the high costs of eradication may be deemed necessary. However, eradication programs are not without risks. For example, an economic analysis of the proposed eradication of the boll weevil from the southern United States predicted that the eradication of the pest would cause the displacement of cotton from the area (Taylor & Lacewell, 1977). In this scenario, increased cotton production due to eradication of the pest would cause prices to fall, forcing production to move to the west where it is more efficient. In the case of the ill-fated pink bollworm eradication project in the desert regions of southern California, early termination of the crop was available as an alternative, but it was not favored by farmers because they did not pay for the full cost of the eradication program or the environmental costs of high pesticide use, and yields were lower. Only resistance to insecticides in pesticide-induced pests made them reconsider alternatives such as short-season cottons and conservation of natural control agents.

JUSTIFYING THE NEED FOR BIOLOGICAL CONTROL

Why then do we feel the need to make economic justifications for biological control? Why has biological control not been more widely supported worldwide? Economists would call it market failure, because the users of pesticides do not pay for long-term consequences of pesticide use and hence may ignore environmentally safer alternatives (Regev, 1984). However, there are also problems of perception, for as Day (1981) assessed in his review of the acceptance of biological control as an alternative for control of alfalfa weevil in the northeastern United States: "At first, the general opinion was that biological insect control was outmoded, because it had not been effective in the east in decades, and it was not likely to be competitive with synthetic insecticides or the newer man-made chemicals such as pheromones, chemosterilants, attractants and hormones." In short, biological control was not perceived as competitive with newer technologies and it was not considered modern. The overselling of bioengineering solutions for crop protection by industry, government, and universities can also be added to the list of reasons why classical biological control is not currently strongly supported. Finally, once established, biological control agents cannot be marshaled for private profit except via augmentative releases.

Often the damage of important pests may not be obvious to funding agencies, or grower groups may not be sufficiently organized to provide the funding. For example, the Egyptian alfalfa weevil in California is a very serious pest not only in alfalfa, but more importantly in pasturelands where it depletes the nitrogen-fixing medics and clovers. In 1974, their feeding damage resulted in $2.40 to $9.59 reduction in fat lamb production (or $5.00 reduction in beef production) and $1.00 to $1.50 reduction in fixed nitrogen per acre per year, in addition to spraying costs of $2.50/acre/year plus materials (Monty Bell, University of California Cooperative Extension, personal communication); note that the external social costs have not been included in this analysis. These losses averaged over the vast expanse of grazing land in California and other western states add to an enormous sum. However, despite the economic significance of this pest, funding for this project has proved elusive, greatly hindering biological control efforts. In contrast, funding for the biological control of the ice plant scales in California was rapid because its damage was readily visible along the freeways; and the California Department of Transportation, which funded the project, had ready access to funds from gasoline taxes.

In developed countries, the advocates of biological control, compared to those promoting predominantly the use of chemical pesticides, are far fewer in number, generally have sparser resources, and have a more difficult educational task. It requires tremendous educational skills, financial re-sources, and personal dedication to effectively convey the necessary information to enable farmers to make educated decisions about pest control. The processes of biological control are not visible to the majority of agriculturalists, and with rare exception its benefits become part of the complicated biology that is subsumed in the business of crop production and is quickly forgotten by old and new clients alike. On rare occasions, the biological and economic success was so dramatic, as occurred with Klamath weed in California, that the generation four decades later knew the history of its control.

The problem is also increasing in developing countries as modern agrotechnology displaces traditional methods, and they too become dependent on pesticides for the control of pests. To combat this problem, the United Nations-sponsored project on rice in Southeast Asia headed by P. E. Kenmore set as its goal the training of 6 million rice farmers to recognize the organisms responsible for the natural control of rice pests.

In short, perceptions of the seriousness of a pest control problem often determine whether an environmentally sound alternative is selected. Of course, this is too simple a statement of the problem, but it does provide the general direction on how to solve it.

BIOLOGICAL CONTROL AS AN ALTERNATIVE TO PESTICIDE USE

In a free-market economy, individual farmers make their own pest control decisions, and purveyors of alternatives such as pesticides have the right to market them (albeit with some restrictions). Under such a system, the perceptions of the problem by farmers and the marketing skills of those proposing alternative solutions often dictate how well biological control is adopted in the field.

In evaluating the effectiveness of chemical control or augmentative release of natural enemies, economists traditionally look at the balance of revenues [$B(x)$, the value of the increase in yield attributable to using x units of the control measure (pesticide or augmentation)] minus the out-of-pocket cost [$C(x)$] of using x units of the control measure. Only infrequently are the social costs [$S(x)$] associated with the control measure included. For augmentative releases of natural enemies and biological control, $S(x)$ is usually zero. The benefit function is usually assumed to be concave from below and the cost per unit of x constant (Fig. 1A). From the farmer perspective, the net benefit (π) function should be

$$\pi = B(x) - C(x).$$

The solution to this function with maximum net benefit occurs when $dB/dx = dC/dx$; hence the optimal quantity of x to use is x_1 when $S(x)$ is excluded, but is x_2 when included (see Fig. 1 B). If the cost per unit of x used increases with

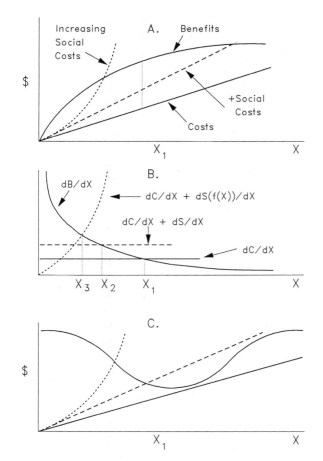

FIGURE 1 Evaluation of effectiveness of chemical control: (A) Without effective biological control, (B) the marginal solutions and (C) with effective biological control (note the change in the benefit function).

x, costs rise rapidly and less pesticide (x_3) is optimal. Unfortunately, the social or external costs of pesticides in terms of pollution, health, and environmental effects are seldom included in the farmer's calculations because farmers usually have no economic incentive to consider them.

In sharp contrast, augmentative releases of natural enemies also engender ongoing costs, but they are environmentally safe and may be more economical than pesticide use. Prime examples of the successful use of this method are the highly satisfactory control of pests in sugarcane in much of Latin America (Bennett, 1969), and in citrus orchards in the Fillmore District of California (van den Bosch et al., 1982).

Conservation of natural enemies ($x = o$) for control of pests such as *Lygus* spp. bugs on cotton in the San Joaquin Valley in California and in other crops elsewhere (DeBach, 1974) often yields superior economic benefits than does control with insecticides (Falcon et al., 1971). In such a case the ill-advised use of chemical pesticides (x) may change the benefit function resulting in additional pest control costs and, at times, lower yields (Fig. 1C; after Gutierrez et al., 1979). With naturally occurring biological control

and economically viable classical biological control (*BC*), the costs of other pest control tactics and social costs often become zero, and the whole of society obtains the maximum benefits: the natural and biological controls supplant other methods of control and solve the problem permanently. In such cases biological control should be favored, as the equation for private profit becomes

$$B(BC) - C(BC) > B(x) - C(x) > B(x) - C(x) - S(x).$$

PERCEIVED RISK

Despite effective natural control, farmers may still perceive a high risk of pest outbreak and may apply cheap pesticides as perceived insurance (or assurance) against risk of pests such as *Lygus* spp. in cotton, but in paying the premium they may become locked in a treadmill of pesticide use so colorfully described by van den Bosch (1978). DeBach (1974) called pesticides "ecological narcotics" because of their effect of suppressing problems temporarily, but causing addiction to their continued use. Regev (1984) showed that this was a classic case of market failure and similarly refers to the addiction to pesticides, and concluded that in general the root of the problem is that pesticides are preferred because the social costs are not paid by the users.

Economists use the concept of utility to evaluate perception of risk (Regev, 1984), but we use a more intuitive approach here. Two concepts enter the picture when one wishes to analyze the reliance of farmers on pesticides: the first is a measure of the mean and variance of profits and the second is the perception of risk (i.e., a subjective probability distribution of profits). In this concept, the higher the variance, the higher is the perceived risk.

If we have effective natural control (e.g., California San Joaquin Valley cotton), risk-averse farmers might still perceive the distribution of profits with and without pesticides to be as illustrated in Fig. 2A. The real distributions of profits with pesticide use and in the absence of resistance to pesticides are likely to be as shown in Fig. 2B, while the presence of resistance would greatly alter the real distributions (see Fig. 2C). Here we assume the same variance, but quite likely the variances (real risk) would increase where resistance is a factor. Obviously, if risk-averse farmers think that despite the same average profit, the variation in profit is lower using pesticides, they will no doubt choose to control pests using them. If farmers are more informed about all the issues, they may still judge the distribution more favorable using pesticides (see Fig. 2) because they have no incentive to assume responsibility for the social costs. The decision might not be so certain in the latter cases, if increases in pesticide costs cause a significant shift in the perception of risk involved in the various control alternatives. One hoped for outcome might be that natural controls are increasingly preferred. If resistance occurs,

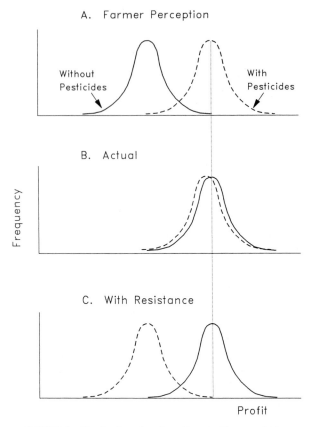

A. Farmer Perception

Without
Pesticides

With
Pesticides

B. Actual

Frequency

C. With Resistance

Profit

FIGURE 2 Distribution of profits with and without pesticides.

farmers soon learn that preserving natural enemies in the field is risk reducing and in some cases may be an option to bankruptcy. In cases of complete biological control, the mean profits may be greatly increased because pesticides would no longer be required, yields would be near maximum, and variance of yield would be narrowed. Numerous examples of the latter exist.

In summary, how farmers perceive risk is important and determines how much they are willing to pay for pest control to minimize that risk (a measure of risk aversion). Adding the social cost of pesticide use to the cost of pesticides narrows the gap between unrealistically perceived risk and real risk to profits. Taxing pesticide users to fund biological control efforts is a socially responsible way to fund permanent solutions for pest problems, but in California such funds go into the coffers of some state bureaucracy.

CONCLUSION

The abundant empirical evidence (see Table 1) shows that biological control as practiced by experts is among the most cost-effective methods of pest control. Because of its

highly positive social and economic benefits, biological control often should be the first pest control tactic explored. However, we also feel that Howarth's admonitions are not without merit—biological control workers must be cautious when introducing exotic organisms—a point that argues against granting license too widely for such introductions. Biological control is a serious endeavor for professionals—it cannot become a bandwagon for enthusiasts having little of the formal training and understanding of the basis of the discipline.

In pest control, the rights of society and the environment are increasingly in conflict with private profit. Classical biological control and other forms of natural control, plus other environmentally and economically sound methods must step in to bridge this vexing gap. Of this set, biological control has the best record and a considerable untapped potential for future benefit.

Acknowledgment

Special thanks are extended to Uri Regev who made suggestions for improving the text.

References

Adkisson, P. L. (1972). The integrated control of insect pests of cotton. Proceedings of the Tall Timbers Conference on Ecology and Animal Control Habitat Management Tallahassee, Florida, 4, 175–188.

Andres, L. A. (1977). The economics of biological control of weeds. Aquatic Botany, 3, 111–123.

Bennett, F. D. (1969). Tachinid flies as biological control agents for sugarcane moth borers. In J. R. Williams, J. R. Metcalfe, R. W. Mungomery, & R. Mathes (Eds.), Pests of sugar cane (pp. 117–18). New York: Elsevier.

Burrows, T. M., Sevacherian, V., Browning, H., & Baritelle, J. (1982). History and cost of the pink bollworm (Lepidoptera: Gelechiidae) in the Imperial Valley. Bulletin of the Entomological Society of America, 28, 286–90.

Caltagirone, L. E., & Huffaker, C. B. (1980). Benefits and risks of using predators and parasites for controlling pests. Ecological Bulletin (Stockholm), 31, 103–109.

Clausen, C. P. (Ed.). (1978). Introduced parasites and predators of arthropod pests: A world review. Agricultural handbook 480. Washington, DC: U.S. Department of Agriculture.

Cullen, J. M. (1985). Bringing the cost benefit analysis of biological control of *Chondrilla juncea* up to date. In E. S. Delfosse (Ed.), Proceedings, Sixth International Symposium on Biological Control of Weeds (pp. 145–142). Agriculture Canada, Vancouver, Canada. August 19–25, 1984.

Cullen, J. M., & Whitten, M. J. (1995). Economics of classical biological control. In H. M. Hokkanen and J. M. Lynch (Eds.), Biological control. Benefits and risks (pp. 270–276). Cambridge: Cambridge University Press.

Day, W. H. (1981). Biological control of alfalfa weevil in northeastern United States. In G. C. Papavizas (Ed.), Biological control in crop production. BARC Symposium No. 5 (pp. 361–374). Totowa, NJ: Allenheld, Osmun.

Dean, H. A., Schuster, M. F., Bolling, J. C., & Riherd, P. T. (1979).

Complete biological control of *Antonina graminis* in Texas with *Neodusmetia sangwani* (a classic example). Bulletin of the Entomological Society of American, 25(4), 262–267

DeBach, P. (1964). Biological control of insect pests and weeds. Reinhold, New York.

DeBach, P. (1971). The use of imported natural enemies in insect pest management. Proceedings of the Tall Timbers Conference on Ecology and Animal Control Habitat Management, Tallahassee, Florida, 3, 211–32.

DeBach, P. (1974). Biological control by natural enemies. London: Cambridge University.

DeBach, P., Dietrick, E. J., Fleschner, C. A., & Fisher, T. W. (1950). Periodic colonization of *Aphytis* for control of the California red scale. Preliminary tests, 1949. Journal of Economic Entomology, 43, 783–802.

Doutt, R. L. (1964). The historical development of biological control. In P. DeBach (Ed.), Biological control of insect pests and weeds (pp. 21–42). New York: Reinhold.

Ehler, L. E., & Andrés, L. A. (1983). Biological control: Exotic natural enemies to control exotic pests. In C. L. Wilson & Graham, C. L. (Eds.), Exotic plant pests and North American agriculture (pp. 295–418). New York: Academic Press.

Ehler, L. E., & van den Bosch, R. (1974). An analysis of the natural biological control of *Trichoplusia ni* (Lepidoptera: Noctuidae) on cotton in California. Canadian Entomologist, 106, 1067–1073.

Ehler, L. E., Eveleens, K. G., & van den Bosch, R. (1973). An evaluation of some natural enemies of cabbage looper in cotton in California. Environmental Entomology, 2, 1009–1015.

Eveleens, K. G., van den Bosch, R., & Ehler, L. E. (1973). Secondary outbreak induction of beet armyworm by experimental insecticide application in cotton in California. Environmental Entomology, 2, 497–503.

Falcon, L. A., van den Bosch, R., Gallagher, J., & Davidson, A. (1971). Investigation on the pest status of *Lygus hesperus* in cotton in central California. Journal of Economic Entomology, 64, 56–61.

FAO. (1987). Production yearbook 1986: Rome. Vol. 40. Rome: UN, FAO.

Garcia, R., Caltagirone, L. E., & Gutierrez, A. P. (1988). Comments on a redefinition of biological control. Roundtable. BioScience, 38, 692–694.

Gould, F. (1986). Simulation models for predicting durability of insect-resistant germ plasm: A deterministic diploid, two-locus model forum. Environmental Entomology, 15, 1–10.

Greathead, D. J. (1995). Benefits and risks of classical biological control. In H. M. Hokkanen & J. M. Lynch (Eds.). Biological control: Benefits and risks (pp. 53–63). Cambridge: Cambridge University Press.

Gutierrez, A. P., Wang, Y., & Regev, U. (1979). An optimization model for *Lygus hesperus* (Heteroptera: Miridae) damage in cotton: The economic threshold revisited. Canadian Entomologist, 111, 41–54.

Gutierrez, A. P., Neuenschwander, P., & Van Alphen, J. J. M. (1993). Factors affecting the biological control of cassava mealybug by exotic parasitoids: A ratio- dependent supply-demand driven model. Journal of Applied Ecology, 30, 706–721.

Gutierrez, A. P., Neuenschwander, P., Schulthess, F., Baumgaertner, J. U., Wermelinger, B., & Loehr, B. (1988a). Analysis of biological control of cassava pests in Africa. II. Cassava mealybug *Phenococcus manihoti*. Journal of Applied Ecology, 25, 921–940.

Gutierrez, A. P., Wermelinger, B., Shulthess, F., Baumgaertner, J. U., Herren, H. R., & Ellis, C. K. (1988b). Analysis of biological control of cassava pests in Africa. I. Simulation of carbon, nitrogen and water dynamics in cassava. Journal of Applied Ecology, 25, 901–920.

Gutierrez, A. P., Yaninek, J. S., Wermelinger, B., Herren, H. R., & Ellis, C. K. (1988c). Analysis of the biological control of cassava pests in Africa. III. Cassava green mite *Mononychellus tanajoa*. Journal of Applied Ecology, 25, 941–950.

Harris, P. (1979). Cost of biological control of weeds by insects in Canada. Weed Science, 27(2), 242–250.

Hassan, S. A. (1982). Mass production and utilization of *Trichogramma*. III. Results of some research projects related to the practical use in the Federal Republic of Germany. First International Symposium l'INRA *Trichogramma*, Antibes, France. Collques de l'INRA, 9, 213–218.

Headley, J. C., & Hoy, M. A. (1987). Benefit/cost analysis on integrated mite management program for almonds. Journal of Economic Entomology, 80(3), 555–559.

Herren, H. R., Neuenschwander, P., Hennessey, R. D., & Hammond, W. N. O. (1987). Introduction and dispersal of *Epidinocarsis lopezi* (Hym., Encyrtidae) an exotic parasitoid of the cassava mealybug *Phenococcus manihoti* (Hom., Pseudococcidae), in Africa. Agricultural Ecosystems of the Environment, 19, 131–134.

Howarth, F. G. (1983). Classical biocontrol: Panacea or Pandora's box. Proceedings of the Hawaiian Entomological Society, 24, 239–244.

Hoy, M. A., Barnett, W. W., Rell, W. D., Castro, D., Cahn, D. & Hendricks, L. C. (1982). Large scale releases of pesticide-resistant spider mite predators. California Agriculture, 36, 8–10.

Hoy, M. A., Barnett, W. W., Hendricks, L. C., Castro, D., Cahn, D., & Bentley, W. J. (1984). Managing spider mites in almonds with pesticide-resistant predators. California Agriculture, 38, 18–20.

Huffaker, C.B. (1957). Fundamentals of biological control of weeds. Hilgardia, 27, 101–157.

Huffaker, C. B., & Caltagirone, L. E. (1986). The impact of biological control on the development of the Pacific. Agricultural Ecosystems and the Environment, 15, 95–107.

Huffaker, C. B., & Kennett, C. E. (1953). Developments toward biological control of cyclamen mite on strawberries in California. Journal of Economic Entomologist, 46, 802–812.

Huffaker, C. B., & Kennett, C. E. (1966). Biological control of *Parlatoria oleae* (Colvee) through the compensatory action of two introduced parasites. Hilgardia, 37(9), 283–335.

Huffaker, C. B., Simmonds, F. J., & Laing, J. E. (1976). Theoretical and empirical basis of biological control. In C. B. Huffaker & P. S. Messenger (Eds.), Theory and practice of biological control (pp. 41–78). New York: Academic Press.

Hussey, N. W. (1970). Some economic considerations in the future development of biological control. In Technical economics of crop protection and pest control (Monograph 36SCI pp. 109–118). London: Society of Chemical Industry.

Hussey, N. W. (1985). Whitefly control by parasites. In N. W. Hussey & N. Scopes (Eds.), Biological pest control—The Glasshouse Experience (pp. 104–115). Ithaca, NY: Cornell University Press.

Hussey, N. W., & Scopes, N. (Eds.). (1985). Biological pest control—The glasshouse experience. Ithaca, NY: Cornell University Press.

Kenmore, P. E. (1980). Ecology and outbreaks of a tropical insect pest of the green revolution, the rice brown planthopper, Nilaparvata lugens (Stål). Unpublished doctoral dissertation, University of California, Berkeley.

Kenmore, P. E., Carino, F. O., Perez, C. A., Dyck, V. A., & Gutierrez, A. P. (1986). Population regulation of the rice brown planthopper (*Nilaparvata lugens* [Stål]) within rice fields in the Philippines. Journal of Plant Protection of Tropics, 1, 19–37.

Lamas, J. M. (1980). Control de los insectos-plaga del algodonero en El Peru—Esquema de la planificacion de una campaña de control integrado y su problematica. Revista Peruana Entomologia, 23, 1–6.

Li, L. Y. (1982). *Trichogramma* sp. and their utilization in the People's Republic of China. First International Symposium on *Trichogramma*, Antibes, France. Collques l'INRA, 9, 23–29.

Louda, S. M., D, Kendall, J, Connor, & Simberloff, D. (1997). Ecological effects of an insect introduced for the biological control of weeds. Science, 277, 1088–1090.

Luck, R. F., Shepard, B. M., & Kenmore, P. E. (1988). Experimental methods for evaluating arthropod natural enemies. Annual Review of Entomology, 33, 367–391.

National Academy of Sciences (1987). Report of the research briefing panel on biological control in managed ecosystems, R. J. Cook, Chair. Washington, DC: National Academy Press.

Neuenschwander, P., Hammond, W. N. O., Gutierrez, A. P., Cudjoe, A. R., Baumgaertner, J. U., & Regev, U. (1989). Impact assessment of the biological control of the cassava mealybug, Phenacoccus manihoti Matile Ferrero (Hemiptera: Pseudococcidae) by the introduced parasitoid Epidinocarsis lopezi (DeSantis) (Hymenoptera: Encyrtidae). Bulletin of Entomological Research, 79, 579–594.

Pak, G. A. (1988). Selection of Trichogramma for inundative control. Unpublished doctoral dissertation, Agricultural University Wageningen, Netherlands.

Pimentel, D. (1995). Biotechnology: Environmental impact of introducing crops and biocontrol agents in North American agriculture. In H. M. Hokkanen and J. M. Lynch (Eds.), Biological control: Benefits and risks (pp. 13–29). Cambridge: Cambridge University Press.

Regev, U. (1984). An economic analysis of man's addiction to pesticides. In G. R. Conway (Ed.), Pest and pathogen control: Strategic, tactical, and policy models (pp. 441–453). New York: John Wiley & Sons.

Reichelderfer, K. H. (1979). Economic feasibility of a biological control technology: Using a parasitic wasp, Pediobius foveolatus, to manage Mexican bean beetle on soybean. (Agriculture Economics Rep. No. 430). Washington, DC: U.S. Department of Agriculture ESCS.

Reichelderfer, K. H. (1981). Economic feasibility of biological control of crop pests. In G. C. Papavizas (Ed.), Biological control in crop production, BARC Symposium No. 5 (pp. 403–417). Totowa, NJ: Allenheld, Osmun.

Reichelderfer, K. H. (1985). Factor affecting the economic feasibility of the biological control of weeds. In E. S. Delfosse (Ed.), Proceedings of the Sixth International Symposium on Biological Control of Weeds, Vancouver, Canada, August 19–25, 1984.

Room, P. M., Harley, K. L. S., Forno, I. W., & Sands, D. P. A. (1981). Successful biological control of the floating weed salvinia. Nature (London), 294, 78–80.

Simmonds, F. J. (1967). The economics of biological control. Journal of the Royal Society of Arts, 115, 880–898

Stenseth, C. (1985a). Whitefly and its parasite Encarsia formosa. In N. W. Hussey & N. Scopes (Eds.), Biological pest control—the glasshouse experience (pp. 30–33). Ithaca, NY: Cornell University Press.

Stenseth, C. (1985b). Red spider mite control by Phytoseiulus in Northern Europe. In N. W. Hussey & N. Scopes (Eds.), Biological pest control—the glasshouse experience (pp. 119–124). Ithaca, NY: Cornell University Press.

Tassan, R. L., Hagen, K. S., & Cassidy, D. V. (1982). Imported natural enemies established against ice plant scales in California. California Agriculture, 36, 16–17.

Taylor, C. R., & Lacewell, R. D. (1977). Boll weevil control strategies: Regional benefits and costs. Southern Journal of Agricultural Economics, 9, 124–135.

Turner, C. E. (1985). Conflicting interests and biological control of weeds. Proceedings of the Sixth International Symposium on Biological Control of Weeds (pp. 203–225). Vancouver, Canada, 1984.

van den Bosch, R. (1978). The pesticide conspiracy. New York: Doubleday.

van den Bosch, R., Messenger, P. S., & Gutierrez, A. P. (1982). An introduction to biological control. New York: Plenum.

von Arx, R., Baumgaertner, J., & Delucchi, V. (1983). A model to simulate the population dynamics of Bemisia tabaci Genn. (Stern., Aleyrodidae) on cotton in the Sudan Gezira. Zeitschrift fuer Angewandte Entomologie, 96, 341–363.

11

Periodic Release and Manipulation of Natural Enemies

GARY W. ELZEN and EDGAR G. KING

Subtropical Agricultural Research Center
U.S. Department of Agriculture
Agricultural Research Service
Weslaco, TX

INTRODUCTION

The development of insecticide-resistant populations of pests, evidence for resurgence on a worldwide basis (Waage, 1989), and environmental concerns have led to increased research in the area of alternative strategies to pesticides for control of damaging pest populations. Classical biological control and the successes in that area are important, and provide background and encouragement for increased efforts in the manipulation of natural enemies (Huffaker, 1974; Greathead, 1986). Methods for manipulation of natural enemies for the purposes of reducing herbivore populations and alleviating plant stress include, but are not limited to, conservation and augmentation (this chapter), habitat management (Chapter 14), and genetic manipulation (Chapter 12). In this chapter we explore the possible avenues that are available, or may be developed, for manipulation of natural enemies to improve their effectiveness in controlling pests, and we will consider the natural enemies in the context of a rational management program that explores the appropriate use of all available insect control methods.

AUGMENTATION

Augmentation of natural enemies to effect suppression of pest populations has received considerable attention and has been reviewed by Beglyarov and Smetnik (1977), Huffaker *et al.* (1977), Knipling (1977), Ridgway and Vinson (1977), Shumakov (1977), Starter and Ridgway (1977), King *et al.* (1984, 1985a), and van Lenteren (1986). Behavior of released parasitoids and predators (Weseloh, 1984) and their effectiveness have additionally been treated (Vin-

son, 1975; Nordlund *et al.,* 1981a, 1981b, 1985; Lewis & Nordlund, 1985).

The concepts of inundative and inoculative releases of entomophages were first described by DeBach and Hagen (1964). Inundative releases rely mainly on the agents released, not their progeny, to effect control, whereas inoculative releases rely on an increase of the initial natural-enemy populations that will suppress subsequent pest generations. Inundative releases may be timed so that both immediate control and subsequent progeny production contribute to pest suppression (Li, 1984). Augmentative or inundative releases have been described by such terms as supplemental releases, strategic releases, programmed releases, seasonal colonization, periodic colonization, and compensatory releases (Ridgway *et al.,* 1977; King *et al.* 1984, 1985a). Examples demonstrating the feasibility of controlling pests by augmentative releases of entomophages are given in Table 1. The major proportion of successfully introduced natural enemies are parasitoids (DeBach, 1964; Laing & Hamai, 1976; van Lenteren, 1986). Greathead (1986) noted that parasitoids have been established and achieved satisfactory control about three times more often than predators have. Individual case studies were presented by King *et al.* (1985a). Steps by which augmentation may be effected are shown in Fig. 1.

Trichogramma spp.

Control of lepidopterous pests by mass rearing and release of *Trichogramma* spp. has occurred for many decades. The pioneering research of Howard and Fiske (1911) and Flanders (1929, 1930) stimulated research with *Trichogramma* spp., and a number of successes in reducing insect populations by augmentation with *Trichogramma* have

TABLE 1   Parasitoid and Predator Species in United States Where Feasibility Has Been Shown in Augmentative Releases

Entomophage	Target pest	Commodity	Reference
Amblyseius californicus (McGregor)	Two-spotted spider mite	Strawberry	Oatman *et al.,* 1977
Aphidius smithi Sharma & Subba Rao	Pea aphid	Pea	Halfill & Featherston, 1973
A. melinus DeBach	California red scale	Citrus	DeBach & Hagen, 1964
Bracon spp.	Pink bollworm	Cotton	Bryan *et al.,* 1973, Legner & Medved, 1979
Agathis (Brasus) diversa Muesebeck	Oriental fruit moth	Peach	Brunson & Allen, 1944
Catolaccus grandis (Burks)	Boll weevil	Cotton	King *et al.,* 1985
Chelonus spp.	Pink bollworm	Cotton	Bryan *et al.,* 1973; Legner & Medved, 1979
Chrysopa carnea Stephens	Several	Several crops	Ables & Ridgway, 1981
Cocinella septempunctata L.	Potato aphid, green peach aphid	Potato	Shands *et al.,* 1972
Cryptolaemus montrouzieri Mulsant	Citrus mealybug	Citrus	DeBach & Hagen, 1964
Diaeretiella rapae McIntosh	Aphid	Potato	Shands *et al.,* 1975
Encarsia formosa Gahan	Greenhouse whitefly	Greenhouse crops	Vet *et al.,* 1980
Hippodamia convergens Guérin-Méneville	Aphid	Pea	DeBach & Hagen, 1964; Cooke, 1963
Leptomastix dactylopii Howard	Mealybug	Citrus	Fisher, 1963
Lixophaga diatraeae (Townsend)	Sugarcane borer	Sugarcane	King *et al.,* 1981
Jalysus spinosus (Say)	Tobacco budworm	Tobacco	Ridgway *et al.,* 1977
Aphid parasites	Greenbug, pea aphid	Sorghum, pea	Ridgway *et al.,* 1977
Lysiphlebus testaceipes (Cresson)	Greenbug	Sorghum	Starks *et al.,* 1976
Macrocentrus ancylivorus Rohwer	Oriental fruit moth	Peach	Brunson & Allen, 1944
Microterys flavus (Howard)	Brown soft scale	Citrus	Hart, 1972
Phytoseiulus persimilis Athias-Henriot	Two-spotted spider mite	Strawberry	Oatman *et al.,* 1968, 1976, 1977
Polistes spp.	Tobacco budworm, Lepidopterous larvae	Cotton	Fye, 1972
Pediobius foveolatus (Crawford)	Mexican bean beetle	Soybean	Ridgway *et al.,* 1977
Predaceous mites	Several mite pests	Fruit, greenhouse crops	Ridgway *et al.,* 1977 Chant, 1959
Fly parasites	Filth fly	Man, other animals	Brunetti, 1981
Praon spp.	Aphid	Potato	Shands *et al.,* 1975
Stethorus picipes Casey	Avocado brown mite	Avocado	McMurtry *et al.,* 1969
Trichogramma pretiosum Riley	Bollworm, tobacco budworm	Cotton	Stinner *et al.,* 1974 Ables *et al.,* 1981, King *et al.,* 1985b
T. minutum Riley	Tomato fruit worm, cabbage looper, tomato hornworm	Tomato	Oatman & Platner, 1978
	Imported Cabbageworm, cabbage looper	Cabbage	Parker, 1971
	Sugarcane borer	Sugarcane	Jaynes & Bynum, 1941
T. cacoecia Marchal	Codling moth	Apple	Dolphin *et al.,* 1972
T. nubiale Ertle & Davis	European corn borer	Corn	Kanour & Burbutis, 1984
Typhlodromus spp.	Cyclamen mite	Strawberry	Huffaker & Kennet, 1956
	McDaniel spider mite	Apple	Croft & McMurtry, 1972

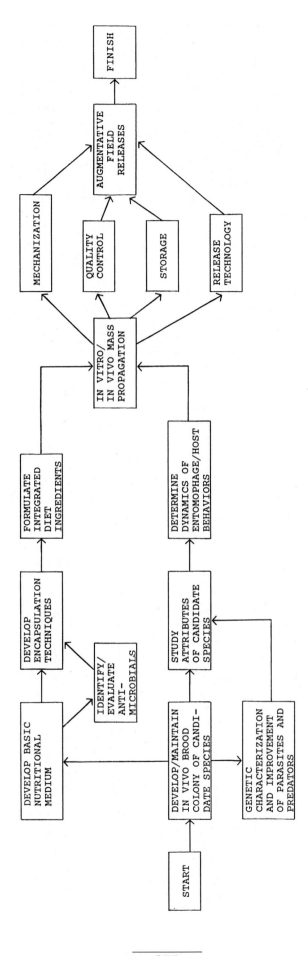

FIGURE 1   Steps for effecting augmentation.

255

been reported. Reviews on the use of *Trichogramma* were presented by Ridgway and Vinson (1977), Ridgway *et al.* (1977) for the Western Hemisphere, by Huffaker (1977) for China, by Beglyarov and Smetnik (1977) for the area of the former USSR, and by King *et al.* (1985b) for augmentation in cotton. As of 1985, *Trichogramma* spp. were the entomophagous arthropods most widely used for augmentation in the world (King *et al.*, 1985b). Hassan (1982) reported 65 to 93% reduction in larval infestations of the European corn borer following *Trichogramma* releases during the late 1970s in Germany [see also Bigler (1983)]. Voronin and Grinberg (1981) reported reduction of pest species of *Loxostege* spp., *Agrotis* spp., and *Ostrinia* spp. following *Trichogramma* releases. In China a significant reduction in populations was reported for *Ostrinia* spp., *Heliothis* spp., and *Cnaphalocrocis* spp., and crop damage was also reduced (Li, 1984).

Oatman and Platner (1985) found that two common lepidopterous pests of avocado in southern California, *Amorbia cuneana* Walsingham and the omnivorous looper *Sabulodes aegrotata* (Guenée), could be effectively controlled by releases of 50,000 *T. platneri* Nagarkatti in each of four uniformly spaced trees per acre. At least three weekly releases were required for control of *S. aegrotata,* while two were necessary for *A. cuneana*. Reduction in insect numbers and crop damage in several agroecosystems was also reported from China and the former USSR (Voronin & Grinberg, 1981; Li, 1984). Oatman and Platner (1971, 1978) demonstrated the technical feasibility of augmenting *T. pretiosum* Riley populations to reduce damage in tomato caused by the tomato fruit worm [*Helicoverpa zea* (Boddie)], cabbage looper [*Trichoplusia ni* (Hübner)], and *Manduca* spp., although these authors found that chemical control was necessary for pests not attacked by *T. pretiosum*. Oatman *et al.* (1983) reported studies conducted to develop an integrated control program for the tomato fruit worm and other lepidopterous pests on summer plantings of fresh-market tomatoes in southern California in 1978 and 1979. A regimen of twice weekly applications of Dipel® (delta-endotoxin of *Bacillus thuringiensis* Berliner var. *kurstaki*), plus twice-weekly releases of *T. pretiosum,* was compared with weekly applications of methomyl. There were no significant differences in fruit yield or size between the two treatment programs. Methomyl adversely affected predator populations, host eggs, and egg parasitization by *T. pretiosum,* whereas Dipel® did not.

Trichogramma spp. have been utilized in the Netherlands for biological control of lepidopteran pests. Two approaches were taken, the selection of "best" species and strains of *Trichogramma* (van Lenteren *et al.,* 1982) and studies of the manipulation of *Trichogramma* behavior. The first approach has been studied extensively, especially in *Brassica* spp. (Pak *et al.,* 1988). Inundative releases of *Trichogramma* were feasible for control of *Mamestra brassica* (Linnaeus) on Brussels sprouts, but control was not very effective at low host densities (van der Schaff *et al.,* 1984). Glas *et al.* (1981) also reported reduction in larval infestation of *Plutella xylostella* (Linnaeus) in cabbage crops. Several years' results of experimental releases of *Trichogramma* were summarized by van Heiningen *et al.* (1985). The second approach to *Trichogramma* manipulation involved examination of semiochemical-mediated behavior. These studies indicated that kairomones and volatile substances released by adult female hosts (sex pheromones) were important in foraging behavior of *Trichogramma* sp. (Noldus & van Lenteren, 1983; Noldus *et al.,* 1988a, 1988b). Preintroductory evaluation as outlined by Wackers *et al.* (1987) may improve prospects for success of augmentative release of strains of *Trichogramma* spp.

Although control can be achieved by augmentation with *Trichogramma* spp. under certain circumstances, variable results and insufficient pest control have also been reported (King *et al.,* 1985b). *Trichogramma pretiosum* was tested in augmentative releases in Arkansas in 1981 and 1982 and in North Carolina in 1983 for management of *Helicoverpa zea* and *Heliothis virescens* (Fabricius) in cotton. These releases failed to provide adequate control in 1981 and 1982. However, in 1983, cotton fields treated by seven augmentative releases of *T. pretiosum* at 306,000 emerged adults per hectare per release yielded significantly more cotton than check fields (not insecticide treated). Insecticidal control fields yielded more cotton than did check or *T. pretiosum* release fields, leading these researchers to conclude that management of *Heliothis* spp. in cotton by augmentative releases of *T. pretiosum* was not economically feasible at the time (King *et al.,* 1985b).

Consistent results in the use of *Trichogramma* necessitate the release of large numbers under controlled conditions to obtain consistent results. In addition to numbers released, the effectiveness of *Trichogramma* may also be influenced by (1) density or phenology of the pest, (2) species or strain of *Trichogramma* released, (3) vigor of the parasitoid released, (4) method of distribution, (5) crop phenology, (6) number of other biological agents present, and (7) proximity to crops receiving insecticide use and drift of insecticides into *Trichogramma* release areas (King *et al.,* 1985b). *Trichogramma* spp. appear highly susceptible to most chemical insecticides; and lethal effects may result from direct exposure to spray application, drift, or posttreatment contact with pesticide residues on foliage (Bull & Coleman, 1985). Chemical insecticides may have contributed to inconsistency in results obtained in Arkansas and North Carolina from augmentation efforts (King *et al.,* 1985b).

King *et al.* (1985b) cited costs for release of *T. pretiosum* to control *Heliothis* spp. at $7.68/ha per application. This cost compares well with the cost of a commonly used pyrethroid, fenvalerate, at $16.18/ha when applied at

0.11 kg/ha. The pyrethroid needs to be applied only once for every two to three parasitoid applications. Nevertheless, the development of resistance to pyrethroids and other classes of insecticides by *Heliothis* (Luttrell *et al.*, 1987; Elzen *et al.*, 1990; Elzen, 1991; Elzen *et al.*, 1992) and to dimethoate by *Lygus lineolaris* (Palisot de Beauvois) (Snodgrass & Scott, 1988), and in general possible development of resistance, make augmentation attractive.

In 1988, E. G. King hypothesized that some of our most intractable pests, such as the boll weevil, could be controlled by augmentative releases of selective parasitoids. He further hypothesized that failure to become established, as in the case of exotic parasitoids of the boll weevil, was not critical in an augmentative release program. In fact, he concluded that population densities of boll weevils tolerated by cotton growers, in season, are so low that they could not support naturally occurring parasitoid populations. These hypotheses were documented in two U.S. Department of Agriculture (USDA). Agricultural Research Service CRIS work projects. These hypotheses were further outlined by Knipling (1992) who developed a theoretical model postulating the suppressive effects of a host-specific parasitoid released against the boll weevil.

The selective parasitoid *Catolaccus grandis* (Burks) (Hymenoptera: Pteromalidae) was selected as the lead candidate for release against the boll weevil in the 1989 ARS work planning session. Previous attempts to import and establish *C. grandis* on the boll weevil were unsuccessful, but it was demonstrated that the parasitoid was preadapted to the in-season biiotic and abiotic environment of the cotton agroecosystem (Cate *et al.*, 1990; Johnson *et al.*, 1973). *Catolaccus grandis* effectively searched for boll weevil-infested squares on the ground and on the plant.

The technical feasibility for suppressing the boll weevil by inoculative/augmentative releases of insectary-produced *C. grandis* (Morales-Ramos *et al.*, 1992; Roberson & Harsh, 1993) was demonstrated in 1992 (Summy *et al.*, 1992, 1995) and in 1993 (Morales-Ramos *et al.*, 1994; Summy *et al.*, 1994). Morales-Ramos *et al.* (1993) reported on a mathematical model stimulating the effects of releases of *C. grandis* on the boll weevil. This model theoretically demonstrated that high rates of parasitism can eliminate in-field reproduction by the boll weevil and preclude the need for subsequent insecticide treatments. King *et al.* (1993) outlined a strategy for integrating augmentation into short-season cotton production. Summy *et al.* (1994) experimentally demonstrated this integration in 1993, but chemical insecticides still were not applied for early season insect pests, viz., invading overwintered boll weevils and cotton fleahoppers. Consequently, there remained a numerical, though not statistical, difference in lint yield between the integrated pest management (IPM)-treated control fields and the parasitoid-release fields. Also, the experimental cotton fields were not managed as commercial cotton fields.

Therefore, additional experiments were conducted to demonstrate that the boll weevil could be suppressed to sub-economic levels in commercially grown cotton by the aid of inoculative/augmentative releases of *C. grandis* during the boll weevil F_1 and F_2 generations as immatures in squares on the plant and on the ground.

Subsequently, King *et al.* (1995) reported the successful feasibility of large-scale rearing of *C. grandis* on boll weevil third instars reared on artificial diet and inoculative/augmentative releases in commercially managed cotton fields in South Texas cotton fields to suppress boll weevils. Up to 90% mortality of boll weevil third instars and pupae was recorded in three parasitoid-release fields, but mortality was substantially less in three IPM-treated control fields. Survival of boll weevil third instars and pupae during the F_2 and F_3 generations was nearly nondetectable in two of the parasitoid-release fields and was reduced in the third parasitoid-release field. Lint yield was significantly higher in the parasitoid-release fields (mean = 1047 lb lint/acre) than in the IPM-treated control (mean = 929 lb lint/acre), yet chemical insecticide applications were reduced from 11 to 5. Expansion of *C. grandis* from field-by-field early-season augmentative releases to early-season area wide releases is projected to be substantially more powerful as a population suppressant, preempting the development of damaging boll weevil populations.

Evaluating Natural Enemies as Candidates for Augmentation

Effective manipulation of natural enemies requires a thorough knowledge of their biology and host associations. Background information may be gained through laboratory studies. However, it is imperative that such data be followed by extensive biological studies in the field. A comparison of laboratory, field cage, and field studies would provide useful information that could be used to predict the impact of the natural enemy on pest populations. Ecological studies may help to determine the influence of habitat, biotypes, or semiochemicals on parasitoid foraging; and may provide helpful information for determining in which habitat to release natural enemies. Suggested steps to accomplish these goals are found in Fig. 2.

The level of control achieved in an augmentation program may be a function of a myriad of interacting factors (Huffaker *et al.*, 1977). Primary factors include the availability of hosts and host–parasitoid synchrony; and conditions of weather during release of natural enemies involving the effects of environmental factors on foraging, influence of habitat type, chemical pesticide usage, and fitness of laboratory-reared parasitoids. Thus, experiments must be conducted that will determine specific situations in which biological control by augmentation may work. All parame-

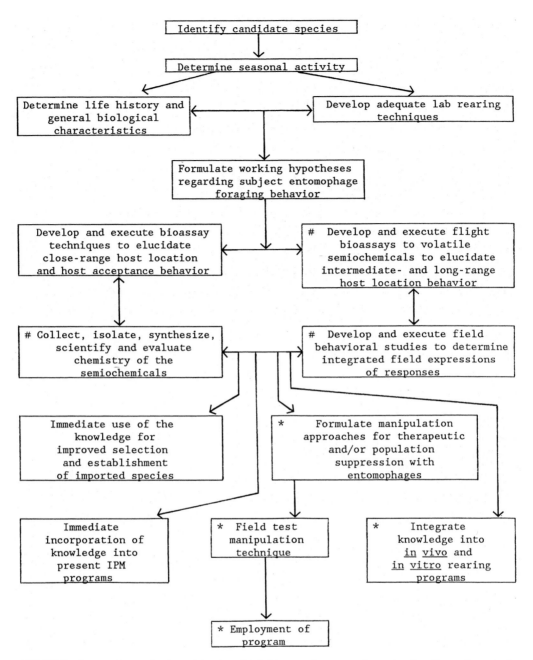

FIGURE 2 Suggested steps for determining in which habitat to release natural enemies. # = Research emphasis (future progress depends on results of these studies); * = long term (>5 years).

ters, biotic and abiotic, should be explored in evaluating augmentation results. Augmentation of natural enemies of row-crop pests may be implemented only after considerable effort has been expended to prove the feasibility of this approach. Thereafter, the efficiency and financial benefits must be determined. Reliance on use of entomophagous arthropods to control pests should be limited to those situations where scientifically, environmentally, and economically sound procedures are available. In addition, funda-

mental knowledge of search rates, functional response, and efficiency could significantly add to the predictability of success in augmentation efforts. We will now give examples of experiments that may aid in evaluating natural enemies as candidates for release.

The wind tunnel flight chamber is a useful tool for examining flight responses and foraging patterns of parasitoids. Most designs presently in use are similar to the wind tunnel of Miller and Roelofs (1978) used for studying moth flight. Nettles (1979, 1980) examined responses of *Eucelatoria* spp. to volatiles from the host and host habitat and suggested the use of *Eucelatoria* attractants to increase parasitoid populations in the vicinity of *Heliothis* hosts. Drost *et al.* (1986) examined the flight behavior mediated by airborne semiochemicals in *Microplitis croceipes* and emphasized the importance of preflight conditioning to the plant–host complex on positive searching responses of *M. croceipes*. *Microplitis croceipes* reared on hosts fed cowpea-seedling leaves instead of artificial diet had an increased percentage of oriented flights to volatiles from cowpea seedling–*H. zea* complex in a flight tunnel. In addition, the increased response was much stronger after adult females had searched a fresh plant–*H. zea* complex (Drost *et. al.*, 1988). Kaas *et al.* (1990) demonstrated learning by *M. croceipes* using a wind tunnel and various "reward" regimens. The flight responses of *M. croceipes* to host and nonhost plants (Navasero & Elzen, 1989) and spring and summer host plants of *Heliothis* spp. (Kaas *et al.*, 1992) have also been reported. Elzen *et al.* (1986) found higher innate searching by *Campoletis sonorensis* (Cameron) on glanded versus glandless varieties of cotton (*Gossypium* spp.). It was implied that volatile chemicals present in the glanded varieties had a positive effect on parasitoid foraging in the wind tunnel that was not produced by glandless cottons or Old World species. Herard *et al.* (1988) conducted experiments with *M. demolitor* that were similar to those of Drost *et al.* (1986). Herard *et al.* (1987) described rearing methods suitable for semiochemical studies. The wind tunnel flight chamber has been further refined with the development of a novel system for injection of semiochemical volatiles directly into the moving air (Zanen *et al.*, 1989). Wind tunnels may aid in efforts to address questions on host habitat location and host location by parasitoids and may provide insights to allow manipulation of parasitoid behavior. Wind tunnels, in addition, are ideal for the early isolation of semiochemicals and for the use in bioassay-directed fractionation and confirmation of synthetic chemical activity.

Parasitism and predation can vary spatially due to variation in parasitoid host plant detection, search rate, or retention of parasitoids on host plants. Host-plant species and stage, host density, and weather are likely to affect all three processes. Parasitism can vary temporally because of variation in host detection, search, retention, parasitoid natality, and mortality. Research designed to gather information to predict distribution of parasitism across host plants under varying conditions could yield important information on the population dynamics of hosts and parasitoids. These predictions are crucial to rational conservation and augmentation of parasitoids. As an example of these dynamic interactions, the search rate of *M. croceipes* in field cages was higher on *Gossypium hirsutum* Linnaeus in summer than on *Geranium dissectum* Linnaeus in spring (Hopper & King, 1986).

However, this difference may arise from different temperatures and not from different host-plant species. In field cages, *M. croceipes* parasitized more hosts on *G. hirsutum* than on *Phaseolus vulgaris* (Linnaeus) and more hosts on *P. vulgaris* (Linnaeus) than on *Lycopersicon esculentum* Mill. (Mueller, 1983). However, it is unclear whether these dissimilarities arose from differences in host-plant attraction or in search rate. Kaas *et al.* (1993) studied the behavioral time budget and periodicity of *M. croceipes* in field cages with *H. virescens* on spring host plants. Results from this study suggested that *M. croceipes* was more efficient on clover than on geranium. Attraction of *C. sonorensis* varies with host-plant species (Elzen *et al.*, 1983) and cotton variety (Elzen *et al.*, 1986), and the attraction to cotton correlated with volatile chemical profile of the varieties (Elzen *et al.*, 1984, 1985). The host-plant species on which *H. zea* has been feeding affects the response of *M. croceipes* to nonvolatile kairomones. Variation in parasitism found in host-plant surveys may arise from variation in attraction or retention of wasps by semiochemicals directly or indirectly derived from the host plants. Studies of host habitat preference may provide clues to the optimal habitat and conditions in which to release parasitoids in augmentation programs.

Screening of the biological characteristics of natural enemies for biological control has been advocated by Sabelis and Dicke (1985) and van Lenteren (1986). An example of a predator currently used in Dutch orchards is *Typhlodromus pyri* Scheuten. As noted by Dicke (1988), despite the use of *T. pyri*, the biology of this predatory mite has not been extensively studied. Based on response of *T. pyri* to volatile kairomones, it was later determined that *T. pyri* prefers the European red mite (*Panonychus ulmi* [Koch]) to *Aculus schlechtendali* (Nalepa), the apple rust mite. A study involving electrophoretic diet analysis confirmed this (Dicke & Dejong, 1988).

Field Evaluations and Assessment

Field experiments may provide insights into the efficiency of a particular parasitoid. Search rate of *M. croceipes* for *H. zea* and *H. virescens* larvae did not depend on host density in field cages (Hopper & King, 1986). However, *M. croceipes* parasitized a higher proportion of *Heliothis* larvae in plots where host density was high than in plots where host density was low (K. R. Hopper, personal communication). Parasitoid aggregation, not increased search rate, caused the increased parasitism at high host density. Further data suggest the potential utilization of *M. croceipes* females at a broad age range for *Heliothis* control (Navasero & Elzen, 1992). Females held for as long as 9 days in the absence of host material in the laboratory remained efficient in parasitizing *H. virescens* larvae. Parasitoids have been shown to aggregate in areas of high host density in laboratory experiments (Hassell, 1971; Murdie & Hassell, 1973; Hubbard & Cook, 1978; T'Hart *et al.,* 1978; Collins *et al.,* 1981; Waage, 1983). Host-plant species vary in attraction and suitability for *Heliothis,* which can cause variation in larval density, and the spatial variations in *Heliothis* parasitism observed in field surveys may in part result from parasitoid aggregation at high host densities. Several parasitoid species are more attracted to plants on which hosts have fed than they are to undamaged plants (Thorpe & Caudle, 1938; Monteith, 1955, 1964; Arthur, 1962; Madden, 1970), and mechanically damaged plants may be searched more by parasitoids (Vinson, 1975). Damaged terminals of *G. hirsutum* attracted more *C. sonorensis* than undamaged terminals (Elzen *et al.,* 1983). *Microplitis croceipes* is attracted to wind-borne odor of *H. virescens* frass and larvae (Elzen *et al.,* 1987a); and *M. croceipes* responds to nonvolatile kairomones produced by *H. zea* (Jones *et al.,* 1971; Gross *et al.,* 1975), *H. virescens,* and *H. subflexa* (Lewis & Jones, 1971). A thorough understanding of parasitoid–host interactions may aid augmentation efforts.

Theoretically, the means to increase the suppressive effect of a natural enemy can be found by increasing their numbers through propagation and release. The rate of establishment may be increased by releasing larger numbers (Beirne, 1975; Ehler & Hall, 1982) but this is not always the case (Greathead, 1971). Field evaluations must provide the data necessary for defining the number of predators or parasitoids required for release per unit area, and this along with mass production technology will determine the economic feasibility of the approach.

Introduced natural enemies, with actual or projected use to augmentation of natural populations, should be evaluated by the criteria proposed by Luck *et al.* (1988). Experimental evaluation through life table analysis, examination of percentage of parasitism, key factor analysis, and use of simulation models may provide insights into the probability of success. Evaluation of natural enemies may include introduction and augmentation; and techniques using cages and barriers, removal of natural enemies, prey enrichment, direct observation, and biochemical evidence of natural-enemy feeding, in quantifiable experiments to gauge the impact of the natural enemies (Luck *et al.,* 1988).

CONSERVATION

One important issue affecting the successful manipulation and conservation of natural enemies is the use of pesticides. The effects of pesticides on beneficial insects are considered in detail in Chapters 13 and 24 and also have been reviewed by Croft and Brown (1975), Croft and Morse (1979), and Croft (1990). Recommendations for changing control practices to preserve natural enemies were addressed by Plapp and Bull (1989). Insecticide use in cotton and the value of predators and parasites for managing *Heliothis* were reviewed by King (1986).

Numerous studies have documented the detrimental effects of pesticides on beneficials, and it may be important to note that an underlying problem in practical implementation of augmentation is the use of pesticides. Unexpected problems may be encountered, even from pesticide drift, so that basic toxicological studies may be necessary to determine if the natural enemies intended for use in augmentation have some degree of tolerance. The significance of insecticidal effects on beneficial insects is reflected in the resurgence of primary and secondary insect pests in heavily sprayed monocultures (Huffaker, 1974). Actions of insecticides on beneficial insects include not only those causing direct mortality but also those acting in indirect ways (Hoy & Dahlsten, 1984). The sublethal effects of pesticides on parasitoids were reviewed by Elzen (1989). No shortage of information exists in describing the ways in which insecticides can affect natural enemies and upset natural control in agroecosystems. First, there are the obvious direct lethal actions: broad-spectrum insecticides, such as organophosphates, are especially selective against entomophages. Because beneficials have more specific enzymes evolved for handling the toxins of their hosts, they are much more susceptible to broad-spectrum insecticides than are their hosts, which have a myriad of plant chemicals with which to contend (Krieger *et al.,* 1971). The resurgence of primary pests and of previously innocuous secondary pests has been reported widely where insecticides selectively destroy natural enemies (Michelbacher *et al.,* 1946; Doutt, 1948; DeBach & Bartlett, 1951; Lingren & Ridgway, 1967). For example, azinphosmethyl, a broad-spectrum organophosphate, selectively destroyed beneficials in apple orchards (Falcon, 1971). Conversely, chloridimeform was found less toxic than some other insecticides to several species of natural enemies (Plapp & Vinson, 1977; Plapp & Bull, 1978).

Elzen *et al.* (in press) has documented the toxicological responses of *C. grandis* to insecticides. Females from three strains of *C. grandis* were tested in a glass vial bioassay (Plapp *et al.,* 1989). Two *in vivo*-reared strains were tested (i.e., the In Vivo strain and the Sinaloa strain). The In Vivo stock was originally imported by J. R. Cate in 1986 from the states of Yucatán, Tabasco, and Oaxaca in Mexico, and has been maintained in the laboratory for 12 years. The Sinaloa strain was collected from cotton near Culiacán, Sinaloa, Mexico in 1996. An *in vivo*-reared strain was tested, (i.e., the In Vitro strain). This may be the only example where insecticides were tested on an *in vitro*-reared parasitoid. Concentration–mortality studies were performed with dimethoate, oxamyl, acephate, malathion, azinphos-methyl, cyfluthrin, methyl parathion, spinosad, and fipronil. Spinosad (Dow-Elanco, Indianapolis, IN) and fipronil (Rhône-Poulenc, Research Triangle Park, NC) are newer insecticides with novel modes of action. Compared with endosulfan, one of the least toxic insecticides to *C. grandis,* methyl parathion was 439-fold more toxic at the lethal dose-50 (LC_{50}). Azinphos-methyl and spinosad, like endosulfan, were lower in toxicity than were other insecticides tested. Fipronil, which is targeted toward boll weevil, was moderately toxic to the In Vivo and Sinaloa strains of *C. grandis,* however, it was 40-fold more toxic than endosulfan to the In Vitro strain. At least one-half of the insecticides tested would be detrimental to *C. grandis* in the field. Methyl parathion use, even in the vicinity of areas where augmentative releases of *C. grandis* are made, would be highly detrimental.

The effect of insecticides on the behavior and reproduction of natural enemies can be dramatic. For instance, Press *et al.* (1981) found that permethrin and pyrethrin reduced the number of adult *Bracon hebetor* Say produced when parental females were exposed to hosts and insecticides simultaneously. Similarly, topical application of carbaryl to female *B. hebetor* resulted in reduced numbers of eggs that develop from vitellogenic oocytes, and resorption of mature ova (Grosch, 1975). Residues of pyrethrin significantly reduced parasitism rates of *Trichogramma pretiosum* on *H. zea* eggs (Jacobs *et al.*, 1984). Formamidines are especially recognized for their ability to disrupt pest mating, reproduction, and feeding behavior (Knowles, 1982; O'Brien *et al.*, 1985). Parasitized hosts have been found to be more susceptible to insecticides than are nonparasitized hosts, thus preventing normal development of immature parasitoids. *Lymantria dispar* (Linnaeus) larvae parasitized by *Apanteles melanoscelus* (Ratzburg) were significantly more susceptible to carbaryl than were nonparasitized larvae, and more time was required for surviving parasitoids to develop (Ahmad & Forgash, 1976). Fix and Plapp (1983) found that *H. virescens* larvae parasitized by *C. nigriceps* were 14.2 times more susceptible to methyl parathion, and 2.5 times more susceptible to permethrin than were unparasitized larvae.

Insecticides that reduce hosts to low densities may, of course, indirectly reduce the abundance of natural enemies. Numbers of *Orius insidiosus* (Say) feeding on cotton leaf perforator *Bacculatrix thruberiella* Bush steadily declined when populations of the leaf perforator were reduced by chlordimeform sprays (Lingren & Wolfenbarger, 1976). While direct mortality is the most severe way that insecticides can have an impact, insecticides such as formamidines may become more important in certain pest management strategies due to the property of suppressing pest populations behaviorally and physiologically at low (sublethal) doses. Low doses of chlordimeform significantly decreased fecundity and egg viability of adult female bollworms, prevented moths from separating after mating (Phillips, 1971), and reduced fecundity in the cotton aphid (*Aphis gossypii* Glover) (Ikeyama & Maekawa, 1973) and in cattle ticks

(*Boophilus microplus* Canestrini) (Mansingh & Rawlins, 1979). Feeding behavior was upset by chlordimeform in larval tobacco cutworms (*Spodoptera litura* Fabricius) (Antoniosus & Saito, 1981), in armyworms (*Leucania separata* Walker) (Wantanabe & Fukami, 1977), and in cockroaches (*Periplaneta americana* Linnaeus). The effects of chlordimeform vary with species (Matsumura & Beeman, 1982) and there is much selectivity among species and stages for actual acute toxicity; some insects are very sensitive, and others are immune, as in the case of adult boll weevils (Wolfenbarger *et al.*, 1973). Additional evidence for the suitability of chlordimeform for some pest management strategies was given by Plapp and Vinson (1977), who found that chlordimeform was approximately 100 times less toxic to the parasitoid *C. sonorensis* than were organophosphates similar to azinphos-methyl. By controlling pests at sublethal doses, the problem of pest resistance may be lessened. Dittrich (1966) found that chlordimeform was effective against some pests that had already become resistant to organophosphate and carbamate insecticides.

The parasitoid *M. croceipes* has been commonly found in cotton in the southeastern United States and is a major parasitoid species (King *et al.*, 1985a), and due to the apparent tolerance of this parasitoid to some commonly used insecticides (King *et al.*, 1985a; Powell *et al.*, 1986; Bull *et al.*, 1987) augmentation releases of *M. croceipes* are anticipated in the future (King *et al.*, 1984). This parasitoid was exposed to insecticides commonly used in cotton using a spray table (Elzen *et al.*, 1987b). Direct treatment with the pyrethroid fenvalerate, a mixture of the formamidine chlordimeform plus fenvalerate, and the carbamate thiodicarb resulted in nearly 100% survival of both sexes at both the lowest and highest field rates recommended for these insecticides. The organophosphate acephate and the carbamate methomyl were extremely toxic to adult *M. croceipes,* and caused 100% mortality at the lowest recommended field rates. In another study, parasitoids were exposed to pesticide-treated foliage and mortality was observed after 24 h. The fenvalerate–chlordimeform mixture caused 10 to 23% mortality, with thiodicarb causing a similar percentage of mortality, whereas methomyl caused significantly high mortality, ranging from 23 to 70%. It is likely that the use of thiodicarb as an ovicide and larvicide for *Heliothis* control will increase in the future because of resistance to pyrethroids. Fortunately, *M. croceipes* appears to be relatively tolerant to this insecticide (Elzen *et al.*, 1989).

MONITORING

Monitoring, as well as sampling natural enemies that are indigenous or released in augmentation programs, may provide useful information for models to predict the impact of entomophages on reducing herbivore-induced damage or plant stress. Modeling of population interactions requires accurate tools to determine absolute densities of beneficials and pests. Monitoring beneficial insect populations, particularly parasitoids, may be complicated by factors such as lack of a stable sex ratio, movement (because females must forage for often patchily distributed hosts), weather factors, and lack of synchrony with host populations. Nevertheless, monitoring methods to evaluate parasitoid populations have been proposed. The most reasonable approach to this problem would involve estimating population numbers from captures of males or females in traps baited with an appropriate attractant, such as sex pheromone. Methods for isolating sex pheromones (Golub & Weatherston, 1984), as well as bioassay-directed fractionation, identification (Heath & Tumlinson, 1984), and synthesis (Sonnet 1984), have been reported in detail. These methods have been adapted for identification of parasitoid pheromones (Eller et al., 1984; Elzen & Powell, 1988). Powell (1986) suggested that monitoring systems be explored using some volatile host or host habitat attractant to trap female beneficials, thereby capturing the agent responsible for parasitism and perhaps obviating any problems that may arise from an unstable sex ratio.

No system has been developed whereby a parasitoid has been monitored with sex pheromone for decision making in an agricultural crop. Although enormous effort has been expended in the field of insect sex pheromones, few studies have resulted in identification of parasitoid sex pheromones. Robacker and Hendry (1977) identified neral and geranial from female *Itoplectis conquisitor* (Say), and demonstrated that these chemicals were attractive to males in the laboratory. Eller et al. (1984) identified and demonstrated the field effectiveness of ethyl palmitoleate, female sex pheromone of *Syndipnus rubiginosus* Walley, a parasitoid of the yellowhead spruce sawfly [*Pikonema alaskensis* (Rohwer)]. Work is underway to correlate field catches of the parasitoid with sawfly parasitism and to incorporate this information into populational models.

Powell and King (1984) demonstrated in the field that males of *M. croceipes* were attracted to virgin females. Diurnal activity of males and females was found to differ; hence knowledge of parasitoid activity periods would be important in developing techniques for sampling parasitoid populations in the field. Scentry® wing traps (Albany International) were more effective in capturing *M. croceipes* males than were Pherocon II® traps (Zoecon). Studies in unsprayed cotton in 1984 revealed that wing traps baited with live virgin females could be used to estimate parasitoid populations (J. E. Powell, personal communication). *Microplitis croceipes* mating behavior and sex pheromone response have been reported (Elzen & Powell, 1988), and a tentative identification of the female-produced sex pheromone has been made (Elzen & Williams, unpublished).

Insect control guidelines often recognize the impact of natural-enemy populations on pest populations (Rude, 1984). However, explicit instructions for using natural enemies in decision making are generally lacking, and where present are used with reservation. Two exceptions are (1) Michelbacher and Smith (1943) recommended that insecticide control decisions in alfalfa for *Colias eurytheme* Boisduval be made only after determining that the number of *Apanteles medicaginis* Muesebeck present was incapable of maintaining the pest under control, and (2) Croft and McMurtry (1972) reported on a decision-making index for predicting the probability of adequate control of a phytophagous mite that would occur depending on the predator to prey ratio per apple leaf. Powell and Zhang (1983) found that pheromone traps baited with live virgin females attracted large numbers of male cereal aphid parasitoids when placed in cereal fields. *Aphidius rhapalosiphi* De Stef and *Praon volucre* Haliday males were caught in separate traps baited with females. Monitoring may be useful in this situation to achieve maximum impact when the parasitoid to aphid ratio is particularly high, especially during early season (Powell, 1986).

REARING

Efficient and cost-effective methods of rearing entomophagous arthropods must be developed if augmentative releases are to be feasible (Beirne, 1974). The importance of adequate and cost-efficient culture methods is particularly evident in greenhouse programs (van Lenteren, 1986). In addition, large numbers of beneficial insects may be employed in greenhouse, field cage, and laboratory studies. Literally thousands of beneficial organisms, available at somewhat unpredictable times, may be required for commercial augmentation releases. Finney and Fisher (1964) discuss problems associated with the culture of entomophagous insects. Considerable attention has been devoted to development of techniques to produce quality insects in large numbers (Smith, 1966; King & Leppla, 1984; Anderson & Leppla, 1992). The genetic implications of long-term laboratory rearing of insects are addressed elsewhere (Mackauer, 1972, 1976; Bouletreau, 1986). Powell and Hartley (1987) described techniques for producing large numbers of parasitoids efficiently. These authors adapted a multicellular host-rearing tray technique (Hartley et al., 1982) to rear *M. croceipes* and several other species of parasitoids. Techniques reduced parasitoid harvest time by one-half, and simultaneous release of nearly 17,000 wasps was possible using low-temperature storage. Morales-Ramos et al. (1992) reported methods for mass rearing of *C. grandis;* and since then, they have described a new system to more efficiently mass rear *C. grandis* using a novel cage system (Morales-Ramos et al., 1997). Powell and Hartley (1987) noted several factors that were important for maintaining this large-scale rearing program and that may be applicable to other programs: (1) continuous host supply; (2) use of environmental chambers to alter developmental rates of hosts and parasitoids; (3) constant appropriate environmental conditions; (4) sanitary rearing conditions with flash sterilization of diet; (5) use of laminar flow hoods; (6) autoclaving reusable supplies; (7) disinfecting work areas; (8) acid or antibiotics in water or food; and (9) adequate technical support, space, and equipment. The quality of mass-reared insects should also be assessed (Boller & Chambers, 1977). Beneficial insects that have been reared for augmentative releases by USDA workers are shown in Table 1.

IN VITRO REARING

Commercialization of augmentation may be practical only for selected organisms for which suitable diets and storage methods are developed. Artificial rearing may increase practical use of augmentation for widespread pest control. Twenty-two species of natural enemy have been reared *in vitro*. Nine species of Hymenoptera (one ectoparasite, four pupal parasitoid species, and four species of *Trichogramma*) and three species of Diptera have been cultured with varying success (King *et al.*, 1984). Predators, notably *Chrysopa carnea* Stephens, have been reared on artificial diets (Vanderzant, 1973; Martin *et al.*, 1978). While there have been numerous successes in oviposition stimulant identification or partial rearing (Nettles & Burks, 1975; Nettles, 1982), definitive development of a feasible *in vitro*-rearing system for beneficials has yet to be developed. Nowadays, most parasitoids are expensive to rear, and the costs involved would preclude mass rearing. Although considerable advances have been made in *in vivo* rearing, the advances have not been achieved with *in vitro* rearing to such an extent. The work of Wu *et al.* (1982) illustrated an instance in which a completely synthetic artificial host egg was produced that contains no insect derivatives, and supports *Trichogramma* oviposition and development. Greany *et al.* (1984) suggested that mass rearing of *Trichogramma* using completely artificial hosts would become economical within the year 1989, but this prediction was not realized.

The previously mentioned boll weevil ectoparasitoid *C. grandis* has been successfully reared on artificial diet (Guerra & Martinez, 1994; Guerra *et al.*, 1993). Methods have been developed to consistently produce *C. grandis* of higher quality (Rojas *et al.* 1996) and at lower cost (Rojas *et al.*, 1997). *Catolaccus grandis* reared by the methods of Rojas *et al.* (1996) have been evaluated in laboratory and field studies in Texas (Morales-Ramos *et al.*, 1998). Females reared on boll weevils had a higher pupal weight and fecundity than females reared *in vitro*, but *in vitro*-reared

females exhibited significantly higher survival during the period of most intensive reproductive activity. The movement searching capacity, and survival under field conditions of the *in vitro*- and *in vivo*-reared *C. grandis* were compared. Dispersal ability and searching capacity was not significantly different with a 30-m radius from a release point. Nevertheless, no significant difference in boll weevil mortality induced by parasitism was recorded between the two methods. The use of artificial diets is a promising method for mass propagating *C. grandis*.

Hymenopterous larval endoparasitoids have not been successfully reared to the adult stage on artificial diet. However, *Cotesia marginiventris* (Cresson) and *M. croceipes* have been reared on artificial media through the first instar (P. Greany, personal communication). Larval endoparasitoids have evolved complex mechanisms that interact with the host's internal dynamics and organs without damaging this environment or causing untimely demise of the host. The workings of these interacting factors must be understood for *in vitro* rearing of larval endoparasitoids to become a reality. Developments in artificial rearing of beneficial insects on artificial diet may allow us to produce sufficient numbers of beneficials to feasibly implement the further evaluation of natural enemies as biological control agents.

Parasitoids that have been reared to adults on artificial media include the larval ectoparasitoid *Exeristes roborator* (Fabricius) (Thompson 1982), endoparasitoids of eggs [*T. pretiosum* (Hoffman *et al.*, 1975) and *T. dendrolimi* (Wu *et al.* 1982)], larvae [*Lixophaga diatreae* (Towns) (Grenier *et al.*, 1978) and *Eucelatoria bryani* Sabrosky (Nettles *et al.*, 1980)], and pupae [*Brachymeria lasus* (Walker) (Thompson, 1983), *Pachycrepoideus vinclemia* Rondani (Thompson *et al.*, 1983), *Itoplectis conquisitor* (Say) (House, 1978), *Pteromalus puparum* Linnaeus (Hoffman & Ignoffo, 1974), and *C. grandis* (Rojas *et al.*, 1996)].

CONCLUSIONS

Recommendations for enhancing biological control using manipulative techniques are as follows: (1) modify chemical control practices to preserve natural enemies, (2) develop selective pesticides favoring arthropod natural enemies, (3) integrate biological control with cultural practices and host-plant resistance, (4) develop methodology for estimating abundance and impact of natural enemies, (5) integrate the natural enemy component into crop−pest models, (6) identify and colonize the best biotypes of predators and parasites including those of foreign origin, (7) increase efforts genetic improvement of natural enemies, (8) manipulate natural enemies with behavior-modifying substances, and (9) develop *in vitro*-rearing techniques for natural enemies. After these criteria are studied, it may become apparent that a particular augmentation effort has the chance for success. At that time an entire, rational pest management package would need to be developed and a pilot program could be implemented. Ultimately, the utilization of natural enemies in pest control will require a series of basic and applied research steps. This process may be arduous, but the final outcome is well worth the effort: pest suppression and management at reduced monetary costs and at reduced impact to the environment.

Acknowledgments

We thank P. J. Elzen (USDA, ARS, Weslaco, TX) for major contributions to the section on conservation and data on the sublethal effects of pesticides. M. G. Rojas (USDA, ARS, Weslaco, TX) provided much information on *C. grandis*. Preparation of this manuscript was supported in part by USDA Competitive Grant No. 87-CRCR-1-2473 to GWE. Mention of a trade name, proprietary product, or specific equipment does not constitute a guarantee or warranty by the USDA and does not imply its approval to the exclusion of other products that may be suitable.

References

Ables, J. R., & Ridgway, R. L. (1981). Augmentation of entomophagous arthropods to control pest insects and mites. In G. C. Papvisas (Ed.); Biological control in crop protection (pp. 273–303). Totowa, NJ: Allenheld, Osman.

Ables, J. R., Vinson, S. B., & Ellis, J. S. (1981). Host discrimination by *Chelonus insularis* (Hymenoptera: scolionidae) and *Trichogramma pretiosum* (Hymenoptera: Trichogrammatidae). Entomophaga, 26, 149–156.

Ahmad, S., & Forgash, A. J. (1976). Toxicity of carbaryl to Gypsy moth larvae parasitized by *Apanteles melanoscelus*. Environmental Entomology, 5, 1183–1186.

Anderson, T. E., & Leppla, N. C. (1992). Advances in insect rearing for research and pest management. Westview Studies in insect biology. Westview Press: Oxford & IBH Pub. Co., Boulder, Co.

Antoniosus, A. G., & Saito, I. (1981). Mode of action of antifeeding compounds in the larvae of the tobacco cutworm *Spodoptera litura* (Fabricius). I. Antifeeding activities of chlordimeform and some plant diterpenes. Applied Entomology and Zoology, 16, 328–334.

Arthur, A. P. (1962). Influence of host tree on abundance of *Itoplectis conquisitor*. (Say) (Hymenoptera: Ichneumonidae), a polyphagous parasite of European pine shoot moth, *Rhyacionia buoliana* (Schiff.) (Lepidoptera: Oleuthreutidae). Canadian Entomologist, 94, 337–347.

Beirne, B. P. (1974). Status of biological control procedures that involve parasites and predators. In F. G. Maxwell & F. A. Harris (Eds.), Proceedings of the Summer Institute on Biological Control of Plant Insects and Diseases (pp. 69–76). Jackson: University Press of Mississippi.

Beirne, B. P. (1975). Biological control attempts by introductions against pest insects in the field in Canada. Canadian Entomologist, 107, 225–236.

Beglyarov, G. A., & Smetnik, A. I. (1977). Seasonal colonization of entomophages in the U.S.S.R. In R. L. Ridgway & S. B. Vinson (Eds.), Biological control by augmentation of natural enemies (pp. 283–328). New York: Plenum Press.

Bigler, F. (1983). Erfahrungen bei der biologischen Bekanpfung des Maiszunslers mit *Trichogramma*—Schlupfwespen in des Scheveiz. Mitteilungen fuer die Schweizerische Landwirtschaft, 31(1/2), 14–22.

Boller, E. F., & Chambers, D. L. (1997). Quality aspects of mass-reared insects. In R. L. Ridgway & S. B. Vinson (Eds.), Biological control by augmentation of natural enemies (pp. 219–235). New York: Plenum Press.

Bouletreau, M. (1986). The genetic and coevolutionary interactions between parasitoids and their hosts. In J. Waage & D. Greathead (Eds.), Insect parasitoids (pp. 169–200). New York: Academic Press.

Brunetti, K. M. (1981). Suppliers of beneficial organisms in North America. Sacramento, CA: California Department of Food and Agriculture Biological Control Services Program.

Brunson, M. H., & Allen, A. W. (1994). Mass liberation of parasites for immediate reduction of oriental fruit moth injury to ripe peaches. Journal of Economic Entomology, 37, 411–416.

Bryan, D. E., Fye R. E., Jackson, C. G. & Patana, R. (1973). Release of parasites for suppression of pink bollworms in Arizona (USDA-ARS W-7). Washington, DC: U.S. Department of Agriculture.

Bull, D. L., & Coleman R. J., (1985). Effect of pesticides on *Trichogramma* spp. Southwestern Entomology, 8, (Suppl.), 156–168.

Bull, D. L., Pryor, N. W., & King, E. G., Jr. (1987). Pharmacodynamics of different insecticides in *Microplitis croceipes* (Hymenoptera: Braconidae), a parasite of lepidopteran larvae. Journal of Economic Entomology, 80, 739–749.

Cate, J. R., Krauter, P. C., & Godfrey, K. E. (1990). Pests of cotton. In D. H. Habect, F. D. Bennett, & J. H. Frank (Eds.), Classical biological control in the southern United States (Southern Cooperation Service Bulletin 335).

Chant, D. A. (1959). Phytoseiid mites (Acarina: Phytoseiidae). I. Bionomics of seven species in southeastern England. Canadian Entomologist, 91(12) (Suppl.), 1–44.

Collins, M. D., Ward, S. A., & Dixon, A. F. G. (1981). Handling time and the functional response of *Aphelinus thompsoni*, a predator and parasite of the aphid *Preponsiphum platanoidis*. Journal of Animal Ecology, 50, 479–487.

Cooke, W. C. (1963). Ecology of the pea aphid in the Blue Mountain area of eastern Washington and Oregon (USDA Tech. Bulletin 1287). Washington, DC: U.S. Department of Agriculture.

Croft, B. A. (1990). Arthropod biological control agents and pesticides. New York: John Wiley & Sons.

Croft, B. A., & Brown, A. W. A. (1975). Responses of arthropod natural enemies to insecticides. Annual Review of Entomology, 20, 285–325.

Croft, B. A., & McMurtry, J. A. (1972). Minimum releases of *Typhlodromus occidentalis* to control *Tetranychus mcdanieli* on apple. Journal of Economic Entomology, 65, 188–191.

Croft, B. A., & Morse, J. (1979). Research advances on pesticide resistance in natural enemies. Entomophaga, 24, 3–11.

DeBach, P. (Ed.). (1964). Successes, trends, and future possibilities. In Biological control of insect pests and weeds (pp. 673–713). London: Chapman & Hall.

DeBach, P., & Bartlett, B. (1951). Effects of insecticides on biological control of insect pests of citrus. Journal of Economic Entomology, 44, 372–383.

DeBach, P., & Hagen, K. S. (1964). Manipulation of entomophagous species. In P. DeBach (Ed.), Biological control of insect pests and weeds (pp. 429–455). London: Chapman & Hall.

Dicke, M. (1988). Prey preference of the phytoseiid mite *Typhlodromus pyri*. I. Response to volatile kairomones. Experimental and Applied Acarology (Northwood), 4, 1–13.

Dicke, M., & DeJong, M. (1988) Prey preference of the phytoseiid mite *Typhlodromus pyri*. II. Electrophoretic diet analysis. Experimental and Applied Acarology (Northwood), 4, 15–25.

Dittrich, V. (1966). N-(2-methyl-4-chlorophenyl)-N', N'-dimethylformamidine (C-8514/Schering 36268) evaluated as an acaricide. Journal of Economic Entomology, 59, 889–893.

Dolphin, R. E., Clevelena, M. L., Mouzin, L. E., & Morrison, R. K. (1972). Releases of *Trichogramma minutum* and *T. cacoeciae* in an apple orchard and the effects on populations of codling moths. Environmental Entomology, 1, 481–484.

Doutt, R. L. (1948). Effects of codling moth sprays on natural control of the Baker mealybug. Journal of Economic Entomology, 41, 116–117.

Drost, Y. C., Lewis, W. J., Zanen, P. O., & Keller, M. A. (1986). Beneficial arthropod behavior mediated by airborne semiochemicals. I. Flight behavior and influence of preflight handling of *Microplitis croceipes* (Cresson). Journal of Chemical Ecology, 12, 1247–1262.

Drost, Y. C., Lewis, W. J., & Tumlinson, J. H. (1988). Beneficial arthropod behavior mediated by airborne semiochemicals. V. Influence of rearing method, host plant, and adult experience on host-searching of *Microplitis croceipes* (Cresson), a larval parasitoid of *Heliothis*. Journal Chemical Ecology, 14, 1607–1616.

Ehler, L. E., & Hall, R. W. (1982). Evidence for competitive exclusions of introduced natural enemies in biological control. Environmental Entomology, 11, 1–4.

Eller, F. J., Bartelt, R. J., Jones, R. L., & Kulman, H. M. (1984). Ethyl (Z)-9-hexadecanoate. A sex pheromone of *Syndipnus rubiginosus,* a sawfly parasitoid. Journal of Chemical Ecology, 10, 291–300.

Elzen, G. W. (1989) Sublethal effects of pesticides on beneficial parasitoids. In P. C. Jepson (Ed.), Pesticides and non-target invertebrates (pp. 129–150). United Kingdom: Intercept, Ltd.

Elzen, G. W. (1991). Pyrethroid resistance and carbamate tolerance in a field population of tobacco budworm in the Mississippi Delta. Southwestern Entomology, (Suppl. 15), 27–31.

Elzen, G. W., & Powell, J. E. (1988). Mating behavior and sex pheromone response of the *Heliothis* parasitoid *Microplitis croceipes* (pp. 257–260). In Proceedings, Beltwide Cotton Conference. National Cotton Council, Memphis, TN.

Elzen, G. W., O'Brien, P. J., & Powell, J. E. (1989). Toxic and behavioral effects of selected insecticides on parasitoid *Microplitis croceipes.* Entomophaga, 34, 87–94.

Elzen, G. W., O'Brien, P. J., & Snodgrass, G. L. (1990). Toxicity of various classes of insecticides to pyrethroid-resistant *Heliothis virescens* larvae. Southwestern Entomology, 15, 33–38.

Elzen, G. W., Williams, H. J., & Vinson, S. B. (1983). Response by the parasitoid *Campoletis sonorensis* (Hymenoptera: Ichneumonidae) to chemicals (synomones) in plants: Implications for host habitat location. Environmental Entomology, 12, 1873–1877.

Elzen, G. W., Williams, H. J., & Vinson, S. D. (1984). Isolation and identification of cotton synomones mediating searching behavior by parasitoid *Campoletis sonorensis.* Journal of Chemical Ecology, 10, 1251–1264.

Elzen, G. W., Williams, H. J., & Vinson, S. D. (1986). Wind tunnel flight responses by parasitoid *Campoletis sonorensis* to cotton cultivars and lines. Entomologia Experimentalis et Applicata, 43, 285–289.

Elzen, G. W., Williams, H. J., Bell, A. A., Stipanovic, R. D., & Vinson, S. B. (1985). Quantification of volatile terpenes of glanded and glandless *Gossypium hirsutum* Linnaeus cultivars and lines by gas chromatography. Journal of Agricultural and Food Chemistry, 33, 1079–1082.

Elzen, G. W., Williams, H. J., Vinson, S. B., & Powell, J. E. (1987a). Comparative flight behavior of parasitoids *Campoletis sonorensis* and *Microplitis croceipes.* Entomologia Experimentalis et Applicata, 45, 175–180.

Elzen, G. W., O'Brien, P. J., Snodgrass, G. L., & Powell, J. E. (1987b). Susceptibility of the parasitoid *Microplitis croceipes* (Hymenoptera: Braconidae) to field rates of selected cotton insecticides. Entomophaga, 32, 545–550.

Elzen, G. W., Leonard, B. R., Graves, J. B., Burris, E., & Micinski, S. (1992). Resistance to pyrethroid, carbamate, and organophosphate insecticides in field populations of tobacco budworm (Lepidoptera: Noctuidae) in 1990. Journal of Economic Entomology, 85, 2064–2072.

Elzen, G. W., Rojas, M. G., Elzen, P. J., King, E. G., & Barcenas, N. M. (in press). Toxicology of *Catolaccus grandis.* In Memorias de Simposio Binacional Mexico-USA, Control biologico del picudo del algodonero mediante liberaciones inoculativas/inundativas del ectoparasito Catolaccus grandis (Burks) (Hymenoptera: Pteromalidae).

Falcon, L. A. (1971). Microbial control as a tool in integrated control programs. In C. B. Huffaker (Ed.), Biological control (pp. 346–364). New York: Plenum Press.

Finney, G. L., & Fisher, T. W. (1964). Culture of entomophagous insects and their hosts. In P. DeBach (Ed.), Biological control of insect pests and weeds (pp. 328–355). London: Chapman & Hall.

Fisher, T. W. (1963). Mass culture of Cryptolaemus and Leptomastix—natural enemies of citrus mealybug (California Agricultural Experiment Station Bulletin 797).

Fix, L. A., & Plapp, F. W., Jr. (1983). Effects of parasitism on the susceptibility of the tobacco budworm (Lepidoptera: Noctuidae) to methyl parathion and permethrin. Environmental Entomology, 12, 976–978.

Flanders, S. E. (1929). The mass production of *Trichogramma minutum* Riley and observations on the natural and artificial parasitism of the codling moth egg. Transcripts, Fourth International Congress on Entomology, 2, 110–130.

Flanders, S. E. (1930). Mass production of egg parasites of the genus *Trichogramma.* Hilgardia, 4, 464–501.

Fye, R. E. (1972). Manipulation of *Polistes exclamans arizonensis.* Environmental Entomology, 1, 55–57.

Gauld, I. D. (1986). Taxonomy, its limitations and its role in understanding parasitoid biology. In J. Waage & D. Greathead (Eds.), Insect parasitoids (pp. 1–21). New York: Academic Press.

Glas, P. C., Smits, P. H., Vlaming, V., & van Lenteren, J. C. (1981). Biological control of lepidopteran pests in cabbage crops by means of inundative releases of *Trichogramma* species: A combination of field and laboratory experiments. Mededelingen van de Faculteit Landbouwwetenschappen Rijksuniversiteit, Gent, 46, 487–497.

Golub, M. A., & Weatherston, I. (1984). Techniques for extracting and collecting sex pheromones from live insects and from artificial sources. In H. E. Hummel & T. A. Miller (Eds.), Techniques in pheromone research (pp. 223–285). New York: Springer-Verlag.

Greany, P. D., Vinson, S. B., & Lewis, W. J. (1984). Finding new opportunities for biological control. Bioscience, 34, 690–696.

Greathead, D. J. (1971). A review of biological control in the Ethiopian region (CIBC Tech. Communications 5).

Greathead, D. J. (1986). Parasitoids in classical biological control. In J. Waage & D. Greathead (Eds.), Insect parasitoids (pp. 289–318). New York: Academic Press.

Grenier, S., Bonnet, G., Delobel, B., & Laviolette, P. (1978). Development en milieu artificiel du parasitoide *Lixophaga diatraeae* (Towns) (Diptera: Tachinidae). Obtention de l'imago a partir de l'oeuf. Comptes Rendu de l' Academie Sciences Paris, 287, 535–538.

Grosch, D. S. (1975). Reproductive performance of *Bracon hebetor* after sublethal doses of carbaryl. Journal Economic Entomology, 68, 659–662.

Gross, H. R., Jr., Lewis, W. J., & Jones, R. L. (1975). Kairomones and their use in management of entomophagous insects. III. Stimulation of *Trichogramma archaeae, T. pretiosum,* and *Microplitis croceipes* with host-seeking stimuli at time of release to improve their efficiency. Journal of Chemical Ecology, 1, 431–438.

Guerra, A. A., & Martinez, S. (1994). An in vitro rearing system for the propagation of the ectoparasitoid *Catolaccas grandis.* Entomologia Experimentalis et Applicata, 72, 11–16.

Guerra, A. A., Robacker, K. M., & Martinez, S. (1993). In vitro rearing of *Bracon mellitor* and *Catolaccus grandis* with artificial diets devoid of insect components. Entomolgia Experimentalis et Applicata, 68, 303–307.

Halfill, J. E., & Featherston, P. E. (1973). Inundative releases of *Aphidius smithi* against *Acyrthosiphon pisum.* Environmental Entomology, 2, 469–472.

Hart, R. E. (1972). Compensatory releases of *Microterys flavus* as a biological control agent against brown soft scale. Environmental Entomology, 1, 414–419.

Hartley, G. G., King, E. G., Brewer, F. D., & Gantt, C. W. (1982). Rearing of *Heliothis* sterile hybrid with a multicellular larval rearing container and pupal harvesting. Journal of Economic Entomology, 75, 7–10.

Hassan, S. A. (1982). Mass production and utilization of *Trichogramma*. III. Results of some research projects related to the practical use in the Federal Republic of Germany. Les *Trichogramma*, Colloques de l'INRA, 9, 213–218.

Hassell, M. P. (1971). Mutual interference between searching insect parasites. Journal of Animal Ecology, 40, 473–486.

Heath, R. R., & Tumlinson, J. H. (1984). Techniques for purifying, analyzing, and identifying pheromones. In H. E. Hummel & T. A. Miller (Eds.), Techniques in pheromone research (pp. 287–322). New York: Springer-Verlag.

van Heiningen, T. G., Pak, G. A., Hassan, S. A., & van Lenteren, J. C. (1985). Four years' results of experimental releases of *Trichogramma* egg parasites against lepidopteran pests in cabbage. Mededelingen van de Faculteit Landbouwwetenschappen Rijksuniversiteit Gent, 50, 379–388.

Herard, F., Keller, M. A., & Lewis, W. J. (1987). Rearing *Micropletis demolitor* (Wilkinson) in the laboratory for use in studies of semiochemical mediated searching behavior. Journal of Entomological Science, 23, 105–111.

Herard, F., Lewis, W. J. & Keller, M. A. (1988). Beneficial arthropod behavior mediated by airborne semiochemicals. IV. Influence of host diet on host-oriented flight chamber responses of *Microplitis demolitor* (Wilkinson). Journal of Chemical Ecology, 14, 1597–1606.

Hoffman, J. D., & Ignoffo, C. M. (1974). Growth of *Pteromalus puparum* in a semi-synthetic medium. Annals of Entomological Society of America, 67, 524–525.

Hoffman, J. D., Ignoffo, C. M., & Dickerson, W. A. (1975). *In vitro* rearing of the endoparasitic wasp, *Trichogramma pretiosum*. Annals of Entomological Society of America, 68, 335–336.

Hopper, K. R., & King, E. G. (1986). Linear functional response of *Microplitis croceipes* (Hymenoptera: Braconidae) to variation in *Heliothis* spp. (Lepidoptera: Noctuidae) density in the field. Environmental Entomology, 15, 476–480.

House, H. C. (1978). An artificial host: Encapsulated synthetic medium for *in vitro* oviposition and rearing the endoparasitoid *Itoplectis conquisitor* (Hymenoptera: Ichneumonidae). Economic Entomology, 110, 331–333.

Howard, L. O., & Fiske, W. F. (1911). The importation into the United States of the parasites of the gypsy moth and the browntail moth (Bureau of Entomology Bulletin 91) Washington DC: U.S. Department of Agriculture.

Hoy, J. B., & Dahlsten, D. L. (1984). Effects of malathion and Staley's bait on the behavior and survival of parasitic Hymenoptera. Environmental Entomology, 13(6), 1483–1486.

Hubbard, S. F., & Cook, R. M. (1978). Optimal foraging by a parasitoid wasp. Journal of Animal Ecology, 47, 593–564.

Huffaker, C. B. (1974). Biological control. New York: Plenum Press.

Huffaker, C. B. (1977). Augmentation of natural enemies in the People's Republic of China. In R. L. Ridgway & S. B. Vinson (Eds.), Biological control by augmentation of natural enemies (pp. 329–340). New York: Plenum Press.

Huffaker, C. B., & Kennett, C. E. (1956). Experimental studies on predation: Predation and cyclamen mite populations on strawberries in California. Hilgardia, 26, 191–222.

Huffaker, C. B., Rabb, R. L., & Logen, J. A. (1977). Some aspects of population dynamics relative to augmentation of natural enemy action. In R. L. Ridgway & S. B. Vinson (Eds.), Biological control by augmentation of natural enemies (pp. 3–38). New York: Plenum Press.

Ikeyama, M., & Maekawa, S. (1973). Development of Spanone for control of rice stem borers. Pesticide Information, 14, 19–22.

Jacobs, R. J., Kouskolekas, C. A., & Gross, H. R. (1984). Effects of permethrin and endosulfan residues on *Trichogramma pretiosum*, an egg parasitoid of *Heliothis zea*. Environmental Entomology, 13, 355–358.

Jaynes, H. A., & Bynum, E. K. (1941). Experiments with *Trichogramma minutum* Riley as a control of the sugarcane borer in Louisiana (USDA Tech. Bulletin 743). Washington, DC: U.S. Department of Agriculture.

Johnson, W. L., Cross, W. H., McGovern, W. L., and Mitchell, H. C. (1973). Biology of *Heterolaccus grandis* in a laboratory culture and its potential as an introduced parasite of the boll weevil in the United States. Environmental Entomology, 2, 112–118.

Jones, R. L., Lewis, W. J., Bowman, M. C., Beroza, M., & Bierl, B. A. (1971). Host seeking stimulant for parasite of corn earworm: Isolation, identification, and synthesis. Science, 173, 842–843.

Kaas, J. P., Elzen, G. W., & Ramaswamy, S. B. (1990). Learning in *Microplitis croceipes* (Cresson) (Hym., Bracondiae). Journal of Applied Entomology, 109, 268–273.

Kaas, J. P., Elzen, G. W., & Ramaswamy, S. B. (1992). Flight responses in a wind tunnel of the parasitoid *Microplitis croceipes* to three spring and three summer host plants of *Heliothis* spp. Entomologia Experimentalis et Applicata, 63, 207–212.

Kaas, J. P., Ramaswamy, S. B., & Elzen, G. W. (1993). Behavioral time budget and periodicity exhibited by *Microplitis croceipes* in field cages with *Heliothis virescens* on spring host plants. Entomophaga, 38, 143–154.

Kanour, W. W., & Burbutis, P. (1984). *Trichogramma nubilale* (Hymenoptera: Trichogrammatidae) field releases in corn and a hypothetical model for control of European corn borer (Lepidoptera: Pyralidae). Journal of Economic Entomology, 77, 103–107.

King, E. G., & Leppla, N. C. (1984). Advances and challenges in insect rearing. New Orleans, LA: Agricultural Research Service, U.S. Department of Agriculture.

King, E. G., & Morrison, R. K. (1984). Some systems for production of eight entomophagous arthropods. In E. G. King & N. C. Leppla (Eds.), Advances and challenges in insect rearing (pp. 206–222). New Orleans, LA: U.S. Department of Agriculture ARS, Southern Region.

King, E. G., Powell, J. E., & Smith, J. W. (1981). Prospects for utilization of parasites and predators for management of Heliothis spp. Proceedings, Workshop on *Heliothis* Management, Patancheu, A. P., India: November 15–20, 1981.

King, E. G., Ridgway, R. L., & Hartstack, A. L., (1984). Propagation and release of entomophagous arthropods for control by augmentation. In P. L. Adkisson & Ma. Shinjun (Eds.), Proceedings, Chinese Academy of Science—U.S. National Academy of Science Joint Symposium on Biological Control of Insects. Beijing, China: Science Press.

King, E. G., Hopper, K. R., & Powell, J. E. (1985a). Analysis of systems for biological control of crop arthropod pests in the U.S. by augmentation of predators and parasites. In M. A. Hoy & D. C. Herzog (Eds.), Biological control in agricultural IPM systems (pp. 201–227). New York: Academic Press.

King, E. G., Bull, D. L., Bouse, L. F., & Phillips, J. R. (1985b). Biological control of bollworm and tobacco budworm in cotton by augmentative releases of Trichogramma (Suppl.) Southwestern Entomology, November 8, 1985. College Station, TX: Southwestern Entomological Society Press.

King, E. G., Hartley, G. G., Martin, D. F., Smith, J. W., Summers, T. E., & Jackson, R. D. (1979). Production of tachinid Lixophaga diatraeae on its natural host, the sugarcane borer, and on an unnatural host, the greater wax moth. (SEA, Advances in Agricultural Techniques, Southern Series No. 3). Washington, DC: U.S. Department of Agriculture.

King, E. G., Summy, K. R., Morales-Ramos, J. A., & Coleman, R. J. (1993). Integration of boll weevil biological control by inoculative/augmentative releases of the parasite Catolaccus grandis in short-season cotton (pp. 915–921). In Proceedings of the Beltwide Cotton Product Research Conference. National Cotton Council, Memphis, TN.

King, E. G., Coleman, R. J., Wood, L., Wendel, L., Greenberg, S., &

Scott, A. W. (1995). Suppression of boll weevil in commercial cotton by augmentative releases of a wasp parasite, Catolaccus grandis (pp. 26–30). In Proceedings of the Beltwide Cotton Product Research Conference National Cotton Council, Memphis, TN.

Knipling, E. F. (1977). The theoretical basis for augmentation of natural enemies, pp. In R. L. Ridgway & S. B. Vinson (Eds.), Biological control by augmentation of natural enemies (pp. 79–124). New York: Plenum Press.

Knipling, E. F. (1992). Principles of insect parasitism analyzed from new perspectives. Practical implications for regulating insect populations by biological means. Washington, DC: U.S. Department of Agriculture, ARS.

Knowles, C. W. (1982). Structure-activity relationship among amidine acaricides and insecticides. In J. R. Coats (Eds.), Insecticide mode of action (pp. 243–277). New York: Academic Press.

Krieger, R. L., Feeny, P. P., & Wilkinson, C. F. (1971). Detoxication enzymes in the guts of caterpillars: An evolutionary answer to plant defenses. Science, 172, 579–581.

Laing, J. E., & Hamai, J. (1976). Biological control of insect pests and weeds by imported parasites, predators, and pathogens. In C. B. Huffaker & P. S. Messenger (Eds.), Theory and practice of biological control (pp. 42–80). New York: Academic Press.

Lawson, F. R., Rabb, R. L., Guthrie, F. E., & Boweny, L. G. (1961). Studies of an integrated control system for hornworms on tobacco. Journal of Economic Entomology, 540, 93–97.

Legner, E. F., & Medved, R. A. (1979). Influence of parasitic Hymenoptera on the regulation of pink bollworm, Pectinophora gossypiella, on cotton in the lower Colorado desert. Environmental Entomology, 8, 922–930.

Lewis, W. J., & Jones, R. L. (1971). Substance that stimulates host-seeking by Microplitis croceipes (Hymenoptera: Braconidae), a parasite of Heliothis species. Annals of the Entomological Society of America, 64, 471–473.

Lewis, W. J., & Nordlund, D. A. (1985). Behavior modifying chemicals to enhance natural enemy effectiveness. In M. A Hoy & D. C Herzog (Eds.), Biological control in agricultural IPM systems (pp. 89–100). New York: Academic Press.

Li, L. (1984). Research and utilization of Trichogramma in China. In P. L. Adkisson & M. Shijun (Eds.), Proceedings, Chinese Academy of Science—U.S. National Academy of Science Joint Symposium on Biological Control of Insects (pp. 204–223). Beijing, China: Science Press.

Lingren, P. D., & Ridgway, R. L. (1967). Toxicity of five insecticides to several insect predators. Journal of Economic Entomology, 60, 1639–1641.

Lingren, P. D., & Wolfenbarger, D. A. (1976). Competition between Trichogramma pretiosum and Orius insidiosus for caged tobacco budworms on cotton treated with chlordimeform sprays. Environmental Entomology, 5, 1049–1052.

Luck, R. F., Shepard, B. M., & Kenmore, P. E. (1988). Experimental methods for evaluating arthropod natural enemies. Annual Review of Entomology, 33, 367–391.

Luttrell, R. G., Roush, R. T., Ale, A., Mink, J. S., Reid, M. R., & Snodgrass, G. L. (1987). Pyrethroid resistance in field populations of Heliothis virescens (Lepidoptera: Noctuidae) in Mississippi in 1986. Journal of Economic Entomology, 80, 985–989.

Mackauer, M. (1972). Genetic aspects of insect control. Entomophaga, 17, 27–48.

Mackauer, M. (1976). Genetic problems in the production of biological control agents. Annual Review of Entomology, 21, 369–385.

Madden, J. L. (1970). Physiological aspects of host tree favourability for the weedwasp, Sirex noctillo F. Proceedings of the Ecological Society of Australia, 3, 147–149.

Mansingh, A., & Rawlins, S. C. (1979). Inhibition of oviposition it the

cattle tick Boophilus microplus by certain acaricides. Pesticide Science, 10, 485–494.

Martin, P. B., Ridgway, R. L., & Schultze, C. E. (1978). Physical and biological evaluation of an encapsulated diet for rearing Chrysopa carnea. Florida, Entomologist, 61, 145, 152.

Matsumura, F., & Beeman, R. W. (1982). Toxic and behavioral effects of chlordimeform on the American cockroach Periplaneta americana. In J. R. Coats (Ed.), Insecticide mode of action (pp. 229–242). New York: Academic Press.

McMurtry, J. A., Johnson, H. G., & Scriven, G. T. (1969). Experiments to determine effects of mass release of Stethorus picipes on the level of infestations of the avocado brown mite. Journal of Economic Entomology, 62, 1216–1221.

Michelbacher, A. E., & Smith, R. F. (1943). Some natural factors limiting the abundance of the alfalfa butterfly. Hilgardia, 15, 369–397.

Michelbacher, A. E., Swanson, C., & Middlekauff, W. W. (1946). Increase in the populations of Leucanium pruinosum on English walnuts following applications of DDT sprays. Journal of Economic Entomology, 39, 812–813.

Miller, J. R., & Roelofs, W. L. (1978). Sustained flight tunnel for measuring insect response to wind borne sex pheromones. Journal of Chemical Ecology, 4, 187–198.

Monteith, L. G. (1955). Host preference of Drino bohemica Mesn. (Diptera: Tachinidae), with particular reference to olfactory response. Canadian Entomologist, 87, 509–530.

Monteith, L. G. (1964). Influence of the health of the food plant of the host on host-finding by tachinid parasites. Canadian Entomologist, 96, 147.

Morales-Ramos, J. A., Summy, K. R., Raulston, J., Cate, J. R., & King, E. G. (1992). Feasibility of mass-rearing Catolaccus grandis (pp. 723–726). In Proceedings of the Beltwide Cotton Production Research Conference. National Cotton Council, Memphis, TN.

Morales-Ramos, J. A., King, E. G., & Summy, K. R. (1993). Magnitude and timing of Catolaccus grandis releases against the cotton boll weevil aided by a simulation model (pp. 915–922). In Proceedings of the Beltwide Cotton Product Research Conference. National Cotton Council, Memphis, TN.

Morales-Ramos, J. A., Rojas, M. G., Roberson, C. J., Jones, R., King, E. G., & Brazzel, J. R. (1994). Suppression of the boll weevil first generation by augmentative releases of Catolaccus grandis in Aliceville, AL (pp. 958–965). In Proceedings of the Beltwide Cotton Production Research Conference. National Cotton Council, Memphis, TN.

Morales-Ramos, J. A., Rojas, M. G., & King, E. G. (1997). Mass propagation of Catolaccus grandis in support of large scale area suppression of the boll weevil in South Texas cotton (pp. 1199–1200). In Proceedings of the Beltwide Cotton Product Research Conference. National Cotton Council, Memphis, TN.

Morales-Ramos, J. A., Rojas, M. G., Coleman, R. J., & King, E. G. (1998). Potential use of in vitro reared Catolaccus grandis (Hymenoptera: Pteromalidae) for biological control of the boll weevil (Coleoptera: Curculionidae). Journal of Economic Entomology, 91, 101–109.

Morrison, R. K., & King, E. G. (1977). Mass production of natural enemies. In R. L. Ridgway & S. B. Vinson (Eds.), Biological control by augmentation of natural enemies (pp. 173–217). New York: Plenum Press.

Morrison, R. K., Jones, S. L., & Lopez, J. D. (1978). A unified system for the production and preparation of Trichogramma pretiosum for field release. Southwestern Entomology, 3, 62–68.

Mueller, T. F. (1983). The effect of plants on the host relations of a specialist parasitoid of Heliothis larvae. Entomologia Experimentalis et Applicata, 34, 78–84.

Murdie, G., & Hassell, M. P. (1973). Food distribution, searching success, and predatory-prey models. In R. W. Horns (Ed.), Mathematical theory of the dynamics of biological populations (pp. 87–101). New York: Academic Press.

Navasero, R. C., & Elzen, G. W. (1989). Responses to *Microplitis crocei- pes* to host and nonhost plants of *Heliothis virescens* in a wind tunnel. Entomologia Experimentalis et Applicata, 53, 57–63.

Navasero, R. C., & Elzen, G. W. (1992). Influence of maternal age and host deprivation on egg production and parasitization by *Microplitis croceipes* (Hym., Braconidae). Entomophaga, 37, 37–44.

Nettles, W. C., Jr. (1982). Contact stimulants from *Heliothis virescens* that influence the behavior of females of the tachinid *Eucelatoria bryani*. Journal of Chemical Ecology, 8, 1183–1191.

Nettles, W. C., Jr. (1979). *Eucelatoria* spp. females: Factors influencing response to cotton and okra plants. Environmental Entomology, 8, 619–623.

Nettles, W. C., Jr. (1980). Adult *Eucelatoria* spp.: Response to volatiles from cotton and okra plants and leaves and from larvae of *Heliothis virescens, Spodoptera eridania,* and *Estigmene acrea.* Environmental Entomology, 9, 759–763.

Nettles, W. C., Jr., & Burks, M. L. (1975). A substance from *Heliothis virescens* larvae stimulating larviposition by females of the tachinid, *Archytas marmoratus.* Journal Insect Physiology, 21, 965–978.

Nettles, W. C., Jr., Wilson, C. M., & Ziser, S. W. (1980). A diet and methods for the *in vitro* rearing of the tachinid, *Eucelatoria* spp. Annals of the Entomological Society of America, 73, 180–184.

Noldus, L. P. J. J., & van Lenteren, J. C. (1983). Kairomonal effects on searching for eggs of *Pieris brassicae, Pieris rapae,* and *Mamestra brassicae* of the parasite *Trichogramma evanescens* Westwood. Mededelingen van de Faculteit Landbouwweten schappen Rijksuniversiteit Gent, 48(2), 183–194.

Noldus, L. P. J. J., Lewis, W. J., Tumlinson, J. H., & van Lenteren, J. C. (1988a). Olfactometer and wind tunnel experiments on the role of sex pheromones of noctuid moths in the foraging behaviours of *Trichogramma* spp. In *Trichogramma* and other egg parasites (Guangzhou, November 10–15, 1986). Colloques de l' INRA, 43, 223–238.

Noldus, L. P. J. J., Buser, J. H. M., & Vet, L. E. M. (1988b). Volatile semiochemicals in host-community location by egg parasitoids. In Insect parasitoids. Proceedings, Third European Workshop (Lyon, September 8–10, 1987). Colloques de l' INRA, 48, 19–20.

Nordlund, D. A., Chelfant, R. B., & Lewis, W. J. (1985). Response of *Trichogramma pretiosum* females to extracts of two plants attacked by *Heliothis zea.* Agricultural Ecosystems Environment, 12, 127–133.

Nordlund, D. A., Jones, R. L., & Lewis, W. J. (1981a). Semiochemicals: Their role in pest control. New York: John Wiley & Son.

Nordlund, D. A., Lewis, W. J., Goss, H. R., & Beevers, M. (1981b). Kairomones and their use for management of entomophagous insects. XII. The stimulatory effects of host eggs and the importance of host egg density to the effective use of kairomones for *Trichogramma pretiosum* Riley. Journal of Chemical Ecology, 7, 909–917.

Oatman, E. R., & Platner, G. (1971). Biological control of the tomato fruitworm, cabbage looper, and hornworms on processing tomatoes in southern California, using mass releases of *Trichogramma pretiosum.* Journal of Economic Entomology, 64, 501–506.

Oatman, E. R., & Platner, G. (1978). Effect of mass releases of *Trichogramma pretiosum* against lepidopterous pests on processing tomatoes in southern California, with notes on host egg population trends. Journal of Economic Entomology, 71, 896–900.

Oatman, E. R., & Platner, G. (1985). Biological control of two avocado pests (pp. 21–23). California Agriculture.

Oatman, E. R., Gilstrap, F. E., & Voth, V. (1976). Effect of different release rates of *Phytoseiulus persimilis* (Acarina: Phytoseiidae) on the twospotted spider mite on strawberry in southern California. Entomophaga, 21, 269–273.

Oatman, E. R., McMurtry, J. A., & Voth, V. (1968). Suppression of the twospotted spider mite on strawberry with mass releases of *Phytoseiulus persimilis.* Journal of Economic Entomology, 61, 1517–1521.

Oatman, E. R., McMurtry, J. A., Gilstrap, F. E., & Voth, V. (1977). Effect of releases of *Amblyseius californica, Phytoseiulus persimilis,* and

Typhlodromus occidentalis on the twospotted spider mite on strawberry in southern California. Journal of Economic Entomology, 70, 45–47.

Oatman, E. R., Wyman, J. A., Van Steenwyk, R. A., & Johnson, M. W. (1983). Integrated control of the tomato fruitworm (Lepidoptera: Noctuidae) and other lepidopterous pests on fresh-market tomatoes in southern California. Journal of Economic Entomology, 76, 1363–1369.

O'Brien, P. J., Elzen, G. W., & Vinson, S. B. (1985). Toxicity of azinphosmethyl and chlordimeform to parasitoid *Bracon mellitor* (Hymenoptera: Braconidae): Lethal and reproductive effects. Environmental Entomology, 14, 891–894.

Pak, G. A., Noldus, L. P. J. J., van Alebeek, F. A. N., & van Lenteren, J. C. (1988). The use of *Trichogramma* egg parasites in the inundative biological control of lepidopterous pests of cabbage in The Netherlands. Ecological Bulletins, 39, 111–113.

Parker, F. D. (1971). Manipulation of pest populations by manipulating densities of both hosts and parasites through periodic releases. In C. B. Huffaker (Ed.), Biological control (pp. 365–376). New York: Plenum Press.

Phillips, J. R. (1971). Bollworm control with chlorphenamidine. Arkansas Farm Research, 4, 9.

Plapp, F. W., Jr., & Bull, D. L. (1978). Toxicity and selectivity of some insecticides to *Chrysopa carnea*, a predator of the tobacco budworm. Environmental Entomology, 8, 431–434.

Plapp, F. W., Jr., & Bull, D. L. (1989). Modifying chemical control practices to preserve natural enemies. In E. G. King & R. D. Jackson (Eds.), Increasing the effectiveness of natural enemies. (pp. 537–546). Proceedings of the Workshop on Biological Control of Heliothis. New Delhi, India: Rekha Printers.

Plapp, F. W., Jr., & Vinson, S. B. (1977). Comparative toxicities of some insecticides to the tobacco budworm and its ichneumonid parasite *Campoletis sonorensis*. Environmental Entomology, 6, 381–384.

Plapp, F. W., Frisbie, R. E., & Jackman, J. A. (1989). Monitoring for pyrethroid resistance in Heliothis spp. in Texas in 1988 (pp. 347–348). In Proc. Beltwide Cotton Product Research Conference. National Cotton Council, Memphis, TN.

Powell, J. E., & Hartley, G. G. (1987). Rearing *Microplitis croceipes* (Hymenoptera: Braconidae) and other parasitoids of Noctuidae on multicellular host-rearing trays. Journal of Economic Entomology, 80, 968–971.

Powell, J. E., & King, E. G. (1984). Behavior of adult *Microplitis croceipes* (Hymenoptera: Braconidae) and parasitism of *Heliothis* spp. (Lepidoptera: Noctuidae) host larvae in cotton. Environmental Entomology, 13, 272–277.

Powell, J. E., King, E. G., Jr., & Jany, C. S. (1986). Toxicity of insecticides to adult *Microplitis croceipes* (Hymenoptera: Braconidae). Journal of Economic Entomology, 79, 1343–1346.

Powell, W. (1986). Enhancing parasitoid activity in crops. In J. Waage & D. Greathead (Eds.), Insect parasitoids (pp. 319–340). New York: Academic Press.

Powell, W., & Zhang, Z. L. (1983). The reactions of two cereal aphid parasitoids, *Aphidius uzbekistanicus* and *A. ervi* to host aphids and their food plants. Physiological Entomology, 8, 439–443.

Press, J. W., Flaherty, B. R., & McDonald, L. L. (1981). Survival and reproduction of *Bracon hebetor* on insecticide-treated *Ephestia cautella* larvae. Journal of the Georgia Entomological Society, 16, 231–234.

Ridgway, R. L., & Vinson, S. B. (1977). Biological control by augmentation of natural enemies. New York: Plenum Press.

Ridgway, R. L., King, E. G., & Carrillo, J. L. (1977). Augmentation of natural enemies for control of plant pests in the Western Hemisphere. In R. L. Ridgway & S. B. Vinson (eds.), Biological control by augmentation of natural enemies (pp. 379–416). New York: Plenum Press.

Robacker, D. C., & Hendry, L. B. (1977). Neral and geranial: Components of the sex pheromone of the parasitic wasp, *Itoplectis conquisitor*. Journal Chemical Ecology, 3, 563–577.

Roberson, J. L., & Harsh, D. K. (1993). Mechanized production processes to encapsulate boll weevil larvae (Anthonomus grandis) for mass production of Catolaccus grandis (pp. 922–923). In Proceedings of the Beltwide Cotton Product Research Conference. National Cotton Council, Memphis, TN.

Rojas, M. G., Morales-Ramos, J. A., & King, E. G. (1996). In vitro rearing of the boll weevil (Coleoptera: Curculionidae) ectoparasitoid *Catolaccus grandis* (Hymenoptera: Pteromalidae) on meridic diets. Journal of Economic Entomology, 89, 1095–1104.

Rojas, M. G., Morales-Ramos, J. A., & King, E. G. (1997). Reduction of cost of mass propagating Catolaccus grandis by the use of artificial diet (pp. 1197–1199). In Proc. Beltwide Cotton Product Research Conference. National Cotton Council, Memphis, TN.

Rude, P. (1984). Integrated pest management for cotton in the western region of the United States. Oakland: University of California Press.

Sabelis, M. W., & Dicke, M. (1985). Long-range dispersal and searching behavior. In W. Helle & M. W. Sabelis (Eds.), Spider mites. Their biology, natural enemies and control (pp. 141–160). Amsterdam: Elsevier Scientific.

Shands, N. A., Simpson, G. W., & Gordon, C. C. (1972). Insect predators for controlling aphids on potatoes. V, Numbers of eggs and schedules for introducing them in large field cages. Journal of Economic Entomology, 65, 810–817.

Shands, N. A., Simpson, G. W., & Simpson, B. A. (1975). Evaluations of field introductions of two insect parasites (Hymenoptera: Braconidae) for controlling potato-infesting aphids. Environmental Entomology, 4, 499–503.

Shumakov, E. M. (1977). Ecological principles associated with augmentation of natural enemies. In R. L. Ridgway & S. B. Vinson (Eds.), Biological control by augmentation of natural enemies (pp. 39–78). New York: Plenum Press.

Smith, C. N. (1966). Insect colonization and mass production. New York: Academic Press.

Snodgrass, G. L., & Scott, W. P. (1988). Tolerance of the tarnished plant bug to dimethoate and acephate in different areas of the Mississippi Delta (pp. 294–295). Proceedings of the Beltwide Cotton Product Research Conference. National Cotton Council, Memphis, TN.

Sonnet, P. E. (1984). Tabulations of selected methods of syntheses that are frequently employed for insect sex pheromones, emphasizing the literature of 1977–1982. In H. E. Hummel & T. A. Miller (Eds.), Techniques in pheromone research (pp. 371–403). New York: Springer-Verlag.

Starks, K. J., Burton, R. L., Teetes, G. L., & Wood, E. A. (1976). Release of parasitoids to control greenbugs on sorghum (USDA-ARS-S-91). Washington, DC: U.S. Department of Agriculture.

Starter, N. H., & Ridgway, R. L. (1977). Economic and social considerations for the utilization of augmentation of natural enemies. In R. L. Ridgway & S. B. Vinson (Eds.), Biological control by augmentation of natural enemies (pp. 431–450). New York: Plenum Press.

Stinner, R. E., Ridgway, R. L., Coppedge, J. R., Morrison, R. K., & Dikerson, W. A. (1974). Parasitism of *Heliothis* eggs after field releases of *Trichogramma pretiosum* in cotton. Environmental Entomology, 3, 492–500.

Summy, K. R., Morales-Ramos, J. A., & King, E. G. (1992). Ecology and potential impact of *Catolaccus grandis* (Burks) on boll weevil infestations in the lower Rio Grande Valley. Southwestern Entomology, 17, 279–288.

Summy, K. R., Morales-Ramos, J. A., King, E. G., Wolfenbarger, D. A., Coleman, R. J., & Greenberg, S. M. (1994). Integration of boll weevil parasite augmentation into the short-season cotton production system of the Lower Rio Grande Valley (pp. 953–958). In Proceedings of the

Beltwide Cotton Product Research Conference. National Cotton Council, Memphis, TN.

Summy, K. R., Morales-Ramos, J. A., & King, E. G. (1995). Suppression of boll weevil infestations on South Texas cotton by augmentative releases of the exotic parasite *Catolaccus grandis* (Hymenoptera: Pteromalidae). Biological Control, 5, 523–529.

T'Hart, J. T., DeJonge, J., Colle, C., Dicke, M., van Lenteren, J. C., & Ramakers, P. (1978). Host selection, host discrimination, and functional response of *Aphidius maticaria* Holiday (Hymenoptera: Braconidae), a parasite of the green peach aphid, *Myzus persicae* (Salz.). Mededelingen van de Faculteit Landbouwwetenschappen Rijksuniversiteit Gent, 43, 441–453.

Thompson, S. N. (1982). *Exeristes roborator:* Quantitative determination of *in vitro* larval growth rates in synthetic media with different glucose concentrations. Experimental Parasitology, 54, 229–234.

Thompson, S. N. (1983). Larval growth of the insect parasite *Brachymeria lasus* reared *in vitro*. Journal of Parasitology, 69, 425–427.

Thompson, S. N., Bedner, L., & Nadel, H. (1983). Artificial culture of the insect parasite *Pachycrepoides vindemiae*. Entomologia Experimentalis et Applicata, 33, 121–122.

Thorpe, W. A., & Caudle, H. B. (1938). A study of the olfactory response of insect parasites to the food plant of their host. Parasitology, 30, 523–528.

van der Schaaf, P. A., Kaskens, J. W. M., Kole, M., Noldus, L. P. J., & Pak, G. A. (1984). Experimental releases of two strains of *Trichogramma* spp. against lepidopteran pests in a Brussels sprouts field crop in The Netherlands. Mededelingen van de Faculteit Landbouwwetenschappen, Rijksuniversiteit Gent, 49(3a), 803–813.

van Heiningen, T. G., Pak, G. A., Hassan, S. A., & van Lenteren, J. C. (1985). Four years' results of experimental releases of *Trichogramma* egg parasites against lepidopteran pests in cabbage. Mededelingen van de Faculteit Landbouwwetenschappen Rijksuniversiteit Gent, 50, 379–388.

Vanderzant, E. S. (1973). Improvements in the rearing diet for *Chrysopa carnea* and the amino acid requirements for growth. Journal of Economic Entomology, 66, 336–338.

van Lenteren, J. C. (1986). Parasitoids in the greenhouse: Successes with seasonal inoculative release systems. In J. K. Waage & D. J. Greathead (Eds.), Insect parasitoids (pp. 341–374). New York: Academic Press.

van Lenteren, J. C, Glas, P. C. G., & Smits, P. H. (1982). Evaluation of control capabilities of Trichogramma and results of laboratory and field research on Trichogramma in The Netherlands. Les Trichogrammes (pp. 257–268). Antibes, France: Ed. INRA Publ.

Vet, L. E. M., van Lenteren, J. C., & Woets, J. (1980). The parasite-host relationship between *Encarsia formosa* (Hymenoptera: Aphelinidae) and *Trialeurodes vaporariorum* (Homoptera: Aleyrodidae). IX. A review of the biological control of the greenhouse whitefly with suggestions for future research. Zeitschrift fuer Angewandte Entomologie, 90, 26–51.

Vinson, S. B. (1975). Biochemical coevolution between parasitoids and their hosts. In P. W. Price (Ed.), Evolutionary strategies of parasitic insects and mites (pp. 14–48). New York: Plenum Press.

Vinson, S. B. (1977). Behavioral chemicals in the augmentation of natural enemies. In R. L. Ridgway & S. B. Vinson (Eds.), Biological control by augmentation of natural enemies (pp. 237–279). New York: Plenum Press.

Voronin, K. E., & Grinberg, A. M. (1981). The current status and prospects of *Trichogramma* utilization in the USSR. In J. R. Coulson (Ed.), First Proceedings on Joint American-Soviet Conference on Use of Beneficial Organisms in the Control of Crop Pests (pp. 49–51). College Park, MD: Entomological Society of America.

Waage, J. K. (1983). Aggregation in field parasitoid populations: Foraging time allocation by a population of *Diadagma* (Hymenoptera: Ichneumonidae). Ecological Entomology, 8, 447–453.

Waage, J. K. (1989). The population ecology of pest-pesticide-natural enemy interactions. In P. C. Jepson (Ed.), Pesticides and non-target invertebrates (pp. 81–93). United Kingdom: Intercept., Ltd.

Wackers, F. L., de Groot, I. J. M., Noldus, L. P. J. J., & Hassan, S. A. (1987). Measuring host preference of Trichogramma egg parasites: An evaluation of direct an indirect methods. Mededelingen van de Faculteit Landbouwwetenschappen Rijksuniversiteit, Gent, 52, 339–348.

Wantanabe, H., & Fukami, J. (1977). Stimulating action of chlordimeform and desmethyl chlordimeform on motor discharge of armyworm, Leucania separata Walker (Lepidoptera: Noctuidae). Journal of Pest Science, 2, 297–301.

Weseloh, R. M. (1984). Behavior of parasites and predators: Influences on manipulative strategies and effectiveness. In P. L. Adkisson & Ma Shijun (Eds.), Proceedings, Chinese Academy of Science—U.S. National Academy of Science Joint Symposium on Biological Control of Insects (pp. 168–180). Beijing, China: Science Press.

Wolfenbarger, D. A., Cantu, E., Lingren, P. D., & Guerra, A. A. (1973). Activity of chlordimeform-HCL and chlordimeform against arthropods attacking cotton. Journal of Economic Entomology, 67, 445–446.

Wu, Z. J., Qin, Li, P. X., Chang, Z. P., & Liu, T. M. (1982). Culturing Trichogramma dendrolimi in vitro with artificial media devoid of insect material. Acta Entomologica Sinica, 25, 128–135.

Zanen, P. O., Lewis, W. J., Cardé, R. T., & Mullinix, B. G. (1989). Beneficial arthropod behavior mediated by airborne semiochemicals. VI. Flight responses of Microplitis croceipes (Cresson), a braconid endoparasitoid of Heliothis spp., to varying olfactory stimulus conditions created with a turbulent jet. Journal of Chemical Ecology, 15, 141–168.

12

Genetic Improvement and Other Genetic Considerations for Improving the Efficacy and Success Rate of Biological Control

M. J. WHITTEN

FAO Intercountry Programme on IPM
in South and Southeast Asia
P.O. Box 3700
MCPO 1277
Makati City, The Philippines

MARJORIE A. HOY

Department of Entomology and Nematology
P.O. Box 110620
University of Florida
Gainesville, Florida

INTRODUCTION

The case in favor of biological control is overwhelming, as a means of managing weeds, pests, and diseases; and in terms of cost effectiveness, environmental acceptability, and safety record. This conclusion is well illustrated by numerous examples detailed in other chapters of this book. Despite this grand record, biological control is but one of a number of approaches available for the management of pests. In this context, it has been attempted for less than 5% of the 5000 odd arthropod pests (Rosen, 1985), and probably for an even smaller percentage of nematodes (van Gundy, 1985), weeds (Bernays, 1985; Charudattan, 1985) and plant pathogens (Baker, 1985; Lindow, 1985; Martin *et al.,* 1985; Schroth & Hancock, 1985). In a balanced perspective, biological control is neither an instant nor a general panacea; and the well-known and spectacular successes, as measured by partial or complete control of the target pests, must be considered against the fact that a substantial proportion of attempts against arthropods have met with limited or no success (Rosen, 1985).

The general objective of this chapter is to identify possible genetic reasons for the low success rate and, more specifically, to explore the proposition that the success rate can be raised by deliberate attempts to improve the efficacy of arthropod natural enemies by some form of artificial selection or genetic manipulation. Genetic improvement of natural enemies of other pests, including pathogens of arthropods (Faulkner & Boucias, 1985; Luthy, 1986; Aizawa, 1987; Yoder *et al.,* 1987; and Chapters 18 and 21 this volume), entomogenous nematodes (Gaugler, 1987), or plant pathogens (Napoli & Staskawicz, 1985; Lindow *et*

al., 1989), will not be discussed here. Other genetic considerations are occasionally raised concerning either the ranking of target arthropod or weed species for biological control or the choice among natural enemies for effecting control. Where these genetic considerations claim to impinge on the efficacy of biological control, the merits of the arguments are addressed in this chapter.

The concept of genetic improvement of arthropod natural enemies is not new. Entomologists have considered selective breeding of biological control agents since early this century [for reviews see DeBach (1958), Sailer (1961), Hoy (1979, 1985b), and Hopper, *et al.* (1995)]. However, achievements in genetic improvement of arthropod natural enemies by 1971 were sufficiently unimpressive to elicit the pithy but fair assessment by Messenger and van den Bosch (1971), "Artificial selection of natural enemies . . . has been considered by many, attempted by few, and, unfortunately, proved practicable in terms of improving biological control by no one." By 1971, selection for improved host finding, changes in host preference, improved climatic tolerance, increased fecundity, and insecticide resistance had been conducted on seven parasitoid species (Table 1). However, it is not clear that any of the selected natural enemies were actually genetically improved, or that their efficacy had been demonstrated in the field. In the only case where a selected strain [*Dahlbominus fuscipennis* (Zetterstedt)] was apparently released into the field, it is not known whether the quality of the strain was restored through the elimination of diseases or deleterious effects of inbreeding, or whether quality was improved over that of the wild strain (Wilkes, 1942, 1947). Since 1971, at least 19 parasitoid and predator species have been selected

TABLE 1   Genetic Improvement of Arthropod Natural Enemies Conducted Prior to 1971

Trait	Species	References
Host finding	*Trichogramma minutum* Riley	Urquijo, 1956; Landaluz, 1950
Host preference	*Horogenes molestae* (Uchida)	Allen, 1954
	Paratheresia claripalpis (Wilkes)	Box, 1956
Climatic tolerance	*Dahlbolminus fuscipennis* (Zetterstedt)	Wilkes, 1942
	Aphytis lingnanensis Compere	White *et al.,* 1970
Increased fecundity	*Dahlbominus fuscipennis*	Wilkes, 1947
Insecticide resistance	*Macrocentrus ancylivorus* Rohwer	Pielou & Glasser, 1952; Robertson, 1957
	Bracon mellitor Say	Adams & Cross, 1967

in the laboratory for enhanced fecundity, host synchrony, nondiapause, pesticide resistance, temperature tolerance, or thelytoky (Table 2); and some of these natural enemies have been successfully evaluated in the greenhouse or the field. Even so, many questions remain about genetic improvement of arthropod natural enemies.

In considering genetic aspects of success and failure in biological control, and in assessing the opportunity for genetic improvement of natural enemies, it might be helpful if we recognize three situations or categories to which we can assign past, present, and proposed biological control projects. We will provide arguments that the opportunity for genetic improvement is low and should receive low priority against alternative strategies for improving the suc-

cess rate in category 1 cases and is high and perhaps should receive greater emphasis in category 2 cases, while the issue currently represents a considerable intellectual and experimental challenge in category 3 cases.

In each category there are other genetic considerations that can be pertinent to our choice of target pest or candidate natural enemy and that can assist decision making on contentious issues such as (1) the appropriate level of emphasis on genetic improvement of a natural enemy versus renewed field exploration, and (2) the stage at which a project should be deemed a failure and should be abandoned. Failure to appreciate existence of the three categories and the genetic implications for projects from each may be partly responsible for some of the present misunder-

TABLE 2   Genetic Improvement of Arthropod Natural Enemies since 1971

Trait	Species	References
Fecundity	*Dahlbominus fulginosus*	Szmidt, 1972
	Phytoseiulus persimilis Athias-Henriot	Voroshilov & Kolmakova, 1977
	Trichogramma fasciatum	Ram & Sharma, 1977
	T. brassicae Bezdenko	Pintureau, 1991
Nondiapause	*Aphidoletes aphidimyza* Rondani	Gilkeson & Hill, 1986
	Metaseiulus occidentalis (Nesbitt)	Hoy, 1984
Pesticide resistance	*Amblyseius fallacis* Garman	Strickler & Croft, 1982
	A. finlandicus (Oudemans)	Kostiainen & Hoy, 1994, 1995
	A. nicholsi Ehara & Lee	Huang *et al.,* 1987
	Aphytis holoxanthus DeBach	Havron *et al.,* 1991a
	A. lingnanensis Compere	Havron *et al.,* 1991b
	A. melinus DeBach	Rosenheim & Hoy, 1988; Spollen & Hoy, 1992c, 1993a
	Chrysoperla carnea Stephens	Grafton-Cardwell & Hoy, 1986
	Metaseiulus occidentalis	Hoy & Knop, 1981; Roush & Hoy, 1981a,1981b; Hoy, 1984; Hoy & Ouyang, 1989
	P. persimilis	Schulten & van de Klashorst, 1974, Avella *et al.,* 1985; Fournier *et al.,* 1987a,
	T. japonicum	Xu *et al.,* 1986
	Trioxys pallidus Haliday	Hoy & Cave, 1988, 1989; Hoy *et al.,* 1989, 1990; Edwards & Hoy, 1993, 1995
	Typhlodromus pyri Scheuten	Markwick, 1986; Suckling *et al.,* 1988
Synchrony	*Cotesia melanoscela* (Ratzeburg)	Hoy, 1975a, 1975b; Weseloh, 1986
Temperature tolerance	*P. persimilis*	Voroshilov, 1979
Thelytoky	*Muscidifurax raptor* Girault & Sanders	Legner, 1987

standing and controversy about the role and scope of genetics and genetic improvement in biological control. First we define the three categories and then we consider the genetic ramifications in each category for improving the success rate.

CATEGORY 1: BIOLOGICAL CONTROL BY RESTORING A NATURAL BALANCE

This category embraces instances where a plant, an animal, or a pathogen has been introduced to a new territory free of its traditional natural enemies, and where environmental and biotic conditions are favorable for its successful propagation. If the species increases in abundance above some nuisance threshold value and becomes a pest, then classical biological control should be considered as a management option. Category 1 situations, therefore, reflect the fact that the newly acquired or enhanced pest status of the introduced organism is essentially due to the lack of effective natural enemies in the new territory, instead of being a direct consequence of some environmental, management, or other biotic factor(s). Category 1 cases probably represent the typical position for the major exotic weeds in Australia, New Zealand, Africa, and North America. Category 1 also includes significant insect pests of coniferous and broad-leaved forests, woodlands, and range or pasture grasslands. Many exotic arthropod and vertebrate pests in Australasia fit this category.

Many significant arthropod pests of intensive agriculture and horticulture, whether exotic or native, now experience a more artificial environment. The habitat may be lacking some critical component for potential natural enemies, such as nesting or hibernation sites resulting from extensive monoculture, or it may contain residues of pesticides or other toxic agrochemicals. Consequently, the direct introduction of natural enemies in such cases, without regard to the novel nature of the environment, could prove to be ineffectual.

Some of the significant successes in classical biological control relate to category 1 pest organisms. The list would include aquatic weeds such as alligator weed [*Alternanthera philoxeroides* (Martius) Grisebach] (Julien, 1981), water hyacinth [*Eichhornia crassipes* (Mart) Solms-Laubach] (Wright, 1981; Beshir & Bennett, 1985; Cofrancesco *et al.,* 1985), and *Salvinia molesta* Mitchell (Room *et al.,* 1985; Thomas & Room, 1985; Chapter 34 this volume); perennial weeds such as prickly pear spp. (*Opuntia* spp.) in Australia (Dodd, 1940), Klamath weed (or St. Johnswort, *Hypericum perforatum* Linnaeus), and ragwort (*Senecio jacobaea* Linnaeus) in Australia and in North America (Andres *et al.,* 1976); *Cordia macrostachya* (Jacquin) Roemer & J. A. Schultes in Mauritius (Simmonds *et al.,* 1976); and, less

impressively, annual weeds such as emex (*Emex spinosa* Linnaeus and *E. australis* Steinh) in Hawaii (Andres *et al.,* 1976). Comprehensive reviews of biological control of weeds can be found in Goeden (1978) and Wapshere *et al.* (1989).

Category 1 invertebrate pests include notable successes in coniferous forests such as *Sirex noctilio* (Fabricius), *Coleophora laricella* (Hübner), *Diprion hercyniae* (Hartig), *D. similis* (Hartig), *Neodiprion sertifer* (Geoffrey), and *Rhyacionia frustrana* (Busck) (Turnock *et al.,* 1976); in broad-leaved forests such as satin and oriental moth [*Stilpnotia salicis* (Linnaeus) and *Anomala orientalis* Water], birch leaf mining sawfly [*Heterarthrus nemoratus* (Fallen)] in North America, oak leaf miner (*Lithocolletis messaniella* Zell.) in New Zealand, and *Eucalyptus* snout beetle (*Gonipterus scutellatus* Gyllenhal)—a serious pest of introduced *Eucalyptus* spp.—in South Africa; and in the tropics and subtropics (Waters *et al.,* 1976) pustule scale (*Asterolecanium pustulans* Cockerell) of shrubs and broad-leaved trees.

We should also recognize the pioneering case of the control of the cottony-cushion scale (*Icerya purchasi* Maskell) by the predatory vedalia ladybeetle [*Rodolia cardinalis* (Mulsant)] in 19th century citrus orchards before widespread use of chemicals; the less successful case some years later of the California red scale [*Aonidiella aurantii* (Maskell)] by the endoparasites *Comperiella bifasciata* Howard and *Prospaltella perniciosi* Tower (Rosen, 1985), and the ectoparasitoid *Aphytis* spp. as detailed in Chapter 27; and finally, Rhodes grass scale [*Antonina graminius* (Maskell)] in Texas grasslands (Hagen *et al.,* 1976) and lucerne aphids [spotted alfalfa aphid, *Therioaphis trifolii* (Monell) f. *maculataand* blue green aphid, *Acyrthosiphon kondoi* (Shinji)] in pastures in Australia (Hughes *et al.,* 1987).

Despite such well-known successes, it remains true that approximately 60% of attempts in category 1 instances have met with limited success. Some failures to date might be construed to have a genetic basis but the solution is unlikely to be found though genetics. For example, *Mimosa pigra* is not a weed in its native Central America where it is the host of a large guild of natural enemies and diseases. Many of these natural enemies would not be approved for introduction into Australia where *Mimosa* is a serious weed. The reason is that evidence would be hard to obtain which demonstrates that many of these natural enemies would not attack the diverse *Acacia* flora of Australia. Similar difficulties exist for woody *Acacia* weeds in Australia of southern African origin (e.g., *A. nilotica*) or for *Acacia* species from Australia that are serious weeds in South Africa. It is unlikely that any genetic measures could tailor the host specificity of potential agents to meet quarantine requirements. We return later to the role and scope of genetics in improving the performance of biological control agents in this category, or in helping to determine the relative suitability of target species for biological control.

CATEGORY 2: BIOLOGICAL CONTROL OF SECONDARY PESTS

This category would include those instances where a pest was traditionally under effective control by its own natural enemies, whether native or introduced, but where control has since failed because the balance has been disturbed by some deliberate and usually identifiable action of man. The most obvious example is the case where pesticides are used to control a major pest in some ecosystem, for example, codling moth (*Laspeyresia pomonella* [Linnaeus]), navel orangeworm (*Amyelois transitella* [Walker]), leaf rollers, oriental fruit moth (*Grapholitha molesta* [Busck]), or apple maggot (*Rhagoletis pomonella* [Walsh]), in deciduous orchards in temperate regions. In these situations, pesticide residues usually cause outbreaks of secondary pests such as tetranychid mites, apple mealybug (*Phenacoccus aceris* Signoret), woolly apple aphid (*Eriosoma lanigerum* [Hausman]), and some scale insects (MacPhee *et al.*, 1976). In such situations, it is generally possible to identify the specific environmental change that has diminished the effectiveness of the natural enemy, viz., the presence of pesticide residues. Consequently, we can define more precisely the "defective" character in the natural enemy requiring artificial selection.

An interesting variation on this theme is the creation of pesticide-induced pests *de novo,* in species often considered to be primary pests. The rice brown plant hopper, *Nilaparvata lugens* Stål, was not a conspicuous pest of rice before the Green Revolution. Its emergence as a major pest was in large measure pesticide-induced (Kenmore *et al.,* 1984). Bentley (1992) and Bentley and Andrews (1991) report similar instances of farmers encountering new pests in Honduras due to pesticide residues in crops. The farmers even conceived of a pesticide conspiracy—the manufacturers having mixed special chemicals that permitted spontaneous generation of pests never seen before. We perceive an ecological instead of a genetic solution to this phenomenon. In the case of the rice brown plant hopper, pesticides are disrupting an ecosystem that contains over 800 species of invertebrates and possibly a higher number of lesser organisms (Settle *et al.,* 1996; Whitten & Settle, in press). It would be inappropriate to describe pesticide susceptibility in this rich fauna as a "defective" character, but more appropriate to describe the technology that encouraged the unnecessary use of pesticides as defective technology. Conservation biological control aimed at maintaining the diversity of organisms that buffer against herbivore outbreaks is the preferred option, instead of selecting for resistance to

pesticides, even in key natural enemies in the tropical rice ecosystem (Whitten & Settle, in press).

The effectiveness of natural enemies has been restored in a number of category 2 cases by the integration of chemicals with other management practices without resort to genetic improvement (MacPhee *et al.,* 1976; Hull & Beers, 1985). For example, Asquith and Colburn (1971) enhanced the effectiveness of native coccinellid predator *Stethorus punctum* LeConte for controlling *Tetranychus urticae* Koch and *Panonychus ulmi* Koch in Pennsylvania apple orchards by introducing "patchiness" into the distribution of pesticide residues.

However, there have been other instances where the release of pesticide-resistant strains has been the deciding factor in successful suppression of the pest (e.g., control of *T. urticae, T. mcdanieli* McGregor, *T. pacificus* McGregor, or *P. ulmi* in North America, New Zealand, Australia, and elsewhere) (Hoyt & Caltagirone, 1971; Hoy, 1979, 1985b; Croft & Strickler, 1983). The principal successes in genetic improvement have been achieved in category 2 situations. We detail these later and give consideration to future directions for genetic improvement of category 2 cases.

CATEGORY 3: BIOLOGICAL CONTROL OF WEEDS AND PESTS IN NOVEL OR DISTURBED ENVIRONMENTS

Category 3 is more difficult to define. Generally, it relates to situations where a pest, usually an arthropod, has been introduced to a new territory and established without its traditional natural enemies, but where the environment can also be considered artificial or novel from the perspective of a potential natural enemy. For example, the physical environment may differ significantly in terms of temperature, humidity, or daylength patterns from those prevailing in the agent's native habitat. The abundance and distribution in time and space of host plant(s) in the case of an exotic arthropod pest may be atypical, especially for pests of intensive horticulture, greenhouse cultivation, or extensive monoculture. We envisage the new environment being deficient in one or more facets, which disrupt normal behavior and reduce survival of the introduced natural enemy leading to its failure to effect control. Such facets could be pertinent and inimical to certain critical behavioral attributes of the candidate natural enemy such as searching ability or mating and oviposition behavior. Other elements of the habitat such as existence of alternative prey and location of refugia for overwintering, estivating, or diapausing may be inadequate. Many of these environments may also contain residues of synthetic pesticides that might differ in amount or type from that prevailing in the home range of the natural enemy.

Category 3 also would include cases where the pest was introduced without its natural enemies but was not a pest initially, and only gradually increased in pest status during subsequent years. It is difficult to demonstrate, but reasonable to assume that, in some instances of such a delay, genetic change during colonization may have occurred, enabling the invader to exploit its new environment more successfully (Parsons, 1983). Thus, an additional novel element for the natural enemy could be a host that has undergone some genetic modification during the period of establishment. Finally, unlike category 2 cases where pesticide residues are generally the principal cause of secondary pests, pesticide residues may represent a complicating instead of a causal factor of pests in category 3.

One documented example of an unsuccessful attempt at biological control of a weed that revealed it ultimately as a category 3 instead of a category 1 pest is silverleaf nightshade (*Solanum elaeagnifolium* Cavanilles). Wapshere (1988) has argued that *S. elaeagnifolium* and its associated herbivores evolved in the Monterrey region of Mexico. It is a weed of some importance in its native range of Mexico and of southwestern United States. Attempts at biological control of the weed in South Africa and in Argentina have been unsuccessful, and efforts have been considered not worth pursuing in California (Goeden, 1971) or in Australia (Wapshere, 1988). Apart from the fact that several key natural enemies lack adequate specificity for introduction, Wapshere listed four factors to support his conclusion: absence of climatic adaptability to the main infested regions (lacking summer rainfall and winter drought), likely inadequate reservoir populations in climates with summer rainfall, inimical effects of cultivation on the insects occurring on the weed, and absence of any agent damaging to the root system (which is the only living part of the plant from autumn to spring).

This careful study indicates why biological control of silverleaf nightshade is not an attractive proposition in terms of classical biological control and illustrates two important points. It identifies the "shortcomings" in the guild of natural enemies available—in this case, lack of specificity and inability of the guild of natural enemies to cope with a different climatic environment—and therefore helps identify selection goals if a genetic improvement program was considered. It also provides, for the time being, a satisfying explanation as to why biological control of this major weed is not an attractive option and lays the way open for other avenues to be pursued. The uncritical collection, release, and ultimate failure of natural enemies that were found associated with a weed species in its native habitat, without the accompanying research as illustrated by the silverleaf nightshade example, allow neither of these outcomes to be contemplated. Biological control of *Mimosa pigra* in Australia is proving to be another case where the

host is preadapted to its new environment—hot tropical with extreme wet/dry seasons—while the introduced natural enemies from Mexico find these alien conditions inhospitable (P. B. Edwards, personal communications). Weeds like silverleaf nightshade and *M. pigra,* which appeared to some as category 1 cases, are now emerging as category 3 cases.

Several successful biological control programs fit within category 3 where no deliberate genetic selection has been applied to the natural enemy. For example, Florida red scale [*Chrysomphalus aonidum* (Linnaeus)] invaded Israel in about 1910, and had become the most important pest of citrus by the mid-1950s but eventually was brought under effective control by an integrated pest management program (IPM) (Rosen, 1980). Similarly, the olive scale [*Parlatoria oleae* (Colvée)], a polyphagous pest introduced to California from the Middle East in 1934, was brought under effective control during the 1950s by the complementary action of two introduced parasites *Aphytis maculicornis* (Masi) and *Coccophagoides utilis* Doutt (Rosen & DeBach, 1978). Genetic improvement was not attempted or required for these scale pests.

Many of our biological control failures, almost by definition, are likely to be classified as category 3 cases, because we might be tempted to argue that the "ineffectual" natural enemy was genetically defective. This statement assumes that the chosen natural enemy was exerting some controlling influence in its native distribution, an assertion that may be without factual foundation in some instances. However, the critical questions, whose correct answers will give a clue to appropriate remedial action, are

1. Was the "genetic" deficiency caused by some inadequate choice among the guild of natural enemies available (an ecological issue)?
2. Was genetic deficiency caused by some sampling deficiencies during collection of the natural enemy (a technical issue)?
3. Was genetic deficiency caused by deterioration or inappropriate conditioning of the material during importation and mass rearings before release (a genetic or a quality control issue)?
4. Were releases poorly timed or conducted with no suitable hosts (an ecological issue)?
5. Was the genetic makeup of the released material deficient because the natural enemy was released into a novel and inhospitable environment?

The option of genetic improvement only relates to the last cause of failure, and the obvious difficulty for the geneticist will be knowing a genetically determined deficiency was responsible and then learning the nature of the deficiency from the applied biologist.

In some instances, the environmental difference might

be slight or subtle, in which case the task of identifying the pertinent difference could be substantial. Genetic considerations also pertain to the initial questions of sampling and genetic deterioration during laboratory manipulation prior to release, but these matters are not covered here. Again, there could be a divergence of opinion between the geneticist and the entomologist whether the appropriate species, biotype, or best ecoclimatic equivalent had been collected during the exploration phase (Caltagirone, 1985; Hoy, 1985c). This last issue is addressed more fully later.

Traditional classifications of biological control practice bear little relationship to the three categories suggested earlier. For example, a more typical classification is classical (inoculative), conservation, and augmentative (repeated and possibly inundative) (Wapshere *et al.*, 1989). These are operational categories and serve a clear purpose for the practitioner. The three categories we have identified earlier are organized from a genetic perspective; they neither contradict nor complement the more operationally determined groupings. However, it is probable that classical biological control would be the norm for category 1 cases. Classical, inundative, and augmentative releases may be required in category 2 situations and may need to be integrated with other management strategies. Category 3 examples could include all operational classes of biological control. The usefulness of the three categories we have defined depends entirely on their utility in focusing our attention on genetic options that will improve the probability of converting a failure into a partial or complete success.

GENETIC CONSIDERATIONS FOR IMPROVED SUCCESS RATES IN BIOLOGICAL CONTROL

Questions of sampling strategies—in terms of sample size, sex ratio, choice of sampling sites concerning center or margins of distribution, pooling of samples, population size and generation number during the quarantine and mass-rearing phases, environmental and culture conditions during the exploration phase through to field release, and release strategies—are all pertinent to successful establishment and effective control of the target organism. The authors have nothing to add to what is found elsewhere in this publication—or the literature generally—on collection and handling. Roush and Hopper have attempted to develop a genetic framework for laboratory rearing (Roush & Hopper, 1995) and release strategies (Roush, 1990), which could help transform these important phases of biological control from an art form to a science. We note that molecular population genetics is beginning to provide new insights into genetic variation in natural populations (Roush, 1990; Oakeshott & Whitten, 1991; Hoelzel, 1992; Avise, 1994; Hoy, 1994a; Symondson & Liddell, 1996).

LIMITED OPPORTUNITY FOR GENETIC IMPROVEMENT IN CATEGORY 1 CASES OF BIOLOGICAL CONTROL

If reasonable care is taken to prevent genetic deterioration of the material collected, imported, and released, and if quality control is maintained in terms of health, vigor, and size of the cultures destined for release, then failure of a natural enemy in category 1—by definition—is likely to reflect some deficiency in the choice or handling of the species chosen for introduction. In essence, failure in category 1 cases is likely to have an ecological or a taxonomic basis and a solution involving one or more of these disciplines. Remedial action is likely to take the form of searching for alternative or additional natural enemies (including presumed biotypes of the failed introduced natural enemy), or to arise out of research directed toward a better understanding of the biology of the host and its natural enemies.

These points are well illustrated by the history of control of aquatic water weed *Salvinia molesta*. Initial attempts at biological control of *S. molesta* with *Cyrtobagus singularis* Hustache collected from *Salvinia auriculata*. Aublet were unsuccessful (Room *et al.*, 1985). A further search in Brazil located for the first time the native range of *S. molesta* (Forno & Harley, 1979) and yielded what was initially presumed to be another biotype of *C. singularis* (Forno *et al.*, 1983). The feeding habits of the Brazilian biotype were different from *C. singularis* and it promised to be a more effective controlling agent (Sands *et al.*, 1983). This promise was realized with the suppression of *S. molesta* in Queensland (Room *et al.*, 1981), but similar success was not obtained for the extensive mats of the weed congesting the Sepik River system in Papua New Guinea until the nitrogen level of the plant was artificially increased by treatment with urea. One application of the nitrogenous fertilizer enabled the weevil to multiply, causing nitrogen mobilization to the damaged parts of the plants and thereby ensuring a sustained buildup in insect numbers and subsequent widespread destruction of the weed (Thomas & Room, 1985). The Papua New Guinea experience was repeated in Sri Lanka (Doeleman, 1989). By this stage, the Brazilian biotype was recognized and described as the separate species *C. salviniae* Calder & Sands. In a trivial sense, the genetic makeup of the two weed species and the two *Cyrtobagus* spp. was responsible for the initial failure, but genetic considerations had no relevance to the operational issues and the practical decisions that ultimately led to perhaps the most spectacular case of biological control yet witnessed of an exotic aquatic weed.

In a similar manner, the limited role of genetics in category 1 cases can be illustrated by the history of control of California red scale [*Aonidiella aurantii* (Maskell)], which started in 1889. Generalist coccinellid predators were first introduced but their effectiveness was limited by dependence on high prey densities (Rosen, 1985). In the next period, inappropriate endoparasites were chosen from the Orient because of poor taxonomy of *Aonidiella* spp. Other attempts failed because of "cryptic effects of the host plant on the suitability of the scale" when *Cycas* palms were substituted for citrus as host plants for the scale (Compere, 1961; Rosen, 1985). One other promising aphelinid [*Pteroptrix chinensis* (Howard)] failed to establish, again because its biology was not sufficiently understood. In the 1940s, once the biological bases for these failures were better understood, a major improvement in the level of control was achieved with the release of the two endoparasites *Comperiella bifasciata* and *Prospaltella perniciosi*. Finally, misidentification of *Aphytis lingnanensis* Compere and *A. melinus* DeBach as *A. chrysomphali* (Mercet), a species already present in California, delayed "for more than half a century the importation of these two most effective natural enemies of California red scale" (Compere, 1961). It is quite clear from these considerations that there would have been little point in geneticists muddying the waters in the long saga of biological control of this scale pest. This example also highlights the importance of sound taxonomy to biological control.

The failure to recognize foreign populations of *A. lingnanensis* and *A. melinus* as potentially effective natural enemies of the red scale in California simply because they were erroneously labeled *A. chrysomphali* has some interesting implications for biological control theory and practice. According to H Compere (1961), L. O. Howard was responsible for these misidentifications, but Howard also arrived at an erroneous corollary in warning G. Compere (H. Compere's father) that "he was wasting his time and the state's money attempting to introduce parasites into California that already existed here" (Compere, 1961). Such a conclusion implies ignorance of the possibility of significant biological variation within a species. In effect, it presupposes that all individuals within a species are genetically equivalent; otherwise there was little justification for discontinuing evaluation of the "mistaken" *A. chrysomphali* for useful variation.

Given that improved taxonomy revealed the existence of several species distinct from *A. chrysomphali*, the example graphically illustrates our limited capacity to recognize and gauge the efficacy of biotypes, geographic races, or even phenotypic variation within a taxon [Caltagirone, 1985; for recent review see Hopper *et al.* (1993)]. If we are not capable of detecting useful variation between distinct species, it does not augur well in the search for such variation within species. It also represents the obverse of the *Cyrtobagus salviniae* story where some workers suggested that the analysis of efficacy should cease simply because the agent was shown to be a separate species from the ineffec-

tual *C. singularis*. Clearly, it is important for biological control workers to identify and exploit naturally occurring variation in natural enemies, whether intra- or interspecific, before embarking on the long and costly task of laboratory selection to improve relevant traits.

A similar point could be made about the place of genetic considerations in the decision-making process by reference to the history of control of the olive scale (*Parlatoria oleae*), a pest introduced to California in 1934. In this instance, a series of morphologically indistinguishable but biologically distinct strains of *Aphytis maculicornis* collected in 1951 from India to Spain complemented the initial 1949 import from Egypt, with the introduction from Iraq and Iran proving the most effective. This latter strain was subsequently described as *A. paramaculicornis* DeBach & Rosen. The final *coup de grâce* was delivered to the olive scale following the release in 1957 of another aphelinid endoparasitoid (*Coccophagoides utilis*) collected in Pakistan in 1957 (Doutt, 1966). Again this example supports the contention that initial efforts should focus on discovering and introducing, successively, strains, or species of natural enemy from the full complement that are presumed to keep the pest in check in its native environment. Attempts to improve natural enemies by artificial selection before these natural sources are reasonably exhausted are premature.

We can also illustrate the secondary importance of genetic improvement of natural enemies for category 1 terrestrial weeds by reference to the range of thistle species that have become aggressive agricultural weeds outside their natural Palearctic distributions, especially in North America, New Zealand, Australia, and South Africa. In particular, thistle species of the tribe Cynareae (family Asteraceae) belonging to the genera *Cirsium, Carduus, Centaurea,* and *Onopordum* are controlled in their native habitats by an extensive guild of natural enemies that have coevolved with them (Zwölfer, 1988). The rich insect faunas of the Cynareae have been divided by Zwölfer into various functional groups: gall makers, leaf miners, folivorous species, root feeders, stem feeders, and flower head feeders. These herbivores, in turn have their own predators and parasitoids. The important phytophagous taxa include a large number of species representing over 27 genera drawn from the Lepidoptera, Coleoptera, Diptera, Homoptera, Heteroptera, and Hymenoptera. Thus, the choice of agents and timing of their introduction are numerous, and optimal strategies clearly require a comprehensive understanding of the complex community ecology of the array of plant–insect interactions that determine abundance and distribution of thistle species in Europe (Sheppard *et al.,* 1989).

There is no *a priori* reason why effective biological control cannot be achieved for the aggressive agricultural weed species when there is no limitation on the importation of effective natural enemies. However, concerns over compensation by the plant to particular species of herbivores acting in isolation (e.g., producing more seed in response to damage to the apical meristem) or interacting negatively with each other to the detriment of control, might dictate a particular sequence of introduction (A. J. Wapshere, J. M. Cullen, personal communications). In other words, the considerations that influence choice of target species and the sequence of introductions will be of an ecological, biogeographical, and taxonomic nature; and will be dictated by questions of efficacy and specificity but not genetics. Questions of genetic improvement are unlikely to be useful in priority setting for research efforts when dealing with weeds such as thistles and, for that matter, for most arthropod pests where restoring some natural balance is the primary objective.

A similar downplaying of genetic considerations is even more appropriate when issues of genetic variability in the target weed are raised as a factor in determining priorities for biological control projects. For example, Burdon and Marshall (1981) have stated that outbreeding weed species with high levels of genetic variability such as Paterson's curse (*Echium plantagineum* Linnaeus), docks (*Rumex* spp.), and various thistle species are likely to be more refractory to biological control than are selfing species and apomicts such as *Opuntia* spp. and skeleton weed (*Chondrilla juncea* Linnaeus). The validity of this general proposition is challenged in the next section.

Repeated failure after a range of natural enemies has been tried is commonly encountered and may indicate that some novel feature confronts the introduced natural enemy for which its own genetic composition is inadequate. Such cases might be better classified as category 3. This is of practical value in two senses. First, it might focus attention on the offending component(s) of the environment in the regions where the releases have failed. By this process we might identify the genetic defect in the natural enemy that is a critical precondition before any artificial selection is commenced. Second, and perhaps more important, this type of analysis may suggest abandoning the target species as being refractory to biological control in its new environment and focusing attention to targets that may yield better returns from limited research funds. Knowing when to terminate a biological control project, and providing a plausible and acceptable argument in support of such a decision to government or some industry that may have invested considerable resources in the project, are probably that aspects of biological control most poorly handled by biological control scientists. These points are illustrated later when we consider the history of biological control of *C. juncea* in Australia.

A common cause of failure for biological control in category 1 projects, at least for weeds, is failure of otherwise suitable and effective natural enemies to meet specific-

ity requirements. For example, a whole range of insect herbivores that are known to exert a controlling influence on the legume *Mimosa pigra* Linnaeus (Lonsdale & Segura, 1987) in Central America are generalists and therefore are unacceptable for importation to control this aggressive weed in Northern Australia because of damage these agents could cause to native *Acacia* spp. (K. Harley, personal communication). Ideally, we should be able to select for altered specificity, but little is known about the mechanisms that determine host specificity of herbivores and less is known about the genetic basis of these characters or the stability of characters so selected.

The principal conclusion we should draw from this discussion is that category 1 cases of biological control, which are primarily concerned with restoring a natural balance and which include many of the major successes in biological control, have the least dependence on genetic considerations, especially on artificial selection as a tactic to improve the success rate. In general, the track record of success is more likely to be improved by the sustained application of modern concepts of taxonomy, population dynamics, community ecology, and behavior. Of course, genetical tools, especially modern molecular analyses, can assist taxonomists and ecologists, but these techniques should be seen as adjuncts to the latter disciplines and not ends in themselves. Better science in these disciplines, coupled with greater resources, increased focus, and improved international cooperation between countries with major exotic weed and arthropod pest problems should improve prospects for increasing the return on investment in biological control.

RELEVANCE OF GENETIC VARIABILITY IN CATEGORY 1 WEEDS TO THEIR SUITABILITY FOR BIOLOGICAL CONTROL

Probably one of the clearest and most emphatic recommendations by geneticists to practitioners of biological control of weeds over the past decade stems from the study by Burdon and colleagues and their genetic interpretations of these studies (Burdon & Marshall, 1981; Brown & Burdon, 1983; Burdon & Brown, 1986). While these authors restrict their consideration to plants, the arguments advanced are sufficiently general that the temptation could exist to extend the conclusion to animal pests. Because their prescriptions are well formulated but could be harmful to the practice of biological control if erroneous, it is important that their arguments be considered in some detail.

The major conclusions reached by these authors are

1. . . . in the biological control of weedy plants, the potential significance of the population genetic structure

of the target species appears to have been severely underestimated."

2. . . . a significant correlation [exists] between the degree of control achieved and the predominant mode of reproduction of the target plant; asexually reproducing species were effectively controlled significantly more often than sexually reproducing ones."

3. . . . It is argued from this result that the genetic structure of the target species has important implications with respect to the selection of species to be controlled using biological agents" (Burdon & Marshall, 1981).

In discussing the ramifications of the high level of genetic variability and heterozygosity in *E. plantagineum,* Burdon and Brown (1986) conclude, "If this high level of isozyme diversity reflects the diversity likely to be found in other parts of the genome, attempts to achieve substantial biological control may require the use of many different control agents." Elsewhere, Brown and Burdon (1983) state, "Genic and genotype variations presumably allow high levels of biochemical flexibility in populations of *E. plantagineum.* Such flexibility could hamper attempts at biological control." In the same article, the authors state, "With its propensity for high genic and multilocus genotypic diversity, *E. plantagineum* is likely to be more difficult to control by biological means than a genetically uniform weed. It is therefore likely to require a greater input of resources to achieve the same level of control as a uniform weed." Clarke (1979) also affirms the current wisdom that biochemical variations may have relevance to a host's withstanding pest or pathogen pressure. Brown and Burdon (1983) in this context state, "If this proposition is correct, then it is clear that *E. plantagineum* in Australia possesses a high level of biochemical flexibility which any biological control program must endeavor to overcome." In at least one instance, a government agency was proposing to abandon, after considerable financial investment, a biological control project concerning an outbreeding weed on the basis of Burdon's thesis. Thus, we are not simply dealing with an academic exercise but have good reason to assess the validity of this hypothesis.

The tenets of Burdon and colleagues may well be applicable to weed pathogens where it is generally acknowledged that host resistance and pathogen specificity and virulence are often under tight genetic control (Flor, 1956). However, we question its relevance to arthropod natural enemies of weeds.

First, the data on which Burdon and Marshall (1981) base their hypothesis are understandably limited. Of the 45 target weeds listed by them, 24 reproduce asexually, 16 reproduce sexually, and 5 have unknown breeding systems. Of the 24 asexual plant species, 18 belong to the Cactaceae with 17 belonging to the single genus *Opuntia*. All 17 cases involving *Opuntia* spp. are apomicts and in each case bio-

logical control was substantial or complete. Possibly, a substantial number of the *Opuntia* apomicts belong to a single species. Given the success rate and the notoriety of the successes, it is unlikely that any instance exists of an *Opuntia* species being a serious pest that has not been subjected to intensive and sustained biological control. A tendency to prejudge examples as probable successes or failures is illustrated by Commonwealth Scientific and Industrial Research Organization (CSIRO)'s experience when it initiated the biological control project on the apomict weed *Salvinia molesta.* It was criticized for investing resources on a weed that had already been shown to be refractory to biological control by earlier workers (Julien, 1987). Thus, success or failure can be infectious in a statistical sense and introduces a bias into the data set if each episode is being considered as an independent statistic.

A second difficulty arises in the case of the apomictic skeleton weed, *C. juncea.* Control of this weed by *Puccinia chondrillina* Bubak & Sydenham is listed as a success, yet only one of the three Australian cultivars has been subjected to substantial control, and none of the clones that occur in the United States or Argentina had been successfully controlled at the time the analysis was completed. Consequently, one success and not less than five failures would be a more accurate assessment of the success rate for biological control of skeleton weed.

No recognition was made of the type of natural enemy involved—whether it was a pathogen or an arthropod. Quite understandably, no allowance could be made by Burdon and Marshall for the enormous variation in endeavor and competence in the various attempts to control each target pest, or can we assume biological control workers consistently placed on the record the failed attempts. There also is uncertainty about the number of unrecorded successes because of inadequate follow-up evaluations. It is difficult to determine whether the prominence of *Opuntia* species in the successes is due to some biological property of cacti unrelated to their genetic composition. In a similar vein, one could suggest, by some principle of association, that the prominence of scale insects among the major successes of insect pests is a function of male haploidy instead of some aspect of the ecology or the biology of this group of pests. It is not clear why the original failed attempt at biological control of *S. molesta* is not included in Burdon and Marshall's original data set because the failure is recorded at least 14 times in the literature (Julien, 1987). We have already outlined the sequence of events and the nongenetic factors that transformed this failure into a notable success. The *S. molesta* example indicates the degree of caution that needs to be exercised in ascribing success or failure to genetics or to any other specific factor.

In general, the data are inadequate to support Burdon and Marshall's general and unqualified conclusion, ". . . the genetic structure of the target species has important implications with respect to the selection of species to be controlled using biological control agents." This proposition must remain hypothetical; and, because of its implications, opportunity should be sought to determine its validity, especially in relation to choosing weed species where the natural enemies for consideration are likely to be arthropods and not pathogens. Indeed, one prediction of Burdon's thesis is not supported by the available evidence. Burdon and Brown (1986) have demonstrated that *E. plantagineum* in Europe has a gene diversity of 35% and the level of heterozygosity is 29%. Because it is known that *E. plantagineum* is generally not abundant in Europe and is attacked there by a wide range of insects, whereas the plant has virtually no major natural enemy in Australia, we might suppose that the herbivores in Europe are maintaining the high genetic diversity (the average per locus expected heterozygosity when two gametes randomly chosen from the population are compared) and heterozygosity. This observation, as acknowledged by Burdon and Brown, is contrary to expectations and does suggest that *E. plantagineum* in Australia should fit comfortably into category 1 biological control situations. Under relaxed selection pressure in Australia, through the absence of the plant's natural enemies, we would predict lower values for these two genetic parameters. In fact, the gene diversity is 34% and the heterozygosity is 32% in Australia, which are not significantly different from the levels recorded in European populations (Burdon & Brown, 1986). It may well be that other unknown factors are maintaining the high values in Australia. However, the minimal data available to date do not support the thesis that genetic diversity is of major relevance in assessing suitability of weeds to biological control, and it would be premature to rank weeds as candidate targets simply on the criterion of their breeding system.

A more comprehensive analysis of the role of breeding systems in weeds on their amenability to biological control has been made by Chaboudez and Sheppard (1995). Their examination of the data indicate that no convincing case yet exists to link the breeding system of a weed to success or failure in attempts at biological control with invertebrate or pathogenic natural enemies.

IMPROVING EFFICACY OF NATURAL ENEMIES FOR CATEGORY 1 WEEDS

There are instances where knowledge of the genetic system of the weed and its natural enemy has enhanced the prospects for selecting an effective biological control agent, and materially assisted in resource allocation decisions. One case concerns skeleton weed (*Chondrilla juncea*) and its pathogen *P. chondrillina*. This highly successful project pioneered the use of a fungal pathogen in classical biological control of a weed, and thus warrants closer attention.

Chondrilla juncea is an obligate triploid apomict, and based on both morphological and electrophoretic evidence the plant exists as a large number of clonal strains in central and western Europe (Chaboudez, 1989). The origin of these clones remains uncertain, except that it has long been assumed to occur in eastern Europe or southern (former) USSR; they are generally considered to be the center of origin of the genus *Chondrilla*.

Three such clones have been introduced accidentally into Australia since European occupation, and have become major weeds of the wheat–sheep zone with higher summer rainfall (Hull & Groves, 1973). In contrast to the limited success achieved with various arthropod natural enemies (Cullen, 1978), an asexual strain of the rust *P. chondrillina* was showed to be a highly specific and an effective agent against one form of skeleton weed in Australia, the narrow-leaf form providing cumulative savings in excess of Australian $300 million (Marsden *et al.,* 1980). However, continuous attempts by CSIRO scientists since 1970 at its Biological Control Unit in Montpellier, southern France, to find effective strains of *P. chondrillina* for the remaining two forms of *C. juncea* had, until recently, proved unsuccessful. In the intervening period, these two strains, especially the intermediate-leaf form, have continued to spread across the wheat–sheep zones of Australia.

The suggestion was made in the 1960s to use chemical mutagens or irradiation on *P. chondrillina* as a means of increasing genetic variation to provide a basis for artificial selection to alter specificity and virulence of the pathogen. Such a proposal was not supported at the time by quarantine authorities on advice from some plant pathologists in Australia responsible for assessing the safety of pathogens proposed for introduction as biological control agents. Indeed, artificial selection, particularly when it pertains to specificity or pathogenicity, poses a special dilemma for quarantine authorities and the practitioner of biological control. The bureaucrat who approves importation of natural enemies has to countenance the risk of future shifts in either specificity or pathogenicity; in effect, a "conventional" assessment has connotations of typological thinking and the species is presumed to be invariant for such key traits. Induced mutations and artificial selection of such traits emphasize biological variability and do not sit comfortably with the typological thinking forced on government officers by the bureaucratic process. Consequently, the search has continued since 1970 in Europe by CSIRO for naturally occurring variants of *P. chondrillina* that meet quarantine specificity and efficacy requirements.

While there is no known scientific basis for preferring naturally occurring variation as against strains derived from mutagenesis and artificial selection, in reality there is a public perception that is less comfortable with, and therefore more inclined to challenge the safety of, organisms engineered by laboratory manipulations. This difficulty is particularly manifested when the natural enemy, such as the frost-inhibiting bacterium *Pseudomonas syringae* (van Hall), is constructed by recombinant DNA techniques. We deal further with this topic under the heading "genetically engineered natural enemies."

The difficulty with the "natural" approach in the case of *Puccinia chondrillina* has been that the plant pathologist searching for suitable strains in the field has lacked a sound conceptual framework that would permit systematic collection and screening of the range of *P. chondrillina* pathotypes present in European fields. It is even possible that the source populations of the clones of *C. juncea* that have established successfully in Australia and Argentina, and that are continuing to spread in the northwestern United States have since become extinct in Europe. Consequently, suitable strains of *P. chondrillina* simply may no longer exist in natural populations of *C. juncea* in Europe.

A significant conceptual breakthrough on biological control of *C. juncea* was achieved by P. Chaboudez and S. Hasan at the Montpellier laboratory of the CSIRO Division of Entomology. Chaboudez (1989, 1994), using electrophoretic techniques, has been able to show that the triploid apomicts of *C. juncea* appear to originate in western Turkey, presumably through some infrequent hybridization event involving two sexually reproducing diploid species of *Chondrilla*. Not only has the diploid progenitor, not known elsewhere in Europe, been found in western Turkey but also there is a concentration of triploid apomictic clones in the region (Chaboudez, 1994), with clone numbers declining in all directions away from this focus. Chaboudez postulated that "successful" clones move out in easterly and westerly directions from the point of origin, some eventually reaching western Europe, where a steady state of recruitment and extinction of clones occurs. There currently exists perhaps several dozen clones in western Europe, with in excess of 300 clones over wider Europe (Chaboudez & Burden, in press; Cullen, personal communication).

Surprisingly, a somewhat parallel evolutionary story appears to pertain to the pathogen *P. chondrillina.* Sexual stages of the rust are known to occur in Turkey in the general region where apomicts of *C. juncea* arise, but climatic conditions in western Europe are rarely suitable for the sexual stage (Chaboudez & Burdon, in press; Hasan *et al.,* 1995). Consequently, asexuality is the prinicipal or only mode of reproduction for *P. chondrillina* in western Europe. This perception encouraged the suggestion that natural selection for virulent forms of *P. chondrillina* must occur largely in eastern Europe and that the "matched" *C. juncea–P. chondrillina* asexual duet migrates together away from the center of origin (Chaboudez, 1989). It is also possible that some clones of *C. juncea* reach western Europe initially without a corresponding *P. chondrillina* pathotype. However, wind dispersal of genetic variants of *P. chondrillina* generated in eastern Europe might permit se-

lection *in situ* in western Europe. This interpretation obviously permits the possibility that, for some of the weedy clones of skeleton weed, there are no naturally occurring pathotypes of *P. chondrillina* for the biologist to discover. Either way, the scenario outlined for the origin of skeleton weed clones and their rust pathogen has led to two practical programs for the isolation, or indeed the genetic selection, of more appropriate natural enemies (Hasan *et al.,* 1995).

Burdon *et al.* (1980), using electrophoretic variants, had previously confirmed and characterized the three clones that had been identified earlier in Australia by their distinctive leaf morphologies (narrow-, intermediate- and broad-leaved. Chaboudez (1989), using seed instead of leaf tissue, improved the technique appreciably, and analyses by Chaboudez (1989,1994), Chaboudez and Burdon (in press), and Hasan *et al.* (1995) have confirmed the number and distribution of clones in Australia, the United States, and western and central Europe. The search in Europe has focused on regions that may have been the source of colonizers for Australia, the United States, and Argentina. This procedure has permitted a more systematic and comprehensive approach to determining if and where suitable clones of *C. juncea* occur in Europe that match the weed clones in the United States and Australia.

A more experimental approach to isolating suitable pathotypes involved the field planting in Turkey of the two "refractory" apomictic clones from Australia (intermediate and broad-leafed forms) along with the North America forms (Hassan *et al.,* 1995). The Turkey garden was established, as a collaborative project by CSIRO and the U.S. Department of Agriculture (USDA) at a site where sexual reproduction in *P. chondrillina* is presumed to occur in the hope that optimal conditions were provided for the field selection of suitable pathotypes of *P. chondrillina* for biological control of skeleton weed in the United States and Australia. As a result of the two-pronged approach, good matches were found for the U.S. clones (Hasan *et al.,* 1995) and reasonable matches were found for two of the three Australian clones (Chaboudez, 1994). The Argentine clones have not yet been examined in this manner.

Deciding when "enough is enough" is an important issue for the researcher, the funding body, and the farmer. The biological control of skeleton weed was commenced in CSIRO in 1967 and received considerable financial assistance from industry sources (Marsden *et al.,* 1980). However, the decision was made in 1970 to withdraw this assistance shortly before an effective rust was eventually isolated and released in 1971, bringing under spectacular control the narrow-leaf form of skeleton weed. Financial support from the grain industry was subsequently restored in 1977 in the hope that effective rusts would be isolated for the remaining two clones of skeleton weed.

The quandary for both CSIRO and the industry funding source has been when to terminate the program if the search continues to prove fruitless for natural enemies effective on the intermediate- and broad-leaf forms of skeleton weed in Australia. From a management perspective, a stop can reasonably be put to the survey of natural populations of *C. juncea* once the search has proved fruitless among pathotypes collected from the genetically related clones identified during a comprehensive survey. The orderly procedure outlined earlier has assisted in this decision-making process. The more experimental approach of screening for new pathotypes of the fungus yielded positive results for both the United States and Australia, and continued funding for such activities can be easily made on a more rational basis.

GENETIC IMPROVEMENT OF NATURAL ENEMIES OF SECONDARY PESTS (CATEGORY 2)

By far the best documented case for genetic improvement of natural enemies relates to the selection of pesticide resistance in natural enemies of arthropod pests of pome and citrus orchards, especially phytophagous mites, a group of secondary pests that achieved prominence in economic entomology through the disruptive influence of broad-spectrum chemicals (McMurtry *et al.,* 1970). It should help the reader if a brief account of the history of genetic improvement in predatory mites and other natural enemies of secondary pests is given. This is not to suggest that the past dictates the potential of future success. It is probable that many of these early attempts were poorly conceived and badly executed. Instead, such an outline should assist in determining whether genetic improvement has a wider role in improving the success rate of biological control and whether the major effort in genetic improvement should take place in the field or in the laboratory under generally artificial conditions.

The best documented case of a group of species emerging for the first time as serious economic pests following persistent use of broad-spectrum pesticides concerns the phytophagous tetranychid mites and their phytoseiid predators. Indeed, very little was known about their biology, taxonomy, or ecology until widespread use of DDT in the late 1940s pushed the tetranychids into prominence as major pests of orchards and horticulture [for a review see McMurtry (1983)]. Interesting statistics on their key natural enemies, the phytoseiid mites, illustrate this point. By 1951 a total of 20 species of phytoseiids had been described; by 1959 this figure rose to 170; and by 1982 nearly 1000 species had been described (Hoy, 1985b).

Between 1950 and 1970, considerable knowledge had accumulated on the taxonomy, biology, phenetics, and ecology of the tetranychid mites and their natural enemies, especially concerning the abundance and the efficacy of predatory phytoseiid mites (Huffaker *et al.,* 1970;

McMurtry *et al.,* 1970; van de Vrie *et al.,* 1972). Much of the field information was collected in agricultural systems where pesticide residues were widespread and where, in effect, the traditional complement of natural enemies was no longer sufficiently abundant to contain populations of the tetranychid mites. Consequently, the ranking of natural enemies, in terms of their effectiveness under these recently imposed artificial conditions, should be treated with considerable caution, particularly when decisions are being made as to which natural enemy should be chosen as a candidate for some long-term and costly program of genetic improvement. For example, if the objective is to select pesticide resistance in the chosen natural enemy, we really need to be confident that the genetically improved natural enemy, together with some IPM system, will lead to pest suppression and therefore warrant the time, effort, and expenditure devoted to the venture. R. Maddern (personal communication) attempted unsuccessfully to select for pesticide resistance in *Stethorus* in South Australia when subsequent studies showed that predatory mites were the more efficacious natural enemy of *Typhlodromus occidentalis*. It therefore would be comforting, though perhaps not essential, to have knowledge that the natural enemy was the principal, or at least a sufficient, controlling agent of the pest before pesticides disrupted the natural balance. We return to this point later.

Careful ecological studies between 1950 and 1970 of tetranychid problems in North America provided valuable information on predator–prey relationships, including synchronization in time and space of predator and prey, overwintering strategies, impact of pesticide residues, and prey density on predator efficiency (McMurtry, 1983). In particular, studies by Fleschner *et al.* (1955) and Collyer (1958, 1964) allowed assessment of the relative importance of coccinellids of the genus *Stethorus* and various phytoseiids. While *Stethorus* was an important predator, Fleschner *et al.* (1955) concluded that predatory mites were mainly responsible for suppressing *Eotetranychus sexmaculatus* in the western United States. Similarly, Collyer's perturbation experiments indicated that *Typhlodromus pyri* Scheuten was primarily responsible for controlling *Panonychus ulmi* in a range of temperate orchards (Collyer, 1964). These conclusions were validated with the critical observation by Hoyt (1969a, 1969b) that *T.* (= *Metaseiulus* or *Galendromus*) *occidentalis* Nesbitt had developed resistance to azinphosmethyl, the principal chemical used for controlling codling moth in the state of Washington. *Tetranychus mcdanieli* had previously developed resistance to the chemical, hence its emergence as a secondary pest in the 1950s and the subsequent need to include acaricides in spray schedules as long as various primary pests such as codling moth were being controlled chemically. Hoyt recognized that acaricides could now be deleted from the spray schedule in the correct belief that the organophosphate (OP)-resistant predatory mite should be capable of establishing a controlling influence on *T. mcdanieli* similar to that pertaining in pre-pesticide days (Hoyt & Caltagirone, 1971).

The next significant step was taken by Croft and Barnes (1971), who proposed biological control of spider mite in southern California by introducing and establishing OP-resistant strains of *Metaseiulus occidentalis* from the state of Washington. Effective control of spider mite was established and reduced the need for acaricides in southern California. The experience of Croft and Barnes directly influenced the decision in 1971 by CSIRO to introduce the same resistant *Typhlodromus occidentalis* into Australia where a similar success was achieved in apple and pear orchards (Readshaw, 1975).

Successful introduction and establishment of field-derived resistant phytoseiid mites for regaining biological control of a range of phytophagous mites, including *M. occidentalis, T. pyri, Phytoseiulus persimilis* Athias-Henriot, and *Amblyseius fallacis* Garman on grapes, almonds, peaches, roses, orchids, and pome fruits, has been a prominent component of many IPM programs since 1970 (Flaherty & Huffaker, 1970; Flaherty & Hoy, 1972; Croft & Brown, 1975; Croft, 1976; Croft & Strickler, 1983; McMurtry, 1983; Hoy, 1985a). In some cases, it was not necessary to introduce resistant biotypes but instead to exploit the resistance already developed *in situ* using the integration concepts first developed by Hoyt. These various IPM projects using resistant predators amply demonstrate the importance of sound ecological judgment and field competency in devising a practical and an effective IPM package for achieving maximum economic and environmental benefits.

By 1975 pesticide resistances had developed under field conditions in a narrow range of natural enemies, including the predator mites *M. occidentalis* (Nesbitt), *T. pyri* (Scheuten), *A. fallacis* (Garman), *Amblyseius hibisci* (Chant), the coccinellid *Coleomegilla maculata* (DeGeer), the black garbage fly *Ophyra leucostoma* (Wiedemann) (predator of the housefly), and the braconids *Macrocentrus ancylivorus* (Rohwer) and *Bracon mellitor* Say (Croft & Brown, 1975; Hoy, 1987, 1990). Furthermore, Croft and Meyer (1973) increased resistance to azinphos-methyl 300-fold in *A. fallacis* by the experimental treatment of a Michigan apple orchard with five to seven annual applications over a 4-year period. Croft and Meyer (1973) recognized that the benefits from using field-generated, OP-resistant predatory mites could be increased and greater flexibility could be given to the pest manager if resistance to a number of commonly used pesticides were incorporated in a single strain of predator. They recognized that the benefits could be short-lived if the primary pest developed resistance to the prevailing chemical and if a new compound was substituted to which

the predator had no resistance. Consequently, Croft and Meyer (1973) suggested that "it would be helpful if resistant predator strains could be detected in early stages of development or if strains could be pre-selected in the laboratory or field." Croft (1972) expressed pessimism about the efficacy of laboratory selection and argued in favor of concentrating efforts in the field where the experiment could be conducted under more natural conditions, on a scale not available in the laboratory. By using *A. fallacis,* an important predator of spider mites in the eastern United States, and essentially applying laboratory selection to field material where resistance genes had been concentrated by continued exposure to an array of chemicals, Croft and colleagues obtained high levels of resistance to OPs and carbamates, and limited response to synthetic pyrethroids (Croft & Meyer, 1973; Strickler & Croft, 1981).

Hoy and colleagues demonstrated the wider benefits of laboratory selection for improved performance in several ways. First, they extended the range of natural enemies subjected to artificial selection to include the parasitoids *A. melinus* and *Trioxys pallidus* Haliday (Hoy & Cave, 1988, 1991; Rosenheim & Hoy, 1988; Hoy *et al.,* 1991; Spollen & Hoy, 1992a, 1992b, 1992c, 1993a, 1993b) and the common green lacewing *Chrysoperla carnea* (Stephens), a generalist predator (Grafton-Cardwell & Hoy, 1986). Second, Hoy and colleagues endeavored to optimize the benefits of combined laboratory and field selection. Previous workers had recognized increased pesticide resistance among field populations of parasites, including *A. melinus, A. africanus* Quednau, and *Comperiella bifasciata* (Hoy, 1987, 1990). However, where attempts were made to increase resistance levels in natural enemies by laboratory selection, insufficient attention was given to the source populations that were often derived from few individuals or had been maintained in the laboratory for extended periods, which usually resulted in strains that failed to respond to selection or strains with very low levels of resistance [less than two- to threefold increases in lethal concentration-50 (LC_{50})] (Hoy, 1987). Intraspecific variation in levels of pesticide resistance in field populations of *M. occidentalis, Amblyseius finlandicus, Chrysoperla carnea, A. melinus,* and *T. pallidus* were surveyed; and field material was used to found populations for laboratory selections (Hoy & Knop, 1979; Grafton-Cardwell & Hoy, 1985; Rosenheim & Hoy, 1986; Hoy & Cave, 1988; Kostiainen & Hoy, 1994, 1995). This approach provides an indication of the genetic variability relating to the character under selection and, therefore, can be a measure of the likelihood of response to selection. More important, it maximizes the chance of capturing useful genes of major effect in the founding colonies.

A third feature of the artificial selection program developed by Hoy and colleagues has been broadening the range of pesticides covered in the selection program and the successful incorporation of different resistance genes into multiple-resistant strains. For example, a laboratory-constructed carbaryl–OP–sulfur-resistant strain of *M. occidentalis* formed the basis of an integrated mite management program for almonds in California (Hoy, 1984, 1985a, 1991). An economic analysis of the costs–benefits of the program in terms of reduced pesticide usage and of ignoring the equally important environmental advantages, suggested a potential annual return on the cumulative research investment in the range of 280 to 370% each year if the program is widely adopted by almond growers in California (Headley & Hoy, 1987). These benefits are actually being accrued; surveys indicate that approximately 60% of the almond acreage in California relies on the integrated mite management program that incorporates resistant *M. occidentalis.* Reduced pesticide use for control of spider mites in almonds, conservatively estimated, results in US $21 million being saved each year by the California almond industry. This large return on the original research investment indicates that a genetic improvement project can be a highly cost-effective investment in agricultural research. Unfortunately, it is apparently the only benefit–cost analysis of a genetic improvement project to date, so it is not clear whether this economic return on research investment is unique. Genetic selection of phytoseiids has consistently yielded strains with field-usable levels of resistances to pesticides (Table 2). The field efficacy of the carbaryl-resistant strain of *A. melinus* was evaluated in citrus orchards in California and in Israel (Spollen & Hoy, 1993b, 1993c).

The azinphos-methyl-resistant strain of *T. pallidus* is cross resistant to chlorpyrifos, endosulfan, methidathion, and phosalone (Hoy & Cave, 1989). This strain was mass reared and released into commercial walnut blocks in 1988 where it established, and survived applications of azinphos-methyl and methidathion (Hoy *et al.,* 1989, 1990). Azinphos-methyl-resistant parasites were recovered from two release sites in 1990, suggesting that this new strain has established and survived over two winters, and has survived multiple applications of azinphos-methyl (Hoy *et al.,* 1990). Large-scale utilization of the azinphos-methyl-resistant strain in walnut IPM will probably require additional releases (Caprio *et al.,* 1991). The original implementation model, in which the resistant strain replaces the susceptible field strain after selection with azinphos-methyl, was shown to be inappropriate (Caprio & Hoy, 1994, 1995; Edwards & Hoy, 1995), indicating that the population genetics and ecology of each species must be well understood.

Pesticide resistance in a natural enemy is clearly a trait of value or relevance only while the particular chemical is used to control a key pest (although additional benefits could accrue if cross-resistances were present in the resistant natural enemy to related pesticides still in use). Pesticide-resistant natural enemies also are valuable only if the secondary pest has itself developed resistance, if the prey–

host of the improved natural enemy is naturally tolerant of the chemical, or if it escapes its impacts through various conservation techniques. Key pests such as codling moth, apple maggot, red-banded leaf roller, and plum curculio did not develop resistance to OP compounds during the 20 to 30 years when they were widely used, whereas some secondary pests (mites, aphids, and leafhoppers) acquired resistance within 5 to 20 years (Croft & Strickler, 1983). Thus, the time period during which a specific resistance is a desirable trait in a natural enemy is that interval between resistance developing in the secondary and primary pests. This window of opportunity might be short-lived and generally will not be predictable. Because pesticides can be withdrawn for other reasons (unacceptable residues or more effective or economic alternative chemicals), the time period to enjoy benefits from artificial selection for resistance could be shorter still.

With such considerations in mind, Hoy and Knop (1979) proposed selection for resistance to new chemicals that might be introduced for crop protection (e.g., synthetic pyrethroids and avermectin) prior to their use in the field. Initial attempts to select for permethrin resistance in *M. occidentalis* using colonies from California orchards and vineyards were unsuccessful (Hoy, 1985b). Limited polygenic response to permethrin selection ($\times 10$) was subsequently obtained using mites collected from a Washington apple orchard (Hoy & Knop, 1981). Subsequent selection for a higher level of permethrin resistance has proved successful (Ouyang & Hoy, unpublished). Strickler and Croft (1981), who also selected for permethrin resistance in *A. fallacis* in anticipation of its use for control of major apple pests in North America (Croft & Hoyt, 1978), experienced a similar limited polygenic response. Hoy and Conley (1987) obtained a positive, but small, response to selection for resistance to avermectin in *M. occidentalis.*

These preliminary results support the view that it may prove necessary to rely on initial field selection of rare resistance (R) alleles of major effect and then to use laboratory selection to increase the frequency of the R allele, or to combine different resistance genes into the one multiply-resistant strain for mass production and field release. Nevertheless, the valid point is made by Hoy and Knop (1979) and Croft and Strickler (1983) that every endeavor to select resistant natural enemies of secondary pests should be made as early as possible to maximize economic returns on the investment.

An important practical question we must consider is whether the response to artificial selection for pesticide resistance ever represents a *de novo* genetic gain exploiting mutations arising in the course of the experiment, or whether the R alleles always need to be present in appreciable frequencies in the founding material so that selection is simply concentrating this preexisting genetic information. A second important question we need to address is the relevance of the specific selection regime applied to the laboratory strain. It is a common misconception that some characters are by their nature quantitative and under polygenic control while others are qualitative and controlled by one or few loci. There is evidence to suggest that the genetic response, in terms of major genes or polygenes, is more a function of the selection regime than the character itself (Whitten & McKenzie, 1982; McKenzie, 1996). It is for this reason that the response to laboratory-based selection for pesticide resistance often has been polygenic while most field-derived resistances have implicated single major genes (Whitten & McKenzie, 1982; Roush & McKenzie, 1987; McKenzie, 1996). The genetic basis of pesticide resistance in a natural enemy may be immaterial when releases are made into areas devoid of the natural enemy, but it can be critical if releases are to be made into an area containing susceptible individuals (Caprio & Hoy, 1995). Special steps need to be taken to ensure resistance is not rapidly lost through genetic dilution (Hoy, 1985b); in any case, polygenic traits tend to be generally less stable than traits under the control of single genes and revert in the absence of continued selection pressure (Falconer, 1960). By contrast, a population where a major gene is fixed is likely to maintain its resistance status, even in the absence of pesticide pressure (Arnold & Whitten, 1976).

Croft originally drew attention to the notion of using starter material from field populations previously exposed to the particular pesticide for which resistance is being selected. He further advocated liberal additions to the strain undergoing laboratory selections. This advice is supported by Hoy and Knop's (1981) experience of a failed response to permethrin selections in a strain of *M. occidentalis* from California followed by 10-fold increases in resistance when a strain from a Washington apple orchard was pressured in the laboratory with permethrin. The point in question here is not the value of the laboratory phase of the genetic improvement program, because there can be no question of the real genetic gains obtained during the selection procedures, and there is nor other way of combining resistances to unrelated groups of chemicals. Instead, it concerns the critical importance of the starting material and the role of the particular selection regime adopted.

It is generally accepted that single-gene-based resistance in natural enemies is desirable for two reasons: strain stability and higher resistance factors invariably associated with major gene derived resistance. Furthermore, the type of selection regime that usually operates in the field favors major gene response (Whitten & McKenzie, 1982; McKenzie, 1996). Consequently, every opportunity should be taken to exploit this genetic response in the field. From these considerations it follows that the intuitive approach of applying a pesticide in the lethal dose range LD_{50} to LD_{90} to successive generations to maximize selection differential, while providing reasonable numbers of surviving offspring

to initiate the next generation with minimal inbreeding, is likely to exploit polygenic variation; and therefore, this approach is not an efficient selection regime. Instead, it is better to maintain a large unselected population with frequent injections of field material and to apply an LD_{99} to around 50 to 70% of each generation, leaving the balance as the breeding colony. Survivors of the LD_{99} should be progeny tested to determine if they contain a major resistance gene. Positive families should form the basis of a resistant colony with outcrossing to field material to remove any inbreeding or adverse genetic effects of colonization. The LD_{99} for heterozygous- and homozygous-resistant genotypes can then be determined and used to construct strains in which the major resistance gene is fixed. These prescriptions presuppose the natural enemy is easy to rear and is highly fecund. Such practical considerations suggest some compromise between the ideal and realistic selection regimes often will be necessary.

Under the ideal regime outlined, survivors are not retained unless they appear to contain a major resistance gone. In this way, a polygenic response should be largely avoided. Temporary loss of response to artificial selection through the incorporation of less resistant field material should be offset by greater long-term responses to selection along with the less well-defined benefits of reduced inbreeding and unintended genetic change caused by the artificial laboratory environment. On this last point, predatory mites—partly due to their small size and the ease of rearing them in the greenhouse or laboratory—can be reared under conditions that more closely simulate field conditions (Hoy, 1985b), which may make them singularly well-suited to genetic improvement programs. To our knowledge, however, no parasites or predators have been selected for resistance using this scheme.

The general genetic model that underpins our major conclusions concerning response to selection for increased pesticide resistance has so far presupposed the existence of a number of loci, at any one of which a spontaneous mutation would generate initially very rare individuals whose frequency would be increased through the provision of special conditions favoring the rare resistance genotype. An alternative situation that has important theoretical and practical implications for genetic improvement programs concerns cases where resistance is caused by gene amplification [e.g., OP resistance in *Myzus persicae* (Suzler) (Devonshire & Sawicki, 1979), in *Culex fatigans* Wiedemann (Mouches *et. al.*, 1986), and in *Phytoseiulus persimilis* (Fournier *et al.*, 1987a, 1987b)]. In any genetic analyses of resistance due to gene amplification, it will map as a single genetic locus; yet it could appear to involve high mutation rates and would display a gradual response to selection more characteristic of a polygenic trait. How frequently gene amplification provides the basis for pesticide resistance remains to be seen (Mouches *et al.*, 1986; Devonshire & Field, 1991;

Hoy, 1994; C. Mouches, personal communication). We consider these implications further when we discuss the potential of molecular genetics in genetic improvement.

One significant benefit of the work of Croft, Hoy, and colleagues has been to provide valuable insights into why natural enemies are less frequently represented among arthropod species that have developed resistance to pesticides. While it is probable that entomologists are more likely to monitor and record resistance in pests than in natural enemies, the 13+ documented cases of resistant natural enemies as against 281 cases among agricultural pest species (Croft & Strickler, 1983; Hoy, 1990, 1991, 1992c) suggest that natural enemies may have a lower capacity to develop resistance, or there is a less effective opportunity in the field for resistance to develop in natural enemies (Huffaker, 1971). Croft and Strickler (1983) provide theoretical reasons why both hypotheses require further investigation, but the response to field selection by the natural enemy in those special cases where the chemical continues to be applied for the control of the principal pest after the secondary pest has acquired resistance does indicate that some natural enemies can also develop resistance given adequate opportunity.

Although the record demonstrates that genetic improvement of natural enemies is an important tool available to the biological control worker where secondary pests have emerged through the adverse effects of pesticide usage on the efficacy of the natural enemies, we now need to consider whether genetic improvement has a place in category 3 cases of biological control.

GENETIC IMPROVEMENT OPPORTUNITIES WHERE THE NATURAL ENEMY IS CONFRONTED WITH ONE OR MORE NOVEL CONDITIONS IN THE ENVIRONMENT (CATEGORY 3)

This grouping of biological control opportunities covers cases where the natural enemy is confronted with a significantly different environment in one or more physical or biological respects. Virtually by definition, we are dealing with unsuccessful cases and we can presume the inadequate performance of the introduced natural enemy reflects one or more genetically determined phenotypic deficiency. The challenge to the biological control practitioner is to choose between five options

1. Manipulate the conditions of establishment (as occurred with *S. molesta* in the Sepik River, Papua New Guinea (Thomas & Room, 1985) by the single application of urea).

2. Seek further samples or biotypes of the natural enemy and see if their performance is adequate. This tactic also was applied to *Salvinia molesta* and led to the

fortuitous collection of a closely related species that proved effective. Similarly, several parasite species have been introduced successfully over the past 100 years to improve the level of control of California red scale.

3. Seek entirely new groups of natural enemies (e.g., pathogens in the case of weeds where suitable arthropod candidates have been exhausted).

4. Abandon the project in favor of a more susceptible target pest.

5. Finally, attempt to improve the natural enemy by artificial selection.

The latter concept has been proposed on occasions for a range of natural enemies and, with the exception of pesticide resistance and nondiapause in natural enemies of secondary pests, has not been used successfully in the field (see Table 2). Furthermore, the only convincing examples of improved resistance to pesticides usually have entailed extensive prior exposure to the relevant selection pressure under field conditions. Laboratory selection serves to reinforce the process by concentrating the frequency of R alleles or can provide a critical step by enabling novel combinations of different resistance mechanisms. Thus, genetic improvement by selecting for pesticide resistance may represent a special case and could prove an inappropriate model for improving the traits of the type as listed in Table 2. At the outset, we should acknowledge that genetic improvement is a high-risk venture and should only be seriously contemplated after the four alternative options have been canvassed thoroughly.

Probably the most awesome step in genetic improvement of natural enemies is to identify the phenotypic weakness that is responsible for the poor level of control. Even this description of the problem is simplistic and naive. We really are looking at a relationship or an adaptation between the natural enemy, its target, and the total environment in which both exist.

Unfortunately, in most cases the precise cause of failure in a natural enemy simply cannot be determined. It is usually only after a successful agent has been introduced that the real cause of earlier failure can be identified (i.e., the supposed deficiencies in the unsuccessful natural enemy are properly defined). The sequence of events leading to the control of *S. molesta* gave a unique opportunity to pinpoint the critical difference between failure and success. This sequence is worth detailing because it illustrates that the deficiency sometimes can be defined only by a *post factum* reconstruction. The pertinent details on the biological control of *S. molesta* were outlined earlier in this chapter (also see Chapter 29 this volume). Sands and colleagues sought to identify why *Cyrtobagus singularis* had not been successful, whereas another "biotype" from southern Brazil had proved effective. It is particularly interesting to note that some workers argued that the question itself ceased to be of any moment once it was realized that the successful biotype was a distinct species. Again, the specific status of the successful biotype is simply irrelevant in reaching a satisfying explanation of why one taxa succeeds where another fails. Sands and Shotz (1985) nevertheless pursued their investigation and identified the larval feeding behavior of *C. salviniae* as being critical; unlike its conspecific, it caused waterlogging of the damaged tissue and subsequent sinking of extensive mats of *S. molesta.* Careful comparative studies on the biology and feeding behavior of *C. singularis* and *C. salviniae* on *S. molesta* and *S. auriculata* eventually dispelled an initial supposition that *C. salviniae* was more effective because it had coevolved with *S. molesta.* It transpired that *C. salviniae* just happens to be a more aggressive herbivore and a more effective controlling agent for both *S. molesta* and *S. auriculata,* the latter species providing the original collections of *C. singularis* (Sands & Shotz, 1985).

We have noted that the success of this project would have been less certain had the influence of nitrogen levels in reproduction of the herbivore not been appreciated. It is a matter of conjecture as to how many potentially useful agents have failed to establish because of inappropriate conditions at the time of release.

The *S. molesta* example clearly illustrates two important points:

1. There is difficulty in defining in any useful *a priori* way what constitutes a deficiency in a natural enemy. Failure to cause waterlogging was not perceived to be the operational difficulty by those workers in whose hands the early attempts had failed, and in a very real sense it was only deemed to be a "phenotypic character" after the variation was perceived.
2. There is inappropriate use of data to confirm some favored hypothesis (e.g., success in this case being due to coevolution between the plant and its herbivore when, in reality, the data were merely consistent with that hypothesis, among others).

When we ultimately embark on a particular genetic improvement program, we cannot select some trait in the natural enemy in an artificial, constrained laboratory environment without continual reference and assessment in the dynamic field context in which the fruits of selection are ultimately evaluated.

We can illustrate this problem by reference to the celebrated case of artificial selection in *Aphytis lingnanensis* whose limited tolerance to temperature extremes reduced its effectiveness away from coastal California. These observations prompted White *et al.* (1970) to select and to obtain response for improved capacity to deal with temperature extremes. It is unfortunate that the gains of this selection program were not evaluated in the field simply because

another species (*A. melinus*), better adapted to the inland climate of California, had been discovered and released (Rosen, 1985). While this example is frequently used to illustrate the obvious principle that artificial selection should be pursued as a last resort, and only after all reasonable attempts to secure naturally occurring natural enemies have been exhausted, it is equally suitable to illustrate the point that some such efforts are possibly conceptually misguided. It is important to note that *A. lingnanensis* had colonized the interior and intermediate regions of California and presumably was exerting a degree of control over California red scale. For the selected laboratory form to displace the field populations of *A. lingnanensis,* we need to suppose that Darwinian fitness is linked in some rigid way with efficiency as a natural enemy at suppressing the host. A further complication would arise from the uncertain outcome when the relatively few selected individuals are released and interbreed with the more numerous and genetically inappropriate field population of the natural enemy (Caprio & Hoy, 1994, 1995). Admittedly, the extraordinary impact that several dozen "Africanized" queen bees, differing polygenically from a vastly more numerous resident population of honeybees in South and Central America, appears to have had in the space of several decades, suggests caution in predicting the outcome of genetic perturbations in natural populations (Taylor, 1985). There is now unpublished evidence that Africanized queens had been much more widely distributed in Brazil than stated in the literature (Michael Allsopp, personal communication) and thus the impact may not be so surprising. We also note that artificial selection for most complex traits such as broadened temperature tolerance can draw on a very limited genetic base and, more significantly, is unlikely to emulate the myriad ways that selection under field conditions can draw on morphological, physiological, biochemical, and behavioral attributes in stimulating a genetic response. In a formal genetic sense, artificial selection begins with a much reduced genetic base simply because of the limited population size and, in general, fewer gene loci are likely to be implicated in a response because of the limited nature of the laboratory selection challenge.

In summary, the general track record for genetic improvement of natural enemies by laboratory selection remains unimpressive in terms of raising the overall success rate of biological control (see Table 2). Of the natural enemies selected in the laboratory since 1971, as far as we are aware, no efforts were made to evaluate the strains selected for enhanced fecundity (Szmidt, 1972; Ram & Sharma, 1977; Voroshilov & Kolmakova, 1977), temperature tolerance (Voroshilov, 1979), thelytoky (Legner, 1987), host synchrony (Weseloh, 1986), or abamectin resistance (Hoy & Ouyang, 1989). Gilkeson and Hill (1986) found the selected nondiapausing strain of *Aphidoletes aphidimyza* was not needed when they learned to manipulate daylength in the greenhouse and thereby they were able to maintain the wild strain in a reproductive mode. Field and Hoy (1986) found the nondiapausing carbaryl–OP-resistant strain of *Metaseiulus occidentalis* performed well in small plots of roses grown in the greenhouse for cut flower production but the strain was not tested in commercial greenhouse production systems. Grafton-Cardwell and Hoy (1986) evaluated life table attributes of the carbaryl-resistant strain of *Chrysoperla carnea* but the strain was not tested in the field or in the greenhouses. The pyrethroid-resistant strains of *Amblyseius fallacis* and *M. occidentalis* are not being used in orchards in North America, although they were tested in apple and pear orchards, in part because the pyrethroids have not been widely adopted for use in IPM programs (K. Strickler, personal communications; M. Hoy, unpublished). Field evaluations of a carbaryl-resil-resistant strain of *Aphytis melinus* in citrus in Israel (Spollen & Hoy, 1992c; D. Rosen, personal communication) suggest this strain has established. The results of efforts to implement a phosmet-resistant strain of *Amblyseius nicholsi* in the People's Republic of China (Huang *et al.*, 1987) or of an OP-resistant strain of *A. finlandicus* in Finland (Kostiainen & Hoy, 1994, 1995) remain unknown.

The azinphos-methyl-resistant strain of *Trioxys pallidus* apparently has established in California walnut orchards (Hoy *et al.*, 1989, 1990; Edwards & Hoy, 1995). Large-scale utilization of this strain in walnut IPM will require extensive dispersal and replacement of the susceptible population by this strain. Additional releases are desirable to speed up the establishment and dispersal of *T. pallidus* (Caprio *et al.*, 1991; Caprio & Hoy, 1995). The pyrethroid-resistant strain of *T. pyri* has been evaluated in New Zealand apple orchards, and Suckling *et al.* (1988) suggested an IPM program utilizing the pyrethroid-resistant strain was planned.

To date, no genetically manipulated parasites or *insect* predators are fully implemented in biological control or IPM programs. It can be argued that many of the failures to evaluate selected strains reflect inadequate time, effort, or commitment. In some cases, the results of evaluation are simply not yet available because the experiments are in progress. In any case, many of the efforts to genetically improve arthropod natural enemies remain untested.

THE SCOPE OF MOLECULAR BIOLOGY AND GENETIC ENGINEERING FOR IMPROVING THE SUCCESS RATE OF BIOLOGICAL CONTROL

The genetic improvement discussed so far has involved classical breeding techniques, choosing as parents those individuals whose phenotype is in the direction of some desired goal. The most successful programs, and those that have had considerable practical value, have drawn on source populations where significant selection under field conditions had already increased the frequency of genes of major effect, as in the case of pesticide resistance of natural enemies.

Where the gene responsible for the phenotypic shift is known and has been biochemically characterized (e.g., a modified acetyl cholinesterase, phosphatase, or esterase in the case of OP resistance), it is then possible to consider transferring such a desirable gene directly from one insect species to another, avoiding the labor-intensive and sometimes futile selection program. In this section we discuss the potential value of recombinant DNA technology for transforming species by transferring specific genes from one species to another entirely unrelated species. We outline briefly the technology and then consider the benefits and risks associated with genetically engineering specific traits into natural enemies.

The most important need is to have a general method of transferring genes across arthropod species. There is every reason to believe the capability will be widely available over the next few years (Handler & O'Brochta, 1991; Presnail & Hoy, 1992; Hoy, 1994a; Carlson *et al.,* 1995; O'Brochta & Atkinson, 1997; Ashburner *et al.,* 1998). Consequently, attention should be given to identifying suitable genes and regulatory sequences for transfer and organisms that should be targets of proposed genetic engineering projects.

It is now possible to isolate specific genes from an insect, to propagate the gene in a bacterial system, and to study its structure and—in some instances—even its function in its new environment if it has been coupled to an appropriate promoter segment that permits expression. It is possible to reintroduce a cloned gene, regardless of the gene's origin, into the genome of *Drosophila melanogaster* Meigen and to study its expression and effect on the phenotype. The preferred technique for such transformations uses a segment of DNA from *D. melanogaster* called a transposable P-element that encodes a transposase enzyme whose function is to facilitate integration of genes into a chromosome (Rubin & Spradling, 1982; Spradling & Rubin, 1982; Engels, 1989). Transformation of *D. melanogaster* with genes from that or other species is now a routine procedure.

Interest in using the P-element vector to transform other arthropods developed soon after the methods were identified (Cockburn *et al.,* 1984; Beckendorf & Hoy, 1985; Kirschbaum 1985). The genetic modification of the P-element system from *D. melanogaster* by Laski *et al.* (1986) encouraged hopes that a general gene transfer technology had been developed for arthropods. There is no evidence that the P-element can function outside the genus *Drosophila* (O'Brochta & Handler, 1988, 1991; Walker, 1989).

Other vectors are being developed, including *hobo, hermes, Minos,* and *mariner* (Maruyama & Hartl, 1991; O'Brochta *et al.,* 1994, 1996; Loukeris *et al.,* 1995a, 1995b). Furthermore, transformation of an arthropod biological control agent with an active transposable element vector may not be desirable if the biological control agent were to be released into the environment for an IPM program. Certainly, such a recombinant natural enemy would be scrutinized by regulatory agencies with special intensity to resolve the stability of the transgenic line (Hoy, 1990, 1993, 1995, 1996).

Progress in developing gene transfer technologies is rapid and it seems reasonable to assume that these technologies will be widely available to the applied entomologist in the foreseeable future. The technologies under development usually entail manipulation of eggs or early embryonic stages and the development of microinjection procedures. Consequently, their application to specific natural enemies might not be a trivial task, especially for species that are endoparasitoids and difficult to rear in the laboratory. Presnail and Hoy (1992) developed a technique called maternal microinjection that resulted in stable transformation of the phytoseiid mite *Metaseiulus occidentalis.* This technique may be useful for delivering DNA to many different arthropods, because stable transformation of the parasitoid *Cardiochiles diaphaniae* Marsh has been demonstrated (Presnail & Hoy, 1996). The technique allows transmission of injected DNA to multiple eggs of the microinjected females, thus being an efficient method of delivering DNA (Presnail & Hoy, 1994).

The goal of genetic engineering of a particular species would have to be well defined in terms of the phenotype required and the expected benefits would have to be substantial to warrant the time and expense necessary to embark along this "hi-tech" pathway. The prospect of incorporating pesticide resistance into a natural enemy is, by itself, sufficiently attractive to warrant further scrutiny of the technology. It may be possible to obtain pesticide resistance genes from both prokaryotic and eukaryotic species for insertion into beneficial arthropods (Beckendorf & Hoy, 1985). Phillips *et al.* (1990) transferred and expressed an OP insecticide-degrading gene from the prokaryote *Pseudomonas* into *D. melanogaster.* Alternatively, pesticide resistance genes may be identified and cloned from another arthropod for which we have good genetic, biochemical, and cytological information, including *Drosophila* spp., *Lucilia cuprina* (Wiedemann), *Musca domestica* Linnaeus, *Culex* spp., *Aedes* spp., and *Anopheles* spp.

We must remember that pesticide resistance in a natural enemy is only of value while the particular pesticide is in use, so there is an overriding time factor to bear in mind. The resistance mechanism should be dominant or semidominant and not additive or recessive. The reason for this restriction follows from the fact that the inserted resistance gene is in addition to, not a substitute for, the original susceptibility gene that would remain and presumably function normally in the engineered organism. Alternatively, the resistance gene can be inserted into the insect's genome in a targeted and specific manner so that it replaces the susceptible alleles in the genome, but such techniques are not yet available for insects other than *D. melanogaster* (Ashburner *et al.,* 1998).

The prospects over the next few years for genetically engineering other valuable traits into natural enemies are less optimistic. The genetic basis for shifts in traits such as fecundity, longevity, diapause, and sex ratio are generally not yet adequately characterized to be of value to the molecular biologist. Here we encounter a fundamental problem sometimes not appreciated by the advocate of genetic engineering and it addresses the differing operational routes whereby geneticists and molecular biologists approach individual genes. The molecular biologist characterizes a gene by the biochemical function of its product, whereas a geneticist characterizes a gene by the phenotypic effects of mutants of the gene. Thus, to a geneticist one gene might be labeled rosy, yellow, or OP resistant while to the molecular biologist the same gene is labeled xanthine dehydrogenase, tryptophan oxygenase, or acetylcholinesterase, respectively. In practical terms, we need to choose genes for which specific mutations will cause a desirable shift in the phenotype but, unfortunately, we usually only determine such an effect after the event. The set of genes we could therefore be interested in is indeterminate and, unfortunately, often indeterminable. This problem, sometimes called "reverse genetics" (Whitten *et al.,* 1996) serves to highlight the fact that classical breeding selects at the phenotypic level; the underlying genotypic change merely follows and is generally not analyzed. In effect, molecular biologists have to work in reverse. They modify the genotype in the calculated hope that the phenotype will shift in the desired direction and not in any unintended direction as well because of pleiotropy. This may be difficult to achieve, especially where the immediate gene product is only distantly connected to the desired phenotypic shift.

There is less of a problem when the phenotype is directly determined by the protein product of the gene (e.g., a novel toxic neuropeptide produced by a microbial pathogen within an insect host) or by an antifreeze protein in a natural enemy (Rancourt *et al.,* 1987). Similarly, in a beneficial arthropod, if the gene product is an enzyme highly effective at metabolizing a pesticide, the phenotypic shift could be significant, stable, and beneficial. However, even if the gene transfer capabilities are adequate, the likelihood of achieving the intended phenotypic shift in arthropod natural enemies, as well as excluding unintended adverse effects on the phenotype, seems problematic based on our present understanding. This issue is well illustrated with genetically engineered pigs. Elevated levels of bovine growth hormone in transgenic pigs caused the desired changes in weight gain, feed efficiency, and low fat levels, but also induced higher incidence of gastric ulcers, arthritis, dermatitis, and viral disease (Pursel *et al.,* 1989).

The difficulties outlined earlier are probably less for pesticide resistance and, more importantly, they may not apply to microbial natural enemies. The opportunities to genetically engineer viruses, bacteria, and fungi seem considerable and only limited by the ingenuity of the pathologist and applied entomologist [for recent review see Whitten *et al.* (1996); and other chapters in same proceedings). The objective is to cause a foreign gene to express in a host, causing its premature death. Our task is to develop a practical delivery system. For example, we could consider coupling the polyhedrin protein promoter from the baculovirus *Autographa californica,* which has been well characterized, with a structural gene coding either for a neuropeptide that disrupts larval development or behavior or for a protein that metabolizes a steroid hormone or some other molecule whose normal titer is also essential for stable development or normal behavior (Luckow & Summers, 1988). The necessary technology and sufficient genes whose products are rapidly becoming adequately characterized are available for us to develop a broad range of more toxic pathogens. Unfortunately, our knowledge is not yet adequate to modify host specificity in controlled ways, and the uncertainty of stability of such critical characters would warrant considerable caution.

One obvious risk in genetically engineering beneficial insects, and particularly microbial pathogens, emanates from the public perceptions of such procedures. It is not patronizing to suggest that the intelligent layman cannot adequately comprehend the complex way biological knowledge is encoded in DNA and accessed during the life cycle of an organism to regulate its development and behavior. Indeed, biologists do not have sufficient understanding of the interface between genotype and phenotype to quantify the biological ramifications and risk of such manipulations (Hoy *et. al.,* 1997).

The first release of a transgenic strain of the predatory mite *Metaseiulus occidentalis* took place in March 1996 (Hoy *et al.,* 1997). This predator strain contained an inactive molecular marker gene and had undergone extensive laboratory tests to evaluate the potential risks of such a release and to confirm the fitness and stability of the transgenic strain (Li & Hoy, 1996; McDermott & Hoy, 1997; Presnail *et al.,* 1997). This field trial did not include permission to allow the predator strain to establish permanently in the environment, and it remains unclear what criteria would be used to evaluate the risks of permanent releases of transgenic natural enemies into the environment as part of a practical pest management program. There is a distinct prospect that public demand for fail-safe procedures, comprehensive environmental impact statements, and guarantees of success could create serious obstacles for the genetic engineer of natural enemies (Hoy, 1992a, 1992b). If these same criteria are applied with equal rigor to traditional biological control programs, it will add further challenges for practitioners in this important component of pest and weed management (Cullen & Whitten, 1995).

Acknowledgments

This work was supported in part by the University of California Agricultural Experiment Station and Western Regional Project W-84; and the Davies, Fischer, and Eckes Endowment in Biological Control at the University of Florida, Gainesville. This is publication R-06125 from the Institute of Food and Agricultural Sciences at the University of Florida.

References

Adams, C. H., & Cross, W. H. (1967). Insecticide resistance in *Bracon mellitor,* a parasite of the bollweevil. Journal of Economic Entomology, 60, 1016.

Aizawa, K. (1987). Strain improvement of insect pathogens. In K. Maramorosch (Ed.), Biotechnology in invertebrate pathology and cell culture (pp. 3–11). Orlando, FL: Academic Press.

Allen, H. W. (1954). Propagation of *Horogenes molestae,* an Asiatic parasite of the Oriental fruit moth, on the potato tuberworm. Journal of Economic Entomology, 47, 278–281.

Andres, L. A., Davis, C. J., Harris, P., & Wapshere, A. J. (1976). Biological control of weeds. In C. B. Huffaker & P. S. Messenger (Eds.), Theory and practice of biological control (pp. 481–499). New York: Academic Press.

Arnold, J. T. A., & Whitten, M. J. (1976). The genetic basis for organophosphorus resistance in the Australian sheep blowfly, *Lucilia cuprina* (Wiedemann) (Diptera, Calliphoridae). Bulletin of Entomological Research, 66, 561–568.

Ashburner, M., Hoy, M. A., & Peloquin, J. (1998). Transformation of arthropods–research needs and long term prospects. Insect Molecular Biology. 7(3), 201–213.

Asquith, D., & Colburn, P. (1971). Integrated pest management in Pennsylvania apple orchards. Bulletin of the Entomological Society of America, 17, 89–91.

Avella, M., Fournier, D., Pralavorio, M., & Berge, J. B. (1985). Selection pour la resistance a la deltamethrine d'une souche de *Phytoseiulus persimilis* Athias-Henriot. Agronomie, 5(2), 177–180.

Avise, J. C. (1994). Molecular markers, natural history and evolution. New York: Chapman & Hall.

Baker, R. T. (1985). Biological control of plant pathogens: Definitions. In M. A. Hoy & D. C. Herzog (Eds.), Biological control in agricultural IPM systems (pp. 25–39). Orlando, FL: Academic Press.

Beckendorf, S. K., & Hoy, M. A. (1985). Genetic improvement of arthropod natural enemies through selection, hybridization or genetic engineering techniques. In M. A. Hoy & D. C. Herzog (Eds.), Biological control in agricultural IPM systems (pp. 167–187). Orlando, FL: Academic Press.

Bentley, J. W. (1992). Alternatives to pesticides in Central America: applied studies of local knowledge. Culture and Agriculture, 44, 10–13.

Bentley, J. W., & Andrews, K. L. (1991). Pests, peasants and publications: Anthropological and entomological views of an integrated pest management program for small-scale farmers. Human Organisation, 50, 113–124.

Bernays, E. A. (1985). Arthropods for weed control in IPM systems. In M. A. Hoy & D. C. Herzog (Eds.), Biological control in agricultural IPM systems (pp. 373–391). Orlando, FL: Academic Press.

Beshir, M. D., & Bennett, F. D. (1985). Biological control of water hyacinth on the White Nile, Sudan. In E. S. Delfosse (Ed.), Proceedings of the Sixth International Symposium on Biological Control of Weeds (pp. 491–496). August 19–25, 1984, Vancouver, Canada: Agriculture Canada.

Box, H. E. (1956). Battle against Venezuela's cane borer. I. Preliminary investigations and the launching of a general campaign. Sugar, 25–27, 30, 45.

Brown, A. H. D., & Burdon, J. J. (1983). Multilocus diversity in an outbreaking weed, Echium plantagineum L. Australian Journal of Biological Science, 36, 503–509.

Burdon, J. J., & Brown, A. H. D. (1986). Population genetics of Echium plantagineum L.—target weed for biological control. Australian Journal of Biological Science, 39, 369–378.

Burdon, J. J., & Marshall, D. R. (1981). Biological control and the reproductive mode of weeds. Journal of Applied Ecology, 18, 649–658.

Burdon, J. J., Marshall, D. R., & Groves, R. H. (1980). Isozyme variation in Chondrilla juncea L. in Australia. Australian Journal of Botany, 28, 193–198.

Caltagirone, L. E. (1985). Identifying and discriminating among biotypes of parasites and predators. In M. A. Hoy & D. C. Herzog (Eds.), Biological control in agricultural IPM systems (pp. 189–200). Orlando, FL: Academic Press.

Carlson, J., Olson, K., Higgs, S., & Beaty, B. (1995). Molecular genetic manipulation of mosquito vectors. Annual Review of Entomology, 40, 359–388.

Caprio, M. A., Hoy, M. A., & Tabashnik, B. E., (1991). Model for implementing a genetically improved strain of a parasitoid. American Entomologist, 37(4), 232–239.

Caprio, M. A., & Hoy, M. A., (1994). Metapopulation dynamics affect resistance development in a predatory mite. Journal of Economic Entomology, 87, 525–534.

Caprio, M., & Hoy, M. A. (1995). Premating isolation in a simulation model generates frequency-dependent selection and alters establishment rates of resistant natural enemies. Journal of Economic Entomology, 88(2), 205–212.

Chaboudez, P. (1989). Modes de reproduction et variabilité genetique des populations de Chondrilla juncea L.: Implications dans la lutte microbiologique contre cette mauvaise herbe. In édit doctorale thése, Université de Montpellier, Sciences et Techniques du Languedoc.

Chaboudez, P. (1994). Patterns of clonal variation in skeleton weed (Chondrilla juncea), an apomictic species. Australian Botanist, 42, 283–95.

Chaboudez, P., & Burdon, J. (in press). Frequency-dependent selection in a wild plant-pathogen system. Oecologia (Berlin),

Chaboudez, P., & Sheppard, A. W. (1995). Are particular weeds more amenable to biological control?—A reanalysis of mode of reproduction and life history. In Biological control of weeds, E. S. Delfosse, & R. R. Scott (Eds.), Proceedings of the Eighth International Symposium Biological Control of Weeds (pp. 95–102). February 2–7, 1992. Lincoln University, Canterbury, New Zealand, Melbourne, Australia: CSIRO Publications.

Charudattan, R. (1985). The use of natural and genetically altered strains of pathogens for weed control. In M. A. Hoy & D. C. Herzog (Eds.), Biological control in agricultural IPM systems (pp. 347–372). Orlando, FL: Academic Press.

Clarke, B. (1979). The evolution of genetic diversity. Proceedings of the Royal Society of London, Series B, 205, 434–474.

Cockburn, A. F., Howells, A. J., & Whitten, M. J. (1984). Recombinant DNA technology and genetic control of pest insects. Biotechnology Genetic Engineering Review, 2, 69–99.

Cofrancesco, A. F., Stewart, R. M., & Sanders, D. R., (1985). The impact of Neochetina eichorniae (Coleoptera: Curculionidae) on water hyacinth in Louisiana. In E. S. Delfosse (Ed.), Proceedings of the Sixth International Symposium on Biological Control of Weeds (pp. 525–535). August 19–25, 1984, Vancouver, Canada: Agriculture Canada:

Collyer, E. (1958). Some insectary experiments with predaceous mites to determine their effect on the development of Metatetranychus ulmi (Koch) populations. Entomologia Experimentalis et Applicata, 1, 138–146.

Collyer, E. (1964). A summary of experiments to demonstrate the role of Typhlodromus pyri (Scheuten) in the control of Panonychus ulmi Koch in England. Acarologia, 9, 363–371.

Compere, H. (1961). The red scale, Aonidiella aurantii (Mask.), and its insect enemies. Hilgardia 31, 173–278.

Croft, B. A. (1972). Resistant natural enemies in pest management systems. Span, 15, 19–22.

Croft, B. A. (1976). Establishing insecticide-resistant phytoseiid mites in deciduous tree fruit orchards. Entomophaga, 21, 383–399.

Croft, B. A., & Barnes, M. M. (1971). Comparative studies on four strains of Typhlodromus occidentalis. III. Evaluations of releases of insecticide-resistant strains into an apple orchard ecosystem. Journal of Economic Entomology, 64, 845–850.

Croft, B. A., & Brown, A. W. A. (1975). Responses of arthropod natural enemies to insecticides. Annual Review of Entomology, 20, 285–335.

Croft, B. A., & Hoyt, S. C. (1978). Consideration for the use of pyrethroid insecticides for deciduous fruit pest control in the USA. Environmental Entomology, 7, 627–630.

Croft, B. A., & Meyer, R. H. (1973). Carbamate and organophosphorous resistance patterns in populations of Amblyseius fallacis. Environmental Entomology, 2, 691–695.

Croft, B. A., & Strickler, K. (1983). Natural enemy resistance to pesticides. Documentation, characterization, theory and application. In G. P. Georghiou & T. Saito (Eds.), Pest resistance to pesticides (pp. 699–702). New York: Plenum Press.

Cullen, J. M. (1978). Evaluating the success of the program for the biological control of Chondrilla juncea L. In T. E. Freeman (Ed.), Proceedings of the Fourth International Symposium on Biological Control of Weeds (pp. 117–121). August 30–September 2, 1976, Gainesville, FL: University of Florida.

Cullen, J. M., & Whitten, M. J. (1995). Economics of biological control. In H. Hokkanen (Ed.), Proceedings of the OECD Workshop on Benefits and Risks of Introducing Biocontrol Agents (pp. 270–276). Saariselka, Finland: Cambridge University Press.

DeBach, P. (1958). Selective breeding to improve adaptations of parasitic insects. In Proceedings of the 10th International Congress of Entomology, 4, 759–768.

Devonshire, A. L., & Field, L. M. (1991). Gene amplification and insecticide resistance. Annual Review of Entomology, 36, 1–23.

Devonshire, A. L., & Sawicki, R. M. (1979). Insecticide-resistant Myzus persicae as an example of evolution by gene duplication. Nature (London), 280, 140–141.

Dodd, A. P. (1940). The biological campaign against prickly pear. Commonwealth-Prickly Pear Board. Brisbane, Australia, 177 pp.

Doeleman, J. A. (1989). Biological control of Salvinia molesta in Sri Lanka: An assessment of costs and benefits. (Australian Centre for International Agricultural Research (ACIAR) Tech. Rep. 12). Canberra, Australia: ACIAR.

Doutt, R. L. (1966). Studies of two parasites of olive scale, Parlatoria oleae (Colvée). I. A taxonomic analysis of parasitic Hymenoptera reared from Parlatoria oleae (Colvée). Hilgardia, 37, 219–231.

Edwards, D. R., & Hoy, M. A. (1993). Polymorphism in two parasitoids detected using random amplified polymorphic DNA polymerase chain reaction. Biological Control, 3, 243–257.

Edwards, O. R., & Hoy, M. A. (1995). Random amplified polymorphic DNA markers to monitor laboratory-selected, pesticide-resistant Trioxys pallidus (Hymenoptera: Aphidiidae) after release into three

California walnut orchards. Environmental Entomology, 24(3), 487–496.

Engels, W. R. (1989). P elements in Drosophila melanogaster. In Mobile DNA (pp. 437–484). Washington, DC: American Society for Microbiology.

Falconer, D. S. (1960). Introduction to quantitative genetics. Edinburgh: Oliver & Boyd.

Faulkner, P., & Boucias, D. G. (1985). Genetic improvement of insect pathogens: Emphasis on the use of baculoviruses. In M. A. Hoy & D. C. Herzog (Eds.), Biological control in agricultural IPM systems (pp. 263–281). Orlando, FL: Academic Press.

Field, R. P., & Hoy, M. A. (1986). Evaluation of genetically improved strains of Metaseiulus occidentalis (Nesbitt) (Acarina: Phytoseiidae) for integrated control of spider mites on roses in greenhouses. Hilgardia, 54(2), 1–32.

Flaherty, D. L., & Huffaker, C. B. (1970). Biological control of Pacific mites and Willamette mites in San Joaquin Valley vineyards. I. Role of Metaseiulus occidentalis. II. Influence of dispersion patterns of Metaseiulus occidentalis. Hilgardia, 40(1), 267–330.

Flaherty, D. L., & Hoy, M. A. (1971). Biological control of Pacific mites and Willamette mites in San Joaquin Valley vineyards. III. Role of tydeid mites. Researches on Population Ecology 13, 80–96.

Fleschner, C. A., Hall, J. C., & Ricker, D. E. (1955). Natural balance of mite pests in an avocado grove. California Avocado Society Yearbook, 39, 155–162.

Flor, H. H. (1956). The complementary genic systems in flax and flax rust. Advances in Genetics, 8, 29–54.

Forno, I. W., Harley, K. L. S. (1979). The occurrence of Salvinia molesta in Brazil. Aquatic Botany, 6, 185–187.

Forno, I. W., Sands, D. P. A., & Sexton, W. (1983). Distribution, biology and host specificity of Cyrtobagous singularis Hustache (Coleoptera: Curculionidae) for the biological control of Salvinia molesta. Bulletin of Entomological Research, 73, 85–95.

Fournier, D., Pralavorio, M., Trottin-Caudal, Y., Coulon, J., Malezieux, S., & Berge, J. B. (1987a). Selection artificielle pour la resistance au methidathion chez Phytoseiulus persimilis A.-H. Entomophaga, 32(2), 209–219.

Fournier, D., Cuany, A., Pralavorio, M. J., & Berge, J. B. (1987b). Analysis of methidathion resistance mechanisms in Phytoseiulus persimilis A.-H. Pesticide Biochemistry and Physiology, 28, 271–278.

Gaugler, R. (1987). Entomogenous nematodes and their prospects for genetic improvement, In K. Maramorosch (Ed.), Biotechnology in invertebrate pathology and cell culture (pp. 457–484). New York: Academic Press.

Gilkeson, L. A., & Hill, S. B. (1986). Genetic selection for and evaluation of nondiapause lines of predatory midge, Aphidoletes aphidimyza (Rondani) (Diptera: Cecidomyiidae). Canadian Entomologist, 118, 869–879.

Goeden, R. D. (1971). Insect ecology of silverleaf nightshade. Weed Science, 19, 45–51.

Goeden, R. D. (1978). Biological control of weeds. In C. P. Clausen (Ed.), Introduced parasites and predators of arthropod pests and weeds. Agricultural Handbook No. 480 (pp. 357–414). Washington, DC U. S. Department of Agriculture.

Grafton-Cardwell, E. E., & Hoy, M. A. (1985). Intraspecific variability in response to pesticides in the common green lacewing, Chrysoperla carnea (Stephens) (Neuroptera: Chrysopidae). Hilgardia, 53(6), 1–32.

Grafton-Cardwell, E. E., & Hoy, M. A. (1986). Genetic improvement of common green lacewing, Chrysoperla carnea (Neuroptera: Chrysopidae): selection for carbaryl resistance. Environmental Entomology, 15(6), 1130–1136.

Hagen, K. S., Viktorov, G. A., Yasumatsu, K., & Shuster, M. F. (1976). Range, forage and grain crops. In C. B. Huffaker & P. S. Messenger (Eds.), Theory and practice of biological control (pp. 397–442). New York: Academic Press.

Handler, A. M., & O'Brochta, D. A. (1991). Prospects for gene transformation in insects. Annual Review of Entomology 36, 159–183.

Hasan, S., Chaboudez, P., & Espiau, C. (1995). Isozyme patterns and susceptibility of North American forms of Chondrilla juncea to European strains of the rust fungus, Puccinia chondrillina. In E. S. Delfosse & R. R. Scott (Eds.), Eighth International Symposium for Control of Weeds (pp. 367–373). February 2–7, 1992, Lincoln University Canterbury, New Zealand Melbourne, Australia: CSIRO Publications.

Havron, A., Rosen, D., Prag, H., & Rossler, Y. (1991a). Selection for pesticide resistance in Aphytis. I. A. holoxanthus, a parasite of the Florida red scale. Entomologia Experimentalis et Applicata, 61, 221–228.

Havron, A., Kenan, G., & Rosen, D. (1991b). Selection for pesticide resistance in Aphytis. II. A. lingnanensis, a parasite of the California red scale. Entomologia Experimentalis et Applicata 61, 229–235.

Headley, J. C., & Hoy, M. A. (1987). Benefit/cost analysis of an integrated mite management program for almonds. Journal of Economic Entomology, 80(3), 555–559.

Hoelzel, A. R. (Ed.). (1992). Molecular genetic analyses of populations: A practical approach. Oxford: IRL Press, Oxford University Press.

Hopper, K. R., Roush, R. T., & Powell, W. (1993). Management of genetics of biological control introductions. Annual Review of Entomology, 38, 27–51.

Hoy, M. A. (1975a). Forest and laboratory evaluations of hybridized Apanteles melanoscelus (Hym.: Braconidae), a parasitoid of Porthetria dispar (Lep.: Lymantriidae). Entomophaga, 20, 261–268.

Hoy, M. A. (1975b). Hybridization of strains of the gypsy moth parasitoid, Apanteles melanoscelus, and its influence upon diapause. Annals of the Entomological Society of America, 68, 261–264.

Hoy, M. A. (1979). Genetic improvement of insects: Fact or fantasy. Environmental Entomology, 5, 833–839.

Hoy, M. A. (1984). Genetic improvement of a biological control agent: Multiple pesticide resistance and nondiapause in Metaseiulus occidentalis (Nesbitt) (Phytoseiidae). In D. Griffiths & C. Bowman (Eds.), Acarology 6 (Vol. 2, pp. 673–679). Chichester: Ellis Horwood.

Hoy, M. A. (1985a). Almonds (California): Integrated mite management for Californian almond orchards. In W. Helle & M. Sabelis (Eds.), Spider mites, their biology, natural enemies and control (Vol. 2, pp. 299–310). Amsterdam: Elsevier.

Hoy, M. A. (1985b). Recent advances in genetics and genetic improvement of the Phytoseiidae. Annual Review of Entomology, 30, 345–370.

Hoy, M. A. (1985c). Improving establishment of arthropod natural enemies. In M. A. Hoy & D. C. Herzog (Eds.), Biological control in agricultural IPM systems (pp. 151–166). Orlando, FL: Academic Press.

Hoy, M. A. (1987). Developing insecticidal resistance in insect and mite predators and opportunities for gene transfer. In H. M. LeBaron, R. O. Mumma, R. C. Honeycutt, & J. H. Duesing (Eds.), Biotechnology in agriculture chemistry (pp. 125–138). American Chemistry Society Symposium Series 334. Washington, DC: American Chemical Society.

Hoy, M. A. (1990). Pesticide resistance in arthropod natural enemies: Variability and selection responses. In R. T. Roush & B. E. Tabashnik (Eds.), Pesticide resistance in arthropods (pp. 203–236). New York: Chapman & Hall.

Hoy, M. A. (1991). Genetic improvement of phytoseiids: In theory and practice. In F. Dusbabek & V. Bukva (Eds.), Modern acarology (Vol. 1, pp. 175–184). Prague: Academia, and The Hague: SPB Academic Publishers.

Hoy, M. A. (1992a). Commentary: Biological control of arthropods: Genetic engineering and environmental risks. Biological Control, 2, 166–170.

Hoy, M. A. (1992b). Criteria for release of genetically-improved phytoseiids: An examination of the risks associated with release of biological control agents. Experimental and Applied Acarology (Northwood), 14, 393–416.

Hoy, M. A. (1992c). Genetic engineering of predators and parasitoids for pesticide resistance. In I. Denholm, A. L. Devonshire, & D. W. Hollomon (Eds.), Resistance 91, achievements and developments in combating pesticide resistance (pp. 307–324). London: Elsevier Applied Science.

Hoy, M. A. (1993). Transgenic beneficial arthropods for pest management programs: An assessment of their practicality and risks. In R. D. Lumsden & J. Vaughn (Eds.), Pest management: Biologically based technologies (pp. 357–369). Washington, DC: American Chemical Society.

Hoy, M. A. (1994a). Insect molecular genetics. An introduction to principles and applications. San Diego: Academic Press. 546 pp.

Hoy, M. A. (1994b). Transgenic pests and beneficial arthropods for pest management programs: An assessment of their practicality and risks. In D. Rosen, F. D. Bennett, & J. L. Capinera (Eds.), Pest management in the subtropics: Biological control, a Florida perspective (pp. 641–670). Andover, United Kingdom: Intercept.

Hoy, M. A. (1995). Impact of risk analyses on pest management programs employing transgenic arthropods. Parasitology Today, 11(6), 229–232.

Hoy, M. A. (1996). Novel arthropod biological control agents. In G. J. Persley (Ed.), Biotechnology and integrated pest management. Proceedings of a Bellagio Conference on Biotechnology for Integrated Pest Management (pp. 164–185). October 1993, Lake Como, Italy, Wallingford: CAB International.

Hoy, M. A., & Cave, F. E. (1988). Guthion-resistant strain of walnut aphid parasite. California Agriculture, 42(4), 4–6.

Hoy, M. A., & Cave, F. E. (1989). Toxicity of pesticides used on walnuts to a wild and azinphosmethyl-resistant strain of Trioxys pallidus (Hymenoptera: Aphidiidae). Journal of Economic Entomology, 82, 1585–1592.

Hoy, M. A., & Cave, F. E. (1991). Genetic improvement of a parasitoid: Response of Trioxys pallidus to laboratory selection with azinphosmethyl. Biocontrol Science and Technology, 1, 31–41.

Hoy, M. A., & Conley, J. (1987). Selection for abamectin resistance in Tetranychus urticae and T. pacificus (Acari: Tetranychidae). Journal of Economic Entomology, 80(1), 221–225.

Hoy, M. A., & Knop, N. F. (1979). Studies on pesticide resistance in the phytoseiid Metaseiulus occidentalis in California. In R. Rodriguez (Ed.), Recent advances in acarology (pp. 89–94). New York: Academic Press.

Hoy, M. A., & Knop, N. F. (1981). Selection for and genetic analysis of permethrin resistance in Metaseiulus occidentalis: Genetic improvement of a biological control agent. Entomologia Experimentalis et Applicata, 30, 10–18.

Hoy, M. A., & Ouyang, Y. L. (1989). Selection of the western predatory mite, Metaseiulus occidentalis (Acari: Phytoseiidae), for resistance to abamectin. Journal Economic Entomology, 82, 35–40.

Hoy, M. A., Cave, F. E., Beede, R., Grant, J., Krueger, W., & Olson, W. (1989). Guthion-resistant walnut aphid parasite. Release, dispersal, and recovery in orchards. California Agriculture, 43(5), 21–23.

Hoy, M. A., Cave, F. E., Beede, R., Grant, J., Krueger, W., & Olson, W. (1990). Release, dispersal, and recovery of a laboratory-selected strain of the walnut aphid parasite Trioxys pallidus (Hymenoptera: Aphidiidae) resistant to azinphosmethyl. Journal of Economic Entomology, 83, 89–96.

Hoy, M. A., Gaskalla, R. D., Capinera, J. L., & Keierleber, C. (1997). Forum: Laboratory containment of transgenic arthropods. American Entomologist, 43(4), 206–209, 255–256.

Hoyt, S. C. (1969a). Integrated chemical control of insects and biological control of mites on apple in Washington. Journal of Economic Entomology, 62, 74–86.

Hoyt, S. C. (1969b). Population studies of five mite species on apple in Washington. Proceedings of the Second International Congress on Acarology (pp. 117–133). Sutton-Bonningon, England, 1967. Budapest: Akademie Kiado.

Hoyt, S. C., & Caltagirone, L. E. (1971). The developing programs of integrated control of pests of apples in Washington and peaches in California. In C. B. Huffaker (Ed.), Biological control (pp. 395–421). New York: Plenum Press.

Huang, M. D., Xiong, J. J., & Du, T. Y. (1987). The selection for and genetical analysis of phosmet resistance in Amblyseius nicholsi. Acta Entomologica Sinica, 30(2), 133–139.

Huffaker, C. B. (1971). The ecology of pesticide interference with insect populations. In J. E. Swift (Ed.), Agricultural chemicals: Harmony or discord for food, people and the environment (pp. 92–104). Berkeley: University of California Division of Agricultural Science.

Huffaker, C. B., van de Vrie, M., & McMurtry, J. A. (1970). Ecology of tetranychid mites and their natural enemies: A review. II. Tetranychid populations and their possible control by predators: An evaluation. Hilgardia, 40(11), 391–458.

Hughes, P. D., Woolcock, L. T., Roberts, J. A., & Hughes, M. A. (1987). Biological control of the spotted alfalfa aphid, Therioaphis trifolii F. maculata on lucerne crops in Australia, by the introduced parasitic hymenopteran Trioxys complanatus. Journal of Applied Ecology, 24, 515–537.

Hull, L. A., & Beers, E. H. (1985). Ecological selectivity: Modifying chemical control practices to preserve natural enemies. In M. A. Hoy & D. C. Herzog (Eds.), Biological control in agricultural IPM systems (pp. 103–122). Orlando, FL: Academic Press.

Hull, V. J., & Groves, R. H. (1973). Variation in Chondrilla juncea L. in south-eastern Australia. Australian Journal of Botany, 21, 113–135.

Julien, M. H. (1981). Control of aquatic Alternanthena philoxeroides in Australia: Another success for Agasicles hygrophila. In E. S. Delfosse (Ed.), Proceedings of the Fifth International Symposium on Biological Control of Weeds (pp. 583–588). July 22–27, 1980. Melbourne, Australia: CSIRO.

Julien, M. H. (1998). (Ed.). Biological Control of weeds: a world catalogue of agents and their targets. 4th ed.

Kenmore, P. E., Carino, F. O., Perez, C. A., Dyck, V. A., & Gutierrez, A. P. (1984). Population regulation of the rice brown planthopper (Nilaparvata lugens Stål) within rice fields in the Philippines Journal of Plant Protection of Tropics, 1(1), 19–37.

Kirschbaum, J. B. (1985). Potential implication of genetic engineering and other biotechnologies to insect control. Annual Review of Entomology, 30, 51–70.

Kostiainen, T., & Hoy, M. A. (1994). Genetic improvement of Amblyseius finlandicus (Acari: Phytoseiidae): Laboratory selection for resistance to azinphosmethyl and dimethoate. Experimental and Applied Acarology, 18, 469–484.

Kostiainen, T., & Hoy, M. A. (1995). Laboratory evaluation of a laboratory-selected OP-resistant strain of Amblyseius finlandicus (Acari: Phytoseiidae) for possible use in Finnish apple orchards. Biocontrol Science and Technology, 5(3), 297–311.

Landaluz, P. U. (1950). Aplicacion de la genetica al aumento de la eficacia del Trichogramma minutum en la luch biologica. Boletin Patologia Vegetal Entomologia Agricola, 18(1–2), 1–12.

Laski, F. A., Rio, D. C., & Rubin, G. M. (1986). Tissue specificity of Drosophila P element transposition is regulated by the level of mRNA splicing. Cell, 44, 7–19.

Legner, E. F. (1987). Transfer of thelytoky to arrhenotokous Muscidifurax raptor Girault & Sanders (Hymenoptera: Pteromalidae). Canadian Entomologist, 119, 265–271.

Li, J., & Hoy, M. A. (1996). Adaptability and efficacy of transgenic and wild-type Metaseiulus occidentalis (Acari: Phytoseiidae) compared as part of a risk assessment. Experimental and Applied Acarology (Northwood), 20, 563–574.

Lindow, S. E. (1985). Foliar antagonists: Status and prospects. In M. A. Hoy & D. C. Herzog (Eds.), Biological control in agricultural IPM systems (pp. 395–413). Orlando, FL: Academic Press.

Lindow, S. E., Panopoulos, N. J., & McFarland, B. L. (1989). Genetic engineering of bacteria from managed and natural habitats. Science, 244, 1300–1307.

Lonsdale, W. M., & Segura, R. (1987). A demographic comparison of native and introduced populations of Mimosa pigra. In Proceedings of the Eight Australian Weeds Conference (pp. 163–166). Sydney.

Loukeris, T. G., Arca, B., Livaderas, I., Dialektaki, G., & Savakis, C. (1995a). Introduction of the transposable element Minos into the germ line of Drosophila melanogaster. Proceedings of the National Academy of Sciences, USA, 92, 9485–9489.

Loukeris, T. G., Livadaras, I., Arca, B., Zabalou, S., & Savakis, C. (1995b). Gene transfer into the medfly, Ceratitis capitata, with a Drosophila hydei transposable element. Science, 270, 2002–2005.

Luckow, V. A., & Summers, M. D. (1988). Trends in the development of baculovirus expression vectors. Biotechnology, 5, 47–55.

Luthy, P. (1986). Genetics and aspects of genetic manipulation of Bacillus thuringiensis. Mitteilungen ausder Biologischen Bundesanstalt fuer Landund -Forstwirtschaft, Berlin–Dahlem, 233, 97–110.

MacPhee, A. W., Caltagirone, L. E., van de Vrie, M., & Collyer, E. (1976). Biological control of pests of temperate fruits and nuts. In C. B. Huffaker & P. S. Messenger (Eds.), Theory and practice of biological control (pp. 337–358). New York: Academic Press.

Markwick, N. G. (1986). Detecting variability and selecting for pesticide resistance in two species of phytoseiid mites. Entomophaga, 31, 225–236.

Marsden, J. S., Martin, G. E., Parham, D. J., Ridsdill-Smith, T. J., & Johnston, B. G. (1980). Returns on Australian agricultural research. The joint Industries Assistance Commission—CSIRO benefit-cost study of the CSIRO Division of Entomology, Canberra, Australia.

Martin, S. B., Abawi, G. S., & Hoch, H. C. (1985). Biological control of soilborne pathogens with antagonists. In M. A. Hoy & D. C. Herzog (Eds.), Biological control in agricultural IPM systems (pp. 433–454). Orlando, FL: Academic Press.

Maruyama, K., & Hartl, D. L. (1991). Evidence for interspecific transfer of the transposable element mariner between Drosophila and Zaprionus. Journal of Molecular Evolution, 33, 514–524.

McDermott, G. J., & Hoy, M. A. (1997). Persistence and containment of Metaseiulus occidentalis (Acari: Phytoseiidae) in Florida: Risk assessment for possible releases of transgenic strains. Florida Entomologist, 80, 42–53.

McKenzie, J. A. (1996). Ecological and evolutionary aspects of insecticide resistance. Austin, TX: R. G. Landes and Academic Press.

McMurtry, J. A. (1983). Phytoseiid predators in orchard systems: A classical biological control success story. In M. A. Hoy, G. L. Cunningham, & L. Knutson (Eds.), Biological control of pests by mites (Special Publication 3304, pp. 21–26). Berkeley, CA: University of California.

McMurtry, J. A., Huffaker, C. B., & van de Vrie, M. (1970). Ecology of tetranychid mites and their natural enemies: A review. I. Tetranychid enemies: Their biological characters and the impact of spray practices. Hilgardia, 40(11), 331–390.

Messenger, P. S., & van den Bosch, R. (1971). The adaptability of introduced biological control agents. In C. B. Huffaker (Ed.), Biological control (pp. 68–92). New York: Plenum.

Mouches, C., Pasteur, N., Berge, J. B., Hyrien, O., Raymond, M., & de Saint Vincent, B. R., (1986). Amplification of an esterase gene is responsible for insecticide resistance in a California Culex mosquito. Science, 233, 778–780.

Napoli, C., & Staskawicz, B. (1985). Molecular genetics of biological control agents of plant pathogens: Status and prospects. In M. A. Hoy & D. C. Herzog (Eds.), Biological control in agricultural IPM systems (pp. 455–463). Orlando, FL: Academic Press.

Oakeshott, J., & Whitten, M. (Eds.). (1991). Molecular approaches to pure and applied entomology. New York: Springer-Verlag.

O'Brochta, D. A., & Atkinson, P. W. (1997). Recent developments in transgenic insect technology. Parasitology Today, 13, 99–104.

O'Brochta, D. A., & Handler, A. M. (1988). Mobility of P elements in drosophilids and nondrosophilids. Proceedings of the National Academy of Sciences, USA, 85, 6052–6056.

O'Brochta, D. A., Warren, W. D., Saville, K. J., & Atkinson, P. W. (1994). Interplasmid transposition of Drosophila hobo elements in non-drosophilid insects. Molecular General Genetics, 244, 9–14.

O'Brochta, D. A., Warren, W. D., Saville, K. J., & Atkinson, P. W. (1996). Hermes, a functional non-drosophilid insect gene vector from Musca domestica. Genetics, 142, 907–914.

Parsons, P. A. (1983). The evolutionary biology of colonising species. Cambridge: Cambridge University Press.

Phillips, J. P., Xin, J. H., Kirby, K., Milne, C. P., Jr., Krell, P., & Wild, J. R. (1990). Transfer and expression of an organophosphate insecticide-degrading gene from Pseudomonas in Drosophila melanogaster. Proceedings of the National Academy of the Sciences, USA, 87, 8155–8159.

Pielou, D. P., & Glasser, R. F. (1952). Selection for DDT resistance in a beneficial insect parasite. Science, 115, 117.

Pintureau, B. (1991). Selection de deux caracteres chez une espece de Trichogrammes, efficacite parasitaire des souches obtenues (Hym.: Trichogrammatidae). Agronomie, 11, 593–602.

Presnail, J. K., & Hoy, M. A. (1992). Stable genetic transformation of a beneficial arthropod, Metaseiulus occidentalis (Acari: Phytoseiidae), by a microinjection technique. Proceedings of the National Academy of the Sciences, USA, 89, 7732–7736.

Presnail, J. K., & Hoy, M. A. (1994). Transmission of injected DNA sequences to multiple eggs of Metaseiulus occidentalis and Amblyseius finlandicus (Acari: Phytoseiidae) following maternal microinjection. Experimental and Applied Acarology (Northwood), 18, 319–330.

Presnail, J. K., & Hoy, M. A. (1996). Maternal microinjection of the endoparasitoid Cardiochiles diaphaniae (Hymenoptera: Braconidae). Annals of the Entomological Society of America, 89(4), 576–580.

Presnail, J. K., Jeyaprakash, A., Li, J., & Hoy, M. A. (1997). Genetic analysis of four lines of Metaseiulus occidentalis (Nesbitt) (Acari: Phytoseiidae) transformed by maternal microinjection. Annals of the Entomological Society of America, 90, 237–245.

Pursel, V. G., Pinkert, C. A., Miller, K. F., Bolt, D. J., Campbell, R. G., & Palmiter, R. D. (1989). Genetic engineering of livestock. Science, 244, 1281–1288.

Ram, A., & Sharma, A. K. (1977). Selective breeding for improving the fecundity and sex ratio of Trichogramma fasciatum (Perkins) (Trichogrammatidae: Hymenoptera) an egg parasite of lepidopterous hosts. Entomologia, 2, 133–137.

Rancourt, D. E., Walker, V. K., & Davies, P. L. (1987). Antifreeze protein gene expression in transgenic Drosophila. Molecular and Cell Biology, 7, 2188–2195.

Readshaw, J. L. (1975). Biological control of orchard mites in Australia with an insecticide-resistant predator. Journal of the Australian Institute of Agricultural Science, 41, 213–214.

Robertson, J. G. (1957). Changes in resistance to DDT in Macrocentrus ancylivorus Rohw. (Hymenoptera: Braconidae). Canadian Journal of Zoology, 35, 629–633.

Room, R. M., Harley, H. L. S., Forno, I. W., & Sands, D. P. (1981). Successful biological control of the floating weed salvinia. Nature (London), 292, 78–80.

Room, P. M., Sands, D. P. A., Forno, I. W., Taylor, M. F. J., & Julien, M. H. (1985). A summary of research into biological control of salvinia in Australia. In E. S. Delfosse (Ed.), Proceedings of the Sixth International Symposium on Biological Control of Weeds (pp. 543–549) August 19–25, 1984, Vancouver, Canada: Agriculture Canada.

Rosen, D. (1980). Integrated control of citrus pests in Israel. In Proceeding of the International Symposium of IOBC/WPRS, Integrated Control in Agriculture and Forestry (pp. 289–292). 1979, Vienna, Austria.

Rosen, D. (1985). Biological control. In G. A. Kerkut & L. I. Gilbert (Eds.), Insect physiology, biochemistry and pharmacology. Comprehensive (Vol. 12, pp. 413–465). Oxford: Pergamon Press.

Rosen, D., & Debach, P. (1978). Diaspididae. In C.P. Clausen (Ed.), Introduced parasites and predators of arthropod pests and weeds: A world review. Agricultural Handbook 480 (pp. 78–128). Washington, DC: U.S. Department of Agriculture.

Rosenheim, J. A., & Hoy, M. A., (1986). Intraspecific variation in levels of pesticide resistance in field populations of a parasitoid, Aphytis melinus (Hymenoptera: Aphelinidae): The role of past selection pressures. Journal of Economic Entomology, 79, 1161–1173.

Rosenheim, J. A., & Hoy, M. A. (1988). Genetic improvement of a parasitoid biological control agent: Artificial selection for insecticide resistance in Aphytis melinus (Hymenoptera: Aphelinidae). Journal of Economic Entomology, 81, 1539–1550.

Roush, R. T. (1990). Genetic variation in natural enemies: Critical issues for colonisation in biological control. In M. Mackauer, L. E. Ehler, & J. Roland (Eds.), Critical issues in biological control (pp. 263–288). Hants, United Kingdom: Intercept.

Roush, R. T., & Hopper, K. R. (1995). Use of single family lines to preserve genetic variation in laboratory colonies. Annals of the Entomological Society of America, 88(6), 713–717.

Roush, R. T., & Hoy, M. A. (1981a). Genetic improvement of Metaseiulus occidentalis: Selection with methomyl, dimethoate, and carbaryl and genetic analysis of carbaryl resistance. Journal of Economic Entomology, 74, 138–141.

Roush, R. T., & Hoy, M. A. (1981b). Laboratory, glasshouse, and field studies of artificially selected carbaryl resistance in Metaseiulus occidentalis. Journal of Economic Entomology, 74, 142–147.

Roush, R. T., & McKenzie, J. A. (1987). Ecological genetics of insecticide and acaricide resistance. Annual, Review of Entomology, 32, 361–380.

Rubin, G. M., Spradling, A. C. (1982). Genetic transformation of Drosophila with transposable element vectors. Science, 218, 348–353.

Sailer, R. I. (1961). Possibilities for genetic improvement of beneficial insects. In Germ plasm resources (pp. 295–303). Washington, DC: American Association for the Advancement of Science (AAAS).

Sands, D. P. A., & Shotz, M. (1985). Control or no control: A comparison of the feeding strategies of two salvinia weevils. In E. S. Delfosse (Ed.), Proceedings of the Sixth International Symposium on Biological Control of Weeds (pp. 551–556). August 19–25, 1984, Vancouver, Canada: Agriculture Canada.

Sands, D. P. A., Schotz, M., & Bourne, A. S. (1983). The feeding characteristics and development of larvae of a salvinia weevil Cyrtobagous sp. Entomologia Experimentalis et Applicata, 34, 291–296.

Schroth, M. N., & Hancock, J. G. (1985). Soil antagonists in IPM systems. In M. A. Hoy & D. C. Herzog (Eds.), Biological control in agricultural IPM systems (pp. 425–431). Orlando, FL: Academic Press.

Schulten, G. G. M., & van de Klashorst, G. (1974). Genetics of resistance to parathion and demeton-S-methyl in Phytoseiulus persimilis A.-H. (Acari: Phytoseiidae). (pp. 519–524). In Proceedings of the Fourth International Congress of Acarology, 1974. Budapest: Akademiai Kiado.

Settle, W. H., Ariawan, H., Astuti, E. T., Cahyana, W., Hakim, A. L., & Hindayana, D. (1996). Managing tropical rice pests through conservation of generalist natural enemies and alternative prey. Ecology, 77(7), 1975–1988.

Sheppard, A. W., Cullen, J. M., Aeschlimann, J.-P., Sagliocco, J.-L., & Vitou, J. (1989). The importance of insect herbivores relative to other limiting factors on weed population dynamics: A case study of Carduus nutans L. In E. S. Delfosse (Ed.), Proceedings of the Seventh International Symposium on Biological Control of Weeds (pp. 227–234). March 6–11, 1988. Rome, Italy: 1st Sper. Patol. Veg. (MAF).

Simmonds, F. J., J. M., Franz, & Sailer, R. I. (1976). History of biological control. In C. B. Huffaker & P. S. Messenger (Eds.), Theory and practice of biological control (pp. 17–39). New York: Academic Press.

Spollen, K. M., & Hoy, M. A. (1992a). Carbaryl resistance in a laboratory-selected strain of Aphytis melinus DeBach (Hymenoptera: Aphelinidae): Mode of inheritance and implications for implementation in citrus IPM. Biological Control and Theories of Applied Pest Management, 2, 211–217.

Spollen, K. M., & Hoy, M. A. (1992b). Genetic improvement of an arthropod natural enemy: Relative fitness of a carbaryl-resistant strain of the California red scale parasite Aphytis melinus DeBach. Biological Control, 2, 87–94.

Spollen, K. M., & Hoy, M. A. (1992c). Resistance to carbaryl in populations of Aphytis melinus DeBach (Hymenoptera: Aphelinidae): Implications for California citrus IPM. Biological Control and Theories of Applied Pest Management, 2, 201–210.

Spollen, K. M., & Hoy, M. A. (1993a). Laboratory-selected California red scale parasite is resistant to Sevin. California Agriculture, 47(1), 16–19.

Spollen, K. M., & Hoy, M. A. (1993b). Residual toxicity of five citrus pesticides to acarbaryl-resistant and a wild strain of the California red scale parasite Aphytis melinus DeBach (Hymenoptera: Aphelinidae). Journal of Economic Entomology, 86(2), 195–204.

Spradling, A. C., & Rubin, G. M., (1982). Transposition of cloned P elements into Drosophila germ line chromosomes. Science, 218, 341–347.

Strickler, K. A., & Croft, B. A. (1981). Variation in permethrin and azinphosmethyl resistance in populations of Amblyseius fallacis (Acarina: Phytoseiidae). Environmental Entomology, 10(2), 233–236.

Strickler, K. A., & Croft, B. A. (1982). Selection for permethrin resistance in the predatory mite Amblyseius fallacis. Entomologia Experimentalis et Applicata, 31, 339–345.

Suckling, D. M., Walker, J. T. S., Shaw, P. W., Markwick, N. G., & Wearing, C. H. (1988). Management of resistance in horticultural pests and beneficial species in New Zealand. Pesticide Science, 23, 157–164.

Symondson, W. O. C., & Liddell, J. E. (Eds.). (1996). The ecology of agricultural pests: Biochemical approaches. London: Chapman & Hall.

Szmidt, A. (1972). Studies on the efficiency of various strains of the parasite Dahlbominus fuscipennis (Zett.) (Hymenoptera, Chalcidoidea) under natural conditions. Ekologia Polska, 20, 299–313.

Taylor, O. R. (1985). African bees: Potential impact in the United States. Bulletin of the Entomological Society of America, 31(4), 15–24.

Thomas, P. A., & Room, M. (1985). Towards biological control of salvinia in Papua New Guinea. In E. S. Delfosse (Ed.), Proceedings of the Sixth International Symposium on Biological Control of Weeds (pp. 567–574). August 19–25, 1984, Vancouver, Canada: Agriculture Canada.

Turnock, W. J., Taylor, K. L., Schroder, D., & Dahlsten, D. L. (1976). Control; Biological control of pests of coniferous forests. In C. B. Huffaker & P. S. Messenger (Eds.), Theory and practice of biological control (pp. 17–39; 289–311). New York: Academic Press.

Urquijo, P. (1956). Selection des estirpes de Trichogramma minutum Riley de maxima effectividad parasitaria. Boletin Patologia Vegetal Entomologia Agricola, 14, 199–216.

van de Vrie, M., McMurtry, J. A., & Huffaker, C. B. (1972). Ecology of tetranychid mites and their natural enemies: A review. III. Biology, ecology and pest status and host-plant relations of tetranychids. Hilgardia, 41(13), 343–432.

van Gundy, S. D. (1985). Biological control of nematodes: Status and prospects in agricultural IPM systems. In M. A. Hoy & D. C. Herzog (Eds.), Biological control in agricultural IPM systems (pp. 467–478). Orlando, FL: Academic Press.

Voroshilov, N. V. (1979). Heat-resistant lines of the mite Phytoseiulus persimilis A.-H. Genetika, 15(1), 70–76.

Voroshilov, N. V., & Kolmakova, L. I. (1977). Heritability of fertility in a hybrid population of *Phytoseiulus*. Genetika, 13(8), 1495–1496.

Walker, V. K. (1989). Gene transfer in insects. In K. Maramorosch & G. H. Sato (Eds.), Advances in cell culture (Vol. 7, pp. 87–124). New York: Academic Press.

Wapshere, A. J. (1988). Prospects for the biological control of silver-leaf nightshade, *Solanum elaeagnifolium,* in Australia. Australian Journal of Agricultural Research, 39, 187–97.

Wapshere, A. J., Delfosse, E. S., & Cullen, J. S. (1989). Recent developments in biological control of weeds. Crop Protection, 8, 27–250.

Waters, W. E., Drooz, A. T., & Pschorn-Walcher, H. (1976). Biological control of pests of broad-leaved forests and woodlands. In C. B. Huffaker & P. S. Messenger (Eds.), Theory and practice of biological control (pp. 313–336). New York: Academic Press.

Weseloh, R. M. (1986). Artificial selection for host suitability and development length of the gypsy moth (Lepidoptera: Lymantriidae) parasite, *Cotesia melanoscela* (Hymenoptera: Braconidae). Journal of Economic Entomology, 79, 1212–1216.

White, E. B., DeBach, P., & Garber, M. J. (1970). Artificial selection for genetic adaptation to temperature extremes in *Aphytis lingnanensis* Compere (Hymenoptera: Aphelinidae). Hilgardia, 40, 161–192.

Whitten, M. J. (1986). Molecular biology—its relevance to pure and applied entomology. Annals of the Entomological Society of America, 79(5), 766–772.

Whitten, M. J., & McKenzie, J. A. (1982). The genetic basis for pesticide resistance. In K. E. Lee (Ed.), Proceedings of the Third Australasian Conference on Grassland Invertebrate Ecology (pp. 1–16). Adelaide, November 30–December 4, 1981. Adelaide: S.A. Government Printer.

Whitten, M. J., Jefferson, R. A., & Dall, D. (1996). Needs and opportunities. In G. J. Persley (Ed.), Biotechnology and integrated pest management. Proceedings of a Bellagio Conference on Biotechnology for Integrated Pest Management (pp. 1–36). October 1993, Lake Como, Italy, Wallingford: CAB International.

Whitten, M. J., & Settle, W. H. (in press). The role of small scale farmers in preserving the link between biodiversity and sustainable agriculture. In Frontiers in biology: The challenges of biodiversity, biotechnology and sustainable agriculture, Proceedings of the 26th General Assembly IUBS. Taipei, Taiwan November 17–22, 1997, Taipei: Academia Sinica.

Wilkes, A. (1942). The influence of selection on the preferendum of a chalcid (*Microplectron fuscipennis* Zett.) and its significance in the biological control of an insect pest. Proceedings of the Royal Society of London, Series B, 130, 400–415.

Wilkes, A. (1947). The effects of selective breeding on the laboratory propagation of insect parasites. Proceedings of the Royal Society of London, Series B, 134, 227–245.

Wright, A. D. (1981). Biological control of water hyacinth in Australia. In E. S. Delfosse (Ed.), Proceedings of the Fifth International Symposium on Biological Control of Weeds (pp. 529–535). July 22–27, 1980, Melbourne, Australia: CSIRO.

Xu, X., Li, K. H., Li, Y. F., Moon, Z., & Li, L. Y. (1986). Culture of resistant strain of *Trichogramma japonicum* to pesticides. Natural Enemies of Insects, 8(3), 150.

Yoder, O. C., Weltring, K., Turgeon, B. G., Garber, R. C., & Vanetten, H. D. (1987). Prospects for development of molecular technology for fungal insect pathogens. In K. Maramorosch (Ed.), Biotechnology in invertebrate pathology and cell culture (pp. 197–218). New York: Academic Press.

Zwölfer, H. (1988). Evolutionary and ecological relationships of the insect fauna of thistles. Annual Review of Entomology, 33, 103–122.

13

Enhanced Biological Control through Pesticide Selectivity

MARSHALL W. JOHNSON

Department of Entomology
University of Hawaii at Manoa
Honolulu, Hawaii

BRUCE E. TABASHNIK

Department of Entomology
University of Arizona
Tucson, Arizona

INTRODUCTION

Complete biological control of crop pests and subsequent elimination of pesticide use has not always been achieved in agroecosystems. In multipest systems, successful biological control of individual species often requires development of integrated pest management (IPM) programs to control those pest species not adequately checked by natural enemies (Wilson & Huffaker, 1976). Usually, the resulting management schemes include pesticides. However, biological and chemical controls are often incompatible because pesticides severely reduce natural-enemy populations. Thus, insecticide applications can disrupt biological control and may cause outbreaks of secondary pests previously suppressed by natural enemies (Bartlett, 1964). The detrimental effects of pesticides are compounded when pest species develop pesticide resistance but natural enemies do not (Chapter 24). Any pesticide unfavorable to parasites and predators may be expected to show decreased effectiveness with prolonged use (Nicholson, 1940).

Numerous reviews and articles have been published on the impact of pesticides on biological control (Nicholson, 1940; Ripper, 1956; Bartlett, 1964; Metcalf, 1986; Croft, 1990a, 1990b; Messing & Croft, 1990a, 1990b; Hardin *et al.*, 1995; van Emden & Peakall, 1996), insecticide resistance in natural enemies (Croft & Brown, 1975; Croft & Strickler, 1983; Theiling & Croft, 1988; Hoy, 1990; Johnson & Tabashnik, 1994; Chapter 24 of this volume), assay techniques to evaluate pesticide effects on natural enemies (Hassan, 1985; 1989; Croft, 1990c; Wright & Verkerk, 1995), and ecological and physiological pesticide selectivity (Newsom *et al.,* 1976; Hull & Beers, 1985; Mullin & Croft, 1985; Poehling, 1989; Croft, 1990d, 1990e). In this

chapter, we discuss direct and indirect effects of pesticides on natural enemies, the consequences of disrupting biological control, and approaches to reduce the negative impact of chemicals on natural-enemy populations.

EFFECTS OF PESTICIDES ON NATURAL ENEMIES

We classify pesticide effects on natural enemies into direct and indirect effects. Direct effects include short- and long-term impacts on natural enemies due to direct contact with pesticides or pesticide residues. Indirect effects are those in which the impact of the pesticide is mediated through the natural enemy's host or prey (Waage, 1989). Indirect effects may be caused by reduction of host or prey populations that serve as food sources for natural enemies (Powell *et al.*, 1985), a change in the host or prey distribution (e.g., less clumped) (Waage, 1989), and ingestion of pesticide-contaminated prey or hosts (Goos, 1973).

Direct Effects

Short-Term Mortality

The most immediate effect of pesticides on natural enemies is short-term (up to 24 h after contact) mortality. Short-term mortality has been studied by exposing individuals from a natural-enemy population to one or more levels of pesticide, using both laboratory and field application techniques (e.g., Croft & Brown, 1975; Plapp & Vinson, 1977; Bellows *et al.*, 1985; Wright & Verkerk, 1995). If a series of doses or concentrations is used, then the population response can be quantified as the dose (LD) or concentration (LC) that is lethal to a designated proportion of the population, such as 50 or 90% (LD$_{90}$). Alternatively, some tests determine natural-enemy mortality caused by a single concentration of toxicant, such as the field application rate (Hassan, 1985) or the LC$_{50}$ for the pest population attacked by the natural enemy (Fabellar & Heinrichs, 1984).

Croft (1977, 1990f) discusses several biological factors that influence the way in which toxins cause relatively immediate mortality in natural enemies, including individual weight, size and sex, developmental stage, starvation and nutritional effects, diapause state, circadian rhythm, and behavior. These factors also influence the responses of pest species to pesticides. When developing assays to measure natural enemy susceptibility to pesticides, these factors should be considered to reduce variation in test results (Busvine, 1971; Hassan, 1977, 1985). Croft (1990c) advocates that research peer groups establish standardized susceptibility tests for evaluating pesticide impact on specific natural enemies.

Developmental stage greatly influences the impact of the pesticide on a natural enemy. Usually adult natural enemies are assayed (Hassan et al., 1987). However, effects of pesticides on parasitoid immature stages may differ significantly from effects on adults (Croft & Brown, 1975). Mode of development can directly influence the impact of pesticides on parasitoids in the field. For example, immature stages of endoparasitoids can be protected in the bodies of their hosts. Studies conducted on egg parasitoids in the genus *Trichogramma* have shown that many pesticides do not penetrate the chorion of eggs of their lepidopterous hosts (Plewka et al. 1975; Bull & Coleman, 1985). Thus, immatures of some *Trichogramma* spp. can survive applications of pesticides when protected in the host egg. Later instars of *Encarsia formosa* Gahan, a parasitoid of the greenhouse whitefly [*Trialeurodes vaporariorum* (Westwood)] are protected within their host's body from direct applications of bioresmethrin (Delorme et al. 1985). Mature pupae of the gregarious eulophid endoparasitoid *Pediobius foveolatus* (Crawford) in mummified Mexican bean beetle larvae can withstand applications of several organophosphate (OP) and carbamate insecticides but young *P. foveolatus* pupae cannot (Flanders et al., 1984). Immature ectoparasitoids may be protected by galleries, mines, and tunnels that their hosts inhabit. Larval stages of ectoparasitoids (e.g., *Diglyphus* spp.) of agromyzid leaf miners may survive insecticide applications due to the protection afforded by the leaf mine (Chandler, 1985).

Behavior of natural enemies, particularly searching behavior, may greatly affect their response to pesticides. The highly mobile searching behavior of natural enemies might be expected to increase their contact with insecticide residues in treated habitats and thus increase their mortality relative to sedentary pests (Croft & Brown, 1975; Flanders et al., 1984; Powell et al., 1985; Waage et al., 1985). There

is little evidence available to support or refute this idea. Studies do show that sublethal residues of some pesticides are repellent to natural enemies, thereby reducing their searching times on treated surfaces (Campbell et al., 1991; Longley & Jepson, 1996). Increased activity (e.g., locomotion, grooming) due to repellency of sublethal residues may result in a natural enemy accumulating additional residues (Jepson et al., 1990).

As with pest populations, some pesticides are highly toxic to a particular natural-enemy population whereas other pesticides are not (Hassan et al., 1987; Theiling & Croft, 1988, 1990b). Sharma and Adlakha (1981) determined the dosage response of the coccinellid predator *Coccinella septempunctata* Linnaeus subjected to spray applications of 24 compounds including chlorinated hydrocarbons, OPs and a carbamate. Based on LC$_{90}$'s, the chlorinated hydrocarbons (endosulfan, toxaphene, DDT) were significantly less toxic to the coccinellid than many of the OPs (naled, malathion, diazinon) and the carbamate carbaryl. In another case (Croft & Brown, 1975), 15 compounds were negligibly toxic to the predatory mite *Amblyseius fallacis* (Garman) at concentrations ≥0.03% but 16 other compounds were highly toxic at concentrations levels ≤0.03%. Residual toxicity of the pyrethroid permethrin to *Trichogramma pretiosum* Riley adults was significantly less than methyl parathion (OP) and methomyl (carbamate) (Bull & Coleman, 1985). *Trichogramma pretiosum* adults displayed intermediate susceptibility to the OPs phosmet and dimethoate. Similarly, the residual toxicities of the pyrethroids fenvalerate and permethrin were less than that of the carbamates methomyl and oxamyl to the eulophid *Diglyphus begini* (Ashmead), a parasitoid of agromyzid leafminers in the genus *Liriomyza* (Rathman et al., 1990). Susceptibility varied among the five *D. begini* populations sampled, with populations inhabiting areas of intense pesticide usage exhibiting higher resistance to the compounds assayed. Responses of populations of the California red scale parasitoid *Aphytis melinus* DeBach varied among five pesticides tested (Rosenheim & Hoy, 1986).

Predators are usually more tolerant of pesticides than are parasitoids (Croft & Brown, 1975; Theiling & Croft, 1988). For example, Rajakulendran and Plapp (1982a) showed that five pyrethroid insecticides (cypermethrin, phenothrin, tralomethrin, fluvalinate, and flucythrinate) were more toxic to the braconid *Campoletis sonorensis* (Carlson) than to the common green lacewing, *Chrysoperla* (= *Chrysopa*) *carnea* Stephens. Bellows et al. (1985) determined the residual toxicities of acephate, dimethoate, and formetenate hydrochloride to three natural enemies of the citrus thrips, *Scirtothrips citri* (Moulton). Natural enemies assayed were the mealybug destroyer, *Cryptolaemus montrouzieri* Mulsant, the parasitoid *Aphytis melinus,* and the predatory mite *Eusieus stipulatus* Athias-Henriot. Susceptibilities to pesticide residues varied, with LC$_{50}$'s for *E. stipulatus* < *A. melinus*

< *C. montrouzieri.* The mealybug destroyer was especially tolerant to formetenate hydrochloride with an LC$_{50}$ approximately 230-fold greater than that for *A. melinus*. A portion of this difference may be attributed to simple dose–weight responses; a *C. montrouzieri* adult is more than 100 times the weight of an *A. melinus* female.

Long-Term, Sublethal Effects

Relatively little attention has been directed toward long-term, sublethal effects of pesticides on natural enemies (Hassan, 1985; Hassan *et al.,* 1987). Wright & Verkerk (1995) point out that standard laboratory assays may not fully document sublethal effects of pesticides, thereby underestimating their full impact on field populations of natural enemies. We consider long-term, sublethal effects to be those that occur more than 24 h after a natural enemy is exposed to a pesticide. Sublethal residues may affect those natural enemies that survive pesticide applications, those that emerge as adults from protected situations, or those that disperse into previously treated areas where residues exist.

Sublethal doses of pesticides can have positive or negative effects on natural enemies (Croft, 1977; Elzen, 1989; Messing & Croft, 1990a; Wright & Verkerk, 1995). Negative effects are more commonly reported and usually outweigh positive effects. Positive effects include increased fecundity (Fleschner & Scriven, 1957; Attallah & Newsom, 1966), enhanced parasitoid efficiency (Irving & Wyatt, 1973), increased mobility (Dempster, 1968; Critchley, 1972), and reduced developmental periods (Adams, 1960; Lawrence *et al.,* 1973). Biological parameters detrimentally affected include daily fecundity, total progeny production, viability, and predation, or parasitism behavior (Table 1).

Reductions in progeny production as a result of decreased adult longevity have been reported in one predator and four parasitoid species (see Table 1). Reduced daily oviposition rates have been reported in five predator and four parasitoid species (see Table 1). In *Bracon hebetor* Say, reduced daily fecundity was a result of physiological impairment of the reproductive system (Grosch, 1970). Bartlett (1964) suggested that the impaired ability of natural enemies to cling to and traverse residue-contaminated surfaces indicated many possible unexplored ways in which natural-enemy functions may be inhibited by pesticides without apparent direct destruction. Disruption of normal behavior has been recorded in several natural enemies and includes loss of the ability to recognize hosts (Flanders, 1943), loss of coordination (Grosch, 1970), reduction of predation efficiency (Wiedl, 1977), temporary paralysis (Kiritani & Kawahara, 1973; Grafton-Cardwell & Hoy, 1985), termination of feeding (Dempster, 1968), and repellency from treated hosts–prey or habitats (Irving & Wyatt,

1973; Jackson & Ford, 1973; Jiu & Waage, 1990; Campbell *et al.,* 1991; Longley & Jepson, 1996; Umoru *et al.,* 1996).

Because searching behavior is an important factor in the ability of a natural enemy to regulate its prey or host at low densities (Huffaker *et al.,* 1976), impairment of this behavior could affect the effectiveness of efficient natural enemies. Perera (1982) demonstrated reduced rates of whitefly parasitism and significant changes in the functional response of *Encarsia formosa* when the parasitoid was provided whitefly hosts on *Phaseolus vulgaris* (Auth) leaves treated with sublethal doses of resmethrin. A typical type II functional response (Holling, 1959) was obtained in the absence of pesticides as compared with a type III (sigmoid) functional response when pesticides were present. On leaves with a uniform pesticide residue (27.6×10^{-7} ml/cm^2), parasitoid searching began only when host densities were ca. 80 third instar larvae per leaf as compared with lower densities on untreated leaves. Total parasitism rates were higher when residues were in a discrete deposit pattern as compared with a uniform pattern. Later studies on *Trichogramma* spp. (Campbell *et al.,* 1991), the aphid parasitoids *Aphidius rhopalosiphi* DeStefani-Perez (Longley & Jepson, 1996; Borgemeister *et al.,* 1993) and *Diaeretiella rapae* McIntosh (Jiu & Waage, 1990; Umoru *et al.,* 1996), and the coccinellid *Coccinella septempunctata* (Wiles & Jepson, 1994) demonstrated that these natural enemies were repelled by sublethal residues of certain pesticides (e.g., deltamethrin, esfenvalerate, malathion, permethrin, pirimicarb), altering their normal behavior and reducing the time they spent searching for hosts or prey in treated areas.

Other effects produced by sublethal doses of pesticides include increased developmental times (Wiedl, 1977), decreased production of F_1 female progeny (O'Brien *et al.,* 1985; Rosenheim & Hoy, 1988a), reduced survival of F_1 progeny (Attallah & Newsom, 1966; Ascerno *et al.,* 1980), and production of deformed F_1 progeny (Ascerno *et al.,* 1980). Due to the many possible sublethal effects of pesticides, the lack of significant natural-enemy mortality following a pesticide application in the field may not reflect the entire impact of the toxin on the natural-enemy population (Wright & Verkerk, 1995).

Indirect Effects

Reduction of Host Populations

Perhaps the greatest detriment to natural-enemy populations other than acute mortality from pesticides is the reduction in population density of hosts that serve as their food sources (Powell *et al.,* 1985). Systemic pesticides have been suggested as being a potential solution to the high acute mortality of natural enemies associated with most conventional pesticide applications. However, even when systemics are used, problems may occur when pest densities are

M. W. Johnson and B. E. Tabashnik

TABLE 1   Biological Parameters of Some Natural Enemies Affected Detrimentally by
Sublethal Exposure to Pesticides[a]

Natural enemy	Family	Pesticide	Reference
		Fecundity[b]	
Amblyseius fallacis	Phytoseiidae	Benomyl	Nakashima & Croft, 1975
A. swirskii	Phytoseiidae	Chlorobenzilate	Swirski *et al.,* 1967
Aphytis holoxanthus	Aphelinidae	Formothion Dimethoate Endosulfan Ethion Sulfur (WP) Sulfur (dust)	Rosen, 1967
Biosteres longicaudatus	Braconidae	Diflubenzuron	Lawrence, 1981
Bracon hebetor	Braconidae	Heptachlor	Grosch, 1970
B. hebetor	Bracondiae	Carbaryl	Grosch, 1975
B. hebetor	Braconidae	Permethrin	Press *et al.,* 1981
B. mellitor	Braconidae	Azinphos-methyl Chlordimeform	O'Brien *et al.,* 1985
Chrysoperla carnea	Chrysopidae	Fenvalerate Permethrin	Grafton-Cardwell & Hoy, 1985
Coleomegilla maculata	Coccinellidae	Toxaphene–DDT	Attallah & Newsom, 1966
Phytoseiulus persimilis	Phytoseiidae	Lindane Pyrethrin and piperonyl butoxide	Schulten *et al.,* 1976
P. persimilis	Phytoseiidae	Dichlofluanid Pirimicarb Pyrazophos	Hassan, 1982
		Total Progeny Production[c]	
A. melinus	Aphelinidae	Dimethoate Malathion Methidathion Chlorpyrifos	Rosenheim & Hoy, 1988
B. mellitor	Braconidae	Pyrethrin Permethrin	Press *et al.,* 1981
Eulophus pennicornis	Eulophidae	Hexaflumuron Teflubenzuron	Butaye & Degheele, 1995
Menochilus sexmaculatus	Coccinellidae	Malathion	Parker *et al.,* 1976
Microctonus aethiopoides	Braconidae	Carbofuran Methidathion Methoxychlor Methyl parathion	Dumbre & Hower, 1976
Trichogramma evanescens	Trichogrammatidae	Metasystox	Plewka *et al.,* 1975
		Viability[d]	
A. melinus	Aphelinidae	Malathion	Abdelrahman, 1973
A. melinus	Aphelinidae	Dimethoate Malathion Methidathion Chlorpyrifos	Rosenheim & Hoy, 1988a
B. hebetor	Braconidae	Heptachlor	Grosch, 1970
B. hebetor	Braconidae	Carbaryl	Grosch & Hoffman, 1973
C. maculata	Coccinellidae	DDT Endrin Toxaphene Toxaphene–DDT	Attallah & Newsom, 1966

(*continues*)

TABLE 1 (*continued*)

Natural enemy	Family	Pesticide	Reference
Diaeretiella rapae	Aphidiidae	Methomyl	Hsieh & Allen, 1986
Harpalus rufipes	Carabidae	DDT	Dempster, 1968
Menochilus sexmaculatus	Coccinellidae	Malathion	Parker *et al.*, 1976
Microctonus aethiopoides	Braconidae	Hydroprene	Ascerno *et al.*, 1980
M. aethiopoides	Braconidae	Carbofuran Methidathion Methoxychlor Methyl parathion	Dumbre & Hower, 1976
Phytoseiulus persimilis	Phytoseiidae	Benomyl	Parr & Binns, 1971
T. evanescens	Trichogrammatidae	Metasystox	Kot & Plewka, 1970

Predation or Parasitism Behavior[e]

Natural enemy	Family	Pesticide	Reference
Aphidius rhopalosiphi	Braconidae	Demeton-*S*-methyl Fenvalerate Deltamethrin	Borgemeister *et al.*, 1993 Longley & Jepson, 1996
Biosteres longicaudatus	Braconidae	Diflubenzuron	Lawrence, 1981
Chrysoperla carnea	Chrysopidae	Fenvalerate Permethrin	Grafton-Cardwell & Hoy, 1985
Coccinella septempunctata	Coccinellidae	Deltamethrin	Wiles & Jepson, 1994
Bracon hebetor	Braconidae	Heptachlor	Grosch, 1970
D. rapae	Braconidae	Malathion Permethrin Pirimicarb Pirimicarb	Jiu & Waage, 1990 Umoru *et al.*, 1996
Encarsia formosa	Aphelinidae	Benomyl Dichlofluanid Tetradifon Lindane	Irving & Wyatt, 1973
E. formosa	Aphelinidae	Resmethrin	Perera, 1982
H. rufipes	Carabidae	DDT	Dempster, 1968
Metaphycus helvolus	Encyrtidae	Sulfur	Flanders, 1943
Lycosa pseudoannulata	Aranea	Cartap	Kirtitani & Kawahara, 1973
P. persimilis	Phytoseiidae	Captan Malathion	Jackson & Ford, 1973
Toxorhynchites brevipalpis	Culicidae	Dieldrin	Wiedl, 1977
Trichogramma spp.	Trichogrammatidae	Esfenvalerate	Campell *et al.*, 1991

Developmental Rates

Natural enemy	Family	Pesticide	Reference
Toxorhynchites brevipalpis	Culicidae	Dieldrin	Wiedl, 1977

Sex Ratio of F_1 Generation

Natural enemy	Family	Pesticide	Reference
Aphytis melinus	Aphelinidae	Chlorpyrifos	Rosenheim & Hoy, 1988
B. mellitor	Braconidae	Azinphos-methyl Chlordimeform	O'Brien *et al.*, 1985
E. formosa	Aphelinidae	Benomyl Dichlofluanid Tetradifon Lindane	Irving & Wyatt, 1973

Longevity of F_1 Generation

Natural enemy	Family	Pesticide	Reference
Coleomegilla maculata	Coccinellidae	Toxaphene–DDT	Attallah & Newsom, 1966
Microctonus aethiopoides	Braconidae	Hydroprene	Ascerno *et al.*, 1980

(*continues*)

TABLE 1 (*continued*)

Natural enemy	Family	Pesticide	Reference
		Morphological Deformities in F_1 Generation	
M. aethiopoides	Braconidae	Hydroprene	Ascerno *et al.*, 1980

[a] Sublethal effects considered only when treated individuals live more than 24 h after treatment with pesticide.
[b] Number of eggs deposited per unit time.
[c] Product of total fecundity and longevity.
[d] Mating or longevity (>24 h postexposure to toxin).
[e] Ability to disperse and search.

reduced to such low levels that natural enemy populations are decimated by lack of prey or hosts or they are forced to emigrate (Boyce, 1936; Bartlett, 1964). This may be particularly serious when natural enemies are highly host specific (Heathcote, 1963). These actions result in natural enemy to pest ratios that favor the numerical increase of the pest. This problem has been suggested as a reason why natural enemies do not develop pesticide resistance as rapidly as their hosts (Georghiou, 1972; May & Dobson, 1986; Tabashnik, 1986; Chapter 24).

Ingestion of Pesticide-Contaminated Hosts or Prey

Mortality of coccinellid predators such as *Menochilus sexmaculatus* (Fabricius), *Coccinella undecimpunctata* Linnaeus, and *Scymnus syriacus* Le Conte has been reported following consumption of aphids poisoned with malathion (Satpathy *et al.*, 1968) and demeton (Ahmed, 1955). Herbert and Harper (1986) reported mortality of *Geocoris punctipes* (Say) nymphs that fed on *Helicoverpa zea* (Boddie) larvae injected with beta-exotoxin of *Bacillus thuringiensis* Berliner. When adult females of *Bracon hebetor* were exposed to larvae of the almond moth, *Ephestia cautella* (Walker), that had been treated topically with various insecticide dosages, parasitoid mortality increased with dosage (Press *et al.*, 1981). Mortality or sedation occurred in all individuals of the jumping spider *Phidippus audax* (Hentz) that ate live third-instar *H. zea* larvae previously fed a wheat germ diet containing 40 ppm abamectin (Roach & Morse, 1988). Emerging adult endoparasitoids can suffer mortality from insecticide residues contacted as they bore through contaminated host integument, egg chorion, or scale covers (Plewka *et al.*, 1975; Flanders *et al.*, 1984; Delorme *et al.*, 1985; Cohen *et al.*, 1988).

CONSEQUENCES OF DISRUPTING BIOLOGICAL CONTROL

Unilateral chemical control of arthropod pests can result in population explosions of both target and nontarget orga-

nisms due to various reasons including hormoligosis (Ferro, 1987; Risch, 1987), pesticide resistance (Georgihou, 1986; Metcalf, 1986), and the destruction of natural enemies (Ripper, 1956; Bartlett, 1964; Reynolds, 1971; Metcalf, 1986). The disruptive effects of pesticides on natural-enemy populations have been recognized for more than 100 years; Comstock (1880) noted that pyrethrum sprays against scale insects did more harm than good by destroying beneficial insects. Pest outbreaks resulting from the destruction of natural enemies are classified as *pest resurgences* and *secondary pest outbreaks* (pest upsets) (Barlett, 1964; van den Bosch *et al.*, 1982). Numerous arthropod species including scale insects, aphids, whiteflies, leafminers, lepidopterans, tetranychid mites, and eriophyid mites often undergo resurgences and secondary pest outbreaks (Ripper, 1956; Reynolds, 1971). Resurgences and secondary pest outbreaks have been reported on such crops as cotton, rice, deciduous fruits, and citrus (Metcalf, 1986).

Resurgences

Resurgences are characterized by a rapid numerical increase of a targeted pest population after use of a broad-spectrum pesticide. Although biological control of the target pest may be only partly effective, the absence of biological control following pesticide treatment results in increased problems (van den Bosch *et al.*, 1982). However, pesticide-induced reductions of natural-enemy populations may not always be the cause. Hardin *et al.* (1995) emphasize that efforts should be made to demonstrate a cause and effect relationship between any particular mechanism and resurgence. Conservation of natural enemies may conflict with the control of populations of major pests. This situation is frequently encountered when early-season pesticide applications disrupt biological control and commit the grower to chemical control for the remainder of the season (Havron *et al.*, 1987). Examples from 14 crop systems including more than 23 pests (and associated natural enemies) that have exhibited or have high potential for pest resurgences are listed in Tables 2 and 3. Pesticide resistance

TABLE 2   Examples of Ecological Selectivity by Modification of a Pesticide to Reduce Potential Pest Resurgences

Crop	Pest	Natural enemy	Pesticide used	Reference
Reduction of Pesticide Dosage				
Alfalfa	*Therioaphis maculata*	*Praon palitans* *Trioxys utilis* Various predators	Demeton	Stern *et al.,* 1959
Apples	*Panonychus ulmi*	*Stethorus punctum*	Cyhexatin	Asquith *et al.,* 1980
Cereals	Aphids	Predators	Pirimicarb	Poehling & Dehne, 1984
Forest	*Bupalus piniaria*	*Cratichneumon nigritarius*	Tetrachlorvinphos Diflubenzuron	Wainhouse, 1987
Peas	*Acyrthosiphon pisum*	*Coccinella septempunctata* *Syrphus vitripennis* *Chrysopa carnea*	Pirimicarb	Hellpap & Schmutterer, 1982
Sorghum	*Schizaphis graminum*	Coccinelids	Various materials	Teetes *et al.,* 1973
Soybeans	*Plathypena scabra*	Geocorids, spiders	Carbaryl, methomyl	Turnipseed, 1972
Soybeans	*Helicoverpa zea*	*Geocoris punctipes* *Nabis* spp.	Carbaryl	Turnipseed, 1972
Tobacco	*Manduca sexta*	*Polistes* spp.	TDE, endrin	Lawson *et al.,* 1961
Use of Nonpersistent Pesticides				
Alfalfa	*T. maculata*	*P. palitans* *T. utilis*	Mevinphos	Stern & van den Bosch, 1959
Alfalfa	*Hypera postica*	*Bathyplectes curculionis*	Methyl parathion	Wilson & Armbrust, 1970

in the primary pest and resurgence of that pest can be tightly coupled and may work synergistically (Hardin *et al.,* 1995). However, Hardin *et al.* argue that insecticide resistance is not a mechanism underlying resurgence, but it may magnify the level of resurgence.

Resurgences have been reported in natural ecosystems as well agroecosystems (Turnock *et al.,* 1976). Decimation of natural enemies of the European spruce sawfly, *Diprion hercyniae* (Hartig), caused by applications of DDT over a 3-year period in New Brunswick, Canada, resulted in a resurgence of the sawfly when pesticide applications were terminated (Neilson *et al.,* 1971). In southwest Texas, insecticidal control efforts were associated with increased populations of the southern pine beetle, *Dendroctonus frontalis* Zimmerman, due in part to elimination of predators (Vitre, 1971).

Secondary Pest Outbreaks

Secondary pest outbreaks, or pest upsets, are characterized by the rapid numerical increase to pest status of a nontargeted, phytophagous arthropod population after use of a broad-spectrum pesticide applied for control of another pest in the ecosystem (van den Bosch *et al.,* 1982). The nontarget pest is usually assumed to be under substantial or complete biological control until the pesticide reduces natural-enemy effectiveness. Examples from 10 crop systems including more than 14 pests (and associated natural ene-

mies) that have exhibited or have potential for pest upsets are listed in Table 4.

Pesticides inducing pest outbreaks are generally those that are exceptionally toxic to the biological control agents involved but not to the pests (Bartlett, 1964). The most serious pest upsets are caused when pesticides decimate natural enemies without significant destruction of the host and persisting residues or repeated applications restrict subsequent natural enemy activity. In fresh-market tomatoes biological control of the agromyzid leafminer *Liriomyza sativae* Blanchard was disrupted by methomyl applications for the tomato fruit worm, *H. zea,* and the tomato pinworm, *Keiferia lycopersciella* (Walsingham), which attack the fruit (Johnson *et al.,* 1980a, 1980b). Weekly pesticide applications for continuous protection of the fruit resulted in secondary outbreaks of the leaf miner due to destruction of its effective natural enemies.

Indirect evidence of secondary pest outbreaks has been recorded in natural ecosystems. A pest upset of pine needle scale, *Phenaecaspis pinifoliae* (Fitch), on lodgepole and Jeffrey pines followed a malathion fogging program directed at mosquitoes in California (Luck & Dahlsten, 1975). Following 24 applications of malathion bait sprays to eradicate the Mediterranean fruit fly, *Ceratitis capitata* (Wiedemann), in northern California, a 90-fold increase was observed in an endemic gall midge *Rhopalomyia california* Felt infesting coyote brush, *Baccharis pilularis* DC (ssp. *consanguinea*) (Ehler *et al.,* 1984).

TABLE 3 Ecological Selectivity by Modification of Pesticide Application Methods to Reduce Potential Pest Resurgences

Crop	Pest	Natural enemy	Pesticide used or application method or target site	Reference
Temporal Discrimination				
Alfalfa	*Hypera postica*	*Bathyplectes curculionis*	Carbofuran	Davis & Hoyt, 1979
Alfalfa	*H. postica*	*B. curculionis*	Methyl parathion	Wilson & Armbrust, 1970
Apples	*Phyllonorycter blancardella*	*Apanteles ornigis*	Oxamyl	Weires *et al.*, 1982
Apples	*Panonychus ulmi*	*Amblyseius fallacis*	Various materials	Croft & Brown, 1975
Apples	*Lymantria dispar*	*Apanteles melanoscelus*	Dimilin	Granett *et al.*, 1976
Grapes	*Erythroneura elegantula*	*Anagrus epos*	Various materials	AliNiazee & Oatman, 1979
Hops	*Phorodon humuli*	*Anthocoris nemoralis*	Dimefox, mephosfolan	Solomon, 1987
Pears	*Psylla pyricola*	*A. nemoralis*	Pyrethroids	Solomon, 1987
Tomatoes	*Trialeurodes vaporariorum*	*Encarsia formosa*	Oxamyl	Way, 1986
Wheat	*Eurygaster integriceps*	*Trissolcus grandis*	Various materials	Novozhilov, 1976
Vegetables	*Pieris brassicae*	*Apanteles glomeratus*	Various materials	Novozhilov, 1976
Vegetables	*Plutella cruciferarum*	*Chorogenes* spp.	Various materials	Novozhilov, 1976
Wheat	*Sitobion avenae*	Various predators	Benomyl	Powell *et al.*, 1985
Habitat Discrimination				
Apples	*Tetranychus mcdanieli*	*Typhlodromus occidentalis*	Tree periphery	Hoyt, 1969
Apples	*Panonychus ulmi*	*Amblyseius fallacis*	Tree periphery	Metcalf, 1972
Beans	*Liriomyza trifolii*	*Oenonogastra microrhopalae*	Granulars in soil	Oetting, 1985
Forest	*Bupalus piniaria*	*Cratichneumon nigritarius*	Outbreak epicenters	Wainhouse, 1987
Hops	*Phorodon humuli*	*Anthocoris nemoralis*	Soil drench	Solomon, 1987
Tobacco	*Manduca sexta*	*Polistes* spp.	Upper plant	Lawson *et al.*, 1961
Tomatoes	*Trialeurodes vaporariorum*	*E. formosa*	Soil drench	Way, 1986
Habitat Partitionment				
Citrus	*Lepidosaphes beckii*	*Aphytis lepidosaphes*	Strip treatment	Debach & Landi, 1961

Secondary pest outbreaks involve nontarget pests usually under substantial or complete biological control until pesticide applications reduce natural-enemy effectiveness. Here again, possible correlations between the effectiveness of a natural enemy based on its searching behavior and the potential to accumulate residues while searching must be considered. Even low levels of pesticide in the agroecosystem may be great enough to impair the searching abilities of the most effective natural enemies, which would lead to an increase in the nontarget pest. Bartlett (1964) suggested that pest outbreaks following treatments are often important indicators of the presence of efficient natural enemies. However, he pointed out that following an upset by natural-enemy elimination, normally less efficient or sometimes even rare species often gain temporary superiority by earlier ingress and may maintain a numerical advantage for long periods. The earliest influx of natural enemies may consist of the most mobile and seasonally abundant forms, or sometimes those species (frequently coccinellid predators) that are ordinarily attracted to high host densities. He sug-

gested caution in predicting natural-enemy efficiency strictly on temporary abundance, particularly following recovery from a pesticide upset.

TECHNIQUES TO REDUCE THE NEGATIVE IMPACT OF CHEMICALS ON NATURAL ENEMIES

Crop Monitoring and Economic Thresholds

Probably the best method for reducing the overall negative impact of chemicals on natural enemies is to avoid pesticide applications altogether. When pesticides are used in the management of pests they should be applied only when necessary and economically justifiable (Higley & Pedigo, 1996a). Decisions concerning the need for or suitability of a pesticide application require knowledge of the pest's impact on the crop and some method of monitoring the pest to determine when it may cause problems (Stern *et*

TABLE 4   Ecological Selectivity by Modification of Pesticide Application Methods or Pesticide to Reduce Potential Secondary Pest Upsets

Crop	Primary pest	Secondary pest	Natural enemy	Pesticide used or application method or target site	References
Temporal Discrimination					
Alfalfa	*Hypera postica*	Various species	Various species	Carbofuran	Davis & Hoyt, 1979
Apples	*Dysaphis plantaginea*	*Panonychus ulmi*	*Stethorus punctum*	Chloropyrifos Methidathion	Hull & Starner, 1983b
Apples	Various pests	*P. ulmi*	*Amblyseius fallacis*	Diazinon Demeton	Croft, 1975
Apples	*Platynota idaeusalis*	*P. ulmi*	*Stethorus punctum*	Fenvalerate	Hull *et al.*, 1985b
Walnuts	*Cydia pomonella*	*Chromaphis juglandicola*	*Trioxys pallidus*	Various materials	Riedl *et al.*, 1979
Habitat Discrimination					
Apples	Various pests	*P. ulmi*	*A. fallacis*	Tree canopy	Croft, 1975
Cotton	*Pectinophora gossypiella*	*Helicoverpa zea*	*Orius* spp.	Lower plant canopy	Davis & Hoyt, 1979
Cotton	*Anthonomus grandis*	*H. zea*	Various species	Pheromone/trap crop	Lloyd *et al.*, 1972
Cotton	*Lygus* spp.	*H. zea* *Spodoptera exigua* *Trichoplusia ni*	Various species Various species Various species	Safflower trap crop	Mueller & Stern, 1974
Cucumbers	*Thrips tabaci*	*Trialeurodes vaporariorum* *Tetranychus urticae*	*Encarsia formosa* *Phytoseiulus persimilis*	Below plants	Pickford, 1983
Peanuts	*Elasmopalpus lignosellus*	Various species	Various species	Granulars in soil	Adkisson, 1972
Peanuts	*E. lignosellus*	Various species	Various species	Directed spray	Adkisson, 1972
Vegetables	Fungus species	*Trialeurodes vaporariorum* *Myzus persicae*	*Verticillium lecanii*	Soil application	Way, 1986
Watermelon	*Dacus cucurbitae*	*Liriomyza* spp.	*Chrysonotomyia punctriventris* *Ganaspidium utilis*	Trap crop	Johnson, 1987
Habitat Partitionment					
Apples	*Platynota idaeusalis*	*P. ulmi*	*Stethorus punctum*	Alternate-row middle	Hull *et al.*, 1983
Citrus	*Lepidosaphes beckii*	*Aonidiella aurantii*	*Aphytis* spp	Strip treatment	DeBach *et al.*, 1955
Reduction of Pesticide Dosage					
Apples	*C. pomonella*	*Tetranychus mcdanieli*	*Typhlodromus occidentalis*	Azinphos-methyl	Hoyt, 1969
Apples	*P. idaeusalis*	*P. ulmi*	*S. punctum*	Various materials	Hull & Beers, 1985
Pears	*C. pomonella*	Various mite species	*T. occidentalis*	Azinphos-methyl	Westigard, 1971
Selective Insecticides with Reduced Rates of Broad-Spectrum Compounds					
Apples	*P. idaeusalis*	*P. ulmi*	*S. punctum*	Azinphos-methyl Methyl parathion (E)	Hull & Starner, 1983a

al., 1959). On watermelon in Hawaii, *Liriomyza* spp. are present in plantings during more than 95% of the growth cycle but densities surpass treatment levels less than 7% of the time (Johnson *et al.*, 1989). In comparison, melon aphid, *Aphis gossypii* Glover, is present less than 80% of the time, but densities may surpass treatment levels on more than 35% of sample dates.

An understanding of the effectiveness of natural enemies is essential to avoid applying pesticides when biological control is adequate (Wilson, 1985). Unfortunately, few monitoring systems have been developed that consider the influence of natural enemies (Flaherty *et al.*, 1981; Li, 1982; Tanigoshi *et al.*, 1983; Wilson *et al.*, 1984). One of the first systems developed was used to predict the necessity of acaricide treatments for European red mite suppression on apple. Treatment decisions are based on the relative densities of the spider mite and its predator *A. fallacis* (Croft, 1977). Using predator to prey ratios, the probability of biological control within a week's time is estimated; acaricide applications are recommended only when biological control is estimated to be inadequate. The impact of pesticide treatments on natural enemies of secondary pests should also be considered when decisions are made to treat for primary pests. Pest management manuals produced for alfalfa hay (Flint, 1985c), almonds (Flint, 1985b), cotton (Flint, 1984), tomatoes (Flint, 1985a), and walnuts (Flint, 1987) advise growers of various natural enemies that should be conserved to avoid secondary upsets.

Several parameters must be considered to permit a pesticide to be used in the most efficient and least disruptive manner. One must first define the biological target (Hull & Beers, 1985). Prerequisites for defining the target are (1) identification of all pests and associated natural enemies that may be affected by pesticide applications in a given habitat, (2) knowledge of the behavior and microhabitats of targeted pests and associated natural enemies, (3) knowledge of the dosage responses of pests and associated natural enemies to available pesticides, and (4) knowledge of all feasible application methods for each compound. This information will provide the basic foundation for maximizing benefits derived from pesticide selectivity. After the biological target is defined, one must then determine options available for the most efficient use of a pesticide with minimal nontarget effects. Vorley & Dittrich (1994) state that the "IPM fit" of a pesticide can be defined by the level of control provided for at least one target pest, the selectivity to key natural enemies, and the risk of resurgence of secondary pests.

Pesticide Selectivity

Bartlett (1964) defined pesticide selectivity as the capacity of a pesticide treatment to spare natural enemies while destroying the target pest. Selectivity differs from specificity, which is the capacity of a compound to cause extraordinarily high mortality in a particular species. Specificity may be considered as an extreme type of selectivity. Selectivity is a relative measure expressed as the natural enemy to pest ratio produced by one pesticide treatment compared with the ratio produced by a reference pesticide treatment (Bartlett, 1964).

Many authors divide pesticide selectivity into physiological selectivity and ecological selectivity based on the mechanism by which a pesticide treatment causes differential mortality to pests and their natural enemies (Metcalf, 1972, 1982; Watson, 1975; Newsom et al., 1976; Hull & Beers, 1985). Ecological selectivity results from differential exposure of pests and natural enemies to a pesticide. Physiological selectivity results from physiological differences in susceptibilies of pests and associated natural enemies to a pesticide as discussed later. Such physiological differences in the natural enemy may be inherent or may be a product of laboratory or field selection. Strategies used to develop selective techniques of pesticide use are dependent on the system under consideration.

Physiological Pesticide Selectivity

Development of physiologically selective compounds is reviewed by Mullin and Croft (1985) and Croft (1990d). Effective use of physiologically selective compounds to enhance biological control is dependent on several factors including the type of pests (primary, secondary) targeted for control and relative levels of pesticide resistance in nontargeted pest and natural-enemy species. When pest resurgence is the primary consideration, the major objective of physiological selectivity is to manipulate the natural enemy to pest ratio to favor the natural enemies by differential mortality of the pest. However, high pest mortality is counterproductive if surviving natural enemies starve or emigrate from the treated crop in search of prey or hosts (Smith, 1970; Tabashnik, 1986). Such loss of natural enemies may also produce a result similar to pest resurgence. Therefore, sucessful prevention of pest resurgences using physiological selectivity requires that a significant portion of the pest population survives the treatment. The pest density that can be tolerated is a function of many factors including the economic threshold, densities of pest and natural enemies following the pesticide treatment, numerical responses of the pest and natural enemies, host crop, and relevant environmental parameters (temperature, humidity, etc.). Few current pest management recommendations include intentional retention of pest populations following treatment.

When secondary pest outbreaks are the foremost concern and the primary pest lacks effective natural enemies, survival of the primary pest after treatment is less important than a potential resurgence problem. In situations where the secondary pest exhibits pesticide resistance to a wide range of compounds, it will generally be preferable to use compounds with high specificity to control the primary pest. This tactic avoids destruction of the secondary pest's natural enemies and reduces further development of resistance in the secondary pest. Unfortunately, current industrial development and production of selective compounds is limited due to small market demand, developmental costs, and potential investment losses if targeted species develop pesticide resistance (Newsom et al., 1976; van Emden & Peakall, 1996). If a mechanism had been established in the 1970s to foster development of selective pesticides by assisting manufacturers to meet regulatory requirements, perhaps selective compounds would be more widely used in agriculture now (National Research Council, 1996).

Efforts have been made by the Working Group on Pesticides and Beneficial Organisms, International Organization for Biological Control, West Palearctic Regional Section (IOBC/WPRS), to identify registered pesticides (acaricides, fungicides, insecticides, herbicides) and plant growth regulators that have selective properties when used at recommended field rates (Franz et al., 1980; Hassan et *al.* 1983, 1987, 1991). Hassan (1985, 1989) reviewed standard assay techniques for analyzing the side effects of pesticides on biological control agents including parasitoids, predators, and a fungal pathogen. These assay techniques are used by the IOBC/WPRS Working Group to identify physiologically selective pesticides. Because no single assay provides sufficient information, a series of sequential tests are used to analyze fully the potential detrimental effects of pesticides. These tests include an initial laboratory test, semifield tests, and a field test (Hassan, 1985). Initial laboratory assays use fresh, dry pesticide residues deposited on glass, leaf, or soil surfaces. Compounds found harmless to specific biological control agents in the laboratory test are not subjected to further testing. Semifield tests include an initial toxicity test and a residue persistence test. Initial toxicity tests consist of applying pesticides to natural materials in the field (protected from direct exposure to sunlight and precipitation) or under simulated field conditions in an environmental chamber. In initial toxicity tests, natural enemies are exposed to treated plants or soil, after which mortality and beneficial capacity are assessed. Pesticide impact on a natural enemy's beneficial capacity is defined as the reduction in parasitism (for parasitoids) or oviposition (for predators) of exposed insects compared with untreated controls. In field tests, effects of field-applied pesticides on beneficials are determined by exposing natural enemies to weathered residues on treated substrates. Field tests are designed according to the crop inhabited by the natural enemy and the timing and number of pesticide treatments used in standard agricultural practices.

Compared with most conventional assays used for pest and beneficial species (Theiling & Croft, 1988), the IOBC/WPRS Working Group tests differ significantly with respect to pesticide exposure rates, exposure periods, and assessment of pesticide impact on the beneficial capacity of the natural enemy. Exposure rates are based on recommended commercial application rates. Exposure periods normally range from one to several weeks (Hassan, 1985) as compared with the short exposure periods (24 to 48 h) used in most conventional assays to determine lethal concentrations or dosages (Theiling & Croft, 1988). Estimates of acute mortality do not provide a complete assessment of the negative impact of pesticides on biological control agents due to the effects of sublethal dosages on longevity, fecundity, and behavior (Wright & Verkerk, 1995). Pesticides are categorized depending on reductions in beneficial capacity of the natural enemy: < 50% = harmless; 50 to 79% = slightly harmful; 80 to 90% = moderately harmful; and > 99% = harmful (Hassan, 1985).

The IOBC/WPRS Working Group has evaluated more than 60 compounds involving 19 natural enemy species (Franz *et al.,* 1980; Hassan *et al.,* 1983, 1987). Twenty-five compounds have been identified as relatively selective to at least one or more natural enemies based on their toxicity

and persistence (Hassan *et al.,* 1987). The IOBC/WPRS methods may provide the best way to identify currently available physiologically selective compounds for use in pest management programs.

Pesticide selectivity has been achieved predominantly by using insecticides to which biological control agents are naturally tolerant (Mullin & Croft, 1985). This approach is restricted by the limited commercial availability of selective compounds (Newsom *et al.,* 1976; Mullin & Croft, 1985). Some negative impacts of insecticides on beneficials may be reduced by laboratory selection for insecticide resistance to those compounds not naturally selective for natural enemies (Croft & Morse, 1979; Hoy, 1985; Johnson & Tabashnik, 1994; Chapter 12).

Laboratory Selection for Resistance

Laboratory selection for increased pesticide resistance in natural enemies has long been advocated (DeBach, 1958) but only relatively recently has been successful with arthropod predators (Hoy, 1987). Phytoseiid mites among this group include *Phytoseiulus persimilis* Athias-Henriot, *Amblyseius fallacis, A. nicholsi* Ehara & Lee, and *Typhlodromus (Metaseiulus) occidentalis* (Nesbitt) (Hoy, 1987; Huang *et al.,* 1987). Probably the most successful results of laboratory selection have occurred in *T. occidentalis* resistance to carbaryl, methomyl, dimethoate, permethrin, and fenvalerate (Hoy, 1985). Selected strains of *T. occidentalis* are now commercially produced for mass releases in almond and apple orchards (Headley & Hoy, 1987). An insect predator, the common green lacewing, *Chrysoperla carnea,* also has been selected for carbaryl resistance (Grafton-Cardwell & Hoy, 1986).

Studies conducted using parasitoids in selection programs have had limited success compared with that for predators (Hoy, 1987; Johnson & Tabashnik, 1994). Several attempts have been made to artificially generate resistant parasitoids (Pielou & Glasser, 1952; Robertson, 1957; Adams & Cross, 1967; Abdelrahman, 1973; Havron, 1983; Hsieh, 1984; Delorme *et al.,* 1985). Although moderate responses to selection have been obtained, in most cases the selection response has not been adequate to incorporate the selected strain into an IPM program. Selection conducted on colonies established from field populations of *Aphytis melinus,* a parasitoid of the California red scale exhibiting low levels of resistance to various insecticides, produced a parasitoid strain approximately 20 times more resistant to carbaryl than did a standard susceptible colony (Rosenheim & Hoy, 1988a, 1988b). Azinphos-methyl resistance was increased 7.5-fold in the parasitoid *Trioxys pallidus* Haliday, which attacks the walnut aphid, *Chromaphis juglandicola* Kaltenbach, in California (Hoy & Cave, 1988). Efforts are currently underway to implement laboratory-selected strains of *A. melinus* (Rosenheim & Hoy,

1988b; Spollen & Hoy, 1992) and *T. pallidus* (Hoy & Cave, 1988; Hoy *et al.*, 1990, 1991) in commercial IPM programs, but results are still preliminary. More detailed discussion of genetic improvement of natural enemies is provided (Chapter 12).

Ecological Selectivity

Due to the limited number of physiologically selective compounds available, greater potential for achieving pesticide selectivity appears to be through the development and use of ecologically selective strategies (Davis & Hoyt, 1979). Ecological selectivity can be accomplished by modification of either the pesticide or the delivery methods (Hull & Beers, 1985).

Ecological selectivity is achieved by reducing the detrimental nontarget effects of broad-spectrum (not physiologically selective) pesticides and thereby conserving natural-enemy populations. Many changes required to make compounds more ecologically selective must be accomplished during the development or manufacturing phases of the materials. Properties that may be altered include degree of systemic action, formulation, encapsulation, and overall persistence (Newsom *et al.*, 1976; Graham-Bryce, 1987). Growers' abilities to employ ecologically selective tactics through alteration of the physical and chemical properties of the toxicant and its carriers are limited by the number of selective products marketed by the chemical industry. However, growers can modify the dosage or application rates. Nonpersistent compounds also can be used to limit the impact of pesticide residues on natural-enemy populations.

Reduction of Application Rates

Perhaps the most readily accomplished modification of a material is the reduction of the application rate. Because broad-spectrum compounds are biologically active against a wide range of organisms, label rates may be higher than required for effective control of some pests. Turnipseed *et al.* (1974) reported that reduced rates of carbaryl and methyl parathion effectively controlled bean leaf beetle [*Cerotoma trifurcata* (Forster)], velvet bean caterpillar (*Anticarsia gemmatalis* Hübner), green cloverworm [*Plathypena scabra* (Fabricius)] and corn earworm [*Helicoverpa zea* (Boddie)] infesting soybeans. Although reduced rates may provide only partial control of a primary pest, benefits may be derived from conservation of the natural enemies. Pest resurgences can be reduced if treatments for the primary pest are properly timed, target sites are restricted, and optimal application tactics are employed (Way, 1986).

If the primary pest has no effective natural enemies and its economic threshold is extremely low, then it may not be feasible to reduce pesticide dosage to conserve the natural enemies of secondary pests. In this situation, primary pest control may be possible if applications are timed to coin-cide with the presence of highly susceptible life stages of pests with discrete generations (see Table 4). Hull and Beers (1985) reported that insecticide applications for the tufted apple bud moth, *Platynota idaeusalis* (Walker), on apple could be reduced approximately 75% below label dosage rates if properly timed. This reduction in pesticide rate allowed *Stethorus punctum* (Le Conte) to control the European red mite.

Selectivity can also be achieved by combining reduced dosages of broad-spectrum materials with physiologically selective compounds or synergists. Hull and Starner (1983a) demonstrated effective control of *P. idaeusalis* on apple when using either encapsulated methyl parathion or azinphos-methyl in combination with low rates of permethrin or fenvalerate (see Table 4). These combinations suppressed the leafroller without destroying populations of *S. punctum* necessary for European red mite conrol. Laboratory studies showed that when the larvicide *Bacillus thuringiensis* was combined with reduced dosages of binapacryl, tricyclohexyltin hydroxide, or fentin hydroxide, selective mortality of the diamondback moth, *Plutella xylostella* (Linnaeus), was achieved with no toxicity to its parasitoid *Thyraeella collaris* Gravely (Hamilton & Attia, 1977). Pyrethroids tend to be more toxic for *Helicoverpa zea* than for the common green lacewing, *Chrysoperla* (*Chrysopa*) *carnea* (Rajakulendran & Plapp, 1982a). When cypermethrin, tralomethin, fluvalinate, and flucythrinate were each combined with chlordimeform at a ratio of 1 part pesticide to 10 parts chlordimeform, the selectivity was significantly increased for *H. zea* as compared with the pyrethroids alone (Rajakulendran & Plapp, 1982b). More research is needed on the potential value of reducing application rates of broad-spectrum compounds alone and in combination with physiologically selective compounds or synergists to reduce their impact on natural enemies (Hull & Beers, 1985).

Nonpersistent Pesticides

Persistence of a given pesticide is partially a function of the rate applied (Busvine, 1971). Persistent pesticide residues can reduce pest and natural-enemy populations to low levels. Residues of carbaryl and carbophenothion applied to soybeans caused high mortality of Mexican bean beetle, *Epilachna varivestis* Mulsant, 21 days following treatment (Flanders *et al.*, 1984). High pest mortality can indirectly reduce natural-enemy efficacy if surviving natural enemies starve or emigrate from the treated crop habitat in search of hosts. Compounds that have long-term residual activity are usually detrimental to natural enemies that disperse into an area shortly after treatment or survive an initial pesticide treatment in a protected stage or refuge (Bartlett, 1964; Flanders *et al.*, 1984). When applied at field rates, temephos and triazophos residues require approximately 21 days to lose their activity against *Aphytis africanus* Quednau that attacks the California red scale, *Aonidiella aurantii* (Maskell) (Schoones & Giliomee, 1984). Acephate residues on citrus caused mortalities in excess of 50% for more than 14 days in the predatory mite *Eusieus stipulatus* Athias-Henriot, the California red scale parasitoid *Aphytis melinus,* and the mealybug destroyer *Crytolaemus montrouzieri* (Bellows *et al.,* 1985). Parasitism of European corn borer [*Ostrinia nubilalis* (Hübner)] eggs by *Trichogramma nubilale* Ertle & Davis was reduced more than 20 days following carbaryl treatment on green pepper (Tipping & Burbutis, 1983). Compounds that maintain residual activity for long periods, relative to the life cycles of affected natural enemies, are not appropriate for integrated programs.

Many investigations on the impacts of pesticides on natural enemies have focused on the immediate or contact toxicity of the pesticide (Theiling & Croft, 1988). Unfortunately, little quantitative information exists describing the relationship between residue dissipation and postapplication survival of natural enemies (Bellows *et al.,* 1985). As stated previously, current work by the IOBC/WPRS Working Group on Pesticides and Beneficial Organisms involves screening of the long-term effects of pesticide residues on natural enemies (Hassan, 1985). Residues are considered harmful to parasitoids when parasitism rates are reduced more than 30%. In approximately 64% of the natural-enemy–pesticide combinations examined by the Working Group, pesticides applied at standard rates were highly toxic more than 16 days after application (Hassan *et al.,* 1987). In tests with the whitefly parasitoid *Encarsia formosa,* amitraz (insecticide), tetradifon (acaricide), and chinomethionate (fungicide) had harmful effects on initial contact but minimal residual effects 5 days after application (Hassan *et al.,* 1987). Pirimicarb in granular form was highly toxic on contact with *Trichogramma* spp., but harmless 5 days after application (Hassan *et al.,* 1987). Singh (1989) lists several "relatively safe" pesticides used in integrated control programs.

Modification of Application Methods

Great progress in reducing the negative impacts of pesticides on natural enemies has been achieved through the modification of delivery or application methods. Strategies employed include temporal discrimination, habitat discrimination, and habitat partition. In temporal and habitat discrimination, pesticides are applied when the natural enemies are absent or in low numbers due to either seasonal or cyclic occurrence (*temporal asynchrony:* Croft, 1990e) or to habitat preferences (*spatial asynchrony:* Croft, 1990e), respectively. Habitat partition involves the restriction of pesticide applications to limited sections of a planting or orchard to establish refuges in which natural enemies are protected.

Temporal Discrimination   Pest resurgences and upsets can be avoided by applying pesticides when biological control agents do not occupy the habitats or crop systems commonly shared with the targeted pest (see Tables 3 and 4). This is practiced in temperate climates where many pests establish populations in crop hosts in the spring prior to associated natural enemies (Davis & Hoyt, 1979; Powell *et al.,* 1985; Solomon, 1987). Bartlett (1964) suggested that pesticide treatments should be eliminated when natural enemies are present and that pest destruction should be maximized when natural enemies are absent or less efficient. However, drastic reductions in prey or host populations prior to immigration of natural enemies may hamper later biological control of targeted pests. Hoyt (1969) reported that prebloom applications of oxythioquinox to control *Tetranychus mcdanieli* McGregor on apple caused such low prey levels that its predator *Typhlodromus occidentalis* suffered mortality due to starvation and did not increase to an effective density until late in the season. This led to economically significant populations of *Tetranychus mcdanieli.*

Natural enemies that share the targeted pest's habitat at the time of application might avoid mortality if they are inactive or are in a relatively resistant stage (Davis & Hoyt, 1979). Conservation of the braconid *Apanteles ornigis* Weed can be achieved by applying oxamyl treatments for spotted tentiform leafminer on apple prior to parasitoid emergence in the spring (Weires *et al.,* 1982). Granett *et al.* (1976) reported that early-season treatments of diflubenzuron for gypsy moth control on apple inhibited the emergence of the endoparasitoid *A. melanoscelus* (Ratzeburg) from moth hosts due to the presence of susceptible stages at application time. However, later season treatments selectively killed gypsy moth larvae, but allowed survival of mature parasitoids ready to emerge.

Habitat Discrimination   When natural enemies are present in a crop, it is often possible to conserve their populations by directing pesticide applications to those habitats that the pests occupy preferentially compared with the biological agents (see Tables 3 and 4). Pest resurgences involving tetranychid mites on deciduous fruit crops have been avoided by applying pesticides only to the peripheral areas of the tree canopy and thereby preserving the phytoseiid predators commonly found deep inside the canopy (Hoyt, 1969; Metcalf, 1972). Pesticide treatments frequently can be limited to those vertical or horizontal strata of a plant in which the highest ratio of targeted pests to natural enemies exists. Endrin and TDE applications directed at tobacco hornworm [*Manduca sexta* (Johannson)] larvae feeding on the upper portions of tobacco plants conserved *Polistes* spp. wasps that prey on *M. sexta* larvae on the lower tobacco leaves (Lawson *et al.,* 1961). Treatments for the pink bollworm, *Pectinophora gossypiella* (Saunders), directed at the lower leaf canopy of cotton allow survival of anthocorid predators (*Orius* spp.) that attack *Heliothis* eggs and larvae (Davis & Hoyt, 1979). When high infestations of *Thrips palmi* Karny on watermelon are limited to the vine tips, oxamyl applications may be directed there, thus conserving parasitoids of *Liriomyza* leafminers that are found predominatly in the basal foliage canopy (Johnson, 1986; Lynch & Johnson, 1987).

Polyphagous pest species often have highly mobile adult stages that move among adjacent plant hosts, both domestic and wild (Nishida & Bess, 1957). This behavior can be exploited to reduce natural-enemy mortality by directing pesticide applications at trap crops. Parasitoids that attack *Liriomyza* leafminers infesting watermelon foliage can be suppressed by insecticides applied directly to the crop for control of melon fly, *Dacus cucurbitae* Coquillet. However, corn (*Zea mays* Bonaf.) borders planted adjacent to watermelon attract melon flies seeking a resting spot. Once in the corn trap crop, melon fly control is achieved when flies ingest malathion-treated bait. This reduces amounts of pesticides applied directly to watermelon and thereby conserves the leafminer parasitoids (Johnson, 1987).

Habitat discrimination can be achieved using baits and treated seeds because target pests usually seek the treated substrate and mortality occurs following contact or ingestion (Newsom *et al.*, 1976). Due to the low chemical dosages required, natural enemies are spared. Newsom *et al.* (1976) reviewed the advantages of seed treatments versus broadcast applications of pesticides. These included cost reduction as a function of insecticide rates used, no additional costs for application methods or equipment, reduced environmental pollution, minimal contact with nontarget species, and reduced selection pressures for the development of pesticide resistance. Pest species controlled using seed treatments include seed maggots, rootworms (Newsom *et al.*, 1976), wireworms (Metcalf, 1982), aphids (Reynolds *et al.*, 1957), and thrips (Reynolds, 1958).

Baits have been used extensively to control pest species of ants such as the imported fire ants, *Solenopsis* spp. (Anonymous, 1972). Baits are removed from the soil surface rapidly by ants and there is thus limited exposure to nontarget species (Newsom *et al.*, 1976). Unfortunately, some nontarget effects have been recorded with the use of mirex for ant control (Harris & Burns, 1972). Other insects controlled by the use of baits include tephritid fruit flies (Steiner *et al.*, 1965, 1970), leaf-cutting bees (Graham-Bryce, 1987), grasshoppers, cutworms, armyworms, and crickets (Metcalf *et al.*, 1962).

Unique approaches to habitat discrimination may be developed based on the physical properties of a compound or its formulation. Systemic materials offer great potential for the control of foliage pests without exposing natural enemies to chemical residues on the plant surface. Suppression of greenhouse whitefly on small tomato plants can be achieved with oxamyl applied as a soil drench without directly affecting subsequent releases of the whitefly paras-

itoid *Encarsia formosa* (Way, 1986) (see Table 3). Soil drenches of dimefox or mephosfolan can be used on hops for early-season control of the damson hop aphid, *Phorodon humuli* (Schrank), without disrupting later biological control provided by the anthocorid *Anthocoris nemoralis* (Linnaeus) (Solomon, 1987). Aldicarb, oxamyl, and disulfoton granules applied to potting media for control of *Liriomyza trifolii* (Burgess) on beans did not reduce parasitism by the braconid *Oenongastra microrhopalae* (Ashmead) (Oetting, 1985). Adkisson (1972) reported that use of granular insecticides in peanuts for control of the lesser cornstalk borer, *Elasmopalpus lignosellus* (Zeller), was effective and reduced potential problems with the approximately 40 secondary pests on peanuts.

Habitat discrimination has great potential for conserving natural enemies, but this tactic requires a good understanding of the habitat preferences of pests and natural enemies. In addition, variation in habitat preferences correlated with plant phenology and seasonal changes must be determined. This is clearly demonstrated by the seasonal movement patterns of the predatory mite *Amblyseius fallacis* in apple orchards in Michigan. Spring populations (April to early June) of the predator are found in the ground cover beneath apple trees where they prey on the two-spotted mite, *Tetranychus urticae* Koch. Around mid-June, *A. fallacis* enters the tree canopy to attack the European red mite, *Panonychus ulmi* (Koch), and other tetranychid species (Croft, 1975). Early-season populations of the predator may be conserved if insecticide sprays for San Jose scale, oyster shell scale, apple aphid, rosy apple aphid, and woolly apple aphid are limited to the apple tree canopy prior to the movement of *A. fallacis* into the apple foliage canopy. Understanding the habitat preferences of *A. fallacis* has been a key component in apple pest management.

Habitat Partitionment　Temporal and habitat discrimination cannot be used when targeted pests and natural enemies are present in identical or highly overlapping habitats or when it is impossible to limit pesticide application to specific habitats. However, it may be possible to establish refuges in a crop in which natural enemies may be protected from pesticide applications. This can be done either by using spot treatments (Bartlett, 1964) or by partitioning a crop into sections and applying treatments to a limited number of the total sections at any one time (Hull & Beers, 1985). Both methods can help to prevent pest resurgences and secondary pest upsets.

Many pests (aphids, tetranychid mites) have clumped distributions within their host crops (Southwood, 1978). This can lead to high pest densities confined to small areas of a total planting or orchard. Natural enemies in a planting or an orchard may be conserved by using spot treatments in which pesticide applications are limited to areas where the pest has surpassed the economic threshold and the pest to natural enemy ratio is unfavorable.

Conservation of natural enemies by partitioning a crop into uniformly distributed sections and treating only a portion of the sections at any one time has been used successfully in citrus and deciduous fruit orchards (Hull & Beers, 1985). This method, referred to as *strip treatment,* was used in citrus to reduce pest resurgences of the purple scale, *Lepidosaphes beckii* (Newman) (DeBach & Landi, 1961) (see Table 3) and secondary pest upsets of the California red scale, *Aonidiella aurantii* (DeBach *et al.,* 1955) (see Table 4). Orchard rows were divided into groups of three pairs; each first pair in each group was treated with oil sprays for purple scale control. After 6 months the second pairs were treated, and 6 months later the third pairs were treated. This treatment pattern reduced total natural-enemy mortality and allowed safe dispersal of parasitoids of *L. beckii* and *A. aurantii* from untreated to treated sections after oil residues dissipated. On apples in Pennsylvania, more than 95% of all pesticides are applied using the *alternate-row middle method* (Lewis & Hickey, 1964; Hull *et al.,* 1983). Pesticides are applied to every other orchard row pair on a 7- to 9-day cycle. This method is effective in reducing secondary upsets of the European red mite (Hull *et al.,* 1983) (see Table 4). Although most common in orchard crops, this technique has good potential in crops with relatively long growth cycles such as alfalfa and cotton.

CONCLUSIONS

Significant reductions in the the negative impact of chemicals on natural enemies would be achieved if pesticides were applied only when necessary. This requires development of economic thresholds and sampling programs along with evaluations of biological control effectiveness. Economic thresholds should be a cornerstone of every IPM program (Higley & Pedigo, 1996a). However, relatively few thresholds exist compared with the number of crop pests for which chemicals are applied (Poston *et al.,* 1983; Peterson, 1996). More studies are needed on yield responses of plants to arthropod injury so that economic thresholds (warning thresholds, action thresholds, density treatment levels) can be developed to reduce unnecessary pesticide use. Implementation of economic thresholds will require monitoring programs that estimate population densities of both the pest and its natural enemies (Higley & Pedigo, 1996b).

When pesticides are applied, their direct and indirect effects can disrupt biological control. Acute mortality is a well-documented direct effect of pesticides on natural enemies. Estimates of acute mortality alone, however, do not provide a complete assessment of the direct impact of pesticides on natural enemies because longer term, sublethal effects can also be important. Sublethal effects of pesticides can have a negative impact on natural enemy searching behavior. This is significant because efficient searching behavior is essential to the effectiveness of biological control agents. In addition, natural enemy populations can suffer indirect effects of pesticides through ingestion of contaminated hosts or prey and reduction of host and prey populations needed for parasitism and food, respectively. Of these factors, short-term mortality and reduction in host or prey populations probably have the greatest impact.

Due to cumulative uptake of toxins, biological control agents that are highly mobile and efficient searchers may be at a disadvantage compared with their less mobile competitors when pesticide residues are present. Studies are needed to determine if acquisition of pesticide residues in treated habitats by natural enemies is correlated with their mobility and searching efficiency. Data now exist that show pesticide-induced repellency can reduce searching efficiency (Campbell *et al.,* 1991; Longley & Jepson, 1996). A fuller understanding of these phenomena are essential for integration of biological and chemical controls.

The IOBC/WPRS Working Group on Pesticides and Beneficial Organisms, together with other workers, has made significant progress in identifying available compounds that can be used in integrated programs. Registration of pesticides should include tests on their impact on natural enemies commonly found in the crop systems where materials will be applied. Resulting information should be provided on pesticide labels. Without this information, one cannot make intelligent pesticide selections. In addition to the type of information that the IOBC/WPRS Working Group has provided, data are needed on the impact of pesticides on natural enemy populations from different areas. Studies have shown that natural enemy responses to pesticides vary with locality. This information would allow a grower to select the least disruptive pesticide for use in integrated programs and would also provide information on the ability of natural enemies to develop pesticide resistance.

Greater efforts are needed to develop additional ecologically selective techniques in the areas of temporal and habitat discrimination and habitat partitionment. This will require more basic studies of the ecology of pests and the associated natural enemies to fully utilize ecological selectivity.

Significant progress has been made in the integration of chemical and biological controls over the last 30 years. However, progress to date is insufficient considering the wide range and number of crop systems that still rely heavily on chemicals for pest suppression. The need to further develop strategies to reduce the pesticide impact on natural enemies will increase as problems with pesticide resistance increase and as more attention is directed toward the reduction of chemical use in agriculture.

Acknowledgments

The thoughtful comments of J. Rosenheim and L. LeBeck were appreciated. This work was supported by USDA grants HAW00945 and HAW00949. This is Journal Series No. 3342 of the Hawaii Institute of Tropical Agriculture and Human Resources.

References

Abdelrahman, I. (1973). Toxicity of malathion to the natural enemies of California red scale, Aonidiella aurantii (Mask.) (Hemiptera: Diaspididae). Australian Journal of Agriculture Research, 24, 119–33.

Adams, J. B. (1960). Effects of spraying 2,4-D amine on coccinellid larvae. Canadian Journal of Zoology, 38, 285–288.

Adams, C. H., & Cross, W. H. (1967). Insecticide resistance in Bracon mellitor, a parasite of the boll weevil. Journal of Economic Entomology, 60, 1016–1020.

Adkisson, P. L. (1972). Use of cultural practices in insect pest management. In Implementing practical pest management strategies (pp. 37–50). Proceedings of the National Extension Insect-Pest Management Workshop, Purdue University, Lafayette, IN, March 14–16, 1972.

Ahmed, M. K. (1955). Comparative effect of systox and schradan on some predators of aphids in Egypt. Journal of Economic Entomology, 48, 530–532.

AliNiazee, M. T., & Oatman, E. R. (1979). Pest management programs. In D. W. Davis, S. C. Hoyt, J. A. McMurtry, & M. T. AliNiazee (Eds.), Biological control and insect pest management. (Division of Agricultural Science Publication 4096, pp. 80–88). Oakland: University of California.

Anonymous. (1972). Report of the Mirex Advisory Committee to William D. Ruckelshaus, Administrator, Washington, DC: Environmental Protection Agency (Revised March 1, 1972).

Ascerno, M. E., Smilowitz, Z., & Hower, A. A. (1980). Effects of the insect growth regulator hydroprene on diapausing Microctonus aethiopoides, a parasite of the alfalfa weevil. Environmental Entomology, 9, 262–264.

Asquith, D., Croft, B. A., Hoyt, S. C., Glass, E. H., & Rice, R. E. (1980): The systems approach and general accomplishments toward better insect control in pome and stone fruits. In C. B. Huffaker (Ed.), New technology of pest control (pp. 249–317). New York: John Wiley & Sons.

Attalah, Y. H., & Newsom, L. D. (1966). Ecological and nutritional studies on Coleomegilla maculata De Geer (Coleoptera: Coccinellidae). III. The effect of DDT, toxaphene, and endrin on the reproductive and survival potentials. Journal of Economic Entomology, 59, 1181–1187.

Bartlett, B. R. (1964). Integration of chemical and biological control. In P. DeBach (Ed.), Biological control of insect pests and weeds (pp. 489–511). London: Chapman & Hall.

Bellows, T. S., Jr., Morse, J. G., Hadjidemetriou, D. G., & Iwata, Y. (1985). Residual toxicity of 4 insecticides used for control of citrus thrips Scirtothrips citri (Thysanoptera: Thripidae) on 3 beneficial species in a citrus agroecosystem. Journal of Economic Entomology, 78, 681–686.

Borgemeister, C., Poehling, H.-M., Dinter, A., & Holler, C. (1993). Effects of insecticides on life history parameters of the aphid parasitoid Aphidius rhopalosiphi (Hym: Aphidiidae). Entomophaga, 38, 245–255.

Boyce, A. M. (1936). The citrus red mite Paratetranychus citri McG. in California and its control. Journal of Economic Entomology, 29, 125–130.

Bull, D. L., & Coleman, R. J. (1985). Effects of pesticides on Trichogramma spp. Southwestern Entomology, 8 (Suppl.), 156–168.

Busvine, J. R. (1971). A critical review of the techniques for testing insecticides (2nd ed.). Slough, United Kingdom: Commonwealth Agricultural Bureau.

Butaye, L., & Degheele, D. (1995). Benzoylphenyl ureas effect on growth and development of Eulophus pennicornis (Hymenoptera: Eulophidae), a larval ectoparasite of the cabbage moth (Lepidoptera: Noctuidae). Journal of Economic Entomology, 88, 600–605.

Campbell, C. D., Walgenbach, J. F., & Kenneday, G. G. (1991). Effect of parasitoids on Lepidoptera pests in insecticide-treated and untreated tomatoes in western North Carolina. Journal of Economic Entomology, 84, 1662–1667.

Chandler, L. D. (1985). Response of Liriomyza trifolii to selected insecticides with notes on hymenopterous parasites. Southwestern Entomology, 10, 228–235.

Cohen, E., Podoler, H., & El-Hamlauwi, M. (1988). Effect of malathion-bait mixture on two parasitoids of the Florida red scale, Chrysomphalus aonidum (L.). Crop Protection 7, 91–95.

Comstock, J. H. (1880). Introduction of the 1880 report of the U.S. Department of Agriculture, (pp. 289–290). Washington, DC: U.S. Department of Agriculture.

Critchley, B. R. (1972). A laboratory study of the effects of some soil-applied organo-phosphorus pesticides on Carabidae (Coleoptera). Bulletin of Entomological Research, 62, 229–242.

Croft, B. A. (1975). Integrated control of apple mites. (Extension Service Bulletin E-825). Michigan State University. East Lansing, MI.

Croft, B. A. (1977). Susceptibility surveillance to pesticides among arthropod natural enemies: Modes of uptake and basic responses. Zeitschrift fuer Pflanzenkrankheiten und Pflanzenschutz, 84, 140–157.

Croft, B. A. (1990a). Natural enemies and pesticides: An historical overview. In B. A. Croft (Ed.), Arthropod biological control agents and pesticides (pp. 3–15). New York: John Wiley & Sons.

Croft, B. A. (1990b). Pesticide effects on natural arthropod enemies: A database summary. In B. A. Croft (Ed.), Arthropod biological control agents and pesticides (pp. 17–46). New York: John Wiley & Sons.

Croft, B. A. (1990c). Standardized assessment methods. In B. A. Croft (Ed.), Arthropod biological control agents and pesticides (pp. 101–126). New York: John Wiley and Sons.

Croft, B. A. (1990d). Physiological selectivity. In B. A. Croft (Ed.), Arthropod biological control agents and pesticides (pp. 221–245). New York: John Wiley & Sons.

Croft, B. A. (1990e). Ecological selectivity. In B. A. Croft (Ed.), Arthropod biological control agents and pesticides (pp. 247–267). New York: John Wiley & Sons.

Croft, B. A. (1990f). Factors affecting susceptibility. In B. A. Croft (Ed.), Arthropod biological control agents and pesticides (pp. 71–100). New York: John Wiley & Sons.

Croft, B. A., & Brown, A. W. A. (1975). Response of arthropod natural enemies to insecticides. Annual Review of Entomology, 20, 285–355.

Croft, B. A., & Morse, J. G. (1979). Research advances on pesticide resistance in natural enemies. Entomophaga, 24, 3–11.

Croft, B. A., & Strickler, K. (1983). Natural enemy resistance to pesticides: Documentation, characterization, theory and application. In G. P. Georghiou & T. Saito (Eds.), Pest resistance to pesticides (pp. 669–702). New York: Plenum Press.

Davis, D. W., & Hoyt, S. C. (1979). Selective pesticides. In D. W. Davis, S. C. Hoyt, J. A. McMurtry, & M. T. AliNiazee (Eds.), Biological control and insect pest management. (Division of Agricultural Science Publication 4096, Berkeley: University of California.

DeBach, P. (1958). Selective breeding to improve adaptations of parasitic insects. Proceedings of the 10th International Congress of Entomology, Montreal (1956), 4, 759–768.

DeBach, P., & Landi, J. (1961). The introduced scale parasite, *Aphytis lepidosaphes* Compere, and a method of integrating chemical with biological control. Hilgardia, 14, 459–497.

DeBach, P., Landi, J., & White, E. B. (1955). Biological control of red scale. California Citrograph, 40, 254, 271–272, 274–275.

Delorme, R., Berthier, A., & Auge, D. (1985). The toxicity of two pyrethroids to *Encarsia formosa* and its host *Trialeurodes vaporariorum* prospecting for a resistant strain of the parasite. Pesticide Science, 16, 332–336.

Dempster, J. P. (1968). The sublethal effect of DDT on the rate of feeding by the ground-beetle *Harpalus rufipes*. Entomologia Experimentalis et Applicata, 11, 51–54.

Dumbre, R. B., & Hower, A. A. (1976). Sublethal effects of insecticides on the alfalfa weevil parasites *Microctonus aethiopoides*. Environmental Entomology, 5, 683–687.

Ehler, L. E., Endicott, P. C., Hertlein, M. B., & Alvarado-Rodriquez, B. (1984). Medfly, *Ceratitis capitata*, eradication in California: Impact of malathion-bait sprays on an endemic gall midge, *Rhopalomyia californica* and its parasitoids. Entomologia Experimentalis et Applicata, 36, 201–208.

Elzen, G. W. (1989). Sublethal effects of pesticides on beneficial parasitoids. In P. C. Jepson (Ed.), Pesticides and non-target invertebrates (pp. 129–150). Wimborne, Dorset, United Kingdom: Intercept.

Fabellar, L. T., & Heinrichs, E. A. (1984). Toxicity of insecticides to predators of rice brown planthoppers, *Nilaparvata lugens* (Stål) (Homoptera: Delphacidae). Environmental Entomology, 13, 832–837.

Ferro, D. N. (1987). Insect pest outbreaks in agroecosystems. In P. Barbosa & J. C. Schultz (Eds.), Insect outbreaks (pp. 195–215). San Diego: Academic Press.

Flaherty, D. L., Hoy, M. A., & Lynn, C. D. (1981). Spider mites. In D. L. Flaherty, F. L. Jenson, A. N. Kasimatis, H. Kido, & W. J. Moller (Eds.), Grape pest management (Agricultural Science Publication 4105 (pp. 111–125). Berkeley: University of California, Press.

Flanders, S. E. (1943). The susceptibility of parasitic Hymenoptera to sulfur. Journal of Economic Entomology, 36, 469.

Flanders, R. V., Bledsoe, L. W., & Edwards, C. R. (1984). Effects of insecticides on *Pediobius foveolatus* (Hymenoptera: Eulophidae), a parasitoid of the Mexican bean beetle (Coleoptera: Coccinellidae). Environmental Entomology, 13, 902–906.

Fleshner, C. A., & Scriven, G. T. (1957). Effect of soil-type and DDT on ovipositional responses of *Chrysopa californica* (Coq.) on lemon. Journal of Economic Entomology, 50, 221–222.

Flint, M. L. (1984). Integrated pest management for cotton in the western region of the United States. Division of Agriculture and Natural Resources Publication 3305). University of California. Oakland, CA.

Flint, M. L. (1985a). Integrated pest management for tomatoes. (Division of Agriculture and Natural Resources Publication 3274). University of California. Oakland, CA.

Flint, M. L. (1985b). Integrated pest management for almonds. (Division of Agriculture and Natural Resources Publication 3308). University of California. Oakland, CA.

Flint, M. L. (1985c). Integrated pest management for alfalfa hay. (Division of Agriculture and Natural Resources Publication 3312). University of California. Oakland, CA.

Flint, M. L. (1987). Integrated pest management for walnuts (2nd ed.). (Division of Agriculture and Natural Resources Publication, 3270). University of California. Oakland, CA.

Franz, J. M., Bogenschutz, H., Hassan, S. A., Huang, P., Naton, E., & Suter, H. (1980). Results of a joint pesticide test programme by the working group: Pesticides and beneficial arthropods. Entomophaga, 25, 231–236.

Georghiou, G. P. (1972). The evolution of resistance to pesticides. Annual Review of Ecology and Systematics, 3, 133–168.

Georghiou, G. P. (1986). The magnitude of the resistance problem. In Pesticide resistance: Strategies and tactics for management (pp. 14–43). Washington, DC: National Academy of Science Press.

Goos, M. (1973). Influence of aphicides used in sugar-beet plantations on arthropods. II. Studies on arachnids—Arachnoidea. Polski Pismo Entomologiczne, 43, 851–859.

Grafton-Cardwell, E. E., & Hoy, M. A. (1985). Short-term effects of permethrin and fenvalerate on oviposition by *Chrysoperla carnea* (Neuroptera: Chrysopidae). Journal of Economic Entomology, 78, 955–959.

Grafton-Cardwell, E. E., & Hoy, M. A. (1986). Genetic improvement of common green lacewing, *Chrysoperla carnea* (Neuroptera: Chrysopidae): Selection for carbaryl resistance. Environmental Entomology, 15, 1130–1136.

Graham-Bryce, I. J. (1987). Chemical methods. In A. J. Burn, T. H. Coaker, & P. C. Jepson (Eds.), Integrated pest management (pp. 113–159). London: Academic Press.

Granett, J., Dubar, D. M., & Weseloh, R. M. (1976). Gypsy moth control with dimilin sprays timed to minimize effects on parasite *Apantales melanoscelus*. Journal of Economic Entomology, 69, 403–404.

Grosch, D. S. (1970). Reproductive performance of a braconid after heptachlor poisoning. Journal of Economic Entomology, 63, 1348–1349.

Grosch, D. S. (1975). Reproductive performance of *Bracon hebetor* after sublethal doses of carbaryl. Journal of Economic Entomology, 68, 659–662.

Grosch, D. S., & Hoffman, A. C. (1973). The vulnerability of specific cells in the oogenetic sequence of *Bracon hebetor* Say to some degradation products of carbamate pesticides. Environmental Entomology, 2, 1029–1032.

Hamilton, J. T., & Attia, F. I. (1977). Effects of mixtures of *Bacillus thuringiensis* and pesticides on *Plutella xylostella* and the parasite *Thyraeella collaris*. Journal of Economic Entomology, 70, 146–147.

Hardin, M. R., Benrey, B., Coll, M., Lamp, W. O., Roderick, G. K., & Barbosa, P. (1995). Arthropod pest resurgences: An overview of potential mechanisms. Crop Protection, 14, 3–18.

Harris, W. G., & Burns, E. C. (1972). Predation on the lone star tick by the imported fire ant. Environmental Entomology, 1, 362–365.

Hassan, S. A. (1977). Standardized techniques for testing side-effects of pesticides on beneficial arthropods in the laboratory. Zeitschrift fur Pflanzenkrankheiten und Pflanzenschutz, 84, 158–163.

Hassan, S. A. (1982). Relative tolerance of three different strains of the predatory mite *Phytoseiulus persimilis* A.-H. (Acari: Phytoseiidae) to 11 pesticides used on glasshouse crops. Zeitschrift Angewandte Entomologie, 93, 55–63.

Hassan, S. A. (1985). Standard methods to test the side-effects of pesticides on natural enemies of insects and mites developed by the IOBC/WPRS Working Group Pesticides and Beneficial Organisms. EPPO Bulletin, 15, 214–255.

Hassan, S. A. (1989). Testing methodology and the concept of the IOBC/WPRS working group. In P. C. Jepson (Ed.), Pesticides and non-target invertebrates (pp. 1–18). Wimborne, Dorset, United Kingdom: Intercept.

Hassan, S. A., Bigler, F., Bogenschutz, H., Brown, J. U., Firth, S. I., & Huang, P. (1983). Results of the second joint pesticide testing programme by the IOBC/WPRS Working Group Pesticides and Beneficial Arthropods. Zeitschrift Angewandte Entomologie, 95, 151–158.

Hassan, S. A., Bigler, F., Bogenschutz, H., Boller, E., Brun, J., & Calis, J. N. M. (1991). Results of the fifth joint pesticide testing programme carried out by the IOBC/WPRS–Working Group Pesticides and Beneficial Organisms. Entomophaga, 36, 55–67.

Hassan, S. A., Albert, R., Bigler, F., Blaisinger, P., Bogenschuetz, H., & Boller, E. (1987). Results of the third joint pesticide testing programme by the IOBC/WPRS Working Group Pesticides and Beneficial Arthropods. Journal of Applied Entomology, 103, 92–107.

Havron, A. (1983). Studies toward selection of Aphytis wasps for pesticide resistance. Unpublished doctoral dissertation, Hebrew University of Jerusalem, Rehovot, Israel.

Havron, A., Rosen, D., & Rossler, Y. (1987). A test method for pesticide tolerance in minute parasitic Hymenoptera. Entomophaga, 32, 83–95.

Headley, J. C., & Hoy, M. A. (1987). Benefit/cost analysis of an integrated mite management program for almonds. Journal of Economic Entomology, 80, 555–559.

Heathcote, G. D. (1963). The effect of coccinellids on aphids infesting insecticide-treated sugar-beet. Plant Pathology, 12, 80–83.

Hellpap, C., & Schmutterer, H. (1982). Studies on the effect of low doses of pirimor on Acyrthosiphon pisum and natural enemies. Zeitschrift Angewandte Entomologie, 94, 246–252.

Herbert, D. A., & Harper, J. D. (1986). Bioassays of a beta-exotoxin of Bacillus thuringiensis against Geocoris punctipes (Hemiptera: Lygaeidae). Journal of Economic Entomology, 79, 592–595.

Higley, L. G., & Pedigo, L. P. (1996a). The EIL concept. In L. G. Higley & L. P. Pedigo (Eds.), Economic thresholds for integrated pest management (pp. 9–21). Lincoln, NE: University of Nebraska Press.

Higley, L. G., & Pedigo, L. P. (1996b). Afterword: Pest science at a crossroads. In L. G. Higley & L. P. Pedigo (Eds.), Economic thresholds for integrated pest management (pp. 291–295). Lincoln, NE: University of Nebraska Press.

Holling, C. S. (1959). Some characteristics of simple types of predation and parasitism. Canadian Entomologist, 91, 385–398.

Hoy, M. A. (1985). Recent advances in genetics and genetic improvement of the Phytoseiidae. Annual Review of Entomology, 30, 345–370.

Hoy, M. A. (1987). Developing insecticide resistance in insect and mite predators and opportunities for gene transfer. In H. M. LeBaron, R. O. Mumma, R. C. Honeycutt, & J. H. Duesing (Eds.), Biotechnology in agricultural chemistry. ACS Symposium Series No. 334 (pp. 125–138). Washington, DC: American Chemical Society.

Hoy, M. A. (1990). Pesticide resistance in arthropod natural enemies: Varability and selection responses. In R. T Roush & B. E. Tabashnik (Eds.), Pesticide resistance in arthropods (pp. 203–236). London: Chapman & Hall.

Hoy, M. A., & Cave, F. E. (1988). Guthion-resistant strain of walnut aphid parasite. California Agriculture, 42, 4.

Hoy, M. A., Cave, F. E., Beede, R. H., Grant, J., Krueger, W. H., & Olson, W. H. (1990). Release, Dispersal, and recovery of a laboratory-selected strain of the walnut aphid parasite Trioxys pallidus (Hymenoptera: Aphidiidae) resistant to azinphosmethyl. Journal of Economic Entomology, 83, 89–96.

Hoy, M. A., Cave, F. E., & Caprio, M. A. (1991). Guthion-resistant parasite ready for implementation in walnuts. California Agriculture, 45, 29–31.

Hoyt, S. C. (1969). Integrated chemical control of insects and biological control of mites on apples in Washington. Journal of Economic Entomology, 62, 74–86.

Hsieh, C.-Y. (1984). Effects of insecticides on Diaeretiella rapae (McIntosh) with emphasis on bioassay techniques for aphid parasitoids. Unpublished doctoral dissertation, University of California, Berkeley.

Hsieh, C. Y., & Allen, W. W. (1986). Effects of insecticides on emergence, survival, longevity, and fecundity of the parasitoid Diaeretiella rapae (Hymenoptera: Aphidiidae) from mummified Myzus persicae (Homoptera: Aphididae). Journal of Economic Entomology, 79, 1599–1602.

Huang, M.-D., Xiong, J.-J., & Du, T.-Y. (1987). The selection for and genetical analysis of phosmet resistance in Amblyseius nicholsi. Acta Entomologica Sinica, 30, 133–139.

Huffaker, C. B., Simmonds, F. J., & Laing, J. E. (1976). The theoretical and empirical basis of biological control. In C. B. Huffaker & P. S. Messenger (Eds.), Theory and practice of biological control (pp. 41–78). New York: Academic Press.

Hull, L. A., & Beers, E. H. (1985). Ecological selectivity: Modifying chemical control practices to preserve natural enemies. In M. S. Hoy & D. C. Herzog (Eds.), Biological control in agricultural IPM systems (pp. 103–122). Orlando, FL: Academic Press.

Hull, L. A., & Starner, V. R. (1983a). Impact of four synthetic pyrethroids on major natural enemies and pests of apple in Pennsylvania. Journal of Economic Entomology, 76, 122–130.

Hull, L. A., & Starner, V. R. (1983b). Effectiveness of insecticide applications timed to correspond with the development of rosy apple aphid (Homoptera: Aphidiidae) on apple. Journal Economic Entomology, 76, 594–598.

Hull, L. A., Beers, E. H., & Meagher, R. L., Jr. (1985a). Impact of selective use of the synthetic pyrethoid fenvalerate on apple pests and natural enemies in large-orchard trials. Journal of Economic Entomology, 78, 163–168.

Hull, L. A., Beers, E. H., & Meagher, R. L., Jr. (1985b). Integration of biological and chemical control tactics for apple pests through selective timing and choice of synthetic pyrethroid insecticides. Journal of Economic Entomology, 78, 714–721.

Hull, L. A., Hickey, K. D., & Kanour, W. W. (1983). Pesticide usage patterns and associated pest damage in commercial apple orchards of Pennsylvania. Journal of Economic Entomology, 577–583.

Irving, S. N., & Wyatt, I. J. (1973). Effects of sublethal doses of pesticides on the oviposition behavior of Encarsia formosa. Annals of Applied Biology, 75, 57–62.

Jackson, G. J., & Ford, J. B. (1973). The feeding behavior of Phytoseiulus persimilis (Acarina: Phytoseiidae), particularly as affected by certain pesticides. Annals of Applied Biology, 75, 165–171.

Jepson, P. C., Chaudry, A. G., Salt, D. W., Ford, M. G., Efe, E., & Chowdhury, A. B. N. U. (1990). A reductionist approach towards short-term hazard analysis for terrestrial invertebrates exposed to pesticides. Functional Ecology, 4, 339–348.

Jiu, G. D., & Waage, J. K. (1990). The effect of insecticides on the distribution of foraging parasitoids Diaeretiella rapae (Hym: Braconidae) on plants. Entomophaga, 35, 49–56.

Johnson, M. W. (1986). Population trends of a newly introduced species, Thrips palmi (Thysanoptera: Thripidae), on commercial watermelon plantings in Hawaii. Journal of Economic Entomology, 79, 718–720.

Johnson, M. W. (1987). Parasitization of Liriomyza spp. (Diptera: Agromyzidae) infesting commercial watermelon plantings in Hawaii. Journal of Economic Entomology, 80, 56–61.

Johnson, M. W., & Tabashnik, B. E. (1994). Laboratory selection for pesticide resistance in natural enemies. (pp. 91–105). In S. K. Narang, A. C. Bartlett, & R. M. Faust (Eds.), Applications of genetics to arthropods of biological control significance. Boca Raton, FL: CRC Press.

Johnson, M. W., Oatman, E. R., & Wyman, J. A. (1980a). Effects of insecticides on populations of the vegetable leafminer and associated parasites on summer pole tomatoes. Journal of Economic Entomology, 73, 61–66.

Johnson, M. W., Oatman, E. R., & Wyman, J. A. (1980b). Effects of insecticides on populations of the vegetable leafminer and associated parasites on fall pole tomatoes. Journal of Economic Entomology, 73, 67–71.

Johnson, M. W., Mau, R. F. L., Martinez, A. P., & Fukuda, S. (1989). Foliar pests of watermelon. Tropical Pest Management, 35, 90–96.

Kiritani, K., & Kawahara, S. (1973). Food-chain toxicity of granular formulations of insecticides to a predator, Lycosa pseudoannulata, of Nephotettix cincticeps. Botyu-Kagaku, 38, 69–75.

Kot, J., & Plewka, T. (1970). The influence of metasystox on different stages of the development of Trichogramma evanescens Westw. Tagunasberichte Deutsche Akademie der Landwirtschaft wissenschaften zu Berlin, 110, 185–192.

Lawrence, P. O. (1981). Developmental and reproductive biologies of the parasitic wasp, Biosteres longicaudatus, reared on hosts treated with a chitin synthesis inhibitor. Insect Science Applications, 1, 403–406.

Lawrence, P. O., Kerr, S. H., & Whitcomb, W. H. (1973). *Chrysopa rufilabris:* Effect of selected pesticides on duration of third larval stadium, pupal stage and adult survival. Environmental Entomology, 2, 477–480.

Lawson, F. R., Rabb, R. L., Guthrie, F. E., & Bowery, T. G. (1961). Studies of an integrated control system for hornworms in tobacco. Journal of Economic Entomology, 54, 93–97.

Lewis, F. H., & Hickey, K. D. (1964). Pesticide application from one side on deciduous fruit trees. Pennsylvania Fruit News, 43, 13–24.

Li, L.-Y. (1982). Integrated rice insect pest control in the Guangdong Province of China. Entomophaga, 27, 81–88.

Lloyd, E. P., Scot, W. P., Shaunak, K. K., Tingle, F. C., & Davich, T. B. (1972). A modified trapping system for suppressing low-density populations of overwintered boll weevils. Journal of Economic Entomology, 65, 1141–1147.

Longley, M., & Jepson, P. C. (1996). Effects of honeydew and insecticide residues on the distribution of foraging aphid parasitoids under glasshouse and field conditions. Entomologia Experimentalis et Applicata, 81, 259–269.

Luck, R. F., & Dahlsten, D. L. (1975). Natural decline of a pine needle scale (*Chionaspis pinifoliae* [Fitch]), outbreak at south Lake Tahoe, California following cessation of adult mosquito control with malathion. Ecology, 56, 893–904.

Lynch, J. A., & Johnson, M. W. (1987). Stratified sampling of *Liriomyza* spp. (Diptera: Agromyzidae) and associated hymenopterous parasites on watermelon. Journal of Economic Entomology, 80, 1254–1261.

May, R. M., & Dobson, A. P. (1986). Population dynamics and the rate of evolution of pesticide resistance. In Pesticide resistance: Strategies and tactics for management (pp. 170–193). Washington, DC: National Academy of Science Press.

Messing, R., & Croft, B. A. (1990a). Sublethal influences. In B. A. Croft (Ed.), Arthropod biological control agents and pesticides (pp. 157–183). New York: John Wiley & Sons.

Messing, R., & Croft, B. A. (1990b). Sublethal influences. In B. A. Croft (Ed.), Arthropod biological control agents and pesticides (pp. 185–218). New York: John Wiley & Sons.

Metcalf, R. L. (1972). Selective use of insecticides in pest management. In Implementing practical pest management strategies (pp. 74–91). Proceedings of the National Extension Insect-Pest Management Workshop. Purdue University, Lafayette, IN, March 14–16, 1972.

Messing, R., & Croft, B. A. (1982). Insecticides in pest management. In R. L. Metcalf & W. H. Luckman (Eds.), Introduction to insect pest management (pp. 217–277). New York: John Wiley & Sons.

Metcalf, R. L. (1986). The ecology of insecticides and the chemical control of insects. In M. Kogan (Ed.), Ecological theory and integrated pest management in practice (pp. 251–297). New York: John Wiley & Sons.

Metcalf, C. L., Flint, W. P., & Metcalf, R. L. (1962). Destructive and useful insects—their habitats and control. New York: McGraw-Hill.

Mueller, A. J., & Stern, V. M. (1974). Timing of pesticide treatments on safflower to prevent *Lygus* from dispersing to cotton. Journal of Economic Entomology, 67, 77–80.

Mullin, C. A., & Croft, B. A. (1985). An update on development of selective pesticides favoring arthropod natural enemies. In M. A. Hoy & D. C. Herzog (Eds.), Biological control in agricultural IPM systems (pp. 123–150). Orlando, FL: Academic Press.

Nakashima, M. J., & Croft, B. A. (1975). Toxicity of benomyl to the life stages of *Amblyseius fallacis*. Journal of Economic Entomology, 67, 675–677.

National Research Council. (1996). Ecologically based pest management: New solutions for a new century. Committee on Pest and Pathogen Control through Management of Biological Control Agents and Enhanced Cycles and Natural Processes, Board of Agriculture. Washington, DC: National Academy Press.

Neilson, M. M., Martineau, R., & Rose, A. H. (1971). *Diprion hercyniae* (Hartig), European spruce sawfly (Hymenoptera: Diprionidae). Commonwealth Institute of Biological Control Technical Communications, 4, 136–143.

Newsom, L. D., Smith, R. F., & Whitcomb, W. H. (1976). Selective pesticides and selective use of pesticides. In C. B. Huffaker & P. S. Messenger (Eds.), Theory and practice of biological control (pp. 565–591). New York: Academic Press.

Nicholson, A. J. (1940). Indirect effects of spray practice on pest populations. Proceedings, Sieberte Internationaler Kongress für Entomologie, Berlin, Germany 1938.

Nishida, T., & Bess, H. A. (1957). Studies on the ecology and control of the melon fly, Dacus (Strumeta) cucurbitae Coquillett (Diptera: Tephritidae). (Hawaiian Agricultural Experimental Station Bulletin No. 34). University of Hawaii, Honolulu, Hawaii.

Novozhilov, K. (1976). Biological principles for the rational use of pesticides in the integrated control of agricultural pests. In Proceedings of the U.S.–U.S.S.R. Symposium: Integrated control of the arthropod, disease and weed pests of cotton, grain sorghum and deciduous fruit (pp. 185–193). Lubbock, TX, September 28–October 1, 1975. Publication MP-1276, Texas Agricultural Experiment Station, Texas A & M University, College Station, Texas.

O'Brien, P. J., Elzen, G. W., & Vinson, S. B. (1985). Toxicity of azinphosmethyl and chlordimeform to parasitoid *Bracon mellitor* (Hymenoptera: Braconidae): Lethal and reproductive effects. Environmental Entomology, 14, 891–894.

Oetting, R. D. (1985). Effects of insecticides applied to potting media on *Oenonogastra microrhopalae* (Ashmead) parasitization of *Liriomyza trifolii* (Burgess). Journal of Entomological Science, 20, 405–410.

Parker, B. L., Ming, N. S., Peng, T. S., & Singh, G. (1976). The effect of malathion on fecundity, longevity, and geotropism of *Menochilus sexmaculatus*. Environmental Entomology, 5, 495–501.

Parr, W. J., & Binns, E. S. (1971). Integrated control of red spider mite (Report of the Greenhouse Crops Research Institute 1970, pp. 119–121). Littlehampton, England.

Perera, P. A. C. R. (1982). Some effects of insecticide deposit patterns on the parasitism of *Trialeurodes vaporariorum* by *Encarsia formosa*. Annals of Applied Biology, 101, 239–244.

Peterson, R. K. D. (1996). The status of economic-decision-level development. In L. G. Higley & L. P. Pedigo (Eds.), Economic thresholds for integrated pest management (pp. 151–178). Lincoln, NE: University of Nebraska Press.

Pickford, R. J. J. (1983). Experiences in the control of Thrips tabaci in England (Sting No. 6, September 1983, p. 8). Newsletter on biological control in glasshouses, Naaldwijk: Glasshouse Crops Research Experimental Station.

Pielou, D. P., & Glasser, R. F. (1952). Selection for DDT resistance in a beneficial insect parasite. Science, 115, 117–118.

Plapp, F. W., Jr., & Vinson, S. B. (1977). Comparative toxicities of some insecticides to the tobacco budworm and its ichneumonid parasite. Environmental Entomology, 6, 381–384.

Plewka, T., Kot, J., & Krukierek, T. (1975). Effect of insecticides on the longevity and fecundity of *Trichogramma evanescens* Westw. (Hymenoptera: Trichogrammatidae). Polish Ecological Studies (PECTDR), 1, 197–210.

Poehling, H. M. (1989). Selective application strategies for insecticides in agricultural crops. In P. C. Jepson (Ed.), Pesticides and non-target invertebrates (pp. 151–175). Wimborne, Dorset, United Kingdom: Intercept.

Poehling, H. M., & Dehne, H. W. (1984). Untersuchungen zum Auftreten von Getreideblattlausen an winterweizen unter prektischjen Anbaubedingungen. II. Einfluss von Insektizidbehandlungen auf Blattlauspopulation und Nutzarthropoden. Medelingen van de Faculteit Landbouwwetenschappen, Rijksuniversiteit Gent, 49, 657–665.

Poston, F. L., Pedigo, L. P., & Welch, S. M. (1983). Economic injury levels: Reality or practicality. Bulletin of the Entomological Society of America, 29, 49–53.

Powell, W., Dean, G. J., & Bardner, R. (1985). Effects of pirimicarb, dimethoate and benomyl on natural enemies of cereal aphids in winter wheat. Annals of Applied Biology, 106, 235–242.

Press, J. W., Flaherty, B. R., & McDonald, L. L. (1981). Survival and reproduction of Bracon hebetor on insecticide-treated Ephestia cautella larvae. Journal of Georgia Entomological Society, 16, 227–231.

Rajakulendran, S. V., & Plapp, F. W., Jr. (1982a). Comparative toxicities of 5 synthetic pyrethroids to the tobacco budworm, Heliothis virescens (Lepidoptera: Noctuidae) and ichneumonid parasite Campoletis sonorensis and a predator Chrysopa carnea. Journal Economic Entomology, 75, 769–772.

Rajakulendran, S. V., & Plapp, F. W., Jr. (1982b). Synergism of five pyrethroids by chlordimeform against the tobacco budworm (Lepidoptera: Noctuidae) and a predator, Chrysopa carnea (Neuroptera: Chrysopidae). Journal of Economic Entomology, 75, 1089–1092.

Rathman, R., Johnson, M. W., Rosenheim, J. A., & Tabashnik, B. E. (1990). Carbamate and pyrethroid resistance in the leafminer parasitoid Diglyphus begini (Hymenoptera: Eulophidae). Journal of Economic Entomology, 83, 2153–2158.

Reynolds, H. T. (1958). Research advances in seed and soil treatment with systemic and non-systemic insecticides. Advances in Pest Control Research, 2, 135–182.

Reynolds, H. T. (1971). A world review of the problem of insect population upsets and resurgences caused by pesticide chemicals. In J. E. Swift (Ed.), Agricultural chemicals—harmony or discord for food, people, environment. (pp. 108–112). University of California Division of Agriculture Science. Oakland.

Reynolds, H. T., Fukuto, T. R., Metcalf, R. L., & March, R. B. (1957). Seed treatment of field crops with systemic insecticides. Journal of Economic Entomology, 50, 527–539.

Riedl, H., Barnes, M. M., & Davis, C. S. (1979). Walnut pest management; historical perpective and present status. In D. J. Boethel & R. D. Eikenbary (Eds.), Pest management programs for deciduous tree fruits and nuts (pp. 15–80). New York: Plenum Press.

Ripper, W. E. (1956). Effect of pesticides on balance of arthropod populations. Annual Review of Entomology, 1, 403–438.

Risch, S. J. (1987). Agricultural ecology and insect outbreaks. In Barbosa & J. C. Schultz (Eds.), Insect outbreaks (pp. 217–238). San Diego: Academic Press.

Roach, S. H., & Morse, R. F. (1988). Effects of abamectin on Phidippus audax (Hentz) (Araneae: Salticidae) when ingested from prey. Journal of Entomology and Science, 23, 112–116.

Robertson, J. G. (1957). Changes in resistance to DDT in Macrocentrus ancylivorus Rohw. (Hymenoptera: Braconidae). Canadian Journal of Zoology, 35, 629–633.

Rosen, D. (1967). Effect of commercial pesticides on the fecundity and survival of Aphytis holoxanthus (Hymenoptera: Aphelinidae). Israel Journal of Agriculture, 17, 47–52.

Rosenheim, J. A., & Hoy, M. A. (1986). Intraspecific variation in levels of pesticide variation in field populations of a parasitoid, Aphytis melinus (Hymenoptera: Aphelinidae): The role of past selection pressures. Journal of Economic Entomology, 79, 1161–1173.

Rosenheim, J. A., & Hoy, M. A. (1988a). Sublethal effects of pesticides on the parasitoid Aphytis melinus (Hymenoptera: Aphelinidae). Journal of Economic Entomology, 81, 476–483.

Rosenheim, J. A., & Hoy, M. A. (1988b). Genetic improvement of a parasitoid biological control agent: Artificial selection for insecticide resistance in Aphytis melinus (Hymenoptera: Aphelinidae). Journal of Economic Entomology, 81, 1539–1550.

Satpathy, J. M., Padhi, G. K., & Dutta, D. N. (1968). Toxicity of eight insecticides to the coccinellid predator Chilomenes sexmaculata F. Indian Journal of Entomology, 27, 72–75.

Schoones, J., & Giliomee, J. H. (1984). The residual toxicity of field-weathered thripicide and spray oil residues on citrus leaves to Aphytis

sp., a parasitoid of the citrus red scale, Aonidiella aurantii (Mask). Citrus Subtropical Fruit Journal, 602, 6–8.

Schulten, G. G. M., van de Klashorst, G., & Russell, V. M. (1976). Resistance of Phytoseiulus persimilis A. H. (Acari: Phytoseiidae) to some insecticides. Zeitschrift fuer Angewandte Entomologie, 80, 337–341.

Sharma, H. C., & Adlakha, R. L. (1981). Selective toxicity of some insecticides to the adults of ladybird beetle, Coccinella septempunctata, and cabbage aphid, Brevicoryne brassicae. Indian Journal of Entomology, 43, 92–99.

Singh, S. P. (1989). Achievements of AICRP on biological control of crop pests and weeds (Bulletin No. 2). Bangalore 560-024: Indian Council of Agricultural Research.

Smith, R. F. (1970). Pesticides: Their use and limitations in pest management. In R. L. Rabb & F. E. Guthrie (Eds.); Concepts of pest management (pp. 103–113). Raleigh, NC: North Carolina State University.

Solomon, M. G. (1987). Fruit and hops. In A. J. Burn, T. H. Coaker, & P. C. Jepson (Eds.), Integrated pest management (pp. 329–360). London: Academic Press.

Southwood, T. R. E. (1978). Ecological methods with particular reference to the study of insect populations. London: Chapman & Hall.

Spollen, K. M., & Hoy, M. A. (1992). Resistance to carbaryl in populations of the parasite Aphytis melinus DeBach (Hymenoptera: Aphelinidae): Implications for California citrus IPM. Biological Control, 2, 203–210.

Steiner, L. F., Mitchell, W. C., Harris, E. J., Kozuma, T. T., & Fujimoto, M. S. (1965). Oriental fruit fly eradication by male annihilation. Journal of Economic Entomology, 58, 961–964.

Steiner, L. F., Hart, W. G., Harris, E. J., Cunningham, R. T., Ohinata, K., & Tamakhi, D. C. (1970). Eradication of the oriental fruit fly from the Mariana Islands by the methods of male annihilation and sterile insect release. Journal of Economic Entomology, 63, 131–135.

Stern, V. M., & van den Bosch, R. (1959). The integration of chemical and biological control of the spotted alfalfa aphid. II. Field experiments on the effects of pesticides. Hilgardia, 29, 103–130.

Stern, V. M., Smith, R. F., van den Bosch, R., & Hagen, K. S. (1959). The integrated control concept. Hilgardia, 29, 81–101.

Swirski, E., Amitai, S., & Dorzia, N. (1967). Field and laboratory trials on the toxicity of some pesticides to predacious mites (Acarina: Phytoseiidae). Ktavim, 17, 149–159.

Tabashnik, B. E. (1986). Evolution of pesticide resistance in predator/prey systems. Bulletin of the Entomological Society of America, 32, 156–161.

Tanigoshi, L. K., Hoyt, S. C., & Croft, B. A. (1983). Basic biological and management components for mite pests and their natural enemies. In B. A. Croft & S. C. Hoyt (Eds.), Integrated management of insect pests of pome and stone fruits (pp. 153–202). New York: Wiley-Interscience.

Teetes, G. L., Brothers, G. W., & Ward, C. R. (1973). Insecticide screening for greenbug control and effect on certain beneficial insects (PR-3166). Texas Agricultural Experimental Station, Texas A & M University, College Station, Texas.

Theiling, K. M., & Croft, B. A. (1988). Pesticide effects on arthropod natural enemies: A database summary. Agricultural Ecosystems of the Environment, 21, 191–218.

Tipping, P. W., & Burbutis, P. P. (1983). Some effects of pesticide residues on Trichogramma nubilale (Hymenoptera: Trichogrammatidae). Journal of Economic Entomology, 76, 892–896.

Turnipseed, S. G. (1972). Management of insect pests of soybeans. Proceedings, Tall Timbers Conference on Ecology and Animal Control Habitat Management, 4, 189–203.

Turnipseed, S. G., Todd, J. W., Greene, G. L., & Bass, M. H. (1974). Minimum rates of insecticides on soybeans: Mexican bean beetle, green cloverworm, corn earworm and velvetbean caterpillar. Journal of Economic Entomology, 67, 287–291.

Turnock, W. J., Taylor, K. L., Schroder, D., & Dahlsten, D. L. (1976).

Biological control of pests of coniferous forests. In C. B. Huffaker & P. S. Messenger (Eds.) (pp. 289–311). Theory and practice of biological control. New York: Academic Press.

Umoru, P. A., Powell, W., & Clark, S. J. (1996). Effect of pirimicarb on the foraging behavior of *Diaeretiella rapae* (Hymenoptera: Braconidae) on host-free and infested oilseed rape plants. Bulletin of Entomological Research, 86, 193–201.

van den Bosch, R., Messenger, P. S., & Gutierrez, A. P. (1982). An introduction to biological control. New York: Plenum Press.

van Emden, H. F., Peakall, D. B. (1996). Beyond silent spring. London: Chapman & Hall.

Vitre, J. P. (1971). Silviculture and the management of bark beetle pests. Proceedings, Tall Timbers Conference on Ecology and Animal Control Habitat Management, 3, 155–168.

Vorley, W. T., & Dittrich, V. (1994). Integrated pest management and resistance management systems. Review of Environmental Contamination and Toxicology, 139, 179–193.

Waage, J. K. (1989). The population ecology of pest-pesticide-natural enemy interactions. In P. C. Jepson (Ed.), Pesticides and non-target invertebrates (pp. 81–93). Wimborne, Dorset, United Kingdom: Intercept.

Waage, J. K., Hassell, M. P., & Godfray, H. C. J. (1985). The dynamics of pest-parasitoid-insecticide interactions. Journal of Applied Ecology, 22, 825–838.

Wainhouse, D. (1987). Forests. In A. J. Burn, T. H. Coaker, & P. C. Jepson (Eds.), Integrated pest management (pp. 361–401). London: Academic Press.

Watson, T. F. (1975). Practical consideration in use of selective insecticides against major crop pests. In J. C. Street (Ed.), Pesticide selectivity (pp. 47–65). New York: Marcel Dekker.

Way, M. J. (1986). The role of biological control in integrated plant protection. In J. M. Franz (Ed.), Biological plant and health protection (pp. 289–303). Stuttgart: Gustav Fischer Verlag.

Weires, R. W., Leeper, J. R., Reissig, W. H., & Lienk, S. E. (1982). Toxicity of several insecticides to the spotted tentiform leafminer (Lepidoptera: Gracillariidae) and its parasite, *Apanteles ornigis*. Journal of Economic Entomology, 75, 680–684.

Westigard, P. H. (1971). Integrated control of spider mites on pear. Journal of Economic Entomology, 64, 496–501.

Wiedl, S. C. (1977). The effects of sublethal concentrations of dieldrin on the predatory efficiency of *Toxorhynchites brevipalpis*. Environmental Entomology, 6, 709–711.

Wiles, J. A., & Jepson, P. C. (1994). Sub-lethal effects of deltamethrin residues on the within-crop behavior and distribution of *Coccinella septempunctata*. Entomologia Experimentalis et Applicata, 72, 33–45.

Wilson, F., & Huffaker, C. B. (1976). The philosophy, scope, and importance of biological control. In C. B. Huffaker & P. S. Messenger (Eds.), Theory and practice of biological control (pp. 3–15). New York: Academic Press.

Wilson, L. T. (1985). Estimating the abundance and impact of arthropod natural enemies in IPM systems. In M. A. Hoy & D. C. Herzog (Eds.), Biological control in agricultural IPM systems (pp. 303–322). Orlando, FL: Academic Press.

Wilson, L. T., Hoy, M. A., Zalom, F. G., & Smilanick, J. M. (1984). Sampling mites on almonds. I. The within-tree distribution and clumping pattern of mites with comments on predator-prey interactions. Hilgardia, 52, 1–13.

Wilson, M. C., & Armbrust, E. J. (1970). Approach to integrated control of the alfalfa weevil. Journal of Economic Entomology, 63, 554–557.

Wright, D. J., & Verkerk, R. H. J. (1995). Integration of chemical and biological control systems for arthropods: Evaluation in a multitrophic context. Pesticide Science, 44, 207–218.

14

Environmental Management to Enhance Biological Control in Agroecosystems

D. K. LETOURNEAU

Department of Environmental Studies
University of California
Santa Cruz, California

M. A. ALTIERI

Division of Insect Biology
University of California
Berkeley, California

INTRODUCTION

The efficacy of parasitoids and predators as agents of biological control depends on conditions set by production technologies such as varietal development, cropping systems, tillage practices, and chemical inputs. Yet trends in agriculture over the last several decades, which include decreasing environmental heterogeneity, increasing fertilizer and pesticide input, increasing mechanization, and decreasing genetic diversity (USDA, 1973; Bottrell, 1980; Whitham, 1983; Altieri & Anderson, 1986; Levins, 1986; Giampietro, 1997), create agricultural environments that pose serious impediments to pest population regulation by natural enemies. In particular, drastic reductions in plant biodiversity and the resulting epidemic effects can adversely affect ecosystem function with further consequences on agricultural productivity and sustainability (Fig. 1). The development and current acceptance of integrated pest management (IPM) programs, the cumulative restrictions and bans on various pesticides, and the growing concerns about pesticide contamination and increased production costs are among the reasons for increasing research efforts toward long-term alternatives and supplements to chemical insecticides. Since van den Bosch and Telford (1964) wrote their pioneering chapter describing habitat modifications to encourage satisfactory biological control of insect pests in agroecosystems, numerous articles have appeared documenting the importance of manipulating the environmental properties of crop fields to make them more favorable to natural enemies and less amenable for insect pests (van Emden & Williams, 1974; Price & Waldbauer, 1975; Perrin, 1980; Cromartie, 1981; Thresh, 1981; Altieri & Letourneau, 1982, 1984; Risch *et al.,* 1983; Herzog &

Funderburk, 1985; van Emden, 1988; Andow, 1991; Altieri *et al.,* 1993; Tonhasca & Byrne, 1994; Altieri, 1994; Wratten & Van Emden, 1995). In the last 20 years, the majority of efforts have been directed at analyzing the effects of reduced tillage and vegetational diversification of agroecosystems. The effects of pest resistant varieties on the suitability of hosts or prey for natural enemies have also been studied. Research is targeting transgenic plants (Johnson & Gould, 1992; Johnson, 1997; Riggin-Bucci & Gould, 1997) to evaluate their compatibility with biological control agents. Research on other types of cultural manipulation such as strip harvesting, trap cropping, and use of nests or artificial shelter has been minimal, with the exception of advances in the use of food sprays (Hagen, 1986) and kairomones (Lewis *et al.,* 1976; Nordlund *et al.,* 1981a, 1981b, 1988; Lewis & Martin, 1990; Metcalf, 1994) to enhance the activity of specific natural enemies.

A virtual explosion of literature reporting the effects of multiple cropping systems on insect dynamics has emerged (Root, 1973; Litsinger & Moody, 1976; Perrin, 1977; Altieri *et al.,* 1978; Perrin & Phillips, 1978; Bach, 1980a, 1980b; Risch, 1980, 1981; Andow, 1983b; Letourneau & Altieri, 1983; Altieri & Liebman, 1986; Andow, 1991; Parajulee *et al.,* 1997; Skovgard & Pats, 1996). This accumulated information provides a basis for designing crop systems with vegetational attributes that enhance reproduction, survival, and efficacy of natural enemies. Unfortunately, because agricultural land use is dictated mainly by economic forces, pest control plans are seldom made on the basis of habitat management. Economies of scale that dominate the agricultural sector in developed countries allow farmers to reduce unit production costs by increasing farm size and by becoming more specialized. For this rea-

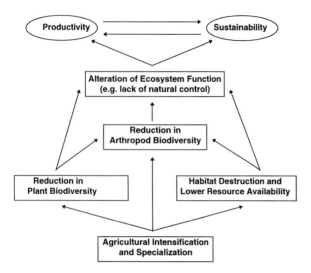

FIGURE 1 Lines of influence between intensification, arthropod biodiversity, and properties of agricultural ecosystems.

son, environmental manipulation strategies with demonstrated effectiveness in experimental systems, such as cotton/alfalfa strip cropping for management of *Lygus* spp. in cotton (Stern *et al.*, 1964), or the use of *Rubus* sp. plantings and now *Prunus* sp. around vineyards for conservation of grape leafhopper parasitoids (Doutt & Nakata, 1973) have not been adopted on a regional scale. We recognize that the prevailing political and economic context of modern farming does not support the maintenance of landscape diversity. This has been one of the main obstacles to the implementation of many of the alternative strategies reviewed in this chapter. However, increasing interests in sustainability (Francis *et al.*, 1990; National Research Council, 1991) and biodiversity (LaSalle & Gauld, 1993; Marc & Canard 1997; Stinner *et al.*, 1997), coupled with global marketing strategies, have resulted in distinct changes in policy recommendations that have begun to encourage agricultural diversification on different scales in the 1990s (e.g., Barghouti *et al.*, 1992; Holt *et al.*, 1995).

Rabb *et al.* (1976) characterize the effective environment of an organism as (1) weather, (2) food, (3) habitat (shelter, nests), and (4) other organisms. Thus, environmental management for biological control concerns the functional environment: the physical and biotic elements that directly or indirectly impact survival, migration, reproduction, feeding, and behaviors associated with these processes. Although pest populations can be controlled directly through cultural control methods that modify the habitat, we do not offer a thorough review of this topic. Instead, our objectives are to discuss conservation (maintenance of natural-enemy abundance and diversity) and enhancement (via increased immigration, tenure, longevity, fertility, and efficiency) strategies that can be used to manipulate natural enemies in

agroecosystems. In general, habitat management regimes to increase natural-enemy effectiveness are directed at: (1) enhancing habitat suitability for immigration and host finding, (2) provision of alternative prey/hosts at times when the pest is scarce, (3) provision of supplementary food (food sprays, pollen and nectar for predators and parasitoids), (4) provision of refugia (for mating or overwintering), and (5) maintenance of noneconomic levels of the pest or alternative hosts over extended periods to ensure continued survival of natural enemies. Examples of cropping techniques that enhance a particular parasitoid population through effects on some of the preceding processes are shown in Table 1, and have been reviewed by Powell (1986). Useful approaches borrowed from conservation biology are reviewed in Letourneau (1998).

The manipulation of entomophagous insects must be approached from several levels, from agroecosystem processes to ecophysiological features of individual organisms. The range of manipulatable elements and their degree of flexibility depend on characteristics of the agroecosystem; these linkages will be described in a section on agricultural habitats. A review of research on modes of environmental management will summarize current knowledge on tactics of input, removal, or adjustments of biotic and physical/chemical elements of the agroecosystem by manual or mechanical means. This will be followed by a discussion of the biological mechanisms involved in natural-enemy responses to cultural practices and a section on relevant ecological theory to stimulate further research on this topic. Thus, we consider the role, methods, and future directions of environmental management as a prophylactic control strategy (*sensu* Vandermeer & Andow, 1986) in a variety of agroecosystems.

CLASSIFICATION OF AGROECOSYSTEMS

Each region has a unique set of agroecosystems that result from climate, topography, soil, economic relations, social structure, and history. Various farming system features can be modified and some can be expected to affect the dynamics of insect populations (Table 2). The agroecosystems of a region often include both commercial and local-use agricultures that rely on technology to different extents depending on the availability of land, capital, and labor. Some technologies in modern systems aim at efficient land use (reliance on agrochemical inputs), while others reduce labor (mechanical inputs). In contrast, resource-poor farmers usually adopt low technology, labor-intensive practices that optimize production efficiency and recycle scarce resources (Mattson *et al.*, 1984; Altieri, 1989).

Common, area-wide environmental management techniques are difficult to design and implement given the differences in climate, agricultural products, and economic

TABLE 1   Examples of Environmental Manipulation Resulting in Increased Parasitism of Specific Insect Pests in a Range of Cropping Systems

Environmental management practice	Parasitoid(s) affected	Pest regulated	Cropping system or environmental diversity element
Use of kairomones	*Trichogramma* spp.	*Heliothis zea* (Boddie)	Soybeans, tomatoes, corn
	Microplitis croceipes (Cresson) (Hymenoptera: Braconidae)	*H. zea*	Soybeans, tomatoes, corn
	Cardiochiles nigriceps Vier. (Hymenoptera: Braconidae)	*H. virescens* (F.)	Soybeans, tomatoes, corn
Provision of alternative hosts	*Macrocentrus* spp. (Hymenoptera: Braconidae)	*Cydia molesta* (Busck) (Lepidoptera: Olethreutidae)	Weeds in peaches
	Lydella grisescens R.-D. (Diptera: Tachinidae)	*Ostrinia nubilalis* (Hübner) (Lepidoptera: Pyralidae)	Ragweeds in maize
	Horogenes spp.	*Plutella xylostella*	*Crataegus* in cabbage
	Anagrus epos (Girault) (Hymenoptera: Mymaridae)	*Erythroneura elegantula* Osborn (Homoptera: Cicadellidae)	Blackberries in vineyards
	Eurytoma tylodermatis Ashm. (Hymenoptera: Eurytomidae)	*Anthonomus grandis* Boheman (Coleoptera: Curculionidae)	Ragweeds in cotton
	Emersonella niveipes Girault (Hymenoptera: Eulophidae)	*Chelymorpha cassidea* (F.) (Coleoptera: Chrysomelidae)	Morning glories in sweet potatoes
	Lysiphlebus testaceipes Cress. (Hymenoptera: Aphidiidae)	*Schizaphis graminum* (Rodani) (Homoptera: Aphididae)	Sunflowers near sorghum
Provision of food other than host	*Tiphia popilliavora* Rohwer (Hymenoptera: Tiphiidae)	*Phyllophaga* spp. (Coleoptera: Scarabeidae)	Nectar from weeds; honeydew from scale insects in various crops
		Lachnosterna spp. (Coleoptera: Scarabeidae)	
	Aphelinus mali (Haldeman) (Hymenoptera: Aphelinidae)	Aphids	Nectar from the honeyplants *Phacelia* and *Eryngium* in apples
	Cotesia medicaginis (Hymenoptera: Braconidae)	*Colias philodice* Godart (Lepidoptera: Pieridae)	Nectar from weeds; honeydew from aphids in alfalfa
	Aphytis proclia (Walker) (Hymenoptera: Aphelinidae)	*Quadraspidiotus perniciosus* (Comstock) (Homoptera: Diaspididae)	Nectar from the honeyplant *Phacelia tanacetifolia* Benth. (Hydrophyllaceae) in orchards
	Various species	*Malacosoma americanum* (F.) (Lepidoptera: Lasiocampidae)	Nectar from weeds in apples
		Laspeyresia (Cydia) pomonella (Busck) (Lepidoptera: Olethuridae)	
	Lixophaga sphenophori (Villeneuve) (Diptera: Tachinidae)	*Rhabdoscelus obscurus* (Boisduval) (Coleoptera: Curculionidae)	Nectar from *Euphorbia* spp. weeds in sugarcane
Adoption of special cropping practices	*Macrocentrus ancylivorus* (Hymenoptera: Braconidae)	*Ancylis comptana* (Froelich) (Lepidoptera: Olethreutidae) *Cydia molesta*	Intercropping strawberries and peaches
	Bathyplectes curculionis (Thomson) (Hymenoptera: Ichneumonidae)	*Hypera postica* (Gyllenhal) (Coleoptera: Curculionidae)	Time of first cutting of alfalfa
	Aphidiids	Aphids	Strip harvesting of alfalfa
	Lysiphlebus testaceipes	*S. graminum*	Resistant cultivars of barley and soybeans
	Anagrus sp.	*Empoasca kraemeri* Ross & Moore (Coleoptera: Cicadellidae)	Mixed cropping maize and beans
	Trichogramma spp. (Hymenoptera: Trichogrammatidae)	*Heliothis zea*	Mixed cropping maize and soybeans

After Cromartie, 1981; Altieri & Letourneau, 1982; and Powell, 1986.

TABLE 2   Agroecosystem Determinants for the Type of
Agriculture in a Given Region

Type of determinants	Factors	Manipulation amenability[a]
Physical	Radiation	
	Temperature	
	Rainfall	
	Irrigation	X
	Soil conditions	X
	Slope	
	Land availability	
Biological	Insect pests and natural enemies	X
	Weed communities	X
	Plant and animal diseases	
	Soil biota[b]	
	Background natural vegetation	X
	Photosynthetic efficiency	
	Cropping patterns	X
	Crop rotation	X
Socioeconomic	Population density	
	Social organization	X
	Economic (prices, markets, capital and credit availability)	X
	Technical assistance	X
	Cultivation implements	
	Degree of commercialization	X
	Labor availability	X
Cultural	Traditional knowledge	
	Ideology	
	Gender issues	
	Historical events	

[a] Factor amenable to manipulation through habitat management and expected to directly or indirectly affect the impact of arthropod natural enemies as biological control agents.

[b] As influenced by fertilizing regimes, composting, etc.

and political structure of each agricultural system. Many farming systems are in transition, with changes forced by shifting resource needs, unequal resource availability, environmental degradation, economic growth or stagnation, political change, etc. Strategies amenable to labor-intensive operations will be radically different from those designed for mechanized, large-scale operations. Specialization and concentration of crops are the most important factors limiting the application of many environmental management options for a particular region.

All agroecosystems are dynamic and subjected to different levels of management so that the crop arrangements in time and space are continually changing in the face of biological, cultural, socioeconomic, and environmental factors. Such landscape variations determine the degree of spatial and temporal heterogeneity characteristic of agricultural regions, which in turn conditions the type of biodiversity present, in ways that may or may not benefit the pest protection of particular agroecosystems. Thus, one of

the main challenges facing agroecologists today is identifying the types of biodiversity assemblages (either at the field or at the regional level) that will yield desirable agricultural results (i.e., pest regulation). This challenge can only be met by further analyzing the relationship between vegetation diversification and population dynamics of herbivore and natural-enemy species, in light of what the unique environment and entomofauna of each are and the diversity and complexity of local agricultural systems.

According to Vandermeer and Perfecto (1995), two distinct components of biodiversity can be recognized in agroecosystems. The first component, planned biodiversity, is the biodiversity that is associated with the crops and livestock purposely included in the agroecosystem by the farmer, and that will vary depending on management inputs and crop spatial–temporal arrangements. The second component, associated biodiversity, includes all soil floras and faunas, herbivores, carnivores, decomposers, etc. that colonize the agroecosystem from surrounding environments and that will thrive in the agroecosystem depending on its management and structure. The relationship of both biodiversity components is illustrated in Fig. 2. Planned biodiversity has a direct function, as illustrated by the bold arrow connecting the planned biodiversity box with the ecosystem function box. Associated biodiversity also has a function, but it is mediated through planned biodiversity. Thus, planned biodiversity also has an indirect function, illustrated by the dotted arrow in the figure, which is realized through its influence on the associated biodiversity. For example, the trees in an agroforestry system create shade, which makes

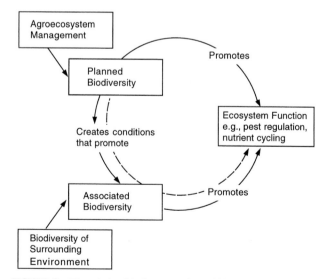

FIGURE 2   The relationship between planned biodiversity (that which the farmer determines, based on management of the agroecosystem) and associated biodiversity (that which colonizes the agroecosystem after it has been set up by the farmer) and how the two promote ecosystem function. (Modified from Vandermeer and Perfecto, 1995.)

it possible to grow only sun-intolerant crops. Thus, the direct function of this second species (the trees) is to create shade. Yet along with the trees might come small wasps that seek out the nectar in the tree's flowers. These wasps may in turn be the natural parasitoids of pests that normally attack the crops. The wasps are part of the associated biodiversity. The trees create shade (direct function) and attract wasps (indirect function) (Vandermeer & Perfecto, 1995).

The key is to identify the type of biodiversity that is desirable to maintain and/or to enhance the implementation of ecological services, and then to determine the best practices that will encourage the desired biodiversity components. As shown in Fig. 4, there are many agricultural practices and designs that have the potential to enhance functional biodiversity, and others that negatively affect it. The idea is to apply the best management practices to enhance and/or to regenerate the kind of biodiversity that can subsidize the sustainability of agroecosystems by providing ecological services such as biological pest control but also nutrient cycling, water and soil conservation, and so on.

Although each farm is unique, farms can be classified by type of agriculture or agroecosystem. Functional grouping is essential for devising area-wide habitat management strategies. Five criteria can be used to classify agroecosystems in a region: (1) the types of crop and livestock; (2) the methods used to grow the crops and produce the stock; (3) the relative intensity of use of labor, capital, and organization, as well as the resulting output of product; (4) the

disposal of the products for consumption (whether used for subsistence or supplement on the farm or sold for cash or other goods); and (5) the structures used to facilitate farming operations (Norman, 1979).

Based on these criteria, it is possible to recognize seven main types of agricultural systems in the world (Grigg, 1974):

1. Shifting cultivation systems
2. Semi-permanent rain-fed cropping systems
3. Permanent rain-fed cropping systems
4. Arable irrigation systems
5. Perennial crop systems
6. Grazing systems
7. Systems regulated by farming practices (alternating arable cropping and sown pasture)

Systems 4 and 5 have evolved into habitats that are much simpler in form and poorer in species than the others, which can be considered more diversified, permanent, and less disturbed, therefore inherently containing elements of natural pest control (Fig. 3). Clearly then, modern systems require more radical modifications of their structure to approach a more diversified, less disturbed state. Figure 4 provides a series of diversification options for such systems. (Alfalfa is treated here as a quasi-perennial crop.) If it is argued that such modifications are not possible in large-scale agriculture due to technical or economic factors, then this point provides a strong conservative argument in favor of small, multiple-use farms.

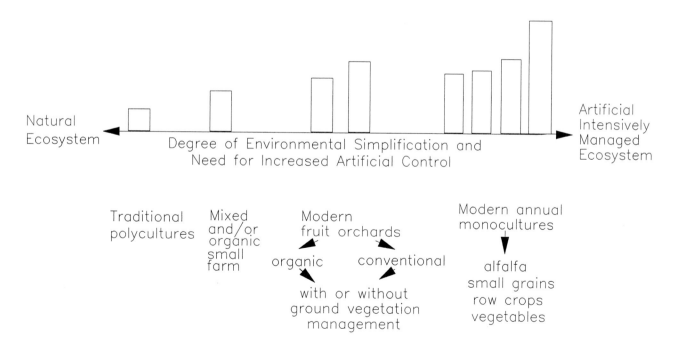

FIGURE 3   Management intensity continuum between ecosystems and agricultural ecosystems in relation to the need for artificial pest control measures.

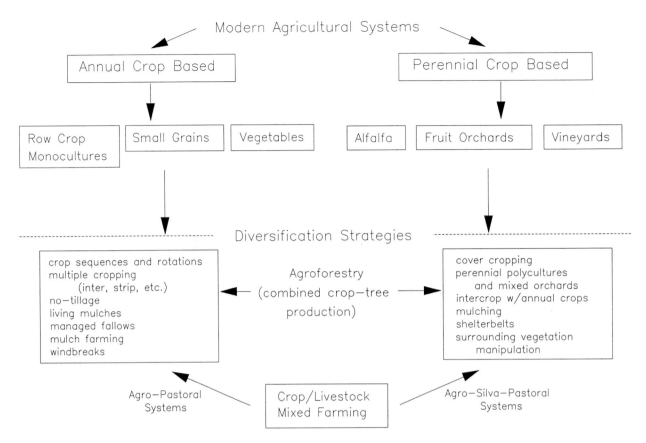

FIGURE 4   Plant diversification strategies in annual and perennial crop systems.

MODES OF ENVIRONMENTAL MANAGEMENT

Vegetational designs across appropriate levels of scale comprise an obvious form of environmental management. At the regional level, landscape vegetation mosaics influence the distribution of food and shelter resources, and, therefore, the colonization patterns of insects (Andow, 1983b; Ryszkowski *et al.,* 1993). On a smaller scale, herbivores and their natural enemies respond to localized patterns of plant spacing, plant structure, and plant species (or varietal) diversity. Environmental components and their management in agroecosystems have three main dimensions: temporal, spatial, and biological (Fig. 5). Other means of biotic management through inundative releases and classical biological control are discussed elsewhere (Chapter 11). Mechanical modes of environmental management, such as cultivating, mowing, and harvesting, affect the structure and permanence of the habitat and thus the life processes of insects in agroecosystems. Chemical inputs, such as the periodic application of water, fertilizers, behavior-modifying agents, and pesticides, affect rates of growth and survival of pests and natural-enemy populations. The impact of pesticides will be treated here only in terms of drift effects (see Chapters 13 and 24). Biotic, physical,

chemical, and mechanical manipulations are imposed on agroecosystems often as a means to achieve objectives unrelated to insect pest management. Nevertheless, the range of environmental manipulations designed for higher yields is sufficiently broad to permit incorporation of tactics that simultaneously improve pest control (see Table 2).

Vegetation Management

Frequently disturbed monocultures favor the rapid colonization and growth of herbivore populations; initial conditions of enemy-free space and high abundance of pests further reduce the ability of natural enemies to regulate them (Price, 1981). In most farming situations these negative factors can be minimized or eliminated by providing continuity of vegetation (and the associated food and shelter) in time and space, thus aiding natural enemies (Table 3).

Many studies document direct behavioral and physiological effects of plants on natural enemies (van Emden, 1965; Leius, 1967; Campbell & Duffey, 1979; Nettles, 1979; Altieri *et al.,* 1981; Boethel & Eikenbary, 1986; De la Cruz & Litsinger, 1986; Elzen *et al.,* 1986; Letourneau, 1987, 1990; Dicke & Sabelis, 1988; Panizzi, 1988; Barker &

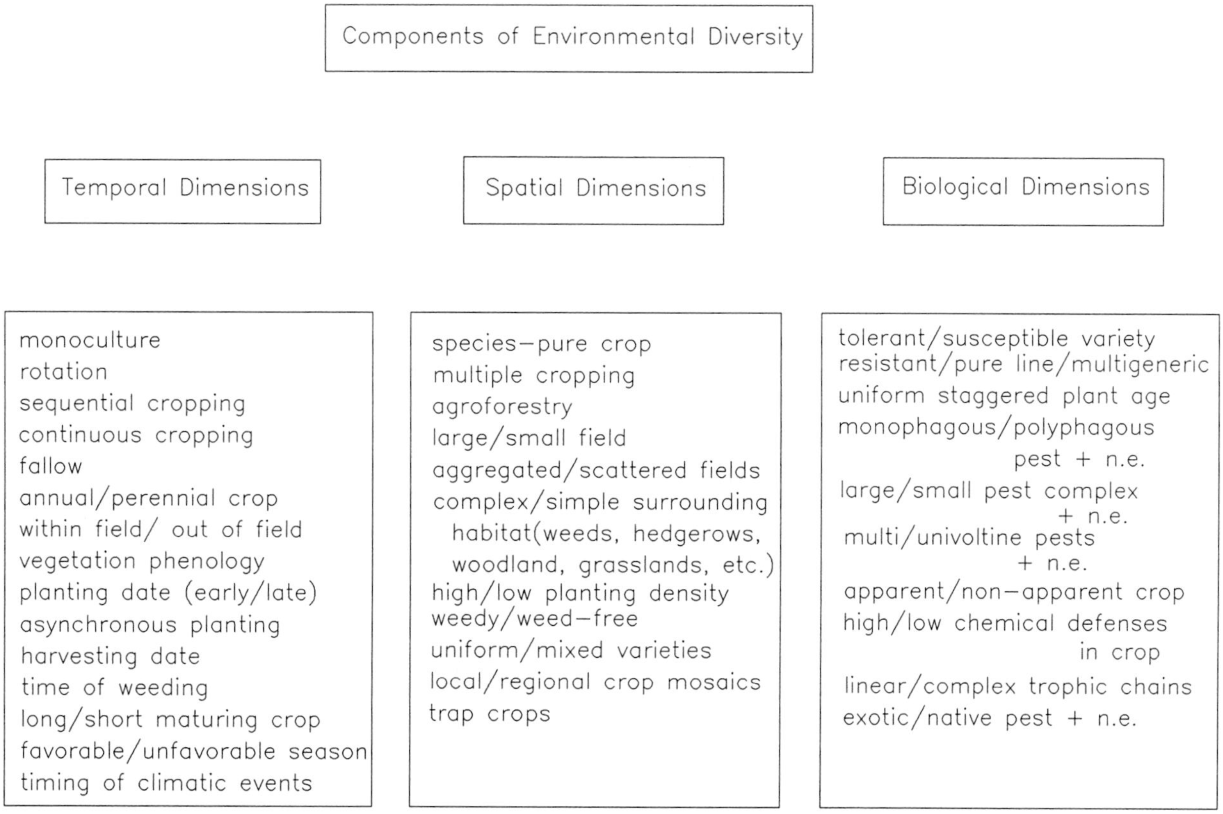

```
┌─────────────────────────────────────────────┐
│        Components of Environmental Diversity  │
└─────────────────────────────────────────────┘
```

```
┌─────────────────────┐   ┌──────────────────┐   ┌──────────────────────┐
│ Temporal Dimensions │   │ Spatial Dimensions│   │ Biological Dimensions │
└─────────────────────┘   └──────────────────┘   └──────────────────────┘
```

Temporal Dimensions	Spatial Dimensions	Biological Dimensions
monoculture	species—pure crop	tolerant/susceptible variety
rotation	multiple cropping	resistant/pure line/multigeneric
sequential cropping	agroforestry	uniform staggered plant age
continuous cropping	large/small field	monophagous/polyphagous
fallow	aggregated/scattered fields	pest + n.e.
annual/perennial crop	complex/simple surrounding	large/small pest complex
within field/ out of field	habitat(weeds, hedgerows,	+ n.e.
vegetation phenology	woodland, grasslands, etc.)	multi/univoltine pests
planting date (early/late)	high/low planting density	+ n.e.
asynchronous planting	weedy/weed—free	apparent/non—apparent crop
harvesting date	uniform/mixed varieties	high/low chemical defenses
time of weeding	local/regional crop mosaics	in crop
long/short maturing crop	trap crops	linear/complex trophic chains
favorable/unfavorable season		exotic/native pest + n.e.
timing of climatic events		

FIGURE 5   Temporal, spatial, and biotic components of environmental diversification of agricultural ecosystems (n.e. = natural enemies of arthropod pests).

Addison, 1997). Some entomophagous insects are more abundant in the presence of particular plants even in the absence of hosts or prey, or they are attracted to or arrested by chemicals released by the herbivore's host plant or other associated plants (Monteith, 1960; Shahjahan, 1974; Nettles, 1979; Martin *et al.*, 1990; Geitzenauer & Bernays, 1996; Stapel *et al.*, 1997). Parasitism of a pest can be higher on some crops than on others (Read *et al.*, 1970; Martin *et al.*, 1976; Nordlund *et al.*, 1985; Johnson & Hara, 1987; Gerard, 1989; Lewis & Gross, 1989; Idris & Grafius, 1996; Bautista & Harris, 1996), and different parasitoids can be associated with particular crops or crop habitats even if the same host is attacked (Lewis & Gross, 1989; Felland, 1990; van den Berg *et al.*, 1990).

Noncrop plants within and around fields can be used to benefit biological control agents (Altieri & Whitcomb, 1979a, 1979b; Barney *et al.*, 1984; Norris, 1986; Bugg & Wilson, 1989; Kiss *et al.*, 1997). Rapidly colonizing, fast-growing plants offer many important requisites for natural enemies—such as alternative prey or hosts, pollen or nectar—as well as microhabitats that are not available in large-scale, weed-free monocultures (van Emden, 1965; Doutt & Nakata, 1973; Cerutti *et al.*, 1989; Idris & Grafius, 1995),

but these interactions can be difficult to define precisely and to implement in control programs (Flaherty *et al.*, 1985). Outbreaks of certain types of crop pests are more likely to occur in weed-free fields than in weed-diversified crop systems (Dempster, 1969; Flaherty, 1969; Root, 1973; Smith, 1976a; Altieri *et al.*, 1977; Ofuya, 1989). Crop fields with dense weed cover and high diversity usually have more predaceous arthropods than do weed-free fields (Pimentel, 1961; Dempster, 1969; Flaherty, 1969; Pollard, 1971; Root, 1973; Smith, 1976b; Speight & Lawton, 1976; Bottenberg *et al.*, 1990; Wyss, 1995; Schellhorn & Sork, 1997). Carabidae (Dempster, 1969; Speight & Lawton, 1976; Thiele, 1977), Syrphidae (Pollard, 1971; Smith, 1976b; Cowgill, 1989), and Coccinellidae (Bombosch, 1966; Perrin, 1975) are abundant in weed-diversified systems. Relevant examples of cropping systems in which the presence of specific weeds has enhanced biological control of particular pests are given in Table 3. The potential for managing weeds as useful components of agroecosystems is great, but not all weeds promote biological control (Powell *et al.*, 1986) even if entomophagous insects are more abundant in the system (Gaylor & Foster, 1987; Bugg *et al.*, 1991).

TABLE 3   Selected Examples of Crop–Noncrop or Crop–Crop Combinations That Enhanced the
Biological Control of Specific Crop Pests

Crop	Weed species or intercrop	Pest(s) regulated	Factor(s) involved	References
Apples	*Phacelia* sp. and *Eryngium* sp.	San Jose scale [*Quadraspidiotus perniciosus* (Comstock)] and aphids	Increased abundance and activity of parasitic wasps (*Aphelinus mali* and *Aphytis proclia*)	Telenga, 1958
Apples	Natural weed complex	*Dysaphis plantaginea* Pass. and *Aphis pomi* (DeGeer)	Increased abundance of aphidophagous predators	Wyss, 1995
Apples	Natural weed complex	Tent caterpillar (*Malacosoma americanum*) and codling moth [*Laspeyresia pomonella* (L.)] (Lepidoptera: Olethreutidae)	Increased activity and abundance of parasitic wasps	Leius, 1967
Beans	Natural weed complex	*Epilachna varivestis* L.	Increased predation on eggs	Andow, 1990
Brassica	Beans	*Brevicoryne brassicae* and *Delia brassicae*	Higher predation and disruption of oviposition behavior	Tukahirwa & Coaker, 1982
Brussels sprouts	Natural weed complex	Imported cabbage butterfly [*Pieris rapae* (L.)] (Lepidoptera: Pieridae) and aphids [*Brevicoryne brassicae* (L.)] (Homoptera: Aphididae)	Alteration of crop background and increase of predator colonization	Smith, 1976b; Dempster, 1969
Brussels sprouts	*Spergula arvensis* L. (Caryophyllaceae)	*Mamestra brassicae* L. (Lepidoptera: Noctuidae), *Evergestis forficalis* L. (Lepidoptera: Pyralidae), *B. brassicae*	Increase of predators and interference with colonization	Theunissen & den Ouden, 1980
Brussels sprouts	*Brassica kaber* (DC.) (Brassicaceae)	*B. brassicae*	Attraction of syphids [*Allograpta obliqua* (day)] (Diptera: Syrphidae) to *B. Kaber* flowers	Altieri, 1984
Cabbage	*Crataegus* sp.	Diamondback moth [*Plutella maculipennis* (Curtis)] (Lepidoptera: Yponomeutidae)	Provision of alternate hosts for parasitic wasps (*Horogenes* sp.)	van Emden, 1965
Cabbage	White and red clover	*Erioischia brassicae,* cabbage aphids, and imported cabbage butterfly (*Pieris rapae*)	Interference with colonization and increase of ground beetles	Dempster & Coaker, 1974
Cacao	Coconut	*Helopeltis theobromae* Miller	Provision of nest sites for ants [*Oecophylla smaragdina* (F.) and *Dolichoderus thoracicus* (Smith) (Hymenoptera: Formicidae)]	Way & Khoo, 1991
Cole crops	Quick-flowering mustards	Cabbageworms (*Peiris* spp.)	Increased activity of parasitic wasps [*Cotesia glomerata* (L.) Hymenoptera: Braconidae]	NAS, 1969
Cole crops	—	*B. brassicae*	Increased abundance of predators	Tukahirwa & Coaker, 1982
Collards	*Amaranthus retroflexus* L. (Amaranthaceae), *Chenopodium album, Xanthium strumarium* L. (Asteraceae)	Green peach aphid (*Myzus persicae*)	Increased abundance of predators (*Chrysoperla carnea,* Coccinellidae, Syrphidae)	Horn, 1981
Corn	Beans	Leafhoppers (*Empoasca kraemeri*), leaf beetle (*Diabrotica balteata* LeConte) (Coleoptera: Chrysomelidae) and fall armyworm [*Spodoptera frugiperda* (Smith)] (Lepidoptera: Noctuidae)	Increase in beneficial insects and interference with colonization	Altieri *et al.,* 1978

(continues)

TABLE 3 (*continued*)

Crop	Weed species or intercrop	Pest(s) regulated	Factor(s) involved	References
Corn	Giant ragweed	European corn borer (*Ostrinia nubialis*)	Provision of alternate hosts for the tachnid parasite, *Lydella grisescens*	Syme, 1975
Corn	Sweet potatoes	Leaf beetles (*Diabrotica* spp.) and leafhoppers *(Agallia lingula)*	Increase in parasitic wasps	Risch, 1979
Cotton	Alfalfa	Plant bugs (*Lygus hesperus* and *L. elisus*)	Prevention of emigration and synchrony in the relationship between pests and natural enemies	Marcovitch, 1935
Cotton	Alfalfa, maize and soybeans	Corn earworm *(H. zea)* and cabbage looper [*Trichoplusia ni* (Hübner)] (Lepidoptera: Noctuidae)	Increased abundance of predators	DeLoach, 1970
Cotton	Forage cowpeas	Boll weevil *(Anthomonus grandis)*	Population increase of parasitic wasps (*Eurytoma* sp.)	Marcovitch, 1935
Cotton	Ragweed and *Rumex crispus*	*Heliothis* spp.	Increased populations of predators	Smith & Reynolds, 1972
Cotton	Sorghum or maize	Corn earworm *(H. zea)*	Increased abundance of predators	DeLoach, 1970
Peaches	Strawberries	Strawberry leaf roller *(Ancylis comptana)* and Oriental fruit moth *(Grapholita molesta)*	Population increase of parasites (*Macrocentrus ancylivora, Microbracon gelechise,* and *Lixophaga variabilis*)	
Peanuts	Maize	Corn borer (*Ostrinia furnacalis* Guenée) (Lepidoptera: Pyralidae)	Abundance of spiders [*Lycosa* sp. (Araneae: Lycosidae)]	Raros, 1973
Sesame	Cotton	*Heliothis* spp.	Increase of beneficial insects and trap cropping	Laster & Furr, 1972
Squash	Maize	*Diaphania hyalinata* (L.) (Lepidoptera: Pyralidae)	Habitat more suitable for parasitoids/alternate hosts for *Trichogramma* spp.	Letourneau, 1987
Soybeans	*Cassia obtusifolia* L. (Fabaceae)	*Nezara viridula* (L.) (Heteroptera: Pentatomidae), *Anticarsia gemmatalis* (Lepidoptera: Noctuidae)	Increased abundance of predators	Altieri *et al.,* 1981
Sugar beet	*Matricaria chamomilla* L., *Lamium purpureum* L. and others	*Aphis fabae* L., *Atomaria linearis* Stephen, *Diplopoda* spp.	Increased abundance of parasites (Hymenoptera, carabids and spiders)	Bosch, 1987
Sugarcane	*Euphorbia* spp. weeds	Sugarcane weevil [*Rhabdoscelus obscurus* (Boisduval)]	Provision of nectar and pollen for the parasite *Lixophaga sphenophori*	Topham & Beardsley, 1975
Sweet potatoes	Morning glory (*Ipomoea asarifolia* Roem et Schult) (Convolvulaceae)	Argus tortoise beetle *(Chelymorpha cassidea)*	Provision of alternate hosts for the parasite *Emersonella* sp.	Carroll, 1978
Winter wheat	Grassland	Cereal aphids	Colonization source for *Aphidius rhopalosiphi* DeStefani Perez females	Vorley & Wratten, 1987

Leius (1967) found that the presence of wildflowers in apple orchards resulted in an 18-fold increase in parasitism of tent caterpillar pupae over nonweedy orchards; parasitism of tent caterpillar eggs increased fourfold, and parasitism of codling moth larvae increased fivefold. A cover crop of bell beans [*Vicia faba* Linnaeus (Fabaceae)] in rain-fed apple orchards in northern California decreased infestation by the codling moth. This lower moth infestation was correlated significantly with increased numbers of predators (Aranae, Coccinellidae, Syrphidae, and Chrysopidae) on the trees (Altieri & Schmidt, 1985). Similar results were obtained by Dickler (1978) in Germany. In New Jersey peach orchards, control of the oriental fruit moth increased in the presence of ragweed (*Ambrosia* sp.), smartweed (*Polygonum* sp.), lambsquarter (*Chenopodium album* Linnaeus) (Chenopodiaceae), and goldenrod (*Solidago* sp.). These weeds provided alternate hosts for the braconid parasite *Macrocentrus ancylivorus* Rohwer (Bobb, 1939). O'Connor (1950) recommended the use of a cover crop in coconut groves in the Solomon Islands to improve the biological control of coreid pests by the ant *Oecophylla smaragdina subnitida* Emery. In Ghana, coconut served this purpose by providing enough shade for cocoa to support high populations of the ant *O. longinoda* Latreille, which kept the cocoa crop free from cocoa caspids (Leston, 1973). Annual crops diversified with cover crops also suffer less damage. For example, Brust *et al.* (1986) reported dramatically higher predation rates of Lepidoptera larvae—black cutworms (*Agrotis ipsilon* [Hufnagel]), armyworms (*Pseudaletia unipunctata* Haworth), stalk borers (*Papaipema nebris* [Guenée]), and European corn borers [*Ostrinia nubilalis* (Hübner)]—tethered to corn sown into a grass–legume mixture than to corn in monoculture. Carabid beetles were more abundant in that living mulch system and were among the larval predators in both systems.

Temporal refugia for predators can result in increased densities early in the season whether they be provided by the crop itself (e.g., De la Cruz & Litsinger, 1986; Coli & Ciurlino, 1990), by other crops (e.g., Murphy *et al.*, 1996), or by permanent noncrop vegetation (Chiverton, 1989; Oraze *et al.*, 1989; Thomas, 1989; Zhuang, 1989; Wratten & Thomas, 1990).

The roles of vegetational mosaics and corridors in the conservation of natural enemies are undoubtedly important, but are difficult phenomena to investigate experimentally. A number of studies have shown that natural enemies move from noncrop habitats to crop fields, but relative vagility of natural enemies and scale of plot designs constrain our abilities to assess landscape-level movement patterns (van Emden, 1988; Hansson, 1991). For example, Rodenhouse *et al.* (1992) found that predaceous arthropods in soybean were more abundant if the plots contained noncrop corridors. In contrast, Pavuk and Barrett (1993) found no difference in larval parasitism of the green clover worm, *Plathy-*

pena scabra (Fabricius) (Lepidoptera: Noctuidae), among these treatments. Landis and Haas (1992) found, in a larger scale comparison, higher parasitism levels of European corn borer, *Ostrinia nubilalis* (Hübner), near wooded sites. Our abilities to assess natural-enemy movement patterns, and to design landscapes to encourage their early arrival and sustained presence in crop fields depend on large-scale empirical studies now feasible using Geographic Information Systems (GIS) technologies (Liebhold *et al.*, 1993).

In summary, design of vegetation management strategies must include knowledge and consideration of (1) crop management in time and space, (2) composition and abundance of noncrop vegetation within and around fields, (3) soil type, (4) surrounding environment, and (5) type and intensity of management. The response of insect populations to environmental manipulation depends on their degree of association with one or more of the vegetational components of the system. Extension of the cropping period, planning temporal or spatial cropping sequences, inclusion of "insectary" plants, or maintenance of movement corridors may allow naturally occurring biological control agents to sustain higher population levels on alternative hosts or prey and to persist in the agricultural environment throughout the year.

Because farming systems in a region are managed over a range of energy inputs, levels of crop diversity, and successional stages, variations in insect dynamics are likely to occur that are difficult to predict. However, based on current ecological and agronomic theory, low pest potentials may be expected in agroecosystems that exhibit the following characteristics:

1. They have high crop diversity through mixtures in time and space (Cromartie, 1981; Altieri & Letourneau, 1982; Risch *et al.*, 1983; but see Andow & Risch, 1985; Nafus & Schreiner, 1986; Altieri, 1994; Coll & Bottrell, 1995).

2. There is discontinuity of monoculture in time through rotations, use of short maturing varieties, use of crop-free preferred host-free periods, etc. (Stern, 1981; Lashomb & Ng, 1984; Desender & Alderweireldt, 1990).

3. There are small, scattered fields creating a structural mosaic of adjoining crops, and uncultivated land potentially providing shelter and alternative food for natural enemies (van Emden, 1965; Altieri & Letourneau, 1982). Pests also may proliferate in these environments depending on plant species composition (Altieri & Letourneau, 1984; Collins & Johnson, 1985; Levine, 1985; Slosser *et al.*, 1985; Lasack & Pedigo, 1986; Mena-Covarrubias *et al.*, 1996; Panizzi, 1997). However, the presence of low levels of pest populations or alternative hosts may be necessary to maintain natural enemies in the area.

4. They have a dominant perennial crop component. Orchards are considered more stable, as permanent ecosys-

tems, than are annual crop systems. Because orchards suffer less disturbance and are characterized by greater structural diversity, possibilities for the establishment of biological control agents are generally higher, especially if floral undergrowth diversity is encouraged (Huffaker & Messenger, 1976; Altieri & Schmidt, 1985).

5. They have high crop densities in the presence of tolerable levels of weed background (Shahjahan & Streams, 1973; Altieri *et al.,* 1977; Sprenkel *et al.,* 1979; Andow, 1983a; Mayse, 1983; Buschman *et al.,* 1984; Ali & Reagan, 1985).

6. There is high genetic diversity resulting from the use of variety mixtures or several lines of the same crop (Perrin, 1977; Whitham, 1983; Gould, 1986; Altieri & Schmidt, 1987).

All the preceding generalizations can serve in the planning of a vegetation management strategy in agroecosystems; however, they must take into account local variations in climate, geography, crops, local vegetation, pest complexes, etc., which might change the potential for pest development under some vegetation management conditions. The selection of component plant species also can be critical (e.g., Bugg *et al.,* 1987; Naranjo, 1987; van den Berg *et al.,* 1990). Systematic studies on the "quality" of plant diversification with respect to the abundance and efficiency of natural enemies are needed. While 59% of the 116 species of entomophages in documented studies reviewed by Andow (1986) exhibited increased abundances when plant species were added to the system, 10% decreased in abundance and 20% were either not affected by vegetational diversity or were variable, sometimes increasing and other times decreasing. Gold *et al.* (1991), Perfecto and Sediles (1992), and Toko *et al.* (1996) found no significant difference in the abundance of specific entomophagous arthropods in response to crop diversification. Similarly, parasitization rates of bean flies were constant despite the addition of maize or other bean varieties in Malawi (Letourneau, 1994). Nafus and Schreiner (1986) found lower parasitism rates in corn intercropped with squash. The addition of squash decreases the abundance of the coccinellid *Coleomegilla maculata* (De Geer) on squash because of a nonuniform distribution of prey (Andow & Risch, 1985). The anthocorid *Orius tristicolor* (White), however, another generalist predator, is more abundant on squash when corn is incorporated. In this case, plant architecture and nonuniform distribution of prey are beneficial (Letourneau 1990). Perfecto *et al.* (1986) found that plant density and diversity interact negatively to determine ground beetle emigration rates. Mechanistic studies to determine the underlying elements of plant mixtures that enhance or disrupt colonization and population growth of natural enemies will allow more precise planning of cropping schemes and will increase the chances of a beneficial effect beyond current levels.

Mechanical Crop Management

Mechanical manipulations of the environment vary along two important parameters, severity of disturbance and frequency of disturbance. Across a range of agroecosystem types, low-input, perennial systems, for example, present a contrast to mechanized, annual crop production systems (Fig. 6). Yet slight modifications in cultural practices for sowing, maintaining, and harvesting annual crops can effect substantial changes in natural-enemy populations, which bring them nearer to those observed in less disturbed perennial counterparts (Barfield & Gerber, 1979; Arkin & Taylor, 1981; Blumberg & Crossley, 1983; Herzog & Funderburk, 1985).

Cultivation and Tillage

Current trends in tillage practices reflect attempts to limit mechanical disturbance of the soil. The objectives in emphasizing surface tillage and no tillage as alternatives to plow tillage are to control soil erosion, enhance crop performance, and use energy more efficiently (Sprague, 1986). Tillage practices have profound effects on the relative densities of pests and natural enemies in a crop field. Minimum tillage systems can conserve and enhance natural enemies of important crop pests (House & All, 1981; Luff, 1982; Blumberg & Crossley, 1983). However, All and Musick (1986) suggest that each crop–pest–enemy interaction must be considered independently. Table 4 demonstrates this point by depicting significant but often divergent responses within the Carabidae, a prominent predaceous group in tillage response studies.

Deep plowing, surface disking, and other manipulations of the soil can affect ground-dwelling arthropods, whether they inhabit the soil consistently or intermittently (e.g., Funderburk *et al.,* 1988; House, 1989a, 1989b; Stinner &

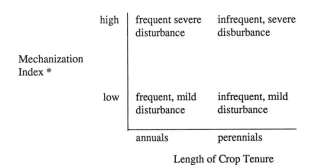

FIGURE 6 Several factors determine the level and frequency of disturbance. The intensity of mechanization and crop tenure are among the most important. Within the constraints of crop production, the use of modified cultivation and strategies will favor various natural enemies. *Determined as a function of plowing depth and compaction.

TABLE 4   Response by Carabid Beetles to Tillage Practices and Cover Cropping in Annual and Perennial Agroecosystems

Tillage practice	Crop	Carabid taxa	Response	References
No tillage	Corn	*Amara* spp., *Pterostichus* spp., and *Amphasia* spp.	Higher population density	Brust *et al.*, 1985
Mowed cover crop	Corn	*Pterostichus* spp. and *Scarites* spp.	Higher abundance than in herbicide-killed cover crop in early season	Laub and Luna, 1992
No tillage	Alfalfa	Carabids	No significant difference in abundance or diversity	Barney & Pass, 1986
Conservation tillage	Soybeans	Carabids	Higher species richness and abundance	House & All, 1981
No tillage	Soybeans	Carabids (9 genera)	Higher species richness, abundance, and biomass	House & Stinner, 1983
No tillage	Wheat, soybeans, corn	*Harpalus* spp. and *Amara* spp.	Higher abundance	House, 1989
No tillage	Apples	*Anisodactylus senctaecrucis* F., *Stenopophus comma* F., *Harpalus aeneus* F., *Amara* spp., and *Bradycellus supestris* Say carabid larvae	Higher abundance in natural vegetation plots; diversity not significantly different	Holliday & Hagley, 1984
Various practices	Spring wheat	40 species	Cropping system and tillage regimes change community composition	Weiss *et al.*, 1990
No tillage	Apples	Carabids (especially *Harpalus* sp.)	Higher abundance in weedy plots than in clean individual plots	Altieri *et al.*, 1985
Cover crop	Apples	*Pterostichus* spp., *Amara* spp.	Lower abundance with cover crop of mixed grass and bell bean than with clean cultivation	Altieri & Schmidt, 1985, 1986
		Agonum spp. *Platynus* spp. *Microlestes* spp.		

House, 1990). The extent of direct mortality depends on their distribution with respect to soil depth and their phenologies. Less directly but potentially as important are effects caused by the removal of resources and natural enemies associated with living undergrowth and plant residues. Few studies have addressed the mechanisms of population and community response to soil manipulations. Therefore, effects due to mechanical disturbance and secondary effects due to weed removal are often confounded. It is difficult to assess experimentally the impact of natural enemies on crop pests in these systems; and the causal links between tillage practices, numbers of natural enemies, and level of biological control have been demonstrated in only a few cases, as discussed later [see Risch *et al.* (1983) and Letourneau (1987) for similar discussions on the paucity of mechanism studies in intercropping experiments].

Brust *et al.* (1985) showed that significantly higher densities of carabids, including *Amara* spp., *Pterostichus* spp., and *Amphasia* spp., in no-tillage systems were the major factor in reducing black cutworm damage significantly below the level achieved in conventional corn systems. Their conclusions were based on comparing treatments with and without soil insecticides. Carcamo (1995) showed that carabid population responses to tillage practices can depend on the species, but found that species diversity and evenness were higher in reduced tillage barley than in conventionally tilled fields. Variable responses were also found for different species of carabids under different tillage intensities by Clark *et al.* (1997). Other studies show that herbivore damage is reduced in no-tillage fields despite similar predator abundance in tilled and no-till fields. For example, reduced rootworm (*Diabrotica* spp.) damage to corn in no-tillage fields compared with plowed fields reflected lower herbivore densities (Stinner *et al.*, 1986). Although spider density was highest in no-tillage systems, predators in general did not exhibit higher densities. In such cases efficiency instead of abundance of predators and parasitoids may be enhanced; the vegetative component may provide alternative resources to entomophages. For example, Foster and Ruesink (1984) showed that the flowering weeds associated with reduced tillage in corn are important nectar sources that increase survival and fecundity of the braconid *Meteorus rubens* (Nees), an important parasitoid of the black cutworm.

Ants are generalist predators sensitive to tillage practices in agroecosystems (Risch & Carroll, 1982). Altieri and Schmidt (1984) report greater species richness, abundance, and predation pressure in uncultivated orchard systems than in those cultivated twice in 6 weeks. Lack of nest disturbance and habitat suitability due to vegetational cover may have been important causes of greater ant abundance. Similar results were predicted for a highly effective predator of bollworm [*Heliothis zea* (Boddie)] *Iridomyrmex pruinosis* (Rogers), in Arkansas cotton fields (Kirkton, 1970) based on field observations. Carroll and Risch (1983) and Letourneau (1983) sampled ant activity in lowland tropical Mexico where farming practices are in transition between slash-and-burn and mechanized cropping practices. The number of ant species at tuna baits in maize fields was similar whether they had been plowed or sown into slash (20 to 23 spp.). However, in central Texas, spring plowing decreased and species richness (as detected with surface and underground baits) from 12 species to 2 species. Among the species that were no longer present at baits (after plowing) were species that prey on *Solenopsis invicta* Buren queens (C. R. Carroll, personal communication 1985).

Pests also can be suppressed directly by plowing the soil (Watson & Larsen, 1968) and burying stubble (Holmes, 1982). For example, Talkington and Berry (1986) significantly reduced the adult emergence of *Fumibotys fumalis* (Guenée), a pyralid moth pest, in peppermint fields by burying the prepupae into the soil; tillage depth was directly correlated with level of reduction. In locations where natural enemies are not effective, deep burial of infested stubble with the conventional moldboard plow may be necessary to effect control of a pest (Umeozor *et al.*, 1985). The study by Telenga and Zhigaev (1959) on *Bothynoderes punctiventris* Germer, the beet weevil, however, shows how differential effects on pests and their natural enemies can be achieved through carefully planned tillage practices. Although deep plowing action destroyed over 90% of the weevil eggs, surface tillage with a disk increased the survival of a parasitoid on those eggs, which ultimately caused a greater level of pest suppression than did deep plowing. Disk-harrowed or plowed plots produced substantially fewer parasitoids of *Meligethus* sp. pollen beetles than did fallow fields or plots of rape that had had direct drilling of winter wheat (Nilsson, 1985). The effects of these practices on parasitism rates were not studied, but a regional use of direct drilling was recommended.

A more subtle effect of plowing involves its impact on weed composition in the field. Studies in northern Florida (Altieri & Whitcomb, 1979a, 1979b) have shown that weed-species composition changes markedly with date of plowing. December plowing stimulated populations of goldenrod (*Solidago altissima* Linnaeus) (Campanulaceae) and also 58 predator species that feed on the aphids (*Uroleucon* spp.) and other herbivores associated with this weed.

By plowing the land in October, camphor weed populations were enhanced along with the 30 predator species associated with camphor weed herbivores.

Mowing, Harvesting, and Weeding

Pruning and mowing crops often results in (1) the arthropod movement from cut plant material, and (2) a period of new growth. These factors can have important consequences on the quality and synchrony of natural-enemy activity. Weeding can also stimulate crop colonization by associated arthropods. The extent to which movement will occur depends on distance and arthropod mobility, but some weeding practices leave associated arthropods intact and promote such movement. When patches of stinging nettle (*Urtica dioica* Linnaeus) (Urticaceae) are cut in May or June, predators (mainly Coccinellidae) are forced to move into crop fields (Perrin, 1975). Similarly, cutting the grass–weed cover has been reported to drive Coccinellidae onto trees in orchards of southeastern Czechoslovakia (Hodek, 1973). Gange and Llewellyn (1989) manipulated the movement of both males and females of the predatory black-kneed capsid[*Blepharidopteris angulatus* (Fallén)] from alder windbreaks to orchards by timing tree pruning with peak aphid densities on *Alnus*.

Some of the best examples of natural-enemy movement prompted by vegetation cutting were shown in alfalfa strip-cutting systems. Van den Bosch and Stern (1969) compared densities of several predators, including *Geocoris pallens* Stål, *Nabis americoferus* Caryon, *Orius tristicolor* White, *Chrysoperla carnea* (Stephens), and *Hippodamia* spp. in strip cut and solid cut fields. Movement out of the field was uncommon even for these mobile predators in strip cut fields; most of them moved onto adjacent plants, so that on a field-wide basis these predators were conserved. Strip cutting also reduced mortality of *Aphidius smithi* Sharma & Rao by providing shelter from adverse physical conditions and host scarcity. Host availability for the parasitoid *Cotesia medicaginis* (Muesebeck) in alfalfa is altered through a different mechanism; oviposition rates of the alfalfa butterfly *Colias philodice eurytheme* Godart peak on new growth following harvest. This causes periodicity in the availability of early instar larvae. Strip cropping can spread the vulnerable stages more evenly over time, and thus can favor the maintenance of *A. medicaginis* populations over the season.

The use of fire to prepare land for cropping by "slash and burn" or by reducing crop residue affects resident natural enemies and incoming colonists. Burning of old, fallow vegetation in a tropical slash-and-burn system decreased ant abundance and foraging activity for more than 4 months (Saks & Carroll, 1980). Although fire has been used as a tool for direct control of pests (Komareck, 1970), generalizations on its effect on natural enemies are not possible. An isolated study showed that controlled burning increased spider and ant densities and biomass due to increased food supply for herbivores in the form of succulent plant growth after the burn (Hurst, 1970). Little is known about specific taxa, and much less is known about effects of fire on predator to prey ratios, arthropod community structure, and habitat suitability.

Chemical Inputs to the System

Fertilizer, irrigation, and pesticide treatments are important components of production strategies in modern agricultural and silvicultural systems. Although the influence of water and fertilizer applications on herbivores is complex (Scriber, 1984; Louda, 1986), fertilizer and herbivory levels may be causally related through changes in plant quality or phenology that affect the dynamics of predator–prey and host–parasitoid interactions. Pesticides, however, have direct detrimental effects on natural enemies and therefore their use in environmental management must be limited to situations where they are timed carefully or selectively applied. Novel advances in the use of behavior-modifying chemicals (Lewis & Nordlund, 1985; Lewis & Martin, 1990) provide powerful new tools for the manipulation of biological control agents. Pollutants can also disturb biological control of crop pests and are discussed later.

Fertilizer

Soil amendments can have a profound effect on the physiological condition of a crop plant. Consequences of these changes for pest management depend on soil variability; the growth, developmental, and biochemical responses of the plant; the direct effects of these changes on herbivores; and the secondary impact on natural enemies. A considerable amount of work has been done on herbivore response to fertilizers that increase nitrogen (N) levels in plants. Mattson (1980) suggests that foliage N level is a major regulator of herbivory rate. Although increased plant quality (higher N) commonly leads to improved performance by individual insects (e.g., growth rates, survival, and fecundity), general statements on the direct responses of herbivores to N fertilizer are not possible due to the array of responses by different species (Scriber, 1984; Letourneau, 1997). Other experiments on links between soil amendments and pest management concern effects on the pest via their response to resistant and susceptible varieties under conditions of different sources or levels of calcium, magnesium, nitrogen, phosphorus (P), potassium (K), or sulfur (Kindler & Staples, 1970; Manuwoto & Scriber, 1985; Culliney & Pimentel, 1986; Shaw et al., 1986; Davidson & Potter, 1995). The environment of a natural enemy is affected by soil amendments through changes in plant quality, as well as by concomitant changes in the herbivores.

Direct effects of fertilizer on biological control are not well documented. Many herbivores exhibit marked increases in population growth on nitrogen-enriched hosts. Concern emerges for the ability of natural enemies to track their prey/hosts under these conditions, but biological control of mites was maintained among apple treated with three levels of nitrogen fertilizer (Huffaker et al., 1970). Although the fecundity of the herbivore Panonychus ulmi (Koch) increased with N level up to a fourfold increase when Amblyseius potentillae (Garman) predators were not present, the predators were able to compensate for most of the increase in prey reproduction. In contrast, fertilized (NPK) cotton plots exhibited higher levels of Heliothis zea than did control plots despite significantly higher population densities of Hippodamia convergens Guérin-Méneville, Coleomegilla maculata langi Timberlake, and Orius insidiosus (Say) in fertilized cotton (Adkisson, 1958). Chiang (1970) demonstrated that fertilized midwestern corn fields (at 50 tons of manure per acre) had significantly fewer (half) corn rootworms than did unfertilized control plots. Although ground beetles and spiders were not affected, the populations of predaceous and herbivorous mites were three times higher in manure treatment plots. Through three seasons of field and laboratory experiments, Chiang concluded that predation by mites accounted for a 20% control of corn rootworm under natural field conditions and a 63% control when manure was applied.

Other effects of fertilizers on natural enemies may be predicted based on the combined information of related studies. For example, the parasitoid Diaretiella rapae (McIntosh) attacks the green peach aphid Myzus persicae (Sulzer) more readily when the aphid is associated with Brassica spp. (Read et al., 1970). The mustard oils in crucifers serve as attractants for the wasp. It has also been shown-demonstrated that some glucosinolates are inversely related to N fertilizer level (Wolfson, 1980). Therefore, soil fertility may have dramatic effects on pest control by constraining the production of semiochemicals that play an important role in mediating interactions between plants, herbivores, and natural enemies.

Synchrony of predators and their prey can be modified by frequency and levels of fertilizer applications. For example, low nutritive quality of host plants may cause immature herbivores to develop more slowly, thus increasing their availability to natural enemies (Feeny, 1976; Moran & Hamilton, 1980; Price et al., 1980). For example, a predaceous pentatomid bug regulated more efficiently Mexican bean beetles on nutritionally poor plants than it did on highly nutritional ones (Price, 1986). Benrey and Denno (1997) showed that Pieris rapae larvae that developed slowly had higher parasitism rates by Cotesia glomerata than did larvae that grew more rapidly on the same host-plant species. If fertilizer increases the nutritive value of crops, then pests may grow more rapidly or feed more efficiently, thus passing through windows of vulnerability more often (see Bernays, 1997). Host-plant phenology can also be driven by fertilizer inputs. Hogg (1986) suggests that the timing of square availability was one factor influencing predation and parasitism rates of *H. zea* in cotton. Although his data are taken from early- and late-season peaks in square production, and not from changes caused by soil fertility, they are indicative of indirect effects of plant quality that could be influenced by fertilization.

Fertilizer-caused changes in nutritive quality of host plants may indirectly affect the survival and reproduction of natural enemies by determining prey quality. Although direct examples of fertilizer effects have not been determined experimentally, nitrogen content is an important aspect of prey quality. For example, nitrogen content may be responsible for higher egg production by *H. convergens* when fed apterous instead of alate green peach aphids (Wipperfürth *et al.,* 1987). Analogous effects may occur with prey of different quality due to host-plant condition. Zhody (1976) found that size, fecundity, and longevity of the aphelinid wasp *Aphelinus asychis* (Walker) was dependent on the food composition of its host *Myzus persicae.* However, low-quality food can also impair the ability of a host to encapsulate the parasitoid (El-Shazley, 1972a, 1972b). Nutrients in the host plant also can modify toxic effects to parasitoids (Duffey & Bloem, 1986) and can influence their sex ratio (Greenblatt & Barbosa, 1981). Host size is often an important determinant of egg fertilization by ovipositing females (Charnov, 1982). The sex ratio of *Diadegma insulare* (Cresson) reared from larvae of *Plutella xylostella* Linnaeus from field plots over a wide range of N fertilizer inputs showed a strong tendency for female bias in heavily fertilized plots (Fox *et al.,* 1990), and Jansson *et al.* (1991) found increased parasitism rates in high N fields. Rates of host feeding by *Encarsia formosa* Gahan on whiteflies were higher on fertilized host plants (Bentz *et al.,* 1996).

Although studies on direct effects of nitrogen on crop architecture and subsequent effects on searching efficiency are not available, some studies suggest that these interactions can occur. Soil nutrient levels can influence plant size, leaf area, canopy closure, and crop architecture. These conditions define searching arenas for natural enemies (Kemp & Moody, 1984). Predator–prey or parasitoid–host contact rates are a function of habitat preference, searching area, prey density, and dispersion patterns. For example, Fye and Larsen (1969) found that the searching efficiency of *Trichogramma* spp. was dependent on structural complexity. Hutchison and Pitre (1983), however, did not find such an effect for *Geocoris punctipes* (Say) on *H. zea.* Shady conditions resulting from overgrowing vegetation reduced parasitism levels of *Pieris* spp. by *Cotesia glomerata* (Linnaeus) (Sato & Ohsaki, 1987).

In addition to direct effects of fertilizer on host-plant

growth form, levels of key chemical constituents in the soil can indirectly affect natural-enemy activity by influencing weed composition in a field. Fields in Alabama, with low soil potassium, were dominated by buckhorn plantain (*Plantago lanceolata* Linnaeus) (Plantaginaceae) and curly dock (*Rumex crispus* Linnaeus) (Polygonaceae), whereas fields with low soil phosphorus were dominated by showy crotalaria (*Crotalaria spectabilis* Roth.) (Fabaceae), morning glory (*Ipomoea purpurea* Roth.) (Convolvulaceae), sicklepod (*Cassia obtusifolia* Linnaeus) (Fabaceae), *Geranium carolinianum* Linnaeus (Geraniaceae), and coffee senna (*Cassia occidentalis* Linnaeus) (Fabaceae) (Hoveland *et al.,* 1976). Soil pH can influence the growth of certain weeds. For example, weeds of the genus *Pteridium* occur on acid soils while *Cressa* sp. inhabits only alkaline soils. Other species (many Compositae and Polygonaceae) grow in saline soils (Anonymous, 1969).

Water

Flooding fields, draining land, and irrigating crops with furrow, drip, or sprinkler methods can affect plant quality and relative humidity (RH) at field level and are all manipulations that can affect biological control on a local scale. The barren desert valleys of southeastern California are suitable habitat for the predaceous earwig *Labidura riparia* (Pallas) due to favorable conditions produced with the development of irrigation agriculture (van den Bosch & Telford, 1964). Expansion of arable land through irrigation in the developing regions of Africa, Asia, and South and Central America may have global effects on the distribution of natural enemies.

As in the case of fertilizer research, much experimental work on the effects of plant stress from water conditions has targeted herbivores (e.g., Miles *et al.,* 1982; Bernays & Lewis, 1986; Louda, 1986). Water availability can affect palatability, feeding duration, developmental time, migration, survival, and fecundity of plant feeders. Thus, many important effects of water conditions on natural enemies are indirect and are mediated through changes in host/prey abundance and dispersion or through qualitative changes similar to those discussed in the previous section (fertilizer). For example, rape plants under drought conditions had increased proline levels and an associated shift in the balance of free amino acids (Miles *et al.,* 1982). Cabbage aphids reached adulthood much faster on these stressed plants. Availability of suitable hosts for parasitoids may be decreased both temporally (duration of vulnerable stages) and absolutely (if parasitoids require slower development than the host) if plants are not well watered.

Direct effects of water include mortality during irrigation and impacts of relative humidity. Ferro and Southwick (1984) and Ferro *et al.* (1979) review the importance of RH on small arthropods. Crop architecture and watering regimes cause large deviations from ambient temperature and humidity levels (Ferro & Southwick, 1984) within foliage boundary layer microhabitats. Irrigation of soybean caused a substantial (4 to 9°C) decrease in canopy temperature and a 16% increase in RH 15 cm above the ground (Downey & Caviness, 1973). Prolonged periods of such irrigation effects can have important consequences for natural enemies because developmental time, and thus population growth and synchrony, are driven by temperature and RH. The tachinid *Eucelatoria armigera* (Coquillett) completes development at different rates depending on temperature and host species (Jackson *et al.*, 1969). Holmes *et al.* (1963) showed that parasitism levels of the wheat stem sawfly by *Bracon cephi* (Gahan) were enhanced by soil moisture and temperature levels that slow plant ripening.

Early initiation of pivotal irrigation in Mississippi soybean fields caused rapid canopy closure that allowed establishment of *Nabis* spp. to be synchronized with the presence of lepidopteran larvae whereas *Nabis* spp. levels increased too late in the season in dryland soybeans (Felland & Pitre, 1991). Leigh *et al.* (1974) also reported highest population levels of *Geocoris pallens* Stål, *G. punctipes* (Say), and *Orius tristicolor* in dense canopies encouraged by both irrigation and close spacing of a cotton crop.

Force and Messenger (1964) showed that a few degrees dramatically affect changes of the innate capacity for increase (*r*) in parasitoids under laboratory conditions. *Cotesia medicaginis* reaches its maximum longevity at 55% RH; longevity decreased sharply at levels above and below that value (Allen & Smith, 1958). This was not deemed an important factor, however, in determining parasitism levels of *Colias* spp. larvae in alfalfa. However, it is known that scale parasitoids in arid citrus groves require irrigated conditions for effective biological control (DeBach, 1958b). The vertical profile and general microclimate depend not only on water inputs but also on mulch practices, row direction, windbreaks, and crop spacing (Hatfield, 1982), all of which are manipulable. Apparently the severity of effects caused by drought conditions depends on many factors, including the availability of free water and nectar in the habitat. Bartlett (1964) found that caged *Microterys flavus* (Howard) adults were able to function well at extremely low RH if provided with a honey–water mixture. It may be possible to favor mobile natural enemies at the expense of their hosts with the provision of water sources during drought conditions.

Use of Semiochemicals

The discovery that the behavior of certain parasitic insects is influenced by chemicals produced by their hosts triggered considerable interest in the use of semiochemicals for manipulating predators and parasitoids in the field, especially for aggregating and/or retaining released parasi-

toids in target locations (Gross, 1981). The various opportunities for, and limitations of, manipulating natural enemies using semiochemicals have been reviewed by Vinson (1977); Nordlund *et al.* (1981a, 1981b, 1988); Powell (1986); Hagen (1986); and Lewis and Martin (1990). According to Lewis *et al.* (1976), host or prey selection is perhaps the most important step in the searching behavior of entomophagous insects that can be manipulated to improve biological control. The use of semiochemicals should be directed at (1) increasing effective establishment of imported species, (2) improving performance and uniform distribution of released species throughout the target area, and (3) optimizing abundance and performance of naturally occurring entomophages (Greenblatt & Lewis, 1983). Based on the environmental factors that influence search behavior, it is possible to devise three main habitat management techniques using semiochemicals:

1. Directing strategies at improving habitat characteristics: use of semiochemicals to make crops more attractive, or to define a more complex mosaic of local search areas (Altieri *et al.*, 1981, but see Gardner & van Lenteren, 1986)
2. Enhancing host plant characteristics: breeding programs directed at improving chemical attractiveness of crops or crops with extrafloral nectaries
3. mimicking high-pest densities through application of diatomaceous earth or artificial eggs impregnated with kairomones (Gross, 1981)

Lewis and Martin (1990) encourage research on multitactic semiochemical packages that combine mating, oviposition, and feeding deterrents with formulations to increase parasitism of crop pests. Advances in the chemical mediation of these interactions (e.g., Williams *et al.*, 1988), the effects of parasitoid adaptation and learning (Lewis *et al.*, 1990; Kester & Barbosa, 1991; Turlings *et al.*, 1993), and an evaluation of the potential for resistance to develop (e.g., Turlings & Tumlinson, 1991) will allow wise application of these habitat management techniques.

Pesticide Drift

Aerial application of insecticides inevitably results in low-level inputs to nontarget areas. Indeed, half the material applied to a field under ideal conditions can drift a considerable distance downwind (Ware *et al.*, 1970). Is inadvertent disruption of natural enemies a common consequence of dilute pesticide inputs downwind of a target field? Although a great deal is known about the effects of direct spraying of various insecticides on natural enemies, now including community level effects (e.g., Croft & Slone 1997), very little experimental work has been conducted to determine effects of low-level inputs (see Chapters 13 and 24). Biological control will be disrupted given sufficient frequency, intensity, and toxicity of sprays (Ridgway *et al.*, 1976; Riehl *et al.*, 1980) even if dosages are sublethal (e.g., De Cock *et al.*, 1996).

However, Risch *et al.* (1986) found that the ratio of natural enemies to herbivores was increased by low, drift-level concentrations of carbaryl. Furthermore, they found that arthropod abundance dropped significantly more in an old field than it did in a corn monoculture. Their results suggest that: (1) low concentrations of insecticides have different effects on herbivores and natural enemies depending on whether the nontarget habitat is a crop field or a field of natural vegetation that serves as a source of colonizers, and (2) these impacts cannot be predicted from a knowledge of effects at high concentrations.

Chemical drift can be minimized by spraying when winds are less than 2 m/sec, using adjuvants, formulating inert emulsions, and using larger droplet sizes (Gebhardt, 1981). Windbreaks surrounding field and regional-wide spray synchrony are forms of cooperative efforts for drift reduction of the effects of low-level pesticide applications.

Although not a problem of drift, purposeful sprays of herbicides to crop fields can have nontarget effects similar to low-level insecticide application. Baker *et al.* (1985) have shown that *Orius* spp. and *Nabis* spp. densities were decreased by monosodium methanearsenate, but the abundance of spiders, *Geocoris* spp., Hymenoptera, and coccinellids was not. Herbicide application can also modify weed species composition in agricultural fields and thereby affect natural enemies (see section on the effects of vegetational diversity).

Dust and Pollution Effects

Dust and pollutants of many types can influence the efficacy of predators and parasitoids as biological control agents. Environmental management then includes consideration of placement of the sources and control of pollutant influx with respect to agricultural fields. It has long been known that some pest outbreaks are caused or exacerbated by dust-fall on crop foliage. Bartlett (1951) found that many "inert" dusts (as described on the pesticide manufacturer's label) rapidly killed *Aphytis chrysomphali* (Mercet) and *Metaphycus luteolus* (Timberlake). Experimental tests by DeBach (1958a) demonstrated an increase in California red scale populations on citrus trees in response to the application of road dust to plants [and see references in van den Bosch & Telford (1964)]. Mechanisms may be mechanical interference or desiccation (Edmunds, 1973). It is also possible that leaf temperature, which can be raised 2 to 4°C by dust cover (Eller, 1977), is a relevant factor. Planned placement of roads and timing of cultivation can reduce the level of dust-fall on crops. Strawberry growers in California profit from daily or twice-daily watering of roadways through the reduction in losses from mites. Ap-

parently, the dust inhibits biological control of phytophagous mites by predaceous species.

The influx of gaseous air pollutants is more difficult to detect and to control. Sulphur dioxide is a common effluent that has documented, negative effects on a variety of organisms (Petters & Mettus, 1982), including honeybees (Ginevan *et al.*, 1980). However, acute exposure of *Bracon hebetor* (Say) females to SO_2 in air causes no reduction of their fertility and fecundity. Petters and Mettus (1982) suggest that damage to parasitic wasps may accrue in the earlier stages, or behavioral avoidance of contaminated areas may explain reports of lower parasitoid and higher herbivore levels near sources of SO_2 pollution than on crops distant from the course. Melanic morphs of the generalist coccinellid predator *Adalia bipunctata* (Linnaeus) occur disproportionately often in the vicinity of coal-processing plants in the United Kingdom. Although earlier investigators suggested a mechanism involving selective toxicity of air pollutants, Muggleton *et al.* (1975) attribute the differences to sunshine levels (negatively correlated with smoke). Whether the coloration of such predators affects their efficiency as biological control agents is not known. Other sources of contamination include auto traffic, drainage from selenium-rich soils (Gerling, 1984), and ozone (Trumble *et al.*, 1987). The literature on these contaminants addresses direct effects on herbivores, but little is known about effects on natural enemies. We could speculate about indirect effects based on current knowledge. For example, the impact of ozone depletion on ultraviolet (UV) radiation, which interacts with plant secondary compounds to slow larval development (Trumble *et al.*, 1991), may result in increased exposure time of hosts to parasitoids. Lead as a contaminant from auto exhaust, now regulated in some areas, has been reported to concentrate in higher trophic levels (Price *et al.*, 1974). Any such concentration of toxic compounds potentially sets up conditions that favor pests.

Some pollutants are inadvertently added to the crop field with soil amendments, such as sludge, manure, and chemical fertilizer (Wong, 1985). Suggestions to use sewage sludge as a nutrient source for crops constitute a clever solution to both waste management and soil fertilization. However, Culliney *et al.* (1986) found a general response of low arthropod diversity when sludge containing heavy metals and toxic chemicals was applied to cole crops. This type of response is common in communities under pollution stress.

MECHANISMS OF ENEMY ENHANCEMENTS

An examination of host-selection processes of entomophagous insects, including host or prey habitat location, host or prey location, and host or prey acceptance (Vinson, 1981), can provide insights into the biological mechanisms by which environmental management enhances biological control. The purposeful design of crop habitats for effective biological control requires a detailed understanding of such mechanisms through basic research. Two initial guidelines among the complex set of interactions between the environment and natural enemies are useful for consideration: (1) the effective environment must be redefined for each phase of host or prey search, and (2) enemies differ in their response to environmental management schemes.

During migration and habitat location, the effective environment may be the local area, a regional landscape, or a series of distant habitat patches with long-distance travel between them. The interplay of colonizer source location, wind patterns, vegetation texture, and host or prey density becomes important on a large scale. Maximum levels of natural control require at the onset both sufficient numbers of natural enemies and temporal synchrony of these invasions. Regional environmental management for enhancing the success of habitat location by natural enemies then would focus on the arrangement of colonizer sources in relation to target sites of potential pest problems as well as on the timing of enemy colonization. Rabb (1978) addressed these needs when he criticized the propensity of single commodity, closed system approaches to pest management in research and decision making as deficient for problems that demand attention to "large unit ecosystem heterogeneity."

Dispersal ranges of natural enemies vary. Migration often occurs in high currents along paths of turbulent convection. Even weak flying insects can disperse over long distances and across wide areas by exploiting the ephemeral but very structured nature of air movement (Wellington, 1983). For example, robust hosts and minute parasitoids can exhibit coupled displacement in long-distance migration, as shown by the Australian plague locust, *Chortoicetes terminifera* Walker and its egg parasitoid, *Scelio fulgidus,* Crawford, which disperse independently on wind currents to the same location (Farrow, 1981). Cumulative numbers over a growing season, however, may be irrelevant if immigration rates of natural enemies are very slow in relation to rising levels of the pest (Doutt & Nakata, 1973; Letourneau & Altieri, 1983; Williams, 1984). Information on source constitution, phenology, and flight patterns then are necessary to design and manage regional-scale agroecosystems for optimal biological control. Flight capacity studies and mathematical models to describe movement patterns based on continuous diffusion or discrete random walk equations have focused on predicting dispersal and migration of herbivores (Okubo 1980; Stinner *et al.,* 1983, 1986). Detailed biological information (Duelli, 1980) coupled with predictive models of natural-enemy movement may aid in predicting synchrony. In some cases, synchronies are difficult to achieve because local species are adapted to exploit natural conditions of prey or host phenologies. For exam-

ple, coccinellid beetles in California estivate during times of prey availability; irrigated crops provide a continuous food supply that was not available before agricultural expansion occurred (Hagen, 1962).

During host or prey location, factors such as the physical texture of plant surfaces, chemical and structural attributes of plants, microclimatic conditions, and patch heterogeneity come into play. Flaherty (1969) showed enhancement in the control of herbivorous mites on grape vines with Johnson grass ground cover. The grass acted as a source of predaceous mites. In the prey location study of Flaherty (1969) and in the habitat location phase study of Doutt and Nakata (1973), the cumulative total number of natural enemies was not as important as the temporal synchrony with growing herbivore populations. Coll and Bottrell (1996) found that plant structural characteristics were important determinants of parasitism levels in monocultures versus mixed crop stands of the Mexican bean beetle. During host or prey acceptance and predation or parasitism, environmental factors operate indirectly through their effects on host or prey behavior and on host or prey quality, and alter levels of vulnerability of natural enemies to mortality factors. Examples of the mechanisms of host or prey selection on all levels of enemy behavior are provided in Table 5.

Cutting across these levels of behavioral organization are activities other than those directly associated with predation or parasitism such as migration to overwintering sites, mating, and acquisition and use of resources other than the primary prey or hosts. Therefore, the interdependence and variability of resource needs and factors such as proximity and availability of resources in time become vital aspects of the environment. These are factors of "habitat suitability" for natural enemies; a reduction of the relative energy expenditure needed, in a particular environment, to fulfill the resource needs of a particular parasitoid or predator will increase its efficiency as a biological control agent.

Conservation of natural enemies through habitat management techniques adapted to the prevailing agronomic schemes then can be of great benefit to farmers. Small changes in agricultural practices may increase natural-enemy populations or enhance efficacy. However, predators and parasitoids are extremely diverse and each family represents a particular range of responses to environmental modification as well as a range of effectiveness as biological control agents (e.g., McMurtry & Croft, 1997). Several examples of habitat management techniques that have been reported to increase the effectiveness and abundance of specific predator groups are given in Table 6.

Although each of the management practices discussed in this chapter can, as single factors, affect the abundance, richness, and efficacy of natural enemies, they also interact in ways that can be complementary, disruptive, or simply unpredictable. Brust (1991) found that tillage practices, organic matter, and irrigation regimes interacted to determine

TABLE 5   Mechanisms of Predator or Prey Efficacy in Suppressing Pests, Including Long- and
Short-Range Host or Prey Selection

Enemy	Herbivore	Plant	Mechanism	Reference
Host/Prey Habitat Location				
Chrysoperla carnea	Aphids	Cotton	Detection of caryophyllene from foliage; detection of indole acetaldehyde from honeydew and phenology-specific plant volatile by gravid females	Flint *et al.,* 1979; van Emden & Hagen, 1976; Hagen 1986 (and references within)
Exeristes ruficollis Grav. (Hymenoptera: Ichneumonidae)	Pine shoot moth	Cotton	Attraction to pinene and oil of pine as gravid female	Thrope & Caudle, 1938
Diaretiella rapae	Cabbage aphid	Crucifers	Attraction to allyl isothiocyanate; increased parasitism	Read *et al.,* 1970; Titayaavan & Altieri, 1990
Trichogramma spp.	*Heliothis zea*	Soybeans	Response of gravid females to volatiles of *Amaranthus retroflexus*	Altieri *et al.,* 1981
Campoletis sonorensis (Cam.) (Hymenoptera: Ichneumonidae)	*H. virescens*	Cotton, sorghum	Plant volatiles attract parasitoid to host habitat	Elzen *et al.,* 1983
Collops vittatus Say (Coleoptera: Melyridae)	Aphids	Cotton	Male response to caryophyllene in foliage	Flint *et al.,* 1981
Microplitis demolitor (Hymenoptera: Braconidae)	*Pseudoplusia includens*	Soybeans	7- and 8-Carbon hydrocarbons, aldehydes and ketones attract parasitoid to plant	Ramachandran & Norris, 1991
T. minutum Riley and *T. pretiosum* Riley (Trichogrammatidae)	*H. virescens*	Soybeans and natural vegetation	Discrimination preference for hosts at different heights	Flanders, 1937; Thorpe, 1985
Leptopilina clavipes (Hartig) (Hymenoptera: Eucoilidae)	*Drosophila* sp.	Fungus	Attracted to odor of mushrooms at stage of decay most likely to contain young maggots	Vet, 1983
Trichopoda pennipes (F.) (Diptera: Tachinidae)	Green stink bugs	—	Attracted to sex pheromones of hosts	Harris & Todd, 1980
Trichogramma spp. and *Chelonus* sp. (Hymenoptera: Trichogrammatidae & Braconidae)	Pink bollworm	—	Intense searching response to hexane soluble chemical in moth scales	Jones *et al.,* 1973; Chiri & Legner, 1983; and see Wang & Zong (1991)
T. brassicae	*Ostrinia nubilalis* Hbn. and *Mamestra brassicae* L. (Lepidoptera: Pyralidae & Noctuidae)	—	Egg extracts cause upwind locomotion	Renou *et al.,* 1992
Coccinellid adults	Aphids	—	Arrest response to simple sugars in honeydew	Hagen, 1986
Aphidoletes aphidimya (Rondani) (Diptera: Cecidomyiidae)	Aphid	—	Oviposition stimulus from honeydew	Wilbert, 1977
Amblyseius fallacis (Garman) (Acari: Phytoseiidae) and *Phytoseiulus macropilis* (Banks) (Acari: Phytoseiidae)	*Tetranychus urticae* Koch (Acari: Tetranychidae)	—	Attraction/arrestment by prey kairomones	Hislop & Prokopy, 1981
Amblyseius limonicus Garman & McGregor, *A. californicus* (McGregor), *A. anonymus* Chant & Baker, *Cydrodromella pilosa* (Chant) (Acari: Phytoseiidae)	*Monochychellus tanajoa* (Bondar) (Acari: Tetranychidae)	Cassava	Attraction to mite-infested leaves	Janssen *et al.,* 1990
Coleomegilla maculata and *Geocoris punctipes*	*H. zea* and *Spodoptera frugiperda*	Corn	Aggregation responses to aqueous larval homogenates	Gross *et al.,* 1985

(continues)

TABLE 5 (*continued*)

Enemy	Herbivore	Plant	Mechanism	Reference
Platycheirus spp. and *Syrphus* spp. (Diptera: Syrphidae)	*Brevicoryne brassicae*	*Brassica* spp.	Visual detection of aphid (or colored paint) aggregations stimulated oviposition	Chandler, 1968
Anthocorids	*B. brassicae*	Brussels sprouts (*Brassica oleracea gemmifera* (Brassicaceae)	Deterred by waxy leaves	Way & Murdie, 1965
Encarsia formosa Gahan (Hymenoptera: Aphelinidae)	*Trialurodes vaporariorum* (Westwood) (Homoptera: Aleurodidae)	Hirsute cucumber varieties	Trichome mat hinders movement and searching efficiency	Hulspas-Jordaan & van Lenteren, 1978
Trichogramma pretiosum and *Chrysopa rufilabris* Burm. (Neuroptera: Chrysopidae)	*H. zea*	Hirsute cotton varieties	Increased trichome density reduced ability for small predators to locate and destroy host–prey	Treacy *et al.*, 1985
Predaceous mites	Herbivorous mites	*Ficus carica* L. (Moraceae)	Glandular hairs correlated with increased or decreased rate of prey consumption depending on predator species	Rasmy, 1977; Rasmy & Elbanhawy, 1974
Bessa harveyi (Townsend) (Diptera: Tachinidae)	Larch sawfly	Larch	Nonhost odors reduce host finding	Montieth, 1960
Adalia bipunctata	Aphids	Radishes and Chinese cabbage	Nonglandular hairs caused larvae to change direction more frequently and thereby encounter prey more frequently	Shah, 1981
Coccinellids	Aphids	Hirsute potato varieties	Reduction in adult searching time; larval mobility impaired; but these effects were attenuated in the field	Obrycki & Tauber, 1984
Coleomegilla maculata	*Ostrinia nubilalis*	Corn	Presence of nonhost plants increased searching area and decreased predation efficiency	Risch *et al.*, 1982
Phytoseiulus persimilis	*Tetranychus urticae*	Lima bean *Phaseolus lunatus*	Damaged host plant produces volatiles attractive to predators	Dicke *et al.*, 1990
Orius tristicolor	*Frankliniella occidentalis* (Pergande) (Thysanoptera: Thripidae)	Squash	Presence of other plant species enhanced early colonization of predators and improved synchrony	Letourneau & Altieri, 1983; Letourneau, 1990
Metaseiulus occidentalis (Nesbitt) (Acari: Phytoseiidae)	*Eotetranychus willamettei* (Ewing) (Acari: Tetranychidae)	Grapes	Close proximity of colonizer source allows improved synchrony	Flaherty, 1969
Peristenus pseudopallipes (Loan) (Hymenoptera: Braconidae)	*Lygus lineolaris* (Palisot de Beauvois) (Heteroptera: Miridae)	—	Odor of flowering *Erigeron* spp. attracted parasitoids	Shahjahan, 1974
Predaceous mites	Tetranychid mites	Apples and peaches	Possible role of pubescence in determining predator–prey contact rates on different plants	Putnam & Herne, 1966
Cotesia marginiventris (Cresson) (Hymenoptera: Braconidae)	*Spodoptera exigua* (Hübner) (Lepidoptera: Noctuidae)	Corn seedlings	Volatiles from damaged leaves attract female wasps	Turlings *et al.*, 1991
Host/Prey Acceptance and Suitability				
Hyposoter exiguae (Viereck) (Ichneumonidae)	*H. zea*	Tomatoes	Increased tomatine in host-plant reduced adult parasitoid longevity, size, and survival	Campbell & Duffey, 1979

(*continues*)

TABLE 5 (*continued*)

Enemy	Herbivore	Plant	Mechanism	Reference
Geocoris punctipes	*Anticarsia gemmatalis*	Soybeans	Predators attacking prey on resistant varieties had increased mortality rates	Rogers & Sullivan, 1986
Eulophid parasitoids	*Liriomyza* spp. (Diptera: Agromyzidae)	Various crops	Match of effective parasitoid species with preferred hosts to increase parasitism levels in different crop habitats	Johnson & Hara, 1987
Functional and Numerical Response				
M. occidentalis	*Tetranychus urticae*	Strawberries	Prey fecundity increased with N level in plants, but functional and numerical response of predator compensated at least 80%	Hamai & Huffaker, 1978
Amblyseius hibisci (Chant) (Phytoseiidae)	*Oligonychus punicae* (Hirst) (Acari: Tetranychidae)	—	Numerical increases in predator population due to pollen availability	McMurtry & Scriven, 1966, 1968

levels of predaceous arthropod activity in peanut agroecosystems; predators were not important regulators of *Diabrotica undecimpunctata howardi* Barber (southern corn rootworm) in conventionally tilled, dry farmed systems, but decreased tillage or increased irrigation or both enhanced most predators. Henze and Sengonca (1992) showed how crop rotation schemes can influence weed composition and crop to weed ratios, which affect aphid densities, and aphid to coccinellid ratios. Increased coccinellid densities may respond to stand density, with different species preferring dense plantings [e.g., *Propylea quatrodecimpuncta* (Linnaeus)] and more sparse plant stands (e.g., *Coccinella* spp.) (Honek, 1983). Comparative studies of different suites of management practices such as in organic versus conventional production systems (Andow & Hidaka, 1989; Kromp, 1989; Pfiffner, 1990; Drinkwater *et al.,* 1995) provide integrative, system level effects on natural enemies in crop habitats.

THEORETICAL ASPECTS OF ENVIRONMENTAL MANAGEMENT

Enemy-Free Space

Perhaps the most general level of theory to guide habitat management for biological control is that of ecological and/or evolutionary escape from predators and parasitoids. Price (1981) acknowledges, in his theory of enemy-free space, that pest irruption is a likely consequence of agricultural practices that foster the spatial and temporal isolation of herbivores from their natural enemies (Price, 1981; Altieri & Letourneau, 1982, 1984; Risch, 1987). Bernays and Graham (1988) postulate that the evolution of diet specialization in insect herbivores could be due to the escape of specialists from their natural enemies. Denno *et al.* (1990) offer supporting evidence for willow beetles. Pest introduction to a novel environment is a classic example of relatively enemy-free conditions when specialized consumers on herbivores are left behind. Temporary release of pests also occurs under conditions of insecticide-caused pest resurgence and secondary pest outbreaks. Evolutionary changes in native crop pests (host shifts) constitute yet another process that may result in a reduction of predation or parasitism pressure. It has been our objective in this chapter to review ways to reduce enemy-free space through environmental manipulations. It is our contention that theory and basic research contribute vital knowledge for designing and integrating methods to enhance, instead of to mitigate, the ecological factors that cause herbivores "to be caught between the devil" (natural enemies) "and the deep blue sea" (host-plant defense/nutrient deficiency) (Lawton & McNeill, 1979).

Island Biogeographic Theory

The insular nature of cultivated areas has motivated several analogies concerncrops as islands available for colonization by arthropods (Price & Waldbauer, 1975; Strong, 1979; Simberloff, 1986). The development of arthropod communities in crops has been analyzed using MacArthur and Wilson's (1967) theory of island biogeography, which allows the prediction of colonization rates and mortality/emigration rates, on a comparative basis, with respect to crop area, distance from the sources of colonizers, and crop longevity (assuming that the system has aspects of equilibrium). The species composition, structure, and abundance

TABLE 6   Examples of Environmental Manipulation Affecting Efficiency and Abundance of Particular Predator Families

Predator family	Environmental management practice(s)	References
	Neuroptera	
Chrysopidae	Mowing of lucerne to force movement to adjacent crops	van den Bosch & Telford, 1964
	Protein-based food sprays in cotton and alfalfa	Hagen, 1986
	Interplanting or adjacent plantings of sorghum with cotton and cabbage	Perrin, 1977
	Placing hollow fibers containing caryophyllene in alfalfa	Hagen, 1986
	Spraying artificial honeydew	Hagen, 1986
Hemerobiidae	Protein and sugar sprays in artichoke	Neuenschwander & Hagen, 1980
	Hemiptera	
Anthocoridae	Planting of legumes (crimson clover, vetch), barley, sorghum, and corn as early sources of *Orius insidiosus*	Whitcomb, personal communication
	Presence of weeds facilitates dispersal of anthocorids to aphid-infested Brussels sprouts	Smith, 1976b
	Intercropping maize for control of thrips on squash	Letourneau & Altieri, 1983; Letourneau, 1990
	Maintaining *Lagopsis supina* as a dominant weed in apple orchards supports early sources of *O. sauteri*	Yu-hua *et al.,* 1997
	Corn, alfalfa, or thistles growing adjacent to apple orchards	McCaffrey & Horsburgh, 1986
	Presence of flowering willows adjacent to apple trees	Solomon, 1981
Lygaeidae	Addition of chopped sunflower seeds on sugar beet for enhancement of *Geocoris* spp.	Tamaki & Weeks, 1972
	Drilled planted soybeans	Ferguson *et al.,* 1984
	Uses of cotton varieties with extrafloral nectarines	de Lima & Leigh, 1984
	Selection of hirsute varieties preferred by *G. punctipes* for oviposition	Naranjo, 1987
Pentatomidae	Provision of supplemental plant feeding (potato foliage for *Podisus maculiventris*) (Say) Hamiptera: Pentatomidae	Ruberson *et al.,* 1986
Nabidae	No-till practices and repeat cropping of soybeans	Hammond & Stinner, 1987
	Coleoptera	
Carabidae	Double cropping barley/wheat and soybeans	Ferguson & McPherson, 1985
	No-tillage agricultural practice	House & Stinner, 1983
	Corn intercropped into a grass–legume mixture	Brust *et al.,* 1985
	Presence of weeds in cereal fields	Speight & Lawton, 1976
	Presence of chrysomelid on several weeds as alternate hosts of *Lebia* beetles	Hemenway & Whitcomb, 1967
	Cover cropping in apple orchards	Altieri & Schmidt, 1986
	Intercropping with white and red clover	Dempster & Coaker, 1974
	Presence of grassy border strips, hedges, wooded areas for survival of spring-breeding beetles	Wallin, 1985
Coccinellidae	Spraying of sucrose solution, artificial honeydew and protein-based food sprays	Hagen, 1986
	Strip cropping and cutting alfalfa in cotton fields	Schlinger & Dietrick, 1960
	Proximity of forest edges or hedgerows to crop fields	van Emden, 1965
	Cutting stinging nettle (*Urtica dioica*) or grass weed cover to force beetle movement to crops	Perrin, 1975; Hodek, 1973
	Increasing corn planting density	Risch *et al.,* 1982
	Double cropped barley–soybeans	Ferguson *et al.,* 1984
	Diptera	
Syrphidae	Occurrence of flowering plants (pollen sources) in the vicinity of crops or provision of flowers (wild mustard) within crops	van Emden, 1965, Altieri & Schmidt, 1986
	Spraying of Wheast® plus sugar or honey	Hagen, 1986
	Hymenoptera	
Formicidae	Use of sugar sources such as extrafloral nectaries and crushed sugarcane, to attract ants	Carroll & Risch, 1983
	Moving colonies to sites of high pest concentration	Carroll & Risch, 1983
	Providing adequate harborage for ground nests of beneficial ants	Buren & Whitcomb, 1977
	Growing annual crops adjacent to permanent ant habitat	Leston, 1973
	Use of shade trees supportive of *Oecophylla* spp.	Leston, 1973
	Use of cover crops in orchards and groves	van den Bosch & Telford, 1964
Vespidae	Provision of artificial structures for nesting of *Polistes annularis* (L.) (Hymenoptera: Vespidae) in cotton and tobacco and of *P. fuscatus* (F.) in cabbage	Lawson *et al.,* 1961; Gould & Jeanne, 1984

(continues)

TABLE 6 (*continued*)

Predator family	Environmental management practice(s)	References
Aranae (spiders)	Maintenance of weed cover in sugarcane	Ali & Reagan, 1985
	Crop fields surrounded by perennial, less disturbed habitats	Reichert & Lockley, 1984
	Cropping systems with lower soil disturbance (no tillage, mulching, orchards, etc.)	Riechert & Lockley, 1984
Acarina (mites)	Addition of pollen to enhance *Amblyseius hibisci* (Chant) in avocados	McMurtry & Scriven, 1966
	Dusting of cattail pollen (*Typha* sp.) to enhance tydeid mites in grapes as alternate prey for *Metaseiulus occidentalis*	Flaherty *et al.,* 1971
	Weed ground cover supportive of two-spotted mites as alternate prey for *M. occidentalis*	Flaherty *et al.,* 1971

of arthropods colonizing a crop field, however, are the result of highly dynamic processes; and the assumption of equilibrium is often inappropriate. Nevertheless, some predictions from the theory seem to hold.

For example, species richness is positively correlated to size on oceanic islands. Similarly, in mainland communities, the number of herbivores associated with a plant is a positive function of the local area planned for, or covered by, that species (Strong, 1979). Larger host islands probably collect more individuals by random probability of encounter. Further, patch detection by dispersers (apparent) may increase with size. Whatever the cause, the effect of an increase in the number of herbivores with an increase in size is important for consideration in pest management strategies. However, any increase in species diversity must be defined by the proportion in each trophic level, at least, and if possible, by the component species' biologies before it can be analyzed in terms of pest management potentials. MacArthur and Wilson's (1967) model treats all members of a species source pool as equivalent colonizers. Clearly, the application of this theory to dynamic and temporary crop islands requires the consideration not only of the number of species and pattern of occurrence, but also of the order and strength of colonizer establishment by trophic level [Altieri & Letourneau, 1984; Robinson & Dickerson, 1987; Thomas & Janzen, 1989; and see Kadmon & Pulliam (1993)].

Extinction rates depend on resource availability in the system. Because the plants are supplied to the system, or "reset" at certain intervals (Levins & Wilson, 1980), the resource base may be more predictable for herbivores, at least early in the season. The immigration rates of natural enemies to large expanses of monoculture may be similarly increased, though spread from the edges may be slow, and thus favor the development of herbivore populations. The equilibrium theory of biogeography does not allow for comparisons of single, large crop fields versus a network of

several small fields of the same total area; yet the contrasting designs are likely to differ in terms of suitability for biological control (Price, 1976).

Although the body of theory based on island community development suggests questions and organizes thought on crop design, the impediments to its application lie in four areas: (1) the frequent disturbance of most crop fields reduces the rigor and applicability of equilibrium models; (2) the few current empirical data available on diversity, size, and distance relationships do not constitute a sufficient basis for environmental design recommendations (Simberloff, 1986); (3) the theory does not distinguish pests and beneficials [see Stenseth (1981)]; and (4) the economic impact of "changing island size" will be viewed as exceedingly risky until demands for more certainty in the theory are met (Simberloff, 1986). Liss *et al.* (1986), however, presented a modification of the MacArthur and Wilson model that incorporates colonizer source composition and changes in "island" habitats over time. They call for further research toward understanding the organization of species pools as sources of arthropod communities in agroecosystems. Kruess and Tscharntke's (1994) study of clover weevils, distance of "islands" from source pools, and parasitism rates demonstrates the kind of studies needed to evaluate the applicability of aspects of island biogeography to agricultural settings.

Consumer Dynamics

Once the natural enemies are within the habitat of their prey or hosts, studies of consumer dynamics become important in predicting the outcome of their interactions. Traditionally, trophic interaction studies in manipulated as well as in natural systems have concentrated on two trophic levels (i.e., plant–herbivore, predator–prey, host–parasite). Many data (Clark & Dallwitz, 1975; Mattson & Addy, 1975; Podoler & Rogers, 1975; Morrow, 1977; Gilbert,

1978; McClure, 1980; Kareiva, 1982; Murdoch *et al.*, 1995) and much theory (Murdoch & Oaten, 1975; Hassell, 1978; May & Anderson, 1978; Clark & Holling, 1979; Murdoch, 1979) on consumer dynamics in two-level systems demonstrate the regulation of populations at the lower trophic level (plant, prey, or host) by natural enemies. Conversely, natural enemies have been ineffective in other cases, at least in the prevailing prey densities (Southwood & Comins, 1976; Strong *et al.*, 1984; Walker *et al.*, 1984). The effectiveness of natural enemies as regulators of herbivore populations depends not only on behavioral and developmental responses of individual enemies and on responses of the entire population to changes in prey or host densities (Murdoch, 1971; Murdoch & Oaten, 1975; Fox & Murdoch, 1978), but also on variation in plant parameters (density, secondary compounds, and associated plant species). The ability of natural enemies to regulate the herbivores depends on the herbivore population's intrinsic growth rate (R_m) (e.g., Getz, 1996) and size distribution (e.g., Murdoch *et al.*, 1997), which in turn reflects the quality of the plant diet. Even small changes in R_m caused by slight differences in plant quality (variety, secondary chemistry, and nutrients) may determine whether parasites or predators can control the herbivore populations (Lawton & McNeill, 1979; Price *et al.*, 1980). The effectiveness of regulation also reflects subtle differences in the timing and distribution of population events in both predator and prey populations (Hassell, 1978; May & Anderson, 1978). Theory and data on interactions involving three trophic levels in a complex habitat are ultimately more suitable as a basis for applying consumer-resource theory to environmental management strategies (Duffey & Bloem, 1986; Price, 1986; Letourneau, 1988; Sabelis *et al.*, 1991).

Thus, the goal of such prophylactic measures of pest control is to avoid the provision of enemy-free space in agricultural environments, and instead to present pests simultaneously with deleterious effects (mortality) caused by their natural enemies and with selectively defensive or suboptimal properties of their food plants. Therefore, studying systems as communities of at least three trophic levels will contribute an understanding of complex interactions that is different from that likely to be gained purely as a by-product of results from two-level studies (Orr & Boethel, 1986).

Effects of Vegetational Diversity

Root (1973) proposed two hypotheses to explain the tendency for low herbivore abundance in diverse vegetation. The "resource concentration" hypothesis predicts that many herbivores, especially those with a narrow host range, are more likely to find, survive, and reproduce on hosts that are in pure or nearly pure stands. "The enemies" hypothesis incorporates the third trophic level. Root (1973) predicted that complex vegetation would provide more resources for

natural enemies (alternate hosts, refugia, and nectar and pollen); as a result, herbivore irruption would be rapidly checked by a higher diversity and an abundance of enemies.

Sheehan (1986) extends the resource concentration concept to predict that specialist enemies will respond to mixed vegetation differently, and probably less favorably, than will generalist predators and parasitoids, because of the importance of alternate prey for generalists. The designation of host–prey specialization categories, however, tends to rely on only one aspect of the resource spectrum of parasitoids and predators (Letourneau, 1987). A range of species (or even individual) characteristics, such as relative vagility, resource needs, and habitat location cues may determine the response of parasitoids and predators to vegetational diversity. Theoretical concepts of connectivity (*sensu* Pimm, 1982) and guilds (Hawkins & MacMahon, 1989) may be important in guiding the design of habitats that optimize natural-enemy conservation. For example, the inclusion of insectary plants in a monoculture adds a functional guild of nectar and pollen producers, which then increases the actual number of trophic linkages in the system. Russell (1989) and Corbett and Plant (1993) suggest that some of our experimental results underestimate the impact of mixed vegetation on enemy enhancement, simply because of mismatched experiment scales with respect to relevant enemy movement patterns [but see Ramert & Ekbom (1996)].

The maintenance of heterogeneity within an agroecosystem may also affect the success of establishment of imported biological control agents. The debate over the degree to which ultimate levels of regulation are attained by single- versus multi-species releases in classical biological control projects continues (Turnbull, 1967; van den Bosch, 1968; Ehler & Hall, 1982; Van Driesche & Bellows, 1996), but analyses of environmental factors as raw materials or as constraints are rarely considered (Beirne, 1985). Certainly factors such as species richness, climatic gradients, and disturbance levels are important in assessing the susceptibility of large-scale communities to biological invasion (Fox & Fox, 1986).

Optimal Foraging

Implicit in environmental management for biological control are considerations of optimal foraging theory. During the host–prey selection process, natural enemies exhibit a chain of responses to stimuli. Several are exemplified in Table 5 as mechanisms with potential for manipulation. The objectives of biological control are to exploit natural processes that allow maximum prey encounter and foraging rates by natural enemies while, in most cases, achieving their persistence in the agroecosystem. Thus, this body of theory is useful for predicting enhancement mechanisms and for evaluating the consequences of under- and overexploitation.

The aggregation of foraging parasitoids in patches of

higher host density has been a critical feature thought to be responsible for successful biological control (Beddington *et al.*, 1978; May & Hassell, 1981; Murdoch & Briggs, 1996). Models of optimal patch use predict predation–parasitism levels between patches, based on host–prey densities (Cook & Hubbard, 1977; Waage, 1979; Iwasa *et al.*, 1984), but the power of these models varies and results depend on the particular spatial scale examined (Hopper *et al.*, 1991; Ives *et al.*, 1993). Murdoch *et al.* (1985) examined the importance of this searching behavior using the successful olive scale *Aphytis paramaculicornis* DeBach & Rosen and *Coccophagoides utilis.* Doutt system, and found little evidence that these parasitoids aggregate in areas of high pest density. Waage (1983) found that *Diadegma* spp. attacking *Plutella xylostella* (Linnaeus) aggregated in patches with greater host density; yet the proportion of hosts parasitized at high host densities was not greater. Roland (1986) found similar results with *Cyzenis albicans:* egg distribution does not affect the level of parasitism. Krivan (1996) showed how prey switching within the context of optimal foraging can lead to the persistence of predators and prey. However, field results can be complicated by community level effects (e.g., Moran & Hurd, 1997).

Predictive models then can be used to clarify the mechanisms involved in enemy behavior and their importance. It may be possible to take advantage of the "simple rules" that foragers use for decisions on how long to remain in a patch, which hosts or prey to seek and to accept, when and where they will oviposit, and (especially for hymenopterous parasitoids) what the sex ratio will be. If these "decisions" are made in response to environmental cues, then they are potentially manipulable in the field (Kareiva & Odell, 1987). Dicke *et al.* (1985) found that searching eucoilid parasitoids remained longer in a patch with moderately higher kairomone concentrations regardless of the actual density of *Drosophila melanogaster* Meigen, its host. The opposite effect could occur if prey or prey cues are difficult to find in a particular cropping system [see Spencer *et al.*, (1996)]. Theoretically, if resource patch boundaries are vague (Schmidt & Brown, 1996), actual biological control could be reduced even if prey are aggregated. Charnov and Skinner (1985) call for careful reflection of both the proximate causes of such responses and the evolutionary causes as complementary approaches that enhance theory and application.

It is also relevant to consider the ultimate population effects on natural enemies given habitat manipulations that exploit behavioral cues and maximize prey reduction. Fewer studies direct such attention to predator fitness. One example shows that although juvenile mantids exhibit a strong type II functional response, such behavior quickly increases beyond the maximum gain in characteristics related to fitness (Hurd & Rathet, 1986). We are not aware of studies addressing ultimate effects of enemy recruitment through enhanced habitat location cues at times of low prey

abundance [but Murdoch *et al.* (1985) do examine local extinction of prey; and theoretical considerations of local extinctions, spatial patchiness, and regional stability of host–parasitoid interactions (Hassell *et al.*, 1991) may help predict these effects]. Thus, the definition of "optimum" may initially depend on maximum influx and efficiency of natural enemies on a local scale, but the ultimate benefits to farmers may depend on large-scale and long-term effects on populations of predators and parasitoids. Thus, enemy response to environmental manipulation will benefit through life table studies over many generations (Hassell, 1986), optimal foraging models that include long-term population changes, and metapopulation models that consider spatial dynamics relative to interpatch heterogeneity (Taylor, 1991; Murdoch *et al.*, 1992; Letourneau, 1998).

CONCLUSIONS

Several manipulable components of agroecosystems provide the raw material for planned habitat management. Research on vegetation management, consumer dynamics, use of semiochemicals, and multitrophic level models of agricultural systems represent some of many advances providing direction. Renewed interest in sustainable agriculture as a response to growing concerns about ground water contamination, soil erosion, occupational health and consumer health, and safety set the stage for long-term planning for pest management. Implementation of habitat management for biological control is a viable option not only as an important component of IPM programs but also as an assurance of constancy of production under low-input and resource-conserving agricultural management.

Data and practical experience reported in this chapter provide ample evidence that it is possible to stabilize insect communities in agroecosystems by designing and constructing vegetational architectures that support populations of natural enemies or that have direct deterrent effects on pest herbivores. What is difficult is that each agricultural situation must be assessed separately, because herbivore–enemy interactions will vary significantly depending on insect species, location and size of the field, plant composition, surrounding vegetation, and cultural management. One can only hope to elucidate the ecological principles governing arthropod dynamics in complex systems, but the biodiversity designs necessary to achieve herbivore regulation will depend on the agroecological conditions and socioeconomic restrictions of each area.

Acknowledgment

Discussions with P. Barbosa, L. R. Fox, and K. Kester improved the quality of this chapter. We appreciate the comments of T. W. Fisher, G. Gordh, and T. S. Bellows, who carefully reviewed earlier versions of the manuscript. Many thanks to M. Hemler, R. Krach, E. Letourneau, and L.

Schmidt for excellent technical assistance. D. K. Letourneau was supported by USDA Competitive Grant No. 85-CRCR-11590 (to D. K. L. and L. R. Fox) and by UC/Santa Cruz Social Sciences Divisional Research Grants.

References

Adkisson, P. L. (1958). The influence of fertilizer applications on populations of *Heliothis zea* (Boddie), and certain insect predators. Journal of Economic Entomology, 51, 757–759.

Ali, A. D., & Reagan, T. E. (1985). Vegetation manipulation impact on predator and prey populations in Louisiana sugarcane ecosystems. Journal of Economic Entomology, 78, 1409–1414.

All, J. N., & Musick, G. J. (1986). Management of vertebrate and invertebrate pests. In M. A. Sprague & G. B. Triplett (Eds.), No-tillage and surface tillage agriculture (pp. 347–388). New York: John Wiley & Sons.

Allen, W. W., & Smith, R. F. (1958). Some factors influencing the efficiency of *Apanteles medicaginis* Muesebeck (Hymenoptera: Braconidae) as a parasite of the alfalfa caterpillar *Colias philodice eurytheme* Boisduval. Hilgardia, 28, 1–42.

Altieri, M. A. (1984). Patterns of insect diversity in monocultures and polycultures of brussels sprouts. Protection Ecology, 6, 227–232.

Altieri, M. A. (1989). Agroecology: a new research and development paradigm for world agriculture. Agriculture, Ecosystems, and Environment, 27, 37–46.

Altieri, M. A. (1994). Biodiversity and pest management in agroecosystems. New York: Food Products Press.

Altieri, M. A., & Anderson, M. K. (1986). An ecological basis for the development of alternative agricultural systems for small farmers in the third world. American Journal of Alternative Agriculture, 1, 30–38.

Altieri, M. A., & Letourneau, D. K. (1982). Vegetation management and biological control in agroecosystems. Crop Protection, 1(4), 405–430.

Altieri, M. A., & Letourneau, D. K. (1984). Vegetation diversity and insect pest outbreaks. CRC Critical Reviews in Plant Science, 2, 131–169.

Altieri, M. A., & Liebman, M. (1986). Insect, weed, and plant disease management in multiple cropping systems. In C. A. Francis (Ed.), Multiple cropping systems (pp. 183–218). New York: Macmillan.

Altieri, M. A., & Schmidt, L. L. (1984). Abundance patterns and foraging activity of ant communities in abandoned, organic and commercial apple orchards in northern California. Agriculture, Ecosystems, and Environment, 11, 3441–3452.

Altieri, M. A., & Schmidt, L. L. (1985). Cover crop manipulation in northern California apple orchards and vineyards: effects on arthropod communities. Biological Agriculture and Horticulture, 3, 1–24.

Altieri, M. A., & Schmidt, L. L. (1986). Population trends, distribution patterns, and feeding preferences of flea beetles (*Phyllotreta cruciferae* Goeze) in collard-wild mustard mixtures. Crop Protection, 5(3), 170–175.

Altieri, M. A., & Schmidt, L. L. (1987). Mixing cultivars of broccoli reduced populations of the cabbage aphid *Breviocoryne brassicae* (Linnaeus). California Agriculture, 41(11/12), 24–26.

Altieri, M. A. (1994). Biodiversity and pest management in agroecosystems. New York: Food Products Press.

Altieri, M. A., & Whitcomb, W. H. (1979a). Manipulation of insect populations through seasonal disturbance of weed communities. Protection Ecology, 1, 185–202.

Altieri, M. A., & Whitcomb, W. H. (1979b). The potential use of weeds in the manipulation of beneficial insects. HortScience, 14, 12–18.

Altieri, M. A., Cure, J. R., & Garcia, M. A. (1993). The role and enhancement of parasitic Hymenoptera biodiversity in agroecosystems. In J. LaSalle & I. D. Gauld (Eds.), Hymenoptera and biodiversity (pp. 257–275). United Kingdom: CAB International.

Altieri, M. A., Schoonhoven, A., & Doll, J. D. (1977). The ecological role of weeds in insect pest management systems: A review illustrated with

bean (*Phaseolus vulgaris* L.) cropping systems. Proceedings of the National Academy of Science, 23, 185–206.

Altieri, M. A., Wilson, R. C., & Schmidt, L. L. (1985). The effects of living mulches and weed cover on the dynamics of foliage and soil arthropod communities in three crop systems. Crop Protection, 4, 201–213.

Altieri, M. A., Francis, C. A., Schoonhoven, A., & Doll, J. (1978). Insect prevalence in bean (*Phaseolus vulgaris*) and maize (*Zea mays*) polycultural systems. Field Crops Research, 1, 33–49.

Altieri, M. A., Lewis, W. J., Nordlund, D. A., Gueldner, R. C., & Todd, J. W. (1981). Chemical interactions between plants and *Trichogramma* sp. wasps in Georgia soybean fields. Protection Ecology, 3, 259–263.

Andow, D. (1983a). Plant diversity and insect populations: Interactions among beans, weeds, and insects. Unpublished doctoral dissertation, Cornell University, Ithaca, NY.

Andow, D. (1983b). The extent of monoculture and its effects on insect pest populations with particular reference to wheat and cotton. Agriculture, Ecosystems, and Environment, 9, 25–35.

Andow, D. (1986). Plant diversification and insect population control in agroecosystems. In D. Pimentel (Ed.), Some aspects of integrated pest management (pp. 277–348). Ithaca, NY: Department of Entomology, Cornell University.

Andow, D. (1990). Population dynamics of an insect herbivore in simple and diverse habitats. Ecology, 72(3), 1006–1017.

Andow, D. (1991). Vegetational diversity and arthropod population response. Annual Review of Entomology, 36, 561–586.

Andow, D. A., & Hidaka, K. (1989). Experimental natural history of sustainable agriculture: Syndromes of production. Agriculture, Ecosystems, and Environment, 27, 447–462.

Andow, D. A., & Risch, S. J. (1985). Predation in diversified agroecosystems: Relations between a coccinellid predator *Coleomegilla maculata* and its food. Journal of Applied Ecology, 22, 357–372.

Arkin, G. F., & Taylor, H. M. (Eds.). (1981). Modifying the root environment to reduce crop stress. St. Joseph, MI: American Society of Agricultural Engineering.

Bach, C. E. (1980a). Effects of plant density and diversity on the population dynamics of a specialist herbivore, the striped cucumber beetle, *Acalymma vittatta* (Fab.). Ecology, 61, 1515–1530.

Bach, C. E. (1980b). Effects of plant diversity and time of colonization on an herbivore-plant interaction. Oecologia (Berlin), 44, 319–326.

Baker, R. S., Laster, M. L., & Kitten, W. F. (1985). Effects of the herbicide monosodium methanearsonate on insect and spider populations in cotton fields. Journal of Economic Entomology, 78(6), 1481–1484.

Barbosa, P., & Letourneau, D. K. (Eds.). (1989). Novel aspects of insect-plant interactions. New York: John Wiley & Sons.

Barfield, B. J., & Gerber, J. F. (Eds.). (1979). Modification of the aerial environment of plants. St. Joseph, MI: American Society of Agricultural Engineering.

Barghouti, S., Garlsus, L., & Unali, D. (Eds.). (1992). Trends in agricultural diversification: Regional perspectives (World Bank Tech. Paper No. 180). Washington, DC: World Bank.

Barker, G. M., & Addison, P. J. (1997). Clavicipitaceous endophytic infection in ryegrass influences attack rate of the parasitoid *Microctonus hyperodae* (Hymenoptera: Braconidae, Euphorinae) in *Listronotus bonatiensis* (Coleoptera: Curculionidae). Environmental Entomology, 26, 416–420.

Barney, R. J., & Pass, B. C. (1986). Ground beetle (Coleoptera: Carabidae) populations in Kentucky alfalfa and influence of tillage. Journal of Economic Entomology, 70, 511–517.

Barney, R. J., Lamp, W. O., Ambrust, E. J., & Kapusta, G. (1984). Insect predator community and its response to weed management in spring-planted alfalfa. Protection Ecology, 6, 23–33.

Bartlett, B. R. (1951). The action of certain "inert" dust materials on parasitic Hymenoptera. Journal of Economic Entomology, 44, 891–896.

Bartlett, B. R. (1964). Patterns in the host-feeding habit of adult parasitic Hymenoptera. Annals of the Entomological Society of America, 57, 344–350.

Bautista, R. C., & Harris, E. J. (1996). Effect of fruit substrates on parasitization of Tephritid fruit flies (Diptera) by the parasitoid *Biosteres arisanus* (Hymenoptera: Braconidae). Environmental Entomology, 25, 470–475.

Beddington, J. R., Free, C. A., & Lawton, J. M. (1978). Characteristics of successful natural enemies in models of biological control of insect pests. Nature (London), 273(5663), 513–519.

Beirne, B. P. (1985). Avoidable obstacles to colonization in classical biological control of insects. Canadian Journal of Zoology, 63, 743–747.

Benrey, B., & Denno, R. F. (1997). The slow-growth-high-mortality hypothesis: A test using the cabbage butterfly. Ecology, 78(4), 987–999.

Bentz, J., Reeves, J. III, Barbosa, P., & Francis, B. (1996). The effect of nitrogen fertilizer applied to *Euphorfia pulcherrima* on the parasitization of *Bemisia argentifolii* by the parasitoid *Encarsia* formosa. Entomologia Experimentalis et Applicata, 78, 105–110.

Bernays, E. A. (1997). Feeding by lepidopteran larvae is dangerous. Ecological Entomology, 22, 121–123.

Bernays, E. A. & Graham, M. (1988). On the evolution of host specificity in phytophagous insects. Ecology, 69, 886–892.

Bernays, E. A., & Lewis, A. C. (1986). The effect of wilting on palatability of plants to *Schistocerca gregaria,* the desert locust. Oecologia (Berlin), 70, 132–135.

Blumberg, A. Y., & Crossley, D. A., Jr. (1983). Comparison of soil surface: Arthropod populations in conventional tillage, no-tillage and old field systems. Agro-Ecosystems, 8, 247–253.

Bobb, M. L. (1939). Parasites of the oriental fruit moths in Virginia. Journal of Economic Entomology, 32, 605–609.

Boethel, D. J., & Eikenbary, R. D. (Eds.). (1986). Interactions of plant resistance and parasitoids and predators of insects. Chichester, United Kingdom: Ellis Horwood.

Bombosch, S. (1966). Occurrence of enemies on different weeds with aphids. In I. Hodek (Ed.), Ecology of aphidophagus insects (pp. 177–179). Prague: Academic Printing House.

Bosch, J. (1987). The influence of some dominant weeds on beneficial and pest arthropods in a sugarbeet field. Zeitschrift fur Pflanzenkrankheiten und Pflanzenschutz, 94, 398–408.

Bottenberg, H., Litsinger, J. A., Barrion, A. T., & Kenmore, P. E. (1990). Presence of tungro vectors and their natural enemies in different rice habitats in Malaysia. Agriculture, Ecosystems, and Environment, 31, 1–15.

Bottrell, D. R. (1980). Integrated pest management. Council on Environmental Quality. Washington, DC: U.S. Government Printing Office.

Brust, G. E. (1991). Soil moisture, no-tillage and predator effects on southern corn rootworm survival in peanut agroecosystems. Entomologia Experimentalis et Applicata, 58, 109–121.

Brust, G. E., Stinner, B. R., & McCartney, D. A. (1985). Tillage and soil insecticide effects on predator-black cutworm (Lepidoptera: Noctuidae) interactions in corn agroecosystems. Journal of Economic Entomology, 78, 1389–1392.

Brust, G. E., Stinner, B. R., & McCartney, D. A. (1986). Predation by soil inhabiting arthropods in intercropped and monoculture agroecosystems. Agriculture, Ecosystems, and Environment, 18, 145–154.

Bugg, R. L., & Wilson, L. T. (1989). *Ammi visnaga* (L.) Lamarck (Apiaceae): Associated beneficial insects and implications for biological control, with emphasis on the bell-pepper agroecosystem. Biological Agriculture and Horticulture, 6, 241–268.

Bugg, R. L., Ehler, L. E., & Wilson, L. T. (1987). Effect of common knotweed (*Polygonum aviculare*) on abundance and efficiency of insect predators of crop pests. Hilgardia, 55, 52 pp.

Bugg, R. L., Wackers, F. L., Brunson, K. E., Dutcher, J. D., & Phatak, S. C. (1991). Cool-season cover crops relay intercropped with cantaloupe: Influence on a generalist predator, *Geocrois punctipes* (Hemiptera: Lygaeidae). Journal of Economic Entomology, 84, 408–416.

Buren, W. F., & Whitcomb, W. H. (1977). Ants of citrus: Some considerations. Proceedings, International Society of Citriculture, 2, 496–498.

Buschman, L. L., Pitre, H. N., & Hodges, H. F. (1984). Soybean cultural practices: Effects on populations of geocorids, nabids, and other soybean arthropods. Environmental Entomology, 13, 305–317.

Campbell, B. C., & Duffey, S. S. (1979). Tomatine and parasitic wasps: Potential incompatibility of plant antibiosis with biological control. Science, 205, 700–702.

Carcamo, H. A. (1995). Effect of tillage on ground beetles (Coleoptera: Carabidae): a farm-scale study in central Alberta. Canadian Entomologist, 127, 631–639.

Carroll, C. R. (1978). Beetles, parasitoids and tropical morning glories: A study in host discrimination. Economic Entomology, 3, 79–86.

Carroll, C. R., & Risch, S. J. (1983). Tropical annual cropping systems: Ant ecology. Environmental Management, 7, 51–57.

Cerutti, F., Delucchi, V., Baumgartner, J., & Rubli, D. (1989). Research on the vineyard ecosystem in Tessin. II. Colonization of grapevines by the cicadellid *Empoasca vitis* Goethe (Hom., Cicadellidae, Typhlocybinae) and of its parasitoid *Anagrus atomus* Haliday (Hym., Mymaridae), and the importance of the surrounding flora. Mitteilungen der Schweizerischen Entomologischen Gesellschaft, 62, 253–267.

Chandler, A. E. F. (1968). The relationship between aphid infestations and oviposition by aphidophagons Syrphidae (Diptera). Annals of Applied Biology, 61, 425–434.

Charnov, E. L. (1982). The theory of sex allocation. Princeton, NJ: Princeton University Press.

Charnov, E. L., & Skinner, S. W. (1985). Complementary approaches to the understanding of parasitoid oviposition decisions. Environmental Entomology, 14, 383–391.

Chiang, H. C. (1970). Effects of manure applications and mite predation on corn rootworm populations in Minnesota. Journal of Economic Entomology, 63, 934–936.

Chiri, A. A., & Legner, E. F. (1983). Field applications of host-searching kairomones to enhance parasitization of the pink bollworm (Lepidoptera: Gelechiidae). Journal of Economic Entomology, 76, 254–255.

Chiverton, P. A. (1989). The creation of within-field overwintering sites for natural enemies of cereal aphids. Proceedings of the Brighton Crop Protection Conference, Weeds (Vol. 3, pp. 1093–1096). Brighton, United Kingdom: November 20–23, 1989. Farnham, Surrey: British Crop Protection Council.

Clark, L. R., & Dallwitz, M. J. (1975). The life system of *Cardiaspina albitextura* (Psyllidae), 1950–1974. Australian Journal of Zoology, 23, 523–561.

Clark, M. S., Gage, S. H., & Spence, J. R. (1997). Habitats and management associated with common ground beetles (Coleoptera: Carbidae) in a Michigan agricultural landscape. Environmental Entomology, 26, 519–527.

Clark, W. C., & Holling, C. S. (1979). Process models, equilibrium structures, and population dynamics: On the formulation and testing of realistic theory in ecology. Fortschritte der Zoologie, 25, 29–52.

Coli, W. M., & Ciurlino, R. (1990). Interaction of weeds and apple pests. Proceedings, Annual Meeting, Massachusetts Fruit Growers' Association, 96, 52–58.

Coll, M., & Bottrell, D. G. (1995). Predator-prey association in monocultures and dicultures—effect of maize and bean vegetation. Agriculture, Ecosystems and Environment, 54, 115–125.

Coll, M., & Bottrell, D. G. (1996). Movement of an insect parasitoid in simple and diverse plant assemblages. Ecological Entomology, 21(2), 141–149.

Collins, F. L., & Johnson, S. J. (1985). Reproductive response of caged adult velvetbean caterpillar and soybean looper to the presence of weeds. Agriculture, Ecosystems, and Environment, 14, 139–149.

Cook, R. M., & Hubbard, S. F. (1977). Adaptive searching strategies in insect parasites. Journal of Animal Ecology, 46, 115–125.

Corbett, A., & Plant, R. E. (1993). Role of movement in the response of natural enemies to agroecysystem diversification: A theoretical evaluation. Entomological Society of America, 22, 519–531.

Cowgill, S. (1989). The role of non-crop habitats on hoverfly (Diptera: Syrphidae) foraging on arable land. Proceedings of the Brighton Crop Protection Conference, Weeds (Vol. 3, pp. 1103–1108). Brighton, United Kingdom: November 20–23, 1989. Farnham, Surrey: British Crop Protection Council.

Croft, B. A., & Slone, D. H. (1997). Equilibrium densities of European red mite (Acari: Tetranychidae) after exposure to three levels of predaceous mite diversity on apple. Environmental Entomology, 26, 391–399.

Cromartie, W. J. (1981). The environmental control of insects using crop diversity. In D. Pimentel (Ed.), CRC handbook of pest management in agriculture (Vol. 1, pp. 223–250). CRC Handbook Series in Agriculture, CRC Press, Boca Raton, FL:

Culliney, T. W., & Pimentel, D. (1986). Ecological effects of organic agricultural practices on insect populations. Agriculture, Ecosystems, and Environment, 15, 253–266.

Culliney, T. W., Pimentel, D., & Lisk, D. J. (1986). Impact of chemically contaminated sewage sludge on the collard arthropod community. Environmental Entomology, 15, 826–833.

Davidson, A. W., & Potter, D. A. (1995). Response of plant-feeding, predatory, and soil-inhabiting invertebrates to Acremonium endophyte and nitrogen fertilization in tall fescue turf. Journal Economic Entomology, 88, 367–379.

DeBach, P. (1958a). Application of ecological information to control citrus pests in California. Proceedings, 10th International Congress on Entomology, 3, 187–194.

DeBach, P. (1958b). The role of weather and entomophagous species in the natural control of insect populations. Journal of Economic Entomology, 51, 474–484.

De Cock, A., DeClercq, P., Tirry, L., & Degheele, O. (1996). Toxicity of Diafenthiuron and imidacloprid to the predatory bug Podisus maculiventris (Heteroptera: pentatomidae). Environmental Entomology, 25, 476–480.

De la Cruz, C. G., & Litsinger, J. A. (1986). Effect of ratoon rice crop on populations of green leafhopper Nephotettix virescens, brown planthopper Nilaparvata lugens, whitebacked planthopper Sogatella furcifera, and their predators. International Rice Research Newsletter, 11, 25–26.

DeLima, J. O. G., & Leigh, T. F. (1984). Effect of cotton genotypes in the western bigeyed bug (Heteroptera: Miridae). Journal of Economic Entomology, 77, 898–902.

DeLoach, C. J. (1970). The effect of habitat diversity on predation. Proceedings, Tall Timbers Conference on Ecological Animal Control by Habitat Management, 2, 223–241.

Dempster, J. P. (1969). Some effects of weed control on the numbers of the small cabbage white (Pieris rapae L.) on brussels sprouts. Journal of Applied Ecology, 6, 339–345.

Dempster, J. P., & Coaker, T. H. (1974). Diversification of crop ecosystems as a means of controlling pests. In D. P. Jones & M. E. Solomon (Eds.), Biology in pest and disease control (pp. 106–114). New York: John Wiley & Sons.

Denno, R. F., & Larsson, S., & Olmstead, K. L. (1990). Role of enemy-free space and plant quality in host-plant selection by willow beetles. Ecology, 71(1), 124–137.

Desender, K., & Alderweireldt, M. (1990). The carabid fauna of maize fields under different rotation regimes. Mededelingen van de Faculteit Landbouwwetenschappen, Rijksuniversiteit Gent, 55(2b), 493–500.

Dicke, M., & Sabelis, M. W. (1988). How plants obtain predatory mites as bodyguards. Netherlands Journal of Zoology, 38, 148–165.

Dicke, M., van Lenteren, M. C., Boskamp, J. G. F., & van Voorst, R. (1985). Intensification and elongation of host searching in Leptopilina heterotoma (Thomson) (Hymenoptera: Eucoilidae) through a kairomone produced by Drosophila melanogaster. Journal of Chemical Ecology, 11, 125–136.

Dicke, M., Sabelis, M. W., Takabayashi, J., Bruin, J., & Posthumus, M. A. (1990). Plant structures of manipulating predator-prey interactions through allelochemicals: Prospects for application in pest control. Journal of Chemical Ecology, 16, 3091–3118.

Dickler, E. (1978). Influence of beneficial arthropods on the codling moth in an orchard with green covered and clean cultivated soil (pp. 16–18). In Summaries of papers presented at the Joint FAO/IAEA and IOBC/WPRS research coordination meetings, Heidelberg, Germany, Berlin: Paul Parey.

Drinkwater, L. E., Workneh, F., Letourneau, D. K., van Bruggen, A. H. C., & Shennan, C. (in review). Fundamental differences between conventional and organic tomato agroecosystems in California. Ecological Applications, 5, 1098–1112.

Doutt, R. L., & Nakata, J. (1973). The Rubus leafhopper and its egg parasitoid: An endemic biotic system useful in grape pest management. Environmental Entomology, 2, 381–386.

Downey, D. A., & Caviness, C. E. (1973). Temperature, humidity and light studies in soybean canopies (Bulletin 784). Agricultural Experimental Station, University of Arkansas.

Duelli, P. (1980). Adaptive and appetitive flight in the green lacewing, Chrysopa carnea. Ecological Entomology, 5, 213–220.

Duffey, S. S., & Bloem, K. A. (1986). Plant defense-herbivore-parasite interactions and biological control. In M. Kogan (Ed.), Ecological theory and integrated pest management practice (pp. 135–184). New York: John Wiley & Sons.

Edmunds, G. F., Jr. (1973). Ecology of black pine leaf scale (Homoptera: Diaspididae). Environmental Entomology, 2, 765–777.

Ehler, L. E., & Hall, R. W. (1982). Evidence for competitive exclusion of introduced natural enemies in biological control. Environmental Entomology, 11, 1–4.

Eller, B. M. (1977). Road dust induced increased leaf temperature. Environmental Pollution, 13, 99–107.

El-Shazly, N. Z. (1972a). Der Einfluss assere Faktoren auf die hämocytare Abwehrreaktion von Neomyzus circumflexus (Buck.) (Homoptera: Aphididae). Zeitschrift fuer Agnewandte Entomologie, 70, 414–436.

El-Shazly, N. Z. (1972b). Der Einfluss von Ernahrung und alterbder Muttertieres auf die hämocytare Abwehrreaktion von Neomyzus circumflexus (Buck.), Entomophaga, 17, 203–209.

Elzen, G. E., Williams, H. J., & Vinson, S. B. (1983). Response by the parasitoid Campoletis sonorensis (Hymenoptera: Ichneumonidae) to chemicals (synomones) in plants: Implications for host habitat location. Environmental Entomology, 12, 1873–1877.

Elzen, G. E., Williams, H. J., & Vinson, S. B. (1986). Wind tunnel flight responses by hymenopterous prarasitoid Campoletis sonorensis to cotton cultivars and lines. Entomologia Experimentalis et Applicata, 42, 285–289.

Farrow, R. A. (1981). Aerial dispersal of Scelio fulgidus (Hym.: Scelionidae), parasite of eggs of locusts and grasshoppers (Orth.: Acrididae). Entomophaga, 26, 349–355.

Feeny, P. P. (1976). Plant apparency and chemical defense. Recent Advances in Phytochemistry, 10, 1–40.

Felland, C. M. (1990). Habitat-specific parasitism of the stalk borer (Lepidoptera: Noctuidae) in northern Ohio. Environmental Entomology, 19, 162–166.

Felland, C. M., & Pitre, H. N. (1991). Diversity and density of foliage-inhabiting arthropods in irrigated and dryland soybean in Mississippi. Environmental Entomology, 20, 498–506.

Ferguson, H. J., & McPherson, R. M. (1985). Abundance and diversity of adult Carabidae in four soybean cropping systems in Virginia. Journal of the Entomological Society of America, 20, 163–171.

Ferguson, H. J., McPherson, R. M., & Allen, W. A. (1984). Effect of four soybean cropping systems on the abundance of foliage-inhabiting insect predators. Environmental Entomology, 13, 1105–1112.

Ferro, D. N., & Southwick, E. E. (1984). Microclimates of small arthropods: Estimating humidity within the leaf boundary layer. Environmental Entomology, 13, 926–929.

Ferro, D. N., Chapman, R. B., & Penman, D. R. (1979). Observations on insect microclimate and insect pest management. Environmental Entomology, 8, 1000–1003.

Flaherty, D. (1969). Ecosystems trophic complexity and Willamette mite Eotetranychus willamettei (Acarina: Tetranychidae) densities. Ecology, 50, 911–916.

Flaherty, D., Lynn, C., Jensen, F., & Hoy, M. (1971). Influence of environment and cultural practices on spider mite abundance in southern San Joaquin Thompson seedless vineyards. California Agriculture, 25(11), 6–8.

Flaherty, D. L., Wilson, L. T., Stern, W. M., & Kido, H. (1985). Biological control in San Joaquin Valley vineyards. In M. A. Hoy & D. C. Herzog (Eds.), Biological control in agricultural IPM systems (pp. 501–520). San Diego: Academic Press.

Flanders, S. E. (1937). Habitat selection by Trichogramma. Annals of the Entomological Society of America, 30, 208–210.

Flint, H. M., Merkle, J. R., & Sledge, M. (1981). Attraction of male Collops vittatus in the field by caryophylline alcohol. Environmental Entomology, 10, 301–304.

Flint, H. M., Salter, S. S., & Walters, S. (1979). Caryophyllene: An attractant for the green lacewing. Environmental Entomology, 8, 1123–1125.

Force, D. C., & Messenger, P. S. (1964). Duration of development, generation time, and longevity of three hymenopterous parasites of Therioaphis maculata, reared at various constant temperatures. Annals of the Entomological Society of America, 54(4), 405–413.

Foster, M. A., & Ruesink, W. G. (1984). Influence of flowering weeds associated with reduced tillage in corn on a black cutworm (Lepidoptera: Noctuidae) parasitoid, Metorus rubens (Nees). Environmental Entomology, 13, 664–668.

Fox, M. D., & Fox, B. J. (1986). The susceptibility of natural communities to invasion. In R. H. Groves & J. J. Burdon (Eds.), Ecology of biological invasions (pp. 57–66). Sydney, Australia: Cambridge University Press.

Fox, L. R., & Murdoch, W. W. (1978). Effects of feeding history of short-term and long-term functional responses in Notonecta hoffmanni. Journal of Animal Ecology, 47, 945–959.

Fox, L. R., Letourneau, D. K., Eisenbach, J., & van Nouhuys, S. (1990). Parasitism rates and sex ratios of a parasitic wasp: Effects of herbivore and host plant quality. Oecologia (Berlin), 83, 414–419.

Francis, G. A., Flora, C. B., & King, L. D. (Eds.). (1990). Sustainable agriculture in the temperate zones. New York: John Wiley & Sons.

Funderburk, J. E., Wright, D. L., & Teare, I. D. (1988). Preplant tillage effects on population dynamics of soybean insect predators. Crop Science, 28, 973–976.

Fye, R. E., & Larsen, D. J. (1969). Preliminary evaluation of Trichogramma minutum as a released regulator of lepidopterous pests of cotton. Journal of Economic Entomology, 62, 1291–1296.

Gange, A. C., & Llewellyn, M. (1989). Factors affecting orchard colonisation by the black-kneed capsid (Blepharidopterus angulatus (Hemiptera: Miridae)) from alder windbreaks. Annals of Applied Biology, 114, 221–230.

Gardner, S. M., & van Lenteren, J. C. (1986). Characterisation of the arrestment responses of Trichogramma evanescens. Oecologia (Berlin), 68, 265–270.

Gaylor, M. J., & Foster, M. A. (1987). Cotton pest management in the southeastern United States influenced by conservation tillage practices. In G. J. House & B. R. Stinner (Eds.), Arthropods in conservation tillage systems. Miscellaneous Publications of the Entomological Society of America, 65, 29–34.

Gebhardt, M. R. (1981). Methods of pesticide application. In D. Pimentel (Ed.), Handbook of pest management in agriculture (Vol. 2, pp. 87–102). Boca Raton, FL: CRC Press.

Gerard, P. J. (1989). Influence of egg depth in host plants on parasitism of Scolypopa australis (Homoptera: Ricaniidae) by Centrodora scolypopae (Hymenoptera: Aphelinidae). New Zealand Entomology, 12, 30–34.

Geitzenauer, H. L., & Bernays, E. A. (1996). Plant effects on prey choice by a vespid wasp, polistes arizonensis. Ecological Entomology, 21(3), 227–234.

Gerling, C. A. (1984). Selenium in agriculture and the environment. Agriculture, Ecosystems, and Environment, 11, 37–65.

Getz, W. M. (1996). A hypothesis regarding the abruptness of density dependence and the growth rate of populations. Ecology, 77(7), 2014–2026.

Gilbert, L. E. (1978). Development of theory in the analysis of insect plant interactions. In Horn et al. (Eds.), Analysis of ecological systems (pp. 116–154). Columbus, OH: Ohio State University Press.

Ginevan, M. E., Lane, D. D., & Greenberg, L. (1980). Ambient air concentration of sulfur dioxide affects flight activity in bees. Proceedings of the National Academy of Sciences, USA, 77, 5631–5633.

Gold, C. S., Altieri, M. A., & Bellotti, A. C. (1991). Survivorship of the cassava whiteflies Aleurotrachelus socialis and Trialeurodes variabilis (Homoptera: Aleyrodidae) under different cropping systems in Colombia. Crop Protection, 10, 305–309.

Gould, F. (1986). Simulation models for predicting durability of insect-resistant germ plasm: Hessian fly (Diptera: Cecidomyiidae) in resistant winter wheat. Environmental Entomology, 15(1), 11–23.

Gould, W. P., & Jeanne, R. L. (1984). Polistes wasps (Hymenoptera: Vespidae) as control agents for lepidopterous cabbage pests. Environmental Entomology, 13(1), 150–156.

Greenblatt, J. A., & Barbosa, P. (1981). Effect of host's diet on two pupal parasitoids of the Gypsy moth: Brachymeria intermedia (Nees) and Coccygominus turionellae (L.). Journal of Applied Ecology, 18, 1–10.

Greenblatt, J. A., & Lewis, W. J. (1983). Chemical environment manipulation for pest insects control. Environmental Management, 7, 35–41.

Grigg, D. B. (1974). The agricultural systems of the world. London: Cambridge University Press.

Gross, H. R. (1981). Employment of kairomones in the management of parasitoids. In D. A. Nordlund, R. L. Jones, & W. J. Lewis (Eds.), Semiochemicals: Their role in pest control (pp. 137–150). New York: John Wiley & Sons.

Gross, J. R., Jr., Pair, S. D., & R. D. Jackson, (1985). Aggregation response of adult predators to larval homogenates of Heliothis zea and Spodoptera frugiperda (Lepidoptera: Noctuidae) in whorl-state cotton. Environmental Entomology, 14, 360–364.

Hagen, K. S. (1962). Biology and ecology of predacious Coccinellidae. Annual Review of Entomology, 7, 289–326.

Hagen, K. S. (1986). Ecosystem analysis: Plant cultivars (HPR), entomophagous species and food supplements. In D. J. Boethel & R. D. Eikenbary (Eds.), Interactions of plant resistance and parasitoids and predators of insects (pp. 151–197). Chichester, United Kingdom: Ellis Horwood.

Hamai, J., & Huffaker, C. B. (1978). Potential of predation by Metaseiulus occidentalis in compensating for increased, nutritionally induced, power of increase of Tetranychus urticae. Entomophaga, 23, 225–237.

Hammond, R. B., & Stinner, B. R. (1987). Soybean foliage insects in conservation tillage systems: Effects of tillage, previous cropping history, and soil insecticide application. Environmental Entomology, 16, 524–531.

Hansson, L. (1991). Dispersal and connectivity in metapopulations. Biological Journal of the Linnean Society, 42, 89–103.

Harris, V. E., & Todd, J. W. (1980). Male-mediated aggregation of male, female and 5th-instar green stink bugs and concomitant attraction of a tachinid parasite, Trichopoda pennipes. Entomologia Experimentalis et Applicata, 27, 117–126.

Hassell, M. P. (1978). The dynamics or arthropod predator-prey systems. Princeton, NJ: Princeton University Press.

Hassell, M. P. (1986). Parasitoids and population regulation. In J. Waage & D. Greathead (Eds.), Insect parasitoids (pp. 202–224). New York: Academic Press.

Hassell, M. P., Comins, H. N., & May, R. M. (1991). Spatial structure and chaos in insect population dynamics. Nature (London), 353, 255–258.

Hatfield, J. L. (1982). Modification of the microclimate via management. In J. L. Hatfield & I. J. Thomason (Eds.), Biometerology in integrated pest management (pp. 147–170). New York: Academic Press.

Hawkins, C. P., & MacMahon, J. A. (1989). Guilds: The multiple meanings of a concept. Annual Review of Entomology, 34, 423–451.

Hemenway, R., & Whitcomb, W. H. (1967). Ground beetles of the genus Lebia latreilla in Arkansas (Coleoptera: Carabidae): Ecology and geographic distribution. Arkansas Academy of Science Proceedings, 21, 15–20.

Henze, M., & Sengonca, C. (1992). Influence of different rotations on population development of cereal aphids and associated predators in winter wheat. Gesunde Pflanzen, 44, 122–125.

Herzog, D. C., & Funderburk, J. E. (1985). Plant resistance and cultural practice interactions with biological control. In M. A. Hoy & D. C. Herzog (Eds.), Biological control in agricultural IPM systems (pp. 67–88). Orlando, FL: Academic Press.

Hislop, R. G., & Prokopy, R. J. (1981). Mite predator responses to prey and predator-emitted stimuli. Journal of Chemical Ecology, 7, 895–904.

Hodek, I. (1973). Biology of Coccinellidae. Prague: Academic Publishing.

Hogg, D. B. (1986). Interaction between crop phenology and natural enemies: Evidence from a study of Heliothis population dynamics on cotton. In D. J. Boethel & R. D. Eikenbary (Eds.), Interactions of plant resistance and parasitoids and predators of insects (pp. 98–124). Chichester, United Kingdom: Ellis Horwood.

Holliday, N. J., & Hagley, E. A. C. (1984). The effect of sod type on the occurrence of ground beetles (Coleoptera: Carabidae) in a pest management apple orchard. Canadian Entomologist, 116, 165–171.

Holmes, N. D. (1982). Population dynamics of the wheat stem sawfly, Cephus cinctus (Hymenoptera: Cephidae), in wheat. Canadian Entomologist, 114, 775–788.

Holmes, N. D., Nelson, W. A., Peterson, L. K., & Farstad, C. W. (1963). Causes of variations in effectiveness of Bracon cephi (Gahan) (Hymenoptera: Braconidae) as a parasite of the wheat stem sawfly. Canadian Entomologist, 95 (2), 113–126.

Holt, R. D., Debinski, D. M., Diffendorfer, J. E., Martinko, E. A., Robinson, G. R., & Ward, G. C. (1995). Perspectives from an experimental study of habitat fragmentation in an agroeco-system. In D. M. Glen, M. P. Greaves, & H. M. Anderson (Eds.), Ecology and integrated farming systems (pp. 147–175). Bristol, UK.

Honek, A. (1983). Factors affecting the distribution of larvae of aphid predators (Col., Coccinellidae and Dipt., Syrphidae) in cereal stands. Zeitschrift fur Angewandte Entomologie, 95, 336–345.

Hopper, K. R., Powell, J. E., & King, E. G. (1991). Spatial density dependence in parasitism of Heliothis virescens (Lepidoptera: Noctuidae) by Microplitis croceipes (Hymenoptera: Braconidae) in the field. Environmental Entomology, 20, 292–302.

Horn, D. J. (1981). Effects of weedy backgrounds on colonization of collards by green peach aphid, Myzus persicae, and its major predators. Environmental Entomology, 10, 285–289.

House, G. J. (1989a). Soil arthropods from weed and crop roots of an agroecosystem in a wheat-soybean-corn rotation: Impact of tillage and herbicides. Agriculture, Ecosystems, and Environment, 25, 233–244.

House, G. J. (1989b). No-tillage and legume cover cropping in corn agroecosystems: Effects on soil arthropods. Acta Phytopathologica et Entomologica Hungarica, 24, 99–104.

House, G. J., & All, J. N. (1981). Carabid beetles in soybean agroecosystems. Environmental Entomology, 10, 194–196.

House, G. J., & Stinner, B. R. (1983). Arthropods in no-tillage soybean agroecosystems: Community composition and ecosystem interactions. Environmental Management, 7, 23–28.

Hoveland, C. S., Buchanan, G. A., & Harris, M. C. (1976). Response of weeds to soil phosphorus and potassium. Weed Science, 24, 144–201.

Huffaker, C. B., & Messenger, P. S. (1976). Theory and practice of biological control. New York: Academic Press.

Huffaker, C. B., Van der Vrie, M., & McMurtry, J. A. (1970). Ecology of tetranychid mites and their natural enemies: A review. II. Tetranychid populations and their possible control by predators: An evaluation. Hilgardia, 40, 391–458.

Hulspas-Jordaan, P. M., & van Lenteren, J. C. (1978). The relationship between host-plant leaf structure and parasitization efficiency of the parasitic wasp, Encarsia formosa Gahan (Hymenoptera: Aphelinidae). Mededelingen van de Faculteit Landbouwwetenschappen, Rijksuniversiteit Gent, 43, 431–440.

Hurd, L. E., & Rathet, I. H. (1986). Functional response and success in juvenile mantids. Ecology, 67, 163–167.

Hurst, G. A. (1970). The effects of controlled burning arthropod density and biomass in relation to bobwhite quail brood habitat on a right-of-way. Proceedings, Tall Timbers Conference on Ecological Animal Control by Habitat Management, 2, 173–183.

Hutchison, W. D., & Pitre, H. N. (1983). Predation of Heliothis virescens (Lepidoptera: Noctuidae) eggs by Geocoris punctipes (Hemiptera: Lygaeidae) adults on cotton. Environmental Entomology, 12, 1652–1656.

Idris, A. B., & Grafius, E. (1995). Wildflowers as nectar sources for Diadegma insulare (Hymenoptera: Ichneumonidae), a parasitoid of Diamondback moth (Lepidoptera: Yponomeutidae). Environmental Entomology, 24, 1726–1735.

Idris, A. B., & Grafius, E. (1996). Effects of wild and cultivated host plants on oviposition, survival, and development of Diamondback moth (Lepidoptera: Plutellidae) and its parasitoid Dindegma insulare (Hymenoptera: ichneumonidae). Environmental Entomology, 25, 825–833.

Ives, A. R., Kareiva, P., & Perry, R. (1993). Response of a predator to variation in prey density at 3 hierarchical scales—Lady beetles feeding on aphids. Ecology, 74(7), 1929–1938.

Iwasa, Y., Suzuki, Y., & Matsuda, H. (1984). Theory of oviposition strategy of parasitoids. I. Effect of mortality and limited egg number. Theoretical Population Biology, 26, 205–227.

Jackson, C. G., Bryan, D. E., & Patana, R. (1969). Laboratory studies of Eucelatoria armigera, a tachinid parasite of Heliothis spp. Journal of Economic Entomology, 62, 907–909.

Janssen, A., Hofker, C. D., Braun, A. R., Mesa, N., Sabelis, M. W., & Bellotti, A. C. (1990). Preselecting predatory mites for biological control: The use of an olfactometer. Bulletin of Entomological Research, 80, 177–181.

Jansson, R. K., Leibee, G. L., Sanchez, C. A., & Lecrone, S. H. (1991). Effects of nitrogen and foliar biomass on population parameters of cabbage insects. Entomologia Experimentalis et Applicata, 61, 7–16.

Johnson, M. T. (1997). Interaction of resistant plants and wasp parasitoids of tobacco budworm (Lepidoptera: Nocruidae). Environmental Entomology, 26(2), 208–214.

Johnson, M. T., & Gould, F. (1992). Interaction of genetically engineered host plant resistance and natural enemies of Heliothis virescens (Lepidoptera: Noctuidae) in tobacco. Environmental Entomology, 21(3), 586–597.

Johnson, M. W., & Hara, A. H. (1987). Influence of host crop of parasitoids (Hymenoptera) of Liriomyza spp. (Diptera: Agromyzidae). Environmental Entomology, 16, 339–344.

Jones, R. L., Beroza, W. J., Bierl, M. B. A., & Sparks, A. N. (1973). Host seeking stimulants (kairomones) for the egg parasite Trichogramma evanescens. Environmental Entomology, 2(4), 593–596.

Kadmon, R., & Pulliam, H. R. (1993). Island biogeography: effect of geographical isolation on species composition. Ecology, 74, 977–981.

Kareiva, P. (1982). Experimental and mathematical analyses of herbivore movement: Quantifying the influence of plant spacing and quality on foraging discrimination. Ecological Monographs, 52, 261–282.

Kareiva, P., & Odell, G. (1987). Swarms of predators exhibit "preytaxis" if individual predators are area-restricted search. American Naturalist, 130, 233–270.

Kemp, W. P., & Moody, U. L. (1984). Relationships between regional soils and foliage characteristics and western spruce budworm (Lepidoptera: Torticidae) outbreak frequency. Environmental Entomology, 13, 1291–1297.

Kester, K. M., & Barbosa, P. (1991). Behavioral and ecological constraints imposed by plants on insect parasitoids: Implications for biological control. Biological Control, 1, 94–106.

Kindler, S. D., & Staples, R. (1970). Nutrients and the reaction of two alfalfa clones to the spotted alfalfa aphid. Journal of Economic Entomology, 63(3), 939–940.

Kirkton, R. M. (1970). Habitat management and its effects on populations of Polistes and Iridomyrmex. Proceedings, Tall Timbers Conference on Ecological Animal Control Habitat Management, 2, 243–246.

Kiss, J., Penksza, K., Toth, F., & Kadar, F. (1997). Evaluation of fields and field margins in nature production capacity with special regard to plant protection. Agriculture, Ecosystems, and Environment, 63(2–3), 227–232.

Komareck, E. V., Sr. (1970). Insect control-fire for habitat management. Proceedings, Tall Timbers Conference on Ecological Animal Control Habitat Management, 2, 157–171.

Krivan, V. (1996). Optimal foraging and predatorprey dynamics. Theoretical Population Biology, 49(3), 265–290.

Kromp, B. (1989). Carabid beetle communities (Carabidae, Coleoptera) in biologically and conventionally farmed agroecosystems. Agriculture, Ecosystems, and Environment, 27, 241–251.

Kruess, A., & Tscharntke, T. (1994). Habitat fragmentation, species loss, and biological control. Science, 264, 1581–1584.

Landis, D. A., & Haas, M. J. (1992). Influence of landscape structure on abundance and within-field distribution of European corn borer (Lepidoptera: Pyralidae) larval parasitoids in Michigan. Environmental Entomology, 21, 409–416.

Lasack, P. M., & Pedigo, L. P. (1986). Movement of stalk borer larvae (Lepidoptera: Noctuidae) from noncrop areas into corn. Journal of Economic Entomology, 79, 1697–1702.

LaSalle, J., & Gauld, I. D. (Eds.). (1993). Hymenoptera and biodiversity. United Kingdom: CAB International.

Lashomb, J. H., & Ng, Y.-S. (1984). Colonization by Colorado potato beetles, Leptinotarsa decemlineata (Say) (Coleoptera: Chrysomelidae), in rotated and nonrotated potato fields. Environmental Entomology, 13, 1352–1356.

Laster, M. L., & Furr, R. E. (1972). Heliothis populations in cotton-sesame interplantings. Journal of Economic Entomology, 65, 1524–1525.

Laub, C. A., & Luna, J. M. (1992). Winter cover crop suppression practices and natural enemies of armyworm (Lepidoptera: Noctuidae) in no-till corn. Environmental Entomology, 21, 41–49.

Lawson, F. R., Rabb, R. L., Guthrie, R. E., & Bowery, T. G. (1961). Studies of an integrated control system for hornworms in tobacco. Journal of Economic Entomology, 54, 93–97.

Lawton, J. H., & McNeill, S. (1979). Between the devil and the deep blue sea: On the problem of being a herbivore. In R. M. Anderson, B. D. Turner, & L. R. Taylor (Eds), Population dynamics (pp. 223–244). Oxford: Blackwell Scientific.

Leigh, T. F., Grimes, D. W., Dickens, W. L., & Jackson, C. E. (1974). Planting pattern, plant population, irrigation and insect interactions in cotton. Environmental Entomology, 3(3), 492–496.

Leius, K. (1967). Influence of wild flowers on parasitism of tent caterpillar and codling moth. Canadian Entomologist, 99, 444–446.

Leston, D. (1973). The ant mosaic-tropical tree crops and the limiting of pests and diseases. Proceedings of the National Academy of Science, 19, 311–341.

Letourneau, D. K. (1987). The enemies hypothesis: Tritrophic interactions and vegetational diversity in tropical agroecosystems. Ecology, 68, 1616–1622.

Letourneau, D. K. (1988). Conceptual framework of three-trophic-level interactions. In P. Barbosa & D. K. Letourneau (Eds.), Novel aspects of insect-plant interactions (pp. 1–9). New York; John Wiley & Sons.

Letourneau, D. K. (1990). Mechanisms of predator accumulation in a mixed crop system. Ecological Entomology, 15, 63–69.

Letourneau, D. K. (1994). Beanfly, management practices, and biological control in Malawian subsistence agriculture. Agriculture, Ecosystems, and Environment, 50, 103–111.

Letourneau, D. K. (1997). Plant-arthropod interactions in agroecosystems, pp. 239–290. In L. Jackson (Ed.), Ecology in agriculture. San Diego; Academic Press.

Letourneau, D. K. (1998). Conservation biology: lessons for conserving natural enemies. In P. Barbosa (Ed.), Conservation biological control. San Diego: Academic Press.

Letourneau, D. K., & Altieri, M. A. (1983). Abundance patterns of a predator Orius tristicolor (Hemiptera: Anthocoridae), and its prey, Frankliniella occidentalis (Thysanoptera: Thripidae): Habitat attraction in polycultures versus monocultures. Environmental Entomology, 122, 1464–1469.

Levine, E. (1985). Oviposition by the stalk borer, Papaipema nebris (Lepidoptera: Noctuidae), on weeds, plant debris, and cover crops in cage tests. Journal of Economic Entomology, 78, 65–68.

Levins, R. (1986). Perspectives in integrated pest management: From an industrial to an ecological model of pest management. In M. Kogan (Ed.); Ecological theory and integrated pest management practice (pp. 1–18) New York: John Wiley & Sons.

Levins, R., & Wilson, M. (1980). Ecological theory and pest management. Annual Review of Entomology, 25, 7–29.

Lewis, W. J., & Gross, H. R. (1989). Comparative studies on field performance of Heliothis larval parasitoids Microplitis croceipes and Cardiochiles nigriceps at varying densities and under selected host plant conditions. Florida Entomologist, 72, 6–14.

Lewis, W. J., & Martin, W. R., Jr. (1990). Semiochemicals for use with parasitoids: Status and future. Journal of Chemical Ecology, 16, 3067–3089.

Lewis, W. J., & Nordlund, D. A. (1985). Behavior-modifying chemicals to enhance natural enemy effectiveness. In M. A. Hoy & D. C. Herzog (Eds.), Biological control in IPM systems (pp. 89–10). San Diego; Academic Press.

Lewis, W. J., Vet, L. E. M., Tumlinson, J. H., van Lenteren, J. C., & Papaj, D. R. (1990). Variations in parasitoid foraging behavior: Essential element of a sound biological control theory. Environmental Entomology, 19, 1183–1193.

Liebhold, A. M., Rossi, R. E., & Kemp, W. P. (1993). Geostatistics and geographic information systems in applied insect ecology. Annual Review of Entomology, 38, 303–327.

Liss, W. J., Gut, L. J., Westigard, P. H., & Warren, C. E. (1986). Perspectives on arthropod community structure, organization, and development in agricultural crops. Annual Review of Entomology, 31, 455–478.

Litsinger, J. A., & Moody, K. (1976). Integrated pest management in multiple cropping systems. In G. B. Triplett, P. A. Sanchez, & R. I. Papendick (Eds.), Multiple cropping (ASA Special Publication No. 27, (pp. 293–316). Madison WI: American Society of Agronomy.

Louda, S. M. (1986). Insect herbivory in response to root-cutting and flooding stress on native crucifer under field conditions. Acta Oecologica, 7, 37–53.

Luff, M. L. (1982). Population dynamics of Carabidae. Annals Applied Biology, 101(1), 164–170.

MacArthur, R. H., & Wilson, E. O. (1967). The theory of island biogeography. Princeton, NJ: Princeton University Press.

Manuwoto, S., & Scriber, J. M. (1985). Differential effects of nitrogen fertilization of three corn genotypes on biomass and nitrogen utilization by the southern armyworm, Spodoptera eridania. Agriculture, Ecosystems, and Environment, 14(1/2), 25–40.

Marc, P., & Canard, A. (1997). Maintaining spider biodiversity in agroecosystems as a tool in pest control. Agriculture, Ecosystems, and Environment, 63(2–3), 229–235.

Marcovitch, S. (1935). Experimental evidence on the value of strip cropping as a method for the natural control of injurious insects, with special reference to plant lice. Journal of Economic Entomology, 28, 62–70.

Martin, P. B., Lingren, P. D., Greene, G. L., & Ridgway, R. L. (1976). Parasitization of two species of Plusiinae and Heliothis spp. after release of Trichogramma pretiosum in seven crops. Environmental Entomology, 5, 991.

Martin, W. R., Jr., Nordlund, D. A., & Nettles, W. C., Jr. (1990). Response of parasitoid Eucelatoria bryani to selected plant material in an olfactometer. Journal of Chemical Ecology, 16, 499–508.

Mattson, P. C., Altieri, M. A., & Gagne, W. C. (1984). Modification of small farmer practice for better pest management. Annual Review of Entomology, 29, 303–402.

Mattson, W. J., Jr. (1980). Herbivory in relation to plant nitrogen content. Annual Review of Ecology and Systematics, 11, 119–167.

Mattson, W. J., Jr., & Addy, N. D. (1975). Phytophagous insects as regulators of forest primary production. Science, 190, 515–522.

May, R. M., & Anderson, R. M. (1978). Regulation and stability of host-parasite interactions. II. Destabilizing processes. Journal of Animal Ecology, 47, 279–267.

May, R. M., & Hassell, M. P. (1981). The dynamics of multiparasitoid-host interactions. America Naturalist, 117, 234–261.

Mayse, M. A. (1983). Cultural control in field crop fields: Habitat management technique. Environmental Management, 7, 15.

McCaffrey, J. P., & Horsburgh, R. L. (1986). Functional response of Orius insidiosus (Hemiptera: Anthocoridae) to the European red mite, Panonychus ulmi (Acari: Tetranychidae) at different constant temperatures. Environmental Entomology, 15, 532–535.

McClure, M. S. (1980). Foliar nitrogen: A basis for host suitability for elongate hemlock seeds, Fiorinia externa (Homoptera: Diaspididae). Ecology, 61, 72–79.

McMurtry, J. A., & Croft, B. A. (1997). Life-styles of phytoseiid mites and their roles in biological control. Annual Review of Entomology, 42, 291–321.

McMurtry, J. A., & Scriven, G. T. (1966). The influence of pollen and prey density on the number of prey consumed by Amblyseius hibisci (Chant) (Acarina: Phytoseiidae). Annals of the Entomological Society of America, 59, 147–149.

McMurtry, J. A., & Scriven, G. T. (1968). Studies on predator-prey interactions between Amblyseius hibisci and Oligonychus punicae: Effects of host-plant conditioning and limited quantities of alternate food. Annals of the Entomological Society of America, 61, 393–397.

Mena-Covarrubias, J., Drummon, F. A., & Maynes, D. L. (1996). Population dynamics of the Colorado potato beetle (Coleoptera: chrysomelidae) on horsenettle in Michigan. Environmental Entomology, 25, 68–77.

Metcalf, R. L. (1994). Role of kairomones in integrated pest management. Phytoparasitica, 22, 275–279.

Miles, P. W., Aspinall, D., & Rosenberg, L. (1982). Performance of the cabbage aphid, Brevicoryne brassicae (Linnaeus), on water-stressed rape plants, in relation to changes in their chemical composition. Australian Journal of Zoology, 30, 337–345.

Monteith, L. G. (1960). Influence of plants other than the food plants of their host on host-finding by tachinid parasites. Canadian Entomologist, 92, 641–652.

Moran, N., & Hamilton, W. D. (1980). Low nutritive quality as defense against herbivores. Journal of Theoretical Biology, 86, 247–254.

Moran, M. D., & Hurd, L. E., (1997). Relieving food limitation reduces survivorship of a generalist predator. Ecology, 78(4), 1266–1270.

Morrow, P. A. (1977). The significance of phytophagous insects in the Eucalyptus forests of Australia. In W. J. Mattson (Ed.), The role of arthropods in forest ecosystems (pp. 19–29). New York: Springer-Verlag.

Muggleton, J., Lonsdale, D., & Benham, B. R. (1975). Melanism in Adalia bipunctata Linnaeus (Col.: Coccinellidae) and its relationship to atmospheric pollution. Journal of Applied Ecology, 12, 451–464.

Murdoch, W. W. (1971). The developmental response of predators to changes in prey density. Ecology, 52, 133–137.

Murdoch, W. W. (1979). Predation and the dynamics of prey populations. Fortschritte der Zoologie, 25, 295–310.

Murdoch, W. W., & Briggs, C. J. (1996). Theory for biological control—recent developments. Ecology, 77(7), 2001–2013.

Murdoch, W. W., & Oaten, A. (1975). Predation and population stability. Advances in Ecological Research, 9, 2–131.

Murdoch, W. W., Chesson, J., & Chesson, P. L. (1985). Biological control in theory and practice. American Naturalist, 125, 344–366.

Murdoch, W. W., Briggs, C. J., Nisbet, R. M., Gurney, W. S. C., & Stewart-Oaten, A. (1992). Aggregation and stability in metapopulation models. American Naturalist, 140, 41–58.

Murdoch, W. W., Briggs, C. J., & Nisbet, R. M. (1997). Dynamical effects of host size—and parasitoid state—dependent attacks by parasitoids. Journal of Animal Ecology, 66, 542.

Murdock, W. W., Luck, R. F., Swarbrick, S. L., Walde, S., Yu, D. S. & Reeve, J. D. (1995). Regulation of an insect population under biological control. Ecology, 76, 206–217.

Murphy, B. C., Rosenheim, J. A., & Granett, J. (1996). Habitat diversification for improving biological control: abundance of Anagrus epos (Hymenoptera: mymaridae) in grape vineyards. Environmental Entomology, 25, 495–504.

Nafus, D., & Schreiner, I. (1986). Intercropping maize and sweet potatoes. Effects on parasitization of Ostrinia furnacalis eggs by Trichogramma chilonis. Agriculture, Ecosystems, and Environment, 15, 189–200.

Naranjo, S. E. (1987). Observations on Geocoris punctipes (Hemiptera: Lygaeidae) oviposition site preferences. Florida Entomologist, 70, 173–175.

National Academy of Sciences (NAS). (1969). Principles of plant and animal control: Vol. 3, Insect-pest management and control (pp. 100–169). Washington, DC: National Academy of Sciences.

National Research Council. (1991). Sustainable agriculture research and education in the field. Washington, DC: National Academy Press.

Nettles, W. C. (1979). Eucelatoria sp. females: Factors influencing responses to cotton and okra plants. Environmental Entomology, 8, 619–623.

Neuenschwander, P., & Hagen, K. S. (1980). Role of the predator, Hemerobius pacificus, in a non-insecticide treated artichoke field. Environmental Entomology, 9, 492–495.

Nilsson, C. (1985). Impact of ploughing on emergence of pollen beetle parasitoids after hibernation. Zeitschrift fuer Angewandte Entomologie, 100, 302–308.

Nordlund, D. A., Jones, R. L., & Lewis, W. J. (1981a). Semiochemicals: Their role in pest control. New York: John Wiley & Sons.

Nordlund, D. A., Lewis, W. J., & Gross, H. R. (1981b). Elucidation and employment of semiochemicals in the manipulation of entomophagous insects. In E. R. Mitchell (Ed.), Management of insect pests with semiochemicals: Concepts and practice (pp. 463–475). New York: Plenum Press.

Nordlund D. A., Lewis, W. J., & Altieri, M. A. (1988). Influences of plant produced allelochemicals on the host and prey selection behavior of entomophagous insects. In P. Barbosa & D. K. Letourneau (Eds.), Novel aspects of insect-plant interactions. New York: John Wiley & Sons.

Norman, M. J. T. (1979). Annual cropping systems in the tropics: An introduction. Gainesville, FL: University Press.

Norris, R. F. (1986). Weeds and integrated pest management systems. HortScience, 21, 402–410.

Obrycki, J. J., & Tauber, M. J. (1984). Natural enemy activity on glandular pubescent potato plants in the greenhouse: An unreliable predictor of effects in the fields. Environmental Entomology, 13, 679–683.

O'Connor, B. A. (1950). Premature nutfall of coconuts in the British

Solomon Islands Protectorate. Agricultural Journal (Fiji Department of Agriculture) 21 (1–2), 21–42.

Ofuya, T. I. (1989). Effects of weeds on colonization of cowpea by Aphis craccivora Koch (Homoptera: Aphididae) and its major predators in Nigeria. Tropical Pest Management, 35, 403–405.

Okubo, A. (1980). Diffusion and ecological problems: Mathematical models. New York: Springer-Verlag.

Oraze, M. J., Grigarick, A. A., & Smith, K. A. (1989). Population ecology of Pardosa ramulosa (Araneae, Lycosidae) in flooded rice field of northern California. Journal of Arachnology, 17, 163–170.

Orr, D. B., & Boethel, J. (1986). Influence of plant antibiosis through four trophic levels. Oecologia (Berlin), 70, 242–249.

Panizzi, A. R. (1988). Parasitism by Eutrichopodopsis nitens (Diptera: Tachinidae) of Nezara viridula (Hemiptera: Pentatomidae) on different host plants. Centro Nacional de Pesquisa de Soja, EMBRAPA, 36, 82–83.

Parajulee, M. N., Montandon, R., Slosser, J. E. (1997). Relay intercropping to enhance abundance of insect predators of cotton aphid (Aphis gossypii Glover) in Texas cotton. International Journal of Pest Management, 43, 227–232.

Pavuk, D. M., & Barrett, G. W. (1993). Influence of successional and grassy corridors on parasitism of Plathypena scabra (F.) (Lepidoptera: Noctuidae) larvae in soybean agroecosystems. Environmental Entomology, 22, 541–546.

Perfecto, I., & Sediles, A. (1992). Vegetational diversity, ants (Hymenoptera; Formicidae), and herbivorous pests in a Neotropical agroecosystem. Environmental Entomology, 21: 61–67.

Perfecto, I., Horwith, B., Vandermeer, J., Schultz, B., McGuiness, H., & Dos Santos, A. (1986). Effects of plant diversity and density on the emigration rate of two ground beetles, Harpalus pennsylvanicus and Evarthrus sodalis (Coleoptera: Carabidae), in a system of tomatoes and beans. Environmental Entomology, 15, 1028–1031.

Perrin, R. M. (1975). The role of the perennial stinging nettle Urtica dioica, as a reservoir of beneficial natural enemies. Annals of Applied Biology, 81, 289–297.

Perrin, R. M. (1977). Pest management in multiple cropping systems. Agro-Ecosystems, 3, 93–118.

Perrin, R. M. (1980). The role of environmental diversity in crop protection. Protection Ecology, 2, 77–114.

Perrin, R. M., & Phillips, M. L. (1978). Some effects of mixed cropping on the population dynamics of insect pests. Entomologia Experimentalis et Applicata, 24, 385–393.

Petters, R. M., & Mettus, R. V. (1982). Reproductive performance of Bracon hebetor females following acute exposure to SO$_2$ in air. Environmental Pollution, Series A, 27, 155–163.

Pfiffner, L. (1990). Effects of different farming systems on the presence of epigeal arthropods, in particular of carabids (Col., Carabidae), in winter wheat plots. Mitteilungen den Schweizerischen Entomologischen Gesellschaft, 63, 63–75 (in German).

Pimentel, D. (1961). Species diversity and insect population outbreaks. Annals of the Entomological Society of America, 54, 76–86.

Pimm, S. L. (1982). Food webs. New York: Chapman & Hall.

Podoler, M., & Rogers, D. (1975). A new method for the identification of key-factors from life-table data. Journal of Animal Ecology, 44, 85–114.

Pollard, D. G. (1971). Hedges. VI. Habitat diversity and crop pests—a study of Brevicoryne brassicae and its syrphid predators. Journal of Applied Ecology, 8, 751–780.

Powell, W. (1986). Enhancing parasitoid activity in crops. In J. Waage & D. Greathead (Eds.), Insect parasitoids (pp. 319–335). London: Academic Press.

Powell, W., Dean, G. J., & Wilding, N. (1986). The influence of weeds on aphid-specific natural enemies in winter wheat. Crop Protection, 4, 182–189.

Price, P. W. (1976). Colonization of crops by arthropods: Non-equilibrium

communities in soybean fields. Environmental Entomology, 5, 605–611.

Price, P. W. (1981). Relevance of ecological concepts to practical biological control. In G. C. Papavizas et al. (Eds.), Biological control in crop production (pp. 3–19). Beltsville Symposium on Agricultural Research, No. 5. Totowa, New Jersey: Allanheld, Osmum, & Co. Publishers.

Price, P. W. (1986). Ecological aspects of host plant resistance and biological control: Interaction among three trophic levels. In D. J. Boethel & R. D. Eikenbary (Eds.), Interactions of plant resistance and parasitoids and predators of insects (pp. 11–30). Chichester, United Kingdom: Ellis Horwood.

Price, P. W., & Waldbauer, G. P. (1975). Ecological aspects of pest management. In R. L. Metcalf & W. H. Luckman (Eds.), Introduction to insect pest management (2nd ed.,(pp. 33–68). New York: Wiley-Interscience.

Price, P. W., Rathke, B. J., & Gentry, D. A. (1974). Lead in terrestrial arthropods: Evidence for biological concentration. Environmental Entomology, 3, 370–372.

Price, P. W., Bouton, C. E., Gross, P., McPheron, B. A., Thompson, J. N., & Weis, A. E. (1980). Interactions among three trophic levels: Influence of plants on interactions between insect herbivores and natural enemies. Annual Review of Ecology and Systematics, 11, 41–65.

Putman, W. L., & Herne, D. C. (1966). The role of predators and other biotic factors in regulating the population density of phytophagous mites in Ontario peach orchards. Canadian Entomologist, 98, 808–820.

Rabb, R. L. (1978). A sharp focus on insect populations and pest management from a wide-area view. Bulletin of the Entomological Society of America, 24, 55–61.

Rabb, R. L., Stinner, R. E., & van den Bosch, R. (1976). Conservation and augmentation of natural enemies. In C. B. Huffaker & P. S. Messenger (Eds.), Theory and practice of biological control (pp. 233–254). New York: Academic Press.

Ramachandran, R., & Norris, D. M. (1991). Volatiles mediating plant-herbivore-natural enemy interactions: Electroantennogram responses of soybean looper, Pseudoplusia includens, and a parasitoid, Microplitis demolitor, to green leaf volatiles. Journal of Chemical Ecology 17, 1665–1690.

Ramert, B., & Ekbom, B. (1996). Intercropping as a management strategy against carrot rust fly (Diptera, psilidae)—A test of enemies and resource concentration hypotheses. Environmental Entomology, 25(5), 1092–1100.

Raros, R. S. (1973). Prospects and problems of integrated pest control in multiple cropping. IRRI Saturday Seminar, Los Baños, The Philippines.

Rasmy, A. H. (1977). Predatory efficiency and biology of the predatory mite Amblyseius gossipi (Acarina: Phytoseiidae) as affected by physical surfaces of the host plant. Entomophaga, 2, 421–423.

Rasmy, A. H., & Elbanhawy, E. M. (1974). Behaviour and bionomics of the predatory mite Phytoseius plumifer (Acarina: Phytoseiidae) as affected by physical surfaces of the host plant. Entomophaga, 19, 255–257.

Read, D. P., Feeny, P. P., & Root, R. B. (1970). Habitat selection by the aphid parasite Diaeretiella rapae (Hymenoptera: Braconidae) and hyperparasite Charips brassicae (Hymenoptera: Cynipidae). Canadian Entomologist, 102, 1567–1578.

Renou, M., Nagnan, P., Berthier, A., & Durier, C. (1992). Identification of compounds from the eggs of Ostrinia nubilalis and Mamestra brassicae having kairomone activity on Trichogramma brassicae. Entomologia Experimentalis et Applicata, 63, 291–303.

Ridgway, R. L., King, E. G., & Carrillo, J. L. (1976). Augmentation of natural enemies for control of plant pests in the western hemisphere. In R. L. Ridgway, & S. B. Vinson (Eds.), Biological control by augmentation of natural enemies (pp. 379–416). New York: Plenum Press.

Riechert, S. E., & Lockley, T. (1984). Spiders as biological control agents. Annual Review of Entomology, 29, 299–320.

Riehl, L. A., Brooks, R. F., McCoy, C. W., Fisher, T. W., & Dean, H. A. (1980). Accomplishments toward improving integrated pest management for citrus. In C. B. Huffaker (Ed.), New technology of pest control (pp. 319–363). New York: John Wiley, & Sons.

Riggin-Bucci, T. M., & Gould, F. (1997). Impact of intraplot mixtures of toxic and nontoxic plants on population dynamics of diamondback moth (Lepidoptera: Plutellidae) and its natural enemies. Journal of Economic Entomology, 90(2), 241–251.

Risch, S. J. (1979). A comparison, by sweep sampling, of the insect found from corn and sweet potato monocultures and dicultures in Costa Rica. Oecologia (Berlin), 42, 195–211.

Risch, S. J. (1980). The population dynamics of several herbivorous beetles in a tropical agroecosystem: The effect of intercropping corn, beans and squash in Costa Rica. Journal of Applied Ecology, 17, 593–612.

Risch, S. J. (1981). Insect herbivore abundance in tropical monocultures and polycultures: An experimental test of two hypotheses. Ecology, 62, 1325–1340.

Risch, S. J. (1987). Agricultural ecology and insect outbreaks. In P. Barbosa & J. C. Schultz (Eds.), Insect outbreaks (pp. 217–233). New York: Academic Press.

Risch, S. J., & Carroll, C. R. (1982). The ecological role of ants in two Mexican agroecosystems. Oecologia (Berlin), 55, 114–119.

Risch, S. J., Andow, D., & Altieri, M. A. (1983). Agroecosystem diversity and pest control: Data, tentative conclusions, and new directions. Environmental Entomology, 12, 625–629.

Risch, S. J., Pimentel, D., & Grover, H. (1986). Corn monoculture versus old field: Effects of low levels of insecticides. Ecology, 67, 505–515.

Risch, S. J., Wrubel, R., & Andow, D. (1982). Foraging by a predaceous beetle, Coleomegilla maculata (Coleoptera: Coccinellidae), in a polyculture: Effects of plant density and diversity. Environmental Entomology, 11, 949–950.

Robinson, J. V., & Dickerson, J. E. (1987). Does invasion sequence affect community structure? Ecology, 68, 587–595.

Rodenhouse, N. L., Barrett, G. W., Zimmerman, D. M., & Kemp, J. C. (1992). Effects of uncultivated corridors on arthropod abundances and crop yields in soybean agroecosystems. Agriculture, Ecosystems, and Environment, 38, 179–191.

Rogers, D. J., & Sullivan, M. J. (1986). Nymphal performance of Geocoris punctipes (Hemiptera: Lygaeidae) on pest-resistant soybeans. Environmental Entomology, 15, 1032–1036.

Roland, J. (1986). Parasitism of winter moth in British Columbia during build-up of its parasitoid Cyzenis albicans: Attack rate on oak v. apple. Journal of Animal Ecology, 55, 215–234.

Root, R. B. (1973). Organization of plant-arthropod association in simple and diverse habitats: The fauna of collards (Brassica oleraceae). Ecological Monographs, 43, 95–124.

Ruberson, J. R., Tauber, M. J., & Tauber, C. A. (1986). Plant feeding by Podisus maculiventris (Hymenoptera: Pentatomidae): Effect on survival, development, and preoviposition period. Environmental Entomology, 15, 894–897.

Russell, E. P. (1989). Enemies hypothesis: A review of the effect of vegetational diversity on predatory insects and parasitoids. Environmental Entomology, 18, 590–599.

Ryszkowski, L., Karg, J., Margalit, G., Paoletti, M. G., & Zlotin, R. (1993). Above ground insect biomass in agricultural landscapes of Europe. In R. G. H. Bunce, L. Ryszkowski, & M. G. Paoletti (Eds.), Landscape ecology and agroecosystems (pp. 71–82). Boca Raton: Lewis Publishers.

Sabelis, M. W., Diekmann, O., & Jansen, V. A. A. (1991). Metapopulation persistence despite local extinction: Predator-prey patch models of the Lotka-Volterra type. Biological Journal of the Linnean Society, 42, 267–283.

Saks, M. E., & Carroll, C. R. (1980). Ant foraging activity in tropical agroecosystems. Agro-Ecosystems, 6, 177–188.

Sato, Y., & Ohsaki, N. (1987). Host-habitat location by Apanteles glomeratus and effects of food-plant on host-parasitism. Ecological Entomology, 12, 291–297.

Schellhorn, N. A., & Sork, V. L. (1997). The impact of weed diversity on insect population dynamics and crop yield in collards, Brassica oleraceae (Brassicaceae). Oecologia, 111(2), 233–240.

Schlinger, E. I., & Dietrick, E. J. (1960). Biological control of insect pests aided by strip-farming alfalfa in experimental programs. California Agriculture, 14, 15.

Schmidt, K. A., & Brown, J. S. (1996). Patch assessment in fox squirrels: the role of resource density, patch size, and patch boundaries. American Naturalist, 147, 360–380.

Scriber, J. M. (1984). Nitrogen nutrition of plants and insect invasion. In R. D. Hauck (Ed.), Nitrogen in crop production (pp. 441–460). American Society of Agronomy.

Shah, M. A. (1981). The influence of plant surfaces on the searching behavior of coccinellid larvae. Entomologia Experimentalis et Applicata, 31, 377–380.

Shahjahan, M. (1974). Erigeron flowers as food and attractive odor source for Peristenus pseudopallipes a braconid parasitoid of the tarnished plant bug. Environmental Entomology, 3, 69–72.

Shahjahan, M., & Streams, A. S. (1973). Plant effects on host-finding by Leiophron pseudopallipes (Hymenoptera: Braconidae), a parasitoid of the tarnished plant bug. Environmental Entomology, 21, 921–925.

Shaw, M. C., Wilson, M. C., & Rhykerd, C. L. (1986). Influence of phosphorous and potassium fertilization on damage to alfalfa, Medicago sativa Linnaeus, by the alfalfa weevil, Hypera postica (Gyllenhall) and potato leafhopper, Empoasca fabae (Harris). Crop Protection, 5, 245–249.

Sheehan, W. (1986). Response by specialist and generalist natural enemies to agroecosystem diversification: A selective review. Environmental Entomology, 15, 456–461.

Simberloff, D. (1986). Island biogeographic theory and integrated pest management. In M. Kogan (Ed.), Ecological theory and integrated pest management practice (pp. 19–35). New York: John Wiley, & Sons.

Skovgard, H., & Pats, P. (1996). Effects of intercropping on maize stemborers and their natural enemies. Bulletin of Entomological Research, 86(5), 599–607.

Slosser, J. E., Jacoby, P. W., & Price, J. R. (1985). Management of sand shinnery oak for control of the boll weevil (Coleoptera: Curculionidae) in the Texas rolling plains. Journal of Economic Entomology, 78, 383–389.

Smith, J. G. (1976a). Influence of crop background on natural enemies of aphids on brussels sprouts. Annals of Applied Biology, 83, 15–29.

Smith, J. G. (1976b). Influence of crop background on aphids and other phytophagous insects on brussel sprouts. Annals of Applied Biology, 83, 1–13.

Smith, R. F., & Reynolds, H. T. (1972). Effects of manipulation of cotton agro-ecosystems on insect populations. In M. T. Farvar & J. P. Milton (Eds.), The careless technology: Ecology and international development (pp. 373–406). Garden City, NY: Natural History Press.

Solomon, M. G. (1981). Windbreaks as a source of orchard pests and predators. In J. M. Thresh (Ed), Pests, pathogens and vegetation (p. 273). Boston, MA: Pitman.

Southwood, T. R. E., & Comins, H. N. (1976). A synoptic population model. Journal of Animal Ecology, 45, 949–965.

Speight, H. R., & Lawton, J. H. (1976). The influence of weed cover on the mortality imposed on artificial prey by predatory ground beetles in cereal fields. Oecologia (Berlin), 23, 211–233.

Spencer, H. G., Kennedy, M., Gray, R. D. (1996). Perceptual constraints on optimal foraging—the effects of variation among foragers. Evolutionary Ecology, 10, 331–339.

Sprague, M. A. (1986). Overview. In M. A Sprague & G. B. Tiplett (Eds.),

No-tillage and surface-tillage agriculture (pp. 1–18). New York: John Wiley & Sons.

Sprenkel, R. K., Brooks, W. M., van Duyn, J. W., & Deitz, L. L. (1979). The effects of three cultural variables on the incidence of *Nomuroea rileyi,* phytophagous Lepidoptera, and their predators on soybeans. Environmental Entomology, 8, 337–339.

Stapel, J. O., Cortesero, A. M., De Moraes, C. M., Tumlinson, J. H., & Lewis, W. J. (1997). Extrafloral nectar, honeydew, and sucrose effects on searching behavior and efficiency of *Microplitis croceipes* (Hymenoptera: Braconidae) in cotton. Environmental Entomology, 26, 617–623.

Stenseth, N. C. (1981). How to control pest species: Application of models from the theory of island biogeography in formulating pest control strategies. Journal of Applied Ecology, 18, 773–794.

Stern, V. M. (1981). Environmental control of insects using trap crops, sanitation, prevention and harvesting. In D. Pimentel (Ed.), CRC handbook of pest management in agriculture (Vol. 1, pp. 199–207). Boca Raton, FL: CRC Press.

Stern, V. M., van den Bosch, R., & Leigh, T. F. (1964). Strip cutting of alfalfa for *Lygus* bug control. California Agriculture, 18, 4–6.

Stinner, B. R., & House, G. J. (1990). Arthropods and other invertebrates in conservation-tillage agriculture. Annual Review of Entomology, 35, 299–318.

Stinner, R. E., Saks, M., & Dohse, L. (1986). Modeling of agricultural pest displacement. In W. Danthanarayana (Ed.), Insect flight: Dispersal and migration (pp. 235–241). New York: Springer-Verlag.

Stinner, R. E., Barfield, C. S., Stimac, J. L., & Dohse, L. (1983). Dispersal movement of insect pests. Annual Review of Entomology, 28, 319–335.

Stinner, D. H., Stinner, B. R., & Martsoslf, E. (1997). Biodiversity as an organizing principle in agroecosystems management: case studies of holistic resources management practitioners in the USA. Agriculture, Ecosystems, and Environment, 62, 199–213.

Strong, D. R. (1979). Biogeographical dynamics of insect-host plant communities. Annual Review of Entomology, 24, 89–119.

Strong, D. R., Lawton, J. H., & Southwood, T. R. E. (1984). Insects on plants. Oxford: Blackwell Scientific.

Syme, P. D. (1975). The effects of flowers on the longevity and fecundity of two native parasites of the European pine shoot moth in Ontario. Environmental Entomology, 4, 337–346.

Talkington, M. L., & Berry, R. E. (1986). Influence of tillage in peppermint on *Fumibotys fumalis* (Lepidoptera: Pyralidae), common groundsel, *Senicio vulgaris,* and soil chemical components. Journal of Economic Entomology, 79, 1590–1594.

Tamaki, G., & Weeks, R. W. (1972). Biology and ecology of two predators, *Geocoris pallens* Stål and *G. fullatus* (Say) (Tech. Bulletin 1446). Washington, DC: U.S. Department of Agriculture.

Taylor, A. D. (1991). Studying metapopulation effects in predator-prey systems. Biological Journal of the Linnean Society, 42, 305–323.

Telenga, N. A. (1958). Biological method of pest control in crops and forest plants in the USSR (pp. 1–15). In Report of the Soviet delegation, Ninth International Conference on Quarantine Plant Protection, Moscow.

Telenga, N. A., & Zhigaev, G. N. (1959). The influence of different soil cultivation on reproduction of *Caenocrepis bothynoderis,* an egg parasite of beet weevil. Nauchnye Trudy Ukrainskogo Nauchno-Issledovatel'skogo Instituta Zashchity Rastenievodstva, 8, 68–75.

Theuinissen, J., & den Ouden, H. (1980). Effects of intercropping with *Spergula arvensis* on pests of Brussels sprouts. Entomologia Experimentalis et Applicata, 27, 260–268.

Thiele, H. V. (1977). Quantitative investigations on the distribution of carabids. In Carabid beetles in their environment. Berlin: Springer-Verlag.

Thomas, C. D., & Janzen, D. H. (1989). Predator-herbivore interactions and the escape of isolated plants from phytophagous insects. Oikos, 55, 291–298.

Thomas, M. B. (1989). The creation of island habitats to enhance populations of beneficial insects (Vol. 3, 1097–1102). Proceedings of the Brighton Crop Protection Conference, Weeds, Brighton, United Kingdom: November 20–23, 1989. Farnham, Surrey: British Crop Protection Council.

Thorpe, K. W. (1985). Effects of height and habitat type on egg parasitism by *Trichogramma minutum* and *T. pretiosum* (Hymenoptera: Trichogrammatidae). Agriculture, Ecosystems, and Environment, 12, 117–126.

Thorpe, W. H., & Caudle, H. B. (1938). A study of the olfactory responses of insect parasites to the food plant of their host. Parasitology, 30, 523–528.

Thresh, J. M. (1981). Pests, pathogens and vegetation: The role of weeds and wild plants in the ecology of crop pests and diseases. Boston, MA: Pitman.

Titayavan, M., & Altieri, M. A. (1990). Synomone-mediated interactions between the parasitoid *Diaeretiella rapae* and *Brevicoryne brassicae* under field conditions. Entomophaga, 35, 499–507.

Toko, M., Yaninek, J. S., & O'Neil, R. J. (1996). Response of *Mononychellus tanajoa* (Acari: tetranychidae) to cropping systems, cultivars, and past interventions. Environmental Entomology, 25, 237–249.

Tonhasca, A., Jr., & Bryne, D. N. (1994). The effects of crop diversification on herbivorous insects: a meta-analysis approach. Ecological Entomology, 19, 239–244.

Topham, M., & Beardsley, J. W. (1975). An influence of nectar source plants on the New Guinea sugarcane weevil parasite, *Lixophaga sphenophori* (Villeneuve). Proceedings of the Hawaiian Entomological Society, 22, 145–155.

Treacy, M. F., Zummo, G. R., & Benedict, J. H. (1985). Interactions of host-plant resistance in cotton with predators and parasites. Agriculture, Ecosystems, and Environment, 13, 151–157.

Trumble, J. T., Hare, J. D., Musselman, R. C., & McCool, P. M. (1987). Ozone-induced changes in host-plant suitability: Interactions of *Keiferia lycopersicella* and *Lycopersicon esculentum.* Journal of Chemical Ecology, 13, 203–218.

Trumble, J. T., Moar, W. J., Brewer, M. J., & Carson, W. G. (1991). Imapct of UV radiation on activity of linear furanocoumarins and *Bacillus thuringiensis* var. *kurstaki* against *Spodoptera exigua:* Implications for tritrophic interactions. Journal of Chemical Ecology, 17, 973–987.

Tukahirwa, E. M., & Coaker, T. H. (1982). Effect of mixed cropping on some insect pests of Brassicas: Reduced *Brevicoryne brassicae* infestations and influences of epigeal predators and the disturbance of oviposition behaviour in *Delia brassicae.* Entomologia Experimentalis et Applicata, 32, 129–140.

Turlings, T. C. J., & Tumlinson, J. H. (1991). Do parasitoids use herbivore-induced plant chemical defenses to locate hosts? Florida Entomologist, 74, 42–50.

Turlings, T. C. J., Tumlinson, J. H., Eller, F. J., & Lewis, W. J. (1991). Larval-damaged plants: Source of volatile synomones that guide the parasitoid *Cotesia marginiventris* to the micro-habitat of its hosts. Entomologia Experimental et Applicata, 58, 75–82.

Turlings, T. C., Wäckers, F. L., Vet, L. E. M., Lewis, W. J., & Tumlinson, J. H. (1993). Learning of host-finding cues by hymenopterous parasitoids. In D. R. Papaj & A. C. Lewis (Eds.), Insect learning (pp. 51–78). New York: Chapman & Hall.

Turnbull, A. L. (1967). Population dynamics of exotic insects. Bulletin of the Entomological Society of America, 13, 333–337.

Umeozor, O. C., van Duyn, J. W., Bradley, J. R., Jr. & Kennedy, G. G. (1985). Comparison of the effect of minimum-tillage treatments on the overwintering emergence of European corn borer (Lepidoptera: Pyralidae) in cornfields. Journal of Economic Entomology, 78, 937–939.

USDA. (1973). Monoculture in agriculture: Extent, causes and problems. Report of the task force on spatial heterogeneity in agricultural landscapes and enterprises. Washington, DC: U.S. Department of Agriculture.

van den Berg, H., Nyambo, B. T., & Waage, J. K. (1990). Parasitism of *Helicoverpa armigua* (Lepidoptera: Noctuidae) in Tanzania: Analysis of parasitoid-crop associations. Environmental Entomology, 19, 1141–1145.

van den Bosch, R. (1968). Comments on population dynamics of exotic insects. Bulletin of the Entomological Society of America, 14, 112–115.

van den Bosch, R., & Stern, V. M. (1969). The effect of harvesting practices on insect populations in alfalfa. In Proceedings, Tall Timbers Conference on Ecological Animal Control by Habitat Management (Vol. 1, pp. 47–54). Tallahassee, FL: Tall Timbers Research Station.

van den Bosch, R., & Telford, A. D. (1964). Environmental modification and biological control. In P. DeBach (Ed.), Biological control of insect pests and weeds (pp. 459–488). New York: Reinhold.

Vandermeer, J., & Andow, D. A. (1986). Prophylactic and responsive components of an integrated pest management program. Journal of Economic Entomology, 79, 299–302.

Vandermeer, J., & Perfecto, I. (1995). Breakfast of biodiversity: The truth about rain forest destruction. Oakland, CA: Food First Books.

Van Driesche, R. G., & Bellows, T. S. Jr. (1996). Biological control. New York: Chapman & Hall.

van Emden, H. F. (1965). The role of uncultivated land in the biology of crop pests and beneficial insects. Scientific Horticulture, 17, 121–136.

van Emden, H. F. (1988). The potential for managing indigenous natural enemies of aphids on field crops. In Biological control of pests, pathogens and weeds: Developments and prospects. Philosophical Transactions of the Royal Society of London, 318, 183–201, 1189.

van Emden, H. F., & Hagen, K. S. (1976). Olfactory reactions of the green lacewing, *Chrysopa carnea,* to tryptophan and certain breakdown points. Environmental Entomology, 5, 469–473.

van Emden, H. F., & Williams, G. F. (1974). Insect stability and diversity in agroecosystems. Annual Review of Entomology, 19, 455–475.

Vet, L. E. M. (1983). Host habitat location through olfactory cues by *Leptopilina clavipe* (Hartig) (Hym.: Encoilidae), a parasite of fungivorous *Drosophila:* The influence of conditioning. Netherlands Journal of Zoology, 33, 225–248.

Vinson, S. B. (1977). Behavioural chemicals in the augmentation of natural enemies. In R. L. Ridgway & S. B. Vinson (Eds.), Biological control by augmentation of natural enemies (pp. 237–279). New York: Plenum Press.

Vinson, S.-B. (1981). Habitat location. In D. A. Nordlund, R. L. Jones, & W. J. Lewis (Eds.), Semiochemicals: Their role in pest control (pp. 51–77). New York: John Wiley & Sons.

Vorley, V. T., & Wratten, S. D. (1987). Migration of parasitoids (Hymenoptera: Braconidae) of cereal aphids (Hemiptera: Aphididae) between grassland, early-sown cereals and late-sown cereals in southern England. Bulletin of Entomological Research, 77, 555–568.

Waage, J. K. (1979). Foraging in patchily-distributed hosts by the parasitoid, *Nemeritis canescens.* Journal of Animal Ecology, 48, 353–371.

Waage, J. K. (1983). Aggregation in field parasitoid populations: Foraging time allocation by a population of *Diadegma* (Hymenoptera: Ichneumonidae). Ecological Entomology, 8, 447–453.

Walker, G. P., Nault, L. R., & Simonet, D. E. (1984). Natural mortality factors acting on potato aphid (*Macrosiphum euphorbiae*) populations in processing-tomato fields in Ohio. Environmental Entomology, 13, 724–732.

Wallin, H. (1985). Spatial and temporal distribution of some abundant carabid beetles (Coleoptera: Carabidae) in cereal fields and adjacent habitats. Pedobiologia, 28, 19–34.

Wang, J. J., & Zong, L. B. (1991). A study on host-seeking kairomone for *Trichogramma confusum* Wiggiani. Colloques de l'INRA, 56, 93–96.

Ware, G. W., Cahill, W. P., Gerhardt, P. O., & Witt, J. M. (1970). Pesticide drift IV. On-target deposits from aerial application of insecticides. Journal of Economic Entomology, 63, 1982–1983.

Watson, T. F., & Larsen, W. E. (1968). Effects of winter cultural practices on the pink bollworm in Arizona. Journal of Economic Entomology, 61, 1041–1044.

Way, M. J., & Khoo, K. C. (1991). Colony dispersion and nesting habits of the ants, *Dolichoderus thoracicus* and *Decophylla smaragdina* (Hymenoptera: Formicidae), in relation to their success as biological control agents on cocoa. Bulletin of Entomological Research, 81, 341–350.

Way, M. J., & Murdie, G. (1965). An example of varietal variations in resistance of Brussels sprouts. Annals of Applied Biology, 56, 326–328.

Weiss, M. J., Balsbaugh, E. U., Jr., French, E. W., & Hoag, B. K. (1990). Influence of tillage management and cropping system on ground beetle (Coleoptera: Carabidae) fauna in the Northern Great Plains. Environmental Entomology, 19, 1388–1391.

Wellington, W. S. (1983). Biometeorology of dispersal. Bulletin of the Entomological Society of America, 29, 24–29.

Whitham, T. G. (1983). Host manipulation by parasites: Within-plant variation as a defense against rapidly evolving pests. In R. F. Denno & M. S. McClure (Eds.), Variable plants and herbivores in natural and managed systems (pp. 15–42). New York: Academic Press.

Wilbert, H. (1977). Der Honigtan als Reizund Engergiequelle fur Entomophage Insekten. Apidologie, 8, 393–400 (in German: English summary).

Williams, D. W. (1984). Ecology of blackberry-leafhopper-parasite system and its relevance to California grape agroecosystems. Hilgardia, 52, 1–33.

Williams, H. J., Elzen, G. W., & Vinson, S. B. (1988). Parasitoid-host-plant interactions, emphasizing cotton (*Gossypium*). In P. Barbosa & D. K. Letourneau (Eds.), Novel aspects of insect-plant interactions (pp. 171–200). New York: John Wiley & Sons.

Wipperfürth, T., Hagen, K. S., & Mittler, T. E. (1987). Egg production by the coccinellid *Hippodaemia convergens* fed on two morphs of the green peach aphid, *Myzus persicae.* Entomologia Experimentalis et Applicata, 44, 195–198.

Wolfson, J. L. (1980). Oviposition response of *Pieris rapae* to environmentally induced variation in *Brassica nigra.* Entomologia Experimentalis et Applicata, 27, 223–232.

Wong, M. H. (1985). Heavy metal contamination of soils and crops from auto traffic, sewage sludge, pig manure, and chemical fertilizer. Agriculture, Ecosystems, and Environment, 13, 139–149.

Wratten, S. D., & Thomas, M. B. (1990). Environmental manipulation for the encouragement of natural enemies of pests. Monograph, British Crop Protection Council, 45, 87–92.

Wratten, S. D., & van Emden, H. F. (1995). Habitat management for enhanced activity of natural enemies of insect pests. In D. M. Glen, M. P. Greaves, & H. M. Anderson (Eds.), Ecology and integrated farming systems (pp. 117–145). New York: Wiley.

Wyss, E. (1995). The effects of weed strips on aphids and aphidophagous predators in an apple orchard. Entomologia Experimentalis et Applicata, 75, 43–49.

Yu-hua, Y., Yi, Y., Xiang-ge, D., & Bai-ge, Z. (1997). Conservation and augmentation of natural enemies in pest management of Chinese apple orchards. Agriculture, Ecosystems, and Environment, 62, 253–260.

Zhody, N. Z. M. (1976). On the effect of food of *Myzus persicae* Sulz. on the hymenopterous parasite *Aphelinus asychis* Walker. Oecologia (Berlin), 26, 185–191.

15

Biology of Parasitic Hymenoptera

G. GORDH

USDA, Agricultural Research Service
Kika de la Garza Subtropical Agricultural
Research Center
2301 S. International Blvd.
Weslaco, Texas 78596

E. F. LEGNER

Department of Entomology
University of California
Riverside, California

L. E. CALTAGIRONE

Division of Biological Control
University of California
Berkeley, California

INTRODUCTION

Ecological and evolutionary aspects of parasitoid biology have become focal points of entomological research during the past 20 years to an extent unparalleled in the development of entomology. Here we review the principal features of parasitic Hymenoptera anatomy, development, biology, and reproduction as these features relate to biological control. Available space does not permit treatment of all parasitic insects. Parasitoids are the group of natural enemies of greatest importance in biological control of pests. Of 1193 species of parasitoids and predators included in the world review of biological control programs by Clausen (1978), 907 (76.03%) were parasitoids; of these, 765 (84.34%) were Hymenoptera, 125 (13.79%) were Diptera, and 17 (1.43%) belonged to other groups. Our discussion will emphasize the parasitic Hymenoptera because of their numerical dominance and their importance in biological control.

Definitions

Parasites are organisms that live in or on, and at the expense of, other organisms (hosts), each individual parasite completing all or a major part of its life cycle in or on one individual of the host. This is a form of a somewhat prolonged symbiosis involving at least two unrelated species. One symbiont (the parasite) lives at the expense of the other symbiont (the host). The parasite provides no benefit to the host, which is weakened or otherwise damaged, and is eventually destroyed, or rendered unable to reproduce.

For insects developing in this manner the larva is always the parasitic form. These insects have been called protelean parasites (Caullery, 1950; Sellier, 1959; Askew, 1971) to distinguish them from other groups of organisms that develop parasitically. The term "parasitoid" was proposed earlier (Reuter, 1913), and in the past 15 years this term has gained widespread acceptance among ecologically and ethologically oriented workers.

Several criteria are used to distinguish parasitoids from other parasitic animals. These include (1) the parasitic behavior is expressed only during the larval stage; (2) the adult stage is free living [exceptions to (1) and (2): adult female of some Strepsiptera]; (3) the parasitoid larva typically kills and consumes one host; (4) the body size of the parasitoid approximates that of the host; (5) the parasitoid life cycle is relatively simple; (6) the parasitoid shares relatively close taxonomic affinity with hosts (insects on insects); (7) parasitoids display a reproductive capacity between true parasites and free living forms (Doutt, 1959). Exceptions to each criterion are found among parasitic Hymenoptera.

While these criteria provide a useful framework for instruction, they may inaccurately portray the fundamental nature of insect parasitism. Our perception of insect parasitism must be altered as the biology and the development of parasitic insects become better documented. Indeed, parasitic insects display a prodigious array of progenitive strategies (Clausen, 1940; Askew, 1971; Price, 1977). They develop internally (endoparasitoids), externally (ectoparasitoids), or initially in one area and finally in the other (Shaw, 1983). Solitary parasitism is a condition in which a parasitoid larva completes development in a one-to-one relationship with its host; supernumerary eggs or larvae are eliminated. In contrast, gregarious parasitism involves the development of more than one individual in one host. Some species of parasitoids are obligatorily solitary or gregarious, while others are facultatively so.

Beyond this basic architecture for parasitism we see numerous variations on themes that are influenced by habitat, adult female anatomy, oviposition behavior, host taxa, host biology, and numerous other factors (Price, 1980). This necessitates special terms for these kinds of interactions. For instance, the term idiobiont was proposed for parasitoids that kill immediately or shortly after initial parasitization, permanently impairing or preventing further development of the hosts (Askew & Shaw, 1986). Typically idiobionts are ectoparasitic, attack concealed hosts, and express a broad host spectrum (generalists). In contrast, the term "koinobiont" was proposed for parasitoids that do not kill rapidly, permanently impairing or preventing further host development. Typically koinobionts are endoparasitic, attack exposed hosts, and reveal a limited host range (specialists). Some koinobionts may induce their hosts to consume more of their food (Hunter & Stoner, 1975) while others cause the reverse response (Doutt, 1963). Exceptions to this classification are easy to find.

When females of more than one species oviposit in a host individual, the condition is termed multiple parasitism. Facultative multiple parasitism is the association of more than one species of parasite on a host simultaneously. Obligatory multiple parasitism is a very rarely encountered phenomenon, and one whose functional significance is not clearly established (Salt, 1968). The condition in which a parasitoid prefers to distribute her progeny in hosts already parasitized by a different species is called cleptoparasitism. Examples are discussed by Arthur et al. (1964), Piel and Covillard (1933), Spradberry (1969), and Syme (1969). Cleptoparasitism is considered as an evolved advantageous multiparasitic strategy by Gauld and Bolton (1988).

Superparasitism is common in parasitic Hymenoptera, and has been defined in several ways: (1) one female allocating more eggs to a host than can successfully develop to maturity; (2) several conspecific females allocating more eggs to a host than can successfully develop to maturity; and (3) several females, not necessarily conspecific, allocating more eggs to a host than can successfully develop to maturity. The latter definition includes the original concept of superparasitism as defined by Howard and Fiske (1911). Each definition views the phenomenon in a different way, with disparate effects on the fitness of ovipositing females. As generally accepted, superparasitism is the condition in which an individual host receives more parasitoids of the same species than can develop to maturity, regardless of whether the parasitoids (eggs or larvae) were distributed by a single female or by several females. Various aspects of

superparasitism involving single females and several females of the same species have been studied by van Dijken and Waage (1987).

Superparasitism and the great majority of cases of multiple parasitism are temporary conditions; eventually the supernumerary individuals are eliminated. The incidence of each condition is dependent on the number of parasitoid progenies to be distributed and the number of susceptible hosts available.

Hyperparasitism is a multilevel trophic system in which one species develops as a parasitoid of another parasitoid. The species attacking the free-living organisms (herbivore or predator) is called a primary parasitoid; the primary parasitoid is attacked by a secondary or hyperparasitoid. Hyperparasitism is rare in the Coleoptera and Diptera, but reaches elaborate development in the Parasitica (Gordh, 1981; Sullivan, 1987).

Hyperparasitic development takes many forms including facultative hyperparasitism, obligatory hyperparasitism, adelphoparasitism, and tertiary hyperparasitism. Facultative hyperparasites can complete feeding and development as a primary parasite or use a primary parasite as a host. Obligatory hyperparasites must complete feeding and development using a primary parasite as host.

Important Anatomical Features of Parasitic Hymenoptera

Parasitic insects are common in the Holometabola; they are found in the Hymenoptera, Diptera, Strepsiptera, Coleoptera, and Lepidoptera. Hymenoptera are the most important group of insects from the viewpoint of applied biological control. The Hymenoptera include about 125,000 nominal species, but many parasitic species await description and the order will be shown substantially larger. Traditional classifications include the suborders Symphyta (= Chalastogastra: sawflies, wood wasps) and Apocrita (= Clistogastra: bees, wasps, ants). Few species of Symphyta are parasitic, and Symphyta assume a minor position in parasitism because nearly all species are phytophagous. The Apocrita are sometimes subdivided into two infraorders: the Parasitica and the Aculeata. Alternative classifications reflect several infraorders (Rasnitsyn, 1988).

Certain anatomical features have undoubtedly contributed to the evolutionary success of the apocritan Hymenoptera. Important features include an appendicular ovipositor, a constricted waist (petiole), and accessory gland secretions. The great majority of these parasitoids exhibit the ability to provide for their progeny. Together, these features have made parasitism a highly successful lifestyle and consequently have focused attention on apocritan Hymenoptera as an important group in applied biological control. The pertinence of each factor is explained next.

Appendicular Ovipositor

The Symphyta and the Parasitica are among the few Holometabola with a lepismatid-like ovipositor (E. Smith, 1970). The significance of this tubular, elongated, egg-laying structure as an adaptation for parasitism cannot be overemphasized. This type of ovipositor permits precise placement of the egg in habitats that other insects cannot reach without elaborate anatomical modifications involving other regions of the body (Rasnitsyn, 1968).

Accessory Gland Secretions

These secretions are associated with the reproductive system of most Insecta. They serve to lubricate the egg, provide a substrate for fungal growth, induce gall formation, and constitute venoms for the subduction of prey and host. Use of glandular secretions against potential hosts is a cardinal landmark in the evolution of parasitism by Hymenoptera.

Constricted Waist (Petiolate Abdomen)

The Aculeata and the Parasitica display a constriction between the thorax and the abdomen. The constriction forms a small, ringlike second abdominal segment, termed the petiole. This constriction promotes abdominal flexibility, which enables the adult to sting hosts and prey into paralysis and also permits egg deposition in confined spaces.

Progeny Provisioning

Ancestral Hymenoptera presumably displayed a phytophagous larval stage (Iwata, 1976). This is seen today in Symphyta females that place their eggs in plant tissue. The behavioral transition from placing an egg in plant tissue to the present condition in which an apocritan female places an egg in or on a host must have occurred early in the evolution of parasitic habits.

PARASITIC HYMENOPTERA IMPORTANT IN BIOLOGICAL CONTROL

The large number of taxa involved, and the comparatively small number of those that have been used in biological control suggest the enormous possibilities that this group offers for pest control. Here we highlight briefly significant features of biology for the most noteworthy families.

The Parasitica

The Parasitica form the largest division of the Hymenoptera, and the group that is most important to biological control. Biologically, the Parasitica primarily are composed of species that are parasitic, but some groups include gall formers and other phytophages.

Ichneumonoidea

This superfamily is among the oldest groups of Parasitica and has a fossil record extending into the Cretaceous (Rasnitsyn, 1980). Ichneumonidae are the largest family in this superfamily, and the host spectrum is broad with focus on Holometabola. All ichneumonids are parasitic but they do not attack Mecoptera, Siphonaptera, or Strepsiptera. Larvae, pupae, and immatures in cocoons are preferred. Adults are often associated with moist habitats and extensive ground cover. Braconidae are related to Ichneumonidae. All species are primary parasites, but host associations have not been established for most species (Matthews, 1974). Current information suggests that braconids display an exceptionally broad host range, mostly Holometabola, but do not attack Trichoptera, Mecoptera, or Siphonaptera. One subfamily, the Aphidiinae, generally regarded as important in biological control, includes about 300 species, all of which are primary internal parasites of aphids. They are sometimes regarded as a distinct family near the Euphorinae (van Achterberg, 1984).

Ceraphronoidea

This superfamily includes two extant families, Ceraphronidae and Megaspilidae, which early classifications placed in the Proctotrupoidea (Masner & Dessart, 1967; Dessart, 1986). Ceraphronidae are very poorly studied, but species apparently develop as endoparasites of larval Diptera, such as Cecidomyiidae. Pupation occurs inside the mature larval integument. Some species attack Thysanoptera, Lepidoptera, and Neuroptera (Chrysopidae, Coniopterygidae). Megaspilidae develop as ectoparasites of diverse taxa. They are hyperparasites of aphidiids on Aphididae, or primary parasites of Coccidae, Mecoptera, Neuroptera, and Diptera. Some myrmecophiles probably attack Diptera.

Evanioidea

This group is of questionable development and composition (Townes, 1950; Crosskey, 1951) and members have been placed among the Ichneumonoidea and Proctotrupoidea in some classifications. Three included families are Evaniidae, Aulacidae, and Gasteruptiidae. Evaniidae are widespread, are predominantly tropical, and under domestic conditions are typically encountered around drains and on windows. All species apparently are endoparasites of cockroach oothecae (Townes, 1949; Brown, 1973). Aulacidae are cosmopolitan, solitary egg–larval endoparasites of wood-boring Coleoptera and Hymenoptera (Hedicke, 1939; Oehlke, 1983; Benoit, 1984). Gasteruptiidae are widespread. Adults visit flowers and rotting logs in search of hosts that include bees and wasps. They show transitional behavior between cleptoparasitism and ectoparasitism.

Trigonaloidea

This superfamily represents a lineage whose members have been placed in many superfamilies (Townes, 1956). The Trigonaloidea includes one extant family, the cosmopolitan Trigonalidae, and one fossil family, the Ichneumonomimidae. Biological studies are limited (Raff, 1934; Cooper, 1954). Species are hyperparasites of larval Hymenoptera and Tachinidae; some develop as primary endoparasites on Symphyta in Australia. Adult females oviposit on vegetation. Frequently several thousand eggs are discharged during one ovipositional episode; the eggs are consumed by larval Symphyta or Lepidoptera and hatch, and the parasitoid larva penetrates the host's hemocoel.

Chalcidoidea

This superfamily is among the numerically largest, biologically most diverse and geologically oldest of the Parasitica (Yoshimoto, 1975; Gordh, 1979; Gibson, 1986). Chalcidoids feed as primary and secondary parasitoids, inquilines, gall inducers, and seed eaters. Development may be internal, external, solitary, or gregarious. Chalcidoids attack all stages of their hosts, including the adult (LaSalle, 1990), and may serve as hosts to chalcidoids, and the host spectrum extends from spiders and ticks to aculeate Hymenoptera. They are among the most important groups for applied biological control (DeBach & Schlinger, 1964). Here we briefly treat families most conspicuous in biological control attempts.

Aphelinidae are primary or hyperparasitoids that develop externally or internally. The biology of many species is unusual and males of some species are hyperparasitic (heteronomous) (Broodryck & Doutt, 1966; Kennett *et al.*, 1966; Walter, 1983; Viggiani, 1984). The higher classification of aphelinids is relatively well studied (Yashnosh, 1976, 1983; Hayat, 1983). Most species attack sternorrhynchous Homoptera and several species are important in biological control of scale insects (Rosen & DeBach, 1979) and whiteflies (Greathead, 1986).

Chalcididae are primary and hyperparasitoids of Lepidoptera and to lesser extent Diptera, Coleoptera, and other Hymenoptera (Burks, 1940; Boucek, 1952, 1974). As hyperparasitoids they attack Tachinidae developing as primary parasitoids of Hymenoptera and Lepidoptera. Development typically is solitary as endoparasitoids of the last instar larva and pupa. Some species have been used in biological control.

Encyrtidae are all endoparasitoids, and predominantly primary parasitoids; some species are hyperparasitic (Gordh, 1979, 1981). The host spectrum of Encyrtidae is broad, but most species are associated with Homoptera (Tachikawa, 1981). Encyrtids are important in biological control of mealybugs and scale insects. Polyembryony is known in many species (see later).

Eulophidae (including Elasminae) are predominantly primary parasitoids; some species are hyperparasitoids and a few are phytophagous (Boucek & Askew, 1968). Development is solitary or gregarious, internal or external. The host spectrum of eulophids is extensive, and includes all immature stages, often attacking concealed Lepidoptera and Diptera. They are important in some biological control programs involving sawflies on pine, and leaf miners.

Eurytomidae are typically solitary, rarely gregarious, ectoparasitoids; some species are endoparasitic of cecidogenic insects. The first instar is hymenopteriform, often with five pairs of spiracles. The host spectrum and feeding strategies of eurytomids are diverse: many species are carnivorous, developing as parasitoids of gall makers or other phytophages in seeds and stems, or as parasitoids of eggs (tree crickets); others are phytophagous, developing in seeds, or in galls induced by other insects.

Mymaridae developed early in the history of the Apocrita (Yoshimoto, 1975). All species develop as internal parasitoids of insect eggs (Burks, 1979a; Huber, 1986). Typically larvae are solitary and prefer Auchenorrhyncha whose eggs are concealed within plant tissues, under bark, and in soil. Pupation occurs inside the host egg. Some species have been successful in biological control programs (Clausen, 1978).

Pteromalidae are a large, biologically diverse, taxonomically difficult group of chalcidoids (Graham, 1969). Egg shape, size, and number are highly variable. Up to 700 eggs per female have been reported. The first instar larva is hymenopteriform, having 13 segments with head and mandibles sometimes large. Ectoparasitic larvae exhibit an open respiratory system, while that of endoparasitic forms is closed. They are solitary or gregarious, but typically they are ectoparasitoids. Their feeding spectrum is very diverse: their hosts are typically Holometabola larvae and pupae in stems, leaf mines, galls, and similar habitats; some species are larval–pupal parasitoids; some are predaceous on cecidomyiid larvae or coccoid and delphacid eggs; some are gall inducers and gall feeders.

Torymidae develop either as gall inducers, phytophages of seed endosperm, primary ectoparasitoids of gall inducers, ectoparasitoids of aculeate Hymenoptera, or hyperparasitoids (Grissell, 1979; Narendran, 1984). The torymid egg is of variable shape, typically sausagelike. The larva is hymenopteriform; those of parasitic species often have spines, while those of phytophagous species are spineless.

Trichogrammatidae are primary, solitary, or gregarious internal parasitoids of insect eggs. Some taxa are among the smallest insects (*Megaphragma* ca. 0.20 mm long) (Doutt & Viggiani, 1968; Burks, 1979b). The larvae undergo hypermetamorphic development: the first instar larva is sacciform or mymariform; the last instar is segmented, robust, and spineless. The family displays a broad host spectrum that includes principal orders of Holometabola, Hemiptera, and Thysanoptera. Trichogrammatidae are used extensively in biological control programs. More information on applied ecology has been published on *Trichogramma* than on any other comparable group of Parasitica.

Proctotrupoidea

The Proctotrupoidea have a fossil record in the Jurassic, which suggests that they are older than the Chalcidoidea. They are more abundant than chalcidoids in the oldest amber deposits (Cretaceous). Taxonomically, the extant fauna of Proctotrupoidea is substantially smaller than the Chalcidoidea. Family-level classification has been reviewed and tentative phylogenies have been proposed (Kozlov, 1970). All species are primary parasitoids and resemble Chalcidoidea; both groups have small body size and reduced wing venation. Proctotrupoids have not been used extensively in biological control, but some groups are important in controlling pests. Some species exert a negative influence when attacking predators (e.g., Heloridae, Townes, 1977).

Diapriidae are internal parasitoids of Diptera pupae within puparia and frequently are encountered in damp habitats (Nauman, 1987). Some species are myrmecophilous, probably attacking inquilines. Gregarious development has been reported, but the biology of most species has not been studied in detail. Females oviposit into either host pupa or host larva.

Platygasteridae are usually univoltine, gregarious, internal parasitoids. Hosts include gall-forming Cecidomyiidae, mealybugs, and whiteflies. Females oviposit in eggs or early instar host larvae. Oviposition is site specific (e.g., the eggs may be laid either in the nerve cord, in the gut, or in the brain). Development is completed within the mature larva. Some species exhibit hymenopteriform larvae; many species are hypermetamorphic. Polyembryony occurs in the family (Marchal, 1906; Ivanova-Kasas, 1972). Some reports suggest species of platygasterids are useful in biological control (Clausen, 1978; Jeon *et al.*, 1985; Lee *et al.*, 1985).

Proctotrupidae (Serphidae) develop as solitary or gregarious internal parasitoids of Coleoptera or Diptera. Adults are taken in damp, concealed habitats under bark, leaf litter, and fungal gardens. The polypodeiform first instar is qui-

escent. When the host is ready to pupate, the larval parasitoid develops rapidly. Proctotrupoids pupate externally in rows on the host's venter, with the apex of the gaster remaining attached to the host. A cocoon is not formed (Townes & Townes, 1981).

Scelionidae are solitary, internal egg parasitoids of Hemiptera, Lepidoptera, Mantodea, and Aranaea. The family is the largest and taxonomically among the best studied proctotrupoids (Masner, 1976, 1980; Galloway & Austin, 1984). Many species appear host-family specific and seem to prefer eggs in aggregations. Egg marking with semiochemicals has been noted in several species. Phoresy is widespread and probably common (Orr *et al.*, 1986). The first instar larva is teleaform. Several species show strongly female-biased sex ratios. Males of most species are protandrous and combative, and sibling mating appears common.

Cynipoidea

The group is large and poorly studied compared with other Parasitica (Konigsman, 1978). Current estimates indicate about 30% of the species are phytophagous (Nordlander, 1984). Cynipoids branched early from generalized parasitic Hymenoptera and Nordlander believes Cynipoidea are primitively parasitic. Fossil Cynipoidea have been taken in Taimyrian and Canadian amber, which makes them as old as chalcidoids (Rasnitsyn & Kovaley, 1988).

Eucoilidae are endoparasitic in schizophorous Diptera larvae. The family was subordinated within the Cynipidae in early classifications. Females oviposit in young host larvae and development proceeds through three to five instars. Pupation occurs within a host puparium. Hypermetamorphosis is probably universal with the first instar eucoiliform, with thoracic lobes and caudal projection for emergence from the egg. The respiratory system gradually develops, with the first instar lacking spiracles, while the fifth instar exhibits eight pairs.

Figitidae (including Anacharitinae) are solitary, primary internal parasitoids of Neuroptera and Diptera. Little is known of their biology (Miller & Lambdin, 1985). The egg is laid in the neuropteran hemocoel and hatches, and the larva feeds internally until the host spins its cocoon. Next the figitid migrates outside and feeds externally. Figitids overwinter as diapausing larvae within the host's body. Among dipteran hosts, Syrphidae and saprophagous cyclorrhaphans seem preferred.

Ibaliidae are solitary egg–larval parasitoids. The most detailed account of development and biology is provided by Chrystal (1930). Females locate eggs or hosts boring in wood and oviposit into the egg or early larva of Siricidae. First instar larva is polypodeiform; later larval stages are hymenopteriform. Early larval development is internal while later development is external. The fourth instar larva has large mandibles but does not feed. Pupation occurs in the host's tunnel. One species is used in Tasmania and in New Zealand to control siricids in pines (Kerrich, 1973).

Aculeata

Despite their large numbers, widespread geographic distribution, and diverse biologies, the parasitic aculeate Hymenoptera are not a highly important group in applied biological control. Parasitism in this group is found in the more primitive taxa. Their hosts are insects of limited economic significance. Higher forms behave as predators. Complex parasitic relationships are found among social bees and ants, but these are beyond this discussion.

Chrysidoidea

Chrysidoids are generalized Aculeata and sometimes they are regarded as the most primitive superfamily, apparently near the Pompilioidea. Fossil dryinids are found in Taimyrian and Canadian amber. All chrysidoids are parasitic, and nearly all are primary parasitoids. Their host associations are cohesive at the family level, each of many genera attacking a family or a group of related families.

Dryinidae are another group with a fossil record dating from the Cretaceous (Ponomarenko, 1981). Species attack auchenorrhynchous Homoptera (Cicadelloidea, Fulgoroidea). A number of female dryinids exhibit chelate forelegs with which they capture their host. Before oviposition they temporarily paralyze the nymphal host (Olmi, 1984a, 1984b). Female dryinids sometimes host feed. The first instar larva is sacciform, but later stages are hymenopteriform. A thalacium is formed from cast larval exuviae (as in the Rhopalosomatidae). Dryinidae apparently lack Malpighian tubules. Polyembryony and thelytoky have been reported in a few species (Ivanova-Kasas, 1972; Ponomarenko, 1975).

Chrysididae are parasitoids of Hymenoptera and phasmid eggs. Host associations reflect the higher classification of the Chrysididae: primitive subfamilies attack sawflies and phasmid eggs, while derived ones attack aculeate Hymenoptera (Bohart & Kimsey, 1982). Cleptoparasitism is expressed in forms that feed on the host and food in its cell. First instar chrysidid larvae are planidiform and typically attack the prepupal stage of the host. Complex female behavior of derived lineages is noted in host attack: she gnaws into the host cocoon to oviposit or gnaws into mud cells and then into the host cocoon inside. Females sometimes seal holes with saliva. Thanatotic behavior is reported in many species.

Bethylidae are primary external parasitoids of Coleoptera and Lepidoptera larvae (Evans, 1964, 1978). Females search for hosts, often in concealed habitats, and sting them into paralysis. Host paralysis is temporary or permanent, depending on the species. Typically oviposition occurs at the site of encounter, but a few species transport the host to a protected place, or construct a cell for it before ovipositing. The egg is hymenopteriform and is "glued" to the host. Development of the larvae is gregarious. Bethylids are not host specific, but species in certain genera are associated predominantly with certain host families. Females adjust the number of eggs laid on a host based on the host size; if mated, they also control the sex ratio (female bias).

PRINCIPAL ASPECTS OF PARASITOID BIOLOGY

Developmental Stages

The Egg

Parasitic insect eggs are variable in size and shape. This variability is a consequence of the enormous number of taxa involved and the developmental requirements of the insect embryos contained in these eggs. The variability in size and shape partially reflects a compromise between needs of the developing embryo and problems associated with oviposition. The nomenclature and early literature associated with parasitoid eggs were characterized by Clausen (1940), were reviewed by Hagen (1964), and are summarized here. Research on egg anatomy during the past 20 years has shown that shape alone is not diagnostic and that unrelated taxa share identical shapes (Hinton, 1969). With the application of electron microscopy we now see that chorion morphology, eggshell complexity, micropylar position, number, and configuration are important features that must be described, studied, and understood. The number of taxa for which information must be collected is very large. The eggs of Hymenoptera are briefly reviewed here.

The hymenopteriform egg is the hypothetical ancestral form or the generalized type found in Hymenoptera. Typically the egg is sausagelike with rounded poles and a body several times longer than wide. The acuminate egg is typically long and narrow, and generally is adapted for extrusion from the long ovipositor of parasitic Hymenoptera that attack insects that induce galls or live in galleries and tunnels. This egg is found in some Ichneumonoidea and Chalcidoidea. The stalked egg is elongate with a constricted stalklike projection from one or both poles. Stalk length is variable, sometimes corkscrew shaped, and often several times longer than the remainder of the egg. Stalked eggs are found in most superfamilies of parasitic Hymenoptera: Chalcidoidea (most families), Chrysidoidea, Cynipoidea, Evaniioidea, Ichneumonoidea, and Proctotrupoidea (most families). The pedicellate egg is an apparent variation of the stalked egg in which one end is modified to anchor the egg to its host. Most pedicellate eggs are deposited externally on the host, but some are internal and attached to the host via the ventral surface of the egg. The pedicel may originate from the stalk, from the body of the egg, or from a modified micropylar structure. This form is distributed among Parasitica, including Chalcidoidea and Ichneumonoidea. The encyrtiform egg is unusual in that it changes shape after oviposition. Within the ovary it is shaped as two spheres connected by a stalk. After oviposition one sphere collapses, all of the cytoplasm being accumulated in the main body of the egg, which then appears stalked. Encyrtiform eggs are always deposited inside the body of the host, with the collapsed sphere projecting from the stalk outside the host's body. An aeroscopic plate, used for embryonic and larval respiration, usually is found on the stalk and sometimes projects onto the body of the egg. This type of egg is characteristic of Encyrtidae (Maple, 1947) and Tanaostigmatidae (LaSalle & LeBeck, 1983).

The Embryo

Embryology concerns the development of the individual organism, typically within the egg (Counce, 1961). A mature egg is distinguished by the animal pole, close to which the nucleus is found, and by the vegetal pole that is opposite the animal pole. The yolk is accumulated in the vegetal pole (Balinsky, 1970). The animal pole is more active in physiological exchange during oogenesis. The internal portion is fluidlike and differs from the surface.

Mature eggs undergo "aging" among some species, resorption in others, and a combination of both in others. Aged eggs of Hymenoptera may be deposited or resorbed. Embryos develop into male or female progenies, depending on the kind of parthenogenesis. In some cases eggs may hatch within the mother, which kills her. The state of meiosis at oviposition varies (Baerends & van Roon, 1950; Bronskill, 1959).

Cleavage is the subdivision of the one-celled egg into smaller units, called blastomeres, by mitotic divisions. Each division results in a reduction in size of the ensuing blastomeres. The total mass of living substance available at the start is not increased appreciably when cleavage is complete. The ova of most arthropods are centrolecithal (i.e., the yolk is massed centrally and surrounded by a peripheral shell of cytoplasm). Cleavage occurs only in the peripheral region and is termed superficial. Some endoparasitic Hymenoptera have little yolk (isolecithal) and show total cleavage (Bronskill, 1959, 1960). Yolk components are called deutoplasm and include protein yolk bodies, lipid yolk bodies, and glycogen particles. Some chalcidoids apparently lack yolk. Cleavage usually begins soon after the egg is laid, but exceptions occur and some eggs hatch inside the mother. The duration of cleavage varies, but generally it is finished after 6 to 8 h at 23°C.

Gastrulation is the process through which three germ layers are formed: the ectoderm, mesoderm, and entoderm. Germ layers produce body organs and appendages. Gastrulation occurs in diverse ways among Hymenoptera, and

may differ among species of one family. The duration apparently ranges from 1 to 12 h. Segmentation occurs early in development of some Hymenoptera, and later in others. Two embryonic membranes, the serosa and the amnion, usually envelop the insect embryo, but in Hymenoptera, one or both may be rudimentary or absent. Embryonic envelopes function in protection and nutrition, and usually occur well developed in species with little yolk. Eggs with little yolk are usually minute when deposited in the host. Then, probably by osmosis or active absorption of host fluid, they gradually become larger (Imms, 1931; Simmonds, 1947c). Expanding eggs of this type have been called hydropic (Flanders, 1942).

In those species in which the trophamnion is broken and cells of the membrane move in the host's hemolymph, these cells increase in size proportional to the growth of the larval parasitoid (Hagen, 1964). They become greatly enlarged while retaining their trophic function because the larva feeds on these cells (Jackson, 1928; O. Smith, 1952). Host nutrition influences the development of these cells and, through them, the development of the parasitoid larva. Some membranes persist, covering the larva. For example, the chorion may remain intact until after first larval ecdysis (Flanders, 1964) or the chorion may be shed early and the developing embryo is enveloped by a trophic membrane that remains without disintegrating until after the parasitoid larva sets itself free in the host (Caltagirone, 1964). The embryology of an endoparasitic hymenopteran, *Coccygomimus turionellae,* is discussed in detail by Bronskill (1959).

Egg orientation is similar in all Hymenoptera studied and follows Hallez's (1886) law of orientation within the polytrophic ovariole. The anterior pole is directed toward the head of the parent female. However, during oviposition, the posterior pole emerges first. The dorsal, ventral, and lateral sides vary within the same individual. The embryo remains in the original cephalocaudal axis during the entire development, but just before eclosion it rotates 180 degrees on the longitudinal axis.

Eclosion from the egg usually occurs when histogenesis is complete. Exceptions occur and first instar larvae of some internal parasitoids are precociously emerged embryos (Imms, 1931). Eclosion is not always immediate, and some eggs may overwinter in the completely incubated condition. Eclosion by parasitoids may require very high relative humidity (Gerling & Legner, 1968). Specific host organs may serve as oviposition sites, and chorions may be coated to avoid encapsulation in the host (King *et al.,* 1968).

Polyembryony is a form of asexual reproduction in which many embryos develop from repeated division within one egg (Marchal, 1898, 1904). It has been reported in several groups of insects (Silvestri, 1906, 1923, 1937; Kornhauser, 1919; Leiby & Hill, 1923, 1924; Leiby, 1929; Parker, 1931; Daniel, 1932; Paillot, 1937; Doutt, 1947,

1952; Ivanova-Kasas, 1972). Among parasitic Hymenoptera, polyembryony is known in Braconidae, Encyrtidae, Platygasteridae, and Dryinidae; and is best documented in the Encyrtidae (Doutt, 1947; Cruz, 1981, 1986a, 1986b).

Generation time in polyembryony varies from several weeks to a year. Embryo development begins just as in monoembryony. Polar nuclei, however, do not enter directly into the blastula stage, but produce an embryonic membrane (trophamnion) that surrounds the developing embryo-like area. The trophamnion extracts nutrients from the host hemolymph (Koscielska, 1962, 1963). The embryo then divides into small groups of cells (morulae) enclosed by the trophamnion. The trophamnion then changes into a chainlike structure with the morulae arranged in a row or branching cluster. This finally breaks up and separate embryos are formed. The number of embryos from a haploid egg is less than that from a diploid egg.

The Larva

Larval Hymenoptera typically consist of a head, three thoracic segments, and 9 to 10 abdominal segments. Apocrita lack thoracic or abdominal appendages, except in Proctotrupoidea and Cynipoidea (see later). Frequently larvae show five instars, but variation is common. Determining the primitive condition is difficult because only a few taxa have been studied. The number of mandible exuviae is used as evidence for determining instar numbers, but dimorphism may occur within one larval instar, and sexual dimorphism in the larvae can be striking. The anatomy of final instar parasitic Hymenoptera larvae has been described (Short, 1978; Finlayson & Hagen, 1979; Gillespie & Finlayson, 1983).

Conventional larval development is seen in many Hymenoptera. The eruciform larva is shaped like a caterpillar with a well-developed head capsule, thoracic legs, and abdominal prolegs. The Symphyta express this larval type that may represent the ancestral condition for Hymenoptera. The hymenopteriform larva represents the generalized larval form seen in the Apocrita. The body is spindle shaped, without thoracic legs, featureless with pale to translucent integument, and head capsule weakly developed or absent. The mandibulate larva has a large, sclerotized head, large falcate mandibles, and body-tapered posteriad. The caudate larva has a body with long, segmented, flexible caudal appendages. The function of caudal appendages is not established; sometimes they are reduced in later instars and lost in the last instar. The caudate form is displayed by some endoparasitic ichneumonid larvae. The vesiculate larva has the proctodaeum everted, and displays short caudal appendages with vesicles at the bases. This larva is seen in some Ichneumonidae and Braconidae. The mymariform larva displays a conical anterior process and the abdomen of some species is segmented. This larval form is found in Mymaridae and Trichogrammatidae. The sacciform larva is ovoid, featureless, and without segmentation. It is found in Dryinidae and Trichogrammatidae. The polypodeiform larva is segmented with paired, short, flexible projections from thoracic and abdominal segments. It is found in Cynipoidea and Proctotrupoidea.

Many Parasitica are hypermetamorphic and larvae change in shape during successive instars as a normal consequence of development. Several first instar morphs are diagnostic of some Hymenoptera families. Thus, the teleaform larva is seen in Scelionidae: the first instar is not segmented, weakly cephalized with prominent protuberances or curved hooks cephalad, and the body has a caudal process with girdles or rings of setae around the abdomen. The cyclopoid larva is seen in Platygasteridae: the first instar shows a large swollen cephalothorax, very large, falcate mandibles and bifurcate caudal processes. The planidiform larva is seen in the Perilampidae and in a few Ichneumonidae: the migratory first instar is legless and somewhat flattened, and often displays strongly sclerotized, imbricated integumental sclerites and spinelike locomotory processes. The form is often confused with the triungulin (Heraty & Darling, 1984). The eucoiliform larva is seen in Eucoilidae, Charipidae, and Figitidae: the first instar displays three pairs of long thoracic appendages but lacks the cephalic process and girdle of setae of the teleaform larva. Subsequent instars are polypodeiform.

Defense from and evasion of predators are important to parasitoid larvae. First instars use various methods of locomotion from moving sluglike to jumping. Fast locomotion is typical of species that lay their eggs a distance from the host (Clausen, 1976). Exposed larvae are protected by spines and sclerites. Internal parasitoid larvae often show large or well-developed mandibles (Salt, 1961), presumably used in defense. Also, cytolitic enzymes, secreted by one parasitoid to kill another, have been suggested as a method of eliminating competitors (Spencer, 1926; Thompson & Parker, 1930; O. Smith, 1952; Salt, 1961; Gerling & Legner, 1968). Identification of the toxic compound or compounds has been difficult.

Methods of eliminating competitors can be sophisticated. For instance, physiological suppression of supernumerary parasites is common among Parasitica (Fisher, 1971). Typically, young larvae are eliminated without physical contact between competitors. Nevertheless, mechanisms of suppression are poorly understood. Encapsulation of moribund parasite eggs and larvae by host phagocytes may result in death of supernumerary individuals via anoxia. Alternatively, older parasitic larvae may consume nutrients essential for development of young larvae and death of the younger individuals results from inanition.

First-instar larvae show diversity in respiration (Clausen, 1950). Endophagous larvae respire through the integument or obtain air from outside the host through tubelike structures (a membranous cocoon attached to the host tracheae). The final instar may possess a different spiracle arrangement and number while early instars may lack spiracles (Hagen, 1964).

The prepupal stage begins when the last larval instar stops feeding, voids meconia, and shows little movement. Although this is often referred to as a resting stage, the animal is active physiologically. The onset of prepupal formation may vary among individuals developed from eggs deposited during a short period (ca. 24 h) or it can occur synchronously (Gerling & Legner, 1968; Legner, 1969a). Connection between midgut and hindgut begins when the last larval instar is fully fed, and is completed during the prepupal stage. The prepupal stage lasts less than 24 h; meconia are eliminated as pellets or encased in a peritrophic sac (Gerling & Legner, 1968).

The Pupa

The pupal stage of parasitic Hymenoptera has been superficially studied, and most work is limited to line drawings or photographs of the habitus. Most hymenopteran parasitoids do not spin cocoons. However, fully fed endophagous larvae can construct membranous cocoons while immersed in the host fluids. Similar cocoonlike structures are found between gregarious (polyembryonic) pupae. The duration of the pupal stage can be variable or remarkably equal among the progeny of one female per day (Legner, 1969a).

Typically, the progenies of one ovipositional episode all exit the host immediately after eclosion from the pupa, or they may remain inside for variable lengths of time depending on when the adult bites through the encasing host (Legner, 1969a).

Oogenesis–Ovisorption–Spermatogenesis

Oogenesis in arrhenotokous Hymenoptera resembles diploid–diploid organisms, but spermatogenesis is highly modified (Crozier, 1975, 1977). Male Hymenoptera are hemizygous, and spermatogenesis is modified to ensure that a balanced set of chromosomes is transmitted via the sperm. The first division is abortive, with no karyokinesis (Crozier, 1975). In most Hymenoptera, the sperm of a haploid male are genetically identical.

Parasitica ovarioles have been characterized as proovigenic or synovigenic. Both conditions are generally viewed as species characteristics. However, the distinction is probably artificial, and when all parasitic Hymenoptera have been studied, we will see a continuum between conditions termed "proovigenic" and "synovigenic." Most eggs are laid by neonate females and the oviposition period of an adult female is usually short. Synovigenic species generally synchronize oogenesis with oviposition and the oviposition

period of an adult female is relatively long. Conventional wisdom holds that synovigenic species are more effective in biological control because they are longer lived and can reproduce at lower densities of the host population. This has yet to be demonstrated. Synovigenic species are further divided into subgroups where ovulation is internally or externally induced.

Ovisorption is "the capacity of the follicle cells to dissolve and absorb the oocyte" (DeWilde, 1964). Several factors influence this phenomenon in which vitellogenesis is interrupted and the oocyte dies in the follicle. Follicle cells cease to participate in egg formation; they may divide and absorb the dead oocyte. Their nuclei become pycnocytic, and the cells break down and are absorbed through the ovarian sheaths (King, 1963; Richards & King, 1967). From a biological control viewpoint, ovisorption is important in parasitoids where effectiveness as natural enemies depends in part on their conservation of reproductive material, correlated with a high searching capacity (Doutt, 1964).

Ovisorption has been reported in several orders of insects (Pfeiffer, 1945; Highnam et al., 1963). When certain Parasitica that ovulate yolk-replete eggs are deprived of hosts, the process of oogenesis and ovisorption occurs synchronously. This enables a female to deposit newly formed, viable eggs after a period of inhibited oviposition (Flanders, 1935b, 1942; Medler, 1962; Hopkins & King, 1964, 1966; King & Richards, 1968a; King & Ratcliffe, 1969). Under other conditions, inviable and viable eggs may be indiscriminately deposited in a host (Gerling & Legner, 1968). Partially absorbed, viable eggs may produce diploid males (Flanders, 1943b), embryonic starvation (Flanders, 1957, 1959b), or change the normal sex ratio (King, 1962).

Ovisorption occurs when conditions are unfavorable for mature eggs (Flanders, 1942; Edwards, 1954b; LaBergrie, 1959; Phipps, 1966) or when parasitism has taken place (Palm, 1948). Host feeding, oviposition, and ovisorption are closely related processes that affect fecundity, longevity, and host-killing capacity of the parasitoid (DeBach, 1943b; Legner & Gerling, 1967; Legner & Thompson, 1977).

Storage of mature eggs in muscular oviducts of hydropic species is correlated with the ability to discharge a large number of eggs quickly during one insertion of the ovipositor, or a large number of eggs singly if hosts are available. Rapid egg deposition probably is responsible for the high number of braconid species yielding a preponderance of male progeny (Clausen, 1940).

Anhydropic species with short oviducts ovulate when environmental conditions are favorable for immediate egg deposition. The rate of oviposition may be governed by the number of ovarioles (Clausen, 1940). In gregarious species the number of eggs deposited per host may be influenced by the number of mature eggs in the ovarioles, which at least in one case is regulated by a polygenic system (Legner, 1987b, 1988a, 1989b). Because ovisorption may preclude ovulation in some anhydropic species, the responsiveness to oviposition stimuli may be a function of the frequency of oviposition (Flanders, 1942; Legner & Gerling, 1967; Gerling & Legner, 1968). Females of anhydropic species may lose the ability to respond to oviposition stimuli if deprived of hosts for long periods (Jackson, 1937). Fecundity is sometimes lowered after ovisorption has occurred (King, 1962).

Oviposition Strategy and Progeny Number

Anatomy of the female reproductive system has been studied in some Parasitica (Iwata, 1959a, 1959b; 1960a, 1960b, 1962; Copeland & King, 1971, 1972a, 1972b, 1972c; Copeland, 1976). Parasitoids have developed ovipositional strategies that are essential for optimizing lifetime performance (Cook & Hubbard, 1977; Hubbard & Cook, 1978; Waage, 1979; Iwasa *et al.*, 1984). Some Parasitica can restrain oviposition, but egg retention lasts only a few days because intrinsic pressure of egg accumulation forces oviposition (Lloyd, 1938; Simmonds, 1956). Adult Parasitica possess a high inherent fecundity, are long lived, and actively search for hosts. Within a generation they can increase progeny production in response to rising host densities (Flanders, 1935a; DeBach & Smith, 1941; Burnett, 1951). This increase can be induced through mechanical and sensory processes (Madden & Pimentel, 1965; Legner *et al.*, 1966; Legner, 1967a, 1967b). Contributions have been made concerning the behavioral mannerisms whereby this acceleration becomes possible (Wylie, 1965a, 1966a, 1966b).

The relationship between parasitoid density and effective rate of reproduction has been studied (DeBach & Smith, 1947; Burnett, 1953, 1956, 1958a, 1958b, 1967; Bouletreau & David, 1967; Legner, 1967b). Factors involved in determining the effective rate include temperature, humidity, host size, adult parasitoid food, and larval competition.

Temperature and humidity influence parasitoid efficiency and oviposition. Low temperature can reduce the oviposition capacity and act with host density to reduce the number of hosts contacted by the parasitoid (Burnett, 1951). In some species, the greatest influence of temperature results from its differential effect on fecundity and rate of development of the host and parasitoid (Burnett, 1949). In other species, an increase in temperature combines with an increase in host density to induce a greater percentage of parasitoid emergence during one parasitoid generation (Burnett, 1951). This illustrates the importance of optimum temperature for maximum host destruction. Parallel results were obtained in a field experiment (Burnett, 1951). Greenhouse studies show that with an increase in temperature,

efficiency of the parasitoid increases and percentage of parasitism rises (Burnett, 1953). Humidity influences the oviposition rate of parasitoids (Martin, 1946; Martin & Finney, 1946). Higher humidities generally promote longer adult longevities (Legner & Gerling, 1967; Olton & Legner, 1974).

Parasitoid females estimate host size, and those of gregarious species adjust the number of eggs laid in response to that estimate, larger hosts receiving more eggs (Edwards, 1954a). The ability to estimate host size and to adjust clutch size in response to that estimate appears widespread in bethylids (Legner *et al.*, 1966; Gordh *et al.*, 1983; Gordh & Medved, 1986). An optimal size for ectophagous parasitoids is manifest in species attacking synanthropic flies (Legner, 1969a).

Adult parasitoids often distribute their progeny in such a way as to avoid competition. In superparasitized hosts supernumerary eggs and larvae die (Salt, 1961). Gregarious parasitoids discriminate among hosts, although the ability may be imperfect or temporary. Under some conditions, females refrain from laying supernumerary eggs; under other conditions (e.g., scarce healthy hosts) restraint fails and females oviposit into parasitized hosts. Thus, for lack or failure of discrimination, or by breakdown of restraint, superparasitism and thereby competition occur. Four modes of competition are usually recognized: (1) physical attack, (2) physiological suppression, (3) injury, and (4) selective starvation.

Supernumerary larvae of gregarious parasitoids are not always eliminated as early as solitary parasitoids. Often, final instar, gregarious larvae are found dead. Death of weaker competitors from food shortage has usually been implied. Adults dwarfed from severe competition also support this idea (Wylie, 1965b). Suffocation has been shown in some examples. Larval competition can be reduced by the female producing unfertilized eggs when superparasitism is possible (Holmes, 1972). Male larvae can compete effectively under crowded conditions (Wylie, 1966b).

Sex Regulation

Sex-ratio theory has been dominated by research on vertebrate animals since Darwin developed the principles of natural selection. Conventional wisdom held that sex ratios typically fluctuated around 1:1 (Fisher, 1930). Later work with Parasitica shows that sex ratios of parasitic insects often deviate from 1:1 (Hamilton, 1967). Apparent from this research is that adult female wasps control the sex of their offspring and that numerous factors shape proximate and ultimate sex ratios in parasitic organisms.

Insect sperm stored in the spermatheca retain their vitality for many years (Lillie, 1919), and are quiescent, except when females are in contact with oviposition sites (Flanders, 1939). The sperm at the opening of the spermatheca must be exposed to an activating agent before they can migrate through the spermathecal duct to the oviduct where the egg is fertilized. The agent apparently originates in the spermathecal gland, which presumably secretes a fluid that is slightly alkaline (Flanders, 1946a, 1946b; Lensky & Schindler, 1967). The capacity of the spermathecal gland, adjusted to the rate of egg deposition, is probably an important factor in determining the sex ratio (Flanders, 1947, 1950).

Three anatomical types of spermathecae have been described in Hymenoptera (Flanders, 1939, 1956). Type I is found in the honeybee, and in *Typhia.* The spermathecal duct can be bent into a valve at its juncture with the capsule and several sperm can be discharged simultaneously. Glandular fluids activate and transport sperm and regulate the number of sperm released on each egg as it passes along the oviduct. The spermathecal gland empties into the sperm capsule instead of the spermathecal duct as in other species. Type II is found in ichneumonids and braconids. The spermathecal gland empties into the lumen of the spermathecal duct. The short exit passage from the capsule to the spermathecal duct is narrow, limiting the rate of sperm passage. The gland is voluminous in ichneumonids; braconids have a smaller gland that is accompanied by a contractile reservoir. Type III occurs in chalcidoids. The spermathecal gland is very small and secretions only activate sperm. The spermatozoa in these species are extraordinarily long.

Hymenoptera regulate the sex ratio of their offspring (Crew, 1965; Hamilton, 1967; Charnov, 1979; Charnov *et al.,* 1981; Green *et al.,* 1982). Factors involving the adult female can influence fertilization of an egg by arrhenotokous Hymenoptera, and thus change the sex ratio. Some factors are female age (Wilkes, 1963; Legner & Gerling, 1967; Legner, 1985b), mating response changing with age (Crandell, 1939), sperm depletion (Gordh & DeBach, 1976; Nadel & Luck, 1985), rate of oviposition (Legner, 1967a, 1967b), delaying or interrupting oviposition (Wilkes, 1963; Legner & Gerling, 1967; Legner, 1985b), host density (Flanders, 1939), changing parasite density (Wylie 1965a, 1965c, 1966b), host size (Holdaway & Smith, 1932; Ullyett, 1936; Taylor, 1937; Legner, 1969a), multiple copulation by females (Flanders, 1946b), female discrimination (Flanders, 1939; Legner & Gerling, 1967; Werren, 1984), and ratio of different hosts (Flanders, 1959a). The haplodiploidic sex-determining mechanism in Hymenoptera results in the whole genetic complement carried by the male subjected to natural selection; thereby deleterious mutations are eliminated (Gauld, 1984). Thus, sibmating, which is common in those species exhibiting female-biased sex ratios, is not deleterious.

Temperature extremes also influence sex ratio in Parasitica (Wilson & Woolcock, 1960; Flanders, 1965; Bowen & Stern, 1966; Quezada *et al.* 1973; Gordh & Lacey, 1976). High temperatures can influence the sex ratio through sterilizing effects during postembryonic development (Wilkes, 1959). Sex-ratio upsets can be induced by low temperature (Schread & Garman, 1933, 1934; Lund, 1938; DeBach & Argyriou, 1966; DeBach & Rao, 1968). Fertility of parasitoids subjected to low temperatures during development may be adversely affected (van Steenburgh, 1934; Anderson, 1935). Larvae stored at 15°C may not change the sex ratio of their offspring, but pupae stored at this temperature can produce a greater proportion of male progeny (Hanna, 1935).

Chalcidoid larvae stored near 0°C can sustain a greater male mortality, causing a predominantly female sex ratio (DeBach, 1943a; DeBach & Argyriou, 1966; DeBach & Rao, 1968). Moursi (1946) reviewed cases where low temperature especially seemed to produce sex-ratio changes in parasitoids. He considered as causes: (1) inadequate stimulation of the spermathecal gland, (2) depletion of spermathecal secretions, and (3) failure of spermathecal nerves and muscles to function or synchronize the discharge of sperm with the expulsion of eggs through the oviduct. Prolonged exposure of males to low, nonlethal temperatures may induce gonad malnutrition in mature larvae and pupae (Flanders, 1938). Some third-instar chalcidoid larvae stored at low temperatures transform into adults with changed fecundities that produce a preponderance of female progeny (Legner, 1967a). Prolonged storage of mature chalcidoid larvae at low temperature does not influence the sex ratio of surviving adults (Legner, 1976).

Males are produced by deuterotokous Hymenoptera in populations that would otherwise be 100% females and thelytokous (E. L. Smith, 1941; Flanders, 1945; Bowen & Stern, 1966; Legner, 1985b). These males can be produced after periods of hot weather (E. L. Smith, 1941; Gordh & Lacey, 1976) or during oviposition at high temperature (Legner, 1985a).

Extranuclear substances, such as microorganisms and/or chemicals, are suspected to be involved in the expression of thelytoky. The manner in which thelytoky becomes incorporated in a hybrid arrhenotokous population of *Muscidifurax raptor* Girault & Sanders (Legner, 1987a, 1987c; 1988b), and the switch from thelytokous to arrhenotokous reproduction in *Trichogramma* spp. after treatment with antibiotics (Stouthamer & Luck, 1988; Stouthamer, 1989) implicate microorganisms. Thus, thelytoky may actually originate as an infection and may be accompanied by reduced parasitoid vigor (Legner & Gerling, 1967; Legner, 1988c). Sex ratio can be shaped by differential mortality of immature parasitoids (Jenni, 1951; Wilkes, 1963). In multiparasitism, the first individual present usually survives. Often, mortality among gregarious larvae involves females more than males (Vandel, 1932; Grosch, 1948). Superparasitism and subsequent larval competition can reduce female progeny (Kanungo, 1955). Conceivably, this mechanism may contribute to the stability of the system by avoiding over-exploitation of the host through reducing the number of females in the next generation. However, the proportion of females can be higher as a result of an increase in superparasitism in mass culture units (Martin & Finney, 1946).

The stage or the species of hot can influence parasitoid sex ratio. Thus, a thelytokous braconid produces only female progenies when oviposition occurs in beetle larvae, but males are produced when eggs are laid in adult beetles (Kunckel & Langlois, 1891). One sawfly feeding on alder is usually unisexual, but closely related species feeding on birch are bisexual (Bischoff, 1927). One aphelinid is unisexual when reproducing on San Jose scale growing on peach trees, and bisexual when reproducing on San Jose scale growing on melon (Flanders, 1944a). A gall-forming eurytomid is unisexual on one variety of *Acacia* and is bisexual on another (Flanders, 1945).

Microorganisms (viruses, bacteria, spiroplasmas) can alter sex ratios by selective elimination of immature males or females (Werren *et al.,* 1981, 1986; Skinner, 1982, 1985; Vinson & Stoltz, 1986). Stoltz and Vinson (1977) and Stoltz *et al.* (1976) found viruses in the calyx epithelial cells of endoparasitoids; Fleming and Summers (1986) found them in the lumen of the oviduct. These viruses were passed from parent to offspring, males transmitting viral DNA to females during copulation (Stoltz *et al.,* 1986).

Host Feeding

Host feeding is a phenomenon in which female parasitoids wound hosts with the ovipositor or mandibles and consume hemolymph at the wound. This behavior affects the biology of the parasitoid and the dynamics of populations in many ways. Marchal (1905) concluded that the ovipositor was used more often for host feeding than for oviposition. Host feeding prolongs the life of the female and supplies protein needed for oogenesis (King & Hopkins, 1963). Neonate adult synovigenic Hymenoptera may not have mature eggs and host feeding may provide nutrients necessary for egg development. Some females with mature eggs in the ovaries will not oviposit until after host feeding (Moursi, 1946). Other females will oviposit immediately after emergence, but if hosts are withheld during middle age, host feeding is required for additional oviposition. Adults of some species do not contain mature eggs at the pupal stage, but oviposition occurs before host feeding (Flanders, 1936).

Host feeding among parasitoids has not been sufficiently studied, but it has been reported in numerous species in 18 families (Jervis & Kidd, 1986); it may be important in applied biological control (Van Driesche *et al.,* 1987). It is unknown in proovigenic species, in some species whose males and females develop in unrelated hosts, and in species where yolk-deficient eggs are stored in the oviducts. In some cases, the published record may be erroneous. Host feeding has been reported in many species of bethylids, but this is questioned (Gordh, unpublished).

The close association of host feeding and oviposition in many Parasitica probably indicates that the habit of ovipositing in other insects evolved with host feeding by adults. Adult predacious habits preceded parasitic oviposition. Bartlett (1964) reasoned that the widespread occurrence of predation by adults of 20 families of parasitic Hymenoptera provides little evidence of the evolutionary pathways through which predation by adult parasitoids might have developed. In the Ichneumonoidea, predation by adults is commonly encountered in the form of host feeding. The habit appears more widespread among the Ichneumonidae than among other Parasitica, and the adults of some species completely consume their hosts.

Host-feeding patterns in Chalcidoidea are not consistent. In some genera of Pteromalidae, Eulophidae, and Aphelinidae it is the rule. Host feeding is prominent in encyrtid genera such as *Metaphycus* and *Microterys,* but it is conspicuously absent in some species of these genera. Host feeding appears sporadic within the Eupelmidae and Eurytomidae, and appears infrequently in the Trichogrammatidae. Predation by adult Cynipoidea is poorly represented, and seems best expressed by Figitidae. Predation by adult Proctotrupoidea has been noted in Scelionidae. Although a few cases are known where specific stages of hosts are preferred, there usually is less specificity shown in host feeding than in ovipositional attack (Bartlett, 1964).

Host feeding has direct effects on the host population, and some parasitoids kill a significant number of hosts by feeding directly on them. Host feeding by parasitoids is often associated with mutilation of the host. This is detrimental to parasitoid reproduction because associated species are affected (Flanders, 1951). Early workers felt that host feeding may work against the regulation of host densities by destroying hosts inhabited by immature parasitoids (Hartley, 1922; Flanders, 1953b). However, host destruction through host feeding is important in natural mortality of synanthropic Diptera (Legner, 1979b). Within the Parasitica, host feeding is not necessary to obtain amino acids because they are found in honeydew and plant nectaries (Zoebelein, 1956a, 1956b). Nevertheless, diet in general is influential on fecundity (Bracken, 1969).

The effects of host feeding on host regulation also have been considered (Flanders, 1953b). Parasitism is more effective than predation at low population densities. Protein requirements of the parasitoid are minimal. Eggs produced by a parasitoid, but not deposited, are absorbed and the egg material is used to prolong life (Flanders, 1950, 1953b). Higher host populations are needed to maintain host-feeding species. However, host-feeding habits of adult parasitoids appear valuable in reducing high host populations and may be an advantage in periodic inundative releases.

Host feeding affects oogenesis and ovisorption. When deprived of hosts, many Parasitica resorb the mature eggs (Flanders, 1935b; Grosch, 1948; Grosch *et al.,* 1955). Experimental studies with starvation and diets of honey and host blood have demonstrated subtle differences in longevity, oogenesis, and egg resorption (Edwards, 1954b; Gerling & Legner, 1968).

Diapause in Parasitoids

Diapause is a state of arrested growth or reproduction of many hibernating or estivating arthropods. Diapause differs from quiescence. For instance, eggs of some parasitoids deposited in hosts hatch, but the parasitoid larva do not develop further until the host is about to form the puparium. Sometimes this behavior is regarded as a form of quiescence instead of diapause. Borderline cases occur, but physiological mechanisms can be recognized in diapausing insects that are absent from quiescent insects (Tauber & Tauber, 1976). The arrest is facultative (environmental stimuli direct the organism to continue or to terminate development) or the arrest is obligatory. In facultative and obligatory diapause, control of development is exercised by the endocrine system (Beck, 1968).

The causes of diapause in the Parasitica are not simple. Individuals may enter diapause at a time when the environment is favorable to continuous development and increase of the species. Older females can decrease the percentage of their progeny entering diapause. Females in which development had passed through diapause produce a much lower incidence of diapause in their progeny. Prolonging an adult female's life by change in diet causes an increase in progeny diapause. Diapause incidence can increase if unnatural hosts are used as larval food. The physiological state of the parent female before and during oviposition can influence the proportion of her progenies that enter diapause (Flanders, 1944b, 1972; Simmonds, 1946, 1947b, 1948; Schneidermann & Horwitz, 1958; Saunders, 1962, 1965, 1966a, 1966b).

The effects of temperature on diapause are variable (Saunders, 1967) and temperature may affect the induction of diapause through photoperiodic influences (Sullivan & Wallace, 1967). High temperatures tend to avert diapause in long-day species; low temperatures may avert diapause in some cases. Temperatures are important in determining whether photoperiod can act. Temperature and photoperiod may act differently on different developmental stages to cause diapause (Eskafi & Legner, 1974).

Parasitoid development can depend on the physiological state of the host, or parasitoids can have their own photoperiodic responses. Pupal diapause can be determined by the photoperiod applied during the larval stage (DeWilde, 1962; Danilevskii, 1965). Photoperiod can induce diapause in the parasitoid while the host remains active (Geyspitz &

Kyao, 1953; Maslennikova, 1958). Some photoperiodic responses of the host and parasitoid are difficult to separate. Some endoparasitoids diapause only when in hosts that are themselves in diapause (Doutt, 1959).

Photoperiod is an important factor in intraspecific geographic differentiation and in shaping evolution. Photoperiodic response in local populations of an insect species may differ according to the latitude at which the species occurs without being accompanied by distinguishing morphological features. These populations may differ in the intensity of response, in the effect of temperature on the response, and in the critical photoperiod. At higher latitudes populations tend to become univoltine and to exhibit obligatory diapause. Moreover, photoperiod-induced diapause tends to be more intense in populations of high latitudes. Two seasonal forms exist, the long day and the short day. Seasons exert their influence according to the particular form.

Doutt (1959) felt that the intervention of diapause in some stage of the life of a parasitoid may be essential for synchronization of development between host and parasitoid. Flanders (1944b) considered diapause in parasitoids as adaptive in that it delays development until the host attains the stage presumably most suitable for the nutritional requirements of the parasitoid. Simmonds (1946, 1947b, 1948), however, attributed diapause to a physiological maladjustment during development. He did not feel diapause was adaptive to enable a species to survive a period unfavorable to further growth. Andrewartha (1952) and Hodek (1965) found that diapause in the adult stage may take the form of a failure to mature eggs or sperm and may be manifested by an extended preovipositional period. The preoviposition period of the adult could be prolonged by subjecting feeding larvae to lower temperatures. Walker and Pimentel (1966) have found a correlation between longevity and diapause.

Genetics

Electrophoretic techniques have revolutionized genetic studies of parasitic Hymenoptera. Electrophoresis has established low levels of variability in cultured stock compared with recently collected wild cultures (Legner, 1979a; Unruh *et al.,* 1983). The stock used to start colonies of some parasitoids possessed only a fraction of the gene pool of the parental wild populations (Unruh *et al.,* 1983). However, Hymenoptera generally show lower genetic variability than do other insect orders (Crozier, 1971, 1975; Metcalf *et al.,* 1975). Small laboratory colonies lose large amounts of behavioral variability after a few generations in culture (Legner, 1979a; Unruh *et al.,* 1983).

Today we recognize that the aspects of parasitoid genetics are more complex than suspected. Studies on the pteromalid genus *Muscidifurax* show extranuclear phases in the inheritance of polygenes for quantitative reproductive traits

(Legner, 1989a). Males of this wasp can change the female's oviposition phenotype upon mating, by transferring an unknown substance capable of regulating such change. This change is passed to the offspring. It appears as if a proportion of the genes in the female are phenotypically plastic and can change expression under the influence of substances in the male seminal fluid. The intensity of this response is different depending on the genetic composition of both the male and the female. Full expression occurs in the F_1 female (Legner, 1988a, 1989a).

Behavioral changes after mating are permanent as revealed by increases or decreases in the response maintained for 16 days following a particular mating (Legner, 1987b, 1989b). There is no change back to the original behavior following a second mating with a male of the opposite genotype (Legner, 1989a). Speculations on the nature of the substance in the male seminal fluid have included microorganisms, accessory gland fluids, and such behavior-modifying chemicals as prostaglandins (Legner, 1987b, 1988a). Because inheritance of such genes seems to occur in a stepwise manner, the entire process has been termed "accretive inheritance" (Legner, 1988b, 1989a).

The inheritance scheme in the pteromalid *Muscidifurax* is fundamentally important to a wider understanding of genetics of Hymenoptera. The kinds of genes capable of phenotypic alteration in mated females before being inherited by their progenies have been termed "wary genes" because they, or precursors of them, are tested in the environment in an attenuated manner before inheritance by the offspring (chromosomal inheritance) (Legner, 1985b, 1989a). We do not know whether such genes possess chemical precursors capable of changing the female's phenotype or are inherited extranuclearly after mating.

The ability of the male substance to switch a polygene locus "on" or "off" in the mother suggests that a locus may exist in two states: active and inactive. In *Muscidifurax* at least eight loci are thought to be active in determining multiple oviposition behavior (Legner, 1991).

In the process of hybridization, wary genes may serve to quicken the pace of evolution by allowing natural selection for both nonlethal undesirable and desirable characteristics to act in the parental generation. Wary genes that are detrimental to the hybrid population thus might be more prone to elimination and beneficial ones may be expressed in the mother before the appearance of her active progeny (Legner, 1987b, 1988a). If wary genes occur more generally in Hymenoptera, their presence might partially account for the rapid evolution thought to have occurred in certain groups (Hartl, 1972; Gordh, 1975, 1979).

Selective Breeding

Increasing the average number of progenies through selective breeding is seen as a possible benefit to biological control. Studies of Parasitica have been conducted with this objective in mind (Wilkes, 1942, 1947; Allen, 1954). Several generations are required for the desired effects, which may include increased fecundity as well as reduced larval mortality. Hybridization techniques may be useful for increasing the fecundity and the host destructive capacity of parasitic insects (Legner, 1972, 1988b). However, the choice of parental race is very important for success (Legner, 1988b), and should probably be restricted to cohorts from similar climatic zones, because negative heterosis could result (Legner, 1972).

Insemination by old males of some chalcidoids is not always successful and females inseminated by them produce fewer female progenies (Wilkes, 1963; Gordh & DeBach, 1976). In some Parasitica the female may be less fecund after copulation, possibly because she exercises greater discrimination in host selection. Some Parasitica oviposit as readily before copulation as they do afterward. Sex ratios in these species are determined partly by the amount of oviposition prior to mating. In aphelinids where oviposition instincts are permanently changed by copulation, male production is obligatory before copulation and is facultative afterward. Insemination has a profound and irreversible effect on behavior of some Parasitica, and heritable traits for fecundity and other reproductive behavior are expressed in the mother immediately after mating at an intensity dictated by the male's genome through an extranuclear phase of inheritance (Legner, 1987b, 1988a).

ECOLOGY OF PARASITIC HYMENOPTERA

The interaction of parasitoid and host is behaviorally complex and ecologically multifaceted. Laing (1937) proposed three steps in the attack activity of a parasitoid: (1) attraction to the host habitat, (2) attraction to host individuals in the habitat, and (3) acceptance or rejection of the host. Flanders (1937a) concurrently proposed a fourth step in the process. His steps were (1) host habitat finding, (2) host finding, (3) host acceptance, and (4) host suitability. A fifth component, host regulation, is important from a biological control standpoint (Vinson, 1975b). Restatements of the procedures in host selection include those by Salt (1935, 1958), Thorpe and Caudle (1938), Doutt (1964), Hodek (1966), Weseloh (1981), and Vinson (1976, 1984).

Habitat Location

Davis (1896) first observed, without indicating the cause, that some plants such as *Nicotiana, Pelargonium, Datura,* and *Eucalyptus* were repellent to the aphelinid *Encarsia formosa* Gahan, a whitefly parasitoid. Later, Picard and Rabaud (1914) observed that many parasitic Hymenoptera attack species in several orders, provided the hosts feed on one species of food plant. Thus, natural enemies are often attracted to a habitat instead of directly to a host in the habitat. Recognizing this concept is important to biological control (Kulman & Hodson, 1961; Hagen *et al.,* 1970). Numerous environmental factors affect parasitoids to bring about this attraction. These include leaf structure (Milliron, 1940; Thompson, 1951; Putnam & Herne, 1966; Downing & Moilliet, 1967), bud size (Graham & Baumhoffer, 1927; Arthur, 1962), plant variety (Franklin & Holdaway, 1966), host-plant odor (Altieri *et al.,* 1982; Vet & Bakker, 1985; Vet & van Alphen, 1985), plant infestation (Flanders, 1940), soil type (Fleschner & Scriven, 1957), leaf exudations (Rabb & Bradley, 1968), honeydew (Milliron, 1940), fungal odor (Madden, 1968; Greany *et al.,* 1977) odor of host food (Laing, 1937; Thorpe & Jones, 1937; Thorpe & Caudle, 1938; Edwards, 1954a), excreta of the host insect (Flanders, 1935a), host pupation depths (Ullyett, 1949), visual stimuli of the plant (Chandler, 1966, 1967), height of host location (McLeod, 1951; Thorpe, 1985), repellent effects of the plant (Davis, 1896; Speyer, 1929), host insect dioecy and habitat (Stary, 1964), and presence of alternative hosts (Eikenbary & Rogers, 1973). Food of a host may positively or negatively affect its parasitoids (Simmonds, 1944; Smith, 1957; Lawson, 1959; Hodek, 1966).

Host Location

After the host habitat is located, hosts are located by random and directed search (Ullyett, 1943, 1947; Doutt, 1959). Olfaction is widely used by parasitoids in locating hosts (Bouchard & Cloutier, 1985; Dicke *et al.,* 1985; van Alphen & Vet, 1986) and this behavior may be acquired (Vet, 1983, 1985). Parasitoids seem more attracted to higher densities of host and to certain patterns of host distribution (Legner, 1967b, 1969b). Host trail odors may facilitate searching (Price, 1970). Other olfactory stimuli exist (Vet & Bakker, 1985; Vet & van Alphen, 1985), and some physical host characteristics affect host selection (Weseloh, 1969, 1971a, 1971b, 1972; Weseloh & Bartlett; 1971; Wilson *et al.,* 1974).

Adding kairomones to a habitat can enable some parasitoids to locate their hosts more efficiently [Jones *et al.,* 1971; Gross *et al.,* 1975; Altieri *et al.,* 1982; cf. Gardner & van Lenteren (1986)]. For example, *Trichogramma* spp. respond to chemical extracts of host moth scales and some braconids respond to extracts of host larval frass (Jones *et al.,* 1973; Beevers *et al.,* 1981; Noldus & van Lenteren, 1983, 1985). Synthesis of kairomones may enhance biological control on a larger scale (Vinson, 1968, 1975a, 1975b, 1976; Lewis *et al.,* 1971, 1972; Weseloh, 1974). In some cases, however, kairomones may function to confuse parasitoids into lesser efficiency (DeBach, 1944; Chiri & Legner, 1982, 1983).

Host Acceptance

After the parasitoid has located the host, decisions must be made by the female parasitoid as to whether the host is acceptable for oviposition. Numerous cues suggest the host may be suitable for attack, including host odor, host size, host location, shape and motion, and host physiology (Bryden & Bishop, 1945; Arthur *et al.,* 1969; Legner & Olton, 1969; Legner & Thompson, 1977). Salt (1935) termed host acceptance a "psychological selection." Huffaker [in Doutt (1959)] suggested it be called "ethological selection." A few species of parasitoids are host specific, but most attack several hosts in nature. However, no parasitoid is indiscriminate. Under natural conditions, a parasitoid will attack only a few of the species on which development is possible.

Superparasitism is widespread in the Parasitica. Host discrimination (the ability to identify parasitized hosts) is intimately associated with superparasitism. The relationship between discrimination and superparasitism has been considered (van Lenteren *et al.,* 1978; Bakker *et. al.,* 1985) and many parasitoids discriminate parasitized hosts and avoid superparasitism. Flanders (1951) indicated that a "spoor effect" may be present. Chemoreceptors on the ovipositor that mediate female oviposition behavior are probably widespread in the Parasitica (Narayanan & Chaudhuri, 1955; Wylie, 1958; Greany & Oatman, 1972; Hawke *et al.,* 1973; Domenichini, 1978; Van Veen & Van Wijk, 1985). An aversion to superparasitism is expressed by many parasitoids ready to oviposit. Biologists interpret this aversion as a mechanism to conserve eggs and increase efficiency in searching for hosts. From the viewpoint of applied biological control, superparasitism is important and mathematical models have been developed to address parasite distribution and the avoidance of superparasitism (Bakker *et al.,* 1972; Rogers, 1975; Griffiths, 1977; Narendran, 1985).

The ability to recognize parasitized hosts and selectively oviposit upon them must be important for hyperparasites. This is an aspect of host selection that has not been studied. Indeed, discrimination by hyperparasites must be complex because hyperparasitism has been derived independently several times in many distantly related lineages (Gordh, 1981). Beddington and Hammond (1977) developed a mathematical model for this system to analyze the implications of hyperparasitism for biological control.

Tertiary hyperparasitism is a condition in which hyperparasites attack one another. Tertiary hyperparasitism has been divided into interspecific tertiary hyperparasitism (allohyperparasitism) (Sullivan, 1972; Matejko & Sullivan, 1984) and intraspecific tertiary hyperparasitism (autohyperparasitism) (Bennett & Sullivan, 1978; Levine & Sullivan, 1983). Biologically, tertiary hyperparasitism is a scheme of development that has rarely been documented in the field. It seems to arise from intensive competition, particularly in systems in which primary and secondary parasitoids are adelphoparasitic (Zinna, 1962). Its significance may be restricted to the laboratory, or to unusual circumstances best exemplified in aphid or scale parasites.

Parasitoid Attack and Host Defenses

Hosts do not necessarily represent passive resources placed at the disposal of the parasitoids (Doutt, 1963). At the individual level, physical interaction of parasitoid and host can be dynamic. A host may ward off a parasitoid externally before oviposition has occurred (Hinton, 1955; Cole, 1957, 1959), or may overcome the parasitoid internally after oviposition (Salt, 1935, 1955a, 1955b, 1957, 1960, 1963a, 1963b, 1965, 1966, 1967, 1968; Walker, 1959; Hadorn & Walker, 1960; Salt & van de Bosch, 1967; Streams, 1968). Internal defense reactions involve cellular (encapsulation and melanization) and humoral phenomena (D. Griffiths, 1960, 1961; K. Griffiths, 1969). Encapsulation is the formation of a cyst by host cells around foreign objects. Encapsulation occurs in epidermal, tracheal, gut, muscle, and nervous tissue. Hemolymphic capsules are formed by the hemocytes congregating and differentiating into two layers. Cells of the inner layer form connective tissue fibers. Opinions differ concerning whether these inner cells form a true syncytium and whether connective fibers are formed directly from the cell cytoplasm or are secreted by them (Schneider, 1950, 1952; Muldrew, 1953; Griffiths, 1960; Petersen, 1962; van den Bosch, 1964; Nappi & Streams, 1970).

Melanization reactions involve the deposition of pigment around a parasitoid. It is thought that melanization is associated with encapsulation and is essentially a cellular phenomenon. The melanin formed is derived from tyrosine by way of phenolase reactions. Apparently, the substrate and enzyme are physically separated within certain blood cells normally, but injury causes the reaction to proceed.

Humoral reactions are poorly understood and two viewpoints have been expressed concerning the role of encapsulation in defense. One view holds that hemocytes play a primary role in causing the death of living eggs and larvae of parasitoids. The other view argues that humoral phenomena cause immunity, and hemocytes merely act as scavengers. The first viewpoint is favored because living parasitoids have been found encapsulated.

Host defense responses vary with the species of the host and parasitoid involved. In general, different hosts utilize different defense mechanisms against the same parasitoid, and different parasitoids cause similar defense reactions in the same host. Temperature, superparasitism, and multiple parasitism also affect immune responses. Parasitoids also may simply starve to death in a host that is unsuitable for its development (Flanders, 1937b)

Host Suitability

Oviposition by a parasitoid is not necessarily an index of host suitability. The attractiveness of the host to the adult parasitoid is often independent of its suitability for larval parasitoid development. Muldrew (1953) suggested that a once-susceptible host population (that probably contained a few resistant individuals) may become totally resistant to parasitoid attack. Alternatively, some host-plant species may provide a superior nutritional source for host insects that results in higher survivorship by parasitoids (Mueller, 1983). In contrast, the host plant may confer on the host insect an immunity to parasitization (E. L. Smith, 1949; Flanders, 1953a; J. Smith, 1957).

In a slightly different context, there is widespread observation and very recent unpublished experimental evidence that citrus trees that have received treatments of DDT or other insecticides change their nutritional value to favor pest insect species. Scale insects, for example, are stimulated to reproduce and grow at a faster rate. The effect of chemical pesticides and fertilizers on plants and their phytophages (insects, mites) was called "trophobiosis" by Chaboussou (1966, 1971). Parasitoids are also eliminated by the treatments so that host increase is unchecked for some time following a treatment. The so-called "DDT check method" (DeBach, 1955) to exclude the activity of natural enemies, therefore, may give distorted information on the actual value of the parasitoids and predators eliminated, because the host plants are artificially stimulated.

Partial resistance of the plant to a pest may render the parasites of the pest more effective as controlling agents (van Emden & Wearing, 1965; van Emden, 1986).

SUMMARY

At least 65,000 species of Hymenoptera are parasitoids, but only 765 species have been used in biological control programs worldwide. In this chapter we have reviewed the various major taxonomic groups and the unique adaptations (morphological, physiological, behavioral, and ecological) that make it possible for them to exploit a diversity of hosts in diverse environments. Among the other taxa that include parasitoids—Diptera, Coleoptera, Strepsiptera, and Lepidoptera—the Hymenoptera stand alone in the numerous species that are parasitoids. Their plasticity as a group is exemplified by their different mode of embryonic development, by interactions (trophic and otherwise) with their hosts, by the sex-determining mechanism, and by the intricate diversity of host selection processes. These and related topics are summarily reviewed based on a large list of publications that is by no means exhaustive.

References

Allen, H. W. (1954). Propagation of *Horogenes molestae,* an Asiatic parasite of the oriental fruit moth, on potato tuberworm. Journal of Economic Entomology, 47, 278–281.

Altieri, M. A., Annamalai, S., Katiyai, K. P., & Flath, R. A. (1982). Effects of plant extracts on the rates of parasitization of *Anagasta kuehniella* (Lep.: Pyralidae) eggs by *Trichogramma pretiosum* (Hym.: Trichogrammatidae) under greenhouse conditions. Entomophaga, 27, 431–438.

Anderson, R. L. (1935). Offspring obtained from females reared at different temperatures in *Habrobracon.* American Naturalist, 69, 183–187.

Andrewartha, H. G. (1952). Diapause in relation to the ecology of insects. Biological Review of Cambridge Philosophical Society, 27, 50–107.

Arthur, A. P. (1962). Influence of host tree on abundance of *Itoplectis conquisitor* (Say) (Hymenoptera: Ichneumonidae), a polyphagous parasite of the European pine shoot moth, *Ryacionia buoliana* (Schiff.) (Lepidoptera: Olethreutidae). Canadian Entomologist, 94, 337–347.

Arthur, A. P., Hedgekak, B. M., & Rollins, L. (1969). Component of the host haemolymph that induces oviposition in a parasitic insect. Nature (London), 223, 966–967.

Arthur, A. P., Stainer, R., & Turnbull, A. L. (1964). The interaction between *Orgilus obscurator* (Nees) and *Temelucha interruptor* (Grav.) (Hymenoptera: Ichneumonidae), parasites of the pine shoot moth *Rhyacionia buoliana* (Schiff.) (Lepidoptera: Olethreutidae). Canadian Entomologist, 96, 1030–1034.

Askew, R. R. (1971). Parasitic insects. New York: American Elsevier.

Askew, R. R., & Shaw, M. R. (1986). Parasitoid communities: Their size, structure and development. In J. Waage & D. Greathead (Eds.), Insect parasitoids (pp. 225–264). London: Academic Press.

Baerends, G. P., & van Roon, J. M. (1950). Embryological and ecological investigations on the development of the egg of *Ammophila campestris* Jur. Tijdschrift voor Entomologie, 92, 53–112.

Bakker, K., Eijsackers, H. J. P., van Lenteren, J. C., & Meelis, E. (1972). Some models describing the distribution of eggs of the parasite *Pseudeucoila bocheii* (Hymenoptera: Cynipidae) over its hosts, larvae of *Drosophila melanogaster.* Oecologia (Berlin), 10, 29–57.

Bakker, K., van Alphen, J. J. M., van Batenburg, F. H. D., van der Hoeven, N., Nell, H. W., & van Strien-van Liempt, W. T. F. H. (1985). The function of host discrimination and superparasitization in parasitoids. Oecologia (Berlin), 67, 572–576.

Balinsky, B. I. (1970). An introduction to embryology. Philadelphia: Saunders.

Bartlett, B. R. (1964). Patterns in the host-feeding habits of adult parasitic Hymenoptera. Annals of the Entomological Society of America, 57, 344–350.

Beck, S. D. (1968). Insect photoperiodism. London: Academic Press.

Beddington, J. R., & Hammond, P. S. (1977). On the dynamics of host-parasite-hyperparasite interactions. Journal of Animal Ecology, 46, 811–821.

Beevers, M., Lewis, W. J., Gross, H. R., Jr., & Nordlund, D. A. (1981). Kairomones and their use for management of entomophagous insects. X. Laboratory studies on manipulation of host-finding behavior of *Trichogramma pretiosum* Riley with a kairomone extracted from *Heliothis zea* (Boddie) moth scales. Journal of Chemical Ecology, 7, 635–648.

Bennett, A. W., & Sullivan, D. J. (1978). Defensive behavior against tertiary parasitism by the larva of *Dendrocerus carpenteri,* an aphid hyperparasitoid. Journal of the New York Entomological Society, 86, 153–160.

Benoit, P. L. G. (1984). Aulacidae, famille nouvelle pour la faune de l'Afrique tropicale (Hymenoptera). Rev. Zool. Afr, 98, 799–803.

Bischoff, H. (1927). Biologie der Hymenopteren. Julius Springer, Berlin. 598 pp.

Bohart, R. M., & Kimsey, L. S. (1982). A synopsis of the Chrysididae in America North of Mexico. Memoirs of American Entomological Institute, 33, 1–266.

Boucek, Z. (1952). The first revision of the European species of the family Chalcididae (Hymenoptera). Acta Entomologica Musei Nationalis Pragae, 27, 1–108.

Boucek, Z. (1974). On some Chalcididae and Pteromalidae (Hymenoptera), with descriptions of new genera and species from Africa and one species from Asia. Journal of the Entomological Society of South Africa, 37, 327–343.

Boucek, Z., & Askew, R. R. (1968). Hymenoptera, Chalcidoidea, Palearctic Eulophidae (Excl. Tetrastichinae). Index of entomophagous insects. Paris: Le Francois.

Bouchard, Y., & Cloutier, C. (1985). Role of olfaction in host finding by aphid parasitoid *Aphidius nigripes* (Hymenoptera: Aphidiidae). Journal of Chemical Ecology, 11, 801–808.

Bouletreau, M., & David, J. (1967). Influence of the density of the larval population on the size of adults, duration of development and frequency of diapause in *Pteromalus puparum* L. (Hymenoptera, Chalcididae). Entomophaga, 12, 187–197.

Bowen, W. R., & Stern, V. M. (1966). Effect of temperature on the production of males and sexual mosaics in a uniparental race of *Trichogramma semifumatum* (Hymenoptera: Trichogrammatidae). Annals of the Entomological Society of America, 59, 823–834.

Bracken, G. K. (1969). Effects of dietary amino acids, salts and protein starvation on fecundity of the parasitoid *Exeristes comstockii* (Hymenoptera: Ichneumonidae). Canadian Entomologist, 101, 92–96.

Bronskill, J. F. (1959). Embryology of *Pimpla turionellae* (L.) (Hymenoptera: Ichneumonidae). Canadian Journal of Zoology, 37, 655–688.

Bronskill, J. F. (1960). The capsule and its relation to the embryogenesis of the ichneumonid parasitoid *Mesoleius tenthredinis* Morl. in the larch sawfly, *Pristiphora erichsonii* (Htg.) (Hymenoptera: Tenthredinidae). Canadian Journal of Zoology, 38, 769–775.

Broodryk, S. W., & Doutt, R. L. (1966). The biology of *Coccophagoides utilis* Doutt (Hymenoptera: Aphelinidae). Hilgardia, 37, 233–254.

Brown, V. K. (1973). The biology and development of *Brachygaster minutus* Olivier (Hymenoptera: Evaniidae), a parasite of the oothecae of *Ectobius* spp. (Dictyoptera: Blattidae). Journal of Natural History, 7, 665–674.

Bryden, J. W., & Bishop, M. W. H. (1945). *Perilitus coccinellae* (Hym., Braconidae) in Cambridgeshire. Entomological Monographs, 81, 51–52.

Burks, B. D. (1940). Revision of the chalcid flies of the tribe Chalcidini in America north of Mexico. Proceedings of the United States National Museum, 88, 237–354.

Burks, B. D. (1979a). Mymaridae. In K. V. Krombein *et al.* (Eds.), Catalog of Hymenoptera in America north of Mexico (Vol. 1 pp. 1022–1033). Washington DC: Smithsonian Institute Press.

Burks, B. D. (1979b). Trichogrammatidae. In K. V. Krombein *et al.* (Eds.), Catalog of Hymenoptera in America north of Mexico (Vol. 1, pp. 1033–1043). Washington, DC: Smithsonian Institute Press.

Burnett, T. (1949). The effect of temperature on an insect host-parasite population. Ecology, 30, 113–134.

Burnett, T. (1951). Effects of temperature and host density on the rate of increase of an insect parasite. American Naturalist, 85, 337–352.

Burnett, T. (1953). Effect of temperature and parasite density on the rate of increase of an insect parasite. Ecology, 34, 322–328.

Burnett, T. (1956). Effects of natural temperature on oviposition of various numbers of an insect parasite (Hymenoptera, Chalcididae, Tenthredinidae). Annals of the Entomological Society of America, 49, 55–59.

Burnett, T. (1958a). Effect of host distribution on the reproduction of *Encarsia formosa* Gahan (Hymenoptera: Chalcidoidea). Canadian Entomologist, 90, 179–191.

Burnett, T. (1958b). Effect of area of search on reproduction of *Encarsia formosa* Gahan (Hymenoptera: Chalcidoidea). Canadian Entomologist, 90, 225–229.

Burnett, T. (1967). Aspects of the interaction between a chalcid parasite and its aleurodid host. Canadian Journal of Zoology, 45, 539–578.

Caltagirone, L. E. (1964). Notes on the biology, parasites and inquilines of *Pontania pacifica* (Hymenoptera: Tenthredinidae), a leaf-gall incitant on *Salix lasiolepis*. Annals of the Entomological Society of America, 57, 279–291.

Caullery, M. (1950). Le parasitisme et la symbiose. Paris: Doin, edit.

Chaboussou, F. (1966). Nouveaux aspects de la phytiatrie et de la phytopharmacie. Le phenomene de la trophobiose. Proceedings FAO Symposium of Integrated Pest Control, October 11–15, 1965, Rome, 1, 33–61.

Chaboussou, F. (1971). Le conditionnement physiologique de la vigne et la multiplication des Cicadelles. Revue de Zoologie Agricole et de Pathologie Vegetable, 70(3), 1–66.

Chandler, A. E. F. (1966). Some aspects of host plant selection in aphidophagous Syrphidae. In Proceedings, Symposium on Ecology of Aphidophagous Insects, Liblice 1965 (pp. 113–115). Prague: Academia.

Chandler, A. E. F. (1967). Oviposition responses by aphidophagous Syrphidae (Diptera). Nature (London), 213, 736.

Charnov, E. L. (1979). The genetical evolution of patterns of sexuality. Darwinian fitness. American Naturalist, 113, 465–480.

Charnov, E. L., Los-Den Hartugh, R. L., Jones, T., & van den Assem, J. (1981). Sex ratio evolution in a variable environment. Nature (London), 289, 27–33.

Chiri, A. A., & Legner, E. F. (1982). Host-searching kairomones alter behavior of *Chelonus* sp. nr. *curvimaculatus,* a hymenopterous parasite of the pink bollworm, *Pectinophora gossypiella* (Saunders). Environmental Entomology, 11, 452–455.

Chiri, A. A., & Legner, E. F. (1983). Field applications of host-searching kairomones to enhance parasitization of the pink bollworm (Lepidoptera: Gelechiidae). Journal of Economic Entomology, 76, 254–255.

Chrystal, R. N. (1930). Studies of the *Sirex* parasites. The biology and post-embryonic development of *Ibalia leucospoides* Hochenw. (Hymenoptera–Cynipoidea). Oxford Forestry Memoirs, 11, 1–63.

Clausen, C. P. (1929). (1940). Entomophagous insects. New York: McGraw-Hill.

Clausen, C. P. (1950). Respiratory adaptations in the immature stages of parasitic insects. Arthropoda, 1, 197–224.

Clausen, C. P. (1976). Phoresy among entomophagous insects. Annual Review of Entomology, 21, 343–368.

Clausen, C. P. (Ed.). (1978). Introduced parasites and predators of arthropod pests and weeds: A world review. Agriculture Handbook No. 480. Washington, DC: U.S. Department of Agriculture.

Cole, L. R. (1957). The biology of four species of Ichneumonidae parasitic on *Tortrix viridana* L. Proceedings of the Royal Entomological Society of London, Series C, 22 48–49.

Cole, L. R. (1959). On the defenses of lepidopterous pupae in relation to the oviposition behavior of certain Ichneumonidae. Journal of Lepidopterists' Society, 13, 1–10.

Cook, R. M., & Hubbard, S. F. (1977). Adaptive searching strategies in insect parasites. Journal of Animal Ecology, 46, 115–125.

Cooper, K. W. (1954). Biology of eumenine wasps. IV. A trigonalid wasp parasitic on *Rygchium rugosum* (Saussure). Proceedings of the Entomological Society of Washington, 56, 280–288.

Copeland, M. J. W. (1976). Female reproductive system of the Aphelinidae (Hymenoptera: Chalcidoidea). International Journal of Insect Morphology and Embryology, 5, 151–166.

Copeland, J. J. W., & King, P. E. (1971). The structure and possible function of the reproductive system in some Eulophidae and Tetracampidae. Entomologist, 104, 4–28.

Copeland, J. J. W., & King, P. E. (1972a). The structure of the female

reproductive system in the Pteromalidae (Chalcidoidea: Hymenoptera). Entomologist, 105, 77–96.

Copeland, J. J. W., & King, P. E. (1972b). The structure of the female reproductive system in the Torymidae (Hymenoptera: Chalcidoidea). Transactions of the Royal Entomological Society of London, 124, 191–212.

Copeland, J. J. W., & King, P. E. (1972c). The structure of the female reproductive system in the Eurytomidae (Chalcidoidea: Hymenoptera). Journal of Zoology, 166, 185–212.

Counce, S. J. (1961). The analysis of insect embryogenesis. Annual Review of Entomology, 6, 295–312.

Crandell, H. A. (1939). The biology of *Pachycrepoideus dubius* Ashmead (Hymenoptera), a pteromalid parasite of *Piophila casei* Linne (Diptera). Annals of the Entomological Society of America, 32, 632–654.

Crew, F. A. E. (1965). Sex determination. New York: Dover.

Crosskey, R. W. (1951). The morphology, taxonomy, and biology of the Evanioidea (Hymenoptera). Transactions of the Royal Entomological Society of London, 102, 247–301.

Crozier, R. H. (1971). Heterozygosity and sex determination in haplodiploidy. American Naturalist, 105, 399–412.

Crozier, R. H. (1975). Animal cytogenetics, 3. Insecta, 7. Hymenoptera. Berlin: Gebruder Borntraeger.

Crozier, R. H. (1977). Evolutionary genetics of the Hymenoptera. Annual Review of Entomology, 22, 263–288.

Cruz, Y. P. (1981). A sterile defender morph in a polyembryonic hymenopterous parasite. Nature (London), 294, 446–447.

Cruz, Y. P. (1986a). The defender role of the precocious larvae of *Copidosomopsis tanytmema* Caltagirone (Encyrtidae, Hymenoptera). Journal of Experimental Zoology, 237, 309–318.

Cruz, Y. P. (1986b). Development of the polyembryonic parasite *Copidosomopsis tanytmema* (Hymenoptera: Encyrtidae). Annals of the Entomological Society of America, 79, 121–127.

Daniel, D. M. (1932). Macrocentrus ancylivorus Rohwer, a polyembryonic braconid parasite of the oriental fruit moth (Tech. Bulletin 187). New York Agricultural Experimental Station.

Danilevskii, A. S. (1965). Photoperiodism and seasonal development of insects. Edinburgh: Oliver & Boyd (original in Russian, 1961).

Davis, G. C. (1896). Pests of house and ornamental plants. Michigan Agricultural Experimental Station Bulletin, 2, 3–45.

DeBach, P. (1943a). The effect of low storage temperature on reproduction in certain parasitic Hymenoptera. Pan-Pacific Entomologist, 19, 112–119.

Debach, P. (1943b). The importance of host-feeding by adult parasites in the reduction of host populations. Journal of Economic Entomology, 36, 647–658.

DeBach, P. (1944). Environmental contamination by an insect parasite and the effect on host selection. Annals of the Entomological Society of America, 37, 70–74.

DeBach, P. (1955). Validity of the insecticidal check method as a measure of the effectiveness of natural enemies of diaspine scale insects. Journal of Economic Entomology, 44, 373–383.

DeBach, P., & Argyriou, L. C. (1966). Effects of short duration suboptimum pre-oviposition temperatures on progeny production and sex ratio in species of *Aphytis* (Hymenoptera: Aphelinidae). Researches on Population Ecology, 8, 69–77.

DeBach, P., & Rao, S. V. (1968). Transformation of inseminated females of *Aphytis lingnanensis* into factitious virgins by low-temperature treatment. Annals of the Entomological Society of America, 61, 332–337.

DeBach, P., & Schlinger, E. (Eds.). (1964). Biological control of insect pests and weeds. London: Chapman & Hall.

DeBach, P., & Smith, H. S. (1941). The effect of host density on the rate of reproduction of entomophagous parasites. Journal of Economic Entomology, 34, 741–745.

DeBach, P., & Smith, H. S. (1947). Effects of parasite population density on rate of change of host and parasite populations. Ecology, 28, 290–298.

Dessart, P. (1986). Tableau dichotomique des genres de Ceraphronoidea (Hymenoptera) avec commentaires et noucelles especes. Frustula Entomologica, 7–8, 307–372.

DeWilde, J. (1962). Photoperiodism in insects and mites. Annual Review of Entomology, 7, 1–26.

DeWilde, J. (1964). Reproduction. In M. Rockstein (Ed.), The physiology of Insecta (pp. 10–58). London: Academic Press.

Dicke, M., van Lenteren, J. C., Boskamp, G. J. F., & van Voorst, R. (1985). Intensification and prolongation of host searching in Leptopilina heterotoma (Thomson) (Hymenoptera: Eucoilidae) through a kairomone produced by Drosophila melanogaster. Journal of Chemical Ecology, 2, 125–136.

Domenichini, G. (1978). Some structures of various Hymenoptera, Chalcidoidea and their functions. Bollettino di Zoologia Agraria e Bachicoltura, Ser. 2, 14, 29–93.

Doutt, R. L. (1947). Polyembryony in Copidosoma koehleri Blanchard. American Naturalist, 81, 435–453.

Doutt, R. L. (1952). The teratoid larva of polyembryonic Encyrtidae (Hymenoptera). Canadian Entomologist, 84, 247–250.

Doutt, R. L. (1959). The biology of parasitic Hymenoptera. Annual Review of Entomology, 4, 141–182.

Doutt, R. L. (1963). Pathologies caused by insect parasites. In E. A. Steinhaus (Ed.), Insect pathology: An advanced treatise (Vol. 2, pp. 393–422). New York: Academic Press.

Doutt, R. L. (1964). Biological characteristics of entomophagous adults. In P. DeBach & E. Schlinger (Eds.), Biological control of insect pests and weeds (2nd ed., p. 152) London: Chapman & Hall.

Doutt, R. L., & Viggiani, G. (1968). The classification of the Trichogrammatidae (Hymenoptera: Chalcidoidea). Proceedings of the California Academy of Science, (Ser. 4), 35, 447–586.

Downing, R. S., & Moilliet, I. K. (1967). Relative densities of predacious and phytophagous mites on three varieties of apple trees. Canadian Entomologist, 99, 738–741.

Edwards, R. L. (1954a). The host-finding and oviposition behaviour of Mormoniella vitripennis (Walker) (Hym.: Pteromalidae), a parasite of muscoid flies. Behaviour, 7, 88–112.

Edwards, R. L. (1954b). The effect of diet on egg maturation and resorption in Mormoniella vitripennis (Hymenoptera, Pteromalidae). Quarterly Review of Microbiological Science, 95, 459–468.

Eikenbary, R. D., & Rogers, C. E. (1973). Importance of alternate hosts in establishment of introduced parasites. Proceedings of the Tall Timbers Conference on the Ecology of Animal Control by Habitat Management, 5, 119–133.

Eskafi, F. M., & Legner, E. F. (1974). Fecundity, development and diapause in Hexacola sp. near websteri, a parasite of Hippelates eye gnats. Annals of the Entomological Society of America, 67, 769–771.

Evans, H. E. (1964). A synopsis of the American Bethylidae (Hymenoptera, Aculeata). Museum of Comparative Zoology Bulletin, 132, 1–222.

Evans, H. E. (1978). The Bethylidae of America north of Mexico. Memoirs of the American Entomology Institute, 27, 332 pp.

Finlayson, J. T., & Hagen, K. S. (1979). Final instar larvae of parasitic Hymenoptera (Pest Mangement Paper 10). Burnaby, B.C. Canada: Simon Fraser University.

Fisher, R. A. (1930). The genetical theory of natural selection. Oxford: Clarendon Press.

Fisher, R. C. (1971). Aspects of the physiology of endoparasitic Hymenoptera. Biological Review, 46, 243–278.

Flanders, S. E. (1935a). Effect of host density on parasitism. Journal of Economic Entomology, 28, 898–900.

Flanders, S. E. (1935b). An apparent correlation between the feeding habits of certain pteromalids and the condition of their ovarian follicles. Annals of the Entomological Society of America, 28, 438–444.

Flanders, S. E. (1936). Japanese species of Tetrastichus parasitic on eggs of Galerucella xanthomelaena. Journal of Economic Entomology, 29, 1024–1025.

Flanders, S. E. (1937a). Habitat selection by Trichogramma. Annals of the Entomological Society of America, 30, 208–210.

Flanders, S. E. (1937b). Starvation of developing parasites as an explanation of immunity. Journal of Economic Entomology, 30, 970–971.

Flanders, S. E. (1938). The effect of cold storage on the reproduction of parasitic Hymenoptera. Journal of Economic Entomology, 31, 633–634.

Flanders, S. E. (1939). Environmental control of sex in hymenopterous insects. Annals of Entomological Society of America, 32, 11–26.

Flanders, S. E. (1940). Environmental resistance to the establishment of parasitic Hymenoptera. Annals of the Entomological Society of America, 33, 245–253.

Flanders, S. E. (1942). Oosorption and ovulation in relation to oviposition in the parasitic Hymenoptera. Annals of the Entomological Society of America, 35, 251–266.

Flanders, S. E. (1943b). The role of mating in the reproduction of parasitic Hymenoptera. Journal Economic Entomology, 36, 802–803.

Flanders, S. E. (1944a). Observations on Prospaltella perniciosi and its mass production. Journal of Economic Entomology, 37, 105.

Flanders, S. E. (1944b). Diapause in parasitic Hymenoptera. Journal of Economic Entomology, 37, 408–411.

Flanders, S. E. (1945). Uniparentalism in the Hymenoptera and its relation to polyploidy. Science, 100, 168–169.

Flanders, S. E. (1946a). The mechanism of sex control of the honey bee. Journal of Economic Entomology, 39, 379–380.

Flanders, S. E. (1946b). Control of sex and sex-limited polymorphism in the Hymenoptera. Quarterly Review of Biology, 21, 135–143.

Flanders, S. E. (1947). Elements of host discovery exemplified by parasitic Hymenoptera. Ecology, 28, 299–309.

Flanders, S. E. (1950). Regulation of ovulation and egg disposal in the parasitic Hymenoptera. Canadian Entomologist, 82, 134–140.

Flanders, S. E. (1951). Mass culture of California red scale and its golden chalcid parasites. Hilgardia, 21, 1–42.

Flanders, S. E. (1953a). Variations in susceptibility of citrus-infesting coccids to parasitization. Journal of Economic Entomology, 46, 266–269.

Flanders, S. E. (1953b). Predatism by the adult hymenopterous parasite and its role in biological control. Journal of Economic Entomology, 46, 541–544.

Flanders, S. E. (1956). The mechanisms of sex ratio regulation in the parasitic Hymenoptera. Insect Society, 3, 325–334.

Flanders, S. E. (1957). Ovigenic-ovisorptic cycle in the economy of the honey bee. Science Monthly, 85, 176–177.

Flanders, S. E. (1959a). Differential host relations of the sexes in parasitic Hymenoptera. Entomologia Experimentalis et Applicata, 2, 125–142.

Flanders, S. E. (1959b). Embryonic starvation, an explanation of the defective honey bee egg. Journal of Economic Entomology, 52, 166–167.

Flanders, S. E. (1964). Dual ontogeny of the male Coccophagus gurneyi Comp. (Hymenoptera: Aphelinidae): A phenotypic phenomenon. Nature (London), 204, 944–946.

Flanders, S. E. (1965). On the sexuality and sex ratios of hymenopterous populations. American Naturalist, 99, 489–494.

Flanders, S. E. (1972). The duality of imaginal diapause inception in pteromalids parasitic on Hypera postica. Annals of the Entomological Society of America, 65, 105–108.

Fleming, J. G. W., & Summers, M. D. (1986). Campoletis sonorensis endoparasitic wasps contain forms of C. sonorensis virus DNA suggestive of integrated and extrachromosomal polydnavirus DNAs. Journal of Virology, 57, 552–562.

Fleschner, C. A., & Scriven, G. T. (1957). Effect of soil-type and DDT on ovipositional response of *Chrysopa californica* (Coq.) on lemon trees. Journal of Economic Entomology, 50, 221–222.

Franklin, R. T., & Holdaway, F. G. (1966). A relationship of the plant to parasitism of European corn borer by the tachinid parasite *Lydella grisescens*. Journal of Economic Entomology, 59, 440–441.

Galloway, I. D., & Austin, A. D. (1984). Revision of the Scelioninae (Hymenoptera:Scelionidae) in Australia. Australian Journal of Zoology (Suppl. Series) 99, 1–138

Gardner, S. M., & van Lenteren, J. C. (1986). Characterisation of the arrestment responses of *Trichogramma evanescens*. Oecologia (Berlin), 68, 265–270.

Gauld, I. D. (1984). An introduction to the Ichneumonidae of Australia. London: British Museum (Natural History).

Gauld, I. D., & Bolton, B. (1988). The Hymenoptera. London: British Museum (Natural History) & Oxford University Press.

Gerling, D., & Legner, E. F. (1968). Developmental history and reproduction of *Spalangia cameroni* Perkins (Hymenoptera: Spalangiidae), parasite of synanthropic flies. Annals of the Entomological Society of America, 61, 1436–1443.

Geyspitz, K. F., & Kyao, I. I. (1953). The influence of the length of illumination on the development of certain braconids (Hymenoptera). Entomolgicheskoe Obozrenie, 33, 32–35.

Gibson, G. A. P. (1986). Evidence for monophyly and relationships of Chalcidoidea, Mymaridae and Mymarommatidae (Hymenoptera: Terebrantes). Canadian Entomologist, 118, 205–240.

Gillespie, D. R., & Finlayson, T. (1983). Classification of the final instar of the Ichneumoninae (Hymenoptera: Ichneumonidae). Memoirs of the Entomological Society of Canada, 124.

Gordh, G. (1975). Some evolutionary trends in the Chalcidoidea (Hymenoptera) with particular reference to host preference. Journal of the New York Entomological Society, 83, 279–280.

Gordh, G. (1976). *Goniozus gallicola* Fouts, a parasite of moth larvae, with notes on other bethylids (Hymenoptera: Bethylidae; Lepidoptera: Gelechiidae) (Tech. Bulletin 1524) Washington, DC: U.S. Department of Agriculture.

Gordh, G. (1979). Superfamily Chalcidoidea. In K. V. Krombein *et al.*(Eds.), Catalog of Hymenoptera in America north of Mexico (Vol. 1, pp. 743–748). Washington, DC: Smithsonian Institute Press.

Gordh, G. (1981). The phenomenon of insect hyperparasitism and its taxonomic occurrence in the Insecta. In M. Klein (Ed.), The role of hyperparasitism in biological control: A symposium (pp. 10–18). Berkeley: University of California Division of Agricultural Science.

Gordh, G., & DeBach, P. (1976). Male inseminative potential in *Aphytis lingnanensis* (Hymenoptera: Aphelinidae). Canadian Entomologist, 108, 583–589.

Gordh, G., & Lacey, L. (1976). Biological studies of *Plagiomerus diaspidis* Crawford, a primary internal parasite of diaspidid scale insects (Hymenoptera: Encyrtidae; Homoptera: Dispididae). Proceedings of the Entomological Society of Washington, 78, 132–144.

Gordh, G., & Medved, R. E. (1986). Biological notes on *Goniozus pakmanus* Gordh (Hymenoptera: Bethylidae), a parasite of pink bollworm, *Pectinophora gossypiella* (Saunders) (Lepidoptera: Gelechiidae). Journal of the Kansas Entomological Society, 59, 723–734.

Gordh, G., Wooley, J. B., & Medved, R. E. (1983). Biological studies on *Goniozus legneri* Gordh (Hymenoptera: Bethylidae), a primary external parasite of the navel orangeworm, *Amyelois transitella* (Walker) and pink bollworm, *Pectinophora gossypiella* (Saunders) (Lepidoptera, Pyralidae, Gelechiidae). Contributions of the American Entomological Institute (Ann Arbor), 20, 433–468.

Graham, M. R. W. de V. (1969). The Pteromalidae of northwestern Europe (Hymenoptera: Chalcidoidea). Bulletin of the British Museum (Natural History) (Suppl. 16). 908 pp.

Graham, S. A., & Baumhofer, L. G. (1927). The pine tip moth in the Nebraska National Forest. Journal of Agricultural Research, 35, 323–333.

Greany, P. D., & Oatman, E. R. (1972). Demonstration of host discrimination in the parasite *Orgilus lepidus* (Hymenoptera: Braconidae). Annals of the Entomological Society of America, 65, 375–376.

Greany, P. D., Tumlinson, J. H., Chambers, D. L., & Boush, G. M. (1977). Chemically mediated host finding by *Biosteres (Opius) longicaudatus*, a parasitoid of tephretitid fruit fly larvae. Journal of Chemical Ecology, 3, 189–195.

Greathead, D. J. (1986). Parasitoids in classical biological control. In J. Waage & D. Greathead (Eds.), Insect parasitoids. New York: Academic Press.

Green, R. F., Gordh, G., & Hawkins, B. A. (1982). Precise sex ratios in highly inbred parasitic wasps. American Naturalist, 120, 653–665.

Griffiths, D. C. (1960). Immunity of aphids to insect parasites. Nature (London) 187, 346.

Griffiths, D. C. (1961). The development of *Monoctonus paludum* Marshall (Hym., Braconidae) in *Nasonovia ribis-nigri* on lettuce, and immunity reactions in other lettuce aphids. Bulletin of Entomological Research, 52, 147–163.

Griffiths, D. C. (1977). Avoidance-modified generalised distributions and their application to studies of superparasitism. Biometrics, 33, 103–112.

Griffiths, K. J. (1969). Development and diapause in *Pleolophus basizonus* (Hymenoptera: Ichneumonidae). Canadian Entomologist, 101, 907–914.

Grissell, E. E. (1979). Family: Torymidae. In K. V. Krombein *et al.* (eds.), Catalog of Hymenoptera in America north of Mexico (Vol. 1, pp. 748–768). Washington, DC: Smithsonian Institute Press.

Grosch, D. S. (1948). Growth in *Habrobracon*. Growth, 12, 243–254.

Grosch, D. S., LaChance, L. E., & Sullivan, R. L. (1955). Notes on the feeding preferences of *Habrobracon* adults (*Microbracon hebetor* [Say]: Hymenoptera: Braconidae). Annals of the Entomological Society of America, 48, 415–416.

Gross, H. R., Jr., Lewis, W. J., Jones, R. L., & Nordlund, D. A. (1975). Kairomones and their use for management of entomophagous insects. III. Stimulation of *Trichogramma achaeae, T. pretiosum*, and *Microplitis croceipes* with host-seeking stimuli at time of release to improve their efficiency. Journal of Chemical Ecology, 1, 431–438.

Hadorn, K., & Walker, I. (1960). *Drosophila* and *Pseudocoila*. I. Selection experiments on increasing the defense reaction of the host. Revue Suisse de Zoologie, 67, 216–225.

Hagen, K. S. (1964). Developmental stages of parasites. In P. H. DeBach & E. Schlinger (Eds.), Biological control of insect pests and weeds (pp. 168–246). London: Chapman & Hall.

Hagen, K. S., Sawall, E. F., Jr. & Tussan, R. L. (1970). The use of food sprays to increase effectiveness of entomophagous insects. Proceedings of the Tall Timbers Conference on the Ecology of Animal Control by Habitat Management, 2, 59–81.

Hallez, P. (1886). Loi de l'orientation de l'embryon chez les insectes. Compte Rendu, 103, 606–608.

Hamilton, W. D. (1967). Extraordinary sex ratios. Science, 156, 477–488.

Hanna, A. D. (1935). Fertility and toleration of low temperature in *Euchalcidia caryobori* Hanna (Hymenoptera, Chalcidinae). Bulletin of Entomological Research, 26, 315–322.

Hartl, D. L. (1972). A fundamental theorem of natural selection for sex linkage or arrhenotoky. American Naturalist, 106, 516–524.

Hartley, E. A. (1922). Some bionomics of *Aphelinus semiflavus* (Howard), chalcid parasite of aphids. Ohio Journal of Science, 22, 209–236.

Hawke, S. D., Farley, R. D., & Greany, P. D. (1973). The fine structure of sense organs in the ovipositor of the parasitic wasp, *Orgilus lepidus* Muesebeck. Tissue Cell, 5, 171–184.

Hayat, M. (1983). The genera of Aphelinidae (Hymenoptera) of the world. Systematic Entomology, 8, 63–102.

Hedicke, H. (1939). Aulacidae. Catalogus Hymenopterorum, 10, 3–28.

Heraty, J. M., & Darling, D. C. (1984). Comparative morphology of the planidial larvae of Eucharitidae and Perilampidae (Hymenoptera: Chalcidoidea). Systematic Entomology, 9, 309–328.

Highnam, F. C., Lusis, O., & Hill, L. (1963). Factors affecting oocyte resorption in the desert locusts Schistocerca gregaria (Forskal). Journal of Insect Physiology, 9, 827–837.

Hinton, H. E. (1955). Protective devices of endopterygote pupae. Transactions of the Society for British Entomology, 12, 49–92.

Hinton, H. E. (1969). Respiratory systems of insect egg shells. Annual Review of Entomology, 14, 343–368.

Hodek, I. (1965). Several types of induction and completion of adult diapause. Proceedings, 12th International Congress of Entomology, (1964), 431–432.

Hodek, I. (1966). Food ecology of aphidophagous Coccinellidae. In Proceedings of the Symposium on Ecology of Aphidophagous Insects (pp. 23–30). Prague: Academia.

Holdaway, F. G., & Smith, N. F. (1932). A relation between size of host puparia and sex ratio of Alysia manducator Pantzer. Australian Journal of Experimental Biology and Medical Science, 10, 247–259.

Holmes, H. B. (1972). Genetic evidence for fewer progeny and a higher percent males when Nasonia vitripennis oviposits in previously parasitized hosts. Entomophaga, 17, 79–88.

Hopkins, C. R., & King, P. E. (1964). Egg resorption in Nasonia vitripennis (Walker) (Hymenoptera, Pteromalidae). Proceedings of the Royal Entomological Society of London, Series A, 39, 101–107.

Hopkins, C. R., & King, P. E. (1966). An electron-microscopical and histochemical study of the oocyte periphery in Bombus terrestris during vitellogenesis. Journal of Cell Science, 1, 201–216.

Howard, L. O., & Fiske, W. F. (1911). The importation into the United States of the parasites of the gypsy moth and the brown tail moth. USDA Bureau of Entomology Bulletin, 91, 1–312.

Hubbard, S. F., & Cook, R. M. (1978). Optimal foraging by parasitoid wasps. Journal of Animal Ecology, 47, 593–604.

Huber, J. T. (1986). Systematics, biology and hosts of the Mymaridae and Mymarommatidae (Insecta: Hymenoptera). Entomography, 4, 185–243.

Hunter, K. W., Jr., & A. Stoner. (1975). Copidosoma truncatellum: Effects of parasitization on food consumption of larval Trichoplusia ni. Environmental Entomology, 4, 381–382.

Imms, A. D. (1931). Recent advances in entomology. Philadelphia: Blakiston & Sons.

Ivanova-Kasas, O. M. (1972). Polyembryony in insects. In S. J. Counce & C. H. Waddington (Eds.), Development systems: Insects (Vol. 1, pp. 243–271). New York: Academic Press.

Iwasa, Y., Suzuki, Y., & Matzuda, H. (1984). Theory of oviposition strategy of parasitoids. I. Effect of mortality and limited egg number. Theoretical Population Biology, 26, 205–227.

Iwata, K. (1959a). The comparative anatomy of the ovary in Hymenoptera. IV. Proctotrupoidea and Agriotypidae (Ichneumonidae) with descriptions of ovarian eggs. Kontyu, 27, 18–20.

Iwata, K. (1959b). The comparative anatomy of the ovary in Hymenoptera. III. Braconidae (Inc. Aphidiidae). Kontyu, 27, 231–238.

Iwata, K. (1960a). The comparative anatomy of the ovary in Hymenoptera. V. Ichneumonidae. Acta Hymenopterol, 1, 115–169.

Iwata, K. (1960b). The comparative anatomy of the ovary in Hymenoptera. Supplement of Aculeata with descriptions of ovarian eggs of certain species. Acta Hymenopterol, 1, 205–211.

Iwata, K. (1962). The comparative anatomy of the ovary in Hymenoptera. VI. Chalcidoidea with description of ovarian eggs. Acta Hymenopterol, 1, 383–391.

Iwata, K. (1976). Evolution of instinct. Comparative ethology of Hymenoptera. New Delhi: Amerind.

Jackson, D. J. (1928). The biology of Dinocampus (Perilitus) rutilus Nees, a braconid parasite of Sitoni linesta L. I. Proceedings of the Zoological Society of London, 1928, 597–630.

Jackson, D. J. (1937). Host-selection in Pimpla examinator F. (Hymenoptera). Proceedings of the Royal Entomological Society of London, Series A, 12, 81–91.

Jenni, W. (1951). Beitrag zuur Morphologie und Biologie der Cynipide Pseudeucoila bochei Weld, eines Larvenparasiten von Drosophila melanogaster Meig. Acta Zoologica, 32, 177–254.

Jeon, M.-J., Lee, B.-Y., Ko, J.-H., Miura, T., & Hirashima, Y. (1985). Ecology of Platygaster matsutama and Inostemma seoulis (Hymenoptera: Platygasteridae), egg-larval parasites of the pine needle gall midge, Thecodiplosis japonensis (Diptera, Cecidomyiidae). Esakia, 23, 131–143.

Jervis, M. A., & Kidd, N. A. C. (1986). Host-feeding strategies in hymenopteran parasitoids. Biological Review, 61, 395–434.

Jones, R. L., Lewis, W. J., Bowman, M. C., Beroza, B., & Bierle, B. A. (1971). Host-seeking stimulant for parasite of corn earworm: Isolation, identification and synthesis. Science, 1973, 872–873.

Jones, R. L., Lewis, W. J., Bowman, M. C., Beroza, B., & Bierle, B. A. (1973). Host seeking stimulants (kairomones) for the egg parasite Trichogramma evanescens. Environmental Entomology, 2, 593–596.

Kanungo, K. (1955). Effects of superparasitism on sex-ratio and mortality. Current Science, 24, 59–60.

Kennett, C. E., Huffaker, C. B., & Finney, G. L. (1966). The role of an autoparasitic aphelinid, Coccophagoides utilis Doutt, in the control of Parlatoria oleae (Colvee). Hilgardia, 37, 255–282.

Kerrich, G. J. (1973). On the taxonomy of some forms of Ibalia Latreille (Hymenoptera: Cynipoidea) associated with conifers. Zoological Journal of the Linnean Society, 53(1), 65–79.

King, P. E. (1962). The effect of resorbing eggs upon the sex ratio of the offspring in Nasonia vitripennis (Hymenoptera, Pteromalidae). Journal of Experimental Biology, 39, 161–165.

King, P. E. (1963). The rate of egg resorption in Nasonia vitripennis (Walker) (Hymenoptera: Pteromalidae) deprived of hosts. Proceedings of the Royal Entomological Society of London, 38, 98–100.

King, P. E., & Hopkins, C. R. (1963). Length of life of the sexes in Nasonia vitripennis (Walker) (Hymenoptera: Pteromalidae) under conditions of starvation. Journal of Experimental Biology, 40, 751–761.

King, P. E., & Ratcliffe, N. A. (1969). The structure and possible mode of functioning of the female reproductive system in Nasonia vitripennis (Hym.: Pteromalidae). Journal of Zoology, 157, 319–344.

King, P. E., & Richards, J. G. (1968). Oosorption in Nasonia vitripennis (Hymenoptera: Pteromalidae). Journal of Zoology, 154, 495–516.

King, P. E., Richards, J. G., & Copeland, M. J. W. (1968). The structure of the chorion and its possible significance during oviposition in Nasonia vitripennis (Walker) (Hymenoptera: Pteromalidae) and other chalcids. Proceedings of the Royal Entomological Society of London, Series A, 43, 13–20.

Konigsmann, E. (1978). Das phylogenetische System der Hymenoptera. III. "Terebrantes" (Unterordnung Apocrita). Deutsche Entomologische Zeitschrift, 23(1/3), 1–55.

Kornhauser, S. J. (1919). The sexual characteristics of the membracid Thelia bimaculata (Fab.). I. External changes induced by Aphelopus theliae (Gahan). Journal of Morphology, 32, 531–635.

Koscielska, M. C. (1962). Investigation upon trophamnion of Ageniaspis fuscicollis Dalm. Studia Societatis Scientiarum Torunensis, 6(7), 1–9 (in Polish with English summary).

Koscielska, M. C. (1963). Investigations on polyembryony in Ageniaspis fuscicollis Dalm. (Chalcidoidea, Hymenoptera). Zoologica Poloniae, 13(3/4), 255–276.

Kozlov, M. A. (1970). Supergeneric groups of the Proctotrupoidea (Hymenoptera). Entomological Review, 49, 115–127.

Kulman, H. M., & Hodson, A. C. (1961). Parasites of the jack pine budworm, Choristoneura pineus, with special reference to parasitism at particular locations. Journal of Economic Entomology, 54, 221–224.

Kunckel d'Herculais, J., & Langois, C. (1891). Moeurs et metamorphoses de *Perilitus brevicollis* Haliday Hymenoptere Braconide parasite de l'Altise de la vigne en Algerie. Annales de la Societe Entomologique de France, 60, 457–466.

LaBergrie, V. (1959). Embryonic diapause in the unlaid egg of parasitic Hymenoptera. Comptes Rendus Hebdomadaires des Seances, de l'Academie des Sciences, Paris, Serie D, 249, 2115–2117.

Laing, J. (1937). Host-finding by insect parasites. I. Observations on the finding of hosts by *Alysia manducator, Mormoniella vitripennis* and *Trichogramma evanescens*. Journal of Animal Ecology, 6, 298–317.

LaSalle, J. (1990). A new genus and species of Tetrastichinae (Hymenoptera:Eulophidae) parasitic on the coffee berry borer, *Hypothenemus hampei* (Ferrari)(Coleoptera: Scolytidae). Bulletin of Entomological Research, 80, 7–10.

LaSalle, J., & LeBeck, L. M. (1983). The occurrence of encyrtiform eggs in the Tanaostigmatidae (Hymenoptera: Chalcidoidea). Proceedings of the Entomological Society of Washington, 85(2), 397–398.

Lawson, F. R. (1959). The natural enemies of the hornworms on tobacco (Lepidoptera: Sphingidae). Annals of Entomological Society of America, 52, 741–755.

Lee, B.-Y., Ko, J.-H., Choi, B.-H., Jeon, M.-J., Miura, T., & Hirashima, Y. (1985). Utilization of proctotrupoid wasps in Korea for control of the pine needle gall midge, *Thecodiplosis japonensis* (Diptera, Cecidomyiidae). Esakia, 23, 145–150.

Legner, E. F. (1967a). Two exotic strains of *Spalangia drosophilae* merit consideration in biological control of *Hippelates collusor* (Diptera: Chloropidae). Annals of the Entomological Society of America, 60, 458–462.

Legner, E. F. (1967b). Behavior changes the reproduction of *Spalangia cameroni, S. endius, Muscidifurax raptor,* and *Nasonia vitripennis* (Hymenoptera: Pteromalidae) at increasing fly host densities. Annals of the Entomological Society of America, 60, 819–826.

Legner, E. F. (1969a). Adult emergence interval and reproduction in parasitic Hymenoptera influenced by host size and density. Annals of the Entomological Society of America, 62, 220–226.

Legner, E. F. (1969b). Distribution pattern of hosts and parasitization by *Spalangia drosophilae* (Hymenoptera: Pteromalidae). Canadian Entomologist, 101, 551–557.

Legner, E. F. (1972). Observations on hybridization and heterosis in parasitoids of synanthropic flies. Annals of the Entomological Society of America, 65, 254–263.

Legner, E. F. (1976). Low storage temperature effects on the reproductive potential of three parasites of *Musca domestica*. Annals of the Entomological Society of America, 69, 435–441.

Legner, E. F. (1979a). Prolonged culture and inbreeding effects on reproductive rates of two pteromalid parasites of muscoid flies. Annals of the Entomological Society of America, 72, 114–118.

Legner, E. F. (1979b). The relationship between host destruction and parasite reproductive potential in *Muscidifurax raptor, M. zaraptor* and *Spalangia endius* (Chalcidoidea: Pteromalidae). Entomophaga, 24, 145–152.

Legner, E. F. (1985a). Effects of scheduled high temperature on male production in thelytokous *Muscidifurax uniraptor* (Hymenoptera: Pteromalidae). Canadian Entomologist, 117, 383–389.

Legner, E. F. (1985b). Natural and induced sex ratio changes in populations of thelytokous *Muscidifurax uniraptor* (Hymenoptera: Pteromalidae). Annals of the Entomological Society of America, 78, 398–402.

Legner, E. F. (1987a). Transfer of thelytoky to arrhenotokous *Muscidifurax raptor* (Hymenoptera:Pteromalidae). Canadian Entomologist, 119, 2651–2671.

Legner, E. F. (1987b). Inheritance of gregarious and solitary oviposition in *Muscidifurax raptorellus* (Kogan & Legner) (Hymenoptera: Pteromalidae). Canadian Entomologist, 119, 791–808.

Legner, E. F. (1987c). Pattern of thelytoky acquisition in *Muscidifurax*

raptor (Girault & Sanders) (Hymenoptera: Pteromalidae). Bull. Soc. Vect. Ecol., 12, 1–11.

Legner, E. F. (1988a). *Muscidifurax raptorellus* (Hymenoptera: Pteromalidae) females exhibit post-mating oviposition behavior typical of the male genome. Annals of the Entomological Society of America, 81, 522–527.

Legner, E. F. (1988b). Hybridization in principal parasitoids of synanthropic Diptera: The genus *Muscidifurax* (Hymenoptera: Pteromalidae). Hilgardia, 56, 1–36.

Legner, E. F. (1988c). Studies of four thelytokous Puerto Rican isolates of *Muscidifurax uniraptor* [Hymenoptera: Pteromalidae]. Entomophaga, 33, 269–280.

Legner, E. F. (1989a). Wary genes and accretive inheritance in Hymenoptera. Annals of the Entomological Society of America, 82, 245–249.

Legner, E. F. (1989b). Phenotypic expressions of polygenes in *Muscidifurax raptorellus* [Hym.:Pteromalidae], a synanthropic fly parasitoid. Entomophaga, 34, 523–530.

Legner, E. F. (1991). Estimations of the number of active loci, dominance and heritability in polygenic inheritance of gregarious behavior in *Muscidifurax raptorellus* Kogan & Legner [Hymenoptera: Pteromalidae]. Entomophaga, 36(1), 1–18.

Legner, E. F., & Gerling, D. (1967). Host-feeding and oviposition on *Musca domestica* by *Spalangia cameroni, Nasonia vitripennis* and *Muscidifurax raptor* (Hymenoptera: Pteromalidae) influences their longevity and fecundity. Annals of the Entomological Society of America, 60, 678–691.

Legner, E. F., & Olton, G. S. (1969). Migrations of *Hippelates collusor* larvae from moisture and trophic stimuli and their encounter by *Trybliographa* parasites. Annals of the Entomological Society of America, 62, 136–141.

Legner, E. F., & Thompson, S. N. (1977). Effects of the parental host on host selection, reproductive potential, survival and fecundity of the egg-larval parasitoid, *Chelonus* sp. nr. *curvimaculatus* Cameron, reared on *Pectinophora gossypiella* (Saunders) and *Phthorimaea operculella* (Zeller). Entomophaga, 22, 75–84.

Legner, E. F., Bay, E. C., & Medved, R. A. (1966). Behavior of three native pupal parasites of *Hippelates collusor* in controlled systems. Annals of the Entomological Society of America, 59, 977–984.

Leiby, R. W. (1929). Polyembryony in insects. Transactions Fourth International Contr. Entomology, 2, 873–887.

Leiby, R. W., & Hill, C. C. (1923). The twinning and monembryonic development of *Platygaster hiemalis*, a parasite of the Hessian fly. Journal of Agricultural Research, 25, 337–350.

Leiby, R. W., & Hill, C. C. (1924). The polyembryonic development of *Platygaster vernalis*. Journal of Agricultural Research, 28, 829–840.

Lensky, Y., & Schindler, H. (1967). Mortality and reversible inactivation of honeybee spermatozoa *in vivo* and *in vitro*. Annales del Abeille, 10, 5–16.

Levine, L., & Sullivan, D. J. (1983). Intraspecific tertiary parasitoidism in *Asaphes lucens* (Hymenoptera: Pteromalidae), an aphid hyperparasitoid. Canadian Entomologist, 115, 1653–1658.

Lewis, W. J., Jones, R. L., & Sparks, A. N. (1972). A host-seeking stimulant for the egg parasite *Trichogramma evanescens:* Its source and a demonstration of its laboratory and field activity. Annals of the Entomological Society of America, 65, 1087–1089.

Lewis, W. J., Sparks, A. N., & Redlinger, L. M. (1971). Moth odor: A method of host-finding by *Trichogramma evanescens*. Journal of Economic Entomology, 64, 557–558.

Lillie, F. R. (1919). Problems of fertilization. Chicago: University of Chicago Science Series.

Lloyd, D. C. (1938). A study of some factors governing the choice of hosts and distribution of progeny by the chalcid *Ooencyrtus kuwanae* Howard. Philosophical Transactions of the Royal Society of London Series B, 229, 275–322.

Lund, H. O. (1938). Studies on longevity and productivity in *Trichogramma evanescens*. Journal of Agricultural Research, 56, 421–439.

Madden, J. (1968). Behavioral responses of parasites to the symbiotic fungus associated with *Sirex nictilio* F. Nature (London), 218, 189–190.

Madden, J. L., & Pimentel, D. (1965). Density and spatial relationships between a wasp parasite and its housefly host. Canadian Entomologist, 97, 1031–1037.

Maple, J. D. (1947). The eggs and first instar larvae of Encyrtidae and their morphological adaptations for respiration. University of California, Berkeley, Publications in Entomology, 8(2), 25–122.

Marchal, P. (1898). Le cycle evolutif de l'*Encyrtus fusicollis*. Bulletin de la Societe Entomologique de France, 109–111.

Marchal, P. (1904). Recherches sur la biologie et le developpement de hymenopteres parasites. I. La polyembryonie specifique ou germinogenie. Archives de Zoologie Experimentale et Generale, 2, 257–335.

Marchal, P. (1905). Observations biologiques sur un parasite de la galeruque de L'orme, le *Tetrastichus xanthomelaenae* (Rond.). Bulletin de la Societe Entomologique de France, (1905), 64–68.

Marchal, P. (1906). Recherches sur la biologie et le developpement des Hymenopteres parasites. Les *Platygasters*. Archives de Zoologie Experimentale et Generale, 4, Serie 4, 485–640.

Martin, C. H. (1946). The effect of humidity on the oviposition activity of *Macrocentrus ancylivorus*. Journal of Economic Entomology, 39, 419.

Martin, C. H., & Finney, G. L. (1946). Control of the sex ratio in *Macrocentrus ancylivorus*. Journal of Economic Entomology, 39, 269–272.

Maslennikova, V. A. (1958). On the conditions determining the diapause in the parasitic Hymenoptera, *Apanteles glomeratus* L. (Braconidae) and *Pteromalus puparum* (Chalcididae). Reviews in Entomololgy, 37, 538–545.

Masner, L. (1976). Revisionary notes and keys to world genera of Scelionidae (Hymenoptera: Proctotrupoidea) Memoirs of the Entomological of Society of Canada, 97, 87, pp.

Masner, L. (1980). Key to genera of Scelionidae of the Holarctic region, with descriptions of new genera and species (Hymenoptera: Proctotrupoidea). Memoirs of the Entomological Society of Canada, 113, 54 pp.

Masner, L., & Dessart, P. (1967). La reclassification des categories taxonomiques superieures des Ceraphronoidea (Hymenoptera). Bulletin de l'Institut Royal, des Sciences Naturelles de Belgique, 43/22. 1–33.

Matejko, I., & Sullivan, D. J. (1984). Interspecific tertiary parasitoidism between two aphid hyperparasitoids: *Dendrocerus carpenteri* and *Alloxysta megourae* (Hymenoptera: Megaspilidae & Cynipidae). Journal of Washington Academy of Science, 74, 31–38.

Matthews, R. W. (1974). Biology of Braconidae. Annual Review of Entomology, 19, 15–32.

McLeod, J. H. (1951). Notes on the lodgepole needle miner, *Recurvaria milleri* Busck. (Lep.: Gelechiidae), and its parasites in western North America. Canadian Entomologist, 83, 295–301.

Medler, J. T. (1962). Development and absorption of eggs in bumblebees (Hymenoptera, Apidae). Canadian Entomologist, 94, 825–833.

Metcalf, R. A., Marlin, J. C., & Whitt, G. S. (1975). Low levels of genetic heterozygosity in Hymenoptera. Nature (London), 257, 792–794.

Miller, G. L., & Lambdin, P. L. (1985). Observations on *Anacharis melanoneura* (Hymenoptera: Figitidae), a parasite of *Hemerobius stigma* (Neuroptera: Hemerobiidae). Entomological News, 96, 93–97.

Milliron, H. E. (1940). A study of some factors affecting the efficiency of *Encarsia formosa* Gahan, an aphelinid parasite of the greenhouse whitefly, *Trialeurodes vaporarium* (Westw.). Michigan Agricultural Experimental Station Technical Bulletin, 173, 1–23.

Moursi, A. A. (1946). The effect of temperature on development and reproduction of *Mormoniella vitripennis* (Walker). Bulletin de la Société Fouad ler D'Entomologie, 30, 39–61.

Mueller, T. F. (1983). The effect of plants on the host relations of a specialist parasitoid of *Heliothis* larvae. Entomologia Experimentalis et Applicata, 34, 78–84.

Muldrew, J. A. (1953). The natural immunity of the larch sawfly (*Pristiphora erichsonii* [Htg.]) to the introduced parasite *Mesoleius tenthre-*

dinis Morley, in Manitoba and Saskatchewan. Canadian Journal of Zoology, 31, 313–332.

Nadel, H., & Luck, R. F. (1985). Span of female emergence and male sperm depletion in the female-biased, quasi-gregarious parasitoid, *Pachycrepoideus vindemiae* (Hymenoptera: Pteromalidae). Annals of the Entomological Society of America, 78, 410–414.

Nappi, A. J., & Streams, F. A. (1970). Abortive development of the cynipid parasite *Pseudeucoila bochei* (Hymenoptera) in species of the *Drosophila melanica* group. Annals of the Entomological Society of America, 63, 321–327.

Narayanan, E. S., & Chaudhuri, R. P. (1955). Studies on *Stenobracon deesae* (Cam.), a parasite of certain lepidopterous borers of graminaceous crops in India. Bulletin of Entomological Research, 45, 647–659.

Narendran, T. C. (1984). Chalcids and sawflies associated with plant galls. In T. N. Ananthakrishnan (Ed.), Biology of gall insects (pp. 273–303). New Delhi: Oxford & IBH Publishing.

Narendran, T. C. (1985). An analysis of the superparasitic behaviour and host discrimination of chalcid wasps (Hymenoptera: Chalcidoidea). Proceedings of the Indian Academy of Sciences (Animal Science), 94, 325–331.

Naumann, I. D. (1987). The Ambositriinae (Hymenoptera: Diapriidae) of Melanesia. Invertebrate Taxonomy, 1, 439–471.

Noldus, L. P., & van Lenteren, J. C. (1983). Kairomonal effects on host searching of *Pieris brassicae, Pieris rapae* and *Mamestra brassicae* of the parasite *Trichogramma evanescens* Westwood. Mededelingen van de Faculteit Landbouwwetenschappen Rijks universiteit Gent, 48(2), 183–194.

Noldus, L. P., & van Lenteren, J. C. (1985). Kairomones for the egg parasite *Trichogramma evanescens* Westwood. I. Effect of volatile substances released by two of its hosts, *Pieris brassicae* L. and *Mamestra brassicae* L. Journal of Chemical Ecology, 2, 781–791.

Nordlander, G. (1984). Vad vet vi om parasitiska Cynipoidea? Entomologisk Tidskrift, 105, 36–40.

Oehlke, J. (1983). Revision der europaischen Aulacidae (Hymenoptera–Evanioidea). Beitraege zur Entomologie, 33, 439–447.

Olmi, M. (1984a). A revision of the Dryinidae (Hymenoptera). Memoirs of the American Entomological Institute, 37, 1–946.

Olmi, M. (1984b). A revision of the Dryinidae (Hymenoptera). Memoirs of the American Entomological Institute, 37, 947–1913.

Olton, G. S. &, Legner, E. F. (1974). Biology of *Tachinaephagus zealandicus*. (Hymenoptera: Encyrtidae), parasitoid of synanthropic Diptera. Canadian Entomologist, 106, 785–800.

Orr, D. B., Russin, J. S., & Boethel, D. J. (1986). Reproductive biology and behavior of *Telenomus calvus* (Hymenoptera: Scelionidae), a phoretic egg parasitoid of *Podisus maculiventris* (Hymenoptera: Pentatomidae). Canadian Entomologist, 118, 1063–1072.

Paillot, A. (1937). Sur le developpement polyembryonaire d'*Amicroplus collaris* Spin., parasite des chenilles d'*Euxoa segetum* Schiff. Comptes Rendu de l'Academic Sciences Paris, 204, 810–812.

Palm, N. E. (1948). Normal and pathological histology of the ovaries in *Bombus terrestris*. (Hymenoptera). Opuscula Entomologica, (Suppl. 7), 1–101.

Parker, H. L. (1931). *Macrocentrus gifuensis* Ashmead, a polyembryonic braconid parasite in the European corn borer (Tech. Bulletin 230 pp. 1–62) Washington, DC: U. S. Department of Agriculture.

Petersen, G. (1962). Haemocytare Abwehrresktion des Wirtes gegen endoparasitische Insekten und ihre Bedeutung fur die Biologische Bekampfung. Bericht uber die 9 Wandersammlung Deutscher Entomologen 6-8V, 1961 (pp. 179–195) Berlin.

Pfeiffer, I. W. (1945). Effect of the corpora allata on the metabolism of adult female grasshoppers. Journal of Experimental Zoology, 99, 183–233.

Phipps, J. (1966). Ovulation and oocyte resorption in Acridoidea (Orthoptera). Proceedings of the Royal Entomological Society of London, Series A, 41, 78–87.

Picard, F., & Rabaud, E. (1914). Sur le parasitism externe des Braconids

(Hym.). Bulletin of the Entomological Society of France, 19, 266–269.

Piel, O., & Covillard, J. (1933). Contribution à l'étude du *Monema flavescens* Wkr. and its parasites (Lepidoptera, Heterogeneidae). Lingnan Scientific Journal, 12 (Suppl.), 173–202.

Ponomarenko, N. G. (1975). Characteristics of larval development in the Dryinidae (Hymenoptera). Entomological Review, 54, 36–39.

Ponomarenko, N. G. (1981). New Dryinidae (Hymenoptera) from the late Cretaceous of the Taymyr and Canada. Paleontology Journal, 1981(1), 115–120.

Price, P. W. (1970). Trail odors: Recognition by insects parasitic in cocoons. Science, 170, 546–547.

Price, P. W. (1977). General concepts on the evolutionary biology of parasites. Evolution, 31, 405–420.

Price, P. W. (1980). Evolutionary biology of parasites. Princeton, NJ: Princeton University Press.

Putnam, W. L., & Herne, D. H. (1966). The role of predators and other biotic agents in regulating the population density of phytophagous mites in Ontario peach orchards. Canadian Entomologist, 98, 808–820.

Quezada, J. R., DeBach, P. H., & Rosen, D. (1973). Biological and taxonomic studies of *Signiphora borinquensis,* new species (Hymenoptera: Signiphoridae), a primary parasite of diaspine scales. Hilgardia, 41, 543–604.

Rabb, R. L., & Bradley, J. R. (1968). The influence of host plant on parasitism of eggs of the tobacco hornworm. Journal of Economic Entomology, 61, 1249–1252.

Raff, J. W. (1934). Observations on saw-flies of the genus *Perga,* with notes on some reared primary parasites of the families Trigonalidae, Ichneumonidae, and Tachinidae. Proceedings of the Royal Society of Victoria, 47, 54–77.

Rasnitsyn, A. P. (1968). Evolution of the function of the ovipositor in relation to the origin of parasitism in Hymenoptera. Entomological Review, 47, 35–40.

Rasnitsyn, A. P. (1980). Origin and evolution of Hymenoptera. Transactions of Paleontological Institute and Academy of Science, USSR, 174, 1–192 (in Russian).

Rasnitsyn, A. P. (1988). An outline of evolution of the hymenopterous insects (Order Vespida). Oriental Insects, 22, 115–145.

Rasnitsyn, A. P., & Kovalev, O. V. (1988). The oldest Cynipoidea (Hymenoptera: Archaeocynipidae) from the Early Cretaceous of Transbaicalia. Vestnik Zoologii, 1, 18–21 (in Russian).

Reuter, O. M. (1913). Lebensgewohnheiten und Instinkte der Insekten. Berlin: Friedlander.

Richards, J. G., & King, P. E. (1967). Chorion and vitelline membranes and their role in resorbing eggs of the Hymenoptera. Nature (London), 214, 601–602.

Rogers, D. (1975). A model for avoidance of superparasitism by solitary insect parasitoids. Journal of Animal Ecology, 44, 623–628.

Rosen, D., & DeBach, P. (1979). Species of *Aphytis* of the world (Hymenoptera: Aphelinidae). Series Entomologica, 17, 1–801.

Salt, G. (1935). Experimental studies in insect parasitism. III. Host selection. Proceedings of the Royal Society of London, 117, Series B, 413–435.

Salt, G. (1955a). Experimental studies in insect parasitism. VII. Host reactions following artificial parasitization. Proceedings of the Royal Society of London, Series 144, 380–398.

Salt, G. (1955b). Experimental studies in insect parasitism. IX. The reactions of a stick insect to an alien parasite. Proceedings of the Royal Society of London, Series B, 146, 93–108.

Salt, G. (1957). Experimental studies in insect parasitism. X. The reactions of some entopterygote insects to an alien parasite. Proceedings of the Royal Society of London, Series B, 147, 167–184.

Salt, G. (1958). Parasitic behaviour and the control of insect pests. Endeavour, 17, 145–148.

Salt, G. (1960). Experimental studies in insect parasitism. XI. The haemocytic reaction of a caterpillar under varied conditions. Proceedings of the Royal Society of London, Series B, 151, 446–467.

Salt, G. (1961). Competition among insect parasitoids. In Mechanisms in biological competition. Symposium of the Society of Experimental Biology, 15, 96–119.

Salt, G. (1963a). The defense reactions of insects to metazoan parasites. Parasitology, 53, 527–642.

Salt, G. (1963b). Experimental studies in insect parasitism. XII. The reactions of six exopterygote insects to an alien parasite. Journal of Insect Physiology, 9, 647–669.

Salt, G. (1965). Experimental studies in insect parasitism. XIII. The haemocytic reaction of a caterpillar to eggs of its habitual parasite. Proceedings of the Royal Society of London, Series B, 162, 303–318.

Salt, G. (1966). Experimental studies in insect parasitism. XIV. The haemocytic reaction of a caterpillar to larvae of its habitual parasite. Proceedings of the Royal Society of London, Series B, 165, 151–178.

Salt, G. (1967). Cellular defense mechanisms in insects. Federal Proceedings, 26, 1671–1674.

Salt, G. (1968). The resistance of insect parasitoids to the defense reactions of their hosts. Biological Review, 43, 200–232.

Salt, G. & van den Bosch, R. (1967). The defense reactions of three species of *Hypera* (Coleoptera, Curculionidae) to an ichneumon wasp. Journal of Invertebrate Pathology, 9, 164–177.

Saunders, D. S. (1962). The effect of age of female *Nasonia vitripennis* (Walker) (Hymenoptera, Pteromalidae) upon the incidence of larval diapause. Journal of Insect Physiology, 8, 309–318.

Saunders, D. S. (1965). Larval diapause of maternal origin: Induction of diapause in *Nasonia vitripennis* (Walker) (Hymenoptera: Pteromalidae). Journal of Experimental Biology, 42, 495–508.

Saunders, D. S. (1966a). Larval diapause of maternal origin. II. The effect of photoperiod and temperature on *Nasonia vitripennis.* Journal of Insect Physiology, 12, 569–581.

Saunders, D. S. (1966b). Larval diapause of maternal origin. III. The effect of host storage on *Nasonia vitripennis.* Journal of Insect Physiology, 12, 899–908.

Saunders, D. S. (1967). Time measurement in insect photoperiodism: Reversal of photoperiodic effect by chilling. Science, 156(3778), 1126–1127.

Schneider, F. (1950). Die Abwehrreaktion des Insektenblutes und ihre Beeinflussung durch die Parasiten. Vierteljahresschrift der Naturforschenden Gesellschaft in Zuerich, 95, 22–44.

Schneider, E. (1952). Einige physiologische Beziehungen zwischen Syrphidenlarven und ihren Parasiten. Zeitschrift fuer Angewandte Entomologie, 33, 150–162.

Schneidermann, H. A., & Horwitz, J. (1958). The induction and termination of facultative diapause in the chalcid wasps *Mormoniella vitripennis* (Walker) and *Tritneptis klugii* (Ratzeburg). Journal of Experimental Biology, 35, 520–551.

Schread, J. C., & Garman, P. (1933). Studies on parasites of the oriental fruit moth. I. *Trichogramma.* Bulletin of Connecticut Agricultural Experimental Station, 353, 691–756.

Schread, J. C., & Garman, P. (1934). Some effects of refrigeration on the biology of *Trichogramma* in artificial rearing. Journal of New York Entomological Society, 42, 263–283.

Sellier, R. (1959). Les insectes utiles. Paris: Payot.

Shaw, M. R. (1983). On evolution of endoparasitism: The biology of some genera of Rogadinae (Braconidae). Contributions of the American Entomological Institute, 20, 307–328.

Short, J. R. T. (1978). The final larval instars of the Ichneumonidae. Memoirs of the American Entomological Institute, 25, 508 pp.

Silvestri, F. (1906). Contribuzioni alla conoscenza biologica degli imenotterio parassiti. I. Biologia del *Litomastix truncatellus* (Dalm.). Bollettino del Laboratorio di Zoologia Generale e Agraria della R. Scuola Superiore d'Agricoltura in Portici, 1, 17–64.

Silvestri, F. (1923). Contribuzioni alla conoscenza dei Tortricidi delle querce. Bollettino del Laboratorio di Zoologia Generale e Agraria della R. Scuola Superiore d'Agricultura in Portici, 17, 41–107.

Silvestri, F. (1937). Insect polyembryony and its general biological aspects. Bulletin of the Museum of Comparative Zoology, 81, 469–498.

Simmonds, F. J. (1946). A factor affecting diapause in hymenopterous parasites. Bulletin of Entomological Research, 37, 95–97.

Simmonds, F. J. (1947a). Some factors influencing diapause. Canadian Entomologist, 79, 226–232.

Simmonds, F. J. (1947b). The biology of the parasites of Loxostege sticticalis L. in North America—Meteorus loxostegei Vier. (Braconidae, Meteorinae). Bulletin of Entomological, Research, 38, 373–379.

Simmonds, F. J. (1947c). The biology of the parasites of Loxostege sticticalis L., in North America—Bracon vulgaris (Cress.) (Braconidae, Agathinae). Bulletin of Entomological Research, 38, 145–155.

Simmonds, F. J. (1948). The influence of maternal physiology on the incidence of diapause. Philosophical Transactions of the Royal Society of London, Series B, 233, 385–414.

Simmonds, F. J. (1956). Superparasitism by Spalangia drosophilae Ashm. Bulletin of Entomological Research, 47, 361–376.

Simmonds, H. W. (1944). The effect of the host fruit upon the scale Aonidiella aurantii Mask. in relation to its parasite Comperiella bifasciata How. Journal of the Australian Institute of Agriculture and Science, 10, 38–39.

Skinner, S. W. (1982). Maternally inherited sex ratio in the parasitoid wasp Nasonia vitripennis. Science, 215, 1133–1134.

Skinner, S. W. (1985). Son-killer: A third extrachromosomal factor affecting the sex-ratio in the parasitoid wasp, Nasonia (= Mormoniella) vitripennis. Genetics, 109, 745–759.

Smith, E. L. (1970). Evolutionary morphology of the external insect genitalia. II. Hymenoptera. Annals of the Entomological Society of America, 63, 1–27.

Smith, E. L. (1941). Racial segregation in insect populations and its significance in applied entomology. Journal of Economic Entomology, 34, 1–13.

Smith, E. L. (1949). A race of Comperiella bifasciata successfully parasitizes California red scale. Journal of Economic Entomology, 35, 809–812.

Smith, J. M. (1957). Effects of the food plant of California red scale, Aonidiella aurantii (Mask.) on reproduction of its hymenopterous parasites. Canadian Entomologist, 89, 219–230.

Smith, O. J. (1952). Biology and behavior of Microctonus vittatae Muesebeck (Braconidae). University of California, Berkeley, Publications in Entomology, 9, 315–344.

Spencer, H. (1926). Biology of parasites and hyperparasites of aphids. Annals of the Entomological Society of America, 19, 119–153.

Speyer, E. R. (1929). The greenhouse whitefly. Journal of the Royal Horticultural Society, 54, 181–192.

Spradberry, J. P. (1969). The biology of Pseudorhyssa sternata Merill (Hym., Ichneumonidae), a cleptoparasite of siricid woodwasps. Bulletin of Entomological Research, 59, 291–297.

Stary, P. (1964). Food specificity in the Aphidiidae (Hymenoptera). Entomophaga, 9, 92–99.

Stoltz, D. B., & Vinson, S. B. (1977). Baculovirus-like particles in the reproductive tracts of female parasitoid wasps. II. The genus Apanteles. Canadian Journal of Microbiology, 23, 28–37.

Stoltz, D. B., Guzo, D., & Cook, D. (1986). Studies on polydnavirus transmission. Virology, 155, 120–131.

Stoltz, D. B., Vinson, S. B., & Mackinnon, E. A. (1976). Baculovirus-like particles in the reproductive tracts of female parasitoid wasps. Canadian Journal of Microbiology, 22, 1013–1023.

Stouthamer, R. (1989). Causes of thelyotoky and crossing incompatibility in several Trichogramma species (Hymenoptera: Trichogrammatidae). Unpublished doctoral dissertation, University of California, Riverside.

Stouthamer, R., & Luck, R. F. (1988). Microorganisms implicated as a cause of thelytokous parthenogenesis in Trichogramma spp. (p. 256). Proceedings of the 18th International Congress on Entomology, Vancouver, B.C., Canada. July 3–9, 1988.

Streams, F. A. (1968). Defense reactions of Drosophila species (Diptera: Drosophilidae) to the parasite Pseudeucoila bochei (Hymenoptera: Cynipidae). Annals of the Entomological Society of America, 61, 158–164.

Sullivan, C. R., & Wallace, D. R. (1967). Interaction of temperature and photoperiod in the induction of prolonged diapause in Neodiprion sertifer (Geoffroy). Canadian Entomologist, 99, 834–850.

Sullivan, D. J. (1972). Comparative behavior and competition between two aphid hyperparasites: Alloxysta victrix and Asaphes californicus (Hymenoptera: Cynipidae and Pteromalidae). Environmental Entomologist, 1, 234–244.

Sullivan, D. J. (1987). Insect hyperparasitism. Annual Review of Entomology, 32, 49–70.

Syme, P. D. (1969). Interaction between Pristomerus sp. and Orgilus obscurator, two parasites of the European pine shoot moth. Bi-Monthly Research Notes, 25(4), 30–31. Forestry Research Laboratory, Sault Ste. Marie, Ont., Canada.

Tachikawa, T. (1981). Hosts of encyrtid genera in the world (Hymenoptera: Chalcidoidea). Memoirs of the College of Agriculture Ehime University, 25, 85–110.

Tauber, M. J., & Tauber, C. A. (1976). Insect seasonality: Diapause maintenance, termination and post-diapause development. Annual Review of Entomology, 21, 81–107.

Taylor, T. H. C. (1937). The biological control of an insect in Fiji. An account of the coconut leaf-mining beetle and its parasite complex. London: Imperial Institute of Entomology.

Thompson, W. R. (1951). The specificity of host relations in predacious insects. Canadian Entomologist, 83, 262–269.

Thompson, W. R., & Parker, H. L. (1930). The morphology and biology of Eulimneria crassifemur, an important parasite of the European corn borer. Journal of Agricultural Research, 40, 321–345.

Thorpe, K. W. (1985). Effects of height and habitat type on egg parasitism by Trichogramma minutum and T. pretiosum (Hymenoptera: Trichogrammatidae). Agricultural Ecosystems in the Environment, 12, 117–126.

Thorpe, W. H., & Caudle, H. B. (1938). A study of the olfactory responses of insect parasites to the food plant of their host. Parasitology, 30, 523–528.

Thorpe, W. H., & Jones, F. G. (1937). Olfactory conditioning in a parasitic insect and its relation to the problem of host selection. Proceedings of the Royal Society, Series B, 124, 56–81.

Townes, H. K. (1949). The Nearctic species of Evaniidae (Hymenoptera). Proceedings of the United States National Museum, 99, 525–539.

Townes, H. K. (1950). The Nearctic species of Gasteruptiidae (Hymenoptera). Proceedings of the United States National Museum, 100, 85–145.

Townes, H. K. (1956). The Nearctic species of trigonalid wasps. Proceedings of the United States National Museum, 106, 295–300.

Townes, H. K. (1977). A revision of the Heloridae (Hymenoptera). Contributions of the American Entomological Institute, 15, 1–12.

Townes, H., & Townes, M. (1981). A revision of the Serphidae (Hymenoptera). Memoirs of the American Entomological Institute, 32, 1–541.

Ullyett, G. C. (1936). Host selection by Microplectron fuscipennis Zett. (Hymenoptera, Chalcididae). Proceedings of the Royal Society of London, Series B, 120, 253–291.

Ullyett, G. C. (1943). Some aspects of parasitism in field populations of Plutella maculipennis Curt. Journal of the Entomological Society of South Africa, 6, 65–80.

Ullyett, G. C. (1947). Mortality factors in populations of Plutella maculi-

pennis Curtis (Lep. Tineidae), and their relation to the problems of control. Union of South Africa, Department of Agriculture and Entomology Memoirs, 2, 77–202.

Ullyett, G. C. (1949). Pupation habits of sheep blowflies in relation to parasitism by *Mormoniella vitripennis* Wlk. (Hym.: Pteromalidae). Bulletin of Entomological Research, 40, 533–537.

Unruh, T. R., White, W., Gonzalez, D., Gordh, G., & Luck, R. F. (1983). Heterozygosity and effective size in laboratory populations of *Aphidius ervi* (Hymenoptera: Aphidiidae). Entomophaga, 28, 245–258.

van Achterberg, C. (1984). Essay on the phylogeny of Braconidae (Hymenoptera: Ichneumonidae). Tijdschift voor Entomologie, 105, 41–58.

van Alphen, J. J. M., & Vet, L. E. M. (1986). An evolutionary approach to host finding and selection. In J. Waage & D. Greathead (Eds.), Insect parasitoids (pp. 23–61). London: Academic Press.

Vandel, A. (1932). Le sexe des parasites depend-il du nombre d' individus renferme, dans la meme hote? (pp. 245–252). Paris: Societe Entomologique France, Paris Livre Centenaire.

Van den Bosch, R. (1964). Encapsulation of the eggs of *Bathyplectes curculionis* (Thomson) (Hymenoptera: Ichneumonidae) in larvae of *Hypera brunneipennis* (Boheman) and *Hypera postica* (Gyllenhal) (Coleoptera: Curculionidae). Journal of Insect Pathology, 6, 343–367.

van Dijken, M., & Waage, J. K. (1987). Self and conspecific superparasitism by the egg parasitoid *Trichogramma evanescens*. Entomologia Experimentalis et Applicata, 43, 183–192.

van Driesche, R. G., Bellotti, A., Herrera, C. J., & Castello, J. A. (1987). Host feeding and ovipositor insertion as sources of mortality in the mealybug *Phenacoccus herreni* caused by two encyrtids, *Epidinocarsis diversicornis* and *Acerophagus coccois*. Entomologia Experimentalis et Applicata, 44, 97–100.

van Emden, H. F. (1986). The interaction of plant resistance and natural enemies: Effects on populations of sucking insects. In D. J. Boethel & R. D. Eikenbary (Eds.), Interactions of plant resistance and parasitoids and predators of insects (pp. 138–150). New York: Wiley & Sons.

Van Emden, H. F., & Wearing, C. H. (1965). The role of the aphid host plant in delaying economic damage levels in crops. Annals of Applied Biology, 56, 323–324.

van Lenteren, J. C., Bakker, K., & van Alphen, J. J. M. (1978). How to analyze host discrimination. Ecology and Entomology, 3, 71–75.

Van Steenburgh, W. E. (1934). *Trichogramma minutum* Riley as a parasite of the oriental fruit moth, (*Laspeyresia molesta* Busck) in Ontario. Canadian Journal of Research, 10, 287–314.

Van Veen, J. C., & Van Wijk, M. L. E. (1985). The unique structure and functions of the ovipositor of the non-paralyzing ectoparasitoid *Colpoclypeus florus* Walk. (Hym., Eulophidae) with special reference to antennal sensilla and immature stages. Zeitschrift fuer Angewandte Entomologie, 99, 511–531.

Vet, L. E. M. (1983). Host-habitat location through olfactory cues by *Leptopilina clavipes* (Hartig) (Hym.: Eucoilidae), a parasitoid of fungivorous *Drosophila*: The influence of conditioning. Netherlands Journal of Zoology, 33, 225–248.

Vet, L. E. M. (1985). (1985). Response to kairomones by some alysiine and eucoilid parasitoid species (Hymenoptera). Netherlands Journal of Zoology, 35, 486–496.

Vet, L. E. M., & Bakker, K. (1985). A comparative functional approach to the host detection behaviour of parasitic wasps. II. A quantitative study on eight eucoilid species. Oikos, 44, 487–498.

Vet, L. E. M., & van Alphen, J. J. M. (1985). A comparative functional approach to the host detection behaviour of parasitic wasps. I. A qualitative study on Eucoilidae and Alysiinae. Oikos, 44, 478–486.

Viggiani, G. (1984). Bionomics of the Aphelinidae. Annual Review of Entomology, 29, 257–276.

Vinson, S. B. (1968). Source of a substance in *Heliothis virescens* (Lepidoptera: Noctuidae) that elicits a searching response in its habitual parasite, *Cardiochiles nigriceps* (Hymenoptera: Braconidae). Annals of the Entomological Society of America, 61, 8–10.

Vinson, S. B. (1975a). Source of a material in the tobacco budworm which initiates host-searching by the egg-larval parasitoid, *Chelonus texanus*. Annals of the Entomological Society of America, 68, 381–384.

Vinson, S. B. (1975b). Biochemical coevolution between parasitoids and their hosts. In P. W. Price (Ed.), Evolutionary strategies of parasitic insects and mites (pp. 14–48). New York: Plenum Press.

Vinson, S. B. (1976). Host selection by insect parasitoids. Annual Review of Entomology, 21, 109–133.

Vinson, S. B. (1984). Parasitoid-host relationships. In W. J. Bell & R. Carde (Eds.), Chemical ecology of insects (pp. 205–236). London: Chapman & Hall.

Vinson, S. B., & Stoltz, D. B. (1986). Cross-protection experiments with two parasitoid (Hymenoptera: Ichneumonidae) viruses. Annals of the Entomological Society of America, 79, 216–218.

Waage, J. (1979). Foraging for patchily distributed hosts by the parasitoid, *Nemeritis canescens*. Journal of Animal Behavior, 48, 353–371.

Walker, I. (1959). Die Abwehrreaktion des Wirtes *Drosophila melanogaster* gegen die zoophage Cynipide *Pseudeucoila bochei* Weld. Revue Suisse de, Zoologie, 66, 569–632.

Walker, I., & Pimentel, D. (1966). Correlation between longevity and incidence of diapause in *Nasonia vitripennis* Walker (Hymenoptera, Pteromalidae). Gerontologia, 12, 89–98.

Walter, G. H. (1983). "Divergent male ontogenies" in Aphelinidae (Hymenoptera: Chalcidoidea): A simplified classification and a suggested evolutionary sequence. Biological Journal of the Linnean Society, 19, 63–82.

Werren, J. H. (1984). Brood size and sex ratio regulation in the parasitic wasp *Nasonia vitripennis* (Walker) (Hymenoptera: Pteromalidae). Netherlands Journal of Zoology, 34, 123–143.

Werren, J. H., Skinner, S. W., & Charnov, E. L. (1981). Paternal inheritance of a daughterless sex ratio factor. Nature (London), 293, 467–468.

Werren, J. H., Skinner, S. W., & Huger, A. M. (1986). Male-killing bacteria in a parasitic wasp. Science, 231, 990–992.

Weseloh, R. M. (1969). Biology of *Cheiloneurus noxius,* with emphasis on host relationships and oviposition behavior. Annals of the Entomological Society of America, 62, 299–305.

Weseloh, R. M. (1971a). Influence of host deprivation and physical host characteristics on host selection behavior of the hyperparasite *Cheiloneurus noxius* (Hymenoptera: Encyrtidae). Annals of the Entomological Society of America, 64, 580–586.

Weseloh, R. M. (1971b). Influence of primary (parasite) hosts on host selection of the hyperparasite *Cheiloneurus noxius* (Hymenoptera: Encyrtidae). Annals of the Entomological Society of America, 64, 1233–1236.

Weseloh, R. M. (1972). Sense organs of the hyperparasite *Cheiloneurus noxius* (Hymenoptera: Encyrtidae) important in host selection processes. Annals of the Entomological Society of America, 65, 41–46.

Weseloh, R. M. (1974). Host recognition by the gypsy moth larval parasitoid, *Apanteles melanoscelus*. Annals of the Entomological Society of America, 67, 583–587.

Weseloh, R. M. (1981). Host location by parasitoids. In D. A. Nordlund, R. L. Jones, & W. J. Lewis (Eds.), Semiochemicals: Their role in pest control (pp. 79–95) New York: John Wiley & Sons.

Weseloh, R. M., & Bartlett, B. R. (1971). Influence of chemical characteristics of the secondary scale host on host selection behavior of the hyperparasite *Cheiloneurus noxius* (Hymenoptera: Encyrtidae). Annals of the Entomological Society of America, 64, 1259–1264.

Wilkes, A. (1942). The influence of selection on the preferendum of a chalcid (*Microplectron fuscipennis* Zett.) and its significance in the biological control of an insect pest. Proceedings of the Royal Society of London, Series B, 130, 400–415.

Wilkes, A. (1947). The effects of selective breeding on the laboratory propagation of insect parasites. Proceedings of the Royal Entomological Society of London, Series B, 134, 227–245.

Wilkes, A. (1959). Effects of high temperature during post embryonic development on the sex ratio of an arrhenotokous insect *Dahlbominus fuliginosus* (Nees) (Hymenoptera: Eulophidae). Canadian Journal of Genetics and Cytology, 1, 102–109.

Wilkes, A. (1963). Environmental causes of variation in the sex ratio of an arrhenotokous insect, *Dahlbominus fuliginosus* (Nees) (Hymenoptera: Eulophidae). Canadian Entomologist, 95, 183–202.

Wilson, D. D., Ridgway, R. L., & Vinson, S. B. (1974). Host acceptance and oviposition behavior of the parasitoid *Campoletis sonorensis* (Hymenoptera: Ichneumonidae). Annals of the Entomological Society of America, 67, 271–274.

Wilson, F., & Woolcock, L. T. (1960). Temperature determination of sex in a parthenogenetic parasite, *Ooencyrtus submetallicus* (Howard) (Hymenoptera: Encyrtidae). Australian Journal of Zoology, 8, 153–169.

Wylie, H. G. (1958). Factor that affect host finding by *Nasonia vitripennis* (Walk.) (Hymenoptera: Pteromalidae). Canadian Entomologist, 90, 597–608.

Wylie, H. G. (1965a). Discrimination between parasitized and unparasitized house fly pupae by females of *Nasonia vitripennis* (Walk.) (Hymenoptera: Pteromalidae). Canadian Entomologist, 97, 279–286.

Wylie, H. G. (1965b). Effects of superparasitism on *Nasonia vitripennis* (Hymenoptera: Pteromalidae). Canadian Entomologist, 97, 326–331.

Wylie, H. G. (1965c). Some factors that reduce the reproductive rate of *Nasonia vitripennis* (Walk.) at high adult population densities. Canadian Entomologist, 97 970–977.

Wylie, H. G. (1966a). Survival and reproduction of *Nasonia vitripennis* (Walk.) at different host population densities. Canadian Entomologist, 98, 275–281.

Wylie, H. G. (1966b) Some mechanisms that affect the sex ratio of *Nasonia vitripennis* (Walk.) (Hymenoptera: Pteromalidae) reared from superparasitized fly pupae. Canadian Entomologist, 98, 645–653.

Yashnosh, V. A. (1976). Classification of the parasitic Hymenoptera of the family Aphelinidae (Chalcidoidea). Entomological Review, 55, 114–120.

Yashnosh, V. A. (1983). Review of the world genera of Aphelinidae (Hymenoptera). I. Key to the genera. Entomological Review, 62, 145–159.

Yoshimoto, C. M. (1975). Cretaceous chalcidoid fossils from Canadian amber. Canadian Entomologist, 107, 499–528.

Zinna, G. (1962). Ricerche sugli insetti entomofagi. III. Specializzazione entomoparassitica negli Aphelinidae: Interdependenze biocenotiche tra due specie associate. Studio morfologico, etologico e fisiologico del *Coccophagoides similis* (Masi) e *Azotus matritensis* Mercet. Bollettino del Laboratorio di Entomologia Agravia 'Fillippo Silvestri' Portici, 20, 73–184.

Zoebelei, G. Z. (1956a). Der Honigtau als Nahrung der Insekten. Teil I. Zeitschrift fuer Angewandte Entomologie, 38, 369–416.

Zoebelein, G. Z. (1956b). Der Honigtau als Nahrung der Insekten. II. Zeitschrift fuer Angewandte Entomologie, 39, 129–167.

16

Terrestrial Arthropod Predators of Insect and Mite Pests

K. S. HAGEN,* N. J. MILLS,† G. GORDH,‡ and J. A. MCMURTRY§

*Division of Biological Control, University of California, 1050 San Pablo Avenue, Albany, CA 94706
†Insect Biology, Wellman Hall, University of California, Berkeley, CA 94720-3112
‡Department of Entomology, University of Queensland, St. Lucia, Queensland 4072, Australia
§Department of Entomology, University of California, Riverside, CA 92521

INTRODUCTION

Terrestial arthropod predators are important natural enemies. However, they often are overlooked and their impact underestimated in both natural and managed ecosystems due to a number of factors. For example, in contrast to all other natural-enemy groups, the duration of the interaction between a predator and each prey item is so brief that direct estimation of a predator's impact in the field is frequently impossible [see Wratten (1987), Luck *et al.* (1988) and Stuart and Greenstone (1990) for a discussion of methods for the indirect assessment of predation]. This brief duration of interaction, the tremendous overall diversity of the group, their frequent nocturnal activity, their more restrictive rearing requirements, and many other characteristics, all pose greater obstacles for the detailed investigation in comparison with other natural-enemy groups.

The majority of predators are carnivorous throughout their life cycle, though in some groups, predation is confined to either juvenile (e.g., Syrphidae, *Chrysoperla*) or adult stages (e.g., Asilidae, Empidae). An interesting variation on this theme is found among some Miridae. The vulnerable first nymphal instar is often phytophagous but rapidly shifts to a predatory diet as it increases in size and searching capacity. There is also considerable variation in the number of prey consumed by predators, forming a continuum that extends from the near parasitoid characteristics of the vedalia beetle through to the voracious appetite of *Chrysoperla carnea* (Stephens).

Conceptually, the action of predators is fundamentally different and more complex than that of parasitoids (Sabelis, 1992). At the individual level, a typical predator must search for prey at each successive stage of its development and the success of this search determines the individual's growth and survival. Thus, both age structure and energetics are essential elements of predation. This was initially considered by Hassell *et al.* (1976) and Beddington *et al.* (1976) but more elegantly incorporated into a conceptual framework by Gutierrez *et al.* (1981, 1990), where predator satiation (Mills,1982a) and predator–prey life ratio (Sabelis, 1992) are the important factors influencing the impact of predation. At the population level, predators are typically less specialized and are seldom confined to attacking a single prey species. Thus, predator population development will often be determined by the abundance and characteristics of several prey species—the conquences of these factors have yet to be fully explored.

The less specialized nature of predators and their requirement for a greater prey abundance to ensure oviposition and survival have generally led to their being considered less suitable than parasitoids in classical biological control (Huffaker *et al.,* 1971; Huffaker *et al.,* 1976). Conversely, predators can attack a wider range of prey stages than parasitoids and can continue to attack prey throughout their development. This means that predators have a distinct advantage over parasitoids in that they do not need to be synchronized with a single susceptible stage in the life cycle of their prey. The historical record, in fact, indicates that predators are no more difficult to establish (Hall & Ehler, 1979), and do not perform any less effectively than parasitoids in achieving complete success in the control of target pests (Hall *et al.*, 1980). Hokkanen (1985) also noted that predators have been more effective against certain groups of pests (such as the Coleoptera) and in annual cropping systems compared with more stable habitats. They have also been particularly effective against those pests that have no parasitoids, such as the Adelgidae and Tetranychidae.

The use of predators in classical biological control may continue to fall behind that of parasitoids due to increasing environmental concerns over importations of less specialized natural enemies. However, their importance in augmentation and conservation is likely to increase in the future as pest management becomes less reliant on the use of synthetic insecticides.

Predaceous arthropods are represented in nearly all taxa that make up the subphyla Chelicerata, Uniramia, and Crustacea, which compose the phylum Arthropoda (Hagen, 1987; New, 1991; Sabelis, 1992). We are concerned here only with the predaceous taxa that belong to the Chelicerata and Uniramia.

SUBPHYLUM CHELICERATA

The Chelicerata is composed of three classes of over 63,000 known species. The most important class with regard to applied biological control is the Arachnida, divided into 13 orders. We will consider only the two most important, the Araneae (spiders) and Acariformes (mites).

Class Arachnida

Subclass Araneae

Of the 87 families that belong to Araneae, 10 contain species that have been involved in biological control studies, but thus far there is no evidence that spiders as individual species can achieve regulation of a pest insect or mite population. Spiders are considered to have a stabilizing influence on species.

Although spiders are ubiquitous, are entirely carnivorous, and prey predominantly on insects, their effect on pest populations is poorly known. This probably can be attributed to several factors, including the secretive habits of many species, the difficulty of evaluating the impact of generalist predators on a given pest species, and the general lack of knowledge of and/or interest in spiders.

Moreover, spiders possess few of the biological traits usually considered important for natural-enemy effectiveness (Riechert & Lockley, 1984). They have long life cycles relative to those of insect or mite pests (usually 1 year or one season), thus they cannot exert a within-season numerical response to increasing pest densities. Spiders do not have distinct preferences for a given prey species, and switching to a more abundant food source seems to be a rare occurrence (Riechert & Lockley, 1984). Also, most spiders exhibit territorial behavior; therefore, a marked density-dependent response through immigration is unlikely.

Nevertheless, some studies suggest that spiders can be a significant mortality factor acting on some pest species, and have certain inherent traits as potentially effective predators. Some spiders have been shown to exhibit a sigmoidal or density-dependent functional response, characterized by a high plateau (Riechert & Lockley, 1984; Provencher & Coderre, 1987; Riechert, 1992). This high plateau (where no further increase in attacks occurs in response to increasing prey density) can be attributed to the characteristic of spiders killing many times more prey than they actually consume. Theoretically, a stable predator–prey interaction can be maintained by a sigmoidal functional response alone, without a numerical response of the natural enemy (Hassell, 1978). This, along with a self-limitation factor in the dynamics of polyphagous predators like spiders, may contribute to regulation of pest populations at low levels (Riechert & Lockley, 1984; Riechert, 1992). Some selectivity also occurs because of such factors as location in the microhabitat, size limitations of the prey, characteristics of the web, foraging behavior, and palatability of the prey. The impact of generalist arthropod predators was demonstrated by preserving or removing spiders in an experimental vegetable garden; by mulching and/or providing flowering plants, the spider assemblages were increased and insect damage was reduced (Riechert & Bishop, 1990). In this study where spiders were removed from manipulated plots, insect damage increased, proving that spider predation was the critical cause.

Families of Spiders Having Potential Importance in Biological Control

Although there have been numerous surveys of the spider faunas in different agricultural habitats and laboratory studies to determine feeding by spiders on various pest species, there have been relatively few attempts to evaluate the impact of spiders on pest populations [see reviews by Riechert and Lockley (1984), Nyffeler and Benz (1987), and Young & Edwards (1990)]. A complete listing of references is not attempted here.

Theridiidae and Linyphiidae   The Theridiidae and Linyphiidae are two families of web builders (scattered, irregular strands and sheet webs, respectively) prevalent in many agroecosystems, such as field crops. Although various species have been observed to feed on insect pest species, there are few data available concerning their impact on pest populations. A *Theridion* species represented 63% of the spider population on avocado foliage in Israel (Mansour *et al.*, 1985). Branches from which spiders were removed showed damage from feeding of geometrid larvae, while no damage occurred on branches on which spiders were undisturbed. Prey consumption by a linyphiid, *Lepthyphantes tenuis* (Blackwall), followed a type II functional response, increasing as the density of wheat aphids rose and then leveling out, but the numbers of aphids killed and not eaten had a simple linear correlation with prey density (Mansour & Heimbach, 1993).

Micryphantidae The Micryphantidae are mainly small spiders (rarely more than 2 mm) that could easily be overlooked as potential natural enemies. Life table studies indicated that micryphantids were an important mortality factor in the life cycle of *Spodoptera litura* (Fabricius) by preying on the newly hatched larvae and disturbing the larval colonization of taro leaves (Yamanaka *et al.,* 1973; Nakasuji *et al.,* 1973). Micryphantids were also observed on avocado leaves, feeding on spider mite *Oligonychus punicae* (Hirst) (McMurtry & Johnson, 1966). *Erigone atra* (Blackwall) demonstrated similar responses to grain aphid densities as the linyphiid *L. tenuis* (Mansour & Heimbach, 1993).

Araneidae and Tetragnathidae Orb weavers in the families Araneidae and Tetragnathidae are also frequently mentioned as comprising a large proportion of the spider faunas in some ecosystems. Their abundance and estimated predation pressure on insect populations in grassland ecosystems was pointed out by Nyffeler and Benz (1987). Orb weavers are the dominant group in some crops, such as cotton and soybeans, in the United States (Whitcomb & Bell, 1964; LeSar & Unzicker, 1978; Nyffeler *et al.,* 1989; Young & Edwards, 1990). In general, these spiders are limited to drifting, flying, or hopping prey ensnared in the webs. There is no conclusive evidence that they are significant biological control agents on crops, but experimental studies are lacking.

Clubionidae There is some evidence that the Clubionidae are significant predators of lepidopteran pests under some conditions. This family consists of hunting spiders that are primarily nocturnal and commonly construct silken tubes for retreats. Bishop and Blood (1981) showed direct numerical relationships between clubiionid, *Chiracanthium diversum* Koch and oxyopid *Oxyopes mundulus* Koch with larval populations of *Heliothis* on cotton in Australia. The former species was considered more important because of its numerical dominance, its densities sometimes being nearly as high as *Heliothis* spp. Although these spiders were considered incapable of controlling the pests by themselves, their role as part of the predator complex was considered important, as was their apparent ability to regulate small numbers of the larvae present between peaks in *Heliothis* activity. *Chiracanthium mildei* Koch was the dominant species of spider on apple trees in Israel in an experiment in which spiders were removed from three to six apple trees and egg masses of *S. littoralis* (Boisduval) were added (Mansour *et al.,* 1980a). *Spodoptera littoralis* larvae and leaf damage were observed 5 days later on the trees from which spiders were removed, while on the control trees no live larvae or damaged leaves were observed. In addition to causing direct mortality, these spiders also were observed to interfere with the colonization and aggregation behavior of the larvae (Mansour *et al.,* 1981). *Chiracanthium mildei* also showed a functional response to increasing densities of 4-day-old *Spodoptera* larvae (Mansour *et al.,* 1980b). The response by individual spiders starved 4 days leveled off at about 180 larvae consumed in an 18-h. period at prey densities of 250 and 300 larvae per cage. It has been pointed out that spiders may consume considerably more prey in the laboratory than would be available in the field (Nyffeler & Benz, 1987).

Lycosidae The Lycosidae, or wolf spiders, are primarily ground-inhabiting hunters that will also forage on low-growing vegetation. Coates (1974) considered *Pardosa crassipalpus* (Purcell) to be one of the most important natural enemies of mite *Tetranychus cinnabarinus* (Boisduval) on strawberries in South Africa. Because of its nocturnal habits, it previously had gone unnoticed. An intensive quantitative evaluation of spider predation was conducted in rice fields in Japan by Kiritani *et al.* (1972). A team of observers recorded all predation events by spiders twice weekly at different times of the day and formulae were derived to estimate the total predation per day of the prevalent spiders on the green rice leafhopper, *Nephotedtex cincticeps* Uhler, and the brown rice plant hopper, *Nilaparvata lugens* (Stål), as well as on other spiders and miscellaneous prey. It was estimated that the two hopper species comprised at least 80% of the diet of *Lycosa pseudoannulata* Boes et Str. The percentage of each of the two species of prey in the spider's diet varied according to their relative abundance in a given season. An orb weaver in the Tetragnathidae and a micryphantid also utilized the same prey (up to 50% of their diet). Further quantitative studies on *L. pseudoannulata* indicated that this predator is important in reducing the density of the green rice leafhopper. *Lycosa* has two and a partial third generations and its population growth fits a logistic curve (Kiritani & Kakiya, 1975).

Lycosa spp. were considered second in importance to ants in predation on the cattle tick *Boophilus microplus* (Can.) in Queensland, Australia (Wilkinson, 1970). An estimated 50% of the engorged females were killed by these two groups of predators. More nocturnally active lycosid species survive pesticide treatments compared with diel-active species that are exposed in cotton fields during applications (Hayes & Lockley, 1990).

Oxyopidae The Oxyopidae is another family of hunting spiders common on low-growing plants. *Oxyopus* species occur worldwide in agroecosystems (Young & Lockley, 1985; Young & Edwards, 1990). *Oxyopus salticus* Hentz can either search actively or assume an ambush behavior (Young & Lockley, 1986). These workers found that predation success on *Lygus lineolaris* (Palisot de Beauvois) was inversely related to prey size. They suggested that early-season *Lygus* nymphs of any size may suffer high levels of predation by the large immature and adult spiders present during that time, while mainly small nymphs would be preyed on by the predominantly small immature spiders present later in the season.

Studies by Nyffeler *et al.* (1987b) demonstrated that *O. salticus* is a generalist predator. Its diet in a Texas cotton field included *Solenopsis invicta* Buren (red imported fire ant) (22%), leafhoppers (17.2%), dipterans (15.6%), aphids (14%), and spiders (14%). Predaceous arthropods, including fire ants, made up 42% of the spider's diet. The major cotton pests (boll weevils, cotton fleahopper, and *Heliothis* spp.) were not numerous. *Oxyopus salticus* killed an estimated 4.5% of the arthropods present per week. Although the green lynx spider, *Peucetia viridens* (Hentz), fed on some of the major cotton pests, over 50% of its diet consisted of beneficial arthropods (Nyffeler *et al.*, 1987a). In an experimental plot in which 45,000 *O. sertatus* Koch were released, the population density of cecidomyiid *Contarina inouyei* Mani was reduced to ca. 25% of the level of the control plot in a cryptomeria forest in Japan (Kayashima, 1961).

Salticidae   Species in the Salticidae (jumping spiders) are common in agroecosystems and prey on various pests (Riechert & Lockley, 1984; Young & Edwards, 1990). These spiders generally are diurnal and hunt their prey visually. In greenhouse experiments, Muniappan and Chada (1970) and Horner (1972) showed that *Phidippus audex* (Hentz) and *Metaphidippus galathea* (Walck), respectively, suppressed populations of the greenbug on seeding plants. However, because only one prey and one predator species were present in the experimental systems, it is invalid to conclude these predators are important under field conditions. Functional responses of *Marpissa tigrina* Tikader suggested that this species is a highly efficient predator of a psyllid on citrus in India (Sanda, 1991), but 18 species of spiders, mainly salticids, preying on citrus psyllids in South Africa did not appear to directly influence the populations of the psylla to any great extent (Van den Berg *et al.*, 1992).

Thomisidae   The Thomisidae (crab spiders) frequently are mentioned as potentially important predators of crop pests (Riechert & Lockley, 1984; Young & Edwards, 1990), but no quantitative studies on predator–prey interactions have been conducted. Putman and Herne (1966), using chromatographic methods, found that spiders, the most abundant of which was *Pliodromus praelustrus* Keyserling, showed a functional response to increasing densities of spider mites [*Panonychus ulmi* (Koch) and *Bryobia*

rubrioculus (Scheuten) (= *arboreus*)] in a peach orchard. Feeding occurred even at very low densities. They concluded that spiders are part of the complex of minor predators, which aids the major predators in maintaining the density of *P. ulmi* at low levels.

Conclusions

There is no evidence that spiders, as individual species, can achieve regulation of a pest insect or mite population. It is more likely that the assemblage of spider species contribute a stabilizing influence on pest species (Riechert & Lockley, 1984; Riechert & Bishop, 1990). Therefore, laboratory or greenhouse experiments investigating the effect of single species of spiders on single species of pests are of questionable relevance to the field situation. A potential adverse effect of spiders on biological control is that of destruction of more effective specific parasitoids and predators or of competition with these natural enemies for prey (Whitcomb & Bell, 1964; Nyffeler *et al.*, 1987a, 1987b). On the other hand, spiders could have a buffering effect on large population fluctuations by stabilizing numbers of the specific enemy, thereby preventing local annihilation of both the pest and the enemy.

Although there is a paucity of information on the importance of spiders in biological control, the evidence is sufficiently encouraging to warrant further study and to attempt to preserve them in the ecosystem.

Subclass Acari

Order Parasitiformes

In the subclass Acari, there are 331 known families of which at least 20 include predaceous species, but only 4 families in the suborder Mesostigmata and 7 in the suborder Prostigmata are discussed here.

Suborder Mesostigmata (= Gamasida)

Phytoseiidae   The Phytoseiidae, containing over 1000 described species, are essentially ubiquitous inhabitants on higher plants, including both annual and perennial agroecosystems. The importance of phytoseiids in the biological control of phytophagous mites has been shown in various situations (e.g., McMurtry, 1982; McMurtry & Croft, 1997).

Life Histories   Phytoseiids go through developmental stages typical of most gamasid mites, (i.e., egg, hexapod larva, octapod protonymph and deutonymph, and adult). Generation times generally require 6 to 10 days at warm temperatures (usually less than that of their prey, such as tetranychid mites and thrips) (McMurtry *et al.*, 1970; Sabelis, 1985). Development can be continuous in warm climates, while in temperate climates species overwinter as inseminated adult females. A reproductive diapause, regulated by photoperiod and influenced also by temperature, has been found in some species.

Except for a few known thelytokous species, phytoseiid females must mate to reproduce. Mating usually occurs as soon as the female molts to the adult stage. The male is attracted at close range by a female sex pheromone (Rock *et al.,* 1976; Hoy & Smilanick, 1979). Many species require multiple matings to produce their full complement of eggs. However, a single mating is sufficient for some species, especially among those that have a high reproductive potential and strongly female-biased sex ratio (Bounfour & McMurtry, 1987). In the species investigated, all the eggs are fertilized; however, during embryo development, some eggs lose the paternal set of chromosomes, with the haploid eggs presumably developing into males and the diploid eggs presumably developing into females. This phenomenon has been termed parahaploidy (Hoy, 1979) or pseudo-arrhenotoky (Schulten, 1985).

Fecundity of *Phytoseiulus* and some *Neoseiulus* species average over 60 eggs per female, with a rate of up to 4 eggs per day at the peak in the oviposition period (McMurtry & Rodriguez, 1987). In general, those species having a high fecundity rate have a rapid developmental period as well, resulting in a high intrinsic rate of increase (r_m) (Sabelis, 1985). Such species (e.g., *P. persimilis* Athias-Henriot) are usually specialized predators closely associated with *Tetranychus* species developing in patchily distributed colonies with dense webbing. These predators have been most widely used in applied biological control of spider mites. However, many species have lower rates of reproduction (1 to 2 eggs per day) and development, with consequently lower intrinsic rates of increase (Tanigoshi, 1982). Such species tend to be more general feeders that utilize foods such as pollen, small insects, and eriophyid mites as well as tetranychid mites. Thus, these phytoseiids seem to be "k-selected" species.

Predator Attributes Although phytoseiids are best known as predators of spider mites (which certainly reflects their prime importance in applied biological control), relatively few species are dependent on this kind of prey for food; and the majority of species are probably general feeders that subsist on a variety of foods, including pollen (McMurtry & Rodriguez, 1987; McMurtry & Croft, 1997). Therefore, species present in a given ecosystem should be investigated for their potential as predators of other pests in addition to spider mites. For example, certain phytoseiids have been found to be important natural enemies of thrips, including *Euseius tularensis* Congdon on *Scirtothrips citri* (Moulton) (Tanigoshi *et al.,* 1985) and *E. addoensis* on *S. aurantii* Faure (Grout & Richards 1992), respectively, and *N. barkeri* (Hughes) on *Thrips tabaci* Lindeman (Ramakers, 1980).

Phytoseiids have also been recorded attacking other groups of phytophagous mites. There is one documented case of control of a tarsonemid mite (cyclamen mite on strawberries in California) (Huffaker & Kennett, 1956). Several species of phytoseiids were able to develop and reproduce on the tarsonemid *Polyphagotarsonemus latus* (Banks) (McMurtry *et al.,* 1984). Control of eriophid mites has not been demonstrated, but these mites can be an important alternate food for phytoseiids, which may result in stabilizing spider mite populations at lower levels than would occur without this source of food (Hoyt, 1969).

Specialized feeding on tetranychid mites has evolved, apparently independently, in several unrelated groups of phytoseiids, including *Phytoseiulus, Neoseiulus, Galendromus* (= *occidentalis* group of *Typhlodromus*) (McMurtry, 1982; McMurtry & Croft, 1997). Species in these three groups are usually associated with *Tetranychus* species that form colonies with dense webbing. The long mediodorsal setae possessed by these phytoseiids may facilitate their movement through the webbing, which may entangle species in other genera (Sabelis & Bakker, 1992).

It has been demonstrated that some of these species show attraction or arrestment responses to kairomones present on mites, webbing, or feces of certain species of spider mites (Sabelis & Dicke, 1985). The predators *Phytoseiulus persimilis, Galendromus occidentalis* (Nesbitt), and *Amblyseius potentillae* (Garman) showed a strong positive response to kairomones from a preferred prey species, but showed little or no response to a species in a different genus of spider mite with which it is not associated under field conditions. However, *P. persimilis* responded to kairomones of both a favorable prey species, *T. urticae* Koch, and an unfavorable one, *T. evansi* Baker & Pritchard (de Moraes & McMurtry, 1985).

The capacity for attacking and feeding on tetranychid prey varies with the phytoseiid species and is determined, at least in part, by the size and reproductive capacity of the predator. *Phytoseiulus persimilis,* a relatively large and highly fecund species, consumes 10 to 12 spider mite eggs during nymphal development and 14 to 22 eggs per day per ovipositing female (Takafuji & Chant, 1976; Sabelis, 1981). Ovipositing females of *P. longipes* Evans, given single stages of *T. pacificus* McGregor, fed on 31 eggs, 30 larvae, 23 protonymphs, 15 deutonymphs, or 4 adult female prey per day (Badii, 1981). These rates are lower in smaller species, such as *G. occidentalis* (Laing, 1968; Sabelis, 1981). The prey consumption capacity for phytoseiids is thus much lower than for the common insect predators of spider mites, such as *Stethorus* or *Scolothrips* species. An advantage to this characteristic is that phytoseiids can develop and reproduce at low prey densities compared with the larger predators. This, along with a short generation time, can result in a numerical response to prey increase with a shorter lag time even when the pest mite numbers are still relatively low.

Some Genera and Species Groups and Their Possible Importance Phytoseiulus is a small genus in which all known species are specialized predators of spider mites (mainly *Tetranychus* species) that develop in dense colonies with much webbing. *Phytoseiulus persimilis* has a lower oviposition rate on other genera of spider mites, such as *Panonychus* or *Oligonychus* (Ashihara *et al.*, 1978), while *Phytoseiulus longipes* showed no reproduction on *Panonychus citri* (McGregor) and only a low rate on *Oligonychus punicae* (Badii & McMurtry, 1984). *Phytoseiulus persimilis* has been studied and used in biological control programs more than any other phytoseiid. Its use has been mainly for mite control in greenhouses and to some extent on strawberries outdoors (McMurtry, 1982; McMurtry & Croft, 1997).

Galendromus is a genus whose species show a close association with tetranychid mites. *Galendromus occidentalis* is the major predator of various *Tetranychus* species on deciduous tree crops and grapes in western North America, and *G. helveolus* Chant (= *floridanus*) appears to be an important predator of *Eotetranychus sexmaculatus* (Riley) in Florida (McMurtry, 1982). The species studied do not show a close association with spider mites having less patchy distributions and producing relatively little webbing. For example, *G. occidentalis* does not appear effective in controlling *Panonychus ulmi* (Hoyt, 1969) and other species in this category. Moreover, it does not respond to chemical cues from that mite as it does to cues from *Tetranychus* species (Hoy & Smilanick, 1981; Sabelis & van de Baan, 1983). Unlike the *Phytoseiulus* species, those in the *G. occidentalis* group that have been studied can subsist and even increase on other types of prey, such as eriophyid and tydeid mites.

Typhlodromus includes the widely distributed *T. pyri* Scheuten, common on deciduous fruit crops, and also some species on subtropical crops (such as *T. athiasae* Porath & Swirski, *T. phialatus* Athias-Henriot, and *T. exhilaratus* Ragusa). These species are more general in their food habits, because they have been reported to reproduce on pollen, on eriophyid and tenuipalpid mites, and on tetranychids (Chant, 1959; Swirski *et al.*, 1967b). Field studies in several countries indicate that *T. pyri* can be important in maintaining *Panonychus ulmi* and sometimes *Tetranychus uricae* at low densities, although it may not be capable of keeping pace with high densities of this mite (McMurtry, 1982; Nyrop, 1988; Croft *et al.*, 1992).

Neoseiulus (*sensu* Muma *et al.*, 1970) is a genus that contains various potentially valuable predators of tetranychid mites, including *N. fallacis* (Garman), *N. californicus* (McGregor), *N. chilenensis* (Dosse), *N. bibens* (Blommers), and *N. longispinosus* (Evans) (McMurtry, 1982; McMurtry & Croft, 1997). These mites usually are found in close association with *Tetranychus* species and all have a high reproductive capacity when feeding on *Tetranychus* (Ma &

Laing, 1973; Blommers, 1976; Ball, 1980). *Neoseiulus fallacis* has an impact on *P. ulmi* as well as *Tetranychus* species on deciduous fruit trees (Croft & McGroarty, 1973).

Other species of *Neoseiulus* are more generalized in their food habits, feeding on other small arthropods as well as pollen. *Neoseiulus barkeri* and *N. cucumeris* are used for thrips control on greenhouse-grown vegetables. Because they subsist on a variety of foods, they are sometimes released even before the pest thrips are detected (van Rijn & Sabelis, 1990).

Amblyseius sensu stricto is a large genus that is well represented in more humid regions and seems to consist largely of species that utilize a variety of food types. *Amblyseius potentillae,* which has the potential of controlling *P. ulmi* on deciduous fruit trees in Italy [where it is referred to as *A. andersoni* (Chant) and in the Netherlands as *A. eharai* Amitai & Swirski], and *A. herbicolus* (Chant) (= *deleoni* Muma & Denmark) are considered potentially effective predators of *P. citri* in Japan and Australia (Tanaka & Kashio, 1977; Beattie, 1978). Swirski *et al.* (1967a) showed that *A. swirskii* Athias could utilize many types of mites and insects found on citrus trees in Israel.

Euseius is a common genus in tropical and subtropical regions of the world. All species that have been studied reproduce readily on pollen and in some cases pollen seems to be a more favorable food than animal prey (McMurtry & Rodriguez, 1987). *Euseius* species frequently are the most abundant phytoseiids on citrus and other subtropical tree crops. Although seasonal population peaks sometimes are correlated with the abundance of pollen from the host tree or other nearby sources, studies have indicated that various species have some impact on *Panonychus* species (either citrus red mite or European red mite) in such countries as the Netherlands, South Africa, Australia, Chile, and the United States (McMurtry, 1982; McMurtry & Croft, 1997). Their potential as predators of thrips and other small insects also deserves further investigation. The importance of such "generalist" phytoseiids seems to be in stabilizing pest populations at low densities rather than in suppression of populations that have increased to higher numbers (McMurtry, 1992).

Ascidae This is a large and diverse family, common in such habitats as soil, humus, litter, foliage, bark, flowers, and nests of vertebrates. The food habits of most species are unknown; most are thought to be predaceous on such groups as nematodes and small arthropods but some may be pollen or fungus feeders (Lindquist & Evans, 1965).

Species in the genus *Blattisocius* have some potential for biological control of certain pest insects in stored products. *Blattisocius tarsalis* (Berlese) is a cosmopolitan species that preys on eggs of many insect species, especially Lepidoptera in stored grains and in laboratory cultures (Darst & King, 1969; Haines, 1981). This species has a short, non-feeding larval stage (+ 24 h) and predaceous protonymphal and deutonymphal stages that consume a mean of 3.7 eggs of *Ephestia cautella* (Walker). The generation time (egg–egg) is as short as 7 days at 27°C. Adult female mites consumed ca. 2.6 *Ephestia* eggs per day and laid 1.6 eggs per day with a mean total of 26 eggs per female. The intrinsic rate of increase was high in relation to that of the moth. The developmental and oviposition rates were much lower on *Tribolium* eggs (Haines, 1981). Adult female mites are phoretic on moths. The preferred location for attachment on *Plodia interpunctella* (Hübner) is at the base of the wings or on the dorsum of the abdomen, where up to 14 mites per moth have been observed (Darst & King, 1969). Experimental studies by Flanders and Badgley (1960, 1963) and White and Huffaker (1969) showed that *B. tarsalis* was the major factor contributing to regulation of *Anagasta kuehniella* (Zeller) in laboratory ecosystems. Graham (1970) and Haines (1981) concluded that this mite can exert effective control of *E. cautella* in the absence of insecticides, and that they can utilize eggs of *Tribolium* as an alternate food source.

Laelapidae This is a diverse family of mites ranging from generalist predator species to facultative parasites to obligate parasites on warm-blooded animals. They are found commonly in such habitats as manure, litter and soil, stored grain, and nests of various vertebrates and insects (Krantz, 1978). Evidence that laelapid mites may have a suppressive effect on pest insect populations was given by Chaing (1970). Corn plots in Minnesota that received manure applications showed higher populations of predaceous mites and lower populations of the corn rootworm, *Diabrotica longicornis* (Say), than did nonmanured plots. Laboratory observations confirmed that a *Androlaelaps* sp. and a *Stratiolaelaps* sp. fed on eggs and larvae of corn rootworm. Chaing concluded that the predaceous mites accounted for a 19.7% control of rootworms under natural conditions and a 63.0% control, in the manured plots.

Some *Hypoaspis* species are associated with phytophagous dynastine scarab beetles and are known to feed on beetle eggs. Some *Hypoaspis* mites, considered important mortality factors of *Oryctes* beetles, were introduced to Tokelau Island for attempted biological control of *O. rhinoceros* Linnaeus, a serious pest of coconuts (Swan, 1974).

Macrochelidae Mites of this family are common inhabitants of humus, litter, or other habitats at or near the ground level. A few occur in such places as nests of birds, mammals, bees, ants, or even on the fur of animals (Krantz, 1978). *Macrocheles* is the largest genus (over 700 species) and the one containing species of potential value in biological control of noxious flies. Most *Macrocheles* species have phoretic adult females, commonly on muscoid flies or scarabeiid dung beetles (Krantz, 1978). These species prey

on the eggs and small larvae of muscoid flies, and may have preferences for certain species. For example, *M. muscaedomesticae* (Scopoli) prefers housefly eggs to those of *Stomoxys calcitrans* (Linnaeus) or *Fannia canicularis* (Linnaeus) (Kinn, 1966; Singh *et al.,* 1966); it also feeds on nematodes (Rodriguez *et al.,* 1962). Both mated or unmated females are phoretic, and arrhenotoky apparently is the general mode of reproduction. Therefore, all mature females carried to a new dung habitat by a scarabeiid beetle or fly can begin laying eggs immediately after arrival (Axtell 1969). Most *Macrocheles* studied have a high reproductive potential, with the ability to complete a generation in as short a time as 4 days, or about 3 generations to 1 of the fly. Some species produce over 150 progeny in 10 to 12 days. Six of seven species treated had a higher intrinsic rate of increase (R_m) than that of *Musca domestica* (Linnaeus) at several temperatures (over 0.9 R_m in some instances) (Cicolani, 1979). Evidence by Jalil and Rodriguez (1970) suggests that *Macrocheles muscaedomesticae* responds to chemical cues from the fly eggs as prey and from the adult flies as carriers.

Various studies have shown that *M. muscaedomesticae* can reduce fly populations by 60 to 90%. Krantz (1983) reviewed this work and listed some limiting factors in the successful use of these mites in biological control of flies in poultry or dairy farms, and suggested some points on which further study is needed. It was pointed out that certain *Macrocheles* species not occurring in domestic habitats might be more effective than *M. muscaedomesticae*.

Macrocheles species also have been investigated for control of flies that breed in dung pads in pastures and rangelands. Because of the ephemeral nature and wide scattering of dung pads in these situations, it is not surprising that the macrochelid mites in such habitats have evolved close phoretic relationships with insects, primarily scarabeiid dung beetles. More than 180 known species of *Macrocheles* are associated with dung beetles, which are themselves major competitors with flies for dung as a food resource (Krantz, 1983). Extensive studies have been conducted in Australia on the biological control potential of *Macrocheles* species on bush fly *Musca vetustissima* Walker. Experiments with the indigenous *Macrocheles glaber* showed that bush fly breeding could be controlled by as few as 50 mites in a 1000-ml dung pad. The mites are phoretic on several species of native dung beetles, but they must arrive at the dung pad soon after the flies have laid their eggs. If the eggs escape predation, the larvae are able to escape by burrowing into the wet dung. Thus, the daily flight activity of the beetles plays an important role (Wallace *et al.,* 1979).

Macrocheles peregrinus Krantz was introduced from Africa and released in northern Australia, where indigenous species of the *M. glaber* group do not occur. The high reproductive potential of this species (life cycle of 3 days at 27°C) was considered a factor in its rapid establishment, and the fact that it is phoretic on many dung beetle species, including some introduced from Africa to effect fly control, was thought to be important in its rapid and extensive dispersal (Wallace & Holm, 1983). By early 1985 (less than 5 years after its introduction), it had spread to within 50 km of the northern range of the *M. glaber-perglaber* complex (Halliday & Holm, 1985). Krantz' (1983) review of the biological control potential of macrochelids against flies in pastoral settings points out some of the aspects on which more information is needed, including systematics, feeding habits, prey preferences, relationships with the beetle carriers (specificity, time of flight, etc.), and interactions with other mites.

Uropodidae These mites are common inhabitants in such ecosystems as litter, detritus, rotting wood, stored products, and manure. They are phoretic in the deutonymphal stage, commonly on insects, to which they attach by means of an anal pedicel. The feeding habits of most species are unknown, but the family apparently includes feeders on such things as fungi, organic detritus, nematodes, and insect eggs and larvae (Krantz, 1978).

Fuscuropoda vegetans (DeGeer) has been studied as a potential biological control agent of muscoid flies in poultry facilities. This species has a relatively long life cycle (24 days plus 7 to 11 days preoviposition) compared with macrochelid mites but its oviposition rate was lower on housefly eggs than on nematodes (4 versus 7 eggs per day) (Jalil & Rodriguez, 1970). Control of housefly populations was demonstrated experimentally (O'Donnell & Axtell, 1965; Rodriguez *et al.*, 1970). The latter authors suggested that *F. vegetans* complemented *Macrocheles* in biological control of fly populations, because the uropodids prefer larvae and the macrochelids prefer eggs.

Order Acariformes

Suborder Prostigmata (= Actinedida)

Tydeidae These mites are common foliage-inhabiting species on a wide variety of crops. They develop through the egg and three nymphal stages to the adult in as short a time as 9 days at 30°C. *Homepronematus anconai* (Baker) has a fecundity of over 40 eggs per female and a fairly high intrinsic rate of increase (0.278 at 30°C) (Knop & Hoy, 1983). There are conflicting opinions on whether foliage-inhabiting tydeids are predaceous, phytophagous, pollenophagous, or saprophagous. Various species probably can be placed in any of these categories. The literature on feeding habits of Tydeidae has been reviewed by Hessein and Perring (1986). These authors also conducted a study demonstrating the impact of tydeid predation on a pest population. On both individual tomato leaflets and tomato seedlings, *H. anconai* prevented the tomato russet mite, *Aculops lycopercicae* (Massee), from reaching damaging numbers. The ty-

deid also fed and reproduced on pollen alone (Knop & Hoy, 1983; Hessein & Perring, 1986).

Bdellidae Bdellids are medium- to large-sized mites (up to 4 mm) commonly occurring on low-growing vegetation, in soil or humus, and, in some cases, on trees or under bark. Prey include such small arthropods as Collembola, aphids, larvae of Diptera, and mites in the Tetranychidae and Penthaleidae (Alberti, 1973; Krantz, 1978). The prey is captured with the aid of silken threads, which also serve to anchor it to the substrate while the predator sucks out the body fluids (Alberti, 1973).

Some species of bdellids are active only during the cool season and estivate in the egg stage during the summer, their phenology being synchronized with such pests as Collembola and mites of the family Penthaleidae and the genus *Bryobia* (Tetranychidae) (Snetsinger, 1956; Wallace, 1967; Soliman & Mohamed, 1972). *Neomolgus aegyptiacus* Soliman consumed ca. 70 Collembola during the larval and three nymphal stages over a 22-day developmental period at 17 to 19°C (Soliman & Mohamad, 1972). This species was always observed in association with Collembola.

Bdellids were demonstrated to be important biological control agents of the collembolan *Sminthurus viridis* Linnaeus (lucerne flea), a pest of alfalfa in Australia. Studies by Wallace (1967) indicated that no outbreaks of *S. viridis* occurred when numbers of *Bdellodes lapidaria* Kramer exceeded 20 per square meters. A classical biological control program involved the introduction of *N. capillatus* (Kramer) from the Mediterranean area and its establishment in more arid parts of Australia where *B. lapidaria* did not occur (Wallace, 1974; Ireson & Webb, 1995).

Anystidae The "whirligig" mites have been observed feeding on such prey as aphids, thrips, leafhoppers, scale crawlers, and spider mites. They have been evaluated as biological control agents of collembolans and penthaleid mites (Michael, 1995).

Cheyletidae Cheyletids are predators that assume an "ambush" mode characterized by lack of motion, extended legs, and palpi abducted at a wide angle. The prey is grasped with the palpi and paralyzed within a few seconds (Wharton & Arlian, 1972). Various species occur in orchards, where they prey on such arthropods as scale crawlers and different mite species, including tetranychids and phytoseiids. They are considered to be of little importance in spider mite control because of low reproductive potential and prey consumption capacity (Laing & Knop, 1983). They also have been observed to feed on mites in the Tenuipalpidae (Haramoto, 1969) and Eriophyidae (Sternlicht, 1970), but no information is available on whether they have a significant impact on populations of these families of mites.

Some cheyletids prey on scale insects, usually by hiding under an empty scale or in other protected places on leaves or twigs, where they attack wandering crawlers (Gerson *et al.*, 1990). *Chetotogenes ornatus* (Canestrini & Fanzago) was reported to be the dominant acarine predator of chaff scale, *Parlatoria pergandii* Comstock, on the bark of citrus trees in Israel (Avidov *et al.*, 1968). It showed a density-dependent response to scale populations and peaked in late summer at the "critical, ebb period of the chaff scale's populations." It was considered "an enemy of some importance" on the chaff scale, but laboratory studies showed a relatively slow development rate, low fecundity, and only two generations per year.

On the other hand, there is evidence that cheyletids have potential for controlling acarid grain mites in stored grains. *Cheyletus eruditus* (Schrank) is often found in stored grains in sufficient numbers to suggest that it is an important predator (Norris, 1958; Solomon, 1969; Jeffrey, 1976; Bruce, 1983). The potential for manipulating *C. eruditus* for biological control of acarid mites was demonstrated by Pulpán and Verner (1965) in Czechoslovakia. They found that if the predator is present before the beginning of July and the infestation rate has not exceeded 1000 mites per 500 g of grain, spontaneous control can occur if there is one predator to 75 to 100 pest mites. They also showed possibilities of transferring predators from granaries where they are abundant to those where they are scarce, and they described a method whereby up to 25,000 mites could be collected in 1 h. Zdarkova (1986) reports additional work on production and utilization of *C. eruditus* for control of stored product mites in Czechoslovakia.

Although *C. eruditus* does not have a rapid rate of development or a high prey consumption capacity (Beer & Dailey, 1956; Solomon, 1969), fecundity is fairly high (up to 130 eggs per female) and all progeny are usually female (Summers & Witt, 1972; Hughes, 1976). It also has the ability to survive and find prey at very low prey densities. However, as the predator density increases and the prey density decreases, cannibalism becomes a major factor (Solomon, 1969; Burnett, 1977).

Stigmaeidae Species in the genera *Zetzellia* and *Agistemus* have been studied as potential predators of phytophagous mites and insects on crop plants. These mites are similar in size to, or slightly smaller than, tetranychid mites and develop through egg, larval, protonymph, and deutonymph stages to the adult. *Zetzellia mali* (Ewing) is common in deciduous orchards in North America. It occurs most frequently along the midrib on the undersides of the leaves and feeds on all stages of spider mites, eriophyid rust mites, and eggs of small insects, as well as on pollen. *Zetzellia mali's* rate of development and reproductive potential are lower than those of most phytoseiid mites, and thus the species is classed as a "k-selected" instead of an

"r-selected" predator. Therefore, its value in biological control is questionable (Laing & Knop, 1983; Santos & Laing, 1985).

Agistemus species have similar biologies and general feeding habits as *Zetzellia* species, and they sometimes reach high population densities on foliage. Collyer (1964) observed *A. longisetus* Gonzalez in numbers as high as 100 per leaf in some orchards and considered it a potentially valuable predator of tetranychid mites in the genera *Panonychus*, as well as of eriophyid and tenuipalpid mites. Collyer reported higher rates of fecundity and development than have been observed for most stigmaeids. Laing and Knop (1983) and Santos and Laing (1985) reviewed the food sources, host plant, and geographic ranges, as well as potential value of Stigmaeidae in biological control.

On the basis of laboratory and field studies, Oomen (1982) considered *Agistemus* species to be the most effective predators of the tenuipalpid mite, *Brevipalpus phoenicus* (Geijskes), on tea in Indonesia. Tenuipalpids have a lower reproductive potential than tetranychids, so the stigmaeids could be more effective predators of this family.

Erythraeidae and Trombidiidae The subcohort Parasitengona includes several families of moderate to very large, reddish mites (from 500 to over 10,000 species) that are general predators of small arthropods. Although the efficacy of these mites may be limited by slow developmental rates (1 to 3 generations per year), they could act as stabilizing factors in the ecosystem. The Erythraeidae and Trombidiidae are characterized by active larvae (many are parasitic on insects), quiescent protonymphs, predaceous deutonymphs, quiescent tritonymphs, and predaceous adults. These are largely ground-inhabiting species, although some (such as the erythraeid genus *Balaustium*) are arboreal, feeding on various soft-bodied insects and mites in orchards. Virtually no quantitative studies have been done on the effects of the predation by these families of mites on pest insect and mite populations. Various associations and potentials for biological control are reviewed by Lindquist (1983), Laing and Knop (1983), Eikwort (1983), and Welbourn (1983).

SUBPHYLUM UNIRAMIA

The subphylum Uniramia, which is made up of animals that belong to the classes Myriapoda and Hexapoda, contains the majority of invertebrate predators that have been used in biological control programs. Apparently, there have been no species in the Myriapoda that have been used directly to control pests; thus, the Hexapoda has been the source of most species used in biological control, and these predators have come from the subclass Pterygota (winged insects). The five other orders in Hexapoda, which formerly fell under the taxon "Aptera" (wingless insects), thus far have been of little importance to applied biological control.

Class Hexapoda

Subclass Pterygota

Of the 29 orders that make up the subclass Pterygota, 9 orders contain families of terrestrial predators: only the families that comprise species that have been involved in biological control programs are discussed here. In general, as the insect taxa advanced phylogentically, the more specialized their prey spectra became. All foraging strategies of predators, including random searching, hunting, ambush, or trapping, are exploited in the higher pterygote orders.

Order Orthoptera

Most families in Orthoptera are composed of phytophagous species, but there are a few among the over 22,000 species that are predaceous. The food habits of Orthoptera can be determined by examining mandibles and feculae, even of dead specimens (Gangwere, 1969).

Gryllotalpidae   Dissections of the alimentary tract of two freshly killed mole cricket species in the genus *Scapteriscus* showed that *S. acletus* Rehn & Hebard was primarily carnivorous while *S. vicinus* Scudder was herbivorous (Matheny, 1981).

Gryllidae   Among its nine subfamilies only the Oecanthinae (tree crickets), Nemobiinae (ground crickets), and Trigonidiinae are considered predaceous enough to offset their feeding on flowers and leaves. Even the common pestiferous field cricket, *Gryllus assimilis* (Fabricius), can be predaceous; it was recorded by Smith (1959) to have destroyed up to 50% of the egg pods of a grasshopper species in Canada. Ground cricket *Pictonemubius ambitiosus* (Scudder) was recorded to have eaten radioactive labeled velvet bean caterpillar eggs placed in a Florida soybean field (Buschman *et al.*, 1977). *Metioche vittaticollis* (Stål), a trigonidiinine found commonly in Asian rice fields, preyed on eggs of five rice pest insect species under cage conditions, but did not feed on rice seedlings; however, it could be reared on shredded carrots (Rubia & Shepard, 1987).

Tettigoniidae   Species in the Listrosclinae are predaceous (Key, 1970a), and some Conocephalinae appear in biological control literature. The tropical Asian *Conocephalus longipennis* (Dehaan) consumes eggs of two moth species in Malaysian rice fields and is considered beneficial even though it also feeds on flowers and young grains; it can reach an estimated density of about 20,000 per hectare and it may make an important contribution to pest control

in rice fields during the first half of the growing season (Manley, 1985). In Philippine rice fields, this species destroyed from 17 to 65% of the rice yellow stem borer egg masses (Pantua & Litsinger, 1984). Unspecified tettigoniids were considered to have consumed the largest proportion of medium to large velvet bean caterpillar larvae in Florida soybean field study plots (Godfrey *et al.*, 1989). Two species of predaceous katydids prey on brood in unguarded *Polistes* wasp nests in Central America (O'Donnell, 1993).

Order Mantodea

This is the only hemimetabolous order other than Odonata composed entirely of predaceous species. Both nymphs and adults are generalist predators. The 1800 or so known species are distributed mostly in the tropics and make up 8 families or 1 family, Mantidae, with 34 subfamilies (Brues *et al.*, 1954; Key, 1970b). Most of the praying mantids are cryptically colored to match the background where they sit and wait for prey. Some species appear as lichens on bark; others are vividly colored as the flowers they sit in to ambush visiting floral insects and some species resemble in color and form flowers that appear to attract insects (Wickler, 1968).

There have been extensive studies made on mantid prey recognition (Rilling *et al.*, 1959), prey capture (Mittelstaedt, 1957; Roeder, 1963), and response to prey density (Holling, 1966; Mook & Davies, 1966; Hurd *et al.*, 1978; Eisenberg *et al.*, 1981; Matsura & Nakamura, 1981; Hurd & Rathet, 1986; Paradise & Stamp, 1991). The biologies of only a few species have been described. *Tenodera sinensis* (Saussure) and *Mantis religiosa* (Linnaeus) are now the most widely distributed temperate species. The temperate species are univoltine and overwinter in the egg stage.

The eggs are placed in cases, oothecae, which can contain several hundred eggs each (Eisenberg & Hurd, 1990), but as the adults of *T. sinensis* age, the size of the egg cases and numbers of nymphs hatching decrease; 200 nymphs hatch from the first case, but only 25 hatch from the sixth (Butler, 1966). *Tenodera sinensis* deposits its oothecae on dead, upright plant stems, while the oothecae of *M. religiosa*, whose adults forage closer to the ground, were found in dense grasses (Eisenberg & Hurd, 1990). *Litaneutria minor* (Scudder) is a ground dweller but deposits its egg cases on low shrubs (Cannings, 1987). Eggs hatch during the spring in temperate climates, and the nymphs molt 3 to 12 times before attaining the adult stage.

The role of mantids as biological control agents is still uncertain even though many U.S. gardeners and organic growers purchase egg cases from eastern suppliers who collect *T. sinensis* and *M. religosa* oothecae from fields during winter. In China, oothecae are either collected or reared for release in various agricultural crops.

Hurd and Eisenberg (1984) conducted replicated releases of *T. sinensis* oothecae in a seminatural field community of low vegetation to determine the impact of releases on herbivore biomass (dry weight) and abundance. The results from this experiment and a previous one led Hurd (1985) to conclude there may be some agricultural situations in which some mantids will prove to be effective biocontrol agents, but his information and results from releases in more or less natural communities do not support this notion.

A review on the use of mantids in China as biological control agents in certain agricultural crops by Olkowski and Thiers (1990) is somewhat more positive. However, the few articles available in Chinese describing pest suppression in the field by mantids (mainly using *Hierudula patellifera* Serville and *T. sinensis*) do not provide enough detail to draw definitive conclusions on the use of mantids in agricultural crops, but under certain conditions there appeared to be suppression of pests such as aphids. Ge (1986) reported many instances of storing and releasing field-collected oothecae, mainly in cotton. He also obtained oothecae for release from mantids reared on artificial diets. The artificial diet used by Ge contained mostly pig liver, aphid powder, sugar, and water mixed with a little yeast and agar.

Three mantid species have now been accidentally introduced into North America: *T. sinensis* from Asia, *M. religiosa* from Europe, and *Iris oratoria* (Linnaeus), which was introduced from the Mediterranean area into southern California in the early 1950s (Gurney, 1955); *I. oratoria* has become quite common in San Joaquin Valley cotton fields.

Order Dermaptera

The 1860 or so species of earwigs in the world are divided into seven families that are mainly found in the tropics and subtropics (Steinmann, 1989). In California, only 1 of the 21 species present is considered native to the state (Langston & Powell, 1975). It appears that many species are predaceous, and some of the cosmopolitan omnivorous species that can be pestivorous have been used as biological control agents.

Remarkably little is known about the biology of most species in the Dermaptera. Of the few species that have been studied, eggs are deposited in groups either in the soil, under stones, beneath bark of trees, or in sheaths of leaves. The eggs are groomed and protected by the mother. Besides several genera of ectoparasitic earwigs of mammals that practice viviparous reproduction, in Hawaii predaceous species *Sphingolabis hawaiiensis* (Bormans) removes the thin embryonic membrane at the time the egg emerges (Marucci, 1955). The mother feeds the fully formed first- and second-instar nymphs bits of prey. Apparently, feeding young nymphs by the mother is common among the earwigs that have been studied. There are four to seven instars and with the molting of each instar there is an increase in antennal segments (Brindle, 1987).

Earwigs are mainly nocturnal and remain close to the soil, though some species are diurnal and move from plant to plant, rarely going to the ground. At least some of the obligatory predatory species use their caudal forceps to capture and hold prey while they feed.

Terry (1905) was one of the first to recognize the importance of earwigs as predators of insect pests. He observed three species feeding on leafhoppers and psyllids in Hawaiian sugarcane fields. The few predaceous species that have been recorded occur in four families.

Carcinophoridae   Three species in this family have been observed to be predaceous during at least part of their life cycle. *Euborellia annulipes* (Lucas), a cosmopolitan wingless species, is omnivorous but prefers insect food, often feeding almost exclusively on cane leafhoppers. It is nocturnal, remaining hidden in leaf sheaths or under fallen trees or stones (Terry, 1905). According to Bharadwaj (1966), it feeds on *Cosmopolites sordidus* (Germar) in Jamaica and on larvae of *Sesamia inferens* Walker infesting sugarcane, but also attacks stored vegetables and roots of greenhouse vegetables; it is a possible carrier of endoparasites of fowls, and invades households in the U.S. Gulf states.

The west African earwig, *Euborellia cincticollis* (Gerstaecker), which has been reported from California, has a biology similar to that of *E. annulipes*. It feeds on aphids and other arthropods as well as plants in the laboratory; in California it feeds on cantaloupes (Knabke & Grigarick, 1971). *Anisolabis eteronoma* Borelli fed on tephritid larvae in rotting fruit on the soil—the nymphs fed on fruit fly larvae developed faster than those fed on fruit alone (Marucci, 1955).

Chelisochidae   The black earwig, *Chelisoches morio* (Fabricius), is an Oriental, Australian, and Afrotropical diurnal species, which was considered to be an important predatory earwig in the agricultural areas of the Hawaiian Islands (Terry, 1905). Eggs are deposited in a heap in the sheaths of large-leaved succulent plants. It is a general predator and has been recorded attacking *Brontispa* spp. (Coleoptera: Chrysomelidae) mining coconut fronds in Micronesia and New Caledonia (Waterhouse & Norris, 1987).

Forficulidae   *Anechura harmandi* (Burr) was the most important and numerous predator of eggs and young larvae of the herbivorous lady beetle *Henosepilachna niponica* (Lewis) in central Japan (Ohgushi, 1986).

Doru taeniatum (Dohrn), a Neotropical species that now occurs in California, Texas, and Florida, was observed eating eggs of the velvet bean caterpillar, *Anticarsia gemmatalis* Hübner, in Florida soybean fields and became radioactive after having preyed on labeled *A. gemmatalis* eggs placed in the field (Buschman *et al.*, 1977). This nocturnal earwig is common in Central America, occurring primarily on large grasses and feeding on eggs and small larvae of *Spodoptera frugiperda* (Smith) on corn; it also can reproduce on grass pollens. It spends most of its time on plants and deposits its eggs in leaf sheaths, and is inactive during the Honduran dry season, hiding in leaf sheaths; thus, Jones *et al.* (1987) suggest that it may not be an important predator of *S. frugiperda* during this period. Jones *et al.* (1988) described the biology and produced a life table of *D. taeniatum* based on laboratory studies.

Forficula auricularia Linnaeus, the European earwig, is an omnivore; aphids appear to be a preferred prey (Skuhravy, 1960; Asgari, 1966). It can be reared rather easily (Lamb & Wellington, 1974). Buxton and Madge (1976) and Madge and Buxton (1976) found *F. auricularia* to be a successful predator of the damson-hop aphid, *Phorodon humuli* (Shrank), in the laboratory; in the field, Campbell (1978) found the European earwig and two *Anthocoris* spp. were the most abundant predators when the damson-hop aphid populations declined (damage to the plants by earwigs was rarely seen). From the numbers of earwigs trapped over a 2-month period, it was estimated that the earwigs ate over 50,000 aphids per day. The European earwig eats aphids in various field crops and is an efficient predator of cereal aphids in low aphid population densities (Sunderland & Vickerman, 1980; Sunderland, 1988). Because pear psylla increased when *F. auricularia* nymphs were adversely affected by codling moth sprays, this indicated that the earwig preyed on the psyllids (Sauphanor *et al.*, 1993).

In Washington State, *F. auricularia* was reared on dog food in the laboratory and then released on prebearing apple trees provided with artificial retreats at a rate of five or six adults per tree (Carroll & Hoyt, 1984). This resulted in a decline in 3 weeks from over 500 *Aphis pomi* DeGeer per tree to less than 50, compared with over 3000 per tree in trees free of earwigs (Carroll & Hoyt, 1984). However, using similar methods on bearing apple trees infested with *A. pomi* failed to slow aphid population growth (Carroll *et al.*, 1985). Serological assays of predators feeding on *A. pomi* in Ontario apple orchards indicated among the several polyphagous arthropods that of a total of 79 *F. auricularia* collected, 44 were seropositive, which indicated that the earwig is a potentially important predator of *A. pomi* (Hagley & Allen, 1990).

Labiduridae *Labidura riparia* (Pallas), the striped earwig, is a worldwide species in the tropical–subtemperate regions. It is a nocturnal generalist predator that is considered an important potential biological control agent because it rarely feeds on plants and has become common in California and in the southern states (Schlinger *et al.*, 1959; Price & Shepard, 1977; Kharboutli & Mack, 1993). It is the most studied predaceous earwig.

This species uses its caudal forceps to catch and hold prey while feeding. Eggs are deposited in the ground, are brooded over, and then are moved by the mother. There are six instars and the species can be reared on a commercial cat food (Shepard *et al.*, 1973). In cultivated areas it is mostly found associated with field crops.

In cotton, *L. riparia* was reported to feed on eggs and larvae of *S. litura* in Egypt (Bishara, 1934), but actually the prey was *S. littoralis* according to Ammar and Farrag (1974), who studied the biology and behavior of *L. riparia* in Egypt. In soybeans, Buschman *et al.* (1977) believed that *L. riparia* was responsible for most of the predation of the velvet bean caterpillar eggs and larvae among the 20 different predators that they observed during 1973 in Florida; in 1982, *Labidura riparia* ate the greatest proportion of these small larvae on the ground, while in 1981 the ant *Pheidole morrisi* Forel was the major consumer of all the velvet bean caterpillar larvae in Florida soybean plots (Godfrey *et al.*, 1989) and similarly, in 78 Florida citrus groves, Tryon (1986) found that on and in the soil beneath citrus trees ants were the dominant predator in 74 groves and *L. riparia* was dominant in 4 where the ants were not present; in Brazilian sugarcane fields, the earwig is an important predator of *Diatraea saccharalis* (Fabricius) eggs and first-instar larvae (Bueno & Berti-Filho, 1987). In Florida, studies at night in corn indicated that there was an unusually low adult male to female ratio of *L. riparia* on the plants: Waddill (1978) suggested that the odd sex ratio may be explained by the fact that females required more food than males. In the laboratory studies, Waddill (1978) provided *Diabrotica balteata* LeConte as prey. In Florida cabbage fields, where pupae of *Trichoplusia ni* (Hübner) were placed on the undersurface of cabbage leaves, *L. riparia* and sparrows consumed from 4 to 90% of the pupae over a 3-day period (Strandberg, 1981a). Because the earwigs were nocturnal and the birds were diurnal in their feeding activity, Waddill calculated that the earwigs consumed an overall average of 42% of the placed pupae. Earwigs were active the entire year but fewer were caught during winter months (Strandberg, 1981b).

Labiidae *Sphingolabis hawaiiensis* occurs from the Philippines to New Guinea, in the New Hebrides, and in the Hawaiian Islands. In Hawaii, this earwig was found by Marucci (1955) to prey on Oriental fruit fly larvae in rotten fruit and in the soil but did not eat fruit fly puparia. The earwig attacks frontally by using its mandibles and caudal forceps. The adults provided their first two instars with fragments of fruit fly larvae that the adults had killed and only partially consumed. The adults kill more prey than they eat.

Order Thysanoptera

The families Aeolothripidae and Thripidae in the suborder Terebrantia, and Phlaeothripidae in the suborder Tubulifera, contain species that are predaceous, mainly on small insects (such as thrips, aphid nymphs, scales, or insect eggs) and on the eggs and immature stages of mites. A few, such as *Franklinothrips* species, may be obligate predators, but most appear to be partially or even primarily phytophagous. There are no obvious modifications in the mouthparts or other parts of the body for the predatory mode of life (Lewis, 1973). Palmer *et al.* (1989) present an excellent introduction to economically important Thysanoptera, outlining the diversity and biology of thrips and giving a key for identification.

Aeolothripidae   Thrips in this family are generally fast moving, with slender legs. A small hook on the second tarsus may aid these predators in grasping prey (Lewis, 1973).

Predaceous *Aeolothrips* species have mainly been reported as thrips predators (Lewis, 1973; Ananthakrishnan, 1984). *Aeolothrips intermedius* Bagnall developed more rapidly on thrips larvae (average: 12 days larva-adult at 24°C) than on tetranychid mites (average: 19 days); development was not completed on aleyrodids or psyllids and no feeding occurred on aphids. Adults needed to feed on flowers to attain sexual maturity. (Bournier *et al.,* 1979). This Palearctic species has three or four generations per year in southern France (Bournier *et al.,* 1978).

Observations on *Franklinothrips* species indicate that they are obligate predators, primarily on other thrips, Lewis, 1973; Ananthakrishnan, 1984). *Franklinothrips tenuicornis* Uzel and *F. vespiformis* Crawford prey on thrips on various tropical crops in the Western Hemisphere. These species were not considered sufficiently abundant to affect thrips control on cacao in Trinidad or on avocado in California (Callan, 1943; McMurtry, 1961).

Aleurodothrips fasciapennes (Franklin) completed development on whiteflies and spider mites, but the highest percentage reached maturity on armored scales. In Florida, significant correlations occurred between numbers of thrips and the purple scale, *Lepidosaphes beckii* (Neuman) (Selhime *et al.,* 1963). A short life cycle permits the completion of 1.5 to 2 generations for every generation of the purple scale. Both laboratory and field studies suggested that this species is a factor in the biological control of purple scale. The impact of this thrips on purple scale is reviewed by Palmer and Mound (1990).

Thripidae   *Scolothrips* species are predators of tetranychid mites and possibly are the most specialized of the predaceous thrips. *Scolothrips sexmaculatus* (Pergande) is a North American species that is associated mainly with *Tetranychus* species characterized by dense colonies with extensive webbing. Both larvae and adults of this species showed a tendency to be arrested by spider mite webbing, after which penetration into the silken network of the prey colony occurred. When mite eggs were replaced on clean leaves without webbing, the thrips rarely preyed on them, but continued searching (Gilstrap & Oatman, 1976). These

workers observed that any stage of *T. pacificus* could be utilized as prey, although small thrips larvae had some difficulty in capturing large mites. The front tarsi were used to hold and manipulate the prey during feeding. Apparently, mites were not perceived until contact was made. *Scolothrips sexmaculatus* has a relatively rapid rate of development (under 10 days at 35°C). The intrinsic rate of increase is comparable to, or even exceeds, that of some phytoseiid mites associated with *Tetranychus* mites. For example, values for *S. sexmaculatus* of 0.26 at 30°C (Coville & Allen, 1977) and 0.304 at 35°C (Gilstrap & Oatman, 1976) compare favorably with those of *Galandromus occidentalis* (Tanigoshi *et al.,* 1975), an effective predator of spider mites on many crops.

Scolothrips species are more voracious than phytoseiid mites, but consume fewer prey than some other specialized insect predators of mites e.g., *Stethorus* spp. in the Coccinellidae). Larvae of both *S. sexmaculatus* and *S. longicornis* Priesner consume 40 to 60 mite eggs during development under various temperature conditions, while ovipositing adults consume 80 to 100 eggs per day at 35°C. At the same temperature, the thrips females laid an average of 14 to 16 eggs per day. *Scolothrips sexmaculatus* had a fecundity of well over 200 eggs per female over a wide range of temperatures (Gilstrap & Oatman, 1976; Gerlach & Sengonca, 1986). Other *Scolothrips* species known to be associated with tetranychid mites are listed in Lewis (1973), Ananthakrishnan (1984), and Chazeau (1985).

Some thrips considered pests and assumed to be strictly phytophagous have been found to be predaceous. Insecticide applications formerly were made to control *F. occidentalis* (Pergande) on cotton in California. However, various field observations of this thrips feeding on *Tetranychus* species prompted further investigations. Marcano-Brito (1980) showed that *F. occidentalis* could complete development on mite eggs alone compared with eggs plus a cotton leaf, with the egg consumption during larval development averaging 182 and 72, respectively. Trichilo and Leigh (1986) demonstrated a functional response to increasing egg densities by both larval and adult thrips, but predation was reduced by spider mite webbing. Pollen and nectar were also utilized as foods. Gonzalez and Wilson (1982) and Pickett *et al.* (1988) considered *F. occidentalis* an important member of the beneficial arthropod complex on cotton, for its effect not only as a predator of spider mites but also as a food source for other predators, such as *Orius tristicolor* White.

Other phytophagous species, such as *F. tritici* (Fitch) and *Thrips tabaci,* also have been observed to feed on tetranychid mites (Lewis, 1973; Boykin *et al.,* 1984).

Phlaeothripidae   Predaceous species in this family utilize various kinds of prey and nonprey foods. They have been investigated mainly concerning their roles as predators of spider mites. *Haplothrips faurei* Hood, a common species in North American orchards, develops and reproduces on spider mites and on moth eggs and to a lesser extent on pollen and on eriophyid mites. Feeding on the mite *Panonychus ulmi* was confined mainly to the egg stage, averaging 143 eggs consumed during a developmental period of 8 to 10 days at 24°C (Putman, 1965). Ovipositing females consumed 43.6 eggs per day and laid 3.3 eggs per day. This species was considered most important in reducing the winter egg population of *P. ulmi* by preying on late summer and autumn populations on peach trees. Chromatograms of thrips collected during this time indicated that nearly 100% had fed on spider mites. In a December sample of twigs, an estimated 76% of the winter eggs of *P. ulmi* were destroyed by the thrips (Putman, 1965; Putman & Herne, 1966).

Haplothrips victoriensis Bagnall is considered a major factor in maintaining *Tetranychus urticae* at low levels on seed alfalfa in South Australia. If the ratio of thrips to motile mites went below 1:10, such as after an insecticidal treatment, mite populations could increase rapidly (Bailey & Caon, 1986). Larvae completed development on either spider mite eggs or alfalfa flowers (presumably the pollen), with faster development on the latter. Adults survived only a short time without flowers, although they ate mite eggs. Some *Haplothrips* spp. prey on diaspine scales (Palmer & Mound, 1990).

These *Haplothrips* species seem to be opportunist feeders whose numbers probably are not closely correlated with any single prey species. Some species in this genus are considered crop pests (Ananthakrishnan, 1984).

Leptothrips mali (Fitch) is common in North American deciduous fruit orchards. This species attacks eggs and motile forms of spider mites, grasping the prey with its front legs (Parrella *et al.*, 1982). Development on eggs of *Panonychus ulmi* was completed in 19 days at 29.4°C, but required 23 days on pollen. Fecundity of 13 eggs per female in laboratory-reared thrips was lower than for field-collected ones, suggesting that additional nutrition, possibly from apple rust mite, was utilized in the field. Parrella *et al.* (1982) concluded that this species was not able to maintain mite populations at low levels, but it was a factor, along with the coccinellid *Stethorus punctum* LeConte, in reducing populations from higher levels.

Order Heteroptera

Although the majority of the terrestrial species of the Heteroptera are plant feeders, there are thousands that are predatory. In North America alone, there are 3834 species in 677 genera listed, along with many references to biologies of predaceous species, in a catalog of Heteroptera edited by Henry and Froeschner (1988). A readily usable key to genera was written by Slater and Baranowski (1978), and Yonke (1991) prepared a key to immature true bugs, along with biological notes. A general coverage of biology of all families is found in Miller (1971).

The original feeding habit and habitat of Heteroptera are thought to be phytophagous and terrestrial by Sweet (1979), but Cobben (1978) believes that carnivory was the original feeding strategy and their primitive lifestyle was semiaquatic. Cohen (1990), studying the feeding adaptations of five predaceous species in four different families of Heteroptera, found that the toothed mandibular stylets penetrate, rasp, and cut prey tissues; and the maxillary stylets form salivary and food canals that deliver saliva and remove liquified prey contents from within. From this work, Cohen concluded that the morphological, physiological (protolytic enzymes), and behavioral adaptations permit a concerted predigestion of prey from within and allow the predators to feed efficiently on prey that are equal to, or greater than, them in size. However, even the predaceous species feed on plant tissues or plant secretions (i.e., Chu, 1969; Stoner, 1972; Tamaki & Weeks, 1972a; Naranjo & Stimac, 1987). Anthocoridae, Miridae, and Nabidae oviposit into plant tissue and will oviposit into Parafilm®-wrapped blocks of water-soaked cellulose sponge; the Parafilm® has to be stretched for anthocorids but not for nabids (Shimizu & Hagen, 1967).

Heteropteran predators are active in many agricultural crops: in alfalfa fields (Smith & Hagen, 1956; Rakickas & Watson, 1974; Benedict & Cothran, 1975; Neuenschwander *et al.*, 1975; Guppy, 1986; Braman & Yeargan, 1990; O'Neil, 1992); in apple orchards (Collyer, 1953a, 1953b; Lord, 1971; Fauvel, 1974; Ràcz, 1988; Arnoldi *et al.*, 1991, 1992); in citrus trees (Van den Berg & Bedford, 1988); in cocoa fields (Entwistle, 1972); in cotton fields (Whitcomb & Bell, 1964; van den Bosch & Hagen, 1966; Ridgway *et al.*, 1967; Falcon *et al.*, 1968; van den Bosch *et al.*, 1969; Irwin *et al.*, 1974; Ehler & van den Bosch, 1974; Wilson & Gutierrez, 1980; Gonzalez *et al.*, 1982; Schuster & Calderon, 1986; Gravena & Pazetto, 1987; Gravena & Da Cunha, 1991; Kapadia & Puri, 1991; Snodgrass, 1991); in coffee fields (LePelley, 1968); in corn fields (Davis *et al.*, 1933; Knutson & Gilstrap, 1989); in pears (Herard 1986); in potato fields (Biever & Chauvin, 1992a; Hough-Goldstein & Whalen, 1993); in rice fields (Kenmore, 1980; Yasumatsu *et al.*, 1981; Napompeth, 1982; Chang & Oka, 1984; Manti, 1991); in soybean fields (Dumas *et al.*, 1964; Irwin & Shepard, 1980; McCarty *et al.*, 1980; Richman *et al.*, 1980; Isenhour & Yeargan, 1981, 1982; Buschman *et al.*, 1984; O'Neil & Stimac, 1988; Godfrey *et al.*, 1989); in sugarcane fields (William, 1931; Guagliumi, 1973); in sugar beet fields (Tamaki & Weeks, 1973; & Long, 1978); and in strawberries and vegetables mostly in greenhouses (Gilkeson *et al.*, 1990; Steiner & Tellier, 1990; Villevielle & Millot, 1991; Gillespie & Quiring, 1992; Higgins, 1992; Van de Veire & Degheele, 1993).

The general importance of Heteroptera in applied biological control was reviewed by Clausen (1940), Sweetman (1958), and Carayon (1961a). Biological control literature mostly deals with Anthocoridae, Berytidae, Lygaeidae, Miridae, Nabidae, Pentatomidae, Phymatidae, and Reduviidae.

Anthocoridae   About 400 species of minute pirate bugs or flower bugs are distributed in nearly all parts of the world; in North America north of Mexico, there are about 90 species recognized (Kelton, 1978; Henry, 1988). Hodgson and Aveling (1988) presented an excellent review on the biology and ecology of anthocorids; most are generalist predators, but there is some plant feeding by most species while some feed on pollen and a few feed on blood. The predaceous species, after contacting the prey, feed by extending their rostra directly forward and everting the stylets beyond the tips of the labral sheaths; in contrast, the mirids retract their labial segments to expose the stylets when feeding (Anderson, 1962b). Eggs are usually deposited horizontally into plant tissue with only the circular whitish caps (opercula) visible at an oblique angle on the plant surface. The structure of the opercular surface varies between species (Sands, 1957). When eggs are deposited into a leaf blade or a petiole, an elongate bulge often is visible on the plant surface; eggs can also be deposited in bark crevices. If the plant tissue becomes dry before the eggs hatch, the embryos are usually killed. There are five instars and the nymphs usually feed on the same prey as adults, with younger nymphs feeding on smaller prey stages. Common prey are mites, thrips, aphids, psyllids, and lepidopteran eggs and small larvae. In temperate regions, there are usually one to three generations and most species overwinter as adults. Short photoperiods induce reproductive diapause in some species (Parker, 1975; Kingsley & Harrington, 1982; Ruberson *et al.*, 1991).

Anthocorinae   The greatest importance of anthocorids is as endemic natural enemies attacking agricultural pests. In Chu's (1969) review, 40 references to anthocorids feeding on about 20 species of Lepidoptera are listed, mostly *Orius* species. Other anthocorid genera that include species important to natural and applied biological control are *Anthocoris, Lyctocoris, Montandoniola, and Xylocoris* (Hodgson & Aveling, 1988).

The biologies of about a dozen *Anthocoris* spp. have been investigated. Their range of prey is generally wide, but the range of plants inhabited depends on the species. Anderson (1962a) studied the bionomics of six *Anthocoris* spp. in England and found two of these species on 12 different tree species; one was on 11, two were on 7, and one was found on only 1 tree species. Thus, the latter *Anthocoris* species is restricted to feeding on prey that occurs on one plant species.

The polyphagous *A. nemorum* (Linnaeus) is the most ubiquitous European species. Collyer (1967), using 30 literature references, records *A. nemorum* associated with 35 different arthropod prey in 20 families of insects and 5 mite families on 17 different plants; Hill (1957) recorded and observed this predator associated with over 33 plant species. Prey of *A. nemorum,* other than pear psylla, noted in the literature since 1965, were listed by Herard (1986), and included 34 insect species in four orders and 5 mite species on at least 16 different plants. There were five references to *A. nemorum* feeding on pear psylla in Europe. As Anderson (1962c) and Herard and Chen (1985) point out, although a diversity of prey is accepted, there are great differences in the nutritive values for the predator. In an olfactometer, *A. nemorum* was attracted to certain plant leaf volatiles (tomato, willow) and reacted to, and distinguished between, odors coming from certain tetranychids but not to aphid species (Dwumfour, 1992).

The efficiency of *A. nemorum* to capture prey aphids increased as the predator to prey ratio increased (Evans, 1976a). This anthocorid was unable to survive on sycamore trees unless there were 5 small sycamore aphids per 10 cm^2 of leaf, while *A. confusus* Reuter required 14 aphids per same leaf area (Dixon & Russel, 1972). Russel (1970) suggests that under natural conditions, the number of anthocorids that reach maturity is dependent on the number of young aphids available during the early stages of the anthocorid's development. If leaf surfaces are hairy, more thorough searching occurs but less surface area is covered compared with leaves with waxy surfaces (Lauenstein, 1980). *Anthocoris nemorum* and *A. nemoralis* (Fabricius) were the most abundant aphidophagous insects consuming hop aphids on hops (Campbell, 1977). Because *A. nemoralis* develops faster than *A. nemorum,* they ate fewer hop aphids. It appears that *A. nemorum* may have more of an impact on aphids and mites than on the myriad other prey it attacks. It has been collected in France and sent to the western United States for release against *Psylla pyricola* Forester, but it did not become established (Herard, 1986).

Anthocoris nemoralis, a common Palearctic species, was accidentally introduced into eastern Canada, and later successfully was introduced into British Columbia from Europe in 1963 to control pear psylla (McMullen & Jong, 1967). It has made its way to California where it occurs commonly feeding on several exotic psyllids that attack ornamental plants (Hagen & Dreistadt, 1990). Although *A. nemoralis* is a generalist predator, its prey and plant habitat range is narrower than *A. nemorum.* Herard (1986) listed 11 insect and 5 mite species as prey of *A. nemoralis* in addition to pear psylla. There are five European references to its feeding on pear psylla, and it is considered to be one of the most important predators of *Psylla pyri* (Linnaeus) and *P. pyricola.* In England, *A. nemoralis* appeared to have a strong density-dependent relationship with *P. pyricola* populations (Hodgson & Mustafa, 1984). In eastern Canada, it is considered the predominant predator of pear psylla (Hagley & Simpson, 1983). The searching patterns of *A.*

nemoralis correlated well with the distribution of pear psylla on pear leaves (Brunner & Burts, 1975). Dempster (1968) notes that while predation by *A. nemoralis* and *A. sarothamni* Douglas & Scott (psyllid predator on scotch broom) on *Arytainia* psyllids was density dependent, they were incapable of controlling the numbers of their prey because of the variability in contemporaneous mortality caused by other predators.

Anthocoris antevolens White, a Nearctic predaceous species commonly found on deciduous trees and shrubs, is transcontinental in Canada and the northern United States. It preys on psyllids and aphids and is considered an important predator of pear psylla (Anderson, 1962b; Watson & Wilde, 1963; Madsen & Wong, 1964; Nickel *et al.*, 1965; Kelton, 1978).

Anthocoris melanocerus Reuter, a northwest Nearctic species, has a similar biology to *A. antevolens,* but it also occurs on herbaceous plants; many of the same authors mentioned under *A. antevolens* also studied *A. melanocerus.* Based on laboratory cage studies, Tamaki and Weeks (1968) found that *A. melanocerus* suppressed *Myzus persicae* (Sulzer) on sugar beets below control populations; however, on broccoli the predator was only temporarily suppressive. In cages, egg cannibalism of the predator was as high as 95%.

Anthocoris confusus, another common European species, has a similar biology to *A. nemoralis* (Hill, 1965) but is somewhat less ubiquitous; it is more of a tree-dwelling species. In England, *A. confusus* appears to feed mainly on callaphidid aphids on four major host trees (Anderson, 1962a) including the sycamore aphid *Drephanosiphum platanoides* (Schrank) (Russell, 1970; Dixon & Russel, 1972). The prey suitable for nymphal development may not be suitable for egg production because *A. confusus* will not reproduce on *Aphis fabae* Scopoli because this aphid induces a reproductive diapause (Anderson, 1962a). Evans (1976a) obtained a high numerical response in *A. confusus* when prey density of *Aulacorthum circumflexus* (Buckton) increased from one up to eight adult aphids per day; above this density, there was no increase in reproductive rate. Also, there can be mutual interference between the adult anthocorids that leads to a density-dependent decrease in egg production due mainly to an increased tendency for adults to migrate out of areas of high predator density (Evans, 1976d). As the age of the plants increased, the numbers of eggs deposited decreased considerably; plant topography can strongly influence the search pattern (Evans, 1976b, 1976c).

The biology of *Montandoniola moraguesi* (Puton), also in the Anthocorinae, is similar to that of *Anthocoris* spp., but differs in that it apparently only feeds on Phlaeothripidae, including the Cuban laurel thrips. *Gynaikothrips ficorum* (Marchal) (Tawfik & Nagui, 1965; Funasaki, 1966). During 1964, the Cuban laurel thrips was found infesting Chinese banyan trees, *Ficus retusa* Linnaeus, at the Honolulu International Airport. It quickly spread to the other Hawaiian Islands within a year, resulting in defoliated trees. *Montandoniola moraguesi* was obtained from the Philippines and introduced to Oahu and Kauai Islands in 1964 (Davis & Krauss, 1965), and in one year the predator spread considerably and reduced the thrips populations on five islands (Funasaki, 1966). It also has been observed to prey on *Liothrips urichi* Karny, which was introduced into the Hawaiian Islands to control the weed *Clidemia hirta* (Linnaeus) D. Don (Funasaki *et al.,* 1988).

Orius is another economically important genus in the subfamily Anthocorinae with 21 species recognized in the Western Hemisphere (Herring, 1966). Its species are smaller than *Anthocoris* but have similar biologies. Carayon and Steffan (1959) reviewed the feeding habits of 13 species of *Orius* and concluded that most species are polyphagous and all probably suck plant juices. The most common prey for the majority were thrips, followed by Homoptera, lepidopteran eggs and small larvae, and mites. *Orius pallidicornis* (Reuter) apparently feeds exclusively on cucurbit pollen (Carayon & Steffan, 1959). Feeding on plant liquids helps sustain *Orius* when prey are scarce (Kiman & Yeargan, 1985), but can be a detriment if a crop is treated with systemic pesticides because the toxins kill the predators when they ingest the contaminated plant sap (Ridgway *et al.,* 1967).

There is much written about the four or five species of *Orius* that are involved in natural and biological control of many pests in a diversity of crops. Ryerson and Stone (1979) list 164 selected references to the bionomics for *O. insidiosus* (Say) and *O. tristicolor* alone.

Orius insidiosus was mainly a midwestern–eastern North American species, but is now transcontinental and has spread into Central and South America. In agriculture, it is essentially associated with corn, soybeans, and to a lesser extent apple trees. Barber (1936) found it on a wide variety of plants but particularly on corn; it could develop as nymphs when exposed only to corn silks but no reproduction occurred. Dicke and Jarvis (1962) found corn pollen to be important in building up populations of the predators. Eggs are deposited into moist corn silks and nymphs feed on the silks and insects, particularly *Heliothis zea* (Boddie), when they are older. The life cycle is about 24 days during the summer (Barber, 1936). Nymphal development was completed on *Acer* pollen plus green beans, but not on green beans alone; development was faster on diets containing arthropod prey than on diets lacking prey. Also, Kiman and Yeargan (1985) found an adult diet of *H. virescens* (Fabricius) eggs resulted in higher fecundity for *O. insidiosus* than for spider mites or a thrips species.

The association of *O. insidiosus* with corn is apparently long-standing because it is attracted to volatiles (hexane extracts) of corn silks (Reid & Lampman, 1989). The synamone brings *O. insidiosus* to corn silks where eggs of *H.*

zea, as well as those of a more recent prey, the European corn borer. *Ostrinia nubilalis* (Häbner), are commonly deposited (Reid, 1991). Corn earworms and fall armyworms feeding on resistant corn foliage or meridic diets containing resistant corn silks were subject to increased mortality by *Orius insidiosus* (Isenhour *et al.,* 1989). When densities of *O. insidiosus* and spiders averaged two to five per plant, survival of small southwestern corn borers on whorl and tassel stage corn was reduced (Knutson & Gilstrap, 1989).

The insidious flower bug was by far the most abundant predator encountered in Arkansas grain sorghum fields; it was sampled as a predator of *H. zea* to determine if predator to prey ratios could be established (Steward *et al.,* 1991). On soybeans, *O. insidiosus* commonly feeds on soybean thrips larvae occurring mainly on trifoliate leaves, which accounted for a concentration of the *Orius* eggs in the upper parts of the plant and caused a decline in the thrips populations (Isenhour & Yeargan, 1981). Crop phenology is also an important factor. In strip-planted soybean and corn in Kentucky, populations of *O. insidiosus* peaked on corn during the period of silking and pollen shed, while predator density was greater on early planted soybeans compared to June planted soybeans (Isenhour & Yeargan, 1981). Corn and alfalfa adjacent to apple orchards in Virginia apparently served as alternate sites for *O. insidiosus* populations throughout the growing season (McCaffrey & Horsburgh, 1986b). These authors determined that thistles also harbor a large number of predators, and calculated temperature threshholds for development (McCaffrey & Horsburgh, 1986b) and functional response of *O. insidiosus* to the European red mite at different temperatures (McCaffrey & Horsburgh, 1986a).

Orius insidiosus was successfully introduced into the Hawaiian Islands against *H. zea* in a 1951 classical biological control program (Funasaki *et al.,* 1988).

Orius minutus (Linnaeus) is a common European species that was probably introduced into British Columbia with the importations of shrubs and herbaceous plants (Kelton, 1978, Lattin *et al.,* 1989). It has a similar biology to *Anthocoris* and other *Orius* spp. in that it is a general predator and feeds to some extent on plants. Herard (1986) lists 18 recent references of *O. minutus* feeding on mites, thrips, Diptera, Lepidoptera, Coleoptera, coccids, aleyrodids, and aphids; it is one of six *Orius* species that has been recorded feeding on pear psylla. Spider mites, particularly the European red mite, are common prey of *O. minutus* (Collyer, 1953a, 1953b; Berker, 1958). Niemczyk (1978b) found that one *O. minutus* per 5 to 10 apple leaves can keep the European red mite at a low level if the mites do not exceed two to five mites per leaf prior to the appearance of the predator. *Orius minutus* and *O. vicinus* Ribaut could reduce the European red mite and other mites in vineyards in the absence of Phytoseiidae (Duso & Girolami, 1983). Niemczyk (1978a) reared many generations in the laboratory using eggs of the Angoumois grain moth, *Sitotroga cerealella* (Olivier), which was a better food than Mediterranean flour moth eggs, the European red mite, the twospotted spider mite, or the green apple aphid.

Orius majusculus (Reuter) can be mass cultured on *A. kuehniella* and geranium leaves (Alauzet *et al.,* 1992); and and *O. laevigatus* (Fieber) are released against psyllids, aphids, and mites but are mostly used against thrips (Rudolf *et al.,* 1993).

Orius sauteri (Poppius), an Asian species, is mass-reared in China for field release, but we are not certain which pests are the targets. It has been reared from eggs to adults on aphids, mites, thrips, rice moth, pollen, and artificial eggs (Zhou & Wang, 1989). Soybean sprouts served as the best ovipositional medium; a mean of 52 eggs was obtained per predator on one bean plant seedling (Zhou *et al.,* 1991a). *Orius insidiosus* has been released in greenhouses to control thrips (Van de Veire & Degheele, 1993).

Orius tristicolor resembles *O. insidiosus* in its size and biology, but it is more widespread in North America (Anderson, 1962b; Kelton, 1978). It is commonly found in alfalfa, cotton, and other low crops as well as in flowering heads of many herbaceous plants. Early studies on the biology of *O. tristicolor* involved using spider mites as prey (Butler, 1966; Askari & Stern, 1972a, 1972b), but its preferred prey is probably thrips for it is most commonly associated with thrips in the field. A diet of pollen plus thrips was found most suitable for development of *O. tristicolor* nymphs, but adult longevity and total egg deposition increased when adults were fed green bean plus pollen plus thrips (Salas-Aguilar & Ehler, 1977). Stoltz and Stern's (1978) results from laboratory feeding studies of *O. tristicolor* exposed to western flower thrips supported the correlation of field data associating the two species. Letourneau and Altieri (1983) described the interactions of *O. tristicolor* and the western flower thrips in squash monocultures and polycultures of squash, corn, and cowpea; and suggest that the determining factor for differential colonization by *Orius* in monocultures and polycultures of squash was the attraction of *Orius* to the crop habitat during its prey location process. Once in the habitat the alarm pheromone of the western flower thrips nymphs becomes a prey-finding kairomone for *O. tristicolor* (Teerling *et al.,* 1993).

The importance of *O. tristicolor* as a mortality factor of aphids is not clear. In Goodarzy and Davis' (1958) study, it is considered to be one of the most important predators of *Therioaphis maculata* Busk., the spotted alfalfa aphid, in Utah. However, Neuenschwander *et al.* (1975) show only a minor affect on pea aphid and spotted alfalfa aphid populations in California alfalfa fields. In Idaho potatoes, *O. tristicolor* abundance was strongly correlated with the presence of thrips, but did not appear to be the cause of green peach aphid decline (Hollingsworth & Bishop, 1982). The searching behavior of *O. tristicolor* is similar to many other predators: once it is in the habitat of the prey, predators search more rapidly with a slow turning rate when starved, but after contacting prey, the searching speed decreases and the turning rate increases (Shields & Watson, 1980). Light intensity can also influence turning ratios of starved females of *O. tristicolor*: under direct current (DC)-powered incandescent light there was no effect on activity levels or turning ratios, but under alternating current (AC) powered incandescent light, the distances traveled were greater regardless of light level (Shields, 1980). Under 12-h photoperiods, *O. tristicolor* ate twice as many mite eggs as when held in the absence of light (Askari & Stern, 1972b).

In applied biological control, the capacity for thrips control by *Orius* on squash was demonstrated by Letourneau and Altieri (1983) using predator inclusion–exclusion cage experiments. It is released with phytoseiids to control western flower thirps attacking vegetables in greenhouses (Gillespie & Quiring, 1992; Cloutier & Johnson, 1993). *Orius tristicolor* was released twice in Hawaiian classical biological control programs: in 1962 against the microlepidopteran *Ithome cancolorella* Chambers that attacked kiwi, and in 1964 against the Cuban laurel thrips (Davis & Krauss, 1963, 1965), but apparently *O. tristicolor* never became established (Clausen, 1978).

Orius tantillus (Motschulsky) is considered to have great potential as a predator of *Thrips palmi* Karny in the Philippines (Mituda & Calilung, 1989).

Lyctocorinae The subfamily Lyctocorinae contains species that are commonly found under bark, in stored grain, and in leaf or straw litter. The few species involved in natural and biological control literature are mainly in two genera.

Lyctocoris beneficus (Hiura) occurs in Japan and Korea, and its biology has been excellently studied and described by Chu (1969). Because this anthocorid attacks overwintering larvae and newly emerging adults of the striped rice stalk borer, *Chilo suppressalis* (Walker), in straw piles and in barns, it was considered a natural enemy that could be reared and possibly used in augmentative releases against the stalk borer. Chu (1969) found that *L. beneficus* fed on 45 of the 73 different species in nine insect orders that he exposed to the anthocorid. This generalist predator, however, prefers a habitat of somewhat dry rice paddy straw. The best insectary diet to mass rear the predator was *Tribolium* larvae, but Chu (1969) also made some progress in developing an artificial diet for the bug; furthermore, he calculated growth and reproductive energetic values of *L. beneficus*. This predator did not have any influence on borer populations in paddy stubble because it could not attack free-crawling borers. Thus, it is limited to preying on borers infesting straw piles.

Lyctocoris campestris (Fabricius) is a worldwide generalist predator of stored product insects that produces best growth and reproductive responses when exposed to grain moths instead of to common stored product-infesting beetles (Parajulee & Phillips, 1993).

Lyctocoris elongatus (Reuter) and *Scoloposcelis flavicornis* Reuter (= *S. mississippensis* Drake & Harris) are predators of bark beetles attacking standing and felled pines. Schmitt and Goyer (1983) determined the developmental times and described the immature stages of both anthocorids.

Xylocoris flavipes (Reuter), a cosmopolitan species called the warehouse pirate bug, is commonly found associated with stored product insect pests. This association prompted Jay *et al.* (1968) to explore the possibility of its being a biological control agent of stored product pests, and their first tests showed that the anthocorid successfully inhibited population buildup of four species of beetles that attack stored grain. Populations of *Tribolium* were significantly reduced in stored peanuts by releasing various densities of *X. flavipes* (Press *et al.*, 1975) and it also controlled moths infesting stored peanuts (Brower & Mullen, 1990). Also, the saw-toothed grain beetle populations infesting shelled corn were reduced 95% by releases of the predator (Arbogast, 1976). Increases of residual populations of almond moth, saw-toothed grain beetle, and *T. castaeum* (Herbst) in small quantities of rolled oats were curtailed by releasing *X. flavipes* (LeCato *et al.*, 1977). In laboratory studies, LeCato and Arbogast (1979) effectively reduced populations of Angoumois grain moth in small quantities of grain because the predator killed the prey before it entered the grain to cause damage. *Xylocoris flavipes* is more effective in coarsely particulated media than in fine media (Press *et al.*, 1978). Wen and Deng (1989) found that a *Xylocoris* sp. in China gave effective control of *T. confusum* Jacquelin duVal in stored wheat grain but not in wheat flour. Releasing parasitoids of almond moth eggs and larvae was compatible with the predation of *X. flavipes* (Brower & Press, 1988; Press, 1989). The release of 50 pairs of *X. flavipes* into empty corn bins with infested residues of grain with stored product insect pests resulted in good reduction of smaller pest species but not larger moth larvae; thus Brower and Press (1992) believe by releasing specific parasitoids for control of the moths and primary grain pests, the whole pest complex might be greatly reduced before the newly harvested grain is brought to the bins for storage.

Xylocoris galactinus (Fieber), a cosmopolitan generalist predator, is found associated with stored grains, but prefers moist habitats; thus, it is also commonly found in manure heaps, in hot beds, and under bark. Chu (1969) made an extensive study of its biology. He exposed *X. galactinus* to over 60 different species and the predator fed on 43 species of insects in 22 families scattered in nine orders; he also had some success rearing it on artificial diets and was able to mass-rear it on *T. castaneum* larvae infesting cracked corn. *Xylocoris galactinus* has five generations in Japan compared with three for *L. beneficus*.

Berytidae   The stilt bugs are essentially phytophagous (Froeschner & Henry, 1988), but Elsey and Stinner (1971) and Elsey (1972a, 1973) found *Jalysus spinosus* (Say) to be predaceous on eggs of *H. virescens* and *Manduca sexta* (Linnaeus) on tobacco plants, and at times it was the most numerous predator of *Heliothis* eggs in North Carolina tobacco fields (Elsey, 1972a). This predator can only prey on eggs, dead insects, or slow-moving insects such as aphids, and cannot penetrate the serosa of 1- or 2-day-old *M. sexta* eggs (Elsey, 1972b).

Lygaeidae   This is the second largest family in the Heteroptera, with about 3000 species in North America. Ashlock and Slater (1988) list 322 species in 85 genera and include references to their biologies. Apparently only the subfamily Geocorinae contains predaceous species, and only 4 or 5 species of the 25 species in *Geocoris* that occur in North America north of Mexico have been involved in the biological control studies. In Readio and Sweet's (1982) key and discussions of the species of Geocorinae that occur in the United States east of the 100 meridian, four of the five important species are covered.

Geocoris spp. are generalist predators that require some feeding on plants for optimal development and reproduction (York, 1944; Sweet, 1960; Ridgway & Jones, 1968; Stoner, 1970; Tamaki & Weeks, 1972b; Naranjo & Stimac, 1985, 1987; Bugg *et al.* 1987, 1991). Readio and Sweet (1982) show a prey list of *Geocoris* species in the United States, and Tamaki and Weeks (1972b) list prey of *Geocoris* species from the world literature. The prey are mostly Heteroptera, Homoptera, Coleoptera, Lepidoptera, and Acarina. Unlike the Anthocoridae, Nabidae, and Miridae; *Geocoris* spp. deposit their oval eggs singly, either horizontally on plant surfaces or in soil duff. Eggs are often parasitized by scelionids. The nymphs usually attack the same type prey as the adults. They locate prey visually, but use their antennae to find plant food or possibly the correct habitat (Readio & Sweet, 1982).

Geocoris atricolor Montandon is a western North American species found in alfalfa and cotton fields. It was reared in the laboratory on potato tuber moth larvae and string beans by Dunbar and Bacon (1972a, 1972b), who determined its developmental time for eggs and nymphs and fecundity under six different temperatures. *Geocoris atricolor* was more similar in its responses to various temperatures to *G. pallens* Stål than to *G. punctipes* (Say), which developed and oviposited best at the lower temperatures tested.

Geocoris bullatus (Say) occurs across the northern United States in sandy soil areas and readily feeds on seeds even in the presence of prey (Sweet, 1960). It did not lay eggs on a diet of seeds and beans or of pea aphids and beans unless sugar beet leaves were provided (Tamaki & Weeks, 1972b). In the northwest United States, it is usually abundant in forage and vegetable crops along with *G. pallens* and became most abundant (25 per square foot) in orchard floor vegetation, particularly under peach trees infested with *Myzus persicae*. *Geocoris bullatus* reduced the number of green peach aphids on fallen leaves returning to each tree by more than half, thus, an average of more than 10,000 more oviparae returned to the trees where predators were excluded from the orchard floor (Tamaki, 1972). Under laboratory conditions, Tamaki and Weeks (1972a) found coccinellids were the most effective in reducing high densities of aphids, but the omnivorous nabids and geocorids were more effective at low aphid densities because they could sustain themselves on fewer prey. They also found that in a mixture of caterpillar and aphid prey, *G. bullatus* attacked only a few caterpillars; thus, the caterpillars interfered only slightly with predation on the aphids. In a greenhouse study, *G. bullatus* was released in cages with *M. persicae* infesting sugar beets and suppressed the aphids so there was only a 9-fold increase in 28 days compared with a 2000-fold increase in cages without *Geocoris* (Tamaki & Weeks, 1972b). By providing chopped sunflower seeds as a supplemental food in sugar beet plots, the number of *Geocoris* eggs laid on sugar beet plants was doubled compared with controls (Tamaki & Weeks, 1972b). *Geocoris bullatus* also preys on *Lygus* spp., and Chow *et al.* (1983) determined the functional response of the predator to three different densities of *Lygus* nymphs at three different temperatures.

Geocoris pallens occurs throughout the western United States east to central Texas, mostly in open, disturbed habitats and agricultural fields (Readio & Sweet, 1982). It has a similar biology to *G. bullatus* but is more southern in its distribution. Butler (1966) and Dunbar and Bacon (1972a) determined its egg and nymphal development at several different temperatures. Tamaki and Weeks (1972b) obtained similar responses when they tested progeny of overwintered *G. pallens*. However, when they tested progeny from first generation adults, there was a consistent 4°C difference over the entire regression line in the rate of egg development; the developmental threshold of the eggs from the overwintered adults was about 10°C and those from the first generation was about 15°C.

In field studies in cotton, *G. pallens* was an effective predator of *H. zea* larvae (van den Bosch *et al.*, 1969), *Lygus* bugs (Leigh & Gonzalez, 1976), cabbage looper (Ehler *et al.*, 1973), pink bollworm (Irwin *et al.*, 1974), and mites (Gonzalez *et al.*, 1982); it also suppressed *M. persicae* on sugar beets (Tamaki & Weeks, 1973).

Because certain pesticides, including herbicides and defoliants, besides insecticides can reduce *Geocoris* populations, the choice of more sparing pesticides is valid. Yokoyama *et al.* (1984) determined the acute toxicity of 22 pesticides, noting those materials that are particularly suitable for use in cotton.

Geocoris punctipes occurs across the southern two-thirds of the United States and extends southward into Colombia. It is commonly found associated with agricultural crops such as alfalfa, cotton, lettuce, peanuts, soybeans, and sugar beets (Readio & Sweet, 1982). It appears to be the most studied geocorine species among the natural enemies of agricultural pests. It has a similar biology to *G. pallens,* but because it is one of the largest *Geocoris* species, it consumes more prey. Butler (1966) determined the number of bollworm eggs consumed by adults and nymphs of *G. punctipes,* the number of different size lygus bug nymphs consumed per individual adult, and number of first-instar lygus bugs consumed by individual *G. punctipes* nymphs of different ages in 24 h (1967). The number of soybean looper eggs consumed daily at four different temperatures was determined by Crocker *et al.* (1975). At 25°C in the laboratory, the developmental time of *G. punctipes* eggs was about 10 days for a total nymphal development of about 27 days, and the preoviposition period was about 5 days with a mean fecundity of 178 eggs; the diet was coddled fifth-instar larvae of the beet armyworm and a section of green string bean (Champlain & Sholdt, 1967a). Champlain and Sholdt (1967b) also determined the temperature range for the development of the immature stages.

After testing eight different diets, Dunbar and Bacon (1972b) concluded that insect eggs were preferred by *G. punctipes,* even though complete nymphal development occurred on pea aphids with green beans and two-spotted spider mites and green beans. No adults were produced on pea aphids alone, and no eggs were deposited with a diet of the spider mites and green beans. With green beans (provided as a water source) and fresh *Heliothis* eggs, *G. punctipes* produced significantly more eggs with higher fertility than when fed fresh *Sitotroga cerealella* eggs; frozen eggs could be useful for temporary rearing even though fecundity was significantly lower than on fresh eggs (Lopez *et al.,* 1987). Cohen (1984) measured the food consumption and utilization from first-instar nymphs to adults of *G. punctipes* fed eggs of *H. virescens* and found that food absorption efficiency was 95.2% and growth efficiency was 52.9%; and he also determined the ingestion efficiency and protein consumption by *G. punctipes* fed pea aphids (Cohen, 1989). In addition, Cohen improved a method of encapsulating a semidefined artificial diet in wax that was fed successfully to *G. punctipes, G. pallens,* and several other predaceous insects (Cohen, 1983). Cohen further simplified an artificial diet for *G. punctipes* by making a paste of fresh beef liver, fat ground beef, sugar solution, wheat germ, fresh spinach, coupled with a salt solution that was squeezed onto sheets of stretched Parafilm® and rolled into

a sausagelike cylinder; on this diet. *G. punctipes* was reared for at least three generations (Cohen, 1985). It has been cultured on the artificial diet for over 50 generations, and Hagler and Cohen (1990) incorporated vertebrate-derived antigens into the diet and determined that simulated prey ingestion could be detected up to 48 h.

Crop plants and weeds can have direct and indirect influences on the predation performance of *Geocoris,* as well as on any of the predaceous hemipterans that at times are herbivorous. Application of systemic insecticides to cotton showed not only the importance of natural populations of *Geocoris, Nabis,* and *Orius* predators in "regulating" *Heliothis* spp. populations but also the fact that the heteropteran predators fed on cotton because they were reduced in numbers by systemic soil treatments that had minimal impact on *Chrysopa* larvae and hymenopteran parasitoids (Ridgway *et al.,* 1967). York (1944) had already demostrated that *Geocoris* fed on plants as a water source but required leafhopper eggs or nymphs for complete development. Stoner (1970) tested many plants and found *G. punctipes* could live over 20 days on certain seeds, about 9 days on cotton leaves, 12 days on dandelion pollen, and 3.5 days on water alone, but required arthropod prey for normal development and fecundity. Big-eyed bugs were also found to use cotton extrafloral nectar as an alternate food (Yokoyama, 1978). *Geocoris punctipes* did feed more on nectared cotton without *H. virescens* eggs; however, when *Heliothis* eggs were present, there was no difference in feeding on nectared versus nectarless plants (Thead *et al.,* 1985).

Sprenkel *et al.* (1979) found the planting dates, row widths, and to a degree the seeding rate of soybeans influenced populations of three lepidopteran pests and the abundance of *G. punctipes, Nabis* spp., and spiders; by planting early, predation and parasitization of the lepidopteran pests were increased.

There also can be interactions between predaceous species such as spider *Oxyopes salticus* and *Geocoris* spp. that prefer *Heliothis* spp. eggs but will feed on the spider's eggs. The big-eyed bug adults ate all eggs or emerging spiderlings from a single unguarded egg sac, and the adult *O. salticus* females fed on more active prey such as larger *Geocoris* nymphs and adults (Guillebeau & All, 1989). These and other results suggest that inundation of fields with *Geocoris* spp. is unlikely to reduce *O. salticus* populations, but predation by increased *O. salticus* populations could decrease geocorid populations. Also, the intraspecific competition between *G. punctipes* for *H. zea* first-instar larvae significantly reduced the number of larvae consumed per individual and suggested to Guillebeau and All (1990) that exceeding an optimal density of *Geocoris* in the field could reduce the biological control of *H. zea.*

Releases of *G. punctipes* in field cages in cotton showed that about 252,000 adults per acre were ineffective, but a release of 630,000 per acre reduced the egg and larval populations of *H. virescens* by 88% (Lingren *et al.,* 1968). The number of *H. virescens* eggs consumed per *G. punctipes* by intermediate predator densities (10,756 to 16,134 per ha) increased significantly as prey density increased (16,134 to 48,402 per ha), but the percentage of eggs consumed remained constant (1.0 to 2.0% per predator per 48 h), suggesting a type I functional response (Hutchison & Pitre, 1983). In field cages, *G. punctipes* and *Nabis* spp. could significantly reduce the Mexican bean beetle populations by preying on eggs and first-instar larvae of the beetles on soybeans (Waddill & Shepard, 1974a). When *G. punctipes* were released in cages over alfalfa infested with *Bemisia tabaci* (Gennadius), 40% of the whiteflies were destroyed (Becker *et al.,* 1992).

Geocoris ochropterus (Fieber), an important general predator of crop pests in India, was reared on frozen ant pupae for potential release against tea insect pests (Mukhopadhyay & Sannigrahi, 1993).

Miridae   This is the largest family of the Heteroptera; Henry and Wheeler (1988) believe the number of species will reach 20,000 in the world faunas. Most known species are phytophagous, but scattered in the Miridae are at least 25 genera containing species that are mostly generalized predators. There are 98 mirid species that are common to North America and the Old World including some predaceous species (Wheeler & Henry, 1992). Some of the most distinctive mirid pest species are at times predaceous, and Schuh (1974) and Wheeler (1976) believe ancestral mirids were generalized predators. Mirids usually deposit their eggs in plant stems and in living or dead twigs, but some coccophagous species deposit their eggs beneath old female scale coverings (Wheeler & Henry, 1978b). There are five nymphal instars and the young usually feed on the same plant or prey as the adult. In temperate climates there is usually one generation, but mirids can have overlapping broods. Overwintering usually occurs in the egg stage and, at times, in the adult stage (Wheeler & Henry, 1992).

Some predaceous species have played important roles in classical and natural biological control. Natural control of the European red mite by mirids was observed by Gilliatt (1935) in Nova Scotia apple orchards; when nicotine sprays were used in the orchard, the nymphal stages of mirids were quickly killed and populations of the mite made rapid increases. He also tested other pesticides that were less harmful to the predators. Certain pesticides on pear trees reduce predaceous mirids, which allows pear psylla populations to increase (Westigard, 1973). Descriptions of the various stages and biologies of many mirid species found in European orchards are nicely covered by Fauvel (1974). Only a few genera that contain predaceous species will be discussed here. The predaceous genera fall under many

different higher taxa, but it appears that the Isometopinae, Deraeocorinae (at least the tribe Deraeocorini), and many genera in the subfamily Orthotylinae contain most of the predaceous species.

Blepharidopterus angulatus (Fallén), the black-kneed capsid, is a Palearctic species that was probably inadvertently introduced into North America after 1935. It occurs in Europe, the former USSR, North Africa, and Canada (ranging from British Columbia to Nova Scotia) (Henry & Wheeler, 1988). Collyer (1952) found this mirid to be most numerous in commercial fruit orchards in England infested with the European red mite, *Panonychus ulmi,* where it is an important predator of the mite (Muir, 1966). It also attacks other mites, aphids, and thrips. An adult female *B. angulatus* can consume 4000 adult female mites; nymphs and adults were observed to suck leaves and fruitlets but no damage to plant tissue occurred (Collyer, 1952). *Blepharidopterus angulatus* also feeds on aphids and leafhoppers. Glen (1973, 1975) made extensive studies of the searching, food requirements, and energetics of the mirid exposed mainly to lime aphids and constructed a model to predict prey densities required for optimal development. Under field conditions, first-instar *B. angulatus* should require a prey density about nine times greater than that required in later instars.

Campylomma verbasci (Meyer-Dür) is a European species that was inadvertently introduced into North America where it now occurs across the continent in the northern United States and southern Canada. It is a parthenogenetic species and an important predator of pear psylla in British Columbia (McMullen & Jong, 1970) and France (Fauvel & Atger, 1981); it preys on eggs of the European red mite on apple in Nova Scotia (Gilliatt, 1935; Lord, 1971), England (Collyer, 1953a), and Poland, but alone it could stop the increase of neither mites nor aphids in Poland (Niemczyk, 1978c). In Ontario apple orchards, serological assays revealed that the mirid preyed on *Aphis pomi* (Hagley & Allen, 1990). In the laboratory it ate two *A. pomi* a day compared with two larger mirids, *Lepidopsallus minisculus* Knight (which ate over six aphids) and *Hyaliodes vitripennis* Say (which consumed over three aphids), and all three mirids ate tetranychid mites (Arnoldi *et al.,* 1992). Both *C. verbasci* and *Atractotomus mali* (Meyer-Dür) can be a pest of apples depending on the host cultivar and prey availability (Wheeler & Henry, 1992).

Campyloneura virgula (Herrick-Shaeffer) is another parthenogenetic Palearctic polyphagous species, which was recently discovered in British Columbia and now extends down the West Coast into northern California. It is predaceous on small arthropods found on a variety of trees and shrubs (Lattin & Stonedahl, 1984). It overwinters in the egg stage in the British Isles, but as an adult on the continent of Europe (Southwood & Leston, 1959). Along with anthocorids, it displayed an ability to regulate pear psylla populations in untreated pear orchards in France (Bouyjou *et al.*, 1984).

Ceratocapsus modestus (Uhler) may be important in the control of leaf gall-forming phylloxera (Wheeler & Henry, 1978a). Carvalho *et al.* (1983) summarized the predaceous habits of this genus and described many new South American species.

The genus *Cyrtorhinus* is composed of valuable species that show a marked preference for particular plants and their associated delphacid plant hoppers, but a species may turn to related plants and plant hoppers in the vicinity of its regular associated habitat and prey eggs. The genus at one time contained a group of species that have distinct large sinuous arolia between the claws and another group with only two very fine parallel small setae between the claws (Usinger, 1939). When Carvalho and Southwood (1955) revised the genus, they gave the generic name *Tytthus* to the latter group; it also contains some very important natural enemies and will be discussed later.

Cyrtorhinus fulvus Knight, a Pacific Island species, prefers taro as a plant host, where it oviposits in leaf petioles and the nymphs and adults feed on the eggs of the taro leafhopper, *Tarophagus proserpina* (Kirkoldy) (Matsumoto & Nishida, 1966). It was purposely introduced from the Philippines to Hawaii in 1938, where it quickly became established and controlled the leafhopper (Fullaway, 1940); later the predator was sent to seven Pacific islands and became established and provided good control in Fiji, Guam, Ponape, and Samoa (Waterhouse & Norris, 1987). By removing the predators by hand every other day in Hawaiian plots, Matsumoto and Nishida (1966) obtained a significant increase in taro leafhopper populations compared with plots where *C. fulvus* was not removed.

Cyrtorhinus lividipennis Reuter is an Asian species that also occurs on Guam and the Philippine Islands (Chiu, 1979; Yasumatsu *et al.*, 1981). Initial introductions into the Hawaiian Islands to control the corn delphacid *Peregrinus maidis* (Ashmead) failed, but later it not only became established but also provided effective control of *P. maidis* and has shown an adaptation to dry corn agroecosystems (Liquido & Nishida, 1985b, 1985c). In Philippine rice plots, predators such as large spiders, coccinellid beetles, dragonflies, damsel flies, lizards, and toads were removed by hand, while plant hoppers, small spiders, and *C. lividipennis* were removed with mouth aspirators by Hsieh and Dyck (1975). These authors concluded that spiders and *C. lividipennis* played an important role in regulating the rice green leafhopper, *Nephotettix virescens* (Distant). The brown plant hopper, *Nilaparvata lugens*, is also a prey of *C. lividipennis* (Chua & Mikil, 1989; Manti, 1991). The mirid eats all stages of the brown plant hopper and its functional response to both eggs and first-instar nymphs was determined by

Sivapragasam and Asma (1985). It also attacks the brown plant hopper on rice in Guam (Usinger, 1939), and prevented increases of this plant hopper in drilled, but not transplanted, rice fields in Fiji (Hinckley, 1963). In the Philippines, Kenmore (1980) considered predation by generalist natural enemies a more significant source of brown plant hopper mortality than that of parasitism. He found spiders, veliid *Microvelia,* and mesoveliid *Mesovelia,* which feed on prey falling on the water surface, to be apparently important in suppressing brow plant hopper populations. In the event that inoculative releases of *C. lividipennis* became necessary to control delphacids, Liquido and Nishida (1985a) developed methods of mass-culturing the mirid on a factitious prey, eggs of the Mediterranean fruit fly. The body dimensions of the predators fed on the fruit fly eggs were equal to those that received its normal prey, the corn leafhopper; the duration of eggs and of nymphal instars was not significantly different and fecundity was the same, but longevity of the mirid fed its natural prey was much longer.

The genus *Deraeocoris* is composed of some 70 species in North America and was last revised by Knight (1921). At that time he called attention to some species that were quite specific to woolly aphids, *Eriosoma* spp., which in their nymphal stages resembled the aphids because of a covering of white, powdery, flocculent material. Furthermore, Knight believed there were also phytophagous species in the genus because they were associated with certain plant species. Today, however, it appears that most species are general predators (Henry & Wheeler, 1988).

Deraeocoris lutescens (Shilling) is a general predator in European orchards; it is partially phytophagous and overwinters in the adult stage (Collyer, 1953b). Collyer (1953b) and Fauvel (1974) briefly describe its stages and biology. Herard (1986) cites references on its occurrence in Europe and its association with pear psylla.

Deraeocoris nebulosus (Uhler) is a generalist predator that occurs throughout most of the United States and Canada on more than 50 species of ornamental trees and shrubs (Wheeler *et al.*, 1975). These authors studied its biology by exposing the mirid mainly to the oak lace bug as prey, because the predator could be important in suppressing populations of lace bugs and other pests on ornamentals.

Deraeocoris ruber (Linnaeus) is distributed widely in Europe, but is more common in southern Europe and has spread to the eastern United States and South America (Henry & Wheeler, 1988). It is found on many trees, bushes, and herbs, where it feeds on various insects and mites, but prefers feeding on prey larger than the red spider mite (Collyer, 1953b). It is a common predator of aphids on hazel in Italy, overwinters in the egg stage (Viggiani, 1971), and is an occasional predator of *Psylla pyri* (Herard, 1986).

Diaphnocoris provancheri (Buryae) (= *pellucida* Uhler) occurs across North America on many trees and shrubs and is predaceous on aphids (Kelton, 1980). It is also an effective predator of *P. ulmi* on apple trees in Nova Scotia where it occurs with three other mirid species (Lord, 1971). In the same orchards, MacLellan (1962) found what was probably *D. provancheri* preferably attacking fresh codling moth eggs, unlike the three other mirid species present, but all the mirids attacked young codling moth larvae.

Heterotoma meriopterum (Scopoli) is another Palearctic species inadvertently introduced into North America where it occurs from British Columbia to Nova Scotia and New York (Henry & Wheeler, 1988). Fauvel (1974) illustrates some of its stages. Herard (1986) found it to be one of the most common mirids in French pear orchards, also occurring on hawthorn and nettle where it feeds on psyllids and aphids. On pear trees, eggs are deposited in young twigs and overwinter there, but normal and unavoidable pruning of pear trees in winter intensely depressed predator populations in the orchard; thus hawthorn trees can be useful for mirid survival and retention near the orchards; also, the pear prunings can be saved for mirid emergence (Herard, 1986). Collyer (1953b) found it mainly predaceous on *P. ulmi* and less so on aphids in apple trees in England.

In the subfamily Isometopinae, *Corticoris signatus* (Heidemann) and *Myiomma cixiformis* (Uhler) prey on obscure scale *Melanaspis obscura* (Comstock) infesting oaks in Pennsylvania (Wheeler & Henry, 1978b). The predaceous habit of this subfamily on diaspine scales was first reported by Hesse (1947), who noted the nymphs and adults of *Letaba bedfordi* Hesse preying on and reducing heavy infestations of citrus red scale in South Africa. Wheeler and Henry (1978a) review the feeding habits reported in the literature for the subfamily and describe the biology and illustrate nymphs of several species.

Lygus spp., the notorious pests on many agricultural crops, are at times predaceous. Wheeler (1976) lists 20 arthropod species in five insect orders as prey. However, lygus also can serve as prey to at least four predatory heteropteran families. Whalon and Parker (1978) employed an immunological microdouble-diffusion technique to determine the presence of apple-infesting *L. lineolaris* Palisot de Beauvois in predator gut contents, using antibodies developed in rabbits.

Macrolophus caliginosus Wagner is considered a good candidate for release in greenhouses because it is a polyphagous predator that has been used against whiteflies (Malausa *et al.,* 1987). It would be more effective against *Myzus persicae* than *Tetranychus urticae* based on preference and functional responses of *M. caliginosus* to both prey (Foglar *et al.,* 1990).

The genus *Phytocoris* is one of the largest in the Miridae, but the habits of most species are poorly known. The nymphs and adults of some *Phytocoris* spp. are concealed as they crouch in the crevices of the bark, apparently awaiting prey (Knight, 1921). Some *Phytocoris* spp. in western North America are of Palearctic origin (Stonedahl,1983).

Spanagonicus albofasciatus (Reuter), a widely distributed New World species, has mixed reviews as a predator. Wene and Sheets (1962) considered it a pest in Arizona cotton fields, but Butler (1965) tested it in the laboratory and found it to eat spider mite eggs; hyaline plant bug eggs; cotton aphids; bollworm eggs and young larvae; lygus bug eggs; and eggs, larvae, pupae, and adults of the banded whitefly. Thus, Butler considered *S. albofasciatus* (Reuter) a potentially important predator in Arizona cotton fields. Its stages and development were studied by Musa and Butler (1967). In Arizona, it is found commonly on 12 different plants in 10 different plant families (Stoner, 1965). In Florida soybean fields, it was observed to feed on *Heliothis zea* and *Pseudoplusia includens* (Walker) eggs by Neal *et al.* (1972) and on velvet bean caterpillar eggs in the same area by Buschman *et al.* (1977).

Stethoconus japonicus Schumacher, an adventive predator of the azalea lacebug, is the first host-specific lacebug predator established in the Western Hemisphere (Henry *et al.,* 1986). It showed the highest biological control potential for species of *Stephanitis* lacebugs, but also preys on hawthorn lacebugs. The biology of this oligophagous mirid is described by Neal *et al.* (1991) and Neal and Haldemann (1992). The biology of *Stethoconus praefectus* (Distant), which attacks *Stephanitis typicus* Distant, a pest of coconut palms, was studied by Mathen and Kurian (1972).

Tytthus mundulus (Breddin) occurs naturally in Australia, Java, the Philippines, and Fiji Islands (Usinger, 1939). A search for more natural enemies of the sugarcane leafhopper, *Perkinsiella saccharicida* Kirkaldy, for release in the Hawaiian Islands was made in the Queensland area of Australia by F.A.G. Muir in 1920, where he discovered *T. mundulus* feeding on the embedded leafhopper eggs. He returned to the Hawaiian Islands with a few of the mirids that were carefully studied before release in July 1920. Another lot of *T. mundulus* was collected in Fiji by Pemberton and these were also released in Hawaii during 1920. By 1923, the sugarcane leafhopper was no longer a problem (Pemberton, 1948). There can be 10 generations of the predator per year (Swezey, 1936). It often consumes the leafhopper eggs and then its own eggs in the same sites, but asynchrony of corn leafhopper in time and place explains why *T. mundulus* is not an effective predator of *Peregrinus maidis* infesting corn (Verma, 1955). It was also successfully introduced into Jamaica and Mauritius (Williams, 1957). In South Africa, *T. mundulus* imported from Mauritius did not become established against the green leaf sucker, *Numicia viridis* Muir, but the inadvertently introduced *T. parviceps* (Reuter) apparently did become established (Carnegie & Harris, 1969). Both of these *Tytthus* spp. are cultured and released in South Africa (Greathead, 1971). Releases of *T. mundulus* were made in Florida in 1984 but establishment is not yet confirmed (Henry & Wheeler, 1988). Attempts were made to introduce *T. mundulus* against the sugarcane leafhopper in Ecuador several times but the material died in shipment in 1977. Bianchi (1980) hand carried the predator and made releases in Ecuador, but its establishment there is questionable. In transit, *T. mundulus* nymphs can be fed housefly eggs inserted into leaf tissue (Stevens, 1975).

Nabidae The damsel bug family contains 31 genera and about 380 species—all known species are predaceous (Lattin, 1989). The subgeneric taxa of *Nabis* (*Reduviolus* and *Tropiconabis*) are not used here. The bionomics of Nabidae have been reviewed by Lattin (1989). Eggs and oviposition into plant tissue are similar to that of the Anthocoridae. There are usually five instars but only four in *Himacerus apterus* Fabricius. This species also differs from most nabids (which apparently overwinter as adults) in that it overwinters in Europe in the egg stage (Koschel, 1971). Most adults of *Nabis alternatus* Parshley, *N. americoferus* Carayon, and *N. capsiformis* (Germar) enter an adult reproductive diapause state during late fall; and at least *N. alternatus* can have up to five generations in Arizona (Stoner *et al.,* 1975). In alfalfa fields in Kentucky, *N. americoferus* has three generations, while in soybeans there is only one (Braman & Yeargan 1990), and in Ontario alfalfa fields there are two (Guppy, 1986).

Nabids are considered polyphagous or generalist predators, but some may feed on plants. This was first suspected when the heteropteran predators were differentially killed compared with mandibulate predators in cotton treated with systemic insecticides (Ridgway *et al.,* 1967). Laboratory tests demonstrated that *N. americoferus* and *G. pallens* feed on the cotton terminals (Ridgway & Jones, 1968). Furthermore, Stoner (1972) found that first instars of *N. alternatus, N. americoferus,* and *N. capsiformis* can be sustained on pollens, certain seeds, and legumes, but no nymphal development occurs (compared with geocorines where some development does occur). Nabids fed on cotton leaves alone survived about as long as those exposed only to water. Stoner concluded from comparing the responses of *Geocoris* and *Nabis* with plant feeding that *Nabis* spp. are more predaceous than *Geocoris* spp. One should be aware that plant feeding, though minor, can involve plant pathogen transmission, because *N. alternatus* possibly transmits a yeast plant pathogen (Burgess *et al.,* 1983).

Nabid prey range is wide among the species studied. Several genera within the subfamily Prostemmatinae are predators of soil-dwelling lygaeid species of the Blissinae, Geocorinae, and Rhyparochromidae (Lattin, 1989). There are at least six *Nabis* spp. that have received attention as important natural enemies of agricultural crop pests, and a *Himacerus* species is considered to be a common predator of forest pests. It appears that no nabids have been intentionally introduced as a natural enemy; however, *Anaptus major* (Costa), a European species, was introduced into North America, probably in ship ballasts (Lattin, 1989). *Himacerus apterus* is a common Palearctic species that is predominantly brachyapterous in the adult stage (Péricart, 1987). In early nymphal instars, it lives on low plants feeding on mites and small insects, and later moves to trees where it feeds on various defoliators (Koschel, 1971). Koschel (1971) made an extensive study of *H. apterus* that was prompted by observations on its predation of various defoliating moth and tenthredinid larvae infesting pines; it was considered as a candidate predator that could be mass-cultured and released in forests. However, Koschel concluded it would be too difficult to mass-culture because *H. apterus* only deposits about 20 eggs per female and is very cannibalistic.

Nabis alternatus is a transcontinental North American species (Henry & Lattin, 1988). Taylor (1949) found *N. alternatus* to be the most numerous insect predator in Utah alfalfa fields, and laboratory studies showed that up to 200 eggs per female were deposited into alfalfa stems. Nymphs consumed an average of 80 pea aphids or 29 lygus bugs to mature; they also fed on thrips and on first- and second-instar grasshopper nymphs, but died when exposed to alfalfa weevil larvae (Taylor, 1949). An accurate count of the number of *Lygus* nymphs consumed and required by *N. alternatus* was made by Perkins and Watson (1972). They found that early-stage nabids successfully preyed on the first 3 lygus instars and only the larger nabid instars were easily able to attack all lygus instars; first-instar *N. alternatus* require about 7 first-instar lygus nymphs and a fifth-instar nabid required about 91 first-instar nymphs for its development to adult, or a total of about 158 first-instar lygus nymphs for all 5 nabid instars compared with about 57 third-instar nymphs (Perkins & Watson, 1972). *Nabis alternatus* also is an effective predator of aphids. The "predator power" of *N. alternatus* was determined by exposing the nabids to green peach aphids on sugar beet plants in field cages under three different temperatures (Tamaki *et al.,* 1981). When nabids insert the eggs into alfalfa stems, only about 5% will hatch if the stems are allowed to dry, while over 55% will hatch if the alfalfa continues to grow (Richards & Harper, 1978). Predation of noctuid eggs and young larvae by *N. alternatus* was demonstrated by Barry *et al.* (1974). They used the nabid alone and in comparison with *Geocoris punctipes* and *Chrysoperla carnea* in field cages over soybeans infested with *H. zea* and *Trichoplusia ni;* the nabid was the only predator that consistently resulted in significantly greater regulation of the pest populations—defoliation, pod damage, and *H. zea* numbers were all significantly reduced. *Nabis alternatus* also is able to grasp flea beetle adults with its front legs and to feed on them (Burgess, 1982). Bjegovic (1968) observed *N. feroides* Remane catching flea beetles and paralyzing them by injecting saliva; this nabid is also considered to be an important predator of cereal leaf beetle eggs.

Nabis americoferus is the most common nabid in North America and was called *N. ferus* (Linnaeus) in Nearctic literature until Carayon (1961b) distinguished differences between the two species. Developmental rates, consumption rates, and reproductive capacity vary with temperature as well as with diet, and temperature thresholds and degree days required for certain developmental stages can vary with the geographic origin of insect populations. This was pointed out by Braman and Yeargan (1988) where the thermal unit requirements calculated for nymphal development of a population of *N. americoferus* from Ontario (Guppy, 1986) were nearly 20% greater than for a Kentucky population, though thermal unit requirements for egg stages were similar (Braman *et al.*, 1984). Nymphs fed larvae of the alfalfa blotch leaf miner developed faster than those fed pea aphids (Guppy 1986), and when Bramon and Yeargan (1988) used green bean pods instead of soybean seedlings as an ovipositional substrate, egg production was often lowered in the three nabid species studied. Pea aphids were preferred three times over potato leafhopper nymphs (Flinn *et al.*, 1985). Egg-to-weight relationships suggested that *N. americoferus* was more efficient at producing eggs than was *N. roseipennis* Reuter (O'Neil, 1992).

Evaluation of predation by *N. americoferus* in field cages has been conducted against several lepidopteran pests and aphids. In cotton, released nabids appeared to perform more effectively against *H. zea* than do geocorids or chrysopids (van den Bosch *et al.*, 1969). Releases of a coccinellid, a geocorid, and *N. americoferus* singly and in combination were made against the green peach aphid, the zebra caterpillar, and the bertha armyworm on caged sugar beets (Tamaki & Weeks, 1972a). The release of *Coccinella transversoguttata* Feldermann alone and in combination with the other predators was effective in reducing the aphid population. The nabid and *G. punctipes* were also effective against the aphids but only for about 1 week. The nabids alone and in some combinations were superior to all other treatments in reducing the populations of the noctuid species when these larvae were small. Among other predators in cotton, *N. americoferus* showed potential as a predator of the pink bollworm (Irwin *et al.*, 1974) and the cabbage looper (Ehler *et al.*, 1973).

Nabis capsiformis is a pantropical species and the most widespread nabid known; human activities may have had little or no role in its dispersal because its ability to disperse through flight is well established (Lattin, 1989). In Mississippi, *N. capsiformis* deposited about 105 eggs per female at 26 to 28°C (Hormchan *et al.*, 1976), and about 200 eggs per female at 28°C (Hormchan *et al.*, 1976); in Queensland,

Australia, females deposited about 200 eggs (Samson & Blood, 1979). A life table was produced at both locations. In Australia, Samson and Blood (1980) compared the voracity and searching ability of *N. capsiformis* with *Chrysopa signata* Schneider and *Micromus tasmaniae* (Walker) as predators of *Heliothis* pests associated with cotton; the adult female *N. capsiformis* and third-stage *C. signata* were the most voracious predatory stages while the *Micromus* consumed few *H. punctigera* Wallengren eggs or young larvae. The searching efficiency of second-instar *C. signata* was more efficient than third-instar *N. capsiformis,* while third-instar *M. tasmaniae* attacked very few eggs (Samson & Blood, 1980). In soybean fields in the southeast United States, *N. capsiformis* is one of five species of *Nabis* present in fields by early July and remains abundant during the growing season (Funderburk & Mack, 1989); in Florida soybean fields, *N. capsiformis* was the major consumer of small velvet bean caterpillar larvae, *Anticarsia gemmatalis* Hübner (Godfrey *et al.*, 1989). Buschman *et al.* (1977) also found in Florida soybean fields that chrysopids, spiders, and three *Nabis* spp. were the most frequently recorded consumers of radioactive eggs of the velvet bean caterpillar.

Nabis ferus is a common Palearctic species (Carayon, 1961b). It occurs commonly in grasses, bushes, and low crops. Herard (1986) lists prey of *N. ferus* noted in recent literature; the list included aphids on many plants, lygus in strawberries, noctuid eggs and larvae in cotton, Colorado potato beetle larvae, dipteran larvae, and thrips; the majority of these references are in Russian.

Nabis roseipennis is widespread east of the Mississippi River but also extends westward through Colorado into British Columbia (Slater & Baronowski, 1978). It was one of the first nabids whose biology was studied. Mundinger (1922) described the immature stages and gave details on oviposition into grass stalks: the nabid first probed near the top of the stalk with her proboscis and then deposited an egg into the stalk. Isenhour and Yeargan (1982) found *N. roseipennis* most often oviposited in median leaflet petioles of soybeans and noted stratification of oviposition sites along the vertical aspect of the soybean plant by *Orius* and *Nabis*. In Kentucky soybean fields, *N. roseipennis* produced the greatest number of nymphs compared with *N. americoferus* and *N. rufusculus* Reuter; all three species were also present in alfalfa fields (Braman & Yeargan, 1990).

Like other previous *Nabis* spp. mentioned, *N. roseipennis* is a generalist. It is a common predator of lepidopteran eggs and larvae such as the green clover worm in the laboratory and soybeans (Sloderbeck & Yeargan, 1983; Yeargan & Braman, 1989), soybean looper in soybeans (Richman *et al.*, 1980), *H. zea* on cotton (Donahoe & Pitre, 1977), and *H. virescens* on cotton (Nadgauda & Pitre, 1987). In the laboratory, the nymphs and adults survived and matured satisfactorily on small *Heliothis* larvae, but development was slow compared with feeding on larger larvae, which accelerated development and maintained a high growth potential (Nadgauda & Pitre, 1987). Predation rates on *Lygus* nymphs were determined in the laboratory at different temperatures by Nadgauda and Pitre (1986). Predatory rates of *N. roseipennis* were also determined by laboratory feedings of potato leafhopper nymphs and adults, while peak populations of four *Nabis* spp. monitored in alfalfa fields coincided with the peak density of the leafhopper (Rensner *et al.*, 1983). Although the nabid attacked third-instar larvae of the boll weevil, it would not attack third-instar Mexican bean beetle larvae (but did eat its eggs) (Wiedenmann & O'Neill, 1990b).

Nabis rufusculus is very similar to *N. roseipennis* in morphology, biology, and distribution (Mundinger, 1922). It occurs in Kentucky alfalfa and soybean fields. In soybeans it has one generation per year: sweep sampling recovered fewer nymphal nabids in comparison with suction or shake sampling, and thus was less satisfactory in assessing population trends (Braman & Yeargan, 1990).

Pentatomidae   This is one of the largest families of Heteroptera with over 2500 described species (Miller, 1971). The family contains many phytophagous pests, but there are also a few important predaceous species that have been involved in biological control projects. The predaceous species all belong to the subfamily Asopinae, which has over 300 species in 27 genera in the New World (Thomas, 1992); very few, if any, are prey specific, and the first-instar nymphs often feed on plant liquids (Esselbaugh, 1949; Froeschner, 1988). Ishihara (1941) reviewed the feeding habits of predaceous Heteroptera, particularly the Pentatomidae; and in his table of 27 asopine species in 20 genera, 18 are listed as preying on Lepidoptera (mainly larvae), 15 on coleopteran larvae, 7 on hymenopteran larvae, 6 on Hemiptera, 3 on Diptera, and 1 on a neuropteran larva. Most of the species listed by Ishihara fed on species in more than one insect order. Over 90 insect species have been reported as prey of *Podisus maculiventris* (Say) (McPherson, 1980).

Predaceous pentatomids deposit clusters of barrel-shaped eggs with elongate chorionic processes that ring the opercular cap on plant surfaces (Esselbaugh, 1949). When the nymphs hatch, they often remain clinging to their eggshells until the first molt. There are five nymphal instars and each instar can be of different color. Saini (1985, 1988) has published color photographs of all stages of 19 (including 5 predaceous) pentatomid species occurring in Argentina— some of which range into North America. When a prey is detected, the nymphs and adults extend their broad rostra directly forward in an attempt to pierce it and inject paralyzing enzymes, but the success of the attack and subduing the victim varies between the prey species (Marston *et al.*, 1978) and the age of the predator (Morris, 1963). Overwintering occurs in either the egg or adult stage depending on

the species. Predators that were fed less frequently laid fewer eggs, less often and later in life (Wiedenmann & O'Neil, 1990a).

The following asopine species are some that have been studied as biological control agents, have been considered as potential natural enemies, or have been imported to control specific introduced pests.

Andrallus spinidens (Fabricius) is a circumtropical nonspecific predator of lepidopteran larvae. In rice fields in West Malaysia, where its abundance coincided with outbreaks of *Melantis leda* (Linnaeus), Manley (1982) concluded that the predator was of limited importance at low prey densities but of much greater value at outbreak densities because of the short life cycle and continuous feeding habit of the predator. In India, *A. spinidens* was observed feeding on the noctuid *Rivala* sp. infesting soybeans by Singh and Singh (1989), who also studied its biology.

Apateticus crocatus (Uhler) is a North American species and is considered an important predator of *Malacosoma* spp. and *Ellopia somniaria* Hulst caterpillars in Victoria, British Columbia The fifth instar refused animal food, preferring plant juices (Downes, 1921). *Apateticus cynicus* (Say), a North American species that is univoltine, overwinters in the egg stage and produces few eggs; thus, it is not considered an outstanding predator (Jones & Coppel, 1963).

Eocanthecona furcellata (Wolf), an Asian and Pacific Island species, was released in Florida against the Colorado potato beetle and the eastern tent caterpillar in 1981, but by 1988 recovery had not yet been reported (Froeschner, 1988). Li *et al.* (1988) developed an artificial diet for this generalist predator. The functional response of the predator with *Lotoia lepida* (Cramer) as prey was measured in India (Senrayan, 1988). There is greater larval survival when exposed to leaves and prey than on prey alone (Senrayan, 1991). *Eocanthecona furcellata* can be reared on frozen moth larvae (Yasuda & Wakamura, 1992).

Euthyrhynchus floridanus (Linnaeus) ranges from the southeastern United States to Brazil. A biology of this nonspecific predator was determined by feeding it *Agrotis ipsilon* (Hufnagel); its developmental period was longer than any other asopine (Oetting & Yonke, 1975). It is found in soybean fields and overwinters in the adult stage. It was reared on larvae of *Galleria mellonella* (Linnaeus) but also fed on *H. virescens* larvae and on all stages of *Epilachna varivestis* Mulsant (Ables, 1975). Like *Alcaeorrhynchus grandis* (Dallas), the egg stage is prolonged in *E. floridanus*, being 19 days in the former and 16 days for the latter compared with about 5 days for *P. maculiventris*, when reared under similar conditions (Richman & Whitcomb, 1978).

Oechalia schellembergii (Guerin-Menville), an asopine species occurring in Australian alfalfa and cotton fields, is a natural enemy of *H. punctigera*. First-instar nymphs molted to the second instar if alfalfa stems were present but not if their own eggs or eggs of *H. punctigera* only were available, and the second instar did not develop to the third instar on alfalfa stems alone (Awan, 1987). Awan *et al.* (1989) found that female *O. schellembergii* took relatively less time to locate wandering fourth-instar larvae of *H. punctigera* than stationary larvae and that movement of prey increases attack frequency. They also reported the predators responded positively to prey larval frass; predators with blocked vision and olfaction searched sporadically with extended rostra and used their front tarsi to locate prey. It is interesting that *Trissolcus basalis* (Wollaston), the parasitoid successfully introduced against *Nezara viridula* (Linnaeus), also parasitized the eggs of *O. schellembergii* as well as another asopine predator (*Cermatalus nasalis* Westwood) in alfalfa fields in South Australia (Awan, 1989).

Perillus bioculatus (Fabricius), a widely distributed Nearctic species, has received much attention as a natural enemy of the Colorado potato beetle, *Leptinotarsa decimlineata* (Say), particularly in Europe where the beetle has invaded. Tamaki and Butt (1978) constructed life tables and determined the feeding potential of the different life stages of the predator and the amount of potato leaf foliage consumed by the larvae and adults of the beetle; thus, given the number of life stages of the predators and defoliators, the amount of predation and defoliation can be predicted. The predator overwinters as an adult and feeds on eggs and larvae of the potato beetle, and deposits about 100 eggs per month. The first instar feeds on the plant while later nymphal stages are essentially predaceous and complete development in about a month after consuming over 400 beetle eggs (Knight, 1923; Landis, 1937). Frozen potato beetle larvae can be used to rear the predator nymphs but not egg-laying females; however, permanent rearings can be made using frozen pupae (Franz, 1967). The predator was introduced into France (Trouvelot, 1932), Germany (Franz & Szmidt, 1960), Hungary (Jermy, 1962), Russia (Beglyarov & Smetnik, 1977), Czechoslovakia (Tadic, 1964), and Italy (Tremblay & Zouliamis, 1968), but apparently never became permanently established in Europe (Szmidt & Wegorek, 1967; Beglyarov & Smetnik, 1977; Jermy, 1980) because it could not overwinter successfully: thus, periodic colonization of the predator is called for (Szmidt & Wegorek, 1967). In small-plot field studies in Delaware, releases of *P. bioculatus* in conjunction with bacterial insecticides provided significant control of the beetles for at least 2 weeks, suppressing larval populations by 76% compared with plots treated only with bacterial insecticide (Hough-Goldstein & Keil, 1991). Release of nymphs at either 1.6 per row-meter or 9.8 per row-meter significantly reduced the number of *L. decimlineata* larvae in small plots, and the number of beetle egg masses was significantly reduced by the higher release rate or by the lower release rate in combination with a bacterial spray

(Hough-Goldstein & Whalen, 1993). Biever and Chauvin (1992a) obtained 30 and 62% reduction of the Colorado potato beetle populations by releasing one to three *P. bioculatus* per plant, respectively, and the same authors (1992b) obtained best results by releasing the predator 4 days after potato plant emergence.

Picromerus bidens (Linnaeus) is a common Palearctic asopine species that was inadvertently introduced into North America about 50 years ago and is now found in Quebec and the northeastern United States (Javahery, 1986; Froeschner, 1988). Lariviere and Larochelle (1989) made a world review of the distribution and bionomics of the predator. It is univoltine and overwinters primarily in the egg stage. *Picromerus bidens* occurs on a wide diversity of plants; in Europe it has been recorded on over 30 plant species and over 50 in North America. Its prey range is also broad because it has been reported attacking over 35 different insect species in six orders. If there is any preference for certain prey, it may be defoliating Symphyta of the Hymenoptera. According to Lariviere and Larochelle (1989), many authors have indicated the importance of this predator in reducing various insect pest populations, particularly tenthredinids and diprionids. Mayne and Breny (1948) found *P. bidens* not to be a good control agent of the Colorado potato beetle in Europe because the predator did not disperse easily in the fields and was too much of a generalist predator, while Javahery (1986) noted that its low rate of reproduction was another weakness. There are some early references to *P. bidens* being an effective predator of bedbugs (Clausen, 1940).

Podisus spp. have attracted much attention of biological control researchers. The genus is widespread in the New World, and most species are generalist predators. Coppel and Jones (1962) compared the biologies of four *Podisus* spp. associated with the introduced sawfly *Diprion similis* (Hartig) in Wisconsin and found their biologies quite similar. One of the species was *P. maculiventris*.

Podisus maculiventris occurs transcontinentally in North America and has been exported to Europe and Korea (Froeschner, 1988). It also was sent to China in 1983 from the United States, but as of 1987 no recoveries had been made (Wang & Gong, 1987). It overwinters as an adult in deciduous woods and their borders (Jones & Sullivan, 1981). Among the 90 prey species listed by McPherson (1980) that were taken from the literature, the main prey consisted of about 50 species of Lepidoptera, 17 of Coleoptera, and 6 of Symphyta (Hymenoptera). Both prey defense by large webworms and age of the adult *P. maculiventris* had important effects on the functional response curves (Morris, 1963). Mukerji and LeRoux (1969a, 1969b, 1969c) exposed different size wax moth larvae to different stages of *P. maculiventus* and measured resulting functional responses. They found that prey size was an important component of the predation process for all stages of the predator. These authors also studied the energetics of *P. maculiventris* using wax moth larvae as prey. Warren and Wallis (1971) fed frozen fall armyworm and tent caterpillar larvae and live wax moth larvae to *P. maculiventris* and determined duration from egg to adult at four different temperatures—at the lowest (50°F) no development occurred. From one field-collected female, 1174 eggs were laid in 126 days. Comparable sized larvae of five different lepidopteran species that attack soybeans were exposed on soybean plants to fifth-instar *P. maculiventus* nymphs, but no significant effect on the ability of the predator to contact any prey of the larvae was found; however, once contact was made, the behavior of the prey larvae significantly affected the ability of the predator to subdue them (Marston *et al.*, 1978). Marston *et al.* (1978) concluded that in pest management programs involving releases of *P. maculiventris,* the mortality of prey larvae will vary between species (e.g., *H. zea*), with mortality of about 60% being considered intermediate of the five pest larval species tested. Lopez *et al.* (1976) found third-instar nymphs and adults of *P. maculiventris* were the most efficient predators of third-instar larvae of *H. zea* and *H. virescens* compared with a coccinellid, a chrysopid, and a geocorid species. Releases of *P. maculiventris* adults at a rate of 100,000 per acre in field cages over cotton resulted in a substantial reduction of *H. virescens* larvae (Lopez *et al.*, 1976). Releases of adult predators in greenhouse cotton naturally infested with *Peridroma saucia* (Huebner) larvae resulted in 75% of the cutworm larvae being consumed within 48 h (Ables & McCommas, 1982).

Podisus maculiventris that was contaminated with *Trichoplusia ni* single-embedded nuclear polyhedrosis virus was released in field cages containing cabbage looper larvae and initiated virus infections and epizootics in 3 to 4 weeks (Biever *et al.*, 1982).

There are some insecticides that are compatible with *P. maculiventris*. The predator is generally more susceptible to organophosphorus and carbamate insecticides than to the velvet bean caterpillar, fall armyworm, or corn earworm, but is more tolerant of certain pyrethroids (Yu, 1988).

Coleopteran prey, such as *Tenebrio molitor* Linnaeus, were exposed as small and large larvae at different intervals to *P. maculiventris* in the laboratory to determine the influence of low prey inputs on survival and reproduction; the predators provided with prey survived regardless of feeding interval, but those predators fed less frequently and laid fewer eggs less often and later in life (Wiedenmann & O'Neil, 1990a). The predator survives long periods without prey and adjusts its weight loss by reducing reproductive effort when prey becomes scarce (O'Neil & Wiedemann, 1990; Legaspi & O'Neil, 1993). Larger *P. maculiventris* females laid eggs at younger ages and more frequently than smaller females (Evans, 1982a), but smaller species of asopines, like *Perillus circumcinctus* Stål that is a specialist

predator on certain chrysomelid larvae, exploit the prey populations more effectively than the generalist *Podisus maculiventris*. Evans (1982b) concludes the key difference in the habits of the two predators is in timing of reproduction. Well-timed oviposition by specialist females ensures that offspring mature when prey are easily harvested. In contrast, the poor timing (late oviposition) by the larger generalist female results in most offpsring facing a prey shortage of chrysomelid larvae in the system investigated. A major attempt to introduce *P. maculiventris* into France to control the introduced Colorado potato beetle was made in the 1930s, but the predator did not become established. Couturier (1938) believed that in France the relative dryness of the summer months was the chief obstacle to the establishment of the bug. However, Drummond *et al.* (1984) evaluated Colorado potato beetle larvae as prey compared with larvae of *Galleria mellonella, Epilachna varivestis,* and *Malacosoma americanium* (Fabricius), finding that this potato beetle was a suboptimal prey because the survival of *P. maculiventris* decreased and its developmental time increased. In the former USSR, releases of second and third instar *P. maculiventris* in fields at predator to prey ratios of 1:10 to 1:30 (amounting to 210,000 predators perhectare) in early potatoes resulted in good control of the Colorado potato beetle, and three releases of 100,000 predators perhectare on eggplants during the summer amounted to 90 to 100% control and a yield of 37,400 kg compared with 4,500 kg in untreated control fields (Khloptseva, 1991).

To increase the retention of released *P. maculiventris* adults against *E. varivestis* in snap bean plots, the adult female predators were dewinged; significantly fewer *E. varivestis* larvae were found on the plants treated with released predators over a 21-day period and a higher yield of snap beans was obtained. Lambdin and Baker (1986) obtained the reduction by releasing 5 or 10 adults per 3.1-m plots. A release of five *P. maculiventris* per cage significantly reduced the 50 *E. varivestis* fourth instars placed on soybeans in small field cages, but Waddill and Shepard (1975) found that *P. maculiventris* did not reach adulthood when fed only eggs or first instars of *E. varivestis,* but the predator matured when fed later instars or pupae of the beetle. These authors also obtained the usual functional response curves. In caged field soybeans infested with *E. varivestis,* O'Neil (1988) estimated the area searched by *P. maculiventris* by calculating the number of prey attacked and the number of prey available for attack and leaf area. As leaf area increased, the predator searched a larger area. When O'Neil (1989) compared the functional response of *P. maculiventris* in both the laboratory and field, the low number of attacks in the field suggested that the influence of handling time was trivial and that predation was associated primarily with the predator's ability to find prey in the soybean canopy. Applying the results of the laboratory study to a field setting is difficult because of the differences in the relative magnitudes of prey density and corresponding attack rates between laboratory and field results (O'Neil, 1989), but a more detailed search model was developed based on varying *E. varivestis* larvae as prey exposed on different size field-caged plants to adult *P. maculiventris* (Wiedenmann & O'Neil, 1992).

Male *P. maculiventris* produce a long-range aggregation pheromone that was identified and synthesized by Aldrich *et al.* (1984). A similar type of pheromone was found in *P. fretus* Olsen by Aldrich *et al.* (1986). In both cases, the aggregation pheromones became kairomones because four species of natural enemies were attracted to the artificial pheromones (Aldrich, 1988).

Life history studies have been made of nearly all the North American *Podisus* spp., and their biologies are similar to *P. maculiventris*. However, the distribution and prey range does vary somewhat between species. *Podisus modestus* (Dallas) is a northern arboreal forest species that preys on sawfly defoliators of conifers (Coppel & Jones, 1962; Tostowaryk, 1971). *Podisus placidus* Uhler is a midwest to eastern species that preys mainly on lepidopteran larvae and sawfly larvae (Coppel & Jones, 1962; Oetting & Yonke, 1971), and was considered for introduction into Yugoslavia to control the fall webworm (Tadic, 1964). *Podisus sagitta* Fabricius occurs from the southern United States to Brazil and it naturally feeds on *Epilachna* larvae in Mexico. DeClereq and Degheele (1990) reared it on wax moth and beet armyworm larvae, and like most asopines, it took about 1 month to develop from egg to adult and to deposit about 600 eggs; lower threshold temperatures for egg and nymphal development were estimated as 13.3 and 12.2°C for *P. sagitta*, and for *P. maculiventris*, 10.7 and 11.7°C, respectively (DeClereq & Degheele, 1992a). These same authors (1992b) reared both predators on a bovine meat diet. *Podisus serieventris* Uhler ranges from the midwestern to eastern United States and from British Columbia to Nova Scotia in Canada (Froeschner, 1988). Its population increased and decreased in conjunction with an outbreak of black-headed budworm on conifers in Nova Scotia (Prebble, 1933). Coppel and Jones (1962) collected it from pines infested with sawflies.

In Brazil, entomologists are evaluating the role of asopine species as predators of defoliators in various crops and forests. Among the natural enemies of geometrid larvae infesting eucalyptus trees, *Alcaeorrynchus grandis*, *P. connexivus* Bergroth, *P. nigrolimbatus* Spinola, and *Thynacanta marginata* (Dallas) are involved (Berti Filho & Fraga, 1987; Zanuncio *et al.*, 1989). In Brazil, *T. marginata* is common in soybean fields, *P. maculiventris* is found in cotton fields, and *A. grandis* is found in cassava fields (Grazia & Hildebrand, 1987). In Argentina soybean fields, four predaceous pentatomids (*P. nigrolimbatus*, *P. nigrispinus* Dallas, *Oplomus cruentus* (Burmeister), and *Stiretrus*

decastigma Herrich-Schaeffer) are found and their eggs can be distinguished from eight phytophagous pentatomids found in soybean fields (Saini, 1984); and the biology and morphology of the latter two species are described by Saini (1989). *Podisus nigrolimbatus* is also a predator of *Pyrrhalta* (= *Xanthogaleruca*) *luteola* (Muller), which defoliates elms in Argentina (Artola *et al.*, 1983). Goncalves (1990) determined predatory capacities of *Podisus connexivus* and *P. nigrolimbatus* in the laboratory and reviewed the literature dealing mainly with Neotropical asopine species.

Stiretrus anchorago (Fabricius) ranges from Kansas in the west to the east coast of the United States and south to Panama; its color becomes darker to the south and is called variety *fimbriatus* (Froeschner, 1988). Its immature stages were described by Oetting and Yonke (1971). This species may be more narrow in its prey associations compared with most known asopine species. At least compared with *P. maculiventris*, *S. anchorago* is more selective of the prey it eats (Waddill & Shepard, 1975). It has been reported to be the most numerous pentatomid attacking *E. varivestis* in Mexico; in the laboratory it feeds voraciously on *E. varivestis* but is reluctant to feed on *Galleria mellonella*. It feeds on alfalfa weevil larvae, *Hypera postica* (Gyllenhal), but even though it completes its development to adult, the predator does not oviposit (Richman, 1977). Waddill and Shepard (1974b) determined the duration of the *S. anchorago* stages at three different constant temperatures and like *P. maculiventris*, it completes its life cycle in about 1 month (Richman & Whitcomb, 1978). Also like *P. maculiventris*, the males produce an aggregation pheromone that attracts both sexes and nymphs, and the pheromone has been identified and synthesized (Kochansky *et al.*, 1989).

Phymatidae The biology of this family is similar to the Asopinae in that both deposit eggs in groups on plant surfaces and have five nymphal instars. The biology of only one species has been studied in detail. Balduf wrote 10 articles on the biology and quantitative food habits of *Phymata americana* Melin, a common North American species. Based on his recording of over 830 different insect prey species of this predator, which sits and waits in fresh flowers that possess nectar and pollen, Balduf (1948) made a summary of his studies and concluded that the phymatid attacked beneficials as well as pests.

Reduviidae This is one of the larger families of Heteroptera. Most known species are predaceous, mostly on insects, but some suck blood from birds and mammals. The biology and ecology of many subfamilies are poorly known (Miller, 1971). Readio (1927) is still one of the best general references to the biology of the Nearctic species. Most species appear to be generalist predators in their habitats, which can be restrictive and thus narrows the prey range.

They kill their prey by injecting saliva when piercing the body of the prey with their stylets, and practice preoral digestion of the prey's internal structures, whose liquefaction permits ingestion (Cohen, 1993). Motion and form of the prey are the main stimuli for attack by some reduviids (Parker, 1971), and motion plus a kairomone is necessary to induce attack of others [e.g., *Phonoctonus fasciatus* (Palisot de Beauvois) and *P. subimpictus* Stål preying on the cotton stainer *Dysdercus superstitiosus* (Fabricius) in Nigeria (Parker, 1972)]. The species of Emsinae have raptorial front legs and are essentially ambush-type predators, but some are highly specialized predators of spiders and use the spider webs to find their prey (Snoddy *et al.,* 1976). Apparently, some species in the Apiomerini smear plant resins on their front tibiae to help capture prey (Miller, 1971). Some reduviids prey on termites: *Salyavata variegata* Amyot & Serville nymphs coat themselves with termite nest crumbs and debris for protection, and also use spent carcasses of freshly killed *Nasutitermes* species as bait to attract more termites (McMahan, 1983).

Reduviid eggs are deposited singly or in groups and may be partially covered with a glutinous substance by the female. Inoue (1986) found that the ovipositional sites for *Agriosphodrus dohrni* Signoret are selected based on the architectural structure of plants (in cherry trees) and height above ground. These sites provide protection from storms in Japan and hibernation sites for nymphs. The females actively select a density-dependent ovipositional site. There are five nymphal instars. Nearctic species may have several generations a year but usually there is only one, and overwintering can occur in any stage depending on the species (Readio, 1927). Readio (1927) summarized the beneficial effects of Reduviidae, believing that the family is of considerable importance even though he knew of only a few cases where a harmful insect was controlled solely by reduviids. Because reduviids are mainly generalist predators, have low fecundity, and are usually monovoltine, their importance as natural enemies of insect pests is limited.

There are 31 subfamilies of Reduviidae recognized in the world faunas (Davis, 1969; Miller, 1971), and in North America there are 10 subfamilies (Froeschner, 1988). At least three contain species that have been involved in applied biological control research. These are the Harpactorinae, Peiratinae, and Reduviinae.

Harpactorinae   This is the largest subfamily and contains most species that have been studied as natural enemies. Most of these species are in the genera *Apiomerus, Arilus, Montina, Sinea, Sycanus,* and *Zelus.*

Apiomerus crassipes (Fabricius) adults, like many phymatids, visit flowers in search of prey; therefore, usually more beneficial insects are prey than are pests (Swadener & Yonke, 1973; Bouseman, 1976). Nymphs have rarely been observed in the field but were reared in the laboratory.

The tibiae of the forelegs have long setae coated with a sticky substance produced by secreting glands of the tibiae—the substance did not come from plant resins because plants were not involved in the rearing (Swadener & Yonke, 1973). An adult of *A. crassipes* was observed feeding on the Florida harvester ant in the field and fed on ants and on *Diabrotica* adults in the laboratory (Morrill, 1975). *Apiomeris spissipes* (Say) adults prefer Coleoptera and Diptera but little is known about what nymphs feed on in the field. Readio (1927) found nymphs under rocks. Cade *et al.* (1978) observed *A. spissipes* capturing harvester ants along their foraging trails and at the periphery of their nests; about 29% of the *Pogonomyrmex* nests examined in Texas had reduviids present.

Arilus cristatus (Linnaeus), the wheel bug, is one of the largest and most easily recognized assassin bugs and has a painful bite if handled carelessly. This generalist predator ranges from Illinois in the West to the East coast. It is commonly found in fruit and nut trees and also in flowers (Readio, 1927; Underhill, 1954) and soybean fields (McPherson *et al.,* 1982).

Montina confusa (Stål) is commonly found feeding on lepidopteran defoliators in Brazilian eucalyptus forests. Its biology and morphology were examined in the laboratory by rearing it on *Galleria mellonella* (Bueno & Berti Filho, 1984a, 1984b).

The genus *Sinea* is mainly Nearctic but there are a few Neotropical species. Of the 11 North American species (Readio, 1927; Froeschner, 1988), the biologies of only 3 have been studied in detail. *Sinea confusa* Caudel occurs in the southwest and ranges to Central America. It is the most common reduviid found in Arizona crops (Werner & Butler, 1957). Duration of development of eggs and nymphs was determined at five different constant temperatures by Butler (1966), and Fye (1979) measured the prey preferred and quantities eaten of *Lygus hesperus* Knight, *Aphis gossypii, Heliothis zea, Trichoplusia ni, Pectinophora gossypiella,* and *Bucculatrix thurberiella* Busck; and looked at the searching ability in petri dishes and on artificial cotton plants. Fye (1979) concluded that *S. confusa* is an indiscriminate feeder and because of the small numbers of adults usually present in cotton fields, their overall impact, even though they feed on a large number of cotton pests, is limited. Generally the feeding by the predators in their various instars was proportionate to the size of the prey and predator. *Sinea diadema* (Fabricus) is one of the most common North American reduviids, and is distributed throughout most of the United States (Slater & Baranowski, 1978). Balduf and Slater (1943) added to the biology of that contributed by Readio (1927). In Illinois it overwinters in the egg stage and completes two generations a year. In addition to the 14 prey species recorded in the literature, Balduf and Slater (1943) observed 19 additional prey species attacked. Balduf made extensive studies on weights of prey (adult

Drosophila flies) ingested compared with weights gained by *S. diadema* nymphs, and found that assimilation is inversely proportional to the amount of food ingested, while elimination (fecal wastes) is directly proportional to the quantity of prey ingested (Balduf, 1950). *Sinea spinipes* (Herrich-Schaeffer) occurs from Colorado and Texas in the West to the East Coast; its biology was extensively investigated by Readio (1927). It overwinters as an adult and has one generation a year. Its nymph preys on nymphs of phytophagous pentatomids, fall webworm larvae, and other caterpillars in fruit, nut, and forest trees.

Sycanus, an Asiatic genus, contains several species that have been utilized in applied biological control. According to Rao *et al.* (1971), *S. leucomesus* Walker was introduced into tea plantations in the Cameron Highlands of Malaysia from the lowlands in 1940 to prey on two mirid pests, *Helopeltis bradyi* Waterhouse and *H. cinchonae* Mann. The predator became established along with two other introduced reduviids, *Isyndus heros* (Fabricius) and *Cosmolestes picticeps* Stål. Simultaneously, two other reduviids, *Euagorus plagiatus* Burmeister and *Rhinocoris marginalis* Thunberg, were introduced but did not become established; and three endemic reduviids in the highlands, *Endochus cameronicus* Miller, *Isyndus,* sp., and *Euagorus* sp., were reared and released. The degree of control of the tea mirid pests by the reduviids is uncertain (Rao *et al.,* 1971). *Sycanus indagator* (Walker) was imported from India and its feeding behavior was studied in the laboratory (Bass & Shepard, 1974). It was released in field cages in Florida soybean fields as a predator of *Pseudoplusia includens* and in field cages over cabbage plants infested with *T. ni.* Greene and Shepard (1974) concluded from laboratory and field studies that survival and larval consumption were too low to suggest success of the predator in Florida cabbage or soybean fields. It apparently did not become established (M. Shepard, personal communications, 1992).

The genus *Zelus* is distributed throughout the tropical and temperate areas of the Western Hemisphere; and there are nine species found in the United States, Canada, and northern Mexico (Hart, 1986). The three species that have received the most attention are *Z. luridus* Stål, *Z. renardii* Kolenati, and *Z. tetracanthus* Stål.

Zelus luridus is the valid name for the species referred to as *Z. exsanguis* Uhler in the biological literature; the latter species occurs only south of the United States, except perhaps southern Texas (Hart, 1986). *Zelus luridus* occurs in Arizona and Colorado in the West and extends to the East Coast. In Kansas, it overwinters as a nymph curled up in a leaf on the ground in some protected place and has one generation per year (Readio, 1927). West and DeLong (1955) found large nymphs overwintering in Ontario, Canada, and observed that the bug did not feed on plants but was a generalist predator of aphids and large tent caterpillars. These authors believed cannibalism was probably a

necessity for the first-instar nymphs. After feeding on caterpillars infected with microsporidia, *Zelus* excreted viable spores that infected caterpillars feeding on contaminated foliage (Kaya, 1979). *Zelus renardii* is found in the southwestern United States and extends south to Guatemala, but also has been inadvertently introduced to the Hawaiian and Philippine Islands, Johnston Island, and Samoa (Hart, 1986). It is a generalist predator with the nymphs commonly feeding on aphids, leafhoppers, and thrips, while the adults will eat any arthropod that they can catch and hold. Furthermore, *Z. renardii* or *Nabis* spp. were observed to severely reduce survivorship of lacewing larvae (Rosenheim *et al.,* 1993). Butler (1966) determined duration of stages influenced by five different temperatures, and Mbata *et al.* (1987) studied its reproductive behavior. Fye (1979) researched the same biotic factors of *Z. renardii* that he investigated for *Sinea confusa* and found he could not rear first-instar *Z. renardii* on bollworm eggs, but Lingren *et al.* (1968) observed adults to consume about 42 bollworm eggs or 75 first instars per day. In Hawaii, because *Z. renardii* had been involved in reducing sugarcane leafhopper populations at times, it is called the leafhopper assassin bug (Zimmerman, 1948). *Zelus tetracanthus* (= *socius* Uhler) ranges from southern Canada to Brazil and Paraguay (Hart, 1986). As common as this assassin bug is, apparently not much is known about its biology; however, Butler (1966) did determine duration of eggs and nymphal stages at five different temperatures.

Peiratinae This subfamily is composed of three genera in North America (*Melanolestes, Rasahus,* and *Sirthenia*) and embraces only a few species. These species are nocturnal and mainly ground dwellers but are attracted to lights and can inflict a painful bite. Their biologies are poorly known (Readio 1927). Apparently, the best known biology is of *M. picipes* (Herrich-Schaeffer), which was described by Readio (1927). He exposed many different insects that are found under rocks to adult *M. picipes* but the assassin bug only ate scarab larvae; Readio believed the nymphs also may be restricted to such prey. Eggs of the predator are deposited under rocks.

Reduviinae This subfamily contains many nocturnal species, and their biologies are poorly known. One species, however, has been introduced into many tropical places. *Platymerus laevicollis* Distant (= *P. rhadamanthus* Gerst), an east African species, is a generalist predator, but because its habitat is more or less restricted to crowns of palms and its adults attack adult *Oryctes* spp., it was considered a potential natural enemy of *O. rhinoceros* (Linnaeus) (Vanderplank, 1958). It was first sent to India from Zanzibar, and later to Mauritius, Pakistan, Malaya, and 10 Pacific islands (Greathead, 1971; Rao *et al.,* 1971) but apparently became established only in the Laccadive Islands (Waterhouse & Norris, 1987). It has a painful bite if handled carelessly (Miller, 1971).

Superorder Neuropteroidea

The 4500 or so species of lacewings in the world are divided into three orders. Because Megaloptera are aquatic in their larval stages, the alder and dobson flies will not be discussed here. All larvae of Neuropteroidea are apparently predaceous. A key to the larvae of all families of Neuropteroidea is nicely presented by Gepp (1984a), and Gepp (1990) illustrates the eggs of lacewing families. The food habits of adult Neuropteroidea are mostly predatory but some feed only on honeydew nectar or pollen. The monograph by Killington (1936, 1937) on British Neuroptera, the extensive revision by Aspöck *et al.* (1980) of the Neuroptera of Europe, and the reviews found in the neuropterology series edited by Gepp *et al.* (1984, 1986) and by Mansell and Aspöck (1990) provide the bases for knowing this important group of insects. Also, the review of the biology of Neuroptera by Balduf (1939) is an extensive and valuable reference. The two orders discussed here are the Raphidioptera and Neuroptera (= Planipennia).

Order Raphidioptera

Aspöck (1986a) reviews the present knowledge of snake flies of the world, noting that the distribution of the two families Raphidiidae and Inocelliidae is essentially Holarctic. The adults and larvae are mandibulate general predators living mostly in arboreal habitats. The elongate eggs are usually deposited in groups in bark crevices, and the flatish larvae search for arthropods found under loose bark or on trunks and branches of trees, though some larvae live in the soil. The adults hunt for the smaller insects associated with trees and bushes. The generation time is usually 2 or 3 years. Tauber (1991b) reviews the biology and presents a key to the larvae of both families.

Aspöck (1986a) mentions the unclarified economic significance of Raphidioptera and points out that these insects may act as effective predators of phytophagous arthropods living in and on forest and fruit trees and believes these predators could be introduced into areas where snake flies do not occur naturally, particularly in the Southern Hemisphere. About 90 years ago, a *Raphidia* sp. from California was sent to New Zealand to control the codling moth, but it did not become established (Clausen, 1978).

Inocelliidae Little biological information is known about this family that contains five genera and 17 species, though Aspöck (1986a) states that the species in this family are arboreal and have longer larval periods requiring 3 years to complete their development. Aspöck *et al.* (1974) and Tauber (1991a) discuss details on larval biology and morphology.

Raphidiidae The 26 genera of this family embrace 151 species that live in forests and orchards (with some larvae found in the soil) (Aspöck *et al.*, 1974, 1975; Aspöck 1986a). Woglum and McGregor (1958) describe the biology and morphology of *Agulla bractea* Carpenter and the biology of *A. astuta* (Banks) in 1959. The larvae of both species were observed in the field preying on black scale eggs and crawlers on citrus trees in southern California. The adult snake flies were also collected in the citrus trees. The biologies of both *Agulla* spp. were found to be similar. The adults were fed black scale crawlers and aphids and deposited batches of eggs in about a week; first-instar larvae did not feed, and there were 10 or 11 instars. Development could be completed in 1 year if larvae were exposed to the cool temperatures during winter. If kept under warm conditions, the larvae did not pupate. Some larvae completed development in 3 years if finally exposed to a cold period. All stages and the larval morphology are described for both species (Woglum & McGregor, 1958, 1959).

Order Neuroptera

The 4000 or so species of lacewings in the order Neuroptera compose 17 families that until recently were placed in the suborder Planipennia (Aspöck *et al.*, 1980; Gepp, 1984a; New, 1989; Tauber, 1991a). Their eggs are broadly oval with a conspicuous micropylar knob and often are placed at the end of silken threads. The eggs are deposited singly or in groups, usually near prey. Most families have campodieform larvae with three instars. All larvae are predaceous and have sucking and piercing mouthparts; they pupate in silken cocoons. Most adults are predaceous, while quite a few feed on honeydew, nectar, and/or pollen; and a few do not feed at all (Withycombe, 1923; Balduf, 1939; Clausen, 1940; Sweetman, 1958; New, 1975). The Neuroptera that are predaceous on aphids are reviewed by New (1988b; those that are predaceous on coccids, by Drea (1990a). The Neuroptera used to control arthropods in apple orchards are reviewed by Principi and Canard (1974), and Szabo and Szentkiralyi (1981); and in pear orchards, by Herard (1986).

The main families that contain species involved in applied biological control are Chrysopidae, Coniopterygidae, and Hemerobiidae. Berothidae is included here because some of its species are predaceous on termites, and are potentially useful as biological control agents.

Berothidae The Berothidae family has 78 species of beaded lacewings in eight genera that are mainly tropical with only one genus occurring in Europe and one genus, in North America. Aspöck (1986b) presented the state of knowledge of the family, including the biology of the only species known in any detail, *Lomamyia latipennis* Carpenter. It was reared on *Zootermopsis angusticollis* Hagen and the three larval instars described by Tauber and Tauber (1968). Johnson and Hagen (1981) readily reared it on *Reticulitermes hesperus* Banks and described the use of an aggressive allomone by the first- and third-instar larvae to paralyze the termites before they consumed their prey; the second instar is quiescent in the termite galleries. Eggs are glued to each other as a cluster at the end of a single silken stalk and are deposited on termite-infested logs. The adult is able to produce and deposit eggs on a diet of honey alone (K. S. Hagen, unpublished data). Brushwein (1987) presented data on the life history, adult longevity and fertility, habitat, and seasonal occurrence of *L. hamata* (Walker) associated with the eastern subterranean termite *R. flavipes* (Kollar). This eastern U.S. predator has a biology similar to *L. latipennis*. Brushwein (1987) reared *L. hamata* larvae on live and freezer-stored termites. Tauber (1991a) reviewed the form and habits of the immature berothids.

Chrysopidae   There are about 1200 species of green lacewings recognized in the world today distributed among nearly 75 genera and 11 subgenera belonging to three extant subfamilies: Nothochrysinae, Apochrysinae, and Chrysopinae (Brooks & Barnard, 1990). The largest subfamily is the Chrysopinae, which contains most of the species involved in applied biological control.

The biology of the Chrysopidae is summarized in Canard *et al.* (1984). Canard and Principi (1984) summarize the development of 19 Chrysopidae species based on 25 references. The length of the preimaginal developmental period is about 10 days in the larval stage, and another 10 days in the cocoon. The biology of most chrysopid species is completely unknown but of those studied, eggs are deposited atop a silken filament (except for *Anomalochrysa* spp., which deposit eggs on the sides on leaf surfaces), either as a single unit or as multiples in clusters (Duelli, 1984b; Gepp, 1984b). All larvae are predaceous and have three instars. They are either elongate and naked or more oval and carry debris on their dorsal hairs (Gepp, 1984b). Many chrysopid larvae are generalist predators, feeding on whatever prey found in the habitat where the eggs are deposited. Like most predators, the search is intensified by short turnings when prey is contacted (Bänsch, 1966). A newly hatched first-instar *Chrysoperla carnea* larva can travel 214 m in 15 h before it dies of starvation (Fleschner, 1950). Some chrysopid larvae are highly specific to their selection of prey e.g., *Chrysopa slossonae* Banks larvae have only been found associated with the woolly alder aphid, *Prociphilus tesselatus* (Fitch) (Tauber & Tauber, 1987a; Milbrath *et al.*, 1993), and *Italochrysa italica* (Rossi) larvae feed only on early stages of the ant *Cremastogaster scutellaris* (Oliver) (Principi, 1946). The quantity and quality of larval prey have an influence on developmental time and adult fecundity—indeed, certain aphids eaten by larvae can cause sterility in the males that emerge (Ca-

nard, 1970a). Larvae of the common chrysopid species are usually more easily identified to species than are adults. Larvae of common North American species have been described by Smith (1922) and Tauber (1974, 1991a), and Japanese species, by Tsukaguchi (1978); references to descriptions of all known larvae are cited by Gepp (1984a, pp. 215–217).

There can be one to several generations per year (Tauber & Tauber, 1982, 1986, 1987b). Overwintering occurs in all stages except the egg, and the stage in which the chrysopid overwinters is characteristic for genera and subgenera (Canard & Principi, 1984; Tauber, 1991a; Principi, 1992; Canard & Grimal, 1993).

Adult chrysopids, depending on the genus or subgenus, either are predaceous or feed on honeydew, nectar, or pollen (Hagen, 1950; Hagen *et al.*, 1970b; Sheldon & MacLeod, 1971; Principi & Canard, 1984). It now appears that many genera of Chrysopidae are not predaceous in their adult stage (Johnson, 1982; Principi & Canard, 1984; Brooks & Barnard, 1990), and many of the nonpredaceous species harbor a mutualistic yeast (*Torulopsis* sp.), which is spread by trophallaxis between adults (Hagen *et al.*, 1970b; Johnson, 1982). The role of the yeast apparently is to synthesize missing essential amino acids that are lacking in honeydew (Hagen & Tassan, 1972; Hagen, 1986). The species in the genus *Chrysopa sensu stricto* are all predaceous in the adult stage. The tracheal trunks associated with the foregut crop are thinner than those of *Chrysoperla* spp., which do not have predaceous adults (Hagen *et al.*, 1970b; Principi & Canard, 1984). There are also differences in the form of mandibles between predaceous and nonpredaceous species (Ickert 1968).

Mating attraction can involve vibrations, mainly via abdominal jerking in both sexes; specific duet courtship songs in some chrysopid species are transmitted from abdominal movements through the legs to a leaf or a thin substrate (Henry, 1983). There are glands in *Chrysopa sensu stricto* that are suspected to emit sex pheromones (Henry, 1984). Some *Chrysoperla* spp. will mate when fed sugar and water, but others require protein in the diet of at least the females to mate; similarly, some *Chrysopa* spp. will mate only on sugar and water, while others require aphids (Tauber & Tauber, 1974). In any case, protein plus carbohydrate is required to obtain egg production and oviposition. Enzymatic protein hydrolysates (particularly of yeasts) plus sugar or honey are highly effective as artificial honeydew (Hagen, 1950; Hagen & Tassan, 1966). However, *Chrysoperla carnea* adults obtain enough transferred metabolites from larval feeding to produce from 20 to 30 eggs in their lifetime when fed on honey or sugar and water alone (Hagen & Tassan, 1966). Rousset (1984) reviewed 16 references concerning the reproductive responses of 12 different chrysopid species exposed to different adult diets. Rousset found the preoviposition period varied from 3.9 to

13.2 days, fecundity reached as high as 960 eggs, and fertility reached as high as 99%.

The four genera of Chrysopidae that contain species involved in biological and integrated control projects are *Brinckochrysa, Chrysopa, Chrysoperla,* and *Mallada* (Doutt & Hagen, 1949; Hassan, 1974; Ridgway & Kinzer, 1974; Shuvakhina, 1975; New, 1975, 1984a, 1991; Beglyarov & Smetnik, 1977; Ridgway & Vinson, 1977; Ables *et al.,* 1979; Hagen & Bishop, 1979; Principi, 1984; Ridgway & Murphy, 1984; Tulisalo, 1984a, 1984b; Barclay 1990; Khloptseva, 1991; Tauber & Tauber, 1993). Very little has been done in classical biological control; most research has been conducted in the augmentation and conservation of chrysopids.

The 16 species in the genus *Brinkochrysa* are mostly concentrated in the Afrotropics. Their larvae are not trash carriers (Brooks & Barnard, 1990). Apparently *B. scelestes* (Banks) has been the only species in the genus used in a biological control project. This chrysopid was mass-cultured using 1000 eggs of the rice moth, *Corcyra cephalonica* (Stainton), per larva (Patel *et al.,* 1988). The adults are not predaceous; most eggs were obtained from females exposed to a honey and castor pollen diet (Krishnamoorthy, 1984). Releases at the rate of six second-instar larvae per tobacco plant infested with *Myzus persicae* reduced the aphid populations nearly 80% in the field (Rao & Chandra, 1984). The chrysopid larvae were not hindered by the glandular trichomes (or their secretions) on the tobacco plants. The chrysopid also oviposits readily on cotton and green gram in India, where larvae are considered important predators of *Helicoverpa armigera, Spodoptera litura,* and aphids (Patel & Vyas, 1985). It is also an effective predator of woolly apple aphid on apple in India (Thakur *et al.,* 1988).

The genus *Chrysopa sensu stricto* comprises some 54 Holarctic species with 27 species occurring in the western Palearctic region, 15 in the eastern Palearctic, and 4 in the Nearctic region (Brooks & Barnard, 1990). New (1980), considering the genus in a broad sense, provides a key for 25 *Chrysopa* species in Australia. However, none of these are actually in *Chrysopa sensu stricto;* most are in *Mallada* and *Chrysoperla* (Brooks & Barnard, 1990). Larvae of *Chrysopa* spp. *sensu stricto* are not trash carriers and overwinter (in diapause) in cocoons. The adults are predaceous. About six species are commonly referred to in biological control literature.

Chrysopa formosa Brauer is a widespread Palearctic species found from Europe to Asia on shrubs, in orchards, and in deciduous forests (Séméria, 1984). The eggs are deposited singly or widely spaced in groups of 2 to 10 on undersides of leaves of herbs and deciduous trees; Tsukaguchi (1978) found eggs on 17 different shrubs and trees infested with 17 different aphid species in Japan. This species was reared in large numbers in the laboratory by feeding larvae

S. cerealella eggs that could be stored at −8 to −18°C for up to a month without harm. Adults were fed aphids and yeast hydrolysate with honey, under a long photoperiod (Kowalska, 1976). Also, adults and larvae can be reared on lyophilized drone honeybee pupae powder (Niijima & Matsuka, 1990). Because of the high egg production and low larval cannibalism, Babrikova (1978) considered it a good candidate as a biological control agent. In 1980, he reduced *Myzus persicae* populations on tobacco by releasing about eight larvae per plant in Bulgaria. Lyon (1986) released *C. formosa, C. perla* (Linnaeus), and *C. carnea* in greenhouses to control aphids on tomatoes, sweet peppers, and eggplants at the rate of 10 second-instar larvae per plant each week. Lyon also released *Aphelinus asychis* Walker along with the chrysopids. This complex of natural enemies controlled the aphids better on green peppers than on tomatoes because young chrysopid larvae were trapped and killed by the sticky hairs on tomato leaves.

Chrysopa kulingensis Navas, a predator of aphids and soft scales in Chinese pine forests, was mass reared as larvae on honeybee larvae and *Corcyra cephalonica* eggs. A release of this chrysopid accounted for a significant decrease in *Matsucoccus* populations and a 20% increase in the predator population 35 days after release compared with the control (Wang & Hu, 1987).

Chrysopa nigricornis Burmeister is a common large Nearctic tree and shrub species (Smith, 1922; Toschi, 1965). It occurs in peach orchards where the larvae prey on aphids and possibly oriental fruit moth eggs (Putman, 1932). At times *C. nigricornis* adults are highly attracted to terpinyl acetate in oriental fruit moth bait pans hung in peach trees (Caltagirone, 1969). The egg clusters of this chrysopid are commonly seen in walnut orchards and on cottonwood trees infested with gall aphids, as well as on cotton plants infested with aphids (van den Bosch & Hagen, 1966).

Chrysopa oculata Say is a common multivoltine Nearctic species that ranges from southern Canada to northern Mexico. Adults emerge in the spring and reproduction and development occurs throughout the summer and autumn until the first frost (Tauber *et al.,* 1987). The induction and duration of diapause is positively related to the latitudinal origin of the population, which is correlated to photoperiodicity of the region (Nechols *et al.,* 1987). *Chrysopa oculata* deposits eggs singly on peach leaves, where a larva can destroy over 600 *Grapholita molesta* (Busck) eggs or about eight larvae during its development (Briand, 1931). In Texas cotton fields both larvae and adults feed on *Aphis gossypii* Glover; the mature larvae apparently enter the soil to pupate (Burke & Martin, 1956). In alfalfa hay fields, *C. oculata* feeds mainly on pea aphids and somewhat on alfalfa weevil larvae and ignores leafhoppers and plant bug nymphs (Lavallee & Shaw, 1969). In Pennsylvania grape vineyards *C. oculata* is the most common lacewing, and its larvae were observed to prey on grape phylloxera on wild grape vines (Jubb & Masteller, 1977). The most suitable prey for *C. oculata* in corn fields was *Rhopalosiphum maidis* (Fitch); if switched from aphids to stalk borer [*Papaipema nebris* (Guenée)] eggs while in third instar, more than 80% of the lacewing larvae died. Aphids were also superior to a diet of *Ostrinia nubilalis* or *Agrotis ipsilon* eggs (although *C. carnea* larvae did well), but larvae of both lacewing species died when given first-instar corn borer larvae. It appears that *C. occulata* larvae are not the generalist predators that *C. carnea* larvae are (Obrycki *et al.*, 1989).

Chrysopa pallens (Rambur) (= *semptempunctata* Wesmael) (Brooks & Barnard, 1990) is a mainly arboreal species widely distributed through the Palearctic region (Principi, 1940; Hölzel, 1984). In Japan, Tsukaguchi (1978) found egg clusters on 19 different plants (mostly trees) infested with 16 different aphid species. He reared and described all larval instars and obtained about 3200 eggs from a female held for about 90 days. Larvae and adults of *C. pallens* were reared for six generations on pulverized lyophilyzed drone honeybee brood and produced as many as 2228 eggs per female held for 166 days (Okada *et al.*, 1974). In China, Cai *et al.* (1985) reared *C. pallens* larvae and adults on artificial diets not containing insect parts for seven generations. Niijima (1989) formulated a chemically defined diet for the adult and larvae, and determined the amino acid requirements for larval development (1993a) and vitamin requirements for larval development (1993b). Its functional responses to cotton bollworms and cotton aphids were found to follow Holling's type II (Huang *et al.*, 1990). *Chrysopa pallens* is an important predator of aphids on apples in Japan. The larvae can survive being exposed to chitin synthesis-inhibiting insecticides, but died at the prepupal stage in their cocoons (Kawashima, 1988). In (former) USSR greenhouses, a release of second-instar larvae at a predator to prey ratio of 1:5 or 1:10 controlled aphids on cucumbers. Releases of adult females at 1:50 resulted in an 82 to 86% reduction in aphid numbers in 3 days. The larvae resulting from the released adults completely destroyed the remaining aphids by the ninth day after the initial release (Ushchekov, 1976).

Chrysopa perla is a Palearctic species that occurs commonly across the northern European countries (Greve, 1984). In the Mediterranean region it is limited to cooler and wetter areas (Séméria, 1984). Apparently *C. perla* is not found in China (Yang, 1988), and only rarely is collected in Japan (Tsukaguchi, 1978). This species is associated with many different trees and shrubs (Zeleny, 1984). However, a diet of certain aphid species can limit the reproductive success of the adult while allowing larval development [e.g., *Brevicoryne brassicae* (Linnaeus) somewhat prolongs developmental time on *C. perla* larva, and when fed to adults results in lower progeny production levels,

while a diet of *Megoura viciae* Buckton reared on *Vicia sativa* (Linnaeus) virtually precluded progeny production (Canard, 1970a 1970b)]. Detailed studies on the influence of food on *C. perla* reproduction are described by Philippe (1972). The Australian *Apertochrysa edwardsi* Banks (as *Chrysopa edwardsi*) was satisfactorily reared as larvae and adults on *B. brassicae* and on acacia psyllid *Acizza acaciaebaileyanae* (Froggatt) (New, 1981). Both prey species are thick and waxy in the nymphal stages, which led New to conclude that *A. edwardsi* larvae and adults are adapted to feed on "waxy" prey. Therefore, this may be an effective species for importation against this type of prey.

Chrysopa perla and six other chrysopid species were reared on one feeding of 50 to 100 mg of frozen *Sitotroga cerealella* eggs per larva by Kawalska (1976). By releasing 4800 *C. perla* eggs in about 40 m^2 of asparagus infested with *Myzus persicae* at a rate of 1 egg per 25 aphids, Kawalska (1976) was able to keep the asparagus free of aphids for 3 months. Weekly releases of *C. perla* and *C. formosa* were made alone and in association with *Aphelinus asychis* Walker against *M. persicae* and *Macrosiphum euphorbiae* (Thomas) infesting eggplants in a greenhouse, which resulted in control similar to that obtained in another greenhouse where pirimicarb was used (Lyon, 1979). In the former USSR, egg and larval releases of *C. perla* successfully controlled *Aphis gossypii* on cucumbers in greenhouses and were also effective in the field against *Myzus persicae* on kale and *Aulacorthum circumflexus* (Buckton) on ornamental plants (Ushchekov, 1989).

The genus *Chrysoperla* contains about 50 species worldwide (Brooks & Barnard, 1990). This genus was once considered a subgenus of *Chrysopa*, but distinctly differs from it morphologically and even more so biologically. The adults of *Chrysoperla* spp. are not predaceous, but feed on honeydew and pollen that are required for egg production (Hagen, 1950; Hagen *et al.*, 1970b; Rousset, 1984); thus far all of the *Chrysoperla* spp. adults studied have contained symbiotic yeasts in their esophageal diverticulae (Hagen *et al.*, 1970b; Johnson, 1982; Principi & Canard, 1984). All species overwinter as adults (Principi, 1992). Eggs are deposited singly or loosely in groups. Larvae are generalist predators but not trash carriers (Balduf, 1939; Principi & Canard, 1984; Adams & Penny, 1985).

The utilization of *Chrysoperla* (principally three species) in biological control has been through augmentation and conservation—the several attempts to purposely introduce them to exotic locations have failed. Because at least some *Chrysoperla* spp. are honeydew feeders, they have been manipulated by applying artificial honeydew in crops to attract adults and induce oviposition where insect damage is anticipated.

Chrysoperla carnea is apparently a Holarctic species. In the world literature, it has gone under at least 20 different names (Brooks & Barnard, 1990). In the Nearctic region it has been called: *C. californica* (Coquillett), *C. downesi* (Smith), *C. mohave* (Banks), *C. plorabunda* (Fitch) as well as a few other names that have not appeared in biological control literature. There are geographic variations in the courtship song of Nearctic *C. carnea* that have led some workers to suspect that this group may include several cryptic species. In their world list, Brooks and Barnard (1990) recognize *C. plorabunda* (Fitch) as a separate species from *C. carnea,* based apparently on Henry's (1983) observation that a Swiss *C. carnea* population did not recognize the courtship song of an American *C. carnea* population and failed to mate, which led him to resurrect *C. plorabunda* from synonomy. Henry and Wells (1990) found additional populations whose songs varied from that of the *C. plorabunda* population in New York (the type locality for *C. plorabunda*). However, the Taubers (1972, 1986, 1987b) have demonstrated that geographic variations in diapause response in *C. carnea* are also under polygenic control, and they (1986) resynonymized *C. plorabunda* under *C. carnea,* recommending that no name changes be implemented until a thorough taxonomic study—which considers intra- and interpopulation variation in morphological, behavioral, and ecological trials of the entire *C. carnea* species complex—is completed. Additional synonyms of *C. carnea* include *C. vulgaris* (Schneider) from Europe and *C. sinica* (Tjeder), *C. nipponicus* (Okamoto), and *C. kurisakiana* (Okamoto) from Asia (Brooks & Barnard, 1990).

Chrysoperla carnea larvae are generalist predators of soft-bodied insects and spider mites. Over 70 different prey species in five insect orders have been recorded in 90 references cited by Balduf (1939) and Herard (1986). The prey are mostly Homoptera, with aphids on low-growing crops and vegetation predominating; however, psyllids, mealybugs, leafhoppers, and whiteflies are also readily consumed. Butler and Henneberry (1988) found that *C. carnea* larvae ate immature stages of *Bemisia tabaci* (Gennadius) on cotton in the laboratory. Interestingly, they also found that the presence of lacewing larvae inhibited both visitation and oviposition by *Bemisia* adults, even after the larvae had left the leaves, and suggested the repelling volatile semiochemical or semiochemicals be identified for possible uses as a control measure. Among the recorded lepidopteran prey, noctuids (*Heliothis* spp.) on cotton were the most often mentioned. *Chrysoperla carnea* larvae exposed to *Heliothis* eggs on artificial cotton plants searched mostly the top half of the "plants" where older lacewing larvae consumed 3 to 6 eggs per day; the time spent handling each *Heliothis* egg did not affect the rate of successful search (Butler & May, 1971). *Chrysoperla carnea* larvae are stimulated into a prey-seeking behavior by kairomones found in moth scales of *H. zea* that are associated with moth oviposition sites (Lewis *et al.,* 1977) and acceptance of the *H. zea* eggs is

stimulated by an additional kairomone, probably in the accessory gland secretion, which is associated with egg deposition (Nordlund *et al.,* 1977). Chemically defined diets have been formulated for *C. carnea* larvae (Hasegawa *et al.,* 1989; Yazlovetsky, 1992).

Adults of *C. carnea* feed on honeydew (Hagen, 1950; Hagen & Tassan, 1972) and pollen (Sundby, 1967; Sheldon & MacLeod, 1971), but the biotype *C. mohave* Banks (a synonym of *C. carnea*) is an estivating strain that eats aphids, honeydew, and probably pollen (C. Tauber & M. Tauber, 1973; M. Tauber & C. Tauber 1975). *Chrysoperla carnea* adults search for food and oviposition sites nocturnally, and respond anemochemotactically to volatile chemicals emanating not only from the food source (honeydew) (Hagen *et al.,* 1970a, 1976b; Duelli, 1980, 1984a), but also from the habitat (the plant involved) (Flint *et al.* 1979; Hagen, 1986). The kairomone that attracts *C. carnea* to honeydew comes from tryptophan (Hagen *et al.,* 1976b), and laboratory olfactometer studies with tethered *C. carnea* females, at least 2 days old, demonstrated that they respond positively in flight toward breakdown chemical volatiles of tryptophan, particularly indole acetaldehyde (van Emden & Hagen, 1976). After antennal contact with honeydew or hydrolized tryptophan in water, *C. carnea* females showed a reduced mean walking speed, and an increased mean turning frequency and turning angle (McEwen *et al.,* 1993b). The synomone that attracts *C. carnea* to a cotton habitat is the terpene caryophyllene, a common volatile coming from cotton leaves (Flint *et al.,* 1979). Adults fly downwind at night but turn upwind when they detect honeydew; however, it appears that *C. carnea* adults respond to the kairomone only when a synomone (plant volatile) is received simultaneously (Hagen, 1986). This type of behavior must be considered when attempting to use artificial honeydew to attract *C. carnea* (and perhaps other *Chrysoperla* spp.) because a synomone may not be present because of species and age variation of crop plants.

Of all the Neuroptera, *C. carnea* has been the most commonly used species in applied biological control, mainly through augmentation and conservation tactics. It rarely has been involved in classical biological control programs, but at least two attempts were made to introduce *C. carnea* into New Zealand. One introduction involved importation of eggs from California. This failed even though adults were successively reared in the laboratory and fed a standard complete diet of dry yeasts and sugar (Hagen & Tassan, 1970). The failure to become established was probably due to the absence of yeast symbiotes—these are apparently not present in the egg stage and would not have been previously present in New Zealand due to the absence of any honeydew or pollen feeding chrysopids (Hagen & Tassan, 1972). A similar result was noted when the cocoons of several chrysopid species were imported from India. Progeny were successfully reared in the insectary, but establishment was never achieved (Rao *et al.,* 1971). However, *C. comanche, C. carnea,* and *Mallada basalis* (Walker), which all "require" yeast symbiotes, became established in the Hawaiian Islands (Adams, 1963), suggesting that they invaded the islands as adults carrying their symbiotic yeasts.

Augmentation of *C. carnea* in field and greenhouse situations to control insects and mite pests has been conducted both by releasing eggs and larvae (Hassan, 1974; Beglyarov & Smitnik, 1977; Ridgway *et al.,* 1977; Stinner, 1977; Ables & Ridgway, 1981; Ridgway & Murphy, 1984; Barclay, 1990; Khloptseva, 1991; Nordlund & Morrison, 1992) and by attracting and feeding wild adult lacewings to promote egg production and oviposition (Hagen *et al.,* 1970a; Hagen & Hale, 1974; Hagen & Bishop, 1979; Ridgway & Murphy, 1984; Barclay, 1990).

Larvae of *C. carnea* can be reared on a variety of insects, but lepidopterans are the most widely used hosts. These include eggs and coddled larvae of potato tuberworms, *Phthorimaea operculella* (Zeller) (Finney, 1948, 1950); eggs of Angoumois grain moth, *Sitotroga cerealella* (Shuvakhina, 1969; Ridgway *et al.,* 1970; Beglyarov & Ushchekov, 1974; Hassan, 1975; Makarenko, 1975; Morrison *et al.,* 1975; Nordland & Morrison, 1992); frozen eggs (Kawalska, 1976), all moth stages (Tulisalo, 1984a), and larvae of Indian meal moth, *Plodia interpunctella,* and *Mamestra brassicae.* (Linnaeus) (Hassan, 1975); eggs of Mediterranean flour moth, *Anagasta kuehniella* (Pasqualini, 1975, Zheng *et al.,* 1993a); and eggs of the wax moth, *Galleria mellonella* (Sgobba & Zibordi, 1984). McEwen *et al.* (1993a) obtained greater survival and faster larval development by mixing *A. kuehniella* eggs in a sugar solution containing yeast hydrolysate (artificial honeydew). The quantity and quality of these hosts can influence the fecundity and longevity of the resulting adult lacewings (Makarenko, 1975; Pasqualini, 1975; Zheng *et al.,* 1993b).

Other prey used to culture *C. carnea* larvae have included green peach aphids cultured on green pepper and spring rape (Tulisalo & Korpela, 1973), and eggs of the eucalyptus borer, *Phoracantha semipunctata* Fabricius (Neumark, 1952).

Lacewing larvae have also been reared on artificial diets (Niijima & Matsuka, 1990). These will undoubtedly be the preferred diets in the future, especially if diet encapsulation methods are further developed (see Chapter 7).

Chrysoperla spp. are much easier to mass-culture than *Chrysopa* spp. because they readily produce high numbers of eggs on artificial diets. The best diets are those that contain an enzymatic protein hydrolysate of yeast, sugar, or honey, either mixed or provided separately (Finney, 1950; Hagen, 1950; Hagen & Tassan, 1966). Another effective adult lacewing diet is yeast *Saccharomyces fragilis* cultured on whey and mixed with sugar or honey (Hagen & Tassan, 1970; Zhou *et al.,* 1981; Nordland & Morrison, 1992). On

these diets a female *C. carnea* can produce 500 to 1000 eggs in about 30 days.

Because there is deterioration of *C. carnea* when mass-cultured over time, Jones *et al.* (1978) recommend not going beyond six generations before renewing the insectary stock, or even by the fourth generation as suggested by Niijima & Matsuka (1990). Diapausing adults can be stored at 5°C for over 30 weeks and still be reproductively effective for mass production (Tauber *et al.,* 1993), and eggs can be stored at 8°C for up to 20 days without loss of viability (Osman & Selman, 1993).

The first field releases of *Chrysoperla carnea* were made against mealybug *Pseudococcus obscurus* Essig infesting pears that were being sprayed with DDT to control the codling moth in the Santa Clara Valley, California. In the unsprayed control blocks, it became clear that the mealybug was being controlled by *C. carnea* (Doutt, 1948). There were no parasitoids reared from the mealybugs, and the only predators present were lacewing larvae. This situation led to mass-culturing *C. carnea* and to subsequently releasing egg; although the adult lacewings were killed by the DDT sprays, eggs and larvae were not harmed. Three releases of 250 eggs each per tree reduced the mealybug infestation rate of the fruit down to 12%, while 58 to 68% of the pears were infested in the DDT-treated blocks, where no lacewing releases were made (Doutt & Hagen, 1949, 1950). The pear growers were going to build an insectary to mass-culture *C. carnea* and integrate lacewing egg releases with the DDT spray treatments; however, when parathion replaced DDT, the new insecticide also controlled the mealybug and so the insectary project was dropped.

It was not until the 1960s before *C. carnea* was again utilized in the field. The target pests were bollworms (*Heliothis* spp.), because lacewing larvae appeared to have an impact on these populations, particularly in cotton fields. Releases of *C. carnea* were first conducted in field cages, which demonstrated that *C. carnea* larvae could reduce bollworm densities. Ridgway and Jones (1968, 1969) obtained 76 to 99% reduction of *H. virescens* by releasing first-instar larvae against bollworms when there were few alternate prey available. Ridway and Jones (1968) also found *H. zea* was reduced on cotton from 74 to 90% with predator to prey ratios from 2.5:1 to 30:1. These release rates ranged from 62,000 per hectare to 741,000 per hectare. Lingren *et al.* (1968) got a 96% reduction of *H. virescens* in cages at a release rate of about 1 million *C. carnea* larvae per hectare. Van den Bosch *et al.* (1969) reduced caged *H. zea* larvae on cotton from 46 to 84% using predator to prey ratios of 1:4 to 1:3, respectively. A release rate of about 200,000 per hectare of 1- to 4-day-old *C. carnea* larvae resulted in 30% less damage to bolls by *H. virescens* in field cages (Lopez *et al.,* 1976). Stark and Whitford (1987) conducted a series of third-instar *C. carnea* releases on caged cotton plants with 20, 40, 60, or 80 *H.*

virescens eggs per cage. At the lowest bollworm density (predator to prey of 2.8:1) 55% of the eggs were attacked, while 42% were attacked at the highest density (predator to prey ratio of 0.8:1), producing a type II functional response. From these studies, they calculated that *C. carnea* larvae had an average search rate of 1.08 × 10m^2 ha/predator-day (or 0.11 row-meter/predator-day). The calculation of such search rates can be useful in determining the optimal numbers for field releases. Only a few releases of *C. carnea* have been made in open cotton fields against bollworms. A 96% reduction of bollworm occurred at a release rate of 721,200 first and second-instar larvae per hectare, but no reduction of bollworms occurred when only adult lacewings were released (Ridgway & Jones, 1969). Gurbanov (1982) recorded a 100% reduction of *H. armigera* eggs in cotton at a 1:1 release ratio but only a 50% reduction of bollworm larvae; he also obtained over 95% reduction of aphids, thrips, and spider mites from these releases. In soybeans, releases of 20, 60, or 120 *C. carnea* eggs in cages where 1000 *H. zea* eggs and 500 *Trichoplusia ni* eggs were placed did not reduce defoliation; Barry *et al.* (1974) suggested that predation of lacewing eggs by *Orius insidiosus* that penetrated the cages over soy plants explained the ineffectiveness of *C. carnea* in their tests. Another lepidopteran, pink bollworm, was reduced from 33 to 80% by releasing third-instar *C. carnea* larvae from 1 to 12 per cotton plants (Irwin *et al.*, 1974).

Field releases against aphids began on vegetables. Shands *et al.* (1972b) released at the rate of about 50,000 to 100,000 *C. carnea* per hectare and obtained 42 to 89% reduction of *Myzus persicae,* but only a 22 to 34% reduction of *Macrosiphum euphorbiae* on potatoes. Adashkevich and Kuzina (1974), using ratios of 1:5 and 1:10 second-instar *C. carnea* larvae, got no reduction, but at a 1:2 ratio using third instars obtained 50% reduction of aphids on potatoes. Releases at a ratio of one second instar to five *M. persicae* on vegetables (eggplants, peppers, tomatoes) accounted for reductions of aphids from 43 to 97% (Begley-arov & Smetnik, 1977). *Aulacorthum solani* (Kaltenbach) was reduced 90% on tomatoes using a 1:5 predator to prey ratio (Ter-Simonjan *et al.,* 1982); they also obtained a 94% reduction of *Aphis gossypii* on cucumbers. A net release of 625 *C. carnea* eggs per tulip tree infested with *Illinoia liriodendri* (Monell) did not control the aphid; predation on released eggs by the Argentine ant precluded potential control (Dreistadt *et al.,* 1986). *Aphis pomi* was reduced significantly on dwarf apple trees at a release rate of 335,000 *C. carnea* eggs per hectare, a ratio of about 1:18 (Hagley, 1989). Inoculative releases of first-instar *C. carnea* (= *C. sinica*) in Chinese wheat fields 1 month before spring harvest yielded the highest number of adults, which then spread to other crops later in the season (Zhou *et al.,* 1991b).

The density of the variegated grape leafhopper, *Erythro-neura variabilis* Beamer, was reduced by up to 35% after releases of 3000 to 8000 hatching *C. carnea* eggs against each of the two broods of leafhoppers. These results suggest that lacewing releases can reduce initial leafhopper density up to 25 per leaf below the economic injury level of 15 per leaf (Daane *et al.,* 1993).

Releases of *C. carnea* against the Colorado potato beetle have given variable results. No control was achieved by releasing eggs at a 1:1 ratio, but an 86 to 95% reduction of the pest populations on potatoes was found when second- and third-instar *C. carnea* larvae were released at 1:1 to 1:5, at 1:10, and at 1:20 ratios by Adashkevich and Kuzina (1974) and Shuvakhina (1975). On eggplants, the Colorado potato beetle was reduced from 74 to 100% with the highest predation occurring when releasing *C. carnea* larvae at 1:1 to 1:5 ratios (Beglyarov & Smetnik, 1977). A release of one *C. carnea* larva against 80 *Tetranychus urticae* in cotton reduced the spider mites by 99.6% (Ishankulieva, 1979).

There has also been some success controlling spider mites by releasing chrysopids. On apple trees, the European red mite was reduced when one *C. carnea* larva was released for every 10 to 25 leaves (Miszezak & Niemczyk, 1978). On citrus trees, a release of one first-instar *C. carnea* (as *C. sinica*) to 100 to 200 *Panonychus citri* reduced the mites by 90% in 5 days (Peng, 1985). Peng (1988) also made field releases of *C. shansiensis* (Kuwayama) against red spider mites on citrus.

Releases of *C. carnea* eggs or larvae had controlled a variety of insect and spider mite pests of crops and ornamentals grown in greenhouses (Tulisalo, 1984b). Scopes (1969) was able to control *Myzus persicae* on chrysanthemums by releasing 1-day-old lacewing larvae at a predator to aphid ratio or 1:50, while third-instar larvae were effective at a ratio of 1:200. However, the potential to control *Aphis gossypii* on cucumbers was limited due to the hairiness of the leaf, which inhibits the locomotion of chrysopid larvae (Scopes, 1969). Aphids infesting celery, eggplants, lettuce, oats, parsley, peppers, strawberries, and tomatoes have been controlled by releasing eggs or larvae of *C. carnea* (Bondarenko & Moiseev, 1972; Beglyarov & Ushchekov, 1974; Tulisalo & Tuovinen, 1975; Kawalska, 1976; Rautapaa, 1977; Tulisalo *et al.,* 1977; Hassan, 1978; Celli *et al.,* 1986). *Tetranychus urticae* infesting peaches grown on a trellis system in a protected environment were controlled by placing 1000 or 1500 *C. carnea* eggs per tree on foliage when the spider mites became active on foliage (Hagley & Miles, 1987). Rautapaa (1977) had to release about 15 to 20 times more eggs as larvae to achieve the same decrease of *Rhopalosiphum padi* on oats under greenhouse conditions.

Chrysopid eggs have usually been released in the field manually, either still attached to the ovipositional substrate used in the insectary (such as paper cards, cloth, or rice hulls), or destalked. The latter can be scattered freely or sprinkled on a 10% sucrose or 1% methyl cellulose solution and then sprayed on foliage (Doutt & Hagen, 1950; Ridgway & Jones, 1969; Shands *et al.,* 1972a; Barry *et al.,* 1974; Jones & Ridgway, 1976; Peng, 1986). Larvae from destalked eggs developed faster than those from stalked eggs (Sengonca & Schimmel, 1993). There has also been considerable interest in mechanical application techniques for chrysopid eggs (McEwen, 1996; Daane & Yokota, 1997; Gardner & Giles, 1996, 1997; Sengonca & Loechte, 1997).

The use of food sprays in the field can augment natural chrysopid populations by arrestment or attraction (Hagen *et al.,* 1970a; New, 1975; Hagen & Bishop, 1979; Ridgway & Murphy, 1984; Hagen, 1986; Barclay, 1990; Evans & Swallow, 1993; Ehler *et al.,* 1997). It is possible to retain and arrest foraging lacewing adults by simply spraying a sugar solution (10% sucrose in water) or molasses on the crop (Ewert & Chiang, 1966; Carlson & Chiang, 1973). Chrysopids are not actually attracted to the sugar sprays, but will stay in the field when they contact the sugar; thus they can be concentrated in sugar-sprayed areas (Hagen *et al.,* 1970a). *Chrysoperla* spp. can be attracted to artificial honeydew and stimulated to oviposit even in the absence of aphids or other honeydew-excreting insects. By simulating a high aphid population level by spraying artificial honeydew that contain sugar and protein, it is possible to induce chrysopids to oviposit before aphids increase, thus controlling other potential pests that are not honeydew producers such as noctuids, thrips, and spider mites.

Pea and spotted alfalfa aphid populations on alfalfa attained only half the population densities reached in plots not receiving protein food sprays (Hagen *et al.,* 1970a). Similarly, *Myzus persicae* populations on pepper plants were slowed and reached only half of the populations on unsprayed peppers (Hagen & Hale, 1974). *Myzus persicae, Lygus hesperus,* and *Autographa californica* (Speyer) numbers were reduced in potato fields by spraying Wheast® (yeast grown on whey) in a sugar solution that increased *C. carnea* abundance and oviposition (Ben Saad & Bishop, 1976). *Heliothis zea* eggs, larvae, and damaged bolls were reduced in plots sprayed with Wheast® and sugar (Hagen *et al.,* 1970a; Kinzer & Jones, 1973; Denver, 1974). However, Butler and Ritchie (1971) did not obtain an increase in *C. carnea* oviposition in cotton when sprayed with Wheast® plus honey and glycerin. A yeast plus sugar spray applied to grape vineyards did promote an increase in one case but not in a second trial (White & Jubb, 1980). Duelli (1987) got attraction but did not obtain an increase of *C. carnea* oviposition in Switzerland when he applied Wheast® plus sugar on alfalfa, apple, or corn, while Dean and Satasook (1983) in England obtained attraction of *C. carnea* to tryptophan-impregnated sticky traps but not to caryophyllene. Application of artificial honeydew will not be of any value

if there is already honeydew present (Hagen *et al.,* 1970a; Tassan *et al.,* 1979).

When applying food sprays, both a protein source (yeasts) and sugar must be applied. Shands *et al.* (1972b) and Hagley and Simpson (1981) applied protein without sugar on potatoes and apples and failed to obtain an increase of *C. carnea.* The artificial honeydew should be applied in swaths across the prevailing evening winds and on crops that are releasing natural volatiles (synomones); otherwise, *C. carnea* adults may not be attracted to tryptophan volatiles (Hagen, 1986).

Shelters were provided for hibernating *C. carnea* adults by placing boxes in field crops or even in fallow fields as fall approached; mortality during hibernation did not exceed 2% (Sengonca & Henze, 1992).

Another advantage of *C. carnea* is that it has a rather broad tolerance to many insecticides, particularly in its larval stage (Bartlett, 1964b; Grafton-Cardwell & Hoy, 1985). As noted earlier, lacewing eggs and larvae survived DDT treatments and controlled mealybugs in pear orchards (Doutt & Hagen, 1950). Fortunately, *C. carnea* larvae are also tolerant of pyrethroid insecticides because they produce pyrethroid esterase or esterases that allow them to detoxify these insecticides (Ishaaya & Casida, 1981). The tolerance to certain insecticides varies geographically among *C. carnea* populations. *Chrysoperla carnea* from areas in California with a history of heavy pesticides usage are more tolerant to certain insecticides compared with populations occurring in areas where less pesticides were used (Grafton-Cardwell & Hoy, 1985).

In conclusion, since *C. carnea* is such a common and readily available generalist predator, it can be utilized in biological and integrated control programs against a variety of insect and spider mite pests in several ways, including mass culture and periodic colonization (although this is rather expensive today); application of behavioral chemicals and artificial food to retain, attract, and induce oviposition; and use of selective insecticides to spare naturally occurring lacewing populations.

Chrysoperla externa (Hagen [= *lanata* (Banks)] ranges over all Latin America and north into Florida and South Carolina and was inadvertently introduced into the Hawaiian Islands (Adams, 1963; Adams & Penny, 1985). Because it is commonly found in agricultural habitats and preys on a wide range of insect pest species, its biology has been investigated and found to be similar to *C. carnea* (Ru *et al.,* 1975; Crouzel & Botto, 1977; Aun, 1986; Núnez, 1988; Barclay, 1990). However, little research was conducted on ways to augment *C. externa* populations until Barclay (1990) tried both releasing *C. externa* eggs and using supplementary food sprays to manipulate lacewing adults in Nicaraguan soybean and cornfields.

Barclay reared the larvae on frozen *Sitotroga cerealella* eggs in individual cells, and used Wheast® plus honey as a paste to feed the adults. Groups of 100 lacewing eggs were cut from paper taken from the oviposition units (eggs could be held at 10°C for up to 10 days). Two releases were made in soybeans and one, in tasselling corn that amounted to release rates from 360,000 to 760,000 *C. externa* eggs per hectare. No significant increases in *C. externa* larval populations were observed in either crop; there was no significant effect on the three targeted noctuid species in soybeans (heavy rains may have reduced the number of lacewing larvae directly or indirectly through increasing fungal attack on lacewing eggs, while many eggs were also parasitized by *Trichogramma* spp.). There were significantly fewer *H. zea* larvae found in the corn release plots compared with the control 1 week after release. Even though heavy rains occurred midway through the test that reduced the number of lacewing larvae, Barclay (1990) considered there had been enough time for the eggs to hatch and the larvae to have had an impact on *H. zea* before the storm.

Chysoperla externa adults were also manipulated in the field using Wheast®–sugar food sprays (Barclay, 1990). Chrysopid adults increased from 2.3- to 6-fold in food-sprayed corn field plots compared with check plots 24 h following each treatment application; significantly more chrysopid eggs were found in the Wheast® + sugar plots compared with plots where sugar alone was sprayed and there were more than three times as many freshly deposited green lacewing eggs found in the Wheast®–sugar treatments compared with controls.

Significantly fewer *C. externa* eggs were found on cotton plants intercropped with corn and weeds than were found in cotton monocultures (Schultz, 1988); Schultz suggests the reduction in egg numbers is due to a corresponding decrease in number of prey (aphids) and increased presence of other predators such as earwigs.

Chrysoperla rufilabris (Burmeister) is widespread in the United States but is virtually absent in the dry southwestern states and most abundant in the more humid southeastern United States (Tauber & Tauber, 1983). Under relatively humid conditions (75% RH), *C. carnea* and *C. rufilabris* are similar in almost all aspects of the life history traits that the Taubers (1983) measured. However, under low (35% RH) and moderate humidity (55% RH), *C. rufilabris* had a prolonged preoviposition period, a reduced fecundity, an increased preimaginal mortality, and a slower developmental rate. The larvae have a broad prey range. In southeastern United States cotton fields, *C. rufilabris* larvae are considered important predators of spider mites (McGregor & McDonough, 1917) and bollworms (Ewing & Ivy, 1943; Lingren *et al.*, 1968). Fewer bollworm eggs were destroyed on pilose cotton than on smooth leaf and hirsute cotton; from functional response studies of the lacewing larvae, Treacy *et al.* (1987) demonstrated that cotton-leaf trichomes are mechanical barriers that reduce the mobility and predatory ability of *C. rufilabris*. In Florida (Buschman *et al.*,

1977) and South Carolina (Shepard *et al.*, 1974; McCarty *et al.*, 1980) relatively few chrysopid larvae were found in soybean fields, but they did consume noctuid eggs labeled with ^{32}P. Aphids, such as the cotton aphid (Ewing & Ivy, 1943; Burke & Martin, 1956), the pecan aphid (Elkarmi *et al.*, 1987), and citrus aphid (Hydorn & Whitcomb, 1972), are common prey. Large numbers of oriental fruit moth eggs were observed to be destroyed by *C. rufilabris* larvae in Vineland, Canada peach trees (Putman, 1932). Putman (1956) also observed that *C. rufilabris* became less abundant than *C. carnea* when DDT was used in Ontario peach orchards—tests showed larvae of *C. rufilabris* larvae were more sensitive to the insecticide. Lawrence (1974) found *C. rufilabris* larvae had various degrees of sensitivity to four common insecticides used to control insect pests in Florida citrus.

Because *C. rufilabris* showed promise for manipulation on cultivated crops, methods of culturing and influence of diets in both larvae and adult lacewings were studied (Hydorn & Whitcomb, 1972, 1979; Ru *et al.*, 1976; Elkarmi *et al.*, 1987; Nordlund & Morrison, 1992). It was determined that *C. rufilabris* larvae readily prey on eggs and young larvae of the Colorado potato beetle in the laboratory, while in field cages, 84% population reduction was obtained with a release rate of 80,940 *C. rufilabris* larvae per hectare (Nordlund *et al.*, 1991). However, these authors concluded that such a rate is not practical unless the cost to mass-culture lacewings is reduced. Larvae of *C. rufilabris* feed voraciously on *Bemisia tabaci* eggs and nymphs on the lower (under) surface of *Hibiscus* leaves; treatments in a greenhouse with five larvae per *Hibiscus* plant produced both marketable and unmarketable plants, and all plants in the 25 or 50 lacewing larvae per plant treatments remained healthy and marketable (Breene *et al.*, 1992).

The genus *Mallada* (= *Anisochrysa*) is the largest of the Chrysopidae with at least 122 described species that are worldwide but absent from much of the Neotropics (Brooks & Barnard, 1990). They are trash carriers and at least some diapause as a third larval instar (Principi, 1956; 1992; Principi *et al.*, 1990). Of this genus, only *M. boninensis* Okomoto and *M. basalis* have been studied in any detail as natural enemies in agricultural crops. In an unsprayed Rhodesian cotton field, 81% of the fertile chrysopid eggs sampled by Brettell (1979) were *M. boninensis*, one of three chrysopid species present. The adults are not predaceous but a larva can destroy a total of 297 *H. armigera* eggs. This species is widely distributed in the Old World and is most common in Asia; Brettell (1979) suspects that it was carried to southern Africa on citrus imported from China some 200 years ago. He also found that of the four widely used pesticides in cotton, DDT was the most toxic to the larvae. This is interesting because *C. carnea* larvae are not susceptible to DDT (Bartlett, 1964b). Among the 10 chrysopid species collected in Guanazhou, China, *M. boninensis* was one of three dominant species found (Wei *et al.*, 1987). In Taiwan, Lee and Shih (1982) reared larvae of *M. boninensis* on nymphs of the psyllid *Paurocephala psylloptera* Crawford; the developmental time was somewhat longer than when the larvae were reared on *Corcyra cephalonica* eggs. They developed a life table for the chrysopid; and by using field cages, they found an optimal ratio of four first-instar *M. boninensis* larvae per an initial prey biomass index of 158 suppressed psyllid populations within 4 to 5 days. In Japan during the summer, predation by *M. boninensis* is considered a major mortality factor of *Ceroplastes japonicus* Green on *Euonymus* (Miyanoshita & Kawai, 1992); and in India, it readily attacks the grape mealybug *Maconellicoccus hirsutus* (Green) (Mani & Krishnamoorthy, 1989; Mani, 1990).

Mallada basalis larvae were reared on *C. cephalonica* eggs; 13 adults were obtained from each milliliter of host eggs (Chen *et al.*, 1993). Eggs of *M. basalis* placed in citrus trees resulted in a reduction of *Panonychus citri* (Wu, 1992).

Among the 11 species of *Mallada* in Australia, *M. signata* (Schneider) is a predator of *Heliothis* spp. in Queensland cotton. Its voracity and searching ability were studied in the laboratory by Samson & Blood (1980), who found the numbers of *H. punctigera* eggs and early instars attacked per day in petri dishes increased exponentially with each successive larval instar of the chrysopid. *Mallada signata* larval searching success and consumption of *Heliothis* were greater than those reported for some other chrysopids [e.g., *C. carnea* (Samson & Blood, 1980)]. Because *Mallada* spp. are similar to *Chrysoperla* spp. (not predaceous in the adult stage and attracted to protein hydrolysates [Neuenschwander *et al.*, 1981)] and are often common in wild refuges as well as in crops (Pantaleoni & Sproccati, 1987), they may prove to be effective predators in crops if artificial honeydew is applied in anticipation of arthropod pest problems.

Coniopterygidae According to Meinander (1990), there are 423 recognized species of dusty wings in the world today; Meinander (1972) revised the family and included references to the few biologies that are known in the family; and in 1986 he mapped the distribution of the 132 species belonging to the 10 genera present in the Americas (Meinander, 1986). Johnson (1980) reviewed the biology and produced a key to the North American genera of the Coniopterygidae, and revised the genus *Aleuropteryx*.

The dusty wings are the smallest of all the Neuroptera, but some of the species present in agricultural and landscape habitats are important predators of spider mites, aphids, coccids, and whiteflies, and these species account for the known biologies (Quayle, 1912; Arrow, 1917; Withycombe, 1923, 1924; Collyer, 1951; Fleschner, 1952; Fleschner & Ricker, 1953; Badgley *et al.*, 1955; Muma, 1967;

Ward, 1970; Henry, 1976; Agekyan, 1978; Johnson, 1980; Wheeler, 1981; Rousset, 1984; Drea, 1990a; Tauber, 1991a).

The adults and larvae are predaceous, often feeding on the same prey, but adults also feed on honeydew. The adults consume the whole prey with mandibulate mouthparts while the fusiform larvae suck their prey dry using piercing mouthparts; both adults and larvae detect their prey tactilely. Many dusty wings are associated with one tree or shrub species, indicating they may be restricted to specific prey. The eggs are usually laid singly on leaf surfaces or in bark crevices. The larvae usually undergo three instars (Meinander, 1972), but four instars have been observed in *Conwentzia californica* Meinander by Quayle (1912), in *Heteroconis picticornis* (Banks) by Badgley *et al.* (1955), and in *Semidalis vicina* (Hagen) by Muma (1967). The flatish cocoons have two or three layers of silk with the innermost being the thickest and are often fastened to the undersides of leaves. Some species overwinter within the cocoon as prepupae, while others overwinter as free larvae.

The Coniopterygidae are divided into two subfamilies: Aleuropteryginae and Coniopteryginae. There are few species in each subfamily that have been considered important to applied biological control.

Aleuropteryginine larvae have antennae and labial palpi of about equal length with long jaws that project from beneath the labrum. *Aleuropteryx juniperi* Ohm is a Palearctic species that was inadvertently introduced into Pennsylvania, probably on juniper nursery stock from Europe, in the early 1900s (Henry, 1976). By 1980, it had spread to Maryland, New Jersey, New York, and West Virginia (Wheeler, 1981). It occurs mainly on *Juniperus* spp., and in England feeds on juniper scale, *Carulapsis juniperi* (Bouché) (Ward, 1970). This was also observed by Henry (1976) in Pennsylvania, and he also found it to feed on *C. minima* (Targioni-Tozzetti) on juniper. Henry (1976) reared *A. juniperi* in the laboratory on *C. minima*. *Aleuropteryx juniperi* usually lays its eggs singly on the underside of juniper needles. The larvae feed at the base of the adult scale but penetrate scale crawlers and nymphs. The adults feed on all stages of minute cypress scale and its honeydew (Henry, 1976). Stimmel (1979) mentions that *A. juniperi* reduced *C. minima* population more than any other natural enemy in experimental juniper plantings.

Heteroconis picticornis (= *Spiloconis picticornis* Banks) was introduced from Hong Kong in 1953. Several thousand were released against spider mites and scale insects infesting citrus and avocado groves in southern California by Badgley *et al.* (1955), but it did not become established. These researchers described the predator's biology. The flat or scalelike eggs take the shape of the contour of the surface upon which they are laid. The larvae pass through four instars in 16 to 32 days, and feed only on diaspine or lecanine scale insects, preferring crawlers and first- and second-instar scale. The fourth instar selects a crevice or crack in which to spin its three-layered silken cocoon. The adults feed on both spider mites (including citrus red mite and citrus rust mite) and scale insects. They lived longest (63 days) on scale crawlers, honey, and water (Badgley *et al.*, 1955).

Hemerobiidae Among the nearly 550 described species in 40 genera of brown lacewings in the world (which now includes the former family Sympherobiidae), about 50% are in two genera: *Hemerobius* and *Micromus* (New, 1988b; Monserrat, 1990). Oswald (1993) recognizes only 25 genera in the world and provides a key to the genera. The hemerobiids are predaceous in both adult and larval stages, feeding mainly on aphids, with evidently fewer alternate food sources than chrysopids (Killington, 1936; Balduf, 1939; New, 1975). Because the adults do not need to rely on honeydew for egg production, they can produce and deposit their eggs at lower aphid densities than do chrysopids (Neuenschwander *et al.*, 1975). Even though brown lacewing larvae and adults are rather polyphagous, different species appear to be tied to certain habitats, [e.g., conifers, broad-leaved trees, or low vegetation (New, 1975; 1988a; Monserrat & Marin, 1996)].

The most interesting biological aspect of many hemerobiids is the remarkable low temperature thresholds of all stages—even lower than most of their prey species. Thus, the potential for using these predators in temperate climates during cool periods is great (Neuenschwander, 1975, 1976; New, 1975, 1984b, 1988a; Canard, 1997).

Eggs are deposited singly on their sides; the conspicuous projecting polar micropylar knob distinguishes a hemerobiid egg from a syrphid egg. Larvae undergo three instars, do not carry trash, and can be recognized from chrysopid larvae by their smaller heads that jerk from side to side when moving. Pupation occurs in a loosely woven silken cocoon near the feeding areas. Hemerobiids have been used in biological control in classical, in augmentative, and in conservation programs. The species involved have been mainly in the genera *Hemerobius* or *Micromus,* but the biologies of some lesser known genera have been described [e.g., *Boriomyia,* which is aphidophagous (Laffranque & Canard, 1975); *Drepanacra,* which feeds mainly on psyllids (New, 1984b, 1988a); and *Sympherobius,* which is predaceous on mealybugs, has been mass-cultured and released against mealybugs attacking citrus (Bodenheimer, 1951) and in Europe it feeds on aphids and European red mite in apple orchards (Principi & Canard, 1974)].

Hemerobius is a widely distributed genus of about 220 species occurring in Africa, Asia, Australia, Europe, and North and South America (Laidlaw, 1936; Carpenter, 1940; Nakahara, 1965; Kevan & Klimaszewski, 1987; New 1988a; Monserrat, 1990). Several species are Holarctic.

Hemerobius humilinus Linnaeus is a Holarctic species, but it is rarely found west of the Rocky Mountains in the United States, whereas in Canada it ranges westward to British Columbia. In the eastern United States it is found in Pennsylvania vineyards (Jubb & Mastellar, 1977) and in apple orchards preying on aphids (Holdsworth, 1970a); in the laboratory its larvae consumed both codling moth eggs and first-instar larvae (Holdsworth, 1970b). In Europe, *H. humilinus* is 1 of 11 species of hemerobiids occurring in apple orchards (Szabo & Szentkiralyi, 1981), where it consumes aphids and European red mites (Principi & Canard, 1974); in China, it occurs in cotton fields (Chao & Chang, 1978).

Hemerobius pacificus Banks is a common species west of the Rocky Mountains in the Nearctic region (Carpenter, 1940; Nakahara, 1965). In California, it mainly occurs along the coast where it is active as a general predator in low crops during the winter and early spring (Neuenschwander *et al.*, 1975). Its temperature threshold for development is 0.4, 4.1, and 0.6°C for egg, larva, and pupa, respectively (Neuenschwander, 1975); and has a threshold of 4.4°C for reproduction (Neuenschwander, 1976). Neuenschwander and Hagen (1980) released 20 *H. pacificus* eggs per plant weekly in artichoke plots from October through the end of April. They demonstrated that *Hemerobius* populations had a great impact on overwintering green peach aphid populations, and even though first-instar larvae of the relatively rare artichoke plume moth, *Platyptilia carduidactyla* (Riley), served as a secondary food source, the impact of the lacewing was measurable at least at higher the moth densities. In plots treated with protein plus sugar sprays, the densities of *H. pacificus* indicated arrestment of adults, and eggs were about twice to four times higher in treated plots, resulting in lower green peach aphid and artichoke plume moth populations.

Hemerobius stigma Stephens was found to be an effective predator of the balsam woolly aphid, *Adelges piceae* (Ratzeburg), in England (Laidlaw, 1936). About 5000 eggs of *H. stigma* and a few eggs of *H. nitidulus* Fabricius collected in England were sent to eastern Canada where *A. piceae* had invaded during the early 1900s (McGugan & Coppel, 1962; Mitchell & Wright, 1967). It was concluded that *H. stigma* and *H. nitidulus* did not become established, based on the assumption that the brown lacewings could not survive the winters (McGugan & Coppel, 1962). However, a common native species, *H. stigmaterus,* Fitch, was synonymized with *H. stigma* by Tjeder (1960); and Garland (1978a) points out the implications of the taxonomic problem, indicating that *H. stigma* was already in the New World. However, Nakahara (1965) considered both names to be valid and recommended withholding judgment utilizing studies of the Old World populations as compared with the variation in the New World populations. Garland (1978b) exposed eggs of *H. stigma* originating in Quebec, Canada, to overnight temperatures as low as −3.8°C and found they survived, indicating that eggs of this species survive occasional frosts.

Micromus has a nearly cosmopolitan distribution and includes from 95 to 150 specific names (Klimaszewski & Kevan, 1988; New, 1988a; Monserrat, 1990; Oswald, 1993). *Micromus* spp. have been used in classical biological control programs as aphidophagous predators in a variety of crops and also have been manipulated in the field and in greenhouses (New, 1975, 1988b; Khloptseva, 1991).

Micromus timidus Hagen (= *vinaceus* Gerstaecker) was imported from cane fields in north Queensland, Australia, in 1919 and became established in the Hawaiian Islands (Williams, 1931; New, 1988a).

Micromus angulatus Stephens, a Holarctic aphidophagous species, can deposit as many as 2000 eggs and in France overwinters as an adult with its hibernal diapause being induced by short photoperiods (Miermont & Canard, 1975). In the former USSR, *M. angulatus* lives in colonies of aphids on soybeans, clover, maize, and cabbage plants near forest fringes and has two to three generations per year; it is mass-cultured in greenhouses using bean and pea aphids grown on horse beans as prey for adults and larvae. It successfully controls aphids on cucumbers in greenhouses at predator (eggs) to prey ratios of 1:1 or 1:3 and when released every 7 to 9 days (Khloptseva, 1991). Apparently there is no cannibalism on pupae when larvae of *M. angulatus* are reared on aphids (Stelzl & Hassan, 1992).

Micromus posticus (Walker) occurs from Texas to Florida and north into Canada (Carpenter, 1940). It preys mainly on aphids infesting citrus groves, pecan orchards, vineyards, alfalfa, cotton fields, and potato fields [see references in Miller & Cave (1987)]. Descriptions of immature stages and life table data are presented by Miller and Cave (1987); they found that in unsprayed Alabama cotton fields only a percentage of *M. posticus* larvae were parasitized by an ichneumonid and the figitid *Anacharis melanoneura* Ashmead. In Florida citrus groves as well as in the laboratory, Selhime and Kanavel (1968) found *M. posticus* larvae were not parasitized by *Anacharis* sp; however, *M. subanticus* (Walker) larvae in citrus were heavily parasitized, which suggested that the rarity of *M. subanticus* in citrus groves was the result of parasitism.

Micromus tasmaniae is a common Australian and New Zealand species that preys on aphids in alfalfa (Milne & Bishop, 1987); cotton aphids (Samson & Blood, 1979, 1980); rose aphids (Maelzer, 1977); potato aphids (Hussein, 1986); and aphids, Collembola, and mites on strawberries [determined by Butcher *et al.* (1988) using immunological testing for proteins]. New (1984b) found *M. tasmaniae* commonly associated with psyllids on acacias. Samson and Blood (1979) found *M. tasmaniae* had a lower voracity for *Heliothis* eggs and larvae than did chrysopid larvae or nabids. Temperature thresholds for development were determined by Samson and Blood (1979) and by New (1984b).

Eggs of *M. tasmaniae* suspended in a 0.03% xanthan gum solution (Hussein, 1984) sprayed on potato foliage yielded enough larvae to reduce *Myzus persicae* populations by 70% compared with the aphid abundance on untreated potato plants, while the yield of potatoes from untreated plants was reduced by 40% (Hussein, 1986). The first lacewing eggs were sprayed on potato plots at the time alate *M. persicae* were migrating into the potato fields and then applied twice weekly over a period of 4 weeks.

The genus *Nusalala* consists of 20 species found in North and South America (Oswald, 1993). In Brazil, *N. uruguaya* (Navas) larvae completed development on two aphid species but all died at the end of the first instar when fed *Toxoptera citricidus* (Kirkaldy) (Souza *et al.*, 1990). The larvae are fast moving, only feed on live aphids, and exhibit cannibalism at low food availability (Souza & Ciociola, 1995).

Order Coleoptera

Coleoptera comprise the greatest number of described species (ca. 300,000) of any order, and occur in all habitats except the depths of salty seas. Crowson (1981) reviewed the morphology and biology of Coleoptera including a chapter on predaceous beetles. Balduf (1935) extensively covered the literature on the biologies of entomophagous Coleoptera up to 1933. Keys to larvae of families and many subfamilies of the order have been written by Klausnitzer (1978) and Lawrence (1991a), but little is known about the ecological importance of the many species that live in soil and in water. Even the prey spectra of many predaceous families are poorly known. However, some of the most spectacular successes in classical biological control have involved the introduction and establishment of predaceous Coleoptera, and the roles that various endemic entomophagous Coleoptera play in the natural biological control of their prey are slowly being revealed.

In a survey of the food habits of beetles, Townsend Glover (1868) listed only 11 families that contained predaceous species. Today about a third of the families (54) are known to contain predaceous species (Hagen *et al.*, 1976a), including 47 terrestrial families (Hagen, 1987). Both larvae and adults are predatory in most families. Of the four suborders, Archostemata, Myxophaga, Adephaga, and Polyphaga, only the latter two contain species that have been involved in biological control studies.

Suborder Adephaga

Of the 10 families that compose the Adephaga, four are terrestrial (Geadephaga): Rhysodidae, Paussidae, Cicindelidae, and Carabidae. The rhysodids apparently are not predaceous. The paussids, which are associated with ants, will not be discussed. The Carabidae is by far the most important family in Adephaga involved in biological control investigations.

The six aquatic families of Adephaga (Hydradephaga), Haliplidae, Amphizoidae, Hygrobiidae, Noteridae, Dytiscidae, and Gyrinidae are all essentially predatory except the Haliplidae, but little is known about their impact on their prey populations.

Cicindelidae Both cicindelid adults and larvae are predaceous. The adults are active hunters, while the larvae sit and wait for their prey. The family is considered a subfamily of Carabidae by some coleopterists, but the larvae of tiger beetles differ greatly from carabids in their morphology and habits. The following comments have largely been taken from Balduf (1935), Thiele (1977), and Pearson (1988).

Adults of the fast-moving, typical, diurnal tiger beetles, *Cicindela* spp., are usually maculate and metallic with bulgy eyes, and are commonly seen in sandy localities either near water or inland. The tropical genera, *Collyris* and *Tricondyla,* are arboreal and have tarsal pads on all legs in both sexes. The diurnal species use their eyes to locate prey. The flightless, nocturnal, black, carabid-like *Amblycheila* and *Omus* species have relatively flat, small eyes and probably rely on contact to attack their prey. The adults of the nocturnal South America *Megacephala fulgida fulgida* Klug evidently locate mole crickets by phonotaxis, orienting toward cricket songs (Fowler, 1987). The size of prey captured is related to the size of the tiger beetle. The adults seize their prey by their large sickle-shaped mandibles, which are armed on the inner side with several large teeth and a molarlike tooth near the base. They ingest the fluids of their prey and use preoral digestion (Evans, 1965). Pearson and Mury (1979) found a close correlation between mandible length and median prey size captured in 13 *Cicindela* spp. found at edges of ponds and in 6 grassland species in southeastern Arizona. Similar correlations have been found elsewhere in the world. The adults are general predators and attack insects, arachnids, and crustacea within certain size ranges. However, some larger species respond to aggregations of smaller prey.

The diurnally active adults hide in soil during the night and may hibernate in rather long burrows in the soil. The terrestrial species dig a shallow hole in the soil and deposit eggs singly, while the arboreal species deposit eggs singly in tunnels made in twigs or bark.

The larvae sit and wait for prey in burrows made in the soil or in wood. Their circular, camouflaged heads and pronota form closings to the entrances of their burrows. The larvae have two or three pairs of hooks on the tergum of the fifth abdominal segment that is used to anchor them vertically. When a prey comes within range, the larva strikes quickly, throwing its front and middle legs out clear of the burrow; its abdomen remains inside. The prey is seized by large, sickle-shaped, dentate mandibles and is consumed after the larva retreats into the burrow. Like the adults, the larvae of tiger beetles appear to be general predators of certain size prey.

There are three larval instars. Generally a 2-year period is required to complete development, but development can range from 1 to 4 years depending on the species. Pupation occurs within their larval burrows, which are plugged by the mature larva. Larvae of many species hibernate within their burrows.

The impact of cicindelids on regulation of agricultural insect pests is poorly known, but several articles that are now in press cite Fowler's (1987) note concerning the predatory behavior of *Megacephala fulgida,* which clearly implicates this nocturnal species as a primary biological control agent in the regulation of early instar populations of *Scaptericus* mole crickets in Brazil. Pearson (1988) mentions a personal communication from J. H. Frank, who observed *M. virginica* (Linnaeus) as a major predator of introduced mole crickets in southeastern United States.

Carabidae Of this huge family of approximately 40,000 described species of ground beetles, the biologies of less than 100 species have been investigated, mostly western European species (Ball, 1978). As common as carabids are, their effects on pest populations are poorly known, and only a few species have been utilized in classical biological control programs.

Since Balduf's (1935) synopsis of carabid biologies, there had been no exhaustive reviews until Thiele's (1977) compilation of his personal investigations together with the existing knowledge of carabid biology and ecology, which resulted in a book on the physiological and behavioral adaptations of carabids in relation to their environments. Thiele's book; the proceedings of symposia dealing with carabids edited by Erwin *et al.* (1979), Brandmayr *et al.* (1983), and Den Boer *et al.* (1986), papers given at the sixth European Carabidologist's meeting published in the *Acta Phytopathologica Entomologica Hungaricae, 22* (1–4), 1987; and Luff's (1987) review on polyphagous ground beetles in agriculture provided the main bases for the following comments.

Carabid food habits range from being ectoparasitoids on specific insect hosts, to obligatorily carnivores, herbivores, and omnivores. Forbes (1883) analyzed the gut contents of 175 adult specimens representing 20 carabid genera and found that 57% of the food was of animal origin (of that percentage, 36% insects; and 27% mollusks, earthworms, myriapods, and arachnids); the vegetable matter eaten, which accounted for 43%, was fungi and pollen. Similar studies, reviewed by Thiele (1977), show that intestinal contents of a species varies through the year, and one cannot generalize about the food habits of sympatric congeneric species (e.g., one *Harpalus* sp. diet is 50% animal matter while another *Harpalus* sp. in the same area is nearly totally phytophagous). Apparently, it is quite common for carabids to feed partly on vegetable matter (Johnson & Cameron, 1969). Although *Pterostichus, Calathus,* and *Agonum* spp. at times eat plant materials, they feed mainly on insects such as ants, aphids, and lepidopteran larvae (Thiele, 1977). In his fine review, Allen (1979) not only lists the food habits of 156 North American carabid species taken from literature, but also lists 107 North American species that have been studied ecologically or biologically. The latest review on carabid food habits by Larochelle (1990) is exhaustive. Based on 1290 references, he concludes 775 species are exclusively carnivorous, 618 species are partially or exclusively entomophagous, 85 species are exclusively phytophagous, and 206 species are omnivorous. Larochelle states in general, "Carabids are opportunistic feeders, depending on available food. It is, therefore, presumed that following more comprehensive studies, carabids will appear more omnivorous."

The oligophagous predators are found more among the primitive carabids such as Cychrini (*Cychrus, Scaphinotus, Brennus*), which feed mainly on slugs and snails (Larochelle, 1972; Greene, 1975; Digweed, 1993) and the Carabini (*Carabus* and *Calosoma*). Many species of *Carabus* feed on mollusks and earthworms. Some species have long mandibles in both their adult and larval forms and appear cychrine-like, feeding mainly on mollusks (Sturani, 1962; Malausa *et al.,* 1983). *Abax parallelepipedus* (Piller & Mitterpacher) and *Pterostichus madidus* (Fabricius) also prey on slugs (Asteraki, 1993; Symondson & Liddell, 1993). *Calosoma* species are commonly called caterpillar hunters, and they do largely feed on lepidopteran larvae (Burgess, 1911; Burgess & Collins, 1917; Price & Shepard, 1978; Wallin, 1991). Young (1984) observed that *C. sayi* DeJean adults in the laboratory did not consume the very small or the very large lepidopteran larvae but they would consume living and dead non-lepidopteran insects. *Pasimachus elongatus* LeConte of the Scaritini fed on six species of Coleoptera as adults and on ants and termites found under cow chips as larvae; in the laboratory 23 adult orthopterans and 18 beetle species were eaten (Cress & Lawson, 1971). *Notiophilus biguttatus* Fabricius (Notiophilini) adults and larvae feed extensively on Collembola. Adults use their large eyes to hunt their prey during the day (Bauer, 1975). Metabolic rate and reproduction of this species are strongly correlated with the rate of Collembola taken in as prey (Ernsting & Issaks, 1987). As the carabid tribes become more derived, it appears that the feeding trends and structures become more omnivorous.

The form of the feeding apparatus of adult carabids, particularly the mandibles, offers a clue as to the feeding habits. Zhavoronkova (1969) found distinct correlations between the nature of feeding by carabids and the structure of their mandibles and proventriculi, and she distinguished

three main feeding categories: obligate zoophages, predominant zoophages, and predominant phytophages. Evans and Forsythe (1985), after comparing many external and internal structures associated with a diversity of carabid feeding habits, concluded that although parts of the feeding mechanism directly reflect the nature of the food of different group beetles, the mechanism as a whole reflects the method of feeding instead of the specific nature of the food taken. The mechanism also appears to be focused toward three main feeding methods: fluid feeding (in which preoral digestion is important), fragmentary feeding, and mixed feeding. Tips of mandibles can wear down with age (Houston, 1981; Wallin, 1988); thus the mandibles of fluid feeders can become so worn as to appear like mandibles of a mixed feeder of phytophage. Evans and Forsythe (1985) and Forsythe (1987) suggest that the head shape in carabids may reflect locomotory adaptations more frequently than feeding adaptation, and the form of the carabid body and the shape and articulation of legs reflect their habits more often than their habitats. According to Crowson's (1982) book review of *Life Forms of Carabidae,* edited by Sharova and Ghilarov (1981) (in Russian), Sharova composed a dichotomous key to the various life forms of adult and larval Carabidae based on food types, locomotory categories, and habitats coupled with morphological characters such as adult and larval mandibles and types of legs, larval urogomphi, general body form, and so on.

The habitats of carabids are diverse and are located and returned to by kinetic responses to gradients of abiotic factors (Thiele, 1977). However, Evans (1983, 1988) reports that certain carabids locate their habitats by responding to kairomones or synomones emanating from their habitats. Evans found that *Bembidion obtusidens* Fall adults were attracted to volatile metabolites of blue-green algae growing in the habitats of the beetles and their prey. Wheater (1989) found some carabids detected prey visually, while others, particularly of the tribe Pterostichini, relied on olfactory cues.

There are few data on the length of carabid preoviposition periods. Not ovipositing during the year they emerge as adults may be common with the larger carabids, like *Calosoma* and *Carabus* that have either estival or hibernal dormancy periods. In some cases, oviposition does not occur until the third year after emergence (Dusaussoy, 1963). Reproduction in some species is regulated by the action of day-length photoperiods on juvenile hormone levels (Ferenz, 1986). Some species lay a single batch of eggs per season (Luff, 1973), while others lay eggs continuously at a relatively constant rate (van Dijk, 1979; Baars & van Dijk, 1984). However, the quantity of food ingested and temperature is directly tied to egg production (van Dijk, 1983). Based on years of pitfall records, Den Boer (1986) has determined the net production (*R*) of 64 carabid species.

A simulation energetics model involving ingestion rate of prey resulting in egg production rate was built by Mols (1987), which involves the use of a feedback mechanism associating gut capacity with ovarial size—the more eggs that are in the ovaries, the less the gut capacity, which inhibits feeding via reaction of the stretch receptors in the abdomen.

Carabids have been divided into spring breeding and fall breeding species depending on which season they lay their eggs. The spring breeders hibernate or enter diapause as adults, reproduce from spring to early summer, and then usually die. The fall breeders reproduce in autumn, usually hibernate as larvae, and estivate as adults. Thiele (1977) lists 36 spring breeding species and 21 fall breeding species. Of these, about 60% are nocturnally active and 22% predominantly active during the day, while 18% were variable.

Eggs are usually deposited singly in the soil. However, Thiele (1977) lists 14 species in the Pterostichini that have been observed to dig a "nest" in soil or rotting wood and to deposit a group of eggs that the female remains with until the larvae hatch.

Carabid larvae predominantly have three instars but some species have only two, and a few species have been reported to have four and five instars. Most larvae have similar type mandibles, with a single cutting edge with a retinaculum but without a suctorial tube or channel, and usually have prominent urogomphi (Thompson, 1979). The larvae of many species are predaceous. Others are phytophagous or omnivorous and some are ectoparasitoids, but the feeding habits of larvae are only known for relatively few species; of the 156 published feeding records of North American carabids listed by Allen (1979), 67 refer to larval feeding. Carabid larvae often kill more prey than they can consume. Once fully grown, the larvae form cells in the soil or rotting wood and pupate. The pupal period is about 1 week during the summer.

Large, showy carabids were among the first beetles observed to be predaceous on caterpillars and manipulated to biologically control insect pests in Europe. In the early 1900s, *Calosoma sycophanta* (Linnaeus) was successfully introduced into the United States from Europe to control gypsy moth (Burgess, 1911; Howard & Fiske, 1911). This success stimulated the first detailed studies on carabid biology. Burgess and Collins (1917) described the biologies of over 30 *Calosoma* spp. Not until the 1960s did research on carabid biology resurge widely (Kulman, 1974; Thiele, 1977), and not until the 1980s were carabids considered again for use as biological control agents (Luff, 1987)

Thiele (1977) concluded from his experiences and a review of carabid research that even though there is little hope of combating pests biologically with ground beetles, their role in natural control of many invertebrates in certain ecosystems should be considered when developing integrated control programs and carabids could be imported in certain classical biological control programs.

One of the few classical biological control attempts involving carabids was the introduction of *C. sycophanta* from Europe (mostly Switzerland) between 1905 and 1910. Of the 4000 *Calosoma* imported, 67% were released against the gypsy moth in the New England states, which resulted in *C. sycophanta* becoming established (Burgess, 1911). As Clausen (1956) stated, evaluating a predator of this type is difficult, but the abundance and general distribution of *C. sycophanta* in the areas heavily infested by the gypsy moth warrant the belief that it is one of the most important of the imported natural enemies of this pest as well as other introduced forest lepidopteran pests. From his behavioral and ecological field studies in New England on the abundance and movement of *C. sycophanta* in relation to gypsy moth populations, Weseloh (1985a, 1985b, 1987, 1990, 1993, 1997) correlated that a relatively low number of adult beetles can have a substantial impact on the moth populations. On the average, 40% of pupae present in his study area were destroyed by *Calosoma* larvae. According to Burgess (1911), a larva can kill over 50 fully grown gypsy moth larvae during a 2-week period, and the adults (which live for 2 to 4 years) in one season consume on the average about 150 large larvae. The slow numerical response of *C. sycophanta* is perhaps the result of its prolonged preoviposition period. Dusaussoy (1963) found that the adult *C. sycophanta* has a hibernal diapause and eggs are rarely produced the first year. Because none of our native *Calosoma* species are as arboreal in larval stages as *C. sycophanta* (Gidaspow, 1959), there is little competition from other predators for gypsy moth larvae and pupae.

Luff (1987) reviewed the literature dealing with carabids associated with agricultural insect pests that had concentrated on aphids in north European cereal fields. Since then, carabid research on other pests in a variety of crops has expanded. In alfalfa, carabids had some impact on aphids and weevils (Lester & Morrill, 1989), and the search pattern of *Calosoma affine* Chaudoir was altered by high populations of *Colias* and *Spodoptera* caterpillar populations (Wallin, 1991). In apple orchards, carabids and other ground-dwelling arthropod predators were found to prey on apple maggot larvae and pupae (Allen & Hagley, 1990). Eight carabid species taken from pitfall traps in an apple orchard gave positive serological reactions to antiserum against codling moth larvae (Hagley & Allen, 1988); and *Anisodactylus californicus* Dejean, *Harpalus pensylvanicus* DeGeer, and three *Pterostichus* species were determined to be capable of killing and consuming mature codling moth larvae (Riddick & Mills, 1994). *Pterostichus* species were found to dominate the carabid community in an unsprayed apple orchard in California (Riddick & Mills, 1996) and predation of tethered codling moth larvae was estimated to be 60% per night during the first generation of the codling moth in spring (Riddick & Mills, 1994). Bean fields managed by a low-input cropping system had a different complex of carabid faunas compared with fields managed by conventional methods, and the complex differed from year-to-year in both systems (Fan *et al.*, 1993). Chen and Willson (1996) documented the carabid community in soybean fields and Tonhasca (1993) reported variable response of carabid numbers in tilled versus nontilled soybeans grown both as a monoculture or an intercrop with corn. The cabbage root fly is susceptible to carabid predation (Varis, 1989; Finch & Elliot, 1992; Finch, 1996).

In cereal fields (mostly barley and wheat) infested with aphids, carabids varied in number and species depending on cultural and pesticide applications (Chiverton, 1987, 1988; Helenius, 1990; Basedow *et al.* 1991; Chiverton & Sothern, 1991; Anderson, 1992; Ekbom *et al.*, 1992; Lys *et al.*, 1992; Wallin *et al.*, 1992; Helenius *et al.*, 1995), but there is some evidence suggesting that aphids may not be an important food for many of the generalist carabids in cereal fields (Bilde & Toft, 1997). In corn fields, large carabids did not return to paraquat- and glyphosate-treated field areas until about 28 days after application; thus, lower rates of predation of early-season lepidopteran pests by these carabid species may occur in no-tillage corn fields that utilize herbicides (Brust, 1990). Low tillage and the presence of broadleaf weeds also increase the activity density of carabid beetles in corn fields, although the weeds adversely affected corn yield (Pavuk *et al.*, 1997). From four genera 14 carabid species were identified by immunoelectro-osmophoresis to have fed on wheat midge larvae (Floate *et al.*, 1990). In potato fields, 19 carabid species were collected in pitfall traps, but were generally lower in sprayed than in unsprayed plots for a few days after application. There was a differential feeding by carabids of aphids on plants and those that fall to the ground (Dixon & McKinlay, 1992). Kromp (1990) found up to 48 carabid species in potato fields, and the total carabids and number of species were always higher in biological (organic) farmed fields than in conventionally farmed plots; he named four species that can be considered bioindicators for biologically farmed agroecosystems. Hance (1987) placed three species of carabids in sugar beet field cages with different levels of *Aphis fabae*. He found that all three species produced important reductions in aphid populations and that carabid density was a major factor in predator efficacy. In Michigan onion fields, a *Bembidion* species preys on eggs and first instars of the onion maggot. In field cage studies, this species reduced maggot numbers by up to 57%; thus, it may be an important biocontrol agent (Grafius & Warner, 1989). The largest of four carabid species tested, *Pterostichus melanarius* (Illinger), consumed the greatest number

of carrot weevil eggs laid on carrot leaves (Zhao *et al.*, 1990).

Suborder Polyphaga

Among the 44 terrestrial families of suborder Polyphaga that contain predatory species, 14 have been involved in biological control studies of arthropod pests. That is not to imply that the other families do not contain important natural enemies, because all the predaceous species of Polyphaga have some influence in their ecosystems.

Cleridae   Cleridae is a large, mainly tropical family of brightly colored beetles. The best known genera are *Enoclerus* and *Thanasimus,* which are predators of scolytid bark beetles. The majority of species appear to be subcortical predators of boring beetles on woody plants, but *Phyllobaenus* is generally associated with Lepidoptera (Wingfield & Warren, 1968; Farrar, 1985) and *Trichodes* and *Lecontella* are associated with Hymenoptera (Krombein, 1967; Bitner, 1972). Other genera, such as *Necrobia* and *Corynetes,* are associated with carcasses and skins and can be pests of stored meats. Detailed biological information on clerids is largely confined to the better known species that are scolytid predators, although an example of a *Phyllobaenus* sp. parasitizing an ant–plant mutualism in the tropics (Letourneau, 1990) provides an interesting, if aberrant, example of clerid biology.

Thanasimus species occur throughout the Holarctic region and are important predators of *Dendroctonus, Ips,* and *Scolytus* bark beetles (Mills, 1983). *Enoclerus* species are also important predators of *Dendroctonus* bark beetles in western North America (Dahlsten, 1982). Adult clerids feed on adult scolytids as they aggregate to attack the host trees. Eggs are laid under the bark scales and the larvae enter the scolytid galleries to feed on the developing brood. Larger larvae are able to bore through the bark, but may also walk across the bark surface to move between galleries. Pupation occurs in cells in the bark toward the base of the trees. Clerids can overwinter as adults, as larvae, and as prepupae in pupal cells.

The consumption of scolytid adults by adult clerids varies from 0.5 to 2.5 individuals per day (Berryman, 1966b; Turnbow *et al.,* 1978; Schlup, 1987). Mean adult longevity for *E. lecontei* (Wolcott) is 70 days (Berryman, 1966b), for *Thanasimus dubius* (Fabricius) is 117 days (Mizell & Nebeker, 1982a), and for *T. formicarius* (Linnaeus) is 127 days (Schlup 1987). Mean fecundity of *E. lecontei* is 515 with individuals able to produce up to 1000 eggs (Berryman, 1966a) and that of *T. formicarius* is 1439 (Schlup, 1987). There appear to be three larval instars, although this may be variable because from two to five instars have been reported in the literature. The duration of development from egg to adult is 61 days for *E. lecontei* (Berryman, 1966b) and 70 days for *T. formicarius* (Schlup, 1987). The predation potential of the larval stages has been estimated as 44 *I. typographus* Linnaeus larvae for *T. formicarius* (Mills, 1985) and 12 *D. brevicomis* LeConte larvae for *E. lecontei* (Berryman 1966b). The adults and larvae are cannibalistic but can readily be reared in the laboratory (Lawson & Morgan, 1992). Adults require scolytid food for oviposition (freeze-stored scolytids are often acceptable) and larvae can be fed on almost any animal-based food such as stored product insect larvae (Mills, unpublished).

Clerid beetles use olfactory cues in host location (Davis *et al.*, 1983; Payne *et al.*, 1984). For example, both sexes of *T. dubius* are attracted by components of scolytid aggregation pheromones: frontalin used by *Dendroctonus frontalis* Zimmermann; ipsdienol used by *Ips* species; and alpha-pinene, a host tree volatile (Mizell & Nebeker, 1982b; Lindgren, 1993). There is geographic variability in the enantiomeric composition of ipsdienol in *I. pini* (Say) populations in North America and this is matched by variation in the response of local populations of *T. dubius* to enantiomeric mixtures of ipsdienol, which results in some prey populations escaping predation (Raffa & Klepzig, 1989; Herms *et al.*, 1991). Given a choice of several prey, *T. dubius* preferred *I. arulus* (Eichhoff) over *D. frontalis* (Mizell & Nebeker, 1982a).

Clerid beetles are often dominant members of the natural-enemy complexes of scolytids and must contribute significantly to the natural control of these pests. For example, Reeve (1997) provides evidence that populations of *D. frontalis* decline when the abundance of *T. dubius* is high. Similarly, *T. formicarius* has been estimated to cause from 5 to 53% mortality of *I. typographus* brood (Mills, 1985) and from 48 to 52% mortality of *Tomicus piniperda* (Linnaeus) brood (Gidaszewski, 1974). Similarly, Weslian and Regnander (1992) report a 45% reduction in *I. typographus* brood emergence from bolts caged with *Thanasimus formicarius*. The Palearctic *T. formicarius* and the Nearctic *T. dubius* are the only clerid species to have been used in classical biological control against bark beetles [although another clerid, *Callimerus arcufer* Chapin, was introduced into Fiji by Tothill *et al.* (1930) against the coconut moth without becoming established]. Unsuccessful releases of *T. formicarius* were made against *D. frontalis* in North America in 1894 and in 1980 (Miller *et al.*, 1987), but apparently *T. dubius* has become established in Australia against *I. grandicollis* (Eichhoff) (Berisford & Dahlsten, 1989; Lawson & Morgan, 1992). Both Berryman (1967) and Moeck and Safranyik (1984) discuss means of preserving and augmenting native clerid predators of scolytids.

Coccinellidae The 3500 or so lady beetles in the world are divided into seven subfamilies: Chilocorinae, Coccidulinae, Coccinellinae, Epilachninae, Hyperaspinae, Lithophilinae, and Scymninae. Most species are predaceous, except those in the Epilachninae (which are all phytophagous) and in the tribe Psylloborini of the Coccinellinae, which contains mostly mycophagous species; the feeding habits of the Lithophilinae are apparently unknown. The predaceous species often consume pollen or extrafloral nectar and honeydew when their prey are scarce (Carter & Dixon, 1984; Hagen, 1986). The bionomics of many coccinellids are similar at the tribal level and have been reviewed by Balduf (1935), Clausen (1940), Hagen (1962, 1974), and Hodek (1967, 1973, 1986). The taxonomic revisions of the Coccinellidae of Japan (Sasaji, 1971) and North America (Gordon, 1985) and revisions of Coccinellinae of the Palearctic and Oriental regions (Iablokoff-Khnzorian, 1982) and Australia (Pope, 1988) discuss the known prey range of various taxa. The prey spectra are fairly similar, at least at the generic level if not at the tribal level (Schilder & Schilder, 1928). In almost all predaceous species, the adult female must feed on its prey to produce eggs.

The form and placement of eggs are also similar at the generic level. There are almost always four larval instars that generally have many similar morphological characteristics at the tribal level. The adults and larvae of the predaceous species have mandibles with one or two teeth at the apex while the mandibles of Epilachninae and Psylloborini have three or more teeth apically (Savoiskaya & Klausnitzer, 1973; LeSage, 1991; Samways *et al.*, 1997). Many larvae in the Scymnini, Ortaliini, Hyperaspini, Coccidulini, Novini, Cryptognathini, Azyini, and Telsimiini have conspicuous waxy coatings and filaments (Pope, 1979). However, within genera such as *Scymnus* or *Hyperaspis*, there are some species that do not produce waxy filaments. Larvae feed on the same prey as adults, but early larval instars may not be able to penetrate the integument of the larger prey. Pupation occurs near the larval feeding site. Basically there are three types of pupae: those that pupate within their last larval exuviae (skin) that envelops the pupae except for a dorsal split (Chilocorinae, Novini); those where the last larval exuviae covers only the ventral surface of the body including the head (Hyperaspini); and those where the last larval exuviae is pushed back and covers the anal end of the pupae, which stands more erect on the substrate (most other species) (Phuoc & Stehr, 1974).

Many coccinellids in the temperate zones are univoltine, usually being active in the summer and dormant during the winter. However, in the Mediterranean-type climatic zones, the reproductive activity is limited to the spring, thus their dormancy spanning the estival, autumnal, and hibernal periods with no oviposition for at least 9 months; during dry summers many species are bivoltine, being active and producing young in the spring and fall and being dormant during the summer and winter. In the subtropics and tropics, coccinellids may have continuous generations. Dormancy occurs in the adult stage, usually as a reproductive diapause that can be induced by nutrition, photoperiod, temperature, or combinations of these (Hodek, 1973). *Hippodamia convergens* Guerin-Meneville is facultative in responding to adverse conditions and has exhibited all the preceding voltinisms in California (Hagen, 1962, 1974). Warm temperatures (>18°C) can offset the influence of short photoperiods during the fall and winter on *H. convergens* adults, which permits reproduction if enough aphids are available (K.S. Hagen, unpublished data). *Coccinella septempunctata* Linnaeus is more sensitive to photoperiods and is univoltine in northern latitudes, intermediate in middle latitudes, and multivoltine in southern latitudes. Thus, it is more obligatory in responding to physical conditions instead of biotic factors in its diapause induction (Hodek, 1973). In some coccinellids diapause is not expressed, [e.g., *Rodolia cardinalis* Mulsant, which just slows its development when temperatures are low (Quezada & DeBach, 1973)].

The searching sequence adult coccinellids use in locating their prey and ovipositional site is similar to that of other natural enemies and is composed of searching for suitable habitats (plants), searching on plants for prey, and capturing and accepting prey (Hodek, 1993). However, some captured prey may not be suitable for sustenance and reproduction and may even be detrimental (Hattingh & Samways, 1992). Apparently, vision plays a dominant role in adult coccinellids locating habitats because there are only a few cases where synomones (plant volatiles that influence host selection of entomophagous insects) have been shown to be involved [e.g., Kesten (1969) found *Anatis ocellata* (Linnaeus) adults more attracted to pine needle volatiles]. There also seems to be a response to the height of plants; thus general plant habitats such as trees (conifers or deciduous), shrubs, or herbs (Hodek, 1973, 1993; Mills, 1981) can be recognized as sources of essential or alternate prey.

The searching behavior of the adult coccinellids can be influenced by the variations in architecture and morphology of different plant species, and even between plant varieties. The search rate and success of finding psyllid nymphs by adult *Curinus coeruleus* (Mulsant) are influenced by the width of subleaflets of the leucaena trees (Da Silva *et al.,* 1992). Adults of *Coccinella septempunctata* Linnaeus and *H.* (*Adonia*) *variegata* (Goeze) searching a normal pea variety, fell off the plant nearly twice as frequently as they did on a leafless variety (Kareiva & Sahakian, 1990). A comparative study of the foraging of coccinellids on crucifers indicated a much greater influence of plant architecture on the rate of predation of cabbage aphids than of coccinellid species (Grevstad & Klepetka, 1992). The differences in the abundance and type of pubescence present can have an effect on coccinellids as well as on other natural enemies in their success of controlling arthropod pests; smaller cocci-

nellid species and younger instars are particularly sensitive (Obrycki, 1986). On glabrous leaves with a slippery wax layer, larvae could move only along the edge or on narrow protruding veins (Shah, 1982).

Adult prey searching on plants appears to be random until contact with a prey is made, although phototactic and geotactic responses may be involved. However, there is now some evidence that chemotactic and vision responses are also important. Colburn and Asquith (1970) found that *Stethorus punctum* (LeConte) adults were preferentially attracted to samples with European red mites in an olfactometer, indicating a distinct (volatile) kairomone is involved in detecting the prey. Obata (1986) used permeable gauze bags and nonpermeable transparent plastic bags to determine that aphidophagous *Harmonia axyridis* (Pallas) adults were attracted visually and olfactorily to both aphids and leaves before actual contact occurs. Visual and olfactory cues were found to be important for host location by adult *Chilocorus nigritus* (Fabricius), although not for larvae (Hattingh & Samways, 1995). *Diomus hennesseyi* Fürsch (a small scymnine) and *Exochomus flaviventris* Mader (a medium-sized chilocorine) are predators of the cassava mealybug in Africa; females of these two species spent more time searching on cassava leaves previously infested with cassava or citrus mealybugs than on clean cassava leaves, indicating that the wax and/or honeydew (both products of the mealybug) contained a contact kairomone; wax from the cassava mealybug arrested both coccinellid species, and honeydew excreted by cassava mealybug arrested *Exochomus* and inexperienced *Diomus* (van den Meiracker *et al.,* 1990), and Heidari and Copland (1993) concluded that mealybug honeydew was both a food and an arrestant for *Cryptolaemus montrouzieri* (Mulsant). Vision was shown to be involved in *C. septempunctata* adults; they detected prey (pea aphids) and a silver foil dummy about 1 cm away prior to physical contact (Stubbs, 1980). Nakamuta (1984), using a video recorder, determined that adults of *C. septempunctata bruckii* Mulsant visually oriented toward prey at a distance of 7 mm, and 8 mm away from a dummy aphid under light conditions; under dark conditions, they did not orient toward prey even at 2 mm, but quickly responded to prey on contact. Maxillae are involved in contact and capture of prey; amputation of the maxillary palpi decreased the number of prey captured or consumed (Nakamuta & Saito, 1985).

Once a prey has been contacted or eaten, an adult coccinellid, like many other hunting predators, changes its searching behavior from a direct or smoothly curved search path to a convoluted complex path (marked by slower speeds and increased number of turns); thus, the beetle tends to stay in the vicinity of the site where the first prey was consumed. This foraging behavior was observed for *Hippodamia convergens* adults by Rowlands *et al.* (1978).

For *C. septempunctata brucki* adults, the switch from an extensive search to a concentrated one is stimulated by the contact with a prey instead of the consumption of it (Nakamuta, 1986). The search for prey by adults of some *Hippodamia, Coccinella,* and *Coleomegilla* spp. can be arrested and concentrated in the field by the presence of artificial diets (containing sugars) that are sprayed on crops (Ewert & Chiang, 1966; Hagen *et al.,* 1970a). Leaf extract and wood extract from *Berberis vulgaris* (Linnaeus) enhanced oviposition of *C. septempunctata* when sprayed on cherry trees (Shah, 1983).

In the field, the number of aphidophagous coccinellids present is correlated with aphid population densities (e.g., Niemczyk & Dixon, 1988; Ferran *et al.,* 1991; Ives *et al.,* 1993). However, the numerical response of aphidophagous coccinellids is probably limited by satiation (Mills, 1982b), egg cannibalism (Mills, 1982b), reduced oviposition in the presence of conspecific larvae (Hemptinne *et al.,* 1992), and optimal foraging across ephemeral patches of prey (Kindlmann & Dixon, 1993).

It has long been known that coccinellid larvae conduct the same searching behavior as that described earlier for the adults, but often the forelegs as well as the head are involved in detecting prey (Fleschner, 1950; Banks, 1954, 1957; Dixon, 1959; Kaddou, 1960; Bänsch, 1964, 1966; Storch, 1976; Hunter, 1978; Stubbs, 1980; Carter & Dixon, 1982, 1984; Obata, 1986). Direct contact with a prey is not always necessary before attack because fourth-instar larvae of *C. septempunctata* can detect an aphid 0.7 mm away (Stubbs, 1980). In the presence of honeydew, larvae spend longer periods searching in the restricted search area than if honeydew were not present (Carter & Dixon, 1984; Heidari & Copland, 1993). A *C. septempunctata* larva can detect a chemical marker that it secreted on plants previously searched unsuccessfully, thus reducing the retracing of its own path (Marks, 1977). The ability to capture aphids varies with the relative sizes of predator and prey and the various escape strategies of the latter, including maintaining a low profile (flat sessile aphids), walking away or dropping from the plant on sighting a coccinellid (larger predators are more conspicuous), kicking or pulling away if an appendage is grasped, and "waxing" the predator with paralyzing substances secreted from the siphunculi (Dixon, 1958, 1970; Brown, 1972a; Wratten, 1976; Hajek & Dahlsten, 1987).

Some prey species are not only nutritionally unsuitable but also distasteful or even toxic to coccinellids, while some prey that are suitable for some species are not for others (Blackman, 1965; Iperti, 1965; Hodek, 1966b, 1973; Hukusima & Kamei, 1970; Hukusima & Komada, 1972; Rogers *et al.,* 1972; Hämäläinen *et al.,* 1975; Mills, 1981; Morales & Burandt, 1985; Obrycki & Orr, 1990; Hattingh & Samways, 1992; Hodek, 1993). An aphid species can be toxic in the spring but not in the fall (Takeda *et al.,* 1964).

Different morphs of the green peach aphid fed on by *H. convergens* influenced levels of egg production (Wipperfürth *et al.,* 1987). When the greenbug from two different resistant cultivars of resistant sorghums were fed to *H. convergens* larvae, the larval–pupal survival was reduced and the developmental time was extended compared with larvae fed greenbugs reared on a susceptible sorghum hybrid (Rice & Wilde, 1989). Only three generations at most of *Harmonia conformis* (Boisduval) could be reared on aphids, but when fed acacia psyllids, continuous generations were produced (Leeper & Beardsley, 1976). This may explain why several earlier introductions against several different aphid pests in the Hawaiian Islands failed. Larvae of the aphidophagous *H. axyridis* continuously reared on grain moth eggs or pea aphids led to conditioning in search behavior: only larvae that ate the same prey as they were reared on adopted intensive searching movements after feeding, and Ettifouri and Ferran (1993) concluded that using larvae reared on substitute prey in biological control systems may decrease the efficiency of released predators. Some plants synthesize toxins (e.g., glycoside) that are sequestered by certain aphids and can deter coccinellids from eating them (Hodek, 1966b; Nishida & Fukami, 1989). Colorful aphids are often not consumed by lady beetles (Johnson, 1907). When *Hippodamia* (*Adonia*) *variegata* larvae feed on *A. nerii* Fonscolombe infesting *Cionura erecta* (Linnaeus) (Asclepiadaceae), the resulting coccinellid adults are either brachypterous or apterous; if these deformed adults are fed aphids from a different species of host plant, the progenies are normal (Pasteels, 1978). Sequestering toxins from plants is not limited to aphids— coccids such as the cottony-cushion scale are toxic to vedalia beetles when the scales are infesting certain *Spartium* or *Genista* species (Balachowsky & Molinari, 1930; Poutiers, 1930). Plant toxins may have been involved in Thompson's (1951) failure to establish 13 coccinellid species to control two diaspine scales attacking Bermuda cedar. All species developed in cloth bags covering scales on tree branches; however, when the coccinellids were liberated, only one species became established, indicating that the coccinellids were more fastidious than formerly believed.

Biological Control with Coccinellidae Historically, a few species of coccinellids played key roles in the development of the three tactics now employed as part of the biological control strategy. The successful introduction of the vedalia beetle, *Rodolia cardinalis,* into California from Australia in 1888 controlled the cottony-cushion scale in 1 year, and provided the "breakthrough" for the development of the classical biological control tactic (Essig, 1931; Doutt, 1958, 1964; Caltagirone & Doutt, 1989). Since then, there have been many different exotic coccinellid species introduced against various homopterous pests around the world, but relatively few have become established. More coccidophagous species have been established than aphidophagous ones.

The failure of another lady beetle (the mealybug destroyer, *Cryptolaemus montrouzieri*), which was unable to survive inland California winters after its introduction from Australia, sparked the research to develop the first augmentation tactics by mass-producing the predator for release in citrus groves infested with mealybugs each spring (Smith & Armitage, 1920). During the 1920s and 1930s there were 16 insectaries that mass-cultured the mealybug destroyer for periodic colonization, being the first modern system that formed the basis of the augmentation tactic of biological control. There are still a few insectaries producing *Cryptolaemus*, and it remains one of the few coccinellid species that is mass-cultured commercially (Fisher, 1963).

Several native *Hippodamia* spp. responded to the spotted alfalfa aphid, (*Therioaphis maculata* Buckton) invasion in California, a phenomenon that led to development of the California integrated control concept during the late 1950s (Stern *et al.,* 1959). Before any parasitoids were widely established and spotted alfalfa aphid-resistant plantings became widespread, native predators essentially controlled this aphid during the spring and fall, although they were usually not sufficiently abundant during the summer to provide control. Therefore, by applying only a single application of a selective aphicide (instead of the accepted 3 to 10 treatments of broad-spectrum insecticides), *Hippodamia* and other generalist predators (mainly Heteroptera) survived, resulting in predator to prey ratios that gave control during the summer (Stern *et al.,* 1959; Smith & Hagen, 1966; Neuenschwander *et al.,* 1975). This integration of chemical and biological control agents is a conservation tactic commonly used today in biological control. The role of coccinellids in biological control of general insect and arthropod pests has been reviewed by Essig (1931), Iperti (1961), Hodek (1966a, 1967, 1970, 1973, 1986), and Hagen (1962, 1974); in control of aphids, by Iperti (1965), Hagen and van den Bosch (1968), van Emden (1972), Frazer and Gilbert (1976), and Frazer (1988a, 1988b); and in control of diaspine scales, by Drea and Gordon (1990).

Among the Coccinellidae there is a striking contrast between coccidophagous species that have frequently been successfully used in biological control and aphidophagous species that have in general been unsuccessful in biological control. From field observations on *Adalia bipunctata* (Linnaeus), Mills (1982b) suggested two factors that might account for this pattern. First, coccidophagous species tend to be small and feed continuously, while the aphidophagous species include some of the largest coccinellids that are characterized by long periods of inactivity due to satiation. Second, the aphidophagous species lay their eggs in batches on the surface of the host plant where they are subject to density-dependent cannibalism, whereas the coccidophagous species lay their eggs singly or in small groups often concealed beneath the prey where they escape the constraints of cannibalism. Hemptinne *et al.* (1992) considered that for progeny to be able to complete their development, aphidophagous coccinellids must synchronize oviposition with the early development of ephemeral aphid populations. They provided evidence that *A. bipunctata* avoids oviposition at later stages of aphid population development through an inhibition response to the presence of conspecific larvae, and argue that this response, in addition to satiation and cannibalism, may limit the effectiveness of aphidophagous coccinellids in biological control. However, Merlin *et al.* (1996) demonstrated a similar oviposition deterrent response for the coccidophagous species *C. montrouzieri*. Dixon *et al.* (1997) considered the success of coccidophagous coccinellids in biological control to be determined by their smaller generation time ratio, the generation time of the coccinellids relative to that of their prey. They provided supporting evidence from a comparative analysis of generation times of coccinellids, aphids, and coccids, indicating that although aphidophagous species have shorter generation times than coccidophagous species, the telescoping of aphid generations leads to a mean generation time ratio of 2.4 for aphidophagous species in contrast to 0.6 for coccidophagous species.

The coccinellid species that have been utilized in biological control are from the subfamilies Chilocorinae, Coccidulinae, Coccinellinae, Hyperaspinae, and Scymninae.

Chilocorinae is a subfamily divided into three tribes, the Chilocorini, Platynaspini, and Telsimiini. The Chilocorini has over 200 species in 18 genera with representatives in all parts of the world (Chapin, 1965b). Most species are coccid or mealybug feeders and often deposit single eggs under scale coverings. Five genera contain species that have been utilized in biological control: *Chilocorus, Cladis, Curinus, Exochomus,* and *Halmus* (which included some species formerly placed in *Orcus*).

Chilocorus is a large genus of about 70 species that mainly occur in Palearctic and Ethiopian regions and prey on 11 different genera of diaspine scales. Several species have been successfully introduced to control scales. *Chilocorus bipustulatus* (Linnaeus) from Israel was established in California against *Parlatoria oleae* (Colvee) (Huffaker & Doutt, 1965), and a strain from Iran gave substantial to complete biological control of *P. blanchardi* Targioni in Mauritania and Morocco (Iperti & Laudeho, 1969; Iperti *et al.,* 1970). It also became established in Niger (Stansly, 1984). After *C. cacti* (Linnaeus) from Cuba became established in Puerto Rico and caused the reported extinction of *Pseudaulacaspis pentagona* (Targioni-Tozzetti), the predator disappeared for years (Wolcott, 1960), but it still contributes to control of the scale (Cock, 1985). *Chilocorus circumdatus* (Schonherr) was introduced from South China in 1895 by Koebele into the Hawaiian Islands where it became established and feeds on coccids and diaspine scales (Leeper, 1976). *Chilocorus kuwanae* Silvestri has been introduced into the United States and China several times since 1895 (as *C. similis* Rossi or *C. kuwanae*), but has become established only in the Santa Barbara area (Smith, 1965). *Chilocorus kuwanae* from Korea was introduced into the Washington, DC area in 1984 against the euonymus scale, *Unaspis euonymi* (Comstock). After 3 years, the predator had greatly increased in number and drastically had reduced the scale populations on test plants (Drea & Carlson, 1987; Hendrickson *et al.*, 1991). *Chilocorus nigritis* (Fabricius) is an economically important biological control agent of red scale on citrus in southern Africa (Samways, 1984, 1989; Samways & Wilson, 1988). *Chilocorus nigritis* is apparently endemic to Reunion, Aldabra, or Madagascar, was inadvertently introduced into Kenya (Greathead & Pope, 1977) and southern Africa, and now has been successfully colonized in eastern Africa by utilizing nontarget scale insects grown on bamboo near citrus orchards. *Chilocorus bipustulatus* and *C. infernalis* Mulsant have been introduced into South Africa (Hattingh & Samways, 1991, 1992) and New Zealand (Hill *et al.*, 1993). *Chilocorus nigritis* has been released and established on islands in the Indian Ocean (Greathead, 1971; Rao *et al.*, 1971) and in the Hawaiian Islands (Leeper, 1976), and is established in Brazil (Samways, 1984) and in Oman (Kinaway, 1991).

The genus *Cladis* has only a single species, *C. nitidula* (Fabricius), which occurs on various West Indian islands and has been reported from Buenos Aires (Chapin, 1965b). This species was introduced into Puerto Rico against the coconut scale, but it was more effective as a predator of *Asterlecanium bambusae* (Boisduval) (Cock, 1985). *Curinus* is also a monotypic genus whose sole representative, *C. coeruleus* Mulsant, is endemic to the Neotropics, but has been widely distributed in the Pacific area against the leucaena psyllid (Waterhouse & Norris, 1987).

Exochomus metallicus (Korschefsky) and *E. quadripustulatus* (Linnaeus) have been introduced against coccids and successfully established in California. The former species is from Eritrea, Ethiopia, and is found only in the Santa Barbara area (Gordon, 1985); the latter species is found along the California coast on a variety of scale insects (Essig, 1931; Gordon, 1985). In Africa, *E. flaviventris* Mader has become one of the most common endemic coccinellids to prey on the exotic cassava mealybug (Kanika-Kiamfu *et al.*, 1993).

Halmus chalybeus (Boisduval) was sent to many locations in California from Australia in 1892 and into the Hawaiian Islands in 1874 by Koebele as *Orcus chalybeus* (Essig, 1931); it became established in the Santa Barbara area of California (Gordon, 1985) and in several Hawaiian Islands.

Eguis platycephalus Mulsant principally a predator of

bamboo scales, was established in Puerto Rico from Cuba (Cock, 1985). The tribe Telsimiini contains a few species in *Telsimia*, which are from Asia and Oceania. *Telsimia nitidae* Chapin was introduced from Guam to the Hawaiian Islands in 1936; in the Hawaiian Islands it preys on coconut scale and *Pinnaspis buxi* (Bouche) (Leeper, 1976).

Coccidulinae is a subfamily that contains some of the most important species used in classical biological control. Among its genera are *Rhizobius* and *Rodolia* of worldwide fame, and also *Azya* and *Pseudoazya*, which include a few species that have been successfully imported and established.

Azyini is primarily a Neotropical tribe composed of two genera, *Azya* with 14 species and *Pseudoazya* with 6 species (Gordon, 1980). The members of Azyini are all "scale" predators (armored, soft, and mealybugs). They deposit single eggs horizontally on or near their prey. *Azya oribgera* Mulsant was introduced into the Hawaiian Islands from Mexico by Koebele in 1908 and has been reported to feed on *Coccus viridis* (Green) and *Ferrisiana virgata* (Cockerell) (Leeper, 1976). *Azya oribgera* also is established on Guam (Chapin, 1965a) and was accidentally introduced into Florida (Woodruff & Sailer, 1977). It became established in Bermuda in 1957 and achieved good control of *Pulvinaria psidii* Maskell and apparently maintains itself in spite of the heavy predation on the coccinellids by tree lizards (Cock, 1985). *Azya luteipes* Mulsant is difficult to distinguish from *A. orbigera* and other species with two dorsal spots except by examination of male genitalia. It occurs naturally along the coastal region of South America from Bahia, Brazil, to Argentina [the references to it occurring in Florida and the Hawaiian Islands are actually *A. orbigera* (Gordon 1980)], and it was collected in Brazil and introduced into Puerto Rico in 1939 to prey on bamboo scale (Bartlett, 1940). Possibly unrecorded or accidental introductions were made into Surinam, Guiana, and Dominica (Gordon, 1980). *Pseudoazya trinitatis* (Marshall) is native to northern South America and probably Trinidad, but was introduced and established in Puerto Rico, the Virgin Islands, and possibly Saipan; the introductions into Florida, Fiji, Vaté Island, and New Hebrides failed (Gordon, 1980). Although it was introduced mainly as a predator of the coconut scale in Puerto Rico, it has become most effective against *Pseudaulacaspis pentagona* (Targioni-Tozzetti), *A. bombusae*, and other scales (Cock, 1985). It has also been recorded as a predator of coconut scale in Grenada, Barbados, and Jamaica; and was established on the latter two islands from shipments from Trinidad (Cock, 1985).

The tribe Coccidulini comprises about 170 species in some 28 genera; about 100 species are Australasian (Pope, 1979). *Rhizobius* spp., which prey mostly on coccids, is the only genus in the tribe that has been utilized thus far in classical biological control. *Rhizobius forestieri* (Mulsant) almost always was misidentified as *R. ventralis* (Erichson) after its introduction into California from Australia in 1889 by Koebele (Pope, 1981). *Rhizobius forestieri* readily preyed on black scale in California (Bartlett, 1978), and has been recorded to attack the same scale, four mealybug species, an aphid species, and *Spodoptera* eggs in Hawaii after its introduction from Australia (via California) by Koebele in 1894 (Leeper, 1976). It became established in Italy and Sicily against *Chrysomphalus dityospermi* (Morgan) in 1908 by Silvestri, and in South Africa and New Zealand effectively attacks scales on *Araucaria* and Norfolk Island pine (Bartlett, 1978; Pope, 1981, Cameron *et al.,* 1989). *Rhizobius forestieri* has been successfully established in Greece as a predator of black scale on olive (Katsoyannos, 1984), and in 1987 it became established on an island off southern France and effectively reduced black scale on clementine trees (Iperti *et al.,* 1989). Richards (1981) lists 14 coccid species as prey of the adult, 9 of which are prey of adult beetles and larvae in Australia. *Rhizobius ventralis* is more specific; it voraciously feeds on *Eriococcus coriaceus* Maskell and rejects a variety of coccid pest species. This narrow prey specificity, coupled with an inability to either produce more generations than their prey or adjust to new habitats, were some of the reasons for the failure or partial failure of *R. ventralis* and *R. forestieri* as biological control agents in other parts of the world (Richards, 1981). The original description of *R. lophanthae* (Blaisdell) (formerly *Lindorus lophanthae*) was based on specimens collected in California (where it had been introduced and established in 1892 as a predator of black scale), but it is mainly a predator of diaspine scales in its native Australia and where it has been established elsewhere (Essig, 1931; Greathead, 1973). It was introduced into the Hawaiian Islands from California in 1894 where it has been recorded to prey on five diaspine species and black scale (Leeper, 1976). It has become established in Argentina and Italy on white peach scale (Waterhouse & Norris, 1987). It became established in Bermuda in 1949, but in spite of its heavy predation on Bermuda cedar scales, the cedar trees continued to decline; it also attacked palmetto scale (Bennett & Hughes, 1959). Among the many coccinellids imported into South Africa in 1900, only *R. lophanthae* became established and became an important addition to the predator complex on red scale (Greathead, 1971). During the 1920s it also was introduced into Mauritius where it attacked coconut scale, but because it was not very effective, other natural enemies were sought and imported after 1937 (Greathead, 1971). Actually, *R. lophanthae* has become established in every exotic location where it has been released except perhaps Chile (Greathead, 1973). In Peru it was introduced as a predator of citrus round scale, *Selanaspidus articulatus* (Morgan), but it also readily fed on citrus white scale, *Pinnaspis aspidistrae* (Sing) (a female *R. lophanthae* eats an average of 450 scales) (Marin, 1983). *Rhizobius pulchellus* Montrouzier from New Caledonia was accidentally

introduced to Vanuatu (New Hebrides) in 1964, where serious outbreaks of coconut scale in coconut plantations were occurring on Vaté Island; the predator quickly reduced the scale population, while *Cryptognatha nodiceps* Marshall, which successfully controlled coconut scale in the Fiji Islands, failed (Cochereau, 1969). Waterhouse and Norris (1987) consider *R. satelles* Blackburn the species involved in the preceding project, and note that *R. satelles* has been incorrectly referred to, or misidentified as, *Lindorus lophanthae, R. lophanthae, R. pulchellus,* and *R. nigrovatus*. It is common in Australia, New Zealand, New Caledonia, and South Africa; and has become widely dispersed in Vanuatu. *Rhizobius satelles* was introduced to Fiji, Tahiti, and Wallis Island, where it became successfully established against *A. destructor* (Waterhouse & Norris, 1987).

The tribe Noviini has four genera and includes 73 species that are mainly found in the African–Asian–Australian regions (Gordon, 1972; Pope, 1979). Their main prey are Margarodidae, principally *Icerya* spp. *Rodolia* has been the genus chiefly involved in biological control projects. The vedalia beetle, *R. cardinalis* Mulsant, is famous for saving the citrus industry in California during the late 1880s (Essig, 1931; Doutt, 1958; DeBach, 1964; Caltagirone & Doutt, 1989). Ever since its introduction into California from Australia, it has given complete control, with only an occasional upset of the vedalia due to extreme winters or detrimental pesticide use (Bartlett, 1978). The vedalia was imported by 57 different countries between 1889 and 1958, ranging from temperate or desert to tropical climates, and in 55 of these countries it has been reported not only to become established but also to produce good control of the cottony–cushion scale (Bartlett, 1978). In Bermuda, the vedalia introduction did not give complete control of *I. purchasi* because it was relatively inactive in the cooler months; and a significant increase in the scale population occurred during this period, but the introduction of *Cryptochaetum iceryae* (Diptera: Cryptochaetidae), an endemic Australian parasitoid, into Bermuda successfully controlled the scale during the cooler periods (Bennett & Hughes, 1959). Similarly, along the cooler northern California coast, *C. iceryae* is more effective than the vedalia in controlling the scale (Quezada & DeBach, 1973), and the same situation occurs along the cooler coast in South Australia (Prasad, 1989). The vedalia preys on *I. purchasi* on nearly all the plants the scale infests, but the scale is relatively immune to attack when it infests certain species in the genera *Ulex, Spartium,* and *Genista* (Poutiers, 1930; Bodenheimer, 1951; Priore, 1963); and is not readily attacked on *Coerulus laurifolius* (Quezada & DeBach, 1973). *Icerya purchasi* has been recorded attacking 40 species in 16 different plant families in just the Hawaiian Islands (Hale, 1970). The biological attributes that make *R. cardinalis* such an effective predator of *I. purchasi* in warm climates are a short generation time (three to one of the scale), a narrow prey specificity, a good voracity, and a high searching capacity (Bodenheimer, 1951; Quezada & DeBach, 1973). *Icerya aegyptiaca* (Douglas), an Asian species, was successfully controlled in Egypt by importing only six individual beetles that became established. *Icerya aegyptiaca* and *I. seychellarum* (Westwood) have also been controlled in the mountainous islands of Micronesia mainly by the introduction of *R. pumila* Weise, but the predator has been less successful on atolls, for after dramatically reducing populations of fluted scales it then became extinct (Schreiner, 1989). *Icerya monterratensis* Riley & Howard in Puerto Rico and in Ecuador and *I. palmeri* Riley & Howard in Chile were also successfully controlled by introducing *R. cardinalis* (Bartlett, 1978). *Rodolia iceryae* Jensen from South Africa did not become established when released in Israel (Mendel & Blumberg, 1991), but is considered to have potential for inoculative releases against *I. pattersoni* Newstead on coffee in Kenya (Kairo & Murphy, 1995).

Coccinellinae is a subfamily that at present is made up of six tribes: Bulaeini, Coccinellini, Discotomini, Psylloborini, Singhikalini, and Tytthaspini. Only the tribe Coccinellini in the Coccinellinae contains species that are of importance to applied biological control, because Bulaeini (*Bulaea*) is essentially phytophagous (with multidentate mandibles apically like Psylloborini), Psylloborini and Tytthaspini are mycophagous, and Discotomini and Singhikalini apparently have not been involved in any applied biological control programs. Coccinellini today consists of 86 genera with over 600 species that are essentially predaceous; most are aphidophagous in the Northern Hemisphere, but it appears that in the Australian, Ethiopian, and Neotropical regions the genera have species more closely associated with other Homoptera, particularly psyllids and coccids.

Most species deposit their eggs vertically in groups near their prey. On hatching, larvae remain clustered together on their eggshells for up to 1 day and often attack the unhatched eggs. This sibling cannibalism is often beneficial for the one egg eaten (which may not have hatched), nearly doubles the life of a newly hatched larva, and prolongs the search for aphids (Banks, 1956; Dixon, 1959; Brown 1972b; Dimetry, 1974; Kawai, 1978; Ng, 1986). Two or more coccinellid species have frequently been recorded from the same habitat, attacking the same aphid colonies, which suggested to Evans (1991) that there is little or no difference between intra-and interspecific interactions among coccinellids when two similarly sized individuals co-occur on a host plant. However, both Mills (1982b) and Osawa (1989, 1993) documented substantial cannibalism in the field, and Agarwala and Dixon (1992) demonstrated asymmetrical interspecific predation in the laboratory between *Adalia* and *Coccinella* species. The species diversity of immature aphidophagous coccinellids did not differ among South Dakota row crops but did vary among years within a 13-year period (Kieckhefer *et al.*, 1992).

At least 15 genera have been utilized in biological control projects. In recent years there has been greater success in introducing and establishing aphidophagous coccinellids in the New World.

Adalia bipunctata, a Palearctic species, has become widely distributed both intentionally and adventively in the temperate regions of the world and now occurs on all continents (Iablokoff-Khnzorian, 1982; Gordon, 1985; Pope, 1988). It was introduced and established in Chile in 1940 and now is found in Argentina (Zuniga, 1985). It is an arboreal predator of many different aphids [67 different species listed by Gordon (1985)] but also preys on psyllids and coccids (Herard, 1986). In France, releases of five adults per apple tree at 15-day intervals controlled *Dysaphis plantaginea* (Passerini) (Remaudiere *et al.*, 1973); and in Poland, *A. bipunctata* adults were able to suppress apple aphids at a ratio of up to 1:250 aphid nymphs, and the *Adalia* larvae at a ratio of 1 predator to 60 nymphs (Niemczyk *et al.*, 1974). On sycamore trees, there has to be more than two young sycamore aphids per 100 cm^2 of leaves for *A. bipunctata* to survive (Dixon, 1970). Ovipositing *A. bipunctata* laid eggs only when there were sufficient lime aphids on lime trees for survival of the first-instar larvae (11 small aphids per 100 cm^2 of leaf) but did not lay proportionately more eggs at higher densities (Wratten, 1973; Mills, 1982b). Ants can reduce the effectiveness of *A. bipunctata* controlling aphids on peach trees (Kreiter & Iperti, 1986). In France (Iperti & Kreiter, 1986) and Finland (Hämäläinen, 1976), *A. bipunctata* has one generation a year in the field, but in the New York area it has two or three (Obrycki *et al.*, 1983). Continuous generations could be cultured in the laboratory at photoperiods of at least 14h light:10h dark at 23°C for the New York populations and at LD 18:6 at 25°C for the Finland population.

Aiolocaria hexaspilota (Hope), an Asian species, was observed to feed on chrysomelid eggs and larvae infesting willow in Japan (Timberlake, 1943). It was introduced to the western (former) USSR to control the Colorado potato beetle and other chrysomelids [e.g., *Melasoma populi* (Linnaeus), but apparently without any great success (Iablokoff-Khnzorian, 1982)].

Aphidecta obliterata (Linnaeus) is a European species predaceous on adelgids (Iablokoff-Khnzorian, 1982). After the balsam woolly aphid, *Adelges piceae*, was accidentally introduced into North America about 1900, it became a serious pest of true firs in Canada and the United States (Amman, 1966; Clark *et al.*, 1971). Among the many different predaceous insects introduced against *A. piceae*, *Aphidecta obliterata* finally became established in Newfoundland associated with *Adelges cooleyi* (Gillette). Studies indicated that predation is evident only when there are more than 25 *A. cooleyi* per 4-in. twig; thus, even though it is established in Newfoundland, it is ineffective because it does not prey at low levels of *A. piceae* populations (Clark *et al.*, 1971). *Aphidecta obliterata* was introduced into North Carolina from Germany in 1960 and became established against *Adelges piceae,* but no attempts were made to determine its effectiveness in the field (Amman, 1966; Gordon, 1985).

Bothrocalvia pupillata (Swartz) occurs in India, Indonesia, and China (Iablokoff-Khnzorian, 1982). It was introduced into the Hawaiian Islands from Hong Kong in 1895 by Koebele (as *Coelophora pupillata*) where it now preys on *Toxoptera citricidus* (Kirkaldy), *Thoracaphis fici* (Takahashe), and *Selenothrips rubricinctus* (Giard) (Leeper, 1976). It also preys on certain psyllids and aleyrodids in the Hawaiian Islands (Funasaki, 1988).

Cleobora mellyi (Mulsant), an Australian species that is considered by some taxonomists to be in the tribe Psylloborini, is polyphagous, feeding on aphids, coccids, psyllids, and chrysomelids. It is considered an important predator of *Chrysophtharta bimaculata* (Olivier), a chrysomelid defoliator of eucalypts in Tasmania (Mensah & Madden, 1994), and has been established against a eucalypt-defoliating chrysomelid, *Paropsis charybdis* Stål, in New Zealand (Cameron *et al.*, 1989). However, *Cleobora mellyi* probably is more naturally a predator of psyllids on acacias (Pope, 1988).

Coccinella septempunctata Linnaeus (C-7) is the most common Palearctic coccinellid species. It was intentionally introduced into North America several times from 1956 to 1971, but these releases were unsuccessful (Angalet *et al.*, 1979), including releases of eggs and larvae (from a stock originally collected near Paris, France) in Maine potato fields over a 5-year period where the adults overwintered in screen cages in grasslands (Shands *et al.*, 1972d). However, a population was discovered in New Jersey in 1973 that may have resulted from an accidental introduction (Angalet & Jacques, 1975). Since 1973, C-7 has been released in every state and in southern Canada and has become widely established east of the Rocky Mountains but only marginally in the West (Gordon, 1985; Schaefer *et al.*, 1987; Gordon & Vandenberg, 1991). C-7 is a polyphagous predator but is essentially aphidophagous. Iablokoff-Khnzorian (1982) reviewed its biology and lists as food sources over 60 aphid species (including several aphids that are toxic), thrips, aleyrodids, psyllids, leafhoppers, coccids, moths, beetles (particularly some chrysomelids), pollen, and nectar. Herard (1986) noted from literature (over 50 references since 1978) that C-7 was recorded preying on 25 different aphid species infesting some 27 different crops as well as various other insect groups, including psyllids. Cultivated and uncultivated habitats in midwestern United States play an important role in supporting populations of *C. septempunctata* (Maredia *et al.*, 1992). The subspecies *C. septempunctata brucki* Mulsant occurs in Asia and was introduced

into the Hawaiian Islands from Okinawa in 1958 (Leeper, 1976). In Japan, it has an that estival diapause that is controlled by long days (LD 18:6 photoperiod) but its hibernal dormancy is not induced by short days (LD 12:12); these responses are opposite those of European C-7 populations (Hodek *et al.*, 1984). Releases of C-7 have been made to control aphids in greenhouses and potato fields. Hämäläinen (1977) released larvae of both C-7 and *Adalia bipunctata* at intervals of a few weeks on chrysanthemums and sweet peppers infested with *Myzus persicae,* which prevented aphid population growth and decreased aphid numbers. One C-7 first-instar larva per 50 aphids was effective on chrysanthemums while one *A. bipunctata* larva per 5 to 10 aphids on sweet peppers resulted in control, but larvae of neither coccinellid species were effective in controlling *Macrosiphum rosae* on roses. Shands and Simpson (1972) applied a spray mixture of C-7 eggs in water and 0.25% agar on potato plants to control aphids; none of several treatments provided satisfactory control of aphids, but larvae resulting from early-season applications of eggs gave the best control of the aphids. In earlier experiments, Shands *et al.* (1972b, 1972c) obtained 30 to 50% control of potato aphids by introducing eggs and young larvae of C-7. C-7 larvae have an impact on aphids infesting cereal crops in Finland (Rautapaa, 1972), and Ferran *et al.* (1986b) derived a formula that calculated the biomass of aphids consumed by each C-7 larva (regardless of instar) per square meter of wheat in France. Olszak and Niemczyk (1986) placed two C-7 adults (with hind wings removed) in caged small apple trees with either 10 or 20 *Aphis pomi* per tree; he noted 100% aphid mortality within 6 days for 10 aphids and 8 days for 20 aphids.

Coccinella undecimpunctata Linnaeus, a Eurasian species, was introduced into New Zealand in 1874 (Miller *et al.*, 1936; Cameron *et al.*, 1989). Since 1959, it has been found in Western Australia, and since 1970, in Tasmania (Pope, 1988). In North America, it was first discovered in 1912 in Massachusetts; now it is found in the northeastern United States, southern Canada, British Columbia, and the Seattle area of Washington State without any intentional action by man (Wheeler & Hoebeke, 1981; Gordon, 1985). It is recorded to prey on at least 12 aphid species, most of which are pests of agricultural crops (Iablokoff-Khnzorian, 1982).

Coelophora inaequalis (Fabricius) occurs in Australia, China, Indonesia, Malaysia, the Mariana Islands, Fiji, New Zealand, and the Philippines (Timberlake, 1943; Pope, 1988). It was introduced into the Hawaiian Islands from Australia, Ceylon, and China in 1894 by Koebele to control *Melanaphis sacchari* (Zehntner) (Leeper, 1976). It has been recorded to prey on nine aphid species and a psyllid in Hawaii (Funasaki *et al.*, 1988). *Coelophora inaequalis* was also successfully introduced into New Zealand (date unknown) (Pope, 1988) and into Puerto Rico from Hawaii to prey on *Sipha flava* (Forbes) (Clausen, 1978), and may be established in Florida (Gordon, 1985).

Coleomegilla maculata (DeGeer) occurs from Canada to Argentina and has been divided into at least eight subspecies by Timberlake (1943). Three subspecies occur in North America (Gordon, 1985). Even though *C. maculata lengi* Timberlake is aphidophagous, it can complete its larval development when fed only corn pollen (Smith, 1960). It readily feeds on European corn borer eggs (Andow & Risch, 1985; Phoofolo & Obrycki, 1997) and does not prefer aphids over Colorado potato beetle larvae; in field cage studies, *C. maculata* adults significantly reduces populations of Colorado potato beetle eggs and small larvae (Groden *et al.*, 1990). Adult *C. maculata* aggregate in maize and cotton plots when sprayed with a yeast–sugar mixture (Nichols & Neel, 1977). Of four coccinellid species tested for aphid control under greenhouse conditions, Gurney and Hussey (1970) found *C. maculata* and *Cycloneda sanguinea* (Linnaeus) to be the most voracious in feeding on aphids, but *Coleomegilla maculata* larvae, because of glandular hairs on cucumber foliage, could not remain long enough to be effective. One second- or third-instar *Cycloneda sanguinea* larva released against 20 *Aphis gossypii* on greenhouse cucumbers almost eliminated the aphids after 16 and 18 days, respectively (Hussey & Bravenboer, 1971).

Harmonia axyridis (Pallas), an Asian arboreal aphid-feeding species, was released in California, the Hawaiian Islands, eastern and southern United States, and Nova Scotia (Timberlake, 1943; Iablokoff-Khnzorian, 1982; Gordon, 1985; McClure, 1987; Tedders & Schaefer, 1994); France (Ongagna *et al.*, 1993); and Greece (Katsoyannos *et al.*, 1997). It has been recovered in Louisiana and Mississippi (Chapin & Brou, 1991), in the southeastern states (Gordon & Vandenberg, 1991; Tedders & Schaefer, 1994), and in the western states (Dreistadt *et al.*, 1995; Lamana & Miller, 1996). Chapin and Brou (1991) listed over 14 aphid species recorded as prey, and McClure (1987) reared *H. axyridis* on a margarodid scale. A life table was developed by Osawa (1992). Hukusima (1971) released 30 adult *H. axyridis* three times in one season in apple tree plots treated with various selective pesticides; the addition of the predators held aphids at lower levels over a longer period compared with the control plots. The cotton aphid was controlled in Chinese cotton fields by a predator to prey ratio of 1:100 to 120; interplanting maize with cotton increased coccinellid populations compared with cotton alone (Dong, 1988). In France, Ferran *et al.* (1996) used releases of 50 larger larvae per four rose bushes to reduce rose aphids to comparable levels achieved with insecticides. Similarly, Trouve *et al.* (1997) found that an early release of 50 second- and third-instar larvae per plant reduced aphid abundance on hops by 72%, keeping them below the insecticide treatment threshold.

Harmonia conformis (Boisduval) is widely distributed in Australia, Tasmania, and New Zealand (Pope, 1988; Cameron *et al.*, 1989). It has been released in California and Hawaii (often under the name *Leis conformis*) several times as an enemy of aphids. Koebele introduced it into Hawaii 1894 and 1904, but it only became established in 1973 when it was released against the acacia psyllid, *Acizzia uncatoides* (Ferris & Klyver), and dramatically reduced the pest population on the big island of Hawaii (Leeper, 1976; Leeper & Beardsley, 1976). Thousands of *H. conformis* were also released in coastal California against the acacia psyllid, but the species never became established (Gordon, 1985). In Australia, the predator feeds mostly on acacia psyllids and only switches to aphids when psyllids are scarce (Hales, 1979).

Harmonia dimidiata (Fabricius), an Asian species (Iablokoff-Khnzorian, 1982), was collected in south China and released in California in 1925 as *L. conformis* against citrus aphids, but it never became established. However, a shipment of beetles from California to Florida in 1925 resulted in establishment there (Clausen, 1978; Gordon, 1985).

Harmonia quadripunctata (Pontopiddian) is an Old World species associated mostly with conifers (Iablokoff-Khnzorian, 1982). It never was intentially introduced to North America, but Vandenberg (1990) discovered scattered specimens in various museums dating back as early as 1924. The present distribution in North America appears to be in an area near the Hudson River in the vicinity of the New Jersey–New York border. A key to the three North American *Harmonia* spp. is shown in Gordon and Vandenberg (1991).

Hippodamia (Hippodamia) convergens Guerin is the most abundant and widespread *Hippodamia* species in North America (Gordon, 1985). *Hippodamia* spp. appear to be polyphagous, based on the variety of arthropods recorded as prey in the 195 references cited by Vanndell and Storch (1972). However, aphids are the main prey of *H. convergens,* and, in fact, are an obligatory requirement for optimal fecundity. *Hippodamia convergens* prefers low vegetation and is primarily univoltine in western North America where it migrates to mountains to spend 9 months (June to February) in aggregations. However, if aphid populations are high enough in late spring (more than two pea aphids per stem in alfalfa), the newly emerged *H. convergens* adults will not readily migrate to the mountains, but will remain and produce subsequent generations in the valleys (Smith & Hagen, 1966; Hagen, 1974; Neuenschwander *et al.*, 1975). Also, artificial diets containing hydrolyzed yeast and sugar applied on crops (alfalfa) in the valley when aphid populations are low can arrest, retain, and induce oviposition (Hagen *et al.*, 1970a). They remain reproductively dormant while in aggregations but mate just before leaving aggregation sites in February. They fly skyward to a flight ceiling of 13°C and are blown westward to the valleys by northeasterly winds (Hagen, 1962). The migratory flight from the overwintering sites in the mountains is necessary before *H. convergens* begins to search for aphids. The propensity to fly is associated with hormonal control and is correlated with ovarian development (Rankin & Rankin, 1980a, 1980b; Davis & Kirkland, 1982). Thus, relatively few *H. convergens* collected from overwintered aggregations and released will remain and immediately search for aphids; the beetles will react as they would when leaving their overwintering sites (Davidson, 1924; DeBach & Hagen, 1964; Kieckhefer & Olson, 1974). The few (ca. 10%) overwintered beetles that do not fly far when released are probably weakened by the parasitism of *Dinocampus coccinellae* (Shrank) (Ruzicka & Hagen, 1985). However, insectary-reared adults, or adults from aggregation sites that have been allowed to fly for 7 to 10 days before release, disperse more slowly and can provide short-term reductions of 25 to 84% in aphid populations (Dreistadt & Flint, 1996). If the elytra of *H. convergens* are glued together so they cannot fly, they can reduce green peach aphid numbers in sugar beet plots four times more than in control plots; they will also deposit more eggs than in plots where beetles able to fly were released (Tamaki & Weeks, 1973). By removing the hind wings of *H. convergens,* the adults ate 22 times more *Heliothis zea* eggs than beetles free to fly (Ignoffo *et al.,* 1977). Regulation of greenbugs in Texas high plains grain sorghum is largely due to the suppressive action of *Hippodamia convergens* and *H. sinuata* (Mulsant); the impact of the *Hippodamia* spp. on greenbugs was determined by exclusion and introduction of beetles (Kring *et al.,* 1985). *Hippodamia convergens* has also been utilized in classical biological control. It was sent to at least 11 countries for release, but only became established in Peru in 1937 (Clausen, 1978), and later in Chile and Argentina (Zuniga, 1985) and Venezuela (Szumkowski, 1961). After five attempts to establish it in the Hawaiian Islands, it finally was found in 1964, but only at the higher altitudes of Maui and Hawaii (Leeper, 1976).

Hippodamia (Adonia) variegata (Goeze) is a Palearctic aphidophagous species absent in Japan, but extends south to India and in Africa to Kenya (Iablokoff-Khnzorian, 1982); it also has been reported in South Africa (Greathead, 1971). This species has been introduced into North America many times between 1957 and 1981 without successful establishment (Gordon, 1985; Gordon & Vandenberg, 1991); however, in 1984, Gordon (1987) reported *H. variegata* to be established in Quebec, apparently as an adventive species, and since then, it has been recorded from the eastern United States and the maritime provinces of Canada (Wheeler, 1993; Wheeler & Stoops, 1996). This Canadian stock of *H. variegata,* as well as populations from France, India, Morocco, and the former USSR, has been cultured and released widely against the Russian wheat aphid in the United States but apparently has not become established in the new release areas (Gordon, 1985; Schaefer & Dysart,

1988; Gordon & Vandenberg, 1991). However, *H. variegata* from South Africa not only became established in Chile in 1975, but also now occurs in Argentina and Brazil; in these countries, it has played an important role along with imported parasitoids in a very successful biological control program against several introduced cereal aphids (Zuniga, 1990).

Hippodamia (Semiadalia) undecimnotata (Schneider) is another Palearctic species that is aphidophagous, but it also readily preys on alfalfa weevil larvae in the (former) USSR (Iablokoff-Khnzorian, 1982). It has been cultured and released in the western United States against the Russian wheat aphid, but thus far no establishment has been recorded (Gordon & Vandenberg, 1991).

Lemnia biplagiata (Swartz), an Oriental species, was introduced into Taiwan from Indochina in 1925 (Clausen, 1978), and Iablokoff-Khnzorian (1982) mentions that it was introduced from Vietnam in 1927, but was already in Taiwan. Semyanov and Bereznaja (1988) concluded that this coccinellid is suitable for use as a biological agent for controlling aphid populations in greenhouses. *Lemnia saucia* Mulsant (= *calypso* in China, *L. mouhoti* in Laos, and *L. swinhoei* in Taiwan), which occurs in India, Japan, and the Philippines, was also collected in Indochina, released into Taiwan against aphids on sugarcane, and became established there (Rao *et al.,* 1971; Clausen, 1978).

Menochilus sexmaculatus (Fabricius), an Oriental species common in forests and gardens, is largely aphidophagous but also preys on psyllids, mealybugs, coccids, and spider mites (Iablokoff-Khnzorian, 1982; Cartwright *et al.,* 1977). It was imported from Pakistan and released for 2 years against the greenbug in Oklahoma; it reproduced and remained in the field, but apparently could not overwinter there (Cartwright *et al.,* 1977).

Oenopia conglobata (Linnaeus) is a Palearctic aphidophagous species but also feeds on coccids and immature chrysomelids, as well as on pollen and nectar (Iablokoff-Khnzorian, 1982; Herard, 1986). It was released several times in North America from 1957 through 1982 (Gordon, 1985) but had not become established by 1990; however, a Russian strain is still being released in the western United States against the Russian wheat aphid (Gordon & Vandenberg, 1991).

Olla v-nigrum (Mulsant) (= *abdominalis*) is a New World species extending from Canada to Argentina. It is primarily arboreal and in North America preys particularly on aphids infesting *Juglans* trees (Timberlake, 1943; Sluss, 1967; Gordon, 1985; Vandenberg, 1992); in the tropics, psyllids are its main prey (Vandenberg, 1992). Extensive biological studies of *O. v-nigrum* were made in France before releasing it in Europe, and it was compared with some other coccinellids such as *Adalia bipunctata, Harmonia axyridis, H. conformis,* and *Hippodamia convergens.* It was concluded that *O. v-nigrum* would be a valuable addition to the Mediterranean region as an aphid predator (Kreiter & Iperti, 1984; Iperti & Kreiter, 1986), while Ferran *et al.* (1986a) concluded that *Harmonia axyridis* and *O. v-nigram* could be used together. In the Hawaiian Islands, *O. v-nigrum* was introduced from Mexico in 1908 (Leeper, 1976); in these islands it is recorded to prey on aphids, psyllids, and aleyrodids (Funasaki *et al.*, 1988) and is an important predator of the leucaena psyllid, *Heteropsylla cubana* (Crawford) (Waterhouse & Norris, 1987). *Olla v-nigrum* was introduced into Guam sometime before 1952 (Chapin, 1965a) and was introduced from Tahiti to New Caledonia in 1987 against *H. cubana,* which had invaded that area around 1985. By 1989, *O. v-nigrum* was present in most parts of New Caledonia and was an important predator of the psyllid (Chazeau *et al.,* 1991). These authors developed a life table for the coccinellid and found the psyllid to be a highly suitable prey, though toxic to other polyphagous predators.

Propylea quatuordecimpunctata (Linnaeus), a common aphidophagous Palearctic species, has been recorded to prey on at least 20 species of aphids, mainly on low vegetation in the Old World (Iablokoff-Khnzorian, 1982; Herard, 1986). Before releasing it in Oklahoma, seven aphid species associated with small grains were tested as prey in the laboratory. The beetles survived on all seven species, but the reproductive level varied (Rogers *et al.,* 1972). Attempts to establish it in the United States from 1971 to 1982 were unsuccessful (Gordon, 1985). However, in 1968, it was discovered in Quebec (Chantal, 1972), predating any intentional releases in North America. Later it was found from the vicinity of Montreal, south along the St. Lawrence River to northern New York, Maine, Vermont, and Massachusetts; all these were considered as unintentional establishments (Schaefer & Dysart, 1988; Wheeler, 1990, 1993; Gordon & Vandenberg, 1991). Populations of *P. quatuordecimpunctata* from Canada, France, and Turkey did not differ significantly in either life table parameters or in critical thresholds for diapause induction (Obrycki *et al.,* 1993).

Hyperaspinae is a subfamily composed of approximately 500 species divided into 14 genera (Pope, 1979). Duverger (1983) elevated the tribe Hyperaspini to subfamily status, removing it from the subfamily Scymninae, with most species belonging to *Cyra, Hyperaspis,* or *Brachycantha.* According to R. D. Gordon (personal communication) of the U.S. National Museum most South American species formerly described under the taxa *Hyperaspis* or *Cleothera* in reality belong to Mulsant's genus *Cyra.* The majority of the hyperaspine species attack the more or less sedentary homopterans such as coccids and mealybugs. Gordon (1985) lists 75 different species of prey that El-Ali (1972) recorded associated with 60 hyperaspine species (mostly *Cyra* and *Hyperaspis*). *Brachycantha* spp. apparently prey on coccids, mealybugs, and aphids that infest plant roots and are attended by ants (El-Ali, 1972).

The eggs of *Hyperaspis* are usually flattened and deposited singly near their prey in irregularities of the bark or in growth rings of twigs but not under scale coverings of coccids (El-Ali, 1972). However, *Cy. notata* (Mulsant) (as *Cleothera notata*) and two other *Cyra* spp. deposit ovoid-shaped eggs inside the ovisacs of mealybugs; their larvae feed on the eggs and pupate at the base of infested cassava plants (Sullivan *et al.,* 1991), but there are other species that pupate on leaves. There are usually four larval instars, but McKenzie (1932) reported that the fall generation of *H. lateralis* Mulsant passes through only three instars (the spring generation passes through four). The few hyperaspine species that have been utilized in applied biological control belong to either *Cyra* or *Hyperaspis.*

Cyra notata and *C. raynevali* Mulsant (as *Hyperaspis* or *Cleothera*) are Neotropical species successfully introduced from South America and established in Africa against the cassava mealybug *Phenacoccus manihoti* Matile-Ferro (Kiyinidau & Fabres, 1987; Löhr & Varela, 1990; Herren & Neuenschwander, 1991; Reyd *et al.,* 1991; Kanika-Kianfu *et al.,* 1992). The biology of *Cyra notata* (= *H. notata* Mulsant) has been elucidated by Sullivan *et al.* (1991) in Columbia and by Dreyer *et al.* (1997a, 1997b) in Nigeria. The biology of *C. raynevali* (= *H. jucunda* Mulsant) was described by She *et al.* (1984) in southern Nigeria and by Kiyindou and Fabres (1987) and Reyd *et al.* (1991) (the latter utilizing *Planococcus citri* Risso as a factitious prey) in the Congo. *Cyra pantherina* (Fürsch) [= *Cyra jocosa* (Mulsant)] occurs from Mexico to Brazil. It was introduced into the Hawaiian Islands from Mexico in 1907 by Koebele as a predator of *Orthezia insignis* Browne, and later was sent from Hawaii to Kenya, where it was reared in large numbers to control *O. insignis* attacking jacaranda trees; the establishment of the predator in Tanzania and Uganda saved the jacarandas and *Orthezia* is no longer a problem (Greathead, 1971). It has also been successful in the control of the same scale on the island of St. Helena (Booth *et al.,* 1995). *Cyra trilineata* (Mulsant) keeps the pink sugarcane mealybug, *Saccharicoccus sacchari* (Cockerell), in check in Venezuela (Box, 1950) and Barbados (Coles, 1964), and also occurs in French Guiana. *Cyra trilineata* (as *Hyperaspis*) was released and established on several Caribbean Islands, where it contributes to the control of *S. sacchari* (Cock, 1985). The coccinellid also was sent to Puerto Rico in 1934 and 1935, but apparently did not become established; however, Wolcott (1937) believed that it became established because the sugarcane mealybug population was reduced and no outbreaks had occurred by 1937.

Hyperaspis binotata (Say) is a common eastern North American species (Gordon, 1985). Arboreal soft scales are its main prey; it is an effective natural enemy of cottony maple scale [*Pulvinaria innumerobilis* (Rathvon)], terrapin scale [*Mesolecanium nigrofasciatum* (Pergande)] (Simanton, 1916), and pine tortoise scale [*Toumeyella parvicornis* (Cockerell)] on jack and Scotch pine (McIntyre, 1960). A number of *Hyperaspis* spp. have been successfully introduced into new areas. *Hyperaspis campestris* Herbst in the former USSR is a successful biological control agent of many scale insects on grapes, citrus, and other subtropical plants. In (former) USSR tea plantations, *Pulvinaria flocifera* (Westwood) can be a serious pest, but releases of about 400 adult *H. campestris* per acre reduced the pest population below the economic level in 2 years (Bogdanova, 1956). *Hyperaspis conviva* Casey, an eastern North American species (Gordon, 1985), can control pine tortoise scale on jack pine in Manitoba, but the scale populations persist and increase if the scale is attended by *Formica obscuripes* Forel (Bradley, 1973). *Hyperaspis maindroni* Sicard from New Guinea became established in India to prey on the hibiscus mealybug (Rao *et al.*, 1971). It was either *H. limbalis* Casey, which occurs at San Diego, California, or *H. inflexa*, which occurs from California to New Jersey, that was introduced into the Hawaiian Islands from California by Koebele in 1906 (Leeper, 1976). These two *Hyperaspis* spp. can only be identified by examining the male genitalia (Gordon, 1985); thus it is not certain which species was introduced and became established to prey on mealybugs. It is recorded to attack *Trionymus insularis* Ehrhorn (Leeper, 1976; Funasaki *et al.*, 1988). Besides several *Cyra* spp. from South America introduced against the cassava mealybug in Africa, there are 32 endemic species of Coccinellidae recorded among the 120 arthropods associated with *P. manihoti*, including five *Hyperaspis* spp. (Neuenschwander *et al.*, 1987). *Hyperaspis senegalensis hottentotta* Mulsant has been the most closely studied endemic *Hyperaspis* species preying on the cassava mealybug (Fabres & Kiyindou, 1985; Kiyindou *et al.*, 1990). It was introduced into California and persisted only 1 or 2 years as a predator of ice plant scales. *Hyperaspis signata* (Olivier), a southeastern species (Gordon, 1985), has a similar biology to *H. binotata* and is an important natural enemy of the pine tortoise scale. It was introduced into Michigan from Minnesota in early 1930 and by October only 8 trees in a plantation of 690 had scale present, compared with 78 trees in May (Orr & Hall, 1931). *Hyperaspis silvestrii* Weise, a Mexican species, was introduced and established in the Hawaiian Islands in 1922 as a predator of avocado mealybug and the coconut mealybug (Leeper, 1976) and in the Philippines on the pineapple mealybug (Bartlett, 1978).

Scymninae is a subfamily composed of seven tribes, with the Cryptognathini, Scymnini, and Stethorini being the main ones that include species that have been utilized in applied biological control. These small pubescent coccinellids are rather prey specific (at least in their larval stages) to the generic level.

The tribe Cryptognathini is composed of approximately 50 species in eight genera, but only a few species in the genera *Cryptognatha* and *Pentilia*, which are Neotropical (Pope, 1979), will be discussed here. Their oval eggs are deposited beneath the covering of the scale insect after the scale body has been eaten by the coccinellid adult. The larvae bear long, white, waxy tufts that project from their sides. When pupating, the larval skin is pushed downward and forms a base upon which the naked pupa rests (Taylor, 1935; Pope, 1979).

Cryptognatha nodiceps Marshall is endemic to Trinidad and Guyana and preys on diaspine scales. It was imported into the Fiji Islands from Trinidad in 1928 to control the coconut scale, *Aspidiotus destructor* (Signoret). It was a remarkably spectacular success. In only 9 months after the first releases, the scale was under control in all the more important islands of Fiji; and in a further 9 months, the scale not only was controlled on every island of Fiji but also was difficult to detect (Taylor, 1935). A similar outstanding example of classical biological control was achieved in Principe by introducing *C. nodiceps* against the coconut scale (Simmonds, 1960). According to Greathead (1971) and Simmonds (1960), it was estimated that the production of copra in Principe Island had declined from 1400 tons per year before the scale invasion to 500 tons at the height of the infestation, amounting to an annual loss of $165,000. The problem was eliminated at a cost of $10,000 for the importation and release of *C. nodiceps*. During the 1930s, *C. nodiceps* was successfully introduced into Florida, Puerto Rico, the Dominican Republic, St. Kitts, and Nevis (Cock, 1985; Gordon, 1985; Waterhouse & Norris, 1987). However, *C. nodiceps* failed to control the coconut scale when it was introduced into Vanuatu (New Hebrides), but the scale was controlled by *Rhizobius satelles* (Chazeau, 1979; Waterhouse & Norris, 1987).

The 11 *Pentilia* spp. range from Mexico to Argentina. *Pentilia castanea* Mulsant was imported from Trinidad and was established in the Dominican Republic and Puerto Rico to prey on the coconut scale (Bartlett, 1940; Cock, 1985; Waterhouse & Norris, 1987).

Scymnini is the largest tribe in the Coccinellidae with over 600 described species, of which over 570 belong to *Diomus, Nephus*, or *Scymnus* (Whitehead, 1967; Pope, 1979; Gordon, 1985). Among the 22 other genera, only *Clitostethus, Cryptolaemus, Pseudoscymnus*, and *Scymnodes* include species to be mentioned here.

Clitostethus spp. were considered to occur only in the Old World, but Fürsch (1987) synonymized the New World genus *Nephaspis* Casey under *Clitostethus* Weise. *Clitostethus* larvae and adults are essentially predators of all stages of whiteflies. The beetles deposit their oval pale eggs horizontally among the aleyrodid nymphs on the underside of leaves. The larvae apparently do not produce a waxy covering, threads, or plates but are often coated by wax secreted by the whitefly nymphs; the pupae are naked and found on the leaves where the larvae had fed (Agekyan, 1978; Yoshida & Mau, 1985; Bellows *et al.*, 1992a, 1992b). *Clitostethus arcuatus* (Rossi), a Palearctic species, preys on several different aleyrodid species including the greenhouse whitefly. It moves easily on a glabrous plant surface, but larvae become trapped by foliar pubescence (Ricci & Cappelletti, 1988). In 1990 it was released in California from Israel to prey on the ash whitefly, *Siphoninus phillyreae* (Haliday), and has become established in coastal southern California (Bellows *et al.*, 1992a, 1992b). *Clitostethus arcuatus* from Madeira became established in Cuba, preying on whitefly attacking citrus and fiddlewood (Cock, 1985). *Clitostethus bicolor* (Gordon) (as *Nephaspis*) was sent to Hawaii in 1979 from Trinidad and became established against the spiraling whitefly, *Aleurodicus dispersus* Russell (Funasaki *et al.*, 1988; Waterhouse & Norris, 1989). Later shipments of *C. bicolor* were made from Hawaii and released against *A. dispersus* in Fiji and American Samoa, where it also became established (Waterhouse & Norris, 1989). *Clitostethus oculatus* (Blatchley) (as *Nephaspis amnicola* Wingo), which occurs from the southeastern United States (Gordon, 1985), the Caribbean, and Central America, was imported from Trinidad and Honduras and became established in Hawaii as an effective predator when populations of four different whitefly species are high (Kumashiro *et al.*, 1983; Yoshida & Mau, 1985; Waterhouse & Norris, 1989); it was also established on Guam and American Samoa (Waterhouse & Norris, 1989).

The genus *Cryptolaemus* comprises seven species that are indigenous to the Australasian zoogeographic region (Booth & Pope, 1986). The mealybug destroyer, *C. montrouzieri montrouzieri* Mulsant, is the subspecies being utilized today, mainly as an augmentative agent to control mealybugs in warm climates and greenhouses. They deposit their small, oval, lemon-yellow colored eggs usually into the cottony egg sac of mealybugs, or at least in the vicinity of mealybugs. The larvae are covered with white cottony wax filaments and when mature they frequently move downward and pupate within their dorsally split larval skins on or about the base of the plant (Essig, 1931; Smith & Armitage, 1931; Bodenheimer, 1951; Fisher, 1963). Larval and pupal development and reproduction were faster and greater when *C. montrouzieri* was exposed to *Planococcus citri* on *Coleus* leaves than to *P. citri* on oleander; *Aphis nerii* on oleander was a much poorer prey than the mealybugs (Averbeck & Haddock, 1984).

The mealybug destroyer was introduced into California in 1891 and 1892 from Sydney, Australia by Koebele, but it could only overwinter in California along the southern California coast (Clausen, 1915; Smith & Armitage, 1920; Essig, 1931). Because it initially showed great promise as a predator of mealybugs infesting citrus, Smith and Armitage (1920, 1931) developed inexpensive techniques to mass-culture both mealybugs and predators, and many commercial insectaries arose in southern California to produce the mealybug destroyer for the citrus growers (Fisher, 1963). However, the successful introductions of various mealybug

parasitoids led to control of mealybugs without annual releases of *Cryptolaemus;* thus, a number of insectaries discontinued producing the predator. Even today, some *C. montrouzieri* are mass-cultured for use in greenhouses (Witcomb, 1940; Doutt, 1951, 1952; Hussey & Scopes, 1985). *Cryptolaemus m. montrouzieri* has been imported into over 40 countries (Bartlett, 1978; Booth & Pope, 1986) including Hawaii, where it preys on at least 17 species of mealybugs (Leeper, 1976). In India, it is released against the grapevine mealybug (Babu & Azam, 1988; Srinivasan *et al.*, 1989; Mani, 1990), but obtained poor establishment against the coffee mealybug (Reddy *et al.*, 1991). In a few countries, it failed to become established permanently, mainly because the winters were too cold. Bartlett (1974) intentionally sought out biotypes of the beetles from the coldest areas of Australia and collected two populations that were propagated and colonized in California to broaden the genetic base for cold hardiness of the species. As a result, we now have *C. m. montrouzieri* persisting in the San Francisco Bay area of California where the original biotype could not persist (K. S. Hagen, unpublished). Thus, reintroducing *C. m. montrouzieri*, utilizing this new biotype, into the areas where the original biotype had failed to become established should be considered.

Diomus is primarily a Neotropical genus, but there are a few species scattered in all zoogeographic regions (Gordon, 1985; Xiong-Fei & Gordon, 1986). The biology is known for only a few species. *Diomus debilis* (LeConte) lays elliptical flattened eggs on leaves and preys on mealybugs that infest salt marsh grasses (Whitehead, 1967). *Diomus* larvae have no apparent waxy secretions and they pupate on the plants where the larvae had fed. Mealybugs are prey for several *Diomus* spp. (Whitehead, 1967; Leeper, 1976; Meyerdirk, 1983; Hennessey & Muaka, 1987; Löhr *et al.*, 1990; Herren & Neuenschwander, 1991), as well as aphids (Leeper, 1976; Bishop & Blood, 1978), psyllids (Leeper, 1976; Pinnock *et al.*, 1978; Leeper & Beardsley, 1976), and spider mites in Brazilian apple orchards (Lorenzato *et al.*, 1986).

Diomus notescens (Blackburn) was introduced into Hawaii from Australia by Koebele in 1894, where it preys on *Nipaecoccus vastator* (Maskell), *P. citri*, and *Toxoptera citricidus* (Leeper, 1976; Funasaki *et al.* 1988). In Australia *D. notescens* is considered an effective predator of *Aphis gossypii* on cotton (Bishop & Blood, 1978). *Diomus pumilio* Weise was collected in southeastern Australia in 1971 feeding on acacia psylla, *Acizza uncatoides*, which was rare. The predator was sent to California where it was cultured and released against *A. uncatoides* infesting acacias adorning the freeway landscape along the coast. *Diomus pumilio* was found to be widespread from San Francisco south to San Diego by the middle 1970s and has eliminated the need for insecticide treatments since then, saving the State Department of California Transportation $55,000 per year in the Bay area alone (Pinnock *et al.*,

1978). Although another survey has confirmed the reduced abundance of the psyllid, *Anthocoris nemoralis* is now the dominant predator instead of *D. pumilio* (Dreistadt & Hagen, 1994). It was also sent to Hawaii as an enemy of the acacia psylla that was attacking koa trees, but it failed to become established; however, another coccinellid, *Harmonia conformis,* dramatically controlled the psyllid (Leeper, 1976; Leeper & Beardsley, 1976). Several *Diomus* spp. were collected in South America and sent to tropical Africa to prey on the cassava mealybug, *Phenacoccus manihoti* (Löhr *et al.,* 1990). At least one *Diomus* sp. became established widely in Africa (Herren & Neuenschwander, 1991). *Diomus hennesseyi* Fürsch, an endemic African species, also attacks the cassava mealybug among some 30 other coccinellid species (Fürsch, 1987; Neuenschwander *et al.,* 1987; Kanika-Kiamfu *et al.,* 1992).

Nephus is composed of seven subgenera, but the few species that concern us fall under *Nephus (Nephus), Nephus (Sidis),* and *Nephus (Scymnobius).* The *Nephus* spp. that have been studied are all predators of mealybugs and their larvae have conspicuous waxy filaments (Whitehead, 1967). *Nephus (Scymnobius) bilucernarius* (Mulsant) was established in Hawaii from Mexico against the pineapple mealybug, *Dysmicoccus brevipes* (Cockerell) in 1930; it also preys on *Nipaecoccus vastator* (Leeper, 1976; Funasaki *et al.,* 1988). *Nephus (N.) roepkei* (Fluiter) was imported to Hawaii from Asia several times and became established in the early 1990s, where it preys on three mealybug species and eggs of *Spodoptera mauritis* (Boisduval) (Funasaki *et al.,* 1988). *Nephus (Sidis) binaevatus* Mulsant is endemic to South Africa and was introduced into southern California in 1922 as an enemy of the grape mealybug [*Pseudococcus maritimus* (Ehrhorn)] on grapes, and the citrophilius mealybug [*P. calceolariae* (Maskell)] on citrus, where it became established as an effective predator on citrus (Smith, 1923; Essig, 1931). Parasitoids introduced later controlled the mealybugs, reducing the importance of *N. bineavatus* (Bartlett, 1978); presently, it is found commonly attacking mealybugs on ornamentals in coastal California. *Nephus (Sidis) reunioni* Fürsch was imported to the former USSR from Isle of Réunion to combat citrus mealybugs, mainly in the Transcaucasia and Crimea; culturing methods and optimal temperatures were determined to provide material for release (Izhevsky & Orlinsky, 1988).

Pseudoscymnus anomalus Chapin (endemic to Micronesia) was imported from Guam to Hawaii in 1970 as an enemy of the coconut scale (Leeper, 1976) and was also successfully established in American Samoa and Vanuatu (Waterhouse & Norris, 1987). These authors consider that the species is specific to the coconut scale. *Pseudoscymnus tsugae* Sasaji & McClure has been described from Japan and is considered a potential biological control agent for hemlock woolly adelgid in the eastern United States (Sasaji & McClure, 1997).

Some *Scymnodes* species are aphid feeders, others are predaceous on coccids, and at least one species eats ants; all are endemic to Australia and Melanesia (Pope & Lawrence, 1990). *Scymnodes (Apolinus) lividigaster* Mulsant has spread naturally to New Zealand and several South Pacific Island groups, and was intentionally introduced into the Hawaiian Islands by Koebele in 1894 as an aphid predator (Leeper, 1976; Pope & Lawrence, 1990), where it proved to be of considerable value against the sugarcane aphid (Pemberton, 1948).

The genus *Scymnus* is composed of five subgenera but only two of these concern us today—*Scymnus (Scymnus)* and *S. (Pullus).* The known larvae of both subgenera have waxy coverings with varying lengths of filaments, except the first instars that show no wax (Whitehead, 1967; Pope, 1979). The main prey of the larvae are aphids and adelgids (Gordon, 1976).

There have been 17 species in the taxon *Scymnus* brought to the continental United States, but thus far only 2 have become established, 1 intentionally and the other accidentally. A few exotic species have become established in the Hawaiian Islands but there are no data on the introductions (Leeper, 1976).

Scymnus (S.) frontalis (Fabricius), an Old World species, was collected in Turkey and is currently being cultured and released in North America against the Russian wheat aphid, *Diuraphis noxia* (Mordvillco). It can be recognized from other North American *Scymnus (Scymnus)* species by taxonomic characters described by Gordon and Vandenberg (1991). Its temperature thresholds for development and reproduction are about 5 to 10°C higher than that of *D. noxia* (Naranjo *et al.,* 1990). *Scymnus (Pullus) impexus* Mulsant was introduced into Canada and the United States from Germany several times from 1951 to 1960 for control of the balsam woolly adelgid, but it has become established only in Oregon (Gordon, 1985). It also was exported from Germany into southeast England against adelgids and became established there (Greathead, 1976). A classical biological study of *S. (P.) impexus* was made by Delucchi (1954). *Scymnus (P.) suturalis* Thunberg is a European species that was inadvertently introduced to the eastern United States (Pennsylvania and New York), possibly via ornamental conifer nursery stock from Europe (Gordon, 1982). Its closest relative is *S. (P.) coniferarum* Crotch, which is an aphid predator on conifers in the western United States, and Gordon (1982) describes the characteristics differentiating the two species. *Scymnus (P.) uncinatus* Sicard from Mexico was introduced and established in Hawaii in 1922 for control of pineapple mealybugs (Leeper, 1976).

Stethorini is a tribe of one genus containing 65 species that are found in most parts of the world (Gordon & Chapin, 1983). There are three subgenera (Pang & Mao, 1979), but because only a few species have been placed in their respective subgenera, these divisions will not always be noted here. *Stethorus* spp. are obligate predators of spider mites, and from most records of Kapur (1948) and all records listed by Gordon and Chapin (1983), it appeared to Gordon that the spider mite prey may be restricted to the family Tetranychidae. Adult *Stethorus* have been observed to feed at extrafloral nectaries and on aphids and their honeydew, but none produced eggs (Putman, 1955).

Stethorus spp. deposit broadly ovoid eggs horizontally on the underside of leaves, usually among mite colonies (Putman, 1955; Colburn & Asquith, 1971; Raros & Haramoto, 1974). Oviposition is reduced on plants with pubescent leaves, because hooked trichomes may kill or impede the movement of larvae (Gunthart, 1945; Putman, 1955; Raros & Haramoto, 1974). The larvae are free of waxy coverings and feed on mite eggs and nymphs by extraoral digestion (Fleschner, 1950; Collyer, 1953a; Robinson, 1953; Putman, 1955; Raros & Haramoto, 1974); a larva eats about 200 tetranychids (ca. 45% eggs, 33% nymphs, 20% adults) (Raros & Haramoto, 1974). Pupation occurs at the feeding site; the pupa is "naked," pushing its last larval skin posteriorly to its base. An adult usually consumes the entire prey, and in 5 days eats about 400 tetranychids (ca. 40% eggs, 35% larvae and nymphs, and 22% adults) (Raros & Haramoto, 1974). Female beetles preferred mite eggs; and when satiated, they spent approximately 45% of their time searching, 14% feeding, and 40% resting (Houck, 1991). In another species, the females showed no significant preference for any life stage of a tetranychid (Tanigoshi & McMurtry, 1977). Oviposition begins when the European red mite, *Panonychus ulmi* (Koch), population ranged from 1 to 2 mites per apple leaf (Hull *et al.*, 1977a, 1977b). Daily rates of prey consumption of ovipositing females may exceed 40 adult spider mites (McMurtry *et al.*, 1970). Tanaka (1966) found that 50 to 100 eggs or 15 to 17 adults of *P. citri* were needed per day for a female to oviposit. A mean of about 4 eggs can be deposited per day over 55 days with a total of 300 to 400 eggs in a lifetime (Tanigoshi & McMurtry, 1977).

Stethorus spp. are important natural enemies of spider mites, particularly the Tetranychidae in the field (Kapur, 1948; McMurtry *et al.*, 1970). The impact of *Stethorus* spp. on tetranychids has been indicated by the rapid increase of mites after certain pesticide treatments on agricultural crops (Gilliatt, 1935; Readshaw, 1975; Field, 1979; Tanigoshi *et al.*, 1983; Welch *et al.*, 1983; Hull & Beers, 1985) and by hand removal of predators from mite-infested trees (Fleschner, 1950, 1958). The potential for these coccinellids to control spider mites in orchards or greenhouses was demonstrated by a rapid decrease of tetranychid populations after releasing *Stethorus* adults (Hussey & Huffaker, 1976).

The use of *Stethorus* in the applied biological control of spider mites has been mainly through conservation (by utilizing selective pesticides) and augmentation. The results from classical biological attempts have been poor indeed (Huffaker *et al.*, 1976). Of the 12 *Stethorus* spp. released in California from 1900 to 1978, Gordon (1985) found no record of establishment. However, one species has been discovered in California, indicating successful establishment (K. S. Hagen, unpublished observations).

Stethorus (Stethorus) punctillum Weise is a Palearctic species that was accidentally introduced into North America and discovered in Massachusetts and Ontario in 1950. This species is now known to occur in eastern North America from Massachusetts to Wisconsin and in the West in Oregon, Washington, and Idaho (Gordon, 1985). *Stethorus (Stethorus) punctillum* is reported to have reduced mite populations in Russian cotton fields and European vineyards (Kapur, 1948). It is one of the most effective predators attacking *P. ulmi* infesting Ontario peach orchards (Putman & Herne, 1966). In Europe, it shows a degree of specificity to fruit tree mites, notably *P. ulmi* and *Tetranychus urticae*. Wheeler *et al.* (1973) made collections of mites from more than 200 ornamental plant species grown in Pennsylvania, and *S. punctillum* was found associated almost exclusively with conifers infested with the spruce spider mite, *Oligonychus ununguis* (Jacobi).

Stethorus (Stethorus) punctum punctum (LeConte) is a native species to North America and closely resembles *S. punctillum*. It ranges from southeastern Canada to North Carolina and west to Montana (Gordon, 1985). It is commonly a predator of *P. ulmi* in apple orchards. Its searching habits and functional responses have been studied by Colburn and Asquith (1971) and Hull *et al.* (1976, 1977a, 1977b). Welch *et al.* (1983) reviewed and compared three pest–predator models; two involved phytoseiid predators and the third, *S. punctum*, which included *T. ulmi* in Pennsylvania apple orchards [originally constructed by Mowery *et al.* (1975)]. Based on weekly sampling of trees, the number of *S. punctum* needed to control the mites was determined and the predicted number of mites per leaf was close to the actual field counts (Welch *et al.*, 1983). Adults overwinter in the ground cover and soil around the base of the trunk of apple trees (Felland & Hull 1996). *Stethorus (S.) punctum picipes* Casey is a western North American species from Idaho to British Columbia south to southern California (Gordon, 1985). It feeds on tetranychids infesting cotton and citrus (Kapur, 1948), and is an effective predator of the avocado brown mite. *Oligonychus punicae* (Hirst), and citrus red mite, *Panonychus citri* (McGregor) (Fleschner, 1950, 1958; McMurtry *et al.*, 1969; Tanigoshi & McMurtry, 1977). Multiple releases of *S. picipes* against the avocado brown mite at a rate of 400 to 500 adult beetles per tree resulted in an earlier increase of *S. picipes* in release areas, a lower peak population of the mite, and a lower percentage of damaged leaves in the release plots compared with check plots receiving no *Stethorus*. On caged avocado seedlings infested with avocado brown mite, introducing first-instar larvae of *S. picipes* 2 weeks after the initial prey infestation (prey to predator ratio averaged about 69:1) resulted in the development, maturation, and production of immature *Stethorus* offspring and initial suppression of *O. punicae* populations. Subsequently, overexploitation of prey occurred, with resultant starvation of the predators, and the prey population increased (Tanigoshi & McMurtry, 1977). Tanigoshi and McMurtry (1977) considered *S. p. picipes* to be a high-density predator but evidence (Congdon *et al.,* 1993) suggests that it is capable of detecting and responding to very low-density populations of *Tetranychus urticae* in raspberries.

Stethorus (Parastethorus) histrio Chazeau is a widespread and common species in Australia and is also found on Réunion Island (Indian Ocean) and New Caledonia (Pacific Ocean) (Houston, 1980); in the Western Hemisphere it occurs in Chile and the Yucatan of Mexico (Gordon & Chapin, 1983). In Australia, Britton and Lee (1972) used the name *S. nigripes* Kapur, and it was introduced into California in 1974 and 1978 under this epithet (Gordon, 1985) but was never recovered until 1991 when K. S. Hagen (unpublished observations) found it established in Albany, California, preying on *Tetranychus urticae* on almond and citrus and on *P. ulmi* on citrus and *Rhamnus californica* Eschscholtz. At the same time that *S. (P.) histrio* was being released in California, *S. nigripes* Kapur (under the name *S. loxtoni* Britton & Lee) was also being released, but in far greater numbers and over a greater area. Richardson (1977) made a detailed study of all aspects of *S. nigripes* (calling it *S. loxtoni*) and argued that of all the Australian *Stethorus* species, this would be the species with the greatest potential of becoming established in California against *T. urticae*. Readshaw (1975) also shared this view, but thus far of all the *Stethorus* species introduced, only *S. histrio* has become established in California.

Sticholotidinae is a subfamily of eight tribes, five of which contain species used in applied biological control. Four tribes are natural enemies of armored scales (Diaspididae): Cephaloscymnini, Microweiseini, Sticholotidini, and Sukunahikonini (Drea & Gordon, 1990). The species in the tribe Serangiini are mainly predators of whiteflies (Gordon, 1977; Clausen, 1978; Waterhouse & Norris, 1989; Onillon, 1990).

The bluntly oval eggs of the armored scale predators (*Coccidophilus, Dilponis, Microweisea, Sticholotis,* and some *Pharoscymnus* spp.) are deposited either as a single unit or as groups beneath the dorsal scale coverings of killed scales (Flanders, 1936; Muma, 1956; Brettell, 1964; Ghani & Ahmad, 1966; Ahmad & Ghani, 1971).

The eggs of *D. inconspicuus* Pope have single slender stalks that usually project from under the scale covering (Brettell, 1964). The whitefly predators (*Catana, Delphastus,* and *Serangium* spp.) deposit their elongate oval, translucent eggs on the underside of leaves among the whitefly immatures (Clausen & Berry, 1932; Muma, 1956; Timofeyeva & Hoang, 1978). Larvae of Sticholotidinae do not pro-

duce waxy coverings (LeSage, 1991). Pupation occurs near larval feeding sites; pupae are exarate with all appendages free and the last larval skin pushed down posteriorly (Phuoc & Stehr, 1974). The tropical and subtropical species are multivoltine, but *Coccidophilus marginata* (LeConte) is univoltine in Canada, and oddly it overwinters in the third larval instar (Martel & Sharma, 1970).

The tribe Microweiseini is represented by three genera that contain a few species of interest to us today. *Coccidophilus citricola* Brethes is a predator of red and purple scales in subtropical and temperate South America. It was introduced into California from Brazil in 1934 and 1935 as a predator of diaspine scales infesting citrus trees; it became established but disappeared a few years later (Flanders, 1936; Clausen, 1978). A mass-culturing method was developed for *C. citricola* in Chile and the beetle was introduced into northern Chile to prey on diaspine scales attacking olive and citrus, but it is not known if it still persists there (Aguilera *et al.,* 1984). *Coccidophilus marginata* occurs in the northeastern United States and southeastern Canada where it is an important enemy of the pine needle scale (Martel & Sharma, 1970). *Coccidophilus atronitens* Carey is a counterpart of *C. marginata,* preying on the same pest but in western North America (Dahlsten *et al.,* 1969). *Microweisea coccidivora* (Ashmead) is a southeastern United States species that is associated with diaspine scales on citrus in Florida (Muma, 1956). *Microweisea misella* (LeConte), a widespread species occurring from southern Canada to Florida and Mexico, was introduced into Bermuda to prey on Bermuda cedar scales in 1951; it became abundant but did not solve the problem because the trees continued to die (Bennett & Hughes, 1959).

The tribe Serangiini contains nine genera, but we shall only mention a few species of whitefly predators belonging to three: *Catana, Delphastus,* and *Serangium. Catana clauseni* Chapin (as a *Cryptognatha* sp.) was collected in Malaysia feeding on citrus blackfly, *Aleurcanthus woglumi* Ashby, and sent to Cuba in 1930 against that same species. It became established and was still in Cuba in 1936 but declined in effect and numbers because of the great reduction of the blackfly by introduced parasitoids (Clausen & Berry, 1932; Clausen, 1978; Cock, 1985). Rees (1948) describes the larvae of *C. clauseni* and how to distinguish it from the larvae of *Delphastus pusillus* (LeConte). *Delphastus pusillus* was successfully imported into the Hawaiian Islands from Trinidad in 1980 as a predator of whiteflies, including the spiraling whitefly, *Aleurodicus dispersus* Russell, and from there to American Samoa in 1984, where it also became established against *A. dispersus* (Funasaki *et al.,* 1988; Waterhouse & Norris, 1989). *Delphastus pusillus* naturally occurs across the southern half of the United States (Gordon, 1985). In Florida, it is a common predator of whiteflies attacking citrus trees (Muma, 1956). Adult females live 60.5 days at 28°C and development time from oviposition to adult emergence is 21.0 days with the consumption of 977.5 eggs of *Bemisia tabaci* (Hoelmer *et al.*, 1993). In California, it is being cultured and released against the recent invading silverleaf whitefly, *Bemisia argentifolii* Bellows & Perring; *D. pusillus* is considered effective because it has a voracious feeding capacity, a high reproductive potential, a positive response to olfactory cues from whitefly-infested leaves, and the ability to discriminate and avoid feeding on parasitized whitefly nymphs (Parrella *et al.*, 1992: Hoelmer *et al.*, 1994; Heinz & Nelson, 1996). *Serangium maculigerum* Blackburn was introduced into Hawaii by Koebele in 1894 and now is found attacking several introduced whiteflies in Oahu (Leeper, 1976; Funasaki *et al.*, 1988). *Serangium parcesetosum* Sicard was first found in India, where it is a predator of citrus whitefly. Timofeyeva and Hoang (1978) were unable to find any specific natural enemies of the whitefly in citrus plantations along the Caucasian coast of the Black Sea, so they imported the predator from India for study and release. The coccinellid produced four generations per year, destroyed up to 90% of the citrus whitefly population, and spread to other plantations; it overwintered successfully in the (former) USSR for at least 2 years. *Serangium parcesetosum* has been studied as a potential biological control agent of the silverleaf whitefly in the United States. This species is more voracious than *D. pusillus*, with a lifetime consumption of whitefly eggs and immatures estimated at just over 2500 at 30°C and nearly 5000 at 20°C (Legaspi *et al.*, 1996). The impact of natural enemies on control of whiteflies is reviewed by Onillon (1990).

The tribe Sticholotidini has 30 genera but only 2 concern us today, *Pharoscymnus* and *Sticholotis*, though there are apparently many new species to be described and studied in the tribe (many of which are distinguished by their reduced hind wings). The few feeding records of the species of this tribe indicate that diaspids and eriococcids are prey. *Pharoscymnus* is an Old World genus. *Pharoscymnus horni* Weise, a Ceylonese species, was cultured along with *Sticholotis madagassa* (Weise) and *Chilocorus nigritus* from Mauritius and was released at a rate of 800 adults per acre of sugarcane in Karnataka, India, for the control of the sugarcane scale, *Melanaspis glomerata* Green; the predators became established and the scale populations were reduced by 10 to 48% (Ansari *et al.*, 1989). *Pharoscymnus numidicus* (Pic) is endemic to North Africa and the Middle East and is one of the most important predators of the date scale, *Parlatoria blanchardi* Targioni-Tozzetti, in Israel (Kehat, 1967). In Morocco, *P. numidicus*, *P. setulosus* Chevrolet, and *P. ovoideus* Sicard are predators of the date scale on palms (Smirnoff, 1957). Because of the near eradication of these predators due to insecticide use, interest in them was rekindled, with the aim of reestablishing them by releasing imported predators. *Pharoseymnus numidicus* was reestablished in all the date plantations in the Arava area of

Israel (Kehat *et al.*, 1974). Thus, whenever well-adapted predators are introduced under favorable climatic conditions, scale populations can be reduced to tolerable levels if no, or selective, pesticides are used (Benassy, 1990). *Sticholotis madagassa* Weise is not an important predator of *Aulacaspis tegalensis* (Zehntner) on sugarcane in Mauritius but is in Tanzania and Réunion Island (Williams & Greathead, 1990). *Sticholotis ruficeps* Weise, an Asian species, somehow invaded Hawaii, where it feeds on *Eriococcus araucariae* Maskell and *Pinnaspis buxi* (Bouchè) (Leeper, 1976).

Derodontidae   This is another small family of beetles that would have remained obscure if it were not for one species, *Laricobius erichsonii* Rosenhaur, which was introduced from Europe to Canada and the United States to control a European invader, the balsam woolly aphid, *Adelges piceae* (Ratzeburg) (Franz, 1958). The biology of the other two genera in Derodontidae is unknown, but a few species have been reported to live in slime molds under bark and in shelf fungi.

Among the seven predaceous insects that have been introduced and established against *A. piceae* in North America, *L. erichsonii* appears to be the only one exerting some control in certain areas in Canada (Franz, 1958) and it is considered the major controlling agent in Central Europe.

Histeridae   The biology of only a dozen of the nearly 4000 described species in this family is known in detail. However, some species have been utilized in biological control programs (Rao *et al.*, 1971; Clausen, 1978; Waterhouse & Norris, 1987). It appears that histerids are carnivorous in both their adult and larval stages. Many genera are associated with carrion and dung, where they prey mainly on immature stages of Diptera and other coprophagous and sarcophagous insects that often occur in burrows and nests. There are genera that contain species associated with trees and rotting vegetation and prey on xylophagous insects such as bark beetles, weevils, and some immature scarabs. Species in the subfamily Hetarinae live in ant or termite nests.

Eggs are usually deposited singly. The species associated with manure place them in the soil beneath cattle dung. There are three instars and pupation is in the soil or in manure within a pupal "cocoon," but the histerids in rotten wood do not make a cocoon. In temperate climates there is usually one generation a year, but some species can complete their development in about 1 month and usually overwinter as an adult (Balduf, 1935; Hinton, 1944; Nikitskiy, 1976; Morgan *et al.*, 1983; Summerlin *et al.*, 1989).

A few histerid species have been imported to control filth flies. In 1909, *Atholus bimaculatus* (Linnaeus) (= *Hister*) was established in the Hawaiian Islands when nine individuals were imported from Germany for control of the horn fly (Legner, 1978). It also attacks houseflies in manure (Legner, 1978). Later, mostly in the 1950s, eight more histerid species were imported into Hawaii to attack the horn fly, of which four became established, including *Pachylister caffer* (Erichson) (Legner, 1978; Funasaki *et al*, 1988). A *P. caffer* first-instar larva can consume over 40 horn fly puparia, and a second-instar larva, over 325 (Summerlin *et al.*, 1989). *Hister nomas* Erichson and *P. caffer* were introduced into California against houseflies and stable flies in 1969 but apparently did not become established (Legner, 1978). *Pachylister chinensis* Quensel, the most active predator of filth flies in manure in the South Pacific, became established in Fiji in 1936. It came from Indonesia and was credited for reducing housefly populations by Simmonds (1958), but Bornemissza (1968) believed other factors explain the reduction of flies in Fiji. *Pachylister chinensis* also became established in New Britain, New Hebrides, Samoa, and the Solomon Islands, but not in several other Pacific islands including the Hawaiian Islands. In California, Legner and Olton (1968) estimate that more than 95% of the destruction of fly populations in manure is caused by parasitoids and predators that attack immature stages of flies. *Carcinops pumilio* (Erichson) is a major predator of fly eggs and larvae in poultry manure in the United States (Geden & Axtell, 1988; Rueda & Axtell, 1997) and a simulation model of its predation on housefly immature stages was developed by Wilhoit *et al.* (1991).

Plasius javanus Erichson was introduced from Java to Fiji in 1913 and 1914 to control the banana weevil, *Cosmopolites sordidus,* where the histerid reduced the weevil damage on one island. It was also introduced into the Cook Islands, Guam, Palau, and Tahiti, where it became established and reduced the damage on some islands. The histerid larvae can devour 30 or more weevil larvae per day over a 4- to 6-week period (Jepson, 1914; Rao *et al.,* 1971; Clausen, 1978). Another species, *Oxysternus maximus* (Linnaeus), was found to be useful in the control of the palm weevil *Rhynchophorus palmarum* (Linnaeus), according to Hinton (1944).

Teretriosoma nigrescens Lewis was introduced from Mexico to west Africa in 1991 for control of the greater grain borer, *Prostephanus truncatus* (Horn) in maize stores. The adult predators respond to the male-produced aggregation pheromone of *P. truncatus* and their seasonal flight patterns follow that of the host both in Mexico (Tigar *et al.,* 1994) and in Togo (Borgemeister *et al.,* 1997). In comparative before and after release studies in Togo, Richter *et al.* (1997) document a reduction in grain damage from 63% before the release of *T. nigrescens* to 21% following release.

Among the many natural enemies associated with *Oryctes rhinoceros*, there are 11 histerid species in four genera in various tropical countries. Of the few histerid species introduced into eight different countries in the Pacific re-

gion, only *Pachylister chinensis* became established (in four islands) while the many introductions of various *Hololepta* spp. failed completely (Waterhouse & Norris, 1987).

Nitidulidae   Nitidulids have one of the largest feeding repertories among the Coleoptera (Lawrence, 1991b). Species of *Epuraea, Glischrochilus,* and *Pityaphagus* have been recorded as predators of scolytid eggs and larvae, but Lawrence (1991b) thinks it is doubtful that any of these are obligate predators. However, species in the Cybocephalinae are predaceous on diaspine scales (Drea, 1990b) and whiteflies (Kirejtshuk *et al.,* 1997). Some 150 species in the subfamily have been described, and some systematists considered them to constitute a family (Blumberg, 1973). Drea (1990b) lists 29 genera of diaspine scales that have been recorded as hosts of cybocephalines, so prey specificity needs to be determined before introducing any cybocephaline against this group (Blumberg & Swirski, 1974). The adults will feed on various arthropods other than diaspine scales such as moth, lecanine, and tetranychid eggs, but no oviposition occurs and survival is variable [oviposition did occur on *Bemisia tabaci* (Gennadius)]; the cybocephaline larvae were even more discriminating (Blumberg & Swirski, 1974).

A few species have been imported to control diaspine scales. Heavy infestations of the date palm scale, *Parlatoria blanchardi* (Targioni-Tozzetti), were killing date palms in Algeria; a coccinellid and *Cybocephalus nigriceps palmarium* Peyerimhoff from southern Oran (Algeria) were imported in the early 1920s, became established, and controlled the scale (Balachowsky, 1928). During the early 1940s, date palms in Moroccan oases were being killed by the date palm scale, and again the introduction of a coccinellid and the cybocephaline gave control where the predators became established (Smirnoff, 1954, 1957). During the early 1930s, a *Cybocephalus* sp. was introduced into California from China against California red scale; 10 years later it was found to be established in southern California (Rosen & DeBach, 1978). A *Cybocephalus* sp. from Trinidad was established in Bermuda by 1955 and contributes to the control of *Pseudaulacaspis pentagona* (Bennett & Hughes, 1959). *Cybocephalus nipponicus* Endrody-Younga, or a closely related species, was introduced from South Korea into the eastern United States for use against *Unaspis euonymi* (Comstock) infesting euonymus. It appears to be established but its impact on the scale is not evident (Drea, 1990b).

Drea (1990b) concluded from reviewing the literature that in general *Cybocephalus* spp. alone do not appear to be important in the control of diaspine scales.

Rhizophagidae   The biology of this small family is poorly known. Worldwide there are 51 described species (Bousquet, 1990). It appears that both larval and adult *Rhi-*

zophagus spp. are monophagous predators on certain species of bark beetles, both larval and adult. Several species in the genus *Rhizophagus* have been recorded or used in biological control of certain bark beetles. The Russians first observed *R. grandis* Gyllenhal to be an important predator of *Dendroctonus micans* (Kugelann), which attacks spruce and some other conifers (Kobakhidze, 1965). In old spruce stands, the bark beetle is not an important pest, but as spruce stands expanded to the West, serious *D. micans* outbreaks occurred in new plantations (Kobakhidze, 1965). Because *R. grandis* is considered the main factor for regulating bark beetle populations, Belgian, English, and French entomologists have developed a mass-rearing method for the predator and made inoculative releases of *R. grandis* in England and France to prevent outbreaks of *D. micans* (Grégoire *et. al.,* 1985; Grégoire, 1988; Evans & King, 1989; Evans & Fielding, 1996). Orientation and oviposition of *R. grandis* are linked by simple oxygenated monoterpenes to its prey, *D. micans* (Grégoire *et al.,* 1992; Wyatt *et al.,* 1993). *Rhizophagus grandis* was introduced into the United States, where inoculative releases have been made against *D. terebrans* Oliver (Moser, 1989).

Two species of *Rhizophagus* were introduced from Europe into Canada and New Zealand against bark beetles in the 1930s but apparently never became established (Clausen, 1978).

Staphylinidae   There are about 27,000 species of rove beetles in 1,500 genera (Frank & Curtis, 1979), but the biologies of only a relatively few species of Staphylinidae are known. Larvae live in the same habitat as adults, mostly concealed in soil, decaying plant material, animal dung, or ant and termite nests; a few species are arboreal. Rove beetles are fungivorous, algivorous, saprophagous, predaceous, or parasitic (either insect parasitoids or external parasitoids on vertebrates). Most species have three larval instars, and pupation usually occurs in cells in the larval substrate; Steninae, some Paederinae, and Aleocharinae spin silken cocoons (Balduf, 1935; Frank, 1991).

Predaceous species are found in Aleocharinae, Omaliinae, Osoriinae, Oxyporinae, Oxytelinae, Quediinae, Paederinae, Staphylininae, Tachyporinae, and Xantholininae. Many of the predaceous species are found associated with animal dung. Among 100 staphylinid species found in cattle dung in South Africa, the 79 predaceous species were from five subfamilies (Aleocharinae, Oxytelinae, Paederinae, Staphylininae, and Xantholininae) and the 19 coprophagous species were in the Oxytelinae (Davis *et al.,* 1988). Hunter *et al.* (1991) found 21 species in 11 genera in the same five subfamilies (except for Xantholininae) associated with dung in Texas; Tachyporinae was represented in Texas but not in the South Africa dung study. Some staphylinids in the preceding subfamilies have been utilized in applied biological control.

Aleocharinae is a huge subfamily that is poorly known taxonomically and biologically. At least 72 genera of the Aleocharinae are predaceous on ants (Seevers, 1965); however, species in several genera have been used as biological control agents. All *Aleochara* spp., where biologies have been studied, are solitary ectoparasitoids of cyclorrhaphous dipteran pupae as larvae, but predators of eggs and larvae of the latter as adults (Balduf, 1935; Moore & Legner, 1971; Peschke & Fuldner, 1977; Klimaszewski, 1984). Klimaszewski (1984) recognizes seven subgenera in *Aleochara.* Several *Aleochara* spp. were introduced into North America from Europe, and some species have been mass-cultured and periodically released against dipteran rootworms. *Aleochara (Coprochara) bilineata* Gyllenhal is a Palearctic species that was apparently accidentally introduced into North America with its rootworm host puparia, *Delia* (= *Hylemya*) spp., and was reintroduced intentionally into Canada in the 1940s because it was thought that the *Aleochara* already present was an endemic species (Clausen, 1978; Muona, 1984). *Aleochara bilineata* attacks the pupae of the cabbage, onion, seed corn, and turnip (*Delia*) rootworms as well as on the spinach leaf miner and housefly (Anderson *et al.,* 1983; Klimaszewski, 1984). Although parasitism of cabbage root fly in Canada is density dependent and can be as high as 94%, the introduction of *A. bilineata* is not considered to be sufficiently effective in reducing the abundance of this pest (Turnock *et al.,* 1995). The insect growth inhibitor diflubenzuron will kill *D. radicum* (Linnaeus) but is not toxic to *A. bilineata* eggs or larvae (Gordon & Cornect, 1986). This species can be mass-cultured (Whistlecraft *et al.,* 1985), and the Russians use mass culturing for releasing 20,000 to 40,000 beetles per hectare at the beginning of the egg-laying period of *D. radicum* and obtain full protection from the cabbage maggot (Beglyarov & Smetnik, 1977). Only 3% of marked *A. bilineata* were recovered in a release garden (Tomlin *et al.,* 1992).

Aleochara (Xenochara) tristis Gravenhorst is a Palearctic species introduced to North America from France as a natural enemy of the face fly, *Musca autumnalis* DeGeer (Drea, 1966). It was released in New Jersey, Pennsylvania, Maryland, and Nebraska, with no recoveries by 1969 (Legner, 1978). However, in a revision of North American *Aleochara,* Klimaszewski (1984) found specimens from Quebec, Canada, and Del Norte County of California; thus, it appears to be established in North America. *Aleochara (X.) taeniata* Erichson was introduced into the United States from Jamaica as a natural enemy of the housefly (White & Legner, 1966) and is now in California and Arizona (Legner, 1978; Klimaszewski, 1984). *Aleochara (A.) curtula* (Goeze), a rootworm and housefly predator–parasitoid, was introduced to North America from Europe. It now occurs in the eastern United States and Canada, while its presence in British Columbia may have resulted from a second introduction (Peschke & Fuldner, 1977; Klimaszewski, 1984).

Oligota, another aleocharine genus, contains species that

are predaceous as adults and larvae on tiny arboreal arthropods, but apparently pupate in the soil; at least *O. oviformis* (Casey), a Holarctic species that preys on spider mites infesting avocado and citrus trees in California, would not pupate until exposed to soil (Badgley & Fleschner, 1956; Moore *et al.*, 1975). It also is an important predator of tetranychid mites on citrus, pear, tea, and mulberry in Taiwan (Lo *et al.*, 1990), and preys on *Oligonychus* mite on maize in Mexico; temperature threshold for the predator development is 12.5 and 15.4°C for the prey (Quinones *et al.*, 1987). In European apple orchards, *O. flavicornis* Boisduval not only preys on tetranychid mites, but also feeds on thrips, whiteflies, and aphid eggs (Steiner, 1974). *Oligota minuta* Cameron, a common predator of cassava green mite in the Neotropics, was sent from Trinidad to Kenya in 1977 and 1983 but did not become established; in 1983, it was sent to West Africa (Cock, 1985). In Japan, Shimoda *et al.* (1993, 1994a, 1994b, 1997) have studied the biology and seasonal abundance of *O. kashmirica benefica* Naomi, an important predator of *Tetranychus urticae.*

Atheta, another aleocharine genus of over 100 subgenera, has to be mentioned because the biology of at least one species is known in some detail. *Atheta coriaria* (Kraatz) adults and larvae consume eggs and early larval instars of several nitidulid beetles and houseflies (Miller & Williams, 1983).

Paederinae is a fairly large subfamily composed of species found under bark, under stones, near water, and in agricultural habitats. *Paederus fuscipes* Curtis, an Old World species, is found with 18 other staphylinids in croplands of eastern China (Lu & Zhu, 1984). It is an important predator of several major rice pests, consuming up to 49 first-instar paddy stem borers and up to 20 first- or second-instar plant hoppers per day; its functional response was determined as type II (Luo *et al.*, 1989). Its importance in controlling leafhopper population densities in Malaysian rice fields is considered to be underestimated (Manley, 1977). *Paederus fuscipes* in sugarcane fields showed an aggregated distribution pattern based on eight different types of measurement (Lin & Gu, 1987). However, *Paederus* species are unlikely to be used in biological control as almost all species in all life stages contain pederin, a hemolymph toxin that causes blistering dermatitis (Morsy *et al.*, 1996).

Subfamily Staphylininae includes the largest species of the family. Several genera (including *Ocypus*) are placed in the genus *Staphylinus* by some taxonomists. In this broad interpretation, seven European *Staphylinus* species have been introduced inadvertently into North America (Newton, 1987). Most of these species were first collected near coastal port cities, indicating they (as well as many European carabids) were brought in the dirt used as ship's ballast from Europe. *Ocypus olens* Müller, one of these introduced staphylinids, now occurs from southern California north to Oregon, and it is considered a potential agent for biological control of the brown garden snail, *Helix aspersa* Müller, in California (Orth *et al.*, 1975). In gardens where *O. olens* is found, there are fewer snails than in gardens where the staphylinid is absent (Fisher *et al.* 1976). In Europe it was one of the first predaceous insects to be manipulated, and around 1840 Boisgirand destroyed earwigs by placing *O. olens* in his own garden (Howard & Fiske, 1911). Nield (1976) studied the biology and searching behavior in the laboratory and in the field. He described the eggs of *O. olens* as hard shelled (as are other *Staphylinus, Philonthus,* and *Quedius* spp.), and thus differ from the soft-shelled carabid eggs; larvae ate earthworms, blowfly puparia, and *Tenebrio* larvae but not blowfly maggots or slugs. Like other predaceous insects, *O. olens* responds to prey on contact (Wheater, 1989), were more active when starved, and showed a greater rate of turning when fed (Wheater, 1991).

The genus *Creophilus* is considered to be in the subfamily Xanthopyginae by some taxonomists. Among the 14 widely distributed *Creophilus* spp. in the world (Moore & Legner, 1971), *C. erythocephalus* (Linnaeus) was successfully introduced from Australia into the Hawaiian Islands in 1921 to prey on the hornfly, *Siphona irritans* (Linnaeus), and houseflies and stable flies (Pemberton, 1948). Greene (1996) provides details of laboratory-rearing methods for *C. maxillosus* (Linnaeus), a predator of filth flies. *Philonthus* is a large, cosmopolitan genus in the Staphylininae and includes species having a variety of different habitats and prey. Fox and MacLellan (1956) did a precipitin test to determine that some *Philonthus* prey on wireworms, *Agriotes sputator* (Linnaeus), in Canada. Adults of *P. theveneti* Horn were observed to feed on houseflies and stable fly eggs in a New Brunswick feedlot by Campbell and Hermanussen (1974). They exposed the staphylinid to populations of three filth fly species in the laboratory. Stable flies were reduced by 76% and houseflies, by 37 to 41%, but no reduction was obtained in the face fly populations; thus, repeating releases of *P. theveneti* was considered as a possible regulating technique against stable flies. *Philonthus politus* Linnaeus and *P. sordidus* Gravenhorst in California (Legner & Olten, 1968), *P. cruentatus* Guelin and *P. flavolimbatus* Erichson in Texas (Hunter *et al.*, 1989), and *P. americanus* Erichson in Nebraska (Seymour & Campbell, 1993), are all common predators of houseflies and stable flies. Harris and Oliver (1979), using a ratio of 1 adult *P. flavolimbatus* to 10 horn fly eggs, obtained over 99% elimination of producing horn fly adults in the laboratory, but simulated field tests using the same ratio gave only a 72% reduction. *Philonthus cognatus* (Stephens) showed a preference for grain aphids over fungal food (Dennis *et al.*, 1991), but only climbs wheat plants for flight (Dennis & Sotherton, 1994).

Subfamily Tachyporinae contains some genera that are mainly mycetophagous. Little is known about the biology of *Tachyporus,* but the adults and larvae of all species are almost certainly predaceous (Campbell, 1979). However, within the genus there are different degrees of entomophagy and mycophagy. When Dennis *et al.* (1991) dissected guts and made field observations of foraging behavior of *T. hypornum* (Fabricius), *T. chrysomelinus* (Linnaeus), and *T. obtusus* (Linnaeus)—all known to prey on the grain aphid *Sitobion avenae* (Fabricius)—they found differences in mycophagy between species. *Tachyporus hypornum* showed a preference for mildew conidia over aphids, while *T. obtusus* preferred aphids over mildew. Of the species that were abundant in cereals, *T. chrysomelinus* females showed a greater preference for aphids than males did. All species consumed mildew conidia, but in choice experiments, aphid consumption increased for female *T. chrysomelinus* and third-instar *Tachyporus* larvae and decreased for *T. hypornum* adults and first- and second-instar larvae; thus, mycophagy can limit aphid predation and influence numerical response (Dennis *et al.,* 1991). Gut dissection should accompany the use of enzyme-linked immunosorbent assay (ELISA) in determining prey spectra and relative quantities of food eaten by polyphagous predators (Sunderland *et al.,* 1987). Thomas *et al.* (1992) and Dennis *et al.* (1994) believe that field boundary habitats could be exploited to provide suitable overwintering sites.

Subfamily Xantholininae has common species and plays an important role as predators in a wide range of habitats (Smetana, 1982), but only a few have been studied in any detail. The genus *Nudobius* is Holarctic and Ethiopian in its distribution, and apparently all species live under bark of dead or dying coniferous trees (Smetana, 1982). *Nudobius pugetances* Casey, a western North American species, preys wholly on secondary insects under the bark, which limits its usefulness as a biological control agent (Struble, 1930). *Thyreocephalus albertsi* (Fauvel) was collected in the Philippine Islands and was introduced into the Hawaiian Islands as a predator of the Oriental fruit fly, *Dacus dorsalis* Hendel. The adult staphylinids search fallen rotting fruit for fruit fly larvae, and a female can destroy 500 to 600 larvae during its life (Marucci & Clancy, 1952). There is no record of its establishment in Hawaii.

Order Diptera

The Diptera order is divided into three suborders: Nematocera, Brachycera and Cyclorrhapha. A large number of dipteran families (about 30%) are predaceous at some stage during their life cycle (Hagen, 1987). Of the 21 nematoceran families, 4 include at least some predators, as do 10 of the 15 brachyceran families and 15 of the 62 cyclorrhaphan families. Some, such as Empidae, Asilidae, and Scatophagidae, are notable predators as adults but are either scavengers or carnivores in the soil during the larval stage. In other families, such as Cecidomyiidae, Syrphidae, and Chamaemyiidae, the larval stage is actively predatory while the adult stage feeds on pollen, nectar, or honeydew. There are only a few families in which both adult and larval stages are predatory, such as Dolichopodidae, and in these cases the prey used by adults and larvae are notably different. The prey of larval dipteran predators include aquatic invertebrates, soil invertebrates, Homoptera and other small arthropods, wood-boring beetles, Orthoptera eggs, and mollusks, while the prey of adults include a wide variety of small arthropods.

Although many of the predaceous Diptera may play an important role in the natural regulation of prey populations, only those that have been more actively considered as biological control agents will be discussed here.

Cecidomyiidae Cecidomyiidae is a large family of very small, fragile insects that are readily recognized by their long, hairy, beadlike antennae and reduced wing venation. The majority of these gall midges are true gall-forming species, some of which are notorious pests; however, the family also includes some important predators of spider mites, Homoptera, and other Cecidomyiidae (Nijveldt, 1969).

A number of cecidomyiid genera, including *Coccodiplosis, Dicrodiplosis,* and *Vincentodiplosis,* appear to have specialized on coccoid prey (Harris, 1968, 1990). There is little detailed information on their biology and impact on prey populations, but preliminary observations (Harris, 1968) indicate that they are likely to be of importance as natural controls. The genera *Arthrocnodax, Feltiella* [revised by Gagne (1995)] and *Therodiplosis* contain a number of species that are predatory on spider mites (Chazeau, 1985), and *Lestodiplosis* contains species predatory on a wide range of prey, including other cecidomyiids. McMurtry *et al.* (1970) reported *A. occidentalis* Felt to be a good predator, consuming 380 mites over a period of 17 days.

The aphid predators are the most well-known group of predatory midges and have received the greatest attention in biological control. The genera *Aphidoletes* (three species) and *Monobremia* (one species) feed exclusively on aphids (Harris, 1973). *Aphidoletes aphidimyza* (Rondani) is the best known species of the group with a distribution throughout the Northern Hemisphere. There are many accounts of its biology (e.g., Harris, 1973; Markkula *et al.,* 1979; Nijveldt, 1988). Adults emerge from pupae in the soil and are nocturnal in habit, living for up to 14 days, with a fecundity of up to 100 eggs. The orange eggs (0.3×0.1 mm) are laid singly or in clusters on plants near aphid colonies. The eggs hatch in 3 to 4 days and the first-instar larvae immediately attack aphids, paralyzing them with an injected venom (Laurema *et al.,* 1986). Larval development is completed in 7 to 14 days during which time they can kill from 5 to 10 large aphids or up to 40 to 80 small aphids. Pupation occurs in silken cocoons in the soil. Adults emerge after 1 to 3 weeks, depending on temperature, and the life cycle is often completed in about 3 weeks [280.2 degree days above a threshold temperature of 6.2°C

(Havelka, 1980) and 284 degree days above 9.2°C (Raworth, 1984)].

Aphidoletes aphidimyza appears able to distinguish between different aphid densities, laying eggs in proportion to the density of *Myzus persicae* on brussels sprouts (El-Titi, 1973) and *Aphis pomi,* on apples (Stewart & Walde, 1997). Aphid honeydew is used as a food by females (Uygun, 1971) and may also be used in prey location (Havelka & Syrovatka, 1991). Honeydew is also attractive to the newly hatched larvae (Wilbert, 1977) and this may help them to find the aphid colony. Adult females can distinguish different host plants. Mansour (1975) reported that after rearing three generations on *M. persicae* on turnip root cabbage, more eggs were laid on this host plant than on brussels sprouts.

There are several features of the biology of *A. aphidimyza* that enhance its potential as a biological control agent. El-Titi (1973) has demonstrated that oviposition is density dependent, although only weakly so under field conditions (Stewart & Walde, 1997). It also tends to kill more aphid prey than it consumes (Uygun, 1971), which could result in density-dependent prey mortality. The range of aphid prey accepted is broad (Harris, 1973) and covers all greenhouse aphid pests (Markkula *et al.,* 1979), enabling it to be used against a variety of aphid pests on different greenhouse crops. Of the aphid predators, *A. aphidimyza* is the only one that can maintain populations throughout the season under glass (Ramakers, 1988). The midge is very easy to rear (Markkula & Tiittanen, 1976, 1985) and has been produced commercially both in Europe and in Canada since the early 1980s. The pupae in cocoons are very resilient and provide an ideal stage for transportation and release in the crop. The ability of *A. aphidimyza* to withstand cold storage at 5°C for up to 8 months has been demonstrated (Gilkeson, 1990). Cold storage is an important attribute for the commercialization of a biological control agent and this advancement is likely to lead to more widespread use of *A. aphidimyza* for the biological control of aphids, not only under glass but also in small-scale plantings of high-density field or orchard crops (e.g., Bouchard *et al.,* 1988).

There has been considerable interest in the potential of *A. aphidimyza* as a biological control agent for aphids in greenhouses in the former USSR (Ushchekov, 1975; Storozhkov *et al.,* 1981), Europe (Markkula & Tiittanen, 1980; van Lenteren & Woets, 1988), China (Cheng *et al.,* 1992), and North America (Meadow *et al.,* 1985; Gilkeson & Hill, 1987). Two to three applications of pupae at 7- to 10-day intervals, using release rates of 1 predator to 200 aphids, kept *Aphis gossypii* in check on cucumber in the former USSR (Ushchekov, 1975). In contrast, in Finland (Markkula & Tiittanen, 1980) and in the United States (Meadow *et al.,* 1985) release rates of one predator to three aphids at 14-day intervals were required to control *M. persicae* on peppers and tomatoes. A comparison of release rates for control of *M. persicae* on peppers was made by Gilkeson

and Hill (1987), who found that a ratio of 1 predator to 10 aphids gave the best results. It seems likely that both host-plant characteristics including size and surface morphology and the degree of aggregation of the target aphid could affect the performance of *Aphidoletes aphidimyza* in different greenhouse crops and could necessitate further examination of required release rates.

Syrphidae Syrphidae, or hover fly, is a large family of Diptera particularly well known for striking mimicry of bees and wasps. The feeding habits of the family are diverse but the subfamily Syrphinae are important predators of aphids and other Homoptera (Gilbert, 1981; Chambers, 1988), and occasionally of chrysomelid leaf beetles (Rank & Smiley, 1994); and the tribe Pipizini of the Eristalinae are important predators of gall-forming aphids (Rojo & Marcos-Garcia, 1997). The predaceous syrphids can be divided into two groups, obligate predators and facultative predators (Chandler, 1968a; Rotheray, 1983). The majority of species belong to the former group, while some *Melanostoma* species and perhaps almost all *Platycheirus* species are facultative predators and have the ability to make use of rotting plant material to complete their development.

Adult females emerge with an undeveloped reproductive system and require a protein source to mature eggs. The principal food of adult syrphids includes pollen and nectar (Gilbert, 1981; Hickman *et al.,* 1995). Mating has rarely been observed in syrphids and often takes place on the wing. A few species such as *Syrphus ribesii* (Linnaeus) form male swarms. The fecundity of *Epistrophe nitidicollis* (Meigen) was estimated to be 126 (Tinkeu, 1995), but that of other syrphids can be as large as 500 to 1500 eggs, as for *Metasyrphus corollae* (Fabricius) (Benestad, 1970; Scott & Barlow, 1984). The white eggs (1.0 × 0.5 mm) are generally laid singly on the underside of leaves supporting aphid colonies. The identity of the eggs of some species can be determined from their surface sculpture (Chandler, 1968b). The duration of the egg stage is about 5 days. The hover fly larvae are active between dusk and dawn; their nocturnal habits make them less conspicuous than other aphid predators. They have sucking mouthparts and often lift the aphid prey from the plant surface while feeding. Prey consumption rates in the laboratory vary from a low of 135 aphids for *Platycheirus clypeatus* (Meigen) (Bankowska *et al.,* 1978) to as many as 550 for *Scaeva pyrastri* (Linnaeus) (Wnuk & Fuchs, 1977). Larval development in *Episyrphus balteatus* (DeGeer) can be completed in as little as 10 days but more frequently extends to about 40 days under field conditions (Ankersmit *et al.* 1986). Pupation occurs on the plant or in the leaf litter and adults emerge after 1 to 3 weeks.

The majority of syrphid predators are multivoltine, although some are univoltine, and overwinter as final instar larvae, pupae, or adults. The range of prey consumed by predaceous syrphid larvae can be extensive. Ruzicka (1975) reported that the polyphagous *M. corollae* can successfully develop on 12 different aphids but that the development rate and adult size [which can affect subsequent fecundity (Scott & Barlow 1984)] vary between prey species. One species, *Cavariella theobaldi* (Bragg), was found to be toxic to the syrphid. Syrphids are also attacked by a wide range of parasitoids (Rotheray, 1984).

The oviposition behavior of aphidophagous syrphids is influenced by a number of olfactory, visual, and mechanical cues (Chandler, 1968a; Chambers, 1988). In *M. corollae*, oviposition is stimulated by olfactory cues originating from honeydew or from aphid cornicle secretions (Bombosch & Volk, 1966; Hagen *et al.*, 1970a; Ben Saad & Bishop, 1976; Budenberg & Powell, 1992). The ovipositor appears to have sensilla responding to honeydew components (Hood Henderson, 1982), whereas antennal sensilla respond to green leaf volatiles (Hood Henderson & Wellington, 1982). For *E. balteatus,* females land on ears of wheat more frequently if contaminated with honeydew and the number of eggs laid increased with honeydew concentration (Budenberg & Powell, 1992). *Metasyrphus corollae* also appears to prefer vertical surfaces (Sanders, 1980) and darker surfaces (Sanders, 1981) for oviposition. Cannibalism of eggs by third-instar larvae of *E. balteatus* is common under laboratory conditions (Branquart *et al.,* 1997), but the extent of cannibalism in the field is unknown. Oviposition by *Platycheirus* and some *Melanostoma* species, the facultative predators, differs from that of the obligate predators in that eggs are frequently laid away from aphid colonies (Chandler, 1968b). These eggs are laid in small groups in a row and there is some evidence that the young larvae are cannibalistic.

Syrphids have occasionally been used in classical biological control but to date all have failed to establish in the target regions, perhaps due to a lack of mating [see Waage *et al.* (1984)] in released material. The impact of syrphids in the context of natural control has seldom been assessed in detail and their role may have been underestimated as a result of their nocturnal activity. Bugg (1993) concludes that it is difficult to demonstrate the effect of field margins and flowers, as adult food sources, on the effectiveness of syrphid predation. However, observations on *E. balteatus* suggest that it selectively feeds on noncrop plants around the margins of cereal fields in England (Cowgill *et al.,* 1993a) and that natural control of cereal aphids could be enhanced by modified management of field boundaries (Cowgill *et al.,* 1993b), particularly using *Phacelia tanacetifolia* as a border plant (Hickman & Wratten, 1996)

Three different techniques have been used to study syrphids: field cages for predator exclusion, laboratory cage studies, and modeling. Predator exclusion studies (Tamaki & Weeks, 1973; Chambers *et al.,* 1983; Hopper *et al.,* 1995) indicate the value of the aphidophagous predator complex as a whole but do not permit the separation of the effects of syrphid from coccinellid predation. Laboratory cages were used by Wnuk (1977) to demonstrate that *E. balteatus* can eliminate colonies of *Aphis pomi* within a few days with predator to prey ratios of from 1:50 to 1:200. Predation within cages is undoubtedly greater than under field conditions and may indicate the potential of syrphids for use in greenhouses. Some preliminary experiments using *M. corollae* for the control of *A. gossypii* on cucumber in greenhouses (Chambers, 1986) have demonstrated promise but currently the cecidomyiid *Aphidoletes aphidimyza* is favored due to its ability to produce self-perpetuating populations and its ease of rearing (see earlier). Bombosch (1963) first suggested that the impact of syrphids could be evaluated by modeling the predator feeding potential. This approach has been extended by Tamaki *et al.* (1974) and Chambers and Adams (1986) but these models remain dependent on how applicable laboratory data on predator feeding potential relate to predator power in the field. For example, Tenhumberg (1995) estimated the feeding potential of *E. balteatus* in field cages to be only half that found in laboratory studies.

Chamaemyiidae Chamaemyiidae is a small family of predatory silver-colored flies specializing on coccoid and aphidoid prey. Several species have been successfully used for the classical biological control of adelgid pests. It is surprising, therefore, that the biology and the range of prey attacked by this group of predators are in general poorly known. Detailed information is restricted to only a few species.

Of some 12 genera worldwide (McAlpine, 1960; Tanasijtschuk, 1993), the majority of species belong to the rather large genus *Leucopis*. The subgenus *Neoleucopis* contains species that are predators of adelgids (McAlpine, 1971; McLean, 1992), especially *Adelges, Pineus,* and *Sacciphantes*. The subgenus *Leucopis* also contains some adelgid specialists (e.g., *L. argenticollis* Zetterstedt and *L. astonea* McAlpine) as well as the genera *Cremifania* (Brown & Clark, 1956) and *Lipoleucopis* (Wilson, 1938). Other species in the subgenus *Leucopis* are predators of aphids (Tawfik, 1965; Sandhu & Kaushal, 1975; Tracewski, 1983; Tanasijtschuk, 1996), phylloxerids (Stevenson, 1967), mealy-bugs (Sluss & Foote, 1971, 1973), or soft scales (N. J. Mills, unpublished).

The adelgid predators can be univoltine, such as *Lipoleucopis praecox* (Meigen) (Wilson, 1938), but are more typically bivoltine, such as *Leucopis tapiae* Blanchard (Eichhorn *et al.,* 1972), *L. pinicola* Malloch (Sluss & Foote, 1973), and *L. obscura* Haliday (Brown & Clark, 1956). In contrast, the aphid and scale predators appear to be multivoltine, as is *L. verticalis* Malloch (Sluss & Foote, 1971), *L. conciliata* McAlpine & Tanasijtschuk (Tanasijtschuk, 1959), and *Chamaemyia polystigma* Meigen (Sluss & Foote, 1973). The life cycle is similar for all species and is represented here by *L. pinicola* (Sluss & Foote, 1973). The adults emerge from puparia in the bark or soil and have a preoviposition period of about 18 days and a fecundity of about 40 eggs. Adults live about 25 to 30 days and lay small groups of white eggs (0.4 × 0.1 mm) in association with colonies of their prey. Eggs hatch after 8 to 10 days and the duration of the three larval instars is about 3 weeks under laboratory conditions. The larvae move away from the prey colonies to pupate and the pupal period lasts about 2 weeks in the laboratory. Methods for rearing *Leucopis* species have been developed by Gaimari and Turner (1996a).

These chamaemyiid predators appear to be relatively restricted in the range of prey that they attack. The subgenus *Neoleucopis* is restricted to adelgids and within this group there are species specializing on *Adelges, Pineus,* or *Sacciphantes* prey (McAlpine, 1971). Greathead (1995) considers the Palearctic species *L. obscura* and *L. atratula* (Ratzeburg) to be associated with *Adelges* on fir, whereas *L. tapiae, L. atrifacies* Aldrich, *L. nigraluna* McAlpine, and *L. manii* McAlpine are asoociated with *Pineus* on pine. Among the aphid and coccid specialists, *Leucopis conciliata* and *L. verticalis* are known only in association with a few closely related prey species (Sluss & Foote, 1971; McAlpine & Tanasijtschuk, 1972). The larvae of the aphid predators *L. gaimarii* Tanasijtshuk and *L. ninae* Tanasijts-huk are voracious predators consuming about 100 aphids before completing their development (Gaimari & Turner, 1996b). The prey location mechanisms of Chamaemyiidae have not been studied.

Several *Leucopis* species and *Cremifania nigrocellulata* Czerny have been used in the classical biological control of *Adelges piceae* in North America; *Pineus pini* (Macquart) in Hawaii; and *P. laevis* (Maskell) in Australia, New Zealand, and Chile (Mills, 1990). Three *Leucopis* species and *C. nigrocellulata* have been established against *A. piceae* in North America but control has not been achieved (Clark *et al.,* 1971). In contrast, *L. tapiae* in New Zealand (Zondag & Nuttall, 1989), *L. obscura* in Chile (Zuniga, 1985), and both *L. tapiae* (originally misidentified as *L. obscura*) and *L. nigraluna* in Hawaii (Culliney *et al.,* 1988; Greathead, 1995) have provided successful control of *Pineus* species. In addition, the Palearctic aphid predator *L. ninae* has been introduced to the western United States against the Russian wheat aphid (Tanasijtschuk, 1996).

Dolichopodidae   This is a large family comprising more than 2000 species of small, bristly flies that are often metallic green or blue. The adults are predatory on small, soft-bodied insects and mites, and the larvae are probably all carnivorous. Dolichopodid species are found in a wide variety of habitats but the best known predators are those of the *Medetera* genus that prey on the larvae and pupae of wood-boring insects.

Medetera is a distinctive genus of more than 250 described species and their association with bark beetles has attracted attention in both the Nearctic and Palearctic regions (Bickel, 1985). Most *Medetera* larvae are subcortical predators where they have been recorded preying on scolytids (DeLeon, 1935; Nuorteva, 1959; Beaver, 1966; Nicolai, 1995), *Pissodes* weevils (Deyrup, 1978), and other bark-inhabiting insects.

Medetera aldrichii Wheeler has been studied in the most detail (DeLeon, 1935; Schmid, 1971; Nagel & Fitzgerald, 1975). Eggs are laid under bark scales and the hatching larvae enter the bark beetle galleries to feed on the prey. Full-grown larvae overwinter before pupation and adult emergence occurs in the spring. Although Kishi (1969) reported *Medetera* larvae to be scavengers instead of predators, other studies (DeLeon, 1935; Beaver, 1966; Schmid, 1971; Nagel & Fitzgerald, 1975) demonstrate that they are voracious predators consuming from 6 to 15 prey during their larval development.

Medetera species appear to be able to make use of the aggregation pheromones of their bark beetle prey to locate suitable host trees for the development of their progeny. Various *Medetera* species have been attracted both to naturally infested tree bolts and to traps baited with synthetic aggregation pheromones (Williamson, 1971; Fitzgerald & Nagel, 1972; Borden, 1982; Mills & Schlup, 1989). Nagel and Fitzgerald (1975) also found that the orientation of the newly hatched larvae of *M. aldrichii* to scolytid entrance holes was stimulated both by the host tree volatile D-α-pinene and an extract of the frass of *Dendroctonus pseudotsugae* Hopkins. *Medetera* species appear not to be particularly specialized in relation to their bark beetle prey and have never been used in a biological control program. Moeck and Safranyik (1984), however, considered that the observed abundance and predation potential of *Medetera* species would make them good candidates for augmentative biological control of bark beetle pests.

Lonchaeidae   Little is known of this moderately sized family of flies though in many cases the larval stages are considered to be scavengers on rotting vegetation or dung. A notable exception, however, is *Lonchaea corticis* Taylor, an active predator of the prepupae and pupae of the subcortical weevil, *Pissodes strobi* (Peck), that attacks the leading shoots of young pines and Sitka spruce (Alfaro & Borden, 1980; Hulme, 1989).

The young larvae develop as scavengers in the larval galleries of the weevil, becoming predatory only when the prey have completed their larval development and formed a pupal cell under the bark. Each *L. corticis* larva can then consume about three pupae (Alfaro & Border, 1980), which results in a significant reduction in overwintering populations of the weevil (Hulme, 1990). One of the most effective means of controlling the weevil is to remove attacked leaders during winter, and techniques have been developed to use cold hardiness as a means of conserving the natural enemies from cut leaders (Hulme *et al.,* 1986).

Sciomyzidae The sciomyzid, or marsh fly, family is composed of about 600 species living in all zoogeographic regions of the world. It was not until 1953 that Berg (1953) first reported an association of sciomyzids with snails and other mollusks. Since then, the family has become one of the better known biologically, in view of its importance in natural and biological control (Knutson, 1976).

The feeding behavior of the sciomyzids varies from parasitism, through predation to saprophagy, with many intermediate forms. Most of the predaceous larvae are aquatic and feed on freshwater snails occurring at or near the water surface. Slugs and terrestrial or semiaquatic snails are the hosts of the parasitic species. *Dictya* species provide an example of the biology of predaceous sciomyzids (Valley & Berg, 1977). Adults live for up to 100 days and generally have a preoviposition period of less than 3 weeks. The fecundity of females extends to several hundred eggs (white, 0.7 × 0.2 mm); these tend to be laid in rows on vegetation and the larvae hatch after 2 to 4 days. The prey include a variety of aquatic pulmonate snails. Larvae may feed for over an hour on a single prey but are seldom found in the field in contact with their prey. The larvae are able to swim across the surface of water and attack their prey by crawling between the shell and mantle. From 12 to 53 snails are killed by a single larva during its development. Puparia are formed on the water surface and adults emerge after 7 to 9 days.

The biology of Nearctic *Tetanocera* species that are predators of semi-aqautic snails is provided by Foote (1996a, 1996b). Vala *et al.* (1995) describe the life history and larval biology of the Afrotropical *Sepedon trichrooscelis* Speiser. Gormally (1988) examines the influence of temperature on snail predation in *Illione albiseta* (Scopoli), and McLaughlin and Dame (1989) provide notes on rearing *D. floridensis* Steyskal.

There are relatively few sciomyzids that are known to attack slugs (Reidenbach *et al.,* 1989). Trelka and Foote (1970) provide details of the biology of three Nearctic *Tetanocera* species that spend the first two instars as parasitoids and the final instar as predators of slugs. The univoltine *Euthycera cribrata* (Rondani) is a European species that lays 400 to 600 eggs after a preoviposition period of several months (Reidenbach *et al.,* 1989). Larvae sit and wait for a passing slug or snail and then burrow into the host, feeding first as a parasitoid and then as a predator or saprophage once the host dies. Each host is used for 4 to 11 days and the larvae attack 15 to 25 slugs over a period of 2 to 3 months of larval development.

In a survey of sciomyzids attacking terrestrial helicid snails in the Mediterranean, Coupland (1996a) found six species. The majority of species were univoltine, with a preoviposition period as long as 72 days and a larval development period of up to 6 months. Most species were associated with forest and scrubland, although the multivoltine

Pherbellia cinerella (Fallen) is associated with pastures (Coupland & Baker, 1995).

The predators of aquatic pulmonate snails appear to have a broad host range. In contrast, the parasitoid species are more restricted in host range, the *Tetanocera* species being confined to a single genus of slugs or even to a single species of *Deroceras* slugs in the case of *T. plebeia* (Loew) and *T. valida* (Loew) (Trelka & Foote, 1970). Prey specificity among the predators and saprophages of terrestrial snails appears to vary from being very broad to quite specific (Coupland *et al.,* 1994; Foote, 1996a). Little is known of the prey location mechanisms in the Sciomyzidae, but Coupland (1996b) demonstrated that the feces of terrestrial snails induced oviposition by *Pherbellia cinerella,* and both feces and mucus stimulated the search of first-instar larvae.

Sciomyzids have been used several times as potential biological control agents of aquatic snails (Mead, 1979; Godan, 1983). For example, in 1959 the Nicaraguan *Sepedon macropus* Walker was imported into Hawaii (previously devoid of Sciomyzidae) against *Lymnaea ollula* Gould, the principle vector of the giant liver fluke. The flies established and became abundant but their impact on the snail vector has not been adequately studied, although the incidence of the liver fluke has declined (Chock *et al.,* 1961; Knutson, 1976). Marharaj *et al.* (1992) have considered the potential of *S. jonesi* Barraclough (= *scapularis* Adams) for biological control of *Bulinus africanus* (Krauss), an important vector of bilharzia blood flukes in Africa.

There have been no introductions of sciomyzids for the biological control of terrestrial snails. However, there has been considerable interest in the potential of European species for the control of four exotic species of helicid snails in Australia (Coupland *et al.,* 1994; Coupland & Baker, 1995; Coupland, 1996a).

Order Hymenoptera

The Hymenoptera constitute one of the largest and oldest orders of holometabolous insects. More than 120,000 species have been described, but the actual number of species is considerably higher. In terms of geologic age, the Hymenoptera have a fossil record referable to the Triassic. Taxonomically, the Hymenoptera are divided into the suborders Symphyta and Apocrita. The Apocrita are divided into the divisions Parasitica and Aculeata. Collectively, Symphyta, Parasitica, and Aculeata represent well-recognized lineages whose relationships to one another are not clearly established.

Hymenoptera are important for understanding entomophagous habits because they are numerically large and biologically diverse, with all forms of phytophagy, predation, and parasitism expressed within each lineage. Traditional definitions of parasitism and predation are provided elsewhere in this book. These definitions fail when examining details of the biology of many Hymenoptera because we see complex life histories with some species utilizing different feeding strategies between the larval and adult stages. For instance, the larva may express characteristics of a parasitoid while the adult shows signs of predation. Indeed, this problem is reflected in numerous articles that cover "host records" and "prey selection" in discussing a species.

Hagen (1987) lists the families of Hymenoptera in which predation has been reported. A modification of that list is provided in Table 1. From the standpoint of feeding behavior, predaceous Hymenoptera adults also consume pollen and take nutrition from floral and extrafloral nectaries. Much of the information concerning predation by Hymenoptera is anecdotal and the role of plant food is not clear. Nevertheless, at the family level trends in the evolution of predation are apparent.

The impact of predatory behavior by Hymenoptera on economic pests has not been appreciated because it is difficult to measure. Many species feed casually on other insects and the magnitude of the impact of this predation has not been appreciated in terms of biological control. However, Hagen *et al.* (1976a) discuss the biology of predators as it relates to biological control.

Anatomical features that have been adapted for predation are important in understanding the position of Hymenoptera as predaceous insects. Mandibulate mouthparts and an appendicular ovipositor are anatomical features common to many Hymenoptera, and are essential for the implementation of predatory behavior. Mandibles are used by many insects to consume prey, but careful observation must be exercised to distinguish between malaxation and feeding. For instance, many reports of adult parasitoids host feeding exist but these must be confirmed to establish genuine predation.

Predaceous Hymenoptera feed on all stages of development, but most often they attack the immature stages of their victims. The egg stage is vulnerable to attack by predators and sometimes species that are usually classified as parasitic behave as predators. For instance, many groups of Parasitica attack the egg stage. When a developing larva consumes one egg, it is generally regarded as parasitic; when a developing larva consumes more than one egg, it is generally regarded as an egg predator. We see this in some Pteromalidae (*Scutellista*), Scelionidae (*Baeus*), and Eurytomidae (*Desantisca* spp.). Larvae, nymphs, and adults may have methods of defense that are somewhat helpful in deterring predation by Hymenoptera. The pupae of some insects have gin traps, which may help deter predation or parasitism. However, the pupa is generally more vulnerable than the nymph, larva, or adult.

Symphyta

The Symphyta (wood wasps, horntails, and sawflies) are cosmopolitan in distribution with about 10,000

TABLE 1   Families of Hymenoptera with Predatory Behavior

Family	Predaceous Stage	Taxa of Prey
Symphyta		
Tenthredinidae	Adult	Coleoptera, Diptera, Homoptera
Parasitica		
Ichneumonidae	Adult, larva	Spider eggs, bee eggs, larvae
Braconidae	Adult, larva	Diptera, Lepidoptera
Evaniidae	Larva	Dictyoptera eggs
Gasteruptiidae	Larva	Hymenoptera larvae
Aphelinidae	Adult	Scale insects
Encyrtidae	Adult, larva	Scale insect eggs
Eulophidae	Adult, larva	Lepidoptera and Coleoptera larvae, eggs of many insects and some spiders
Eupelmidae	Adult, larva	Eggs of many insects
Eurytomidae	Larva	Orthoptera, Homoptera eggs
Pteromalidae	Adult, larva	Coccidoidea eggs
Scelionidae	Adult	Mantid adult
Torymidae	Larva	Mantid eggs
Trichogrammatidae	Adult	Lepidoptera eggs
Aculeata		
Chrysididae	Adult, larva	Hymenoptera and their prey
Bethylidae	Adult	Coleoptera, Lepidoptera
Dryinidae	Adult	Homoptera
Mutillidae	Adult	Hymenoptera larvae
Tiphiidae	Adult	Scarabaeidae larvae
Thynnidae	Adult	Scarabaeidae larvae
Pompilidae	Adult	Spiders
Formicidae	Adult, larva	Many arthropods, including insects
Eumenidae	Adult, larva	Lepidoptera, Coleoptera larvae
Masaridae	Adult, larva	
Vespidae	Adult, larva	Lepidoptera, Coleoptera larvae
Sphecidae	Adult, larva	Many insects

After Hagen, K. S. (1987). In F. Slansky, Jr. & J. G. Rodriguez (Eds.). *Nutritional ecology of insects, mites, and spiders.* (pp. 533–577). New York: John Wiley & Sons.

nominal species. Symphyta are the most primitive Hymenoptera based on adult and larval morphology. Xyelidae are among the earliest fossils and possess features not present in other Hymenoptera (Rasnitsyn, 1980). Some contemporary taxonomists believe that the Symphyta are a paraphyletic group because they possess anatomical features common to all Hymenoptera. Hymenopterists have not thoroughly explored the relationship between taxonomic characters and biological characteristics. Thus, we cannot infer that biological features observed in closely related groups did not evolve more than once.

The most conspicuous adult feature that separates most Symphyta from Apocrita is the broadly attached, sessile metasoma (gaster) to the mesosoma, though some groups of Parasitica also display a relatively broadly attached mesosoma and metasoma. The appendicular ovipositor is well-developed in Symphyta and Apocrita. In sawflies it is used to saw plant tissue, but in the more highly evolved Hymenoptera it serves several additional functions that are discussed elsewhere.

Larval Symphyta differ from Apocrita in that the former displays a well-developed head capsule with strong mandibles and resembles caterpillars. The larval thorax of most sawflies has three pairs of legs and the abdomen holds six to eight pairs of prolegs, providing the larva feeds externally. There are no crochets on the prolegs of sawflies as seen in caterpillars. Conventional wisdom views the Symphyta as phytophagous. The larval stages of most Symphyta are phytophagous but a few (Orussidae) are parasitic. Clausen (1940) reports that the adults of some tenthredinids are predaceous.

Parasitica

Aspects of predation among so-called parasitic Hymenoptera (division Parasitica) have been discussed by Jervis and Kidd (1986) and Heimpel and Collier (1996). The consumption of host body tissues or fluids by an adult parasitoid is an aspect of biology that may be viewed as a form of predation. For lack of a better term, this phenomenon has been called "host feeding" by entomologists. Bartlett (1964a) provided the first comprehensive review of the phenomenon in parasitic Hymenoptera. Among parasitic Hymenoptera both sexes typically take food as adults, but host feeding seems restricted to adult females. The victim is not necessarily killed, and the role that moribund hosts play in population dynamics has not been examined. In a different context, victims may not be used as a host for progeny development. That is, many adult parasitic Hymenoptera feed on body fluids of potential hosts without laying eggs in or upon these potential hosts. Host feeding appears widespread among the Apocrita. Jervis and Kidd (1986) report that more than 140 species in 17 families of Hymenoptera engage in host feeding.

Host feeding can affect the female wasp in several ways, including physical maintenance, longevity, and fecundity. The relationship between physical maintenance and longevity is intimate and difficult to separate and few studies have attempted to do this. From the female parasitoid's perspective, host feeding many be an alternative to starvation. During the process of starvation some wasps, such as the ichneumonid *Pimpla turionellae* (Linnaeus), may undergo the loss of flight musculature (Sandlan, 1979).

Several studies of parasitic Hymenoptera in many families have shown that fecundity is highest in females that have been provided host hemolymph. Flanders (1950) called neonate parasitic Hymenoptera females synovigenic when they did not possess a full complement of eggs ready for oviposition. Flanders contrasted this condition with so-called proovigenic parasitoids that emerge as adults with a full complement of eggs ready for oviposition. Flanders hypothesized that synovigenic female parasitoids must consume nutrients that are necessary for the development of their eggs. Feeding on host body fluids is one source of nutrition that may provide necessary nutrients for egg development in synovigenic parasitoids. While this characterization is conceptually appealing, we should observe that there are few rigorous data that could be used to support this dichotomy of reproductive capacity or physiology. However, one study of host feeding in *Aphytis melinus* DeBach clearly shows that while host hemolymph does not support the longevity of parasitoid adults, the typical host-feeding pattern of one host per day contributes nearly four eggs per host feed to their fecundity (Heimpel *et al.*, 1997).

Female parasitoids use various anatomical techniques to feed on an insect that can be viewed as a potential host. To feed upon a host, some species use the ovipositor to inflict a wound from which hemolymph wells. The fluid is imbibed by the female parasitoid. Other species construct a feeding tube with the ovipositor. The parasitoid thrusts the ovipositor into the body of the host and hemolymph coagulates around the ovipositor, thereby forming a tube. When the tube is completed, the female parasitoid applies her mouthparts to the apex of the tube and feeds on host fluids via capillary action (Fulton, 1933). Some species of parasitoids use their mandibles to mutilate the host and feed on host body fluids at the wound. Other species consume host tissues in addition to body fluids (Waloff, 1974).

Several types of host-feeding behavior have been outlined by Jervis and Kidd (1986). Concurrent host feeding involves females that oviposit and host feed on the same individual; nonconcurrent host feeding involves females that do not oviposit on the same individual used for host feeding. Destructive host feeding involves the death of the victim on which the parasitoid feeds; nondestructive host feeding relates to victims that survive the host-feeding process. Destructive feeding is usually nonconcurrent because the victim is not suitable as an ovipositional resource (a host). The impact of nondestructive host feeding may be even more difficult to document under field conditions, and thus more difficult to assess in terms of usefulness in biological control. Nondestructive feeding may be nonconcurrent or concurrent. During the lifetime of an adult parasitoid we can see autogenous and anautogenous host-feeding behavior. Autogenous species are capable of laying a complement of eggs without host feeding. In such species, host feeding may be facultative. Anautogenous species must host feed to provide nutrition for the egg that is otherwise unavailable; in such species, host feeding is obligatory.

Host feeding may be related to egg resorption. Many parasitic wasps resorb eggs that are not passed from the ovariole in a timely manner. Studies by Antolin and Williams (1989) on *Muscidifurax zaraptor* Kogan & Legner, a synovigenic pupal parasitoid of houseflies, and by Collier (1995) on *A. melinus,* a synovigenic nymphal ectoparasitoid of California red scale, report that both parasitoids resorb eggs when starved and that host feeding is necessary for continued egg maturation.

Early specialists in biological control apparently regarded host feeding as an ancillary aspect of parasitoid behavior. Nevertheless, a few studies suggest that host feeding can result in a significant level of mortality (DeBach, 1943). Gadd and Fonseka (1945) note that *Neoplectrus maculatus* Ferriere kills and consumes one nettle grub for each three that are used as hosts. Gadd *et al.* (1946) suggest that *Platyplectrus natadae* Ferriere kills one nettle grub for each four that are used as hosts and *P. taprobanes* (Gadd) kills one nettle grub for each six that are used as hosts. However, theoretical evidence suggests that the inclusion of host feeding in a host–parasitoid model leads to an increase in the equilibrium density of a host population (Mills & Getz, 1996). In an effort to resolve the conflict of between conventional wisdom and theory, Jervis *et al.* (1996) analyzed the biological control record of parasitoid introductions against Homoptera. They found that destructive host feeders were superior in both rates of establishment and success, suggesting that other attributes, such as searching ability, may be linked to host feeding in parasitoids.

Aculeata

Predation by adult aculeates represents a very different phenomenon than the host feeding observed within the parasitic Hymenoptera. A feature almost universally present in aculeates is the sting, with its attendant glandular secretions. Only females possess the sting. A few groups of Aculeata, such as some bees and ants, have lost the sting. However, for the bulk of the Hymenoptera, the sting represents an obvious anatomical adaptation that enables the bee, wasp, or ant to defend itself from predation and/or to immobilize prey.

Chrysidoidea   are regarded as the most primitive aculeate Hymenoptera. Six families are members of the Chrysidoidea: Bethylidae, Chrysididae, Dryinidae, Embolemidae, Sclerogibbidae, and Scolebythidae. The Sierolomorphidae have sometimes been placed within the Chrysidoidea in some classifications, but their true affinities lie with the Scolioidea. All families of Chrysidoidea are viewed as

parasitic, but adults of some groups show predaceous characteristics. For instance, adult Dryinidae commonly feed on hosts, which may be nonconcurrent (Waloff, 1974; Chua *et al.,* 1984). Dryinidae are remarkable among parasitic Hymenoptera in that the adults are larger than the hosts that they parasitize. This fact may be attributed to the prolonged feeding time without apparent harm to the host. A consequence of this feeding behavior is that smaller species of hosts can be utilized by relatively large-bodied dryinids (Ponomarenko, 1975).

Bethylidae host feeding by adult females is more sporadic. Epyrinae, such as *Laelius pedatus* (Say), *Holepyris glabratus* (Fabricius), and *Sclerodermus* spp., often host feed (Bridwell, 1919, 1920). *Sclerodermus* species do not pierce the integument of the host, but feed on hemolymph exuded through the cuticle during malaxation (Bridwell, 1920). Female *Sclerodermus* also feed on hosts on which parasitoid larvae are developing and will feed on their own eggs and larvae; Bridwell also reports larvae feeding on their sibling larvae, but indicates that he has observed no maternal behavior toward larval progeny. Reports of host feeding have been made for some Bethylidae of the genus *Goniozus.* Many species of *Goniozus* malaxate their hosts but do not host feed (G. Gordh, personal observation). The account by Ranaweera (1950) leaves little doubt that host feeding was observed in a wasp he called *G. montanus,* and numerous reports of host feeding by *G. nephatidis* make it unlikely that the phenomenon does not occur in that species (Rao & Cherian, 1928; Jayaratnam, 1941).

Formicoidea   are regarded as among the most highly evolved Hymenoptera, in part because they display sociality. Sociality combined with predatory habits can represent a quantum leap above the impact expressed by conventional predation. Predation by ants is expressed in most subfamilies of ants. Hölldobler and Wilson (1990) summarized the genera of ants involved in prey specialization. Specialized prey include arthropod eggs, centipedes, millipedes, apterygotes, termites, beetles, moth larvae, and other Hymenoptera. Some ants are generalist perdators. As a consequence of interspecific fights, some species of ants carry other species of ants (enemies) back to the nest to be used as food for the colony. In other ants the predator–prey relationship is more intimate and cannibalism occurs (Driessen *et al.,* 1984).

Ants are behaviorally and biologically complex, and they express predatory behavior in many ways. Adaptations for predatory behavior are more apparent in ants than in other insects. Ants possess a sting that provides an effective weapon for the injection of offensive chemicals that neutralize prey. Mandibles are well-adapted structures that can be used to dismember prey. Behavior in ants is complex and workers frequently cooperate in the retrieval of prey.

The importance of ants in biological control is often

complex and sometimes contradictory (Way, 1992). To accurately assess the impact of ants in biological control, we must correctly identify the ants involved. Next, we must determine whether the ant species is endemic, adventive, or introduced. We must also determine whether the ant apecies is herbivorous, predaceous, or omnivorous. Finally, we must determine whether the ant is operating under natural or applied biological control.

Ants apparently exert significant natural biological control in some forest habitats (Finnegan, 1974). Some of the general attributes that contribute to the effectiveness of ants as predators includes long-term colony existence, long hunting seasons, large populations of workers in a colony, large areas covered by workers, elaborate behaviors involving worker recruitment, and nonspecificity with regard to prey life stage. Red wood ants of the *Formica rufa* group include 8 species in Europe and 15 species in North America. In Europe, red wood ants have been noted as effective in the control of forest pests (Gösswald, 1951; Ruppertshofen, 1955).

The usefulness of red wood ants is not universally accepted and some reservations have been expressed concerning their efficacy in forested habitats (Adlung, 1966). Inaccurate taxonomic identifications have contributed to the failure to appreciate the role of some species in the *F. rufa* group (Finnegan, 1971). Three species (*Formica lugubris* Zetterstedt, *F. polyctena* [Förster], and *F. aquilonia* Yarrow) are effective in Europe. One species, *F. lugubris,* was successfully imported into Canada from Italy for control of forest pests in Quebec (Finnegan, 1975). Another species, *F. obscuripes* Forel, was successfully moved from Manitoba to Quebec (Finnegan, 1977). The overall effectiveness of these insects in Canada remains to be established, but introduction and exclusion studies of *F. neoclara* (Emery) in pear orchards showed the importance of this ant as a predator of pear psylla (Paulson & Akre, 1992a, 1992b). *Formica yessensis* Forrel has been demonstrated to be an important natural enemy of the pine caterpillar *Dendrolimus spectabilis* Butler in Korea (Kim & Kim, 1973; Kim & Choi, 1976; Kim & Murakami, 1978), and the potential of *Formica* species as predators of gypsy moth has been evaluated in the United States (Weseloh, 1996).

Ants are not universally recognized as effective applied biological control agents in orchards, forests, and ornamental trees. Haney (1988) provides a world survey of the role of ants in citrus pest management. The citrus ant, *Oecophylla smaragdina* (Fabricius), is an effective predator for the control of insect pests. Huang and Yang (1987) document the use of citrus ants in China and note that the earliest written record of their application dates from 304 A.D. Leston (1973) discusses the use of citrus ants on other crops outside China. *Oecophylla smaragdina* is a so-called weaver ant, a species that bind leaves and twigs together with silk (Hölldobler & Wilson, 1977). The ants inhabit their arboreal nests during the night and feed on insects on the tree during the day. Ants colonize adjacent trees with the aid of bamboo bridges placed between the canopies of the trees by farmers. The effectiveness of these ants in the control of pests such as soft scales and mealybugs has been questioned (Groff & Howard, 1924; Hoffman, 1936; Chen, 1962). However, improved cultural practices and fostering of other natural enemies improve the effectiveness of *O. smaragdina* in the control of orchard pests (Yang, 1982, 1984; Yang *et al.,* 1983, 1984; Dejean, 1991). The negative effect of pesticides on the action of *O. smaragdina* on citrus in China has been observed (Chen, 1985), and the elimination of chemicals in orchards further improves the effectiveness of *O. smaragdina.* Studies also indicate the potential of *O. smaragdina* in the control of *Helopeltis pernicialis* Stonedahl, Malipatil and Houston, a new pest of cashew in Australia (Peng *et al.,* 1997).

Other species of predatory ants may also exert a positive or negative impact on applied biological control of tree pests. The Argentine ant, *Iridomyrmex humilis* (Mayr), has been observed to remove 98% of the eggs of the chrysopid *Chrysoperla carnea* that were placed in street trees to control the tulip tree aphid *Illinoia liriodendri* (Dreistadt *et al.,* 1986). In contrast, foraging *F. hemorrohoidalis* Emery ants were believed to kill 95% of the pupae of western spruce budworm, *Choristoneura occidentalis* Freeman, in north central Washington (Campbell & Torgersen, 1982), but carpenter ants are of limited importance as predators against spruce budworm in northwest Ontario (Sanders & Pang, 1992).

Ants may act as predators of pests in annual field crops. Some studies in the United States show that the red imported fire ant, *Solenopsis invicta,* attacks pests of cotton including cotton fleahopper (Breene *et al.,* 1990), boll weevil (Sterling *et al.,* 1984), and tobacco budworm (McDaniel & Sterling, 1982; Nuessly & Sterling, 1994). This ant is an introduced predator from South America that is currently widespread in the cotton-growing region of the southeastern United States. While *S. invicta* may be beneficial for some pests of cotton (Breene, 1991), it may not be suitable against aphids on cotton because the ant protects aphids from other predators (Showler & Reagan, 1987), takes honeydew from aphids, and is predaceous on aphids that have been parasitized (Vinson & Scarborough, 1991). The red imported fire ant is also regarded as a significant predator of the sugarcane borer, *Diatraea saccharalis,* in Louisiana (Reagan *et al.,* 1972, Ali & Reagan, 1985) and of the fall armyworm in sweet sorghum (Fuller *et al.,* 1997). In general, this species of ant is regarded as highly noxious and millions of dollars a year are spent on its control in the United States. Kirk (1981) noted that *Lasius neoniger* Emery may be an effective predator of corn rootworm larvae (*Diabrotica* spp.) in South Dakota.

A few studies in tropical regions show the potential for ants as biological control agents. Way *et al.* (1989) found that a complex of ant species in Sri Lanka contribute to the control of coconut caterpillar, *Opisina arenosella* Walker, through egg predation, and Löhr (1992) correlated decrease of coreid bug damage to coconuts with an increase in ants. Eskafi and Kolbe (1990) note that *S. geminata* (Fabricius) can eliminate up to 25% of the medfly larvae, *Ceratitis capitata* (Wiedemann), falling from coffee and citrus in Guatemala. Wong *et al.* (1984) show that *Iridomyrmex humilis* serves a similar role in controlling medfly larvae, pupae, and teneral adults in Hawaii. Perfecto (1991) provides data that suggest *S. geminata* and *Pheidole radowszkoskii* Mayr may be helpful in the control of fall armyworm *Spodoptera frugiperda*, feeding on corn grown in Nicaragua. Endemic ants in Ghana can reduce the effectiveness of an introduced parasitoid of the cassava mealybug (Cudjoe *et al.*, 1993).

Pompiloidea are considered by some taxonomists as transitional between sphecoids and scolioids. All three superfamilies are fossorial and are active burrowers. The superfamily includes the Pompilidae and Rhopalosomatidae. The Rhopalosomatidae are sometimes placed in the Scolioidea, but are considered within the Pompiloidea here. The Rhopalosomatidae are parasitoids of Gryllidae, but little is known of their biology (Gurney, 1953). The larva apparently forms a thalacium on the lateral aspect of the abdomen of crickets. Pompilidae are biologically distinctive in that the larvae all use spiders as food. The female wasp stings a spider and transports it to the nest. The adult female provisions a cell with one spider. After the nest has been provisioned, the female uses the apex of the gaster to tamp earth or mud over the entrance. Some *Dipogon, Priocnemis,* and Ageniellini construct the cell before hunting. This is probably a derived condition. Most female pompilids prepare the nest after the spider has been located and paralyzed, while other aculeates prepare the nest before searching for hosts. Predation, if it may be regarded as such, is restricted to adult pompilids. Williams (1956) reports that species of *Pepsis* feed on hemolymph of spiders that have been paralyzed and subsequently will be used as hosts for larval development. Concurrent feeding may be common in pompilids.

Scolioidea consists of several families, including Mutillidae, Sapygidae, Scoliidae, Sierolomorphidae, and Tiphiidae. Predation among adult Mutillidae does not occur. Gregarious parasitism is unusual by Mutillidae larvae because they are voracious, and consume all available food. Presumably cannibalism, a form of predation, occurs in Mutillidae when more than one mutillid larva attempts to develop on a host (Brothers, 1984). Sapygids are all parasitic on other aculeate Hymenoptera; we know of no records of predation. Scoliids are exclusively solitary parasitoids of Coleoptera larvae. We know nothing of the biology of Sierolomorphidae. Tiphiidae adults are fossorial wasps and the larvae develop as parasitoids of some families of

Coleoptera, usually Scarabaeidae, Tenebrionidae, and Cicindelidae or sometimes Curculionidae. One genus, *Diamma,* attacks mole crickets. Tiphiids are also regarded as parasitic.

Sphecoidea includes the Sphecidae, which is recognized here in the sense of Bohart and Menke (1976) containing 11 subfamilies of solitary wasps that provision their nests with other insects and are considered beneficial. Prey specificity is not at the subfamily level, but is more at the tribal level. Feeding habits represent a continuum ranging from predation to parasitism. The larger sphecids, like Ampulicinae that capture cockroaches and Sphecinae that capture spiders or Orthoptera, are actually external parasitoids because they completely develop feeding on a single victim, while the sphecids that provision each cell with several small prey are indeed predators. Pemphredoninae insects provision their nests with Homoptera or, in a few cases, with Thysanoptera or Collembola. Astatinae provides Hemiptera for its larvae. In Larrinae, the tribe Larrini provisions nests chiefly with crickets and the tribe Tachytini, chiefly with grasshoppers. Crabroninae provides flies as prey. Nyssoninae is a large subfamily with six tribes, each having their own prey spectra but mostly Homoptera (chiefly leafhoppers) and flies. However, *Sphecius speciosus* (Drury), in the tribe Gorytini, is the cicada killer whose larvae are external parasitoids of cicadas, while other nyssonines are cleptoparasitoids of other sphecids. Philanthinae is ground nesting and provides aculeate Hymenoptera or Coleoptera as prey (Evans & Eberhard, 1970; Bohart & Menke, 1976). The best known Philanthinae are the bee wolves. *Philanthus* spp. provision their larvae with bees and wasps; thus, they are not beneficial but have a fascinating foraging behavior (Evans & O'Neil, 1988).

In an intensive field study of several British pemphredonine species in the genus *Passaloecus* that provision their nests with aphids, Corbet and Backhouse (1975) concluded that solitary aculeates would not make good biological control agents for use in British field crops. These wasps have several disadvantages that preclude them from being effective natural enemies: they require high temperatures to be active, and have a slow numerical response to changes in prey density due to low fecundity and long generation time coupled with the emigration time from infested crops to their nests; finding and developing suitable nesting sites also require time and energy. Perhaps these disadvantages extend to all predaceous solitary aculeates. Corbet and Backhouse (1975), however, suggest possible use of pemphredonids in greenhouses.

Vespoidea is considered to have only the family Vespidae, which comprises some 30 genera, seven tribes, and six subfamilies: Euparaguinae, Masarinae, Eumeninae, Stenogastrinae, Vespinae, and Polistinae (Carpenter, 1982). The primitive subfamilies Masarinae and Eumeninae are solitary; the former store pollen and nectar as food for their larvae and the latter provide their larvae with paralyzed or dismembered arthropods. The Stenogastrinae are subsocial because a feedball (evidently produced in the gut of the female) is placed in the cell before the egg hatches and later the larva is fed progressively with masticated insects and spiders. The other subfamilies are social wasps and their larvae are fed masticated arthropods and often also nectar (Evans & Eberhard, 1970; Ross & Matthews, 1991).

Among the social vespoid wasps, the Polistinae subfamily has received the most attention by biological control workers—*Polistes* has been the main genus involved. Most of the more than 150 *Polistes* spp. known live in the tropics and are mainly caterpillar hunters (Evans & Eberhard, 1970; Reeve, 1991). *Polistes* spp. adapt to using man-made structures to build their paper nests that allows relocation, higher density, and protection from spraying of pesticides. Rabb and Lawson (1957) found foraging of *Polistes* spp. in North Carolina to be influenced by caterpillar abundance in tobacco and soybean fields and the number of natural nesting sites. Fewer hornworms were killed in tobacco fields when there was an abundance of green clover worms in soybean fields. By erecting wooden structure shelters around tobacco fields, *P. annularis* (Linnaeus) were induced to nest in the shelters, and these wasps reduced hornworms 60% although they had no effect on tobacco budworms (Lawson *et al.*, 1961).

Rabb *et al.* (1976) and Gillaspy (1979) concluded from their own attempts and those in the literature that increasing *Polistes* numbers to control lepidopteran larvae was not practical. Gould and Jeanne (1984) obtained mixed results in the control of *Pieris rapae* (Linnaeus) infesting cabbage plants by transplanting *P. fuscatus* (Fabricius) colonies in wooden nesting boxes. An average nest of 5.6 wasps was sufficient to reduce *Pieris* populations by up to five larvae (44%) on each of approximately 210 cabbage plants per season, while the availability of alternate prey influenced the degree of control of a target pest (Gould & Jeanne, 1984). Morimoto (1960) obtained increased predation of *Pieris* larvae in cabbage fields where he placed artificial nests for *Polistes*. In the temperate zone, nesting sites placed in sun-warmed sites will encourage larger and faster developed broods (Jeanne & Morgan, 1992). By placing *P. antennalis* Perez nests in cotton fields at the rate of 100 wasps per mu (0.67 ha), Li *et al.* (1984) obtained effective control of noctuid larvae.

In the tropics the jackspaniard wasp, *Polistes cinctus cinctus* Lepeletier [in literature, *P. annularis* (Linnaeus), *P. canadensis* (Linnaeus)] was highly regarded on St. Vincent Island because it allowed cotton to be grown without pesticides, while in other Caribbean islands there were great losses to the cotton leaf worm *Alabama argillacea* (Hübner) and this led to the exportation of *P. c. cinctus* to many of the Caribbean islands. Also on St. Vincent, the farmers

erected sheds in the fields for the wasp to nest in (Myers, 1931). The jackspaniard wasp also readily utilized the arrowroot leaf roller as an alternate host on St. Vincent Island (Cock, 1985). In Brazil and Peru, even though two *Polistes* spp. are not deliberately manipulated for biological control of the coffee leaf miner, *Perileucoptera coffeella* (Guerin-Meneville), they are recognized as important predators of this pest (Enriquez *et al.*, 1976; Reis & Souza, 1996). In Brazil, Parra *et al.* (1977) and Tozatti and Gravena (1988) have reported that predation by vespids can control outbreaks of the coffee leaf miner. The Indians in Amazonia, who grow 10 different crops without pesticides in cleared areas of 2 to 4 ha in the forest, owe much of the insect pest control to 10 different social wasps, including four *Polistes* spp. that nest in forest edges (Raw, 1988).

The large New World tropical genus *Mischocyttarus* has species with biologies similar to *Polistes* (Gadagkar, 1991) and workers have investigated the foraging behavior of *M. flavitarsus* (Saussere) in greenhouses (Cornelius, 1991). Even though considered a generalist predator, this wasp species was deterred by aposomatic prey species (Bernays, 1988). This genus is manipulatable and has a wide prey range, making it a good potential as an augmentable biological control agent.

CONCLUSIONS

The broad ecological role of predation has been discussed by Huffaker (1970); and the potential of predators in pest control has been presented by Hagen *et al.* (1976a), Luff (1983), and Wratten (1987). The current review of the major taxa of arthropod predators makes it clear that there are a variety of examples in which predators play a decisive role in the dynamics of phytophagous arthropod populations, through importation, augmentation, or natural control.

The introduction of the vedalia beetle into southern California against the cottony-cushion scale in 1888 provides the earliest and most spectacular example of the potential of using predators for the classical biological control of newly invading pests. Numerous other examples have been documented in the earlier sections of this chapter. The historical record indicates that while predators have been used far less frequently than parasitoids for the classical biological control of arthropod pests, their overall rates of establishment and success are equal to those of parasitoids (Hall & Ehler, 1979; Hall *et al.*, 1980), and, in fact, they may perform better than parasitoids in field crops and on plantations (Hokkanen, 1985). Predators used in the classical biological control of arthropod pests have in most cases been specialists; there is little justification for the importation of more polyphagous predators. Preadapted guilds of polyphagous predators are indigenous to most regions of the world and their introduction to other target areas can only be justified in the absence of such guilds, such as depauperate island faunas or where indigenous predators fail to adapt successfully to exotic target crops. In both cases, such importations should be considered with the utmost caution and specialist predators should be used whenever possible.

Predator augmentation dates back as far as 1916, when a commercial insectary was initiated for the rearing of *Cryptolaemus montrouzieri* for inoculative seasonal control of mealybugs in the citrus orchards of southern California (Hagen, 1974). Since then, many different predators have been successfully augmented for the control of pests in protected cropping (Hussey & Scopes, 1985), but there remain few examples of successful field augmentation. Augmentation of the predaceous mite *Phytoseiulus persimilis* has been widely adopted for the control of the two-spotted mite in California and Florida strawberry fields (Grossman, 1989; Decou, 1994). This example illustrates the potential for increased integration of field augmentation of predators into pest management strategies in the future, particularly in row crops, but also highlights the necessity to place greater emphasis on the development of improved economically viable rearing techniques. As we move from an era of introduced pests to one of indigenous pests through the successful implementation of classical biological control and quarantine restrictions, augmentation is likely to become a more frequently practiced form of biological control (Mills, 1991; Parrella *et al.,* 1992).

Gilliatt (1935) was one of the first to appreciate the role of insect predators in the natural control of potential pests in a study of the European red mite in apple orchards in Nova Scotia. This work led to the development of the concept of integrated pest management (Pickett, 1959; Stern *et al.,* 1959) and the wider acceptance of the importance of generalist, as well as specialist, predators as a major factor in the prevention of pest outbreaks (Luff, 1983; Whitcomb & Godfrey, 1991; Riechert, 1992). More than 1000 species of arthropods have been documented from alfalfa fields in California (Smith & van den Bosch, 1967) and over 1000 predaceous arthropods, in soybean fields in Florida (Whitcomb, 1974). The conservation of this tremendous diversity of predators in field crops is one of the most important challenges for the future and for the development of sustainable agriculture. However, we need a far greater understanding of how microclimate, plant spacing, alternative food sources, and integrated controls affect not only the abundance of predators in field crops but also their impact on potential pests, before conservation will achieve the degree of success seen with other strategies of biological control.

Acknowledgment

The assistance of Robert Zuparko during the final draft preparation of this chapter is gratefully acknowledged.

References

Ables, J. R. (1975). Notes on the biology of the predacious pentatomid *Euthyrhynchus floridanas* (L.). Journal of the Georgia Entomological Society, 10, 353–356.

Ables, J. R., & McCommas, D. W., Jr. (1982). Efficacy of *Podisus maculiventris* as a predator of variegated cutworm on greenhouse cotton. Journal of the Georgia Entomological Society, 17, 204–206.

Ables, J. R., & Ridgway, R. L. (1981). Augmentation of entomophagous arthropods to control insect pests and mites. In G. C. Papavizas (Ed.), Biological control in crop production (b). Beltsville Symposium on Agricultural Research (pp. 273–303). Granada: Allanheld, Osmun.

Ables, J. R., Reeves, B. G., Morrison, R. K., Kinzer, R. E., Jones, S. L., & Ridgway, R. L. (1979). Methods for the field releases of insect parasites and predators. Transactions of the American Society of Agricultural Engineers, 18, 58–62.

Adams, P. A. (1963). Taxonomy of Hawaiian *Chrysopa* (Neuroptera: Chrysopidae). Proceedings of the Hawaiian Entomological Socieyt, 18, 221–223.

Adams, P. A., & Penny, N. D. (1985). Neuroptera of the Amazon Basin. IIa. Introduction and Chrysopini. Acta Amazonica, 15, 413–479.

Adashkevich, B. P., & Kuzina, N. P. (1974). Chrysopids in vegetable crops. Zashchita Rastenii, 9, 28–29.

Adlung, K. G. (1966). A critical evaluation of the European research on use of red wood ants (*Formica rufa* group) for the protection of forests against harmful insects. Zeitschriff fuer Angewandte Entomologie, 57, 167–189.

Agarwala, B. K., & Dixon, A. F. G. (1992). Laboratory study of cannibalism and interspecific predation in ladybirds. Ecological Entomology, 17, 303–309.

Agekyan, N. G. (1978). *Clitostethus arcuatus* Rossi (Coleoptera, Coccinellidae), a predator of the citrus whitefly in Adzharia. Entomological Review, 56, 22–23.

Aguilera, A., Mendoza, R., Vargas, H., & Diaz, G. (1984). Nuevos aportes sobre la actividad depredadora de *Coccidophilus citricola* Brethes (Coleoptera: Coccinellidae). Idesia (Chile), 8, 47–54.

Ahmad, R., & Ghani, M. A. (1971). The biology of *Sticholotis marginalis* Kapur (Col.: Coccinellidae) CIBC Technical Bulletin, 14, 91–95.

Alauzet, C., Dargagnon, D., & Hatte, M. (1992). Production d'un Heteroptere predateur: *Orius majusculus* (Het.: Anthocoridae). Entomophaga, 37, 249–252.

Alberti, G. (1973). Ehrnahrungsbiologie und Spinnvermogen der Schnabelmilben (Bdellidae, Trombidiformes). Zeitschrift fuer Morphologie der Tiere, 76, 285–388.

Aldrich, J. R. (1988). Chemical ecology of the Heteroptera. Annual Review of Entomology, 33, 211–238.

Aldrich, J. R., Kochansky, J. P., & Abrams, C. B. (1984). Attractant for a beneficial insect and its parasitoids: Pheromone of the predatory spined soldier bug, *Podisus maculiventris* (Hemiptera: Pentatomidae). Environmental Entomology, 13, 1031–1036.

Aldrich, J. R., Lusby, W. R., & Kochansky, J. P. (1986). Identification of a new predaceous stink bug pheromone and its attractiveness to the eastern yellow jacket. Experientia, 42, 583–585.

Alfaro, R. I., & Borden, J. H. (1980). Predation by *Lonchaea corticis* (Diptera: Lonchaeidae) on the white pine weevil, *Pissodes strobi* (Coleoptera: Curculionidae). Canadian Entomologist, 112, 1259–1270.

Ali, A. D., & Reagan, T. E. (1985). Vegetation manipulation impact on predator and prey populations in Louisiana sugarcane ecosystems. Journal of Economic Entomology, 78, 1409–1414.

Allen, R. T. (1979). The occurrence and importance of ground beetles in agricultural and surrounding habitats. In T. L. Erwin *et al.* (Eds.), Carabid beetles: Their evolution, natural history and classification (pp. 485–505). The Hague: Dr. W. Junk.

Allen, W. R., & Hagley, E. A. (1990). Epigeal arthropods as predators of mature larvae and pupae of the apple maggot (Diptera: Tephritidae). Environmental Entomology, 19, 309–312.

Amman, G. D. (1966). *Aphidecta obliterata* (Coleoptera: Coccinellidae), an introduced predator of the balsam woolly aphid. *Chermes piceae* (Homoptera: Chermidae), established in North Carolina. Journal of Economic Entomology, 59, 506–509.

Ammar, E. D., & Farrag, S. M. (1974). Studies on the behavior and biology of the earwig *L. riparia* Pallas (Dermaptera, Labiduridae). Zeitschrift fuer Angewandte Entomologie, 7, 189–196.

Ananthakrishnan, T. N. (1984). Bioecology of thrips. Oak Park, MI: Indira Publishing House.

Anderson, A. (1992). Effects of fenvalerate and esfenvalerate on carabid and staphylinid species in spring barley fields. Norwegian Journal of Agricultural Sciences, 6, 411–417.

Anderson, A., Hansen, A. G., Rydland, N., & Oyre, T. G. (1983). Carabidae and Staphylinidae (Col.), a predator of eggs of the turnip root fly *Delia floralis* Fallen (Diptera: Anthomyiidae) in cage experiments. Zeitschrift fuer Angewandte Entomologie, 95, 499–506.

Anderson, N. H. (1962a). Bionomics of six species of *Anthocoris* (Heteroptera: Anthocoridae) in England. Transactions of the Royal Entomological Society of London, 114, 67–95.

Anderson, N. H. (1962b). Anthocoridae of the Pacific Northwest with notes on distributions, life histories and habitat (Heteroptera). Canadian Entomologist, 94, 1325–1334.

Anderson, N. H. (1962c). Growth and fecundity of *Anthocoris* spp. reared on various prey (Heteroptera: Anthocoridae). Entomologia Experimentalis et Applicata, 5, 40–52.

Andow, D. A., & Risch, S. J. (1985). Predation in diversified agroecosystems: Relations between a coccinellid predator *Coleomegilla maculata* and its food. Journal of Applied Ecology, 22, 357–372.

Angalet, G. W., & Jacques, R. L. (1975). The establishment of *Coccinella septempunctata* L. in the continental United States. United States Department of Agriculture Cooperative Economic Insect Report, 25, 883–884.

Angalet, G. W., Tropp, J. M., & Eggert, A. N. (1979). *Coccinella septempunctata* in the United States recolonizations and notes on its ecology. Environmental Entomology, 8, 846–901.

Ankersmit, G. W., Dijkman, H., Keuning, N. J., Mertens, H., Sins, A., & Tacoma, H. M. (1986). *Episyrphus balteatus* as a predator of the aphid *Sitobon avenae* on winter wheat. Entomologia Experimentalis et Applicata, 42, 271–277.

Ansari, M. A., Pawar, A. D. Murthy, K. R. K., & Ahmed, S. N. (1989). Sugarcane scale, *Melanaspis glomerata* Green and its biocontrol prospects in Karnataka. Plant Protection Bulletin (Iaridaba), 41, 21–23.

Antolin, M. F., & Williams, R. L. (1989). Host feeding and egg production in *Muscidifurax zaraptor* (Hymenoptera: Pteromalidae). Florida Entomologist, 72, 129–134.

Arbogast, R. T. (1976). Suppression of *Oryzaephilus surmamensis* (L.) (Coleoptera Cucujidae) on shelled corn by the predator *Xylocoris flavipes* (Reuter) (Hemiptera, Anthocoridae). Journal of the Georgia Entomological Society, 11, 67–71.

Arnoldi, D., Stewart, R. K., & Boivin, G. (1991). Field survey and laboratory evaluation of the predator complex of *Lygus lineolaris* and *Lygocoris communis* (Hemiptera: Miridae) in apple orchards. Journal of Economic Entomology, 84, 830–836.

Arnoldi, D., Stewart, R. K., & Boivin, G. (1992). Predatory mirids of the green apple aphid *Aphis pomi*, the two-spotted spidermite *Tetranychus urticae* and the European red mite *Panonychus ulmi* in apple orchards in Quebec. Entomophaga, 37, 283–292.

Arrow, G. J. (1917). The life-history of *Conwentzia psociformis*. Entomological Monthly Magazine, 53, 254–257.

Artola, J. A., Garcia, M. F., & Dicindio, S. E. (1983). Bioecologia de *Podisus nigrolimbatus* Spinola (Heteroptera: Pentatomidae), predator de *Pyrrhalta luteola* (Muller) (Coleoptera: Chrysomelidae). IDIA, No. 401/404, 25–33.

Asgari, A. (1966). Untersuchungen Uber die im Raum Stuttgard-Hohenheim als wichstigste Pradatoren der grunen Apfelblattlaus (*Aphidula pomi* DeG.) auftreden Arthropoden. Zeitschrift fuer Angewandte Zoologie, 53, 35–93.

Ashihara, W., Hamamura, T., & Shinkaji, N. (1978). Feeding, reproduction, and development of *Phytoseiulus persimilis* Athias-Henriot (Acarina: Phytoseiidae) on various food substances. Bulletin of Fruit Tree Research Station of Japan, E-2, 91–98.

Ashlock, P. D., & Slater, A. (1988). Family Lygaeidae. In T. J. Henry & R. C. Froeschner (Eds.), Catalog of Heteroptera or true bugs (pp. 167–245). Leiden: E. J. Brill.

Askari, A., & Stern, V. M. (1972a). Biology and feeding habits of *Orius tristicolor* (Hemiptera: Anthocoridae). Annals of the Entomological Society of America, 65, 96–100.

Askari, A., & Stern, V. M. (1972b). Effect of temperature and photoperiod on *Orius tristicolor* feeding on *Tetranychus pacificus*. Journal of Economic Entomology, 65, 132–135.

Aspöck, H. (1986a). The Raphidioptera of the world: A review of present knowledge. In J. Gepp, H. Aspöck, & H. Hölzel (Eds.), Recent research in Neuropterology (pp. 15–29). Grasz, Austria: Gepp.

Aspöck, H., Aspöck, U., & Hölzel, H. (1980). Die Neuropteren Europeas, 2 vols. Krefeld: Goecke & Evers.

Aspöck, H., Aspöck, U., & Rausch, H. (1975). Raphidiopteren-Larven als Bodenbewohner (Insecta, Neuropteroidea) Zeitschrift fuer Angewandte Zoologie, 62, 361–375.

Aspöck, H., Rausch, H., & Aspöck, U. (1974). Untersuchungen uber die Okologie der Rhaphidiopteren Mittleuropas (Insecta, Neuropteroidea). Zeitschrift fuer Angewandte Entomologie, 76, 1–30.

Aspöck, U. (1986b). The present state of knowledge of the family Berothidae (Neuropteroidea: Planipennia). In J. Gepp, H. Aspöck, & H. Hölzel (Eds.). Recent research in neuropterology (pp. 87–101). Graz, Austria: Gepp.

Asteraki, E. J. (1993). The potential of carabid beetles to control slugs in grass/clover swards. Entomophaga, 38, 193–198.

Aun, V. (1986). Aspectos da biologia de *Chrysoperla externa* (Hagen) (Neuroptera: Chrysopidae). Dissertation Escola Superior de Agricultura, Universidade de São Paulo, Piracicaba, Brazil.

Averbeck, J. K., & Haddock, J. D. (1984). The effect of prey diet on predation efficiency of the coccinellid, *Cryptolaemus montrouzieri* (Mulsant). Proceedings of the Indiana Academy of Science, 93, 211–212.

Avidov, Z., Blumberg, D., & Gerson, U. (1968). *Cheletogenes ornatus* (Acarina: Chyletidae), a predator of the chaff scale on citrus in Israel. Israel Journal of Entomology, 3, 77–93.

Awan, M. S. (1987). Plant feeding by first stages of predacious pentatomids *Oechalia schellenbergii* (Gurin-Meneville) and *Cermatulus nasalis* (Westwood), a useful adaptation (pp. 113–116). Proceedings 5th Pakistani Congress of Zoology, University of Karachi, November 8–11, 1986.

Awan, M. S. (1989). Parasitism of two endemic predacious stink bugs by the introduced egg parasitoid *Trissolcus basalis* (Wallaston) (Hymenoptera: Scelionidae). Annales de la Societe Entomologique de France, 25, 119–120.

Awan, M. S., Wilson, L. T., & Hoffman, M. P. (1989). Prey location by *Oechalia schellenbergii*. Entomologia Experimentalis et Applicata, 51, 225–231.

Axtell, R. C. (1969). Macrochelidae (Acarina: Mesostigmata) as biological control agents for synanthropic flies. In Proceedings, 2nd International Congress on Acarology, 1967 (pp. 401–416). Budapest: Akademie Kaido.

Baars, M. A., & van Dijk, Th. S. (1984). Population dynamics of two carabid beetles at a Dutch heathland. II. Egg production and survival in relation to density. Journal of Animal Ecology, 58, 389–400.

Babrikova, T. (1978). Some bioecological characteristics of *Chrysopa formosa* Br. (Chrysopidae: Neuroptera). Rastenier'dni Naut, 15, 114–

119 (from (1979). Review of Applied Entomology, Series A, 67, 4098).

Babu, T. R., & Azam, K. M. (1988). Effect of low holding temperature during pupal instar on adult emergence, pre-oviposition and fecundity of *Cryptolaemus montrouzieri* Mulsant (Coleoptera: Coccinellidae). Insect Science Applications, 9, 175–177.

Badgley, M. E., & Fleschner, C. A. (1956). Biology of *Oligota oviformis* Casey (Coleoptera: Staphylinidae). Annals of the Entomological Society of America, 49, 501–502.

Badgley, M. E., Fleschner, C. A., & Hall, J. C. (1955). The biology of *Spiloconis picticornis* Banks (Neuroptera: Coniopterygidae). Psyche, 62, 75–81.

Badii, M. H. (1981). Experiments on the dynamics of predation of *Phytoseiulus longipes* on the prey. *Tetranychus pacificus* Acarina: Phytoseiidae, Tetranychidae). Unpublished doctoral dissertation, University of California, Riverside.

Badii, M. H., & J. A. McMurtry, (1984). Life history of and life table parameters for *Phytoseiulus longipes* with comparative studies of *P. persimilis* and *Typhlodromus occidentalis* (Acari: Phytoseiidae). Acarologia, 25, 111–123.

Bailey, P., & Caon, G. (1986). Predation on twospotted mite, *Tetranychus urticae* Koch (Acarina: Tetranychidae), by *Haplothrips victoriensis* Bagnall (Thysanoptera: Phlaeothripidae) and *Stethorus nigripes* Kapur (Coleoptera: Coccinelliidae) on seed lucerne crops in South Australia. Australian Journal of Zoology, 34, 515–525.

Balachowsky, A. (1928). Observations biologique sur les parasites des coccides du Nord-African. Annales des Epiphyties (Paris), 14, 280–312.

Balachowsky, W. V., & Molinari, L. (1930). L'extension de la cochenille australene (*Icerya purchasi* Mask.) en France et de son predateur *Novius cardinalis* Annales des Epiphyties (Paris), 16, 1–24.

Balduf, W. V. (1935). The bionomics of entomophagous Coleoptera. New York: John S. Swift.

Balduf, W. V. (1939). The bionomics of entomophagous insects, II. New York: John S. Swift.

Balduf, W. V. (1948). A summary of studies on the ambush bug *Phymata pennsylvanica americana* Melin (Phymatidae, Hemiptera). Illinois Academy of Science Transactions, 41, 101–106.

Balduf, W. V. (1950). Utilization of food by *Sinea diadema* (Fabr.) (Reduviidae, Hemiptera). Annals of the Entomological Society of America, 43, 354–360.

Balduf, W. V., & Slater, J. S. (1943). Additions to the bionomics of *Sinea diadema* (Fabr.) (Reduviidae, Hemiptera). Proceedings of the Entomological Society of Washington, 45, 11–18.

Ball, G. E. (1978). A book review of H-U. Thiele's 1977 carabid beetles in their environments. Science, 201, 704–705.

Ball, J. C. (1980). Development, fecundity, and prey consumption of four species of predacious mites (Phytoseiidae) at two constant temperatures. Environmental Entomology, 9, 298–303.

Bankowska, R., Mikolajczyk, W., Palmowska, J., & Trojan, P. (1978). Aphid-aphidophage community in alfalfa cultures (*Medicago sativa* L.) in Poland. III. Abundance regulation of *Acyrthosiphon pisum* (Harr.) in a chain of oligophagous predators. Annals of Zoology, 34, 39–77.

Banks, C. J. (1954). The searching behaviour of coccinellid larvae. Animal Behavior, 2, 37–38.

Banks, C. J. (1956). Observations on the behaviour and mortality in Coccinellidae before dispersal from egg shells. Proceedings of the Royal Entomological Society of London, Series A, 31, 56–60.

Banks, C. J. (1957). The behaviour of individual coccinellid larvae on plants. British Journal of Animal Behavior, 5, 12–24.

Bänsch, R. (1964). Vergleichende Untersuchungen zur Biologie und zum Beutefangverhalten aphidivorer Coccinelliden, Chrysopiden und Syrphiden. Zoologische Jahrbuecher Syst. 91, 271–340.

Bänsch, R. (1966). On prey-seeking behaviour of aphidophagous insects. In I. Hodek (Ed.), Ecology of aphidophagous insects (pp. 123–128). Prague: Academia.

Barber, G. W. (1936). *Orius insidiosus* (Say), an important natural enemy of the corn ear worm (Technical Bulletin No. 504). Washington, DC: U.S. Department of Agriculture.

Barclay, W. W. (1990). Role of *Chrysoperla externa* as a biological control agent of insect pests in Mesoamerican agricultural habitats. Unpublished masters thesis, Entomology Department, University of California, Berkeley.

Barry, R. M., Hatchett, J. H., & Jackson, R. D. (1974). Cage studies with predators of the cabbage looper, *Trichoplusia ni,* and corn earworm, *Heliothis zea,* in soybeans. Journal of the Georgia Entomological Society, 9, 71–78.

Bartlett, B. R. (1964a). Patterns in the host-feeding habits of adult Hymenoptera. Annals of the Entomological Society of America, 57, 344–350.

Bartlett, B. R. (1964b). Toxicity of some pesticides to eggs, larvae, and adults of the green lacewing, *Chrysopa carnea.* Journal of Economic Entomology, 57, 366–369.

Bartlett, B. R. (1974). Introduction into California of cold-tolerant biotypes of the mealybug predator, *Cryptolaemus montrouzieri,* and laboratory procedures for testing natural enemies for cold-hardiness. Environmental Entomology, 3, 553–556.

Bartlett, B. R. (1978). Margarodidae, Ortheziidae, Pseudococcidae. In C. P. Clausen (Ed.), Introduced parasites and predators of arthropod pests and weeds: A world review. Agricultural Handbook No. 480 (pp. 132–170) Washington, DC: U.S. Department of Agriculture.

Bartlett, K. A. (1940). The collection in Trinidad and southern Brazil of coccinellids predatory on bamboo scales. Proceedings of the Sixth Pacific Science Congress, 4, 339–343.

Basedow, T., Braun, C., Leuhr, A., Nauman, J., Norgall, T., & Yanes, G. Y. (1991). Abundance, biomass and species number of epigeal predatory arthropods in fields of winter wheat and beets at different levels of intensity: Differences and their reasons: Results of three intensity levels in Hess (Germany). 1985–1988. Zoologische Jahrbuecher, Abteilung fuer Systematik Oekologie und Geographie de Tiere, 118, 87–116.

Bass, J. A., & Shepard, M. (1974). Predation by *Sycanus indagnator* on larvae of *Galleria mellonella* and *Spodoptera frugiperda.* Entomologia Experimentalis et Applicata, 17, 143–148.

Bauer, T. (1975). Zur Biologie und Autökologie von *Notiophilus biguttatus* F. and *Bembidion foraminosum* Strm. als Bewohner-ökologisch extremer Standorte Zum Lebeneformtyp des visuell jagenden Raubers unter den Laufkafern (II). Zoologischer Anzeiger, 194, 305–318.

Beattie, G. A. C. (1978). Biological control of citrus mites in New South Wales. In P.R. Larey (Ed.), International Society of Citriculture, 1978, (pp. 156–158). DeLeon Springs, FL: Painter Printing.

Beaver, R. A. (1966). The biology and immature stages of two species of *Medetera* (Diptera: Dolichopodidae) associated with the bark beetle *Scolytus scolytus* (F.). Proceedings of the Entomological Society of London, 41, 145–190.

Becker, H., Corliss, J., DeQuattro, J., Gerrietts, M., Senft, D., Stanley, D. (1992). Get the whitefly swatters fast! Agricultural Research, 40, 4–13.

Beddington, J. R., Hassell, M. P., & Lawton, J. H. (1976). The components of arthropod predation, II. The predator rate of increase. Journal of Animal Ecology, 45, 165–186.

Beer, R. E., & Dailey, D. T. (1956). Biological and systematic studies of two species of cheyletid mites, with a description of a new species. University of Kansas Science Bulletin, 38, 393–436.

Beglyarov, G. A., & Smetnik, A. I. (1977). Seasonal colonization of entomophages in the U.S.S.R. In R. L. Ridgway & S. B. Vinson (Eds.), Biological control by augmentation of natural enemies (pp. 283–328). New York: Plenum Press.

Beglyarov, G. A., & Ushchekov, A. T. (1974). Experimentation and outlook for the use of chrysopids. Zashchita Rastenii, 9, 25–27.

Bellows, T. S., Jr., Paine, T. D., & Gerling, D. (1992a). Development, survival, longevity, and fecundity of *Clitostethus arcuatus* (Coleoptera: Coccinellidae) in *Siphoninus phillyreae* (Homoptera: Aleyrodidae) in the laboratory. Environmental Entomology, 21, 659–663.

Bellows, T. S., Paine, T. D., Gould, J. R., Bezark, L. G., & Ball, J. C. (1992b). Biological control of ash whitefly: A success in progress. California Agriculture, 46, 24–28.

Benassy, C. (1990). Date palm. In D. Rosen (Ed.), Armored scale insects, their biology, natural enemies and control (Vol. B, pp. 585–591). Amsterdam: Elsevier.

Benedict, J. H., & Cothran, W. R. (1975). A faunistic survey of Hemiptera-Heteroptera found in northern California hay alfalfa. Annals of the Entomological Society of America, 68, 897–900.

Benestad, E. (1970). Food consumption at various temperature conditions in larvae of *Syrphus corollae* (Fabr.) (Diptera, Syrphidae). Norsk Entomologisk Tidsskrift, 17, 87–91.

Bennett, F. D., & Hughes, I. W. (1959). Biological control of insect pests in Bermuda. Bulletin of Entomological Research, 50, 423–436.

Ben Saad, A. A., & Bishop, G. W. (1976). Effect of artificial honeydew on insect communities in potato fields. Environmental Entomology, 5, 453–457.

Berg, C. O. (1953). Sciomyzid larvae (Diptera) that feed on snails. Journal of Parasitology, 39, 630–636.

Berisford, C. W., & Dahlsten, D. L. (1989). Biological control of *Ips grandicollis* (Eichhoff) (Coleoptera: Scolytidae) in Australia—A preliminary evaluation. In D. L. Kulhary & M. C. Miller (Eds.), Potential for biological control of *Dendroctonus* and *Ips* bark beetles (pp. 81–93). Nacogdoches, TX: S. F. Austin State University,

Berker, J. (1958). Die natürliche Feinde der Tetranychiden. Zeitschrift fuer Angewandte Entomologie, 43, 15–172.

Bernays, E. A. (1988). Host specificity in phytophageous insects: Selection pressure from generalist predators. Entomologia Experimentalis et Applicata, 49, 131–140.

Berryman, A. A. (1966a). Factors influencing oviposition, and the effect of temperature on development and survival of *Enoclerus lecontei* (Wolcott) eggs. Canadian Entomologist, 98, 579–585.

Berryman, A. A. (1966b). Studies on the behavior and development of *Enoclerus lecontei* (Wolcott), a predator of the western pine beetle. Canadian Entomologist, 98, 519–526.

Berryman, A. A. (1967). Preservation and augmentation of insect predators of the western pine beetle. Journal of Forestry, 65, 260–262.

Berti Filho, E., & Fraga, A. I. A. (1987). Inimigos naturais para o controle de lepidopteros desfolhadores de *Eucalyptus* sp. Brasil Florestal, Rio de Janeiro, 62, 18–22.

Bharadwaj, R. K. (1966). Observations on the bionomics of *Euborell iaannulipes* (Dermaptera: Labiduridae). Annals of the Entomological Society of America, 59, 441–450.

Bianchi, F. A. (1980). *Perkinsiella saccharicida* Kirkaldy and *Tytthus mundulas* Breddin. Proceedings of the Hawaiian Entomological Society, 23, 186–187.

Bickel, D. J. (1985). A revision of the Nearctic *Medetera* (Diptera: Dolichopodidae). (Tech. Bulletin 1692). Washington, DC: U.S. Department of Agriculture.

Biever, K. D., & Chauvin, R. L. (1992a). Suppression of the Colorado potato beetle (Coleoptera: Chrysomelidae) with augmentative releases of predaceous stinkbugs (Hemiptera: Pentatomidae). Journal of Economic Entomology, 85, 720–726.

Biever, K. D., Chauvin, R. L. (1992b). Timing of infestation by the Colorado potato beetle (Coleoptera: Chrysomelidae) on suppressive effect of field released stinkbugs (Hemiptera: Pentatomidae) in Washington. Environmental Entomology, 21, 1212–1219.

Biever, K. D., Andrews, P. L., & Andrews, P. A. (1982). Use of a predator, *Podisus maculiventris,* to distribute virus and initiate epizootics. Journal of Economic Entomology, 75, 150–152.

Bilde, T., & Toft, S. (1997). Consumption by carabid beetles of three cereal aphid species relative to other prey types. Entomophaga, 42, 21–32.

Bishara, I. (1934). The cotton worm *Prodenia litura* F. in Egypt. Society Royal Entomology d'Egypte Bulletin, 18, 288–420.

Bishop, A. L., & Blood, P. R. B. (1978). Temporal distribution and abundance of the coccinellid complex as related to aphid populations on cotton in southeast Queensland. Australian Journal of Zoology, 26, 153–158.

Bishop, A. L., & Blood, P. R. B. (1981). Interactions between natural populations of spiders and pests in cotton and their importance to cotton production in Southeastern Queensland. General Applied Entomology, 13, 98–104.

Bitner, R. M. (1972). Predation by the larvae of *Lecontella cancellata* (LeConte) (Coleoptera: Cleridae) on seven species of aculeate Hymenoptera. Entomology News, 83, 23–26.

Bjegovic, P. (1968). Some biological characteristics of the damsel bug, *Nabis feroides* Rm. (Hemiptera, Nabidae) and its part in regulating population dynamics of the cereal leaf beetle, *Lema melanopa* L. Zastita Bilja, 19, 235–246.

Blackman, R. L. (1965). Studies on specificity in Coccinellidae. Annals of Applied Biology, 56, 336–338.

Blommers, L. (1976). Capacities for increase and predation in *Amblyseius bibens* (Acarina: Phytoseiidae). Zeitschrift fuer Angewandte Entomologie, 81, 225–244.

Blumberg, D. (1973). Survey and distribution of Cybocephalidae (Coleoptera) in Israel. Entomophaga, 18, 125–131.

Blumberg, D., & Swirski, E. (1974). The development and reproduction of cybocephalid beetles on various foods. Entomophaga, 19, 437–443.

Bodenheimer, F. S. (1951). Citrus entomology. Dr. W. Junk, The Hague, Netherlands.

Bohart, R. M. & Menke, A. S. (1976). Sphecid wasps of the world. Berkeley: University of California Press.

Bogdanova, N. L. (1965). *Hyperaspis campestris* Herbst (Coleoptera: Coccinellidae) as destroyer of *Chloropulvinaria floccifera* Westw. (Homoptera: Coccidoidea). Entomologicheskoe Obozrenie, 35, 311–322.

Bombosch, S. (1963). Untersuchungen zue Vermehrung von *Aphis fabae* Scop, in Samenrubenbestanden unter besonderer Berucksichtigung der Schwebfliegen (Diptera: Syrphidae). Zeitschift fuer Angewandte Entomologie, 52, 105–141.

Bombosch, S., & Volk, S. (1966). Selection of oviposition site by *Syrphus corollae* F. In I. Hodek (Ed.), Ecology of aphidophagous insects (pp. 117–119). Prague: Academia.

Bondarenko, N. V., & Moiseev, E. G. (1972). Evaluation of the effectiveness of chrysopids in the control of aphids. Zashchita Rastenii, 17, 19–20.

Booth, R. G., & Pope, R. D. (1986). A review of the genus *Cryptolaemus* (Coleoptera: Coccinellidae) with particular reference to the species resembling *C. montrouzieri* Mulsant. Bulletin Entomological Research 76, 701–717.

Booth, R. G., Cross, A. E., Fowler, S. V., & Shaw, R. H. (1995). The biology and taxonomy of *Hyperaspis pantherina* (Coleoptera: Coccinellidae) and the classical biological control of its prey, *Orthezia insignis* (Homoptera: Ortheziidae). Bulletin of Entomological Research, 85, 307–314.

Borden, J. H. (1982). Aggregation pheromones. In J. B. Mitton & K. B. Sturgeon (Eds.), Bark beetles in North American conifers (pp. 74–139). Austin: University Texas Press.

Borgemeister, C., Meikle, W. G., Scholz, D., Adda, C., Degbey, P., & Markham, R. H. (1997). Seasonal and weather factors influencing the annual flight cycle of *Prostephanus truncatus* (Coleoptera: Bostrichidae) and its predator *Teretriosoma nigrescens* (Coleoptera: Histeridae) in Benin. Bulletin Entomological Research, 87, 239–246.

Bornemissza, G. F. (1968). Studies on the histerid beetle *Pachylister chinensis* in Fiji, and its possible value in the control of buffalo fly in Australia. Australian Journal of Zoology, 16, 673–688.

Bouchard, D., Hill, S. B., & Pilon, J. G. (1988). Control of green apple aphid populations in an orchard achieved by releasing adults of *Aphidoletes aphidimyza* Rondani (Diptera: Cecidomyiidae). In E. Niemczyk & A. F. G. Dixon (Eds.), Ecology and effectivenss of Aphidophaga (pp. 257–260). The Hague: Academia.

Bounfour, M., & McMurtry, J. A. (1987). Biology and ecology of *Euseius scutalis* (Athias-Henriot) (Acarina: Phytoseiidae). Hilgardia, 55, 1–23.

Bournier, A., Lacasa, A., & Pivot, Y. (1978). Biologie d'un thrips prédateur *Aeolothrips intermedius* (Thys.: Aeolothripidae). Entomophaga, 23, 403–410.

Bournier, A., Lacasa, A., & Pivot, Y. (1979). Régime alimentaire d'un thrips prédateur *Aeolothrips intermedius* (Thys.: Aeolothripidae). Entomophaga, 24, 353–361.

Bouseman, J. K. (1976). Biological observations on *Apiomerus crassipes* (F.) (Hemiptera: Reduviidae). Pan-Pacific Entomologist, 52, 178–179.

Bousquet, Y. (1990). A review of the North American species of *Rhizophagus* Herbst and a revision of the Nearctic members of the subgenus *Anomophagus* Reitter (Coleoptera: Rhizophagidae). Canadian Entomology, 122, 131–171.

Bouyjou, B., Canard, M., & Xuan, N. T. (1984). Analyse par battage des principux predators et proies potentielle en verger de poiriers nontraite. IOBC WPRS Bulletin, 7, 148–166.

Box, H. E. (1950). The more important insect pests of sugarcane in northern Venezuela. Proceedings of the Hawiian Entomological Society, 14, 41–51.

Boykin, L. W., Campbell, W. V., & Beute, M. K. (1984). Effect of pesticides on *Neozygites floridana* (Entomophthorales: Entomophthoraceae) and arthropod predators attacking the twospotted spider mite (Acari: Tetranychidae) in North Carolina peanut fields. Journal Economic Entomology, 77, 969–975.

Bradley, G. A. (1973). Effect of *Formica obscuripes* (Hymenoptera: Formicidae) on the predator-prey relationship between *Hyperaspis congressis* (Coleoptera: Coccinellidae) and *Toumeyella numismaticum* (Homoptera: Coccidae). Canadian Entomologist, 105, 1113–1118.

Braman, S. K., & Yeargan, K. V. (1988). Comparison of developmental and reproductive rates of *Nabis americoferus, N. roseipennis,* and *N. rufusculus* (Hemiptera: Nabidae). Annals of the Entomological Society of America, 81, 923–930.

Braman, S. K., & Yeargan, K. V. (1990). Phenology and abundance of *Nabis americoferus, N. roseipennis,* and *N. rufusculus* (Hemiptera: Nabidae) and their parasitoids in alfalfa and soybean. Journal of Economic Entomology, 83, 823–830.

Braman, S. K., Sloderbeck, P. E., & Yeargan, K. V. (1984). Effects of temperature on the development and survival of *Nabis americoferus* and *N. roseipennis.* Annals of the Entomological Society of America, 77, 592–596.

Brandmayr, P., DenBoer, P. J., & Weber, F. (Eds.). (1983). Biology of carabids: The synthesis of field study and laboratory experiment. Report Fourth Symposium European Carabidologists. Wageningen: Central Agricultural Publication Document (Pudoc).

Branquart, E., Hemptinne, J.-L., Bauffe, C., & Benfekih, L. (1997). Cannibalism in *Episyrphus balteatus* (Dipt.: Syrphidae). Entomophaga, 42, 145–152.

Breene, R. G. (1991). Control of cotton pests by red imported fire ants. IPM Practitioner, 13, 1–4.

Breene, R. G., Sterling, W. L., & Nyffeler, M. (1990). Efficacy of spider and ant predators on the cotton fleahopper (Hemiptera: Miridae). Entomophaga, 35, 393–401.

Breene, R. G., Meagher, R. L., Jr., Nordlund, D. A., & Wang, Y. T. (1992). Biological control of *Bemisia tabaci* (Homoptera: Aleyrodi-

dae) in a greenhouse using *Chrysoperla rufilabris* (Neuroptera: Chrysopidae). Biological Control, 2, 9–14.

Brettell, J. H. (1964). Biology of *Diloponis inconspicuus* Pope (Coleoptera: Coccinellidae), a predator of citrus red scale, with notes on the feeding behaviour of other scale predators. Journal of Entomological Society of South Africa, 27, 19–28.

Brettell, J. H. (1979). Green lacewings (Neuroptera: Chrysopidae) of cotton fields in central Rhodesia. I. Biology of *Chrysopa boninensis* Okomoto and toxicity of certain insecticides to the larva. Rhodesian Journal of Agricultural Research, 17, 141–150.

Briand, L. J. (1931). Notes on *Chrysopa oculata* Say and its relation to the Oriental peach moth (*Laspeyresia molesta* Busck.) infestation in 1930. Canadian Entomologist, 63, 123–126.

Bridwell, J. C. (1919). Some notes on Hawaiian and other Bethylidae (Hymenoptera) with descriptions of new species. Proceedings of the Hawaiian Entomological Society, 4, 21–38.

Bridwell, J. C. (1920). Some notes on Hawaiian and other Bethylidae (Hymenoptera) with the description of a new genus and species. 2nd paper. Proceedings of the Hawaiian Entomological Society, 4, 291–314.

Brindle, A. (1987). Dermaptera. In F. W. Stehr (Ed.), Immature insects (pp. 171–178). Dubuque, IA: Kendall/Hunt.

Britton, E. B., & Lee, B. (1972). *Stethorus loxtoni* sp. n. (Coleoptera: Coccinellidae), a newly-discovered predator of the two-spotted mite. Journal of Australian Entomological Society, 11, 55–60.

Brooks, S. J., & Barnard, P. C. (1990). The green lacewings of the world: A generic review (Neuroptera: Chrysopidae). Bulletin of British Museum (Natural History) Entomology, 59, 117–286.

Brothers, D. J. (1984). Gregarious parasitoidism in Australian Mutillidae (Hymenoptera). Australian Entomology Magazine, 11, 8–10.

Brower, J. H., & Mullen, M. A. (1990). Effects of *Xylocoris flavipes* (Hem.: Anthocoridae) releases on moth populations in experimental peanut storage. Journal of Entomological Science, 25, 268–276.

Brower, J. H., & Press, J. W. (1988). Interactions between the egg parasite *Trichogramma pretiosum* (Hym.: Trichogrammatidae) and a predator, *Xylocoris flavipes.* (Hem.: Anthocoridae) of the almond moth, *Cadra cautella,* (Lep.: Pyralidae). Journal of Entomological Science, 23, 342–349.

Brower, J. H., & Press, J. W. (1992). Suppression of residual populations of stored-product pests in empty corn bins by releasing the predator *Xylocoris flavipes* (Reuter). Biological Control, 2, 66–72.

Brown, H. D. (1972a). Predaceous behaviour of four species of Coccinellidae (Coleoptera) associated with the wheat aphid, *Schizaphis graminum,* (Rondani), in South Africa. Transactions of the Royal Entomological Society of London, 124, 21–36.

Brown, H. D. (1972b). The behaviour of newly hatched coccinellid larvae (Coleoptera: Coccinellidae). Journal of the Entomological Society of South Africa, 35, 149–157.

Brown, N. R., & Clark, R. C. (1956). Studies of predators of the balsam woolly aphid *Adelges piceae* (Ratz) (Homoptera: Adelgidae). I. Field identification of *Neoleucopis obscura* (Hal.), *Leucopina americana* (Mall.) and *Cremifania nigrocellulata* (Z.) (Diptera, Chamaemyiidae). Canadian Entomologist, 88, 272–279.

Bruce, W. A. (1983). Mites as biological control agents of stored products pests. In M. A. Hoy, G. L. Cunningham, & L. Knutson (Eds.), Biological control of pests by mites Special Publication 3304, (pp. 74–78). Berkeley: DANR, University of California.

Brues, C. T., Melander, A. L., & Carpenter, F. M. (1954). Classification of insects. Museum of Comparative Zoology, Cambridge, Mass.

Brunner, J. F., & Burts, E. C. (1975). Searching behavior and growth rates of *Anthocoris nemoralis* (Hemiptera: Anthocoridae), a predator of the pear psylla, *Psylla pyricola.* Annals of the Entomological Society of America, 68, 311–315.

Brushwein, J. R. (1987). Bionomics of *Lomoimyia hamata* (Neuroptera:

Berothidae). Annals of the Entomological Society of America, 80, 671–679.

Brust, G. E. (1990). Direct and indirect effects of four herbicides on the activity of carabid beetles (Coleoptera: Carabidae). Pesticide Science, 30, 309–320.

Budenberg, W. J., & Powell, W. (1992). The role of honeydew as an ovipositional stimulant for two species of syrphids. Entomologia Experimentalis et Applicata, 64, 57–61.

Bueno, V. H. P., & Berti Filho, E. (1984a). *Montina confusa* (Stål 1859) (Hemiptera, Reduviidae, Zelinae): I. Aspectos biologicos. Revista Brasileira de Entomologia, 28, 345–353.

Bueno, V. H. P., & Berti Filho, E. (1984b). *Montina confusa* (Stål 1859) (Hemiptera Reduviidae, Zelinae). II. Aspectos morphologicos de ninfas e adultos. Revista Brasileira de Entomologia, 28, 355–364.

Bueno, V. H. P., & Berti Filho, E. (1987). Consume e longevidade de adulttos de *Labidura ripara* (Pallas) (Dermaptera, Labiduridade) em ovos de *Diatraea saccharalis* (F.) (Lepidoptera: Pyralidae). Turialba, San Jose, 37, 365–368.

Bugg, R. L. (1993). Habitat manipulation to enhance the effectiveness of aphidophagous hover flies (Diptera: Syrphidae). Sust. Agric. 5, 12–15.

Bugg, R. L., Ehler, L. E., & Wilson, L. T. (1987). Effect of common knotweed (*Polygonum aviculare*) on abundance and efficiency of insect predators of crop pests. Hilgardia, 55, 1–53.

Bugg, R. L., Wickers, F. L., Brunson, K. E., Dutcher, J. D., & Phatak, S. C. (1991). Cool-season cover crops relay intercropped cantaloupe: Influence on a generalist predator, *Geocoris punctipes* (Hemiptera: Lygaeidae). Journal of Economic Entomology, 84, 408–416.

Burgess, A. F. (1911). *Calosoma sycophanta. Its life history, behavior, and successful colonization in New England* (Bulletin No. 101). Washington, DC: U.S. Department of Agriculture Bureau of Entomology.

Burgess, A. F., & Collins, C. W. (1917). The genus *Calosoma* (Bulletin No. 417). Washington, DC: U.S. Department of Agriculture.

Burgess, L. (1982). Predation on adults of the flea beetle *Phyllotreta cruciferae* by the western damsel bug, *Nabis alternatus* (Hemiptera: Nabidae). Canadian Entomologist, 114, 763–764.

Burgess, L., Dueck, J., & McKenzie, D. L. (1983). Insect vectors of yeast *Nematospora coryli* in mustard, *Brassica juncea*, crops in southern Saskatchewan. Canadian Entomologist, 115, 25–30.

Burke, H. R., & Martin, D. F. (1956). The biology of three chrysopid predators of the cotton aphid. Journal of Economic Entomology, 49, 698–700.

Burnett, T. (1977). Biological models of two acarine predators of the grain *Acarus siro* L. Canadian Journal of Zoology, 55, 1312–1323.

Buschman, L. L., Pitre, H. N., & Hodges, H. F. (1984). Soybean cultural practices: Effects on populations of geocorids, nabids, and other soybean arthropods. Environmental Entomology, 13, 305–317.

Buschman, L. L., Whitcomb, W. H., Hemenway, R. C., Mays, D. L., Ru, N., & Leppla, H. C. (1977). Predators of velvetbean caterpillar eggs in Florida soybeans. Environmental Entomology, 6, 403–407.

Butcher, M. R., Penman, D. R., & Scott, R. R. (1988). Field predation of two-spotted spider mite in a New Zealand strawberry crop. Entomophaga, 33, 173–183.

Butler, G. D., Jr. (1965). *Spanagonicus albofasciatus* as an insect and mite predator (Hemiptera: Miridae). Journal of Kansas Entomological Society, 38, 70–75.

Butler, G. D., Jr. (1966). Development of several predaceous Hemiptera in relation to temperature. Journal of Economic Entomology, 59, 1306–1307.

Butler, G. D., Jr., & Henneberry, T. J. (1988). Laboratory studies of *Chrysoperla carnea* predation on *Bemisia tabaci*. Southwestern Entomologist, 13, 165–170.

Butler, G. D., Jr., & May, C. J. (1971). Laboratory studies of the searching capacity of larvae of *Chrysopa carnea* for eggs of *Heliothis* spp. Journal of Economic Entomology, 64, 1459–1461.

Butler, G. D., & Ritchie, P. L., Jr. (1971). Feed Wheast and the abundance and fecundity of *Chrysopa carnea*. Journal of Economic Entomology, 64, 933–934.

Butler, L. (1966). Oviposition in the Chinese mantid. *Tenodera aridifolia sinensis* (Saussure) (Orthoptera: Mantidae). Journal of Georgia Entomological Society, 1, 5–7.

Buxton, J. H., & Madge, D. S. (1976). The evaluation of the European earwig (*Forficula auricularia*) as a predator of the damson-hop aphid (*Phorodon humuli*). I. Feeding experiments. Entomologia Experimentalis et Applicata, 19, 109–114.

Cade, W. H., Simpson, P. H., & Breland, O. P. (1978). *Apiomerus spissipes* (Hemiptera: Reduviidae): A predator of harvester ants in Texas. Southwestern Entomologist, 3, 195–197.

Cai, C. R., Zhang, X. D., & Zhao, J. Z. (1985). Tentative studies on the artificial diets of larvae and adults of *Chrysopa septempunctata* (Neuroptera: Chrysopidae). Natural Enemy Insects, 7, 125–128 (in Chinese, English summary).

Callan, E. M. (1943). Natural enemies of the cacao thrips. Bulletin of Entomological Research, 34, 313–321.

Caltagirone, L. E. (1969). Terpenylacetate bait attracts *Chrysopa* adults. Journal of Economic Entomology, 62, 1237.

Caltagirone, L. E., & Doutt, R. L., (1989). The history of the vedalia beetle importation to California and its impact on the development of biological control. Annual Review of Entomology, 34, 1–16.

Cameron, P. J., Hill, R. L., Barn, J., & Thomas, W. P. (1989). A review of biological control of invertebrate pests and weeds in New Zealand 1874 to 1987. (CIBC Technical Communication No. 10). CAB International, Wallingford, UK.

Campbell, C. A. M. (1977). A laboratory evaluation of *Anthocoris nemorum* and *A. nemoralis* (Hem.: Anthocoridae) as predators of *Phorodon humuli* (Hom.: Aphididae). Entomophaga, 22, 309–314.

Campbell, C. A. M. (1978). Regulation of the damson-hop aphid [(*Phorodon humuli* (Shrank)] on hops (*Humulus lupulus* L.) by predators. Journal of Horticultural Science, 53, 235–242.

Campbell, J. B., & Hermanussen, J. F. (1974). *Philonthus theveneti*: Life history and predatory habits against stable flies, house flies, and face flies under laboratory conditions. Environmental Entomology, 3, 356–358.

Campbell, J. M. (1979). A revision of the genus *Tachyporus* Gravenhorst (Coleoptera: Staphylinidae) of North and Central America. Memoirs of the Entomological Society of Canada, 109, 1–93.

Campbell, R. W., & Torgersen, T. R. (1982). Some effects of predaceous ants on western spruce budworm pupae in north central Washington. Environmental Entomology, 11, 111–114.

Canard, M. (1970a). Sterlite d'origine alimentaire chez le male d'un predateur aphidiphage *Chrysopa perla* (L.) (Insectes, Neuropteres). Comptes Rendus Hebdomadaires des Seances de l'Academie des Sciences, Series D, 271, 1097–1099.

Canard, M. (1970b). Incidences de la valeur alimentaire de divers pucerons (Homoptera: Aphidae) sur le potentiel de multiplication de *Chrysopa perla* (L.) (Neuroptera, Chrysopidae). Annals of the Zoological Ecology of Animal, 2, 345–355.

Canard, M. (1997). Can lacewings feed on pests in winter? (Neur.: Chrysopidae and Hemerobiidae). Entomophaga, 42, 113–117.

Canard, M., & Grimal, A. (1993). Multiple action of photoperiod on diapause in the green lacewing *Mallada picteti* (McLachlan) (Neuroptera: Chrysopidae). Bollettino dell'Istituto di Entomologia "Guido Grandi" della Universita degli Studi di Bologna, 47, 235–245.

Canard, M., & Principi, M. M. (1984). Development of Chrysopidae. In M. Canard, Y. Semeria, & T. R. New (Eds.), Biology of Chrysopidae (pp. 57–74). The Hague: Dr. W. Junk.

Canard, M., Semeria, Y., & New, T. R. (Eds.). (1984). Biology of Chrysopidae. The Hague: Dr. W. Junk.

Cannings, R. A. (1987). The ground mantis, *Litaneutria minor* (Dictoptera: Mantidae) in British Columbia. Journal of Entomological Society of British Columbia, 84, 64–65.

Carayon, J. (1961a). Quelque remarques sur les Hemipteres–Heteropteres: Leur importance comme insectes auxiliaires et les possibilites de leur utilisation dans la lutte biologique. Entomophaga, 6, 133–141.

Carayon, J. (1961b). Valeur systematique des voies ectodermiques de l'appaieil genital femelle chez les Hemipteres Nabidae. Bulletin du Museúm National d'Histoire Naturelle, 33, 183–196.

Carayon, J., & Steffan, J. R. (1959). Observations sur le regime alimentaiie des Orius et particalierement d'Orius pallidicornis (Reuter) (Heteroptera: Anthocoridae). Cahiers des Natural. Bull. Nat. Parisiens n.s. 15, 53–64.

Carlson, R. E., & Chiang, H. C. (1973). Reduction of an Ostrina nubilalis population by predatory insects attracted by sucrose sprays. Entomophaga, 18, 205–211.

Carnegie, A. J. M., & Harris, R. H. G. (1969). The introduction of mirid egg predators (Tytthus spp.) into South Africa. Proceedings of the South African Sugar Technology Association, 43, 113–116.

Carpenter, F. M. (1940). A revision of the Nearctic Hemerobiidae. Berothidae, Sisyridae, Polystaechotidae and Dilardiae (Neuroptera). Proceedings of the American Academy Arts Science, 74, 193–280.

Carpenter, J. M. (1982). The phylogenetic relationships and natural classification of the Vespoidea (Hymenoptera). Systematics and Entomology, 7, 11–38.

Carroll, D. P., & Hoyt, S. C. (1984). Augmentation of European earwigs (Dermaptera: Forficulidae) for biological control of apple aphid (Homoptera: Aphididae) in an apple orchard. Journal of Economic Entomology, 77, 738–740.

Carroll, D. P., Walker, J. T. S., & Hoyt, S. C. (1985). European earwigs (Dermaptera: Forficulidae) fail to control apple aphids on bearing apple trees and woolly apple aphids (Homoptera: Aphididae) in apple root stock stool beds. Journal of Economic Entomology, 78, 972–974.

Carter, M. C., & Dixon, A. F. G. (1982). Habitat quality and the foraging behaviour of coccinellid laravae. Journal of Animal Ecology, 51, 865–878.

Carter, M. C., & Dixon, A. F. G. (1984). Honeydew: An arrestant stimulus for coccinellids. Ecological Entomology, 9, 383–387.

Cartwright, B. O., Eikenbary, R. D., Johnson, J. W., Johnson, T. N., Farris, T. H., & Morrison, R. D. (1977). Field release and dispersal of Menochilus sexmaculatus, an imported predator of greenbug, Schizaphis graminum. Environmental Entomology, 6, 699–704.

Carvalho, J. C. M., & Southwood, T. R. E. (1955). Revisao do complexo Cyrtorhinus Fieber-Mecamma Fieber (Hemiptera: Heteroptera). Boletim do Museu Paraense Emilio Goeldi, 11, 7–72.

Carvalho, J. C. M., Fontes, A. V., & Henry, T. J. (1983). Taxonomy of the South American species of Ceratocaspsus, with descriptions of 45 new species (Hemiptera: Miridae). United States Technical Bulletin, 1676, 1–58.

Celli, G., Corazza, L., Nicoli, G., Burchi, C., Cornale, R., & Benuzzi, M. (1986). Lotta biologica con Chrysoperla carnea Steph. (Neuroptera: Chrysopidae) Agli Afidi della fragola in serra. Due anni di esperienze. Istituto di Entomologia "G. Grandi," Universita degli Studi, Bologna. ATTI Giornate Fitopatologische, 1, 93–102.

Chaing, H. C. (1970). Effects of manure applications and mite predation on corn rootworm populations in Minnesota. Journal of Economic Entomology, 63, 934–936.

Chambers, R. J. (1986). Preliminary experiments on the potential of hoverflies (Dipt.: Syrphidae) for the control of aphids under glass. Entomophaga, 31, 197–204.

Chambers, R. J. (1988). Syrphidae. In A. K. Minks & P. Harrewijn (Eds.), Aphids, their biology, natural enemies, and control (Vol. 2B, pp. 259–270). Amsterdam: Elsevier.

Chambers, R. J., & Adams, T. H. L. (1986). Quantification of the impact of hoverflies (Diptera: Syrphidae) on cereal aphids in winter wheat: An analysis of field populations. Journal of Applied Ecology, 23, 895–904.

Chambers, R. J., Sunderland, K. D., Wyatt, I. J., & Vickerman, G. P. (1983). The effects of predator exclusion and caging on cereal aphids in winter wheat. Journal of Applied Ecology, 209, 209–224.

Champlain, R. A., & Sholdt, L. L. (1967a). Life history of Geocoris punctipes (Hemiptera: Lygaeidae) in the laboratory. Annals of the Entomological Society of America, 60, 881–883.

Champlain, R. A., & Sholdt, L. L. (1967b). Temperature range for development of immature stages of Geocoris punctipes (Hemiptera: Lygaeidae). Annals of the Entomological Society of America, 609, 881–883.

Chandler, A. E. F. (1968a). Some host-plant factors affecting oviposition by aphidophagous Syrphidae (Diptera). Annals of Applied Biology, 61, 415–423.

Chandler, A. E. F. (1968b). A preliminary key to the eggs of some of the commoner aphidophagous syrphidae (Diptera) occurring in Britain. Transactions of the Royal Entomological Society of London, 120, 199–210.

Chang, S. J., & Oka, H. I. (1984). Attributes of a hopper-predator community in a rice field. Agricultural Ecosystems in the Environment, 12, 73–78.

Chant, D. A. (1959). Phytoseiid mites (Acarina: Phytoseiidae). I. Bionomics of seven species in southeastern England. II. A taxonomic review of the family Phytoseiidae, with descriptions of thirty-eight new species. Canadian Entomologist, 91(Suppl. 12).

Chantal, C. (1972). Additions a la faune coleopterique du Quebec. Nature Canada, 99, 243–244.

Chao, C. C., & Chang, S. J. (1978). Population fluctuations of green lace wings in cotton fields. Acta Entomologica Sinica, 21, 271–278 (in Chinese); (1979). Review of Applied Entomology, Series A, 5092, 67, 618.

Chapin, E. A. (1965a). Coleoptera, Coccinellidae. Insects of Micronesia. Bishop Museum of Honolulu, Hawaii, 16, 189–254.

Chapin, E. A. (1965b). The genera of the Chilocorini (Coleoptera, Coccinellidae). Bulletin of the Museum of Comparative Zoology, 133, 227–271.

Chapin, J. B., & Brou, V. A. (1991). Harmonia axyridis (Pallas), the third species of the genus to be found in the United States. Proceedings of the Entomological Society of Washington, 93, 630–635.

Chazeau, J. (1979). La lutte biologique contre la cochenille transparente du cocotier, Temnaspidiotis destructor (Signoret). ORSTOM Series in Biology, 44, 11–22.

Chazeau, J. (1985). Predaceous insects. In W. Helle & M. W. Sabelis (Eds.), Spider mites, their biology, natural enemies and control (Vol. 1B, pp. 211–246). Amsterdam: Elsevier Science.

Chazeau, J., Bouye, E., & Larbogne, L. (1991). Cycle de development et table de vie d'Olla v-nigrum (Col.: Coccinellidae) ennemi naturel d'Heteropsylla cubana (Hom.: Psyllidae) introduit in Nouvelle-Caledonie. Entomophaga, 36, 275–285.

Chen, S. (1962). The oldest practice of biological control: The culture and efficacy of Oecophila smaragdina Fabr. in orange orchards. Acta Entomologica Sinica, 11, 401–407 (in Chinese).

Chen, S. (1985). Integrated pest management especially by natural control in citrus orchards. Natural Enemy Insects, 7, 223–302.

Chen, S. M., Cheng, W. Y., & Weng, Z. T. (1993). Non-partition rearing of the green lacewing Mallada basalis (Neuroptera: Chrysopidae). Report of the Taiwan Sugar Research Institute, 141, 25–33.

Chen, Z. Z., & Willson, H. R. (1996). Species composition and seasonal distribution of carabids (Coleoptera: Carabidae) in an Ohio soybean field. Journal of Kansas Entomological Society, 69, 310–316.

Cheng, H. K., Zhao, J. H., Xie, M., Wei, S. X., Song, X. P., & Wang, J. Z. (1992). Tests on the effect of releasing Aphidoletes aphidimyza (Dip.: Cecidomyiidae) to control the aphid Myzus persicae, in greenhouses and plastic tunnels. Chinese Journal of Biological Control, 8, 97–100.

Chiu, S. C. (1979). Biological control of the brown planthopper. International Rice Communications News, 15, 33–36.

Chiverton, P. A. (1987). Predation of *Rhopalosiphum padi* (Homoptera: Aphididae) by polyphagous predatory arthropods during the aphids' pre-peak period in spring barley. Annals of Applied Biology, 111, 257–270.

Chiverton, P. A. (1988). Searching behavior and cereal aphid consumption by *Bembidion lampros* and *Pterostichus cupreus* in relation to temperature and prey density. Entomologia Experimentalis et Applicata, 47, 173–182.

Chiverton, P. A., & Sotherton, N. W. (1991). The effects on beneficial arthropods of the exclusion of herbicides from cereal crop edges. Journal of Applied Ecology, 28, 1027–1039.

Chock, Q. C., Davis, C. J., & Chong, M. (1961). *Sepedon macropus* (Diptera: Sciomyzidae) introduced into Hawaii as a control for the liver fluke snail, *Lymnaea ollula*. Journal of Economic Entomology, 54, 1–4.

Chow, T., Long, G. E., & Tamaki, G. (1983). Effects of temperature and hunger on the functional response of *Geocoris bullatus* (Say) (Hemiptera: Lygaeidae) to *Lygus* spp. (Hemiptera: Miridae) density. Environmental Entomology, 12, 1332–1338.

Chu, Y. I. (1969). On the biomics of *Lyctocoris beneficus* (Hiara) and *Xylocoris galactinus* (Fleber) (Anthocoridae Heteroplera), Journal of the Faculty of Agriculture, Kyushu University, 15, 1–136.

Chua, T. H., & Mikil, E. (1989). Effects of prey number and stage on the biology of *Cyrtorhinus lividipennis* (Hemiptera: Miridae): A predator of *Nilaparvata lugens* (Homoptera: Delphacidae). Environmental Entomology, 18, 251–255.

Chua, T. H., Dyck, V. A., & Pena, N. B. (1984). Functional response and searching efficiency in *Pseudogonatopus flavifemur* Esaki & Hash. (Hymenoptera: Dryinidae), a parasite of rice planthoppers. Researches on Population Ecology (Kyoto), 26, 74–83.

Cicolani, B. (1979). The intrinsic rate of natural increase in dung macrochelid mites. predators of *Musca domestica* eggs. Bollettino di Zoologia, 46, 171–178.

Clark, R. C., Greenbank, D. D., Bryant, D. G., & Harris, J. W. E. (1971). *Adelges piceae* (Ratz.), balsam woolly aphid (Homoptera: Adelgidae). CIBC Technical Communications, 4, 113–125.

Clausen, C. P. (1915). Mealybugs on citrus trees. University of California Agricultural Experimental Station Bulletin, 258, 19–48.

Clausen, C. P. (1940). Entomophagous insects. New York: McGraw-Hill.

Clausen, C. P. (1956). Biological control of insect pests in the continental United States (Technical Bulletin No. 1139). Washington, DC: U.S. Department of Agriculture.

Clausen, C. P. (1978). (Ed.). Introduced parasites and predators of arthropod pests and weeds: A world review Agricultural Handbook No. 480. Washington, DC: U.S. Department of Agriculture.

Clausen, C. P., & Berry, P. A. (1932). The citrus blackfly in Asia, and the importation of its natural enemies into tropical America (Technical Bulletin No. 320). Washington, DC: U.S. Department of Agriculture.

Cloutier, C., & Johnson, S. E. (1993). Predation by *Orius tristicolor* (Hemiptera: Anthocoridae) on *Phytoseiulus persimilis* (Acarina: Phytoseiidae): Testing for compatibility between biocontrol agents. Environmental Entomology, 22, 477–482.

Coates, T. J. D. (1974). The influence of some natural enemies and pesticides on various populations of *Tetranychus cinnabarinus* (Boisduval), *T. lombardinii* Baker & Pritchard and *T. ludeni* Zacker (Acari: Tetranychidae) with aspects of their biologies (Entomological Memoir 42 Rep.). South African Department of Agriculture. Entomological Memoir, Department of Agriculture, Republic of South africa, No. 42, pp. 1–40.

Cobben, R. H. (1978). Evolutionary trends in Heteroptera. II. Mouthpart structures and feeding strategies. Agricultural Research Reports, Wageningen (1978), 5, 1–407.

Cochereau, P. (1969). Controle biologique d'*Aspidiotus destructor* Signoret (Homoptera, Diaspinae) dan l'ile Vate' (Nouvelles Hebrides)

aumoyen de *Rhizobius pulchellus* Montrouzier (Coleoptera: Coccinellidae). ORSTOM, Series in Biology, 8, 57–100.

Cock, M. J. W. (Ed.). (1985). A review of biological control of pests in the commonwealth Caribbean and Bermuda up to 1982. London: Commonwealth Agricultural Bureau.

Cohen, A. C. (1983). Improved method of encapsulating artificial diet for rearing predators of harmful insects. Journal of Economic Entomology, 76, 951–959.

Cohen, A. C. (1984). Food consumption, food utilization and metabolic rates of *Geocoris punctipes* (Het: Lygaeidae) fed *Heliothis virescens* (Lep.: Noctuidae) eggs. Entomophaga, 29, 361–367.

Cohen, A. C. (1985). Simple method for rearing the insect predator *Geocoris punctipes* (Heteroptera: Lygaeidae) on a meat diet. Journal of Economic Entomology, 78, 1173–1175.

Cohen, A. C. (1989). Ingestion efficiency and protein consumption by a heteropteran predator. Annals of the Entomological Society of America, 82, 495–499.

Cohen, A. C. (1990). Feeding adaptations of some predaceous Hemiptera. Annals of the Entomological Society of America, 83, 1215–1223.

Cohen, A. C. (1993). Organization of digestion and preliminary characterization of salivary trypsin-like enzymes in a predaceous heteropteran, *Zelus renardii*. Journal of Insect Physiology, 39, 823–829.

Colburn, R., & Asquith, D. (1970). A cage used to study finding of a host by the ladybird beetle, *Stethorus punctum*. Journal of Economic Entomology, 63, 1376–1377.

Colburn, R., Asquith, D. (1971). Observations on the morphology and biology of the ladybird beetle *Stethorus punctum*. Annals of the Entomological Society of America, 64, 1217–1221.

Coles, L. W. (1964). Collection of predaceous lady beetle. *Hyperaspis trilineata*, in Barbados, and shipment to Hawaii. Journal of Economic Entomology, 57, 768.

Collier, T. R. (1995). Host feeding, egg maturation, resorption, and longevity in the parasitoid *Aphytis melinus* (Hymenoptera: Aphelinidae). Annals of the Entomological Society of America, 88, 206–214.

Collyer, E. (1951). The separation of *Conwentzia pineticola* End. from *Conwentzia psociformis* (Curt.), and notes on their biology. Bulletin of Entomological Research, 42, 555–564.

Collyer, E. (1952). Biology of some predatory insects and mites associated with the fruit tree red spider mite (*Metatetranychus ulmi* Koch) in southeastern England. Journal of Horticultural Science, 27, 117–129.

Collyer, E. (1953a). Biology of some predatory insects and mites associated with the fruit tree red spider mite (*Metatetranychus ulmi* Koch) in southeastern England. II. Some important predators of the mite. Journal of Horticultural Science, 28, 85–97.

Collyer, E. (1953b). Biology of some predatory insects and mites associated with the fruit tree red spider mite (*Metatetranychus ulmi* Koch) in southeastern England. III. Further predators of the mite. Journal of Horticultural Science, 28, 98–113.

Collyer, E. (1964). Phytophagous mites and their predators in New Zealand orchards. New Zealand Journal of Agricultural Research, 7, 551–568.

Collyer, E. (1967). On the ecology of *Anthocoris nemorum* (L.) (Hemiptera-Heteroptera). Proceedings of the Royal Entomological Society London, Series A, 42, 107–118.

Congdon, B. D., Shanks, C. H., & Antonelli, A. L. (1993). Population interaction between *Stethorus punctum picipes* (Coleoptera: Coccinellidae) and *Tetranychus urticae* (Acari: Tetranychidae) in red raspberries at low predator and prey densities. Environmental Entomology, 22, 1302–1307.

Coppel, H. C., & Jones, P. A. (1962). Bionomics of *Podisus* spp. associated with the introduced pine sawfly, *Diprion similis* (Htg.) in Wisconsin. Wisconsin Academy of Science, Arts, and Letters, 51, 31–56.

Corbet, S. A., & Backhouse, M. (1975). Aphid-hunting wasps: A field study of *Passaloecus*. Transactions of the Royal Entomological Society of London, 127, 11–30.

Cornelius, M. L. (1991). Tritrophic level interactions: Effects of host plant traits on two predators of lepidopteran larvae, the Argentine ant, *Iridomyrmex humilis* and the paper wasp *Mischocyttarus flavitarsis*. Unpublished doctoral dissertation, University of California, Berkeley.

Coupland, J. B. (1996a): The biological control of helicid snail pests in Australia: Surveys, screening and potential agents. In I. F. Henderson (Ed.), Slug and snail pests in agriculture. British Crop Protection Council Symposium Proceedings, 66, 255–261.

Coupland, J. B. (1996b). Influence of snail feces and mucus on oviposition and larval behavior of *Pherbellia cinerella* (Diptera: Sciomyzidae). Journal of Chemical Ecology, 22, 183–189.

Coupland, J., & Baker, G. (1995). The potential of several species of terrestrial Sciomyzidae as biological control agents of pest helicid snails in Australia. Crop Protection, 14, 573–576.

Coupland, J., Espiau, B. A., & Baker, G. (1994). Seasonality, longevity, host choice, and infection efficiency of *Salticella fasciata* (Diptera: Sciomyzidae), a candidate for the biological control of pest helicid snails. Biological Control, 4, 32–37.

Couturier, A. (1938). Contribution o l'etude biologique de *Podisus maculiventris* Say. Predateur Americain du doryphore. Annales des Epiphyties et de Phytogenetique, 4, 95–165.

Coville, P. L., & Allen, W. W. (1977). Life table and feeding habits of *Scolothrips sexmaculatus* (Thysanoptera: Thripidae). Annals of the Entomological Society of America, 64, 1217–1221.

Cowgill, S. E., Wratten, S. D., & Sotherton, N. W. (1993a). The selective use of floral resources by the hoverfly *Episyrphus balteatus* (Diptera: Syrphidae) on farmland. Annals of Applied Biology, 122, 223–231.

Cowgill, S. E., Wratten, S. D., & Sotherton, N. W. (1993b). The effect of weeds on the numbers of hoverfly (Diptera: Syrphidae) adults and the distribution and composition of their eggs in winter wheat. Annals of Applied Biology, 123, 499–515.

Cress, D. C., & Lawson, F. A. (1971). A life history of *Pasimachus elongatus*. Journal of Kansas Entomological Society, 30, 142–144.

Crocker, R. L., Whitcomb, W. H., & Ray, R. M. (1975). Effects of sex, developmental stage, and temperature on predation by *G. punctipes*. Environmental Entomology, 4, 531–534.

Croft, B. A., & McGroarty, D. L. (1973). A model study of acaricide resistance, spider mite outbreaks, and biological control patterns in Michigan apple orchards. Environmental Entomology, 2, 633–638.

Croft, B. A., MacRae, I. V., & Currans, K. G. (1992). Factors affecting biological control of apple mites by mixed populations of *Metaseiulus occidentalis* and *Typhlodromus pyri*. Experimental and Applied Acarology (Northwood), 14, 343–355.

Crouzel, I. S. de, & Botto, E. N. (1977). Cicio de vida de *Chrysopa lanata lanata* (Banks) y algunas observaciones biologicas en condiciones de laboratorio. Revista de Investigaciones Agropecuarias INTA, Buenos Aires, Serie 5, Patologia Vegetal, 13, 1–14.

Crowson, R. A. (1981). The biology of the Coleoptera. London: Academic Press.

Crowson, R. A. (1982). Another view about life forms in ground beetles (Coleoptera: Carabidae). Entomologia Generalis, 7, 379–380.

Cudjoe, A. R., Neuenschwander, P., & Copland, M. J. W. (1993). Interference by ants in biological control of the cassava mealybug *Phenacoccus manihoti* (Hemiptera: Pseudococcidae) in Ghana. Bulletin of Entomological Research, 83, 15–22.

Culliney, T. W., Beardsley, J. W., & Drea, J. J. (1988). Population regulation of the Eurasian pine adelgid (Homoptera; Adelgidae) in Hawaii. Journal of Economic Entomology, 81, 142–147.

Daane, K. M., & Yokota, G. Y. (1997). Release strategies affect survival and distribution of green lacewings (Neuroptera: Chrysopidae) in augmentation programs. Environmental Entomology, 26, 455–464.

Daane, K. M., Yokota, G. Y., Rasmussen, Y. D., Zheng, Y., & Hagen, K. S. (1993). Effectiveness of leafhopper control varies with lacewing release methods. California Agriculture, 47, 19–23.

Dahlsten, D. L. (1982). Relationships between bark beetles and their natural enemies. In J. B. Mitton & K. B. Sturgeon (Eds.), Bark beetles in North American confiers (pp. 140–182). Austin: University of Texas Press.

Dahlsten, D. L., Garcia, R., Prine, J. E., & Hunt, R. (1969). Insect problems in forest recreation areas. California Agriculture, 23, 4–6.

Darst, P. H., & King, E. W. (1969). Biology of *Melichares tarsalis* in association with *Plodia interpunctella*. Annals of the Entomological Society of America, 62, 747–749.

Da Silva, P. G., Hagen, K. S., & Gutierrez, A. P. (1992). Functional response of *Curinus coeruleus* (Col.: Coccinellidae) to *Heteropsylla cubana* (Hom.: Psyllidae) on artificial and natural substrate. Entomophaga, 37, 555–564.

Davidson, W. M. (1924). Observations and experiments on dispersion of the convergent lady-beetle (*Hippodamia convergens* Guerin) in California. Transactions of the American Entomological Society, 50, 163–175.

Davis, A. L. V., Doube, B. M., & McLennon, P. D. (1988). Habitat associations and seasonal abundance of coprophilous Coleoptera (Staphylinidae, Hydrophilidae and Histeridae) in the Hluhluwe region of South Africa. Bulletin of Entomological Research, 78, 425–434.

Davis, C. J., & Krauss, N. L. H. (1963). Recent introductions for biological control in Hawaii. VIII. Proceedings of the Hawaiian Entomological Society, 18, 245–249.

Davis, C. J., & Krauss, N. L. H. (1965). Recent introductions for biological control in Hawaii. X. Proceedings of the Hawaiian Entomological Society, 19, 87–91.

Davis, E. G., Horton, J. R., Gable, C. H., Walter, E. V., Blanchard, R. A., & Heinrich, C. (1933). The southwestern corn borer (Technical Bulletin 388). Washington, DC: U.S. Department of Agriculture.

Davis, H. G., George, D. A., McDonough, L. M., Tamaki, G., & Burditt, A. K. (1983). Checkered flower beetle (Coleoptera: Cleridae) attractant: Development of an effective bait. Journal of Economic Entomology, 76, 674–675.

Davis, J. R., & Kirkland, R. L. (1982). Physiological and environmental factors related to the dispersal flight of the convergent lady beetle, *Hippodamia convergens* Guerin-Meneville. Journal of Kansas Entomological Society, 55, 187–196.

Davis, N. (1969). Contribution to the morphology and phylogeny of the Reduvioidea. IV. The Harpactorid complex. Annals of the Entomological Society of America, 62, 74–94.

Dean, G. J., & Satasook, C. (1983). Response of *Chrysoperla carnea* (Stephens) (Neuroptera: Chrysopidae) to some potential attractants. Bulletin of Entomological Research, 73, 619–624.

DeBach, P. (1943). The importance of host-feeding by adult parasites in the reduction of host populations. Journal of Economic Entomology, 36, 647–658.

DeBach, P. (Ed.). (1964). Biological control of insect pests and weeds. London: Chapman & Hall.

DeBach, P., & Hagen, K. S. (1964). Manipulation of entomophagous species. In P. DeBach (Ed.), Biological control of insect pests and weeds (pp. 429–458). London: Chapman & Hall.

De Clereq, P., & Degheele, D. (1990). Description and life history of the predatory bug *Podisus sagitta* (Fab.) (Hemiptera: Pentatomidae). Canadian Entomologist, 122, 1149–1156.

De Clereq, P., & Degheele, D. (1992a). Development and survival of *Podisus maculiventris* (Say) and *Podisus sagitta* (Fab.) (Heteroptera: Pentatomidae) at various constant temperatures. Canadian Entomologist, 124, 125–133.

De Clereq, P., & Degheele, D. (1992b). A meat-based diet for rearing the predatory stinkbugs *Podisus maculiventris* and *Podisus sagitta* (Het.: Pentatomidae). Entomophaga, 37, 149–157.

Decou, G. C. (1994). Biological control of the two-spotted spider mite (Acarina: Tetranychidae) on commercial strawberries in Florida with

Phytoseiulus persimilis (Acarina: Phytoseiidae). Florida Entomologist, 77, 33–41.

Dejean, A. (1991). Adaptation d'*Oecophylla longioda* (Formicidae: Formicinae) aux variations spatio-temporelles de la densite de proies. Entomophaga, 36, 29–54.

De Leon, D. (1935). A study of *Medetera aldrichii* Wh. (Diptera–Dolichopodidae), a predator of the mountain pine beetle (*Dendroctonus monticolae* Hopk., Col.–Scolytidae). Entomologica Americana, 15, 55–91.

Delucchi, V. (1954). *Pullus impexus* (Muls.) (Coleoptera, Coccinellidae), a predator of *Adeleges piceae* (Ratz.) (Hemiptera, Adelgidae), with notes on its parasites. Bulletin Entomological Research, 45, 243–278.

de Moraes, G. J., & McMurtry, J. A. (1985). Chemically mediated arrestment of the predaceous mite *Phytoseiulus persimilis* by extracts of *Tetranychus evansi* and *Tetranychus urticae*. Experimental and Applied Acarology (Northwood), 1, 127–138.

Dempster, J. P. (1968). Intra-specific competition and dispersal: As exemplified by a psyllid and its anthocorid predator. In T. R. E. Southwood (Ed.), Insect abundance (pp. 8–17). Oxford: Blackwell.

Den Boer, P. J. (1986). What can carabid beetles tell us about dynamics of populations? In P. J. Den Boer, M. L. Luff, D. Mossakowski, & F. Weber (Eds.), Carabid beetles—their adaptations and dynamics (pp. 313–330). Stuttgart: Gustov Fischer.

Den Boer, P. J., Luff, M. L., Mossakowski, D., & Weber, F. (Eds.). (1986). Carabid beetles—their adaptations and dynamics. Stuttgart: Gustov Fischer.

Dennis, P., & Sotherton, N. W. (1994). Behavioural aspects of staphylinid beetles that limit their aphid feeding potential in cereal crops. Pedobiologia, 38, 222–237.

Dennis, P., Thomas, M. B., & Sotherton, N. W. (1994). Structural features of field boundaries which influence the overwintering densities of beneficial arthropod predators. Journal of Applied Ecology, 31, 361–370.

Dennis, P., Wratten, S. D., & Sotherton, N. W. (1991). Mycophagy as a factor limiting predation of aphids (Hemiptera: Aphididae) by staphylinid beetles (Coleoptera: Staphylinidae) in cereals. Bulletin of Entomological Research, 81, 25–31.

Denver, C. (1974). The use of biological control in an Arkansas farming operation. Proceedings, Tall Timbers Conference on Ecological Animal Control and Habitat Management, Tallahassee, 5, 61–65.

Deyrup, M. A. (1978). Notes on the biology of *Pissodes fasciatus* LeConte and its insect associates (Coleoptera: Curculionidae). Pan-Pacific Entomologist, 54, 103–106.

Dicke, F. F., & Jarvis, J. L. (1962). The habits and seasonal abundances of *Orius insidiosus* (Say) on corn. Journal of Kansas Entomological Society, 35, 339–344.

Digweed, S. C. (1993). Selection of terrestrial gastropod prey by cychrine and pterostichine ground beetles (Coleoptera: Carabidae). Canadian Entomologist, 125, 463–472.

Dimetry, N. Z. (1974). The consequences of egg cannibalism in *Adalia bipunctata* (Coleoptera: Coccinellidae). Entomophaga, 19, 445–451.

Dixon, A. F. G. (1958). The escape responses shown by certain aphids to the presence of the coccinellid *Adalia decempunctata* (L.). Transactions of the Royal Entomological Society of London, 10, 319–334.

Dixon, A. F. G. (1959). An experimental study of the searching behaviour of the predatory coccinellid beetle *Adalia decempunctata* (L.). Journal of Animal Ecology, 28, 259–281.

Dixon, A. F. G. (1970). Factors limiting the effectiveness of the coccinellid beetle, *Adalia bipunctata* (L.), as a predator of the sycamore aphid, *Drepanosiphum platanoides* (Schr.). Journal of Animal Ecology, 39, 739–751.

Dixon, A. F. G., & Russel, R. J. (1972). The effectiveness of *Anthocoris nemorum* and *A. confusus* (Hemiptera: Anthocoridae) as predators of the sycamore aphid, *Drepanosiphum platanoides*. II. Searching behav-

ior and the incidence of predation in the field. Entomologia Experimentalis et Applicata, 15, 35–50.

Dixon, P. L., & McKinlay, R. G. (1992). Pitfall catches of and aphid predation by *Pterostichus melanarius* and *Pterostichus madidus* in insecticide treated and untreated potatoes. Entomologia Experimentalis et Applicata, 64, 63–72.

Dixon, A. F. G., Hemptinne, J.-L., & Kindlmann, P. (1997). Effectiveness of ladybirds as biological control agents: Patterns and processes. Entomophaga, 42, 71–83.

Donahoe, M. C., & Pitre, H. H. (1977). *Reduviolus roseipennis* behavior and effectiveness in reducing numbers of *Heliothis zea* on cotton. Environmental Entomology, 6, 872–876.

Dong, Y. G. (1988). Trials on the control of *Aphis gossypii* Glover with *Coccinella axyridis* Pallas. Zhejiang Agricultural Science, 3, 135–139.

Doutt, R. L. (1948). Effect of codling moth sprays on natural control of the Baker mealybug. Journal of Economic Entomology, 41, 116.

Doutt, R. L. (1951). Biological control of mealybugs infesting commercial greenhouse gardenias. Journal of Economic Entomology, 44, 37–40.

Doutt, R. L. (1952). Biological control of *Planococcus citri* on commercial greenhouse *Stephanotis*. Journal of Economic Entomology, 45, 342.

Doutt, R. L. (1958). Vice, virtue, and the vedalia. Bulletin of the Entomological Society of America, 4, 119–123.

Doutt, R. L. (1964). The historical development of biological control. In P. DeBach (Ed.), Biological control of insect pests and weeds (pp. 21–42). London: Chapman & Hall.

Doutt, R. L., & Hagen, K. S. (1949). Periodic colonization of *Chrysopa californica* as a possible control of mealybugs. Journal of Economic Entomology, 42, 560.

Doutt, R. L., & Hagen, K. S. (1950). Biological control measures applied against *Pseudococcus maritimus* on pears. Journal of Economic Entomology, 43, 94–96.

Downes, W. (1921). The life history of *Apateticus crocatus* Uhl. Proceedings of Entomological Society of British Columbia, 16, 21–27.

Drea, J. J. (1966). Studies of *Aleochara tristis* (Coleoptera: Staphylinidae), a natural enemy of the face fly. Journal of Economic Entomology, 59, 1368–1373.

Drea, J. J. (1990a). Neuroptera. In D. Rosen (Ed.), Armored scale insects, their biology, natural enemies and control (Vol. B pp. 51–59). Amsterdam: Elsevier.

Drea, J. J. (1990b). Other Coleoptera, In D. Rosen (Ed.), Armored scale insects, their biology, natural enemies and control (Vol. B, pp. 41–49). Amsterdam: Elsevier.

Drea, J. J., & Carlson, R. W. (1987). The establishment of *Chilocorus kuwanae* (Coleoptera: Coccinellidae) in eastern United States. Proceedings of the Entomological Society of Washington, 89, 821–824.

Drea, J. J., & Gordon, R. D. (1990). Predators. In D. Rosen (Ed.), Armored scale insects, their biology, natural enemies and control (Vol. B, pp. 19–39). Amsterdam: Elsevier.

Dreistadt, S. H., & Flint, M. L. (1996). Melon aphid (Homoptera: Aphididae) control by inundative convergent lady beetle (Coleoptera: Coccinellidae) release on chrysanthemum. Environmental Entomology, 25, 688–697.

Dreistadt, S. H., & Hagen, K. S. (1994). Classical biological control of the acacia psyllid, *Acizzia uncatoides* (Homoptera: Psyllidae), and predator-prey-plant interactions in the San Francisco Bay area. Biological Control, 4, 319–327.

Dreistadt, S. H., Hagen, K. S., & Bezark, L. G. (1995). *Harmonia axyridis* (Pallas) (Coleoptera: Coccinellidae), first western United States record for this Asiatic lady beetle. Pan-Pacific Entomologist, 71, 135–136.

Dreistadt, S. H., Hagen, K. S., & Dahlsten, D. L. (1986). Predation by *Iridomyrmex humilis* (Hym.: Formicidae) on eggs of *Chrysoperla carnea* (Neu.: Chrysopidae) released for inundative control of *Illinoia liriodendri* (Hom.: Aphididae) infesting *Liriodendron tulipifera*. Entomophaga, 31, 397–400.

Dreyer, B. S., Baumgartner, J., Neuenschwander, P., & Dorn, S. (1997a). The functional responses to two *Hyperaspis notata* strains to their prey, the cassava mealybug *Phenacoccus manihoti.* Mitteilungen der Schweizerischen Entomologischen Gesellschaft, 70, 21–28.

Dreyer, B. S., Neuenschwander, P., Bouyjou, B., Baumgartner, J., & Dorn, S. (1997b). The influence of temperature on the life table of *Hyperaspis notata.* Entomologia Experimentalis et Applicata, 84, 85–92.

Driessen, G. J. J., Van Raalte, A. T., & DeBruyn, G. J. (1984). Cannibalism in the red wood ant, *Formica polyctena* (Hymenoptera: Formicidae). Oecologia (Berlin), 63, 13–22.

Drummond, F. A., James, R. L., Casagrande, R. A., & Faubert, H. (1984). Development and survival of *Podisus maculiventris* (Say) (Hemiptera: Pentatomidae), a predator of the Colorado potato beetle (Coleoptera: Chrysomelidae). Environmental Entomology, 13, 1283–1286.

Duelli, P. (1980). Adaptive dispersal and appetitive flight in the green lacewing *Chrysopa carnea.* Ecological Entomology, 5, 213–220.

Duelli, P. (1984a). Flight, dispersal, migration. In M. Canard, Y. Semeria, & T. R. New (Eds.), Biology of Chrysopidae (pp. 110–116). The Hague: Dr. W. Junk.

Duelli, P. (1984b). Oviposition. In M. Canard, Y. Semeria, & T. R. New (Eds.). Biology of Chrysopidae (pp. 129–133). The Hague: Dr. W. Junk.

Duelli, P. (1987). The influence of food on the oviposition-site selection in a predatory and a honeydew-feeding lacewing species (Planipennia: Chrysopidae). Neuroptera International, 4, 205–210.

Dumas, B. A., Boyer, W. P., & Whitcomb, W. H. (1964). Effect of various factors on surveys of predaceous insects in soybean. Journal of Kansas Entomological Society, 37, 192–201.

Dunbar, D. M., & Bacon, O. G. (1972a). Influence of temperature on development and reproduction of *Geocoris atricolor, G. pallens,* and *G. punctipes* (Heteroptera: Lygaeidae) from California. Environmental Entomology, 1, 596–599.

Dunbar, D. M., Bacon O. G. (1972b). Feeding, development and reproduction of *Geocoris punctipes* (Heteroptera: Lygaeidae) on eight diets. Annals of the Entomological Society America, 65, 892–895.

Dusaussoy, G. (1963). Observations sur le comportement de *Calosoma sycophanta* L. en elevage. Revue de Pathologie Vegetale et Entomologie Agricole de France, 42, 53–65.

Duso, C., & Girolami, V. (1983). Role of Anthocoridae in control of *Panonychus ulmi* (Koch) in vineyards. Bulletino dell' Instituto di Entomologia "Guido Grandi" Universita degli Studi di Bologna, 37, 157–169.

Duverger, C. (1983). Contribution a connaissance des Coccinellidae d'Iran. Nouvelle. Reviews in Entomology, 13, 73–93.

Dwumfour, E. F. (1992). Volatile substances evoking orientation in predatory flowerbug *Anthocoris nemorum* (Heteroptera: Anthocoridae). Bulletin of Entomological Research, 82, 465–469.

Ehler, L. E., & van den Bosch, R. (1974). An analysis of natural biological control of *Trichoplusia ni* (Lepidoptera: Noctuidae). Canadian Entomologist, 106, 1067–1073.

Ehler, L. E., Eveleens, K. G., & van den Bosch, R. (1973). An evaluation of some natural enemies of cabbage looper on cotton in California. Environmental Entomology, 2, 1009–1015.

Ehler, L. E., Long, R. F., Kinsey, M. G., & Kelly, S. K. (1997). Potential for augmentative biological control of black bean aphid in California sugarbeet. Entomophaga, 42, 241–256.

Eichhorn, O., Pschorn-Walcher, H., & Schroder, D. (1972). Bekampfung verschleppter Forstschadlinge. Anzeiger fuer Shaedlingskunde und Pflanzenschutz, (1971) 44, 145–152.

Eikwort, G. C. (1983). Potential use of mites as biological control agents of leaf-feeding insects. In M. A. Hoy, G. L. Cunningham, & L. Knutson (Eds.), Biological control of pests by mites (Special Publication) 3304, pp. 42–52). Berkeley: DANR, University of California.

Eisenberg, R. M., & Hurd, L. E., (1990). Egg dispersion in two species of praying mantids (Mantodea: Mantidae). Proceedings of the Entomological Society of Washington, 92, 808–810.

Eisenberg, R. M., Hurd, L. E., & Bartley, J. A. (1981). Ecological consequences of food limitation for adult mantids *Tenodera sinensis* Saussure. American Midland Naturalist, 106, 209–218.

Ekbom, B. S., Wiktelius, S., & Chiverton, P. A. (1992). Can polyphagous predators control the bird cherry-oat aphid *(Rhopalsiphum padi)* in spring cereals? A simulation study. Entomologia Experimentalis et Applicata, 65, 215–223.

El-Ali, A. A. (1972). A biosystematic study of Hyperaspini of California with emphasis on immature stages. Unpublished doctoral dissertation. University of California, Berkeley.

Elkarmi, L. A., Harris, M. K., & Morrison, R. K. (1987). Laboratory rearing of *Chrysoperla rufilabris* (Burmeister), a predator of insect pests of pecans. Southwestern Entomologist, 12, 73–78.

Elsey, K. D. (1972a). Predator of eggs of *Heliothis* spp. on tobacco. Environmental Entomology, 1, 433–437.

Elsey, K. D. (1972b). Defenses of eggs of *Manduca sexta* against predation by *Jalysus spinosus.* Annals of the Entomological Society of America, 65, 896–897.

Elsey, K. D. (1973). *Jalysus spinosus:* Spring biology and factors that influence occurrence of the predator on tobacco in North Carolina. Environmental Entomology, 2, 421–425.

Elsey, K. D., & Stinner, R. E. (1971). Biology of *Jalysus spinosus,* an insect predator found on tobacco. Annals of the Entomological Society of America, 64, 779–783.

El Titi, A. (1973). Einflusse von Beutedichte und Morphologie der Wirtspflanze auf die Eiablage von *Aphidoletes aphidimyza* (Rond.) (Diptera: Itonididae). Zeitschrift fuer Angewandte Entomologie, 72, 400–415.

Enriquez, E., Bejarano, S., & Vila, V. (1976). Observaciones sobre avispas predatoras de *Leucoptera coffeella* Guer-Men. en el central y sur del Peru. Revista Peruana de Entomologia, 18, 82–83.

Entwistle, P. F. (1972). Pests of cocoa. London: Longman.

Ernsting, G., Issaks, J. A. (1987). Effects of food intake and temperature on energy budget parameters in *Notiophilus biguttatus.* Acta Phytopathogica Entomologica Hungicae, 22, 135–145.

Erwin, T. L., Ball, G. E., Whitehead, D. R., Halpern, A. L. (Eds.). (1979). Carabid beetles: Their evolution, natural history and classification. The Hague: Dr. W. Junk.

Eskafi, F. M., & Kolbe, M. M. (1990). Predation on larval and pupal *Ceratitis capitata* (Diptera: Tephritidae) by the ant *Solenopsis geminata* (Hymenoptera: Formicidae) and other predators in Guatemala. Environmental Entomology, 19, 148–153.

Esselbaugh, C. O. (1949). Notes on the bionomics of some midwestern Pentatomidae. Entomologica Americana, 28, 1–69.

Essig, E. O. (1931). A History of entomology. New York: Macmillan.

Ettifouri, M., & Ferran, A. (1993). Influence of larval rearing diet on the intensive searching behaviour of *Harmonia axyridis* (Col.: Coccinellidae) larvae. Entomophaga, 38, 51–59.

Evans, E. W. (1982a). Consequences of body size for fecundity in the predatory stinkbug, *Podisus maculiventus* (Hemiptera: Pentatomidae). Annals of the Entomological Society of America, 75, 418–420.

Evans, E. W. (1982b). Timing of reproduction by predatory stinkbugs (Hemiptera: Pentatomidae): Patterns and consequences for a generalist and a specialist. Ecology, 63, 147–158.

Evans, E. W. (1991). Intra versus interspecific interactions of lady beetles (Coleoptera: Coccinellidae) attacking aphids. Oecologia (Berlin), 87, 401–408.

Evans, E. W., & Swallow, J. G. (1993). Numerical responses of natural enemies to artificial honeydew in Utah alfalfa. Environmental Entomology, 22, 1392–1401.

Evans, H. E., & O'Neill, K. M. (1988). The natural history and behavior of North American beewolves. Ithaca, NY: Comstock.

Evans, H. F. (1976a). The role of predator-prey size ratio in determining the efficiency of capture by *Anthocoris nemorum* and the escape reactions of its prey, *Acyrthosiphon pisum*. Ecological Entomology, 1, 85–90.

Evans, H. F. (1976b). The effect of prey density and host plant characteristics on oviposition and fertility in *Anthocoris confusus* (Reuter). Ecological Entomology, 1, 157–161.

Evans, H. F. (1976c). The searching behavior of *Anthocoris confusus* (Reuter) in relation to prey density and plot surface topography. Ecological Entomology, 1, 163–169.

Evans, H. F. (1976d). Mutual interference between predatory anthocorids. Ecological Entomology, 1, 283–286.

Evans, H. F., & Fielding, N. J. (1996). Restoring the natural balance: Biological control of *Dendroctonus micans* in Great Britain. In J. K. Waage (Ed.), Biological control introductions: Opportunities for improved crop production (pp. 47–57). Farnham, United Kingdom: British Crop Protection Council.

Evans, H. F., & King, C. J. (1989). Biological control of *Dendroctonus micans* (Coleoptera: Scolytidae). Potential for biological control of *Dendroctonus* and *Ips* bark beetles. In D. L. Kulhavy & M. C. Miller (Eds), British experience of rearing and release of *Rhizophagus grandis* (Coleoptera: Rhizophagidae) (pp. 109–128). Nocogdoches, TX: S. F. Austin State University.

Evans, H. F., & West Eberhard, M. J. (1970). The wasps. Ann Arbor: University of Michigan Press.

Evans, M. E. G. (1965). The feeding method of *Cicindela hybrida* L. Proceedings of the Royal Entomological Society of London, 40, 61–66.

Evans, M. E. G., & Forsythe, T. G. (1985). Feeding mechanisms and their variation in form of some adult ground-beetles (Coleoptera: Caraboidea). Journal of Zoology, 206, 113–143.

Evans, W. G. (1983). Habitat selection in the Carabidae. Coleopterists' Bulletin, 37, 164–67.

Evans, W. G. (1988). Chemically mediated habitat recognition in shore insects. Journal of Chemical Ecology, 14, 1441–1454.

Ewert, M. A., & Chiang, H. C. (1966). Dispersal of three species of coccinellids in corn fields. Canadian Entomologist, 98, 999–1003.

Ewing, K. P., & Ivy, E. E. (1943). Some factors influencing bollworm populations and damage. Journal of Economic Entomology, 36, 602–606.

Fabres, G., & Kiyindou, A. (1985). Comparison du potentiel biotique de deux coccinelles (*Exochomus flariventris* et *Hyperaspis senegalensis hottentotta*, Col., Coccinellidae) predatrices de *Phenacoccus manihoti* (Hom., Pseudococcidae) au Congo. Acta Oecol. Oecol. Appl. 6, 339–348.

Falcon, L. A., van den Bosch, R., Ferris, C. A., Stromberg, L. K., Etzel, L. K., & Stinner, R. E. (1968). A comparison of season-long cotton-pest-control programs in California during 1966. Journal Economic Entomology, 61, 633–642.

Fan, Y., Liebman, M., Groden, E., & Alford, A. R. (1993). Abundance of carabid beetles and other ground dwelling arthropods in conventional versus low-input bean cropping systems. Agriculture, Ecosystems and Environment, 43, 127–139.

Farrar, R. R. (1985). A new record of the clerid, *Phyllobaenus pubescens* (LeConte) (Coleoptera), as a predator of *Heliothis zea* (Boddie) (Lepidoptera: Noctuidae) on cotton. Coleopterists' Bulletin, 39, 33–34.

Fauvel, G. (1974). Les heteropteres predateurs en verger. In L. Brader (Ed.), Les organismes auxiliares en verger de pommiers (IOBC WPRS Brochure No. 3, (pp. 125–150).

Fauvel, G., & Atger, P. (1981). Etude de l'evolution des insectes auxiliaires et de leurs relations avec le psylle du poirier (*Psylla pyri* L.) et l'acorien rouge (*Panonychus ulmi* [Koch]) dans deux vergers du Sud-Est de la France en 1979. Agronomie, 1, 813–820.

Felland, C. M., & Hull, L. A. (1996). Overwintering of *Stethorus punctum punctum* (Coleoptera: Coccinellidae) in apple orchard ground cover. Environmental Entomology, 25, 972–976.

Ferenz, H. J. (1986). Photoperiodic regulation of juvenile hormone and reproduction in carabid beetle, *Pterostichus unigrita*. In P. J. Den Boer, M. L. Luff, D. Mossakowski, & F. Weber (Eds.), Carabid beetles—their adaptations and dynamics (pp. 113–123). Stuttgart: Gustov Fischer.

Ferran, A., Iperti, G., Lapchin, L., & Rabasse, J. M. (1991). La localisation, le comportement et les relations "prore-predateur" chez *Coccinella septempunctata* dans un champ de ble. Entomophaga, 36, 213–225.

Ferran, A., Iperti, G., Kreiter, S., Quilici, S., & Shanderl, H. (1986a). Preliminary results of a study of the potentials of some aphidophagous coccinellids for use in biological conrol. In I. Hodek (Ed.), Ecology of Aphidophaga 2 (pp. 479–484). Prague: Academia.

Ferran, A., Iperti, G., Lapchin, L., Lyon, J. P., & Rabasse, J. M. (1986b). The efficiency of aphidophagous predators in cereal crops: A new approach. In I. Hodek (Ed.), Ecology of Aphidophaga 2 (pp. 423–428). Prague: Academia.

Ferran, A., Niknam, H., Kabiri, F., Picart, J.-L., De Herce, C., & Brun, J. (1996). The use of *Harmonia axyridis* larvae (Coleoptera: Coccinellidae) against *Macrosiphum rosae* (Hemiptera: Sternorrhyncha: Aphididae) on rose bushes. European Journal of Entomology, 93, 59–67.

Field, R. P. (1979). Integrated pest control in Victorian peach orchards: the role of *Stethorus* spp. (Coleoptera: Coccinellidae). Journal of the Australian Entomological Society, 18, 315–322.

Finch, S. (1996). Effect of beetle size on predation of cabbage root fly eggs by ground beetles. Entomologia Experimental et Applicata, 81, 199–206.

Finch, S., & Elliott, M. S. (1992). Carabidae as potential biological agents for controlling infestations of cabbage root fly. Phytoparasitica, 20, 67–75.

Finnegan, R. J. (1971). An appraisal of indigenous ants as limiting agents of forest pests in Quebec. Canadian Entomologist, 103, 1489–1493.

Finnegan, R. J. (1974). Ants as predators of forest pests. Entomophaga, 7, 53–59.

Finnegan, R. J. (1975). Introduction of a predacious red wood ant, *Formica lugubris* (Hymenoptera: Formicidae), from Italy to eastern Canada. Canadian Entomologist, 107, 1271–1274.

Finnegan, R. J. (1977). Establishment of a predacious red wood ant, *Formica obscuripes* (Hymenoptera: Formicidae), from Manitoba to Eastern Canada. Canadian Entomologist, 109, 1145–1148.

Finney, G. L. (1948). Culturing *Chrysopa californica* and obtaining eggs for field distribution. Journal of Economic Entomology, 41, 719–721.

Finney, G. L. (1950). Mass-culturing *Chrysopa californica* to obtain eggs for field distribution. Journal of Economic Entomology, 43, 97–100.

Fisher, T. W. (1963). Mass culture of *Cryptolaemus* and *Leptomastix*, natural enemies of citrus mealybug (Agricultural Experimental Station Bulletin 797). Berkeley, University of California.

Fisher, T. W., Moore, I., Legner, E. F., & Orth, R. E. (1976). *Ocypus olens:* A predator of brown garden snail. California Agriculture, 30, 20–21.

Fitzgerald, T. D., & Nagel, W. P. (1972). Oviposition and larval bark-surface orientation of *Medetera aldrichii* (Diptera: Dolichopodidae): Response to a prey-liberated plant terpine. Annals of the Entomological Society of America, 63, 328–330.

Flanders, S. E. (1936). *Coccidophilus citricola* Brethes, a predator enemy of red and purple scales. Journal of Economic Entomology, 29, 1023–1024.

Flanders, S. E. (1950). Regulation of ovulation and egg disposal in the parasitic Hymenoptera. Canadian Entomologist, 82, 134–140.

Flanders, S. E., & Badgley, M. E. (1960). A host-parasitic interaction conditioned by predation. Ecology, 41, 363–365.

Flanders, S. E., & Badgley, M. E. (1963). Predator-prey interactions in self-balanced laboratory populations. Hilgardia, 35, 145–183.

Fleschner, C. A. (1950). Studies on searching capacity of the larvae of three predators of the citrus red mite. Hilgardia, 20, 233–265.

Fleschner, C. A. (1952). Dusty wings on citrus. California Agriculture, 6, 4.

Fleschner, C. A. (1958). Field approach to population studies of tetrany-chid mites on citrus and avocado in California. Proceedings of the 10th International Congress of Entomology, 2, 669–674.

Fleschner, C. A., & Ricker, D. W. (1953). Food habits of coniopterygids on citrus in southern California. Journal of Economic Entomology, 46, 458–461.

Flinn, P. W., Honer, A. A., & Taylor, R. A. J. (1985). Preference of Reduviolus americoferus (Hemiptera: Nabidae) for potato leafhopper nymphs and pea aphids. Canadian Entomologist, 117, 1503–1508.

Flint, H. M., Salter, S. S., & Walters, S. (1979). Caryophyllene: An attractant for the green lacewing. Environmental Entomology, 8, 1123–1125.

Floate, K. D., Doane, J. F., & Gillott, D. (1990). Carabid predators of the wheat midge (Diptera: Cecidomyiidae) in Saskatchewan (Canada). Environmental Entomology, 19, 1503–1511.

Foglar, H., Malausa, J. C., & Wajnberg, E. (1990). The functional response and preference of Macrolophus caliginosus (Heteroptera: Miridae) for two of its prey: Myzus persicae and Tetranychus urticae. Entomophaga, 35, 465–474.

Foote, B. A. (1996a). Biology and immature stages of snail-killing flies belonging to the genus Tetanocera (Insecta: Diptera: Sciomyzidae). I. Introduction and life histories of predators of shoreline snails. Annals of Carnegie Museum, 65, 1–12.

Foote, B. A. (1996b). Biology and immature stages of snail-killing flies belonging to the genus Tetanocera (Insecta: Diptera: Sciomyzidae). II: Life histories of predators of snails of the family Succineidae. Annals of Carnegie Museum, 65, 153–166.

Forbes, S. A. (1883). The food relations of the Carabidae and Coccinellidae. Illinois State Laboratory Natural History Bulletin, 6, 33–63.

Forsythe, T. G. (1987). Ground beetles: Form and function. Antenna, 11, 57–61.

Fowler, H. G. (1987). Predatory behavior of Megacephala fulgida (Coleoptera: Cicindelidae). Coleopterists' Bulletin, 41, 407–408.

Fox, C. J. S., & MacLellan, C. R. (1956). Some Carabidae and Staphylinidae shown to feed on a wireworm. Agriotes sputator (L.), by the precipitin test. Canadian Entomologist, 88, 228–231.

Frank, J. H. (1991). Staphylinidae. In F. W. Stehr (Ed.), Immature insects (Vol. 2, pp. 341–352). Dubuque, IA: Kendal/Hunt.

Frank, J. H., & Curtis, G. A. (1979). Trend lines and number of species of Staphylinidae. Coleopterists' Bulletin, 33, 133–139.

Franz, J. M. (1958). Studies on Laricobius erichsonii Rosen (Coleoptera: Derodontidae), a predator on chermesids. Entomophaga, 3, 109–196.

Franz, J. M. (1967). Beobachtungen uber das verhalten der raubwanze Perillus bioculatus (F.) (Pentatomidae) gegen uber ihrer beute Leptinotarsa decimlineata (Say) (Chrysomelidae). Zeitschrift fuer Pflanzenkrankheiten und Pflanzenschutz, 74, 1–13.

Franz, J., & Szmidt, A. (1960). Beobachtungen Beim zuchten von Perillus bioculatus (F.) (Heterop., Pentatomidae), einem aus Nordamerika importierten rauber des kartoffelkafers. Entomophaga, 5, 87–110.

Frazer, B. D. (1988a). Coccinellidae. In A. K. Minks & P. Harrewijn (Eds.), Aphids: Their biology, natural enemies and control (Vol. B, pp. 231–247). Amsterdam: Elsevier.

Frazer, B. D. (1988b). Predators. In A. K. Minks & P. Harrewijn (Eds.), Aphids: Their biology, natural enemies and control (Vol. B, pp. 217–230). Amsterdam: Elsevier.

Frazer, B. D., & Gilbert, N. (1976). Coccinellids and aphids: A quantitative study of the impact of adult ladybirds (Coleoptera: Coccinellidae) preying on field populations of pea aphids (Homoptera: Aphididae). Journal of the Entomological Society of British Columbia, 73, 33–56.

Froeschner, R. C. (1988). Family Pentatomidae Leach, 1815, The stinkbugs. In T. J. Henry & R. C. Froeschner (Eds.), Catalog of Heteroptera or true bugs of Canada and continental United States (pp. 544–597). Leiden: E. J. Brill.

Froeschner, R. C., & Henry, T. J. (1988). Family Berytidae. In T. J. Henry & R. C. Froeschner (Eds.), Catalog of Heteroptera, or true bugs, of

Canada and the Continental United States (pp. 56–60). Leiden: E. J. Brill.

Fullaway, D. T. (1940). An account of the reduction of the immigrant taro leafhopper (Megameles proserpina) population to insignificant numbers by the introduction and establishment of the egg-sucking bug Cyrtorhinus fulvus. Sixth Pacific Science Congress Proceedings, 4, 345–346.

Fuller, B. W., Reagan, T. E., Flynn, J. L., & Boetel, M. A. (1997). Predation on fall armyworm (Lepidoptera: Noctuidae) in sweet sorghum. Journal of Agricultural Entomology, 14, 151–155.

Fulton, B. B. (1933). Notes on Habrocytus cerealellae, a parasite of Angoumois grain moth. Annals of the Entomological Society of America, 26, 536–553.

Funasaki, G. Y. (1966). Studies on the life cycle and propagation technique of Montandoniola maraguesi (Puton) (Heteroptera: Anthocoridae). Proceedings of the Hawaiian Entomological Society, 19, 209–211.

Funasaki, G. Y., Lai, P. Y., Nakahara, L. M., Beardsley, J. W., & Ota, A. K. (1988). A review of biological control introductions in Hawaii: 1890 to 1985. Proceedings of the Hawaiian Entomological Society, 28, 105–160.

Funderburk, J. E., & Mack, T. P. (1989). Seasonal abundance and dispersion patterns of damsel bugs (Hemiptera: Nabidae) in Alabama and Florida soybean fields. Journal of Entomological Science, 24, 9–15.

Fürsch, H. (1987). Neue afrikanische Scymnini-Arten (Coleoptera: Coccinellidae) als fresfeinde von manihot-schadlingen. Revue de Zoologie Afrcaines, 100, 387–394.

Fye, R. E. (1979). Cotton insect populations. Development and impact of predators and other mortality factors (Technical Bulletin No. 1592). Washington, DC: U.S. Department of Agriculture.

Gadagkar, R. (1991). Belongaster, Mischocyttarus, Parapolybia, and independent founding Ropalidia. In K. G. Ross & R. W. Matthews (Eds.), The social biology of wasps (pp. 149–190). Ithaca, Comstock NY: Cornell University Press.

Gadd, C. H., & Fonseka, W. T. (1945). Neoplectrus maculatus Ferriere— a predator and parasite of Natada mararia Mo. and other nettlegrubs. Ceylon Journal of Science, Section B: Zoology, 23, 9–18.

Gadd, C. H., Fonseka, W. T., & Ranaweera, D. J. W. (1946). Parasites of tea nettlegrubs with special reference to Platyplectrus natadae Ferriere and Autoplectrus taprobanes Gadd. Ceylon Journal of Science, Section B: Zoology, 23, 81–94.

Gagne, R. J. (1995). Revision of Tetranychid (Acarina) mite predators of the genus Feltiella (Diptera: Cecidomyiidae). Annals of the Entomological Society of America, 88, 16–30.

Gaimari, S. D., & Turner, W. J. (1996a). Methods for rearing aphidophagous Leucopis spp. (Diptera: Chamaemyiidae). Journal of Kansas Entomological Society, 69, 363–369.

Gaimari, S. D., & Turner, W. J. (1996b). Larval feeding and development of Leucopis ninae Tanasijtshuk and two populations of Leucopis gaimarii Tanasijtshuk (Diptera: Chamaemyiidae) on Russian wheat aphid, Diuraphis noxia (Mordvilko) (Homoptera: Aphididae), in Washington. Proceedings of the Entomological Society of Washington, 98, 667–676.

Gangwere, S. K. (1969). A combined short-cut technique to study of food selection in Orthopteroidea. Turtox News, 47, 121–125.

Gardner, J., & Giles, K. (1996). Handling and environmental effects on viability of mechanically dispensed green lacewing eggs. Biological Control, 7, 245–250.

Gardner, J., & Giles, K. (1997). Mechanical distribution of Chrysoperla rufilabris and Trichogramma pretiosum: Survival and uniformity of discharge after spray dispersal in an aqueous suspension. Biological Control, 8, 138–142.

Garland, J. A. (1978a). Reinterpretation of information on exotic brown lacewings (Neuroptera: Hemerobiidae) used in biocontrol programmes in Canada. Manitoba Entomologist, 12, 25–28.

Garland, J. A. (1978b). Observations on survival of eggs of Hemerobius *stigma* (Neuroptera: Hemerobiidae) following exposure to frost. Manitoba Entomologist, 12, 61–62.

Ge, D. (1986). Rearing and release of mantids for controlling cottonpests. Natural Enemy Insects, 8, 200–204 (in Chinese).

Geden, C. J., & Axtell, R. C. (1988). Predation by *Carcinops pumilio* (Coleoptera: Histeridae) and *Macrocheles muscaedomesticae* (Acarina: Macrochelidae) on the house fly (Diptera: Muscidae): Functional response, effects of temperature, and availability of alternate prey. Environmental Entomologist, 17, 739–744.

Gepp, J. (1984a). Erforschungsstand der Neuropteron—Larven der Erde. In J. Gepp, H. Aspöck, & H. Hölzel (Eds.), Progress in worlds neuropterology (pp. 183–239). Graz, Austria: Gepp.

Gepp, J. (1984b). Morphology and anatomy of the preimaginal stages of Chrysopidae: A short survey. In M. Canard, Y. Semeria, & T. R. New (Eds.), Biology of Chrysopidae (pp. 9–19). The Hague: Dr. W. Junk.

Gepp, J. (1990). An illustrated review of egg morphology in the families of Neuroptera (Insects: Neuroteroidea). In M. W. Mansell & H. Aspöck (Eds.), Proceedings, Third International Symposium on Neuropterology (pp. 131–149). Pretoria: Department of Agricultural Development.

Gepp, J., Aspöck, H., & Hölzel H. (Eds.). (1984). Progress in worlds neuropterology. Proceedings, First International Symposium on Neuropterology. Graz, Austria: Gepp.

Gepp, J., Aspöck, H., & Hölzel, H. (Eds.). (1986). Recent research in Neuropterology. Proceedings, Second International Symposium on Neuropterology. Graz, Austria: Gepp.

Gerlach, S., & Sengonca, C. (1986). Frassaktivität and Wirksamkeit des räuberischen Thrips, *Scolothrips longicornis* Priesner (Thysanoptera, Thripidae). Journal of Applied Entomology, 101, 444–452.

Gerson, U., O'Connor, B. M., & Houck, M. A. (1990). Acari. In D. Rosen (Ed.), Armored scale insects, their biology, natural enemies and control (Vol. B, pp. 77–97). Amsterdam: Elsevier.

Ghani, M. A., & Ahmad, R. (1966). Biology of *Pharoscymnus flexibilis* Muls. (Col.: Coccinellidae). CIBC Technical Bulletin, 7, 107–111.

Gidaspow, T. (1959). North American caterpillar hunters of the genera *Calosoma* and *Callisthenes* (Coleoptera: Carabidae). Bulletin of American Museum of Natural History, 116, 229–343.

Gidaszewski, A. (1974). Analysis of the occurrence and vigor of *Tomicus piniperda* (L.) and *T. minor* (Hertg.) in forest stands in the Wielkopolski National Park during the years 1969–1970. Polskie Pismo Entomologiczne, 44, 789–815.

Gilbert, F. S. (1981). Foraging ecology of hoverflies: Morphology of the mouthparts in relation to feeding on nectar and pollen in some common urban species. Ecological Entomology, 6, 245–262.

Gilkeson, L. A. (1990). Cold storage of the predatory midge *Aphidoletes aphidimyza* (Diptera: Cecidomyiidae). Journal of Economic Entomology, 83, 965–970.

Gilkeson, L. A., & Hill, S. B. (1987). Release rates for control of green peach aphid (Homoptera: Aphididae) by the predatory midge *Aphidoletes aphidimyza* (Diptera: Cecidomyiidae) under winter greenhouse conditions. Journal of Economic Entomology, 80, 147–150.

Gilkeson, L. A., Morewood, W. D., & Elliott, D. E. (1990). Current status of biological control of thrips in Canadian greenhouses with *Amblyseius cucumeris* and *Orius tristicolor*. IOBC WPRS Bulletin, 13 (5), 75–76.

Gillaspy, J. E. (1979). Management of *Polistes* wasps for caterpillar predation. Southwestern Entomologist, 4, 334–350.

Gillespie, D. R., & Quiring, D. J. M. (1992). Competition between *Orius tristicolor* (White) (Hemiptera: Anthocoridae) and *Amblyseius cucumeris* (Oudemans) (Acari: Phytoseiidae) feeding on *Frankliniella occidentalis* (Pergande) (Thysanoptera: Thripidae). Canadian Entomologist, 124, 1123–1128.

Gilliatt, F. C. (1935). Some predators of the European red mite, *Paratetranychus pilosus* C&F, in Nova Scotia. Canadian Journal of Research, 13, 19–38.

Gilstrap, F. E., & Oatman, E. R. (1976). The bionomics of *Scolothrips*

sexmaculatus (Pergande) (Thysanoptera: Thripidae), an insect predator of spider mites. Hilgardia, 44, 27–59.

Glen, D. M. (1973). The food requirements of *Blepharidopterus angulatus* (Heteroptera: Miridae) as a predator of the lime aphid, *Eucallipterus tiliae*. Entomologia Experimentalis et Applicata, 16, 255–267.

Glen, D. M. (1975). Searching behaviour and prey-density requirements of *Blepharidopterus angulatus* (Fall.) (Heteroptera: Miridae) as a predator of the lime aphid, *Eucallipterus tiliae* (L.) and leafhopper, *Alnetoidea alneti* (Dahlbom). Journal of Animal Ecology, 44, 85–114.

Glover, T. (1868). The food and habits of beetles (Annual Report of U.S. Commercial Agriculture for 1868, pp. 78–117). Washington, DC: U.S. Department of Agriculture.

Godan, D. (1983). Pest slugs and snails. Biology and control. Berlin: Springer-Verlag.

Godfrey, K. E., Whitcomb, W. H., & Stimac, J. L. (1989). Arthropod predators of velvetbean caterpillar, *Anticarsia gemmatalis* Hübner (Lepidoptera: Noctuidae), eggs and larvae. Environmental Entomology, 18, 118–123.

Goncalves, L. (1990). Biologia e capacidade predatorla de *Podisus nigrolimbatus* Spinola, 1832, *Podisus connexivas* Bergroth, 1891 (Hemiptera: Pentatomidae: Asopinae) em condicoes de laboratorio. Unpublished doctoral dissertation Escola Superios de Agricultura de Larvas, M. G. Brasil.

Gonzalez, D., & Wilson, L. T (1982). A food-web approach to economic thresholds: A sequence of pests/predaceous arthropods on California cotton. Entomophaga, 27 (special issue), 31–43.

Gonzalez, D., Patterson, B. R., Leigh, T. F., & Wilson, L. T. (1982). Mites: A primary food source for two predators in San Joaquin Valley cotton. California Agriculture, 26, 18–20.

Goodarzy, K., & Davis, D. W. (1958). Natural enemies of the spotted alfalfa aphids in Utah. Journal of Economic Entomology, 51, 612–616.

Gordon, R. D. (1972). The tribe Noviini in the New World (Coleoptera: Coccinellidae) Journal of Washington Academy of Science, 62, 23–31.

Gordon, R. D. (1976). The Scymnini (Coleoptera: Coccinellidae) of the United States and Canada: Key to genera and revision of *Scymnus, Nephus* and *Diomus*. Bulletin of Buffalo Society Natural Science, 28, 362.

Gordon, R. D. (1977). Classification and phylogeny of the New World Sticholotidinae (Coccinellidae). Coleopterists' Bulletin, 31, 185–228.

Gordon, R. D. (1980). The tribe Azyini (Coleoptera: Coccinellidae): Historical review and taxonomic revision. Transactions of the American Entomological Society, 106, 149–203.

Gordon, R. D. (1982). An Old World species of *Scymnus (Pullus)* established in Pennsylvania and New York. Proceedings of the Entomological Society of Washington, 84, 250–255.

Gordon, R. D. (1985). The Coccinellidae (Coleoptera) of America north of Mexico. Journal of the New York Entomological Society, 93, 1–912.

Gordon, R. D. (1987). The first North American records of *Hippodamia variegata* (Goeze) (Coleoptera: Coccinellidae). Journal of the New York Entomological Society, 95, 307–309.

Gordon, R. D., & Chapin, E. A. (1983). A revision of the New World species of *Stethorus* Weise. Transactions of the American Entomological Society, 109, 229–276.

Gordon, R., & Cornect, M. (1986). Toxicity of the insect growth regulator diflubenzuron to the rove beetle *Aleochara bilineata*, a parasitoid and predator of cabbage maggot *Delia radicum*. Entomologia Experimentalis et Applicata, 42, 179–185.

Gordon, R. D., & Vandenberg, N. (1991). Field guide to recently introduced species of Coccinellidae (Coleoptera) in North America, with a revised key to North American genera of Coccinellini. Proceedings of the Entomological Society of Washington, 93, 845–864.

Gormally, M. J. (1988). Temperature and the biology and predation of *Ilione albiseta* (Diptera: Sciomyzidae). Potential biological control agent of liver fluke. Hydrobiologia, 166, 239–246.

Gösswald, K. (1951). Die rote Waldameise im Dienste der Waldhygiene: Forstwirtschaftliche Bedeutung, Nutzung, Lebensweise, Zucht, Vermeharung und Schutz. Lüneburg: Metta Kinau.

Gould, W. P., & Jeanne, R. L. (1984). *Polistes* wasps (Hymenoptera: Vespidae) as control agents for lepidopterous cabbage pests. Environmental Entomology, 13, 150–156.

Grafius, E., & Warner, F. W. (1989). Predation by *Bembidion quadrimaculatum* (Coleoptera: Carabidae) on *Delia antiqua* (Diptera: Anthomyiidae). Environmental Entomology, 18, 1056–1059.

Grafton-Cardwell, E. E., & Hoy, M. A. (1985). Intraspecific variability in response to pesticides in the common green lacewing, *Chrysoperla carnea* (Stephens) (Neuroptera: Chrysopidae). Hilgardia, 53, 1–32.

Graham, W. M. (1970). Warehouse ecology studies of bagged maize in Kenya. II. Ecological observations on an infestation by *Ephestia (Cadra) cautella* (Walker) (Lepidoptera: Phycitidae). Journal of Stored Products Research, 6, 157–167.

Gravena, S., & DaCunha, H. F. (1991). Predation of cotton leafworm first instar larvae, *Alabama argillacea* (Lep.: Noctuidae). Entomophaga, 36, 481–491.

Gravena, S., & Pazetto, J. A. (1987). Predation and parasitization of cotton leafworm eggs, *Alabama argillacea* (Lep.: Noctuidae). Entomophaga, 32, 241–248.

Grazia, J., & Hildebrand, R. (1987). Hemipteros predadores insectos. In Encontio Sul-Brasileiro de controle biologico de pragas 1 (pp. 21–37). Passo Fundo:

Greathead, D. J. (1971). A review of biological control in the Ethiopian region (CIBC Technical Communications No. 5).

Greathead, D. J. (1973). A review of introductions of *Lindorus lophanthae* (Blaisd.) (Col.: Coccinellidae) against hard scales (Diaspididae). CIBC Technical Bulletin, 16, 29–33.

Greathead, D. J. (1976). A review of biological control in western and southern Europe (CIBC Technical Communications No. 7).

Greathead, D. J. (1995). The *Leucopis* spp. (Diptera: Chamaemyiidae) introduced for biological control of *Pineus* sp. (Homoptera: Adelgidae) in Hawaii: Implications for biological control of *Pineus ?boerneri* in Africa. Entomologist, 114, 83–90.

Greathead, D. J., & Pope, R. D. (1977). Studies on the biology and taxonomy of some *Chilocorus* spp. (Coleoptera: Coccinellidae) preying on *Aulacaspis* spp. (Hemiptera: Diaspididae) in East Africa, with the description of a new species. Bulletin of Entomological Research, 67, 259–270.

Greene, A. (1975). Biology of the five species of Cychrini (Coleoptera: Carabidae) in the steppe region of southeastern Washington. Melanderia, 19, 1–43.

Greene, G. L. (1996). Rearing techniques for *Creophilus maxillosus* (Coleoptera: Staphylinidae), a predator of fly larvae in cattle feedlots. Journal of Economic Entomology, 89, 848–851.

Greene, G. L., & Shepard, M. (1974). Biological studies of a predator, *Sycanus indagator*. II. Field survival and predation potential. Florida Entomologist, 57, 33–38.

Grégoire, J. C. (1988). The greater European spruce beetle. In A. A. Berryman (Ed.), Population dynamics of forest insects (pp. 456–478). New York: Plenum.

Grégoire, J. C., Merlin, J., Pasteels, J. M., Jaffuel, R., Vouland, G., & Schvester, D. (1985). Biocontrol of *Dendroctonus micans* by *Rhizophagus grandis* Gyll. (Col., Rhizophagidae) in the Massif Central (France). A first appraisal of the mass rearing and release methods. Zeitschrift fuer Angewandte Entomologie, 99, 182–190.

Grégoire, J. C., Covillien, D., Krebber, R., Konig, W. A., Meyer, H., & Franke, W. (1992). Orientation of *Rhizophagus grandis* (Coleoptera: Rhizophagidae) to oxygenated monoterpenes in a species-specific predator-prey relationship. Chemoecology, 3, 14–18.

Greve, L. (1984). Chrysopid distribution in northern latitudes. In M. Canard, Y. Semeria, & T. R. New (Eds.), Biology of Chrysopidae (pp. 180–186). The Hague: Dr. W. Junk.

Grevstad, F. S., & Klepetka, B. W. (1992). The influence of plant architecture on the foraging efficiencies of a suite of ladybird beetles feeding on aphids. Oecologia (Berlin) 92, 399–404.

Groden, E., Drummond, F. A., Casagrande, R. A., & Haynes, D. L. (1990). *Coleomegilla maculata* (Coleoptera: Coccinellidae): Its predation upon the Colorado potato beetle (Coleoptera: Chrysomelidae) and its incidence in potatoes and surrounding crops. Journal of Economic Entomology, 83, 1306–1315.

Groff, G. W., & Howard, C. W. (1924). The cultured citrus ant of South China. Lingnan Agricultural Review, 2, 108–114.

Grossman, J. (1989). Update: Strawberry IPM features biological and mechanical control. IPM Practitioner, 11, 1–4.

Grout, T. G., & Richards, G. I. (1992). *Euseius addoensis addoensis,* an effective predator of citrus thrips, *Scirtothrips aurantii,* in the eastern Cape province of South Africa. Experimental and Applied Acarology (Northwood), 15, 1–13.

Guagliumi, P. (1973). *Pragas da cana-de-acucar* (Colec. Can. No. 10 Inst. Acucar Alcool. Serv. Document). Brasil.

Guillebeau, L. P., & All, J. N. (1989). *Geocoris* spp. (Hemiptera: Lygaeidae) and the striped lynx spider (Araneae: Oxyopidae): Cross predation and prey preferences. Journal of Economic Entomology, 82, 1106–1110.

Guillebeau, L. P., & All, J. N. (1990). Big-eyed bugs (Hemiptera: Lygaeidae) and the striped lynx spider (Araneae: Oxyopidae): Intra- and interspecific interference on predation of first instar corn earworm (Lepidoptera: Noctuidae). Journal of Entomological Science, 25, 30–33.

Gunthart, E. (1945). Ueber Spinnmilben und deren naturlichen Feinde. Mitteilungen Schweizerischen Entomologischen Gesellschaft, 19, 279–308.

Guppy, J. C. (1986). Bionomics of the damsel bug, *Nabis americoferus* Carayon (Hemiptera: Nabidae), a predator of the alfalfa blotch leafminer (Diptera: Agromyzidae). Canadian Entomologist, 118, 745–751.

Gurbanov, G. G. (1982). Effectiveness of the use of the common lacewing (*Chrysoperla carnea* (Stevens)) in the control of sucking pests and cotton moth on cotton. Izvestiya Akademii Nauk Azerbaidzhonskoi SSR, Biologicheskikh Nauk, 2, 92–96 (in Russian); (1985). Review of Applied Entomology, 73, 933.

Gurney, A. B. (1953). Notes on the biology and immature stages of a cricket parasite of the genus *Rhopalosoma*. Proceedings of the U.S. National Museum, 103, 19–34.

Gurney, A. B. (1955). Further notes on *Iris oratoria* in California. Pan-Pacific Entomologist, 31, 67–72.

Gurney, B., & Hussey, N. W. (1970). Evaluation of some coccinellid species for biological control of aphids in protected cropping. Annals of Applied Biology, 65, 451–458.

Gutierrez, A. P., Baumgaertner, J. U., & Hagen, K. S. (1981). A conceptual model for growth, development and reproduction in the ladybird beetle, *Hippodamia convergens* (Coleoptera: Coccinellidae). Canadian Entomologist, 113, 21–33.

Gutierrez, A. P., Hagen, K. S., & Ellis, C. K. (1990). Evaluating the impact of natural enemies: A multitrophic perspective. In M. Mackauer, L. E. Ehler, & J. Roland (Eds.), Critical issues in biological control (pp. 81–109). Andover: Intercept.

Hagen, K. S. (1950). Fecundity of *Chrysopa californica* as affected by synthetic food. Journal of Economic Entomology, 43, 101–104.

Hagen, K. S. (1962). Biology and ecology of predaceous Coccinellidae. Annual Review of Entomology, 7, 289–326.

Hagen, K. S. (1974). The significance of predaceous Coccinellidae in biological and integrated control of insects. Entomophaga Mem. H. S., 7, 35–44.

Hagen, K. S. (1986). Ecosystem analysis: Plant cultivars (HPR), entomophagous species and food supplements. In D. J. Boethal & R. D. Eikenbary (Eds.), Interactions of plant resistance and parasitoids and predators of insects (pp. 151–197). New York: John Wiley & Sons.

Hagen, K. S. (1987). Nutritional ecology of terrestrial insect predators. In

F. Slansky, Jr. & J. G. Rodriguez (Eds.), Nutritional ecology of insects, mites, and spiders (pp. 533–577). New York: John Wiley & Sons.

Hagen, K. S., & Bishop, G. W. (1979). Use of supplemental foods and behavioral chemicals to increase the effectiveness of natural enemies. In D. W. Davis *et al.* (Eds.), Biological control and insect pest management Division of Agricultural Science Publication 4096, (pp. 49–60). Berkeley: University of California.

Hagen, K. S., & Dreistadt, S. H. (1990). First California record for *Anthocoris nemoralis* (Fabr.) (Hemiptera: Anthocoridae), a predator important in the biological control of psyllids (Homoptera: Psyllidae). Pan-Pacific Entomologist, 66, 323–324.

Hagen, K. S., & Hale, R. (1974). Increasing natural enemies through use of supplementary feeding and non-target prey. In F. G. Maxwell & F. A. Harris (Eds.), Proceedings summer institute on biological control of plant insects and diseases (pp. 170–181). Mississippi: Jackson University Press.

Hagen, K. S., & Tassan, R. L. (1966). The influence of protein hydrolysates of yeast and chemically defined diet upon the fecundity of *Chrysopa carnea* Steph. Vest. Csl. Spol. Zool, 30, 219–227.

Hagen, K. S., & Tassan, R. L. (1970). The influence of food wheast and related *Saccharomyces fragilis* yeast products on the fecundity of *Chrysopa carnea* (Neuroptera: Chrysopidae). Canadian Entomologist, 102, 806–811.

Hagen, K. S., & Tassan, R. L. (1972). Exploring nutritional roles of extracellular symbiotes on the reproduction of honeydew feeding adult chrysopids and tephritids. In J. G. Rodriguez (Ed.), Insect and mite nutrition (pp. 323–351). Amsterdam: North Holland.

Hagen, K. S., & van den Bosch, R. (1968). Impact of pathogens, parasites, and predators on aphids. Annual Review of Entomology, 13, 335–384.

Hagen, K. S., Sawall, E. F., & Tassan, R. L. (1970a). The use of food sprays to increase effectiveness of entomophagous insects. Proceedings, Tall Timbers Conference on Ecological Animal Control and Habitat Management, Tallahassee, 2, 59–81.

Hagen, K. S., Tassan, R. L., & Sawall, E., Jr. (1970b). Some ecophysiological relationships between certain *Chrysopa* honeydews and yeasts. Bollettin del Laboratori Entomologica Agricultura "Filippo Silvestri," Portici, 28, 113–34.

Hagen, K. S., Bombosch, S., & McMurtry, J. A. (1976a). The biology and impact of predators. In C. B. Huffaker & P. S. Messenger (Eds.), Theory and practice of biological control (pp. 93–142). New York: Academic Press.

Hagen, K. S., Greany, P., Sawall, E. F., Jr. & Tassan, R. L. (1976b). Tryptophan in artificial honeydews as a source of an attractant for adult *Chrysopa carnea*. Environmental Entomology, 5, 458–468.

Hagler, J. R., & Cohen, A. (1990). Effects of time and temperature on digestion of purified antigen by *Geocoris punctipes* (Hemiptera: Lygaeidae) reared on artificial diet. Annals of the Entomological Society of America, 83, 1177–1180.

Hagley, E. A. C. (1989). Release of *Chrysoperla carnea* Stephens (Neuroptera: Chrysopidae) for control of the green apple aphid *Aphis pomi* De Geer (Homoptera: Aphididae). Canadian Entomologist, 121, 309–314.

Hagley, E. A. C., & Allen, W. R. (1988). Ground beetles (Coleoptera: Carabidae) as predators of the codling moth, *Cydia pomonella* (L.) (Lepidoptera: Tortricidae). Canadian Entomologist, 120, 917–926.

Hagley, E. A. C., & Allen, W. R. (1990). The green apple aphid, *Aphis pomi* De Geer (Hemiptera: Aphididae), as prey of polyphagous arthropod predators in Ontario. Canadian Entomologist, 122, 1221–1228.

Hagley, E. A. C., & Miles, N. (1987). Release of *Chrysoperla carnea* Stephens (Neuroptera: Chrysopidae) for control of *Tetranychus urticae* Koch (Acarina: Tetranychidae) on peach grown in a protected environment structure. Canadian Entomologist, 119, 205–206.

Hagley, E. A. C., & Simpson, M. (1981). Effect of food sprays on numbers of predators in an apple orchard. Canadian Entomologist, 113, 75–77.

Hagley, E. A. C., & Simpson, M. (1983). Effect of insecticides on predators of the pear psylla, *Psylla pyricola* (Hem.: Psyllidae), in Ontario. Canadian Entomologist, 115, 1409–1414.

Haines, C. P. (1981). Laboratory studies on the role of an egg predator, *Blattisocius tarsalis* (Berlese) (Acari: Ascidae), in relation to the natural control of *Ephestia cautella* (Walker) (Lepidoptera: Pyralidae) in warehouses. Bulletin of Entomological Research, 71, 555–574.

Hajek, A. E., & Dahlsten, D. L. (1987). Behavioral interactions between three birch aphid species and *Adalia bipunctata* larvae. Entomologia Experimental et Applicata, 45, 81–87.

Hale, L. (1970). Biology of *Icerya purchasi* and its natural enemies in Hawaii. Proceedings of the Hawaiian Entomological Society, 20, 533–550.

Hales, D. (1979). Population dynamics of *Harmonia conformis* (Boisd.) (Coleoptera: Coccinellidae) on acacia. General Applied Entomology, 11, 3–8.

Hall, R. W., & Ehler, L. E. (1979). Rate of establishment of natural enemies in classical biological control. Bulletin of Entomological Society of America, 25, 280–282.

Hall, R. W., Ehler, L. E., & Bisabri-Ershadi, B. (1980). Rate of success in classical biological control of arthropods. Bulletin of Entomological Society of America, 26, 111–114.

Halliday, R. B., & Holm, E. (1985). Experimental taxonomy of Australian mites in the *Macrocheles glaber* group (Acarina: Macrochelidae). Experimental and Applied Acarology (Northwood), 1, 277–286.

Hämäläinen, M. (1976). Rearing the univoltine ladybeetles, *Coccinella septempunctata* and *Adalia bipunctata* (Col., Coccinellidae), all year around in the laboratory. Annales Agricultura Fenniae, 15, 66–71.

Hämäläinen, M. (1977). Control of aphids on sweet peppers, chrysanthemums and roses in small greenhouses using ladybeetles *Coccinella septempunctata* and *Adalia bipunctata*. Annales Agricultura Fenniae, 16, 117–131.

Hämäläinen, M., Markkula, M., & Ray, T. (1975). Fecundity and larval voracity of four lady beetle species (Col., Coccinellidae). Annales Entomologici Fennici, 41, 124–127.

Hance, T. (1987). Predation impact of carabids at different population densities on *Aphis fabae* development in sugarbeet. Pedobiologia, 30, 251–262.

Haney, P. B. (1988). Identification, ecology and control of the ants in citrus: A world survey. Proceedings, Sixth International Citrus Congress, 3, 1227–1251.

Haramoto, F. (1969). Biology and control of *Brevipalpus phoenicis* (Geijskes) (Acarina: Tenuipalpidae). Hawaiian Agricultural Experimental Station Technical Bulletin, 68, 1–63.

Harris, K. M. (1968). A systematic revision and biological review of the cecidomyiid predators (Diptera: Cecidomyiidae) on world Coccoidea (Hemiptera: Homoptera). Transactions of Royal Entomological Society of London, 119, 401–494.

Harris, K. M. (1973). Aphidophagous Cecidomyiidae (Diptera): Taxonomy, biology and assessments of field populations. Bulletin of Entomological Research, 63, 305–325.

Harris, K. M. (1990). Cecidomyiidae and other Diptera. In D. Rosen (Ed.), Armored scale insects, their biology, natural enemies and control (Vol. B, pp. 61–66). Amsterdam: Elsevier.

Harris, R. L., & Oliver, L. M. (1979). Predation of *Philonthus flavolimbatus* on horn fly. Environmental Entomology, 8, 259–260.

Hart, E. R. (1986). Genus *Zelus* Fabricius in the United States, Canada, and Northern Mexico (Hemiptera: Reduviidae). Annals of the Entomological Society of America, 79, 535–548.

Hasegawa, M., Niijima, K., & Matsuka, M. (1989). Rearing *Chrysoperla carnea* (Neuroptera: Chrysopidae) on chemically defined diets. Applied Entomology and Zoology, 24, 96–102.

Hassan, S. A. (1974). Die Massenzucht und Verwendung von *Chrysopa*— Arten (Neuroptera: Chrysopidae) zur Bekampfung von Schadinsekten. Zeitschrift fuer Pflanzenkrankheiten Pflanzenschutz, 81, 620–637.

Hassan, S. A. (1975). Uber die Massenzucht von *Chrysoperla carnea* Steph. (Neuroptera: Chrysopidae). Zeitschrift fuer Angewandte Entomologie, 79, 310–319.

Hassan, S. A. (1978). Releases of *Chrysopa carnea* Steph. to control *Myzus persicae* (Sulzer) on eggplant in small greenhouse plots. Zeitschrift Pflanzenkrankheiten Pflanzenpathologie Pflanzenschutz, 85, 118–123.

Hassell, M. P. (1978). The dynamics of arthropod predator-prey systems. Princeton, NJ: Princeton University Press.

Hassell, M. P., Lawton, J. H., & Beddington, J. R. (1976). The components of arthropod predation. I. The prey death rate. Journal of Animal Ecology, 45, 135–164.

Hattingh, V., & Samways, M. J. (1991). Determination of the most effective method for field establishment of biocontrol agents of the genus *Chilocorus* (Coleoptera: Coccinellidae). Bulletin of Entomological Research, 81, 169–174.

Hattingh, V., & Samways, M. J. (1992). Prey choice and substitution in *Chilocorus* spp. (Coleoptera: Coccinellidae). Bulletin of Entomological Research, 82, 327–334.

Hattingh, V., & Samways, M. J. (1995). Visual and olfactory location of biotypes, prey patches, and individual prey by the ladybeetle *Chilocorus nigritus*. Entomologia Experimentalis et Applicata, 75, 87–98.

Havelka, J. (1980). Effect of temperature on the developmental rate of preimaginal stages of *Aphidoletes aphidimyza* (Diptera, Cecidomyliidae). Entomologia Experimentalis et Applicata, 27, 83–90.

Havelka, J., & Syrovatka, O. (1991). Stimulatory effect of the honeydew of several aphid species on females of the gall midge *Aphidoletes aphidimyza* (Rondani) (Diptera: Cecidomyiidae): Electroantennograph studies. Journal of Applied Entomology, 112, 341–344.

Hayes, J. L., & Lockley, T. C. (1990). Prey and nocturnal activity of wolf spiders (Araneae: Lycosidae) in cotton fields in the delta region of Mississippi. Environmental Entomology, 19, 1512–1518.

Heidari, M., & Copland, M. J. W. (1993). Honeydew: A food resource or arrestant for the mealybug predator *Cryptolaemus montrouzieri*. Entomophaga, 38, 63–68.

Heimpel, G. E., & Collier, T. R. (1996). The evolution of host-feeding behaviour in insect parasitoids. Biological Review, 71, 373–400.

Heimpel, G. E., Rosenheim, J. A., & Kattari, D. (1997). Adult feeding and lifetime reproductive success in the parasitoid *Aphytis melinus*. Entomologia Experimentalis et Applicata, 83, 305–315.

Heinz, K. M., & Nelson, J. M. (1996). Interspecific interactions among natural enemies of *Bemisia* in an inundative biological control program. Biological Control, 6, 384–393.

Helenius, J. (1990). Conventional and organic cropping systems at Suitia (Finland). VI. Insect populations in barley. Journal of Agricultural Science of Finland, 62, 349–356.

Helenius, J., Holopainen, J., Muhojoki, M., Pokki, P., Tolonen, T., & Venalainen, A. (1995). Effect of undersowing and green manuring on abundance of ground beetles (Coleoptera, Carabidae) in cereals. Acta Zoologica Fennica, 196, 156–159.

Hemptinne J.-L., Dixon, A. F. G., & Coffin, J. (1992). Attack strategy of ladybird beetles (Coccinellidae): factors shaping their numerical response. Oecologia (Berlin), 90, 238–245.

Hendrickson, R. M., Jr., Drea, J. J., & Rose, M. (1991). A distribution and establishment program for *Chilocorus kuwanae* (Silvestri) (Coleoptera: Coccinellidae) in the USA. Proceedings of the Entomological Society of Washington, 93, 197–200.

Hennessey, R. D., & Muaka, T. (1987). Field biology of the cassava mealybug, *Phenacoccus manihoti,* and its natural enemies in Zaire. Insect Science Applications, 8, 899–903.

Henry, C. S. (1983). Acoustic recognition of sibling species within the holarctic lacewing (*Chrysoperla carnea*) (Neuroptera: Chrysopidae). Systematics and Entomology, 8, 293–301.

Henry, C. S. (1984). The sexual behavior of certain lacewings. In M. Canard, Y. Semeria, & T. R. New (Eds.), Biology of Chrysopidae (pp. 101–110). The Hague: Dr. W. Junk.

Henry, C. S., & Wells, M. M. (1990). Geographical variation in the song of *Chrysoperla plorabunda* (Neuroptera: Chrysopidae) in North America. Annals of the Entomological Society of America, 83, 312–325.

Henry, T. J. (1976). *Aleuropteryx juniperi:* A European scale predator established in Nroth America (Neuroptera: Coniopterygidae). Proceedings of the Entomological Society of Washington, 78, 195–201.

Henry, T. J. (1988). Anthocoridae. In T. J. Henry & R. C. Foreschner (Eds.), Catalog of the Heteroptera, or true bugs, of Canada and the continental United States (pp. 12–28). Leiden: E. J. Brill.

Henry, T. J., & Froeschner, R. (Eds.). (1988). Catalog of the Heteroptera, or true bugs, of Canada and the continental United States. Leiden: E. J. Brill.

Henry, T. J., & Lattin, J. D. (1988). Family Nabidae. In T. J. Henry & R. C. Froeschner (Eds.), Catalog of the Heteroptera, or true bugs, of Canada and the continental United States (pp. 508–520). Lieden: E. J. Brill.

Henry, T. J., & Wheeler, A. G., Jr. (1988). Family Miridae Hahn, 1833. The plant bugs. In T. J. Henry & R. C. Froeschner (Eds.), Catalog of the Heteroptera, or true bugs, of Canada and continental United States (pp. 251–507). Leiden: E. J. Brill.

Henry, T. J., Neal, J. W., Jr., & Gott, K. M. (1986). *Stethoconus japonicus* (Heteroptera: Miridae): A predator of *Stephanitis* lace bugs newly discovered in the United States, promising in the biocontrol of azalea lace bug (Heteroptera: Tingidae). Proceedings of the Entomological Society of Washington, 88, 722–730.

Herard, F. (1986). Annotated list of the entomophagous complex associated with pear psylla, *Psylla pyri* (L.) (Hom.: Psyllidae) in France. Agronomie, 6, 1–34.

Herard, F., & Chen, K. (1985). Ecology of *Anthocoris nemoram* (L.) (Het: Anthocoridae) and evaluation of its potential effectiveness for biological control of pear psylla. Agronomie, 5, 855–863.

Herms, D. A., Haack, R. A., & Ayres, B. D. (1991). Variation in semiochemical-mediated prey-predator interaction: *Ips pini* (Scolytidae) and *Thanasimus dubius* (Cleridae). Journal of Chemical Ecology, 17, 515–524.

Herren, H. R., & Neuenschwander, P. (1991). Biological control of cassava pests in Africa. Annual Review of Entomology, 36, 257–283.

Herring, J. L. (1966). The genus *Orius* of the Western Hemisphere (Hemiptera: Anthocoridae). Annals of the Entomological Society of America, 59, 1093–1109.

Hesse, A. J. (1947). A remarkable new dimorphic isometopid and two other new species of Hemiptera predaceous upon the red scale of citrus. Journal of the Entomological Society of South Africa 10, 1–45.

Hessein, N. A., & Perring, T. M. (1986). Feeding habits of the Tydeidae with evidence of *Homeopronematus anconai* (Acari: Tydeidae) predation on *Aculops lycopercici* (Acari: Eriophyidae). International Journal of Acarology, 12, 215–221.

Hickman, J. M., & Wratten, S. D. (1996). Use of *Phacelia tanacetifolia* strips to enhance biological control of aphids by hoverfly larvae in cereal fields. Journal of Economic Entomology, 89, 832–840.

Hickman, J. M., Lovei, G. L., & Wratten, S. D. (1995). Pollen feeding by adults of the hoverfly *Melanostoma fasciatum* (Diptera: Syrphidae). New Zealand Journal of Zoology, 22, 387–392.

Higgins, C. J. (1992). Western flower thrips (Thysanoptera: Thripidae) in greenhouses: Population dynamics, distribution on plants, and associations with predators. Journal of Economic Entomology, 85, 1891–1903.

Hill, A. R. (1957). The biology of *Anthocoris nemorum* (L.) in Scotland (Hemiptera: Anthocoridae). Transactions of the Royal Entomological Society of London, 109, 379–394.

Hill, A. R. (1965). The bionomics and ecology of *Anthocoris confusus* Reuter in Scotland. I. The adult and egg production. Transactions of the Society of British Entomology, 16, 245–256.

Hill, M. G., Allan, D. J., Henderson, R. C., & Charles, J. G. (1993). Introduction of armoured scale predators and establishment of the predatory mite *Hemisarcoptes coccophagus* (Acari: Hemisarcoptidae) on latania scale, *Hemiberlesia lantaniae* (Homoptera: Diaspididae) in kiwifruit. Bulletin of Entomology Research, 83, 369–376.

Hinckley, A. D. (1963). Ecology and control of rice planthoppers in Fiji. Bulletin of Entomology Research, 54, 467–481.

Hinton, H. E. (1944). The Histeridae associated with stored products. Bulletin of Entomological Research, 35, 309–338.

Hodek, I. (Ed.). (1966a). Ecology of aphidophagous insects. Prague: Academia.

Hodek, I. (Ed.). (1966b). The food ecology of aphidophagous Coccinellidae. In I. Hodek (Ed.), Ecology of aphidophagous insects (pp. 23–30). Prague: Academia.

Hodek, I. (Ed.). (1967). Bionomics and ecology of predaceous Coccinellidae. Annual Review of Entomology, 12, 79–104.

Hodek, I. (Ed.). (1970). Coccinellids and the modern pest management. BioScience, 20, 543–552.

Hodek, I. (Ed.). (1973). Biology of Coccinellidae. Prague: Academia Czechoslovak Academy of Science.

Hodek, I. (Ed.). (1986). Ecology of Aphidophaga (Vol. 2) Prague: Academia.

Hodek, I. (Ed.). (1993). Habitat and food specificity in aphidophagous predators. Biocontrol Science and Technology, 3, 91–100.

Hodek, I., Okuda, T., & Hodkova, M. (1984). Reverse photoperiodic responses in two subspecies of *Coccinella septempunctata* L. Zool. Jb. Syst., 111, 439–448.

Hodgson, C., & Aveling, C. (1988). Anthocoridae. In A. K. Minks & P Horrewijn (Eds.), Aphids, their biology, natural enemies and control (Vol. B, pp. 279–292). Amsterdam: Elsevier.

Hodgson, C. J., & Mustafa, T. M. (1984). Aspects of chemical and biological control of *Psylla pyricola* Forster in England. IOBC WPRS Bulletin, 7, 330–353.

Hoelmer, K. A., Osborne, L. S., & Yokomi, R. K. (1993). Reproduction and feeding behavior of *Delphastus pusillus* (Coleoptera: Coccinellidae), a predator of *Bemisia tabaci* (Homoptera: Aleyrodidae). Journal of Economic Entomology, 86, 322–329.

Hoelmer, K. A., Osborne, L. S., & Yokomi, R. K. (1994). Interactions of the whitefly predator *Delphastus pusillus* (Coleoptera: Coccinellidae) with parasitized sweetpotato whitefly (Homoptera: Aleyrodidae). Environmental Entomology, 23, 136–139.

Hoffman, W. E. (1936). Notes on citrus insects. Lingnan Agricultural Journal, 2, 215–218.

Hokkanen, H. M. T. (1985). Success in classical biological control. CRC Critical Reviews in Plant Science, 3, 35–72.

Holdsworth, R. P., Jr. (1970a). Aphids and aphid enemies: Effect of integrated control in an Ohio apple orchard. Journal of Economic Entomology, 63, 530–555.

Holdsworth, R. P., Jr. (1970b). Codling moth control as part of an integrated program in Ohio. Journal of Economic Entomology, 63, 894–897.

Hölldobler, B., & Wilson, E. O. (1977). Weaver ants. Scientific American, 237, 146–154.

Hölldobler, B., Wilson, E. O. (1990). The ants. Cambridge: Belknap Press/Harvard.

Holling, C. S. (1966). The functional response of invertebrate predators to prey density. Memoirs of Entomological Society of Canada, 48, 86.

Hollingsworth, C. S., & Bishop, G. W. (1982). *Orius tristicolor* (Heteroptera: Anthocondae) as a predator of *Myzus persicae* (Homoptera: Aphididae) on potatoes. Environmental Entomology, 11, 1046–1048.

Hölzel, H. (1984). Chrysopidae of the Palearctic Region—a review. In J.

Gepp, H. Aspöck, & H. Hölzel (Eds.), Progress in world's neuropterology (pp. 61–68). Austria, Graz: Gepp.

Hood Henderson, D. E. (1982). Fine structure and neurophysiology of a gustatory sensillum on the ovipositors of *Metasyrphus venablesi* and *Eupeodes volucris* (Diptera: Syrphidae). Canadian Journal of Zoology, 60, 3187–3195.

Hood Henderson, D. E., & Wellington, D. E. (1982). Antennal sensilla of some aphidophagous Syrphidae (Diptera): The fine structure and electro-antennogramme study. Canadian Journal of Zoology, 60, 3172–3186.

Hopper, K. R., Aidara, S., Agret, S., Cabal, J., Coutinot, D., & Dabire, R. (1995). Natural enemy impact on the abundance of *Diuraphis noxia* (Homoptera: Aphididae) in wheat in southern France. Environmental Entomology, 24, 402–408.

Hormchan, P., Schuster, M. F., & Hepner, L. W. (1976). Biology of *Tropiconabis capsiformis*. Annals of the Entomological Society of America, 69, 1016–1018.

Horner, N. V. (1972). *Metaphidippus galathea* as a possible biological control agent. Journal of Kansas Entomological Society, 45, 324–327.

Houck, M. A. (1991). Time and resource partitioning in *Stethorus punctum* (Coleoptera: Coccinellidae). Environmental Entomology, 20, 494–497.

Hough-Goldstein, J., & Keil, C. B. (1991). Prospects for integrated control of the Colorado potato beetle (Coleoptera: Chrysomelidae) using *Perillus bioculatus* (Hemiptera: Pentatomidae) and various pesticides. Journal of Economic Entomology, 84, 1645–1651.

Hough-Goldstein, J., & Whalen, S. (1993). Inundative release of predatory stink bugs for control of Colorado potato beetle. Biological Control 3, 343–347.

Houston, K. J. (1980). A revision of the Australian species of *Stethorus* Weise (Coleoptera: Coccinellidae). Journal of Australian Entomological Society, 19, 81–91.

Houston, W. W. K. (1981). The life cycles and age of *Carabus glabratus* Paykull and *C. problematicus* Herbst (Col.: Carabidae) on moorland in northern England. Ecology and Entomology, 6, 263–271.

Howard, L. O., & Fiske, W. F. (1911). The importation into the United States of the parasites of the gypsy moth and the brown-tail moth. (Bureau of Entomology Bulletin No. 91). Washington, DC: U.S. Department of Agriculture.

Hoy, M. A. (1979). Parahaploidy of the "arrhenotokous" predator, *Metaseiulus occidentalis* (Acarina: Phytoseiidae), demonstrated by XX-irradiation of males. Entomologia Experimentalis et Applicata, 26, 97–104.

Hoy, M. A., & Smilanick, J. M. (1979). A sex pheromone produced by immature and adult females of the predatory mite, *Metaseiulus occidentalis*, Acarina: Phytoseiidae. Entomologia Experimentalis et Applicata, 26, 291–300.

Hoy, M. A., & Smilanick, J. M. (1981). Non-random prey location by the phytoseiid predator *Metaseiulus occidentalis*: Differential responses to several spider mite species. Entomologia Experimentalis et Applicata, 29, 241–253.

Hoyt, S. C. (1969). Population studies of five mite species on apple in Washington. In G. O. Evans (Ed.), Proceedings of the Second International Congress on Acarology, Sutton-Bonington, England, 1967 (pp. 117–133). Budapest: Akademie Kiado.

Hsieh, C., & Dyck, V. A. (1975). Influence of predators on population density of the rice green leafhoppers. Plant Protection Bulletin (Taiwan), 17, 346–352.

Huang, H., Yian, J. Z., & Li, D. Q. (1990). Predation model of *Chrysopa septempunctata* on cotton insect pests. Natural Enemy Insects, 12, 7–12 (in Chinese with English abstract).

Huang, H. T., & Yang, P. (1987). The ancient cultures citrus ant. BioScience, 37 (9), 665–671.

Huffaker, C. B. (1970). The phenomenon of predation and its roles in nature. In P. J. Den Boer & G. R. Gradwell (Eds.), Proceedings, Advanced Study Institute, Dynamic Populations, Oosterbeek (pp. 327–343).

Huffaker, C. B., & Doutt, R. L. (1965). Establishment of the coccinellid *Chilocorus bipustulatus* Linnaeus, in California olive groves. Pan-Pacific Entomologist, 41, 61–63.

Huffaker, C. B., & Kenneth, C. E. (1956). Experimental studies on predation: Predation and cyclamen-mite populations on strawberries in California. Hilgardia, 26, 191–222.

Huffaker, C. B., Messenger, P. S., & DeBach, P. (1971). The natural enemy component in natural control and the theory of biological control. In C. B. Huffaker (Ed.), Biological control (pp. 16–67). New York: Plenum Press.

Huffaker, C. B., Simmonds, F. J., & Laing, J. E. (1976). The theoretical and empirical basis of biological control. In C. B. Huffaker & P. S. Messenger (Eds.), Theory and practice of biological control (pp. 42–78). New York: Academic Press.

Hughes, A. M. (1976). The mites of stored food and houses. (Ministry of Agriculture Fisheries & Food Technical Bulletin 9). London: Her Majesty's Stationery Office.

Hukusima, S. (1971). Simultaneous suppression of major phytophagous arthropods in apple orchards by combination spray programs with release of *Harmonia axyridis* Pallas. Research Bulletin of Faculty of Agriculture, Gifu University, 31, 113–135.

Hukusima, S., & Kamei, M. (1970). Effects of various species of aphids as food on development, fecundity and longevity of *Harmonia axyridis* Pallas (Coleoptera: Coccinellidae). Research Bulletin of Faculty of Agriculture, Gifu University, 29, 53–66.

Hukusima, S., & Komada, N. (1972). Effects of several species of aphids on development and nutrition of *Propylea japonica* Thunberg (Coleoptera: Coccinellidae). Proceedings of Kansas Plant Protection Society, 14, 7–13.

Hull, L. A., & Beers, E. H. (1985). Ecological selectivity: Modifying chemical control practices to preserve natural enemies. In M. A. Hoy & D. C. Herzog (Eds.), Biological control in agricultural IPM systems (pp. 103–122). Orlando, FL: Academic Press.

Hull, L. A., Asquith, D., & Mowery, P. D. (1976). Distribution of *Stethorus punctum* in relation to densities of the European red mite. Environmental Entomology 5, 337–342.

Hull, L. A., Asquith, D., & Mowery, P. D. (1977a). The functional responses of *Stethorus punctum* to densities of the European red mite. Environmental Entomology, 6, 85–90.

Hull, L. A., Asquith, D., & Mowery, P. D. (1977b). The mite searching ability of *Stethorus punctum* within an apple orchard. Environmental Entomology, 6, 684–688.

Hulme, M. A. (1989). Laboratory assessment of predation by *Lonchaea corticis* (Diptera: Lonchaeidae) on *Pissodes strobi* (Coleoptera: Curculionidae). Environmental Entomology, 18, 1011–1014.

Hulme, M. A. (1990). Field assessment of predation by *Lonchaea corticis* (Diptera: Lonchaeidae) on *Pissodes strobi* (Coleoptera: Curculionidae) in *Picea sitchensis*. Environmental Entomology, 19, 54–58.

Hulme, M., Dawson, A. F., & Harris, J. W. E. (1986). Exploiting cold-hardiness to separate *Pissodes strobi* (Peck) (Coleoptera: Curculionidae) from associated insects in leaders of *Picea sitchensis* (Bong.) Carr. Canadian Entomologist, 118, 1115–1122.

Hunter, J. S., III, Bay, D. E., & Fincher, G. T. (1989). Laboratory and field observations on the life history and habitats of *Philonthus cruentatus* and *Philonthus flavolimbatus*. Southwestern Entomologist, 14, 41–47.

Hunter, J. S., III, Fincher, G. T., Bay, D. E., & Beerwinkle, K. R. (1991). Seasonal distribution and diel flight activity of Staphylinidae (Coleoptera) in open and wooded pasture in east central Texas. Journal of Kansas Entomological Society, 64, 163–173.

Hunter, K. W. (1978). Searching behavior of *Hippodamia convergens* larvae (Coccinellidae: Coleoptera). Psyche, 85, 249–253.

Hurd, L. E. (1985). Ecological considerations of mantids as biocontrol agents. Antenna, 9, 19–22.

Hurd, L. E., & Eisenberg, R. M. (1984). Experimental density manipulations of the predator *Tenodera sinensis* (Orthoptera: Mantidae) in an old-field community. II. The influence of mantids on arthropod community structure. Journal of Animal Ecology, 53, 955–967.

Hurd, L. E., & Rathet, L. H. (1986). Functional response and success in juvenile mantids. Ecology, 67, 163–167.

Hurd, L. E., Eisenberg, R. M., & Washburn, S. O. (1978). Effects of experimentally manipulated density on field populations of the Chinese mantis (*Tenodera aridifolia sinensis* Saussure). American Midland Naturalist, 99, 58–64.

Hussein, M. Y. (1984). A spray technique for mass release of eggs of *Micromus tasmaniae* Walker (Neuroptera: Hemerobiidae). Crop Protection, 3, 369–378.

Hussein, M. Y. (1986). Biological control of aphids on potatoes by inundative releases of predators. In M. Y. Hussein & A. G. Ibrahim (Eds.), Biological control in the tropics (pp. 137–147). Pertanian, Malaysia: Penerbit University.

Hussey, N. W., & Bravenboer, L. (1971). Control of pests in glasshouse culture by introduction of natural enemies. In C. B. Huffaker (Ed.). Biological control (pp. 195–216). New York: Plenum.

Hussey, N. W., & Huffaker, C. B. (1976). Spider mites. In V. L. Delucchi (Ed.), Studies in biological control (pp. 178–228). London: Cambridge University Press.

Hussey, N. W., & Scopes, N. (Eds.). (1985). Biological pest control, the glasshouse experience. New York: Cornell University Press.

Hutchison, W. D., & Pitre, H. N. (1983). Predation of *Heliothis virescens* (Lepidoptera: Noctuidae) eggs by *Geocoris punctipes* (Hemiptera: Lygaeidae) adults on cotton. Environmental Entomology, 12, 1652–1656.

Hydorn, S. B., & Whitcomb, W. H. (1972). Effects of parental age at oviposition on progeny of *Chrysopa rufilabris*. Florida Entomologist, 55, 79–85.

Hydorn, S. B., & Whitcomb, W. H. (1979). Effects of larval diet on *Chrysopa rufilabris*. Florida Entomologist, 62, 293–298.

Iablokoff-Khnzorian, S. M. (1982). Les Coccinelles Coleopteres—Coccinellidae Triba Coccinellini des regions Palearctique et Orientale. Paris: Boubeo.

Ickert, G. (1968). Beitrage zur Biologie einheimischer Chrysopiden. (Planipennia, Chrysopidae). Entomologische Abhandlungen, 36, 123–192.

Ignoffo, C. M., Garcia, C., Dickerson, W. A., Schmidt, G. T., & Biever, K. D. (1977). Imprisonment of entomophages to increase effectiveness: evaluation of a concept. Journal of Economic Entomology, 70, 292–294.

Inoue, H. (1986). Habitat use by refuging predator, *Agriosphodrus dohrni* Signoret. Nymphal microhabitat suitability and density dependent microhabitat suitability and density dependent microhabitat selection by ovipositing females. Researches on Population Ecology (Kyoto), 28, 321–332.

Iperti, G. (1961). Les coccinelles leut utilisation en agriculture. Revue de Zoologie Agricole et Appliquee, 1–6, 1–28.

Iperti, G. (1965). Contribution a l'etude de la specificite chez les principales coccinelles aphidophages des Alpes—maritimes et des basses—Alpes. Entomophaga, 10, 159–178.

Iperti, G., & Kreiter, S. (1986). Two aphidophagous coccinellids: How to use their complementary biological potentialities for better biological control. In I. Hodek (Ed.), Ecology of Aphidophaga (Vol 2, pp. 475–478). Prague: Academia.

Iperti, G., & Laudeho, Y. (1969). Les entomophages de *Parlatoria blanchardi* Targ., dans les palmerales de l'adrar Mauritanien 1. Annales de Zoologie—Ecologie Animale, 1, 17–30.

Iperti, G., Giuge, L., & Roger, J. P. (1989). Installation de *Rhyzobius Forestieri* (Col.: Coccinellidae) surl'Ile de Parquerolles. Entomophaga, 34, 365–372.

Iperti, G., Laudého, Y., Brun, J., & deJanvry, E. C. (1970). Entomophages de *Parlatoria blanchardi* Targ. les palmeraies de l'adrar Mauritanien. Annales de Zoologie—Ecologie Animale, 2, 617–638.

Ireson, J. E., & Webb, W. R. (1995). Effectiveness of *Neomulgus capillatus* (Acarina: Bdellidae) as a predator of *Sminthurus viridis* (Collembola: Sminthuridae) in northwestern Tasmania. Journal of the Australian Entomological Society, 34, 237–240.

Irwin, M. E., & Shepard, M. (1980). Sampling predaceous Hemiptera on soybean. In M. Kogan & D. C. Herzog (Eds.), Sampling methods in soybean entomology (pp. 503–531). New York: Springer-Verlag.

Irwin, M. E., Gill, R. W., & Gonzalez, D. (1974). Field-cage studies of native egg predators of the pink bollworm in southern California cotton. Journal of Economic Entomology, 67, 193–196.

Isenhour, D. J., & Yeargan, K. V. (1981). Effect of crop phenology on *Orius insidiosus* populations on strip-cropped soybean and corn. Journal of Georgia Entomological Society, 16, 310–322.

Isenhour, D. J., & Yeargan, K. V. (1982). Oviposition sites of *Orius insidiosus* (Say) and *Nabis* spp. in soybean. (Hemiptera: Anthocoridae and Nabidae). Journal of Kansas Entomological Society, 55, 65–72.

Isenhour, D. J., Wiseman, B. R., & Layton, R. C. (1989). Enhanced predation by *Orius insidiosus* (Hemiptera: Anthocoridae) on larvae of *Heliothis zea* and *Spodoptera frugiperda* (Lepidoptera: Noctuidae) caused by prey feeding on resistant corn genotypes. Environmental Entomology, 18, 418–422.

Ishaaya, I., & Casida, J. E. (1981). Pyrethroid esterase(s) may contribute to natural pyrethroid tolerance of larva of the common green lacewing. Environmental Entomology, 10, 681–684.

Ishankulieva, T. (1979). The use of larvae of the common lacewing in the control of pests of cotton. Izvestiya Akademiya Nauk Turkmenska SSR, Biol, 1, 17–21 (in Russian); (1980). Review of Applied Entomology, Series A: Agricultural, 68, 3840.

Ishihara, T. (1941). Hemiptera-Heteroptera as the natural enemies for controlling the pests. Botany and Zoology, Theoretical Applications, 9, 223–229 (in Japanese).

Ives, A. R., Kareiva, P., & Perry, R. (1993). Response of a predator to variation in prey density at three hierarchical scales: Lady beetles feeding on aphids. Ecology, 74, 1229–1238.

Izhevsky, S. S., & Orlinsky, A. D. (1988). Life history of the imported *Scymnus reunioni* (Coleoptera: Coccinellidae) predator of mealybugs. Entomophaga, 33, 101–114.

Jalil, M., & Rodriquez, J. G. (1970). Biology of and odor perception by *Fuscuropoda vegetans* (Acarina: Uropodidae), a predator of the housefly. Annals of the Entomological Society of America, 63, 935–938.

Javahery, M. (1986). Biology and ecology of *Picromerus bidens* (Hemiptera: Pentatomidae) in southeastern Canada. Entomological News, 97, 87–98.

Jay, E., Davis, R., & Brown, S. (1968). Studies on the predaceous habits of *Xylocoris flavipes* (Reuter) (Hemiptera: Anthocoridae). Journal of Georgia Entomological Society, 3, 126–130.

Jayaratnam, T. J. (1941). The bethylid parasite (*Perisierola nephantidis* M.) of the coconut caterpillar (*Nephantis serinopa* Meyr.). Tropical Agriculture, (Agricultural Journal of Ceylon), 97, 115–127.

Jeanne, R. L., & Morgan, R. C. (1992). The influence of temperature on nest site choice and reproductive strategy in temperate zone *Polistes* wasps. Ecological Entomology, 17, 135–141.

Jeffrey, I. G. (1976). A survey of the mite fauna of Scottish farms. Journal of Stored Product Research, 12, 149–156.

Jepson, F. P. (1914). A mission to Java in quest of natural enemies for a coleopterous pest of bananas. Bulletin of Department of Agriculture of Fiji, 7, 1–18.

Jermy, T. (1962). Preliminary observations on the natural enemies of *Perillus bioculatus* F. in Hungary (Hemiptera, Pentatomidae). Folia Entomologica Hungarica, (N.S.), 15, 17–23.

Jermy, T. (1980). The introduction of *Perillus bioculatus* into Europe to control the Colorado beetle. Bull. Org. Eur. Mediterr. Prot. Plant, 10, 475–479.

Jervis, M. A., & Kidd, N. A. C. (1986). Host-feeding strategies in hymenopteran parasitoids. Biological Review, 61, 395–434.

Jervis, M. A., Hawkins, B. A., & Kidd, N. A. C. (1996). The usefulness of destructive host feeding parasitoids in classical biological control: Theory and observation conflict. Ecological Entomology, 21, 41–46.

Johnson, J. B. (1982). Bionomics of some symbiote using Chrysopidae (Insecta: Neuroptera) from western United States. Unpublished doctoral dissertation, University California. Berkeley.

Johnson, J. B., & Hagen, K. S. (1981). A neuropterous larva uses an allomone to attack termites. Nature (London), 289, 506–507.

Johnson, N. E., & Cameron, R. S. (1969). Phytophagous ground beetles. Annals of the Entomological Society of America, 62, 909–924.

Johnson, R. H. (1907). Economic notes on aphids and coccinellids. Entomological News, 13, 171–174.

Johnson, V. (1980). Review of the Coniopterygidae (Neuroptera) of North America with a revision of the genus *Aleuropteryx*. Psyche, 87, 259–298.

Jones, P. A., & Coppel, A. C. (1963). Immature stages and biology of *Apateticus cynicus* (Say) (Hemiptera: Pentatomidae). Canadian Entomologist, 95, 770–779.

Jones, R. W., Gilstrap, F. E., & Andrews, K. L. (1987). Activities and plant associations of the earwig, *Doru taeniatum*, in a crop-weed habitat. Southwestern Entomologist, 12, 107–118.

Jones, R. W., Gilstrap, F. E., & Andrews, K. L. (1988). Biology and life tables for the predaceous earwig, *Doru taeniatum* (Derm., Forticulidae). Entomophaga, 33, 43–54.

Jones, S. L., & Ridgway, R. L. (1976). Development of methods of field distribution of eggs of the insect predator *Chrysoperla carnea* Stephens. Washington, DC: U.S. Department of Agriculture, Agricultural Research Services.

Jones, S. L., Kinzer, R. E., Bull, D. L., Ables, J. R., & Ridgway, R. L. (1978). Deterioration of *Chrysopa carnea* in mass culture. Annals of the Entomological Society of America, 71, 160–162.

Jones, W. A., Jr., & Sullivan, M. J. (1981). Overwintering habitats, spring emergence patterns and winter mortality of some South Carolina Hemiptera. Environmental Entomology, 10, 409–414.

Jubb, G. L., & Masteller, E. C. (1977). Survey of arthropods in grape vineyards of Erie County, Pennsylvania: Neuroptera. Environmental Entomology, 6, 419–428.

Kaddou, I. K. (1960). The feeding behavior of *Hippodamia quinquesignata* (Kirby) larvae. University of California Publication of Entomology, 16, 181–232.

Kairo, M. T. K., & Murphy, S. T. (1995). The life history of *Rodolia iceryae* Janson (Col., Coccinellidae) and the potential for use in inoculative releases against *Icerya pattersoni* Newstead (Hom., Margarodidae) on coffee. Journal of Applied Entomology, 119, 487–491.

Kanika-Kiamfu, J., Iberti, G., & Brun, J. (1993). Etude de la consommation alimentaire d'*Exochomus flaviventris* (Col.: Coccinellidae) predateur de la cochenille du manioc *Phenacoccus manihoti* (Hom.: Pseudococcidae). Entomophaga, 38, 291–298.

Kanika-Kiamfu, J., Kiyindou, A., Brun, J., & Iperti, G. (1992). Comparison des potentialites biologiques de trois coccinelles predatrices de la cochenille farineuse du manioc *Phenacoccus manihoti* (Hom.: Pseudococcidae). Entomophaga, 37, 277–282.

Kapadia, M. N., & Puri, S. N. (1991). Biology and comparative predation efficacy of three heteropteran species recorded as predators of *Bemisia tabaci* in Maharashtra. Entomophaga, 36, 555–559.

Kapur, A. P. (1948). On the Old World species of the genus *Stethorus* Weise (Coleoptera: Coccinellidae). Bulletin of Entomological Research, 39, 297–320.

Kareiva, P., & Sahakian, R. (1990). Tritrophic effects of a simple architectural mutation in pea plants. Nature (London), 345, 433–434.

Katsoyannos, P. (1984). The establishment of *Rhyzobius forestieri* (Col.: Coccinellidae) in Greece and its efficiency as an auxiliary control agent against a heavy infestation of *Saissetia oleae* (Hom.: Coccidae). Entomophaga, 29, 387–397.

Katsoyannos, P., Kontodimas, D. C., Stathas, G. J., & Tsartsalis, C. T. (1997). Establishment of *Harmonia axyridis* on citrus and some data on its phenology in Greece. Phytoparasitica, 25, 183–191.

Kawai, A. (1978). Sibling cannibalism in the first instar larvae of *Harmonia axyridis* Pallas (Coleoptera, Coccinellidae). Kontyu, Tokyo, 46, 14–19.

Kawalska, T. (1976). Mass rearing and possible uses of Chrysopidae against aphids in glasshouses. IOBC WPRS Bulletin, 1976, 80–85.

Kawashima, K. (1988). Effects of chitin synthesis inhibitors on the four-spotted lacewing, *Chrysopa septempunctata* Wesmael. Annual Report of the Society Plant Protection Japan, 39, 246–247 (in Japanese); (1990). Review of Applied Entomology, Series A, 78, 5492.

Kaya, H. K. (1979). Microsporidian spores: Retention of infectivity after passage through gut of the assassin bug, *Zelus exsanguis* (Stål). Proceedings of the Hawaiian Entomological Society, 23, 91–94.

Kayashima, I. (1961). Study of the lynx spider, *Oxyopes sertatus* L. Koch, for biological control of the crytomerian leaf fly *Contarina inouyei* Mani. Review of Applied Entomology, Series A, 51, 413.

Kehat, M. (1967). Studies on the biology and ecology of *Pharoscymnus numidieus* (Coccinellidae) an important predator of the dark palm scale *Parlatoria blanchardi*. Annales de la Societe Entomologique de France (N.S.) 3, 1053–1065.

Kehat, M., Swirski, E., Blumberg, D., & Greenberg, S. (1974). Integrated control of date palm pests in Israel. Phytoparasitica, 2, 141–149.

Kelton, L. A. (1978). The insects and arachnids of Canada. IV. The Anthocoridae of Canada and Alaska (Heteroptera: Anthocoridae) (Branch Publication 1639). Canadian Department of Agricultural Research.

Kelton, L. A. (1980). Description of a new species of *Parthenicus* Reuter, new records of Holarctic Orthotylini in Canada, and new synonymy for *Diaphnocoris pellucida* (Heteroptera: Miridae). Canadian Entomologist, 112, 341–344.

Kenmore, P. E. (1980). Ecology and outbreaks of a tropical insect pest of the green revolution, the rice brown planthopper, *Nilaparvata lugens* (Stål). Unpublished doctoral dissertation, University of California, Berkeley.

Kesten, U. (1969). Zur morphologie und biologie von *Anatis ocellata* (L.) (Coleoptera: Coccinellidae). Zeitschrift fuer Angewandte Entomologie, 63, 412–445.

Kevan, D. K. M., & Klimaszewski, J. (1987). The Hemerobiidae of Canada and Alaska. Genus *Hemerobius* L. G. Italia Entomologia, 3(16), 305–369.

Key, K. H. L. (1970a). Orthoptera. The insects of Australia (pp. 323–347). Melbourne: Melbourne University Press.

Key, K. H. L. (1970b). Mantodea. The insects of Australia (pp. 294–301). Melbourne: Melbourne University Press.

Kharboutli, M. S., & Mack, T. P. (1993). Effect of temperature, humidity, and prey density on feeding rate of the striped earwig (Dermaptera: Labiduridae). Environment Entomology, 22, 1134–1139.

Khloptseva, R. I. (1991). The use of entomophages in biological pest control in the USSR. Biocontrol News Information, 12, 243–246.

Kieckhefer, R. W., Elliott, N. C., & Beck, D. A. (1992). Aphidophagous coccinellids in alfalfa, small grains, and maize in eastern South Dakota. Great Lakes Entomologist, 25, 15–23.

Kieckhefer, R. W., & Olson, G. A. (1974). Dispersal of marked adult coccinellids from crops in South Dakota. Journal of Economic Entomology, 67, 52–54.

Killington, F. J. (1936, 1937). A monograph of the British Neuroptera (Vol. 1, Vol. 2, Ray Soc. No. 123). London: Ray Society, London, UK

Kim, C. H., & Choi, J. S. (1976). Studies on the biological control of pine caterpillar (*Dendrolimus spectabilis* Butler) by red wood ant (*Formica rufa* var. *yessolensis* Forel). Korean Journal of Plant Protection, 15, 7–16.

Kim, C. H., & Kim, J. M. (1973). Studies on red wood ant (*Formica* sp.) for the control of pine caterpillar (*Dendrolimus spectabilis* Butler). Korean Journal of Plant Protection, 12, 109–114.

Kim, C. H., & Murakami, Y. (1978). Studies on the biological control of pine caterpillar (*Dendrolimus spectabilis* Butler) by red wood ant (*Formica rufa truncicola* var. *yessoensis* Forel). Journal of the Institute Agricultural Research Utilities Gyeongsang University, 12, 91–123 (in Korean, English summary).

Kiman, Z. B., & Yeargan, K. V. (1985). Development and reproduction of the predator *Orius insidiosus* (Hemiptera: Anthocoridae) reared on diets of selected plant material and arthropod prey. Annals of the Entomological Society of America, 78, 464–467.

Kinaway, M. M. (1991). Biological control of the coconut scale insect (*Aspidiotus destructor* Sign, Homoptera: Diaspididae) in the southern region of Oman (Dhofar). Tropical Pest Management, 37, 387–389.

Kindlmann, P., & Dixon, A. F. G. (1993). Optimal foraging in ladybird beetles (Coleoptera Coccinellidae) and its consequences for their use in biological control. European Journal of Entomology, 90, 443–450.

Kingsley, P. C., Harrington, B. J. (1982). Factors influencing termination of reproductive diapause in *Orius insidiosus* (Hemiptera: Anthocoridae). Environmental Entomology, 11, 461–462.

Kinn, D. N. (1966). Predation by the mite, *Macrocheles muscaedomesticae* (Acarina: Macrochelidae) on three species of flies. Journal of Medical Entomology, 3, 155–158.

Kinzer, R. C., & Jones, S. L. (1973). Manipulation of *Chrysopa* sp. with a protein supplement and sugar spray. Folia Entomologica Mexicana, 25–26, 43.

Kirejtshuk, A. G., James, D. G., & Heffer, R. (1997). Description and biology of a new species of *Cybocephalus* Erichson (Coleoptera: Nitidulidae), a predator of Australian citrus whitefly. Australian Journal of Entomology, 36, 81–86.

Kiritani, K., & Kakiya, N. (1975). An analysis of the predator-prey system in the paddy field. Researches on Population Ecology (Kyoto), 17, 29–38.

Kiritani, K., Kawahara, S., Sasaba, T., & Nakasuji, F. (1972). Quantitative evaluation of predation by spiders on the green rice leafhopper, *Nephotettix cincticeps* Uhler, by a sight-count method. Researches on Population Ecology (Kyoto), 13, 187–200.

Kirk, V. M. (1981). Corn rootworm: Population reduction associated with the ant, *Lasius neoniger*. Environmental Entomology, 10, 966–967.

Kishi, Y. (1969). A study of the ability of *Medetera* sp. (Diptera: Dolichopodidae) to prey upon the bark and wood boring Coleoptera. Applied Entomology and Zoology, 4, 177–184.

Kiyindou, A., & Fabres, G. (1987). Etude de la capacite d'accroissement chez *Hyperaspis raynerali* (Col.: Coccinellidae) predateur introduct au Congo pour la regulation des populations de *Phenacoccus manihoti* (Hom.: Pseudococcidae). Entomophaga, 32, 181–189.

Kiyindou, A., LeRü, B., & Fabres, G. (1990). Influence de la nature et de l'abondance des proies sur l'augmentation des effectifs de deux coccinelles prédatrices de la cochenille du manioc au Congo. Entomophaga, 35, 611–620.

Klausnitzer, B. (1978). Ordung Coleoptera (Larven). The Hague: Dr. W. Junk.

Klimaszewski, J. (1984). A revision of the genus *Aleochara* Gravenhorst of America north of Mexico (Coleoptera: Staphylinidae, Aleocharinae). Memoirs of the Entomological Society of Canada, 129, 3–211.

Klimaszewski, J., & Kevan, D. K. M. (1988). The brown lacewing flies of Canada and Alaska (Neuroptera: Hemerobiidae), III. The genus *Micromus* Rambur. G. Italia Entomologia, 4 (19), 31–76.

Knabke, J. J., & Grigarick, A. A. (1971). Biology of the African earwig,

Euborellia cincticollis (Gerstaecker), in California and comparative notes on *Euborellia annulipes* (Lucas). Hilgardia, 41, 157–194.

Knight, H. H. (1921). Monograph of North American species of *Deraeocoris*—Heteroptera: Miridae. University of Minnesota Agricultural Experimental Station Technical Bulletin, 1, 76–210.

Knight, H. H. (1923). Studies on the life history and biology of *Perillus bioculatus* Fabricius, including observations on the nature of the color pattern (19th Rep State Entomology of Minnesota 1922, pp. 50–96).

Knop, N. F., & Hoy, M. A. (1983). Biology of a tydeid mite, *Homeopronematus anconai* (n. comb.) (Acari: Tydeidae), important in San Joaquin Valley vineyards. Hilgardia, 51, 1–30.

Knutson, A. E., & Gilstrap, F. E. (1989). Direct evaluation of natural enemies of the southwestern corn borer (Lepidoptera: Pyralidae) in Texas corn. Environmental Entomology, 18, 732–739.

Knutson, L. (1976). Sciomyzid flies. Another approach to biological control of snail-borne diseases. Insect World Digest, 3, 13–18.

Kobakhidze, D. N. (1965). Some results and prospects of the utilization of beneficial entomophagous insects in the control of insect pests in Georgian SSR (USSR). Entomophaga, 10, 323–330.

Kochansky, J., Aldrich, J. R., & Lusby, N. R. (1989). Synthesis and pheromonal activity of 6, 10, 13-trimethy (-1-tetradecanol) for predatory stink bug, *Stiretus anchorago* (Heteroptera: Pentatomidae). Journal of Chemical Ecology, 15, 1717–1728.

Koschel, H. (1971). Zur kenntnis der raubwanze *Himacerus apterus* F. (Heteroptera, Nabidae). Zeitschrift fuer Angewandte Entomologie, 69, 1–24, 113–137.

Kowalska, T. (1976). Mass rearing and possible uses of Chrysopidae against aphids in glasshouses. IOBC WPRS Bulletin, 4, 80–85.

Krantz, G. W. (1978). A manual of acarology (2nd ed.). Corvallis: Oregon State University Bookstores.

Krantz, G. W. (1983). Mites as biological control agents of dung-breeding flies, with special reference to the Macrochelidae. In M. A. Hoy, G. L. Cunningham, & L. Knutson (Eds.), Biological control of pests by mites (Special Publication 3304, pp. 91–102). Berkeley: DANR, University of California.

Kreiter, S., & Iperti, G. (1984). Etude des potentialites biologiques et ecologiques d'un predateur aphidiphage *Olla v-nigrum* Muls. (Coleoptera, Coccinellidae) en vue de son introduction en France. 109 Congr. Nat. Soc. Scivartes, Dijon, Sci Fasc., 11, 275–282.

Kreiter, S., & Iperti, G. (1986). Effectiveness of *Adalia bipunctata* against aphids in a peach orchard with special reference to ant/aphid relationships. In I. Hodek (Ed.), Ecology of Aphidophaga 2 (pp. 537–543). Prague: Academia.

Kring, T. J., Gilstrap, F. E., & Michels, G. J., Jr. (1985). Role of indigenous coccinellids in regulating greenbugs (Homoptera: Aphididae) on Texas grain sorghum. Journal of Economic Entomology, 78, 269–273.

Krishnamoorthy, A. (1984). Influence of adult diet on the fecundity and survival of the predator, *Chrysopa scelestes* (Neur.: Chrysopidae). Entomophaga, 29, 445–450.

Krombein, K. V. (1967). Trap-nesting wasps and bees: Life histories, nests, and associates. Washington, DC: Smithsonian Press.

Kromp, B. (1990). Carabid beetles (Coleoptera: Carabidae) as bioindicators in biological and conventional farming in Austrian potato fields. Biological Fertility of Soils, 9, 182–187.

Kulman, H. M. (1974). Comparative ecology of North American Carabidae with special reference to biological control. Entomophaga, Mem. H.S., 7, 61–70.

Kumashiro, B. R., Lai, P. Y., Funasaki, G. Y., & Teramoto, K. K. (1983). Efficacy of *Nephaspis amnicola* and *Encarsia haitiensis* in controlling *Aleurodicus dispasus* in Hawaii. Proceedings of the Hawaiian Entomological Society, 24, 261–269.

Laffranque, J. P., & Canard, M. (1975). Biologie du predateur Aphidiphage *Boriomyia subnebulosa* (Stephens) (Neuroptera, Hemerobiidae):

Etudes au laboratoire et dans les conditions hivernaleks du sud-ouest de la France. Annales de Zoologie—Ecologie Animale, 7, 331–343.

Laidlaw, W. B. R. (1936). The brown lacewing flies (Hemerobiidae): Their importance as controls of *Adelges cooleyi* Gillette. Entomology Monthly Magazine, 72, 164–174.

Laing, J. E. (1968). Life history and life table of *Phytoseiulus persimilis* Athias-Henriot. Acarologia, 10, 578–588.

Laing, J. E., & Knop, N. F. (1983). Potential use of predaceous mites other than Phytoseiidae for biological control of orchard mites. In M. A. Hoy, G. L. Cunningham, & L. Knutson (Eds.), Biological control of pests by mites (Special Publication 3304, pp. 28–35). Berkeley: DANR, University of California.

Lamana, M. L., & Miller, J. C. (1996). Field observations on *Harmonia axyridis* Pallas (Coleoptera: Coccinellidae) in Oregon. Biological Control, 6, 232–237.

Lamb, R. J., & Wellington, W. G. (1974). Techniques for studying the behavior and ecology of the European earwig, *Forficula auricularia* (Dermaptera: Forficulidae). Canadian Entomologist, 106, 881–888.

Lambdin, P. L., & Baker, A. M. (1986). Evaluation of dewinged spined soldier bugs. *Podisus maculiventris* (Say), for longevity and suppression of the Mexican bean beetle, *Epilachna varivestis* Mulsant, on snapbeans. Journal of Entomological Science, 21, 263–266.

Landis, B. J. (1937). Insect hosts and nymphal development of *Podisus maculiventris* (Say) and *Perillus bioculatus* F. (Hemiptera: Pentatomidae). Ohio Journal of Science, 37, 252–259.

Langston, R. L., & Powell, J. A. (1975). The earwigs of California (Order Dermaptera). Bulletin of the California Insect Survey, 20, 1–25.

Lariviere, M. C., & Larochelle, A. (1989). *Picromerus bidens* (Heteroptera: Pentatomidae) in North America, with a world review of distribution and bionomics. Entomological News, 100, 133–146.

Larochelle, A. (1972). Notes on the food of Cychrini (Coleoptera: Carabidae). Great Lakes Entomologist, 5, 81–83.

Larochelle, A. (1990). The food of carabid beetles. (Suppl. 5). Canada: Fabreries.

Lattin, J. D. (1989). Bionomics of the Nabidae. Annual Review of Entomology, 34, 383–400.

Lattin, J. D., & Stonedahl, G. M. (1984). *Campyloneura virgula*, a predacious Miridae not previously recorded from the United States (Hemiptera). Pan-Pacific Entomologist, 60, 4–7.

Lattin, J. D., Asquith, A., & Booth, S. (1989). *Orius minutus* (Linnaeus) in North America (Hemiptera: Anthocoridae). Journal of the New York Entomological Society, 97, 409–416.

Lauenstein, G. (1980). Zum Suchverhalten von *Anthocoris nemorum* L. (Het.: Anthocoridae). Zeitschrift fuer Angewandte Entomologie, 89, 428–442.

Laurema, S., Husberg, G.-B., & Markkula, M. (1986). Composition and functions of the salivary gland of the larva of the aphidmidge *Aphidoletes aphidimyza*. In I. Hodek (Ed.), Ecology of Aphidophaga (pp. 113–118). Prague: Academia.

Lavallee, A. G., & Shaw, F. R. (1969). Preferences of golden-eye lacewing larvae for pea aphids, leafhopper, and plant bug nymphs and alfalfa weevil larvae. Journal of Economic Entomology, 62, 1228–1229.

Lawrence, J. F. (1991a). Key to the families and many subfamilies of Coleoptera larvae (Worldwide). In F. Stehr (Ed.), Immature insects (Vol. 2, pp. 184–298). Dubuque, IA: Kendall/Hunt.

Lawrence, J. F. (1991b). Nitidulidae (Cucujoidea). In F. W. Stehr (Ed.), Immature insects. (Vol. 2, pp. 456–460). Dubuque, IA: Kendall/Hunt.

Lawrence, P. O. (1974). Susceptibility of *Chrysopa rufilabris* to selected insecticides and miticides. Environmental Entomology, 3, 146–150.

Lawson, F. R., Rabb, R. L., Guthrie, F. E., & Bowery, T. G. (1961). Studies of an integrated control system for hornworms on tobacco. Journal of Economic Entomology, 54, 93–97.

Lawson, S., & Morgan, F. D. (1992). Rearing of two predators, *Thanasimas dubius* and *Temnochila virescens,* for the biological control of *Ips grandicollis* in Australia. Entomologia Experimentalis et Applicata, 65, 225–233.

LeCato, G. L., & Arbogast, R. T. (1979). Functional response of *Xylocoris flavipes* to Angoumois grain moth and influence of predation on regulation of laboratory populations. Journal of Economic Entomology, 72, 847–849.

LeCato, G. L., Collins, J. M., & Arbogast, R. T. (1977). Reduction of residual populations of stored-product insects by *Xylocoris flavipes* (Hemiptera: Anthocoridae). Journal of Kansas Entomological Society, 50, 84–88.

Lee, S. J., & Shih, C. I. T. (1982). Biology, predation and field cage release of *Chrysopa boninensis* Okamoto on *Pouraephala psylloptera* Crawford and *Corcyra cephalonica* Stanton. Journal of Agriculture and Forestry, 31, 129–144.

Leeper, J. R. (1976). A review of the Hawaiian Coccinellidae. Proceedings of the Hawaiian Entomological Society, 22, 279–306.

Leeper, J. R., & Beardsley, J. W., Jr. (1976). The biological control of *Psylla uncatoides* (Ferris & Klyver) (Homoptera: Psyllidae) on Hawaii. Proceedings of the Hawaiian Entomological Society, 22, 307–321.

Legaspi, J. C., & O'Neil, R. J. (1993). Life history of *Podius maculiventris* given low numbers of *Epilachna varivestis* as prey. Environmental Entomology, 22, 1192–1200.

Legaspi, J. C., Legaspi, B. C., Meagher, R. L., & Ciomperlik, M. A. (1996). Evaluation of *Serangium parcesetosum* (Coleoptera: Coccinellidae) as a biological control agent of the silverleaf whitefly (Homoptera: Aleyrodidae). Environmental Entomology, 25, 1421–1427.

Legner, E. F. (1978). Natural enemies imported in California for the biological control of face fly, *Musca autumnalis* DeGeer and horn fly, *Haematobia irritans* (L.) (pp. 77–99). Proceedings of the 46th Annual Conference on the California Mosquito Vector Control Association, January 29–February 1, 1978. Fresno, CA.

Legner, E. F., & Olton, G. S. (1968). The biological method and integrated control of house and stable flies in California. California Agriculture, 22, 2–4.

Leigh, T. F., & Gonzalez, D. (1976). Field cage evaluation of predators for control of *Lygus hesperus* Knight on cotton. Environmental Entomology, 5, 948–952.

LePelley, R. H. (1968). Pests of coffee. London: Longmans.

LeSage, L. (1991). Coccinellidae. In F. Stehr (Ed.), Immature insects (Vol. 2, pp. 485–494). Dubuque, IA: Kendall/Hunt.

LeSar, C. D., & Unzicker, J. D. (1978). Life history, habitats, and prey preferences of *Tetragnatha laboriosa* (Araneae: Tetragnathidae). Environmental Entomology, 7, 879–884.

Lester, D. G., & Morrill, W. L. (1989). Activity density of ground beetles (Coleoptera: Carabidae) in alfalfa and sainfoin. Journal of Agricultural Entomology, 6, 71–76.

Leston, D. (1973). The ant mosaic-tropical tree crops and the limiting of pests and diseases. Pest Articles and News Summaries, 19, 311–341.

Letourneau, D. K. (1990). Code of ant-plant mutualism broken by parasite. Science, 248, 215–217.

Letourneau, D. K., & Altieri, M. A. (1983). Abundance patterns of a predator, *Orius tristicclor* (Hemiptera: Anthocoridae), and its prey, *Frankliniella occidentalis* (Thysanoptera: Thripidae): Habitat attraction in polycultures versus monocultures. Environmental Entomology, 12, 1464–1469.

Lewis, T. (1973). Thrips, their biology, ecology and economic importance. London: Academic Press.

Lewis, W. J., Nordlund, D. A., Gross, H. R., Jr., Jones, R. L. & Jones, S. L. (1977). Kairomones and their use for the management of entomophagous insects. V. Moth scales as a stimulus for predation of *Heliothis zea* (Boddie) eggs by *Chrysoperla carnea* Stephens larvae. Journal of Chemical Ecology, 3, 483–487.

Li, L., Guo, M., Wu, H., & Chen, H. (1988). An artificial diet for *Eocanthecona furcellata*. Chinese Journal of Biological Control, 4, 45–49.

Li, T. S., Li, C. Z., & Wei, J. E. (1984). Biology of *Polistes antennalis* Perez and its use in the control of lepidopteran insects in cotton fields. Natural Enemy Insects, 6, 101–103.

Lin, Y., & Gu, D. (1987). On the spatial distribution pattern of adult of *Paederus fuscipes* in the sugarcane fields and its applications. Natural Enemy Insects, 9, 63–69.

Lindgren, B. S. (1993). Attraction of douglas fir beetle, spruce beetle and a bark beetle predator (Coleoptera: Scolytidae and Cleridae) to enantiomers of frontalin. Journal of Entomological Society of British Columbia, 89, 13–17.

Lindquist, E. E. (1983). Some thoughts on the potential use of mites in biological control, including a modified concept of "parasitoids." In M. A. Hoy, G. L. Cunningham, & L. Knutson (Eds.), Biological control of pests by mites (Special Publication 3304, pp. 12–20). Berkeley: DANR, University of California.

Lindquist E. E., & Evans, G. O. (1965). Taxonomic concepts in the Ascidae, with a modified setal nomenclautre for the isiosoma of the Gamasina (Acarina: Mesostigmata). Memoirs of the Entomological Society of Canada, 47, 1–64.

Lingren, P. D., Ridgway, R. L., & Jones, S. L. (1968). Consumption by several common arthropod predators of eggs and larvae of two *Heliothis* species that attack cotton. Annals of the Entomological Society of America, 61, 613–618.

Liquido, N. J., & Nishida, T. (1985a). Population parameters of *Cyrtorhinus lividipennis* Reuter (Heteroptera: Miridae) reared on eggs of natural and factitious prey. Proceedings of the Hawaiian Entomological Society, 25, 87–92.

Liquido, N. J., & Nishida, T. (1985b). Observations on some aspects of the biology of *Cyrtorhinus lividipennis* Reuter (Heteroptera: Miridae). Proceedings of the Hawaiian Entomological Society, 25, 95–101.

Liquido, N. J., & Nishida, T. (1985c). Variation in number of instars, longevity, and fecundity of *Cyrtorhinus lividipennis* Reuter (Hemiptera: Miridae). Annals of the Entomological Society of America, 78, 459–463.

Lo, K. C., Lee, W. T., Wu, T. K., & Ho, C. C. (1990). Use of predators for controlling spider mites (Acarina, Tetranychidae) in Taiwan, China. Tukuba-gun, Japan: National Agricultural Research Center.

Löhr, B. (1992). The pugnacious ant, *Anoplolepus custodiens* (Hymenoptera: Formicidae) and its beneficial effect on coconut production in Tanzania. Bulletin of Entomological Research, 82, 213–218.

Löhr, B. Santos, & Varela, A. M. (1990). Exploration for natural enemies of cassava mealybug *Phenacoccus manihoti* (Homoptera: Pseudococcidae) in South America for the biological control of this introduced pest in Africa. Bulletin of Entomological Research, 80, 417–426.

Lopez, J. D., Jr., House, V. S., & Morrison, R. K. (1987). Suitability of frozen *Sitotroga ceraelella* (Olivier) eggs for temporary rearing of *Geocoris punctipes*. Southwestern Entomologist, 12, 223–228.

Lopez, J. D., Jr., Ridgway, R. L., & Pinnell, R. E. (1976). Comparative efficacy of four insect predators of the bollworm and tobacco budworm. Environmental Entomology, 5, 1160–1164.

Lord, F. T. (1971). Laboratory tests to compare the predatory value of six mirid species in each stage of development against the winter eggs of the European red mite. *Panonychus ulmi* (Acari: Tetranychidae). Canadian Entomologist, 103, 1663–1669.

Lorenzato, D., Grellmann, E. O., Chouene, E. C., & Meyer-Cachapuz, L. M. (1986). Flutuacao populacional de acaros fitoagos eseus predadores associados a cultura da macieira (*Malus domestica* Bork) e efeitos dos controles quimico e biologico. Agronomia–Sulriograndense, 22, 215–242.

Lu, Z., & Zhu, J. (1984). Notes on predaceous rove beetles from cropland in Jiangsu province. Natural Enemy Insects, 6, 20–27.

Luck, R. F., Shepard, B. M., & Kenmore, P. E. (1988). Experimental methods for evaluation of arthropod natural enemies. Annual Review of Entomology, 33, 367–392.

Luff, M. L. (1973). The annual activity pattern and life cycle of *Pterostichus madidus* (F.) (Col.: Carabidae). Entomologica Scandinavica, 4, 259–273.

Luff, M. L. (1983). The potential of predators for pest control. Agricultural Ecosystems in the Environment, 10, 159–181.

Luff, M. L. (1987). Biology of polyphagous ground beetles in agriculture. Agricultural Zoology Review, 2, 237–278.

Luo, X., Zhuo, W., & Wang, Y. (1989). Preliminary study on predatory effect of *Paederus fuscipes* Curtis (Coleop.: Staphylinidae) Natural Enemy Insects, 11 (10), 12–19.

Lyon, J. P. (1979). Lachers experimentaux de chrysopes et d'Hymenopteres parasites sur puceravis en series d'aubergines. Annales de Zoologie—Ecologie Animale, 11, 51–65.

Lyon, J. P. (1986). Use of aphidophagous and polyphagous beneficial insects in greenhouses. In I. Hodek (Ed.), Ecology of Aphidophaga (pp. 471–474). Dordrecht: Dr. W. Junk.

Lys, J. A., & Nentwig, W. (1992). Augmentation of beneficial arthropods by strip-management. IV. Surface activity, movements and activity density of abundant carabid beetles in a cereal field. Oecologia (Berlin), 92, 373–382.

Ma, W.-L., & Laing, J. W. (1973). Biology, potential for increase and prey consumption of *Amblyseius chilenensis* (Dosse) (Acarina: Phytoseiidae). Entomophaga, 18, 47–60.

MacLellan, C. R. (1962). Mortality of codling moth eggs and young larvae in an integrated control orchard. Canadian Entomologist, 94, 655–666.

Madge, D. S., & Buxton, J. H. (1976). The evaluation of the European earwig (*Forficula auricularia*) as a predator of the damson-hop aphid (*Phorodon humali.* II. Choice of prey. Entomologia Experimentalis et Applicata, 19, 221–226.

Maelzer, D. A. (1977). The biology and main causes of changes in numbers of the rose aphid, *Macrosiphum rosae* (L.) on cultivated roses in South Australia. Australian Journal of Zoology, 25, 269–284.

Madsen, H. F., & Wong, T. T. Y. (1964). Effects of predators on control of pear psylla. California Agriculture, 18, 2–3.

Maharaj, R., Appleton, C. C., & Miller, R. M. (1992). Snail predation by larvae of *Sepedon scapularis* Adams (Diptera: Sciomyzidae), a potential biocontrol agent of snail intermediate hosts of schistosomiasis in South Africa. Medical Veterinary Entomology, 6, 183–187.

Makarenko, G. N. (1975). Quality effect of the *Sitotroga cerealella* eggs on the breeding of golden-eyed fly. In G. V. Guseva & V. A. Scheretilnikova (Eds.), Biological methods for plant protection (Vol. 4, pp. 298–308). Proceedings, All-Union Scientific Research Institute of Plant Protection, Leningrad (in Russian).

Malausa, J. C., Drescher, J., & Franco, E. (1987). Perspectives for the use of a predaceous bug *Macrolophus caliginosus* Wagner (Heteroptera: Miridae) on glasshouse-crops. IOBC WPRS Bulletin, 10, 106–107.

Malausa, J. C., Raviglione, M. C., Boggio, F., Bovo, P. G., & Cossutta, F. (1983). I. Carabus olympiae sella dell'alta Valle Sessera. France: Pro Natura Biellese.

Mani, M. (1990). Rid the grape-vine of mealybug. Indian Horticulture, 35, 28–29.

Mani, M., & Krishnamoorthy, A. (1989). Feeding potential and development of green lacewing *Mallada boninensis* (Okamoto) on the grape mealybug, *Maconellicoccus hirsutus* (Green). Entomon, 14, 19–20.

Manley, G. V. (1977). *Paederus fuscipes* (Col.: Staphylinidae), a predator of rice fields in West Malaysia. Entomophaga, 22, 47–59.

Manley, G. V. (1982). Biology and life history of the rice field predator *Andrallus spinidens* F. (Hemiptera: Pentatomidae). Entomological News, 93, 19–24.

Manley, G. V. (1985). The predation status of *Conocephalus longipennis* (Orthoptera: Tettigoniidae) in rice fields of west Malaysia. Entomological News, 96, 167–170.

Mansell, M. W., & Aspöck, H. (Eds.). (1990). Advance in neuropterology.

Proceedings, Third International Symposium on Neuropterology. Pretoria: Department of Agricultural Development.

Mansour, F., & Heimbach, U. (1993). Evaluation of lycosid, micryphantid and linyphid spiders as predators of *Rhopalosiphum padi* (Hom.: Aphididae) and their functional response to prey density-laboratory experiments. Entomophaga, 38, 79–87.

Mansour, F., Rosen, D., Shulov, A., & Plaut, H. N. (1980a). Evaluation of spiders as biological control agents of *Spodoptera littoralis* larvae on apple in Israel. Acta Oecol., Oecol. Appl. 1, 225–232.

Mansour, F., Rosen, D., & Shulov, A. (1980b). Functional response of the spider *Chiracanthium mildei* (Arachnida: Clubionidae) to prey density. Entomophaga, 25, 313–316.

Mansour, F., Rosen, D., & Shulov, A. (1981). Disturbing effect of a spider on larval aggregations of *Spodoptera littoralis*. Entomologia Experimentalis et Applicata, 29, 234–237.

Mansour, F., Wysoki, M., & Whitcomb, W. H. (1985). Spiders inhabiting avocado orchards and their role as natural enemies of *Boarmia selenaria* Schiff (Lepidoptera: Geometridae) larvae in Israel. Acta Oecol., Oecol. Appl., 6, 315–321.

Mansour, M. H. (1975). The role of plants as a factor affecting oviposition by *Aphidoletes aphidimyza* (Dipt.: Cecidomyiidae). Entomologia Experimentalis et Applicata, 18, 173–179.

Manti, I. (1991). Mirid predation on brown planthopper eggs. International Rice Research Newsletter, 16, 24–25.

Marcano-Brito, R. (1980). Factors affecting the distribution and abundance of 3 species of *Tetranychus* spider mites on cotton and the effect of their damage on transpiration and photosynthesis. Unpublished doctoral dissertation, University of California, Riverside.

Maredia, K. M., Gage, S. H., Landis, D. A., & Scriber, J. M. (1992). Habitat use patterns by the seven-spotted lady beetle (Coleoptera: Coccinellidae) in a diverse agricultural landscape. Biological Control, 2, 159–165.

Marin, L. R. (1983). Biologia capacidad predacion de *Lindorus lophanthae* (Blais.) (Col. Coccinellidae) predator de *Pinnaspis aspidistrae* (Sing.) (Hom.: Diaspididae). Revista Peruana de Entomologia, 26, 63–66.

Markkula, M., & Tiittanen, K. (1976). A method for mass rearing of *Aphidoletes aphidimyza*. IOBC WPRS Bulletin, 4, 183–184.

Markkula, M., & Tiittanen, K. (1980). Biological control of pests in glasshouses in Finland. The situation today and in the future. IOBC WPRS Bulletin, 3, 127–133.

Markkula, M., & Tiittanen, K. (1985). Biology of the midge *Aphidoletes* and its potential for biological control. In N. W. Hussey & N. Scopes (Eds.), Biological control pest control, the glasshouse experience (pp. 74–81). Ithaca, NY: Cornell University Press.

Markkula, M., Tiittanen, K., Hämäläinen, M., & Forsberg, A. (1979). The aphid midge *Aphidoletes aphidimyza* (Diptera, Cecidomyiidae) and its use in biological control of aphids. Annales Entomologici Fennici, 45, 89–98.

Marks, R. J. (1977). Laboratory studies of plant searching behaviour by *Coccinella septempunctata* L. larvae. Bulletin of Entomological Research, 67, 235–241.

Marston, N. L., Schmidt, G. T., Bierer, K. D., & Dickerson, W. A. (1978). Reaction of five species of soybean caterpillars to attack by the predator, *Podisus maculiventris*. Environmental, Entomology, 7, 53–56.

Marucci, P. E. (1955). Notes on the predatory habits and life cycle of two Hawaiian earwigs. Proceedings of the Hawaiian Entomological Society, 15, 565–569.

Marucci, P. E., & Clancy, D. W. (1952). The biology and laboratory culture of *Thyreocephalus albertisi* (Fauvel) in Hawaii. Proceedings of the Hawaiian Entomological Society, 14, 525–532.

Martel, P., & Sharma, M. L. (1970). Quelques notes sur *Microweisra marginata* (LeConte), (Coleoptera: Coccinellidae), predateur de la cochenille du pin, *Phenacapsis pinifoliae* (Fitch). Annales de la Societe Entomologique de France, 15, 61–65.

Mathen, K., & Kurian, C. (1972). Description, life-history and habits of *Stethoconus praefectus* (Distant) (Heteroptera: Miridae), predacious on *Stephanitis typicus* Distant (Heteroptera: Tingidae), a pest of coconut palm. Indian Journal of Agricultural Science, 42, 255–262.

Matheny, E. L., Jr. (1981). Contrasting feeding habits of pest mole cricket species. Journal of Economic Entomology, 74, 444–445.

Matsumoto, B. M., & Nishida, T. (1966). Predator-prey investigations on the taro leafhopper and its egg predator. Hawaii Agricultural Experiment Station Technical Bulletin, 64, 1–32.

Matsura, T., & K. Nakamura. (1981). Effects of prey density on mutual interference among nymphs of a mantis, *Paratenodera angustipennis* (S.). Japanese Journal of Ecology, 31, 221–223.

Mayne, R., & Breny, R. (1948). *Picromerus bidens* L.: Morphologie Biologie. Determination de sa valeur d'utilisation dans la lutte biologique contre le doryphore de la pomme de terre—La valeur economique antidoryphorique des Asopines indigenes belges. Parasitica, 4 (4), 189–224.

Mbata, K. S., Hart, E. R., & Lewis, R. E. (1987). Reproductive behavior in *Zelus renardi* Kolenti 1857 (Hemiptera; Reduviidae). Iowa State Journal of Research, 62, 261–266.

McAlpine, J. F. (1960). A new speices of *Leucopis (Leucopella)* from Chile and a key to the world genera and subgenera of Chamaemyiidae (Diptera). Canadian Entomologist, 92, 51–58.

McAlpine, J. F. (1971). A revision of the subgenus *Neoleucopis* (Diptera: Chamaemyiidae). Canadian Entomologist, 103, 1851–1874.

McAlpine, J. F., & Tanasijtschuk, V. N. (1972). Identity of *Leucopis argenticollis* and a new species (Diptera: Chamaemyiidae). Canadian Entomologist, 104, 1865–1875.

McCaffrey, J. P., & Horsburgh, R. L. (1986a). Functional response of *Orius insidiosus* (Hemiptera: Anthocoridae) to European red mite, *Panonychus ulmi* (Acari: Tetranychidae), at different constant temperatures. Environmental Entomology, 15, 532–535.

McCaffrey, J. P., & Horsburgh, R. L. (1986b). Biology of *Orius insidiosus,* a predator in Virginia apple orchards. Environmental Entomology, 15, 984–988.

McCarty, M. T., Ridgway, R. L., & Turnipseed, S. G. (1980). Identification of predaceous arthropods in soybeans by using autoradiography. Environmental Entomology 9, 199–203.

McClure, M. S. (1987). Potential of the Asian predator, *Harmonia axyridis* Pallas (Coleoptera: Coccinellidae), to control *Matsucoccus resinosae* Bean & Godwin (Homoptera: Margarodidae) in the United States. Environmental Entomology, 16, 224–230.

McDaniel, S. G., & Sterling, W. L. (1982). Predation of *Heliothis virescens* (F.) eggs on cotton in east Texas. Environmental Entomology, 11, 60–66.

McEwen, P. K. (1996). Viability of green lacewing *Chrysoperla carnea* Steph. (Neuropt., Chrysopidae) eggs stored in potential spray media, and subsequent effects on survival of first instar larvae. Journal of Applied Entomology, 120, 171–173.

McEwen, P. K., Jervis, M. A., & Kidd, N. A. C. (1993a). Influence of artificial honeydew on larval development and survival in *Chrysoperla carnea* (Neur., Chrysopidae). Entomophaga, 38, 241–244.

McEwen, P. K., Clow, S., Jervis, M. A., Kidd, N. A. C. (1993b). Alteration in searching behaviour of adult female green lacewings *Chrysoperla carnea* (Neur.: Chrysopidae) following contact with honeydew of the black scale *Saissetia oleae* (Hom.: Coccidae) and solutions containing acidhydrolysed L-tryptophan. Entomophaga, 38, 347–354.

McGregor, E. A., & McDonough, F. L. (1917). The red spider on cotton. (USDA Bulletin No. 416, pp. 1–72). Washington, DC: U.S. Department of Agriculture.

McGugan, B. M., & Coppel, H. C. (1962). Biological control of forest insects, 1910–1958. CIBC Technical Communications, 2, 35–216.

McIntyre, T. (1960). Natural factors control the pine tortoise scale in Northeast. Journal of Economic Entomology, 53, 325.

McKenzie, H. L. (1932). The biology and feeding habits of *Hyperaspis lateralis* Mulsant (Coleoptera-Coccinellidae). University of California Publications in Entomology, 6, 9–20.

McLaughlin, R. E., & Dame, D. A. (1989). Rearing *Dictya floridensis* (Diptera: Sciomyzidae) in a continuously producing colony and evaluation of larvae food sources. Journal of Medical Entomology, 26, 522–527.

McLean, I. F. G. (1992). Behaviour of larval and adult *Leucopis* (Diptera: Chamaemyiidae). British Journal of Entomology and Natural History, 5, 35–36.

McMahan, E. (1983). Adaptations, feeding preferences, and biometrics of a termite-baiting assassin bug (Hemiptera: Reduviidae). Annals of Entomological Society of America, 76, 483–486.

McMullen, R. D., & Jong, C. (1967). New records and discussion of predators of the pear psylla, *Psylla pyricola* Forster, in British Columbia. Journal of Entomological Society of British Columbia, 64, 35–40.

McMullen, R. D., & Jong, C. (1970). The biology and influence of pesticides on *Campylomma verbasci* (Heteroptera: Miridae). Canadian Entomologists, 102, 1390–1394.

McMurtry, J. A. (1961). Current research on biological control of insect and mite pests. California Avocado Society Yearbook, 45, 104–106.

McMurtry, J. A. (1982). The use of phytoseiids for biological control: Progress and future prospects. In M. A. Hoy (Ed.), Recent advances in knowledge of the Phytoseiidae (Publication 3284, pp. 23–48). Berkeley: University of California Press.

McMurtry, J. A. (1992). Dynamics and potential impact of 'generalist' phytoseiids in agroecosystems and possibilities for establishment of exotic species. Experimental and Applied Acarology (Northwood), 14, 371–382.

McMurtry, J. A., & Croft, B. A. (1987). Life-styles of phytoseiid mites and their roles in biological control. Annual Review of Entomology, 42, 291–321.

McMurtry, J. A., & Johnson, H. G. (1966). An ecological study of the spider mite *Oligonychus punicae* (Hirst) and its natural enemies. Hilgardi, 37, 363–402.

McMurtry, J. A., & Rodriquez, J. G. (1987). Nutritional ecology of phytoseiid mites. In F. Slansky & J. G. Rodriquez (Eds.), Nutritional ecology of insects, mites and spiders (pp. 609–644). New York: John Wiley & Sons.

McMurtry, J. A., Badii, M. H., & Johnson, H. G. (1984). The broad mite, *Polyphagotarsonemus latus,* as a potential prey for phytoseiid mites in California. Entomophaga, 29, 83–86.

McMurtry, J. A., Huffacker, C. B., & van de Vrie, M. (1970). Ecology of tetranychid mites and their natural enemies: A review. I. Tetranychid enemies: Their biological characters and the impact of spray practices. Hilgardia, 40, 331–390.

McMurtry, J. A., Johnson, H. C., & Scriven, G. T. (1969). Experiments to determine effects of mass releases of *Stethorus picipes* on the level of infestation of the avocado brown mite. Journal of Economic Entomology, 62, 1216–1221.

McPherson, J. E. (1980). A list of prey species of *Podisus maculiventris* (Hemiptera: Pentatomidae). Great Lakes Entomologist, 13, 17–24.

McPherson, R. M., Smith, J. C., & Allen, W. A. (1982). Incidence of arthropod predators in different soybean cropping systems. Environmental Entomology, 11, 685–689.

Mead, A. R. (1979). Economic malacology with particular reference to *Achatina fulica*. In V. Fetter & J. Peake (Eds.), Pulmonates (Vol. 2B, pp. 1–150). London: Academic Press.

Meadow, R. H., Kelly, W. C., & Shelton, A. M. (1985). Evolution of *Aphidoletes aphidimyza* (Dip.: Cecidomyiidae) for control of *Myzus persicae* (Hom.: Aphididae) in greenhouse and field experiments in the United States. Entomophaga, 30, 385–392.

Meinander, M. (1972). A revision of the family Coniopterygidae (Planipennia). Acta Zoologica Fennica, 136, 1–357.

Meinander, M. (1986). The Coniopterygidae of America (Neuroptera). In J. Gepp, H. Aspöck, & H. Hölzel (Eds.), Recent research in neuropterology (pp. 31–43). Lanach, Austria: J. Gepp.

Meinander, M. (1990). The Coniopterygidae (Neuroptera: Planipennia). A check-list of the species of the World, descriptions of new species and other new data. Acta Zoologica Fennica, 189, 1–95.

Mendel, Z., & Blumberg, D. (1991). Colonization trials with *Cryptochaetum iceryae* and *Rodolia iceryae* for improved biological control of *Icerya purchasi* in Israel. Biological Control, 1, 68–74.

Mensah, R. K., & Madden, J. L. (1994). Conservation of two predator species for biological control of *Chrysophtharta bimaculata* (Col.: Chrysomelidae) in Tasmanian forests. Entomophaga, 39, 71–83.

Merlin, J., Lemaitre, O., & Grégoire, J. C. (1996). Chemical cues produced by conspecific larvae deter oviposition by the coccidophagous ladybird beetle, *Cryptolaemus montrouzieri*. Entomologia Experimentalis et Applicata, 79, 147–151.

Meyerdirk, D. E. (1983). Biology of *Diomus flavifrons* (Blackburn)(Coleoptera: Coccinellidae), a citrus mealybug predator. Environmental Entomology, 12, 1275–1277.

Michael, P. (1995). Biological control of redlegged earth mite and lucerne flea by the predators *Anystis wallacei* and *Neomolgus capillatus*. Plant Protection Quarterly, 10, 55–57.

Miermont, Y., & Canard, M. (1975). Biologie da predateur aphidiphage *Eumicromus angulatus* (Neur.: Hemerobiidae): Etudes au laboratoire et observations dans le sud-ouest de la France. Entomophaga, 20, 179–191.

Milbrath, L. R., Tauber, M. J., & Tauber, C. A. (1993). Prey specificity in *Chrysopa*: On interspecific comparison of larval feeding and defensive behavior. Ecology, 74, 1384–1393.

Miller, D., Clark, A. F. & Dumbleton, L. D. (1936). Biological control of noxious insects and weeds in New Zealand. New Zealand Journal of Science and Technology, 8, 579–593.

Miller, G. L., & Cave, R. D. (1987). Bionomics of *Micromus posticus* (Walker) (Neuroptera: Hemerobiidae) with descriptions of immature stages. Proceedings of the Entomological Society of Washington, 89, 776–789.

Miller, K. V., & Williams, R. N. (1983). Biology and host preference of *Atheta coriaria* (Coleoptera: Staphylinidae), an egg predator of Nitidulidae and Muscidae. Annals of the Entomological Society of America, 76, 158–161.

Miller, M. C., Moser, J. C., McGregor, M., Grégoire, J.-C., Baisier, M., & Dahlsten, D. L. (1987). Potential for biological control of native North American *Dendroctonus* beetles (Coleoptera: Scolytidae). Annals of the Entomological Society of America, 80, 417–428.

Miller, N. C. E. (1971). The biology of the Heteroptera. United Kingdom: E. W. Classey.

Mills, N. J. (1981). Essential and alternative foods for some British Coccinellidae (Coleoptera). Entomologist's Gazette, 32, 197–202.

Mills, N. J. (1982a). Satiation and the functional response: A test of a new model. Ecology and Entomology, 7, 305–315.

Mills, N. J. (1982b). Voracity, cannibalism and coccinellid predation. Annals of Applied Biology, 101, 144–148.

Mills, N. J. (1983). The natural enemies of scolytids infesting conifer bark in Europe in relation to the biological control of *Dendroctonus* spp. in Canada. Biocontrol News Information, 4, 305–328.

Mills, N. J. (1985). Some observations on the role of predation in the natural regulation of *Ips typographus* populations. Zeitschrift fuer Angewandte Entomologie, 99, 209–215.

Mills, N. J. (1990). Biological control of forest aphid pests in Africa. Bulletin of Entomological Research, 80, 31–36.

Mills, N. J. (1991). Biological control, a century of pest management. Bulletin of Entomological Research, 80, 359–362.

Mills, N. J., & Getz, W. M. (1996). Modelling the biological control of

insect pests: A review of host-parasitoid models. Ecology Modeleling, 92, 121–143.

Mills, N. J., & Schlup, J. (1989). The natural enemies of *Ips typographus* in central Europe: Impact and potential use in biological control. In D. L. Kulhavy & M. C. Miller (Eds.), Potential for biological control of *Dendroctonus* and *Ips* bark beetles (pp. 131–146). Texas: Austin State University.

Milne, W. M., & Bishop, A. L. (1987). The role of predators and parasites in the natural regulation of lucerne aphids in eastern Australia. Journal of Applied Ecology, 24, 893–905.

Miszezak, M., & Niemczyk, E. (1978). Green lacewing, *Chrysoperla carnea* (Neuroptera: Chrysopidae), as a predator of European mite, *Panonychus ulmi* Koch, on apple trees. II. The effectiveness of *Chrysoperla carnea* larvae in control of *Panonychus ulmi* Koch. Fruit Science Report, 4, 21–31 (Skierniewice, Poland).

Mitchell, R. G., & Wright, K. H. (1967). Foreign predator introductions for control of the balsam woolly aphid in Pacific Northwest. Journal of Economic Entomology, 60, 140–147.

Mittelstaedt, H. (1957). Prey capture in mantids: Recent advances in invertebrate physiology (pp. 51–71). Eugene: University of Oregon Publishing.

Mituda, E. C., & Calilung, V. J. (1989). Biology of *Orius tantillus* (Motshulsky) (Hemiptera: Anthocoridae) and its predatory capacity against *Thrips palmi* Karny (Thysanoptera: Thripidae) on watermelon. Philippine Agriculture, 72, 165–184.

Miyanoshita, A., & Kawan, S. (1992). Influence of predation by *Mallada boninensis* (Okamoto) (Neuroptera: Chrysopidae) and autumn movement of female adults on survival of *Cerplastes japonicus* Green (Homoptera: Coccidae): A model experiment with cage. Japanese Journal of Applied Entomology and Zoology, 36, 196–199.

Mizell, R. F., & Nebeker, T. E. (1982a). Preference and oviposition rates of adult *Thanasimus dubius* on three prey species. Environmental Entomology, 11, 139–143.

Mizell, R. F., & Nebeker, T. E. (1982b). Response of the clerid predator *Thanasimus dubius* (F.) to bark beetle pheromones and tree volatiles in a wind tunnel. Journal of Chemical Ecology, 10, 177–187.

Moeck, H. A., & Safranyik, L. (1984). Assessment of predator and parasitoid control of bark beetles. (Information Report, BC-X-248). Canadian Forestry Service.

Mols, P. J. M. (1987). Hunger in relation to searching behaviour, predation and egg production of the carabid beetle *Pterostichus coerulescens* L. Results of simulation. Acta Phytopathologica et Entomologica Hungarica, 22, 187–206.

Monserrat, V. J. (1990). A systematic checklist of the Hemerobiidae of the World (Insecta: Neuroptera). In M. W. Mansell & H. Aspöck (Eds.), Advances in neuropterology. Proceedings of the 3rd International Symposium on Neuropterology (pp. 215–262). Pretoria: Department of Agricultural Development.

Monserrat, V. J., & Marin, F. (1996). Plant substrate specificity of Iberian Hemerobiidae (Insecta: Neuroptera). Journal of Natural History, 30, 775–787.

Mook, L. J., & Davies, D. M. (1966). The European praying mantis (*Mantis religiosa* L.) as a predator of the red-legged grasshopper (*Melanoplus femurrubrum* (DeGeer). Canadian Entomologist, 98, 913–918.

Moore, I., & Legner, E. F. (1971). Host records of parasitic staphylinids of the genus *Aleochara* in America (Coleoptera: Staphylinidae). Annals of Entomological Society of America, 64, 1184–1185.

Moore, I., Legner, E. F., & Badgley, M. E. (1975). Description of the developmental stages of the mite predator, *Oligota oviformis* Casey, with notes on the asemeterium and its glands (Coleoptera: Staphylinidae). Psyche, 82, 181–188.

Morales, J., & Burandt, C. L., Jr. (1985). Interactions between *Cycloneda sanguinea* and the brown citrus aphid: Adult feeding and larval mortality. Environmental Entomology, 14, 520–522.

Morgan, P. B., Patterson, R. S., & Weidhaas, D. E. (1983). A life history study of *Carcinops pumilio* Erichson (Coleoptera: Histeridae). Journal of Georgia Entomological Society, 18, 353–358.

Morimoto, R. (1960). *Polistes* wasps as natural enemies of agricultural and forest pests. II. Scientific Bulletin of the Faculty of Agriculture, Kyushu University, 18, 117–132.

Morrill, W. L. (1975). An unusual predator of the Florida harvester ant. Journal of Georgia Entomological Society, 10, 50–51.

Morris, R. F. (1963). The effect of predator age and prey defense on the functional response of *Podisus maculiventris* Say to the density of *Hyphantria cuniea* Drury. Canadian Entomologist, 95, 1009–1020.

Morrison, R. K., House, V. S., & Ridgway, R. L. (1975). Improved rearing unit for larvae of a common green lacewing. Journal of Economic Entomology, 68, 821–822.

Morsy, T. A., Arafa, M. A. S., Younis, T. A., & Mahmoud, I. A. (1996). Studies on *Paederus alfierii* Koch (Coleoptera: Staphylinidae) with special reference to the medical importance. Journal of Egyptian Society of Parasitology, 26, 337–351.

Moser, J. C. (1989). Inoculative release of an exotic predator for the biological control of the black turpentine beetle. In D. L. Kulhavy & M. C. Miller (Eds.), Potential for biological control of *Dendroctonus* and *Ips* bark beetles (pp. 189–200). Nacogdoches, TX: S. F. Austin State University.

Mowery, P. D., Asquith, D., & Bode, W. M. (1975). Computer simulation for predicting the number of *Stethorus punctum* needed to control the European red mite in Pennsylvania apple trees. Journal of Economic Entomology, 68, 250–254.

Muir, R. C. (1966). The effect of sprays on the fauna of apple trees. IV. The recolonization of orchards plots by the predatory mirid *Blepharidopterus angulatus* and its effect on populations of *Panonychus ulmi*. Journal of Applied Ecology, 3, 269–276.

Mukerji, M. K., & LeRoux, E. J. (1969a). The effect of predator age on the functional response of *Podisus maculiventus* to the prey size of *Galleria mellonella*. Canadian Entomologist, 101, 314–327.

Mukerji, M. K., & LeRoux, E. J. (1969b). A quantitative study of food consumption and growth of *Podisus maculiventris* (Hemiptera: Pentatomidae). Canadian Entomologist, 101, 387–403.

Mukerji, M. K., & LeRoux, E. J. (1969c). Study on energetics of *Podisus maculiventris* (Hemiptera: Pentatomidae). Canadian Entomologist, 101, 449–460.

Mukhopadhyay, A., & Sannigrahi, S. (1993). Rearing success of a polyphagous predator *Geocoris ochropterus* (Hem.: Lygaeidae) on preserved ant pupae of *Oecophylla smaragdina*. Entomophaga, 38, 219–224.

Muma, M. H. (1956). Life cycles of four species of ladybeetles. Florida Entomologist, 39, 115–118.

Muma, M. H. (1967). Biological notes on *Coniopteryx vicina* (Neuroptera: Coniopterygidae). Florida Entomologist, 50, 285–293.

Muma, M. H., Denmark, H. A., & DeLeon, D. (1970). Phytoseiidae of Florida. Arthropods of Florida and neighboring land areas, volume 6. Division of Plant Industry, Gainesville, FL.

Mundinger, F. G. (1922). The life history of two species of Nabidae (Hemip.: Heterop.) (Technical Publication 16). New York State College of Forestry, 22, 149–167.

Muniappan, R., & Chada, H. L. (1970). Biological control of the greenbug by the spider *Phidippus audax*. Journal of Economic Entomology, 63, 1712.

Muona, J. (1984). Review of Palearctic Aleocharinae also occurring in North America (Coleoptera: Staphylinidae). Entomoligica Scandinavica, 15, 227–231.

Musa, M. S., & Butler, G. B., Jr. (1967). The stages of *Spanogonicus albofasciatus* and their development. Journal Kansas Entomological Society, 40, 596–600.

Myers, J. G. (1931). A preliminary report on the investigation into the biological control of West Indian insect pests. London: His Majesty's Stationery Office.

Nadgauda, D., & Pitre, H. N. (1986). Effects of temperature on feeding, development, fecundity, and longevity of *Nabis roseipennis* (Hemiptera: Nabidae) fed tobacco budworm (Lepidoptera: Noctuidae) larvae and tarnished plant bug (Hemiptera: Miridae) nymphs. Environmental Entomology, 15, 536–539.

Nadgauda, D., & Pitre, H. N. (1987). Feeding and development of *Nabis roseipennis* (Hemiptera: Nabidae) on *Heliothis virescens* (Lepidoptera: Noctuidae) larvae. Journal of Entomological Science, 21, 45–50.

Nagel, W. P., & Fitzgerald, T. D. (1975). *Medetera aldrichii* larval feeding behavior and prey consumption (Dipt.: Dolichopodidae). Entomophaga, 20, 121–127.

Nakahara, W. (1965). Contributions to the knowledge of the Hemerobiidae of western North America (Neuroptera). Proceedings of United States National Museum, 116, 205–222.

Nakamuta, K. (1984). Visual orientation of a lady beetle, *Coccinella septempunctata* L. (Coleoptera: Coccinellidae), toward its prey. Applied Entomology and Zoology, 19, 82–86.

Nakamuta, K. (1986). Area-concentrated search in the lady beetle, *Coccinella septempunctata:* Eliciting cue and giving up time. In I. Hodek (Ed.), Ecology of Aphidophaga (Vol. 2, pp. 83–88). The Hague: Dr. W. Junk.

Nakamuta, K., & Saito, T. (1985). Recognition of aphid prey by the lady beetle, *Coccinella septempunctata bruckii* Mulsant (Coleoptera: Coccinellidae). Applied Entomology and Zoology, 20, 479–483.

Nakasuji, F., Yamanaka, K., & Kiritani, K. (1973). The disturbing effect of micryphantid spiders on the larval aggregations of the tobacco cutworm *Spodoptera litura*. Kontyu, 41, 220–227.

Napompeth, B. (1982). Biological control research and development in Thailand. (pp. 301–323). Proceedings, International Conference of Plant Protection in the Tropics, 1982, Malaysia.

Naranjo, S. E., & Stimac, J. L. (1985). Development, survival, and reproduction of *Geocoris punctipes* (Hemiptera: Lygaeidae): Effects of plant feeding on soybean and associated weeds. Environmental Entomology, 14, 523–530.

Naranjo, S. E., & Stimac, J. L. (1987). Plant influences of predation and oviposition by *Geocoris punctipes* (Hemiptera: Lygaeidae) in soybean. Environmental Entomology, 167, 182–189.

Naranjo, S. E., & Gibson, R. L., & Walgenbach, D. D. (1990). Development, survival and reproduction of *Scymnus frontalis* (Coleoptera: Coccinellidae), an imported predator of Russian wheat aphid, at four fluctuating temperatures. Annals of Entomological Society of America, 83, 527–531.

Neal, J. W., Jr., & Haldemann, R. L. (1992). Regulation of seasonal egg hatch by plant phenology in *Stethoconus japonicus* (Heteroptera: Miridae), a specialist predator of *Stephanitis pyrioides* (Heteroptera: Tingidae). Environmental Entomology, 21, 793–798.

Neal, J. W., Jr., Haldemann, R. H., & Henry, T. J. (1991). Biological control potential of a Japanese plant bug *Stethoconus japonicus* (Heteroptera: Miridae), an adventive predator of the azalea lace bug (Heteroptera: Tingidae). Annals of the Entomological Society of America, 84, 287–293.

Neal, T. M., Greene, G. L., Mead, F. W. & Whitecomb, W. H. (1972). *Spanogonicus albofasciatus* (Hemiptera: Miridae): A predator in Florida soybeans. Florida Entomologist, 55, 247–250.

Nechols, J. R., Tauber, M. J., & Tauber, C. A. (1987). Geographical variability in ecophysiological traits controlling dormancy in *Chrysopa oculata* (Neuroptera: Chrysopidae). Journal of Insect Physiology, 33, 627–633.

Neuenschwander, P. (1975). Influence of temperature and humidity on the immature stages of *Hemerobius pacificus*. Environmental Entomology, 4, 215–220.

Neuenschwander, P. (1976). Biology of adult *Hemerobius pacificus*. Environmental Entomology, 5, 96–100.

Neuenschwander, P., & Hagen, K. S. (1980). Role of the predator *Hemerobius pacificus* in a non-insecticide treated artichoke field. Environmental Entomology, 9, 492–495.

Neuenschwander, P. M., Canard, M., & Michalakis, S. (1981). The attractivity of protein hydrolysate baited McPhail traps to different chrysopid and hemerobiid species (Neuroptera) in a Creton olive orchard. Annales de la Societe Entomologique de France (N.S.), 17, 213–220.

Neuenschwander, P., Hagen, K. S., & Smith, R. F. (1975). Predation on aphids in California's alfalfa fields. Hilgardia, 43, 53–78.

Neuenschwander, P., Hennessey, R. D., & Herren, H. R. (1987). Food web of insects associated with cassava mealybug, *Phenacoccus manihoti* Matile-Ferrero (Hemiptera: Pseudococcidae), and its introduced parasitoid, *Epidinocarsis lopezi* (DeSantis) (Hymenoptera: Encyrtidae), in Africa. Bulletin of Entomological Research, 77, 177–189.

Neumark, S. (1952). *Chrysoperla carnea* St. and its enemies in Israel (No. 1). Ilanoth: Forest Research Station.

New, T. R. (1975). The biology of Chrysopidae and Hemerobiidae (Neuroptera) with reference to their usage as biocontrol agents: A review. Transactions of the Royal Entomological Society of London, 127, 115–140.

New, T. R. (1980). A revision of the Australian Chrysopidae (Insecta: Neuroptera). Australian Journal of Zoology, (Suppl.) Series No. 77, 143.

New, T. R. (1981). Aspects of the biology of *Chrysopa edwardsi* Banks (Neuroptera: Chrysopidae) near Melbourne, Australia. Neuroptera International, 1, 165–174.

New, T. R. (1984a). Chrysopidae: Ecology on field crops. In M. Canard, Y. Semeria, & T. R. New (Eds.), Biology of Chrysopidae (pp. 160–167). Boston: Dr. W. Junk.

New, T. R. (1984b). Comparative biology of some Australian Hemerobiidae. In J. Gepp *et al.* (Eds.), Progress in world's neuropterology (pp. 153–166). Graz, Austria: Gepp.

New, T. R. (1988a). A revision of the Australian Hemerobiidae. Invertebrate Taxonomy, 2, 339–411.

New, T. R. (1988b). Neuroptera. In A. K. Minks & P. Harrewijn (Eds.), Aphids, their biology, natural enemies and control (Vol. B, pp. 249–258). Amsterdam: Elsevier.

New, T. R. (1989). *Planipennia* (lacewings). Handbook of Zoology (Berlin) IV Insecta, 30, 1–132.

New, T. R. (1991). Insects as predators. Kensington N.S.W., Australia: New South Wales University Press.

Newton, A. F. (1987). Four *Staphylinus (sensu lato)* species new to North America, with notes on other introduced species (Coleoptera: Staphylinidae). Coleopterists' Bulletin, 41, 381–384.

Ng, S. M. (1986). Egg mortality of four species of aphidophagous Coccinellidae in Malaysia. In I. Hodek (Ed.), Ecology of Aphidophaga (pp. 77–81). Prague: Academia.

Nichols, P. R., & Neel, W. W. (1977). The use of food Wheast as a supplemental food for *Coleomegilla maculata* (DeGeer) (Coleoptera: Coccinellidae) in the field. Southwestern Entomology, 2, 102–105.

Nickel, J. L., Shimizu, J. T., & Wong, T. T. Y. (1965). Studies on natural control of pear psylla in California. Journal of Economic Entomology, 58, 970–976.

Nicolai, V. (1995). The impact of *Medetera dendrobaena* Kowarz (Dipt., Dolichopodidae) on bark beetles. Journal of Applied Entomology, 119, 161–166.

Nield, C. E. (1976). Aspects of the biology of *Staphylinus olens* (Muller), Britain's largest staphylinid beetle. Ecological Entomology, 1, 117–126.

Niemczyk, E. (1978a). *Orius minutus* (L.) (Heteroptera: Anthocoridae): The occurrence in apple orchards, biology and effect of different food on development. Bull. Entomol. Pologne, 48, 203–209.

Niemczyk, E. (1978b). The role of *Orius minutus* (L.) in controlling the European mite—*Panonychus ulmi* Koch on young apple trees. Bull. Entomol. Pologne, 48, 211–219.

Niemczyk, E. (1978c). *Campylomma verbasci* Mey-Dur (Heteroptera, Miridae) as a predator of aphids and mites in apple orchards. Bull. Entomol. Pologne, 48, 221–235.

Niemczyk, E., & Dixon, A. F. G. (Eds.). (1988). Ecology and effectiveness of aphidophaga. The Hague: SPB Academic.

Niemczyk, E., Olszak, R., Miszczak, M., & Bakowski, G. (1974). Effectiveness of some predaceous insects in the control of phytophagous mites and aphids on apple trees (Annual Rep. 1974). Poland: Research Institute of Pomology.

Niijima, K. (1988). Cecidomyiidae. In A. K. Minks & P. Harrewijn (Eds.), Aphids, their biology, natural enemies, and control (Vol. 2B, pp. 271–277). Amsterdam: Elsevier.

Niijima, K. (1989). Nutritional studies on an aphidophagous chrysopid, *Chrysopa septempunctata* Wesmael. I. Chemically-defined diets and general nutritional requirements. Bulletin of the Faculty of Agriculture, Tamagawa University, 29, 22–30.

Niijima, K. (1993a). Nutritional studies on an aphidophagous chrysopid, *Chrysopa septempunctata* Wesmael (Neuroptera: Chrysopidae). II. Amino acid requirements for larval development. Applied Entomology and Zoology, 28, 81–87.

Niijima, K. (1993b). Nutritional studies on an aphidophagous chrysopid, *Chrysopa septempunctata* Wesmael (Neuroptera: Chrysopidae). III. Vitamin requirement for larval development. Applied Entomology and Zoology, 28, 89–95.

Niijima, K., & Matsuka, M. (1990). Artificial diets for the mass production of chrysopids (Neuroptera). FFTC Book Series, 40, 190–198.

Nijveldt, W. (1969). Gall midges of economic importance: Vol. 8. Gall midges—miscellaneous. London: Crosby Lockwood.

Nikitskiy, N. B. (1976). Larval morphology and mode of life of Histeridae (Coleoptera) in bark-beetle passages. Entomological Review, 4, 102–111.

Nishida, R., & Fukami, H. (1989). Host plant iridoid-based chemical defense of an aphid, *Acyrthosiphon nipponicus,* against ladybird beetles. Journal of Chemical Ecology, 15, 1837–1845.

Nordlund, D. A., & Morrison, R. K. (1992). Mass rearing of *Chrysoperla* species. In T. E. Anderson & H. C. Leppla (Eds.), Advances in insect rearing for research and pest management (pp. 427–439). Boulder, CO: Westview.

Nordlund, D. A., Vacek, D. C., & Ferro, D. N. (1991). Predation of Colorado potato beetle (Coleoptera: Chrysomelidae) eggs and larvae by *Chrysoperla rufilabris* (Neuroptera: Chrysopidae) larvae in the laboratory and field cages. Journal of Entomological Society, 26, 443–449.

Nordlund, D. A., Lewis, W. J., Jones, R. L., Gross, H. R., Jr. & Hagen, K. S. (1977). Kairomones and their use for management of entomophagous insects. VI. An examination of the kairomones for the predator *Chrysopa carnea* Stephens at the oviposition sites of *Heliothis zea* (Boddie). Journal of Chemical Ecology, 3, 507–511.

Norris, J. D. (1958). Observations on the control of mite infestations in stored wheat by *Cheyletus* spp. (Acarina, Cheyletidae). Annals of Applied Biology, 46, 411–422.

Nuessly, G. S., & Sterling, W. L. (1994). Mortality of *Helicoverpa zea* (Lepidoptera: Noctuidae) eggs in cotton as a function of oviposition sites, predator species, and desiccation. Environmental Entomology, 23, 1189–1202.

Núñez, E. (1988). Ciclo biologico y crianza de *Chrysoperla externa* y *Ceraeochrysa cincta* (Neuroptera: Chrysopidae). Revista Peruana de Entomologia, 31, 76–82.

Nuorteva, M. (1959). Untersuchungen über einige in den Frassbildern der Borkenkäfer lebends *Medetera*—Arten (Dipt., Dolichopodidae). Annales Entomologici Fennici, 25, 192–210.

Nyffeler, M., & Benz, G. (1987). Spiders in natural pest control: A review. Journal of Applied Entomology, 103, 321–339.

Nyffeler, M., Dean, D. A., & Sterling, W. L. (1987a). Predation by green lynx spider, Peucetia viridins (Araneae: Oxyopidae), inhabiting cotton and woolly croton plants in East Texas. Environmental Entomology, 16, 355–359.

Nyffeler, M., Dean, D. A., & Sterling, W. L. (1987b). Evaluation of the importance of the striped lynx spider, Oxyopes salticus (Araneae: Oxyopidae), as a predator in Texas cotton. Environmental Entomology, 16, 1114–1123.

Nyffler, M., Dean, D. A., & Sterling, W. L. (1989). Prey selection and predatory importance of orb-weaving spiders (Araneae: Araneidae) in Texas cotton. Environmental Entomology, 18, 373–380.

Nyrop, J. P. (1988). Spatial dynamics of an acarine predator-prey system: Typhlodromus pyri (Acari: Phytoseiidae) preying on Panonychus ulmi (Acari: Tetranychidae). Environmental Entomology, 17, 1019–1031.

Obata, S. (1986). Mechanisms of prey finding in aphidophagous ladybird beetle, Harmonia axyridis (Coleoptera: Coccinellidae). Entomophaga, 31, 303–311.

Obrycki, J. J. (1986). The influence of foliar pubescence on entomophagous species, In D. J. Boethel & R. D. Eikenbary (Eds.), Interactions of plant resistance and parasitoids and predators of insects (pp. 61–83). Chichester: E. Horwood.

Obrycki, J. J., & Orr, C. J. (1990). Suitability of three prey species for Nearctic populations of Coccinella septempunctata, Hippodamia variegata, and Propyleaquatuor decimpunctata (Coleoptera: Coccinellidae). Journal of Economic Entomology, 83, 1292–1297.

Obrycki, J. J., Hamid, M. N., Sajap, A. S., & Lewis, L. C. (1989). Suitability of corn insect pests for development and survival of Chrysoperla carnea and Chrysoperla oculata (Neuroptera: Chrysopidae). Environmental Entomology, 18, 1126–1130.

Obrycki, J. J., Tauber, M. J., Tauber, C. A., & Gollands, B. (1983). Environmental control of the seasonal life cycle of Adalia bipunctata (Coleoptera: Coccinellidae). Environmental Entomology, 12, 416–421.

Obrycki, J. S., Orr, D. B., Orr, C. J., Wallendorf, M., & Flanders, R. V. (1993). Comparative development and reproductive biology of three populations of Propylea quatuordecimpunctata (Coleoptera: Coccinellidae). Biological Control, 3, 27–33.

O'Donnell, A. E., & Axtell, R. C. (1965). Predation by Fuscuropoda vegetans (Acarina: Uropodidae) on the house fly (Musca domestica). Annals of the Entomological Society of America, 58, 403–404.

O'Donnell, S. (1993). Interactions of predaceous katydids (Orthoptera: Tettigoniidae) with Neotropical social wasps (Hymenoptera: Vespidae): Are wasps a defense mechanism or prey? Entomological News, 104, 39–42.

Oetting, R. D., & Yonke, T. R. (1971). Immature stages and biology of Podisus placidus and Stiretras fimbriatus (Hemiptera: Pentatomidae). Canadian Entomologist, 103, 1505–1516.

Oetting, R. D., Yonke, T. R. (1975). Immature stages and notes on the biology of Euthyrhynchus floridanus (L.) (Hemiptera: Pentatomidae). Annals of the Entomological Society of America, 68, 659–662.

Ohgushi, T. (1986). Population dynamics of an herbivorous lady beetle, Henosepilachna niponica, in a seasonal environment. Journal of Animal Ecology, 55, 861–879.

Okada, I., Matsuka, M., & Tani, M. (1974). Rearing a green lacewing, Chrysopa septempunctata Wesmael on pulverized drone honeybee brood. Bulletin of the Faculty of Agriculture, Tamagaina University, 14, 26–32.

Olkowski, W., & Thiers, P. (1990). Promising work with praying mantids in Chinese biological control. IPM Practitioner, 12, 6–9.

Olszak, R. W., & Niemczyk, E. (1986). Predaceous Coccinellidae associated with aphids in apple orchards. Ekologia Polski, 34, 711–721.

O'Neil, R. J. (1988). Predation by Podisus maculiventris (Say) on Mexican bean beetle, Epilachna varvestis Mulsant, in Indiana soybeans. Canadian Entomologist, 120, 161–166.

O'Neil, R. J. (1989). Comparison of laboratory and field measurements of functional response of Podisus maculiventris (Heteroptera: Pentatomidae). Journal of Kansas Entomological Society, 62, 148–155.

O'Neil, R. J. (1992). Body weight and reproductive status of two nabid species (Heteroptera: Nabidae) in Indiana. Environmental Entomology, 21, 191–196.

O'Neil, R. J., & Stimac, J. L. (1988). Measurement and analysis of arthropod predation on velvetbean caterpillar, Anticarsia gemmatalis (Lepidoptera: Noctuidae), in soybeans. Environmental Entomology, 17, 821–826.

O'Neil, R. J., & Wiedenmann, R. N. (1990). Body weight of Podisus maculiventris (Say) under various feeding regimens. Canadian Entomologist, 122, 285–294.

Ongagna, P., Guige, L., Iperti, G., & Ferran, A. (1993). Life cycle of Harmonia axyridis (Col.: Coccinellidae) in its area of introduction: South-eastern France. Entomophaga, 38, 125–128.

Onillon, J. C. (1990). The use of natural enemies for the biological control of white files. In D. Gerling (Ed.), Whiteflies: Their bionomics, pest status and management (pp. 287–313). Andover, United Kingdom: Intercept.

Oomen, P. A. (1982). Studies on population dynamics of the scarlet mite Brevipalpus phoenicis, a pest of tea in Indonesia. Mededelingen Landbouwhogeshool Wageningen, 82, 1–85.

Orr, L. W., & Hall, R. C. (1931). An experiment in direct biotic control of a scale insect on pine. Journal of Economic Entomology, 24, 1087–1089.

Orth, R. E., Moore, T., Fisher, T. W., & Legner, E. F. (1975). A rove beetle, Ocypus olens, with potential for biological control of the brown garden snail, Helix aspersa, in California, including a key to the Nearctic species of Ocypus. Canadian Entomologist, 107, 111–116.

Osawa, N. (1989). Sibling and non-sibling cannibalism by larvae of a lady beetle Harmonia axyridis Pallas (Coleoptera: Coccinellidae) in the field. Researches on Population Ecology (Kyoto), 31, 153–160.

Osawa, N. (1992). A life table of the ladybird beetle Harmonia axyridis Pallas (Coleoptera: Coccinellidae) in relation to aphid abundance. Japanese Journal of Entomology, 60, 575–579.

Osawa, N. (1993). Population field studies of the aphidophagous ladybird beetle Harmonia axyridis (Coleoptera: Coccinellidae): Lifetables and key factor analysis. Researches on Population Ecology, 35, 335–348.

Osman, J. Z., & Selman, B. J. (1993). Storage of Chrysoperla carnea Steph. (Neuroptera: Chrysopidae) eggs and pupae. Journal of Applied Entomology, 115, 420–424.

Oswald, J. D. (1993). Revision and cladistic analysis of the world genera of the family Hemerobiidae (Insecta: Neuroptera). Journal of the New York Entomological Society, 101, 143–299.

Palmer, J. M., & Mound, L. A. (1990). Thysanoptera. In D. Rosen (Ed.), Armored scale insects, their biology, natural enemies and control (Vol. B, pp. 67–76). Amsterdam: Elsevier.

Palmer, J. M., Mound, L. A., & du Heaume, G. J. (1989). 2. Thysanoptera. London: CAB International Institute of Entomology, British Museum.

Pantaleoni, R. A., & Sproccati, M. (1987). I. Neurotteri delle colture agrarie: Studi preliminari circa l'influenza di siepi ed altre aree non coltivate sulle populazioni di Crisopidi. Bollettino dell, Istituto Entomologia "Guido Grandi" Universita degli Studi di Bologna, 42, 193–203.

Pantua, P. C., & Litsinger, J. A. (1984). A meadow grasshopper Conocephalus longipennis (Orthoptera: Tettigoniidae) predator of rice yellow stem borer egg masses. International Rice Research Newsletter, 9, 13.

Paradise, C. J., & Stamp, N. E. (1991). Abundant prey can alleviate previous adverse effects on growth of juvenile praying mantids (Orthoptera: Mantidae). Annals of the Entomological Society of America, 84, 396–406.

Parajulee, M. N., & Phillips, T. W. (1993). Effects of prey species on development and reproduction of the predator *Lytocoris campestris* (Heteroptera: Anthocoridae). Environmental Entomology, 22, 1035–1042.

Parker, A. H. (1971). The predatory and reproductive behaviour of *Vestula lineaticeps* (Sign.) (Hem., Reduviidae). Bulletin of Entomological Research, 61, 119–124.

Parker, A. H. (1972). The predatory and sexual behaviour of *Phonoctonus fasciatus* (P. & B.) and *P. subimpictus* Stål (Hem., Reduviidae). Bulletin of Entomological Research, 62, 139–150.

Parker, N. J. B. (1975). An investigation of reproductive diapause in two British populations of *Anthocoris nemorum* (Hemiptera: Anthocoridae). Journal of Entomology Series A: General Entomology, 49, 173–178.

Parra, J. R., Goncalves, P. W., Gravena, S., & Marconato, A. R. (1977). Parasitos e predadores do bicho-mineiro do coffeeiro *Perileucoptera coffeella* (Guerin-Meneville, 1842) en São Paulo. Anais Sociedade Entomologica do Brasil, 6, 138–143.

Parrella, M. P., Heinz, K. M., & Nunney, L. (1992). Biological control through augmentative releases of natural enemies: A strategy whose time has come. American Entomologist, 28, 172–179.

Parrella, M. P., Rowe, D. J., & Horsburg, R. L. (1982). Biology of *Leptothrips mali*, a common predator in Virginia apple orchards. Annals of the Entomological Society of America, 75, 130–135.

Parrella, M. P., Bellows, T. S., Gill, R. J., Brown, J. K., & Heinz, K. M. (1992). Sweet potato whitefly: Prospects for biological control. California Agriculture, 46, 25–26.

Pasqualini, E. (1975). Prove di allevamento in ambiente condizionato di *Chrysoperla carnea* Steph. (Neuroptera: Chrysopidae). Bolletino Ism. Entomologia Universita Bologna, 32, 291–304.

Pasteels, J. M. (1978). Apterous and brachypterous coccinellids at the end of the food chain, *Cionura erecta* (Asclepiadaceae)—*Aphis nerii*. Entomologia Experimentalis et Applicata, 24, 379–384.

Patel, K. G., & Vyas, H. N. (1985). Ovipositional site preference by green lacewing *Chrysopa (Chrysoperla) sceletes* Banks on cotton and green grain. Gujarat Agricultural University Research Journal, 10, 79–80.

Patel, K. G., Yadav, D. N., & Patel, R. C. (1988). Improvement in mass rearing technique of green lacewing *Chrysopa sceletes* Banks (Neuroptera: Chrysopidae). Gujarat Agricultural University Research Journal, 14, 1–4.

Paulson, G. S., & Akre, R. D. (1992a). Evaluating the effectiveness of ants as biological control agents of pear psylla (Homoptera: Psyllidae). Journal of Economic Entomology, 85, 70–73.

Paulson, G. S., & Akre, R. D. (1992b). Introducing ants (Hymenoptera: Formicidae) into pear orchards for the control of pear psylla, *Cacopsylla pyricola* (Foerster) (Homoptera: Psyllidae). Journal of Agricultural Entomology, 9, 37–39.

Pavuk, D. M., Purrington, F. F., Williams, C. E., & Stinner, B. R. (1997). Ground beetle (Coleoptera: Carabidae) activity density and community composition in vegetationally diverse corn agroecosystems. American Midland Naturalist, 138, 14–28.

Payne, T. L., Dickens, J. C., & Richerson, J. V. (1984). Insect predator-prey coevolution via enantiomeric specificity in a kairomone-pheromone system. Journal of Chemical Ecology, 10, 487–492.

Pearson, D. L. (1988). Tiger beetles. Annual Review of Entomology, 33, 123–147.

Pearson, D. L., & Mury, E. J. (1979). Character divergence and convergence among tiger beetles (Coleoptera: Cicindelidae). Ecology, 60, 557–566.

Pemberton, C. E. (1948). History of the entomology department of experiment stations, H.S.P.A. 1904–1945. Hawaii Planters' Board, 52, 53–90.

Peng, R. K., Christian, K., & Gibb, K. (1997). Control threshold analysis for the tea mosquito bug, *Helopeltis pernicialis* (Hemiptera: Miridae) and preliminary results concerning the efficiency of control by the green ant. *Oecophylla smaragdina*. International Journal of Pest Management, 43, 233–237.

Peng, Y. (1985). Field release of *Chrysopa sinica* as a strategy in the integrated control of *Panonychus citri*. Chinese Journal of Biology, Control, 1, 2–7 (in Chinese with English summary).

Peng, Y. (1986). Sprays of *Chrysopa* eggs for pest control. Chinese Journal of Biological Control, 2, 39.

Peng, Y. (1988). Mass rearing and field release of *Chrysopa shansiensis* against citrus red spider mites. Chinese Journal of Biology Control, 4, 137 (in Chinese).

Perfecto, I. (1991). Ants (Hymenoptera: Formicidae) as natural control agents of pests in irrigated maize in Nicaragua. Journal of Economic Entomology, 84, 65–70.

Péricart, J. (1987). Hémiptères Nabidae d'Europe Occidentale de du Magheb (Vol. 71). France: Faune.

Perkins, P. V., & Watson, T. F. (1972). *Nabis alternatus* as a predator of *Lygus hesperus*. Annals of the Entomological Society of America, 65, 625–629.

Peschke, K., & Fuldner, D. (1977). Ubersicht und neue Untersuchungen Zurlebensweise der parasitoidon Aleocharinae (Coleoptera: Staphylinidae). Zool. Jb. Syst., 104, 242–262.

Philippe, R. (1972). Biologie de la reproduction de *Chrysopa perla* (L.) (Neuroptera, Chrysopidae) en fonction de l'alimentation maginale. Annales de Zoologie—Ecologica Animale, 4, 213–227.

Phoofolo, M. W., & Obrycki, J. J. (1997). Comparative prey suitability of *Ostrinia nubilalis* eggs and *Acyrthosiphon pisum* for *Coleomegilla maculata*. Biological Control, 9, 167–172.

Phuoc, D. T., & Stehr, F. W. (1974). Morphology and taxonomy of known pupae of Coccinellidae (Coleoptera) of North America, with a discussion of phylogenetic relationships. Contributions of American Entomological Institute, Ann Arbor, 10, 1–125.

Pickett, A. D. (1959). Utilization of native parasites and predators. Journal Economic Entomology, 52, 1103–1105.

Pickett, C. H., Wilson, L. T., & Gonzalez, D. (1988). Population dynamics and within-plant distribution of the western flower thrips (Thysanoptera: Thripidae), an early-season predator of spider mites infesting cotton. Environmental Entomology, 17, 551–559.

Pinnock, D. E., Hagen, K. S., Cassidy, D. V., Brand, R. J., Milstead, J. E., & Tassan, R. L. (1978). Integrated pest management in highway landscapes. California Agriculture, 32, 33–34.

Ponomarenko, N. G. (1975). Characteristics of larval development in the family Dryinidae (Hymenoptera). Entomological Review, 54, 36–39.

Pope, R. D. (1979). Wax production by coccinellid larvae (Coleoptera). Systematic Entomology, 4, 171–196.

Pope, R. D. (1981). "*Rhizobius ventralis*" (Coleoptera: Coccinellidae), its constituent species and their taxonomy and historical roles in biological control. Bulletin of Entomological Research, 71, 19–31.

Pope, R. D. (1988). A revision of the Australian Coccinellidae (Coleoptera). I. Subfamily Coccinellinae. Invertebrate Taxonomy, 2, 633–735.

Pope, R. D., & Lawrence, J. F. (1990). A review of *Scymnodes* Blackburn, with the description of a new Australian species and its larva (Coleoptera: Coccinellidae). Systematic Entomology, 15, 241–252.

Poutiers, R. (1930). Sur le comportement de *Novius cardinalis* vis-a-vis de certains alcaloides. Comptes Rendus de la Societe de Biologie, 103, 1023–1025.

Prasad, Y. K. (1989). The role of natural enemies in controlling *Icerya purchasi* in South Australia. Entomophaga, 34, 391–395.

Prebble, M. L. (1933). The biology of *Podisus serieventris* Uhler, in Cape Breton, Nova Scotia. Canadian Journal of Research, 9, 1–30.

Press, J. W. (1989). Compatibility of *Xylocoris flavipes* (Hemiptera: Anthocoridae) and *Venturia canescens* for (Hymenoptera: Ichneumonidae) suppression of the almond moth, *Cadra cautella* (Lepidoptera: Pyralidae). Journal of Entomological Science, 24, 156–160.

Press, J. W., Flaherty, B. R., & Arbogast, R. T. (1975). Control of the red flour beetle, *Tribolium castaneum*, in a warehouse by a predaceous bug, *Xylocoris flavipes*. Journal of the Georgia Entomological Society, 10, 76–78.

Press, J. W., LeCato, G. L., & Flaherty, B. R. (1978). Influence of media particle size on the distribution of the predaceous bug *Xylocoris flavipes*. Journal of the Georgia Entomological Society, 13, 275–278.

Price, J. F., & Shepard, M. (1977). Striped earwig, *Labidura riparia*, colonization of soybean fields and response to insecticides. Environmental Entomology, 6, 679–683.

Price, J. F., & Shepard, M. (1978). *Calosoma sayi:* Seasonal history and response to insecticides in soybeans. Environmental Entomology, 7, 358–363.

Principi, M. M. (1940). Contributi allo studio dei neurotteri italiani. I. *Chrysopa septempunctata* Wesm. e *Chrysopa flavifrons* Brauer. Bolletino dell'Istituto di Entomologia della Universita di Bologna, 12, 63–144.

Principi, M. M. (1946). Contributi allo studio dei neurotteri italiani. IV. *Nothochrysa italia* Rossi. Bolletino dell'Istituto di Entomologia Universita Bologna, 15, 85–102.

Principi, M. M. (1956). Contributi allo studio dei neurotteri italiani. XIII. Studio morphologico, etologico e sistematico di un gruppo omogeneo di specie del gen. *Chrysopa* Leach (*C. flavifrons* Brauer, *prasima* Burn. e *clathrata* Schn.). Bolletino dell'Istituto di Entomologia Universita Bologna, 21, 319–410.

Principi, M. M. (1984). I. Neurotteri crisopidi e le possibilita della loro utilizzaione in lotta biologica e in lotta integrata. Bolletino dell'Istituto di Entomologia della Universita degli Studi di Bologna, 38, 231–262.

Principi, M. M. (1992). Insect diapause and its occurrence in some species of Chrysopidae (Insecta: Neuroptera) as a result of photoperiodic influence. Bulletino dell'Istituto di Entomologia "Guido Grandi" della Universita degli Studi di Bologna, 46, 1–30.

Principi, M., & Canard, M. (1974). Neuropteres. In Les organismes auxiliaires en verger de pommiers. (OBC WPRS Brochure No. 3, pp. 151–162).

Principi, M., & Canard, M. (1984). Feeding habits. In M. Canard, Y. Semeria, & T. R. New (Eds.), Biology of Chrysopidae (pp. 76–92). The Hague: Dr. W. Junk.

Principi, M. M., Memmi, M., & Sgobba, D. (1990). Influenza della temperatura sulla diapausa larvale di *Mallada flavifrons* (Brauer) (Neuroptera: Chrysopidae). Bollettino dell'Istituto di Entomologia "Guido Grandi" dell'Universita degli Studi di Bologna, 44, 37–55.

Priore, R. (1963). Studio morpho-biologico sulla *Rodolia cardinalis* Muls. Bollettino Entomologica Agricultura "Filippo Silvestri," Portici, 21, 62–198.

Provencher, L., & Coderre, D. (1987). Functional responses and switching of *Tetragnatha laboriosa* Hentz (Araneae: Tetragnathidae) and *Clubiona pikei* Gertsh. (Araneae: Clubionidae) for the aphids *Rhopalosiphum maidis* Fitch and *Rhopalosiphum padi* (L.) (Homoptera: Aphididae). Environmental Entomology, 16, 1305–1309.

Pulpán, J., & Verner, P. H. (1965). Control of tyroglyphoid mites in stored grain by the predatory mite *Cheyletus eruditus* (Schrank). Canadian Journal of Zoology, 43, 417–432.

Putman, W. L. (1932). Chrysopids as a factor in the natural control of the Oriental fruit moth. Canadian Entomologist, 64, 124–126.

Putman, W. L. (1955). Bionomics of *Stethorus punctillum* Weise (Coleoptera: Coccinellidae) in Ontario. Canadian Entomologist, 87, 9–33.

Putman, W. L. (1956). Differences in susceptibility of two species of *Chrysopa* (Neuroptera: Chrysopidae) to DDT. Canadian Entomologist, 88, 520.

Putman, W. L. (1965). The predaceous thrips *Haplothrips faurei* Hood (Thysanoptera: Phleothripidae) in Ontario peach orchards. Canadian Entomologist, 97, 9–33.

Putman, W. L., & Herne, D. H. C. (1966). The role of predators and other biotic agents in regulating the population density of phytophagous mites in Ontario peach orchards. Canadian Entomologist, 98, 808–820.

Quayle, H. J. (1912). Red spiders and mites of citrus trees. University of California Agricultural Station Bulletin, 234, 481–530.

Quezada, J. R., & DeBach, P. (1973). Bioecological and population studies of the cottony-cushion scale, *Icerya purchasi* Mask., and its natural enemies, *Rodolia cardinalis* Mul. and *Cryptochaetum iceryae* Will, in Southern California. Hilgardia, 41, 631–688.

Quinones, F. J., Bravo, H., Vera, J., & Carrillo, J. L. (1987). Requerimientos termicos para el desarollo de *Oligota oviformis* Casey (Coleoptera: Staphylinidae) y supresa *Oligonychus mexicanus* McGregor & Ortega (Acariformes: Tetranychidae). Agrociencia, Mexico, 67, 125–136.

Rabb, R. L., & Lawson, F. R. (1957). Some factors influencing the predation of *Polistes* wasps on tobacco hornworm. Journal of Economic Entomology, 50, 778–784.

Rabb, R. L., Stinner, R. E., & van den Bosch, R. (1976). Conservation and augmentation of natural enemies. In C. B. Huffaker & P. S. Messenger (Eds.), Theory and practice of biological control (pp. 233–254). New York: Academic Press.

Ràcz, V. (1988). The association of the predatory bug *Atractotomus mali* Mey.-D. (Heteroptera: Miridae) with aphids on apple in Hungary. In E. Niemczyk & A. F. G. Dixon (Eds.), Ecology and effectiveness of aphidophaga (pp. 43–46). The Hague: SPB Academic.

Raffa, K. F., & Klepzig, K. D. (1989). Chiral escape of bark beetles from predators responding to a bark beetle pheromone. Oecologia (Berlin), 80, 566–569.

Rakickas, R. J., & Watson, T. F. (1974). Population trends of *Lygus* spp. and selected predators in strip-cut alfalfa. Environmental Entomology, 3, 781–784.

Ramakers, P. M. J. (1980). Biological control of *Thrips tabaci* (Thysanoptera: Thripidae) with *Amblyseius* sp. (Acari: Phytoseiidae). IOBC WPRS Bulletin, 3(3), 203–208.

Ramakers, P. M. J. (1988). Biological control in greenhouses. In A. K. Minks & P. Harrewijn (Eds.), Aphids, their biology, natural enemies, and control (Vol. 2C, pp. 199–208). Amsterdam: Elsevier.

Ranaweera, D. J. W. (1950). Observations on the life-history of *Bethylus distigma* Motschulsky (*Goniozus montanus* Kieff) (Insecta, Hymenoptera, Bethylidae). Zoological Society of India Journal, 2, 127–132.

Rank, N. E., & Smiley, J. T. (1994). Host-plant effects on *Parasyrphus melanderi* (Diptera: Syrphidae) feeding on a willow leaf beetle *Chrysomela aeneicollis* (Coleoptera: Chrysomelidae). Ecological Entomology, 19, 31–38.

Rankin, M. A., & Rankin, S. M. (1980a). Some factors affecting the presumed migratory, flight activity of the convergent lady, beetle, *Hippodamia convergens* (Coccinellidae: Coleoptera). Biological Bulletin (Woods Hole, Massachusetts), 158, 356–369.

Rankin, M. A., & Rankin, S. M. (1980b). The hormonal control of migratory flight behaviour in the convergent lady beetle. *Hippodamia convergens*. Physiological Entomology, 5, 175–182.

Rao, R. S. N., & Chandra, I. J. (1984). *Brinckochrysa scelestes* (Neur.: Chrysopidae) as a predator of *Myzus persicae* (Hom.: Aphididae) on tobacco. Entomophaga, 29, 283–285.

Rao, R. Y., & Cherian, M. C. (1928). Notes on the life history and habits of *Parasierola* sp., the bethylid parasite of *Nephantis serinopa* Meyr. Madras Presidency, Department of Agriculture Yearbook, 1927, 11–22.

Rao, V. P., Ghani, M. A., Sankaran, T., & Mathur, K. C. (1971). A review of the biological control of insects and other pests in southeast Asia and the Pacific region (CIBC Technical Communications No. 6, Commonwealth Agricultural Bureaux, Farnham Royal, UK).

Raros, E. S., & Haramoto, F. H. (1974). Biology of *Stethorus siphonulus* Kapur (Coccinellidae: Coleoptera), a predator of spider mites, in Hawaii. Proceedings of the Hawaiian Entomological Society, 31, 457–465.

Rasnitsyn, A. P. (1980). The origin and evolution of hymenopteran insects. Trudy Paleontologicheskogo Instituta, 179, 1–180 (in Russian).

Rautapaa, J. (1972). The importance of *Coccinella septempunctata* L. (Col., Coccinellidae) in controlling cereal aphids, and the effect of aphids on the yield and quality of barley. Annales Agriculturae Fenniae, 11, 424–436.

Rautapaa, J. (1977). Evaluation of predator-prey ratio using *Chrysoperla carnea* Steph. in control of *Rhopalosiphum padi* (L.). Annales Agriculturae Fenniae, 16, 103–109.

Raw, A. (1988). Social wasps (Hymenoptera: Vespidae) and insect pests of crops of the Surui and Cinta Larga indians in Rondonia, Brazil. The Entomologist, 107, 104–109.

Raworth, D. A. (1984). Population dynamics of the cabbage aphid, *Brevicoryne brassicae* (Homoptera: Aphididae) at Vancouver, British Columbia. IV. Predation by *Aphidoletes aphidimyza* (Diptera: Cecidomyiidae). Canadian Entomologist, 116, 889–893.

Readio, J., & Sweet, M. H. (1982). A review of the Geocorinae of the United States east of the 100 meridian (Hemiptera: Lygaeidae). Entomological Society of America Miscellaneous Publications, 12, 1–91.

Readio, P. A. (1927). Studies on the biology of the Reduviidae of America north of Mexico. University of Kansas Science Bulletin, 17, 5–291.

Readshaw, J. L. (1975). The ecology of tetranychid mites in Australian orchards. Journal of Applied Ecology, 12, 473–495.

Reagan, T. E., Coburn, G., & Hensley, S. D. (1972). Effects of Mirex on the arthropod fauna of a Louisiana sugarcane field. Environmental Entomology, 1, 588–591.

Reddy, K. B., Sreedharan, K., & Bhat, P. K. (1991). Effect of rate of prey, *Planococcus citri* (Risso) on the fecundity of mealybug predator, *Cryptolaemus montrouzieri* Mulsant. Journal of Coffee Research, 21, 149–150.

Rees, B. E. (1948). The larva of *Catana clauseni* Chapin, and its comparison with the larva of *Delphastus* (Lec.). Proceedings of the Entomological Society of Washington, 50, 231–234.

Reeve, H. K. (1991). *Polistes*. In K. G. Ross & R. W. Matthews (Eds.), The social biology of wasps (pp. 99–148). Ithaca, NY: Comstock/Cornell University Press.

Reeve, J. D. (1997). Predation and bark beetle dynamics. Oecologia (Berlin), 112, 48–54.

Reid, C. D. (1991). Ability of *Orius insidiosus* (Hemiptera: Anthocoridae) to search for and attack European corn borer and corn earworm eggs on corn. Journal of Economic Entomology, 84, 83–86.

Reid, C. D., & Lampman R. L. (1989). Olfactory responses of *Orius insidiosus* (Hemiptera: Anthocoridae) to volatiles of corn silks. Journal of Chemical Ecology, 15, 1109–1115.

Reidenbach, J. M., Vala, J. C., & Ghamizi, M. (1989). The slug-killing Sciomyzidae (Diptera): Potential agents in the biological control of crop pest molluscs. In I. F. Henderson (Ed.), Slugs and snails in world agriculture. British Crop Protection Council Monograph, 41, 273–280.

Reis, P. R., & Souza, J. C. (1996). Integrated management of the leaf miner, *Perileucoptera coffeella* (Guerin-Meneville) (Lepidoptera: Lyonetiidae), and its effects on coffee yield. Anais Sociedade Entomologica do Brasil, 25, 77–82.

Remaudiere, G., Iperti, G., Leclant, F., Lyon, J. P., & Michel, M. F. (1973). Biologie et ecologie des aphids et de leurs enemies naturales. Application a la lutte integee en vergers. Entomophaga Mem. H. S. 6, 1–34.

Rensner, P. E., Lamp, W. O., Barney, R. J., & Armbrust, E. J. (1983). Feeding tests of *Nabis roseipennis* (Hemiptera: Nabidae) on potato leafhopper, *Empoasca fabae* (Homoptera: Cicadellidae) and their movement into spring-planted alfalfa. Journal of Kansas Entomological Society, 56, 446–450.

Reyd, G., Gery, R., Ferran, A., Iperti, G., & Brun, J. (1991). Etude de la consommation alimentaire d'*Hyperaspis raynevali* (Col.: Coccinellidae) predateur de la cochenille farineuse du manioc *Phenacoccus manihoti* (Hom.: Pseudococcidae). Entomophaga, 36, 161–171.

Ricci, C., & Cappelletti, G. (1988). Relationship between some morphological structures and locomotion of *Clitostethus arcuatus* Rossi (Coleoptera: Coccinellidae), a whitefly predator. Frustula Entomologica, 11, 195–202.

Rice, M. E., & Wilde, G. E. (1989). Antibiosis effect of sorghum on convergent lady beetle (Coleoptera: Coccinellidae), a third-trophic

level predator of the greenbug (Homoptera: Aphididae). Journal of Economic Entomology, 82, 570–573.

Richards, A. M. (1981). *Rhizobius ventralis* (Erichson) and *R. forestieri* (Mulsant) (Coleoptera: Coccinellidae), their biology and value for scale insect control. Bulletin of Entomological Research, 71, 31–46.

Richards, L. A., & Harper, A. M. (1978). Oviposition by *Nabis alternatus* (Hemiptera: Nabidae) in alfalfa. Canadian Entomologist, 110, 1359–1362.

Richardson, N. L. (1977). Biology of *Stethorus loxtoni* Britton & Lee (Coleoptera: Coccinellidae) and its potential as a predator of *Tetranychus urticae* Koch (Acarina: Tetranychidae) in California. u-Unpublished doctoral dissertation, Berkeley: University of California.

Richman, D. B. (1977). Predation on alfalfa weevil, *Hypera postica* (Gyllenhal), by *Stiretrus anchorago* (F.) (Hemiptera, Pentatomidae). Florida Entomologist, 60, 192.

Richman, D. B., & Whitcomb, W. H. (1978). Comparative lifecycles of four species of predatory stinkbugs (Hemiptera: Pentatomidae). Florida Entomologist, 61, 113–119.

Richman, D. B., Hemenway, R. C., Jr. & Whitcomb, W. H. (1980). Field evaluation of predators of the soybean looper. *Pseudoplusia includens* (Lepidoptera: Noctuidae). Environmental Entomology, 9, 315–317.

Richter, J., Biliwa, A., Helbig, J., & Henning-Helbig, S. (1997). Impact of *Teretriosoma nigrescens* Lewis (Coleoptera: Histeridae) on *Prostephanus truncatus* (Horn) (Coleoptera: Bostrichidae) and losses in traditional maize stores in southern Togo. Journal of Stored Products Research, 33, 137–142.

Riddick, E. W., & Mills, N. J. (1994). Potential of adult carabids (Coleoptera: Carabidae) as predators of fifth-instar codling moth (Lepidoptera: Tortricidae) in apple orchards in California. Environmental Entomology, 23, 1338–1345.

Riddick, E. W., & Mills, N. J. (1996). *Pterostichus* beetles dominate the carabid assemblage in an unsprayed orchard in Sonoma County, California. Pan-Pacific Entomologist, 72, 213–219.

Ridgway, R. L., & S. L. Jones, (1968). Plant feeding by *Geocoris pallens* and *Nabis americoferus*. Annals of the Entomological Society of America, 61, 232–233.

Ridgway, R. L., & Jones, S. L. (1969). Inundative releases of *Chrysopa carnea* for control of *Heliothis* on cotton. Journal of Economic Entomology, 62, 177–180.

Ridgway, R. L., & Kinzer, R. E. (1974). Chrysopids as predators of crop pests. Entomophaga, 7, 45–51.

Ridgway, R. L., & Murphy, W. L. (1984). Biological control in the field. In M. Canard, Y. Semeria, & T. R. New (Eds.), Biology of Chrysopidae (pp. 220–238). The Hague: Dr. W. Junk.

Ridgway, R. L., & Vinson S. B., (Eds.). (1977). Biological control by augmentation of natural enemies. Insect and mite control with parasites and predators. New York: Plenum Press.

Ridgway, R. L., King, E. G., & Carrillo, J. L., (1977). Augmentation of natural enemies for control of plant pests in the Western Hemisphere. In R. L. Ridgway & S. B. Vinson (Eds.), Biological control by augmentation of natural enemies (pp. 379–416). New York: Plenum Press.

Ridgway, R. L., Morrison, R. K., & Badgley, M. (1970). Mass rearing a green lacewing. Journal of Economic Entomology, 63, 834–836.

Ridgway, R. L., Lingren, P. D., Cowan, C. B., Jr., & Davis, J. W. (1967). Populations of arthropod predators and *Heliothis* spp. after applications of systemic insecticides to cotton. Journal of Economic Entomology, 60, 1012–1016.

Riechert, S. E. (1992). Spiders as representative "sit-and-wait" predators. In M. J. Crawley (Ed.), Natural enemies (pp. 313–328). London: Blackwell Science.

Riechert, S. E., & Bishop, L. (1990). Prey control by an assemblage of generalist predators in garden test systems. Ecology, 71, 1441–1450.

Riechert, S. E., & Lockley, T. (1984). Spiders as biological control agents. Annual Review of Entomology, 29, 299–320.

Rilling, S., Mittelstaedt, H., & Roeder, K. D. (1959). Prey recognition in the praying mantis. Behavior, 14, 164–184.

Rock, G. C., Monroe, R. J., & Yeargan, D. R. (1976). Demonstration of a sex pheromone in the predaceous mite Neoseiulus fallacis. Environmental Entomology, 5, 264–266.

Rodriquez, J. G., Singh, P. & Taylor, B. (1970). Manure mites and their role in fly control. Journal of Medical Entomology, 7, 335–341.

Rodriquez, J. G., Wade, C. F., & Wells, C. N. (1962). Nematodes as a natural food for Macrocheles muscaedomesticae (Acarina: Macrochelidae), a predator of the housefly egg. Annals of the Entomological Society of America, 55, 507–511.

Roeder, K. D. (1963). Nerve cells and insect behavior. Cambrige: Harvard University Press.

Rogers, C. E., Jackson, H. B., & Eikenbary, R. D. (1972). Responses of an imported coccinellid, Propylea 14-punctata, to aphids associated with small grains in Oklahoma. Environmental Entomology, 1, 198–202.

Rojo, S., & Marcos-Garcia, M. A. (1997). Syrphid predators (Dipt.: Syrphidae) of gall forming aphids (Hom.: Aphididae) in Mediterranean areas: Implications for biological control of fruit trees pests. Entomophaga 42, 269–276.

Rosen, D., & DeBach, P. (1978). Diaspididae. In C. P. Clausen (Ed.), Introduced parasites and predators of arthropod pests and weeds: A World review. Agriculture Handbook No. 480 (pp. 78–128). Washington, DC: U.S. Department of Agriculture.

Rosenheim, J. A., Wilhoit, L. R., & Armer, C. A. (1993). Influence of intraguild predation among generalist insect predators on the suppression of an herbivore population. Oecologia (Berlin), 96, 439–449.

Ross, K. G., & Matthews, R. W. (1991). The social biology of wasps. Ithaca, NY: Comstock/Cornell University Press.

Rotheray, G. E. (1983). Feeding behaviour of Syrphus ribesii and Melanostoma scalare on Aphis fabae. Entomologia Experimentalist et Applicata, 34, 148–154.

Rotheray, G. E. (1984). Host relations, life cycles and multiparasitism in some parasitoids of aphidophagous Syrphidae (Diptera). Ecological Entomology, 9, 303–310.

Rousset, A. (1984). Reproductive physiology and fecundity. In M. Canard, Y. Semeria, & T. R. New (Eds.), Biology of Chrysopidae (pp. 116–129). The Hague: Dr. W. Junk.

Rowlands, M. L., Rowlands, J., & Chapin, J. W. (1978). Prey searching behavior in adults of Hippodamia convergens (Coleoptera: Coccinellidae). Journal of the Georgia Entomological Society, 18, 309–315.

Ru, N., Whitcomb, W. H., & Murphey, M. (1976). Culturing of Chrysopa rufilabris (Neuroptera: Chrysopidae). Florida Entomologist, 59, 21–25.

Ru, N., Whitcomb, W. H., Murphey, M., & Carlysle, T. C. (1975). Biology of Chrysopa lanata (Neuroptera: Chrysopidae). Annals of the Entomological Society of America, 68, 187–190.

Ruberson, J. R., Bush, L., & Kring, T. J. (1991). Photoperiodic effect on diapause induction and development in the predator Orius insidiosus (Heteroptera: Anthocoridae). Environmental Entomology, 20, 786–789.

Rubia, E. G., & Shepard, B. M. (1987). Biology of Metiochle vittaticollis (Stål) (Orthoptera: Gryllidae), a predator of rice pests. Bulletin of Entomological Research, 77, 669–676.

Rudolf, E., Malausa, J. C., Millot, P., & Pralavorio, R. (1993). Influence des basses temperatures sures potentialites biologiques d'Orius laevigatus et d'Orius majusculus (Het.; Anthocoridae). Entomophaga, 38, 317–325.

Rueda, L. M., & Axtell, R. C. (1997). Arthropods in litter of poultry (broiler chicken and turkey) houses. Journal of Agricultural Entomology, 14, 81–91.

Ruppertshofen, H. (1955). Biologischer Forstschutz durch Vögel und rote Waldameisen im Stadtwald Mölla im Verlauf von 250 Jahren. Waldhygiene, 1, 45.

Russel, R. J. (1970). The effectiveness of Anthocoris nemorum and A. confusus (Hemiptera: Anthocoridae) as predators of the sycamore aphid, Drepanosiphum platanoides, I. The number of aphids consumed during development. Entomologia Experimentalis et Applicata, 13, 194–207.

Ruzicka, A. (1975). The effects of various aphids as larval prey on the development of Metasyrphus corollae (Dipt.: Syrphidae). Entomophaga, 20, 393–402.

Ruzicka, Z., & Hagen, K. S. (1985). Impact of parasitism on migratory flight performance in females of Hippodamia convergens (Coleoptera). Acta Entomologica Bohemoslovaca, 82, 401–406.

Ryerson, S. A., & Stone, J. D. (1979). A selected bibliography of two species of Orius: The minute pirate bug, Orius tristicolor, and Orius insidiosus (Heteroptera: Anthocoridae). Entomological Society of America Bulletin, 25, 131–135.

Sabelis, M. W. (1981). Biological control of two-spotted spider mites using phytoseiid predators. Modelling the predator-prey interactions at the individual level (Agricultural Research Rep. 910). Wageningen, The Netherlands: PUDOC.

Sabelis, M. W. (1985). Development. In W. Helle & M. W. Sabelis (Eds.), Spider mites: Their biology, natural enemies and control (Vol. 1B, pp. 43–53). Amsterdam: Elsevier Science.

Sabelis, M. W. (1992). Predatory arthropods. In M. J. Crawley (Ed.), Natural enemies (pp. 225–264). London: Blackwell Science.

Sabelis, M. W., & Bakker, F. M. (1992). How predatory mites cope with the web of their tetranychid prey: A functional view on dorsal chaetotaxy in the Phytoseiidae. Experimental and Applied Acarology (Northwood), 16, 203–225.

Sabelis, M. W., & Dicke, M. (1985). Long-range dispersal and searching behaviour. In W. Helle & M. W. Sabelis (Eds.), Spider mites: Their biology, natural enemies and control (Vol. 1B, pp. 141–160). Amsterdam: Elsevier Science.

Sabelis, M., & van de Baan, H. E. (1983). Location of distant spider mite colonies by phytoseiid predators: Demonstration of specific kairomones emitted by Tetranychus urticae and Panonychus ulmi. Entomologia Experimentalis et Applicata, 33, 303–314.

Saini, E. D. (1984). Identificacion de los huevos de pentatomidos (Heteroptera) encontrados en cultivos de soja. IDIA, Nos. 425–428, 79–84.

Saini, E. D. (1985, 1988). Identifiction practica de pentatomidos perjudicial y beneficos (Vol. 1, Vol. 2) Argentina: Instituto, Patologia Vegetale INIA-Castelar.

Saini, E. D. (1989). Aspectos morphologicas y biologicos de Stiretrus deastigma Herr-Schaf. y Oplomus cruentus (Burm.), pentomidos (Heteroptera) predadores. Revista Sociedad Entomologica Argentina, 48, 53–65.

Salas-Aguilar, J., & Ehler, L. E. (1977). Feeding habits of Orius tristicolor. Annals of the Entomological Society of America, 70, 60–62.

Samson, P. R., & Blood, P. R. B. (1979). Biology and temperature relationships of Chrysopa sp., Micromus tasmaniae and Nabis capsiformis. Entomologia Experimentalis et Applicata, 25, 253–259.

Samson, P. R., & Blood, P. R. B. (1980). Voracity and searching ability of Chrysopa signata (Neuroptera: Chrysopidae), Micromus tasmaniae (Neuroptera: Hemerobiidae) and Tropiconabis capsiformis (Hemiptera: Nabidae). Australian Journal of Zoology, 28, 515–580.

Samways, M. J. (1984). Biology and economic value of the scale predator Chilocorus nigritis (F.) (Coccinellidae). Biocontrol News Information, 5, 91–105.

Samways, M. J. (1989). Climate diagrams and biological control: An example from areography of the lady bird Chilocorus nigritis (Fabricius, 1978) (Insecta, Coleoptera, Coccinellidae). Journal of Biogeography, 16, 345–351.

Samways, M. J., & Wilson, J. (1988). Aspects of feeding behavior of Chilocorus nigritus (F.) (Col.: Coccinellidae) relative to its effectiveness as a biological control agent. Journal of Applied Entomology, 106, 177–182.

Samways, M. J., Osborn, R., & Saunders, T. L. (1997). Mandible form relative to the main food type in ladybirds (Coleoptera: Coccinellidae). Biocontrol Science and Technology, 7, 275–286.

Sanda, G. L. (1991). Mode of hunting and functional response of the spider *Marpissa tigrina* Tikader (Salticidae: Arachnida) to the density of its prey, *Diaphorina citri*. Entomon, 16, 279–282.

Sanders, C. J., & Pang, A. (1992). Carpenter ants as predators of spruce budworm in the boreal forest of northwestern Ontario. Canadian Entomologist, 124, 1093–1100.

Sanders, W. (1980). Das Eiablageverhalten der Schwebfliege *Syrphus corollae* Fabr. in Abhangigkeit von der Raumlichen Lage der Blattlauskolonie. Zeitschrift fuer Angewandte Zoologie, 67, 35–46.

Sanders, W. (1981). Der Einfluss von weissen und schwarzen Flachen auf das Verhalten der Schwebfliege *Syrphus corollae* Fabr. Zeitschrift fuer Angewandte Zoologie, 68, 307–314.

Sandhu, G. S., & Kaushal, K. K. (1975). Biological studies and host range of *Leucopis* sp. (Chamaemyiidae: Diptera), a predatory fly on aphids in the Punjab. Indian Journal of Entomology, 37, 185–187.

Sandlan, K. P. (1979). Host-feeding and its effects of the physiology and behavior of the ichenumonid parasitoid *Coccygomomus turionellae*. Physiological Entomology, 4, 383–392.

Sands, W. A. (1957). The immature stages of some British Anthocoridae (Hemiptera). Transactions of the Royal Entomological Society of London, 109, 295–310.

Santos, M., & Laing, J. E. (1985). Stigmaeid predators. In W. Helle & M. W. Sabelis (Eds.), Spider mites: Their biology, natural enemies and control (Vol. 1B, pp. 197–203). Amsterdam: Elsevier Science.

Sasaji, H. (1971). Fauna Japonica: Coccinellidae (Insecta: Coleoptera). Tokyo, Japan: Academic Press.

Sasaji, H., & McClure, M. S. (1997). Description and distribution of *Pseudoscymnus tsugae* sp. nov. (Coleoptera: Coccinellidae), an important predator of hemlock woolly adelgid in Japan. Annals of the Entomological Society of America, 90, 563–568.

Sauphanor, B., Chabrol, L., Faivre D'Arcier, F., Sureau, F., & Lenfant, C. (1993). Side effects of deflubenzuron on pear psylla predator: *Forficula auricularia*. Entomophaga, 38, 163–174.

Savoiskaya, G. I., & Klausnitzer, B. (1973). Morphology and taxonomy of the larvae with keys for their identification. In I. Hodek (Ed.), Biology of Coccinellidae (pp. 36–55). Prague: Academia.

Schaefer, P. W., & Dysart, R. J. (1988). Palearctic aphidophagous coccinellids in North America. In E. Niemczyk & A. F. G. Dixon (Eds). Ecology and effectiveness of aphidophaga (pp. 99–103). The Hague: S.P.B. Academic.

Schaefer, P. W., Dysart, R. J., & Specht, H. B. (1987). North American distribution of *Coccinella septempunctata* (Coleoptera: Coccinellidae) and its mass appearance in coastal Delaware. Environmental Entomology, 16, 368–373.

Schilder, F. A., & Schilder, M. (1928). Die Nahrung der Coccinelliden und ibre Beziehung zur Verwandtschaft der Arten, Arb. Biol. Reichs. Anstalt Land-u Forstw, 16, 213–282.

Schlinger, E. I., van den Bosch, R., & Dietrick, E. J. (1959). Biological notes on the predaceous earwig *Labidura riparia* (Pallas), a recent immigrant to California (Dermaptera: Labiduridae). Journal of Economic Entomology, 52, 247–249.

Schlup, J. (1987). Investigations on the population dynamics of Ips typographus L. and on the biology of its main predator Thanasimus formicarius L., Unpublished diploma thesis, Zoology Institute, University of Bern, Switzerland.

Schmid, J. M. (1971). *Medetera aldrichii* (Diptera: Dolichopodidae) in the Black Hills. II. Biology and densities of the immature stages. Canadian Entomologist, 103, 848–853.

Schmitt, J. J., & Goyer, R. A. (1983). Laboratory development and de-

scription of immature stages of *Scoloposcelis mississippansis* Drake & Harris and *Lyctocoris elongatus* (Reuter) (Hemiptera: Anthocoridae) predators of southern pine bark beetles (Coleoptera: Scolytidae). Annals of Entomological Society of America, 76, 868–872.

Schreiner, I. (1989). Biological control introductions in the Caroline and Marshall Islands. Proceedings of the Hawaiian Entomological Society, 29, 57–69.

Schuh, R. T. (1974). The Orthylinae and Phylinae (Hemiptera: Miridae) of South Africa with a phylogentic analysis of the ant-mimetic tribes of the two subfamilies for the world. Entomologica Americana, 47, 1–332.

Schulten, G. G. M. (1985). Mating. In W. Helle & M. W. Sabelis (Eds.), Spider mites: Their biology, natural enemies and control (Vol. 1B, pp. 55–65). Amsterdam: Elsevier Science.

Schultz, B. B. (1988). Reduced oviposition by green lacewings (Neuroptera: Chrysopidae) on cotton intercropped with corn, beans or weeds in Nicaragua. Environmental Entomology, 17, 229–232.

Schuster, M. F., & Calderon, M. (1986). Interactions of host plant resistant genotypes and beneficial insects in cotton ecosystems. In D. J. Boethel & R. D. Eikenberg (Eds.), Interactions of plant resistance and parasitoids and predators of insects (pp. 84–97). Berkshire: Ellis Horwood.

Scopes, N. E. A. (1969). The potential of *Chrysoperla carnea* as a biological control agent of *Myzus persicae* on glasshouse chrysanthemums. Annals of the Applied Biology, 64, 433–439.

Scott, S. M., & Barlow, C. A. (1984). Effect of prey availability during development on the reproductive output of *Metasyrphus corollae* (Diptera: Syrphidae). Environmental Entomology, 13, 669–674.

Seevers, C. H. (1965). The systematics, evolution and zoogeography of staphylinid beetles associated with army ants. Fiediana, Zoology, 47, 138–351.

Selhime, A. G., & Kanavel, R. F. (1968). Life cycle and parasitism of *Micromus posticus* and *M. subanticus* in Florida. Annals of the Entomological Society of America, 61, 1212–1215.

Selhime, A. G., Muma, M. H., & Clancey, D. W. (1963). Biological, chemical, and ecological studies on the predatory thrips, *Aleurodothrips fasciapennis* in Florida citrus groves. Annals of the Entomological Society of America, 56, 709–712.

Séméria, Y. (1984). Savannah: Mediterranean climates. In M. Canard, Y. Séméria, & T. R. New (Eds.), Biology of Chrysopidae (pp. 167–180). The Hague: Dr. W. Junk.

Semyanov, V. P., & Bereznaja, E. B. (1988). Biology and prospects of using Vietnam's lady beetle *Lemnia biplagiata* (Swartz) for control of aphids in greenhouses. In E. Niemczyk & A. F. G. Dixon (Eds.), Ecology and effectiveness of aphidophaga (pp. 267–269). The Hague: SPB Academic.

Sengonca, C., & Henze, M. (1992). Conservation and enhancement of *Chrysoperla carnea* (Stephens) (Neuroptera, *Chrysoperla*) in the field by providing hibernation shelters. Journal of Applied Entomology, 114, 497–501.

Sengonca, C., & Loechte, C. (1997). Development of a spray and atomizer technique for applying eggs of *Chrysoperla carnea* (Stephens) in the field for biological control of aphids. Zeitschrift fuer Pflanzenkrankheiten und Pflanzenschutz, 104, 214–221.

Sengonca, C., & Schimmel, C. (1993). A comparative study on *Chrysoperla carnea* (Steph.) eggs with and without stalk with semisynthetic diet and with aphids fed culture with regard to field application. Anzeiger fuer Schaedlingskunde, Pflanzen- und Umweltschutz, 66, 81–84.

Senrayan, R. (1988). Functional response of *Eocanthecona furrellata* (Wolff.) (Heteroptera: Pentatomidae) in relation to prey density and defense with reference to its prey *Latoia lepida* (Cramer) (Lepidoptera: Limacodidae). Proceedings of the Indian Academy of Science, Animal Science, 97, 339–345.

Senrayan, R. (1991). Plant feeding by *Eocantheconga furcellata* (Wolff.)

(Heteroptera: Asopinae): Effect on development survival and reproduction. Phytophaga (Madras), 31, 103–108.

Seymour, R. C., & Campbell, J. B. (1993). Predators and parasitoids of house flies and stable flies (Diptera: Muscidae) in cattle confinements in West Central Nebraska. Environmental Entomology, 22, 212–219.

Sgobba, D., & Zibordi, L. (1984). Prove di allevamento massala di larvae di Chrysoperla carnea (Steph.) (Neuroptera: Chrysopidae): Propos le peril contenmento del cannibalismo. Bollettino dell'Istituto di Entomologia Universita Bologna, 39, 1–16.

Shah, M. A. (1982). The influence of plant surfaces on searching behaviour of coccinellid larvae. Entomologia Experimentalis et Applicata, 31, 377–380.

Shah, M. A. (1983). A stimulant in Berberis vulgaris inducing oviposition in coccinellids. Entomologia Experimental et Applicata, 33, 119–121.

Shands, W. A., & Simpson, G. W. (1972). Insect predators for controlling aphids on potatoes. II. A pilot test of spraying eggs of predators on potatoes separated by bare fallow land. Journal of Economic Entomology, 65, 1383–1387.

Shands, W. A., Gordon, C. C., & Simpson, G. W. (1972a). Insect predators for controlling aphids on potatoes. VI. Development of a spray technique for applying eggs in the field. Journal of Economic Entomology, 65, 1099–1103.

Shands, W. A., Simpson, G. W., & Brunson, M. H. (1972b). Insect predators for controlling aphids on potatoes. I. In small plots. Journal of Economic Entomology, 65, 511–514.

Shands, W. A., Simpson, G. W., & Gordon, C. C. (1972c). Insect predators for controlling aphids on potatoes. V. Numbers of eggs and schedules for introducing them in large field cages. Journal of Economic Entomology, 65, 810–817.

Shands, W. A., Simpson, G. W., & Storch, R. H. (1972d). Insect predators for controlling aphids on potatoes. IX. Winter survival of Coccinella species in field cages over grassland in northeastern Maine. Journal of Economic Entomology, 65, 1392–1396.

Sharova, I. Ch., & Ghilarov, M. S. (Eds.). (1981). Life forms of carabides. Moscow: Nauka (in Russian).

She, H. D. N., Odebiyi, J. A., & Herren, H. R. (1984). The biology of Hyperaspis jucunda (Col.: Coccinellidae), an exotic predator of the cassava mealybug Phenacoccus manihoti (Hom.: Pseudococcidae) in southern Nigeria. Entomophaga, 29, 87–93.

Sheldon, J. K., & MacLeod, E. G. (1971). Studies on the biology of the Chrysopidae. II. The feeding behaviour of the adult of Chrysopa carnea (Neuroptera). Psyche, 78, 107–121.

Shepard, M., Carner, G. R., & Turnipseed, S. G. (1974). Seasonal abundance of predaceous arthropods in soybeans. Environmental Entomology, 3, 985–988.

Shepard, M., Waddill, V., & Kloft, W. (1973). Biology of the predaceous earwig Labidura riparia (Dermaptera: Labiduridae). Annals of the Entomological Society of America, 66: 837–841.

Shields, E. J. (1980). Locomotory activity of Orius tristicolor under various intensities of flickering and non-flickering light. Annals of Entomological Society of America, 73, 74–77.

Shields, E. J., & Watson, T. F. (1980). Searching behavior of female Orius tristicolor. Annals of the Entomological Society of America, 73, 533–535.

Shimizu, J. T., & Hagen, K. S. (1967). An artificial ovipositional site for some Heteroptera that insert their eggs into plant tissue. Annals of the Entomological Society of America, 60, 1115–1116.

Shimoda, T., Shinkaji, N., & Amano, H. (1993). Seasonal occurrence of Oligota-kashmirica-benefica Naomi (Coleoptera: Staphylinidae) on arrow root and effect of prey consumption rate on development and oviposition. Japanese Journal of Applied Entomology and Zoology, 37, 75–82.

Shimoda, T., Shinkaji, N., & Amano, H. (1994a). Oviposition behavior of Oligota kashmirica benefica Naomi (Coleoptera: Staphylinidae) I.

Adaptive significance of egg-covering behavior by adult females. Japanese Journal of Applied Entomology and Zoology, 38, 1–6.

Shimoda, T., Shinkaji, N., & Amano, H. (1994b). Oviposition behavior of Oligota kashmirica benefica Naomi (Coleoptera: Staphylinidae). II. Preference of oviposition sites a leguminous arrowroot. Japanese Journal of Applied Entomology and Zoology, 38, 65–70.

Shimoda, T., Shinkaji, N., & Amano, H. (1997). Prey stage preference and feeding behaviour of Oligota kashmirica benefica (Coleoptera: Staphylinidae), an insect predator of the spider mite Tetranychus urticae (Acari: Tetranychidae). Experimental and Applied Acarology (Northwood), 21, 665–675.

Showler, A. T., & Reagan, T. E. (1987). Ecological interactions of the red imported ant in the southeastern United States. Journal of Entomological Science, (Suppl), 52–64.

Shuvakhina, E. Y. (1969). Rearing Chrysoperla carnea Steph. and C. septempunctata Wesm. on a laboratory scale. Zashchita Rastenii, 3, 82–86.

Shuvakhina, E. Y. (1975). Common golden-eyed fly as a predator of Colorado potato beetle and the possibility of its application on potato. In G. V. Guseva & V. A. Shcheretilnikova (Eds.), Biological methods for plant protection (pp. 284–297). Proceedings of All-Union Science Research Institute Plant Protection, Leningrad (in Russian).

Simanton, F. L. (1916). Hyperaspis binotata, a predatory enemy of the terrapin scale. Journal of Agricultural Research, 6, 197–203.

Simmonds, F. J. (1960). Biological control of coconut scale, Aspidiotus destructor Sign., in Principe, Portugese West Africa. Bulletin of Entomological Research, 51, 223–237.

Simmonds, H. W. (1958). The housefly problem in Fiji and Samoa. South Pacific Quarterly Bulletin, 8, 29–30.

Singh, K. J., & Singh, O. P. (1989). Biology of a pentatomid predator, Andrallus spiniders (Fab.) on Rivula sp., a pest of soybean in Madhya Pradesh. Journal of Insect Science, 2, 134–138.

Singh, P., King, W. E., & Rodriquez, J. G. (1966). Biological control of muscids as influenced by host preference of Macrocheles muscaedomesticae (Acarina: Macrochelidae). Journal of Medical Entomology, 3, 78–81.

Sivapragasam, A., & Asma, A. (1985). Development and reproduction of the mirid bug, Cyrtorhinus lividipennis (Heteroptera: Miridae) and its functional response to the brown plant hopper. Applied Entomology and Zoology, 20, 373–379.

Skuhravy, J. O. (1960). Die Nahrung des Ohrwurms (Forficula auricularia L.) in den Feldkulturen. Casopis Ceskeho Spol. Entomol. 57, 329–339.

Slater, J. A., & Baranowski, R. M. (1978). How to know the true bugs (Hemiptera-Heteroptera). Dubuque, IA: Wm. C. Brown.

Sloderbeck, P. E., & Yeargan, K. V. (1983). Comparison of Nabis americoferus and N. roseipennis (Hemiptera: Nabidae) as predators of the green cloverworm (Lepidoptera: Noctuidae). Environmental Entomology, 12, 161–165.

Sluss, R. R. (1967). Population dynamics of walnut aphid Chromaphis juglandicola (Kalt.) in northern California. Ecology, 48, 41–58.

Sluss, T. P., & Foote, B. A. (1971). Biology and immature stages of Leucopis verticalis (Diptera: Chamaemyiidae). Canadian Entomologist, 103, 1427–1434.

Sluss, T. P., & Foote, B. A. (1973). Biology and immature stages of Leucopis pinicola and Chamaemyia polystigma (Diptera: Chamaemyiidae). Canadian Entomologist, 105, 1443–1452.

Smetana, A. (1982). Revision of the subfamily Xantholininae of America north of Mexico (Coleoptera: Staphylinidae). Memoirs of the Entomological Society of Canada, 120, 1–389.

Smirnoff, W. A. (1954). Les Cybocephalus (Col., Cybocephalidae) d'Afrique du Nord, predateurs de Parlatoria blanchardi Targ. (Homoptera: Coccoidea) parasite du palmier-dattier. Revue de Pathologie Vegetale, 33, 84–101.

Smirnoff, W. A. (1957). La cochenille du palmier dattier (*Parlatoria blanchardi* Targ.) en Afrique du Nord. Comportement importance economique, predateurs et lutte biologique. Entomophaga, 11, 1–98.

Smith, B. C. (1960). A technique for rearing coccinellid beetles on dry foods, and influence of various pollens on development of *Coleomegilla maculata lengi* Timb. (Coleoptera: Coccinellidae). Canadian Journal of Zoology, 38, 1047–1049.

Smith, D. S. (1959). Note on destruction of grasshopper eggs by the field cricket *Acheta assimilis* luctuosus (Serville) (Orthoptera: Gryllidae). Canadian Entomologist, 91, 127.

Smith, H. S. (1923). The successful introduction and establishment of the ladybird, *Scymnus binaevatus* Mulsant, in California. Journal of Economic Entomology, 16, 516–518.

Smith, H. S., & Armitage, H. M. (1920). Biological control of mealybugs in California. California State Department of Agricultural Monthly Bulletin, 9, 104–164.

Smith, H. S., & Armitage, H. M. (1931). The biological control of mealybugs attacking citrus (Agricultural Bulletin 509). Berkeley: University of California.

Smith, R. C. (1922). The biology of Chrysopidae. Cornell University Agricultural Experimental Station Memoirs, 58, 1287–1372.

Smith, R. F., & Hagen, K. S. (1956). Predators of the spotted alfalfa aphid. California Agriculture, 10, 8–10.

Smith, R. F., & Hagen, K. S. (1966). Natural regulation of alfalfa aphids in California. In I. Hodek (Ed.), Ecology of aphidophagous insects (pp. 297–319). Prague: Academia.

Smith, R. F., & van den Bosch, R. (1967). Integrated control. In W. W. Kilgore & R. L. Doutt (Eds.), Pest control (pp. 295–340). New York: Academic Press.

Smith, S. G. (1965). *Chilocorus similis* Rossi (Coleoptera: Coccinellidae) disinterment and case history. Science, 148, 1614–1616.

Snetsinger, R. (1956). Biology of *Bdella depressa*, a predaceous mite. Journal of Economic Entomology, 49, 745–746.

Snoddy, E. L., Humphreys, W. J., & Blum, M. S. (1976). Observations on the behavior and morphology of the spider predator, *Stenolemus lanipes* (Hemiptera: Reduviidae). Journal of the Georgia Entomological Society, 11, 55–58.

Snodgrass, G. L. (1991). *Deraeocoris nebulosus* (Heteroptera: Miridae): Little known predator in cotton in the Mississippi Delta. Florida Entomologist, 74, 340–344.

Soliman, Z. R., & Mohamed, M. I. (1972). On the development and biology of the predaceous mite, *Neomolgus aegyptiacus* Sol. (Acarina: Bdellidae). Zeitschrift fuer Angewandte Entomologie, 71, 90–95.

Solomon, M. E. (1969). Experiments on predator-prey interactions of storage mites. Acarologia, 10, 484–503.

Southwood, T. R. E., & Leston, D. (1959). Land and water bugs of the British Isles. London: Fredrick Warner.

Souza, B., & Ciociola, A. I. (1995). Behavioral aspects of *Nusalala uruguaya* (Navas) (Neuroptera: Hemerobiidae) in the laboratory. Anais Sociedade Entomologica do, Brasil, 24, 173–175.

Souza, B., Matioli, J. C., & Ciociola, A. I. (1990). Biologia comparada de *Nusalala uruguaya* (Navas) (Neuroptera: Hemerobiidae) alimentata con diferentes especies de afideos. I. Fase de larva. Anaisda Escola Superior de Agricultura "Luiz de Queiroz" Universidade de Sao Paulo, 47, 283–300.

Sprenkel, R. K., Brooks, W. M., van Duyn, J. W., & Deitz, L. L. (1979). The effects of three cultural variables on the incidence of *Nomuraea rileyi*, phytophagous Lepidoptera, and their predation on soybeans. Environmental Entomology, 8, 334–339.

Srinivasan, T. R., & Babu, P. C. S. (1989). Field evaluation of *Cryptolaemus montrouzieri* Mulsant, the coccinellid predator against grapevine mealybug *Maconellicoccus hirsutus* (Green). South Indian Horticulture, 37, 50–51.

Stansly, P. A. (1984). Introduction and evaluation of *Chilocorus bipustulatus* (Col.: Coccinellidae) for control of *Parlatoria blanchardi* (Hom.: Diaspididae) in date groves of Niger. Entomophaga, 29, 29–39.

Stark, S. B., & Whitford, F. (1987). Functional response of *Chrysoperla carnea* (Neur.: Chrysopidae) larvae feeding on *Heliothis virescens* (Lep.: Noctuidae) eggs on cotton in field cages. Entomophaga, 32, 521–527.

Steiner, H. (1974). Carabidae, Staphylinidae, Cantharidae. In L. Brader (Ed.), Les organismes auxiliaries en verger de pommiers. (Brochure No. 3, pp. 123–124). Wageningen: IOBC WPRS.

Steiner, M. Y., & Tellier, A. J. (1990). Western flower thrips, *Frankliniella occidentalis* (Pergande), in greenhouse cucumbers in Alberta, Canada. IOBC WPRS Bulletin, 13, 202–205.

Steinmann, H. (1989). World catalogue of Dermaptera. Dordrecht: Kluwer Academic.

Stelzl, M., & Hassan, S. A. (1992). Culturing of *Micromus angulatus* (Steph.) (Neuropteroidea, Hemerobiidae), a new candidate beneficial insect for the biological control of insect pests in greenhouses. Journal of Applied Entomology, 114, 32–37.

Sterling, W. L., Dean, D. A., Fillman, D. A., & Jones, D. (1984). Naturally-occurring biological control of the boll weevil (Col., Curculionidae). Entomophaga, 29, 1–9.

Stern, V. M., Smith, R. F., van den Bosch, R., & Hagen, K. S. (1959). The integration of chemical and biological control of the spotted alfalfa aphid. Hilgardia, 29, 81–101.

Sternlicht, M. (1970). Contribution to the biology of the citrus bud mite, *Aceria sheldoni* (Ewing). Annals of Applied Biology, 65, 221–230.

Stevens, G. S. (1975). Transportation and culture of *Tytthus mundulus*. Journal of Economic Entomology, 68, 753–754.

Stevenson, A. B. (1967). *Leucopis simplex* (Diptera: Chamaemyiidae) and other species occurring in galls of *Phylloxera vitifoliae* (Homoptera: Phylloxeridae) in Ontario. Canadian Entomologist, 99, 815–820.

Steward, V. B., Yearian, W. C., & Kring, T. J. (1991). Sampling of *Heliothis zea* (Lepidoptera: Noctuidae) and grain sorghum panicles by a modification of the beat-bucket method. Journal of Economic Entomology, 84, 1095–1099.

Stewart, H. C., & Walde, S. J. (1997). The dynamics of *Aphis pomi* de Geer (Homoptera: Aphididae) and its predator, *Aphidoletes aphidimyza* (Rondani) (Diptera: Cecidomyiidae), on apple in Nova Scotia. Canadian Entomologist, 129, 627–636.

Stimmel, J. F. (1979). Seasonal history and distribution of *Carulaspis minima* (Targ.-Tozz.) in Pennsylvania (Homoptera: Viaspididae). Proceedings of the Entomological Society of Washington, 81, 222–229.

Stinner, R. E. (1977). Efficacy of inundative releases. Annual Review of Entomology, 22, 515–531.

Stoltz, R. L., & Stern, V. M. (1978). The longevity and fecundity of *Orius tristicolor* when introduced to increasing numbers of the prey *Frankliniella occidentalis,* Environmental Entomology, 7, 197–198.

Stonedahl, G. M. (1983). New records for Palearctic *Phytocoris* in western North America (Hemiptera: Miridae). Proceedings of the Entomological Society Washington, 85, 463–471.

Stoner, A. (1965). Host plants of *Spanogonicus albofasciatus* in Arizona. Journal of Economic Entomology, 58, 322–324.

Stoner, A. (1970). Plant feeding by a predaceous insect, *Geocoris punctipes.* Journal of Economic Entomology, 63, 1911–1915.

Stoner, A. (1972). Plant feeding by *Nabis,* a predaceous genus. Environmental Entomology, 1, 557–558.

Stoner, A., Metcalfe, A. M., & Weeks, R. E. (1975). Seasonal distribution, reproduction diapause, and parasitization of three *Nabis* spp. in southern Arizona. Environmental Entomology, 4, 211–214.

Storch, R. H. (1976). Prey detection by 4th stage *Coccinella transverso-guttata* (Coleoptera: Coccinellidae). Animal Behavior, 24, 690–693.

Storozhkov, Y. V., Moiseev, E. E., & Bondarenko, N. V. (1981). A test of biological control of cucumbers in a large greenhouse. Zashchita Rastenii, 1, 28–29 (in Russian).

Strandberg, J. O. (1981a). Predation of cabbage looper. *Trichoplusia ni*, pupae by the striped earwig, *Labidura riparia*, and two bird species. Environmental Entomology, 10, 712–715.

Strandberg, J. O. (1981b). Activity and abundance of the earwig, *Labidura riparia*, in a winter cabbage production system. Environmental Entomology, 10, 701–704.

Struble, G. R. (1930). The biology of certain Coleoptera associated with bark beetles in western yellow pine. University of California Publications in Entomology, 5, 105–134.

Stuart, M. K., & Greenstone, M. H. (1990). Beyond ELISA: A rapid, sensitive, specific immunodot assay for identification of predator stomach contents. Annals of the Entomological Society of America, 83, 1101–1107.

Stubbs, M. (1980). Another look at prey detection by coccinellids. Ecological Entomology, 5, 179–182.

Sturani, M. (1962). Osservazioni e ricerche biologiche sul genere *Carabus* Linnaeus (*sensa lato*). Memorie della Societa Entomologica Italiana, 41, 85–202.

Sullivan. D, J., Castillo, J. A., & Bellotti, A. C. (1991). Comparative biology of six species of coccinellid beetles (Coleoptera: Coccinellidae) predaceous on mealybug, *Phenacoccus herreni* (Homoptera: Pseudococcidae), a pest of cassava in Columbia, South America. Environmental Entomology, 20, 685–689.

Summerlin, J. W., Fincher, G. T., Roth, J. P., & Petersen, H. D. (1989). Laboratory studies on the life cycle and prey relationships of *Pachylister caffer* Erichson (Coleoptera: Histeridae). Journal of Entomological Science, 24, 329–338.

Summers, F. M., & Witt, R. L. (1972). Nesting behavior of *Cheyletus eruditus* (Acarina: Cheyletidae). Pan-Pacific Entomologist, 48, 261–269.

Sundby, R. A. (1967). Influences of food on the fecundity of *Chrysoperla carnea* Stephens. Entomophaga, 12, 475–479.

Sunderland, K. D. (1988). Carabidae and other invertebrates. In A. K. Minks & P. Harrewijn (Eds.), Aphids, their biology, natural enemies and control (Vol. B, pp. 293–310). Amsterdam: Elsevier.

Sunderland, K. D., & Vickerman, G. P. (1980). Aphid feeding by some polyphagous predators in relation to aphid density in cereal fields. Journal of Applied Ecology, 17, 389–396.

Sunderland, K. D., Crook, N. E., Stacey, D. L., & Fuller, B. T. (1987). A study of feeding by polyphagous predators on cereal aphids using ELISA and gut dissection. Journal of Applied Ecology, 24, 907–933.

Swadener, S. O., & Yonke, T. R. (1973). Immature stages and biology of *Apiomerus crassipes* (Hemiptera: Reduviidae). Annals of the Entomological Society of America, 66, 188–196.

Swan, D. I. (1974). A review of work on predators, parasites and pathogens for the control of *Oryctes rhinoceros* (L.) (Coleoptera: Scarabaeidae) in the Pacific area (CIBC Miscellaneous Publication No. 7).

Sweet, M. H. (1960). The seedbugs; a contribution to the feeding habits of Lygaeidae (Hemiptera-Heteroptera). Annals of the Entomological Society of America, 53, 317–321.

Sweet, M. H. (1979). On the original feeding habits of the Hemiptera (Insecta). Annals of Entomological Society of America, 72, 575–579.

Sweetman, H. L. (1958). The principles of biological control. Dubuque, IA: Wm. C. Brown.

Swezey, O. H. (1936). Biological control of the sugar cane leafhopper in Hawaii. Bulletin of the Experimental Station Hawaiian Sugar Planters Association, 21, 57–101.

Swirski, E., Amitai, S., & Dorzia, N., (1967a). Laboratory studies on the feeding, development and reproduction of the predacious mites *Ambly-seius rubini* Swirski & Amitai and *Amblyseius swirskii* Athias (Acarina, Phytoseiidae) on various kinds of food substances. Israel Journal of Agricultural Research, 17, 101–119.

Swirski, E. S., Amitai, S., & Dorzia, N. (1967b). Laboratory studies on the feeding, development and oviposition of the predaceous mite *Typhlodromus athiasae* P. & S. (Acarina: Phytoseiidae) on various kinds of food substances. Israel Journal of Agricultural Research, 17, 213–228.

Symondson, W. O. C., & Liddell, J. E. (1993). The detection of predation by *Abax parallelepipedus* and *Pterostichus madidus* (Coleoptera: Carabidae) on Mollusca using a quantitative ELISA. Bulletin of Entomological Research, 83, 641–647.

Szabo, S., & Szentkiralyi, F. (1981). Communities of Chrysopidae and Hemerobiidae (Neuroptera) in some apple-orchiards. Acta Phytopathologica Academiae Scientiarum Hungaricae, 16, 157–169.

Szmidt, A., & Wegorek, W. (1967). Population dynamicsche wirkung von *Perillus bioculatus* (Fabr.) (Het. Pentatomidae) auf den Kartoffelkafer. Entomophaga, 12, 403–408.

Szumkowski, W. (1961). Aparicon de un coccinellido predator nuevo para Venezuala. Agronomia Tropical (Venezuala), 11, 33–37.

Tadic, M. (1964). Possibilities for introducing into Yugoslavia *Podisus placidus* Uhl., an American predator of the fall army webworm. Journal of Science and Agricultural Research, 16, 43–52.

Takafuji, A., & Chant, D. A. (1976). Comparative studies of two species of predaceous mites (Acarina: Phytoseiidae) with special reference to their responses to the density of their prey. Researches on Population Ecology (Kyoto), 17, 255–310.

Takeda, S., Hukusima, S., & Yamada, S. (1964). Seasonal abundance of coccinellid beetles. Research Bulletin of the Faculty of Agriculture, Gifu University, 19, 55–63.

Tamaki, G. (1972). The biology of *Geocoris bullatus* inhabiting orchard floors and its impact on *Myzus persicae* on peaches. Environmental Entomology, 1, 559–565.

Tamaki, G., & Butt, B. A. (1978). Impact of *Perillus bioculatus* on the Colorado potato beetle and plant damage (Technical Bulletin 1581). Washington, DC: U.S. Department of Agriculture.

Tamaki, G., & Long, G. E. (1978). Predator complex of the green peach aphid on sugarbeets: Expansion of the predator power and efficacy model. Environmental Entomology, 7, 835–842.

Tamaki, G., & Weeks, R. E. (1968). *Anthocoris malanocerus* as a predator of the green peach aphid on sugar beets and broccoli. Annals of the Entomological Society of America, 61, 579–584.

Tamaki, G., & Weeks, R. E. (1972a). Efficiency of three predators, *Geocoris bullatus*, *Nabis americoferus*, and *Coccinella transversoguttata*, used alone or in combination against three prey species, *Myzus persicae*, *Ceramica picta*, and *Mamestra configurata*, in a greenhouse study. Environmental Entomology, 1, 258–263.

Tamaki, G., & Weeks, R. E. (1972b). Biology and ecology of two predators, *Geocoris pallens* Stål and *G. bullatus* (Say) (Technical Bulletin No. 1446). Washington, DC: U.S. Department of Agriculture.

Tamaki, G., & Weeks, R. E. (1973). The impact of predators on populations of green peach aphids on field-grown sugarbeets. Environmental Entomology, 2, 345–349.

Tamaki, G., McGuire, J. U., & Turner, J. E. (1974). Predator power and efficacy: A model to evaluate their impact. Environmental Entomology, 3, 625–630.

Tamaki, G., Weiss, M. A., & Long, G. E. (1981). Evaluation of plant density and temperature in predator-prey interactions in field cages. Environmental Entomology, 10, 716–720.

Tanaka, M. (1966). Fundamental studies on the utilization of natural enemies in the citrus grove in Japan. I. The bionomics of natural enemies of the most serious pests. II. *Stethorus japonicus* Kamiya (Coccinellidae) a predator of the citrus red mite, *Panonychus citri* (McGregor). Bulletin of Horticulture Research Station (Japan), Series D, 4, 22–42 (in Japanese; English summary).

Tanaka, M., & Kashio, T. (1977). Biological studies on *Amblyseius largoensis* (Muma) (Acarina: Tetranychidae). Bulletin of Fruit Tree Research Station, Series D (Kuchinotsu), 1, 49–67.

Tanasijtschuk, V. N. (1959). Species of the genus *Leucopis* Mg. (Diptera, Chamaemyiidae) from the Crimea. Entomological Review, 38, 829–844.

Tanasijtschuk, V. N. (1993). Morphological differences and phylogenetic relationships of genera of Chamaemyiidae Diptera. Entomological Review, 72, 66–100.

Tanasijtschuk, V. N. (1996). Two species of *Leucopis* Meigen (Diptera: Chamaemyiidae) predacious on the Russian wheat aphid, *Diuraphis noxia* (Mordvilko) (Homoptera: Aphididae), in North America. Proceedings of the Entomological Society of Washington, 98, 640–646.

Tanigoshi, L. K. (1982). Advances in knowledge of the Phytoseiidae. In M. A. Hoy (Ed.), Recent advances in knowledge of the Phytoseiidae (Publication 3284, pp. 1–22). Berkeley: University of California Press.

Tanigoshi, L. K., & McMurtry, J. A. (1977). The dynamics of predation of *Stethorus picipes* (Coleoptera: Coccinellidae) and *Typhlodromus floridanus* on the prey *Oligonychus punicae* (Acarina: Phytoseiidae, Tetranychidae). Hilgardia, 45, 237–288.

Tanigoshi, L. K., Hoyt, S. C., & Croft, B. A. (1983). Basic biology and management components for mite pests and their natural enemies. In B. A. Croft & S. C. Hoyt (Eds.), Integrated management of insect pests of pome and stone fruits (pp. 153–202). New York: Wiley-Interscience.

Tanigoshi, L. K., Fargerlund, J., Nishio-Wong, J. Y., & Griffiths, H. J. (1985). Biological control of citrus thrips, *Scirtothrips citri*, in southern California citrus groves. Environmental Entomology, 14, 733–741.

Tanigoshi, L. K., Hoyt, S. C., Browne, R. W., & Logan, J. A. (1975). Influence of temperature on population increase of *Metaseiulus occidentalis* (Acarina: Phytoseiidae). Annals of the Entomological Society of America, 68, 979–986.

Tassan, R. L., Hagen, K. S., & Sawall, E. F., Jr. (1979). The influence of field food sprays on the egg production rate of *Chrysoperla carnea*. Environmental Entomology, 8, 81–85.

Tauber, C. A. (1974). Systematics of North American chrysopid larvae: *Chrysopa carnea* group (Neuroptera). Canadian Entomologist, 106, 1113–1153.

Tauber, C. A. (1991a). Order Neuroptera. In F. W. Stehr (Ed.), Immature insects (Vol. 2, pp. 126–143). Dubuque, IA: Kendall/Hunt.

Tauber, C. A. (1991b). Order Raphidioptera. In F. W. Stehr (Ed.), Immature insects (Vol. 2, pp. 123–125). Dubuque, IA: Kendall/Hunt.

Tauber, C. A., & Tauber, M. J. (1968). *Lomamyia latipennis* (Neuroptera: Berothidae) life history and larval descriptions. Canadian Entomologist, 100, 623–629.

Tauber, C. A., & Tauber, M. J. (1973). Diversification and secondary integradation of two *Chrysoperla carnea* strains (Neuroptera: Chrysopidae). Canadian Entomologist, 105, 1153–1167.

Tauber, C. A., & Tauber, M. J. (1982). Evolution of seasonal adaptations and life history traits in *Chrysopa*: response to diverse selective pressures. In H. Dingle & J. P. Hegmann (Eds.), Evolution and genetics of life histories (pp. 51–72). New York: Springer-Verlag.

Tauber, C. A., & Tauber, M. J. (1986). Ecophysiological responses in life-history evolution: Evidence for their importance in a geographically widespread insect species complex. Canadian Journal of Zoology, 64, 875–884.

Tauber, C. A., & Tauber, M. J. (1987a). Food specificity in predacious insects: A comparative ecophysiological and genetic study. Evolutionary Ecology, 1, 175–186.

Tauber, C. A., & Tauber, M. J. (1987b). Inheritance of seasonal cycles in *Chrysoperla* (Insecta: Neuroptera). Genetic Research, 49, 215–223.

Tauber, C. A., Tauber, M. J., & Nechols, J. R. (1987). Thermal requirements for development in *Chrysopa oculata*: A geographically stable trait. Ecology, 68, 1479–1487.

Tauber, M. J., & Tauber, C. A. (1972). Geographic variation in critical photoperiod and in diapause intensity of *Chrysoperla carnea* (Neuroptera). Journal of Insect Physiology, 18, 25–29.

Tauber, M. J., & Tauber, C. A. (1974). Dietary influence on reproduction in both sexes of five predacious species (Neuroptera). Canadian Entomologist, 106, 921–925.

Tauber, M. J., & Tauber, C. A. (1975). Criteria for selecting *Chrysopa carnea* biotypes for biological control: Adult dietary requirements. Canadian Entomologist, 107, 589–595.

Tauber, M. J., & Tauber, C. A. (1983). Life history traits of *Chrysopa carnea* and *Chrysopa rufilabris* (Neuroptera: Chrysopidae): Influence of humidity. Annals of the Entomological Society of America, 76, 282–285.

Tauber, M. J., & Tauber, C. A. (1993). Adaptations to temporal variation in habitats: Categorizing, predicting, and influencing their evaluation in agroecosystems. In K. C. Kim & B. A. McPheron (Eds.), Evolution if insect pests/ patterns of variation (pp. 103–127). New York: John Wiley & Sons.

Tauber, M. J., Tauber, C. A., & Gardescu, S. (1993). Prolonged storage of *Chrysoperla carnea* (Neuroptera: Chrysopidae). Environmental Entomology, 22, 843–848.

Tawfik, M. F. S. (1965). The life-history of *Leucopis puncticornis aphidivora* Rand. (Diptera: Octhiphilidae). Bulletin de la Societe Entomologique de Egypte, 49, 55–57.

Tawfik, M. F. S., & Nagui, A., (1965). The biology of *Montandoniella moraguesi* Puton, a predator of *Gynaikothrips ficorum ficorum* Marchal in Egypt. Bulletin de la Societe Entomologique de Egypte, 49, 195–200.

Taylor, E. J. (1949). A life history of *Nabis alternatus*. Journal of Economic Entomology, 42, 991.

Taylor, T. H. C. (1935). The campaign against *Aspidiotus destructor* Sign., in Fiji. Bulletin of Entomological Research, 26, 1–102.

Tedders, W. L., & Schaefer, P. W. (1994). Release and establishment of *Harmonia axyridis* (Coleoptera: Coccinellidae) in the southeastern United States. Entomological News, 105, 228–243.

Teerling, D. R., Gillespie, D. R., & Borden, J. H. (1993). Utilization of western flower thrips alarm pheromone as a prey-finding kairomone by predators. Canadian Entomologist, 125, 431–437.

Tenhumberg, B. (1995). Estimating predatory efficiency of *Episyrphus balteatus* (Diptera: Syrphidae) in cereal fields. Environmental Entomology 24, 687–691.

Terry, F. W. (1905). Leaf-hoppers and their natural enemies, V. Forticulidae, Syrphidae and Hemerobiidae. Hawaiian Sugar Planters Association Bulletin, 1, 162–181.

Ter-Simonjan, L. G., Bushtshik, T. N., & Baskresenskaja, V. N. (1982). Biological control in vegetable plantings. Acta Entomologica Fennica, 40, 33–35.

Thakur, J. W., Pawar, A. D., & Rawat, U. S. (1988). Observations on the correlation between population density of apple woolly aphid and its natural enemies and their effectiveness in Kulla Valley (H.P.). Plant Protection Bulletin of India, 40, 13–15.

Thead, L. G., Pitre, H. N., & Kellogg, T. F. (1985). Feeding behavior of adult *Geocoris punctipes* (Say) (Hemiptera: Lygaeidae) on nectaried and nectariless cotton. Environmental Entomology, 14, 134–137.

Thiele, H. U. (1977). Carabid beetles in their environments. Springer-Verlag.

Thomas, D. B. (1992). Taxonomic synopsis of the asopine Pentatomidae (Heteroptera) of the Western Hemisphere. Entomological Society of America Thomas Say Foundation, 16, 1–156.

Thomas, M. B., Sotherton, N. W., Coombes, D. S., & Wratten, S. D. (1992). Habitat factors influencing the distribution of polyphagous predatory insects between field boundaries. Annals of Applied Biology, 120, 197–202.

Thompson, R. G. (1979). Larvae of North American Carabidae with a key to the tribes. In T. L. Erwin *et al.* (Eds.), Carabid beetles: Their evolution, natural history, and classification (pp. 205–291). The Hague: Dr. W. Junk.

Thompson, W. R. (1951). The specificity of host relations in predaceous insects. Canadian Entomologist, 83, 262–269.

Tigar, B. J., Osborne, P. E., Key, G. E., Flores, S. M. E., & Vazquez, A. M. (1994). Distribution and abundance of *Prostephanus truncatus* (Coleoptera: Bostrichidae) and its predator *Teretriosoma nigrescens* (Coleoptera: Histeridae) in Mexico. Bulletin of Entomological Research, 84, 555–565.

Timberlake, P. H. (1943). The Coccinellidae of ladybeetles of the Koebele collection-I. Hawaiian Planters Record, 47, 1–67.

Timofeyeva, T. V., & Hoang, D. N. (1978). Morphology and biology of the Indian ladybird *Serangium parcesetosum* Sicard (Coleoptera, Coccinellidae) predacious on the citrus whitefly in Adzaria. Entomological Review, 57, 210–214 (translated from Russian).

Tinkeu, L. N. (1995). Oviposition and hatching of the eggs of *Epistrophe nitidicollis* (Diptera: Syrphidae). Mededelingen Faculteit Landbouwkundige en Toegepaste Biologische Wetenschappen Universiteit Gent, 60, 619–623.

Tjeder, B. (1960). Neuroptera from New Foundland, Miquelon, and Labrador. Opuscula Entomologica, 25, 146–149.

Tomlin, A. D., McLeod, D. G. R., Moore, L. V., Whistlecraft, J. W., Miller, J. J., & Tolman, J. H. (1992). Dispersal of *Aleochara bilineata* (Col.: Saphylinidae) following inundative releases in urban gardens. Entomophaga, 37, 55–63.

Tonhasca, A. J. (1993). Carabid beetle assemblage under diversified agroecysystems. Entomologia Experimentalis et Applicata, 68, 279–285.

Toschi, C. A. (1965). The taxonomy, life histories, and mating behavior of the green lacewings of Strawberry Canyon (Neuroptera: Chrysopidae). Hilgardia, 36, 391–431.

Tostowaryk, W. (1971). Life history and behavior of *Podisus modestus* (Hemiptera: Pentatomidae) in boreal forest in Quebec. Canadian Entomologist, 103, 662–674.

Tothill, J. D., Taylor, T. H. C., & Paine, R. W. (1930). The coconut moth in Fiji. London: Imperial Bureau of Entomology.

Tozatti, G., & Gravena, S. (1988). Fatores naturais de mortalidade de *Perileucoptera coffeella* (Guerin-Meneville) (Lepidoptra: Lyonetiidae), em cafe, Jaboticabal. Cientifica Jaboticabal, 16, 179–187.

Tracewski, K. T. (1983). Description of the immature stages of *Leucopis* sp. nr. *albipuncta* (Diptera: Chamaemyiidae) and their role as predators of the apple aphid, *Aphis pomi* (Homoptera: Aphididae). Canadian Entomologist, 115, 735–742.

Treacy, M. F., Benedict, J. H., Lopez, J. D., & Morrison, R. K. (1987). Functional response of a predator (Neuroptera: Chrysopidae) to bollworm (Lepidoptera: Noctuidae) eggs on smooth leaf, hirsute, and pilose cottons. Journal of Economic Entomology, 86, 376–379.

Trelka, D. G., & Foote, B. A. (1970). Biology of slug-killing *Tetanocera* (Diptera: Sciomyzidae). Annals of the Entomological Society of America, 63, 877–895.

Tremblay, E., & Zouliamis, N. (1968). Dati conclusivi sull introduzione sulla biologis ed impigo del *Perillus bioculatus* (Fab.) (Heteroptera: Pentatomidae) nell'Itala meridenale. Bollettino del Laboratori Entomologica Agricultura "Filippo Silvestri," Portici, 26, 99–122.

Trichilo, P. J., & Leigh, T. F. (1986). Predation on spider mites by the western flower thrips *Frankliniella occidentalis* (Thysanoptera: Thripidae). Environmental Entomology, 15, 821–825.

Trouve, C., Ledee, S., Ferran, A., & Brun, J. (1997). Biological control of the damson-hop aphid, *Phorodon humuli* (Hom.: Aphididae), using the ladybeetle *Harmonia axyridis* (Col.: Coccinellidae). Entomophaga, 42, 57–62.

Trouvelot, B. (1932). Researches sur les parasites et predateurs attaquant le doryphore en Amerique du nord et envol en France des premeres colonies des especis les plus avactives. Annales de Epiphyties, 17, 408–445.

Tryon, E. H., Jr. (1986). The striped earwig, and ant predators of sugar cane root stock borer in Florida citrus. Florida Entomologist, 69, 336–343.

Tsukaguchi, S. (1978). Descriptions of the larvae of *Chrysopa* Leach (Neuroptera, Chrysopidae) of Japan. Kontyu, 46, 99–122.

Tulisalo, U. (1984a). Mass rearing techniques. In M Canard, Y. Semeria, & T. R. New (Eds.), Biology of Chrysopidae (pp. 213–220). The Hague: Dr. W. Junk.

Tulisalo, U. (1984b). Biological control in the greenhouse. In M. Canard, Y. Semeria, & T. R. New (Eds.), Biology of Chrysopidae (pp. 228–233). The Hague: Dr. W. Junk.

Tulisalo, U., & Korpela, S. (1973). Mass rearing of the green lacewing (*Chrysoperla carnea* Steph.). Annales Entomologici Fennici, 39, 143–144.

Tulisalo, U., & Tuovinen, T. (1975). The green lacewing, *Chrysoperla carnea* Steph. (Neuroptera: Chrysopidae), used to control the green peach aphid, *Myzus persicae* Sulz., and the potato aphid, *Macrosiphum euphorbiae* Thomas (Homoptera: Aphididae), on greenhouse green peppers. Annales Entomologici Fennici, 41, 94–102.

Tulisalo, U., Tuovinen, T., & Kurppa, S. (1977). Biological control of aphids with *Chrysoperla carnea* on parsley and green pepper in the greenhouse. Annales Entomologici Fennici, 43, 97–100.

Turnbow, R. H., Franklin, R. T., & Nagel, W. P. (1978). Prey consumption and longevity of adult *Thanasimus dubius*. Environmental Entomology, 7, 695–697.

Turnock, W. J., Boivin G., & Whistlecraft, J. W., (1995). Parasitism of overwintering puparia of the cabbage maggot, *Delia radicum* (L.) (Diptera: Anthomyiidae), in the relation to host density and weather factors. Canadian Entomologist, 127, 535– 542.

Underhill, R. A. (1954). Habits and life history of *Arilus cristatus* (Hemiptera: Reduviidae). Walla Walla College Publications, Department of Biological Science, 11, 1–15.

Ushchekov, A. T. (1975). A predacious cecidomyiid in glasshouses. Zashchita Rastenii 1975 (9), 21–22.

Ushchekov, A. T. (1976). *Chrysopa septempunctata* Wesm. in glasshouses. Zashchita Rastenii, 10, 16–17 (in Russian).

Ushchekov, A. T. (1989). *Chrysopa perla* for aphid control. Zashchita Rastenii, 11, 20–22 (in Russian).

Usinger, R. L. (1939). Distribution and host relationships of *Cyrtorhinus* (Hemiptera: Miridae). Proceedings of the Hawaiian Entomological Society, 10, 271–273.

Uygun, V. D. (1971). Der Einfluss der Nahrungsmenge auf Fruchtbarkeit und Lebensdauer von *Aphidoletes aphidimyza* (Rond.) (Diptera: Itonidae). Zeitschrift fuer Angewandte Entomologie, 69, 234–258.

Vala, J.-C., Casc, C., Gbedjissi, G., & Dossou, C. (1995). Life history, immature stages and sensory receptors of *Sepedon (Parasepedon) trichrooscelis* an Afrotropical snail-killing fly (Diptera: Sciomyzidae). Journal of Natural History, 29, 1005–1014.

Valley, K., & Berg, C. O. (1977). Biology, immature stages and new species of snail-killing Diptera of the genus *Dictya* (Sciomyzidae). Search, 7, 1–44.

Van den Berg, E. C., & Bedford, G. (1988). Notes on the biology, feeding behaviour and importance of *Menida lythrodes* (Hemiptera: Pentatomidae), predator of two important citrus pests. Phytophylactica, 20, 15–16.

Van den Berg, M. A., Dippensaar-Shoemaz, A. S., Deacon, V. E., & Anderson, S. H. (1992). Interactions between citrus psylla, *Trioza erytreae* (Hem.: Triozidae), and spiders in an unsprayed citrus orchard in the Transvaal Lowveld. Entomophaga, 37, 599–608.

Vandenberg, N. J. (1990). First North American records for *Harmonia quadripunctata* (Pontopiddian) (Coleoptera: Coccinellidae), a lady beetle native to the Palearctic. Proceedings of the Entomological Society of Washington, 92, 407–410.

Vandenberg, N. J. (1992). Revision of the New World lady beetles of the genus *Olla* and description of a new allied genus (Coleoptera: Coccinellidae). Annals of the Entomological Society of America, 85, 370–392.

van den Bosch, R., & Hagen, K. S. (1966). Predaceous and parasitic arthropods in California cotton fields. California Agricultural Experimental Station Bulletin, 820, 1–32.

van den Bosch, R., Leigh, T. F., Gonzalez, D., & Stinner, R. E. (1969). Cage studies on predators of the bollworm in cotton. Journal of Economic Entomology, 62, 1486–1489.

van den Meiracker, R. A. F., Hammond, W. N. O., & van Alphen, J. J. M. (1990). The role of kairomones in prey finding by *Diomus* sp. and *Exochomus* sp., two coccinellid predators of the cassava mealybug, *Phenacoccus manihoti*. Entomologia Experimentalis et Applicata, 56, 209–217.

Vanderplank, F. L. (1958). The assassin bug, *Platymerus rhadamanthus* Gerst (Hemiptera: Reduviidae), a useful predator of the rhinoceros beetles *Oryctes boas* (F.) and *Oryctes monoceros* (Oliv.) (Coleoptera: Scarabaeidae). Journal of Entomological Society of South Africa, 21, 309–314.

Van de Veire, M., & Degheele, R. D. (1993). Control of western flowr thrips, *Frankliniella occidentalis* with the predator *Orius insidiosus* on sweet peppers. IOBC WPRS Bulletin, 16(2), 185–188.

van Dijk, Th. S. (1979). On the relationship between reproduction, age and survival in two carabid beetles: *Calathus melanocephalus* L. and *Peterostichus coerulescens* L. (Coleoptera, Carabidae). Oecologia (Berlin), 40, 63–80.

van Dijk, Th. S. (1983). The influence of food and temperature on the amount of reproduction in carabid beetles. How to translate the results of laboratory experiments into reality of the field. In P. Brandmayr *et al.* (Eds.), The synthesis of field study and laboratory experiment (pp. 105–123). (Rep. 4th Symposium of European carabidolgists). Wageningen: Central Agricultural Publication Document (Pudoc).

van Emden, H. F. (Ed.). (1972). Aphid technology. London: Academic Press.

van Emden, H. F., & Hagen, K. S. (1976). Olfactory reactions of the green lacewing *Chrysopa carnea* to tryptophan and certain breakdown products. Environmental Entomology, 5, 469–473.

van Lenteren, J., & Woets, J. (1988). Biological and integrated control in greenhouses. Annual Review of Entomology, 33, 239–269.

Vanndell, W. L., & Storch, R. H. (1972). Food lists of *Hippodamia* (Coleoptera: Coccinellidae). University of Maine Agricultural Experimental Station Technical Bulletin, 55, 19.

Van Rijn, P. C. J., & Sabelis, M. W. (1990). Pollen availability and its effect on the maintenance of populations of *Amblyseius cucumeris*, a predator of thrips. International Symposium on Crop Protection. Mededelingen Faculteit Landbouwwetenschappen Rijksuniversteit Gent, 55, 335–342.

Varis, A. L. (1989). Cabbage field Carabidae and their role as natural enemies of *Delia radicum* and *Delia floralis* (Diptera: Anthomyiidae). Acta Entomologica Fennica, 53, 61–64.

Verma, J. S. (1955). Biological studies to explain the failure of *Cyrtorhinus mundulas* (Breddin) as an egg-predator of *Peregrinus maidis* (Ashmead) in Hawaii. Proceedings of the Hawaiian Entomological Society, 15, 623–634.

Viggiani, G. (1971). Osservazioni biologiche sul mirid predator *Deraeocoris ruber* (L.) (Rhynchota, Heteroptera). Bollettino del Laboratori Entomologia Agrario Filippo Silvestri, 29, 270–286.

Villevieille, M., & Millot, P. (1991). Lutte biologique contre *Frankliniella occidentalis* avec *Orius laevigatus* sur frasier. IOBC WPRS Bulletin, 14, 57–64.

Vinson, S. B., & Scarborough, T. A. (1991). Interactions between *Solenopsis invicta* (Hymenoptera: Formicidae), *Rhopalosiphum maidis* (Homoptera: Aphididae), and the parasitoid *Lysiphlebus testaceipes*

Cresson (Hymenoptera: Aphidiidae). Annals of the Entomological Society of America, 84, 158–164.

Waage, J. K., Carl, K. P., Mills, N. J., & Greathead, D. J. (1984). Rearing entomophagous insects. In P. Singh & R. F. Moore (Eds.), Handbook of insect rearing (Vol. 1, pp. 45–66) Amsterdam: Elsevier.

Waddill, V. H. (1978). Sexual differences in foraging on corn of adult *Labidura riparia* (Derm.: Labiduridae). Entomophaga, 23, 339–342.

Waddill, V., & Shepard, M. (1974a). Potential of *Geocoris punctipes* (Hemiptera: Lygaeidae) and *Nabis* spp. (Hemiptera: Nabidae) as predators of *Epilachna varivestis* (Coleoptera: Coccinellidae). Entomophaga, 19, 421–426.

Waddill, V., & Shepard, M. (1974b). Biology of predaceous stinkbug, *Stiretrus anchorogo* (Hemiptera: Pentatomidae). Florida Entomologist, 57, 249–253.

Waddill, V., & Shepard, M. (1975). A comparison of predation by the pentatomids, *Podisus maculiventris* (Say) and *Stiretrus anchorago* (F.) on the Mexican bean beetle, *Epilachna varivestis* Mulsant. Annals of the Entomological, Society of America, 68, 1023–1027.

Wallace, M. M. H. (1967). The ecology of *Sminthurus viridis* (L.) (Collembola). I. Processes influencing numbers in pastures in Western Australia. Australian Journal of Zoology, 15, 1173–1205.

Wallace, M. M. H. (1974). An attempt to extend the biological control of *Sminthurus viridis* (Collembola) to new areas in Australia by introducing a predatory mite, *Neomolgus cappillatus* (Bdellidae). Australian Journal of Zoology, 22, 519–529.

Wallace, M. M. H., & Holm, E. (1983). Establishment and dispersal of the introduced predatory mite, *Macrocheles peregrinus* Krantz in Australia. Journal of the Australian Entomological Society, 22, 345–348.

Wallace, M. M. H., Tyndale-Biscoe, M., & Holm, E. (1979). The influence of *Macrocheles glaber* on the breeding of the Australian bushfly, *Musca vetusstissima* in cow dung. In J. G. Rodriquez (Ed.), Recent advances in Acarology (Vol. 2, pp. 217–222). New York: Academic Press.

Wallin, H. (1988). Mandible wear in the carabid beetle *Pterostichus melanarius* in relation to diet and burrowing behavior. Entomologia Experimentalis et Applicata, 48, 43–50.

Wallin, H. (1991). Movement patterns and foraging tactics of a caterpillar hunter inhabiting alfalfa fields. Functional Ecology, 5, 740–749.

Wallin, H., Chiverton, P. A., Ekbom, B. S., & Borg, A. (1992). Diet, fecundity, and egg size in some polyphagous predatory carabid beetles. Entomologia Experimentalis et Applicata, 65, 129–140.

Waloff, N. (1974). Biology and behaviour of some species of Dryinidae (Hymenoptera). Journal of Entomology, Series A: Physiology & Behavior, 49, 97–109.

Wang, L. Y., & Hu, H. L. (1987). Studies of *Chrysopa kulingensis* Navas. Natural Enemy Insects, 9, 25–28 (in Chinese).

Wang, Y., & Gong, X.-Ch. (1987). Notes on the introduced predator. *Podisus maculiventris* in Anhui and Beijing. Chinese Journal of Biological Control, 3, 81–83.

Ward, L. K. (1970). *Aleuropteryx juniperi* Ohm. (Neuroptera: Coniopterygidae) new to Britain feeding on *Carulaspis juniperi* Bouché (Hemiptera: Diaspidae). Entomologist's Monthly Magazine, 106, 74–78.

Warren, L. O., & Wallis, G. (1971). Biology of the spined soldier bug, *Podisus maculiventris* (Hemiptera: Pentatomidae). Journal of the Georgia Entomological Society, 6, 109–116.

Waterhouse, D. F., & Norris, K. R. (1987). Biological control Pacific prospects. Melbourne: Inkata Press.

Waterhouse, D. E., & Norris, K. R. (1989). Biological control Pacific prospects—Supplement 1. No. 12 Canberra: Australian Centre International Agricultural Research Manager.

Watson, T. K., & Wilde, W. H. A. (1963). Laboratory and field observations on two predators of the pear psylla in British Columbia. Canadian Entomologist, 95, 435–438.

Way, M. J. (1992). Role of ants in pest management. Annual Review of Entomology, 37, 479–504.

Way, M. J., Cammell, M. E., Bolton, B., & Kanagaratnam, P. (1989). Ants (Hymenoptera: Formicidae) as egg predators of coconut pests, especially in relation to biological control of the coconut caterpillar, Opisina arenosella Walker (Lepidoptera; Xyloryctidae), in Sri Lanka. Bulletin of Entomological Research, 79, 219–233.

Wei, D., Huang, B., Guo, Z., Chen, X., Luo, Q., & Liang, W. (1987). Species and habits of green lacewings in Guangzhou. Natural Enemy Insects, 9, 21–24 (in Chinese).

Welbourn, W. C. (1983). Potential use of trombidioid and erythracoid mites as biological control agents of insects. In M. A. Hoy, G. L. Cunningham, & L. Knutson (Eds), Biological control of pests by mites, (Spec. Publication 3304, pp. 103–104. Berkeley: DANR, University of California.

Welch, S. M., Logan, J. A., & Mowery, P. D. (1983). Biological control models for management of phytophagous mites. In B. A. Croft & S. C. Hoyt (Eds.), Integrated management of insect pests of pome and stone fruits (pp. 309–341). New York: Wiley-Interscience.

Wen, B., & Deng, W. (1989). A simulation study on the control effect of Tribolium confusum (Col.: Tenebrionidae) by Xylocoris sp. (Hem.: Anthocoridae). Chinese Journal of Biological Control, 5, 6–8 (in Chinese with English summary).

Wene, G. P., & Sheets, L. W. (1962). Relationship of predatory and injurious insects in cotton fields in the Salt River Valley area of Arizona. Journal of Economic Entomology, 55, 395–398.

Werner, F. G., & Butler, G. D., Jr. (1957). The reduviids and nabids associated with Arizona crops. Arizona Agricultural Experimental Station Technical Bulletin, 133, 11.

Weseloh, R. M. (1985a). Predation by Calosoma sycophanta L. evidence for a large impact on gypsy moth, Lymantria dispar L. pupae. Canadian Entomologist, 117, 1117–1126.

Weseloh, R. M. (1985). Changes in population size, dispersal behavior, and reproduction of Calosoma sycophanta associated with changes in gypsy moth, Lymantria dispar abundance. Environmental Entomology, 14, 370–377.

Weseloh, R. M. (1987). Emigration and spatial dispersion of the gypsy moth predator Calosoma sycophanta. Entomologia Experimentalis et Applicata, 44, 187–93.

Weseloh, R. M. (1990). Experimental forest releases of Calosoma sycophanta (Coleoptera: Carabidae) against the gypsy moth. Journal of Economic Entomology, 83, 2229–2234.

Weseloh, R. M. (1993). Behavior of the gypsy moth predator, Calosoma sycophanta (Coleoptera: Carabidae), as influenced by time of day and reproductive status. Canadian Entomologist, 125, 887–894.

Weseloh, R. M. (1996). Effect of supplemental foods on foraging behavior of forest ants in Connecticut. Environmental Entomology, 25, 848–852.

Weseloh, R. M. (1997). Orientation of Calosoma sycophanta L. (Coleoptera: Carabidae) in forests: Insights from visual responses to objects. Canadian Entomologist, 129, 347–354.

Weslian, J., & Regnander, J. (1992). The influence of natural enemies on brood production in Ips typographus (Col.: Scolytidae) with special reference to egg-laying and predation by Thanasimus formicarius (Col.: Cleridae). Entomophaga, 37, 333–342.

West, A. S., & DeLong, B. (1955). Notes on the biology and laboratory rearing of a predatory insect, Zelus exsanguis (Stål) (Hemiptera: Reduviidae). Report of the Entomological Society of Ontario, 86, 97–101.

Westigard, P. H. (1973). The biology and effect of pesticides on Deraeocoris brevis piceatus (Heteroptera: Miridae). Canadian Entomologist, 105, 1105–1111.

Whalon, M. E., & Parker, B. L. (1978). Immunological identification of tarnished plant bug predators. Annals of Entomological Society of America, 71, 453–456.

Wharton, G. W., & Arlian, L. G. (1972). Predatory behavior of the mite Cheyletus aversor. Animal Behavior, 20, 719–723.

Wheater, C. P. (1989). Prey detection of some predatory Coleoptera (Carabidae & Staphylinidae). Journal of Zoology, 218, 171–185.

Wheater, C. P. (1991). Effect of starvation on locomotor activity in some predaceous Coleoptera (Carabidae, Staphylinidae). Coleopterists' Bulletin, 45, 371–378.

Wheeler, A. G., Jr. (1976). Lygus bugs as facultative predators. In D. R. Scott & L. E. O'Keefe (Eds.), Lygus bug: Host plant interaction (pp. 28–35). Moscow, ID: University Press of Idaho.

Wheeler, A. G., Jr. (1981). Updated distribution of Aleuropteryx juniperi (Neuroptera: Coniopterygidae), a predator of scale insects on ornamental juniper. Proceedings of the Entomological Society of Washington, 83, 173.

Wheeler, A. G., Jr. (1990). Propylea quatuordecimpunctata: Additional U.S. records of an adventive lady beetle (Coleoptera: Coccinellidae). Entomological News 101, 164–166.

Wheeler, A. G., Jr. (1993). Establishment of Hippodamia variegata and new records of Propylea quatuordecimpunctata (Coleoptera: Coccinellidae) in the eastern United States. Entomological News, 104, 102–110.

Wheeler, A. G., Jr., & Henry, T. J. (1978a). Ceratocapsus modestus (Hemiptera: Miridae), a predator of grape phylloxera: Seasonal history and description of fifth instar. Melsheimer Entomology Series, No 25, 6–10.

Wheeler, A. G., Jr., & Henry, T. J. (1978b). Isometopinae (Hemiptera: Miridae) in Pennsylvania: Biology and description of fifth instars, with observations of predation on obscure scale. Annals of the Entomological Society of America, 71, 607–614.

Wheeler, A. G., Jr., & Henry, T. J. (1992). A synthesis of the Holarctic Miridae (Heteroptera): Distribution, biology, and origin with emphasis on North America. Entomological Society of America Thomas Say Foundation Monograph, 15, 306.

Wheeler, A. G., Jr., & Hoebeke, E. R. (1981). A revised distribution of Coccinella undecimpunctata L. in eastern and western North America (Coleoptera: Coccinellidae). Coleopterists' Bulletin, 35, 213–216.

Wheeler, A. G., Jr., & Stoops, C. A. (1996). Status and spread of the Palearctic lady beetles Hippodamia variegata and Propylea quatuordecimpunctata (Coleoptera: Coccinellidae) in Pennsylvania, 1993–1995. Entomological News, 107, 291–298.

Wheeler, A. G., Jr., Colburn, R. B., & Lehman, R. D. (1973). Stethorus punctillum associated with spruce spidermite on ornamentals. Environmental Entomology, 2, 718–720.

Wheeler, A. G., Jr., Stinner, B. R., & Henry, T. J. (1975). Biology and nymphal stages of Deraeocoris nebulosus (Hemiptera: Miridae), a predator of arthropod pests on ornamentals. Annals of Entomological Society of America, 68, 1063–1068.

Whistlecraft, J. W., Harris, C. R., Tolman, J. H., & Tomlin, A. D. (1985). Mass rearing technique for Aleochara bilineata (Coleoptera: Staphylinidae). Journal of Economic Entomology, 78, 995–997.

Whitcomb, W. H. (1974). Natural populations of entomophagous arthropods and their effect on the agroecosystem. In F. G. Maxwell & F. A. Harris (Eds.), Proceedings, Summer Institute of Biological Control of Plant Insects and Diseases (pp. 150–169). Jackson: University Press of Mississippi.

Whitcomb, W. H., & Bell, K. (1964). Predaceous insects, spiders and mites of Arkansas cotton fields. Bulletin of Arkansas University, 690, 1–83.

Whitcomb, W. H., & Godfrey, K. E. (1991). The use of predators in insect control. (pp. 215–241). In D. Pimentel (Ed.), Handbook of pest management in agriculture (2nd ed., Vol. 2). Boca Raton, FL: CRC Press.

White, E. B., & Legner, E. F. (1966). Notes on the life history of Aleochara taeniata, a staphylinid parasite of the house fly, Musca domestica. Annals of the Entomological Society America, 59, 573–577.

White, E. G., & Huffaker, C. B. (1969). Regulatory process and population cyclicity in laboratory populations of Anagasta kuehniella (Zeller) (Lepidoptera: Phycitidae). I. Competition for food and predation. Researches on Population Ecology, 11, 57–83.

White, T. D., & Jubb, G. L., Jr. (1980). Potential of food sprays for augmenting green lacewing populations in vineyards. Melsheimer Entomology Series, 29, 35–42.

Whitehead, V. B. (1967). The validity of the higher taxonomic categories of tribe Scymnini (Coleoptera: Coccinellidae). Unpublished doctoral dissertation, University of California. Berkeley.

Wickler, W. (1968). Mimicry in plants and animals. [Translated from German by R. D. Martin.] New York: McGraw-Hill.

Wiedenmann, R. N., & O'Neil, R. J. (1990a). Effects of low rates of predation on selected life-history characteristics of *Podisus maculiventris* (Say) (Heteroptera: Pentatomidae). Canadian Entomologist, 122, 271–283.

Wiedenmann, R. N., & O'Neil, R. J. (1990b). Response of *Nabis roseipennis* (Heteroptera: Nabidae) to larvae of Mexican bean beetles *Epilachna varivestis* (Col.: Coccinellidae). Entomophaga, 35, 449–458.

Wiedenmann, R. N., & O'Neil, R. J. (1992). Searching strategy of the predator *Podisus maculiventis* (Jay) (Heteroptera: Pentatomidae). Environmental Entomology, 21, 1–9.

Wilbert, H. (1977). Der Honigtau als Reiz- und Energiequelle für Entomophage Insekten. Aphidologie, 8, 393–400.

Wilhoit, L. R., Stinner, R. E., & Axtell, R. C. (1991). CARMOD: A simulation model for *Carcinops pumilio* (Coleoptera: Histeridae) population dynamics and predation on immature stages of houseflies (Diptera: Muscidae). Environmental Entomology, 20, 1079–1088.

Wilkinson, P. R. (1970). Factors affecting the distribution and abundance of the cattle tick in Australia: Observations and hypotheses. Acarologia, 12, 492–508.

Williams, F. X. (1931). The insects and other invertebrates of Hawaiian sugarcane fields. Hawaiian Sugar Planters' Association.

Williams, F. X. (1956). Life history studies of *Pepsis* and *Hemipepsis* wasps in California (Hymenoptera, Pompilidae). Annals of the Entomological Society of America, 49, 447–466.

Williams, J. R. (1957). The sugar-cane Delphacidae and their natural enemies in Mauritius. Transactions of the Royal Entomological Society of London, 109, 65–110.

Williams, J. R., & Greathead, D. J. (1990). Sugarcane. In D. Rosen (Ed.), Armored scale insects: Their biology, natural enemies and control (Vol. B, (pp. 563–578). Amsterdam: Elsevier.

Williamson, D. L. (1971). Olfactory discernment of prey by *Medetera bistriata* (Diptera: Dolichopodidae). Annals of the Entomological Society of America, 64, 586–589.

Wilson, F. (1938). Notes on the insect enemies of chermes with particular reference to *Pineus pini* Koch and *P. strobi* Hartig. Bulletin of Entomological Research, 29, 337–389.

Wilson, L. T., & Gutierrez, A. P. (1980). Within-plant distribution of predators on cotton. *Gossypium hirsutum:* Comments on sampling and predator efficiencies. Hilgardia, 48, 3–11.

Wingfield, M. T., & Warren, L. O. (1968). Notes on the biology of a clerid beetle *Phyllobaenus singularis* (Wolc.). Journal of Georgia Entomological Society, 3, 41–44.

Wipperfürth, T., Hagen, K. S., & Mittler, T. E. (1987). Egg production by the coccinellid *Hippodamia convergens* fed on two morphs of the green peach aphid, *Myzus persicae*. Entomologia Experimentalis et Applicata, 44, 195–198.

Witcomb, W. D. (1940). Biological control of mealybugs in greenhouses. Massachusetts Agricultural Experimental Station Bulletin, 375, 1–22.

Withycombe, C. L. (1923). Notes on the biology of some British Neuroptera. Transactions of the Royal Entomological Society of London, 1922, 51–594.

Withycombe, C. L. (1924). Notes on the economic value of the Neuroptera with special reference to the Coniopterygidae. Annals of Applied Biology, 11, 112–125.

Wnuk, A. (1977). The natural enemy *Episyrphus balteatus* (Deg.) (Diptera, Syrphidae) limiting *Aphis pomi* Deg. (Hom.: Aphididae). Polski Pismo Entomologiczne, 47, 455–460.

Wnuk, A., & Fuchs, R. (1977). Observations on the effectiveness of the limitation of *Brevicoryne brassicae* (L.) by Syrphidae. Polski Pismo Entomologiczne, 47, 147–156.

Woglum, R. S., & McGregor, E. A. (1958). Observations on the life history and morphology of *Agulla bractea* Carpenter (Neuroptera: Raphidiodea: Raphidiidae). Annals of Entomological Society of America, 51, 129–141.

Woglum, R. S., & McGregor, E. A. (1959). Observations on the life history and morphology of *Agulla astuta* (Banks) (Neuroptera: Raphidiodea: Raphidiidae). Annals of the Entomological Society of America, 52, 489–502.

Wolcott, G. N. (1937). Annual report of fiscal year 1935–1936 of Entomology (Report of Agricultural Experimental Station, pp. 53–64). Rio Piedras: University of Puerto Rico.

Wolcott, G. N. (1960). Efficiency of lady beetles (Coccinellidae: Coleoptera) in insect control. Journal of Agricultural University of Puerto Rico, 44, 166–172.

Wong, T. T., McInnis, K. O., Nishimoto, J. I., Ota, A. K., & Chang, U. S. C. (1984). Predation of the Mediterranean fruit fly (Diptera: Tephritidae) by the Argentine ant (Hymenoptera: Formicidae) in Hawaii. Journal of Economic Entomology, 77, 1454–1458.

Woodruff, R. E., & Sailer, R. I. (1977). Establishment of the genus *Azya* in the United States (Coleoptera: Coccinellidae). Florida Department of Agriculture. (Entomological Circular 176, pp. 1–2).

Wratten, S. D. (1973). The effectiveness of the coccinellid beetle, *Adalia bipunctata* (L.), as a predator of the lime aphid, *Eucallipterus tiliae* (L.). Journal of Animal Ecology, 42, 785–802.

Wratten, S. D. (1976). Searching by *Adalia bipunctata* (L.) (Coleoptera: Coccinellidae) and escape behavior of its aphid and cicadellid prey on lime (*Tilia vulgaris* Hayne). Ecological Entomology, 1, 139–142.

Wratten, S. D. (1987). The effectiveness of native natural enemies. In A. J. Burn, T. H. Cooker, & P. C. Jepson (Eds.), Integrated pest management (pp. 89–112). London: Academic Press.

Wu, T. K. (1992). Feasibility of controlling citrus red mite, *Panonychus citri* (Acarina: Tetranychidae) by green lacewing *Mallada basalis* (Neuroptera: Chrysopidae). Chinese Journal of Entomology, 12, 81–89.

Wyatt, T. D., Phillips, A. D. G., & Grégoire, I.-C. (1993). Turbulence trees and semiochemicals wind-tunnel orientation of the predator *Rhizophagus-grandis* to its bark beetle prey *Dendroctonus-micans*. Physiological Entomology, 18, 204–210.

Xiong-Fei, P., & Gordon, R. D. (1986). The Scymnini (Coleoptera: Coccinellidae) of China. Coleopterists Bulletin, 40, 157–199.

Yamanaka, K., Nakasuji, F., & Kiritani, K. (1973). Life tables of the tobacco cutworm *Spodoptera litura* and the evaluation of effectiveness of natural enemies. Japanese Journal of Applied Entomology and Zoology, 16, 205–214.

Yang, C. (1988). The significance of the Chinese resource of chrysopids to biological pest control and the World's lacewing fauna. Chinese Journal of Biological Control, 3, 131–136 (in Chinese).

Yang, P. (1982). A preliminary study of the biology of *Oecophylla smaragdina* Fabr. and its use in the control of citrus pests. Acta Scientiarum Naturalium Universitatis Sunyatseni, 3, 102–105 (in Chinese).

Yang, P. (1984). The growth and development of colonies of *Oecophylla smaragdina* Fabr. Natural Enemy Insects, 6, 240–243 (in Chinese).

Yang, P., Chen, S., & Zheng, J. (1983). The yellow citrus ant in Fujian. Natural Enemy Insects, 5, 59–60 (in Chinese).

Yang, P., Chen, S., & Zheng, J. (1984). Benefits in the use of ants for pest control. Agriculture Fujian, 10, 38 (in Chinese).

Yasuda, T., & Wakamura, S. (1992). Rearing of predatory stink bug, *Eocanthecona furcellata* (Wolff) (Heteroptera: Pentatomidae), on frozen larvae of *Spodoptera litura* (Fabricius) (Lepidoptera: Noctuidae). Applied Entomology and Zoology, 27, 303–305.

Yasumatsu, K., Wongski, T., Tirawat, C., Wongsiri, N., & Lewvanich, A. (1981). Contributions to the development of integrated rice pest control in Thailand. Japanese International Cooperative Agency.

Yazlovetsky, I. G. (1992). Development of artificial diets for entomophagous insects by understanding their nutrition and digestion. In T. E. Anderson & N. C. Leppla (Eds.), Advances in insect rearing for research and pest management (pp. 41–61). Boulder, CO: Westview Press.

Yeargan, K. V., & Braman, S. K. (1989). Comparative behavioral studies of indigenous hemipteran predators and hymenopterous parasites of the green cloverworm (Lepidoptera: Noctuidae). Journal of Kansas Entomological Society, 62, 156–163.

Yokoyama, V. Y. (1978). Relation of seasonal changes in extrafloral nectar and foliar protein and arthropod populations in cotton. Environmental Entomology, 7, 799–802.

Yokoyama, V. Y., Pritchard, J., & Dowell, R. V. (1984). Laboratory toxicity of pesticides to Geocoris pallens (Hemiptera: Lygaeidae), a predator in California cotton. Journal of Economic Entomology, 77, 10–15.

Yonke, T. R. (1991). Order Hemiptera. In F. W. Stehr (Ed.), Immature insects (Vol. 2, pp. 22–65). Dubuque, IA: Kendall/Hunt.

York, G. T. (1944). Food studies of Geocoris spp., predators of beet leafhopper. Journal of Economic Entomology, 37, 25–29.

Yoshida, H. A., & Mau, R. F. L. (1985). Life history and feeding behavior of Nephaspis amnicola Wingo. Proceedings of the Hawaiian Entomological Society, 25, 155–160.

Young, O. P. (1984). Prey of adult Calosoma sayi (Coleoptera, Carabidae). Journal of the Georgia Entomological Society, 19, 503–507.

Young, O. P., & Edwards, G. B. (1990). Spiders in United States field crops and their potential effect on crop pests. Journal of Arachnology, 18, 1–27.

Young, O. P., & Lockley, T. C. (1985). The striped lynx spider, Oxyopes salticus (Araneae: Oxyopidae) in agroecosystems. Entomophaga, 30, 329–346.

Young, O. P., & Lockley, T. C. (1986). Predation of the striped lynx spider, Oxyopes salticus (Araneae: Oxyopidae) on the tarnished plant bug, Lygus lineolaris (Heteroptera: Miridae): A laboratory evaluation. Annals of the Entomological Society of America, 79, 879–883.

Yu, S. J. (1988). Selectivity of insecticides to the spined soldier bug (Heteroptera: Pentatomidae) and its lepidopterous prey. Journal of Economic Entomology, 81, 119–122.

Zanuncio, J. C., Malheiros, R. R., Zanuncio, T. V., & Padua, R. L. A. (1989). Hemipteras predadoros de lagartus desfolhadoras de Eucalyptus spp. Congress of Brasilian Entomology (Belo Horizonte), 12, 465.

Zdarkova, E. (1986). Mass rearing of the predator Cheyletus eruditus (Schrank) (Acarina: Cheyletidae) for biological control of acarid mites infesting stored products. Crop Protection, 5, 122–124.

Zeleny, J. (1984). Chrysopid occurrence in west palearctic temperate forests and derived biotypes. In M. Canard et al. (Eds.) Biology of chrysopidae (pp. 151–160). The Hague: Dr. W. Junk.

Zhao, D. X., Boivin, G., & Stewart, R. K. (1990). Consumption of carrot weevil. Listronotus oregonensis (Coleoptera: Curculionidae) by four species of carabids on host plants in the laboratory. Entomophaga, 35, 57–60.

Zhavoronkova, T. N. (1969). Certain structural peculiarities of the Carabidae (Coleoptera) in relation to their feeding habits. Entomological Review, 48, 462–471.

Zheng, Y., Hagen, K. S., Daane, K. M., & Mittler, T. E. (1993a). Influence of larval dietary supply on food consumption, food utilization efficiency, growth and development of the lacewing Chrysoperla carnea. Entomologia Experimentalis et Applicata, 67, 1–7.

Zheng, Y., Daane, K. M., Hagen, K. S., & Mittler, T. E. (1993b). Influence of larval food consumption on fecundity of the lacewing, Chrysoperla carnea (Stephens) (Neuroptera: Chrysopidae). Entomologia Experimentalis et Applicata, 67, 9–14.

Zhou, W., & Wang, R. (1989). Rearing of Orius sauteri (Hem.: Anthocoridae) with natural and artificial diets. Chinese Journal of Biological Control 5, 9–12 (in Chinese).

Zhou, W., Wang, R., & Qiu, S. (1991a). Use of soybean sprouts as the oviposition material in mass rearing of Orius sauteri (Het.: Anthocoridae). Chinese Journal of Biological Control, 7, 7–9 (in Chinese with English summary).

Zhou, W., Wang, R., & Qui, S. (1991b). Field studies on survival of Chrysoperla sinica (Neu.: Chrysopidae) mass reared and inoculatively released in wheat fields in northern China. Chinese Journal of Biological Control, 7, 97–100.

Zhou, W., Liu, Z., Cheng, W., & Qui, S. (1981). Study on cultivation of adult Chrysopa sinica with powder artificial diet. Acta Phytopathologica Sinica, 7, 2–3.

Zimmerman, E. C. (1948). Insects of Hawaii: Vol. 3, Heteroptera. Honolulu: University of Hawaii Press.

Zondag, R., & Nuttall, M. J. (1989). Pineus laevis Maskell, pine twig chermes or pine woolly aphid (Hemiptera: Adelgidae). In A review of invertebrate pests and weeds in New Zealand 1874 to 1987. CIBC Technical Communications, 10, 295–297.

Zuniga, E. (1985). Eighty years of biological control in Chile. Historical review and evaluation of projects undertaken (1903–1983). Agricultural Technology, 45, 175–183.

Zuniga, E. (1990). Biological control of cereal aphids in the southern core of South America, In R. K. Campbell & R. D. Eikenbary (Eds.), Aphid-plant interactions (pp. 362–367). Amsterdam: Elsevier Science.

CHAPTER

17

Arthropods and Vertebrates
in Biological Control of Plants

T. S. BELLOWS

Department of Entomology
University of California
Riverside, California

D. H. HEADRICK

Crop Science Department
California Polytechnic State University
San Luis Obispo, California

INTRODUCTION

Biological control of plants using insect and other metazoan herbivores has a long history of research and application throughout the world. Some of the world's worst weed pests, both terrestrial and aquatic, have been controlled biologically with herbivores. In many of these cases, biological control has been the only possible option, for example, in cases where weeds become naturalized over millions of hectares of native grasslands or colonize natural waterways. Although cultural and chemical control cannot address the presence of a huge, self-reproducing population over vast areas, the use of natural control agents has often proved beneficial and effective.

In the context of "use" of herbivorous insects, mites, nematodes, and vertebrates, we are almost exclusively concerned with the introduction of a herbivore to control either an adventive or a native plant pest. In such work, herbivores become candidate "agents" for use against a particular target weed. As this field has developed, the majority of cases have involved the discovery and release of herbivores into the environment to become naturally reproducing populations; the use of large numbers of mass-reared individuals, as is sometimes done with parasitoids or predators against other insect pests, has found only very limited application in the control of weeds. Thus, we are concerned here with the ecological context of introducing new herbivore species (through the usual channels of strict quarantine and evaluation, see Chapters 5 and 6) into environments as permanent members of the new community, with the aim that their presence will reduce the damage caused by a weedy plant. In most cases, the target weed will be adventive, or introduced, into the target environment. The work

of biological control of weeds with invertebrates thus involves three broad fields of endeavor: research into plant biology and ecology, herbivore biology and ecology, and herbivore–plant interactions.

In this volume there are two additional chapters that address biological control of weedy plants. Chapter 34 reviews programs against both aquatic and terrestrial weeds from a programmatic viewpoint. Chapter 35 reviews the use of plant pathogens as agents against weedy plants. This chapter reviews the use of invertebrates and vertebrates in plant biological control programs and discusses the taxa that have been used in these programs. We touch briefly on ecological attributes of biological control agents, and review procedures for ensuring the ecological safety of candidate natural enemies.

EFFECTS OF HERBIVORY ON WEEDY PLANTS

Herbivory is widely common among invertebrates, and most plants have a number of herbivores, both generalists and in some cases specialists, feeding on them (Strong *et al.,* 1984). Plants are, quite clearly, capable of sustaining considerable herbivory without undue damage to their own survival and reproduction. How is it, then, that herbivorous insects can cause sufficient change in plant population dynamics that they can control a plant population?

Biological control of plants differs in some important ways from control of invertebrate pests or pathogens. While predation and parasitism of invertebrate pests are direct causes of mortality in the target population, plant death rarely occurs from the attack of a single or a few herbivores. In contrast, a variety of mechanisms are involved in the interactions between invertebrate herbivores and the plants that they feed on. Together these processes can lead to the demise of a plant population; even when in isolation, they do not cause the death of individual plants.

The effects of herbivorous arthropods on plants occur at two levels: changes in the growth and reproduction of individual plants, and changes in the dynamics of the target plant population. In a great deal of literature the effects of herbivory on individual plants have been explored, but the quantitative impact of herbivores on plant populations has been reported in very few studies (Crawley, 1989). In terms of practical biological control, the matters of interest are the overall impact of herbivores on the dynamics of plant populations. Nonetheless, an appreciation of the component effects on individual plants sheds important light on the ecological interplay between plants and their herbivores. We will consider here how five different impacts of herbivores on plants can lead to plant population suppression and control.

Reduced Reproduction or Recruitment to the Plant Population

Seed- and flower-feeding insects are often used in introduction programs against weedy plants. Herbivory of the reproductive structures can reduce fruit and seed set, and can in some cases reduce recruitment to the seedling stage of the populations. However, the relationship between seed production and seedling recruitment is often not linear. While seed production might be limiting for some plant populations, for many plant populations, including annual, biennial, and long-lived woody species, seed production vastly outmatches seedling recruitment. Sites for seedling establishment are often the resource that limits this stage of population growth (Andersen, 1989). In ecological context, this is a situation where the impact of mortality in one stage (the seed) is reduced in significance if a mortality or resource limiting a subsequent stage (seedling establishment) is density dependent. The presence of a limited number of seedling establishment sites acts as a density-dependent buffer once the seed supply exceeds the supply of favorable germination sites.

Nonetheless, seedling recruitment of some plant species has been reduced by the introduction of seed- or flower-feeding insects. Seed production of the tree *Sesbania punicea* (Cavanille) Bentham was reduced by 98% following the introduction of the apionid *Trichapon lativentre* (Bequin Billecocq) (Hoffman & Moran, 1992). The addition of the weevil *Rhyssomatus marginatus* Fåhraeus further destroyed 84% of the produced seeds, and the two agents together arrested nearly completely the reproduction of the weed (Hoffman & Moran, 1992). An in-depth discussion of

the significance of seed-attacking insects is given by Neser & Kluge (1986).

Direct Mortality of Plants

Herbivorous arthropods may occasionally kill their host plants directly. Plant death in such cases is usually caused either by consumption of whole plants, often by an unusually large population of herbivores, or by extensive disruption of plant physiological systems such as water transport. Plant death can also occur through repeated defoliation accompanied by destruction of plant carbohydrate reserves (preventing regrowth). In other cases, effects of herbivory act together with additional agents or environmental stresses to cause plant death indirectly.

Indirect Mortality of Plants

Herbivory can reduce existing plant biomass or cause redirection of resources in plant tissues through many mechanisms (including defoliation, removal of plant sap, destruction of roots, and galling of plant tissues). Such herbivory often leads to reduction in stored reserves in a plant, and reduces the rate of plant growth or regrowth in the presence of herbivory. The relationship between injury from herbivores and individual plant performance (such as total seed set or growth rate) is often linear (Crawley, 1989). At moderate levels of herbivory, however, many plants are able to compensate partially or fully for losses (Trumble *et al.*, 1993), so the relationship between the proportion of plants actually dying and the amount of herbivorous injury may be nonlinear, and is governed by a damage threshold (Harris, 1986). This threshold is the level or amount of biomass that must be removed before the plants cannot compensate and consequently incur an increased risk of death. Many species of forage grasses, for example, remain healthy with annual losses to herbivory of 40% of their biomass, but when herbivory removes more than 50%, grass stands decline (Harris, 1986).

Plant response to herbivory varies not only with the amount of biomass consumed or removed, but also with the season in which herbivory takes place. In addition, plant response will vary with the specific tissues that are attacked; particularly when the tissues attacked are also those used by the plant for storage of carbohydrate reserves. For plant species in which reserves are stored in foliage, defoliation is much more likely to result in plant death than for species in which storage occurs in other structures, such as the cambium or roots. Seasonal timing of herbivory also is important in determining the outcome of plant–herbivore interactions. Defoliation of musk thistle, *Carduus thoermeri* Weinmann, at the rosette stage caused mortality, but similar levels of defoliation after development of the fruiting stalk had little effect on the plants (Cartwright & Kok, 1990).

Galling of plant tissues has its greatest effect when tissues are attacked early in development (Harris, 1986). The plant reacts as if the resources used for gall formation had been used for normal tissue growth and little compensatory growth takes place. Extensive galling can act as a nutrient sink, reducing other vegetative growth and, if galling occurs in the reproductive tissues, reducing reproduction (Dennell, 1988).

Interactions with Other Stresses

Because herbivory is fundamentally a loss of biomass or nutrients to a plant, the degree of herbivory that a plant can stand is often a function of what other factors are present, timing of the herbivory, and environmental conditions affecting the plant.

Multiple Herbivores

A series of introduced herbivores that either act sequentially through the year, or that attack different plant tissues, can be valuable in suppressing plant growth and reproduction. In parts of Hawaii, for example, control of *Lantana camara* Linnaeus was achieved by the introduction of both a tingid (*Teleonomia scrupulosa* Stål) that annually defoliated the plants and additional agents that attacked the plant during the winter months (Andres & Goeden, 1971). Seed production of *S. punicea* in South Africa was virtually eliminated by the combined action of two seed-feeding insects (Hoffman & Moran, 1992). Tansy ragwort (*Senecio jacobaea* Linnaeus) in the northwestern United States was controlled by the complimentary actions of two herbivores, cinnabar moth [*Tyria jacobaeae* (Linnaeus)] and a flea beetle [*Longitarsus jacobaeae* (Waterhouse)] (McEvoy *et al.,* 1990).

Abiotic Stress

Defoliated plants tend to respond to injury by regrowth, drawing on stored reserves of carbohydrates and other nutrients. New foliage is then used by the plant to rebuild carbohydrate reserves. Plants that have used stored reserves for regrowth can be more vulnerable to abiotic stresses, such as winter cold or summer drought, until reserves have been restored. If a defoliation–refoliation cycle occurs shortly before periods of drought or other stress, the impact of this environmental stress on the plant can be more severe than would ordinarily occur. Defoliation of tansy ragwort by the cinnabar moth reduces root reserves and makes the plants more susceptible to death caused by frost (Harris *et al.,* 1978).

Plant Disease

Herbivory may increase plant susceptibility to disease. Limb death in *Mimosa pigra* Linnaeus in Australia, caused by the pathogen *Botryodiplodia theobromae* Patouillard, was apparently stimulated by a stem-boring moth introduced against this weed (Wilson & Pitkethley, 1992).

Plant Competition

In nearly every community, plants compete for resources: light, water, nutrients, germination sites, or canopy growth space. Grasses are important competitors of the rangeland weed *Centaurea maculosa* Lamark, reducing survival of rosettes by 83% over control sites lacking grasses (Müller-Schärer, 1991). The competitive ability of wheat against the weed *Amsinckia intermedia* von Fischer and Meyer was sharply increased in the presence of the flower gall nematode *Anguina amsinckiae* (Steiner & Scott) Thorne.

The ability of weeds to compensate for herbivory by increased growth is greatest when competition is absent and when nutrients and water are not limiting. As plant competition increases, weeds are less able to regrow following attack by herbivores. Competition may act over time such that areas left bare by weed plant death, if quickly filled by other more desirable plants, are not available as sites favorable for germination and establishment of weed seedlings.

TAXA USED IN BIOLOGICAL CONTROL

Julien (1992) has provided a world catalog of agents imported and released in various countries for control of adventive weedy plants. Some 259 species of invertebrates have been employed, predominantly insects (254 species) and some mites (5 species). Releases of these species have resulted in establishment at one or more locations for 62% (161) of the species used. Control (at least partial, including cases in which the agent was part of the complex of herbivores that eventually resulted in control) of the target weed was achieved in at least one location by 65 species (25% of all species released). A large number of families and species have been employed, and a summary of their taxonomic groupings is presented in Tables 1 and 2. Care should be taken in interpreting the information in these tables as definitive concerning the relative merits of any particular group, because this table is only a historical record and not a test of hypotheses relating to the suitability of one or more natural-enemy groups. Included in Julien's (1992) review and in these tables are many species whose releases are relatively recent and still under evaluation.

The taxa employed in such programs include insects,

TABLE 1 A Summary of Invertebrates Intentionally Moved between Countries for Biological Control of Adventive Weeds

Order	No. of families employed	No. of species employed	Establishments (no. of species)	Successes (no. of species)
Coleoptera	7	109	66 (61%)	33 (30%)
Lepidoptera	21	82	46 (56%)	15 (18%)
Hemiptera	8	19	15 (79%)	8 (42%)
Diptera	6	35	25 (71%)	4 (11%)
Thysanoptera	2	4	2 (50%)	1 (25%)
Hymenoptera	3	4	3 (75%)	2 (50%)
Orthoptera	1	1	1 (100%)	0 (0%)
Acarina	3	5	3 (60%)	2 (40%)
Fungi	4	8	8 (100%)	5 (63%)
Nematodes	1	1	1 (100%)	0 (0%)
Overall	56	268	170 (63%)	70 (26%)

Data from Julien, M. H. (Ed.) (1992).
Biological control of weeds: A world catalogue of agents and their target weeds. Wallingford, United Kingdom: CAB International.

mites, nematodes, fish, and pathogens (see Chapter 35 for pathogens used in weed biological control). Among these taxa, which groups of herbivores emerge as important for weed control depends in large part on which plant species become pests, and which groups feed on those plants. For these reasons, the number of families or species recognized as important may increase with time if additional weeds become targets of such programs.

Invertebrates

From the historical record, we note that of 65 species of herbivorous insects and mites that have provided control of their target weeds, 40 are included in six families of insects: three families of Coleoptera; two, of Lepidoptera; and one, of Homoptera. These include Chrysomelidae (12 species), Curculionidae (14), Cerambycidae (4), Pyralidae (3), Arctiidae (3), and Dactylopidae (4). Relatively few generalities about the biologies of these groups are possible at the family level because within most families of insects diverse biologies are exhibited by member species. Broader descriptions of herbivorous insect families are given by Arnett (1968), Borror *et al.* (1989), and CSIRO (1970). Insect–plant interactions are discussed by Strong *et al.* (1984). Van Driesche and Bellows (1996) provide a slightly fuller overview of the biologies of the major families of herbivores used in biological control.

Coleoptera

Chrysomelidae

Members of this family feed on foliage and flowers as adults. Larvae feed on foliage, are leaf miners, or are stem

or root borers. Some 39 species of chrysomelids have been used for biological control and 12 have successfully controlled their target pests in at least one location. *Agasicles hygrophila* Selman and Vogt has been used successfully in Australia and other countries (Julien, 1981) for aquatic forms of alligator weed, *Alternanthera philoxeroides* (Martius) Grisebach, a plant native to South America. The beetle is multivoltine with as many as five generations per year in Argentina, and adults lay an average of more than 1100 eggs during their 7-week adult life span (Clausen, 1978). The terrestrial weed *S. jacobaea* was controlled by two herbivores, one of which was the chrysomelid *L. jacobaeae*.

Curculionidae

Larvae of this family either feed externally on foliage, or feed internally in fruits or stems. Of 36 species of curculionids that have been used in biological control programs, 14 have provided control of their target weeds. These include the weevil *Rhinocyllus conicus* (Frölich), which controlled nodding thistle, (*Carduus nutans* Linnaeus) in Canada (Harris, 1984); *Neohydronomus affinis* Hustache, which control water lettuce (*Pistia stratiotes* Linnaeus) in several locations including Florida (Dray *et al.,* 1990); and *Microlarinus lypriformis* (Wollaston), which contributed to partial controlled of puncture vine (*Tribulus terrestris* Linnaeus) in the southwestern United States and Hawaii (Huffaker *et al.,* 1983).

Cerambycidae and Buprestidae

Larvae of these families typically bore in the stems of woody plants. Adults often feed on flowers. Seventeen species of cerambycids or buprestids have been used in biolog-

TABLE 2   A Summary of the Historical Record of Movement of Groups of Invertebrates and Fungi between Countries for Control of Adventive Weeds

Order	No. of species employed	Establishments (no. of species)	Successes (no. of species)
Coleoptera	109	66 (61%)	33 (30%)
Chrysomelidae	39	21	12
Curculionidae	36	24	14
Cerambycidae	14	9	4
Apionidae	9	3	1
Bruchidae	7	6	2
Buprestidae	3	3	1
Anthribidae	1	0	0
Lepidoptera	82	46 (56%)	15 (18%)
Pyralidae	23	11	3
Noctuidae	10	4	1
Tortricidae	9	5	2
Gelechiidae	5	3	0
Arctiidae	4	3	3
Gracillariidae	4	3	1
Pterophoridae	3	3	2
Lyonetidae	3	2	1
Cochylidae	3	2	0
Geometridae	3	2	0
Sesiidae	3	1	0
Aegeriidae	2	0	0
Coleophoridae	2	2	0
Oecophoridae	1	1	0
Pterolonchidae	1	0	0
Carposinidae	1	1	1
Heliodinidae	1	1	1
Sphingidae	1	1	0
Lycaenidae	1	1	0
Dioptidae	1	0	0
Hepialidae	1	0	0
Hemiptera	19	15 (79%)	8 (42%)
Dactylopiidae	6	5	4
Tingidae	5	3	1
Coreidae	3	2	1
Pseudococcidae	1	1	1
Miridae	1	1	0
Aphidae	1	1	0
Psyllidae	1	1	1
Delphacidae	1	1	0
Diptera	35	25 (71%)	4 (11%)
Tephritidae	17	11	2
Cecidomyiidae	6	5	1
Agromyzidae	5	4	1
Anthomyiidae	4	3	0
Ephydridae	2	2	0
Syrphidae	1	0	0
Thysanoptera	4	2 (50%)	1 (25%)
Phlaeothripidae	3	2	1
Thripidae	1	0	0
Hymenoptera	4	3 (75%)	2 (50%)
Tenthredinidae	2	1	0
Eurytomidae	1	1	1
Pteromalidae	1	1	1
Orthoptera	1	1 (100%)	0 (0%)
Pauliniidae	1	1	0

(continues)

TABLE 2 *(continued)*

Order	No. of species employed	Establishments (no. of species)	Successes (no. of species)
Acarina	5	3 (60%)	2 (40%)
Eriophyidae	3	2	1
Tetranychidae	1	1	1
Galumnidae	1	0	0
Fungi	8	8 (100%)	5 (63%)
Uredinales	5	5	3
Hyphomycetes	1	1	0
Ustilaginales	1	1	1
Coelomycetes	1	1	1
Nematodes	1	1 (100%)	0 (0%)
Anguinidae	1	1	0
Overall	56	268	170 (63%)

After Van Driesche, R. G., & Bellows, T. S. (1996). *Biological control.* New York: Chapman & Hall; data from Julien, M. H. (Ed.). (1992). *Biological control of weeds: A world catalogue of agents and target weeds.* Wallingford, United Kingdom: CAB International.

ical control projects and four have provided some control of their target weed species.

Bruchidae

The larvae of this family typically develop inside seeds, especially of legumes. While seven species of bruchids have been employed in biological control projects and some have caused dramatic reduction of seeds and seedlings, in only one case has this herbivory resulted in suppression of the target weed. In South Africa, two species of seed-feeding bruchids are important in biological control of *Prosopis* spp., a plant that has some desirable characteristics but has become an invasive weed. By limiting seed production through biological agents, the plant can be preserved but its reproduction severely restricted, reducing its invasiveness and pest status (Zimmerman, 1991).

Lepidoptera

Pyralidae

Species in this third largest lepidopteran family are usually small, with the adults appearing as small, delicate moths with a scaled proboscis. Larval feeding habits are quite varied and species may be foliage feeders, borers, or feeders on stored products, including beeswax. One of the most famous members of this family in the history of biological control is the Phycitinae species *Cactoblastis cactorum* (Bergroth), which successfully controlled prickly pear cacti (*Opuntia* spp.) in Australia. Of 23 pyralid species employed in biological control, three have suppressed their target weeds.

Noctuidae

This is the largest family in the Lepidoptera. The adults are heavy-bodied, night-flying moths; the larvae are usually smooth bodied and dull in color. Most feed on foliage, but some are stem borers or feed on fruits. Of the 10 species employed as biological control agents one has controlled its target weed.

Tortricidae

This is one of the largest families of microlepidoptera. The larvae are foliage feeders, most often of perennial plants. They may tie or roll leaves and often bore into plant parts. Of the nine species employed in biological control, two have controlled their target weeds.

Arctiidae

This is a moderate-sized family with four rather distinct subfamilies (the Pericopinae, Lithosiinae, Arctiinae, and Ctenuchinae), which are given family status by some authors. Subfamily Arctiinae, or tiger moths, comprise most of the species in the family. Of the four species that have been used as biological control agents, three have successfully suppressed their target weeds. *Tyria jacobaeae* was responsible in part for control of tansy ragwort, *S. jacobaea,* in Oregon (McEvoy *et al.,* 1990).

Gracillariidae

Members of this family are minute moths whose larvae are leaf miners. Mines are typically blotch in form and the leaf is often folded. Many mines may occur per leaf. Pupation often takes place in the mine or on the surface of the mined leaf. One of the four species employed for biological control suppressed its host.

Hemiptera

Dactylopiidae

These insects resemble mealybugs (Pseudococcidae) in appearance and habits. Dactylopiids feed on cacti and while the number of species in the family is not large, the family has played an important role in successful biological control of several species of cacti. Four of the six species used have provided control of their target cacti (Goeden *et al.,* 1967).

Tingidae

Nymphs and adults feed together on the same host plant by sucking plant fluids from leaves and other tissues, causing the foliage to become bronzed; and, when bug densities are high, plants die. Some species are relatively specific in their choice of host plants, which may be either herbaceous or woody species. One species, *Teleonemia scrupulosa,* played an important role in the control of lantana *(Lantana camara),* a woody ornamental of great importance as a pasture pest in tropical areas around the world (Julien, 1992).

Coreidae

Most coreids are phytophagous. Some species feed on seeds and others on foliage. Some species have enlarged hind tibia or femurs. Many species produce noticeable odors. One of the three species employed in biological control has controlled its host plant.

Diptera

Cecidomyiidae

These are small delicate flies with long legs and long antennae. About two-thirds of known species form galls and most of the remaining species feed on decaying vegetation. Galls are formed on all parts of plants, including buds, leaves, flowers, and stems. One of the six species employed for biological control has been proved successful.

Tephritidae

The members of this family are small to medium-sized flies, usually with patterned wings. Larvae feed inside plant tissues, tunneling in seed heads, forming galls, or feeding in fruits. A few species are leaf miners. Seventeen species have been used in biological control projects, mainly species that feed in seed heads of thistles, knapweeds, and other plants, or species that form galls. Establishment has occurred in a substantial proportion of cases. Two species have successfully controlled their target weeds.

Agromyzidae

The larvae of most species of Agromyzidae are leaf miners. Species in the family attack a wide range of plants and are found in most habitats. Of the five species employed for biological control, one has been successful.

Thysanoptera

Phlaeothripidae

Some species in this family are herbivorous. Of the three species employed for biological control, one has suppressed its target weed. *Liothrips urichi* Karmy controlled *Clidemia hirta* (Linnaeus) in some habitats in Fiji (Rao, 1971).

Hymenoptera

Tenthredinidae

This is the most common group of sawflies, and most species are small or moderate in size. The larvae are external feeders on foliage. Typically there is one generation a year. Most species feed on woody shrubs or trees. A few species are gall makers or leaf miners. Of the two species that have been employed, neither has been successful.

Eurytomidae

These are minute, darkly-colored wasps. They vary in their biologies. Some species are parasitic, while others feed in seeds or make galls. The single species of this family that has been employed for biological control suppressed its target weed.

Pteromalidae

The Pteromalidae comprise a diverse group of Hymenoptera, which includes some gall formers. One of these, *Trichilogaster acaciaelongifoliae* Froggat, has been important in the biological control of *Acacia longifolia* (Andrews) Willdenow in South Africa (Dennill & Donnelly, 1991). The wasp is univoltine, and females oviposit in floral buds of the target plant, leading to galling of these tissues. Extensive attack can occur, and attacked buds do not produce inflorescences or seeds. Attack by the wasp has resulted in 30% mortality of adult trees in some locations. The final impact of this wasp is still under evaluation, but the insect appears promising as a control agent of *A. longifolia.*

Acarina

Eriophyidae

These are extremely small mites (about 0.15 mm in length) that feed on plant tissues. Some species cause galls, while others feed externally on plant tissues and thereby discolor fruits or other plant parts. The potential biological control uses of the family are reviewed by Gerson and Smiley (1990), who note that eriophyids, while slow acting, are often highly host specific. Of the three species that have been employed for biological control, one has suppressed its target weed.

Tetranychidae

Spider mites, which are larger than eriophyids, feed on plant foliage but do not cause galls; many species produce visible webbing. Gerson and Smiley (1990) summarize past biological control uses of the family, which have been extremely limited. The only target weed against which a spider mite (*Tetranychus lintearius* Dufour) has been deliberately introduced has been gorse (*Ulex europaeus* Linnaeus) in New Zealand (Hill *et al.,* 1991), and evaluation of this program is continuing.

Nematodes

A number of plant-parasitic nematodes in the Tylenchina, especially the family Anguinidae, induce galls on plant foliage. The taxonomy of this group is reviewed by Siddiqi (1986). Some nematodes in this group are resistant to dehydration, a feature that enhances their survival. A number of species in the group have been considered for use in biological control through augmentation (Parker, 1991). One species has been introduced against an adventive plant (Julien, 1992).

Vertebrates

A variety of domestic animals such as goats, sheep, and geese are used to some degree for terrestrial weed control; such uses are usually in the context of cultural regimes, where intensive foraging is used to reduce unwanted vegetation, including weedy species. In the case of aquatic weeds, fish have been used to reduce and manage intermittent or perennial weed problems. A total of 11 fish species have been introduced into various countries to suppress aquatic vegetation (Julien, 1992). These occur in three families, Cyprinidae (which include various carp species), Cichlidae, and Osphronemidae (Table 3).

All these species are generalist herbivores that suppress whole communities of aquatic macrophytes. As such, these species are limited in their application to enclosed water bodies, such as irrigation ditches or ponds, where total or near total weed suppression is desired; and the fish population can be contained, thus avoiding introduction of nonnative fish to waterways and other natural breeding grounds of native fish. The potential for such fish to affect natural systems, including both macrophytic and native fish communities, is high. Each introduction must be considered carefully, taking into account the potential for subsequent spread to other water bodies by flood or casual relocation by people. In some instances sterile hybrids or sterile triploids are used to minimize the risk of establishing breeding

TABLE 3   Species of Fish Introduced into One or More Countries for the Control of Aquatic Plants

Cyprinidae
   Aristichthys nobilis (Richardson) (big head)
   Ctenopharyngodon idella (Cuvier & Valenciennes) (grass carp, white amur)
   Hypophthalmichthys molitrix (Cuvier & Valenciennes) (silver carp)
   Puntius javanicus (Bleeker)
Cichlidae
   Oreochromis aureus (Steinbachner)
   O. mossambicus (Peters)
   O. niloticus (Linnaeus)
   Tilapia macrochir (Boulenger)
   T. melanopleura Dumeril
   T. zillii (Gervais)
Osphronemidae
   Osphronemus goramy Lacepedes (giant gourami)

Data from Julien, M. H. (Ed.). (1992). *Biological control of weeds: A world catalogue of agents and their target weeds.* Wallingford, United Kingdom: CAB International.

populations of introduced fish (Anderson, 1993; Van Driesche & Bellows, 1996).

ATTRIBUTES AND SAFETY OF CONTROL AGENTS

The desirable attributes of weed biological control agents vary depending on the ecological situation into which the agents are placed. Our awareness of this diversity of desirable attributes has matured over the nearly 100 years that such work has taken place. While it is tempting to point to past programs and to analyze the suitability of agents in the light of today's ecological understandings, such critiques are less useful than might be surmised, because decisions made early in the development of this science might not be made in similar situations today. Instead, we turn to current, conservative practice in the conduct of biological control programs to define attributes desirable among natural enemies of weedy plants.

Candidate natural enemies, when introduced into a new environment, face the usual challenges of colonizing that environment, securing resources for growth and development, locating mates, reproducing, and surviving seasonal weather patterns and attack by extant natural enemies. Having accomplished this, they are expected to disperse in the environment (preferably throughout the adventive range of the weed), persist, reproduce, and reduce the population density of the weed to a noneconomic status. In most cases, it is also desirable that in their new environment the herbivore has little or no impact on any nontarget species of plant. Weed biological control agents are sought to be ecologically widely benign to nontarget plants, while being sufficiently damaging to the target plant to be effective in controlling it.

While at first this may seem a daunting prospect, in many cases the ecological setting of the target weed permits considerable flexibility in achieving these goals. For example, weeds may be taxonomically isolated in their new environment. In such settings, weed-feeding agents that might feed on a few related plants in their native home would, if introduced into the weed's new adventive range, find themselves with only a single host-plant species present in this new range. In such circumstances, these agents are highly specific to the target, because their other normal hosts are lacking. In other cases, adventive weeds may be related to extant plants in the new environment. In these cases, more work may be involved in finding herbivores sufficiently selective that their impact on native vegetation will be at acceptably low levels.

Host Specificity

Issues concerning host specificity are always situation specific. Generalizations are not possible, because the setting in which each weedy plant is addressed will be unique. Nonetheless, we find throughout the work of plant biological control a generally high degree of conservatism concerning what natural enemies are released into an environment, with emphasis placed on herbivores sufficiently specific that risk of impact to nontarget species is minimized. By and large, this work has been successful in avoiding widespread nontarget impacts, and only in rare cases have nontarget impacts been· recorded as significant (Van Driesche & Bellows, 1996).

Clearly defining host specificity, as it might be realized in a natural setting, is a difficult undertaking, in part because laboratory research will rarely mimic field behavior of any particular herbivore. The best solution for evaluating the potential host range of a candidate natural enemy is to study it in detail in its native home, defining its ecologically realized host range, and determining the impact it has both on the target weed and on nontarget plants (Clement & Sobhian, 1991; Baudoin et al., 1993). When such work is not possible, then weed biological control agents are evaluated in special quarantine facilities, where their biology and host range can be tested against both target and nontarget plants, prior to any planned release into the environment. While evaluating the results of laboratory trials is difficult because the behavior of many herbivores is modified by confinement, it is usually the most conservative assessment of herbivore impact. When a candidate herbivore is shown not to feed on non-target plants in forced settings in the laboratory, there is a high likelihood that it will not feed on them in natural settings. Wapshere (1989)

and Harley and Forno (1992) discuss interpreting the results of host-specificity trials.

Wapshere (1989) formalized an important and informative "reverse-order" testing procedure for evaluating host range of herbivores (Fig. 1). By reasoning that host-plant use by insects must involve two steps, both acceptance by the adult for oviposition and suitability for immature development, Wapshere (1989) proposed that host-range testing should start with forced feeding trials of immature stages of the candidate herbivore. Plants on which the immatures can successfully develop to fertile adults are then possible candidates for inclusion in the host range of the herbivore. These plants are further tested by isolating adults of the herbivore with them, and by determining whether oviposition occurs. If either larval maturation fails, or if adults refuse to oviposit on the plant species, then it is extremely

HERBIVORE HOST RANGE TESTING

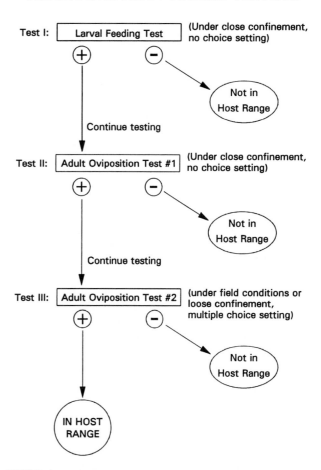

FIGURE 1 "Reverse-order" testing sequence for estimating host ranges of herbivorous arthropods. [After Wapshere, A. J. (1989). *Comparing methods of selecting effective biocontrol agents for weeds* (pp. 557–560). In Anonymous, Proceedings of the First International Weed Control Congress, Melbourne, Australia.]

unlikely that in natural settings the plant would ever be utilized in any significant way. If both larval development and oviposition in confinement take place, an additional adult oviposition test is conducted under field conditions or loose confinement with multiple species of plants. In some cases, an insect will oviposit on a plant in close confinement that it does not oviposit on when offered the same plant with other plant species (Buckingham *et al.,* 1989; Thompson & Habeck, 1989; Wilson & Garcia, 1992). If the adult does locate and oviposit on the plant species in this mixed-species setting, the plant is likely to be in the ecological host range of this herbivore. If the plant is avoided or ignored in this "choice" test, then it likely is not in the herbivore's host range in natural settings.

Once estimates of the host range have been made for a particular candidate herbivore, the degree of host specificity required for a particular situation must also be determined. While few herbivores are monophagous on a single plant species, many are oligophagous, feeding on only a few related species of plants. Such oligophagous agents may be judged suitable for a program if they do not attack any vital economic or native plants, or when it is clear that their impact on native and economic plants will be minimal. Polyphagous herbivores are not usually employed in biological control programs, but may be successful in some cases where the target weed is the only plant in the agent's host range present in the new environment. One setting in which polyphagous herbivores are used is in contained waterways, such as irrigation canals, where control of macrophytic aquatic weeds is affected by herbivorous fish.

Other Attributes

While the degree of host specificity is a vital and often primary focus of the evaluation of candidate herbivores, several other features of herbivore ecology are likely important in the success of any particular agent. In reviewing biological programs against weedy plants, it has not been possible to identify specific characteristics that are closely correlated with success or establishment. However, some features of herbivore biology are more common among successful agents. Of all the released insect species that have resulted in successful establishment, in 60% of the cases species where both the immature stages and the adults damage the host have been involved; in 37%, organisms where only the immature stages damage the plant have been involved; and in 3%, insects where only the adult damages the plant have been involved. Hence, a close association between the plant and the entire life cycle appears to be a common feature of herbivores established against weed pests.

Mobility and dispersal capabilities are clearly important in expanding an herbivore's geographic range to become coincident with the weed's range. Not all stages need be motile, however, because there are examples of biological control agents that have successfully established, dispersed, and controlled weeds that have had sessile stages. Some degree of seasonal synchrony between the herbivore and the target weed is necessary, and diapause cues, if obligatory, must be present for such synchrony to be effective. Ecological similarity between the target region and the area of origin of the herbivore often can help ensure that such climatic and seasonal parameters will not be limiting. Finally, we often see introduced herbivores become attacked by existing parasites or predators; thus, the reproductive potential of the herbivore in its new home must be sufficient for it to overcome natural mortality factors to become an effective control agent.

Additional characteristics have been proposed as likely to be important to a successful weed-feeding agent. A good agent should not be self-limiting, but should be tolerant of high densities of its own species so that it will be able to become sufficiently dense to damage its host. Species that are limited in their country of origin by natural enemies (parasitoids, predators, diseases) may be very effective in a new region provided that the natural enemies can be eliminated during the introduction process and that there are not species in the new location likely to take their place. A candidate agent should complement existing stresses placed on the weed by other herbivores in the system (unless it is expected to replace an existing, but ineffective agent). Thus, weed biological control programs often consider herbivores that feed on different portions of the plants as complementary to one another.

AGENT SELECTION

Agent selection is both art and science, and requires a certain degree of synthesis of both agent biology and target plant biology. It has not been possible to accurately predict which agents, of any potential list of candidate species, comprise the set of herbivores that will provide control of a target weed (Van Driesche & Bellows, 1996). However, because of the intense scrutiny such agents undergo, it is often possible to place candidate agents into such categories as "possible" or "unlikely"; and because of the high costs of evaluating and screening potential agents, such sifting or sorting is highly desirable. Where possible, the clearest and simplest way of selecting herbivores is to use species that have controlled the target weed in other countries with similar climates. Thus, the use of *Cyrtobagous salviniae* Calders and Sands to control *Salvinia molesta* D.S. Mitchell in South Africa was an obvious choice following its successful use earlier in Australia and Papua New Guinea against this weed (Thomas & Room, 1986; Cilliers, 1991).

In the absence of a prior program to draw from, there are three approaches to agent preliminary selection or screening (Wapshere, 1992). These include studies in the weed's native range to identify herbivores having the greatest impact there (while making appropriate allowance for parasitoids and other factors that might be different in the new target area for the agent) (Wapshere, 1974). A second method is to study the historical record of past projects (Julien, 1992) and to identify agents or features of successful agents that have been used against related weeds. Finally, in the absence of direct information, agents might be scored based on possession of particular attributes they (or their plant hosts) possess that seem, based on theoretical considerations, to be important for successful biological control (Harris, 1973; Goeden, 1983). Caution must always be employed in such cases, however, because biological performance of herbivores can be both unexpected and unexplainable based on preliminary, laboratory, or solely theoretical considerations. While such concepts cannot comprise a detailed guide to selecting agents, these concepts, together with the biological attributes discussed earlier, may serve as a foundation for sorting agents into at least priority rankings for evaluation.

The number of organisms that are obtained for evaluation as biological control agents against weeds is always greater than the number finally released, and these are greater than the number that become established. The filter of quarantine is a crucial part of this work, and the conservative approach that has been the standard of this science, to only release herbivores after careful scrutiny and review, has served widely to protect our environments from herbivores with undesirable attributes, while allowing effective control of weeds that had previously devastated natural and agronomic systems. The use of invertebrates and vertebrates as effective control agents has been and will continue to be a crucial component of our efforts to manage invasive and noxious weeds.

References

Andersen, A. N. (1989). How important is seed predation to recruitment in stable populations of long-lied perennials? Oecologia (Berlin), 81, 310–315.

Anderson, L. W. J. (1993). Progress and promise in management of aquatic-site vegetation using biological and biotechological approaches. In R. D. Lumsden & J. L. Vaughn (Eds.), Pest management: Biologically based technologies (pp. 241–249). Washington, DC: American Chemical Society.

Andres, L. A., & Goeden, R. D. (1971). The biological control of weeds by introduced natural enemies. In C. B. Huffaker (Ed.), Biological Control (pp. 143–164). New York: Plenum Press.

Arnett, R. H. (1968). The beetles of the United States (a manual for identification). Ann Arbor, MI: American Entomological Institute.

Baudoin, A. B. A. M., Abad, R. G., Kok, L. T., & Bruckart, W. L. (1993). Field evaluation of Puccinia carduorum for biological control of musk thistle. Biological Control, 3, 53–60.

Borror, D. J., Triplehorn, C. A., & Johnson, N. F. (1989). An Introduction to the Study of Insects (6th ed.). Philadelphia: Saunders College.

Buckingham, G. R., Okrah, E. A., & Thomas, M. C. (1989). Laboratory host range tests with Hydrellia pakistanae (Diptera: Ephydridae), an agent for biological control of Hydrilla verticillata (Hydrocharitaceae). Environmental Entomology, 18, 164–171.

Cartwright, B., & Kok, L. T. (1990). Feeding by Cassida rubiginosa (Coleoptera: Chrysomelidae) and the effects of defoliation on growth of musk thistles. Journal of Entomological Science, 25, 538–547.

Cilliers, C. J. (1991). Biolgical control of water fern, Salvina molesta (Salviniaceae), in South Africa. Agriculture, Ecosystems and Environment, 37, 219–224.

Clausen, C. P. (Ed.). (1978). Introduced parasites and predators of arthropod pests and weeds: A world review. Agricultural Handbook No. 480. Washington, DC: U.S. Department of Agriculture.

Clement, S. L., & Sobhian, R. (1991). Host-use patterns of capitulum feeding insects of yellow starthistle: Results from a garden plot in Greece. Environmental Entomology, 20, 724–730.

Crawley, M. J. (1989). Insect herbivores and plant population dynamics. Annual Review of Entomology, 34, 531–564.

CSIRO. (1970). [Corporate author]. The insects of Australia. A textbook for students and research workers. Carlton, Victoria: Melbourne University Press.

Dennill, G. B., & Donnelly, D. (1991). Biological control of Acacia longifolia and related weed species (Fabaceae) in South Africa. Agriculture, Ecosystems and Environment, 37, 115–135.

Dennill, G. B. (1988). Why a gall former can be a good biocontrol agent: The gall wasp Trichilogaster acaciaelongifoliae and the weed Acacia longifolia. Ecological Entomology, 13, 1–9.

Dray, F. A., Jr., Center, T. D., Habeck, D. H., Thompson, C. R., Cofrancesco, A. F., & Balciunas, J. K. (1990). Release and establishment in the southeastern United States of Neohydronomus affinis (Coleoptera: Curculionidae), an herbivore of water lettuce. Environmental Entomology, 19, 799–802.

Gerson, U., & Smiley, R. L. (1990). Acarine biocontrol agents, an illustrated key and manual. New York: Chapman & Hall.

Goeden, R. D., Fleshner, C. A., & Ricker, D. (1967). Biological control of prockly pear cacti on Santa Cruz Island, California. Hilgardia, 38, 579–606.

Goeden, R. D. (1983). Critique and revision of Harris' scoring system for selection of insect agents in biological control of weeds. Protection Ecology, 5, 287–301.

Harley, K. L. S., & Forno, I. W. (1992). Biological control of weeds, a handbook for practitioners and students. Melbourne, Australia: Inkata Press.

Harris, P. (1973). The selection of effective agents for the biological control of weeds. Canadian Entomologist, 105, 1495–1503.

Harris, P. (1986). Biological control of weeds. In Franz, J. M. (Ed.), Biological plant and health protection, biological control of plant pests and of vectors of human and animal diseases (pp. 123–138). International symposium of the Akademie der Wissenschaften und der Literatur, Mainz, November 15–17, 1984, Mainz and Darmstadt, Germany.

Harris, P. (1984). Carduus nutans L., nodding thistle and C. acanthoides L., plumeless thistle (Compositae). In J. S. Kelleher & M. A. Hulme (Eds.). Biological control programmes against insects and weeds in Canada 1969–1980 (pp. 115–126). London: CAB International.

Harris, P., Wilkinson, A. T. S., Thompson, L. S., & Neary, M. (1978). Interaction between the cinnabar moth, Tyria jacobaeae L. (Lep.: Arctiidae) and ragwort Senecio jacobaea L. (Compositiae) in Canada (pp. 174–180). In Anonymous, Proceedings, Fourth International Symposium on Biological Control of Weeds, Gainesville, FL.

Hill, R. L., Gindell, J. M., Winks, C. J., Sheat, J. J., & Hayes, L. M. (1991). Establishment of gorse spider mite as a control agent for gorse (pp. 31–34). Proceedings of the 44th New Zealand Weed and Pest Control Conference. Wellington, N.Z.

Hoffmann, J. H., & Moran, V. C. (1992). Oviposition patterns and the supplementary role of a seed-feeding weevil, Rhyssomatus marginatus (Coleoptera: Curculionidae), in the biological control of a perennial leguminous weed, *Sesbania punicea.* Bulletin of Entomological Research, 82, 343–347.

Huffaker, C. B., Hamai, J., & Nowierski, R. M. (1983). Biological conrol of puncturevine, *Tribulus terrestris,* in California after twenty years of activity of introduced weevils. Entomophaga, 28, 387–400.

Julien, M. H. (1981). Control of aquatic Alternanthera philoxeroides in Australia; another success for Agasicles hygrophila (pp. 583–588). Proceedings of the Fifth International Symposium on Biological Control of Weeds.

Julien, M. H. (Ed.). (1992). Biological control of weeds: A world catalogue of agents and their target weeds. Wallingford, United Kingdom: CAB International.

McEvoy, P. B., Cox, C. S., James, R. R., & Rudd, N. T. (1990). Ecological mechanisms underlying successful biological weed control: Field experiments with ragweed Senecio jacobaea (pp. 55–66). In Delfosse E. (Ed.), Proceedings of the Seventh International Symposium on Biological Control of Weeds, Rome, Italy.

Müller-Schärer, H. (1991). The impact of root herbivory as a function of plant density and competition: Survival, growth and fecundity of *Centaurea maculosa* in field plots. Journal of Applied Ecology, 28, 759–776.

Neser, S., & Kluge, R. L. (1986). The importance of seed-attacking agents in the biological control of invasive alien plants. In I. A. Macdonald, W., F. J. Kruger, & A. A. Ferrar (Eds.), The Ecology and Management of Biological Invasions in Southern Africa (pp. 285–293). Cape Town, South Africa: Oxford University Press.

Parker, P. E. (1991). Nematodes as biological control agents of weeds. In D. O. TeBeest, (Ed.), Microbial control of weeds (pp. 58–68). New York: Chapman & Hall.

Rao, U. P. (1971). Biological control of pests in Fiji (Miscellaneous Publication 2). Slough, U.K.: Commonwealth Institute of Biological Control.

Siddiqi, M. R. (1986). Tylenchida parasites of plants and insects. Slough, United Kingdom: CAB International, Commonwealth Institute of Parasitology.

Strong, D. R., Lawton, J. H., & Southwood, T. R. E. (1984). Insects on plants: Community patterns and mechanisms. Cambridge, MA: Harvard University Press.

Thomas, P. A., & Room, P. M. (1986). Taxonomy and control of *Salvinia molesta.* Nature (London), 320, 581–584.

Thompson, C. R., & Habeck, D. H. (1989). Host specificity and biology of the weevil *Neohydronomus affiinis* [Coleoptera: Curculionidae] a biological control agent of *Pistia stratiotes.* Entomophaga, 34, 299–306.

Trumble, J. T., Kolodny-Hirsch, D. M., & Ting, I. P. (1993). Plant compensation for arthropod herbivory. Annual Review of Entomology, 38, 93–119.

Van Driesche, R. G., & Bellows, T. S. (1996). Biological control. New York: Chapman & Hall.

Wapshere, A. J. (1974). Towards a science of biological control of weeds (pp. 3–12). In Anonymous, Proceedings of the Third International Symposium on Biological Control of Weeds, Montpellier, France, 1973.

Wapshere, A. J. (1989). A testing sequence for reducing the rejection of potential biological control agents for weeds. Annals of Applied Biology, 114, 515–526.

Wapshere, A. J. (1992). Comparing methods of selecting effective biocontrol agents for weeds (pp. 557–560). In Anonymous, Proceedings of the First International Weed Control Congress, Melbourne, Australia:

Wilson, B. W., & Garcia, C. A. (1992). Host specificity and biology of *Heteropsylla spinulosa* [Hom.: Psyllidae] introduced into Australia and Western Samoa for the biological control of *Mimosa invisa.* Entomophaga, 37, 293–299.

Wilson, C. G., & Pitkethley, R. N. (1992). *Botryodiplodia* die-back of *Mimosa pigra,* a noxious weed in northern Australia. Plant Pathology, 41, 777–779.

Zimmermann, H. G. (1991). Biological control of mesquite, Prosopis spp. (Fabaceae), in South Africa. Agriculture, Ecosystems and Environment, 37, 175–186.

18

A Perspective on Pathogens as Biological Control Agents for Insect Pests

B. A. FEDERICI

Department of Entomology
University of California
Riverside, California

INTRODUCTION

Synthetic chemical insecticides have been the principal component of most insect control programs since the discovery of the insecticidal properties of DDT just prior to World War II. They continue to be used extensively due primarily to their high efficacy and relatively low cost, but also because most are easy to produce and use, are persistent, and have a broad spectrum of activity that includes many different types of insects. This combination of properties led to widespread and heavy use soon after discovery throughout the world to control agricultural insect pests and insect vectors of human and animal diseases. There can be little doubt that chemical insecticides have been of great value in reducing economic damage caused by insects and by the prevalence of many plant and animal diseases transmitted by insects. However, their extensive use has resulted in a variety of problems, the most important of which include the development of resistance in many pest populations, the elevation to major pest status of relatively minor insect pests resulting from increased use of insecticides to control resistant pests, the elimination of natural enemies and disruption of natural ecosystems, and the fear on the part of the public that continued use of chemical insecticides will further contaminate the environment—particularly the food and water supply—with persistent and harmful chemicals. New chemical insecticides are much more expensive to develop, and this, plus pressures to reduce the use of chemicals in insect control programs, has led to increased emphasis on the development of alternatives control agents and strategies. These include biological control agents, cultural control (also referred to as environmental management), use of pheromones, increased development of insect-resistant plants, and better integration of nonchem-

ical agents and tactics into integrated pest management (IPM) programs.

Like all other organisms, insects are susceptible to a variety of diseases caused by pathogens. Many of these pathogens cause diseases that are acute and fatal, and therefore can be important short-term regulators of insect populations. As a result, insect pathogens are currently the subject of a considerable research effort aimed at developing the most effective pathogens as biological control agents for insects. There has been interest in using pathogens to control insects for well over 100 years, and several are currently used quite successfully, but interest in their further development is now greater than ever due to the search for more environmentally friendly alternatives to chemical insecticides.

The pathogens that cause disease in insects fall into four main groups: viruses, bacteria, fungi, and protozoa. In the present chapter, after a brief historical review and a preliminary description of strategies for using pathogens in control programs, each of these groups will be described in terms of the biological properties that determine their utility as agents for controlling insects. In addition to the pathogens, insect-parasitic nematodes will also be treated in this chapter because many of the same principles apply to the use of these organisms as biological control agents.

A great deal has been written about the biology of insect pathogens and their use as control agents, including many excellent reviews, but few are actually used on a widespread basis as operational control agents. Thus, instead of reviewing this literature again, the intent is to summarize for nonspecialists (i.e., those outside the fields of insect pathology and microbial control) the key biological features of the major pathogens developed or considered for development as control agents and to provide a critical evaluation of why these agents are not more widely used in insect control. The emphasis will be on a description and an assessment of the attributes and limitations of the different pathogen types based on the numerous studies conducted in this field over the past three decades. The most appropriate uses for each type will be identified as well as the obstacles that must be overcome to have these agents used more widely in insect control. In addition, it will be shown how knowledge of the molecular biology of certain insect pathogens, particularly nuclear polyhedrosis viruses (NPVs) and *Bacillus thuringiensis* (B.t.), and the techniques of recombinant DNA technology are being employed to improve the insecticidal properties of these pathogens. References to the primary literature and reviews will be given where pertinent for those interested in a wider range of examples and more detailed information. The principal pathogens and nematodes that will be discussed in this chapter are listed in Table 1 along with references to more specialized articles.

Historical Background

Serious interest in using pathogens to control insect pests can be traced back well over 100 years. According to Steinhaus (1975), several workers during the last century, including A. Bassi in Italy, V. Auduoin in France, and later J. LeConte in the United States and E. Metchnikoff in Russia, suggested that pathogens might prove to be effective agents for controlling crop pests. However, none of these early suggestions led to the successful use of pathogens to control insects. In the present century, renewed interest in using pathogens for insect control followed the successful use of other natural enemies such as predaceous beetles and parasitic wasps. Control of the cottony-cushion scale, *Icerya purchasi* Maskell, by the vedalia beetle, *Rodolia cardinalis* (Mulsant), in the United States led to the establishment of permanent research programs by governmental agencies in this and several other countries aimed at using natural enemies, including pathogens, as pest control agents (Paillot, 1933; Steinhaus, 1975). Two notable early successes emerged from these programs: the development of *Bacillus popilliae* Dutky as a control agent for the Japanese beetle, *Popillia japonica* Newman, in the United States (White & Dutky, 1942), and the use of the NPV of the European spruce sawfly, *Gilpinia hercyniae* (Hartig), as a classical biological control agent in Canada (Balch & Bird, 1944).

As part of an expanded research effort in the United States, Professor Harry S. Smith at the University of California hired E. A. Steinhaus shortly after the end of World War II to develop pathogens of insects as biological control agents (Steinhaus, 1963, 1975). Steinhaus, who had been trained as a bacteriologist, was well aware of the success attained with *Bacillus popilliae* and the *G. hercyniae* NPV; and soon after his appointment he began research in earnest aimed at using insect pathogens, particularly bacteria and viruses, as control agents. Though certainly other individu-

als were important to the development of insect pathology and microbial control [see Cameron (1973)], the modern era of these two closely intertwined disciplines is due largely to the extraordinary vision, scholarship, and energy of Steinhaus. Steinhaus resurrected bacterium *Bacillus thuringiensis* Berliner as a potential control agent, and his studies along with those of Hannay (1953) and Angus (1954) were pivotal points in its ultimate commercialization and successful use as a microbial insecticide. In addition, through his early studies of the NPV of the alfalfa caterpillar, *Colias eurytheme* Boisduval, Steinhaus did much to focus attention on the potential of viruses as microbial insecticides. His studies of viral and bacterial diseases of insects and other pioneering efforts in the field of insect pathology reinvigorated studies of fungi and protozoa as control agents, and were important to the development of the international fields of insect pathology and microbial control as we know them today.

Current Usage in Perspective

Despite the rather extensive body of literature on insect pathogens and microbial control, only one pathogen, (B.t.), is used routinely in a variety of control programs and has proved to be a commercial success in industrialized countries. This is also the pathogen in which current interest concerning further development and use is the highest. Before discussing the different pathogen types, it is worthwhile to consider why this is so. Such an assessment identifies the key features required for a pathogen to be successful as an insect control agent. To do this requires a few definitions concerning the different ways in which pathogens are used or considered for use in insect control programs, and a preliminary view (to be expanded on later) of the performance expectations by which the potential of insect pathogens and their degree of success are evaluated.

In general, the guiding principle here is one of economics; the pathogens that will be developed and used are those that are the most *cost-effective*, in either the short or long term. Though this may be obvious, it is all too often overlooked in much of the literature on the use of pathogens as control agents. The effectiveness of a control strategy based on an individual pathogen will always be compared with that obtained with other available control strategies, including those based on other pathogens and chemical insecticides. Also of importance, cost-effectiveness can vary with such important factors as crop, season, species complex to be controlled, geographic location, governmental regulations affecting registration and use, and status of economic development in the country where the pathogen is used. The latter two are particularly important because the costs of development and the use of a pathogen in developing countries, especially for viruses and fungi, are much less than in highly industrialized countries. This is because the costs of material and labor for production are low, many of

TABLE 1   Principal Pathogens and Nematodes Developed or Considered for Use as Biological
Control Agents for Insect Pests and Vectors of Disease

Pathogen	Major targets	References[a]
Viruses		
Baculoviruses		
Nuclear polyhedrosis viruses	Caterpillars, sawfly larvae	Federici, 1998
		Granados & Federici, 1986
		Adams & Bonami, 1991
		Hunter-Fujita *et al.,* 1998
		Miller, 1997; Treacy, 1998
Granulosis viruses	Caterpillars	Tweeten *et al.,* 1981
Nonoccluded (*Oryctes* virus)	Scarab larvae	Bedford, 1981
		Zelazny *et al.,* 1992
Cytoplasmic polyhedrosis viruses	Caterpillars	Aruga & Tanada, 1971;
		Katigitri, 1981
Entomopoxviruses	Grasshoppers, scarab larvae	Granados, 1981; Arif, 1984
Iridoviruses	Mosquitoes	Anthony & Comps, 1991
Bacteria		
Bacillus popilliae	Scarab larvae	Klein, 1981; Klein, 1997
B. thuringiensis	Caterpillars	Navon, 1993
		Baum *et al.,* 1998
	Mosquito and blackfly larvae	de Barjac & Sutherland, 1990;
		Becker & Margalit, 1993
		Jenkins, 1998
	Bettle larvae	Keller & Langenbruch, 1990
B. sphaericus	Mosquito larvae	de Barjac & Sutherland, 1990
		Baumann *et al.,* 1991
		Charles *et al.,* 1996
Fungi		
Mastigomycotina		
Coelomomyces species	Mosquito larvae	Federici, 1981; Couch &
		Bland, 1985
Lagenidium giganteum	Mosquito larvae	Federici, 1981
Zygomycotina		
Entomophthoraceous fungi	Aphids, caterpillars, beetles,	Wilding, 1981
	grasshoppers	Humber, 1989
Deuteromycotina		
Beauveria bassiana	Bettle larvae, caterpillars,	McCoy *et al.,* 1988
	grasshoppers	Wraight and Carruthers, 1998
Metarhizium anisopliae	Beetle larvae, leafhoppers,	McCoy *et al.,* 1988
	spittlebugs, cockroaches	
Verticillium lecani	Aphids, whiteflies	Hall, 1981;
		Hall & Paperiok, 1982
Microsporidia		
Nosema species	Grasshoppers, mosquito	Brooks, 1988; Henry, 1991
	larvae, beetles	
Vairimorpha necatrix	Caterpillars	Maddox *et al.,* 1981;
		Brooks, 1988
Nematodes		
Mermithids	Mosquito larvae	Petersen, 1982
Steinernematids, Heterorhabditids	Caterpillars, beetle larvae,	Gaugler & Kaya, 1990
	mole crickets	Kaya & Gaugler, 1993
		Grewal and Georgis, 1998

[a] Review articles and books that provide extensive coverage of the biology, use, and control potential of pathogens and nematodes along with references to relevant primary literature. The textbook *Insect Pathology* by Tanada and Kaya (1993), and *Manual of Techniques in Insect Pathology* edited by Lacey (1997) also provide useful information and numerous references to the primary literature on insect pathogens and nematodes.

the farms are smaller and less mechanized, and the registration process is cheaper and less cumbersome. For these reasons, caution must be exercised in drawing conclusions about the cost-effectiveness of a pathogen based on studies carried out in developing countries and then applying these results to prospective use in developed countries. Alternatively, experience gained from the greater use of pathogens in developing countries may provide valuable insights into how to integrate pathogens into IPM programs in developed countries. Nevertheless, though pathogens (especially viruses) are used more widely in developing countries such as China and India, even in these, usage is marginal in comparison to chemical insecticides.

In most cropping systems, pest populations contain individuals infected with pathogens capable of causing both acute (e.g., nuclear polyhedrosis viruses) and chronic (e.g., microsporidia) diseases. The periodic spread of these pathogens throughout the pest population, especially when the population becomes large and economically important, often leads to devastating epizootics that reduce the pest population to below economic thresholds for years afterwards (Fuxa & Tanada, 1987). Typically, however, the pest population does not crash until there has been substantial economic damage. This point is well illustrated by the outbreaks and declines of many lepidopteran forest pests such as those of the gypsy moth, *Lymantria dispar,* and the spruce budworm, *Choristoneura fumiferana.* Because modeling of these epizootics is not sufficiently precise at present to predict accurately their occurrence and extent (Anderson, 1982), monitoring pathogens within the population is not considered a reliable method of using pathogens as control agents. Thus, programs aimed at using pathogens for pest control generally involve strategies in which pathogens are deliberately introduced against pest populations for short- or long-term control.

In view of the preceding considerations, the most cost-effective use of a pathogen, as with other biological control agents, is as a classical biological control agent (DeBach & Rosen, 1991). In such cases, introduction leads to large-scale outbreaks of disease (epizootics) and establishment of the pathogen in the target population. Within a few years, the pest population is reduced below the economic threshold on a permanent basis. Unfortunately, though pathogens may establish and become endemic in a target population within a few years of introduction, there are very few examples where establishment has resulted in a classical biological control success. The best example in insects is the use of an NPV to control the European spruce sawfly in Canada, as noted earlier (Balch & Bird, 1944). Seasonal introduction is another way in which pathogens can be used. For this type of use, the pathogen is released into the pest population at the beginning of the season or early in the development of the pest population. The pathogen reduces the population below the economic threshold, but

typically only for one or a few seasons. The protozoan *Nosema locustae* Canning has been used in such a strategy to reduce grasshopper populations over a period of several weeks (Henry, 1990). Again, however, this tactic has been of only limited utility.

The most common strategy is to use pathogens as microbial insecticides, and because this strategy has been quite successful with bacteria and viruses, it will likely be of greater use in the future. In this case, a formulation of the pathogen is applied against a target pest on a periodic basis, as needed, much as in the case of chemical insecticides (Federici, 1991; Moscardi, 1999). Depending on the target pest, applications may often be fewer than those required with a chemical insecticide because the pathogens are quite specific and typically do not kill predatory and parasitic insects. Thus, these natural enemies remain in the ecosystem to slow the increase in the pest population after the initial mortality caused by the microbial insecticide. Moreover, the reproduction of the pathogen in the target insect, as in the cases of viruses, adds to the amount of the pathogen in the crop environment, and this can extend control and improve cost-effectiveness. In pests with only one or a few generations per season, a single application can yield effective season-long control where a combination of these factors is in operation.

Another important issue affecting the use of pathogens is performance expectations. In most cases, a successful pathogen is one that can reduce the pest or vector population to below an economic or transmission threshold routinely and reliably at a cost that is economical in proportion to the value of the crop or impact of the disease. Clearly, pathogens that are effective as classical biological control agents, or on a seasonal introduction basis, would be the most cost-effective. However, few pathogens have been used successfully in these strategies. Often where there are claims of success, cost-effectiveness has not been clearly demonstrated. As a result, the efficacy of most pathogens is evaluated in terms of their utility as microbial insecticides. Since World War II, expectations for the efficacy of any control strategy have been very high, having been set largely by the performance of chemical insecticides. Traditionally, these have been fast-acting, broad-spectrum control agents with substantial residual activity that are relatively inexpensive, in addition to bring easy to produce, formulate, and use. Thus, under most circumstances pathogens are evaluated on the basis of how they compare with chemical insecticides, in particular, with how quickly they kill the target insect and what the cost is. This leads to what might be called the microbial control agent paradox—the advantages of microbial insecticides can also be disadvantages—a paradox that is important because of its economic consequences. Ironically, it is now generally agreed that the two properties of many chemical insecticides originally considered their best attributes (i.e., a broad spectrum of activity and significant residual activity) are actually detrimental. These properties are largely responsible for the destruction of natural-enemy populations and the development of insecticide resistance, and if such environmental costs are calculated, the cost-effectiveness of chemical insecticides is lower than calculated (Higley & Wintersteen, 1992). On the other hand, insect pathogens typically have a narrow spectrum of activity and relatively poor residual activity. Though these are now considered beneficial properties, they have discouraged interest by industry in the development of many potentially useful pathogens because of the relatively high costs of development and registration in comparison with the likely return on investment. This is a relative argument. The costs for development and registration of a microbial insecticide are actually much cheaper in developed countries than for a chemical insecticide (e.g., approximately a million dollars for a microbial compared with at least several million dollars for most chemicals in the United States, Europe, and Japan). This is because many pathogens with potential for development as microbial insecticides have already been identified. Nevertheless, a company still must see the potential for making its investment pay off within a few years. For many pathogens that are highly specific in their host range, this is simply not possible given the current regulatory environment and market size in most industrialized countries.

Whether on a small or a large scale, industrial interest is extremely important because industry will be the source of pathogens to be used as microbial insecticides. This is particularly true in developed countries where the farms tend to be large. In fact, most farmers, whether large or small, want a reliable supply of control agents. Though they are willing to change cultural practices, because of numerous other responsibilities, farmers are generally unwilling to serve as the manufacturers of their own microbial insecticides.

In light of the economics of pesticide development, it is easy to see why *B.t.* has been the most widely developed and used insect pathogen. In essence, it compares favorably with chemical insecticides in many crop and forest systems where lepidopterous insects are the key or major pests. It is fast acting; is relatively inexpensive; and is easy to produce, formulate, and use. Its residual activity is low, and none of the subspecies have a broad host range in comparison with most chemical insecticides. Yet the range of insects it controls provided, and continues to provide, a market large enough to justify commercial development.

Overall, in most pest control situations the prospects for the use of pathogens is best where they will be used as microbial insecticides, to be used one or more times as needed during the pest season. Under these circumstances, the availability, reliability, and cost-effectiveness of microbial insecticides become paramount. These issues will be returned to again at the close of this chapter, but should be

kept in mind as the properties of the various pathogen groups and our ability to improve their efficacy are discussed.

VIRUSES

All viruses are obligate intracellular parasites, and as such must be grown in living hosts. With insect viruses, this means either in insects (i.e., *in vivo*) or in cultured insect cells (*in vitro*); no methods exist for growing viruses on "artificial" media. Most of the viruses that occur in insects belong to one of the major taxonomic groups listed in Table 1. In general, these viruses are divided into two broad non-taxonomic categories, occluded viruses and nonoccluded viruses (Payne & Kelly, 1981; Tanada & Kaya, 1993; Hunter-Fujita *et al.*, 1998). Occluded viruses are so named because after formation in infected cells, the mature virus particles (virions) are occluded within a protein matrix, forming paracrystalline bodies that are generically referred to as either inclusion or occlusion bodies (Fig. 1). In nonoccluded viruses, the virions occur freely or occasionally form paracrystalline arrays of virions that are also known as inclusion bodies; these, however, have no occlusion body protein interspersed among the virions. More detailed information on the biological properties of insect viruses can be found in works by Adams and Bonami (1991), Tanada and Kaya (1993), and Miller (1997).

Biological Properties Related to Insect Control

A short description of the most important biological properties is given here for the four most common virus types to set the stage for a more detailed discussion of occluded baculoviruses, which as a group have received the most attention as control agents.

Iridoviruses

Iridoviruses are nonoccluded viruses with a linear double-stranded DNA genome that produce large, icosahedral virions (125 to 200 nm) that replicate in the cytoplasm of a wide range of tissues in infected hosts, causing disease that is generally fatal. Virions can form paracrystalline arrays in infected tissues, imparting an iridescent hue to infected hosts, from which the name of this virus group is derived. Over 30 types are known, and all have been extremely difficult to transmit *per os*. The host range of each type appears to be quite narrow based on natural occurrence in host field populations. Prevalence and mortality rates in natural populations of host insects are typically less than 1% (Kelly & Robertson,1973; Tanada & Kaya,1993).

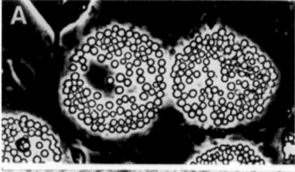

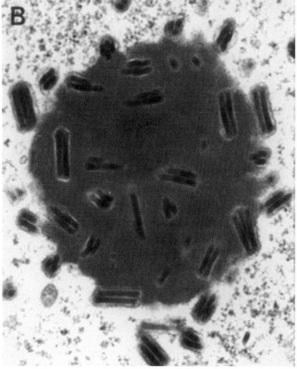

FIGURE 1 Nuclear polyhedrosis virus polyhedra. (A) A wet mount preparation viewed with phase microscopy showing refractile polyhedra in two infected nuclei. (B) Electron micrograph through a single polyhedron showing the enveloped rod-shaped virions, characteristic of NPVs, occluded within the polyhedral matrix. On ingestion, this matrix dissolves in the insect midgut, and the virions invade the host via midgut microvilli.

Cytoplasmic Polyhedrosis Viruses

Cytoplasmic polyhedrosis viruses (CPVs) are occluded double-stranded RNA viruses with a multipartite genome that replicate and form large (ca 0.5 to 2 μm) polyhedral to spherical occlusion bodies in the cytoplasm of midgut epithelial cells, causing a chronic disease. Infection in early instars retards growth and development, extending the larval phase by several weeks, and in many cases is ultimately fatal. This virus type is relatively common among lepidopterous insects and dipterous insects of the suborder Nematocera (e.g., mosquitoes, blackflies, midges). Isolates that

have been studied in any detail are generally easy to transmit *per os* to the original host and other species of the same order, and thus the host range of this virus type is probably the broadest among the insect viruses (Aruga & Tanada, 1971; Katagiri, 1981; Tanada & Kaya, 1993).

Entomopoxviruses

Entomopoxviruses are occluded double-stranded DNA viruses that produce large virions (150 × 300 nm) that replicate in the cytoplasm of a wide range of tissues in most hosts, causing an acute fatal disease. Occlusion bodies, depending on the isolate, vary from being ovoidal to spindle shaped and generally occlude 100 or more virions. These viruses have been most commonly reported from coleopterous insects, from which there are over 20 isolates, but are also known from lepidopterous, dipterous (midges), and orthopterous (grasshoppers) insects. This virus type is easily transmitted *per os,* though the experimental host range of individual isolates, where tested, is relatively narrow, generally being restricted to closely related species (Granados, 1981; Arif, 1984).

Occluded Baculoviruses

There are two types of occluded baculoviruses, NPVs and granulosis viruses (GVs). Both NPVs and GVs are highly infectious *per os,* and in some insect species these viruses cause widespread epizootics periodically that result in significant (>90%) declines in larval populations.

NPVs are known from a wide range of insect orders as well as from Crustacea (shrimp), but by far have been most commonly reported from lepidopterous insects, from which well over a 500 isolates are known (Tanada & Kaya, 1993; Volkman *et al.,* 1995). Many of these are different viruses (i.e., viral species). NPVs replicate in the nuclei of cells, generally causing an acute fatal disease (see Fig. 1). The virions are large (80 to 200 × 280 nm) and consist of one or more rod-shaped nucleocapsids with a double-stranded circular DNA genome enclosed in an envelope. The occlusion bodies of NPVs are referred to commonly as polyhedra because they are typically polyhedral in shape. Polyhedra are large (ca 0.5 to 2 μm), and form in the nuclei, where each occludes as many as several hundred virions. NPVs of lepidopterous insects infect a range of host tissues, but those of other orders are typically restricted to the midgut epithelium (Federici, 1993, 1997). Some NPVs have a very narrow host range, and may only replicate efficiently in a single species, whereas others, such as the AcMNPV [i.e., the NPV of the alfalfa looper, *Autographa californica* (Speyer)] have a relatively broad host range and are capable of infecting species from different genera (Granados & Federici, 1986; Blissard & Rohrmann, 1990).

GVs, of which there are now over 100 isolates, are

closely related to NPVs but differ from the latter in several important respects (Tweeten *et al.,* 1981; Federici, 1997; Volkman *et al.,* 1995). The virions of GVs are similar to those of NPVs but contain only one nucelocapsid per envelope. GVs are only known from lepidopterous insects. Like NPVs, they initially replicate in the cell nucleus, but replication involves early lysis of the nucleus (as virions begin to assemble), which in NPVs only occurs after most polyhedra have formed. After the nucleus lyses, GV replication continues throughout the cell, which now consists of a mixture of cytoplasm and nucleoplasm. When completely assembled, the virions are occluded individually in small (200 × 600 nm) occlusion bodies referred as granules. Many GVs primarily infect the fat body, whereas others have a broader tissue tropism and replicate throughout the epidermis, tracheal matrix, and fat body. One, the GV of the grape leaf skeletonizer, *Harrisina brillians* Barnes & McDonnough, is unusual in that it only replicates in the midgut epithelium (Federici & Stern, 1990).

Use of Viruses as Insect Control Agents

The best example of the use of a virus as an insect control agent is the NPV of the European spruce sawfly, *G. hercyniae,* as a classical biological control agent to control this important forest pest (Balch & Bird, 1944; Cunningham & Entwistle, 1981). The European spruce sawfly was introduced into eastern Canada from northern Europe around the turn of the century and was a severe forest pest by the 1930s. Hymenopteran parasites were introduced from Europe in the mid-1930s as part of a biological control effort, and inadvertently along with these came the NPV, which was first detected in 1936. Natural epizootics caused by the virus began in 1938, by which time the sawfly had spread over 31,000 km^2. Most sawfly populations were reduced to below economic threshold levels by 1943, and remain under natural control today, the control being effected by a combination of the NPV (which accounts for more than 90% of the control) and the parasites.

While viruses, particularly NPVs, are frequently associated with rapid declines in the populations of important lepidopterous and hymenopterous (sawfly) pests, the *G. hercyniae* NPV is the only true example of a virus that has been effective as a classical biological control agent. Another baculovirus, the "nonoccluded" baculovirus of the palm rhinoceros beetle *(Oryctes rhinoceros)* has been a quasi-classical biological control success in that once introduced into populations it can yield control for several years, but ultimately it dissipates and must be reapplied against the pest population (Bedford, 1981; Zelazny *et al.,* 1992). Moreover, aumentative seasonal introductions, though possibly effective, have only rarely been effective and are not well documented. Thus, the control potential of most viruses is best evaluated by assessing their utility as microbial

insecticides. From this standpoint, iridoviruses are essentially useless because of their poor infectivity *per os.* CPVs are not much better because, though highly infectious *per os,* the disease they cause is chronic (Aruga & Tanada, 1971; Payne, 1981; Tanada & Kaya, 1993; Kunimi, 1998). CPVs have, however, been useful in some situations, such as for suppression of the pine caterpillar, *Dendrolimus spectabilis,* in Japan. The interested reader is referred to Katagiri (1981) for a review of these viruses as control agents. Entomopoxviruses have not yet been developed as control agents for any insect. The control potential and procedures for development of these viruses are similar to those described earlier for entomopoxviruses, and later below for NPVs; therefore, they will not be discussed here in any detail. The interested reader is referred to the publications by Granados (1981) and Arif (1984) for additional information about entomopoxviruses.

Nuclear Polyhedrosis Viruses as Conventional Microbial Insecticides

The viruses most commonly used or considered for development as microbial insecticides in industrialized as well as less developed countries are the NPVs. The reasons for this are that NPVs are often common in, and easily isolated from, pest populations, production in their hosts is straightforward and inexpensive, and technology for formulation and application is relatively simple and adaptable to standard pesticide application methods. Most NPVs, however, are narrow in their host range, infecting only a few closely related species. Furthermore, though several can be grown *in vitro* in small to moderate volumes (ca. 20 to 300-liter cell cultures), no fermentation technology currently exists for their mass production on a commercial scale that would permit repeated applications to hundreds of thousands of acres, which is possible with B.t. and chemical insecticides. These two key limitations have been significant disincentives for the commercial development of NPVs, especially in industrialized countries. The large chemical and pharmaceutical companies that might be expected to take an interest in NPVs have, with rare exception, demonstrated little interest in producing a product that must be grown in living caterpillars or sawfly larvae, and that in most cases has a small market. The production technology is more suitable for "cottage industries" in either industrialized or developing countries. However, in industrialized countries, where all pathogens used as insecticides must be registered by one or more governmental agencies, the regulatory procedures and associated costs of development—conservatively estimated at $1,000,000—impede development by smaller companies such as those involved in the production of predators and parasites.

Despite these drawbacks, several NPVs have been registered as microbial insecticides though the market size for

TABLE 2 Major Viruses Used or under Development for Control of Insect Pests[a]

Target pest	Virus	Crop or habitat	Product name	Producer
Registered or in use				
Caterpillars				
Anticarsia gemmatalis	NPV	Soybeans	None	Various/local (Brazil)
Adoxophyes orana	GV	Fruit orchards	Capex	Andermatt Biocontrol (Switzerland)
Cydia pomonella	GV	Apples, walnuts	Madex	Andermatt Biocontrol
Helicoverpa zea	NPV	Cotton, vegetables	Gemstar	Thermo-Trilogy (United States)
Lymantria dispar	NPV	Decidous forests	Gypcheck Disparvirus	Thermo-Trilogy Canadian Forest Service
Mamestra brassicae	NPV	Vegetables	Mamestrin	Caliope (France)
Orgyia pseudotsugata	NPV	Douglas fir forests	TM Biocontrol-1	Thermo-Trilogy
Spodoptera littoralis	NPV	Cotton	Spodopterin	Caliope
S. exigua	NPV	Vegetables	Spod-X	Thermo-Trilogy
Sawfly larvae				
Gilpinia hercyniae	NPV	Spruce forests		
Neodiprion sertifer	NPV	Pine forests	Neocheck-S Sentifervirus	
N. lecontei	NPV	Pine forests	Lecontivirus	Canadian Forest Service
Under development				
Caterpillars				
Autographa californica	NPV	Vegetables	Gusano	Thermo-Trilogy
S. exigua	NPV	Vegetables	Spod-X	Thermo-Trilogy
Cydia pomonella	GV	Apples, walnuts	CYD-X	Thermo-Trilogy
Plodia interpunctella	GV	Stored nuts	None	U.S. Department of Agriculture

[a] See Shah and Goettel (1998) or the website "www.sipweb.org" for a recent list of pathogen products.

most is quite small (Table 2). Whether registered or not, several are used in many less developed countries, particularly for control of lepidopteran pests of field and vegetable crops (Entwistle, 1998). Moreover, over the past decade there has been renewed interest in developing NPVs because recombinant DNA technology offers potential for improving the efficacy of these viruses. For economic reasons, as in the past, this interest is still restricted largely to NPVs of major lepidopteran pests and a few key sawflies. The high host specificity of NPVs makes them of most use where a single insect species is the only pest or at least a key one of a particular crop. Utility is increased if the target insect is not very sensitive to B.t., and is resistant to chemicals; or if the latter are too costly, environmentally unacceptable, or unavailable. The viruses listed in Tables 2 and 3 all meet, or at one time met, these criteria. Before dis-

TABLE 3 Comparative Efficacy of Representative Occluded Viruses[a]

Virus	Occlusion body yield per larva	Occlusion body application rate per hectare	LEs per hectare[a]
Anticarsia gemmatalis NPV	2×10^9	1.5×10^{11}	50
Helicoverpa zea NPV	1×10^{10}	1×10^{12}	100
Lymantria dispar NPV	2×10^9	5×10^{11}	250
Orgyia pseudotsugata NPV	2×10^9	3×10^{11}	125
Spodoptera exigua NPV	2×10^9	1×10^{12}	500
Neodiprion sertifer NPV	1×10^9	1×10^{11}	200
Cydia pomonella GV	1×10^{13}	2×10^{15}	300

[a] Amount needed to reduce crop loss or pest populations by approximately 80 to 95%.

cussing the possibilities for improvement through genetic engineering, it is worthwhile to summarize with a few examples the use of conventional (i.e., unengineered) NPVs and GVs as microbial insecticides and to consider further their advantages and limitations.

Production and Formulation

The NPVs are mass-produced in larval hosts grown on an artificial diet or vegetation from natural host plants, and in the latter case can be done in the field (Ignoffo, 1973; Moscardi, 1999). All viruses currently used in control programs are produced in larvae. To maximize production, larvae are infected *per os* at an advanced stage of development, such as during the late fourth instar, and are reared either as groups, or as individuals for species that are cannibalistic. After ingesting the virus, occlusion bodies dissolve in the alkaline midgut, releasing virions. In lepidopterans, the virus first invades midgut epithelial cells where, during the first 24 h of infection, it undergoes an initial colonizing phase of replication in the nuclei of these cells. No occlusion bodies are produced in these nuclei, but instead progeny virions migrate through the basement membrane, and invade and colonize almost all other tissues of the host. In these, virus replication results in virions that are occluded in polyhedra. Maximum production of polyhedra occurs in tissues that are the most nutrient rich and metabolically active such as the fat body, epidermis, and tracheal matrix. This definitive phase of viral disease occurs over a period of 5 to 10 days, and represents several cycles of replication as the virus spreads throughout the tissues and invades most host cells. The actual length of the disease depends on several factors including the host and viral species, the larval instar at the time of infection, the amount of inoculum, and the temperature. Near the end of the disease, after most polyhedra have formed, the nuclei lyse. As more and more nuclei lyse, the larva eventually dies; after this the body "liquefies," releasing literally billions of polyhedra. In commercial production, larvae may be harvested prior to liquefaction to keep bacteria, which quickly colonize dead larvae, at a low level in the final product. Alternatively, antibiotics can be added to the diet to keep bacterial counts low. After the larval production phase is complete, the larvae are collected and formulated. Formulation varies considerably, and depends on how the virus will be used, but both liquid concentrates and wettable powders have been developed that usually can be applied with conventional equipment (Ignoffo, 1973; Shapiro, 1986; Moscardi, 1999).

The production of lepidopteran GVs and sawfly NPVs is similar to that described for lepidopteran NPVs [see Cunningham and Entwistle (1981) and Shapiro (1986)]. As noted previously, however, the sawfly NPVs differ from the lepidopteran NPVs in that the former only replicate and form polyhedra in midgut epithelial cells. Polyhedral yields

therefore are lower than those obtained with lepidopteran NPVs.

Use of NPVs and GVs

The extent to which conventional NPVs can be useful as microbial insecticides depends on several factors that include the relative importance of the target pest in the pest complex attacking a crop, the amount of virus that must be used to control the pest in both the short term and long term (persistence and carryover), the value of the crop, and the cost and availability of alternative control measures (Federici, 1998). NPVs are ideal candidates for use where a single lepidopteran species is the major pest for most of the growing season on a crop with a high cash value where other available pest control methods are not cost-effective. Thus, NPVs that have been developed and used are those effective against species of *Helicoverpa, Heliothis,* and *Spodoptera* on such crops as cotton, corn, and sorghum; and even more so on tomatoes, strawberries, and flowers such as chrysanthemums, and *Anticarsia gemmatalis* in soybeans. The cost-effectiveness of these viruses is determined by the amount of virus that must be applied and the frequency of application necessary to keep the pest below an economic threshold. As noted earlier, this will vary with the virus, pest, and crop; and more importantly, among different countries. The amount of virus required is best assessed in terms of larval equivalents (LEs) (Ignoffo, 1973) necessary to achieve effective control. Illustrative examples are given in Table 3. The number of LEs required to obtain effective control is a critical component in the determination of cost-effectiveness due to the cost of labor and materials that go into virus production. This can range from 50 LEs per hectare per treatment using the *A. gemmatalis* NPV (AgNPV) to control the velvet bean caterpillar on soybeans to 500 LEs for the *S. exigua* multiple nucleocapsids per virion envelope (SeMNPV) on lettuce and chrysanthemums. Moreover, the number of LEs required to control a specific pest can vary with the crop due to differences in insect feeding behavior and in crop characteristics (such as phenology, leaf chemistry, and plant structure). For example, whereas 500 LEs per hectare may be required to control *S. exigua* on lettuce, the value may be as high as 1000 LEs on alfalfa or as low as 200 LEs on tomatoes. Clearly, the use of the SeMNPV to control the beet armyworm on tomatoes is much more cost-effective than its use would be on alfalfa. This type of economic evaluation is essential for determining if a specific NPV merits commercial development as well as for determining if it will be useful against a specific insect on a particular crop.

Because the actual use of NPVs and GVs as microbial insecticides in insect control programs is very limited in developed countries and is not well documented in most developing countries with the exception of Brazil, it is not possible to know the amount of hectares treated. The largest documented program in the world is the use of the AgNPV against this pest on soybeans in Brazil, where the hectares treated exceed 1 million (2.2 million acres) annually (Moscardi, 1999). Based on estimates collated by Entwistle (1998) and Moscardi (1999), this single program probably exceeds the total area currently treated worldwide with all other NPVs and GVs. These viruses may be more widely used than indicated here, but this information is not readily available in the published literature.

The AgNPV/soybean program and a few others are briefly described here to provide examples of viruses used in control programs. A wider range of examples is provided in reviews by Hunter-Fujita *et al.* (1998) and Moscardi (1999), as well as in earlier reviews by Fuxa (1990), Shapiro (1986), Shapiro and Robertson (1992), Payne (1982, 1988), and Pinnock (1975), which remain worthy reading.

AgNPV for Control of *A. gemmatalis* on Soybeans   The use of the AgNPV to control the velvet bean caterpillar, *A. gemmatalis,* was initiated in the early 1980s and grew by the mid-1990s to a program of approximately 1 million hectares (2.2 million acres) of soybeans (Moscardi, 1999). The virus is applied at a rate of 50 LEs per hectare (1.5×10^{11} polyhedra) when populations are in the first three instars, and when there are 20 or fewer larvae per row-meter of plants. Only one application is made per growing season. The cost to the farmer is about $1.50 per hectare, which is cheaper than chemical insecticides, and the level of control is comparable. The program is coordinated by the state agency, Embrapa, which awards contracts for the production of virus to five private companies. Most of the virus is produced in field plots, where yields of AgNPV-infected larvae have grown to 35 metric tons per year (Moscardi, 1999).

The success of this program is due to several factors, a primary one being that the virus only has to be applied once per season. In addition, soybeans can tolerate a significant amount of defoliation without affects on yield, and labor costs for the production of the virus are quite low.

Helicoverpa zea NPV (HzNPV) for Control of the Heliothis and Helicoverpa Species   The HzNPV was the first virus commercialized for use on a large scale (Ignoffo, 1973), and was sold originally by Sandoz, Inc. (which has merged with Ciba-Geigy to become Novartis) as Elcar. It was was developed for use against *Heliothis virescens* and *Helicoverpa zea* in cotton, but was not a commecial success due to the advent of the synthetic pyrethroids, which were cheaper and highly effective. This virus is still used in many countries such as China and India to control various members of the *Heliothis/Helicoverpa* complex on crops such as cotton, corn, and sorghum. Total usage worldwide

is probably in the range of 200,000 to 300,000 ha, and the rates of application are in the range of 100 to 200 LE per hectare. Typically, several applications are required per growing season, with the rate and frequency depending on the crop and size of the pest population.

Though one of the most effective NPVs, this virus has not been used much in developed countries because it was not being produced commercially on a large scale. Novartis has sold its microbial insecticide business to Thermo-Trilogy, and the HzNPV is now again available under the product name Gemstar.

***Spodoptera* NPVs for Control of *Spodoptera* Species**   The *Spodoptera* complex consists of important species that attack a wide variety of field and vegetable crops. The most improtant pests include *S. exigua, S. frugiperda, S. litura,* and *S. littoralis.* Each of these is attacked by an NPV that is fairly specific, and isolates of several of these have been developed in different regions of the world for *Spodoptera* control. For vegetable crops, the SeNPV has been developed for control of *S. exigua* on vegetable and ornamental crops. Thermo-Triology currently produces this virus under the trade name Spod-X, and the product is used to control *S. exigua* on chrysanthemumns in greenhouses in the Netherlands, and on vegetable crops in Thailand, with application rates typically ranging from 100 to 300 LE per hectare depending on the crop and insect pressure. It is also being evaluated for use on cotton in the United States. Total area treated at present is quite small, amounting to less than 10,000 ha worldwide. The *S. frugiperda* NPV (SfNPV) has been developed for control of this pest in maize in Latin America, where annual usage is currently in the range of 20,000 ha (Moscardi, 1999). The *S. litura* and *S. littoralis* NPVs (Entwistle, 1998) are primarily used for control of these pests on cotton in Middle Eastern countries and Asia, but accurate figures on their use are not available.

GVs for Insect Control   The granulosis viruses of codling moth, *Cydia pomonella* (CpGV), the potato tuberworm, *Phthorimaea operculella* (PoGV), and the cabbageworm, *Pieris rapae* (PrGV), are used to varying degrees to control these pests. The CpGV is registered for use against codling moth larvae on apples in Canada, the United States, and several European countires including Switzerland and Germany. The virus is applied at rates varying from 1 to 2.5×10^{13} granules per hectare, at a cost of $50 to $80 per application (see Table 3). In cool areas at high altitudes such as in Switzerland, the moth has only one or two generations per year, and thus the virus is cost-effective. In California, however, populations may have to be treated once a week, thereby decreasing the cost-effectiveness of the virus (Cunningham, 1998). Commerical preparations currently on the market include Carpovirusine (France), Madex (Switzerland), and Granusal (Germany), with sales worldwide scheduled to reach 60,000 ha in 1998 (Moscardi, 1998).

The PoGV is used in many developing countries to protect potatoes that are stored aboveground after harvest. Under direction of Center International for Potato Research (CIP, Lima, Peru), the virus is produced by "cottage industries" in countries such as Peru, Morroco, Libya, China, and Indonesia; and is typically applied to potatoes as a dust. The PrGV has been used for many years in mainland China to control *P. rapae* on vegetable crops (Entwistle, 1998). The virus, for use in either dusts or wettable powders, is produced in larvae in field plots of cabbage and lettuce, and then is harvested as the infected larvae begin to die. The area treated with the virus is not known, but probably is at least several tens of thousands of hectares per year.

Limitations

Viruses are not widely used at present in industrialized countries because products based on them are of very limited availability. Moreover, effective and competitively priced alternatives are readily available. Among these are new and more specific chemical insecticides such as the pyrroles, Fipronil and Imidocloprid, and existing and new products based on B.t., including transgenic plants. Furthermore, viruses still have what are considered key limitations. These include relatively slow speed of kill, narrow spectrum of activity (host range), little residual activity, and lack of a cost-effective system for mass production *in vitro*. On the other hand, these limitations have not inhibited the development and use of viruses in developing countries where NPVs and even a few GVs are used, especially on field and vegetable crops, in China, India, and Brazil, as well as in many smaller countries in Latin America, Africa, and Southeast Asia (Hunter-Fujita *et al.*, 1998; Zhang, 1981; Peng *et. al.*, 1988, 1991; McKinley *et al.*, 1989, Moscardi, 1989, 1990, 1999). The reasons for this are that chemical insecticides in many of these countries are expensive, their widespread and heavy use has resulted in resistance, labor costs for virus production *in vivo* are low, production technology is simple, and registration for use of viruses either is not required or is easily obtained.

If pressure continues to reduce the use of synthetic chemical insecticides, viruses will receive increased attention, particularly for the control of lepidopterous insects for which no other effective control agents such as B.t. and parasites exist. This should lead to increased efforts to develop and use conventional viruses in IPM programs in both industrialized and developing countries. In addition to conventional viruses, the development of recombinant DNA technology (i.e., genetic engineering) offers promise for improving viral efficacy by reducing or eliminating some of their major disadvantages.

Use of Recombinant DNA Technology to Improve Nuclear Polyhedrosis Viruses

The two most significant limitations of conventional viral insecticides are the relatively slow speed of kill and the narrow host range. Recombinant DNA technology is being used to overcome both of these limitations (Miller, 1988; Maeda, 1989; Hawtin & Possee, 1992; Black *et al.*, 1997; Treacy, 1998). When a conventional virus is used against an insect population (e.g., against noctuid larvae), the population typically consists of a mixture of instars and may contain more than one economically important species. A virulent NPV sprayed at an appropriate rate will kill the first and second, and in some cases the third instars, within 2 to 4 days. However, the more advanced instars may live for a week or more, causing further damage to the crop. Thus, current approaches to improving the efficacy of viral insecticides are aimed at developing broad-spectrum viruses that will cause a cessation of larval feeding within 24 to 48 h of infection, either by death or by paralysis. This is being done by deleting genes that delay death from viruses, and by engineering viruses to express genes encoding enzymes or peptide hormones that disrupt larval metabolism, or peptide neurotoxins that paralyze or kill the insect directly. Because it already has a broader host range than most occluded baculoviruses and can be genetically manipulated with ease in several cell lines, the *Autographa Californica* MNPV (AcMNPV) is the virus that has been the subject of most of the engineering studies to date.

The virus is engineered by deleting the gene encoding EGT (ecdysteroid UDP-glucosyltransferase) and/or by adding one or two genes encoding toxins under the control of strong viral promoters. By using this strategy, genes for juvenile hormone esterase (Hammock *et al.*, 1990), B.t. endotoxins (Martin *et al.*, 1990; Merryweather *et al.*, 1990; Pang *et al.*, 1992), insecticidal neurotoxins from the straw itch mite (*Pyemotes tritici*), scorpions (*Androctonus australis*, *Buthus epeus*, and *Leirus quinquestriatus lebraeus*), and spiders (*Agelenopsis aperta*, *Diguetia canites*, *Tegenaria agnestis*) have been engineered into the AcMNPV (McCutchen *et al.*, 1991; Tomalski & Miller, 1991; Black *et al.*, 1997; Treacy, 1998). Of these, the most promising results have been obtained with viruses producing neurotoxins, where the time between feeding and paralysis has been reduced by as much as 40% in comparison to wild-type AcMNPV (Treacy, 1998).

Engineering viruses to express insecticidal proteins in many cases could also result in an expanded host range. This is because in many lepidopteran host species that do not develop a patent disease when infected by conventional viruses, there can be limited viral replication, with these less susceptible hosts developing a mild disease and surviving infection. However, the same hosts infected by an engineered virus that expresses a potent insecticidal protein will likely succumb because the virus does not need to replicate extensively to paralyze or kill the larva.

BACTERIA

Bacteria are relatively simple unicellular microorganisms that lack internal organelles such as a nucleus and mitochondria and which reproduce by binary fission. With a few exceptions, most of those used as microbial insecticides grow readily on a wide variety of inexpensive substrates, a characteristic that greatly facilitates their mass production. The overwhelming majority of the bacteria currently used or under development as microbial control agents for insects are spore-forming members of the bacterial family Bacillaceae, and belong to the genus *Bacillus*. These insect-pathogenic bacilli occur in healthy and diseased insects, but also occur in, and can be isolated from, many other habitats including insect frass, soil, plants, granaries, and aquatic environments.

Biological Properties in Relation to Insect Control

Bacteria, especially the various subspecies of *Bacillus thuringiensis* Berliner, have been and will continue for the foreseeable future to be the insect pathogens used most widely and successfully in insect control. The reasons for this are that they are easy to mass-produce, formulate, and use in large-scale operational control programs; kill target insects quickly (≤ 48 h); have a spectrum of activity that includes many economically important pests; and are much safer than most synthetic chemical pesticides for nontarget organisms and the environment.

There are two major types of bacteria that are used in insect control: (1) those that cause fatal infectious diseases and (2) those that kill insects primarily through the action of insecticidal toxins. An example of the first type is *Bacillus popilliae* Dutky, a bacterium that infects and kills coleopteran larvae, particularly soil-inhabiting members of the family Scarabaeidae (Klein, 1997). The second type is exemplified by B.t., a species that produces toxins—both protein endotoxins and nucleotide exotoxins—capable of killing insects whether or not they are directly associated with the bacterium. Molecular biologists have made extensive use of the latter property by inserting the genes encoding various endotoxins into other microorganisms and plants, thereby making them insecticidal (Fischhoff *et al.,* 1987; Vaeck *et al.,* 1987; Perlak *et al.,* 1990; Baum *et al.,* 1998; Jenkins, 1998). The use of B.t. endotoxins in bac-

terial insecticides and transgenic plants has become so important that this bacterial species is treated separately in Chapter 21, and will be discussed only briefly here. Therefore, after a short summary of the main properties of B.t., the main subject of this section will be three other bacteria that are used as biological control agents of insect, *B. popilliae, B. sphaericus,* and *Serratia entomophila.*

Bacillus thuringiensis

The species *Bacillus thuringiensis* Berliner is actually a complex of bacterial subspecies all of which are characterized by the production of a parasporal body during sporulation. This parasporal body contains one or more proteins in a crystalline form (Fig. 2), and many of these are highly toxic to certain species of insects. In the insecticidal isolates, the toxins are known as endotoxins and often occur in the parasporal body as protoxins that after ingestion are activated by proteolysis in the gut. The activated toxins destroy midgut epithelial cells, killing sensitive insects within a day or two of ingestion. In insect species only moderately sensitive to the toxins, such as *Spodoptera* species, the spore contributes to the activity of the bacterium. A primary reason for the success of B.t. is that it is easily grown on a large scale (e.g., 50,000-liter batches), in simple, cheap, readily available media.

The most widely used B.t. is the HD1 isolate of *B. thuringiensis* subsp. *kurstaki* (B.t.k.), an isolate that produces four major endotoxin proteins packaged into the crystalline parasporal body (see Fig. 2 B). This isolate is used as the active ingredient in numerous commercially available bacterial insecticides used to control lepidopterous pests in field and vegetable crops, and forests (Navon, 1993; van Frankenhuyzen, 1993). Another successful B.t. is the ONR60A isolate of *B. thuringiensis* subsp. *israelensis* (B.t.i.), which is highly toxic to the larvae of many mosquito and blackfly species. This isolate also produces a parasporal body that contains four major endotoxins (see Fig. 2 B), but these are different than those that occur in B.t.k. Several commercial products based on B.t.i. are available and are used to control both nuisance and vector mosquitoes and blackflies (Federici *et al.,* 1990). A third isolate of B.t. that has been developed commercially is the DSM2803 isolate of *B. thuringiensis* subsp. *morrisoni* (strain tenebrionis). This isolate produces a cuboidal parasporal body toxic to many coleopterous insects, and is used commercially to control several beetle pests.

All the preceding isolates are essentially used as bacterial insecticides, applied as needed. A wide variety of formulations are available including emulsifiable concentrates, wettable powders, and granules for use against different pests in a variety of habitats. On a worldwide basis, millions of hectares are treated annually with products based on B.t. Estimates indicate the worldwide market is about

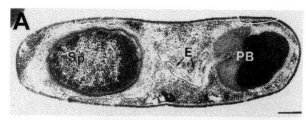

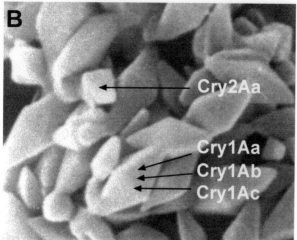

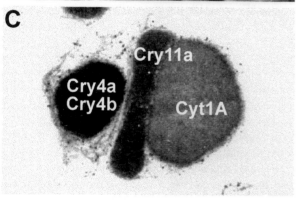

FIGURE 2   Sporulated cells of *Bacillus thuringiensis* (B.t.). (A) An electron micrograph of a sporulating cell of *B.t.* subsp. *israelensis*. The spherical structure is a spore adjacent to which is a parasporal body. The parasporal body contains protein endotoxins responsible for most of the insecticidal properties of this insecticidal bacterium. (B) Parasporal bodies typical of most isolates of *B. thuringiensis* subsp. *kurstaki* (B.t.k.), used widely to control caterpillar pests. The bipyramidal crystals contain three endotoxin proteins (Cry1Aa, Cry1Ab, and Cry1Ac), and the cuboidal crystal an additional toxin (Cry2A). This toxin complexity accounts for the broad spectrum of activity of many isolates of B.t.k. (C) A parasporal body of *B. thuringiensis* subsp. *israelensis* (B.t.i.) used widely to control the larvae of mosquitoes and blackflies. This parasporal body is also composed of four major endotoxins, a large semispherical inclusion containing Cyt1Aa, a dense spehrical body that apparently contains the Cry4Aa and Cry4Ba proteins, and a bar-shaped body that contains Cry11Aa. The endotoxin inclusions of this subspecies are held together by an envelope of unknown composition (arrowheads). This parasporal body has the highest specific toxicity of known B.t. species, and this is due to synergistic interactions between the Cyt1Aa and Cry proteins as well as synergistic interactions among the Cry proteins. B.t. endotoxins act by destroying the insect midgut epithelium (stomach).

$80 to $100 million and growing. Though most use is currently as a bacterial insecticide, plants have been engineered to produce B.t. proteins for resistance to insects, and this use could surpass the use of B.t. insecticides in the next century. For a more thorough treatment of B.t., the interested reader is referred to Chapter 21.

Biology and Use of *Bacillus popilliae* for Scarab Control

The milky diseases of scarabs, which were first discovered over 50 years ago (Dutky, 1937, 1963), are caused by *B. popilliae* and the closely related *B. lentimorbus*. The term "milky disease" is derived from the opaque white color that characterizes diseased larvae and results from the accumulation of sporulating bacteria in the hemolymph. The disease is initiated when grubs feeding on the roots of grasses or other plants ingest the bacterial spores. The spores germinate in the midgut and vegetative cells invade the midgut epithelium where they grow and reproduce, changing in form as they progress toward invasion of the hemocoel (Splittstoesser *et al.,* 1978). After passing through the basement membrane of the midgut, the bacteria colonize the hemolymph over a period of several weeks and sporulate, reaching populations of 10^8 cells per milliliter. The disease is fatal, providing that the larvae ingest a sufficient number of spores early in their development. Dead larvae in essence become foci of spores that can serve as a source of infection for up to 30 years (Klein, 1981; 1997).

One of the drawbacks of *B. popilliae* and its close relatives is that despite decades of research, suitable media for their growth and mass production *in vitro* have not been developed. This has inhibited both research and large-scale commercial development of *B. popilliae*. As has been done for several decades, the technical material (i.e., spores) that are the basis of commercial formulations are produced in living, field-collected scarab larvae. Nevertheless, a small but steady market remains for *B. popilliae* in the United States due to the serious problems caused by scarab larvae, such as damage to turf grass by larvae of the Japanese beetle, *Popillia japonica*.

To produce *B. popilliae,* scarab larvae are collected from field populations, injected with spores, and held in environmental chambers for from 1 to 2 weeks until the bacteria have sporulated and killed most of the larvae. Powdered formulations of the bacterial spores are then made by drying and grinding the larvae. These preparations are applied to soils infested with grubs at rates of about 1 kg of formulation per hectare. Though treatment can be expensive, averaging $220 per hectare in the United States, a single treatment will typically last for 10 years, making the use of *B. popilliae* cost-effective.

When *B. popilliae* sporulates, it produces a parasporal body; studies have shown that this body contains a Cry protein similar to the endotoxins produced by *B. thurin-giensis*.

Biology and Use of *Bacillus sphaericus* for Mosquito Control

Since the mid-1960s, it has been known that many isolates of *B. sphaericus* Neide were toxic to certain mosquito species. Over the past three decades, three isolates have been evaluated for their mosquito control potential, 1593 from Indonesia, 2297 from Sri Lanka, and 2362 from Nigeria. The 1593 and 2297 isolates were obtained from soil and water samples at mosquito breeding sites, whereas 1593 was isolated from a dead adult blackfly. *Bacillus sphaericus* (B.s.) is active primarily against mosquitoes belonging to the genera *Culex* and *Anopheles*, and exhibits only minor toxicity to most *Aedes* species and blackflies (Baumann *et al.*, 1991; Charles *et al.*, 1996). Of these isolates, 2362 is the most toxic to the widest range of mosquito species. Its toxicity is sufficiently high that it has become commercialized, and is available under the product name Vectolux from Abbott Laboratories.

The toxicity of B.s. like B.t. is due to protein endotoxins that are produced during sporulation and assembled into parasporal bodies. B.s. is unusual in that the main toxin is a binary toxin. The most toxic strains encode two proteins, one of 51.4 kDa and another of 41.9 kDa (Baumann *et al.*, 1988). These are proteolytically activated in the mosquito midgut to release peptides of 43 and 39 kDa, respectively, which associate to form the binary toxin, with the former protein apparently constituting the binding domain and the latter, the toxin domain. Like B.t.i., the target of action is the midgut epithelium. The toxins bind to microvilli, causing hypertrophy and lysis of cells.

In addition to the binary toxin, other toxins (Mtx toxins) are produced during vegetative growth, but the binary toxin is the most active and important from the standpoint of the activity of Vectolux.

Biology and Use of *Serratia entomophila* for Scarab Control

A novel bacterium named *Serratia entomophila* causes amber disease in the grass grub, *Costelystra zealandica*, an important pest of pastures in New Zealand, and has been developed as a biological control agent for this pest (Jackson *et al.*, 1992). This bacterium adheres to the chitinous initma of the foregut, were it grows extensively, eventually causing the larvae to develop an amber color and resulting in death. The bacterium is easily grown and mass-produced *in vitro*, and can now be grown to densities as high as 4×10^{10} cell per milliliter. Successful mass production of *S. entomophila* led to its rapid commercialization. It is now used to treat infested pastures in New Zealand at a rate of

1 liter of product per hectare. Liquid formulations of this living non-spore-forming bacterium are applied with subsurface application equipment. The rapid development and commercialization of the bacterium, even though the use is rather restricted, shows how microbials can be successful in niche markets where there are few alternatives, and where mass production methods—the most critical factor—are available.

FUNGI

Fungi constitute a large and diverse group of eukaryotic organisms distinguished from others by the presence of a cell wall, as in plants, but lacking in chloroplasts and thus the ability to carry out photosynthesis. Fungi live as either saprophytes or parasites of plants and animals, and for growth require organic food obtained by absorption from the substrates on which they live. The vegetative phase is known as a thallus, and can be either unicellular (as in yeasts) or multicellular and filamentous, forming a mycelium, the latter being characteristic for most of the fungi that attack insects. During vegetative growth, the mycelium consists primarily of hyphae that may be septate or nonseptate, and these grow throughout the substrate to acquire nutrients. Reproduction can be sexual or asexual; and during this phase the mycelium produces specialized structures such as motile spores, sporangia, and conidia, which are typically the agents by which fungi infect insects. Fungi usually grow best under wet or moist conditions, and those that are saprophytic as well as many of the parasitic species are easily cultured on artificial media.

The fungi are divided into five major subdivisions and these reflect the evolution of the biology of fungi from aquatic to terrestrial habitats. For example, species of the genera *Coelomomyces* and *Lagenidium* (subdivision Mastigomycotina) are aquatic and produce motile zoospores during reproduction, whereas members of the genera *Metarhizium* and *Beauveria* (subdivision Deuteromycotina) are terrestrial and reproduce and disseminate via nonmotile conidia (Humber, 1997).

Biological Properties in Relation to Insect Control

Unlike most other pathogens, fungi usually infect insects by active penetration through the cuticle, a trait that makes them attractive as candidates for control of insects with sucking mouthparts. The typical life cycle begins when a spore, either a motile spore or a conidium, lands on the cuticle of an insect. Soon after, under suitable conditions, the spore germinates, producing a germ tube that grows and penetrates down through the cuticle into the hemocoel (St.

Leger, 1992). Once in the hemolymph, the fungus colonizes the insect. Hyphal bodies bud off from the penetrant hyphae and either continue to grow and divide in a yeastlike manner or to elongate, forming hyphae that grow throughout the insect body. Complete colonization of the body typically requires 7 to 10 days, after which the insect dies. Some fungi produce peptide toxins during vegetative growth, and in these strains death can occur within 48 h. Subsequently, if conditions are favorable, which generally means an ambient relative humidity of greater than 90% in the immediate vicinity of the dead insect, the mycelium will form reproductive structures and spores, thereby completing the life cycle. Depending on the type of fungus and species, these will be produced either internally or externally; and can be motile spores, resistant spores, sporangia, or conidia.

From the standpoint of microbial control, the ability of fungi to infect insects via the cuticle gives them a significant advantage over viruses, bacteria, and protozoa. If they could be effectively developed, they would be useful against the wide range of important insect pests with sucking mouthparts such as aphids, leafhoppers, whiteflies, and scales.

Fungi as Insect Control Agents

Fungi are one of the most common types of pathogens observed causing disease in insects in the field. Moreover, outbreaks of fungal diseases under favorable conditions often lead to spectacular epizootics that decimate populations of specific insects over areas as large as several hundred square kilometers [see Carruthers and Soper (1987) for a review and Andreadis and Weseloh (1990) for an example]. As a result, there has been interest in using fungi to control insects for well over a century, with the first efforts employing *Metarhizium anisopliae* (Metchnikoff) to control the wheat cockchafer, *Anisoplia austriaca* Hubst, in Russia (Krassilstschik, 1888; Steinhaus, 1949). Though there have been numerous attempts since then to develop fungi as commercial microbial insecticides, very few of these efforts have met with success. Thus, at present there are less than a handful of commercially accessible fungal insecticides available for use in industrialized countries, and true commercial success has been elusive to date. The biotechnology firm Mycogen* in San Diego, California, was founded in the 1980s to exploit the biological control potential of insecticidal fungi, but now concentrates almost exclusively on the development of transgenic plants. On the other hand, imperfect fungi such as *M. anisopliae* and *Beauveria bassiana* (Balsamo) have been produced and used in developing countries (e.g., in Brazil and China) using cottage industry technology such as that used to produce viruses. A

*Recently purchased by Dow Agrosciences Inc.

quasi-commercial product "Boverin" has also been developed and used in Russia for control of the Colorado potato beetle, but has been ineffective in the United States. Current efforts to find alternatives to chemical insecticides have intensified research on fungi, with the aim being to find new isolates or to improve existing strains through molecular genetic manipulation that will be more successful as either classical biological control agents or mycoinsecticides. The principal problem has been a lack of cost-effective production methods. Mycotech, Inc., a small biotechnology firm in Montana, has developed a novel fermentation strategy to produce fungi such as *B. bassiana* for insect control. Products produced by this method are now coming to market and are undergoing evaluation for control of root weevils, aphids and whiteflies in greenhouses, and for control of outbreak locust populations (Wraight and Carruthers, 1998).

The literature on the biology and potential use of many species of fungi is extensive and is covered very well in the book by McCoy *et al.* (1988), to which readers interested in a much more thorough coverage than can be given here are referred. The shorter reviews by Ferron (1978), Hall and Papierok (1982), and McCoy (1990) are also recommended. My intent here is to briefly summarize the critical features of the biology of fungi that have been or continue to be serious candidates for use in insect control, and through the use of selected examples, illustrate the advantages and disadvantages of these as biological control agents. The major types of fungi considered for insect control are listed in Table 4.

Aquatic Fungi

Coelomomyces and Lagenidium

Two types of aquatic fungi that attack mosquito larvae have been studied for use as biological control agents, species of *Coelomomyces* (class Chytridiomycetes; order Blastocladiales) and *Lagenidium giganteum* (class Oomycetes; order Lagenidiales).

The genus *Coelomomyces* is composed of over 70 species of obligately parasitic fungi that have a complex life cycle involving an alternation of sexual (gametophytic) and asexual (sporophytic) generations (Couch & Bland, 1985; Whisler, 1985). In all species studied to date, the sexual phase parasitizes a microcrustacean host, typically a copepod, whereas the asexual generation develops, with rare exception, in mosquito larvae. In the life cycle, a biflagellate zygospore invades the hemocoel of a mosquito larva where it produces a sporophyte that colonizes the body and forms resistant sporangia. The larva dies and subsequently the sporangia undergo meiosis, producing uniflagellate meiospores that invade the hemocoel of a copepod host, where a gametophyte develops. At maturation, the gametophyte cleaves, forming thousands of uniflagellate gametes.

TABLE 4   Major Types of Fungal Pathogens Considered for Microbial Control[a]

Fungal subdivision	Genus	Infective stage	Target
Mastigomycotina			
Class Chytridiomycetes			
	Coelomomyces	Zoospores	Mosquito larvae
Class Oomycetes			
	Lagenidium	Zoospores	Mosquito larvae
Zygomycotina			
Class Zygomycetes			
	Conidiobolus	Conidia	Aphids
	Entomophaga	Conidia	Grasshoppers, caterpillars
	Zoophthora	Conidia	Aphids, caterpillars, beetles
Deuteromycotina			
Class Hyphomycetes			
	Beauveria	Conidia	Beetles, caterpillars, plant hoppers
	Metarhizium	Conidia	Plant and leafhoppers, stinkbug
	Nomurea	Conidia	Caterpillars
	Paecilomyces	Conidia	Whiteflies, leafhopper
	Verticillium	Conidia	Aphids, whiteflies, scales

[a]From Roberts, D. W., & Humber, R. A. (1981). *The biology of conidial fungi* (Vol. 2, pp. 201–236). New York: Academic Press.

Cleavage results in death of the copepod and in escape of the gametes, which fuse and form biflagellate zygospores that seek out another mosquito host, completing the life cycle. The life cycles of these fungi are highly adapted to those of their hosts. Moreover, as obligate parasites these fungi are very fastidious in their nutritional requirements, and as a result no species of *Coelomomyces* has been cultured *in vitro.*

Coelomomyces is the largest genus of insect-parasitic fungi, and has been reported worldwide from numerous mosquito species, many of which are vectors of important diseases such as malaria and filariasis. In some of these species, *Anopheles gambiae* in Africa for example, epizootics caused by *Coelomomyces* kill greater than 95% of the larval populations in some areas (Couch & Umphlett, 1963; Chapman, 1985). Such epizootics led to efforts to develop several species as biological control agents during the past three decades (Federici, 1981). However, these efforts have largely been discontinued due to the discovery that the life cycle required a second host for completion, the inability to culture these fungi *in vitro,* and the development of B.t.i. as a bacterial larvicide for mosquitoes.

Though the difficulties encountered with *Coelomomyces* make it unlikely this fungus will ever be developed as a biological control agent, there is still considerable interest in *Lagenidium giganteum.* This oomycete fungus has two important advantages over *Coelomomyces:* it is easily cultured on artificial media and it does not require an alternate host (Federici, 1981). In the life cycle, a motile zoospore invades a mosquito larva through the cuticle. Once within

the hemocoel, the fungus colonizes the body over a period of 2 to 3 days, producing an extensive mycelium consisting largely of nonseptate hyphae. Toward the end of growth, the hyphae become septate, and out of each segment an exit tube forms that grows back out through the cuticle and forms zoosporangia at the tip. Zoospores quickly differentiate in these, exiting out through an apical pore to seek out a new substrate. In addition to this asexual cycle, thick-walled resistant sexual oospores can also be formed within the mosquito cadaver.

Techniques have been developed to produce both zoosporangia and oospores *in vitro,* and methods are currently being developed to modify existing technology so that the fungus can be mass-produced. Several years of field trials in California and North Carolina have shown that the zoosporangia are too fragile for routine use in operational control programs. The oospore, however, is quite stable though germination remains unpredictable. Nevertheless, field results indicate that germination of even a small percentage of oospores can result in the initiation of epizootics that lead to season-long mosquito control (Kerwin & Washino, 1987). Although several technical problems related to mass production and formulation remain to be overcome, *L. giganteum* remains a promising candidate for successful commercial development. Its principal advantage over B.t.i. is that if effective formulations can be developed, it appears that in many habitats only a single application would be required, at most, per season. Even less frequent applications may be possible in some habitats because evidence suggests that the oospores can overwinter, initiating epizootics the following seasons. The extent to which this occurs and can be relied on for effective mosquito control remains to be determined.

A small biotechnology firm, AgraQuest of Sacramento, California, has begun producing a product based on *L. giganteum* with the trade name of Lagenex. The product has not been in use long enough to access either its operational efficacy or its commercial success.

In addition to *Coelomomyces* and *L. giganteum,* the aquatic hyphomycete fungi *Culicinomyces clavosporus* Couch and *Tolypocladium cylindrosporum* have been considered for mosquito control (Federici, 1981; Soares & Pinnock, 1984); however, high production costs, lack of clear and cost-effective control in the field, and the advent of B.t.i. have eliminated these fungi as serious candidates for development as microbial control agents.

Terrestrial Fungi

Fungi that have received the most attention for use in biological control are terrestrial fungi, with most emphasis being placed on the development of selected species of hyphomycetes such as *M. anisopliae* and *B. bassiana* for use as microbial insecticides. In addition, the more specific and nutritionally fastidious entomophthoraceous fungi continue to receive attention, but for their potential use as classical biological control agents instead of as microbial insecticides. Representative examples of these terrestrial fungi are discussed later.

Order Entomophthorales

Entomophthorales comprises a large order of zygomycete fungi that contains numerous genera, many species of which are commonly found parasitizing insects and other arthropods. These fungi routinely cause localized and in some cases widespread epizootics in populations of hemipterous and homopterous insects, particularly aphids and leafhoppers, but also in other types of insects such as grasshoppers, flies, beetle larvae, and caterpillars. In addition, a few species of the genus *Conidiobolus* are able to cause mycoses in some mammals, including humans (Humber, 1989; Humber *et al.,* 1989). Apart from these few species, most of the entomophthoraceous fungi are highly specific, obligate parasites of insects and therefore pose no threat to nontarget organisms when used for biological control. As in the case of *Coelomomyces,* however, the complex nutritional requirements, which have made mass production *in vitro* impossible to date, and the high degree of host specificity make these fungi poor candidates for development as microbial insecticides. Moreover, the conidia are very fragile, providing a challenge to formulation, and the resistant spores, like the oospores of *L giganteum,* are difficult to germinate in a predictable manner. Nevertheless, there is

evidence that these fungi can provide effective insect control through modifying cultural practices in crop production where these fungi occur naturally and through the introduction of foreign strains and species (i.e., a classical biological control approach).

The general aspects of the infection process and colonization of the host by entomophthoraceous fungi are similar to those described earlier for other fungi. They differ from other groups primarily in the types of reproductive structures formed and the specific details of their life cycles [see MacLeod (1963)]. At the end of the infection process, if external conditions are favorable (high humidity), the mycelium generates conidiophores that typically grow out of the insect carcass and form fragile primary conidia at their tips. Shortly after formation, in the most unique feature of these fungi, these conidia are physically discharged (i.e., shot off to land on another substrate, i.e., insect). This results in a "halo" of conidia on the substrate around a dead insect. In species of insects that live in groups or develop to high population densities, this is a very effective method of dissemination, and it is easy to see how it can result in epizootics. If the primary conidium fails to land on a suitable substrate, it can generate a secondary conidium that will repeat the process, which can continue for one or two more times until the conidial reserves are spent. In addition to this type of spore, these fungi can form thick-walled resistant zygospore (sexual) and azygospores (asexual) within the insect that can survive in the cadaver or soils and can germinate months or years later.

The most important genera found attacking insects in the field are *Conidiobolus* (aphids), *Erynia* (aphids), *Entomophthora* (aphids), *Zoophthora* (aphids, caterpillars, beetles), and *Entomophaga* (grasshoppers, caterpillars). Though many species of these cause epizootics and have received considerable study, none really appears to have much potential for development as a commercial microbial insecticide (Wilding, 1981). On the other hand, cultural control, classical biological control, and environmental monitoring methods continue to show promise for using entomophthoraceous fungi for insect control. For example, the introduction of *Erynia radicans* from Israel into Australia to control the spotted alfalfa aphid, *Therioaphis maculata* (Buckton), has been a classical biological control success (Milner *et al.,* 1982). By using a cultural method in which the first cutting of alfalfa was moved forward in the growing season to concentrate hosts under the windrows when warm and moist, Brown and Nordin (1982) were able to enhance the transmission of *E. phytonomi* among larvae of the alfalfa weevil, *Hypera postica* (Gyllenhal). This resulted in a significant outbreak of disease that reduced the weevil population, saving farmers an estimated $40 per year. In the former Soviet Union, Voronina (1971) characterized zones based on moisture and temperature and was able to develop a model that predicted the occurrence of epizootics by a species of *Entomophthora* in populations of the pea aphid, *Acrythosiphon pisum* (Harris), thereby reducing the overall level of insecticide usage. The success of such tactics obviously requires a sound understanding of the ecology and epizootiology of these fungi.

Another example of apparent classical biological control is the natural outbreaks of *Entomophaga miamiaga* in larval populations of the gypsy moth, *Lymantria dispar,* an important pest of decidous forests, throughout several states comprising the middle Atlantic and New England regions of the United States (Andreadis & Weseloh, 1990; Hajek *et al.,* 1990; Hajek *et al.,* 1996). These outbreaks of *E. miamiaga* have reduced larval populations to below economic thresholds, and the fungus is spreading westward naturally and with human assistance to gypsy moth populations established in other states. The source of this fungus is Japan, though it is not clear when the fungus causing present outbreaks of disease was introduced into the United States. The fungus was purposely introduced into the United States around the turn of the century, but apparently did not become established. Then in the late 1980s, outbreaks of *E. miamiaga* began to occur in Connecticut and New York, and later in Virginia. In areas were it has become established, providing there is sufficient rainfall, the fungus appears to be capable of keeping the gypsy moth population below defoliation levels. It will require another 10 years of evaluation to determine if this is a valid case of classical biological control by a fungus.

Class Hyphomycetes

The hyphomycete fungi belong to the fungal subdivision Deuteromycotina (imperfect fungi), a grouping erected to accommodate fungi for which the sexual phase (perfect state) has been lost or remains unknown. This group contains the fungal species that most workers consider to have the best potential for development as microbial insecticides, *Beauveria bassiana* and *Metarhizium anisopliae,* the agents of the white and green muscardine diseases of insects respectively. Unlike the fungi discussed earlier, these two species have very broad host ranges and probably are capable of infecting insects of most orders. This broad host range is a concern from the standpoint of safety to nontarget organisms such as predatory and parasitic insects, though natural occurrence in these insects is apparently rare. Aside from this, where these fungi are mass-produced for insect control, workers often develop allergic reactions to the conidia as a result of being exposed to them for long periods during the production process.

With respect to the general life cycle of these fungi, the process of invasion, colonization of the insect body, and formation of conidiophores and conidia is similar to that described earlier for the other fungi (Roberts & Humber, 1981). During invasion and colonization, some fungal spe-

cies produce peptide toxins that quicken host death. The infectious agent is the conidium (Fig. 3), and the taxonomy for the hyphomycetes is based primarily on the morphology of the reproductive structures, particularly the conidiophores and conidia (Samson, 1981; Humber, 1997). Most of the hyphomycete fungi used or under development grow well on a variety of artificial media, and this attribute and their ability to infect insects via the cuticle are the two traits that favor commercial development. In the cottage industry commercial operations in Brazil, China, and the former USSR, solid or semisolid substrates are used for production in which the primary ingredients are grain or grain hulls.

In general, the development of *B. bassiana* and *M. anisopliae* is being targeted for control of insects that live in cooler and moist environments, such as beetle larvae in soil and plant hoppers on rice, though the former species is also being evaluated against whiteflies in greenhouses and grass-

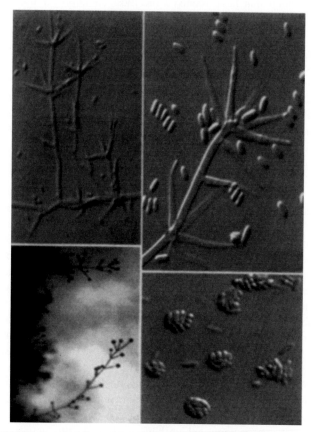

FIGURE 3 Typical reproductive structures of Deuteromycete (imperfect) fungi. The conidia visible as free conidia and conidial clusters in the two panels on the right are the principal infective units. When these come in contact with an insect host, they germinate and penetrate into the body forming a mycelium that colonizes the insect over a period of several days. When conditions are appropriate, typically meaning high relative humidity, hyphae penetrate back out through the cuticle producing conidiophores, the visible branched structures in the panels on the left, which form reproductive conidia at their tips (Humber, 1997).

hoppers, especially locusts, in field crops. In addition to these two species, several species with much narrower host ranges are considered to have potential for development, including *Paecilomyces fumoso-rosea* (for whiteflies), *Verticillium lecanii* (for aphids and whiteflies in greenhouses), *Hirsutella thompsonii* (for mites), and *Nomurea rileyi* (for noctuid caterpillars).

With these apparent advantages, the question arises as to why none of the hyphomycete fungi have been commercially successful as microbial insecticides in developed countries. There are several reasons related to their biological properties. First and foremost is that the production of conidia or mycelial fragments that are used as the active ingredient of formulations is not cost-effective in that too much material is required to achieve an acceptable level of control. Typical figures range from 2 to 3 kg of dried mycelia or conidia (10^{13}) per hectare of treatment for well-exposed insect populations such as plant hoppers in rice fields (Table 5). For cryptic insects such as beetle larvae in soil, the figure is approximately 10-fold higher. At best, yields of mycelia and conidia on solid or semisolid substrates are about 4 to 10% of the weight of the substrate on which they are grown. This means it requires from 5 kg (for 10^{13} conidia) to well over 50 kg of grain/grain hulls to produce enough active ingredient to treat 1 ha with a species such as *M. anisopliae*. In addition to the problem of inefficient yields, formulations are bulky and preservation of fungal viability beyond a few months is low due to the fragile nature of the conidia. In the case of mosquito and blackfly control, similar constraints apply. In addition, the discovery of cost-effective strains of B.t. and B.s. (Table 6) has generally eliminated imperfect fungi as well as many other microorganisms for consideration as biological control agents for these important nuisance and vector insects.

Despite these problems, which have hampered the development of fungi over the past 30 years in developed countries, several products did come to market including Mycar (*H. thompsonii*) for citrus rust mite control and Mycotal

TABLE 5   Comparison of the Efficiency of Production Relative to Application Rates for Representative Pathogens Evaluated as Microbial Insecticides for Control of Agricultural Pests

Agent	LC$_{90}$ per hectare	Production yield	Medium per hectare[a]
For caterpillars			
Bacteria			
Bacillus thuringiensis	300 g powder[b]	10 g/liter	30 liters
Virus			
Heliothis zea NPV	6×10^{11} polyhedra	6×10^{10} polyhedra per larvae	100 larvae
Spodoptera exigua NPV	1×10^{12} polyhedra	2×10^9 polyhedra per larva	500 larvae
Protozoa			
Vairimorpha necatrix	1×10^{12} spores	1.7×10^9 spores per larvae	600 larvae
Fungi			
Metarhizium anisopliae	1×10^{14} conidia	2×10^9 conidia per g rice	50 kg rice
Beauveria bassiana	1×10^{14} conidia	2×10^{10} conidia per g grain	5 kg grain
Nematode			
Steinernema carpocapsae	1×10^{10} dauer larvae	10^8/2 liters	400 liters
For grasshoppers			
Microsporidia			
Nosema locustae	2.5×10^9 spores[c]	6.4×10^9 spores per grasshopper	1 grasshopper
Fungi			
B. bassiana	1×10^{14} conidia	2×10^{10} conidia per g grain	5 kg grain
For spittlebugs			
Fungi			
M. anisopliae	1×10^{10} conidia[d]	2×10^9 conidia per g rice	5 kg rice

[a]Refers to the amount of liquid or solid substrate needed to produce the amount of technical material required to achieve population reductions of 90% for 1 ha.

[b]Technical powder; refers to the unformulated dried solids obtained after fermentation. Of this weight, only approximately 25% consists of insecticidal proteins in *B. t.*, and 5 to 10% in *B. s.*; the rest is fermentation solids, bacterial spores, and remnants of bacterial cells.

[c]For control of grasshoppers (*Melanoplus sanguinipes* and others).

[d]For control of the spittlebug, *Mahanarva posticata*, on rangeland. At this rate, only 65% of the population is controlled, but this provides sufficient crop protection to be assessed as cost-effective (in Brazil).

TABLE 6   Comparison of the Efficiency of Production Relative to Application Rates for
Representative Bacterial and Fungal Pathogens Evaluated as Microbial Insecticides for Control of
Mosquito Larvae

Agent	LC$_{90}$ per hectare	Production yield	Medium per hectare[a]
Bacteria			
Bacillus thuringiensis subsp. *israelensis* (B.t.)	200 g Technical powder[b]	10 g/liter	20 liters
B. sphaericus (B.s.) 2362	100 g Technical powder[b]	10 g/liter	10 liters
Fungi			
Culicinomyces clavosporus	10^{14} conidia	10^{11} conidia per liter	1000 liters
Metarhizium anisopliae	2 kg Conidia	1 kg/25 kg rice	50 kg/rice

[a] Refers to the amount of liquid or solid substrate needed to produce the amount of technical material required to achieve population reductions of 90% for 1 ha. Production on media used commercially is in most cases more cost-effective than these figures indicate.

[b] Technical powder refers to the unformulated dried solids obtained after fermentation. Of this weight, only approximately 25% consists of insecticidal proteins in B.t., and 5 to 10% in B.s.; the rest is fermentation solids, bacterial spores, and remnants of bacterial cells.

(V. lecanii) for use against aphids in greenhouses. However, these products were not efficacious, and turned out to be commercial failures. This situation has changed somewhat over the past few years in that there is renewed interest in developing various strains of *B. bassiana* for control of whiteflies and locusts. Mycotech with its proprietary production system and Thermo-Trilogy are developing new products based on this fungus with the goal being to make *B. bassiana* an operational microbial insecticide. The current focus is to develop this fungus for use under favorable circumstances, (e.g., against insects with high levels of insecticide resistance and/or sucking insects attacking a cash crop of high value in greenhouses). A few substantiated successes are needed to clearly demonstrate the potential of this pathogen group.

In developing countries, *B. bassiana* and *M. anisopliae* have been used in some crops with considerable success. For example, in China *B. bassiana* has been used to control the European corn borer, *Ostrinia nubilalis*, in maize. The fungus is produced in large covered pits on maize stalks. In Brazil, a preparation of *M. anisopliae* known as Metaquino has been used for many years to control the spittlebug, *Mahanarva posticata*, on sugarcane plantations and in pastures. Fungal conidia are produced in sealed plastic bags on rice. Figures indicate that as many as 50,000 ha are treated annually, and reductions in spittlebug populations are sufficient to keep populations below damaging levels [see McCoy *et al.* (1988)]. In the South Pacific, *Metarhizium anisopliae* has also been used to assist control of the rhinoceros beetle, *Orycetes rhinocerous*, a serious pests of coconut palms. Application of conidia at a rate of 50 g/m^2 of soil yielded 80% larval mortality and improved coconut yields by 25%. While these are examples of local successes, their applicability to agricultural production in developed countries is questionable.

PROTOZOA

Protozoa is a general term applied to a large and diverse group of eukaryotic unicellular motile microorganisms that belong to what is now known as the kingdom Protista (Levine *et al.,* 1980). Members of this kingdom can be free-living and saprophytic, commensal, symbiotic, or parasitic. The cell contains a variety of organelles, but no cell wall, and cells vary greatly in size and shape among different species. Feeding is by ingestion or more typically by adsorption, and vegetative reproduction is by binary or multiple fission. Both asexual and sexual reproduction occur and the latter, often useful for taxonomy, can be very complex. Many protozoa produce a resistant spore stage that is also used in taxonomy. The kingdom is divided into a series of phyla based primarily on the mode of locomotion and structure of locomotory organelles, and includes the Sarcomastigophora (flagellates and amoebae), Apicomplexa (sporozoa), Microspora (microsporidia), Acetospora (haplosporidia), and Ciliophora (ciliates). Some types of protozoa, such as the free-living amoebae and ciliates, are easily cultured *in vitro,* whereas many of the obligate intracellular parasites have not yet been grown outside of cells.

Biological Properties in Relation to Insect Control

As might be expected from such a large and diverse group of organisms, many species of protozoans are associated with insects and the biology of these associations covers the gamut from being symbiotic to parasitic. Those that are parasitic have the general feature of causing diseases that are chronic. Many of the parasitic types, especially the microsporidia, build up slowly in insect populations, eventually causing epizootics that lead to rapid declines in populations of specific species (Brooks, 1988). These epizootics attracted the interest of workers in the possibility of using protozoa to control pest insects, and over the past several decades there have been numerous studies aimed at evaluating this potential. In general, these studies have demonstrated that protozoa hold little potential for use as fast-acting microbial insecticides because of the chronic nature of the diseases they cause and the lack of commercially suitable methods for mass production. However, as in the case of the entomophthoraceous fungi, the possibility exists that protozoans, particularly microsporidia, may be useful as classical biological control agents or in intermediate to long-range pest population management strategies. Clear examples of the effectiveness of such strategies remain to be demonstrated.

The life cycles and biologies that occur among the various types of protozoa that attack insects are too diverse in relation to their pest control potential for even a few to be covered here. Instead, the group with the most potential—the microsporidia—will be described in terms of their general biology and possible use in insect control. For more detailed coverage of the many types of protozoa that attack insects, the excellent publications by Canning (1982), Henry (1990), and Maddox (1987); and, in particular, the treatise by Brooks (1988) are highly recommended.

General Biology of Microsporidia

The microsporidia (phylum Microspora) are the most common and best studied of the protozoans that cause important diseases of insects. Well over 800 species are known, and most of these have been described from insects (Brooks, 1988; Tanada & Kaya, 1993). Microsporidia have been most commonly described from insects of the orders Coleoptera, Lepidoptera, Diptera, and Orthoptera, but are also known from other orders and probably occur in all. The epizootics in insect populations caused by protozoa are usually due to microsporidia.

All microsporidia are obligate intracellular parasites and are unusual in that they lack mitochondria. In addition, they produce spores that are distinguished from the spores of all other known types of organisms by the presence of a polar filament (Fig. 4), a long coiled tube within the spore used to infect hosts with the sporoplasm (Vavra, 1976).

The typical microsporidian life cycle begins with the ingestion of the spore by a susceptible insect. Once within the midgut, the polar filament everts, rapidly injecting the sporoplasm into host tissue. The sporoplasm is unicellular but may be uni- or binucleate. On entry into the cytoplasm of a host cell (e.g., the fat body in many species of insects) the sporoplasm forms a plasmodium (meront) that undergoes numerous cycles of vegetative growth (merogony). During these cycles, the cells multiply extensively, dividing

FIGURE 4 Representative microsporidian spores. (A) A wet mount preparation of spores viewed with phase contrast microscopy. (B) A transmission electron micrograph through a microsporidian spore. The arrows point to cross sections through the polar filament, which is used to inject the infectious sporoplasm into host tissues. Typically, several million to as many as several billion spores are produced per host. The usual mode of infection is by ingestion, after which the spores extrude the polar filament, injecting the microsporidian sporoplasm into midgut epithelial cells, or directly into tissues such as the fat body.

by binary or multiple fission and spreading to other cells; and in many species, the cells spread to other tissues of the host. After several mergonic cycles, the microsporidian undergoes sporulation. This consists of two major phases, sporogony—a terminal reproductive division committed to sporulation—and spore morphogenesis. In the sexual phase of reproduction, meiosis occurs early during sporogony. The spores, which in general measure several microns in diameter and length, have a thick wall and are highly refractile when viewed by phase microscopy. The disease often lasts for several weeks, during which spores accumulate in the tissues of infected hosts, yielding billions per individual.

Microsporidian systematics is based on the size and structure of the spores, life cycles, and host associations. In addition to transmission by ingestion, many microsporidia are transmitted vertically from adult females to larvae via the egg (transovarially). With respect to host range, some species are species specific whereas others occur in many species of the same family or order, and some can be transmitted to insects of different orders.

Microsporidia as Biological Control Agents

Naturally occurring epizootics caused by microsporidia are periodically very effective in significantly reducing insect pest populations. The problem is that these epizootics cannot be predicted with any degree of accuracy or relied on for adequate control, even though many of the conditions that facilitate their occurrence are known. A classic example of this is the epizootics caused by *Nosema pyrausta* (Paillot) in populations of the European corn borer, *O. nubilalis* (Maddox, 1987). These are useful when they occur, but insufficient to be relied on because they often occur too late to prevent economic damage. Thus, efforts have been directed toward developing methods for amplifying spore loads in the field through inundative releases, in essence using microsporidia as microbial insecticides.

As obligate intracellular parasites that lack mitochondria, microsporidia cannot be grown on artificial media. Several species have been grown, however, in established insect cell lines, though this is not practical for field use. For field application, whether for microbial insecticide trials or introductions into populations, spores are grown in living hosts. With such methods the yield can be quite high (10^9 to 10^{10} spores per host). These yields in terms of the number of larvae that must be grown to treat a hectare and infect most of the target population are comparable to the requirements for NPVs. Thus, if the microsporidia could cause acute diseases, they would be on an equal footing with many of the NPVs. However, the diseases are chronic, and even if a high percentage of the target pest population is infected,

there all too often is little, if any, crop protection. In fact, if advanced instars such as the third and fourth are treated, the larvae may live longer and cause greater crop damage than if the fields were left untreated. Thus, microsporidia are not useful as microbial insecticides.

The dimorphic species *Vairimorpha necatrix,* which has a broad host range among noctuid larvae, provides a good example of the problems encountered when microsporidia are used as microbial insecticides. In some trials on tobacco, Fuxa and Brooks (1979) found that application of spores at a rate of 10^{12} per acre reduced feeding damage as effectively as *Bacillus thuringiensis* (Dipel). More typically, rates as high as 10^{13} spores per acre (1000 LEs) did not give adequate control. Moreover, when the same rate was used against *Helicoverpa zea* on soybeans, although 99% of the larvae were infected, there was little reduction in larval density or feeding damage.

Another species, *N. algerae,* has been extensively evaluated as a control agent for larvae of anopheline mosquitoes. Field trials showed that applications of 2×10^{12} spores per hectare could reduce populations by as much as 86% (Anthony *et al.,* 1978). However, this high rate of application, which relied on spores produced in larvae, was not economical. In addition, soon after these trials cost-effective bacterial insecticides based on B.t.i. became commercially available.

A more positive example is the use of *N. locustae* Canning against acridid grasshoppers, in which this microsporidian is used as a population management tool instead of as a microbial insecticide [see Henry (1981; 1990)]. *Nosema locustae* has a very broad host range that includes more than 100 species of grasshoppers. Application of 2.5×10^9 spores per hectare reduced grasshopper densities by 50% from 28 to 30 days after application. In addition, population reductions carried over into the next generation because of a decrease in the fecundity of grasshoopers not killed directly. This is a very low rate in terms of production costs because the yield of spores from the grasshopper *Melanoplus differentialis* averaged 6.4×10^9 spores per grasshopper (i.e., a little more than a sixth of a "grasshopper equivalent" per acre). This microsporidian is the only protozoan registered by the Environmental Protection Agency (EPA) for use in insect control. Nevertheless, despite these positive characteristics, *N. locustae* has not been well accepted by farmers, who apparently prefer to see the grasshoppers controlled more quickly.

There is now a general realization that microsporidia and other protozoans have virtually no potential for use as microbial insecticides. They may, however, be useful as population management tools. Studies are currently in progress aimed at evaluating the effects of introducing microsporidia from European populations of the gypsy moth, *L. dispar,* into North American populations.

NEMATODES

Nematodes comprise a very large group of diverse but relatively simple multicellular eukaryotic organisms belonging to the phylum Nematoda (roundworms). As with other large groups of organisms, there is considerable variation in the biologies of different species within the phylum. Many species are free-living, whereas others are either facultative or obligate parasites. In general, nematodes are bilaterally symmetrical, elongate, and vermiform; taper at both ends; and are covered by a cuticle that they must molt to progress in development. Most species have a stylet plus specialized feeding glands and an alimentary tract. Life stages include an egg, several juvenile stages (larvae), and adults. The latter, depending on the species, can be sexually dimorphic as well as hermaphroditic. Nematodes that attack insects vary in size from being less than 1 mm in length to as large as 30 cm. Those visible with the unaided eye are typically white in color.

Biological Properties in Relation to Insect Control

Over the past 50 years, many different types of nematodes attacking insects have been examined with respect to their potential for development as biological control agents. The groups that have received the most study are (1) obligately parasitic mermithids (family Mermithidae), which have been evaluated for mosquito control; (2) facultatively parasitic steinernematids (family Steinernematidae) and heterorhabditids (family Heterorhabditidae), which have been evaluated for control of insects breeding in cryptic habitats (e.g., in soil, or within trees); and (3) facultatively parasitic neotylenchid (family Neotylenchidae) nematodes of the genus *Deladenus* that have been developed for control of *Sirex* wood wasps. Those that have been effective or appear to hold potential for further development are the facultative parasites, and this is due largely to the development of techniques for their mass production on artificial media. An important regulatory advantage that the nematodes possess in the United States over microbial pathogens is that the nematodes do not have to be registered as insecticides by the EPA.

Selected examples are provided later to illustrate both the advantages and problems encountered in developing and using nematodes as biological control agents. For more information of the biology and systematics of insect-parasitic nematodes, as well as a much more in-depth coverage of their potential as biological control agents, the interested reader is referred to the reviews by Poinar (1979), Petersen (1982), Gaugler and Kaya (1990), Kaya and Gaugler (1993), and Grewal and Georgis (1998).

Nematodes as Biological Control Agents

Mermithid Nematodes

Mermithid nematodes are among the largest nematodes attacking insects, and the adult females typically measure from 5 to 20 cm or more in length. In parasitized insects with a translucent cuticle, such as the larvae of many aquatic insects, the advanced stages of developing nematodes can often be observed within the hemocoel where they appear as long, thin, white worms. Mermithids are obligate parasites and have been reported from many different orders of insects as well as from other arthropods such as spiders and crustaceans. However, the only ones seriously considered for use as biological control agents are the species *Romanomermis culicivorax* and *R. iyengari,* which are capable of parasitizing many species of mosquito larvae (Petersen, 1982). The life cycle of *R. culicivorax,* which is quite typical of mermithids, will be used to illustrate the properties that resulted in studies of the biological control potential of this nematode type.

The females of *R. culicivorax* are found in the wet soil at the bottom of aquatic habitats in which mosquitoes breed. Here, after mating, the females lay thousands of eggs. The embryo develops into a first-instar juvenile over a period of about 1 week, and then molts to a second-stage juvenile while still within the egg. This second stage (preparasite) then hatches out and swims to the surface of the pond where it seeks out and, with the aid of its stylet, invades early instar mosquito larvae via the cuticle. Once within the hemocoel, the immature larva grows over a period of 7 to 10 days by absorbing nutrients through its cuticle, and molts once during this period. On completion of this parasitic phase, the third-stage juvenile punctures its way back out through the cuticle of its host, thereby killing the mosquito; then it drops to the bottom of the pond where it matures without feeding over another 7 to 10 days, and molts to the adult stage. The adults then mate and the females lay eggs, completing the life cycle.

By using several different species of colonized mosquitoes and simple rearing techniques, Petersen (1982) and colleagues were able to develop mass-rearing techniques for *R. culicivorax,* and conducted a wide range of trials in which they evaluated this nematode by applying the preparasitic stage in the laboratory and in the field. In the field studies, rates of parasitism (= mortality) of anopheline larvae as high as 85% were obtained when preparasites were applied at a rate of 2400/m^2. Even higher levels of parasitism (96%) were reported by Levy and Miller (1977) when they applied preparasites at a rate of 3600/m^2 against floodwater mosquitoes in Florida.

When introduced into suitable habitats, *R. culicivorax* can recycle, providing some level of control in subsequent generations of mosquitoes, but this level is typically not sufficient to interrupt the vector potential of mosquitoes or to alleviate the annoyance problem. In addition, both laboratory and field studies showed that *R. culicivorax* was quite sensitive to chloride ions; and the methods developed for mass production, storage, shipment, and use turned out not to be cost-effective, particularly in comparison with commercial formulations of the bacterial insecticide B.t.i., which was discovered while the development and evaluation of *R. culicivorax* were underway. As a result of its limitations and the availability of cheaper and more effective alternative biological methods of control, there is little interest today in the further development of *R. culicivorax,* or other mermithids for that matter, as biological control agents.

Steinernematid and Heterorhabditid Nematodes

The steinernematids and heterorhabditids are small (less than 1 to 3 mm) terrestrial nematodes that occur most commonly in nature as parasites of soil-inhabiting insects. They have relatively simple and typical nematode life cycles that aside from the egg include four larval stages and the adult. These nematodes are unusual, however, in that they have established a mutualistic relationship with bacteria that they harbor within their alimentary tracts, and these bacteria actually kill the insects rather quickly after the nematode invades the insect body. The bacteria have evolved specific relationships with individual species of nematodes. For example, the bacterial species *Xenorhabdis nematophilis* is associated with steinernematid *Steinernema carpocapsae,* whereas *Photorhabdis luminescens* is associated with heterorhabditid *Heterorhabditis bacteriophora* (Kaya & Gaugler, 1993). In addition to this unique relationship with symbiotic bacteria, these nematodes also produce an unusual quasi-resistant larval stage—the so-called "dauer larva"—the insect-infective stage of these nematodes, which is actually the third-instar juvenile surrounded by the molted cuticle of the second stage (Fig. 5).

In the life cycle of these nematodes, the dauer larva seeks out and infects an insect via the mouth and anus, or through the spiracles. Once within the hemocoel, the nematode begins to feed on hemolymph; while doing so it defecates, releasing the symbiotic bacteria. These quickly colonize the insect, killing it within 1 to 3 days. The nematodes feed on the bacteria and tissues of the dead larva, maturing and undergoing from two to three generations within the dead insect's body over a period of 1 to 2 weeks. The final generation results in the formation of thousands of dauer larvae that leave the cadaver in search of a new host.

Mass production of these nematodes has been greatly facilitated by the ease with which the symbiotic bacteria and nematodes can be grown on a variety of artificial media *in vitro.* Fermentation-type mass production of dauer lar-

FIGURE 5 Infective juveniles (dauer larvae) of the steinernematid nematode, *Steinernema carpocapsae* (Courtesy of Robin Bedding; Lacey and Brooks, 1997). These infective stages are motile, and gain access to hosts by being ingested, or by penetrating into the host body via the anus or spiracles. Once in the hemocoel, they begin feeding and defecate symbiotic bacteria that then multiply, typically killing the host within 48 h. The progeny nematodes feed on these bacteria as well as putrefying host tissues.

vae, the stage that is used for insect control, is now feasible in vessels with capacities of 15,000 liters and greater, with the yield of dauers for steinernematids being in the range of 10^5/ml (Friedman, 1990).

Though isolated in nature primarily from soil-inhabiting insects, the experimental insect host range of these nematodes is quite broad and includes more than 200 species of beetle, lepidopteran, and orthopteran pests as well as other insects such as cockroaches. However, field trials have shown that the nematodes do not survive well in environments that are not protected and moist. Formulations of dauer larvae, which must be kept wet and cool between production, formulation, and use, perform best when used against insects breeding in cryptic habitats, such as in soil or shaded and moist environments. The best results have been obtained against beetle grubs or caterpillars in stalks, tree trunks, or soil (including potted plants in horticultural operations), where applications of dauers at rates of 1 to 7×10^9 dauers per hectare have yielded levels of control ranging from 70 to 99% (Georgis & Hague, 1991).

From an economic standpoint, these nematodes have demonstrated enough potential that they are now produced and marketed by several small companies in the United States and Europe (Georgis & Hague, 1991; Grewal & Georgis, 1998). The costs range from 30 to 100% greater than the cost of comparable control obtained with chemical insecticides. However, by considering the diminishing availability of many chemical insecticides, the presence of resistance in some of the target pests, and the need for biological control agents in organically grown vegetable crops, the nematodes are approaching cost-effectiveness in several crop/habitat situations. Further improvements in mass production and formulation should improve their prospects for use in the future, at least in selected habitats.

Neotylenchid Nematodes of the Genus *Deladenus*

Although an old story, the use of *Deladenus siricidicola* to control *Sirex* wood wasps in pine forests in Tasmania and Victoria, Australia, is worth summarizing briefly because it illustrates how unique situations can arise where a biological control agent without mass-market potential can be developed and used on a small scale to bring about cost-effective control where even chemical insecticides have been ineffective.

Neotylenchid nematodes of the genus *Deladenus* are facultative parasites of wood wasps of the family Siricidae, and *D. siricidicola* is the species that has been used the most successfully in biological control (Bedding, 1974). There are two phases in the life cycle of *D. siricidicola,* a free-living phase that feeds on fungi that grow in the trees and are transmitted by the wasps; and a parasitic phase that invades wasp larvae and develops in larvae and adults, destroying eggs in the adult females. The latter phase consists of females that first develop as free-living nematodes, and then invade the wasp larvae to complete their development as parasites. The free-living phase can be mass-produced by feeding the nematodes on the fungi grown *in vitro*. They can then be inoculated into holes bored in trees in infested areas in forests. By using this technique, Bedding (1974) was able to achieve rates of parasitism of *Sirex* wasps as high as 90% over a 4-year period, and reduced the number of trees killed by the wasps from 200 to none over a 4-year period in a 1000-acre forest in Tasmania.

DISCUSSION AND CONCLUSIONS

Research on the use of pathogens to control insects has been in progress now for more than a century. It is worthwhile to look back over this period to see what lessons have been learned and to assess the implications of this knowledge for the development of more effective and environmentally sound pest control agents and strategies for the future. Three principal conclusions emerge from the studies conducted to date with pathogens. The first is that classical biological control using pathogens has resulted only rarely in effective pest control. There have been numerous purposeful introductions of pathogens as well as accidental introductions (e.g., with parasitic wasps), but other than the NPV of the European spruce sawfly in Canada, and possibly the fungus *E. miamiaga* against the gypsy moth in the United States, none of these have provided effective long-term pest control. Thus, considering the large number of insect pests on a worldwide basis, classical biological control with pathogens is unlikely to be an effective strategy in the future against most pests.

The second major conclusion is that regardless of whether a pathogen is used as part of an augmentative release program or as a microbial insecticide, a cost-effective and reliable supply of the pathogen must be available on a routine basis. This is particularly true in situations where farms tend to be large and highly mechanized, or where a pest extends over large geographic areas, as in the case of many forest pests. Thus, it is extremely important that methods be available for the mass production, formulation, distribution, and use of products of standardized efficacy. Lack of methods for mass production of cost-effective formulations of pathogens has been, and will continue to be, the largest obstacle to the more widespread development and use of pathogens as insect control agents. The initial acceptance and subsequent success of B.t. have been due largely to the ease with which various subspecies of this bacterium can be mass-produced *in vitro* on a variety of inexpensive media. Data relevant to assessing the importance of the relationship between mass production, efficacy, and product success are presented in Table 5 for agricultural pests, and in Table 6 for mosquito larvae. From these data it can be seen that ease of mass production and relatively inexpensive substrates were important characteristics of the pathogens considered the most successful as biological insecticides.

The third major conclusion to emerge from the studies of the last century is that the biological properties and high selectivity of most pathogens, in marked contrast to synthetic chemical insecticides, make most pathogen types much more suitable for development against particular pests and pest groups. None are truly broad spectrum in the sense of many chemical insecticides, and thus the economic incentives for development of any pathogen as a microbial insecticide are much less than for chemicals. Aside from overcoming technical obstacles, their more widespread use in the future, therefore, will be dependent on the development of niche markets and determination of how pathogens and nematodes can best be incorporated into IPM programs.

Each of these conclusions has important implications for future research and development efforts, and therefore they will be discussed in greater detail in an attempt to further assess in a realistic manner the potential that pathogens and nematodes hold for use as biological control agents or microbial insecticides, and to identify specific obstacles that must be overcome to optimize the use of these environmentally compatible control agents.

Classical Biological Control

Though classical biological control strategies based on pathogens will not be widely used in the future, there are some pathogen types that offer more hope than others. For example, when moisture is sufficient, many of the entomophthoraceous fungi cause seasonal epizootics in their native habitats. These fungi are often overlooked in foreign exploration programs aimed at introducing biological control agents to control nonindigenous pests. As noted earlier, it is highly unlikely these fungi will ever be useful as microbial insecticides; however, because they are relatively specific, they should be included in foreign exploration programs for natural enemies of pests attacking both annual and perennial crops. There is no evidence to support their use against disease vectors such as mosquitoes and blackflies, or household pests such as cockroaches and termites. Essentially the same can be said for the microsporidia. However, the chronic nature of the disease caused by these pathogen types makes it probable they will only be useful against important pests of perennial crops, such as the gypsy moth, with population cycles of several years. Moreover, there is some evidence that microsporida could be useful against ants such as the imported fire ants, *Solenopsis richteri* and *S. invicta* (Briano *et al.*, 1996; Williams *et al.*, 1998). In situations where population crashes are caused by existing pathogens, the introduction and establishment of even a single pathogen could provide additional stress that might slow the growth of the pest population, or cause it to crash sooner.

The possibility exists that an occasional virus or nematode may also be useful as a classical biological control agent, because there is at least one example of each having been successful in the past. In general these agents will be of little value in providing long-term pest control. Interestingly, despite the success of B.t. as a microbial insecticide, there is no indication that this bacterium or any other has ever served as an effective natural enemy of an insect pest. Thus, the potential for using bacteria as classical biological control agents is virtually nil.

Though current prospects are bleak for classical biological control, there is always the possibility that in the more distant future we will be able to genetically engineer pathogens to serve as classical biological control agents once we determine the mechanisms that account for success of the few that have been used effectively in such a strategy. The NPV of sawfly *G. hercyniae* serves as an excellent example. Unlike the NPVs that attack caterpillars, the NPVs of sawflies replicate exclusively and extensively in the midgut epithelium where they form large numbers of polyhedra. Instead of being "trapped" in the body until death as in the case of the lepidopteran NPVs, many polyhedra are passed with feces; thus, soon after infection the larva becomes a source of fresh virus for horizontal transmission. This trait along with the infection of the nucleus and gregarious behavior of *Gilpinia hercyniae* probably all play a role in the success of this virus (see Federici & Stern, 1990). However, engineering NPVs of lepidopteran pests to produce occluded virions (i.e., polyhedra) in the midgut epithelium, at least for those species with gregarious behavior, might be an effective strategy. It has been argued that such uses will lead to the development of resistance in the pest population to such a strategy. While theoretically true, it must be noted that the *G. hercyniae* NPV remains effective as a classical biological control agent to this day. Such an engineering strategy is not possible at present because the genetic basis for the control of NPV polyhedra formation in different tissues is not known.

Mass Production and Cost-Effectiveness

The direct relevance of having a cost-effective method of mass production to the success of a microbial insecticide becomes apparent from a comparison of the technologies currently available for producing different control agents with respect to the quantity of the control agent that must be used to achieve an acceptable level of control. Though information for commercial products is proprietary, the estimates of production efficacy presented in Tables 5 and 6, along with information on product characteristics in Table 7 illustrate the relative advantages and disadvantages of different pathogens and nematodes. A useful figure to calculate here that relates directly to cost is the amount of medium or substrate that must be used to produce a sufficient amount of the control agent to obtain effective control per hectare. By comparing different pathogens, it can be seen that bacteria and viruses require the least amount of substrate where the target level of control is a 90% reduction of the pest population. This leads to the question of why viral microbial insecticides have not been as successful as bacteria. The reasons are that viruses are much more specific than *B.t.,* thereby limiting market size. With the exception of the old Sandoz Crop Protection unit (now Thermo-Trilogy), the companies that produce and market insecticides have traditionally not been interested in products that must be produced *in vivo*. Thus, without even considering postproduction features such as harvesting,

TABLE 7   Product Characteristics for Different Types of Pathogens and Nematodes used as Insecticides[a]

Characteristic	Viruses	Bacteria	Fungi	Microsporidia	Nematodes
Broad host/target range	+ +	+ + + +	+ + + + + +	+ +	+ + + +
Mass production					
In vitro	+	+ + + + + +	+ + + +	+	+ + + + +
In vivo	+ + + +	N/A	N/A	+ + +	N/A
Efficacy per unit weight	+ + + + + +	+ + + + +	+ +	+	+ + +
Safety to nontargets	+ + + + + +	+ + + + +	+ +	+ + +	+ + + +
Formulation	+ + + + +	+ + + + +	+ +	+ + +	+ + + +
Storage	+ + + + +	+ + + + + +	+ +	+ +	+ + +
Residual activity	+ +	+	+	+	+ + +

[a]Based on the following representative pathogens developed or under development for each type: viruses, nuclear polyhedrosis viruses (NPV); bacteria, *Bacillus thuringiensis;* fungi, *Beauveria bassiana;* microsporidia, *Nosema locustae;* and nematodes, *Steinernema carpocapsae.*

formulation, and stability, it can been seen that bacteria, because of their low cost of production per "control unit" *and* ease of production, are the most cost-effective. In addition, because the principal active ingredients are spores and "nonliving" insecticidal proteins, they have been easier than other pathogens to harvest, formulate, and stabilize.

Future Utility of Different Pathogen Types

The extent to which pathogens will be used in the future is dependent on a variety of intrinsic and extrinsic factors and the extent to which these change or can be modified. The intrinsic factors vary from one group to another but in general include suitability for mass production, mode of action, spectrum of activity, and other biological features of these control agents listed in Table 7. The extrinsic factors include such components as the availability of synthetic chemical insecticides, governmental regulations, and identification of appropriate roles conventional or genetically improved pathogens can play in IPM programs. The prospects for more widespread use and the obstacles that must be overcome are discussed next for each type of control

agent. The probable future utility for each type of agent against the most important groups of insects is summarized in Table 8.

Bacteria

Most emphasis will continue to be on the range of insecticidal proteins produced by B.t., although these may be complemented by insecticidal proteins being discovered in other bacteria such as *Clostidium bifermentans* (Delecluse *et al.,* 1995) and the mutualistic bacteria associated with the steinernematid and heterorhabditid nematodes. The three major thrusts of this research will be to isolate improved strains (more toxic to recalcitrant species, novel host spectra) from nature, to determine the mode of action at the molecular level, and to more extensively use these proteins in transgenic plants. Determination of the mode of action at the molecular level should make it possible to develop microbial insecticides based on B.t. that are active against pests with chewing mouthparts (grasshoppers, cockroaches, termites, etc.) or chewing and sponging mouthparts (houseflies, tephritid fruit flies, etc.) for which no effective B.t.s

TABLE 8   Summary of Utility of the Different Types of Pathogens for Use as Microbial Insecticides against Different Types of Insect Pests

Insect group	Viruses	Bacteria	Fungi	Microsporidia	Nematodes
Lepidoptera	+ + +	+ + + + + +	+	+	+ +
Coleoptera	+	+ + +	+ +	+	+ + + +
Diptera	−	+ + + +	+	+	+
Homoptera	−	−	+ + +	−	+
Hemiptera	−	−	+ + +	−	+
Orthoptera	+	−	+ +	+ +	+ + +
Hymenoptera	+ + +	−	+	+ +	+

are currently available. Isolation or development of new types of protein insecticides based on B.t. proteins presents considerable opportunity because the technology already exists for their mass production, formulation, and application; and the regulatory procedures in most countries, with the possible exception of new regulations being developed by the European Economic Community, facilitate registration. Moreover, improved genetic engineering techniques and our rapidly expanding knowledge of plant molecular biology offer opportunities for attacking both chewing and sucking insects with insecticidal proteins directly through the plant. For the latter, the development of resistance management strategies is paramount (Tabashnik, 1994; Gould, 1998).

The research on B.t. as well as that on viruses, with their emphasis on the midgut epithelium, is also important in that it identifies this tissue as a source of novel target sites. Aside from the specific mode of action of B.t. proteins, the various other receptors on the midgut epithelium such as transmembrane transport systems (ion channels and enzymes) become targets for which inactivating and inhibiting insecticidal peptides can be designed to be delivered by conventional methods or through transgenic plants.

Viruses

If the availability of chemical insecticides declines, and significant resistance emerges to B.t. whether in the form of bacterial insecticides or transgenic plants, the market for conventional viruses will expand for insects pests for which viruses are available and for which other control measures are ineffective. These are big "if's." Availability of viruses restricts this market largely to the use of NPVs and GVs against lepidopterous pests; to a much lesser extent, the use of NPVs against sawfly larvae; and possibly EPVs against grasshoppers, locusts, and certain beetle pests. For the NPVs used against lepidopterous insects in developed countries, the two key technical obstacles to more widespread use are increasing the speed of kill or paralysis and the development of cost-effective methods for mass production *in vitro*. Because these are technological problems, it should be possible to overcome them through various engineering strategies, whether by direct engineering of the virus to improve efficacy or through development of improved fermentation systems. The costs for registration of a conventional virus in developed countries are reasonable (ca. $1 million in the United States) for viruses targeted against important pests. Viruses will not be developed for small niche markets, however, without significant exemption from the existing, but largely unnecessary, requirements for toxicological studies.

An economic risk that exists for the development of viral insecticides, because most are targeted against lepidopterous pests, is that the need for them will be eliminated by the growing use of transgenic plants, the development of new strains of B.t. (either naturally occurring or genetically engineered), or the development of new types of chemical insecticides.

In developing countries, where the procedures for registration are more reasonable considering the exemplary safety record of viral insecticides, the use of viruses will continue to expand as resistance to chemical insecticides increases and as new chemicals become less available or too expensive.

Fungi

As noted earlier, fungi have an important advantage over other pathogens in that the infectious unit, the zoospore or conidium, can penetrate directly through the cuticle. Despite this advantage, no fungus to date has been a commercial success in developed countries. This is due largely to a lack of cost-effective methods for mass production of fungi, but also because the conidia have been somewhat difficult to formulate and stabilize. The latter are technological problems and thus should be possible to overcome. Mass production is also a technological problem, but its resolution so far has been elusive. Overcoming this key obstacle with one or more of the hyphomycete fungi such as *Beauveria bassiana* or *Metarhizium anisopliae,* and a subsequent demonstration that these pathogens can provide reliable cost-effective control of one or a few key beetle or homopterous or hemipterous pests remain worthwhile goals because their realization would stimulate interest in further development of this group. The availability of fungal microbial insecticides that could effectively control such pests as aphids, whiteflies, thrips, and mites would be welcome additions and add much needed balance to IPM programs.

It is now possible to transform insect-pathogenic fungi, and thus the possibility exists for improving efficacy using recombinant DNA technology. For example, the genes for enzymes involved in penetration of the insect cuticle could be amplified, making a fungus more infectious (St. Leger, 1992; St. Leger *et al.,* 1996). However, given the relatively broad host range of many fungi, any microbial insecticides with such properties will have to be evaluated very carefully prior to field testing and further development.

Microsporidia

The chronic nature of the diseases caused by microsporidia and the lack of suitable methods for their mass production leave almost no hope that they will prove to be useful in the future as microbial insecticides. In some cases, where short-term control is not needed (e.g., grasshoppers on rangeland), they may be useful in long-term population management strategies. Their other possible uses, though marginal as noted earlier, may be as additional stress factors in classical biological control strategies (Anderson, 1982).

Nematodes

The major types of nematodes evaluated to date are the mermithids, steinernematids, and heterorhabditis, the latter two being quite similar in life cycle and potential. The fastidious nutritional requirements of the mermithids make it highly unlikely they will ever be practical or cost effective as control agents for any pests. In contrast to this, both the steinernematids and heterorhabditis show considerable potential for further development, especially against cryptic insects breeding in moist or wet habitats, such as soil- or tree-inhabiting insects, which are difficult to control with other technologies. The potential for the use of these nematodes has increased considerably over the past few years due to the development of improved methods for their mass production *in vitro*. Though costs are still high in comparison with other control strategies, further improvements in mass production methods and methods for formulating dauer larvae, as well as identification of new strains, should improve the utility of these nematodes as control agents. They also have an advantage over pathogens, as noted earlier, in that as "macroorganisms," like parasites, they do not have to be registered as insecticides, at least in the United States.

In summary, while not constituting a major control strategy, pathogens and nematodes have been useful in the past, and will play a greater role in the future, but primarily as microbial insecticides. They will not be as broadly effective as many of the chemical insecticides, but data indicate they are more environmentally compatible; and in the long term, when the cost to the environment is considered, they may be more cost-effective and therefore may be more appropriate for use. Greater use of these environmentally compatible agents will be facilitated by improvements in their efficacy, production, and formulation; by moderation of governmental regulations concerning registration; and by more thorough delineation of the role pathogens and nematodes can play in IPM programs.

References

Adams, J. R., & Bonami, J. R. (Eds.), (1991). Atlas of invertebrate viruses. Boca Raton, FL: CRC Press.

Anderson, R. M. (1982). Theoretical basis for the use of pathogens as biological control agents of pest species. Parasitology, 84, 3–33.

Andreadis, T. G., & Weseloh, R. M. (1990). Discovery of *Entomophaga maimaiga* in North American gypsy moth, *Lymantria dispar*. Proceedings of the National Academy of Sciences, USA, 87, 2461–2465.

Angus, T. A. (1954). A bacterial toxin paralyzing silkworm larvae. Nature (London), 173, 545.

Anthony, D. W., Savage, K. E., Hazard, E. I., Avery, S. W., Boston, M. D., & Oldacre, S. W. (1978). Field tests with *Nosema algerae* Varva and Undeen (Microsporida, Nosematidae) against *Anopheles albimanus* Weidemann in Panama. Miscellaneous Publications of the Entomological Society of America, 11, 17–26.

Anthony, D. W., & Comps, M. (1991). Iridoviridae. In J. R. Adams & J.

R. Bonauri (Eds.), Atlas of invertebrate viruses pp. 55–86. Boca Raton, FL: CRC Press.

Arif, B. M. (1984). The entomopoxviruses. Advances in Virus Research, 29, 195–214.

Aruga, H., & Tanada, Y. (1971). The cytoplasmic-polyhedrosis virus of the silkworm. Japan: University of Tokyo Press.

Balch, R. E., & Bird, F. T. (1944). A disease of the European spruce sawfly, *Gilpinia hercyniae* [Htg.] and its place in natural control. Scientific Agriculture, 25, 65–80.

Baum, J. A., Johnson, T. B., & Carlton, B. C. (1998). *Bacillus thuringiensis:* natural and recombinant bioinsecticide products. In F. R. Hall & J. J. Menn (Eds.), Biopesticides: use and delivery (pp. 189–209). Totowa: Humana Press.

Baumann, L., Broadwell, A. H., & Baumann, P. (1988). Sequence analysis of the mosquitocidal toxin genes encoding 51.4- and 41.9-kilodalton proteins from *Bacillus sphaericus* 2362 and 2297. Journal of Bacteriology, 170, 2045–2050.

Baumann, P., Clark, M. A., Baumann, L., & Broadwell, A. H. (1991). *Bacillus sphaericus* as a mosquito pathogen: Properties of the organism and its toxins. Microbiology Reviews, 55, 425–436.

Becker, N., & Margalit, J. (1993). Use of *Bacillus thuringiensis* subsp. *israelensis* against mosquitoes and blackflies. In P. F. Entwistle, J. S. Cory, M. J. Bailey, & S. Higgs (Eds.), Bacillus thuringiensis, An Environmental Biopesticide: Theory and Practice (pp. 147–170). Chichester: John Wiley & Sons.

Bedding, R. A. (1974). Five new species of *Deladenus* (Neotylenchidae) entomophagous-mycetophagous nematodes parasitic in siricid woodwasps. Nematologica, 20, 204–225.

Bedford, G. O. (1981). Control of the rhinoceros beetle by baculovirus. In H. D. Burges (Ed.), Microbial control of pests and plant diseases 1970–1980 (pp. 409–426). London: Academic Press.

Black, B. C., Brennan, L. A., Dierks, P. M., & Gard, I. E. (1997). Commercialization of baculoviral insecticides. In L. K. Miller (Ed.), The Baculoviruses (pp. 341–387). New York: Plenum Press.

Blissard, G. W., & Rohrmann, G. F. (1990). Baculovirus diversity and molecular biology. Annual Review of Entomology, 35, 127–155.

Briano, J. A., Patterson, R. S., & Cordo, H. A. (1996). The black imported fire ant, *Solenopsis richteri*, infected with *Thelohania solenopsae*: intracolonial prevalence of infection and evidence for transovarial transmission. Journal of Invertebrate Pathology, 67, 178–179.

Brooks, W. M. (1988). Entomogenous protozoa. In C. M. Ignoffo & N. B. Mandava (Eds.), Handbook of natural pesticides: Vol. 5. Microbial pesticides, Part A, Entomogenous protozoa and fungi (pp. 1–149). Boca Raton, FL: CRC Press.

Brown, G. C., & Nordin, G. L. (1982). Alfalfa crop management augmented by managing diseases of insect pests. In R. E. Frisbie (Ed.), Consortium for integrated pest management success stories (USDA-CSRS/EPA project report, pp. 23–30).

Cameron, J. W. M. (1973). Insect pathology. In R. F. Smith, T. H. Mittler, & C. N. Smith (Eds.), History of entomology (pp. 285–306). Palo Alto, CA: Annual Reviews.

Canning, E. U. (1982). An evaluation of protozoal characteristics in relation to the biological control of pests. Parasitology, 84, 119–149.

Carruthers, R. I., & Soper, R. S. (1987). Fungal diseases. In J. R. Fuxa & Y. Tanada (Eds.), Epizootiology of insect diseases (pp. 357–416). New York: John Wiley & Sons.

Chapman, H. C. (1985). Ecology and use of *Coelomomyces* in biological control: A review. In J. N. Couch & C. E. Bland (Eds.), The genus Coelomomyces (pp. 361–368). New York: Academic Press.

Charles, J.-F., Nielsen-LeRoux, C., & Delecluse, A. (1996). *Bacillus sphaericus* toxins: Molecular biology and mode of action. Annual Review of Entomology, 41, 451–472.

Couch, J. N., & Bland, C. E. (1985). Taxonomy. In J. N. Couch & C. E. Bland (Eds.), The genus Coelomomyces (pp. 81–297). New York: Academic Press.

Couch, J. N., & Umphlett, C. J. (1963). *Coelomomyces* infections. In E. A. Steinhaus (Ed.), Insect pathology: An advanced treatise (Vol. 2, pp. 99, 149–188). New York: Academic Press.

Cunningham, J. C., & Entwistle, P. F. (1981). Control of sawflies by baculovirus. In H. D. Burges (Ed.), Microbial control of pests and plant diseases, 1970–1980 (pp. 379–407). London: Academic Press.

Cunningham, J. T. (1998). North America. In F. R. Hunter-Fujita, P. F. Entwistle, H. F. Evans, & N. E. Crook (Eds.), Insect viruses and pest management (pp. 313–331). Chichester: John Wiley & Sons.

Davidson, E. W., & Yousten, A. A. (1990). The mosquito larval toxin of *Bacillus sphaericus*. In H. de Barjac & D. Sutherland (Eds.), Bacterial control of mosquitoes and blackflies; biochemistry, genetics, and applications of Bacillus thuringiensis and Bacillus sphaericus (pp. 237–255). New Brunswick, NJ: Rutgers University Press.

DeBach, P., & Rosen, P. (1991). Biological control by natural enemies. Cambridge: Cambridge University Press.

de Barjac, H. (1978). Une nouvelle variete de *Bacillus thuringiensis* tres toxic pour les mostiques: *B. thuringiensis* var. *israelensis* serotype 14. Comptes Rendu de l'Academie Sciences Paris, 286D, 797–800.

de Barjac, H., & Sutherland, D. (Eds.). (1990). Bacterial control of mosquitoes and blackflies; biochemistry, genetics, and applications of Bacillus thuringiensis and Bacillus sphaericus. New Brunswick, NJ: Rutgers University Press.

Dutky, S. R. (1937). Investigation of the diseases of the immature stages of the Japanese beetle. Unpublished doctoral dissertation, Rutgers University, New Brunswick, NJ.

Dutky, S. R. (1963). The milky diseases. In E. A. Steinhaus (Ed.), Insect pathology: An advanced treatise (Vol. 2, pp. 75–115). New York: Academic Press.

Entwistle, P. E. (1998). A world survey of virus control of insect pests. In F. R. Hunter-Fujita, P. F. Entwistle, H. F. Evans, & N. E. Crook (Eds.), Insect viruses and pest management (pp. 189–200). Chichester: John Wiley & Sons.

Federici, B. A. (1993). Viral pathobiology in relation to insect control. In N. E. Beckage, S. N. Thompson, & B. A. Federici (Eds.), Parasites and pathogens of insect: Vol. 2. Pathogens (pp. 81–101). San Diego: Academic Press.

Federici, B. A. (1997). Baculovirus pathogenesis. In L. K. Miller (Ed.), The Baculoviruses (pp. 33–59). New York: Plenum Press.

Federici, B. A. (1981). Mosquito control by the fungi *Culicinomyces, Lagenidium,* and *Coelomomyces*. In H. D. Burges (Eds.), Microbial control of pests and plant diseases 1970–1980 (pp. 555–572). London: Academic Press.

Federici, B. A. (1991). Microbial insecticides. Pesticide Outlook, 2, 22–28.

Federici, B. A. (1998). Naturally occurring baculoviruses for insect pest control. In F. R. Hall & J. J. Menn (Eds.), Biopesticides: use and delivery (pp. 301–320). Totowa: Humana Press.

Federici, B. A., & Stern, V. M. (1990). Replication and occlusion of a granulosis virus in larval and adult midgut epithelium of the western grapeleaf skeletonizer, *Harrisina brilliana*. Journal of Invertebrate Pathology, 56, 401–414.

Federici, B. A., Luthy, P., & Ibarra, J. E. (1990). The parasporal body of BTI: Structure, protein composition, and toxicity. In H. de Barjac & D. Sutherland (Eds.), Bacterial control of mosquitoes and blackflies; biochemistry, genetics, and applications of Bacillus thuringiensis and Bacillus sphaericus (pp. 16–44). New Brunswick, NJ: Rutgers University Press.

Ferron, P. (1978). Biological control of insect pests by entomogenous fungi. Annual Review of Entomology, 23, 409–442.

Fischhoff, D. A., Bowdisch, K. S., Perlak, F. J., Marrone, P. G., McCormick, S. H., & Niedermeyer, J. G. (1987). Insect tolerant transgenic tomato plants. Bio/Technology, 5, 807–813.

Friedman, M. J. (1990). Commercial production and development. In R. Gaugler & H. K. Kaya (Eds.), Entomopathogenic nematodes in biological control (pp. 139–172). Boca Raton, FL: CRC Press.

Fuxa, J. R. (1990). New directions for insect control with baculoviruses. In R. Baker & P. Dunn (Eds.), New directions in biological control (pp. 97–113). New York: Alan R. Liss.

Fuxa, J. R., & Tanada, Y. (Eds.). (1987). Epizootiology of insect diseases. New York: John Wiley & Sons.

Fuxa, J. R., & Brooks, W. M. (1979). Effects of *Vairimorpha necatrix* in sprays and cornmeal on *Heliothis* species in tobacco, soybeans, and cotton. Journal of Economic Entomology, 72, 462–467.

Gaugler, R., & Kaya, H. K. (Eds.). (1990). Entomopathogenic nematodes in biological control. Boca Raton, FL: CRC Press.

Georgis, R., & Hague, N. G. M. (1991). Nematodes as biological insecticides. Pesticide Outlook, 2, 29–32.

Gould, F. (1998). Susceptibility of transgenic insecticidal cultivars: Integrating pest genetics and ecology. Annual Review of Entomology, 43, 701–726.

Granados, R. R. (1981). Entomopoxvirus infections in insects. In E. W. Davidson (Ed.), Pathogenesis of invertebrate microbial diseases (pp. 101–126). Totowa, NJ: Allanheld, Osman.

Granados, R. R., & Federici, B. A. (Eds.). (1986). The biology of baculoviruses: Vol. 1. Biological properties and molecular biology; Vol. 2. Practical application for insect control. Boca Raton, FL: CRC Press.

Grewall, P., & Georgis, R. (1998). Entomopathogenic nematodes. In F. R. Hall & J. J. Menn (Eds.), Biopesticides: use and delivery (pp. 271–299). Totowa: Humana Press.

Hajek, A. E., Humber, R. A., Elkinton, J. S., May, B., Walsh, S. R. A., & Silver, J. C. (1990). Allozyme and restriction fragment length polymorphism analyses confirm *Entomophaga miamiaga* responsible for 1989 epizootics in North American gypsy moth populations. Proceedings of the National Academy of Sciences, USA, 87, 6979–6982.

Hajek, A. E., Elkinton, J. S., & Witcosky, J. J. (1996). Introduction and spread of the fungal pathogen *Entomophaga miamiaga* (Zygomycetes: Entomophthorales) along the leading edge of gypsy moth (Lepidoptera:Lymantriidae) spread. Environmental Entomology, 25, 1235–1247.

Hall, R. A., & Papierok, B. (1982). Fungi as biological control agents of arthropods of agricultural and medical importance. Parasitology, 84, 205–240.

Hammock, B. D., Bonning, B. C., Possee, R. D., Hanzlik, T. N., & Maeda, S. (1990). Expression and effects of juvenile hormone esterase in a baculovirus vector. Nature (London), 344, 458–461.

Hannay, C. L. (1953). Crystalline inclusions in aerobic spore-forming bacteria. Nature (London), 172, 1004–1006.

Hawtin, R. E., & Possee, R. D. (1992). Genetic manipulation of the baculovirus genome for insect pest control. In N. E. Beckage, S. N. Thompson, & B. A. Federici (Eds.), Pathogens and parasites of insects. Orlando, FL: Academic Press.

Henry, J. E. (1981). Natural and applied control of insects by protozoa. Annual Review of Entomology, 26, 49–73.

Henry, J. E. (1990). Control of insects by protozoa. In R. Baker & P. Dunn (Eds.), New directions in biological control (pp. 161–176). New York: Alan R. Liss.

Higley, L. G., & Wintersteen, W. K. (1992). A novel approach to environmental risk assessment of pests as a basis for incorporating environmental costs into economic injury levels. American Entomologist, 38, 34–39.

Humber, R. A. (1989). Synopsis of a revised classification for the Entomophthorales (Zygomycotina). Mycotaxon, 34, 441–460.

Humber, R. A. (1997). Fungi: identification. In L. A. Lacey (Ed.), Manual of techniques in insect pathology (pp. 153–185). San Diego: Academic Press.

Humber, R. A., Brown, C. C., & Kornegay, R. W. (1989). Equine zygomycosis caused by *Conidiobolus lamprauges*. Journal of Clinical Microbiology, 27, 573–576.

Hunter-Fujita, F. R., Entwistle, P. E., Evans, H. F., & Crook, N. E. (Eds.). (1998). Insect viruses and pest management. Chichester: John Wiley & Sons.

Ignoffo, C. M. (1973). Development of a viral insecticide: Concept to commercialization. Experimental Parasitology, 33, 380–406.

Jackson, T. A., Pearson, J. F., O'Callaghan, M. O., Mahanty, H. K., & Willocks, M. J. (1992). Pathogen to product-development of Serratia entomophila (Enterobacteriaceae) as a commercial biological control agent for the New Zealand grass grub (Costelystra zealandica). In T. A. Jackson & T. R. Glare (Eds.), Use of pathogens in scarab pest management (pp. 191–198). Andover: Intercept.

Jenkins, J. N. (1998). Transgenic plants expressing toxins from Bacillus thuringiensis. In F. R. Hall & J. J. Menn (Eds.), Biopesticides: use and delivery (pp. 221–232). Totowa: Humana Press.

Katagiri, K. (1981). Pest control by cytoplasmic polyhedrosis viruses. In H. D. Burges (Ed.), Microbial control of pests and plant diseases 1970–1980 (pp. 433–440). London: Academic Press.

Kaya, H. K., & Gaugler, R. (1993). Entomopathogenic nematodes. Annual Review of Entomology, 38, 181–206.

Keller, B., & Langenbruch, G.-A. (1993). Control of coleopteran pests by Bacillus thuringiensis. In P. F. Entwistle, J. S. Cory, M. J. Bailey, & S. Higgs (Eds.), An environmental biopesticide: Theory and practice Bacillus thuringiensis, (pp. 171–191). Chichester: John Wiley & Sons.

Kelly, D. C., & Robertson, J. S. (1973). Icosahedral cytoplasmic deoxyriboviruses. Journal of General Virology, 20, 17–41.

Kerwin, J. L., & Washino, R. K. (1987). Ground and aerial application of asexual stage of Lagenidium giganteum for control of mosquitoes associated with rice culture in the Central valley of California. Journal of the American Mosquito Control Association, 3, 59–64.

Klein, M. G. (1981). Advances in the use of Bacillus popilliae for pest control. In H. D. Burges (Ed.), Microbiol control of pests and plant diseases 1970–1980 (pp. 183–192). London: Academic Press.

Klein, M. G. (1997). Bacteria of soil-inhabiting insects. In L. A. Lacey (Ed.), Manual of techniques in insect pathology (pp. 101–116). San Diego: Academic Press.

Krassilstscshik, I. M. (1888). La production industriale des parasites vegetaux pour la destruction des insectes nuisibles. Bulletin des Sciences France, 19, 461–472.

Kunimi, Y. (1998). Japan. In F. R. Hunter-Fujita, P. F. Entwistle, H. F. Evans, & N. E. Crook (Eds.), Insect viruses and pest management (pp. 269–279). Chichester: John Wiley & Sons.

Lacey, L. A. (1997). Manual of techniques in insect pathology. San Diego: Academic Press.

Lacey, L. A., & Brooks, W. M. (1997). Initial handling and diagnosis of diseased insects. In L. A. Lacey (Ed.), Manual of techniques in insect pathology (pp. 1–15). San Diego: Academic Press.

Levine, N. D., Corlis, J. O., Cox, F. E. G., Deroux, G., Grain, J., & Honigberg, B. M. (1980). A newly revised classification of the Protozoa. Journal of Protozoology, 27, 37–58.

Levy, R., & Miller, T. W., Jr. (1977). Experimental release of Romanomermis culicivorax (Mermithidae: Nematoda) to control mosquitoes breeding in southwest Florida. Mosquito News 37, 483–486.

MacLeod, D. M. (1963). Entomophthorales infections. In E. A. Steinhaus (Ed.), Insect pathology: An advanced treatise (Vol. 2, pp. 189–231). New York: Academic Press.

Maddox, J. V. (1987). Protozoan diseases. In J. R. Fuxa & Y. Tanada (Eds.). Epizootiology of insect diseases (pp. 417–452). New York: John Wiley & Sons.

Maddox, J. V., Brooks, W. M., & Fuxa, J. R. (1981). Variamorpha necatrix a pathogen of agricultural pests: Potential for pest control. In H. D. Burges (Ed.), Microbial control of pests and plant diseases 1970–1980 (pp. 587–594). London: Academic Press.

Maeda, S. (1989). Expression of foreign genes in insects using baculovirus vectors. Annual Review of Entomology, 34, 351–372.

Martens, J. W. M., Honee, G., Zuidema, D., Van Lent, J. W. M., Visser, B., & Vlak, J. M. (1990). Insecticidal activity of a bacterial crystal protein expressed by a recombinant baculovirus in insect cells. Applied Environmental Microbiology, 56, 2764–2770.

McCoy, C. W. (1990). Entomogenous fungi as microbial insecticides. In R. Baker & P. Dunn (Eds.), New directions in biological control (pp. 139–159). New York: Alan R. Liss.

McCoy, C. W., Samson, R. A., & Boucias, D. G. (1988). Entomogenous fungi. In C. M. Ignoffo & N. B. Mandava (Eds.), Handbook of natural pesticides: Vol. 5. Microbial pesticides, Part A, Entomogenous protozoa and fungi (pp. 151–236). Boca Raton, FL: CRC Press.

McCutchen, B. F., Choudary, P. V., Crenshaw, R., Maddox, D., Karmita, S. G., & Palekar, N. (1991). Development of a recombinant baculovirus expressing an insect active neurotoxin: Potential for pest control. Bio/Technology, 9, 848–852.

McKinley, D. J., Moawad, G., Jones, K. A., Grzywacz, D., & Turner, C. (1989). The development of nuclear polyhedrosis virus for the control of Spodoptera littoralis (Boisd.) in cotton. In M. B. Green & D. J. de B. Lyon (Eds.), Pest management in cotton (pp. 93–100). Chichester: Ellis Horwood.

Merryweather, A. T., Weyer, U., Harris, M. P. G., Hirst, M., Booth, T., & Possee, R. D. (1990). Construction of genetically engineered baculovirus insecticides containing the Bacillus thuringiensis kurstaki HD-73 delta endotoxin. Journal of General Virology, 71, 1535–1544.

Miller, L. K. (1988). Baculoviruses as gene expression vectors. Annual Review of Microbiology, 42, 177–199.

Miller, L. K. (Ed.), (1997). The Baculoviruses New York: Plenum Press.

Milner, R. J., Soper, R. S., & Lutton, G. G. (1982). Field release of an Israeli strain of the fungus Zoophthora radicans for biological control of Therioaphis trifolii f. maculata. Journal of the Australian Entomological Society, 21, 113–118.

Moscardi, F. (1999). Assessment of the application of baculoviruses for control of lepidoptera. Annual Review of Entomology, 44, 257–289.

Moscardi, F. (1989). The use of viruses for pest control in Brazil: The case of the nuclear polyhedrosis virus of the soybean caterpillar, Anticarsia gemmatalis. Memorias do Instituto Oswaldo Cruz, 4, 51–56.

Moscardi, F. (1990). Development and use of soybean caterpillar baculovirus in Brazil. In D. E. Pinnock (Ed.), Proceedings of the Fifth International Colloquium on Invertebrate Pathology and Microbial Control (pp. 184–187). Adelaide: Society for Invertebrate Pathology.

Navon, A. (1993). Control of lepidopteran pests with Bacillus thuringiensis. (pp. 125–146). In P. F. Entwistle, J. S. Cory, M. J. Bailey, & S. Higgs (Eds.), Bacillus thuringiensis, an environmental biopesticide: Theory and practice. Chichester: John Wiley & Sons.

Paillot, A. (1933). L'infection chez les insectes. Trevoux, France: G. Patisier.

Pang, Y., Frutos, R., & Federici, B. A. (1992). Synthesis and toxicity of full-length and truncated bacterial CryIVD mosquitocidal proteins expressed in lepidopteran cells using a baculovirus vector. Journal of General Virology, 73, 89–101.

Payne, C. C. (1981). Cytoplasmic polyhedrosis viruses. In E. W. Davidson (Ed.), Pathogenesis of invertebrate microbial diseases (pp. 61–100). Totowa, NJ: Allanheld, Osman.

Payne, C. C. (1982). Insect viruses as control agents. Parasitology, 84, 35–77.

Payne, C. C. (1988). Pathogens for the control of insects: Where next? Philosophical Transactions of the Royal Society of London, Series B, 318, 225–248.

Payne, C. C., & Kelly, D. C. (1981). Identification of insect and mite viruses. In H. D. Burges (Ed.), Microbial control of pests and plant diseases, 1970–1980 (pp. 61–91). London: Academic Press.

Peng, H., Li, C., & Zhe, Y. (1988). Application of viral insecticide to controlling Buzura suppressaria in Tung tree. Chinese Bulletin of Biological Control, 4, 187–193.

Peng, H., Xie, T., Feng, J., Zhang, Y., & Liu, Y. (1991). Study on new viral pesticide with high effect and without environment pollution in China (pp. 680–683). Proceedings, International Symposium on Tea Science, Shizuoka, Japan.

Perlak, F. J., Deaton, R. W., Armstrong, T. A., Fuchs, R. L., Sims, S. R., & Greenplate, J. T. (1990). Insect resistant cotton plants. Bio/Technology, 8, 939–943.

Petersen, J. J. (1982). Current status of nematodes for the biological control of insects. Parasitology, 84, 177–204.

Pinnock, D. E. (1975). Pest populations and virus dosage in relation to crop productivity. In M. Summers, R. Engler, L. A. Falcon, & P. V. Vail (Eds.), Baculoviruses for insect pest control: Safety considerations (pp. 145–154). Washington, DC: American Society for Microbiology.

Poinar, G. O., Jr. (1979). Nematodes for the biological control of insects. Boca Raton, FL: CRC Press.

Roberts, D. W., & Humber, R. A. (1981). Entomogenous fungi. In G. T. Cole & B. Kendrick (Eds.), The biology of conidial fungi (Vol. 2, pp. 201–236). New York: Academic Press.

Samson, R. A. (1981). Identification of entomopathogenic Deuteromycetes. In H. D. Burges (Ed.), Microbial control of pests and plant diseases 1970–1980 (pp. 93–106). London: Academic Press.

Shah, P., & Goettel, M. S. (1998). Directory of Microbial Products and Services. Gainesville: Society for Invertebrate Pathology Press.

Shapiro, M. (1986). In vivo production of baculoviruses. In R. R. Granados & B. A. Federici (Eds.), The biology of baculoviruses (Vol. 2, pp. 31–61). Boca Raton, FL: CRC Press.

Shapiro, M., & Robertson, J. L. (1992). Enhancement of gypsy moth (Lepidoptera: Lymantriidae) baculovirus activity by optical brighteners. Journal of Economic Entomology, 85, 1120–1124.

Soares, G. G., Jr., & Pinnock, D. E. (1984). Effect of temperature on germination, growth, and infectivity of the mosquito pathogen Tolypocladium cylindrosporum (Deuteromycotina; Hyphomycetes). Journal of Invertebrate Pathology, 43, 242–246.

Splittstoesser, G. M., Kawanishi, C. Y., & Tashiro, H. (1978). Infection of the European chafer, Amphimallon majalis, by Bacillus popilliae: Light and electron microscope observations. Journal of Invertebrate Pathology, 31, 84–90.

St. Leger, R. J. (1992). Biology and mechanisms of insect-cuticle invasion by Deuteromycete fungal pathogens. In N. E. Beckage, S. N. Thompson, & B. A. Federici (Eds.), Pathogens and parasites of insects. Orlando, FL: Academic Press.

St. Leger, R. J., Lokesh, J., Bidochka, M. J., & Roberts, D. W. (1996). Construction of an improved mycoinsecticide overexpressing a toxic protease. Proceedings of the National Academy of Sciences U.S.A., 93, 6349–6354.

Steinhaus, E. A. (1949). Principles of insect pathology. New York: McGraw-Hill.

Steinhaus, E. A. (1963). Introduction. In E. A. Steinhaus (Ed.), Insect pathology: An advanced treatise (Vol. 1, pp. 1–17). New York: Academic Press.

Steinhaus, E. A. (1975). Disease in a minor chord. Columbus: Ohio State University Press.

Tabashnik, B. E. (1994). Evolution of resistance to Bacillus thuringiensis. 1994. Annual Review of Entomology, 39, 47–79.

Tanada, Y., & Kaya, H. K. (1993). Insect pathology. San Diego: Academic Press.

Tomalski, M. D., & Miller, L. K. (1991). Insect paralysis by baculovirus-mediated expression of a mite neurotoxin gene. Nature (London). 352, 82–85.

Treacy, M. F. (1998). Recombinant baculoviruses. In F. R. Hall & J. J. Menn (Eds.), Biopesticides: use and delivery (pp. 321–340). Totowa: Humana Press.

Tweeten, K. A., Bulla, L. A., & Consigli, R. A. (1981). Applied and molecular aspects of insect granulosis viruses. Microbiology Review, 45, 379–408.

Vaeck, M., Reynearts, A., Hofte, H., Jansens, S., De Beuckeleer, M., & Dean, C. (1987). Trangenic plants protected from insect attack. Nature (London), 328, 33–37.

van Frankenhuyzen, K. (1993). The challenge of Bacillus thuringiensis. In P. F. Entwistle, J. S. Cory, M. J. Bailey, & S. Higgs (Eds.), Bacillus thuringiensis, an environmental biopesticide: theory and practice (pp. 1–35). Chichester: J. Wiley & Sons.

Vavra, J. (1976). Structure of the microsporidia. In L. A. Bulla & T. C. Cheng (Eds.), Comparative pathobiology (Vol. 2, pp. 1–85). New York: Plenum Press.

Volkman, L. E., Blissard, G. W., Friesen, P., Keddie, B. A., Possee, R., & Theilmann, D. A. (1995). Family Baculoviridae. In F. A. Murphy, C. M. Fauquet, D. H. L. Bishop, S. A. Ghabrial, A. W. Jarvis, & G. P. Martelli. (Eds.), Virus taxonomy. Classification and nomenclature of viruses. Sixth Report of the International Committee on Taxonomy of Viruses (pp. 104–113). Wien: Springer.

Voronina, E. G. (1971). Entomophthorosis epizootics of the pea aphid, Acrythosiphon pisum. Entomological Review (USSP), 4, 444–458.

Whisler, H. C. (1985). Life history of species of Coelomomyces. In J. N. Couch & C. E. Bland (Eds.), The genus Coelomomyces (pp. 1–22). New York: Academic Press.

White, R. T., & Dutky, S. R. (1942). Cooperative distribution of organisms causing milky disease of Japanese beetle grubs. Journal of Economic Entomology, 35, 679–682.

Wilding, N. (1981). Pest control by Entomophthorales. In H. D. Burges (Ed.), Microbial control of pests and plant diseases 1970–1980 (pp. 539–554). London: Academic Press.

Williams, D. F., Knue, G. J., & Becnel, J. J. (1998). Discovery of Thelohania solenopsae from the red imported fire ant, Solenopsis invicta, in the United States. Journal of Invertebrate Pathology, 71, 175–176.

Wraight, S. P., & Carruthers, R. I. (1998). Production, delivery, and use of mycoinsecticides for control of insect pests in field crops. In F. R. Hall & J. J. Menn (Eds.), Biopesticides: use and delivery (pp. 233–269). Totowa: Humana Press.

Zelazny, B., Lolong, A., & Pattang, B. (1992). Oryctes rhinocerous (Coleoptera: Scarabaeidae) populations suppressed by a baculovirus. Journal of Invertebrate Pathology, 59, 61–68.

Zhang, G. (1981). The production and application of the nuclear polyhedrosis virus of Heliothis armigera in biological control. Acta Phytopathology Sinica, 8, 235–240.

19

Cross-Protection and Systemic Acquired Resistance for Control of Plant Diseases

J. ALLAN DODDS

Department of Plant Pathology
University of California
Riverside, California

INTRODUCTION

An axiom of plant pathology is that most plant species are resistant to most pathogens, leading plant pathologists to focus on those interactions that lead to disease between genetically susceptible hosts and their pathogens. Once the susceptible reaction is known and described, a comparison with related varieties that are genetically resistant to the pathogen then follows. Genetic analysis next occurs in an attempt to define genes for resistance in the resistant host, and genes for avirulence/virulence in the pathogen. A great deal of contemporary research is done with such systems on the assumption that a full understanding of what is commonly known as the gene-for-gene interaction will be needed if we are ever to unravel the host-pathogen interaction (Hammond-Kosack & Jones, 1997; Staskawicz *et al.,* 1995).

One of the more surprising observations in plant pathology is that a fully susceptible host can become resistant to a virulent pathogen without the introduction of a gene for resistance either through conventional breeding or by plant transformation. This is not such a feat of magic in mammalian pathology, where a complex multicomponent immune system can be called on to mount a defense in a susceptible individual on infection by a virulent pathogen. While susceptible plants lack such a system, which depends so much on a circulating blood system, they do possess the ability to respond to pathogen attack with an array of biochemical, physiological, and anatomical changes, all of which appear to be aimed at containing the pathogen. This chapter deals with two such responses, cross-protection and systemic acquired resistance (SAR).

COMPARISON OF CROSS-PROTECTION AND SYSTEMIC ACQUIRED RESISTANCE

A priority is to make clear the differences that separate these two responses. Cross-protection is considered to take place when the presence of an established pathogen prevents a related pathogen from fully expressing its disease-causing potential. Much of the literature on cross-protection deals with viruses but other pathogens types can do this (Chakraborty & Gupta, 1995; Freeman & Rodriguez, 1993; Louter & Edgington, 1990). In practice, a mild strain of a virus is used to infect a plant that may well remain almost symptomless. The mild strain does, however, spread throughout the plant. A strain capable of causing severe disease is then inoculated a few days later at the same or a distant site (an upper younger but already infected leaf), 3 days often being sufficient. The control for the experiment is a plant that was not previously infected with the mild strain, and this plant will soon develop symptoms of infection with the severe strain. The experimental plant does not develop these symptoms, either in the leaf inoculated with the severe strain if local symptoms are expected or in even younger leaves that should develop symptoms of systemic infection. When the challenge inoculation is performed with a virus that is unrelated to the mild virus used to first inoculate the plant, no protection is seen. Cross-protection has been reviewed extensively by Fulton (1982, 1986) and Hamilton (1980).

SAR occurs when a plant is inoculated with a virulent pathogen against which it is able to mount a resistance reaction, usually in the form of a necrotic hypersensitive reaction (HR) like a local lesion. When that plant is later inoculated with a pathogen to which it is normally susceptible (related or not to the first pathogen), infection can be greatly limited or prevented. This increase in resistance in a normally susceptible leaf takes place in leaves far away from the originally inoculated leaf, and therefore in the absence of the first pathogen, which was restricted to the site of inoculation by the HR. The increased resistance can show surprisingly little specificity; the inducing pathogen could be a virus that causes local lesions; and the induced resistance could be active against a fungal, bacterial, or nematode pathogen. SAR is also called induced resistance and has been reviewed extensively, including excellent reviews by Kessman *et al.* (1994) and Sticher *et al.* (1997).

There are two main reasons why cross-protection and SAR are commonly considered together. The first is the observation that both cause a susceptible plant to become resistant, and the second is that a delay of a few days is needed between the first inoculation (often called the protective inoculation) and inoculation with the second pathogen (often called the challenge inoculation), or else the response is not seen. The major differences between these two phenomena include the specificity with respect to the pathogen against which resistance is active, which is high for cross-protection (same or related pathogen to the one used for protection) and is low for SAR (active against diverse pathogens); and the need for the protective pathogen to be present in tissue with increased resistance, which is a requirement for cross-protection and is not needed for SAR. These differences are important and point to very different mechanisms, even if the consequence is similar, and for this reason these topics will be treated separately. Before doing so it is worth pointing out that these two phenomena could both be at work at the same time in a plant if the cross-protective strain was able not only to deploy its own specific mechanisms at sites of interaction with the challenge strain, but also to trigger any of the host responses that are now known to characterize SAR.

SYSTEMIC ACQUIRED RESISTANCE

SAR is a very active research area with many review articles (Hunt & Ryals, 1996; Kessman *et al.*, 1994a, 1994b; Ryals *et al.*, 1996; Schneider *et al.*, 1996; Sticher *et al.*, 1997) and their availability dictates that this topic will not be considered in detail here, in part because the use of SAR in the field as a biological control has not reached the level that can be documented for cross-protection. The general features of SAR were described earlier, though it may be worth noting that the distant acquisition of resistance at sites away from the site of induction applies to roots as well as to leaves. Also, plants in the field may well be exposed to repeated biological interactions such that they are in a higher state of SAR than are comparable experimental plants maintained in greenhouse conditions.

For SAR to work, it is necessary to propose a signal that is produced at the site of induction, where the HR response is seen, and which is translocated to the distant sites where SAR is later detected. Salicylic acid (SA) seems to have some properties that could make it a candidate as a signal, but it may not be the only signal (Sticher *et al.*, 1997). The signal or signals can prime the plant to respond quickly to future attacks by pathogens. Part of this priming activity may be the induction of systemically distributed pathogenesis related (PR) proteins, known as SAR proteins, which are being characterized, as are the SAR genes. While attack by a pathogen is the primary method for activating SAR, it is also possible to duplicate much of the SAR biology with certain chemicals applied to plants, these include SA, jasmonates, and even some synthetic compounds, opening the door for chemical control based on a principle discovered as a biological control. Biological and chemical induction, signaling, and subsequent involvement of SAR proteins in defense against challenge infections are all active research areas and the review articles listed should be consulted for detailed information.

CROSS-PROTECTION

The remainder of this chapter will deal with viral cross-protection, which currently is not nearly such an active area of research as SAR, but which has several examples of actual deployment as a field biological control method. The current lack of activity in cross-protection is explained in part by the strides made using viral genes and genome sequences as transgenes for pathogen-derived resistance (see Chapter 20). This approach has the promise of providing much of the benefits of cross-protection for disease management without the major inherent risks, primarily the avoidance of using a "live vaccine."

One of the first descriptions of cross-protection was by McKinney (1929) who showed that severe yellow symptoms of one strain of tobacco mosaic virus (TMV) did not appear in inoculated plants preinfected with a more normal strain that produced a green mosaic. This became a useful tool during the period when most plant virology was conducted in the greenhouse, because the detection of cross-protection between a characterized and an uncharacterized virus isolate could be used to infer relatedness (i.e., the new virus was probably a strain of the characterized virus). The discovery that a plant or even all plants of a given variety may be naturally infected with a mild strain (an undetected latent infection) can also be deduced if such plants fail to show symptoms when inoculated with a greenhouse-type strain (Fernow, 1967).

Mechanisms of Cross-Protection

Mechanistic studies of cross-protection have used carefully designed experimental approaches (Dodds, 1990). It has been demonstrated that a challenge virus (e.g., TMV) can actually accumulate in the infected plant without expressing the symptoms that would normally appear with a similar level of expression in a nonprotected plant (Cassels & Herrick, 1977). Alternatively, the challenge virus (e.g., cucumber mosaic virus, CMV) can fail to accumulate, providing a more easily understood basis for lack of symptoms of the challenge virus (Dodds, 1982; Dodds *et al.*, 1985). An important observation on cross-protection made with both TMV (Sherwood & Fulton, 1982) and CMV (Dodds *et al.*, 1985) is that protection in the inoculated leaf can be overcome by using viral RNA instead of virus particles as the challenge inoculum, implying that the coat protein (CP) is directly involved in at least the early steps of initial establishment of the challenge virus in the challenge inoculated leaf. Additional evidence for this has been provided by Sherwood (1987b). Viroids, which lack any viral proteins including CPs, are also capable of exhibiting cross-protection (Duran-Vila & Semancik, 1990; Fernow, 1967; Khoury *et al.*, 1988; Niblett *et al.*, 1978; Palukaitis & Zaitlin, 1984; Singh *et al.*, 1989, 1990, 1993), and thus the mechanism just suggested cannot be active for this class of pathogens. This must also be the case for strains of TMV that lack the ability to make functional CP but still can act as protective strains in cross-protection experiments (Sarkar & Smitamana, 1981; Gerber & Sarkar, 1989). Breakdown of cross-protection in parts of individual plants (Rezende & Sherwood, 1991), and host effects on the level of activity (Rezende *et al.*, 1992; Singh & Singh, 1995) have also been described.

If a virus genome had only a few base differences from another, even a single base difference, would these two variants cross-protect? Because RNA viruses exist as a quasi species, or population of nearly identical molecules with minor sequence variability caused by copy errors during replication, this is a hard question to answer. One attempt to address this question was with satellite tobacco mosaic virus (STMV). Two cDNA clones were captured with 5 single base differences out of the 1059 nucleotides that constitute the genome. These were sufficient to make it possible to distinguish the presence or absence of either strain by the use of a fingerprinting technique know as RNase protection assay (RPA). These two variants were capable of reciprocal cross-protection (Kurath & Dodds, 1994). This is an interesting result because it provides experimental evidence to support the hypothesis that cross-protection could be occurring at all times within a singly infected plant and could lead to segregation of minor variants in different tissue types or tissue patches following founder effects that caused such variants to dominate in a single cell or tissue piece that later divided to become such a patch. Ideas like this have been proposed to partially explain the basis for mosaic symptoms in plants (Mathews, 1991).

An important question to ask about viral cross-protection is, "Are some viral genes not involved?", or "Which viral genes are involved?" An approach to this problem is to use a second unrelated virus as a viral vector to deliver a single viral gene of the protecting virus to a plant and then to see if its presence and activity are sufficient to mimic the type of cross-protection measured when the entire infectious virus is used as the protecting strain (Culver, 1996). The study involves cross-protection with TMV and the viral vector was potato virus X (PVX), genetically engineered to express TMV CP from its genome. Unmodified PVX did not prevent the lethal necrosis in *Nicotiana benthamiana* caused by TMV, but the modified PVX did offer good protection. The role of CP in TMV cross-protection was therefore confirmed, but the author also showed that if the CP of TMV was not expressed (no CP made but its RNA sequence is still present) there was still limited protection. This result helps rationalize previous reports that strains of TMV that do not make a functional CP can still cross-protect (Sarkar & Smitamana, 1981; Gerber & Sarkar, 1989). This approach of using viral vectors to deliver single cross-protecting genes could lead to additional understanding of the mechanisms that underlie cross-protection, which remain unclear, but are probably multifactorial as this study suggests.

In partial recognition of the probable complexity underlying the phenomenon, several mechanisms have been proposed to explain cross-protection including (1) the CP of the protecting virus acts to encapsidated challenge virus strain RNA preventing its expression, (2) the challenge virus cannot uncoat itself and thus fails to initiate an infection, (3) the replicative RNAs from the protecting virus hybridize with RNAs of the challenge virus and render them inactive, (4) the protecting virus depletes host metabolites and/or occupies host structures that are needed by the challenge virus to establish an infection (Dodds *et al.*, 1985; Mathews, 1991; Palukaitis & Zaitlin, 1984; Ponz & Bruening, 1986; Shalla & Peterson, 1978; Sherwood, 1987a; Sherwood & Fulton, 1982; Zaitlin, 1976).

Practical Application and Biological Control

The increased understanding of the dynamics and mechanisms of cross-protection has probably served as a catalyst to better justify the methods used for selection and deployment of mild strains in field situations. This is done with the hope that their deliberate introduction and dissemination will provide a level of disease management against more severe field strains through the activity of cross-protection. The remainder of this chapter will summarize some of these case histories, and draws on previous review articles that emphasized this practical aspect of cross-protection (Fletcher, 1978; Gonsalves & Garnsey, 1989; Hamilton, 1985; Lecoq, 1998).

Concerns Raised by Cross-Protection

There has always been a degree of reluctance to use cross-protection as a biological control in field situations. This is because several problems can be imagined, some based on the reality that the use of cross-protection inevitably leads to the deliberate widespread distribution of a "live vaccine." A first concern is the problem of nontarget crops that may also be susceptible to the virus. A strain selected for mildness on the target crop may not be mild on nontarget crops, and its deliberate distribution in the target crop as a live vaccine would increase its inoculum potential compared with earlier times when the target crop was not deliberately 100% infected at the time of planting. A second concern is the stability of the mild strain in the target crop. Viral pathogenicity can hinge on small numbers of nucleotide changes in single viral genes, and the possibility of a selected mild strain mutating to a more virulent form needs to be evaluated. A third concern is the longevity of the protective principle, knowing that mild strains may lose their ability to protect, either because of their own genetic drift or because of the different challenge strains that may be encountered from location to location and from year to year. A fourth concern is interaction with unrelated viruses or other pathogens, where the worst case would be a strong synergistic reaction between the mild virus and the second pathogen. Another scenario that could be imagined in such a mixed infection involves the possible encapsidation of the genome of a virus in the capsid protein of a second virus in a mixed infection, leading to an extension of vector transmission of one of the viruses (Bourdin & Lecoq, 1991). A fifth concern is that increasing the distribution of a virus by the widespread inclusion of a mild strain wherever the crop is grown puts an unusually high incidence of the virus into an agricultural ecosystem in which virus evolution is presumed to take place. Interactions such as recombination between the virus with other strains, related viruses, and even unrelated viruses could lead to the evolution of new viruses.

Selection and Deployment of Mild Protective Strains

These concerns notwithstanding, cross-protection is proposed for field use when a severe strain of a virus is clearly limiting the possibility of continuing to cultivate a crop, and when no tolerant or resistant varieties are available. A good example of "no place to turn" is the problem associated with severe stem-pitting forms of citrus tristeza virus (CTV) such as the capoa bonito strain found in parts of Brazil's sweet orange industry (Muller, 1980; Muller et al., 1988). When a field strain is really severe, then the task of selecting a strain that is sufficiently "mild" becomes relative and quite different from the criteria that might be applied for a greenhouse study of cross-protection, such as those described earlier. The strain selected for use may have virulence factors that would not be tolerated in an industry with few virus problems, but that are still preferable compared with the level of destruction anticipated for unchecked widespread attack by the severe field strain. Other criteria besides mildness may also figure in the selection of a suitable protective strain, not the least of which is that it be protective, hopefully against a wide range of strains. In a study of 100 strains of CTV, none were found to be protective against a set of isolates that were severe (Roistacher & Dodds, 1993), so mildness alone is not predictive of protectiveness, and extensive cross-protection trials will have to be carried out to select a protective strain.

The selection of a suitable mild strain is by no means easy. The major tools available include the chance encounter of a suitable isolate in a field plant showing no disease (Hughes & Ollenu, 1994; Costa & Muller, 1980; Muller & Costa, 1987); a similar discovery during the course of field and greenhouse observations on parts of an individual plant (Lecoq et al., 1991), the deliberate use of mutagenesis and the selection of variants following such procedures as cold or heat treatment of infected plants (Kosaka & Fukinushi, 1993; Oshima, 1975), or host passage through alternate host plants (Roistacher et al., 1988). Regardless of the method, the time to develop a suitable mild protective strain will be measured in years and not months, especially if consideration is given to additional desirable phenotypes (Lecoq, 1998). Chance discovery of a natural mild strain would seem to be a good method, because the strain should have the benefits of natural selection. To discover such a strain it may take a true epidemic of a severe strain, and the near loss of an entire industry before "survivors" such as a standout citrus tree in a grove of severely declining trees can be identified. Loss of an industry is a large price to pay for the discovery of a cross-protective strain.

Given a choice between several mild, protective strains, then other useful properties would include a high capacity to be invasive in the host. In perennial crops this may include the capacity to recover after being suppressed by climatic extremes. Another would be genetic stability so as to avoid the loss of protectiveness, or the acquisition of virulence. Again, the suppression and recovery of virus titer in perennial crops actually favors the selection of mutants in new flush growth. Vector transmission properties need also be considered, and conservatively it would be wise to select (or engineer) the mild protective strain to be nontransmissible, because this would limit the unregulated distribution to nontarget crops, in the event that the mild strains have problems.

How to actually deploy the mild, protective strain presents some special problems. For annual crops the model of tomato mosaic (Broadbent, 1976) provides some useful pointers. The mild protective strain was mechanically inoculated using a spray device to greenhouse tomato seedlings by government extension agents using inoculum from well-characterized stocks to ensure trueness to type. Lecoq (1998) discusses these aspects more fully. In perennial crops, trueness to type is more likely to take the form of a new improved variety, one with the protective strain already present, from which propagation can be made. The benefits of a certification program for nursery mother trees then can be used to maximize trueness to type.

Examples of Field Biological Control

One of the main reasons for including a chapter on cross-protection is that this biological control method has been practiced in the field on several crops with some degree of success. Examples include tomato and tomato mosaic virus (ToMV), cucurbits and zucchini yellow mosaic virus (ZYMV), papaya and papaya ringspot virus (PRV), and citrus and CTV. These will be discussed and other examples are noted for which there are only a few studies (Alrefai & Korben, 1995; Huss et al., 1989; Howell & Mink, 1988; Hughes & Ollenu, 1994; Kosaka & Fukunishi, 1994; Kristic et al., 1995; Rezende & Pacheco, 1998; Singh & Singh, 1995; Wen et al., 1991). There is also a special situation involving the presence of a satellite RNA in the inoculum along with a mild strain of CMV (Gallitelli et al., 1991; Tousignant & Kaper, 1997; Wu et al., 1989). The main effect here seems to be the attenuating effect of the satellite RNA on the predicted disease symptoms of CMV, and as such differs somewhat from cross-protection, but is an interesting use of satellites. It is being used extensively in the field in China (Tien & Wu, 1991) and Japan (Iwaki et al., 1986; Sayama et al., 1993). A similar interaction has been noted with potato spindle tuber viroid (Montasser et al., 1991).

Citrus tristeza Virus

The use of mild strain cross-protection has been practiced on a large scale in sweet orange in Sao Paulo State, Brazil (Costa & Muller, 1980; Muller, 1980; Muller & Costa, 1987; Muller et al., 1988), and in grapefruit in South Africa (VanVuuren et al., 1993). In both cases the major varieties were becoming nonproductive because of the effects of stem-pitting strains of CTV, which is transmitted by aphids in a semipersistent manner. The selection of protective strains in both cases was from stand-out field trees, found to contain mild protective isolates. Following trials and comparisons with candidate strains, a selected strain was then introduced into mother trees known to be free of other viruses, and propagations from these trees

were then used whenever groves were replanted or new groves were established. This required considerable management of nursery industries. Tens of millions of trees have been preimmunized with selected protective strains in both countries and the most severe problems associated with stem pitting are for the most part minimized. Lessons from both countries are that breakdown of cross-protection in some individual preimmunized trees is to be expected, the mild strain selected initially may not function well in all areas, and it may not be the strain of choice in all varieties. The Brazilian situation demonstrates that cross-protection has allowed a country to continue growing its major sweet orange type (the Pera sweet orange, which is unusually sensitive to CTV-induced stem pitting) instead of having to switch to a more tolerant sweet orange variety. Cross-protection against CTV continues to be used on a wide scale in both Brazil and South Africa.

Papaya ringspot Virus

This is another aphid-transmitted virus (nonpersistent transmission) and is being managed in Hawaii and in Taiwan by mild strain cross-protection, though the development of transgenic resistant lines may replace this as an approach (Wang et al., 1987; Yeh et al., 1988; Gonsalvez & Garnsey, 1989; Tennant et al., 1994). In contrast to sweet orange, where the effect of CTV is on fruit yield and general tree quality, PRV causes fruit blemishes that make the fruit unmarketable. Another major difference is the use of mutagenesis and greenhouse selections as the method used to obtain a mild protective strain (Yeh & Gonsalves, 1984). The initial work was done with Hawaii isolates tested in Hawaii, with promising results. When the same mild strain was used in Taiwan on a much larger scale, results were not as good, suggesting that there can be no shortcuts in developing mild strains. Each country will probably have to expend the time and energy to select strains optimized for local use against prevailing severe strains.

Tomato Mosaic Virus

Annual crops would seem to be a good target for cross-protection because the technology does not need to be used all the time, but only when needed. A good example of this is the use of mild protective strains of ToMV in tomatoes grown under glass in northern Europe and elsewhere (Broadbent, 1976; Fletcher, 1978; Oshima, 1975). As with papaya, the effect of the severe strains can be on fruit yield and quality in the form of severe distortion. The mild strain used was selected by mutation followed by biological purification through single local lesions (Rast, 1972). It took several years of selection and testing of candidate strains before the final strain, MII-16, was chosen and approved for field use. The containment of the industry in greenhouses was a significant aid to the use of a mechanical inoculum in a controlled manner (sprayed on seedlings) under government supervision. This technology served the industry well in the 1970s but was eventually replaced by the introduction of new resistant cultivars, which makes another point about cross-protection, which is that it can be a good stopgap measure when traditional approaches can no longer be relied on. It can buy time to address the traditional approach. Lecoq (1998) gives a detailed account of the use of this approach in France.

Zucchini Yellow Mosaic Virus

The ZYMV is a major nonpersistent aphid transmitted viral pathogen of cucurbits that first made its appearance in the early 1980s. It causes severe distortion of fruits of melons, squashes, and cucumbers. No resistant varieties are known, and so cross-protection is a possible approach to disease management of this very visible viral disease. A fortuitous observation of an individual melon plant led to the discovery of a side branch with mild symptoms on a plant with overall severe symptoms. This became the source of a mild strain that has been used to demonstrate good cross-protection in experimental field situations in several countries (Lecoq, 1998; Lecoq *et al.*, 1991; Perring *et al.*, 1995; Spence *et al.*, 1996; Walkey, 1992; Walkey *et al.*, 1992; Wang *et al.*, 1991).

CONCLUSIONS

An important point to reemphasize is that SAR and cross-protection are such different phenomena that they are best treated separately, as was done in this chapter. SAR is rapidly moving out of the realm of biological control and into chemical control as chemicals are found that can mimic the response to attack by pathogens. Both phenomena are also being "challenged" as biological controls by the substitution of transgenic plants for the early steps that up until now have required the application of a biological agent as the protective inoculum.

Cross-protection is not a biological control method to choose without an understanding of the major commitment to research, development, and extension that will be needed before it can be used with effectiveness and on a wide scale. The examples given all represent situations where very severe pathogens were present and became a serious problem and where there were no alternative management practices to fall back on, at least not when action needed to be taken. The need to select or create mild strains, to test these for their relative protective ability, to determine a means to disseminate or apply a true-to-type mild protective strain, and to monitor its effectiveness against a variety of field challenge strains in different sites and at different seasons and years makes cross-protection a long-term commitment. Because the first selected mild protective strain may lose effectiveness, it is important to maintain a program that can select new strains for later use including

special situations (e.g., other varieties besides the major crop). (There are indications that grapefruit, lemons, limes, etc. each may need its own strain of CTV different from the one used for sweet orange, for example.) For some crops, including citrus, it may be a long time before transgenic plants with pathogen-derived resistance are available. For others, such as tomato, cucurbits, and papaya, this may happen sooner. Cross-protection can therefore fill a technology gap while waiting for transgenics to arrive, and can be deployed again should new technologies fail.

References

Alrefai, R. H., & Korban, S. S. (1995). Cross protection against virus disease in fruit trees. Fruit Varieties Journal, 49, 21–30.

Bourdin, D., & Lecoq, H. (1991). Evidence that heteroencapsidation between two potyviruses is involved in aphid transmission of a nontransmissible isolate from mixed infections. Phytopathology, 81, 1459–1464.

Broadbent, L. (1976). Epidemiology and control of tomato mosaic virus. Annual Review of Phytopathy, 14, 75–96.

Cassels A. C., & Herrick, C. C. (1977). Cross protection between severe and mild strains of tobacco mosaic virus in doubly infected tomato plants. Virology, 78, 252–260.

Chakraborty A, & Gupta, P. K. S. (1995). Factors affecting cross protection of Fusarium wilt of pigeon pea by soilborne nonpathogenic fungi. Phytoparasitica, 23, 323–334.

Costa A. S., & Muller, G. W. (1980). Tristeza control by cross protection: A US-Brazil cooperative success. Plant Disease, 64, 538–541.

Culver, J. N. (1996). Tobamovirus cross protection using a potexvirus vector. Virology, 226, 228–235.

Dodds, J. A. (1982). Cross protection and interference between electrophoretically distinct strains of cucumber mosaic virus in tomato. Virology, 118, 235–240.

Dodds, J. A. (1990). Approaches to studying viral cross protection. In New directions in biological control: Alternatives for suppressing agricultural pests and diseases (pp. 213–221). New York: Alan R. Liss.

Dodds, J. A., Lee, S. Q., & Tiffany, M. (1985). Cross protection between strains of cucumber mosaic virus: Effect of host, and type of inoculum on accumulation of virions and double stranded RNA of the challenge virus. Virology, 144, 301–309.

Duran-Vila, N., & Semancik, J. S. (1990). Variations in the cross protection effect between 2 strains of citrus exocortis viroid. Annals of Applied Biology, 117, 367–377.

Fernow, K. H. (1967). Tomato as a test plant for detection of mild strains of potato spindle tuber virus. Phytopathology, 57, 1347–1352.

Fletcher, J. T. (1978). The use of avirulent virus strains to protect plants against the effects of virulent strains. Annals of Applied Biology, 89, 110–114.

Freeman, S., & Rodriguez, R. J. (1993). Genetic conversion of a fungal pathogen to a nonpathogenic endophytic mutualist. Science, 260, 75–78.

Fulton, R. W. (1982). The protective effects of systemic virus infection. In R. K. S. Wood (Ed.), Active defense mechanisms in plants, NATO Advanced Study Institute Series (pp. 231–245). New York: Plenum Press.

Fulton, R. W. (1986). Practices and precautions in the use of cross protection for plant virus disease control. Annual Review of Phytopathology, 24, 67–81.

Gallitelli D., Vovlas, C., Martelli, G., & Montasser, M. S. (1991). Satellite-mediated protection of tomato against cucumber mosaic virus. II. Field test under natural epidemic conditions in southern Italy. Plant Disease, 75, 93–95.

Gerber, M., & Sarkar, S. (1989). The coat protein of tobacco mosaic virus does not play a significant role for cross protection. Journal of Phytopathology, 124, 323–331.

Gonsalves, D., & Garnsey, S. M. (1989). Cross-protection techniques for control of plant virus diseases in the tropics. Plant Disease, 73, 592–596.

Hamilton, R. I. (1980). Defenses triggered by previous invaders: Viruses. In J. G. Horsfall & E. B. Cowling (Eds.), Plant disease (Vol. 5, pp. 279–303). Academic Press. New York:

Hamilton, R. I. (1985). Using plant viruses for disease control. HortScience, 20, 848–852.

Hammond-Kosack, K. E., & Jones, J. D. G. (1997). Plant disease resistance genes. Annual Review of Plant Physiology and Plant Molecular Biology, 48, 575–607.

Howell W. E., & Mink, G. I. (1988). Natural spread of cherry rugose mosaic virus and two prunus necrotic ringspot virus biotypes in a central Washington sweet cherry orchard. Plant Disease, 72, 636–640.

Hughes J. D., & Ollennu, L. A. A. (1994). Mild strain protection of cocoa in Ghana against cocoa swollen shoot virus—a review. Plant Pathology, 43, 442–457.

Hunt, M. D., & Ryals, J. A. (1996). Systemic acquired resistance signal transduction. Critical Reviews in Plant Science, 15, 583–606.

Huss, B., Walter, B., & Fuchs, M. (1989). Cross protection between arabis mosaic virus and grapevine fanleaf virus in Chenopodium quinoa. Annals of Applied Biology, 114, 45–60.

Iwaki, M., Zenbayashi, R., Hanada, K., Shibukawa, S., & Hiroshi, T. (1986). Control of cucumber mosaic virus (CMV) disease of tomato by application of an attenuated strain of CMV. Annals of the Phytopathological Society of Japan, 52, 745–751.

Kessmann, H., Staub, T., Hofmann, C., Maetzke, T., Herzog, J., & Ward, E. (1994a). Induction of systemic acquired resistance in plants by chemicals. Annual Review of Phytopathology, 32, 439–459.

Kessmann, H., Staub, T., Ligon, J., OOstendorp, M., & Ryals, J. (1994b). Activation of systemic acquired disease resistance in plants. European Journal of Plant Pathology, 100, 359–369.

Khoury, J., Singh, R. P., Boucher, A., & Coombs, D. H. (1988). Concentration and distribution of mild and severe strains of potato spindle tuber viroid in cross protected tomato plants. Phytopathology, 78, 1331–1336.

Kosaka, Y., & Fukinushi, T. (1993). Attenuated isolates of soybean mosaic virus derived at low temperature. Plant Disease, 77, 882–886.

Kosaka Y., & Fukunishi, T. (1994). Application of cross-protection to the control of black soybean mosaic disease. Plant Disease, 78, 339–341.

Krstic, B., Ford, R. E., Shukla, D. D., & Tosic, M. (1995). Cross-protection studies between strains of sugarcane mosaic, maize dwarf mosaic, johnsongrass mosaic, and sorghum mosaic potyviruses. Plant Disease, 79, 135–138.

Kurath, G., & Dodds, J. A. (1994). Satellite tobacco mosaic virus sequence variants with only five nucleotide differences can interfere with each other in a cross protection-like phenomenon in plants. Virology, 202, 1065–1069.

Lecoq, H. (1998). Control of plant virus diseases by cross protection. In A. Hadidi, R. K. Khetarpal, & H. Koganezawa (Eds.), Plant virus disease control (pp. 33–40). St. Paul: APS Press.

Lecoq, H., Lemaire, J. M., & Wipfscheibel, C. (1991). Control of zucchini yellow mosaic virus in squash by cross protection. Plant Disease, 75, 208–211.

Louter J. H., & Edgington, L. V. (1990). Indications of cross-protection against Fusarium crown and root rot of tomato. Canadian Journal of Plant Pathology, 12, 283–288.

Mathews, R. E. F. (1991). Plant virology (3rd Ed.). San Diego: Academic Press.

McKinney, H. H. (1929). Mosaic diseases in the Canary Islands, West Africa and Gibralta. Journal of Agricultural Research, 39, 557–578.

Montasser M. S., Kaper, J. M., & Owens, R. A. (1991). Report of potential biological control of potato spindle tuber viroid disease by virus-satellite combination. Plant Disease, 75, 319–319.

Muller, G. W. (1980). Use of mild strains of citrus tristeza virus (CTV) to reestablish commercial production of "Pera" sweet orange in Sao Paulo, Brazil. Proceedings of the Florida State Horticultural Society, 93, 62–64.

Muller, G. W., Costa, A. S., Castro, J. L., & Guirado, N. (1988). Results from preimmunization tests to control the capoa bonito strain of tristeza (pp. 82–85). Proceedings of 10th Conference of International Organization Citrus Virology, Valencia, Spain.

Muller, G. W., & Costa, A. S. (1987). Search for outstanding plants in tristeza infected citrus orchards: The best approach to control the disease by preimmunization. Phytophylactica, 19, 197–198.

Niblett, C. L., Dickson, E., Fernow, K., Horst, R. K. & Zaitlin, M. (1978). Cross protection among four viroids. Virology, 91, 198–203.

Oshima, N. (1975). The control of tomato mosaic virus disease with attenuated virus of tomato strain of TMV. Review of Plant Protection Research, 8, 126–135.

Palukaitis, P., & Zaitlin, M. (1984). A model to explain the "cross protection" phenomenon shown by plant viruses and viroids. In T. Kosuge & E. W. Nester (Eds.), Plant microbe interactions: Molecular and genetic perspectives (pp. 213–222). New York: Macmillan.

Perring, T. M., Farrar, C. A., Blua, M. J., & Wang, H. L. (1995). Cross protection of cantaloupe with a mild strain of zucchini yellow mosaic virus—effectiveness and application. Crop Protection, 14, 601–606.

Ponz, F., & G. Bruening, G. (1986). Mechanisms of resistance to plant viruses. Annual Review of Phytopathology, 24, 355–381.

Rast, A., & Th, B. (1972). MII-16, an artificial symptomless mutant of tobacco mosaic virus for seedling inoculation of tomato crops. Netherlands Journal of Plant Pathology, 78, 110–112.

Rezende, J. A. M., & Sherwood, J. L. (1991). Breakdown of cross protection between strains of tobacco mosaic virus due to susceptibility of dark green areas to superinfection. Phytopathology, 81, 1490–1496.

Rezende, J. A. M., Urban, L., Sherwood, J. L., & Melcher, U. (1992). Host effect on cross protection between 2 strains of tobacco mosaic virus. Journal of Phytopathology, 136, 147–153.

Rezende, J. A. M., & Pacheco, D. A. (1998). Control of papaya ringspot virus-type W in zucchini squash by cross-protection in Brazil. Plant Disease, 82, 171–175.

Roistacher C. N., Dodds, J. A., & Bash, J. A. (1988). Cross protection against citrus tristeza virus seedling yellows and stem pitting viruses by protective isolates developed in greenhouse plants (pp. 91–100). Proceedings of the 10th Conference of International Organization of Citrus Virology, Valencia, Spain.

Roistacher, C. N., & Dodds, J. A. (1993). Failure of 100 'mild' citrus tristeza virus isolates from California to cross protect against a challenge by severe sweet orange stem pitting isolates (pp. 100–106). Proceedings of the 12th Conference of International Organization of Citrus Virology,

Ryals, J., Neuenschwander, U., Willits, M., Molina, A., Steiner, H. Y., et al. (1996). Systemic acquired resistance. Plant Cell, 8, 1899–1919.

Sarkar, S., & Smitamana, P. (1981). A proteinless mutant of tobacco mosaic virus: Evidence against the role of a viral coat protein for interference. Molecular General Genetics, 184, 158–159.

Sayama, H., Sato, T., Kominato, M., & Natsuaki, T. (1993). Field testing of a satellite containing attenuated strain of cucumber mosaic virus for tomato protection in Japan. Phytopathology, 83, 405–410.

Schneider, M., Schweizer, P., Meuwly, P., & Metraux, J. P. (1996). Systemic acquired resistance in plants. International Journal of Cytology, 168, 303–340.

Shalla T. A., & Peterson, L. J. (1978). Studies on the mechanism of viral cross protection. Phytopathology, 68, 1681–1683.

Sherwoood, J. L. (1987a). Mechanisms of cross-protection between plant virus strains. In D. Evered & S. Harnett (Eds.), Plant resistance to viruses (pp. 136–150). Chichester: John Wiley & Sons.

Sherwood, J. L. (1987b). Demonstration of the specific involvement of coat protein in tobacco mosaic virus (TMV) cross protection using a TMV coat protein mutant. Journal of Phytopathology, 118, 358–362.

Sherwood, J. L., & Fulton, R. W. (1982). The specific involvement of coat protein in tobacco mosaic virus cross protection. Virology, 119, 150–158.

Singh, M., & Singh, R. P. (1995). Host dependent cross protection between PVYN, PVYO, and PVA in potato cultivars and Solanum brachycarpum. Canadian Journal of Plant Pathology, 17, 82–86.

Singh R. P., Boucher, A., & Somerville, T. H. (1990). Cross protection with strains of potato spindle tuber viroid in the potato plant and other solanaceous hosts. Phytopathology, 80, 246–250.

Singh R. P., Boucher, A., & Somerville, T. H. (1993). Interactions between a mild and a severe strain of potato spindle tuber viroid in doubly infected potato plants. American Potato Journal, 70, 85–92.

Singh, R. P., Khoury, J., Boucher, A., & Somerville, T. H. (1989). Characteristics of cross-protection with potato spindle tuber viroid strains in tomato plants. Canadian Journal of Plant Pathology, 11, 263–267.

Spence, N. J., Mead, A., Miller, A., & Shaw, E. D. (1996). The effect on yield in courgette and marrow of the mild strain of zucchini yellow mosaic virus used for cross-protection. Annals of Applied Biology, 129, 247–259.

Staskawicz, B. J., Ausubel, F. M., Baker, B., Ellis, J. G., & Jones, J. D. G. (1995). Molecular genetics of plant disease resistance. Science, 268, 661–667.

Sticher, L., Mauch-Mani, B., & Metraux, J. P. (1997). Systemic acquired resistance. Annual Review of Phytopathology, 35, 235–271.

Tennant P. F., Gonsalves, C, Ling, K. S., Fitch, M., Manshardt, R., & Slightom, J. L., (1994). Differential protection against papaya ringspot virus isolates in coat protein gene transgenic papaya and classically cross-protected papaya. Phytopathology, 84, 1359–1366.

Tien, P., & Wu, G. S. (1991). Satellite RNA for the biological control of plant disease. Advances in Virus Research, 39, 321–339.

Tousignant, M. E., & Kaper, J. M. (1997). Control of cucumber mosaic virus using viral satellites. In G. O. Boland & L. D. Kuykendall (Eds.),

Plant microbe interactions and biological control (pp. 283–295). New York: Marcel Dekker.

VanVuuren, S. P., Collins, R. P., & DaGraca, J. V. (1993). Evaluation of citrus tristeza virus isolates for cross protection of grapefruit in South Africa. Plant Disease, 77, 24–28.

Walkey, D. G. A. (1992). Zucchini yellow mosaic virus—control by mild strain protection. Phytoparasitica, 20, 99–103.

Walkey, D. G. A, Lecoq, H., Collier, R., & Dobson, S. (1992). Studies on the control of zucchini yellow mosaic virus in courgettes by mild strain protection. Plant Pathology, 41, 762–771.

Wang, H. L., Yeh, S. D., Chiu, R. J., & Gonsalves, D. (1987). Effectiveness of cross protection by mild mutants of papaya ringspot virus for control of ringspot disease of papaya in Taiwan. Plant Disease, 71, 491–497.

Wang, H. L., Gonsalves, D., Provvidenti, R., & Lecoq, H. L. (1991). Effectiveness of cross protection by a mild strain of zucchini yellow mosaic virus in cucumber, melon, and squash. Plant Disease, 75, 203–207.

Wen, F., Lister, R. M., & Fattouh, F. A. (1991). Cross-protection among strains of barley yellow dwarf virus. Journal of General Virology, 72, 791–799.

Wu, G. S., Kang, L. Y., & Tien, P. (1989). The effect of satellite RNA on cross protection among cucumber mosaic virus strains. Annals of Applied Biology, 114, 489–496.

Yeh, S. D., & Gonsalves, D. (1984). Evaluation of induced mutants of papaya ringspot virus for control by cross protection. Phytopathology, 74, 1086–1091.

Yeh, S. D., Gonsalves, D., Wang, H. L., Namba, R., & Chiu, R. J. (1988). Control of papaya ringspot virus by cross protection. Plant Disease, 72, 375–380.

Zaitlin, M. (1976). Viral cross protection: More understanding is needed. Phytopathology, 66, 382–383.

20

Genetic Mechanisms for Engineering Host Resistance to Plant Viruses

BRET COOPER

Division of Plant Biology
Department of Cell Biology
The Scripps Research Institute
10550 North Torrey Pines Road, BCC 206
La Jolla, California 92037

INTRODUCTION

As lucidly mentioned in Chapter 19, the mechanism behind the phenomenon of viral cross-protection was one of the several pressing interests of plant virologists in the late 1970s and early 1980s. One theory held that the coat protein (CP) of the protecting virus played a role in cross-protection activity, although other mechanisms were not dismissed. The possibility that viral RNA mediated cross-protection could not be definitely excluded. By 1985, researchers were hopeful that their advances in plant genetic transformation techniques would finally enable them to elucidate the role of CP in cross-protection.

By expressing the CP gene from tobacco mosaic tobamovirus (TMV) in transgenic tobacco, Powell-Abel *et al.* (1986) validated that CP could confer cross-protection-like resistance to TMV in transgenic plants. Unfortunately, the mechanism behind classical viral cross-protection activity remained unresolved despite the experiments. However, the discovery of this new form of virus resistance, called CP-mediated protection, brought forth a viable alternative to cross-protection—one that does not have the side effects associated with an infectious cross-protecting virus.

There have been a number of successful attempts at developing CP-mediated protection to other viruses since the inception of the new technology. This science has also stimulated the exploration of other viral genes useful for conferring virus resistance in transgenic plants. What have followed these pioneering studies are academic attempts at creating novel virus protection strategies for transgenic plants and detailed analyses to the mechanisms behind these approaches. Thorough investigations of the durability of the resistance and evaluations for agricultural application, which relate to the mechanism of engineered resistance, have emanated from the idea that a greater understanding of the technologies will allow for greater protection of plants from viruses.

ENGINEERED VIRAL RESISTANCE STRATEGIES

Until the discovery of plant genetic transformation techniques, viral resistance traits could only be bred in cultivars, just as were any of the traits typically exploited for crop improvement. Genetic transformation now allows for the introduction to plants exogenous genes that confer virus resistance. Most of the scope of this review comprises pathogen-derived resistance developed via genetic transformation. Pathogen-derived resistance is aptly named because it describes any strategy in which a gene from a pathogen directly confers resistance to the host in which it is expressed. CP-mediated resistance is one such example. Pathogen-derived resistance might appear to be paradoxical, but the reasons behind the workings of pathogen-derived resistance are quickly becoming more understandable. The theorized conception of expressing genes from pathogens in transgenic organisms is occasionally attributed to Sanford and Johnston (1985) when pathogen-derived resistance is reviewed. Although correct in their prediction that defective or differentially expressed pathogen-derived genes in transgenic plants could adversely affect an incoming virus, pathogen-derived resistance in the form of cross-protection was already a recognized and an applied strategy used for protecting plants (see Chapter 19).

Pathogen-derived resistance can be distinguished from structural resistance mechanisms that include physiological defenses of a plant, such as the cutin that protects an epidermis, or anatomical traits that have been exploited by plant breeders to increase a plant's ability to ward off pests. Also distinguishable is genetic resistance that includes resistance genes (R genes), some of which mediate a hypersensitive defense reaction to pathogens. These genes have been used in the traditional breeding of plants for resistance, although the cloning of these R genes will enable developers to cross the species barrier using genetic transformation as part of an engineered resistance strategy. Other engineered strategies that are not pathogen-derived but will be useful in the control of plant viruses, such as the expression of antiviral inhibitors, typically exploit biochemistry to impede the spread, to block the transmission, or to interfere with the replication or translation of a virus. The plant-derived antiviral inhibitors may be implicated in certain aspects of the broadly conceived nonhost viral resistance, the observation that a wide range of plants cannot be infected by certain viruses probably because the virus is unable to replicate (the definition of immunity) or spread through the host. Although nonhost resistance usually plays an insignificant role in plant breeding, an understanding of the finer aspects of nonhost resistance may make for compelling plant virology in the next century and offer insights to engineering the broadest forms of plant virus resistance.

The nonvirologist needs to be aware of some factors before he peruses this review as reference. The transgenic plants described here express a gene of interest that confers degrees of resistance to plant viruses. In some of the experiments that are described, transgenic plants that do not express or contain the gene are used as controls. Transgenic controls, or vector controls as they are sometimes called, seem to be just as susceptible to virus as nontransgenic controls. Some researchers have even done away with using vector controls and use nontransgenic controls exclusively. Therefore, although some controls are transgenic plants, all controls described here are distinguishable from the transgenic plants expressing genes that confer resistance.

Many times, researchers evaluate resistance by a reduction in visual symptoms. Whereas this is one acceptable measure of disease incidence, and usually accounts for most reports describing "percentages of plants with symptoms," it is not necessarily an accurate report of a plant's absolute resistance to virus. A resistant plant that does not develop symptoms by these visual measures could still accumulate high titers of virus. This distinction can have ramifications, certainly if one is trying to reduce the viral load in the field, which can be an effective means of virus control in itself.

Virologists typically distinguish titers of virus inocula in mass–volume concentrations. Whereas this allows for consistency when one researcher is working with a single virus, concentrations are not sufficient to compare the infectivity between two different viruses. Concentration and infectivity can be independent of each other. One different way to relate the infectivity of viruses is to determine their inocula potentials on local lesion hosts. Not all viruses have common local lesion hosts, however. Unfortunately, many researchers neglect to evaluate viral inocula potentials. Therefore, the concentration data in this chapter is for reference; concentrations may mean very little from one virus, or one study, to another.

CP-Mediated Protection

Since its inception, CP-mediated protection has become the method of choice for protecting transgenic plants against viruses because the technology can be used against a wide range of plant viruses. Scientifically, only the inability to transform a great many plant species impedes the application of this transgenic technology. However, that will quickly be overcome as methods of transformation are refined. The mechanistic action of CP-mediated protection is best understood for the TMV model system. In 1986, Beachy's group described the regeneration of transgenic tobacco plants that expressed the TMV CP gene and accumulated the transgenic protein (Powell-Abel et al., 1986). Their transgenic plants inoculated with 0.25 mg/ml TMV develop symptoms at a slower rate than control plants. Whereas control plants exhibit mosaic and vein clearing symptoms 3 to 4 days postinoculation, anywhere between 50 and 80% fewer TMV CP plants develop symptoms at this time. Some TMV CP plants slowly develop symptoms over several days after control plants are completely infected. However, some transgenic TMV CP plants remain uninfected after challenge. Continued evaluation of the durability of the protection showed that it is not stable at high-growth temperatures (Nejidat & Beachy, 1989).

CP-mediated protection retains similarities to cross-protection. Like cross-protection, TMV CP-mediated protection breaks down under high concentrations of challenge virus (Powell-Abel et al., 1986) and there seems to be a modest degree of broad-spectrum activity to related tobamoviruses (Nejidat & Beachy, 1990). Although the mechanistic action of cross-protection can be linked to either the RNA or the CP of the protecting virus, the activity of TMV CP-mediated protection corresponds to the presence and amounts of accumulated transgenic CP and not the CP mRNA (Powell et al., 1990). Furthermore, TMV CP works alone in mediating protection and not by inducing inherent plant resistance mechanisms such as those triggered by the pathogenesis related (PR) proteins of the plant (Carr et al., 1989).

An early event in the infection cycle of the incoming virus, such as the uncoating of virus that must precede viral replication, seems to be inhibited by transgenic CP because TMV RNA inoculum breaks through the protection (Powell-Abel *et al.*, 1986; Register & Beachy, 1988). This idea is further substantiated by the fact that TMV CP-coated mRNA is translated less efficiently in transgenic CP protoplasts than the naive mRNA is, suggesting once again that the transgenic CP inhibits the disassembly of the virion (Osbourn *et al.*, 1989); [also see Wu *et al.* (1990)]. Nevertheless, there may be some interference in a later step involving replication and during local and systemic spread (Osbourn *et al.*, 1989; Clark *et al.*, 1990; Wisniewski *et al.*, 1990). At least initially, the protection mechanism is favored when either the transgenic CP actually interferes with or blocks an uncoating site or its initial presence shifts kinetics favoring disassembly to those favoring assembly (Register & Nelson, 1992).

TMV CP mutants were tested to gauge the effects on the assembly and disassembly processes with regard to CP-mediated protection. TMV hybrid viruses expressing the CP of the distantly related sunn−hemp mosaic tobamovirus (SHMV) overcame the resistance phenotype of TMV CP plants, once again illustrating the importance of the transgenic CP being able to interfere with a similarly coated virion (Clark *et al.*, 1995b). Furthermore, when mutations that result in the disruption of CP−CP interactions are introduced to the CP genes that are subsequently expressed in transgenic plants, those plants are more susceptible to TMV infection compared with plants that express wild-type CP or CP mutants with strong CP−CP interactions (Bendahmane *et al.*, 1997; Clark *et al.*, 1995a). Thus, the ability of the transgenic CP to interact with the viral CP is requisite for initiating protection.

CP-mediated protection has been successfully employed against the aphid-transmissible and seed-transmissible alfalfa mosaic alfamovirus (AlMV). AlMV differs from TMV in that it has a tripartite genome instead of a monopartite genome, and three of its virions are bacilliform particles and one is spheroidal. Unlike TMV, AlMV requires CP to initiate replication (Bol *et al.*, 1971; Meshi *et al.*, 1987). Amazingly, the differences between the two viruses mean little with respect to CP-mediated protection activity. Transgenic tobacco accumulating the AlMV CP exhibit at least a 3-day delay in symptom appearance on the inoculated leaves, have fewer lesions, and accumulate lower amounts of virus compared with controls (Tumer *et al.*, 1987). Like TMV CP plants, some AlMV CP plants have a delay in symptom development and some completely escape infection after inoculation with AlMV. The fact that replication is not enhanced by the presence of the AlMV CP suggests that, again, the transgenic CP interferes with an early step in viral infection. Also like TMV-CP mediated protection, AlMV CP-mediated protection breaks down if the challenge inoculum concentration is high (Tumer *et al.*, 1987), and AlMV CP plants are not resistant to unrelated, heterologous viruses (Loesch-Fries *et al.*, 1987). Likewise, the levels of AlMV CP accumulation in the plant directly correlate with levels of protection, and CP mRNA does not directly mediate protection (van Dun *et al.*, 1987; van Dun *et al.*, 1988).

AlMV CP-mediated protection differs from TMV-CP mediated protection in that certain lines of AlMV CP plants resist the infection of AlMV RNA (Tumer *et al.*, 1991). Other experiments suggest that AlMV CP-mediated protection breaks down if the challenge inoculum is RNA (Loesch-Fries *et al.*, 1987; van Dun *et al.*, 1987). It is not known if these opposing observations are explained by the CP genes from the strains of AlMV used for transformation, the differential CP accumulation levels in the plants, and the viruses tested; or if the RNA-protecting CP interferes with a later step in infection.

Worth mentioning is the CP-mediated protection obtained with the CP of cucumber mosaic cucumovirus (CMV). CMV is similar to AlMV in its genome organization, although CMV virions retain perfect icosahedral symmetry. Anywhere from 70 to 100% of the transgenic plants accumulating CMV CP escape infection of CMV at concentrations of 5 or 25 μg/ml (Cuozzo *et al.*, 1988). At least in some cases, CMV CP-mediated protection does not break down with high virus concentrations (Cuozzo *et al.*, 1988; Nakajima *et al.*, 1993), which makes it unique among the previously discussed CP-mediated protection mechanisms for TMV or AlMV. Additionally, the extent of CMV CP-mediated protection does not necessarily depend on amount of accumulating CMV CP in the transgenic plants (Namba *et al.*, 1991). These same authors also have shown that CPs derived from some CMV strains offer broader ranges of protection (Namba *et al.*, 1991). Thus, the derivative strain is an important consideration for anyone wanting to develop virus-resistant plants using a CP.

RNA-Mediated Protection

Stark and Beachy (1989) first described plants expressing the CP of soybean mosaic potyvirus (SMV) that are protected against heterologous potyvirus infection. Potyviruses are distinct from tobamoviruses in that their monopartite genomes translate as single polyproteins. Homozygous tobacco plants accumulating high amounts of SMV CP are completely resistant to 10 μg/ml potato virus Y potyvirus (PVY) or 5 μg/ml tobacco etch potyvirus (TEV), but the lower expressing CP heterozygotes only show a 3- to 5-day delay in symptoms. In contrast to the relation between potyvirus resistance and high levels of accumulated CP shown by Stark and Beachy (1989), Lawson *et al.* (1990) showed that potato plants expressing the lowest amounts of PVY CP are the most resistant.

To get closer to understanding the disparity between resistance and CP levels, Lindbo and Dougherty (1992a) constructed a transgene that expresses nontranslatable TEV CP mRNA. In contrast to Powell *et al.* (1990) who showed that the nontranslatable TMV CP gene confers no resistance, Lindbo and Dougherty (1992a; 1992b) effectively demonstrated that tobacco plants expressing the untranslatable TEV CP gene cannot become infected with TEV and that replication of incoming virus is almost completely inhibited. The authors proposed that the transgenic TEV mRNA mediated plant protection to TEV. They defined other observations that characterized this RNA-mediated protection including specificity to the derivative virus, protection independent of inoculum concentration, and reduction of viral replication (Lindbo & Dougherty, 1992a, 1992b). Similar observations bolster the theory (van der Vlugt *et al.*, 1992; Kollar *et al.*, 1993b; Farinelli & Malnoe, 1993).

Lindbo and Dougherty (1992a) concurrently observed that some of the transgenic tobacco plants accumulating TEV CP that are susceptible to TEV infection lose their symptoms several weeks after inoculation. They conclusively found that recovered tissues are virus free, express lower levels of transgene RNA compared with the noninoculated transgenic plants, and cannot be superinfected with TEV even though the tissue is susceptible to other viruses (Lindbo *et al.*, 1993).

Their emergent theory suggests that mechanisms of protection are related for the transgenic plants accumulating TEV CP and the plants expressing the untranslatable TEV CP gene (Lindbo *et al.*, 1993). Once the threshold of a particular RNA species is reached or surpassed due to the presence of TEV RNA, an inherent cellular posttranscriptional control mechanism is activated (Lindbo *et al.*, 1993; Smith *et al.*, 1994). Overly abundant RNAs, mRNAs from the plant, and viral RNAs are subsequently targeted within the cell and eliminated. In other words, plants accumulating TEV CP recover from symptoms and remain at a state of hyperawareness that prevents superinfection once the incoming TEV RNA and homologous transgenic TEV RNA reach a threshold for the activation of the posttranscriptional control (Dougherty *et al.*, 1994). Plants expressing the untranslatable TEV CP genes are possibly resistant to initial infection because they are already induced to a point of hyperawareness to the viral RNA.

The RNA-mediated virus resistance and the underlying inherent cellular mechanism that regulates gene expression posttranscriptionally appear to be similar to the gene-silencing effects of cosuppression, which, among other things, influence the flower color patterns in plants (Jorgensen, 1995). To investigate this relation between RNA-mediated resistance and gene silencing, Ratcliff *et al.*, (1997) inoculated *Nicotiana clevelandii* with tomato black ring nepovirus (TBRV) and noticed that the plants naturally recover from their symptoms and then remain resistant to superinfection by TBRV of the same strain but not others. The authors then inoculated recovered leaves with a potato virus X potexvirus (PVX) viral recombinant harboring an RNA

sequence of TBRV and showed that the recovered leaves are resistant to the recombinant PVX but not to wild-type PVX. Concurrently, Covey *et al.* (1997) offered proof that there is a direct suppression of virus replication due to an increase in RNA degradation in recovered kohlrabi leaves initially inoculated with cauliflower mosaic caulimovirus. Together, their results demonstrate that the RNA of the virus is targeted for degradation in recovered leaves by a cellular gene-silencing mechanism that also serves as a virus resistance mechanism. Therefore, gene silencing is the best explanation for the virus resistance seen in transgenic plants expressing only the mRNAs of plant virus CPs. Gene silencing may also explain why the dark green tissues of a plant infected with a virus that induces mosaic symptoms are resistant to superinfection (Atkinson & Matthews, 1970; Ratcliff *et al.*, 1997), and may offer an alternative explanation for the mode of action for the phenomenon of cross-protection (Lindbo *et. al.*, 1993).

Transgenic plants expressing CP genes of potato leafroll luteovirus (PLRV), tomato spotted wilt tospovirus (TSWV) or PVX resist the infection of the respective viruses (Kawchuk *et al.*, 1991; MacKenzie & Ellis, 1992; Lawson *et al.*, 1990; Hemenway *et al.*, 1988). The resistance from untranslatable genes (Kawchuk *et al.*, 1991; de Haan *et al.*, 1992), a suppression in viral replication (Pang *et al.*, 1993), a lack of correlation between CP accumulation levels and protection (MacKenzie & Ellis, 1992), and a gene silencing from transgenes (Angell & Baulcombe, 1997) suggest that RNA-mediated protection mechanisms are activated in these systems. However, the CP also may offer protection because broad-spectrum resistance has been demonstrated in transgenic plants expressing TSWV and PVX CP genes (Vaira *et al.*, 1995; Prins *et al.*, 1995; Spillane *et al.*, 1997). Likewise, there are several examples in which transgenic plants accumulating potyvirus CP exhibit broad-spectrum resistance activity to heterologous potyviruses (Stark & Beachy, 1989; Ling *et al.*, 1991; Namba *et al.*, 1992; Fang & Grumet, 1993). It is more difficult to explain the broad-spectrum resistance seen in these examples with RNA-mediated protection because that mechanism is usually strain specific, although it is not dismissed. It remains to be determined if CP transgenes from nepoviruses, carlaviruses, or other virus groups confer CP-mediated or RNA-mediated protection or both (Brault *et al.*, 1993; MacKenzie & Tremaine, 1990).

Replicase-Mediated Protection

Viral replicases as functional proteins or mutants thereof have been employed as pathogen-derived resistance agents in transgenic plants. The replicase itself is requisite for viral replication, thus making the viral replication process a primary target of attack if one were trying to devise a scheme to engineer viral resistance for plants. Most of the 5′ end of the TMV genome encodes the 126-kDa component of the replicase and the 183-kDa read-through protein extended from the 126-kDa component (Dawson & Lehto, 1990). The difference between the two genes is the 3' end of the 183-kDa open reading frame, which in itself encodes a potential 54-kDa product that is difficult to detect in infected plants. Expressed in transgenic tobacco, the 54-kDa gene gives levels of resistance that surpass TMV CP-mediated protection (Golemboski *et al.,* 1990). The 54-kDa plants are resistant to TMV at a concentration of 500 μg/ml and are resistant to TMV RNA infection for at least 48 days. Like CP-mediated protection, replication is suppressed by the undetectable 54-kDa protein and not its transcript (Carr & Zaitlin, 1991; Carr *et al.,* 1992). The transgenic plants are also resistant to the closely related TMV mutant YSI/1 but not to the distantly related TMV-U2, TMV L, or unrelated CMV, suggesting that sequence specificity is a requirement for resistance. Similar results have been shown with plants expressing the pea early browning tobravirus putative 54-kDa truncated read-through replicase gene (MacFarlane & Davies, 1992). However, some plants expressing the 54-kDa truncated read-through replicase gene from pepper mild mosaic mottle tobamovirus develop resistance only after recovering from initial susceptibility (Tenllado *et al.,* 1995), an observation previously associated with RNA-mediated protection. This implies that RNA-mediated protection might serve as a secondary form of resistance if the initial resistance from the replicase protein product is not sufficient.

In the mentioned examples, the resistance gene used is only part of the whole replicase gene. Conceivably, the read-through replicase protein acts as a dominant negative mutant that can interfere with the activity of the wild-type replicase produced by the natural infection (Herskowitz, 1987). Theoretically, not only would the read-through products of certain viruses act in a dominant negative mutant manner, but also would the loss-of-function replicase genes act the same. In practice, the modified and truncated replicase from CMV RNA 2 gives transgenic tobacco plants expressing the construct resistance to high levels of CMV or CMV RNA (Anderson *et al.,* 1992). Plants from these experiments remain without symptoms for at least 120 days. The transgenic plants are also resistant to most CMV strains from the same subgroup but not to CMV strains from the alternative subgroup, other cucumoviruses, PVX, TMV, or TEV (Zaitlin *et al.,* 1994). Additionally, resistance does not break down at high temperatures as in transgenic plants accumulating CP (Anderson *et al.,* 1992). The hard-to-detect mutant replicase protein probably inhibits the replication of viral infection in inoculated leaves of the transgenic plants, but the systemic movement may also be inhibited by the defective replicase (Carr *et al.,* 1994; Gal-On *et al.,* 1994; Zaitlin *et al.,* 1994; Hellwald & Palukaitis, 1995). The use of engineered defective replicase genes is

not limited to CMV. Transgenic tobacco plants expressing a copy of a PVX nonfunctional replicase gene are also resistant to high levels of PVX inoculum (Longstaff *et al.,* 1993; Braun & Hemenway, 1992).

There is some consternation whether the functional full-length replicase genes can confer resistance or only loss-of-function products are efficacious. Initial attempts to obtain resistant plants expressing full-length AlMV or brome mosaic bromovirus (BMV) replicase genes failed (Taschner *et al.,* 1991; Mori *et al.,* 1992). In fact, the transgenic plants are able to support the replication of respective AlMV or BMV RNAs dependent on the replicase function. However, there are a few reports where full-length replicase genes were successfully used to obtain resistance. Some lines of plants accumulating only very low levels of PVX replicase are highly resistant to PVX inoculated at concentrations of 5 μg/ml, although others do not reduce virus accumulation at all (Braun & Hemenway, 1992). Likewise, replicase-mediated protection to PVY was achieved using the full-length gene (Audy *et al.,* 1994).

There might be an explanation for the disparity between the abilities of some full-length genes to complement infection and others to confer resistance. Some lines of transgenic tobacco plants transformed with the full-length replicase gene of TMV are not resistant to TMV infection (Donson *et al.,* 1993). However, some highly resistant lines exhibit an unusually broad range of resistance (for replicase-mediated protection) to other viruses including tobacco mild green mosaic tobamovirus, TMV-U5, green tomato atypical tobamovirus, and ribgrass mosaic tobamovirus. These resistant tobacco plants contain an IS10-like transposable element inserted in the replicase gene that causes the premature termination of the replicase and results in a potentially truncated dominant-negative mutant protein. This suggests that inadvertent mutations in the full-length replicase genes in some of the other systems cause the gene to confer resistance instead of complementing viral replication. Nevertheless, loss of function does not explain every situation. Transgenic plants expressing the cymbidium ringspot tombusvirus full-length replicase gene are resistant even though there is no mutation in the transgene to make it act as a dominant-negative mutant (Rubino & Russo, 1995). The authors suggest that an RNA-mediated protection mechanism confers resistance.

Defective Movement Protein-Mediated Protection

Viral encapsidation and replication processes tend to be highly adapted for individual viruses. Thus, the breadth of CP and replicase-mediated protection is limited to viruses related to those from which the transgenes originated. A mechanism that disrupts a function that many unrelated viruses share could confer multivirus resistance. Most viruses encode movement proteins (MPs) that potentiate the cell-to-cell transport of the viruses in infected hosts (Deom *et al.*, 1992). Because some viruses can complement the movement of heterologous viruses that are, by themselves, unable to spread within a plant, MPs of different viruses may have functional compatibility (Atabekov & Taliansky, 1990; Giesman-Cookmeyer *et al.*, 1995; Cooper *et al.*, 1996; Rao *et al.*, 1998). Therefore, it should be possible to inhibit the spread of many viruses by interfering with a function shared by viral MPs. If such a function could be interrupted in transgenic plants, it might result in resistance to one or more viral diseases.

Unlike the expressed CP genes, functional MPs from expressed transgenes of TMV and tobacco rattle tobravirus (TRV) do not protect the plant (Deom *et al.*, 1987; Angenent *et al.*, 1990). In fact, transgenic TMV MP and other transgenic MPs derived from cucumoviruses, alfamoviruses, and dianthoviruses remain functional by modifying plasmodesmata and/or complementing the spread of movement-defective viruses (Deom *et al.*, 1987; Wolf *et al.*, 1989; Kaplan *et al.*, 1995; Vaquero *et al.*, 1994; Poirson *et al.*, 1993; Fujiwara *et al.*, 1993). In at least one documented case, the wild-type TMV MP transgene increases the virulence of challenge viruses instead of conferring resistance (Cooper *et al.*, 1995). This is not always the case, however, because transgenic tobacco plants expressing a functional TMV MP exhibit a delay in the infection of PVX and vice versa (Ares *et al.*, 1998).

As an alternative to the fully functional MPs, defective, dysfunctional, or nonfunctional MPs can be expressed in transgenic plants to act as dominant negative mutants that interfere with or block the sites of activity of functional wild-type MP (Herskowitz, 1987; Deom *et al.*, 1992; von Arnim & Stanley, 1992). Malyshenko *et al.* (1993) regenerated transgenic tobacco plants expressing a gene encoding a temperature-sensitive MP of TMV. Plants held at the high temperature that renders the transgenic MP dysfunctional are protected against TMV infection (3 μg/ml). Protection is observed as a decreased amount of virus accumulation in inoculated and systemic leaves. Virus replication is not inhibited in protoplasts at the elevated temperature. Therefore, the presence of a transgenic MP that is dysfunctional at the high temperature inhibits TMV movement in the inoculated leaf.

Malyshenko *et al.* (1993) also regenerated transgenic tobacco expressing the BMV MP gene. BMV cannot infect tobacco, but it can replicate in tobacco protoplasts (Sakai *et al.*, 1983) meaning that the BMV MP might be nonfunctional in tobacco. Transgenic plants accumulating BMV MP exhibit a reduced accumulation of TMV in inoculated and systemic leaves, but TMV replicates to normal levels in the protoplasts from these plants. Apparently, the BMV MP

interferes with the activity of the TMV MP and prevents the spread of infection.

Lapidot *et al.* (1993) obtained plant protection with an engineered dysfunctional TMV MP. Gafny *et al.* (1992) had previously showed that deleting amino acid codons 3, 4, and 5 from the TMV MP gene made the MP unable to support viral movement. When this mutated gene is expressed by tobacco plants, less dysfunctional MP (dMP) is detected in the cell wall subcellular fractions compared with the amount of wild-type MP in transgenic plants, and the size exclusion limits of the plasmodesmata of the dMP transgenics are only partially modified (Lapidot *et al.*, 1993). The dMP transgenic local lesion hosts exhibit smaller and fewer lesions, and the dMP systemic hosts exhibit a 14-day delay in the appearance of symptoms when inoculated with TMV (0.25 μg/ml); TMV RNA; and two other tobamoviruses, tobacco mild green mosaic and sunn-hemp mosaic virus. Less virus accumulates in the upper leaves of dMP transgenic plants compared with controls. Histochemical analysis and the use of a TMV clone expressing a marker gene clearly demonstrate that neither the numbers of infection sites are reduced nor viral replication is inhibited, but that cell-to-cell movement is blocked or delayed. These dMP transgenic plants eventually become infected, and this implies that the wild-type MP can compete with the dMP for active sites in the cell.

Continued studies of the TMV dMP plants have showed that the transgene confers broad-spectrum protection to taxonomically distinct AlMV, CMV, TRV, tobacco ringspot nepovirus (TRSV), and peanut chlorotic streak caulimovirus (Cooper *et al.*, 1995). The TMV dMP also delays the infection of tomato mosaic, U5, Ob and Cg tobamoviruses, and tomato spotted wilt tospovirus; and the protection to TMV does not break down at elevated growth temperatures (Cooper, *et al.*, unpublished results). Interestingly, the mechanism of resistance to the heterologous viruses in these plants is different from the mechanism of resistance to tobamoviruses whose local spread is restricted. The local spread of the heterologous viruses does not appear adversely affected in the inoculated leaves; there is a significant delay in the appearance in systemic symptoms and virus accumulation in the upper leaves. Again, it appears that given sufficient opportunity the MP from the challenge virus can overtake the protective activity of the transgenic dMP. This, and the fact that transgenic plants expressing wild-type MPs are susceptible to infection, suggest that the dMP confers protection and not the RNA transcript. Although these plants are not as resistant to TMV as TMV CP or replicase plants, dMP protection methods offer the possibility of attaining multivirus protection.

Beck *et al.* (1994) regenerated *Nicotiana benthamiana* expressing a defective 13-kDa gene from the white clover mosaic potexvirus (WClMV) triple-gene block, which has a role in the local spread of potexviruses. These plants are protected against three strains of WClMV, WClMV RNA, PVX, narcissus mosaic potexvirus, and potato virus S carlavirus (PVS). The plants are not resistant to TMV. Protection is observed as a delay in the appearance of symptoms in a population of plants and as a reduction in the accumulation of viral RNA in the inoculated leaves of the plants. Although virus protection is achieved with the defective 13-kDa gene, Beck *et al.* (1994) did not show that the defective protein inhibits viral movement, and the 13-kDa protein was not detected in the plants because of a lack of an antibody.

Other Viral Genes Tested

Other virus-derived genes have been tested for their abilities to confer transgenic plants with virus protection. Maiti *et al.* (1993) showed that a potyvirus proteinase gene was useful for making plants less susceptible to infection. In this particular example, the NIa gene from the tobacco vein mottling potyvirus (TVMV) was expressed in transgenic tobacco. NIa is the multifunctional protein with domains for the virus genome-linked protein (VPg) and the proteinase. Transgenic plants accumulating NIa are protected against TVMV, but not to the taxonomically related potyviruses TEV or PVY. Although the NIa plants do not exhibit the broader protection, they are resistant to high concentrations of TVMV, as high as 50 μg/ml, whereas the TVMV CP protection breaks down at TVMV concentrations of 10 μg/ml. NIa expressed *in planta* might interfere with the virus-encoded VPg functions, or it may alter processing of the viral polyprotein. However, plants expressing related constructs exhibit recovery, suggesting that other components of a potyvirus genome besides the CP can confer RNA-mediated protection (Swaney *et al.*, 1995). Plants expressing the TVMV cylindrical inclusion (CI) protein gene are not resistant to TVMV (Maiti *et al.*, 1993).

Antisense RNA

Another form of pathogen-derived resistance is antisense RNA protection. Antisense applications are finding their way into many arenas of science in and outside plant pathology. The simple principle behind the theory is that the antisense RNA will hybridize to the sense RNA message of importance and inhibit sense strand translation, transport, or stability within the cell. Medical researchers are exploring antisense technology to inhibit human immunodeficiency virus (HIV) replication or to regulate gene expression in cancer cells. Plant breeders have used antisense technology to create tomatoes that ripen slowly.

For virus resistance, one could imagine designing antisense constructs for any of the number of essential viral genes including CP, MP, replicase, or proteinase. In theory,

inhibition of the gene or open reading frame that encodes an essential function would stop the procession of infection. Antisense protection has not showed as much promise as CP-mediated protection or replicase-mediated protection; thus, antisense modes of action are typically not explored. Often the failures might be pinpointed to the targeted gene. It might be a poor idea to design antisense constructs to RNAs expressed from subgenomic promoters because these RNAs are expressed at high levels, higher than the antisense RNA expressed from the constitutive promoter in the plant. Antisense will probably work best when trying to regulate lower levels of RNA expression. With the use of antisense technology there have been limited successes in protecting plants against viruses, mainly because antisense gene suppression is usually overcome.

One of the first reports of antisense protection for plant viruses came from Hemenway *et al.* (1988). Tobacco plants expressing an antisense RNA of the PVX CP are delayed in the accumulation of PVX (70 to 80% reduction) in inoculated leaves at inoculum concentrations of 0.05 μg/ml but not 0.5 μg/ml. The virus reduction is not seen in the systemic leaves because all the antisense plants develop symptoms at the normal rate. In contrast to plants accumulating PVX CP (sense plants), the antisense plants are much less resistant. The protection in the PVX CP plants may be a result of an RNA-mediated protection mechanism. Because the antisense plants are not as resistant as CP plants, an RNA-mediated protection mechanism can possibly be disregarded as one of the mechanisms of action for the antisense plants. Perhaps the low levels of antisense RNA expressed by the plant are sufficient to inhibit early stages of CP translation via hybridization, whereas the inhibitory effects of hybridization are overcome as larger amounts of viral RNA replicate.

Powell *et al.* (1989) investigated similar antisense protection mechanisms with TMV antisense RNA to the CP or to the CP plus the 3′ end of TMV. Tobacco plants accumulating antisense transcripts to the CP plus the 3′ end are protected from infection by low levels of TMV (0.01 to 0.05 μg/ml), but the plants accumulating the antisense CP sequence alone are not. Protection is manifested as an escape from infection and not a delay in symptoms. On the other hand, CP plus the 3′ end antisense plants exhibit only a delay in symptoms when inoculated with 4 μg/ml TMV RNA. These antisense plants are not as resistant as the TMV CP plants, but differ in that they can delay infections caused by viral RNA. The authors have postulated that the transgenic 3′ end inhibits replicase synthesis of the viral negative strand, but that this inhibition is overcome once increased levels of inoculum out-compete the inhibiting transgenic antisense RNA.

Resistance in *N. benthamiana* to high concentrations of bean yellow mosaic potyvirus (BYMV; 100 μg/ml) has been achieved by using a similar antisense construct to the BYMV 3' end (Hammond & Kamo, 1995). Negative strand synthesis might also be inhibited by the antisense RNA, but these transgenic plants exhibit recovery phenotypes, which implicates a prevailing gene-silencing-type mechanism. Nelson *et al.* (1993) achieved protection by expressing in transgenic tobacco the antisense RNA to 51 nucleotide sequences at the 5' end of TMV. Their results suggest that it is less important for the antisense sequence to bind to a certain gene than it is for it to bind to a region of RNA that influences the replication or expression of that particular gene.

Worthy of mention is the high level of protection seen in tobacco plants transformed with an antisense gene of the CP from TEV (Lindbo & Dougherty, 1992a). Plants with the antisense gene resist TEV infection at 5 μg/ml but are not as resistant as plants expressing the nontranslatable CP gene in the sense direction. Even better protection was obtained using the antisense to a TSWV MP (Prins *et al.*, 1997). RNA-mediated protection mechanisms may be activated in these cases, but other resistance mechanisms cannot be discounted.

Defective Interfering RNA/Satellite RNA

The vein of pathogen-derived resistance has continued with the expression of defective interfering (DI) RNAs and/or satellite (Sat) RNAs in transgenic plants. DI RNAs, first identified with tomato bushy stunt tombusvirus (TBSV), are by-products of the recombination of aborted viral replicons and are dependent on the progenitor virus for replication and encapsidation (Hillman *et al.*, 1987). This particular TBSV DI RNA, when associated in an infection with TBSV, attenuates symptoms probably by depressing viral replication of the progenitor virus by out-competing for essential replication materials. It follows that any DI RNA that attenuates the symptoms of a virus could be expressed in plants as a means of viral disease protection. Stanley *et al.* (1990) expressed a DI DNA in *N. benthamiana* that was able to delay and reduce symptoms of African cassava mosaic geminivirus. Similarly, DI RNA of cymbidium ringspot tombusvirus prevents stem necrosis, but not milder disease symptoms, in transgenic *N. benthamiana* challenged with RNA transcripts of cymbidium ringspot virus (Kollar *et al.*, 1993a). At the time of its discovery, DI RNA-mediated protection was touted over CP-mediated protection because there would be no undesirable or consumable protein in the DI RNA transgenic plants compared with CP plants. Moreover, DI RNA plants resist infection by viral RNA, one of the downfalls of CP-mediated protection. Obviously the disadvantages of DI RNA-mediated protection had to be reconciled. Natural DI RNAs are uncommon, so protecting DIs would have to be engineered, thus creating an obstacle for virologists. There could be more serious consequences if a transgenic DI RNA mutated and resulted in debilitating disease symptoms on infection (Li *et al.*, 1989).

Some viruses have Sat RNAs that ameliorate symptoms. Sat RNAs are considered parasites of viruses. They are self-replicating molecules that encode no known proteins and have no similarity to their associated viruses but do depend on the virus for encapsidation and spread. The expression of the TBSV Sat RNA or the expression of the CMV Sat RNA in transgenic plants results in a marked reduction in symptoms and replication when the plants are challenged with the respective viruses (Gerlach *et al.*, 1987; Harrison *et al.*, 1987). Again touted as a viable alternative to CP-mediated protection, Sat RNA-mediated protection has lost popularity since few viruses have associated Sat RNAs that can be used for protection and since Sleat and Palukaitis (1992) showed that Sat RNAs that have increased virulence can differ from benign Sat RNAs by a single nucleotide.

R Genes

Natural disease *R* genes give plants the innate ability to repel the attack of pathogens. The *R* genes, when activated by pathogens, initiate a cascade of plant defense–signal–induction reactions resulting in a hypersensitive response to the pathogen (Hammond-Kosack & Jones, 1996). The reactions include a rapid oxidative burst at the site of infection, a toxic accumulation of oxidative intermediates, and a tissue necrosis. The pathogen ceases to spread and the plant develops an induced state of resistance due to systemic acquired resistance (SAR) (Dangl *et al.*, 1996; see also Chapter 19). Plant breeders and geneticists traditionally and successfully have used natural *R* genes to ward off agricultural diseases. However, the sexual compatibility of plants and the specificity of resistance to a particular pathogen have limited the successfulness of any particular disease *R* gene. Plant genetic transformation and recombinant DNA techniques can allow for the crossing of species barriers and broaden the usefulness of disease *R* genes from heterologous plants. The advantage to using natural *R* genes over pathogen-derived genes in an engineered resistance strategy is that plant-derived genes are not exotic and will stimulate much less public anxiety with their usage. The disadvantage is that these *R* genes might also require interactions with other genes that are species specific to function properly. An *R* gene that functions in one plant species might not function in another in the absence of those secondary genes needed to trigger the essential signal pathways for a resistance response.

Of the handful of plant disease *R* genes that have been cloned, only one confers virus resistance (Whitham *et al.*, 1994; Bent, 1996). The *N* gene, originally from *Nicotiana glutinosa*, confers hypersensitive resistance to most tobamoviruses (Holmes, 1938). After cloning the *N* gene in 1994, Whitham *et al.* (1996) were then able to show that *N* is also capable of conferring hypersensitive resistance to TMV in transgenic tomato. Soon, other genes will be cloned and used to confer resistance in heterologous plants. Detailed genetic maps exist or are being created for *Rx,* which gives resistance to PVX in potato (Baulcombe *et al.,* 1994); *TTR1,* a gene from *Arabidopsis thaliana* that confers tolerance to TRSV (Lee *et al.,* 1996); *L2,* which gives resistance to most tobamoviruses in pepper (de la Cruz *et al.,* 1997); and *Sw-5,* a gene from tomato that gives resistance to tospoviruses (Brommonschenkel & Tanksley, 1997):

Pathogenesis-Related Proteins

Along with *R* genes that are typically specific to a certain virus or single class of viruses, plants contain genes that have a broader spectrum of activity to many unrelated viruses and even other types of pathogens. Most easily recognized are genes encoding the pathogenesis-related (PR) proteins that accumulate in the presence of disease infection, regardless of the infectious agent (Rigden & Coutts, 1988). These proteins can be chitinases, heat-shock proteins, proteinase inhibitors, cellulases, and salicylic acid-inducible proteins that are part of an overall disease signaling system (Yang & Klessig, 1996). Noticeably, the direct functions of the PR proteins, aside from possibly inducing SAR or inhibiting proteinases, would have little effect on viral infections. Cloned *Pr-1, GRP,* and *PR-S* genes constitutively expressed in transgenic plants give the tobacco no additional protection to TMV or AIMV (Linthorst *et al.,* 1989). *Pr-1b* constitutively expressed in tobacco gives the transgenic plants no advantage over TMV infection (Cutt *et al.,* 1989). Possibly, PR protein genes will never confer as much resistance as genes that are pathogen derived. Notwithstanding, the understanding of their roles in horizontal plant resistance will offer much insight to creating or improving broad-spectrum resistance strategies to a variety of pathogens.

Antiviral Agents

The other approaches to combating viral diseases in transgenic plants typically involve the use of antiviral agents. These compounds resemble those being used to treat human ailments such as the reverse transcriptase inhibitors or proteinase inhibitors effective against HIV, or monoclonal antibodies that can specifically target cancer cells. Researchers have constructed ribozymes, RNA molecules with enzymatic activity, to specifically cleave other RNA molecules (Haseloff & Gerlach, 1988). Lamb and Hay (1990) adapted this idea for plant virus protection. They created ribozymes that cleave consensus sequences in PLRV RNA. These ribozymes and others like them could be used to confer protection to viruses when expressed in transgenic plants.

Another agent, melittin, distinguished as a peptide for its ability to reduce HIV-1 production *in vitro* (Wachinger *et al.,* 1992) can reduce TMV infection by 90% (Marcos *et al.,* 1995). The compound binds to TMV RNA, thus possibly antagonizing CP/RNA interactions. It is unknown if melittin-like agents would be effective in conferring resistance in transgenic plants, but it is certainly possible. It is also plausible that other antiviral agents can be developed to inhibit viral infection.

Naturally occurring, or biological, antiviral agents have been tested for their efficacy against plant viruses. One particular agent is $2'-5'$ oligoadenylate synthetase which, when induced by interferon in mammals, polymerizes ATP chains via $2'-5'$ phosphodiester bonds (Samuel, 1991). These $2'-5'$ adenosine triphosphate (ATP) chains activate RNase L, which degrades viral RNAs. When $2'-5'$ oligoadenylate synthetase is expressed in transgenic potato plants, the plants exhibit a 7-day delay in symptom development after challenge with PVX (Truve *et al.,* 1993). Similar plants are also resistant to PVS and tolerant to PVY (Truve *et al.,* 1994).

The pokeweed antiviral protein (PAP), a naturally occurring antiviral agent, has been investigated for efficacy as well. Depending on the host and the amount of accumulating PAP, tobacco and potato transgenic for PAP have degrees of broad-spectrum protection to PVX, PVY, and CMV (Lodge *et al.,* 1993). Because of its similarity to ribosome-inhibiting proteins from other plant species that also have antiviral activity, PAP possibly acts by preventing translation of viral RNA (Stirpe *et al.,* 1992).

Antibodies expressed in plants may be useful for the control of plant viruses. Tavladoraki *et al.* (1993) developed antibodies with specific activity to artichoke mottled crinkle tombusvirus (AMCV), isolated the genes that encode the variable domains of the heavy and light chains for the particular immunoglobulin, created a single-chain antibody gene by combining the two former genes, and expressed this construct in transgenic *N. benthamiana.* Plants accumulating the antibody exhibit a 5- to 14-day delay in symptoms compared with controls when challenged with 4 μg/ml AMCV, a 1000-fold higher concentration than the minimum necessary to induce a systemic infection. There is a slight reduction of virus in the inoculated leaves but there is an eightfold reduction of virus in the upper leaves. The transgenic plants are not resistant to CMV. The mechanism behind the protection is unclear. The antibody may interfere with the uncoating or reassembly of the virus because it is CP specific.

TRANSGENIC PLANT PERFORMANCE

Transgenic plants with any of the previously mentioned genetic mechanisms of resistance must be durable enough to resist natural infections in the field and must have traits that make them economically viable crops. These characteristics have to be carefully selected after virus-resistant transgenic plants are created. The ability of a resistance mechanism to confer resistance to viruses transmitted by vectors is paramount. Oftentimes, in laboratories (sometimes in field trials) transgenic plants are challenged by hand (or air brush). This type of inoculation is not common in the wild. Natural virus vectors can constantly and persistently transmit viruses and put plants under indefinite amounts of challenge pressure. Thus, the one leaf, one virus, one concentration, and single inoculation procedure that dominates current laboratory challenge experiments may not be an accurate measure of the resilience of transgenic plants that are protected against viruses transmitted by vectors. Transgenic plants not only must have the abilities to resist viruses, but also must retain agronomic traits that engender them desirable commercial cultivars. Sometimes the accumulation of transgenic protein or its biochemical activity within the plant cells adversely affects the plants. Therefore, the overall qualities of the plants for agricultural production must be tested under natural growth conditions in the field.

Virus Transmission

Invertebrate vectors transmit most of the plant viruses that cause severe diseases. Reasonably, these virus problems can be tackled by eliminating or controlling the vectors. Environmental regulations, such as the impending worldwide ban on methyl bromide, and the cost of insecticides, among others outlined in previous chapters, limit a grower's ability to directly control the vector. Despite environmental concerns for topical pesticides, genetic transformation technologies may greatly assist in controlling viral diseases transmitted by insect vectors. For example, plants can be transformed to directly control vectors. As discussed in Chapter 21, *Cry* genes from *Bacillus thuringiensis* (B.t.) are useful for controlling Coleopteran species and, consequently, the comoviruses and sobemoviruses they can transmit. Alternatively, insects could be transformed to be unable to transmit viruses. Dengue fever antisense genes expressed in mosquitoes result in reduced transmission of the virus (Olson *et al.,* 1996). Perhaps this technology can be used to control the transmission of plant viruses by aphids.

For engineered resistance to be useful for transgenic plants, plants must be able to resist vector inoculation of viruses. A few laboratory studies document the protection of transgenic plants to viruses transmitted by vectors. Transgenic potato accumulating both the PVX and PVY CP are protected against each individual virus and against the mixed infection when challenged by hand (Lawson *et al.,* 1990). One particular line from their study is protected against PVY when inoculated with viruliferous green peach

aphids; other lines protected against mechanical inoculation do not resist virus infection by aphid inoculation. Their report provides two important pieces of information. First, an RNA-mediated mechanism of protection, a likely explanation for the potyvirus resistance, is sufficient to protect plants from the aphid transmission of viruses. Second, plants have to be screened for the best levels of protection against aphid transmission of viruses, despite the known levels of protection to virus infection from mechanical inoculation.

Replicase-mediated protection mechanisms and CP-mediated protection mechanisms are also sufficient to confer protection from a virus transmitted by aphids. Plants expressing a truncated RNA 2 replicase gene from CMV-Fny (subgroup I) are resistant to CMV subgroup I strains transmitted by aphids but not to subgroup II strains (Zaitlin *et al.,* 1994). Significantly fewer transgenic tobacco plants accumulating the CP of the nonaphid-transmissible CMV-C (subgroup I) become infected when inoculated by aphid with CMV-WL (subgroup II) or CMV-Chi (subgroup I; Quemada *et al.,* 1991).

Resistance to virus transmission by other vectors has also been tested in transgenic plants. de Haan *et al.* (1992) have demonstrated that RNA-mediated protection is sufficient to give tobacco high levels of protection to TSWV transmitted by thrips, and Reavy *et al.* (1995) found that transgenic *N. benthamiana* accumulating the CP of potato mop-top furovirus are protected against mechanical and fungal inoculation. Rice plants accumulating CP of rice stripe tenuivirus exhibit a 75% reduction in disease symptoms when challenged with viruliferous plant hoppers (Hayakawa *et al.,* 1992).

As of yet, transgenic tobacco accumulating TRV CPs that are resistant to mechanical inoculation are not resistant to inoculation by viruliferous trichodorid nematodes (van Dun & Bol, 1988; Ploeg *et al.,* 1993). This may be a result of reduced expression of CP in roots by the constitutive promoter. Root-specific promoters may be necessary to enhance expression levels. Nematodes may also inject destabilized particles that can overcome CP-mediated protection, but this has yet to be determined.

Field Trials

The step from laboratory tests to field trials can be a great leap for some scientists who have no desire to work an acre of land. The chasm diminishes once there is sufficient economic push to test engineered resistance for commercially viable crops. Field trials are true measuring sticks for efficacy and are required before any commercial claims can be made for the levels of resistance. Theoretically, transgenic plants with moderate levels of resistance that delay replication or virus spread could be sufficient for agricultural use. A delay in infection may be sufficient to allow a crop to bear quality fruit, or may be all that is necessary to break an infection cycle. However, proven field immunity most likely will be the selling point for the transgenic plants, especially if perennial crops are ever to be protected by engineered resistance.

Hundreds of field trials of virus-resistant transgenic plants have been performed over the years. It should be noted, however, that the results from many field trials are not published for a variety of commercial or scientific reasons. A few are discussed here to show the efficacy of the engineered resistance mechanism under agricultural conditions. In general, CP-mediated (RNA-mediated in some cases?) and replicase-mediated mechanisms (Kaniewski *et al.*, 1995a) can confer field immunity to transgenic plants, provided the cultivars and lines with the best traits are selected (Kaniewski & Thomas, 1997).

One successful field trial assayed the resistance of transgenic potatoes expressing a PVY-N CP gene (Malnoe *et al.*, 1994). Transgenic plants were planted next to infected plants in an aphid-infested field. By week eight, all nontransgenic bait plants were infected, but none of the PVY CP plants were infected with PVY-N 15 weeks after planting and only 23% were infected with PVY-O. Of the clones propagated from tubers of transgenic plants 28% were infected with PVY-O 6 weeks after their planting. These tests show not only that the transgenic potatoes are resistant to initial infection but also that the resistance reduces the ability of the virus to be transmitted by vegetative propagation.

Transgenic cucumbers accumulating the CMV-C CP were evaluated for resistance under natural field conditions (Gonsalves *et al.*, 1992). Only 8% (average) of the transgenic plants planted for a whole growing season, in the presence of known CMV sources and aphids, had CMV symptoms compared with 86% of the susceptible controls and 3% of a cultivar bred for CMV resistance. Enzyme-linked immunosorbent assays (ELISAs) revealed that a greater number of the transgenic- and nontransgenic-resistant plants accumulated viruses, indicating that there is only a suppression of symptoms in some plants. The authors report a slight advantage in total fruit yield from the transgenic plants, but this is not a conclusive report for yield advantages for CP plants because symptomatic nontransgenic plants can yield large amounts of diseased and undesirable fruit.

Transgenic potatoes expressing dual CPs of PLRV strains were tested for their ability to resist PLRV transmitted by aphids in the field (Thomas *et al.*, 1997). There were significant delays of infection and symptom development in some transgenic plants. However, at the end of the two 12-week experiments, there was only a 17% reduction in the number of plants infected in one experiment and no reduction in the other. Although resistance to primary infection was unremarkable, the infected transgenic potatoes were capable of reducing secondary spread and transmission to adjacent plants. The authors remind transgenic plant developers that initial screens for virus resistance may exclude plants with phenotypes that are useful for controlling secondary infections and transmission, which is another legitimate way of controlling viral diseases in the field.

Effects of Engineered Resistance Genes on Plants

The effects of transgene expression in plants can be unnoticeable, negligible, insignificant, or deleterious. Whenever a foreign gene is expressed in a plant, there is potential for a transgenic protein–plant protein interaction that consequently results in a new plant phenotype. For virus-derived transgenes, the transgenic viral protein should interact with the same plant proteins as the analogous protein produced by the infecting virus. Surprisingly, the expression of single virus-derived genes in plants does not cause the plants to have the corresponding diseaselike symptoms. Transgenic plants expressing viral CPs can look like nontransgenic plants that are not infected. However, some transgenic plants have phenotypes that differ from their nontransgenic parents. Although phenotype differences may be a result of the transformation event in some cases, phenotype differences also are a result of transgene expression. Potato plants expressing the CP gene of PVX have smaller leaves, stunted inflorescence, changes in light sprout and tuber shape, and lower yields (Jongedijk *et al.*, 1992; Malnoe *et al.*, 1994). Transgenic tobacco plants accumulating high levels of PAP are severely stunted in growth and their leaves are mottled (Lodge *et al.*, 1993). Plants expressing the PVX MP are stunted and their leaves are chlorotic (Ares *et al.*, 1998). The size-exclusion limits of plasmodesmata are modified in transgenic plants accumulating TMV MP or dMP (Wolf *et al.*, 1989; Lapidot *et al.*, 1993). In addition, transgenic plants accumulating TMV MP have reduced carbon partitioning to their roots (Lucas *et al.*, 1993). Because it is usually impossible to predict how the transgene or the recombination event will alter the phenotype of the transgenic plant, careful selection of the transformants and crossing to other plant lines is necessary to ensure that the plants retain other useful agronomic traits (Jongedijk *et al.*, 1992; Thomas & Kaniewski, 1997).

REGULATION AND RISK ASSESSMENT

Currently, transgenic plants and their products are regulated in the United States under the authority of the U.S. Department of Agriculture (USDA), the Food and Drug Administration (FDA), and the Environmental Protection

Agency (EPA). The FDA is responsible for the safety of commercial foods, except for meat and poultry, and has the ability to regulate any food additives and nonpesticide food substances. The USDA regulates meat and poultry products while an important subdivision, the Animal and Plant Health Inspection Service (APHIS), can control the release and spread of agricultural diseases and pests. The FDA has jurisdiction over any pesticide and can create tolerances for pesticide residues in food. The scope of these three agencies is sufficient to control all aspects of commerce for any plant genetically transformed to resist viral infection. Therefore, it is important to consider the regulatory framework that controls the release of transgenic plants because it can dictate which particular engineered resistance mechanisms can be commercially developed.

In 1992, the FDA issued a policy statement concerning clarification of its jurisdiction over fruits, vegetables, grains, or plant by-products generated from genetically transformed plants. The FDA maintains that genetically engineered foods will be treated no differently than traditionally produced foods regulated under the Federal Food, Drug, and Cosmetic Act (FFDCA). The FDA believes that food regulations should be based on food characteristics and food components and not on the method by which the foods were created (FDA, 1992). Under FDA guidelines, transgenically produced or induced substances that do not exceed amounts currently consumed in nontransgenic foods would be unlikely to need FDA premarket review. Arguably, because plant viruses have always been present in food sources, virus-resistant transgenic plants expressing virus-derived transgenes should not be subjected to FDA review as long as the expressed viral proteins do not exceed amounts of viral proteins currently found in nontransgenic foods. Virus-resistant plants with nonpathogen-derived engineered resistance mechanisms may face governmental scrutiny, especially if the mechanisms significantly alter the chemical composition of the plant.

Shortly after the FDA proposed its regulations for transgenic foods, the USDA released its proposal to regulate the commercial release of transgenic plants (USDA, 1992, 1993). The USDA claims that the jurisdiction given by the Federal Plant Pest Act grants APHIS the ability to regulate plants transformed with the DNA from any regulated plant pest if the plants can then directly or indirectly cause disease in other plants. APHIS is concerned with the introduction of genes from disease-causing agents to plants with the ability to outcross to other crops or weedy relatives. Outcrossing is a legitimate concern that will have to be addressed as transgenic plants are released. Crop-to-weed gene flow has been proved to exist between johnsongrass and sorghum (Arriola & Ellstrand, 1996) and transgenes for herbicide resistance in oilseed rape can outcross to a weedy relative (Mikkelsen *et al.*, 1996). Consequently, the only exemptions from USDA regulation are corn, cotton, potato,

soybean, tobacco, tomato, or any other plant that has negligible ability to outcross; contains stable, integrated, and characterized genetic material that does not produce an infectious entity; is not toxic to nontarget organisms; does not encode products for pharmaceutical use; and does not contain sequences from animal or human pathogens. Also exempt from regulation are those types of transgenic plants encoding plant virus elements that are noncoding regulatory regions, sense or antisense CPs, or any other plant virus antisense constructs derived from viruses that are endemic to the agricultural areas where the transgenic plants will be planted.

The EPA has assumed the most convoluted and controversial authority over transgenic plants that resist the infection of diseases and the damage of pests. The EPA has jurisdiction over pesticides and their risks to the environment under the Federal Insecticide, Fungicide, and Rodenticide Act (FIFRA), and over pesticide residues with respect to their consumption under FFDCA and the Food Quality Protection Act (FQPA; EPA, 1994; 1997). The EPA is concerned with the production of pesticides by transgenic plants and their regulatory status under the law. In proposing policy to regulate these plants and to promulgate their jurisdiction, the EPA defined pesticides produced in transgenic plants as plant-pesticides. In accordance with the FIFRA, plant-pesticides must be demonstrated not to have unreasonably adverse effects on the environment. These plant-pesticides will be exempt from regulation if they are derived from closely related plants, if they act primarily by affecting the plant, or if they are CPs from plant viruses. In accordance with the FFDCA as amended by the FQPA, the use of plant-pesticides must not result in increased dietary exposures. Plant-pesticides will be exempt from a requirement of a tolerance if the plant-pesticide genes are derived from plants that are sexually compatible to the transgenics, if the plant-pesticides are derived from antisense genes, or if the plant-pesticides are plant virus CPs or segments thereof. The EPA believes that most plant-pesticides, given the breadth of their policy, will be exempt from regulation. Although an across-the-board exemption was proposed, exemptions are currently made on a case-by-case basis. According to the EPA, only the plant viral CPs of watermelon mosaic potyvirus and zucchini yellow mosaic potyvirus, papaya ringspot potyvirus, CMV, and PLRV are exempt from the requirement of a tolerance at this time.

The new EPA classification and definition of plant-pesticide has created scientific disdain. A consortium of scientists and representatives from 11 scientific societies including the American Phytopathological Society, the American Society of Agronomy, and the Institute of Food Technologists have prepared a report protesting the plant-pesticide classification and the proposed EPA regulations (Beachy *et al.*, 1996). Among other points, the group believes that genetic mechanisms for disease and pest resistance are not the equivalent of chemical pesticides, and should not be classified as such. Technically, all natural plants, by the EPA definition, could be construed as plant-pesticides because they have inherent genetic mechanisms to resist pests. The group believes that transgenic plants are sufficiently regulated by the FDA and APHIS and that any regulatory oversight should focus on plant traits and not the traits created by a particular method. Most of all, the group is concerned with the economic impact of the proposed EPA rules because the designation of a plant as a pesticide betrays scientific authority stating that the transgenic plants are safe for consumption. The group also believes that the rules discourage small companies from developing publicly unacceptable or regulation-burdened transgenic plants. Small companies are already thwarted by the EPA policy and are hesitant to develop plant disease resistance technology (Sanford, 1997).

There has been scientific concern about the environmental release of transgenic plants expressing pathogen-derived genes. Some of the risks that may or may not be associated with CP-mediated protection, for example, are the potential for new viruses to be formed through transencapsidation or recombination, and the potential for increased seed transmission. The potential for viral nucleic acids to become transencapsidated has been demonstrated experimentally (Candelier-Harvey & Hull, 1993). However, because transencapsidation occurs naturally without the aid of transgenic plants (Creamer & Falk, 1990), the risk associated with transgenic plants is low. As for the potential for genomic recombination, there are no examples that demonstrate a significant risk. One particular study shows that a disabled virus can rescue its regulatory genomic end from a transgenic plant that expresses that end (Greene & Allison, 1994). However, this recombination event occurs under high-selection pressures. It is difficult to imagine such pressures occurring in nature that do not already produce recombination events. As for seed transmission, it is not a well-characterized process. One study attributes seed transmission to the maternal genes of a plant (Wang & Maule, 1994). While the viral CP is certainly an important part in any type of viral transmission, there is no reason to believe that it alone could increase the risk of virus seed transmission in transgenic plants greater than the level of risk that already occurs naturally. In fact, all these risks of increased viral transmission might be lowered if plants resist certain viral infections.

The EPA has suggested that exemptions should be extended to include all plant viral proteins and the viral genetic sequences that encode them that are used in virus resistance strategies. There should be no such broad exemption. Transgenic wild-type functional MPs cause synergistic infections in the presence of other unrelated viruses (Cooper et al., 1995). Replicase genes can function in plants and support viral infection (Taschner et al., 1991). Real dangers

in increased disease severity could occur in wild or weedy relatives if an MP or replicase functioned in those plants because of outcrossing. MPs and replicases that are proved to be nonfunctional will eliminate the potential for synergistic viral infections in other hosts. Actually, any wild-type viral gene expressed in a transgenic plant is or can produce an active biological molecule that plays a role in disease infection. Viral genes, including DI-RNAs or Sat RNA's, should be carefully evaluated as protective agents before they are practically used in agriculture. At this time, a complete knowledge of the biological activity of any one viral protein is not within our grasp. There should be no haphazard release of molecularly manipulated active biological agents without a sound reasoning or a scientific commitment to their understanding.

Some risks do exist with the use of pathogen-derived resistance mechanisms. These risks do not preclude using the technology, however. Engineered resistance genes will increase the selection pressure on viruses to overcome protection. What happens if we are overwhelmed with resistance-breaking viruses? Should there be government regulation to prevent this from occurring? These issues are being considered, but most likely will have to be dealt with once they occur. In the meantime, growers need to understand that they should only deploy pathogen-derived resistance as part of an overall integrated pest management strategy. Engineered resistance will not be a cure for ever-persistent plant diseases.

PROSPECTS

The discovery of CP-mediated resistance remains important for its far-reaching impact on the plant sciences. The technology was one of the firsts to verify that inheritable plant traits can be manipulated at a molecular level for the improvement of agricultural production. It advanced concepts in designing alternative strategies for transgenic disease resistance and essentially brought forth a new age of plant pathology. Today there is a variety of methods of choice for protecting plants against viruses. Yet, no one method is perfect for all applications. Replicase genes tend to give the greatest amount of protection, but can they pass the muster of government regulation? Can CP genes give enough protection if they are used instead? Is RNA-mediated protection the method of choice because plants can recover if they become infected? On the other hand, is broad-spectrum activity of dMP-mediated protection preferable if only modest resistance is needed? These are just some of the questions that will have to be addressed as research for these protection mechanisms is continued, funded, or brought to market.

Antiviral mechanisms of the future should confer immunity to multiple pathogens. Field immunity has been achieved for PLRV in potatoes expressing PLRV replicase genes and for CMV in tomatoes expressing CMV CP genes (Kaniewski *et al.,* 1995a, 1995b). One way to improve broad-spectrum resistance is to express multiple CPs in plants (Lawson *et al.,* 1990; Prins *et al.,* 1995; Marcos & Beachy, 1997), use defective MP-mediated resistance strategies (Beck *et al.,* 1994; Cooper *et al.,* 1995), or apply alternative antiviral strategies (Lodge *et al.,* 1993; Truve *et al.,* 1994). Despite these achievements, no single engineered resistance mechanism has been developed that confers immunity to a broad spectrum of heterologous viruses. A combination of several engineered resistance mechanisms might be necessary to achieve this goal. More than likely, genes encoding plant-derived antiviral inhibitors of mutual functions of viral replication or cell-to-cell spread will be used to confer broad-spectrum immunity or to relegate infections to a subliminal nature.

Acknowledgments

I thank Drs. Christoph Reichel and Paloma Ma's for their valuable input and criticism. I am grateful to Dr. Brian Federici for financial support during the writing of this chapter and to Dr. Roger Beachy whose laboratory was where the work took place.

References

Anderson, J. M., Palukaitis, P., & Zaitlin, M. (1992). A defective replicase gene induces resistance to cucumber mosaic virus in transgenic tobacco plants. Proceedings of the National Academy of Sciences, USA, 89, 8759–8763.

Angell, S. M., & Baulcombe, D. C. (1997). Consistent gene silencing in transgenic plants expressing a replicating potato virus X RNA. EMBO Journal, 16, 3675–3684.

Angenent, G. C., van den Ouweland, J. M. W., & Bol, J. F. (1990). Susceptibility to virus infection of transgenic tobacco plants expressing structural and nonstructural genes of tobacco rattle virus. Virology, 175, 191–198.

Ares, X., Calamante, G., Cabral, S., Lodge, J., Hemenway, P., Beachy, R. N. (1998). Transgenic plants expressing potato virus X ORF2 protein (p24) are resistant to tobacco mosaic virus and Ob tobamoviruses. Journal of Virology, 72, 731–738.

Arriola, P. E., & Ellstrand, N. C. (1996). Crop-to-weed gene flow in the genus *Sorghum* (Poaceae): Spontaneous interspecific hybridization between johnsongrass, *Sorghum halepense,* and crop sorghum, *S. bicolor.* American Journal of Botany, 83, 1153–1160.

Atabekov, J. G., & Taliansky, M. E. (1990). Expression of a plant virus-coded transport function by different viral genomes. Advances in Virus Research, 38, 201–248.

Atkinson, P. H., & Matthews, R. E. F. (1970). On the origin of dark green tissue in tobacco leaves infected with the tobacco mosaic virus. Virology, 40, 344–356.

Audy, P., Palukaitis, P., Slack, S. A., & Zaitlin, M. (1994). Replicase-mediated resistance to potato virus Y in transgenic tobacco plants. Molecular Plant-Microbe Interactions, 7, 15–22.

Baulcombe, D., Gilbert, J., Goulden, M., Kohm, B., & Cruz, S. S. (1994). Molecular biology of resistance to potato virus X in potato. Biochemical Society Symposia, 60, 207–218.

Beachy, R. N., Cantrell, R. P., Chitwood, D. J., Cook, R. J., Cregan, P.

B., & Day, P. R. Gantt, D. G. Gilchrist, G. G. Kennedy, T. J. Ng, C.O. Qualset, J. S. Thenell, S. A. Tolin & A. K. Vidaver. (1996). Appropriate oversight for plants with inherited traits for resistance to pests. Institute of Food Technologists. Chicago, IL.

Beck, D. L., van Dolleweerd, C. J., Lough, T. J., Balmori, E., Voot, D. M., Andersen, M. T., (1994). Disruption of virus movement confers broad-spectrum resistance against systemic infection by plant viruses with a triple gene block. Proceedings of the National Academy of Sciences, USA, 91, 10310–10314.

Bendahmane, M., Fitchen, J. H., Zhang, G., & Beachy, R. N. (1997). Studies of coat protein-mediated resistance to tobacco mosaic tobamovirus: Correlation between assembly of mutant coat proteins and resistance. Journal of Virology, 71, 7942–7950.

Bent, A. F. (1996). Plant disease resistance genes: Function meets structure. Plant Cell, 8, 1757–1771.

Bol, J. F., van Vloten-Doting, L., & Jaspars, E.M.S. (1971). A functional equivalence of top component aRNA and coat protein in the initiation of infection by alfalfa mosaic virus. Virology, 46, 73–85.

Brault, V., Candresse, T., le Gall, O., Delbos, R. P., Lanneau, M., & Dunez, J. (1993). Genetically engineered resistance against grapevine chrome mosaic nepovirus. Plant Molecular Biology, 21, 89–97.

Braun, C. J., & Hemenway, C. L. (1992). Expression of amino-terminal portions or full-length viral replicase genes in transgenic plants confers resistance to potato virus X infection. Plant Cell, 4, 735–744.

Brommonschenkel, S. H., & Tanksley, S. D. (1997). Map-based cloning of the tomato genomic region that spans the *Sw-5* tospovirus resistance gene in tomato. Molecular & General Genetics, 256, 121–126.

Candelier-Harvey, P., & Hull, R. (1993). Cucumber mosaic virus genome is encapsidated in alfalfa mosaic virus coat protein expressed in transgenic tobacco plants. Transgenic Research, 2, 277–285.

Carr, J. P., & Zaitlin, M. (1991). Resistance in transgenic tobacco plants expressing a nonstructural gene sequence of tobacco mosaic virus is a consequence of markedly reduced virus replication. Molecular Plant-Microbe Interactions, 4, 579–585.

Carr, J. P., Gal-On, A., Palukaitis, P., & Zaitlin, M. (1994). Replicase-mediated resistance to cucumber mosaic virus in transgenic plants involves suppression of both virus replication in the inoculated leaves and long-distance movement. Virology, 199, 439–447.

Carr, J. P., Marsh, L. E., Lomonossoff, G. P., Sekiya, M. E., & Zaitlin, M. (1992). Resistance to tobacco mosaic virus induced by the 54-kDa gene sequence requires expression of the 54-kDa protein. Molecular Plant-Microbe Interactions, 5, 397–404.

Carr, J. P., Beachy, R. N., & Klessig, D. F. (1989). Are the PR1 proteins of tobacco involved in genetically engineered resistance to TMV? Virology, 169, 470–473.

Clark, G. C., Fitchen, J., Nejidat, A., Deom, C. M., & Beachy, R. N. (1995b). Studies of coat protein-mediated resistance to tobacco mosaic virus (TMV). II. Challenge by a mutant with altered virion surface does not overcome resistance conferred by TMV coat protein. Journal of General Virology, 76, 2613–2617.

Clark, G. C., Fitchen, J. H., & Beachy, R. N. (1995a). Studies of coat protein-mediated resistance to TMV. I. The PM2 assembly defective mutant confers resistance to TMV. Virology, 208, 485–491.

Clark, W. G., Register, J. C., Nejidat, A., Eichholtz, D. A., Sanders, P. R., & Fraley, R. T. (1990). Tissue-specific expression of the TMV coat protein in transgenic tobacco plants affects the level of coat protein-mediated virus protection. Virology, 179, 640–647.

Cooper, B., Schmitz, I., Rao, A. L. N., Beachy, R. N., & Dodds, J. A. (1996). Cell-to-cell transport of movement-defective cucumber mosaic and tobacco mosaic viruses in transgenic plants expressing heterologous movement protein genes. Virology, 216, 208–213.

Cooper, B., Lapidot, M., Heick, J. A., Dodds, J. A., & Beachy, R. N. (1995). A defective movement protein of TMV in transgenic plants confers resistance to multiple viruses whereas the functional analog increases susceptibility. Virology, 206, 307–313.

Covey, S. N., Al-Kaff, N. S., Langara, A., & Turner, D. S. (1997). Plants combat infection by gene silencing. Nature (London), 385, 781–782.

Creamer, R., & Falk, B. (1990). Direct detection of transcapsidated barley yellow dwarf luteoviruses in doubly infected plants. Journal of General Virology, 71, 211–217.

Cuozzo, M., O'Connell, K. M., Kaniewski, W., Fang, R.-X., Chua, N.-H., & Tumer, N. E. (1988). Viral protection in transgenic tobacco plants expressing the cucumber mosaic virus coat protein or its antisense RNA. Bio/Technology, 6, 549–557.

Cutt, J. R., Harpster, M. H., Dixon, D. C., Carr, J. P., Dunsmuir, P., & Klessig, D. F. (1989). Disease response to tobacco mosaic virus in transgenic tobacco plants that constitutively express the pathogenesis-related PR1b gene. Virology, 173, 89–97.

Dangl, J. L., Dietrich, R. A., & Richberg, M. H. (1996). Death don't have no mercy: Cell death programs in plant-microbe interactions. Plant Cell, 8, 1793–1807.

Dawson, W. O., & Lehto, K. M. (1990). Regulation of tobamovirus gene expression. Advances in Virus Research, 38, 307–342.

de Haan, P., Gielen, J. J., Prins, M., Wijkamp, I. G., van Schepen, A., & Peters, D. (1992). Characterization of RNA-mediated resistance to tomato spotted wilt virus in transgenic tobacco plants. Bio/Technology, 10, 1133–1137.

de la Cruz, A., Lopez, L., Tenllado, F., Diaz-Ruiz, J. R., Sanz, A. I., & Vaquero, C. (1997). The coat protein is required for the elicitation of the Capsicum L2 gene-mediated resistance against the tobamoviruses. Molecular Plant-Microbe Interactions, 10, 107–113.

Deom, C. M., Lapidot, M., & Beachy, R. N. (1992). Plant virus movement proteins. Cell, 69, 221–224.

Deom, C. M., Oliver, M. J., & Beachy, R. N. (1987). The 30-kilodalton gene product of tobacco mosaic virus potentiates virus movement. Science, 237, 389–394.

Donson, J., Kearney, C. M., Turpen, T. H., Khan, I. A., Kurath, G., & Turpen, A. M. (1993). Broad resistance to tobamoviruses is mediated by a modified tobacco mosaic virus replicase transgene. Molecular Plant-Microbe Interactions, 6, 635–642.

Dougherty, W. G., Lindbo, J. A., Smith, H. A., Parks, T. D., Swaney, S., & Proebsting, W. M. (1994). RNA-mediated virus resistance in transgenic plants: Exploitation of a cellular pathway possibly involved in RNA degradation. Molecular Plant-Microbe Interactions, 7, 544–552.

EPA. (1994). Proposed policy; plant-pesticides subject to the Federal Insecticide, Fungicide, and Rodenticide Act and the Federal Food, Drug, and Cosmetic Act. Federal Register, 59, 60496–60518.

EPA. (1997). Plant-pesticides; supplemental notice of proposed rulemaking. Federal Register, 62, 27132–27155.

Fang, G., & Grumet, R. (1993). Genetic engineering of potyvirus resistance using constructs derived from the zucchini yellow mosaic virus coat protein gene. Molecular Plant-Microbe Interactions, 6, 358–367.

Farinelli, L., & Malnoe, P. (1993). Coat protein gene-mediated resistance to potato virus Y in tobacco: Examination of the resistance mechanisms—is the transgenic coat protein required for protection? Molecular Plant-Microbe Interactions, 6, 284–292.

FDA. (1992). Statement of policy: Foods derived from new plant varieties. Federal Register 57, 22984–23005.

Fujiwara, T., Giesman-Cookmeyer, D., Ding, B., Lommel, S. A., & Lucas, W. J. (1993). Cell-to-cell trafficking of macromolecules through plasmodesmata potentiated by the red clover necrotic mosaic virus movement protein. Plant Cell, 5, 1783–1794.

Gafny, R., Lapidot, M., Berna, A., Holt, C. A., Deom, C. M., & Beachy, R. N. (1992). Effects of terminal deletion mutations on function of the movement protein of tobacco mosaic virus. Virology, 187, 499–507.

Gal-On, A., Kaplan, I., Roossink, M. J., & Palukaitis, P. (1994). The kinetics of infection of zucchini squash by cucumber mosaic virus indicate a function for RNA 1 in virus movement. Virology, 205, 280–289.

Gerlach, W. L., Llewllyn, D., & Haseloff, J. (1987). Construction of a plant disease resistance gene from the satellite RNA of tobacco ringspot virus. Nature (London), 328, 802–805.

Giesman-Cookmeyer, D., Silver, S., Vaehongs, A. A., Lommel, S. A., & Deom, C. M. (1995). Tobamovirus and dianthovirus movement proteins are functionally homologous. Virology, 213, 38–45.

Golemboski, D. B., Lomonossoff, G. P., & Zaitlin, M. (1990). Plants transformed with a tobacco mosaic virus nonstructural gene sequence are resistant to the virus. Proceedings of the National Academy of Sciences, USA, 87, 6311–6315.

Gonsalves, D., Chee, P., Provvidenti, R., Seem, R., & Slightom, J. L. (1992). Comparison of coat protein-mediated and genetically-derived resistance in cucumbers to infection by cucumber mosaic virus under field conditions with natural challenge inoculation by vectors. Bio/Technology, 10, 1562–1570.

Greene, A. E., & Allison, R. F. (1994). Recombination between viral RNA and transgenic plant transcripts. Science, 263, 1423–1425.

Hammond, J., & Kamo, K. K. (1995). Effective resistance to potyvirus infection conferred by expression of antisense RNA in transgenic plants. Molecular Plant-Microbe Interactions, 8, 674–682.

Hammond-Kosack, K. E., & Jones, J. D. G. (1996). Resistance gene-dependent plant defense responses. Plant Cell, 8, 1773–1791.

Harrison, B. D., Mayo, M. A., & Baulcombe, D. C. (1987). Virus resistance in transgenic plants that express cucumber mosaic virus satellite RNA. Nature (London), 328, 799–802.

Haseloff, J., & Gerlach, W. L. (1988). Simple RNA enzymes with new and highly specific endoribonuclease activities. Nature (London), 334, 585–591.

Hayakawa, T., Zhu, Y., Itoh, K., Kimura, Y., Izawa, T., & Shimamoto, K. (1992). Genetically engineered rice resistant to rice stripe virus, an insect-transmitted virus. Proceedings of the National Academy of Sciences, USA, 89, 9865–9869.

Hellwald, K. H., & Palukaitis, P. (1995). Viral RNA as a potential target for two independent mechanisms of replicase-mediated resistance against cucumber mosaic virus. Cell, 83, 937–946.

Hemenway, C., Fang, R.-X., Kaniewski, W. K., Chua, N.-H., & Tumer, N. E. (1988). Analysis of the mechanism of protection in transgenic plants expressing the potato virus X coat protein or its antisense RNA. EMBO Journal, 7, 1273–1280.

Herskowitz, I. (1987). Functional inactivation of genes by dominant negative mutations. Nature (London), 329, 219–222.

Hillman, B. I., Carrington, J. C., & Morris, T. J. (1987). A defective interfering RNA that contains a mosaic of a plant virus genome. Cell, 51, 427–433.

Holmes, F. O. (1938). Inheritance of resistance to tobacco-mosaic disease in tobacco. Phytopathology, 28, 553–561.

Jongedijk, E., de Schutter, A. A., Stolte, T., van den Elzen, P. J., & Cornelissen, B. J. (1992). Increased resistance to potato virus X and preservation of cultivar properties in transgenic potato under field conditions. Bio/Technology, 10, 422–429.

Jorgensen, R. A. (1995). Cosuppression, flower color patterns, and metastable gene expression states. Science, 268, 686–691.

Kaniewski, W., Lawson, C., Loveless, J., Thomas, P., Mowry, T., & Reed, G. (1995a). Expression of potato leafroll virus (PLRV) replicase genes in Russet Burbank potatoes provide field immunity to PLRV. In M. Manka (Ed.), Environmental biotic factors in integrated plant disease control (pp. 289–292). Poznan: The Polish Phytopathological Society.

Kaniewski, W., Mitsky, T., & Loveless, J. (1995b). Highly resistant transgenic tomatoes expressing cucumber mosaic virus (CMV) coat protein (CP) genes. In M. Manka (Ed), Environmental biotic factors in integrated plant disease control (pp. 293–296). Poznan: The Polish Phytopathological Society.

Kaniewski, W. K., & Thomas, P. E. (1997). Field testing resistance of transgenic plants. In G. D. Foster & S. C. Taylor (Eds.), Methods in molecular biology: Vol. 81. Plant virology protocols: From virus isolation to transgenic resistance (pp. 497–508). Totowa, NJ: Humana Press.

Kaplan, I. B., Shintaku, M. H., Li, Q., Zhang, L., Marsh, L. E., & Palukaitis, P. (1995). Complementation of virus movement in transgenic tobacco expressing the cucumber mosaic virus 3a gene. Virology, 209, 188–199.

Kawchuk, L. M., Martin, R. R., & McPherson, J. (1991). Sense and antisense RNA-mediated resistance to potato leafroll virus in Russet Burbank potato plants. Molecular Plant-Microbe Interactions, 4, 247–253.

Kollar, A., Dalmay, T., & Burgyan, J. (1993a). Defective-interfering RNA-mediated resistance against cymbidium ringspo tombusvirus in transgenic plants. Virology, 193, 313–318.

Kollar, A., Thole, V., Dalmay, T., Salamon, P., & Balazs, E. (1993b). Efficient pathogen-derived resistance induced by integrated potato virus Y coat protein gene in tobacco. Biochimie, 75, 623–629.

Lamb, J. W., & Hay, R. T. (1990). Ribozymes that cleave potato leafroll virus RNA within the coat protein and polymerase genes. Journal of General Virology, 71, 2257–2264.

Lapidot, M., Gafny, R., Ding, B., Wolf, S., Lucas, W. J., & Beachy, R. N. (1993). A dysfunctional movement protein of tobacco mosaic virus that partially modifies the plasmodesmata and limits virus spread in transgenic plants. Plant Journal, 4, 959–970.

Lawson, C., Kaniewski, W., Haley, L., Rozman, R., Newell, C., & Sanders, P. & (1990). Engineering resistance to mixed virus infection in a commercial potato cultivar: Resistance to potato virus X and potato virus Y in transgenic Russet Burbank. Bio/Technology, 8, 127–134.

Lee, J. M., Hartman, G. L., Domier, L. L., & Bent, A. F. (1996). Identification and map location of TTR1, a single locus in Arabidopsis thaliana that confers tolerance to tobacco ringspot nepovirus. Molecular Plant-Microbe Interactions, 9, 729–735.

Li, X. H., Heaton, L. A., Morris, J, & Simon, A. E. (1989). Turnip crinkle defective interfering RNAs intensify viral symptoms and are generated de novo. Proceedings of the National Academy of Sciences, USA, 86, 9173–9177.

Lindbo, J. A., & Dougherty, W. G. (1992a). Pathogen-derived resistance to a potyvirus: Immune and resistant phenotypes in transgenic tobacco expressing altered forms of a potyvirus coat protein nucleotide sequence. Molecular Plant-Microbe Interactions, 5, 144–153.

Lindbo, J. A., & Dougherty, W. G. (1992b). Untranslatable transcripts of the tobacco etch virus coat protein gene sequence can interfere with tobacco etch virus replication in transgenic plants and protoplasts. Virology, 189, 725–733.

Lindbo, J. A., Silva-Rosales, L., Proebsting, W. M., & Dougherty, W. G. (1993). Induction of a highly specific antiviral state in transgenic plants: Implications for regulation of gene expression and virus resistance. Plant Cell, 5, 1749–1759.

Ling, K., Namba, S., Gonsalves, C., Slightom, J. L., & Gonsalves, D. (1991). Protection against detrimental effects of potyvirus infection in transgenic tobacco plants expressing the papaya ringspot virus coat protein gene. Bio/Technology, 9, 752–758.

Linthorst, H. J., Meuwissen, R. L., Kauffmann, S., & Bol, J. F. (1989). Constitutive expression of pathogenesis-related proteins PR-1, GRP, and PR-S in tobacco has no effect on virus infection. Plant Cell, 1, 285–291.

Lodge, J. K., Kaniewski, W. K., & Tumer, N. E. (1993). Broad-spectrum virus resistance in transgenic plants expressing pokeweed antiviral protein. Proceedings of the National Academy of Sciences, USA, 90, 7089–7093.

Loesch-Fries, L. S., Merlo, D., Zinnen, T., Burhop, L., Hill, K., & Krahn, K. (1987). Expression of alfalfa mosaic virus RNA 4 in transgenic plants confers virus resistance. EMBO Journal, 6, 1845–1851.

Longstaff, M., Brigneti, G., Boccard, F., Chapman, S., & Baulcombe, D.

(1993). Extreme resistance to potato virus X infection in plants expressing a modified component of the putative viral replicase. EMBO Journal, 12, 379–386.

Lucas, W. J., Olesinski, A., Hull, R. J., Haudenshield, J. S., Deom, C. M., & Beachy, R. N. (1993). Influence of the tobacco mosaic virus 30-kDa movement protein on carbon metabolism and photosynthate partitioning in transgenic tobacco plants. Planta, 190, 88–96.

MacFarlane, S. A., & Davies, J. W. (1992). Plants transformed with a region of the 201-kilodalton replicase gene from pea early browning virus RNA1 are resistant to virus infection. Proceedings of the National Academy of Sciences, USA, 89, 5829–5833.

MacKenzie, D. J., & Tremaine, J. H. (1990). Transgenic Nicotiana debneyii expressing viral coat protein are resistant to potato virus S infection. Journal of General Virology, 71, 2167–2170.

MacKenzie, D. J., & Ellis, P. J. (1992). Resistance to tomato spotted wilt virus infection in transgenic tobacco expressing the viral nucleocapsid gene. Molecular Plant-Microbe Interactions, 5, 34–40.

Maiti, I. B., Murphy, J. F., Shaw, J. G., & Hunt, A. G. (1993). Plants that express a potyvirus proteinase gene are resistant to virus infection. Proceedings of the National Academy of Sciences, USA, 90, 6110–6114.

Malnoe, P., Farinelli, L., Collet, G. F., & Reust, W. (1994). Small-scale field tests with transgenic potato, cv. Bintje, to test resistance to primary and secondary infections with potato virus Y. Plant Molecular Biology, 25, 963–975.

Malyshenko, S. I., Kondakova, O. A., Nazarova, Ju. V., Kaplan, I. B., Taliansky, M. E., & Atabekov, J. G. (1993). Reduction of tobacco mosaic virus accumulation in transgenic plants producing non-functional viral transport proteins: Journal of General Virology, 74, 1149–1156.

Marcos, J. F., & Beachy, R. N. (1997). Transgenic accumulation of two plant virus coat proteins on single self-processing polypeptide. Journal of General Virology, 78, 1771–1778.

Marcos, J. F., Beachy, R. N., Houghten, R. A., Blondelle, S. E., & Perez-Paya, E. (1995). Inhibition of a plant virus infection by analogs of melittin. Proceedings of the National Academy of Sciences, USA, 92, 12466–12469.

Meshi, T., Watanabe, Y., Saito, T., Sugimoto, A., Maeda, T., & Okada, Y. (1987). Function of the 30 kDa protein of tobacco mosaic virus: Involvement in cell-to-cell movement and dispensability for replication. EMBO Journal, 6, 2557–2563.

Mikkelsen, T. R., Andersen, B., & Jorgensen, R. B. (1996). The risk of crop transgene spread. Nature (London), 380, 31.

Mori, M., Mise, K., Okuno, T., & Furusawa, I. (1992). Expression of brome mosaic virus-encoded replicase genes in transgenic tobacco plants. Journal of General Virology, 73, 169–172.

Nakajima, M., Hayakawa, T., Nakamura, I., & Suzuki, M. (1993). Protection against cucumber mosaic virus (CMV) strains O and Y and chrysanthemum mild mottle virus in transgenic tobacco plants expressing CMV-O coat protein. Journal of General Virology, 74, 319–322.

Namba, S., Ling, K., Gonsalves, C., Gonsalves, D., & Slightom, J. L. (1991). Expression of the gene encoding the coat protein of cucumber mosaic virus (CMV) strain WL appears to provide protection to tobacco plants against infection by several different CMV strains. Gene, 107, 181–188.

Namba, S., Ling, K., Gonsalves, C., Slightom, J. L., & Gonsalves, D. (1992). Protection of transgenic plants expressing the coat protein gene of watermelon mosaic virus II or zucchini yellow mosaic virus against six potyviruses. Phytopathology, 82, 940–946.

Nejidat, A., & Beachy, R. N., (1989). Decreased levels of TMV coat protein in transgenic tobacco plants at elevated temperatures reduce resistance to TMV infection. Virology, 173, 531–538.

Nejidat, A., & Beachy, R. N. (1990). Transgenic tobacco plants expressing a coat protein gene of tobacco mosaic virus are resistant to some other tobamoviruses. Molecular Plant-Microbe Interactions, 3, 247–251.

Nelson, A., Roth, D. A., & Johnson, J. D. (1993). Tobacco mosaic virus infection of transgenic *Nicotiana tabacum* plants is inhibited by antisense constructs directed at the 5′ region of viral RNA. Gene, 127, 227–232.

Olson, K. E., Higgs, S., Gaines, P. J., Powers, A. M., Davis, B. S., & Kamrud, K. I., (1996). Genetically engineered resistance to Dengue-2 virus transmission in mosquitoes. Science, 272, 884–886.

Osbourn, J. K., Watts, J. W., Beachy, R. N., & Wilson, T. M. (1989). Evidence that nucleocapsid disassembly and a later step in virus replication are inhibited in transgenic tobacco protoplasts expressing TMV coat protein. Virology, 172, 370–373.

Pang S. Z., Slightom, J. L., & Gonsalves, D., (1993). Different mechanisms protect transgenic tobacco against tomato spotted wilt and impatiens necrotic spot tospoviruses. Bio/Technology, 11, 819–824.

Ploeg, A. T., Mathis, A., Bol, J. F., Brown, D. J., & Robinson, D. J. (1993). Susceptibility of transgenic tobacco plants expressing tobacco rattle virus coat protein to nematode-transmitted and mechanically inoculated tobacco rattle virus. Journal of General Virology, 74, 2709–2715.

Poirson, A., Turner, A. P., Giovane, C., Berna, A., Roberts, K., & Godefroy-Colburn, T., (1993). Effect of the alfalfa mosaic virus movement protein expressed in transgenic plants on the permeability of plasmodesmata. Journal of General Virology, 74, 2459–2461.

Powell, P. A., Stark, D. M., Sanders, P. R., & Beachy, R. N. (1989). Protection against tobacco mosaic virus in transgenic plants that express tobacco mosaic virus antisense RNA. Proceedings of the National Academy of Sciences, USA, 86, 6949–6952.

Powell, P. A., Sanders, P. R., Tumer, N., Fraley, R. T., & Beachy, R. N. (1990). Protection against tobacco mosaic virus infection in transgenic plants requires accumulation of coat protein rather that coat protein RNA sequences. Virology, 175, 124–130.

Powell-Abel, P., Nelson, R. S., De, B., Hoffmann, N., Rogers, S. G., & Fraley, R. T., (1986). Delay of disease development in transgenic plants that express the tobacco mosaic virus coat protein gene. Science, 232, 738–743.

Prins, M., Kikkert, M., Ismayadi, C., de Graauw, W., de Haan, P., & Goldbach, R. (1997). Characterization of RNA-mediated resistance to tomato spotted wilt virus in transgenic tobacco plants expressing NS(M) gene sequences. Plant Molecular Biology, 33, 234–243.

Prins, M., de Haan, P., Luyten, R., van Veller, M., van Grinsfen, M. Q., & Goldbach, R. (1995). Broad resistance to tospoviruses in transgenic tobacco plants expressing three tospoviral nucleoprotein gene sequences. Molecular Plant-Microbe Interactions, 8, 85–91.

Quemada, H. D., Gonsalves, D., & Slightom, J. L. (1991). Expression of coat protein gene from cucumber mosaic virus strain C in tobacco: Protection against infections by CMV strains transmitted mechanically or by aphids. Phytopathology, 81, 794–802.

Rao, A. L. N., Cooper, B., & Deom, C. M. (1998). Defective movement of viruses in the *Bromoviridae* is differentially complemented in *Nicotiana benthamiana* expressing tobamovirus or dianthovirus movement proteins. Phytopathology. 88, 666–672.

Ratcliff, F., Harrison, B. D., & Baulcombe, D. C. (1997). A similarity between viral defense and gene silencing in plants. Science, 276, 1558–1560.

Reavy, B., Arif, M., Kashiwazaki, S., Webster, K. D., & Barker, H. (1995). Immunity to potato mop-top virus in *Nicotiana benthamiana* plants expressing the coat protein gene is effective against fungal inoculation of the virus. Molecular Plant-Microbe Interactions, 8, 286–291.

Register, J. C., & Beachy, R. N. (1988). Resistance to TMV in transgenic plants results from interference with an early event in infection. Virology, 166, 524–32.

Register, J. C., & Nelson, R. S. (1992). Early events in plant virus infection: Relationships with genetically engineered protection and host gene resistance. Seminars in Virology 3, 441–451.

Rigden, J., & Coutts, R. (1988). Pathogenesis-related proteins in plants. Trends in Genetics, 4, 87–89.

Rubino, L., & Russo, M. (1995). Characterization of resistance to cymbidium ringspot virus in transgenic plants expressing a full-length viral replicase gene. Virology, 212, 240–243.

Sakai, F., Dawson, J. R. O., & Watts, J. W. (1983). Interference in infections of tobacco protoplasts with two bromoviruses. Journal of General Virology, 64, 1347–1354.

Samuel, C. E. (1991). Antiviral actions of interferon. Interferon-regulated cellular proteins and their surprisingly selective antiviral activities. Virology, 183, 1–11.

Sanford, J. (1997). Allowing small companies to develop and market biopesticides. Genetic Engineering News, 17, 4.

Sanford, J. C., & Johnston, S. A. (1985). The concept of parasite-derived resistance—deriving resistance genes from the parasite's own genome. Journal of Theoretical Biology, 113, 395–405.

Sleat, D. E., & Palukaitis, P. (1992). A single nucleotide change within a plant virus satellite RNA alters the host specificity of disease induction. Plant Journal, 2, 43–49.

Smith, H. A., Swaney, S. L., Parks, T. D., Wernsman, E. A., & Dougherty, W. G. (1994). Transgenic plant virus resistance mediated by untranslatable sense RNAs: Expression, regulation, and fate of nonessential RNAs. Plant Cell, 6, 1441–1453.

Spillane, C., Verchot, J., Kavanagh, T. A., & Baulcombe, D. C. (1997). Concurrent suppression of virus replication and rescue of movement-defective virus in transgenic plants expressing the coat protein of potato virus X. Virology, 236, 76–84.

Stanley, J., Frischmuth, T., & Ellwood, S. (1990). Defective viral DNA ameliorates symptoms of geminivirus infection in transgenic plants. Proceedings of the National Academy of Sciences, USA, 87, 6291–6295.

Stark, D. M., & Beachy, R. N. (1989). Protection against potyvirus infection in transgenic plants: Evidence for broad spectrum resistance. Bio/Technology, 7, 1257–1262.

Stirpe, F., Barbieri, L., Batelli, M. G., Soria, M., & Lappi, D. A. (1992). Ribosome-inactivating proteins in plants: Present status and future prospects. Bio/Technology, 10, 405–412.

Swaney, S., Powers, H., Goodwin, J., Rosales, L. S., & Dougherty, W. G. (1995). RNA-mediated resistance with nonstructural genes from the tobacco etch virus genome. Molecular Plant-Microbe Interactions, 8, 1004–1011.

Taschner, P. E., van der Kuyl, A. C., Neeleman, L., & Bol, J. F. (1991). Replication of an incomplete alfalfa mosaic virus genome in plants transformed with viral replicase genes. Virology, 181, 445–450.

Tavladoraki, P., Benvenuto, E., Trinca, S., De Martinis, D., Cattaneo, A., & Galeffi, P. (1993). Transgenic plants expressing a functional single-chain Fv antibody are specifically protected from virus attack. Nature (London), 366, 469–472.

Tenllado, F., Garcia-Luque, I., Serra, M. T., & Diaz-Ruiz, J. R. (1995). *Nicotiana benthamiana* plants transformed with the 54-kDa region of the pepper mild mottle tobamovirus replicase gene exhibit two types of resistance responses against viral infection. Virology, 211, 170–183.

Thomas, P. E., & Kaniewski, W. K. (1997). Agronomic performance of transgenic plants. In G. D. Foster & S. C. Taylor (Eds.), Methods in molecular biology: Vol. 81. Plant virology protocols: From virus isolation to transgenic resistance (pp. 509–520). Totowa, NJ: Humana Press.

Thomas, P. E., Kaniewski, W. K., & Lawson, E. C. (1997). Reduced field spread of potato leafroll virus in potatoes transformed with the potato leafroll virus coat protein gene. Plant Disease, 81, 1447–1453.

Truve, E., Aaspollu, A., Honkanen, J., Puska, R., Mehto, M., & Hassi, A. (1993). Transgenic potato plants expressing mammalian 2′–5′ oligoadenylate synthetase are protected from potato virus X infection under field conditions. Bio/Technology, 11, 1048–1052.

Truve, E., Kelve, M., Aaspollu, A., Kuusksalu, A., Seppanen, P., & Saarma, M. (1994). Principles and background for the construction of transgenic plants displaying multiple virus resistance. Archives of Virology, (Suppl.), 9 41–50.

Tumer, N. E., O'Connell, K. M., Nelson, R. S., Sanders, P. R., Beachy, R. N., & Fraley, R. T. (1987). Expression of alfalfa mosaic virus coat protein gene confers cross-protection in transgenic tobacco and tomato plants. EMBO Journal, 6, 1181–1188.

Tumer, N. E., Kaniewski, W., Haley, L., Gehrke, L., Lodge, J. K., & Sanders, P. (1991): The second amino acid of alfalfa mosaic virus coat protein is critical for coat protein-mediated protection. Proceedings of the National Academy of Sciences, USA, 88, 2331–2335.

USDA. (1992). Genetically engineered organisms and products; notification procedures for the introduction of certain regulated articles; and petition for nonregulated status. Federal Register, 57, 53036–53053.

USDA. (1993). Genetically engineered organisms and products; notification procedures for the introduction of certain regulated articles. Federal Register, 58, 17044–17059.

Vaira, A. M., Semeria, L., Crespi, S., Lisa, V., Allavena, A., & Accotto, G. P. (1995). Resistance to tospoviruses in Nicotiana benthamiana transformed with the N gene of tomato spotted wilt virus: Correlation between transgene expression and protection in primary transformants. Molecular Plant-Microbe Interactions, 8, 66–73.

van der Vlugt, R. A., Rutier, R. K., & Goldbach, R. (1992). Evidence for sense RNA-mediated protection to PVY[N] in tobacco plants transformed with the viral coat protein cistron. Plant Molecular Biology, 20, 631–639.

van Dun, C. M. P, & Bol, J. F. (1988). Transgenic tobacco plants accumulating tobacco rattle virus coat protein resist infection with tobacco rattle virus and pea early browning virus. Virology, 167, 649–652.

van Dun, C. M. P., Overduin, B., van Vloten-Doting, L., & Bol, J. F. (1988). Transgenic tobacco expressing tobacco streak virus or mutated alfalfa mosaic virus coat protein does not cross-protect against alfalfa mosaic virus infection. Virology, 164, 383–389.

van Dun, C. M. P., Bol, J. F., & van Vlotten-Doting, L. (1987). Expression of alfalfa mosaic virus and tobacco rattle virus coat protein genes in transgenic tobacco plants. Virology, 159, 299–305.

Vaquero, C., Turner, A. P., Demangeat, G., Sanz, A., Serra, M. T., & Roberts, K. (1994). The 3a protein from cucumber-mosaic virus increases the gating capacity of plasmodesmata in transgenic tobacco plants. Journal of General Virology, 75, 3193–3197.

von Arnim, A., & Stanley, J. (1992). Inhibition of African cassava mosaic virus systemic infection by a movement protein from the related geminivirus tomato golden mosaic virus. Virology, 187, 555–564.

Wachinger, M., Saermark, T., & Erfle, V. (1992). Influence of amphipathic peptides on the HIV-1 production in persistently infected T lymphoma cells. FEBS Letters, 309, 235–241.

Wang, D. W., & Maule, A. J. (1994). A model for seed transmission of a plant virus—genetic and structural analysis of pea embryo invasion by pea seed-borne mosaic virus. Plant Cell, 6, 777–787.

Whitham, S., McCormick, S., & Baker, B. (1996). The N gene of tobacco confers resistance to tobacco mosaic virus in transgenic tomato. Proceedings of the National Academy of Sciences, USA, 93, 8776–8781.

Whitham, S., Dinesh-Kumar, S. P., Choi, D., Hehl, R., Corr, C., & Baker, B. (1994). The product of the tobacco mosaic virus resistance gene N: Similarity to Toll and the Interleukin-1 receptor. Cell, 78, 1101–1115.

Wisniewski, L. A., Powell, P. A., Nelson, R. S., & Beachy, R. N. (1990). Local and systemic spread of tobacco mosaic virus in transgenic tobacco. Plant Cell, 2, 559–567.

Wolf, S., Deom, C. M., Beachy, R. N., & Lucas, W. J. (1989). Movement protein of tobacco mosaic virus modifies plasmodesmatal size exclusion limit. Science, 246, 377–379.

Wu, X. J., Beachy, R. N., Wilson, T. M., & Shaw, J. G. (1990). Inhibition of uncoating of tobacco mosaic virus particles in protoplasts from transgenic tobacco plants that express the viral coat protein gene. Virology, 179, 893–895.

Yang, Y., & Klessig, D. F. (1996). Isolation and characterization of a tobacco mosaic virus-inducible myb oncogene homolog from tobacco. Proceedings of the National Academy of Sciences, USA, 93, 14972–14977.

Zaitlin, M., Anderson, J. M., Perry, K. L., Zhang, L., & Palukaitis, P. (1994). Specificity of replicase-mediated resistance to cucumber mosaic virus. Virology, 201, 200–205.

21

Bacillus thuringiensis in Biological Control

B. A. FEDERICI

Department of Entomology and
Interdepartmental Graduate Programs in Genetics and Microbiology
University of California
Riverside, California

INTRODUCTION

The insecticidal bacterium *Bacillus thuringiensis* (B.t.) is the most successful commercial biological control agent of insect pests. Moreover, aside from the parasites and predators that have been effective in classical biological programs, it is the most widely used biological control agent in the world. B.t. is currently used in many countries to suppress numerous lepidopteran and coleopteran pests of forests, vegetables, and field crops; and to control the larvae of many species of vector and nuisance mosquitoes and blackflies. The principal reasons for the success of B.t. include the high efficacy of its insecticidal proteins, the existence of a diversity of proteins that are effective against a range of important pests, its relative safety to nontarget insect predators and parasites, its ease of mass production at a relatively low cost, and its adaptability to conventional formulation and application technology. In addition, knowledge of B.t. and plant molecular biology, recombinant DNA technology, and plant transformation techniques have been used to develop insect-resistant B.t.-transgenic varieties of major crops such as cotton, corn, potatoes, and rice, creating a new multibillion dollar industry. New types of insecticidal B.t. proteins are being discovered routinely, and other advances in knowledge about B.t. proteins and plant molecular biology indicate the use of these proteins will be extended in the future to control a greater range of invertebrate pests. Thus, among the various types of insect pathogens, parasites, and predators, B.t. has excellent prospects for more widespread use in future insect control programs. An advantage of these programs is that the successful use of B.t. markedly reduces the use of synthetic chemical insecticides.

Whereas it is widely known in the pest control field that B.t. kills insects through the action of protein toxins, specific knowledge of its complexity, diversity, and mode of action as well as its potential for further development is often lacking. Due to the present and growing importance of B.t., my purpose in this chapter therefore is to provide an overview of the basic biology of B.t. and how it is being used in biological control, both in the form of bacterial insecticides and transgenic plants. The chapter is designed primarily to provide information for non-B.t. specialists. After a general discussion of the basic biology and systematics of B.t., I focus on the mode of action of the key types of insecticidal proteins and the composition of the most common isolates used in commercial formulations of bacterial insecticides. I then cover the use of selected B.t. proteins in recombinant bacteria and transgenic plants, and close with short sections on resistance and the prospects for further use of B.t. in the future.

GENERAL BIOLOGY OF BACILLUS THURINGIENSIS

B.t. is a spore-forming bacterium that can be readily isolated on simple media such as nutrient agar from a variety of habitats including soil, water, plants, grain dust, dead insects, and insect feces. Compared with other insect pathogens, its life cycle is simple. When nutrients are sufficient for growth, the spore germinates producing a vegetative cell that grows and reproduces by binary fission. The bacterium continues to multiply until one or more nutrients, such as sugars, certain amino acids, and oxygen, become insufficient for continued vegetative growth. Under these conditions, the bacterium sporulates producing a spore and parasporal body, the latter composed primarily of insecticidal protein toxins (Fig. 1).

Though B.t. can be isolated from many environmental sources and is typically referred to as a "soil bacterium," it has several features indicating that its principal ecological niche is insects (Federici, 1993; Meadows, 1993). The original isolations of B.t., for example, were made from diseased caterpillars (Ishiwata, 1901; Berliner, 1915), and these remain a good source of isolates. More importantly, B.t. produces a range of toxins and toxin synergists that are very effective at killing certain species of insects, especially larvae of lepidopterous insects, providing a rich substrate for B.t. reproduction. The principal toxins are the Cry and Cyt protein δ-endotoxins, but many isolates of B.t. also produce the β-exotoxin, the synergist zwittermycin, and enzymes such as phospholipases that enhance the activity of the δ-endotoxins. In addition, the spore itself can synergize the activity of Cry proteins in some insects. In many insect species, especially grain-feeding lepidopterans, the bacterium reproduces to very high levels after insect death, with millions of spores being produced per cadaver (Fig. 2). Thus, B.t. ecology and reproductive biology suggest that its toxins and toxin synergists evolved to debilitate or kill directly a range of insect species, thereby providing a substrate for reproduction of this bacterium (Federici, 1993). An interesting aspect of B.t. general biology is that unlike most other insect pathogens, with the possible exception of certain grain-feeding lepidopterans, B.t. does not cause natural epizootics in insect populations.

Bacillus thuringiensis as a Bacterial Species

The species B.t. was first isolated from diseased larvae of the silkworm, *Bombyx mori,* in Japan by Ishiwata (1901). It was not officially described, however, until it was re-isolated by Berliner (1915) from diseased larvae of the Mediterranean flour moth, *Anagasta kuehniella,* in Thuringia, Germany, hence the derivation of species name *thurin-*

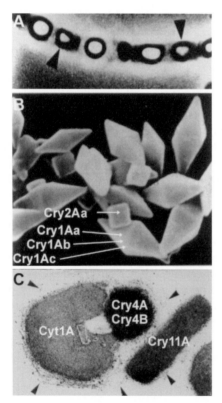

FIGURE 1 Sporulating cells and endotoxin-containing parasporal bodies of *Bacillus thuringiensis.* (A) Phase-contrast micrograph of a wet-mount preparation of sporulating B.t. cells illustrating ovoidal spores and adjacent parasporal bodies (arrowheads). (B) Scanning electron micrograph of the bipyramidal and cuboidal endotoxin inclusions characteristic of the HD1 isolate of *B. thuringiensis* subsp. *kurstaki* (H3a3b). Commercially, this is the most successful strain of B.t. with more than 100 formulations marketed in different countries around the world to control caterpillar pests. (C) Transmission electron micrograph of the parasporal body of the ONR60A isolate of *B. thuringiensis* subsp. *israelensis* (H14) used widely to control the larvae of mosquitoes and blackflies. Each crystalline inclusion contains a different endotoxin. The arrowheads point to the parasporal envelope that holds the inclusions together.

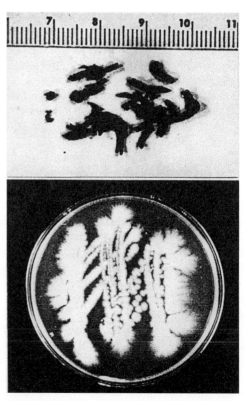

FIGURE 2 Occurrence of *Bacillus thuringiensis* in larvae of the navel orangeworm, *Amyelois transitella.* The top panel shows larvae killed by *B. thuringiensis* subsp. *aizawai* (H7) during an epizootic in a laboratory colony fed on grain (wheat). The bottom panel shows a pure culture of the *B. thuringiensis* subsp. *aizawai* strain isolated from a small piece of fat body from a dead larva. These larvae contain as many as 10^8 spores per cadaver.

giensis. Though commonly referred to in the singular as B.t., *B. thuringiensis* as currently recognized is actually a complex of subspecies, all of which are characterized by the production of a parasporal body during sporulation. The parasporal body is the principal characteristic used to differentiate this species from the closely related species, *B. cereus,* and other bacilli. This parasporal body contains one or more proteins, typically as crystalline inclusions, and most of these are highly toxic to one or more species of insects (Fig. 3). The toxins are known as endotoxins, and occur in the parasporal body as protoxins, which after ingestion dissolve and are converted to active toxins through cleavage by proteolytic enzymes in the insect gut. The activated toxins bind to the midgut microvillar membrane in sensitive insects, lyse the cells, and destroy much of the midgut epithelium, causing insect death. Thus, B.t. endotoxins are stomach poisons selective for insects and certain other invertebrates. At present, there are more than 60 subspecies of B.t. (Table 1), distinguished from one another on the basis of immunological differences in flagellar (H antigen) serotype (de Barjac & Franchon, 1990). Each subspecific name corresponds with a specific H antigen number. For example, *B. thuringiensis* subspecies *kurstaki* is H3a3b,

whereas *B. thuringiensis* subspecies *morrisoni* is H8a8b. In the literature, the term "variety" is also used in place of subspecies, as is occasionally the term "strain." Because the H antigen serotype-subspecific name often does not correlate with insecticidal properties, acronyms and numbers are often used to designate specific isolates, especially those with important insecticidal properties. For example, HD1 (isolate number 1 from Howard Dulmage) is used to designate a specific isolate of *B. thuringiensis* subspecies *kurstaki* (H3a3b) that produces four major endotoxin proteins and has a broad spectrum of activity against lepidopteran pests. Another isolate of *B. thuringiensis* subsp. *kurstaki* (H3a3b) is HD73. This produces only a single endotoxin protein and as a result has a much narrower spectrum of activity against insects than does HD1. Historically, HD1 was the first B.t. isolate developed commercially for the control of lepidopterous pests, and remains the most widely used in commercial products today. This isolate has also been the source of much of our knowledge of B.t. genetics and molecular biology as well as endotoxin genes used in transgenic bacteria and plants.

Despite the large number of B.t. subspecies that have been described, the taxonomic validity of these as well as maintaining B.t. as a species separate from *B. cereus* has been in question for many years. At the species level, the

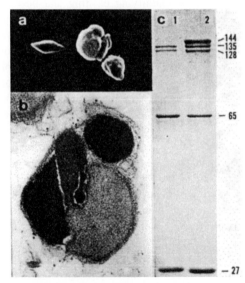

FIGURE 3 Endotoxin complexity in a parasporal body of *Bacillus thuringiensis.* (a) and (b) show, respectively, scanning and transmission electron micrographs of the parasporal body of *B. thuringiensis* subsp. *morrisoni* (H8a8b), isolate PG-14, a mosquitocidal isolate of this subspecies. In (c1), the protein composition of *B. thuringiensis* subspecies *israelensis* parasporal body is compared with that of the PG-14 parsporal body (c2). Note that the latter isolate contains the four major proteins characteristic of most mosquitocidal isolates of B.t. (Cry4A, 135 kDa; Cry4B, 128 kDa; Cry11A 65 kDa; and Cyt1A, 27 kDa), and in addition a protein of 144 kDa. The four mosquitocidal proteins are encoded by a large transmissible plasmid of 125 kb, whereas the 144-kDa protein is encoded on a different plasmid. Isolates of B.t. with a broad spectrum of activity against insect species typcially contain four or more endotoxin proteins in the parasporal body.

TABLE 1 Subspecies/Serovarieties of *Bacillus thuringiensis*[a]

H antigen[b]	Subspecies/ serovariety[c, d]	H antigen	Subspecies/ serovariety
1	*thuringiensis*	13	*pakistani*
2	*finitimus*	**14**	***israelensis***
3a	*alesti*	15	*dakota*
3a3b	***kurstaki***	16	*indiana*
4a4b	*sotto*	17	*tohokuensis*
4a4c	*kenyae*	18	*kumamotoensis*
5a5b	***galleriae*[e]**	19	*tochigiensis*
5a5c	*canadensis*	20a20b	*yunnanensis*
6	*entomocidus*	20a20c	*pondicheriensis*
7	***aizawai***	21	*colermi*
8a8b	***morrisoni***	22	*shandongiensis*
8a8c	*ostriniae*	23	*japonensis*
8b8d	*nigeriensis*	24	*neoleonensis*
9	*tolworthi*	25	*coreanensis*
10	*darmstadiensis*	26	*silo*
11a11b	*toumanoffi*	27[f]	*mexicanensis*
11a11c	*kyushuensis*		
12	*thompsoni*		

[a]From de Barjac, H. & Franchon, E. (1990). *Entomophaga, 35,* 233–240.

[b]Flagellar antigen used to prepare antibodies.

[c]Also referred to in the literature as "variety" and occasionally "strain."

[d]H antigens and subspecies in bold indicate they are used in commercial products.

[e]Used in the former Soviet Union and People's Republic of China.

[f]See Thiery & Frachon (1997) for a more complete list.

primary problem is that the only phenotypic character that clearly differentiates *B. cereus* from B.t. is the parasporal body synthesized by the latter species during sporulation (Baumann *et al.,* 1987). The information for parasporal body production is encoded on large transmissible plasmids in most B.t. subspecies [see Hofte and Whiteley (1989) for review]. When these plasmids are lost naturally or are cured from B.t. subspecies by growing cells at 42°C, no parasporal body is produced. Cured B.t. strains that hosted these plasmids cannot be reliably distinguished from *B. cereus.* At the subspecies level, phenotypic biochemical differences among many are minor, indicating that many subspecies may not be valid using accepted standards of differentiating bacterial subspecies. In other cases, the differences in biochemical properties and insecticidal activity are so significant that some subspecies could be viewed as distinct species.

Molecular Systematics of *B. thuringiensis*

Attempts to clarify the relationships between B.t. and *B. cereus,* and among subspecies of B.t., have used the methods of molecular systematics. These studies have provided strong evidence that B.t. and *B. cereus* are so closely related that they could be considered the same species. For example, in a study of small subunit ribosomal RNA sequences, Ash *et al.* (1991) found that *B. cereus* and B.t. fell into the same clade, and were much more closely related to each other than to 48 of the 51 *Bacillus* species they studied. Even more definitive evidence for the close relationship of B.t. and *B. cereus* was found by Carlson and Kolsto (1993), who used DNA–DNA hybridization techniques and chromosome mapping to study the relationship of these species. Under stringent hybridization conditions they found that 19 random *B. cereus* probes hybridized strongly to *B. t.* subsp. *thuringiensis* (HD2) fragments, but did not hybridize with or gave only weak signals to fragments of *B. subtilis* under conditions of low stringency. Moreover, they found that when the *Not*I chromosomal map of *B. t.* subsp. *thuringiensis* was compared with that of four different strains of *B. cereus,* the B.t. map was more similar to two of the *B. cereus* strains than these were to the two other *B. cereus* strains.

With respect to the relationships among different subspecies, Carlson and Kolsto (1993) also compared the *Not*I chromosomal fragment patterns of 10 different B.t. subspecies and found that they differed markedly. In another study using DNA fragment patterns, Priest *et al.* (1994) found considerable variation within and among the restriction fragment length polymorphisms (RFLPs) of ribosomal gene clusters of several different B.t. subspecies. Yet they also found that in subspecies *B. thuringiensis subsp. israelensis* and *B. thuringiensis subsp. aizawai,* the RFLPs for each were so unique and characteristic among different isolates that they could be used to assign isolates to these subspe-

cies. Care must be taken, however, in interpreting the differences in chomosomal and ribosomal gene clusters to mean that these subspecies actually differ much from one another because even minor differences in large DNA sequences can lead to major differences in RFLPs that do not correspond with significant biological differences.

From the standpoint of B.t. being used as an insecticide, taxonomic studies, especially flagellar serotyping, have aided isolate classification but have proved unreliable as accurate predictors of insecticidal activity. Perhaps the best example of this is found in *B. thuringiensis* subsp. *morrisoni* (H8a8b), which includes isolates active against lepidopteran (isolate HD12), coleopteran (isolate DSM2803 and others), or dipteran (isolate PG14) insects. Another example of this is found in the occurrence of the 125-kb plasmid that encodes mosquitocidal Cry4 and Cyt1 proteins among numerous subspecies of B.t. including *israelensis, morrisoni, entomocidus, kenyae,* and *thompsoni* (Lopez-Meza *et al.,* 1996; Thiery *et al.,* 1996).

The absence of an absolute correlation between subspecies/serotype and insecticidal activity continues to make it essential that the toxicity spectrum of new isolates be determined through bioassays, and ultimately be characterized in terms of the nucleotide sequences of genes encoded and expressed by individual B.t. isolates. Other methods have been examined for classifying existing and new isolates, including typing isolates with batteries of monoclonal antibodies to known endotoxins or using the polymerase chain reaction to identify genes carried by isolates. However, these have not proved better than H antigen serotyping for identification and classification of isolates. Moreover, though useful for identifying new toxins, these techniques carry the risk of missing unknown proteins with novel host spectra.

To summarize the current status of B.t. systematics, most molecular evidence is in agreement with more classical biochemical and physiological studies that indicate B.t. and *B. cereus* are the same species, and that the latter becomes the former when it acquires one or more plasmids that express genes for insecticidal proteins. Nevertheless, maintaining B.t. as a separate species has practical value because of its insecticidal properties, as does dividing B.t. isolates into subspecies based on flagellar antigens. The latter are useful in cataloging the more than 20,000 isolates of B.t. collected to date. However, to understand the insecticidal properties of an isolate, regardless of its subspecies/serotype or other designation, knowledge of the insecticidal protein genes encoded and expressed is required.

INSECTICIDAL PROTEINS OF *BACILLUS THURINGIENSIS*

There are two principal active components in commercial preparations of B.t., the spore and parasporal body. For most insect pests, the parasporal body, which consists of one or more insecticidal toxin proteins, accounts for most of the formulation's activity, including initial paralysis followed by death. The lethal effects of these proteins have been known since their discovery in the 1950s (Hannay, 1953; Angus, 1954). They are generally referred to as δ-endotoxins, "δ" designating a particular class of toxins, and endotoxin referring to their localization within the bacterial cell after production, as opposed to being secreted. In the early 1980s, shortly after the development of recombinant DNA techniques, it was discovered that B.t. δ-endotoxins were encoded by genes carried on plasmids. This discovery led quickly to a major research effort in many laboratories around the world aimed at understanding the genetic and molecular biology of these toxins. This effort resulted in cloning and sequencing of numerous B.t. genes, over 100 to date, and characterization of the toxicity of the proteins each gene encodes (Crickwore *et al.,* 1998). Pertinent information derived from these studies through 1988 was summarized in the excellent review by Hofte and Whiteley (1989). At the time Hofte and Whiteley wrote their review, a wide variety of confusing names and acronyms were being used to refer to B.t. endotoxin genes and proteins. Computer analyses showed that the nucleotide sequences of most of these genes were quite similar. To standardize the terminology, Hofte and Whiteley (1989) proposed a simplified nomenclature for naming all insecticidal B.t. genes and proteins. In this nomenclature, the proteins are referred to as Cry and Cyt proteins. Though modified by Crickmore *et al.* (1998), this system is still in use today, and is described later along with its modifications and additions.

Cry and Cyt Protein Nomenclature

The Cry and Cyt nomenclature developed by Hofte and Whiteley (1989) was originally based on the spectrum of activity of the proteins as well as their size and apparent relatedness as deduced from nucleotide and amino acid sequence data, and protein gel analyses. At that time, with the exception of a 27-kDa cytolytic protein gene from *B. thuringiensis* subsp. *israelensis,* all genes appeared to be related, and probably derived from the same ancestral gene. Thus Hofte and Whiteley termed these "*cry*" (for crystal) genes and the proteins they encode "Cry" proteins. This designation was followed by a Roman numeral that indicates pathotype (I and II for toxicity to lepidopterans, III for toxicity to coleopterans, and IV for toxicity to dipterans), followed by an uppercase letter indicating the chronological order in which genes with significant differences in nucleotide sequences were described. The I and II for lepidopteran-toxic proteins also indicate size differences, with the I referring to proteins with a mass in the range of 130 kDa, and the II designating those with a mass from 65 to 70 kDa. Some epithets also include a lowercase letter in parentheses, which indicates minor differences in the nucle-

otide sequence within gene or protein type. Thus, CryIA referred to a 130-kDa protein toxic to lepidopterous insects for which the first gene (*cryIA*) was sequenced, whereas CryIVD referred to a 72-kDa protein with mosquitocidal activity for which the encoding gene was the fourth from this pathotype sequenced. Though not perfect, this system was preferable to the chaos that existed before it was proposed.

The 27-kDa CytA protein first isolated from *B. thuringiensis* subsp. *israelensis* differs from other B.t. proteins not only in its smaller size, but also in that it is highly cytolytic to a wide range of cell types *in vitro,* including those of vertebrates [see Federici *et al.* (1990) for reviews]. In addition, this shares no apparent relatedness with Cry proteins. Because of these differences and its broad cytolytic activity, Hofte and Whiteley (1989) referred to this as the CytA protein encoded by the *cytA* gene.

In their revision of B.t. gene nomenclature, Hofte and Whiteley (1989) listed 38 published gene sequences that encoded 13 different Cry proteins and the single CytA protein. Since their publication, the number of B.t. Cry endotoxins has more than tripled, and several new *cyt* genes have also been described (Crickmore *et al.,* 1998). As more and more *cry* genes were sequenced and analyzed, it was decided to name genes based on their relatedness as determined primarily from the degree of their deduced amino acid identity. As a result, the nomenclature developed by Hofte and Whiteley (1989) has been modified in the following manner. The *cry* and *cyt* descriptors have been maintained, but the Roman numerals have been replaced with Arabic numbers to indicate major relationships (90% identity), with higher degrees of identity being indicated by following uppercase letters (95% identity), and minor variations of these alleles being designated by lowercase letters, with the parentheses around the latter eliminated. Thus, for example, what was CryIA(c) is now Cry1Ac (and the corresponding gene, *cry1Ac*), a relatively minor change. However, for some genes and proteins the changes are greater. For example, CryIVD does not cluster with the earlier CryIV (now Cry4) proteins, and thus is now the taxon Cry11Aa. The previous and new nomenclature for representative examples of the most commonly studied and used endotoxin proteins is presented in Table 2. The complete list of currently recognized B.t. genes and proteins and references to the literature describing these can be obtained from the following website:

http://www.biols.susx.ac.uk
/Home/Neil_Crickmore/Bt/index.html.

Though the new designations supposedly carry no specific information concerning insecticidal spectrum, because the numbers have been maintained for many of the genes listed by Hofte and Whiteley (1989), and because a high degree of correlation between relatedness and insecticidal spectrum remains, primary insecticidal activity can often be inferred.

To illustrate this, Cry1 still refers to lepidopteran toxicity; Cry2, to lepidopteran toxicity and in some, dipteran activity; Cry3, to coleopteran toxicity; and Cry4, to dipteran toxicity.

Mode of Action and Structure of Cry Proteins

The spore can play an important role in the pathogenicity of B.t. to certain insect species, but the parasporal body causes the rapid paralysis and ultimate death of most target species [for references see Huber and Luthy (1981), Aronson *et al.* (1986), Hofte and Whiteley (1989), Moar *et al.* (1989)]. The parasporal body consists of one or more Cry δ-endotoxins, and though the three-dimensional structure of three of these is now known from X-ray crystallographic studies, the mode of action of these toxins has not been resolved at the molecular level. Therefore, I will first review our knowledge of the mode of action of Cry endotoxins derived from early and later studies, and then consider the three-dimensional structure and insight provided into the mode of action.

Mode of Action

In the typical B.t., *B. thuringiensis* subsp. *kurstaki,* for example, the parasporal body dissolves after ingestion encountering the alkaline (pH 8 to 10) juices of the midgut. Dissolution requires the reduction of disulfide bridges that stabilize the Cry molecules in the parasporal crystal (Aronson, 1993). Most Cry toxins are actually protoxins of about 130 to 140 kDa (e.g., Cry1 and Cry4 proteins) from which an active toxin "core" in the range of 60 to 70 kDa is released in the midgut by proteolytic cleavage from the C-terminal half of the molecule (Fig. 4). These activated toxin molecules pass through the peritrophic membrane and bind to specific receptors on the apical microvillar brush border membrane of midgut epithelial cells, which lies just outside the peritrophic membrane. Binding is an essential step in the intoxication process, and in susceptible insects the toxicity of a particular B.t. protein is correlated with the number of specific binding sites (i.e., receptors) on microvilli and with the affinity of the B.t. molecules for these sites (Hofmann *et al.,* 1988; Van Rie *et al.* 1989, 1990). However, binding by itself, even high-affinity binding, does not

TABLE 2   Nomenclature for Representative Insecticidal Proteins and Their Encoding Genes from *Bacillus thuringiensis*

Old nomenclature[a]		New nomenclature[b]		Insect spectrum	Number of amino acids	Mass (kDa)
Gene	Protein	Gene	Protein			
cryIA(a)	CryIA(a)	*cry1Aa*	Cry1Aa	Lepidoptera	1176	133.2
cryIA(b)	CryIA(b)	*cry1Ab*	Cry1Ab	Lepidoptera	1155	131.0
cryIA(c)	CryIA(c)	*cry1Ac*	Cry1Ac	Lepidoptera	1178	133.3
cryIB	CryIB	*cry1Ba*	Cry1Ba	Lepidoptera	1207	138.0
cryIC	CryIC	*cry1Ca*	Cry1Ca	Lepidoptera	1189	134.8
cryID	CryID	*cry1Da*	Cry1Da	Lepidoptera	1165	132.5
cryIIA	CryIIA	*cry2Aa*	Cry2Aa	Lepidoptera/Diptera	633	70.9
cryIIB	CryIIB	*cry2Ab*	Cry2Ab	Lepidoptera	633	70.8
cryIIIA	CryIIIA	*cry3Aa*	Cry3Aa	Coleoptera	644	73.1
cryIIIB	CryIIIB	*cry3Ba*	Cry3Ba	Coleoptera	649	74.2
cryIVA	CryIVA	*cry4Aa*	Cry4Aa	Diptera[c]	1180	134.4
cryIVB	CryIVB	*cry4Ba*	Cry4Ba	Diptera	1136	127.8
cryIVC	CryIVC	*cry10Aa*	Cry10Aa	Diptera	675	77.8
cryIVD	CryIVD	*cry11Aa*	Cry11Aa	Diptera	643	72.4
jeg80	Jeg80	*cry11Ba*	Cry11Ba	Diptera		80
cytA	CytA	*cyt1Aa*	Cyt1Aa	Diptera/others[d]	248	27.4
cytB	CytB	*cyt2Aa*	Cyt2Aa	Diptera		

[a]From Hofte, H., & Whiteley, H. R. (1989). *Microbiological Review, 53,* 242–255.

[b]For a complete list of crystalline endotoxin proteins see the following website:
http://www.biols.susx.ac.uk/Home/Neil_Crickmore/Bt/index.html

These proteins are protoxins that are activated *in vivo* by proteolytic cleavage in the insect midgut after ingestion. The Cry proteins are cleaved to form activated toxins in the range of 60 to 65 kDa, with most of the protein (ca. 600 amino acids) in the 130-kDa size range, proteins being cleaved from the C terminus. A small amount of cleavage also occurs at the N terminus of all Cry proteins.

[c]Cry4 and Cry11 proteins are the most toxic known proteins to dipteran insects, and have only been reported to have significant activity against members of the suborder Nematocera (e.g., insects such as mosquitoes, blackflies, chironomid midges, psychodid flies, and crane flies).

[d]*In vitro,* the CytA protein has been shown to be cytolytic for a wide range of cell types including those from invertebrates as well as vertebrates. Proteolysis yields a cytolytic protein of 25 kDa.

always lead to toxicity, indicating that insertion and probably postinsertional processing in the midgut membrane are required to obtain toxicity. For example, in three different studies it has been demonstrated that the Cry1Ac toxin can bind with high affinity to microvillar membrane vesicles from *Lymantria dispar* (Wolfersberger, 1990), *Spodoptera frugiperda* (Garczynski *et al.,* 1991) and *Heliothis virescens* (Gould *et al.,* 1992), but with little or no subsequent toxicity. This indicates that insertion and likely postinsertional processing of Cry proteins are essential to intoxication.

In highly sensitive insect species, the microvilli lose their characteristic structure within minutes of toxin insertion, and the cells become vacuolated and begin to swell (Huber & Luthy, 1981; Luthy & Ebersold, 1981). This swelling continues until the cells lyse and slough from the basement membrane of the midgut epithelium. As more and more cells slough, the alkaline gut juices begin to leak into the hemocoel where, as a result, the hemolymph pH rises by a half unit or more. This causes the paralysis and eventual death of the insect (Heimpel & Angus, 1963; Heimpel, 1967).

Though this general picture of the mode of action has been known for many years, the actual process of intoxication at the molecular level remains unresolved, especially the series of events that occurs after the toxin binds to the receptor and inserts into the microvillar membrane. There

is good evidence for an influx of cations, especially potassium, and water into the columnar cells, which continues until cell lysis. To explain this influx, it has been proposed that activated Cry molecules insert into the apical microvillar membrane forming transmembrane pores that are cation-selective at the alkaline pH that exists in the lepidopteran midgut (see Wolfersberger, 1989; Knowles & Ellar, 1987; Knowles & Dow, 1994; and Schwartz *et al.,* 1993). For pores to form, it is postulated that during membrane insertion, the molecules undergo conformational changes that permit insertion. After insertion, the toxin molecules aggregate to form a pore composed of six toxin molecules. As more and more Cry molecules enter the membrane, additional pores form, accelerating the influx of cations and water, followed by cell hypertrophy and death. A more detailed discussion of the mode of action of Cry toxins and the data supporting various pore-formation models can be found in the reviews by Gill *et al.* (1992), Knowles and Ellar (1987), and Schnepf *et al.* (1998).

Structure

Analysis of B.t. *cry* gene sequences conducted toward the end of the 1980s showed that the active portion of Cry toxin molecules (i.e., essentially amino acids 30 to 630 for most Cry molecules) contained five blocks of conserved amino acids distributed along the molecule, and a highly variable region within the C-terminal half (Hofte & Whiteley, 1989). The variable region was thought to be responsible for the insect spectrum of activity, and experimental evidence from recombinant DNA studies was obtained in support of this hypothesis. For example, by swapping the highly variable regions of Cry1Aa and Cry1Ac, the insect spectrum of these molecules could be reversed (Ge *et al.,* 1989). In addition, binding studies showed that in most cases the degree of sensitivity of an insect to a particular Cry molecule was directly correlated with the number of high-affinity binding sites on the midgut microvillar membrane (Hofmann *et al.,* 1988; Van Rie *et al.,* 1989). Studies over the past decade have built upon this foundation and considerably improved our knowledge of B.t. protein molecular biology as a result of determination of the crystal structure of three Cry proteins (Cry3A, Cry1Aa, and Cry2A), identification of regions on the Cry molecule involved in midgut binding and specificity, and identification of several glycoprotein receptors for B.t. molecules on the insect microvillar membrane.

The structure of the Cry3A molecule was the first solved, and similarities in conserved amino acid blocks among Cry toxins and conservation in hydrophobicity indicate this structure is likely a good general model for Cry toxins. Thus, the information discussed later combines interpretations from the crystal structure of the Cry3A molecule with recombinant DNA experiments on Cry1 molecules.

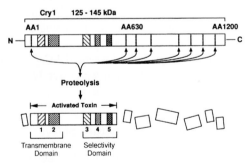

FIGURE 4   Schematic illustration of the structure and activation of typical Cry toxins of *Bacillus thuringiensis*. Cry1 and Cry2 toxins are typically active against lepidopteran insects, and the latter occur naturally at about half the size of the former. When ingested by a larva, midgut proteases activate the protoxin, releasing the active toxic peptide by progressive cleavage of peptide fragments from both the N and C termini. Similar processing occurs to activate Cry3 (toxic to coleopteran insects) and Cry4 (toxic to nematoceran dipterans) toxins in the midgut of sensitive insects. AA indicates amino acid residues, whereas the arrows indicate protease cleavage sites. The numbers and hashed boxes indicate the approximate location of the five converged blocks of amino acids that are present in many Cry toxins. [After Hofte H., & Whitelely, H. R. (1989). *Microbiological Reviews, 53,* 242–255].

The solution of the crystal structure of the Cry3A molecule (Li *et al.,* 1991) showed that this protein is basically wedge shaped and consists of three domains (Fig. 5). Domain I is composed of amino acids 1 to 290 and contains a hydrophobic seven-helix amphipathic bundle, with six helices surrounding a central helix. This domain contains contains all the first conserved block and a major portion of the second conserved block of amino acids noted earlier. Theoretical computer models of the helix bundle of this domain show that after insertion and rearrangment, aggregations of six of these domains could form a pore through the microvillar membrane (Li *et al.,* 1991). Domain II extends from amino acids 291 to 500 and contains three antiparallel β sheets around a hydrophobic core. This domain contains most of the hypervariable region and most of conserved blocks three and four. The crystal structure of the molecule together with recombinant DNA experiments and binding studies indicates that the three extended loop structures in the β sheets are responsible for initial recognition and binding of the toxin to binding sites on the microvillar membrane (Lee *et al.,* 1992). Domain III is composed of amino acids 501 to 644 and consists of two antiparallel β sheets, within which is found the remainder of conserved block number three along with blocks four and five. The three-dimensional structure resolved by Li *et al.* (1991) indicates this domain is involved in maintaining the structural integrity of the molecule, and site-directed mutagenesis studies of conserved amino acid block 5 in the Cry1Aa molecule suggest this domain may also play a role in pore formation (Chen *et al.,* 1993). Though the structure of the molecule, experiments, and computer modeling indi-

cate binding is attributed primarily to the hypervariable region of this domain, studies by Wu and Aronson (1992) also show that positively charged amino acids in domain I are important to binding and toxicity. Moreover, domain III has also been involved in receptor binding (DeMaagd *et al.,* 1996).

An important aspect of understanding B.t. molecular biology and determining its mode of action at the molecular level is identification of the proteins to which the toxin molecules bind on the microvillar membrane. Sangadala *et al.* (1994) have contributed to this area with the identification of the first two proteins in the midgut of *Manduca sexta* to which the Cry1Ac toxin binds. They identified an aminopeptidase of 120 kDa as a major binding protein, and a second protein of 65 kDa, an alkaline phosphatase, as a minor binding protein. More recently, a cadherin-like Z10-kDa protein in *M. sexta* had been identified as a receptor for the Cry1Ab toxin (Francis & Bulla, 1997). These findings are important because they indicate that the initial recognition signal on the midgut of a sensitive insect is not an ion channel, but instead one or more "housekeeping" enzymes that extend from the microvillar membrane into the gut lumen. Thus, if these observations hold for other Cry proteins, the midgut receptors are then essentially docking proteins for the toxins, indicating the toxins do not directly affect ion permeability by binding to ion channels.

These studies of the structure of B.t. toxins and their binding properties have identified key regions of the molecule responsible for binding and putatively the first binding proteins on the insect midgut epithelium. While further studies are certainly needed to more clearly define the specificity of binding and binding sites, the most enigmatic area of B.t. molecular biology with respect to its mode of action is how the toxin causes toxicity after binding to the membrane. Understanding this should provide information that will enable the development of insecticidal Cry proteins for use against insects against which we currently have none, such as cockroaches and grasshoppers.

Mode of Action and Structure of Cyt Proteins

Four Cyt proteins are currently recognized, Cyt1A, Cyt2A, Cyt1B, and Cyt2B (Crickmore *et al.,* 1998), though most of our knowledge of these proteins is based on studies of Cyt1A and Cyt2A, and the crystal structure of the latter toxin. These proteins are highly hydrophobic and all have a mass in the range of 24 to 28 kDa. They share no significant amino acid sequence identify with Cry proteins, and are thus unrelated. The first Cyt protein, Cyt1A, was identified as a component of the parasporal body of *B. thuringiensis* subsp. *israelensis.* Initially, it was thought that Cyt1A played little role in the toxicity of this subspecies to mosquito and blackfly larvae, but there is now general agreement that not only is it toxic to these and related flies

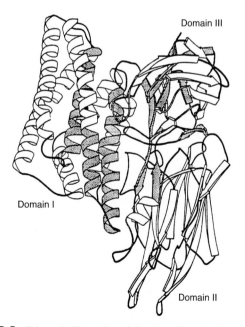

FIGURE 5 Schematic illustration of the three-dimentional structure of the Cry3Aa molecule (from Knowles & Dow, 1994).

belonging to the dipteran suborder Nematocera, but also it is important as a synergist of Cry4 and Cry11 toxins, and as a preventative against developing insect resistance. The Cyt proteins derive their name from being cytolytic to a wide range of invertebrate and vertebrate cells *in vitro* (Thomas & Eller, 1983). While they are rare in comparison to Cry proteins, since the early work on Cyt1A in *B. thuringiensis* subsp. *israelensis,* Cyt proteins have been reported from mosquitocidal isolates of *B. thuringiensis* subsp. *kyushuensis* (H11a11c), *B. thuringiensis* subsp. *morrisoni* (H8a8b), *B. thuringiensis* subsp. *medellin* (H30), and *B. thuringiensis* subsp. *jegatheson* (H28a28c).

Mode of Action

As in the case of Cry proteins, Cyt proteins are synthesized during sporulation as protoxins and are assembled into crystalline inclusions that make up a portion of the parasporal body. They are always associated with Cry proteins, but do not cocrystallize with these, instead forming a separate inclusion. The shape of inclusion formed by Cyt proteins varies from hemispherical to angular because they assemble in a spherical parasporal body simultaneously along with inclusions formed by the other Cry proteins with which they occur. However, when produced alone in a recombinant B.t., the Cyt1A protein forms a bipyramidal crystal with 12 faces (Wu & Federici, 1993).

Based on studies of Cyt1A, after ingestion the 27.3-kDa protoxin molecules dissolve from the crystal under the alkaline conditions present in the midgut of susceptible insects such as mosquito larvae. Proteolytic enzymes then activate the protoxin by cleaving amino acids at both the C and N terminus, releasing an active toxin of 24 kDa. This molecule passes through the peritrophic membrane and then inserts into the microvillar brush border membrane of midgut epithelial cells. Unlike Cry molecules, it appears that Cyt proteins do not require a glycoprotein receptor for binding. Cyt proteins have a high affinity for the lipid portion of the membrane, specifically for unsaturated phospholipids such as cholesterol, phospholipidcholine, and sphigomyelin (Thomas & Ellar, 1983). After binding to the microvillar membrane, the cells hypertrophy and subsequently lyse.

The specific mode of action of Cyt toxins at the molecular level is not known. The leading hypothesis is that Cyt molecules insert into the microvillar membrane and then assemble into clusters of as many as 18 molecules to form pores that act much like those prosposed for Cry toxins (Koni & Ellar, 1994; Gill *et al.,* 1992). Lysis then results from the influx of cations and water. However, evidence also has been provided that Cyt molecules do not form pores, but instead act like detergents, binding to and perturbing the lipid bilayer and disrupting the structural arrangement of membrane proteins (Butko *et al.,* 1996, 1997), which would also cause lysis.

Unlike Cry proteins, Cyt proteins exhibit the same cytolytic activity *in vitro* against a wide range of invertebrate and vertebrate cell types, though at higher concentrations. However, toxicity is not observed when ingested *in vivo* by most insects tested, including larvae of lepidopterous species, nonnematoceran flies such as houseflies and fruit flies, and vertebrates such as mice. The mechanism of this *in vivo* specificity is not known, but may involve specific combinations of unsaturated phospholipids, or an unknown protein receptor. Though Cyt proteins have been thought to be only toxic to nematocerous dipterans, it has been reported that at least one beetle species, *Chrysomela scripta,* is sensitive to the Cyt1A protein (Federici & Bauer, 1998).

An important property of Cyt proteins, based primarily on studies of Cyt1A, is the ability to synergize the toxicity of the Cry proteins with which they occur. For example, combination of Cyt1A with the Cry4A, Cry4B, or Cry11A proteins with which it occurs in the parasporal body of *B. thuringiensis* subsp. *israelensis* results in toxicities three- to fivefold higher than the specific toxicity of Cyt1A or any of the Cry proteins alone (Wu & Chang, 1985; Ibarra & Federici, 1986; Wu *et al.,* 1995; Crickmore *et al.,* 1996). The mechanism of this synergism is not known but probably involves cooperativity of the Cyt and Cry proteins in binding to and/or inserting into the microvillar membrane. Relevant to this is the finding that combination of Cyt1A with Cry4 or Cry11A proteins enables the latter to overcome high levels of resistance to them in mosquito *Culex quinquefasciatus* (Wirth *et al.,* 1997).

Structure

The Cyt class of *B. thuringiensis* endotoxins are 24–28 kDa in mass, and occur primarily in subspecies toxic to nematoceran dipterans. The crystal structure for Cyt2A has been solved, and based on sequence similarities among Cyt proteins, it is assumed all have a similar structure. In contrast to the three-domain structure of activated Cry toxins, the Cyt2A molecule is a single domain, consisting of a β-sheet core wrapped in two outer layers of α-helix hairpins (Li *et al.,* 1996). Owing to the latitude of correlating structure with function, the structure of Cyt2A can be used to support either the pore-forming or detergent-like mode of action for Cyt proteins.

MAJOR B.T. PATHOTYPES

The numerous isolates of B.t. that have been screened for insecticidal activity can be divided among three major pathotypes, those that exhibit toxicity to (1) lepidopterous, (2) dipterous, or (3) coleopterous insects. By far most isolates and subspecies fall into the first category, which from the turn of the century until the mid-1970s was the only pathotype known. The first isolate exhibiting substantial toxicity to dipterous insects, mainly to the larvae of nematocerous dipterans such as mosquitoes and blackflies, was the ONR 60A isolate of *B. thuringiensis* subsp. *israelensis* (H 14) discovered in Israel in 1976 (Goldberg & Margalit, 1977). Several years later, the first B.t. isolate with high toxicity to coleopterous insects, the "*tenebrionis*" isolate of *B. thuringiensis* subsp. *morrisoni* (H 8a8b), was discovered in Germany (Krieg *et al.,* 1983). Since these discoveries, isolates of these and other subspecies have been found that exhibit substantial toxicity to dipterous or coleopterous insects, but isolates with toxicity to lepidopterous insects remain the most common pathotype known. Isolates of B.t. active against nematodes have also been reported (Feitelson *et al.,* 1992).

Parasporal Body Shape, Complexity, and Pathotype

In general, the shape of the parasporal body is a good but not absolute indicator of an isolate's pathotype. For example, most isolates of B.t. produce a large bipyramidal parasporal crystal (0.5×1 μm) that is almost always only toxic to lepidopterous insects (Heimpel & Angus, 1963; Moar *et al.,* 1989). In isolates of B.t. active against lepidopterous insects, such as the HD1 isolate of *B. thuringiensis* subsp. *kurstaki* (H 3a3b), the bipyramidal crystal may be accompanied by a smaller cuboidal crystal toxic to both lepidopterous and dipterous insects (Yamamoto & McLaughlin, 1981). Others, such as the ONR60A isolate of *B. thuringiensis* subsp. *israelensis* (H 14) and the PG-14 isolate of *B. thuringiensis* subsp. *morrisoni* (H 8a8b), produce spherical parasporal bodies (0.7 to 1 μm) that are toxic primarily to nematocerous dipterans (e.g., mosquito and blackfly larvae; Federici *et al.* 1990), whereas the tenebrionis strain of *B. thuringiensis* subsp. *morrisoni* (H 8a8b) produces a thin, cuboidal crystal that is toxic only to certain species of coleopterans (Krieg *et al.,* 1987; Keller & Langenbruch, 1993).

The degree of protein complexity within the parasporal body of different isolates can vary considerably (Fig. 6). Single crystals can be composed of a single type of protein molecule or a mixture of as many as three. In addition, a single parasporal body may be composed of two (e.g., certain isolates of *B. thuringiensis* subsp. *kurstaki*) to four (e.g., *B. thuringiensis* subsp. *israelensis*) crystals (Hofte & Whiteley, 1989; Federici *et al.,* 1990). An example of a simple crystal is that of the HD-73 isolate of *B. thuringiensis* subsp. *kurstaki*. This isolate only encodes and produces the Cry1Ac protein that forms a typical bipyramidal crystal during sporulation. The related HD1 isolate of the same subspecies, however, carries at least five *cry* genes (*cry1Aa, cry1Ab, cry1Ac, cry2A,* and *cry2B*) and produces at least four of these encoded proteins (Cry1Aa, Cry1Ab, Cry1Ac, and Cry2A) during sporulation. The three Cry1A proteins cocrystallize, forming a single bipyramidal crystal, whereas

the smaller Cry2A protein forms the associated cuboidal inclusion. Another important subspecies of B.t. is *B. thuringiensis* subsp. *aizawai* (H 7). Typically, isolates of this subspecies produce a single bipyramidal crystal per cell in which Cry1Aa, Cry1Ab, Cry1C, and Cry1D have cocrystallized. The most complicated situation occurs in *B. thuringiensis* subsp. *israelensis,* where two Cry4 proteins, Cry11A, and two Cyt proteins—Cyt1A being the dominant one—crystallize into three different inclusion types that are bound together in a fibrous envelope (Federici *et al.,* 1990).

COMMERCIAL PRODUCTS BASED ON *BACILLUS THURINGIENSIS*

The first commercial products based on *B. thuringiensis* consisted of sporulated, lysed cells of the fermented natural isolates described above, formulated as powders or liquid emulsifiable concentrates. Products based on these isolates remain the most common types of products in use today (Table 3). However, in an effort to improve upon these products, studies of the molecular biology and genetics of *B. thuringiensis* led to the development of several new types of products in which B.t. genes have been manipulated in different ways. These products include bacterial insecticides based on transconjugate or recombinant bacteria, and insect-resistant transgenic plants. In transconjugate bacteria, also referred to as transconjugants, plasmids encoding *cry* genes that do not occur together naturally have been combined in a comon host cell, usually a derivative of *B. thuringiensis* subsp. *kurstaki*. In recombinant bacteria, one or more cloned *cry* genes are expressed in either *B. thuringiensis* or another bacterium such as *Pseudomonas flourescens,* to improve the insecticidal properties of the products. Impovements consist of increased activity and stability, or expanded pest spectra achieved by combining *cry* genes which encode proteins that target pests such as those belonging to the *Spodoptera* complex, which are difficult to control with products based on the HD1 isolate of *B. thuringiensis* subsp. *kurstaki*. Lastly, in transgenic plants, usually after optimizing gene codon usage for expression in plants, one or more *cry* genes have introduced into commercial crops such as cotton, corn, soybeans, and potatoes to control certain major lepidopteran and coleopteran pests that feed on these crops.

Products Based on Natural Isolates of *B. thuringiensis*

The isolates of the subspecies described above serve as the basis for most of the commerical bacterial insecticides used in pest control throughout the world. Whereas the proteins in the parasporal bodies of these isolates are the principle insecticidal components, the high insecticidal activity and broad insect spectrum of the HD1 isolate of *B.*

thuringiensis subsp. *kurstaki* and the ONR60A isolate of *B. thuringiensis* subsp. *israelensis* are not due to the proteins alone, but instead to synergistic interactions among the proteins, proteins and spores, or other molecules produced during fermentation. The isolates and strains used in commercial products from the simplest, *B. thuringiensis* subsp. *morrisoni* strain tenebrionis (DSM 2803), to the most complex, *B. thuringiensis* subsp. *israelensis* (ONR60A), are described below. It should be realized that in most cases these strains served as the starting materials for the selection of strains that are actually used in commercial products. Once an isolate with good insecticidal properties is identified, it is typically subjected to additional selection and screening in the laboratory to develop strains with consistent commercial properties such as high toxicity, stability, high fermentation biomass yield, and good shelf life. Most of the products based on B.t. isolates and strains are applied at a rate $<$ to 1 lb or 1 quart/acre.

Bacillus thuringiensis subsp. *morrisoni* (H8a8b), Isolate DSM 2803

The DSM 2803 isolate of *B. thuringiensis* subsp. *morrisoni*, strain tenebrionis, was originally isolated from diseased larvae of beetle *Tenebrio molitor* (Krieg *et al.,* 1983). Studies of this isolate showed that it produced a single thin rectangular crystal composed of a 67-kDa Cry3A protein toxic to many species of coleopterous insects (Keller & Langenbruch, 1993). Subsequent studies showed that the gene encoding the protein encodes a toxin of 73.1 kDa, but that the protein was processed before being packaged into the crystal. In the 15 years since the discovery of the DSM 2803, numerous other isolates of B.t.-producing Cry3 proteins have been obtained from soil and grain dust samples, and currently four different types of Cry3 proteins (Cry3A–D) are recognized.

Because of the need to control many coleopterous pests, especially the Colorado potato beetle, *Leptinotarsa decemlineata,* the DSM 2803 isolate and several others where quickly developed into commercial bacterial insecticides such as MTRAK[R], Trident[R], and Novodor[R] (Table 3). In addition, the *cry3A* gene was used to construct transgenic potatoes (NewLeaf[R], Monsanto Company), resistant to the Colorado potato beetle. While several of these products initially sold quite well, the advent of synthetic chemical insecticide Admire[R] (imidocloprid) has greatly reduced the market, at least at present, for bacterial insecticides based on Cry3 proteins.

Bacillus thuringiensis subsp. *kurstaki* (H3a3b), Isolates HD1 and NRD12

Probably the most common subspecies of B.t. isolated from nature is *B. thuringiensis* subsp. *kurstaki*. As noted

earlier, the parasporal body of this subspecies can vary considerably in complexity by having from only one protein to as many as four. The strains used in the first commercial formulations marketed over 30 years ago, and still commonly used today in products such as Dipel[R], Foray[R], and Thuricide[R] are those derived from the HD1 isolate (see Table 3). These products are widely used to control caterpillar pests of vegetable and field crops, ornamentals, and forests (Navon, 1993; van Frankenhuyzen, 1993). HD1 has several properties that contribute to its ongoing success. First, the parasporal body consists of four Cry proteins, Cry1Aa (133.2 kDa), Cry1Ab (131 kDa), Cry1Ac (132.3 kDa), and Cry2Aa (70.9 kDa). Though these vary in their insect spectrum and specific toxicity (Table 4), together they give this isolate a broad spectrum of activity against a wide range of caterpillar species attacking field crops (cotton, corn, and soybeans), vegetables (tomatoes, broccoli, lettuce, and cabbage), fruit (strawberries, grapes, and peaches), and forests (deciduous and fir trees). This protein complexity probably also accounts for the lack of economically important resistance after more than 30 years of use in all but a very few species of lepidopterans, the notable exception being larvae of the diamondback moth, *Plutella xylostella* (Tabashnik, 1994).

Second, in addition to the parasporal body proteins, HD1 produces several components that synergize or potentiate the toxicity of the Cry proteins. These include the synergist zwittermycin which is an organic molecule, and the spore (Moar *et al.;* 1989; Miyasono *et al.,* 1994). The mechanism of zwittermycin is unknown. The spore is a synergist in the broad sense of the word, and has its greatest effect against insects with moderate or low sensitivity to the Cry toxins, such as larvae of gypsy moth (*Lymantria dispar*), the diamondback moth (*P. xylostella*), and the beet armyworm (*Spodoptera exigua*). In larvae of these species, after an initial intoxication resulting from activation of the Cry toxins in the midgut, the spore germinates and produces enzymes such as phospholipases and proteases. In addition, insecticidal proteins produced during vegetative growth have been identified (Estruch *et al.,* 1996). These contribute to the permeabilization and lysis of the midgut epithelial cells.

Third, many isolates of *B. thuringiensis* subsp. *kurstaki* produce the β-exotoxin, an adenine nucleotide that acts as a competitive inhibitor of DNA-dependent RNA polymerase [see Lecadet and de Barjac (1981), and Sebesta *et al.* (1981) for reviews]. It occurs widely in many natural isolates of B.t. and in many countries it is not allowed in commercial B.t. preparations because it can be teratogenic for mammals. However, it is permitted in formulations of B.t. in Finland and in certain African countries where it is used to control filth breeding flies, and it is also used in Russia against insects such as the Colorado potato beetle. Though β-exotoxin has received relatively little attention recently, it does provide another interesting example of

TABLE 3 Representative Examples of Commercially Available Microbial Insecticides Based on
Bacillus thuringiensis[a]

Target pest	B.t. subspecies	Crop or habitat	Product[b, c]	Producer[d, e]
Caterpillars (Lepidoptera)	*B. thuringiensis kurstaki* (H3a3b)	Vegetables	Biobit	Abbott Laboratories
			Condor[f]	Ecogen
			Cutlass[f]	Ecogen
			Dipel	Abbott Laboratories
			Thuricide	Thermo-Trilogy
			Javelin	Thermo-Trilogy
			Toarow CT	Toagosei Chemical
		Forestry	Dipel	Abbott Laboratories
			Foray	Abbott Laboratories
			Thuricide	Thermo-Trilogy
	B. thuringiensis aizawai (H7)	Vegetables	Xentari	Abbott Laboratories
		Beehives	Certan[g]	Thermo-Trilogy
		Flowers	Clorbac	Abbott Laboratories
Beetles (Coleoptera)	*B. thuringiensis morrisoni*[h](H8a8b)	Vegetables	Foil[f]	Ecogen
			MTRAK	Mycogen
			Novodor	Abbott Laboratories
			Trident	Thermo-Trilogy
Mosquitoes and blackflies (Diptera)	*B. thuringiensis israelensis* (H 14)	Breeding waters	Vectobac	Abbott Laboratories
			Teknar	Thermo-Trilogy
			Skeetal	Abbott Laboratories

[a] From Shah and Goettel (1999) and technical bulletins from manufacturers.

[b] Most of these are general trade names that apply to a variety of different types of formulations such as flowable emulsifiable concentrates, wettable powders, and dusts. At present, over 100 different B.t. products are marketed throughout the world, with worldwide sales estimated at $100 million per annum.

[c] The target species for vegetables crops are numerous lepidopteran and coleopteran pests, among the most important of which are *Trichoplusia ni* (cabbage looper), *Heliocoverpa zea* (corn earworm), *Spodoptera exigua* (beet armyworm), *Plutella xylostella* (diamondback moth), *Ostrinia nubilalis* (European corn borer), and *Leptinotarsa decemlineata* (Colorado potato beetle). Major forest pests include the lepidopterans *Lymantria dispar* (gypsy moth) and *Choristoneura fumiferana* (spruce budworm).

[d] Addresses can be found in Shah and Goettel (1999), or the website www.sipwebiorg.

[e] This list is not meant to be exhaustive, but instead to provide the name of major present and upcoming manufacturers. There are many smaller companies in industrialized and developing countries that also produce a variety of B.t. products.

[f] Transconjugate bacteria.

[g] For control of larvae of the wax moth, *Galleria mellonela*.

[h] The name *"tenebrionis"* has been used to describe pathotypes of this subspecies/serotype toxic to coleopteran insects.

synergism between different potential components of B.t. Because it occurs widely in natural isolates, Dubois (1986) examined its potential to synergize the activity of B.t. spore–crystal mixtures. By using a derivative of the HD1 isolate of *B. thuringiensis* subsp. *kurstaki,* he found that the LC_{50} of this isolate against the gypsy moth larvae was reduced significantly by the addition of 0.01% β-exotoxin. Though synergism could be detected earlier, by day 11 postfeeding, larvae treated with only the β-exotoxin or B.t. preparation had suffered only 20% mortality, whereas mortality was greater than 75% in those treated with a combination of the two.

As in the case of the spores, the mechanism of synergism between the β-exotoxin and other B.t. components is not well defined. Based on its mode of action, one possibility is that the β-exotoxin interferes with protein synthesis in midgut cells, especially differentiating regenerative cells that require a high level of protein synthesis, and by doing so it impedes the larva's ability to repair damage to the midgut epithelium. This would hasten the action of endotoxins and spore germination.

Another isolate of *B. thuringiensis* subsp. *kurstaki* currently used in several commercial preparations is NRD12 isolated by Norman Dubois. This isolate, which has the same Cry toxin composition as the HD1 isolate (Moar *et al.,* 1989) was originally reported to have much better insecticidal activity against species of the genus *Spodoptera.* However, it was later demonstrated that much of this activity was due to low levels of β-exotoxin.

Bacillus thuringiensis subsp. *aizawai* (H7)

Although the HD1 isolate of *B. thuringiensis* subsp. *kurstaki* is the most widely used an successful isolate of B.t.,

TABLE 4   Toxicity of B.t. Cry Proteins to First Instars of Three
Lepidopteran Pest Species[a]

| Cry protein | LC$_{50}$ in ng/cm^2 of diet | | |
	Tobacco hornworm	Tobacco budworm	Cotton leaf worm
Cry1Aa	5.2	90	>1350
Cry1Ab	8.6	10	>1350
Cry1Ac	5.3	1.6	>1350
Cry1C	>128	>256	104

[a]Tobacco hornworm *(Manduca sexta)*, tobacco budworm *(Heliothis virescens)*, cotton leaf worm *(Spodoptera littoralis)*. From Hofte and Whiteley (1989).

products based on it are not always effective at economical rates in controlling certain noctuid pests, especially species of *Spodoptera* such as the beet armyworm (*S. exigua*), the fall armyworm (*S. frugiperda*), and cotton and sorghum pests that occur in other regions of the world such as *S. litura* and *S. littoralis*. This is due to the relative lack of sensitivity of species in this genus to the Cry1A proteins produced by HD1 (see Table 4). The search for isolates of B.t. that would be more effective against *Spodoptera* species led to the discovery of several isolates of *B. thuringiensis* subsp. *aizawai* that are more effective than HD1. These isolates typically produce a single bipyramidal crystal per cell that contains a complex of four proteins, Cry1Aa (133.2 kDa), Cry1Ab (131 kDa), Cry1C (134.8 kDa), and Cry1D (132.5). These isolates are the basis for the product Xentari[R] (see Table 3) recommended for use against *Spodoptera* species (as well as many others, including populations of the diamondback moth, *P. xylostella*, resistant to HD1 preparations), and Certan[R] recommended for control of the waxmoth, *Galleria mellonella*, in beehives (see Table 3). The Cry1C protein is especially important to the increased activity of *Bothuringiensis* subsp. *aizawai* against the beet armyworm and diamondback moth.

Bacillus thuringiensis subsp. *israelensis* (H14), isolate ONR60A

The second most widely used isolate of B.t. used in insect control is the ONR60A isolate of *B. thuringiensis* subsp. *israelensis*, which is used in many different regions of the world primarily for the control the larvae of nuisance and vector mosquito and blackfly species (Mulla, 1990; Becker & Ludwig, 1993; Becker & Margalit, 1993). Though used primarily against mosquito and blackfly larvae, this isolate is toxic to all species that have been tested of the dipteran suborder Nematocera (the "long-horned" flies), which includes flies such as the mushrooms flies

(family Sciaridae), crane flies (family Tipulidae), and midges (family Chironomidae).

The parasporal body of this isolate, which has an LC$_{50}$ in the range of 10 to 15 ng/ml of water against fourth instars of *Aedes* and *Culex* mosquitoes, is the most toxic per unit weight of all known B.t. isolates. This high toxicity and broad spectrum among nematocerous dipterans are due a complex of four major proteins, Cry4A (134 kDa), Cry4B (128 kDa), Cry11A (72 kDa), and Cyt1A (27.3 kDa). These are localized within three different inclusion types within the parasporal body, each of which is individually enveloped in a fibrous envelope of unknown composition, several layers of which also surround the entire parasporal body. The apparent function of this envelope is to hold the different toxin inclusions together to increase the probability that all toxins will enter the midgut simultaneously. After ingestion and dissolution in the midgut of sensitive insects, these proteins interact synergistically to bring about the high toxicity characteristic of this isolate. In addition to the major proteins, lesser amounts of related Cyt1A and Cry11 proteins also occur in the parasporal body. All these toxin proteins are encoded on a large plasmid of approximately 125 kb. Although this plasmid was initially identified in the ONR60A isolate of *B. thuringiensis* subsp. *israelensis*, it has been reported to occur quite widely in other mosquitocidal subspecies of B.t. including subsp. *thompsoni, entomocidus, kenyae,* and *morrisoni* (Ibarra & Federici, 1986; Ragni *et al.*, 1995; Lopez-Meza *et al.*, 1996).

An unusual feature of the Cry and Cyt toxins of *B. thuringiensis* subsp. *israelensis*, and other subspecies carrying the same 125-kb plasmid, is the ability of these proteins to interact synergistically. Though synergism was initially reported between combinations of the Cyt1A and Cry4 or Cry11A proteins (Wu & Chang, 1985; Ibarra & Federici, 1986; Chilcott & Ellar, 1988), it was later shown that combinations of Cry4 and/or Cry11A proteins are also synergistic (Crickmore *et al.*, 1995; Poncet *et al.*, 1995). As noted earlier, these synergistic interactions, an example of which is provided in Table 5, account for the broad nematocerous dipteran spectrum and high toxicity of *B. thuringiensis* subsp. *israelensis* to mosquito and blackfly larvae. In addition, laboratory studies by Georghiou and Wirth (1997) suggest that the Cyt1A protein delays the development of resistance to the Cry4 and Cry11A proteins. In a series of experiments using various combinations of these toxins against larvae of *Culex quinquefasciatus*, toxin combinations containing Cyt1A led to only a 3.2-fold level of resistance after 28 generations, whereas the levels of resistance in combinations of Cry toxins lacking Cyt1A ranged from 90- to 900-fold or more.

Based on the high efficacy of *B. thuringiensis* subsp. *israelensis*, a range of commercial products based on the ONR60A isolate were quickly developed shortly after its discovery (Mulla, 1990; Becker and Margalit 1993). Rep-

TABLE 5   Examples of Cyt1A: Cry11A Synergism against First Instars of the Mosquito *Aedes aegypti*

Protein	Ratio	LC$_{50}$ (ng/ml)[a]	Range (ng/ml)[b]	Slope
CytA	—	60a	45–104	1.96
Cry11A	—	85b	60–200	<0.9
CytA + Cry11A	1:1	14.8c	10–22	1.54
CytA + Cry11A	3:1	14.8c	10–22	1.64
CytA + Cry11A	1:3	13.7c	7–22	1.32
CytA + Cry11A	1:10	11.2c	6–18	1.44
CytA + Cry11A	10:1	15.2c	12–19	2.2
Parasporal body	—	2.5d	0.5–4.4	2.7

[a]Numbers followed by the same letter indicate no significant difference at the 95% confidence level.

[b]Confidence limits of 95%.

resentative examples products still marketed in the United States, Europe, and other countries are shown in Table 3. In the United States and Europe, these products are used primarily for the control of nuisance mosquitoes and black-flies, and in some situations, chironomid midges. However, the most important use of *B. thuringiensis* subsp. *israelensis* from a medical perspective has been in the World Health Organization's Onchocerciasis Control Program in West Africa. Formulations of *B. thuringiensis* subsp. *israelensis,* mainly Teknar[R] and Vectobac[R], have been used in the dry season since the mid-1980s to control blackfly larvae that had begun to develop resistance to the chemical insecticide Abate (Guillet *et al.,* 1990). The use of products based on *B. thuringiensis* subsp. *israelensis* preserved the Onchocerciasis Control Program, allowing the return of farmers to many river valleys that had become largely uninhabitable due to the prevalence of onchocerciasis.

Products based on *B. thuringiensis* subsp. *israelensis* have been in use now for well over a decade, and no significant resistance has been reported in field populations of mosquitoes, blackflies, or chironomids. In the Rhine Valley in Germany, where *B. thuringiensis* subsp. *israelensis* is used against the floodwater mosquito (*Aedes vexans*) for more than 10 years, specific tests for the development of resistance have indicated no decrease in the sensitivity of mosquito populations (Becker & Ludwig, 1993). This lack of resistance is probably due to two major factors, the complexity of the *B. thuringiensis* subsp. *israelensis* parasporal body as described earlier, and the treatment of only selected areas along the Rhine. Because *A. vexans* is capable of moving several miles from its origin, the it latter permits the seasonal and year-to-year mixing of treated and untreated mosquito populations (natural refugia), thereby reducing the buildup of resistance genes in the treated populations.

Products Based on Transconjugate or Recombinant Bacteria

Transconjugate Bacteria

Transconjugate bacteria used in commercial products are based on various strains of *B. thuringiensis* as the host cell. Typically, the *cry* gene complement of a recipient strain is increased by transforming the strain with a plasmid carrying a *cry1* or *cry3* gene. The introduced plasmid is used to amplify copies of genes already in the strain, for example *cry1Ac,* to increase the yield of a particular protein, or to introduce new *cry* genes to expand the spectrum of activity. Examples of products on the market based on transconjugate strains include Cutlass®, Condor®, and Foil®. In the first two, a strain of *B. thuringiensisis* subsp. *kurstaki* has been enhanced by transformation with a plasmid from *B. thuringiensis* subsp. *aizawai* that contains a *cry1A* gene to increase toxicity against certain lepidoptera. In Foil®, a plasmid bearing a *cry3A* gene was introduced into a strain of *B. thuringiensis* subsp. *kurstaki* that produces Cry1Ac, yielding a strain toxic to both lepidopteran and coleopteran insects (Baum *et al.,* 1998).

Recombinant Bacteria

During this decade, several bacterial strains constructed using recombinant DNA technology have been registered in the United States. The first products registered were developed by Mycogen (San Diego, CA), and used *Pseudomonas fluorescens* as the host strain. The bacterium was engineered to produce large amounts of wild-type or recombinant Cry proteins, after which the cell wall was chemically fixed around the crystal to provide protection from ultraviolet light. Examples of products and the protein(s) they contain include MVP®, which contains a Cry1Ac-Cry1Ab chimera toxic to lepidopteran insects; MTRAK®, which contains Cry3A toxic to coleopteran insects; and MATTCH®, which contains Cry1Ac and Cry1C, toxic to lepidopteran insects (this product is a physical mixture of two *P. flourscens* strains, each producing one of the toxins).

Ecogen (Langhorne, PA) pioneered a strategy in which strains of *B. thuringiensis* were engineered to produce more complex mixtures of toxins using recombinant plasmids. Current products on the market are CRYMAX®, which contains Cry1Ac, and Cry2A, and is toxic to lepidopteran insects; Lepinox®, which contains Cry1Aa, Cry1Ac, a Cry1Ac-Cry1F chimera, and Cry2A, toxic to lepidopteran insects and designed especially for the *Spodoptera* complex; and Raven®, which contains Cry1Aa, Cry3A and

Cry3Ba, and is toxic to both lepidopteran and coleopteran insects (Baum *et al.,* 1998).

in the field is a significant threat when pest populations are placed under intensive selection pressure.

RESISTANCE TO *BACILLUS THURINGIENSIS* INSECTICIDES

Though resistance to B.t. products among insect species under field conditions has been rare, laboratory studies show that insects are capable of developing high levels of resistance to one or more Cry proteins. Under laboratory selection, for example, populations of the Indian meal moth (*Plodia interpunctella*) developed levels of resistance ranging from 75- to 250-fold to Cry1Aa, Cry1Ab, Cry1Ac, Cry2A, and Cry1C. In addition, under heavy selection pressure in the laboratory, populations of mosquitoes (*C. quinquefasciatus*), beetles (the Colorado potato beetle and the cottonwood leaf beetle, *Chrysomela scripta*), and tobacco budworm (*Heliothis virescens*) all developed levels of resistance ranging from several 100- to several 1000-fold to the Cry toxins against which they were selected (see Tabashnik, 1994; Bauer, 1995 for reviews). Rotation of Cry proteins in bacterial insecticides is a potentially useful tactic for managing resistance to individual Cry proteins. However, because most Cry proteins are related, the potential for cross-resistance remains a major problem. In fact, high levels of cross-resistance among Cry proteins has already been demonstrated in the laboratory in populations of the tobacco budworm.

Whereas the preceding results were obtained using laboratory models, resistance in the field has become a major problem in the diamondback moth in Hawaii, Japan, the Philippines, and Florida, where populations were treated heavily and frequently (weekly or more) with products based on the HD1 isolate of *B. thuringiensis* subsp. *kurstaki.* In Hawaiian populations, because of cross-resistance, resistance extended to include Cry1Aa, Cry1Ab, Cry1Ac, Cry1F, and Cry1J, whereas certain Floridian populations were resistant to Cry1Aa, Cry1Ab, and Cry1Ac. These results make it clear that even with complex products containing mixtures of Cry endotoxins and synergists, resistance

TRANSGENIC *BACILLUS THURINGIENSIS* CROPS

The ability to clone B.t. endotoxin genes led quickly to the development of the first transgenic B.t. plants in the mid-1980s. Since then, many major crops that suffer substantial economic damage from caterpillar and beetle pests have been genetically engineered to produce Cry proteins to control these insect pests, though only a few of these are available commercially (see Jenkins, 1998; Schulet *et al.,* 1998 for recent reviews). In the United States, the available crops are cotton, corn, and potatoes (Table 6). Numerous others under development include rice, soybeans, broccoli, lettuce, walnuts, apples, and alfalfa. During the 1998 growing season, approximately 12 million acres of B.t.-corn were grown in the United States and 2.8 acres of B.t.-cotton. Several tens of thousands of acres of B.t.-potatoes were also grown. These amounts are expected to grow considerably over the next few years as more companies commercialize B.t.-crops. Moreover, if these initial crops prove to be economic successes, most minor crops will eventually be engineered to produce Cry and other proteins for control of their major invertebrate pests. A list of B.t.-transgenic crops currently under development can be found on the United States Department of Agriculture's website, hhtp://www.aphis.usda.gov/bbep/bp/.

Resistance Development Concerns

There has been concern about the use of B.t. transgenic crops ever since the concept was first developed. Initially, environmentalists raised concerns over the safety to vertebrates of eating products containing Cry proteins. However, because safety studies demonstrated that vertebrate stomach juices rapidly inactivate Cry proteins, the emphasis has shifted to concern over the development of resistance to Cry proteins in insect populations. Were this to occur, the value of microbial insecticides based on B.t. Cry proteins

TABLE 6   Insecticidal Protein Composition of B.t. Bacterial Insecticides versus Transgenic Plants

B.t. subspecies (isolate)	Cry and Cyt proteins							
	Cry1Aa	Cry1Ab	Cry1Ac	Cry2Aa	Cry4A	Cry4B	Cry11A	Cyt1A
Bacillus thuringiensis kurstaki (HD1)+	+	+	+	+				
B. thuringiensis israelensis (60A)					+	+	+	+
B.t. cotton			+					
B.t. corn			+					
B.t. corn		+						

could be greatly diminished because of the documented capacity, even under field conditions, of insect populations to develop cross-resistance. The potential for the development of resistance to Cry proteins is of concern not only to environmentalists, but also to most scientists in academia, government, and B.t. insecticide and transgenic plant industries. With respect to the latter, failure of B.t. crops in the field after more than a decade of development would result in significant economic losses to these companies.

The threat of resistance is real and easy to understand. Basically, the first generation of transgenic crops is based on plants that produce only a single Cry protein, and thus these lack the complexity described earlier of conventional B.t.-based bacterial insecticides. For example, current lines of B.t. cotton produce the Cry1Ac protein, and are targeted to control the most important cotton pest, the tobacco budworm, in the southeastern United States. This protein was selected for engineering into cotton because it is the most toxic to the tobacco budworm (see Table 4). Similarly, in the case of corn, most lines have been engineered to produce the Cry1Ab toxin to control the European corn borer (*Ostrinia nubilalis*), and in potatoes, to produce the Cry3A toxin to control the Colorado potato beetle.

In addition to the lack of complexity in comparison to bacterial insecticides, each of these toxins is produced continually by the plant. While perhaps an advantage from the standpoint of saving application costs, continuous production of toxin places the insect population under heavy selection pressure. Actually, over the past two growing seasons, studies of B.t. cotton in the United States and Australia have shown that Cry1Ac production in the plant decreases over the growing season, in some cases leading to sublethal doses for larvae near the end of the season. This is of particular concern because it is known from experience with conventional chemical insecticides that exposure of populations to sublethal doses of a toxin permits the survival of individuals that are heterozygous for toxin resistance genes, allowing these genes to build up in the population.

Because first generation B.t. crops generally target only a single pest species, another problem they present is lack of adequate control of insects not very sensitive to the toxin produced by the crop. It is important to realize that lack of or low sensitivity is not resistance, but nevertheless can lead to control problems and resistance to other toxins. For example, as can be seen from Table 4, the Cry1Ac protein is not very toxic to species of armyworms (*Spodoptera* species) or bollworms (*Helicoverpa* species). Already problems have been encountered in the United States and Australia in B.t. cotton with bollworms. In the United States, corn earworm (*H. zea*) populations invaded limited acreages of B.t. cotton in Texas, virtually destroying the crop because of their high tolerance to Cry1Ac. In 1998 in Australia, where the principal cotton pests are bollworms (*H. armigera* and *H. punctigera*), which are also quite tolerant

to Cry1Ac, the efficacy of the B.t. cotton lasted for only approximately half the season. Applications of other pesticides, including B.t.-based insecticides, had to be made to control these pests. The lack of control resulted not only from the lower sensitivity of the bollworms to Cry1Ac, but from the levels of toxin decreasing in the cotton plants as the growing season progresses. A danger of decreasing toxin levels is that this could prime target populations for the development of resistance to other Cry proteins to which these populations may initially be quite sensitive.

Bacillus thuringiensis Resistance Management in Transgenic Crops

The possibility of resistance to transgenic crops has prompted the development of a variety of conceptual strategies for managing resistance (McGaughey & Whalon, 1992; Tabashnik, 1994; Gould, 1998). The most prominent of these are listed in Table 7. Most of these, for example, pyramiding various toxin genes within plants, use of tissue-specific toxin production, and induced toxin synthesis in which toxin is only synthesized after an insect begins to feed, are years away from field deployment or commercial availability. In the meantime, resistance management relies primarily on the refuge strategy. In this strategy, some percentage of the crop, usually 4 to 30%, must consist of non-B.t. plants.

The value of the non-B.t. plants is to maintain a high percentage of susceptible insects (i.e., frequency of susceptible genes) in the target population. When moths lay eggs on B.t. plants, a high percentage of the first instars that feed on these plants will die, because this is the stage most sensitive to the toxin. However, a high percentage of the larvae that emerge and feed on non-B.t. plants will survive, and theoretically mate with adults that survived growth on B.t. plants. The latter survivors presumably survived because they were heterozygous or homozygous for resistance. They should constitute a low percentage of the mating population, and by mating with insects not selected for

TABLE 7   Key Strategies for Managing Resistance to B.t. Crops[a]

Protein pyramiding	Multiple Cry genes
	Cry plus other insecticidal genes
Protein synthesis	Consitutive
	Tissue specific
	Chloroplast specific
	Inducible
Field tactics	Refuges
	spatial, temporal
	Cry gene crop rotation

[a]Modified from McGaughey, W. H., & Whalon, M. E. (1992). *Science, 258,* 1451–1455.

resistance, the percentage of resistance genes is diluted and remains low in the target population.

To delay resistance, the refuge strategy is being used in B.t. cotton and B.t. corn in the United States. B.t. cotton has been planted for 3 years, but it is not yet possible to assess the success of this strategy. During the first 3 years, there has been no confirmed evidence of resistance. However, aside from the refuges planted in the crops, many surrounding non-B.t. crops and noncrop plants provide refuges for susceptible insects, which then contribute to the dilution of resistance genes in the target population gene pool. The true test of this strategy will come when large contiguous areas, comprising square miles of B.t. crops producing Cry genes, are planted successively for several years.

While these experiments in nature are ongoing, more sophisticated plant engineering strategies are being used to engineer resistance management strategies directly into the plants. In addition to pyramiding Cry genes, mixtures of insecticidal proteins with different modes of action will be engineered into plants. Already several types of proteins to meet these needs have been identified, including non-Cry proteins from B.t. insecticidal non-B.t. toxins, lectins, and enzymes selectively toxic to specific insects (Jenkins, 1998; Schuler *et al.*, 1998).

Other Concerns about *Bacillus thuringiensis* Crops

Concerns have been raised about the safety of B.t. proteins, and therefore B.t. crops, to nontarget insects, especially the predators and parasites used as biological control agents. It must be realized that B.t. kills the target pest, but even B.t. insecticides also cause mortality in certain nontarget insects. B.t. products used to control caterpillar pests such as gypsy moth and spruce budworm larvae in forests, for example, will also kill certain species of nontarget lepidopteran larvae in these habitats. The mortality caused in the nontarget populations must be kept in perspective and viewed in the context of the relative risk of using B.t. in comparison with using available synthetic chemical insecticides. The latter typically have a much broader spectrum of toxicity, and will kill pest and nontarget insects belonging to a wide range of insect orders. Because all B.t. proteins have a very restricted spectrum of activity, each usually only exhibiting toxicity to insects of a single order (Lepidoptera, Diptera, or Coleoptera), the use of B.t. proteins is much more environmentally compatible than is the use of chemical insecticides. B.t. proteins may persist in the environment, when used in either insecticides or B.t. crops, but their toxic effects on nontarget predators and parasites are low and temporary, being reduced even further after the crop is harvested.

It should also be obvious that as a control agent targeted to kill insect pests, B.t. crops will also reduce the predator and parasite populations, particularly the latter, that depend on the target insect for their reproduction. Again, the mortality caused by B.t. in these nontarget populations must be kept in perspective. Most crop production, especially that of field, vegetable, and fruit crops, occurs in monocultures that are not natural ecological habitats. The crop uniformity characteristic of these monocultures permits pest insects as well as their predators and parasites to build up into populations that are much larger than those that would occur in more diverse natural habitats. In this context, the use of B.t. crops has a neutral effect in that it reduces predators and parasite populations that would not occur if it were not for the unnatural presence of a large crop habitat and concomitantly large host pest populations that the predators and parasites use as a resource. Moreover, even in the unnatural ecological habitat of a crop monoculture, the use of B.t. because of its greater specificity, will have less impact overall on the predator and parasite populations than will the use of broad-spectrum chemical insecticides.

References

Angus, T. A. (1954). A bacterial toxin paralyzing silkworm larvae. Nature (London), 173, 545–546.

Aronson, A. I. (1993). The two faces of *Bacillus thuringiensis*: insecticidal proteins and post-exponential survival. Molecular Microbiology, 7, 489–496.

Ash, C., Farrow, J. A. E., Wallbanks, S., & Collins, M. D. (1991). Phylogenetic heterogeneity of the genus *Bacillus* revealed by comparative analysis of small-subunit-ribosomal RNA sequences. Letters in Applied Microbiology, 13, 202–206.

Bauer, L. S. (1990). Resistance: a threat to the insecticidal crystal proteins of *Bacillus thuringiensis*. Florida Entomologist, 78, 414–443.

Baum, J. A., Johnson, T. B., & Carlton, B. C. (1998). *Bacillus thuringiensis*: natural and recombinant bioinsecticide products. In F. R. Hall & J. J. Menn (Eds.), Biopesticides: use and delivery (pp. 189–209). Totowa: Humana Press.

Baumann, L., Okamoto, K., Unterman, B. M., Lynch, M. J., & Baumann, P. (1984). Phenotypic characterization of *Bacillus thuringiensis* (Berliner) and *B. cereus* (Frankland & Frankland). Journal of Invertebrate Pathology, 44, 329–341.

Becker, N., & Ludwig, M. (1993). Investigations on possible resistance in *Aedes vexan* after a 10-year application of *Bacillus thuringiensis israelensis*. Journal of the American Mosquito Control Association, 9, 221–224.

Becker, N., & Margalit, J. (1993). Use of *Bacillus thuringiensis israelensis* against mosquitoes and blackflies. In P. F. Entwistle, J. S. Cory, M. J. Bailey, & S. Higgs (Eds.), Bacillus thuringiensis, an environmental biopesticide: theory and practice (pp. 147–170). London: J. Wiley & Sons.

Berliner, E. (1915). Ueber die schlaffsucht der *Ephestia kuhniella* und *Bac. thuringiensis* n. sp. Z. Allg. Entomologie, 2, 21–56.

Butko, P., Huang, F., Pusztai-Carey, M., & Surewicz, W. K. (1996). Membrane permeabilization induced by cytolytic delta-endotoxin CytA from *Bacillus thuringiensis* var. *israelensis*. Biochemistry, 35, 11355–11360.

Butko, P., Huang, F., Pusztai-Carey, M., & Surewicz, W. K. (1997). Interaction of the delta-endotoxin CytA from *Bacillus thuringiensis* var. *israelensis* with lipid membranes. Biochemistry, 36, 12862–12868.

Carlson, C. R., & Kolsto, A.-B. (1993). A complete physical map of a *Bacillus thuringiensis* chromosome. Journal of Bacteriology, 175, 1053–1060.

Chen, X. J., Lee, M. K., & Dean, D. H. (1993). Site-directed mutations in a highly conserved region of *Bacillus thuringiensis* δ-endotoxin affect inhibition of short circuit current across *Bombyx mori* midguts. Proceedings of the National Academy of Sciences, USA, 90, 9041–9045.

Chilcott, C. N., & Ellar, D. J. (1988). Comparative toxicity of *Bacillus thuringiensis* var. *israelensis* crystal proteins *in vivo* and *in vitro*. Journal of General Microbiology, 134, 2551–2558.

Crickmore, N., Bone, E. J., Williams, J. A., & Ellar, D. J. (1995). Contribution of the individual components of the delta-endotoxin crystal to the mosquitocidal activity of *Bacillus thuringiensis* subsp. *israelensis*. FEMS Microbiology Letters, 131, 249–254.

Crickmore, N., Zeigler, D. R., Feitelson, J., Schnepf, E., Van Rie, J., & Lereclus, D., (1998). Revision of the nomenclature for the *Bacillus thuringiensis* pesticidal crystal proteins. Microbiology and Molecular Biology Reviews, 62, 807–813.

de Barjac, H., & Franchon, E. (1990). Classification of *Bacillus thuringiensis* strains. Entomophaga, 35, 233–240.

De Maagd, R. A., Kwa, M. S. G., van der Klei, H., Yamamoto, T., Schipper, B., Vlak, J. M., Stiekema, W. J., & Bosch, D. (1996). Domain III substitution in *Bacillus thuringiensis* CryIA(b) results in superior toxicity for *Spodoptera exigua* and altered membrane protein recognition. Applied and Environmental Microbiology, 62, 1537–1543.

Dubois, N. R. (1986). Synergism between β-exotoxin and *Bacillus thuringiensis* subspecies *kurstaki* (HD-1) in gypsy moth, *Lymantria dispar*, larvae. Journal of Invertebrate Pathology, 48, 146–151.

Estruch, J. J., Warren, G. W., Mullins, M. A., Nye, G. J., Craig, J. A., & Koziel, M. G. (1996). Vip3A, a novel *Bacillus thuringiensis* vegetative insecticidal protein with a wide spectrum of activities against lepidopteran insects. Proceedings of the National Academy of Sciences USA, 93, 5389–5304.

Federici, B. A. (1994). *Bacillus thuringiensis*: Biology, application, and prospects for further development. In R. J. Akhurst (Ed.), Proceedings of the Second Canberra Meeting on *Bacillus thuringiensis* (pp. 1–15). Canberra: CPN Publications Pty, Ltd.

Federici, B. A., & Bauer, L. S. (1998). Cyt1Aa protein of *Bacillus thuringiensis* is toxic to the cottonwood leaf beetle, *Chrysomela scripta*, and suppresses high levels of resistance to Cry3Aa. Applied and Environmental Microbiology, 64, 4368–4371.

Federici, B. A., Luthy, P., & Ibarra, J. E. (1990). The parasporal body of BTI: structure, protein composition, and toxicity. In H. de Barjac & D. Sutherland (Eds.), Bacterial control of mosquitoes and blackflies; biochemistry, genetics, and applications of *Bacillus thuringiensis* and *Bacillus sphaericus* (pp. 16–44). New Brunswick: Rutgers University Press.

Francis, B. R., & Bulla, Jr., L. A. (1997). Further characterization of BT-R₁, the cadherin-like receptor for Cry1Ab toxin in tobacco hornworm (*Manduca sexta*) midguts. Insect Biochemistry and Molecular Biology, 27, 541–550.

Feitelson, J. S., Payne, J., & Kim, L. (1992). *Bacillus thuringiensis*: Insects and beyond. Bio/Technology, 10, 271–276.

Garczynski, S. F., Crim, J. W., & Adang, M. J. (1991). Identification of putative brush border membrane-binding molecules specific to *Bacillus thuringiensis* δ-endotoxin by protein blot analysis. Applied and Environmental Microbiology, 57, 2816–2820.

Ge, A. Z., Shivarova, N. I., & Dean, D. H. (1989). Location of the *Bombyx mori* specificity domain on a *Bacillus thuringiensis* δ-endotoxin protein. Proceedings of the National Academy of Sciences, USA, 86, 4037–4041.

Georghiou, G. P., & Wirth, M. C. (1997). Influence of exposure to single versus multiple toxins of *Bacillus thuringiensis* subsp. *israelensis* in development of resistance in the mosquito *Culex quinquefasciatus* (Diptera: Culicidae). Applied and Environmental Microbiology, 63, 1095–1101.

Gill, S. S., Cowles, E. A., & Pietrantonio. (1992). The mode of action of *Bacillus thuringiensis* endotoxins. Annual Review of Entomology, 37, 615–636.

Goldberg, L. J., & Margalit, J. (1977). A bacterial spore demonstrating rapid larvicidal activity against *Anopheles sergentii, Uranotaenia unguiculata, Culex univitattus, Aedes aegypti,* and *Culex pipiens*. Mosquito News, 37, 355–358.

Gould, F. (1998). Sustainability of transgenic insecticidal cultivars: integrating pest genetics and ecology. Annual Review of Entomology, 43, 701–726.

Gould, F., Martinez-Ramirez, A., Anderson, A., Ferre, J., Silva, F. J., & Moar, W. J. (1992). Broad-spectrum resistance to *Bacillus thuringiensis* toxins in *Heliothis virescens*. Proceedings of the National Academy of Sciences, USA, 89, 7986–7990.

Guillet, P., Kurtak, D. C., Philippon, B., & Meyer, R. (1990). Use of *Bacillus thuringiensis israelensis* for Onchocerciasis control in West Africa. In H. de Barjac & D. Sutherland (Eds.), Bacterial control of mosquitoes and blackflies; biochemistry, genetics, and applications of *Bacillus thuringiensis* and *Bacillus sphaericus* (pp. 187–201). New Brunswick: Rutgers University Press.

Hannay, C. L. (1953). Crystalline inclusions in aerobic spore-forming bacteria. Nature (London), 172, 1004–1006.

Heimpel, A. M. (1967). A critical review of *Bacillus thuringiensis* var. *thuringiensis* Berliner and other crystalliferous bacteria. Annual Review of Entomology, 12, 287–322.

Heimpel, A. M., & Angus, T. A. (1963). Diseases caused by certain sporeforming bacteria. In E. A. Steinhaus (Ed.), Insect pathology: an advanced treatise, (Vol. 2, pp. 21–73). New York: Academic Press.

Hofmann, C., Vanderbruggen, H., Hofte, H., Van Rie, J., Jansen, S., & Van Mallaert, H. (1988). Specificity of *Bacillus thuringiensis* δ-endotoxins is correlated with the presence of high affinity binding sites in the brush border membrane of target insect midguts. Proceedings of the National Academy of Sciences, USA, 85, 7844–7888.

Hofte, H., & Whiteley, H. R. (1989). Insecticidal crystal proteins of *Bacillus thuringiensis*. Microbiological Reviews, 53, 242–255.

Huber, H. E., & Luthy, P. (1981). Bacillus thuringiensis delta-endotoxin: composition and activation. In E. W. Davidson (Ed.), Pathogenesis of invertebrate microbial diseases (pp. 209–234). Totowa: Allanheld, Osman & Co.

Ibarra, J. E., & Federici, B. A. (1986). Isolation of a relatively non-toxic 65-kilodalton protein inclusion from the parasporal body of *Bacillus thuringiensis* subsp. *israelensis*. Journal of Bacteriology, 165, 527–533.

Ishawata, S. (1901). On a type of severe flacherie (sotto disease). Dainihon Sanshi Kaiho, 114, 1–5.

Jenkins, J. J. (1998). Transgenic plants expressing toxins from *Bacillus thuringiensis*. In F. R. Hall & J. J. Menn (Eds.), Biopesticides: use and delivery (pp. 211–232). Totowa: Humana Press.

Keller, B., & Langenbruch, G. A. (1993). Control of coleopteran pests by Bacillus thuringiensis. In P. F. Entwistle, J. S. Cory, M. J. Bailey, & S. Higgs (Eds.), *Bacillus thuringiensis,* an environmental biopesticide: theory and practice (pp. 171–191). London: J. Wiley & Sons.

Knowles, B. H., & Dow, J. A. T. (1994). The crystal δ-endotoxins of *Bacillus thuringiensis*: Models for their mechanism of action on the insect gut. BioEssays, 15, 469–476.

Knowles, B. H., & Ellar, D. J. (1987). Colloid-osmotic lysis is a general feature of the mechanism of action of *Bacillus thuringiensis* δ-endotoxins with different insect specificity. Biochemica Biophysica Acta, 924, 509–518.

Koni, P. A., & Ellar, D. J. (1993). Cloning and characterization of a novel *Bacillus thuringiensis* cytolytic toxin δ-endotoxin. Journal of Molecular Biology, 229, 319–327.

Krieg, A., Huger, A., Langenbruch, G., & Schnetter, W. (1983). *Bacillus thuringiensis* var. *tenebrionis*: a new pathotype effective against larvae of Coleoptera. Journal of Applied Entomology, 96, 500–508.

Krieg, A., Schnetter, W., Huger, A. M., & Langenbruch, G. A. (1987). *Bacillus thuringiensis* subsp. *tenebrionis,* Strain BI 256-82: a third pathotype within the H-Serotype 8a:8b. Systematic and Applied Microbiology, 9, 138–141.

Lecadet, M.-M., & Debarjac, H. (1981). *Bacillus thuringiensis* beta-exotoxin. In E. W. Davidson (Ed.), Pathogenesis of invertebrate microbial diseases (pp. 293–321). Totowa: Allanheld, Osman & Co.

Lee, M. L., Milne, R. E., Ge, A., & Dean, D. H. (1992). Location of a *Bombyx mori* receptor binding region on a *Bacillus thuringiensis* δ-endotoxin. Journal of Biological Chemistry, 267, 3115–3121.

Li, J., Carroll, J., & Ellar, D. J. (1991). Crystal structure of insecticical δ-endotoxin from *Bacillus thuringiensis* at 2.5 Å resolution. Nature (London), 353, 815–821.

Li, J., Koni, P. A., & Ellar, D. J. (1996). Structure of the mosquitocidal δ-endotoxin CytB from *Bacillus thuringiensis* p. *kyushuensis* and implications for membrane pore formation. Journal of Molecular Biology, 257, 129–152.

Lopez-Meza, J. E., Federici, B. A., Poehner, W. J., Martinez-Castillo, M., & Ibarra, J. E. (1995). Highly mosquitocidal isolates of *Bacillus thuringiensis* subsp. *kenyae* and *entomocidus* from Mexico. Biochemical Systematics and Ecology, 23, 461–468.

Luthy, P., & Ebersold, H. R. (1981). *Bacillus thuringiensis* delta-endotoxin: histopathology and molecular mode of action. In E. W. Davidson (Ed.), Pathogenesis of invertebrate microbial diseases (pp. 235–267). Totowa: Allanheld, Osman & Co.

Meadows, M. P. (1993). *Bacillus thuringiensis* in the environment: ecology and risk assessment. In P. F. Entwistle, J. S. Cory, M. J. Bailey, & S. Higgs (Eds.), Bacillus thuringiensis, an environmental biopesticide: theory and practice (pp. 193–220). London: J. Wiley & Sons.

McGaughey, W. H., & Whalon, M. E. (1992). Managing insect resistance to *Bacillus thuringiensis* toxins. Science, 258, 1451–1455.

Miyasono, M., Inagaki, S., Yamamoto, M., Ohba, K., Ishiguro, T., Takeda, R., & Hayashi, Y. (1994). Enhancement of δ-endotoxin activity by toxin-free spore of *Bacillus thuringiensis* against the diamondback moth, *Plutella xylostella*. Journal of Invertebrate Pathology, 63, 111–112.

Moar, W. J., Trumble, J. T., & Federici, B. A. (1989). Comparative toxicity of spores and crystals from the NRD-12 and HD-1 strains of *Bacillus thuringiensis* subsp. *kurstaki* to neonate beet armyworm (Lepidoptera: Noctuidae). Journal of Economic Entomology, 82, 1593–1603.

Mulla, M. S. (1990). Activity, field efficacy, and use of *Bacillus thuringiensis israelensis* against mosquitoes. In H. de Barjac & D. Sutherland (Eds.), Bacterial control of mosquitoes and blackflies; biochemistry, genetics, and applications of Bacillus thuringiensis and Bacillus sphaericus (pp. 134–160). New Brunswick: Rutgers University Press.

Navon, A. (1993). Control of lepidopteran pests with *Bacillus thuringiensis*. In P. F. Entwistle, J. S. Cory, M. J. Bailey, & S. Higgs (Eds.), Bacillus thuringiensis, an environmental biopesticide: theory and practice (pp. 125–146). London: J. Wiley & Sons.

Poncet, S., Delecluse, A., Klier, A., & Rapoport, G. (1995). Evaluation of synergistic interactions among CryIVA, CryIVB and CryIVD toxic components of *Bacillus thuringiensis* subsp. *israelensis* crystals. Journal of Invertebrate Pathology, 66, 131–133.

Priest, F. G., Kaji, D. A., Rosato, Y. B., & Canhos, V. P. (1994). Characterization of *Bacillus thuringiensis* and related bacteria by robosomal RNA gene restriction fragment length polymorphisms. Microbiology, 140, 1015–1022.

Ragni, A., Thiery, I., & Delecluse, A. (1995). Characterization of six highly mosquitocidal *Bacillus thuringiesis* strains that do not belong to H-14 serotype. Current Microbiology, 31, 1–7.

Sangadala, S., Walters, F. W., English, L. H., & Adang, M. J. (1994). A mixture of *Manduca sexta* aminopeptidase and alkaline phosphatase enhances *Bacillus thuringiensis* insecticidal CryIA(c) toxin binding and $^{86}Rb^{Ipl}$-K^+ efflux *in vitro*. Journal of Biological Chemistry, 269, 10088–10092.

Schnepf, E., Crickmore, N., Van Rie, J., Lereclus, D., Baum, J., & Feitelson, J. (1998). *Bacillus thuringiensis* and its pesticidal proteins. Microbiology and Molecular Biology Reviews, 62, 775–806.

Schnepf, H. E., & Whiteley, H. R. (1981). Cloning and expression of the *Bacillus thuringiensis* crystal protein gene in *Escherichia coli*. Proceedings of the National Academy of Sciences USA, 78, 2893–2897.

Schuler, T. H., Poppy, G. M., Kerry, B. R., & Denholm, I. (1998). Insect-resistant transgenic plants. Trends in Biotechnology, 16, 168–175.

Schwartz, J.-L., Garneau, Savaria, D., Masson, L., Brousseau, R., & Rousseau, E. (1993). Lepidopteran-specific crystal toxins from *Bacillus thuringiensis* form cation- and anion-selective channels in planar lipid bilayers. Journal of Membrane Biology, 132, 53–62.

Sebesta, K., Farkas, J., Horska, K., & Vankova, J. (1981). Thuringiensin, the beta-exotoxin of Bacillus thuringiensis. In H. D. Burges (Ed.), Microbial control of pests and plant diseases, 1970–1980 (pp. 249–281). London: Academic Press.

Shah, P. A., & Goettel, M. S. (1999). Directory of microbial control products and services. Gainesville: Society for Invertebrate Pathology Press.

Smith, R. A., & Couche, G. A. (1991). The phylloplane as a source of *Bacillus thuringiensis* variants. Applied and Environmental Microbiology, 57, 311–315.

Tabashnik, B. E. (1994). Evolution of resistance to *Bacillus thuringiensis*. Annual Review of Entomology, 39, 47–79.

Thiery, I., & Franchon, E. (1997). Bacteria: identification, isolation, culture and preservation of entomopathogenic bacteria. In L. A. Lacey (Ed.), Manual of techniques in insect pathology (pp. 55–77). San Diego: Academic Press.

Thomas, W. E., & Ellar, D. J. (1983). *Bacillus thuringiensis* var. *israelensis* crystal delta-endotoxin: effects on insects and mammalian cells *in vitro* and *in vivo*. Journal of Cell Science, 60, 181–197.

Thomas, W. E., & Ellar, D. J. (1983). Mechanism of action of *Bacillus thuringiensis* var. *israelensis* insecticidal δ-endotoxin. FEBS Letters, 154, 362–368.

van Frankenhuyzen, K. (1993). The challenge of *Bacillus thuringiensis*. In P. F. Entwistle, J. S. Cory, M. J. Bailey, & S. Higgs (Eds.), Bacillus thuringiensis, an environmental biopesticide: theory and practice (pp. 1–35). London: J. Wiley & Sons.

van Frankenhuyzen, K. (1993). The challenge of *Bacillus thuringiensis*. In P. F. Entwistle, J. S. Cory, M. J. Bailey, & S. Higgs (Eds.), Bacillus thuringiensis, an environmental biopesticide; theory and practice (pp. 1–35). Chichester: J. Wiley & Sons.

Van Rie, J., Jansens, S., Hofte, H., Degheele, D., & Van Mellaert, H. (1989). Specificity of *Bacillus thuringiensis* δ-endotoxins: Importance of specific receptors on the brush border membrane of the midgut of target insects. European Journal of Biochemistry, 186, 239–247.

Van Rie, J., Jansens, S., Hofte, H., Degheele, D., & Van Mallaert, H. (1990). Receptors on the brush border membrane of the insect midgut as determinants of the specificity of *Bacillus thuringiensis* delta-endotoxins. Applied and Environmental Microbiology, 56, 1378–1385.

Wirth, M. C., Georghiou, G. P., & Federici, B. A. (1997). CytA enables CryIV endotoxins of *Bacillus thuringiensis* to overcome high levels of CryIV resistance in the mosquito, *Culex quinquefaciatus*. Proceedings of the National Academy of Sciences USA, 94, 10536–10540.

Wolfersberger, M. G. (1990). The toxicity of two *Bacillus thuringiensis* delta δ-endotoxins to gypsy moth larvae is inversely related to the affinity binding cites on the brush border membranes for toxins. Experientia, 46, 475–477.

Wu, D., & Aronson, A. I. (1992). Localized mutagenesis defines regions of the *Bacillus thuringiensis* δ-endotoxin involved in toxicity and specificity. Journal of Biological Chemistry, 267, 2311–2317.

Wu, D., & Chang, F. N. (1985). Synergism in mosquitocidal activity of 26 and 65 kDa protein from *Bacillus thuringiensis* subsp. *israelensis* crystal. FEBS Letters, 190, 232–236.

Wu, D., & Federici, B. A. (1993). A 20-kilodalton protein preserves cell viability and enhances CytA crystal formation during sporulation in *Bacillus thuringiensis*. Journal of Bacteriology, 175, 5276–5280.

Wu, D., Johnson, J. J., & Federici, B. A. (1994). Synergism of mosquitocidal toxicity between CytA and CryIVD proteins using inclusions produced from cloned genes of *Bacillus thuringiensis*. Molecular Microbiology, 13, 965–972.

Yamamoto, T., & McLaughlin, R. E. (1981). Isolation of a protein from the parasporal crystal of *Bacillus thuringiensis* var. *kurstaki* toxic to the mosquito larva, *Aedes taeniorhynchus*. Biochemical and Biophysical Research Communications, 103, 414–421.

Nutrition of Entomophagous Insects and Other Arthropods

S. N. THOMPSON

Department of Entomology
University of California
Riverside, California

K. S. HAGEN

Center for Biological Control
University of California,
Berkeley
Albany, California

INTRODUCTION

Significant advances have been made in our knowledge of the nutrition of entomophagous insects and arthropods since this topic was discussed by Doutt (1964) and Hagen (1964). Many nutritional factors important or essential for growth, development, and reproduction have been identified, and success has been achieved at feeding and rearing several entomophagous species in the absence of their natural food. This success has confirmed that the nutrition of these organisms represents a complex and often tritrophic interaction of physiological, behavioral, and ecological factors involving the entomophage, its host or prey, and the host or prey food source (Price *et al.*, 1980; Barbosa *et al.*, 1982; Price, 1986; Hare, 1992). The interaction is often considered the purview of nutritional ecology (Slansky, 1982, 1986; Hagen, 1987; Slansky & Rodriguez, 1987; Vinson & Barbosa, 1987).

Host Effects on the Biological Attributes of Parasitoids

The complexity of parasitoid nutrition was alluded to by Salt (1941), who stated that ". . . far from being a purely passive victim, obliterated without a trace, the host is often able to impress its mark . . . upon the insect parasitoid that destroys it. There can be no doubt that the host may bequeath to its parasite an important and sometimes striking legacy of morphological, physiological, and behavioristic characters." Early studies demonstrated that the host influences the growth and survival of the developing parasite as well as the sex ratio, fecundity, longevity, and vigor of the

adult wasp (Flanders, 1935; Clausen, 1939; Salt, 1941). These various host effects have subsequently been described and investigated by many others (Arthur & Wylie, 1959; Wylie, 1967; Nozato, 1969; Sandlan, 1979a, 1979b; Vinson & Iwantsch, 1980; Charnov *et al.*, 1981; Charnov, 1982; Luck *et al.*, 1982; Werren, 1984; Mackauer, 1986; Opp & Luck, 1986; Strand, 1986; Waage, 1986; Mackauer & Sequeira, 1993; King, 1987, 1989; Kazmer & Luck, 1995; Mackauer *et al.*, 1997).

Host and Parasitoid Size and the Effects of Parasitoid Size on Fitness

A common observation concerns the relationship between host biomass and size of solitary parasitoids. Larger parasitoids generally develop from larger hosts. The relationship has been described for parasitoid species that attack every host developmental stage, but applies more generally to parasitoids of host eggs and pupae where host size is fixed, that is, with idiobiontic parasites (Askew & Shaw, 1986). Idiobionts are generally but not always ectoparasitic in habit (see Godfray, 1994; Quickie, 1997). The relationship holds when a parasitoid is reared on different host species of variable size as well as when reared on different sized individuals of the same metamorphic stage of a single host species (Salt, 1940; Jowyk & Smilowitz, 1978; Mellini & Campadelli, 1982; Sandlan, 1982; Mellini & Beccari, 1984; Harvey *et al.*, 1994, 1995).

Bai (1986) compared the size of adult *Trichogramma pretiosum* Riley reared on the eggs of five host species and demonstrated a direct correlation between parasitoid size and volume of the host egg from which it emerged. The effects of host size on quality of *T. pretiosum* were investigated by Bai *et al.* (1992) who reported that female wasps emerging from natural hosts were larger, more fecund, and longer lived than those reared from fictitious hosts. Parasitoid size was related to the size of the host egg, even when the wasp was gregarious, and in the latter case parasitoid size was related to the total number that emerged. The authors concluded that females of *T. pretiosum* measure host size and allocate eggs in a manner that produces a specific size distribution of females wasps that emerge from natural hosts, and suggested that size-related components of fitness are under strong selection pressure.

The size–fitness hypothesis in adult *T. pretiosum* under field conditions was examined by Kazmer and Luck (1995). They observed that mate location by males and host location by females increased with size in smaller parasitoids but were unaffected by size in larger individuals. Average wasp fitness as indicated by the preceding parameters, however, did not increase linearly with size. Moreover, parasitoid size was not a reliable predictor of individual fitness. Instead, consistent size–fitness relationships were evident only when fitness parameters were averaged over many genotypes and environments.

Bigler *et al.* (1987) and Hohmann *et al.* (1988) demonstrated that adult females of *T. maidis* Pintureau & Voegele and *T. platneri* Nagarkatti, respectively, also were less fecund and shorter lived when reared on the smaller eggs of natural hosts.

Host size may also influence the size of koinobiontic parasites, those that develop on hosts that continue to feed and to grow. Koinobionts are most commonly endoparasites (see Godfray, 1994; Quickie, 1997). Liu (1985) demonstrated a relationship between host and parasitoid size by comparing the size of the solitary parasitoid *Aphidius sonchi* Marshall when reared on different sized instars of its aphid host, *Hyperomyzus lactucae* (Linnaeus). Parasitoids reared from first-instar hosts were smaller than those reared on third instars. Moreover, those parasitoids reared on the larger hosts had more eggs at emergence. Parasitoid survival and development times, however, were equivalent on first- and third-instar hosts. Mackauer and Kambhampati (1988) later examined the effects of extremely small host size on parasitism by *A. smithi* Sharma & Subba Rao. When parasitizing adults of *Acyrthosiphon pisum* (Harris), female parasitoids deposited eggs in the hemocoel of the host and occasionally in aphid embryos still developing within the host's ovarioles. Parasitoids that developed in host embryos were much smaller and less fecund than those that developed on the body tissues of the host. The longevity of both groups, however, was similar.

The effect of host nutrition on growth and development of *Venturia canescens* (= *Nemeritis*) (Gravenhorst) was examined by Harvey *et al.* (1995). Parasites reared on third-instar *Plodia interpunctella* Hübner that were nutrient deficient due to partial starvation early during parasitism had significantly longer development times and displayed greater mortality than parasites reared on continuously well-fed hosts. Parasitoids reared on nutrient-deficient third-instar hosts, however, were not different in size from those raised on well-fed hosts. Wasps reared from fifth-instar hosts that were maintained at high density were significantly smaller and developed faster than those reared from hosts maintained at low density. Host starvation experiments with parasitized third-instar hosts established that parasitoid survivorship and size increased with the length of host access to food, while parasite development time decreased with increased host access to food. The authors concluded that host growth was essential for successful parasite development when smaller early-instar hosts were parasitized and that successful development may be influenced by host-feeding rate and food quality.

Visser (1994) examined the size–fitness relationship of *Aphaereta minuta* (Nees) and observed that larger females had more and larger eggs, lived longer, and searched for hosts more efficiently than did smaller parasitoids. When these various biological characters were considered together, fitness increased linearly with size. The author further demonstrated that the size–fitness relationship held under field conditions.

Positive linear relationships between host and parasitoid size often are not evident for koinobiontic species developing on immature host stages that continue to feed. Sequeira and Mackauer (1992a), for example, reported that the biomass of *Aphidius ervi* Haliday larvae reared on different sized nymphal instars of the aphid host *Acyrthosiphon pisum* varied in a nonlinear fashion with host biomass. Maximum parasite biomass and development time to eclosion of the adult parasitoid varied with host age at the time of parasitization. The authors proposed that host quality is not a linear function of host size in parasitoids developing in feeding and growing host stages. Additional studies suggested that parasitoids developing in hosts below a certain size threshold maximize fitness by balancing biomass or growth with development time (Sequeira & Mackauer, 1992b). Sequeira and Mackauer (1994) subsequently demonstrated that fecundity and reproductive success of *Aphidius ervi* also varied in a nonlinear manner with parasitoid biomass and host size. The interactions of life history parameters, including development time, biomass, and size, considered as adaptations to resource utilization in koinobiontic and idiobiontic parasitoids were further discussed by Mackauer *et al.* (1997).

Although adult size of gregarious parasitoids is generally decreased as the number of developing parasites increases (Salt, 1940; Waage & Ng, 1984; Bai *et al.*, 1992), a correlation between total parasitoid biomass and/or numbers with host size has been demonstrated with some gregarious larval parasitoids (Wylie, 1965; Bouletreau, 1971; Thurston

& Fox, 1972; Opp & Luck, 1986; Hawkins & Smith, 1986). Opp and Luck (1986) reported results of host size selection studies with the gregarious parasitoids *Aphytis melinus* DeBach and *A. lingnanensis* Compere. Adult females of both species selected small individuals of their host, *Aonidiella aurantii* (Maskell), for depositing male eggs and large hosts for depositing female eggs. Indeed, there was a limit to host size below which no female eggs were deposited. No difference related to host size was observed in the mortality of developing parasitoids. There was, however, a positive correlation between host size and that of emerging female parasitoids, and large females were more fecund. At the same time there was an upper limit to female parasitoid pupal size, which suggested that progeny gained no additional advantage in reproductive success when reared on larger hosts. The authors concluded that this may account for the facultative gregarious character of these parasitoid species on large hosts. The mechanisms by which gregarious populations moderate their development relative to host size have been described by Beckage and Riddiford (1983).

Effects of Host Nutrition on Parasitoid Attributes

Evidence that nutritional factors influence the success of parasitoids is provided from studies on parasitoids reared from hosts maintained on different food sources (Vinson, 1976 1981). Smith (1957) observed differences in larval mortality and adult size, sex ratio, and reproductive rate of several parasitoid species when reared on *A. aurantii* maintained on different food plants. *Habrolepis rouxi* Compere displayed limited mortality on *A. aurantii* maintained on citrus, but 100% mortality was noted on hosts maintained on sago palm. Similar results were obtained with *Comperiella bifasciata* Howard. Pimentel (1966) and Altahtawy *et al.* (1976) demonstrated differences in parasitoid fecundity and longevity depending on host food source. Oviposition and longevity of *Apoanagyrus lopezi* (De Santis) were significantly affected by the food plant of its host *Phenacoccus manihoti* Matile-Ferrero (Sousissi & Ru, 1997). Because the effects on the parasitoid were related to the levels of plant resistance to the host, the latter authors suggested that secondary plant chemicals within the host may have been responsible.

Allelochemicals or plant secondary compounds, including plant toxins, or toxic substances produced by the host insect from plant-derived natural products, often impact the effects of nutrition on parasite characters [see Herzog and Furderburk (1985) and Hare (1992)]; and their potential effects should not be overlooked when evaluating the results of studies employing natural foods. *Hyposoter exiguae* (Viereck) was adversely affected by tomatine sequestered by *Helicoverpa* (= *Heliothis*) *zea* (Boddie) feeding on to-

matoes (Campbell & Duffey, 1979). Subsequent investigations suggested that the toxic effects were due to sterol binding and the authors demonstrated that the effects of tomatine could be alleviated by supplementing the diet of the host with additional phytosterol (Campbell & Duffey, 1981). Thurston and Fox (1972) reported that nicotine influenced the emergence of *Cotesia* (= *Apanteles*) *congregata* (Say) when reared on *Manduca sexta* (Linnaeus) maintained on tobacco, and Barbosa *et al.* (1991) further observed that nicotine caused significant mortality of the parasite but had little if any effect on the host. Later studies suggested that nicotine in the host tissues may act by mediating the availability of nutrients to developing *C. congregata* and/or may affect assimilation of nutrients by the parasites themselves (Bentz & Barbosa, 1992). Similar but more severe effects of nicotine were reported for *Hyposoter annulipes* (Cresson) (Barbosa *et al.*, 1986). Rutin and hordenine in the diet of *M. sexta* also had negative effects on the development of *C. congregata,* but the effects of these chemicals on the parasite paralleled those on the host (Barbosa *et al.*, 1991).

Reitz and Trumble (1997) reported increased mortality of the parasite *Archytas marmoratus* (Townsend) when reared on *Spodoptera exigua* (Hübner) fed three furanocoumarins, secondary plant metabolites extracted from *Apium* spp. Again, the effect of these chemicals on the parasite was directly mediated through their effects on the host. Benn *et al.* (1979) reported that *Microplitis* sp. accumulated pyrrolizidine alkaloids from its host *Nyctemera annulata* Boisduval, which in turn were sequestered from the plant food, *Senecio spathulatus* A. Rich. The authors did not report significant detrimental effects on parasitoid life characters but suggested that these secondary plant chemicals may protect the host and parasitoid from predation.

The effects of host-contained antifeedants on parasite development and subsequent quality of adult parasitoids were discussed by Weseloh (1984) and Jermy (1990).

Other investigators have attempted to examine the effects of nutritional quality on parasitoids. Zohdy (1976) observed that *Aphelinus asychis* Walker display longer larval development time and decreased adult longevity when reared on *Myzus persicae* (Sulzer) maintained on defined diets deficient in sucrose or iron. These effects appeared to be related to decreased host size rather than to any difference in the nutritional quality of the host. Survival of *Aphaereta pallipes* (Say) was affected by the balance of amino acids and glucose in the artificial diet used for rearing its host, *Agria* (= *Pseudosarcophaga*) *housei* [= *affinis* (Fallen)] Shewell (House & Barlow, 1961). Nadarajan and Jayaraj (1975) observed differences in larval development and adult size, fecundity, and sex ratio of *Tetrastichus israeli* (Mani & Kurian) when reared on several different host species. The authors correlated these differences to the total level of essential amino acids in host tissues. Although parasitoids reared from some host species with high levels of essential amino acids were larger and longer lived, the results were variable, as were the specific amino acid compositions of the different hosts.

Host Nutrition and Parasitoid Behavior

While parasitoid fecundity and sex ratio are correlated with host size and nutritional factors, behavior is also involved in determining reproductive success (Vet & Dickie, 1992). Several studies have demonstrated that larval diet may influence parasitoid behavior. Vet (1983) reported that the response of adult *Leptopilina clavipes* (Hartig), a parasite of *Drosophila* spp., to chemicals or kairomones emanated from yeast was increased when the parasite was reared on hosts fed yeast. Similar results demonstrating effects of host diet on the behavior of adult parasitoids have been reported for several other parasitoid species (Herard *et al.* 1988; Sheehann & Shelton, 1989). Thus, nutrition and behavior interact in a complex fashion to affect quality attributes.

Salt (1940) demonstrated that adults of *Trichogramma evanescens* Westwood display behavioral dimorphism related to host size. Large females obtained from large hosts failed to oviposit in small hosts. Small females, however, accepted hosts of all sizes. Male wing development was markedly influenced by host size. The latter effect was also noted for the parasitoid *Gelis corruptor* (Forester) (Salt, 1952). Sandlan (1982) reported that adult females of *Coccygomimus* (= *Pimpla*) *turionellae* (Linnaeus) lacked obvious morphological and behavioral polymorphism, but large females experienced difficulty handling and ovipositing in smaller hosts. Conversely, small females were more efficient in attacking small hosts. Fecundity was strongly influenced by longevity, with the greatest longevity reported for larger individuals reared from large hosts.

Effects of Prey Species on the Biological Attributes of Predators

In contrast to the extensive literature on parasitoids, fewer studies have been conducted on effects of various natural foods on the biological character of predators. Putnam's (1932, 1937) investigations with *Chrysoperla rufilabris* Burmeister were among the first to demonstrate that larval diet had a significant effect on predator development and survival. The observations that prey species affected juvenile growth and mortality of *C. rufilabris* were later confirmed by Hydorn and Whitcomb (1979). Muma (1957) demonstrated that the larval development time and mortality, and adult longevity of *Chrysopa lateralis* (Guerin) were significantly affected by the species of prey fed on by predatory larvae. Muma suggested that this was due to differences in the nutritional value of the prey. Adult fe-

cundity of *Chrysoperla* (= *Chrysopa*) *carnea* (Stephens) was markedly affected by the availability of different foods (Sundby, 1966, 1967). Atwal and Sethi (1963) reported that *Coccinella septempunctata* Linnaeus attained a greater weight when feeding on *Lipaphis erysimi* (Kaltenbach) than on two other aphid species, and demonstrated that *L. erysimi* had significantly higher protein levels. Smith (1965) observed that 10 coccinellid species fed dried, powdered aphids, grew larger and faster when feeding on *Acyrthosiphon pisum* and *Rhopalosiphum maidis* (Fitch) than on *Aphis fabae* Scopoli.

Later studies further confirmed the importance of food on development and the biological character of predators. *Hippodamia sinuata* Mulsant developed faster when exclusively fed *R. maidis* compared with a diet of *Schizaphis graminum* (Rondani) (Michels & Behle, 1991). Elliott *et al.* (1994) reported that *Cycloneda ancoralis* (Germar) larvae also showed considerable differences in development time as well as final adult size when fed four different aphid species. Although the number of prey eaten differed between prey species, the amount of food eaten was not positively correlated with development time or predator size. Thus, differences in nutritional quality between prey species were probably important factors for predator growth and development.

Investigations on the predaceous mite *Euseius tularensis* (Congdon) have demonstrated that oviposition on citrus trees was significantly decreased when trees were fertilized at lower rates (Grafton-Cardwell & Ouyang, 1996). Decreasing concentrations of nitrogen and manganese in leaves were correlated with decreased oviposition, demonstrating a direct relationship between predator reproduction and nutritional status of citrus trees.

The preceding discussion serves to illustrate that nutritional factors influence the biological attributes of entomophagous species and to indicate the complexity of the biological interactions affecting their nutrition. This chapter does not attempt a fuller synthesis of the nutritional ecology of these organisms, but instead focuses on feeding, nutritional requirements, and nutritional physiology, as well as on the current state of technology aimed at artificial rearing. Behavioral and ecological considerations are discussed only when they have direct relevance for nutrition.

FOOD UTILIZATION

Food Utilization by Parasitoids

High efficiencies of food utilization have often been predicted for parasitoids. The larval stages consume food of high nutritional content, are relatively inactive within the host, and have a limited food supply, all of which may contribute to selection for efficient food utilization (Fisher,

1971, 1981; Slansky & Scriber, 1985); see also Wiegert and Petersen (1983).

Food utilization by parasitoid larvae was reviewed by Slansky and Scriber (1985). The hymenopterous species described were *Coccygomimus* (= *Pimpla*) *instigator* (Fabricius) and *Pteromalus puparum* (Linnaeus) (Chlodny, 1968); *Gelis macrurus* (Thomson) and *Hidryta frater* (Cresson) (= *sordidus*) (Edgar, 1971); *Brachymeria intermedia* (Nees) and *C. turionellae* (Greenblatt *et al.*, 1982); *Diadromus pulchellus* Wesmael (Rojas-Rousse & Kalmes, 1978); *Trypargilum* (= *Trypoxylon*) *politum* (Say) (Cross *et al.*, 1978); *Phanerotoma flavitestacea* Fischer (Hawlitzky & Mainguet, 1976); *Venturia* (= *Nemeritis*) *canescens* (Gravenhorst) (Fisher, 1968); *Cidaphus alarius* Gravenhorst and *Phynadeuon dumetorum* Gravenhorst (Varley, 1961); and *Cotesia* (= *Apanteles*) *glomerata* (Linnaeus) (Slansky, 1978). The mean net conversion efficiency [i.e., proportion of assimilated food converted to body mass, see Petrusewicz (1967), Calow (1977), Hagen *et al.* (1984)] of these species varied widely from 11 to 62%, with a mean of 37% that was equal to or less than that for many groups of insect herbivores and detritivores. Studies by Cameron and Redfern (1974) suggested that conversion efficiencies for *Eurytoma tibialis* Boheman and *Habrocytus elevatus* (Walker) were at the high end of this range. Net conversion efficiencies of parasitoids may not be exceptionally high because selection has occurred for rapid, instead of efficient growth (Slansky, 1986). The well-known inverse relationship between growth efficiency and assimilation (Welch, 1968) may also be important. In contrast to net conversion efficiency, the preceding parasitoids had relatively high percentages of assimilation (i.e., percentage of ingested food that is assimilated), ranging from 55 to 94%, with a mean of 67%, compared with means of 40 to 50% for most herbivores and detritivores.

The highest nutritional efficiencies for a parasitoid were reported by Howell & Fisher (1977) for the ichneumonid *V. canescens*. Based on caloric values, the authors developed the energy budget shown in Fig. 1. The larval stage had a 65% net conversion efficiency and 95% assimilation when maintained on the host *Anagasta* (= *Ephestia*) *kuehniella* (Zeller). Net conversion efficiency to the adult was 20%.

Several workers have calculated the proportion of food-host available that is consumed by the parasitoid and converted to parasitoid biomass. These "exploitation" indices vary from 3 to 80% (Slansky, 1986). Larval *V. canescens* consumed 90% of its host's biomass and converted 55%, but there was no clear correlation between host size and parasitoid size or biomass conversion (Howell & Fisher, 1977).

Food Utilization by Predators

Based on similar reasoning to that outlined earlier for parasitoids, food utilization by predators also may be hy-

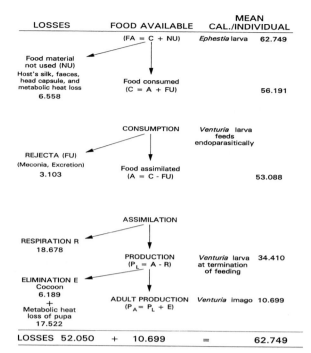

FIGURE 1 Nutritional energy budget for *Venturia* (*Nemeritis*) *canescens* (Gravenhorst) feeding on the host, *Anagasta* (= *Ephestia*) *kuehniella* (Zeller). FA = total food available for consumption by the parasitized host larva; NU = uneaten food, silk, fecal matter, and metabolic heat loss; C = food eaten or consumed; A = food absorbed or assimilated following digestion; FU = unabsorbed food passed through the gut and rejected; P = growth or productivity; R = respiration; E = eliminated energy including cocoon and metabolic heat loss; P_L = productivity or biomass of the larval parasite; P_A = productivity or biomass of the adult parasitoid. [From Howell, J., & Fisher, R.C. (1977). *Ecology and Entomology, 2,* 143–151.]

pothesized as highly efficient. This is particularly the case with predators that "sit and wait" (Lawton, 1971), thus avoiding metabolic expenditure in search for food. Slansky and Scriber (1985) reviewed studies on food utilization of 11 predaceous insects. Predators had similar net conversion efficiencies ranging from 4 to 64%, with a mean of 34%, but higher assimilation efficiencies, ranging from 37 to 98% with a mean of 86%, than those of the parasitoids discussed earlier.

Zheng *et al.* (1993a) reported that *Chrysoperla carnea* exhibited gross conversion efficiencies (i.e., proportion of ingested food converted to body mass) ranging from 40 to 60%, depending on the developmental stage and level of food consumption. Conversion efficiency was greater in later instars and generally increased as food supply decreased.

Cohen (1984, 1989a) examined food utilization by the predator *Geocoris punctipes* (Say) when reared from first-instar nymphs to adults on eggs of *Heliothis virescens* (Fabricius). Assimilation efficiency was approximately 95%;

gross conversion efficiency, 53%; and net conversion efficiency, 55%. *Geocoris punctipes* reared on an artificial diet (Cohen, 1981 1989b) displayed significantly lower assimilation and conversion efficiency than did predators reared on host eggs (Cohen & Urias, 1988).

Food consumption by predators is greatly influenced by prey density, distribution of prey developmental stages available, and predator satiation (McMurty & Rodriguez, 1987). "Exploitation indices," therefore, are more difficult to evaluate in predator–prey interactions.

DEVELOPMENTAL NUTRITIONAL REQUIREMENTS

General Nutritional Requirements of Insects

The qualitative nutritional requirements of insects, determined principally by deletion studies using defined and deficient artificial diets, have been outlined previously (Dadd, 1973, 1977, 1985; Friend & Dadd, 1982; Hagen *et al.,* 1984). All insects have similar requirements for approximately 30 chemicals including protein and/or 10 essential amino acids (arginine, histidine, isoleucine, leucine, lysine, methionine, phenylalanine, threonine, tryptophan, and valine); B-vitamin complex (biotin, folic acid, nicotinic acid, pantothenic acid, pyridoxine, riboflavin, and thiamin); and other water soluble growth factors (including choline and inositol), certain fat soluble vitamins, cholesterol or a structurally similar phytosterol, a polyunsaturated fatty acid, minerals, and an energy source (usually provided by simple or complex carbohydrates and/or lipids). Quantitative requirements, however, may differ considerably and often can be related to natural feeding habits (Dadd, 1985).

Requirements for Development of Hymenopteran Parasitoids

The nutritional requirements of entomophagous insects are similar, and similar to those of nonentomophagous species. House (1977) referred to this common feature of insect nutrition as the "rule of sameness" (House, 1966a, 1974). The rule has been confirmed by studies with parasitic and predaceous insects (Table 1). In assessing the essentiality of nutrients, it is important to note that most studies were conducted by rearing a single generation on a synthetic or semisynthetic diet (see later subsections on *in vitro* culture of parasitoids and *in vitro* culture of predators, Table 4). Some investigations overlooked the potential contribution of nutrients stored within the egg. Stored nutrients may support limited development and, in the case of trace nutrients, supply a sufficient quantity to ensure development of one generation. Studies with the parasitoids *Itoplectis conquisitor* (Say) (Yazgan, 1972) and *Exeristes roborator* (Fa-

bricius) (Thompson, 1981a), for example, demonstrated partial larval development on diets lacking various essential amino acids and B-complex vitamins. Numerous studies have demonstrated that entomophagous insects have no distinctive or unusual qualitative nutritional requirements (House, 1977; Thompson, 1981a, 1981b, 1981c, 1981d, 1986a; Grenier *et al.,* 1994; Vinson, 1994).

The nutritional value of foodstuff goes far beyond basic qualitative content. The quantitative balance of different nutrients is a critical and often dominant factor determining dietary acceptability and suitability (House, 1969, 1974). The predominant foods of both parasitic and predaceous insects are of animal origin, thus being generally high in protein content and low in carbohydrate and fat (House, 1977). Studies with several parasitic species have demonstrated a requirement for diets high in protein and/or amino acids (Thompson, 1986a). For example, at the 6% amino acid level, *E. roborator* completed larval development without glucose and/or fatty acids (Thompson, 1976a). Glucose, however, markedly improved survival when the amino acid level was reduced to 3%, and at 1% amino acid no development occurred without the carbohydrate. Thompson (1982) further characterized the quantitative effects of dietary carbohydrate on larval growth of *E. roborator*. Similar effects of amino acid level on larval development were reported by Yazgan (1972) for *I. conquisitor*. Adult eclosion was reduced by dietary amino acid levels below 6% and by deletion of glucose. Fatty acids were only marginally beneficial in enhancing growth and development rates of both species (Thompson, 1976a, 1977; Yazgan, 1972). A polyunsaturated fatty acid, however, was required in small amounts. Adult *I. conquisitor* (Yazgan, 1972) and *E. roborator* (Thompson, 1981a) displayed crumpled wings and/or bent ovipositors without a polyunsaturated fatty acid in the larval diet. Linolenic acid alleviated these deformities in *I. conquisitor,* and linoleic and linolenic acids were supplied together in the case of *E. roborator* (S. N. Thompson, unpublished).

The effect of nutritional balance on larval growth of *Brachymeria lasus* (Walker) was described by Thompson (1983a). Media containing 0 to 10% glucose with 2% amino acids, and 1 to 8% amino acids with or without 2% glucose were tested. All media contained 15% albumin and 2.5% lipids. Weight gain increased on diets containing 2% glucose when the amino acid level was increased from 1 to 4%, but was reduced at the higher amino acid levels. Similar effects of varying the amino acid level were obtained with diets lacking glucose, but the overall weight gain was less than that observed with diets containing glucose. On diets containing 2% amino acids, weight gain increased dramatically when glucose was increased from 0.5 to 4%, but decreased sharply at higher glucose levels. Growth rates on these diets were generally in the range of 150 to 200 mg/g·day. The maximum rate, 260 mg/g·day, was ob-

TABLE 1 Qualitative Developmental Nutritional Requirements of Entomophagous Insects

		Nutritional requirement[a]								
		Amino acids		Lipids		Vitamins				
	Taxa	10 Essentials/others	Carbohydrates	Cholesterol	Fatty acids[b]	B complex	Fat soluble	Nucleic acid	Minerals	References
Parasitoids										
Order Diptera										
Agria housei	A	B[c]	B	A[d]	B	A	R[e]	B[f]	A	House, 1974, 1977
Eucelatoria bryani	A[g]	A[h]B[i]	—	—	—	—	—	—	—	Nettles, 1986a, 1987a
Hymenoptera										
Brachymeria intermedia	A	—	—	A	B	—	—	—	—	Thompson, 1981a, 1981b
B. lasus	A[g]	—	A[j]	A[k]	B	—	—	—	—	Thompson, 1981b, 1981c, 1983a, 1983b
B. ovata (Say)	A	—	—	—	—	—	—	—	—	Thompson, 1981b
Exeristes roborator	A	B[l]	B[m]	A	A[n]B[b]	A[n]	NR	NR	—	Thompson, 1976a, 1976b, 1977, 1982
Itoplectis conquisitor	A	—	A[o]B[m]	—	A[n]B	A	NR	NR	—	Yazgan, 1972
Pachycrepoideus vindemiae	A	—	—	A[k]	B	—	—	—	—	Thompson, 1981b, 1981c
Predators										
Order Neuroptera										
Chrysopa septempunctata	A	—	—	—	—	A	NR	—	—	Niijima, 1993a, 1993b
Chrysoperla carnea	A	B[p]	—	A[q]	A[q]	—	—	—	—	Vanderzant, 1973
Coleoptera										
Harmonia axyridis	—	—	—	—	—	—	—	—	A	Matsuka & Takahashi, 1977
Hippodamia convergens		A/B[r]	—	—	—	—	—	—	—	Racioppi *et al.*, 1981

[a] A = Absolute or essential requirement without which the insect fails to grow or develop; B = beneficial for improving growth; R = required for reproduction; NR = not required.

[b] The form in which the fatty acids were included was often important. For example, free fatty acids emulsified in Tween® derivatives were nontoxic, but otherwise C16-18 fatty acids were highly toxic to *E. roborator* (Thompson, 1977), as well as to *I. conquisitor* (Yazgan, 1972). Triglycerides, however, were nontoxic.

[c] Alanine, glycine, and tyrosine are beneficial for larval growth and development.

[d] Cholesterol, 7-dehydrocholesterol, β-sitosterol, and stigmosterol were utilized.

[e] Vitamin E was required for reproduction and Vitamin A improved larval development rate.

[f] Nucleotides were beneficial but nucleosides and individual purines and pyrimidines had no effect.

[g] Both protein and free amino acids were essential for complete development.

[h] Asparagine was essential for development during the larval stage. Proline was "semiessential" and very few pupae survived in its presence.

[i] Tyrosine and glycine improved pupal and adult survival.

[j] Glucose was essential for completion of larval development and pupation.

[k] Cholesterol, cholestanol, and β-sitosterol were utilized.

[l] Larvae failed to survive on a synthetic medium containing only the 10 essential amino acids, or on media containing the essentials plus single nonessential amino acids. A mixture of protein (albumin) and free amino acids produced best results.

[m] Glucose was important for larval growth, development, and survival at low amino acid levels.

[n] Polyunsaturated fatty acid was required.

[o] Glucose or fatty acids were essential for satisfactory pupal development and survival.

[p] Aspartate, cysteine, glutamate, glycine, proline, serine, and tyrosine were very beneficial for completion of larval development and successful pupation.

[q] In the absence of soybean oil and lecithin, larval survival was reduced and pupation was negligible. The exact lipid constituents necessary, however, were not determined.

[r] In the absence of tryptophan and cystine, adults displayed deformities of the tibia and tarsi.

tained on a medium containing 2% glucose and 2% amino acids. The effects of nutrient balance were closely related to the osmolality of the artificial medium (Thompson, 1983b) (see later subsection on the importance of dietary osmotic pressure to successful *in vitro* culture).

Requirements for Development of Dipteran Parasitoids

Similar quantitative requirements to those of hymenopteran parasitoids were demonstrated by House (1966b) for the dipteran *Agria housei*. Maximum growth and survival, however, were achieved when all nutrients were increased proportionally over the levels in a basal diet that contained 2.25% amino acids, 0.05% salts, 1.16% lipids, and 2.25% miscellaneous ingredients, including glucose, ribonucleic acid, vitamins, and agar. When amino acid level alone was increased, survival was reduced. On a diet containing nutrient levels equivalent to pork liver, that is, 20% amino acids, 4% glucose, 3.5% lipids, 2% salts, and 0.75% ribonucleic acid, survival was greater than 80%. House (1967, 1970) subsequently demonstrated that the relative balance of amino acids and glucose was critical in determining growth or development; moreover, *A. housei* larvae selected diets for feeding on the basis of nutrient balance.

A requirement for asparagine by the parasitoid *Eucelatoria bryani* Sabrosky (see Table 1; Nettles, 1986a) and the absence of a requirement for a polyunsaturated fatty acid by *A. housei* (House & Barlow, 1960) were consistent with findings for non-parasitic dipteran insects (Dadd, 1977).

Requirements for Development of Predators

Fewer studies have been conducted to establish the basic qualitative and quantitative nutritional requirements of predators. The amino acid requirements for larval development of *Chrysopa pallens* (= *septempunctata*) (Rambur) were described by Niijima (1993a). The studies employed a chemically defined diet composed of 23 amino acids, sucrose, cholesterol, six fatty acids, five organic acids, 11 minerals, and 17 vitamins (Hasegawa *et al.,* 1989; Niijima, 1989). The 10 usually essential amino acids were required for larval development beyond the first instar, but several amino acids could be deleted from the diet without adverse effects. Decreasing the total amino acid level from approximately 40% resulted in a gradual increase in larval development time. Later, Niijima (1993b) reported the essentiality for B vitamins as well as for choline. Nicotinic and pantothenic acids were essential for larval development. *Chrysopa pallens,* however, could complete larval development of a single generation in the individual absence of most other B vitamins, although at reduced yields and rates. Other water-soluble vitamins including ascorbate and car-

nitine were also required over successive generations. Fat-soluble vitamins were not required.

Evaluation of Developmental Nutritional Requirements

Most quantitative nutritional studies with parasitoids have evaluated the effects of nutritional balance using univariate or monofactorial analyses. Grenier *et al.* (1986) emphasized that such an approach has severe limitations because it ignores potential interactions between nutrients, including ". . . additivity, competitivity, antagonism or synergy." As a result, interpretation of effects of nutrient variation aimed at medium optimization is made difficult. The authors suggested that nutritional studies be designed and analyzed in a multidimensional fashion that accounts for interactions between all nutrients and several biological criteria.

Bonnot (1986, 1988) reported the results of a canonical correlation analysis of 46 experiments with the dipterous parasitoid *Lixophaga diatraeae* (Townsend) reared on an artificial medium. Canonical analysis constructs maximum correlations between all linear combinations of variables within sets, in this case between growth and development, and dietary parameters. Because the method may generate biologically meaningless correlations, accurate interpretation requires insightful knowledge of biological correspondence between variables. The concentrations of 30 medium components were varied and the effects on nine developmental criteria were determined; nine linear correlations were obtained, three with correlation coefficients greater than 0.95 (Table 2 and Figure 2).

Bonnot (personal communication) provided the following explanation of the correlations shown in Figure 2 in relation to the nutrient composition and developmental criteria listed in Table 2. Correlation 1 demonstrated that development time (5 to 8) through all stages, from the fast stadium to adult eclosion, was related to the levels of dietary proteins (1 to 7) and amino acids (15 to 29). The relationship varied between instars and developmental stages and was most critical during the third larval stadium (7) when maximum growth occurred in preparation for pupation. The correlation was negative for the first stadium (5) and the pupal stage (8) and positive for the second (6) and third (7) stadia. The second and third correlations demonstrated the importance of nonnutritional factors, including surface tension and texture provided by agar (11) and lecithin (13). The contribution of these components to the physical properties of the medium was supportive in correlation 3 but opposite in correlation 2. Again, the effect of nutrient balance differed with the developmental stadia. In the case of correlation 2, the cumulative development time to the third stadium (6) was more strongly correlated than

TABLE 2   Developmental Criteria and Nutrient Composition Employed for Investigation of *in vitro* Culture of *Lixophaga diatraeae* (Townsend).

Developmental criteria	Nutrients tested	
1. Percentage of first-instar survival	1. Ovalbumin	16. Threonine
2. Percentage of second-instar survival	2. Bovine serum albumin	17. Serine
3. Percentage of third-instar survival and pupation	3. Soy protein	18. Glutamate
4. Percentage of pupal survival	4. Denatured bromelain (tyrosine-rich protein extracted from pineapple)	19. Proline
5. Development time (days) to the second stadium		20. Glycine
6. Cumulative development time to the third stadium	5. Lactalbumin	21. Alanine
7. Cumulative development time to the pupal stage	6. Casein	22. Valine
8. Cumulative development time to adult eclosion	7. Gelatin	23. Methionine
9. Maximum weight	8. Fibrin	24. Isoleucine
	9. Organic acids	25. Leucine
	10. Glycogen	26. Tyrosine + phenylalanine
	11. Agar	27. Lysine
	12. Inorganic salts	28. Histidine
	13. Lecithin	29. Arginine
	14. Fat	30. Cysteine
	15. Aspartate	

a After Bonnot G., 1986. *Compte Rendu de Colloque du CNERNA, Paris,* 227–240.

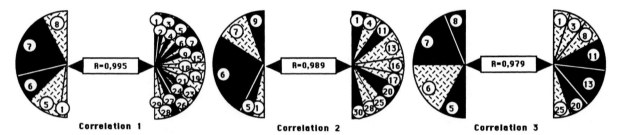

FIGURE 2   Diagrammatic representation of canonical correlations between nutrient composition and developmental criteria during in vitro culture of *Lixophaga diatraeae* (Townsend) on effects of nutrient balance on development, as shown in Table 2. Sector sizes are proportional to the correlation coefficient between the concerned original and canonical variables of the same set. Solid sectors are positively correlated and hatched sectors, negatively correlated

was the development time to the second stadium (5). Cumulative development time to the pupal stage was negatively correlated, but maximum weight (9) was positively correlated. Thus, increased agar (11) and/or lowered lecithin (13) levels increased maximum weight (9) but slowed development in the third stadium (7). Correlation 2 also suggested that nutrient conditions favoring maximum weight caused increased mortality (1) during the first stadium. Correlation 3 was strongly negative for cumulative development time to the third stadium (6).

Developmental Nutrition, Environmental Conditions and Biological Characteristics of Parasitoids

Studies with parasitoid *A. housei* have demonstrated that optimal nutritional balance can be influenced by environ-

mental factors. House (1966b) demonstrated that the effects of dietary glucose level on larval survival and development could be modulated by temperature. The nutritive value of the basal medium (see earlier subsection on requirements for development of dipterous parasitoids; House, 1966a) was increased by increasing the temperature from 20 to 25 and 30°C at glucose levels between 0 and 1.5%. At higher glucose levels, larval survival and development were reduced with increasing temperature. Later, House (1972) successfully formulated two media of different composition whose superiority for larval growth and development of *A. housei* was reversed at two different temperatures, 15 and 30°C. House (1966b) pointed out that such nutritional effects might have ecological significance by affecting the host range of insects. It was suggested that in establishing host range, an insect might be affected differently if the nutrient composition of its food were uniform and the temperature varied within the range, rather than if the temperature were uniform and the composition of food were variable. Alternatively, the insect might not be affected if variation in food composition were accompanied by compensatory changes in temperature. Thus, an insect species that attacks a particular foodstuff in a region with a specific temperature might, if introduced into another area with a different temperature, adapt to a different food source whose nutrient composition is favored at the new temperature.

Little is known of the effects of developmental nutrition on the behavior of parasitoid larvae aside from measurements related to growth and development rate. Of interest, therefore, was the report by Veerman et al. (1985) that a photoperiodic response by Cotesia glomerata was influenced by the carotenoid content of its host's diet. Vitamin A was essential for photoperiodic induction of diapause and the authors suggested that this vitamin or a derivative functions as a photoreceptor pigment.

Importance of Nonnutritional Factors in Feeding and Nutrition

Nonnutritional factors are intimately and intrinsically involved in food acceptance and intake. These include physical properties such as form and texture, but also nonnutritive chemicals that elicit specific behavioral and/or physiological responses essential for finding and accepting foodstuff, and in some cases for initiating behaviors associated with the feeding process itself (Bernays & Simpson, 1982; Bernays, 1985). Although such factors have been best documented in phytophagous insects, they also play a role in the biology of entomophagous insects and will likely be of importance in the development of continuous in vitro culture. Bernays (1997) suggested that predation on lepidopterous larvae may be stimulated by a variety of cues emanating from feeding insects, including sounds of mandible activity and/or odors released from damaged foliage.

FEEDING AND NUTRITION OF ADULT ENTOMOPHAGOUS SPECIES

Supplemental Foods and Performance

The adults of many entomophagous species feed and are routinely fed in the laboratory to optimize life span. A source of carbohydrate is the most common food employed. Hohmann et al. (1988) reported that adult female Trichogramma platneri Nagarkatti were short lived and only realized about half their potential fecundity if honey was not available. Similar results were later reported by McDougall and Mills (1997) for this species and by Bai et al. (1992)

with Trichogramma pretiosum Riley. Availability of honey during the first day following emergence increased adult longevity of T. pretiosum sixfold. Morales-Ramos et al. (1996a) demonstrated that female Catolaccus grandis (Burks) displayed significantly increased longevity when fed a glucose–fructose solution. Adult female Brachymeria lasus failed to develop eggs unless sucrose was provided (Mao & Kunimi, 1994). Similar findings have been reported for predators (see later section on current applications). McMurtry and Scriven (1966) observed increased survival of the predaceous mites Amblyseius limonicus Garman & McGregor, A. hibisci (Chant), Typhlodromus occidentalis Nesbitt and T. rickeri Chant when fed sucrose or molasses.

The significance of adult feeding in nature was often overlooked by early biological control specialists. Bierne (1962) suggested that many biological control attempts failed as a result, and other authors have since reiterated the potential importance of supplemental foods for enhancing the efficiency of biological control agents (Powell, 1986; van Emden, 1990; Kidd & Jervis, 1989; Jervis & Kidd, 1991; Jervis et al., 1993, 1996) (see later subsection on host feeding by hymenopterous parasitoids).

One of the earliest field studies indicating the importance of adult feeding for improving parasitoid performance was that of Leius (1967a), who reported a relationship between the natural abundance and variety of wild flowers in apple orchards and the incidence of parasitism of Malacosoma americanum (Fabricius) and Laspeyresia (= Carpocapsa) pomonella (Linnaeus) by hymenopteran parasitoids I. conquisitor, Apophua (= Glypta) simplicipes (Cresson), Scambus hispae (Harris), Telonomus sp., Ooencyrtus clisiocampae (Ashmead), and Eupelmus spongipartus Foerster. Eighteen times as many M. americanum pupae, four times as many M. americanum eggs, and five times as many L. pomonella eggs were parasitized in orchards with an undergrowth of wild flowers when compared with orchards lacking similar flora. Similar results demonstrating the importance of feeding on wild flowers were reported by Idris and Grafius (1995) with the parasitoid Diadegma insulare (Cresson). Those authors demonstrated how seasonal availability and distribution of eight flower species affected parasitoid success. Lingren and Lukefar (1977) demonstrated that adult Campoletis sonorensis (Cameron), a parasitoid that feeds on the extrafloral nectar of cotton, lived significantly longer when exposed to extrafloral nectaried cotton than nectariless cotton and, moreover, that parasitism of hosts was higher on the nectaried form of cotton.

Foster and Ruesink (1984) reported that the parasitoid Meteorus rubens (Nees von Esenbeck) exhibited increased longevity and fecundity when exposed to five species of flowering weeds associated with reduced tillage of corn.

Similar results implying the importance of natural foods

have been reported for predators (Muma, 1957). Adjei-Maafo and Wilson (1983) demonstrated that 15 categories of entomophagous arthropods, including the predators *Deraeocoris signatus* (Distant), *Geocoris lubra* (Kirkaldy), *Nabis capsiformis* Germar, *Chrysoperla* spp., *Laius bellalus* Guérin, *Coccinella repanda* (Thunberg), and *Verania frenata* Erichson, were present at densities two to three times higher on nectaried versus nonnectaried cotton. Several workers have observed increased mite predator densities in environments rich in pollen (Kennett *et al.*, 1979; Grout & Richards, 1990). Hemptinne and Desprets (1986) reported that following hibernation *Adalia bipunctata* (Linnaeus) feeds on pollens as an alternate food, which allows the predators to lay eggs as soon as prey become available. Fecundity of the predaceous tydeid mite *Homeopronematus anconai* (Baker) was dramatically increased when the normal food prey, *Aculops lycopersici* (Massee), was supplemented with pollen of *Typha latifolia* Linnaeus (Hessein & Perring, 1988). Although semiochemicals (see later section on chemical cues influencing the behavior of entomophagous species) contribute by attracting these insects to the plant, the nutrition provided by nectars and pollens undoubtedly plays an important role.

Plant Foods

Leius (1960) reviewed the early literature describing how adult parasitoids feed from flowers and other plant parts. In most cases insects fed on floral and extrafloral nectars as well as pollens. Although knowledge of the specific nutritional requirements of adult entomophagous insects is limited, much data are available on the chemical and nutritional composition of these plant products. Floral nectars contain up to 75% by weight of simple sugars, mainly sucrose, fructose, and glucose (Baker & Baker, 1983), but considerable qualitative and quantitative differences exist between plant species. Free amino acids are also abundant in nectars, although most nectars do not contain all 10 essential amino acids. Small amounts of proteins, lipids, dextrins, and vitamins that are nutritionally beneficial are also present. The composition of extrafloral nectars is equally complex, if not more so, as that of floral nectars (Baker *et al.*, 1978). Pollens have a complex composition of small molecular nutrients and many pollens have high levels of free amino acids (Barbier, 1970; Stanley & Linskens, 1974). In comparison with nectars, pollens generally have higher levels of protein, lipid, and polysaccharides. Pollens and nectars together can provide a complete diet for the successful growth, development, and reproduction of many insects. Smith (1961) demonstrated that predator *Coleomegilla maculata lengi* Timberlake can complete larval development on pollen alone. Several phytoseid mite predators readily consume pollen and nectar and often display

higher rates of oviposition when fed these foods instead of prey (McMurtry & Rodriguez, 1987). Thus, when prey are scarce, plant products play a critical role in maintaining predators (Hodek, 1973, 1996). The complex ecological and evolutionary interactions between plant flowers, nectars, and pollens, and various insect groups were discussed by Hagen (1986a).

The plant-feeding habits of three hymenopteran species, *I. conquisitor*, *Scambus buolianae* (Hartig), and *Orgilus obscurator* (Nees), were examined by Leius (1960). The attractiveness of the flowers of wild mustard, white sweet clover, wild parsnip, silky milkweed, and annual sow thistle was tested. Except for annual sow thistle, *I. conquisitor* was attracted to and fed from all these flowers, and wild parsnip was most attractive. Similar results were obtained with *S. buolianae*. *Orgilus obscurator* was attracted to and fed on wild parsnip only, but further tests demonstrated that this parasitoid would feed on other umbelliferous plant flowers, including those of wild carrot and water hemlock. The nutritive value of various pollens for fecundity and longevity of *S. buolianae* was described by Leius (1963). *Itoplectis conquisitor* and *S. buolianae* also accepted various seminatural foods, including honey, sucrose solution with or without plant pollens, and raisins.

The plant feeding behavior of *O. obscurator* was examined by Syme (1975), who reported a broad range of food plants, including species from five families. It was demonstrated that the adult parasitoids may emerge prior to the availability of the insect host, and Syme (1977) suggested that a variety of plant species be made available as food to ensure sufficient longevity of the adult female for adequate pest control.

Investigations with *Diadegma insulare* demonstrated that this parasitoid feeds on a variety of wild flowers and that longevity and fecundity were influenced markedly by the specific flora present (Idris & Grafius, 1995). The greatest benefit was observed with feeding on wild mustard, yellow rocket, and wild carrot, all species that produce nectar. Moreover, longevity and fecundity were correlated with the flower corolla opening diameter.

The flower-visiting habits of hymenopteran parasitoids were reviewed and further described by Jervis *et al.* (1993).

Studies with predaceous phytoseiid mites have also demonstrated the importance of plants as alternate food sources. Tanigoshi *et al.* (1983) reported that the facultative predator *Euseius tularensis* (Congdon) (= *E. hibisci*) was able to complete immature development and to produce viable offspring when fed ice plant pollen alone, and Congdon and McMurtry (1988) later observed that this mite predator preferred to consume ice plant pollen instead of several prey species, including citrus thrips and citrus red mites. Zhao and McMurtry (1990) subsequently reported that *E. tularensis* reproduced more rapidly when feeding on ice plant pollen than when feeding on a diet of prey. Ouyang *et al.* (1992), however, compared the pollen of several plant species as food sources for *E. tularensis;* and demonstrated that development, survivorship, and reproduction varied markedly with the pollen source.

Host Feeding by Hymenopteran Parasitoids

In addition to feeding on plants and plant products, many parasitoids, particularly idiobiontic species, feed on potential hosts [see Godfray (1994) and Quickie (1997)]. Host feeding by koinobiontic parasitoids is less common. Adult female hymenopterans puncture or damage host larvae or pupae and feed on the hemolymph and/or internal tissues. Kidd and Jervis (1989) estimated that as much as one-third of the world's parasitoid faunas, or as many as 100,000 species, host feed. Some parasitoids may kill more host individuals by host feeding (including ovipositor probing followed by host rejection) than by parasitism (DeBach, 1943, 1954; Kidd & Jervis, 1989; Jervis & Kidd, 1991). Legner (1979) emphasized that consideration of a parasitoid's host destructive capacity was important to correctly evaluate the impact of periodic inundative field releases on pest populations. Others also suggested that host-feeding behavior was an important consideration for assessing the potential of a biological control agent (Greathead, 1986; Yamamura & Yano, 1988; Huffaker & Gutierrez, 1990). Kidd and Jervis (1989, 1991b) discussed the significance of host feeding on parasitoid–host population dynamics and outlined implications for biological control. Despite the widespread occurrence of host feeding by parasites, Kidd and Jervis concluded that in most cases host feeding would have little effect on host population dynamics, and similar conclusions were reached by Briggs *et al.* (1995).

Jervis *et al.* (1996) compared the performance in past biological control programs of host-feeding parasitoid species that were destructive versus those that were nondestructive. In the case of nondestructive feeding the host survives, while destructive feeding behavior results in host death. They reported that nondestructive species were much more successful as biological control agents, but suggested that the difference between the two parasitoid feeding patterns may be influenced by the availability of nonhost food sources for nondestructive species.

Leius (1962, 1967b) demonstrated the importance of feeding habits to fecundity of hymenopteran *S. buolianae.* Egg production was reduced by two-thirds and longevity, by one-third when females were permitted to host feed intermittently or were deprived after 15 days of age. Eggs were not laid if females were deprived for 20 days. The effects of feeding on host body fluids—in conjunction with honey, pollen, and raisin—on fecundity and longevity of *S. buolianae* and *I. conquisitor* were examined by Leius (1961a) and Leius (1961b). Maximum fecundity and lon-gevity of both species were obtained when host fluids and seminatural foods were provided together. Host feeding, however, was essential and *S. buolianae* did not lay eggs when deprived of host hemolymph or tissues.

Bartlett (1964) examined the host-feeding behavior of encyrtid parasitoid *Microterys flavus* (Howard), and was among the first to suggest a direct relationship between host-feeding behavior and nutrition. Bartlett hypothesized that host feeding developed coincidentally with depletion of eggs and suggested that host mutilation was a reflection of "frustrated" host feeding when the host failed to bleed readily. Host feeding by *M. flavus* was usually displayed following egg laying, and oviposition resumed after host feeding. After reviewing this "predatory" habit for adults from 20 families of parasitic Hymenoptera, Bartlett concluded that this behavior was indicative of the necessity for dietary supplementation of some ubiquitous substance or substances required by many diverse species. Bartlett also reported that a food supplement of enzymatic yeast and soy hydrolysate with honey satisfied the nutrient requirements for sustaining reproductive activity in *M. flavus.*

Proovigenic and Synovigenic Parasitoids

Parasitic hymenopterans are often catagorized as proovigenic or synovigenic (Flanders, 1950; Quickie, 1997; see also Chapter 15). Females of proovigenic species complete oogenesis prior to, or shortly after, emergence, and lay eggs over a relatively short period of time primarily on larval stages of their host. Host feeding is only important for ensuring that the female lives long enough to deposit all her eggs. Proovigeny is most common among koinobiontic endoparasites. In contrast, females of synovigenic species eclose with only a fraction of their total egg complement as mature eggs. Synovigenic parasitoids generally are idiobionts and attack primarily host eggs and pupae. They are longer lived than are proovigenic species and produce eggs throughout their adult lives. To sustain oogenesis the females of many synovigenic species require additional nutrient. Based on egg type, Dowell (1978) described two types of synovigenic parasitoids. The first group produce large anhydropic or yolk-rich eggs that contain sufficient nutrient for completion of embryonic development prior to oviposition. Parasitoids that produce anhydropic eggs host feed to obtain sufficient nutrition for sustaining egg production. Other synovigenic species produce hydropic or yolk-deficient eggs. Embryonic development in hydropic eggs occurs in the host following oviposition. In this case the adult does not require additional nutrient to support egg development and has no requirement to host feed. Quickie (1997) provided a detailed outline of the parasite life history traits associated with koinobiontic and idiobiontic species, including proovigeny and synovigeny. As will be evident from the discussion later, the proovigenic–synovigenic and idiobiont–koinobiont distinctions are not entirely clear when parasitoid host-feeding strategies are considered.

Patterns of Host Feeding by Hymenopteran Parasitoids

Jervis and Kidd (1986) reviewed the feeding behavior of 140 hymenopteran parasitoids and concluded that host feeding was important for egg production, while nonhost foods were important for maintenance and longevity. Four types of host feeding were distinguished. Concurrent feeding occurred when the female used the same host individual for feeding and oviposition. Feeding was referred to as nonconcurrent if the female used different host individuals for feeding and oviposition. Either feeding habit may be nondestructive or destructive. In the latter case, probing and/or mutilation as well as feeding may contribute to death. Destructive feeding was most often nonconcurrent because it usually rendered a host unsuitable for oviposition. The distribution of feeding habits among species for which complete information was available was nonconcurrent/destructive (42 species), concurrent/nondestructive (11 species), concurrent/nondestructive and nonconcurrent/destructive (8 species), and concurrent/nondestructive and nonconcurrent/nondestructive (3 species). Feeding patterns, however, were not mutually exclusive, and parasitoids differed in their lifetime and diurnal patterns of feeding. It was concluded that concurrent/nondestructive feeding was most likely when hosts were readily available and that destructive feeding was advantageous when host density was low. Jervis et al. (1992) reviewed and outlined analytical methods for determining the dietary range of adult parasitoids.

The effect of host size on parasitoid host-feeding pattern has been examined in several studies. When given a choice, parasitoids generally feed on small hosts and oviposit on larger hosts (Bartlett, 1964; van Alphen, 1980; Godfray, 1994). This is often reflected by feeding on earlier host stages and by ovipositing on later and larger stages. Kidd and Jervis (1991a, 1991b) concluded that this feeding pattern has the advantages of reduced handling time on early host stages, reduced waste of progeny from mortality due to limited food resources, and reduced progeny mortality from the parent's host-feeding activity.

The preceding issues were further investigated by Heimpel and Rosenheim (1995, 1996) in their studies of effects of host-feeding pattern on reproductive success in *Aphytis melinus* DeBach. The authors characterized the role of host feeding as a dynamic balance between current and future reproduction. Hosts used for oviposition represent current reproduction while those used for feeding represent an investment toward future reproduction. The influences of numerous factors including egg load, diet, age, experience, and host size on host-feeding pattern were examined. Later,

Heimpel *et al.* (1997) distinguished the roles of host and nonhost feeding for the same parasitoid. They demonstrated that host feeding increased longevity and fecundity but only when a nonhost source of sugar was also available. Both sugar and host feeding were necessary for maximum lifetime reproductive success. Similar results demonstrating the essentiality of both host and nonhost feeding for optimal fecundity were reported by Morales-Ramos *et al.* (1996a) for the parasitoid *Catolaccus grandis*.

Models of Host Feeding by Hymenopteran Parasitoids

Several models were deduced by Jervis and Kidd (1986) to assess how the energetic demands and constraints on a parasitoid affect its host-feeding strategy. One model predicted the feeding strategy for maximizing egg production of a single synovigenic female. The energetic needs of the parasitoid were arithmetically expressed as follows:

$$N_f = (r + s + oE)/h$$

and

$$E = [N_a \cdot h - (r + s)]/h + o$$

where *r, s,* and *o* = energy required for maintenance, host location, and production and oviposition of an individual egg, respectively:

h = net energy gained from feeding on a single host

E = total number of eggs oviposited

N_f = number of hosts required to be fed on to meet energy demands

$N_a = N_f + E$ = number of hosts attacked

The relationship defines the optimal allocation of feeding and oviposition at different host densities (Fig. 3). Figure 4 illustrates the influence of host availability on oviposition when *h, r,* and *o* vary. The authors assumed that the percentage of hosts used for oviposton increases as the host density increases, but emphasized that the predictions of this simplistic model cannot explain the diversity of host-feeding behavior among parasitic Hymenoptera. Several potential constraints were discussed. Moreover, the model assumed a constant longevity that the authors pointed out would be a disadvantage for the parasitoid. Subsequently, Kid and Jervis (1989) described more sophisticated models of host feeding that did consider variable longevity.

The value of host-feeding models for predicting parasitoid population stability and persistence was discussed by Jervis and Kidd (1991). The authors emphasized that while host feeding in itself does not influence stability, stability is affected by certain characteristics of synovigeny, namely,

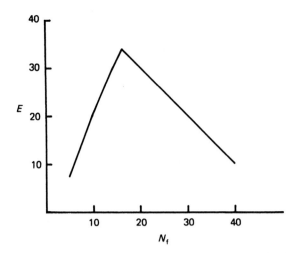

FIGURE 3 Relationship between the number of eggs a parasitoid deposits during its lifetime (E) and the number of hosts fed upon (N_f). Below the optimum, egg numbers are constrained by the relationship $E = [N_f - (r + s)]/o$. Above the optimum, egg numbers are limited by the relationship $E = N_a - N_f$. Indicated values were calculated from $N_a = 50$, h = 5, $r + s = 10$ and $o = 2$ (see text for further explanation). [From Jervis, M. A., & Kidd, A. C., (1986). *Biological Review, 61, 395–434.*]

model and two additional models; the first assumed that the parasitoid feeds solely to obtain energy to mature eggs and the second assumed that the parasitoid feeds to obtain energy for both longevity and egg maturation. Their results supported the last model and based on the available information on patterns of host feeding and oviposition by parasitoids they concluded that parasitoids host feed both for longevity and egg maturation.

Several other authors have further discussed the importance of host-feeding strategies for parasioid success and have reviewed the various host-feeding models (Yamamura & Yano, 1988; Huffaker & Gutierrez, 1990; Godfray, 1994; Heimpel & Rosenheim, 1995; Heimpel & Collier, 1996).

egg limitation, egg resorption, and variable longevity, that collectively constitute a significant destabilizing mechanism. They further explained that synovigeny determines a parasitoid's lifetime searching efficiency through variable longevity and fecundity and, by egg resorption, its survival during periods of host scarcity. When hosts are scarce, parasitoids survive by using energy obtained from egg resorption, but at the expense of oviposition. Even at low host densities, host encounter rates remain relatively high, but host feeding causes host densities to continue decreasing. When egg resorption and host-feeding attacks can no longer sustain the parasitoids, they die without reproducing. Thus, host-feeding behavior and synovigeny are inextricably linked and synovigeny is a critically important component that must be considered when developing host-feeding models.

Houston *et al.* (1991) employed a proovigenic model to examine the effect of host encounter rate on host-feeding behavior and to determine the extent of energy reserves when a parasitoid would be expected to change from host feeding to ovipositon. This model assumed that the sole purpose of host feeding is to obtain energy to maximize fecundity by increasing host-searching time. These authors reported that a parasitoid will oviposit if energy reserves are maintained above a critical value and will host feed to avoid starvation when energy reserves fall below the critical level. Increased host encounter rates are expected to increase the critical energy level. Chan and Godfray (1993) describe similar modeling strategies for determining the function of host feeding. They employed the proovigenic

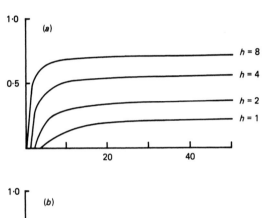

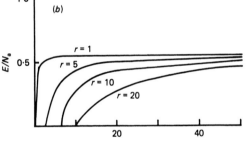

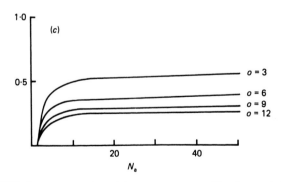

FIGURE 4 The proportion of host encounters resulting in parasitoid oviposition (E/N_a) in relation to the number of encounters (N_a): (a) with different amounts of energy extracted (h), (b) with different maintenance costs (r), and (c) with different egg production and oviposition costs (o) (see text for further explanation). [From Jervis, M. A., & Kidd, A. C. (1986). *Biological Review, 61, 395–434.*]

Host Feeding by Dipteran Parasitoids

Host feeding occurs among dipteran parasitoids but is not as common as in the Hymenoptera (Clausen, 1940). Shahjahan (1968) suggested that host feeding by tachinid parasitoids may increase longevity and fecundity and Nettles (1987b) reported that fecundity was prolonged by feeding *Eucelatoria bryani* host hemolymph compared with a sucrose solution. The latter effect could not be mimicked by substituting a solution of free amino acids or bovine serum albumin. Adult *Trichopoda giacomellii* Berthold displayed significantly greatly longevity and fecundity when fed raisins compared with water (Coombs, 1997), and this author suggested that the availability of suitable foods may be essential for successful establishment of parasitoids in the field for biological control.

Honeydew Feeding by Adult Entomophagous Species

Honeydew, the excretion of various homopteran insects, is a food source for many adult entomophagous insects. Neuropterans of the genus *Chrysoperla* and other genera with nonpredaceous adults feed actively on honeydew, nectar, and pollen (Principi & Canard, 1984). Although honeydew does not contain all the essential amino acids, Hagen and Tassan (1970, 1972) demonstrated that yeast symbiotes residing in the gut can provide the missing amino acids to some non-predaceous stages. Predatory neuropterous adults also feed on honeydew, but reproductive activity ensues only after prey are eaten (Hagen & Sawall, 1970; Hagen, 1986a). Similar results have been reported with coccinellid predators (Hagen, 1962). Heidari and Copland (1993) reported that honeydew not only was an important food source for last-instar larvae and adults of the coccinellid predator *Cryptolaemus montrouzieri* (Mulsant), but also acted as an arrestant causing increased prey-searching time. The phytoseid mite predator *Amblyseius largoensis* (Muma) can complete development on honeydew alone (Kamburov, 1971). McMurtry and Scriven (1964) reported that the predaceous mite *Euseius hibisci* displayed reduced mortality and reduced preoviposition time, and increased oviposition when fed honeydew with prey compared with prey alone.

An important component of honeydew feeding by many predators appears to be the attractiveness of tryptophan breakdown products (van Emden & Hagen, 1976; Dean & Satasook, 1983). Tryptophan, therefore, is an important component of "artificial honeydew" employed for supplemental feeding of predators (see later section on current applications).

Dipterans and hymenopteran parasitoids have also been observed feeding on honeydew (Clausen, 1940; Zoebelein, 1956; Idoine & Ferro, 1988). The potential importance of honeydew as a supplementary food was suggested early on by Clausen *et al.* (1933) who observed adult female *Tiphia matura* Allen & Jaynes migrating long distances from the location of their host to feed on honeydew. Ichneumonids of the genus *Rhyssa* appear to be dependent on honeydew for maintaining the longevity necessary to parasitize and adequately control populations of *Sirex* (Hocking, 1967). The nutritional value of honeydews appears to vary with the homopteran source. Wilbert (1977) demonstrated considerable differences in longevity among several hymenopterous parasitoids fed aphid or coccid honeydew.

Nutritional Requirements of Adult Entomophagous Species

The nutritional requirements of adult entomophagous insects remain obscure. Some requirements of the ichneumonid parasitoid *Exeristes comstockii* (Cresson) were examined by Bracken (1965, 1966, 1969). Adult females fed an artificial medium containing amino acids, sucrose, fatty acids, cholesterol, vitamins, and inorganic salts produced eggs at a rate equivalent to that of individuals fed *Galleria mellonella* (Linnaeus) larvae and sucrose. Egg production was reduced or eliminated when amino acids, sucrose, vitamins or salts were deleted. Sucrose, pantothenic acid, folic acid, and thiamin were all essential for egg laying.

The nutritional requirements of adult predators are also unknown. Numerous seminatural diets, however, have been successfully developed for maintaining chrysopid spp. and various adult coccinellids (see next section). These studies suggest that predators require a complete and well-balanced diet to ensure maximum longevity and reproductive potential. Tauber and Tauber (1974) examined the influence of diet on reproduction of several predaceous chysopids and observed significant differences in their dietary requirements for mating and oviposition. Adult females of *Chrysoperla externa* (= *lanata*) (Hagen), which are not predaceous, required a protein–carbohydrate diet to successfully mate, while nonpredaceous males and females of *C. downesi* (Banks) only required a diet of sugar and water for mating. The predaceous adults of *Chrysopa nigricornis* Burmeister required aphid prey to mate, but those of *C. quadripunctata* Burmeister and *Sympherobius amiculus* (Fitch) successfully mated when fed sugar and water alone. All species required both prey and protein–carbohydrate diet for sustained oviposition, but females of *C. quadripunctata* and *S. amiculus* initiated oviposition without prey. Nutritional data for the adults of several chrysopid species were summarized by Roussett (1984), and effects of various diets on fecundity were summarized by Hagen (1986b).

TABLE 3 Parasitoid Species Reared *in vitro*[a]

Taxa	Natural host stage	Medium Condition[b]	Type[c]	Source of eggs or larvae	Development	References
Order Diptera						
Family Sarcophagidae						
Agria housei	Lepidopteran pupae	Xenic	Oligidic	Dissection of the female parasitoid	Many generations	House & Traer, 1948
		Axenic	Meridic	Dissection of the female parasitoid	First-instar larva to fecund adult	House, 1954
Boettacheria (= Sarcophaga) cimbicis (Townsend)	Hymenoptera	Axenic	Meridic/holidic	Dissection of the female parasitoid	First-instar larva to adult[d]	House, 1966c
		Xenic	Oligidic	—	First-instar larva to adult	Shannon, 1923
Kellymia kellyi	Orthopteran nymphs/adults	Xenic	Oligidic	—	40 Generations	Smith, 1958
Sarcophaga aldrici	Lepidopteran pupae	Xenic	Oligidic	Larviposition directly onto artificial medium	Many generations	Arthur & Coppel, 1953
S. sarracenioides Aldrich	Hymenopteran pupae	Xenic	Oligidic	—	First-instar larva to adult	Shannon, 1923
Family Tachinidae						
Archytas marmoratus	Lepidopteran larvae/pupae	Axenic	Oligidic/meridic[e]	Dissection from *Helicoverpa zea*	Second instar to pupa	Bratti, 1993, 1994a; Farneti et al., 1997
Eucelatoria bryani	Lepidopteran larvae	Axenic	Meridic, holidic	Dissection from *Heliothis virescens*	First-instar larva to adult	Nettles et al., 1980; Nettles, 1986b
			Oligidic/meridic[e]	Dissection from *Galleria mellonella*		Bratti & Nettles, 1992; Bratti & D'Amelio, 1994
Exorista larvarum	Lepidopteran larvae	Axenic	Oligidic	Dissection from *G. mellonella*	Egg to adult	Mellini et al., 1993b, 1993c; Mellini & Campadelli, 1994a, 1994b, Bratti et al., 1995; Bratti & Coulibaly, 1995
			Oligidic/meridic[e]			
E. sorbillans	Lepidopteran larvae	Axenic	Oligidic[e]	Surface of *Bombyx mori*	Egg to adult	Watanabe & Mistuhashi, 1995
Lixophaga diatraeae	Lepidopteran larvae	Axenic	Meridic	Dissection of the female parasitoid	Embryo and first-instar larva to adult	Grenier et al., 1978; Grenier, 1979
Palexorista laxa	Lepidopteran larvae	Axenic	Oligidic[e] Oligidic/meridic[e]	Dissection from *Helicoverpa zea*	First-instar larva to adult	Bratti & Nettles, 1988; Bratti & Nettles, 1994
Parasetigena agilus (= silvestres) Robineau-Desvoidy		Xenic	Meridic/holidic		Slight growth of first-instar larva	Prell, 1915
Phryxe caudata	Lepidopteran larvae	Axenic	Meridic	Dissection of the female parasitoid	Embryo and first- to last-instar larva	Grenier et al., 1974, 1975; Grenier, 1979
Pseudogonia rufifrons	Lepidopteran larvae	Axenic	Oligidic[e]	Dissection from *G. mellonella*	Second-instar larva to adult	Baronio & Sehnal, 1980
Pseudoperichaeta nigrolineata	Lepidopteran larvae	Axenic	Oligidic[e] Oligidic/meridic	Dissection of the female parasitoid	First-instar larva to adult[f]	Bratti, 1989, 1990a,b; Fanti & Bratti, 1991
			Meridic		First- to second-instar larva	Grenier, 1988
Order Hymenoptera						
Family Ichneumonidae						
Campoletis distincta (= perdistinctus) (Provancher)	Lepidopteran larvae	Axenic	Oligidic	—	Egg to pupa	Vanderzant (cf. Thompson, 1981a)

(continues)

609

TABLE 3 *(continued)*

Taxa	Natural host stage	Medium Condition[b]	Type[c]	Source of eggs or larvae	Development	References
C. sonorensis	Lepidopteran larvae	Axenic	Holidic/meridic[f]	Dissection from *Heliothis virescens*	Egg to third-instar larva	Hu & Vinson, 1997a,b, 1998
Diapetimorpha introita	Lepidopteran pupae	Axenic	Oligidic		Egg to fecund adult	Greany & Carpenter, 1996
Exeristes roborator	Lepidopteran larvae	Axenic	Meridic/holidic	Surface of *Pectinophora gossypiella* following oviposition	Egg to fecund adult	Carpenter & Greany, 1997; Thompson, 1975, 1976
Itoplectis conquisitor	Lepidopteran pupae	Axenic	Meridic	Dissection from *G. mellonella* following oviposition	Egg to adult	Yazgan & House, 1970
I. conquisitor	Lepidopteran pupae	Axenic	Meridic/holidic	Direct oviposition into artificial medium	Egg to fecund adult	Yazgan, 1972
I. conquisitor	Lepidopteran pupae	Axenic	Meridic		Egg to adult	House, 1978
Coccygomimus (= *Pimpla*) *turionellae*	Lepidopteran pupae	Axenic	Oligidic	Dissection from *G. mellonella*	Egg to adult	Bronskill & House, 1957
Venturia canescens	Lepidopteran larvae	Axenic	Oligidic[e]	Dissection of *Ephestia kuehniella*	Egg to pupa	Nakahara et al., 1997; Ohbayashi et al., 1994; Yamamoto et al., 1997
Family Braconidae						
Bracon mellitor	Coleopteran larvae	Axenic	Oligidic/meridic[e]	Removed from host larvae following oviposition	Egg to adult	Guerra, 1992; Guerra et al., 1993
Cardiochiles nigriceps	Lepidopteran larvae	Axenic	Oligidic/meridic	Dissection from *H. virescens*	Egg to second-instar larva	Pennocchio et al., 1992
Cotesia (= *Apanteles*) *marginiventris*	Lepidopteran larvae	Axenic	Oligidic[e]/meridic	Dissection from *H. zea*	Egg to first-instar larva	Greany, 1980, 1981, 1986
Diachasmimorpha (= *Biosteres*) *longicaudatus*	Dipteran larvae/pupae	Axenic	Oligidic	Dissection from *Anastrepha suspensa*	Egg to pupa[f]	Lawrence, 1991
Lysiphlebus fabarum	Hymenopteran adults	Axenic	Oligidic/meridic	Dissection from *Aphis fabae* Scopoli	First-instar larva to adult	Rotundo et al., 1988
Microplitis croceipes	Lepidopteran larvae	Axenic	Oligidic/meridic[e]	Dissection from *H. zea*	Egg to first-instar larva	Greany, 1986
Family Chalcididae						
Brachymeria intermedia	Lepidopteran pupae	Axenic	Meridic / Oligidic[e]	Dissection from *G. mellonella*	Egg to last-instar larva / Egg to adult	Thompson, 1980; Dindo, 1990; Dindo & Campadelli, 1992; Dindo et al., 1997a, 1997b
B. lasus	Lepidopteran pupae	Axenic	Meridic	Dissection from *G. mellonella*	Egg to fecund adult	Thompson, 1981b, 1981c, 1981d, 1983a, 1983b
B. ovata	Lepidopteran pupae	Axenic	Meridic	Dissection from *G. mellonella*	Egg to last instar	Thompson, 1981b
Family Encyrtidae						
Ageniaspis fuscicollis	Lepidopteran eggs and larvae	Axenic	Meridic	Dissection from field-collected *Hyponomeuta* sp. eggs	Egg to second-instar larva	Nenon, 1972a, 1972b
Ooencyrtus pityocampae	Lepidopteran eggs	Axenic	Oligidic/meridic	Oviposition directly into artificial medium	Egg to fecund adult	Masutti et al., 1992
Family Eulophidae						
Anastatus japonicus	Lepidopteran eggs	Axenic	Oligidic[e]	Oviposition directly into artificial medium	Egg to adult	Li, 1989
Tetrastichus schoenobii	Lepidopteran eggs and larvae	Axenic	Meridic[e]	Dissection from host	Egg to fecund adult	Ding et al., 1980a

TABLE 3 (continued)

Species	Host stage	Rearing[b]	Diet[c]	Method	Development	References
Family Pteromalidae						
Catolaccus grandis	Coleopteran larvae	Axenic	Oligidic/meridic[e]	Removed from host larvae following oviposition	Egg to adult	Guerra, 1992; Guerra et al., 1993; Rojas et al., 1996a
Pachycrepoideus vindemiae	Dipteran pupae	Axenic	Meridic[e]	Dissection from Pieris brassicae	Egg to adult	Thompson et al., 1983
Pteromalus puparum	Lepidopteran pupae	Axenic?	Oligidic[e]	Dissection from P. brassicae	Egg to pupa	Bouletreau, 1968
P. puparum	Lepidopteran pupae	Axenic?	Oligidic[e]	Dissection from P. rapae	Egg to adult	Bouletreau, 1972
P. puparum	Lepidopteran pupae	Axenic?	Oligidic[e]	Dissection from P. rapae	Egg to fecund adult	Hoffman et al., 1973
P. puparum	Lepidopteran pupae	Axenic?	Oligidic	Dissection from P. rapae	Eggs to fecund adult	Hoffman & Ignoffo, 1974
Family Scelionidae						
Telenomus heliothidis	Lepidopteran eggs	Xenic	Oligidic[e]	Dissection from Heliothis virescens	Egg to fecund adult	Strand et al., 1988
Trissolcus basalis	Hemipteran eggs	Axenic	Oligidic[e]	Dissection from host eggs	Egg to third-instar larvae	Volkoff et al., 1992
Tetrastichus schoenobii	Lepidopteran larvae	Axenic	Oligidic[e]	Dissection from adult parasitoid	Egg to adult	Ding et al., 1980a
Family Trichogrammatidae						
Trichogramma brassicae	Lepidopteran eggs	Axenic	Oligidic	Oviposition directly into artificial medium	Egg to adult	Grenier & Liu, 1990
T. californicum	Lepidopteran eggs	Xenic	Oligidic	Oviposition directly into artificial medium	Egg to adult	Xie et al., 1997
			Oligidic	Oviposition directly into artificial medium	Egg to second-instar larva	Rajendram, 1978b
T. chilonis	Lepidopteran eggs	Axenic	Oligidic[e]	Oviposition directly into artificial medium	Egg to adult	Liu et al., 1995a, 1995b
T. confusum	Lepidopteran eggs	—	Oligidic/meridic	Oviposition directly into artificial medium	Egg to adult	Liu & Wu, 1982
T. dendrolimi	Lepidopteran eggs	—	Oligidic[e]	Oviposition directly into artificial medium	Egg to prepupa	Guan et al., 1978
		Axenic	Oligidic[e]	Oviposition directly into artificial medium	Egg to adult	Liu et al., 1979
		Axenic	Oligidic[e]	Oviposition directly into artificial medium	Egg to adult	Liu et al., 1995a, 1995b
T. dendrolimi		Axenic	Oligidic/meridic	Oviposition directly into artificial medium	Egg to adult	CRGHP, China, 1979
		Axenic	Oligidic	Oviposition directly into artificial medium	35 Generations	Wu et al., 1980, 1982
		Axenic	Oligidic	Oviposition directly into artificial medium	61 Generations	Gao et al., 1982
		Axenic	Oligidic[e]	Oviposition directly into artificial medium	Egg to adult	Dai, 1985
T. galloi	Lepidopteran eggs	Axenic	Oligidic[e]	Dissection from H. virescens or Heliocoverpa perva	Egg to adult	Consoli & Parra, 1997
T. japonicum	Lepidopteran eggs	Axenic	Oligidic	Oviposition directly into artificial medium	Egg to adult	Liu et al., 1983
		Axenic	Oligidic[e]	Oviposition directly into artificial medium	Egg to adult	Liu et al., 1995a,b
T. maidis	Lepidopteran eggs	Axenic	Oligidic[e]	Oviposition directly into artificial medium	Egg to adult	Grenier & Bonnot, 1988
T. minutum	Lepidopteran eggs	Axenic	Oligidic[e]	Oviposition directly into artificial medium	Egg to adult	Nordlund et al., 1997
		Axenic	Oligidic[e]	Oviposition directly into artificial medium	Egg to adult	Xie et al., 1997
			Oligidic[e]	Oviposition directly into artificial medium	Egg to adult	Liu et al., 1995a,b
T. nubilalae	Lepidopteran eggs	Axenic	Oligidic[e]	Oviposition directly into artificial medium	Egg to adult	Liu et al., 1983
T. pretiosum	Lepidopteran eggs	Axenic	Oligidic[e]	Dissection from Trichoplusia ni	Egg to fecund adult	Hoffman et al., 1975
			Oligidic[e]	Oviposition directly into artificial medium	Egg to adult	Liu & Wu, 1982; Xie et al., 1986b
				Dissection from Heliothis virescens	Egg to fecund adult	Strand & Vinson, 1985
				Dissection from H. virescens or Helicoverpa zea	Egg to adult	Consoli & Parra, 1997

[a] Table modified and updated from Hagen, K. S. (1986b). In M. Y. Hussein & A. G. Ibrahim (Eds.), *Biological control in the tropics: Proceedings of the first regional symposium on biological control* (pp. 35–85). Malaysia: Penerbit Universiti Pertanian.

[b] Axenic = reared under aseptic (sterile) conditions or on diets containing antibiotics; xenic = reared with unknown number of associated microorganisms and without antimicrobial compounds.

[c] Holidic = media with all ingredients known in chemical structure; meridic = media with holidic base to which is added at least one substance or preparation of unknown structure of uncertain purity; oligidic = media in which crude organic materials supply most dietary requirements.

[d] Fecund adults were subsequently obtained by inclusion of vitamin E.

[e] Contains insect-derived components such as host hemolymph, other tissues or tissue homogenates, and/or extracts.

[f] Larval molting was stimulated by or required 20-hydroxyecdysone.

IN VITRO CULTURE OF PARASITOIDS

The possibility of rearing entomophagous insects *in vitro,* that is, on artificial food in the absence of the host, has long fascinated biological control experts. Simmonds (1944) stated that ". . . the breeding of parasites for biological control . . . suggests the possibility of devising purely artificial nutritive media. . . . If this could be done, it might prove of great practical importance." Traditional rearing of parasitoids and predators often involves the tedious task of maintaining large colonies of the host species as well as the plant or other food source of the host (Waage *et al.,* 1985). *In vitro* culture offers a simple alternative for mass-rearing programs (Mellini, 1978, 1991; Greany *et al.,* 1984; Bratti, 1990b; Grenier *et al.,* 1994; Senft, 1997). *In vitro* culture also enables dietary and nutritional manipulations for fundamental studies of nutrition and biochemistry. Greany *et al.* (1984) outlined some of the benefits of *in vitro* culture as a tool for studying parasitoids, including the facilitation of behavioral investigations and clarification of the influence of diet and inheritance on various biological characteristics such as fecundity, vigor, selectivity, searching efficiency, and environmental tolerance.

Physiological Factors of Importance to Successful Development of Parasitoids

The physiological and metabolic adaptations exhibited by many insect parasitoids (see Chapter 7 for definitions) in relation to their parasitic way of life are of critical importance for successful *in vitro* culture (Mellini, 1975a; Thompson, 1981a; Grenier *et al.,* 1986; Campadelli & Dindo, 1987). The parasitoid–host relationship is often, but incorrectly, thought to lack the complex physiological interactions typical of the host associations of other metazoan parasites (Thompson, 1985, 1986a; Dindo, 1987; Mellini, 1993). The immature stages of many parasitoids, particularly koinobiontic endoparasitoids that parasitize the larval stages of their hosts, are truly parasitic and such parasite–host relationships are characterized by extensive physiological and biochemical interactions between the parasitoid and the host (Beckage, 1985; Thompson, 1985, 1986a, 1986b, Lawrence, 1986a, 1990; Lawrence & Lanzrein, 1993). Many of these interactions are intimately associated with nutrition and successful development of the parasitoid (Beckage & Riddiford, 1983; Thompson, 1986a). Thompson (1981b) stated ". . . it should not be expected that the essential nutritional requirements of metazoan parasitic animals will differ from related 'free-living' forms, but rather that differences will appear in the physiological and metabolic adaptations that enable the former to obtain their nutrition from a living source. . . ." Mellini (1975b, 1978, 1983) and Grenier *et al.* (1986) discussed the poten-

tial importance of the host endocrine system and of hormonal interaction in *in vitro* culture (see later subsection on the role of hormones in *in vitro* culture of parasitoids). Nonhormonal host growth factors also appear to play important roles in regulating parasite development (Greany, 1986; Nettles, 1990; Ferkovich & Oberlander, 1991a). The extent to which parasite–host physiological interactions need to be considered in the successful development of *in vitro* culture remains to be established and will undoubtedly vary with the parasitoid species under study. Although our knowledge of parasitoid physiology and biochemistry is still limited, remarkable success has been achieved to date at *in vitro* culture (Table 3) and defined artificial diets have been formulated for several species (Table 4).

Artificial Diets and *in vitro* Culture of Dipteran Parasitoids

Early studies on rearing parasitoids apart from their hosts utilized a variety of natural foodstuffs, including fish and liver products. House and Traer (1948) reared the sarcophagid dipteran *Agria housei* for many generations on a diet of salmon and liver. In contrast to 38% pupation among larvae reared on the host, *Choristoneura fumiferana* (Clemens), 88% of larvae pupated when reared on the artificial medium. The related parasitoid, *Sarcophaga aldrichi* Parker, was also reared on this medium as well as on liver alone (Arthur & Coppel, 1953). Subsequently, Coppel *et al.* (1959) maintained *A. housei* in the laboratory on fresh pork liver. Approximately 1000 *A. housei* larvae were reared on one-half pound of sliced liver and were not affected by putrification of the tissue. Smith (1958) maintained *Kellymyia kellyi* (Aldrich) for 40 generations on pork liver and also was successful in rearing larvae on a mixture of powdered milk, powdered egg, and brewer's yeast moistened with water to form a thick paste.

The first chemically defined medium for rearing a parasitoid was developed by House (1954) for *A. housei.* The diet was prepared aseptically and consisted of 19 amino acids, ribonucleic acid, dextrose, inorganic salts, B vitamins, choline, and inositol, gelled with agar. Of the 84% of larvae reaching the third instar, 60% pupated and 32% of the pupae emerged as adults. The medium was later refined and many of the developmental nutritional requirements of *A. housei* were determined (House, 1977). House (1966c) established that vitamin E was necessary for reproduction and inclusion of this vitamin enabled continuous *in vitro* culture of *A. housei.*

In contrast to the preceding species, other dipteran parasitoids have proved more difficult to culture in vitro. Many of these have specialized physiological adaptations associated with parasitism that sarcophagid parasitoids lack. Tachinid parasitoids, for example, have relatively high

TABLE 4   Composition of Defined Artificial Media Formulated for *in vitro* Culture of Selected Parasitoid Species

Lixophaga diatraeae (Townsend) (Diptera: Tachinidae)[a]

Component	mg/100 ml	Component	mg/100 ml	Component	mg/100 ml	Component	mg/100 ml
Amino nitrogen	6,827	Amino nitrogen		Organic acids		Lipids	2,270 or 3,270
Asparagine	201	Enzymatic hydrolysates		Malic acid	50	Cholesterol	150
Threonine	78	Lactalbumin	2,604	α-Ketoglutaric acid	10	Lecithin (egg)	500
Serine	57	Ovalbumin	715	Succinic acid	5	Choline–HCl	20
Glutamic acid	186	Soy Protein	916	Fumaric acid	5	Corn oil	1,600 or 2,600
Proline	109	Casein	41	Citric acid	10	Carbohydrate	3,000
Glycine	91	Inorganic salts	808.9453	Pyruvic acid	10	Glycogen	3,000
Alanine	169	NaCl	78	Vitamins	18.6275	Miscellaneous	825 or 925
Valine	104	KCl	100	Ascorbic acid	4.1	Glutathione	10
Methionine	10	$MgSO_4 \cdot 7H_2O$	417	Pantothenic acid	1.1	ATP	65
Isoleucine	77	$CaCl_2 \cdot H_2O$	180	Nicotinic acid	2.2	RNA	50
Leucine	29	KI	0.02	Riboflavin	0.5	Agarose	700 or 800
Tyrosine	89	$Co(CH_3CO_2)_2 \cdot 4H_2O$	0.02	Pyridoxine	0.15	Properties	
Phenylalanine	62	$Mn(CH_3CO_2)_2 \cdot 4H_2O$	5.50	Folic acid	$27 \cdot 10^{-3}$		400 mOs
Lysine–HCl	175	$Cu(CH_3CO_2)_2 \cdot H_2O$	0.20	Biotin	$10 \cdot 10^{-3}$		6.5 pH
Histidine–HCl·H_2O	260	$Fe(NO_3)_3 \cdot 9H_2O$	8.90	Hiamine	$37 \cdot 10^{-3}$	Rearing conditions	
Tryptophan	40	$Zn(NO_3)_2 \cdot 6H_2O$	2.30	Cyanocobalamin	$35 \cdot 10^{-4}$		22.5 $\pm$ 5°C
Phosphoethanolamine	20	$MoCl_3$	$28 \cdot 10^{-4}$	Retinol	2.0		85 $\pm$ 5 Relative humidity
Arginine–HCl	176	SeO_2	$7 \cdot 10^{-4}$	α-Tocopherol	2.0		
Cysteine	10	$VaCl_3$	$18 \cdot 10^{-4}$				
Gelatin	608	H_2PO_4	17				

(continues)

TABLE 4 (continued)

Brachymeria lasus (Walker) (Hymenoptera: Chalcididae)[b]

	mg/100 ml		mg/100 ml		mg/100 ml
Amino nitrogen	15,960	Inorganic salts	298	Lipids	2,700
Bovine serum albumin	15,000	$CaCl_2$	15	Cholesterol	10
Alanine	50	$CoCl_2 \cdot 6H_2O$	2.5	Intralipid®	
Arginine–HCl	40	$FeCl_3 \cdot 6H_2O$	10	Soybean oil	2,000
Aspartate	100	$ZnCl_2$	2.5	Glycerol	450
Cysteine–HCl	10	K_2HPO_2	180	Egg yolk phospholipid	240
Glutamic acid	100	$Na_2HPO_4 \cdot 7H_2O$	25		
Glycine	50	$MgSO_4 \cdot 7H_2O$	60		
Histidine · 2HCl	80	$CuSO_4 \cdot 5H_2O$	2.5	Carbohydrate	4,000
Hydroxyproline	40	$MnSO_4 \cdot H_2O$	0.5	Glucose	4,000
Isoleucine	40				
Leucine	70		μg/100 ml		
Lysine · 2HCl	70	Vitamins	223,680		
Methionine	20	Biotin	60	Properties	
Phenylalanine	40	Choline chloride	200,000	525 mOs	
Proline	40	Cyanocobalamine	20	pH 6.5	
Serine	40	Folic acid	100	Rearing conditions	
Threonine	40	Inositol	15,000	25°C	
Tryptophan	10	Nicotinamide	5,000	Dark	
Valine	120	Pantothenic acid	2,000		
		Pyridoxine–HCl	300		
		Riboflavin	1,000		
		Thiamin	200		

(continues)

TABLE 4 (continued)

Exeristes roborator (Fabricius) (Hymenoptera: Ichneumonidae)[c]

	mg/100 ml		mg/100 ml		mg/100 ml
Amino nitrogen	6,000	Inorganic salts	606	Lipids	496
Alanine	300	$CaCl_2$	30	Cholesterol	50
Arginine–HCl	240	$CoCl_2 \cdot 6H_2O$	5	Linoleic acid	50
Aspartic acid	600	$FeCl_3 \cdot H_2O$	20	Linolenic acid	50
Cysteine–HCl·H_2O	60	$ZnCl_2$	5	Palmitic acid	123
Glutamic acid	600	K_2HPO_4	370	Palmitoleic acid	25
Glycine	300	$Na_2HPO_4 \cdot 7H_2O$	50	Oelic acid	123
Histidine dihydrochloride	480	$MgSO_4 \cdot 7H_2O$	120	Stearic acid	25
Hydroxyproline	240	$CuSO \cdot 5H_2O$	5		
Isoleucine	240	$MnSO_4 \cdot H_2O$	1	Carbohydrate	2,000
Leucine	420			Glucose	2,000
Lysine dihydrochloride	420	Vitamins	223.68		
Methionine	120	*d*-Biotin	0.06	Miscellaneous	5,000
Phenylalanine	240	Choline cloride	200.00	Sephadex® LH20	5,000
Proline	240	Cyanocobalamine	0.02		
Serine	240	Folic acid	0.10	Properties	
Threonine	240	*i*-Inositol	15.00	pH 7	
Tryptophan	60	Nicotinamide	5.00	Rearing conditions	
Valine	720	D-Pantothenic acid (hemi Ca)	2.00	23°C	
		Pyridoxine–HCl	0.30	Dark	
		Riboflavin	1.00		
		Thiamine–HCl	0.20		

(continues)

TABLE 4 *(continued)*

Eucelatoria bryani (Sabrosky) (Diptera: Tachinidae)[d]

	mg/100 ml		mg/100 ml		mg/100 ml		mg/100 ml
Amino nitrogen	3,983	Inorganic salts		Vitamins	19.25	Miscellaneous	1,906
Asparagine	170	NaCl	261	Thiamin	0.1	Adenine	6
Threonine	70	KCl	59	Riboflavin	1.0	Thymine	6
Serine	50	$MgSO_4 \cdot 7H_2O$	123	Pyridoxine	0.1	Guanine	6
Glutamine	200	$MgCl_2 \cdot 6H_2O$	319	Biotin	0.05	Cytosine	3
Proline	50	KH_2PO_4	40	Ascorbic acid	5.0	Uracil	6
Glycine	80	K_2HPO_4	51	Nicotinic acid	2.0	ATP	20
Alanine	100	KI	0.02	Calcium pantothenate	4.0	Glutathione	10
Valine	100	$Co(CH_3CO_2)_2 \cdot 4H_2O$	0.02	Folic acid	1.0	Calcium phosphorylcholine	129
Methionine	30	$Mn(CH_3CO_2)_2 \cdot 4H_2O$	5.5	Inositol	5.0	Phosphoethanolamine	5
Cysteine	8	$Cu(CH_3CO_2)_2 \cdot H_2O$	0.2	Retinol	0.5	Yeastolate®	200
Isoleucine	80	$Fe(NO_3)_3 \cdot 9H_2O$	8.9	α-Tocopherol	0.5	Kanamycin	5
Leucine	100	$Zn(NO_3)_2 \cdot 6H_2O$	2.0	Lipids	210.2	Penicillin	10
Tyrosine	70	Organic acids	330	Distearoyl phosphatidyl choline	0.1	Agar	1,500
Phenylalanine	70	Malic acid	30	Dimyristoyl phosphatidyl choline	0.1	Properties	
β-Alanine	10	α-Ketoglutaric acid	45	Tween 80	150	6.7–6.8 pH	
Lysine	150	Fumaric acid	50	Cholesterol	10	Rearing conditions	
Histidine	75	Potassium acetate	83	Triolein	50	29°C	
Arginine	125	Citric acid	25	Carbohydrates	1,500	Photoperiod 14/10 h	
Tryptophan	40	Calcium succinate	67	Trehalose	1,000	Light/dark	
γ-Aminobutyric acid	5	Pyruvic acid	30	Glucose	50	50–90% Variable relative humidity (see text for explanation)	
Lactalbumin hydrolyzate	900						
Bactopeptone®	500						
Bovine serum albumin	1,000						

(continues)

TABLE 4 *(continued)*

Trichogramma pretiosum (Riley) (Hymenoptera: Trichogrammatidae)[e]

	mg/100 ml		mg/100 ml		mg/100 ml
Amino nitrogen	20,818	Inorganic salts	197,051	Myo-inositol	384
Bovine serum albumin	20,000	NaH_2PO_4,H_2O	70,000	Nicotinic acid	10
β-alanine	18	KCL	72,000	Calcium pantothenate	50
Arginine·HCl	48	$MgSO_4$ (anhydrous)	55,000	Pyridoxine·HCl	31
Aspartic acid	78	$(NH_4)MO_7O_{24}·H_2O$	2	Riboflavin	30
Asparagine (anhydrous)	78	$CoCl_2·6H_2O$	2	Thiamine·HCl	10
Cystine·2HCl	6	$CuCl_2·2H_2O$	3	γ-Aminobenzoic acid	20
Glutamic acid	90	$MnCl_2·4H_2O$	12	Nicotinamide	160
Glutamine	60	$ZnCl_2$	2	β-Carotene	1,000
Glycine	12	$FeSO_4·6H_2O$	30	Lipids	2,790
Histidine	12	Organic acids		Cholesterol	100
Hydroxyproline	80	Malic acid	5.1	Intralipid	
Isoleucine	45	α-Ketoglutaric acid	3	Soybean oil	2,000
Lysine·HCl	42	Succinic acid	2	Glycerol	450
Methionine	60	Fumaric acid	0.3	Egg yolk phospholipid	240
Phenylalanine	60	Vitamins	50,951	Carbohydrates	6,000
Proline	30	d-Biotin	23	Glucose	150
DL-Serine	24	Choline chloride	49,200	Sucrose	990
Threonine	12	Cyanocobalamine	26	Maltose	60
Tryptophan	6	Folic acid	7	Trehalose	4,800
Tyrosine	15				
Valine	42				

Manduca sexta hemolymph[f] variable concentrations

Properties
- 460 mOs (with 30% hemolymph)
- ph 6.6

Rearing conditions
- 27°C dark
- humidity adjusted during rearing

[a] From Grenier, S., et al. (1978). *Comptes Rendu de l'Academie Sciences Paris, Series D, 287,* 535–538.
[b] From Thompson, S. N. (1981d). *Experimental Parasitology, 32,* 414–418.
[c] From Thompson, S. N. (1975). *Annals of the Entomological Society of America, 68,* 220–226.
[d] From Nettles, W. C. (1986b). *Environmental Entomology, 15,* 1111–1115.
[e] From Strand, M. R., & Vinson, S. B. (1985). *Entomologia Experimentalis et Applicata, 39,* 203–209.
[f] Required for pupation.

respiratory rates (Ziser & Nettles, 1979; Bonnot *et al.,* 1984) and during or immediately following the first stadium the larvae form a direct connection to the host's tracheal system (Keilen, 1944; Fisher, 1971). First-instar larvae of the parasitoid *Eucelatoria bryani* attach to the host's tracheal system 12 h after hatching, and respiratory considerations proved to be critical for the development of *in vitro* culture methods (Nettles *et al.,* 1980). During initial studies, first-instar larvae dissected from the host were placed directly in a liquid artificial diet. Following the first stadium, larvae were transferred to diets gelled with agar, thus exposing larvae directly to atmospheric oxygen. Improvements in the methods allowed development without transfer. Powdered artificial diet containing 1.5% agar was preconditioned by maintaining it at 50% relative humidity for 24 h. The diet was then poured into petri dishes and held at 90% relative humidity. Young larvae dissected from the host 18 to 24 h after larviposition fed on the liquid diet covering the surface of the gelled medium. This was consistent with the normal feeding habit of first-instar larvae that feed on and develop in the host's hemolymph. As the liquid was slowly absorbed by the agar gel, the surface of the gelled medium dried and larvae were exposed to the atmosphere.

The artificial medium for rearing *E. bryani* was composed of mixtures of organic acids, amino acids, nucleic acid bases, B and fat-soluble vitamins, phospholipids and derivatives, as well as adenosine triphosphate (ATP), lactalbumin hydrolysate, Bactopeptone®, Yeastolate®, albumin, cholesterol, triolein, glucose, and trehalose (Table 4). When reared as described earlier, larvae developed at a rate equivalent to that of larvae reared on the host, *Heliothis virescens,* and 13% developed into adults with a sex ratio of approximately 1:2 (males to females). Adults were fecund but produced fewer progeny than host-reared insects. The medium was later refined and simplified (Nettles, 1986a) and some of the basic developmental nutritional requirements of *E. bryani* were determined (see earlier subsection on requirements for development of dipterous parasitoids). Nettles (1986b) tested the nutritive value of albumin or soy flour additives and reported greatly increased adult yields. Insects reared on this medium were more fecund (W. C. Nettles, personal communication). Further improvements in the *in vitro*-rearing method were outlined by Bratti and Nettles (1992).

The use of tissue culture media in diets for *in vitro* rearing of *E. bryani* was described by Bratti and D'Amelio (1994). Several tissue culture media including TC-100 M, TNM-FG and Schneider's medium supported only limited parasite development, even when supplemented with additional nutrients including fetal bovine serum and egg yolk. The IPL-52B M based medium, however, did support reasonable development with approximately 57% completing larval development, 37% pupal yield, and 23% adult emer-

gence. Addition of *G. mellonella* pupal extract improved the results with the other tissue culture media and yielded results equal to those with IPL 52B M without the insect additive. Among the pupal homogenate-supplemented media, the medium based on TNM-FH produced the best overall results. Parasitoids reared on all tissue culture media supplemented with pupal extract were fecund. The authors suggested that the low level of free amino acids in some of the tissue culture media may have limited growth and development.

Bratti and Nettles (1994) later compared the development of *E. bryani* with that of another tachinid parasite, *Palexorista laxa* (Curran), when maintained on the meridic diet of Nettles (1986b) supplemented with egg yolk but devoid of insect components, or on a diet of *H. zea* pupal homogenate. As in the preceding experiments with *E. bryani, P. laxa* larvae were dissected from hosts 18 to 24 h following larviposition. The percentage of yield of adult *E. bryani* reared on the semidefined diet was significantly greater than that of *P. laxa,* but on the pupal homogenate a significantly greater yield was obtained for *P. laxa.* The yield for *P. laxa* reared on pupal homogenate, approximately 90%, was the highest reported to that date for a tachinid parasitoid fed nonliving host material. *Palexorista laxa* was reared earlier *in vitro* on *Manduca sexta* hemolymph containing soy residue as a thickening agent (Bratti & Nettles, 1988).

Larval development of *Phryxe caudata* Rondani to the third instar was obtained using a liquid artificial diet described by Grenier *et al.* (1974). In contrast to the results of Nettles *et al.* (1980) with *E. bryani,* development of *P. caudata* was not improved by rearing larvae on gelled diets (Grenier *et al.,* 1975). Nettles *et al.* (1980) suggested that this may have resulted from the slower development rate and respiratory requirements of the latter species when reared *in vitro.* The importance of respiratory requirements of *P. caudata* was reported by Bonnot (1986).

Grenier *et al.* (1978) described the *in vitro* culture of *Lixophaga diatraeae,* the first tachinid successfully cultured from the first-instar larvae to adult on an artificial medium. The medium was composed of organic acids, amino acids, B and fat-soluble vitamins, gelatin, enzymatic hydrolysates (of lactalbumin, ovalbumin, soy protein, and caesin), ATP, lecithin, cholesterol; and was gelled with agarose (see Table 4). Adult insects were fecund and their offspring developed normally in the host, *G. mellonella* (S. Grenier, personal communication). Dietary osmolality was a critical factor for successful development of *P. caudata* and *L. diatraeae* and could not exceed 450 mOs/kg (Grenier *et al.,* 1986) (see later subsection on the importance of dietary osmotic pressure to successful *in vitro* culture).

Embryonic development of *P. caudata* and *L. diatraeae* on artificial media was examined by Grenier (1979). Uteri containing newly fertilized eggs were dissected from adult female parasitoids and placed on an agarose-gelled medium similar to that described earlier for rearing larvae. The yield of larvae was equivalent to that observed *in vivo* on host material, and was much greater than that obtained with a liquid diet. Respiratory requirements appeared to be critical for successful *in vitro* embryogenesis of these tachinid species.

Attempts at *in vitro* culture of *Pseudogonia rufifrons* Wied. (= *Gonia cinerascens* Rondani) have been described by several authors (Bratti & Monti, 1988; Bratti, 1989, 1990a, 1990b; Fanti & Bratti, 1991; Mellini *et al.*, 1993a). Complete larval development and pupation were achieved by Fanti and Bratti (1991) on the diet described by Nettles *et al.* (1980) for *E. bryani*, supplemented with *Galleria mellonella* pupal hemolymph and 20-hydroxyecdysone. Adult yield was correlated with the time the eggs and larvae spent in the host prior to dissection and *in vitro* culture. The complex behavioral and physiological interactions of this parasitoid with its host were outlined by Mellini and Coulibaly (1991).

Host pupal homogenate and hemolymph were also successfully employed for *in vitro* maintenance of the tachinids *Archytas marmoratus* (Townsend) and *Lydella thompsoni* Hert. (Bratti & Costantini, 1991; Bratti, 1993). Growth and development, however, were poor. Pupation of *A. marmoratus* cultured from the first instar was achieved with the diet of Nettles *et al.* (1980) supplemented with *H. zea* pupal homogenate. Limited larval growth and development of *A. marmoratus* was later reported on diets containing veal homogenate and *G. mellonella* pupal extract (Farneti *et al.*, 1997).

Successful *in vitro* culture of the tachinid *Exorista larvarum* Linnaeus on oligidic media was reported by Mellini *et al.* (1993b). Diets were composed of bovine serum, extracts or homogenates of *G. mellonella* pupae, and unspecified "additives"; and were gelled with agar. Pupal yields from eggs were as high as 90% and approximately 70% of these developed into adults. Development time was similar to that obtained on host insects. Further study examined the effects of modifying the qualitative and quantitative composition of the diet and demonstrated that cotton could replace agar as a support medium. (Mellini *et al.*, 1993c). Mellini and Campadelli (1994a, 1994b) reported that powdered yeast extract and/or egg yolk could replace *G. mellonella* larval or pupal homogenates, and *E. larvarum* was successfully cultured without insect additives. Pupae reared from the artificial medium were of similar size to those reared on *G. mellonella* and adult parasitoids were fecund. Further improvements optimized the dietary and nutritional requirements for *in vitro* culture (Mellini & Campadelli, 1995, 1996a). Ultimately, a diet composed of 74% skim milk, 8% yeast, 16% egg yolk, 4% saccharose, and 0.01% ascorbate produced a pupal yield of 52% and an adult

emergence of approximately 80% from parasitoid eggs removed from the surface of host larvae.

Mellini and Campadelli (1996b, 1996c) compared *in vitro* and *in vivo* culture of *E. lavarum* and discussed the potential for mass culture of this parasitoid species. There were few substantial differences between parasitoids reared *in vitro* and those reared from the host, *G. mellonella*. Pupal yields were almost the same on the diet as on the host. Under comparable conditions where insects were reared individually, the utilization of the artificial diet was approximately 20%, while utilization on the host was about 28%. Depending on the amount of artificial diet available, however, the weight of the pupae could be as much as 50% greater than that of pupae obtained with the host. Emergence rates based on numbers of pupae were approximately 84% on the host and 88% on the diet. Despite the success of the *in vitro* culture system, the authors pointed out that the host was still necessary for obtaining parasitoid eggs and that addition of small amounts of *G. mellonella* pupal homogenate to the artificial medium greatly increased the weight of parasitoid pupae. Moreover, the authors emphasized that successful rearing of *E. larvarum* may reflect the relatively simple host relationship of this idiobiontic species. *Exorista larvarum* forms primary integumental respiratory funnels and larvae displayed similar behavior on the gelled diet as in the host.

Several attempts have been made at adapting various tissue culture media for rearing *E. larvarum*, and similar results to those obtained for *Eucelatoria bryani* (Bratti & D'Amelio, 1994) were initially reported by Bratti and Coulibaly (1995). With *Exorista larvarum*, TMM-FH supported the best development, but addition of *G. mellonella* pupal extract significantly increased pupal yield and weight as well as adult emergence. Parasitoids cultured *in vitro* were fecund and readily parasitized *G. mellonella* larvae. Those reared on Schneider's medium supplemented with pupal extract displayed the greatest yield of pupae and adult progeny. Later, Bratti *et al.* (1995) reported rearing five consecutive generations on a diet of TNM-FH with 10% chicken egg yolk and lacking any insect additives.

Partial success at culture of *E. sorbillans* using MGM 450 tissue culture medium supplemented with *Bombyx mori* hemolymph was reported by Mitsuhashi and Oshiki (1993). Watanabe and Mitsuhashi (1995) observed that when *B. mori* hemolymph was mixed with wheat bran or modified melon fly diet, a few adult *E. sorbillans* were obtained. Later, Mitsuhashi (1996) demonstrated complete development to the adult stage on a diet composed of MGM 450 with 20% *B. mori* hemolymph, 20% fetal bovine serum, 2% peptone, and a small aliquot of powdered milk. Addition of 20-hydroxyecdysone to a hemolymph free medium improved molting from the first to second stadium, but had no effect when added to a hemolymph fortified medium (see later subsection on the role of hormones in *in vitro* culture of parasitoids).

A comparison of the *in vitro* development of several tachinid parasitoids including *Eucelatoria bryani, Exorista larvarum, Palexorista laxa, Pseudogonia rufifrons, A. marmoratus,* and *L. thompsoni* was made by Bratti and Campadelli (1993) and Bratti (1994a, 1994b).

Artificial Diets and *in vitro* Culture of Hymenopteran Parasitoids

Simmonds (1944) was among the first to attempt rearing hymenopteran parasitoids *in vitro*. Larvae of three ichneumonid ectoparasitoids were maintained for extended periods on raw beef and gelatin. Slight growth but no development ensued. Subsequently, Bronskill and House (1957) were successful in rearing *Coccygomimus turionellae* on a thick slurry of pork liver in 0.8% saline. The autoclaved pork homogenate was dispensed into sterile test tubes and surface-sterilized eggs dissected from host pupae were transferred onto the medium. Mature larvae were placed in gelatin capsules for pupation and 7% of the eggs became adults. By comparison, 50% of parasitoid adults were obtained from parasitized *G. mellonella*.

Yazgan and House (1970) reported the culture of the closely related ichneumonid *Itoplectis conquisitor* on a diet similar to that formulated for *Agria housei* (House, 1977). The first holidic diet for rearing a hymenopterous parasitoid *in vitro* was described by Yazgan (1972) for this species. The diet contained mixtures of amino acids, fatty acids, B vitamins, fat-soluble vitamins, lipogenic growth factors, glucose, and RNA; and was gelled with agar (see Table 4). The medium was ground into a viscous slurry. Parasitoid eggs dissected from the host were placed directly on the ground medium. Development from the egg to fecund adult was obtained but development time was twice that observed on the host, *G. mellonella.*

Thompson (1975) reared the parasitoid *Exeristes roborator* on a diet with a similar nutrient composition (see Table 4), but in contrast to *I. conquisitor, E. roborator* larvae would not tolerate direct contact with gelled media. Exposure to atmospheric oxygen was important for successful *in vitro* culture of *E. roborator* and success was achieved by retaining suspensions of the liquid diet in lipophilic Sephadex® LH-20 gel filtration medium. Mortality, size, and development time of *E. roborator* reared *in vitro* were similar to those of individuals reared on *Pectinophora gossypiella* (Saunders). Yazgan (1972) and Thompson (1976a, 1976b) determined many of the developmental nutritional requirements of *I. conquisitor* and *E. roborator* (see earlier subsection on requirements for development of hymenopteran parasitoids).

Greany and Carpenter (1996) reported on the *in vitro* culture of the ichneumonid ectoparasitoid *Diapetimorpha introita* (Cresson) maintained on an oligidic medium encapsulated in Parafilm®. The diet was composed of a defined insect culture medium supplemented with glutamine, ground bovine liver, and chicken egg yolk. Although the specific composition was not reported, the complete medium contained approximately 6% carbohydrate, 13% protein, 10% fat, 1% inorganics, and 70% water. *Diapetimorpha introita* individuals reared *in vitro* were somewhat smaller and had a longer development time than parasitoids reared on the host *Spodoptera frugiperda* (Carpenter & Greany, 1998). Adult parasitoids reared on the artificial diet, however, were fecund and displayed longevity comparable to that of individuals reared on the host. Greany and Carpenter (1996) also reported successful culture of the ectoparasitoid *Cryptus albitarsus* (Cresson) on this diet.

Chemically defined diets for rearing various chalcid parasitoids of the genus *Brachymeria* were described by Thompson (1980, 1981d) (see Table 4). Complete development of *B. lasus* from egg to adult at rates approaching those observed in the host, *G. mellonella,* were achieved on diets containing heat-denatured albumin, amino acids, glucose, B vitamins, inorganic salts, lipogenic growth factors, and Intralipid®. The latter, a phospholipid emulsion of soybean oil, was essential for complete development. Larvae were reared from eggs dissected from host pupae immediately following oviposition, and parasitoids were cultured individually in the wells of microtissue culture plates. Larval development time was approximately twice as long on the synthetic medium when compared with that on the host, *G. mellonella,* and approximately 80% of the larvae became adults. Although fat-soluble vitamins were not necessary for complete development, adult *B. lasus* lacked the characteristic yellow coloration on the femur if vitamin A was not added (S. N. Thompson, unpublished).

Later studies on *B. intermedia* attempted the development of meridic and oligidic diets for assessing the potential for *in vitro* mass culture of this parasitoid. Initially, only a few adults were successfully cultured on *G. mellonella* pupal extract (Dindo, 1990), or on pupal extract with a beef-based commercial baby food (Dindo & Campadelli, 1992). Later, however, Dindo *et al.* (1994) and Dindo (1995) reported successful *in vitro* culture of this species on bovine and chicken-based commercial foods supplemented with 10 to 20% pupal homogenate. Adult parasitoids were fecund (Dindo, 1995; Dindo *et al.,* 1995). Subsequently, *B. intermedia* was reared on a diet of 85% veal homogenate, 10% egg yolk, 2.5% wheat germ, and 2.5% yeast extract (Dindo *et al.,* 1997a, 1997b). The adult yield was approximately 23% and adults were fecund.

In vitro culture of a pteromalid, *Pteromalus puparum,* was described by Bouletreau (1968, 1972), who achieved complete development on host hemolymph in hanging-drop slide mounts, and similar results were reported by Hoffman *et al.* (1973). Hoffman and Ignoffo (1974) reported limited success with an artificial medium containing yeast hydrolysate, fetal bovine serum, and Grace's tissue culture medium.

Guerra (1992) reported rearing another pteromalid, *Catolaccus grandis* (Burks), as well as the braconid *Bracon mellitor* Say, *in vitro* on an artificial diet similar to that described by Xie *et al.* (1986b) for *Trichogramma pretiosum,* and composed of chicken egg yolk, milk, and hemolymph from various lepidopterous larvae (see later section on *in vitro* culture of *Trichogramma*). Approximately 25% of the eggs of these species reached the adult stage. The sex ratio of *in vitro* reared parasitoids was 1:1 and adults of both species mated and laid viable eggs on *Anthonomus grandis* Boheman larvae. Complete *in vitro* development of *B. mellitor* and *C. grandis* on artificial diets without host hemolymph was later described by Guerra *et al.* (1993). The medium was composed of amino acids, glucose, vitamins, fatty acids, cholesterol, inorganic salts, and egg yolk. The concentration of agar was critical for optimal success. Development from egg to adult required approximately 16 days for each species and was similar to the time observed on the normal host. Approximately 50% adult emergence was obtained, which was also similar to eclosion levels on host insects. Adult morphology and mating appeared normal. Later modifications and improvements in the diet significantly increased the pupal weight, longevity and fecundity of adult females (Rojas *et al.,* 1996a, 1996b). The amino acids histidine, proline, and glutamic acid, which together accounted for approximately 30% of the total amino acid component, were critical for development and fecundity. The most suitable meridic diet contained a mixture of canola and olive oils in addition to egg yolk. Guerra *et al.* (1994) described various oviposition stimulants for *C. grandis,* but artificial oviposition directly into the artificial medium through a Parafilm® shield was described by Rojas *et al.* (1996a).

Several other parasitoids, including *Anastatus japonicus* Ashmead, *Telenomus dendrolimusi, Dibrachys cavus* (Walker), *Habrobracon hebetor* (Say), and *B. greeni* Ashmead, also have been reared *in vitro* on host hemolymph or chicken egg yolk, milk, and hemolymph diets (Li *et al.,* 1988; Li, 1989, 1992). In the case of *A. japonicus,* media were encapsulated in artificial eggs composed of a polymeric film through which the adult parasitoids oviposited. Complete development was obtained. On diets containing hemolymph from the pupae of *Antheraea pernyi* or *Philosamia cynthia ricini* Donovan hemolymph, longevity and fecundity of artificially reared females were greater than those of parasitoids reared on host silkworm eggs (Liu *et al.,* 1988).

Ding *et al.* (1980a) reared the tetrastichid egg parasitoid

Tetrastichus schoenobii Ferriere on modified Gardiner's tissue culture medium supplemented with egg yolk, milk and hemolymph from *A. pernyi* Guérin-Méneville. Approximately 60% of the parasitoids completed development to the adult stage. No deformities were evident and *in vitro*-reared insects displayed normal fecundity. Masutti *et al.* (1991) reared another egg parasitoid, the encyrtid *Ooencyrtus pityocampae* (Mercet), on a semisynthetic diet without hemolymph or other insect components. The diet contained egg yolk, casein and yeast hydrolysates, trehalose, ribonucleic acid, inorganic salts, vitamins, and cholesterol. Mortality of the later instars was high and only a few adult female parasitoids were obtained. The time for complete development was similar to that observed for host-reared parasitoids, and the adults were fecund. Factors influencing oviposition by this species into artificial eggs were outlined by Masutti *et al.* (1992).

In vitro studies with egg yolk and milk diets containing hemolymph from last-instar *M. sexta* were conducted on the scelionid egg parasitoid *Trissolcus basalis* (Woll.) by Volkoff *et al.* (1992). In this case parasitoid eggs hatched and larvae completed development, but no pupation was observed. If larvae were reared in the host to the second instar prior to *in vitro* culture, pupation and adult emergence were obtained.

In vitro culture of embryos of the polyembryonic parasitoid *Copidosoma floridanum* was reported by Iwabuchi (1996). Successful development to the morula stage in parasitoid eggs dissected from host eggs was obtained on MGM 450 supplemented with up to 5% bovine serum albumin. Addition of juvenile hormone promoted the development of polymorulae.

Silver and Nappi (1986) reported partial *in vitro* culture of the egg parasitoid *Leptopilina heterotoma.* Parasitoid eggs dissected from host eggs 30 h following oviposition completed embryonic development and hatched, and the larvae molted to the second stage when maintained on Shield's and Sang's tissue culture medium.

In vitro Culture of Koinobiontic Endoparasitoids

Greany (1980, 1981) reported studies on the *in vitro* embryonic development of braconid *Cotesia* (= *Apanteles*) *marginiventris* (Cresson) maintained in Grace's tissue culture medium supplemented with fetal bovine serum, bovine serum albumin, and whole egg ultrafiltrate. Insects were reared from the embryonic germ band stage to mature first-instar larvae on this diet cocultured with host fat body tissue. Similar results were reported for *in vitro* development of *Microplitis croceipes* (Cresson) (Greany, 1986). The importance of protein nutrition for successful culture was emphasized and protein secretion by the fat body was implicated as a factor explaining the essentiality of this tissue for successful embryonic development. Greany *et al.* (1990) isolated and characterized a host hemolymph protein required for egg development. Subsequently, Ferkovich *et al.* (1991) described the *in vitro* stimulation of *M. croceipes* egg development by a fat body cell line, and Ferkovich and Oberlander (1991b) partially characterized factors that induce germ band formation and hatching.

Pennacchio *et al.* (1992) reported *in vitro* rearing the braconid *Cardiochiles nigriceps* Viereck from the postgerm band egg to second-instar larva. Complete larval development, however, was not obtained. The medium was composed of mixtures of amino acids, carbohydrates, salts, and vitamins supplemented with bovine albumin, lactalbumin hydrolysate, fetal bovine serum, milk, and egg yolk.

Campoletis sonorensis was cultured *in vitro* from egg to third-instar larvae by Hu and Vinson (1997a). Several tissue culture media were tested and TNM-FH proved to be the most suitable for egg hatching and further larval development. The complete medium was composed of TNM-FH and 20% fetal bovine serum supplemented with mixtures of various amino acids, sugars, and proteins. Egg yolk improved development. Molting was stimulated by 20-hydroxyecdysone included at levels of 10^{-4} to $10^{-3}\%$, alone or together with 10^{-5} to $5 \times 10^{-5}\%$ juvenile hormone (see later subsection on the role of hormones in *in vitro* culture of parasitoids). Further investigations demonstrated that insect additives, including hemolymph, fat body, and epidermal cell preparations, increased larval growth rate, but development past the third instar was not obtained (Hu & Vinson, 1997b). When the culture methods were modified to increase exposure of the developing larvae to air and to decrease the water content of the medium, development to mature fifth-instar larvae was obtained (Hu & Vinson, 1998).

In vitro culture of the ichneumonid endoparasite *Venturia canescens* was reported by Ohbayashi *et al.* (1994). Parasitoid eggs were removed from host larvae at the pregerm band embryonic stage. Several tissue culture media were tested for further egg development. In Schneider's medium supplemented with 10% fetal bovine serum parasitoid embryos completed development to the germ band stage but failed to hatch. Embryos in eggs cultured in MGM 450, however, completed germ band development and hatched, although at slower and lower rates than did eggs maintained in host larvae. In this case coculture with host fat body had no effect on parasitoid development. Hatched larvae survived for a few days but grew little. When MGM 450 medium was supplemented with an equal amount of *G. mellonella* pupal extract, pregerm band eggs completed embryonic and larval development and pupated (Nakahara *et al.*, 1997). Adult eclosion, however, was not obtained. Because the growth-promoting effect of the insect additive was destroyed by heating or by passage through a 100-kD

molecular sieve, the authors concluded that a protein was likely responsible for the enhanced parasitoid development. First-, second-, or third-instar parasitoid larvae dissected from the host also developed to the pupal stage in the MGM 450 medium but failed to eclose (Yamamoto *et al.*, 1997). A few adults were obtained when fourth-instar larvae were cultured.

In vitro Culture of Trichogramma

The *in vitro* culture of a *Trichogramma* species, *T. pretiosum,* was first achieved by Hoffman *et al.* (1975) following earlier less successful attempts by Rajendram (1978a) to rear *T. californicum* Nagaraja & Nagarkatti on modified Grace's tissue culture medium encapsulated in paraffin droplets. Complete development of *T. pretiosum* was obtained on filter paper disks soaked in sterile *Heliothis zea* hemolymph. *In vitro* culture to the adult stage required approximately 25% more time than was observed on the host, *Trichoplusia ni* (Hübner). Although most adults did not fully expand their wings, they mated and laid eggs without difficulty. Progeny from eggs of *in vitro*-cultured parasitoids had a sex ratio of 1.2:1 males to females when reared on host eggs. Hoffman *et al.* (1975) also reported development to the prepupal stage on a semisynthetic artificial diet similar to that described for *Pteromalus puparum* (Hoffman & Ignoffo, 1974) (see earlier subsection on artificial diets and *in vitro* culture of hymenopterous parasitoids) but supplemented with wheat germ oil. Strand and Vinson (1985) obtained complete *in vitro* culture of *Trichogramma pretiosum* on an artificial medium supplemented with up to 40% *Manduca sexta* hemolymph (see Table 4). Hemolymph was necessary to induce pupation. Survival to the adult stage was 70% and the sex ratio of adults was approximately 1:2 males to females (M. R. Strand, personal communication). Xie *et al.* (1986a) also reported that host hemolymph was required for pupation of *T. pretiosum* and, moreover, that factors in the host egg influenced adult emergence. Irie *et al.* (1987) subsequently reported that the requirement of host hemolymph for the complete *in vitro* development was due to the presence of specific factors that were extractable in 76% ethanol. Further purification of the pupation factor by traditional chromatographic methods demonstrated the presence of two active carbohydrate-containing components. The authors suggested that the pupation factor had characteristics of a sugar or low-molecular weight-peptide.

Consoli *et al.* (1993) and Consoli and Parra (1995, 1997) reported rearing *T. pretiosum* on a diet composed of 70% *H. zea* larval hemolymph, 20% egg yolk, 5% fetal bovine serum, and 5% *H. virescens* egg homogenate. Another species, *T. galloi* Zucchi, was reared on a similar medium but lacking insect egg homogenate. Development time of both species was slightly longer than observed on host material and adults reared *in vitro* displayed significantly lower fecundity when compared with parasitoids reared from host eggs.

Grenier and Bonnot (1988) reported successful artificial culture of *T. pretiosum* as well as *T. maidis* on artificial diets composed of egg yolk and host hemolymph encapsulated in an artificial polyethylene membrane. Subsequent studies with *T. brassicae* Bezdenko demonstrated that oviposition into artificial eggs was stimulated by treating the external surface with organic solvents that appeared to alter the physical properties of the membrane (Grenier *et al.*, 1993).

Successful *in vitro* culture of *T. dendrolimi* Matsumura in hanging-drop mounts containing hemolymph of *Antheraea pernyi* was described by Guan *et al.* (1978). Little success, however, was achieved with semidefined diets containing a variety of nutrient additives. Liu *et al.* (1979) reported success with hanging-drop mounts containing media with *A. pernyi* or *Attacus cynthia* (Drury) hemolymph, egg yolk, bovine milk, organic acids, and procine serum. Experiments were also conducted on *T. japonicum* Ashmead, *T. australicum* Girault, and *T. evanescens,* but the extent of development was not reported.

Wu *et al.* (1980, 1982) and Wu and Qin (1982a) obtained the successful culture of *T. dendrolimi* to the adult stage on semisynthetic media without host hemolymph but containing egg yolk, chicken embryo fluid, bovine milk, and peptone. Only 16% of the eggs completed development, however; and most adults obtained were females and of poor "vitality." Nevertheless, the results suggested that in contrast to *T. pretiosum,* the *in vitro* culture of *T. dendrolimi* may not require host factors (Xie *et al.,* 1986a; Irie *et al.,* 1987). Liu and Wu (1982) reported on *in vitro* culture of *T. dendrolimi, T. confusum* Viggiani, and *T. pretiosum* on a medium of Grace's tissue culture medium, yeast hydrolysate, fetal calf serum, chicken embryo extract, bovine milk, and chicken egg yolk. Adults were less viable than normal and displayed abnormal wing development. Additional studies on the culture of *T. dendrolimi* on media without insect components were described by Grenier *et al.* (1995). Diets were composed of casein, casein hydrolysate, yeast hydrolysate, egg yolk, milk, inorganics, and vitamins. In these investigations greater viability was reported and 80 to 90% of the enclosed adults were normal.

The Cooperative Research Group of Hubei Province (CRGHP) in China has conducted extensive studies on the complete *in vitro* culture of *T. dendrolimi* in hemolymph-based artificial media encapsulated in artificial eggs into which the adult females oviposited (CRGHP, 1979). Gao *et al.* (1982) reported rearing 35 continuous generations of this species in hanging-drop mounts of this artificial medium. Additional studies on artificial mass culture of *T. japonicum* as well as *T. nubilalae* Ertle & Davis were reported by Liu *et al.* (1983). Liu *et al.* (1995a) reported

rearing five generations of *T. minutum* Riley on artificial host eggs.

Xie *et al.* (1997) reported the *in vitro* culture of *T. minutum* and *T. brassicae* from egg to adult on a culture medium devoid of insect additives. The diet was composed of Grace's tissue culture medium, egg yolk, ultracentrifuged egg yolk, milk, and yeast extract. The results were similar to those described earlier for *T. confusum, T. dendrolimi,* and *T. pretiosum* (Wu *et al.,* 1980, 1982; Wu & Qin, 1982a; Liu & Wu, 1982) when reared without insect components. Adult yields were poor and less than half of the adults were viable and able to spread their wings.

The quality of *T. minutum* reared on a diet composed of 7% yeast extract, a commercial amino acid product, nonfat dried milk, egg yolk and chicken embryo extract, and *M. sexta* egg liquid was described by Nordlund *et al.* (1997). Parasitoids were reared for 10 generations. Compared with parasitoids reared on *H. zea* eggs, the artificially cultured insects had longer development times and a higher proportion of deformed females. Adult yield and sex ratio of parasitoids reared *in vivo* and *in vitro* were comparable. Moreover, *in vitro*-cultured parasitoids were generally larger, displayed greater longevity, and were capable of parasitizing more host eggs.

Role of Hormones in the *in vitro* Culture of Parasitoids

Several studies have attempted to determine the effects of hormone supplementation on parasitoid development *in vitro*. Nenon (1972a, 1972b) demonstrated that host hormones greatly increased *in vitro* survival of developing embryos and larvae of the encyrtid parasite *Ageniaspis fuscicollis* (Dalman). The parasitoid was maintained on a diet of chicken embryo extract, equine serum, and beef peptone. Ecdysteroid or juvenile hormone added to the medium had little effect, but when included together, resulted in nearly 100% survival to the second instar.

Juvenile hormone was demonstrated to be an important stimulus for formation of polymorulae prior to larval differentiation during *in vitro* embryonic development of another polyembryonic parasite, *Copidosoma floridanum,* in tissue culture medium (Iwabuchi, 1995, 1996). Juvenile hormones I, II, or II at concentrations of 0.01 ng/ml to 1 μg/ml stimulated polymorula proliferation in a dose-dependent manner. The analogues farnesyl methylether and methoprene also stimulated polymorula formation when included at 1 μg/ml. The roles of juvenile hormone and 20-hydoxyecdysone in the development of *C. floridanum* were described by Strand *et al.* (1991a, 1991b) and Baehrecke *et al.* (1993). Those studies established that embryonic development and larval differentiation or morphogenesis of this parasite are closely synchronized with hormonally mediated events during the final two instars of the host; moreover, they indicated that both juvenile hormone and 20-hydroxyecdysone play important roles in regulating parasite development.

The tachinid *Pseudogonia rufifrons* also depends on its host's endocrine system for growth and development (Baronio & Sehnal, 1980; Mellini & Coulibaly, 1991), and Fanti (1990) reported that molting from the first- to second-larval stadium *in vitro* required 20-hydroxyecdysone. Development from the second-instar larva to the adult stage occurred in the absence of hormones. Molting of the tachinids *Pseudoperichaeta nigrolineata* Walker (Grenier, 1988) and *Exorista sorbillans* (Mitsuhashi, 1996) from the first- to second-instar larva was also stimulated by 20-hydroxyecdysone.

Lawrence (1991) demonstrated that the hymenopteran parasite *Diachasmimorpha* (= *Biosteres*) *longicaudatus* Ashmead, a parasite whose development is also synchronized with that of its host (Lawrence, 1982, 1986b), required 20-hydroxyecdysone for successful *in vitro* culture. In this case, the hormone was required throughout larval development. Ecdysone was ineffective in stimulating molting. Larval development of the endoparasite *Campoletis sonorensis* was also stimulated by 20-hydroxyecdysone (Hu & Vinson, 1997a). Although juvenile hormone alone was ineffective, when provided together with 20-hydroxyecdysone, molting was stimulated beyond that effect observed for 20-hydroxyecdysone alone.

In contrast to the preceding results, 20-hydroxyecdysone failed to stimulate development of *Brachymeria intermedia in vitro* (Thompson, 1980). Greany (1980, 1981) reported that 20-hydroxyecdysone inhibited egg hatching of *Cotesia marginiventris;* and ecdysone, 20-hydroxyecdysone, and juvenile hormone analogue hydroprene had no effect on larval growth or development. The deleterious effect of 20-hydroxyecdysone, however, could be overcome by simultaneous application of hydroprene.

Future study of the effects of host hormones *in vitro* culture systems will require careful and detailed experimental design. Hormones act in a complex and often synergistic fashion, and the timing of their application, as well as the method of exposure, will likely prove to be critical in assessing their potential use in *in vitro* system [see Lawrence (1991)].

Role of Teratocytes in the Nutrition of Parasitoids

Teratocytes, cells derived from the embryonic membrane of the parasitoid egg, are released into the host hemocoel at the time of egg hatching (Vinson, 1970; Vinson & Iwantsch, 1980; Dahlman & Vinson, 1993). Although their significance is not completely understood, it has been suggested that they play a role in parasite nutrition (Dahlman,

1990). Sluss (1968) demonstrated that the teratocytes of *Perilitus coccinellae* (Schrank) increased in volume severalfold in the coccinellid host and were subsequently eaten by the developing parasitoid larvae. Greany (1980) observed that teratocytes present in artificial culture medium for *C. marginiventris* caused dissociation of cocultured fat body and suggested that teratocytes may facilitate larval growth. Rotundo *et al.* (1988), however, obtained complete larval development of braconid *Lysiphlebus fabarum* (Marshall) on a similar artificial diet lacking fat body or teratocytes.

A role for teratocytes in the successful *in vitro* culture of the egg parasitoid *Telenomus heliothidis* Ashmead was demonstrated by Strand *et al.* (1988). Embryonic development of *T. heliothidis* was achieved in Hinks TNH-FH medium containing 30% *M. sexta* hemolymph. Mature embryos were transferred to a medium containing 40% *M. sexta* hemolymph, egg yolk, milk, and trehalose. Development to the adult stage required 1 day longer than on the host, *H. virescens,* and 42% of the larvae became adults. The sex ratio was approximately 1:1 males to females (M. R. Strand, personal communication). The presence of teratocytes had no effect on larval development to the third instar. However, when teratocytes were removed from the medium during larval development, pupation was markedly reduced and the development time of parasitoids that completed development was increased. Based on observation of the medium, the authors concluded that the teratocytes aided larval feeding by dispersing the particulate material and solubilizing nutrients. Strand *et al.* (1986) suggested that teratocytes of *T. heliothidis* aided in decomposition and necrosis of the tissues of the host, *H. virescens,* partially due to the release of lytic enzymes. Thus, their function *in vitro* may parallel that occurring during the development of the parasitoid in the host.

Importance of Dietary Osmotic Pressure to Successful *in vitro* Culture of Parasitoids

Parasitoids such as the idiobionts *Itoplectis conquisitor* and *Exeristes roborator* are very tolerant of high osmotic pressures. Artificial diets that supported *in vitro* culture of those species (Yazgan, 1972; Thompson, 1975) had osmolalities approaching 2000 mOs/kg (S. N. Thompson, unpublished). In contrast, osmotic pressure was a critical factor in formulating artificial media for *B. lasus* (Thompson, 1983b). The effect of both carbohydrate and amino acid levels was similar and appeared to be closely related to osmolality (Fig. 5). The optimal range of osmotic pressure in the artificial diets was 550 to 700 mOs/kg, which is was considerably greater than the 350 to 450 mOs/kg of host hemolymph and tissues. The optimal range observed for the artificial media probably did not represent the actual opti-

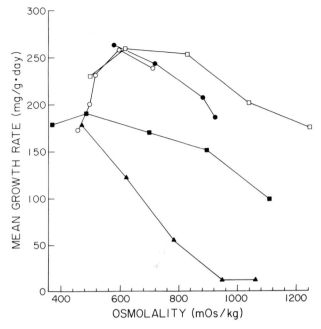

FIGURE 5 The relationship of osmolality to mean growth rate of *Brachymeria lasus* (Walker) reared *in vitro* on chemically defined diets containing various levels of glucose and free amino acids. o = D-glucose levels of 0.5, 1, 2, 4, and 6% with 2% amino acids; ● = D-glucose levels of 4, 6, 8, and 10% with 2% amino acids; □ = amino acid levels of 1, 2, 4, 6, and 8% with 2% D-glucose; ■ = amino acid levels of 1, 2, 4, 6, and 8% in the absence of D-glucose; ▲ = L-glucose levels of 0, 2, 4, 6, and 7.5% with 2% amino acids and without D-glucose. Mean growth rate (G):

$$G = \frac{\Delta_w}{W_e \cdot t}, \text{ (after Thompson, 1983b)}$$

where Δ_w = gain in dry weight over the entire developmental period (μg); t = developmental time (days); and W_e = mean exponential wet weight during the developmental period (mg).

mal osmolality for larval development but instead the point at which the detrimental effects of osmotic pressure exceeded the beneficial effects of increased nutrient level. To test this hypothesis, Thompson supplemented a diet containing 2% amino acids with nutritionally inert L-glucose and observed a sharp decline in growth rate as dietary osmolality increased (see Fig. 3). In contrast, supplementing the medium with D-glucose enhanced growth rate throughout the same range of osmotic pressure up to approximately 700 mOs/kg. Thus, the optimal dietary osmolality is likely lower than 600 mOs, the osmolality of the optimal diet tested containing 2% amino acids and 2% glucose.

Thompson (1981c) was unsuccessful in attaining complete development of pteromalid parasitoid *Pachycrepoideus vindemiae* (Rondani) on an artificial medium similar to that used for successful *in vitro* culture of *Brachymeria* spp. However, when the amino acid component was replaced with a mixture of the corresponding polymerized

amino acids and the osmolality was reduced to approximately 390 mOs/kg, development from the egg to adult stage was obtained (Thompson *et al.*, 1983).

Although various *Trichogramma* species have been cultured in the range of 300 to 460 mOs/kg (Grennier & Bonnot, 1988; Strand & Vinson, 1985), 320 mOs/kg appears to be optimal and most species fail to hatch and develop at osmolalities greater than 450 (Grenier, 1994). Tachinids *Phryxe caudata* and *Lixophaga diatraeae* also failed to develop at higher osmolalities (Grenier et al., 1986).

Studies on the development of chemically defined diets clearly demonstrate that the importance of dietary osmotic pressure varies between parasitoid species.

IN VITRO CULTURE OF PREDATORS

In vitro Culture of Coccinellids

Studies on artificial rearing of predators have emphasized maintenance of the adult stage for maximizing egg production instead of complete *in vitro* culture (see later section on current applications). The larval stages of many predators are the preferred biological control agent, and eggs and larvae produced by adults are placed directly in the field. Nevertheless, considerable effort has been aimed at complete artificial culture of several species (Table 5). Among the first species reared artificially from egg to adult was the coccinellid *Coleomegilla maculata* (DeGeer) (Szumkowski, 1952). Adults fed voraciously on raw liver or meat and were maintained on these foods for months in the absence of prey. Larval survival, however, was poor on meat products alone and only 38% became adults. Supplementation with vitamins resulted in approximately 86% of the larvae reaching the adult stage. Oviposition and egg viability were greatly increased by addition of vitamin E to the adult diet. The culture methods were later refined and a diet of fresh yeast and glucose supported larval development (Szumkowski, 1961a, 1961b).

Smith (1965, 1966) successfully reared several coccinellid species, including *C. maculata*, on dried aphids supplemented with pollen. Success was also achieved on a diet of 40% brewer's yeast and 55% sucrose with supplements of cholesterol, inorganic salts, RNA, wheat germ oil, and vitamins. Adults were fed the same diet supplemented with powdered liver. Subsequently, Attallah and Newsom (1966) reared eight generations of *C. maculata* on a defined diet of casein, sucrose, wheat germ, soybean hydrolysate, glycogen, butter fat, a liver factor, corn oil, brewer's yeast, dextrose, cotton leaf extract (supplying carotenoids and sterols), ascorbate, inorganic salts, vitamins, and agar. Adults reared *in vitro* were fecund and mating was stimulated by addition of vitamin E to the diet. The medium failed to support growth of *Coccinella novemnotata* Herbst, *Cyclo-*

TABLE 5　Predaceous Insect Species Reared *in vitro*

Taxa	Medium[a]	Development	References
Order Coleoptera			
Family Coccinellidae			
Adalia bipunctata	Oligidic	Partial larval	Hawkes, 1920
	Oligidic/meridic	Egg to fecund adult	Kariluoto, 1980
Aiolocaria hexaspilota (= *mirabilis*) (Hope)	Oligidic[b]	Egg to adult	Matsuka *et al.*, 1972
Calosoma sycophanta	Oligidic	Egg to adult	Greany & Carpenter, 1996
Chilocorus kuwanae	Oligidic[b]	Egg to adult	Matsuka *et al.*, 1972
Coccinella septempunctata	Oligidic[b]	Egg to adult	Okada *et al.*, 1971
	Oligidic[b]	Egg to adult	Tanaka & Maeta, 1965
	Oligidic/meridic	Egg to fecund adult	Kariluoto, 1980
	Oligidic[b]	Egg to fecund adult	Smirnoff, 1958
C. transversoguttata Falderman	Oligidic/meridic	Egg to adult	Kariluoto, 1980
Coleomegilla maculata	Oligidic	Egg to fecund adult	Szumkowski, 1952, 1961a, 1961b
	Oligidic/meridic	Egg to fecund adult	Attallah & Newsom, 1966
	Oligidic/meridic	Egg to fecund adult	Kariluoto *et al.*, 1976; Kariluoto, 1980
	Oligidic	Partial larval development	Greany & Carpenter, 1996
Crytolaemus montrouzieri	Oligidic[b]	Egg to adult	Chumakova, 1962
Harmonia axyridis	Oligidic[b]	16 Generations	Okada *et al.*, 1971a; Matsuka *et al.*, 1972
Hippodamia convergens	Merdic/oligidic	Egg to fecund adult	Racioppi *et al.*, 1981
Menochilus sexmaculatus	Oligidic[b]	3 Generations	Matsuka *et al.*, 1972
Olla v-nigrum (= *abdominalis*)	Oligidic/meridic	Egg to fecund adult	Bashir, 1973
Propylea japonica	Oligidic[b]	Egg to adult	Matsuka *et al.*, 1972
P. quatuordecim (= *punctata*) (Linneaus)	Oligidic/meridic	Egg to adult	Kariluoto, 1980
Rodolia cardinalis	Oligidic[b]	Egg to fecund adult	Smirnoff, 1958
Chilocorus bipustulatus	Oligidic[b]	Egg to fecund adult	Smirnoff, 1958
Clitostethus arcuatus	Oligidic[b]	Egg to fecund adult	Smirnoff, 1958
Exochomus anchorifer	Oligidic[b]	Egg to fecund adult	Smirnoff, 1958
E. nigromaculatus	Oligidic[b]	Egg to fecund adult	Smirnoff, 1958
E. quedripostulatus	Oligidic[b]	Egg to fecund adult	Smirnoff, 1958
Oenopia (= *Harmonia*) *conglobata*	Oligidic[b]	Egg to fecund adult	Smirnoff, 1958
O. doublieri	Oligidic[b]	Egg to fecund adult	Smirnoff, 1958
Mycetaea tafiletica	Oligidic[b]	Egg to fecund adult	Smirnoff, 1958
Pharoscymnus numidicus	Oligidic[b]	Egg to fecund adult	Smirnoff, 1958
P. ovoideus	Oligidic[b]	Egg to fecund adult	Smirnoff, 1958
Psyllobora (= *Thea*) *vigintiduopunctata*	Oligidic[b]	Egg to fecund adult	Smirnoff, 1958
Rhizobius litura	Oligidic[b]	Egg to fecund adult	Smirnoff, 1958
R. lophantae	Oligidic[b]	Egg to fecund adult	Smirnoff, 1958
Scymnus kiesenwetteri	Oligidic[b]	Egg to fecund adult	Smirnoff, 1958
S. pallidivestis	Oligidic[b]	Egg to fecund adult	Smirnoff, 1958
S. suturalis	Oligidic[b]	Egg to fecund adult	Smirnoff, 1958
Stethorus punctillum	Oligidic[b]	Egg to fecund adult	Smirnoff, 1958
Order Neuroptera			
Family Chrysopidae			
Chrysoperla carnea	Oligidic/meridic	Egg to fecund adult	Hagen & Tassan, 1965
	Oligidic/meridic	3 Generations	Hassan & Hagen, 1978
	Holidic	Newly hatched larvae to fecund adults	Hasegawa *et al.*, 1989
Chrysoperla sinica	Oligidic	Egg to fecund adult	Cai *et al.*, 1983
	Oligidic	Egg to fecund adult	Zhou & Zhang, 1983
Order Hemiptera			
Family Lygaeidae			
Geocoris punctipes	Oligidic[b]	Egg to fecund adult	Dunbar & Bacon, 1972
	Oligidic/meridic	Egg to fecund adult	Cohen, 1981, 1983, 1985
	Oligidic	Egg to fecund adult	Greany & Carpenter, 1996
			Carpenter & Greany, 1997
Family Miridae			
Macrolophus caliginosus	Holidic	Egg to adult	Grenier *et al.*, 1989

(continued)

TABLE 5 *(Continued)*

Taxa	Medium[a]	Development	References
Order Hemiptera *continued*			
Family Pentatomidae			
Lyctocoris campestris	Oligidic	Egg to fecund adult	Greany & Carpenter, 1996
Podisus maculoventris	Oligidic	Egg to fecund adult	De Clercq & Degheele, 1992
	Oligidic	Egg to fecund adult	Greany & Carpenter, 1996
P. sagitta	Oligidic	Egg to fecund adult	De Clercq & Degheele, 1992
Perillus bioculatus	Oligidic	Egg to fecund adult	Greany & Carpenter, 1996
Xylocoris flavipes	Oligidic	Egg to fecund adult	Greany & Carpenter, 1996

[a]Holidic = media with all ingredients known in chemical structure; meridic = media with holidic base to which is added at least one substance or preparation of unknown structure of uncertain purity; oligidic = media in which crude organic materials supply most dietary requirements.

[b]Contains insect-derived components such as host hemolymph, other tissues or tissue homogenates, and/or extracts.

neda spp., *Hippodamia convergens* Guérin and *Olla v-nigrum* (= *abdominalis*) (Mulsant). The latter species, however, was successfully cultured *in vitro* by Bashir (1973). Maximum egg production by adults required inclusion of vitamin E in the larval diet. In contrast to the results of Szumkowski (1952), supplementation of the adult diet alone was insufficient for optimal egg production.

Smirnoff (1958) reported successful *in vitro* culture of several coccinellid species including *Psyllobora* (= *Thea*) *virgintiduopunctata* (Linnaeus), *Chrysopa septempunctata*, *Oenopia* (= *Harmonia*) *doublieri* (Mulsant), *O.* (= *Harmonia*) *conglobata* (Linnaeus), *Rhizobius lophanthae* (Blaisdell), *R. litura* (Fabricius), *Rodolia cardinalis* (Mulsant), *Exochomus anchorifer* Allard, *E. quadripustulatus* (Linnaeus), *E. nigromaculatus* Erhorn, *Scymnus suturalis* Thunberg, *S. pallidivestis* Mulsant, *S. kiesenwetteri* Mulsant, *Stethorus punctillum* Weise, *Chilocorus bipustulatus* (Linnaeus), *Clitostethus arcuatus* Rossi, *Pharoscymnus numidicus* Pie, *P. ovoideus* Sicard, and *Mycetaea tafilaletica* Smirnoff (Endomychidae). The diet was composed of sucrose, honey, royal jelly, alfalfa flour, yeast, and agar supplemented with dried pulverized prey. Rearing of larvae of a few species was improved by addition of beef jelly. All species developed more rapidly and lived longer on the artificial diet compared with insects reared under natural conditions. Adults were active, mated readily, and were fecund. Tanaka and Maeta (1965) reared *Harmonia axyridis* (Pallas), *Chrysopa septempunctata*, and *Chilocorus kuwanae*. Silvestri on the diet of Smirnoff (1958) as well as on a variety of other artificial media. Successful artificial culture of all three species was obtained but adults failed to lay eggs. Chumakova (1962) reared *Crytolaemus montrouzieri* on similar crude diets supplemented with dried prey.

Partial *in vitro* development of *Coleomegilla maculata* and complete *in vitro* culture of *Calosoma sycophanta* (Linnaeus) were reported by Greany and Carpenter (1996) on an encapsulated oligidic medium containing ground liver, egg yolk, and glutamine. Fecundity of *C. sycophanta* was

much lower than that of predators reared on prey (P. D. Greany, personal communication).

Considerable success was achieved by Okada *et al.* (1971a, 1971b, 1972) and Matsuka *et al.* (1972) at rearing *H. axyridis* on diets containing powdered larvae and pupae of drone honeybee (*Apis mellifera* Linnaeus). Sixteen generations of *H. axyridis* and three generations of *Menochilus sexmaculatus* (Fabricius) were cultured *in vitro*. An improved rearing method was later described by Okada and Matsuka (1973) and adapted by Matsuka *et al.* (1982) for maintaining adult *R. cardinalis*. Adults of *Chilocorus rubidus* Hope, *Scymnus hilaris* Motschulsky, *S. otohime* Kamiya, *Vibidia duodecimguttata* Poda and *Stethorus japonicus* Kamiya could also be maintained on the diet but did not lay eggs (Matsuka *et al.*, 1972). Niijima *et al.* (1986) described the use of drone honeybee powder for rearing several coccinelleds including *Adalia bipunctata*, *Anatis halonis* Lewis, *Coccinella explanata* Miyatake, *C. septempunctata*, *Coccinula crotchi* (Lewis), *Eocaria muiri* Timberlake, *H. axyridis*, *H. yedoensis* Takizawa, *Hippodamia convergens*, *H. tredecimpunctata* Linnaeus, *Lemnia beplagiata* (Swartz), *M. sexmaculatus*, *Propylea japonica* (Thunberg), *S. hilaris*, and *S. otohime*. Results were variable, but 11, 16, and 25 successive generations of *E. muiri*, *Harmonia axyridis*, and *M. sexmaculatus*, respectively, were cultured from the egg to adult stage. Larval development, adult longevity, and fecundity were reported to be good.

Fractionation of honeybee powder was described by Matsuka and Okada (1975), who concluded that the active factor stimulating predator growth was unstable but nonproteinaceous. Further attempts to analyze bee powder were described by Niijima *et al.* (1977). Based on these analytical results, Niijima *et al.* (1986) formulated several chemically defined diets for rearing *H. axyridis*. Larvae developed from the first to third instars on a diet containing 18 amino acids, sucrose, cholesterol, 10 vitamins, and 6 minerals.

Extensive *in vitro* rearing studies on *Adalia bipunctata* were described by Kariluoto *et al.* (1976). Approximately

60 variations of seven artificial diets were tested (Table 6 and Fig. 6). These contained varying amounts of wheat germ, brewer's yeast, casein, cotton leaf extract, egg yolk, sucrose, liver fractions, honey, glycogen, soybean hydrolysate, butter fat, corn oil, dextrose, amino acids, inorganic salts, ascorbate, choline, vitamin E, and antibiotics. On the best diets 60 to 80% of larvae became adults but development time was slower than on the natural prey, and adult weight was lower (see Table 6 and Fig. 6). The medium was modified by Kariluoto (1978) and the levels of antibiotics, including sorbic acid and methyl-*o*-hydroxybenzoate, were found to be critical for successful development. Kariluoto (1980) subsequently obtained fecund adults of *A. bipunctata* and *Chrysopa septempunctata* reared *in vitro*. Two other species of coccinellids were also successfully reared from egg to adult stage.

Racioppi *et al.* (1981) developed an oligidic diet supplemented with liver extracts for rearing *Hippodamia convergens* Guérin *in vitro*. The diet contained corn, soy, milk, casein and soy hydrolysates, wheat germ oil, sucrose, yeast, honey, tryptophan, cystine, B vitamins, and other water-soluble growth factors. Four generations were cultured on the oligidic diet. *In vitro*-cultured insects were smaller than insects reared on aphids, and although copulation was observed, no oviposition occurred and egg production was absent.

In vitro Culture of Chrysopa and Chrysoperla

Early attempts to culture *Chrysoperla* spp. *in vitro* met with limited success, but complete culture of *Chrysopa carnea* was finally achieved by Hagen and Tassan (1965), who encapsulated a liquid medium in paraffin droplets. The diet was composed of enzymatic yeast protein hydrolysate, casein hydrolysate, ascorbate, choline, and fructose. Adults were fecund but development time from the egg to the adult was approximately twice that of insects reared on aphids. Vanderzant (1969) successfully cultured *C. carnea* for seven generations on pieces of cellulose sponge soaked in enzymatic casein and soy hydrolysates, fructose, inorganic salts, lecithin, cholesterol, vitamins, inositol, choline, and ascorbate. Although development on this semidefined diet was slow, 50 to 65% of larvae became adults compared with 85% when reared on insect eggs. Hassan and Hagen (1978) reported rearing three generations of *C. carnea* on an artificial diet of honey, sucrose, yeast flakes, casein and yeast enzymatic hydrolysates, and egg yolk. Development time and pupal weights were similar to those of insects on eggs of *Sitotroga cerealella* (Olivier). Martin *et al.* (1978) further evaluated encapsulated diets for rearing *C. carnea*.

Chemically defined artificial diets for rearing *C. carnea* were described by Hasegawa *et al.* (1989) and Niijima (1989). Those authors demonstrated that the ratio of sugar to amino acids was an important factor for larval develop-

ment. A carbon to nitrogen ratio between 3:7 and 8:2 sustained development, but diets with a ratio less than 2:8 or greater than 9:1 were inadequate. The best diet contained mixtures of amino acids, fatty acids, sugars, organic acids, vitamins, and minerals. Approximately 65% of the eggs that hatched on this medium reached the adult stage, and adult females were fecund. A second generation was also reared and adults produced viable eggs. This chemically defined diet was later used for rearing *C. pallens* and the amino acid and vitamin requirements for development of this species were reported (Niijima, 1993a, 1993b) (see earlier subsection on (Requirements for development of predators).

Ye *et al.* (1979) successfully cultured *Chrysoperla sinica* (Tjeder) for 10 generations on a medium of egg, brewer's yeast, honey, sucrose, and ascorbate. Adults were fed powdered liver, brewer's yeast, and honey. This species was also reared on an encapsulated medium of soybean and beef hydrolysates, brewer's yeast, egg yolk, honey, sucrose, linoleic acid, and ascorbate (Cai *et al.* 1983; Zhou & Zhang, 1983).

Artificial mass culture of *C. carnea* was reported by Yazlovetsky & Nepomnyashchaya (1981) following the development of a suitable artificial medium for supporting larval development (Nepomnyashchaya *et al.*, 1979). The microencapsulated medium was composed of casein hydrolysate, wheat germ extract, brewer's yeast extract, sucrose, soybean oil, lecithin, cholesterol, choline, and ascorbate. Another microencapsulation technique for mass-producing artificial eggs for *C. carnea* was described by Morrison *et al.* (1975).

The importance of larval nutrition on the performance of adult *C. carnea* was described by Zheng *et al.* (1993a, 1993b). Larvae that were provided abundant food in the form of *Anagasta kuehniella* eggs developed faster; and the adults had a shorter preoviposition period, deposited eggs longer, and had significantly greater fecundity than adults reared from larvae whose food supply was restricted. Moreover, the low overall fecundity of adults arising from poorly fed larvae was not overcome by unrestricted feeding of adults on an artificial diet of whey yeast, enzymatic protein hydrolysate of brewer's yeast, and honey. Similar studies on this species were reported by McEwen *et al.* (1993a, 1996). When a diet of *A. kuehniella* was supplemented with an "artificial honeydew" composed of yeast autolysate, sugar, and water, larval development was faster and adult yield was greater than was observed with insects reared on host eggs alone. Adults fed the artificial diet during larval development were able to produce eggs without a proteinaceous meal.

In vitro Culture of Hemipterans

Several diets have been described for rearing and maintenance of the hemipteran predator *Geocoris punctipes*

TABLE 6 Composition of Eight Diets for Rearing the Predator *Adalia bipunctata* (Linnaeus)[a,b]

Diet components		1	2	3	4	5	6	7	8[c]
Water	ml	200.0	200.0	200.0	220.0	220.0	220.0	220.0	200.0
Wheat germ	g	2.0	2.0	2.0	2.0	5.0	5.0	5.0	4.0
Brewer's yeast	g	2.0	2.0	2.0	2.0	5.0	10.0	7.0	1.0
Casein	g	1.5	1.5	1.5	1.5	—	—	—	6.0
Carrot–lipid–casein[d]	g	3.0	3.0	3.0	3.0	3.0	3.0	3.0	—
Cottonleaf extract		—	—	—	—	—	—	—	1.0
Egg yolk	g	—	20.0	20.0	—	—	—	—	—
Whole egg	g	—	—	—	50.0	50.0	50.0	50.0	—
Sucrose	g	7.0	7.0	7.0	7.0	7.0	7.0	7.0	3.0
Liver fractions	g	—	—	—	—	—	—	—	6.0
Beef liver	g	20.0	20.0	50.0	50.0	50.0	50.0	50.0	—
Honey	g	10.0	10.0	10.0	10.0	10.0	10.0	10.0	—
Glycogen	g	—	—	—	—	—	—	—	2.0
Soybean hydrolysate	g	—	—	—	—	—	—	—	3.0
Butter fat	g	—	—	—	—	—	—	—	2.0
Corn oil	ml	—	—	—	—	—	—	—	2.0
Dextrose	g	—	—	—	—	—	—	—	1.0
Amino acid solution	ml	—	—	—	—	—	20.0	—	—
Salt mixture	g	1.0	1.0	1.0	1.0	1.0	1.0	1.0	1.25
Ascorbic acid	g	0.5	0.5	0.5	0.5	1.0	1.0	1.0	0.5
Vitamin stock	ml	3.0	3.0	3.0	3.0	3.0	3.0	3.0	3.0
Choline chloride (15% solution in water)	ml	1.5	1.5	1.5	1.5	1.5	1.5	1.5	—
DL-α-Tocopherol, Merck	ml	0.1	0.1	0.1	0.1	0.1	0.1	0.1	(0.2)
Nipagin (15% solution in 95% EtOH)	ml	1.5	1.5	1.5	1.5	1.5	1.5	1.5	0.2g
Aureomycin	g	0.8	0.8	0.8	0.3	0.3	0.3	0.3	0.06
Sorbic acid	g	0.2	0.2	0.2	0.2	0.2	0.2	0.2	0.2
Agar	g	4.0	4.0	4.0	4.0	3.7	3.7	3.7	4.0
Total growth	g	258	278	308	358	362	387	264	237

[a] See effects in Figure 6.

[b] From Kariluoto, K. T., *et al.* (1976). *Annales Entomologici Fennici, 42*, 91–97.

[c] From Atallah, Y. H., & Newsom, L. D. (1966). *Journal of Economic Entomology, 59*, 1173–1179.

[d] Carrot lipids, extracted from 1 kg of washed and macerated carrots with chloroform–methanol (1:2). The extract still containing the solvent was added to 100 g of casein and evaporated to dryness under vacuum. The carrol–lipid–casein mixture was prepared monthly and stored at −18°C.

(Dunbar & Bacon, 1972). These media, however, were supplemented with insects. Cohen (1981) reported the successful *in vitro* culture of *G. punctipes* from first-stage nymphs to adults on encapsulated semidefined diets. Six media containing yeast and casein hydrolysates, cholesterol, sucrose, agar, inorganic salts, corn oil, lecithin, phenylalanine, and a vitamin mixture were formulated and encapsulated in a variety of materials. The latter included mixtures of polybutene 32 (Chevron), dental impression wax (Sybron/Kerr), Vaseline®, Epoline C-16 (Eastman), Sunoco (Sun Oil), candelilla wax (Frank B. Ross, Inc.) and Paraplast®. The best results were obtained with vitamin-enriched medium encapsulated in a mixture of 5% polybutene 32 and 95% dental impression wax. Growth and development of *G. punctipes in vitro* was far superior to that observed with insects reared on *Spodoptera exigua* (Hübner). The percentage of nymphs reaching the adult stage and adult survival were also significantly greater on the artificial diet. Subsequently, Cohen (1983) described modifications of media content, preparation, and encapsulation; and was successful in rearing two generations of *G. punctipes. Geocoris pallens* Stål, *H. convergens, H. axyridis,* and *Nabis* spp. also fed successfully on the encapsulated medium. In all tests, superior results were obtained on medium encapsulated with 30% polybutene 32 and 70% dental wax.

Cohen (1985) described a diet composed of equal parts of fresh ground beef and beef liver supplemented with sucrose for continuous rearing of *G. punctipes.* The ingredients were blended into a paste and small aliquots wrapped in stretched Parafilm® were presented to developing nymphs for feeding. Twelve generations were successfully cultured, and artificially reared predators displayed greater fecundity and adult weight than did individuals reared on insect eggs and coddled larvae (Cohen & Urias, 1986).

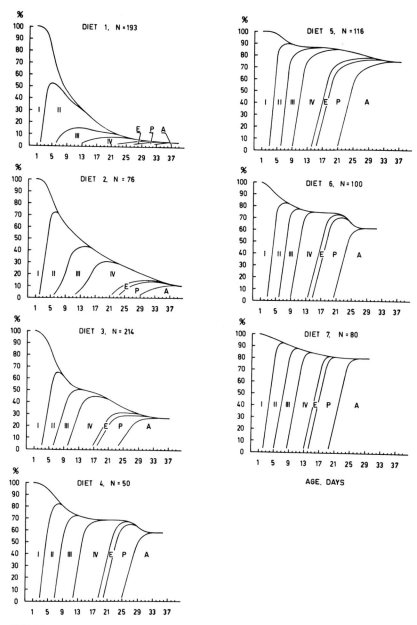

FIGURE 6 Effects of nutrient balance of eight diets shown in Table 6 on development of the Coccinellid predator *Adalia bipunctata* (Linnaeus). Roman numerals represent larval instars. E = prepupa, P = pupa, and A = adult. [From Kariluoto, K. T., *et al.* (1976). *Annales Entomologici Fennici, 42* 91–97.]

Development, however, was slower on the artificial diet. To date over 160 generations of *G. punctipes* have been continually reared on this diet (A. C. Cohen, personal communication).

Meat-based diets have also been formulated for rearing *Podisus maculiventris* (Say) and *P. sagitta* (Fabricius) (DeClercq & Degheele, 1992).

Greany and Carpenter (1996) reported the successful *in vitro* culture of several predateous hemipterans including *G. punctipes, Lyctocoris campestris* (Fabricius), *Perillus bioculatus* (Say), *P. maculiventris,* and *Xylocoris flavipes*

(Reuter) on an encapsulated oligidic diet without insect components. The diet, initially developed for rearing the parasitoid *Diapetimorpha introita* (see earlier subsection on artificial diets and *in vitro* culture of hymenopteran parasitoids), supported egg to adult development. Fecundity of the predators reared on the artificial diets, however, was considerably less than that observed with insects reared on prey (P. D. Greany, personal communication).

Artificial diets for complete *in vitro* culture of the mirid predator *Macrolophus caliginosus* Wagner were described by Grenier *et al.* (1989). The diets contained organic acids, sugars, vitamins, amino acids, ribonucleic acid, albumin, and inorganic salts; and were presented to predators in small bags. Survival of insects reared on the best diet was similar to that of predators reared on eggs of *Anagasta* (*Ephestia*) *kuehniella,* with 21% of the eggs developing into adults. If the artificial diet was supplemented with vegetable material, 62% of the predators reached the adult stage.

In vitro Culture of Predaceous Mites

Few attempts have been made to develop *in vitro* culture methods for predaceous mites. *Amblyseius limonicus* and *A. hibisci* were reared from egg to adult on a diet of yeast hydrolysate, sucrose, or molasses (McMurtry & Scriven, 1966). However, development rate was low and mortality high. Later, Shehata and Weismann (1972) tested an artificial diet for development of the predator *Phytoseiulus persimilis* Athias-Henriot. The diet was composed of pangamin extract, honey, ascorbic acid, amino acids, and B vitamins; and was sandwiched in a thin layer between two pieces of stretched Parafilm®. Newly hatched larvae developed to the adult stage in approximately 3 days. Artificially cultured individuals laid viable eggs when fed *Tetranychus urticae* Koch.

Kennett and Hamai (1980) successfully cultured seven species of predaceous mites, including *A. largoensis, A. limonicus, A. hibisci, Typhloseiopsis arboreus* (Chant), *Metaseiulus pomoides* Schuster & Pritchard, *Typhlodromus pyri* Scheuten, and *Iphiseius degenerans* (Berlese), on an artificial media similar to that described by Hassan and Hagen (1978). The diet was composed of honey, sucrose, yeast flakes, yeast and casein hydrolysates, and egg yolk, and was encapsulated in paraffin droplets. All the preceeding species oviposited in the presence of the artificial diet, but at much lower levels than those observed with natural foods including pollen and prey. Eggs produced by predators reared on artificial diet were viable and subsequent development proceeded normally on the artificial food. Development time of all species to the adult stage was similar to that observed with natural food.

CONTINUOUS ARTIFICIAL MASS CULTURE

A principal goal of studies on *in vitro* culture of entomophagous insects is continuous artificial mass culture with exclusion of the host insect. Achieving this goal requires careful consideration of factors that otherwise would not be considered of direct importance to nutrition. Culture of a parasitoid requires the direct deposition of the eggs or larvae onto an artificial substrate. Artificial food must be acceptable for feeding by all stages of a predator. Behavioral considerations, therefore, may be critical for the successful continuous culture of many entomophagous species. Suc-

cessful *in vitro* culture achieved thus far reflects the level of complexity of the behavioral interactions between parasitoid and host or predator and prey. The first success with parasitoids was achieved with the sarcophagid dipterans, many of which readily oviposit and develop on carrion. Arthur and Coppel (1953) and Smith (1958) reported rearing *Sarcophaga aldrici* and *Kellymyia Kellyi*, respectively, for many generations on fish and liver. H. L. House (personal communication) reared *Agria housei* continuously for 756 generations on pork liver. In contrast to these species, however, the behavioral interaction between many parasitoids and their hosts are complex and diverse, involving numerous physical and chemical cues that initiate specific behavioral patterns leading to oviposition.

Behavior and in vitro Culture

Host selection and successful parasitism is a multistep process involving host habitat location, host location, host acceptance, host suitability, and host regulation (Doutt, 1959; Vinson, 1976, 1984; Tumlinson *et al.,* 1993; Vinson *et al.,* 1998; Chapter 15). Many factors, particularly those that influence host acceptance, are critical for continuous culture. Multiple events leading to successful oviposition, including examination of the host, thrusting with the ovipositor, insertion, and finally oviposition (Schmidt, 1974) may each be stimulated by different chemical as well as physical cues (Arthur, 1981; Vinson, 1984). These cues may be associated with the host species, the plant, or other food source of the host; or may result from interactions involving both the host and its food (Vinson, 1975; Price *et al.,* 1980; Barbosa 1988).

The importance of physical factors associated with the host's food plant was demonstrated by Mellini *et al.* (1980) to be essential for successful oviposition and parasitism by *Pseudogonia rufifrons*. This tachinid parasitoid deposits microtype eggs on the leaves of certain plants, and host larvae become infected by ingesting the parasitoid eggs. Leaf color, shape, size, thickness, and reflectivity were among the factors influencing oviposition. The authors constructed polished, thin, yellow, oval-pointed artificial bee's wax leaves, 2 to 7 cm², on which large numbers of parasitized eggs were deposited. The parasitoid readily developed in *Galleria mellonella* after the host fed on the artificial leaves. Complex combinations of physical cues, including size, shape, color, texture, and movement have been demonstrated to influence the oviposition behavior of numerous parasitoids (Arthur, 1981; Jones, 1981; Nordlund *et al.,* 1981).

The ovipostion behavior of predators is also affected by the characteristics of their host plants. Evans (1976) reported that females of *Anthocoris confusus* (Reuter) distinguish between turgid and flaccid foliage when searching for oviposition sites.

Chemical Cues Influencing the Behavior of Entomophagous Species

Chemical factors or semiochemicals are known to play important roles in both parasitoid–host (Arthur, 1981; Jones, 1981; Vinson, 1984; Noldus, 1989; Lewis & Martin, 1990; Vet & Dicke, 1992; Vinson *et al.*, 1998) and predator–prey interactions (Greany & Hagen, 1981; Jones, 1981; Hagen, 1986a, 1987; Vet & Dicke, 1992). The involvement of chemicals in host acceptance and oviposition by parasitoids is well documented. Several investigations indicate that parasitoid response to semiochemicals can be influenced markedly by experience (Lewis & Tumlinson, 1988; Vet & Groenewold, 1990). Such adaptive changes have been demonstrated to affect host acceptance (Kaiser *et al.*, 1989), and associative learning may prove to be invaluable as a tool for modifying parasitoid performance when used in conjunction with *in vitro* culture.

During predator–prey relationships, kairomones produced by the prey may serve as attractants, arrestants, and/ or phagostimulants (Dicke *et al.*, 1990). *Chrysopa carnea* adults are attracted to a variety of chemicals such as tryptophan by-products formed from honeydew (Hagen *et al.*, 1976). Allelochemicals from plants may also affect predator behavior. The predatory bug *A. nemorum* has been reported to orient relative to volatile substances emitted by the host plant (Dwumfour, 1992). Constant *et al.* (1996) observed increased egg laying by the mirid *Macrolophus caliginosus* Wagner on damp dental cotton sprayed with an ethanol extract of *Inula viscosa* (Linneaus), the predator's natural host plant. Feeding attractants and deterrents produced by prey also affect the behavior of mite predators (McMurtry & Rodriquez, 1987).

Attempts to utilize behavior-modifying chemicals in the development of continuous artificial culture are restricted to a few species. *Itoplectis conquisitor* accepts a host and oviposits following detection of specific components of host hemolymph during probing with the ovipositor (Arthur *et al.*, 1969). The parasitoid oviposited into host hemolymph placed on paraffin tubes, and the authors fractionated the hemolymph and the isolated component or components that induced oviposition. The active fraction was colorless and water soluble and gave a strong reaction to ninhydrin and Folin–phenol reagents. Moreover, it had a molecular weight of approximately 7 kD and was nondializable and heat stable. It was concluded that the oviposition stimulant was proteinaceous and, Arthur *et al.* (1973) were successful in stimulating similar oviposition activity with a variety of amino acid mixtures containing trehalose and/or $MgCl_2$. The best results were obtained with a mixture of serine (0.5 *M*), arginine (0.05 *M*), leucine (0.065 *M*), and $MgCl_2$ (0.025 *M*); and the ovipositional activity observed dramatically exceeded that stimulated by host hemolymph. House (1978) developed a synthetic, artificial host composed of an artificial diet encapsulated in paraffin. The diet, based on that described by Yazgan (1972), was composed of casein, gelatin, amino acids, inorganic salts, glycogen, lipids, glucose, trehalose, water- and fat-soluble vitamins, and agar. The female parasitoid readily accepted and oviposited into the artificial host, and the first successful complete artificial culture of a hymenopterous parasitoid was achieved. The single adult obtained was a male of comparable size to that reared by Yazgan (1972).

Considerable knowledge is available describing how chemical and other factors influence adult reproductive capacity of several parasitoid species for which *in vitro* culture methods have been developed. For example, larviposition by *Eucelatoria bryani* is known to be stimulated by kairomones emanating from the host's cuticle, and female adults carefully and repeatedly examine artificial hosts coated with cuticular extracts (Burks & Nettles, 1978). Tucker and Leonard (1977) extracted a kairomone from the pupae of *Lymantria dispar* (Linnaeus) that appeared responsible for ovipositional behavior by *Brachymeria intermedia*. *Tetrastichus schoenobii* was induced to oviposit in artificial eggs coated with host scales (Ding *et al.*, 1980b). Oviposition can also be induced by chemical factors not normally associated with the host or the parasitoid's environment. Xie *et al.* (1997), for example, reported that Elmer's® glues served as both arrestants and probing and oviposition enhancers for several *Trichogramma* species. Although continuous artificial culture has been attempted with only a few species, the method may be applicable.

Continuous *in vitro* Culture of *Trichogramma*

No parasitoid species or group of species has received as much attention from biological control workers as species of the genus *Trichogramma*. More extensive efforts to develop continuous artificial culture have been made with *Trichogramma* spp. than have been made with other parasitoids, and the Chinese have been particularly active in developing this technology (Fig. 7; Li, 1993). Many aspects of the ovipositional behavior of this genus were described by Salt (1934, 1940) in his classical studies on *T. evanescens;* see Fisher (1986). Studies demonstrate the importance of physical (Rajendram & Hagen, 1974) and chemical factors, including kairomones (Nordlund *et al.*, 1985; Kaiser *et al.*, 1989; Renou *et al.*, 1992), for eliciting oviposition. Rajendram (1978a, 1978b) obtained artificial oviposition by *T. californicum* into physiological saline or Neisenheimer's salt solution encapsulated in paraffin. Nettles *et al.* (1982, 1983) subsequently reported that a dilute solution of KCl and $MgSO_4$ induced oviposition by *T. pretiosum* into artificial wax eggs (Nettles *et al.*, 1984). Wu and Qin (1982b) reported that leucine, phenylalanine, and/or isoleucine stimulated oviposition by *T. dendrolimi* into artificial eggs. With a complete mixture of all three amino acids, 600, 400, and 320 mg/100 ml, respectively, adult females

FIGURE 7   Continuous artificial mass culture of *Trichogramma* spp. in China. (a) through (f) Courtesy of Li Li-ying, Director, Guangdon Entomological Institute, China, 1994. (a) Mechanical press and template to making artificial eggs; (b) dispensing medium into artificial eggs; (c) petri plates containing artificial eggs for oviposition; (d) developing larvae; (e) pupae; (f) *in vitro*-cultured parasitoids ovipositing into artificial media.

laid more eggs than when insect hemolymph was used. The ovipositional behavior of several additional species of *Trichogramma* and implications for continuous culture were described by Li *et al.* (1989).

Morrison *et al.* (1983) were successful at developing an inexpensive synthetic membrane as an alternative for paraffin through which *T. pretiosum* would oviposit. The silicone–polycarbonate copolymer was clear and highly elastic, and adult females oviposited through the surface into an ovipositional stimulant at rates comparable to those obtained with host eggs. The use of polyethylene as an alternative to wax for producing artificial eggs for oviposition by *T. dendrolimi* was described by the CRGHP of China (1985), and for oviposition by *T. maidis* by Grenier and Bonnot (1988).

The potential for large-scale continuous artificial culture of *T. pretiosum* was demonstrated by Xie *et al.* (1986b). Based on studies described earlier on artificial media (Xie *et al.* 1986a) and influence and interaction of media ingredients on ovipositional behavior (Nettles *et al.* 1985), three *in vitro* culture methods were developed. These methods utilized microtiter tissue culture plates, multiple-drop rearing in petri plates, and flooded petri plate rearing. The basic

diet was composed of 50% heat-treated insect hemolymph, 25% egg yolk, and 15 g/100 ml dried milk suspension. Each of the methods supported the rearing of large populations of parasitoid larvae. By using 10-cm flooded petri plates and the basic medium supplemented with the egg contents of *Manduca sexta,* approximately 10,000 larvae could be reared with adult emergence as high as 75%. Microbial contamination and subsequent loss of entire petri plates were major obstacles with this method, but the authors reported successful use of several antibiotics for reducing losses. The most promising method for mass production was the multiple-drop petri plate technique. Grenier and Liu (1990) tested a variety of fungicides for controlling contamination in artificial diets for *Trichogramma.*

Dai (1985) described the continuous culture of 61 generations of *T. dendrolimi* on egg yolk, milk, and hemolymph diets. The percentage of parasitism by artificially reared parasitoids remained at 100%, but emergence of adult parasitoids was reduced from 89 to 46% over successive generations. The proportion of females remained approximately 75 to 80%. Additional studies on continuous culture of *Trichogramma* spp. were reported by Li (1989, 1993), Li *et al.* (1988), and Liu *et al.* (1995a).

Liu *et al.* (1995b) described an automated mass production system employed for rearing *T. chilonis, T. dendrolimi, T. japonicum, T. minutum,* and *T. pretiosum.* The rearing method generates sufficient artificial host eggs each hour to produce 6 to 7 million individual *Trichogramma.*

Performance of *in vitro*-Cultured Entomophagous Insects

Although widespread use of continuous *in vitro* culture for mass-rearing predators and parasitoids is still in the future, several field trials with *in vitro*-reared *Trichogramma* have been made. Continuous artificial mass culture of *T. dendrolimi* was described by Li (1982) and Dai (1985). Gao *et al.* (1982) reported that field release of *in vitro*-reared parasitoids resulted in 93% parasitism of *Heliothis armigera* (Hübner) eggs in cotton. Additional field trials with *Trichogramma* were described by Liu *et al.* (1995a).

Liu *et al.* (1988) examined the efficacy of *in vitro*-cultured *Anastatus* sp. Parasitoids reared on an artificial diet readily attacked the hemipterous host *Tessaratoma papillosa* Drury growing on litchi trees in pots or orchards. Approximately 90% of host eggs were parasitized. Han (1988) compared the performance of *in vitro*-cultured *A. japonicus* with parasitoids reared on eggs of *A. pernyi.* When field released at a rate of 1.25×10^5 per hectare, artificially reared parasitoids attacked approximately 86% of the host *T. papillosa* compared with 75% for parasitoids reared *in vivo.*

The effectiveness of augmentative release of *in vitro*-reared *Trichogramma* spp. against a variety of insect pests was described by Dai (1988). *Trichogramma dendrolimi,* released at a rate of 1.5×10^5 per hectare against *H. armigera* and 3.0×10^5 per hectare against *Dendrolimus punctata,* parasitized 90 and 83%, respectively, of these host species. These results compared with 88 and 42%, respectively, for parasitoids reared on host eggs. Released at a rate of 1.5×10^5 per hectare against *Chilo sacchariphagus* and *C. infuscatellus,* artificially cultured *T. confusum* parasitized 90 and 61% of hosts, respectively, compared with 93 and 67%, respectively, by parasitoids reared *in vivo.* The significance of these results relative to economic thresholds of pest damage was not discussed in any of the preceding studies.

During the 8 years preceding 1994, over 1 billion *Trichogramma* spp. and 10 million *Anastatus* spp. were reared on artificial host eggs and released throughout China on 1333-ha sugarcane fields and 1500 litchi trees (Liu *et al.* 1995a, 1995b). Control of stem borer eggs was 75 to 93% by *T. chilonis* reared on artificial host eggs, compared with 67 to 83% by this parasitoid reared on silkworm eggs and 12 to 34% by chemical control. Parasitism of *Tessaratoma*

papollosa eggs by *Anastatus* spp. was 89 to 96% compared with 9% pest control by chemical application.

In vitro-cultured *Catolaccus grandis* have been tested in the field by Morales-Ramos *et al.* (1996b). Parasitism of *Anthonomus grandis* by parasitoids cultured *in vitro* was significantly less than that obtained with parasitoids reared on host larvae.

Yazlovetsky and Nepomnyashchaya (1981) compared the effectiveness of *Chrysopa carnea* reared on an artificial diet with predators reared on *Sitrotroga cerealella,* and reported them equally effective as predators of *Myzus persicae.* The pattern of prey selection by the hemipteran *G. punctipes* reared *in vitro* for several years has been equal to that of predators reared in the field (Hagler & Cohen, 1991).

Major concerns of programs aimed at artificial mass culture of entomophagous insects are the evaluation and the maintenance of insect quality. Quality is ultimately expressed as efficacy in the field. It represents the expression of biological properties critical to the success of the insect as a biological control agent. Quality requirements and assessment of entomophagous insects were discussed by Neuenschwander *et al.* (1989), Bigler (1989), Hopper *et al.* (1993), Cohen and Staten (1993), and Bennett (1993). The topics of quality and quality control, however, are beyond the scope of the present chapter, and the reader is referred to Leppla and Ashley (1989) and Chapter 7.

CURRENT APPLICATIONS

Application of nutrition in biological control programs is largely restricted to use of food and food supplements to enhance the activity and effectiveness of entomophagous insects in the field (Hagen & Hale, 1974; Hagen & Bishop, 1979; Greenblatt & Lewis, 1983; Hagen, 1986a; Gross, 1987). Such use is necessitated by a lack of synchrony between natural enemies and their hosts and/or isolation of entomophagous insects from the natural environment that normally supplies alternate food sources such as nectars and honeydews (Hagen, 1986a). These factors frequently occur in crop monocultures and may intensify following pesticide application. The importance of nutritional supplements for adult parasitoids and predators is well established. Studies with species of *Trichogramma* demonstrated that fecundity and longevity were increased markedly by feeding adult insects (Anunciada & Voegele, 1982; Bai, 1986), and future use of feeding prior to or following release in the field may have a significant effect on their success as biological control agents. Few studies on the effects of feeding parasitoids on field performance are available. Temerak (1976) reported spraying honey solution on sorghum stalks during winter to provide supplementary food to *Bracon brevicornis* Wesmal in the absence of pollen, honeydew, and nectar. Parasitoid cocoons significantly increased after spraying and the prevalence of hosts was decreased. Despite field trials employing kairomones for attracting and stimulating host searching by *Trichogramma* spp. (Lewis *et al.*, 1979, 1982), no attempt has been made to use kairomones in combination with supplemental foods to maintain parasitoid populations when host numbers are low.

Successful use of supplementary food sprays has been reported in biological control studies employing predaceous insects. For example, Ewert and Chiang (1966) sprayed sucrose solutions on corn to aggregate coccinellid and chrysopid adults. Increased predator density and reproductive activity significantly lowered aphid populations. Carlson and Chiang (1973) reported increased numbers of *Chrysopa* and/or *Chrysoperla* sp. and *Glischiochilus quadrisignatus* (Say) in corn sprayed with sugar or molasses solutions. Increased predation resulted in significant reductions in the population of *Ostrinia nubilalis* (Hübner).

Studies on sugar-sprayed alfalfa plots were described by Hagen *et al.* (1976). *Chrysopa carnea* and *Hippodamia* sp. were arrested in large numbers following spraying during periods of low host populations. Within 24 h the population of coccinellid adults increased 20-fold and that of *C. carnea*, 200-fold. Populations of *Lygus* spp. also increased after application of sugar sprays (Lindquist & Sorenson, 1970). Hagen *et al.* (1971) concluded that sucrose was an arrestant for adult *Lygus* and coccinellids. *Chrysopa carnea* and coccinellids, as well as *Lygus*, were attracted to potato plants sprayed with honey, suggesting a critical role for volatile components (Ben Saad & Bishop, 1976a, 1976b).

The addition of semiochemicals to supplemental foods for *C. carnea* has proved highly beneficial. The complex interactions of semiochemicals and food in influencing the behavior of *C. carnea* were described by Hagen and Bishop (1979). These authors reported that the adult responds to a volatile signal, that is, a synomone, from plant habitats in which prey are located and then is attracted to the prey by tryptophan breakdown products from honeydew (van Emden & Hagen, 1976; Dean & Satassok, 1983). The specific behavioral and flight patterns exhibited by *C. carnea* in response to these interactions were described by Duelli (1980). The habitat synomone affecting the behavior of *C. carnea* in cotton was reported by Flint *et al.* (1979) to be caryophyllene, but several chemicals from other plants also displayed synomone activity for this species (Hagen, 1986b).

Hagen *et al.* (1971) and Tassan *et al.* (1979) were successful in attracting *C. carnea* adults to alfalfa fields by applying artificial honeydew composed of various yeast products (brewer's yeast/Wheast®) and sugar. Although the specific synomone of alfalfa is unknown, application of caryophyllene together with the kairomone from tryptophan markedly improved attraction of *C. carnea* during the onset of flowering (Hagen, 1986a). Application of artificial

honeydew was also successful for aggregating *Hippodamia* spp., as well as other coccinellids and predators. Subsequent trials employing Wheast® in combination with sucrose and honey or molasses applied to various crops were successful in manipulating *C. carnea* populations (Hagen & Hale, 1974). To retain *C. carnea* adults after attracting them, sugar must be present in the food spray.

Later studies demonstrated the value of artificial honeydews containing tryptophan, sucrose, and yeast as supplemental foods for improving performance of *C. carnea* in olive groves (Liber & Niccoli, 1988; McEwen *et al.*, 1993a, 1993b, 1996). McEwen *et al.* (1994) also reported that tryptophan sprays were effective in concentrating predators in the olive tree canopy.

In contrast to the preceding results, Butler and Ritchie (1971) reported that *C. carnea* adults were attracted to Wheast® and sugar mixtures sprayed on cotton but no increase in egg deposition was noted. Similar studies demonstrated inconsistent oviposition in grape plots (White & Jubb, 1980). No attraction of chrysopid adults occurred in treated apple orchards (Hagley & Simpson, 1981) or in potato fields when Wheast® only was applied. Dean and Satasook (1983) outlined reasons why food sprays might not be practical in control programs for cereal aphids in England. Variable abundance of the univoltine *C. carnea* populations, low plant growth form, and development of sooty mold on plants were all suggested as factors affecting the field performance of *C. carnea*. Duelli (1987) failed to observe an increase in oviposition by chrysopids when artificial honeydews were applied to alfalfa, corn, sunflowers, and plum orchards. Duelli suggested that the different responses of sibling species of *C. carnea* in Europe and North America may be a reflection of behavioral differences.

The industrial *in vitro* production of *T. dendrolimi* and *T. chilonis* Ishii for biological control in China was outlined by Dai *et al.* (1991). At that time 132,000 and 237,600 artificial host eggs were being produced per hour. Daily production amounted to 30 to 43 million *T. dendrolimi* and 17 to 23 million *T. chilonis* at a cost of approximately 50% that of mass culture on silkworm eggs. Mass release was conducted for control of several lepidopterous pests in the provinces of Hubei, Guangdong, Guangxi, and Henan; and rates of parasitism consistently exceeded 80%.

CONCLUSIONS AND FUTURE DIRECTIONS

Nutrition and Dietetics

"Nutrition is about nourishment; that is, it is the action or processes of transforming substances found in foodstuff into body materials and energy to do all the things attributed to life. Nutritional requirements depend on the synthetic abilities of the organism and the basis is genetical. Therefore, through nutrition we have a direct and essential connection between an environmental factor, foodstuff and the vital processes of the insect organism" (House, 1977). Knowledge of the nutrition and nutritional requirements of entomophagous insects has improved significantly over the past three decades. Most nutritionally related research with insects, however, has not been aimed at developing a basic understanding of their nutrition, but instead has been directed largely at the development of successful insect rearing. The field of insect "nutrition" has been dominated by efforts to examine feeding and to develop artificial diets. These topics are really the subject of dietetics (Beck, 1972). They are critically important for insect rearing, but have provided a limited understanding of insect nutrition, aside from establishing qualitative nutritional requirements.

The qualitative nutritional requirements of insects, including entomophagous forms, are remarkably similar despite the diversity of their feeding habits (Beck, 1972; Dadd, 1973; Hagen, 1986b). Although our knowledge of dietetics and nutrition has advanced, practical application of nutritional principles to insect rearing in support of biological control programs is lacking. Rearing each insect species has become a unique challenge because our meager knowledge of nutrition fails to provide a broad and sound basis for approaching insect husbandry (House, 1977). In the case of entomophagous species, foodstuff is in a constant state of qualitative and quantitative flux, and almost nothing is known of the quantitative nutritional requirements for the various life stages and physiological functions of these insects. Future investigations should emphasize the quantitative nutritional requirements for growth and development, oogenesis, and other activities and processes of particular importance to the biological control application desired. Although energy budgets for development of some entomophagous species have been calculated, additional and more detailed studies are needed. As discussed earlier, the requirements for many nutrients are often dependent on the presence and concentration of others, and correct nutrient balance often may be critical for successful nutrition. Several species of parasitoids and predators for which artificial diets have been successfully developed can serve as models for *in vitro* investigation on the quantitative requirements for specific nutrients. House (1977) was successful in examining many aspects of the quantitative nutrition of *Agria housei* in this manner. Thompson (1976a, 1982) used a defined artificial medium to examine the quantitative requirements for supporting larval growth of the parasitoid *Exeristes roborator*.

Dependence of Parasitoid Development on Host Physiology

The development of many parasitoids is intimately associated with the physiology of their hosts, as discussed earlier. Many studies suggest that changes in the host's physiology following parasitism are adaptive for the parasitoid, ensuring its successful development (Vinson & Iwantsch, 1980; Thompson, 1985, 1986a, 1993). Slansky (1986) outlined how parasitoids overcome potential nutrient constraints by altering their host's behavior and physiology. Changes in the composition of the host's internal milieu likely have significant nutritional consequences for the parasitoid (Grenier *et al.*, 1986; Thompson, 1983c, 1986a). Endocrine interactions appear of critical importance to successful parasitoid development. Developmental synchrony between many larval endoparasitoids and their hosts is well known (Beckage, 1985; Lawrence, 1991), and strongly suggests that the host's hormones and endocrine physiology influence parasitoid development (Lawrence, 1986a; Lawrence & Lanzrein, 1993). It is important that additional study of the endocrine interactions between parasitoids and their hosts be initiated to characterize the involvement of host hormones in parasitoid development. Investigation is also necessary to identify and characterize growth factors critical to parasite survival and development (Ferkovich & Oberlander, 1991a).

Nutritional Ecology of Entomophagous Species

Nutritional and physiological investigations are essential in achieving successful culture of entomophagous insects. The use of these insects to augment biological control programs, however, also requires careful consideration of ecological factors and of the relationship between a parasitoid's or predator's nutrition and its success as a biological control agent. Slansky (1986), Hagen (1987), and Vinson and Barbosa (1987) discussed the importance of ecological considerations in the nutrition of entomophagous insects. They emphasized that the behavior and regulatory physiology of insects are in a state of continuous flux in response to food supply, and that nutrition can be fully understood only by considering the insect's nutritional ecology. In the case of entomophagous species, both the ecology of the parasitoid or predator and that of the host and prey need to be understood.

Genetic Basis and Modification of Nutrition

Nutritional and dietary requirements are genetically based and genetic manipulation holds promise as a means for modifying the nutrition of entomophagous insects. Studies by Chabora and Chabora (1971) suggested that nutritional content varies between strains of insects. They demonstrated that the yields of parasitoids *Nasonia vitripennis* (Walker) and *Muscidifurax raptor* Girault & Saunders were significantly increased when they were reared on a hybrid of two strains of the host *Musca domestica* Linnaeus. Thus, it may be possible to select for nutritional characteristics in a host and a prey. Collins (1984) discussed the artificial selection of desired characteristics in insect rearing. Genetic selection of the nutritional character of host and prey species might provide a valuable tool for studying the nutrition of parasitoids and predators.

Rousch (1979), Hoy (1979, 1986, 1990, 1992), and Caprio *et al.* (1991) outlined the potential for genetic improvement of parasitoids and predators used in pest management programs. Most studies have employed genetic selection to increase field effectiveness and included selection for sex ratio, host-finding ability, host preference, insecticide resistance, and improved climatic tolerance (Hoy, 1976; Chapter 12). Implicit to any genetic improvement program is a requirement for maintenance of vigor and vitality reflected, in part, by longevity and fecundity (Bartlett, 1993). Because the latter characteristics are intimately associated with nutrition, as discussed earlier, many previous genetic programs may, in fact, have involved selection for nutritionally related traits. The diversity of biological traits previously selected strongly suggests the possibility that genetic selection of nutritional and dietary requirements can be successful for modifying the nutrition of entomophagous species.

Genetic Engineering of Entomophagous Species

Advances in molecular biology (recombinant DNA technology) further suggest the possibility for genetic manipulation of the nutrition of entomophagous insects (Thompson, 1990; Chapter 4). Genetic engineering, that is, incorporation of foreign or *in vitro* altered genes for expression of desirable traits by an organism, has advanced considerably (Beckendorf & Hoy, 1985; Whitten, 1989; Crampton, 1992). Heilmann *et al.* (1993) discussed the potential of genetic engineering for improving the performance of entomophagous insects. The technology, however, is still in its infancy, and methods for genetic transformation of eukaryotic organisms are limited at the present time (Schmidt, 1990; Finnegan, 1992). Nevertheless, stable genetic transformation of the phytoseiid mite predators *Metaseiulus occidentalis* (Nesbitt) (Presnail & Hoy, 1992; Presnail *et al.*, 1997) and *Amblyseius finlandicus* (Oudemans) (Presnail & Hoy, 1994) and the braconid parasitoid *Cardiochiles diaphaniae* (Marsh) (Oudenmans) (Presnail & Hoy, 1996) has been achieved by microinjection, and Hoy (1993, 1995) and McDermott and Hoy (1997) have discussed the use of transgenic beneficial arthropods in pest management.

The future holds considerable promise for dramatic advances in our understanding the nutritional and dietary requirements of entomophagous insects. That knowledge will play an important role in the development of efficient rearing programs for parasitoids and predators.

Acknowledgments

The authors acknowledge N. E. Beckage, University of California, Riverside, California; W. C. Nettles, USDA-ARS, College Station, Texas; and A. C. Cohen, USDA-ARS, Tucson, Arizona, for their comments and suggestions during the early stages of preparing this chapter.

We also express our gratitude to P. D. Greany, USDA-ARS, Gainesville, Florida; and S. Grenier, Institut National des Sciences Appliquees, Villeurbanne Cedex, France, for their assistance during preparation of later drafts of the manuscript.

Finally, the senior author is grateful to Dr. Catherine Tauber, Cornell University, Ithaca, New York for clarifying the taxonomy of *Chrysopa* and *Chrysoperla*.

References

Adjei-Maafo, I. K., & Wilson, L. T. (1983). Factors affecting the relative abundance of arthropods on nectaried and nectariless cotton. Environmental Entomology, 12, 349–352.

Altahtawy, M. M., Hammad, S. M., & Hegazi, E. M. (1976). Studies on the dependence of *Microplitis rufiventris* Kok (Hym.: Braconidae) parasitizing *Spodoptera littoralis* (Boisduval) on its own food as well as its host. Zeitschrift fuer Angewandte Entomologie, 81, 3–13.

Anunciada, L., & Voegele, J. (1982). L'importance de la nourriture dans le potentiel biotique de *Trichogramma maidis* Pintureau et Voegele et *T. nagarkattii* Voegele et Pintureau (Hym.: Trichogrammatidae) et l'oosorption dans les femelles en contention ovarienne. Les Trichogrammes. Colloques de 1'INRA, 9, 79–84.

Arthur, A. P. (1981). Host acceptance by parasitoids. In D. A. Nordlund, R. L. Jones, & W. J. Lewis (Eds.), Semiochemicals: Their role in pest control (pp. 97–120). New York: John Wiley & Sons.

Arthur, A. P., & Coppel, H. C. (1953). Studies on dipterous parasites of the spruce budworm, *Choristoneura fumiferana* (Clemens) (Lepidoptera: Tortricidae). I. *Sarcophaga aldrichi* Parker (Diptera: Sarcophagidae). Canadian Journal of Zoology, 31, 374–391.

Arthur, A. P., & Wylie, H. G. (1959). Effects of host size on sex ratio, development time and size of *Pimpla turionellae* (Linnaeus) (Hymenoptera: Ichneumonidae). Entomophaga, 4, 297–301.

Arthur, A. P., Batsch, W. W., & Rollins, L. (1969). Component of the host haemolymph that induces oviposition in a parasitic insect. Nature (London), 223, 966–967.

Arthur, A. P., Hegdekar, B. M., & Batsch, W. W. (1973). A chemically defined, synthetic medium that induces oviposition in the parasite *Itoplectis conquisitor* (Hymenoptera: Ichneumonidae). Canadian Entomologist, 105, 787–793.

Askew, R. R., & Shaw, M. R. (1986). Parasitoid communities: Their size, structure and development. In J. Waage & D. Greathead (Eds.), Insect parasitoids (pp. 225–264). London: Academic Press.

Attallah, Y. H., & Newsom, L. D. (1966). Ecological and nutritional studies on *Coleomegilla maculata* De Geer (Coleoptera: Coccinellidae). I. The development of an artificial diet and a laboratory rearing technique. Journal of Economic Entomology, 59, 1173–1179.

Atwal, A. S., & Sethi, S. L. (1963). Biochemical basis for the food preference of a predator beetle. Current Science, 32, 511–512.

Baehrecke, E. H., Aiken, J. M., Dover, B. A., & Strand, M. R. (1993). Ecdysteroid induction of embryonic morphogenesis in a parasitic wasp. Developmental Biology, 158, 275–287.

Bai, B. (1986). *Trichogramma* quality. Host effect on quality attributes of *Trichogramma pretiosum* Riley (Hymenoptera: Trichogrammatidae) and host discrimination by *Copidosoma truncatellum* Dalman (Hymenoptera: Encyrtidae). Unpublished master's thesis, University of California, Riverside.

Bai, B., Luck, R. F., Forster, L., Stephens, B., & Janssen, J. A. M. (1992). The effect of host size on quality attributes of the egg parasitoid,

Trichogramma pretiosum. Entomologia Experimentalis et Applicata, 64, 37–48.

Baker, H. G., & Baker, I. (1983): A brief historical review of the chemistry of floral nectar. In B. Bentley & T. Elias (Eds.), The biology of nectaries (pp. 126–152). New York: Columbia University Press.

Baker, H. G., Opler, P. A., & Baker, I. (1978). A comparison of amino acid complements of floral and extrafloral nectars. Botanical Gazette (Chicago), 139, 322–332.

Barbier, M. (1970). Chemistry and biochemistry of pollens. Progress in Phytochemistry, 2, 1–34.

Barbosa, P. (1988). Natural enemies and herbivore–plant interactions, influence of plant allelochemicals and host specificity. In P. Barbosa & D. K. Letourneau (Eds.), Novel aspects of insect-plant interactions (pp. 201–229). New York: John Wiley & Sons.

Barbosa, P., Gross, P., & Kemper, J. (1991). Influence of plant allelochemicals on the tobacco hornworm and its parasitoid, *Cotesia congregata.* Ecology, 72, 1567–1575.

Barbosa, P., Saunders, J. A., & Waldvogel, M. (1982). Plant-mediated variation in herbivore suitability and parasitoid fitness. In J. H. Visser & A. K. Minks (Eds.), Proceedings of the Fifth International Symposium on Insect–Plant Relationships (pp. 63–71). Wageningen: Pudoc.

Barbosa, P., Saunders, J. A., Kemper, J., Trumbule, R., Olechno, J., & Martinat, P. (1986). Plant allelochemicals and insect parasitoids: Effects of nicotine on *Cotesia congretata* (Say) (Hymenoptera: Bracondiae) and *Hyposoter annulipes* (Cresson) (Hymenoptera: Ichneumonidae). Journal of Chemical Ecology, 12, 1319–1327.

Baronio, P., & Sehnal, F. (1980). Dependence of the parasitoid *Gonia cinerascens* on the hormones of its lepidopterous hosts. Journal of Insect Physiology, 26, 619–626.

Bartlett, A. C. (1993). Maintaining genetic diversity in laboratory colonies of parasites and predators. In S. K. Narang, A. C. Bartlett, & R. M. Faust (Eds.), Applications of genetics to arthropods of biological control significance (pp. 133–145). Boca Raton, FL: CRC Press.

Bartlett, B. R. (1964). Patterns in the host-feeding habit of adult parasitic Hymenoptera. Annals of the Entomological Society of America, 57, 344–350.

Bashir, M. O. (1973). Effect of nutrition on development and reproduction of aphidophagous coccinellids with special reference to *Olla abdominalis* (Say). Unpublished doctoral dissertation, University of California, Berkeley.

Beck, S. D. (1972). Nutrition, adaptation and environment. In J. G. Rodriguez (Ed.), Insect and mite nutrition (pp. 1–6). Amsterdam: North-Holland.

Beckage, N. E. (1985). Endocrine interactions between endoparasitic insects and their hosts. Annual Review of Entomology, 30, 371–413.

Beckage, N. E., & Riddiford, L. M. (1983). Growth and development of the endoparasitic wasp *Apanteles congregatus* dependence on host nutritional status and parasite load. Physiology and Entomology, 8, 231–241.

Beckendorf, S. K., & Hoy, M. A. (1985). Genetic improvement of arthropod natural enemies through selection, hybridization or genetic engineering techniques. In Biological control in agricultural IPM systems (pp. 167–187). New York: Academic Press.

Benn, M., DeGrave, J., Gnanasunderam, C., & Hutchins, R. (1979). Host-plant pyrrolizidine alkaloids in *Nyctemera annulata* Boisduval: Their persistence through the life-cycle and transfer to a parasite. Experientia, 35, 731–732.

Bennett, F. D. (1993). Increasing genetic diversity for release of parasites and predators. In S. K. Narang, A. C. Bartlett, & R. M. Faust (Eds.), Applications of genetics to arthropods of biological control significance (pp. 160–166). Boca Raton, FL: CRC Press.

Bentz, J. A., & Barbosa, P. (1992). Effects of dietary nicotine and partial starvation of tobacco hornworm, *Manduca sexta,* on the survival and development of the parasitoid *Cotesia congregata.* Entomologia Experimentalis et Applicata, 65, 241–245.

Ben Saad, A. A., & Bishop, G. W. (1976a). Attraction of insects to potato plants through use of artificial honeydews and aphid juice. Entomophaga, 21, 49–57.

Ben Saad, A. A., & Bishop, G. W. (1976b). Effect of artificial honeydews on insect communities in potato fields. Environmental Entomology, 5, 453–457.

Bernays, E. A. (1985). Regulation of feeding behavior. In G. A. Kerkut & L. I. Gilbert (Eds.), Comprehensive insect physiology, biochemistry and pharmacology (Vol 4, pp. 1–32). New York: Pergamon Press.

Bernays, E. A. (1997). Feeding by lepidopteran larvae is dangerous. Ecological Entomology, 22, 121–123.

Bernays, E. A., & Simpson, S. J. (1982). Control of food intake. Advances in Insect Physiology, 16, 59–118.

Bierne, B. (1962). Trends in applied biological control of insects. Annual Review of Entomology, 7, 387–400.

Bigler, F. (1989). Quality assessment and control in entomophagous insects used for biological control. Journal of Applied Entomology, 108, 390–400.

Bigler, F., Meyer, A., & Bosshartied, S. (1987). Quality assessment in *Trichogramma maidis* Pintureau et Voegele reared from eggs of the factitious hosts *Ephestia kuehniella* Zell. and *Sitotroga cerealella* (Oliver). Journal of Applied Entomology, 104, 340–353.

Bonnot, G. (1986). Les particularités de la nutrition des insectes parasites. In C. L. Legner (Ed.), La nutrition de crustacés et des insectes. Compte Rendu du Colloque du CNERNA, Paris, 227–240.

Bonnot, G. (1988). Multivariate analysis of nutritional experiments with parasitoids. Colloques de l'INRA, 48, 81–83.

Bonnot, G., Delobel, B., & Grenier, S. (1984). Elevage, croissance et développement de *Phryxe caudata* (Diptera, Tachinidae) sur son hôte de substitution *Galleria mellonella* (Lepidoptera, Pyralidae) et sur milieu artificiel. Bulletin de la Societe Linnéenne de Lyon, 53, 313–320.

Bouletreau, M. (1968). Premiers résultats de l'élevage des larves d'un Hyménoptère chalcidien (*Pteromalus puparum*) sur hémolymphe de Lépidoptère. Entomophaga, 13, 217–222.

Bouletreau, M. (1971). Croissance larvaire et utilisation de l'hôte chez *Pteromalus puparum* (Hym.: Chalc.): Influence de la densité de population. Annales de Zoologie—Ecologie Animale, 3, 305–318.

Bouletreau, M. (1972). Développement et croissance larvaires en conditions semi-artificielles et artificielles chez un Hyménoptère entomophage: *Pteromalus puparum* L. (Chalc.). Entomophaga, 17, 265–273.

Bracken, G. K. (1965). Effects of dietary components on fecundity of the parasitoid *Exeristes comstockii* (Cresson) (Hymenoptera: Ichneumonidae). Canadian Entomologist, 97, 1037–1041.

Bracken, G. K. (1966). Role of ten dietary vitamins on fecundity of the parasitoid *Exeristes comstockii* (Cresson) (Hymenoptera: Ichneumonidae). Canadian Entomologist, 98, 918–922.

Bracken, G. K. (1969). Effects of dietary amino acids, salts and protein starvation on fecundity of the parasitoid *Exeristes comstockii* (Hymenoptera: Ichneumonidae). Canadian Entomologist, 101, 91–96.

Bratti, A. (1989). Allevamento *in vitro* di *Pseudosonia rufifrons* Wied. in estratti di omogeneizzato di crisalidi di *Galleria mellonella* L. e su diete meridiche. Bollettino dell'Istituto di Entomologia "Guido Grandi" della Universita degli Studi di Bologna, 44, 11–22.

Bratti, A. (1990a). Allevamento *in vitro* di *Pseudogonia rufifrons* Wied. in estratti di omogeneizzato di *Galleria mellonella* L. e su diete meridiche. Bollettino dell'Istituto di Entomologia "Guido Grandi" della Universita degli Studi di Bologna, 44, 11–22.

Bratti, A. (1990b). Tecniche di allevamento *in vitro* per gli stadi larvali di insetti entomofagi parassitoidi. Bollettino dell, Istituto di Entomologia "Guido Grandi" della Universita degli Studi di Bologna, 44, 169–221.

Bratti, A. (1993). *In vitro* rearing of *Lydella thompsoni* Hert. and *Archytas maromratus* (Town.) larval stages: Preliminary results. Bollettino dell, Istituto di Entomologia "Guido Grandi" della Universita degli Studi di Bologna, 48, 93–100.

Bratti, A. (1994a). Principi generali per l'allestimento di diete artificiali per gli stadi larvali dei Ditteri Tachinidi e nuovi contributi nel campo della sperimentazione. In G. Vioggiani (Ed.), M.A.F.-Convegno "Lotta biologica," Acireale 1991 (pp. 41–53). Roma: Istituto Sperimentale Patologia Vegetale.

Bratti, A. (1994b). Una dieta di base per l'allevameto in vitro di tre species Ditteri Tachinidi: Palexorista laxa (Curran), Eucelatoria bryani Sab. ed Exorista larvarum (L.) (pp. 705–706). Atti XVII Congresso Nazionale Italiana Entomologia, Udine, Giugno 13–18, 1994.

Bratti, A., & Campadelli, G. (1993). Comparison of insect-material in a meridic diet for Exorista larvarum L. (Dipt. Tachinidae) in vitro rearing. Bollettino dell'Istituto di Entomologia "Guido Grandi" della Universita degli Studi di Bologna, 48, 59–65.

Bratti, A., & Costantiti, W. (1991). Effetti di nuove diete artificiali dell'ospite sulla coppia ospite parassita Galleria mellonella L. (Lep. Galleriidae)-Archytas marmoratus (Town.) (Dopt. Tachinidae). Bollettino dell'Istituto di Entomologia "Guido Grandi" della Universita degli Studi di Bologna, 46, 49–62.

Bratti, A., & Coulibaly, A. K. (1995). In vitro rearing of Exorista larvarum on tissue culture-based diets. Entomologia Experimental et Applicata, 74, 47–53.

Bratti, A., & D'Amelio, L. (1994). In vitro rearing of Eucelatoria bryani Sab. (Diptera Tachinidae) on tissue culture-based diets. Bollettino dell'Istituto di Entomologia "Guido Grandi" della Universita degli Studi di Bologna, 48, 109–114.

Bratti, A., & Monti, M. (1988). Allevamento in vitro delle larve di Pseudogonia rufifrons Wied. (Dipt. Tachinidae) su omogeneizzato di crisalidi di Galleria mellonella L. (Lep. Galleriidae). Bollettino dell'Istituto di Entomologia "Guido Grandi" della Universita degli Studi di Bologna, 43, 115–126.

Bratti, A., & Nettles, W. C. (1988). In vitro rearing of Palexorista laxa (Curran) (Diptera Tachinidae) on haemolymaph-based diets. Bollettino dell'Istituto di Entomologia "Guido Grandi" della Universita degli Studi di Bologna, 43, 25–30.

Bratti, A., & Nettles, W. C. (1992). In vitro rearing of Eucelatoria bryani: Improvements and evaluation of factors affecting efficiency. Entomologia Experimentalis et Applicata, 63, 213–219.

Bratti, A., & Nettles, W. C. (1994). Comparative growth and development in vitro of Eucelatoria bryani Sab. and Palexorista laxa (Curran) (Diptera Tachinidae) fed a meridic diet and a diet of Helicoverpa zea (Boddie) (Lepidoptera Noctuidae) pupae. Bollettino dell'Instituto di Entomologia "Guido Grandi" della Universita degli Studi di Bologna, 49, 119–129.

Bratti, A., Campadelli, G., & Mariani, M. (1995). In vitro rearing of Exorista larvarum (L.) on diet without insect components. Bollettino dell'Istituto di Entomologia "Guido Grandi" della Universita degli Studi di Bologna, 49, 225–236.

Briggs, C. J., Nisbet, R. M., Murdoch, W. W., Collier, T. R., & Metz, J. A. J. (1995). Dynamic effects of host-feeding in parasitoids. Journal of Animal Ecology, 64, 403–416.

Bronskill, J., & House, H. L. (1957). Notes on rearing a pupal endoparasite, Pimpla turionellae (Linnaeus) (Hymenoptera: Ichneumonidae), on unnatural food. Canadian Entomologist, 89, 483.

Burks, M. L., & Nettles, W. C., Jr. (1978). Eucelatoria sp.: Effects of cuticular extracts from Heliothis virescens and other factors on oviposition. Environmental Entomology, 7, 897–900.

Butler, G. D., & Ritchie, P. L. (1971). Wheast® and the abundance and fecundity of Chrysopa carnea. Journal of Economic Entomology, 64, 933–934.

Cai, C., Zhang, X., & Zhao, J. (1983). Studies on the artificial diet for rearing larvae of Chrysopa sinica Tjeder. Natural Enemy Insects, 5(2), 82–85.

Calow, P. (1977). Conversion efficiencies in heterotrophic organisms. Biological Reviews, 52, 385–409.

Cameron, R. A. D., & Redfern, M. (1974). A simple study in ecological energetics using a gall-fly and its insect parasites. Journal of Biological Education, 8, 75–82.

Campadelli, G., & Dindo, M. L. (1987). Recenti progressi nello studio delle diete artificiali per l'allevamento di insetti entomologi parassiti. Bollettino dell'Istituto di Entomologia "Guido Grandi" della Universita degli Studi di Bologna, 42, 101–118.

Campbell, B. C., & Duffey, S. S. (1979). Tomatine and parasitic wasps—potential incompatibility of plant antibiosis with biological control. Science, 205, 700–702.

Campbell, B. C., & Duffey, S. S. (1981). Alleviation of tomatine-induced toxicity to the parasitoid, Hyposoter exiguae, by phytosterols in the diet of the host, Heliothis zea. Journal of Chemical Ecology, 7, 927–946.

Caprio, M. A., Hoy, M. A., & Tabashnik, B. E. (1991). Model for implementing a genetically improved strain of a parasitoid. American Entomologist, 37, 232–239.

Carlson, R. E., & Chiang, H. C. (1973). Reduction of Ostrinia nubilalis population by predatory insects attracted by sucrose sprays. Entomophaga, 18, 205–211.

Carpenter, J. E., & Greany, P. D. (1998). Comparative development and performance of artificially-reared versus host-reared Diapetimorpha introita (Cresson) (Hymenoptera: Ichneumonidae) wasps. Biological Control, 11, 203–208.

Chabora, P. C., & Chabora, A. J. (1971). Effects of an interpopulation hybrid host on parasite population dynamics. Annals of the Entomological Society of America, 64, 558–562.

Chan, M. S., & Godfray, H. C. J. (1993). Host-feeding strategies of parasitoid wasps. Evolutionary Ecology, 7, 593–604.

Charnov, E. L. (1982). The theory of sex allocation. Princeton, NJ: Princeton University Press.

Charnov, E. L., Hartogh Los-den, R. L., Jones, W. T., & van den Assem, J. (1981). Sex ratio evolution in a variable environment. Nature (London), 289, 27–33.

Chlodny, J. (1968). Evaluation of some parameters of the individual energy budget of the larvae of Pteromalus puparum (Linnaeus) (Pteromalidae) and Pimpla instigator (Fabricius) (Ichneumonidae). Ekologia Polska (Series A), 16, 505–512.

Chumakova, B. M. (1962). Significance of individual food components for the vital activity of mature predatory and parasitic insects. Voprosy Ekologii Kievsk, 8, 133–134.

Clausen, C. P. (1939). The effect of host size upon the sex ratio of hymenopterous parasites and its relation to methods of rearing and colonization. Journal of the New York Entomological Society, 47, 1–9.

Clausen, C. P. (1940). Entomophagous insects. New York: McGraw-Hill.

Clausen, C. P., Jaynes, H. A., & Gardner, T. R. (1933). Further investigations of the parasites of Popillia japonica in the far east. USDA Technical Bulletin, 366, 68.

Cohen, A. C. (1981). An artificial diet for Geocoris punctipes. Southwestern Entomology, 6, 109–113.

Cohen, A. C. (1983). Improved method of encapsulating artificial diet for rearing predators of harmful insects. Journal of Economic Entomology, 76, 957–959.

Cohen, A. C. (1984). Food consumption, food utilization, and metabolic rates of Geocoris punctipes (Het.: Lygaeidae) fed Heliothis virescens (Lep.: Noctuidae) eggs. Entomophaga, 29, 361–367.

Cohen, A. C. (1985). Simple method for rearing the insect predator Geocoris punctipes (Heteroptera: Lygaeidae) on a meat diet. Journal of Economic Entomology, 78, 1173–1175.

Cohen, A. C. (1989a). Ingestion efficiency and protein consumption by a heteropteran predator. Annals of the Entomological Society of America, 82, 495–499.

Cohen, A. C. (1989b). Using a systematic approach to develop artificial diets for predators. In T. E. Anderson & N. C. Leppla (Eds.), Advances in insect rearing for research and pest management (pp. 77–91). Boulder, CO: Westview Press.

Cohen, A. C., & Staten, R. T. (1993). Long-term culturing and quality assessment of predatory big-eyed bugs, *Geocoris punctipes*. In S. K. Narang, A. C. Bartlett, & R. M. Faust (Eds.), Applications of genetics to arthropods of biological control significance (pp. 235–254). Boca Raton, FL: CRC Press.

Cohen, A. C., & Urias, N. M. (1986). Meat-based artificial diets for *Geocoris punctipes* (Say). Southwestern Entomology, 11, 171–176.

Cohen, A. C., & Urias, N. M. (1988). Food utilization and egestion rates of the predator *Geocoris punctipes* (Hemiptera: Heteroptera) fed artificial diets with rutin. Journal of Entomological Science, 23, 174–179.

Collins, A. M. (1984). Artificial selection of desired characteristics in insects. In E. G. King & N. C. Leppla (Eds.), Advances and challenges in insect rearing. Washington, DC: USDA-ARS, U.S. Government Printing Office.

Congdon, B. D., & McMurtry, J. A. (1988). Prey selectivity in *Euseius tularensis* [Acari: Phytoseiidae]. Entomophaga, 33, 281–287.

Consoli, R. L., & Parra, J. R. P. (1995). Parasitism capacity of *Trichogramma galloi* Zucchi and *T. pretiosum* Riley reared *in vitro* and *in vivo*. In *Trichogramma* and other egg parasitoids. Colloques de l'INRA, 73, 149–154.

Consoli, R. L., & Parra, J. R. P. (1997). Development of an oligidic diet for *in vitro* rearing of *Trichogramma galloi* Zucchi and *Trichogramma pretiosum* Riley. Biological Control, 8, 172–176.

Consoli, F. L., Parra, J. R. P., Monteiro, R. C., & Zucchi, R. A. (1993). Avancos na criacao de *Trichogramma galloi* Zucchi, *in vitro*. 14th Brazilian Congress of Entomology, (p. 356). Piracicaba: SEB.

Constant, B., Grenier, S., & Bonnot, G. (1996). Artificial substrate for egg laying and embryonic development by the predatory bug *Macrolophus caliginosus*. (Heteroptera: Miridae). Biological Control, 7, 140–147.

Coombs, M. T. (1997). Influence of adult food deprivation and body size on fecundity and longevity of *Trichopoda giacomellii:* A South American parasitoid of *Nezara viridula*. Biological Control, 8, 119–123.

Cooperative Research Group of Hubei Province (CRGHP)China. (1979). Studies on the artificial host egg of the endoparasitoid wasp *Trichogramma*. Acta Entomologica Sinica, 22, 301–309.

Cooperative Research Group of Hubei Province (CRGHP) China. (1985). Study on artificial host egg-EII for *Trichogramma*. Journal of Wuhan University, 4, 1–10.

Coppel, H. C., House, H. L., & Maw, M. G. (1959). Studies on dipterous parasites of the spruce budworm, *Choristoneura fumiferana* (Clemens) (Lepidoptera: Tortricidae). VII. *Agria affinis* (Fall.) (Diptera: Sarcophagidae). Canadian Journal of Zoology, 37, 817–830.

Crampton, J. M. (1992). Potential applications of molecular biology in entomology, In J. M. Crampton & P. Eggleston (Eds.), Insect molecular science, (pp. 3–20). New York: Academic Press.

Cross, E. A., Mostafa, A. E.-S., Bauman, T. R., & Lancaster, I. J. (1978). Some aspects of energy transfer between the organ-pipe mud dauber *Trypoxylon politum* and its araneid spider prey. Environmental Entomology, 7, 647–752.

Dadd, R. H. (1973). Insect nutrition: Current developments and metabolic implications. Annual Review of Entomology, 18, 381–420.

Dadd, R. H. (1977). Qualitative requirements and utilization of nutrients: Insects. In M. Rechcigl (Ed.), Handbook series in nutrition and food: Vol. 1, Section D. nutritional requirements (pp. 305–346). Cleveland: CRC Press.

Dadd, R. H. (1985). Nutrition: Organisms. In G. A. Kerkut & L. I. Gilbert (Eds.), Comprehensive insect physiology, biochemistry and pharmacology (Vol. 4, pp. 313–390). New York: Pergamon Press.

Dahlman, D. L. (1990). Evaluation of teratocyte functions: An overview. Archives of Insect Biochemistry and Physiology, 13, 29–39.

Dahlman, D. L., & Vinson, S. B. (1993). Teratocytes: Developmental and biochemical charactistics. In N. E. Beckage, S. N. Thompson, & B. A. Federici (Eds.), Parasites and pathogens of insects (pp. 145–165). New York: Academic Press.

Dai, K. J. (1985). Study on artificial host egg-EII for *Trichogramma*. Journal of Wuhan University, 4, 1–10.

Dai, K. J. (1988). Research and utilization of artificial host egg for propagation of parasitoid *Trichogramma*. Colloques de l'IRNA, 43, 311–318.

Dai, K. J., Ma, Z. J., Zhang, L. W., Cao, A. H., Zhang, Q. X., & Xu, L. (1991). Research on technology of industrial production of the artificial host egg of *Trichogramma*. In *Trichogramma* and other egg parasitoids. Colloques de l'IRNA, 56, 137–139.

Dean, G. J., & Satasook, C. (1983). Response of *Chrysopa carnea* (Stephens) (Neuroptera: Chrysopidae) to some potential attractants. Bulletin of Entomological Research, 73, 619–624.

DeBach, P. (1943). The importance of host feeding by adult parasites in the reduction of host populations. Journal of Economic Entomology, 36, 647–658.

DeBach, P. (1954). Relative efficacy of the red scale parasites *Aphytis chrysomphali* Mercet and *Aphytis* "A" on citrus trees in southern California. Bullettino Lab. Zool. Gen. Agric. (Filippo Silvestri), Portici, 33, 135–151.

De Clercq, P., & Degheele, D. (1992). A meat-based diet for rearing the predatory stinkbugs *Podisus maculiventris* and *Podisus sagitta*. Entomophaga, 37, 194–195.

Dicke, M., Sabelis, M. W., Takabayashi, J., Bruin, J., & Posthumus, M. A. (1990). Plant strategies of manipulating predator-prey interactions through allelochemicals: Prospects for application in pest control. Journal of Chemical Ecology, 16, 3091–3118.

Dindo, M. L. (1987). Effetti indotti da parassitoidi imenotteri nei loro ospiti. Bollettino dell'Istituto di Entomologia "Guido Grandi" della Universita degli Studi di Bologna, 42, 1–46.

Dindo, M. L. (1990). Alcune osservazioni sulla biologia di *Brachymeria intermedia* (Nees) (Hym. Chalcididae *in vivo* e *in vitro*. Bollettino dell' Istituto di Entomologia "Guido Grandi" della Universita degli Studi di Bologna, 44, 221–232.

Dindo, M. L. (1995). Possibilities of culturing *Brachmeria intermedia* (Nees) (Hym Chalcididae), a solitary pupal gypsy moth parasitoid, on artificial diets. In P. Luciano (Ed.), Integrated protection in cork-oak forests (Vol. 18, pp. 95–99) Avignon, France: IOBC.

Dindo, M. L., & Campadelli, G. (1992). Preliminary studies on the artificial culture of *Brachymeria intermedia* (Nees) (Hym Chalcididae) on oligidic diets. Bollettino dell' Istituto di Entomologia "Guido Grandi" della Universita degli Studi di Bologna, 46, 93–99.

Dindo, M. L., Farneti, R., & Gardenghi, G. (1997a). Artificial culture of the pupal parasitoid *Brachymeria intermedia* (Nees) (Hymenoptera Chalcididae) on oligic diets. Boln. Asoc. Esp. Entomol., 21, 11–25.

Dindo, M. L. Gardenghi, G., & Grasso, M. (1995). Notes on the anatomy and histology of the female reproductive system of *Brachymeria intermedia* (Nees) (Hymenoptera Chalcididae) reared *in vivo* and *in vitro*. Bollettino dell' Istituto di Entomologia "Guido Grandi" della Universita degli Studi di Bologna, 50, 5–13.

Dindo, M. L., Sama, C., & Farneti, R. (1994). Allevamento in vitro di unm endoparassitoide pupale, Brachymeria intermedia (Nees) (Hymenoptera: Charlcididae) (pp. 639–641). Atti XVII Congresso Nazionale Italiana, Entomologia. Udine, Giugno 13–18, 1994.

Dindo, M. L., Sama, C., Fanti, P., & Farneti, R. (1997b). *In vitro* rearing of the pupal parasitoid *Brachymeria intermedia* (Hym: Chalcididae) on artificial diets with and without host components. Entomophaga, 42, 415–423.

Ding, D. C., Qiu, H. G., & Hwang, C. B. (1980a). In vitro rearing of an egg-parasitoid Tetrastichus schoenobii (Hymenoptera: Tetrastichidae) (pp. 55–58). Shanghai: Contr. Shanghai Institute of Entomology.

Ding, D. C. Zhang, T. P., & Zhong, Y. K. (1980b). Studies on Tetrastichus schoenobii Ferriere (Hymenoptera: Tetrastichidae): Oviposition into artificial media (pp. 59–61). Shanghai: Contr. Shanghai Institute of Entomolgy.

Doutt, R. L. (1959). The biology of parasitic Hymenoptera. Annual Review of Entomology, 4, 161–182.

Doutt, R. L. (1964). Biological characteristics of entomophagous adults. In P. DeBach (Ed.), Biological control of insects, pests and weeds (pp. 145–167). London: Chapman & Hall.

Dowell, R. (1978). Ovary structure and reproductive biologies of larval parasitoids of the alfalfa weevil (Coleoptera: Curculionidae). Canadian Entomologist, 110, 507–512.

Duelli, P. (1980). Adaptive dispersal and appetitive flight in the green lacewing, Chrysopa carnea. Ecology and Entomology, 5, 213–220.

Duelli, P. (1987). The influence of food on the oviposition site selection in a predatory and a honeydew-feeding lacewing species (Planipennia: Chrysopidae). Neuroptera International, 4, 205–210.

Dunbar, D. M., & Bacon, O. G. (1972). Feeding, development and production of Geocoris punctipes (Heteroptera: Lygaeidae) on eight diets. Annals of Entomological Society of America, 65, 892–895.

Dwumfour, E. F. (1992). Volatile substances evoking orientation in the predatory flowerbug Anthocoris nemorum (Heteroptera: Anthocoridae). Bulletin of Entomological Research, 82, 465–469.

Edgar, W. D. (1971). Aspects of the ecology and energetics of the egg sac parasites of the wolf spider Pardosa lugubris (Walckenaer). Oecologia (Berlin), 7, 155–163.

Elliott, N. C., French, B. W., Michels, G. J., & Reed, D. K. (1994). Influence of four apid prey species on development, survival and adult size of Cycloneda ancoralis. Southwestern Entomology, 19, 57–61.

Evans, H. F. (1976). The effect of prey density and host plant characteristics on oviposition and fertility in Anthocoris confusus (Reuter). Ecological Entomology, 1, 157–161.

Ewert, M. A., & Chiang, H. C. (1966). Dispersal of three species of coccinellid in corn fields. Canadian Entomologist, 98, 999–1003.

Fanti, P. (1990). Fattori ormonali inducenti la prima muta laravale del parassitoide Pseudogonia rufifrons Wied. (Diptera: Tachinidae) in substrati di crescita in vivo e in vitro. Bollettino dell' Istituto di Entomologia "Guido Grandi" della Universita degli Studi di Bologna, 45, 47–59.

Fanti, P., & Bratti, A. (1991). In vitro rearing of the larval stages of the parasitoid Pseudogonia rufifrons Wied. (Diptera: Tachinidae): Preliminary results. Redia, 74, 449–452.

Farneti, R., Dindo, M. L., & Cristiani, G., (1997). In vitro rearing of the larval-pupal parasitoid Archytas marmoratus (Townsend) (Diptera: Tachinidae) on oligidic diets: Preliminary results. Bollettino dell' Istituto di Entomologia "Guido Grandi" della Universita degli Studi di Bologna, 51, 53–61.

Ferkovich, S. M., & Oberlander, H. (1991a). Growth factors in invertebrate in vitro culture. In Vitro Cellular and Developmental Biology, 27A: 483–486.

Ferkovich, S. M., & Oberlander, H. (1991b). Stimulation of endoparasitoid egg development by a fat body cell line: Activity and characterization of factors that induce germ ban formation and hatching (pp. 181–187). Proceedings, Eighth International Conference on Invertebrate and Fish Tissue Culture, Columbia, MO.

Ferkovich, S. M., Dillard, C., & Oberlander, H. (1991). Stimulation of embryonic development in Microplitis croceipes (Braconidae) in cell culture media preconditioned with a fat body cell line derived from a nonpermissive host, gypsy moth, Lymantria dispar. Archives of Insect Biochemistry and Physiology, 18, 169–175.

Finnegan, D. J. (1992). Transposable elements and their biological consequences in Drosophila and other insects. In J. M. Crampton and P. Eggleston (Eds.), Insect molecular science (pp. 35–48) New York: Academic Press.

Fisher, R. C. (1968). Conversion studies in parasitic Hymenoptera. Proceedings, 13th International Congress of Entomology, 1, 376–377.

Fisher, R. C. (1971). Aspects of the physiology of endoparasitic Hymenoptera. Biological Reviews, 46, 243–278.

Fisher, R. C. (1981). Efficiency of parasites in assimilating host tissues into their own. Parasitology, 82, 33–34.

Fisher, R. C. (1986). George Salt and the development of experimental insect parasitology. Journal of Insect Physiology, 32, 249–253.

Flanders, S. E. (1935). Host influence on the prolificacy and size of Trichogramma. Pan-Pacific Entomologis, 11, 175–177.

Flanders, S. E. (1950). Regulation of ovulation and egg disposal in the parasitic Hymenoptera. Canadian Entomologist, 82, 134–140.

Flint, H. M., Salter, S. S., & Walters, S. (1979). Caryophyllene: An attractant for the green lacewing. Environmental Entomology, 8, 1123–1125.

Foster, M. A., & Ruesink, W. G. (1984). Influence of flowering weeds associated with reduced tillage in corn on a black cutworm (Lepidoptera: Noctuidae) parasitoid, Meteorus rubens (Nees von Esenback). Environmental Entomology, 13, 664–668.

Friend, W. G., & Dadd, R. H. (1982). Insect nutrition—a comparative perspective, In H. H. Draper (Ed.), Advances in nutritional research (Vol. 4, pp. 205–247). New York: Plenum.

Gao, Y. G., Dai, K. J., Shong, L. C., et al. (1982). Studies on the artificial host egg for Trichogramma, In INRA Publ. (Ed.), Les Trichogrammes Antibes (France), April 20–23, 1982. Colloques de l'INRA, 9, 181.

Godfray, H. C. J. (1994). Parasitoids: Behavioral and evolutionary ecology. Princeton, NJ: Princeton University Press.

Grafton-Cardwell, E. E., & Ouyang, Y. (1996). Influence of citrus leaf nutrition on survivorship, sex ratio, and reproduction of Euseius tularensis (Acari: Phytoseiidae). Environmental, Entomology, 25, 1020–1025.

Greany, P. (1980). Growth and development of an insect parasitoid in vitro. American Zoologist, 20, 946.

Greany, P. (1981). Culture of hymenopteran endoparasites in vitro. In Vitro, 17, 230.

Greany, P. (1986). In vitro culture of hymenopterous larval endoparasitoids. Journal of Insect Physiology, 32, 409–419.

Greany, P., & Carpenter, J. E. (1996). Culture medium for parasitic and predaceous insects. U.S. Patent 08, 692, 565 (docket no. 0010.96. pending).

Greany, P., & Hagen, K. S. (1981). Prey selection. In E. A. Nordlund, R. L. Jones, & W. J. Lewis (Eds.), Semiochemicals: Their role in pest control (pp. 121–135). New York: John Wiley & Sons.

Greany, P., Vinson, S. B., & Lewis, W. J. (1984). Insect parasitoids: Finding new opportunities for biological control. Bioscience, 34, 690–696.

Greany, P. D., Clark, W. R., Ferkovich, S. M., Law, J. H. & Ryan, R. O. (1990). Isolation and characterization of a host hemolymph protein required for development of the eggs of the endoparasite, Microplitis croceipes. In H. H. Hagedorn, J. G. Hildebrand, M. G. Kidwell, & J. H. Law (Eds.), Molecular insect science (p. 306). New York: Plenum Press.

Greathead, D. J. (1986). Parasitoids in classical biological control. In J. K. Waage & D. J. Greathead (Eds.), Insect parasitoids, 13th Symposium of the Royal Entomological Society of London (pp. 289–318). London: Academic Press.

Greenblatt, J. A., & Lewis, W. J. (1983). Chemical environmental manipulation for pest insect control. Environmental Management, 7, 35–41.

Greenblatt, J. A., Barbosa, P., & Montgomery, M. E. (1982). Host's diet effects on nitrogen utilization efficiency for two parasitoid species: Brachymeria intermedia and Coccygomimus turionellae. Physiology and Entomology, 7, 263–267.

Grenier, S. (1979). Developpement embryonnaire in vitro, en milieu artificiel defini de deux parasitoides ovolarvipares, Phryxe caudata et Lixophaga diatraeae (Diptera, Tachinidae). Entomologia Experimentalis et Applicata, 26, 13–23.

Grenier, S. (1988). Developmental relationships between the tachinid parasitoid *Pseudoperichaeta nigrolineata* and two host species—hormonal implications. Colloques de l'INRA, 48, 87–89.

Grenier, S. (1994). Rearing of *Trichogramma* and other egg parasitoids on artificial diets. In E. Wajnberg & S. A. Hassan, Biological control with egg parasitoids (pp. 73–92). Wallingford, United Kingdom: CAB International.

Grenier, S., & Bonnot, G. (1988). Development of *Trichogramma dendrolimi* and *T. maidis* (Hymenoptera: Trichogrammatidae) in artificial media and artificial host eggs—in *Trichogramma* and other egg parasites. Colloques l'INRA, 43, 319–326.

Grenier, S., & Liu, W. H. (1990). Antifungals: Mold control and safe levels in artificial media for *Trichogramma* (Hymenoptera, Trichogrammatidae). Entomophaga, 35, 283–291.

Grenier, S., Bonnot, G., & Delobel, B. (1974). Définition et mise au point de milieux artificiels pour l'élevage *in vitro* de *Phryxe caudata* Rond. (Diptera, Tachinidae). I. Survie du parasitoïde sur milieux dont la composition est basée sur celle de l'hémolymphe de l'hôte. Annales de Zoologie—Ecologie Animale, 6, 511–520.

Grenier, S., Bonnot, G., & Delobel, B. (1975). Définition et mise au point de milieux artificiels pour l'élevage *in vitro* de *Phryxe caudata* Rond. (Diptera, Tachinidae). II. Croissance et mues larvaires du parasitoïde en milieux définis. Annales de Zoologie—Ecologie, Animale, 7, 13–25.

Grenier, S., Bonnot, G., & Delobel, B. (1986). Physiological considerations of importance to the success of *in vitro* culture: An overview. Journal of Insect Physiology, 32, 403–408.

Grenier, S., Greany, P. D., & Cohen, A. C. (1994). Potential for mass release of insect parasitoids and predators through development of artificial culture techniques. In D. Rosen, F. D. Bennett, & J. L. Capinera (Eds.), Pest management in the subtropics: Biological control—A Florida perspective (pp. 181–205). Andover: Intercept.

Grenier, S., Veith, V., & Renou, M. (1993). some factors stimulating oviposition by the oophagous parasitoid *Trichogramma brassicae* bezd. (Hym., Trichogrammatidae) in artificial host eggs. Journal of Applied Entomology, 115, 66–76.

Grenier, S., Bonnot, G., Delobel, B., & Laviolette, P. (1978). Développement en milieu artificiel du parasitoïde *Lixophaga diatraeae* (Towns.) (Diptera, Tachinidae). Obtention de l'imago à partir de l'ouef. Comptes Rendu de l'academie Sciences Paris, Series D, 287, 535–538.

Grenier, S., Guillaud, J., Delobel, B., & Bonnot, G. (1989). Nutrition et elevage du predateur polyphage *Macrolophus caliginosus* (Heteroptera, Miridae) sur milieu artificiels. Entomophaga, 34, 77–86.

Grenier, S., Yang, H., Guillaud, J., & Chapelle, L. (1995). Comparative development and biochemical analyses of *Trichogramma* (Hymenoptera: Trichogrammatidae) grown in artificial media with hemolymph or devoid of insect components. Comparative Biochemistry and Physiology [Part] B, 111, 83–90.

Gross, H. R. (1987). Conservation and enhancement of entomophagous insects—a perspective. Journal of Entomological Science, 22, 97–105.

Grout, T. G., & Richards, G. I. (1990). The influence of windbreak species on citrus thrips (Thysanoptera: Thripidae) populations and their damage to South African citrus orchards. Journal of the Entomological Society of South Africa, 53, 151–157.

Guan, X., Wu, Z., Wu, T., & Feng, H. (1978). Studies on rearing *Trichogramma dendrolimi in vitro*. Acta Entomologica Sinica, 21, 221–226.

Guerra, A. A. (1992). *In vitro* rearing of *Bracon mellitor* and *Catolaccus grandis* with different insect hemolymph-based diets. Southwestern Entomology, 17, 123–126.

Guerra, A. A., Martinez, S., & Del Rio, H. S. (1994). Natural and synthetic oviposition stimulants for *Catolaccus grandis* (Burks) females. Journal of Chemical Ecology, 20, 1583–1594.

Guerra, A. A., Robacker, K. M., & Martinez, S. (1993). *In vitro* rearing of *Bracon mellitor* and *Catolaccus grandis* with artificial diets devoid of insect components. Entomologia Experimentalis et Applicata, 68, 303–307.

Hagen, K. S. (1962). Biology and ecology of predaceous Coccinellidae. Annual Review of Entomology, 7, 289–326.

Hagen, K. S. (1964). Nutrition of entomophagous insects and their hosts. In P. DeBach (Ed.), Biological control of insect pests and weeds (pp. 356–380). London: Chapman & Hall.

Hagen, K. S. (1986a). Ecosystem analysis: Plant cultivars (HPR), entomophagous species and food supplements. In D. J. Boethel & R. D. Eikenbary (Eds.), Interactions of plant resistance and parasitoids and predators of insects (pp. 151–197). New York: Halsted Press.

Hagen, K. S. (1986b). Dietary requirements for mass rearing of natural enemies for use in biological control. In M. Y. Hussein & A. G. Ibrahim (eds.), Biological control in the tropics: Proceedings of the first regional symposium on biological control (pp. 35–85). Malaysia: Penerbit Universiti Pertanian.

Hagen, K. S. (1987). Nutritional ecology of terrestrial predators. In F. Slansky & J. G. Rodriguez (Eds.), Nutritional ecology of insects, mites, spiders, and related invertebrates (pp. 533–577). New York: John Wiley & Sons.

Hagen, K. S., & Bishop, G. W. (1979). Use of supplemental foods and behavioral chemicals to increase the effectiveness of natural enemies. In D. W. Davis, S. C. Hoyt, J. A. McMurtry, & M. T. Ali Niazee (Eds.), Biological control and insect pest management. (Agricultural Science Publication 4096, (pp. 49–60). Berkeley: University of California.

Hagen, K. S., & Hale, R. (1974). Increasing natural enemies through use of supplementary feeding and non-target prey. In F. G. Maxwell & F. A. Harris (Eds.), Proceedings of the summer institute on biological control of plant insects and diseases (pp. 170–181). Mississippi: University Press.

Hagen, K. S., & Sawall, E. F. (1970). Some ecophysiological relationships between certain *Chrysopa*, honeydews and yeasts. Bollettino Laboratori Entomologia Agricultura Portici, 28, 113–134.

Hagen, K. S., & Tassan, R. L. (1965). A method of providing artificial diets to *Chrysopa* larvae. Journal of Economic Entomology, 58, 999–1000.

Hagen, K. S., Tassan, R. L. (1970). The influence of food Wheast and related *Saccharomyces fragilis* yeast products on the fecundity of *Chrysopa carnea* (Neuroptera: Chrysopidae). Canadian Entomologist, 102, 806–811.

Hagen, K. S., & Tassan, R. L. (1972). Exploring nutritional roles of extracellular symbiotes on the reproduction of honeydew feeding adult chryosopids and tephritids. In J. G. Rodriguez (Ed.), Insect and mite nutrition (pp. 323–351). Amsterdam: North-Holland.

Hagen, K. S., Dadd, R. H., & Reese, J. (1984). The food of insects. (pp. 79–112). In C. B. Huffaker & R. L. Rabb (Eds.), Ecological entomology. New York: John Wiley & Sons.

Hagen, K. S., Sawall, E. F., Jr., & Tassan, R. L. (1971). The use of food sprays to increase effectiveness of entomophagous insects. Proceedings, Tall Timbers Conference on Ecological Animal Control by Habitat Management, 2, 59–81.

Hagen, K. S., Greany, P., Sawall, E. F., Jr., & Tassan, R. L. (1976). Tryptophan in artificial honeydews as a source of an attractant for adult *Chrysopa carnea*. Environmental Entomology, 5, 458–468.

Hagler, J. R., & Cohen, A. C. (1991). Prey selection by *in vitro*- and field-reared *Geocoris punctipes*. Entomologia Experimentalis et Applicata, 59, 201–205.

Hagley, E. A. C., & Simpson, C. M. (1981). Effect of food sprays on numbers of predators in an apple orchard. Canadian Entomologist, 113, 75–77.

Han, S. (1988). *In vitro* rearing *Anastatus japonicus* Ashmead (Hym., Eupelmidae) for controlling litchi stink bug, *Tessaratoma papillosa*

Drury (Hemiptera, Pentatomidae). Natural Enemy Insects, 10, 170–173.

Hare, D. J. (1992). Effects of plant variation on herbivore-natural enemy interactions. In R. S. Fritz & E. L. L. Simms (Eds.), Plant resistance to herbivores and pathogens: Ecology, evolution, and genetics (pp. 278–298). Chicago: University of Chicago Press.

Harvey, J. A., Harvey, I. F., & Thompson, D. J. (1994). Flexible larval growth allows use of a range of host sizes by a parasitoid wasp. Ecology, 75, 1420–1428.

Harvey, J. A., Harvey, I. F., & Thompson, D. J. (1995). The effect of host nutrition on growth and development of the parasitoid wasp Venturia canescens. Entomologia Experimentalis et Applicata, 75, 213–220.

Hasegawa, M., Niijima, K., & Matsuka, M. (1989). Rearing Chrysoperla carnea (Neuroptera: Chrysopidae) on chemically defined diets. Applied Entomology and Zoology, 24, 96–102.

Hassan, S. A., & Hagen, K. S. (1978). A new artificial diet for rearing Chrysopa carnea larvae (Neuroptera, Chrysopidae). Zeitschrift fuer Angewandte Entomologie, 86, 315–320.

Hawkes, O. A. M. (1920). Observations on the life-history, biology, and genetics of the lady-bird beetle, Adalia bipunctata (Linnaeus). Proceedings of the Zoological Society (London), 29, 475–490.

Hawkins, B.A., & Smith, J. W., (1986). Rhaconotus roslinensis (Hymenoptera: Braconidae), a candidate for biological control of stalkboring sugarcane pests (Lepidoptera: Pyralidae): Development, life tables, and intraspecific competition. Annals of the Entomological Society of America, 79, 905–911.

Hawlitzky, W., & Mainguet, A. M. (1976). Analyse quantitative des lipides, des substances azotées et du glycogène chez la larve d'un insecte parasite ovolarvaire Phanerotoma flavitestacea (Hymenoptera: Braconidae). Entomologia Experimentalis et Applicata, 20, 43–55.

Heidari, M., & Copland, M. J. W. (1993). Honeydew: A food resource or arrestant for the mealybug predator Cryptolaemus montrouzieri. Entomophaga, 38, 63–68.

Heilmann, L. J., DeVault, J. D., Leopold, R. L., & Narang, S. K. (1993). Improvement of natural enemies for biological control: A genetic engineering approach. In S. K. Narang, A. C. Bartlett, & R. M. Faust (Eds.), Applications of genetics to arthropods of biological control significance (pp. 167–189). Boca Raton, FL: CRC Press.

Heimpel, G. E., & Rosenheim, J. A. (1995). Dynamic host feeding by the parasitoid Aphytis melinus: The balance between current and future reproduction. Journal of Animal Ecology, 64, 153–167.

Heimpel, G. E., & Rosenheim, J. A. (1996). Egg limitation, host quality, and dynamic behavior by a parasitoid in the field. Ecology, 77, 2410–2420.

Heimpel, G. E., Rosenheim, J. A., & Kattari, D. (1997). Adult feeding and lifetime reproductive success in the parasitoid Aphytis melinus. Entomologia Experimentalis et Applicata, 83, 305–315.

Heimpel, G. E., & Collier, T. R. (1996). The evolution of host-feeding behaviour in insect parasitoids. Biological Reviews, 71, 373–400.

Hemptinne, J. L., & Desprets, A. (1986). Pollen as a spring food for Adalia bipunctata. In I. Hodele (Ed.), Ecology of Apidophaga (pp. 29–35). Prague: Academia.

Herard, F. M., Keller, A., Lewis, W. J., & Tumlinson, J. H. (1988). Beneficial arthropod behavior mediated by airborne semiochemicals. IV. Influence of host diet on host-oriented flight chamber responses of Microplitis demolitor Wilkinson. Journal of Chemical Ecology, 14, 1597–1606.

Herzog, D. C., & Funderburk, J. E. (1985). Plant resistance and cultural practice interactions with biological control. In M. A. Hoy & C. C. Hertzog (Eds.), Biological control in agricultural IPM systems (pp. 67–88). Princeton, NJ: Princeton University Press.

Hessein, N A., & Perring, T. M. (1988). The importance of alternate foods for the mite Homeopronematus anconae (Acari: Tydeidae). Annals of the Entomological Society of America, 81, 488–492.

Hocking, H. (1967). The influence of food on longevity and oviposition in Rhyssa persuasoria (Linnaeus) (Hymenoptera: Ichneumonidae). Journal of Australian Entomological Society, 6, 83–88.

Hodek, I. (1973). Biology of Coccinellidae. Prague: W. Junk, Academia.

Hodek, I. (1996). Ecology of Coccinellidae. Dordrecht: Kluwer.

Hoffman, J. D., & Ignoffo, C. M. (1974). Growth of Pteromalus puparum in a semisynthetic medium. Annals of the Entomological Society of America, 67, 524–525.

Hoffman, J. D., Ignoffo, C. M., & Dickerson, W. A. (1975). In vitro rearing of the endoparasitic wasp, Trichogramma pretiosum. Annals of the Entomological Society of America, 68, 335–336.

Hoffman, J. D., Ignoffo, C. M., & Long, S. H. (1973). In vitro cultivation of an endoparasitic wasp, Pteromalus puparium. Annals of the Entomological Society of America, 66, 633–634.

Hohmann, C. L., Luck, R. F., & Oatman, E. R. (1988). A comparison of longevity and fecundity of adult Trichogramma platneri (Hymenoptera: Trichogrammatidae) reared from the eggs of the cabbage looper and the angoumois grain moth, with and without access to honey. Journal of Economic Entomology, 81, 1307–1312.

Hopper, K. R., Roush, R. T., & Powell, W. (1993). Management of genetics of biological-control introductions. Annual Review of Entomology, 38, 27–51.

House, H. L. (1954). Nutritional studies with Pseudosarcophaga affinis (Fall.), a dipterous parasite of the spruce budworm, Choristoneura fumiferana (Clemens). I. A chemically defined medium and aseptic-culture technique. Canadian Journal of Zoology, 32, 331–341.

House, H. L. (1966a). The role of nutritional principles in biological control. Canadian Entomologist, 98, 1121–1134.

House, H. L. (1966b). Effects of varying the ratio between the amino acids and other nutrients in conjunction with a salt mixture on the fly Agria affinis (Fall.). Journal of Insect Physiology, 12, 299–310.

House, H. L. (1966c). Effects of vitamins E and A on growth and development, and the necessity of vitamin E for reproduction in the parasitoid Agria affinis (Fallen) (Diptera: Sarcophagidae). Journal of Insect Physiology, 12, 409–417.

House, H. L. (1967). The role of nutritional factors in food selection and preference as related to larval nutrition of an insect, Pseudosarcophaga affinis (Diptera: Sarcophagidae), on synthetic diets. Canadian Entomologist, 99, 1310–1321.

House, H. L. (1969). Effects of different proportions of nutrients on insects. Entomologia Experimentalis et Applicata, 12, 651–669.

House, H. L. (1970). Choice of food by larvae of the fly, Agria affinis, related to dietary proportions of nutrients. Journal of Insect Physiology, 16, 2041–2050.

House, H. L. (1972). Inversion in the order of food superiority between temperatures affected by nutrient balance in the fly larva Agria housei (Diptera: Sarcophagidae). Canadian Entomologist, 104, 1559–1564.

House, H. L. (1974). Nutrition. In M. Rockstein (Ed.), The physiology of Insecta (Vol. 5, pp. 1–62). New York: Academic Press.

House, H. L. (1977). Nutrition of natural enemies. In R. L. Ridgway & S. B. Vinson (Eds.), Biological control by augmentation of natural enemies (pp. 151–182). New York: Plenum.

House, H. L. (1978). An artificial host: Encapsulated synthetic medium for in vitro oviposition and rearing the endoparasitoid Itoplectis conquisitor (Hymenoptera: Ichneumonidae). Canadian Entomologist, 110, 331–333.

House, H. L., & Barlow, J. S. (1960). Effects of oleic and other fatty acids on the growth rate of Agria affinis (Fall.) (Diptera: Sarcophagidae). Journal of Nutrition, 72, 409–414.

House, H. L., & Barlow, J. S. (1961). Effects of different diets of a host, Agria affinis (Fall.) (Diptera: Sarcophagidae), on the development of a parasitoid, Aphaereta pallipes (Say) (Hymenoptera: Braconidae). Canadian Entomologist, 93, 1041–1044.

House, H. L., & Traer, M. G. (1948). An artificial food for rearing

Pseudosarcophaga affinis (Fall.), a parasite of the spruce budworm *Choristoneura fumiferana* (Clemens) (79th Annual Report, (pp. 1–4). Entomological Society of Ontario.

Houston, A. I., McNamara, J. M., & Godfray, H. C. J. (1991). The effect of variability on host feeding and reproductive success in parasitoids. Bulletin of Mathematical Biology, 54, 465–478.

Howell, J., & Fisher, R. C. (1977). Food conversion efficiency of a parasitic wasp, *Nemeritis canescens*. Ecology and Entomology, 2, 143–151.

Hoy, M. A. (1976). Genetic improvement of insects: Fact or fantasy. Environmental Entomology, 5, 833–839.

Hoy, M. A. (1979). The potential for genetic improvement of predators for pest management programs. In M. A. Hoy & J. J. McKelvey (Eds.), Genetics in relation to insect management (Working papers, pp. 106–115). New York: The Rockefeller Foundation.

Hoy, M. A. (1986). Use of genetic improvement in biological control. Agriculture Ecosystems and Environment, 15, 109–119.

Hoy, M. A. (1990). Genetic improvement of arthropod natural enemies: Becoming a conventional tactic? In R. Baker & P. Dunn (Eds.), New directions in biological control: UCLA Symposium on Molecular and Cellular Biology (Vol. 112, pp. 405–417). New York: Alan R. Liss.

Hoy, M. A. (1992). Criteria for release of genetically-improved phytoseiids: An examination of the risks associated with release of biological control agents. Experimental and Applied Acarology (Northwood), 14, 393–416.

Hoy, M. A. (1993). Transgenic beneficial arthropods for pest management programs: An assessment of their practicality and risks. In R. D. Lumsden & J. Vaughn (Eds.), Pest management: Biologically based technologies (pp. 357–369). Washington, DC: American Chemical Society.

Hoy, M. A. (1995). Impact of risk analyses on pest-management programs employing transgenic arthropods. Parasitology Today, 11, 229–232.

Hu, J. S., & Vinson, S. B. (1997a). *In vitro* rearing of *Campoletis sonorensis* (Hymenoptera: Ichneumonidae), a larval endoparasitoid of *Heliothis virescens* (Lepidoptera: Noctuidae) from egg to third instar in an artificial medium devoid of insect sources. Entomologia Experimentalis et Applicata, 85, 263–273.

Hu, J. S., & Vinson, S. B. (1997b). *In vitro* development of *Campoletis sonorensis* [Hym:Ichneumonidae], a larval endoparasitoid of *Heliothis virescens* [Lep: Noctuidae] in an artificial medium with insect sources from egg to third larval instar. Entomophaga, 42, 375–385.

Hu, J. S., & Vinson, S. B. (1998). The *in vitro* development from egg to prepupa of *Campoletis sonorensis* (Hymenoptera: Ichneumonidae) in an artificial medium: Importance of physical factors. Journal of Insect Physiology, 44, 455–462.

Huffaker, C. B., & Gutierrez, A. P. (1990). Evaluation of efficiency of natural enemies in biological control. In D. Rosen (Ed.); World crop pests: Armoured scale insects—their biology, natural enemies and control (pp. 473–495). Amsterdam: Elsevier.

Hydorn, S. B., & Whitcomb, W. H. (1979). Effects of larval diet on *Chrysopa rufilabris*. Florida Entomologist, 62, 293–298.

Idris, A. B., & Grafius, E. (1995). Wildflowers as nectar sources for *Diadegma insulare* (Hymenoptera: Ichneumonidae), a parasitoid of diamondback moth (Lepidoptera: Yponomeutidae). Biological Control, 24, 1726–1735.

Idoine, K., & Ferro, D. N. (1988). Aphid honeydew as a carbohydrate source for *Edovum puttleri* (Hymenoptera: Eulophidae). Environmental Entomology, 17, 941–944.

Irie, K., Xie, Z.-N., Nettles, W. C., Jr., Morrison, R. K., Chen, A. C., & Holman, G. M. (1987). The partial purification of a *Trichogramma pretiosum* pupation factor from hemolymph of *Manduca sexta*. Insect Biochemistry, 17, 269–275.

Iwabuchi, K. (1995). Effect of juvenile hormone on the embryogenesis of a polyembryonic wasp, *Copidosoma floridanum, in vitro*. In Vitro Cellular and Developmental Biology, 31, 803–805.

Iwabuchi, K. (1996). *In vitro* culture of polyembryonic wasps. Proceedings, Ninth International Conference of Invertebrate Cell Culture, (pp. 127–133). Society for *In vitro* Biology.

Jermy, T. (1990). Prospects of antifeedant approach to pest control—a critical review. Journal of Chemical Ecology, 16, 3151–3166.

Jervis, M. A., & Kidd, A. C. (1986). Host-feeding strategies in hymenopteran parasitoids. Biological Reviews, 61, 395–434.

Jervis, M. A., & Kidd, A. C. (1991). The dynamic significance of host-feeding by insect parasitoids—what modellers ought to consider. Oikos, 62, 97–99.

Jervis, M. A., Kidd, N. A. C., & Walton, M. (1992). A review of methods for determining dietary range in adult parasitoids. Entomophaga, 37, 565–574.

Jervis, M. A., Kidd, N. A. C., Fitton, M. G., Huddleston, T., & Dawah, H. A. (1993). Flower visiting by hymenopteran parasitoids. Journal of Natural History, 27, 67–105.

Jervis, M. A., Hawkins, B. A., & Kidd, A. C. (1996). The usefulness of destructive host feeding parasitoids in classical biological control: Theory and observation conflict. Ecology and Entomology, 21, 41–46.

Jones, R. L. (1981). Chemistry of semiochemicals involved in parasitoid–host and predator–prey relationships. In D. A. Nordlund, R. L. Jones, & W. J. Lewis. (Eds.), Semiochemicals: Their role in pest control (pp. 239–250). New York: John Wiley & Sons.

Jowyk, E. A., & Smilowitz, Z. (1978). A comparison of growth and developmental rates of the parasite *Hyposoter exiguae* reared from two instars of its host, *Trichoplusia ni*. Annals of the Entomological Society of America, 71, 467–472.

Kaiser, L., Pham-Delegue, H., Backchine, E., & Masson, C. (1989). Olfactory responses of *Trichogramma maidis* Pint. et Voeg.: Effects of chemical cues and behavioral plasticity. Journal of Insect Behavior, 2, 701–712.

Kamburov, S. S. (1971). Feeding, development and reproduction of *Amblyseius largoensis* on various food substances. Journal of Economic Entomology, 64, 643–648.

Kazmer, D. J., & Luck, R. F. (1995). Field tests of the size-fitness hypothesis in the egg parasitoid *Trichogramma pretiosum*. Ecology, 76, 412–425.

Kariluoto, K. T. (1978). Optimum levels of sorbic acid and methyl-*p*-hydroxy-benzoate in an artificial diet for *Adalia bipunctata* (Coleoptera, Coccinellidae) larvae. Annales Entomologici Fennici, 44, 94–97.

Kariluoto, K. T. (1980). Survival and fecundity of *Adalia bipunctata* (Coleoptera, Coccinellidae) and some other predatory insect species on an artificial diet and a natural prey. Annales Entomologici Fennici, 46, 101–106.

Kariluoto, K. T., Junnikkala, E., & Markkula, M. (1976). Attempts at rearing *Adalia bipunctata* L. (Col., Coccinellidae) on different artificial diets. Annales Entomologici, Fennici, 42, 91–97.

Keilin, D. (1944). Respiratory systems and respiratory adaptations in larvae and pupae of Diptera. Parasitology, 36, 1–66.

Kennett, C. E., & Hamai, J. (1980). Oviposition and development in predaceous mites fed with artificial and natural diets (Acari: Phytoseiidae). Entomologia Experimentalis et Applicata, 28, 116–122.

Kennett, C. E., Flaherty, D. L., & Hoffman, R. W. (1979). Effect of windborne pollens in the population dynamics of *Amblyseius hibisci* (Acarina: Phytoseiidae). Entomophaga, 24, 83–98.

Kidd, N. A. C., & Jervis, M. A. (1989). The effects of host-feeding behaviour on the dynamics of parasitoid-host interactions, and the implications for biological control. Researches on Population Ecology (Kyoto), 31, 235–274.

Kidd, N. A. C., & Jervis, M. A. (1991a). Host-feeding and oviposition strategies of parasitoids in relation to host stage. Researches on Population Ecology (Kyoto), 33, 13–28.

Kidd, N. A. C., & Jervis, M. A. (1991b). Host-feeding and oviposition by parasitoids in relation to host stage: Consequences for parasitoid-host population dynamics. Researches on Population Ecology (Kyoto), 33, 87–99.

King, B. H. (1987). Offspring sex ratios in parasitoid wasps. Quarterly Review of Biology, 62, 367–396.

King, B. H. (1989). Host-size-dependent sex ratios among parasitoid wasps: Does host growth matter? Oecologia (Berlin), 78, 420–426.

Lawrence, P. O. (1982). Biosteres longicaudatus: Developmental dependence on host (Anastrepha suspensa) physiology. Experimental Parasitology, 53, 396–405.

Lawrence, P. O. (1986a). Host-parasite hormonal interactions: An overview. Journal of Insect Physiology, 32, 295–298.

Lawrence, P. O. (1986b). The role of 20-hydroxyecdysone in the moulting of an endoparasitoid, Biosteres longicaudatus. Journal of Insect Physiology, 32, 329–337.

Lawrence, P. O. (1990). The biochemical and physiological effects of insect hosts on the development and ecology of their insect parasites—an overview. Archives of Insect Biochemistry and Physiology, 13, 217–228.

Lawrence, P. O. (1991). Hormonal effects on insects and other endoparasites in vitro. In Vitro Cellular and Developmental Biology, Part A, 27, 487–496.

Lawrence, P. O., & Lanzrein, B. (1993). Hormonal interactions between insect endoparasites and their host insects. In N. E. Beckage, S. N. Thompson, & B. A. Federici (Eds.), Parasites and pathogens of insects (Vol. 1, pp. 59–86). New York: Academic Press.

Lawton, J. H. (1971). Ecological energetics studies on larvae of the damselfly Pyrrhosoma nymphula (Sulzer) (Odonata: Zygoptera). Journal of Animal Ecology, 40, 385–423.

Legner, E. F. (1979). The relationship between host destruction and parasite reproductive potential in Muscidifurax raptor, M. zaraptor and Spalangia endius (Chalcidoidea: Pteromalidae). Entomophaga, 24, 145–152.

Leius, K. (1960). Attractiveness of different foods and flowers to the adults of some hymenopterous parasites. Canadian Entomologist, 92, 369–376.

Leius, K. (1961a). Influence of various foods on fecundity and longevity of adults of Scambus buolianae (Hartig) (Hymenoptera: Ichneumonidae). Canadian Entomologist, 93, 1079–1084.

Leius, K. (1961b). Influence of food on fecundity and longevity of adults of Itoplectis conquisitor (Say) (Hymenoptera: Ichneumonidae). Canadian Entomologist, 93, 771–780.

Leius, K. (1962). Effects of body fluids of various host larvae on fecundity of females of Scambus buolianae (Hartig) (Hymenoptera: Ichneumonidae). Canadian Entomologist, 94, 1078–1082.

Leius, K. (1963). Effects of pollen on fecundity and longevity of adult Scambus buolianae (Hartig) (Hymenoptera: Ichneumonidae). Canadian Entomologist, 95, 202–207.

Leius, K. (1967a). Influence of wild flowers on parasitism of tent caterpillar and codling moth. Canadian Entomologist, 99, 444–446.

Leius, K. (1967b). Food sources and preferences of adults of a parasite, Scambus buolianae (Hym.: Ich.), and their consequences. Canadian Entomologist, 99, 865–871.

Leppla, N. C., & Ashley, T. R. (1989). Quality control in insect mass production: A review and model. Bulletin of the Entomological Society of America, 35, 33–44.

Lewis, W. J., & Martin, W. R. (1990). Semiochemicals for use in biological control: Status and future. Journal of Chemical Ecology, 16, 3067–3085.

Lewis, W. J., & Tumlinson, J. H. (1988). Host detection by chemically mediated associative learning in a parasitic wasp. Nature (London). 331, 257–259.

Lewis, W. J., Beevers, M., Nordlund, D. A., Cross, H. R., & Hagen, K. S. (1979). Kairomones and their use for management of entomophagous insects. IX. Investigations of various kairomone-treatment patterns for Trichogramma spp. Journal of Chemical Ecology, 5, 673–680.

Lewis, W. J., Nordlund, D. A., Gueldner, R. C., Teel, P. E. A., & Tumlin-

son, J. H. (1982). Kairomones and their use for management of entomophagous insects. XII. Kairomone activity for Trichogramma spp. of abdominal tips, feces, and a synthetic pheromone blend of Heliothis zea (Boddie) moths. Journal of Chemical Ecology, 8, 1323–1332.

Li, L. Y. (1982). Trichogramma sp. and their utilization in People's Republic of China. In Les Trichogrammes. Colloques de l'INRA, 9, 23–29.

Li, L. Y. (1989). Mass production of Trichogramma spp. and Anastatus japonicus Ashmead with artificial diets in China. In The use of natural enemies to control agricultural pests. Japan: Food & Fertilizer Tech. Center/ASPAC, National Research Center.

Li, L. Y. (1992). In vitro rearing of parasitoids of insect pests in China. Korean Journal of Applied Entomology, 31, 241–246.

Li, L. Y. (1993). In vitro rearing of egg parasitoids. Proceedings of the Division of Natural Enemies of Insect Pests, China: Guangdong Entomological Institute (1979–1993).

Li, L. Y., Liu, W. H., Chen, C. S., Han, S. T., & Shin, J. C. (1988). In vitro rearing of Trichogramma sp. and Anastatus sp. in artificial "eggs" and the methods of mass production. In Trichogramma and other egg parasites. Colloques de l'INRA, 43, 339–352.

Li, L. Y., Chen, Q. X., & Liu, W. H. (1989). Oviposition behavior of twelve species of Trichogramma and its influence on the efficiency of rearing them in vitro. Natural Enemy Insects, 11, 31–35.

Liber, H., & Niccoli, A. (1988). Observations on the effectiveness of an attractant food spray in increasing chrysopid predation on Prays oleae (Bern.) eggs. Redia, 71, 467–482.

Lindquist, R. K., & Sorenson, E. L. (1970). Interrelationships among aphids, tarnished plant bugs and alfalfas. Journal of Economic Entomology, 63, 192–195.

Lingren, P. D., & Lukefar, M. J. (1977). Effects of nectariless cotton on caged populations of Campoletis sonorensis. Environmental Entomology, 6, 586–588.

Liu, J. F., Liu, Z. C., Wang, C. X., Yang, W. H., & Li, D. S. (1995a) Study on rearing Trichogramma minutum on artificial host eggs. In Trichogramma and other egg parasitoids. Colloques de l'INRA, 73, 161–162.

Liu, S. (1985). Development, adult size and fecundity of Aphidius sonchi in two instars of its aphid host, Hyperomyzus lactucae. Entomologia Experimentalis et Applicata, 37, 41–48.

Liu, W., & Wu, Z. (1982). Recent results in rearing Trichogramma in vitro with the artificial media devoid of insect additives. Acta Entomologica Sinica, 25, 160–163.

Liu, Z. C., Liu, J. F., Wang, C. X., Yang, W. H., & Li, D. S. (1995b). Mechanized production of artificial egg for mass-rearing of parasitic wasps. In Trichogramma and other egg parasitoids. Colloques de l'INRA, 73, 163–164.

Liu, W., Xie, Z. N., Xiao, G. F., Zhou, Z. F., Ou Yang, D. H., et al. (1979). Rearing of Trichogramma dendrolimi in artificial diets. Acta Phytologica Sinica, 6, 17–24.

Liu, W. H., Zhou, Y. F., Chen, C. X., Han, S. C., & Li, L. Y. (1983). In vitro rearing Trichogramma japonicum and Trichogramma nubilalae. Natural Enemy Insects, 2, 15–17.

Liu, Z.-C., Wang, Z. Y., Sun, Y. R., Liu, J. F., & Yang, W.-H. (1988). Studies on culturing Anastatus sp., a parasitoid of Litchi stink bug, with artificial host eggs. In Trichogramma and other egg parasites. Colloques de l'INRA 43, 353–360.

Luck, R. F., Podoler, H., & Kfir, R. (1982). Host selection and egg allocation behavior by Aphytis melinus and A. lingnanensis: Comparison of two facultatively gregarious parasitoids. Ecology and Entomology, 7, 397–408.

Mackauer, M. (1986). Growth and developmental interactions in some aphids and their hymenopterous parasites. Journal of Insect Physiology, 32, 275–280.

Mackauer, M., & Kambhampati, S. (1988). Parasitism of aphid embryos by Aphidius smithi: Some effects of extremely small size. Entomologia Experimentalis et Applicata, 49, 167–173.

Mackauer, M., & Sequeira, R. (1993). Patterns of development in insect parasites. In N. E. Beckage, S. N. Thompson, & B. A. Federici (Eds.), Parasites and pathogens of insects (Vol. 1, pp. 1–23). New York: Academic Press.

Mackauer, M., Sequeira, R., & Otto, M. (1997). Growth and development in parasitoid wasps: Adaptation to variable resources. In Dettner, et al. (Eds.), Ecological studies (Vol. 1, pp. 191–203) Berlin: Springer-Verlag.

Mao, H., & Kunimi, Y. (1994). Longevity and fecundity of Brachmeria lasus (Walker) (Hymenoptera: Chalcidoidea), a pupal parasitoid of the oriental tea tortrix, Homona magnanima Diakonoff (Lepidoptera: Tortricidae) under laboratory conditions. Applied Entomology and Zoology, 29, 237–243.

Martin, P. B., Ridgway, R. L., & Schuetze, C. E. (1978). Physical and biological evaluations of an encapsulated diet for rearing Chrysopa carnea. Florida Entomologist, 61, 145–152.

Masutti, L., Slavazza, A., & Battisti, A. (1992). Oviposition of Ooencyrtus pityocampae (Mercet) into artificial eggs (Hym., Encyrtidae). Redia, 74, 457–462.

Masutti, M., Battisti, A., Milani, M., & Zanata, M. (1991). First success in the in vitro rearing of Ooencyrtus pityocampae (Mercet) (Hym.: Encyrtidae). Redia, 75, 227–232.

Matsuka, M., & Okada, I. (1975). Nutritional studies on an aphidophagous coccinellid, Harmonia axyridis. I. Examination of artificial diets for the larval growth with special reference to drone honeybee powder. Bulletin of the Faculty of Agricriculture, Tamagawa University, 15, 1–9.

Matsuka, M., & Takahashi, S. (1977). Nutritional studies of an aphidophagous coccinellid Harmonia axyridis. II. Significance of minerals for larval growth. Applied Entomology and Zoology, 12, 325–329.

Matsuka, M., Watanabe, M., & Niijima, K. (1982). Longevity and oviposition of vedalia beetles on artificial diets. Environmental Entomology, 11, 816–819.

Matsuka, M., Shimotori, D., Senzaki, T., & Okada, I. (1972). Rearing some coccinellids on pulverized drone honeybee brood. Bulletin of the Faculty of Agricriculture, Tamagawa University, 12, 28–38.

McDermott, G. J., & Hoy, M. A. (1997). Persistence and containment of Metaseiulus occidentalis (Acari: Phytoseiidae) in Florida: Risk assessment for possible releases of transgenic strains. Florida Entomologist, 80, 42–53.

McDougall, S. J., & Mills, N. J. (1997). The influence of host, temperature and food sources on the longevity of Trichogramma platneri. Entomologia Experimentalis et Applicata, 83, 195–203.

McEwen, P. K., Jervis, M. A., & Kidd, N. A. C. (1993a). Influence of artificial honeydew on the larval development and surivial of the green lacewing, Chrysoperla carnea. Entomophaga, 38, 241–244.

McEwen, P. K., Jervis, M. A., & Kidd, N. A. C. (1993b). The effect on olive moth Prays oleae population levels of applying artificial food to olive trees (pp. 361–368). A.N.P.P. Third International Conference on Pests in Agriculture, Montpellier.

McEwen, P. K., Jervis, M. A., & Kidd, N. A. C. (1994). Use of a sprayed L-tryptophan solution to concentrate numbers of the green lacewing Chrysoperla carnea in olive tree canopy. Entomologia Experimentalis et Applicata, 70, 97–99.

McEwen, P. K., Jervis, M. A., & Kidd, N. A. C. (1996). The influence of an artificial food supplement on larval and adult performance in the green lacewing Chrysoperla carnea (Stephens). International Journal of Pest Management 42, 25–27.

McMurtry, J. A., & Rodriguez, J. G. (1987). Nutritional ecology of phytoseiid mites. In F. Slansky & J. G. Rodriguez (Eds.), Nutritional ecology of insects, mites, spiders, and related invertebrates (pp. 609–644). New York: John Wiley & Sons.

McMurtry, J. A., & Scriven, G. T. (1964). Studies on feeding, reproduction and development of Amblyseius hibisci (Acarina: Phytoseiidae) on various food substances. Annals of the Entomological Society of America, 57, 649–655.

McMurtry, J. A., & Scriven, G. T. (1966). Effects of artificial foods on reproduction and development of four species of phytoseiid mites. Annals of the Entomological Society of America, 59, 267–269.

Mellini, E. (1975a). Possibilità di allevamento di insetti entomofagi parasiti su diete arfificiali. Bolletino dell'Istituto di Entomologia "Guido Grandi" della Universita degli Studi di Bologna, 32, 257–290.

Mellini, E. (1975b). Studi sui ditteri larvevoridi XXV. Sul determinismo ormonale delle influenze esercitate dagli ospiti sui loro parassiti. Bollettino dell'Istituto di Entomologia "Guido Grandi" della Universita degli Studi di Bologna, 31, 165–203.

Mellini, E. (1978). Moderni problemi di entomoparassitologia (pp. 263–292). Estratto dagli Atti XI Congresso Nazionale Italiano di Entomologia Portici-Sorrento.

Mellini, E. (1983). L'ipotesi della dominazione ormonale, esercitata dagli ospiti sui parassitoidi, alla luce delle recenti scopterte nella endocrinologia degli insetti. Bollettino dell'Istituto di Entomologia "Guido Grandi" della Universita degli Studi di Bologna, 38, 135–166.

Mellini, E. (1991). Artificial diets for the rearing of parasitical entomophagous insects. Alma Mater Studiorum. Universita degli Studi di Bologna, 7, 187–216.

Mellini, E. (1993). Saggio breve sulla entomofagia degli insetti. Bollettino dell'Istituto di Entomologia "Guido Grandi" della Universita degli Studi di Bologna, 47, 179–221.

Mellini, E., & Beccari, G. (1984). Relazioni tra dimensioni degli ospiti e percentuali di parassitizzazione nella coppia ospite-parassita Galleria mellonella L.-Gonia cinerascens Rondani. Bollettino dell'Istituto di Entomologia "Guido Grandi" dela Universita degli Studi di Bologna, 38, 71–88.

Mellini, E., & Campadelli, G. (1982). Potenziale megetico del parassitoide Gonia cinerascens Rond. misurato sull'ospite di sostituzione Galleria mellonella L. Memorie della Societa Entomologica Italiana, 60, 239–252.

Mellini, E., & Campadelli, G. (1994a). Qualitative improvements in the composition of oligidic diets for the parasitoid Exorista larvarum (L.). Bollettino dell'Istituto di Entomologia "Guido Grandi" della Universita degli Studi di Bologna, 49, 187–196.

Mellini, E., & Campadelli, G. (1994b). Further simplification in the composition of oligidic diets for the parasitoid Exorista larvarum (L.) Bollettino dell'Istituto di Entomologia "Guido Grandi" della Universita degli Studi di Bologna, 49, 211–223.

Mellini, E., & Campadelli, G. (1995). Formulas for "inexpensive" artificial diets for the parasitoid Exorista larvarum (L). Bollettino dell'Istituto di Entomologia "Guido Grandi" della Universita degli Studi di Bologna, 50, 95–106.

Mellini, E., & Campadelli, G. (1996a). Latest results in the rearing of the parasitoid Exorista larvarum (L.) on oligidic diets. Bollettino dell'Istituto di Entomologia "Guido Grandi" della Universita degli Studi di Bologna, 50, 143–153.

Mellini, E., & Campadelli, G. (1996b). A first overall comparison between the in vitro and in vivo production of the parasitoid Exorista larvarum (L.). Bollettino dell'Istituto di Entomologia "Guido Grandi" della Universita degli Studi di Bologna, 50, 183–199.

Mellini, E., & Campadelli, G. (1996c). Actual possibilities of mass production of the parasitoid Exorista larvarum (L.) (Diptera: Tachinidae) on oligidic diets. Bollettino dell'Istituto di Entomologia "Guido Grandi" della Universita degli Studi di Bologna, 50, 233–241.

Mellini, E., & Coulibaly, A. K. (1991). Un decennio di sperimentazione sul sistema ospite-parassita Galleria mellonella L.—Pseudogonia rufifrons Wied.: Sintesi dei risultati. Bollettino dell'Istituto di Entomologia "Guido Grandi" della Universita degli Studi di Bologna, 45, 191–249.

Mellini, E., Campadelli, G., & Dindo, M. L. (1993a). Possibile impiego di plasma bovino nell'allestimento di diete artificiali per le larve del parassitoide Pseudogonia rufifrons Wied. G. Viggiani (Ed.), Convegno "lotta biologica" acireale 1991 (pp. 145–150).

Mellini, E., Campadelli, G., & Dindo, M. L. (1993b). Artificial culture of the parasitoid *Exorista larvarum* L. (Dipt. Tachinidae) on bovine serum-based diets. Bollettino dell'Istituto di Entomologia "Guido Grandi" della Universita degli Studi di Bologna, 47, 223–231.

Mellini, E., Campadelli, G., & Dindo, M. L. (1993c). Artificial culture of the parasitoid *Exorista larvarum* L. (Dipt. Tachinidae) on oligidic media: Improvements of techniques. Bollettino dell'Istituto di Entomologia "Guido Grandi" della Universita degli Studi di Bologna 48, 1–10.

Mellini, E., Malagoli, M., & Ruggeri, L. (1980). Substrati artificiali per l'ovideposizione dell'entomoparassita *Gonia cinerascens* Rondani (Diptera: Larvaevoridae) in cattività. Bollettino dell'Istituto di Entomologia "Guido Grandi" della Universita degli Studi di Bologna, 35, 127–156.

Michels, G. J., & Behle, R. W. (1991). Effects of two prey species on the development of *Hippodamia sinuata* (Coleoptera: Coccinellidae) larvae at constant temperatures. Journal of Economic of Entomology, 84, 1480–1484.

Mitsuhashi, J. (1996). *In vitro* rearing of tachinid maggots on artificial diets. Proceedings Ninth International Conference of Invertebrate Cell Culture (pp. 134–141). Society for *In vitro* Biology.

Mitsuhashi, J., T., & Oshiki, T. (1993). Preliminary attempts to rear an endoparasitic fly, *Exorista sorbillans* (Diptera, Tachinidae) *in vitro*. Japanese Journal of Entomology, 61, 459–464.

Morales-Ramos, J. A., Rojas, M. G., & King, E. G. (1996a). Significance of adult nutrition and oviposition experience on longevity and attainment of full fecundity of *Catolaccus grandis* (Hymenoptera: Pteromalidae). Annals of the Entomological Society of America, 89, 555–563.

Morales-Romas, J. A., Rojas, M. G., Coleman, R. J., Greenberg, S. M., Summy, K. R., & King, E. G. (1996b). Comparison of *in vivo* versus *in vitro*-reared *Catolaccus grandis* in the field. In Proceedings of the Beltwide Cotton Conference (pp. 1099–1104). Memphis: National Cotton Council of America.

Morrison, R. K., House, V. S., & Ridgway, R. L. (1975). Improved rearing unit for larvae of a common lacewing. Journal of Economic Entomology, 68, 821–822.

Morrison, R. K., Nettles, W. C., Ball, D., & Vinson, S. B. (1983). Successful oviposition by *Trichogramma pretiosum* through a synthetic membrane. Southwestern Entomology, 8, 248–251.

Muma, M. H. (1957). Effects of larval nutrition on the life cycle, size, coloration, and longevity of *Chrysopa lateralis* Guer. Florida Entomologist, 40, 5–9.

Nadarajan, L., & Jayaraj S. (1975). Influence of various hosts on the development and reproduction of pupal parasite, *Tetrastichus israeli* M. & K. (Eulophidae: Hymenoptera). Current Science, 44, 458–460.

Nakahara, Y., Iwabuchi, K., & Mitsuhashi, J. (1997). *In vitro* rearing of the larval endoparasitoid, *Venturia canescens* (Granvenhorst) (Hymenoptera: Ichneumonidae). III. Growth promoting ability of extract of *Galleria mellonella* pupae. Applied Entomology and Zoology, 32, 91–99.

Nenon, J.-P. (1972a). Culture *in vitro* des embryons d'un Hyménoptère endoparasitè polyembryonnaire: *Ageniaspis fuscicollis* (= *Encyrtus fuscicollis*). Rôle des hormones de synthèse. Comptes Rendus de l'Academie Sciences Paris, 274D, 3299–3302.

Nenon, J.-P. (1972b). Culture *in vitro* des larves d'un Hyménoptère endoparasite polyembryonnaire: *Ageniaspis fuscicollis*. Rôle des hormones de synthèse. Comptes Rendus de l'Academie Sciences Paris, 274D, 3409–3412.

Nepomnyashchaya, A., Maintcher, E., & Yazlovetsky, I. (1979). New approach to elaboration of artificial nutritive diets for mass rearing of entomophagous insects. Optimization of nutritive diets by the simplex lattice method (pp. 29–35). Kishinev, Shtiintsa: Biochemistry and Physiology Institute.

Nettles, W. C. (1986a). Asparagine: A host chemical essential for the growth and development of *Eucelatoria bryani*, a tachinid parasitoid of *Heliothis* spp. Comparative Biochemistry and Physiology [Part A], 85, 697–701.

Nettles, W. C. (1986b). Effects of soy flour, bovine serum albumin and three amino acid mixtures on growth and development of *Eucelatoria bryani* (Diptera: Tachinidae) rearing on artificial diets. Environmental Entomology, 15, 1111–1115.

Nettles, W. C. (1987a). Amino acid requirements for growth and development of the tachinid, *Eucelatoria bryani*. Comparative Biochemistry and Physiology [Part A], 86, 349–354.

Nettles, W. C. (1987b). *Eucelatoria bryani* (Diptera: Tachinidae): Effect on fecundity of feeding on hosts. Environmental Entomology, 16, 437–440.

Nettles, W. C. (1990). *In vitro* rearing of parasitoids: Role of host factors in nutrition. Archives of Insect Biochemistry and Physiology, 13, 167–175.

Nettles, W. C., Wilson, C. M., & Ziser, S. W. (1980). A diet and methods for the *in vitro* rearing of the tachinid, *Eucelatoria* sp. Annals of the Entomological Society of America, 73, 180–184.

Nettles, W. C., Morrison, R. K., Xie, Z. N., Ball, D., Shenkir, C. A., & Vinson, S. B. (1982). Synergistic action of potassium chloride and magnesium sulfate on parasitoid wasp oviposition. Science, 218, 164–166.

Nettles, W. C. Morrison, R. K., Xie, Z. N., Ball, D., Shenkir, C. A., & Vinson, S. B. (1983). Effect of cation, anions and salt concentrations on oviposition by *Trichogramma pretiosum* in wax eggs. Entomologia Experimentalis et Applicata, 33, 283–289.

Nettles, W. C. Morrison, R. K., Xie, Z. N., Ball, D., Shenkir, C. A., & Vinson, S. B. (1984). Ovipositional stimulant for *Trichogramma* spp. U.S. Patent 4,484,539.

Nettles, W. C. Morrison, R. K., Xie, Z. N., Ball, D., Shenkir, C. A., & Vinson, S. B. (1985). Effect of artificial diet media, glucose, protein hydrolysates, and other factors on oviposition in wax eggs by *Trichogramma pretiosum*. Entomologia Experimentalis et Applicata, 38, 121–129.

Neuenschwander, P., Haug, T., Ajounu, O., Davis, H., Akinwumi, B., & Madojemu, E. (1989). Quality requirements in natural enemies used for inoculative release: Practical experience from a successful biological control programme. Journal of Applied Entomology, 108, 409–420.

Niijima, K. (1989). Nutritional studies on an aphidophagous chrysopid, *Chrysopa septempunctata* Wesmael I. Chemically-defined diets and general nutritional requirements. Bulletin of the Faculty of Agriculture, Tamagawa University, 29, 22–30.

Niijima, K. (1993a). Nutritional studies on an aphidophagous chrysopid, *Chrysopa septempunctata* Wesmael (Neuroptera: Chrysopidae). II. Amino acid requirement for larval development. Applied Entomology and Zoology, 28, 81–87.

Niijima, K. (1993b). Nutritional studies on an aphidophagous chrysopid, *Chrysopa septempunctata* Wesmael (Neuroptera: Chrysopidae). III. Vitamin requirement for larval development. Applied Entomology and Zoology, 28, 89–95.

Niijima, K., Matsuka, M., & Okada, I. (1986). Artificial diets for an aphidophagous coccinellid, *Harmonia axyridis*, and its nutrition. In I. Hodek (Ed.), Ecology of aphidophaga (pp. 37–50). Prague: Academia.

Niijima, K., Nishimura, R., & Matsuka, M. (1977). Nutritional studies of an aphidophagous coccinellid, *Harmonia axyridis*. III. Rearing of larvae using a chemically defined diet and fraction of drone honeybee powder. Bulletin of the Faculty of Agriculture, Tamagawa University, 17, 45–51.

Noldus, L. P. J. J. (1989). Semiochemicals, foraging behavior and quality of entomophagous insects for biological control. Journal of Applied Entomology, 108, 425–451.

Nordlund, D. A., Chalfant, R. B., & Lewis, W. J. (1985). Response of

Trichogramma pretiosum females to extracts of two plants attacked by *Heliothis zea.* Agriculture Ecosystems and Environment, 12, 127–133.

Nordlund, D. A., Jones, R. L., & Lewis, W. J. (Eds.). (1981). Semiochemicals: Their role in pest control. New York: John Wiley & Sons.

Nordlund, D. A., Wu, Z. X., & Greenberg, S. M. (1997). *In vitro* rearing of *Trichogramma minutum* Riley (Hymenoptera: Trichogrammatidae) for ten generations, with quality assessment comparisons of *in vitro* and *in vivo* reared adults. Biological Control, 9, 201–207.

Nozato, K. (1969). The effect of host size on the sex ratio of *Itoplectis cristatae* Momoi (Hymenoptera: Ichneumonidae), a pupal parasite of the Japanese pine shoot moth, *Petrova* (= *Evetria*) *cristata* (Walsingham) (Lepidoptera: Olethreutidae). Kontyu, 37, 134–146.

Ohbayashi, T., Iwabuchi, K., & Mitsuhashi, J. (1994). *In vitro* rearing of a larval endoparasitoid, *Venturia canescens* (Gravenhorst) (Hymenoptera: Ichneumonidae). I. Embryonic development. Applied Entomology and Zoology, 29, 123–126.

Okada, I., & Matsuka, M. (1973). Artificial rearing of *Harmonia axyridis* on pulverized drone honeybee brood. Environmental Entomology, 2, 301–302.

Okada, I., Hoshiba, H., & Maruoka, T. (1971a). An artificial rearing of a coccinellid beetle, *Harmonia axyridis* (Pallas), on drone honeybee brood. Bulletin of the Faculty of Agriculture, Tamagawa University, 11, 91–97.

Okada, I., Matsuka, M., & Hoshiba, H. (1971b). Utilization of drone larvae and pupae as semiartificial diet for aphidophagous insects. Apiacta, 6, 119–120.

Okada, I., Hoshiba, H., & Maehava, T. (1972). An artificial rearing of a coccinellid beetle, *Harmonia axyridis* Pallas, on pulverized drone honeybee brood. Bulletin of the Faculty of Agriculture, Tamagawa University, 12, 39–47.

Opp, S. B., & Luck, R. F. (1986). Effects of host size on selected fitness components of *Aphytis melinus* and *A. lingnanensis* (Hymenoptera: Aphelinidae). Annals of the Entomological Society of America, 79, 700–704.

Ouyang, Y., Grafton-Cardwell, E. E., & Bugg, R. L. (1992). Effects of various pollens on development, survivorship, and reproduction of *Euseius tularensis* (Acari: Phytoseiidae). Environmental Entomology, 21, 1371–1376.

Pennacchio, F., Vinson, S. B., & Tremblay, E. (1992). Preliminary results on *in vitro* rearing of the endoparasitoid *Cardiochiles nigriceps* from egg to second instar. Entomologia Experimentalis et Applicata, 64, 209–216.

Petrusewicz, K. (1967). Concepts in studies on the secondary productivity of terrestrial ecosystems. In K. Petrusewicz (Ed.), Secondary productivity of terrestrial ecosystems (Vol. 1, pp. 17–49). Warsaw: Panstwowe Wydawnictwo Naukowe.

Pimentel, D., (1966). Wasp parasite (*Nasonia vitripennis*) survival on its fly host (*Musca domestica*) reared on various foods. Annals of the Entomological Society of America, 59, 1031–1038.

Powell, W. (1986). Enhancing parasitoid activity in crops. In J. Waage & D. Greathead (Eds.), Insect parasitoids (pp. 319–340). London: Academic Press.

Prell, H. (1915). Zur Biologie der Tachinen *Parasitiogena segregata* Rdi und *Panzeria rudis* Fall. Zeitschrift fuer Angew Entomologie, 2, 57–148.

Presnail, J. K., & Hoy, M. A. (1992). Stable genetic transformation of a beneficial arthropod, *Metaseiulus occidentalis* (Acari: Phytoseiidae), by a microinjection technique. Proceedings of the National Academy of Sciences, USA, 89, 7732–7736.

Presnail, J. K., & Hoy, M. A. (1994). Transmission of injected DNA sequences to multiple eggs of *Metaseiulus occidentalis* and *Amblyseius finlandicus* (Acari: Phytodeiidae) following maternal microinjection. Experimental and Applied Acarology (Northwood), 18, 319–330.

Presnail, J. K., Hoy, M. A. (1996). Maternal microinjection of the endo-

parasitoid *Cardiochiles diaphaniae* (Hymenoptera: Braconidae). Annals of the Entomological Society of America, 89, 576–580.

Presnail, J. K., Jeyaprakash, A., Li, J. B., & Hoy, M. A. (1997). Genetic analysis of four lines of *Metaseiulus occidentalis* (Acari: Phytoseiidae) transformed by maternal microinjection. Annals of the Entomological Society of America, 90, 237–245.

Price, P. W. (1986). Ecological aspects of host plant resistance and biological control: Interactions among three trophic levels. In Interactions of plant resistance and parasitoids and predators of insects (pp. 11–30). New York: Halsted.

Price, P. W., Bouton, C. E., Gross, P., McPheron, B. A., Thompson, J. N., & Weis, A. E. (1980). Interactions among three tropic levels: Influence of plant interactions between insect herbivores and natural enemies. Annual Review of Ecology and Systematics, 11, 41–65.

Principi, M. M., & Canard, M. (1984). Feeding habits. In M. Canard, Y. Séméria, & T. R. New (Eds.), Biology of chrysopidae (pp. 76–92). The Hague: W. Junk.

Putnam, W. L. (1932). Chrysopids as a factor in the natural control of the oriental fruit moth. Canadian Entomologist, 64, 121–126.

Putnam, W. L. (1937). Biological notes on the Chrysopidae. Canadian Journal of Research Section D, Zoological Science, 15, 29–37.

Quickie, D. L. J. (1997). Parasitic wasps. New York: Chapman & Hall.

Racioppi, J. V., Burton, R. L., & Eikenbary, R. (1981). The effects of various oligidic synthetic diets on the growth of *Hippodamia convergens.* Entomologia Experimentalis et Applicata, 30, 68–72.

Rajendram, G. F. (1978a). Oviposition behavior of *Trichogramma californicum* on artificial substrates. Annals of the Entomological Society of America, 71, 92–94.

Rajendram, G. F. (1978b). Some factors affecting oviposition of *Trichogramma californicum* (Hymenoptera: Trichogrammatidae) in artificial media. Canadian Entomologist, 110, 345–352.

Rajendram, G. F., & Hagen, K. S. (1974). *Trichogramma* oviposition into artificial substrates. Environmental Entomology, 3, 399–401.

Reitz, S. R., & Trumble, J. T. (1997). Effects of linear furanocoumarins on the herbivore *Spodoptera exigua* and the parasitoid *Archytas marmoratus:* Host quality and parasitoid success. Entomologia Experimentalis et Applicata, 84, 9–16.

Renou, M., Nagnan, P., Berthier, A., & Durier, C. (1992). Identification of compounds from the eggs of *Ostrinia nubilalis* and *Mamestra brassicae* having kairomone activity on *Trichogramma brassicae.* Entomologia Experimentalis et Applicata 63, 291–303.

Rojas, M. G., Morales-Ramos, J. A., & King, E. G. (1996a). *In vitro* rearing of the boll weevil (Coleoptera: Curculionidae) ectoparasitoid *Catolaccus grandis* (Hymenoptera: Pteromalidae) on meridic diets. Journal of Economic Entomology, 89, 1095–1104.

Rojas, M. G., Morales-Ramos, J. A., & King, E. G. (1996b). Laboratory evaluation of first, second, fifth, and tenth *in vitro*-reared generations of *Catolaccus grandis.* In Proceedings of the Beltwide Cotton Conference (pp. 1104–1107). Memphis: National Cotton Council of America.

Rojas-Rousse, D., & Kalmes, R. (1978). The development of male *Diadromus pulchellus* (Hymenoptera: Ichneumonidae) in the pupae of *Acrolepiopsis assectella* (Lepidoptera: Plutellidae): Comparison of assimilation and energy losses under two temperature regimes. Environmental Entomology, 7, 469–481.

Rotundo, G., Cavalloro, R., & Tremblay, E. (1988). *In vitro* rearing of *Lysiphlebus fabarum* (Hym.: Braconidae). Entomophaga, 33, 261–267.

Rousch, R. T. (1979). Genetic improvement of parasites. In M. A. Hoy & J. J. McKelvey (Eds.), Genetics in relation to insect management (Working papers, (pp. 97–105). New York: The Rockefeller Foundation.

Roussett, A. (1984). Reproductive physiology and fecundity. In M. Canard, Y. Séméria, & T. R. New (Eds.), Biology of Chrysopidae (pp. 116–129) The Hague: W. Junk.

Salt, G. (1934). Experimental studies in insect parasitism. II. Superparasitism. Proceedings of the Royal Entomological Society of London, Series B, 114, 455–476.

Salt, G. (1940). Experimental studies in insect parasitism. VII. The effects of different hosts on the parasite *Trichogramma* Westw. (Hym.: Chalcidoidea). Proceedings of the Royal Entomological Society of London, Series A, 15, A 81–124.

Salt, G. (1941). The effects of hosts upon their insect parasites. Biological Review, 16, 239–264.

Salt, G. (1952). Trimorphism in the ichneumonid *Gelis corruptor*. Quarterly Journal Microscopic Science, 93, 453–475.

Sandlan, K. P. (1979a). Sex ratio in *Coccygomimus turionellae* Linnaeus (Hymenoptera: Ichneumonidae) and its ecological implications. Ecological Entomology, 41, 365–378.

Sandlan, K. P. (1979b). Host-feeding and its effects on the physiology and behavior of the ichneumonid parasitoid, *Coccygomimus turionellae*. Physiology and Entomology, 4, 383–392.

Sandlan, K. P. (1982). Host suitability and its effects on parasitoid biology in *Coccygomimus turionellae* (Hymenoptera: Ichneumonidae). Annals of the Entomological Society of America, 75, 217–221.

Schmidt, G. T. (1974). Host acceptance behavior of *Campoletis sonorensis* toward *Heliothis zea*. Annals of the Entomological Society of America, 67, 835–844.

Schmidt, R. R. (1990). Investigation of mechanisms: The key to successful use of biotechnology. In R. Baker & P. Dunn (Eds.), New directions in biological control: UCLA Symposium on Molecular and Cellular Biology (Vol. 112, pp. 1–22). New York: Alan R. Liss.

Senft, D. (1997). Mass reared insects get fast-food. Agricultural Research, 45, 4–7.

Sequeira, R., & Mackauer, M. (1992a). Nutritional ecology of an insect host-parasitoid association: The pea aphid-*Aphidius ervi* system. Ecology, 73, 183–189.

Sequeira, R., & Mackauer, M. (1992b). Covariance of adult size and development time in the parasitoid wasp *Aphidius ervi* in relation to the size of its host, *Acyrthosiphon pisum*. Evolutionary Ecology, 6, 34–44.

Sequeira, R., & Mackauer, M. (1994). Variation in selected life-history parameters of the parasitoid wasp, *Aphidius ervi*: Influence of host developmental stage. Entomologia Experimentalis et Applicata, 71, 15–22.

Shahjahan, M. (1968). Effect of diet on the longevity and fecundity of the adults of the tachnid parasite *Trichopoda pennipes pilipes*. Journal of Economic Entomology, 61, 1102–1103.

Shannon, R. C. (1923). Rearing dipterous larvae on nutrient agar. Proceedings of the Entomological Society of Washington, 25, 103–104.

Shehata, K. K., & Weismann, L. (1972). Rearing the predaceous mite *Phytoseiulus persimilis* Athias-Henriot on artificial diet (Acarina: Phytoseiidae). Biologia (Bratislava), 27, 609–615.

Sheehan, W., & Shelton, A. M. (1989). The role of experience in plant foraging by the aphid parasitoid *Diaretiella rapae* (Hymenoptera: Aphidiidae). Journal of Insect Behavior, 2, 743–759.

Silver, M. J., & Nappi, A. J. (1986). *In vitro* study of physiological suppression of supernumerary parasites by the endoparasitic wasp *Leptopilina heterotoma*. Journal of Parasitology, 72, 405–409.

Simmonds, F. J. (1994). The propagation of insect parasites on unnatural hosts. Bulletin of Entomological Research, 35, 219–226.

Slansky, F., Jr. (1978). Utilization of energy and nitrogen by larvae of the imported cabbageworm, *Pieris rapae,* as affected by parasitism by *Apanteles glomeratus*. Environmental Entomology, 7, 179–185.

Slansky, F., Jr. (1982). Toward a nutritional ecology of insects. In J. H. Viser & A. K. Minks (Eds.), Proceedings of the Fifth International Symposium on Insect–Plant Relationships (pp. 253–259) Wageningen: Pudoc.

Slansky, F., Jr. (1986). Nutritional ecology of endoparasitic insects and their hosts: An overview. Journal of Insect Physiology, 32, 255–261.

Slansky, F., Jr., & Rodriguez, J.G. (1987). Nutritional ecology of insects, mites, spiders, and related invertebrates: An overview. In F. Slansky & J. G. Rodriguez (Eds.), Nutritional ecology of insects, mites, spiders, and related invertebrates (pp. 1–69) New York: John Wiley & Sons.

Slansky, F., Jr., & Scriber, J. M. (1985). Food consumption and utilization. In G. A. Kerkut & L. I. Gilbert (Eds.), Comprehensive insect physiology, biochemistry and pharmacology (Vol. 4, pp. 87–163). New York: Pergamon.

Sluss, R. (1968). Behavioral and anatomical responses of the convergent lady beetle to parasitism by *Perilitus coccinellae* (Schrank) (Hymenoptera: Braconidae). Journal of Invertebrate Pathology, 10, 9–27.

Smirnoff, W. A. (1958). An artificial diet for rearing coccinellid beetles. Canadian Entomologist, 90, 563–565.

Smith, B. C. (1961). Results of rearing some coccinellid (Coleoptera: Coccinellidae) larvae on various pollens. Proceedings of the Entomological Society of Ontario, 91, 270–271.

Smith, B. G. (1965). Growth and development of coccinellid larvae on dry foods (Coleoptera: Coccinellidae). Canadian Entomologist, 97, 760–768.

Smith, B. G. (1966). Effect of food on some aphidophagous Coccinillidae. In I. Hodek (Ed.), Ecology of aphidophagous insects (pp. 75–81). Prague: Academia Publishing House, Czechoslovakia Academy of Science.

Smith, J. M. (1957). Effects of the food plant of California red scale *Aonidiella aurantii* (Mask.) on reproduction of its hymenopterous parasites. Canadian Entomologist, 89, 219–230.

Smith, R. W. (1958). Parasites of nymphal and adult grasshoppers (Orthoptera: Acrididae) in Western Canada. Canadian Journal of Zoology, 36, 217–262.

Souissi, R., & Le Ru, B. (1997). Effect of host plants on fecundity and development of *Apoanagyrus lopezi,* an endoparasitoid of the cassava mealybug *Phenacoccus manihoti*. Entomologia Experimentalis et Applicata, 82, 235–238.

Stanley, R. G., & Linskens, H. F. (1974). Pollen—biology, biochemistry, management. Berlin: Springer-Verlag.

Strand, M. R. (1986). The physiological interactions of parasitoids with their hosts and their influence on reproductive strategies. In J. Waage & D. Greathead (Eds.), Insect parasitoids (pp. 97–136). London: Academic Press.

Strand, M. R., & Vinson, S. B. (1985). *In vitro* culture of *Trichogramma pretiosum* on an artificial medium. Entomologia Experimentalis et Applicata, 39, 203–209.

Strand, M. R., Baehrecke, E. H., & Wong, E. A. (1991a) The role of host endocrine factors in the development of polyembryonic parasitoids. Biological Control, 1, 144–152.

Strand, M. R., Goodman, W. G., & Baehrecke, E. H. & (1991b). The juvenile hormone titer of *Trichoplusia ni* and its potential role in embryogenesis of the polyembryonic wasp *Copidosoma floridanum*. Insect Biochemistry, 21, 205–214.

Strand, M. R., Meola, S. M., & Vinson, S. B. (1986). Correlating pathological symptoms in *Heliothis virescens* eggs with development of the parasitoid *Telenomus heliothidis*. Journal of Insect Physiology, 32, 389–402.

Strand, M. R., Meola, S. M., Nettles, W. C., & Xie, Z. N. (1988). *In vitro* culture of the egg parasitoid *Telenomus heliothidis:* The role of teratocytes and medium consumption in development. Entomologia Experimentalis et Applicata, 46, 71–78.

Sundby, R. A. (1966). A comparative study of the efficiency of three predatory insects *Coccinella septempunctata* L. [Coleoptera, Coccinellidae] *Chrysopa carnea* St. [Neuroptera, Chrysopidae] and *Syrphus ribesii* L. [Diptera, Syrphidae] at two different temperatures. Entomophaga, 11, 395–404.

Sundby, R. A. (1967). Influence of food on the fecundity of *Chrysopa carnea* Stephens [Neuroptera, Chrysopidae]. Entomophaga, 12, 475–479.

Syme, P. D. (1975). The effects of flowers on the longevity and fecundity of two native parasites of the European pine shoot moth in Ontario. Environmental Entomology, 4, 337–346.

Syme, R. D. (1977). Observations on the longevity and fecundity of Orgilus obscurator (Hymenoptera: Bracondiae) and the effects of certain foods on longevity. Canadian Entomologist, 109, 995–1000.

Szumkowski, W. (1952). Observations on Coccinellidae. II. Experimental rearing of Coleomegilla on a non-insect diet. Transactions, Ninth International Congress of Entomology, 1, 781–785.

Szumkowski, W. (1961a). Dietas sin insectos vivos para la cria de Coleomegilla maculata Deg. (Coccinellidae, Coleoptera). Agronomía Tropical (Maracay, Venezuela), 10, 149–154.

Szumkowski, W. (1961b). Aparicion de un coccinelido predator nuevo para Venezuela. Agronomía Tropical (Maracay, Venezuela), 11, 33–37.

Tanaka, M., & Maeta, Y. (1965). Rearing of some predacious coccinellid beetles by the artificial diets. Bulletin of the Horticultural Research Station of Japan, Series D, 3, 17–35.

Tanigoshi, L. K., Nishio-Wong, J. Y., & Fargerlund, J. (1983). Greenhouse and laboratory studies of Euseius hibisci (Chant) (Acarina: Phytoseiidae), a natural enemy of the citrus thrips, Scirtothrips citri (Moulton) (Thysanoptera: Thripidae). Environmental Entomology, 12, 1298–1302.

Tassan, R. L., Hagen, K. S., & Sawall, E. F. (1979). The influence of field food sprays on the egg production rate of Chrysopa carnea. Annals of the Entomological Society of America, 8, 81–85.

Tauber, M. J., & Tauber, C. A. (1974). Dietary influence on reproduction in both sexes of five predaceous species (Neuroptera). Canadian Entomologist, 106, 921–925.

Temerak, S. A. (1976). Studies on certain mortality factors affecting distribution and abundance of sugarcane borers in upper Egypt. Unpublished doctoral dissertation, Assiut University, Egypt.

Thompson, S. N. (1975). Defined meridic and holidic diets and aseptic feeding procedures for artificially rearing the ectoparasitoid Exeristes roborator (Fabricius). Annals of the Entomological Society of America, 68, 220–226.

Thompson, S. N. (1976a). Effects of dietary amino acid level and nutritional balance on larval survival and development of the parasite Exeristes roborator. Annals of the Entomological Society of America, 69, 835–838.

Thompson, S. N. (1976b). The amino acid requirements for larval development of the hymenopterous parasitoid. Exeristes roborator Fabricius (Hymenoptera: Ichneumonidae). Comparative Biochemistry and Physiology, 53A, 211–213.

Thompson, S. N. (1977). Lipid nutrition during larval development of the parasitic wasp, Exeristes. Journal of Insect Physiology, 23, 579–583.

Thompson, S. N. (1980). Artificial culture techniques for rearing larvae of the chalcidoid parasite, Brachymeria intermedia. Entomologia Experimentalis et Applicata, 27, 133–143.

Thompson, S. N. (1981a). The nutrition of parasitic Hymenoptera. Proceedings, Ninth International Conference of Plant Protection, 1, 93–96.

Thompson, S. N. (1981b). Essential amino acid requirements of four species of parasitic Hymenoptera. Comparative Biochemistry and Physiology Part A, 69, 173–174.

Thompson, S. N. (1981c). Brachymeria lasus and Pachycrepoideus vindemiae: Sterol requirements during larval growth of two hymenopterous insect parasites reared in vitro on chemically defined diets. Experimental Parasitology, 51, 220–235.

Thompson, S. N. (1981d). Brachymeria lasus: Culture in vitro of a chalcid insect parasite. Experimental Parasitology, 32, 414–418.

Thompson, S. N. (1982). Exeristes roborator: Quantitative determination of in vitro larval growth rates in synthetic media with different glucose concentrations. Experimental Parasitology, 54, 229–234.

Thompson, S. N. (1983a). Larval growth of the insect parasite Brachymeria lasus reared in vitro. Journal of Parasitology, 69, 425–427.

Thompson, S. N. (1983b). Brachymeria lasus: Effects of nutrient level on in vitro larval growth of a chalcid insect parasite. Experimental Parasitology, 55, 312–319.

Thompson, S. N. (1983c). Metabolic and physiological effects of metazoan endoparasites on their host species. Comparative Biochemistry and Physiology Part B, 74B, 183–211.

Thompson, S. N. (1985). Metabolic integration during the host associations of multicellular animal endoparasites. Comparative Biochemistry and Physiology [Part B], 81, 21–42.

Thompson, S. N. (1986a). Nutrition and in vitro culture of insect parasitoids. Annual Review of Entomology, 31, 197–219.

Thompson, S. N. (1986b). The metabolism of insect parasites (parasitoids): An overview. Journal Insect Physiology, 32, 421–423.

Thompson, S. N. (1990). Nutritional considerations in propagation of entomophagous species. In R. Baker & P. Dunn (Eds.), New directions in biological control. UCLA Symposia on Molecular and Cellular Biology (Vol. 112, pp. 389–404). New York: Alan R. Liss.

Thompson, S. N. (1993). Redirection of host metabolism and effects on parasite nutrition. In N. E. Beckage, S. N. Thompson, & B. A. Federici (Eds.), Parasites and pathogens of insects (Vol. 1, pp. 125–144). New York: Academic Press.

Thompson, S. N., Bednar, L., & Nadel, H. (1983). Artificial culture of the insect parasite Pachycrepoideus vindemiae. Entomologia Experimentalis et Applicata, 33, 121–122.

Thurston, R., & Fox, P. M. (1972). Inhibition by nicotine of emergence of Apanteles congregatus from its host, the tobacco hornworm. Annals of the Entomological Society of America, 65, 547–550.

Tucker, J. E., & Leonard, D. E. (1977). The role of kairomones in host recognition and host acceptance behavior of the parasite, Brachymeria intermedia. Environmental Entomology, 6, 527–531.

Tumlinson, J. H., Lewis, W. J., & Vet, L. E. M. (1993). How parasitic wasps find their hosts. Scientific American, 226, 100–106.

Vanderzant, E. S. (1969). An artificial diet for larvae and adults of Chrysopa carnea, an insect predator of crop pests. Journal of Economic Entomology, 62, 256–257.

Vanderzant, E. S. (1973). Improvements in the rearing diet for Chrysopa carnea and the amino acid requirements for growth. Journal of Economic Entomology, 66, 336–338.

van Alphen, J. J. M. (1980). Aspects of the foraging behavior of Tetrastichus asparagi Crawford and Tetrastichus spec. (Eulophidae), gregarious egg parasitoids of the asparagus beetles Crioceris asparagi L. and C. duodecimpunctata L. (Chrysomelidae). I. Host species selection, host stage selection and host discrimination. Netherlands Journal of Zoology, 30, 307–325.

van Emden, H. F. (1990). Plant diversity and natural enemy efficiency in agroecosystems. In Mackauer, L. E. Ehler, & J. Roland (Eds.), Critical issues in biological control (pp. 63–80). Andover: Intercept.

van Emden, H. F., & Hagen, K. S. (1976). Olfactory reactions of the green lacewing, Chrysopa carnea to tryptophan and certain breakdown products. Environmental Entomology, 5, 469–473.

Varley, G. C. (1961). Conversion rates in hyperparasitic insects. Proceedings of the Royal Entomological Society of London, Series C, 26, 11.

Veerman, A., Slagt, M. E., Alderlieste, M. F. J., & Veenendaal, R. L. (1985). Photoperiodic induction of diapause in an insect is vitamin A dependent. Experientia, 41, 1194–1195.

Vet, L. E. M. (1983). Host-habitat location through olfactory cues by Leptopilina clavipes (Hartig) (Hym.: Eucoilidae), a parasitoid of fungivorous Drosophila: The influence of conditioning. Netherlands Journal of Zool. 33, 225–248.

Vet, L. E. M., & Dicke, M. (1992). Ecology of infochemical use by natural enemies in aa tritrophic context. Annual Review of Entomology, 37, 141–172.

Vet, L. E. M., & Groenewold, A. W. (1990). Semiochemicals and learning in parasitoids. Journal of Chemical Ecology, 16, 3119–3135.

Vinson, S. B. (1970). Development and possible functions of teratocytes in the host-parasite association. Journal Invertebrate Pathology, 16, 93–101.

Vinson, S. B. (1975). Biochemical coevolution between parasitoids and their hosts. In P. Price (Ed.), Evolutionary strategies of parasitic insects and mites (pp. 14–48). New York: Plenum.

Vinson, S. B. (1976). Host selection by insect parasitoids. Annual Review of Entomology, 21, 109–133.

Vinson, S. B. (1981). Habitat location. In D. A. Nordlund, R. L. Jones, & W. J. Lewis (Eds.), Semiochemicals, their role in pest control (pp. 51–77). New York: John Wiley & Sons.

Vinson, S. B. (1984). Parasitoid-host relationship. In W. J. Bell & R. T. Cardé (Eds.), Chemical ecology of insects (pp. 205–233). London: Chapman & Hall.

Vinson, S. B. (1994). Parasitoid in vitro rearing: Successes and challenges. In J. P. R. Ochieng-Odero (Ed.), Techniques of insect rearing for the development of integrated pest and vector management strategies (Vol. 1, pp. 49–104). Nairobi: ICIPE.

Vinson, S. B., & Barbosa, P. (1987). Interrelationships of nutritional ecology of parasitoids. In F. Slansky & J. G. Rodriguez (Eds.), Nutritional ecology of insects, mites, spiders, and related invertebrates (pp. 673–695). New York: John Wiley & Sons.

Vinson, S. B., Bin, F., & Vet, L. E. M. (1998). Critical issues in host selection by insect parasitoids. Biological Control, 11, 77–78.

Vinson, S. B., & Iwantsch, G. F. (1980). Host suitability for insect parasitoids. Annual Review of Entomology, 25, 397–419.

Visser, M. E. (1994). The importance of being large: The relationship between size and fitness in females of the parasitoid Aphaereta minuta (Hymenoptera: Braconidae). Journal of Animal Ecology, 63, 963–978.

Volkoff, N., Vinson, S. B., Wu, Z. X., & Nettles, W. C. (1992). In vitro rearing of Trissolcus basalis (Hym., Scelionidae), an egg parasitoid of Nezara viridula (Hem., Pentatomidae). Entomophaga, 37, 141–148.

Waage, J. K. (1986). Family planning in parasitoids: Adaptive patterns of progeny and sex allocation. In J. K. Waage & D. Greathead (Eds.), Insect parasitoids (pp. 63–96). New York: Academic Press.

Waage, J. K., & Ng, S. M. (1984). The reproductive strategy of a parasitic wasp. I. Optimal progeny and sex allocation in Trichogramma evanescens. Journal of Animal Ecology, 53, 401–415.

Waage, J. K., Carl, K. P., Mills, N. J., & Greathead, D. J. (1985). Rearing entomophagous insects. In P. Singh & R. F. Moore (Eds.), Handbook of insect rearing (pp. 45–66). Amsterdam: Elsevier.

Watanabe, M., & Mitsuhashi, J. (1995). In vitro rearing of an endoparasitic fly, Exorista sorbillans (Diptera: Tachinidae). Applied Entomology and Zoology, 30, 319–325.

Welch, H. E. (1968). Relationships between assimilation efficiencies and growth efficiencies for aquatic consumers. Ecology, 49, 755–759.

Werren, J. H. (1984). A model for sex ratio selection in parasitic wasps: Local mate competition and host quality effects. Netherlands Journal of Zoology, 34, 81–96.

Weseloh, R. M. (1984). Effects of the feeding inhibitor plictran and low Bacillus thuringiensis Berliner dose on Lymantria dispar (L.) (Lepidoptera: Lymantriidae): Implication for Cotesia melanoscelus (Ratzburg) (Hymenoptera: Braconidae). Environmental Entomology, 13, 1371–1376.

White, T. D., & Jubb, G. L. (1980). Potential of food sprays for augmenting green lacewing populations in vineyards (Melsheimer Entomology Series 29, pp. 35–42).

Whitten, M. J. (1989). The relevance of molecular biology to pure and applied entomology. Entomologia Experimentalis et Applicata, 53, 1–16.

Wiegert, R. G., & Petersen, C. E. (1983). Energy transfer in insects. Annual Review of Entomology, 28, 455–486.

Wilbert, H. (1977). Der honigtau als reizund engergiequelle fur entomophage insekten. Apidologie, 8, 393–400.

Wu, Z. X., & Qin, J. (1982a). Culturing Trichogramma dendrolimi in vitro with media devoid of insect materials (pp. 431–435). Proceedings of the Chinese Academy of Science—United States National Academy of Science Joint Symposium on the Biological Control of Insects.

Wu, Z. X., & Qin, J. (1982b). Ovipositional response of Trichogramma dendrolimi to the chemical contents of artificial eggs. Acta Entomologica Sinica, 25, 363–372.

Wu, Z., Zhang, Z., Li, T., & Liu, D. (1980). Artificial media devoid of insect additives for rearing larvae of the endoparasitoid wasp Trichogramma. Acta Entomologica Sinica, 23, 232.

Wu, Z., Qin, J., Li, T. X., Chang, Z. P., & Liu, T. M. (1982). Culturing Trichogramma in vitro with artificial media devoid of insect materials. Acta Entomologica Sinica, 25, 128–135.

Wylie, H. G. (1965). Effects of superparasitism on Nasonia vitripennis (Walker) (Hymenoptera: Pteromalidae). Canadian Entomologist, 97, 326–331.

Wylie, H. G. (1967). Some effects of host size on Nasonia vitripennis and Muscidifurax raptor (Hymenoptera: Pteromalidae). Canadian Entomologist, 99, 742–748.

Xie, Z. N., Nettles, W. C., Saldana, G., & Nordlund, D. A. (1997a). Elmer's School Glue and Elmer's Glue All: Arrestants and probing/oviposition enhancers for Trichogramma spp. Entomologia Experimentalis et Applicata, 82, 115–118.

Xie, Z.-N., Nettles, W. C., Jr., Morrison, R. K., Irie, K., & Vinson, S. B. (1986a). Effect of ovipositional stimulants and diets on the growth and development of Trichogramma pretiosum in vitro. Entomologia Experimentalis et Applicata, 42, 119–124.

Xie, Z.-N., Nettles, W. C., Jr., Morrison, R. K., Irie, K., & Vinson, S. B. (1986b). Three methods for the in vitro culture of Trichogramma pretiosum Riley. Journal of Entomological Science, 21, 133–138.

Xie, Z. N., Wu, Z. X., Nettles, W. C., Saldana, G., & Nordlund, D. A. (1997b). In vitro culture of Trichogramma spp. on artificial diets containing yeast extract and ultracentrifuged chicken egg yolk but devoid of insect components. Biological Control, 8, 107–110.

Yamamoto, Y., Ohori, M., Ohbayashi, T., Iwabuchi, K., & Mitsuhashi, J. (1997). In vitro rearing of the larval endoparasitoid, Venturia canescens (Gravenhorst) (Hymenoptera: Ichneumonidae). II. Larval development. Applied Entomology and Zoology, 32, 256–258.

Yamamura, N., & Yano, E. (1988). A simple model of host-parasitoid interaction with host-feeding. Researches on Population Ecology (Kyoto), 30, 353–369.

Yazgan, S. (1972). A chemically defined synthetic diet and larval nutritional requirements of the endoparasitoid Itoplectis conquisitor (Hymenoptera). Journal of Insect Physiology, 18, 2123–2141.

Yazgan, S., & House, H. L. (1970). An hymenopterous insect, the parasitoid Itoplectis conquisitor, reared axenically on a chemically defined diet. Canadian Entomologist, 102, 1304–1306.

Yazlovetsky, I. G., & Nepomnyashchaya, A. M. (1981). Essai d'elevage massif de Chrysopa carnea sur milieux artificiels micro-encapsules et de son application dans la lutte contre les pucerons en serre. Lutte Biologique et Irtégrée Coutre les Pucerons. Colloque Franco-Soviétique, Rennes (pp. 51–58). September 26–27, 1979, Paris, France: Institue National de la Recherche Agronomique.

Ye, Z., Han, Y., Wang, D., Liang, S., & Li, S. (1979). Studies on the artificial diets of larvae and adults of Chrysopa sinica Tjeder. Acta Phytologica Sinica, 6, 11–16.

Zhao, Z., & McMurtry, J. A. (1990). Development and reproduction of three Euseius (Acari: Phytoseiidae) species in the presence and absence of supplementary foods. Experimental and Applied Acarology (Northwood), 8, 233–242.

Zheng, Y., Hagen, K. S., Daane, K. M., & Mittler, T. E. (1993a). Influence of larval dietary supply on the food consumption, food utilization efficiency, growth and development of the lacewing *Chrysoperla carnea.* Entomologia Experimentalis et Applicata, 67, 1–7.

Zheng, Y., Daane, K. M., Hagen, K. S., & Mittler, T. E. (1993b). Influence of larval food consumption on the fecundity of the lacewing *Chrysoperla carnea.* Entomologia Experimentalis et Applicata, 67, 9–14.

Zhou, W., & Zhang, X. (1983). Preliminary study on the use of encapsulated diet for rearing *Chrysopa sinica.* Acta Phytologica Sinica, 10(3), 161–165.

Ziser, S., & Nettles, W. C. (1979). The rate of oxygen consumption by *Eucelatoria* sp. in relation to larval development and temperate. Annals of the Entomological Society of America, 72, 540–543.

Zoebelein, G. (1956). Der Honigtau als Nahrung der Insekten. Zeitschriff fuer Angewandte Entomologie, Part I, 38, 369–416; Part II, 39, 129–167.

Zohdy, N. M. (1976). On the effect of the food of *Myzus persicae* Sulzer on the hymenopterous parasite *Aphelinus asychis* Walker. Oecologia (Berlin), 26, 185–191.

23

Sex Ratio and Quality in the Culturing of Parasitic Hymenoptera

A Genetic and Evolutionary Perspective

R. F. LUCK

Department of Entomology
University of California
Riverside, California

L. NUNNEY

Department of Biology
University of California
Riverside, California

R. STOUTHAMER

Department of Entomology
Agricultural University
6700 EH Wageningen, the Netherlands

INTRODUCTION

Species of parasitic Hymenoptera are used extensively in biological control as agents for the permanent suppression of a pest or as insectary-reared, augmentative agents for the seasonal suppression of a pest. Both approaches require the agent to be cultured but the methods used to culture them often affect their quality, especially that of females. Poorly producing cultures frequently manifest detrimental sex-ratio changes (van Dijken *et al.,* 1993), occasionally becoming entirely male with the consequent extinction of the culture (e.g., Platner & Oatman, 1972). These sex-ratio changes arise from the interactions between conditions within the culture and factors such as sex determination, sex allocation, haplodiploidy, sex-ratio distorting elements, or unusual life histories of the parasitoids (Luck *et al.,* 1992; Godfray, 1994), for example, heteronomous parasitism (Walter, 1983; Godfray & Waage, 1990). In this chapter we review some of these processes and factors as they impinge on the productivity of parasitoid cultures. In the first half of the chapter we discuss sex determination, arrhenotoky, deuterotoky, thelytoky, sex-ratio distorting factors (including genomic and cytoplasmic factors) and heteronomous (auto-) parasitism. In the second half of the chapter we discuss sex allocation and sex-ratio theory including factors that determine the number of eggs allocated to a host (i.e., clutch size).

SEX DETERMINATION

Most parasitic Hymenoptera are haplodiploid. Males are normally haploid with one set of maternally derived chromosomes; females are diploid with two sets of chromosomes, one maternally derived and one paternally derived. Under this form of inheritance, fertilized eggs become female and unfertilized eggs become male. Females control fertilization because after mating they store sperm in the spermatheca, and they determine the sex of the offspring at oviposition by regulating the sperms' access to the egg.

The haplodiploid genetic system in which males arise from unfertilized eggs is a form of parthenogenesis known as arrhenotoky. Hymenoptera can reproduce parthenogenetically in two other modes: deuterotoky and thelytoky. The distinction between these modes is based on the sex of the offspring arising from unfertilized eggs (Winkler, 1920; Suomalainen *et al.,* 1987). In the two less common forms of parthenogenesis (deuterotoky and thelytoky) *all* offspring arise from unfertilized eggs. They differ in that deuterotokous females produce offspring of *both* sexes whereas thelytokous females produce *only* daughters. However, the distinction between deuterotoky and thelytoky has become increasingly blurred because several strains or species, initially thought to be thelytokous, when reexamined, have been found to produce an occasional male. Although these males have usually been considered non-functional (White, 1984), examples have been found where they mate and pass their genes to subsequent generations, which then reproduce thelytokously [e.g., *Aphytis mytilaspidis* (LeBaron) (Rössler & DeBach, 1972, 1973) and *Trichogramma* species (Luck *et al.,* 1992; Stouthamer & Kazmer, 1994)]. Thus, the distinction between deuterotoky and thelytoky has become ambiguous and the term "thelytoky" has been frequently applied to cases that are more appropriately termed "deuterotoky." For purposes of this review, we do not make a distinction between deuterotoky and thelytoky. Incidence of thelytoky among species of parasitic Hymenoptera and a familial classification based on mode of reproduction are indicated in Tables 1 and 2, respectively.

Diploid males are found occasionally (Whiting, 1943) and this has raised the question of how sex is determined in the Hymenoptera. Three models have been proposed to explain the usual association of sex with ploidy (Crozier, 1977; Luck *et al.*, 1992; Cook, 1993b): (1) a single-locus, multiple-allele (SLMA) model (Whiting, 1943); (2) a multiple-locus, multiple-allele (MLMA) model (Snell, 1935; Crozier, 1971, 1975, 1977); and (3) a genetic balance model (da Cuhna & Kerr, 1957; Kerr & Nielsen, 1967). The SLMA model proposes that a single locus with several alleles is responsible for determining the sex of an offspring. If the offspring is heterozygous at the sex-determining locus (i.e., it is diploid and received a different sex-determining allele from each parent), the offspring is female, but if it is homozygous (i.e., it is diploid and received the same sex-determining allele from each parent), the offspring is male. Haploid offspring are always male because they are hemizygous and therefore can never carry two sex-determining alleles. The multiple-locus, multiple-allele model extends and subsumes the single-locus model

TABLE 1 Number of Species per Family and Superfamily from Which Thelytoky has been Reported

	Family	Superfamily
Tenthredinoidea		90
Ichneumonoidea		32
Braconidae	15	
Aphidiidae	6	
Ichneumonidae	11	
Chalcidoidea		121
Torymidae	3	
Pteromalidae	3	
Eurytomidae	9	
Leucospidae	1	
Eupelmidae	2	
Encyrtidae	23	
Aphelinidae	38	
Signiphoridae	4	
Eulophidae	9	
Mymaridae	9	
Trichogrammatidae	20	
Cynipoidea		5[a]
Pelicinoidea		1
Proctotrupoidea		5
Bethyloidea		6
Bethylidae	1	
Dryinidae	5	
Apoidea		6
Formicidae		4

[a] Not included are >2000 spp. of Cynipidae with cyclic thelytoky (see text).

TABLE 2 Classification of Parasitic Hymenoptera by Family and Mode of Reproduction[a,b]

Family	A	A + T	T	?	%T
Aphelinidae	16	1	8	0	35
Encyrtidae	24	1	4	7	14
Eulophidae	8	1	3	7	20
Ichneumonidae	3	1	1	14	10
Braconidae	19	0	1	26	2
Mymaridae	3	0	2	1	33
Totals	73	4	19	55	15

[a] A = arrhenotokous, T = thelytokous, ? = unknown mode of reproduction.

[b] Based on all parasitic Hymenoptera species whose life history is described in Clausen (1978).

to include several sex-determining loci (Crozier, 1977; Cook, 1993b). With this model, diploid offspring that are heterozygous at *one or more* loci are female. Diploids that are homozygous at *all* sex-determining loci are males. Again, haploid offspring are always males because they are hemizygous. The SLMA model is therefore a special case of the MLMA model. Cook (1993b) refers to the SLMA and MLMA collectively as complementary sex determination (CSD).

In contrast to the SLMA and MLMA models, the genetic balance model proposes that the sex of an offspring is determined by the balance between the effects of the female-determining loci and the male-determining loci (da Cuhna & Kerr, 1957; Kerr & Nielsen, 1967). It is assumed that the male-determining loci (and not the female-determining loci) are subject to dosage compensation (i.e., two copies have the same net effect as one copy). In the haploid offspring, the strength of the male-determining loci outweighs that of the female-determining loci. These individuals are male. However, in diploid offspring, the cumulative effects of the female-determining loci outweigh the non-cumulative effects of the male-determining loci. These individuals are female.

Both the SLMA and MLMA model predict the appearance of diploid males when hymenopteran populations are inbred (brother–sister matings) whereas the genetic balance model does not. In contrast to the MLMA model, the SLMA model predicts that diploid males occur after one or two generations of inbreeding whereas several to many generations of continuous inbreeding are required before diploid males appear if the MLMA model holds. The absence of diploid males following inbreeding, however, cannot be taken as evidence that the SLMA and MLMA models are inapplicable because homozygosity at some sex-determining loci may be lethal, causing the diploid males to die as eggs or larvae (Crozier, 1971, 1975, 1977).

The immature stages of parasitoids are usually not apparent and hidden and, thus, this mortality may go unnoticed. However, sex linkage is analogous to haplodiploidy, and we do not know of any examples of sex-linked alleles that are not both homozygous lethal and hemizygous lethal.

The generality of the SLMA or MLMA models as an explanation for sex determination in Hymenoptera species can be tested by continuously inbreeding individuals and testing their progeny for the presence of diploid males. Diploid males have been documented in several hymenopteran species: the braconid parasitoid *Bracon hebetor* Say (Horn, 1943; P. W. Whiting, 1943; A. R. Whiting, 1961); *B. serinopae* (Clark *et al.,* 1963); *Cotesia rubecula* (Steiner in Stouthamer *et al.,* 1992); *Micropletis croceipes* (Steiner & Teig, 1989); *Bathyplectes curculionis* (Unruh *et al.,* 1984); *Diadromus pulchellus* (Hedderwick *et al.,* 1985); the honeybee, *Apis mellifera* Linnaeus (Woyke, 1963), and a related species, *A. cerana* Fabricius (Hoshiba *et al.,* 1981); a bumblebee, *Bombus atratus* Franklin; several stingless bees, *Melipona quadrifasciata* Lee (Camargo, 1979), *B. terrestris* (Duchateau & Marien, 1995; Hoshiba *et al.,* 1995), *Megachile rotunda* (Frohlich *et al.,* 1990, Mc-Corquodale & Owen, 1994), and *Trigona quadrifasciata* (Tarelho, 1973, cited in Moritz, 1986); and the sawflies, *Neodiprion nigroscutum* Middleton (Smith & Wallace, 1971), *N. pinetum* (Norton) (Wallace in Stouthamer *et al.,* 1992), *Althalia rosae* (Naito & Suzuki, 1991), and *Diprion pini* (Beaudoin *et al.,* 1994). In several of these cases sex determination may depend on two sex-determining loci. Inbred *B. atratus* (mother–son matings) produced diploid males, and Garofalo and Kerr (1975) posited a two-loci model to explain the occurrence of these males in this species. Crozier (1971, 1977) disagreed and argued for a one-locus model. Tarelho (1973, cited in Moritz 1986) reported diploid males in natural colonies of *T. quadrifasciata*. He assumed that the queen mated once. This assumption, when coupled with the ratio of diploid males to workers, suggested that a two-loci model explained sex determination in this species. Smith and Vikki (1978) reported that sex determination in *N. nigroscutum* is consistent with a two-loci, multiple allele model; however, they provided no data to substantiate their conclusion. Although these examples suggest that an MLMA model may explain sex determination in some Hymenoptera species, an unambiguous example remains to be documented.

Several well-documented examples exist that are consistent with the SLMA model. Crossing experiments have demonstrated that the sex of *Bracon hebetor* is controlled by a single locus (Whiting, 1943) with nine known alleles (Horn, 1943; Whiting, 1961). Behavioral studies have indicated this species to be outbreeding in the field (Antolin & Strand, 1992). Similarly, the gender of *Apis mellifera* (Woyke, 1963), *Melipona quadrifasciata* (Camargo 1979),

N. nigroscutum (Smith & Wallace, 1971, but see Smith & Vikki 1978), *Diadromus pulchellus* (Periquet *et al.,* 1993), and *Athalia rosae* (Naito & Suzuki, 1991) are all determined by a single locus with several alleles.

These results show that the genetic balance model (de Cuhna & Kerr, 1957; Kerr & Nielsen, 1967) cannot be the general mode of sex determination in the Hymenoptera species. However, long-term inbreeding experiments and the homozygosity characteristic of many thelytokous parasitoids suggest that the SLMA or MLMA model does not apply generally either. Inbreeding experiments with *Nasonia vitripennis* (Walker) (Schmieder & Whiting, 1947; Skinner & Werren, 1980), *Mellitobia* species (Schmieder & Whiting, 1947), *Muscidifurax raptor* Girault & Sanders (Legner, 1979; Fabritius, 1984), *Leptopilina heterotoma* (Thompson) (Hey & Gargiulo, 1985), and *Goniozus nephantidis* (Cook, 1993a) failed to produce diploid males. Although it can be argued that inbreeding experiments were not continued long enough to create complete homozygosity at all sex-determining loci or that homozygous sex alleles were lethal (Crozier, 1971, 1975, 1977), indirect evidence suggests otherwise. Smith (1941) was the first to point out that many thelytokous species are highly homozygous. In these species normal meiosis occurs and diploidy is restored by a fusion of two of the meiotic products [gamete duplication or terminal fusion (Suomalainen *et al.,* 1987)]. With thelytokous *M. uniraptor* Kogan & Legner (Legner, 1985) and some *Trichogramma* species (Stouthamer & Kazmer, 1994), diploidy is restored by gamete duplication, leading to complete homozygosity in one generation (Stille & Dävring, 1980). With *Aphytis mytilaspidis* it appears to be terminal fusion (Rössler & DeBach, 1973), which also implies homozygosity over time in much of the genome. In both cases an increasing frequency of diploid males is expected if either SLMA or MLMA models apply, but this does not happen. These thelytokous forms produce females generation after generation; thus, the sex locus models (SLMA and MLMA) do not apply. Moreover, Skinner and Werren (1980) sibmated *N. vitripennis* for six generations without the appearance of diploid males or increased mortality. Similarly, Cook (1993a) sibmated *G. nephantidis* for 22 generations without the appearance of diploid males or increased motality. Both of these results eliminate an MLMA model as an explanation for sex determination involving up to 6 loci in the case of *N. vitripennis* and up to 15 loci in the case of *G. nephantidis*. This pattern may be generally the case with species of Chalcidoidea and Cynipoidea, and with Bethylidae. These taxa manifest fewer male-biased sex ratios in cultures than does Ichneumonoidea when inbred (Luck *et al.,* 1992; Stouthamer *et al.,* 1992). Thus, sex determination in Hymenoptera differs among taxonomic groups and has clear implications for culture maintenance in the different taxa.

The inbreeding that occurs with small populations in culture is expected to increase homozygosity. In those taxonomic groups in which sex is determined by polymorphic sex loci, such homozygosity leads to diploid males. The rapidity with which this happens depends on, among other things, (1) the amount of genetic diversity in the sex-determining alleles among the individuals used to initiate the culture [for field example see Ross *et al.,* (1993)] (2) the effective population size [for field example see Kukuk and May (1990)] (3) the number of sex-determining loci involved, and (4) whether diploid males survive and are capable of inseminating females (Stouthamer *et al.,* 1992). Depending on the strain or species, diploid males either survive as well as haploid males and are capable of mating [e.g., *D. pulchellus* (El Agoze *et al.,* 1994; El Agoze & Periquet, 1993) or *Bombus terrestris* (Duchateau & Marien, 1995)], or suffer increased mortality and sterility [e.g., *Bracon hebetor* (Torvik, 1931; Inaba, 1939) or *Neodiprion nigroscutum* (Smith & Wallace, 1971)]. In *B. hebetor,* diploid sperm are unable to penetrate the egg (MacBride, 1946); and in the few cases in which the sperm do fertilize an egg, the resulting triploid females are usually sterile (Torvik, 1931).

Another aspect of genetic variation peculiar to the haplodiploid genetic system is the difference in inbreeding depression manifested by the male versus female genome. The genome can be classified into three functional parts: (1) genes that code for traits expressed in both males and females (e.g., traits such as general metabolism or flight); (2) genes that code for traits expressed only in females (e.g., traits associated with reproduction, pheromone production, fertilization, and mating behavior; and (3) genes that code for traits expressed only in males (e.g., traits associated with male mating behavior. Because of haplodiploidy, these three groups of genes are exposed to different selection regimes. Those traits expressed only in males are exposed to selection each generation in the hemizygous male. Such selection rapidly eliminates deleterious alleles from the genome. However, in each generation a majority of the copies of these genes are carried by the females but they are unexpressed. Thus, these copies are hidden from selection. Those traits expressed in both males and females are also exposed to selection in the male each generation. Thus, deleterious alleles will be eliminated in the male but, unless they are completely dominant, they will remain somewhat hidden in the diploid female as heterozygotes. Thus, the elimination of these alleles occurs more rapidly, although if they are completely recessive, the rate is approximately the same as that of the previous case. In contrast, deleterious, recessive alleles expressed only in the females may remain hidden within the genome at low frequencies for long periods. Such alleles are subjected to selection only when homozygous. Thus, when a population is inbred, females should be affected more than males as the increased homozygosity from inbreeding exposes the recessive, deleterious alleles. Estimates of the percentage of the genome expressed in both males and females, in males only, or in females only are difficult to obtain but the few that exist vary with the taxon (e.g., von Borstel & Smith, 1960; Crozier, 1976).

If cultures used in biological control are subjected to inbreeding, the genetic load (deleterious genes) hidden in the female-limited genome can influence the sex ratio by reducing the production of female offspring. The potential for such an effect is strongest in those species that outbreed [e.g., many Ichneumonoidea (Luck *et al.,* 1992)]. With increased inbreeding there is an increased chance that some of the rare, deleterious mutations in the female-limited genome may become much more common (due to genetic drift). These mutations affect such factors as viability of female embryos, fertility, mating, fertilization, and oviposition; thus, they will differentially affect the production of female offspring. Mortality of female embryos and an increased number of functionally virgin females (those incapable of mating or fertilizing their eggs) would result in an increasing percentage of male offspring in the culture. This differential effect on the production of female offspring along with the occurrence of diploid males may explain the low percentage of female progeny reported in laboratory cultures of several (outbreeding) Ichneumonidae and Braconidae (e.g., Bradley & Burgess, 1934; Simmonds, 1947; Oatman & Platner, 1974; Flanders & Oatman, 1982; Smith *et al.,* 1990) as compared with inbreeding Chalcidoidea (Schmieder & Whiting, 1947; Skinner & Werren, 1980; Fabritius, 1984) and perhaps Eucoilidae (Hey & Gargiulo, 1985). Cultures of parasitoid species that typically outbreed in nature are more likely to go extinct or to produce a low frequency of female offspring if the conditions within the cultures promote inbreeding. Inbreeding effects are exacerbated when cultures are initiated with a few individuals possessing a low diversity of sex-determining alleles or alleles that code for female-specific traits. Bottlenecks, unequal participation of individuals in matings, and unequal production of offspring all increase the rate at which alleles (i.e., genetic diversity) are lost in culture (Unruh *et al.,* 1983; Stouthamer *et al.,* 1992).

SEX-RATIO DISTORTING FACTORS

In addition to the interaction between inbreeding and the sex-determining mechanisms, sex-ratio distortion due to several other factors that alter offspring sex ratios have been detected in parasitic Hymenoptera [reviewed by Ebbert (1992)]. Some of them have the potential to extirpate a culture or to reduce its productivity. Sex-ratio distorters in these cases involve heritable elements within either the genome or the cytoplasm and include such factors as sperm morphology (Wilkes, 1964; Lee & Wilkes, 1965; Wilkes & Lee, 1965), the paternal sex-ratio (PSR) element (Werren *et al.*, 1981, 1987), primary male syndrome (Hunter *et al.*, 1993), maternal sex ratio (Skinner, 1982), son-killer trait (Skinner, 1985; Werren *et al.*, 1986), nonreciprocal cross-incompatibilities (Saul, 1961; Conner & Saul, 1986; Richardson *et al.*, 1987; Breeuwer & Werren, 1990), and thelytoky (Stouthamer *et al.*, 1990a, 1990b, 1993; Zchori-Fein *et al.*, 1992).

Sperm morphology can affect fertilization rates. Through selection, Wilkes (1964) increased the sex ratio (percentage of males) in the eulophid wasp *Dahlbominus fuscipennis* (Zetterstedt) from the normal 8% males to about 95% males. From crossing experiments between the high and normal sex-ratio lines, the high sex-ratio trait appeared to be genetic and expressed only in males. Males from the high sex-ratio line produced few female offspring when crossed with normal females whereas females from the high sex-ratio line produced normal sex ratios when crossed with males from the normal sex-ratio line. The cause appeared to be the low number of successfully fertilized eggs. Experiments showed that males of the normal sex-ratio strain produced two main types of sperm differing in the direction of the helix on the sperm head (i.e., dextral or sinistral) (Lee & Wilkes, 1965; Wilkes & Lee, 1965). Evidence suggested that the sinistrally coiled sperm was unable to penetrate the vitelline membrane of the egg, thus leaving the "fertilized" egg unfertilized and functionally haploid (Wilkes & Lee, 1965).

In normal chromosomal inheritance, the carriers of traits are located on the chromosomes; however, occasionally supernumary (B) chromosomes occur. One example of this is the parasitic element PSR (Werren *et al.*, 1981, 1987), which has been found in *Nasonia vitripennis*. Males carrying PSR cause the females they inseminate to produce only sons. Sperm carrying PSR fertilize an egg, but the paternal genome subsequently condenses and forms a nonfunctional mass (Reed, 1993; Dobson & Tanouye, 1996). The PSR element remains and is transmitted intact (Beukeboom & Werren, 1993). Thus, the fertilized egg carries the maternal (haploid) set of chromosomes plus the PSR element from the male. Such an egg gives rise to male offspring that carry the PSR trait (and element). When these males mate again with females, only male offspring with the PSR factor arise from the fertilized eggs. The dynamics of this element within a population are largely determined by the percentage of eggs that are fertilized (Skinner, 1987) and the level of population subdivision (Werren & Beukeboom, 1993). As long as the percentage of fertilization is less than 50%, the factor should decrease in frequency in a randomly mating population (Skinner, 1987), but in subdivided populations the PSR frequency may decline even if fertilization percentages are higher than 50% (Werren & Beukeboom, 1993). An additional example of a PSR-like factor is the

primary male syndrome in the aphelinid *Encarsia pergandiella* Howard reported by Hunter *et al.*, (1993).

Another form of non-Mendelian inheritance involves extrachromosomal factors. These factors may be microorganisms, microsporidia, plasmids, and mitochondria. Over the past 20 years several extrachromosomal factors that influence sex ratio have been detected in parasitic Hymenoptera. [However, not all microorganisms found in the ovaries of several species of Hymenoptera are known to affect their wasp hosts, for example, *D. fuscipennis* (Byers & Wilkes, 1970) and *Coccygominus turionellae* (Linnaeus) (Middeldorf & Ruthmann, 1984)].

Intensive studies of *N. vitripennis* have revealed at least three different extrachromosomal factors that distort sex ratios. These include (1) maternal sex ratio (MSR), (2) son-killer trait, and (3) nonreciprocal cross-incompatibilities (NRCI) or cytoplasmic incomptibility (CI). Females carrying the MSR factor produce male offspring when they are virgins, as expected. However, after mating practically all their offspring are female (Skinner, 1982). MSR has a strictly maternal inheritance that is consistent with a hypothesis that it is caused by a microorganism. However, its exact nature remains unknown. A similar pattern of sex allocation has been observed in *Coccophagus lyciminia* (Walker) (Flanders, 1943). Virgin females produce only male offspring whereas mated females produce only female offspring. However, neither the cause nor the mode of inheritance of this trait is known.

Son-killer trait (Skinner, 1985) in *N. vitripennis* is caused by a rod-shaped bacterium (Werren *et al.*, 1986; Gherna *et al.*, 1991). Infection with this bacterium leads to the death in the larval stage of male offspring arising from unfertilized eggs. It does not kill the offspring arising from fertilized eggs (females and PSR males) (Werren *et al.*, 1986). The son-killer bacterium infects many different tissues and its transmission from mother to offspring most likely takes place through the hemolymph of the parasitized host (Huger *et al.*, 1985). In Hymenoptera species no other confirmed cases of the son-killer bacteria are known; however, the symptoms described in a strain of *Caraphractus cinctus* Walker by Jackson (1958) are consistent with a son-killing bacterium. Virgin females of a low sex-ratio strain produced very few male offspring, about 3% of what the normal line produces, whereas mated females from both normal and low sex-ratio lines produced similar numbers of females. Male killing factors similar to those in *N. vitripennis* are known from several Coccinellidae (Matsuka *et al.*, 1975; Kai, 1979; Gotoh, 1982; Gotoh & Niijima, 1986; Hurst *et al.*, 1993, 1994, 1997), where several different bacteria are able to cause male killing; for review see Hurst *et al.* (1997).

In contrast to the factors discussed earlier, (NRCI) only becomes evident in crosses between strains. It is usually due to a cytoplasmic incompatibility in which one strain carries a Rickettsia-like microorganism that the other lacks. Eggs containing microorganisms (*Wolbachia*) [for review see Werren (1997)] are compatible with sperm from both infected and uninfected males, whereas eggs free of microorganisms can only be fertilized successfully by sperm from uninfected males. In Hymenoptera species this trait results in all male offspring in crosses between infected males and uninfected females. The reciprocal crosses in which uninfected males are crossed with infected females result in offspring with a normal sex ratio. Although transmission of this trait appears to be purely through the maternal line (Saul, 1961), it can be acquired by wasps in laboratory cultures (Conner & Saul, 1986), possibly through their hosts. The incompatibility can be removed by antibiotic treatments (Richardson *et al.*, 1987; Breeuwer & Werren, 1990). Microorganism-induced NRCI has only been documented in parasitic Hymenoptera in *Nasonia* spp. (Breeuwer & Werren, 1990). In other species of Hymenoptera [*Leptopilina heteretoma* (= *Pseudocoila bochi*) (Veerkamp, 1980), *Aphidius ervi* Haliday and *A. pulcher* (Mackauer, 1969), and several *Trichogramma* spp. from North America (Nagarkatti & Fazaluddin, 1973; Pinto *et al.*, 1991) and Europe] similar incompatibilities have been found but their cause has not been determined. Such determination is important because the cause is not always extrachromosomal. In *T. deion* Pinto & Oatman an NRCI between two strains appears to be caused by an incompatibility between nuclear genes (Stouthamer *et al.*, 1996).

Microorganisms can have yet another effect: they are associated with thelytoky in several species of Hymenoptera. While not all thelytokous species have a microbial cause (Stouthamer, 1989), bacteria of the genus *Wolbachia* have been found in thelytokous *Trichogramma* spp. and *M. uniraptor* (Stouthamer & Werren, 1993; Stouthamer *et al.*, 1993). In addition, circumstantial evidence suggests that infection with the *Wolbachia* bacteria may be widespread in other thelytokous species such as *E. formosa* (Zchori-Fein *et al.*, 1992), *Apoanagyrus* (= *Epidinocarsis*) *diversicornis* (Howard) (Pijls *et al.*, 1996), and other species (see Stouthamer *et al.*, 1990a; Stouthamer, 1997). For mass rearings, thelytoky has an advantage over arrhenotoky in that all the hosts produced will be used for the production of female wasps. Because only females are the effective biological control agents, the use of thelytokous wasps can result in a cost reduction. An often assumed advantage of thelytokous over arrhenotokous wasps is the higher rate of population increase. This is based on the assumption that thelytokous females will produce equal numbers of offspring as their arrhenotokous conspecifics. However, in the case of the *Wolbachia*-associated thelytoky in *Trichogramma*, there appears to be a substantial cost of the infection on the fecundity of the thelytokous females (Stouthamer & Luck, 1993), although in other species this cost may not be present (Stouthamer *et al.*, 1994; Horjus & Stouthamer, 1995). Aeschlimann (1990) has speculated that

using the thelytokous forms of a wasp species may be more effective for biological control than using the arrhenotokous form. However, which form will be more effective in biological control will depend on the life history of the wasp and the density of the host that needs to be controlled (Stouthamer, 1993).

SEX RATIO IN HETERONOMOUS PARASITOIDS

Species in which males and females parasitize different hosts (e.g., Hunter *et al.*, 1996) or in which the female is a primary parasitoid and the male is a secondary (hyper-) parasitoid are called heteronomous parasitoids (Walter, 1983). Heteronomous parasitoids are found in eight genera of Aphelinidae: *Aneristus, Coccophagus, Euxanthellus, Prococcophagus, Lounsburia, Physcus, Coccophagoides,* and *Encarsia;* and the series of unusual male ontogenies evinced among these genera have been classified by Walter (1983). The best studied cases involve heteronomous hyperparasitism, in which females are primary endoparasitoids and the males are hyperparasitiods developing on a larva or pupa of their own species (typically a female) or of another species of internal parasitoid. The sex ratio in such wasps is constrained by the availability of suitable hosts for either male or female offspring and not merely by the decision of a female to fertilize her eggs. Hunter (1989) found that female *E. pergandiella* Howard preferred to hyperparasitize (lay male eggs) and that the frequency of male or female eggs did not depend on the frequency or abundance of suitable hosts. In contrast, Donaldson and Walter (1991a, 1991b) report that the sex ratio of the heteronomous hyperparasitoid *C. atratus* Compere in the laboratory and in the field depends on the ratio of host types available (i.e., those on which male versus female offspring are produced). Several studies have indicated that the sex ratio is influenced by whether the hosts that are available for the production of sons are conspecific or heterospecific. A greater proportion of sons are produced if heterospecific hosts are available (Avilla & Copland, 1987; Avilla *et al.*, 1991; Williams, 1991). Although such species may preferentially hyperparasitize (= sons), most of the suitable hosts they encounter in the field will generally be unparasitized hosts, leading to female-biased sex ratios (Rice, 1937; Neuffer, 1964; Smith *et al.*, 1964; Hunter, 1989; Hunter,1993).

Godfray and Waage (1990) and Godfray and Hunter (1994) proposed that the sex ratio of heteronomous parasitoids should be strongly influenced by the availability of their host types (i.e., those for sons versus those for daughters). Sons are laid in previously parasitized hosts or they are laid in a different host species [e.g., a lepidopteran egg (Hunter *et al.*, 1996)]. Daughters are laid as primary parasitoids in immature homopterans. They considered two cases, one in which the heteronomous female is limited by the eggs she has available and a second case in which she is limited by the time (= life expectancy) she has available. In the case in which she is egg limited, she should produce equal numbers of sons and daughters as Fisher's (1930) rule suggests. Thus the sex ratio produced under these conditions is expected to be 1:1. In contrast, a time-limited heteronomous parasitoid that searches a habitat that contains both daughter-producing and son-producing hosts should allocate sons and daughters based on whatever host type she encounters. Thus, the time-limited parasitoid is in effect, host limited and the sex ratio produced should reflect the relative abundance of host types that the female encounters. In reality, conditions may often lie between these extremes, with the relative abundance of each host type interacting with a female's degree of egg limitation, causing the female strategy to switch depending on conditions. (Hunter & Godfray, 1995). Experiments to test these predictions resulted in qualitative agreement with the predictions (Hunter & Godfray, 1995).

Heteronomous parasitoids have been important in biological control (e.g., Smith *et al.,* 1964; Huffacker & Kennett, 1966) but care in rearing and release may be necessary. Under laboratory-rearing conditions extreme female-biased sex ratios can be obtained (J. Williams, 1972; Avilla & Copland, 1987; Avilla *et al.,* 1991; Stouthamer & Luck, 1991; T. Williams, 1991). Often, male eggs can be laid only in hosts that contain a parasitoid larva or pupa; therefore, in the laboratory when mated females are exposed to large numbers of unparasitized hosts, they will encounter only hosts suitable for the production of female progeny. Hosts suitable for the production of male progeny will only occur in the culture when the female larvae have reached a suitable stage for hyperparasitism. Removing the parasitized host material before the females can hyperparasitize some previously parasitized hosts or before the hyperparasitoids (males) emerge will extirpate the culture. Similar problems occur when wasps are released in the field for biological control depending on the complex of parasitoid species already present on the host. If other parasitiod species are present, some of these may serve as hosts for the male offspring (hyperparasitoids). If the only hosts available for the production of sons are previously parasitized hosts by conspecific females, then the released females will encounter mainly unparasitized hosts and, therefore, produce mostly female offspring. In this case, to improve colonization success two releases should be made at a location. The second release should be timed so that the female offspring of the first release have reached a suitable stage for the production of sons by females of the second release (Stouthamer & Luck, 1991).

POSTCOPULATION FERTILIZATION LAGS

The absence or low frequency of female offspring in recently mated females (postcopulation lag in fertilization)

has been reported in several Hymenoptera: *Aphidius smithii* Sharma & Subba Rao (Mackauer, 1976), *Chelonus kelliea* Marsh and *C. phthorimaeae* Gahan (Powers & Oatman, 1985), *Orgilus jenniae* Marsh (Flanders & Oatman, 1982), *B. hebetor* (Genieys, 1925), *N. vitripennis* (van den Assem & Feuth-De Bruijn, 1977), and *Campoletis flavicincta* (Ashmead) (Hoelscher & Vinson, 1971). The duration of these lags varies but ranges from 2 to 6 h before the first fertilized eggs are laid. It may also be related to the time that has elapsed between emergence of the female and mating (Hoelscher & Vinson, 1971). Such lags have not been detected in *Apanteles subandinus* Blanchard (Cardona & Oatman, 1975) or *A. dignus* Muesebeck (Cardona & Oatman, 1971). Although the cause of these postcopulation lags is unknown, it may be due to the time it takes the sperm to reach the spermatheca and the filled spermatheca to become functional. In *D. fuscipennis* the movement of the sperm from the vagina to the spermatheca is completed in about 30 min (Wilkes, 1963).

MULTIPLE COPULATIONS

In *Macrocentrus ancylivorus* Rohwer females the number of matings is known to influence the sex ratio. In this species males transfer a spermatophore during copulation (Flanders, 1945). For successful insemination to occur, the spermatophore must be aligned with the duct though which the sperm moves to the spermatheca. Multiple matings result in several spermatophores that are lodged in the vagina, thus impeding the proper alignment of the spermatophores and preventing the sperm from entering the spermatheca (Flanders, 1945). Thus, multiple mating in this parasitoid results in functionally virgin females. Van den Assem and Feuth-De Bruijn (1977) report that *N. vitripennis* females that mate a second time lay a high percentage of male offspring during the first 24 h after mating. Allen *et al.* (1994) reported a low frequency of multiple mating in a field population of *Aphytis melinus* (6%). Higher frequences occurred when females of this species where confined with males under laboratory conditions. The males guarded the females after they inseminated them. If the guarding male was removed at various times after insemination and a second male was allowed to inseminate the female, sperm was transferred to the female. The amount of sperm that was transferred by the second male decreased as a function of the time the first male was allowed to guard the female. Sperm from the second male was used to fertilize fewer eggs than the first male's sperm, but use of sperm from both males appears to be random.

TEMPERATURE AND PHOTOPERIOD

Photoperiod influences the sex ratio of *C. flavicincta* (Ashmead) (Hoelscher & Vinson, 1971) and *Pteromalus puparium* Linnaeus (Bouletreau, 1976). The greatest percentage of female *C. flavicincta* was observed at photoperiods of 12:12 L to D after the F_2 generation in the laboratory culture, and it increased over the six generations of the experiment (generation 1: 17% female; generation 6: 33% female). However, the sex ratio remained male biased, as was typical of field populations (range 57 to 70% male). At all other photoperiods the laboratory culture was more male biased. At 8:16 (L to D) the sex ratio averaged 86% males; at 10:14, it averaged 92% males; at 14:10, it averaged 81% males; and at 16:8, it averaged 88% males. When reared in all dark or in all light, the cultures became 100% male after three generations. With the exception of the 12:12 light regimen, the male bias in the sex ratio increased each generation. The effects of photoperiod on sex ratio in *C. flavicincta* may have been confounded by the occurrence of diploid males that arose from inbreeding. However, this does not explain the sex ratios that occurred at 12:12. This system clearly calls for more research. *Pteromalus puparium* produced 20 to 30% males at a photoperiod of 10:14 L to D and 60 to 70% males at 14:10 L to D. Female fecundity and the proportion of inseminated females were unaffected by photoperiod (Bouletreau, 1976).

Temperature extremes also affect the sex ratio of progeny from parasitic species of Hymenoptera. Exposure of *Dahlbominus fuscipennis* larvae and pupae to high temperatures caused extreme variation in the sex ratio of progeny produced by the resulting females (Wilkes, 1959). Under optimal rearing conditions (23°C), *D. fuginosus* produced 90% daughters. After 5 days postembryonic exposure to 27°C, females produced only 54% daughters; at 29°C, 38% daughters; and at 31°C, 17% daughters. Kfir and Luck (1979) found a similar effect when *Aphytis* spp. were exposed to high temperatures, but one that differed between congeners. Compared with a sex ratio of about 60% female at 27°C, *A. melinus* DeBach adults produced 51% females when they were exposed as pupae to 32°C, and 37% females when they were reared at temperatures that fluctuated between 29 and 32°C (24-h cycle). *Aphytis lingnanensis* Compere was more severely affected than was *A. melinus*. Adult *A. lingnanensis* exposed as pupae to 32°C produced 18% female progeny whereas adults exposed as immatures to temperatures that fluctuated between 29 and 32°C produced 24% females.

Low temperatures also affect the sex ratio produced by parents. DeBach and Rao (1968) reported that adults of *A. lingnanensis* exposed to −1°C were completely sterilized. Sperm were killed in the testes and spermatheca (after mating). Kfir and Luck (1979) also reported a decrease in the percentage of female progeny of *A. lingnanensis* and *A. melinus* exposed to low temperatures. *Aphytis lingnanensis* and *A. melinus* produced 24 and 47% females, respectively, when the adult females were exposed to 2°C for 7 h. before oviposition, and 16 and 24%, respectively, when exposed

to −2°C for 4 h. Shorter periods of exposure to 2°C only marginally affected the sex ratio of these parasitoids. Longer periods of exposure to −2°C resulted in no offspring (*A. lingnanensis*) or few offspring (*A. melinus*), although in the latter case, 9 of the 16 offspring were female. Continuously rearing *C. flavicincta* at a constant temperature of 32 or 18°C resulted in a more male-biased sex ratio than when this species was reared at 27 or 21°C (Hoelscher & Vinson, 1971).

Low-temperature effects can be particularly important if parasitoids are stored at low temperatures to retard development. Both high- and low-temperature effects may come into play when parasitoids are shipped and exposed to temperature extremes while in transit. Problems can be reduced by insulating the cultures while they are in transit to protect them when they are exposed to ambient conditions when on the loading dock or when they are being transferred from one vehicle or aircraft to another. It is at this stage that many parasitoid shipments have met their demise.

SEX ALLOCATION

When a female parasitoid encounters a patch of one or more hosts, she must "decide" which hosts, if any, to parasitize and how many eggs to lay (Waage & Ng, 1984). The number of eggs laid per host by gregarious parasitoids during a visit (= clutch size) is limited in large part by host size (e.g., Klomp & Teerink, 1962; Gordh, 1976; Luck *et al.*, 1982; Schmidt & Smith, 1987; Bai *et al.*, 1992; Hardy *et al.*, 1992; Mayhew & Godfray, 1997). In addition, many species of parasitic Hymenoptera adjust the sex ratio of their offspring at oviposition to environmental conditions such as the presence of other females or the quality of hosts (Charnov, 1979, 1982; King, 1987, 1992; Frank, 1990; Luck *et al.*, 1992; Godfray, 1994). Because genes in a parental generation have two means of entering subsequent generations (i.e., either through sons or through daughters), selection favors the genomes of parents who choose the appropriate mixture of offspring genders that maximize both the number of surviving offspring and their reproductive success. Thus, natural selection favors females who allocate combinations of clutch size and gender that maximize the number of grandchildren (= a female's reproductive success or fitness) (Taylor & Bulmer, 1980; Taylor, 1981; Charnov, 1982; Charnov & Skinner, 1984, 1985; Frank, 1985; Hardy, 1994; Mayhew & Godfray, 1997; Nunney, 1985; Nunney & Luck, 1988; Waage & Godfray, 1985; Godfray, 1986, 1994; Waage, 1986; Waage & Ng, 1984).

In panmictic (randomly mating) populations, natural selection favors parents who invest equally in daughters and sons (Fisher, 1930); if each gender is of equal cost, a 1:1 sex ratio is expected. When population sex ratios deviate from equality, parents who lay more eggs of the less common sex will derive more reproductive success on average. (Their progeny participate in more matings, on average; therefore, more of their genes will be represented in future generations when compared with parents who do not skew their progeny's sex ratio to the uncommon sex.) Thus, frequency-dependent selection maintains the optimal sex ratio.

However, not all populations are panmictic. As Hamilton (1967) first noted, natural selection favors parents who invest more in daughters than in sons when populations are structured as patches (i.e., where hosts occur in patches such as egg masses or loose clusters of eggs), or where several offspring are laid in a single host (= gregarious parasitoids) (van den Assem *et al.,* 1980). Hamilton's (1967) model assumed that (1) all the parental females were equally fecund, (2) the offspring of these females mate randomly before dispersal of the inseminated female offspring, and (3) the mating groups are limited to the offspring emerging from the natal patch.

One prediction made by this model is that the sex ratio (percentages of males) should increase as the number of females contributing offspring to a patch increases (Fig. 1). When a single female is the sole contributor of offspring to a patch, the model predicts that all of the offspring in the patch will be daughters, but to satisfy preceding assumptions (2) and (3), it is necessary that sufficient sons are allocated to ensure that all the daughters are inseminated before they emigrate (Hartl, 1971). The pattern of increasing sex ratio with an increasing number of contributing females has been observed in four families of parasitic wasps:

1. Scelionid wasps, which are solitary egg parasitoids attacking either solitary host eggs or host egg masses of varying size (Viktorov, 1968; Viktorov & Kochetova, 1973; Schwartz & Gerling, 1974; Waage, 1982; van Welzen & Waage, 1987; Strand, 1988; Noda & Hirose, 1989)

2. Pteromalids *Eupteromalus dubius* (Ashmead) (Wylie, 1976); *Muscidifurax raptor* Girault & Sanders (King & Seidel, 1993); *N. vitripennis,* a gregarious pupal parasitoid of cyclorrhaphous flies [Werren, 1980, 1983, 1984; Parker & Orzack, 1985; Orzack & Parker, 1990; King & Skinner, 1991, but see Orzack, 1990, 1992]; *Pachycrepoideus vindemiae* (Rondani), a solitary parasitoid of cyclorrhaphous and drosophilid flies that occur in patches (Nadel, 1985); *Spalangia cameroni* Perkins (King 1989), solitary parasitoids of cyclorrhaphous flies; and *Pteromalus puparium* (Takagi 1985), a gregarious parasitoid of lepidopteran pupae

3. Trichogrammatids *T. japonicum* Ashmead (Kuno, 1962), *T. evanescens* Westwood (Waage & Lane, 1984; van Dijken & Waage, 1987), *T. chilonis* Ishii (Suzuki *et al.,* 1984), *T. lutea* Girault (Kfir, 1982), and *T. pretiosum* Riley (Luck *et al.,* in lit.), all of which are facultatively gregarious parasitoids of lepidopteran eggs

4. Bethylid wasps (Griffiths & Godfray, 1988; Morgan & Cook, 1994).

The initial model proposed by Hamilton (1967), however, must be modified to include the unequal relatedness of sons and daughters to their mothers that results from inbreeding in haplodiploid species (Hamilton, 1979). The sibmating that occurs in the local mating groups results in inbreeding because some of the matings involve sibs sharing the same mother. This inbreeding results in daughters that are more related than sons to their mothers. Sons possess only one set of chromosomes, which they inherit only from their mother. Because their mother possesses two sets of chromosomes, sons always carry exactly 50% of their mother's genes. In contrast, daughters possess two sets of chromosomes, one from their mother and one from their father. When the mother's and father's genetic complements are similar because of inbreeding, the daughters inherit 50% of the mother's genes directly, plus they inherit more maternal genes indirectly through the father because some of his genes are identical by descent to the mother's. Thus, the daughters possess more of the mother's genes than the sons possess. Compared with Hamilton's (1967) original diploid model, this asymmetrical relatedness of daughters and sons to their mother leads to a slower sex-ratio increase (percentage of males) with an increase in the number of contributing females (see Fig. 1). However, under panmixia (i.e., no inbreeding) equal investment in daughters and sons is still favored with haplodiploidy (Hartl & Brown, 1970; Hartl, 1971).

Several factors apparently favor the evolution of biased sex ratios. Hamilton (1967) suggested that it is competition among brothers in a mating group for access to mates that selects for female-biased sex ratios; hence, his notion has been referred to as local mate competition (LMC). Alternatively, Maynard Smith (1978) proposed that sibmating is

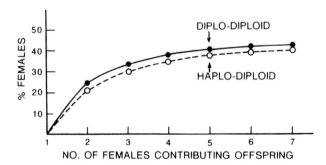

FIGURE 1 The predicted relationship between the number of females contributing offspring to a mating group and the sex ratio of their offspring (percentage of males) with local parental control (LPC) with diplodiploid species (solid line) and haplodiploid species (dashed line). See text for additional explanation.

the factor selecting for female-biased sex ratios. Sibmated daughters pass on twice as many parental genes, and in patches founded by one to several mothers, the opportunity for sibmating is high. If sons mate with their sisters (sibmating) or if they avoid competing with their brothers for mates (LMC), then parental fitness is increased. Thus, females who allocate more resources to daughters are favored if this increases sibmating or reduces competition among brothers for access to mates.

Alternatively, we may assess the influence a mother has over the sex ratio of the mating group to which she contributes offspring. The more influence she has, the more female biased should be her offspring's sex ratio. As Nunney (1985) has pointed out, sibmating and LMC are both the expression of one underlying process: local parental control (LPC). With a fixed clutch size allocated by each female, a decision to lay a female egg increases sibmating but it also is a decision to lay one less male egg, which reduces LMC among sons. LPC (Nunney, 1985) refers to a female's ability to control her fitness by affecting the sex ratio of the mating group to which she contributes offspring. The more influence she has over the sex ratio of the mating group (and hence over the reproductive success of her sons) the more she should invest in daughters. If she is the sole contributor of offspring and these offspring mate only among themselves, then she has complete control of the mating group's sex ratio. As the number of other females contributing offspring to the group increases, each female has a decreasing influence on the sex ratio of the mating group and she should reduce her investment in daughters. Decreasing an individual female's influence to zero approaches panmictic mating conditions that favor equal investment in daughters and sons. The degree of LPC is not only dependent on the number of founding females; for example, LPC decreases if males disperse and successfully invade other mating groups (Nunney & Luck, 1988) or if females remate after dispersal. Thus, by assessing whether factors increase or decrease the mother's influence on the sex ratio of her offspring's mating group, it is easy to propose an expected direction of a sex ratio shift (i.e., more or less investment in daughters).

Regardless of the assumed selective factors involved, theory predicts that female-biased sex ratios are to be expected when offspring arise from one or a few mothers and the offspring mate mostly among themselves (Taylor & Bulmer, 1980; Charnov, 1982; Nunney & Luck, 1988; King, 1992; Luck *et al.*, 1992; Godfray, 1994; Hardy, 1994). Such sex ratios appear to be common where parasitic hymenopterans lay eggs in hosts that occur as egg masses or as clustered hosts, or where the parasitoid is gregarious (e.g., Arthur & Wylie, 1959; Wilkes, 1963; Walker, 1967; Viktorov, 1968; van den Assem, 1971; Holmes, 1972; Viktorov & Kochetova, 1973; Schwartz & Gerling, 1974; Wy-

lie, 1976; Hamilton, 1979; Werren, 1980; 1983; 1984; Waage, 1982; van Alphen & Thunnissen, 1983; van den Assem *et al.*, 1984; Waage & Lane, 1984; Nadel, 1985; Putters & van den Assem, 1985; King, 1987, 1989; Orzack & Parker, 1990; Noda & Hirose, 1989; King & Skinner, 1991; Morgan & Cook, 1994; Mayhew & Godfray, 1997; Luck & Janssen, submitted).

When offspring in a patch arise from a single mother, her reproductive success is proportional to the number of mated daughters that emigrate from that patch (Hartl, 1971; Green *et al.*, 1982; Hardy, 1992). These circumstances favor a mother who lays just enough sons in the patch to ensure that all of her daughters are fertilized (Hartl, 1971). This has been termed precise sex allocation (Green *et al.*, 1982; Hardy, 1992). Thus, in small patches in which only a few offspring are laid (< 8), precise sex allocation implies the allocation of a single son. This pattern contrasts with that arising from a random ($=$ a binomial) allocation of sex (Hartl, 1971; Green *et al.*, 1982; Hardy, 1992). Binomial allocation of sex yields some patches that contain only sons, others containing only daughters, and still others containing a variable mixture of sons and daughters. With LMC, daughters that emerge in patches lacking sons emigrate unmated and may be constrained to lay only sons (Godfray, 1990). In those patches in which more than one son is laid, the sons compete with each other for access to their sisters; hence, the sons are wasted genetic resources that would be better spent as daughters. Consequently, LMC favors precise sex allocation because it reduces the variation in the number of sons per patch, thus maximizing the number of inseminated daughters (Hartl, 1971; Green *et al.*, 1982; Hardy, 1992). This also can be seen in terms of LPC: a mother producing a precise sex ratio maximizes her control over the reproductive fitness of her offspring.

As the number of mothers allocating offspring to a patch increases, an increasing number of the offspring are expected to be sons. The presence of daughters from unrelated mothers in the patch offers additional routes by which a mother's genes can enter the next generation via her sons mating with the other females' daughters (Hamilton, 1967; Taylor & Bulmer, 1980; Charnov, 1982). Precise allocation of sex is still favored because it reduces the variance in the number of sons allocated per clutch (Hardy, 1992; Nagelkerke & Hardy, 1994; Luck *et al.*, in lit.) and maximizes maternal control over their reproductive success (Nunney, 1985). The betylids (Green *et al.*, 1992; Mayhew & Godfray, 1997), and *T. pretiosum* (Luck & Janssen, submitted) manifest precise sex allocation. In the case of *T. pretiosum*, it also manifests precise sex allocation when several mothers are allocating offspring to a patch (Luck & Janssen, submitted).

Sex-ratio adjustments such as these have important implications for culturing parasitoids. In species that adjust their progeny's sex ratio based on the number of mothers allocating offspring to a patch, the presence of other females in a culture (i.e., moderate to high parasitoid–host ratios) under circumstances in which they are exposed to hosts for extended periods likely increases the production of male offspring at the expense of female offspring. Confining the females of some species of parasitoids together prior to or while exposing them to hosts results in less female-biased sex ratios than if they were individually confined with hosts (Viktorov, 1968; Wylie, 1976; Werren, 1980; Orzack & Parker, 1986). Because the production of female offspring is the desired product for field release, the production of male offspring wastes resources. To the extent that is practical, isolating parental females, each to their own patch and limiting the time that hosts are exposed to the females, may increase the production of female offspring. The percentage of females varying their sex ratios range, for example, from a high of 82% females to a low of 50% or less (e.g., Bai *et al.,* 1992).

SEX RATIO AND VARIATION IN FITNESS

With some species of Hymenoptera (principally ectoparasitoids) the size distribution of available hosts influences sex allocation and the optimal sex ratio (Charnov, 1979, 1982; Charnov *et al.,* 1981; Green, 1982; Frank, 1990). When hosts vary in size, natural selection favors parents who allocate sons and daughters to hosts based on host size if the reproductive consequences of being small differs among the sexes (Charnov, 1979, 1982; Charnov *et al.,* 1981; Werren, 1984; King, 1987, 1992; Frank; 1986, 1990). If, for example, a female loses more fitness by laying small daughters than she does by laying small sons, then the female is expected to allocate daughters to the larger hosts and sons to the smaller hosts (Charnov *et al.,* 1981; Charnov, 1982; Green, 1982; Frank, 1990; Godfray, 1994). This expectation assumes that (1) the wasps use a range of host sizes, (2) a single egg is laid per host, (3) the host occurs as a nongrowing stage or is paralyzed when it is parasitized so that all the food available to the developing offspring is contained in the host at the time it is attacked, and (4) the female controls the gender of her offspring via arrhenotoky (as in most parasitic Hymenoptera) (Charnov *et al.,* 1981; Charnov, 1982).

A model developed by Charnov (Charnov *et al.,* 1981; Charnov, 1982) and based on these assumptions predicts an optimal strategy in which only sons are allocated to small hosts and daughters to large hosts. This strategy requires that the parental female manipulates the primary sex ratio. Thus, to test whether Charnov's model is an appropriate explanation for a relationship between host size and an offspring's gender, the sex of the eggs at oviposition must be assessed. Determining the sex ratio at adult emergence

(the secondary sex ratio) is an inappropriate test because this may not reflect the primary sex ratio. For example, daughters are presumed to require more resources than sons to reach pupation weight. Female offspring allocated to small hosts may deplete the available resources and die before maturation (Hare & Luck, 1991). In contrast, the resources present in small hosts may be sufficient for a male wasp to complete its development. Thus, the observation that only males emerge from small hosts may arise from the differential death of females and not from the mother's manipulation of the sex ratio.

Charnov's model also predicts that the shift from sons to daughters is sudden and occurs at a specific host size (Charnov *et al.,* 1981; Charnov, 1982). However, as Charnov points out, in reality, this shift is more likely to be gradual because wasps probably vary in their allocation of sons and daughters because the relative value of being a large son versus a large daughter may differ seasonally or geographically. The model also assumes that the experimentalist classifies the reproductive value of a host in the same way as the wasp and that the wasp classifies the value of the host without error. Both of these assumptions seem unlikely.

A third prediction of Charnov's model has to do with the concept of large versus small hosts. A host is only large or small relative to the size of the other hosts with which it is associated (Charnov *et al.,* 1981; Charnov, 1982). With changes in the size distribution of hosts the model predicts that the size of hosts to which female eggs are allocated will change. If the mean size of hosts decreases, then the proportion of female eggs allocated to a given host size will increase compared with those allocated to the same host size when it is part of a distribution of hosts with a larger mean. Although Charnov's model assumes a panmictic mating structure, Werren (1984) has demonstrated that this assumption is not essential to the evolution of a size-dependent sex allocation pattern. A number of species of parasitic wasps are known to allocate males to small hosts and females to large hosts (Clausen, 1939; Arthur & Wylie, 1959; van den Assem, 1971; Sandlan, 1979; Charnov *et al.,* 1981; Charnov, 1982; van den Assem *et al.,* 1984; King, 1988, 1990; de Jong & van Alphen, 1989; Werren & Simbolotti, 1989; Heinz & Parrella, 1990; Nishimura, 1993; van Dijken *et al.,* 1993; Bokonon-Ganta *et al.,* 1995; Heinz, 1996; Lampson *et al.,* 1996; Bernal *et al.,* 1998); see King (1987) for a review. In solitary parasitoids differential mortality has been eliminated as an explanation for size-dependent gender allocation in several species (Sandlan, 1979; Charnov *et al.,* 1981; van den Assem, *et al.,* 1984; Luck & Podoler, 1985; Opp & Luck, 1986; King, 1987, 1988; Heinz & Parrella, 1990).

Crucial to Charnov's (1982) variation-in-fitness hypothesis is that the fitness consequences of being small differ between the sexes. Females are assumed to lose more fitness than males lose by being small. Van den Assem *et al.* (1989) and Heinz (1991) determined the relative value of small females versus small males in the laboratory. They found that both male and female reproductive success increased with increasing wasp size, but the female's reproductive success increased more rapidly than that of the male. The male was equally effective at courting and inseminating females over a range of sizes whereas the fecundity of females increased with female size. Allen *et al.* (1994) found that larger *A. melinus* males were more likely to mate when large and small males competed for access to a virgin female. Similarly, Kazmer and Luck (1995) found that a greater fraction of the larger than smaller *T. pretiosum* males emerging from hosts parasitized in the field were collected at traps with caged, virgin females. Thus, larger members of both sexes are more successful at finding resources, but larger females have the additional advantage of increased fecundity, walking speed, and dispersal. Bigler *et al.* (1987) determined the percentage of hosts parasitized by small- and intermediate-sized wasps (a direct measure of a female's reproductive success) reared on factitious hosts and released in the field. In each of four releases the intermediate-sized wasps parasitized more hosts than the smaller wasps. The difference in the percentage of hosts parasitized consistently favored the intermediate-sized wasps but the difference was not significant. Large *T. pretiosum* females parasitized significantly more hosts in the field than small females and a greater fraction of the larger than smaller females emerging from hosts parasitized in the field are collected at host patches placed in the field (Kazmer & Luck, 1991, 1995). Similar results were obtained with *Aphaereta minuta* (Visser, 1994) and *Achrysocaroides zwoelferi* (West *et al.*, 1996). Furthermore, in a time-limited ectoparasitoid (koinobiont), large *Asobara tabida* females dispersed further than smaller females did, which increased their probability of encountering a host patch (Ellers *et al.*, 1998). These results imply that larger females were more successful than were smaller females at finding hosts.

These results may be consistent with Charnov's model but a more rigorous test of the differences in relative size-dependent fitness among the sexes is needed. However, the results do have implications for the culturing of parasitoids. Size-related fitness is important in the production of female offspring because this fitness is, in part, the same as parasitoid quality. Thus, in cultures of biological control agents, host quality is important to the production of fit females for release, especially for those used in augmentative biological control. Moreover, the size of the host used in a culture can influence the sex ratio of the wasps produced (Salt, 1935; Klomp & Teerink, 1962, 1967; Bai *et al.*, 1992). For example, use of small diaspidid scale insects to culture *Aphytis melinus* resulted in a high frequency of male wasps and a few small female wasps, whereas large-scale insects produced a strongly female-biased sex ratio (Abdelrahman, 1974).

LOCAL PARENTAL CONTROL AND VARIATION IN FITNESS–HOST SIZE INTERACTIONS

Werren (1984) investigated the combined effects of LPC and host size (quality) on the expected sex ratios of parasitic wasps. He assumed hosts were of two types, large and small. In cases in which female offspring benefit more in fitness than males from large hosts, parents should allocate a greater fraction of sons to the smaller hosts. The expected sex ratio depends on the fraction of the small hosts among the hosts parasitized. The change in sex ratio as this fraction is increased can be characterized in three parts: (1) when few small hosts are available, a Hamiltonian sex ratio is expected; (2) when an intermediate fraction of the hosts available are small, a linearly increasing sex ratio (percentage of males) is expected because males are exclusively produced on the small hosts; and (3) when mostly small hosts are available, the sex ratio (percentage of males) is expected to decline back to the Hamiltonian expectation, because females are increasingly produced from small hosts. The degree to which the sex ratio deviates from the Hamiltonian sex ratio depends on the relative fitness of daughters produced on the small hosts: the less fit the daughter, the greater the sex-ratio increase when an intermediate fraction of parasitized hosts are small, so that when a large number of mothers contribute offspring to a mating group, a male-biased sex ratio is expected. Thus, the average number of females contributing offspring and the relative fitness of daughters produced on small hosts combine to determine the overall sex ratio. In a laboratory test of this model, Werren and Simbolottii (1989) found that the sex ratio (percentage of males) increased as the proportion of small hosts (0.8 mm) increased in *Larophagus distinguendus*. This result contrasts with the predicted result that a Hamiltonian sex ratio is expected as the proportion of small hosts increase (Werren, 1984; Werren & Simbolotti, 1989). However, this response depended on the size of the small hosts offered. If the small hosts were either 1.0 or 1.3 mm in size, then the prediction of the model is borne out; if they were 0.8 mm, it was not. Mayhew and Godfray (1997) found that single males of *Laelius pedatus* were allocated to small hosts, single females were allocated to larger hosts, and gregarious broods were allocated to larger hosts. This implies that a critical host size exists below which a female offspring is of little reproductive value. *Spalangia cameroni* has also been demonstrated to shift its offspring's sex ratio in response to both changing host size and number of other females allocating offspring to other hosts in a patch (King, 1988, 1989).

CLUTCH SIZE AND SEX RATIO

With gregarious parasitoids (species that lay a clutch of several eggs per host), the size of the initial clutch is correlated with host size (Salt, 1935, 1940; Klomp & Teerink, 1962; Gordh, 1976; Luck *et al.,* 1982; Schmidt & Smith, 1987; Bai *et al.,* 1992). A parasitoid measures the size of the host [e.g., *Trichogramma* (Klomp & Teerink, 1962; Schmidt & Smith, 1987)] apparently during its initial transit across the host (Schmidt & Smith, 1987). Usually, additional clutches are allocated to hosts only under conditions of high parasitoid to host ratios (van Alphen & Visser, 1990) and reflect decreasing host quality and host availability. When a single clutch of eggs is allocated to a host, the sex ratio of emerging offspring is frequently female biased and presumed to be so because of LMC (Suzuki & Iwasa, 1980; Werren, 1980; Waage & Lane, 1984). Subsequent clutches allocated to previously parasitized hosts are often less female biased than the first, if their size is equal to or less than that of the first (Werren, 1980; Parker & Orzack, 1985; Orzack & Parker, 1986, 1990). It has been asserted that the sex-ratio shift in the second clutch is due to the second female exerting LPC to influence LMC (Suzuki & Iwasa, 1980; Werren, 1980) but the agreement between the predicted and observed pattern is open to question (Orzack, 1990). Other factors that affect the sex ratio may be involved. The sex-ratio shift manifested by the offspring of gregarious species in superparasitized hosts can result from at least three interacting processes: (1) LPC, (2) differential mortality of daughters versus sons, and (3) changes in the relative fitness of female versus male offspring with size and with difference in the resource requirements of daughters versus sons (Taylor, 1981; Godfray, 1986). Although superparasitism is adaptive under conditions of high parasitoid to host ratios (van Alphen & Visser, 1990), the increased number of eggs allocated to a host increases the competition among the wasp larvae (Klomp & Teerink, 1962; Werren, 1983; van Alphen & Visser, 1990). The offspring emerging from such hosts are smaller (Klomp & Teerink, 1967; Werren, 1983; Strand, 1986; Suzuki *et al.,* 1984; Bai *et al.,* 1992; Hardy *et al.,* 1992) or are more likely to die as immatures (Klomp & Teerink, 1967; Suzuki *et al.,* 1984). If the development of sons requires less resources than daughters, then the allocation of male eggs in previously parasitized hosts would be favored in circumstances of scramble competition arising from superparasitism. Consequently, a smaller female-biased sex ratio among the offspring of a second clutch allocated to a previously parasitized host may be favored because of the asymmetrical effects of one or more factors [see preceding (1) to (3)] on the offspring of the second or subsequent clutches (Taylor, 1981; Godfray, 1986).

Regardless of the reasons for the higher frequency of male wasps emerging from superparasitized hosts, conditions that promote superparasitism in a culture must be minimized. Superparasitized hosts produce smaller females and a higher fraction of male offspring than nonsuperparasitized hosts produce. Furthermore, at high levels of super-

parasitism, mortality of the offspring may ensue (Klomp & Teerink, 1967; Suzuki *et al.,* 1984). To the extent practical, isolating mothers, each with its own hosts, and limiting the time they are exposed to the hosts prevent superparasitism. This design works as well for culturing gregarious parasitoids as it does for solitary species.

SEX RATIO OF POLYEMBRYONIC WASPS

Polyembryony is the development of multiple individuals from a single egg and it occurs in four families of parasitic Hymenoptera: Braconidae, Encyrtidae, Platygasteridae, and Dryinidae (Ivanova-Kasas, 1972). In polyembryonic encyrtids, the initial egg divides and forms a varying number of blastomeres (Strand, 1989a). As cell division continues, blastomeres aggregate into groups of embryonic tissue, each of which is surrounded by serosal tissue and a membrane or membranes. These aggregates, or morulae, are themselves surrounded by a serosal membrane and large groups of them are scattered throughout the host's body. Collectively, the embryonic and serosal tissue are referred to as the polygerm (Strand, 1989a). In *Copidosoma floridanum* (Ashmead), most of the morulae give rise to reproductive larvae (>1000) that consume the host and pupate within the host during the host's final larval instar. However, some of the morulae produce precocious, nonreproductive larvae (<50) that feed minimally and die when the reproductive larvae consume the host. Cruz (1981) suggests that the non-reproductive larvae defend the polygerm from other parasitoid species (interspecific competition).

In laboratory experiments (Strand, 1989b), mated *C. floridanum* females were found to lay either one egg per host (in which case the offspring were either all female or all male) or two eggs per host (in which case 71 out of 72 broods were a mixture of male and female offspring. The number of individuals reared from a host typically ranged between 1000 and 1400 individuals (Grbic *et. al.,* 1992). All-male broods averaged 1063 ± 137.6, all-female broods averaged 1218.5 ± 198.4, and mixed broods averaged 1191.2 ± 240.9 (Strand, 1989b). The emerging (i.e., secondary) sex ratio in the mixed broods was 0.042 ± 0.066 (proportion of males) (Strand, 1989b) or 0.12 ± 0.03 (Grbic *et al.,* 1992) even though the primary sex ratio was 0.5 (from the initial two eggs). In field-collected hosts in Wisconsin, the mean offspring per host was 1385 ± 79.2 and the brood size did not differ among single- or mixed-sexed broods (Grbic *et al.,* 1992). In the field, mixed broods predominate (Strand, 1989a; Grbic *et al.,* 1992). The highly female-biased sex ratio arises from the precocious larvae, which are female, differentially attacking male morulae in hosts with mixed broods. This is thought to arise from the asymmetrical relatedness of brothers and sisters because of the haplodiploid genetic system in Hymenoptera; sisters share fewer genes with their brothers than are shared by brothers with their sisters (Grbic *et al.*, 1992). Mixed broods arise more frequently when *C. floridanum* encounters hosts infrequently, whereas single broods arise when hosts are encountered frequently (Hardy *et al.*, 1993). Laboratory experiments estimated that 95% of the females would be inseminated when the ratio of females to males was 30:1 or a sex ratio of 0.033 (Strand, 1989a). It is on this basis, along with the effect of host encounter rate on the frequency of mixed broods, that Hardy *et al.* (1993) argue that the sex-ratio patterns observed in *C. floridanum* are consistent with LMC/LPC.

SUMMARY

A major objective in the culturing of biological control agents is to provide the conditions that yield a high frequency of high-quality female wasps for release. By quality we mean those attributes that contribute to the agent's ability to parasitize the greatest percentage of hosts in the field. Two factors contribute to quality. The most important appears to be size, which influences fecundity (van dem Assem *et al.*, 1989) and the ability to find hosts in the field (Bigler *et al.*, 1987; Kazmer & Luck, 1991, 1995; Visser, 1994; West *et al.*, 1996; Ellers *et al.*, 1998). The second factor is individual heterozygosity. We have argued that in outbreeding species the effect of inbreeding on female function may be severe (Stouthamer *et al.*, 1992).

Among factors that can be manipulated in culturing parasitic Hymenoptera to achieve low sex ratio (high percentage of females) and high quality are (1) minimizing inbreeding in an outbreeding species (e.g., in Ichneumoidea), (2) isolating mothers with a large number of high-quality hosts to maximize the production of female offspring, (3) limiting exposure time of hosts to parental females to reduce superparasitism (a condition that produces an increased percentage of males and of small females of low quality), and (4) providing hosts for solitary parasitoids that induce the allocation of a high frequency of female eggs (e.g., large hosts). In the special case of heteronomous species, it may be necessary to leave some previously parasitized hosts with the parental females to obtain male offspring. These recommendations arise from a practical understanding of parasitoid life histories and an understanding of the theories of sex allocation and sex determination. Although these theories may be incomplete and have been subjected to limited tests, their predictions have been generally borne out; thus, they are of value in designing culturing conditions for parasitic Hymenoptera. However, clearly additional research is needed. It is also clear from the recent research on sex ratio that a strong and important linkage exists between fundamental research in evolutionary ecology of parasitic Hymenoptera, such as that associated with the theories of sex allocation and sex determination, and the practical application of its results, such as that involved in the culturing of biological control agents. Still, other less predictable factors affect the sex ratios of parasitoids in culture and appear to be serendipitous in their occurrence e.g., sex-distorting genomic and cytoplasmic factors or nonreciprocal cross-incompatibility. Consequently, clear guidelines as to the rearing conditions that might ameliorate or exclude their effects are more difficult to provide.

Acknowledgment

A portion of this effort was funded by National Science Foundation Grant No. BSR 89-21100 (RFL), BARD Grant US-2359-93C (RFL), and a European Union Grant EU-FAIR-CT94-1433 (RS).

References

Abdelrahman, I. (1974). Studies in ovipositional behaviour and control of sex in *Aphytis melinus* DeBach, a parasite of California red scale, *Aonidiella aurantii* (Mask.). Australian Journal of Zoology, 22, 231–247.

Aeschlimann, J. P. (1990). Simultaneous occurrence of thelytoky and bisexuality in hymenopteran species and its implications for the biological control of pests. Entomophaga, 35, 3–5

Allen, G. R., Kazmer, D. J., & Luck, R. F. (1994). Post-copulatory male behaviour, sperm precedence and multiple mating in a solitary parasitoid wasp. Animal Behaviour, 48, 635–644.

Antolin, M. F., & Strand, M. R. (1992). Mating system of *Bracon hebetor* (Hymenoptera: Braconidae). Ecological Entomology, 17, 1–7

Arthur, A. P., & Wylie, H. G. (1959). Effects of host size on the sex ratio, development time and size of *Pimpla turionellae* (Hymenoptera: Ichneumonidae). Entomophaga, 4, 297–301.

Avilla, J., & Copland, M. J. W. (1987). Effects of host stage on the development of the facultative autoparasitoid *Encarsia tricolor* (Hymenoptera: Aphelinidae). Annals of Applied Biology, 110, 381–389.

Avilla, J., Anadon, J., Sarasua, M. J., & Albajes, R. (1991). Egg allocation of the autoparasitoid *Encarsia tricolor* at different relative densities of the primary host (*Trialeurodes vaporariorum*) and two secondary hosts (*Encarsia formosa* and *E. tricolor*). Entomologia Experimentalis et Applicata, 59, 219–227.

Bai, B., Luck, R. F., Forster, L., Janssen, J. A. H., & Stephens, B. (1992). The effect of host size on quality attributes of the egg parasitoid, *Trichogramma pretiosum*. Entomologia Experimentalis et Applicata, 64, 37–48.

Beaudoin, L., Geri, C., Allais, J. P., & Goussard, F. (1994). Influence of consanguinity on the sex-ratio and diapause of *Diprion pini* L. (Hym., Diprionidae) populations. I. Observations on a rearing population. Relation with the sex determinism and consanguinity. Journal of Applied Entomology, 118, 267–280.

Bernal, J. S., Luck, R. F., & Morse, J. G. (1998). Sex ratios in field populations of two parasitoids (Hymenoptera: Chalcidoidea) of *Coccus hesperidum* L. (Homoptera: Coccidae). Oecologia (Berlin).

Beukeboom, L. W., & Werren, J. H. (1993). Transmission and expression of the parasitic paternal sex ratio (PSR) chromosome. Heredity, 70, 437–443.

Bigler, F., Meyer, A., & Bossart, S. (1987). Quality assessment in *Trichogramma maidis* Pintureau et Voegele reared from eggs of the factitious hosts *Ephestia kuehniella* Zell. and *Sitotroga cerealella* (Oliver). Journal of Applied Entomology, 104, 340–353.

Bouletreau, M. (1976). Influence de la photoperiode subie par les adultes sur la sex ratio de la decendance chez *Pteromalus puparum* (Hymenoptera; Chalcididae). Entomologia Experimentalis et Applicata, 19, 197–204.

Bokonon-Ganta, A. H., Neuenschwander, P., van Alphen, J. J. M., & Vos, M. (1995). Host stage selection and sex allocation by *Anagyrus mangicola* (Hymenoptera: Encyrtidae), a parasitoid of the mango mealybug, *Rastrococcus invadens* (Homoptera: Pseudococcidae). Biological Control, 5, 479–486.

Bradley, W. G., & Burgess, E. D. (1934). The biology of *Cremastus flavoorbitalis*, an ichneumonid parasite of the European corn borer (Technical Bulletin, 441). Washington, DC: U.S. Department of Agriculture.

Breeuwer, J. A. J., & Werren, J. H. (1990). Microorganisms associated with chromosome destruction and reproductive isolation between insect species. Nature (London), 346, 558–559.

Byers, J. R., & Wilkes, A. (1970). A rickettsialike organism in *Dahlbominus fuscipennis*: Observations on its occurence and ultrastructure. Canadian Journal of Zoology, 48, 959–964.

Camargo, C. A. (1979). Sex determination in bees. XI. Production of diploid males and sex determination in *Melipona quadrifasicata*. Journal of Apicultural Research, 18, 77–84.

Cardona, C., & Oatman, E. R. (1971). Biology of *Apanteles dignus*. a primary parasite of the tomato pinworm. Annals of the Entomological Society of America, 64, 996–1007.

Cardona, C., Oatman, E. R. (1975). Biology and physical ecology of *Apanteles subandinus* Blanchard (Hymenoptera: Braconidae), with notes on temperature responses of *Apanteles scutellaris* Muesebeck and its host potato tuberworm. Hilgardia, 43, 1–51.

Charnov, E. L. (1979). The genetical evolution of patterns of sexuality: Darwinian fitness. American Naturalist, 113, 465–480.

Charnov, E. L. (1982). The theory of sex allocation. Princeton, NJ: Princeton University Press.

Charnov, E. L. & Skinner, S. W. (1984). Evolution of host selection and clutch size in parasitic wasps. Florida Entomologist, 67, 5–21.

Charnov, E. L., & Skinner, S. W. (1985). Complementary approaches to the understanding of parasitoid ovipositon decisions. Environmental Entomology, 14, 383–391.

Charnov, E. L., Los-den Hartogh, E. L., Jones, W. T., & van den Assem, H. (1981). Sex ratio evolution in a variable environment. Nature (London), 289, 27–33.

Clark, A. M., Bertrand, H. A., & Smith, R. E. (1963). Life span differences between haploid and diploid males of *Habrobracon serinopae* after exposure as adults to X-rays. American Naturalist, 97, 203–208.

Clausen, C. P. (1939). The effect of host size upon the sex ratio of hymenopterous parasites and its relation to methods of rearing and colonizations. Journal of the New York Entomological Society, 47, 1–9.

Conner, G. W., & Saul, G. B. (1986). Acquisition of incompatibility by inbred wild-type stocks of *Mormoniella*. Journal of Heredity, 77, 211–213.

Cook, J. M. (1993a). Experimental tests of sex determination in *Goniozus nephantidis*. Heredity, 71, 130–137.

Cook, J. M. (1993b). Sex determination in the Hymenoptera: A review of models and evidence. Heredity, 71, 421–435.

Crozier, R. H. (1971). Heterozygosity and sex-determination in haplodiploidy. American Naturalist, 105, 399–412.

Crozier, R. H. (1975). Hymenoptera. In B. John (Ed.), Animal cytogenetics: Vol. 3. Insecta 7 (pp. 1–95). Berlin: Gebr. Borntraeger.

Crozier, R. H. (1976). Why male-haploid and sex-limited genetic systems seems to have unusually sex-limited mutational genetic loads. Evolution, 30, 623–624.

Crozier, R. H. (1977). Evolutionary genetics of the Hymenoptera. Annual Review of Entomology, 22, 263–288.

Cruz, Y. P. (1981). A sterile defender morph in a polyembryonic hymenopterous parasite. Nature (London), 294, 446–447.

da Cuhna, A. B., & Kerr, W. E. (1957). A genetical theory to explain sex-determination by arrhenotokous parthenogenesis. Forma et Functio, 1, 33–36.

DeBach, P., & Sudha Rao, V. (1968). Transformation of inseminated females of *A. lingnanensis* into factitious virgins by low temperature treatment. Annals of the Entomological Society of America, 61, 332–337.

de Jong, P. W., & van Alphen, J. J. M. (1989). Host size selection and sex allocation in *Leptomastix dactylopii*, a parasitoid of *Planococcus citri*. Entomologia Experimentalis applicata, 50, 161–169.

Dobson, S., & Tanouya, M. (1996). The paternal sex ratio chromosome induces chromosome loss independently of *Wolbachia* in the wasp *Nasonia vitripennis*. Development, Genes & Evolution, 206, 207–217.

Donaldson, J. S., & Walter, G. H. (1991a). Brood sex ratios of the solitary parasitoid wasp, *Coccophagus atratus*. Ecological Entomology, 16, 25–33.

Donaldson, J. S., & Walter, G. H. (1991b). Host population structure affects field sex ratios of the heteronomous hyperparasitoid, *Coccophagus atratus*. Ecological Entomology, 16, 35–44.

Duchateau, M. J., & Marien, J. (1995). Sexual biology of haploid and diploid males in the bumble bee *Bombus terrestris*. Insectes Sociaux, 42, 255–266.

Ebbert, M. A. (1992). Endosymbiotic sex ratio distorters in insects and mites. In D. L. Wrench & M. A. Ebbert (Eds.), Evolution and diversity of sex ratios in haplodiploid insects and mites (pp. 150–191). London: Chapman & Hall.

El Agoze, M., Drezen, J. M., Renault, S., & Periquet, G. (1994). Analysis of the reproductive potential of diploid males in the wasp *Diadromus pulchellus* (Hymenoptera: Ichneumonidae). Bulletin of Entomological Research, 84, 213–218.

El Agoze, M., & Periquet, G. (1993). Viability of diploid males in the parasitic wasp, *Diadromus pulchellus* (Hym.: Ichneumonidae). Entomophaga, 38, 199–206.

Ellers, J., van Alphen, J. J. M., & Svenster, J. G. (1998). A field study of size-fitness relationships in the parasitoid *Asobara tabida*. Journal of Animal of Ecology, 67, 318–324.

Fabritius, K. (1984). Untersuchungen uber eine Inzucht von *Muscidifurax raptor* unter Laborbedingungen. Entomological Genetics, 9, 237–241.

Fisher, R. A. (1930). The genetical theory of natural selection. Oxford: Oxford University Press.

Flanders, S. E. (1943). The role of mating in the reproduction of parasitic Hymenoptera. Journal of Economic Entomology, 36, 802–803.

Flanders, S. E. (1945). The role of the spermatophore in the mass propagation of *Macrocentrus ancylivorus*. Journal of Economic Entomology, 38, 323–327.

Flanders, R. V., & Oatman, E. R. (1982). Laboratory studies on the biology of *Orgilus jenniae*, a parasitoid of the potato tuberworm, *Phthorimaea operculella*. Hilgardia, 50, 1–33

Frank, S. A. (1985). Hierarchial selection theory and sex ratios. II. On applying theory, and a test with fig wasps. Evolution, 39, 949–964

Frank, S. A. (1986). Hierarchical selection theory and sex ratios. I. General solutions for structured populations. Theoretical Population Biology, 29, 312–342.

Frank, S. A. (1990). Sex allocation theory for birds and mammals. Annual Review of Ecology and Systematics, 21, 13–55.

Frohlich, D. R., Brindley, W. A., Burris, T. E., Youssef, N. N. (1990). Esterase isozymes in a solitary bee, *Megachile rotundata* (Fab.): Characterization, developmental multiplicity, and adult variability. Biochemical Genetics, 28, 347–358.

Garofalo, C. A., & Kerr, W. E. (1975). Sex determination in bees. I. Balance between femaleness and maleness genes in *Bombus atratus*. Genetica, 45, 203–209.

Genieys, P. (1925). *Habrobracon brevicornis* Wesm. Annals of the Entomological Society of America, 18, 143–202.

Gherna, R., Werren, J. H., Weisburg, W., Cote, R., Woese, C. R., Mandelco, L. & Brenner, R. (1991). *Arsenophonus nasoniae,* genus novel, species novel, causative agent of the "sonkiller" trait in the parasitic wasp *Nasonia vitripennis.* International Journal of Systematic Bacteriology, 41, 563–565.

Godfray, H. J. C. (1986). Models for clutch and sex-ratio with sibling interaction. Journal of Theoretical Biology, 30, 215–231.

Godfray, H. J. C. (1990). The causes and consequences of constrained sex allocation in haplodiploid animals. Journal of Evolutionary Biology, 3, 3–17.

Godfray, H. J. C. (1994). Parasitoid behavioral and evolutionary ecology. In J. R. Krebs & T. H. Clutton-Brock (Eds.), Monographs in behavior and ecology. Princeton, NJ: Princeton University.

Godfray, H. J. C., & Hunter, M. S. (1994). Heteronomous hyperparasitoids, sex ratios and adaptations—a reply. Ecological Entomology, 19, 93–95.

Godfray, H. J. C., & Waage, J. K. (1990). The evolution of highly skewed sex ratios in aphelinid wasps. American Naturalist, 136, 715–721.

Gordh, G. (1976). *Goniozus gallicola* Fouts, a parasite of moth larvae, with notes on other bethylids (Hymenoptera: Bethylidae; Lepidoptera: Gelechiidae) (Technical Bulletin 1524) Washington, DC: U.S. Department of Agriculture.

Gotoh, T. (1982). *Harmonia axyridis* (Pallas) (Coleoptera: Coccinellidae). Applied Entomology and Zoology, 17, 319–324.

Gotoh T., & Niijima, K. (1986). Characteristics and agents of abnormal sex ratio in two aphidophagous coccinellid species. In I. Hodek (Ed.), Ecology of Aphidiphaga. Proceedings, 2nd Symposium, Zvikovske Pokhradi, September 2–8, 1984 (pp. 545–550). Boston: Junk.

Grbic, M., Ode, P. J., & Strand, M. R. (1992). Sibling rivalry and brood sex ratios in polyembryonic wasps. Nature (London), 360, 254–256.

Green, R. F. (1982). Optimal foraging and sex ratio in parasitic wasps. Journal of Theoretical Biology, 95, 43–48.

Green, R. F., Gordh, G., & Hawkins, B. A. (1982). Precise sex ratio in highly inbred parasitic wasps. American Naturalist, 120, 653–665.

Griffiths, N., & Godfray, H. J. C. (1988). Local mate competition, sex ratio and clutch size in bethylid wasps. Behavioral Ecology and Sociobiology, 22, 211–217.

Hamilton, W. D. (1967). Extraordinary sex ratios. Science, 156, 477–488.

Hamilton, W. D. (1979). Wingless and fighting males in fig wasps and other insects. In M. S. Blum & M. A. Blum (Eds.), Sexual selection and reproductive competition in insects (pp. 167–220). New York: Academic Press.

Hardy, I. C. W. (1992). Non-binomial sex allocation and brood sex ratio variances in the parasitoid Hymenoptera. Oikos, 65, 143–158.

Hardy, I. C. W. (1994). Sex ratio and mating structure in the parasitoid Hymenoptera. Oikos, 69, 3–20.

Hardy, I. C. W., Griffiths, N. T., & Godfray, H. C. J. (1992). Clutch size in a parasitoid wasp: A manipulation experiment. Journal of Animal Ecology, 61, 121–129.

Hardy, I. C. W., Ode, P. J., & Strand, M. R. (1993). Factors influencing brood sex ratios in polyembryonic Hymenoptera. Oecologia (Berlin), 93, 343–348.

Hare, J. D., & Luck, R. F. (1991). Indirect effects of citrus cultivars on life history parameters of a parasitic wasp. Ecology, 72, 1576–1585.

Hartl, D. L. (1971). Some aspects of natural selection in arrhenotokous populations. American Zoology, 11, 309–325.

Hartl, D. L., & Brown, S. W. (1970). The origin of male haploid genetic systems and their expected sex ratio. Theoretical Population Biology, 1, 165–190.

Hedderwick, M. P., El Agose, M., Garaud, P., & Periquet, G. (1985). Mise en evidence de males heterozygotes chez L'Hymenoptere *Diadromus pulchellus.* Genetics, Selection, Evolution, 17, 303–310.

Heinz, K. M. (1991). Sex-specific reproductive consequences of body size in the solitary ectoparasitoid, *Diglyphus begini.* Evolution, 45, 1511–1515.

Heinz, K. M. (1996). Host size selection and sex allocation behaviour among parasitoid trophic levels. Ecological Entomology, 21, 218–226.

Heinz, K. M., & Parrella, M. P. (1990). The influence of host size on sex ratios in the parasitoid, *Diglyphus begini* (Hymenoptera: Eulophidae). Ecological Entomology, 15, 391–399.

Hey, J., & Gargiulo, M. K. (1985). Sex ratio changes in *Leptopilina heterotoma* in response to inbreeding. Journal of Heredity, 76, 209–211.

Hoelscher, C. E., & Vinson, S. B. (1971). The sex ratio of a hymenopterous parasitoid, *Campoletis perdistinctus,* as affected by photoperiod, mating, and temperature. Annals of the Entomological Society of America, 64, 1373–1376.

Holmes, H. B. (1972). Genetic evidence for fewer progeny and higher percent males when *Nasonia vitripennis* oviposits in previously parasitized hosts. Entomophaga, 17, 79–88.

Horjus, M., & Stouthamer, R. (1995). Does infection with thelytoky-causing *Wolbachia* in the pre-adult and adult life stages influence the adult fecundity of *Trichogramma deion* and *Muscidifurax uniraptor?* Proceedings, Section in Experimental and Applied Entomology of the Netherlands Entomological Society, 6, 35–40.

Horn, A. R. (1943). Proof for multiple allelism of sex differentiating factors in *Habrobracon.* American Naturalist, 77, 539–550.

Hoshiba, H., Okada, I., & Kusanagi, A. (1981). The diploid drone of *Apis cerana japonica* and its chromosomes. Journal of Apicultural Research, 20, 143–147.

Hoshiba, H., Duchateau, M. J., & Velthuis, H. H. W. (1995). Diploid males in the bumble bee *Bombus terrestris* (Hymenoptera). Karyotype analyses of diploid females, diploid males and haploid males. Japanese Journal of Entomology, 63, 203–207.

Huffaker, C. B., & Kennett, C. E. (1966). Studies of two parasites of olive scale, *Parlatoria oleae* (Colvée). IV. Biological control of *Parlatoria oleae* (Colvée) through the compensatory action of two introduced parasites. Hilgardia, 37, 283–335.

Huger, A. M., Skinner, S. W., & Werren, J. H. (1985). Bacterial infections associated with the son-killer trait in the parasitoid wasp *Nasonia vitripennis.* Journal of Invertebrate Pathology, 46, 272–280.

Hunter, M. S. (1989). Sex allocation and egg distribution of an autoparasitoid, *Encarsia pergandiella.* Ecological Entomology, 14, 57–67.

Hunter, M. S. (1993). Sex allocation in a field population of an autoparasitoid. Oecologia (Berlin), 93, 421–428.

Hunter, M. S., & Godfray, H. C. J. (1995). Ecological determinants of sex allocation in an autoparasitoid wasp. Journal of Animal Ecology, 64, 95–106.

Hunter, M. S., Nur, U., & Werren, J. H. (1993). Origin of males by genome loss in an autoparasitoid wasp. Heredity, 70, 162–171.

Hunter, M. S., Rose, M., & Polaszek, A. (1996). Divergent host relationships of males and females in the parasitoid *Encarsia porteri* (Hymenoptera: Aphelinidae). Annals of the Entomological Society of America, 89, 667–675.

Hurst, G. D. D., Majerus, M. E. N., & Walker, L. E. (1993). The importance of the cytoplasmic male killing elements in *Adalia bipunctata.* Heredity, 69, 84–91.

Hurst, G. D. D., Purvis, E. L., Sloggett, J. J., & Majerus, M. E. N. (1994). The effect of infection with male-killing *Rickettsia* on the demography of female *Adalia bipunctata.* Heredity, 73, 309–316.

Hurst, G. D. D., Hurst, L. D., & Majerus, M. E. N. (1997). Cytoplasmic sex-ratio distorters. In O'Neill, S. L., A. A. Hoffman, & J. H. Werren (Eds.). Influential passengers: Inherited microorganisms and arthropod reproduction (pp. 125–154). New York: Oxford University Press.

Inaba, F. (1939). Diploid males and triploid females of the parasitic wasp, *Habrobracon pectinophorae.* Cytologia, 9, 517–523.

Ivanova-Kasas, O. M. (1972). Polyembryony in insects. In S. J. Counce & C. H. Waddington (Eds.), Developmental systems. Insects (Vol. 2, pp. 243–272). New York: Academic Press.

Jackson, D. J. (1958). Observations on the biology of *Caraphractus cinctus,* a parasitoid of the eggs of Dytiscidae. Transactions of the Royal Entomological Society of London, 110, 533–553.

Kai, H. (1979). Maternally inherited sonless abnormal sex-ratio (SR) condition in the lady beetle *Harmonia axyridis.* Acta Genetica Sinica, 6, 296–304.

Kazmer, D. J., & Luck, R. F. (1991). Female body size, fitness and biological control quality: Field experiments with *Trichogramma pretiosum.* In E. Waijnberg & S. B. Vinson (Eds.) Proceedings of the Third International Symposium *Trichogramma.* San Antonio, Texas (pp. 107–110) Paris: INRA.

Kazmer, D. J., & Luck, R. F. (1995). Field tests of the size-fitness hypothesis in the egg parasitoid *Trichogramma pretiosum.* Ecology, 76, 412–425.

Kerr, W. E., & Nielsen, R. A. (1967). Sex determination in bees (Apinae). Journal of Apicultural Research, 6, 3–9.

Kfir, R. (1982). Reproduction characteristics of *Trichogramma brasiliensis* and *T. lutea,* parasitizing eggs of *Heliothis armigua.* Entomologia Experimentalis et Applicata, 32, 249–255.

Kfir, R., & Luck, R. F. (1979). Effects of constant and variable temperature extremes on sex ratio and progeny production by *Aphytis melinus* and *A. lingnanensis.* Ecological Entomology, 4, 335–344.

King, B. H. (1987). Offspring ratios in parasitoid wasps. Quarterly Review of Biology, 62, 367–396.

King, B. H. (1988). Sex-ratio manipulation in response to host size by the parasitoid wasp *Spalangia cameroni:* A laboratory study. Evolution, 42, 1190–1198.

King, B. H. (1989). A test of local mate competition theory with a solitary species of parasitoid wasp, *Spalangia cameroni.* Oikos, 55, 50–54.

King, B. H. (1990). Sex ratio manipulation by the parasitoid wasp *Spalangia cameroni* in response to host age: A test of the host-size model. Evolutionary Ecology, 4, 149–156.

King, B. H. (1992). Sex determination and sex ratio patterns in parasitic Hymenoptera. In D. L. Wrench & M. A. Ebbert (Eds.), Evolution and diversity of sex ratios in haplodiploid insects and mites (pp. 418–441). London: Chapman & Hall.

King, B. H., & Seidl, S. E. (1993). Sex ratio response of the parasitoid wasp *Muscidifurax raptor* to other females. Oecologia (Berlin), 94, 428–433.

King, B. H., & Skinner, S. W. (1991). Proximal mechanisms of the sex ratio and clutch size responses of the wasp *Nasonia vitripennis* to parasitized hosts. Animal Behavior, 42, 23–32.

Klomp, H., & Teerink, B. J. (1962). Host selection and number of eggs per oviposition in the egg parasite, *Trichogramma embryophagum* Htg. Nature (London), 195, 1020–1021.

Klomp, H., & Teerink, B. J. (1967). The significance of oviposition rates in the egg parasite, *Trichogramma embryophagum* Htg. Archives Neerlandaises de Zoologie, 17, 350–375.

Kukuk, P. F., & May, B. (1990). Diploid males in a primitively eusocial bee, *Lasioglossum (Dialictus) zephyrum* (Hymenoptera: Halictidae). Evolution, 44, 1522–1528.

Kuno, E. (1962). The effect of population density on the reproduction of *Trichogramma japonicum* Ashmead (Hymenoptera: Trichogrammatidae). Researches on Population Ecology (Kyoto), 4, 47–59.

Lampson, L. J., Morse, J. G., & Luck, R. F. (1996). Host selection, sex allocation, and host feeding by *Metaphycus helvolus* (Hymenoptera: Encyrtidae) on *Saissetia oleae* (Homoptera: Coccidae) and its effect on parasitoid size, sex and quality. Environmental Entomology, 25, 283–294.

Lee, P., & Wilkes, A. (1965). Polymorphic spermatozoa in the hymenopterous wasp *Dahlbominus.* Science, 147, 1445–1446.

Legner, E. F. (1979). Prolonged culture and inbreeding effects on reproductive rates of two pteromalid parasities of muscoid flies. Annals of the Entomological Society of America, 72, 114–118.

Legner, E. F. (1985). Effects of scheduled high temperature on male production in the thelytokous *Muscidifurax uniraptor.* Canadian Entomologist, 117, 383–389.

Luck, R. F., & Podoler, H. (1985). Competitive exclusion of *Aphytis lingnanensis* by *A. melinus:* Potential role of host size. Ecology, 66, 904–913.

Luck, R. F., & Janssen, J. A. M., Pinto, J. D., & Oatman, E. R. (in lit.). Precise sex allocation and sex ratio shifts by the parasitoid *Trichogramma pretiosum* with increasing foundress numbers. Behavioral Ecology and Sociobiology.

Luck, R. F., Podoler, H., & Kfir, R. (1982). Host selection and egg allocation behavior by *Aphytis melinus* and *A. lingnanensis:* Comparison of two facultatively gregarious parasitoids. Ecological Entomology, 7, 397–408.

Luck, R. F., Stouthamer, R., & Nunney, L. (1992). Sex determination and sex ratio patterns in parasitic Hymenoptera. In D. L. Wrench & M. A. Ebbert (Eds.), Evolution and diversity of sex ratios in haplodiploid insects and mites (pp. 442–476). London: Chapman & Hall.

MacBride, D. H. (1946). Failure of sperm of *Habrobracon* diploid males to penetrate the eggs. Genetics, 31, 224.

McCorquodale, D. B., & Owen, R. E. (1994). Laying sequence, diploid males, and nest usurpation in the leafcutter bee, *Megachile rotundata* (Hymenoptera: Megachilidae). Journal of Insect Behavior, 5, 731–738.

Mackauer, M. (1969). Sexual behavior of and hybridization between three species of *Aphidius* Nees parasitic on the pea aphid. Proceedings of the Entomological Society of Washington, 71, 339–351.

Mackauer, M. (1976). An upper boundary for the sex ratio in a haplodiploid insect. Canadian Entomologist, 108, 1399–1402.

Matsuka, M., Hashi, H., & Okada, I. (1975). Abnormal sex-ratio found in the lady beetle, *Harmonia axyridis* Pallas (Coleoptera: Coccinellidae). Applied Entomology and Zoology, 10, 84–89.

Maynard Smith, J. (1978. The evolution of sex. Cambridge: Cambridge University Press.

Mayhew, P. J., & Godfray, H. C. J. (1997). Mixed sex allocation strategies in a parasitoid wasp. Oecologia (Berlin), 110, 218–221.

Middeldorf, J., & Ruthmann, A. (1984). Yeast-like endosymbionts in an ichneumonid wasp. Zeitschrift für Naturforschung, Teil C, 39, 322–326.

Moritz, R. F. A. (1986). The genetics of bees other than *Apis mellifera.* In T. E. Rinderer (Ed.), Bee genetics and breeding (pp. 121–154). New York: Academic Press.

Morgan, D. J. W., & Cook, J. M. (1994). Extremely precise sex ratios in small clutches of a bethylid wasp. Oikos, 71, 423–430.

Nadel, H. (1985). Sex allocation and the mating system of a female-biased parasitoid, *Pachycrepoideus vindemiae* (Rondani) (Chalcidoidea: Pteromalidae). Unpublished doctoral dissertation, University of California, Riverside.

Nagarkatti, S., & Fazaluddin, M. (1973). Biosystematic studies on *Trichogramma* species. II. Experimental hybridization between some *Trichogramma* spp. from the new world. Systematic Zoology, 22, 103–117.

Nagelkerke, C. J., & Hardy, I. C. W. (1994). The influence of developmental mortality on optimal sex allocation under local mate competition. Behavioral Ecology, 5, 401–411.

Naito, T., & Suzuki, H. (1991). Sex determination in the sawfly, *Athalia rosae ruficornis:* Occurrence of triploid males. Journal of Heredity, 82, 101–104.

Neuffer, G. (1964). Zu den Aussetzenversuchen von *Prospaltella perniciosi* gegen die San Jose Schildlaus in Baden-Wurttemberg. Entomophaga, 9, 131–136.

Nishimura, K. (1993). Oviposition strategy of the parasitic wasp *Dinarmus basalis* (Hymenoptera, Pteromalidae). Evolutionary Ecology, 7, 199–206.

Noda, T., & Hirose, Y. (1989). Male second strategy in the allocation of sexes by the parasitic wasp, *Gryon japonicum*. Oecologia (Berlin), 81, 145–148.

Nunney, L. (1985). Female-biased sex ratios: individual or group selection. Evolution, 39, 349–361.

Nunney, L., & Luck, R. F. (1988). Factors influencing the optimum sex ratio in a structured population. Theoretical Population Biology, 33, 1–30.

Oatman, E. R., & Platner, G. R. (1974). The biology of *Temelucha* sp., *platneri* group, a primary parasite of the potato tuberworm. Annals of the Entomological Society of America, 67, 275–280.

Opp, S. B., & Luck, R. F. (1986). Effects of host size on selected fitness components of *Aphytis melinus* DeBach and *A. lingnanensis* Compere (Hymenoptera: Aphelinidae). Annals of the Entomological Society of America, 79, 700–704.

Orzack, S. H. (1990). The comparative biology of second sex ratio evolution within a natural population of a parasitic wasp, *Nasonia vitripennis*. Genetics, 124, 385–396.

Orzack, S. H. (1992). Sex ratio evolution in parasite wasps. In D. L. Wrench & M. A. Ebbert (Eds.), Evolution and diversity of sex in haplodiploid insects and mites (pp. 477–511). London: Chapman & Hall.

Orzack, S. H., & Parker, E. D., Jr. (1986). Sex ratio control in a parasitic wasp *Nasonia vitripennis*. I. Genetic variation in facultative sex-ratio adjustment. Evolution, 40, 331–340.

Orzack, S. H., & Parker, E. D., Jr. (1990). Genetic variation for sex ratio traits within a natural population of a parasitic wasp, *Nasonia vitripennis*. Genetics, 124, 373–384.

Parker, E. D., & Orzack, S. H., Jr. (1985). Genetic variation for the sex ratio in *Nasonia vitripennis*. Genetics, 110, 93–105.

Periquet, G., Hedderwick, M. P., El Agoze, M., & Poire, M. (1993). Sex determination in the hymenopteran *Diadromus pulchellus*: Validation of the one-locus multi-allele model. Heredity, 70, 420–427.

Pijls, J. W. A. M., van Steenbergen, H. J., & van Alphen, J. J. M. (1996). Asexuality cured: The relations and differences between sexual and asexual *Apoanagyrus diversicornis*. Heredity, 76, 506–513.

Pinto, J. D., Stouthamer, R., Platner, G. R., & Oatman, E. R. (1991). Variation in reproductive compatibility in *Trichogramma* and its taxonomic significance. Annals of the Entomological Society of America, 84, 37–46.

Platner, G. R., & Oatman, E. R. (1972). Techniques for culturing and mass producing parasites of the potato tuberworm. Journal of Economic Entomology, 65, 1336–1338.

Powers, N. R., & Oatman, E. R. (1985). Biology and temperature responses of *Chelonus kellieae* and *Chelonus phthorimaeae* (Hymenoptera: Braconidae) and their host, the potato tuberworm, *Phthorimaea operculella* (Lepidoptera: Gelechiidae). Hilgardia, 52, 1–32.

Putters, F. A., & van den Assem, J. (1985). Precise sex ratio in a parasitic wasp: The result of counting eggs. Behavioral Ecology and Sociobiology, 17, 265–270.

Reed, K. M. (1993). Cytogenetic analysis of the paternal sex ratio chromosome of *Nasonia vitripennis*. Genome, 36, 157–161.

Rice, P. L. (1937). A study of the insect enemies of San Jose scale (*Quadraspidiotus perniciosus*) with special reference to *Prospaltella perniciosi*. Unpublished doctoral thesis, Columbus: Ohio State University.

Richardson, P. N., Holmes, W. P., & Saul, G. B. (1987). The effect of tetracycline on nonreciprocal cross incompatibility in *Nasonia vitripennis*. Journal of Invertebrate Pathology, 50, 176–183.

Ross, K. G., Vargo, E. L., Keller, L. & Trager, J. C. (1993). Effect of a founder event on variation in the genetic sex-determining system of the fire ant *Solenopsis invicta*. Genetics, 135, 843–854.

Rössler, Y., & DeBach, P. (1972). The biosystematic relations between a thelytokous and arrhenotokous form of *Aphytis mytilaspidis*. I. The reproductive relations. Entomophaga, 17, 391–423.

Rössler, Y., & DeBach, P. (1973). Genetic variability in the thelytokous form of *Aphytis mytilaspidis*. Hilgardia, 42, 149–175.

Salt, G. (1935). Experimental studies in insect parasitism. III. Host selection. Proceedings of the Royal Society of London. Series B., 117, 413–435.

Salt, G. (1940). Experimental studies in insect parasitism. VII. The effects of different hosts on the parasite *Trichogramma evanescens* Wesw. (Hym. Chalcidoidea). Proceedings of the Royal Entomological Society of London, Series, 15, 81–85.

Sandlan, K. (1979). Sex ratio regulation in *Coccygomimus turionella* and its ecological implications. Ecological Entomology, 4, 365–378.

Saul, G. B. (1961). An analysis of non-reciprocal cross incompatibility in *Mormoniella vitripennis*. Zeitschrift fuer Vererbungslehre, 92, 28–33.

Schmidt, J. M., & Smith, J. J. B. (1987). The measurement of exposed host volume by the parasitoid, *Trichogramma minutum* and the effects of wasp size. Canadian Journal of Zoology, 65, 2837–2845.

Schmieder, R. G., & Whiting, P. W. (1947). Reproductive economy in the chalcidoid wasp *Melittobia*. Genetics, 32, 29–37.

Schwartz, A., & Gerling, D. (1974). Adult biology of *Telenomus remus* (Hymenoptera: Scelionidae) under laboratory conditions. Entomophaga, 19, 483–492.

Simmonds, F. J. (1947). Improvement of the sex-ratio of a parasite by selection. The Canadian Entomologist, 74, 41–44.

Skinner, S. W. (1982). Maternally inherited sex ratio in the parasitoid wasp, *Nasonia vitripennis*. Science, 215, 1133–1134.

Skinner, S. W. (1985). Son-killer: A third extrachromosomal factor affecting the sex-ratio in the parasitoid wasp, *Nasonia vitripennis*. Genetics, 109, 745–759.

Skinner, S. W. (1987). Paternal transmission of an extrachromosomal factor in a wasp: Evolutionary implications. Heredity, 59, 47–53.

Smith, H. D., Maltby, H. L., & Jimenez, J. J. (1964). Biological control of the citrus blackfly in Mexico (Technical Bulletin 1311). Washington, DC: U.S. Department of Agriculture.

Smith, J. W., Jr., Rodriguez-del-Bosque, L. A., & Agnew, C. W. (1990). Biology of *Mailochia pyralidis* (Hymenoptera: Ichneumonidae), an ectoparasite of *Eoreuma loftini* (Lepidoptera: Pyralidae) from Mexico. Annals of the Entomological Society of America, 83, 961–966.

Smith, S. G. (1941). A new form of spruce sawfly identified by means of its cytology and parthenogenesis. Scientific Agriculture, 21, 245–305.

Smith, S. G., & Vikki, N. (1978). Coleoptera. In B. John (Ed.), Animal cytogenetics: Vol. 3. Insecta 5, Berlin: Gebr. Borntraeger.

Smith, S. G., & Wallace, D. R. (1971). Allelic sex determination in a lower hymenopteran, *Neodiprion nigrosculum* Midd. Canadian Journal of Genetics and Cytology, 13, 617–621.

Snell, G. D. (1935). The determination of sex in *Habrobracon*. Proceedings of the National Academy of Sciences, USA, 21, 446–453.

Steiner, W. W. M., & Teig, D. A. (1989). *Microplitis croceipes*: Genetic characterization and developing insecticide resistant biotypes. The Southwestern Entomologist, 12, 71–87.

Stille, B., & Dävring, L. (1980). Meiosis and reproductive strategy in the parthenogenetic gall wasp *Diplolepis rosae*. Heredity, 92, 353–362.

Stouthamer, R. (1989). Causes of thelytoky and crossing incompatibility in several Trichogramma species. Unpublished doctoral dissertation, University of California, Riverside.

Stouthamer, R. (1993). The use of sexual versus asexual wasps in biological control. Entomophaga, 38, 3–6.

Stouthamer, R. (1997). Wolbachia-induced parthenogenesis. In S. L. O'Neill, A. A. Hoffmann, & J. H. Werren. (Eds.). Influential passengers: Inherited microorganisms and arthropod reproduction (pp. 102–124). New York: Oxford University Press.

Stouthamer, R., & Luck, R. F. (1991). Transition from bisexual to unisexual cultures in *Encarsia perniciosi* Tower (Hymenoptera: Aphelinidae): New data and a reinterpretation. Annals of the Entomological Society of America, 84, 150–157.

Stouthamer, R., & Luck, R. F. (1993). Influence of microbe-associated parthenogenesis on the fecundity of Trichogramma deion and T. pretiosum. Entomologia Experimentalis et Applicata, 67, 183–192.

Stouthamer, R. J., & Werren, J. H. (1993). Microorganisms associated with parthenogenesis in wasps of the genus Trichogramma. Journal of Invertebrate Pathology, 61, 6–9.

Stouthamer, R., Luck, R. F., & Hamilton, W. D. (1990a). Antibiotics cause parthenogenetic Trichogramma (Hymenoptera: Trichogrammatidae) to revert to sex. Proceedings of the National Academy of Sciences, USA, 87, 2424–2427.

Stouthamer, R., Pinto, J. D., Platner, G. R., & Luck, R. F. (1990b). Taxonomic status of thelytokous species of Trichogramma (Hymenoptera: Trichogrammatidae). Annals of the Entomological Society of America, 83, 475–481.

Stouthamer, R. J., Luck, R. F., & Werren, J. H. (1992). Genetics of sex determination and the improvement of biological control using parasitoids. Environmental Entomology, 21, 427–435.

Stouthamer, R. J. J., Breeuwer, A. J., Luck, R. F., & Werren, J. H. (1993). Molecular identification of micro-organisms associated with parthenogenesis. Nature (London), 361, 66–68.

Stouthamer, R., & Kazmer, D. J. (1994). Cytogenetics of microbe-associated parthenogenesis and its consequences for gene flow in Trichogramma wasps. Heredity, 73, 317–327.

Stouthamer, R., Luko, S., & Mak, F. (1994). Influence of parthenogenesis Wolbachia on host fitness. Norwegian Journal of Agricultural Science, 16, (Suppl.) 177–122.

Stouthamer, R., Luck, R. F., Pinto, J. D., Platner, G. D., & Stephens, B. (1996). Non-reciprocal cross-incompatibility in Trichogramma deion. Entomologia Experimentalis et Applicata, 80, 481–489.

Strand, M. (1986). In J. K. Waage & D. Greathead (Eds.), Insect parasitoids. London: Academic Press.

Strand, M. (1988). Variable sex ratio strategy of Telenomus heliothidis (Hymenoptera, Scelionidae): Adaptation to host and conspecific density. Oecologia (Berlin), 77, 219–224.

Strand, M. (1989a). Clutch size, sex ratio and mating by the polyembryonic encyrtid Copidosoma floridanum (Hymenoptera: Encyrtidae). Florida Entomologist, 72, 32–42.

Strand, M. (1989b). Oviposition behavior and progeny allocation of the polyembryonic wasp Copidosoma floridanum (Hymenoptera: Encyrtidae). Journal of Insect Behavior, 2, 355–369.

Suomalainen, E., Saura, A., & Lokki, J. (1987). Cytology and evolution in parthenogenesis. Boca Raton, FL: CRC Press.

Suzuki, Y., & Isawa, Y. (1980). A sex ratio theory of gregarious parasitoids. Researches on Population Ecology (Kyoto), 22, 366–382.

Suzuki, Y., Tsuji, H., & Sasakawa, M. (1984). Sex allocation and effects of superparasitism on secondary sex ratios in the gregarious parasitoid, Trichogramma chilonis. Animal Behaviour, 32, 478–484.

Takagi, M. (1985). The reproductive strategy of the gregarious parasitoid, Pteromalus puparum (Hymenoptera: Pteromalidae). I. Optimal number of eggs in a single host. Oecologia (Berlin), 68, 1–6.

Tarelho, Z. V. S. (1973). Contribuição ao estudo citogenético dos Apoidea. Unpublished doctoral dissertation, Universidade São Paulo, Rebeirao Preto.

Taylor, P. D. (1981). Intra-sex and inter-sex sibling interactions as sex ratio determinants. Nature (London), 291, 64–66.

Taylor, P. D., & Bulmer, M. G. (1980). Local mate competition and sex ratio. Journal of Theoretical Biology, 86, 409–419.

Timberlake, P. H., & Clausen, D. P. (1924). The parasites of Pseudococcus maritimus (Ehrhorn) in California. University of California Technical Bulletin of Entomology, 3, 223–292.

Torvik, M. M. (1931). Genetic evidence for diploidism of biparental males in Habrobracon. Biological Bulletin, 61, 139–156.

Unruh, T. R., Gordh, G., & González, D. (1984). Electrophoretic studies on parasitic Hymenoptera and implications for biological control. In Proceedings of the XVII International Congress on Entomology, Hamburg, Germany.

Unruh, T. R., White, W., González, D., Gordh, G., & Luck, R. F. (1983). Heterozygosity and effective size in laboratory populations of Aphidius ervi (Hym.: Aphidiidae). Entomophaga, 28, 245–258.

van Alphen, J. J. M., & Thunnissen, I. (1983). Host selection and sex allocation by Pachycrepoideus vindemiae Rondani (Pteromalidae) as a facultative hyperparasitoid of Asobara tabida Nees (Braconidae: Alysiinae) and Leptopilina heterotoma (Cynipoidea: Eucolidae). Netherlands Journal of Zoology, 33, 497–514.

van Alphen, J. J. M., & Visser, M. E. (1990). Superparasitism as an adaptive strategy for insect parasitoids. Annual Review of Entomology, 35, 59–79.

van den Assem, J. (1971). Some experiments on sex ratio and sex regulation in the pteromalid Lariophagus distinguendas. Netherlands Journal of Zoology, 21, 373–402.

van den Assem, J, & Feuth-De Bruijn, E. (1977). Second matings and their effect on the sex ratio of the offspring in Nasonia vitripennis. Entomologia Experimentalis, et Applicata, 21, 23–28.

van den Assem, J., Gijswijt, M. J., & Nubel, B. K. (1980). Observations on courtship and mating strategies in a few species of parasitic wasps (Chalcidoidea). Netherlands Journal of Zoology, 30, 208–227.

van den Assem, J., Putters, F. A. & Prins, T. H. (1984). Host quality effects on sex ratio of the parasitic wasp. Anisopteromalus calandrae (Chalcidoidea: Pteromalidae). Netherlands Journal of Zoology, 34, 33–62.

van den Assem, J., van Tersel, J. J. A., & Los-den Hartogh, R. L. (1989). Is being large more important for female than for male parasitic wasps? Behaviour, 108, 160–195.

van Dijken, M. J., & Waage, J. K. (1987). Self and conspecific superparasitism by the egg parasitoid Trichogramma evanescens. Entomologia Experimentalis et Applicata, 43, 183–192.

van Dijken, M. J., van Stratum, P., & van Alphen, J. J. M. (1993). Superparasitism and sex ratio in the solitary parasitoid, Epidinocarsis lopezi. Entomologia Expermentalis et Applicata, 68, 51–58.

van Welzen, C. R. L., & Waage, J. K. (1987). Adaptive responses to local mate competition by the parasitoid, Telenomus remus. Behavioral Ecology and Sociobiology, 21, 359–365.

Veerkamp, F. A. (1980). Behavioural differences between two strains and the hybrids of the wasp Pseudocoila bochei, a parasite of Drosophila melanogaster. Netherlands Journal of Zoology, 30, 431–449.

Viktorov, G. A. (1968). The influence of population density on sex-ratio in Trissolcus grandis Thoms. (Hymenoptera: Scelionidae). Zoologicheskii Zhurnal, 47, 1039–1045.

Viktorov, G. A., & Kochetova, N. I. (1973). Significance of population density to the control of sex ratio in Trissolcus grandis (Hymenoptera: Scelionidae). Zoologicheskii Zhurnal, 50, 1753–1755.

Visser, M. E. (1994). The importance of being large—the relationship between size and fitness in females of the parasitoid Aphaereta minuta (Hymenoptera: Broconidae). Journal of Animal Ecology, 63, 963–978.

von Borstel, R. C., & Smith, P. A. (1960). Haploid intersexes in the wasp Habrobracon. Heredity, 15, 29–34.

Waage, J. K. (1982). Sib-mating and sex ratio strategies in scelionid wasps. Ecological Entomology, 7, 103–112.

Waage, J. K. (1986). Family planning in parasitoids: Adaptive patterns of progeny and sex allocation. In J. K. Waage & D. Greathead (Eds.), Insect parasitoids (pp. 63–95). London: Academic Press.

Waage, J. K., & Godfray, H. J. C. (1985). Reproductive strategies and population ecology of insect parasitoids. In R. M. Sibley & R. H. Smith (Eds.), Behavioural ecology. Ecological consequences of adaptive behaviour (pp. 449–470). Oxford: Blackwell Scientific.

Waage, J. K., & Lane, J. A. (1984). The reproductive strategy of a parasitic wasp. II. Sex allocation and local mate competition in *Trichogramma evanescens.* Journal of Animal Ecology, 53, 417–426.

Waage, J. K., & Ng, S. M. (1984). The reproductive strategy of a parasitic wasp. I. Optimal progeny and sex allocation in *Trichogramma evanescens.* Journal of Animal Ecology, 53, 401–415.

Walker, I. (1967). Effect of population density on the viability and fecundity in *Nasonia vitripennis* Walker (Hymenoptera, Pteromalidae). Ecology, 48, 294–301.

Walter, G. H. (1983). 'Divergent male ontogenies' in Aphelinidae: A simplified classification and a suggested evolutionary sequence. Biological Journal of the Linnean Society, 19, 63–82.

Werren, J. H. (1980). Sex ratio adaptations to local mate competition in a parasitic wasp. Science, 208, 1157–1158.

Werren, J. H. (1983). Sex ratio evolution under local mate competition in a parasitic wasp. Evolution, 37, 116–124.

Werren, J. H. (1984). A model for sex ratio selection in parasitic wasps: Local mate competition and host quality effects. Netherlands Journal of Zoology, 34, 81–96.

Werren, J. H. (1997). Biology of *Wolbachia.* Annual Review of Entomology, 42, 587–609.

Werren, J. H., & Beukeboom, L. W. (1993). Population genetics of a parasitic chromosome: Theoretical analysis of PSR in subdivided populations. American Naturalist, 142, 224–241.

Werren, J. H., & Simbolotti, G. (1989). Combined effects of host quality and local mate competition on sex allocation in *Lariophagus distinguendus.* Evolutionary Ecology, 3, 203–213.

Werren, J. H., Nur, U., & Eickbush, D. (1987). An extrachromosomal factor causing loss of paternal chromosomes. Nature (London), 327, 75–76.

Werren, J. H., Skinner, S. W., & Charnov, E. L. (1981). Paternal inheritance of a daughterless sex ratio factor. Nature (London), 293, 467–468.

Werren, J. H., Skinner, S. W., & Huger, A. M. (1986). Male-killing bacteria in a parasitic wasp. Science, 231, 990–992.

West, S. A., Flanagan, K. E., & Godfray, H. C. J. (1996). The relationship between parasitoid size and fitness in the field a study of *Achrysocarioides zwoelferi* (Hymenoptera: Eulophidae). Ecological Enotomology

White, M. J. D. (1984). Chromosomal mechanisms in animal reproduction. Bolletiro di Zoologia, 51, 1–23.

Whiting, A. R. (1961). Genetics of *Habrobracon.* Advances in Genetics, 10, 295–348.

Whiting, P. W. (1943). Multiple alleles in complementary sex determination of *Habrobracon.* Genetics, 28, 365–382.

Whiting, P. W. (1960). Polyploidy in *Mormoniella.* Genetics, 45, 949–970.

Wilkes, A. (1959). Effects of high temperature during post embryonic development on the sex ratio of an arrhenotokous insect, *Dahlbominus fulginosus* (Nees) (Hymenoptera: Eulophidae). Canadian Journal of Genetics and Cytology, 1, 102–109.

Wilkes, A. (1963). Sperm transfer and utilization by the arrhenotokous wasp *Dahlbominus fuscipennis.* Canadian Entomologist, 97, 647–657.

Wilkes, A. (1964). Inherited male-producing factor in an insect that produces its males from unfertilized eggs. Science, 144, 305–307.

Wilkes, A., & Lee, P. E. (1965). The ultrastructure of dimorphic spermatozoa in the hymenopteran *Dahlbominus fuscipennis.* Canadian Journal of Genetics and Cytology, 7, 609–619.

Williams, J. R. (1972). The biology of *Physcus semintus* and *P. subflavus,* parasites of the sugar cane scale insect *Aulacaspis tegalensis.* Bulletin of Entomological Research, 61, 463–484.

Williams, T. (1991). Host selection and sex ratio in a heteronomous hyperparasitoid. Ecology and Entomology, 16, 377–386.

Winkler, H. K. A. (1920). Verbreitung und Ursache der Parthenogenesis im Pflanzen-und Tierreiche. Jena: Gustav Fisher.

Woyke, J. (1963). Drone larvae from fertilized eggs of the honeybee. Journal of Apicultural Research, 2, 19–24.

Woyke, J. (1979). Sex determinination in *Apis cerana indica.* Journal of Apicultural Research, 18, 122–127.

Wylie, H. G. (1976). Observations on the life history and sex ratio variability of *Eupteromalis dubius* (Hymenoptera: Pteromalidae), a parasitoid of cyclorrhaphous Diptera. Canadian Entomologist, 108, 1267–1274.

Zchori-Fein, E., Rousch, R. T., & Hunter, M. S. (1992). Male production induced by antibiotic treatment in *Encarsia formosa,* an asexual species. Experientia, 48, 102–105.

24

Evolution of Pesticide Resistance
in Natural Enemies

BRUCE E. TABASHNIK

Department of Entomology
University of Arizona
Tucson, Arizona

MARSHALL W. JOHNSON

Department of Entomology
University of Hawaii
Honolulu, Hawaii

INTRODUCTION

Resistance to pesticides in agricultural pests and vectors of human disease is an urgent worldwide problem (NRC, 1986; Roush & Tabashnik, 1990). In contrast, documented examples of pesticide resistance in field populations of natural enemies are relatively rare (Croft, 1990). Beneficial organisms account for fewer than 3% of the 447 species of insects and mites reported as resistant by Georghiou (1986). Thus, reported cases of resistance in pests outnumber those in natural enemies by more than 30/1.

Hypotheses offered to explain this pattern can be put into three general categories: (1) bias in documentation of resistance, (2) differential preadaptation to pesticides, and (3) differences in population ecology (Table 1). The first hypothesis asserts that resistance in a pest is more likely to be observed and reported than resistance in a natural enemy. Hypotheses in the other two categories contend that biological differences between pests and natural enemies cause resistance to evolve more readily in pests than in natural enemies.

We emphasize that the categories are somewhat arbitrary and the hypotheses are not mutually exclusive. Combinations of factors may operate in a particular case and the relative importance of factors may vary among species. In this chapter, we describe the general categories and specific hypotheses, examine relevant empirical evidence and theory, and suggest avenues for further study.

HYPOTHESES AND EVIDENCE

Bias in Documentation

Pesticide resistance is a genetically based, statistically significant increase in the ability of a population to tolerate one or more pesticides. In most cases, resistance is documented with laboratory bioassays showing that a population with a history of extensive exposure to pesticides has a significantly greater LC_{50} or LD_{50} (concentration or dose of pesticide that kills 50%) compared with a conspecific population that has had less exposure to pesticides. One can also document resistance by showing that treatment with a fixed concentration or dose causes significant differences in mortality among conspecific populations (Roush & Miller, 1986; Tabashnik *et al.*, 1987). Evidence of significant increase through time within a population in LC_{50}, LD_{50}, or survival in response to a fixed concentration or dose provides more direct documentation of resistance.

Resistant pest populations can attract immediate attention when pesticide treatments fail to control them. Resistance in natural enemies, however, does not create problems and may go unnoticed. Thus, the bias in documentation hypothesis states that pesticide resistance is more likely to be documented in pests than in natural enemies (Georghiou, 1972; Croft & Brown, 1975).

Croft and Brown (1975) noted that if resistance in natural enemies appears rare due to inadequate documentation, then "systematic testing of samples of natural enemies in heavily treated ecosystems should detect more cases of resistance." Data available on pesticide impact on natural enemies more than doubled from 1970 to 1984 (Theiling & Croft, 1988) and the number of natural-enemy species reported as resistant to one or more pesticides also doubled during the same period (Georghiou, 1972, 1986). The proportion of cases of resistance accounted for by beneficial species, however, remained at about 3% in the 1970 and 1984 surveys (6/225 versus 12/447). These data do not resolve the question; they only show that the cumulative effect of factors contributing to more documented cases of resistance in pests than in natural enemies remained con-

TABLE 1   Hypotheses Proposed to Explain the Scarcity of
Pesticide Resistance in Natural Enemies

Category	Specific hypotheses	References
Documentation bias	—	Georghiou, 1972; Croft & Brown, 1975
Preadaptation	Detoxification enzymes	Croft & Morse, 1979; Croft & Strickler, 1983
	Intrinsic tolerance	Rosenheim & Hoy, 1986
	Genetic variation	Georghiou, 1972
	Fitness cost	This chapter
Population ecology	Food limitation	Huffaker, 1971; Georghiou, 1972
	Life history	Croft, 1982; Tabashnik & Croft, 1985
	Exposure	Croft & Brown 1975
	Genetic systems	This chapter

sistent through time. Cases of resistance were included in the surveys only if resistance was due to field application of pesticides and was sufficient to cause diminished mortality at field application rates (Georghiou, 1981, 1986). Cases of resistance in natural enemies due to laboratory selection are reviewed elsewhere (Croft & Strickler, 1983; Croft, 1990; Hoy, 1990; Johnson & Tabashnik, 1994).

The effect of bias in documentation is difficult to quantify. In principle, one could survey a large number of randomly chosen pest and natural-enemy species and then could compare the frequency and magnitude of pesticide resistance between the two groups. There is, however, no such survey. Resistance studies focus on species expected to show resistance; such samples are not random with respect to probability of detecting resistance. Few studies have made intraspecific comparisons of susceptibility among populations of natural enemies or monitored a natural-enemy population's susceptibility through time to detect resistance. For example, resistance ratios (the LD_{50} or LC_{50} of a potentially resistant natural-enemy population divided by the LD_{50} or LC_{50} of a susceptible strain) were available for only about 1% of the cases in the survey of pesticide effects on natural enemies compiled by Theiling and Croft (1988).

We attempted to assess the possible effects of bias in documentation by reviewing evidence from 1979 to 1995 surveys of pesticide susceptibility in 20 species of natural enemies from various ecosystems (Tables 2 and 3). Cases of resistance to pesticides among arthropod parasites and natural enemies reported before 1979 were reviewed by Croft and Strickler (1983). We included only studies that estimated LC_{50}'s for more than one population, which enabled calculation of the maximum resistance ratio (LC_{50} of the most resistant population divided by LC_{50} of the most

susceptible population). Although LC_{50}'s estimated in laboratory bioassays do not necessarily predict survival in the field (Hoy, 1990), field survival is more likely if the LC_{50} exceeds the recommended field application rate. Thus, when this information was reported in the original publication, we included it in our review (see Tables 2 and 3).

Predatory phytoseiid mites received considerable attention and had higher levels of resistance than most other natural enemies (see Tables 2 and 3). A statistically significant difference in susceptibility between at least one pair of populations was reported in 77% (10/13) of cases for phytoseiids. Maximum resistance ratios for phytoseiids exceeded 10-fold in 40% (8/20) of cases. The maximum LC_{50} was greater than the recommended field application rate in 6 of 12 cases. When data for pesticides that had not been applied to field populations were excluded, 88% of cases showed significant variation in pesticide susceptibility, 50% had maximum resistance ratios greater than 10, and 55% showed maximum LC_{50}'s greater than field rates (see Table 3).

For 15 species of nonphytoseiid natural enemies, 83% (29/35) of cases showed significant variation in susceptibility, maximum resistance ratios exceeded 10 in 23% (10/44) of cases, and 41% (11/27) of cases showed maximum LC_{50}'s greater than field rates. *Chrysoperla carnea* (Stephens) accounted for 5 of 11 cases showing ability to survive field rates of pesticides in nonphytoseiids. The ability to survive field rates in these five cases represents natural tolerance instead of resistance, because LC_{50}'s of the most susceptible populations exceeded recommended field concentrations. In fact, four of the five cases were due to tolerance to pyrethroids that were not registered for field use when the bioassays were done (Grafton-Cardwell & Hoy, 1985).

For hymenopteran species, statistically significant variation in susceptibility among populations was almost universal (93%), but maximum resistance ratios >10 were less common (24%) as were cases where the maximum LC_{50} exceeded the field rate (25%). Notable cases of resistance in Hymenoptera include >10-fold resistance of *Diglyphus begini* (Ashmead) to both carbamates and pyrethroids (Rathman *et al.*, 1990), 66-fold resistance to methidathion in *Comperiella bifasciata* Howard (Schoonees & Giliomee, 1982), and >1200-fold resistance to malathion in *Anisopteromalus calandrae* (Howard) (Baker & Weaver, 1993).

The set of cases reviewed here suggests that statistically significant variation in susceptibility to pesticides among conspecific populations is common in natural enemies. Resistance that may be high enough to confer survival at field application rates appears to be about twice as common in predatory phytoseiid mites compared with that of hymenopteran parasitoids. Phytoseiids may evolve resistance readily because they are exposed to pesticides in all life stages, have limited dispersal, and can subsist on plant materials when prey density is low (Georghiou, 1972; Croft & Brown, 1975). These and other traits of phytoseiids will be discussed in relation to specific hypotheses described later. Hoy (1985) and Croft and van de Baan (1988) consider pesticide resistance in phytoseiids in greater detail.

In summary, available evidence is not sufficient to conclusively refute or support the documentation bias hypothesis. Our review of recent field surveys shows that substantial resistance was found commonly in phytoseiid mites, but was less common in most of the other natural enemies tested. Results from a survey of fruit entomologists suggest that among insects and mites from apple orchards, resistance to azinphosmethyl was nearly as frequent in natural enemies as in pests (Croft, 1982). However, because of the unusually prolonged use of azinphosmethyl and other factors, resistance to azinphosmethyl in apple arthropods may not be indicative of broader trends (Tabashnik & Croft, 1985). Bioassays of several pest and natural-enemy species from two or more populations representing variation in insecticide exposure for a particular crop and region could help to assess the extent of bias in documentation.

Differential Preadaptation

The differential preadaptation hypothesis states that resistance to pesticides evolves more readily in pests than in natural enemies due to differences in responses to pesticides between pests and natural enemies that existed *before* selection for pesticide resistance. Several specific hypotheses fall under the general category of differential preadaptation:

1. Pests evolve resistance more readily because they have better intrinsic detoxification capabilities than natural enemies have (Gordon, 1961; Croft & Morse, 1979; Croft & Strickler, 1983).
2. Pests evolve resistance more readily because they have greater intrinsic tolerance to pesticides than natural enemies have (Rosenheim & Hoy, 1986).
3. Pests evolve resistance more readily because they have more genetic variation in tolerance to pesticides than natural enemies have (Georghiou, 1972).
4. Pests evolve resistance more readily because the fitness cost associated with resistance is lower for pests than for natural enemies.

These four hypotheses are not mutually exclusive. Each hypothesis has two parts: (1) an intrinsic (before pesticide exposure) difference between pests and natural enemies in response to pesticides, which (2) enables pests to evolve resistance more readily than natural enemies evolve it. A test of any preadaptation hypothesis must examine both components. Tests of part (1) require comparisons between susceptible populations of pests and natural enemies. Ide-

ally, such tests should compare populations that have had little or no exposure to pesticides, but such populations may be rare. Thus, attempts to measure intrinsic differences between pests and natural enemies can be confounded by the history of insecticide exposure (Croft & Strickler, 1983).

Tests of part (2) are more difficult because they require examination of how a particular intrinsic difference will influence evolution of resistance. Direct tests would need to determine if the measured intrinsic difference caused differences in rates of evolution. For example, an ideal comparison would contrast the rate of evolution of resistance in two populations of a natural-enemy species that differed only in initial detoxification capability. The preadaptation hypothesis would be supported if the natural-enemy strain with higher initial detoxification capability evolved resistance faster than the other strain evolved it. Further, if intrinsic detoxification capacity is a primary determinant of the rate of resistance evolution, then a natural-enemy strain with initial detoxification capability comparable to that of a pest would be expected to evolve resistance at a rate comparable to the pest.

Indirect tests could compare rates of evolution of resistance between pest and natural-enemy species that had intrinsic differences, but such tests could not restrict the cause of a difference in evolutionary rates to any particular factor. In other words, species differ in many ways and interspecific comparisons do not allow one to determine if a given factor is responsible for differences in the rate of resistance evolution between species. Another type of investigation involves theoretical models. This approach attempts to measure the potential impact of intrinsic differences by comparing projected rates of resistance evolution under various assumptions about initial conditions. In the following section we consider empirical and theoretical evidence relevant to specific preadaptation hypotheses.

Detoxification Enzymes

The origins of this hypothesis are reviewed by Croft and Strickler (1983) and Rosenheim *et al.* (1996). The principal idea is that herbivorous pest insects have evolved enzymes to detoxify plant secondary compounds and thus are better preadapted to detoxify pesticides than are entomophagous natural enemies, which do not eat plants. We review available evidence to determine (1) whether herbivorous pests have higher intrinsic levels of detoxification enzymes than natural enemies have, and (2) whether such differences cause natural enemies to evolve resistance more slowly than pests evolve it.

In the broadest survey to date, Brattsten and Metcalf (1970) used the synergistic ratio of carbaryl with piperonyl butoxide to compare levels of mixed function oxidase (MFO) detoxification enzymes in 53 insect species from

TABLE 2 Pesticide Susceptibility of Field and Stored Grain Populations of Natural Enemies, 1979 to 1995

Species (family)	Location	Habitat	No. of populations tested	Pesticide	Maximum resistance ratio at LC_{50}	Maximum LC_{50} > field rate	References
Acarina							
Amblyseius fallacis (Phytoseiidae)	Michigan, New York	Apple, soybean	12	Azinphosmethyl	34[a]	Yes	Strickler & Croft, 1981
			12	Permethrin	15[a]	No	Croft & Wagner, 1981
			2	SD 57706[b]	46[c]	—	
			2	NCI 85913[b]	2[a]	—	
			2	Fluvalinate	5[a]	—	
			2	ZR 3903[b]	8[a]	—	
			4	Carbaryl	3[c]	Yes	Croft et al., 1982
			8	DDT	>43[a]	Yes	Scott et al., 1983
			5	Methoxychlor	13	—	
Euseius (=Amblyseius) hibisci (Phytoseiidae)	California	Citrus	2	Dimethoate	22	No	Tanigoshi & Congdon, 1983
Metaseiulus (=Typholodromus) occidentalis (Phytoseiidae)	California	Grape	14	Dimethoate	11[a]	Yes	Hoy et al., 1979
			6	Methomyl	3[a]	No	
		Pear, wild blackberry	11	Azinphosmethyl	5[a]	—	Hoy & Knop, 1979
			11	Diazinon	4[a]	—	
			2	Dimethoate	1[c]	—[d]	
Typhlodromus arboreus (Phytoseiidae)	Oregon	Apple	2	Azinphosmethyl	7	No	Croft & Aliniazee, 1983
			2	Carbaryl	2	No	
T. pyri (Phytoseiidae)	England, the Netherlands, New Zealand	Apple	10	Azinphosmethyl	6	Yes	Kapentanakis & Cranham, 1983
			10	Carbaryl	>62	Yes	
			7	Permethrin	2	No	
Araneae							
Chiracanthium mildei (Clubionidae)	Israel	Citrus, cotton	2	Malathion	3	—	Mansour, 1984
Coleoptera							
Xylocoris flavipes (Anthocoridae)	South Carolina	Stored grain	2	Malathion	33[a]	—	Baker & Arbogast, 1995
Diptera							
Aphidoletes aphidimyza[e] (Cecidomyiidae)	Michigan	Apple	14	Azinphosmethyl	3[a]	Yes	Warner & Croft, 1982
Hymenoptera							
Anisopteromalus calandrae (Pteromalidae)	South Carolina	Stored grain	2	Chlorpyrifos-methyl	5[a]	—	Baker & Weaver, 1993
			2	Malathion	>1200	Yes	
			2	Pirimiphos-methyl	7[a]	—	Baker, 1994
			2	Cyfluthrin	2[a]	—	
			2	deltamethrin	2[a]	—	
Aphytis africanus (Aphelinidae)	South Africa	Citrus	2	Methidathion	6[a]	No	Schoones & Giliomee, 1982

Species (Family)	Location	Crop/Host		Pesticide	LC_{50} ratio	Resistance	Reference
A. holoxanthus (Aphelinidae)	Israel	Citrus	7	Malathion	3[a]	No	Havron, 1983
A. melinus (Aphelinidae)	California	Citrus	7	Carbaryl	2[a]	No	Rosenheim & Hoy, 1986
			8	Chlorpyrifos	2[a]	No	
			11	Dimethoate	3[a]	No	
			12	Malathion	8[a]	No	
			9	Methidathion	8[a]	No	
Bracon hebetor (Braconidae)	Georgia, South Carolina	Stored grain	3	Malathion	8[a]	—	Baker et al., 1995
Comperiella bifasciata (Encyrtidae)	South Africa	Citrus	2	Methidathion	66[a]	No	Schoonees & Giliomee, 1982
Diaretiella rapae (Aphidiidae)	California	Cauliflower, wild mustard	2	Methomyl	2	No	Hsieh, 1984
Diglyphus begini (Eulophidae)	Hawaii, California	Bean, tomato	5	Fenvalerate	17[a]	Yes	Rathman et al., 1990
			5	Methomyl	21[a]	No	
			5	Oxamyl	20[a]	No	
			5	Permethrin	13[a]	Yes	
Ganaspidium utilis (Ecucoilidae)	Hawaii, California	Bean, tomato	5	Fenvalerate	3[a]	Yes	Rathman et al., 1995
			5	Malathion	9[a]	No	
			5	Methomyl	3[a]	No	
			5	Oxamyl	2[c]	No	
			5	Permethrin	12[a]	Yes	
Pholetesor ornigis (Braconidae)	Ontario (Canada)	Apple	3	Azinphosmethyl	2[c]	—	Trimble & Pree, 1987[f]
			2	Fenvalerate	2[a]	—	
			2	Methomyl	4[a]	—	
			2	Permethrin	3[a]	—	
Trioxys pallidus (Aphidiidae)	California	Walnut	5	Azinphosmethyl	5	No	Hoy & Cave, 1988
Neuroptera							
Chrysoperla carnea (Chrysopidae)[g]	California	Alfalfa	4	Carbaryl	3[a]/2[c]	No/No	Grafton-Cardwell & Hoy, 1985
			4	Diazinon	2[c]/4	No/No	
			4	Fenvalerate	2/2[c]	Yes/Yes	
			4	Methomyl	3[a]/3[c]	No/No	
			4	Permethrin	5/11	Yes/Yes	
			4	Phosmet	2[c]/16	No/Yes	

[a]Statistically significant difference is between LC_{50}'s of at least one pair of populations (by nonoverlap of 95% fiducial limits).
[b]This is an experimental pyrethroid; see reference for details.
[c]No significant differences are among LC_{50}'s.
[d]Pear populations are not treated with dimethoate.
[e]Bioassays are of eggs.
[f]A subsequent survey detected no decrease in susceptibility from 1987 to 1989 (Trimble et al., 1990).
[g]Bioassays are of adults/first-instar larvae.

TABLE 3 Summary of Pesticide Susceptibility of Field and Stored Grain Populations of Natural Enemies, 1979 to 1995

Group	No. of species	Significant variation[a] (%)[b]	Maximum resistance ratio >10 (%)[b]	Maximum LC$_{50}$ > field rate (%)[b]
Phytoseiid mites	5	77(10/13)	40(8/20)	50(6/12)
Phytoseiid mites (field-used pesticides only)[c]	5	88(7/8)	50(7/14)	55(6/11)
Nonphytoseiids	15	83(29/35)	23(10/44)	41(11/27)
Hymenoptera	11	93(25/27)	24(1/13)	25(5/20)
Chrysoperla carnea	1	29(2/7)	17(2/12)	42(5/12)
Totals	20	81(39/48)	28(18/64)	44(17/39)
Totals (field-used pesticides only)[c]	20	84(36/43)	24(17/58)	45(17/38)

[a] Nonoverlap of 95% fiducial limits occurs for LC$_{50}$'s of at least one pair of populations.

[b] Numbers show the percentage of cases that meet each criterion (numbers of cases are in parentheses).

[c] This excludes cases in which populations were tested versus compounds that had not been applied to them in the field (i.e., pyrethroids tested before field use and dimethoate versus *M. occidentalis* in grapes).

eight orders. Piperonyl butoxide inhibits MFO enzymes that detoxify carbaryl. Thus, the ratio of the LD$_{50}$ (or LC$_{50}$) of carbaryl alone to the LD$_{50}$ (or LC$_{50}$) of carbaryl plus piperonyl butoxide (synergistic ratio) is an indicator of MFO activity. Brattsten and Metcalf (1970) found that a few of the phytophagous species had extraordinarily high synergistic ratios [most notably *Pogonomyrmex barbatus* (F. Smith), the red harvester ant, synergistic ratio >223], but the *median* synergistic ratio for 25 herbivores (3.6) was less than half the median for 15 entomophagous species (8.4). The proportion of species with low synergistic ratios (i.e., below the overall median of 4.8) was significantly greater for herbivores (17/25) than for predators and parasites (3/15) ($p < 0.01$, median test; Fig. 1). The median synergistic ratios for 13 crop pests (2.3) and 27 herbivorous and non-herbivorous pests (4.2) were also significantly lower than the median for entomophagous species ($p < 0.01$ and $p < 0.05$, respectively). These findings contradict part (1) of the preadaptation hypothesis.

Consideration of part (2) of the hypothesis shows that natural enemies that have not been reported as resistant to insecticides had greater synergistic ratios than did several pests that are notorious for their ability to evolve insecticide resistance. For example, the synergistic ratios for six of seven species of parasitic Hymenoptera studied were higher than the ratios for the German cockroach [*Blattella germanica* (Linnaeus)], the corn earworm [*Helicoverpa zea* (Boddie)], the fall armyworm (*Spodoptera frugiperda* J. E.

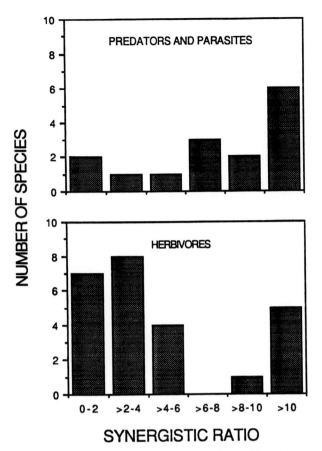

FIGURE 1 Synergistic ratio of carbaryl with piperonyl butoxide as an indicator of mixed function oxidase (MFO) activity in entomophagous and herbivorous insects [Adapted from Brattsten, L. B., & Metcalf, R. L. (1970). *Journal of Economic Entomology, 63,* 1347–1348.]

Smith), and the southern house mosquito (*Culex pipiens quinquefasciatus* Say).

In conclusion, the data of Brattsten and Metcalf (1970) do not support the idea that phytophagous pests have higher levels of detoxification enzymes than natural enemies have. Further, their data suggest that low MFO activity, as indicated by the synergistic ratio of carbaryl to piperonyl butoxide, does not necessarily slow evolution of pesticide resistance. Some pests with low synergistic ratios have evolved resistance whereas some natural enemies with relatively high ratios have not.

The preceding conclusions were later confirmed by findings that piperonyl butoxide affected a predatory mite (*Amblyseius fallacis* Garman) more than it affected its herbivorous prey mite (*Tetranychus urticae* Koch) (Strickler & Croft, 1985). In addition, the synergistic ratios of piperonyl butoxide with propoxur for adults of the ectoparasitic braconid *Oncophanes americanus* (Weed) (14.0) were 8-, 2-, and 12-fold greater than the synergistic ratios for adults, large larvae, and medium larvae, respectively, of its herbivorous tortricid host, *Argyrotaenia citrana* (Fernald) (Croft & Mullin, 1984). The synergistic ratio for large larvae of the ectoparasite (5.5) was more than triple the synergistic ratio for adults or medium larvae of the host, but it was slightly less than the synergistic ratio of large larvae of the host (6.0).

Croft and Mullin (1984) stated that "synergist tests are useful in qualitatively estimating the availability of an MFO detoxification pathway," but they questioned the value of piperonyl butoxide synergist tests as an indicator of MFO activity across diverse arthropod species. Thus, it is important to consider other measures of enzyme activity, such as results from *in vitro* assays.

In vitro inhibition studies and hydrolysis assays showed that predatory lacewing larvae, *C. carnea,* have unusually active esterases that detoxify pyrethroids (Ishaaya & Casida, 1981). Although direct comparative tests were not done, preparations of whole lacewing larvae were more active than larval gut or integument preparations from an herbivore, the cabbage looper, *Trichoplusia ni* (Hübner). Lacewing larvae were chosen for enzymatic tests because they have high natural tolerance to pyrethroids and thus were suspected to have high esterase activity (Plapp & Bull, 1978). Therefore, the lacewing esterase activity data are a biased sample of natural-enemy detoxification enzyme capacity.

Mullin *et al.* (1982) found differences in *in vitro* detoxification enzyme activity between susceptible populations of a predatory mite (*Amblyseius fallacis*) and its herbivorous prey mite (*Tetranychus urticae*). MFO and *trans*-epoxide hydrolase had greater activity in the prey than in the predator, but the opposite was observed for *cis*-epoxide hydrolase and glutathione transferase. Esterase activity was similar in susceptible populations of the two species.

Although the herbivorous mite had greater activity than the predatory mite had for an MFO enzyme (aldrin epoxidase) and for *trans*-epoxide hydrolase, neither of these two enzymes was more active in resistant than in susceptible populations of the herbivore. These data suggest that resistance in the herbivore is not due to elevated levels of aldrin epoxidase or *trans*-epoxide hydrolase. Thus, it appears that intrinsically greater levels of these two enzymes were not responsible for the ability of the herbivorous pest to evolve resistance more readily than its predator evolved it. Conversely, of the five types of enzyme activity measured, only glutathione transferase was greater in resistant prey than in susceptible prey, suggesting that this enzyme could be partly responsible for resistance in the prey species. Contrary to the expectation of the preadaptation hypothesis, however, susceptible predators had more than 10-fold higher activity for this enzyme than susceptible pests had.

The evidence suggests that the detoxification enzyme differences between *A. fallacis* and *T. urticae* reported by Mullin *et al.* (1982) do not support part (2) of the preadaptation hypothesis. It may be inappropriate to generalize from this case because predatory phytoseiid mites are the only group of natural enemies that has readily evolved pesticide resistance. Intrinsic disadvantages that slow evolution of resistance in other natural enemies may be absent or diminished in phytoseiids.

Van de Baan and Croft (1990) compared *in vitro* enzyme activities between the mirid predator *Deraeocoris brevis* Knight and susceptible and resistant strains of its prey *Psylla pyricola* Foerster. They found that the pest had significantly higher esterase activity than did the predator. Further, elevated esterase activity was associated with resistance in the pest. However, glutathione *S*-transferase and P-450 monooxygenase activities were higher in the predator than in the susceptible strain of the pest. The results with esterase are consistent with the preadaptation hypothesis, but the results with the latter two enzyme activities are not.

In a related study of *in vitro* enzyme activity, Croft and Mullin (1984) compared various life stages of a braconid ectoparasite *Oncophanes americanus* (Weed) and its tortricid host, *Argyrotaenia citrana* (Fernald). Based on data from whole-body preparations averaged across life stages, none of the five assays conducted showed significantly lower enzyme activity in the parasitoid compared with that of the host. The parasitoid had three-fold higher *cis*-epoxide hydrolase activity ($p < 0.001$) and aldrin epoxidase activity ($p = 0.06$) than its host had. No significant differences were found between species in *trans*-epoxide hydrolase, glutathione transferase, or esterase activities. Enzyme activities from midgut tissue of late larvae of the host were much higher than the activities of whole-body preparations from any life stage of either species. Midgut tissue of the parasitoid was not assayed, so interspecific comparisons of midgut tissue enzyme activity were not possible. In summary, the data from this parasite–host system do not support part (1) of the preadaptation hypothesis and they do not address part (2) of the hypothesis.

Another *in vitro* study compared adult midgut levels of three types of enzyme between an herbivorous coccinellid, *Epilachna varivestis* Mulsant, and a predatory coccinellid, *Hippodamia convergens* Guerin-Meneville (Mullin, 1985). *Trans*-epoxide hydrolase was virtually identical in the two species (<2% difference), but the predator had 4-fold greater *cis*-epoxide hydrolase activity and 2.5-fold greater esterase activity than the herbivore had. Thus, this comparison between beetles that differ in feeding habit does not support part (1) of the preadaptation hypothesis.

Review of these studies and similar studies of four piercing–sucking herbivores (Mullin, 1985) showed broad overlap in detoxification enzyme levels between herbivores and natural enemies. For example, aldrin epoxidase, *trans*-epoxide hydrolase, and *cis*-epoxide hydrolase were lower in the herbivore *Aphis nerii* Fonscolombe than in any of the three natural enemies tested (*A. fallacis, O. americanus,* and *P. fovelatus*). Whole-body levels of aldrin epoxidase for the herbivores *T. urticae* and *A. citrana* and the three natural enemies tested fell within the range of the three aphid species tested [*A. nerii, Myzus persicae* (Sulzer), and *Macrosiphum euphorbiae* (Thomas)].

In one of the most detailed comparative studies of detoxification abilities, Yu (1987) measured 15 components of enzymatic detoxification capacity in the spined soldier bug [*Podisus maculiventris* (Say), Heteroptera: Pentatomidae] and four of its noctuid prey species [*Heliothis virescens* (Fabricius), *H. zea, Spodoptera frugiperda,* and *Anticarsia gemmatalis* (Hübner)]. *In vitro* assays of midgut tissues from adult predators and final-instar larvae of the lepidopteran prey showed that the predator lacked none of the detoxification enzyme systems that were found in the prey. The prey had greater detoxification capacity in 70% (42/60) of pairwise comparisons between the predator and prey. For 11 of 15 components measured, either 3 or 4 of the prey species had greater activity than the predator had. The reverse was found, however, for cytochrome P-450, microsomal desulfurase, and glutathione transferase toward 1-chloro-2,4-dinitrobenzene (CDNB). In addition, the proportion of the high-spin form of cytochrome P-450, thought to indicate allelochemical-oxidizing capacity (Yu, 1987), was higher in the predator than in any of the prey.

Topical bioassays showed that the predator was generally more susceptible to organophosphorous and carbamate insecticides, but was more tolerant of pyrethroids compared with its prey (Yu, 1988). However, the organophosphate tetrachlorvinphos was much more toxic to two of the prey species (*S. fruigiperda* and *A. gemmatalis*) than to the predator. Although penetration of tetrachlorvinphos was slower for the predator than for *S. frugiperda,* injection tests and acetylcholinesterase inhibition tests implied that the predator's ability to survive tetrachlorvinphos was due to enhanced enzymatic detoxification. Yu (1988) concluded that the predator's high pyrethroid tolerance was probably due to reduced penetration or target site insensitivity.

Yu's (1987) *in vitro* assays showing that the predator generally had lower enzymatic detoxification capacity than its prey support part (1) of the preadaptation hypothesis, but several significant exceptions also were found. Yu's (1988) bioassays show that patterns of susceptibility to insecticides could not be reliably predicted on the basis of enzyme activities. Indeed, higher levels of certain "detoxification" enzymes, such as desulfurase, may actually increase the toxicity of some insecticides (Yu, 1988).

The data from *in vivo* and *in vitro* studies reviewed here do not support the idea that natural enemies have consistently lower levels of all types of detoxification enzymes than pests have, but this does not exclude the possibility of more subtle differences between the two groups. For ex-

ample, an alternative hypothesis is that oxidative detoxification enzymes are more active in pests than in natural enemies, whereas hydrolytic detoxification enzymes are similar in pests and in natural enemies (Plapp & Vinson, 1977). The first part of this hypothesis, however, is contradicted by data from *in vivo* synergism tests and is not strongly supported by data from *in vitro* tests. Aldrin epoxidase data from a predatory mite and its herbivorous prey mite (Mullin *et al.,* 1982) support the hypothesis, but data from whole-body preparations of an ectoparasite and its lepidopterous host show the opposite trend (Croft & Mullin, 1984). Furthermore, intraspecific variation in MFO enzymes among developmental stages, sometimes even within an instar, can be greater than the typical differences found between pests and natural enemies (Gould, 1984 and cited references; Croft & Mullin, 1984).

Hydrolytic enzymes such as esterases and epoxide hydrolases are often equally or more active in natural enemies compared with pests (see preceding text). Yu (1987) found, however, that hydrolytic enzyme activities were lower in a predator than in its prey in 88% (14/16) of comparisons, but oxidative detoxification activities were lower in the predator than in the prey in 59% (19/32) of comparisons. Likewise, van de Baan and Croft (1990) found hydrolytic activity was lower in a predator than in its prey, whereas the opposite was true for oxidative activity.

The ratio of *trans*- to *cis*-epoxide hydrolase is, on average, higher for herbivores than for natural enemies (Mullin & Croft, 1984). The lower *trans* to *cis* ration of natural enemies, however, does not explain a reduced ability to evolve resistance. The pests *Blattella germanica* and *Myzus persicae* Sulz., both well known for their ability to evolve resistance, had lower *trans* to *cis* ratios than six of seven entomophagous species studied (Mullin, 1985). Nonetheless, differences between pests and natural enemies in *trans* to *cis* ratios and other biochemical traits may be useful in designing selective pesticides (Mullin & Croft, 1984, 1985).

In summary, the detoxification enzymes found in pests are also present in natural enemies. Data from *in vivo* synergism tests show that MFO enzyme levels in natural enemies were not consistently lower than those in pests; evidence to date shows the opposite trend. Although *in vitro* enzyme assays have indicated many cases in which pests have higher levels of detoxification enzymes than natural enemies have, the reverse has been reported with nearly equal frequency. Thus, part (1) of the hypothesis generally is not supported. In those cases in which detoxification enzyme levels are higher in pests than in natural enemies, there is little or no evidence indicating that this difference contributes to an ability to evolve resistance more rapidly. Thus, there is little support for part (2) of the hypothesis. Finally, the differential detoxification hypothesis does not address nonmetabolic resistance (such as reduced penetration and target site insensitivity) or widespread resistance in nonherbivorous pests (Tabashnik, 1986).

Intrinsic Tolerance

The idea that natural enemies are intrinsically less tolerant to pesticides than pests are, is in effect a generalized version of the differential detoxification enzyme hypothesis. The concept is the same, but unlike the differential detoxification hypothesis, the mechanism causing the intrinsic difference is unspecified. Tests of this version of the pre-adaptation hypothesis must determine (1) whether pests have higher intrinsic tolerance to pesticides than natural enemies have, and if so, then (2) whether such intrinsic differences in tolerance cause natural enemies to evolve resistance more slowly than pests evolve resistance.

Two types of bias make it difficult to evaluate part (1) of the hypothesis. First, widespread resistance and cross-resistance in pests can make it difficult to assess their intrinsic tolerance. Inclusion of resistant pest populations in surveys of tolerance would tend to inflate measures of tolerance in pests relative to natural enemies. Second, researchers may concentrate efforts on pesticides that natural enemies can tolerate because such compounds are particularly useful in integrated pest management (IPM). These two biases operate in opposite directions and their relative magnitude is not easily determined.

Brattsten and Metcalf (1970) found that some pests were very tolerant to carbaryl [e.g., the ants *Pogonomyrmex barbatus* ($LD_{50} > 5800$ μg/g) and *P. californicus estebanius* Pergande ($LC_{50} = 450$ μg/cm²)]. However, median LD_{50} values did not differ significantly between 22 pests (35.5 μg/g) and 8 entomophagous species (28.8 μg/g). Similarly, medians for carbaryl LC_{50} did not differ significantly between five pests (13.5 μg/cm²) and seven parasites (14.5 μg/cm²).

Surveys of studies that compare the LC_{50} or LD_{50} values within complexes of natural enemies and pests also do not support the hypothesis that natural enemies are intrinsically more susceptible to pesticides than their prey or hosts are (Croft & Brown, 1975; Theiling & Croft, 1988). Croft and Brown (1975) found that natural enemies were more tolerant than their prey or hosts were in 67 of 92 cases in which the same bioassay method was used to compare species. Although predators were usually more tolerant than their prey were (63 of 77 cases), parasitoids were usually less tolerant than their hosts were (11 of 15 cases).

Theiling and Croft (1988) calculated LC_{50} or LD_{50} ratios for 870 cases in which a natural enemy was compared with its prey or host. Ratios ranged widely, yet natural enemies were more tolerant than their prey or hosts were in 57% of the cases. Predators were more tolerant than their prey were, but parasitoids were less tolerant than their hosts were. For 10 of 12 families of natural enemies, including

Braconidae and Aphelinidae, the average ratio for the family showed greater tolerance in the natural enemy. Phytoseiidae natural enemies were as tolerant as their prey were and only Ichneumonidae were less tolerant than their hosts were. The average ratio for 18 of 24 compounds including diazinon, permethrin, parathion, and carbaryl showed that natural enemies were more tolerant than their prey or hosts were.

The preceding results contradict the differential intrinsic tolerance hypothesis. Theiling and Croft (1988) suggest some reasons why their survey may make natural enemies appear more tolerant than they really are. They note that LC_{50} or LD_{50} comparisons between pest and natural-enemy pairs are available for only a small subset of the studies of pesticide impact on natural enemies (fewer than 7% of the cases surveyed by Theiling & Croft, 1988). Natural enemies suspected to be tolerant to pesticides may be more likely to be compared with their prey or hosts in bioassays. Similarly, compounds thought to be more toxic to pests than to natural enemies may be more likely to be tested in comparative studies.

Although some surveys suggest that natural enemies are not always less tolerant to pesticides than pests are, they are less tolerant in many cases (Theiling & Croft, 1988). Therefore, it is useful to consider how reduced intrinsic tolerance might affect evolution of insecticide resistance. If ability to survive field rates of pesticide is a criterion for resistance, then a natural enemy with lower intrinsic tolerance would have to increase its tolerance more substantially to be considered resistant.

Simulations by Tabashnik and Croft (1985) suggest that, under certain conditions, reduced intrinsic tolerance could also slow evolution of resistance as measured by changes in the frequency of a resistance (R) allele. Simulations based on a one-locus model showed that 10- or 100-fold reduction in the LC_{50} of homozygous susceptible (SS) individuals had little impact on projected times for resistance evolution in natural enemies of apple pests. In contrast, however, reducing the LC_{50} of all three presumed genotypes (SS, RS, RR) by 10- or 100-fold greatly slowed resistance evolution in nearly all cases. When the LC_{50}'s of all three genotypes were reduced, heterozygous individuals were rendered functionally recessive, which slowed evolution of resistance (see Curtis *et al.*, 1978; Taylor & Georghiou, 1979; Tabashnik & Croft, 1982).

The assumption of lower LC_{50}'s for all three genotypes implies that an R allele increases the LC_{50} by a fixed multiple. Thus, if the SS individuals are less tolerant, then so are RS and RR individuals. Lowering the LC_{50} of SS individuals without altering the LC_{50}'s of RS or RR individuals assumes that an R allele provides a fixed level of tolerance, regardless of the tolerance of SS individuals. We do not know if either assumption is widely applicable.

In summary, surveys of comparative bioassay studies

suggest that, on average, natural enemies are not intrinsically less tolerant to pesticides than are pests. Such surveys may be biased, however; and there are at least 371 documented cases in which pests are more tolerant than their natural enemies are (Theiling & Croft, 1988). Reduced intrinsic tolerance could retard evolution of resistance in some natural enemies if resistance is defined as ability to survive field concentrations of a pesticide or if R alleles confer a fixed multiple of increased tolerance relative to susceptible individuals. The finding that Phytoseiidae species, which have been frequently recorded as evolving resistance, had low selectivity ratios (LC$_{50}$ of natural enemy divided by LC$_{50}$ of its prey or host) compared with other natural enemies (Theiling & Croft, 1988) suggests that low intrinsic tolerance relative to pests is not a major impediment to evolution of resistance in natural enemies.

The idea that natural enemies are not generally less tolerant to pesticides than pests are differs from widely held perceptions. Such perceptions may be based partly on observations that field applications of pesticides affect natural enemies more than they affect pests. However, disruption of natural-enemy populations by field applications of pesticides may be due to reduction of host or prey populations in addition to direct toxic effects (see food limitation later in this chapter, and Johnson and Tabashnik in Chapter 13). Mathematical models show that if a pest and its natural enemy are equally susceptible to a pesticide, the pesticide will have a more severe impact on the natural-enemy population than on the pest population (Wilson & Bossert, 1971; Tabashnik, 1986). Therefore, pesticide applications can be extremely detrimental to natural-enemy populations, even if the natural enemy's tolerance is similar to that of the pest.

Genetic Variation

The genetic variation hypothesis says that natural enemies evolve resistance more slowly because natural-enemy populations have less genetic variation than pest populations have (Huffaker, 1971; Georghiou, 1972). Surveys of electrophoretic data show that hymenopterans have less variation in allozymes than most other insects have (Hedrick and Parker, 1997). In particular, the expected heterozygosity or average gene diversity (H$_{exp}$) for 13 species of wasps (0.044 ± 0.006 SE) was less than half the H$_{exp}$ for 158 species of Orthoptera, Homoptera, Coleoptera, Lepidoptera, and Diptera (0.120 ± 0.006 SE) (Graur, 1985). These data are consistent with the idea that hymenopteran parasites have less genetic variation than do herbivorous insects.

Although hymenopteran parasites may have less allozymic variation than most other insects, including many pests, this measure of genetic variation may be unrelated to the ability to evolve insecticide resistance. Some pests known for their resistance evolution had high expected heterozygosity (*Heliothis virescens* = 0.389, *H. zea* = 0.327, *Lygus hesperus* = 0.256) but others had very low values (*B. germanica* = 0.015, *M. persicae* = 0.000). The ability of some pests to readily evolve pesticide resistance, even though they display little or no electrophoretic heterozygosity, shows that this type of genetic variation is not a prerequisite for resistance evolution.

Genetic variation in tolerance to pesticides is required for evolution of resistance, but it has rarely been measured (Roush & McKenzie, 1987; Tabashnik & Cushing, 1989; Tabashnik, 1995). Could intrinsic differences in genetic variation influence resistance evolution? Single-locus population genetics theory predicts that the rate of evolution of resistance increases approximately linearly with the logarithm of initial frequency of an R allele (May & Dobson, 1986). Large differences in initial frequency of an R allele have relatively small effects on rates; a 1000-fold difference between initial frequencies of an R allele in a pest (10^{-3}) and a natural enemy (10^{-6}) would lead to rates of resistance evolution twice as fast in the pest as in the natural enemy. Reducing the initial R allele frequency from 10^{-3} to 10^{-4} or 10^{-5} in simulations of a one-locus model had little effect on projected rates of resistance evolution for most of the species of natural enemies of apple pests that were considered (Tabashnik & Croft, 1985).

Most economically significant cases of pesticide resistance are thought to be under monogenic control, but there are also many examples of polygenic resistance (Roush & McKenzie, 1987; Tabashnik & Cushing, 1989). According to quantitative genetic theory, the rate of increase in pesticide tolerance would be directly proportional to the additive genetic variance (Via, 1986). Therefore, large differences in additive genetic variance would have a major impact on resistance evolution.

In laboratory selection for resistance, natural enemies often do not respond readily (Hoy, 1990; Johnson & Tabashnik, 1994), which provides indirect support for the hypothesis that insufficient genetic variation limits resistance evolution in natural enemies. Laboratory selection programs achieved resistance levels sufficient to enable survival at recommended field rates of pesticide in only 22% of studies with parasitoids and in 68% of studies with predators (Johnson & Tabashnik, 1994). Estimates of heritability, which is the proportion of total phenotypic variance accounted for by additive genetic variance, are available for only a tiny fraction of the pesticide selection experiments reported. In this meager sample, the mean realized heritability of resistance for six pests (0.30, n = 14 cases; Omer *et al.*, 1993) was greater than the maximum value reported for three natural enemies (range = −0.002 to 0.27, mean = 0.11, n = 13 cases; Johnson & Tabashnik, 1994).

To summarize, natural enemies (particularly parasitic hymenopterans) may have less allozymic heterozygosity than pests have, but among herbivores this index of genetic variation is not well correlated with the ability to evolve pesticide resistance. The extent of genetic variation in pesticide susceptibility in pest and in natural-enemy populations has not been characterized in a way that enables rigorous comparison between the two groups. However, relatively slow responses of natural enemies to laboratory selection for resistance in some cases suggest that limited genetic variation may sometimes contribute to the failure of natural enemies to evolve resistance in the field.

Fitness Cost

It is often presumed that in the absence of pesticides, a resistant individual is less fit than a susceptible individual is. If this fitness cost of resistance were substantially greater for natural enemies than for pests, it might retard evolution of resistance in natural enemies. Review of data available for pests suggests that the fitness cost is not generally large, but it may depend on the nature of the resistance mechanism (Roush & McKenzie, 1987). Studies of the predators *Metaseiulus occidentalis* and *C. carnea* and the parasitoid *Anisopteromalus calandrae* show little or no fitness cost associated with resistance (Roush & Hoy, 1981; Roush & Plapp, 1982; Grafton-Cardwell & Hoy, 1986; Baker *et al.*, 1998). Likewise, Baker (1995) found that in two hymenopteran parasitoids, resistance to malathion was stable in the absence of exposure to insecticides. Stability of fenvalerate resistance in the leafminer parasitoid *Diglyphus begini* was not less than that in its leafminer host (Spollen *et al.*, 1995). Additional data are needed to evaluate this hypothesis more thoroughly.

Population Ecology

The concept underlying the population ecology hypothesis is that pesticide resistance evolves more readily in pests than in natural enemies because of differences in population ecology. There are several specific hypotheses under the general category of differences in population ecology:

1. Pests evolve resistance more readily because natural enemies suffer from food limitation following insecticide treatments (Huffaker, 1971; Georghiou, 1972)
2. Pests evolve resistance more readily because of differences between pests and natural enemies in life history traits (Croft, 1982; Tabashnik & Croft, 1985)
3. Pests evolve resistance more readily because they are more exposed to pesticides than natural enemies are (Croft & Brown, 1975)
4. Pests evolve resistance more readily because of differences between pests and natural enemies in their genetic systems (e.g., ploidy level)

As with preadaptation hypotheses, each hypothesis has two parts: (1) there is some difference in population ecol-

ogy between pests and natural enemies, and (2) the difference enables pests to evolve resistance more readily than do natural enemies.

Food Limitation

The food limitation hypothesis is based on the population dynamics of interactions between natural enemies and pests (Huffaker, 1971; Georghiou, 1972; Croft & Brown, 1975; Tabashnik, 1986). The idea is that the few resistant pests surviving an initial pesticide treatment will have an abundant food supply (the crop). In contrast, resistant natural enemies surviving treatment will find their food supply (prey or host) severely reduced. Thus, resistance evolves more slowly in natural enemies because they starve, emigrate, or have reduced reproduction following treatments that eliminate much of their food supply (Tabashnik, 1986).

Pesticide treatments can reduce the food supply of natural enemies while leaving the pests' food supply intact. Thus, there is little question that pests and natural enemies differ in the way in which their food supply is affected by pesticides. It is difficult to determine, however, the extent to which natural-enemy populations are limited by the availability of prey or hosts after pesticide treatments. Pesticide applications reduce natural-enemy populations, but in most cases, direct effects of the pesticides on a natural enemy are confounded with indirect effects of the pesticides on the natural enemy's food supply. Further, the effect of food limitation on a natural enemy's ability to evolve resistance cannot be determined easily in the field.

The food limitation hypothesis could be tested directly by contrasting responses to pesticide treatments in a natural-enemy population feeding on a susceptible strain of a pest versus a population of the same natural-enemy feeding on a resistant strain of the pest. The food limitation hypothesis predicts that pesticide resistance will evolve in the latter case, but not in the former. To our knowledge, this type of experiment has not been performed. Here we review other experimental, historical, and theoretical evidence to evaluate the food limitation hypothesis.

If food limitation is a major factor slowing evolution of resistance in natural enemies, then one could predict that

1. Natural enemies will evolve resistance readily when provided with abundant food in artificial selection programs.
2. Natural enemies that use food sources not reduced by pesticides will evolve resistance more readily than do natural enemies that specialize on susceptible pests killed by pesticides. Food sources not reduced by pesticides include plants and pests that are not killed because they are protected from pesticide exposure, naturally tolerant, or resistant.

Successful laboratory selection for pesticide resistance in natural enemies (Croft & Strickler, 1983; Hoy, 1985, 1990; Croft, 1990; Johnson & Tabashnik, 1994) provides some support for the food limitation hypothesis. However, as noted above earlier, laboratory selection may not produce high levels of resistance in natural enemies as readily as in pests. Such differences in the level of resistance achieved in pest and natural-enemy species may be due to intrinsic limitations of natural enemies (see earlier section on differential preadaptation); or to technical problems associated with sampling, rearing, and selecting large numbers of natural enemies.

We are aware of only one study that directly compared evolution of pesticide resistance in a pest and its natural enemy in the laboratory. Morse and Croft (1981) compared responses to selection for resistance to azinphosmethyl in predatory mite *Amblyseius fallacis* and its prey *Tetranychus urticae*. A susceptible strain of the predator initiated from only a few individuals did not evolve resistance after seven selections. In contrast, a composite susceptible strain, initiated with 600 adult females from three predator strains, evolved 80-fold resistance in 22 selections. In a separate, but parallel experiment, a susceptible noncomposite strain of the pest showed a 20-fold increase in LC_{50} in 22 selections. In the only experiment to compare resistance evolution in the predator and in the prey with susceptible strains of both species in contact, the composite strain of the predator evolved 80-fold resistance whereas the pest strain evolved only 5-fold resistance.

These results suggest that the pest did not evolve resistance more readily than its predator, but interpretations are complicated by several factors. First, a noncomposite pest strain evolved resistance more readily than a noncomposite predator strain. Second, even though the proportional increase in LC_{50} was higher for the predator than for the pest in some experiments, the final LC_{50} after selection was always higher for the pest because of the pest's initially higher LC_{50}. Third, the food limitation hypothesis was not tested directly because predators fed on leaf nectaries and survived even in the absence of prey. Thus, it was not possible to determine if lack of food would retard the evolution of resistance in the predator.

The ability of phytoseiid mites to subsist on plant materials and their ability to evolve resistance in the field (Georghiou, 1972; Croft & Brown, 1975) is consistent with the second prediction from the food limitation hypothesis. Also consistent with this prediction is the general pattern that natural enemies usually become resistant only after their prey or host evolves resistance (Georghiou, 1972; Croft & Brown, 1975; Tabashnik & Croft, 1985).

Rosenheim and Hoy (1986) concluded that food limitation was not a key factor affecting resistance evolution in *Aphytis melinus,* an aphelinid parasite of the California red

scale [*Aonidiella aurantii* (Maskell)]. They noted that a nonresistant host population can survive pesticide applications if it is not contacted by treatments or if it is naturally tolerant. Thus, treating the periphery of citrus trees with dimethoate to control citrus thrips or with chlorpyrifos to control lepidopteran pests should not severely reduce populations of California red scale, which are distributed throughout the tree. In addition, California red scale populations are normally tolerant of the relatively low concentrations of dimethoate used to suppress citrus thrips. Therefore, food limitation should not restrict the evolution of resistance to dimethoate or chlorpyrifos in *Aphytis melinus*. Conversely, food limitation should slow evolution of resistance in *A. melinus* to carbaryl, malathion, and methidathion because these insecticides are used to control scales.

Analysis of ranges in LC_{50} values showed that resistance in *A. melinus* to dimethoate and chlorpyrifos was not consistently greater than resistance to carbaryl, malathion, and methidathion. These results support the conclusion that food limitation was not a major determinant of rates of resistance evolution in *A. melinus*. Rosenheim and Hoy (1986) note that this is not a strong test of the hypothesis because many factors other than food limitation (e.g., cross-resistance and variation among insecticides in the duration and extent of use) could have affected the outcome.

The extremely high resistance to malathion found in the parasitoid *Anisopteromalus calandrae* is consistent with the food limitation hypothesis because in this case, the host [*Sitophilus oryzae* (L.)] is protected from exposure to malathion by feeding inside grain kernels during its immature stages (Baker & Weaver, 1993; Baker & Throne, 1995). Substantial resistance to malathion in the warehouse pirate bug, *Xylocoris flavipes* (Reuter), also fits the food limitation hypothesis, because this predator's prey are highly resistant to malathion (Baker & Arbogast, 1995).

Mathematical models have been used to project the potential impact of food limitation on population dynamics and ability to evolve pesticide resistance of natural enemies. According to the Lotka–Volterra equations of predator–prey population growth, equivalent pesticide-caused mortality will suppress a predator population more than that of its prey (Wilson & Bossert, 1971). This occurs because the predator's birth rate and the prey's death rate are proportional to the product of the population sizes of both species. In contrast, the predator's death rate and the prey's birth rate are not affected by the population size of the other species. Thus, a pesticide treatment that kills 90% of predator and prey populations reduces the predator's birth rate and the prey's death rate by a factor of 100, but reduces the predator's death rate and the prey's birth rate only by a factor of 10. More refined models also show that natural-enemy populations are more severely suppressed by pesticides than are pest populations, even though the immediate mortality is similar for both populations (Waage *et al.,* 1985).

Does suppression of natural-enemy populations affect their ability to evolve resistance? May and Dobson (1986) emphasized the general distinction between overcompensating and undercompensating density dependence. Pests generally rebound above their long-term average or equilibrium levels following pesticide treatments and thus show overcompensating density dependence. In contrast, natural enemies recover slowly, showing undercompensating density dependence. Undercompensating density dependence reduces the average population size, thereby increasing the extent to which immigration of susceptible individuals is expected to slow evolution of resistance (Comins, 1977; Taylor & Georghiou, 1979; Tabashnik & Croft, 1982). So, in the presence of immigration by susceptibles, pests with overcompensating density dependence will evolve resistance faster than natural enemies with undercompensating density dependence (May & Dobson, 1986).

Simulation studies of 12 natural enemies of apple orchard pests showed that incorporation of a simplified version of the food limitation hypothesis substantially improved the correspondence between predicted and reported times for resistance evolution (Tabashnik & Croft, 1985). In this analysis, natural enemies were assumed to begin evolving resistance only after their prey or host had become resistant. Although results supported the food limitation hypothesis, this approach oversimplified dynamic ecological and evolutionary processes.

Other simulation studies included the evolutionary potential for resistance in both predator and in prey, as well as coupled predator–prey population dynamics (Tabashnik, 1986). The key assumption of these simulations was that low prey density reduced the predator's rates of consumption, survival, and fecundity. Predator functional response and the effects of food shortage on predator survival and fecundity were based, in part, on experimental data from mites (Dover *et al.,* 1979). The predator and prey were assumed to have equal intrinsic tolerance and equal genetic potential for evolving resistance. However, intensive pesticide use caused rapid resistance evolution in the pest (prey), but suppressed resistance evolution or caused local extinction of the predator. These theoretical results imply that food limitation is sufficient to account for pests' ability to evolve pesticide resistance more readily than natural enemies evolve resistance.

In summary, pesticide treatments reduce the food supply of natural enemies more than that of pests. Severe reductions in food supply can slow resistance evolution in natural enemies. Indirect support for the food limitation hypothesis is provided by the success of laboratory selection programs in which natural enemies are provided abundant food, by

the general trend that natural enemies evolve resistance in the field only when their prey or hosts are resistant or are protected from pesticide exposure, and by theoretical models. Food limitation, however, does not appear to explain resistance evolution in *Aphytis melinus*. Direct tests of this hypothesis are needed.

Life History Traits

Life history traits include number of generations per year, rate and timing of reproduction, survival, developmental rate, and sex ratio. Theoretical work and historical patterns suggested that the rate of resistance evolution increases as reproductive capacity, particularly the number of generations per year, increases (Tabashnik & Croft, 1982, 1985; Georghiou & Taylor, 1986; May & Dobson, 1986). If the number of generations per year is consistently higher for pests than for natural enemies, then this might explain why pests evolve resistance more readily than do natural enemies.

The median number of generations per year, however, is less for pests than for natural enemies (Stiling, 1990). Further, analysis of extensive databases and re-evaluation of relevant theory suggest that there is no simple relationship between generations per year and rate of resistance evolution (Rosenheim & Tabashnik, 1990; 1991, 1993). Thus, available evidence refutes the idea that pests become resistant faster because they have more generations per year than natural enemies have.

Differences between pests and natural enemies in other life history traits may affect resistance evolution. For example, pests generally have higher fecundity (Stiling, 1990) and may maintain larger populations in treated habitats (Croft & Tabashnik, 1990) compared with natural enemies.

Exposure

The rate of resistance evolution is a function of selection intensity, which is determined in part by extent of exposure to pesticides. Croft and Brown (1975) speculated that natural enemies are often less intensively selected than pests are because pesticides are directed at pests; natural enemies contact pesticides only insofar as they occupy the same habitat as their prey or host. However, mobile predators and parasites might pick up more toxicant and thus suffer greater mortality from residual deposits than would sedentary pests in the same habitat (Croft & Brown, 1975; see also Johnson & Tabashnik in Chapter 13).

Evaluation of these hypotheses requires detailed information about exposure to pesticides in the field (e.g., the proportion of the population exposed, levels and duration of exposure in treated habitats, and gene flow between treated and untreated habitats). Because most of this information is not available, much empirical work is needed to assess the effects of pesticide exposure on resistance evolution in pests versus natural enemies.

Genetic Systems

Most evolutionary considerations about pesticide resistance are based on the assumption that organisms are diploid and sexually reproducing, but a variety of genetic systems occur among insects and mites. Differences in genetic systems between pests and natural enemies could influence their relative ability to evolve resistance. For example, hymenopteran insects are haplodiploid. In one modeling study, however, resistance evolved faster under haplodiploidy than under diplodiploidy (Horn & Wadleigh, 1988), which suggests that haplodiploidy does not slow evolution of resistance in hymenopteran parasitoids.

Phytoseiid predatory mites have various genetic systems including thelytoky (females only), parahaploidy (embryos of both sexes are diploid, but males lose one chromosome set during embryonic development), and perhaps true arrhenotoky (unfertilized eggs produce haploid males and fertilized eggs produce diploid females) (Hoy, 1985). Parahaploid phytoseiids such as *Typhlodromus occidentalis* and *Phytoseiulus persimilis* Athias-Henriot may have some advantages of haploidy (exposing haploid individuals to selection) and diploidy (recombination) (Hoy, 1985). This could help to explain why phytoseiids evolve resistance to pesticides more readily than do other natural enemies, but it does not support the idea that the genetic systems of natural enemies retard their resistance evolution.

Related factors that could affect evolution of resistance in pests and in natural enemies are the extent of sexual versus asexual reproduction, inbreeding, and sociality. Social insects would not be expected to evolve resistance readily because they have small effective population size (few reproductives) and their reproducing individuals usually have limited exposure to pesticides. These factors may explain the paucity of documented cases of resistance in social Hymenoptera and Isoptera insects (Georghiou, 1981), many of which are pests. They do not help to explain the lack of resistance in parasitic hymenopterans.

In summary, genetic systems and related factors may influence resistance evolution in pests and natural enemies. It does not appear, however, that there are consistent differences in these traits that would favor resistance evolution in pests compared with that of natural enemies.

Perversity of Nature

The perversity of nature hypothesis states that the worst possible outcome is for pests to evolve resistance more readily than natural enemies do; therefore, this is what occurs. Although compelling, this hypothesis is completely untestable.

SUMMARY AND IMPLICATIONS

The hypotheses proposed to explain the scarcity of documented cases of resistance in natural enemies are not mutually exclusive. The relative importance of each factor may vary among species and pesticides. Several factors may act jointly in some cases. Thus, there may be no single, general explanation. Nonetheless, review of the available evidence casts doubt on some hypotheses, supports others, and indicates potentially productive avenues for further research.

Resistance to pesticides in pest populations is probably more likely to be documented than is natural-enemy resistance, but the magnitude of this effect is difficult to measure. Bioassays comparing interpopulation variation for several pests and natural enemies from a given crop and region could help to assess the influence of bias in documentation.

Available data show that natural enemies possess the detoxification enzyme systems found in pests. The intrinsic levels of detoxification enzymes are not consistently higher for pests than for natural enemies. When pests have higher intrinsic levels of detoxification enzymes than natural enemies have there is little evidence showing that this difference contributed to faster evolution of resistance in the pest. Versions of the preadaptation hypothesis based on intrinsic differences in detoxification ability thus appear to have limited power to explain differences between pests and natural enemies.

Differences in intrinsic pesticide tolerance between pests and natural enemies are difficult to assess. Surveys suggest that intrinsic pesticide tolerance is not consistently higher for pests than for natural enemies, but comparative data are limited and may be biased. Higher intrinsic tolerance in pests could account in part for more documented cases of resistance in pests, particularly if the criteria for resistance include the ability to survive field applications of pesticides. Lack of genetic variation for pesticide tolerance in natural enemies could also retard their evolution of resistance. This facet of the preadaptation hypothesis needs investigation. Similarly, more data are needed to determine whether fitness costs associated with resistance are substantially lower for pests than for natural enemies.

Indirect evidence suggests that differences in population ecology are important in slowing resistance evolution in natural enemies relative to pests. Experiments are needed to assess the extent to which natural-enemy populations are suppressed by food limitation, and how such suppression affects their ability to evolve resistance. Experiments are needed also to measure exposure to pesticides in pests and in natural enemies. Differences between pests and natural enemies in number of generations per year do not explain the relative scarcity of resistance in natural enemies, but differences in other life history traits may be important. Differences in genetic systems and related factors are not likely to explain why pests evolve resistance more readily than do natural enemies.

We think that food limitation due to reduction in host or prey populations by pesticides is a major factor influencing natural-enemy populations. If this is true, then there are some important implications for management. First, many natural enemies should evolve resistance when provided abundant food in artificial selection programs (Hoy, 1985, 1990; Croft, 1990; Johnson & Tabashnik, 1994). Second, intensive pesticide use may disrupt biological control, even if the natural enemies are not susceptible—either because of natural tolerance or resistance (Tabashnik, 1986; see also Johnson and Tabashnik in Chapter 13). Thus, to maintain effective biological control, use of selective pesticides should be sparing and judicious.

Acknowledgments

We are grateful to K. Theiling and B. Croft for providing a preprint of their review paper and references from their extensive natural-enemy database, and to J. Baker for providing reprints and preprints of his exciting work. The thoughtful comments of J. Rosenheim, B. Croft, K. Theiling, H. van de Baan, C. Bach, M. Caprio, A. Moore, and L. Yudin improved the manuscript. We thank N. Finson, D. Horn, T. Sparks, and S. Toba for their assistance. We also thank our colleagues for faithfully citing this chapter during its long dormancy. This work was supported by several USDA grants (USDA grant HAW00947H and grants from the following USDA programs: Special Grants in Tropical and Subtropical Agriculture, Western Regional Pesticide Impact Assessment Program and Integrated Pest Management) and a Fujio Matsuda Scholar Award from the University of Hawaii Foundation.

References

Baker, J. E. (1994). Sensitivities of laboratory and field strains of the parasitoid *Anisopteromalus calandrae* (Hymenoptera: Pteromalidae) and its host, *Sitophilus oryzae* (Coleoptera: Curculionidae), to deltamethrin and cyfluthrin. Journal of Entomological Science, 29, 100–109.

Baker, J. E. (1995). Stability of malathion resistance in two hymenopterous parasitoids. Journal of Economic Entomology, 88, 232–236.

Baker, J. E., & Arbogast, R. T. (1995). Malathion resistance in field strains of the warehouse pirate bug (Heteroptera: Anthocoridae) and a prey species *Tribolium castaneum* (Coleoptera: Tenebrionidae). Journal of Economic Entomology, 88, 241–245.

Baker, J. E., Perez-Mendoza, J., Beeman, R. W., & Throne, J. E. (1998). Fitness of a malathion-resistant strain of the parasitoid *Anisopteromalus calandrae* (Hymenoptera: Pteromalidae). Journal of Economic Entomology, 91, 50–55.

Baker, J. E., & Throne, J. E. (1995). Evaluation of a resistant parasitoid for biological control of weevils in insecticide-treated wheat. Journal of Economic Entomology, 88, 1570–1579.

Baker, J. E., & Weaver, D. K. (1993). Resistance in field strains of the parasitoid *Anisopteromalus calandrae* (Hymenoptera: Pteromalidae) and its host, *Sitophilus oryzae* (Coleoptera: Curculionidae), to malathion, chlorpyrifos-methyl, and pirimiphos-methyl. Biological Control, 3, 233–242.

Baker, J. E., Weaver, D. K., Throne, J. E., & Zettler, J. L. (1995). Resistance to protectant insecticides in two field strains of the stored-product insect parasitoid *Bracon hebetor* (Hymenoptera: Braconidae). Journal of Economic Entomology, 88, 512–519.

Brattsten, L. B., & Metcalf, R. L. (1970). The synergistic ratio of carbaryl with piperonyl butoxide as an indicator of the distribution of multifunction oxidases in the Insecta. Journal of Economic Entomology, 63, 1347–1348.

Comins, H. N. (1977). The development of insecticide resistance in the presence of immigration. Journal Theoretical Biology, 177–197.

Croft, B. A. (1982). Developed resistance to insecticides in apple arthopods: A key to pest control failures and successes in North America. Entomologia Experimentalis et Applicata, 31, 88–110.

Croft, B. A. (Ed.). (1990). Arthropod biological control agents and pesticides. New York: John Wiley & Sons.

Croft, B. A., & Aliniazee, M. T. (1983). Differential resistance to insecticides in *Typhlodromus arboreus* Chant and associated phytoseiid mites of apple in the Willamette Valley, Oregon. Environmental Entomology, 12, 1420–1423.

Croft, B. A., & Brown, A. W. A. (1975). Responses of arthropod natural enemies to insecticides. Annual Review of Entomology, 20, 285–335.

Croft, B. A., & Morse, J. G. (1979). Recent advances on pesticide resistance in natural enemies. Entomophaga, 24, 3–11.

Croft, B. A., & Mullin, C. A. (1984). Comparison of detoxification enzyme systems in *Argyrotaenia citrana* (Lepidoptera: Tortricidae) and the ectoparasite, *Oncophanes americanus* (Hymenoptera: Braconidae). Environmental Entomology, 13, 1330–1335.

Croft, B. A., & Strickler, K. (1983). Natural enemy resistance to pesticides: Documentation, characterization, theory and application. In G. P. Georghiou & T. Saito (Eds.), Pest resistance to pesticides (pp. 669–702). New York: Plenum Press.

Croft, B. A., & Tabashnik, B. E. (1990). Factors affecting resistance. In B. A. Croft (Ed.), Arthropod biological control agents and pesticides (pp. 403–428). New York: John Wiley & Sons.

Croft, B. A., & van de Baan, H. E. (1988). Ecological and genetic factors influencing evolution of pesticide resistance in tetranychid and phytoseiid mites. Experimental Applied Acarology (Northwood), 4, 227–300.

Croft, B. A., & Wagner, S. W. (1981). Selectivity of acaricidal pyrethroids to permethrin-resistant strains of *Amblyseius fallacis*. Journal of Economic Entomology, 74, 703–706.

Croft, B. A., Wagner, S. W., & Scott, J. G. (1982). Multiple- and cross-resistances to insecticides in pyrethroid-resistant strains of the predatory mite, *Amblyseius fallacis*. Environmental Entomology, 11, 161–164.

Curtis, C. F., Cook, L. M., & Wood, R. J. (1978). Selection for and against insecticide resistance and possible methods of inhibiting the evolution of resistance in mosquitoes. Ecological Entomology, 3, 273–287.

Dover, M. J., Croft, B. A., Welch, S. M., & Tummala, R. L. (1979). Biological control of *Panonychus ulmi* (Acarina: Tetranychidae) by *Amblyseius fallacis* (Acarina: Phytoseiidae) on apple: A prey-predator model. Environmental Entomology, 8, 282–292.

Georghiou, G. P. (1972). The evolution of resistance to pesticides. Annual Review of Ecology and Systematics, 3, 133–168.

Georghiou, G. P. (1981). The occurrence of resistance to pesticides in arthropods: An index of cases reported through 1980. Rome: FAO.

Georghiou, G. P. (1986). The magnitude of the resistance problem. In Pesticide resistance: Strategies and tactics for management (pp. 14–43). Washington, DC: National Academy.

Georghiou, G. P., & Taylor, C. E. (1986). Factors influencing the evolution of resistance. In Pesticide resistance: Strategies and tactics for management (pp. 143–146). Washington, DC: National Academy.

Gordon, H. T. (1961). Nutritional factors in insect resistance to chemicals. Annual Review of Entomology, 6, 27–54.

Gould, F. (1984). Mixed function oxidases and herbivore polyphagy: The devil's advocate position. Ecological Entomology, 9, 29–34.

Grafton-Cardwell, E. E., & Hoy, M. A. (1985). Intraspecific variability in response to pesticides in the common green lacewing, Chrysoperla carnea (Stephens) (Neuroptera: Chrysopidae). Hilgardia, 53, 1–32.

Grafton-Cardwell, E. E., & Hoy, M. A. (1986). Genetic improvement of the common green lacewing, Chrysoperla carnea (Neuroptera: Chrysopidae): Selection for carbaryl resistance. Environmental Entomology, 15, 1130–1136.

Graur, D. (1985). Gene diversity in Hymenoptera. Evolution, 39, 190–199.

Havron, A. (1983). Studies toward selection of Aphytis wasps for pesticide resistance. Unpublished doctoral dissertation, Hebrew University of Jerusalem, Rehovot, Israel.

Hedrick, P. W., & Parker, J. D. (1997). Annual Review of Ecology and Systematics, 28, 55–83.

Horn, D. J., & Wadleigh, R. W. (1988). Resistance of aphid natural enemies to insecticides. In P. Harrewijin & A. K. Minks (Eds.), Aphids, their biology, natural enemies, and control (Vol. B, pp. 337–347). Amsterdam: Elsevier.

Hoy, M. A. (1985). Recent advances in genetics and genetic improvement of the Phytoseiidae. Annual Review of Entomology, 30, 345–370.

Hoy, M. A. (1990). Pesticide resistance in arthropod natural enemies: Variability and selection, In R. T. Roush & B. E. Tabashnik (Eds.), Pesticide resistance in arthropods (pp. 203–236). New York: Chapman & Hall.

Hoy, M. A., & Cave, F. E. (1988). Guthion-resistant strain of walnut aphid parasite. California Agriculture, 42, 4–5.

Hoy, M. A., & Knop, N. F. (1979). Studies on pesticide resistance in the phytoseiid Metaseiulus occidentalis in California. In K. G. Rodriguez (Ed.), Recent advances in acarology (Vol. 1, pp. 89–94) New York: Academic Press.

Hoy, M. A., Flaherty, D, Peacock, W, & Culver, D. (1979). Vineyard and laboratory evaluations of methanol, dimethoate, and permethrin for a grape pest management program in the San Joaquin Valley of California. Journal of Economic Entomology, 72, 250–255.

Hsieh, C.-Y. (1984). Effects of insecticides on Diaeretiella rapae (McIntosh) with emphasis on bioassay techniques for aphid parasitoids. Unpublished doctoral dissertation, University of California, Berkeley.

Huffaker, C. B. (1971). The ecology of pesticide interference with insect populations, In J. W. Swift (Ed.), Agricultural chemicals—harmony or discord for food people environment (pp. 92–107). University of California Division of Agricultural Science. Berkeley.

Ishaaya, I., & Casida, J. E. (1981). Pyrethroid esterase(s) may contribute to natural pyrethroid tolerance of larvae of the common green lacewing. Environmental Entomology, 10, 681–684.

Johnson, M. W., & Tabashnik, B. E. (1994). Laboratory selection for pesticide resistance in natural enemies. In S. Karl Narang, A. C. Bartlett, & R. M. Faust (Eds.), Applications of genetics to arthropods of biological control significance (pp. 91–105). Boca Raton, FL: CRC Press.

Kapentanakis, E. G., & Cranham, J. E. (1983). Laboratory evaluation of resistance to pesticides in the phytoseiid predator Typhlodromus pyri from English apple orchards. Annals of Applied Biology, 103, 389–400.

Mansour, F. (1984). A malathion-tolerant strain of the spider Chiracanthium mildei and its response to chloropyrifos. Phytoparasitica, 12, 163–166.

May, R. M., & Dobson, A. P. (1986). Population dynamics and the rate of evolution of pesticide resistance. In Pesticide resistance: Strategies and tactics for management (pp. 170–193). Washington DC: National Academy.

Morse, J. G., & Croft, B. A. (1981). Developed resistance to azinphosmethyl in a predator-prey mite system in greenhouse experiments. Entomophaga, 26, 191–202.

Mullin, C. A. (1985). Detoxification enzyme relationships in arthropods of differing feeding strategies. In P. A. Hedin (Ed.), Bioregulators for pest control (pp. 267–278.) Washington, DC: American Chemical Society.

Mullin, C. A., & Croft, B. A. (1984). Trans-epoxide hydrolase: A key indicator enzyme for herbivory in arthropods. Experientia, 40, 176–178.

Mullin, C. A., & Croft, B. A. (1985). An update on development of selective pesticides favoring arthropod natural enemies. In M. A. Hoy & D. C. Herzog (Eds.), Biological control in agricultural integrated pest management systems (pp. 123–150). New York: Academic Press.

Mullin, C. A., Croft, B. A., Strickler, K., Matsumara, F., & Miller, J. R. (1982). Detoxification enzyme differences between a herbivorous and predatory mite. Science, 217, 1270–1271.

National Research Council (NRC). (1986). Pesticide resistance: Strategies and tactics for management. Washington, DC: National Academy.

Omer, A. D., Tabashnik, B. E., Johnson, M. W., & Leigh, T. F. (1993). Realized heritability of resistance to dicrotophos in greenhouse whitefly. Entomologia Experimentalis et Applicata, 68, 211–217.

Plapp, F. W., Jr., & Bull, D. L. (1978). Toxicity and selectivity of some insecticides to Chrysopa carnea, a predator of the tobacco budworm. Environmental Entomology, 7, 431–434.

Plapp, F. W., Jr., & Vinson, S. B. (1977). Comparative toxicities of some insecticides to the tobacco budworm and its ichneumonid parasite, Campoletis sonorensis. Environmental Entomology, 6, 381–384.

Rathman, R. J., Johnson, M. W., Rosenheim, J. A. & Tabashnik, B. E. (1990). Carbamate and pyrethroid resistance in the leafminer parasitoid Diglyphus begini (Hymenoptera: Eulophidae). Journal of Economic Entomology, 83, 2153–2158.

Rathman, R. J., Johnson, M. W., Tabashnik, B. E., & Purcell, M. (1992). Sexual differences in insecticide susceptibility and synergism with piperonyl butoxide in the leafminer parasitoid Diglyphus begini (Hymenoptera: Eulophidae). Journal of Economic Entomology, 85, 15–20.

Rathman, R. J. Johnson, M. W., Tabashnik, B. E., & Spollen, K. M. (1995). Variation in susceptibility to insecticides in the leafminer parasitoid Ganaspidium utilis (Hymenoptera: Eucoilidae). Journal of Economic Entomology, 88, 475–479.

Rosenheim, J. A., & Hoy, M. A. (1986). Intraspecific variation in levels of pesticide resistance in field populations of a parasitoid, Aphytis melinus (Hymenoptera: Aphelinidae): The role of past selection pressures. Journal of Economic Entomology, 79, 1161–1173.

Rosenheim, J. A., Johnson, M. W., Mau, R. F. L., Welter, S. C., & Tabashnik, B. E. (1996). Biochemical preadaptations, founder events, and the evolution of resistance in arthropods. Journal of Economic Entomology, 89, 263–273.

Rosenheim, J. A., & Tabashnik, B. E. (1990). Evolution of pesticide resistance: Interactions between generation time and genetic, ecological, and operational factors. Journal of Economic Entomology, 83, 1184–1193.

Rosenheim, J. A., & Tabashnik, B. E. (1991). Influence of generation time on the rate of response to selection. American Naturalist, 137, 527–541.

Rosenheim, J. A., & Tabashnik, B. E. (1993). Generation time and evolution. Nature (London), 365, 791–792.

Roush, R. T., & Hoy, M. A. (1981). Laboratory, glasshouse, and field studies of artificially selected carbaryl resistance in Metaseiulus occidentalis. Journal of Economic Entomology, 74, 142–147.

Roush, R. T., & McKenzie, J. A. (1987). Ecological genetics of insecticide and acaricide resistance. Annual Review of Entomology, 32, 361–380.

Roush, R. T., & Miller, G. L. (1986). Considerations for design of insecticide resistance monitoring programs. Journal of Economic Entomology, 79, 293–298.

Roush, R. T., & Plapp, F. W., Jr. (1982). Biochemical genetics of resistance to aryl carbamate insecticides in the predaceous mite, Metaseiulus occidentalis. Journal of Economic Entomology, 75, 708–713.

Roush, R. T., & Tabashnik, B. E. (Eds.). (1990). Pesticide resistance in arthropods. New York: Chapman & Hall.

Schoonees, J., & Giliomee, J. H. (1982). The toxicity of methidathion to parasitoids of red scale, Aonidiella auranti (Hemiptera: Diaspididae). Journal of the Entomological Society of South Africa, 45, 261–273.

Scott, J. G., Croft, B. A., & Wagner, S. W. (1983). Studies on the mechanism of permethrin resistance in Amblyseius fallacis (Acarina: Phytoseiidae) relative to previous insecticide use on apple. Journal of Economic Entomology, 76, 6–10.

Spollen, K. M., Johnson, M. W., & Tabashnik, B. E. (1995). Stability of fenvalerate resistance in the leafminer parasitoid Diglyphus begini (Hymenoptera: Eulophidae). Journal of Economic Entomology, 88, 192–197.

Stiling, P. (1990). Calculating the establishment rates of parasitoids in classical biological control. American Entomologist, 36, 225–229.

Strickler, K., & Croft, B. A. (1981). Variation in permethrin and azinphosmethyl resistance in populations of Amblyseius fallacis (Acarina: Phytoseiidae). Environmental Entomology, 10, 233–236.

Strickler, K., & Croft, B. A. (1985). Comparative rotenone toxicity in the predator, Amblyseius fallacis (Acari: Phytoseidae), and the herbivore, Tetranychus urticae (Acari: Tetranychidae), grown on lima beans and cucumbers. Environmental Entomology, 14, 243–246.

Tabashnik, B. E. (1986). Evolution of pesticide resistance in predator-prey systems. Bulletin of the Entomological Society of America, 32, 156–161.

Tabashnik, B. E. (1995). Insecticide resistance. Trends in Ecology and Evolution, 10, 164–165.

Tabashnik, B. E., & Croft, B. A. (1982). Managing pesticide resistance in crop–arthropod complexes: Interactions between biological and operational factors. Environmental Entomology, 11, 1137–1144.

Tabashnik, B. E., & Croft, B. A. (1985). Evolution of pesticide resistance in apple pests and their natural enemies. Entomophaga, 30, 37–49.

Tabashnik, B. E., & Cushing, N. L. (1989). Quantitative genetic analysis of insecticide resistance: Variation in fenvalerate tolerance in a diamondback moth (Lepidoptera: Plutellidae) population. Journal of Economic Entomology, 82, 5–10.

Tabashnik, B. E., Cushing, N. L., & Johnson, M. W. (1987). Diamondback moth (Lepidoptera: Plutellidae) resistance to insecticides in Hawaii: Intra-island variation and cross-resistance. Journal of Economic Entomology, 80, 1091–1099.

Tanigoshi, L. K., & Congdon, B. D. (1983). Laboratory toxicity of commonly-used pesticides in California citriculture to Euseius hibisci (Chant) (Acarina: Phytoseiidae). Journal of Economic Entomology, 76, 247–250.

Taylor, C. E., & Georghiou, G. P. (1979). Suppression of insecticide resistance by alteration of gene dominance and migration. Journal of Economic Entomology, 72, 105–109.

Theiling, K. M., & Croft, B. A. (1988). Pesticide side-effects on arthropod natural enemies: A database summary. Agriculture Ecosystems and Environment, 21, 191–218.

Trimble, R. M., & Pree, D. J. (1987). Relative toxicity of six insecticides to male and female Pholetesor ornigis (Weed) (Hymenoptera: Braconidae), a parasite of the spotted tentiform leafminer, Phyllonorycter blancardella (Fabr.) (Lepidoptera: Gracillariidae). Canadian Entomologist 119, 153–157.

Trimble, R. M. Pree, D. J., & Vickers, P. M. (1990). Survey for insecticide resistance in some Ontario populations of the apple leafminer parasite, Pholetesor ornigis (Weed) (Hymenoptera: Braconidae). Canadian Entomologist, 122, 969–973.

Van de Baan, H. E., & Croft, B. A. (1990). Factors influencing insecticide resistance in Psylla pyricola (Homoptera: Psyllidae) and susceptibility in the predator Deraeocoris brevis (Heteroptera: Miridae). Environmental Entomology, 19, 1223–1228.

Via, S. (1986). Quantitative genetic models and the evolution of pesticide resistance, In Pesticide resistance: Strategies and tactics for management (pp. 222–235). Washington, DC: National Academy.

Waage, J. K., Hassell, M. P., & Godfray, H. C. J. (1985). The dynamics of pest-parasitoid-insecticide interactions. Journal of Applied Ecology 22, 825–838.

Warner, L. A., & Croft, B. A. (1982). Toxicities of azinphosmethyl and selected orchard pesticides to an aphid predator, Aphidoletes aphidimyza. Journal of Economic Entomology 75, 410–415.

Wilson, E. O., & Bossert, W. H. (1971). A primer of population biology. Sunderland, MA: Sinauer Associates.

Yu, S. J. (1987). Biochemical defense capacity in the spined soldier bug and its lepidopterous prey. Pesticide Biochemistry and Physiology, 28, 216–233.

Yu, S. J. (1988). Selectivity of insecticides to spined soldier bug (Heteroptera: Pentatomidae) and its lepidopterous prey. Journal of Economic Entomology, 81, 119–122.

25

Hypovirulence to Control Fungal Pathogenesis

DENNIS W. FULBRIGHT

Department of Botany and Plant Pathology
Michigan State University
East Lansing, Michigan

INTRODUCTION

Nearly a century ago chestnut blight struck the Eastern forests of North America initiating the most destructive forest epidemic in recorded history (Anagnostakis, 1982). The pathogen, earlier known as *Endothia parastica* and later called *Cryphonectria parasitica*, was unleashed in New York City, having hitchhiked to North America on Japanese chestnut trees, *Castanea crenata*, a species more resistant to infection by the fungal pathogen than North America's native species the American chestnut tree, *Castanea dentata* (Milgroom *et al.,* 1997). The pathogen is a wound-infecting, stem-cankering ascomycete that produces masses of sticky asexual conidia throughout the summer and light, airborne ascospores in the late summer each year until the stem is girdled and the bark is killed (Fig. 1). Because it is not a good saprophyte, where or how *Cryphonectria parasitica* survives after it has killed most of the chestnut trees in a localized area is not entirely clear. It is known to reinfect young sucker sprouts emerging from the root collar of infected or dying individuals and it can be found on certain oak species in some areas (Schwadron, 1995). It is well known that some chestnut trees will sucker from the root collar rapidly producing new shoots that can become reinfected and if they attain much size; the pathogen initiates continuous rounds of infection on the sprouts.

The introduction of chestnut blight through the reckless importation of foreign species from one continent to another taught us early about the accidental importation of harmful species and the biological disaster that can develop if the biological lessons of competition and host–pathogen interactions are ignored. The chestnut blight disaster led to quarantine laws that certainly must have prevented other biological invaders from moving to new islands and continents, but the laws did not stop all invaders as witnessed by the development of other major disease epidemics throughout the world (Scheffer, 1997).

However, another maybe even more important lesson has developed from the chestnut blight epidemic. This lesson is about the natural biological constraints that even the most destructive pathogen must endure to maintain its long-term predominance on a particular host. Not long after the dissemination of chestnut blight into Europe, just prior to World War II, European chestnut trees also began dying rapidly of chestnut blight. However, about 20 years after the pathogen's introduction, sucker sprouts were observed to survive infection; chestnut trees, from these surviving sprouts, began to reclaim the landscape, and once again, chestnut products (wood and nuts) became an important commodity in the Mediterranean region of Europe (Heiniger & Rigling, 1994).

To understand how these European chestnut trees survived meant that more information was needed if we were to understand the constraints faced by the pathogen. Yet, little work was being done on chestnut blight because it was such a devastating disease; many government agencies and researchers had given up on research as a being waste of time and money in the United States and elsewhere. Full credit must be given to early French researchers, including Jean Grente and co-workers who were able to determine the overall mechanism of the European chestnut tree's recovery (Grente & Berthelay-Sauret, 1978). The researchers applied fungal genetic techniques and concepts to the system to understand what might be happening on the surviving trees while pursuing applied biological control experi-

691

(A)

(B)

FIGURE 1   Naturally occurring lethal-type chestnut blight canker expanding through the bark. (A) Before outer bark removed and (B) outer bark removed. Canker probably initiated at the branch scar.

ments to determine if the system could be exploited at nonrecovering blighted locations. They discovered that the chestnut blight pathogen on infected but surviving trees was reduced in aggressiveness, and it appeared that the natural defense mechanisms of the tree were able to stop or slow the infection to the point where the tree could compartmentalize the infection to just the outer bark. This caused scaring and some cambial damage, with the subsequent development of wound tissue and callus; however, overall, the trees remained healthy and productive.

This remarkable research, in terms of deciphering the the biological puzzle presented to the researchers, should be a highlight in biological control treatises. The chestnut blight pathogen in this state of reduced aggression was morphologically distinct from the aggressive form of the pathogen; in culture it was white instead of displaying its usual diagnostic orange pigmentation, was reduced in sporulation, and had a slower growth rate. That Grente and Berthelay-Sauret recognized this new form of the pathogen

as *Endothia parasitica* and not another canker-inducing species or a weak, secondary pathogen infecting stressed trees was a great achievement in itself, but not as great as the next. Subsequently, it was soon discovered that when this less aggressive, morphologically altered strain came into hyphal contact with aggressive strains of the chestnut blight fungus, the traits characteristic of the less aggressive strain transferred to the aggressive strain effectively reducing its virulence in the tree. These strains were not avirulent and did not become nonpathogenic on the chestnut; it was clear that they were still capable of recognizing and infecting the host, but where defense mechanisms of the tree species had earlier been ineffective, they were now effective in reducing canker expansion on the tree. Grente termed this natural control of chestnut blight "hypovirulence" and he proposed that a genetic factor in the cytoplasm of the hypovirulent strain was responsible for the hypovirulent phenotype (Grente & Berthelay-Sauret, 1969). He theorized that this cytoplasmic factor, when transferred to virulent strains from hypovirulent strains during hyphal anastomosis, was responsible for the transmissible nature of the genetic traits associated with hypovirulent strains.

In North America, hypovirulence did not develop 20 years after the pathogen was introduced, or if it did it was not biologically successful and therefore not recognized. By the 1950s, it was estimated that over 3 billion chestnut trees were destroyed by the pathogen or prematurely harvested before blight infection destroyed the trees' value. While initial research in the early part of the century had provided great detail as to the life cycle and movement of the fungus, it did little to save trees. Breeding programs designed to capture the natural resistance in the Asian species' background initially failed to provide a disease-resistant tree with the important characteristics of the American species, such as girth and height. Forest management strategies such as eradication also failed. Except for uninfected trees that had been planted outside of the natural range of the native species, mature, nut-bearing chestnut trees were becoming extremely rare in North America. By the 1960s, the Eastern forests only harbored remnant sprouts growing from the root collars of blight-killed trees, offering hope that one day the tree could be brought back as a co-dominant forest species.

In the United States, research on and scientific interest in chestnut and chestnut blight waned for decades until two events occurred in the 1970s, one growing out of the other. Researchers at the Connecticut Agricultural Experiment Station, a research institute that had maintained an active chestnut tree breeding/research program, discovered Grente's work and began to investigate the cytoplasmic factor and its potential for controlling chestnut blight of American chestnut in North America. Van Alfen *et al.* (1975) soon discovered that the European hypovirulent strains could transfer the hypovirulent phenotype to North

American virulent isolates infecting American chestnut. The American tree responded similarly to the European tree in presenting a defense reaction to the invading hypovirulent strain. This was followed by an astute observation.

Newspaper coverage of the Connecticut research was read by a nature center volunteer near Grand Rapids, Michigan, who, while cross-country skiing, found diseased American chestnut trees with callusing cankers similar to those shown in the newspaper article. She sent the cankers to Connecticut and soon it was confirmed that hypovirulent strains were present in Michigan (Fulbright *et al.*, 1983). The hypovirulent isolates recovered from the callused cankers on the Michigan trees were not typical of the hypovirulent strains isolated in Europe; they were orange instead of white and produced more spores in culture than did European hypovirulent strains (Elliston, 1978). Since then, hypovirulent strains have been found in other regions of North America, including New Jersey (Hillman *et al.*, 1992) Ontario (McKeen, 1995), and other isolated locations (Jaynes & Elliston, 1982), but none are associated with as much chestnut survival or are as geographically widespread as in Michigan.

SEARCHING FOR THE CAUSE OF HYPOVIRULENCE

The cytoplasmic factor associated with hypovirulence in chestnut blight was determined by Day *et al.* (1977) to be double-stranded RNA (dsRNA) molecules. This was not too surprising because the genetic component of nearly all fungal viruses is dsRNA. What was surprising was that there were no virions, icosehedral or other virus-like particles present in the hypovirulent strains. Instead, dsRNA was associated with pleiomorphic vesicles presumably of fungal origin (Dodds, 1980). At first there seemed to be a perfect correlation between the presence of dsRNA in the cytoplasm and the hypovirulent phenotype. Whether dsRNA was moved into a new strain via hyphal fusion or removed by curing agents, the characteristics of hypovirulent strains were gained or lost, respectively. It became apparent that the simple presence of dsRNA could not account for the dramatic culturable and pathogenic differences observed among dsRNA-containing strains. For example, *C. parasitica* strains would lose pigmentation when infected with various dsRNA genomes from Europe, but never lost pigmentation when infected with isolates of Michigan origin. Some hypovirulent strains grew almost as fast as non-dsRNA infected strains, yet still retained a reduction in the ability to establish cankers on the tree host. A few strains regardless of origin were so debilitated in culture that it was a wonder the isolate could have existed in nature at all (Elliston, 1978).

Elliston (1985) was the first to attempt to ascribe some of the obvious cultural differences to the dsRNA infection status of the isolates. He postulated that one of the reasons for phenotypic variation existing in the hypovirulent strains could be due to *E. parasitica* strains being infected with more than one dsRNA genome. Mixed virus infections in one fungal thallus can be difficult to detect. While most of the dsRNA molecules are of different size classes, there is enough overlap in the dsRNA genome size that one dsRNA molecule could easily mask the presence of another on agarose or polyacrylamide gels. Elliston determined through single-conidial isolation techniques that one of the first Michigan hypovirulent strains isolated was, in fact, doubly infected. By effectively segregating the cytoplasm through asexual spore cultures, he was able to detect the singular effect of each dsRNA molecule on the fungal host. One dsRNA genome had a greater effect on the fungal culture morphology than the other had, leading to speculation that different dsRNA genomes have separate, unequal effects, that were responsible for different outcomes of fungal infection on trees. In a similar experiment, where two dsRNAs were both transferred into the same genetic background, Smart and Fulbright (1995) found that these coinfected strains were markedly more reduced in virulence when compared with strains infected with each virus separately. This indicates that if two viruses are released separately in a stand of trees, the fungal strains that become infected with both dsRNA molecules may result in a third type of effect on the pathogen.

Other studies (Fulbright, 1985) suggested that a cytoplasmic genetic factor in addition to dsRNA might be involved in the generation of the hypovirulent phenotype. Another type of hypovirulence first described in Michigan did not appear to be associated with dsRNA (Mahanti *et al.*, 1993). Additional studies indicated that mitochondria dysfunction may play a role in the hypovirulent phenotype of some hypovirulent strains. Transmissible from strain to strain via hyphal fusion, the mitochondrial hypovirulent phenotype is maternally inherited, which delineates it from dsRNA-associated hypovirulence. The only dsRNA known to be maternally inherited in *C. parasitica* has been a dsRNA associated with the mitochondria (Polashock & Hillman, 1994) and genetically engineered transgenic hypovirulent strains (Anagnostakis *et al.*, 1998). In the dsRNA-free hypovirulent strains, most of the respiratory activity is through the alternative oxidase pathway whereby the fungal thallus is cyanide resistant. Many of the strains showing these traits began to deteriorate on subculturing to fresh medium. This process has been observed before in other fungi and is described as senescence.

It was speculated that mitochondrial DNA (mtDNA) mutations might be involved in this form of hypovirulence. To explore this possibility, mutations that cause respiratory defects were induced and selected in Ep155, a stable virulent strain used in most laboratories studying chestnut blight. The study found that a cytoplasmically transmissible, hypovirulent phenotype could be selected (Monteiro-Vitorello, 1995). If such mutants can be selected in *C. parasitica,* it is possible that they might be selected in other fungal pathogen species giving rise to the possibility of inducing a hypovirulent phenotype where dsRNAs are lacking.

HYPOVIRUSES

While mitochondrial hypovirulence appears to play a role in the biological control of chestnut blight in specific stands of American chestnut trees in Michigan and Ontario (McKeen, 1995), the preponderance of hypovirulent strains found in surviving chestnut stands in Europe and North America appears to be dsRNA associated. Much is now known about these naked dsRNA genomes in terms of their structure, function, and genetic relatedness. Studies were initiated on three dsRNA genomes, one from Europe (CHV1-Ep713) (Choi & Nuss, 1902), the second from New Jersey (CHV2-NB58) (Hillman *et al.,* 1992), and a third from Michigan (CHV3-GH2) (Fulbright, 1990). The result was the establishment of the first virus family without structural proteins, Hypoviridae (Hillman *et al.,* 1995). Those dsRNA genomes that (1) reduced virulence, (2) were not associated with viral-encoded coat proteins, and (3) showed some nucleic acid or enzymatic protein motif related to the type dsRNA of the Hypoviridae family were called hypoviruses. The dsRNA from isolates that are obviously hypovirulent but not well characterized are usually referred to as hypovirus-like dsRNAs to separate them from the dsRNAs that have been given species names such as CHV1-Ep713. This is the type species for the virus family and the designation indicates it is *Cryphonectria* hypovirus number 1 from the fungal isolate Ep713.

Hypovirus CHV1-Ep713 has been instrumental in developing our understanding of hypovirulence and the symptoms associated with hypovirulence, that is, how fungal viruses may interfere with important functions of the pathogen such as sporulation, growth rate, and virulence. It was intuitive to believe that a substance that the fungus normally produced was now lacking; and, in fact, studies showed that many potential pathogenicity or virulence compounds including cutinase (Varley *et al.,* 1992), laccase (Rigling, 1995), oxalic acid (Vannini *et al.,* 1993), polygalacturonase (Gao et al. 1996), and others (Kazmierczak *et al.,* 1996) were down regulated or lacking altogether. So far, none of these interesting enzymes or compounds has proved to be an obvious pathogenicity factor in the *C. parasitica/Castanea* pathosystem and the search for such a fungal product continues.

As part of that continuing process, Nuss (1996) and coworkers concentrated their studies on G proteins (guanine nucleotide binding proteins), a group of regulatory proteins that play an essential role in cellular response to environmental stimuli (Gao & Nuss, 1996). One such protein, CPG-1, was found to be reduced in accumulation during hypovirus infection, and strains genetically disrupted for this protein also shared many of the characteristics of hypovirus infection. This work suggested that hypoviruses may suppress or interfere with virulence by way of disrupting one or more of the fungal-signaling components of the G protein pathway.

BIOLOGICAL CONTROL

The most exciting aspect of the natural development of hypovirulence is that it has allowed researchers to focus on the biological constraints fungal pathogens may confront in a living substrate. Hypovirulence, initially found as a naturally occurring biological control of chestnut blight, has now been part of experimental and practical application for almost 40 years. In Europe, chestnut farmers can purchase a tube of "hypovirulence paste" that can be applied to trees displaying the types of cankers usually associated with virulent strains (Heinger & Rigling, 1994). It even can be applied to grafting wounds to prevent virulent strain infection of the healing graft union.

Introducing any biological control into an environment managed toward crop production means that conditions at the introduction foci are generally manipulated toward plant production and away from pathogen or microorganism enhancement. Introducing hypovirulence, a form of the pathogen (weakened albeit, but a pathogen nonetheless), means that certain conditions must be manipulated in favor of the pathogen. The colonization of the hypovirulent fungus must be maintained or sustained in such a way that the pathogen ultimately supports itself and disseminates by natural means. After observing Michigan's American chestnut tree stands for 20 years, I am still surprised to see how well hypovirulence has maintained itself (personal observation). Even in extreme drought (1988 and 1998) and extreme winters (1993 and 1994), hypovirulent strains have not disadvantaged their host trees and have not allowed the trees to escape infection. Signs of the fungus can still be found on most trees in a blighted, surviving chestnut stand. The sustainable nature of naturally occurring hypovirulence is the goal of chestnut blight researchers, a goal that has elluded North American researchers to this point.

The factors limiting successful establishment of hypovirulent strains are not well enumerated or understood. Concerns usually center around the vegetative compatibility with virulent strains presently in the stand or in the surrounding area. It is widely believed that hypovirus transmission is inhibited among strains heteroallelic for veg-

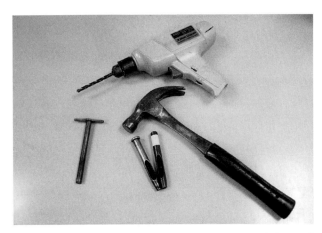

FIGURE 2 Instruments used to make holes in bark around the margin of the virulent cankers have included cordless electric drills, leather punches, and cork borers.

etative incompatibility loci, and these incompatible interactions are responsible for curtailing the rapid dispersal of hypovirulent strains in the natural range of chestnut in Eastern North America (Liu & Milgrom, 1996). These loci appear to have direct roles in blocking nuclear transmission and establishing heterokaryon formation. It appears that only certain vegetative incompatibility loci block hypovirus dispersal while others have no measurable effect.

To bypass this potential problem, at least three different strategies have been employed. First, multiple hypovirulent strains have been introduced to a nonrecovering stand, hopefully representing all or most of the various vegetative compatibility groups in the area (Anagnostakis, 1990). A second strategy is to isolate a representative sample of the *Cryphonectria parasitica* population that is to be treated (Cummings-Carlson *et al.*, 1998). These strains can be tested to determine the number and type of vegetative compatibility groups that are found at the site. Hypoviruses can then be introduced into these strains through hyphal fusion. The newly infected strains can then be reintroduced back to the location from which the strains were recovered. The third strategy used to bypass vegetative compatibility problems involves recombinant DNA techniques whereby integrated full-length complementary DNA (cDNA) could be found in ascospore progeny that included vegetative compatibility groups of the parental types and the recombinant types (Anagnostakis *et al.*, 1998). Generally, dsRNA genomes are excluded from the ascospore, but the integration event allows virus heritability through the ascospore. Integrated full-length cDNA will generate infectious hypovirus in the cytoplasm, allowing the hypovirus access to new vegetative compatibility groups.

Chestnut stands treated using the first two traditional approaches listed earlier are still currently monitored and enough success has been documented at each location that

these strategies are still recommended when deploying hypovirulence (Anagnostakis, 1990; Cummings-Carlson, 1998). The nontraditional genetically engineered strain was also successfully deployed in a test limited by size and time. One isolate, however, was recovered from a superficial canker that represented a different vegetative compatibility type from that of either transgenic parent. This indicates that the dsRNA was able to appear in a new vegetative compatibility type as predicted by the model. The dsRNA was cytoplasmic and was originally generated from the integrated, full-length hypovirus clone (Anagnostakis *et al.*, 1998).

Another important consideration when releasing hypoviruses in chestnut stands involves choosing the proper hypovirus or hypovirus-like dsRNA. If too debilitated, it may reduce the ability of the host fungus to infect the bark of the tree, thereby causing the strain to be lost. The inability of the fungus to sporulate asexually may reduce dissemination of hypovirulence. In a treated stand of American chestnut trees, one particular debilitating dsRNA genome was introduced for three consecutive years, after which it could only be found in 36% of the bark samples assayed. Another hypovirus, one that could be found in a majority of its asexual spores, was released and after 2 years was found in the isolates recovered from 38% of the bark samples tested. The first debilitating hypovirus is still present in the pathogen population but at a reduced frequency (Cummings-Carlson *et al.*, 1998).

TREATING CHESTNUT TREES WITH HYPOVIRULENT STRAINS

To treat chestnut trees, hypovirulent strains traditionally have been placed around the margins of cankers caused by virulent strains. A hole is made with a cork borer or leather punch in the bark down to the cambium at the canker edge (Fig. 2). The hypovirulent inoculum is placed in the hole and taped to prevent the rapid desiccation of the introduced strain. If the hypovirus or mitochondria transfer to the thallus of the virulent strain inciting the canker, a layer of wound periderm or callus will develop at the margins, walling off the fungus (Fig. 3A to C). Effective treatments generally lead to the eventual closing of the canker when periderm completely covers the exposed wood within 2 or 3 years.

The fate of the hypovirulent strain placed around the canker, as well as the fate of the original virulent strain—now hypovirulent, has not been well studied. What appears to be most important is the fate of the hypovirulent-inducing factor, the hypovirus or the mitochondrion. Dissemination through asexual spores requires the development of stroma, which means the hypovirulent strain must maintain enough virulence to continue to grow in the bark (Shain &

FIGURE 3 Hyprovirulent-treated canker. (A) Swelling canker with older infected rhytidome attached; (B) loose bark removed exposing the callus tissue, (C) slices through the callus layer exposing the living tissue at the edge of the canker, and (D) removal of outer bark shows a "break out" where the fungal thallus has broken through the callus layer. Treatment holes around edge of the canker are still visible.

Miller, 1992). If the strain is too debilitated, it will be lost, if it is too virulent, the strain will be ineffective in stopping the girdling of the stem (Taylor *et al.,* 1998).

Many treated cankers end up with "break overs" where at some point around the callus the canker fungus grows over the callus and into healthy tissue (see Fig. 3 D). This is an area that needs to be followed up, because it could be a hypovirulent strain that is continuing to grow, which will mean more hypovirus dissemination through stroma development and asexual spore production. However, it could also mean that parts of the canker-inciting strain did not become infected with the hypovirulent factor, indicating that the current infection may still girdle the stem if not treated again.

By using these techniques, the largest stand of mature (>2000 stems) American chestnut trees left in North America near West Salem, Wisconsin, was continually treated for 6 years—from 1992 to 1997 (Cummings-Carlson *et al.,*

1998). Two different hypovirus-like dsRNAs were used in the treatment. Every canker that could be treated was treated using leather punches and the hypovirulent strain made up in an agar slurry. At the beginning of the treatment, the virulent strain in the stand represented a single compatibility group (another was found later). Many of the cankers at this site are producing callus and it appears that some of the trees have cankers in remission. However, many trees are still dying each year. In 1998, treatment was stopped and the fate of the stand will be left to natural hypovirus dissemination.

References

Anagnostakis, S. L. (1982). Biological control of chestnut blight. Science, 215, 466–471.

Anagnostakis, S. L. (1990). Improved tree condition maintained in two

Connecticut plots after treatments with hypovirulent strains of the chestnut blight fungus. Forest Sciences, 36, 113–124.

Anagnostakis, S. L., Chen, B., Geletka, L. M. & Nuss, D. L. (1998). Field release of transgenic hypovirulent Cryphonectria parasitica strains demonstrates hypovirus transmission to ascospore progeny, persistance and limited dissemintion. Phytopathology, 88, 598–604.

Bissegger, M., Rigling, D, & Heiniger, (1997). Population structure and disease development of Cryphonectria parasitica in European chestnut forests in the presence of natural hypovirulence. Phytopathology, 87, 50–59.

Chen, B., Choi, G. H., & Nuss, D. L. (1994). Attenuation of fungal virulence by synthetic infectious hypovirus transcripts. Science, 26, 1762–1764.

Choi, G. H., & Nuss, D. L. (1992). Hypovirulence of chestnut blight fungus conferred by an infectious viral cDNA. Science, 257, 800–803.

Cummings-Carlson, J., Fulbright, D., MacDonald, W. L., & Milgroom, M. G. (1998). West Salem: A research update. Journal of American Chestnut Foundation, 12, 24–26

Day, P. R., Dodds, J. A., Elliston, J. E., Jaynes, R. A., & Anagnostakis, S. L. (1977). Double-stranded RNA in Endothia parasitica. Phytopathology, 67, 1393–1396.

Dodds, J. A. (1980). Association of type I viral-like dsRNA with club-shaped particles in hypovirulent strains of Endothia parasitica. Virology, 107, 1–12.

Elliston, J. E. (1978). Pathogenicity and sporulation of normal and diseased strains of Endothia parasitica in American chestnut. In W. L. MacDonald, F. C., Cech, & H. C. Smith, (Eds.) Proceedings of the American Chestnut Symposium (pp. 95–100). Morgantown, WV: West Virginia University Books.

Elliston, J. E. (1985). Characteristics of dsRNA-free and dsRNA-containing strains of Endothia parasitica in relation to hypovirulence. Phytopathology, 75, 151–158.

Elliston, J. E. (1985). Further evidence for two cytoplasmic hypovirulence agents in a strain of Endothia parasitica from western Michigan. Phytopathology, 75, 1405–1413.

Fulbright, D. W. (1985). A cytoplasmic hypovirulent strain of Endothia parasitica without double-stranded RNA. Phytopathology (Abstract), 75, 1328.

Fulbright, D. W. (1990). Molecular basis for hypovirulence and its ecological relationships. In R. Baker & P. Dunn (Eds.), New directions in biological control (pp. 693–702). New York: Alan R. Liss.

Fulbright, D. W., Weidlich, W. H., Haufler, K. Z., Thomas, C. S., & Paul, C. P. (1983). Chestnut blight and recovering American chestnut trees in Michigan. Canadian Journal of Botany, 61, 3164–3171.

Gao, S., Choi, G. H., Shain, L., & Nuss, D. L. (1996). Cloning and targeted disruption of enpg-l, encoding of the major in vitro extracellular endopolygalacturonase of the chestnut blight fungus, Cryphonectria parasitica. Applied and Environmental Microbiology, 62, 1984–1990.

Gao, S., & Nuss, D. L. (1996). Distinct roles for two G protein subunits in fungal virulence, morphology, and reproduction revealed by targeted gene disruption. Proceedings of the National Academy of Sciences, USA, 93, 14122–14127.

Grente, J., & Berthelay-Sauret, S. (1969). Pathologie vegetale—L'hypoviruence exclusive, phenomene original en pathologie vegetale. Comptes Rendus de l' Academie Sciences Paris, 268, 2347–2350.

Grente, J., & Berthelay-Sauret, S. (1978). Biological control of chestnut blight in France. In W. L. MacDonald, F. C. Cech, & H. C. Smith (Eds.), Proceedings of the American Chestnut Symposium (pp. 30–34). Morgantown, WV: West Virginia University Books.

Heiniger, U., & Rigling, D. (1994). Biological control of chestnut blight in Europe. Annual Review of Phytopathology, 32, 581–599.

Hillman, B. I., Tian, Y., Bedker, P. J., & Brown, M. P. (1992). A North American hypovirulent isolate of the chestnut blight fungus with a European isolate-related dsRNA. Journal of General Virology, 73, 681–686.

Hillman, B. I., Fulbright, D. W., Nuss, D. L., & Van Alfen, N. K. (1995). Hypoviridae. In F. A. Murphy, et al. (Eds.), Virus taxonomy: Classification and nomenclature of viruses, sixth report of the International Committee on Taxonomy of Viruses. New York: Springer-Verlag.

Jaynes, R. A., & Elliston, J. E.(1982). Hypovirulent isolates of Endothia parasitica associated with large American chestnut trees. Plant Disease, 66, 679–772.

Kazmierczak, P., Pfeiffer, P., Zhang, L., & Van Alfen, N. K. (1996). Transcriptional repression of specific host genes by the mycrovirus Cryphonectria hypovirus 1. Journal of Virology, 70, 1137–1142.

Liu, Y.-C., & Milgroom, M. G. (1996). Correlation between hypovirus transmission and the number of vegetative incompatibility (vic) genes different among isolates from a natural population of Cryphonectria parasitica. Phytopathology, 86, 79–86.

Mahanti, N., Bertrand, H., Monteiro-Vitorello, C., & Fulbright, D. W. (1993). Elevated mitochondrial alternative oxidase activity in dsRNA-free, hypovirulent isolates of Cryphonectria parasitica. Physiology and Molecular Plant Pathology, 42, 455–463.

McKeen, C. D. (1995). Chestnut blight in Ontario: Past and present status. Canadian Journal of Plant Pathology, 17, 295–304.

Milgroom, M. G., Wang, K., Zhou, Y., Lipari, S. E., & Kaneko, S. (1997). Intercontinental population structure of the chestnut blight fungus, Cryphonectria parasitica. Mycologia, 88, 179–190.

Monteiro-Vitorello, C. B., Bell, J. A., Fulbright, D. W., & Bertrand, H. (1995). A cytoplasmically-transmissible hypovirulence phenotype associated with mitochondrial DNA mutations in the chestnut blight fungus Cryphonectria parasitica. Proceedings of the National Academy of Sciences, USA, 92, 5935–5939.

Nuss, D. L. (1996). Using hypoviruses to probe and perturb signal transduction processes underlying fungal pathogenesis. The Plant Cell, 8, 1845–1853.

Polashock, J. J., & Hillman, B. I. (1994). A small mitochondrial double-stranded (ds) RNA element associated with a hypovirulent strain of the chestnut blight fungus and ancestrally related to yeast cytoplasmic T and W dsRNAs. Proceedings of the National Academy of Sciences, USA, 91, 8680–8684.

Rigling, D. (1995). Isolation and characterization of Cryphonectria parasitica mutants that mimic a specific effect of hypovirulence-associated dsRNA on laccase activity. Canadian Journal of Botony, 73, 1655–1661.

Scheffer, R. P. (1997). Natural history of some destructive diseases: Native plants, alien pathogens. In The nature of plant diseases (pp. 81–89). Cambridge University Press.

Schwadron, B. A. (1995). Distribution and persistence of American chestnut sprouts, Castanea dentata [Marsh.] Borkh., in Northeastern Ohio woodlands. Ohio Journal of Science, 95, 281–288.

Shain, L., & Miller, J. B. (1992). Movement of cytoplasmic hypovirulence agents in chestnut blight cankers. Canadian Journal of Botony, 70, 557–561.

Smart, C. D., & Fulbright, D. W. (1995). Characterization of a strain of Cryphonectria parasitica doubly infected with hypovirulence-associated dsRNA viruses. Phytopathology, 85, 491–494.

Taylor, D., Jarosz, A. M., Lenski, R. E., & Fulbright, D. W. (in press). Acquisition of hypovirulence in host-pathogen systems with three trophic levels. American Naturalist.

Vannini, A., Smart, C. D., & Fulbright, D. W. (1993). The comparison of oxalic acid production *in vivo* and *in vitro* by virulent and hypovirulent *Cryphonectria (Endothia) parasitica*. Physiology and Molecular Plant Pathology, 43, 443–451.

Van Alfen, N. K., Jaynes, R. A., Anagnostakis, S. L., & Day, P. R. (1975). Chestnut blight: Biological control by transmissible hypovirulence in *Endothia parasitica*. Science, 189, 890–891.

Varley, D. A., Podila, G. K., & Hiremath, S. T., (1992). Cutinase in *Cryphonectria parasitica,* the chestnut blight fungus: Suppression of cutinase gene expression in isogenic hypovirulent strains containing double-stranded RNAs. Molecular and Cellular Biology, 12, 539–4544.

26

Controlling Soil-Borne Plant Pathogens

T. S. BELLOWS

Department of Entomology
University of California
Riverside, California

INTRODUCTION

Biological control of soil-dwelling plant pathogens has been the focus of numerous studies, and has been treated in several texts and reviews. As in the case of pathogens of aboveground plant parts, management of soil-borne pathogens occurs at the microbial level (Andrews, 1992). Agents used against these pathogens include both competitive antagonists as well as predators and microbial hyperparasites. Biological control agents for pathogens and nematodes function either by destroying the pest organisms (predaceous and mycoparasitic antagonists), or through competitive exclusion. Competitive exclusion takes place when an antagonist occupies or uses resources in a nonpathogenic manner and in so doing excludes pathogenic organisms, thus preventing the disease.

Diseases of roots and subterranean stems are caused by a variety of pathogens. Because of this diversity, the species antagonist to plant pathogens and the mechanisms by which they accomplish their beneficial action are also quite varied. The taxonomic diversity, biology, and use of these antagonists are covered in some detail in several texts and reviews, including Cook and Baker (1983), Parker *et al.* (1985), Lynch (1987), Campbell (1989), Adams (1990), Hornby (1990), Sayre and Walter (1991), Stirling (1991), Tjamos *et al.* (1992), and Cook (1993).

This chapter introduces briefly some of the antagonists of soil-borne pathogens as representative of the broad taxa that are useful in this field, and then discusses ways in which these antagonists can be employed to bring about biological control of soil-borne plant pathogens.

ROOT PATHOGENS AND THEIR BIOLOGICAL CONTROL AGENTS

Microbial Pathogens and Antagonists

Root diseases are caused by a wide variety of fungi and by some bacteria in many crops and plant systems. Biological control agents recognized as significant in suppression of microbial root diseases are largely antagonists that can occupy niches similar to the pathogens and either by natural means or through manipulation outcompete the pathogens in these niches. Antibiotic production is also important in a few cases, as are mycoparasitism and induced resistance.

Antagonists are known among bacteria and fungi. *Streptomyces scabies,* the causative organism of potato scab, is suppressed by naturally occurring populations of *Bacillus subtilis* (Ehrenberg) Cohn and saprotrophic *Streptomyces* spp. Other microorganisms recognized as suppressing fungal diseases include species of *Pseudomonas* and *Bacillus.* Saprotrophic *Fusarium* fungi are able to suppress populations of pathogenic *Fusarium* spp. through competition for nutrients.

There are few well-documented cases of induced resistance for soilborne pathogens, and these are mostly of wilt diseases. Examples of organisms that induce resistance in plants include nonpathogenic strains of *Fusarium* spp., *Verticillium* spp., and *Gaeumannomyces* spp.

Mycoparasitic flora such as *Trichoderma* spp., *Coniothyrium minitans* Campbell, and *Sporidesmium sclerotivorum* Uecker *et al.* can be added to soil against fungal diseases. *Bacillus* spp. and especially *Pseudomonas* spp. are among bacteria that have properties particularly suited to effective suppression of root-infecting pathogens in soil, such as antibiotic production and competition for Fe^{3+} ions. Mycetophagous soil amoebae have also been noted feeding on pathogenic fungi. These amoebae generally require moist conditions in which to function, and may be important in natural control of some fungi.

Mycorrhizae are nonpathogenic fungi associated with roots in some temperate forest trees (and are often essential symbionts for coniferous tree species). Ectomycorrhizae are mostly basidiomycetes that form a sheath over the root, and hyphae spread out into the soil. These fungi have been studied in relation to nutrient uptake, but they also affect root disease. As they completely enclose the root, they change the quantity and quality of exudates reaching the soil, and consequently roots with mycorrhizae have a rhizosphere flora different from roots lacking these these symbionts (Campbell, 1985). In at least one case, the mycorrhizal fungus *Pisolithus tinctorius* (Persoon) Coker & Couch, the symbiont sheath is sufficiently thick that it forms a barrier to infection by such pathogens as *Phytophthora cinnamomi* Rands attacking eucalyptus trees. Other mycorrhizal fungi produce antibiotics effective against *P. cinnamomi* in plate tests. The intentional manipulation of mycorrhizal fungi for disease control has not been widely implemented, but opportunities for selected uses may be possible (Campbell, 1989).

Plant-Parasitic Nematodes

Plant-parasitic nematodes inhabit many soils and attack the roots of plants. They are affected by a range of natural enemies, including bacteria, nematophagous fungi, and predaceous nematodes and arthropods. There is some limited evidence for virus association with nematodes (Loewenberg *et al.*, 1959), but the etiology of these viruses is not well known (Stirling, 1991). The biologies of natural enemies of nematodes have been reviewed by Sayre and Walter (1991) and Stirling (1991).

Bacteria Affecting Plant-Parasitic Nematodes

Several bacterial diseases of nematodes have been reported (Saxena & Mukerji, 1988); other bacteria produce compounds that are detrimental to plant-parasitic nematodes (Stirling, 1991). The most widely studied bacterial pathogen of nematodes are in the genus *Pasteuria.* Early work was focused on *P. penetrans* (Thorne) Sayre & Starr sensu stricto Starr & Sayre. This taxon represents an assemblage of numerous pathotypes and morphotypes, and may represent a complex of species (Starr & Sayre, 1988). This bacterium has been found infecting a large number of nematode species (more than 200 in about 100 genera, Sayre &

Starr, 1988; Stirling, 1991), does not attack other soil organisms, and is the most specific obligate parasite of nematodes known. Its spores attach to and penetrate the nematode cuticle. Most attention has been centered on populations of *P. penetrans* sensu stricto (Starr & Sayre, 1988) that attack root knot nematodes (*Meloidogyne* spp.). The spores of *P. penetrans* germinate a few days after a contaminated nematode begins feeding on a root (Sayre & Wergin, 1977). The bacterium reproduces throughout the entire female body, and the female either may be killed or may mature but produce no eggs. Bacterial spores (about 2 million from each infected nematode, Mankau, 1975) are released when the nematode body decomposes, and they remain free in the soil until contacted by another nematode. They tolerate dry conditions and a wide range of temperatures, and may remain viable in the soil for more than 6 months. Because *P. penetrans* is an obligate parasite, it has not yet been possible to develop *in vitro* culturing techniques for this bacterium. Different populations of the bacterium show varying degrees of specificity to small numbers of nematode species, but the mechanisms and degree of specificity remain to be elucidated (Stirling, 1991). *Pasteuria penetrans* appears responsible for some cases of natural regulation of nematode populations (Sayre & Walter, 1991). Variation in susceptibility of *Meloidogyne* spp. biotypes to *P. penetrans* may contribute to variation in effectiveness (Tzortzakakis *et al.*, 1997).

A few strains of *Bacillus thuringiensis* Berliner are also known to have activity against nematodes, including plant-parasitic species. Zuckerman *et al.* (1993) report efficacy of a strain against *M. incognita* (Kofoid & White) Chitwood, *Rotylenchus reniformis* Linford & Oliveira, and *Pratylenchus penetrans* Cobb in field and in greenhouse trials. The body openings of these nematodes are too small to permit the ingestion or other ingress of the bacterium, and Zuckerman *et al.* (1993) suggest that the mode of action is either a β-exotoxin (Prasad *et al.*, 1972; Ignoffo & Dropkin, 1977) or a δ-endotoxin released following bacterial cell lysis. A strain of *B. thuringiensis* with a nematotoxic δ-endotoxin is the subject of a European patent application by Mycogen Corporation of San Diego, California (Zuckerman *et al.*, 1993).

Fungi Affecting Plant-Parasitic Nematodes

A large group of fungi attack nematodes in the soil (Barron, 1977; Stirling, 1991; Galper *et al.*, 1991). Numerous species have been reported from all types of soils. The taxonomy of the group has been subject to revision, and we use the generic names recognized in Stirling (1991) in this chapter.

Some nematophagous fungi are endoparasitic in nematodes. Among these are genera that reproduce through motile zoospores (e.g., *Catenaria anguillulae* Sorokin, *Lagen-*

idium caudatum Barron, and *Aphanomyces* sp.), which generally appear only weakly pathogenic in healthy nematodes (Stirling, 1991). Other endoparasitic fungi possess adhesive conidia, and the infection process begins when conidia adhere to a nematode's cuticle (e.g., the genera *Verticillium, Drechmeria, Hirsutella,* and *Nematoctonus*). In *Nematoctonus* spp., the germinating spores secrete a nematotoxic compound that causes rapid immobilization and death of nematodes (Giuma *et al.,* 1973). A few species [*Catenaria auxilaris* (Kuhn) Tribe, *Nematophthora gynophila* Kerry & Crump] parasitize adult females or nematode eggs instead of juveniles.

Other fungi capture nematodes through use of special trapping structures, and have been termed "predatory." Among the more common of these fungi are species in such genera as *Monacrosporium, Arthrobotrys,* and *Nematoctonus*. These fungi consist of a sparse mycelium, modified to form organs capable of capturing nematodes. These organs include adhesive structures, such as adhesive hyphae, branches, knobs, or nets (Stirling, 1991). There are also nonadhesive rings, the cells of which expand when touched on their inner surface, constricting the interior of the ring, and trapping nematodes. Most of these fungi are not specific and attack a wide range of nematode species. They are widely distributed (Gray, 1987, 1988) and most are capable of saprotrophic growth, but often appear limited in this phase in the soil. Many soils suppress the growth of these fungi (a condition called soil fungistasis or mycostasis). This is possibly due to two different causes. Mankau (1962) concluded that a water-diffusible substance was responsible for inhibited germination in tests of southern Californian soil. Other studies have indicated increased activity following soil amendments with nutrients (Olthof & Estey, 1966) or organic material (Cooke, 1968), which implies fungistasis may be a result of resource limitation. Following saprotrophic growth, formation of trapping structures occurs, and is apparently stimulated by nematodes (Nordbring-Hertz, 1973; Jansson & Nordbring-Hertz, 1980). Stirling (1991) suggested that this phase of predaceous activity is followed by diversion of resources to reproduction, followed by a relatively dormant phase.

Several species of fungi are facultatively parasitic on nematodes. Of the few of these fungi that are significant pathogens of root knot and cyst nematodes, *Verticillium* spp. are among the most important. These fungi can parasitize nematode eggs, and *V. chlamydosporium* Goddard plays a major role in limiting multiplication of *Heterodera avenae* Wollenweber in English cereal fields (Kerry *et al.,* 1982a, 1982b). *Paecilomyces lilacinus* (Thom) Samson parasitizes eggs of *Meloidogyne incognita* (Jatala *et al.,* 1979) and *Heterodera zeae* Koshy, Swarup & Sethi (Dunn, 1983; Godoy *et al.,* 1983). *Dactylella oviparasitica* Stirling & Mankau, a parasite of *Meloidogyne* eggs, is thought to be at least partly responsible for natural decline of root knot

nematodes in California peach orchards (Stirling *et al.,* 1979).

Predacious Nematodes

Predatory nematodes are found in four main taxonomic groups (Monochilidae, Dorylaimidae, Aphelenchidae, and Diplogasteridae), each with a distinct feeding mechanism and food preferences (Stirling, 1991). The monochilids have a large buccal cavity that bears a large dorsal tooth; all species are predaceous, feeding on protozoa, nematodes, rotifers, and other prey, which may be swallowed whole or pierced and the body contents removed. The dorylaimids are typically larger than their prey and possess a hollow spear that is used either to pierce the body of the prey or to inject enzymes into the food source and to then remove the predigested contents. The group is considered omnivorous, but the feeding habits are known for only a few species (Ferris & Ferris, 1989). Almost all the predatory aphelenchids are in the genus *Seinura*. Although small, they can feed on nematodes larger than themselves by injecting the prey with a rapidly paralyzing toxin via their stylet. The diplogasterids, typically a bacteria-feeding group, have a stoma armed with teeth, and the species with large teeth prey on other nematodes. Species in all these groups are generally omnivorous, feeding on free-living as well as plant-parasitic nematodes. The role of individual species in the population dynamics of plant-parasitic nematodes in the soil has been difficult to quantify, but it is possible that a number of species may act together to produce a significant impact (Stirling, 1991).

Insects and Mite Predators

Several microarthropods in the soil, including mites and Collembola, prey on nematodes, and high predation rates have been recorded *in vitro* (Stirling, 1991). A few genera are obligate predators of nematodes, while other genera are more omnivorous and consume nematodes as well as other foods (Moore *et al.,* 1988; Walter *et al.,* 1988; Sayre & Walter, 1991). The information available suggests that, as a group, microarthropods are probably significant predators on nematodes in some soils and habitats. Limited information about predation rates in soil is available, however, and more work will be necessary to assess the impact of this group on nematode populations.

MICROBIAL AND HABITAT CHARACTERISTICS

Pathogens and Antagonists

Competition for resources in soil environments is vital to the ability of any particular organism to increase in numbers and consequently to reduce the numbers or activity of other organisms, including plant pathogens (Campbell, 1989; Andrews, 1992). As discussed in Chapter 32, microbial competition can be important at two main stages of growth of pathogen populations. First, there may be competition during initial establishment on a fresh resource that previously was not colonized by microorganisms, as is typical for ruderal or *r*-selected species. Second, after initial establishment, there is further competition to secure enough of the limited resources present to permit survival and eventual reproduction. This later competition after population establishment is typical of *K*-selected species. Microorganisms show many traits that may characterize them as particularly adept at either the colonization (*r*-strategist) phase or the subsequent (*K*-strategist) phases of competition (Begon *et al.,* 1986). These concepts represent the endpoints of a continuum, and there are varying degrees of *r*- and *K*-related characteristics in different microbes in various habitats; see Andrews and Harris (1986) for further discussion on these concepts in microbial ecology.

Plant pathogens are spread across this *r−K* range of characteristics, and also vary in other biological characteristics. Opportunistic pathogens are able to attack young, weakened, or predisposed plants, but may be poor competitors (*Botrytis, Pythium,* and *Rhizoctonia*). Other pathogens tolerate environmental stresses well. These organisms often live in situations with few competitors, because few species are able to exist in some stressful environments. Pathogens, such as the *Penicillium* species that cause postharvest rots, produce antibiotics that inhibit competitors. Other species have a very high competitive ability [*Fusarium culmorum* (Smith) Saccardo]. Knowledge of the ecology of a pathogen can be important in determining what biological control strategy might be most effective. Stress-tolerant and competitive species, for example, would require different biological control strategies and agents than would ruderal ones.

Just as plant pathogens vary in *r−K* and other characteristics, antagonists of these pathogens have varied biological attributes. The properties of an effective biological control agent will depend on the setting in which it is intended to function. In many agricultural settings, disturbance makes new resources available to microbes through crop residue burial, cultivation, or planting. A frequent need, therefore, is a control agent that has the characteristics of an *r*-strategist (Campbell, 1989), which can grow quickly and colonize new resources rapidly when faced with minimal nutrient and environmental restrictions. It should function well in disturbed environments and have some means (such as spores) of surviving in the soil, or on the plant near the pathogen inoculum, or near the source or site of infection. Biological control agents that are *r*-strategists are an approximate equivalent of a protectant fungicide, being in place before the pathogen infection cycle can begin. In other programs, such as those directed against a pathogen that has already invaded the plant host, a more competitive species will be required. Finally, a biological control agent may have to be tolerant of abiotic stresses, particularly for use in dry climates.

Soil and the Rhizosphere

Although there is much variation in soil types in different locations, soils are typically rich in microflora, with propagules numbering in the hundreds of thousands per gram of soil (Campbell, 1989). In most soils, growth of microorganisms is carbon limited (Campbell, 1989), either because what carbon is available is not physically accessible or because the microbes do not possess the enzymes necessary to degrade the carbon-containing molecules that are present. An exception to this general limitation is the rhizosphere, the roots and the regions immediately adjacent to them. The rhizosphere contains easily metabolized carbon and nitrogen sources such as amino acids, simple sugars, and other compounds exuded by the roots. Consequently, this region is more favorable than surrounding soil for the support of microflora. While root pathogens and plant-parasitic nematodes may be found growing on or in roots, many microbes in the soil may be dormant because of resource limitations. Because there are many dormant organisms in the soil prepared to take advantage of any favorable period or opportunity, competition for resources in the soil may be significant and may limit the ability to augment beneficial organisms and have them flourish, unless soils are first sterilized to eliminate potential competitors. Consequently, much research surrounding biological control of root diseases and nematodes has centered around identifying soils that are naturally suppressive to particular disease organisms and investigating the microbial components of the soil responsible for the suppression.

CONTROL OF SOIL BORNE PLANT PATHOGENS

Organisms for biological control of soil borne diseases can be used primarily in two ways, either through conservation or through augmentation. Among the diversity of nonpathogenic microbes usually associated with plants, there is substantial opportunity for development of resident species as competitors of, or antagonists to, pathogenic organisms. Management of such antagonistic organisms for biological control can range from treatment of soil to favor the desirable organisms (conservation) through inoculation of soils or plants with specific beneficial microorganisms (augmentation).

Substantial work has been done in characterizing the role of microorganisms in biological control of plant diseases

(Elad, 1986). The biological mechanisms underlying the success of these antagonists may include initial competition for occupancy of inoculation sites, competition for limiting nutrients or minerals, parasitism, and induction of plant defenses.

Conservation

Biological control of plant pathogens through conservation is accomplished either by preserving existing microbes that attack or compete with pathogens or by enhancing conditions for their survival and reproduction at the expense of pathogenic organisms. Conservation is applicable in situations where microorganisms important in limiting disease-causing organisms already occur, primarily in the soil and in plant residues. Microorganisms may be conserved by avoiding practices that negatively affect them (such as soil treatments with fungicides). The soil environment may be enhanced for some beneficial organisms through adding organic matter (including such soil amendments as composted organic matter, manures, cover crops, or crop residue).

Root Diseases

In the case of soil microflora, species employed for biological control of plant pathogens are often competitive antagonists. Adding amendments to soil is one way in which soil microorganisms may be managed to enhance populations of these beneficial organisms. Addition of or-

ganic matter to soils for control of *Streptomyces scabies,* the causative organism of potato scab, is one example. Addition of carbon sources to soil increases general microbial activity that leads to reductions in *S. scabies.* Specifically, *Bacillus subtilis* and saprotrophic species of *Streptomyces* were encouraged by barley, alfalfa or soy meal (Campbell, 1989). Soy meal was also a substrate for antibiotic production against *S. scabies.* A general rise in soil organic matter also gave control of *Phytophthora cinnamomi* in avocado in Australia (Malajczuk, 1979). The addition of more than 10 tons of organic matter per hectare per year led to general increases in numbers of bacteria. Lysis of the hyphae and sporangia of the pathogen were attributed to species of *Pseudomonas, Bacillus,* and *Streptomyces.*

Some soils appear to suppress disease naturally, and may contain antagonistic or antibiotic floras that flourish without the need for amendments. One example of such suppressive soils is the *Fusarium*-suppressive soil in the Chateaurenard District of the Rhone Valley in France (Fig. 1). Here, *Fusarium oxysporum* f. sp. *melonis* Snyder & Hansen is present, but no disease develops when susceptible melon varieties are grown. These soils are suppressive for several other types of *F. oxysporum,* but not to other species or genera of pathogens. The suppressive nature of the soils is clearly biotic, because the soils lose their suppressive ability when steam sterilized, and the suppressive ability can be transferred to other soils. The antagonists principally responsible for this suppression are nonpathogenic strains of *F. oxysporum* and *F. solani* (Martius) Saccardo. The sup-

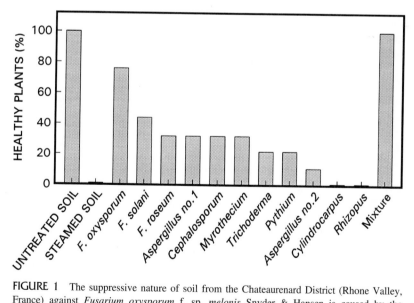

FIGURE 1   The suppressive nature of soil from the Chateaurenard District (Rhone Valley, France) against *Fusarium oxysporum* f. sp. *melonis* Snyder & Hansen is caused by the presence of other nonpathogenic fungi in the soil. Untreated soil and steamed soil with a mixture of the fungi present completely suppress the disease. The species of fungi present vary in their individual ability to suppress the disease, with a nonpathogenic form of *F. oxysporum* Schlechtendal playing an important role. (After Alabouvette *et al.* (1979).

pression appears to be due to fungistasis induced by nutrient limitation. The competing fungi appear to have nearly the same ecological niche as the pathogenic forms, and the saprotrophic forms outcompete the pathogens for limiting resources so that dormant chlamydospores of the pathogen do not germinate in the presence of host root exudates. It may be possible to develop systems for other areas using the antagonists from the Chateaurenard area (Campbell, 1989), although additional research may be necessary to permit their effective operation in different soils. Other soils suppressive to *Fusarium* wilts are also known. There are numerous other examples of suppressive soils (Larkin *et al.*, 1996), although some soils or combinations appear to give somewhat variable results.

Plant-Parasitic Nematodes

There are several reports of substantial natural control (control by natural enemies without intentional manipulation) of plant-parasitic nematodes. Stirling (1991) and Sayre and Walter (1991) review several of these. The natural suppression of the cereal cyst nematode *Heterodera avenae* in cereal cultivation in England (Gair *et al.*, 1969) is one example of such suppression. In this case, populations of the nematode initially increased for the first 2 to 3 years of cultivation and then declined continually during 13 years of continuous cultivation of both oats and barley (a more susceptible crop) (Fig. 2). Four species of nematophagous fungi were present in the soil. The two species principally responsible for nematode suppression were *Nematophthora gynophila* and *Verticillium chlamydosporium*. Both fungi attacked female nematodes, either destroying them or reducing their fecundity. The activity of both fungi during laboratory trials was greatest in wet soils (Kerry *et al.*,

1980). Although natural suppression of the nematode population takes some time to develop in these soils, once established it maintains the population below the economic threshold (Stirling, 1991).

Conserving nematode antagonists (as opposed to directly enhancing their numbers, as discussed later) in soils is a matter that has received relatively little attention. The application of toxins (e.g., insecticides and fungicides) to aerial portions of crops or directly to soils often leads to pesticide activity in the soil. All nematicides are nonselective in their action and hence will kill predatory nematodes (Stirling, 1991). In addition, herbicides have well-documented effects on soil microorganisms (Anderson, 1978) and may well exert some influence on microbial antagonists of nematodes. Insecticides may negatively affect soil microarthropods. Many fungicides are known to be detrimental to nematophagous fungi (Mankau, 1968; Canto-Saenz & Kaltenbach, 1984; Jaffee & McInnis, 1990), but at levels higher than would be expected under normal field practice. Among the fumigant nematicides, ethylene dibromide (EDB) and dibromo-chloro-propene (DBCP) appear nontoxic to the nematode-trapping fungi (Mankau, 1968), and several herbicides were not harmful to *Arthrobotrys* sp. (Cayrol, 1983).

Despite these potentially significant effects on beneficial microfloras and faunas, and the possibility of conserving these organisms by appropriate choice of pesticide, little has emerged to integrate these ideas into normal farming practice. Perhaps because there has been no serious emergence of nematode problems associated with the use of these materials, the status quo is justified. Nonetheless, the opportunities for conserving biologically important agents should be considered in the development of future integrated management programs for plant-parasitic nematodes.

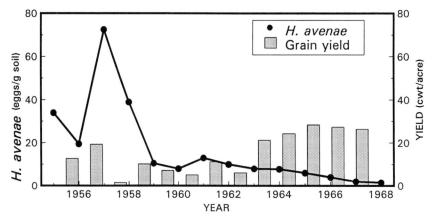

FIGURE 2 Postcropping levels of *Heterodera avenae* Wollenweber and crop yields under continuous grain [oats (*Avena sativa* Linnaeus) 1955 to 1962 and barley (*Hordeum vulgare* Linnaeus) 1963 to 1968] culture in the United Kingdom. [After Gair, R., *et al.* (1969). *Annals of Applied Biology, 63,* 503–512.]

Cultivation practices may also be selected to favor natural enemies of nematodes. Among these are minimal or conservation tillage, which reduced the number of cysts of *H. avenae* on roots and the amount of damage caused by the nematode on wheat in Australia (Roget & Rovira, 1987). Other practices that may affect populations of natural enemies include normal tillage (which adds crop residue to the soil and thus may favor certain beneficial organisms) and crop rotation sequences (Stirling, 1991).

The knowledge that some soils are naturally suppressive to nematodes prompts the question of whether or not the features of these soils can be used to improve biological control. In all documented instances where they have been studied, the suppressive properties of these soils appear to result primarily from the action of one or two specific biological control agents (Stirling, 1991). The suppressiveness requires substantial time to develop, and considerable economic loss might be incurred during the initial seasons of cropping before the suppressive nature has fully developed. Some risk is involved also, because the suppressive nature of the soil may not develop to suitable levels. Careful management of crop variety, particularly concerning susceptibility to nematode damage, during the initial phases of land usage for cropping, is an important part of taking advantage of the potential of these resident natural enemies (Stirling, 1991).

Where soils are not naturally suppressive to nematode populations, they may be manipulated to enhance what natural control agents are present. Most attention in this arena has been given to the addition of organic matter to the soils. Much of the information concerning the effects of these amendments is circumstantial, but the beneficial effects appear widespread. Many different soil amendments have been considered and evaluated, and it is clear that the reduction of plant damage from nematodes following such amendments may occur through a variety of mechanisms (Stirling, 1991).

One of these mechanisms is through general improvement of soil structure and fertility. Addition of crop residue or animal manures increases ion exchange capacity of the soil, chelates micronutrients to make them accessible by the plant, and adds available nitrogen. Grown under such improved conditions, healthy plants are better able than stressed plants to tolerate damage from nematodes. In a second mechanism, certain amendments may directly improve plant resistance to nematodes (Sitaramaiah & Singh, 1974).

Populations of some nematophagous and antagonistic organisms respond directly to soil amendments. Spores of many of fungi fail to germinate in otherwise suitable but nonamended soils (Dobbs & Hinson, 1953) and this soil mycostasis can affect both spores and mycelia (Duddington *et al.,* 1956a, b, 1956b, 1961; Cooke & Satchuthananthavale, 1968) (Table 1). Before predation on nematodes by

TABLE 1   Results of a Field Microplot Experiment with *Heterodera schachtii* Schmidt on Sugar Beets Showing the Effects of Soil Amendments and Nematode-Tapping Fungi on Nematode Populations and Yield

Treatment	Final nematode population/ 100 g soil		Yield (ton/ha)
	Cysts	Eggs	
1. Untreated	467	114	45.4
2. Bran (20 ton/ha)	383	103	56.3
3. *Monacrosporium thaumasia* (Drechsler) de Hoog & van Oorschot mycelium (6.8 ton/ha) at planting	488	144	56.3
4. Treatment 2 + treatment 3	333	94	59.2
5. Treatment 4 + *M. thaumasia* mycelium (6.8 ton/ha) in midseason	363	128	60.1

After Duddington, C.L., *et al.* (1956a). *Nematologica, 1,* 341–343.

these fungi can take place, mycelial growth and trap formation must occur. The addition of organic matter provides a substrate that may stimulate spore germination. Organic amendments stimulate a broad range of soil microorganisms; thus, the effects of amendments on populations of these organisms are complex. There is a general increase in microbial population growth immediately following the addition of organic matter; subsequently as part of the community succession, there is an increase in populations of nematode-trapping fungi (Stirling, 1991). Speigel *et al.* (1988, 1989) concluded that the beneficial effects of chitin amendments resulted from the action of specialized microorganisms.

Augmentation

Biological control of plant pathogens through augmentation is based on mass-culturing antagonistic species and adding them to the cropping system. This is considered augmentation because the organisms used are usually present in the system, but at lower numbers or in locations different from those desired. The purpose of augmentation is to increase the numbers or to modify the distribution of the antagonist in the system. In some cases such organisms are taken from one habitat (e.g., the soil) and augmented in another (e.g., the phyllosphere). There are several examples where such organisms, when moved to a new habitat (e.g., from the soil to the aboveground part of a plant) colonize and serve as successful agents of biological control (Andrews, 1992; Cook, 1993). The activity of augmenting microbial agents is sometimes termed "introduction" in the plant pathology literature, in the sense of "adding" them to the system (Andrews, 1992; Cook, 1993).

Augmentation of antagonists may suppress disease through two different types of mechanisms. The first is when the antagonistic organisms compete, at potential infection sites or zones, with the pathogens themselves. In this approach, the antagonist population is directly responsible for disease suppression. A second approach is inoculation of plants with nonpathogenic organisms that prompt general plant defenses against infection by pathogens (induced resistance, discussed later), and control is achieved through greater resistance of the plant to infection.

Parasitism is another mechanism by which beneficial microorganisms suppress plant pathogens. Some species of *Trichoderma,* for example, attack pathogenic fungi, leading to the lysis of the pathogen. Natural enemies of plant-parasitic nematodes include bacterial diseases and nematophagous and nematopathogenic fungi. These organisms can also be augmented into soil environments to suppress pathogenic organisms.

Augmentation of antagonists and parasites of plant disease organisms can generally be of two types, inoculation and inundation (also see Chapter 32). Inoculative releases consist of small amounts of inoculum, with the intention that the organisms in this inoculum will establish populations of the antagonist that, following subsequent population growth, will then limit the pathogen population. In inundative releases, a large amount of inoculum is applied, with the expectation that control will result directly from this large initial population with limited reliance on subsequent population growth. Biological control of plant pathogens may also rely on a hybrid of these two concepts. A large amount of inoculum may be applied, both to increase the population of the antagonist and to improve its distribution to favor biological control, and antagonism can result from both these applied organisms and the increased population of antagonists resulting from their reproduction. In some cases, augmentation of an antagonist will result in more than one season of control (Junqueira & Gasparotto, 1991; Cook, 1993). More generally, beneficial microorganisms are added seasonally or more frequently.

Root Diseases

The floras with demonstrated antagonistic properties under field conditions include fungi, principally *Trichoderma* spp., and, among bacteria, *Bacillus* spp. and *Pseudomonas* spp.

Among bacteria, species of *Bacillus* are regularly used for biological control of root diseases. Members of the genus have advantages, particularly because they form spores that permit simple storage and long shelf life, and they are relatively easy to inoculate into the soil. The consequence of this biology is, however, that although the inoculum may be present in the soil, it may be in a dormant or resting stage. Species of *Bacillus* have provided good control on some occasions. Capper and Campbell (1986) showed a doubling of wheat yield over wheat plants naturally infected with take-all disease by those also inoculated with *B. pumilus* Meyer & Gottheil. *Bacillus pumilus* and *B. subtilis* were also used to protect wheat from diseases caused by species of *Rhizoctonia* (Merriman *et al.,* 1974). A major difficulty with the use of *Bacillus* spp. is that the control provided is often very variable, with very different results in different locations, or even in different parts of a season in the same location (Campbell, 1989). *Bacillus subtilis* is used as a seed inoculant on cotton and peanuts (*Arachis hypogaea* Linnaeus). Treatment promotes increased root mass, nodulation, and early emergence; and suppression of diseases caused by species of *Rhizoctonia* and *Fusarium.*

Of substantially more promise as antagonists of root diseases are species of *Pseudomonas,* particularly the *P. fluorescens* Migula and *P. putida* (Trevisan) Migula groups (Campbell, 1989). These fluorescent bacteria are easy to grow in the laboratory, are normal inhabitants of the soil, and colonize and grow well when inoculated artificially. They produce a number of antibiotics as well as siderophores. Several have received patents and are marketed commercially for control of root rot in cotton (Campbell, 1989). One example is the use of isolates of a species of *Pseudomonas* as an antagonist of take-all disease of wheat (Weller, 1985). Isolates of *P. fluorescens* from soils showing control of take-all can be applied as seed coats and inoculated into fields suffering from the disease. Such treatments give 10 to 27% yield increases compared with untreated, infected control groups. Evidence points to both siderophore and antibiotic production as important.

Species of the fungal genus *Trichoderma* can be saprotrophic and mycoparasitic and have been used against wilt diseases of tomatoes, melons, cotton, wheat, and chrysanthemums. The antagonists were applied to seeds or incorporated into the planting mix at transplanting in a bran mixture. Although disease did develop, it did so much more slowly than it did in untreated soils, giving from 60 to 83% reduction in disease (Siven & Chet, 1986). The mode of action against *Verticillium albo-atrum* Reinke & Berthier wilt of tomatoes appeared to be antibiosis.

Mycoparasitic *Trichoderma* spp. have also been used successfully against diseases caused by *Rhizoctonia* and *Sclerotium* pathogens. *Sclerotium rolfsii* attacks many crop plants and survives unfavorable periods by forming sclerotia in the soil. Strains of *T. harzanium* that have β-1–3 glucanases, chitinases, and proteases have been isolated. These enzymes permit *T. harzanium* to parasitize the hyphae and sclerotia of the pathogen, invading and causing lysis of the cells. *Trichoderma harzanium* is grown on autoclaved bran or seed, and this material is then mixed with the surface soil (Chet & Henis, 1985). Two other fungi known to parasitize sclerotia are *Coniothyrium minitans*

Campbell and *Sporidesmium sclerotivorum* (Ayers & Adams, 1981).

Sporidesmium sclerotivorum is a hyphomycete that in nature behaves as an obligate parasite of sclerotia of *Botrytis cinerea* and several species of *Sclerotium* (Adams, 1990). This hyphomycete has been studied as an agent against botrytis rot in lettuce, where it shows considerable potential. It can be grown *in vitro* on various carbon sources, and is efficient in converting glucose into mycelium. Spores produced in mass culture are collected, processed, and applied to infected soil; and field tests are promising (Adams, 1990).

Plant-Parasitic Nematodes

Organisms most often considered for augmentation against plant parasitic nematodes are nematophagous or parasitic fungi and bacteria. The bacterial pathogen of nematodes most studied is *Pasteuria penetrans* sensu stricto Starr & Sayre (Starr & Sayre, 1988), which is an obligate parasite of root knot nematodes (*Meloidogyne* spp.) and has not been successfully cultured *in vitro*. This restriction in mass-culturing has limited attempts to test the bacterium's effectiveness (Stirling, 1991). In experimental trials this pathogen has demonstrated potential for controlling *Meloidogyne* spp. (Mankau, 1972; Stirling *et al.*, 1990; Fig. 3), infesting a high proportion of nematodes in soil to which bacterial spores had been added, and in other trials (USDA, 1978) reducing damage to plants in plots containing the bacterium. Observations by Mankau (1975) indicated that populations of the bacterium did not increase rapidly in field soil. The development of a mass production method in which roots containing large numbers of infected *Meloidogyne* spp. females were air-dried and finely ground to

produce an easily handled powder enabled more extensive testing (Stirling & Watchel, 1980). When dried root preparations laden with spores were incorporated into field soil at rates of 212 to 600 mg/kg of soil, the number of juvenile *Meloidogyne javanica* (Treub) Chitwood in the soil and the degree of galling was substantially reduced (Stirling, 1984); other authors have reported similar results (Stirling, 1991). Effective use of such a bacterium through such inundative release would require spore concentrations on the order of 10^5 spores per gram of soil (Stirling *et al.*, 1990). Such quantities could only be produced on a large scale with an efficient *in vitro* culturing method, a problem that has received attention but has not yet yielded a solution (Stirling, 1991). Use in inoculative releases, where smaller numbers of spores are applied, and subsequently a crop tolerant of nematode damage is grown to permit the increase of both nematode and bacterial populations has been suggested (Stirling, 1991). Conserving the bacterium in the presence of nematicides appears possible. Of seven tested nematicides only one showed slight toxicity to the bacterium (USDA, 1978).

The use of *Bacillus thuringiensis* strains with activity against nematodes is also possible. Because these bacteria may be cultured in fermentation media, their mass culture is much simpler than that of *P. penetrans*. Suppression of nematodes was possible through drench applications and through incorporating the bacterium into a methylcellulose seed coat (Zuckerman *et al.*, 1993).

Considerable attention has been given to the nematode-trapping fungi as possible augmentative agents. Mass culture on nutrient media is possible for these fungi. Two cultures of nematophagous *Arthrobotrys* fungi have been developed and tested for addition to soil for specific target environments. Cayrol *et al.* (1978) reported the successful

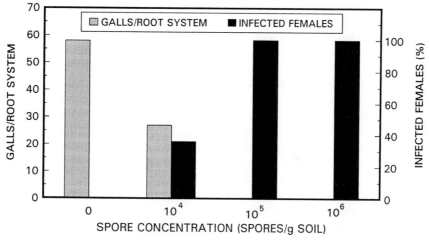

FIGURE 3   Number of galls caused by *Meloidogyne javanica* (Treub) Chitwood [and percentage of females infected with *Pasteuria penetrans* (Thorne) Sayre & Starr sensu stricto Starr & Sayre] after movement as juveniles through 4 cm of soil infested at various densities with *P. penetrans* spores. [After Stirling, G. R. *et al.*, (1990). *Nematologica, 36*, 246–252.]

use of *A. robusta* Cooke & Ellis var. *antipolis,* commercially formulated as Royal 300®, against the mycetophagous nematode *Ditylenchus myceliophagus* Goodey in commercial production of the mushroom *Agaricus bisporus* (Lange) Singer. The nematophagous fungus was seeded simultaneously with *A. bisporus* into mushroom compost, which led to 28% increases in harvest and reduced nematode populations by 40%. The results justified the commercial use of the fungus for nematode control in mushroom culture. Cayrol and Frankowski (1979) reported the use of *Arthrobotrys superba* Corda (Royal 350®) in tomato fields, applied to the soil at a rate of 140 g/m², resulting in protection of the tomatoes and colonization of the soil by the fungus.

Other reports have indicated little efficacy of fungal preparations when added alone to soil (Barron, 1977; Sayre, 1980; Rhoades, 1985). In general, there has been limited success in the use of these agents; see Stirling (1991) for a summary. The fungistatic nature of soil (Mankau, 1962; Cooke & Satchuthananthavale, 1968) may limit the ability of these fungi to grow even when added in substantial numbers to soil.

Many of the predaceous fungi may be unsuited for control of root knot nematodes, *Meloidogyne* spp. Stirling (1991) suggested that *Monacrosporium lysipagum* (Drechsler) Subramanian and *M. ellipsosporum* (Grove) Cooke & Dickinson, which can invade egg masses, may warrant further investigation. The nematode-trapping fungi are likely to be more effective against ectoparasitic nematodes and such species as *Tylenchus semipenetrans* Cobb, where juvenile stages migrate through the rhizosphere. Little attention has been given to testing predaceous fungi against such nematodes.

Fungi that are internal parasites of nematodes have been difficult to culture on nutrient media, and consequently there have been few attempts to use them for augmentative control of nematodes. Alternative mass-culturing techniques may hold some promise (Stirling, 1991). Lackey *et al.* (1993) report the production and formulation on alginate pellets [see also Fravel *et al.* (1985)] of *Hirsutella rhossiliensis* Minter & Brady that, when added to soil, led to transmission to the nematode *Heterodera schachtii* Schmidt and suppressed nematode invasion of roots.

Among the facultatively parasitic fungi that attack nematodes, *Paecilomyces lilacinus* and *V. chlamydosporium* have received the most attention as possible augmentative agents. The results of studies with *P. lilacinus* have been variable, with some studies showing some positive effect of the fungus, while others show little or no effect (Stirling, 1991). The mechanisms leading to the beneficial effect are not clearly elucidated, but may be from metabolic products or effects other than direct parasitism of eggs. Studies have generally involved the addition of fungal preparations to the soil at the rate of 1 to 20 ton/ha, which is likely too great for widespread commercial use. Additions at lower

rates (0.4 ton/ha) in a variety of carriers (alginate pellets, diatomaceous earth, wheat granules) also have resulted in limited beneficial effects (Cabanillas *et al.,* 1989; Stirling, 1991). Tribe (1980) suggested the direct addition of *V. chlamydosporium* to the soil. Kerry (1988) added hyphae and conidia, formulated in sodium alginate pellets or in wheat bran, to soil, and the fungus proliferated in the soil only from granules containing bran. When chlamydospores were used as inoculum, the fungus was able to establish without a food base (De Leij & Kerry, 1991). Of three isolates studied, only one successfully colonized tomato root surfaces. This species apparently has considerable promise, but screening programs will be necessary to identify isolates with characteristics suitable for biological control (Stirling, 1991).

Predaceous microarthropods and predaceous nematodes have evoked considerable interest. Most research has been done in simple microcosms, and there have been no attempts to evaluate augmentative release of these organisms in a field setting. In one experiment, Sharma (1971) found nematode numbers reduced by 50% or more in glass jars inoculated with mites and springtails compared with similar jars containing no predators, but the author pointed out possible causes for the reduction other than simple predation. Experiments with predaceous nematodes have in general failed to demonstrate a measurable impact of the predator (Stirling, 1991). One exception was the reduction of galling by *Meloidogyne incognita* on tomato by predaceous nematodes (Small, 1979). The general suitability of these groups of organisms for inundative release is questionable, because of the potential difficulties in developing technologies for their rearing, packaging, transport, and delivery beneath the soil in a viable state (Stirling, 1991).

Induced Resistance

Induced resistance is a form of disease control that uses the natural defense responses of the plant, which may include production of phytoalexins, additional lignification of cells, and other mechanisms (Horsfall & Cowling, 1980; Bailey, 1986), to defend the plant against pathogenic infection. These plant defenses are induced, or promoted, in the plant prior to exposure to the pathogen by challenging the plant with a nonpathogenic microbe. The plant defenses then limit later infection by the pathogen. The challenging organism employed may be an avirulent strain of the pathogen, a different *forma specialis,* or even a different species. Induced resistance and cross-protection are treated more fully in Chapter 19.

Root Diseases

There are few well-documented cases of induced resistance for soil-borne pathogens, and these are mostly of wilt diseases. Dipping tomato roots in a suspension of *F. oxy-*

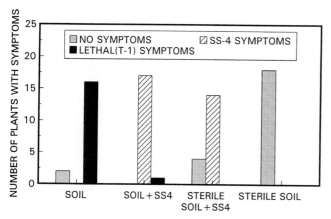

FIGURE 4 Induced resistance with a reduced virulence strain (SS-4) of *Verticillium albo-atrum* Reinke & Berthier in a field of cotton (*Gossipium hirsutum* Linnaeus) infested with a virulent strain (T-1) of the pathogen. [After Schnathorst, W. C., & Mathre, D. E. (1966). *Phytopathology, 56,* 1204–1209.]

sporum f. sp. *dianthi* a few days before likely exposure to the pathogen *F. oxysporum* f. sp. *lycopersici* (Saccardo) Snyder & Hansen conferred protection that lasts a few weeks. Cotton may be protected for 3 months or longer by spraying the roots at transplanting with a mildly pathogenic strain of the disease-causing pathogen *V. albo-atrum* (Schnathorst & Mathre, 1966) (Fig. 4).

The role of some fungi against take-all disease of wheat includes some elements of induced resistance. *Gaeumannomyces graminis* (Saccardo) Arx & Olivier var. *graminis*

grows on grass roots and also has been found on wheat, where it occupies a niche similar to that of the pathogen *G. graminis* var. *tritici* Walker. The antagonist invades the root cortex but not the stele, and is halted by the lignification and suberization of the cortex and stele. Root cells with these chemically changed walls are less susceptible to invasion by the pathogen. Although this interaction produced yield increases in Europe, the strains or species present in the United States did not appear to confer resistance and in Australia there were only slight yield increases (Campbell, 1989). These variable results, while somewhat common for biological control of soil borne pathogens, do not reduce the value of the antagonists where they do work, but instead serve to outline some of the potential challenges in defining the taxonomy, biology, and host-plant relations in this group of organisms.

DEVELOPING AND USING BENEFICIAL SPECIES

Since the first experimental reports of biological control of plant pathogens (Hartley, 1921; Sanford, 1926; Millard & Taylor, 1927; Henry, 1931), substantial potential for microbial control of pathogens has been demonstrated. A number of products or programs have reached the stage of commercial development or availability (Table 2). Products in current use include both those aimed at specialty markets such as control of nematodes on mushroom production and at large-scale markets such as seed treatments for cotton and soybeans.

TABLE 2 Some Commercially Available Antagonists or Products for Soil Plant Pathogens and Plant-Parasitic Nematodes

Organism	Trade name	Target
Targeted against Pathogens		
Bacillus subtilis (Ehrenberg) Cohn	Kodiak, Epic	Seed treatment against *Rhizoctonia* spp., *Pythium* spp., and *Fusarium* spp. root diseases
Pseudomonas cepacia (Burkholder) Palleroni and Holmes	Blue Circle; Intercept	*Rhizoctonia* spp. *Pythium* spp., and *Fusarium* spp. diseases of seedlings
P. fluorescens Migula	Dagger G	*Rhizoctonia* spp. and *Pythium* spp. diseases of seedlings
Fusarium oxysporum Schlechtendal (nonpathogenic strains)	Fusaclean; Biofox C	Diseases from pathogenic strains of *F. oxysporum*
Gliocladium virens Millers, Giddens & Foster	GlioGard	*Rhizoctonia* spp. and *Pythium* spp. diseases of seedings and bedding plants
Mycorrhyzae	Vaminoc	*Botrytis* spp. and *Pythium* sp. diseases
Streptomyces griseovirides Anderson *et al.*	Mycostop	*Alternaria* sp. and *Fusarium* spp. diseases
Trichoderma harzanium/polysporum (Link) Rifai	BINAB	Wood-rot fungi; *Verticillium malthousei* Ware in mushrooms
T. harzianum Rifai	F-Stop; Trichodex; Supravit; T-35; TY	*Heterobasidion annosum;* diseases caused by *Rhizoctonia* spp., *Pythium* spp., *Fusarium* spp. *B. cinerea* Persoon: Fries, and *Sclerotium rolfsii* Saccardo
T. lignorum (Tode) Harz	Trichodermin-3	*Rhizoctonia* spp. and *Fusarium* spp. diseases
Targeted against Nematodes		
Arthrobotrys robusta Cooke & Ellis var. *antipolis*	Royal 300®	*Ditylenchus myceliophagus* Goodey
A. superba Corda	Royal 350®	*Meloidogyne* spp.
Chitin-based amendment	Clandosan®	Plant-parasitic nematodes

Research and development of antagonists of plant pathogens is composed of several steps, as discussed in Chapter 32. These include initial discovery of candidate agents; refinement of our knowledge of their biology, ecology, and mode of action; microcosm and field trials of their efficacy; and large-scale development for commercial production.

The adoption of any biological control agent in commercial agriculture is dependent on its reliability and its availability. Limitations to development of any particular agent include cost of development and size of potential market. Many pesticides for control of plant diseases have a broad spectrum of activity; are applicable in a variety of crops and settings; and may act either prophylactically, therapeutically, or both. Biological controls, in contrast, may have narrow ranges of activity and may work in only a few crops or soil types. While they can often act both prophylactically and therapeutically, their action may take some time to develop.

Microorganisms intended for use as biological control agents must be viewed in a biological instead of a chemical paradigm (Cook, 1993). Although an effective pesticide may work in many places, each place may have unique soil, edaphic, and biological features that limit or enhance the effectiveness of microbial antagonists of pathogens. Consequently, each microbial biological control system may have to make use of locally adapted strains, taking advantage of resident antagonistic floras and faunas and augmenting their effectiveness with additional species or strains, or enhancing resident populations through soil amendments. Although the different strains may use common mechanisms to achieve biological control, competitive abilities adapted to local conditions may be vital to permit the organisms to compete for resources and effectively control pathogens. Hence, antagonists to plant pathogens may have a much narrower market than a chemical pesticide has, and be unattractive for development by major corporations (Andrews, 1992). In this context, it may be appropriate for public institutions such as government experiment stations to undertake the development of such biological controls, in the same way that they take the responsibility for development of new plant varieties (Cook, 1993).

References

Adams, P. B. (1990). The potential of mycoparasites for biological control of plant diseases. Annual Review of Phytopathology, 28, 59–72.

Alabouvette, C., Rouxel, F., & Louvet, J. (1979). Characteristics of Fusarium wilt-suppressive soils and proposals for their utilization in biological control. In B. Schippers & W. Gams (eds.), Soil-borne plant pathogens (pp. 165–182). London: Academic Press.

Anderson, J. R. (1978). Pesticide effects on non-target soil microorganisms. In I. R. Hill & S. J. L. Wright (Eds.), Pesticide microbiology (pp. 313–533). London: Academic Press.

Andrews, J. H. (1992). Biological control in the phyllosphere. Annual Review of Phytopathology, 30, 603–635.

Andrews, J. H., & Harris, R. F. (1986): r- and K- selection and microbial ecology. Advances in Microbial Ecology, 9, 99–147.

Ayers, W. A., & Adams, P. B. (1981). Mycoparasitism and its application to biological control of plant diseases. In G. C. Papavizas (Ed.). Biological control in crop production. Beltesville Symposium in Agricultural Research 5 (pp. 91–105) Granada: Allenheld, Osmum Publishing.

Bailey, J. A. (1986). Biology and molecular biology of plant-pathogen interactions. Berlin: Springer-Verlag.

Barron, G. L. (1977). The nematode-destroying fungi. Topics in mycobiology Vol. I Guelph, Ontario, Canada: Canadian Biological Publications.

Begon, M., Harper, J. L., & Townsend, C. R. (1986). Ecology: Individuals, populations and communities. Oxford, United Kingdom: Blackwell Scientific Publications.

Cabanillas, E., Barker, K. R., & Nelson, L. A. (1989). Survival of Paecilomyces lilacinus in selected carriers and related effects on Meloidogyne incognita on tomato. Journal of Nematology, 21, 164–172.

Campbell, R. (1985). Plant microbiology. London: Edward Arnold.

Campbell, R. (1989). Biological control of microbial plant pathogens. Cambridge, United Kingdom: Cambridge University Press.

Canto-Saenz, M., & Kaltenbach, R. (1984). Effect of some fungicides on Paecilomyces lilacinus (Thom) Samson. Proceedings of the First International Congress of Nematology, Guelph, Canada, 1, 13–14.

Capper, A. L., & Campbell, R. (1986). The effect of artificially inoculated antagonist bacteria on the prevalence of take-all disease of wheat in field experiments. Journal of Applied Bacteriology, 60, 155–160.

Cayrol, J. C. (1983). Lutte biologique contre les Meloidogyne au moyen d'Arthrobotrys irregularis. Revenue de Nématologie, 6, 265–73.

Cayrol, J. C., & Frankowski, J. P. (1979). Une methode de lutte biologique contre les nematodes a galles des racines appartenant au genre Meloidogyne. Revue Horticole, 193, 15–23.

Cayrol, J. C., Frankowski, J. P., Laniece, A., d'Hardemare, G., & Talon, J. P. (1978). Contre les nematodes en champigonniere. Mise au point d'une methode de lutte bologique a l'aide d'un hyphomycete predateur: Arthobotrys robusta souche antipolis (Royal 300). Revue Horticole, 184, 23–30.

Chet, I., & Henis, Y. (1985). Trichoderma as a biocontrol agent against soilborne root pathogens. In C. A. Parker, et al. (Eds.) Ecology and management of soilborne plant pathogens (pp. 110–112) St. Paul, M N: American Phytopathological Society.

Cook, R. J. (1993). Making greater use of introduced microorganisms for biological control of plant pathogens. Annual Review of Phytopathology, 31, 53–80.

Cook, R. J., & Baker, K. F. (1983). The nature and practice of biological control of plant pathogens. St. Paul, MN: American Phytopathological Society.

Cooke, R. C. (1968). Relationships between nematode-destroying fungi and soil-borne phytonematodes. Phytopathology, 58, 909–913.

Cooke, R. C., & Satchuthananthavale, V. E. (1968). Sensitivity to mycostasis of nematode-trapping Hyphomycetes. Transactions of the British Mycological Society, 51, 555–561.

De Leij, F., & Kerry, B. R. (1991). The nematophagous fungus, Verticillium chlamydosporium, as a potential biological control agent for Meloidogyne arenaria. Revue de Nématologie, 14, 157–164.

Dobbs, C. G., & Hinson, W. H. (1953). A widespread fungistasis in soil. Nature (London), 172, 197–199.

Duddington, C. L. Everard, C. O. R., & Duthoit, C. M. G. (1961). Effect of green manuring and a predaceous fungus on cereal root eelworm on oats. Plant Pathology, 10, 108–109.

Duddington, C. L., Jones, F. G. W., & Moriarty, F. (1956a). The effect of predaceous fungus and organic matter upon the soil population of beet eelworm Heterodera schachtii Schm. Nematologica, 1, 344–348.

711

Duddington, C. L., Jones, F. G. W., & Williams, T. D. (1956b). An experiment on the effect of a predaceous fungus upon the soil population of potato root eelworm, *Heterodera rostochiensis* Woll. Nematologica, 1, 341–343.

Dunn, M. T. (1983). *Paecilomyces nostocides,* a new hyphomycete isolated from cysts of *Heterodera zeae.* Mycologia, 75, 179–182.

Elad, Y. (1986). Mechanisms of interactions between rhizosphere microorganisms and soil-borne plant pathogens. In V. Jensen, A. Kjoller, & L. H. Sorensen (Eds.) Microbial communities in soil (pp. 49–60). New York: Elsevier.

Ferris, V. R., & Ferris, J. M. (1989). Why ecologists need systematists: Importance of systematics to ecological research. Journal of Nematology, 21, 308–314.

Fravel, D. R., Marios, J. J., Lumsden, R. D. & Connick, W. J., Jr. (1985). Encapsulation of potential biocontrol agents in an alginate-clay matrix. Phytopathology, 75, 774–777.

Gair, R., Mathias, P. L., & Harvey, P. N. (1969). Studies of cereal nematode populations and cereal yields under continuous or intensive culture. Annals of Applied Biology, 63, 503–512.

Galper, S., Cohn, E., Spiegel, Y., & Chet, I. (1991). A collagenolytic fungus, *Cunninghamella elegans,* for biological control of plant parasitic nematodes. Journal of Nematology, 23, 269–274.

Giuma, A. Y., Hackett, A. M., & Cooke, R. C. (1973). Thermostable nematoxins produced by germinating conidia of some endozoic fungi. Transactions of the British Mycological Society, 60, 49–56.

Godoy, G., Rodriguez-Kabana, R., & Morgan-Jones, G. (1983). Fungal parasites of *Meloidogyne arenaria* in infested soils. II. Effects on microbial populations. Nematropica, 13, 201–213.

Gray, N. F. (1987). Nematophagous fungi with particular reference to their ecology. Biological Reviews, 62, 245–304.

Gray, N. F. (1988). Fungi attacking vermiform nematodes. In G. O. Poinar & H.-B. Jansson (Eds.). Diseases of nematodes Vol 2, (pp. 3–38). Boca Raton, FL: CRC Press.

Hartley, C. (1921). Damping off in forest nurseries. USDA Bulletin, 934, 1–99.

Henry, A. W. (1931). The natural microflora of the soil in relation to the foot rot problem of wheat. Canadian Journal of Research, 4, 69–77.

Hornby, D. (Ed.) (1990). Biological control of soil-borne plant pathogens. Wallingford, United Kingdom: CAB International.

Horsfall, J. G., & Cowling, E. B. (Ed.). (1980). Plant disease: An advanced treatise: Vol. 2. How disease develops in populations. New York: Academic Press.

Ignoffo, C. M., & Dropkin, V. H. (1977). Deleterious effects of the thermostable toxin of *Bacillus thuringiensis* on species of soil-inhabiting, myceliophagous, and plant-parasitic nematodes. Journal of the Kansas Entomological Society, 50, 394–398.

Jaffee, B. A., & McInnis, T. M. (1990). Effects of carbendazim on the nematophagous fungus *Hirsutella rhossiliensis* and the ring nematode. Journal of Nematology, 22, 418–419.

Jansson, H.-B., & Nordbring-Hertz, B. (1980). Interactions between nematophagous fungi and plant parasitic nematodes: Attraction, induction of trap formation and capture. Nematologica, 26, 383–389.

Jatala, P., Kaltenbach, R., & Bocangel, M. (1979). Biological control of *Meloidogyne incognita acrita* and *Globodera pallida* on potatoes. Journal of Nematology, 11, 303.

Junqueira, N. T. V., & Gasparotto, L. (1991). Controle biológico de fungos estromáticos causadores de doenças foliares em Seringueira. In Anonymous. Controle Biológico de Doenças de Plantas (pp. 307–331). Org. W. Bettiol., Brasilia, DF: EMBRAPA.

Kerry, B. R. (1988). Two microorganisms for the biological control of plant parasitic nematodes. Proceedings of the Brighton Crop Protection Conference, 2, 603–607.

Kerry, B. R., Crump, D. H., & Mullen, L. A. (1980). Parasitic fungi, soil moisture and multiplication of the cereal cyst nematode, *Heterodera avenae.* Nematologica, 26, 57–68.

Kerry, B. R., Crump, D. H., & Mullen, L. A. (1982a). Studies of the cereal cyst nematode, *Heterodera avenae* under continuous cereals, 1975–1978. II. Fungal parasitism of nematode eggs and females. Annals of Applied Biology, 100, 489–499.

Kerry, B. R., Crump, D. H., & Mullen, L. A. (1982b). Natural control of the cereal cyst nematode, *Heterodera avenae* Woll., by soil fungi at three sites. Crop Protection, 1, 99–109.

Lackey, B. A., Muldoon, A. E., & Jaffee, B. A. (1993). Alginate pellet formulation of *Hirsutella rhossiliensis* for biological control of plant-parasitic nematodes. Biological Control, 3, 155–160.

Larkin, R. P., Hopkins, D. L., & Martin, F. N. (1996). Suppression of fusarium wilt of watermelon by nonpathogenic *Fusarium oxysporum* and other microorganisms recovered from a disease-suppressive soil. Phytopathology, 86, 812–819.

Loewenberg, J. R., Sullivan, T., & Schuster, M. L. (1959). A viral disease of *Meloidogyne incognita,* the southern root-knot nematode. Nature (London), 184, 1896.

Lynch, J. M. (1987). Biological control within microbial communities of the rhizosphere. In M. Fletcher, T. R. G. Gray, & J. G. Jones (Eds.) Ecology of microbial communities (pp. 55–82). Society of General Microbiology, Symposium 41. Cambridge, United Kingdom: Cambridge University Press.

Malajczuk, N. (1979). Biological suppression of *Phytophthora cinnamomi* in eucalyptus and avocado in Australia. In B. Schippers & W. Gams (Eds.). Soil-Borne Plant Pathogens (pp. 635–652). London: Academic Press.

Mankau, R. (1962). Soil fungistasis and nematophagous fungi. Phytopathology, 52, 611–615.

Mankau, R. (1968). Effects of nematicides on nematode-trapping fungi associated with the citrus nematode. Plant Disease Reporter, 52, 851–855.

Mankau, R. (1972). Utilization of parasites and predators in nematode pest management ecology. Proceedings of the Tall Timbers Conference on Ecology, Animal Control, and Habitat Management, 4, 129–143.

Mankau, R. (1975). *Bacillus penetrans* n. comb. causing a virulent disease of plant-parasitic nematodes. Journal of Invertebrate Pathology, 26, 333–339.

Merriman, R. R., Price, R. D., & Baker, K. F. (1974). The effect of inoculation of seed with antagonists of *Rhizoconia solani* on the growth of wheat. Australian Journal of Agricultural Research, 25, 213–218.

Millard, W. A., & Taylor, C. B. (1927). Antagonism of microorganisms as the controlling factor in the inhibition of scab by green manuring. Annals of Applied Biology, 14, 202–216.

Moore, J. C., Walter, D. E., & Hunt, H. W. (1988). Arthropod regulation of micro- and mesobiota in below-ground detrital food webs. Annual Review of Entomology, 33, 419–439.

Nordbring-Hertz, B. (1973). Nematode-induced morphogenesis in the predacious fungus *Arthrobotrys oligospora.* Physiologica Plantarum, 29, 223–233.

Olthof, T. H. A., & Estey, R. H. (1966). Carbon and nitrogen levels of a medium in relation to growth and nematophagous activity of *Arthrobotrys oligospora* Fres. Nature (London), 209, 1158.

Parker, C. A., Rovira, A. D., Moore, K. J., Wong, P. T. W., & Kollmorgen, J. F. (Eds.). (1985). *Ecology and management of soilborne plant pathogens.* St. Paul, MN: American Phytopathological Society.

Prasad, S. S. S. V., Tilak, K. V. B. R., & Gollakote, K. G. (1972). Role of *Bacillus thuringiensis* var. *thuringiensis* on the larval survivability and egg hatching of *Meloidogyne* spp., the causative agent of root knot disease. Journal of Invertebrate Pathology, 20, 377–378.

Rhoades, H. L. (1985). Comparison of fenamiphos and *Arthrobotrys amerospora* for controlling plant nematodes in central Florida. Nematropica, 15, 1–7.

Roget, D. K., & Rovira, A. D. (1987). A review of the effect of tillage on the cereal cyst nematode. Wheat Research Council of Australia, Workshop Report Series No. 1, 31–35.

Sanford, G. B. (1926). Some factors affecting the pathogenicity of *Actinomyces scabies.* Phytopathology, 16, 525–547.

Saxena, G., & Mukerji, K. G. (1988). Biological control of nematodes. In K. G. Mukerji & K. L. Garg (Eds.). *Biocontrol of plant diseases* (Vol. 1, pp. 113–127). Boca Raton, FL: CRC Press.

Sayre, R. M. (1980). Promising organisms for biocontrol of nematodes. Plant Disease, 64, 526–532.

Sayre, R. M., & Walter, D. E. (1991). Factors affecting the efficacy of natural enemies of nematodes. Annual Review of Phytopathology, 29, 149–166.

Sayre, R. M., & Starr, M. P. (1988). Bacterial diseases and antagonisms of nematodes. In G. O. Poinar & H.-B. Jansson (Eds.). *Diseases of nematodes* (Vol. 1, pp. 69–101). Boca Raton, FL: CRC Press.

Sayre, R. M., & Wergin, W. P. (1977). Bacterial parasite of a plant nematode: Morphology and ultrastructure. Journal of Bacteriology, 129, 1091–1101.

Schnathorst, W. C., & Mathre, D. E. (1966). Cross-protection in cotton with strains of *Verticillium albo-atrum.* Phytopathology, 56, 1204–1209.

Sharma, R. D. (1971). Studies on the plant parasitic nematode *Tylenchorhynchus dubius.* Mededelingen Landbouwwetenschappen Wageningen, 71, 1–154.

Sitaramaiah, K., & Singh, R. S. (1974). The possible effects on *Meloidogyne javanica* of phenolic compounds produced in amended soil. Journal of Nematology, 6, 152.

Siven, A., & Chet, I. (1986). *Trichoderma harzianum:* An effective biocontrol agent of *Fusarium* spp. In V. Jensen, *et al.* (Eds.). Microbial communities in soil (pp. 89–95). London: Elsevier.

Small, R. W. (1979). The effects of predatory nematodes on populations of plant parasitic nematodes in pots. Nematologica, 25, 94–103.

Speigel, Y., Cohn, E., & Chet, I. (1989). Use of chitin for controlling *Heterodera avenae* and *Tylenchulus semipenetrans.* Journal of Nematology, 21, 419–422.

Speigel, Y., Chet, I., Cohn, E., Galper, S., & Sharon, E (1988). Use of chitin for controlling plant parasitic nematodes. III. Influence of temperature on nematicidal effect, mineralization and microbial population buildup. Plant and Soil, 109, 251–256.

Starr, M. P., & Sayre, R. M. (1988). *Pasteuria thornei* sp. nov. and *Pasteuria penetrans* sensu stricto emend., mycelial and endospore-forming bacteria parasitic, respectively, on plant parasitic nematodes of the genera *Pratylenchus* and *Meloidogyne.* Annales de l'Institute Pasteur/Microbiology, 139, 11–31.

Stirling, G. R. (1984). Biological control of *Meloidogyne javanica* with *Bacillus penetrans.* Phytopathology, 74, 55–60.

Stirling, G. R. (1991). Biological control of plant parasitic nematodes: Progress, problems and prospects. Wallingford, United Kingdom: CAB International.

Stirling, G. R., & Watchel, M. F. (1980). Mass production of *Bacillus penetrans* for the biological control of root-knot nematodes. Nematologica, 26, 308–312.

Stirling, G. R., McKenry, M. V., & Mankau, R. (1979). Biological control of root-knot nematodes (*Meloidogyne* spp.) on peach. Phytopathology, 69, 806–809.

Stirling, G. R., Sharma, R. D., & Pery, J. (1990). Attachment of *Pasteuria penetrans* spores to *Meloidogyne javanica* and its effects on infectivity of the nematode. Nematologica, 36, 246–252.

Sutton, J. C., & Peng, G. (1993a). Manipulation and vectoring of biocontrol organisms to manage foliage and fruit diseases in cropping systems. Annual Review of Phytopathology, 31, 473–493.

Tjamos, E. C., Papavizas, G. C., & Cook, R. J. (Eds.). (1992). Biological control of plant disease. New York: Plenum Press.

Tribe, H. T. (1980). Prospects for the biological control of plant parasitic nematodes. Parasitology, 81, 619–639.

Tzortzakakis, E. A., Channer, A. G. D. R., Gowen, S. R., & Ahmed, R. (1997). Studies on the potential use of *Pasturia penetrans* as a biocontrol agent of root-knot nematodes (*Meloidogyne* spp.) Plant Pathology, 46, 44–55.

U.S. Department of Agriculture (USDA). (1978). Biological agents for pest control, status and prospectus. (Stock no. 001-000-03756-1). Washington, DC: U.S. Government Printing Office.

Walter, D. E., Hunt, H. W., & Elliot, E. T. (1988). Guilds or functional groups? An analysis of predatory arthropods from a shortgrass steppe soil. Pedobiologia, 31, 247–260.

Weller, D. M. (1985). Application of fluorescent pseudomonads to control root diseases. In C. A., Parker, *et al.* (Eds.). Ecology and management of soilborne plant pathogens (pp. 137–140). St. Paul, MN: American Phytopathological Society.

Zuckerman, B. M., Dicklow, M. B., & Acosta, N. (1993). A strain of *Bacillus thuringiensis* for the control of plant-parasitic nematodes. Biocontrol Science and Technology, 3, 41–46.

Biological Control in Subtropical and Tropical Crops

C. E. KENNETT

Division of Biological Control
University of California
Albany, California

J. A. MCMURTRY

Department of Entomology
University of California
Riverside, California

J. W. BEARDSLEY

Department of Entomology
University of Hawaii at Manoa
Honolulu, Hawaii

INTRODUCTION

A comprehensive review of biological control efforts against all tropical and subtropical perennial crops would require an entire volume and is obviously beyond the scope of this chapter. What is presented here reflects, to a considerable extent, the research interests and experience of the authors. We have attempted to review biological control efforts against pests of seven commercially important crops, particularly those in which there has been substantial biological control activity during the past 10 to 15 years: citrus, olive, avocado, tea, coffee, coconut, and banana. Citrus, because of its commercial importance worldwide, as well as the long history of biological control efforts against citrus pests, is treated in more detail than are other crops.

CITRUS

Worldwide, some 70 species of insects are considered to be major pests of citrus (Ebeling, 1959). Of these, more than one-third belong to the Homoptera order. Among others, this includes armored scales (Diaspididae), soft scales (Coccidae), fluted scales (Margarodidae), mealybugs (Pseudococcidae), and whiteflies (Aleyrodidae). The most successful biological control projects in citrus have been conducted with examples of these families.

In 1888, the vedalia lady beetle, *Rodolia* (= *Novius*) *cardinalis* (Mulsant), was introduced into California from Australia for control of the cottony-cushion scale, *Icerya purchasi* Maskell, an introduced species that was devastating the citrus industry in southern California (Doutt, 1958).

This undertaking, which resulted in the complete subjugation of the scale, was the first outstanding success in the field of classical biological control. This success generated much interest in the biological method of pest control and, as a consequence, similar efforts against other citrus pests in southern California were soon undertaken. Although few of the many biological control projects in citrus during the succeeding 100 years have been as spectacularly successful as that with the cottony-cushion scale, the number of successful projects far exceeds that on any other crop. Much of the methodology and theory of biological control has been derived from the work on citrus.

Armored Scales: Diaspididae

California Red Scale, *Aonidiella aurantii* (Maskell)

According to Ebeling (1959), California red scale is probably the most important citrus pest in the world. It is less widely distributed than the purple scale, but causes greater damage and is more difficult to control. Of oriental origin, it is the most important pest of citrus in California, South Africa, Australia, New Zealand, and northwestern Mexico, and is a major pest in the eastern Mediterranean Basin, North Africa, and parts of South America (Quayle, 1938).

Between 1889 and 1947, a great variety of exotic natural enemies was introduced into California for control of the red scale. The earliest efforts emphasized introduction of coccinellid predators, of which more than 40 species were imported between 1889 and 1892 (Compere, 1961). Failure of nearly all these to become established resulted in a greater emphasis on parasitoid introductions. Between 1900 and 1908, several species including *Aspidiotiphagus citrinus* (Crawford), *Comperiella bifasciata* Howard, and *Pteroptrix* (= *Casca*) *chinensis* (Howard) were imported from southern China. None of these, however, became established (Compere, 1961; Flanders *et al.*, 1958). During and prior to these efforts, an aphelinid parasitoid of unknown origin had been found to parasitize the red scale in southern California. This species, later identified as *Aphytis chrysomphali* (Mercet), was extensively propagated between 1902 and 1904 and was distributed to orchardists on request, but the early optimism for control of red scale with this parasitoid was not realized (Compere, 1961). Thus, *Aphytis* material encountered during foreign explorations either was disregarded or, if shipped to California, was identified as *A. chrysomphali.*

Further introductions of *C. bifasciata* (all from Japan) were made during 1916 and 1917, 1922, and 1924 (Compere, 1961). In California, the parasitoid failed to reproduce on red scale in the laboratory but was successfully propagated on *Chrysomphalus bifasciculatus* Ferris in 1924. It again failed to become established on red scale, but was later found to reproduce on the yellow scale, *Aonidiella citrina* (Coquillett), and was established on this scale in 1931 (Smith, 1942). It was later determined that the host scales in Japan were *A. taxus* Leonardi and *C. bifasciculatus*, not *A. aurantii* and *C. aonidum* (Linnaeus) as believed.

Between 1918 and 1934, various coccinellid predators were imported from South Africa, China, Australia, and South America but none became established (Compere, 1961). In 1924, an *Aphytis sp.* (presumably *A. lingnanensis* DeBach but misidentified as *A. chrysomphali*) was imported from southern China but did not become established. In 1937, *Habrolepis rouxi* Compere, an encyrtid internal parasitoid of red scale, was imported from South Africa. Widely colonized in southern California, *H. rouxi* became established only in a very limited area (Flanders, 1944a). Though of little value, this species holds the distinction of being the first intentionally introduced parasitoid to become established on red scale in California. In 1940, *Comperiella bifasciata* was again imported, this time on red scale collected in southern China (Smith, 1942), and for the first time was successfully propagated on it in California (Flanders, 1943a). Although establishment remained questionable for several years, *C. bifasciata* eventually spread throughout all but the coastal areas of southern California (DeBach, 1948, DeBach *et al.*, 1955). Studies by Flanders (1944b) and Teran and DeBach (1963) showed that the *C. bifasciata* from *Aonidiella aurantii* in China and that from *A. taxus* and *Chrysomphalus bifasciculatus* in Japan were distinct biological races, each being limited to its preferred host in the field.

Between 1947 and 1949, many shipments of parasitized red scale were sent to California from southern China and Taiwan. From this material only *Aphytis lingnanensis* from

southern China and *Encarsia* (= *Prospaltella*) *perniciosi* (Tower) (a red scale race) from Taiwan became established (DeBach *et al.*, 1950; DeBach 1953; Rosen & DeBach, 1978). *Encarsia perniciosi* was relatively common at first, but its eventual range was limited to the coastal areas whereas *A. lingnanensis* quickly gained dominance and by 1958 had displaced the long-established *A. chrysomphali* except at a few coastal sites (DeBach & Sundby, 1963). Although *A. lingnanensis* alone gave good control of red scale in coastal areas, it was ineffective in interior areas.

Further exploration for new parasitoids during 1956 and 1957 resulted in the importation of several new species (DeBach, 1959), but of these only *A. melinus* DeBach from India and Pakistan became established. The other *Aphytis* species, *A. fisheri* DeBach and *A. proclia* (Walker), as well as *Coccobius* (= *Physcus*) *debachi* (Compere & Annecke) and *Aspidiotiphagus citrinus*, failed to become established. Subsequent importations made between 1960 and 1964 included *A. coheni* DeBach, *A. holoxanthus* DeBach (*ex* Israel, 1960), and *A. africanus* Quednau (*ex* South Africa, 1961), all of which failed to become established (Rosen & DeBach, 1978).

Following colonization in 1958, *Aphytis melinus* quickly displaced *A. lingnanensis* at the original release sites and by 1964 had become dominant on red scale in southern California and had further displaced *A. lingnanensis* although the latter species remained common or even dominant in some coastal areas (DeBach, 1966, 1969). The general decline in red scale abundance in southern California after 1962 was attributed to *A. melinus* and *A. lingnanensis*, with the former being complemented by *Comperiella bifasciata* in the interior and intermediate areas and the latter, by *E. perniciosi* in coastal areas (DeBach, 1965a, 1969). Although substantial to complete control has been attained in many districts, only partial control is provided in others (Rosen & DeBach, 1978). DeBach (1962, 1965b) attributed this variability to the different climatic conditions that occur within the citrus belts of southern California. Kfir and Luck (1979) concluded that the negative effects of extreme temperatures on sex ratios and progeny production in *A. melinus* and *A. lingnanensis* not only contribute to their ineffectiveness in the more extreme climates of southern California but also may partially explain the displacement of *A. lingnanensis* by *A. melinus.*

In the more climatically extreme citrus districts of central and northern California, where only *A. melinus* and *C. bifasciata* have been established, biological control of red scale is, at best, only partially effective, even though both parasitoids have been present for more than 30 years. *Aphytis riyahdi* DeBach, introduced from Saudi Arabia (DeBach, 1977), was colonized in central California, but failed to become established.

In Australia, attempts to develop biological control of red scale have closely paralleled those in California. Beginning about 1902, various indeterminate natural enemies were imported but only one parasitoid, thought to be *A. chrysomphali,* was reported as established in Western Australia (Wilson, 1960). By the 1940s, only one additional species, *Comperiella bifasciata,* had been established. Subsequent importation of *A. melinus* (1961) and *E. perniciosi* (1970) resulted in establishment, but introduction of *A. lingnanensis, A. coheni, A. riyahdi,* and *Habrolepis* sp., sometime after 1977, was unsuccessful (Furness *et al.,* 1983). In the Lower Murray Valley (Victoria and South Australia), the long-established *A. chrysomphali* was found to have been displaced by *A. melinus* about 1972 to 1974 (Furness *et al.,* 1983). Campbell (1976) documented the decline in red scale abundance following colonization of *A. melinus* and Furness *et al.* (1983) credited *A. melinus* and *C. bifasciata* with substantial to complete control of the scale. In Queensland, fully effective control was attributed to *A. lingnanensis* and *C. brifasciata* (Smith, 1978a). While *C. bifasciata* had been introduced into Queensland in 1948, the presence of *A. lingnanensis* apparently was due to natural ecesis, possibly from Victoria, where it had been introduced earlier.

Since 1960, introduction of one or more red scale parasitoids has also been undertaken in France, Cyprus, Greece, Sicily, Israel, Morocco, and South Africa. In Mediterranean Basin countries where *A. melinus* has been established, the parasitoid has, as in Australia, displaced *A. chrysomphali* within a short period of time (DeBach & Argyriou, 1967; Rosen, 1967; Orphanides, 1984). However, displacement of native or introduced congeners by *A. melinus* has not always been the rule. Although *A. melinus* displaced *A. chrysomphali* in Israel, it did not displace the earlier, accidentally introduced *A. coheni* (Rosen, 1967) whereas in South Africa the native *A. africanus* has retained its dominant position despite establishment of *A. melinus, A. lingnanensis,* and *A. coheni* (Bedford, 1968; Annecke & Moran, 1982). In addition to California, Australia, and South Africa, partial to complete control has been reported in Chile (Gonzalez, 1969), Cyprus (Orphanides, 1984), Greece (DeBach & Argyriou, 1967), France (Benassy & Bianchi, 1974), and Argentina (Crouzel *et al.,* 1973). In most cases the improved level of control was due to *A. melinus* but in France *C. bifasciata* alone was responsible.

Florida Red Scale, *Chrysomphalus aonidum* (Linnaeus)

A polyphagous species of oriental origin, Florida red scale is scattered throughout the tropical and subtropical regions of North, Central, and South America; the Mediterranean Basin; South Africa; Australia; Asia; and the Pacific Islands. It has been reported as particularly damaging to citrus in Israel, Egypt, Lebanon, Brazil, Mexico, Texas, and Florida (Quayle, 1938; Bodenheimer, 1951; Ebeling, 1959).

The first attempt to introduce exotic natural enemies of

C. aonidum took place in Israel from 1945 to 1947 when *Comperiella bifasciata* was imported from California. This resulted in failure, but renewed attempts in 1956 and 1957 resulted in the importation of *Aphytis holoxanthus* and *Pteroptrix* (= *Casca*) *smithi* (Compere) from Hong Kong (Rivnay, 1968). Direct colonization of these parasitoids (fewer than 150 *A. holoxanthus* and ca. 120 *P. smithi*) resulted in establishment although *P. smithi* was not recovered in the field until 1960. The rapid natural dispersal of *A. holoxanthus,* aided by human transfer of parasitized scales, resulted in complete control of *Chrysomphalus aonidum* throughout the coastal plain citrus districts within 2 to 3 years (Rivnay, 1968; DeBach *et al.,* 1971). *Pteroptrix smithi* was later found to have dispersed along the coastal plain and while its activity was judged complementary to that of *A. holoxanthus,* studies by Steinberg *et al.* (1986) indicate it is now the dominant species. According to Harpaz and Rosen (1971), biological control of what was previously the most serious pest of citrus in Israel permitted the development of an effective integrated control program in citrus.

Further introductions and establishment of *A. holoxanthus* in Mexico, Florida, South Africa, Australia, and Texas have all met with complete subjugation of *C. aonidum* (Maltby *et al.,* 1968; Selhime *et al.,* 1969; Cilliers, 1971; Smith, 1978b; Annecke & Moran, 1982; Dean, 1982; Bedford, 1989).

Purple Scale, *Cornuaspis* (= *Lepidosaphes*) *beckii* (Newman)

This scale has long been recognized as one of the most destructive scale pests of citrus. Considered indigenous to the Far East, it has been a major pest of citrus in parts of Central and South America, Australia, various Mediterranean countries, South Africa, Hawaii, Florida, Texas, and California (DeBach & Landi, 1961). It is highly specific to citrus, rarely being found on other plants.

In California, where it gained entry in 1889, purple scale became a major pest in the coastal citrus districts. Efforts toward biological control were initiated as early as 1891 when two coccinellid predators, *Orcus chalybeus* (Boisduval) and *Rhizobius setelles* Blackburn [= *lophanthae* (Blaisdell)], were introduced. Both became established, but only *Rhizobius* remains common although it is of no importance in purple scale control (Rosen & DeBach, 1978).

Many years later, and within a short period of time (1948 to 1950), two new species of parasitoids, later described as *Aphytis lepidosaphes* Compere and *Coccobius* (= *Physcus*) *fulvus* (Compere & Annecke), were introduced from the Orient (DeBach & Landi, 1961). Following a lengthy colonization program. *A. lepidosaphes* was found to be established throughout the infested areas. A colonization program for *C. fulvus,* however, resulted in establishment only at one locality (Rosen & DeBach, 1978). Although *A. lepidosaphes* provided only partial control, the rate of scale population increase was greatly reduced (DeBach & Landi, 1961). This degree of control, coupled with a periodic, alternating strip treatment program of oil sprays, provided satisfactory control without disrupting the natural enemies of other citrus pests.

Between 1952 and 1968, *A. lepidosaphes* was shipped to 13 other countries or states, with substantial to complete control of purple scale occurring in most cases (DeBach, 1971). These included Texas (Dean, 1961), Mexico (Maltby *et al.*, 1968), Peru (Herrara, 1964), Chile (Gonzalez, 1969), Greece (DeBach & Argyriou, 1967), France (Benassy *et al.*, 1974), and South Africa (Bedford, 1973; Annecke & Moran, 1982). Concurrent with these efforts, surveys for natural enemies of diaspine scales in many parts of the world (DeBach, 1971) showed that not only was *C. beckii* more widely dispersed outside the Orient than was previously known but also the highly host-specific *A. lepidosaphes* had dispersed to as many, or more, countries (13) than it had been deliberately introduced to by man. According to DeBach, *A. lepidosaphes* appears to be responsible for substantial to complete control in nearly every country where accidental colonization (ecesis) has occurred; see also Rosen (1965) and Fabres (1974). Fortuitous biological control via ecesis has also occurred in other species of *Aphytis* (DeBach, 1971), but the best documentation is that concerning *A. lepidosaphes*.

Yellow Scale, *Aonidiella citrina* (Coquillett)

Like the California red scale, this species is also of oriental origin. Its distribution, however, is less extensive than that of California red scale. In addition to California, Florida, and Texas, it also occurs in Mexico, Australia, New Guinea, Japan, China, India, Pakistan, Afghanistan, Iran, Russia, Turkey, Ethiopia, and Argentina.

In California, *A. citrina* had become a serious pest of citrus by the 1880s. At that time and for many years afterward, it was thought to be a strain of California red scale until McKenzie (1937) discovered the morphological characters that separate the two species. Prior to 1900, several species of parasitoids were found to attack the yellow scale in California. These included *Encarsia* (= *Prospaltella*) *aurantii* (Howard), *Aspidiotiphagus citrinus*, and *Aphytis citrinus* Compere [= *aonidiae* (Mercet)]. These were thought to have been accidentally introduced from Japan; and although the near annihilation of yellow scale in the San Gabriel Valley in 1889 was attributed to two of them (Craw, 1891), a similar decline did not occur in other areas of the state (Flanders, 1956).

Following the discovery that the Japanese race of *Comperiella bifasciata* (imported earlier for control of California red scale) reproduced in yellow scale in the laboratory,

mass propagation and colonization were undertaken (Smith, 1942). In southern California, *C. bifasciata* was credited with effective control within 1 to 3 years following colonization whereas in the more northerly San Joaquin Valley, it did not exert a similar control (Flanders, 1956). In 1948 and 1949, the Chinese race of *C. bifasciata* was colonized in the San Joaquin Valley and was credited with temporary control in at least one orchard (Flanders, 1956). Later, however, DeBach (1955) and Kennett (1973) found that the Japanese race was capable of effective control in the San Joaquin Valley in the absence of disruptive pesticide treatments. That this control was due to the Japanese race and not the Chinese race was indicated by laboratory tests showing that *C. bifasciata* reared from field-collected yellow scale did not reproduce in California red scale (Rosen & DeBach, 1978). Many other parasitoids, including several that attack California red scale or other diaspine scales, were released on yellow scale in California between 1953 and 1967, but of these only *Aphytis melinus* became established on this scale in one county.

According to DeBach *et al.* (1978) the yellow scale had become extinct throughout southern California by 1970. They attributed this to displacement by California red scale, aided by the action of *C. bifasciata*.

Yanone Scale, *Unaspis yanonensis* (Kuwana)

The yanone, or arrowhead, scale is considered indigenous to China (Ebeling, 1959). In Japan, where it was first found to occur in 1907, it became the most serious insect pest of citrus (Clausen, 1927). In 1964, it was reported to be established in southern France (Commeau & Sola, 1964), where it has become a major pest of citrus (Panis, 1982).

Although 18 species of predators and 4 species of parasitoids were recorded as natural enemies of *Unaspis yanonensis* in Japan, none provided economic control (Murakami, 1975). Attempts to introduce natural enemies of yanone scale began in 1955 when *A. lingnanensis* was imported from California. This parasitoid, however, failed to become permanently established (Nishida & Yanai, 1965). Subsequent efforts with *A. lingnanensis*, imported from Hong Kong [where it occurred on *U. citri* (Comstock)], resulted in temporary establishment, but the parasitoid was unable to overwinter (Tanaka & Inoue, 1977). In 1975, a shipment of *U. yanonensis* from Hong Kong (derived from citrus fruits exported from China) yielded several specimens of a new *Aphytis* sp. after receipt in Japan (Tanaka & Inoue, 1977). However, the single live female obtained from this material failed to reproduce on *U. yanonensis*. In 1980, foreign exploration in China resulted in the introduction of two new parasitoids (Nishino & Takagi, 1981). Both of these, *A. yanonensis* DeBach & Rosen and

Coccobius fulvus became established at various localities on Kyushu, Shikoku, and Honshu. Preliminary studies (Furuhashi & Nishino, 1983; Yukinari & Kagawa, 1985; Takagi & Ujiye, 1986; Furuhashi & Ohkubo, 1990) of these parasitoids have indicated a potential for highly effective control of *U. yanonensis*. As with the Florida red scale in Israel, complete control of yanone scale could be the key to development of an integrated control program for citrus pests in Japan.

Preliminary field studies with *A. yanonensis* in France, following introduction in 1984, have given promising results (Benassy & Pinet, 1987).

Dictyospermum Scale, *Chrysomphalus dictyospermi* (Morgan)

This highly polyphagous species, widely distributed in subtropical regions, is the most serious insect pest of citrus in the western Mediterranean Basin (Quayle, 1938). It is a major pest in Greece and Iran but a minor pest in Florida, California, Mexico, and South America.

In Greece, the dictyospermum or Spanish red scale is considered to be a more serious pest of citrus than either California red scale or purple scale is. Although locally existing parasitoids, including *A. chrysomphali*, had long been known to be ineffective, apparently no attempts to introduce exotic species had been made prior to 1962 (DeBach & Argyriou, 1967). In 1962 and 1963, *A. lingnanensis*, *A. coheni*, and *A. melinus* were imported from California for control of both *C. dictyospermi* and *Aonidiella aurantii*. By 1963, *Aphytis melinus* was established on both scales (DeBach & Argyriou, 1967) whereas *A. lingnanensis* and *A. coheni*, although recovered in low numbers, failed to persist. Within 2 years *A. melinus* had displaced *A. chrysomphali* at all original release sites. According to DeBach and Argyriou, the effective rate of dispersal for *A. melinus* was 75 to 100 km/year. By 1968, only occasional specimens of *C. dictyospermi* could be found on the previously scale-encrusted trees at the original release sites (DeBach, 1974).

Efforts to develop biological control of *C. dictyospermi* in Italy were begun in 1908 and in France (Corsica) in 1923 (Greathead, 1976). However, it was only after *A. melinus* was introduced in 1964 (Italy) and 1972 (France) that substantial to complete control was forthcoming (Greathead, 1976; Benassy, 1977; Rosen & DeBach, 1978). In Morocco, complete control was attained following the introduction of *A. melinus* in 1966 (Benassy & Euverte, 1970) whereas natural ecesis of the parasitoid from Greece has given good control in Turkey (Tuncyureck & Oncuer, 1974). The biological control of dictyospermum scale is most remarkable in that highly successful results have been attained with a parasitoid that had never been found with

certainty to occur on this scale in the field prior to its colonization in California and Greece (DeBach, 1974).

Rufous Scale, *Selenaspidis articulatus* (Morgan)

The rufous or West Indies scale is widely distributed in tropical and subtropical regions of the world. Its status as a major pest is, however, generally restricted to citrus in the West Indies and northern South America (Talhouk, 1975).

In Peru, where *S. articulatus* was reported as more damaging to citrus than California red scale was (Wille, 1952; Beingolea, 1969), early investigations indicated little or no control by the resident parasitoid–predator complex. Introduction of *A. lingnanensis* was attempted in 1962 but the parasitoid failed to become established (Beingolea, 1969). However, the conclusion that rufous scale was native to Africa (Ebeling, 1959) led to exploration in Uganda, from which parasitized *S. articulatus* was shipped to Peru. This material yielded a new species of *Aphytis*, later described as *A. roseni* DeBach & Gordh, which was successfully colonized and established (Bartra, 1974). Complete control of rufous scale in the coastal citrus districts was reported by 1975 (Rosen & DeBach, 1978).

Unarmored or Soft Scales: Coccidae

Black Scale, *Saissetia oleae* (Olivier)

The black scale, a highly polyphagous, cosmopolitan species, is of economic importance mainly on citrus and olive. Though believed native to South Africa (DeLotto, 1976), *S. oleae* rarely is found on citrus there or in East Africa, central California, and parts of Mexico. It is most important as a pest of citrus in Australia, southern California, Chile, and numerous countries of the Mediterranean Basin.

In southern California, where *S. oleae* became a major pest prior to 1880, attempts at biological control began in 1891 and 1892 with the introduction of various coccinellid predators from Australia. Of these, only *Rhizobius ventralis* Erichson and *Orcus chalybeus* became established (Essig, 1931). Additional species of *Rhizobius*, *Orcus*, and *Leis* were introduced from Australia between 1892 and 1902 but none became established (Bartlett, 1978). Introduction of hymenopterous parasitoids began with the introduction of the egg "predator" *Scutellista caerulea* (Fonscolombe) (= *S. cyanea* Motschulsky) from South Africa in 1901. This was followed by the introduction of some 30 species of parasitoids and about 10 species of coccinellids over the next six decades (Bartlett, 1969, 1978; Luck, 1981). Of the nine parasitoids that became permanently established, the most important are *Metaphycus helvolus* (Compere), *M. bartletti* Annecke & Mynhardt, *M. lounsburyi* (Howard), *Diversi nervus elegans* Silvestri, and *S. caerulea. Metaphycus lounsburyi* showed promise following establishment in 1918 (Smith 1921), but it was not until 1940, following the introduction of *M. helvolus,* that substantial control of *S. oleae* was attained (Bartlett, 1978). Various facets of the campaign against *S. oleae* that are of particular interest include the emphasis placed on introduction of coccinellid predators following the earlier success with the vedalia beetle against the cottony-cushion scale, and their failure to become established; the apparent displacement of *Moranila californica* (Howard), another chalcidoid egg predator, by *S. caerulea* (Flanders, 1958); failure of the introduced *Coccophagus* species to attack *S. oleae* on citrus, their being confined in most cases to the scale on other host plants (Clausen, 1956); the mistaken introduction of the hyperparasitoid *Quaylea whittieri* Girault, which became notorious for its attack on *M. lounsburyi* but essentially disappeared in later years (Flanders, 1943); the differing degrees of control achieved on univoltine (inland) and bivoltine (coastal) scale populations; the role of host feeding by *M. helvolus* in reducing *S. oleae* populations (DeBach, 1943); the fortuitous near eradication of the nigra scale, *Saissetia nigra* Nietner, by *M. helvolus* (Smith, 1944); and the complete elimination of *Eucalymnatus tesselatus* (Signoret) prior to reaching pest status, also by *M. helvolus* (Bartlett, 1969).

Although the appearance of *S. oleae* as a pest of citrus in Australia was coincidental to that in California, much of the early (1897 to 1905) efforts to develop biological control was centered on movement of native natural enemies, particularly coccinellid predators, among the several states (Wilson, 1960). During this period, numerous parasitoids were introduced from various countries, including *Metaphycus lounsburyi* from South Africa and *Moranila californica* (1902) and *Scutellista caerulea* from California. Together, these were credited with excellent control of *S. oleae* in Western Australia (Wilson, 1960). In 1942, *Metaphycus helvolus* was introduced from California and became established in South Australia. According to Wilson, the general status of *S. oleae* as a minor pest in Australia is due to both presumably native as well as exotic natural enemies, a quite different result from that in California.

In Chile, biological control of *S. oleae* was initiated in 1903 with the introduction of *Rhizobius ventralis,* but major efforts were not undertaken until the 1930s when eight species of parasitoids, including *Metaphycus lounsburyi,* were introduced from California and Peru (Graf Marin & Cortes, 1939; Bartlett, 1978). Of these, only *S. caerulea* became established. Introduction of *M. helvolus,* first undertaken in 1943, was unsuccessful, but a later attempt in 1951 resulted in establishment. Duran (1944) reported that the appearance of *M. lounsburyi* in Chile about 1944 was by ecesis from Peru, where it had been introduced for control of *S. oleae* on olive. Gonzalez (1969) reported that

M. helvolus and *M. lounsburyi* were highly effective against *S. oleae* in areas where the scale is bivoltine but much less so in univoltine areas. *Scutellista caerulea, Metaphycus flavus* (Howard) and *Coccophagus caridei* (Brethes) were also credited with aiding in control.

Attempts to develop biological control of *S. oleae* on citrus in the Mediterranean Basin are of more recent origin. As suggested by Greathead (1976), the increase in severity of *S. oleae* infestations following the advent of synthetic insecticides was possibly responsible for these efforts (see also Rivnay, 1968). Beginning in 1953, with the introduction of several parasitoid species in France, efforts against *S. oleae* have been undertaken in Italy (1960, 1971), Corsica (1971), Iran (1960), Greece (1962, 1968) and Israel (1963, 1972–1978). The early introductions in France were unsuccessful (Wilson, 1976) but later efforts resulted in establishment of *M. helvolus* in 1969 and *M. lounsburyi* in 1976 (Panis, 1979, 1983). The early efforts in Israel also failed to result in establishment of parasitoids (Bartlett, 1978) but subsequent introductions, mostly from South Africa during 1972 to 1977, resulted in establishment of *M. helvolus, M. lounsburyi, M. bartletti,* and *S. caerulea* (Wysoki, 1979).

Red Wax Scale, *Ceroplastes rubens* Maskell

This coccid is an important pest of many economic plants, including citrus, in southern Asia, Japan, Australasia, and numerous islands in the Pacific Ocean. It also occurs in Kenya and the Seychelles Islands. In Japan, where it was first found in 1897, it soon became a serious pest of citrus in Kyushu and later spread to Honshu and Shikoku. Early attempts to establish natural enemies imported from California (1924) and Hawaii (1932–1938) were unsuccessful (Yasumatsu & Tachikawa, 1949). In 1946, however, surveys of parasitoids attacking *C. rubens* revealed the presence of a new, highly effective parasitoid in western Kyushu (Yasumatsu, 1951). This parasitoid, first thought to be *Anicetus annulatus* Timberlake, was later considered to be *A. ceroplastis* Ishii and finally described as a new species, *A. beneficus* Ishii and Yasumatsu, in 1957.

Beginning in 1948, *Anicetus beneficus* was colonized throughout the citrus districts of Honshu and Shikoku, where it gave full economic control within three to four host generations (Yasumatsu, 1953). In Korea, significant reductions in densities of *C. rubens* were achieved within the same time span following introduction of *A. beneficus* in 1975 (Kim *et al.,* 1979). This case of biological control is of particular interest in that it was suggested (Yasumatsu, 1958) that *A. beneficus* was native to Japan and possibly arose as a mutation from *A. ceroplastis,* a closely related species which attacks another species of *Ceroplastes* in Japan. Although such a phenomenon had not been demonstrated in biological control, an alternative hypothesis that

A. beneficus was an accidental introduction also seemed questionable as the parasitoid had not been found to occur naturally outside of Japan. However, the discovery of *A. beneficus* attacking *C. rubens* in China (Jiang & Gu, 1983) appears to indicate that the parasitoid came to Japan from China.

White Wax Scale, *Ceroplastes destructor* Newstead

This wax scale is an important pest of citrus in Australia. It has also been reported from citrus in New Zealand, and occurs over much of East and South Africa, where it is thought to have originated. In Australia, the earliest attempt at biological control was undertaken during the 1930s when some 25 species of parasitoids (collected from indeterminate species of *Ceroplastes* in Kenya and Uganda) were imported (Wilson, 1960). Three of these, *Diversinervus elegans, Bothriophryne ceroplastae* Compere, and *Scutellista caerulea,* were reared successfully on *C. destructor* but only the latter two were released. Neither species, however, became established. Renewed attempts between 1968 and 1973 resulted in the introduction of six parasitoids from South Africa. Four of these became established, of which *Anicetus communis* Annecke and *Paraceraptocerus nyasicus* Compere contributed most to the control of *C. destructor* (Sands *et al.,* 1986). Earlier, Milne (1981) reported a case of complete suppression of *C. destructor,* principally by *A. communis.* The eventual outcome of this project further demonstrates that successful biological control most often lies in obtaining natural enemies that attack the target pest in its known or presumed area of origin.

Citricola Scale, *Coccus pseudomagnoliarum* (Kuwana)

This soft scale became a major pest of citrus in southern California following its discovery there about 1907 (Quayle, 1938). Prior to 1951, the scale was known to occur only in Japan, California, and Arizona. Since then, it has been reported from the former Soviet Union (1951), Iran (1963), Turkey (1971), Greece (1972), Sicily (1973), Italy (1976), and Australia (1978).

Attempts to develop classical biological control of citricola scale have, to date, been limited to California. While none of these efforts have been successful, the results are of particular interest in that this scale is one of the few citrus pests in California where no progress has been made (Flanders & Bartlett, 1964; Kennett, 1988). Between 1922 and 1985, several attempts with both monophagous and oligophagous parasitoids from Japan failed to result in establishment of any of the six species liberated (Gressitt *et al.,* 1954). In southern California, however, citricola scale eventually was controlled by a complex of parasitoids, in-

cluding native, cosmopolitan, and exotic species, the latter having been introduced for control of the black scale, *Saissetia oleae* (Bartlett, 1953; 1969). In central and northern California, where citrus is grown under greater climatic extremes, these same parasitoids have proved to be ineffective, and the scale continues to be an important pest.

Failure of the Japanese parasitoids to become established, particularly the host-specific *Metaphycus orientalis* (Compere) and *Microterys okitsuensis* Compere, appears to be due to their inability to bridge the hot summer period, at which time the univoltine, even-brooded host scale are too small for parasitoid reproduction (Compere, 1924; Flanders & Bartlett, 1964; Kennett, 1988). The recent appearance of citricola scale in countries other than Japan and the United States suggests that classical biological control may yet be possible in central California through introduction of oligophagous parasitoids found to attack the scale in those countries (Oncuer, 1974; Kennett, 1988).

Brown Soft Scale, *Coccus hesperidum* Linnaeus

This soft scale is the most cosmopolitan of the many different coccids that attack citrus. This scale is rarely a major pest of citrus but is prone to quickly reach outbreak status when its natural enemies are disrupted by chemical treatments against other pests. Typical of observations in this regard are those of Beingolea (1969), who stated that while the rarity of *C. hesperidum* in Peruvian citrus orchards was due to its natural enemies (*Metaphycus luteolus* Timberlake and *Coccophagous quaestor* Girault), its potential to become a severe pest became evident following the advent of synthetic pesticides.

Many natural enemies are known to attack *C. hesperidum* throughout the world. In South Africa, for example, this scale is attacked by more than 25 species of chalcidoid wasps (Prinsloo, 1984).

Because of its status as a minor pest of citrus, few efforts have been made to develop classical biological control of *C. hesperidum.* Apparently, the earliest attempts were made during the early 1900s when various, mostly indeterminate, species of parasitoids were introduced into Western Australia (Wilson, 1960). Records of establishment are lacking for most of these, but the scale is generally kept under substantial control by native and cosmopolitan species. In Texas, where *C. hesperidum* became an economic pest during the 1950s, various exotic parasitoids were introduced in 1954 and again during the 1960s but none of the five species involved became established (Dean, 1955; Hart *et al.,* 1969). Failure of resident parasitoids and predators to provide control was attributed to interference by ants (Dean, 1955), but later studies indicated drift of pesticides from adjoining crops as the causative factor because few of the resident parasitoids observed earlier could be found (Hart *et al.,* 1969).

This example of natural biological control is an exception to the usual experience where native or cosmopolitan natural enemies tend to be ineffective against exotic pests.

Fluted Scales: Margarodidae

Cottony-Cushion Scale, *Icerya purchasi* Maskell

Native to Australia, this scale has spread throughout most of the subtropical, tropical, and warmer temperate regions of the world. It has a wide host range and is notorious for its ability to severely debilitate or even kill mature trees. This scale is adapted to a wide range of climates and is a pest of many fruit and shade trees, the most important being citrus, mango, and guava.

Following its discovery on Acacia in northern California about 1868, the cottony-cushion scale spread rapidly and by 1886 was devastating the citrus industry in southern California. Chemical controls were generally ineffective and damage was so great that many orchardists destroyed their trees. This dire situation led state and federal horticultural officials to consider the importation of insect natural enemies as a possible solution to the problem (Doutt, 1958). Consequently, a search for natural enemies of cottony-cushion scale was conducted in Australia in 1888 and 1889. This resulted in the shipment to California of dipterous parasitoid *Cryptochaetum* (= *Lestophonus*) *iceryae* (Williston), and a coccinellid predator *Rodolia* (= *Vedalia*, = *Novius*) *cardinalis* (Mulsant). Both readily became established, and while *C. iceryae* was considered the more promising, its activity was quickly surpassed by the highly voracious *R. cardinalis*. A second lady beetle, *Rodolia koebelei* (Horn), was also introduced from Australia in 1892 and although it became established and persisted for many years, it was eventually displaced by *R. cardinalis*. By late 1889, the scale had been so decimated by the vedalia lady beetle that it was no longer considered even a minor threat to the California citrus industry (Quayle, 1938; Doutt, 1958; DeBach, 1974). While the spectacular control of *I. purchasi* was credited to *R. cardinalis,* the eventual role of *C. iceryae,* as suggested by Smith and Compere (1916), was not recognized until much later. Studies by Quezada and DeBach (1973) showed that *Rodolia* and *Cryptochaetum* acted in concert, with the lady beetle tending to displace the fly in the desert areas and the reverse being the case along the coast, whereas in the intermediate (interior) areas, both tended to be commonly present, with their seasonal relative abundances fluctuating according to the prevailing environmental conditions.

Since the complete suppression of cottony-cushion scale in California, *R. cardinalis* has been introduced into some 57 countries, with successful establishment and good control reported in 55 (DeBach, 1964; Bartlett, 1978). In colder regions, however, the vedalia beetle is often eliminated during winter and recolonization is required to maintain control (Greathead, 1976). In spite of such exceptions, the vedalia beetle has demonstrated an adaptability to a greater range of climatic conditions than any other natural-enemy introduction (Clausen, 1978b). Attempts to introduce *C. iceryae* into other countries have been more limited and less successful, with only 5 of the 10 cases resulting in establishment (Bartlett, 1978).

Other Fluted Scales: *Icerya seychellarum* (Westwood), *I. aegyptiaca* (Douglas), and *I. montserratensis* Riley & Howard

Introductions of *R. cardinalis* in various countries for control of these fluted scales have resulted in their control either by the vedalia beetle alone or in conjunction with other native or introduced species such as *R. pumila* Weise, *R. limbatus* Motschulsky, *R. chermisina* Mulsant, *Cryptochaetum monophlebi* (Skuse), and *C. grandicorne* Rondini (Kuwana, 1922; Rodriguez-Lopez, 1942; Vesey-Fitzgerald, 1953; Beardsley, 1955; Bedford, 1965).

Mealybugs: Pseudococcidae

Citrophilus Mealybug, *Pseudococcus calceolariae* (Maskell)

Presumably native to Australia, *P. calceolariae* first appeared as a pest of citrus in California in 1913. It has since been found in New Zealand, Chile, Italy, southern Russia, South Africa, Sardinia, and the Canary Islands.

In California, the citrophilus mealybug rapidly reached major pest status in the coastal citrus districts, the infestations far surpassing those of the citrus mealybug (Clausen, 1978b). Periodic colonization of the coccinellid, *Cryptolaemus montrouzieri* Mulsant, earlier employed for control of the citrus mealybug, was used extensively against *P. calceolariae* for some years; and, while it provided measurable relief, the need for additional natural enemies became apparent as the mealybug continued to extend its range and increase in severity (Bartlett, 1978).

Although the origin of *P. calceolariae* was unknown, it was deduced by Smith and Compere (1929) to have come from Australia. Foreign exploration, conducted there in 1927 and 1928, resulted in importation of two parasitoids, *Coccophagus gurneyi* Compere and *Hungariella pretiosa* (Timberlake), and several predators. Following colonization of the parasitoids in 1928 and 1929, complete control of *P. calceolariae* was attained within 2 years (Compere & Smith, 1932). Later, Bartlett and Lloyd (1958) found *P. calceolariae* to be extensively attacked by another parasitoid, *Pauridea* (= *Tetracnemoidea*) *peregrina* (Timberlake), which had been introduced in 1934 for control of the long-tailed mealybug. The extremely effective control of *P. calceolariae* in California has been consistently born out not only by the extreme rarity of the mealybug but also by the complete absence of outbreaks following synthetic insecticide applications for other pests (DeBach, 1974; Bartlett, 1978; Clausen, 1978b). According to these authors, the consistency and completeness of this control rivals that of the cottony-cushion scale by *Rodolia cardinalis* and *Cryptochaetum iceryae.*

Effective control of *P. calceolariae* on citrus also has been attained in the former Soviet Union (Prokopenko & Mokrousova, 1963) and Chile (Gonzalez, 1969).

Green's Mealybug, *Pseudococcus citriculus* Green

This mealybug, of widespread distribution in East Asia, is also known from Israel, Hawaii, Brazil, and Paraguay. In Israel, where it was first found in 1937, *P. citriculus* quickly became a serious pest of citrus (Rivnay, 1968). Initially, it was thought to be a closely related species, *P. comstocki* (Kuwana), but concern as to its identity led to extensive studies (taxonomic, host preference), which indicated that it was not *P. comstocki* but instead *P. citriculus* (Bodenheimer, 1951). Surveys during 1939 and 1940 (Rivnay, 1968) showed that while *P. citriculus* was attacked by many native predators and parasitoids, the degree of control was unsatisfactory. In 1939, while the mealybug's identity was yet unresolved, shipments of *P. comstocki* from Japan produced several species of parasitoids, among which *Clausenia purpurea* Ishii and *Allotropa burrelli* Muesbeck were the most abundant (Rivnay, 1942). The latter species failed to reproduce on *P. citriculus,* but *C. purpurea* was readily propagated and colonized. Following this, Rivnay (1946) reported that while resident parasitoids were highly competitive, once *C. purpurea* was introduced it always became the dominant species. Both Bodenheimer (1951) and Gruenberg [see Bodenheimer (1951)] concluded that none of the natural enemies (i.e., species of *Scymnus, Clausenia, Leptomastidea, Leptomastix,* and *Anagyrus*) had become dominant and that effective control would have ensued without the introduction of *C. purpurea.* However, it was subsequently reported (Rosen, 1967) that *C. purpurea* alone was responsible for the complete control of *P. citriculus,* see also DeBach (1974).

Citrus Mealybug, *Planococcus citri* (Risso)

This highly polyphagous mealybug is an important pest of citrus in many parts of the world. In California, it became a serious pest in the coastal citrus districts following its introduction about 1880 (Clausen, 1915). Efforts at biological control began with the introduction of a coccinellid predator *Cryptolaemus montrouzieri* from eastern Australia in 1892. While the beetle at first often decimated heavy infestations of *P. citri,* it was generally ineffective and later was found to persist only in milder climates of the immediate coastal areas (Smith & Armitage, 1920). Attempts to utilize hymenopterous parasitoids for control of *P. citri* began with the introduction of *Leptomastidea abnormis* (Girault) from Sicily in 1914 (Smith, 1917). This species became established but it was only partially effective (Clausen, 1956, 1978b).

A second phase in the campaign against *P. citri* began with the development of a method for mass-rearing *P. citri* on etiolated potato sprouts (Smith & Armitage, 1920, 1931). Utilized at first for the mass culture of *Leptomastidea abnormis,* this technique was so successful that it was then applied to the mass culture of *Cryptolaemus.* Between 1916 and 1930, 16 insectaries for *Cryptolaemus* production were established in the citrus-producing areas (Fisher, 1963). Though this program resulted in the colonization of many millions of *Cryptolaemus* each year, and was highly successful, the number of insectaries declined to only four by 1963. To a great extent, this was due to the successful biological control of the more serious citrophilus mealybug in 1930.

Further attempts to obtain new parasitoids of *P. citri* resulted in the introduction of *Leptomastix dactylopii* Howard from Brazil in 1934 (Compere, 1939) and *Pauridia peregrina* Timberlake from south China in 1950 (Flanders, 1951), but neither of these persisted (Clausen, 1956) in the field. *Pauridia peregrina* has persisted as an insectary pest, often virtually halting the mealybug production program as a result of its attack on first-instar mealybugs. However, mass culture and periodic colonization of these two parasitoids, in addition to *Cryptolaemus,* have been credited, along with the earlier established *L. abnormis,* with reducing the frequency of *P. citri* outbreaks (DeBach & Hagen, 1964). Subsequent introductions of various mealybug parasitoids, including *Anagyrus pseudococci* (Girault), were made between 1950 and 1956 (Bartlett & Lloyd, 1958) but all were unsuccessful.

Results of biological control campaigns against *P. citri* in other countries have been quite variable. In Western Australia, where mealybugs (including *P. citri*) were serious pests of citrus, highly successful control was obtained simply by transference of *Cryptolaemus montrouzieri* from New South Wales (Wilson, 1960). Elsewhere, results following the introduction of *C. montrouzieri* have, in most instances, paralleled those in California, with failure of the beetle to become permanently established being due to its inability to survive winter conditions (Wood, 1963; Rosen, 1967; Argyriou, 1969; Beingolea, 1969; Greathead, 1976). Establishment has, in general, been limited to more tropical regions such as Florida and Hawaii and limited areas in several Mediterranean countries (Greathead, 1976). Mass production and periodic colonization of *Cryptolaemus* have also been practiced with some degree of success in several Mediterranean countries, including Spain (Gomez-Clemente, 1954) and Italy (Mineo, 1967). In Queensland, mass releases of *Leptomastix dactylopii* have been used to augment natural populations during periods of scarcity (Smith *et al.,* 1988). Between 1984 and 1987, this program resulted in the reduction of *P. citri*-infested fruits to an acceptable level by harvest.

Extensive introductions of citrus mealybug parasitoids also have been attempted elsewhere, most often with limited success (Zinna, 1960; Bartlett, 1978; Luck, 1981). However, Beingolea (1969) reported that *Pauridea peregrina* (an accidental introduction) effectively controlled *P. citri* in Peru; and Gonzalez (1969) rated the introduced parasitoids *L. abrnomis, A. pseudococci,* and *L. dactylopii* as the most important natural enemies in controlling *P. citri* in Chile.

Although the many and varied efforts to obtain complete biological control of *P. citri* rarely have succeeded, the program of periodic colonization (augmentation) of natural enemies for control of *P. citri* stands as a landmark in the development of this method of biological pest suppression.

Long-Tailed Mealybug, *Pseudococcus longispinus* (Targioni-Tozzetti)

Cosmopolitan species *P. longispinus* generally occurs throughout tropical and subtropical climates where it attacks a great variety of plants, including citrus, avocado, and mango. Although its origin is unknown, Flanders (1940) suggested, on the basis of extreme scarcity and high parasitism, that *P. longispinus* is native to Australia.

Following a major outbreak of *P. longispinus* on citrus in southern California, a parasitoid introduction program was undertaken in 1933. This was facilitated by previously, but recently acquired knowledge concerning parasitoids of *P. longispinus* in other countries (Flanders, 1940). Consequently, *Anarhopus sydneyensis* Timberlake was introduced from Australia in 1934; *Hungariella peregrina* (Compere), from Brazil in 1935; and *Anagyrus fusciventris* Girault, from Hawaii in 1936. Each of these in turn became established, it later being stated (Flanders, 1940) that *A. sydneyensis* was the dominant species.

Subsequent outbreaks of *P. longispinus* between 1943 and 1945 led to studies (DeBach, 1949; DeBach *et al.,* 1949) showing the mealybug was attacked by six primary parasitoids and eight predators. Of the parasitoids, *Anarhopus sydneyensis* was by far the most abundant, followed by *Hungariella pretiosa, Coccophagus gurneyi,* and *H. peregrina. Anagyrus fusciventris,* however, was not present. Most abundant among the predators were the brown lacewing (*Sympherobius californica* Banks), *Cryptolaemus montrouzieri,* and the green lacewing [*Chrysoperla californica* (Coquillett)]. From these studies it was concluded that

while the parasitoids were possibly more important than the predators during the spring, predators were the major factor in subjugating the mealybug.

Although not unique, this example is unusual in that both exotic and native predators and parasitoids contributed to the suppression of *P. longispinus,* the native predators perhaps being most important overall. Fortuitously, *A. sydneyensis* and *H. peregrina* have provided complete control of *P. longispinus* on avocado in California (Flanders, 1944c).

Similar efforts against *P. longispinus* on citrus in Israel were initiated in 1953 with the introduction of *A. sydneyensis* and *H. peregrina,* but only the latter species became established (Rivnay, 1968).

Whiteflies: Aleyrodidae

Citrus Blackfly, *Aleurocanthus woglumi* Ashby

According to Clausen (1978c) the citrus blackfly is endemic to tropical and subtropical Asia, whence it has spread to East Africa, South Africa, the Seychelles Islands, the West Indies, Central America, northern South America, and the United States (Texas and Florida). Although *A. woglumi* was found in Jamaica in 1913 and quickly spread to Cuba, Panama, and Costa Rica, the initial impetus for biological control did not occur until the late 1920s, following Silvestri's (1926, 1928) publications on aleyrodid parasitoids in Asia. This led to exploration in southern Asia from 1929 to 1931, during which time several parasitoids and predators were sent from Malaya to Cuba (Clausen & Berry, 1932). The first species to be introduced, *Eretmocerus serius* Silvestri, readily became established whereas later attempts with *Encarsia* (= *Prospaltella*) *divergens* (Silvestri) and *E. smithi* (Silvestri) met with failure. The coccinellid predators *Catana clauseni* Chapin and *Scymnus smithianus* Clausen & Berry, both obtained in Sumatra, were colonized and established in 1930; and although the former species at times nearly eradicated the blackfly, it was unable to persist at low prey densities and was eventually supplanted by *Eretmocerus serius.* By 1951, *C. clauseni* could no longer be found in Cuba (Clausen, 1978c). Following the successful control of *A. woglumi* in Cuba, *E. serius* was introduced into Jamaica, the Bahamas, Haiti, Costa Rica, Barbados, the Seychelles, Kenya, and South Africa at various times between 1931 and 1959 with generally outstanding results.

In Mexico, attempts at biological control began in 1938 when *E. serius* was introduced from the Canal Zone. Following its failure to become established, *E. serius* was reintroduced in 1943, this time successfully (Smith *et al.,* 1964). The degree of control, however, was not comparable with that in Cuba and elsewhere in the West Indies, except in areas having continuously high humidity. This led to further exploration in Asia from 1948 to 1950. The first introductions, *Encarsia divergens* and *E. smithi,* sent from Malaya in 1948, failed to become established. Exploration was then moved to India and Pakistan (1949 and 1950), resulting in the shipment of eight parasitoid and two predator species (Smith *et al.,* 1964). Of these, *E. smithi, E. clypealis* (Silvestri), *E. opulenta* (Silvestri), and *Amitus hesperidum* Silvestri became established. Each of these parasitoids demonstrated a capability to control the blackfly under suitable climatic conditions (Clausen, 1978c). While *E. smithi* appeared to be the most promising species and became dominant over *Eretocerus serius,* it subsequently came to be dominated by the other three species. Of these, *A. hesperidum* was found to be adapted to most of the varied climates of Mexico and was the most capable in quickly reducing blackfly infestations. However, as the general abundance of *Aleurocanthus woglumi* declined over time, *Amitus hesperidum* came to be dominated by *Encarsia clypealis* and *E. opulenta.* Flanders (1969) explained the order of dominance on the basis of the parasitoids' reproductive strategies (as they influenced interspecific competition). The spatial distribution of the three dominant species was also found to be influenced by climatic conditions. Thus, *E. opulenta* is the overall most effective species, particularly in the more arid regions, whereas *E. clypealis* is most effective in humid areas and *A. hesperidum* is the least effective, particularly in hot, dry areas. This example provides support for the general practice of multiple species introductions instead of preselection and introduction of a presumed single "best" species. It seems highly unlikely that the degree of control obtained would have occurred if only one species been introduced on the basis of its distribution and apparent effectiveness in India–Pakistan.

Subsequent establishment of *E. opulenta* or *Eretmocerus serius* in other countries, either alone, as in El Salvador (Quezada, 1974), Kenya (Wheatley, 1964), South Africa (Bedford & Thomas, 1965) and Venezuela (Anonymous, 1978); with another species, as in Barbados (Pschorn-Walcher & Bennett, 1967) and Florida (Dowell *et al.,* 1979); or with two other species, as in Texas (Summy *et al.,* 1983), have all resulted in excellent control. In those cases where multiple species were introduced, *Encarsia opulenta* became dominant and some degree of displacement of the other species (*Eretmocerus serius, A. hesperidum,* or *Encarsia clypealis*) followed. In Jamaica, the introduction of *E. opulenta* resulted not only in better control but also in displacement of the long-established *Eretmocerus serius* (van Whervin, 1968).

Woolly Whitefly, *Aleurothrixus floccosus* (Maskell)

Long known as a more or less serious pest of citrus in tropical North America, the Caribbean Islands, and South America, the woolly whitefly rarely had been found outside these regions until 1966 when it was simultaneously reported from southern California, the Cote d'Azur in southern France, and the environs of Malaga in Spain. By the 1970s, it had spread to Portugal, Corsica, Italy (including Sicily), Morocco, and Reunion. More recently it has been reported from Hawaii, Texas, Sardinia, and Egypt.

In 1967, *Amitus spiniferus* Brethes, *Eretmocerus paulistus* Hempel, and an *Encarsia* sp. were introduced into California from Mexico but only *A. spiniferus* became established (DeBach & Warner, 1969). Reintroduction of *Eretmocerus paulistus* from Mexico and Florida in 1968, however, resulted in establishment. Although early results were promising, the work was moved to Baja California, Mexico, in 1970 when a chemical eradication program was undertaken in California. Following the failure of eradication efforts, *A. spiniferus* and *E. paulistus* were reintroduced from Baja California in 1971 along with *Cales noacki* Howard, which had been introduced from Chile in 1970. By 1975, these parasitoids, with *C. noacki* and *A. spiniferus* predominating, had effected complete control of *Aleurothrixus floccosus* in the original area of invasion (San Diego County) and in Baja California (DeBach & Rose, 1976). Surveys in 1983 and 1984 have indicated a good level of control throughout the infested areas of California with only 38% of 566 survey sites showing infestations, most of which were light or spotty (S. C. Warner, personal communication).

In 1970, these parasitoids were introduced at Malaga, Spain. While only *C. noacki* became established, *A. floccosus* came under complete control throughout the province by 1974 (Greathead, 1976). Similar results have occurred following introduction of *C. noacki* in France (1971), Reunion (1976), Portugal (1978), and Sicily (1983); and in Hawaii (1981) and Italy (1982), where both *C. noacki* and *Amitus spiniferus* were introduced. A possible case of natural ecesis may have occurred with *C. noacki* in Morocco, where both the parasitoid and *A. floccosus* were reported to be established in 1973.

This example of classical biological control clearly demonstrates the advantages to be gained through *a priori* knowledge of the natural enemies that attack potentially invasive pest species in their native land or lands. In this case, the many earlier records of parasitoids of *A. floccosus* in North and South America were of inestimable value in expediting natural-enemy introductions shortly after the invasive trend of this pest became apparent.

Orange Spiny Whitefly, *Aleurocanthus spiniferus* (Quaintance)

Of Asian origin, this whitefly is widely distributed throughout much of that region and is at times a serious pest of citrus in certain areas (Clausen, 1978c). As yet, however, it has not exhibited the same degree of invasiveness as have the other major citrus whiteflies although it is now known to occur in East Africa, Mauritius, Guam, and Hawaii.

The first effort to develop biological control of *A. spiniferus* was conducted in Japan, where it had become a serious pest in Kyushu during the early 1920s. In 1925, *Encarsia smithi* and a coccinellid predator *Cryptognatha* sp. were imported from China (Kuwana & Ishii, 1927). Although only *E. smithi* was established, it quickly provided a complete and lasting control (DeBach, 1974). In Guam, where *A. spiniferus* was first recorded in 1951, attempts at biological control were initiated in 1952 with the importation of *E. smithi, E. clypealis, E. opulenta* Silvestri), *Eretmocerus serius,* and *Amitus hesperidum* from Mexico (Peterson, 1955). Of these, only *Encarsia smithi* and *A. hesperidum* became established, but complete control was obtained with *E. smithi* predominating. In Hawaii, discovery of *Aleurocanthus spiniferus* infestations in 1974 was quickly followed by introduction during the same year of *Amitus hesperidum* from Mexico and *E. smithi* from Japan. Both of these parasitoids became established (Nakao & Funasaki, 1976), but the degree of control has not been reported.

Citrus Whitefly, *Dialeurodes citri* (Ashmead)

Of Asian origin, the citrus whitefly has, rather sporadically and over a long period of time, invaded various citrus-producing regions in both hemispheres. In the United States, it was known to be a serious pest of citrus in Florida before 1880 and later became of great concern in other Gulf Coast states. In California, where *D. citri* first appeared in 1907, it has never become a major pest although sporadic minor outbreaks have been observed. In Europe, *D. citri* was first found in France about 1945 or 1946. Its subsequent spread throughout much of the Mediterranean Basin has, to a great extent, coincided with that of the woolly whitefly. It is now also known to occur in Sardinia, Italy, Corsica, Corfu, Turkey, Israel, the former Soviet Union, Japan, India, Pakistan, and Taiwan.

The earliest attempt to use biological control against *D. citri* was undertaken in 1910 when the U.S. Department of Agriculture (USDA) sent R. S. Woglum to Asia in search of natural enemies (Woglum, 1913). Some 16 months later, Woglum returned to Florida with cultures of *Encarsia* (= *Prospaltella*) *lahorensis* (Howard) and coccinellid predator *Serangium flavescens* (Motschulsky), both of which were obtained in what is now Pakistan. Unfortunately, host stages suitable for reproduction were lacking and the stocks were lost. No further attempts to import natural enemies of *D. citri* were made until 1968 when *E. lahorensis* was introduced into California (DeBach & Warner, 1969). Further introductions from India, Japan, Hong Kong, and Flor-

ida resulted in the establishment of an *Encarsia* sp. from India and a coccinellid *Delphastus pusillus* (LeConte) from Florida (Rose & DeBach, 1981).

Studies in California by Rose and DeBach showed that while *E. lahorensis* was capable of rapid suppression of *Dialeurodes citri* and was an efficient searcher at low host densities, it was slow to disperse whereas the *Encarsia* sp. from India, although less responsive to host population increase, had superior dispersal abilities. Following the establishment of *E. lahorensis* in Florida in 1977 (some 66 years after the initial attempt), a massive colonization program was conducted throughout the state (Sailer *et al.,* 1984). Within several years the parasitoid had become established in 59 of the state's 67 counties. The generally marked reductions in *D. citri* abundance, as reported in both Florida and California, suggest a successful conclusion to these biological control campaigns. Substantial results also have been obtained in Italy, where *E. lahorensis* was established in 1975 (Viggiani & Battaglia, 1983). This parasitoid has also been established in Sardinia, Corfu (1977) and Israel (1980), but the results have not been reported. In the former Soviet Union, where *E. lahorensis* failed to become established, some success has been reported with introduced parasitic fungi (*Aschersonia* spp.) and coccinellid *Serangium parcesetosa* Sicard introduced from India (Shenderovskaya, 1976).

Bayberry Whitefly, *Parabemisia myricae* (Kuwana)

This whitefly appears to be the latest in a series of citrus aleyrodids that have spread into various regions of the world in recent decades. Its discovery in California during the 1970s led the introduction of several aleyrodid parasitoids between 1979 and 1980 (Rose & DeBach, 1992). While an *Encarsia* sp. and an *Eretmocerus* sp. were established, and good control was obtained within 4 years, an unintroduced (native ?) species, *Eretmocerus debachi* (Rose & Rosen), became the overall dominant parasitoid.

Psyllids: Psyllidae

Although only two species of psyllids have been reported as pests of citrus, both are of major economic importance in that they transmit greening (likubin) disease in citrus. The first species, *Diaphorina citri* (Kuwayama), is generally distributed over much of Asia and is also known from Brazil, Saudi Arabia, and the islands of Reunion and Mauritius; while on the contrary, the second species, *Trioza erythreae* (Del Guernica), occurs throughout most of eastern Africa, extending from Sudan to South Africa as well as the islands of Madagascar, Mauritius, Reunion, and St. Helena.

In Reunion, where *D. citri* and *T. erythreae* occur together, the former generally is restricted to the low elevation (< 500 m), low rainfall (< 1000 mm) areas, while the latter is more widespread but is best adapted to the areas of higher rainfall, which occur above 500 m. In 1974, two eulophid parasitoids, *Tetrastichus dryi* Waterston and *Psyllaephagus pulvinatus* Waterston, were introduced from South Africa for control of *T. erythreae*. Although only *T. dryi* became established, it alone virtually eliminated the psyllid from citrus groves within 2 to 3 years (Aubert & Quilici, 1983). In 1978, *T. radiatus* Waterston was introduced from India for control of *D. citri* and, as in the previous case, it, along with the accidentally introduced *Diaphorencyrtus aligarhensis* (Shaffee *et al.*), brought about a similar scarcity of the psyllid within 2 years. The complete suppression of these two psyllids in Reunion is unique in that it represents the only case where a serious disease of citrus has been rendered innocuous via classical biological control of its insect vectors. *Tetrastichus radiatus* was subsequently introduced into Taiwan from Reunion in 1984 to supplement the work of two naturally occurring parasitoids, *D. aligarhensis* and *Psyllaephagus* sp. Although it was not readily established (Chiu *et al.*, 1985), subsequent observations (Chien *et al.*, 1988) indicated it had become the most prevalent parasitoid on *D. citri*.

Mites

Many species of phytophagous mites infest citrus throughout the world. The families having pest species include Tetranychidae, Tenuipalpidae, Tarsonemidae, and Eriophyidae (Ebeling, 1959). Biological control of what are probably the two most widespread species is considered here.

Tetranychidae

Citrus Red Mite, Panonychus citri *(McGregor)*

Citrus red mite has a worldwide distribution on citrus. Like citrus, it may be of Asian origin. *Panonychus citri* feeds on leaves and fruit and occasionally on green twigs, causing a bronzing or silvering effect. Severe infestations may cause a partial defoliation of trees (Jeppson *et al.*, 1975). Biological control of citrus red mites is practiced mainly through conservation of indigenous natural enemies (i.e., by avoiding spraying with pesticides toxic to the natural enemies). Beetles in the genus *Stethorus* (Coccinellidae) and *Oligota* (Staphylinidae) are specialized predators of tetranychid mites present in all areas of the world where citrus is grown. Their impact on mite populations usually is thought to be significant only at medium or high mite densities. However, closer examination sometimes has revealed that *Stethorus* species are able to perceive isolated spider mite colonies and begin oviposition when average mite densities are very low (Hull *et al.*, 1977; Haney *et al.*,

1987). Other insectan predators reported attacking citrus red mites include *Saula japonica* Gorham, a coleopteran in the Endomychidae in Japan, neuropterans in the Coniopterygidae, and dipterans in the Cecidomyidae (McMurtry, 1985).

Phytoseiid mites are common predators of citrus red mites. The most prevalent phytoseiids in many areas of the world are various species of *Euseius*. *Euseius* species are "generalist" feeders that reproduce readily on other foods as well as mites, especially pollen, so their abundance is not always correlated with that of mite prey (McMurtry, 1969, 1977, 1985; Kennett *et al.*, 1979; Garcia-Mari *et al.*, 1984; Ferragut *et al.*, 1987). However, most of these studies suggest that *Euseius* species sometimes maintain citrus red mite populations at low levels, either because the predators maintain relatively high densities as a result of utilizing other foods and thus prevent increases of these mites (Kennett *et al.*, 1979; Garcia-Mari *et al.*, 1984), or because they actually show functional and numerical responses to increases of these red mites (McMurtry *et al.*, 1969; Keetch, 1972; Schwartz, 1978; Garcia-Mari *et al.*, 1984). Other phytoseiids considered potentially important predators of citrus red mites include several species in the *Amblyseius largoensis* group, including *A. eharai* Amitai & Swirski in Japan (Tanaka & Kashio, 1977) and *A. herbicolus* (Chant) in Australia (Beattie, 1978), both formerly referred to as *A. deleoni* Muma & Denmark. An unrelated species, *Amblyseius newsami* (Evans), is considered an important predator of citrus red mites in Guangdong Province, China; the species is encouraged by manipulation of a ground cover that provides pollen and moderates temperatures (Huang, 1978, 1981). Phytoseiids that are specialized predators of citrus red mites have not been reported.

A noninclusion virus disease is an additional mortality factor on citrus red mites in California. A survey of 51 orchards in central and southern California indicated infected citrus red mites in 82% of these orchards, with high mite populations frequently being controlled by epizootics of the virus (Shaw *et al.*, 1968). Infected mites can be found also at low population densities of citrus red mites in orchards under chemical pest control as well as unsprayed ones (Shaw & Pierce 1972). Citrus red mite populations usually decline, even before high densities are reached, when samples of adult females show an infection of 10 to 15% and/or when there is a progressive increase in the rate of infection on two or more biweekly sampling dates (McMurtry *et al.*, 1979). In one study, phytoseiid mites were considered secondary in importance to the virus in suppressing citrus red mite populations in the spring (McMurtry *et al.*, 1979). Estimation of the percentage of infection is based on the presence of birefringent crystals in slide-mounted mites examined microscopically under polarized light (Smith & Cressman, 1962). A portable polarizing unit for field detection of diseased mites was developed (Reed *et al.*, 1972).

Improved biological control of citrus red mites has been attempted by the "classical" method of importation, propagation, release, and attempted establishment of exotic predators. Over 20 species have been collected from citrus in different parts of the world, and propagated and colonized in California orchards. Only two of these have become established in California on citrus. *Euseius stipulatus* Athias-Henriot, introduced from the Mediterranean region in 1971, has displaced the native *E. hibisci* (Chant) in many coastal orchards of southern California. Interestingly, *E. stipulatus,* like *E. hibisci,* has very broad feeding habits, although it shows a higher reproductive capacity than *E. hibisci* shows when feeding on citrus red mites. Because *E. stipulatus* rapidly (within two seasons) displaced *E. hibisci,* also in orchards having few citrus red mites, additional factors must favor the exotic species over the native one (McMurtry, 1982, 1989; Friese, 1985). *Typhlodromus rickeri* Chant, introduced from India in 1962, has been recovered in a few coastal orchards and only in relatively low numbers. This species is less of a "generalist" than are the *Euseius* species, being more closely associated with mites and having a higher rate of reproduction on both tetranychid and eriophyid mites (McMurtry & Scriven, 1964b; McMurtry, 1982).

Observations from various parts of the world indicate that most serious problems with citrus red mites are induced by the use of pesticides, usually applied for other citrus pests (McMurtry, 1985). Therefore, the key to managing citrus red mites is to avoid the use of pesticides that destroy predators and/or stimulate increases of this mite.

Eriophyidae

Citrus Rust Mite, Phyllocoptruta oleivora *(Ashmead)*

This mite is considered an important pest of citrus in humid areas throughout the world. Mites damage the fruits by puncturing the epidermal cells of the peel, causing the formation of lignin, which results in a russeting and cracking and a rusty brown coloration on oranges and grapefruit and a silvering of lemons (McCoy & Albrigo, 1975). Various species of insects and mites are known to prey on citrus rust mite, but their role in rust mite control is virtually unknown. Most species of phytoseiid mites probably will prey on rust mites, but the suitability of this prey for phytoseiid reproduction varies with the species. For example, rust mite was not a favorable food for *T. athiasae* Porath & Swirski (Swirski *et al.,* 1967b) and *E. hibisci* (McMurtry & Scriven, 1964a), but promoted medium and high reproduction for *A. swirskii* Athias-Henriot and *T. rickeri,* respectively (McMurtry & Scriven, 1964b; Swirski *et al.,* 1967a). Insects known to feed on citrus rust mite include species in the Coniopterygidae (Neuroptera) (Muma, 1967, 1969; McMurtry, 1977), Cecidomyidae (Diptera), and Phlaeothripidae (Thysanoptera) (van Brussel, 1975).

One of the major mortality factors affecting citrus rust mite in humid regions is the fungus disease *Hirsutella thompsonii* Fisher. The fungus was first reported from an epizootic of citrus rust mite in Florida by Fisher *et al.* (1949). Muma (1955, 1958) presented field data indicating the fungus can be important in the natural control of rust mites. The intensity and duration of the fungus attack was demonstrated to be proportional to rust mite density. The possibility of culturing *H. thompsonii* and applying formulations of it to enhance field infection at lower mite densities for economic control has been investigated since the 1960s, beginning with culturing the fungus on artificial media (McCoy & Kanavel, 1969). Suppression of mites in the field with applications of fragmented mycelia of the fungus was demonstrated by McCoy *et al.,* (1971). Development and testing of commercial formulations (Mycar®) and eventual registration have occurred. Although spring applications of Mycar have not appeared promising, apparently because of unsuitable weather conditions (too dry to promote infection), summer applications significantly reduced rust mite populations in Florida. This mycoacaracide, combined with oil, showed promise for compatible crop protection against both rust mite and greasy spot fungus disease, *Mycosphaerella citri* Whiteside (McCoy & Couch, 1982).

Snails

Helicidae

The brown garden snail. *Helix aspersa* Müller, of European origin, is the most important snail pest in California. Although most commonly a pest of landscape ornamentals and home gardens, this snail has also been an important pest of citrus for many years (Basinger, 1931).

Attempts to develop biological control of *H. aspersa* in California began with the introduction of predaceous snails and beetles during the 1950s and early 1960s. These efforts resulted in establishment of only one species, a staphylinid beetle (Fisher & Orth, 1985). In 1966, however, another (opportunistic) predaceous snail *Rumina decollata* Linnaeus (also of European origin) was found to have invaded California (Fisher, 1966). Subsequent observations suggested that *R. decollata* had been present there for 5 to 10 years before its discovery; later studies showed it to be established in at least 14 counties (Fisher *et al.,* 1980; Fisher & Orth, 1985).

Experimental releases of *R. decollata* in southern California citrus orchards were begun in 1975 and, in most cases, resulted in complete control (displacement) of *Helix* within 4 to 6 years after the initial predator seedings (Fisher & Orth, 1985). These authors estimated that *R. decollata* is now used to control *H. aspersa* in some 20,000 ha of citrus in southern California. Methods for suppression of *H. aspersa* during the period preceding full control by *R. decol-*

lata have been discussed by Fisher *et al.* (1983) and Fisher and Orth (1985).

OLIVE

The common olive, *Olea europea* Linnaeus, is a major crop in most countries of the Mediterranean Basin and the Middle East. Outside these regions, commercial olive production is almost entirely limited to the United States (California) and South America (Peru, Chile, and Argentina), where it is introduced.

In the Western Hemisphere, "classical" biological control of olive pests has been concerned primarily with armored and soft scales that were introduced along with olive or other host plants. In the Eastern Hemisphere, where olive is native, biological control work has dealt almost exclusively with introduced pests such as the olive fly and the black scale (*Saissetia oleae*). Indigenous and cosmopolitan species that are considered major pests, particularly in the Eastern Hemisphere, are often induced pests and few efforts in biological control have been attempted against them.

Olive Pests

Olive Scale, *Parlatoria oleae* (Colvée) (Diaspididae)

This highly polyphagous armored scale became a serious pest of olive, various deciduous fruits, and many ornamentals in California shortly after its discovery there in 1934 (McKenzie, 1952).

Initial attempts toward biological control of *P. oleae* included colonization of an encyrtid parasitoid *Habrolepis rouxi* Compere from 1940 to 1942 and an aphelinid parasitoid *Aphytis* sp. imported from Egypt in 1948. Neither became permanently established. Subsequent efforts, however, resulted in the establishment of *Aphytis paramaculicornis* DeBach & Rosen [= *A. maculicornis* (Masi)], imported from Iran in 1951. Early results were promising (Doutt, 1954), with markedly lower scale densities occurring within 1 to 2 years, but later studies (Huffaker *et al.,* 1962) showed that pest densities often remained above the economic threshold.

Further efforts to improve the parasitoid complex on *P. oleae* were made with the aphelinid *Coccophagoides utilis* Doutt and the encyrtid *Anthemus inconspicuus* Doutt, both introduced from Pakistan in 1957 (Doutt, 1966). Both of these became established but the latter species disappeared about 1961 whereas the former species was not recovered from *P. oleae* until the same year. Subsequent colonization of *C. utilis* from insectary propagation (4,000,000) (Finney, 1966) and field insectary trees (several millions) resulted in widespread establishment in olive and many other hosts of *P. oleae;* within 2 to 3 years the combined effect of *A. paramaculicornis* and *C. utilis* brought about complete con-

trol (<0.5% infested fruit) in olives wherever both parasitoids were present (Huffaker & Kennett, 1966; Kennett *et al.,* 1966). Subsequent studies in olives (Huffaker *et al.,* 1986) indicated a continuing and complete control, with scale densities being even lower than those extant some 15 to 20 years earlier.

Black Scale, *Saissetia oleae* (Olivier) (Coccidae)

This soft scale is the most widely distributed insect pest of olives (Ebeling, 1959). It is a common and, at times serious, pest of olives throughout the Mediterranean region, the Near East, South America, and North America.

Efforts toward biological control of *S. oleae* have been undertaken in all countries where olives are grown commercially. The results generally have been much less successful than those in citrus (Bartlett, 1978). Most of the more promising natural enemies have been tried in various combinations in various countries over a period of many years, but rarely has control been more than partial. Prior to the 1950s, most efforts were centered in California, Chile, Peru, and Argentina (Duran, 1944; Beingolea, 1955), with *Scutellista caerulea, Metaphycus lounsburyi, M. stanleyi* Compere, *M. helvolus,* and *Lecanobius utilis* Compere being used most frequently. Subsequent efforts after 1950, particularly in California (van den Bosch *et al.,* 1955; Bartlett & Medved, 1966), resulted in several new introductions, including *Diversinervus elegans* and *M. bartletti,* but only the latter species became established on *S. oleae* in the major olive districts of central and northern California (Kennett, 1980, 1986). From 1979 to 1985 introductions of black scale parasitoids have resulted in the establishment of several species, notably *Prococcophagus probus* Annecke & Mynhart and *M. zebratus* (Mercet) (Daane *et al.,* 1991).

In the Mediterranean region, where major efforts began during the 1960s and 1970s, most of the preceding species, as well as several accidentally introduced species, have been established in one or more countries (Argyriou & Katsoyannis, 1976; Blumberg & Swirski, 1977; Viggiani & Mazzoni, 1980; Panis, 1983) but, as in California and South America, control generally has been less than satisfactory.

In most cases, failure to obtain meaningful control has been attributed to the univoltine, even-brooded nature of *S. oleae* where host stages suitable for parasitoid reproduction are absent for much of the year.

Olive Fly, *Dacus oleae* (Gmelin) (Tephritidae)

Of African origin, this fruit fly is the key pest of olive throughout the Mediterranean Basin and the Middle East (Clausen, 1978a). Where the fly is not controlled, larval infestation of mature olive fruit may at times reach 100%.

Attempts at biological control of *Dacus oleae* started in 1910 when the braconid parasitoid *Opius africanus* Szepligeti was introduced into Italy from Eritrea (Silvestri, 1914).

However, it failed to become established. Likewise, 13 additional species introduced from East Africa and Australia between 1911 and 1914 failed to establish (Silvestri, 1924). Further efforts in 1914 resulted in the introduction of another braconid, *Opius concolor* Szepligeti, from Tripoli but, while establishment apparently took place, it was not recovered until 1929 (Monastero, 1931). Other early introductions of *O. concolor* were made in France, Greece, and Spain, but all were unsuccessful (Greathead, 1976; Clausen, 1978a). Reintroductions in France and Greece, however, were successful (Delanoue, 1960), but (as in Italy) *O. concolor* never became abundant and did not disperse readily.

In 1958, a program for mass production and release of *O. concolor* was undertaken on the island of Elba (Fenili & Pegazzano, 1962). Relatively high levels of parasitism occurred at first but the parasitoid later became quite rare. Subsequent work in Italy and France (Delanoue, 1960; Monastero & Genduso, 1964; Monastero & Delanoue, 1967; Liotta & Mineo, 1968; Monastero, 1968; Genduso & Ragusa, 1969), using periodic releases of many thousands and in some cases many millions of *O. concolor,* gave economic control of *D. oleae* in certain experiments but not in others. In one case (Monaco, 1969), native parasitoids (chalcidoids) gave a higher level of parasitization than did the mass-released *O. concolor*. Studies on native parasitoids of *D. oleae* in France (Arambourg & Pralavorio, 1974) indicate noneconomic control by the three chalcidoid parasitoids involved. On the basis of these studies, however, it was suggested that one of them, *Eupelmus urozonus* Dalman, be mass-produced for field augmentation (Arambourg & Pralavorio, 1974).

Although certain of the mass release programs with *O. concolor* proved to be economically feasible, this approach to olive fly control has not been pursued in recent years.

Olive Moth, *Prays oleae* (Bernard) (Yponomeutidae)

This moth is generally distributed throughout the Mediterranean Basin where it is considered a major pest of olives.

Many species of natural enemies (principally parasitoids) are known to attack *P. oleae* (Arambourg, 1969; Stavraki, 1967, 1970a) but natural control generally has been ineffective. Greathead (1976) attributed the absence of natural control to the disruptive effect of synthetic pesticides on natural enemies of the moth.

Various components of biological suppression of *Prays oleae* have been investigated, including those on efficacy of indigenous parasitoids (Arambourg, 1969; Campos & Ramos, 1982), augmentation of the most promising native parasitoids via mass production and release (Arambourg, 1970; Arambourg *et al.,* 1970), introduction of parasitoids into countries where not found to occur previously (Stavraki, 1970b), use of the insect pathogen *Bacillus thuringien-sis* Berliner as a supplement to native parasitoids (Panis, 1979; Viggiani, 1981), and use of *B. thuringiensis* (B.t.) alone (Niccoli & Tiberi, 1985). Of these, the most promising results were obtained with the native parasitoid–pathogen treatment studies.

AVOCADO

The avocado, *Persea americana* Miller, is native to southern Mexico and Central America (Williams, 1977; Storey *et al.,* 1986). Extensive commercial production of avocado now occurs in the United States, in several countries bordering the Mediterranean region, and in southern Africa, in addition to several countries in Central and South America. To date, there seems to have been relatively little movement of avocado pests from the area of origin to other countries; instead, the crop is more commonly attacked by certain existing pest species in each area where production occurs.

Published works on biological control research on avocado pests mainly are from California and Israel. These works are therefore the basis for this section.

Avocado Pests

Omnivorous Looper, *Sabulodes aegrotata* (Guenée) (Geometridae)

This apparently indigenous species becomes sporadically abundant on avocado in California. Although feeding injury is confined mainly to the foliage and is, therefore, of minor significance, a heavy infestation of omnivorous looper can result in considerable feeding on the surface of young fruit, causing scarring or pitting (Ebeling, 1959). Indigenous parasitoids reportedly maintain this pest below damaging levels most of the time, and a heavy infestation in a given orchard one season is usually followed by several seasons of low populations (Ebeling, 1959; Oatman *et al.,* 1983; Oatman & Platner, 1985). The most common parasitoids of this insect are *Trichogramma platneri* Nagarkatti on the eggs, and the larval parasitoids *Apanteles caberate* Muesebeck, *Bracon xanthonotus* Ashmead, and *Meteorus tersus* Muesebeck (Oatman *et al.,* 1983).

Experiments by Oatman and Platner (1985) showed that three weekly releases of *T. platneri* of 50,000 in each of four uniformly spaced trees per acre resulted in effective control of omnivorous looper although parasitism was lower than that on the tortricid *Amorbia cuneana* Walsingham at comparable release rates. Because augmentative releases are only occasionally necessary, the adult populations should be monitored with the aid of pheromone traps to determine when and if releases are needed.

Giant Looper, *Boarmia selenaria* (Schiffermüller) (Geometridae)

This species is sometimes injurious to avocados in Israel, especially in situations where insecticide drift of aerial sprays applied to cotton fields has caused upsets. Larger larvae chew large holes in the fruit, often resembling injury caused by rats (Wysoki *et al.*, 1975).

Wysoki and Izhar (1980) list 32 species of natural enemies of *B. selenaria* in Israel, including bacteria, fungi, spiders, mites, insects, and birds. The most abundant were the parasitoids *Apanteles cerialis* Nixon (Braconidae) and two species of tachinid flies. However, parasitism by these species did not reach a high rate until late summer when the main damage to avocados had already occurred. Therefore, control was supplemented by aerial applications of B.t. Because only the young larvae are sensitive to B.t. preparations, a monitoring system was devised utilizing virgin female traps to determine the peaks of activity of the males. Applications made about 2 weeks after these peaks provided a useful complement to natural control of the looper (Izhar *et al.*, 1979; Wysoki *et al.*, 1981).

Trichogramma platneri was introduced from California for potential control of both giant looper and honeydew moth, *Cryptoblabes gnidiella* (Milliére) (Wysoki & Renneh, 1985), but establishment had not been verified.

Western Avocado Leaf roller, *Amorbia cuneana* Walsingham (Tortricidae)

This apparently indigenous species is sporadically injurious to avocados in California. It feeds primarily on leaves, but it also feeds on the fruit surface when it has webbed leaves against the fruit (Ebeling, 1959). In California, 15 species of parasitoids from this pest are recorded (Oatman *et al.*, 1983). The egg parasitoid *Trichogramma platneri* and occasionally the larval parasitoid *Elachertus proteoteratus* Howard were the only common Hymenoptera attacking *A. cuneana*. This pest was most commonly parasitized by tachinid flies, all larval–pupal parasitoids. Oatman and Platner (1985) observed that *T. platneri* apparently prefered eggs of *A. cuneana* to those of the omnivorous looper because they parasitized a larger percentage of this species' eggs in a cluster. Effective control of this pest was achieved by only two weekly releases of 50,000 *T. platneri* on four uniformly spaced trees per acre. Monitoring of moth populations was considered essential to determine if and when supplemental releases were necessary (Oatman & Platner, 1985).

Long-Tailed Mealybug, *Pseudococcus longispinus* (Targioni-Tozzetti) (Pseudococcidae)

This species was once considered a serious pest of avocados in San Diego County, California. In 1941, two parasitoids, *Anarhopus sydneyensis* Timberlake from Australia and *Hungariella peregrina* (Compere) from Brazil and Argentina, previously established on mealybug infestations on dracena in San Diego and on citrus in Downey, respectively, were released on avocado in the Carlsbad area. Both species became established in this area. In less than 2 years after the releases, mealybug populations were greatly reduced, and the biological control of long-tailed mealybug on avocado in southern California was considered a complete success (Flanders, 1944c). Presently, the mealybug occurs throughout the avocado growing areas of California, but only in low numbers (personal observations).

Outbreaks of *P. longispinus* occurred in Israel in avocado orchards adjacent to cotton and citrus being sprayed with nonselective pesticides. This apparently resulted from destruction of natural enemies, especially *H. peregrina*, which had been introduced into Israel in 1954. Following the outbreaks, *H. peregrina* was mass-reared and released in the heavily infested areas. Aerial applications of nonselective pesticides were forbidden within 200 m of avocado orchards. *Pseudococcus longispinus* was again brought under control (Wysoki, 1979; Swirski *et al.*, 1980; Wysoki *et al.*, 1981). An additional parasite, *Anagyrus fusciventris* Girault, was imported from Australia in 1971 and apparently is an additional factor in the biological control of this mealybug (Wysoki, 1979).

Latania Scale, *Hemiberlesia lataniae* (Signoret) (Diaspididae)

Latania scale has been known as a pest of avocado in California since 1928 and at one time a considerable acreage was fumigated or sprayed. Scales are usually most abundant on branches or twigs; although when present in high numbers, they also infest the fruit, so that fruit may be downgraded (Ebeling, 1959). Such noticeable infestations are presently unusual and generally can be attributed to upset situations in which natural enemies are destroyed by pesticides. A relatively large number of predators and parasitoids attack the latania scale in California. One of the most important appears to be coccinellid *Chilocorus stigma* Say, or twice-stabbed lady beetle. The adult beetles are also important in that they transport the phoretic stage (deutonymph) of the mite *Hemisarcoptes malus* (Shimer). This parasite, known from many species of armored scales, feeds on all stages of the scale beneath the covering. Feeding by *Hemisarcoptes* may cause mortality or reduction in longevity or fecundity of the host scale, depending on size of the scale and the number of individuals feeding on it (Gerson & Schneider, 1981). Although these mites have been termed predators, parasites, or parasitoids by various authors, the term "parasite," as used by Lindquist (1983), seems to be the most fitting. Mites in the Cheyletidae [*Cheletomimus berlesei* (Oudemans)] and Camerobiidae families

(*Neophyllobius* sp.) have been observed preying on the crawler stage of latania scale (Ebeling, 1959).

Of the Hymenoptera order, *Aphytis proclia* (Walker) is considered the most important parasitoid of latania scale (Fleschner, 1954; Ebeling, 1959). It has not been determined which species of natural enemy is most important in maintaining latania scale at low levels or whether the collective action of several of these species is responsible.

Bayberry Whitefly, *Parabemisia myricae* (Kuwana)(Aleyrodidae)

This pest was introduced into Israel in 1978, where it spread throughout the country and became a pest of both citrus and avocado. Of four parasitoids imported for biological control of *P. myricae,* the most effective has been *Eretmocerus* sp. from California. This appears to be a successful case of biological control of an imported pest by introduction and establishment of a single species of natural enemy (Swirski *et al.,* 1988).

Greenhouse Thrips, *Heliothrips haemorrhoidalis* (Bouché) (Thripidae)

This thrips has become an increasingly important avocado pest in California. It is a pest mainly on the Hass variety, where it occurs primarily on the fruit, causing brown scarring and downgrading of fruit. The Trichogrammatid egg parasitoid *Megaphragma mymaripenne* Timberlake is not an effective regulating agent, even though over 50% of the thrips egg blisters sometimes show parasite emergence holes (Hessein & McMurtry, 1988). Predators, including the green lacewing, *Chrysoperla carnea* (Stephens), and the thrips *Franklinothrips vespiformis* (Crawford), sometimes may be responsible for significant thrips mortality but their occurrence is inconsistent. Two eulophid larval parasitoids have been introduced to California. *Goetheana parvipennis* Gahan was imported from Trinidad in 1962 (McMurtry & Johnson, 1963) and from the Bahamas in 1983. Neither introduction resulted in establishment of *G. parvipennis.* This parasitoid is known from several species of thrips, especially the red-banded or cacao thrips, *Selenothrips rubrocinctus* (Giard), on which it was established in the West Indies after its introduction from Africa (Dohanian, 1937; Callan, 1943).

Thripobius semiluteus Boucek, introduced from Australia in 1986, appears to be established and spreading in southern California and may offer some promise for control and a reduction in the use of pesticides for greenhouse thrips (McMurtry *et al.,* 1991).

Avocado Brown Mite, *Oligonychus punicae* (Hirst) (Tetranychidae)

Oligonychus punicae, originally described from specimens from India on pomegranate and grape, was first discovered in California in the 1920s and it is also common on that plant in Mexico and Guatemala. Feeding by this species is confined mainly to the upper surfaces of the leaf beginning along the midrib or in depressions in the leaf surface, except in heavy infestations (ca. 80 to 100 adult females or 200 to 300 total post-embryonic stages per leaf), when infestations extend to the lower leaf surfaces and fruits. Such population levels can result in partial defoliation, especially on the "Hass" variety (McMurtry & Johnson, 1966; McMurtry *et al.,* 1969; McMurtry, 1985). Population peaks usually are below these levels and damage is rarely considered sufficiently severe to warrant chemical control.

The coccinellid *Stethorus picipes* Casey is considered the major factor responsible for suppressing *O. punicae* populations after medium to high densities have been attained. Control of the mite is usually effected before severe bronzing occurs, with *S. picipes* showing a numerical response to increases of *O. punicae* when the latter is still at relatively low or medium densities (10 to 20 adult females per leaf). Conversely, in cases where severe bronzing of foliage occurs, there is a delayed increase (longer lag period) of *S. picipes* (McMurtry & Johnson, 1966; McMurtry, 1985). The other major predator, *Euseius hibisci,* usually does not respond soon enough to overtake an increasing *O. punicae* population, possibly because of a lack of tendency to congregate and oviposit on leaves infested with this mite, and the inability to gain access to prey on the upper leaf surface that is protected beneath silken webbing (McMurtry & Johnson, 1966). Releases of *S. picipes* at 200 or 400 beetles per tree significantly lowered peak populations of *O. punicae* and leaf damage, compared with control plots. However, such a practice is not cost-effective because of lack of efficient mass production technology and the irregular need to augment natural control (McMurtry *et al.,* 1969).

Persea Mite, *Oligonychus perseae* (Tuttle, Baker, and Abbatiello)

O. perseae, a recent introduction to California, possibly from Mexico, colonizes the underside of leaves in semicircular "nests" of tightly woven silken webbing. It became a major pest of California avocados in the 1990s (Bender, 1993). Neither *E. hibisci* nor *S. picipes* enter the "nests" of *O. perseae* and they apparently are unable to regulate populations of this pest. *Galendromus annectens* (DeLeon), previously an uncommon species on avocado, became considerably more abundant and widespread in response to high populations of *O. perseae*. Releases of *G. helveolus* from a culture originating in Florida, resulted in establishment of this species. The relative importance of these two *Galendromus* species varies from one area to another and one year to another (Aponte & McMurtry, unpublished data).

Six-Spotted Spider Mite, *Eotetranychus sexmaculatus* (Riley) (Tetranychidae)

This mite develops in colonies on the undersides of avocado leaves, mainly along the midrib or main veins, and produces a chlorosis that is also visible on the upper side of the leaf. Unlike *O. punicae*, *E. sexmaculatus* can cause defoliation at relatively low population densities (Fleschner *et al.*, 1955; Ebeling, 1959). However, numbers usually do not exceed two to three mites per leaf. Maintenance of these low levels is attributed to the phytoseiid mites, *Euseius hibisci* and *Amblyseius limonicus* Garman & McGregor. These predators occupy, and lay their eggs in, the same microhabitat as *Eotetranychus sexmaculatus* (lower leaf surface along the midrib) where they would contact this species more frequently than *O. punicae* (McMurtry, 1985). Natural control of *E. sexmaculatus* was demonstrated experimentally by Fleschner *et al.* (1955). *Eotetranychus sexmaculatus* nearly defoliated a limb from which predators were removed by hand for approximately 3 months, while on adjacent limbs from which predators were not removed, no damage was evident on the foliage.

Galendromus helveolus (Chant) (= *floridanus* Muma)

Galendromus helveolus, a more specialized predator of tetranychid mites, occurs in Mexico, Central America, and Florida. Although it has not been studied on avocado, Muma (1970) credited this phytoseiid with regulating *E. sexmaculatus* populations at low densities on citrus in Florida.

TEA

The tea plant, *Camellia sinensis* (Linnaeus), has been cultivated from ancient times, and is now grown extensively in India, Sri Lanka, Indonesia, China, Japan, the former Soviet Union, and East Africa. The plant is cultivated for the young shoots and it is pruned to a low bush with a flat "plucking table." Arthropod pests of tea include mites in the Tetranychidae, Tenuipalpidae, Tarsonemidae, and Eriophyidae families, scolytid beetles; lepidopterous larvae in the Tortricidae, Limacodidae, and Geometridae families; and coccid scale insects (Cranham, 1966).

Probably the most successful case of classical biological control of a pest on tea involves the tea tortrix, *Harmona coffearia* Nietner, in Sri Lanka (Ceylon). The tortrix was a serious pest of tea from about 1910 to the late 1930s. The parasitoid *Macrocentrus harmonae* Nixon was collected in Java in 1935 and 1936 and liberated in small numbers in Ceylon, where it spread to all tea-growing districts by 1941. A decrease of tea tortrix from high to low numbers was coincident with increases in parasitization by *M. harmonae* from 1937 through 1941 (Gadd, 1941). The tortrix has since remained a minor pest, except for occasional high fluctuations in some areas or outbreaks attributed to destruction of the parasitoid by insecticides (Cranham, 1966).

The major tea pests in the Republic of Georgia (formerly of the Soviet Union) are reportedly controlled largely by native parasitoids and predators. These pests included three introduced scale insects, *Temnaspidiotus destructor* (Signoret) (coconut scale). *Aspidiotus transparens* Kuwana (transparent scale), and *Abgroaspis cyanophili* Signoret (cyanophilis scale); and an indigenous scale. *Chloropulvinaria floccifera* Westwood. Native natural enemies, including parasitoids in the Aphelinidae family and three species of Coccinellidae in combination with releases of 3000 to 5000 individuals per hectare of the imported coccinellid *Rhizobius setellus* Blackburn were effective in reducing infestations of the introduced species to low numbers in tea plantations. *Chloropulvinaria floccifera* was also observed to be reduced to low numbers by native Hymenoptera and Coccinellidae. Releases of *Cryptolaemus* sp. were recommended in plantations where native natural enemies were not exerting the desired control. Control of two lepidopterous pests. *Argyrothaemao pulhellana* Howard and *Sparganothus pilleriana* Schiffermüller, by native natural enemies usually needed to be supplemented by an application of entobacterin-3 (Gaprindashvili, 1975).

Although mites are considered the most serious pest of tea in some areas (Cranham, 1966; Banarjee & Cranham, 1985), little research has been done to evaluate or manipulate indigenous natural enemies or to introduce exotic species. The most detailed study on effects of natural enemies was conducted by Oomen (1982) in Indonesia on tenuipalpid mite *Brevipalpus phoenicus* (Geijskes). Differential predator levels were achieved by a pesticide exclusion method. In DDT-treated plots the most common species of Phytoseiidae were killed, permitting Stigmaeidae and an *Amblyseius* species to develop to high densities, and the pest mite *B. phoenicis* was maintained at considerably lower levels than in untreated plots. It was concluded that competition from the normally prevalent phytoseiids was inhibiting the less abundant but more effective predators of *B. phoenicis*. Oomen suggested that practices should be employed to limit the use of pesticides that are toxic to the Stigmaeidae. Laboratory tests showed the stigmaeid *Agis-*

temus denotatus Gonzalez to have a higher intrinsic rate of increase than *B. phoenicis* had and three stigmaeid species suppressed populations of *B. phoenicus* on excised leaves (Oomen, 1982).

COFFEE

Commercial species of coffee, the major one being *Coffea arabica* Linnaeus, are native to Africa. The coffee plant has been introduced to many tropical areas of the world, especially in Central and South America, where it has become a major export crop. Being a perennial crop, the coffee plant harbors many species of arthropods, over 850 species of insects being listed as feeding on this plant (Le Pelley, 1973). However, a relatively small number of the phytophagous species cause serious problems. The fact that natural enemies do keep many of the potential pests in check is illustrated by the outbreaks of certain species following large-scale applications of broad-spectrum pesticides or natural phenomena such as volcanic dust from eruptions of volcanos.

The most spectacular case of classical biological control of an insect pest of coffee is that of the coffee mealybug, *Planococcus kenyae* (Le Pelley). This insect was introduced from Uganda to Kenya sometime before 1938 and it became a devastating pest in coffee plantations (Le Pelley, 1968). It was initially misidentified, but after it was recognized to be the species indigenous to Uganda, investigations there revealed a complex of parasitoids, of which several were introduced to Kenya in 1938. *Anagyrus* species nr. *kivuensis* Compere quickly became established and in less than 2 years effective biological control of the mealybug in Kenya had been achieved (Le Pelley, 1968). Cases of pest resurgence have occurred following the use of certain broad-spectrum pesticides or from interference by the ant *Pheidole punctulata* Mayr. Thus, ant control measures may be necessary for optimal activity of the parasitoid (Le Pelley, 1973). It was estimated that at least £10 million had been saved by the biological control of coffee mealybug in Kenya (Melville, 1959).

COCONUT

Coconut is a major low elevation subsistence crop throughout the tropics, as well as a plantation crop of importance for production of vegetable oil. Major insect pests of coconut include foliage-feeding Coleoptera, Lepidoptera, and Orthoptera; phloem-feeding Homoptera; and and crown- and trunk-boring Coleoptera. Classical biological control has been used successfully against a number of such pests in many parts of the tropics. Some of the more important pest species that have been subjects of biological control efforts follow.

Important Pests

Coconut Rhinoceros Beetle, *Oryctes rhinoceros* (Linnaeus) (Scarabaeidae)

The coconut rhinoceros beetle is a large, robust scarab, adults of which bore into and feed on meristematic tissue in the growing points of coconut and other palms. Damage is often severe, and a single attack may be sufficient to kill or seriously impair otherwise vigorous trees. The larvae feed in rotting vegetable material such as dead coconut trunks, compost pits, and sawdust piles.

Clausen (1978a) stated that *O. rhinoceros* is native to southeast Asia, Indonesia, and possibly the Philippines and Formosa. It has been spread through commerce to Mauritius in the Indian Ocean, and through the islands of the western Pacific as far east as Fiji, Samoa, and Tonga. In newly invaded areas it has caused major losses due to reduced yields and dead trees. Gressitt (1953) reviewed the distribution and hosts of the beetle and gave a detailed account of its biology and ecology in the Palau Islands. Cumber (1957) summarized data on the outbreak in Western Samoa.

Biological control work against this pest has been reviewed by Clausen (1978a) and by Waterhouse and Norris (1987). Early work in Fiji and Palau involved attempts to colonize entomophagous insects such as *Scolia* spp. and various predaceous Coleoptera (Carabidae, Elateridae, and Histeridae). With the exception of *Scolia ruficornis* Fabricius from Zanzibar, which became established at several release sites, most such introductions apparently failed or provided no appreciable control. *Scolia ruficornis* has been reported to parasitize up to 30% of the grubs (Waterhouse & Norris, 1987), but its effectiveness is limited by the habits of the female wasps that seek hosts only in loose, friable materials, such as compost and sawdust, and ignore infested dead trees, which are a major breeding habitat.

Successful biological control of *O. rhinoceros* is attributed largely to the dissemination of the *Oryctes* baculovirus in those areas of the Indian and Pacific Oceans to which the beetle has spread during this century (Waterhouse & Norris, 1987). Discovered in southeast Asia, the baculovirus was first spread from Malaya to Western Samoa in 1967 (Marschall, 1970). Since then, the virus has been disseminated widely throughout the Pacific region and apparently has produced significant reductions in *Oryctes* populations and damage almost everywhere (Waterhouse & Norris, 1987).

Coconut Stick Insect, *Graeffea crousanii* (Le Gullon) (Phasmidae)

The coconut stick insect appears to be native to the South Pacific region. It occurs from New Caledonia to French Polynesia and at times has caused serious defoliation of coconuts, particularly in Fiji, Samoa, and Tonga (Clausen, 1978a; Waterhouse & Norris, 1987). Eupelmid egg parasites *Paranastatus nigriscutellatus* Eady and *P. verticalis* Eady occur naturally in Fiji and have been utilized in inundative release programs against this pest. One or both of these parasites were successfully introduced into Samoa and Tonga. Tachinid parasites associated with other phasmids in the Solomon Islands were introduced into Fiji during the 1960s, but failed to establish (Waterhouse & Norris, 1987).

Coconut Katydids, *Sexava* spp. (Tettigonaidae)

These katydids are widely distributed in the Pacific region (e.g., Indonesia, Papua-New Guinea, Palau, and Bismark Archipelago). In Indonesia, inundative releases of gregarious encyrtid egg parasite *Leefmansia bicolor* Waterson have been attempted but results are inconclusive (Munaan & Wikardi, 1986). A stylopid *Stictotrema sp.* has also been utilized in biological control efforts against *Sexava* spp. (Clausen, 1978a).

Coconut Scale, *Aspidiotus destructor* Signoret (Diaspididae)

The coconut scale, also called the transparent scale because of the semi-transparent scale covering, is a relatively polyphagous diaspidid that is distributed widely in tropical and subtropical regions. This pest probably originated somewhere in the Old World tropics, because species of the genus *Aspidiotus* are concentrated mainly in lands around the Indian Ocean (Ferris, 1941). During the twentieth century, *A. destructor* has extended its geographic range significantly, resulting in serious outbreaks in many newly infested areas, particularly on oceanic islands. Although primarily a pest of coconut, this scale has caused serious damage to several other tropical crops, such as banana, breadfruit, mango, and papaya. A "subspecies" (more likely a sibling species), *A. destructior rigidus* Reyne, has done major damage to coconuts in Indonesia.

The extensive biological control efforts that have been undertaken against the coconut scale have been summarized by Rosen and DeBach (1978) and by Waterhouse and Norris (1987). In almost every case, initial invasion of a new area by the scale has been followed by buildup of dense populations that caused economic damage. In most such areas the subsequent introduction of natural enemies has been followed by marked reductions in scale populations, and significant declines in economic losses due to scale damage. The detailed article by Taylor (1935) on the history of the scale in Fiji, one of the first insular regions that the scale invaded, provides an excellent example.

A number of predators and parasites have been utilized to combat coconut scale, and it is of interest that species that have been effective in one area against this pest sometimes have been ineffective or have failed to establish in others. For example, the coccinellid beetle *Cryptognatha nodiceps* Marshall, which was given major credit for the successful control of the scale in Fiji, has failed to become established in Tahiti, Hawaii, and Vanuatu. Other coccinellids (e.g., *Chilocorus nigritus* (Fabricius), *Pseudoscymnus anomalus* Chapin, *Rhizobius pulchellus* Montrouzier, *Telesimia nitida* Chapin, and others) have been successfully employed against this scale in several places. Parasitoids (e.g., *Aphytis* spp. and *Aspidiotiphagus* spp.) may be important in some areas (e.g., Hawaii; Beardsley, unpublished). Waterhouse and Norris (1987) have provided a tabular summary of reported biological control introductions against *A. destructor*. Schreiner (1990) reviewed biological control introductions into Micronesia against this pest. Undoubtedly the coconut scale will continue to extend its geographic range, and additional efforts to bring it under biological control will be required.

Red Coconut Scale, *Furcaspis oceanica* Lindinger (Diaspididae)

This armored scale apparently is endemic to one of the volcanic islands of Micronesia, where it is now widely distributed. Its known hosts are restricted to coconut, other palms, and *Pandanus* spp. In most areas indigenous parasites, particularly encyrtid wasp *Anabrolepis oceanica* Doutt, have kept it under reasonably good biological control. Accidental introductions into the Mariana island of Saipan (during the 1940s) and Guam (1960s) resulted in damaging outbreaks. An attempt was made during the 1950s to establish *A. oceanica* on Saipan but follow-up studies are lacking (Doutt, 1950a; Gardner, 1958).

Coconut Hispine Beetles, *Brontispa* spp. and *Promecotheca* spp. (Chrysomelidae)

Hispine beetles of the genera *Brontispa* and *Promecotheca* have caused serious damage to coconuts, particularly young trees, in many parts of the south and western Pacific. Adults and larvae of *Brontispa* spp. feed primarily on the surface of leaflets in young, unexpanded leaves, whereas larvae of *Promecotheca* spp. mine within leaves, and adults feed on the undersides of the distal portions of leaflets. The major pest species of *Brontispa* are *B. longissima* (Gestro) and *B. mariana* Spaeth. The former is probably native to Indonesia and New Guinea, but has been spread eastward through the Solomon Islands, Samoa, and French Polynesia (Waterhouse & Norris, 1987). *Brontispa mariana* is believed to have originated on one of the high islands of the Western Carolines (possibly Yap) and became a pest when introduced into the northern Mariana islands of Saipan and Rota (Clausen, 1978a), and on various atolls in the Caroline Islands. Additionally, *B. palauensis* (Esaki & Chujo) was accidentally introduced into Guam in 1978 (Muniappan, 1980), and *B. chalybeipennis* (Zacher) from Ponape and the Marshall Islands was found established at Honolulu, Hawaii, in 1985 (Beardsley, 1986).

In the genus *Promecotheca, P. coeruleipennis* (Blanchard) (= *reichei* Baly), indigenous to the South Pacific, caused major losses in Fiji until brought under biological control (Taylor, 1937; Clausen, 1978a). *Promecotheca papuana* (Csiki) has been a problem in New Guinea and the Bismarck Archipelago, as was *P. opacicollis* Gestro in the New Hebrides.

Biological control efforts against *Brontispa* species have involved mostly the introduction and spread of a gregarious internal parasitoid of larvae and pupae, *Tetrastichus brontispae* (Ferriere), originally from Indonesia. This natural enemy has been effective in some areas (e.g., Saipan against *B. mariana,* French Polynesia against *B. longissima*), but less so in others (Waterhouse & Norris, 1987; Doutt, 1950). Introductions of several other parasites have been attempted. A nonspecific egg parasitoid. *Trichogrammatoidea nana* (Zehntner), was successfully established in the Solomon Islands and possibly elsewhere, but apparently has not been particularly effective.

Biological control efforts against *Promecotheca* species were first attempted in Fiji during the early 1930s. The successful project there was described in detail by Taylor (1937). Success was attributed to the introduction from Java of a gregarious internal parasitoid of larvae and pupae, *Pediobius parvulus* (Ferriere) (Eulophidae). This parasitoid was subsequently introduced into other areas against *P. papuana* in New Guinea and the Bismarks, and against *P. opacicolis* in the New Hebrides. It was apparently effective in the latter case, but less satisfactory in the former (Clausen, 1978a).

Lepidopterous Pests

Levuana iridescens Bethune-Baker

A number of damaging lepidopterous pests of coconut flowers and foliage have been subjects of biological control attempts. The biological control of the zygaenid moth *Levuana iridescens* Bethune-Baker, which became a serious defoliator of coconut in Fiji during the 1920s, has been considered to be one of the outstanding successes in the Pacific region (Tothill *et al.,* 1930; Clausen, 1978a). The

introduction of tachinid fly *Bessa remota* (Aldrich) from Malaya was given major credit for this success.

The Coconut Flat Moth or Coconut Leaf Miner, *Agonoxena argaula* Meyrick (Agononenidae)

This is a serious but sporadic pest of coconut and other palms that is widely distributed in the South Pacific region and in Hawaii. Related species occur in Micronesia, Indonesia, Papua-New Guinea, and the Bismark and Solomon Islands. Waterhouse and Norris (1987) discussed the pest status and summarized biological control attempts against this pest. Chalcid wasp *Brachymeria agonoxenae* Fullaway was successfully introduced into Hawaii from American Samoa in 1948. Results of this importation have not been formally evaluated, although Lai (1988) lists this species as being under complete biological control.

Tirathaba spp.

The pest status and biological control of the coconut spike moths. *Tirathaba* spp., were summarized by Waterhouse and Norris (1987). Whether or not the flower-feeding larvae of these moths cause economically significant damage has been a matter of debate for many years.

BANANA

Banana Pests

Banana Skipper, *Erionota thrax* (Linnaeus) (Hesperiidae)

This butterfly is native to southeast Asia and Indonesia. During the second half of the twentieth century, it has extended its range into new insular areas, including New Guinea, Mauritius, Guam, and Hawaii, where it has caused substantial damage in both subsistence and commercial banana plantings.

Larvae of *E. thrax* feed in rolled leaf sections on banana and a few closely related plants (e.g., Manila hemp). When abundant, they can cause complete defoliation (Mau *et al.,* 1980). Complete biological control of this pest was achieved in Guam and in Hawaii through the introduction from southeast Asia of two parasitoid wasps, a gregarious larval parasite *Apanteles erionotae* Wilkinson, and an egg parasite *Ooencyrtus erionotae* Ferriere (Mau *et al.,* 1980; Lai, 1988).

Banana Scab Moth, *Lamprosema* (= *Nacoleia*) *octasema* (Meyrick) (Pyralidae)

The larvae of this pyralid feed on flowers and developing fruit of banana. Damage is caused not only by the destruction of young fruit but also more commonly by scarification of the skin of developing fruit. Unsightly scarification results in culling, or serious downgrading of fruit quality. Larvae also feed on flowers of *Heliconia* spp., *Pandanus* spp., and *Nipa* spp. palms.

Lamprosema octasema is known from southeast Asia, Indonesia, northeast Australia, New Guinea, and throughout the South Pacific region as far east as Samoa and Tonga. It is believed to be native to southeast Asia and Indonesia. Apparently, damage has been most severe in Pacific island plantations (Waterhouse & Norris, 1987).

Efforts to locate and introduce natural enemies to combat banana scab moth have been made in Fiji, New Britain, and Western Samoa. Introductions have been attempted with several species of parasitoids, but only the egg–larval braconid *Chelonus striatigenas* Cameron was successfully introduced into Fiji. It has been credited with achieving partial biological control (Rao *et al.,* 1971, Waterhouse & Norris, 1987).

Banana Root Borer, *Cosmopolites sordius* (Germar) (Curculionidae)

This curculionid beetle, also called the banana corm weevil, is a serious pest throughout most of the banana-growing regions of the world. It is presumed to be native to the Indo-Malayan region. Banana and manila hemp appear to be the only confirmed hosts. Damage is caused primarily by the tunneling and feeding activities of the larvae in the rootstock (corms) of the hosts. Weevil activity predisposes banana corms to rot and weakens the root system so that plants are easily wind-thrown.

Numerous attempts have been made to achieve biological control of *C. sordidus* (Clausen, 1978a; Waterhouse & Norris, 1987). The natural enemies that have been utilized are rather general predators and it appears that no parasitoids are presently known that attack it. Predators such as histerid beetle *Plaesius javanus* Erichson and hydrophilid beetle *Dactylosternum hydrophiloides*. Macleary, both from southeast Asia, have been distributed widely to combat *C. sordidus*. Several of these, particularly *P. javanus,* are now established at many locations in the Pacific and Caribbean regions. *Plaesius javanus,* together with *P. laevigatus* Marseul, appears to have reduced the pest status of the weevil in Fiji (Waterhouse & Norris, 1987), but for most sites where introductions have been made, very little information is available on the establishment or effectiveness of these predators, and the weevil still is a serious problem in many places. Waterhouse and Norris (1987) suggested the possibility that parasitic nematodes (such as those of the steinernematid and heterorhabditid groups) may prove to be useful against this pest.

Biological control of cassava pests is treated in Chapter 30.

References

Annecke, D. P., & Moran, V. C. (1982). Insects and mites of cultivated plants in South Africa. South Africa: Butterworth & Co.

Anonymous. (1978). Pest control: The case of the black fly shows the effectiveness of biological control. Noticias Agricolas, 8, 51–52.

Arambourg, Y. (1969). Inventaire de la biocoenose parasitaire de *Prays oleae* dans le Basin Mediterraneen. Entomophaga, 14, 185–194.

Arambourg, Y. (1970). Technique d'elevage et essais experimentaux de lachers de *Chelonus elaeaphilus* Silv. parasite de *Prays oleae* Bern. (teinge de l'olivier). In Colloque franco-sovietique sur l'utilisation des entomophages. Antibes: Annales Zoologie Ecologie Animale no. hors serie Institut Nationale Recherche Agronomie. 70–73, pp. 57–61).

Arambourg, Y., & Pralavorio, R. (1974). The ectophagous chalcidoids (Hym.: Chalcidoidea) parasitizing *Dacus oleae* Gmel. (Dip.: Trypetidae). Annales Institut Phytopathologique Benaki, 11, 30–46.

Arambourg, Y., Pralavorio, R., & Chabot, B. (1970). Possibilities d'elevage d'*Ageniaspis fuscicollis praysincola* Silv. parasite de *Prays oleae* Bern. (Lep.: Hyponomeutidae) sur on hôte de replacement. Annales de Zoologie—Ecologie Animale, 2, 657–658.

Argyriou, L. C. (1969). Biological control of citrus insects in Greece. In H. D. Champan (Ed.), Proceedings, First International Citrus Symposium, Riverside, California, 2, 817–822.

Argyriou, L. C., & Katsoyannis, P. (1976). Establishment of *Metaphycus helvolus* Compere in Kerkyra (Corfu) on *Saissetia oleae* (Olivier). Annales Institut Phypathologique Benaki, 11, 200–208.

Aubert, B., & Quilici, S. (1983). Nouvel quilibre biologique observ la Reunion sur les populations de psyllids l'introduction et l'establissement d'hymenoptères chalcidiens. Fruits, 38, 771–780.

Banarjee, B., & Cranham, J. E. (1985). Tea. In W. Helle & M. W. Sabelis (Eds.), Spider mites, their biology, natural enemies and control (Vol. 1B, pp. 371–374). Amsterdam: Elsevier Science.

Bartlett, B. R. (1953). Natural control of citricola scale in California. Journal of Economic Entomology, 46, 25–28.

Bartlett, B. R. (1969). The biological control campaigns against soft scales and mealybugs on citrus in California. In. H. D. Champan (Ed.), Proceedings, First International Citrus Symposium, Riverside, California, 2, 875–878.

Bartlett, B. R. (1978). Coccidae. In. C. P. Clausen (Ed.), Introduced parasites and predators of arthropod pests and weeds: A world review. Agriculture Handbook No. 480. Washington, DC: U.S. Department of Agriculture.

Bartlett, B. R., & Lloyd, D. C. (1958). Mealybugs attacking citrus in California—a survey of their natural enemies and the release of new parasites and predators. Journal of Economic Entomology, 51, 90–93.

Bartlett, B. R., & Medved, R. A. (1966). The biology of effectiveness of *Diversinervus elegans* (Encyrtidae: Hymenoptera), an imported parasite of lecaniine scale insects in California. Annals of the Entomological Society of America, 59, 974–976.

Bartra, C. E. (1974). Biology of *Selenaspidis articulatus* Morgan and its principal biological control agents. Revista Peruana de Entomologia, 17, 60–68.

Basinger, A. J. (1931). The European brown snail in California (Bulletin 515). University of California Agriculture Experiment Station.

Beardsley, J. W. (1955). Fluted scales and their biological control in United States administered Micronesia. Proceedings of the Hawaiian Entomological Society, 15, 391–399.

Beardsley, J. W. (1986). Notes and exhibitions: *Brontispa chalybeipennis* (Zacher). Proceedings of the Hawaiian Entomological Society, 27, 16.

Beatti, G. A. C. (1978). Biological control of citrus mites in New South Wales (pp. 156–158). In W. Grierson (Ed.), Proceedings of the International Society for Citriculture. Sidney, Australia.

Bedford, E. C. G. (1965). An attempt to control the Seychelles scale,

Icerya seychellarum (Westwood) (Homoptera: Coccidae) in South Africa by introducing *Cryptochaetum monophlebi* Skuse (Diptera: Cryptochaetidae). Journal of the Entomological Society of South Africa, 28, 155–165.

Bedford, E. C. G. (1968). The biological control of red scale *Aonidiella aurantii* (Mask) on citrus in South Africa. Journal of the Entomological Society of South Africa, 31, 17–28.

Bedford, E. C. G. (1973). Biological control proves successful. Citrus and Subtropical Fruit Journal, February, 5–11.

Bedford, E. C. G. (1989). The biological control of circular purple scale, Chrysomphalus aonidum (Linnaeus), on citrus in South Africa (Technical Communication, Department of Agriculture and Water Supplement, South Africa No. 218).

Bedford, E. C. G., & Thomas, E. D. (1965). Biological control of the citrus blackfly, *Aleurocanthus woglumi* (Ashby) (Homoptera: Aleyrodidae) in South Africa. Journal of Entomological Society South Africa, 28, 117–132.

Beingolea, O. (1955). The present status of *S. oleae* on olive in the Yauco and Ilo Valleys. Boletin Trimestad de Experimentacion Agropecuaria. 4, 18–22.

Beingolea, O. (1969). Biological control of citrus pests in Peru. In H. D. Champan (Ed.), Proceedings, First International Citrus Symposium, Riverside, California, 2, 827–838.

Benassy, C. (1977). Notes on parasites of some diaspine scale insects (*Chrysomphalus, Lepidosaphes, Unaspis*). Boletin informative de plagas-Ministerio de Agricultura, Servicio de Defensa Contra Plagas e Inspeccion Fitopatologica, 3, 55–73.

Benassy, C., & Bianchi, H. (1974). Observations sur *Aonidiella aurantii* Mask. et son parasite indigene *Comperiella bifasciata* How. (Hymenoptera. Encyrtidae). Bulletin S. R. O. P., 3, 39–50.

Benassy, C., & Euverte, G. (1970). Note on the action of two·species of *Aphytis* as biological control agents against two citrus coccids, *aonidiella aurantii* Mask and *Chrysomphalus dicryospermi* (Morg.) in Morocco. Annales de Zoologie—Ecologie Animale, 2, 357–372.

Benassy, C., & Pinet, C. (1987). On the introduction into France of *Aphytis yanonensis* DeBach & Rosen (Hym.: Aphelinidae), a parasite of the citrus arrowhead scale: *Unaspis yanonensis* Kuw. (Homopt: Diaspidinae). Comptes Rendus de l'Académie d'Agriculture de France, 73, 33–38.

Benassy, C., Bianchi, H., & E. Franco, (1974). Note sur l'introduction en France d'*Aphytis lepidosaphes* Comp. (Hymenopt., Aphelinidae) parasite de la cochenille virgule des Citrus (*Lepidosaphes beckii* Newm.) (Homopt., Diaspididae). Comptes Rendus de Académie Agriculture de France, 60, 191–196.

Bender, G. (1993). A new mite problem in avocado. California Grower, 16, 8–9.

Blumberg, D., & Swirski, E. (1977). Release and recovery of *Metaphycus* spp. (Hymenoptera: Encyritidae) imported for control of the Mediterranean black scale, *Saissetia oleae* (Olivier) in Israel. Phytoparasitica, 5, 115–118.

Bodenheimer, F. S. (1951). Citrus entomology in the Middle East. The Hague: W. Junk.

Callan, E. McC. (1943). Natural enemies of the cacao thrips. Bulletin of Entomological Research, 34, 313–321.

Campbell, M. M. (1976). Colonization of *Aphytis melinus* DeBach [Hymenoptera: Aphelinidae] in *Aonidiella aurantii* (Mask.) [Hemiptera: Coccidae] on citrus in South Australia. Bulletin of Entomological Research, 65, 659–668.

Campos, M., & Ramos, P. (1982). *Ageniaspis fuscicollis praysincola* Silv. (Hym.: Encyrtidae), a parasite of *Prays oleae* Bern. (Lep.: Hyponomeutidae) in Granada. Boletin Association Espanola Entomologie 4, 63–71.

Chien, C. C., Chiu, S. C., & Ku, S. C. (1998). Biological control of *Diaphorina citri* in Taiwan. Fruits, 44, 401–407.

Chiu, S. C., Lo, K. C., Chien, C. C., Chen, C. C., & Chen, C. F. (1985). A review of the biological control of crop pests in Taiwan (1981–

1984). (Special Publication No. 19, pp. 1–8) Taiwan Agricultural Research Institute.

Cilliers, C. J. (1971). Observations on circular purple scale *Chrysomphalus aonidum* (Linn.), and two introduced parasites in Western Transvaal citrus orchards. Entomophaga, 16, 269–284.

Clausen, C. P. (1915). Mealybugs of citrus trees Bulletin 258, pp.19–48) California Agricultural Experiment Station.

Clausen, C. P. (1927). The citrus insects of Japan (Technical Bulletin 15). Washington, DC: U.S. Department of Agriculture.

Clausen, C. P. (1956). Biological control of insect pests in the continental United States (Technical Bulletin No. 1139) U.S. Department of Agriculture.

Clausen, C. P. (1978a) (Ed.). Introduced parasites and predators of arthropod pests and weeds: A world review. Agriculture Handbook No. 480, Washington, DC: U.S. Department of Agriculture.

Clausen, C. P. (1978b). Biological control of citrus insects. In The citrus industry (Vol IV) Berkeley: University of California Division of Agricultural Science.

Clausen, C. P. (1978c). Aleyrodidae. In C. P. Clausen (Ed.). Introduced parasites and predators of arthropod pests and weeds. Agriculture Handbook. No. 480 (pp. 27–35). Washington, DC: U.S. Department of Agriculture.

Clausen, C. P., & Berry, P. A. (1932). The citrus blackfly in Asia, and the importation of its natural enemies into tropical America (Technical Bulletin No. 320). Washington, DC: U.S. Department of Agriculture.

Commeau, J., & Sola, E. (1964). Une nouvelle cochenille des agrumes sur la Cote d'Azur. Phytoma, 16, 49–50.

Compere, H. (1924). A preliminary report on the parasitic enemies of the citricola scale (*Coccus pseudomagnoliarum* [Kuwana]) with descriptions of two new chalcidoid parasites. Bulletin of Southern California Academy of Science, 24, 113–123.

Compere, H. (1939). Mealybugs and their insect enemies in South America. University of California Publications in Entomology, 7, 57–73.

Compere, H. (1961). The red scale and its natural enemies. Hilgardia, 31, 173–278.

Compere, H., & Smith, H. S. (1932). The control of the citrophilus mealybug, *Pseudococcus gahani,* by Australian parasites. Hilgardia, 6, 585–618.

Cranham, J. E. (1966). Tea pests and their control. Annual Review of Entomology, 11, 491–514.

Craw, A. (1891). Internal parasites discovered in the San Gabriel Valley; recommendations and notes. Bulletin of California State Board of Horticuture, 57, 1–7.

Crouzel, De, I. S., Bimboni, H. G., Zanelli, M., & Botto, E. N. (1973). Lucha biologica contra la "conchinilla roja australiana" *Anonidiella aurantii* (Maskell) (Hom. Diaspididae) en citricos. Patologia Vegetal Serie 5, 10, 251–318.

Cumber, R. A. (1957). The rhinoceros beetle in Western Samoa. (Technical Paper No. 107). South Pacific Commission.

Daane, K. M., Barzman, M. S., Kennett, C. E., & Caltagirone, L. E. (1991). Parasitoids of black scale in California: Establishment of *Proccoccophagus probus* Annecke & Mynhardt and *Coccophagus rusti* Compere (Hymenoptera: Aphelinidae) in olive orchards. Pan-Pacific Entomologist, 67, 99–106.

Dean, H. A. (1955). Factors affecting biological control of scale insects on Texas *Citrus.* Journal of Economic Entomology, 48, 444–447.

Dean, H. A. (1961). *Aphytis lepidosaphes* (Hymenoptera: Chalcidoidea), an introduced parasite of purple scale. Annals of the Entomological Society of America, 54, 918–920.

Dean, H. A. (1982). Reduced pest status of the Florida red scale on Texas citrus associated with *Aphytis holoxanthus.* Journal of Economic Entomology, 75, 147–149.

DeBach, P. (1943). The importance of host feeding by adult parasites in the reduction of host populations. Journal of Economic Entomology, 36, 647–658.

DeBach, P. (1948). The establishment of the Chinese race of *Comperiella bifasciata* on *Aonidiella aurantii* in southern California. Journal of Economic Entomology, 41, 985.

DeBach, P. (1949). Population studies of the long-tailed mealybug and its natural enemies on citrus tress in southern California, 1946. Ecology, 30, 14–25.

DeBach, P. (1953). The establishment in California of an oriental strain of *Prospaltella perniciosi* Tower on the California red scale. Journal of Economic Entomology, 46, 1103.

DeBach, P. (1955). Validity of the insecticidal check method as a measure of the effectiveness of natural enemies of diaspine scale insects. Journal of Economic Entomology, 48, 584–588.

DeBach, P. (1959). New species and strains of *Aphytis* (Hymenoptera, Eulophidae) parasitic on the California red scale, *Aonidiella aurantii* (Mask.) in the Orient. Annals of the Entomological Society of America, 52, 354–362.

DeBach, P. (1962). Ecological adaptation of parasites and competition between parasite species in relation to establishment and success. 11th International Congress of Entomology, 2, 687–690.

DeBach, P. (1964). Successes, trends and future possibilities. In P. DeBach (Ed.), Biological control of insect pests and weeds. New York: Reinhold.

DeBach, P. (1965a). Some biological and ecological phenomena associated with colonizing entomophagous insects. In H. G. Baker & G. L. Stebbins (Eds.), The genetics of colonizing species (pp. 287–306). New York: Academic Press.

DeBach, P. (1965b). Weather and the success of parasites in population regulation. Canadian Entomologist, 97, 848–863.

DeBach, P. (1966). The competitive displacement and coexistence principles. Annual Review of Entomology, 11, 183–212.

DeBach, P. (1969). Biological control of diaspine scale insects on citrus in California. In H. D. Champan (Ed.), Proceedings, First International Citrus Symposium, Riverside, California, 2, 801–815.

DeBach, P. (1971). Fortuitous biological control from ecesis of natural enemies. In Entomological essays to commemorate the retirement of Professor K. Yasumatsu. Tokyo: Hokuryukan.

DeBach, P. (1974). Biological control by natural enemies. London: Cambridge University Press.

DeBach, P. (1977). A newly imported California red scale parasite from Saudi Arabia. California Agriculture, 31(12), 6–7.

DeBach, P., & Argyriou, L. C. (1967). The colonization and success in Greece of some imported *Aphytis* spp. (Hym. Aphelinidae) parasitic on citrus scale insects (Hom. Diaspididae). Entomophaga, 12, 325–342.

DeBach, P., & Hagen, K. S. (1964). The conservation and augmentation of natural enemies. In P. DeBach (Ed.), Biological control of insect pests and weeds (pp. 429–458). New York: Reinhold.

DeBach, P., & Landi, J. (1961). The introduced purple scale parasite. *Aphytis lepidosaphes* Compere, and a method of integrating chemical with biological control. Hilgardia, 31, 459–497.

DeBach, P., & Rose, M. (1976). Biological control of woolly whitefly. California Agriculture, 30 (5), 4–7.

DeBach, P., & Sundby R. A. (1963). Competitive displacement between ecological homologues. Hilgardia, 34, 105–166.

DeBach, P., & Warner, S. C. (1969). Research on biological control of whiteflies. Citrograph, 54, 301–303.

DeBach, P., Fleschner, C. A., & Dietrick, E. J. (1949). Population studies of the long-tailed mealybug and its natural enemies on citrus in southern California, 1947. Journal of Economic Entomology, 42, 777–782.

DeBach, P., Hendrickson, R. M., Jr. & Rose, M. (1978). Competitive displacement: Extinction of the yellow scale, *Aonidiella citrina* (Coq.) (Homoptera: Diaspididae) by its ecological homologue, the California red scale, *Aonidiella aurantii* (Mask.) in southern California. Hilgardia, 46, 1–35.

DeBach, P., Landi, J. H., & White, E. B. (1955). Biological control of red scale. Citrograph, 40, 254, 271–272.

DeBach, P., Rosen, D., & Kennett, C. E. (1971). Biological control of

coccids by introduced natural enemies. In C. B. Huffaker (Ed.), Biological control (pp. 165–194). New York: Plenum Press.

DeBach, P., Dietrick, E. J., Fleschner, C. A., & Fisher, T. W. (1950). Periodic colonization of *Aphytis* for control of the California red scale. Preliminary tests, 1949. Journal of Economic Entomology, 43, 783–802.

Delanoue, P. (1960). Essai d'elevage artificiel permanent d'*Opius concolor* Szpl. sur un hôte intermediare et lachers experimentaux de ce parasite de *Dacus oleae* Gmel. dans les Alps-Maritimes. Fed. Internat. Oleic. 10, 10 pp. Federacion Internationale Oleiculture.

DeLotto, G. (1976). On the black scales of southern Europe (Homoptera: Coccoidea: Coccidae). Journal of the Entomological Society of South Africa, 39, 147–149.

Dohanian, S. M. (1937). Life history of the thrips parasite *Dasyscapus parvipennis* Gahan and the technique for breeding it. Journal of Economic Entomology, 30, 78–80.

Doutt, R. L. (1950a). The parasite complex of *Furcaspis oceanica*. Annals of the Entomological Society of America, 43, 501–507.

Doutt, R. L. (1954). An evaluation of some natural enemies of olive scale. Journal of Economic Entomology, 47, 39–43.

Doutt, R. L. (1958). Vice, virtue and the vedalia. Bulletin of the Entomological Society of America, 4, 119–123.

Doutt, R. L. (1966). A taxonomic analysis of parasitic Hymenoptera reared from *Parlatoria oleae* (Colvée). Hilgardia, 37, 219–231.

Dowell, R. V., Fitzpatrick, G. E., & Reinert, J. A. (1979). Biological control of citrus blackfly in southern Florida. Environmental Entomology, 8, 595–597.

Duran, M. (1944). Un enemigo natural de la Saissetia oleae (Bern.) nuevo para Chile (Agricultural Technique, 4 pp. 255–256). Reading, United Kingdom: Lamport Gilbert & Co.

Ebeling, W. (1959). Subtropical fruit pests. University of California Division of Agricultural Science.

Essig, E. O. (1931). A history of entomology. New York: Macmillan.

Fabres, G. (1974). Contribution l'etude d'*A lepidosaphes* (Hym. Aphelinidae) parasites de *Lepidosaphes beckii* (Hom. Diaspididae) en Nouvelle Caldonie. Annales de la Societe Entomologique de France, 10, 371–379.

Fenili, G. A., & Pegazzano, F. (1962). Experimento di introduzioni in Toscana di l'*Opius siculus* Monastero (= *concolor* Szepl.) parasitta del *Dacus oleae* Gmel. Redia, 47, 172–187.

Ferragut, F., Garcia-Mari, F., Costa-Comelles, J., & Laborda, R. (1987). Influence of food and temperature on development and oviposition of *Euseius stipulatus* and *Typhlodromus phialatus* (Acari: Phytoseiidae). Experimental and Applied Acarology (Northwood), 3, 317–329.

Ferris, G. F. (1941). The genus *Aspidiotus* (Homoptera: Coccoidea: Diaspididae). Microentomology, 6, 33–69.

Finney, G. L. (1966). The culture of *Coccophagoides utilis* Doutt, a parasite of *Parlatoria oleae* (Colvée). Hilgardia, 37, 337–343.

Fisher, F. E., Griffiths, J. T., Jr., & Thompson, W. L. (1949). An epizootic of *Phyllocoptruta oleivora*. Phytopathology, 39, 510–512.

Fisher, T. W. (1963). Mass culture of *Cryptolaemus* and *Leptomastix*—natural enemies of citrus mealybug (Bulletin 797). California Agriculture Experiment Station.

Fisher, T. W. (1966). *Rumina decollata* (Linnaeus, 1758) (Achatinidae) discovered in southern California. Veliger, 9, 16.

Fisher, T. W., & Orth, R. E. (1985). Observations of the snail Rumina decollata Linnaeus, 1758 (Stylommatophora: Subulinidae) with particular reference to its effectiveness in the biological control of Helix aspersa Müller, 1774 (Stylommatophora: Helicidae) in California (Occasional Papers, No. 1). Riverside: Department of Entomology, University of California.

Fisher, T. W., Bailey, J. B., & Sakovich, N. J. (1983). A new approach: Skirt pruning, trunk treatment for snail control. Citrograph, 68, 292–294, 296–297.

Fisher, T. W., Orth, R. E., & Swanson, S. C. (1980). Snail against snail. California Agriculture, 34 (10–11), 18–20.

Flanders, S. E. (1940). Biological control of the long-tailed mealybug, *Pseudococcus longispinus*. Journal of Economic Entomology, 33, 754–759.

Flanders, S. E. (1943). Mass production of the California red scale and its parasite *Comperiella bifasciata*. Journal of Economic Entomology, 36, 233–235.

Flanders, S. E. (1944a). The introduction and establishment of *Habrolepis rouxi* in California. Journal of Economic Entomology, 37, 444–445.

Flanders, S. E. (1944b). Observations on *Comperiella bifasciata,* an endoparasite of diaspine coccids. Annals of the Entomological Society America, 37, 365–371.

Flanders, S. E. (1944c). Control of the long-tailed mealybug on avocados by hymenopterous parasites. Journal of Economic Entomology, 37, 308–309.

Flanders, S. E. (1951). Citrus mealybug. Four new parasites studied in biological control experiments. California Agriculture, 5(7), 11.

Flanders, S. E. (1953). Aphelinid biologies with implications for taxonomy. Annals of the Entomological Society America, 46, 84–94.

Flanders, S. E. (1956). Struggle for existence between red and yellow scale. Citrograph, 41, 396, 398, 400, 402–403.

Flanders, S. E. (1958). *Moranila californica* as a usurped parasite of *Saissetia oleae*. Journal of Economic Entomology, 51, 247–248.

Flanders, S. E. (1969). Herbert Smith's observations on citrus blackfly parasites in India and Mexico and the correlated circumstances. Canadian Entomologist, 101, 467–480.

Flanders, S. E., & Bartlett, B. R. (1964). Observations on two species of *Metaphycus* (Encyrtidae, Hymenoptera) parasitic on citricola scale. Mushi, 38, 39–42.

Flanders, S. E., Gressitt, J. L., & Fisher, T. W. (1958). *Casca chinensis,* an internal parasite of California red scale. Hilgardia, 28, 65–91.

Fleschner, C. A. (1954). Biological control of avocado pests. California Avocado Society Yearbook, 38, 125–129.

Fleschner, C. A., Hall, J. C., & Ricker, D. W. (1955). Natural balance of mite pests in an avocado grove. California Avocado Society Yearbook, 39, 155–162.

Friese, D. D. (1985). Factors influencing competition between E. hibisci and E. stipulatus (Acarina: Phytoseiidae). Unpublished doctoral dissertation, University of California, Riverside, December 1985.

Furness, G. O., Buchanan, G. A., George, R. S., & Richardson, N. L. (1983). A history of the biological and integrated control of red scale, *Aonidiella aurantii* on citrus in the Lower Murray Valley of Australia. Entomophaga, 28, 199–212.

Furuhashi, K., & Nishino, M. (1983). Biological control of arrowhead scale. *Unaspis yanonensis,* by parasitic wasps introduced from the People's Republic of China. Entomophaga, 28, 277–286.

Furuhashi, K., & Ohkubo, N. (1990). Use of parasitic wasps for controlling the arrowhead scale, Unaspis yanonensis (Hom.: Diaspididae), in Japan. In FFTC-NARC International Seminar, Tsukuba Sci. City, Ibaraki, October 1989.

Gadd, C. H. (1941). The control of tea tortrix by its parasite *Macrocentrus homonae*. Tea Quarterly, 14, 93–87.

Gaprindashvili, N. K. (1975). Biological control of the main pests on tea plantations in the Georgian SSR. Eighth International Plant Protection Congress, 1975, 3, 29–33.

Garcia-Mari, F., Ferragut, F., Costa-Comelles, J., & Marzal, C. (1984). Population dynamics of the citrus red mite *Panonychus citri* (McG.) and its predators in Spanish citrus orchards. Proceedings, 1984 International Society of Citriculture. São Paulo, Brazil.

Gardner, T. R. (1958). Biological control of insect and plant pests in the Trust Territory of Guam. Proceedings, 10th International Congress of Entomology 4, 465–469.

Genduso, P., & Ragusa, S. (1969). Lotta biologica artificiale contro la mosca delle olive a mezzo dell' *Opius c. siculus* Mon. in Puglia nel 1968. Bollettino dell'Instituto di Entomologia Agricultura Palermo, 7, 196–216.

Gerson, U., & Schneider, R. (1981). Laboratory and field studies of the mite *Hemisarcoptes coccophagus* Meyer (Astigmata: Hemisarcoptidae), a natural enemy of armored scale insects. Acarologia, 22, 199–208.

Gomez-Clemente, F. (1954). The present situation in the biological control of some citrus scales (*Planococcus citri* and *Icerya purchasi*). Bollettino Patologia Vegetale Entomologia Agricultura, 19, 19–35.

Gonzalez, R. (1969). Biological control of citrus pests in Chile. In H. D. Chapman (Ed.), Proceedings of the First International Citrus Symposium, University of California, Riverside, 2, 839–847.

Graf Marin, A., & Cortes Pena, R. (1939). The introduction of parasites into Chile against insect pests: A summary of the importations and their results. Proceedings, Sixth Pacific Scientific Congress, 4, 351–357.

Greathead, D. J. (1976). A review of biological control in western and southern Europe (Technical Communication No. 7). Slough, United Kingdom: CIBC. Commonwealth Agricultural Bureau, Farnham Royal.

Gressitt, J. L. (1953). The coconut rhinoceros beetle (Oryctes rhinoceros) with particular reference to the Palau Islands (Bulletin 212). Honolulu: B. P. Bishop Museum.

Gressitt, J. L., Flanders, S. E., & Bartlett, B. (1954). Parasites of citricola scale in Japan, and their introduction into California. Pan-Pacific Entomologist, 30, 5–9.

Haney, P. B., Luck, R. F., & Moreno, D. (1987). Increases in densities of the citrus red mite *Panonychus citri* (Acarina: Tetranychidae) in association with the Argentine ant *Iridomyrmex humilis* (Hymenoptera: Formicidae), in southern California citrus. Entomophaga 32:49–57.

Harpaz, I., & Rosen, D. (1971). Development of integrated control programs for crop pests in Israel. In C. B. Huffaker (Ed.), Biological control (chap. 20). New York: Plenum Press.

Hart, W. G., Ingle, S., & Garza, M. (1969). Current status of brown soft scale in citrus groves of the Lower Rio Grande Valley. Annals of the Entomological Society of America, 62, 855–858.

Herrara, J. M. (1964). Ciclos biologicos de los queresas de las citricos en la costa central. Peruana Entomologia, 7, 1–8.

Hessein, N. A., & McMurtry, J. A. (1988). Observations of *Megaphragma mymaripenne* Timberlake (Hymenoptera: Trichogrammatidae), an egg parasite of *Heliothrips haemorrhoidalis* Rouch (Thysanoptera: Thripidae). Pan-Pacific Entomologist, 64, 250–254.

Huang, M. (1978). Studies on the integrated control of the citrus red mite with the predaceous mite as a principal controlling agent. Acta Entomologica Sinica, 21, 261–270.

Huang, M. (1981). Biological control of citrus red mite, *Panonychus citri* (McG.) in Guandong Province. Proceedings, International Society of Citriculture 1981, 2, 643–646.

Huffaker, C. B., & Kennett, C. E. (1966). Biological control of *Parlatoria oleae* (Colvée) through the compensatory action of two introduced parasitoids. Hilgardia, 37, 283–335.

Huffaker, C. B., Kennett, C. E., & Finney, G. L. (1962). Biological control of olive scale, *Parlatoria oleae* (Colvée), in California by imported *Aphytis maculicornis* (Masi) (Hymenoptera: Aphelinidae). Hilgardia, 32, 541–636.

Huffaker, C. B., Kennett, C. E., & Tassan, R. L. (1986). Comparisons of parasitism and densities of *Parlatoria oleae* (1952–1982) in relation to ecological theory. American Naturalist, 128, 379–393.

Hull, L. A., Assquith, D., & Mowery, P. D. (1977). The mite searching ability of *Stethorus punctum* within an apple orchard. Environmental Entomology, 6, 684–688.

Izhar, Y., Wysoki, M., & Gur, L. (1979). The effectiveness of *Bacillus thuringiensis* Berliner on *Boarmia selenaria* Schiff (Lepidoptera: Geometridae) in laboratory tests and field trials. Phytoparasitica, 72, 65–77.

Jeppson, L. R., Baker, E. W., & Keifer, H. H. (1975). Mites injurious to economic plants. Berkeley: University of California Press.

Jiang, H., & Gu, H. G. (1983). An investigation on parasitoids of red wax scale, *Ceroplastes rubens* Maskell. Natural Enemy Insects, 5, 249–250.

Keetch, D. P. (1972). Ecology of the citrus red mite. *Panonychus citri* (McGregor) (Acarina: Tetranychidae) in South Africa III. The influence of the predacious mite, *Amblyseius addoenis* van der Merwe & Ryke. Journal of Entomological Society of South Africa, 35, 69–79.

Kennett, C. E. (1973). Biological control of California red scale and yellow scale in San Joaquin Valley citrus groves—a progress report. Sunkist Newsletter, 432, 1–2.

Kennett, C. E. (1980). Occurrence of *Metaphycus bartletti* Annecke & Mynhardt, a South African parasite of black scale, *Saissetia oleae* (Olivier) in central and northern California (Hymenoptera: Encyrtidae; Homoptera: Coccidae). Pan-Pacific Entomologist, 56, 107–110.

Kennett, C. E. (1986). A survey of the parasitoid complex attacking black scale, *Saissetia oleae* (Olivier), in central and northern California (Hymenoptera: Chalcidoidea; Homoptera: Coccidae). Pan-Pacific Entomologist, 62, 363–369.

Kennett, C. E. (1988). Results of exploration for parasitoids of citricola scale *Coccus pseudomagnoliarum* (Homoptera: Coccidae), in Japan and their introduction in California. Kontyu, 56, 445–457.

Kennett, C. E., Flaherty, D. L. & Hoffman, R. W. (1979). Effect of wind-borne pollens on the population dynamics of *Amblyseius hibisci* (Acarina: Phytoseiidae). Entomophaga, 24, 83–98.

Kennett, C. E., Huffaker, C. B., & Finney, G. L. (1966). The role of an autoparasitic aphelinid, *Coccophagoides utilis* Doutt, in the control of *Parlatoria oleae* (Colvée). Hilgardia, 37, 255–282.

Kfir, R., & Luck, R. F. (1979). Effects of constant and variable temperature extremes on sex ratio and progeny production by *Aphvtis melinus* and *A. lingnanensis* (Hymenoptera: Aphelinidae). Ecological Entomology, 4, 335–344.

Kim, H. S., Moon, D. Y., Park, J. S., Lee, S. C., Lippold, P. C., & Chang, Y. D. (1979). Studies on integrated control of citrus pests. II. Control of ruby scales (*Ceroplastes rubens*) on citrus by introduction of a parasite natural enemy, *Anicetus benificus* (Hymenoptera, Encyrtidae). Korean Journal of Plant Protection, 18, 107–110.

Kuwana, I. (1922). Studies on Japanese Monophlebinae. Contribution II. The genus Icerva (Plant Quarterly Station Bulletin) Japan Department of Agriculture Com. Imp. Yokohama, Japan.

Kuwana, I., & Ishii, T. (1927). On *Prospaltella smithi* Silv., and *Cryptognatha* sp., the enemies of *Aleurocanthus spiniferus*, imported from Canton, China. Journal of Okitsu Horticultural Society, 22, 77–80.

Lai, P. Y. (1988). Biological control, a positive point of view. Proceedings of the Hawaiian Entomological Society, 28, 179–190.

LePelley, R. H. (1968). Pests of coffee. London: Longmans.

LePelley, R. H. (1973). Coffee insects. Annual Review of Entomology, 18, 121–142.

Lindquist, E. (1983). Some thoughts on the potential use of mites in biological control, including a modified concept of "parasitoids." In M. A. Hoy, G. L. Cunningham, & L. Knutson (Eds.), Biological control of pests by mites. Berkeley: University of California Division of Agricultural Natural Resources.

Liotta, G., & Mineo, G. (1968). Lotta biologica artificiale contro la mosca delle olive a mezzo dell'*Opius concolor siculus* Mon. in Siciia nel 1968. Bollettino dell' Istituto di Entomologia Agricultura Palermo, 7, 183–196.

Luck, R. F. (1981). Parasitic insects introduced as biological control agents for arthropod pests. Handbook of pest management in agriculture (Vol. 2, pp. 125–284). Boca Raton, FL: CRC Press.

Maltby, W. L., Jimenez Jimenez, E., & DeBach, P. (1968). Biological control of armored scale insects in Mexico. Journal of Economic Entomology, 61, 1086–1088.

Marschall, K. J. (1970). Introduction of a new virus disease of coconut rhinoceros beetle in Western Samoa. Nature (London), 225, 288–289.

Mau, R. F. L., Kumashiro, M. B., & Teramoto, K. (1980). Biological control of the banana skipper *Pelopidas thrax* (Linnaeus) (Lepidoptera: Hesperiidae) in Hawaii. Proceedings of the Hawaiian Entomological Society, 23, 231–237.

McCoy, C. W., & Albrigo, L. G. (1975). Feeding injury to the orange caused by the citrus rust mite, *Phyllocoptruta oleivora* (Prostigmata: Eriophyidae). Annals of the Entomological Society of America, 68, 289–297.

McCoy, C. W., & Couch, T. L. (1982). Microbial control of the citrus rust mite with the mycoacaricide Mycar. Florida Entomologist, 65, 116–127.

McCoy, C. W., & Kanavel, R. F. (1969). Isolation of *Hirsutella thompsonii* from the citrus rust mite *Phyllocoptruta oleivora,* and its cultivation on various synthetic media. Journal of Invertebrate Pathology, 14, 386–390.

McCoy, C. W., Selhime, A. G., Kanavel, R. F., & Hill, A. J. (1971). Suppression of citrus red mite populations with applications of fragmented mycelia of *Hirsutella* thompsonii. Journal of Invertebrate Pathology, 17, 270–276.

McKenzie, H. L. (1937). Morphological differences distinguishing California red scale, yellow scale and related species. California University Publications in Entomology, 6, 323–336.

McKenzie, H. L. (1952). Scale studies. X. Distribution and biological notes on the olive pariatoria scale *Parlatoria oleae* (Colvée) in California (Homoptera; Coccoidea: Diaspididae). Bulletin of California Department of Agriculture, 41, 127–138.

McMurtry, J. A. (1969). Biological control of citrus red mite in California. In H. D. Chapman (Ed.), Proceedings of the First International Citrus Symposium, University of California, Riverside, 2, 855–862.

McMurtry, J. A. (1977). Biological control of citrus mites. (Vol. 2, pp. 855–862). In W. Grierson (Ed), Proceedings of the International Society of Citriculture. Orlando, Florida: Painter Publishing.

McMurtry, J. A. (1982). The use of phytoseiids for biological control: Progress and future prospects. In M. A. Hoy (Ed.), Recent advances in knowledge of the Phytoseiidae (pp. 23–48). Berkeley: University of California Division of Agricultural Sciences.

McMurtry, J. A. (1985). Avocado. In W. Helle & M. W. Sabelis (Eds.), Spider mites, their biology, natural enemies and control (Vol. 1B, pp. 327–337). Amsterdam: Elsevier Science.

McMurtry, J. A. (1985). Citrus. In W. Helle & M. W. Sabelis (Eds.), Spider mites, their biology, natural enemies and control (Vol. 1B, pp. 339–347). Amsterdam: Elsevier Science.

McMurtry, J. A. (1989). Utilizing natural enemies to control pest mites on citrus and avocado in California, USA. In G. P. Channabasavanna & C. A. Viraklamath (Eds.), Proceedings of the Seventh International Congress of Acarology, 1986 (pp. 325–337). New Delhi: Oxford & IPH Publishing.

McMurtry, J. A., & Johnson, H. G. (1963). Progress report on the introduction of a thrips parasite from the West Indies. California Avocado Society Yearbook, 47, 48–51.

McMurtry, J. A., & Johnson, H. G. (1966). An ecological study of the spider mite *Oligonychus punicae* (Hirst) and its natural enemies. Hilgardia, 37, 363–402.

McMurtry, J. A., & Scriven, G. T. (1964a). Studies on the feeding, reproduction and development of *Amblyseius hibisci* (Acarina: Phytoseiidae) on various food substances. Annals of the Entomological Society of America, 57, 649–655.

McMurtry, J. A., & Scriven, G. T. (1964b). Biology of the predaceous mite *Typhlodromus rickeri* (Acarina: Phytoseiidae). Annals of the Entomological Society of America, 57, 362–367.

McMurtry, J. A., Johnson, H. G., & Scriven, G. T. (1969). Experiments to determine effects of mass releases of *Stethorus picipes* on the level of infestation of the avocado brown mite. Journal of Economic Entomology, 62, 1216–1221.

McMurtry, J. A., Johnson H. G. & Newberger, S. J. (1991). Greenhouse thrips parasitoid established on avocado in California. California Agriculture, 45(60), 31–32.

McMurtry, J. A., Shaw, J. G., & Johnson, H. G. (1979). Citrus red mite populations in relation to virus disease and predaceous mites in southern California. Environmental Entomology, 8, 160–164.

Melville, A. R. (1959). The place of biological control in the modern science of entomology. Kenya Coffee, 24, 81–85.

Milne, W. M. (1981). Insecticidal versus natural control of white wax scale (*Gascardia destructor*) at Kenthurst, N.S.W., during 1972–73. Journal of Australian Entomological Society, 20, 167–170.

Mineo, G. (1967). On *Cryptolaemus montrouzieri* Muls. (Observations on morphology and bionomics). Bolletino dell' Istituto del Entomologia Agricultura Fitopatologia Palermo, 6, 99–143.

Monaco, R. (1969). The action taken against *D. oleae* by *O. concolor* distributed in Apulia in the Gargano olive groves, and by the indigenous parasites in the same habitat. Entomologica, 5, 139–191.

Monastero, S. (1931). Un nuevo parasitto endofago della mosca delle olive trovato in Altanilla Milicia (Sicilia) (Fam. Braconidae, Gen. *Opius*). Atti della Accademia Palermo, 16, 95–201.

Monastero, S. (1968). New large-scale tests on the biological control of the olive fly *(D. oleae)* by means of *O. concolor siculus* in Sicily in 1967. Entomophaga, 13, 251–261.

Monastero, S., & Delanoue, P. (1967). First large-scale test of biological control of *D. oleae* by *O. c. siculus* in Sicily. Entomophaga, 12, 381–398.

Monastero, S., & Genduso, P. (1964). Esperimiento di lotta biologica artificiale controla mosca delle olive (*Dacus oleae* Gmel.) esequiti, nel 1964, nell'Isola di Salina (Eolie). Bollettino Istituto de Entomologia Agricultura Osserv Fitopatologia Palermo, 5, 281–289.

Muma, M. H. (1958). Predators and parasites of citrus mites in Florida. Proceedings, 10th International Congress on Entomology, 4, 633–647.

Muma, M. H. (1967). Biological notes on *Coniopteryx vicina* (Neuroptera: Coniopterygidae). Florida Entomologist, 50, 285–293.

Muma, M. H. (1969). Biological control of various insects and mites on Florida citrus. In H. D. Chapman (Ed.), Proceedings of the First International Citrus Symposium, University of California, Riverside, 2, 863–870.

Muma, M. H. (1970). Natural control petential of *Galendromus floridanus* (Acarina: Phytoseiidae) on Florida citrus trees. Florida Entomologist, 53, 79–88.

Munaan, A., & Wikardi, E. (1986). Towards the biological control of coconut pests in Indonisia. In M. Y. Hussein & A. G. Ibrahim (Eds.), Biological control in the tropics (pp. 149–158). Malaysia: Penerbit Universiti Pertanian.

Muniappan, R. (1980). Biological control of the Palau coconut beetle, *Brontispa palauensis* (Esaki & Chujo), on Guam. Micronesica, 16, 359–360.

Murakami, V. (1975). Biological and ecological studies on the parasites of *Unaspis yanonensis* (Kuwana) in Japan. JIBP Synthesis, 7, 125–131.

Nakao, H. K., & Funasaki, G. Y. (1976). Introductions for biological control. Proceedings of the Hawaiian Entomological Society, 22, 329–331.

Niccoli, A., & Tiberi, R. (1985). Use of *Bacillus thuringiensis* Berliner in the control of injurious insects in field and forest environments. Redia, 68, 305–322.

Nishida, K., & Yanai, S. (1965). Parasitization of *Aphytis lingnanensis* Compere to *U. yanonensis*. Kumamoto Prefecture Experiment Station, 2, 15–19.

Nishino, M., & Takagi, K. (1981). Parasites of *Unaspis yanonensis* (Kuwana) introduced from the People's Republic of China. Plant Protection, 35, 253–256 (in Japanese).

Oatman, E. R., & Platner, G. R. (1985). Biological control of two avocado pests. California Agriculture, 39(11–12), 21–23.

Oatman, E. R., McMurtry, J. A., Waggonner, M., Platner, G. A., & Johnson, H. G. (1983). Parasitization of *Amorbia cuneana* (Lepidoptera: Tortricidae) and *Sabulodes aegrotata* (Lepidoptera: Geometridae) on avocado in southern California. Journal of Economic Entomology, 76, 52–53.

Oncuer, C. (1974). The *Coccus* species (Homoptera: Coccidae) damaging citrus groves in the Aegean region: Studies on their morphological characters, distribution and natural enemies. Bitki Koruma Bulteni (Suppl. 1), 1–51.

Oomen, P. A. (1982). Studies on the population dynamics of the scarlet mite, *Brevipalpus phoenicis,* a pest of tea in Indonesia. Mededelingen Landbouwhogeschool Wageningen, 82 (1), 1–85.

Orphanides, G. M. (1984). Competitive displacement between *Aphytis* spp. (Hym. Aphelinidae) parasites of the California red scale in Cyprus. Entomophaga, 29, 275–281.

Panis, A. (1979). Integrated control in olive cultivation. Informatore Fitopatologica, 29, 27–28.

Panis, A. (1982). Scale insects (Homoptera, Coccoidea, Coccidae) within the framework of integrated control in Mediterranean citrus culture. Revue de Zoologie Agricole et de Pathologie Vegetale, 79, 12–22.

Panis, A. (1983). Biological control of the black scale, *Saissetia oleae* (Olivier) in the context of integrated control in French olive cultivation. Symbioses, 15, 53–74.

Peterson, G. D., Jr. (1955). Biological control of the spiny whitefly in Guam. Journal of Economic Entomology, 48, 681–683.

Prinsloo G. L. (1984). An illustrated guide to the parasitic wasps associated with citrus pests in the Republic of South Africa (Science Bulletin No. 402). Republic of South Africa: Department of Agriculture.

Prokopenko, A. I., & Mokrousova, L. A. (1963). Naturalization of a new parasite. Zashchita Rastenii, 11, 49–50 (in Russian).

Pschorn-Walcher, H., & Bennett, F. D. (1967). The successful biological control of citrus blackfly [*Aleurocanthus woglumi* Ashby] in Barbados, West Indies. PANS, A13, 375–384.

Quayle, H. J. (1938). Insects of citrus and other subtropical fruits. Ithaca, NY: Comstock.

Quezada, J. R. (1974). Biological control of *Aleurocanthus woglumi* (Homoptera: Aleyrodidae) in El Salvador. Entomophaga, 19, 243–254.

Quezada, J. R., & DeBach, P. (1973). Bioecological and population studies of the cottony-cushion scale, *Icerya purchasi* Mask, and its natural enemies, *Rodolia cardinalis* Mul. and *Cryptochaetum iceryae* Will., in southern California. Hilgardia, 41, 631–688.

Rao, V. P., Ghani, M. A., Sankaran, R., & Mathur, K. C. (1971). A review of biological control of insects and other pests in Southeast Asia and the Pacific Region (Technical Communication 6). London: Commonwealth Institute of Biological Control Commonwealth Agriculture Bureau.

Reed, D. K., Rich, J. E., & Shaw, J. G. (1972). A portable apparatus for detection of virus-diseased red mites in the field. Journal of Economic Entomology, 65, 890–891.

Rivnay, E. (1942). *Clausenia purpurea* Ishii. a parasite of *Pseudococcus comstocki* Kuw. introduced into Palestine. Bulletin Societe Fouad 1er Entomologie Egypte, 26, 1–19.

Rivnay, E. (1946). The status of *Clausenia purpurea* Ishii and its competition with other parasites of *Pseudococcus comstocki* Kuw. in Palestine. Bull. Societe Fouad 1er Entomol. Egypte, 30, 11–19.

Rivnay, E. (1968). Biological control of pests in Israel (a review 1905–1965). Israel Journal of Entomology, 3, 1–156.

Rodriguez-Lopez, L. (1942). La icerta plaga de los Citricos. Ecuador Departmento de Agricultura Boletin, 17, 1–10.

Rose, M., & DeBach, P. (1981). Citrus whitefly parasites established in California. California Agriculture, 35 (7–8), 21–23.

Rose, M., & De Bach, P. (1992). Biological control of *Parabemisia myricae* (Kuwana) (Homoptera: Aleyrodidae) in California. Israel Journal of Entomol. 25–26, 73–95.

Rosen, D. (1965). The hymenopterous parasites of citrus armored scales in Israel. Annales of the Entomological Society of America, 58, 388–396.

Rosen, D. (1967). Biological and integrated control of citrus pests in Israel. Journal of Economic Entomology, 60, 1422–1427.

Rosen, D., & DeBach, P. (1978). Diaspididae. In C. P. Clausen (Ed.), Introduced parasites and predators of arthropod pests and weeds. Agriculture Handbook No. 480 (pp. 78–128). Washington, DC: U.S. Department of Agriculture.

Sailer, R. I., Brown, R. E., Munir, B., & Nickerson, J. C. E. (1984). Dissemination of the citrus whitefly (Homoptera: Aleyrodidae) parasite Encarsia lahorensis (Howard) (Hymenoptera: Aphelinidae) and its effectiveness as a control agent in Florida. Bulletin of the Entomological Society of America, 30(2), 36–39.

Sands, D. P. A., Lukins, R. G., & Snowball, G. J. (1986). Agents introduced into Australia for the biological control of Gascardia destructor (Newstead) (Hemiptera: Coccidae). Journal of Ausralian Entomological Society, 25, 51–59.

Schreiner, I. (1990). Biological control introductions into the Caroline and Marshall Islands. Proceedings of the Hawaiian Entomological Society, 29, 57–69.

Schwartz, A. (1978). Citrus red mite. In E. G. C. Bedford (Ed.), Citrus pests in the Republic of South Africa Department of Agriculture. (Technical Service Report of South Africa Science Bulletin 391, pp. 41–46).

Selhime, A. G., Muma, M. H., Simanton, W. A., & McCoy, C. W. (1969). Control of Florida red scale in Florida with the parasite Aphytis holoxanthus. Journal of Economic Entomology, 62, 954–955.

Shaw, J. G., & Pierce, H. D. (1972). Persistence of a viral disease of citrus red mites despite chemical control. Journal of Economic Entomology, 65, 611–620.

Shaw, J. G., Tashiro, H., & Dietrick, E. J. (1968). Infection of the citrus red mite with virus in central and southern California. Journal Economic Entomology, 61, 492–1495.

Shenderovskaya, L. P. (1976). Introduced insect enemies and microorganisms. Zashchita Rastenii, 3, 52–53.

Silvestri, F. (1914). Report of an expedition to Africa in search of the natural enemies of fruit flies (Trypaneidae) (Bulletin 3). Hawaiian Board of Agriculture, Forestry Division of Entomology. Honolulu.

Silvestri, F. (1924). Etat actuel de la lutte contre la mouche de l'olive. 6th Congres Internationale d'Oleiculture, Nice (1923), 48–77.

Silvestri, F. (1926). Descrizione di tre specie di Prospaltella e di una di Encarsia (Hym., Chalcididae) parasite di Aleurocanthus (Aleyrodidae). Eos, 2, 179–189.

Silvestri, F. (1928). Contribuzione alla conoscenza degli Aleurodidae (Insecta: Hemiptera) viveti su citrus in estremo oriente e dei lora parasite. Bollettino del Laboratori Zoologia, Portici, 21, 1–60.

Smith, D. (1978a). Biological control of scale insects on citrus in southeastern Queensland. I. Control of red scale Aonidiella aurantii (Maskell). Journal of Australian Entomological Society, 17, 367–371.

Smith, D. (1978b). Biological control of scale insects on citrus in southeastern Queensland. II. Control of circular black scale Chrysomphalus ficus Ashmead, by the introduced parasite. Aphytis holoxanthus DeBach. Journal of Australian Entomological Society, 73, 373–377.

Smith, D., Papacek, D. F., & Murray, D. A. H. (1988). The use of Leptomastix dactylopii Howard (Hymenoptera: Encyrtidae) to control Planococcus citri (Risso) (Hemiptera: Pseudococcidae) in Queensland citrus orchards. Queensland Journal of Agriculture and Animal Science, 45, 157–164.

Smith, H. D., Maltby, H. L., & Jiménez, J. (1964). Biological control of the citrus blackfly in Mexico (Technical Bulletin No. 1311). Washington, DC: U.S. Department of Agriculture.

Smith, H. S. (1917). On the life history and successful introduction into the United States of the Sicilian mealybug parasite. Journal of Economic Entomology, 10, 262–268.

Smith, H. S. (1921). Biological control of the black scale in California. Bulletin of California Department of Agriculture, 10, 127–137.

Smith, H. S. (1942). A race of Comperiella bifasciata successfully parasitizes California red scale. Journal of Economic Entomology, 35, 809–812.

Smith, H. S., & Armitage, H. M. (1920). Biological control of mealybugs in California. Bulletin of California Department of Agriculture, 9, 103–158. Sacramento, California.

Smith, H. S., Armitage, H. M. (1931). The biological control of mealybugs attacking citrus (Bulletin 509). California Agriculture Experiment Station.

Smith, H. S., & Compere, H. (1916). Observations on the Lestophonus, a dipterous parasite of the cottony cushion scale. Bulletin of California Department of Agriculture, 5, 384–390. Sacramento, California.

Smith, H. S., & Compere, H. (1929). New insect enemies of the citrophilus mealybug from Australia. Bulletin of California Department Agriculture, 18, 214–218.

Smith, K. M., & Cressman, A. W. (1962). Birefringent crystals in virus-diseased citrus red mites. Journal of Insect Pathology, 4, 229–236.

Smith, R. H. (1944). Bionomics and control of the nigra scale, Saissetia nigra. Hilgardia, 16, 225–288.

Stavraki, A. (1970a). Contribution to the inventory of the parasite complex in Greece of some insects injurious to olive. Entomophaga, 15, 225–231.

Stavraki, H. G. (1970b). Donnes preliminaires sur les lachers de Chelonus elaeaphilus (Hym. Braconidae) contre Prays olease (Lep., Hyponomeutidae) dans les oliveraires de Kessariani et Thebes en 1968. Annales Institut Phytopathologique Benaki, 9, 281–287.

Stavraki-Pavlopoulou, H. (1967). Essais preliminaires sur les lachers d' Opius concolor Szepl. (Hymen., Braconidae) parasite due Dacus oleae Gmel. (Dip.: Tephritidae) dans l'ile de Chalki 1965. Annales Institut Phytopathologique Benaki, 8, 23–31.

Steinberg, S., Podoler, H., & Rosen, D. (1986). Biological control of the Florida red scale, Chrysomphalus aonidum (Linnaeus), in Israel by two parasite species: Current status in the coastal plain. Phytoparasitica, 14, 199–204.

Storey, W. B., Bergh, B., & Zentmeyer, G. A. (1986). The origin, indigenous range and dissemination of the avocado. California Avocado Society Yearbook, 70, 127–134.

Summy, K. R., Gilstrap, F. E., Hart, W. G., Caballero, J. M., & Saenz, I. (1983). Biological control of citrus blackfly (Homoptera: Aleyrodidae) in Texas. Environmental Entomology, 12, 782–786.

Swirski, E., Amitai, S., & Dorzia, N. (1967a). Laboratory studies on the feeding, development and reproduction of the predaceous mites Amblyseius rubini Porath & Swirski and Amblyseius swirskii Athias (Acarina: Phytoseiidae) on various kinds of food substances. Israel Journal of Agriculture Research, 17, 101–119.

Swirski, E., Amitai, S., & Dorzia, N. (1967b). Laboratory studies on the feeding, development and oviposition of the predaceous mite Typhlodromus athiasae P. & S. (Acarina: Phytoseiidae) on various kinds of food substances. Israel Journal of Agricultural Research, 17, 213–218.

Swirski, E., Wysoki, M., & Izhar, Y. (1988). Integrated pest management in the avocado orchards of Israel. Applied Agricultural Research, 3(1), 1–7.

Swirski, E., Izhar, Y., Wysoki, M., Gurevitz, E., & Greenberg, S. (1980). Integrated control of the long-tailed mealybug, Pseudococcus longispinus (Hom.: Pseudococcidae) in avocado plantations in Israel. Entomophaga, 25, 415–426.

Takagi, K., & Ujiye, T. (1986). Suppressive effects on the arrowhead scale, Unaspis yanonensis Hemiptera: Diaspididae) of the introduced parasitoids Aphytis yanonensis and Coccobius fulvus (Hymenoptera: Aphelinidae). Bulletin of Fruit Tree Research Station, Series D, 8, 53–64.

Talhouk, A. S. (1975). Citrus pests throughout the world. Citrus Technical Monograph 4, 21–24 (Ciba-Geigy, Basel, Switzerland).

Tanaka, M., & Inoue, K. (1977). Introduction of efficient parasite of the arrowhead scale, *Unaspis yanonensis* (Kuwana) from Hong Kong. Bulletin of Fruit Tree Research Station, Series D, 1, 69–85.

Tanaka, M., & Kashio, T. (1977). Biological studies on *Amblyseius largoensis* Muma (Acarina: Phytoseiidae) as a predator of the citrus red mite *Panonychus citri* (McGregor) (Acarina: Tetranychidae. Bulletin Fruit Research Station. Japan, Series D-1, pp 49–67.

Taylor, T. H. C. (1935). The campaign against *Aspidiotus destructor* in Fiji. Bulletin of Entomological Research, 26, 1–102.

Taylor, T. H. C. (1937). The biological control of an insect pest in Fiji, an account of the coconut leafmining beetle and its parasite complex. London: Imperial Bureau of Entomology, Imperial Agricultural Bureau.

Teran, A. L., & DeBach, P. (1963). Observaciones sobre *Comperiella bifasciata* How. (Hymen., Encyrtidae). Revista Agronomica del Noroeste Argentino, 4, 5–23.

Tothill, J. D., Taylor, T. H. C., & Paine, R. W. (1930). The coconut month in Fiji. A history of its control by means of parasites. London: Imperial Bureau of Entomology, Imperial Agricultural Bureau,

Tuncyureck, M., & Oncuer, C. (1974). Studies on aphelinid parasites and their hosts, citrus diaspine insects, in citrus orchards in the Aegean region. Bulletin OILB SROP, 3, 95–108.

van Brussel, E. W. (1975). Interrelations between citrus rust mite, Hirsutella thompsonii and greasy spot on citrus in Surinam. Wageningen: Pudoc.

van den Bosch, R., Bartlett, B. R., & Flanders, S. E. (1955). A search for natural enemies of lecaniine scale insects in northern Africa for introduction into California. Journal of Economic Entomology, 48, 53–55.

van Whervin, L. W. (1968). The introduction of *Prospaltella opulenta* Silvestri into Jamaica and its competitive displacement of *Eretmocerus serius* Silvestri. PANS, A14, 456–464.

Vesey-Fitzgerald, D. (1953). Review of the biological control of coccids on coconut palms in the Seychelles. Bulletin of Entomological Research, 44, 405–413.

Viggiani, G. (1981). Recent findings on integrated control in olive groves. Informatore Fitopatologico, 31(1/2), 37–43.

Viggiani, G., & Battaglia, D. (1983). Experiments on the biological control of *Dialeurodes citri* (Ashm). using *Encarsia lahorensis* (How.) at fruit-farm level, and present status of the parasite in Campania and other areas (pp. 181–189). XIII Congresso Nazionale Italia Entomologia.

Viggiani, G., & Mazzoni, P. (1980). Preliminary results of the introduction into Italy of *Metaphycus bartletti* Ann. & Myn., a parasite of *Saissetia oleae* (Oliv.) (Vol. 2). XII Congresso Nazionale Italia Entomologia.

Waterhouse, D. F., & Norris, K. R. (1987). Biological control Pacific prospects. Melbourne: Australian Centre International Agricultural Research/Inkata Press.

Wheatley, P. E. (1964). The successful establishment of *Eretmocerus serius* Silv. (Hymenoptera: Eulophidae) in Kenya. East African Agriculture and Forestry Journal, 29, 236.

Wille, J. (1952). Entomologia Agricultura de Peru (2nd ed.). Lima: Ministerio de Agricultura.

Williams, L. O. (1977). The botany of the avocado and its relatives. In J. Sauls (Ed.), Proceedings of the First International Tropical Fruit Short Course—The Avocado (pp. 9–15). Gainesville: University of Florida.

Wilson, F. (1960). A review of the biological control of insects and weeds in Australia and Australian New Guinea (Technical Communication No. 1). Slough, United Kingdom: CIBC. Commonwealth Agricultural Bureau, Farnham Royal.

Woglum, R. S. (1913). Report of a trip to India and the Orient in search of the natural enemies of the citrus white fly (Bureau of Entomology Bulletin 120). Washington, DC: U.S. Department of Agriculture.

Wood, B. J. (1963). Imported and indigenous natural enemies of citrus coccids and aphids in Cyprus, and an assessment of their potential value in integrated control programmes. Entomophaga, 8, 66–82.

Wysoki, M. (1979). Introductions of beneficial insects into Israel by the Institute of Plant Protection Quarantine Laboratory, ARO, during 1971–78. Phytoparasitica, 7, 101–106.

Wysoki, M., & Izhar, Y. (1980). The natural enemies of *Boarmia (Ascotis) selenaria* Schiff. (Lepidoptera: Geometridae) in Israel. Acta Oecol./Oecol. Applic. 1, 283–290.

Wysoki, M., & Renneh, S. (1985). Introduction into Israel of *Trichogramma platneri* Nagarkatti, an egg parasite of Lepidoptera. Phytoparasitica, 13, 139–140.

Wysoki, M., Swirski, E., & Izhar, Y. (1981). Biological control of avocado pests in Israel. Protection Ecology, 3, 25–28.

Wysoki, M., Izhar, Y., Swirski, E., & Greenberg, E. (1975). The giant looper *Boarmia (Ascotis) selenaria* Schiff. (Lepidoptera: Geometridae), a new pest of avocado plantations in Israel. California Avocado Society Yearbook, 58, 71–81.

Yasumatsu, K. (1951). Further investigations on the hymenopterous parasites of *Ceroplastes rubens* in Japan. Journal of Kyushu University Faculty of Agriculture. 10, 1–27.

Yasumatsu, K. (1953). Preliminary investigations on the activity of a Kyushu race of *Anicetus ceroplastis* Ishii which has been liberated against *Ceroplastes rubens* Maskell in various districts in Japan. Science Bulletin of Kyushu University Faculty of Agriculture, 14, 17–26.

Yasumatsu, K. (1958). An interesting case of biological control of *Ceroplastes rubens* Maskell in Japan. Proceedings, 10th International Congress on Entomology, 4, 771–775.

Yasumatsu, K., & Tachikawa, T. (1949). Investigations on the hymenopterous parasites of *Ceroplastes rubens* Maskell in Japan. Journal of Kyushu University Faculty of Agriculture, 9, 99–120.

Yukinari, M., & Kagawa, M. (1985). Hibernation and harmonization with host scale of *Aphytis yanonensis* DeBach & Rosen (Hymenoptera: Aphelinidae), the imported parasitoids of the arrowhead scale, *Unaspis yanonensis* (Hemiptera: Diaspididae), in citrus orchard in Tokushima. Bulletin of Tokushima Horticultural Experiment Station, 13, 7–15 (in Japanese).

Zinna, G. (1960). Esperimenti di lotta biologica controil cotonello degli agrumi (*Pseudococcus citri* (Risso)) nell'Isola di Procida mediante l'impiego di due parassiti esotica, *Pauridia peregrina* Timb. e *Leptomastix dactylopii* How. Bolletino del Laboratori Entomologia Agricultura "Filippo Silvestri" Portici, 28, 257–284.

28

Biological Control in Deciduous Fruit Crops

M. T. AliNiazee and B. A. Croft

Department of Entomology
Oregon State University
Corvallis, Oregon

INTRODUCTION

Biological control is an important component of pest control programs for deciduous fruit crops (apple, pear, peach, and cherry) worldwide (Minks & Gruys, 1980; Croft, 1982a; Hoyt *et al.,* 1983; Blommers, 1994; Prokopy & Croft, 1994). For certain key pests for which almost no damage to the harvested fruit can be tolerated control by predators, parasitoids, and other natural enemies is not sufficiently effective to be used as the sole source of control [codling moth (*Cydia pomonella* Linnaeus), apple maggot (*Rhagoletis pomonella* [Walsh]), cherry fruit flies (*Rhagoletis cingulata* Loew, *R. indifferens* Curran), oriental fruit moth (*Grapholitha molesta* [Busck])]. Pesticides are often necessary for control of these pests. Biological control can be very effective against certain sporadic and many secondary pests (mites, aphids, and leafhoppers). However, sometimes nonselective pesticides used for control of key pests disrupt the balance between sporadic and secondary pests and their natural enemies to the extent that chemical control of these pests becomes essential. In extreme cases, strict chemical control programs may be necessary for a wide diversity of pests from both groups, if appropriate use of selective pesticides is not pursued and no attention is given to natural-enemy conservation.

We discuss biological control of pests of deciduous fruit crops with emphasis primarily on western North American orchard species. Some discussion is made for pests encountered in other production areas and, for some topics, a more global treatment is given (genetic improvement of natural enemies of apple pests for traits of pesticide resistance). Discussion is limited to biological control of arthropod pests by arthropod predators and parasitoids. We do not treat all pests that occur on these crops, but focus on those that play a dominant role in the crop ecosystem and those most commonly studied in recent years.

In our presentation we will consider first the biological control of secondary pests and then consider key and sporadic pests together. Key pests usually attack the fruit directly and occur at damaging levels each year. Sporadic pests feed on the fruit, other parts of tree or both, and are less common than key pests. Similar to sporadic pests, secondary pests tend to be infrequently present unless disturbed by pesticides or other factors, and they usually feed on foliage and other nonfruit parts of the tree; see Croft (1982a) for further discussion and identification of these pest types.

Research on deciduous tree fruit pests and their natural enemies has spanned most areas of biological control including (1) importation and establishment of exotic natural enemies; (2) augmentation of endemic or exotic natural enemies through mass-culturing and periodic release; (3) conservation through cultural manipulation, utilization of natural-enemy food sprays, and selective pesticide development and use; (4) development of genetically modified strains of predators, parasitoids, and pathogens using hybridization and artificial selection techniques; and (5) a combination of these tactics. Examples of these different approaches to biological control will be discussed.

BIOLOGICAL CONTROL OF SECONDARY PESTS

The greatest success achieved in biological control of pests of deciduous fruit crops has been with the natural enemies of secondary pests. Most research has been with endemic predators using selective pesticides and improved cultural methods. Some success involving the selection of pesticide resistant strains of predatory species has been achieved (Hoy *et al.,* 1983; Hoy, 1985; Croft, 1990). By far, the greatest amount of study with this group has involved conservation of natural enemies of plant-feeding mites, aphids, leaf miners, and leafhoppers. Relatively little effort has been made to introduce new exotic natural enemies to control these pests. In part, this is because effective endemic species are already present in most areas. They are capable of providing significant biological control if managed properly. For similar reasons, relatively little augmentation of natural enemies for control of these pests has been attempted. In general, the high costs of rearing and releasing predators and parasitoids preclude extensive use of augmentation procedures in many cases on these crops. The same is not true for certain pathogens of these pests that have been extensively mass-produced and introduced into orchards as sprays [*Bacillus thuringiensis* Berliner (B.t.), and a granulosis virus against codling moth].

Management of Endemic Natural Enemies of Secondary Pests

Spider Mites

Plant-feeding mites, including spider mites of the Tetranychidae family, are some of the most troublesome pests of commercial deciduous fruit crops worldwide (van de Vrie *et al.,* 1972; van de Vrie, 1986). Generally, in noncommerical orchards, where no sprays are used and poor horticultural practices are avoided, these pests, being controlled by large complexes of predatory species, are seldom problems. However, when plants are heavily fertilized, pruned, and sprayed with pesticides that are not selective to natural enemies, these pests can rise to high population levels (e.g., over 100 per leaf) and cause severe damage including reduction of fruit size and numbers, poor fruit coloring, and affects on other features of tree health (van de Vrie *et al.,* 1972). They may kill the tree after several successive years of attack by dense populations (Croft & Hoyt, 1983).

Biological control programs for spider mite pests are based on selective chemical control of other orchard pests, while at the same time allowing for biological control of mites. Predators from several different groups are the primary natural enemies involved (acarines from the Phytoseiidae, Stigmaeidae, or other predatory mite groups; coccinellid beetles from the genus *Stethorus;* various hemipterans; neuropterans from the Chrysopidae and Hemerobiidae families; and certain predaceous thrips, spiders, and others). By far the most widely exploited natural enemies have been the phytoseiid mites and predatory *Stethorus* spp.

Table 1 lists major integrated mite control programs that have been researched and implemented in North America. The mite pests and natural enemies listed are the principal focus of biological control research in each area. Table 2 lists selective acaricides used in management; and the effects of each on the principal natural enemies, their prey (tetranychid mites), and alternate prey (eriophyid mites).

Manipulation of many different arthropods within the species complex can be achieved by selection of various products. By careful monitoring of populations of pest mites, alternate prey, and predators, possibilities for biological control can be forecast based on predator to prey ratios even relatively early in a cycle of pest population increase (Croft, 1982b). When prey levels exceed a desired level, selective acaricides can be used to reduce pest mites without influencing predators and the latter then regulate the pests; see Croft & McGroarty (1977) and Mowery *et al.* (1977) for more specific examples of manipulations with selective acaricides. Studies in several fruit production areas have indicated that use of combinations of predators and selective acaricides can reduce acaricide usage from 50 to 90% of commercial programs based on using nonselective acaricides or insecticides. Also, pesticide resistance development in spider mites can be greatly curtailed where the combination of predators and selective acaricides is employed, instead of just using acaricides alone (Croft *et al.,* 1987; Croft, 1990).

In addition to the development of selective acaricides, a great deal of research has gone into identifying other means to conserve and augument the natural enemies of spider mites and to improve biological control. These practices include conservation of alternative prey species [*Aculus schlechtendali* (Nalepa) as prey for *Metaseiulus occidentalis* (Nesbitt) in apple (Hoyt, 1969)], ground cover manipulations to ensure overwintering survivorship and early dispersal of *Amblyseius fallacis* (Garman) into trees (Croft & McGroarty, 1977), alternate middle-row spraying to conserve *Stethorus punctum* (LeConte) (Hull & Beers, 1985), transferring foliage cuttings of trees infested with predators to sites where *Typhlodromus pyri* (Schenten) are absent (Solomon & Easterbrook, 1984; Blommers, 1994), and manipulation of suitable predatory species based on their feeding behavior and prey exploitation capabilities (Croft & McRae, 1993; McMurtry & Croft, 1997).

Means to exploit insecticide-resistant forms of mite predators have been extensively researched. Resistance features developed in the field and those resulting from genetic manipulation through hybridization and artificial selection have been used on a wide scale (Table 3). In North America, resistant strains of *A. fallacis, T. pyri,* and *S. punctum* have been transferred, released, and manipulated for management purposes (Croft & Strickler, 1983; Croft, 1994). These studies have usually involved interorchard movement of resistant strains. Similar successes with populations of *M. occidentalis* having resistance to more than one pesticide have been accomplished, especially in the western United States (Hoy, 1985; Croft, 1994; Croft, 1990).

Table 3 lists some known cases of insecticide-resistant phytoseiids in which resistance development occurred in the field or was a result of laboratory selection. Natural enemies of deciduous fruits and nut pests have been the

TABLE 1   Major Integrated Mite Control Programs on Apple or Closely Related Tree Fruit Crops in North America

State or province	Principal pests controlled[a]	Natural enemies used[b]	Research references[c]	Implementation references[a]
California	Tu, Tp, Pu	To, Ss, T?	Croft & Barnes, 1971; Rice et al., 1976	Davis, 1978
Colorado	Tu, Pu	To	Quist, 1972	Quist, 1972
Illinois	Pu, Tu	Af	Meyer, 1970, 1974	Meyer, 1981
Massachusetts	Pu, Tu, As	Af, Sp	Hislop et al., 1980	Hauschild et al., 1981; Prokopy et al., 1981
Michigan	Pu, Tu, As	Af, Sp, Ag, Zm	Croft & McGroarty, 1977; Croft, 1983	Croft, 1975; Whalon et al., 1982
Missouri	Pu, Tu	Af	Poe & Enns, 1969	—
New Jersey	Pu, Tu, As	Af, Sp	Swift, 1970	Christ, 1971
New York	Pu, Tu	Af, Tp, Sp	Watve & Lienk, 1977; Wieres et al., 1979	Tette et al., 1979, 1981
North Carolina	Pu, Tu, As	Af, Sp	Rock & Yeagan, 1971; Rock & Apple, 1983	Rock, 1972
Ohio	Pu, Tu, As	Af, Zm, Ag	Holdsworth, 1968	Holdsworth, 1974
Oregon	Pu, Tu, Eu	To, Tp	Zwick, 1972; AliNiazee, 1974; Croft & AliNiazee, 1983; Hadam et al., 1986	Whalon & Croft, 1984
Pennsylvania	Pu, Tu	Sp, Af	Asquith, 1971; Hull et al., 1983	Asquith, 1972; Hull, 1979
Utah	Tu, Tm	To	Davis, 1970; Croft, 1972	Whalon & Croft, 1984
Virginia, West Virginia	Pu, Tu	Af, Sp	Clancy & McAllister, 1968; Parrella et al., 1980, 1981a, 1981b, 1982	WV Extension Service, 1982
Washington	Pu, Tm, As	To, Spp, Zm	Hoyt, 1969a, 1968b	Hoyt et al., 1967, 1970, Hudson et al., 1974
British Columbia	Tm, Pu, As	To, Tp, Zm	Downing & Molliet, 1971	Downing & Arand, 1978 B.C. Ministry of Agriculture, 1982
Nova Scotia	Pu, As	Many spp.	Sanford & Herbert, 1970, Hardman et al., 1995	Anonymous, 1970, N.S. Ministry of Agriculture; 1982; Hardman et al., 1995
Ontario	Pu, Tu, As	Af	—	Hagley et al., 1978
Quebec	Pu, Tu, As	Af	Parent, 1967	Paradis, 1981; Bostanian & Coulombe, 1986

[a] Pests of importance in each area are listed sequentially. Abbreviations: Tm = *Tetranychus mcdanieli*; Pu = *Panonychus ulmi*; AS = *Aculus schlechtendali*; Tu = *T. urticae*; Tp = *T. pacificus*; Eu = *Eotetranychus carpini borealis*.

[b] Natural-enemy importance in each area listed sequentially. Abbreviations: To = *Typhlodromus (Metaseiulus) occidentalis*; Sp = *Stethorus punctum*; Zm = *Zetzellia mali*; Af = *Amblyseius fallacis*; Tp = *T. pyri*; T? = several miscellaneous phytoseiids; Ag = *Agistemus fleschneri*; Ss = *Scolothrips sexmaculatus*; Spp = *Stethorus picipes*.

[c] Only references *not cited* in Croft & Strickler (1983), Whalon & Croft (1984), and Croft & Raush (1988) are included in the reference section of this chapter.

species most commonly studied. As discussed by Croft *et al.* (1990), genetic selection studies have provided some outstanding successes in achieving selective use of pesticides. However, there have been problems in artificially selecting populations for insecticide resistance traits. In many cases where a resistant natural-enemy population is selected in the laboratory, relatively low levels of recessive and polygenically determined resistance traits are maintained in populations. These resistance features sometimes are not stable under field conditions and can be lost by hybridization with native susceptible individuals, thus mak-

ing them difficult to manage in the field (Croft, 1983; Hoy, 1985; Vidal & Kreiter, 1995). Means to select for more stable resistance factors in natural-enemy populations are being studied (Croft *et al.*, 1990).

In many areas of the world, importation, establishment, and management of insecticide-resistant phytoseiid mites have been accomplished and have provided levels of biological control similar to those achieved in North American tree fruit orchards; see reviews in Croft & Strickler (1983), Hoy (1985), Croft (1990), Blommers (1994). The most striking successes have been demonstrated in Australia,

TABLE 2 Selective Acaracides Used in IPM of Spider Mites and Their Toxicity to Various Arthropod Groups[a]

Compound	Arthropod group toxicity[b]				
	Tetranychids	Eriophyids	Phytoseiids	Stigmaeids	Insect predators
Endosulfan	L	H	L	H	L
Ethion	M	H	M	NI	M
Formetanate	H	H	H	H	L
Chlordimiform	H	H	H	H	L
Propargite	H	H	L	H	L
Cyhexatin	H	H	L–M	H	L
Fenbutatin oxide	H	H	L	H	L
Abamectin	H	H	L–M	NI	NI
Hexethiazox	H	L	L	NI	L
Clofentazene	H	L	L	NI	L

[a] After Croft, B.A. (1982b).
[b] Abbreviations: H = high toxicity; M = moderate toxicity; L = low toxicity; NI = no information.

New Zealand, and parts of Europe with *A. fallacis, M. occidentalis, T. pyri,* and to some extent, *P. persimilis* (Table 3; Hoy, 1985; Croft, 1990).

Aphids

Of the major aphid pests of deciduous fruits [including *Aphis pomi* De Geer (apple aphid), *Dysaphis plantaginea* (Passerini) (rosy apple aphid), *Eriosoma lanigerum* (Hausmann) (woolly apple aphid), and *Myzus persicae* (Sulzer) (green peach aphid)], the most extensive potential for biological control is possible with *A. pomi, D. plantaginea,* and *E. lanigerum. Myzus persicae* is a vector of many crop diseases; thus, the threshold for *M. persicae* is very low and biological control of this pest is not usually feasible when vectored diseases are involved. Other cases where biological control of aphids has been highly successful include the walnut aphid *Chromaphis juglandicola* Kaltenbach and the filbert aphid *Myzocallis coryli* (Goetze).

Early work on biological control of aphids of tree fruits emphasized the classical approach of introduction of natural enemies from different areas. For example, the woolly apple aphid parasite, *Aphelinus mali* (Haldemann), which was originally found in eastern North America, was distributed to other parts of the continent from 1921 to 1939 (DeBach, 1964). In addition, *A. mali* was introduced to over 50 countries with establishment occurring in at least 42 countries (Clausen, 1978). The results of these introductions have been variable. Excellent control was noticed in northwestern United States, Australia, British Columbia, New Zealand, Chile, and Uruguay while low to moderate control was recorded in Europe and parts of Asia.

Research has been focused on *A. pomi* by Morse & Croft (1987) and Prokopy & Croft (1994). In Massachusetts, the potential impact of the cecidomyiid predator *Aphidoletes ophidimyza* (Rondani) was measured by estimating critical

predator to prey ratios needed for effective biological control. Adams and Prokopy (1977, 1980) documented resistance in this predatory fly to some of the organophosphate insecticides used in orchards. They also identified several selective pesticides that could be used to conserve *A. aphidimyza.* An aphid to predator ratio of 40:1 or less was necessary to avoid use of selective pesticides in the field. Similar work by Warner and Croft (1982), Morse (1981), and Morse and Croft (1987) focused on identifying selective pesticides and examining predator–prey relationships of *A. aphidimyza* and *A. pomi* in Michigan apple orchards. Warner and Croft (1982) documented more than 12-fold resistance to azinphos-methyl in field populations; and found phosalone, phosphamidon, carbophenothion, pirimicarb, and several fungicides and acaricides to be relatively harmless to this dipteran predator. In modeling studies (Fig. 1) of critical prey to predator ratios for control of *A. pomi* by *A. aphidimyza,* Morse and Croft (1987) found that a value of 70 to 90:1 would provide adequate biological control of this pest before an action threshold level of from 40 to 60 per terminal was reached. They concluded that previously determined critical ratios for predicting biological control success may have been too conservative because they were based on single generational estimates made in laboratory investigations.

Research on biological control of *A. pomi* in Washington State has involved predators such as earwigs (*Forficula auricularia* Linnaeus) and Aphidiidae parasitoids (Carroll & Hoyt, 1984a, 1984b, 1986). The parasitoids *Praon unicum* Smith and *Lysiphlebus testaceipes* (Cresson) played a major role in biological control of this pest in early season in selected orchards, even though they did not successfully develop through their entire life cycle while feeding on *A. pomi.* Alternate winter and summer aphid hosts of these parasitoids occurred on weeds, on grasses, and in nearby peach orchards. *Myzus persicae* was one of the major hosts

TABLE 3 Cases of Insecticide-Resistant Phytoseiid Mites Occurring on Deciduous Tree Fruits and Other Closely Related Agricultural Crops

Species	Crop	Compound	Fold –R	Selection conditions[a]	Country	References[b]
Amblyseius chilenesis	Apple	Azinphos-methyl	2	F	Uruguay	Croft *et al.,* 1976
		Phosmet	10	F		
A. fallacis	Apple	Azinphos-methyl	100–1000	F	MI, NC, U.S.	Motoyama *et al.,* 1970, Croft & Nelson, 1973
		Carbaryl	25–77	F, L[a]	IL, U.S.	Croft *et al.,* 1973
		Diazinon	119	F	MI, U.S.	Croft *et al.,* 1976
		Parathion	103–152	F	MI, NC, U.S.	Motoyama *et al.,* 1970, Croft *et al.,* 1973
		Permethrin	60	L[a]	MI, U.S.	Strickler & Croft 1981
		Permethrin/fenvalerate	—	F	PA, U.S.	Hull & Starner, 1983
A. hibisci	Citrus	Carbaryl	—	F	CA, U.S.	Tanigoshi & Fargerland, 1984
		Dimethoate	22	F	CA, U.S.	Tanigoshi & Fargerland, 1983
		Parathion	—	F	CA, U.S.	Kennett, 1970
			50–70	F	CA, U.S.	Tanigoshi & Fargerland, 1984
A. longispinosus	Citrus	Carbaryl	17	L[a]	Taiwan	Lo *et al.,* 1984, Lo, 1986
		Carbaryl/dimethoate	—			Lo, 1986
		Dimethoate	15			
		Malathion	38			
A. potentillae (= *andersoni*)	Apple	Azinphos-methyl Phenthoate Methyl Tetrachlorvinphos	—	F	Italy	Ivancich-Gambaro, 1975
		Azinphos-methyl Tetrachlorvinphos	—	F	Switzerland	Baillod, 1986
		Parathion	100	F	Switzerland	Anber & Oppenoorth, 1986
		Propoxur	2300	F		
Metaseiulus occidentalis	Apple	Azinphos-methyl	101–104	F	WA, U.S.	Croft & Jeppson, 1970; Ahlstrom & Rock, 1973
	Grape	Dimethoate	11	F, L[a]	CA, U.S.	Hoy *et al.,* 1979
		Methomyl	3	F	CA, U.S.	
	Almond	Methomyl	12	L[a]	CA, U.S.	Roush & Hoy, 1981
	Almond	Benomyl	—	F	CA, U.S.	Hoy *et al.,* 1980
		Bayleton	—	F	CA, U.S.	
	Almond	Propoxur	>100	F	CA, U.S.	Roush & Plapp, 1982
	Apple	Permethrin	10	L[a]	WA, U.S.	Hoy & Knop, 1981
	Apple	Bioresmethrin	—	L[a]	Australia	van de Klashorst, 1984
	Almond	Carbaryl + OP + permethrin	—	L[a]	Several	Hoy, 1984
	Almond	Carbaryl + OP + sulfur	—	L[a]	Several	Hoy, 1984
Typhlodromus abberans	Grape	Parathion	—	F	Italy–Switzerland	Corino *et al.,* 1986
T. arboreus	Apple	Azinphos-methyl	7	F	OR, U.S.	Croft & AliNiazee, 1983
		Carbaryl	3–5	F		
T. pyri	Apple	Azinphos-methyl	10	F	New Zealand	Hoyt, 1972
	Apple		11–42	F	New Zealand	Penman *et al.,* 1976
	Apple		14	F	New Zealand	Wong & Chapman, 1979
	Apple		12–20	F	NY, U.S.	Watve & Lienk, 1976
	Apple		4–6	F	U.K.	Kapetanakis & Cranham, 1983
		Carbaryl	20	F		
		Parathion	50	F		
	Apple	Bromophos	7	F	Netherlands	Ovenmeer & van Zon, 1983
		Carbaryl	50	F		
		Parathion	100	F		

(continues)

TABLE 3 *(continued)*

Species	Crop	Compound	Fold −R	Selection conditions[a]	Country	References[b]
	Apple	Paraoxon	1000	F	Netherlands	van de Baan *et al.,* 1985
		Parathion	100	F		
	Apple	Cypermethrin	3	L[a]	New Zealand	Markwick, 1984
	Apple	Parathion	199–271	F	OR, U.S.	Hadam *et al.,* 1986
		Azinphos-methyl	5–7	F	OR, U.S.	
		Carbaryl	25–27	F	OR, U.S.	
	Apple	Azinphos-methyl	—	F	Switzerland	Baillod *et al.,* 1985
	Grape	Phosmet	—	F		
		Diazinon	—	F		
		Methidathion	—	F		
		Mevinphos	—	F		

[a] Represent genetic selection or genetic improvement attempts.

[b] The complete list of references for this table can be found in Croft & Raush (1988).

on peaches. The relatively insecticide-tolerant earwig *F. auricularia* was an important predator in orchards that often prevented the resurgence of *A. pomi* following treatments with selective pesticides (Carroll & Hoyt, 1984a, 1984b). Furthermore, augmentative releases of this species early in the growing season kept aphids below 50 per tree compared with 2000 to 3000 per tree in orchard plots where earwigs were excluded or where releases were not made but earwigs were not excluded (Carroll & Hoyt, 1984b).

Research on the natural enemies of woolly apple aphid in Washington State has focused on the host-specific aphelinid endoparasitoid *Aphelinus mali* and a complex of generalist predators (Walker, 1985). When predators were excluded, *A. mali* alone was incapable of preventing pest populations from reaching unacceptably high levels. Generalist predators appeared to be the dominant factor reducing aphid densities. The most important species were the coccinellid *Coccinella transversoguttata* Brown in midsum-

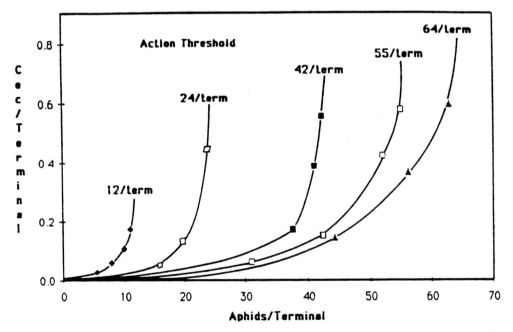

FIGURE 1   Multiple-generation apple aphid/cecidomyiid simulation run with different, initial predator-to-prey ratios and resulting in five different peak aphid densities (12 to 64 aphids per terminal). [From Morse J. G., & Croft, B. A. (1987). *Entomophaga, 32,* 339–356.]

mer, the neuropteran *Chrysoperla nigricornis* (Burmeister), and the mirid *Dareaocoris brevis* (Uhler) throughout mid- and late summer. These and other generalist predators such as *Adalia bipunctata* (Linnaeus) were important natural enemies of woolly apple aphid in Oregon (AliNiazee, unpublished). Other environmental-related factors that influenced the population dynamics of woolly aphids were high overwintering mortality due to cold temperatures, high summer temperatures that increased a late production of aphids, and high levels of emigration (Walker, 1985).

The walnut aphid *Chomaphis juglandicola* (Kaltenbach) is a major pest of walnuts in California. Early studies showed that parasitic wasp *Trioxys pallidus* Haliday was effective in France. This parasitoid was later imported, reared, and released in several counties in California, where it successfully overwintered and established rapidly (Schlinger *et al.,* 1960). The French biotype was highly effective in controlling *C. juglandicola* in the mild coastal counties, but failed in inland areas with high summer temperatures and low humidity. Subsequently, another biotype of *T. pallidus* from Iran, where the climatic conditions are similar to that of central California, was imported and rapidly became established throughout the state (van den Bosch *et al.,* 1970; AliNiazee & Hagen, 1995).

A similar program against filbert aphid *Myzocallis coryli* Goetze resulted in importation, mass production, and release of French and Spanish biotypes of *T. pallidus* in the filbert orchards of the Willamette Valley of Oregon (Messing & AliNiazee, 1989; AliNiazee, 1991; AliNiazee, 1995). Establishment has occurred at nearly 2 dozen sites with rapid colonization occurring at many new sites every year. The parasitoid is now well distributed throughout the hazelnut growing areas of Oregon and Washington (AliNiazee, 1997). The life cycle of the parasitoid appears to be well synchronized with that of the host aphids (Messing, 1986).

Leaf Miners

Biological control research with pests of the genus *Phyllonorycter* has been focused mostly on the complex of endemic parasitoids attacking them, including several species from the hymenopteran Braconidae, Eulophidae, Pteromalidae, and Ichneumonidae (Hagley *et al.,* 1981; Wieres *et al.,* 1982; Hagley, 1985; Drummond *et al.,* 1985; van Driesche *et al.,* 1985; Barrett & Jorgensen, 1986). Leaf miners have reached outbreak levels in almost all areas of North America in the past two decades. This may be attributed to development of pesticide resistance in leaf miners, pesticide susceptability in their natural enemies (Pree *et al.,* 1980, 1986; Trimble & Pree, 1987), and the disruption of natural enemies by use of nonselective pesticides (Wieres *et al.,* 1982; Maier, 1983; Hagley, 1985; van Driesche *et al.,* 1985; Barrett *et al.,* 1995).

Biological control research on leaf miners has documented the major natural enemies present in different fruit-growing areas and measured their impacts on leaf miner populations (e.g., Johnson *et al.,* 1976; Hagley, 1985; Ridgeway & Mahr, 1985; Barrett & Jorgensen, 1986; Van Driesche *et al.,* 1994; Barret *et al.,* 1995). Considerable work to evaluate the effects of commonly used pesticides on the natural enemies of leaf miner species has been undertaken (Hagley *et al.,* 1981; Wieres *et al.,* 1982; van Driesche *et al.,* 1985). Developing means to achieve ecological selectivity of pesticide use by taking advantage of the spatial or temporal asynchrony between the two trophic level groups has been another focus of research (Hagley *et al.,* 1981; Wieres *et al.,* 1982; van Driesche *et al.,* 1985; Maier, 1992). The most advanced application has arisen from use of detailed temperature-dependent phenological models of the emergence of pest and natural enemies [*Sympiesis manylandensis* Girault, *Photesor ornigis* (Weed)] and from identification of the rather narrow biological windows of spray invulnerability for these parasitoids (Drummond *et al.,* 1985).

Leafhoppers

The primary leafhoppers attacking deciduous tree fruits in North America are the white apple leafhopper, *Typhlocyba pomaria* McAtee; and several species less specific to fruit crops, the most common being *Empoasca fabae* (Harris), *Edwardsiana rosae* (Linnaeus), several *Erythroneura* spp., and others (Sayedoleslami 1978; Beers, 1991). Problems with control of leafhopper species have increased on apple, presumably due to resistance to organophosphate insecticides (Croft & Hoyt, 1983). Biological control research on these species has been oriented almost exclusively toward the egg and larval parasitoids that attack these foliage-feeding pests, especially *T. pomaria.*

In Michigan apple orchards, overwintering eggs of *T. pomaria* are principally attacked by the mymarid *Anagrus epos* Girault. Parasitism levels found during studies ranged from 20 to 50%, but were as high as 100% in local areas (Sayedoleslami & Croft, 1980). Distribution studies of both parasitoids and overwintering egg populations of *T. pomaria* indicated good congruence between the pest and natural enemy in all subunits of tree habitat. Also, a density-dependent association between the two species was indicated. Parasitism by nymphal–adult dryinid parasitoid *Aphelopus typhlocyba* Muesebeck was at lower levels as compared with the egg parasitoid (Sayedoleslami, 1978). Croft (1982b) reported that both parasites seemed to tolerate long-used organophosphate insecticides such as azinphos-methyl. In Washington apple orchards, the white apple leafhopper eggs are heavily parasitized by *Anagrus epos* and a high degree of synchronization was found between the life cycle of the host and the parasitoid (Beers, 1991). In New Zealand, the mymarid *Anagrus armatus* provides high levels of parasitism of overwintering apple leafhopper eggs (Teulon & Penman, 1986).

Introductions of Biological Control Agents

Introductions of exotic natural enemies have been undertaken frequently for secondary pests of deciduous fruit crops. Examples from the last two decades for North America are presented in Table 1.

Spider Mites

The transfer and introduction of exotic predators of spider mites from one region within a country or from one area of the world to another for control on deciduous fruit crops has almost exclusively involved insecticide-resistant biotypes of phytoseiid mites (Croft & Strickler, 1983; Hoy, 1985; Nyrop *et al.*, 1994). These imported mites have included species where resistance has developed in field populations following selection for a number of years and also after selection in the laboratory. The most successful cases have been in Australia and New Zealand using mites imported from North American apple orchards and in England using imported *T. pyri* from New Zealand (Croft, 1990).

Pear Psylla, *Psylla pyricola* Förster

Although it is difficult to classify this species as a key or secondary pest (depending on its status in different areas and local natural-enemy abundance), it is usually a secondary, indirect pest of pears. Considerable efforts to locate and import more effective natural enemies for its control in North American orchard systems have been attempted. In 1963, anthocorid *anthocoris nemoralis* (Fabricius) was imported from western Europe into British Columbia orchards and later into western areas of the United States with good establishment reported. It has appeared to be able to withstand the cold temperatures of the region. Several coccinellids, including *Harmonia conformis* Madar, *H. dimidiata* (Mulsant), *A. axyhidis* (Pallas), *Coccinella septempunctata* L., and *Dimus pumilio* Weise, were introduced in the late 1970s into the western United States for control of this pest. Importation and subsequent releases from northern Greece and Yugoslavia of several natural enemies [including biotypes of three predators, *Chrysopa carnea* (Stephens), *Synharmonia conglobata* (Linnaeus), and *Propylea quatourdecimpuncta* (Linnaeus)] and parasitic encyrtid *Prionomitus mitratus* (Dalm) were made in the western United States in the early 1980s (Croft & Bode, 1983). The establishment and benefits of these introductions have not

been reported. *Trechnites psyllae* (Ruschka) has been imported from Greece and released in California and Oregon; however, no establishment has been reported. Future success of biological control against pear psylla may depend on foreign exploration for new beneficials in poorly sampled areas such as Central Asia (Unruh *et al.*, 1995).

BIOLOGICAL CONTROL OF DIRECT, PRIMARY, AND SPORADIC PESTS

Biological control of direct pests of deciduous tree fruits has been difficult to achieve, in part due to their damage to the consumable products and the generally lower tolerance for this damage. However, one of the earliest reported successful cases of biological control came against a direct pest of apples, *Anthonomus* sp. (Coleoptera: Curculionidae), in France by ichneumonid parasitoids of this weevil (Marchal, 1907). Inundative releases and augmentation of naturally occurring beneficial insects for control of key pests have been practiced in tree fruit crops with some success (Table 4; Ridgway & Vinson, 1977; Billioti, 1977; Van Driesche & Carey, 1987).

Numerous attempts have been made to utilize biological control based on foreign exploration and importation of promising agents against the primary and sporadic direct pests of apple, pear, peach, cherry, walnut, and filbert, crops. Many of the introduced natural enemies have been established and in a number of cases seem to suppress pest populations to lower levels under undisturbed conditions. In this section, several of these attempts are reviewed. In general, however, the biological control of direct pests is practiced less frequently on a commercial scale due to a low level of tolerance by consumers to the damage caused by these pests. This forces growers to maintain low pest populations, thus limiting the availability of hosts for reproduction of beneficials. Additionally, the routine use of pesticides is common in most commercial orchards and is a major hindrance to the widespread establishment and persistence of these natural enemies.

Management of Endemic Natural Enemies

Coding Moth [*Cydia pomonella* (Linnaeus)]

This insect is probably the most important pest of the apple worldwide except for isolated areas of Asia and Japan (Putman, 1963; Croft & Riedl, 1991). A native of Eurasia, it was probably accidentally introduced worldwide through shipments of infested fruit. It was first recorded in North America in 1750 (Essig, 1931) and since has spread throughout the continent.

Management of codling moth throughout its range is

TABLE 4   A List of Selected Biological Control Programs Involving Importation, Establishment, Conservation, and Augmentation of Parasitoids against Primary Tree Fruit Pests in North America

Target pest	Areas of use	Type of biological control application[a]	Origin of parasitoid	Success of program (1–5 scale)[b]	References[c]
Operophtera brumata	NW–NE U.S., Canada	I, E, C	Western Europe	5	Embree, 1971; Clausen, 1978; J. C. Miller, unpublished
O. occidentalis	NW U.S.	C	Native	3	J. C. Miller, unpublished; M. T. Ali-Niazee, unpublished
Archips rosanus	NW U.S., Canada	C, A	Native and imported	3	AliNiazee, 1977; AliNiazee, unpublished
Choristoneura rosaceana	NW U.S., Canada	C, A	Native and imported	2	AliNiazee, unpublished
Cydia pomonella	W, E U.S., S. Africa, New Zealand, Peru	I, E	Spain, France, Canada	1	Simmons, 1944; Clausen, 1978
Rhagoletis pomonella	W, E U.S.	I, E, C	Hawaii, native	1	Clausen, 1978; AliNiazee, 1985
R. indifferans	W U.S.	I, E	Hawaii	1	Clausen, 1978
A. argryspila	W Canada	C, A	Native	2	Mayer, 1972
Grapholitha molesta	W U.S., Canada, Australia, Uruguay	I, E	Japan, Korea, Chile, France	2	Clausen, 1978
Anarsia lineatella	W U.S.	I, E	France	2	Clausen, 1978
Cnephasia longana	W U.S.	I, E	France	2	Richter, 1966
Psylla pyricola	W U.S.	I, E	Europe	2	McMullin, 1971; Croft & Bode, 1983

[a] Abbreviations: I = introductions; E = establishment; C = conservation; A = augmentation.
[b] Lowest (1) to highest (5).
[c] Also found in Croft & Strickler (1983), Whalon & Croft (1984), and Croft & Raush (1988)

primarily dependent on the use of chemical pesticides (Barnes, 1959; Madsen & Morgan, 1970). The excessive use of chemicals against this insect has been responsible for development of secondary and sporadic pests in many areas (Croft & Hoyt, 1983), thus adversely affecting several attempted integrated pest management (IPM) programs. Because of this heavy dependency on pesticide use, biological control of codling moth has been neglected for years, although many early reports indicated the usefulness of predators and parasitoids in suppressing populations (Brodie, 1907; Rosenberg, 1934; Boyce, 1941; Simmons, 1944).

Conservation and management of endemic parasitoids such as *Trichogramma* sp. for control of codling moth have been implemented in some areas, including in several countries of Western Europe and parts of North America. The former Soviet Union has been particularly active in using inundative releases of these egg parasitoids to control codling moth on large acreages (Beglyarov & Smetnik, 1977). This research has been focused on development of methods for parasitoid mass production and release and means to increase effectiveness. In Germany, Stein (1960) reported nearly a 50% reduction in damage by codling moth when *Trichogramma* spp. were released against first-generation populations of the moth. In North America, there has been less use of these agents and research has been more limited. Work with *T. platneri* Nagarkatti, an indigenous egg parasitoid of codling moth, has been moderately encouraging (Mills, 1995; Caprile *et al.,* 1994). *Tricho-*

gramma minutum Riley was reported to be an important natural enemy of codling moth in Ontario (Boyce, 1941), where during certain years it provided over 50% parasitism and significantly reduced larval densities.

Other larval parasitoids reported from codling moth that may have potential for management include *Pimpila pterelas* Auctt. and *Ascogaster quadridentata* Wesmael (Simmons, 1944; Putman, 1963). The latter species, which was accidentally introduced from Europe to North America, is the most important larval parasitoid of this pest and causes as much as 25% parasitism in some areas. Other larval parasitoids include *Macrocentrus delicatus* Cresson, *M. instabilis* Muesebeck, *M. ancylivorous* Rohwer, and *Phanerotoma fasciata* Provancher; pupal parasitoids, including *Dibrachys cavus* (Walker), *Eupelmus cyaniceps* Ashmead, *Pimpla annulipes* Brullè, and *Eurytoma* sp., are regarded as generally unimportant (Putman, 1963; Clausen, 1978). A list of the most important natural enemies of codling moth is provided in Table 5.

A number of predaceous insects and birds have been reported to suppress the populations of codling moth in parts of Europe and North America (Putman, 1963; Solomon *et al.,* 1976; Glenn & Milson, 1978). Among these, *Tenebroides corticalis* Melsch, *Solenopsis molesta* (Say) (an ant species), some species of carabids (Riddick & Mills, 1994; Riddick & Mills, 1996a, 1996b, Epstein *et al.,* 1997) and staphylinids, and the spider *Agelena naevia* Walck were found to feed on larvae of codling moth. However,

TABLE 5   Major Natural Enemies of Codling Moth and Their Status as Biological Control Agents

Species	Host stage attacked	Origin	Status (effectiveness)	References
Parasitoids				
Trichogramma platneri	Egg	North America	Partial	Mills, 1995; Unruh, 1996
Trichogramma sp.	Egg	Worldwide	Partial	Boyce, 1941
Ascogaster quadridentata	Egg, larval	Europe, North America	Partial	Clausen, 1978
Liotryphon quadridentatus	Cocoon	Central Asia, introduced to North America	Partial	Mills, 1995
L. caudatus	Cocoon	Central Asia, introduced to North America	Partial	Mills, 1995
Mastrus ridibundus	Cocoon	Central Asia, introduced to North America	Partial	Mills, 1995
Microdus rufipes	Larval	Central Asia, introduced to North America	Partial	Mills, 1995
Pristomerus vulnerator	Larval–pupal	Central Asia, introduced to North America	Partial	Mills, 1995
Predators				
Pterostichus sp.	Larval	North America, others	Minor	Riddick & Mills, 1994; Epstein *et al.,* 1997
Harpalus aeneus	Larval	North America, others	Minor	Epstein *et al.,* 1997
Amara anea	Larval	North America, others	Minor	Epstein *et al.,* 1997
Chrysoperla sp.	Egg	North America, others	Minor	Putman, 1963
Mirids (multiple spp.)	Egg–larval	North America, others	Partial	MacLellan, 1962
Spiders (multiple spp.)	Egg, larval	Worldwide	Partial	Bajwa, 1996
Birds (multiple spp.)	Larval, cocoon	Worldwide	Partial	Solomon *et al.,* 1976

the most effective predators were the woodpeckers *Dendrocopos pubescens* Linnaeus and *D. villosus* Linnaeus, particularly in Nova Scotia (MacLellan, 1959). In Wales, woodpeckers and the blue tits *Paarus caevunlens* Linnaeus and *P. major* Linnaeus were important (Solomon *et al.,* 1976). In Oregon, a large number of predaceous insects, including the mirids *Deraeocoris* sp. and *Phytocoris* sp., are generally present in apple orchards throughout the growing season (AliNiazee, unpublished). MacLellan (1962) reported that a number of mirid species were also important egg and larval predators of codling moth in Nova Scotia.

Early research on augmentation and conservation of biological control agents of codling moth indicated that sprays of lead arsenate and sulfur fungicides inhibited oviposition of the parasitoid *Ascogaster quadridentata* (Cox & Daniel, 1935; Boyce, 1941). Similar reductions of other parasitic and predatory insects and birds caused by organophosphate, carbamate, and pyrethroid insecticides are probably responsible for the failure of natural control of the codling moth in commercial orchards (Croft & Hoyt, 1983). Early-season use of these insecticides at petal fall and first cover in most fruit growing areas for control of pests such as leafrollers, aphids, scales, and other sporadic moth species coincides with the buildup of most parasitic and predaceous insects in commercial orchards. These sprays cause disruption of biological control of primary and secondary pests such as coding moth and leaf rollers. Use of microbial pesticides (such as B.t. and a codling moth granulosis virus), juvenile hormonelike pesticides, and chitin synthesis inhibitors like diflubenzeron (AliNiazee, unpublished) have introduced a new diversity of materials with greater selectivity to a broad range of these natural enemies (Westigard, 1979). Further study of their use and features of selectivity may provide for greater integration of biological controls for the codling moth with chemical suppression tactics.

In addition to early work with mass releases and augmentation of *Trichogramma* spp., the larvae of *Chrysoperla carnea* and *C. rufilabris* Burmeister have been reported to attack eggs of the codling moth (Putman, 1963). Increasing the effectiveness of these species and other generalist predators through use of food sprays has been suggested (Hagen *et al.,* 1971). Furthermore, several species of parasitoids and predaceous insects are not available for purchase in inundative release programs. Although the economics of these control programs have not be carefully documented, they suggest new avenues for achieving biological control of the codling moth.

INTRODUCTION OF EXOTIC BIOLOGICAL CONTROL AGENTS

Codling Moth

Exploration, importation, and introduction of natural enemies against the codling moth has been pursued spordically for nearly a century. This work began in 1904 and 1905, when the ichneumonid *Apistephialtes caudatus* (Ratzeburg) was taken from Spain and released in California (Clausen, 1956a). Although some recoveries were made, establishment did not occur. Likewise, additional releases of these parasitoids during 1935 and 1936 in the central United States failed to result in establishment (Clausen, 1956a). Similarly, bethylid *Goniozus emigratus* (Rohwer), imported from Hawaii and released in California, failed to establish during the early years. However, work by Gordh and Moczar (1990) suggests that this species is present in California. One of the first successful establishments against this pest was with *Ascogaster quadridentata* in Washington, Idaho, and California during the 1940s (Clausen, 1978).

In British Columbia, Canadian scientists released some 6000 *A. quadridentata* from 1934 to 1939 and establishment was rapid. However, similar success with *A. caudatus,* and *Cryptus sexannulatus* Gravenhorst in British Columbia and Ontario from Europe was not achieved. Importation and releases of these three species were also made in South Africa, New Zealand, Australia, Peru, and Pakistan with widespread establishment of *A. quadridentata* and poor or no establishment of the other two species (Clausen 1978). Extensive search has been conducted for the natural enemies of the codling moth in former Soviet Republics, particularly Kazakhstan. A number of species have been imported and some have been released in the apple orchards of Washington and California. Three hymenopterans *Liotryphon caudatus, Mastrus ridibundus,* and *Microdus rufipes* are currently being explored (Mills, 1995; Unruh, 1997).

As noted earlier, results of foreign exploration, importation, and establishment research against the codling moth are mixed. Although *A. quadridentata* is established in several regions and seems to provide good suppression of codling moth in unsprayed situations, its role in commercial sprayed orchards is unknown. *Trichogramma embryophagum* (Hartig) and two other species of *Trichogramma* were introduced into California from the former Soviet Union (Croft & Bode, 1983). Future studies on importation of new species and biotypes of this group of egg parasitoids may lead to biological control of this pest. To this end, foreign exploration in the future should concentrate more in Central Asia, China, Afghanistan, Kashmir, the Swat Valley of Northern Pakistan, and certain areas of Iran where codling moth is thought to be endemic.

Apple Maggot (*Rhagoletis pomonella*)

The apple maggot, *Rhagoletis pomonella,* is a native North American insect. It became a pest of apple in the eastern United States over a century ago by making a host shift from its native hosts, *Crateagus* spp. (Dean & Chapman, 1973). Apple maggot was recorded on the West Coast in Portland, Oregon, in 1979 (AliNiazee & Penrose, 1981), and is now found in at least six western states including Oregon, Washington, California, Utah, Idaho, and Colorado (AliNiazee & Brunner, 1986). This expansion is in addition to its original distribution in the midwestern and eastern United States and Canada (Bush, 1966).

In the eastern United States, larval–pupal parasitoids are important regulating factors for apple maggot populations (Monteith, 1971; Dean & Chapman, 1973; Cameron & Morrison, 1977). Several species of Opiinae braconids have been reported from these areas including *Opius canaliculatus* Gahan and *Diachasma alloeum* (Muesebeck) from Quebec (Rivard, 1967); *Biosteres melleus* (Gahan), *O. canaliculatus,* and *D. alleoum* from New York (Dean & Chapman, 1973); and *B. melleus, Opius lectus* Gahan, *D. alloeum, D. ferrugineum* (Gahan), and *O. downesi* Gahan from Connecticut (Maier, 1981). The opiines had significant effects in reducing apple maggot populations on hawthorn, but their impact on populations of this pest in apple was marginal.

In the western United States, AliNiazee (1985) reported that two opiines, *Opius lectoides* Gahan and *O. downesi,* which commonly attack a closely related species, the snowberry maggot (*Rhagoletis zephyria* Snow), were associated with *R. pomonella* in Oregon, mainly on hawthorn fruit. As many as 60% of the pupae were parasitized by these parasitoids on the native host, while less than 2% on apple were attacked. Both species of *Opius* have fairly short ovipositors that may not be long enough to reach the host larvae in the larger apple fruit. *Diachosma alloeum* has a much longer ovipositor and has been very successful in parasitizing apple maggot larvae in apples in the eastern United States.

No attempt has been made to conserve and augment parasitoids of apple maggot in North America, in part because biological control alone cannot provide economic suppression of this species (AliNiazee *et al.,* 1995). Relatively little is known about the biology of these natural enemies. This has also contributed to the lack of efforts to manipulate their populations for practical purposes. The synchrony of the life cycles of these parasitoids with their apple maggot host seems to be excellent in New York (Dean & Chapman, 1973) and Oregon (AliNiazee, unpublished). Reduced effectiveness of these biological control agents due to insecticide use in commercial orchards may be common. Other parasitoids from apple maggot include *Psilus* sp., *Aphaereta auripes* (Provancher), and several undescribed eulophid species, none of which seem to be important as regulatory agents of this pest.

Several generalist predators, including carabids, ants, spiders, and birds, feed on various stages of the apple maggot (Dean & Chapman, 1973). The mature larval stage and the adult stage after emergence and before flight appear highly vulnerable to predation. However, predators have not provided natural control of this pest. Thus, management of this species through endemic natural enemies will continue to be focused largely on parasitic species.

Exploration and importation of natural enemies from foreign lands were originally thought to hold little promise for control of the apple maggot, mainly because this pest is strictly a North American insect. This conclusion has been reconsidered because closely related fruit fly species are heavily parasitized in other areas of the world (Clausen, 1978). For example, some of the opiines associated with tropical fruit flies were released against the apple maggot during the 1950s. *Opius longicaudatus compensans* (Silvestri) and *O. longicaudatus taiensis* Fullaway were imported from Hawaii and released in West Virginia in 1954 (Clausen, 1978), but these were not established. Other parasitoids for potential introduction include the parasitoids of the exotic European cherry fruit fly, *Rhagoletis cerasi* Linnaeus and the walnut husk fly, *R. completa* Cresson.

Attempting to establish parasitoids from one region and host species to another within North America is another subject that needs further attention. The apple maggot is present in the western United States (AliNiazee & Brunner, 1986), and some of the most effective parasitoids attacking this pest in the eastern states are lacking (AliNiazee, 1985). Therefore, it appears that the introduction of these parasitoids would be a viable option to consider for improving biological control in the West. Attempts should also be made to introduce parasitoids of other closely related species into this area of relatively recent introduction of the apple maggot (AliNiazee, 1985).

Other Fruit Fly Species

The *Rhagoletis* flies are the most important pests of cultivated cherries in North America and Europe. Several species have been the subject of considerable classical biological control efforts, despite an extremely low tolerance for their damage. An infestation rate of less than 0.2% is required for commercial marketing of cherries in the United States. Four species of parasitoids associated with the Oriental fruit fly (*Dacus dorsalis* Hendel) were some of the earliest introductions. They included the braconids *Opius longicaudatus compensans, O. longicaudatus farmosanus* (Fullaway), *O. oophilus* Fullaway, and *O. longicaudatus novacaledonicus* Fullaway. These species were imported from Hawaii and released against *R. indifferens* Curran and *R. fausta* Osten Saken in Oregon and Washington in the early 1950s (Clausen, 1956b). None of these species became established in the region. A parasitoid of *R. cerasi* (the European cherry fruit fly) was imported and released against the eastern cherry fruit fly (*R. cingulata* Loew) from 1959 to 1964 in New Jersey, with no record of establishment. Other species of parasitoids, including *Biostus sub-*

laevis (Gahan), *Coptera occidentalis* Muesebeck, and *Phygadeuon wiesmanni* Sachtleben, are currently being evaluated for release in different areas of North America against several *Rhagoletis* species (K. Hagen, personal communication, October, 1992).

Other Insects

Foreign exploration and importation of natural enemies of other North American fruit pests have been pursued against the Oriental fruit moth [*Grapholitha molesta* (Busck)], the European winter moth (*Operophtera brumata* Linnaeus), and, to a certain extent, against *Archips rosanus* Linnaeus and the omnivorous leaftier, *Cnephasia longana* (Haworth). The latter three species are pests of apples, pears, and cherries while the Oriental fruit moth is a common pest of peaches, prunes, and nectarines.

The search for parasitoids of Oriental fruit moth was conducted in the United States (1920 to 1930), France and Australia (1931 to 1932), and Japan and Korea (1932 to 1933) (Clausen, 1978). Initially, releases of *Macrocentrus ancylivorous* Rohwer, a braconid native to the northeastern United States, were made into parts of Canada during the 1920's and 1930s. This resulted in rapid establishment and a considerable pest population suppression in some areas (LeRoux, 1971). Several parasitoids were imported from Japan and Korea in the late 1930s into California. Initially stocks of *Diadegma molestae* (Uchida) were imported into California and the eastern United States with mixed results. *Phanerotoma molestae* Muesebeck and *Agathis diversa* Muesebeck were imported from northern China in 1950 (Clausen, 1978). However, neither species became established. In the 1940s, *M. ancylivorus* was introduced into California for control of Oriental fruit moth (Finney *et al.,* 1947). It readily became established and provided excellent control of this pest and other closely related lepidopteran species. *Macrocentrus ancylivorus* has been introduced into other countries including Argentina where it has become established (Clausen, 1978). An interesting aspect of this species is its ability to tolerate pesticides. There is good circumstantial evidence that in Canada (and perhaps in parts of the United States) this natural enemy either has natural tolerances or has developed resistances to DDT, parathion, and carbaryl, thus allowing it to fit well into integrated control programs on peach crops in many areas (Dustan & Boyce, 1966).

The leaf roller complexes that occur in most areas of tree fruit production in North America are important pest groups that have received some attention in relation to biological control research; see Croft and Hoyt (1983) and van der Geest and Evenhuis (1991) for species and area involved. *Choristoneura rosaceana* Harris, *Archips rosanus, A. argyrospila* (Walker), *Pandemis limitata* Robinson, *P. pyrusana, Argyrotaenia velutinana* (Walker), and other species can be particular problems (Chapman & Lienk, 1971). In Europe, several of these species, as well as the summer fruit tortrix, *Adoxophyes orana* (von Rösler-stamm), are the principal moth pests, having replaced the codling moth in some areas (Croft & Hull, 1983). Egg parasitoids from the Trichogrammatidae family, in particular, *T. minutum* Riley, parasitize up to 50% of egg masses of these pests (Glass, 1963). In the Netherlands, *T. embry-ophagus cacaoeciae* Marchal has been mass-produced and released for control of *A. orana* with mixed results (Even-huis, 1980). Susceptibility to unfavorable weather and broad-spectrum pesticides were major factors limiting their effectiveness.

Leaf rollers are attacked by many parasitoid species. For example, more than 50 species of parasitic Hymenoptera and several species of dipteran Tachinidae have been reared from the larvae of *A. orana* and other closely related leaf rollers in Europe (Evenhuis, 1980). *Ascogaster rufidens* Wesmael and *A. quadridentata* Wesmael are common egg–larval parasites of this pest in the Netherlands. The eulo-phid *Colpoclypeus florus* (Walker) is an important gregari-ous ectoparasite of older leaf roller larvae (de Jong, 1980). This parasitoid appears to be highly effective against the pandemis and oblique-banded leaf roller in the state of Washington (Nobbs, 1997). In North America, the leaf roll-ers *Archips rosanus, A. argyrospila, Argyrotaenia citrana* (Fernald), and *C. rosaceana* are parasitized by many spe-cies. AliNiazee (1977) found nine species of larval–pupal parasitoids attacking *A. rosanus* from Oregon, while Pro-kopy (1968) found nearly 25 species attacking *A. argyros-pila* in Connecticut. In British Columbia, even larger num-bers of these entomophagous species were noted by Mayer (1973). Ichneumonids and tachinids were the most common groups represented.

Levels of parasitism and control provided by the parasi-toids of leaf rollers range from 1 to 60%, depending on when and where samples were taken (Mayer, 1973; Ali-Niazee, 1977). In commercial orchards, rates are typically low, presumably due to the use of broad-spectrum pesti-cides to which these natural enemies are extremely suscep-tible. In natural or unmanaged environments, parasitism levels may be much higher. For example, Mayer (1973) reported that percentage of parasitism of the leaf roller complex in British Columbia orchards ranged from 19 to 32% on unsprayed apple trees. On the hazelnut (filbert) in Oregon, AliNiazee (1977) found that levels of parasitism ranged from 7 to 48% on these moderately sprayed hosts.

The biological control potential of parasitoid complexes associated with leaf rollers in deciduous fruit pests is of considerable interest to fruit entomologists in light of the increasing pest status and increasing incidences of pesticide resistance among species of this group (Croft & Hull, 1983, 1991). In general, there seems to be a difference of opinion among researchers concerning the importance of biological

control in suppressing leaf roller pests below economic levels. The levels of biological control provided by natural enemies varies from region to region and season to season. For example, in British Columbia, Mayer (1973) concluded that parasitoids alone were incapable of controlling leaf rollers under most circumstances and years. Paradis and LeRoux (1965) felt that larval–pupal parasitoids were in-effective in controlling *A. argyrospila* below economic lev-els in Quebec. In other parts of North America and Europe, the natural-enemy complexes associated with these pests seem to be very important mortality agents. For example, Paradis and LeRoux (1965) found that the parasitoid *Ito-plectis conquisitor* (Say) was acting in a density-dependent manner and provided significant biological control of *A. argyrospila*. The egg parasitoids, particularly *T. minutum,* are very important biological control agents in British Co-lumbia and in Oregon (Mayer, 1973, AliNiazee, unpub-lished). Importation of other egg parasitoid species such as *T. cacoeciae* from Europe (Biliotti, 1977) and further dis-semination of the indigenous species of *Trichogramma* from New York to other tree fruit production areas may contribute even greater levels of biological control against leaf roller pests of orchards.

The European winter moth, *Operophtera brumata,* which is a serious pest of apples in Europe and a pest of hazelnuts, apples, and cherries in Oregon (AliNiazee, 1986), has been the subject of intensive biological control studies throughout its distributional range, and especially since its introduction into North America (Embree, 1971; Roland & Embree, 1995). Introduction and establishment of two parasitoids, including tachinid *Cyzenis albicans* (Fallén) and ichneumonid *Agrypon falveolatum* (Graven-horst), have resulted in an effective biological control pro-gram against this pest in Nova Scotia and surrounding states in the northeastern United States. Efforts to control this pest through biological control introduction in the west-ern regions of North America are still in the early stages of development (see Chapter 29).

SUMMARY OF STATUS OF BIOLOGICAL CONTROL OF DECIDUOUS FRUIT PEST SPECIES

Biological control research and implementation for pests of deciduous tree fruits have for the past quarter century tended to emphasize augmentation and conservation of nat-ural enemies associated with secondary indirect pests of these crops such as mites, aphids, leaf miners, scales, and leafhoppers. Population dynamics and biological control effectiveness of the natural enemies associated with these pests have been studied in some detail. Efforts also have been made to develop selective uses of registered pesticides and/or to identify physiologically selective pesticides for control of pest species (e.g., note the many examples of identification of selective acaricides to date). Use of resistant beneficials has had a major impact on the selective use of pesticides on tree fruit crops, especially with the organophosphate insecticides.

Research on classical biological control of primary direct pests of these crops has been limited, although several outstanding successful cases of biological control have been reported (e.g., winter moth, woolly apple aphid, and others). In the future, a balance between research with conservation and augmentation of native natural enemies versus classical biological control studies of imported pests of these crops should be sought. Unfortunately, it is more difficult to support these more long-term studies of foreign exploration, importation, and establishment of exotics than it is to work with better management of already established forms of predators or parasitoids. Major efforts should be made to emphasize the importance of these more costly but highly rewarding approaches to biological control of pests on these crops and the follow-up evaluations that are necessary to document their value and effectiveness in the future.

References

Adams, R. G., Jr., & Prokopy, R. J. (1977). Apple aphid control through natural enemies. Massachusetts Fruit Notes, 42, 6–10

Adams, R. G., Jr., & Prokopy, R. J. (1980). *Aphidoletes aphidimyza* (Rondani)(Diptera: Cecidomyiidae): An effective predator of the apple aphid (Homoptera: Aphididae) in Massachusetts. Protection Ecology, 2, 27–39.

AliNiazee, M. T. (1974). Role of predatory mite, *Typhlodromus arboreus* in biological control of spider mites on apple in western Oregon. Proceedings of the Fourth International Congress on Acarology (pp. 634–642). Austria: Saalfeldon.

AliNiazee, M. T. (1977). Bionomics and life history of a filbert leafroller, *Archips rosanus* (Lepidoptera: Tortricidae). Annals of the Entomological Society of America, 70, 391–401.

AliNiazee, M. T. (1985). Opine parasitoids (Hymenoptera: Braconidae) of *Rhagoletis pomonella* and *R. zephyria* (Diptera: Tephritidae) in the Willamette Valley, Oregon. Canadian Entomologist, 117, 163–166.

AliNiazee, M. T. (1986). The European winter moth as a pest of filberts: Damage and chemical control. Journal of the Entomological Society of British Columbia, 83, 6–12.

AliNiazee, M. T. (1991). Biological control of the filbert aphid, *Myzocallis coryli* in hazelnut orchards. Proceedings of the Oregon, Washington, British Columbia and Nut Growers Society, 76, 46–53.

AliNiazee, M. T. (1995). The economic, environmental, and sociopolitical impact of biological control. (pp. 47–56). In J. R. Nechols, L. A. Andres, J. W. Beardsley, R. D. Goeden, & C. G. Jackson (Eds.), Biological control in the United States. Accomplishments and benefits of Regional Research Project W-84, 1964–1989. (Publication 3361). University of California, Division of Agriculture and Natural Resources.

AliNiazee, M. T. (1997). Biology, impact and management of *Trixys pallidus* in hazelnut orchards of Oregon. Acta Horticulturae, 445, 477–482

AliNiazee, M. T., & Brunner, J. F. (1986). Apple maggot in the western United States: A review of its establishment and current approaches to management. Journal of Entomological Society of British Columbia, 83, 49–53.

AliNiazee, M. T., & Hagen, K. S. (1995). Walnut aphid. In J. R. Nechols, L. A. Andres, J. W. Beardsley, R. D. Goeden, & C. G. Jackson (Eds.), Biological control in the United States. Accomplishments and benefits of Regional Research Project W-84, 1964–1989 (Publication 3361, Part 1, pp. 140–141). University of California, Division of Agriculture and Natural Resources. Oakland, California

AliNiazee, M. T., & Penrose, R. L. (1981). Apple maggot in Oregon: A possible new threat to the northwest apple industry. Bulletin of the Entomological Society of America, 27, 245–246.

AliNiazee, M. T., Mohammad, A. B., & Jones, V. P. (1995). Apple maggot. In J. R. Nechols, L. A. Andres, J. W. Beardsley, R. D. Goeden, & C. G. Jackson (Eds.), Biological control in the United States. Accomplishments and benefits of Regional Research Project W-84, 1964–1989 (Publication 3361, pp. 221–223). University of California, Division of Agriculture and Natural Resources. Oakland, California

Bajwa, W. I. (1996). Integration of microbial and chemical controls against codling moth, Cydia pomonella, L. Laboratory and field evaluation. Unpublished doctoral thesis, Oregon State University, Corvallis.

Barnes, M. M. (1959). Deciduous fruit insects and their control. Annual Review of Entomology, 4, 343–362.

Barrett, B. A., Hoyt, S. C., & Brunner, J. F. (1995). Western tentiform leafminer. In J. R. Nechols, L. A. Andres, J. W. Beardsley, R. D. Goeden, & C. G. Jackson (Eds.), Biological control in the United States. Accomplishments and benefits of Regional Research Project W-84, 1964–1989 (Publication 3361, pp. 213–216). University of California, Division of Agriculture and Natural Resources, Oakland, California.

Barrett, B. A., & Jorgensen, C. D. (1986). Parasitoids of the Western tentiform leafminer *Plyllonorycter elmaella* (Lepidoptera: Gracillariidae) in Utah apple orchards. Environmental Entomology 15, 635–641.

Beers, E. H. (1991). Control strategies for leafminers and leafhoppers revisited. In K. Williams (Ed.), New directions in tree fruit pest management (pp. 157–167). Yakima, WA: Good Fruit Grower.

Beglyarov, G. A., & Smetnik, A. I. (1977). Seasonal colonization of entomophages in U.S.S.R. In R. L. Ridgway & S. B. Vinson (Eds.), Biological control by augmentation of natural enemies (pp. 283–328). New York: Plenum Press.

Biliotti, E. (1977). Augmentation of natural enemies in western Europe. In R. L. Ridgeway & S. B. Vinson (Eds.), Biological Control by Augmentation of Natural Enemies (pp. 341–347). New York: Plenum Press.

Blommers, L. H. M. (1994). Integrated pest management in European apple orchards. Annual Review of Entomology, 39, 213–242.

Bostanian, J. J., & Coulombe, L. J. (1986). An integrated pest management program for apple orchards in southwest Quebec. Canadian Entomologist, 118, 1131–1142.

Boyce, H. R. (1941). Biological contol of codling moth in Ontario. Annual Report of the Entomological Society of Ontario, 71, 40–44.

Brodie, W. (1907). Parasitism of *Carpocapsa pomonella*. Annual Report of the Entomological Society of Ontario, 37, 5–15.

Bush, G. L. (1966). The taxonomy, cytology and evolution of the genus *Rhagoletis* in North America (Diptera: Tephritidae). Bulletin of the Museum of Comparative Zoology (Harvard), 134, 431–562.

Cameron, P. J., & Morrison, F. O. (1977). Analysis of mortality in the apple maggot, *Rhagoletis pomonella* (Diptera: Tephritidae) in Quebec. Canadian Entomologist, 109, 769–788.

Caprile, J., Klonsky, K., Mills, N., McDougall, S., Micke, W., & VanSteenwyk, R. (1994). Insect damage limits yield, profits of organic apples. California Agriculture, 48, 21–28.

Carroll, D. P., & Hoyt, S. C. (1984a). Natural enemies and their effects on apple aphid colonies on young apple trees in central Washington. Environmental Entomology, 13, 469–481.

Carroll, D. P., & Hoyt, S. C. (1984b). Augmentation of European earwigs

(Dermaptera: Forficulidae) for biological control of apple aphid (Homoptera: Aphidae) in an apple orchard. Journal of Economic Entomology, 77, 738–740.

Carroll, D. P., & Hoyt, S. C. (1986). Hosts and habitats of parasitoids (Hymenoptera: Aphidiidae) implicated in biological control of apple aphid (Homoptera: Aphidae). Environmental Entomology, 15, 1171–1178.

Chapman, P. J., & Lienk, S. E. (1971). Tortricid fauna of apple in New York; including an account of apple's occurence in the state, especially as a naturalized plant (Special Publication). New York State Agricultural Experiment Station. Geneva, NY.

Clausen, C. P. (1956a). Biological control of insect pests in the continental United States (Technical Bulletin 1139). Washington, DC: U.S. Department of Agriculture.

Clausen, C. P. (1956b). Biological control of fruit flies. Journal of Economic Entomology, 49, 176–178.

Clausen, C. P. (Ed). (1978). Introduced parasites and predators of arthropod pests and weeds: A world review. Agriculture Handbook No. 480. Washington, DC: U.S. Department of Agriculture.

Cox, J. A., & Daniel, D. M. (1935). Ascogaster carpocapsae Vier. in relation to arsenical sprays. Journal of Economic Entology, 28, 113–120.

Croft, B. A. (1982a). Apple pest management. In R. L. Metcalf & W. H. Luckmann (Eds.), Introduction to insect pest management (pp. 465–498). New York: Wiley Interscience.

Croft, B. A. (1982b). Developed resistance to insecticides in apple arthropods: A key to pest control failures and successes in North America. Entomologia Experimentalis et Applicata, 31, 88–110.

Croft, B. A. (1983). Status and management of pyrethroid resistance in the predatory mite, Amblyseius fallacis. Great Lakes Entomologist, 16, 17–32.

Croft, B. A. (1990). Arthropod biological control agents and pesticides. New York: Wiley Interscience.

Croft, B. A. (1994). Biological control of apple mites by a phytoseiid mite complex and Zetzellia mali (Acari: Stigmaeidae): Long term effects and impact of azinphosmethyl on colonization by Amblyseius andersoni (Acari: Phytoseiidae). Environmental Entomology, 23, 1317–1325

Croft, B. A., & AliNiazee, M. T. (1983). Differential tolerance or resistance to insecticides in Typhlodromus arboreus Chant and associated phytoseiid mites from apple in the Willamette Valley, Oregon. Journal of Economic Entomology, 12, 1420–1423.

Croft, B. A., & Bode, W. (1983). Tactics for deciduous tree fruit IPM. In B. A. Croft & S. C. Hoyt (Eds.), Integrated management of insect pests of pome and stone fruit insect pests (pp. 219–270). New York: Wiley Interscience.

Croft, B. A., & Hoyt, S. C. (Eds.). (1983). Integrated management of insect pests of pome and stone fruits. New York: Wiley Interscience.

Croft, B. A., & Hull, L. A. (1983). The orchard as an ecosystem. In B. A. Croft & S. C. Hoyt (Eds.), Integrated pest management of insect pests of pome and stone fruits (pp. 19–42). New York: Wiley Interscience.

Croft, B. A., & Hull, L. A. (1991). Chemical control and resistance in tortricid pests of pome and stone fruits. In L. P. S. van der Geest & H. H. Evenhuis (Eds.) World crop pests, Serial 5 (pp. 473–483). Amsterdam: Elsevier.

Croft, B. A., & McGroarty, D. L. (1977). The role of Amblyseius fallacis (Acarina: Phytoseiidae) in Michigan apple orchards (Research Report 333). East Lansing: Michigan State University.

Croft, B. A., & MacRae, I. V. (1993). Biological control of apple mites: Impact of Zetzellia mali (Acari: Stigmaeidae) on Typhlodromus pyri and Metaseiulus occidentalis (Acari: Phytoseiidae). Environmental Entomology, 22, 865–873.

Croft, B. A., & Raush, R. T. (1988). Technical policy issues in management of pesticide resistance in arthropod pest agriculture. Proceedings, AAAs Symposium, Washington, DC, 1987.

Croft, B. A., & Riedl, H. W. (1991). Chemical control and resistance to pesticides of the codling moth, Cydia pomonella. In L. P. S. van der Geest & H. H. Evenhius (Eds), Tortricoid pests (pp. 371–387). Amsterdam: Elsevier.

Croft, B. A., & Strickler, K. (1983). Natural enemy resistance to pesticides: Documentation, characterization, theory and application. In G. P. Georghiou & T. Saito (Eds.), Pest resistance to pesticides (pp. 669–702). New York: Plenum Press.

Croft, B. A., Hoyt, S. C., & Westigard, P. H. (1987). Spider mite management on pome fruits, revisited: Organotin and acaricide resistance management. Journal of Economic Entomology, 80, 304–311.

Croft, B. A., Whalon, M. E., & Roush, R. T. (1990). History, methods and constraints of genetic improvement. In Athropod biological control agents and pesticides (pp. 473–501). New York: Wiley Interscience.

Dean, R. L., & Chapman, P. J. (1973). Bionomics of the apple maggot in eastern New York. Search Agriculture, 3, 1–64.

DeBach, P. (Ed.). (1964). Biological control of insect pests and weeds. New York: Reinhold.

de Jong, D. J. (1980). Tortricids in integrated control in orchards. In A. K. Minks & P. Gruys (Eds.), Integrated control of insect pests in the Netherlands (pp. 19–22). Wageningen: Centre Agricultural Publications and Documents.

Drummond, F. A., van Driesche, R. G., & Logan, P. A. (1985). Model for the temperature-dependent emergence of overwintering Phyllonorycter crataegella (Clemens) (Lepidoptera: Gracillariidae), and its parasitoid, Sympiesis marylandensis Girault (Hymenoptera: Eulophidae). Environmental Entomology, 14, 305–311.

Dustan, A. G., & Boyce, H. R. (1966). Parasitism of the oriental fruitmoth Grapholitha molesta (Busck) (Lepidoptera: Tortricidae) in Ontario, 1956–1965. Entomological Society of Ontario Proceedings, 96, 100–102.

Embree, D. G. (1971). The biological control of the winter moth in eastern Canada by introduced parasites. In C. B. Huffaker (Ed.), Biological control (pp. 217–226). New York: Plenum Press.

Epstein, D., Brown, J., & Zack, R. (1997). Potentially important biological control organisms: Sampling generalist predators in apple orchards. Abstract of the 71st Annual Western Orchard Pest and Disease Management Conference, Portland, Oregon, 71, 35–36.

Essig, E. O. (1931). A history of entomology. New York: Macillan Press.

Evenhuis, H. H. (1980). Relations between insect pests of apple, and their parasites and predators. In A. K. Minks & P. Gruys (Eds.), Integrated control of insect pests in the Netherlands (pp. 33–36). Wageningen: Centre for Agricultural Publications and Documents.

Finney, G. L., Flanders, S. E., & Smith, H. S. (1947). Mass culture of Macrocentrus ancylivorus and its host the potato tuber moth. Hilgardia, 17, 437–483.

Glass, E. H. (1963). Parasitism of the red banded leafroller, Argyrotanenia velutianana by Trichogramma minutum. Annals of the Entomological Society of America, 56, 564.

Glenn, C. A., & Milson, N. F. (1978). Survival of mature larvae of codling moth (Cydia pomonella) on apple trees and ground. Annals of Applied Biology, 90, 133–146.

Gordh, G., & Moczar, L. (1990). A catalog of the world Bethylidae (Hymenoptera: Aculeata). Memoirs of the American Entomological Institute, 46, 1–364.

Hadam, J. J., AliNiazee, M. T., & Croft, B. A. (1986). Phytoseiid mites of major crops in the Willamette Valley, OR and pesticide resistance in Typhlodromus pyri Schulten. Environmental Entomology, 15, 1255–1262.

Hagen, K. S., Sawall, E. F., Jr., & Tassan, R. L. (1971). The use of food sprays to increase effectiveness of entomophagous insects. Proceedings, Tall Timbers Conference, Ecological Animal Control Habitat Management, 2, 59–81.

Hagley, E. A. C. (1985). Parasites recovered from the overwintering generation of the spotted tentiform leafminer, *Phyllonorycter blancardella* (Lepidoptera: Gracillariidae) in pest-management apple orchards in southern Ontario. Canadian Entomologist, 117, 371–374.

Hagley, E. A. C., Pree, D. J., Simpson, C. M., & Hikichi, A. (1981). Toxicity of insecticides to parasites of the spotted tentiform leafminer (Lepidoptera: Gracillariidae). Canadian Entomologist, 113, 899–906.

Hardman, J. M., Smith, R. F., & Bent, E. (1995). Effects of different integrated pest management programs on biological control of mites on apples by predator mites (Acari) in Nova Scotia. Environmental Entomology, 24, 125–142

Hoy, M. A. (1985). Recent advances in genetics and genetic improvement of the Phytoseiidae. Annual Review of Entomology, 30, 345–370.

Hoy, M. A., Westigard, P. H., & Hoyt, S. C. (1985). Release and evaluation of a laboratory selected, pyrethroid resistant strain of the predaceous mite *Metaseiulus occidentalis* in southern Oregon pear orchards and a Washington apple orchard. Journal Economic Entomology, 76, 383–388.

Hoyt, S. C. (1969). Integrated chemical control of insects and biological control of mites on apple in Washington. Journal of Economic Entomology, 62, 74–86.

Hoyt, S. C., Leeper, J. R., Brown, G. C., & Croft, B. A. (1983). Basic biology and management components for insect IPM. In B. A. Croft & S. C. Hoyt (Eds.), Integrated management of insect pests of pome and stone fruits (chap. 5). New York: Wiley Interscience.

Hull, L. A., & Beers, E. H. (1985). Ecological selectivity: Modifying chemical control practices to preserve natural enemies. In M. A. Hoy & D. C. Herzog (Eds.), Biological control in agricultural IPM systems (pp. 103–122). New York: Academic Press.

Johnson, E. F., Lang, J. E., & Trottier, R. (1976). The seasonal occurrence of *Lithocolletis blancardella* (Gracillariidae), and its major natural enemies in Ontario apple orchards. Proceedings of the Entomological Society of Ontario, 107, 31–45.

LeRoux, E. (1971). Biological control attempts on pome fruits (apple and pear) in North America. 1860–1970. Canadian Entomologist, 103, 963–974.

MacLellan, C. R. (1959). Woodpeckers as predators of the codling moth in Nova Scotia. Canadian Entomologist, 91, 673–680.

MacLellan, C. R. (1962). Mortality of codling moth eggs and young larvae in an integrated control orchard. Canadian Entomologist, 94, 655–666.

Madsen, H. F., & Morgan, C. V. G. (1970). Pome fruit pests and their control. Annual Review of Entomology, 15, 295–320.

Maier, C. T. (1981). Parasitoids emerging from puparia of *Rhagoletis pomonella* (Diptera: Tephritidae) infesting hawthorn and apple in Connecticut. Canadian Entomologist, 113, 867–870.

Maier, C. T. (1983). Relative abundance of the spotted tentiform leafminer, *Phyllonorycter blancardella* (F.), and the apple blotch leafminer *P. crataegella* (Clemens (Lepidoptera: Gracillariidae), on sprayed and unsprayed apple trees in Connecticut. Annals of the Entomological Society of America, 76, 992–995.

Maier, C. T. (1992). Seasonal development, flight activity and density of *Sympiesis marylandensis*, a parasitoid of leafmining *Phyllonorycter* spp. in Connecticut apple orchards and forests. Environmental Entomology, 21, 164–172.

Marchal, P. (1907). Utilisation des insectes auxiliaries entomphagous dans la lutte contre les insectes nuisib les a l'agriculture. Annales de l'Institut National Agronomique, 6, 281–354.

Mayer, D. F. (1973). Leafroller (Lepidoptera: Tortricidae) host plant parasite relationships in Okanogan Valley of British Columbia. Unpublished masters thesis, Simon Fraser University, Burnaby, Canada.

McMurthy, J. A., & Croft, B. A. (1997). Life styles of phytoseiid mites and their roles in biological control. Annual Review of Entomology, 42, 291–321.

Messing, R. H. (1986). Biological control of the filbert aphid Myzocallis coryli in western Oregon. Unpublished doctoral thesis, Oregon State University, Corvallis.

Messing, R. H., & AliNiazee, M. T. (1989). Introduction and establish-

ment of *Trioxys pallidus* in Oregon, U.S.A. for control of filbert aphid, *Myzocallis coryli*. Entomophaga, 34, 153–163.

Mills, N. (1995). Potential constraints for biological control of tree fruit pests. Proceedings of the Washington State Horticultural Association, 91, 221–226.

Minks, A. K., & Gruys, P. (eds.). (1980). Integrated control of insect pests in The Netherlands. Wageningen: PUDOC.

Monteith, L. G. (1971). The status of parasites of the apple maggot, *Rhagoletis pomonella* (Diptera: Tephritidae) in Ontario. Canadian Entomologist, 103, 507–512.

Morse, J. G. (1981). Studies on the apple aphid, Aphis pomi, and its cecidomyiid predator, Aphidoletes aphidimyza: Development of a predator-prey model. Unpublished docdoral dissertation, Michigan State University, East Lansing.

Morse, J. G., & Croft, B. A. (1987). Biological control of *Aphis pomi* (Homoptera: Aphididae) by *Aphidoletes aphidimyza* (Diptera: Cecidomiyiidae): A predator-prey model. Entomophaga, 32, 339–356.

Mowery, P. D., Asquith, D., & Hull, L. A. (1977). MITESIM, a computer predictive system for European red mite management. Pennsylvania Fruit News, 56, 64–67.

Nobbs, C. (1997). Alternate host of the gregarious Eulophid ectoparasitoid, *Colpoclypeus florus*. Abstracts of the 71st Annual Western Orchard Pest and Disease Management Conference, Portland, Oregon, 71, 31–32

Nyrop, J. P., Minns, J. C., & Herring, C. P. (1994). Influence of ground cover on dynamics of *Amblyseius fallacis* in New York apple orchards. Agricultural Ecosystems and Environment, 50, 61–72.

Paradis, R. O., & LeRoux, E. J. (1965). Researches sur la biologie et la dynamique des populations natuerelles d'*Archips argyrospilus* (Wlk.) dans de sud-ouest du Quebec. Memoirs of the Entomological Society of Canada, 43, 1–77.

Pree, D. J., Marshall, D. B., & Archibald, D. E. (1986). Resistance to pyrethroid insecticides in the spotted tentiform leafminer, *Phyllonorycter blancardella* (Lepidoptera: Gracillariidae) in southern Ontario. Journal of Economic Entomology, 79, 318–322.

Pree, D. J., Hagley, E. A. C., Simpson, C. M., & Hikichi, A. (1980). Resistance of the spotted tentiform leafminer *Phyllonorycter blancardella* (Lepidoptera: Gracillariidae) to organophosphorus insecticides in southern Ontario. Canadian Entomologist, 112, 469–474.

Prokopy, R. J. (1968). Parasites of the leaf rollers Archips argyrospilus and A. griseus in Connecticut. Journal of Economic Entomology, 61, 348–352.

Prokopy, R. J., & Croft, B. A. (1994). Apple pest management. In R. L. Metcalf & W. H. Luckmann (Eds.), Introduction to pest management, (3rd ed., chap. 13). New York: John Wiley & Sons.

Putman, W. L. (1963). The codling moth, *Carpocapsae promonella* (L.) (Lepidoptera: Tortricidae): A review with special reference to Ontario. Proceedings of the Entomological Society of Ontario, 93, 22–59.

Riddick, E. W., & Mills, N. J. (1994). Potential of adult carabids (Coleoptera: Carabidae) as predators of fifth-instar codling moth (Lepidoptera: Totricidae) in apple orchards in California. Environmental Entomology, 23, 1338–1345.

Riddick, E. W., & Mills, N. J. (1996a). A comparison of the seasonal activity of *Pterostichus* beetles (Coleoptera: Carabidae) in a commercial apple orchard in Sonoma County, California. Pan-Pacific Entomologist, 72, 82–88.

Riddick, E. W., & Mills, N. J. (1996b). *Pterostichus* beetles dominate the Carabid assemblage in an unsprayed orchard in Sonoma County, California. Pan-Pacific Entomologist, 72, 213–219.

Ridgeway N. H., & Mahr, D. L. (1985). Natural enemies of the spotted tentiform leafminer *Phyllonorycter blancardella* (Lepidoptera: Gracillariidae) in sprayed and unsprayed apple orchards in Wisconsin, USA. Environmental Entomology, 14, 459–463.

Ridgway R. L., & Vinson, S. B. (Eds.). (1977). Biological control by augmentation of natural enemies. New York: Plenum Press.

Rivard, I. (1967). *Opius lectus* and *O. alloeus,* larval parasites of the apple maggot, *Rhagoletis pomonella* (Diptera: Tephritidae), in Quebec. Canadian Entomologist, 99, 895–896.

Rock, G. C., & Apple, J. L. (1983). Integrated pest and orchard management systems (IPOMS) for apples in North Carolina (NCS Technical Bulletin 276). North Carolina State Univ. Raleigh.

Roland, J., & Embree, D. G. (1995). Biological control of the winter moth. Annual Review of Entomology, 40, 475–492.

Rosenberg, H. T. (1934). The biology and distribution in France of the larval parasites of *Cydia pomonella*. Bulletin of Entomological Research, 25, 201–256.

Sayedoleslami, E. H. (1978). Aspects of the temporal and spatial coincidence of the white apple leafhopper Typhlkocyba pomaria McAtee, Cicadellidae: Homoptera) and two parasitic Hymenoptera. Unpublished doctoral dissertation, Michigan State University, East Lansing.

Sayedoleslami, H., & Croft, B. A. (1980). Spatial distribution of overwintering eggs of the white apple leafhopper (*Typhlocyba pomaria*) and parasitism by *Anagrus epos*. Environmental Entomology, 9, 624–628.

Schlinger, E. I., Hagen, K. S., & van den Bosch, R. (1960). Imported French parasite of walnut aphid established in California. California Agriculture, 14, 3–4.

Simmons, F. J. (1994). Observations on the parasites of *Cydia pomonella* L. in southern France. Scientific Agriculture, 25, 1–30.

Solomon, M. E., & M. A., Easterbrook, (1984). O. P. resistant *Typhlodromus pyri* for apple orchard mite management. Proceedings, 10th International Congress on Plant Protection, 3, 20–25.

Solomon, M. E., Glen, D. M., Kendall, D. A., & Milsom, N. F. (1976). Predation of overwintering larvae of codling moth (*Cydia pomonella* (L.)) by birds. Journal of Applied Ecology, 13, 341–352.

Stein, W. (1960). Versuche zur biologischen bekampfung des apfelgruttung *Trichogramma*. Entomophaga, 5, 237–247.

Teulon, D. A. J., & Penman, D. R. (1986). Sticky board sampling of leafhoppers in three apple orchards under different management regimes. New Zealand Journal of Agricultural Research, 29, 289–298.

Trimble, R. M., & Pree, D. J. (1987). Relative toxicity of six insecticides to male and female *Pholetesor orgnigis* (Weed) (Hymenoptera: Braconidae), a parasite of the spotted tentiform leafminer, *Phyllonorycter blancardella* (Fabr.) (Lepidoptera: Gracillariidae). Canadian Entomologist, 119, 153–157.

Unruh, T. R., Westigard, P. H., & Hagen, K. S. (1995). Pear psylla. In J. R. Nechols, L. A. Andres, J. W. Beardsley, R. D. Goeden, & C. G. Jackson (Eds.), Biological control in the United States. Accomplishments and benefits of Regional Research Project W-84, 1964–1989 (Publication 3361, pp. 95–100). University of California, Division Agriculture and Natural Resources. Oakland, California.

Unruh, T. R. (1996). Biological control of the codling moth, *Cydia pomonella* (L.) on apple, *Malus domestica* L. Abstracts of the 70th Annual Western Orchard Pest & Disease Management Conference, Portland, Oregon, 70, 21.

Unruh, T. R. (1997). Mass releases of two Eurasia parasitoids of codling moth. Abstracts of the 71st Annual Western Orchard Pest & Disease Management Conference, Portland, Oregon, 71, 41.

van de Bosch, R., Frazer, B. D., Davis, C. S., Messenger, P. S., & Horn, R. (1970). *Trioxys pallidus,* an effective new walnut aphid parasite from Iran. California Agriculture, 24, 8–10.

van der Geest, L. P. S., & Evenhuis, H. H. (1991). Tortricid pests: Their biology, natural enemies and control. Amsterdam: Elsevier.

van de Vrie, M. (1986). Apple. In W. Helle & M. W. Sabelis (Eds.), Spider mites: Their biology, natural enemies and control (Vol 1B, pp. 311–325). Amsterdam: Elsevier.

van de Vrie, M., McMurtry, J. A., & Huffaker, C. B. (1972). Ecology of tetranychid mites and their natural enemies: A review. III. Biology, ecology, and pest status, and host plant relations of tetranychids. Hilgardia, 41, 343–432.

Van Driesche, R. G., & Carey, E. (1987). Opportunities for increased use of biological controls in Massachusetts (Research Bulletin 718). Massachusetts Agricultural Experimental Station. Amherst, Mass.

Van Driesche, R. G., Prokopy, R. J., & Christie, M. (1994). Effects of second-stage IPM practices on parasitism of apple blotch leafminer in Massachusetts apple orchards. Environmental Entomology, 23, 140–146.

Van Driesche, R. G., Marshall Clark, J., Brooks, R. W., & Drummond, F. J. (1985). Comparative toxicity of orchard insecticides to the apple blotch leafminer, *Phyllonorycter crataegella* (Lepidoptera: Gracillariidae), and its eulophid parasitoid, *Sympiesis marylandensis* (Hymenoptera: Eulophidae). Journal of Economic Entomology, 78, 926–932.

Vidal, C., & Kreiter, S. (1995). Resistance to a range of insecticides in the predaceous mite *Typhlodromus pyri* (Acari: Phytoseiidae): Inheritance and physiological mechanisms. Journal of Economic Entomology, 88, 1097–1105.

Walker, J. T. S. (1985). The influence of temperature and natural enemies on population development of woolly apple aphid, *Eriosoma lanigerum* (Hausmann). Unpublished doctoral dissertation, Washington State University, Washington.

Warner, L. A., & Croft, B. A. (1982). Toxicities of azinphosmethyl and selected orchard pesticides to an aphid predator, *Aphidoletes aphidimyza*. Journal of Economic Entomology, 75, 410–415.

Westigard, P. H. (1979). Codling moth: Control on pear with diflubenzuron and effects on nontarget pest and beneficial species. Journal of Economic Entomology, 75, 552–554.

Whalon, M. E., & Croft, B. A. (1984). Apple IPM implementation in North America. Annual Review of Entomology, 29, 435–470.

Wieres, R. W., Leeper, J. R., Ressig, W. H., & Lienk, S. E. (1982). Toxicity of several insecticides to the spotted tentiform leafminer (Lepidoptera: Gracillariidae) and its parasite, *Apanteles ornigis*. Journal of Economic Entomology, 75, 680–684.

Wieres, R. W., McNicholas, F. J., & Smith, G. L. (1979). Integrated mite control in Hudson and Champlain Valley apple orchards. Search Agriculture, 9 1–11.

29

Biological Control of Forest Insects

D. L. DAHLSTEN and N. J. MILLS

Division of Biological Control
University of California
Berkeley, California

INTRODUCTION

Ecological Attributes of Forest Ecosystems

Forest environments present some very different and unique ecological attributes as compared with other environments where biological control programs are attempted. Forests tend to be more complex because they are composed of a diversity of species, ages, intraspecific genetic composition, spacing, and stocking levels (square meters or cubic meters of wood per hectare). Intensively managed forests, even-aged stands, plantations of single and mixed species, and seed orchards are more closely parallel to agriculture but even these stands usually exist in a variety of different stand conditions. It is important to look at some of these ecological attributes in detail because the opportunities for biological control vary depending on the environment and the species involved. To illustrate some of the important ecological differences in forest systems that may determine the success of various biological control tactics to be used, it is useful to contrast forestry with agriculture. Evaluating the outcome of biological control programs is also affected by the ecosystem being studied and these problems will be discussed later.

Forests are a multiple-use resource. This means that in addition to timber production that wildlife, recreation, watershed, and grazing are often components of resource management decisions. In agriculture, the goal of management is to harvest a commodity one or more times per year. In forestry, pest management is further complicated by multiple goals and competing interests of the resource managers as well as the general public, which includes sportsmen, woodcutters, environmentalists, cattlemen, bird watchers, hikers, and such governmental agencies as the Army Corps of Engineers.

Forests tend to be extremely large continuous areas with gradual boundaries. Although the average size of farms has increased in the United States, agricultural units are small by comparison to the vast forested areas. In addition, the changes between agricultural fields are abrupt. Accessibility is also a problem in forested areas. Not only are the biological control opportunities limited by the size of the areas, but also sampling of the pest insects, and therefore a quantitative evaluation of control activity, become very difficult and costly.

The control strategy in forests is also affected by the length of time to harvest. In agriculture, harvesting is an annual event or with some crops there are multiple harvests each year. In forestry, a short period of time to harvest is 20 to 30 years, such as in the southeastern United States. The length of a rotation in other regions is 50 to 100 years. The cost of treating a long-term crop on an annual basis would be impossible economically.

Complexity in an ecological sense is also greater in forest environments than in agricultural ecosystems. Forests vary from single-species plantations to multistoried stands, and plant diversity is greater than in an agricultural field even in the simplest forest stand. The biological control researcher or practitioner must often deal with stands with height (or depth) of up to 200-ft trees, a mixture of age and/or size classes and of stages from seeds to mature trees, a mixture of tree species, numerous canopy levels including herbaceous plants, and different stocking levels or spacing.

Foresters can influence the structure of a forest stand and contain insect pests through their management activities. The type of silvicultural system—even age (clear-cut, shelterwood, or seed tree), or all age—and the various activities such as site preparation, thinning, and herbicide use associated with the cutting system can either create or alleviate pest problems and certainly can affect the natural enemies of potential pest insects. In the most intensively managed forests, genetically selected stock, fertilization, and watering may be part of the operation. However, even the most intensively managed forest cannot match the uniformity of an agricultural system. In agriculture, in most cases a single genotype is selected (hybrids) and often the time of ripening and location of the fruit or vegetable on the plant is controlled to allow for mechanical harvesting. In addition, there is every attempt to keep a crop uniform through careful fertilization and irrigation. The frequency of perturbation is much greater in agriculture and often involves the complete removal of the plants once or several times annually. Harvesting in a forest, even an intensively managed one, may occur every 30 to 100 years. Stand entries for thinning or cutting in an all-age silvicultural system at 5- to 10-year intervals would be considered frequent.

Influence of Forests on Biological Control

Given the ecological attributes of forest ecosystems cited earlier, it is interesting to speculate on how these attributes might affect the potential for biological control. Pschorn-Walcher (1977) has analyzed the influence on the classical approach (importation) but augmentation and conservation are likewise influenced by the structure of forests. The vast, diverse, relatively less perturbed, long-lived, and highly stable in space and time ecosystem confers both advantages and disadvantages for biological control.

The long history and multiple-storied structure of forest stands have made for great floral and faunal diversity and for many coevolved relationships. Diversity confers an advantage for foreign exploration because a large complex of natural enemies should be available from which to choose (Pschorn-Walcher, 1977). These same attributes could make it more difficult for colonization of natural enemies. Theoretically, there would be a greater chance for the introduced natural enemies to be in competition with related native natural enemies, there being a higher probability that relatives would be present in the rich forest fauna. The vastness and diversity create sampling and evaluation problems but less perturbation allows long-term evaluation studies.

The relative uniformity of forested regions is an advantage for the collector because natural-enemy complexes frequently exhibit only minor regional differences (Pschorn-Walcher, 1977). However, any widely distributed pest or a pest introduced in a number of locations in a large forest region would make any colonization program a long-term venture. New pest introductions occurring in one or two locations could be approached with a short-term strategy of natural-enemy colonization. Sampling and evaluation would also be simplified in smaller areas if accessibility did not pose a problem. Accessibility is frequently a major problem in forest pest control projects.

Pschorn-Walcher (1977) contends that the great differences between forest and agroecosystems dictate a different approach to biological control in forestry from agriculture. This is based on the attributes of the forest environment discussed earlier where the less disrupted environment had allowed the development of much more complexity in natural-enemy association. The approach to biological control in agriculture, where there is much less predictability because of continuous perturbation, can be faster using trial-and-error releases until the best natural enemy is found. With forest insects, preintroduction studies are desirable to understand the interrelationships of the various natural enemies and to select the most likely natural enemies for success. Natural-enemy complexes of forest insects can be chosen with a higher degree of predictability for successful introductions and therefore preintroduction studies are justified (Pschorn-Walcher, 1977). Studying the parasitoid complex in some detail provides information on the colonization potential, host density preference, host specificity, host relations, and inbreeding problems of the constituent species. It also allows the identification and elimination of cleptoparasitic species.

Biological Control Strategies in Forestry

There are a wide variety of approaches in biological control including importation, augmentation, and conservation. These approaches have been defined generally and have been discussed in detail in several books (DeBach, 1964a; Huffaker, 1971; Huffaker & Messenger, 1976). In addition to the attributes of the forest as described earlier, the nature of the pest and whether the pest is native or exotic must also be considered before selecting a particular approach.

The major efforts in the biological control of forest insects have been in Canada and the United States and the classical approach of importation has been the most commonly used. This is no doubt because the highest proportion of introduced forest pests occurs in North America (Pschorn-Walcher, 1977). The majority of these insects are lepidopteran and hymenopteran defoliators (sawflies). These relatively large insects appear more commonly as pests in the less perturbed, contiguous forest environment. Because forests are not as intensively managed as agricultural ecosystems, this may also explain why homopterans, which are common subjects for biological control in agriculture, are not common as forest pests.

The problem of evaluating biological control is that it is not entirely clear what is meant by success for programs other than those classified as complete successes. DeBach (1964b) defined "substantial" [intermediate of Laing & Hamai (1976)] and "partial" success, stating that success is relative and that programs are measured in an economic sense. Unfortunately, there is little quantitative information on which to make judgments and evaluate successes in biological control, particularly partial successes, which are very subjective. It appears that in some cases establishment is incorrectly equated with partial successes. The completely successful programs are rather obvious in most cases. The importance of rigorous quantitative evaluation of programs to the growth and development of biological control as a science cannot be overemphasized.

IMPORTATION OF NATURAL ENEMIES

Introduced Pests

By far the most common approach to biological control in forestry has been the importation of natural enemies against introduced pest species or classical biological control (Turnock *et al.,* 1976; Pschorn-Walcher, 1977). This has usually involved colonizing and establishing a relatively small number of natural enemies for control of an exotic pest through direct inoculative releases of freshly imported parasitoids. With a few exceptions, parasitoids have almost always been the natural enemies introduced in forestry.

Some estimates of the numbers of importations of parasitoids and predators and their success of establishment and control are given in Table 1. These data, taken from the International Institute of Biological Control (IIBC) database of worldwide biological control importations (Greathead & Greathead, 1992), indicate that for forest insect pests 78% of importations involve parasitoids (Hymenoptera or Diptera: Tachinidae). Only homopteran pests have attracted substantial importations of predators and while the overall rates of establishment of these two groups of natural enemies are equal, the parasitoids have on average been more than twice as successful in achieving some degree of control of the target forest pests. A number of examples of successful programs using parasitoids are discussed later under the section on case histories.

Approximately 40 species of aphid predators were introduced against the balsam woolly aphid, *Adelges piceae* (Ratzeburg), in an unsuccessful colonization program (Clark *et al.,* 1971). There are no known parasitoids of the balsam woolly aphid. Attempts to introduce predators against bark beetles (Scolytidae) have been made on several occasions. In 1892 and 1893, A. D. Hopkins tried to introduce the clerid *Thanasimus formicarius* (Linnaeus) from Germany to West Virginia for control of the southern pine beetle, *Dendroctonus frontalis* Zimmermann. This introduction was a failure because there have been no recoveries, but is notable because it was the first attempt to import a natural enemy of a forest insect into the United States (Dowden, 1962). Other unsuccessful attempts have been made using *Rhizophagus* spp. (Rhizophagidae) from England both in Quebec, Canada in 1933 and 1934 with one species against the Eastern spruce beetle, *D. rufipennis* (Kirby), and in New Zealand in 1933 with three species against the European bark beetle, *Hylastes ater* (Paykull) (Clausen, 1978). Several species of carabid beetles have been imported for control of the gypsy moth, *Lymantria dispar* (Linnaeus), which is considered by some to be a forest insect and by others to be an urban pest (see later text and Chapter 36). One carabid, in particular, *Calosoma sycophanta* (Linnaeus), has become well established and is credited with having significant impact, at least during the later stages of gypsy moth outbreaks (Weseloh, 1990a).

The importation of red wood ants in North American forests has only been attempted on a few occasions (Finnegan & Smirnoff, 1981). *Formica lugubris* Zetterstedt was imported from Italy in 1971 and 1973 and was introduced into the forests of Quebec (Finnegan, 1975), and *Formica obscuripes* Forel was moved from Manitoba to Quebec in Canada in 1971 and 1972 (Finnega, 1977). The 15 species in the *F. rufa* Linnaeus group in North America are not well known but *F. obscuripes* appeared to have potential and did not occur in the east; therefore, the decision was made to move this species to the east (Finnegan, 1977).

TABLE 1   The Number of Importations of Parasitoids and Predators against Forest Insect Pests, the Percentage of Those Established, and the Percentage of Establishments That Have Achieved at Least Some Degree of Control[a]

	Total	Order of insect pest			
		Lepidoptera	Homoptera	Hymenoptera	Coleoptera
Number of target pests	42	15	10	9	8
Number of importations	360	175	71	93	21
Parasitoids	280	138	6	81	15
% Established	29	30	83	30	73
% Successful establishments	38	32	40	38	64
Predators	80	9	65	0	6
% Established	29	33	29	0	17
% Successful establishments	17	0	21	0	0

[a]Data from BIOCAT, IIBC, Greathead and Greathead, 1992.

The effectiveness of these introductions against defoliators like the Swaine jack pine sawfly and the spruce budworm is unknown, although they have been observed feeding on the spruce budworm and other forest insects (McNeil *et al.,* 1978). The ant populations continue to be encouraged so that eventually they will be well established in a relatively large area. However, ants tend to be polyphagous and may have significant negative environmental impacts, and so are unlikely to be used in any future introductions.

There is only one example of the use of a vertebrate in a forest biological control program. This is the introduction and the colonization of the masked shrew, *Sorex cinereus* Kerr, in Newfoundland, Canada, for control of the larch sawfly, *Pristiphora erichsonii* (Hartig). This was a unique situation in that there were no insectivores and few small fossorial animals on the island of Newfoundland. The shrews were taken from northern New Brunswick, Canada, in 1958 and subsequently released. The shrews also feed on other pest insects and it is tentatively felt that the release was successful (Turnock & Muldrew, 1971) although there is public concern about this introduction.

There has been little use of pathogens in classical biological control programs in forestry but notable exceptions are the accidental introduction of the nuclear polyhedrosis virus (NPV) of the spruce sawfly, *Gilpinia hercyniae* (Hartig), into eastern Canada (McGugan & Coppel, 1962) and the importation of the nematode *Deladenus siricidicola* Bedding for the control of wood wasp *Sirex noctilio* Fabricius in Australia (Bedding & Akhurst, 1974). The fungal pathogen *Entomophaga maimaiga* Humber, Shimazu & Soper, originally introduced from Japan against the gypsy moth in 1911, suddenly reappeared in the northeastern United States in 1989, causing spectacular epizootics (Andreadis & Weseloh, 1990; Hajek *et al.,* 1990; Hajek, 1997). Five species of microsporidia also were introduced against gypsy moth in eastern Maryland from 1986 to 1991 (Jeffords *et al.,* 1988, 1989; Maddox *et al.,* 1992). Experimental releases indicate that microsporidia can be introduced through releases of contaminated egg masses and that successful overwintering occurs with a transovarially transmitted *Nosema* sp.

Mass-rearing and release programs have not been common in the biological control of forest insects, no doubt due to large areas, inaccessibility, and difficulty and logistics in rearing natural enemies of tree-infesting insects. There have been four notable exceptions:

- The propagation of 882 million individuals of eulophid *Dahlbominus fuscipennis* (Zetterstedt) at the Belleville laboratory in Canada for release against *Gilpinia hercyniae* (McGugan & Coppel 1962)
- The propagation of 200 million *D. fuscipennis* by the Maine Forest Service in the United States between 1935

and 1940 for release against *G. hercyniae* (Clausen, 1978)
- The mass rearing and release of several parasitoids of the gypsy moth, *Lymantria dispar,* in the eastern United States (Leonard, 1974)
- The use of nematode *Deladenus siricidicola* against *Sirex noctilio* in Australia (Bedding & Akhurst, 1974)

An alternative to mass rearing is field rearing and redistribution of natural enemies as has been achieved with the parasitoids of the larch casebearer (see later text).

Native Pests

The importation of exotic natural enemies against native insects has potential in the control of forestry pests. This has not been explored thoroughly but with increasing public and political concern about the use of toxic pesticides in the environment more will undoubtedly be done in the future. The importation of exotic natural enemies for the control of native pests is very controversial. Hokkanen and Pimentel (1984, 1989) evaluated this approach of using natural enemies from allied hosts in agricultural, weed, and forest pest control and referred to these as new associations. They concluded that this should be the preferred approach in biological control. This is based on the development of the idea that through genetic feedback mechanisms host–parasitoid systems evolve toward homeostasis and because of this coevolved equilibrium, parasitoids would be limited in their effectiveness as biological control agents (Pimentel, 1961, 1963). It could be reasoned further, then, that generalists would be preferable to specialists in the selection of biological control agents.

The case, however, is not that clear-cut and the most appropriate source of natural enemies should be based on careful evaluation of each specific pest problem. The Pimentel genetic feedback concept is disputed by Huffaker *et al.* (1971), who believe that natural enemies become better adapted through time to control their host; they cite examples of long-standing and effective introduced natural enemies such as the vedalia beetle and *Cryptochaetum* for control of cottony-cushion scale (100 years) and seven other examples to support their case.

Hokkanen and Pimentel (1984) used a chi-square analysis to conclude that successes in biological control were about 75% higher using new associations. Goeden and Kok (1986) disputed the conclusions of Hokkanen and Pimentel (1984) on the biological control of weeds because of a bias in the data toward cactaceous plants, which are not representative of target weeds, and because of inaccuracy with some of the examples. Further, Dahlsten and Whitmore (1989) analyzed the 286 examples of successful biological control used by Hokkanen and Pimentel (1984) and showed a statistically significant advantage for old associations in terms of complete versus intermediate versus partial success. The use of new associations as the preferred method for biological control is further contradicted by the analyses of Hall and Ehler (1979) and Hall *et al.* (1980), who found that the establishment rate of natural enemies was significantly higher for old associations, the complete success of old associations was higher but not statistically significant, and the rate of all successes (complete, partial, etc.) for old associations was significantly higher. Waage (1990) also showed a significant advantage for old over new associations in all categories of success, but pointed out differences in the interpretation of new associations. It is encouraging that there are examples of successful introductions of natural enemies for control of exotic as well as native pests. Both approaches have merit depending on the ecological circumstances where the target pest occurs. Logically this makes good sense. It should be realized, however, that the evaluation of most biological control programs has not been scientifically thorough and that much of the interpretation and, therefore, conclusions concerning success are vague or circumstantial. The controversy is based on the analysis of examples of successes from the literature, principally a review of beneficial insect introductions by Clausen (1978), and the IIBC database (Greathead & Greathead, 1992).

Three native forest tree pests and one introduced shade tree pest that were listed as partial successes for new associations by Hokkanen and Pimentel (1984) were examined more closely. Each of the examples associates a single natural enemy with the pest. The first is *Aprostocetus brevistigma* (Gahan), a gregarious internal parasitoid of the elm leaf beetle, *Xanthogaleruca luteola* (Müller). The beetle was introduced into North America from Europe, but the eulophid parasitoid is native to the northwestern United States and thus is a new association (Clausen, 1978). The introduction of the parasitoid into California against the elm leaf beetle is listed as a partial success (Hokkanen & Pimentel, 1984). By tracing back to DeBach (1964b), to Laing and Hamai (1976), and finally to Clausen (1956), one can perceive the potential for misinterpretation. In the three earlier references two parasitoids are listed, a tachinid *Erynniopsis antennata* Rondani, which was imported from Europe, and thus is an old association, in addition to the eulophid mentioned earlier. Clausen (1956) states that the release of two parasitoids in California was followed by high parasitism rates, particularly by the tachinid. The subjective observation was made that damage was not so general or destructive as prior to establishment and therefore this qualifies as a partial success as defined by DeBach (1964b).

Hokkanen and Pimentel (1984) listed the eulophid as a partial success because it was a new association even though Clausen (1956) had stated that the tachinid was more abundant. In actuality there was no way to quantify the effects of the parasitoid introductions because there were no checks. Clausen (1978) later state, "Infestations of the elm leaf beetle fluctuate widely from year to year; therefore, it has not been possible to evaluate the effect of parasitoids on the pest population." Sampling methods have been developed and population fluctuations and mortality of the elm leaf beetle have been quantified (Clair *et al.*, 1988; Driestadt *et al.*, 1991; Dahlsten *et al.*, 1993). These and other observations indicate that there is still a need for improved control of the elm leaf beetle and that the parasitoids present are not effective.

The other three examples of new associations are from native *Neodiprion* sawflies—a nucleopolyhedrosis virus of *N. lecontei* (Fitch) and two parasitoids from *N. swainei* Middleton, *Exenterus amictorius* Panzer and *Pleolophus basizonus* (Gravenhorst).

With *N. lecontei* it is not clear from the literature (Bird, 1971; Cunningham & DeGroot, 1981) that the virus is a new association; more than likely it is not. In addition, the virus is used and is recommended to be used as a microbial insecticide. There are instances, however, where outbreaks may be destroyed by natural transmission (Bird, 1971). This appears to be a reasonably successful case of biological control in some areas with an old association type organism (i.e., one that is specific to its host).

The two parasitoids of *N. swainei* both attack several sawfly species in Canada and the United States and were introduced from Europe where both attack a number of diprionid sawflies (McGugan & Coppel, 1962). They are generalist sawfly parasitoids and the association with *N. swainei* is a new one. McLeod and Smirnoff (1971) conclude that the introduction of the two species appears to be beneficial but it is not clear what this means. At present, there is more interest in the use of an old association, a *Borrelina* virus (Finnegan & Smirnoff, 1981).

Both of these sawfly species are good examples of efforts against native species and there are others from both the United States (Clausen, 1978) and Canada (McGugan & Coppel, 1962). An extremely successful project using a parasitoid from a host in a different genus in North America against native geometrid moth *Oxydia trychiata* (Guenée) in South America is discussed later in the case history section. Biological control efforts against native species through the importation of exotic natural enemies or by periodic inoculation of native natural enemies have merit as has been pointed out by Carl (1982). An overemphasis of one method over another, such as the "new association" approach as recommended by Hokkanen and Pimentel (1984) prior to a careful evaluation of the target pest, is not a sound pest management strategy. Careful planning and evaluation should be part of any biological control program, particularly where exotic natural-enemy introductions are planned for native species. Examples of such evaluations for Canada are the Douglas fir tussock moth, *Orgyia pseudotsugata* (McDunnough) (Mills & Schoenberg, 1985); spruce budworm, *Choristoneura fumiferana* (Clemens) (Mills 1983a); and bark beetles (Mills, 1983b; Moeck & Safranyik, 1984).

AUGMENTATION OF NATURAL ENEMIES

The effects of natural enemies can be enhanced by various manipulations of the organisms themselves or by alteration of their environment. Although augmentation and conservation can be distinguished theoretically, it is sometimes difficult to distinguish the concepts in practice (Rabb *et al.*, 1976). The two approaches were defined by DeBach (1964c) as the manipulation of natural enemies themselves (augmentation) or their habitat (conservation). Because neither approach has been used extensively in forestry, particularly in the United States and Canada, the majority of the literature deals with examples in agriculture (DeBach & Hagen, 1964; van den Bosch & Telford, 1964; Rabb *et al.*, 1976; Stern *et al.*, 1976; Ridgway *et al.*, 1977). Augmentation of natural enemies involves periodic colonization or inoculation, development of adapted strains by artificial selection, and mass release or inundation (DeBach & Hagen, 1964). The natural enemies used in augmentation include entomopathogens, parasitoids, and predators.

Inoculative attempts have been made with several parasitoids, primarily egg parasitoids, against several forest pests in Germany, Poland, Austria, and Argentina (Turnock *et al.*, 1976). The inoculation of *Rhizophagus grandis* Gyllenhal against *Dendroctonus micans* (Kugelann) in the former Soviet Union, France, and the United Kingdom; and of nematode *Deladenus* against *S. notilio* in Australia are discussed in case histories later. Red wood ants (*Formica* spp.) have been relocated in several countries in Europe and these ants are considered to be effective predators on forest pests. Otto (1967) reviewed these programs and concluded that good results were obtained primarily in pine forests against dipterous and lepidopterous larvae. The ants were less effective against sawflies and ineffective against beetles. Effective protection of coniferous forests using ants has been achieved against five lepidopteran and three sawfly pests in Germany, Switzerland, Italy, the former Soviet Union, Poland, and Czechoslovakia (Otto, 1967; Turnock *et al.*, 1976). Three of eight species in the *Formica rufa* group in Europe are considered to be good biological control agents—*F. lugubris*, *F. polyctena* Forster, and *F. aquilonia* Yarrow.

Based on the success of inoculations in Europe, further study of inoculation is warranted. In general, inoculative releases or periodic colonizations would be best suited for isolated pest infestations, seed orchards, plantations, or other smaller forested units. Inundative releases would be limited to smaller units as well. The use of this technique has been infrequent in forestry but there are a few notable examples. *Trichogramma* spp., small egg parasitoids, are easily mass-reared and are used for inundative releases in agriculture and forestry (DeBach, 1964c). The possibility of mass-rearing *Trichogramma* was first discussed in 1895 and apparently the first inundative release was against a forest pest with thousands of *T. minutum* Riley that were obtained from field-collected eggs and then released against the brown-tail moth, *Euproctis chrysorrhea* (Linnaeus) (Howard & Fiske, 1911; DeBach, 1964c). Inundative releases of *Trichogramma* for control of various forest defoliators have been made in Germany and the former Soviet Union, and scelionid egg parasitoid *Telenomus verticillatus* Kieffer has been used against lasiocampid *Dendrolimus pini* (Linnaeus) in the former Soviet Union (DeBach, 1964c). In China, inundative releases of *Trichogramma* spp. are regularly made against various forest defoliators and this is facilitated by a large, low-cost labor force and numerous cottage industry-type insectaries scattered around the country (McFadden *et al.*, 1981). In Ontario, a program (Smith *et al.*, 1990a) has analyzed the potential of augmentative releases of *T. minitum* against the spruce budworm in experimental 1- to 2-ha plots. Two releases of *T. minutum*, 1 week apart, increased parasitism of sentinel eggs by 14 to 83% and reduced larval populations by 42 to 82%. The most effective release rate was determined to be 12×10^6 females per hectare from field tests (Smith *et al.*, 1990b) and the most effective release strategy was either a single "staggered" release or a double release 4 days apart, 12 days after the appearance of the first budworm egg masses (determined from simulation modeling) (Smith & You, 1990).

An interesting successful effort in Spain against sawfly *Diprion pini* (Linnaeus) involved the collection and storage of 2 tons of sawfly cocoons. The idea was to increase the number of parasitoids where they were most needed and subsequently the cocoons were placed in special emergence boxes that were placed around the infested area. Cocoons exposed to *Dahlbominus fuscipennis* in the laboratory were also added to the boxes. Parasitoid emergence from these boxes contributed about 3 million additional *D. fuscipennis* and ichneumonid *Exenterus oriolus* Hartig to the area. A parasitization rate of nearly 65% was attained (Ceballos & Zarko, 1952; DeBach, 1964c).

There has been increased understanding of the chemical communication among insects. Further, parasitoids and predators use chemical cues (kairomones) to locate their hosts and prey (Vinson, 1984; Haynes & Birch, 1985; Vet & Dicke, 1992; Tumlinson *et al.*, 1993). This has been found to be particularly true for certain bark beetle natural enemies that respond to beetle aggregation pheromones (Borden, 1982, 1985). Much remains to be discovered about the use of kairomones by natural enemies because only a few have been identified so far. There may be great potential for the use of these chemicals in augmentation programs. Kairomones could be used to collect and redistribute natural enemies, to attract natural enemies to infested areas, or to assist in the selection of exotic natural enemies for native hosts.

Mills (1983b) suggests the use of *Dendroctonus* aggregation pheromones as a method of selecting useful European bark beetle egg predators for introduction into Canada. Miller *et al.* (1987) have demonstrated *Thanasimus undulatus* Say (normally associated with *D. pseudotsugae* Hopkins in North America) to exhibit cross-attraction in field tests to other bark beetle pheromones and *R. grandis* (specific to *D. micans* in Europe) to be attracted to the frass of three North American *Dendroctonus* species in the laboratory. In addition, it was found that *Rhizophagus* was stimulated to produce more eggs when exposed to the larval frass of *D. valens* LeConte from California (Gregoire *et al.*, 1991). In assessing the potential for biological control of the mountain pine beetle, *D. ponderosae* Hopkins, Moeck and Safranyik (1984) concluded that inundative releases of native clerid beetles against low level *D. ponderosae* populations offered the best potential for biological control. Kairomones could be used to assist with this type of inundative release to collect and relocate, or to attract predators to selected areas. Exotic natural enemies might be manipulated in this way also after they have been imported and successfully established. There is also evidence that the symbiotic fungi associated with bark beetles may produce kairomones that attract parasitoids of bark beetles, suggesting yet another avenue for natural-enemy manipulation (Dahlsten, 1992).

Insectivorous bird encouragement programs have been used extensively in Europe primarily by providing nest boxes in forests for cavity nesting birds (Bruns, 1960). In California, it has been demonstrated that populations of the mountain chickadee, a ubiquitous insectivorous bird in the forest, can be increased two- to threefold by providing nest boxes (Dahlsten & Copper, 1979). The impact of birds on insect populations is difficult to evaluate but it is speculated that birds operate in an inverse density-dependent manner and their importance would be in preventing outbreaks of forest pests instead of suppressing them (Dahlsten *et al.*, 1990, 1992).

Entomopathogens, as a group of natural enemies, have begun to play a dominant role in biological pest control in forestry within the last decade (Ahmed & Leather, 1994; Lacey & Goettel, 1995). The use of chemical spray treatments over extensive forest areas has become a political issue and public demand has forced the use of the safer, nonpolluting microbial insecticides. The principal entomopathogens used are the bacterium *Bacillus thuringiensis* Berliner (B.t.) and baculoviruses. These control agents have

been tested against a wide variety of forest defoliators in the form of inundative treatments and have the advantage of a much reduced impact on other groups of natural enemies and nontarget organisms, but see Miller (1990).

Microbial insecticides are likely to receive as wide an application in forestry as in agriculture or horticulture for several reasons (Morris *et al.*, 1986). Forest protection is of much greater concern to the general public due to the more extensive areas covered by forest pests. Forest pest problems also tend to involve only single target species instead of a complex of pests, as is often the case in agriculture, which requires the development of only a single microbial product. The forest crop is also better able to withstand the slower action of microbial treatments in comparison with other crops.

Natural epizootics of B.t. do not occur but the bacillus can be produced in large quantities in a fermentor for aerial application. Morris (1982) reviewed the use of B.t. in field applications against forest insects and concluded that deciduous defoliators in general are more susceptible to good control than coniferous defoliators are. This is perhaps surprising, because one of the greatest challenges in the application technology of B.t. is to obtain an even dispersion of the pathogen throughout the forest canopy. The cone-shaped canopy of conifers is more conducive to an even foliage coverage than are the more cylindrical canopies of hardwoods. However, more of the insect pests of conifers are concealed feeders by comparison with hardwood pests and this is also likely to influence the success of B.t. applications. In general, the use of B.t. has been inconsistent in its efficacy against forest defoliators and this is considered by van Frankenhuyzen (1990) to be a result of it having not contact toxicity, having limited residual toxicity, and requiring accurate timing of application.

The spruce budworm, *Choristoneura fumiferana*, in North America and the gypsy moth, both in Europe and North America, have been the main targets of extensive development of B.t. as a means of inundative biological control. Use of B.t. against gypsy moth is discussed later in the section on case histories. The first successes with B.t. were achieved against the spruce budworm and guidelines for the operational use of B.t. in Canada and the United States have been formulated (Morris *et al.*, 1984). Through public demand the use of large-scale conventional chemical insecticide treatments in Canada has been reduced from 98% of the 2.8 million hectares treated in 1981 to an anticipated 26% of 1.9 million hectares requiring protection in 1986 (Morris *et al.*, 1986). The cost of B.t. continues to be greater than that of insecticides; and in an effort to reduce costs and improve efficacy against spruce budworm, attempts are being made to accurately time a single application, instead of applying two applications (Fleming & van Frankenhuyzen, 1992).

Baculoviruses, comprising the NPVs and granulosis viruses (GVs), have been widely tested in field trials against forest insects (Cunningham, 1982). They show a marked degree of specificity for their phytophagous hosts and have no impact on nontarget organisms. Natural epizootics of NPV are often responsible for the termination of outbreaks of major forest insect pests, particularly among the Diprionidae and Lymantriidae families.

Use of NPVs against the gypsy moth is discussed later in the case history section, but diprionid sawflies provide some of the most striking examples of the use of NPVs in biological control (Cunningham & Entwistle, 1981). The virulence of the diprionid NPVs is appreciably greater than that of other host groups (Entwistle, 1983), and the gregarious habit of the diprionid larvae promotes the spread of virus through the larval population.

Virus production cannot be achieved on artificial media, and for sawflies, in contrast to Lepidoptera, which can be reared on artificial diets, foliage-fed host larvae are required for mass production of the virus. Thus, either host larvae must be collected from the field for infection in the laboratory (e.g., Rollinson et al., 1970) or a heavily infested plantation could be sprayed with virus and the infected larvae harvested as they die (e.g., Cunningham & DeGroot, 1981). The periodic inundation of the virus can be conducted either by distribution of host cocoons containing infected eonymphs in forest stands or by more conventional aerial or ground spray machinery. The former method has some potential for *Neodiprion swainei* (Smirnoff, 1962), which has an NPV that spreads rapidly from epicenters, while the latter has been widely used for *N. lecontei* and *N. sertifer* (Geoffroy) (Cunningham & Entwistle, 1981). The NPV of *N. sertifer* has been successfully used in 12 countries and is the most operationally used of the sawfly NPVs. One factor that contributes to this success is the more synchronous hatching of the larvae of *N. sertifer,* as a result of overwintering as eggs instead of as eonymphs, which facilitates the timing of spraying to infect the younger, more susceptible larval instars.

CONSERVATION OF NATURAL ENEMIES

This approach may eventually have its greatest application in the forest environment. The size, complexity, and lack of perturbation in forests lend them to natural-enemy conservation. Although Turnock et al. (1976) state that forest communities are simple, unstable, and prone to wide insect population fluctuation, forests are complex by comparison to agriculture, as discussed earlier.

Conservation of natural enemies should be considered a part of all silvicultural systems and treatments. In addition, there are measures that can be taken directly to conserve natural enemies. By comparison, much more has been done

in agriculture to conserve natural enemies (van den Bosch & Telford, 1964). In agriculture, various cultural methods can be selected to directly influence natural enemies such as strip harvesting and habitat diversification (Stern et al., 1976; Altieri & Letourneau, 1982; Altieri et al., 1993).

Efforts should be made to better understand the influence of forest practices on the natural enemies of pests. While there has been considerable effort to understand the dynamics of the target pests, the natural enemies by comparison have been poorly studied. Knowledge of forestry practices is particularly important at this time because there is a trend to intensive management in forestry (Dahlsten, 1976). Such knowledge would permit the forest manager to select silvicultural methods that would least harm or even encourage natural enemies of various pests. The forest manager can manipulate the age and species composition of the stand as well as the stocking levels (density) and each of these attributes can have an influence on the distribution and abundance of insects and their natural enemies. Changes in forest practice could mean the emergence of new pests, insects about which little is known. Cutting practices conceivably could create secondary pests in much the same manner as chemical insecticides.

Herbicides are being used in increasing amounts in forestry, yet little is known about what the changes in the plant community mean in terms of natural enemies. If flowers of small herbaceous plants are important to natural enemies as Syme (1977) has indicated for a parasitoid of the European pine shoot moth, *Rhyacionia buoliana* (Denis & Schiffermüller), then the potential side effects of herbicides on beneficial insects in forests should be examined more closely.

Judicious use of chemical insecticides is also important as a means of conserving natural enemies. There is in theory considerable naturally occurring biological control and examples are given for forests in the western United States by Hagen et al. (1971). Often the importance of natural enemies is not known until their effect on the host insect or mite is disrupted. Secondary outbreaks have not been well documented in forestry but an extensive outbreak of the spruce spider mite, *Oligonychus ununguis* (Jacobi), following the application of DDT for western budworm control in Montana and Idaho has been documented (Johnson, 1958). In California, an outbreak of the pine needle scale, *Chionaspis pinifoliae* (Fitch), occurred on Jeffrey and lodgepole pines after an area near Lake Tahoe was fogged with Malathion to control adult mosquitoes (Luck & Dahlsten, 1975). The influence of natural enemies was well documented in this study because the collapse of the scale population, after the spraying was halted, occurred over a 3-year period and was demonstrated to be due to a small complex of predators and parasitoids. Other insecticide-induced outbreaks have been documented for the target insects. The elimination of parasitoids and virus diseases of the European spruce sawfly, *Gilpinia hercyniae,* after 3 years of spraying with DDT in New Brunswick, Canada, resulted in an outbreak of the sawfly (Neilson *et al.,* 1971). In Texas, an increase in an infestation of southern pine beetle, *Dendroctonus frontalis,* was felt to be due to the deleterious effects of chemical insecticides on the natural enemies of the bark beetles (Williamson & Vite, 1971). Swezey and Dahlsten (1983) have documented the effects of lindane on the emergence of natural enemies of the western pine beetle, *D. brevicomis* LeConte.

The use of more selective insecticides also can be an important aspect of natural-enemy conservation. Laboratory assays showed that spruce budworm larvae parasitized by *Apanteles fumiferanae* Viereck survive better than unparasitized larvae when treated with B.t. due to a lower feeding rate (Nealis & van Frankenhuyzen, 1990). This result was confirmed in field trials (Nealis *et al.,* 1992; Cadogan *et al.,* 1995), which also indicated that delaying B.t. treatment of budworm from the third- to fourth- to the fourth- through sixth-instar period resulted in a greater conservation of parasitoids and a more significant reduction in budworm density.

The enhancement of the physical environment to encourage natural enemies has great potential as demonstrated by the few attempts that have been made. Parasitoids and vertebrate and invertebrate predators can be benefited by the encouragement of specific plants for food, shelter, and protection from their natural enemies and winter cold (Buckner, 1971; Sailer, 1971). A program was initiated in permanent outbreak areas in Poland in 1958 to increase the effectiveness of natural enemies by applying fertilizers, planting deciduous trees and shrubs, and providing nectar plants for parasitoids and predators (Burzynski, 1970; Koehler, 1970).

In Canada, Syme (1981) has demonstrated that the probability for control of the European pine shoot moth, *Rhyacionia buoliana,* can be enhanced by the presence of wild carrot, *Daucus carota* Linnaeus, in pine plantations. The longevity and fecundity of the most effective introduced parasitoid, the braconid *Orgilus obscurator* (Nees), are increased due to its feeding on the nectar of several flowers including wild carrot (Syme, 1977). Plantation managers can preserve wild carrot in plantations where it occurs or can plant it to encourage *O. obscurator.* Similarly, laboratory studies on the parasitoids of the southern pine beetle, *Dendroctonus frontalis,* indicate the potential of diet supplements to increase longevity and enhance biological control (Mathews & Stephen, 1997).

There are several suggestions in the literature of procedures to conserve the natural enemies of bark beetles. Bedard (1933) recommended examination of infested trees for high degrees of parasitism prior to control in order to conserve parasitoids. The disruption of old infestations of mountain pine beetle in lodgepole pine should be avoided because braconid *Coeloides rufovariegatus* (Provancher) is most abundant in the oldest infestations (DeLeon, 1935). Wind-thrown western white pines should not be disturbed because of the high populations of mountain pine beetle parasitoids (Bedard, 1933). Another method is to leave small-diameter Douglas fir infested with Douglas fir beetle, *D. pseudotsugae,* because the parasitoid *Coeloides vancouverensis* (Dalle Torre) is more abundant in these trees (Ryan & Rudinsky, 1962).

There have been similar recommendations to conserve the predators of bark beetles. Clerid predators of the western pine beetle eventually move to the lower portions of the bole of infested trees and, therefore, it was recommended that the lower section of the trees not be treated with insecticide during control projects (Berryman, 1967). With the southern pine beetle it was found that the clerids emerged later than the bark beetles and it was recommended that infested trees not be removed until after clerid emergence (Moore, 1972).

Several other conservation techniques have been suggested for natural enemies of forest pests. For example, Hulme *et al.* (1986, 1987) have recommended techniques for the conservation of natural enemies of the white pine weevil. Either differential cold hardiness or selected mesh sizes of net bags can be used to separate weevils from parasitoids and predators in attacked leaders.

BIOLOGICAL CONTROL ORGANIZATIONS IN FORESTRY

Biological control of forest insects has been most prevalent in temperate and Mediterranean regions. This primarily reflects the regions of more extensive and actively managed commercial forests and the regions where forest pests have been accidentally introduced by trade and travel.

There are no organizations devoted solely to the biological control of forest insects. However, the Canadian Department of Agriculture Laboratory at Belleville, and an international organization, the CAB IIBC, formerly the Commonwealth Institute of Biological Control, have devoted a portion of their efforts to forest insects.

Normally projects are organized around the target pest and a project group is created involving federal, state or province, and private agencies. The *Sirex* project in Australia and the gypsy moth and European spruce sawfly projects in the United States are good examples. In both Australia and the United States there are government agencies in place that are responsible for importation and quarantine but not specifically for forest insects. These agencies, however, participate whenever a forest insect biological control program is initiated.

Australia

In Australia, commonwealth and state ministers met in 1962 and created a national *Sirex* fund. All states contributed approximately in proportion to their acreage of pine and this was matched by the commonwealth government (Taylor, 1981). Private forest owners also contributed to the *Sirex* fund that was administered by a committee that coordinated research as well as the practical operations. A number of organizations were involved in the program, including the Divisions of Entomology and of Forest Research of Commonwealth Scientific and Industrial Research Organization (CSIRO); the Waite Agricultural Research Institute, University of Adelaide; the Forest Research Institute, New Zealand; Forest Commission of Victoria and Tasmania; and the University of Tasmania. The IIBC also participated in this project. The biological control program against bark beetle *Ips grandicollis* (Eichoff) in Australia from 1981 to 1990 was organized in a similar way, with private timber companies and government agencies participating in the funding. It was nationally funded by the state forest services from 1983 to 1990 (Forest Insect Pest Management Association, 1992).

United States

In the United States, the Animal and Plant Health Inspection Service (APHIS) of the U. S. Department of Agriculture (USDA) is responsible for controlling the importation of living insect material into the United States. Formerly this responsibility resided in other federal organizations, the Federal Entomology Research Branch and the Bureau of Entomology and Plant Quarantine. The forest biological control projects are normally in cooperation with state agencies and other federal agencies such as the Forest Service and the Agricultural Research Service (ARS).

Generally, importation activities are restricted to the federal government agency and ARS Beneficial Insects Research Laboratory maintains quarantine facilities. The states participate in all the other biological control activities. The one exception to the restriction on importation activities is the state of California. California began importation on an independent basis in 1899 and maintains its own quarantine facilities for receiving insects and mites (Clausen, 1956).

Some examples of early cooperative programs were the natural-enemy studies of the gypsy moth and brown-tail moth that were conducted in Europe when the Massachusetts State Board of Agriculture paid the cost of the foreign exploration by federal personnel. In later releases, the New Jersey Department of Agriculture (NJDA) cooperated with APHIS in distributing parasitoids. The gypsy moth biological control program was one of the largest and most extensive projects in the United States. In the biological control program against the European spruce sawfly in the United States, the Canadian Department of Agriculture provided large stocks of parasitoids that were reared and distributed by the Federal Bureau of Entomology and Plant Quarantine and the State of Maine Forest Service. Because there is not a specific agency for forest biological control in the United States, the programs—always cooperative projects between APHIS, ARS, forest insect researchers in the U.S. Forest Service, and entomologists at universities—involve other state and federal agencies in the implementation phase.

Canada

The evolution of biological control programs and laboratories in Canada has been reviewed by Beirne (1973). The first introductions of natural enemies against forest pests were conducted by C. G. Hewitt, the second dominion entomologist, who arranged for parasitoid collections in England for use against the larch sawfly in 1910 to 1913. In 1912, the Natural Control Investigations Laboratory was established under J. D. Tothill in Fredericton to combat the invasion of the brown-tail moth in eastern provinces.

Subsequently biological control programs were transferred to Belleville, Ontario, where the Dominion Parasite Laboratory was established by the Division of Entomology, Canadian Department of Agriculture under the charge of A. B. Baird in 1929. A 40-room, controlled environment quarantine building was added in 1936 and this provided the necessary facilities for the mass-rearing of imported parasitoids that were released on a massive scale against forest pests between 1934 and 1949. The Belleville Laboratory was responsible for the biological control programs against forest insects between 1929 and 1954, covering the period of peak activity (Embree & Pendrel, 1986) and dramatic successes in the use of natural enemies to control introduced forest insects.

Interest in the potential of insect pathogens for the biological control of forest insects developed in the 1940s as a result of the successful introduction of the viral disease of the European spruce sawfly, and this led to the establishment in 1950 of the Insect Pathology Research Institute (now known as the Forest Pest Management Institute) in Sault Ste. Marie, Ontario. Then in 1954, an agreement was made to transfer the responsibility for biological control of forest pests from the Belleville Laboratory to the Forest Insect and Pathology Unit, headed by M. L. Prebble, with the Belleville Laboratory continuing to handle the quarantined importation of natural enemies from abroad.

From this time the biological control projects became dispersed to the regional research centers of the Forestry Service. With the closure of the Belleville Laboratory in 1972, quarantine facilities moved to Ottawa and currently biological control groups are most active, with their own quarantine laboratories, at the Pacific Forest Center in Victoria, British Columbia and the Great Lakes Forest Center in Sault Ste. Marie, Ontario. Canadian perspectives on the biological control of forest insect pests are provided by Hulme (1988), Nealis (1991), and Smith (1993).

International Institute of Biological Control

The CAB International Institute of Biological Control (IIBC) has been active in the foreign exploration for natural enemies of forest insects in Europe for introduction into Canada, New Zealand, and the United States (Greathead, 1980). This institute was established in 1927 as the Farnham House Laboratory in England, with W. R. Thompson as superintendent from 1928. In these early years much of the collection work was conducted in the United Kingdom or from field stations in France and Czechoslovakia.

In 1940, the difficulties of continuing operations in Europe during World War II led to the transfer of the Laboratory to Belleville, Canada, where it was renamed as the Imperial Parasite Service. After 6 years of being housed at the Belleville Laboratory, the headquarters moved to Ottawa and stations were opened in Trinidad, West Indies, and Fontana (near Riverside), California. Activities were resumed in Europe in 1947 with a station near Zurich in Switzerland, which moved to its present location in Delémont in 1962. F. J. Simmonds succeeded W. R. Thompson as director in 1958 and within a few years the headquarters was moved to Trinidad from Canada, where it remained until 1984. The headquarters is now situated at Silwood Park, the field station of the Imperial College of Science and Technology, London University, United Kingdom under the directorship of J. K. Waage, with regional stations in Europe, Africa, Asia, and the Caribbean. The institute has developed BIOCAT, the most complete database available on global biological control introductions of predators and parasitoids (Greathead & Greathead, 1992).

Foreign exploration for parasitoids and predators of forest insects has been a major concern of the institute from its early years as the Farnham House Laboratory to current times at the Pakistan, European, and now U.K. stations, where forestry projects continue.

CASE HISTORIES

Lepidoptera

Larch Casebearer, *Coleophora laricella* Hübner: Coleophoridae

The larch casebearer is native to central Europe and is relatively innocuous in the alpine area on its normal host, *Larix decidua* Mill (Jagsch, 1973). The moth is a defoliator of *Larix* species and becomes a pest in Europe and Asia wherever larch is planted outside of the alpine range. A fairly rich complex of parasitoids is thought to maintain the casebearer at lower densities in its endemic region (Ryan *et al.*, 1987). This insect was probably introduced on nursery stock into North America from Europe and was first found at Northampton, Massachusetts, in 1896 and in Canada at Ottawa in 1905 (Otvos & Quednau, 1981). The moth spread rapidly on tamarack *Larix laricina* (Du Roi) K. Koch in eastern Canada so that by 1947 it was in Newfoundland, the Maritimes, and Ontario; and in the United States in Maine, Michigan, and Wisconsin (McGugan & Coppel, 1962). It is currently widely distributed in the eastern United States and Canada. In 1957, the casebearer was discovered on western larch, *Larix occidentalis* Nuttall, in Idaho (Denton, 1958) and in 1966, in British Columbia (Molnar *et al.*, 1967). It is now widely distributed over the range of western larch including British Columbia, Montana, Idaho, Washington, and Oregon (Clausen, 1978).

The casebearer has one generation per year. The adults begin appearing in late May and lay eggs on either side of the needles. The larvae hatch and burrow directly down into the needles. In the late summer the larvae emerge from the mined needles and form overwintering cases. They feed for a while and then move to branches and twigs to pass the winter. In the early spring the larvae with their cases move and begin feeding on the young buds and foliage. Pupation occurs within the enlarged case, which is commonly attached to a branch on a leaf whorl. The larval feeding, when extensive, causes a loss of growth that is its greatest impact on larch (Ryan *et al.*, 1987).

The biological control program had its beginning in 1928 in western Canada with a request to the Farnham House Laboratory of IIBC for information on the parasitoid complex of the casebearer in Europe (McGugan & Coppel, 1962). Importation and field releases of 5 species of parasitoids occurred in eastern Canada between 1931 and 1939 as follows: 1,037 *Agathis pumila* (Ratzeburg), Braconidae; 29,664 *Chrysocharis laricinellae* (Ratzeburg), Eulophidae; 506 *Cirrospilus pictus* (Nees), Eulophidae; 3,283 *Dicladocerus westwoodii* Westwood, Eulophidae; and 97 *Diadegma laricinellum* (Strobl), Ichneumonidae (Clausen, 1978). All species were subsequently recovered at release sites in Ontario but only two became well established and spread rapidly, *A. pumila* and *Chrysocharis laricinellae*. Between 1942 and 1947, large-scale redistribution releases were made at a number of sites in eastern Canada. The parasitoids were obtained at established colony sites at Millbridge, Ontario (Clausen, 1978). By 1948, populations of the casebearer were low at the original release sites. The parasitoids followed the spread of the casebearer to the west assisted by occasional releases (Ryan *et al.*, 1987). This can be cited as an example of a successful biological control program (Webb & Quednau, 1971).

A separate, extensive parasitoid importation program was also conducted between 1932 and 1936 in the eastern United States in New England and New York (Clausen, 1978). Of the same parasitoids as released in Canada 4 were used in the United States (Clausen, 1978) as follows: 8,141 *A. pumila*, 24,671 *C. laricinellae*, 231 *D. westwoodii*, and 3,580 *D. laricinellum*. Although there is little information to go on, the results were apparently the same in the eastern United States with the establishment of *A. pumila* and *C. laricinellae* followed by high parasitization rates particularly by *A. pumila* (Dowden, 1962). Releases of the two established parasitoids were also made in 1937, 1950, and 1952 in Michigan and Wisconsin.

In the western United States, the first releases of *A. pumila* were made in 1960 with 2360 adult parasitoids that were collected in Rhode Island (Clausen, 1978). These were released at five locations in Idaho. Recoveries were made at three sites in 1962. Between 1964 and 1969, field rearing of *A. pumila* in whole tree cloth cages permitted the release of this parasitoid at 400 sites in Idaho, Montana, Washington, and British Columbia (Ryan *et al.*, 1987). The parasitoid became established and built up at some sites but at other sites either it did not become established or it did not build up. In addition, significant defoliation still occurred throughout much of the area by 1970 and the program was rated as a failure (Turnock *et al.*, 1976; Ryan *et al.*, 1987).

A renewed effort was made to establish biological control of the larch casebearer in the western United States between 1971 and 1985 (Ryan *et al.*, 1987). Six parasitoid species were released (Ryan *et al.*, 1987); their biologies are reported by Ryan (1980). Only two became established: *C. laricinellae* and *A. pumila*. Long-term research plots were set up in the Blue Mountains of Oregon and Washington and both casebearer abundance and parasitism were monitored for 17 years to provide life table data for analysis. From 1971 to 1978, larch casebearer populations increased to over 53 moths per 100 buds, causing severe defoliation (Ryan *et al.*, 1987), but populations declined to under 1 moth per 100 buds in 1988 as levels of parasitism increased. The life table analysis (Ryan, 1983, 1986, 1990, 1997) covers as many as 13 plots and provides a detailed picture of the role of parasitism in the successful biological control of the casebearer in this region. In general, parasitism levels remained low until the late 1970s or early 1980s, and then peaked at more than 50% for *A. pumila* in 10 out of 13 plots and at more than 90% in 4 of these 10 plots. In contrast, levels of parasitism by *C. laricinellae* remained below 30% in almost all plots. Residual mortality, other than parasitism, was the key factor in 11 of the 13 plots prior to parasitoid establishment, but parasitism by *A. pumila* was the key factor in the three most intensively studied plots following parasitoid establishment and the parasitism appeared to act in a delayed density-dependent manner.

In British Columbia, the larch casebearer biological control program was reviewed in 1974 due to the successes in eastern Canada (Otvos & Quednau, 1981). Four parasitoids have been released: *A. pumila, C. laricinellae, D. laricinellum,* and *Dicladocerus japonicus,* Yoshimoto. The story is much the same as with the other release programs—*A. pumila* and *C. laricinellae* have become well established and the other two have not been recovered. It is too early to evaluate the effects of the two parasitoids, but *C. laricinellae* is fairly common in British Columbia and may be responsible for the reduction of larch casebearer and less tree mortality (Otvos & Quednau, 1981).

The larch casebearer is a successful biological control program in both eastern Canada and northwestern United States. It is an example of a classic introduction program with the subsequent redistribution of the parasitoids from areas of establishment to new regions. It is interesting because the two parasitoids complement one another in their action against the casebearer. *Agathis* is extrinsically superior at low host densities and *Chrysocharis* is effective at high host densities. Quednau (1970) hypothesized that *Agathis* can give only partial control on its own and that success is only possible through cooperative interaction with *Chrysocharis*. There has been no success in establishing other parasitoid species. This program also is an example of one where there was a rigorous attempt to evaluate efficacy of the parasitoids (Ryan, 1985, 1986, 1990, 1997; Ryan *et al.*, 1987).

Winter Moth, *Operophtera brumata* (Linnaeus): Geometridae

This polyphagous defoliator of hardwoods is native to most of Europe and parts of Asia, where it is particularly frequent on fruit trees and oak. It was first recognized as an accidental introduction on the south shore of Nova Scotia in 1949 and eventually extended its range to the whole of this region together with small isolated parts of New Brunswick and Prince Edward Island by 1958.

In the first few years after its appearance in Nova Scotia, damage was evident in apple orchards, shade trees, and oak forests. However, at this time hardwoods were not commercially exploited in the province and thus the winter moth was not considered a serious pest (Embree, 1971). Consequently, it was possible to initiate a biological control program instead of a program of insecticide eradication. The general research policy in the early 1950s was directed toward population dynamics of forest insect populations and thus the biological control program was initiated in 1954 with a view to population studies of the host and introduced parasitoids.

Prior to the introduction of parasitoids from Europe, the winter moth fluctuated erratically at high population densities. These fluctuations resulted from the coincidence of hatching of the overwintering eggs and bud burst in early spring (Embree, 1965). This same key mortality factor was also found to be responsible for changes in population levels of winter moth in the United Kingdom (Varley & Gradwell, 1968).

Three tachinid and three ichneumonid parasitoids were obtained in sufficient quantity for introduction into Nova Scotia from Europe. The parasitoids were collected and shipped to Canada by staff of the Belleville Laboratory and the IIBC and field releases were made during the period from 1954 to 1962. These included releases of over 22,000 individuals of the tachinid *Cyzenis albicans* (Fallén) and 2,261 individuals of the ichneumonid *Agrypon flaveolatum* (Gravenhorst), the only two species that became established. *Cyzenis albicans* is very fecund and oviposits micro-type eggs around the edge of damaged foliage where they are ingested by late-instar host larvae. The eggs hatch in the midgut of the host and the larvae bore through the gut wall to develop rapidly after the host has pupated. The tachinid pupates and overwinters within the host pupal case in the ground. The biology of *A. flaveolatum* is similar but it oviposits directly into the host larvae and has larger eggs and much lower fecundity.

Following the establishment of these two parasitoids, parasitism by *C. albicans* increased rapidly to 50% in 1960 and winter moth populations declined from 1961 to 1963, a few years after parasitoid establishment in the region (Embree, 1966, 1991; Roland & Embree, 1995). Parasitism by *A. flaveolatum* increased following the initial decline of the outbreak, and while it may have enhanced the depression of winter moth abundance, population models incorporating aggregation of parasitoid attack indicate that *C. albicans* alone is sufficient to account for successful biological control (Hassell, 1980). Another reanalysis of life table data from both the United Kingdom and Nova Scotia (Roland, 1988; Roland & Embree, 1995) suggests that the decline of the moth may have been due to increased predation of pupae by native species, instead of the direct action of the two parasitoids. Roland suggests that because hosts parasitized by *C. albicans* remain in the soil throughout the winter, they may serve as an important food resource for native predators in early spring and late summer. Life table data indicate a marked increase in pupal predation of winter moth following parasitoid establishment (Embree, 1965), supporting the contention that the role of the parasitoid introductions was indirect and realized through the enhancement of pupal predation.

Between 1972 and 1978, winter moth has been noted in Oregon, Washington, and British Columbia on various hardwood and fruit trees. Both *C. albicans* and *A. flaveolatum* were relocated to these areas between 1979 and 1982 and recoveries were made in many regions (Embree & Otvos, 1984; Kimberling *et al.*, 1986). Populations of winter moth have also declined in British Columbia, where similar life table data have been collected from 1982 to 1990 (Roland, 1988, 1990, 1994; Roland & Embree, 1995). The parasitoids contributed significant levels of mortality, reaching a peak of 80% in 1984 and 1985, and there was a concurrent rise in the level of pupal mortality from 25% in

1982 to 98% in 1987. Because host numbers have stabilized, parasitism (mainly by *C. albicans*) shows a weak delayed density-dependent response (Roland, 1990, 1994), but pupal mortality is strongly and directly density dependent (Roland, 1994).

This program is often considered a good example of biological control in which, in contrast to earlier multiple introduction programs, selective introductions were made. These led to the establishment of a high-host density specialist (*Cyzenis*), with high fecundity to bring about the collapse of an outbreak, and a low-host density specialist (*Agrypon*), that has good searching ability to maintain the collapsed population at a low level of abundance. However, the main reason for the release of a smaller number of parasitoid species was the relatively meager size of collections in Europe, where winter moth abundance was not high at the time. Thus, the only conscious selection process was of parasitoid species obtained in sufficient quantity for meaningful release (Mesnil, 1967), although once the two established parasitoids were becoming effective in the early 1960s, a decision was made to curtail releases of other species (Embree, 1966). The winter moth program in Nova Scotia has been cited as an example in the debate over the mechanism of host control by introduced parasitoids. This debate focuses on whether aggregated attacks by parasitoids (Hassell, 1980) or local extinctions and recolonizations of hosts and parasitoids (Murdoch *et al.*, 1985) account for the long-term stability of winter moth populations. In reality, it appears that pupal predation interacting with introduced parasitoids may provide the stability (Roland, 1994; Roland & Embree, 1995), although this is not a universally accepted view (Embree, 1991).

Winter moth is still considered to occur at outbreak levels of abundance in Nova Scotia apple orchards despite the presence of *C. albicans* (Embree, 1991). There is some evidence that aggregation by *C. albicans* in response to winter moth feeding damage is mediated by borneol, a volatile that occurs in greater abundance in oak than in apple foliage (Roland *et al.*, 1995). However, as in oak woodland, predation appears to be the most important regulating factor in the dynamics of winter moth populations in apples (Pearsall & Walde, 1994).

Colombian Defoliator, *Oxydia trychiata* (Guenée): Geometridae

A successful example of the use of an exotic parasitoid to control a native forest pest was the importation of scelionid egg parasitoid *Telenomus alsophilae* Viereck from North America to Colombia in South America against a geometrid defoliator (Bustillo & Drooz, 1977; Drooz *et al.*, 1977). There are a number of interesting facets to the program because the normal geometrid host of the parasitoid in North America, the fall cankerworm *Alsophila pometaria*

(Harris), is in a different subfamily than is the target pest, *Oxydia trychiata,* in South America. The Colombian geometrid, *O. trychiata,* has a wide distribution extending from Costa Rica to most of the countries in South America. The moth has three generations per year and apparently is capable of normal development on introduced tree species (citrus, coffee, pine, and cypress). There has been an attempt to establish exotic conifer species in Colombia for the production of pulp and paper. This previously unimportant insect became a pest in these pine and cypress plantations (Drooz *et al.,* 1977).

Telenomus alsophilae has several biological attributes that are well worth noting because they may be the key to this unique cross-genus introduction. First of all, its normal host, the fall cankerworm, feeds on several broad-leaved trees but its host in South America feeds on conifers. This indicates that host-plant odors or other differences between conifers and broad-leaved trees are probably not important in host egg finding. There may have been a clue to this because the fall cankerworm feeds on several genera of hardwoods, but they are all deciduous as compared with the conifers. The parasitoid is seemingly easy to handle because changes in photoperiod and lack of cold in the winter did not hinder development (Drooz *et al.,* 1977). This also indicated that the shift to the Southern Hemisphere may not be a problem. The climate of the origin of the parasitoid in Virginia (30°N, elevation 370 m, mean winter temperature 2°C, and mean summer temperature 24°C), compared with that of the release site in Colombia (6°N, 2340 m, temperature range 6 to 26°C with an annual mean of 16°C throughout the year) shows a shift from a temperate to a tropical climate, although the extremes are about the same. The rainfall patterns in the two regions undoubtedly differ too. Again this demonstrates the ecological plasticity of this parasitoid. In addition, the parasitoid is long lived—at least 6 months (Drooz *et al.,* 1977).

An important attribute is the ease with which the parasitoid can be reared, something important to every biological control program. A method to rear the parasitoids continuously has been developed (Drooz *et al.,* 1977). For this the eggs of yet a third geometrid, *Abbottana clemataria* (Smith), were used. The key element here was that *A. clemataria* could be propagated on an artificial diet and could serve as the factitious host.

The egg parasitoid can apparently shift hosts readily because field parasitized eggs of the fall cankerworm were supplemented with fall cankerworm eggs parasitized in the laboratory from parasitoids reared on the factitious host. This is another key attribute.

About 18,000 parasitoids were sent to and released in a pine plantation in Colombia between October and December 1975 (Bustillo & Drooz, 1977; Drooz *et al.,* 1977). Parasitization rates on *O. trychiata* eggs were very high and

by the time the parasitoid had undergone three generations in April of 1976, few adults could be found at normal emergence time. Only 13 egg masses of *O. trychiata* could be found and these were 99% parasitized on the average. By May, the outbreak was confirmed to have been controlled when larvae could not be found in the area (Drooz *et al.,* 1977). They speculate that the parasitoid may be able to maintain itself on any of the four species of *Oxydia* or other geometrids in Colombia.

It remains to be seen if this was a totally unique situation brought about by a parasitoid with extremely flexible and broad biological and ecological attributes. Ecological flexibility and a long adult longevity may be prerequisites for exotic parasitoids to flourish on a different host insect, on a different host plant, and through a change in hemispheres and climatic regions.

European Pine Shoot Moth, *Rhyacionia buoliana* (Denis & Schiffermüller): Tortricidae

The European pine shoot moth occurs throughout Europe and parts of Asia, where it is a major pest of pine plantations. It was first discovered in North America at New York in 1914 and was later also found on imported nursery stock in Canada in 1925. While its distribution extended throughout the northeastern United States and eastern provinces of Canada, British Columbia, and northwestern United States, it was considered an important pest only in the red pine plantations in the northwestern United States and southern Ontario.

In 1927, the IIBC was requested to collect parasitoids in the United Kingdom for introduction into Canada and this led to the release of eight species during the period from 1928 to 1943 and a further five species from material collected in continental Europe between 1954 and 1958 (McGugan & Coppel, 1962). Two additional species were released during the period from 1968 to 1974, one from Germany and one from Argentina (Syme, 1981). A similar program of parasitoid introductions was conducted in the New England states from 1931 to 1937 (Dowden, 1962). This program is another example of the multiple introduction approach used in the biological control of important pests, where emphasis is placed on the need to provide rapid results without the need to resort to more detailed preintroduction studies. Of the 15 species of parasitoids released in New England and in southern Ontario, only three larval parasitoids, the braconid *Orgilus obscurator,* and the ichneumonids *Eulimneria rufifemur* (Thomson) and *Temelucha interruptor* (Gravenhorst), became firmly established. It was not until the early 1960s, however, that *T. interruptor* was recognized as a cleptoparasitoid detrimental to the potential impact of *O. obscurator* (Arthur *et al.,* 1964).

Orgilus obscurator is a specific larval parasitoid with a high fecundity and an efficient host-finding ability. It is able to avoid superparasitism and can be found as the predominant parasitoid in very low-host density situations (Syme, 1977). In contrast, *T. interruptor* is a more general parasitoid of Microlepidoptera and while it also has a high fecundity, it is inefficient at host finding and oviposits most successfully in host larvae previously attacked by *O. obscurator*. Both parasitoids attack young host larvae and only develop further when the host larvae approach maturity. However, the first-instar larva of *T. interruptor* is competitively superior to that of *O. obscurator*, which is killed at an early stage to ensure the successful development of the cleptoparasitoid (Schroeder, 1974).

In general, the parasitoid introduction programs against pine shoot moth in North America are considered unsuccessful. However, isolated reports of high levels of parasitism by *O. obscurator* followed by the collapse of shoot moth populations at Dorcas Bay in Ontario (Syme, 1971) and near Quebec City (Beique, 1960) indicate the potential of this species as a control agent. The occurrence of wild carrot, *Daucus carota*, at Dorcas Bay where parasitism by *O. obscurator* reached 92% prompted further investigations on the influence of this nectar and pollen source on rates of parasitism in Ontario. Syme (1977) demonstrated the beneficial influence of flowers on the longevity and fecundity of *O. obscurator* and was able to show increased rates of parasitism and elimination of pine shoot moth populations when the parasitoid was released into plantations where *D. carota* was plentiful (Syme, 1981).

Thus, this project has been at least partially successful and serves to point out two important considerations in dealing with classical biological control introductions against forest pests. First, larger and more mature parasitoid complexes of forest insects have evolved intricate interactions, including cleptoparasitism, which if undetected in a broad-spectrum, multiple-introduction program, can lead (as in this case) to unfortunate results. In the absence of *T. interruptor*, the better adapted *O. obscurator* is likely to have been a successful control agent. In addition, while *T. interruptor* was the only cleptoparasitoid to become established in North America, others were included in the release program. An opportunity to test the biological control potential of *O. obscurator* alone now exists in Chile, where the European pine shoot moth has invaded the extensive plantations of *Pinus radiata* D. Don (Ciesla, 1993), although in this case the exotic pine may not be as suitable for host location or host attack by the parasitoid. Then second, the availability of suitable food sources such as nectar and pollen in the forest environment may constrain the potential longevity and fecundity of adult parasitoids. While agricultural, horticultural, and urban environments often lack such food sources through cultural practices, the more natural forest environments may also be deficient in adult parasitoid food in regions with poor acidic soils or shaded-out ground cover.

Gypsy Moth, *Lymantria dispar* (Linnaeus): Lymantriidae

The gypsy moth is native to the Palearctic region and is a consistent pest of broadleaf forests in eastern and southern Europe. It was brought to North America and accidentally released in Massachusetts in 1868. Since then, it has become a serious pest of hardwoods throughout the northeastern states and has a continually expanding range that currently extends into Ontario and Quebec and southward into Virginia with isolated infestations in Minnesota and Oregon. In California, it is frequently intercepted during routine inspections by California Department of Food and Agriculture (CDFA) personnel at border entry points.

A biological control program was set up by the USDA Bureau of Entomology in 1905 and extensive foreign exploration for parasitoids and predators has been conducted in Europe, North Africa, and Asia at various intervals (Doane & McManus, 1981). This was the first major classical biological control program against a forest insect (the first attempt in the United States was against a bark beetle in 1892 and 1893 but was not a major forest program) and has been one of the most extensive programs undertaken in the history of biological control.

The gypsy moth project was important for other historical reasons because: (1) insect disease was recognized as an important natural enemy, (2) the sequence theory of natural enemies was proposed first by W. F. Fiske, (3) a number of future important contributors to biological control were trained on the project (H. S. Smith, W. R. Thompson, and W. D. Tothill), (4) sleeve cages and other equipment were invented and techniques were developed that remain in use today, and (5) L. O. Howard and W. F. Fiske developed their ideas on density-dependent and density-independent mortality in insect population dynamics. L. O. Howard, chief of the USDA Bureau of Entomology, was also instrumental in stimulating the Canadian interest in biological control in the early 1900s by making available facilities and scientific assistance from the Melrose Highlands Parasite Laboratory.

The early importations of natural enemies occurred between 1905 and 1914 and 1922 and 1933. While some collecting occurred in Japan, attention was focused on Europe where temporary field laboratories were set up wherever gypsy moth outbreaks were sufficient to permit the rearing of parasitoids from a large number of hosts. Frequent shipments of parasitoids and predators were made to the laboratory at Melrose Highlands and this resulted in the liberation of over 690,000 living insects of more than 45 species during this period (Dowden, 1962). The enormous importation and multiple release program enabled two larval/pupal predators, two egg parasitoids, six larval parasitoids, and one pupal parasitoid to become established in the New England states. The two egg parasitoids also were subject to large-scale rearing releases [*Ooencyrtus kuwanae* (Howard)], or to large-scale relocation releases (*Anastatus disparis* Ruschka). Most establishments occurred rapidly after the initial field releases but the tachinids *Parasetigena silvestris* (Robineau-Desvoidy) and *Exorista larvarum* (Linnaeus) were not recovered until 1937 and 1940, respectively, and the chalcidid *Brachymeria intermedia* (Nees) was only recovered in 1965.

The success of the established parasitoids and predators in controlling the gypsy moth in New England was limited and large-scale aerial applications of DDT were used until the early 1960s. Since 1960, renewed interest in the search for additional natural enemies has led to further exploration in Europe, Morocco, Iran, India, Japan, and Korea (Doane & McManus, 1981). Since 1963, the USDA, ARS Beneficial Insects Research Laboratory (BIRL) has received gypsy moth natural enemies in its quarantine facilities (currently at Newark, DE) and has distributed more than 200,000 individuals of about 60 species to other facilities for culture, study, and field release. From 1963 to 1971, the Gypsy Moth Methods Improvement Laboratory at Otis Air Force Base in Massachusetts was charged with the development of rearing procedures for the imported natural enemies. During this period, in conjunction with the NJDA, about 7 million parasitoids of 17 species were reared and released in the forests of New Jersey and Pennsylvania. From 1971 to 1977, a Gypsy Moth Parasite Distribution Program was established in which the NJDA and the University of Maryland reared an additional 2 million parasitoids of 18 species for distribution throughout the New England states. Since the late 1970s, new parasitoids and a predator collected from the gypsy moth in Japan and Korea and from the Indian gypsy moth, *Lymantria obfuscata* Walker, have been imported and released by the BIRL quarantine (Coulson *et al.,* 1986; Pemberton *et al.,* 1993). More than 100,000 individuals of nine new species or strains have been released in the field in Delaware, Massachusetts, and Pennsylvania.

While much increased knowledge of the biology and rearing methods of the imported parasitoids was gained during this massive program of importation, propagation, and release of parasitoids, it has resulted in the addition of only one pupal parasitoid, *Coccygomimus disparis* (Viereck), to the complex of 10 species established during the initial importation program (Schafer *et al.,* 1989). This has prompted Tallamy (1983) to compare the establishment of gypsy moth parasitoids with island biogeography theory, suggesting that a dynamic equilibrium now exists between further introductions and the extinction of established parasitoids [but see Washburn (1984)]. In the last 30 years, two of the parasitoids that were initially established, *Anastatus disparis* and *Exorista larvarum,* have become very rarely recorded, while two pupal parasitoids, *Brachymeria intermedia* and *C. disparis,* have become established. However, the main reasons for the failure to establish more additional parasitoids in recent years are that almost all species are bivoltine, requiring suitable alternative overwintering hosts, and that many of the parasitoid species released during the 1960s were probably only incidental parasitoids of the gypsy moth in their areas of origin.

In 1980, Forestry Canada and the IIBC initiated a novel approach for the search for new and more effective natural enemies of the gypsy moth. This approach was based on the premise that there is a marked difference in the natural-enemy complexes attacking outbreak versus endemic populations of forest defoliators (Pschorn-Walcher, 1977; Mills, 1990). Because the introduction of natural enemies from outbreak populations in Europe had failed to contain the gypsy moth, the emphasis was shifted to natural enemies of endemic populations. Since 1984, this program (Mills *et al.,* 1986; Mills, 1990) has led to the discovery of the tachinid *Ceranthia samarensis* (Villeneuve) as a low-density parasitoid of gypsy moth in Europe. Methods for the collection, importation, release (Mills & Nealis, 1992; Nealis & Quednau, 1996), and rearing (Quednau, 1993) of this parasitoid have been developed and within-season recoveries and overwintering success have been documented (Nealis & Quednau, 1996). This newly discovered low-density parasitoid offers renewed potential for the biological control of this notorious pest.

The failure of the established natural enemies to control the expanding outbreaks of the gypsy moth encouraged attempts during the 1970s to augment the impact of previously established species. Inoculative releases of small numbers of the predator *Calosoma sycophanta* into building populations of gypsy moth (Weseloh, 1990b) or populations at the leading edge of gypsy moth distribution (Weseloh *et al.,* 1995) increased the level of pupal predation, but did not appear to have an important impact on population growth. Inundative releases of *Cotesia melanoscela* (Ratzeburg) have provided significantly increased rates of parasitism, but have had little impact on foliage protection or egg mass counts for the following generation (Weseloh & Anderson, 1975; Kolodny-Hirsch *et al.,* 1988). In contrast, several other inundative releases of this and other species failed to provide evidence of increased parasitism in comparison with control plots (Doane & McManus, 1981). The success of augmentative parasitoid releases is dependent on an effective synchronization of adult parasitoid activity and susceptible larval instars. Wieber *et al.* (1995) have indicated that this can be achieved, with minimal predation and hyperparasitism, by the release of *C. melanoscela* cocoons in either November or December.

This approach may prove to be rather more successful than timing the release of adult parasitoids within season.

The combined release of parasitoids and pathogens has also been used as a method of augmentation. Wollam and Yendol (1976) demonstrated a synergistic effect of the release of *C. melanoscela* in plots treated with a double application of low concentration (see later text) B.t., over plots treated with each of these natural enemies alone. The resultant reduction in defoliation and subsequent egg mass densities has (Weseloh *et al.*, 1983) been attributed to the retarding effect of B.t. on host larval growth, which exposes the younger larvae to parasitism for a longer time period. A similar effect of *C. melanoscela* in conjunction with viral treatments is unlikely because this parasitoid avoids oviposition in moribund host larvae (Versoi & Yendol, 1982).

Augmentation through use of microbial pathogens alone has been of considerable importance against gypsy moth with significant advances in recent years. Early trials with B.t. in the 1960s were not effective in providing foliage protection. However, the discovery of improved strains (Dubois, 1985b) and successive improvements in formulation and application technology since the late 1970s have led to greater success. The results of aerial applications during the 1970s remained highly variable but a double application of low concentrations [20 billions of international units (BIU) per hectare] was developed and used operationally for the first time in 1980. This also met with limited success. Further experimental work in the 1980s (Dubois, 1985a; Dubois *et al.*, 1988) indicated that the use of a single application at higher concentrations (30 BIU per hectare) and acrylamide stickers could provide good foliage protection and reduce subsequent egg mass densities significantly. This development reduced the cost of B.t. applications and has been used operationally with success on 40 to 70% of the 1.3 to 1.5 million hectares of hardwood forest treated since 1983. Field trials using the Foray 48B strain of B.t. have demonstrated that egg mass densities can be reduced by as much as 99% by a single application high dosage rate of 90 BIU ha^{-1} (Dubois *et al.*, 1993), and by 95 to 96% using either a double or a single application of 50 BIU ha^{-1} (Cunningham *et al.*, 1996a, 1997), indicating that repeated annual applications are no longer necessary to achieve control.

Many field trials have been conducted with virus (NPV) sprays against the gypsy moth in both North America and Europe (Cunningham, 1982). An NPV strain (Hamden standard) isolated from a natural epizootic in Connecticut in 1967 forms the basis of the product "Gypchek" that was registered for use against gypsy moth in North America in 1978. Early trials of the baculovirus produced erratic results and while continued improvements in formulation and application have produced more positive results, they have never been accepted for operational use (Podgwaite, 1985). The main reasons for this are the relatively low virulence

of the virus, its rapid degradation on foliage in the field, and the more recent successes with the use of B.t. However, the development of improved screening formulations to prolong the activity of the virus in the field (Podgwaite *et al.*, 1991, 1992) and use of optical brighteners to enhance virulence and provide ultraviolet protection (Shapiro & Robertson, 1992; Dougherty *et al.*, 1996) have generated renewed interest in the potential of the virus in gypsy moth management. For example, a double application of 5×10^{12} polyhedral inclusion bodies (PIB) per hectare against early-instar larvae provided an 80% reduction in overwintering egg mass densities (Cunningham *et al.*, 1996a), with some evidence of a carryover effect of the virus in larval populations the following year (Cunningham *et al.*, 1996b). Similarly, the inclusion of optical brighteners in the formulation has provided enhanced reduction of egg mass densities in field trials (Webb *et al.*, 1994; Cunningham *et al.*, 1997).

The gypsy moth program has been spectacular both in the scale and the continued enthusiasm with which it has been conducted. The results, though, have been disappointing and serve as a good example of the failure of classical biological control in situations where the introduced pest is also a severe pest in its region of origin. Berryman (1991) argues that in fact the early parasitoid establishments did result in a reduction of the equilibrium density of gypsy moth populations (but not to a level that prevented defoliation). Additionally, it has been found that artificial elevations of local populations through the release of feral or sterile egg masses are rapidly reduced to an undetectable level through parasitism, primarily by the polyphagous tachinid *Compsilura concinnata* (Meigen) (Liebhold & Elkinton, 1989; Elkinton *et al.*, 1999; Gould *et al.*, 1990). Thus, reduction of outbreak populations of the gypsy moth in North America is only likely to be achieved through use of microbial pathogens as biopesticides, whereas endemic densities may be more effectively maintained through the action of generalist parasitoids such as *C. concinnata*, and low-density parasitoids such as *C. samarensis*. Thus, the search for natural enemies in areas where the gypsy moth is not a pest, in nonoutbreak populations or from related nonpest *Lymantria* species, may prove to be a better strategy for classical introductions in the future.

Hymenoptera: Symphyta

European Spruce Sawfly, *Gilpinia hercyniae* (Hartig): Diprionidae

The European spruce sawfly is native to Europe and was first noted as an accidental introduction in Canada in 1922. By 1930, a severe outbreak caused concern in the Gaspe Peninsula and by 1936, the sawfly threatened to devastate the spruce forests of the whole of the eastern boreal region. The sawfly extended its range across all eastern provinces of Canada and into the adjacent United States, causing severe damage over an area of more than 10,000 sq mi (McGugan & Coppel, 1962). A classical biological control program was initiated in 1933. At this time, *G. hercyniae* was not distinguished from the closely related *G. polytomum* (Hartig). The IIBC, then known as the Farnham House Laboratory in the United Kingdom, was enlisted to make large-scale parasitoid collections from *G. polytomum* in Europe. Initial studies revealed that apart from the egg parasitoids, all other parasitoids overwintered in the host cocoon. This simplified parasitoid collections in Europe to procurement of host eggs and cocoons. Between 1932 and 1940, a team of up to 30 people collected over half a million cocoons of *G. polytomum* in Europe for shipment to Canada. In addition, more than half a million eggs and no fewer than 31 million cocoons of other spruce- and pine-feeding sawflies were dispatched to supplement the numbers of the less host-specific parasitoid species available for field release (Morris *et al.*, 1937; Finlayson & Finlayson, 1958). A total of 96 species of primary and secondary parasitoids was obtained from these cocoon collections at the Belleville Laboratory in Canada and a multiple-introduction program involving two egg parasitoids and 25 larval and cocoon parasitoids was conducted from 1933 to 1951. The importation of a wide variety of parasitoids from diverse hosts permitted the inclusion of several pine sawfly pests as additional targets for some of the releases (McGugan & Coppel, 1962).

In 1936, the addition of an elaborate controlled environment quarantine building at the Belleville Laboratory permitted mass rearing of several imported European parasitoids. *Dahlbominus fuscipennis,* a gregarious ectoparasitoid of prepupae, readily attacked cocoons in the laboratory and was selected for a large mass rearing program. The mass rearing peaked in 1940, when 221.5 million *D. fuscipennis* were released. By the end of the program in 1951, 890 million directly imported or laboratory-reared individuals had been liberated in the field (McGugan & Coppel, 1962).

Of the 27 parasitoid species released during the course of this program, only 5 species became established over an appreciable number of generations, although 4 other species were recovered during the years shortly after release. Three of the five species (*D. fuscipennis, Exenterus amictorius,* and *E. confusus* Kerrich) were widely established only during the outbreak and since then have not been recorded from *G. hercyniae*. While *E. amictorius* had little impact, the other two species achieved variable but appreciable levels of parasitism and have been credited with the decline of the outbreak in at least some areas. Two other parasitoids, *Exenterus vellicatus* Cushman (Ichneumonidae) and *Drino bohemica* Mesnil (Tachinidae), never became important until the collapse of the outbreak, but have replaced the three species present during the outbreak to maintain host populations at nondamaging levels.

The European spruce sawfly outbreak began to decline in 1939 and 1940 and this coincided in the southern part of the range with the occurrence of an NPV. This virus is believed to have been accidentally imported and released in Canada with parasitoid material. It spread rapidly to produce epizootics throughout most of the outbreak range; by 1943, host population levels had declined to light infestations. Unlike other diprionid sawflies, *G. hercyniae* larvae are not gregarious, and the rapid spread and subsequent impact of the virus were attributed to its virulence (Bird & Elgee, 1957). Other studies in the United Kingdom, where *G. hercyniae* was accidentally introduced from the continent of Europe in 1968, indicate that birds play an important role in the transmission of the virus, which developed to epizootic levels and brought about the collapse of the outbreak by 1974 (Entwistle, 1976).

The importance of *D. bohemica, E. vellicatus,* and the NPV virus in maintaining the spruce sawfly at low population densities in Canada has been inadvertently demonstrated through chemical spray treatments aimed against spruce budworm. Both in the early 1960s and again in the early 1970s, sawfly population levels increased immediately following the cessation of a 2- to 3-year spray treatment, due to the detrimental effects of the spray on the natural enemies, but declined after several generations as a result of increased parasitism and reappearance of the virus (Neilson *et al.,* 1971; Magasi & Syme, 1981).

This successful program has several interesting features. The success of the accidental introduction of the viral disease provides to date the most outstanding example of the use of a pathogen in classical biological control. Its ability to control the sawfly population in the absence of parasitoids has been demonstrated (Bird & Burk, 1961; Entwistle, 1976) and in Canada it has persisted in the forest environment since the initial introduction despite the low-host densities (Magasi & Syme, 1981). The multiple-introduction program of parasitoids resulted in the establishment of the two more effective and specific species, despite the release of a wide range of potential competitors. However, the continuous and large-scale release of poorly adapted parasitoids, which were later recovered only from other sawfly hosts, was successful in inducing significant levels of mortality prior to the introduction of the virus.

Larch Sawfly, *Pristiphora erichsonii* (Hartig): Tenthredinidae

This sawfly, a comparatively rare insect in Europe, was first recognized to be widely distributed in larch forests throughout the eastern provinces of Canada in 1884. Several short-lived but severe infestations were noted from 1906 to 1916 in which huge quantities of merchantable tamarack (*Larix laricina*) were destroyed (McGugan & Coppel, 1962). Since then, the sawfly has been found throughout the range of larch in North America but remains more important on tamarack than on western larches. There remains some debate as to whether the sawfly was a recent introduction in the late nineteenth century or of much older origin in North America (Ives & Muldrew, 1981). The lack of native parasitoids, however, prompted a classical biological control program to be initiated in 1910; and the introduction of parasitoids was conducted in 1910 to 1913, 1934, and again in 1961 to 1964.

During the early phase of parasitoid introductions (McGugan & Coppel, 1962) collections were made in England and shipped to Canada for quarantine, screening, and direct release of the parasitoids. This led to the establishment of the ichneumonid larval parasitoid, *Mesoleius tenthredinis* Morley, which in Manitoba was found in 20% of sawfly cocoons in 1916 and had parasitized over 80% of the population by 1927 (Criddle, 1928). Tachinid *Zenillia nox* (Hall) was collected in Japan in 1934 by the USDA and was released in both New Brunswick and British Columbia but failed to establish. The success of parasitism by *M. tenthredinis* prompted an extensive relocation program to distribute this parasitoid throughout Canadian larch forests. Rapid establishment was reported with subsequent reduction in sawfly populations and reduced timber losses.

Thus, this appeared to be another example of the success of classical biological control in Canada. However, in the late 1930s, larch sawfly defoliation again became prevalent in Manitoba and because parasitism by *M. tenthredinis* appeared to have dropped to low levels, an additional 75,000 parasitoids from British Columbia were relocated across central Canada. While the parasitoid's range was increased, levels of parasitism remained low due to the heavy encapsulation of the parasitoid eggs by host larvae (Muldrew, 1953). The appearance of a resistant European strain of the sawfly, capable of encapsulating *M. tenthredinis* eggs, appears to have resulted from the parasitoid introduction program in 1913, when imported larch sawfly cocoons were placed directly in the field. The resistant strain has since spread across Canada and into neighboring states of the United States, becoming predominant in most regions (Wong, 1974).

In 1957, renewed efforts were made to obtain additional parasitoids from Europe and Japan, and long-term study plots were set up in Manitoba to evaluate the dynamics of the larch sawfly populations and the impact of introductions. These studies (Ives, 1976) indicated that mortality in the cocoon and adult stages determined population trends and that high water tables and predation by small mammals were largely responsible for the erratic population abundance. The native tachinid, *Bessa harveyi* (Townsend), considered the most important parasitoid in the renewed outbreaks, had little impact.

Eleven parasitoids were collected by the IIBC in Europe and Japan and shipped to Canada between 1959 and 1965.

Of these, five of the more abundant ones were selected for release, being available in sufficient quantity for initial releases of at least 200 adults. In addition, a separate, unique introduction of the masked shrew, *Sorex cinereus,* from New Brunswick to the island of Newfoundland was undertaken in 1958 in an attempt to fill the vacant niche for an insectivore and to increase the cocoon predation of the sawfly. The masked shrew was successfully established, as well as two of the parasitoids. One of these parasitoids was ichneumonid *Olesicampe benefactor* Hinz, which attacked young sawfly larvae. The second was a Bavarian strain of *M. tenthredinis*, which was demonstrated to be only weakly encapsulated by the resistant sawfly strain and was able to pass this attribute on to the progeny of mixed (England × Bavarian) crosses (Turnock & Muldrew, 1971).

Parasitism by *M. tenthredinis* initially increased following the release of the Bavarian strain, but *O. benefactor* became the dominant parasitoid influencing cocoon survival. Parasitism by this latter species at the release point in Manitoba achieved levels of around 90% between 1967 and 1972 (Ives & Muldrew, 1981) and was the dominant factor responsible for the collapse of populations of the sawfly (Ives, 1976). Following the success of *O. benefactor* releases in the late 1960s, this parasitoid was relocated from Manitoba to most other provinces in Canada (Turnock & Muldrew, 1971) as well as to Maine (Embree & Underwood, 1972), Minnesota (Kulman *et al.*, 1974), and Pennsylvania (Drooz *et al.*, 1985).

The impact of the masked shrew on larch sawfly cocoon survival in Newfoundland has not been adequately assessed. Predation of cocoons must certainly have increased but outbreaks continued through the 1960s and 1970s and thus the shrews appear to be unable to prevent larch sawfly populations reaching outbreak densities (Ives & Muldrew, 1981).

Thus, *O. benefactor* appeared to offer the greatest potential for control of larch sawfly populations in Canada. However, in 1966, hyperparasitoid *Mesochorus globulator* Thunberg began to attack *O. benefactor* in Manitoba. This polyphagous hyperparasitoid is common in Europe and also may have been accidentally introduced during the initial 1910 to 1913 parasitoid introductions. It has since spread throughout the region and into Wisconsin, but does not appear to have reached Pennsylvania (Drooz *et al.*, 1985). While hyperparasitism attained very high levels (80 to 90%) in Manitoba during the 1970s, sawfly populations remained low. Control may be achieved by *O. benefactor* despite the occurrence of the hyperparasitoid, although it may be too early to assess the potential of the sawfly to escape parasitism in the future. Any reduction in impact of *O. benefactor* could also allow the Bavarian strain of *Mesoleius tenthredinis* to spread more widely and to play a more dominant role in maintaining host populations at a low level, as it does in Europe.

This project provides further evidence of the value of the more specific and well-adapted parasitoids in classical biological control. As in the case of the European spruce sawfly, while a wide range of parasitoids was released, only the more specific species became established. However, while in the absence of hyperparasitism *O. benefactor* may have been an ideal control agent, its competitive superiority over the Bavarian strain of *M. tenthredinis* may have prevented the latter from establishing and spreading more widely. This, together with the known occurrence of various geographic strains of *M. tenthredinis,* differing in ability to avoid encapsulation by the host, emphasizes the value of detailed studies of parasitoid biologies prior to introduction. In addition, the accidental introduction of a parasitoid-resistant strain of the host and probably also a hyperparasitoid, *Mesochorus globulator,* demonstrates the importance of quarantine handling of imported material in classical biological control programs. The introduction of the masked shrew into Newfoundland is a rare example of the use of a vertebrate predator, which may be justified in this case by filling a vacant niche and increasing soil predation levels in general. The more general feeding habits of such predators, however, make them unsuitable as control agents for specific target prey.

European Wood Wasp, *Sirex noctilio* Fabricius: Siricidae

Biological control attempts against the wood wasp, *Sirex noctilio,* represent one of the very few large programs directed against wood-boring insects. Wood wasps generally are considered secondary pests that attack dead or dying trees. *Sirex noctilio* occurs in Canada and Europe but is most common in the Mediterranean region. It is more or less specific to *Pinus* species (Spradbery & Kirk, 1978), and is unique among European siricids in that it is able to kill standing green trees. Given the right circumstances, as occurred in New Zealand and Australia, this insect was able to cause serious losses in *Pinus radiata* plantations. Monterey pine, *P. radiata,* has a very restricted natural range, occurring in a small area in California. However, a number of years ago this fast-growing conifer was planted in Australia and New Zealand because of a shortage of softwoods in the natural forests. Over the years the acreage devoted to *P. radiata* plantations has increased.

Sirex noctilio was discovered on the North Island of New Zealand around 1900, but it was not until 1927 that it was abundant enough in exotic pine plantations for control to be considered (Taylor, 1981). Abnormally high mortality occurred in *P. radiata* plantations between 1940 and 1949 in New Zealand. *Sirex noctilio* was discovered in southern Tasmania in 1952 and in Victoria in 1961. By this time, *S. noctilio* had caused considerable damage to *P. radiata* plantations in New Zealand (Taylor, 1976).

The biology of *S. noctilio* is especially interesting because the wasp harbors symbiotic fungus *Amylostereum areolatum* (Fr.) Boidin that serves as a kairomone for parasitoids of the wood wasp. In addition, parasitic nematode *Deladenus siricidicola* is totally dependent on the wood wasp and the fungus (Bedding, 1972). Adults of *S. noctilio* emerge from midsummer to late fall and mate in the upper foliage of trees. Females oviposit by drilling holes through the bark into the sapwood of trees that are weakened or damaged, and at the same time introduce the symbiotic fungus (Taylor, 1981). The adults live only a few days in nature and the eggs hatch when the surrounding area has been invaded by the fungus. This occurs after drying because the fungus will not grow otherwise. The first- and second-instar larvae feed exclusively on fungus. The third- and fourth-instar larvae tunnel more deeply in the wood. The late-instar larvae turn back toward the bark to about 5 cm from the bark surface and enter the prepupal stage. Pupation may not occur until the second or third year after hatching, depending on the climate. The adults emerge in approximately 3 weeks following pupation. Each generation emerges over a period of 2 to 3 years; and the proportion of individuals emerging in the first, second, and third year varies by site (Taylor, 1981).

The first biological control program was initiated in New Zealand in 1927 (Taylor, 1981). Between 1929 and 1932, ichneumonid *Rhyssa persuasoria* Linnaeus was introduced and colonized, but the control was not satisfactory (Turnock et al., 1976). A second parasitoid, ibaliid *Ibalia leucospoides* (Hochenwald), was colonized between 1954 and 1958 and control was improved (Zondag, 1959). These two parasitoids were then relocated from New Zealand to Tasmania, *R. persuasoria* in 1957 and *I. leucospoides* in 1959 and 1960.

The large-scale biological control effort did not commence until 1961, following the discovery of *S. noctilio* in Victoria. As mentioned earlier, a national *Sirex* fund was established, involving a consortium of federal, state, and private agencies to provide financial support and a committee to coordinate research and control activities in Victoria (Taylor, 1981). A worldwide search for natural enemies was instituted by the Division of Entomology, CSIRO, in 1962. The search for parasitoids in many parts of the Northern Hemisphere was completed by 1973, and during this 11-year period 21 species of parasitoids were sent to Tasmania for culturing (Taylor, 1976). The strategy was to obtain all the available parasitoids of siricids in conifers and as many strains as possible from different climatic zones, although the major effort was concentrated in the Mediterranean region. This meant that collections were made from siricids in conifers other than *Pinus* and from genera and species other than *Sirex noctilio*. Ten parasitoid species, including many subspecies and geographic races, were released in Tasmania and Victoria. Six species have become established and another species, the ichneumonid *Rhyssa hoferi* Rohwer, is probably established but cannot yet be confirmed (Taylor, 1981). Of the seven species, two are Holarctic (*R. persuasoria* and *I. leucospoides*), two are Palearctic [*I. rufipes drewseni* Borries and the ichneumonid *Odontocolon geniculatus* (Kreichbaumer)], and three are Nearctic [the stephanid *Schlettererius cinctipes* Cresson and the ichneumonids *Megarhyssa nortoni* (Cresson) and *R. hoferi*].

All the species are more or less complementary, although there may be some competition within the guild attacking the larger larvae. The *Ibalia* species attack first- or second-instar siricid larvae and the two species have different emergence times so they do not compete directly. The ichneumonids attack more advanced larvae of their host and there may be differential preference based on tree diameter (Taylor, 1981). *Schlettererius cinctipes* emerges after the peak emergence of the ichneumonids. The other two species are also complementary because *O. geniculatus* is small, emerges in spring, and attacks late-hatching larvae that are still closer to the bark surface. *Rhyssa hoferi* is adapted to drier areas and should do well in areas less favorable for the other rhyssines (Taylor, 1981).

A fortuitous introduction of parasitic nematode *Deladenus siricidicola* was discovered in New Zealand in 1962 by Zondag (1969). This parasite causes the females of *Sirex noctilio* to lay infertile eggs. A search was made for other nematode parasites between 1965 and 1973, but only *D. siricidicola* was found suitable for biological control programs (Bedding & Akhurst, 1974). Several strains of this nematode have been released throughout Tasmania and Victoria and this species is now well established throughout the range of *S. noctilio*. In 1981, it was still being reared and released in parts of Victoria where the siricid density was the greatest (Taylor, 1981). The nematode also affects the reproduction of some female parasitoids (Bedding, 1967). This latter effect apparently does not adversely affect the biological control program. The nematode has led to the reduction of siricid populations to very low levels in some areas.

The *S. noctilio* biological control program is significant for several reasons. A large group of organizations cooperated in putting together a well-funded, extensive worldwide search for parasitoids as well as a research program that looked into many aspects of the tree, *Sirex,* fungus, and parasitoid relationship (Taylor, 1981). As with *G. hercyniae,* there was a fortuitous introduction of the natural enemy (in this case a nematode). *Sirex noctilio* is an example of a Northern Hemisphere insect that became a pest in the Southern Hemisphere on a host tree that was both new to the pest and to the Australasian environment. The biological control program was well planned, with foreign exploration in the Northern Hemisphere extending beyond just *S. noctilio* to include other siricid hosts and colonization of

parasitoids, taking into account both climatic tolerance and host stages attacked.

The program seems to have been a success (Turnock *et al.* 1976). The combination of parasitoids, nematodes, and sound forest management should minimize losses due to *S. noctilio* (Taylor, 1976). There is no doubt that natural enemies have helped to control *S. noctilio* but the final evaluation of the role of the natural enemies in the success of the program may require several more years before the effects of other factors such as site, climatic zone, and annual precipitation can be determined (Taylor, 1981).

Coleoptera

Great European Spruce Beetle, *Dendroctonus micans* (Kugelann): Scolytidae

This bark beetle occurs in coniferous forests from Siberia in the east to central France and the United Kingdom. It is one of only two *Dendroctonus* species occurring in the Palearctic region. *Dendroctonus micans* is primarily pest of spruce, *Picea* spp., but will occasionally attack Scots pine, *Pinus sylvestris* Linnaeus. The bark beetle has been expanding its range southward and westward for many years. Approximately 200,000 ha are currently suffering from outbreaks of *D. micans;* the recently invaded areas include the United Kingdom, France, the Republic of Georgia (formerly of the Soviet Union), and Turkey (Evans, 1985; Gregoire *et al.,* 1989). In the inner parts of its range where the beetle has been established for a long time, populations remain at low densities and it is not a pest. The beetle is only a pest in those areas where it is extending its range.

The European spruce bark beetle differs from the more aggressive North American *Dendroctonus* species in that it attacks its host tree in low numbers, patch-killing the bark. Successive attacks over a period of 5 to 8 years may be necessary to kill a tree except during beetle outbreaks (Gregoire, 1985). The beetle shows kin-mating, has gregarious larvae, and apparently lacks associated fungi that are characteristic of many Scolytidae. *Dendrotonus micans* has very few natural enemies, perhaps due to its unusual biology. By living in thick bark at the base of trees in galleries filled with large quantities of resin the beetles seem to be protected from both competitors and generalist natural enemies (Everaerts *et al.,* 1988).

There is one specific predator, *Rhizophagus grandis,* of the spruce beetle that is very abundant in areas where the bark beetle has been present for long periods of time (i.e., the interior areas of its distribution). This rhizophagid beetle is believed to be responsible for maintaining the low, stable *D. micans* population in these areas (Kobakhidze, 1965; Gregoire, 1976). *Rhizophagus grandis* is the most studied rhizophagid in Europe (Moeck & Safranyik, 1984).

A large-scale biological control program was initiated

against *D. micans* in Soviet Georgia in 1963 (Kobakhidze, 1965). The spruce beetle had extended its range into Georgia following World War II in timber imported from the north. The predator did not follow so an inoculation (augmentation, relocation) program was planned. Large numbers of *R. grandis* were reared on *D. micans* broods in bolts and released as larvae and adults on spruce trees infested by *D. micans* (Kobakhidze *et al.,* 1968). Much of the information on *R. grandis* and the biological control program is in Russian and is not translated, but apparently 200,000 *R. grandis* are produced each year and effective control has been achieved (Gregoire *et. al.,* 1989).

The great European spruce beetle was first observed in the Massif Central in France in the early 1970s when outbreaks began to occur (Gregoire *et al.,* 1985; Gregoire *et al.,* 1989). A program funded by the European Economic Community (EEC) was undertaken in 1983 and involved the cooperation of the Universite Libre de Bruxelles, Belgium; and two French organizations, the Institut National de la Recherche Agronomique (INRA) in Avignon and the Parc National des Cevennes at Florac, where the project is being conducted. The rationale for the program was the same as in Soviet Georgia (i.e., to relocate the predator with the invading *D. mincans*). Initially, the Russian method of rearing was used but later, semiartificial methods were developed (Gregoire *et al.,* 1989). The criteria for the inoculation of *R. grandis* was 50 pairs of adults at the base of each attacked tree if the site contained less than 10 attacked trees per hectare and 500 to 1000 pairs (per tree) if there were more than 10 attacked trees per hectare. A total of 2,350 predators was released in 1983; 8,500, in 1984; 16,350, in 1985; and 41,800, in 1986. The predators are established and the project is currently being evaluated. *Dendrotonus micans* continues to spread into other spruce regions of France and the inoculation of *R. grandis* may need to continue for several years.

The spruce bark beetle was discovered in the United Kingdom in 1982 and surveys indicated that the beetle had been in there since 1972 (Evans, 1985). A program similar to the one in France was initiated in 1983 and an experimental release of 27 pairs of *R. grandis* was made. In 1984, 31,168 *R. grandis* were released in 950 locations throughout the infested area. Subsequent releases of this predator were made in 1985 (39,392) and 1986 (17,604) (Evans & King, 1989; Evans & Fielding, 1996).

This biological control program is significant because it is the only successful biological control project against a scolytid, at least in the Republic of Georgia; the programs in France and the United Kingdom are still being evaluated. In the latter country, the predator is well established and may infest up to 80% of the *D. micans* broods (Fielding, 1992), and evidence is accumulating that *R. grandis* is an effective biological control agent against *D. micans* (Evans & Fielding, 1994, 1996). Another bark beetle program involves the importation of parasitoids and predators from the United States against *Ips grandicollis* in Australia, which is a classical biological control project, but it too is still being evaluated (Berisford & Dahlsten, 1989). Another significant aspect of the *D. micans* project is that it is a good example of the inoculation (augmentation, relocation) technique. A coevolved specific predator is being used and it is a single species inoculation. The predator is attracted to the frass of *D. micans* but laboratory studies showed the predator to be attracted to the frass of three North American *Dendroctonus* species as well (Miller *et al.,* 1987). This suggests the possible use of *R. grandis* as an exotic importation against native North American species of *Dendroctonus*.

References

Ahmed, S. I., & Leather, S. R. (1994). Suitability and potential of entomopathogenic microorganisms for forest pest management, some points for consideration. International Journal of Pest Management, 40, 287–292.

Altieri, M. A., & Letourneau, D. K. (1982). Vegetation management and biological control in agroecosystems. Crop Protection, 1, 405–430.

Altieri, M. A., Cure, J. R., & Garcia, M. A. (1993). The role and enhancement of parasitic Hymenoptera biodiversity in agroecosystems. In J. LaSalle & I. D. Gauld (Eds.), Hymenoptera and biodiversity (pp. 257–275). Wallingford, United King: CAB International.

Andreadis, T. G., & Weseloh, R. M. (1990). Discovery of *Entomophaga maimaiga* in North American gypsy moth, *Lymantria dispar.* Proceedings of the National Academy of Sciences, USA, 87, 2461–2465.

Arthur, A. P., Steiner, J. E. R., & Burnbull, A. L. (1964). The interaction between *Orgilus obscurator* (Nees) (Hymenoptera: Braconidae) and *Temelucha interruptor* (Grav.) (Hymenoptera: Ichneumonidae), parasites of the pine shoot moth, *Rhyacionia buoliana* (Schiff.) (Lepidoptera: Olethreutidae). Canadian Entomologist, 96, 1030–1034.

Bedard, W. D. (1933). Unpublished report in files of the Pacific Southwest Forest and Range Experiment Station, Berkeley: Forest Service, U.S. Department of Agriculture.

Bedding, R. A. (1967). Parasitic and free-living cycles in entomogenous nematodes of the genus *Deladenus.* Nature (London), 214, 174–175.

Bedding, R. A. (1972). Biology of *Deladenus siricidicola* (Neotylenchidae), an entomophagous mycetophagous nematode parasitic in siricid woodwasps. Nematologica, 18, 482–493.

Bedding, R. A., & Akhurst, R. J. (1974). Use of the nematode *Deladenus siricidicola* in the biological control of *Sirex noctilio* in Australia. Journal of the Australian Entomological Society, 13, 129–135.

Beique, R. (1960). The importance of the European pine shoot moth, *Rhyacionia buoliana* (Schiff.) in Quebec City and vicinity. Canadian Entomologist, 92, 858–862.

Beirne, B. P. (1973). Influences on the development and evolution of biological control in Canada. Bulletin of the Entomological Society of Canada, 5, 85–89.

Berisford, C. W., & Dahlsten, D. L. (1989). Biological control of *Ips grandicollis* (Eichoff) (Coleoptera: Scolytidae) in Australia. In D. L. Kulhavy & M. C. Miller (Eds.), Potential for biological control of Dendroctonus and Ips bark beetles (pp. 81–93). Nacogdoches, TX: Center for Applied Studies, School of Forestry, Stephen F. Austin State University.

Berryman, A. A. (1967). Preservation and augmentation of insect predators of the western pine beetle. Journal of Forestry, 65, 260–262.

Berryman, A. A. (1991). The gypsy moth in North America: A case of successful biological control? Trends in Ecology and Evolution 6, 110–111.

Bird, F. T. (1971). *Neodiprion lecontei* (Fitch), red-headed pine sawfly (Hymenoptera: Diprionidae). In Biological control programmes against insects and weeds in Canada, 1959–1968 (Technical Communication No. 4, chap. 41, (pp. 148–150). Commonwealth Institute of Biological Control Commonwealth Agricultural Bureaux, Farnham Royal, England.

Bird, F. T., & Burk, J. M. (1961). Artificially disseminated virus as a factor controlling the European spruce sawfly, *Diprion herycinae* (Htg.), in the absence of introduced parasites. Canadian Entomologist, 92, 228–238.

Bird, F. T., & Elgee, D. E. (1957). A virus disease and introduced parasites as factors controlling the European spruce sawfly, *Diprion hercyinae* (Htg.) in central New Brunswick. Canadian Entomologist, 89, 371–378.

Borden, J. H. (1982). Aggregation pheromones. In J. B. Mitton & K. B. Sturgeon (Eds.), Bark beetles in North American conifers: A system for the study of evolutionary biology (pp. 74–139). Austin, TX: University of Texas Press.

(1985). Aggregation pheromones. In G. A. Kerkut & L. I. Gilbert (Eds.), Comprehensive insect physiology, biochemistry and pharmacology (Vol. 9, pp. 257–285). Oxford, United Kingdom: Pergamon Press.

Bruns, M. (1960). The economic importance of birds in forests. Bird Study, 7, 193–208.

Buckner, C. M. (1971). Vertebrate predators. USDA, Forestry Service Research NE, 194, 21–31.

Burzynski, J. (1970). Biologische Bekampfungs—methoden von Forstschadlingen. Tagungsber. Deutsche Akademie Landwirtschaftwiss Berlin, 100, 37–42 (in German).

Bustillo, A. E., & Drooz, A. T. (1977). Comparative establishment of a Virginia (USA) strain of *Telenomus alsophilae* on *Oxydia trychiata* in Colombia. Journal of Economic Entomology, 70, 767–770.

Cadogan, B. L., Nealis, V. G., & Van Frankenhuyzen, K. (1995). Control of spruce budworm (Lepidoptera: Tortricidae) with *Bacillus thuringiensis* timed to conserve a larval parasitoid. Crop Protection, 14, 31–36.

Carl, K. P. (1982). Biological control of native pests by introduced natural enemies. Biocontrol News Information, 3, 191–200.

Ceballos, G., & Zarko, E. (1952). The biological control of an outbreak of Diprion pini (L.) on Pinus silvestris in the Sierra de Albarracin (pp. 1–38). Madrid: Madrid Instituta España Entomologia (in Spanish).

Ciesla, W. M. (1993). Recent introductions of forest insects and their effects: A global overview. FAO Plant Protection Bulletin, 41, 3–13.

Clair, D. J., Dahlsten, D. L., & Dreistadt, S. H. (1988). Biological control of the elm leaf beetle, *Xanthogaleruca luteola*, in California—a case study. In P. Allen & D. van Dusch (Eds.), Global perspectives on agroecology and sustainable agricultural systems (Vol. 2, pp. 497–504). Santa Cruz: Agroecology Program, University of California.

Clark, R. C., Greenbank, D. O., Bryant, D. G., & Harris, J. W. E. (1971). *Adelges piceae* (Ratz.), balsam woolly aphid (Homoptera: Adelgidae). In Biological control programmes against insects and weeds in Canada, 1959–1968 (Technical Communication, No. 4, pp. 113–127). Commonwealth Institute of Biological Control. Commonwealth Agricultural Bureaux, Farnham Royal, England.

Clausen, C. P. (1956). Biological control of insect pests in the continental United States (Technical Bulletin No. 1139). Washington, DC: U.S. Department of Agriculture.

Clausen, C. R. (1978). (Ed.). Introduced parasites and predators of arthropod pests and weeds: A world review. Agriculture Handbook No. 480. Washington, DC: ARS, U.S. Department of Agriculture.

Coulson, J. R., Fenster, R. W., Schaefer, P. W., Ertle, L. R., Kelleher, J. S., & Rhoads, L. D. (1986). Exploration for and importation of natural enemies of the gypsy moth, *Lymantria dispar* (L.) (Lepidoptera: Lymantriidae), in North America: An update. Proceedings of the Entomological Society Washington 88, 461–475.

Criddle, N. (1928). The introduction and establishment of the larch sawfly parasite, *Mesoleius tenthredinus* Morley, into southern Manitoba. Canadian Entomologist, 60, 51–53.

Cunningham, J. C. (1982). Field trials with baculoviruses: Control of forest insect pests. In E. Kurstak (Ed.), Microbial and viral pesticides (pp. 335–386). New York: Marcel Dekker.

Cunningham, J. C., & DeGroot, P. (1981). *Neodiprion lecontei* (Fitch), redheaded pine sawfly (Hymenoptera: Diprionidae). In J. S. Kelleher & M. A. Hulme (Eds.), Biological control programmes against insects and weeds in Canada, 1969–1980 (pp. 323–329). London: Commonwealth Agriculture Bureau.

Cunningham, J. C., & Entwistle, P. F. (1981). Control of sawflies by baculoviruses, In H. D. Burges (Ed.), Microbial control of pests and plant diseases 1970–1980 (pp. 379–407). New York: Academic Press.

Cunningham, J. C., Brown, K. W., Scarr, T., Fleming, R. A., & Burns, T. (1996b). Aerial spray trials with nuclear polyhedrosis virus and *Bacillus thuringiensis* of gypsy moth (Lepidoptera: Lymantriidae) in 1994. II. Impact one year after application. Proceedings of the Entomological Society of Ontario, 127, 37–43.

Cunningham, J. C., Payne, N. J., Brown, K. W., Fleming, R. A., Burns, T., Mickle, R. E. (1996a). Aerial spray trials with nuclear polyhedrosis virus and *Bacillus thuringiensis* of gypsy moth (Lepidoptera: Lymantriidae) in 1994. I. Impact in the year of application. Proceedings of the Entomological Society of Ontario, 127, 21–35.

Cunningham, J. C., Brown, K. W., Payne, N. J., Mickle, R. E., Grant, G. G., & Fleming, R. A. (1997). Aerial spray trials in 1992 and 1993 against gypsy moth, *Lymantria dispar* (Lepidoptera: Lymantriidae), using nuclear polyhedrosis virus with an without an optical brightener compared to *Bacillus thuringiensis*. Crop Protection, 16, 15–23.

Dahlsten, D. L. (1976). The third forest. Environment, 18, 35–42.

Dahlsten, D. L. (1992). Evidence for microbial production of kairomones for bark beetle parasitoids. In D. C. Allen & L. P. Abrahamson (Tech. Eds.), Proceedings: North American Forest Insect Work Conference, March 25–28, 1991 Denver, Colorado (General Technical Report PNW-GTR-294, p. 30). Portland, Oregon: USDA, Forest Service, Pacific Northwest Station.

Dahlsten, D. L., & Copper, W. A. (1979). The use of nesting boxes to study the biology of the mountain chickadee (*Parus gambeli*) and its impact on selected forest insects. In J. G. Dickson, R. N. Conner, R. R. Fleet, J. C. Kroll, & J. A. Jackson (Eds.), The role of insectivorous birds in forest ecosystems (pp. 217–260). New York: Academic Press.

Dahlsten, D. L., & Whitmore, M. C. (1989). The case for and against the biological control of bark beetles (Coleoptera: Scolytidae). In D. L. Kulhavy & M. C. Miller (Eds.), Potential for biological control of Dendroctonus and Ips bark beetles (pp. 3–19). Nacogdoches, TX: Center for Applied Studies, School of Forestry, Stephen F. Austin State University.

Dahlsten, D. L., Copper, W. A., Rowney, D. L., & Kleintjes, P. K. (1990). Quantifying bird predation of arthropods in forests. In M. L. Morrison, C. J. Ralph, J. Verner, & J. R. Jehl, Jr. (Eds.), Avian foraging: Theory, methodology and applications. Studies in Avian Biology No. 131 (pp. 44–52).

Dahlsten, D. L., Copper, W. A., Rowney, D. L., & Kleintjes, P. K. (1992). Population dynamics of the mountain chickadee in northern California. In D. R. McCullough & R. H. Barrett (Eds.), Wildlife 2001: Populations (pp. 502–510). London: Elsevier Science.

Dahlsten, D. L., Tait, S. M., Rowney, D. L., & Gingg, B. J. (1993). A monitoring system and development of ecologically sound treatments for elm leaf beetle. Journal of Arboriculture, 19(4), 181–186.

DeBach, P. (Ed.). (1964a). Biological control of insect pests and weeds. London: Chapman & Hall.

DeBach, P. (Ed.). (1964b). Successes, trends and future possibilities. In P. DeBach (Ed.), Biological control of insect pests and weeds (pp. 673–713). London: Chapman & Hall.

DeBach, P. (Ed.). (1964c). The scope of biological control. In P. DeBach (Ed.), Biological control of insect pests and weeds (pp. 3–20). London: Chapman & Hall.

DeBach, P., & Hagen, K. S. (1964). Manipulation of entomophagous species. In P. DeBach (Ed.), Biological control of insect pests and weeds (pp. 429–450). London: Chapman & Hall.

DeLeon, D. (1935). The biology of Coeloides dendroctoni Cushman (Hymenoptera: Braconidae) an important parasite of the mountain pine beetle (Dendroctonus monticolae Hopk.). Annals of the Entomological Society of America, 28, 411–424.

Denton, R. E. (1958). The larch casebearer in Idaho—a new defoliator for western forests (Research Note 51). Washington, DC: U.S. Department of Agriculture, Forest Service, Intermountain Forestry Range Experiment Station.

Doane, C. C., & McManus, M. L. (1981). The gypsy moth: Research toward integrated pest management (Technical Bulletin 1584). Washington, DC: USDA Forestry Service.

Dougherty, E. M., Guthrie, K. P., & Shapiro, M. (1996). Optical brighteners provide baculovirus activity enhancement and UV radiation protection. Biological Control, 7, 71–74.

Dowden, P. B. (1962). Parasites and predators of forest insects liberated in the United States through 1960. Agriculture Handbook No. 226. Washington, DC: U.S. Department of Agriculture Forest Service.

Dreistadt, S. H., Dahlsten, D. L., Rowney, D. L., Tait, S. M., Yokota, G. Y., & Copper, W. A. (1991). Treatment of destructive elm leaf beetle should be timed by temperature. California Agriculture, 45 (2), 23–25.

Drooz, A. T., Bustillo, A. E., Fedde, G. F., & Fedde, V. H. (1977). North American egg parasite successfully controls a different host genus in South America. Science, 197, 390–391.

Drooz, A. T., Quimby, J. W., Thompson, L. C., & Kulman, H. M. (1985). Introduction and establishment of Olesicampe benefactor Hinz (Hymenoptera: Ichneumonidae), a parasite of the larch sawfly, Pristiphora erichsonii (Hartig) (Hymenoptera: Tenthredinidae), in Pennsylvania. Environmental Entomology, 14, 420–423.

Dubois, N. R. (1985a). Recent field studies on the use of Bacillus thuringiensis to control the gypsy moth (Lymantria dispar L.). Proceedings of Symposium: Microbial control of spruce budworms and gypsy moths (GTR-NE-100, pp. 83–85). Washington, DC: U.S. Department of Agriculture Forestry Service.

Dubois, N. R. (1985b). Selection of new more potent strains of Bacillus thuringiensis for use against gypsy moth and spruce budworm. Proceedings of Symposium: Microbial control of spruce budworms and gypsy moths (GTR-NE-100, pp. 99–102). Washington, DC: U.S. Department of Agriculture Forestry Service.

Dubois, N. R., Reardon, R. C., & Kolodny-Hirsch, D. M., (1988). Field efficacy of the NRD-12 strain of Bacillus thuringiensis against gypsy moth (Lepidoptera: Lymantriidae). Journal of Economic Entomology, 81, 1672–1677.

Dubois, N. R., Reardon, R. C., & Mierzejewski, K. (1993). Field efficacy and deposit analysis of Bacillus thuringiensis, Foray 48B, against gypsy moth (Lepidoptera: Lymantriidae). Journal of Economic Entomology, 86, 26–33.

Elkinton, J. S., Gould, J. R., Ferguson, C. S., Liebhold, A. M., & Wallner, W. E. (1990). Experimental manipulation of gypsy moth density to assess impact of natural enemies. In A. D. Watt, S. R. Leather, M. D. Hunter, & N. A. C. Kidd (Eds.), Population dynamics of forest insects (pp. 275–287). Andover, United Kingdom: Intercept.

Embree, D. G. (1965). The population dynamics of the winter moth in Nova Scotia, 1954–62. Memoirs of the Entomological Society of Canada, 46, 57.

Embree, D. G. (1966). The role of introduced parasites in the control of the winter moth in Nova Scotia. Canadian Entomologist, 98, 1159–1168.

Embree, D. G. (1971). The biological control of winter moth in Canada by introduced parasites. In C. B. Huffaker (Ed.), Biological control (pp. 217–226). New York: Plenum Press.

Embree, D. G. (1991). The winter moth Operophtera brumata in eastern Canada, 1962–1988. Forest Ecology Magement, 39, 47–54.

Embree, D. G., & Otvos, I. S. (1984). Operophtera brumata (L.) winter moth (Lepidoptera: Geometridae). In J. S. Kelleher & M. A. Hulme (Eds.), Biological control programmes against insects and weeds in Canada, 1969–80 (pp. 353–357). Slough, United Kingdom: Commonwealth Agriculture Bureau.

Embree, D. G., & Pendrel, B. A. (1986). Biological control revisited. Bulletin of the Entomological Society of Canada, 18, 24–28.

Embree, D. G., & Underwood, G. R. (1972). Establishment in Maine, Nova Scotia and New Brunswick of Olesicampe benefactor (Hymenoptera: Ichneumonidae), an introduced ichneumonid parasite of the larch sawfly, Pristiphora erichsonii (Hymenoptera: Tenthredinidae). Canadian Entomologist, 104, 89–96.

Entwistle, P. F. (1976). The development of an epizootic of a nuclear polyhedrosis virus disease in European spruce sawfly Gilpinia hercyinae (pp. 184–188). Proceedings, First International Colloquium on Invertebrate Pathology, Kingston, Canada.

Entwistle, P. F. (1983). Control of insects by virus diseases. Biocontrol News Information, 4, 203–225.

Evans, H. F. (1985). Great spruce bark beetle, Dendroctonus micans: An exotic pest new to Britain. Antenna, 9, 117–121.

Evans, H. F., & Fielding, N. J. (1994). Integrated management of Dendroctonus micans in Great Britain. Forest Ecology and Management, 65, 17–30.

Evans, H. F., & Fielding, N. J. (1996). Restoring the balance: Biological control of Dendroctonus micans in Great Britain. In Biological control introductions—opportunities for improved crop protection. Proceedings, British Crop Protection Council, 68, 47–57.

Evans, H. F., & King, C. J. (1989). Biological control of Dendroctonus micans (Coleoptera: Scolytidae): British experience of rearing and release of Rhizophagus grandis (Coleoptera: Rhizophagidae). In D. L. Kulhavy & M. C. Miller (Eds.), Potential for biological control of Dendroctonus and Ips bark beetles (pp. 109–128). Nacogdoches, TX: Center for Applied Studies, School of Forestry, Stephen F. Austin State University.

Everaerts, C., Gregoire, J.-C., & Merlin, J. (1988). The toxicity of Norway spruce monoterpenes to two bark beetle species and their associates. In W. J. Mattson, J. Levieux, & C. Becnard-Dagen (Eds.), Mechanisms of woody plant defenses against insects (pp. 335–344). New York: Springer-Verlag.

Fielding, N. J. (1992). Rhizophagus grandis as a means of biological control against Dendroctonus micans in Britain (U.K. Research Information Note–Forestry Commission Research Division No. 224). London, H.M.S.O.

Finlayson, L. R., & Finlayson, T. (1958). Notes on parasites of Diprionidae in Europe and Japan and their establishment in Canada on Diprion hercyinae (Htg.) (Hymenoptera: Diprionidae). Canadian Entomologist, 90, 557–563.

Finnegan, R. J. (1975). Introduction of a predaceous red wood ant, Formica lugubris (Hymenoptera: Formicidae) from Italy to eastern Canada. Canadian Entomologist, 107, 1271–1274.

Finnegan, R. J. (1977). Establishment of a predaceous red wood ant, Formica obscuripes (Hymenoptera: Formicidae), from Manitoba to eastern Canada. Canadian Entomologist, 109, 1145–1148.

Finnegan, R. J., & Smirnoff, W. A. (1981). Neodiprion swainei (Middleton), Swaine jack pine sawfly (Hymenoptera: Diprionidae). In J. S. Kelleher & M. A. Hulme (Eds.), Biological control programmes against insects and weeds in Canada, 1969–1980 (pp. 341–348). London: Commonwealth Agriculture Bureau.

Fleming, R. A., & van Frankenhuyzen, K. (1992). Forecasting the efficacy of operational Bacillus thuringiensis Berliner applications against spruce budworm, Choristoneura fumiferana Clemens (Lepidoptera:

Tortricidae), using dose ingestion data: Initial models. Canadian Entomologist, 124, 1101–1113.

Forest Insect Pest Management Association (1992). The biological control of Ips grandicollis (Eichhoff) in Australia. Adelaide, South Australia: South Australia Woods and Forests Department.

Goeden, R. D., & Kok, L. T. (1986). Comments on a proposed "new" approach for selecting agents for the biological control of weeds. Canadian Entomologist, 118, 51–58.

Gould, J. R., Elkinton, J. S., & Wallner, W. E. (1990). Density-dependent suppression of experimentally created gypsy moth, Lymantria dispar (Lepidoptera: Lymantriidae) populations by natural enemies. Journal of Animal Ecology, 59, 213–234.

Greathead, C. J. (1980). Biological control of pests and the contributions of CIBC. Antenna, 4, 88–91.

Greathead, D. J., & Greathead, A. (1992). Biological control of insect pests by insect parasitoids and predators: The BIOCAT database. Biocontrol News Information, 13, 61N–68N.

Gregoire, J.-C. (1976). Note sur deux ennemis naturels de Dendroctonus micans Kug. en Belgique. Bulletin Annales de la Societe Royale Belged Entomologie, 112, 208–212.

Gregoire, J.-C. (1985). Host colonization strategies in Dendroctinus: Larval gregariousness vs. mass attack by adults. In L. Safranyik (Ed.), Proceedings, Meeting of IUFRO Working Parties S2.07-05 and 06. Banff, Canada, September 1983 (pp.147–154). Canadian Forestry Service, Victoria, B.C.

Gregoire, J.-C., Baisier, M., Drumont, A., Dahlsten, D. L., Meyer, H., & Francke, W. (1991). Volatile compounds in the larval frass of Dendroctonus valens and Dendroctonus micans (Coleoptera: Scolytidae) in relation to oviposition in the predator, Rhizophagus grandis (Coleoptera: Rhizophagidae). Journal Chemical Ecology, 17(10), 2003–2019.

Gregoire, J.-C., Baisier, M., Merlin, J., & Naccache, Y. (1989). Interactions between Rhizophagus grandis C: Rhizophagidae) and Dendroctonus micans (Coleoptera: Scolytidae) in the field and the laboratory: Their application for the biological control of D. micans in France. In D. L. Kulhavy & M. C. Miller (Eds.), Potential for biological control of Dendroctonus and Ips bark beetles (pp. 95–108). Nacogdoches, TX: Center for Applied Studies, School of Forestry, Stephen F. Austin State University.

Gregoire, J.-C., Merlin, J., Pasteels, J. M., Jaffuel, R., Vouland, G., & Schvester, D. (1985). Biocontrol of Dendroctonus micans by Rhizophagus grandis Gyll. (Col., Rhizophagidae) in the Massif Central (France). Zeitschriftfuer Angewandte Entomologie, 99, 182–190.

Hagen, K. S., van den Bosch, R., & Dahlsten, D. L. (1971). The importance of naturally-occurring biological control in the western United States. In C. B. Huffaker (Ed.), Biological control (pp. 253–293). New York: Plenum Press.

Hajek, A. E. (1997). Fungal and viral epizootics in gypsy moth (Lepidoptera: Lymantriidae) populations in central New York. Biological Control, 10, 58–68.

Hajek, A. E., Humber, R. A., Elkinton, J. S., May, B., Walsh, S. R. A., & Silver, J. C. (1990). Allozyme and restriction fragment length polymorphism confirm Entomophaga maimaiga responsible for 1989 epizootics in North American gypsy moth populations. Proceedings of the National Academy of Sciences USA, 87, 6979–6982.

Hall, R. W., & Ehler, L. E. (1979). Rate of establishment of natural enemies in classical biological control. Bulletin of the Entomological Society of America, 25, 280–282.

Hall, R. W., Ehler, L. E., & Bisabri-Ershadi, B. (1980). Rate of success in classical biological control of arthropods. Bulletin of the Entomological Society of America, 26, 111–114.

Hassell, M. P. (1980). Foraging strategies, population models and biological control: A case study. Journal of Animal Ecology, 49, 603–628.

Haynes, K. A., Birch, M. C. (1985). The role of other pheromones, allomones and kairomones in the behavioral responses of insects. In G. A. Kerkut & L. T. Gilbert (Eds.), Comprehensive insect physiology, biochemistry and pharmacology (Vol. 9, pp. 225–255). Oxford: Pergamon Press.

Hokkanen, H., & Pimentel, D. (1984). New approach for selecting biological control agents. Canadian Entomologist, 116, 1109–1121.

Hokkanen, H., & Pimental, D. (1989). New associations in biological control: Theory and practice. Canadian Entomologist, 121, 829–840.

Howard, L. O., & Fiske, W. F. (1911). The importation into the United States of the parasites of the gypsy moth and the brown-tail moth. U. S. DA, Bureau of the Entomological Bulletin, 91, 1–312.

Huffaker, C. B. (Ed.). (1971). Biological control. New York: Plenum Press.

Huffaker, C. B., & Messenger, P. S. (Eds.). (1976). Theory and practice of biological control. New York: Academic Press.

Huffaker, C. B., Messenger, P. S., & DeBach, P. (1971). The natural enemy component in natural control and the theory of biological control. In C. B. Huffaker (Ed.), Biological control (pp. 16–67). New York: Plenum Press.

Hulme, M. A. (1988). The recent Canadian record in applied biological control of forest pests. Forestry Chronicle, 64, 27–31.

Hulme, M. A., Dawson, A. F., & Harris, J. W. E. (1986). Exploiting cold-hardiness to separate Pissodes strobi (Peck) (Coleoptera: Curculionidae) from associated insects in the leaders of Picea sitchensis (Bong.) Carr. Canadian Entomologist, 118, 1115–1122.

Hulme, M. A., Harris, J. W. E., & Dawson, A. F. (1987). Exploiting adult girth to separate Pissodes strobi (Peck) (Coleoptera: Curculionidae) from associated insects in the leaders of Picea sitchensis (Bong.) Carr. Canadian Entomologist, 119, 751–753.

Ives, W. G. H. (1976). The dynamics of larch sawfly (Hymenoptera: Tenthredinidae) populations in southeastern Manitoba. Canadian Entomologist, 108, 701–730.

Ives, W. G. H., & Muldrew, J. A. (1981). Pristiphora erichsonii (Hartig), larch sawfly (Hymenoptera: Tenthredinidae). In J. S. Kelleher & M. Hulme (Eds.), Biological control programmes against insects and weeds in Canada 1969–1980 (pp. 369–380). London: Commonwealth Agriculture Bureau.

Jagsch, A. (1973). Populationsdynamik and Parasitenkomplex der Larchenminiermotte, Coleophora laricella Hbn., in naturlichen Verbreitungsgebeiet der Europaischen Larche, Larix decidua Mill. Zeitschrift fuer Angewandte Entomologie, 73, 1–42.

Jeffords, M. R., Maddox, J. V., McManus, M. L., Webb, R. E., & Wieber, A. (1988). Egg contamination as a method for the inoculative release of exotic microsporidia of the gypsy moth. Journal of Invertebrate Pathology, 51, 190–196.

Jeffords, M. R., Maddox, J. V., McManus, M. L., Webb, R. E., & Wieber, A. (1989). Evaluation of the overwintering success of two European microsporidia inoculatively released into gypsy moth populations in Maryland (USA). Journal of Invertebrate Pathology, 53, 235–240.

Johnson, P. C. (1958). Spruce spider mite infestations in northern Rocky Mountain Douglas-fir forests (Intermountain Forestry Range Experiment Station Research Paper, 55). Washington: DC: USDA, Forestry Service.

Kimberling, D. N., Miller, J. C., & Penrose, R. L. (1986). Distribution and parasitism of winter moth, Operophthora brumata (Lepidoptera: Geometridae), in western Oregon. Environmental Entomology, 15, 1042–1046.

Kobakhidze, D. N. (1965). Some results and prospects of the utilization of beneficial entomophagous insects in the control of insect pests in Georgian SSR (USSR). Entomophaga, 10, 323–330.

Kobakhidze, D. N., Tvaradze, M. S., Yashvili, G. V., & Kraveishvili, I. K. (1968). Artificial rearing of Rhizophagus grandis Gyll. for the control of Dendroctonus micans in Georgia. Soobschenie Akademie Nauk Gruzinskoi SSR, 51, 435–440 (in Russian).

Koehler, W. (1970). The theoretical background of the "complex method." Tagungsber. Deutsche Akademie Landwirtschaftwiss Berlin, 110, 31–35 (in German).

Kolodny-Hirsch, D. M., Reardon, R. C., Thorpe, K. W., & Raupp, M. J. (1988). Evaluating the impact of sequential releases of *Cotesia melanoscela* (Hymenoptera: Braconidae) on *Lymantria dispar* (Lepidoptera: Lymantriidae). Environmental Entomology, 17, 403–408.

Kulman, H. M., Thompson, L. C., & Witter, J. A. (1974). Introduction of parasitoids of the larch sawfly in Minnesota. Great Lakes Entomologist, 7, 23–25.

Lacey, L. A., Goettel, M. S. (1995). Current developments in microbial control of insect pests and prospects for the early 21st century. Entomophaga, 40, 3–27.

Laing, J. E., & Hamai, J. (1976). Biological control of insect pests and weeds by imported parasites, predators and pathogens. In C. B. Huffaker & P. S. Messenger (Eds.), Theory and practice of biological control (pp. 685–743). New York: Academic Press.

Leonard, D. E. (1974). Recent developments in ecology and control of the gypsy moth. Annual Review of Entomology, 19, 197–229.

Liebhold, A. M., & Elkinton, J. S. (1989). Elevated parasitism in artificially augmented populations of *Lymantria dispar* (Lepidoptera: Lymantriidae). Environmental Entomology, 18, 986–995.

Luck, R. F., & Dahlsten, D. L. (1975). Natural decline of a pine needle scale (*Chionaspis pinifoliae* [Fitch]) outbreak at South Lake Tahoe, California following cessation of adult mosquito control with malathion. Ecology, 56, 893–904.

Maddox, J. V., McManus, M. L., Jeffords, M. R., & Webb, R. E. (1992). Exotic insect pathogens as classical biological control agents with an emphasis on regulatory consideration. In W. C. Kauffman & J. R. Nechols (Eds.), Selection criteria and ecological consequences of importing natural enemies (pp. 27–39). Lanham, MD: Thomas Say Publications in Entomology/Entomological Society of America.

Magasi, L. P., & Syme, P. D. (1981). *Gilpinia hercyinae* (Hartig), European spruce sawfly (Hymenoptera: Diprionidae). In J. S. Kelleher & M. A. Hulme (Eds.), Biological control programmes against insects and weeds in Canada 1969–1980 (pp. 295–297). London: Commonwealth Agriculture Bureau.

Mathews, P. L., & Stephen, F. M. (1997). Effect of artificial diet on longevity of adult parasitoids of *Dendroctonus frontalis* (Coleoptera: Scolytidae). Environmental Entomology, 26, 961–965.

McFadden, M. W., Dahlsten, D. L., Berisford, C. W., Knight, F. B., & Metterhouse, W. W. (1981). Integrated pest management in China's forests. Journal of Forestry, 79, 723–726, 799.

McGugan, B. M., & Coppel, H. C. (1962). A review of the biological control attempts against insects and weeds in Canada. II. Biological control of forest insects, 1910–1958 (Technical Communication No. 2, pp. 35–216). Commonwealth Institute of Biological Control. Commonwealth Agricultural Bureaux, Farnham Royal, England.

McLeod, J. M., & Smirnoff, W. A. (1971). *Neodiprion swainei* Midd., Swaine jack pine sawfly (Hymenoptera: Diprionidae). In Biological control programmes against insects and weeds in Canada, 1959–1968 (Technical Communication No. 4, chap. 43, pp. 162–167). Commonwealth Institute of Biological Control. Commonwealth Agricultural Bureaux, Farnham Royal, England.

McNeil, J. N., Delisle, J., & Finnegan, R. J. (1978). Seasonal predatory activity of the introduced red wood ant, *Formica lugubris* (Hymenoptera: Formicidae) at Valcartier, Quebec in 1976. Canadian Entomologist, 110, 85–90.

Mesnil, L. P. (1967). History of a success in biological control: The winter moth project in Canada. Technical Bulletin of the Commonwealth Institute of Biological Control, 8, 1–6.

Miller, J. C. (1990). Field assessment of the effects of a microbial pest control agent on nontarget Lepidoptera. American Entomologist, 36, 135–139.

Miller, M. C., Moser, J. C., McGregor, M., Gregoire, J.-C., Baisier, M., & Dahlsten, D. L. (1987). Potential for the biological control of native North American *Dendroctonus* beetles (Coleoptera: Scolytidae). Annals of Entomological Society of America, 80, 417–428.

Mills, N. J. (1983a). Possibilities for the biological control of *Choristoneura fumiferana* (Clemens) using natural enemies from Europe. Biocontrol News Information, 4, 103–125.

Mills, N. J. (1983b). The natural enemies of scolytids infesting conifer bark in Europe in relation to the biological control of *Dendroctonus* spp. in Canada. Biocontrol News Information, 4, 305–328.

Mills, N. J. (1990). Are parasitoids of significance in endemic populations of forest defoliators? Some experimental observations from gypsy moth, *Lymantria dispar* (Lepidoptera: Lymantriidae). In A. D. Watt, S. R. Leather, M. D. Hunter, & N. A. C. Kidd (Eds.), Population dynamics of forest insects (pp. 265–274). Andover, United Kingdom: Intercept.

Mills, N. J., & Nealis, V. G. (1992). European field collections and Canadian releases of *Ceranthia samarensis* (Dipt.: Tachinidae), a parasitoid of the gypsy moth. Entomologa, 37, 181–191.

Mills, N. J., & Schoenberg, F. (1985). Possibilities for the biological control of the Douglas-fir tussock moth, *Orgyia pseudotsugata* (Lymantriidae), in Canada, using natural enemies from Europe. Biocontrol News Information, 6, 7–18.

Mills, N. J., Fischer, P., & Glanz, W.-D. (1986). Host exposure: A technique for the study of gypsy moth larval parasitoids under non-outbreak conditions. Proceedings of the 18th IUFRO World Congress, Division 2, 2, 777–785.

Moeck, H., & Safranyik, L. (1984). Assessment of predator and parasitoid control of bark beetles (Information Report BC-X-248). Canadian Forestry Service, Pacific Forest Research Center. Victoria, B.C.

Molnar, A. C., Harris, J. W. E., & Ross, D. A. (1967). British Columbia Region. Annual Report Forest Insect and Disease Survey—1966 (pp. 108–123). Canadian Forestry Service, Victoria, B.C.

Moore, G. E. (1972). Southern pine beetle mortality in North Carolina caused by parasites and predators. Environmental Entomology, 1, 58–65.

Morris, K. R. S., Cameron, E., & Jepson, W. F. (1937). The insect parasites of the spruce sawfly (*Diprion polytomum* Htg.) in Europe. Bulletin of Entomological Research, 28, 341–393.

Morris, O. N. (1982). Bacteria as pesticides: Forest applications. In E. Kurstak (Ed.), Microbial and viral pesticides (pp. 239–287). New York: Marcel Dekker.

Morris, O. N., Dimond, J. B., & Lewis. F. B. (1984). Guidelines for the operational use of *Bacillus thuringiensis* against the spruce budworm. USDA Handbook No. 621. Washington, DC: U.S. Department of Agriculture.

Morris, O. N., Cunningham, J. C., Finney-Crawley, J. R., Jaynes, R. P., & Kinoshita, A. (1986). Microbial insecticides in Canada: Their registration and use in agriculture, forestry and public and animal health. Bulletin of Entomological Society of Canada, (Suppl. 2), 1–43.

Muldrew, J. A. (1953). The natural immunity of the larch sawfly (*Pristiphora erichsonii* [Htg.]) to the introduced parasite (*Mesoleius tenthredinis* Morley), in Manitoba and Saskatchewan. Canadian Journal Zoology, 31, 313–332.

Murdoch, W. W., Chesson, J., & Chesson, P. L. (1985). Biological control in theory and practice. American Naturalist, 125, 344–366.

Nealis, V. G. (1991). Natural enemies and forest pest management. Forestry Chronicle, 67, 500–505.

Nealis, V. G., & van Frankenhuyzen, K. (1990). Interactions between *Bacillus thuringiensis* Berliner and *Apanteles fumiferanae* Vier. (Hymenoptera: Braconidae), a parasitoid of the spruce budworm, *Choristoneura fumiferana* Clem. (Lepidoptera: Tortricidae). Canadian Entomologist, 122, 585–594.

Nealis, V. G., & Quednau, F. W. (1996). Canadian field releases and overwintering survival of *Ceranthia samarensis* (Villeneuve) (Diptera: Tachinidae) for biological control of the gypsy moth, *Lymantria dispar* (L.) (Lepidoptera: Lymantriidae). Proceedings of the Entomological Society of Ontario, 127, 11–20.

Nealis, V. G., van Frankenhuyzen, K., & Cadogan, B. L. (1992). Conservation of spruce budworm parasitoids following application of *Bacillus thuringiensis* var. *kurstaki* Berliner. Canadian Entomologist, 124, 1085–1092.

Neilson, M. M., Martineau, R., & Rose, A. M. (1971). *Diprion hercyniae* (Hartig), European spruce sawfly (Hymenoptera: Diprionidae). In Biological control programmes against insects and weeds in Canada, 1959–1968 (Technical Communication No. 4, pp. 136–143). Commonwealth Institute of Biological Control. Commonwealth Agricultural Bureaux, Farnham Royal, England.

Otto, D. (1967). The importance of *Formica* colonies in the reduction of important pest insects. A literature review. Waldhygiene, 7, 65–90 (in German).

Otvos, I. S., & Quednau, F. W. (1981). *Coleophorea laricella* (Hübner), larch casebearer (Lepidoptera: Coleophoridae). In J. S. Kelleher & M. A. Hulme (Eds.), Biological control programmes against insects and weeds in Canada, 1969–1980 (pp. 281–284). London: Commonwealth Agriculture Bureau.

Pearsall, I. A., & Walde, S. J. (1994). Parasitism and predation as agents of mortality of winter moth populations in neglected apple orchards in Nova Scotia. Ecological Entomology, 19, 190–198.

Pemberton, R. W., Lee, J. H., Reed, D. K., Carlson, R. W., & Han, H. Y. (1993). Natural enemies of the Asian gypsy moth (Lepidoptera: Lymantriidae) in South Korea. Annals of the Entomological Society of America, 86, 423–440.

Pimentel, D. (1961). Animal population regulation by the genetic feedback mechanism. American Naturalist, 95, 65–79.

Pimentel, D. (1963). Introducing parasites and predators to control native pests. Canadian Entomologist, 95, 785–792.

Podgwaite, J. D. (1985). Gypchek: Past and future strategies for use. Proceedings of the Symposium: Microbial control of spruce budworms and gypsy moths (GTR-NE-100, pp. 91–93). Washington, DC: USDA Forestry Service.

Podgwaite, J. D., Reardon, R. C., Kolodny-Hirsch, D. M., & Walton, G. S. (1991). Efficacy of ground application of the gypsy moth (Lepidoptera: Lymantriidae) nucleopolyhedrosis virus product, Gypchek. Journal of Economic Entomoloy, 84, 440–444.

Podgwaite, J. D., Reardon, R. C., Walton, G. S., Venables, L., & Kolodny-Hirsch, D. M. (1992). Effects of aerially applied Gypchek on gypsy moth (Lepidoptera: Lymantriidae) populations in Maryland woodlots. Journal of Economic Entomology, 85, 1136–1139.

Pschorn-Walcher, H. (1977). Biological control of forest insects. Annual Review of Entomology, 22, 1–22.

Quednau, F. W. (1970). Competition and cooperation between *Chrysocharis laricinellae* and *Agathis pumila* on larch casebearer in Quebec. Canadian Entomologist, 102, 602–612.

Quednau, F. W. (1993). Reproductive biology and laboratory rearing of *Ceranthia samarensis* (Villeneuve) (Diptera: Tachinidae), a parasitoid of gypsy moth, *Lymantria dispar* (L.). Canadian Entomologist, 125, 749–759.

Rabb, R. L., Stinner, R. E., & van den Bosch, R. (1976). Conservation and augmentation of natural enemies. In C. B. Huffaker & P. S. Messenger (Eds.), Theory and practice of biological control (pp. 233–254). New York: Academic Press.

Ridgway, R. L., King, E. G., & Carillo, J. L. (1977). Augmentation of natural enemies for control of plant pests in the western hemisphere. In R. L. Ridgway & S. B. Vinson (Eds.), Biological control by augmentation of natural enemies: Insect and mite control with parasites and predators (pp. 379–416). New York: Plenum Press.

Roland, J. (1988). Decline in winter moth populations in North America: Direct versus indirect effect on introduced parasites. Journal of Animal Ecology, 57, 523–531.

Roland, J. (1990). Interaction of parasitism and predation in the decline of winter moth in Canada. In A. D. Watt, S. R. Leather, M. D. Hunter, & N. A. C. Kidd (Eds.), Population dynamics of forest insects (pp. 289–302). Andover, United Kingdom: Intercept.

Roland, J. (1994). After the decline: What maintains low winter moth

density after successful biological control? Journal of Animal Ecology, 63, 392–398.

Roland, J., & Embree, D. G. (1995). Biological control of the winter moth. Annual Review of Entomology, 40, 475–492.

Roland, J., Denford, K. E., & Jimenez, L. (1995). Borneol as an attractant for *Cyzenis albicans*, a tachinid parasitoid of the winter moth, *Operophtera brumata* L. (Lepidoptera: Geometridae). Canadian Entomologist, 127, 413–421.

Rollinson, W. D., Hubbard, H. B., & Lewis, F. B. (1970). Mass rearing of the European pine sawfly for production of the nuclear polyhedrosis virus. Journal of Economic Entomology, 63, 343–344.

Ryan, R. B. (1980). Rearing methods and biological notes for seven species of European and Japanese parasites of the larch casebearer (Lepidoptera: Coleophoridae). Canadian Entomologist, 112, 1239–1248.

Ryan, R. B. (1983). Population density and dynamics of larch casebearer (Lepidoptera: Coleophoridae) in the Blue Mountains of Oregon and Washington before the build-up of exotic parasites. Canadian Entomologist, 115, 1095–1102.

Ryan, R. B. (1985). A hypothesis for decreasing parasitization of larch casebearer (Lepidoptera: Coleophoridae) on larch foliage by *Agathis pumila*. Canadian Entomologist, 117, 1573–1574.

Ryan, R. B. (1986). Analysis of life tables for the larch casebearer (Lepidoptera: Coleophoreidae) in Oregon. Canadian Entomologist, 118, 1255–1263.

Ryan, R. B. (1990). Evaluation of biological control: Introduced parasites of larch casebearer (Lepidoptera: Coleophoridae) in Oregon. Environmental Entomology, 19, 1873–1881.

Ryan, R. B. (1997). Before and after evaluation of biological control of the larch casebearer (Lepidoptera: Coleophoridae) in the Blue Mountains of Oregon and Washington, 1972–1995. Environmental Entomology, 26, 703–715.

Ryan, R. B., & Rudinsky, J. A. (1962). Biology and habits of the Douglas-fir beetle parasite, *Coeloides brunneri* Viereck in western Oregon. Canadian Entomologist, 94, 748–763.

Ryan, R. B., Tunnock, S., & Ebel, F. W. (1987). The larch casebearer in North America. Journal of Forestry, 85 (7), 33–39.

Sailer, R. I. (1971). Invertebrate predators. USDA, Forestry Service, Research Paper NE, 194, 32–44.

Schafer, P. W., Fuester, R. W., Chianese, R. J., Rhoads, L. D., & Tichenor, R. B. (1989). Introduction and North American establishment of *Coccygomimus disparis* (Hymenoptera: Ichneumonidae), a polyphagous pupal parasite of Lepidoptera, including gypsy moth. Environmental Entomology, 18, 1117–1125.

Schroeder, D. (1974). A study of the interactions between the internal larval parasites of *Rhyacionia buoliana* (Lepidoptera: Olethreutidae). Entomophaga, 19, 145–171.

Shapiro, M., & Robertson, J. L. (1992). Enhancement of gypsy moth (Lepidoptera: Lymantriidae) baculovirus activity by optical brighteners. Journal of Economic Entomology, 85, 1120–1124.

Smirnoff, W. A. (1962). Transovum transmission of virus of *Neodiprion swainei* Middleton (Hymenoptera: Tentheridinidae). Journal of Insect Pathology, 4, 192–200.

Smith, S. M. (1993). Insect parasitoids: A Canadian perspective on their use for biological control of forest insect pests. Phytoprotection, 74, 51–67.

Smith, S. M., & You, M. (1990). A life system simulation model for improving inundative releases of the egg parasite, *Trichogramma minutum* against the spruce budworm. Ecological Modeling, 51, 123–142.

Smith, S. M., Carrow, J. R., & Laing, J. E. (1990a). Inundative release of the egg parasitoid, *Trichogramma minutum* (Hymenoptera: Trichogrammatidae), against forest insect pests such as the spruce budworm, *Choristoneura fumiferana* (Lepidoptera: Tortricidae): The Ontario project 1982–1986. Memoirs of the Entomological Society of Canada, 153, 1–87.

Smith, S. M., Wallace, D. R., House, G., & Meating, J. (1990b). Suppression of spruce budworm populations by *Trichogramma minutum* Riley,

1982–1986. Memoirs of the Entomological Society of Canada, 153, 56–81.

Spradbery, J. P., & Kirk, A. A. (1978). Aspects of the ecology of siricid woodwasps (Hymenoptera: Siricidae) in Europe, North Africa and Turkey with special reference to the biological control of Sirex noctilio F. in Australia. Bulletin of Entomological Research, 68, 341–359.

Stern, V. M., Adkisson, P. L., Beingolea, O. G., & Viktorov, G. A. (1976). Cultural controls. In C. B. Huffaker & P. S. Messenger (Eds.), Theory and practice of biological control (pp. 593–613). New York: Academic Press.

Swezey, S. L., & Dahlsten, D. L. (1983). Effects of remedial application of lindane on emergence of natural enemies of the western pine beetle, Dendroctonus brevicomis (Coleoptera: Scolytidae). Environmental Entomology, 12, 210–214.

Syme, P. D. (1971). Rhyacionia buoliana (Schiff.), European pine shoot moth (Lepidoptera: Olethreutidae). Commonwealth Institute of Biological Control Technical Communication, 4, 194–205.

Syme, P. D. (1977). Observations on the longevity and fecundity of Orgilus obscurator (Hymenoptera: Braconidae) and the effects of certain foods on longevity. Canadian Entomologist, 109, 995–1000.

Syme, P. D. (1981). Rhyacionia buoliana (Schiff.), European pine shoot moth (Lepidoptera: Tortricidae). In J. S. Kelleher & M. A. Hulme (Eds.), Biological control programmes against insects and weeds in Canada, 1969–1980 (pp. 387–394). London: Commonwealth Agriculture Bureau.

Tallamy, D. W. (1983). Equilibrium biogeography and its application to insect host-parasite systems. American Naturalist, 121, 244–254.

Taylor, K. L. (1976). The introduction and establishment of insect parasitoids to control Sirex noctilio in Australia. Entomophaga, 21, 429–440.

Taylor, K. L. (1981). The Sirex woodwasp: Ecology and control of an introduced forest insect, In R. L. Kitching & R. E. Jones (Eds.), The ecology of pests: Some Australian case histories (pp. 231–248). Melbourne, Australia: CSIRO.

Tumlinson, J. H., Turlings, T. C. J., & Lewis, W. J. (1993). Semiochemically mediated foraging behavior in beneficial parasitic insects. Archives of Insect Biochemistry and Physiology, 22, 385–391.

Turnock, W. J., & Muldrew, J. A. (1971). Pristiphora erichsonii (Hartig), larch sawfly (Hymenoptera: Tenthredinidae). In Biological control programmes against insects and weeds in Canada, 1959–1968. (Technical Communication No. 4, pp. 175–194). Commonwealth Institute of Biological Control. Commonwealth Agricultural Bureaux, Farnham Royal, England.

Turnock, W. J., Taylor, K. L., Schroder, D., & Dahlsten, D. L. (1976). Biological control of pests of coniferous forests. In C. B. Huffaker & P. S. Messenger (Eds.), Theory and practice of biological control (pp. 289–311). New York: Academic Press.

van den Bosch, R., & Telford, A. D. (1964). Environmental modification and biological control. In P. DeBach, (Ed.), Biological control of insect pests and weeds (pp. 459–488). London: Chapman & Hall.

van Frankenhuyzen, K. (1990). Development and current status of Bacillus thuringiensis for control of defoliating forest insects. Forestry Chronicle, 66, 498–507.

Varley, A. C., & Gradwell, G. R. (1968). Population models for the winter moth. Symposium of the Royal Entomological Society of London, 4, 132–142.

Versoi, P. L., & Yendol, W. G. (1982). Discrimination by the parasite, Apanteles melanoscelus, between healthy and virus-infected gypsy moth larvae. Environmental Entomology, 11, 42–45.

Vet, L. E. M., & Dicke, M. (1992). Ecology of infochemical use by natural enemies in a tritrophic context. Annual Review of Entomology, 37, 141–172.

Vinson, S. B. (1984). How parasitoids locate their hosts: A case of insect espionage. In T. Lewis (Ed.), Insect communication (pp. 325–348). London: Academic Press.

Waage, J. K. (1990). Ecological theory and the selection of biological control agents, In M. Mackauer, L. E. Ehler, & J. Roland (Eds.), Critical issues in biological control (pp. 135–157). Andover, United Kingdom: Intercept.

Washburn, J. O. (1984). The gypsy moth and its parasites in North America: A A community in equilibrium. American Naturalist, 124, 288–292.

Webb, F. E., & Quednau, F. W. (1971). Coleophora laricella (Hübner), larch casebearer (Lepidoptera: Coleophoridae). In Biological control programmes against insects and weeds in Canada, 1959–1968 (Technical Communication No. 4, pp. 131–136). CAB, Commonwealth Institute of Biological Control. Commonwealth Agricultural Bureaux, Farnham Royal, England.

Webb, R. E., Shapiro, M., Podgwaite, J. D., Ridgway, R. L., Venables, L., & White, G. B. (1994). Effect of optical brighteners on the efficacy of gypsy moth (Lepidoptera: Lymantriidae) nuclear polyhedrosis virus in forest plots with high or low levels of natural virus. Journal of Economic Entomology, 87, 134–143.

Weseloh, R. M. (1990a). Gypsy moth predators: An example of generalist and specialist natural enemies. In A. D. Watt, S. R. Leather, M. D. Hunter, & N. A. C. Kidd (Eds.), Population dynamics of forest insects (pp. 233–243). Andover, United Kingdom: Intercept.

Weseloh, R. M. (1990b). Experimental forest releases of Calosoma sycophanta (Coleoptera: Carabidae) against the gypsy moth. Journal of Economic Entomology, 83, 2229–2234.

Weseloh, R., & Anderson, J. (1975). Inundative release of Apanteles melanoscelus against the gypsy moth. Environmental Entomology, 4, 33–36.

Weseloh, R. M., Andreadis, T. G., Moore, R. E. B., Anderson, J. F., Dubois, N. R., & Lewis, F. B. (1983). Field confirmation of a mechanism causing synergism between Bacillus thuringiensis and the gypsy moth parasitoid, Apanteles melanoscelus. Journal of Invertebrate Pathology, 41, 99–103.

Weseloh, R. M., Bernon, G., Butler, L., Fuester, R., McCullough, D., & Stehr, F. (1995). Releases of Calosoma sycophanta (Coleoptera: Carabidae) near the edge of gypsy moth (Lepidoptera: Lymantriidae) distribution. Environmental Entomology, 24, 1713–1717.

Wieber, A. M., Webb, R. E., Ridgway, R. L., Thorpe, K. W., Reardon, R. C., & Kolodny-Hirsch, D. M. (1995). Effect of seasonal placement of Cotesia melanoscela (Hym.: Braconidae) on its potential for effective augmentative release against Lymantria dispar (Lep.: Lymantriidae). Entomophaga, 40, 281–292.

Williamson, D. L., & Vite, J. P. (1971). Impact of insecticidal control on the southern pine beetle population in east Texas. Journal of Economic Entomology, 64, 1440–1444.

Wollam, J. D., & Yendol, W. G. (1976). Evaluation of Bacillus thuringiensis and a parasitoid for suppression of the gypsy moth. Journal of Economic Entomology, 69, 113–118.

Wong, H. R. (1974). The identification and origin of the larch sawfly, Pristiphora erichsonii (Hymenoptera: Tenthredinidae) in North America. Canadian Entomologist, 106, 1121–1131.

Zondag, R. (1959). Progress report on the establishment in New Zealand of Ibalia leucospoides (Hochenw.) a parasite of Sirex noctilio (F.) (New Zealand Forestry Research Notes 20). Rotorua, N.Z.

Zondag, R. (1969). A nematode infection of Sirex noctilio (F.) in New Zealand. New Zealand Journal of Science, 12, 732–747.

30

Enhancement of Biological Control in Annual Agricultural Environments

MARCOS KOGAN
Integrated Plant Protection Center
Oregon State University
Corvallis, Oregon

DAN GERLING
Department of Zoology
Tel Aviv University
Ramat Aviv, Israel

JOSEPH V. MADDOX
Center of Agricultural Entomology
Illinois Natural History Survey
Champaign, Illinois

INTRODUCTION

Biological control has been most successful when applied to stable perennial agroecosystems (DeBach, 1994; Huffaker & Messenger, 1976; Luck, 1981; Price, 1981; Hokkanen, 1985; Van Driesche & Bellows, 1996). Annual crops, in general, are too unstable to sustain delicate multitrophic-level interactions without major herbivore population fluctuations. When the natural control of potential pest species has been upset by cultural operations and chemical pest control, adequate levels of biological control are difficult to restore. To be sure, many annual crop ecosystems benefit from a high level of natural control, particularly when an ecosystem has not been invaded by exotic pests that require the use of disruptive insecticides (Wilson, 1985). Turnipseed and Kogan (1983) suggested that indigenous natural enemies are important in the regulation of minor phytophagous pests, but it is their impact on the major pests that usually attracts the attention of investigators. When, however, minor pests become major because of imbalances caused by the misuse of insecticides, the door is open to disaster (Reynolds *et al.,* 1982). The explosion of the brown planthopper, *Nilaparvata lugans* (Stål), on rice in the Orient was one example of such a disaster (Teng, 1994). Then it is essential to identify and analyze the regulatory mechanisms of endemic minor pest species.

In this chapter we use a rather broad definition of annual crops because we include, among others, examples from alfalfa, sugarcane and cassava—crops that may have two- or three-year cycles. It might be better to consider the agroecosystems dealt with here as nonpermanent or tran-

sient, with transience ranging from a few weeks, as in some vegetable or truck crops, to a few years. Because these crops are subject to periodic disruption, restoration of natural control of upset herbivore populations is often difficult to achieve—hence the few examples of successful classical biological control of annual crop pests through the release of parasitoids or predators. On the other hand, these same characteristics make transient crops particularly suited for augmentative releases of natural enemies and the use of microbial pesticides or the adoption of methods that enhance the natural epizootics of insect pathogens. These aspects of biological control are emphasized in this chapter.

This chapter is divided into three sections: (a) the ecological characteristics of transient crops and colonization patterns of these crops by herbivores and natural enemies, (b) applied biological control under transient crop conditions (including natural control, introductions of new natural enemies, and augmentation of natural enemies), and (c) the interactions of biological control with other control methods, mainly cultural practices, host-plant resistance, and chemical insecticides.

ECOLOGICAL CHARACTERISTICS AND FAUNAL COLONIZATION PATTERNS

Dynamics of Transient Crop Development

The cycle of a typical transient (e.g., an annual) crop in temperate zones involves soil preparation in late fall and early spring, fertilization, preplant or preemergence applications of herbicide, planting, cultivation, and harvest. In subtropical regions, double- or even multiple-cropping may be possible within the yearly cycle. Rainfall distribution and temperature usually determine optimal planting dates and the length of the growing cycle. In cold, high latitudes, soybean must complete the cycle from planting to harvest in about 90 days. In the subtropics, the use of 140-day varieties is not uncommon (Hinson & Hartwig, 1982). Throughout this cycle, soybean plants accrue biomass at an exponential rate and undergo profound physiological changes. Their total above-ground accumulation of biomass may reach 10 tons (dry matter) per hectare, partitioned throughout the season into vegetative and reproductive structures. As the cycle progresses, growth is accompanied by a parallel increase in architectural and microclimatic complexity within the crop canopy and the underground structures that lead to the diversity and proliferation of potential feeding niches or food resources for colonizing herbivores. The availability of these resources is probably the most important single factor in setting numerical limits on species packing in a given transient crop community. Detailed discussions of the effects of habitat variability and diversity on the patterns of pest colonization and local abundance are found in Kim and McPheron (1993).

The following summary of the dynamics of variation of food resources in a typical annual field crop is based on a soybean model (Kogan, 1981). The exclusively crop-dependent components of a herbivore's feeding niche have functional, spatial, and temporal characteristics. Functional characteristics are determined by the physiology (internal milieu) of the plant and refer to the various plant organs and tissues (roots, shoots, stems, flowers, sheaths, and seeds, as well as mesophyll, parenchyma, sclerenchyma, and vascular tissue) used differentially by various species of herbivores. Spatial characteristics depend on the stratification of the aerial and subterranean volumes of the plant and on the patterns of plants within fields (borders versus center rows or plants in peaks or valleys within fields). Such stratification may cause nutritional variability within and among plants (Denno & McClure, 1985; Slansky & Rodriquez, 1987; Heinrichs, 1988) or subtle but critical variability in microclimate closely related to an insect's ecological preferences. Both functional and spatial characteristics vary in time, resulting in profound differences in plant resources at various phenological stages of development. The most general pattern of the yearly ecological dynamics of a transient crop is an initial, more or less long phase of gradual geometric increase in niche complexity and resource diversity open to herbivore occupancy, followed by a sudden drop in diversity and complexity as plants senesce and the crop reaches harvest maturity.

This generalized pattern of crop dynamics presents a scenario of changing opportunities to potential herbivore colonizers and their complement of natural enemies. The instability of multitrophic interactions under these conditions is one of the major obstacles to classical biological control in transient crops, and thus, we must review relevant concepts on the colonization of transient crops by herbivores and their natural enemies.

Colonization of Transient Crops by Herbivores and Natural Enemies

Transient crops are recolonized yearly by herbivores and their natural enemies. The sources of colonizing species vary according to the crop and may include the agroecosystem encompassing the crop (either a monoculture or a multiple-crop system) and the relative geographic location of interacting agroecosystems. Well-adapted, host-specific, native species that overwinter in or near the crop field, are the first source of colonizers. Corn rootworms, *Diabrotica* spp. (Chrysomelidae), overwinter as eggs and colonize corn plants when the crop is grown without rotation with non-host crops (but see Krysan *et al.*, 1986). In temperate zones, harsh winters usually have a modulating effect on the survival of overwintering native species and thus affect the size of colonizing populations. In the midwestern United States, the bean leaf beetle, *Cerotoma trifurcata* Forster (Chrysomelidae), and the Mexican bean beetle, *Epilachna varivestis* Mulsant (Coccinellidae), overwinter as adults in woodlots surrounding grain legume crop fields. The success of colonization usually depends on the synchronization of the emergence of the overwintered population with the establishment of a host crop in fields adjacent to hibernacula. When spring planting is delayed because of insufficient or excess precipitation, the beetles may lack food or oviposition sites, and colonization may fail. These species usually remain on the crop, however, increasing gradually on succeeding generations if environmental conditions are favorable.

A second group of colonizers are polyphagous species, the populations of which increase on wild plants or on temporarily more attractive crops. These species migrate into a succession of crops as plants reach a preferred stage of growth or as the crops on which they had resided become unsuitable. The corn earworm, *Helicoverpa zea* (Boddie) (Noctuidae), for example, develops on corn early in the season in North Carolina and produces two generations. When second-generation adults emerge, corn is no longer suitable and moths disperse to crops such as cotton, peanut, tomato, and late-planted soybean at bloom. Waves of ovipositing moths are often massive and generate damaging larval populations (Stinner *et al.*, 1977; Kennedy & Margolies, 1985). Also in this category are multivoltine species that arrive in small numbers onto a crop at various times during the season and may or may not become established. If they do become established, their short life cycle and high reproductive rate result in the buildup of populations that may prove damaging. Examples are aphids, whiteflies, leafhoppers, and spider mites.

A third group of colonizers are migrant species that overwinter and reproduce early in the season in regions of subtropical climes. Successive generations expand their geographic range from the overwintering foci, generally following jet stream paths and the availability of suitable hosts (Rabb & Kennedy, 1979; Sparks, 1979).

The island biogeographical or dynamic equilibrium theory has been proposed as a model for the colonization of annual crops by arthropods (Price, 1976; Mayse & Price, 1978; Price & Waldbauer, 1982); however, it has proven of little value in explaining or predicting patterns of colonization of transient crops and its application has been criticized on both theoretical and practical grounds (Rey & McCoy, 1979; Liss *et al.*, 1986; Simberloff, 1986). Although detailed studies on the dynamics of crop colonization under diverse cropping conditions are few, those that exist suggest that the number of colonizing species increases as the crop matures and that a lag occurs between crop colonization by herbivores and subsequent colonization by natural enemies (Price, 1976; Mayse & Price, 1978).

The ability of natural enemies to follow herbivore colonizers closely is an important factor in the regulation of herbivore populations. The availability of prey at an early stage of plant growth may determine the abundance of predators at later stages when other prey species may be present. Anecdotal accounts by soybean researchers in the southern United States (Harper *et al.*, 1983) suggest that the green cloverworm, *Plathypena scabra* (Fabricius) (Noctuidae), an early-season herbivore, is a beneficial species because it serves as prey for predaceous hemipterans. Later in the season, those predators help moderate the population growth of such serious pest species as *Helicoverpa zea*, *Anticarsia gemmatalis* (Hübner), and *Pseudoplusia includens* Walker (Noctuidae). The green cloverworm, however, is a migrant species which usually reaches midwestern soybean fields at critical stages of crop development and therefore poses a potential economic threat in those states. Patterns of regional crop colonization by *H. zea* and *H. virescens* provide another example of the influence of crop developmental stage and relative attractiveness on pest status of those species (Bradley, 1993).

Recruitment of Crop Colonizers

The diversity of the arthropod community associated with annual crops seems to depend mainly on the extent of the area planted to that crop (Strong, 1979). Plant architecture, however, influences the complexity of available feeding niches, and these ultimately determine the complexity and richness of those communities (Lawton, 1978; Kogan, 1981). Whether a crop is introduced or native is also important. Kogan (1981) considered three sources for species recruitment in introduced crops: (1) oligophagous species associated with native plants that have taxonomic (or better

yet, chemical) affinity with the introduced crop; (2) polyphagous species capable of rapidly expanding their host range as new food resources become available or replace previous ones; and (3) oligophagous species that are associated with plants unrelated to the crop and that may undergo gradual host shifts. Native crops have a preponderance of host-specific, coevolved species and a full complement of effective natural enemies. The colonization of introduced hosts by native herbivores that originally fed on plant species closely related to the introduced crop has resulted in some of the most serious pest problems on record. The classic example is the Colorado potato beetle, *Leptinotarsa decemlineata* Say (Chrysomelidae).

In summary, transient crops are recolonized annually by a herbivorous fauna that varies spatially and temporally with the dynamics of the crop, the characteristics of the ecosystem, and the spatial relationship of the crop ecosystem to other adjacent or distant ecosystems. A complement of natural enemies associated with those herbivores usually colonizes the crop after a lag that is determined by the foraging patterns of the natural enemies and the sources of the colonizers. The buildup of natural populations of enemies depends on the availability of suitable prey or hosts. The nature and complexity of this colonizing arthropod fauna depend on whether the crop is native to or introduced into a region. Additionally, the colonizing fauna depends on how long the crop has been under cultivation, increasing exponentially for several growing cycles until it approaches a plateau determined by the area planted to the crop and the complexity of the crop's available feeding niches (asymptotic growth) (Strong, 1974; Lawton, 1978; Kogan, 1981; Strong *et al.*, 1984).

It is this rapidly changing and cyclically perturbed habitat that poses the greatest obstacles to the success of classical biological control in transient crops. Despite the inherent ecological instability of these crops, however, most herbivore populations are effectively regulated by a complement of natural enemies. This regulation is most dramatically demonstrated when natural enemies are inadvertently eliminated by broad-spectrum insecticides (Metcalf, 1986; Kid & Jervis, 1996).

In the following sections, we discuss (a) the role of naturally occurring biological control agents, (b) attempts to use classical biological control in transient crops, (c) the use of augmentative releases, and (d) interactions among biological control and other IPM tactics. Examples are drawn, as much as possible, from the three major categories of biological control agents: parasitoids, predators, and diseases.

NATURAL CONTROL IN TRANSIENT CROPS

Transient crops in most growing regions of the world have a diverse and abundant population of natural control agents as long as fields have not been sterilized by intensive use of broad-spectrum pesticides.

Predators

Extensive and intensive surveys have been conducted using as the target either the crop, particular species or guilds of species within a single crop, or the various crops in a region. Survey methods for natural enemies of crop plant arthropod pests have been reviewed by Powell *et al.* (1996). One of the most extensive surveys of natural enemies of any crop was conducted by Whitcomb and Bell (1964) in Arkansas cotton fields. They identified 600 species of predators representing 45 families of insects, 19 families of spiders, and 4 families of mites. Other extensive surveys were done on spiders on soybean (Florida: Neal, 1974; Illinois: LeSar & Unzicker, 1978). Data from these surveys are compared (Fig. 1A). The number of unique species occurring at each location far exceeded the number of species occurring in common at any two locations combined (Arkansas + Florida, Arkansas + Illinois, Florida + Illinois) or co-occurring at all three locations. The spider community of cotton in Arkansas was far richer than the spider communities of soybean either in Florida or in Illinois. There were about as many species of spiders common to Arkansas cotton fields and Illinois soybean fields as there were to Arkansas cotton fields and Florida soybean fields, but there were three times more species in common in those two comparisons than there were species common to Florida and Illinois soybean fields. Although all three communities had a diverse spider population, the spider community of cotton was much more diverse (Fig. 1A).

Geographic location, however, seems to play a greater role in determining species composition than the quality of the crop matrix. A similar comparison is made among surveys of carabids in Illinois and Iowa corn fields (Dritschilo & Erwin, 1982), in North Carolina soybean fields (Deitz *et al.*, 1976), and in Arkansas cotton fields (Whitcomb & Bell, 1964) (Fig. 1B). In contrast to the spider fauna, the carabids were much more localized. Only one species appeared in all three surveys, and only 18 species (about 20% of the total of 87 species recorded from all three regions) co-occurred in any two agroecosystems. These comparisons suggest that crop communities have a rich fauna of predators and that many species are probably well adapted to local conditions. Although the effectiveness of this predaceous fauna has not been evaluated in detail, resurgences of pests are often attributed to the disruption of the natural control agents by broad-spectrum pesticides (Huffaker & Messenger, 1976; Shepard *et al.*, 1977; Croft, 1990; Mochida *et al.*, 1990; Kidd & Jervis, 1996).

Parasitoids

Assessments of naturally occurring parasitoids are usually based on surveys of individual host species or guilds of hosts. Extensive surveys have been conducted on the parasitoids of some of the major pests of transient crops, e.g., *Helicoverpa zea* and *H. virescens* (Fabricius) (Noctuidae) and *Nezara viridula* (Linnaeus) (Pentatomidae). *Helicoverpa zea* and *H. virescens* are highly polyphagous and have been recorded in the United States from 235 plant species in 36 families. A literature survey of the parasitoids of these two species produced 60 species of Hymenoptera in 6 families (Braconidae, Chalcididae, Eulophidae, Ichneumonidae, Scelionidae, and Trichogrammatidae) and 61 species of Diptera in 4 families (Muscidae, Phoridae, Sarcophagidae, and Tachinidae) (Kogan *et al.*, 1989). The efficacy of natural control agents in cotton in the United States was assessed by Goodenough *et al.* (1986), and King and Coleman (1989) assessed the potential for biological control of *Heliothis* spp. and *Helicoverpa zea* and concluded that natural enemies alone would probably remain incapable of controlling the pests throughout their geographic and host ranges.

A partial host record of *N. viridula* shows that it is also a highly polyphagous species, having been recorded from 44 common cultivated and wild hosts in 18 different plant families (Todd & Herzog, 1980). Jones (1988) surveyed the world literature for records of *N. viridula* parasitoids and found 57 species in 2 dipteran and 5 hymenopteran families. These lists represent a valuable resource for identifying candidates for biological control.

Species guilds, rather than single species, are often the object of detailed studies. Comprehensive studies of parasitoids of lepidopterous caterpillars in soybean in the United States were reviewed in Pitre (1983). Ten primary parasitoids and ten hyperparasitoids were recorded on cereal aphids in Europe (Vorley, 1986). In most cases, extensive surveys of common herbivorous insects of transient crops reveal the presence of a rich associated fauna of natural enemies; however, many of those herbivores remain serious pests. Obviously, qualitative surveys reveal very little about the effectiveness of natural enemies in population regulation. Enrichment of the complement of natural enemies of transient crops through augmentative releases or through classical biological control will be discussed in subsequent sections (see also Hokkanen, 1997).

Entomopathogens

Perhaps the most effective natural control agents of explosive pest populations in transient crops are entomopathogens (Lumsden & Vaughn, 1993; Hajek & Leger, 1994). A good example of efficacy of a fungal pathogen in regu-

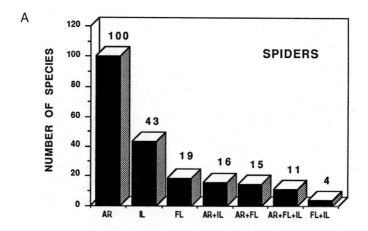

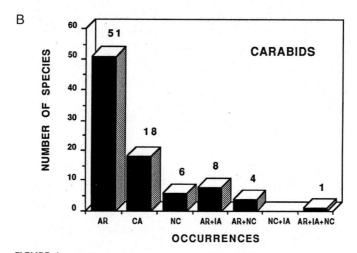

FIGURE 1 (A) Comparison of number of spider species in three agroecosystems: Arkansas cotton, Illinois soybean, and Florida soybean. (B) Comparison of number of carabid species in three agroecosystems: Arkansas cotton, Illinois and Iowa corn, and North Carolina soybean.

lating lepidopterous caterpillar populations is provided by the fungus *Nomuraea rileyi* (Farlow) Samson. *Nomuraea rileyi* is primarily a pathogen of many species of lepidopterous larvae (Ignoffo, 1981). Natural epizootics frequently cause crashes of susceptible host populations. Under favorable environmental conditions, this fungus may be the single most important mortality factor regulating populations of the velvetbean caterpillar, *A. gemmatalis,* in soybean fields in Brazil (Moscardi & Sosa-Gómez, 1992, 1996) and populations of the green cloverworm, *P. scabra,* in soybean in the midwestern United States (Pedigo *et al.,* 1982). The success of the soybean IPM program in Brazil was due, to a large extent, to the correct assessment of natural epizoot-

ics of *N. rileyi* (Kogan *et al.,* 1977; Kogan & Turnipseed, 1987). However, epizootics are often not predictable and occasionally occur too late in the growing season to prevent economic damage to the crop (Ignoffo *et al.,* 1975; Kish & Allen, 1978, Fuxa, 1984). Despite these adverse characteristics of some epizootics, their dramatic nature has caused substantial research to be directed toward using *N. rileyi* as a biological control agent. The strategies used to exacerbate the impact of natural epizootics are discussed under the section on enhancement of the efficiency of natural enemies.

According to Yearian *et al.* (1986), *H. zea* and *H. virescens* on cotton in the United States are infected by many naturally occurring pathogens. The most common are *Nomuraea rileyi* and *Entomophthora* spp. (Fungi), *Nosema heliothidis* and *Varimorpha necatrix* (Microsporida), and the nuclear polyhedrosis viruses of *H. zea* and of *Autographa californica* (Speyer) (Noctuidae). Again, although natural epizootics do occur, they are often inadequate to maintain Heliothini populations below the economic injury level (EIL). Consequently, much effort has been directed at developing manipulative methods to enhance entomopathogen efficacy in the regulation of populations of *Heliothis* spp. on cotton, corn, grain sorghum, and soybean (King & Coleman, 1989).

CLASSICAL BIOLOGICAL CONTROL

Classical biological control of major pests of transient crops has had few spectacular successes. Those that do exist, however, suggest once again that the success of a biological control program cannot be predicted on the basis of assumptions or preconceptions related to the ecological instability of those crops (Hokkanen, 1985). The following case histories illustrate the point.

Nezara viridula (Linnaeus) (Pentatomidae), the Southern Green Stink Bug or Green Vegetable Bug

Based on the polymorphism of *N. viridula* in southeast Asia, this region is considered the center of origin of the species (Yukawa & Kiritani, 1965). *Nezara viridula* is presently found throughout the tropics and subtropics of all continents (Panizzi, 1997). However, Hokkanen (1986) suggested that *N. viridula* is of Ethiopian origin, based on records of polymorphism as well as the number of host-specific parasitoids in the Ethiopian region. Because it is an immigrant pest of many important crops, many attempts to establish parasitoids into newly invaded areas have been made. Programs in Hawaii and Australia have been very successful (Caltagirone, 1981), and programs of importation and release of natural enemies are currently underway in continental Africa (several countries), Argentina, Brazil, Cape Verde, Chile, New Zealand, Taiwan, and the United States (Jones, 1988). The program in Australia provides an insight into the conditions for successful biological control of *N. viridula*.

Nezara viridula was first recorded in Australia in 1913 and has since been the subject of several successful biological control projects, mainly involving colonization of the scelionid egg parasitoid *Trissolcus basalis* (Woll.) imported from Egypt and Pakistan. The early history of the control of *N. viridula* by the importation of natural enemies was recorded by Wilson (1960), Clausen (1978), and Caltagirone (1981). Update of this history and assessment of factors that may have led to the successful control of the pest in Australia are based mainly on Strickland (1981) and Clarke (1990, 1992).

Nezara viridula spread to the Ord Valley (Northeast Western Australia) in 1974, over a decade after the last introduction of parasitoids from Pakistan to other parts of Australia. Within 2 years, it had become a severe pest due to its polyphagous habit that enables *N. viridula* to damage many vegetable and field crops. Damage, for example, was so severe in sorghum that fields had to be abandoned. *Trissolcus basalis* collected in other parts of Australia was mass reared in an insectary and about 44,100 were released in fields in the Ord Valley. The host population began to decline due to parasitism a few months later, and good control was obtained (Strickland, 1981).

Subsequent observations indicated that the parasitoids were usually present regardless of the level of abundance of the host population. Conditions that helped to maintain populations of stinkbugs at low levels and prevented their upsurge following their decline were explained as follows:

1. The prevailing cropping system in the Ord Valley involved diverse plant species that were infested by the stink bug at different population levels. The parasitoids, therefore, were able to move from centers of high host population to centers of low host population, thereby maintaining an overall low equilibrium position throughout the entire spectrum of crops.

2. In addition to *N. viridula*, *T. basalis* attacked several other locally occurring pentatomids and thus had a continuous supply of hosts (Strickland, 1981).

The success of *T. basalis* as the parasitoid of a very mobile and polyphagous pest is attributable to a combination of the characteristics of its own host range and the characteristics of the feeding range of its host species. That combination guarantees an environment that continually provides fresh adult parasitoids capable of keeping the pest at low population levels. Throughout Australia, however, *N. viridula* is a significant pest of grain, legumes, and tomatoes (Clark, 1992), and based on historical data, it was suggested that there is only circumstantial evidence *N. viridula* is under "good" biological control by *T. basalis* (Clarke, 1990).

Since *N. viridula* remains a major pest of many transient crops in most parts of the world, efforts to control it by means of natural enemies continue. According to Jones (1988), African and Asian egg parasitoids in the genera *Trissolcus, Telenomus,* and *Gryon* and six New World tachinid adult parasitoids deserve consideration in biological control programs against *N. viridula*. The tachinid species are *Trichopoda pennipes* (Fabricius), *T. pilipes* (Fabricius), *T. giacomellii* (Blanchard), *T. gustavoi* (Mallea), *Eutrichopodopsis nitens* Blanchard, and *Ectophasiopsis arcuata* (Bigot).

Dacus cucurbitae Coquillet (Tephritidae), the Melon Fly

The melon fly is native to the Indo-Malayan region and was first recorded in Hawaii in 1897. Prior to its entry, cucurbit crops were widely grown for local consumption and some were exported to California. Following the introduction of the fly, growing cantaloupes became impractical and the production of other melons, cucumbers, and tomatoes was seriously curtailed (Nishida & Bess, 1950). Biological control of the melon fly was undertaken by introducing *Biosteres (Opius) fletcheri* (Silv.) (Braconidae) from India. The parasitoids were mass reared in Hawaii and field releases made in 1916 and 1917 resulted in their establishment. Two additional species of *Biosteres, B. longicaudatus watersi* Fullaway from India and *B. angeleti* Fullaway from Borneo, were introduced during 1950 and 1951, respectively (Clausen, 1978).

The 1916 and 1917 releases resulted in a 50% reduction of the melon fly populations and, although the flies were still a pest, melons were again a profitable crop in Hawaii (Fullaway, 1920). In subsequent years, the melon fly again became a severe pest requiring multiple applications of insecticides and generating additional control-related research (Nishida & Bess, 1950). Studies showed that the change in parasitoid efficiency was probably associated with changes in land use and agricultural practices (Newell *et al.,* 1952; Nishida, 1955).

Because melons and other perishable crops are available in the field for only a short period, these plants form an unstable resource to which the biology and life cycle of *D. cucurbitae* are well adapted. Consequently, parasitoids of *D. cucurbitae* must be able to follow the short-lived and localized fly populations throughout their range if efficient control is to be achieved. In Hawaii, control had been possible because of the presence of *Momordica balsamina,* the fruits of which constituted a stable wild host for *D. cucurbitae* and its parasitoids. Changes in agricultural practices and increased land use, however, reduced the areas where *M. balsamina* grew abundantly, thereby reducing the reservoirs of the natural enemies and making it more difficult for the natural enemies to reach the cultivated fields. The main fly population now had its origin in cultivated fruits where parasitization was much lower than in the fruits of *M. balsamina*: 1% for tomatoes, 0–16.5% for melons, and 0.2–6.5% for cucumbers versus 20–37.8% for *M. balsamina* (Nishida, 1955). Thus, a change in the diversity of the habitat proved detrimental to this biological control project.

Oulema melanopus (Linnaeus) (Chrysomelidae), the Cereal Leaf Beetle

Oulema melanopus, a pest of cereals and a native of Europe, was first recorded from Berien County, Michigan, in 1962. According to Haynes and Gage (1981), however, damaging populations in the area were probably present since the latter part of the 1940s. Thus, actual invasion preceded detection by more than 10 years. Expansion of the area infested by the cereal leaf beetle occurred rapidly and the current range extends from Illinois in the west to the New England states in the east, south into the northern ranges of Tennessee and North Carolina, and north and west to Wisconsin. Spread dynamics of the cereal leaf beetle was modeled by Andow *et al.* (1993). Strict interstate quarantines and treatment of potentially infested bales of hay and grain were enforced. Subsequent research discovered that the cereal leaf beetle overwintered under the bark scales of Christmas trees, and certification for the movement of these trees was also required. Eradication efforts covered a period of about 7 years, included extensive areas in Michigan, Indiana, and Illinois, and reached a peak of over 650,000 ha that were blanket-sprayed with carbaryl in 1966. Eradication efforts were abandoned when the spread of the beetle was considered out of control. Moreover, widespread public opposition to the program had been mounted by city dwellers whose cars had paint damage due to the sprayed insecticide.

The beetle has one generation per year and overwinters as unmated adults (Castro *et al.,* 1965). With the spread of the beetle out of control, research was initiated in several areas, including sterile male techniques, behavioral control by means of attractants, and biological control by means of imported natural enemies. Clausen (1978) summarizes the biological control program. Initiated in 1963, the search for natural enemies of the cereal leaf beetle concentrated in France, Italy, and Germany. From 1964 to 1967 five parasitoids were imported, and four became established: *Tetrastichus julis* (Walker) (Eulophidae), *Diaparsis carinifer* (Thomson) and *Lemophagus curtus* Townes (Ichneumonidae), and the egg parasitoid *Anaphes flavipes* (Förster) (Mymaridae) (Haynes & Gage, 1981).

In the absence of efficient natural enemies in the classical biological control sense, researchers turned to the mass release of the egg parasitoid *A. flavipes*. Releases were made in Indiana in 1966, and the parasitoid was recovered at most release sites later in the same season. Since the beetle was not easily reared in the laboratory, cultures of the parasitoid were maintained on beetles collected in the field. These beetles were also used in the screening of wheat, oat, and barley lines and varieties for resistance against the beetle. A parasitoid nursery was established in Niles, Michigan, for the redistribution of parasitoids reared on field-infested populations.

The continuous monitoring of the cereal leaf beetle indicates that populations generally have declined since 1971. Causes for the decline seem to be combinations of such factors as weather-related mortality, mortality due to introduced parasitoids, genetic changes in beetle populations, and changes in overwintering habitat (Haynes & Gage, 1981). *Oulema melanopus* is now a permanent component of the grain ecosystem in north and central North America. Although sporadic outbreaks may require treatment, populations seem to have equilibrated. The cereal leaf beetle story suggests that immigrant pests, after an initial period of explosive expansion, may follow a pattern of adaptation within the agroecosystem that results in an equilibrium state not necessarily detrimental to the crop. Haynes and Gage (1981) sum up the lessons from the cereal leaf beetle invasion as follows:

"Since the discovery of the cereal leaf beetle in North America, research associated with this species is a chronology of how society deals with the introduction of an exotic pest. The numerous faults of the programs that were implemented to control the cereal leaf beetle reflect the priority placed on structural change in the agricultural production system. The initial response was detection, then eradication and containment, followed by an intensive program of host plant resistance and ultimately a biological control effort of questionable success. A great deal of activity and research effort since the early 1960s added much to the understanding of the cereal leaf beetle problem, but both the activity and the research appear to have had minimal impact on the present or final outcome."

This rather pessimistic assessment of the impact of the research effort should not obscure the value of investigating the successes and failures of this biological control effort. The insight gained may be valuable in coping with other immigrant pests.

Hypera postica (Gyllenhal) (Curculionidae), the Alfalfa Weevil

Hypera postica was first found in the United States near Salt Lake City, Utah, in 1904, probably an accidental introduction from Europe (Titus, 1907, 1910). The weevil was confined to 12 western states until 1952, when it was detected in Maryland (Bissell, 1952). From Maryland, it spread rapidly and is now found in the contiguous 48 states.

The weevil has one generation a year and spends the winter as aestivating adults and as eggs. Eggs hatch in the spring about the time alfalfa begins to grow. In the Midwest, larval feeding continues through May, when pupation occurs. After emergence, adults leave the field for available cover, where they undergo summer aestivation. In the fall, adults return to the field and begin laying eggs (Manglitz & App, 1957).

Parasitoids of the alfalfa weevil were first introduced from Europe into the United States in 1911; by 1919 they were well established in many areas of the western United States (Chamberlein, 1924). The parasitic wasp *Bathyplectes curculionis* (Thomson) (Ichneumonidae) is the most widely distributed and most successful introduced parasitoid in the Midwest. During the 1960s and 1970s, both *B.*

curculionis and *B. anurus* (Thomson) were released into Illinois by USDA personnel and are now found in most midwestern populations of the alfalfa weevil (Dysart & Day, 1976). The USDA Animal and Plant Health Inspection Service (APHIS) conducted an extensive survey to determine the level of parasitization of *H. postica* in more than 700 alfalfa-producing counties in the United States (Kingsley *et al.*, 1993). They reported that since 1980, more than 16 million parasitoids were released across the United States as part of the Alfalfa Weevil Biological Control Program. The larval parasitoid *B. anurus* and one of the adult parasitoids, *Microctonus aethiopoides* (Loan), were released in the largest numbers. The survey determined that *B. curculionis* had already spread over most alfalfa-growing areas of the country. Overall levels of parasitization increased in Iowa, Missouri, and Nebraska fields and insecticide usage declined. Less than 3% of 180 fields sampled over the 8-year evaluation survey reached the economic injury level for the alfalfa weevil. The two *Bathyplectis* species exhibited differential success when parasitizing the two strains (eastern and western) of the alfalfa weevil but reached an average of 21% parasitization of the larvae surveyed. *Microctonus aethiopoides* possibly is an important mortality factor helping to maintain alfalfa weevil populations below the economic injury level. Before the first cutting of alfalfa, *M. aethiopoides* killed an average of 26% of adult weevils in the four areas surveyed in eastern United States (Kingsley *et al.*, 1993). Oloumi-Sadeghi *et al.* (1993) suggested that *B. anurus* may have displaced *B. curculionis* in some areas in Illinois.

In 1973, a fungal disease of alfalfa weevil larvae was reported in Ontario, Canada (Harcourt *et al.*, 1974). This fungus was an Entomophthorales similar to the fungus known for many years as an important natural control agent of the cloverleaf weevil, *Hypera punctata* (Arthur) (Arthur, 1886). The taxonomy of the alfalfa weevil fungus, currently called *Zoophthora phytonomi* (Thomson), is not totally resolved, but the fungus of the alfalfa weevil is not the same species as the fungus of the cloverleaf weevil. The alfalfa weevil fungus was not intentionally introduced from Europe and its origin is unknown, but soon after its discovery in Ontario, it spread rapidly and began to cause noticeable alfalfa weevil mortality in many other areas of North America (Muka, 1976; Puttler *et al.*, 1978; Barney *et al.*, 1980; Watson *et al.*, 1981; Los & Allen, 1983; Nordin *et al.*, 1983). This fungus is now the major naturally occurring biological control agent of the alfalfa weevil throughout most of its range (Carruthers & Soper, 1987). A similar, if not identical, fungus causes comparable mortality in *Hypera variabilis* (Herbst) in Israel (Ben Ze'ev & Kenneth, 1982).

The fungus overwinters in the soil as thick-walled resting spores that germinate in the spring and produce germ-conidia. These germ-conidia infect weevil larvae, which at death produce more infectious conidia. Conidia produced by infected larvae are responsible for the horizontal transmission of the disease (Ben Ze'ev & Kenneth, 1982). Younger larvae tend to produce conidia and older larvae tend to produce resting spores (Watson *et al.,* 1980). Brown and Nordin (1982) developed a detailed epizootiological model of this disease and estimated that the first incidence of *Z. phytonomi* occurs in Kentucky after an accumulation of 220–290 degree days. After the first incidence of the disease, the alfalfa weevil population must reach a threshold density that allows sufficient horizontal transmission for an epizootic to develop. Brown and Nordin estimated this threshold to be 1.7 weevil larvae per stem. The mortality rates caused by the fungus are often quite high, commonly between 30 and 70% at the time of peak larval occurrence and often 100% later in the season (Morris, 1985).

The primary problem with this fungus as a biological control agent is that it often appears late relative to currently recommended harvest dates (Armbrust *et al.,* 1985). Brown and Nordin (1982) proposed using computer-directed harvest dates that are earlier than the dates normally recommended. The microenvironment in the windrows promotes an earlier-than-normal epizootic and reduces the need for insecticides.

The appearance of *Z. phytonomi* as a major mortality factor in alfalfa weevil populations after the two parasitoids *B. curculionis* and *B. anurus* were established in the United States poses the question of how these biological control agents will coexist. All three agents attack the larval stage. Approximately 5 days elapses from infection to death in diseased larvae and parasitized larvae die within 10 days. These time periods suggest that an alfalfa weevil larva infected and parasitized simultaneously would probably die from the fungus before the parasitoid completed its development. Detailed laboratory studies have not been conducted, but field studies indicate that the disease has a negative impact on the two parasitoids (Loan, 1981; Los & Allen, 1983; Morris, 1985). It will be interesting to compare the present relationships between fungus and parasitoids with the relationships that will exist after several decades. Economic analyses of an areawide program for the control of the alfalfa weevil with biological control agents showed a highly favorable cost–benefit ratio (White *et al.,* 1995).

Ostrinia nubilalis (Hübner) (Pyralidae), the European Corn Borer

Believed to have been accidentally introduced in shipments of broom corn from Europe, *Ostrinia nubilalis* was first discovered in the United States near Boston, Massachussetts, in 1917 (Caffrey & Worthley,1927) and was reported in northern Illinois in 1939. Its range now includes most of the major corn-growing regions of the United

States. Soon after its introduction, the borer became a devastating pest of corn (maize) in the eastern United States, stimulating an extensive program of foreign exploration for parasitoids. Between 1920 and 1930, almost 3 million parasitoids, representing 24 species, were imported into the United States from Europe and the Orient. In 1962, 6 of these parasitoids were established in populations of corn borer.

Two of the introduced parasitoids, the tachinid *Lydella thompsoni* (Hertig) and the ichneumonid *Eriborus terebrons* (Gravenhorst), commonly parasitized up to 50% of the borers in the Midwest during 1958–1963. However, in the 1960s, European corn borer parasitism by *L. thompsoni* decreased rapidly and few, if any, can now be found in the United States (Hill *et al.,* 1978; Burbutis *et al.,* 1981). Several theories have been proposed to explain the decline of this once-important parasitoid, but the most widely accepted is that the microsporidium *Nosema pyrausta* competitively replaced *L. thompsoni.* Interestingly, the only parasitoid commonly found in the Midwest is the braconid *Macrocentrus grandii* (Goidanich). This parasitoid is infected by *N. pyrausta* and high levels of mortality result (Andreadis, 1980, 1982; Siegel *et al.,* 1986). In Illinois, *M. grandii* parasitized an average of 19.5% of first-generation corn borer larvae but only an average of 5% of second-generation larvae in 1982 and 1983. First-generation borer populations usually have a lower prevalence of *N. pyrausta* than second-generation populations; thus, the parasitoid may avoid the microsporidium by parasitizing primarily first-generation larvae. Corn volatiles are kairomones for *M. grandii,* suggesting that generational differences could also be due to changes in volatile composition as corn matures between the onset of first and second *O. nubilalis* generations (Udayagiri & Jones, 1992). Landis and Haas (1992) suggested that landscape structure could explain differences in levels of parasitization by *E. terebrans* in corn fields in Michigan.

Nosema pyrausta was described by Paillot (1927) from European corn borers collected in France and was first found by Steinhaus (1951) in the United States in larval European corn borers from Iowa, New Jersey, Ohio, and South Dakota. *Nosema pyrausta* was not intentionally introduced into the United States and is thought to have entered via infected parasitoids introduced from Europe. *Nosema pyrausta* now infects corn borers throughout most of its range, and high incidences (up to 100% in specific locations) have been reported from many states (Van Denburg & Burbutis, 1962; Hill & Gary, 1979; Andreadis, 1984; Siegel *et al.,* 1987).

Nosema pyrausta infects most body tissues, and infectious spores are passed in the feces of infected larvae. Horizontal transmission occurs when healthy larvae ingest sufficient numbers of spores, usually in larval tunnels contaminated by frass from infected larvae. Although some disease-induced mortality occurs when larvae are infected by oral ingestion of spores, the most dramatic mortality occurs when *N. pyrausta* is transmitted transovarially from infected females to their progeny (Windels *et al.,* 1976). Transovarially infected larvae experience 30–80% higher mortality than do healthy larvae (Kramer, 1959; Windels *et al.,* 1976; Siegel *et al.,* 1987). Crashes usually occur after several years of rising corn borer populations and when the prevalence of *N. pyrausta* approaches 100%. Since the horizontal transmission of infection in corn borer populations depends on the probability of healthy larvae inhabiting a corn stalk with infected larvae, the initial infection level of transovarially (vertical infection) infected larvae and the larval population density are two of the most important variables affecting infection levels in corn borer populations (Maddox, 1987).

In many areas of the United States *N. pyrausta* is the most important biological mortality factor in corn borer populations. It has little promise as a microbial insecticide because it is widely distributed and epizootics occur naturally.

In some years the fungus *Beauveria bassiana* is an increasingly important mortality factor of the corn borer in Illinois and Iowa (Bing & Lewis, 1991). It will be interesting to observe how fungus, microsporidium, and parasitoids coexist as biological control agents of the corn borer.

Phenacoccus manihoti Matile-Ferrero (Pseudococcidae), the Cassava Mealybug

Cassava (*Manihot esculenta* Crantz) is a root crop that serves as a major energy source for 300–500 million people in approximately 90 countries in tropical regions of the world (Bellotti & Schoonhoven, 1985). Eighty percent of its production is concentrated in Brazil, Indonesia, Zaire, Nigeria, Thailand, and India. Native to tropical South America, the plant was carried by the Portuguese to the Congo basin in Africa in the early 16th century (Cock, 1985). Although the plant is a perennial shrub reproducing vegetatively, cassava roots may be harvested 7–18 months after planting. Roots are harvested by pulling the stems and uprooting the whole plant. Several characteristics of cassava and its production make biological pest control a feasible practice. It is a long-season, long-term crop, usually grown in small acreages under multicropping conditions, thus affording a relatively stable and diverse ecosystem. Furthermore, the economic threshold for pests is usually high because of the capacity to recover from insect injury and the ability of the plant to lose considerable foliage without tuber yield being affected. Moreover, it usually is uneconomical for the small farmer to attempt to reduce pest populations with pesticide applications (Bellotti & van Schoonhoven, 1978; Bellotti *et al.,* 1999).

Several species of mealybugs of the genus *Phenacoccus* have been recorded in association with cassava in South America and Africa. *Phenacoccus gossypii* Townsend & Cockerell, *P. grenadensis* Green & Laing, and *P. madeirensis* Green are polyphagous, but *P. manihoti* Matile-Ferrero seems to be specific to cassava and the only species capable of producing severe distortion of leaves. Another South American species was separated from *P. manihoti* and described as *P. herreni* Cox & Williams (Cox & Williams, 1981). According to Bellotti *et al.* (1985), mealybug damage is a recent phenomenon, but one that is increasing in areas where it had not previously been found. The new pest condition results from an imbalance between the mealybug, the local cassava land race, and the existing natural enemies. The situation was particularly catastrophic in Africa. *Phenacoccus manihoti* was first discovered in Zaire in 1973 and spread into almost all other cassava-growing areas of the continent (Fig. 2). The losses caused by *P. manihoti,* together with another explosive pest, cassava green mite, *Mononychellus* spp., were estimated at $2.0 billion per year, and the pests affected an area of about 5.5 million hectares (Neunschwander *et al.,* 1984; Herren & Neuenschwander, 1991; Herren, 1996; Bellotti *et al.,* 1999).

Attempts to control the mealybug with natural enemies followed the identification of the immigrant species, *P. manihoti,* as the only species causing extensive damage to cassava in Africa (Cox & Williams, 1981). Surveys of native natural enemies associated with *P. manihoti* in Gabon revealed that various guilds have incorporated the immigrant pest in their host or prey range, but none with great efficiency (Boussienguet, 1986). The list included two primary parasitoids, four hyperparasitoids, nine predators, and eight parasitoids of the predatory species. An updated list was prepared by Neuenschwander *et al.* (1987), who also provided a diagram of the trophic interactions of this fauna (Fig. 3). Although both predators and parasitoids are well represented in this fauna, none of the native natural enemies were well adapted to the cassava mealybug (or perhaps to the host plant or to both).

Extensive searches for natural enemies of the cassava mealybug were conducted in South America, the purported center of origin of *P. manihoti* and where it is not considered a serious pest. Between 1977 and 1981, the Commonwealth Institute of Biological Control (CIBC), in collaboration with the International Institute of Tropical Agriculture (IITA), surveyed the tropical areas of central and northern South America and found that the parasitoids *Aenasius vexans* Kerrich, *Apoanagyrus diversicornis* (Howard), and *Anagyrus* spp. seemed to be specific to the cassava mealybug in Brazil, French Guiana, and Guyana (Cox & Williams, 1981). In 1980, a species of *Diomus* (Coccinellidae) was imported and released in experimental fields at IITA (IITA, 1981, 1985), and 1 year later the encyrtid *Apoanagyrus (Epidinocarsis) lopezi* (DeSantis), collected in

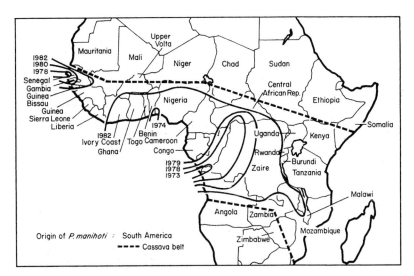

FIGURE 2 Range in Africa of the cassava mealybug *Phenacoccus manihoti* and the establishment of the introduced natural enemy *Epidinocarsis lopezi*.

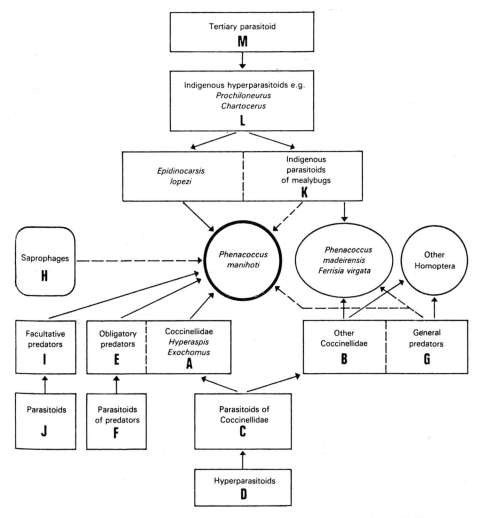

FIGURE 3 Trophic interactions of the complex of entomophagous arthropods associated with *Phenacoccus manihoti* on cassava in the Gabon (after Neuenschwander *et al.*, 1987).

Paraguay by M. Yaseen (CIBC-Trinidad), was imported to Nigeria and released at two sites. The parasitoids were established and recovered from parasitized mealybugs. Because the parasitoid had spread 150 km in 14 months, it seemed to hold good potential for the biological control of the cassava mealybug (Lema & Herren, 1985; Herren, 1996).

The spread of *A. lopezi* was spectacular. By December of 1985, it had become established over 650,000 km² in 13 African countries and a wide range of ecological conditions (Herren *et al.*, 1987; Neuenschwander *et al.*, 1987). Exclusion experiments and continuous monitoring demonstrated the efficiency of the parasitoid in regulating *P. manihoti* populations in Africa (Neuenschwander *et al.*, 1986, 1987). IITA (1985) reported that a significant reduction in population levels of the cassava mealybug had been observed in all regions colonized by *A. lopezi*. In those areas, the mealybug was recorded at populations of 10–20 per terminal cassava shoot. Prior to the establishment of the parasitoid, peak populations in excess of 1500 per shoot were common (IITA, 1985). The successful importation and establishment of *A. lopezi* gave further impetus to the biological control program at IITA, and additional species of parasitoids and predators are being released experimentally with various degrees of success (IITA, 1987b; Herren, 1996). These positive field results contradicted predictions based on laboratory data assessing percentage parasitization that led to the conclusion that *A. lopezi* could not efficiently reduce cassava mealybug populations (Odebiyi & Bokonon-Ganta, 1989). Furthermore, according to Herren and Neuenschwander (1991), despite the considerable positive impact of the cassava mealybug biological control campaign as demonstrated by field evaluations, professional and lay perceptions of this impact have varied greatly from one country to another. Herren and Neuenschwander have identified the following problems in countries that have demonstrable areas under good biological control: (1) The cassava mealybug is still spreading in several countries; consequently, nationwide damage remains high, even if *A. lopezi* has brought the cassava mealybug under control in the first release sites. (2) Within a large area under the umbrella of biological control by *A. lopezi*, individual fields or corners of fields may have comparatively high infestations; most of these infestations seem to be the result of bad farming practices. (3) Biological control activities are free to the farmer and, most often, to the government as well; those activities sometimes lead to the funding of a project or are coupled with food aid to the farmers. Therefore, there is vested interest in declaring cassava mealybug infestations a continued disaster. (4) Ignorance about mechanisms of pest impact and biological control has sometimes led to false expectations: although the cassava mealybug produces noticeable symptoms at the end of the dry season at levels too low to measurably affect yield, memory of the really devastating cassava mealybug infestations encountered before the release of *A. lopezi* is fading (Herren & Neuenschwander, 1991).

Implementation of the biocontrol program ran concurrently with basic biological studies. Examples of such studies are multitrophic interactions in cassava systems (Schulthess *et al.*, 1997; Souissi & Le Ru, 1997), the possible intraspecific larval competition in *A. lopezi* and *Leptomastix dactylopii* (Baaren & van Nenon, 1996), and host discrimination and competition in *Apoanagyrus* spp.(Pijls *et al.*, 1995). These studies should provide the foundation for future developments in this biocontrol system. Other promising natural enemies such as the coccinellid *Hyperaspis raynevali* Mulsant, imported from Guyana into the Congo (Kiyindou & Fabres, 1987), also are being investigated. In addition to these arthropod natural enemies, a report by Le Ru (1986) suggests that the entomophthoraceous fungus *Neozygites fumosa* (Speare) Remaudiere & Keller is perhaps the most significant mortality factor for *P. manihoti* in the Congo.

The success of the biological control program of the cassava mealybug in Africa is perhaps one of the best demonstrations of the potential of this tactic for IPM in transient crops (Fig. 4). Biological control, however, is but one of the tactics being deployed against this and other cassava pests. IITA, CIAT (Centro Internacional de Agricultura Tropical, Palmira, Colombia), and the Brazilian cassava program are actively engaged in the use of plant resistance, cultural control, and the selective use of pesticides in the development of comprehensive IPM systems for this important crop (Cock & Reyes, 1985; Herren, 1996; Belliotti *et al.*, 1999).

Two other important pests of cassava which are receiving biological control emphasis are cassava hornworm and cassava green mite.

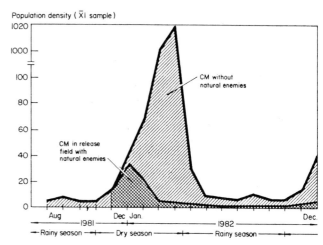

FIGURE 4 Decline in cassava mealybug populations following releases of natural enemies (IITA, 1981–1982).

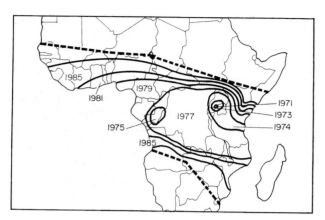

FIGURE 5   Spread of the cassava green mite in sub-Saharan Africa since first detected near Kennjda, Uganda, in 1971 (courtesy IITA, 1987).

Mononychellus tanajoa (Bondar) (Tetranychidae), the Cassava Green Mite

About 50 species of phytophagous mites have been reported on cassava; those in the genera *Tetranychus* and *Mononychellus* are particularly destructive both in South America and in Africa, mainly when they reach highest infestation levels during dry seasons (Bellotti *et al.,* 1982; Mesa & Bellotti, 1987; Herren & Neuenschwander, 1991). The South American species, *Mononychellus tanajoa* (Bondar), known as the cassava green mite, was first detected in east Africa (Uganda) in 1971 (Lyon 1974; Bellotti & Schoonhoven, 1978; Gutierrez, 1987; IITA, 1987a). It

spread rapidly throughout most cassava-growing regions of Africa and reached the northwestern sub-Saharan countries by 1984 (Fig. 5). The cassava green mite seems to be specific to species of *Manihot* and a few other Euphorbiaceae. Yield losses range from 13 to 80%, mainly as a result of defoliation caused by heavy infestations (Théberge, 1985; IITA, 1987a). According to Bellotti and Schoonhoven (1978), several predators have been recorded feeding on cassava mites: coccinellids of the genera *Stethorus, Chilomenes,* and *Verania,* the staphylinid *Oligota minuta* Cameron, the anthocorid *Orius insidiosus* (Say), several species of cecidomyiids and thrips, and the phytoseiid mites *Typhlodromalus limonicus* Garman & McGregor and *T. rapax* (DeLeon). Although it is not clear that any phytoseiids are specialized predators of *Mononychellus,* the phytoseiid mites and *Oligota minuta* seem to be the predominant predators. *Oligota minuta* is more closely associated with *Mononychellus* spp., and *Stethorus* with *Tetranychus* spp. (Bellotti, 1985). Other studies show that 19 species of predaceous mites are present in cassava fields infested by the cassava green mite in Colombia (Bennett & Yassen, 1975; Bellotti *et al.,* 1982; Mesa & Bellotti, 1987).

Efforts are underway to establish a comprehensive biological control program for the cassava green mite complex in Africa. The program involves cooperation among national and international research centers sponsored by IITA, CIBC, and CIAT (Fig. 6). According to this plan, five species of predaceous mites—*Typhlodromalus limonicus, Neoseiulus anonymus* (Chant & Baker), *N. idaeus* Denmark

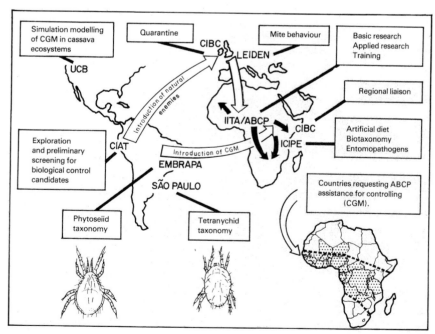

FIGURE 6   Diagram of a network of international collaboration for the biological control of the green cassava mite complex (courtesy IITA, 1987).

& Muma, *Galendromus annectens* (DeLeon), and *Euseius concordis* (Chant)—are mass produced at CIAT, Colombia, on *Mononychellus progresivus* Doreste using a method developed by Mesa and Bellotti (1987). Predaceous mite shipments go through CIBC quarantine in London and are forwarded to Africa (IITA, Nigeria) for field releases. This biological control effort, coupled with the propagation of resistant cassava varieties and cultural control methods, should help to alleviate the impact of the green mites on cassava in Africa (IITA, 1987a; Bellotti *et al.*, 1999). Studies of multitrophic interactions in cassava (Pijls *et al.*, 1995; Schulthess *et al.*, 1997) and additional research on pest–predator interactions should help advance the cassava IPM program in Africa and South America.

INUNDATIVE AND AUGMENTATIVE RELEASES

Using inundative releases of natural enemies to augment their field populations and to improve their ability to control pests has been practiced for many years (King *et al.*, 1985a). Because annual crops are transient and fail to provide a stable environment for continuous occupation by natural enemies, they are particularly suited for the inundative release approach in biological control. By manipulating the kinds of natural enemies, the stage of their development for release, the numbers released, and the time and modes of release, managers must take a much more active role in the control of pests than is required under classical biological control. The role of the pest manager parallels that of the crop manager in the manipulation of the annual crop itself, and thus it opens numerous new possibilities. It also poses new challenges and difficulties.

The success of inundative release programs depends upon some or all of the following: (a) the nature of the crop plant, (b) the developmental stage of the crop plant, (c) the developmental stage of the pest, (d) the absolute density of pest organisms, (e) the quality (host specificity, search capacity, and proper identity) of the natural enemies, (f) the density of the natural enemies, (g) the climate, (h) the complementary or antagonistic effects of other natural enemies, (i) the possibility to integrate the use of natural enemies with other control methods (in particular with insecticides), and (j) the cost of the inundative release program compared with the costs of other methods. In the following section we discuss five examples of inundative or inoculative release programs that represent the range of crop–pest systems and spectrum of natural enemies for which a significant body of information is available. Whereas most research and practical applications of inundative releases used insect parasitoids (Wajnberg & Hassan, 1994), much emphasis has been placed on entomophagous nematodes in recent years (Kaya & Gaugler, 1993).

Use of *Trichogramma* spp. to Control Lepidopterous Pests of Annual Crops

Species of *Trichogramma* are presently the most widely used insects in inundative and augmentative control in the world (Ridgway & Morrison, 1985; Wajnberg & Hassan, 1994; Smith, 1996; Parra & Zucchi, 1997). The area of crops covered by *Trichogramma* releases increases annually and amounted to approximately 11,000,000 ha in the former USSR (Voronin, 1982), 5500 ha in western Europe (Hassan *et al.*, 1986), 355,000 ha in the United States, about 2,000,000 ha in the People's Republic of China (van Lenteren, 1987), and extensive areas in Mexico (Jimenez, 1980). Reports from other countries suggest an expanding pattern of interest and use of *Trichogramma* spp. (e.g., on sugarcane in India) (Sithanantham & Paul, 1989; Mohyuddin, 1991).

Transient crops on which *Trichogramma* is used include rice, millet, various crucifers, sugar beets, sorghum, corn, cassava, cotton, sugarcane, peanut, and some solanaceous vegetable crops (Smith, 1996). However, documentation of the release data and the outcomes of release programs is lacking in many cases, making it difficult to evaluate results and to draw conclusions about the applicability of particular practices to the overall use of *Trichogramma* for biological control. Consequently, we have selected a few of the more prominent examples for detailed analysis.

Cotton

Cotton (*Gossypium* spp.) is a perennial plant grown as a warm-season annual. In the Northern Hemisphere it is typically planted in March or April and, according to the variety, harvested between August and November. The plant is plagued with many pest insects, particularly in the Homoptera, Hemiptera, and Lepidoptera, from germination to picking time (Reynolds *et al.*, 1982; Luttrell *et al.*, 1994). However, the early season, which is typified by vigorous plant growth, is often characterized by a relatively smaller risk of insect damage than is the latter part of the season. The risk differential is due to the relative abundance of natural enemies in the early part of the season (Bar *et al.*, 1979), to the lower susceptibility of plants to damage because no mature fruiting bodies are present and because of their capacity for compensatory growth (Wilson, 1986a), and to the lower numbers and less damaging characteristics of many early-season pests. Consequently, the use of early-season insecticide treatments in cotton is particularly unwarranted and its avoidance may enable growers to extend substantially the insecticide-free period of the crop.

Trichogramma pretiosum (Riley) and *T. australicum* Girault are used in Colombia to control early-season lepidopterous cotton pests (*Alabama argillacea* [Hübner], *Tricho-*

plusia ni [Hübner], *Pseudoplusia includens* Walker, *Sacadodes pyralis* Dyar, and *Heliothis* spp.) (Amaya, 1982). Releases are intended to prevent damage by these pests and to facilitate an insecticide-free period of about 100 days, after which treating against boll weevils (*Anthonomus grandis* Boheman) (S. Orduz, personal communication) is often necessary. The parasitoids are mass reared on *Sitotroga cerealella* Zeller, mostly in local "factories," and released in the field as pupae within host eggs that are glued to 6-cm² cardboard strips, each bearing 3000 eggs, 85% of which are parasitized. Releases start 20–25 days after germination and continue throughout the season. The first three releases are carried out at 5-day intervals to establish an overlapping parasitoid population; subsequent releases are made every 8 days. Each release consists of 20 cards (about 51,000 parasitoids) up to square formation, and 30 cards (about 76,000 parasitoids) thereafter (Amaya, 1982).

Alternatively, releases may start at planting time, when parasitoids are liberated along field margins to kill Lepidoptera that develop on the surrounding vegetation, and continue within the field at approximately weekly intervals for about 3 months. Parasitoids are released as freshly emerged adults, first at the rate of 40,000–50,000 per ha and later 30,000–36,000 per ha. The exact timing of release is determined by field scouting performed twice weekly by growers (S. Orduz, personal communication). When unavoidable insecticide applications have occurred, *Trichogramma* releases are made as soon as 2 days after applications to maintain continuous control (Amaya, 1982).

Inundative programs using *Trichogramma* spp. resulted in a marked reduction of insecticide treatments. In a 15,000-ha area of the Valle del Cauca (Colombia), the number of treatments changed from 20 per year in 1975 to 4 in 1981; on 6000 ha in the northern part of that region, the number was reduced to 1.2 treatments per year. An update on the use of *Trichogramma* in the Valle del Cauca is found in Garcia-Roa (1991).

The success of using *Trichogramma* to control cotton pests in Colombia and in Mexico is due, in part, to the relatively cheap and efficient *Trichogramma* production methods that have been developed. However, emphasis on applied research for the improvement and maintenance of parasitoid quality through continuous selection, the development of parasitoid storage techniques, the accurate determination of the quantities of parasitoids to be released, and the correct timing of the releases (Amaya, 1982; S. Orduz, personal communication) have been crucial to the success of the program.

In the former USSR, races of *T. euproctidis* Girault are used for the control of *Heliothis armigera* Hübner and of cutworms, *Agrotis* sp., in central Asian cotton (Voronin, 1982). For *H. armigera,* three releases of the parasitoids at the rate of 1:1 or 1:2 pest:parasitoid are used, with resulting parasitism of 66.6–90%. Release rates against cutworms are 200,000 per ha three times, once every 5–7 days, when cotton is in the seedling stage. This scheme provides complete protection of the crop. Release thresholds in Tadjikistan against *H. armigera* are 10–20 larvae per 100 plants in mid-stapled varieties and 3–5 larvae per 100 plants in thin-stapled varieties. At these thresholds only 50% parasitization efficiency is sufficient for economic control (Voronin, 1982).

In the People's Republic of China, about 680,000 ha of cotton receive *Trichogramma* releases (Huffaker, 1977). In over 100,000 ha of the Shaanxi Province, control of *Heliothis* is achieved with the release of *T. chilonis* Ishii (= *T. confusum* Viggiani). The parasitoids are applied at the rate of 120,000 per ha in a total of three releases at 3 to 4 day intervals during the F_2 host generation; 75% parasitization is achieved (King *et al.,* 1985b). In the Jiangang farm, *H. armigera* was controlled on 3546 ha yearly between 1975 and 1984 by releasing 414,000 parasitoids per ha. Parasitization reached 45%, with a residual worm density of 4/100 plants. The large amount of data obtained permitted the construction of a reliable model for predicting the parasitizing efficiency of *T. confusum* in cotton fields during the third and fourth host generations (Zhou, 1988).

Large-scale studies on the feasibility of using *Trichogramma* species, especially *T.* nr. *pretiosum* for the control of the bollworms *H. zea* and *H. virescens,* have been conducted in the United States since 1970 (King *et al.,* 1985c). A three-year pilot test was conducted in southeast Arkansas (1981–1982) and North Carolina (1983) to evaluate the feasibility of using *Trichogramma* for controlling *Heliothis* spp. in cotton. King *et al.* (1985b) summarized the project and its achievements and concluded that . . . mean parasitism rate of 47.4% of *Heliothis* spp. by *Trichogramma* nr. *pretiosum* augmented in cotton is insufficient to provide adequate control of *Heliothis*" (King *et al.,* 1985c; King & Coleman, 1989). Explanations for the failure of *Trichogramma* in the United States were presented in contrast with its successes in China, South America, and Mexico. One key reason is the higher production cost of the parasitoids in the United States, especially compared to the low cost and accessibility of insecticides. The low cost of insecticides in the United States also results in lower economic thresholds of *Heliothis,* which in turn promotes numerous insecticide treatments. An additional factor that plays an important role in many of the cotton-growing areas of the world is the frequent need to use insecticides against other pests. Such treatments further disrupt parasitoid activity.

Corn

Corn (*Zea mays*) is an annual crop that, like cotton, is grown during the warm season of the year. Growth cycle from planting to harvest varies from 2 to 5 or 6 months (usually 3–5 months) according to the variety and growing conditions. Corn originated in the Western Hemisphere and spread worldwide shortly after its introduction into the Old World (Aldrich *et al.*, 1975). Since then, it has become a cosmopolitan staple. In addition to the indigenous pest complex of corn, many local insect species have adapted to the crop, and presently each geographic region has both cosmopolitan and local corn pests (Chiang, 1978).

The European corn borer, *Ostrinia nubilalis,* originally fed on unknown hosts but readily moved onto corn, spreading from Europe to reach the status of a severe pest of worldwide importance in countries that have temperate or cold climates (Balachowski, 1951). Along the northern fringes of its distribution (Germany, Switzerland, parts of the former USSR, China, and Canada), the corn borer has only one generation per year. In these regions the European corn borer may be the main or only serious corn pest (Hassan, 1982). The number of generations per year increases at lower latitudes just as the complex of pests associated with corn expands. Consequently, insecticide treatments against the corn borer in its univoltine range not only are expensive and environmentally disruptive but may cause the outbreak of such secondary pests as aphids, pests that would otherwise be controlled by natural enemies (Hassan, 1982). Efforts to control the corn borer by releasing artificially reared *Trichogramma* were first reported from the former USSR (Zimin, 1935) and such efforts have continued (Voegele, 1988). However, commercial efforts to use *Trichogramma* were initiated only during the past decade after successful field trials were carried out in Europe (Bigler, 1986; Voegele, 1988). The number of countries using commercial *Trichogramma* releases rose within a few years from two, the former USSR and the People's Republic of China, to more than ten, including Austria, Bulgaria, Czechoslovakia, Colombia, France, Italy, Germany, Switzerland, the Philippines, Taiwan, and the United States.

The primary reasons that propelled the commercial use of *Trichogramma* as a principal means of corn borer control were

1. Concern over the disadvantages of chemical pesticides as the sole method of corn borer control, including the rising cost of insecticides; the disruption of natural enemies, which resulted in outbreaks of secondary pests; inefficient control due to pest resistance; and environmental contamination

2. Increase in the efficiency of *Trichogramma* production, including cheaper production of factitious hosts and the development of storage techniques (Voegele, 1988) and delivery systems (von der Heyde, 1991)

3. Awareness of the importance of the specific biological characteristics of the parasitoid to be used, leading to the search for more efficient parasitoid species. This search resulted in the discovery of *T. maidis* Pintureau & Voegele and to the development of techniques for quality control and strain improvement (Beglyarov & Smetnik, 1977; Huffaker, 1977; Voronin, 1982; Voegele, 1988; Pintureau, 1991; Pavlik, 1993a). It also led to biological and ecological studies on the various parasitoid species and strains (Bigler *et al.*, 1982; Voronin, 1982; Hawlitzky, 1986; Pavlik, 1993a, 1993b).

4. Ascertaining the requirements for optimal field releases to be carried out, in relation to both host and parasitoid stages, and to the quantity and modes of parasitoid releases (Stengel, 1982; Voronin, 1982; Hassan *et al.*, 1986; Hawlitzky, 1986; Shen *et al.*, 1988).

Some of the largest and best-documented programs using *Trichogramma* for the control of corn borers are summarized in Table 1, where similarities in the overall methodology of *O. nubilalis* control can be seen. Most workers agree that parasitoids must be in the field before, or at the latest, the start of, the European corn borer oviposition wave, and various methods have been devised to accomplish this. Hassan *et al.* (1986) used light traps to detect the first appearance of adult moths, the time at which parasitoids must be dispatched from the mass production plant to reach growers for optimal release. In France, Stengel (1982) and Hawlitzky (1986) reported on a day–degree calculation based on records of the development and flight of the moths since 1963. These data, together with the emergence of moths from caged pupae, are used to determine the onset of oviposition. Economic threshold is reached when 10–12% of the eggs have been laid about 3 weeks after flight initiation. This threshold varies according to climatic conditions and corn variety, ranging from 6% for early and 15% for late varieties.

Parasitoids are released in the field while in the pupal stage within a host egg. These eggs may be glued to cards or loose within perforated capsules from which adults emerge. Parasitoids are most effective during the first 5 postemergence days; thus, young adults should be present in the field during the entire peak oviposition period of the moths. Continuous adult parasitoid presence is achieved through sequential releases, usually 7–10 days apart, or through host eggs that contain parasitoids in different developmental stages; adult wasps should emerge gradually (Hassan, 1982; Bigler, 1986).

Rain and predators may cause mortality of the parasitized eggs released in the field. Egg predation becomes more severe the longer the exposure period. Predation greatly reduces the probability of survival of young parasitoid larvae that remain in the field as long as 1–2 weeks, but there are methods to minimize such mortality factors. In France, Hawlitzky *et al.* (1987) placed parasitized eggs in specially designed perforated capsules 1–3 days before emergence.

TABLE 1 Summary of Mass Release Programs of *Trichogramma* spp. against Corn Borer in Various Countries

Country	*Trichogramma* species	Releases						Lab host species	% Parasitism	Reference
		No./ha./release and (yearly total)	# of releases	Spacing (days)	Method of release timing	# of release points/ha.	Method of release			
Germany	*T. evanescens*	75,000(1) (150,000)	2	7–14	light traps	50	egg cards + saran @ 1500 eggs	*Sitotroga cerealella*	83–87	Hassan *et al.* 1986
		150,000(2) (300,000)	2	7–14					76–89	
Switzerland	*T. maidis*	50,000(1) (150,000)	3	7–10	light traps	50	egg cards	*Ephestia kuehniella*	75–93	Bigler 1986
		150,000(2) (300,000)	2	30?					60–90	
France	*T. maidis*	100,000 (300,000)	3	10	Field emergence cage	up to 100	capsules	*Ephestia kuehniella*	70–904	Hawlitzky *et al.* 1987, Voegele 1988
(former) USSR	*T. euprochidis* *T. evanescens*	30–50,000						*Sitotroga cerealella*	60	Voronin 1982, Beglyarov & Smetnik 1977
Colombia	*T. perkinsi*(5)	76,500	all season	5–5	30 days after planting	30	egg cards	*Sitotroga cerealella*	90	Amayha 1982
P.R.C.	*T. maidis*	100,000–3 million (300–390,000)	3	—	—	75	egg card	—	69.6–83.4	Huffaker 1977
P.R.C.		225–300,000 (675–900,000)	2–3(4)			15	loose in plastic containers	*Corcyra cephalonica*		Coulson *et al.* 1982
Miyun County	*T. dendrolimi* *T. ostriniae*	3–5,000	15			15	capsules	*Antheraea pernyi* *Corcyra cepholonica*	80–90	Coulson *et al.* 1982
India	*T. chilonis*	50,000	all season	10–15	—	—	—	*Corcyra cephalonica*	host populations reduced by 50%	Varma *et al.* 1991
U.S.A.	*T. nubilale*	(4.4 million)	3	—	—	—	—	—	57	Prokrym *et al.* 1992
Philippines	*T. evanescens*	140–160,000	1	—	—	—	—	—	74	Felk 1990

(1) Field corn; (2) sweet corn; (3) 1 generation per year; (4) 2 generations per year; (5) Host is *Diatraea* sp.

In Germany, Hassan (1982) placed egg cartons within a 3 × 6 cm screen "saran" bag as protection against predators and a plastic cover as protection against rain. In the People's Republic of China (Coulson *et al.,* 1982), plastic bags are employed.

The rates of release vary among countries from about 50,000 per ha in the former USSR to about 400,000 in the People's Republic of China. Selection of release rates reflects differences in the availability and quality of the parasitoid and its ability to control the host, as well as differences in the severity of the pest and in the standards of pest control that are required. In addition, univoltine borers require fewer parasitoids than do bivoltine borers. Thus, Bigler (1986) used 150,000 wasps per ha for univoltine borers in parts of Switzerland and twice that amount in regions where the pest is bivoltine.

Quality control of *Trichogramma* spp. is of paramount importance, as can be demonstrated by the reduction of parasitism from 75.2% in 1978 to 18.8% in 1979 in Switzerland, a reduction attributed to the deterioration of the quality of the mass-reared wasps (Bigler *et al.,* 1982). Maintenance of stock quality is usually achieved by rearing at least one annual generation of the parasitoids on *O. nubilalis* eggs (Bigler *et al.,* 1982; Hassan, 1982; Voronin, 1982). Mass-reared stock can also be strengthened by introducing field-collected material, a practice that is very common in the People's Republic of China (Coulson *et al.,* 1982). Voegele (1988) discussed the preservation of stock quality through the retention of original traits and improvement of parasitoids. He postulated that although the criteria for which we should select and breed *Trichogramma* populations are largely unknown, several methods for the retention and improvement of strains can be pursued. These include, in addition to the cyclic return to natural hosts, using isogenic females, manipulating the nutrition of the parasitoids in artificial rearing media, optimizing the host/parasitoid ratio in the culture, manipulating the parasite diapause, using semiochemicals from the plant or from the host insect, and selecting for insecticide resistance. Genes for insecticide resistance as well as genes for response to certain environmental stimuli may also be introduced into the parasitoid populations.

Miscellaneous Crops

In the former USSR, *Trichogramma* spp. have been used to control lepidopterous pests of peas and cabbage. Parasitization of 89–96% of the eggs of *Laspereysia dorsana* Fabricius and 67% of the eggs of *Autographa gamma* Linnaeus attacking peas was achieved following the enrichment of the environment with nectariferous plants (*Phacelia tanecetifolia*). Use of the nectar sources marked an improvement over the 29 and 31% control that had been

obtained without those sources; increased parasitoid longevity was attributed to the availability of the supplementary food. Similar results were obtained in the control of *A. gamma* on cabbage, where improvement was from 50–60% to 80–90% parasitization (Voronin, 1982). Noctuid larvae that infest sugar beet and potato were controlled in the former USSR by releasing 20,000–60,000 parasitoids per ha, resulting in a 60–90% reduction in infestation levels (Beglyarov & Smetnik, 1977). *Trichogramma chilonis* Ishii is being used to control the diamondback moth, *Plutella xylostella* (Linnaeus), on cole crops in Japan (Miura & Kobayashi, 1993) and in Germany (Wuehrer & Hassan, 1993).

Rice pests, especially the rice leaf roller, *Cnaphalocoris medinalis* Guenée, are controlled in the People's Republic of China by five seasonal releases of from 150,000 to 600,000 *T. australicum* per ha, depending upon host density. The resulting parasitism amounts to 80% and the total cost is half that for insecticidal control (Huffaker, 1977). Shen *et al.* (1988) reported successful results with inoculative releases of only 15,000 *T. dendrolimi* Matsumura per ha on seven experimental hectares of rice.

In Colombia, *Trichogramma* is used for the biological control of various crop pests in addition to those on cotton and corn. These include beans and soybeans, where the pests are *Anticarsia gemmatalis* and *Heliothis* sp., and cassava, where the main pest is the sphingid moth *Erinnyis ello* (Linnaeus). Parasitoids are released on egg cards at the rate of 51,000 (20 cards) per ha from 10 days after germination for beans and 76,500 (30 cards) per ha starting 30 days after plant emergence for cassava. Initial releases are spaced 5 days apart; later releases are 8 days apart. No percentages of parasitization were given, but satisfactory control was reported (Amaya, 1982). Sugarcane borers are being controlled through releases of *T. chilonis* in India (Varma *et al.,* 1991).

Smith (1996) suggested that assessment of *Trichogramma* impact as a biocontrol agent should improve with better guidelines for standardizing terminology and measurements. Successful use of *Trichogramma* depends on its fit in the overall IPM program and should not be viewed as a replacement for insecticides (Smith, 1996).

Biological Control of Spider Mites on Transient Crops

Biological control of spider mites has been practiced for over two decades with considerable success (Huffaker *et al.,* 1970); however, most of the work involved greenhouse cultures. Open-air cultures that are attacked by mites are usually either treated with acaricides or efforts are made to avoid using toxicants against insect pests to conserve the natural enemies of the spider mites (Jeppson *et al.*, 1975). The intentional or nonintentional development of resistance to pesticides in predaceous mites has opened new options for successful use of these mites in biological control (Croft, 1990). McMurtry and Croft (1997) suggested a classification of predaceous mites into four categories based on food habits and related biological and morphological traits. Definition of these categories may aid in the assessment of a given species potential as a biological control agent in IPM systems.

The use of natural enemies for the active suppression of spider mites out-of-doors was studied by Oatman *et al.* (1976, 1977a, 1977b, 1981), who used three species of phytoseiid mites, *Amblyseius californicus* (McGregor), *Phytoseiulus persimilis* Athias-Henriot, and *Typhlodromus occidentalis* Nesbit, to suppress *Tetranychus urticae* Koch in California strawberry fields.

These studies showed that *P. persimilis* was the most efficient of the three predator species in suppressing the pest population. The predaceous mite was successfully established in southern California, where it survived in strawberry and lima bean fields as well as on weed species in the genera *Malva, Solanum,* and *Convolvulus.* The weeds acted as reservoirs for the predaceous mites, from which they dispersed to strawberry and lima bean fields during the season. However, these predators were not successful natural control agents, and Oatman *et al.* (1981) showed that spider mites continued to cause serious damage to strawberries in the same regions and that predation by all natural enemies, including phytoseiid mites, ". . . was not a major mortality factor, as *T. urticae* reached extremely high peaks in the untreated check fields."

Commercial use of inundative mite releases in open fields has been practiced in Israel for over a decade and in The Netherlands and France for the past several years. In Israel, spring-growing melons, cantaloupes, and watermelons that are grown in the warm and arid Jordan and Arava Valleys have been subjected each season to attacks by *T. cinnabarinus* (Boisduval) and *T. urticae*. The normal practice of using acaricides against these mites was expensive and, in many cases, insufficient due to the buildup of resistance. The ineffectiveness of chemical controls enabled commercial companies specializing in the culture of *A. persimilis* to introduce inundative mite releases and to establish a biological control program. In this program, fields are surveyed every week from germination on, and predators are released when spider mites are discovered. The release rate is 20,000 predators per ha, or about 1 predator to 10 spider mites (including eggs), when the plants are at the four-leaf stage, and double that amount when plants are bigger and runners have started forming. This release method has the disadvantage of dispersing predators evenly throughout the field, whereas spider mites are usually ag-

gregated in foci. The result is that local flare-ups may occur, and the introduction of additional natural enemies may be required. The problem can be circumvented with preemptive releases of a mixture of 5/1 spider mites/predators in fields not yet infested.

The shift from acaricides to biological control resulted in an average net savings of approximately $300 per ha. Furthermore, growers experienced better yields due to the absence of phytotoxic effects of pesticides and a reduction in soil compaction with the withdrawal of spraying equipment. In addition, aphid attacks that often occurred in conventional control fields were substantially reduced.

Commercial control of *T. urticae* in vegetable crops through the release of *P. persimilis* is gaining acceptance in France and The Netherlands. The system is based on the integration of pesticide treatments against diseases and thrips and on two blanket releases of 4–5 predaceous mites per m². Treatments against thrips with mevinphos are made 2 days before the first mite release about 2–3 weeks after planting. Treatments are accompanied by inspection and monitoring of infestation levels. Infestations usually decline below the economic injury level following the second release. Should the infestation persist, a third release is made in remaining foci. This system is integrated with treatments against *Botrytis,* mildew, and *Pseudoperenospora* and it has been applied successfully to strawberries in France and to strawberries and pickling cucumbers in Holland. As in Israel, the major advantages are healthier and stronger plants that last longer and extend the growing season.

In the Willamette Valley of Oregon, spider mites are economically significant pests of diverse crops, including ornamentals, vegetable and field crops, small fruit, pome fruit, mint, and hop. The complex landscape-level interactions of spider mites and predaceous mites within this cropping system, involving 10–12 crops, and the surrounding natural vegetation have been studied by Croft and coworkers. The system involves not only timely releases of mass-reared predators but also habitat management to enhance natural enemy overwintering and colonization success and the management of resistance in the spider mites to acaricides (Kogan *et al.,* 1999).

Control of the Mexican Bean Beetle, *Epilachna varivestis* Mulsant (Coccinelidae), by Inoculative Releases of an Imported Parasitoid

The following account is based mainly on Jones *et al.* (1983). Importation of the tachinid *Aploymyiopsis epilachnae* (Aldrich) from Mexico during 1922–1923 was the first attempt to control the Mexican bean beetle on common bean (*Phaseolus vulgaris* Linnaeus) (Smyth, 1923). The parasitoid failed to become established despite extensive releases of the flies between 1931 and 1935 in 19 states. Although up to 90% parasitization was attained shortly after releases, the fly could not survive the winter (Landis & Howard, 1940).

The first importations specifically aimed at controlling the Mexican bean beetle on soybean were made in 1966, when two parasitoids of Oriental species of *Epilachna* were brought from India by the U.S. Department of Agriculture (Angalet *et al.*, 1968). The egg parasitoid *Tetrastichus ovulorum* Ferriere (Eulophidae) did not adapt to the new host, but the eulophid *Pediobius foveolatus* (Crawford), a larval–pupal parasitoid, selectively attacked *E. varivestis* but not the larvae of beneficial coccinellids. Although the parasitoid produced various generations within a season, thereby attaining high levels of parasitization, it could not overwinter in the mid-Atlantic states of the United States.

A program of inoculative releases tested in Maryland in 1972 and 1973 was followed by an areawide suppression program in 1974, based on the establishment of nurse crops of common bean (Stevens *et al.*, 1975a, 1975b). Patches of common bean were strategically established early in the growing season in areas adjacent to soybean fields. The Mexican bean beetle was attracted to the bean patches and established healthy colonies that served as breeding hosts for *P. foveolatus* kept over the winter in laboratory colonies (Stevens *et al.*, 1975a). From these patches, the parasitoids readily spread to soybean fields, where levels of parasitization remained between 60 and 90%. The program was implemented in Maryland, Delaware, and Virginia (Schultz & Allen, 1976; Mellors *et al.*, 1983) and has been tested in South Carolina (Shepard & Robinson, 1976). The feasibility of the inoculative releases was discussed by Flanders (1985).

Releases of *P. foveolatus* in central Florida in 1975 and 1976 reduced Mexican bean beetle populations to nearly undetectable levels in commercial fields, although in home gardens, common bean continued to be damaged. The success of the parasitoid in Florida has been attributed to the long growing season that allows up to 10 generations of the parasitoid. In addition, there is an abundance of beggarweed, *Desmodium tortuosum*, a preferred wild host of the Mexican bean beetle that serves as natural inoculum of the parasitoid (Jones *et al.*, 1983).

This program provides one of the few examples of the use of a nurse crop in connection with inoculative releases of a parasitoid originally obtained from a host species different from that of the species targeted for biological control. The economic feasibility of the program has been demonstrated (Reichelderfer & Bender, 1979). Further research focused on strains of *P. foveolatus* imported from Japan (Honshu Island) at latitudes comparable to those in regions of the United States affected by the Mexican bean beetle. Results have failed to show overwintering survival

of these new strains (Jones *et al.*, 1983). An interesting biological trait of *P. foveolatus* is the adjustment of clutch size to reduce the detrimental effect of superparasitism (Hooker & Barrows, 1992).

Besides the use of *P. foveolatus*, research continued on the natural control of *E. varivestis* on both susceptible and resistant soybean cultivars. Some of the most fruitful studies have focused on the generalist predator *Podisus maculiventris* (O'Neil *et al.*, 1996).

Use of Microbial Pesticides

Species of entomopathogenic viruses, bacteria, fungi, and protozoa have been investigated as possible microbial control agents for insect pests of annual crops. Although many of these pathogens have shown promise in field trials, relatively few microbial insecticides are commercially available for use on annual crops, although several are used on a limited scale.

The most widely used microbial insecticide is the spore-forming bacterium *Bacillus thuringiensis*. This bacterium produces a toxic crystal at the time of sporulation that is very active against Lepidoptera but is safe to humans and innocuous to natural enemies. The crystal must be ingested by the insect pest for *B. thuringiensis* to kill the larvae. Burges and Daoust (1986) estimated total annual sales in the United States at $40 million, with most of the insecticide used to control forest Lepidoptera. Since about 50% of all insecticides used in the United States are applied to cotton, one would expect that *B. thuringiensis* would be used extensively on this crop, but this is not the case. Control has been too variable, possibly because *Heliothis* spp., major cotton pests, bore into squares before ingesting enough of the leaf surface to cause mortality. The largest use of *B. thuringiensis* on annual crops has been on vegetables. It was estimated that in 1985 between $5 and $10 million was spent on *B. thuringiensis* for the control of *Plutella xylostella* (Linnaeus), *Artogeia rapae* (Linnaeus), and *Trichoplusia ni* (Hübner).

Several different crop plants, engineered to express the genes coding for the insecticidal proteins produced by *Bacillus thuringiensis*, are now available (Meeusen & Warren, 1989; Stoner, 1996; Waage, 1996). Although there are many advantages to having such plants, their widespread use may hasten the development of resistance to the *B. thuringiensis* toxins (Gould *et al.*, 1992; Gould, 1998). Insect resistance to *B. thuringiensis* has been documented in several insect species (Walton & McGaughey, 1993; Tabashnik *et al.*, 1996; Bartlett *et al.*, 1997), and while in some cases resistance is restricted to a single class of *B. thuringiensis* toxins, in at least one case resistance is not specific (Gould *et al.*, 1992; Tabashnik *et al.*, 1997). In these cases, broad-spectrum resistance could develop in field populations of insects (Roush & Shelton, 1997).

There are no commercial fungal products available for insect pests of annual crops in the United States, but government-sponsored mass production of *Beauveria bassiana* is conducted in the former USSR, primarily for the control of the Colorado potato beetle. Species of *Metarhizium* have been extensively tested for the control of planthoppers in sugarcane and pasture grasses in Brazil. Many additional fungi have been field tested, but commercialization is not imminent for any of the species tested.

With the emphasis on lepidopteran defoliators of soybean, three strategies have been considered in the experimental development of *Nomuraea rileyi* as a biological control agent: (1) microbial insecticides (inundative releases) (Getzen, 1961; Mohamed *et al.*, 1978), (2) induced epizootics (inoculative releases) (Sprenkel & Brooks, 1975; Ignoffo *et al.*, 1976), and (3) manipulation of the ecosystem (Sprenkel *et al.*, 1979).

Nomuraea rileyi will probably never be used extensively as a microbial insecticide. Ignoffo (1981) listed characteristics of *N. rileyi* that limit its success as a microbial insecticide: (1) *N. rileyi* kills slowly, allowing older caterpillars to cause unacceptable damage before dying; (2) it requires free water for germination, growth, and sporulation; (3) it has a temperature range of 15–30°C, and extreme field temperatures may limit its effectiveness; and (4) to be effective, large spore dosages must be directed at young insects.

Because early larval instars are much more susceptible to *N. rileyi* than later instars and naturally occurring epizootics usually take place too late, an attractive concept is the induction of epizootics earlier than their natural occurrence. The basis for this approach is the prophylactic application of conidia or the release of infected larvae to induce an earlier than normal epizootic, thus suppressing larval feeding when plants are most sensitive to feeding injury (Ignoffo, 1981). When acceptable environmental conditions followed the early application of *N. rileyi*, epizootics were induced by the distribution of larval cadavers as well as by conidial application (Sprenkel & Brooks, 1975; Ignoffo *et al.*, 1976). In experimental plots, *N. rileyi* did not spread from treated to untreated plots and Ignoffo *et al.* (1976) concluded that natural spread resulted from limited outward spread of conidia from dead larvae. This conclusion suggests that early applications of *N. rileyi* require complete coverage of the crop and that the inoculation of a field only at several spots is not feasible.

Theoretically, *N. rileyi* epizootics could be improved by altering the conditions of the ecosystem. The two most obvious variables that can be modified are moisture within the plant canopy and age of the host plant. Early planting (Sprenkel *et al.*, 1979) and use of closed canopy plant varieties (Burleigh, 1975) have given limited experimental successes.

The role of *N. rileyi* as a biological control agent remains enigmatic. It is one of the major mortality factors of lepidopterous defoliators of several crops, especially soybeans, but natural epizootics usually occur too late to protect the crop from extensive larval feeding, and inundative releases (microbial insecticides) are probably not practical. Carruthers and Soper (1987) suggest that the approach most likely to enhance natural control by *N. rileyi* is habitat management through cultural practices.

A nuclear polyhedrosis virus (NPV) was isolated from larvae of *Anticarsia gemmatalis* collected in southern Brazil (Carner & Turnipseed, 1977). The virus was imported into the United States, characterized in the laboratory, and tested for infectivity. Dosages as low as 17 polyhedral inclusion bodies (PIBs) per larva were effective, and field tests in small plots at rates of 49 larvae equivalents (LE) per ha significantly reduced populations of *A. gemmatalis*. These preliminary results were confirmed in Florida (Moscardi, 1977), and since the early 1980s, extensive field and laboratory studies have continued in Brazil at the National Soybean Research Center (CNPSoja, Londrina, Parana). The virus (AgNPV) is highly specific to *A. gemmatalis* and is effective at field dosages above 10 LE/ha. Populations are reduced below the economic injury level with a single application of the virus suspension, and mortality reaches 80% at 40 LE/ha (Moscardi, 1983; Moscardi & Sosa-Gomez, 1996). Leaf consumption by diseased larvae was reduced by about 75%, and although the halflife of crude preparations or a purified preparation with a clay adjuvant was either 6 or 7 days, respectively, a single application was sufficient to control the caterpillars.

Large-scale field testing started in Parana during the 1980–1981 and 1981–1982 growing seasons. The AgNVP was applied as a crude preparation at 50 LE/ha when *A. gemmatalis* larvae were less than 1.5 cm long. Applications were made with ground equipment at rates of 100–200 liters of water per ha. Virus used in the field experiments was extracted from batches of 50 cadavers of large caterpillars (>2.5 cm). The dead larvae were macerated in water and filtered through several layers of cheesecloth. The suspension was then transferred to a sprayer tank containing the amount of water needed to cover 1 ha. Infectivity after 4 days was 80%. Experiments were conducted in areas of high incidence of *A. gemmatalis,* and check plots were either treated with standard insecticides or left untreated. In all cases, yields were as high with the virus treatment as they were with insecticides (Moscardi, 1983; Moscardi & Sosa-Gomez, 1996). An estimated 11,000 ha of soybeans was treated with the AgNPV in 1983–1984, and the area was expected to increase to 300,000 ha in 1984–1985 and approached 1,000,000 ha in the 1989–1990 season (Moscardi & Sosa-Gomez 1992, Gazzoni *et al.* 1994).

The baculoviruses (nuclear polyhedrosis viruses) of insects have not enjoyed spectacular success as microbial pesticides of crop plants in the United States. Although some are registered for use as microbial insecticides, none are currently being sold as commercial products. Field trials have been conducted on control of insect pests of annual crops with many different baculoviruses. Many of these trials have produced encouraging results, but the costs of production make large-scale commercialization difficult. Many of these viruses can best be produced in a cottage industry environment and in areas where hand labor is moderate (see the earlier Brazilian example).

Genetically engineered baculoviruses have been developed and many of these have enhanced insecticidal activity. The production of NPVs in tissue culture is also being emphasized (Bonning & Hammock, 1996). The combination of enhanced insecticidal activity and a more economically efficient production method using tissue culture may allow baculoviruses to become commercially available for control of the pests of crop plants in the United States (Wood & Granados, 1991; Wood, 1991; Federici, 1993; Possee *et al.*, 1993; Wiseman & Hamm, 1993; Wood, 1993; Shuler, 1995; Murhammer, 1996).

INTERACTIONS OF BIOLOGICAL CONTROL WITH OTHER IPM TACTICS

More than in any other managed ecosystem, the future of biological control in transient crops rests on a clear perception of the impact on natural enemies of cultural practices, host-plant resistance, and—above all—the use of chemical pesticides. These aspects of IPM have received considerable attention and excellent sources of reference are found in Hoy and Herzog (1985) and Boethel and Eikenbary (1986). The following sections are based on materials in these books. Altieri (1994, 1995) offered excellent insights into the role of habitat management on the enhancement of biological control.

Impact of Cultural Practices on Natural Enemies

Cultural practices may enhance or inhibit colonization of crop fields by natural enemies and their efficiency as mortality factors for the pest complex on the crop. Effects on natural enemies may result from the direct impact of alterations of the micro- or macroenvironment or indirectly through the effect of such alterations on the host plant and the herbivorous host or prey (Herzog & Funderburk, 1985; Andow, 1997). The main cultural practices that affect both host and natural enemy populations are (1) planting and harvest dates, (2) plant density and row width, (3) tillage practices, and (4) water management (Herzog & Funderburk, 1986). Whether herbicides used in weed control in transient crops have an effect on natural enemies is uncertain, but the effect of removing weedy hosts on natural

enemy prey has been studied to some extent (Herzog & Funderburk, 1986; Dutcher, 1993). For example, naturally occurring populations of *Trichogramma* spp. differentially parasitized artificially placed *Helicoverpa zea* eggs on soybean depending on the weed species associated with the soybean. Parasitization was higher on eggs on soybean associated with *Desmodium* or *Croton* than on soybean associated with grasses or in monocultures (Altieri, 1981). Planting dates may affect colonization success of both herbivores and their complement of natural enemies. Predation rates on soybean caterpillars as well as incidence of the fungus *N. rileyi* were higher on early-planted than on late-planted soybean, but when soybean was planted late, as a second crop after small grain, planting in narrow rows and at high seeding rates minimized infection by *N. rileyi* (Sprenkel *et al.*, 1979). Tillage detrimentally affects natural enemies that along with their pest hosts overwinter in the stubble in the field (Carl, 1979; Holmes, 1982; and references in Herzog and Funderburk (1986) for soybean). However, Andow (1992) could not show an effect of three different conservation tillage systems on natural parasitism of *Ostrinia nubilalis* eggs by *Trichogramma* sp., but chewing by predators was higher in chisel-plowed and lowest in no-tillage systems (see also Stinner & House, 1990).

Effect of Plant Resistance on Biological Control

Plant resistance has been considered the IPM tactic most compatible with biological control (Kogan, 1982). However, incompatibilities arise when mechanisms of resistance indiscriminately affect both pests and natural enemies or when natural enemies are indirectly affected through their hosts or prey. Bottrell *et al.* (1998) reviewed the interactions of natural enemies and plant resistance. They stressed that understanding plant–pest–natural enemy interactions is essential to the successful integration of plant resistance with biological control for optimal IPM results. Experimental evidence of incompatibilities was shown in tomato (Duffey & Isman, 1981; Duffey & Bloem, 1986; Duffey *et al.*, 1986). *Helicoverpa zea*, *Spodoptera exigua*, and a common endoparasitic wasp, *Hyposoter exiguae* (Vier.) (Ichneumonidae), were used to demonstrate such chemical incompatibilities. When the caterpillars ingested a diet containing the glycoalkaloid tomatine, the development of the parasitoid was significantly affected or, depending on dosage, the parasitoid was killed (Duffey & Bloem, 1986). Glandular trichomes and their principal constituent, 2-tridecanone, of the insect-resistant wild tomato *Lycopersicon hirsutum* f. *glabratum* had a similarly adverse effect on *Trichogramma pretiosum* (Kashyap *et al.*, 1991). Such studies demonstrate that, depending on the mechanism of resistance, natural enemies can be detrimentally affected. Programs exploiting such mechanisms should weigh the risk of reducing the natural enemy load versus the benefit of the particular resistance trait.

Similar conclusions were drawn by Obrycki (1986), who studied the impact of potato glandular trichomes on *Edovum puttleri* Grissell (Eulophidae), an egg parasitoid of the Colorado potato beetle. Greenhouse studies showed that *E. puttleri* readily parasitizes *L. decemlineata* eggs on *Solanum tuberosum* but that the parasitoid is entrapped in glandular trichomes of *Solanum berthaultii*. On *S. tuberosum*, egg mortality is increased not only due to parasitism but probably also due to host feeding and superparasitism. However, aphid parasitoids that are equally affected by *S. berthaultii* trichomes in the greenhouse were not greatly affected in the field, leading to the conclusion that moderate levels of trichomes and the biological control of potato aphids are not incompatible. It is therefore apparent that both biochemical and physical plant defenses are potentially detrimental to natural enemies. Since behavioral adaptations of parasitoids of insects adapted to resistant lines may occur in nature, identifying such adapted populations would be useful in the search for new sources of natural enemies.

Impact of Pesticides on Natural Enemies

It is important to assess the selectivity of any pesticide to achieve maximum enhancement of natural enemy action in an IPM system. Even the so-called natural insecticides — neem extracts and *Bacillus thuringiensis* — were shown to have a detrimental effect on the parasitism of *Plutella xylostella* eggs by *Trichogramma pretiosum* and *T. principium* (Klemm & Schmutterer, 1993).

The incompatibility between chemical and biological control has been, in fact, the main force behind the evolution of IPM. Because the subject has been so thoroughly debated in the past 40 years (see exhaustive review in Croft, 1990), we merely reemphasize that it is essential to evaluate the impact of natural enemies when assessing the need to apply chemical controls. One criterion for such an assessment is the concept of "inaction levels" of natural enemies proposed by Sterling (1984). Based on this concept, decisions to initiate insecticide sprays should be based not only on the economic injury levels of pests (the action level) but also on the effective population density of natural enemies (inaction levels). This concept has been slow to penetrate IPM programs, but a successful IPM system for soybean in Brazil has essentially rested on the assessment of the level of natural infections of soybean caterpillars by the fungus *Nomuraea rileyi* (Kogan *et al.*, 1977, 1998). Equally important is to identify, augment, or create strains of natural enemies with high levels of resistance to pesticides. The outstanding results obtained with pesticide-resistant predaceous mites were discussed earlier (see also Hoy, 1994).

Strains of *Trichogramma chilonis* were identified in India with differential tolerances to some highly toxic, nonselective insecticides such as methylparathion and quinalphos (Mandal & Somchoudhury, 1992). Use of such tolerant strains may determine the success of an augmentative release program where pesticides must also be used.

CONCLUSION

There are examples of classical biological control of transient crop pests, but the ecological characteristics of these crops often reduce the probability of success in comparison to success rates with perennial, more stable agroecosystems. Successes of inundative and inoculative releases of natural enemies, however, have been numerous, suggesting that the potential for this method of control is yet to be fully realized. As with all other control tactics in the IPM arsenal, biological control of pests of transient crops yields the best results when it is a component, preferably the key component, of the IPM strategy. Biological control when fully integrated with all other control tactics, including the conservative use of selective insecticides, when needed, increases the cost-effectiveness of the pest control system, it is usually well accepted by growers, and it leads to increased confidence in the public about the soundness of an agricultural production operation.

Outbreaks of new pests, native or exotic, are constantly challenging established IPM systems. Outbreaks of *Bemisia argentifolii* in the United States and in other countries have led to a worldwide effort to develop new IPM strategies. It has become apparent that biological control alone is not the solution. Those outbreaks have generated a flurry of new research and what is evolving are comprehensive IPM systems, well adapted to the agroecological characteristics of the affected regions. Even under the rigorously controlled conditions of greenhouses, effectiveness of the whitefly parasitoid, *Encarsia formosa,* is enhanced if integrated with other IPM tactics (Hoddle *et al.*, 1998). Study of the *Bemisia* case history will, likely, become a model for similar events in the future (Gerling & Meyer, 1996).

References

Aldrich, S. R., Scott, W. D., & Leng, E. R. (1975). "Modern Corn Production," 2nd ed. A&L Publications, Champaign, IL.

Altieri, M. A. (1981). Weeds may augment biological control of insects. California Agriculture, 35, 22–24.

Altieri, M. A. (1994). "Biodiversity and Pest Management in Agroecosystems." Food Products Press, New York.

Altieri, M. A. (1995). "Agroecology: The Science of Sustainable Agriculture," 2nd ed. Westview, Boulder, CO.

Amaya, A. M. (1982). Investigacion, utilizacion y resultados obtenidos en diferentes cultivos con el uso de *Trichogramma* en Colombia Sur America. *In* "Les Trichogrammes, ler Symposium International," pp.

201–207. Colloques de l'INRA (Institut National de la Recherche Agronomique), Paris.

Andow, D. A. (1992). Fate of eggs of first generation *Ostrinia nubilalis* (Lepidoptera: Pyralidae) in three conservation tillage systems. Environmental Entomology, 21, 388–393.

Andow, D. A. (1997). "Ecological Interactions and Biological Control." Westview, Boulder, CO.

Andow, D. A., Kareiva, P. M., Levin, S. A., & Okubo, A. (1993). Spread of invading organisms: Patterns of spread. *In* "Evolution of Insect Pests: Patterns of Variation" (K. C. Kim & B. A. McPheron, Eds.), pp. 219–242. Wiley, New York.

Andreadis, T. G. (1980). *Nosema pyrausta* infection in *Macrocentrus grandii*, a braconid parasite of the European corn borer, *Ostrinia nubilalis*. Journal of Invertebrate Pathology, 35, 229–233.

Andreadis, T. G. (1982). Impact of *Nosema pyrausta* on field populations of *Macrocentrus grandii*, an introduced parasite of the European corn borer, *Ostrinia nubilalis*. Journal of Invertebrate Pathology, 39, 298–302.

Andreadis, T. G. (1984). Epizootiology of *Nosema pyrausta* in field populations of the European corn borer (Lepidoptera: Pyralidae). Environmental Entomology, 13, 882–887.

Angalet, G. W., Coles, L. W., & Stewart, J. A. (1968). Two potential parasites of the Mexican bean beetle from India. Environmental Entomology, 10, 782–786.

Armbrust, E. J., Maddox, J. V., & McGuire, M. R. (1985). Controlling alfalfa pests with biological agents. *In* "Integrated Pest Management on Major Agricultural Systems" [Texas Agricultural Experiment Station MP-1616]. (R. E. Frisbie & P. L. Adkisson, Eds.), pp. 424–443.

Arthur, J. C. (1886). A new larval *Entomophthora*. Botanical Gazette, 11, 14.

Baaren, J., & van Nenon, J. P. (1996). Intraspecific larval competition in two solitary parasitoids, *Apoanagyrus (Epidinocarsis) lopezi* and *Leptomastix dactylopii*. Entomologica Experimentalis Applicata, 81, 325–333.

Balachowski, A. S. (1951). "La Lutte Contre les Insectes." Payot, Paris.

Bar, D., Gerling, D., & Rossler, Y. (1979). Bionomics of the principal natural enemies attacking *Heliothis armigera* in cotton fields in Israel. Environmental Entomology, 8, 468–475.

Barney, R. J., Watson, P. L., Black, K., Maddox, J. V., & Armbrust, E. J. (1980). Illinois distribution of the fungus *Entomophthora phytonomi* (Zygomycetes: Entomophthoraceae) in larvae of the alfalfa weevil (Coleoptera: Curculionidae). Great Lakes Entomology, 13, 149–150.

Bartlett, A. C., Dennehy, T. J., & Antilla, L. (1997). An evaluation of resistance to Bt toxins in native populations of the pink bollworm. *In* "Proceedings of the Beltwide Cotton Conferences, Memphis, Tenn," Vol. 2, pp. 885–891. National Cotton Council of America.

Beglyarov, G. A., & Smetnik, A. I. (1977). Seasonal colonization of entomophages in the USSR. *In* "Biological Control by Augmentation of Natural Enemies" (R. L. Ridgway & S. B. Vinson, Eds.), pp. 283–328. Plenum, New York.

Bellotti, A. (1985). Cassava. *In* "Spider Mites, Their Biology, Natural Enemies and Control" (W. Helle & M. W. Sabelis, eds.), Vol. 1B. Elsevier, Amsterdam.

Bellotti, A., & van Schoonhoven, A. (1978). Mite and insect pests of cassava. Annual Review of Entomology, 23, 39–67.

Bellotti, A., & van Schoonhoven, A. (1985). Cassava pests and their control. *In* "Cassava: Research, Production and Utilization" (J. H. Cock & J. A. Reyes, Eds.), pp. 343–392. CIAT, Cali, Colombia.

Bellotti, A., Reyes, J. A., & Guerrero, J. M. (1982). Acaros presentes en al cultivo de la yuca y su control. Guia de Estudio. Ser. 04SC-02-04. CIAT, Cali, Colombia.

Bellotti, A. C., Reyes, J. A., Guerrero, J. M., & Varela, A. M. (1985). The mealybug and cassava green spider mite complex in the Americas: Problems of and potential for biological control. *In* "Cassava: Research, Production and Utilization" (J. H. Cock & J. A. Reyes, Eds.), pp. 393–439. CIAT, Cali, Colombia.

Bellotti, A. C., Smith, L., & Lapointe, S. L. (1999). Recent advances in cassava pest management. Annual Review of Entomology 44, 343–370.

Bennett, F. D., & Yaseen, M. (1975). "Investigation on the Cassava Mite *Mononychellus tanajoa* (Bondar) and Its Natural Enemies in the Neotropics." Commonwealth Institute of Biological Control, Trinidad, W. Indies.

Ben-Ze'ev, I., & Kenneth, R. G. (1982). *Zoophthora phytonomi* and *Conidiobolus osmodes* (Zygomycetes: Entomophthoraceae), two pathogens of *Hypera* species (Coleoptera: Curculionidae) coincidental in time and place. Entomophaga, 25, 171–186.

Bigler, F. (1986). Mass production of *Trichogramma maidis* and its field application against *Ostrinia nubilalis* in Switzerland. Journal of Applied Entomology, 101, 23–29.

Bigler, F., Baldinger, J., & Luisoni, L. (1982). L'impact de la methode d'elevage et de l'hote sur la qualite intrinsique de *Trichogramma evanescens* Westw. *In* "Les Trichogrammes, er Symposium International," pp. 167–180. Colloques de l'INRA (Institut National de la Recherche Agronomique), Paris.

Bing, L. A., & Lewis, L. C. (1991). Suppression of *Ostrinia nubilalis* (Hübner) (Lepidoptera: Pyralidae) by endophytic *Beauveria bassiana* (Balsamo) Vuillemin. Environmental Entomology, 20, 1207–1211.

Bissell, T. L. (1952). United States Bureau Entomology and Quarantine, Cooperative Economic Insect Report, 2, 4.

Boethel, D. J., & Eikenbary, R. D. (1986). "Interactions of Plant Resistance and Parasitoids and Predators of Insects." Ellis Horwood, Chichester.

Bonning, B. C., & Hammock, B. D. (1996). Development of recombinant baculoviruses for insect control. Annual Review of Entomology, 41, 191–210.

Bottrell, D. G., Barbosa, P., & Gould, F. (1998). Manipulating natural enemies by plant variety selection and modification: A realistic strategy? Annual Review of Entomology, 43, 347–367.

Boussienguet, J. (1986). The entomophagous insects of the cassava mealybug, *Phenacoccus manihoti* (Homoptera, Coccoidea, Pseudococcidae) in Gabon: I. Faunistic review and trophic relationships. Annules de la Societe Entomologique de France, 22, 35–44.

Bradley, J. R., Jr. (1993). Influence of habitat on pest status and management of *Heliothis* species on cotton in the southern United States. *In* "Evolution of Insect Pests: Patterns of Variation" (K. C. Kim, & B. A. McPheron, Eds.), pp. 375–391. Wiley, New York.

Brown, G. C., & Nordin, G. L. (1982). An epizootic model of an insect–fungal pathogen system. Bulletin of Mathematical Biology, 44, 731–740.

Burbutis, P. P., Erwin, N., & Ertle, L. R. (1981). Reintroduction and establishment of *Lydella thompsoni* and notes on other parasites of the European corn borer in Delaware. Environmental Entomology, 10, 779–781.

Burges, H. D., & Daoust, R. A. (1986). Current status of the use of bacteria as biocontrol agents. *In* "Fundamental and Applied Aspects of Invertebrate Pathology" (R. A. Samson, J. M. Vlak, & D. Peters, Eds.), pp. 514–517. 4th International Colloquium on Invertebrate Pathology, Wageningen, The Netherlands.

Burleigh, J. G. (1975). Comparison of *Heliothis* spp. larval parasitism and *Spicaria* infection in closed and open canopy cotton varieties. Environmental Entomology, 4, 574–576.

Caffrey, D. J., & Worthley, L. H. (1927). "A Progress Report on the Investigations of the European Corn Borer" [USDA Bulletin No. 1476].

Caltagirone, L. E. (1981). Landmark examples in classical biological control. Annual Review of Entomology, 26, 213–232.

Carl, K. P. (1979). The importance of cultural measures for the biological control of the cereal leaf beetle *Oulema melanopus* (Col.: Chrysomelidae). Mitteilungen der Schweizerischen Entomologischen Gesellschaft, 52, 443.

Carner, G. R., & Turnipseed, S. G. (1977). Potential of a nuclear polyhedrosis virus for control of the velvetbean caterpillar in soybean. Journal of Economic Entomology, 70, 608–610.

Carruthers, R. I., & Soper, R. S. (1987). Fungal diseases. In "Epizootiology of Insect Diseases" (J. R. Fuxa & Y. Tanada, Eds.), pp. 357–416. Wiley, New York.

Castro, T. R., Ruppel, R. F., & Gomulinski, M. S. (1965). Natural history of the cereal leaf beetle in Michigan. Michigan State University Agricultural Experiment Station Quarterly Bulletin 47, 623–653.

Chamberlein, T. R. (1924). "Introduction of Parasites of the Alfalfa Weevil into the United States" [USDA Circular 301].

Chiang, H. C. (1978). Pest management in corn. Annual Review of Entomology, 23, 101–123.

Clarke, A. R. (1990). The control of Nezara viridula (L.) with introduced egg parasitoids in Australia. Australian Journal of Agricultural Research, 41, 1127–1146.

Clarke, A. R. (1992). Current distribution and pest status of Nezara viridula (L.) (Hemiptera: Pentatomidae) in Australia. Journal of the Australian Entomological Society, 31, 289–297.

Clausen, C. P. (1978). "Introduced Parasites and Predators of Arthropod Pests and Weeds: A World Review" [USDA Handbook 480].

Cock, J. H. (1985). Cassava: A basic energy source in the tropics. In "Cassava: Research, Production and Utilization" (J. H. Cock & J. A. Reyes, Eds.), pp. 1–29. CIAT, Cali, Colombia.

Cock, J. H., & Reyes, J. A., Eds., (1985). "Cassava: Research, Production and Utilization, Preliminary Edition." CIAT, Cali, Colombia.

Coulson, J. R., Klassen, W., Cook, R. J., King, E. G., Chiang, H. C., Hagen, K. S., & Yendol, W. G. (1982). "Notes on Biological Control of Pests in China, 1979." USDA, Office of International Cooperation and Development, China Program, Washington, D.C.

Cox, J. M., & Williams, D. J. (1981). An account of cassava mealybugs (Hemiptera: Pseudococcidae) with description of a new species. Bulletin of Entomological Research, 71, 247–258.

Croft, B. A. (1990). "Arthropod Biological Control Agents and Pesticides." Wiley, New York.

DeBach, P., Ed. (1964). "Biological Control of Insect Pests and Weeds." Reinhold, New York.

Deitz, L. L., Van Duyn, J. W., Bradley, J. R., Jr., Rabb, R. L., Brooks, W. M., & Stinner, R. E. (1976). "A Guide to the Identification and Biology of Soybean Arthropods in North Carolina." [North Carolina Agricultural Experiment Station Technical Bulletin 238].

Denno, R. F., & McClure, M. F. (1985). "Variable Plants and Herbivores in Natural and Managed Systems." Academic Press, New York.

Dritschilo, W., & Erwin, T. L. (1982). Responses in abundance and diversity of cornfield carabid communities to difference in farm practices. Ecology, 63, 900–904.

Duffey, S. S., & Bloem, K. A. (1986). Plant defense–parasite–herbivore interactions and biological control. In "Ecological Theory and Integrated Pest Management Practice" (M. Kogan, Ed.), pp. 135–183. Wiley, New York.

Duffey, S. S., & Isman, M. B. (1981). Inhibition of insect larval growth by phenolics in glandular trichomes of tomato leaves. Experientia, 37, 574–576.

Duffey, S. S., Bloem, K. A., & Campbell, B. C. (1986). Consequences of sequestration of plant natural products in plant–insect–parasitoid interactions. In (D. J. Boethel & R. D. Eikenbary, Eds.), pp. 31–60. "Interactions of Plant Resistance and Parasitoids and Predators of Insects," Ellis Horwood, Chichester.

Dutcher, J. D. (1993). Recent examples of conservation of arthropod natural enemies in agriculture. (R. D. Lumsden & J. L. Vaughn, Eds.) pp. 101–108. In "Pest Management: Biologically Based Technologies" [Proceedings of the Beltsville Sympesia XVIII, USDA–ARS] American Chemical Society, Washington, D.C.

Dysart, J. R., & Day, W. H. (1976). Release and recovery of introduced parasites of the alfalfa weevil in eastern North America. Agricultural Research Service, USDA Production Research Report 167.

Federici, B. A. (1993). Viral pathobiology in relation to insect control. In (N. Beckage, N. Thompson, & B. A. Federici, Eds.), "Parasites and Pathogens of Insects", Vol. II, pp. 81–101. Academic Press, San Diego.

Felk, G. (1990). Status of rearing, releasing and establishment of Trichogramma evanescens Westw. for control of Asian corn borer Ostrinia furnicalis Gueneé in the Philippines. In "Integrated Pest Management in Tropical and Subtropical Cropping Systems" 89, pp. 625–637. Deutsche Landwirtschafts-Gesellschaft, Frankfurt, Germany.

Flanders, R. V. (1985). Biological control of the Mexican bean beetle: Potentials for and problems of inoculative releases of Pediobius foveolatus. In "World Soybean Research Conference III: Proceedings," (R. Shibles, Ed.). Westview, Boulder, CO.

Fullaway, D. T. (1920). The melon fly: Its control in Hawaii by a parasite introduced from India. Haw. For. Agric., 17, 101–105.

Fuxa, J. R. (1984). Dispersion and spread of the entomopathogenic fungus Nomuraea rileyi (Moniliales: Moniliaceae) in a soybean field. Environmental Entomology, 13, 252–258.

Garcia-Roa, F. (1991). Effectiveness of Trichogramma spp. in biological control programs in the Cauca Valley, Colombia. Collogues INRA, 56, 197–199.

Gazzoni, D. L., Sosa Gomes, D. R., Moscardi, F., Hoffmann-Campo, C. B., Spalding Correa-Ferreira, S., de Oliveira, L. J., & Corso, I. C. (1994). Insects. In Tropical Soybean: Improvement and Production," pp. 81–108. EMBRAPA-CNPSo, Lodrina, Brazil [FAO Plant Protection Series 27, Rome, Italy].

Gerling, D., & R. T. Meyer, (1996). "Bemisia: 1995—Taxonomy, Biology, Damage, Control and Management." Intercept, Andover, U.K.

Getzen, L. W. (1961). Spicaria rileyi (Farlow) Charles, an entomogenous fungus of Trichoplusia ni (Hübner). Journal at Insect Pathology, 3, 2–10.

Goodenough, L. et al. (1986). Efficacy of entomophagous arthropods. In (S. J. Johnson, E. G. King, & J. Bradley, Jr., Eds.), pp. 75–91. "Theory and Tactics of Heliothis Population Management: 1. Cultural and Biological Control" [Southern Cooperative Series Bulletin, 316].

Gould, F. (1998). Sustainability of transgenic insecticidal cultivars: Integrating pest genetics and ecology. Annual Review of Entomology 43, 701–726.

Gould, F., Martinez-Ramirez, A., Anderson, A., Ferre, J., Silva, F. J., & Moar, W. J. (1992). Broad-spectrum resistance to Bacillus thuringiensis toxins in Heliothis virescens. Proceedings of the National Academic of Sciences, USA, 89, 7986–7990.

Gutierrez, J. (1987). The cassava green mite in Africa: One or two species (Acari: Tetranychidae)? Experimental & Applied Acarology, 3, 163–168.

Hajek, A. E., & St. Leger, R. J. (1994). Interactions between fungal pathogens and insect hosts. Annual Review of Entomology, 39, 293–322.

Harcourt, D. G., Guppy, J. C., MacLeod, D. M., & Tyrrel, D. (1974). The fungus Entomophthora phytonomi pathogenic to the alfalfa weevil, Hypera postica. Canadian Entomologist, 106, 1295–1300.

Harper, J. D., McPherson, R. M., & Shepard, M. (1983). Geographical and seasonal occurrence of parasites, predators and entomopathogens. In "Natural Enemies of Arthropod Pests in Soybean" H. N. Pitre, Ed.), pp. 7–19. [Southern Cooperative Series Bulletin 285].

Hassan, S. A. (1982). Mass-production and utilization of Trichogramma: 3. Results of some research projects related to the practical use in the Federal Republic of Germany. In "Les Trichogrammes, er Symposium International," pp. 214–218. Colloques de l'INRA (Institute National de la Recherche Agronomique), Paris.

Hassan, S. A., Stein, E., Danneman, K., Reichel, W. (1986). Mass-production and utilization of Trichogramma: 8. Optimizing the use to control the European corn borer, Ostrinia nubilalis. Journal of Applied Entomology, 101, 508–515.

Hawlitzky, N. (1986). Etude de la biologie de la pyrale du mais, *Ostrinia nubilalis* Hübner (Lep.: Pyralidae) en region parisienne durant quatre annees et recherche d'elements previsionnels du debut de ponte. Acta Oecologica, Oecologica Applicata, 7, 47–68.

Hawlitzky, N., Stengel, M., Voegele, J., Crouzet, B., & Raynaud, B. (1987). Strategy used in France on the biological control of the European corn borer, *Ostrinia nubilalis* Hübner (Lep.: Pyralidae) by oophagous insects, *Trichogramma maidis* Voeg. et Pint. (Hym.: Trichogrammatidae). *In* "Conference Internationale sur les Ravageurs en Agriculture, Paris."

Haynes, D. L., & Gage, S. H. (1981). The cereal leaf beetle in North America. Annual Review of Entomology, 26, 259–287.

Heinrichs, E. A. (1988). "Plant Stress/Insect Interactions." Wiley, New York.

Herren, H. R. (1996). Cassava and cowpea in Africa. *In* "Biotechnology and Integrated Pest Management" (G. J. Parsley, Ed.) pp. 136–149. CAB International, Wallingford, U.K.

Herren, H. R., & Neuenschwander, P. (1991). Biological control of cassava pests in Africa. Annual Review of Entomology, 36, 257–283.

Herren, H. R., Neuenschwander, P., Hennessey, R. D., & Hammond, W. N. O. (1987). Introduction and dispersal of *Epidinocarsis lopezi* (Hym.: Encyrtidae), an exotic parasitoid of the cassava mealybug, *Phenacoccus manihoti* (Hom.: Pseudococcidae), in Africa. Agriculture Ecosystems, & Environment, 19, 131–144.

Herzog, D. C., & Funderburk, J. E. (1985). Plant resistance and cultural practice interactions with biological control. *In* "Biological Control in Agricultural IPM Systems" (M. A. Hoy & D. C. Herzog, Eds.), pp. 67–88. Academic Press, Orlando, FL.

Herzog, D. C., & Funderburk, J. E. (1986). Ecological bases for habitat management and pest cultural control. *In* "Ecological Theory and Integrated Pest Management Practice" (M. Kogan, Ed.), pp. 217–250. Wiley, New York.

Hill, R. E., & Gary, W. J. (1979). Effects of the microsporidium *Nosema pyrausta* on field populations of European corn borers in Nebraska. Environmental Entomology, 8, 91–95.

Hill, R. E., Carpino, D. P., & Mayo, Z. B. (1978). Insect parasites of the European corn borer, *Ostrinia nubilalis* in Nebraska from 1958-1976. Environmental Entomology, 7, 249–253.

Hinson, K., & Hartwig, E. E. (1982). "Soybean Production in the Tropics." FAO Plant Production and Protection Paper 4 (revised by H. C. Minor).

Hoddle, M. S., Van Driesch, R. G., & Sanderson, J. P. (1998). Biology and use of the whitefly parasitoid Encarsia formosa. Annual Review of Entomology, 43, 645–669.

Hokkanen, H. (1985). Success in classical biological control. CRC Critical Reviews in Plant Sciences, 3, 35–72.

Hokkanen, H. (1986). Polymorphism, parasites, and the native area of *Nezara viridula* (Hemiptera, Pentatomidae). Annales Entomologici Fennici, 52, 28–31.

Hokkanen, H. (1997). Role of biological control and transgenic crops in reducing use of chemical pesticides for crop protection. *In* "Techniques for Reducing Pesticide Use: Economic and Environmental Benefits" (Pimentel, D., Ed.), pp. 103–127. Wiley, New York.

Holmes, N. D. (1982). Population dynamics of the wheat stem sawfly, *Cephus cinctus* (Hymenoptera: Cephidae). Canadian Entomologist, 114, 775–788.

Hooker, M. E., & Barrows, M. (1992). Clutch size reduction and host discrimination in the superparasitizing gregarious endoparasitic wasp *Pediobius foveolatus* (Hymenoptera: Eulophidae). Annals of the Entomological Society of America, 85, 207–213.

Hoy, M. A. (1994). "Insect Molecular Genetics: An Introduction to Principles and Applications." Academic Press, San Diego, CA.

Hoy, M., & Herzog, D. C., Eds., (1985). "Biological Control in Agricultural IPM Systems." Academic Press, Orlando, FL.

Huffaker, C. B. (1977). Augmentation of natural enemies in the People's Republic of China. *In* "Biological Control by Augmentation of Natural Enemies" (R. L. Ridgway & S. B. Vinson, Eds.), pp. 329–339. Plenum, New York.

Huffaker, C. B., & Messenger, P. S., Eds. (1976). "Theory and Practice of Biological Control." Academic Press, New York.

Huffaker, C. B., van de Vrie, M., & McMurtry, J. A. (1970). Ecology of tetranychid mites and their natural enemies: A review. II. Tetranychid populations and their possible control by predators: An evaluation. Hilgardia, 40, 391–458.

Ignoffo, C. M. (1981). The fungus *Nomuraea rileyi* as a microbial insecticide. *In* "Microbial Control of Pests and Plant Diseases 1970–1980," pp. 513–538. Academic Press, London.

Ignoffo, C. M., Marston, N. L., Hostetter, D. L., Puttler, B., & Bell, J. V. (1976). Natural and induced epizootics of *Nomuraea rileyi* in soybean caterpillars. Journal of Invertebrate Pathology, 27, 191–198.

Ignoffo, C. M., Puttler, B., Marston, N. L., Hostetter, D. L., & Dickerson, D. A. (1975). Seasonal incidence of the entomopathogenic fungus *Spicaria rileyi* associated with noctuid pests of soybeans. Journal of Invertebrate Pathology, 25, 135–137.

IITA. (1981). Cassava mealybug: Biological control. *In* "Research Highlights for 1980," pp. 40–43. International Institute of Tropical Agriculture, Ibadan, Nigeria.

IITA. (1985). Dissemination, dispersal, and impact of *E. lopezi*—A natural enemy of the cassava mealybug. *In* "Research Highlights for 1984," pp. 35–39. International Institute of Tropical Agriculture, Ibadan, Nigeria.

IITA. (1987a). Strategies for classical biological control of cassava green mites. *In* "Annual Report and Research Highlights 1986," pp. 112–114. International Institute of Tropical Agriculture, Ibadan, Nigeria.

IITA. (1987b). Update on release, establishment, and impact of *Epidinocarsis lopezi* and other natural enemies of the cassava mealybug. *In* "Annual Report and Research Highlights 1986," pp. 115–118. International Institute of Tropical Agriculture, Ibadan, Nigeria.

Jeppson, L. R., Keifer, H. H., & Baker, E. W. (1975). "Mites Injurious to Economic Plants." Univ. of California Press, Berkeley.

Jimenez, E. (1980). Review of some interesting developments. Plant protection, Mexico. I.O.B.C. Newsletter, 15, 5.

Jones, W. A. (1988). World review of the parasitoids of the southern green stink bug, *Nezara viridula* (Linnaeus) (Heteroptera: Pentatomidae). Annals of the Entomological Society of America, 81, 262–273.

Jones, W. A., Young, S. Y., Shepard, M., & Whitcomb, W. H. (1983). Use of imported natural enemies against insect pests of soybean. *In* "Natural Enemies of Arthropod Pests in Soybean" (H. N. Pitre, Ed.), pp. 63–77. [Southern Cooperative Series Bulletin, 285].

Kashyap, R. K., Kennedy, G. G., & Farrar, R. R., Jr. (1991). Mortality and inhibition of *Helicoverpa zea* egg parasitism rates of *Trichogramma* in relation to trichome/methyl ketone mediated insect resistance of *Lycopersicon hirsutum* f. *glabratum* accession PI 134417. Journal of Chemical Ecology, 17, 2381–2395.

Kaya, H. K., & Gaugler, R. (1993). Entomopathogenic nematodes. Annual Review of Entomology, 38, 181–206.

Kennedy, G. G., & Margolies, D. C. (1985). Mobile arthropod pests: Movement in diversified agroecosystems. Bulletin of the Entomological Society of America, 31, 21–27.

Kidd, N. A. C., & Jervis, M. A. (1996). Population dynamics. *In* "Insect Natural Enemies: Practical Approaches to Their Study and Evaluation" (M. Jervis & N. Kidd, Eds.), pp. 293–374. Chapman & Hall, London.

Kim, K. C., & McPheron, B. A. (1993). "Evolution of Insect Pests: Patters of Variation." Wiley, New York.

King, E. C., & Coleman, R. J. (1989). Potential for biological control of *Heliothis* species. Annual Review of Entomology, 34, 53–75.

King, E. C., Bull, D. L., Bouse, L. F., & Phillips, J. R., Eds., (1985a). Biological control of bollworm and tobacco budworm in cotton by augmentative releases of *Trichogramma*. Southwestern Entomologist, 8, (Suppl.), 1–198.

King, E. C., Bull, D. L., Bouse, L. F., & Phillips, J. R., Eds. (1985b). Introduction: Biological control of *Heliothis* spp. in cotton by augmentative releases of *Trichogramma*. Southwestern Entomologist, 8, (Suppl.), 1–10.

King, E. C., Coleman, R. J., Phillips, J. R., & Dickerson, W. A. (1985c). *Heliothis* spp. and selected natural enemy populations in cotton: A comparison of three insect control programs in Arkansas (1981–82) and North Carolina (1983). Southwestern Entomologist, 8, (Suppl.), 71–98.

Kingsley, P. C., Bryan, M. D., Day, W. H., Burger, T. L., Dysart, R. J., & Schwalbe, C. P. (1993). Alfalfa weevil (Coleoptera: Curculionidae) in biological control: Spreading the benefits. Environmental Entomology, 22, 1234–1250.

Kish, L. P., & Allen, G. E. (1978). "The Biology and Ecology of *Nomuraea rileyi* and a Program for Predicting Its Incidence on *Anticarsia gemmatalis* in Soybean" [Florida Agricultural Experiment Station Bulletin 795,].

Kiyindou, A., & Fabres, G. (1987). Etude de la capacite d'accroissement chez *Hyperaspis raynevali* (Col.: Coccinellidae) predateur introduit au Congo pour la regulation des populations de *Phenacoccus manihoti* (Hom.: Pseudococcidae). Entomophaga, 32, 181–189.

Klemm, U., & Schmutterer, H. (1993). Effects of neem preparations on *Plutella xylostella* (L.) and its natural enemies of the genus *Trichogramma*. Zeitschrift fuer Pflanzenkrankheiten Pflanzenschutz, 100, 113–128.

Kogan, M. (1981). Dynamics of insect adaptations to soybean: Impact of integrated pest management. Environmental Entomology, 10, 363–371.

Kogan, M. (1982). Plant resistance in pest management. *In* "Introduction to Insect Pest Management" (R. L. Metcalf & W. H. Luckmann, Eds.), 2nd ed., pp. 93–134. Wiley, New York.

Kogan, M., & Turnipseed, S. G. (1987). Ecology and management of soybean arthropods. Annual Review of Entomology, 32, 507–538.

Kogan, M., Croft, B. A., & Suthurst, R. F. (1999). Applications of ecology for integrated pest management. *In* "Ecological Entomology" J (A. Gutierrez, Ed.), 2nd ed. Wiley, New York.

Kogan, M., Helm, C. G., Kogan, J., & Brewer, E. (1989). Distribution and economic importance of *Heliothis virescens* and *Helicoverpa zea* in North, Central, and South America, including a listing and assessment of the importance of their natural enemies and host plants. *In* "Proceedings, Workshop on Biological Control of *Heliothis*: Increasing the Effectiveness of Natural Enemies" (E. G. King & R. D. Jackson, Eds.), pp. 240–297. Far Eastern Regional Research Office, USDA, New Delhi, India.

Kogan, M., Turnipseed, S. G., Shepard, M., Oliveira, E. B., & Borgo, A. (1977). Pilot insect pest management program for soybean in southern Brazil. Journal of Economic Entomology, 70, 659–663.

Kramer, J. P. (1959). Some relationships between *Perezia pyraustae* Paillot (Sporozoa: Nosematidae) and *Pyrausta nubilalis* (Hübner) (Lepidoptera: Pyralidae). Journal of Insect Pathology, 1, 25–33.

Krysan, J. L., Foster, D. E., Brason, T. F., Ostlie, K. R., & Cranshaw, W. S. (1986). Two years before the hatch: Rootworms adapt to crop rotation. Bulletin of the Entomological Society of America, 32, 250–253.

Landis, B. J., & Howard, N. F. (1940). *Paradexodes epilachnae*, a tachinid parasite of the Mexican bean beetle. USDA Technical Bulletin, 721.

Landis, D. A., & Haas, M. J. (1992). Influence of landscape structure on the abundance and within-field distribution of European corn borer (Lepidoptera: Pyralidae) larval parasitoids in Michigan. Environmental Entomology, 21, 409–416.

Lawton, J. H. (1978). Host-plant influences on insect diversity: The effects of space and time: "Diversity of Insect Faunas" (L. A. Mound & N. Waloff, Eds.), pp. 105–125. [Symposium, Royal Entomological Society of London 5]. Blackwell, Oxford.

Lema, K. M., & Herren, H. R. (1985). Release and establishment in Nigeria of *Epidinocarsis lopezi*, a parasitoid of the cassava mealybug, *Phenacoccus manihoti*. Entomologica Experimentalis Applicata, 38, 171–176.

Le Ru, B. (1986). Epizootiology of the entomophthoraceous fungus *Neozygites fumosa* in a population of the cassava mealybug, *Phenacoccus manihoti* (Homoptera: Pseudococcidae). Entomophaga, 31, 79–90.

LeSar, C. D., & Unzicker, J. D. (1978). Soybean spiders: Species composition, population densities, and vertical distribution. Illinois Natural History Survey Biology Notes 107.

Liss, W. J., Gut, L. J., Westigard, P. H., & Warren, C. E. (1986). Perspectives on arthropod community structure, organization, and development in agricultural crops. Annual Review of Entomology, 31, 455–478.

Loan, C. (1981). Suppression of the fungi *Zoophthora* spp. by captafol: A technique to study interaction between disease and parasitism in the alfalfa weevil, *Hypera postica* (Coleoptera: Curculionidae). Proceedings of the Entomological Society of Ontario, 112, 81–82.

Los, L. M., & Allen, W. A. (1983). Incidence of *Zoophthora phytonomi* (Zygomycetes: Entomophthorales) in *Hypera postica* (Coleoptera: Curculionidae) larvae in Virginia. Environmental Entomology, 12, 1318–1321.

Luck, R. F. (1981). Parasitic insects introduced as biological control agents for arthropod pest. "Handbook of Pest Management in Agriculture" (D. Pimentel, Ed.), Vol. II, pp. 125–284. CRC Press, Boca Raton, FL.

Lumsden, R. D., & Vaughn, J. L., Eds. (1993). Pest Management: Biologically Based Technologies. Proceedings of the Beltsville Symposium XVIII, USDA–ARS. American Chemical Society, Washington, D.C.

Luttrell, R. G., Fitt, G. P., Ramalho, F. S., & Sugonyaev, E. S. (1994). Cotton pest management: A worldwide perspective. Annual Review of Entomology, 39, 517–526.

Lyon, W. F. (1974). A plant-feeding mite, *Mononychellus tanajoa* (Bondar), new to the African continent threatens cassava (*Manihot esculenta* Crantz) in Uganda. Proceedings of the National Academy of Sciences, USA, 19, 36–37.

Maddox, J. V. (1987). Protozoan diseases. *In* "Epizootiology of Insect Diseases" (J. R. Fuxa & Y. Tanada, Eds.), pp. 417–452. Wiley, & New York.

Mandal, S. K., & Somchoudhury, A. K. (1992). Effect of some important insecticides on the pupal stages of five ecotypes of *Trichogramma chilonis* Ishii. Journal of Entomological Research (New Delhi), 16, 245–250.

Manglitz, G. R., & App, B. A. (1957). Biology and seasonal development of the alfalfa weevil in Maryland. Journal of Economic Entomology, 50, 810–813.

Mayse, M. A., & Price, P. W. (1978). Seasonal development of soybean arthropod communities in east central Illinois. Agroecosystems, 4, 387–405.

McMurtry, J. A., & Croft, B. A. (1997). Life-styles of phytoseiid mites and their roles in biological control. Annual Review of Entomology, 42, 291–321.

Meeusen, R. L., & Warren, G. (1989). Insect control with genetically engineered crops. Annual Review of Entomology, 34, 373–381.

Mellors, W. K., Forrester, O. T., & Schwalbe, C. P. (1983). Parasite production in nurse plots after inoculative release of *Pediobius foveolatus* (Hymenoptera: Eulophidae) against Mexican bean beetle larvae (Coleoptera: Cocinellidae). Journal of Economic Entomology, 76, 1452–1455.

Mesa, N. C., & Bellotti, A. (1987). Biologically controlling destructive cassava mites with Phytoseiidae mites. Cassava Newsletter (CIAT, Colombia), 11, 4–7.

Metcalf, R. L. (1986). The ecology of pesticides and the chemical control of insects. *In* "Ecological Theory and Integrated Pest Management Practice" (M. Kogan, Ed.), pp. 251–297. Wiley, New York.

Miura, K., & Kobayashi, M. (1993). Effect of temperature on the development of *Trichogramma chilonis* Ishii (Hymenoptera: Trichogrammatidae), an egg parasitoid of the diamondback moth. Applied Entomology and Zoology, 28, 393–396.

Mochida, O., Kiritani, K., & Bay-Petersen, J. (1990). "The Use of Natural Enemies to Control Agricultural Pests." Food and Fertilizer Technology Center for the Asia and Pacific Region, Taipei, Taiwan.

Mohamed, A. K. L., Bell, J. V., & Sikorowski, P. P. (1978). Field cage tests with *Nomurea rileyi* against corn earworm larvae on sweet corn. Journal of Economic Entomology, 71, 102–110.

Mohyuddin, A. I. (1991). Utilization of natural enemies for the control of insect pests of sugarcane. Insect Science and Its Application, 12, 19–26.

Morris, M. J. (1985). Influence of the fungal pathogen, *Erynia* sp. (Zygomycetes: Entomophthorales), on larval populations of the alfalfa weevil, *Hypera postica* (Gyllenhal) (Coleoptera: Curculionidae) in Illinois, M.S. Thesis, Univ. of Illinois, Urbana, IL.

Moscardi, F. (1977). Control of *Anticarsia gemmatalis* Hübner on soybean with a baculovirus and selected insecticides and their effect on natural epizootics of the entomogenous fungus *Nomuraea rileyi* (Farlow) Samson, M.S. Thesis, Univ. of Florida, Gainesville, FL.

Moscardi, F. (1983). "Utilização de *Baculovirus anticarsia* para o controle da lagarta da soja, *Anticarsia gemmatalis*" [EMBRAPA/CNPSoja Comun. Tec. 23].

Moscardi, F., & Sosa-Gómez, D. R. (1992). Use of viruses against soybean caterpillars in Brazil. *In* "Pest Management in Soybean," (L. G. Copping, M. B. Green, & R. T. Ress, Eds.), pp. 98–109. Elsevier, London.

Moscardi, F., & Sosa-Gómez, D. R. (1996). *In* "Biotechnology and Integrated Pest Management" (G. J. Persley, Ed), pp. 98–112. CAB International Wallingford, U.K.

Muka, A. A. (1976). A disease of the alfalfa weevil in New York. Proceedings, Forage Insects Research Conference, 18, 28–29.

Murhammer, D. W. (1996). Use of viral insecticides for pest control and production in cell culture. Applied Biochemistry and Biotechnology, 59, 199–200.

Neal, T. M. (1974). Predaceous arthropods in the Florida soybean agroecosystem. M.S. Thesis, Univ. of Florida, Gainesville, FL.

Neuenschwander, P., Haug, T., Herren, H. R., & Madoneju, E. (1984). Root and tuber improvement program. *In* "Biological Control Annual Report for 1983," pp. 114–118. International Institute of Tropical Agriculture, Ibadan, Nigeria.

Neuenschwander, P., Hennessey, R. D., & Herren, H. R. (1987). Food web of insects associated with cassava mealybug, *Phenacoccus manihoti* Matile-Ferrero (Hemiptera: Pseudococcidae), and its introduced parasitoid, *Epidinocarsis lopezi* (De Santis) (Hymenoptera: Encyrtidae), in Africa. Bulletin of Entomological Research, 77, 177–190.

Neuenschwander, P., Schulthess, F., & Madojemu, E. (1986). Experimental evaluation of the efficiency of *Epidinocarsis lopezi*, a parasitoid introduced into Africa against the cassava mealybug *Phenacoccus manihoti*. Entomologica Experimentalis Applicata, 42, 133–138.

Newell, I. M., Mitchell, W. C., & Rathbun, F. L. (1952). Infestation norms for *Dacus cucurbitae* in *Momordica balsamina*, and seasonal differences in activity of the parasite *Opius fletcheri*. Hawaiian Entomological Society Proceedings, 14, 497–508.

Nishida, T. (1955). Natural enemies of the melon fly, *Dacus cucurbitae* Coq. in Hawaii. Annals of the Entomological Society of America, 48, 171–178.

Nishida, T., & Bess, H. A. (1950). Applied ecology in melon fly control. Journal of Economic Entomology, 43, 877–883.

Nordin, G. L., Brown, G. C., & Millstein, J. A. (1983). Epizootic phenology of *Erynia* disease of the alfalfa weevil, *Hypera postica* (Gyllenhal) (Coleoptera: Curculionidae) in Central Kentucky. Environmental Entomology, 12, 1350–1355.

Oatman, E. R., Gilstrap, F. E., & Voth, V. (1976). Effect of different release rates of *Phytoseiulus persimilis* (Acarina: Phytoseidae) on the two-spotted spider mite on strawberry in Southern California. Entomophaga, 21, 269–274.

Oatman, E. R., McMurtry, J. A., Gilstrap, F. E., & Voth, V. (1977a). Effect of releases of *Amblyseius californicus*, *Phytoseiulus persimilis*, and *Typhlodromus occidentalis* on the two-spotted spider mite on strawberry in Southern California. Journal of Economic Entomology, 70, 45–47.

Oatman, E. R., McMurtry, J. A., Gilstrap, F. E., & Voth, V. (1977b). Effect of releases of *Amblyseius californicus* on the two-spotted spider mite on strawberry in Southern California. Journal of Economic Entomology, 70, 638–640.

Oatman, E. R., Wyman, J. A., Browning, H. W., & Voth, V. (1981). Effects of releases and varying infestation levels of the two-spotted spider mite (*Tetranychus urticae*) on strawberry yield in Southern California. Journal of Economic Entomology, 74, 112–115.

Obrycki, J. J. (1986). The influence of foliar pubescence on entomophagous species. *In* "Interactions of Plant Resistance and Parasitoids and Predators of Insects" (D. J. Boethel & R. D. Eikenbary, Eds.), pp. 61–83. Ellis Horwood, Chichester.

Odebiyi, J. A., & Bokonon-Ganta, A. H. (1989). Biology of *Epidinocarsis* (= *Apoanagyrus*) *lopezi* (Hymenoptera: Encyrtidae), an exotic parasite of cassava mealybug, *Phenacoccus manihoti* (Homoptera: Pseudococcidae) in Nigeria. Entomophaga, 31, 251–260.

Oloumi-Sadeghi, H., Steffey, K. L., Roberts, S. J., Maddox, J. V., & Armbrust, E. J. (1993). Distribution and abundance of two alfalfa weevil (Coleoptera: Curculionidae) larval parasitoids in Illinois. Environmental Entomology, 22, 220–225.

O'Neil, R. J., Nagarajan, K., Wiedenmann, R. N., & Legaspi, J. C. (1996). A simulation model of *Podisus maculiventris* (Say) (Heteroptera: Pentatomidae) and Mexican bean beetle, *Epilachna varivestis* (Mulsant) (Coleoptera: Coccinellidae), population dynamics in soybean, Glycine max (L.). Biological Control: Theory and Applied Pest Management, 6, 330–339.

Paillot, A. (1927). Sur deux protozoaires nouveaux parasites des chenilles de *Pyrausta nubilalis* Hb. Comptes Rendus Hebdomadaires des Seances de l'Academie des Sciences, 185, 673–675.

Panizzi, A. R. (1997). Wild hosts of pentatomids: Ecological significance and role in their pest status. Annual Review of Entomology, 42, 99–122.

Parra, J. R. P., & Zucchi, R. A., Eds. (1997). "Trichogramma e o Controle Biológico de Pragas." FEALQ, Piracicaba, Brazil.

Pavlik, J. (1993a). The size of the female and quality assessment of mass-reared *Trichogramma* spp. Entomologica Experimentalis Applicata, 66, 171–177.

Pavlik, J. (1993b). Variability in the host acceptance of European corn borer, *Ostrinia nubilalis* Hbn. (Lep.: Pyralidae) in strains of the egg parasitoid *Trichogramma* spp. (Hym.: Trichogrammatidae). Journal of Applied Entomology, 115, 77–84.

Pedigo, L. P., Bechinski, E. J., & Higgins, R. A. (1982). Partial life tables of the green cloverworm (Lepidoptera: Noctuidae) in soybean and a hypothesis of population dynamics in Iowa. Environmental Entomology, 12, 186–195.

Pijls, J. W. A. M., Kofker, K. D., van Staalduinen, M. J., & van Alphen, J. J. M. (1995). Interspecific host discrimination and competition in *Apoanagyrus (Epidinocarsis) lopezi* and A. (E.) *diversicornis*, parasitoids of the cassava mealybug *Phenacoccus manihoti*. Ecological Entomology, 20, 326–332.

Pintureau, B. (1991). Selection of two characteristics in a *Trichogramma* sp.: Parasitic efficacy of the obtained strains (Hymenoptera: Trichogrammatidae). Agronomie (Paris), 11, 593–602.

Pitre, H., Ed. (1983). "Natural Enemies of Arthropod Pests in Soybean" [Southern Cooperative Series Bulletin 285].

Possee, R. D., Cayley, P. J., Cory, J. S., & Bishop, D. H. L. (1993). Genetically engineered viral insecticides: New insecticides with improved phenotypes. Pesticide Science, 39, 109–115.

Powell, W., Walton, M. P., & Jervis, M. A. (1996). Populations and communities. "Insect Natural Enemies: Practical Approaches to Their Study and Evaluation" (M. Jervis & N. Kidd, Eds.), pp. 223–292. Chapman & Hall, London.

Price, P. W. (1976). Colonization of crops by arthropods: Non-equilibrium communities in soybean fields. Environmental Entomology, 5, 505–611.

Price, P. W. (1981). Relevance of ecological principles to practical biological control. In "Biological Control in Crop Production" (G. C. Papavizas, Ed.), pp. 3–19. [Beltsville Symposia in Agricultural Research, 5]. Allanheld, Osmun Publ., Granada.

Price, P. W., & Waldbauer, G. P. (1982). Ecological aspects of pest management. In "Introduction to Insect Pest Management" (R. L. Metcalf & W. H. Luckmann, Eds.), 2nd ed., pp. 33–68. Wiley, New York.

Prokrym, D. R., Andow, D. A., Ciborowski, J. A., & Sreenivasam, D. D. (1992). Suppression of Ostrinia nubilalis in sweet corn. Entomologica Experimentalis Applicata, 64, 73–85.

Puttler, B., Hostetter, D. L., Long, S. H., & Pinnell, R. E. (1978). Entomophthora phytonomi, a fungal pathogen of the alfalfa weevil in the mid-great plains. Environmental Entomology, 7, 670–671.

Rabb, R. L., & Kennedy, G. G., Eds. (1979). "Movement of Highly Mobile Insects: Concepts and Methodology in Research." Department of Entomology, North Carolina State University, Raleigh, N.C.

Reichelderfer, K., & Bender, F. (1979). Application of a simulative approach to evaluating alternative methods for the control of agricultural pests. American Journal of Agricultural Economics, 61, 258–267.

Rey, J. R., & McCoy, E. D. (1979). Application of island biogeography theory to pests of cultivated crops. Environmental Entomology, 8, 577–582.

Reynolds, H. T., Adkisson, P. L., Smith, R. F., & Frisbie, R. E. (1982). Cotton insect pest management. In "Introduction to Insect Pest Management" (R. L. Metcalf & W. H. Luckmann, Eds.), 2nd ed., pp. 375–441. Wiley, New York.

Ridgway, R. L., & Morrison, R. K. (1985). Worldwide perspective on practical utilization of Trichogramma with special reference to control of Heliothis on cotton. Southwestern Entomologist, 8, (Suppl.), 190–198.

Roush, R. T., & Shelton, A. M. (1997). Assessing the odds: The emergence of resistance to Bt transgenic plants. Nature Biotechnology, 15, 816–817.

Schulthess, F., Neuenschwander, P., & Gounou, S. (1997). Multi-trophic interactions in cassava, Manihot esculenta, cropping systems in the subhumid tropics of West Africa. Agriculture, Ecosystems, & Environment, 66, 211–222.

Schultz, P. B., & Allen, W. A. (1976). Field evaluation of Mexican bean beetle suppression through use of Pediobius foveolatus in Virginia. Vegetable Growers News 31, 1–3.

Shen, X., Wang, K., & Meng, G. (1988). The inoculative release of Trichogramma dendrolimi for controlling corn borer and rice leaf roller. In "Trichogramma and Other Egg Parasites. 2nd International Symposium, Guanzhou, People's Republic of China" (J. Voegele, J. Waage, & J. C. van Lenteren, Eds.), pp. 575–580. [Colloques de l'INRA No. 43].

Shepard, M., & Robinson, J. (1976). Suppression of Mexican bean beetle in soybean by the imported parasite, Pediobius foveolatus. Sumter Area Agriculture Development Project Report 1976, 49–53.

Shepard, M., Carner, G. R., & Turnipseed, S. G. (1977). Colonization and resurgence of insect pests of soybean in response to insecticide and field isolation. Environmental Entomology, 6, 501–506.

Shuler, M. L. (1995). "Baculovirus Expression Systems and Biopesticides." Wiley, New York.

Siegel, J. P., Maddox, J. V., & Ruesink, W. G. (1986). The impact of Nosema pyrausta on a braconid Macrocentrus grandii in central Illinois. Journal of Invertebrate Pathology, 47, 271–276.

Siegel, J. P., Maddox, J. V., & Ruesink, W. G. (1987). Survivorship of the European corn borer, Ostrinia nubilalis (Hübner) (Lepidoptera: Pyralidae) in central Illinois. Environmental Entomology, 16, 1071–1075.

Simberloff, D. (1986). Island biogeographic theory and integrated pest management. In "Ecological Theory and Integrated Pest Management Practice" (M. Kogan, Ed.), pp. 19–35. J Wiley, New York.

Sithanantham, S., & Paul, A. V. N. (1989). Control of Heliothis species (Lep.: Nocutidae) by augmentative releases of predators and parasites in India. In "Increasing the Effectiveness of Natural Enemies. Proceedings of the Workshop on Biological Control of Heliothis" (E. G. King & R. D. Jackson, Eds.), pp. 427–439. Far Eastern Regional Research Office, USDA, New Delhi, India.

Slansky, F., Jr., & Rodriguez, J. G. (1987). "Nutritional Ecology of Insects, Mites, Spiders, and Related Invertebrates." Wiley, New York.

Smith, S. M. (1996). Biological control with Trichogramma: Advances, successes, and potential of their use. Annual Review of Entomology, 41, 375–407.

Smyth, E. G. (1923). A trip to Mexico for parasites of the Mexican bean beetle. Washington Academy of Sciences Journal, 13, 259–260.

Souissi, R., & Le Ru, B. (1997). Effect of host plants on fecundity and development of Apoanagyrus lopezi, an endoparasitoid of the cassava mealybug Phenacoccus manihoti. Entomologica Experimentalis Applicata, 82, 235–238.

Sparks, A. N. (1979). An introduction to the status, current knowledge, and research on movement of selected Lepidoptera in southeastern United States. In "Movement of Highly Mobile Insects: Concepts and Methodology in Research" (R. L. Rabb & G. G. Kennedy, Eds.), pp. 382–383. Department of Entomology, North Carolina State University, Raleigh, N.C.

Sprenkel, R. K., & Brooks, W. M. (1975). Artificial dissemination and epizootic initiation of Nomuraea rileyi, an entomogenous fungus of lepidopterous pests of soybeans. Journal of Economic Entomology, 68, 847–851.

Sprenkel, R. K., Brooks, W. M., Van Duyn, J. W., & Deitz, L. L. (1979). The effects of three cultural variables on the incidence of Nomuraea rileyi, phytophagous Lepidoptera and their predators on soybeans. Environmental Entomology, 8, 334–339.

Steinhaus, E. A. (1951). Report on diagnoses of diseased insects, 1944–50. Hilgardia 20, 629–678.

Stengel, M. (1982). Essai de mise au point de la prevision des egats pour la lutte contre la pyrale du mais (Ostrinia nubilalis) en Alsace (Est de la France). Entomophaga 27, 105–114.

Sterling, W. (1984). "Action and Inaction Levels in Pest Management" [Texas Agricultural Experiment Station Bulletin 1480].

Stevens, L. M., Steinhauer, A. L., & Coulson, J. R. (1975a). Suppression of Mexican bean beetle on soybeans with annual inoculative releases of Pediobius foveolatus. Environmental Entomology, 4, 947–952.

Stevens, L. M., Steinhauer, A. L., & Elden, T. C. (1975b). Laboratory rearing of the Mexican bean beetle and the parasite Pediobius foveolatus, with emphasis on parasite longevity and host–parasite ratios. Environmental Entomology, 4, 953–957.

Stinner, B. R., & House, G. J. (1990). Arthropod and other invertebrates in conservation-tillage agriculture. Annual Review of Entomology, 35, 299–318.

Stinner, R. E., Rabb, R. L., & Bradley, J. R., Jr. (1977). Natural factors operating in the population dynamics of Helicoverpa zea in North Carolina. In "Proceedings, 15th International Congress on Entomology," pp. 622–642. Washington, D.C.

Stoner, K. A. (1996). Plant resistance to insects: A resource available for sustainable agriculture. Biological Agriculture & Horticulture, 13, 7–38.

Strickland, G. R. (1981). Integrating insect control for Ord soybean production. Journal of Agriculture, Western Australia, 22, 81–82.

Strong, D. R. (1974). Rapid asymptotic species accumulation in phytophagous insect communities: The pests of cacao. Science, 185, 1064–1066.

Strong, D. R. (1979). Biogeographical dynamics of insect–host plant communities. Annual Review of Entomology, 24, 89–119.

Strong, D. R., Lawton, J. H., & Southwood, R. (1984). "Insects on Plants: Community Patterns and Mechanisms." Harvard Univ. Press, Cambridge, MA.

Tabashnik, B. E., Groeters, F. R., Finson, N., Liu, Y. B., Johnson, M. W., Heckel, D. G., Luo, K., & Adang, M. J. (1996). Resistance to *Bacillus thuringiensis* in *Plutella xylostella:* The moth heard round the world. *In* "Molecular Genetics and Evolution of Pesticide Resistance," pp. 130–140. American Chemical Society, Washington, D.C.

Tabashnik, B. E., Liu, Y. B., Finson, N., Masson, L., & Heckel, D. G. (1997). One gene in diamondback moth confers resistance to four *Bacillus thuringiensis* toxins. Proceedings of the National Academy of Sciences, USA, 94, 1640–1644.

Teng, P. (1994). Integrated pest management in rice. Experimental Agriculture, 30, 115–137.

Théberge, R. L. (1985). Common African pests and diseases of cassava, yam, sweet potato, and cocoyam. International Institute of Tropical Agriculture, Ibadan, Nigeria.

Titus, E. G. (1907). A new pest on the alfalfa. Desert Farmer, 3, 7.

Titus, E. G. (1910). "The Alfalfa Leaf Weevil" [Utah Agricultural Experiment Station Bulletin 110].

Todd, J. W., & Herzog, D. C. (1980). Sampling phytophagous Pentatomidae on soybean. *In* "Sampling Methods in Soybean Entomology" (M. Kogan & D. C. Herzog, Eds.), pp. 438–478. Springer-Verlag, New York.

Turnipseed, S., & Kogan, M. (1983). Soybean pests and indigenous natural enemies. *In* "Natural Enemies of Arthropod Pests in Soybean" (H. N. Pitre, Ed.), pp. 1–6. [Southern Cooperative Series Bulletin 285].

Udayagiri, S., & Jones, R. L. (1992). Flight behavior of *Macrocentrus grandii* Goidanich (Hymenoptera: Braconidae), a specialist parasitoid of European corn borer (Lepidoptera: Pyralidae): Factors influencing response to corn volatiles. Environmental Entomology, 21, 1448–1456.

Van Denburg, R. S., & Burbutis, P. P. (1962). The host–parasite relationship of the European corn borer. *Ostrinia nubilalis,* and the protozoan, *Perezia pyraustae,* in Delaware. Journal of Economic Entomology, 55, 65–67.

Van Driesche, R. G., & Bellows, T. S., Jr. (1996). "Biological Control." Chapman & Hall, New York.

van Lenteren, J. C. (1987). Environmental manipulation advantageous to natural enemies of pests. *In* "IPM quo vadis. Parasitis Symposium, Geneva" (V. Delucchi, Ed.), pp. 123–163.

Varma, G. C., Rataul, H. S., Shenhmar, M., Singh, S. P., Jalali, S. K. (1991). Role of inundative release of egg parasitoid *Trichogramma chilonis* Ishii in the control of *Chilo auricilius* Dudgeon on sugarcane. Journal of Insect Science, 4, 165–166.

Voegele, J. (1988). Reflections upon the last ten years of research concerning *Trichogramma* (Hym.: Trichogrammatidae), *In* "*Trichogramma* and Other Egg Parasites. Second International Symposium, Guanzhou, People's Republic of China" (J. Voegele, J. Waage, & J. C. van Lenteren, Eds.), pp. 17–29. [Colloques de l'INRA No. 43].

Von der Heyde, J. (1991). Four years of experimental and practical experience with TRICHOCAP. Zeitschrift Pflanzenkrankheiten Pflanzenschutz, 98, 453–456.

Vorley, W. T. (1986). The activity of parasitoids (Hymenoptera: Braconidae) of cereal aphids (Hemiptera: Aphididae) in winter and spring in southern England. Bulletin of Entomological Research, 76, 491–504.

Voronin, K. E. (1982). Biocenotic aspects of *Trichogramma* utilization in integrated plant protection control. *In* "Les Trichogrammes, er Symposium International," pp. 269–274. Colloques de l'INRA (Institut National de la Recherche Agronomique), Paris.

Waage, F. (1996). Integrated pest management and biotechnology: An analysis of their potential integration. *In* "Biotechnology and Integrated Pest Management" (G. J. Persley, Ed.), pp. 37–60. CAB International, Wallingford, U.K.

Wajnberg, E., & Hassan, S. A. (1994). "Biological Control with Egg Parasitoids." CAB International, Wallingford, U.K.

Walton, M. E., & McGaughey, W. H. (1993). Insect resistance to *Bacillus thuringiensis. In* "Advances in Engineered Pesticides" (L. Kim, Ed.), pp. 215–231. New York.

Watson, P. L., Barney, R. J., Maddox, J. V., & Armbrust, E. J. (1981). Sporulation and mode of infection of *Entomophthora phytonomi,* a pathogen of the alfalfa weevil. Environmental Entomology, 10, 305–306.

Whitcomb, W. H., & Bell, K. (1964). Predaceous insects, spiders, and mites of Arkansas cotton fields. Arkansas Agricultural Experiment Station Bulletin, 690, 1–84.

White, J. M., Allen, P. G., Moffitt, L. J., & Kingsley, P. P. (1995). Economic analysis of an areawide program for biological control of the alfalfa weevil. American Journal of Alternative Agriculture, 10, 173–179.

Wilson, L. T. (1985). Estimating the abundance and impact of arthropod natural enemies in IPM systems. *In* "Biological Control in Agricultural IPM Systems" (M. A. Hoy & D. C. Herzog, Eds.), pp. 303–322. Academic Press, New York.

Wilson, L. T. (1986a). Developing economic threshold in cotton. *In* "Integrated Pest Management of Major Agricultural Systems" (R. E. Frisbie & P. L. Adkisson, Eds.), pp. 308–344. [Texas Agricultural Experiment Station MP-1616].

Wilson, L. T. (1986b). The compensatory response of cotton to leaf and fruit damage. Beltwide Cotton Production Research Conference Proceedings (Las Vegas, Nevada), 1986, 149–153.

Wilson, T. (1960). A Review of the Biological Control of Insects and Weeds in Australia and Australian New Guinea. Commonwealth Institute of Biological Control, Technical Communication, 1.

Windels, M. B., Chiang, H. C., & Furgaia, B. (1976). Effects of *Nosema pyrausta* on pupal and adult stages of the European corn borer, *Ostrinia nubilalis.* Journal of Invertebrate Pathology, 27, 239–242.

Wiseman, B. R., & Hamm, J. J. (1993). Nuclear polyhedrosis virus and resistant corn silks enhance mortality of corn earworm (Lepidoptera: Noctuidae) larvae. Biological Control, 3, 337–342.

Wood, H. A. (1991). Development of genetically enhanced baculovirus pesticides. *In* "Biotechnology for Biological Control of Pests and Vectors" (K. Maramorosch, Ed.), pp. 69–76. CRC Press, Boca Raton, FL.

Wood, H. A., & Granados, R. R. (1991). Genetically engineered baculoviruses as agents for pest control. Annual Review of Microbiology, 45, 69–87.

Wuehrer, B. G., & Hassan, S. A. (1993). Selection of effective species/strains of *Trichogramma* (Hym., Trichogrammatidae) to control the diamondback moth *Plutella xylostella* (Lep., Plutellidae). Journal of Applied Entomology, 116, 80–89.

Yearian, W. C., Hamm, J. J., & Carner, G. R. (1986). Efficacy of *Heliothis* pathogens. *In* "Theory and Tactics of *Heliothis* population management: 1. Cultural and Biological Control" (S. J. Johnson, E. G. King, & J. R. Bradley, Jr., Eds.), pp. 92–103. [Southern Cooperative Series Bulletin 316].

Yukawa, J., & Kiritani, K. (1965). Polymorphism in the southern green stink bug. Pacific Insects, 7, 639–642.

Zhou, L. T. (1988). Study on parasitizing efficiency of *Trichogramma confusum* Viggiani in controlling *Heliothis armigera* Hübner and its modelling. *In* "*Trichogramma* and Other Egg Parasites. Second International Symposium, Guanzhou, People's Republic of China" (J. Voegele, J. Waage, & J. C. van Lenteren, Eds.), pp. 641–644. [Colloques de l'INRA No. 43].

Zimin, G. 1935. Trichogrammes dus mais dans la lutte contre la pyrale du mais. Zashchita Rastenii, 1, 69–80.

Glasshouse Environments

MICHAEL P. PARRELLA
Department of Entomology
University of California
Davis, California

LISE STENGÅRD HANSEN
Danish Pest Infestation Laboratory
Ministry of Foods, Agriculture, and Fisheries
Skovbrynet 14, DK-2800 Lyngby, Denmark

JOOP VAN LENTEREN
Department of Entomology
Agricultural University
Wageningen, The Netherlands

INTRODUCTION

The world glasshouse area is estimated to be 280,000 ha; 50,000 ha of this area is covered with glass and 230,000 ha with plastic, with vegetable crops grown on about 65% of this area and ornamentals on 35%. Many of the reasons which make the glasshouse environment ideally suited for the development and reproduction of pests can be used to support the concept that natural enemies can be more easily manipulated in such an environment to bring about successful biological control when contrasted with a field situation. The glasshouse environment provides the natural enemy with food (pests), protection from the environment (frosts, winds, drought, heavy rains, etc.), and a physical barrier to dispersal. Because most pest problems are peculiar to individual glasshouses where infestations stem from contamination by survivors or migrants from earlier crops, the beneficial fauna (i.e., natural enemy) must be manipulated separately for each glasshouse; this manipulation must be based on a firm knowledge of biology and of the pest–beneficial–plant–environment interaction.

Biological control is now used regularly on an estimated area of 15,000 ha, of which the great majority is in vegetable production (van Lenteren, 1995). Sufficient data from the former U.S.S.R. and some Eastern countries (including China) are lacking, so this obviously is an underestimation of the total area worldwide. A historical account of the use of biological control in glasshouses has been described by Greathead (1976), Hussey and Scopes (1985), and Lipa (1985).

A large concentration of glasshouse industry is found in western Europe. Many publications on biological control in glasshouses have originated in The Netherlands and the U.K., where there is a long tradition of research and practical application of biological control in Europe. The following descriptions of use of biological control are based primarily on research conducted in western Europe, though important contributions from other regions of the world are included.

This chapter was originally prepared with a literature cutoff date of 1988. Since 1988, there have been significant changes with respect to biological control in greenhouses. The international movement of pests (*Liriomyza huidobrensis* Blanchard, the pea leafminer; *Frankliniella occidentalis* Pergande, the western flower thrips; *Bemisia argentifolii* Perring & Bellows, the silverleaf whitefly) has dramatically complicated established biological control programs and has prompted a search for effective natural enemies of these pests. Historically, biological control in glasshouses has concentrated primarily on vegetable crops, with little attention toward ornamental production. This has changed dramatically—today there is an international cadre of scientists developing biological control strategies for flower and foliage crops.

This chapter has been modified since its original submission to include some of the high-profile changes since 1988. This has been accomplished by including some general references that the reader can use to gain more specific information.

VEGETABLE CROPS

In glasshouse vegetables, biological control is applied by the seasonal inoculative release method (van Lenteren, 1983, 1986). Limited numbers of parasites or predators are released periodically in short-term crops (6–9 months) to initiate the population buildup of beneficials for control throughout the growing season. With some systems, relatively large numbers of beneficials are released (called an inundative release) to obtain an immediate reduction of the pest population in addition to a long-term effect.

The number of hectares and pest species where biologi-

TABLE 1   World Use of Biological Control in Glasshouses for Various Pest–Beneficial Systems (van Lenteren, 1987; van Lenteren & Woets, 1988)

Year	Number of hectares				
	Tetranychus urticae/ *Phytoseiulus persimilis*	*Trialeurodes vaporariorum/* *Encarsia formosa*	Leafminers[a]/ parasites[d]	Aphids[b]/ *Aphidoletes aphidimyza*	Thrips[c]/ *Amblyseius* spp.
1968	13				
1969	30				
1970	218	115			
1971	110	128			
1972	157	144			
1973	262	286			
1974	241	561			
1975	312	690			
1976	677	1070			
1977	829	1223			
1978	908	1144		3	
1979	978	1233	3	7	1
1980	943	1181	32	7	1
1981	1016	1161	38	9	23
1982	1022	1087	10	7	35
1983	1167	1231	33	7	10
1984	1227	1410	41	13	28
1985	1300	1600	460	13	65
1986	—	—	—	—	140

[a] *Liriomyza bryoniae, L. trifolii, L. sativae,* and *Chromatomyia syngenesiae.*

[b] *Myzus persicae, Aphis gossypii, Macrosiphium euphorbiae,* and *Aulacorthum solani,* among others.

[c] *Thrips tabaci.*

[d] *Opius* spp., *Dacnusa* spp., *Diglyphus* spp., *Chrysocharis* spp.

cal control is practiced increased dramatically in commercial glasshouses from 1968 to 1986 (Table 1) (van Lenteren, 1987; van Lenteren & Woets, 1988). This information is partially updated by van Lenteren (1995).

Two pest–beneficial systems have been used extensively since the late 1960s (use of *Phytoseiulus persimilis* Athias-Henriot (Acari: Phytoseiidae) to control the two-spotted spider mite, *Tetranychus urticae* Koch (Acari: Tetranychidae), and *Encarsia formosa* Gahan (Hymenoptera: Aphelinidae) to control the greenhouse whitefly, *Trialeurodes vaporariorum* (Westwood) (Homoptera: Aleyrodidae)) followed by the more recent development and application of biological control for other glasshouse pests (leafminers, thrips, and aphids). Until recently, biological control was applied almost exclusively on cucumber and tomato, which are by far the largest vegetable crops. Today, biological control is being extended to other vegetable crops such as sweet pepper, eggplant, melon, and strawberries.

FACTORS CONDUCIVE TO THE USE OF BIOLOGICAL CONTROL IN VEGETABLES

In addition to the points mentioned previously in the general description of glasshouse environments, several aspects of vegetable growing are especially conducive to the use of biological control. These factors are responsible for the present widespread use of this method of pest suppression in vegetable crops, and some make biological control more attractive to the grower than chemical control.

Statutory Harvest Interval and Seasonal Labor Schedule (Ramakers, 1980a)   The intensive harvesting of fruit (two to three times a week) is difficult to coordinate with reentry interval restrictions and approved residues inherent in pesticide usage. Frequent applications needed in glasshouses (due to a favorable climate for pest development, pesticide resistance, etc.) create problems for chemical control in vegetable crops.

Higher Risk of Phytotoxicity during Winter Months (van Lenteren *et al.,* 1995)   Young vegetable crops planted in winter are generally less vigorous and especially susceptible to pesticides. This condition is aggravated by the application of carbon dioxide (Hussey & Scopes, 1977) to improve yields. In cucumber crops, yield increases of 20–25% have been found in glasshouses with biological control compared to glasshouses with chemical control (Gould, 1971).

Most Pests Are Indirect Feeders   Pest control in vegetable crops has fairly wide tolerance limits because the pests are primarily confined to the foliage at low to moderate densities. Consequently, greater fluctuations of pest densities can be tolerated because the marketability of the end product, the fruit, is unaffected. However, at high densities, damage to the fruit can occur. The introduction of western flower thrips into vegetable greenhouses has caused major problems because of its tendency to feed on (and deform) young, developing fruit.

Chemical Control Difficult   Due to pesticide resistance or unavailability of registered pesticides, chemical control of several glasshouse pests is difficult or even impossible.

Low Initial Pest Densities at Crop Inception   Annual removal of plants and good sanitation procedures provide ideal starting conditions for biological control, i.e., very low pest densities at the start of the growing season.

Considerable Experience Is Available   Climatic and growing conditions are similar in glasshouses throughout many parts of the world. The accumulation of great amounts of knowledge and considerable experience from practical use help to increase the rate of success.

CONTROL OF THE TWO-SPOTTED SPIDER MITE
Tetranychus urticae WITH *Pytoseiulus persimilis*

Tetranychus urticae is a major pest of glasshouse crops. The world literature on spider mites was reviewed by Hussey and Huffaker (1976).

Infestations of spider mites are very common on cucumber in most countries, whereas the occurrence of these pests on tomato and sweet pepper varies. Mites feed on cell chloroplasts, thereby reducing the photosynthetic capability of the leaf. As the population increases, damaged areas coalesce and the leaves eventually die.

Biological control of *T. urticae* is ideally suited to cucumbers, as the crop can tolerate severe damage (up to 30% of the leaf area) without yield reduction (Hussey & Parr, 1963). This, combined with access to an extremely efficient predator, explains the widespread use of biological control of *T. urticae* (Table 1).

Since the discovery of the predatory mite, *Phytoseiulus persimilis,* by Dosse (1959), its efficiency in glasshouses has been documented by many researchers (Chant, 1961; Bravenboer & Dosse, 1962; Hussey *et al.,* 1965; Legowski, 1966; Gould, 1968; Dixon, 1973; French *et al.,* 1976; Gould, 1977). This came as a welcome solution to increasing problems with acaricide resistance in two-spotted spider mites. Comprehensive bibliographies of *P. persimilis* are available (Pruszynski, 1979; Petitt & Osborne, 1984; Osborne *et al.,* 1985).

Several factors contribute to the success of *P. persimilis*

under diverse glasshouse conditions. At temperatures between 15 and 35°C, the developmental time of *P. persimilis* is shorter than that of *T. urticae*. At 20°C, *P. persimilis* and *T. urticae* increase at a rate of 4.6 and 2.7 times per week, respectively (Scopes, 1985). Bravenboer and Dosse (1962) reported that the optimal temperature for developmental time and reproductive and feeding capacity of *P. persimilis* is between 25 and 30°C. In small-scale experiments, Force (1967) obtained optimal control of *T. urticae* at a constant temperature of 25°C where stable populations were established between low densities of *T. urticae* and *P. persimilis*. This ensured survival of the predator. At 30°C, no regulation of *T. urticae* was found and at 20°C the prey was quickly eradicated. This latter situation should be avoided, as it necessitates further introductions of the predator when new infestations of *T. urticae* develop. However, the greater complexity of the environment in a glasshouse (contrasted with the simple experiments of Force (1967)) facilitates the survival of both species. Stenseth (1979) reported adequate control at temperatures ranging from 15 to 27°C.

Chant (1961) discussed the advantages of *P. persimilis* as follows: (1) high mobility, (2) voracious, (3) dependent on *T. urticae* for food, and (4) avoids leaves without prey. In addition, females of *P. persimilis* do not feed on spider mite eggs and migrate from a leaf when all active prey are eaten. However, this occurs after depositing their own eggs among eggs of *T. urticae,* both of which hatch at approximately the same time. This favors the survival of *P. persimilis* and results in wide distribution of predators.

Phytoseiulus persimilis has demonstrated a remarkable ability to disperse within a glasshouse. According to unpublished data by Hussey and Parr (referred to by Hussey and Bravenboer (1971)), very colony of spider mites in a cucumber glasshouse is associated with a predator 18 days after introducing *P. persimilis* onto one plant in ten. This predaceous mite can traverse 10 tomato plants in 10 days (Hussey & Scopes, 1977). *Phytoseiulus persimilis* is strongly attracted to patches of two-spotted mites by specific kairomones deposited by *T. urticae* on the leaves (Sabelis & van de Baan, 1983). As a result of herbivory, the plants themselves produce specific volatiles, which have been found to repel spider mites as well as attract predators. As the strength of the plant's reaction to a pest infestation varies greatly among crop cultivars, this knowledge can help to explain varying results of biological control and eventually be utilized by selection of cultivars more attractive to predators. A review of herbivore-induced defenses can be found in Dicke (1994/1995). Within a spider mite colony, *P. persimilis* detects its prey by chance contact (Jackson & Ford, 1973), but the predator stays in the colony until all prey are eliminated (Sabelis *et al.,* 1984).

Compared with other natural enemies of spider mites, phytoseiids have relatively low minimum food requirements for development and reproduction; this accounts for their efficiency even at low prey densities (McMurtry *et al.,*

1970). This was demonstrated by Hussey *et al.* (1965), where six weekly inspections of cucumber plants on which *T. urticae* had practically been eliminated by *P. persimilis* revealed no predators. After deliberately reinfesting these plants with *T. urticae*, Hussey *et al.* found that the spider mite population was controlled after 29 days by a few initial predators that had survived on the leaves during the period of low prey density.

A noteworthy feature of *P. persimilis* is the short time in which control can be obtained: 35 days (Chant, 1961); 22–33 days (Hussey *et al.*, 1965); 22 days at a predator: prey ratio of 8:20 (Force, 1967), and within 2 weeks at an initial predator:prey ratio of 1:10 (Stenseth, 1979). This fact is of great importance with respect to grower acceptance of *P. persimilis*.

The importance of the initial density of *T. urticae* for the success of *P. persimilis* cannot be overestimated (Fig. 1) (Hussey *et al.*, 1965). An estimate of the pest density is given by the leaf damage index (LDI) (Hussey & Parr, 1963), which relates the number of mites feeding per leaf to a visual damage rating. When predators are introduced at low LDIs, reduction of the pest population density is achieved before the economic injury level is reached. If plants are damaged to a mean LDI of 1.0 before predator introduction, reduction of the mite population occurs more quickly, but the economic injury level is exceeded. In this scenario, it is possible that the predator population would become so numerous that it would quickly eradicate the mites and then die out from starvation.

Low relative humidity (RH) affects *P. persimilis* adversely. Stenseth (1979) found that survival of the egg stage

of *P. persimilis* dropped from 99.7% at 80% RH to 7.5% at 40% RH (27°C). According to Pralavorio and Almaguel-Rojas (1980), few predators completed their larval development at 50% RH or lower over a range of temperatures. Furthermore, at low relative humidities, adult longevity and fecundity of *P. persimilis* are greatly reduced. This phytoseiid mite avoids excessive heat which normally occurs at the tops of cucumber plants in midsummer. The predators leave the apical foliage and hide beneath the lowest leaves, leaving *T. urticae* free to increase at the upper halves of the plants (Hussey & Scopes, 1977). This problem can be avoided by timing the original introduction of *P. persimilis* to achieve almost complete control of *T. urticae* before warm temperatures occur in the glasshouse (usually before June).

On tomato, biological control of spider mites was less successful for many years. Recently, good results have been obtained with strains of *P. persimilis* specifically selected for use on this crop.

Introduction Methods

Different methods of introducing *P. persimilis* on vegetable crops have been developed and a brief discussion of these methods follows.

Patch Introduction

Predaceous mites are introduced at the site of the initial spider mite infestation as a result of ex-diapausing female *T. urticae*. This is followed by introductions of *P. persimilis* on cucumber plants infested with *T. urticae* on which no predators have been discovered during weekly inspections (sampling). This method has been used on cucumber with success (Gould, 1968, 1970; Stenseth, 1980). This method is not unreasonably time-consuming if the inspections are conducted routinely while other procedures (pruning and picking) are being carried out. Danish cucumber growers spend about 8 h per year per 1000 m² with this method (Hansen *et al.*, 1984a).

Pest-in-First

Cucumber plants are deliberately infested with *T. urticae* immediately after planting. After ca. 10 days, *P. persimilis* is introduced on the same plants. This method gives the most predictable control (Hussey *et al.*, 1965; Legowski, 1966; Gould, 1970; Dixon, 1973). This method was initially adopted as it led to predictable control, but it has subsequently been abandoned.

Simultaneous Introduction

A uniform distribution of *T. urticae* and *P. persimilis* is created either before spider mite infestations are observed

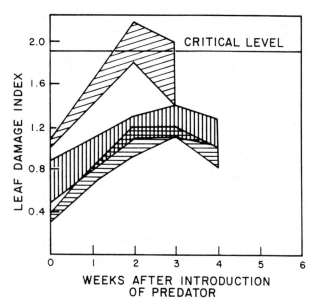

FIGURE 1 Leaf damage when predator was introduced on different host populations. Shaded areas illustrate maximum and minimum degree of damages within the six replicates of each treatment (Hussey *et al.*, 1965).

(Legowski, 1966; Stenseth, 1980) or at the first sign of leaf damage (French *et al.,* 1976; Stenseth, 1980). This method is preferable when large numbers of ex-diapausing females are expected in the glasshouse (Stenseth, 1985).

Blind Release

Early in the season, before *T. urticae* is observed on the plants, *P. persimilis* is introduced by means of a slow-release system, i.e., a container with different stages of the predator and an alternative food source.

CONTROL OF GREENHOUSE WHITEFLY, *Trialeurodes vaporariorum,* USING *Encarsia formosa*

The greenhouse whitefly, *Trialeurodes vaporariorum,* has an enormous host range; this species has been found on plants from 249 genera in 84 plant families (Russel, 1977). A thorough review of the literature on the whitefly pest problem and use of *Encarsia formosa* has been completed (Vet *et al.,* 1980; van Roermund & van Lenteren, 1992a, 1992b). In glasshouses the whitefly is considered a major pest of vegetable crops; it is a most serious pest of tomato and cucumber.

Trialeurodes vaporariorum feeds on the phloem of the plant, but the principal injury arises from the excretion of honeydew by all developmental stages. Honeydew on foliage and fruits gives rise to the growth of sooty molds (*Cladosphaerospermum* spp.), which reduces photosynthesis and hinders respiration (Hussey *et al.,* 1958).

Successful use of the parasite *E. formosa* by Speyer (1927) led to commercial application of this beneficial wasp for the ensuing 25 years in England and other European countries. During this period, no precise recommendations detailing introduction rates and intervals were made, which resulted in unsatisfactory control in many cases. After the advent of synthetic organic pesticides in the 1940s, use of *E. formosa* was largely discontinued.

The development of acaricide resistance in *T. urticae* as early as 1949 (Hussey, 1985) in glasshouses in Europe prompted growers to continue using *P. persimilis* for control. As this resistance problem increased, so did grower dependency on the predaceous mite. Pesticide application for control of *T. vaporariorum* upset biological control of *T. urticae* due to the low chemical tolerance of *P. persimilis.* Consequently, interest in biological control of *T. vaporariorum* was revived in the 1970s. Investigations in England and The Netherlands led to more precise recommendations concerning the use of *E. formosa* (Woets, 1973, 1976, 1978; Parr *et al.,* 1976). The efficiency of this

beneficial wasp is best demonstrated by examining the rapid increase in the area on which *E. formosa* was used during the 1980s (Table 1).

Biological Characteristics

Several factors that contribute to making *E. formosa* extremely attractive as a biological control agent in glasshouses include the following.

Searching Capacity

Hussey and Scopes (1977) found that the parasitoids are arrested when they are in contact with immature whiteflies and honeydew. The parasitoid is not able to locate infested plants from a distance; searching is random at all levels, and even after a host has been found, the search pattern does not change (Noldus & van Lenteren, 1990). The only change observed was that a parasitoid keeps searching much longer on a leaf once a whitefly or honeydew is detected (van Lenteren *et al.,* 1996). This explains the efficiency of *E. formosa* and was further confirmed when the detailed searching behavior of the parasitoid was modeled and related to the population dynamics and spatial distribution of whitefly and parasitoid (van Roermund *et al.,* 1997).

Parasitization Efficiency

Encarsia formosa can discriminate between parasitized and healthy hosts, thus avoiding superparasitism and egg deposition in marginal hosts (van Lenteren *et al.,* 1976). Furthermore, when all hosts encountered on a leaf are parasitized, the wasp quickly migrates in search of other leaves with unparasitized hosts (van Lenteren *et al.,* 1977).

Host Feeding

Encarsia formosa feeds on unparasitized hosts only (Nell *et al.,* 1976) and thus avoids eliminating any of its own offspring. A description of *E. formosa*'s oviposition behavior can be found in van Lenteren *et al.* (1980a).

During the 1970s, increasing energy prices forced growers to reduce average temperature in glasshouses, and research was initiated to find tomato varieties well suited to low temperatures (18°C day, 7°C night). This temperature regime was considered incompatible with the use of *E. formosa,* as the intrinsic rate of natural increase (r_m) of the parasite was believed to be lower than that of *T. vaporariorum* at temperatures below 20°C. Intensive research in The Netherlands, reviewed by van Lenteren and Hulspas-Jordan (1983), revealed that between 12 and 25°C, *E. formosa* exhibits a greater r_m than *T. vaporariorum* (Fig. 2). Furthermore, adult *E. formosa* migrate at temperatures as low as 13°C. Consequently, it was concluded that reducing

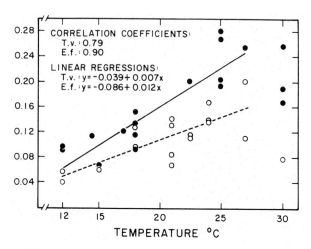

FIGURE 2 Values for intrinsic rate of increase of *T. vaporariorum* (---, ○) and *E. formosa* (—, ●) at different temperatures. The data for 30°C are not used for calculation of the linear regression (van Lenteren & Hulspas-Jordan, 1983).

temperatures in the glasshouse would present no problems for the continued use of *E. formosa*.

Encarsia formosa shows a strong functional response to whitefly on tomato (van Lenteren *et al.,* 1977; van Roermund & van Lenteren, 1993) and successful biological control is regularly achieved on tomato. This was not the case with cucumbers, where several workers have reported difficulties in obtaining good control of *T. vaporariorum* with *E. formosa*. A possible explanation is that cucumber is a better host plant for the greenhouse whitefly (Woets & van Lenteren, 1976) and parasitization by *E. formosa* is less efficient on cucumber (compared with tomato) (van Lenteren *et al.,* 1977). It has been shown that cucumber has longer hairs on the leaves, which retain large amounts of honeydew. This reduces the walking speed of *E. formosa* and the parasite spends more time preening. The solution for obtaining sufficient control of *T. vaporariorum* on cucumber is either to use cultivars with few hairs (van Lenteren *et al.,* 1995) or to apply inundative instead of inoculative releases.

Parr *et al.* (1976) reported that light (sunshine) is a major stimulus of *E. formosa* activity, which may help explain several reports of control failures during winter and early spring when light intensity is low. Humidity also seems to play an important role. Milliron (1940) found the greatest percentage of parasitization at relative humidities between 50 and 70%. According to Ekbom (1977), this also explains the difficulties of using *E. formosa* on cucumber, which is cultivated at a higher relative humidity than tomato. In "patches" with high densities of whitefly and large amounts of honeydew, *E. formosa* does not always exert sufficient control (Ekbom, 1977).

Introduction Methods

Encarsia formosa is easily mass produced at relatively small cost. The parasites are introduced into glasshouses as pupae; this is a resilient stage protected by the larval skin of the whitefly.

Introduction of *E. formosa* when greenhouse whitefly densities are low is of paramount importance for successful control. An initial density of 10 adult *T. vaporariorum* per 100 m² is considered too high (Ekbom, 1977). A primary parasitization rate of 50% or more is necessary to prevent an increase in the population of *T. vaporariorum*. More than ten introduction methods for *E. formosa* have been described (Onillon, 1990) and a brief overview of some of these methods follows.

Pest-in-First

This method involves deliberate infestation of the plants with *T. vaporariorum* followed by several introductions of *E. formosa*. This allows for precise timing of parasite introductions to coincide with development of whitefly third instars, which are preferred by the parasite for oviposition. Reliable control can be obtained with this method (Gould *et al.,* 1975; Parr *et al.,* 1976) but the reluctance of growers to introduce whiteflies into their crops has prevented widespread use of this method (Ekbom, 1977; Stacey, 1977).

Blind Introductions or Periodic Release ("Dribble Method")

Successive, regular introductions of *E. formosa* are made, starting soon after the crop is planted. These introductions are made in anticipation of the greenhouse whitefly occurring in the glasshouse. In some cases, it has been necessary to make five to ten introductions of *E. formosa* to achieve successful control (Gould *et al.,* 1975; Parr *et al.,* 1976). de Lara (1981) reported successful control on tomato after four introductions at 2 week intervals.

A variation of this method involves introductions of *E. formosa* when *T. vaporariorum* is first observed in the glasshouse. Ekbom (1977) recommended two to three releases to obtain satisfactory control, as did Stenseth and Aase (1983). In The Netherlands, four introductions are made, two in each of the first two generations of *T. vaporariorum* (Woets, 1978). This is the primary method used in Sweden and in Denmark, where growers spend 1.5 h per year per 1000 m² glasshouse for inspection of plants for whiteflies and introduction of parasitoids (Hansen *et al.,* 1984a).

"Banker" Plants

Plants with well-established populations of *T. vaporariorum* and *E. formosa* are placed at intervals throughout the glasshouse. Stacey (1977) has reported promising results with this method.

The introduction of a new pest, the sweet potato whitefly, *Bemisia argentifolii* Bellows and Perring, into western Europe in 1987 quickly developed into a threat to the success of biological control in several glasshouse vegetable crops. Attempts to control *B. argentifolii* with *E. formosa* on vegetable crops have led to varying results. *Encarsia formosa* lays fewer eggs, develops more slowly, and has a higher mortality in the immature stages of *B. argentifolii* (Bosclair *et al.*, 1990; Szabo *et al.*, 1993). Regular inundative releases of *E. formosa* have been suggested as the solution (Szabo *et al.*, 1993). New parasitoids and other natural enemies are now being evaluated for control of *B. argentifolii* (Drost *et al.*, 1996; Heinz & Parrella, 1994a, 1994b, Hoddle *et al.*, 1997).

CONTROL OF LEAFMINERS USING PARASITES

Several leafminer species (Diptera: Agromyzidae) are pests of glasshouse crops. In western Europe the tomato leafminer, *Liriomyza bryoniae* Kaltenbach, is found on tomato, cucumber, and melon crops. Until the mid-1970s, this pest rarely occurred in sufficient numbers to acquire pest status. Leafminers usually entered the glasshouses after the month of May and were controlled in most cases by naturally occurring parasites migrating into glasshouses from surrounding outdoor areas (Woets & van der Linden, 1982).

A change in growing substrate (from soil to artificial media) caused growers to abandon their habit of soil disinfection—a major method of eliminating leafminer pupae. Hence, leafminers were able to overwinter in the glasshouses and infest young plants at the start of the next growing season. On mature tomato plants, a relatively heavy infestation (up to 15 mines per leaf adjacent to a truss of swelling fruit (Wardlow, 1985a)) can be tolerated without reduction in yield. However, on a young plant the leafminer larvae often enter the stem and can quickly kill the plant.

Since the late 1970s, biological control of *L. bryoniae* has been investigated primarily in The Netherlands, England, and Sweden. Three common parasites have all given satisfactory control of leafminers: *Dacnusa sibirica* Telenga alone (Nedstam, 1983), *D. sibirica* combined with *Opius pallipes* Wesmael (Hymenoptera: Braconidae) (de Lara, 1981; Woets & van der Linden, 1982; Woets, 1983), or *D. sibirica* combined with *Diglyphus isaea* Walker (Hymenoptera: Eulophidae) (Wardlow, 1984).

The braconid parasites overwinter in the glasshouse, provided that no soil disinfection is carried out. This practice allows for the survival of pupal parasites (Braconidae) and *L. bryoniae* pupae. Dutch growers rely heavily on the effect of these overwintering parasites and each year a sample of the first leafminer generation is examined to determine the degree of parasitization. Based on this, the need to introduce parasites is considered. Overwintering parasites can supply adequate control in tomato glasshouses in The Netherlands (W. J. Ravensberg, personal communication).

Diglyphus isaea often migrates into the glasshouses in July and August and can eradicate the *L. bryoniae* population, possibly due to its host-feeding behavior (Woets & van der Linden, 1985). This species is an ectoparasite on leafminer larvae and is more difficult to handle and transport than the braconids, which are endoparasites and emerge from leafminer pupae.

Hendrikse and Zucchi (1979) and Hendrikse *et al.* (1980) found that both *D. sibirica* and *O. pallipes* have a shorter developmental time, lay more eggs than *L. bryoniae*, and are able to recognize parasitized leafminer larvae.

In tomato crops, endoparasites are introduced as pupae within leafminer puparia when the first *L. bryoniae* larvae are observed in the glasshouse. The number of parasites introduced should be sufficient to obtain a 90% parasitization rate of the second leafminer generation (Wardlow, 1985a). According to Woets and van der Linden (1982), an introduction of *O. pallipes* corresponding in numbers to 3% of the total larvae in the first leafminer generation is necessary to achieve satisfactory control.

In the United States and Canada, other leafminer species, *Liriomyza trifolii* (Burgess) and the vegetable leafminer, *L. sativae* Blanchard, are capable of causing severe problems in glasshouses. The problem with insecticide resistance in *Liriomyza* spp. in the United States is particularly serious (Parrella, 1987). Several researchers have investigated the potential of parasites to control *L. trifolii* (Lindquist & Casey, 1983) and *L. sativae* (McClanahan, 1980) on tomatoes.

In the early 1980s, *L. trifolii* was accidentally introduced into Europe and quickly became established in glasshouses in The Netherlands and in southern France. This was a serious threat to the continued use of biological control in glasshouses, as chemical control of *L. trifolii* was generally incompatible with the beneficials used for control of indigenous leafminers and other pests. Intensive research has led to potential solutions to this problem. In The Netherlands, promising results have been obtained with a parasite, *Chrysocharis parksi* Crawford, introduced from California, in combination with *D. isaea* (Woets & van der Linden, 1985). *Dacnusa sibirica* also has a good effect on this pest (W. J. Ravensberg, personal communication). In southern France, a Mediterranean strain of *D. isaea* is providing good control in many tomato glasshouses. Thus, it seems that the abandonment of biological control in glasshouses was avoided. Another leafminer, *L. huidobrensis,* was introduced into Europe in the late 1980s (van der Linden, 1990). This pest was primarily a problem on leafy and tuberous vegetable crops and IPM programs have targeted these crops (van der Linden, 1993). In general, biological control of leafminers on vegetable crops is achieved relatively easily due to the availability of a range of different parasitoid species suited for different pest species and environmental conditions.

Reviews of the literature on *L. bryoniae* and *L. trifolii,* their biology, parasites, and host plants are available (Parrella & Robb, 1985; Minkenberg & van Lenteren, 1986; Parrella, 1987).

CONTROL OF APHIDS WITH PARASITOIDS AND PREDATORS

Aphids, together with two-spotted spider mites and greenhouse whiteflies, are very serious glasshouse pests. Many genera are represented in glasshouses and some of these are polyphagous, e.g., *Myzus persicae* (Sulzer) (green peach or peach–potato aphid), *Aphis gossypii* Glover (melon or cotton aphid), *Macrosiphum euphorbiae* (Thomas) (potato aphid), and *Aulacorthum solani* Kaltenbach (glasshouse or potato aphid) (Homoptera: Aphididae).

Aphids exhibit extremely rapid reproduction. The four aphid species mentioned previously can increase at the rate of four to eight times per week (20°C) (Rabasse & Wyatt, 1985). Damage caused by aphids is attributed primarily to their sucking plant juices, particularly from young, developing plant tissues, which distorts buds and leaves. Aphids also cause damage by the excretion of large amounts of honeydew.

Despite numerous studies of aphidophagous insects, only a few species have shown potential in glasshouses on a large scale because few natural enemies have the potential to match the reproductive and developmental rates of aphids (van Steenis, 1995). The present discussion will concentrate mostly on two such groups, hymenopteran parasitoids (Braconidae and Eulophidae) and dipteran predators (Cecidomyiidae).

Several species of parasitoids have been investigated as candidates for biological control of aphids in glasshouses. The species which have been studied seriously meet most, if not all, of the requirements essential for an effective natural enemy: high reproductive capacity, short generation time, good dispersal capabilities, and a life cycle well synchronized with their aphid hosts (Mackauer & Way, 1976). Satisfactory control of *M. persicae,* the most important of the glasshouse aphids, has been obtained by means of the parasite *Aphidius matricariae* Haliday (van Lenteren *et al.,* 1980c; Rabasse *et al.,* 1983). *Aphidius matricariae* is well adapted to glasshouse conditions and is often found to be the primary parasite in cases of natural control where parasites have migrated into a glasshouse. Hofsvang and Hågvar

(1982) reported promising results with another parasitoid, *Ephedrus cerasicola* Stary.

Despite these promising results, commercial use of aphid parasites has been practiced on a relatively small scale (van Lenteren, 1985). The outcome is somewhat unpredictable, as the balance between aphids and parasites is often upset by hyperparasites during the early summer (van Lenteren *et al.,* 1980c). Because of their long developmental period and short period of egg laying, a single introduction of parasites will lead to synchronized generations of aphids and parasites, allowing the aphids to increase unchecked between parasite generations. Hussey and Bravenboer (1971) stated that control can only be obtained when the rate of aphid population buildup is suboptimal, e.g., due to crowding or host plant resistance.

The availability of many different species of parasitoids and predators has led to an increase in successful control of aphids; in practice, the grower can choose the optimal species during the initial phase of the growing season and supplement with other species that are better suited to conditions in the glasshouse later on; e.g., control of *Macrosiphum euphorbiae* (Thom.) on tomato can be achieved by *Aphidius ervi* Halliday and/or *Aphelinus abdominalis* Dalman during the initial phase, with additions of *Chrysoperla carnea* (Stephens) (Neuroptera: Chrysopidae) or other predators later (J. Reitzel, personal communication). On sweet pepper, successful control is obtained by the use of the braconid parasitoid *Aphidius colemani* Viereck and the cecidomyid *Aphidoletes aphidimyza* (Diptera) (see later) (van Schelt *et al.,* 1990).

In cucumber, introductions of *Aphidius colemani* starting early in the season may lead to satisfactory control of *Aphis gossypii* Glov. for the first 3 months. Later, or if introductions start during the summer, control may fail (van Steenis & El-Khawass, 1996). In commercial glasshouses, this problem is overcome by introducing other beneficial species during the summer months, either *Aphidoletes aphidimyza* or the coccinelid *Hippodamia convergens* Guérin-Meneville. The latter species prefers relatively high temperatures, well above 20°C, but is extremely efficient under the right conditions due to its voracity (J. Reitzel, personal communication).

A less conspicuous natural enemy of aphids, the cecidomyiid *Aphidoletes aphidimyza* (Rondani), is presently being used commercially to control aphids on vegetable crops in Finland, Denmark, Canada, The Netherlands, the United States, and the former U.S.S.R.

The success of the aphid midge, *A. aphidimyza,* is due to the following traits (Markkula & Tittanen, 1985): (1) it preys on all species of aphids, (2) it exhibits a good functional response to increasing aphid density, (3) it is easily mass produced and survives transportation, (4) a permanent population of *A. aphidimyza* develops in the glasshouse and

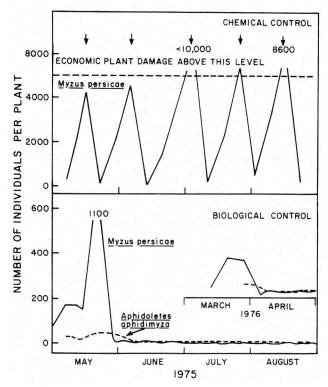

FIGURE 3 Effect of chemical and biological control of *Myzus persicae* on sweet peppers. The experiment was made in the glasshouses of the Agricultural Research Centre in Finland, 1975–1976. In one glasshouse, mevinphos was used when aphids began to damage the plants. In the other, *Aphidoletes aphidimyza* cocoons were applied at a rate of 1 cocoon for 3 aphids. The aphid midges overwintered in the glasshouse although it was not heated during midwinter. They reappeared on the plants the following spring, when heating began. Distribution of a single batch of aphid midge pupae into the soil gave better control than six treatments of mevinphos (Markkula & Tiittanen, 1982).

under favorable conditions it can overwinter and emerge the following spring, and (5) adults are highly mobile and exhibit good searching ability.

Aphidoletes aphidimyza larvae require a minimum of seven *M. persicae* to complete development (Uygun, 1971). With such low food requirements, they are able to survive during periods of prey scarcity. At high aphid densities, this predator can kill up to 10 times the minimum prey requirement.

One drawback to the use of *A. aphidimyza* is the induction of diapause at short day length; the critical day length is 15–17 h, depending on the biotype of the predator. In northern Europe, *A. aphidimyza* gradually disappears from the glasshouse during the months of September and October. If the aphid population has been reduced to a low level at this time, the density will usually remain below the economic injury level for the remainder of the season due to reduced light intensity and temperature in glasshouse at this time (Hansen, 1983). Diapause in *A. aphidimyza* is

facultative and can be prevented by an illumination regime L:D = 16:8. A strain of *A. aphidimyza* with a critical day length of 9 h has been selected (Gilkeson, 1986) and with this strain, use of this predator may not be restricted just to the summer months.

Pupae of *A. aphidimyza* are introduced into the glasshouse when aphids are first observed at rates of 1 pupa per 3 aphids or 2–5 pupae per m² (Markkula *et al.*, 1979). To avoid synchronization of generations, the introduction is repeated after 2–4 weeks. The effect of *A. aphidimyza* on *M. persicae* on sweet pepper is often superior to chemical control (Fig. 3, (Markkula & Tiittanen, 1982)). To overcome the difficulty of detecting the aphids early and, hence, the risk of introducing the predators too late, introductions can be made by means of an open rearing unit or "banker plants," described earlier (Hansen, 1983; van Steenis, 1995).

CONTROL OF ONION THRIPS, *Thrips tabaci,* AND WESTERN FLOWER THRIPS, *Frankliniella occidentalis,* USING PREDATORS

In recent years thrips have changed their status in the glasshouse dramatically; they have gone from a nonpest (incidental) occurring on crops to major pest status. This is especially true on cucumber and sweet pepper crops. The primary reason for the emergence of thrips as pests may be similar to those explaining the rise in the pest status of leafminers—use of artificial growing media with subsequent omissions of soil disinfection. Consequently, thrips are more often present in the glasshouse when a young crop is planted. In addition, there has been a dramatic change in the pattern of pesticide use, which is most obvious in the elimination of frequent applications of broad-spectrum pesticides for the control of *T. urticae*. These sprays prevented thrips from reaching high densities. Due to widespread use of *P. persimilis* for control of two-spotted spider mites, pesticides are applied infrequently, and thrips are left uncontrolled. A possible contributing factor is a greatly drier glasshouse climate resulting from the use of drip irrigation. Furthermore, many growers have shifted to more slow-growing cucumber varieties, which are not able to "outgrow" a thrips infestation.

Historically, the most common pest on vegetable crops in Europe is *Thrips tabaci* Lindeman (Thysanoptera: Thripidae). However, the western flower thrips, *Frankliniella occidentalis* Pergande, which is common in Canada and the United States, has recently been introduced into Europe, where it has become a serious pest in both vegetable and ornamental crops. Thrips feed on plant sap after piercing tissues with their maxillary stylets and mandibles and the tissues around the feeding areas become desiccated, thereby reducing the phytosynthetic area of the leaf. On cucumber a relatively high density of thrips can be tolerated, as reduction in yield or fruit damage has only been found at densities above 25 thrips per leaf (Hansen, 1988).

Chemical control of *T. tabaci* is incompatible with the use of *P. persimilis* for spider mite control. As *T. tabaci* increased in importance, the widespread use of predaceous mites on cucumber crops was threatened. In response to this, intensive research was conducted in several countries to develop biological control methods for thrips.

Ramakers (1980b) reported promising results with native phytoseiid mites, *Amblyseius barkeri* (Hughes) (= *Amblyseius mckenziei* Sch. & Pr.) and *A. cucumeris* (Oud.). This is unusual in that satisfactory biological control can be achieved using a nonspecific predator. Both predaceous mites feed on various arthropods, e.g., *T. urticae* and possibly plant exudates, but they show a pronounced association with thrips. *Amblyseius barkeri* is the main species on cucumber, whereas *A. cucumeris* is more common on sweet pepper (Ramakers, 1980b). Large numbers of these predatory mites can be produced at negligible costs using stored-product mites (Acaridae) as substitute prey (Ramakers & van Lieburg, 1982).

In The Netherlands considerable effort has been directed toward solving the *T. tabaci* problem on sweet pepper. If mixed populations of both predaceous mite species are introduced on sweet pepper, *A. cucumeris* consistently is the dominant species (Ramakers, 1983). *Amblyseius cucumeris* is more difficult to mass rear but seems to give better control on sweet pepper (Ramakers & van Lieburg, 1982). In 1985, *A. cucumeris* was introduced on 68 ha of sweet pepper, and the method utilized in most cases was to introduce predators early in the season and before occurrence of thrips (Klerk & Ramakers, 1986). Since *A. cucumeris* is a nonspecific predator, thrips need not be present at the time of predator introduction. In 83% of the nurseries, control of thrips was completely successful. In 1986, the acreage on which *A. cucumeris* was applied was increased to 140 ha, and the success rate was similar to that obtained in 1985 (Ravensberg & Altena, 1987).

More recently, another predator species, *Amblyseius degenerans* Berlese, has become available for control of thrips on sweet pepper, onion thrips, and the western flower thrips (*Frankliniella occidentalis* (Pergande)). This species is in some aspects considered superior to other phytoseiids available (van Houten & van der Stay, referred to by Ramakers & Voet, 1996). The high price of this predator seems prohibitive to its widespread use and attempts are being made to develop less expensive methods for mass rearing and alternative introduction methods (Ramakers & Voet, 1996). A further possibility is the use of an additional predator, the anthocorid *Orius insidiosus* (Say) (Ramakers, 1993), *Anthocoris nemorum* (Linnaeus), or *Orius majusculus* (Reuter), to complement the activities of the predatory mites at times of high invasion pressure (Jacobsen, 1993).

Predators in the *Amblyseius* genus partly feed on pollen. In cucumber crops more attention must be directed toward establishing them successfully, e.g., by multiple introductions, if they are to be introduced as a preventive measure, as modern cucumber varieties do not produce pollen. The possibilities of using *A. barkeri* on cucumber was investigated in seven and thirteen commercial glasshouses, respectively (Hansen, 1988, 1989). Several introductions of large numbers of predators (300–2000 predators per m^2) were used, after thrips were found on the plants. Satisfactory control was obtained in six and nine glasshouses, respectively (Table 2).

The success of the control seemed to be independent of introductory rates above a minimum of 300–400 predators per m^2. Furthermore, initial thrips densities were of minor significance. Predators established successfully in all glasshouses, and peak predator densities were not significantly lower in cases with unsatisfactory control. The explanation of the variable results may be found in differences in the increase rate of thrips due to climatic factors or variation among cucumber varieties in their suitability as host plants for thrips.

At present, *A. cucumeris* or a combination of this species and *A. barkeri* is used on cucumber. Based on experience from practical use, introductions of predators are now correlated with the thrips density and the actual status of the plant (growth stage, fruiting, amount of leaf damage, etc.) and are coordinated with introductions of predators for spider mite control (J. Reitzel, personal communication). Updates on the biological control of thrips in glasshouses can be found in Anonymous (1990, 1993a, b), Loomans *et al.* (1995), and Jacobsen (1997).

ORNAMENTAL CROPS

Many of the same major and minor pests which attack vegetable crops under glass also attack ornamentals. However, the number of pests attacking ornamentals seems greater, possibly due to the diversity of crops grown under the heading "Ornamentals." For example, in The Netherlands there are about 110 species of cut flowers and 300 species of pot plants and within each there are many cultivars (Fransen, 1993a, 1993b). Strategies developed for using natural enemies in vegetables cannot be directly transferred to an ornamental crop for several reasons. The most important is that an ornamental crop necessitates a very low economic threshold for insects or their damage. This places serious constraints on natural enemies because they must be used in such a way as to maintain the crop virtually pest free. Growers of ornamentals apply pesticides on a regular, calendar basis to their crops on a year-round cycle. Such practices are not conducive to using biological control, especially when considering the broad toxicity spectrum of

TABLE 2 Results of Glasshouse Experiments Using *Amblyseius barkeri* for Biological Control of *Thrips tabaci* on Cucumber (Hansen, 1989)[a]

Glasshouse	Area (m²)	Infested plants[b] (%)	Total number of A.b./m² introduced[c]	Peak number of T.t./leaf ± SE[d]	Peak number of A.b./leaf ± SE[d]	Results[e]	Control lasted weeks[f]
HC 5	1500	3.5	1000	3.6 ± 0.7	0.7 ± 0.2	+	9
HC 8	1500	5.7	700	1.9 ± 0.8	0.2 ± 0.07	+	10
HC 10	1500	11.2	600	2.1 ± 0.6	0.8 ± 0.2	+	10
AN 4	160	35.1	600	12.2 ± 3.9	7.7 ± 1.5	+	17
AN 5	160	36.1	600	5.8 ± 1.5	3.4 ± 0.5	+	19
AN 6	160	22.2	600	2.8 ± 0.8	0.5 ± 0.1	+	19
AN 7	160	36.4	400	3.2 ± 0.9	0.9 ± 0.2	+	17
AN 8	160	15.6	400	1.2 ± 0.4	0.7 ± 0.4	+	17
JP 1	600	3.4	1300	6.4 ± 1.3	2.1 ± 0.3	+	24
JP 2	600	19.2	2000	24.6 ± 5.5	1.2 ± 0.2	−	12
JP 3	600	8.3	1100	9.7 ± 1.7	2.5 ± 0.5	−	10
AP 1	1950	13.0	900	10.1 ± 2.6	1.0 ± 0.3	−	9
JK 6	600	26.5	600	13.3 ± 4.9	0.6 ± 0.2	−	4

[a]Predators were introduced weekly at a rate of 200 predators/m² in most cases. Satisfactory control was defined as situations in which the mean thrips density (all stages except eggs) remained below ca. 10 per leaf throughout the growing season. In glasshouse AN 4, control was considered satisfactory because a high density of predators on the same date quickly reduced the thrips populations to a low level.

[b]Percentage of infested plants at first predator introduction.

[c]Number introduced: all stages. Peak number; all stages except eggs.

[d]T.t.: *Thrips tabaci*; A.b.: *Amblyseius barkeri*.

[e]+, control satisfactory; −, unsatisfactory.

[f]Number of weeks after first predator introduction.

many of the insecticides used. The high value of an ornamental crop (e.g., cut chrysanthemums can be worth about $200,000/ha) (Newman & Parrella, 1986) together with the potentially large losses associated with even moderate insect damage justifies the indiscriminant use of insecticides by many growers.

A general rule of thumb in ornamentals is that the cost of pest control using chemicals (including material and application) is generally <1.0% of the overall cost of producing the crop. This reflects the large production costs involved in growing an ornamental crop. Provided that pesticides remain effective in reducing pest populations, growers will continue to rely on them. It is generally much easier to apply a pesticide to a crop that will provide results rather than dealing with a biological control agent that may or may not be successful. In addition, the beneficial is likely to require several releases in a crop and regular monitoring will probably be necessary to evaluate effectiveness. In this time frame, the quality of the crop remains uncertain.

From a grower's perspective, pesticides are safe, effective, relatively easy to use, inexpensive, and protect a very high value crop from serious pest damage without anxiety to the grower. For biological control to be effective, methods of using beneficials must be developed that will address growers' concerns. A universal problem in ornamentals is that there has been a paucity of biological control trials in these crops. As stated by Hussey and Scopes (1985, p. 114),

"There have been numerous attempts to use biological control on short-term crops but these have not been supported by basic research and so, not surprisingly, most of these introductions failed." Many growers are willing to give biological control a "try," but without specific guidelines which address their crop–pest situation, success is doubtful. Often, when guidelines developed in other crops (tree fruits, greenhouse and field vegetables) are utilized, there are far more failures than successes. After a failure, especially where crop loss is involved, growers are unlikely to try biological control again.

As we progress toward the 21st century, there are factors which actually favor the adoption of biological pest control methods in ornamentals. These will be discussed at the end of this section. Today, there are a few biological control programs in ornamental crops, although there is growing interest among researchers to develop biological control for ornamental crops (Anonymous, 1993b). Great strides have been made in poinsettias (Hoddle *et al.*, 1997), gerbera (Fransen, 1993b), chrysanthemums, and roses. The latter two crops will be discussed in more detail.

Chrysanthemums

Chrysanthemums are one of the major floricultural crops grown throughout the world. There are approximately 2350 ha of chrysanthemums; the major producers, in decreasing order, are Japan, Holland, West Germany, Colombia, and the United States (Anonymous, 1982). Three major methods of production are (1) cut flowers, (2) potted flowering plants, and (3) "garden mums" grown at bedding plant nurseries for eventual transplant into home gardens. Biological control is difficult to achieve in the last two production categories because of the short duration of these crops (<8 weeks) and because the entire plant is sold to the consumer; consequently, even very little damage to any part of the plant is unacceptable. Biological control of aphids was attempted on potted chrysanthemums, but this method has not gained wide acceptance (Scopes, 1970).

Major pest problems on chrysanthemums grown for cut flowers are leafminers, aphids, and thrips. Minor pests include spider mites, mealybugs, plant bugs, and several lepidopteran pests. There have been excellent studies completed where biological control has been evaluated for most of these pests on chrysanthemums. However, there have been few comprehensive studies where biological controls for these pest species were integrated into pest control strategies for chrysanthemum (Scopes & Biggerstaff, 1973; Price *et al.*, 1980; Wardlow, 1985b, 1986; Parrella & Jones, 1987).

Aphids

The major species attacking chrysanthemums include *M. persicae*, *Brachycaudus helichrysi* (Kaltenbach) (leaf-curling plum aphid), *A. gossypii*, and *Macrosiphoniella sanborni* (Gillette) (chrysanthemum aphid) (Homoptera: Aphididae). The last is rarely a problem on chrysanthemums in the United States; it appears to be very sensitive to the broad-spectrum pesticides which are regularly applied to this crop.

Studies with the biological control of aphids have involved examining the potential of coccinellids (Coleoptera), chrysopids (Neuroptera), cecidomyiids, syrphids (Diptera), and fungi (Scopes, 1969; Gurney & Hussey, 1970; Hall & Burgess, 1979; Markkula & Tittanen, 1985; Chambers, 1986).

Predators

Initial studies using predators for aphid control showed that aphids were unevenly distributed on a chrysanthemum crop within a glasshouse. This distribution varies vertically on the plant as well as between varieties for each of the aphid species (Rabasse & Wyatt, 1985). Thus, establishing a uniform density of predators across an entire chrysanthemum crop was difficult and necessitated regular predator releases, which was uneconomical. Success has been obtained with the predator midge *Aphidoletes aphidimyza* Rond. because of its outstanding searching ability and low

cost of mass rearing. The major disadvantage of this predator is low fecundity, but this has been disputed (Gilkeson, 1987). Research with the syrphid fly *Metasyrphis corollae* (Fabricius) (Diptera: Syrphidae), which has higher fecundity and larval voracity that *A. aphidimyza*, has been promising (Chambers, 1986). Problems with this syrphid are that a pollen source is needed to initiate gametogenesis and that both adults and larvae respond insufficiently to low aphid populations. Given the large reproductive potential of *A. gossypii*, damage to chrysanthemum crops may occur. Both the predator midge and syrphid have great potential for aphid control when combined with other control options, such as the use of fungi and parasites.

Parasites

When populations of aphids develop on chrysanthemums, it is very common to find many species of parasites associated with them. However, parasite migration into a greenhouse is insufficient to reduce the damage caused by these pests (Wyatt, 1970). In California, we have observed *Diaretiella rapae* (McIntosh) and *Lysiphlebus* spp. (Hymenoptera: Braconidae) migrating into chrysanthemum greenhouses in response to *M. persicae* populations, but satisfactory control, from the grower's viewpoint, has never been observed. The only feasible way of controlling aphids with parasites is to introduce enough parasites so that significant population increase of the aphids is reduced. This has not been evaluated at the research level although Scopes (1970) tried to establish *Aphidius matricariae* (Hymenoptera: Braconidae) early in the life of a chrysanthemum crop by distributing parasitized aphids on aphid-infested cuttings in the boxes of cuttings prior to planting. There are many commercially produced chrysanthemum cuttings, all of which vary in their suitability as hosts for aphids. Biological control is more feasible on those cultivars which are not good hosts for aphids (Wyatt, 1965).

Fungi

The aphid fungus *Verticillium* (Vertalec®) was widely used to control aphids on chrysanthemums in Europe (Hall, 1985). Vertalec® is primarily used during the months April through September because the pulling of shade cloth during this period increases relative humidity and makes the development of epizootics possible. This fungus is not equally effective against all species of aphids attacking chrysanthemums; a decreasing order of sensitivity is *M. persicae*, *B. helichrysi*, *A. gossypii*, and *M. sanborni*. The registration of Vertalec® in the United States is unlikely; however, the fungus *Beauveria bassiana* has gained U.S. registration (Botani Gard®, Naturalis O®) and has great potential to control aphids, thrips, and whiteflies (Murphy *et al.,* 1998). Compared to *V. lecanii, B. bassiana* is active over a broader range of environmental conditions.

Lepidoptera

Many species of Lepidoptera attack chrysanthemums (Jarrett, 1985) but two of the most important defoliators are *Spodoptera exigua* Hübner (the beet armyworm) (Lepidoptera: Noctuidae) and *Lacanobia oleracea* (the tomato moth) (Lepidoptera: Noctuidae). Control of these pests with insecticides can easily upset biological control of other pests. Research has concentrated on the use of biological insecticides (most notably *Bacillus thuringiensis* Berliner var. *kurstaki*), with special emphasis on formulations and strains that are particularly effective against *Spodoptera* (Dipel 2X® and Javelin®, respectively). Research on a granulosis virus as a control agent of *S. exigua* on chrysanthemums appears promising (Vlak *et al.,* 1982).

Plant Bugs

In unsprayed chrysanthemum crops, both in the United States and in England (Wardlow, 1985b; Jones *et al.,* 1986), *Lygus* spp. (Hemiptera: Miridae) will migrate into glasshouses from adjacent areas. These bugs feed on developing terminals and young buds and can virtually destroy an entire crop. No research on biological control has been attempted on chrysanthemums. However, these pests must be controlled in a way that does not disrupt biological control of other pests. Action must be taken soon after these pests are discovered and options include screening *Lygus* out of chrysanthemum ranges or using a pesticide (e.g., Diazinon) which is less harmful to beneficials than many other effective pesticides.

Mites

The two-spotted spider mite (*Tetranychus urticae* Koch) can be a problem on chrysanthemums, with some cultivars more sensitive than others. Treating boxes of chrysanthemum cuttings with both *T. urticae* and the predator *Phytoseiulus persimilis* Athias-Henriot at the rate of one per plant and one per 50 plants gave excellent control (Scopes & Biggerstaff, 1973). More recently, the recommendation for use of *P. persimilis* is to release predators weekly throughout the life of a chrysanthemum crop at the rate of 10 predators for every 200 plants (Wardlow, 1986). When a spider mite infestation is discovered, release of 10 predators per 10 plants within infested areas is recommended (Wardlow, 1986). The use of diazinon-resistant predators is suggested (Wardlow, 1986) in the event that this material is required for control of other pests.

A review of biological control of *T. urticae* in greenhouses is available (Osborne *et al.,* 1985).

Mealybugs

Historically, the citrus mealybug (*Planococcus citri* [Risso]) (Homoptera: Pseudococcidae) has been a problem on chrysanthemums and biological control has been sought as a solution (Whitcomb, 1940). The predaceous coccinellid *Cryptolaemus montrouzieri* Mulsant (the mealybug destroyer) (Coleoptera: Coccinellidae) has been used successfully on chrysanthemum with releases at the rate of one adult predator for every two plants. This crop has many of the requirements outlined by Whitcomb (1940) that are necessary for successful control of *P. citri* using this predator: appropriate temperature, dense, bushy growth, and the ability to tolerate densities of mealybugs. This last aspect may limit the use of *C. montrouzieri* on commercially grown chrysanthemums (Wardlow, 1986).

The intensity of chemical application on most chrysanthemum crops has all but eliminated the citrus mealybug as a pest. However, as growers reduce sprays and use more biological control, the citrus mealybug is likely to appear more often. In England, results have been promising with releases of one adult *C. montrouzieri* per m². Even at this low rate, the predators are expensive (Wardlow, 1986). Experiments with the coccinellid and the parasite *Leptomastix dactylopii* Howard (Hymenoptera: Encyrtidae) have shown that this combination can successfully control *P. citri* on crotons, *Clivia, Cattleya,* and *Pilea* (Copeland *et al.,* 1985). Research is needed with this combination of natural enemies on chrysanthemums.

Leafminers

Several species of leafminers attack chrysanthemum, but the two most important are *Liriomyza trifolii* (Burgess) and *Chromatomyia syngenesiae* (Hardy) (Diptera: Agromyzidae). The latter species is indigenous to western Europe and was introduced into the United States, probably in the 1980s (Spencer, 1973). Although this species has been shown to be resistant to insecticides (Hussey, 1969), *L. trifolii* is considered to be more tolerant to a wide range of pesticides (Lindquist *et al.,* 1984; Parrella & Keil, 1985). *Liriomyza trifolii* is native to the eastern part of the United States but spread to many chrysanthemum-growing areas of the world in the mid to late 1970s. It is established throughout much of Europe but has been excluded from England (Powell, 1982).

Both leafmining species are extremely important pests of chrysanthemums because they are very difficult to control with standard insecticides. In addition, these materials, which are only partially effective against leafminers, are generally incompatible with other natural enemies. Consequently, biological control of other pests of chrysanthemums can be upset by sprays for leafminer control. Thus, effective biological control of *L. trifolii* and *C. syngenesiae* is critical to a successful biological control effort in chrysanthemums.

In England, where *C. syngenesiae* is the pest species, the use of *Dacnusa* spp. (Hymenoptera: Braconidae) (3 adults/1000 plants) 1 week after planting followed by introduction of *Diglyphus isaea* (Hymenoptera: Eulophidae) (3 adults/1000 plants) 6 weeks after planting has provided good control (Wardlow, 1985b, 1986). With *L. trifolii*, the use of *Diglyphus* spp. has been emphasized, with regular weekly releases of parasites necessary to effect control (Gaviria *et al.*, 1982, Jones *et al.*, 1986). Large data gaps exist in trying to estimate the number of *Diglyphus* needed to effect control given a certain density of *L. trifolii* at the start of a chrysanthemum crop.

Because of the insecticide resistance capability of these leafminers, effective chemical control is difficult to achieve, so more and more growers are considering biological control. However, total reliance on biological control in the absence of other alternative methods (screening out leafminers, isolating rooting areas and mother blocks from production areas, avoidance of continuous cropping in large areas, etc.) will make it more difficult for biological control to be effective. Therefore, the adoption of fully integrated pest management (IPM) strategies by growers is suggested (Price *et al.*, 1980; Wardlow 1985b, 1986; Parrella & Jones, 1987).

IPM for Chrysanthemums

A limiting factor in the development of IPM for chrysanthemums in the United States has been the lack of a biological control program for *L. trifolii* (Price *et al.*, 1980). Research at the University of California at Riverside concentrated on this during 1980–1985 and a successful program was developed (Jones *et al.*, 1986). This biological control program is based on a "window of nonmarketability" in the production of a chrysanthemum crop and targets releases of parasites in the genus *Diglyphus* at a time when the foliage will not be included in the finished cut flower. The length of the flowers sold to the retailer is ca. 12 cm whereas the height of a crop at harvest may exceed 20 cm. Generally, this difference represents foliage produced during the first 6 weeks of crop growth, which either is left behind in the beds to be tilled under in preparation for the next planting or is torn out and discarded.

Parasite releases early in the crop may provide control of *L. trifolii* that exceeds results obtained with chemicals. This depends on the density of leafminers on the crop and population levels in adjacent crops which may be sources of migrating leafminers. The primary objective is to reduce

the population of *L. trifolii* within the first 6 weeks; if this is achieved, then chemical sprays for leafminer control during the rest of the crop may not be needed (Jones *et al.*, 1986). Sampling plans have been developed which rapidly estimate adult and larval populations of *L. trifolii* (Parrella & Jones, 1985; Jones & Parrella, 1986). Therefore, if populations after 6 weeks are still too high, then a pesticide application can be recommended. This is not a failure, because most growers would save thereby six applications of insecticide. In addition, parasite-compatible pesticides may be used (Parrella *et al.*, 1983), although more research is needed in this area.

The use of *Diglyphus* spp. alone to control leafminers may be successful, but other actions directed toward reducing problems with *L. trifolii* will enhance the success. These are outlined in Tables 3 and 4, which present an abbreviated form of an IPM program for chrysanthemums. This IPM program is discussed in detail elsewhere (Parrella & Jones, 1987). In an effort to make biological control of leafminers on chrysanthemums more predictable, a simulation model was developed to estimate the number of parasites needed to effect biological control (Parrella *et al.*, 1992; Heinz *et al.*, 1993). We anticipate that such models can be used as an additional tool in the decision-making process.

Today in the United States a factor limiting the development of IPM on chrysanthemums is lack of a control for the thrips, *Frankliniella occidentalis* Pergande (Thysanoptera: Thripidae), other than chemicals. Only broad-spectrum biocides are effective against thrips, but are incompatible with most other biological control agents.

TABLE 3 Cultural–Physical Pest Management Strategies for Chrysanthemum[a]

Crop stage	Strategy
Mother block	Isolation from production areas; screen out pests
Cuttings	Train workers to spot pests (offer incentives)
Rooting area	Isolate from production areas and mother block; screen out pests
Production ranges	Sterilize soil before planting; avoid excessive nitrogen levels; clean up weeds adjacent to greenhouses; avoid planting leafminer-sensitive varieties near vents and doorways; plant leafminer-sensitive varieties adjacent to each other in central location; avoid continuous cropping; break up large greenhouses into smaller units; remove plant debris (after pinching or stripping of leaves at cutting) from greenhouse immediately

[a] See Parrella & Jones (1987) for details.

TABLE 4 Sampling and Biological Control Strategies for Chrysanthemum Pests[a]

Crop stage	Pest	Strategy	
		Sampling	Biological control
Mother block	*Myzus persicae* Sulzer	Visual searches of growing tips, yellow cards	*Cephalosporium lecanii*
	Liriomyza trifolii (Burgess)	Yellow cards and leaf samples	*Steinernema feltiae*[b]
	Spodoptera exigua Hübner	Pheromone traps and visual search	—
Rooting area	*Spodoptera exigua*	Pheromone traps and visual search	—
	Myzus persicae	Visual searches of growing tips, yellow cards	*Cephalosporium lecanii*
Production ranges	*Liriomyza trifolii*	Yellow cards and leaf samples	*Diglyphus intermedius* (Girault)
			Steinernema feltiae[b]
	Spodoptera exigua	Pheromone traps and visual search	*Bacillus thuringiensis* Berliner
	Lygus spp.	Visual searches of developing buds	—
	Frankliniella occidentalis Pergande	Yellow traps and visual searches of foliage and flowers	—
	Tetranychus urticae Koch	Visual searches of foliage and flowers	*Phytoseiulus persimilis* Athias-Henriot
			Metaseiulus occidentalis (Nesbitt)

[a] See Parrella and Jones (1987) for details.
[b] See Harris (1988) for details.

IPM for Roses

Roses are an important floricultural crop worldwide, with ca. 2900 ha grown in Holland, West Germany, the United States, Italy, France, Japan, and Israel (in decreasing order). Commercial rose growers generally purchase budded stock plants from large propagators and these plants will produce roses for many years. Consequently, unlike other floricultural crops, roses can be considered perennial; through a system of pinching and pruning, the same plant can be productive for 10 years or more. This provides a more stable environment and should be more conducive to biological control efforts. A destabilizing factor is the general susceptibility of roses to many diseases (Hasek, 1980), which requires almost regular applications of fungicides. More information is needed on the compatibility of natural enemies with specific fungicides used in roses.

Major pest problems on roses are two-spotted spider mites, western flower thrips, aphids, and leafrollers. Despite the perennial nature of roses and their inherent suitability for biological control, little research has been done with this crop. Only limited information is available on biological control of mites on roses (Smitely, 1993).

Mites

An integrated pest management program has been developed for roses in France using releases of *P. persimilis* for control of *T. urticae* (Pralavorio *et al.,* 1985). These researchers point out the difficulties associated with using predaceous mites successfully — uneven temperatures in the greenhouse, inconsistent diapause of *T. urticae* in southern areas, and low relative humidity in southern France in spring and summer. This information demonstrates the diversity of climatic conditions that exist within glasshouses from the Mediterranean to northern Europe. Successful biological control will depend on the development of programs in each area; in some areas biological control may not be feasible (van Lenteren *et al.,* 1980b). Broad generalizations or guidelines for the use of *P. persimilis* for *T. urticae* control on roses will likely be unsuitable for many rose-growing areas. In the United States, where conditions vary enormously from New York to California, the use of general guidelines will result in failure at many locations. However, the concept that biological control is less likely to be successful in North America than it is in Europe in glasshouses (Smith & Webb, 1977) is unfounded. If they are to be successful, biological control programs for roses and other crops must be developed specifically for the major growing areas in the United States.

In the United States, *P. persimilis* gave consistent results on roses when an economic threshold of 10 mites per leaflet was established (Boys & Burbutis, 1972). These researchers suggested that a more feasible approach may be to integrate the predator into one overall IPM program. In California, promising results were obtained using *Metaseiulus occidentalis* (Nesbitt) for biological control of *T. urticae* on greenhouse roses. Because of its insecticide resistance capability, this predator was integrated into a pest control program and persisted in greenhouses for more than 2 years (Field & Hoy, 1984).

New Developments to Ensure Successful Application of Biological Control

Over the past two decades the acreage with biological control as well as the number of producers of beneficials has increased (van Lenteren *et al.,* 1997). The successful application of biological control is dependent on the quality of the beneficials. Within the IOBC global working group "Quality Control of Mass-Reared Arthropods," much work has been put into the description of procedures to ensure uniform quality of the natural enemies. This work is being carried out as a collaboration between producers of natural enemies and scientists and has led to the design of guidelines for 20 species of natural enemies (van Lenteren, 1996). These tests relate to product control and will be complemented with field performance tests in the near future. Quality control is closely related to the issues of risks involved with importation of exotic biocontrol agents; a review of the registration procedures existing in the European countries has been compiled by van Lenteren (1997).

Expert systems, or decision-support systems (DSS), have been developed for certain aspects of glasshouse production (climate control, managing production practices, and pest diagnosis). Among the factors favoring the use of DSS in the glasshouse industry is the fact that this sector is technologically highly advanced, with a widespread use of computerized control of environmental conditions and fertilization. Many growers thus possess substantial knowledge of computer systems (Shipp *et al.,* 1996). Recently developed expert systems in this field have included integrated management of pests on specific crops. In The Netherlands use of three or four natural enemies is mandatory if the products are to be marketed using a special IPM label (Ramakers, 1996). To help growers manage this increasingly complex system, a DSS was developed initially for pest management on sweet pepper as a pilot crop. This system was extended to cucumber and tomato crops and a prototype was evaluated in 1995 (Altena, 1996; Ramakers & van der Maas, 1996). In Canada a system aimed at all aspects of cucumber and tomato production was developed and presented for field testing in 1996 (Shipp *et al.,* 1996).

SUMMARY

The tremendous success achieved with biological control in the glasshouse environment has set a very high standard that is difficult for many other segments of agriculture to match. This success has occurred primarily on glasshouse-grown vegetables in Europe and Canada, where there has been outstanding cooperation between research and extension and the necessary funding in the direction of biological control. In the United States, where the glasshouse vegetable industry is small (although increasing), funding for biological control research has been limited. Many scientists will receive considerable grant funds from the chemical industry, which hampers the development of biological control.

The cooperation between research and extension in Europe has led to establishment of several commercial insectaries that produce natural enemies of high quality and that consult with a grower on all aspects of the grower's arthropod problems. The economic advantages have been demonstrated to the growers and they are eager for biological control. The extension effort in biological control has gone beyond the grower and has reached the homeowner (Hansen *et al.,* 1984b); in Denmark biological control agents are available in local supermarkets. A strong research effort continues and is enhanced through cooperation via workshops held ca. every 3–4 years on biological control in glasshouses (IOBC, Western and Eastern Palearctic Section and Nearctic section). At each of these meetings in 1987, 1990, 1993 and 1996 (Anonymous, 1987, 1990, 1993a, 1993b), more than 50 papers were presented. Research is concentrating on thrips in greenhouses, a qualitative assessment of the number of natural enemies that must be released to achieve biological control, and quality control of natural enemies, and there is renewed interest on the use of pathogens (with emphasis on the fungi and nematodes) for biological control of a wide range of greenhouse pests.

It is reassuring to find that many papers at these workshops concentrated on ornamental crops. This, together with a separate conference focusing on biological control in ornamental crops (Anonymous, 1993b), suggests that there is real interest among the researcher–grower continuum to develop biological control strategies. Although comprising ca. 35% of glasshouse production around the world, these crops have been ignored by researchers in Europe and North America in the past. The largest commercial flower crop grown in the world, carnations, has had virtually no studies on biological control. This crop is perennial (ca. 2 years), with spider mites as a major problem. The stability of this system is very positive for biological control. Many examples of isolated studies where biological control was attempted and was successful on an ornamental crop have been published. However, these studies generally do not address mainstream commercial growers who are producing a high-quality crop. The concept of how biological control can fit into the overall production practices of these growers has not been addressed. Until this is done, universal acceptance and the adoption of biological control by growers of ornamentals seem unlikely.

Biological control programs on potted flowering crops and foliage plants (poinsettias, cymbidium orchids, palms, etc.) have really just begun (Anonymous, 1993b; Heinz & Parrella, 1994a) and the future looks very promising.

Increasing problems with insecticide resistance, lack of new or replacement chemicals, tougher pesticide regulations with respect to worker safety, increasing costs of new insecticides, greater responsibility for using a particular pesticide being placed on the grower, and the general urbanization of many areas surrounding ornamental greenhouses demand the use of alternatives to insecticides. If research and extension can meet the challenge, growers will alter their thinking and biological control may become an important part of their pest control program.

Acknowledgments

The authors appreciate the comments made by Richard Lindquist regarding the many drafts of this chapter and the critical reviews of the editors.

References

Altena, K. (1996). What happened to "CAPPA"?: The Koppert experience. Bulletin of the IOBC/WPRS, 19(1), 5–6.

Anonymous. (1982). Aalsmeer flower auction. Flowers . . . unlimited. "International Developments in Floriculture." Vereniging Verenigde Bioemenueilingen Aalsmeer, Aalsmeer, The Netherlands.

Anonymous. (1987). Proceedings, Working Group on Integrated Control in Glasshouses. Bulletin of the IOBC/WPRS, 1987/X/2.

Anonymous. (1990). Proceedings, Working Group on Integrated Control in Glasshouses. Bulletin of the IOBC/WPRS, 1990/XIII/5.

Anonymous. (1993a). Proceedings, Working Group on Integrated Control in Glasshouses. Bulletin of the IOBC/WPRS, 1993/16/2.

Anonymous. (1993b). Proceedings, Working Group on Integrated Control in Glasshouses. Bulletin of the IOBC/WPRS, 1993/16/8.

Anonymous. (1996). Proceedings, Working Group on Integrated Control in Glasshouses. Bulletin of the IOBC/WPRS, 1996/19/1.

Bosclair, J., Bruern, G. J., & van Lenteren, J. C. (1990). Can Bemisia tabaci be controlled with Encarsia formosa? Bulletin of the IOBC/WPRS, 1990/XIII/5, 32–35.

Boys, F. E., & Burbutis, P. B. (1972). Influence of Phytoseiulus persimilis on populations of Tetranychus urticae at the economic threshold on roses. Journal of Economic Entomology, 65, 114–117.

Bravenboer, L., & Dosse, G. (1962). Phytoseiulus riegeli Dosse als Prädator einiger Schadmilben aus der Tetranychus urticae-gruppe. Entomologica Experimentalis et Applicata, 5, 291–304.

Chambers, R. J. (1986). Preliminary experiments on the potential of hoverflies (Dipt.: Syrphidae) for the control of aphids under glass. Entomophaga, 31, 197–204.

Chant, D. A. (1961). An experiment in biological control of Tetranychus telarius (Linnaeus) (Acarina: Tetranychidae) in a greenhouse using the predacious mite Phytoseiulus persimilis Athias-Henriot (Phytoseiidae). Canadian Entomologist, 93, 437–443.

Copeland, M. J., Tingle, C. D., Saynor, M., & Panis, A. (1985). Biology of glasshouse mealybugs and their predators and parasitoids. In "Biological Pest Control: The Glasshouse Experience." (N. W. Hussey & N. E. A. Scopes, Eds.), pp. 82–86. Blandford, Poole, Dorset.

de Lara, M. (1981). Development of biological methods of pest control in the United Kingdom glasshouse industry. In "Proceedings, 1981 British Crop Protection Conference," pp. 599–607.

Dicke, M. (1994/1995). Why do plants talk? Chemoecology, 5/6, 159–165.

Dixon, G. M. (1973). Observations on the use of Phytoseiulus persimilis Athias-Henriot to control Tetranychus urticae Koch on tomatoes. Plant Pathology, 22, 134–138.

Dosse, G. (1959). Uber einige neue Raubmilbenarten (Acar. Phytoseiidae). Pflanzenschutzberichte, 21, 44–66.

Drost, Y. C., Fadi Elmula, A., Posthuma-Doodeman, C. J. A. M., & van Lenteren, J. C. (1996). Development of criteria for evaluation of natural enemies in biological control: Bionomics of different parasitoids of Bemisia argentifolii. Bulletin of the IOBC/WPRS, 1996/19/1, 31–34.

Ekbom, B. (1977). Development of a biological control program for greenhouse whiteflies (Trialeurodes vaporariorum Westwood) using its parasite Encarsia formosa (Gahan) in Sweden. Zeitschrift für angewandte Entomologie, 84, 145–154.

Field, R. P., & Hoy, M. A. (1984). Biological control of spider mites on greenhouse roses. California Agriculture, 38, 29–32.

Force, D. C. (1967). Effect of temperature on biological control of two-spotted spider mites by Phytoseiulus persimilis. Journal of Economic Entomology, 60, 1308–1311.

Fransen, J. J. (1993a). Development of integrated protection in glasshouse ornamentals. Journal of Pesticide Science 36, 329–333.

Fransen, J. J. (1993b). Integrated pest management in glasshouse ornamentals in the Netherlands: A step by step policy. Bulletin IOBC/WPRS, 1993/16/2, 35–38.

French, N., Parr, W. J., Gould, H. J., Williams, J. J., & Simmonds, S. P. (1976). Development of biological methods for the control of Tetranychus urticae on tomatoes using Phytoseiulus persimilis. Annals of Applied Biology, 83, 177–189.

Gaviria, J. D., Gafaro, F., Prieto, A. J., Escobar, J., Garcia, J. H., & Ruiz, H. (1982). Advances en el control integrado del los insectos plagas del cultivo de chrisontemo Chrysanthemum morifolium Ramat & Henfl, en el departmato del Cuaca, Proceedings of the IX Congress de la Sociedad Colombiana de Entomologia.

Gilkeson, L. (1986). Genetic selection for and evaluation of non-diapause lines of predatory midge Aphidoletes aphidimyza (Rondani) (Diptera: Cecidomyiidae). Canadian Entomologist, 118, 869–879.

Gilkeson, L. (1987). A note on fecundity of the aphid predator, Aphidoletes aphidimyza (Rondani) (Diptera: Cecidomyiidae). Canadian Entomologist, 119, 1145–1146.

Gould, H. J. (1968). Observations on the use of a predator to control red spider mite on commercial cucumber nurseries. Plant Pathology, 17, 108–112.

Gould, H. J. (1970). Preliminary studies of an integrated control programme for cucumber pests and an evaluation of methods of introducing Phytoseiulus persimilis Athias-Henriot for the control of Tetranychus urticae Koch. Annals of Applied Biology, 66, 505–513.

Gould, H. J. (1971). Large scale trials of an integrated control programme for cucumber pests on commercial nurseries. Plant Pathology, 20, 149–156.

Gould, H. J. (1977). Biological control of glasshouse whitefly and red spider mite on tomatoes and cucumbers in England and Wales 1975–76. Plant Pathology, 26, 57–60.

Gould, H. J., Parr, W. J., Woodville, H. C., & Simmonds, S. P. (1975). Biological control of glasshouse whitefly (Trialeurodes vaporariorum) on cucumbers. Entomophaga, 20, 285–292.

Greathead, D. J. (1976). A Review of Biological Control in Western and Southern Europe. In "CIBC Technical Communication 7," pp. 52–64. Commonwealth Agricultural Bureau, Slough.

Gurney, B., & Hussey, N. W. (1970). Evaluation of some coccinellid species for the biological control of aphids in protected cropping. Annals of Applied Biology, 65, 451–458.

Hall, R. A. (1985). Aphid control by fungi. In "Biological Pest Control: The Glasshouse Experience" (N. W. Hussey & N. E. A. Scopes, Eds.), pp. 138–141. Blandford, Poole, Dorset.

Hall, R. A., & Burgess, H. D. (1979). Control of aphids in glasshouses with the fungus Verticillium lecanii. Annals of Applied Biology, 93, 235–246.

Hansen, L. S. (1983). Introduction of Aphidoletes aphidimyza (Rond.) (Diptera: Cecidomyiidae) from an open rearing unit for the control of aphids in glasshouses. Bulletin of the IOBC/WPRS, 1983/VI/3, 146–150.

Hansen, L. S. (1988). Control of Thrips tabaci (Thysanoptera: Thripidae) on glasshouse cucumber using large introductions of predatory mites Amblyseius barkeri (Acarina: Phytoseiidae). Entomophaga, 33, 33–42.

Hansen, L. S. (1989). The effect of initial thrips density (Thrips tabaci Lind. (Thysanoptera: Thripidae)) on the control exerted by Amblyseius *barkeri* (Hughes) (Acarina: Phytoseiidae) on glasshouse cucumber. Journal of Applied Entomology, 107, 130–135.

Hansen, L. S., Jakobsen, J., & Reitzel, J. (1984a). Extent and economics of biological control of *Tetranychus urticae* and *Trialeurodes vaporariorum* in Danish glasshouses. EPPO Bulletin, 14, 393–399.

Hansen, L. S., Pedersen, O. C., & Reitzel, J. (1984b). "Skadedyr og Nyttedyr. Håndbog om biologisk bekaempelse i drivhuset." De danske Haveselskaber, Copenhagen.

Harris, M. A. (1988). Current recommendations for leafminer control. *In* "Proceedings, 4th Conference on Insect and Disease Management on Ornamentals" (A.D. Ali *et al.*, Eds.), pp. 138–142. Society of American Florists, Alexandria, VA.

Hasek, R. F. (1980). Roses. *In* "Introduction to Floricuture" (R. A. Larson, Ed.), pp. 81–105. Academic Press, New York.

Heinz, K. M., & Parrella, M. P. (1994a). Biological control of *Bemisia argentifolii* (Gennadius) (Homoptera: Aleyrodidae) infesting *Euphorbia pulcherrima* evaluations of releases of *Encarsia luteola* Howard (Hymenoptera: Aphelinidae) and *Delphastus pusillus* LeConte (Coleoptera: Coccinellidae). Environmental Entomology, 23, 1346–1353.

Heinz, K. M., & Parrella, M. P. (1994b). Poinsettia (*Euphorbia pulcherrima* Willd. ex Klotz) cultivar-mediated differences in performance of five natural enemies of *Bemisia argentifolii* Bellows and Perring, n. sp. (Homoptera: Aleyrodidae). Biological Control, 4, 305–318.

Heinz, K. M., Nunney, L., & Parrella, M. P. (1993). Toward predictable biological control of *Liriomyza trifolii* (Diptera: Agromyzidae) infesting greenhouse cut chrysanthemums. Environmental Entomology, 22, 1217–1233.

Hendrikse, A., & Zucchi, R. (1979). The importance of observing parasite behaviour for the development of biological control of the tomato leafminer (*Liriomyza bryoniae* Kalt.). Mededelingen van de Faculteit Landbouwwetenschappen, Rijksuniversiteit Gent, 44, 107–116.

Hendrikse, A., Zucchi, R., van Lenteren, J. C., & Woets, J. (1980). *Dacnusa sibirica* Telenga and *Opius pallipes* Wesmael (Hym.: Braconidae) in the control of the tomato leafminer *Liriomyza bryoniae* Kalt. Bulletin of the IOBC/WPRS, 1980/III/3, 83–89.

Hoddle, M. S., van Driesche, R. V., Roy, S., Smith, T., Mazzola, M., Lopes, P., & Sanderson, J. (1997). "A Grower's Guide to Using Biological Control for Silverleaf Whitefly on Poinsettias in the Northeast United States. Floral Facts." UMASS Extension, Univ. of Massachusetts, Amherst, MA.

Hofsvang, T., & Hågvar, E. B. (1982). Comparison between the parasitoid *Ephedrus cerasicola* Stary and the predator *Aphidoletes aphidimyza* (Rondani) in the control of *Myzus persicae* Sulzer. Zeitschrift für angewandte Entomologie, 94, 412–419.

Hussey, N. W. (1969). Differences in susceptibility of different strains of chrysanthemum leafminer (*Phytomyza syngenesiae*) to BHC and diazinon. *In* "Proceedings, 5th British Insecticide and Fungicide Conference," pp. 93–97.

Hussey, N. W. (1985). History of biological control in protected culture. Western Europe. *In* "Biological Pest Control: The Glasshouse Experience" (N. W. Hussey & N. E. A. Scopes, Eds.), pp. 11–22. Blandford, Poole, Dorset.

Hussey, N. W., & Bravenboer, L. (1971). Control of pests in glasshouse culture by the introduction of natural enemies. *In* "Biological Control" (C. B. Huffaker, Ed.), pp. 195–216. Plenum, New York.

Hussey, N. W., & Huffaker, C. B. (1976). Spider mites. *In* "Studies in Biological Control" (V. L. Delucchi, Ed.), pp. 179–228. Cambridge Univ. Press, Cambridge, U.K.

Hussey, N. W., & Parr, W. J. (1963). The effect of glasshouse red spider mite (*Tetranychus urticae* Koch) on the yield of cucumbers. Journal of Horticultural Science, 38, 255–263.

Hussey, N. W., & Scopes, N. E. A. (1977). The introduction of natural enemies for pest control in glasshouses: Ecological considerations. *In* "Biological Control by Augmentation of Natural Enemies" (R. L. Ridgway & S. B. Vinson, Eds.), pp. 349–377. Plenum, New York.

Hussey, N. W., & Scopes, N. E. A. (1985). "Biological Pest Control: The Glasshouse Experience." Cornell Univ. Press, Ithaca, NY.

Hussey, N. W., Parr, W. J., & Gould, H. J. (1965). Observations on the control of *Tetranychus urticae* Koch on cucumbers by the predatory mite *Phytoseiulus riegeli* Dosse. Entomologica Experimentalis et Applicata, 8, 271–281.

Hussey, N. W., Parr, W. J., & Gurney, B. (1958). The effect of whitefly populations on the cropping of tomatoes. Glasshouse Crops Research Institute Annual Report, 79–86.

Jackson, G. J., & Ford, J. B. (1973). The feeding behaviour of *Phytoseiulus persimilis* Athias-Henriot (Acarina: Phytoseiidae) particularly as affected by certain pesticides. Annals of Applied Biology, 75, 165–171.

Jacobsen, R. (1993). Control of *Frankliniella occidentalis* with *Orius majusculus:* Experiences during the first full season of commercial use in the U.K. Bulletin of the IOBC/WPRS, 1993/16/2, 81–84.

Jacobsen, R. (1997). Pest management (IPM) in glasshouses. *In* "Thrips as Crop Pests" (T. Lewis, Ed.). CAB International, U.K.

Jarrett, P. (1985). Experience with the selective control of caterpillars using *Bacillus thuringiensis. In* "Biological Pest Control: The Glasshouse Experience" (N. W. Hussey & N. E. A. Scopes, Eds.), pp. 142–144. Blandford, Poole, Dorset.

Jones, V. P., & Parrella, M. P. (1986). Development of sampling strategies for larvae of *Liriomyza trifolii* (Diptera: Agromyzidae) in chrysanthemums. Environmental Entomology, 15, 268–273.

Jones, V. P., Parrella, M. P., & Hodel, D. R. (1986). Biological control of *Liriomyza trifolii* on greenhouse chrysanthemums. California Agriculture, 40, 10–12.

Klerk, M.-L. J. de, & Ramakers, P. M. J. (1986). Monitoring population densities of the phytoseiid predator *Amblyseius cucumeris* and its prey after large scale introductions to control *Thrips tabaci* on sweet pepper. Mededelingen van de Faculteit Landbouwwetenschappen, Rijksuniversiteit Gent, 51, 1045–1048.

Legowski, T. J. (1966). Experiments in predator control of the glasshouse red spider mite on cucumbers. Plant Pathology, 15, 34–41.

Lindquist, R. K., & Casey, M. L. (1983). Introduction of parasites for control of *Liriomyza* leafminers on greenhouse tomato. Bulletin of the IOBC/WPRS, 1983/VI/3, 108–115.

Lindquist, R. K., Casey, M. L., Heylar, N., & Scopes, N. E. A. (1984). Leafminers on greenhouse chrysanthemum: Control of *Chromatomyia syngenesiae* and *Liriomyza trifolii.* Journal of Agricultural Entomology, 3, 256–263.

Lipa, J. J. (1985). History of biological control in protected culture. Eastern Europe. *In* "Biological Pest Control: The Glasshouse Experience" (N. W. Hussey & N. E. A. Scopes, Eds.), pp. 23–33. Blandford, Poole, Dorset.

Loomans, A. J. M., van Lenteren, J. C., Tomasini, M. G., Maini, S., & Riudavets, J. (1995). Biological control of thrips pests. Wageningen Agricultural University Papers 95–1.

Mackauer, M., & Way, M. J. (1976). *Myzus persicae,* an aphid of world importance. *In* "Studies in Biological Control" (V. L. Delucchi, Ed.), pp. 51–117. Cambridge Univ. Press, Cambridge, U.K.

Markkula, M., & Tiittanen, K. (1982). Possibilities of biological and integrated control of pests on vegetables. Acta Entomologici Fennici, 40, 15–23.

Markkula, M., & Tiittanen, K. (1985). Biology of the midge *Aphidoletes Laphidimyza* and its potential for biological control. "Biological Pest Control: The Glasshouse Experience" (N. W. Hussey & N. E. A. Scopes, Eds.), pp. 74–81. Blandford, Poole, Dorset.

Markkula, M., Tittanen, K., Hämälänen, N., & Forsberg, A. (1979). The aphid midge *Aphidoletes aphidimyza* (Diptera, Cecidomyiidae) and its use in biological control of aphids. Annales Entomologici Fennici, 45, 89–98.

McClanahan, R. J. (1980). Biological control of *Liriomyza sativae* on greenhouse tomatoes. Bulletin of the IOBC/WPRS, 1980/III/3, 135–140.

McMurtry, J. A., Huffaker, C. B., & van de Vrie, M. (1970). Tetranychid enemies: Their biological characters and the impact of spray practices. Hilgardia, 40, 331–390.

Milliron, H. E. (1940). A study of some factors affecting the efficiency of Encarsia formosa Gahan, an aphelinid parasite of the greenhouse whitefly, Trialeurodes vaporariorum (Westw.) [Michigan Agricultural Experiment Station Technical Bulletin 173].

Minkenberg, O. P. J. M., & van Lenteren, J. C. (1986). The leafminers Liriomyza bryoniae and L. trifolii (Diptera: Agromyzidae), their parasites and host plants: A review. Wageningen Agricultural University Papers 86–2.

Murphy, B. C., Morisawa, T. A., Newman, J. P., Tjosvold, S. A., & Parrella, M. P. (1998). Fungal pathogen provides control of western flower thrips in greenhouse flowers. California Agriculture, 52, 32–36.

Nedstam, B. (1983). Control of Liriomyza bryoniae Kalt. by Dacnusa sibirica Tel. Bulletin of the IOBC/WPRS, 1983/VI/3, 124–127.

Nell, H. W., Sevenster-van der Lelie, L. A., van Lenteren, J. C., & Woets, J. (1976). The parasite-host relationship between Encarsia formosa (Hymenoptera: Aphelinidae) and Trialeurodes vaporariorum (Homoptera: Aleyrodidae). II. Selection of host stages for oviposition and feeding by the parasite. Zeitschrift für angewandte Entomologie, 81, 372–376.

Newman, J. P., & Parrella, M. P. (1986). A license to kill. Greenhouse Manager, 5, 86–92.

Noldus, L. P. J. J., & van Lenteren, J. C. (1990). Host aggregation and parasitoid behavior: Biological control in a closed system. In "Critical Issues in Biological Control" (M. Mackauer, L. E. Ehler, & J. Roland, Eds.), pp. 229–262. Intercept Ltd., Andover, U.K.

Onillon, J. C. (1990). The use of natural enemies for the biological control of whiteflies. In "Whiteflies: Their Bionomics, Pest Status and Management" (D. Gerling, Ed.), pp. 187–313. Intercept, Ltd., Andover, U.K.

Osborne, L. S., Ehler, L. E., & Nechols, J. R. (1985). Biological Control of the Two-Spotted Spider Mite in Greenhouses. Agricultural Experiment Station, Institute of Food and Agricultural Science, Univ. of Florida Bulletin 853.

Parr, W. J., Gould, H. J., Jessop, N. H., & Ludlam, F. A. B. (1976). Progress towards a biological control programme for glasshouse whitefly (Trialeurodes vaporariorum) on tomatoes. Annals of Applied Biology, 83, 349–363.

Parrella, M. P. (1987). Biology of Liriomyza. Annual Review of Entomology, 32, 201–224.

Parrella, M. P., & Jones, V. P. (1985). Yellow traps as monitoring tools for Liriomyza trifolii (Diptera: Agromyzidae) in chrysanthemum greenhouses. Journal of Economic Entomology, 78, 53–56.

Parrella, M. P., & Jones, V. P. (1987). Development of integrated pest management strategies in floriculture. Bulletin of the Entomological Society of America, 33, 28–34.

Parrella, M. P., & Keil, C. B. (1985). Toxicity of methamidophos to four species of Agromyzidae. Journal of Agricultural Entomology, 2, 234–237.

Parrella, M. P., & Robb, K. L. (1985). Economically important members of the genus Liriomyza Mik: A selected bibliography. Miscellaneous Publications of the Entomological Society of America, 59, 1–26.

Parrella, M. P., Christie, G. D., & Robb, K. L. (1983). Compatibility of insect growth regulators and Chrysocharis parksi (Hymenoptera: Eulophidae) for the control of Liriomyza trifolii (Diptera: Agromyzidae). Journal of Economic Entomology, 76, 949–951.

Parrella, M. P., Heinz, K. M., & Nunney, L. (1992). Biological control through inundative releases of natural enemies: A strategy whose time has come. American Entomologist, 38, 172–179.

Petitt, F. L., & Osborne, L. S. (1984). Selected bibliography of the predacious mite, Phytoseiulus persimilis Athias-Henriot, (Acarina: Phytoseiidae). Bibliographies of the Entomological Society of America, 3, 1–11.

Powell, D. F. (1982). The eradication campaign against American serpentine leafminer, Liriomyza trifolii at Efford Experimental Horticultural Station, Hampshire, England, U.K. Plant Pathology, 30, 195–204.

Pralavorio, M., & Almaguel-Rojas, L. (1980). Influence de la temperature et de l'humidite relative sur le developpement et la reproduction de Phytoseiulus persimilis. Bulletin of the IOBC/WPRS, 1980/III/3, 157–162.

Pralavorio, M., Millot, P., & Fournier, D. (1985). Biological control of greenhouse spider mites in southern France. In "Biological Pest Control: The Glasshouse Experience" (N. W. Hussey & N. E. A. Scopes, Eds.), pp. 125–128. Blandford, Poole, Dorset.

Price, J. F., Overman, A. J., Englehard, A. W., Iverson, M. K., & Yingst, V. W. (1980). Integrated pest management demonstrations in commercial chrysanthemums. Proceedings of the Florida State Horticultural Society, 93, 190–194.

Pruszynski, S. (1979). Phytoseiulus persimilis A.-H. bibliografia. Brochure of the Instytit Ochrony Roslin, ul. Miczurina 20, 60318 Poznan.

Rabasse, J. M., Lafont, J. P., Delpuech, I., & Silvie, P. (1983). Progress in aphid control in protected crops. Bulletin of the IOBC/WPRS, 1983/VI/3, 151–162.

Rabasse, J. M., & Wyatt, I. J. (1985). Biology of aphids and their parasites in greenhouses. In "Biological Pest Control: The Glasshouse Experience" (N. W. Hussey & N. E. A. Scopes, Eds.), pp. 66–73. Blandford, Poole, Dorset.

Ramakers, P. M. J. (1980a). Biological control in Dutch glasshouses; practical application and progress in research. Symposium on Integrated Crop Protection, Valence, June 1980.

Ramakers, P. M. J. (1980b). Biological control of Thrips tabaci (Thysanoptera: Thripidae) with Amblyseius spp. (Acari: Phytoseiidae). Bulletin of the IOBC/WPRS, 1980/III/3, 203–207.

Ramakers, P. M. J. (1983). Mass production and introduction of Amblyseius mckenziei and A. cucumeris. Bulletin of the IOBC/WPRS, 1983/VI/3, 203–206.

Ramakers, P. M. J. (1993). Coexistence of two thrips predators, the anthocorid Orius insidiosus and the phytoseiid Amblyseius cucumeris on sweet pepper. Bulletin of the IOBC/WPRS, 1993/6/2, 133–136.

Ramakers, P. M. J. (1996). Use of natural enemies as indicators for obtaining an IPM label. Bulletin of the IOBC/WPRS, 1996/19/1, 119–122.

Ramakers, P. M. J. & van der Maas, A. A. (1996). Decision support system "CAPPA" for IPM in sweet pepper. Bulletin of the IOBC/WPRS, 1996/19/1, 123–126.

Ramakers, P. M. J., & van Lieburg, M. J. (1982). Start of commercial production and introduction of Amblyseius mckenziei Sch. & Pr. (Acarina: Phytoseiidae) for the control of Thrips tabaci Lind. (Thysanoptera: Thripidae) in glasshouses. Mededelingen van de Faculteit Landbouwwetenschappen, Rijksuniversiteit Gent. 47, 541–545.

Ramakers, P. M. J., & Voet, S. J. P. (1996). Introduction of Amblyseius degenerans for thrips control in sweet peppers with potted castor beans as banker plants. Bulletin of the IOBC/WPRS, 1996/19/1, 127–130.

Ravensberg, W. J., & Altena, K. (1987). Recent developments in the control of thrips in sweet pepper and cucumber. Bulletin of the IOBC/WPRS, 1987/X/2, 160–164.

Russel, L. M. (1977). Hosts and distribution of the greenhouse whitefly, Trialeurodes vaporariorum (Westwood) (Hemiptera: Homoptera: Aleyrodidae). United States Department of Agriculture Cooperative Plant Pest Report, 2, 449–458.

Sabelis, M. W., & van de Baan, H. E. (1983). Location of distant spider mite colonies by phytoseiid predators: Demonstration of specific kairomones emitted by Tetranychus urticae and Panonychus ulmi. Entomologica Experimentalis et Applicata, 33, 303–314.

Sabelis, M. W., Vermaat, J. E., & Groeneveld, A. (1984). Arrestment responses of the predatory mite Phytoseiulus persimilis to steep odour gradients of a kairomone. Physiological Entomology, 9, 437–446.

Scopes, N. E. A. (1969). The potential of Chrysopa carnea as a biological control agent of *Myzus persicae* on glasshouse chrysanthemums. Annals of Applied Biology, 64, 433–439.

Scopes, N. E. A. (1970). Control of *Myzus persicae* on year round chrysanthemums by introducing aphids parasitized by *Aphidius matricariae* into boxes of rooted cuttings. Annals of Applied Biology, 66, 323–327.

Scopes, N. E. A. (1985). Red spider mite and the predator *Phytoseiulus persimilis*. *In* "Biological Pest Control: The Glasshouse Experience" (N. W. Hussey & N. E. A. Scopes, Eds.), pp. 112–115. Blandford, Poole, Dorset.

Scopes, N. E. A., & Biggerstaff, S. M. (1973). Progress towards integrated pest control on year round chrysanthemum. *In* "Proceedings, 7th British Insecticide and Fungicide Conference," pp. 227–234.

Shipp, J. L., Clarke, N. D., Papadopoulos, A. P., Jarvis, W. R., Jewett, T. J., & Khosla, S. K. (1996). Harrow Greenhouse Crop Manager: A decision-support system for integrated management of greenhouse cucumber and tomato. Bulletin of the IOBC/WPRS, 1996/19/1, 155–158.

Smitely, D. (1993). Biocontrol of thrips and mites on roses. Roses Inc. Bull., Sept., pp. 57–63.

Smith, F. F., & Webb, R. E. (1977). Biogeographic and agronomic problems related to the utilization of biocontrol organisms in commercial greenhouses in the continental United States. *In* "Proceedings, XV International Congress on Entomology, Aug. 19–27, 1976, Washington, D.C.," pp. 89–93. ARS-NE-85, 1977.

Spencer, K. A. (1973). "Agromyzidae (Diptera) of Economic Importance" [Series Entomologica 9]. Dr. W. Junk, The Hague.

Speyer, E. R. (1927). An important parasite of the greenhouse whitefly. Bulletin of Entomological Research, 17, 301–308.

Stacey, D. L. (1977). "Banker" plant production of *Encarsia formosa* Gahan and its use in the control of glasshouse whitefly on tomatoes. Plant Pathology, 26, 63–66.

Stenseth, C. (1979). Effect of temperature and humidity on the development of *Phytoseiulus persimilis* Athias-Henriot and its ability to regulate populations of *Tetranychus urticae* Koch (Acarina: Phytoseiidae, Tetranychidae). Entomophaga, 24, 311–317.

Stenseth, C. (1980). Investigation of uniform introduction technique for use of *Phytoseiulus persimilis* for control of *Tetranychus urticae* on cucumber. Meldinger fra Norges Landbrukshøgskole, 59(7).

Stenseth, C. (1985). Red spider mite control by *Phytoseiulus persimilis* in Northern Europe. *In* "Biological Pest Control: The Glasshouse Experience" (N. W. Hussey & N. E. A. Scopes, Eds.), pp. 119–124. Blandford, Poole, Dorset.

Stenseth, C., & Aase, I. (1983). Use of the parasite *Encarsia formosa* (Hym.: Aphelinidae) as part of pest management on cucumbers. Entomophaga, 28, 17–26.

Szabo, P., van Lenteren, J. C., & Huisman, P. W. T. (1993). Development time, survival and fecundity of *Encarsia formosa* on *Bemisia tabaci* and *Trialeurodes vaporariorum*. Bulletin of the IOBC/WPRS, 1993/16/2, 163–168.

Uygun, N. (1971). Der Einfluss der Nahrungsmenge auf Fruchtbarkeit und Lebensdauer von *Aphidoletes aphidimyza* (Rondani 1847) (Diptera: Itonididae). Zeitschrift fuer Angewandte Entomologie, 69, 234–258.

van der Linden, A. (1990). Prospect for the biological control of *Liriomyza huidobrensis* (Blanchard), a new leafminer for Europe. Bulletin of the IOBC/WPRS, 1990/XIII/5, 100–103.

van der Linden, A. (1993). Development of an IPM program in leafy and tuberous crop with *Liriomyza huidobrensis* as a key pest. Bulletin of the IOBC/WPRS, 1993/16/2, 93–95.

van Lenteren, J. C. (1983). The potential of entomophagous parasites for pest control. Agriculture, Ecosystems, & Environment, 10, 143–158.

van Lenteren, J. C. (1985). Data on application areas of biological control in greenhouses. Sting, 8, 17–24.

van Lenteren, J. C. (1986). Parasitoids in the greenhouse: Successes with seasonal inoculative release systems. *In* "Insect Parasitoids" (J. Waage & D. Greathead, Eds.), pp. 342–374. Academic Press, London.

van Lenteren, J. C. (1987). World situation of biological control in greenhouses and factors limiting use of biological control. Bulletin of the IOBC/WPRS, 1987/X/2, 78–81.

van Lenteren, J. C. (1995). Integrated pest management in protected crops. *In* "Integrated Pest Management" (D. Dent, Ed.), pp. 311–343. Chapman & Hall, London.

van Lenteren, J. C. (1996). Quality control tests for natural enemies used in greenhouse biological control. Bulletin of the IOBC/WPRS, 1996/19/1, 83–86.

van Lenteren, J. C. (1997). Benefits and risks of introducing exotic macrobiological control agents into Europe. Bulletin OEPP/EPPO, 27, 15–27.

van Lenteren, J. C., & Hulspas-Jordan, P. M. (1983). Influence of low temperature regimes on the capability of *Encarsia formosa* and other parasites in controlling greenhouse whitefly. Bulletin of the IOBC/WPRS, 1983/VI/3, 54–70.

van Lenteren, J. C., & Woets, J. (1988). Biological and integrated pest control in greenhouses. Annual Review of Entomology, 33, 239–269.

van Lenteren, J. C., Hua, L. Z., & Kamerman, J. W. (1995). The parasite-host relationship between *Encarsia formosa* (Hymenoptera: Aphelinidae) and *Trialeurodes vaporariorum* (Homoptera: Aleyrodidae). XXVI. Leaf hairs reduce the capacity of *Encarsia* to control greenhouse whitefly on cucumber. Journal of Applied Entomology, 199, 553–559.

van Lenteren, J. C., Nell, H. W., Sevenster-van der Lelie, L., & Woets, J. (1976). The parasite-host relationship between *Encarsia formosa* (Hymenoptera: Aphelinidae) and *Trialeurodes vaporariorum* (Homoptera: Aleyrodidae). I. Discrimination between parasitized and unparasitized hosts by the parasite. Zeitschrift für angewandte Entomologie, 81, 377–380.

van Lenteren, J. C., Nell, H. W., & Sevenster-van der Lelie, L. A. (1980a). The parasite-host relationship between *Encarsia formosa* (Hymenoptera: Aphelinidae) and *Trialeurodes vaporariorum* (Homoptera: Aleyrodidae). IV. Oviposition behaviour of the parasite, with aspects of host selection, host discrimination and host feeding. Journal of Applied Entomology, 89, 442–454.

van Lenteren, J. C., Ramakers, P. M. J., & Woets, J. (1980b). World situation of biological control in greenhouses, with special attention to factors limiting application. Mededelingen van de Faculteit Landbouwwetenschappen, Rijksuniversiteit Gent, 45, 537–544.

van Lenteren, J. C., Ramakers, P. M. J., & Woets, J. (1980c). Integrated control of vegetable pests in greenhouses. *In* "Integrated Control of Insect Pests in The Netherlands" (A. K. Minks & P. Gruys, Eds.), pp. 109–118. Pudoc, Wageningen, The Netherlands.

van Lenteren, J. C., Roskam, M. M., & Timmer, R. (1997). Commercial mass production and pricing of organisms for biological control of pests in Europe. Biological Control, 10, 143–149.

van Lenteren, J. C., van Roermund, H. J. W., & Sütterlin, S. (1996). Biological control of greenhouse whitefly (*Trialeurodes vaporariorum*): How does it work? Biological Control, 6, 1–10.

van Lenteren, J. C., Woets, J., van der Poel, N., van Boxtel, W., van de Merondonk, S., van der Kamp, R., Nell, H., & Sevenster-van der Lelie, L. (1977). Biological control of the greenhouse whitefly *Trialeurodes vaporariorum* (Westwood) (Homoptera: Aleyrodidae) by *Encarsia formosa* Gahan (Hymenoptera: Aphelinidae) in Holland: An example of successful applied ecological research. Mededelingen van de Faculteit Landbouwwetenschappen, Rijksuniversiteit Gent, 42, 1333–1343.

van Roermund, H. J. W., & van Lenteren, J. C. (1992a). The parasite-host relationship between *Encarsia formosa* (Hymenoptera: Aphelinidae) and *Trialeurodes vaporariorum* (Homoptera: Aleyrodidae). XXXIV. Life-history parameters of the greenhouse whitefly, *Trialeurodes vaporariorum* as a function of host plant and temperature. Wageningen Agricultural University Papers 92-3, 1–102.

van Roermund, H. J. W., & van Lenteren, J. C. (1992b). The parasite-host relationship between *Encarsia formosa* (Hymenoptera: Aphelinidae) and *Trialeurodes vaporariorum* (Homoptera: Aleyrodidae). XXXV. Life-history parameters of the greenhouse whitefly parasitoid *Encarsia formosa* as a function of host stage and temperature. Wageningen Agricultural University Papers 92-3, 103–147.

van Roermund, H. J. W., & van Lenteren, J. C. (1993). The functional response of the whitefly parasitoid *Encarsia formosa*. Bulletin of the IOBC/WPRS, 1993/16/2, 141–144.

van Roermund, H. J. W., van Lenteren, J. C., & Rabbinge, R. (1997). Biological control of greenhouse whitefly with the parasitoid *Encarsia formosa* on tomato: an individual-based simulation approach. Biological Control, 9, 25–47.

van Schelt, J., Doubma, J. B., & Ravensberg, W. J. (1990). Recent developments in the control of aphids in sweet pepper and cucumber. Bulletin of the IOBC/WPRS, 1990/XIII/5, 190–193.

van Steenis, M. J. (1995). Evaluation of four aphidiine parasitoids for biological control of *Aphis gossypii*. Entomologica Experimentalis Applicata, 75, 151–157.

van Steenis, M. J., & El-Khawass, K. A. M. H. (1996). Different parasitoid introduction schemes determine the success of biological control of *Aphis gossypii*. Bulletin of the IOBC/WPRS, 1996/19/1, 159–162.

Vet, L. E. M., van Lenteren, J. C., & Woets, J. (1980). The parasite-host relationship between *Encarsia formosa* (Hymenoptera: Aphelinidae) and *Trialeurodes vaporariorum* (Homoptera: Aleyrodidae). IX. A review of the biological control of the greenhouse whitefly with suggestions for future research. Zeitschrift für angewandte Entomologie, 90, 26–51.

Vlak, J. M., den Belder, E., Peters, D., & van de Vrie, M. (1982). Bekämpfung eines eingeschleppten Schadlings *Spodoptera exigua* in Gewachshausen it den autochtonen virus. Mededelingen van het Proefstation voor de Bloemisterji te Aalsmeer, No. 81, 1005–1016.

Wardlow, L. R. (1984). Monitoring the activity of tomato leafminer (*Liriomyza trifolii* Kalt.) and its parasites in commercial glasshouses in Southern England. Mededelingen van de Faculteit Landbouwwetenschappen, Rijksuniversiteit Gent, 49, 781–791.

Wardlow, L. R. (1985a). Leafminers and their parasites. *In* "Biological Pest Control: The Glasshouse Experience" (N. W. Hussey & N. E. A. Scopes, Eds.), pp. 61–65. Blandford, Poole, Dorset.

Wardlow, L. R. (1985b). Chrysanthemums. *In* "Biological Pest Control: The Glasshouse Experience" (N. W. Hussey & N. E. A. Scopes, Eds.), pp. 180–185. Blandford, Poole, Dorset.

Wardlow, L. R. (1986). Adapting integrated pest control to work for ornamentals. Grower, 6, 26–29.

Whitcomb, W. D. (1940). "Biological Control of Mealybugs in Greenhouses." Massachusetts Agricultural Experiment Station Bulletin 375.

Woets, J. (1973). Integrated control in vegetables under glass in The Netherlands. Bulletin of the IOBC/WPRS, 73/4, 26–31.

Woets, J. (1976). Progress report on the integrated pest control in glasshouses in Holland. Bulletin of the IOBC/WPRS, 76, 34–38.

Woets, J. (1978). Development of an introduction scheme for *Encarsia formosa* Gahan (Hymenoptera: Aphelinidae) in greenhouse tomatoes to control the greenhouse whitefly *Trialeurodes vaporariorum* (Westwood) (Homoptera: Aleyrodidae). Mededelingen van de Faculteit Landbouwwetenschappen, Rijksuniversiteit Gent, 43, 379–385.

Woets, J. (1983). Observations on *Opius pallipes* Wesmael (Hym.: Braconidae) as a potential candidate for biological control of the tomato leafminer *Liriomyza bryoniae* Kalt. (Dipt.: Agromyzidae) in Dutch greenhouse tomatoes. Bulletin of the IOBC/WPRS, 1983/VI/3, 134–141.

Woets, J., & van Lenteren, J. C. (1976). The parasite-host relationship between *Encarsia formosa* (Hymenoptera: Aphelinidae) and *Trialeurodes vaporariorum* (Homoptera: Aleyrodidae). VI. The influence of the host plant on the greenhouse whitefly and its parasite *Encarsia formosa*. Bulletin of the IOBC/WPRS, 76, 125–137.

Woets, J., & van der Linden, A. (1982). On the occurrence of *Opius pallipes* Wesmael and *Dacnusa sibirica* Telenga (Braconidae) in cases of natural control of the tomato leafminer *Liriomyza bryoniae* Kalt. (Agromyzidae) in some large greenhouses in The Netherlands. Mededelingen van de Faculteit Landbouwwetenschappen, Rijksuniversiteit Gent, 47, 533–540.

Woets, J., & van der Linden, A. (1985). First experiments on *Chrysocharis parksi* Crawford (Hym.: Eulophidae) as a parasite for leafminer control (*Liriomyza* spp.) (Dipt.: Agromyzidae) in European greenhouse tomatoes Mededelingen van de Faculteit Landbouwwetenschappen, Rijksuniversiteit Gent, 50, 763–768.

Wyatt, I. J. (1965). The distribution of *Myzus persicae* (Sulz.) on year-round chrysanthemum. I. Summer season. Annals of Applied Biology, 56, 439–459.

Wyatt, I. J. (1970). The distribution of *Myzus persicae* (Sulz.) on year-round chrysanthemums. II. Winter season: The effect of parasitism by *Aphidius matricariae* Hal. Annals of Applied Biology, 65, 31–41.

32

Foliar, Flower, and Fruit Pathogens

T. S. BELLOWS

Department of Entomology
University of California
Riverside, California

INTRODUCTION

Biological control of plant pathogens is fundamentally a matter of ecological management of a community of organisms. In the case of plant pathogens, however, there are two distinctions from biological control of organisms such as insects and plants. First, management occurs at the microbial level, typically in biological microcosms (leaf surfaces, fruit surfaces, etc.) (Andrews, 1992). Second, biological control agents include competitors as well as parasites. Hyperparasites of plant pathogens function in much the same way as natural enemies (parasitoids) in arthropod systems (by destroying the pest organisms). Competitors function by occupying and using resources in a nonparasitic manner and in so doing exclude pathogenic organisms from colonizing plant tissues. Microbes which negatively affect pathogenic organisms are referred to as antagonists.

Diseases of stems, leaves, flowers, and fruit are caused by a wide variety of pathogens. Because of this diversity, the antagonist species which negatively affect plant pathogens and the mechanisms by which they accomplish their beneficial action are also quite varied. Their biology and taxonomic diversity are covered in several texts and reviews, including Cook and Baker (1983), Fokkema and van den Heuvel (1986), Campbell (1989), Adams (1990), and Tjamos *et al.* (1992). Use of microbes for control of plant pathogens also is covered in these texts and also by Parker *et al.* (1985), Lynch (1987), Wilson and Wisniewski (1989), Adams (1990), Jeffries and Jeger (1990), Andrews (1992), Cook (1993), and Sutton and Peng (1993a).

This chapter will introduce some of the antagonists of plant pathogens as representative of the broad taxa which are important in this field and will then turn to the methods by which these beneficial organisms can be employed in limiting plant diseases.

PATHOGENS AND THEIR BIOLOGICAL CONTROL AGENTS

Stems

Stem diseases produce symptoms which include decay and cankers on forest and orchard trees and wilts such as Dutch elm disease and chestnut blight (caused by the fungus *Cryphonectria parasitica* (Murrill) Barr of Asian origin infecting the American chestnut, *Castanea dentata* [Marsham] Borkjauser). Because the etiologies of stem diseases vary, the taxa involved in biological control also vary. In many stem diseases, the pathogen colonizes a part of the host which initially is relatively free of microorganisms, such as a pruning wound. Successful biological control in such circumstances depends on rapidly colonizing this pristine environment with a nonpathogenic antagonistic competitor. Primary among these organisms are competitively antagonistic fungi, including saprotrophic members of the genera *Fusarium, Cladosporium, Trichoderma,* and *Phanerochaete,* and such antibiotic-producing bacteria as *Bacillus subtilis* and *Agrobacterium* spp. In the case of chestnut blight, hypovirulent strains of the pathogen itself are crucial in bringing about biological control (see also Chapter 25). In this case, hypovirulence is transmitted cytoplasmically to virulent strains already infecting trees, and disease symptoms decline and disappear.

Leaves

The growth of microorganisms on leaves is normally severely restricted by environmental factors (Campbell, 1989). Nutrient levels generally are low on leaf surfaces, and microclimate variables, especially leaf surface moisture, temperature, and irradiation, are often unfavorable for microbial development. In temperate climates and in arid tropical regions, water will be intermittent on leaf surfaces, but may be continually present in humid tropical regions. Temperatures on leaf surfaces exposed to direct radiation may rise to several degrees above ambient. The result of such variation is that microbial floral development on leaf surfaces varies from general scarcity in temperate climes to more extensive microbial films in tropical rain forests (Campbell, 1989).

Microbial surveys of saprotrophs on surfaces of crop plants in temperate conditions yield a diversity of species that are candidates as antagonists of pathogens. Results of such surveys include species of fungi, such as *Aureobasidium pullulans* (de Bary) Arnaud and *Cladosporium* spp., and such yeasts as *Cryptococcus* spp. and *Sporobolomyces* spp. Beneficial bacteria in the phyllosphere include members of such genera as *Erwinia, Pseudomonas, Xanthomonas, Chromobacterium,* and *Klebsiella.* These lists, based on microbial surveys, usually give no indication of activity of the organisms, but this information can be obtained from experimental studies. For example, studies on control of botrytis rot in lettuce, *Lactuca sativa* Linnaeus (Wood, 1951), indicated that several organisms were successful in suppressing the disease when sprayed on lettuce plants, among them *Trichoderma viride* Persoon:Fries, *Pseudomonas* sp., *Streptomyces* sp., and *Fusarium* sp. The microbial composition and biological activity of microbes on leaves can vary with season, position on the top or bottom of the leaf, and location in the plant canopy, depending on the degree of exposure relative to prevailing winds and rain (Campbell, 1989).

Biological control of the black-crust pathogen (*Phyllachora huberi* Hennings) on rubber tree (*Hevea brasiliensis* Müller Argoviensis) foliage is accomplished by the hyperparasites *Cylindrosporium concentricum* Greville and *Dicyma pulvinata* (Berkeley & Curtis) Arx (Junqueira & Gasparotto, 1991). Botrytis leaf spot in onion (*Allium cepa* Linnaeus) was suppressed by *Gliocladium roseum* Link: Bainier (Sutton & Peng, 1993a). Other diseases which can be suppressed by various saprotrophs include powdery mildews, other botrytis rots, and turfgrass diseases (Sutton & Peng, 1993a).

Nonpathogenic species of the fungal genus *Colletotrichum* (Kúc, 1981; Dean & Kúc, 1986) can be used to induce resistance in cucumbers against pathogenic species of the same genus. Inoculation with a nonpathogenic strain of a virus confers protection to plants from pathogenic

strains in many diseases (see also Chapter 19). The bacterium *Bdellovibrio bacteriovorus* Stolp & Starr is a parasite of pathogenic bacteria. Finally, there are numerous parasitic fungi which attack pathogenic fungi (Kranz, 1981). Among those which have been studied in detail, principally as agents against leaf rusts and mildews, are *Sphaerellopsis filum* (Bivona-Bernardi ex Fries) Sutton, *Verticillium lecanii* (Zimmerman) Viegas, and *Ampelomyces quisqualis* Cesati ex Schlechtendal.

Flowers and Fruits

Flowers are ephemeral structures and as such have limited opportunity to become infected. One major disease of flowers which has received attention is fire blight of rosaceous plants, caused by the bacterium *Erwinia amylovora* (Burril) Winslow et al. Biological suppression of the disease has been achieved through use of the nonpathogenic species *Erwinia herbicola* (Lohnis) Dye (Beer *et al.,* 1984; Lindow, 1985), sometimes in combination with *Pseudomonas syringae* van Hall. *Erwinia herbicola* was used successfully by spraying aqueous suspensions of it onto the flowers just before the time of potential infection (Campbell, 1989). The mode of action is primarily competitive exclusion, with the antagonist competing with the pathogen for a growth-limiting resource and possibly other effects such as induced cessation of nectar secretion or accumulation of a host toxin (Wilson & Lindow, 1993).

The fruit diseases addressable through biological control include both pre- and postharvest diseases. One of the first systems developed was against *Botrytis cinerea* Persoon: Fries in vineyards, where sprays with spore suspensions of the antagonist *Trichoderma harzianum* Rifai were effective in suppressing disease incidence. Several organisms, including *Gliocladium roseum, Penicillium* sp., *Trichoderma viride* and *Colletotrichum gloeosporioides* Penzig, were as effective as fungicides in suppressing *B. cinerea* on strawberries (Peng & Sutton, 1991). A number of other examples also have been reported (Sutton & Peng, 1993a).

Postharvest diseases, which can be responsible for 10–50% loss of produce (Wilson & Wisniewski, 1989; Jeffries & Jeger, 1990), include major diseases caused by *Botrytis cinerea, Rhizopus* spp., and other fungi in several crops. Numerous reports deal with suppression of postharvest disease in fruit crops by species of *Penicillium, Bacillus, Trichoderma, Debaryomyces,* and *Pseudomonas* (Campbell, 1989; Wilson & Wisniewski, 1989; Jeffries & Jeger, 1990). The mode of action of many of these is generally antagonism, often through the production of antibiotics which reduce the longevity and germination of spores of pathogens. Some species appear to suppress pathogen growth through nutritional competition or induction of host resistance (Wilson & Wisniewski, 1989). Competitive and parasitic fungi, including *Trichoderma* spp., *Cladosporium her-*

barum (Persoon:Fries) Link, and *Penicillium* spp., can be as effective as commercial fungicides. *Enterobacter cloacae* (Jordan) Hormaeche & Eduards reduces rots by *Rhizopus* spp., but there are hesitations about its use on uncooked food products.

Antagonistic organisms vary both in their innate ability to suppress plant pathogens and in their ability to thrive and compete in different environments. Consideration of organisms for any particular biological control program must take into account these differing capabilities. In addition, the selection of organisms will depend on the approach taken for their use (inoculative augmentation, inundative augmentation, or natural control through conservation).

HABITAT AND AGENT CHARACTERISTICS

Pathogens and Antagonists

To understand the principles that apply to biological control of plant pathogens, we must first consider the ecology of the system at the level of the pathogens and the agents used to control them. Aerial plant surfaces, by and large, present hostile environments to colonizing microbes (Campbell, 1989; Andrews, 1992). Pathogenic microbes attempting to colonize these surfaces may face a number of difficulties, including competition with other, nonpathogenic, microbes.

Various forms of competition in these environments are important for any particular organism to increase in numbers and consequently reduce the numbers or activity of other organisms, including plant pathogens (Campbell, 1989; Andrews, 1992). Microbial competition can be important at two main stages of growth of pathogen populations. First, there may be competition during initial establishment on a fresh resource that was not previously colonized by microorganisms. Second, after initial establishment, there is further competition to secure enough of the limited resources present to permit survival and eventual reproduction. Microorganisms show many traits which may characterize them as particularly adept at either the colonization phase or subsequent phases of competition. Species referred to as *r*-strategists (ruderal species) have a high reproductive capacity. Such species produce so many spores or reproductive bodies that there is a high likelihood that some will be found near any newly available resource. These species are effectively dispersed and establish readily in disturbed habitats or in the presence of noncolonized resources. They are found in disturbed settings where easily decomposable organic matter or plant exudates are found and where initial resource capture is crucial for survival. In contrast to these *r*-strategists, species found in more stable situations face competition for space and limited resources

(Begon *et al.,* 1986). These organisms, termed *K*-strategists, become more dominant as a community matures and becomes more crowded. These concepts form the endpoints of a continuum, and there are varying degrees of *r*- and *K*-related characteristics in different microbes in various habitats (see Andrews and Harris (1986) for further discussion on these concepts in microbial ecology).

Plant pathogens are spread across this *r–K* range of characteristics and vary in other important biological characteristics. There are opportunistic pathogens that are able to attack young, weakened, or predisposed plants but may be poor competitors (*Botrytis, Pythium,* and *Rhizoctonia*). There are pathogens that tolerate environmental stresses. These organisms often live in situations with few competitors, as few species are able to exist in such environments. Some pathogens, such as the *Penicillium* species causing postharvest rots, produce antibiotics that inhibit competitors. Other species have a very high competitive ability (e.g., *Fusarium culmorum* [Smith] Saccardo). It is important to understand the ecology of a target pathogen before one can effectively consider what biological control strategy might be most effective. Stress-tolerant and competitive species, for example, would require different biological control strategies and agents than would ruderal ones.

In the same way that plant pathogens vary in *r–K* and other characteristics, the properties of an effective biological control agent will depend on the setting in which it is intended to function. In many agricultural settings, disturbance makes new resources available to microbes through crop residue burial, cultivation, or planting. A frequent need, therefore, is a control agent that has the characteristics of an *r*-strategist (Campbell, 1989), which can grow quickly and colonize new resources rapidly, with minimal nutrient and environmental restrictions. It should function well in disturbed environments and have some means (such as spores) of surviving in the soil or on the plant near the pathogen inoculum or the source or site of infection. Biological control agents that are *r*-strategists are an approximate equivalent of a protectant fungicide, being in place before the pathogen infection cycle can begin. In other programs, such as those directed against a pathogen which has already invaded the plant host, a more competitive species will be required. Finally, a biological control agent may have to be tolerant of abiotic stresses, particularly for use in dry climates.

Phyllosphere Habitat Characteristics

The phyllosphere is a unique environment in its structure, ecology, nutrient availability, and exposure to climatic factors (Andrews, 1992). Leaves are relatively hostile to microorganisms. They are generally hydrophobic and covered with cutin and wax, which limits the amount of exudate (and hence nutrients) that reaches the leaf surface. These and other factors (such as lack of moisture) impose severe environmental restrictions on microbial growth on leaf surfaces. Fungal pathogens of leaves often enter the leaf tissue very shortly after germination of the pathogen and consequently are protected inside the plant for much of their growth. Bacterial pathogens may multiply on the leaf surface prior to invasion of leaf tissues. Biological control can take place either through general inhibition and competition on the leaf surface prior to invasion of leaf tissues or through suppression of the disease after the pathogen has invaded. Biological control within leaf tissues can occur through one of several mechanisms, including induced resistance in the plant and hyperparasitism of the pathogen.

Woody stems are habitats low in nutrients and often difficult for pathogens to penetrate. Because the wood itself supports very few saprotrophic microorganisms, pathogens colonizing the wood through wounds, dead branches, or roots find very few competitors. Because there are few organisms present, protection of the wood from these decay organisms can be achieved by protecting the relatively small, well-defined wound or branch stub through inoculation (augmentation) with specific microorganisms. Such wounds are initially very low in sugars or other nonstructural carbohydrates, and antagonists such as *Trichoderma* spp. can successfully compete for these limited resources. Many of the organisms used in biological control of stem diseases are employed by applying them directly to stem wounds, where they colonize resources and subsequently exclude pathogenic forms. This initial occupancy by antagonists subsequently limits infection by decay-causing organisms and hence controls the succession of microorganisms in the wood. Of the successful, commercially available biological control products for plant diseases, several are for diseases of woody stems (Campbell, 1989).

BIOLOGICAL CONTROL OF PLANT PATHOGENS

Organisms for biological control of plant disease can be used in various ways, but most attention has been given to their conservation and augmentation rather than the importation and addition of new species as is often done for insect or weed control. This is in part because there is usually a diverse set of microbes already associated with plants. These microbes provide substantial opportunity for development of resident species as competitors or antagonists to pathogenic organisms.

The are several ways in which a conserved or augmented microbial biological control agent can operate against a targeted plant pathogen (Elad, 1986). Among these are (1) competition, (2) induction of plant defenses, and (3) parasitism.

Some agents act through competition for limited resources, and through this competition the growth of the pathogen population is suppressed, reducing the incidence or severity of disease. One important component of competition can be competition for Fe^{3+} ions. These ions are sequestered by chemicals called siderophores, which are produced by many species of plants and microbes. Highly efficient siderophores from nonpathogenic microbes can remove Fe^{3+} ions from the soil, outcompeting siderophores from pathogens and thereby limiting the growth of pathogen populations. Some biological control agents compete through the production of antimicrobial substances such as antibiotics which inhibit the growth of pathogens directly rather than by preemptive consumption of limiting resources.

Another important mechanism limiting infection is the induction of plant defenses against pathogens by nonpathogenic organisms. Cross-protection and induced resistance are mechanisms in which plants are intentionally exposed to certain microbes, thereby conferring in the treated plants some resistance to infection by pathogens. Induced plant defenses may include lignification of cell walls through the addition of chemical cross-linkages in cell wall peptides, which makes the establishment of infection through lysis more difficult; suberification of tissues (where plant cell walls are infiltrated with the fatty substance suberin, making them more corklike); and other general defenses, including production of chitinases and β-1,3-glucanases. These plant defenses then limit later infection by the pathogen. The biological control agent employed may be an avirulent strain of the pathogen, a different forma specialis, or even a different species of microorganism (see also Chapter 19).

Parasitism is a third mechanism by which beneficial microorganisms suppress plant pathogens. Some species of *Trichoderma,* for example, attack pathogenic fungi, leading to the lysis of the pathogen. The yeast *Candida saitoana* Nakase & Suzuki causes severe cytological injury to *B. cinerea* on apple fruit (El-Ghaouth *et al.,* 1998).

Conservation

Control of plant pathogens through conservation of naturally occurring antagonists is accomplished either by preserving existing microbes which attack or compete with pathogens or by enhancing conditions for their survival and reproduction at the expense of pathogenic organisms. Conservation is applicable in situations where microorganisms important in limiting disease-causing organisms already occur, primarily in the soil and plant residues but in some cases also on leaf surfaces. They may be conserved by avoiding practices which negatively affect them (such as soil treatments with fungicides).

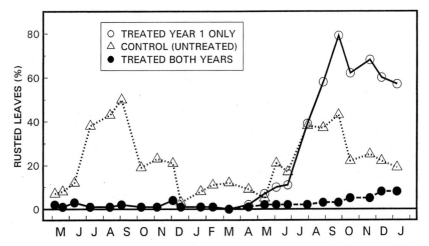

FIGURE 1 The importance of conserving fungi as biological control agents can be demonstrated by the use of fungicides, as in the case of fungicide treatment effects on incidence of rust (*Hemileia vastatrix* Berkeley & Broome) on coffee (*Coffea arabica* Linnaeus) leaves. The untreated control has seasonally variable incidence with maxima of approximately 50%, whereas plots treated in both years have very low incidence. Plots treated only in the first year have low incidence in that year but subsequently have incidences much higher than the untreated control, indicating the presence of microorganisms important in limiting the disease (after Mulinge & Griffiths, 1974).

Conservation of existing flora may be important in limiting the extent of a number of leaf diseases (Campbell, 1989). These effects are often revealed through the use of fungicides which deplete extant fungi, permitting the development of previously unimportant diseases. Fokkema and de Nooij (1981), for example, evaluated the effects of various fungicides on leaf surface saprotrophs that have been used in biological control. Wide-spectrum fungicides permitted almost no growth of saprotrophs, whereas more selective agents permitted some growth of several genera of saprotrophs. In cases where these saprotroph populations play an important role in limiting disease organisms, the application of fungicides would eliminate their contribution to pathogen suppression. One such case is illustrated by Fokkema and de Nooij (1981). Plants treated with benomyl (a systemic fungicide) had fewer saprotrophs and developed more necrotic leaf area when inoculated with *Cochliobolus sativus* (Ito & Kuribayashi) Drechsler ex Dastur than nontreated plants (*C. sativus* is insensitive to benomyl). Another example (Mulinge & Griffiths, 1974) is leaf rust of coffee (*Coffea arabica* Linnaeus), caused by *Hemileia vastatrix* Berkeley & Broome. The disease can be controlled by proper application of fungicide. However, if fungicides are applied in one year and not in the next, the disease is worse on the treated plants than on those which did not receive treatments either year (Fig. 1). The elimination of the saprotrophic flora by the fungicide removes their natural suppressing influence on the disease organisms, resulting in greater disease incidence. Here, careful use of selective fungicides will be crucial to conserving the important antagonistic flora and permitting their beneficial action.

Augmentation

Biological control of plant pathogens through augmentation is based on mass-culturing antagonistic species and adding them to the cropping system. This is considered augmentation of natural enemy populations, because the organisms used are usually present in the system, but at lower numbers or in locations different from those desired. The purpose of augmentation is to increase the numbers or modify the distribution of the antagonist in the system. In some cases such organisms are taken from one habitat (e.g., the soil) and augmented in another (e.g., the phyllosphere). Augmentation of microbial agents is sometimes termed "introduction" in the plant pathology literature, in the sense of "adding" the microbes to the system (Andrews, 1992; Cook, 1993). The organisms introduced, however, are usually found in a local ecosystem and are not "introduced" from another region of the world (in the sense of Chapter 5). There are several examples where such organisms, when moved to a new habitat (e.g., from the soil to the aboveground part of a plant), colonize and serve as successful biological control agents (Andrews, 1992; Cook, 1993).

Augmentation of antagonists of plant disease organisms can generally be of two types, inoculation and inundation. Inoculative releases consist of small amounts of inoculum, with the intention that the organisms in this inoculum will establish populations of the antagonist which will then increase and limit the pathogen population. In inundative releases, a large amount of inoculum is applied, with the expectation that control will result directly from this large initial population, with limited reliance on subsequent population growth. Biological control of plant pathogens may also rely on a hybrid of these two concepts. A large amount of inoculum may be applied, both to increase the population of the antagonist and to improve its distribution to favor biological control, and antagonism can result from both these applied organisms and the increased population of antagonists resulting from their reproduction. Biological control of black-crust (*Phyllachora huberi*) on rubber tree foliage by the hyperparasites *Cylindrosporium concentricum* and *Dicyma pulvinata* (Junqueira & Gasparotto, 1991) is one example of long-term control of a plant pathogen by a single augmentation in an agricultural system (Cook, 1993). In this case, rubber trees were treated with spore suspensions of the antagonists (inundatively), which resulted in control over more than one season. More generally, beneficial microorganisms are added seasonally or more frequently.

Stem Diseases

The control of *Heterobasidion annosum* (Fries) Brefeld, the causative agent of butt rot in conifer stumps, by *Phanerochaete gigantea* (Fries:Fries) Rattan et al. was one of the first commercially available agents for biological control of a plant pathogen (Campbell, 1989). The disease caused by *H. annosum* is primarily a disease of managed plantations. The fungus colonizes freshly cut stumps, invades the dying root system, and can then infect nearby trees through natural root grafts, causing death of the trees. *Heterobasidion annosum,* however, is a poor competitor, and when a stump is intentionally inoculated with *Ph. gigantea* (and usually with chemical nitrogen sources which encourage the growth of the antagonist), the antagonist rapidly colonizes the resource, excluding future attack by the pathogen and even eliminating existing pathogen infection. (Table 1). Very little inoculum is needed on a freshly cut stump, and the shelf life of the pellet formulation is about 2 months at 22°C. The antagonist is able to outcompete *H. annosum* even when the initial inoculum favors the pathogen by as much as 15:1 (Rishbeth, 1963)

The ascomycete fungi *Eutypa armeniaceae* Hansford & Carter and *Nectria galligena* Bresadola in Strass infect apricots and apples, respectively, and cause stem cankers and eventual death of the trees. Pruning wounds in apricots are treated with *Fusarium laterium* Nees:Fries through specially adapted pruning cutters. *Fusarium laterium* produces an antibiotic which inhibits germination and growth of *E. armeniaceae*. When applied, the concentration of the antagonist must be greater than 10^6 conidia/ml. Integrated application which includes a benzimidazole fungicide gives better control than either fungicide or antagonist alone. *Nectria galligena* infection can be reduced through sprays of suspensions of *Bacillus subtilis* or of *Cladosporium cladosporioides* (Fresenius) de Vries. These are not in current commercial use because apples are currently sprayed for *Venturia inaequalis* (Cooke) G. Winter (apple scab) so frequently that *N. galligena* is controlled by those sprays.

Crown gall is a stem disease caused by the bacterium *Agrobacterium tumefaciens* (Smith & Townsend) Conn. It affects both woody and herbaceous plants in 93 families. Infection is typically from the soil, rhizosphere, or pruning tools. Control can be effected by treating plants with a suspension of a related saprotrophic bacterium, *Agrobacterium radiobacter* (Beijerink & van Delden) Conn strain K-84. This strain of the bacterium produces an antibiotic which is taken up by a specific transport system in the pathogen bacterium, which is then killed. The commercially available formulations of this agent are effective primarily against pathogen strains which attack stone fruits, but other bacteria are under investigation for use against strains pathogenic in other crops. This agent has been altered by gene-modifying technology to produce a new strain (strain 1024) which lacks the ability to transfer antibiotic resistance to the target bacterium.

The fungus *Chondrostereum purpureum* (Persoon:Fries) Pouzar infects stems of fruit trees and produces a toxin which leads to a condition known as silverleaf disease. Stems can be inoculated with a species of *Trichoderma*

TABLE 1   Colonization of Scots Pine Stumps (*Pinus sylvestris* Linnaeus) after Inoculation with Various Fungi as Antagonists to *Heterobasidion annosum* (Fries) Brefeld (Which Was Inoculated into All Stumps) (after Rishbeth, 1963)

| | Percentage of mean area of stump section colonized after | | | | | |
| | 10 weeks | | | 6 months | | |
Inoculated Antagonist	Antagonist	Pg[a]	Ha[b]	Antagonist	Pg	Ha
None	—	28	38	—	80	7
Botrytis cinerea Persoon:Fries	5	5	55	0	0	25
Trichoderma viride Persoon:Fries	0	10	65	0	43	40
Leptographium lundbergii Lagerberg & Melin	95	9	5	37	47	0
Phanerochaete gigantea (Fries:Fries) Rattan et al.	80	80	Trace	75	75	0

[a] *Phanerochaete gigantes* (Fries:Fries) Rattan et al.
[b] *Heterobasidion annosum* (Fries) Brefeld.

grown on wooden dowels or prepared as pellets which are inserted into holes bored in the affected stem. Treated stems recover from the disease more rapidly than untreated stems. The *Trichoderma* sp. can be applied to pruning wounds to prevent initial establishment of *C. purpureum*.

Leaf Diseases

Control of leaf diseases at the time of pathogen germination has been shown in the laboratory. This control occurs in the presence of competitive organisms, which may include fungi, yeast, or bacteria. The mode of action in some cases is competition for nutrients which, together with water, are necessary for successful germination and invasion of many pathogens. The germination of *Botrytis* sp., for example, is inhibited by certain bacteria and yeasts (Blakeman & Brodie, 1977). This inhibition is less pronounced when additional nutrients are supplied, indicating that the mechanism is, at least in part, resource competition. Studies on control of botrytis rot in lettuce (Wood, 1951) indicated that several organisms were successful in suppressing the disease when sprayed on lettuce plants, among them species of *Pseudomonas*, *Streptomyces*, *Trichoderma viride*, and *Fusarium*. Peng and Sutton (1991) evaluated 230 isolates of mycelial fungi, yeasts, and bacteria and tested them as antagonists of *B. cinerea* in strawberry in both laboratory and field trials. Several organisms (including members of each taxonomic group tested) were effective, some as effective as captan (a commercial fungicide) (Table 2). Sutton and Peng (1993b) further evaluated *Gliocladium roseum* and determined that the suppression of *B. cinera* by this antagonist was probably a result of competition for leaf substrate. The fungi *Gliocladium roseum* and *Myrothecium verrucaria* (Albertini & Schweinitz) Ditmar were also effective in suppressing *B. cinerea* in black spruce (*Picea mariana* [Miller] Britton Steams Poggenburg) seedlings (Zhang *et al.*, 1994).

Bacteria may also be used to limit frost damage to leaves and blossoms of plants. Certain bacterial species such as *Pseudomonas syringae* and *Erwinia herbicola* serve as nucleation sites on leaves for the formation of ice, and in their presence ice forms soon after temperatures fall below freezing. If these ice-nucleating bacteria are replaced by competitive antagonists (such as certain strains of *Ps. syringae*) that lack the protein that causes ice nucleation, frost is prevented even at temperatures from -2 to $-5°C$ (Lindow, 1985). The protective bacteria, after being applied to the leaves, colonize them for up to 2 months, an interval suitable to protect from frost during the limited season that low temperatures are likely. A naturally occurring, non-ice-nucleating strain of *Ps. fluorescens* is registered in the United States as a commercial product (Frostban B) for suppression of frost damage (Wilson & Lindow, 1994).

Spraying suspensions of propagules, generally at high concentrations, is the principal method for applying biological control agents to foliage (and to flowers), and dusts

TABLE 2 Effects of Various Microorganisms and Captan on Incidence of *Botrytis cinerea* Persoon:Fries in Stamens and Fruits of Strawberry in Field Plots[a]

| Treatment | Incidence of *B. cinerea* (%) | | | |
| | Cambridge plots | | Arkell plots | |
	Stamens	Fruits	Stamens	Fruits
Water check	4 b	35 c	2 b	59 b
Botrytis cinerea Fries:Persoon check	19 a	59 a	16 a	71 a
Captan	5 b	40 b	3 b	56 b
Bacillus sp.	15 a	56 a	12 a	74 a
Cryptococcus laurentii (Kufferath) Skinner	17 a	43 b	11 a	72 a
Rhodotorula glutinis (Fresenius) Harrison	6 b	33 c	12 a	48 c
Alternaria alternata (Fries) Kessler	9 b	51 a	3 b	52 c
Myrothecium verrucaria (Albertini & Schweinitz) Ditmar	15 a	48 b	13 a	60 b
Fusarium graminearum Schwabe	18 a	47 b	16 a	65 a
Fusarium sp.	17 a	46 b	16 a	64 a
Drechslera sp.	16 a	26 c	9 a	56 b
Trichoderma roseum (Persoon) Link	12 a	27 c	13 a	41 c
Epicoccum purpurascens Ehrenberg ex Schlechtendal	6 b	43 b	4 b	51 c
Colletotrichum gloeosporioides Penzig	4 b	44 b	6 b	41 c
Trichoderma viride Persoon:Fries	5 b	38 b	2 b	37 c
Penicillium sp.	4 b	14 d	1 b	38 c
Gliocladium roseum Link:Bainer	4 b	25 c	3 b	43 c

[a] All plots were treated with suspensions of conidia of *B. cinerea*. Several antagonists were as effective as captan in reducing incidence of the pathogen (after Peng & Sutton, 1991) (values in columns followed by the same letter did not differ significantly).

(such as lyophilized bacterial preparations) are also used. Spray methodology has yet to be refined in terms of sprayer characteristics, droplet size, pressures, etc., and other methods of application with greater efficiency may be necessary to effectively target certain plant parts (Sutton & Peng, 1993a).

Flower Diseases

One of the principal diseases of flowers of rosaceous plants is fire blight, which is particularly severe on pear (Campbell, 1989). The causal bacterium, *Erwinia amylovora,* also occurs on leaves and may cause stem cankers. The bacterium is transferred by insects to flowers in the spring from overwintering sites on stem cankers and subsequently from flower to flower. Infection enters the pedicel and from there the stem. Infected flowers and small stems die, and cankers form on other stems. Chemical control is difficult and expensive and is sometimes ineffective because of resistance to copper compounds and streptomycin. Biological control has been effective using *Erwinia herbicola,* sometimes in combination with *Pseudomonas syringae* (Wilson & Lindow, 1993). Suspensions of *E. herbicola* are sprayed onto the flowers just prior to the period of potential infection. The antagonist occupies the same niche as the pathogen, reducing the numbers of *E. amylovora* by competition, and there is also evidence for the production of bacteriocins (chemicals which suppress population growth of related bacteria) by some strains. Control can be good, comparable to that achieved by commercial bactericides, though repeated application of the bacterium is necessary (Isenbeck & Schultz, 1986). Another approach to control is to reduce secondary infections on leaves, which leads to reductions in the overwintering population of the pathogen. This control is achieved by treatment with the antagonists *Ps. syringae* and other bacteria (Lindow, l985). A novel approach to dissemination of the antagonistic bacteria has been evaluated by Thomson *et al.* (1992). These workers mixed *E. herbicola* and *Ps. fluorescens* with pollen in a special apparatus at the entrance to honeybee (*Apis mellifera* Linnaeus) hives. Bees emerging from these hives through the mixtures transmitted the antagonists to the flowers efficiently, although disease control was not evaluated because of absence of disease in the test orchards. Johnson *et al.* (1993) report efffective control of *E. amylovora* using *Ps. fluorescens* strain A506 and *E. herbicola* strain C9-1. Crab apple has been proposed as a model system to further explore the possibilities of biological control of fire blight (Pusey, 1997).

Fruit Diseases

Fruits are subject to attack both by general pathogens (*Botrytis, Rhizopus,* and *Penicillium*) and by a few more specialist pathogens such as *Colletotrichum coffeanum* Noack, the causative agent of coffee berry disease, and *Monilinia* spp. brown rots of rosaceous fruits. Although many of these are controlled by fungicides, *Trichoderma viride* has been shown to limit disease from *Monilinia* spp. Various *Bacillus* spp. also are antagonistic to these fungi through production of antibiotics and by reducing the longevity and germination of spores. Both the bacteria and culture filtrates have been used with some success against these pathogenic fungi, but there has been no commercial development, probably because fungicides used routinely in orchards for control of other diseases give some control of brown rot (Campbell, 1989).

Among the most serious diseases of soft fruits are postharvest rots (Dennis, 1983), especially those caused by *Botrytis cinerea*. Potential for biological control of postharvest diseases has been reviewed by Wilson and Wisniewski (1989). In strawberries, the *B. cinerea* grows saprotrophically on crop debris, and from there infects flowers or fruit. Various species of *Trichoderma* have been evaluated and were as effective as standard fungicides (Tronsmo & Dennis, 1977). The antagonists *Cladosporium herbarum* and *Penicillium* sp. gave excellent results in controlling botrytis rot on tomato (Newhook, 1957). Honeybees have been used to distribute *Gliocladium roseum* to strawberry flowers (Peng *et al.,* 1992) and raspberry (Sutton & Peng, 1993a) to suppress botrytis rot. Selected isolates of *Aureobasidium pullulans* (de Bary) G. Arnaud, *Rhodotorula glutinis* (Fresen.) F. C. Harrison, and *Bacilis subtilis* reduced size and number of postharvest lesions on apple fruit when applied to apple trees in the field late in the growing season (Leibinger *et al.,* 1997).

Induced Resistance

Leaf and Stem Diseases

Induced resistance can control anthracnose diseases caused by *Colletotrichum* spp. (Kúc, 1981; Dean & Kúc, 1986). *Colletotrichum lindemuthianum* (Saccardo & Magnus) Lamson-Scribner causes anthracnose of beans, *Colletotrichum lagenarium* (Passerine) Saccardo causes cucumber anthracnose, and *Cladosporium cucumerinum* Ellis & Arthur causes scab in cucumbers. Inoculation of cucumbers with *Colletotrichum lindemuthianum* (which does not cause disease in cucumbers) makes the plants resistant to both *Colletotrichum lagenarium* and *Cladosporium cucumerinum*. Treatment may be applied to an early leaf and protection subsequently appears in later leaves, even if the initially inoculated leaf is removed. The factor causing resistance travels systemically through the plant. Variations on this approach include inoculating an early leaf with a pathogen, inducing resistance throughout the plant, and then removing the infected leaf. Induced resistance also

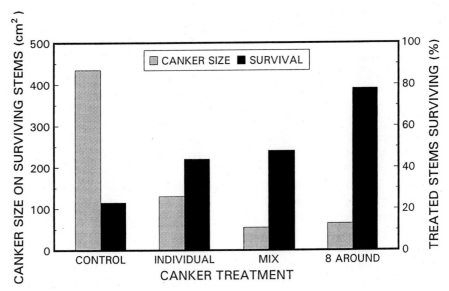

FIGURE 2   Example of hypovirulence in American chestnut (*Castanea dentata* [Marsham] Borkjauser) inoculated with normal (control) and hypovirulent strains of *Cryphonectria parasitica* (Murrill) Barr. "8 around" is eight individual strains placed around a canker (after Jaynes & Elliston, 1980).

occurs in some virus diseases (Thomson, 1958) and may last for years, as in the case of healthy citrus seedlings being inoculated with an avirulent strain of citrus tristeza virus.

Stem rot in carnations, caused by *Fusarium roseum* Link:Fries *'Avenaceum',* can be prevented by inoculating wounds inflicted during propagation with the nonpathogenic *F. roseum 'Gibbosum'.* This inoculation produced a germination inhibitor and also reduced the time needed for the stems to develop resistance to the pathogen. This hastening of resistance was caused by activation of the host's

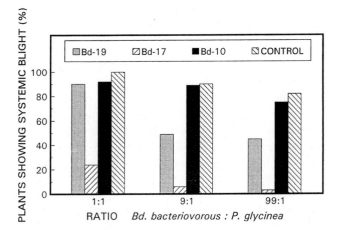

FIGURE 3   Percentage of soybean (*Glycine max* [Linnaeus]) plants showing symptoms of systemic blight after leaves were inoculated with different strains of *Bdellovibrio bacteriovorus* Stolp & Starr mixed in varying ratios with the pathogen *Pseudomonas syringae* pv. *glycinae* (Coeper) Young, Dye & Wilkie (after Scherff, 1973).

defense mechanisms and is another example of induced resistance. Induction of resistance is discussed more fully in Chapter 19.

Hypovirulence

Chestnut blight, caused by *Cryphonectria parasitica,* is controlled by employing naturally occurring hypovirulent strains of the disease pathogen. A number of hypovirulent strains are known, and natural spread of hypovirulent strains in Europe helps to limit damage there (Heiniger & Rigling, 1994). Inoculation of infected trees with a hypovirulent strain leads to reduced canker size and greater stem survival (Fig. 2). In applied programs hypovirulent strains are inoculated into infected trees at the rate of 10 inoculated trees/ha. The hypovirulent strain spreads from these locations and, on contacting more virulent strains, fuses with these strains and exchanges a dsRNA viral element present in the hypovirulent strain (Jaynes & Elliston, 1980; Heiniger & Rigling, 1994; see also Chapter 25). The dsRNA element, which causes hypovirulence, is transferred to the virulent strains and attenuates their effects. Active cankers are eliminated in 10 years (van Alfen, 1982).

Parasitism of Pathogens

Leaf Diseases

Some plant pathogens, including fungi and some bacteria, are known to be attacked by other pathogens. *Bdellovibrio bacteriovorus* is a bacterium that can attack other bacteria by penetrating the cell wall and lysing the host bacterium, subsequently reproducing inside the host. Different strains of *Bd. bacteriovorus* have been examined for virulence against *Pseudomonas syringae* pv. *glycinae* (Coeper) Young, Dye & Wilkie, the cause of soybean blight. By applying *Bd. bacteriovorus* at sufficiently high rates, disease symptoms were reduced more than 95% (Scherff, 1973) (Fig. 3).

Parasites of fungi pathogenic on leaves are numerous (Kranz, 1981), but only a few have been studied in much detail, such as *Sphaerellopsis filum*, *Verticillium lecanii*, and *Ampelomyces quisqualis*. The mycoparasitic fungus typically penetrates the host hypha or spore and kills it. Some of the control may be from the antagonist overgrowing the sporulating pustules of the pathogen and preventing spore release, thus reducing inoculum in the environment, even if the spores are not killed. A typical problem with implementation of these mycoparasitic fungi is that they often do not affect a large proportion of the pathogens unless humidity and temperature are high. Consequently, although much reduction of spore production may take place, there is still sufficient inoculum of the pathogen remaining to cause disease. These mycoparasites often are seen naturally only at high incidences of disease, which is unsuitable for general economic control of the target pathogens. They may have some use as augmented agents in particular systems, either in the tropics or in greenhouses, where environmental conditions are more favorable.

DEVELOPING AND USING BENEFICIAL SPECIES

Growth of our knowledge about biological control of plant diseases has been extensive since the first experimental reports (Hartley, 1921; Sanford, 1926; Millard & Taylor, 1927; Henry, 1931), and substantial potential for microbial control of pathogens has been demonstrated. A number of products or programs have reached the stage of commercial development or availability (Table 3). Products in current use include both those aimed at specialty markets, such as control of certain stem or flower diseases for which chemical control is either unavailable or expensive, and other products aimed at larger scale markets such as seed treatments for cotton and soybeans.

The cycle for research, development, and implementation of antagonists of plant pathogens is composed of several steps. These include initial discovery of candidate agents; refinement of our knowledge of their biology, ecology, and mode of action; microcosm and field trials of their efficacy; and large-scale development for commercial production.

The first challenge in the development of a biological control program is the discovery process. Many microorganisms show potential as antagonists of particular pathogens. Protocols have been proposed to make the process of screening these candidates more efficient (Andrews, 1992; Cook, 1993). The principal difficulties are screening out candidates that are effective only during *in vitro* (agar plate) trials but are not effective in natural settings, and in selecting candidates that can be successfully cultured in large quantities. Following discovery of suitable candidates, research focuses on their mode of action and on factors which may enhance or limit their efficacy in targeted settings (greenhouses, field plots). In addition, experimental fermentation and formulations must be developed for production of materials suitable for use in agricultural settings. Finally, issues of large-scale production and delivery must be addressed. Products for use must be effective on an economical basis, and economies of scale may play an important role in the eventual availability of any organisms or product. Products must have a satisfactory shelf life, and safe and effective methods for application must be discovered or developed (Cook, 1993; Sutton & Peng, 1993a). Such application methods might include sprays of suspensions or dusts, contact application, bee vectoring, and production of antagonists in a crop environment (Sutton & Peng, 1993a).

The adoption of any biological control agent in commercial agriculture is dependent on its reliability and its availability. Limitations to the process of eventual adoption therefore include cost of development and size of potential market. Many pesticides for control of plant diseases have a broad spectrum of activity, are applicable in a variety of crops and settings, and may act either prophylactically, therapeutically, or both. Biological controls, in contrast, often have narrow ranges of activity because of their biology and may work in only a few crops or field settings because of their ecology. Although they can often act both prophylactically and therapeutically, their action may take some time to develop. Hence they may have a much narrower market than a chemical pesticide and be unattractive for development by major corporations (Andrews, 1992). In this context, it may be appropriate for public institutions such as government experiment stations to undertake the development of such biological controls, in the same way that they take the responsibility for development of new plant varieties (Cook, 1993).

Microorganisms intended for use as biological control agents must be viewed in a biological rather than a chemical paradigm (Cook, 1993). Where an effective pesticide may work in many places, each place may have unique soil, edaphic, and biological features which limit or enhance the effectiveness of microbial antagonists of pathogens. Consequently, each microbial biological control system may have to make use of locally adapted strains, taking advantage of resident antagonistic flora and fauna and augmenting their effectiveness with additional species or strains or enhancing resident populations (as through soil amendments or other modifications to the environment or crop).

851

Organism	Trade name	Target
Agrobacterium radiobacter (Beijerink & van Delden) Conn strain K-84	Agtrol; Galltrol; Norbac 84-C	*Agrobacterium tumefaciens* (Smith & Townsend) Conn (crown gall)
Pseudomonas cepacia (Burkholder) Palleroni and Holmes	Blue Circle; Intercept	*Rhizoctonia* spp., *Pythium* spp., and *Fusarium* spp. diseases of seedlings
Pseudomonas fluorescens Migula	Dagger G	*Rhizoctonia* spp. and *Pythium* spp. diseases of seedlings
Coniothyrium minitans Campbell	Coniothyrin	*Sclerotinia sclerotiorum* (Libert) de Bary in sunflower
Fusarium oxysporum Schlechtendal (nonpathogenic strains)	Fusaclean; Biofox C	Diseases froom pathogenic strains of *Fusarium oxysporum*
Gliocladium virens Millers, Giddens & Foster	GlioGard	*Rhizoctonia* spp. and *Pythium* spp. diseases of seedlings and bedding plants
Phanerochaete gigantea (Fries:Fries) Rattan et al.	—	*Heterobasidion annosum* (Fries) Brefeld (butt rot)
Pythium oligandrum Drechsler	Polygandron	*Pythium ultimum* Trow in sugar beet
Streptomyces griseoviridis Anderson et al.	Mycostop	*Alternaria* sp. and *Fusarium* spp. diseases
Trichoderma harzianum polysporum (Link) Rifai	BINAB	Wood rot fungi; *Verticillium malthousei* Ware in mushrooms
Trichoderma harzianum Rifai	F-Stop; Trichodex; Supravit; T-35; TY	*Heterobasidion annosum;* diseases caused by *Rhizoctonia* spp., *Pythium* spp., *Fusarium* spp., *Botrytis cinerea* Persoon:Fries, and *Scleotium rolfsii* Saccardo
Trichoderma lignorum (Tode) Harz	Trichodermin-3	*Rhizoctonia* spp. and *Fusarium* spp. diseases

Although different strains may use common mechanisms to achieve biological control (such as production of antibiotics), competitive abilities adapted to local conditions may be vital to permit the organisms to compete for resources and effectively control pathogens.

References

Adams, P. B. (1990). The potential of mycoparasites for biological control of plant diseases. Annual Review of Phytopathology, 28, 59–72.

Andrews, J. H. (1992). Biological control in the phyllosphere. Annual Review of Phytopathology, 30, 603–635.

Andrews, J. H., & Harris, R. F. (1986). r- and K-selection and microbial ecology. Advances in Microbial Ecology, 9, 99–147.

Bailey, J. A., Ed. (1986). "Biology and Molecular Biology of Plant-Pathogen Interactions." Springer-Verlag, Berlin; New York.

Beer, S. V., Rundle, J. R., & Norielli, J. L. (1984). Recent progress in the development of biological control for fire blight. Acta Horticulturae, 151, 195–201.

Begon, M., Harper, J. L., & Townsend, C. R. (1986). "Ecology: Individuals, Populations and Communities." Blackwell Sci., Oxford.

Blakeman, J. P., & Brodie, I. D. S. (1977). Competition for nutrients between epiphytic microorganisms and germination of spores of plant pathogens on beetroot leaves. Physiological Plant Pathology, 10, 29–42.

Campbell, R. (1989). "Biological Control of Microbial Plant Pathogens." Cambridge Univ. Press, Cambridge, U.K.

Cook, R. J. (1993). Making greater use of introduced microorganisms for biological control of plant pathogens. Annual Review of Phytopathology, 31, 53–80.

Cook, R. J., & Baker, K. F. (1983). "The Nature and Practice of Biological Control of Plant Pathogens." American Phytopathological Society, St. Paul, MN.

Dean, R. A., & Kúc, J. (1986). Induced systemic protection in cucumber: Time of production and movement of the signal. Phytopathology, 76, 966–970.

Dennis, C., Ed. (1983). "Post-Harvest Pathology of Fruits and Vegetables." Academic Press, London.

El-Ghaouth, A., Wilson, C. L., & Wisniewski, M. (1998). Ultrastructural and cytochemical aspects of the biological control of Botrytis cinerea by Candida saitoana in apple fruit. Phytopathology, 88, 282–291.

Elad, Y. (1986). Mechanisms of interactions between rhizosphere microorganisms and soil-borne plant pathogens. In "Microbial Communities in Soil" (V. Jensen, A. Kjoller, & L. H. Sorensen, Eds.), pp. 49–60. Elsevier, New York.

Fokkema, N. J., & de Nooij, M. P. (1981). The effect of fungicides on the microbial balance in the phyllosphere. European and Mediterranean Plant Protection Organization Bulletin, 11, 303–310.

Fokkema, N. J., & van den Heuval, J. (1986). "Microbiology of the Phyllosphere." Cambridge Univ. Press, Cambridge, U.K.

Hartley, C. (1921). Damping off in forest nurseries. USDA Bulletin 934, 1–99.

Heiniger, U., & Rigling, D. (1994). Biological control of chestnut blight in Europe. Annual Review of Phytopathology, 32, 581–599.

Henry, A. W. (1931). The natural microflora of the soil in relation to the foot rot problem of wheat. Canadian Journal of Research, 4, 69–77.

Isenbeck, M., & Schultz, F. A. (1986). Biological control of fireblight (Erwinia amylovora [Burr.] Winslow et al.) on ornamentals. II. Investigation about the mode of action of the antagonistic bacteria. Journal of Phytopathology, 116, 308–314.

Jaynes, R. A., & Elliston, J. E. (1980). Pathogenicity and canker control by mixtures of hypovirulent strains of Endothia parasitica in American chestnut. Phytopathology, 70, 453–456.

Jeffries, P., & Jeger, M. J. (1990). The biological control of postharvest diseases of fruit. Biocontrol News and Information, 11, 333–336.

Johnson, K. B., Stockwell, V. O., McLaughlin, R. J., Sugar, D., Loper, J. E., & Roberts, R. G. (1993). Effect of antagonistic bacteria on establishment of honey bee-disperesed *Erwinia amylovora* in pear blossoms and on fire blight control. Phytopathology, 83, 995–1002.

Junqueira, N. T. V., & Gasparotto, L. (1991). Controle biológico de fungos estromáticos causadores de doenças foliares em Seringueira. *In* "Controle Biológico de Doenças de Plantas," pp. 307–331. Org. W. Bettiol., Brasilia, DF: EMBRAPA.

Kranz, J. (1981). Hyperparasitism of biotrophic fungi. *In* "Microbial Ecology of the Phylloplane" (J. P. Blakeman, Ed.), pp. 327–352. Academic Press, London.

Kúc, J. (1981). Multiple mechanisms, reaction rates and induced resistance in plants. *In* "Plant Disease Control" (R. C. Staples & G. H. Toenniessen, Eds.), pp. 259–284. Wiley, New York.

Leibinger, W., Breuder, B., Hahn, M., & Mendgen, K. (1997). Control of postharvest pathogens and colonization of the apple surface by antagonistic microorganisms in the field. Phytopathology, 87, 1103–1110.

Lindow, S. E. (1985). Integrated control and the role of antibiosis in biological control of fireblight and frost injury. *In* "Biological Control on the Phylloplane" (C. E. Windels & S. E. Lindow, Eds.), pp. 83–115. American Phytopathological Society, St. Paul, MN.

Lynch, J. M. (1987). Biological control within microbial communities of the rhizosphere. *In* "Ecology of Microbial Communities" (M. Fletcher, T. R. G. Gray, & J. G. Jones, Eds.), pp. 55–82 [Society of General Microbiology Symposium 41]. Cambridge Univ. Press, Cambridge, U.K.

Millard, W. A., & Taylor, C. B. (1927). Antagonism of microorganisms as the controlling factor in the inhibition of scab by green manuring. Annals of Applied Biology, 14, 202–216.

Mulinge, S. K., & Griffiths, E. (1974). Effects of fungicides on leaf rust, berry disease, foliation and yield of coffee. Transactions of the British Mycological Society, 62, 495–507.

Newhook, F. J. (1957). The relationship of saprophytic antagonism to control *Botrytis cinerea* Pers. on tomatoes. New Zealand Journal of Science and Technology, Series A, 38, 473–481.

Parker, C. A., Rovira, A. D., Moore, K. J., Wong, P. T. W., & Kollmorgen, J. F., Eds. (1985). "Ecology and Management of Soilborne Plant Pathogens." American Phytopathological Society, St. Paul, MN.

Peng, G., & Sutton, J. C. (1991). Evaluation of microorganisms for biocontrol of *Botrytis cinerea* in strawberry. Canadian Journal of Plant Pathology, 13, 247–257.

Peng, G., Sutton, J. C., & Kevan, P. G. (1992). Effectiveness of honey bees for applying the biocontrol agent *Gliocladium roseum* to strawberry flowers to suppress *Botrytis cinerea*. Canadian Journal of Plant Pathology, 14, 117–129.

Pusey, P. L. (1997). Crab apple blossoms as a model for research on biological control of fire blight. Phytopathology, 87, 1096–1102.

Rishbeth, J. (1963). Stump protection against *Fomes annosus*. III. Inoculation with *Peniophora gigantea*. Annals of Applied Biology 52, 63–77.

Sanford, G. B. (1926). Some factors affecting the pathogenicity of *Actinomyces scabies*. Phytopathology, 16, 525–547.

Scherff, R. H. (1973). Control of bacterial blight of soybean by *Bdellovibrio bacteriovorus*. Phytopathology, 63, 400–402.

Sutton, J. C., & Peng, G. (1993a). Manipulation and vectoring of biocontrol organisms to manage foliage and fruit diseases in cropping systems. Annual Review of Phytopathology, 31, 473–493.

Sutton, J. C., & Peng, G. (1993b). Biocontrol of *Botrytis cinerea* in strawberry leaves. Phytopathology, 83, 615–621.

Thomson, A. D. (1958). Interference between plant viruses. Nature, 181, 1547–1548.

Thomson, S. V., Hanson, D. R., Flint, K. M., & Vandenberg, J. D. (1992). Dissemination of bacteria antagonistic to *Erwinia amulovora* by honey bees. Plant Disease, 76, 1052–1056.

Tjamos, E. C., Papavizas, G. C., & Cook, R. J., Eds. (1992). "Biological Control of Plant Disease." Plenum, New York.

Tronsmo, A., & Dennis, C. (1977). The use of [antagonistic] *Trichoderma* species to control strawberry fruit rots [caused by *Botrytis cinerea, Mucor mucedo*]. Netherlands Journal of Plant Pathology, 83 (Suppl. 1), 449–455.

van Alfen, N. K. (1982). Biology and potential for disease control of hypovirulence of *Endothia parasitica*. Annual Review of Phytopathology, 20, 349–362.

Wilson, C. L., & Wisniewski, M. E. (1989). Biological control of postharvest diseases of fruits and vegetables: An emerging technology. Annual Review of Phytopathology, 27, 425–441.

Wilson, M., & Lindow, S. E. (1993). Interactions between the biological control agent *Pseudomonas fluorescens* A506 and *Erwinia amylovora* in pear blossoms. Ecology and Epidemiology, 83, 117–123.

Wilson, M., & Lindow, S. E. (1994). Ecological similarity and coexistence of epiphytic ice-nucleating (ice$^+$) *Pseudomonas syringae* strains and a non-ice-nucleating (ice$^-$) biological control agent. Applied and Environmental Microbiology, 60, 3128–3137.

Wood, R. K. S. (1951). The control of diseases of lettuce by use of antagonistic organisms. 1. Control of *Botrytis cinerea* Pers. Annals of Applied Biology, 38, 203–216.

Zhang, P. G., Sutton, J. C., & Hopkin, A. A. (1994). Evaluation of microorganisms for biocontrol of *Botrytis-cinerea* in container-grown black spruce seedlings. Canadian Journal of Forest Research, 24, 1312–1316.

Biological Control of Insects and Mites on Grapes

D. L. FLAHERTY

Cooperative Extension, University of California
County Civic Center
Visalia, California

L. T. WILSON

Department of Entomology
Texas A&M University
College Station, Texas

INTRODUCTION

The grape has been cultivated since antiquity, beginning in the region between and to the south of the Black and Caspian seas in Asia Minor. Winkler *et al.* (1974) state

"That region, most botanists agree, is the home of *Vitis vinifera* L., the species from which all cultivated varieties of grapes were derived before the discovery of North America. From Asia Minor culture of the grape spread both west and east. Before 600 B.C. the Phoenicians probably carried wine varieties to Greece, thence to Rome, and on to southern France. No later than the second century A.D. the Romans took the vine to Germany. Probably at even an earlier date, raisin and table grapes were moving around the eastern end of the Mediterranean Sea to the countries of North Africa. The lines of spread of wine varieties differed from those of raisin and table varieties because of differences in custom and religion between the peoples of the northern and southern shores of the Mediterranean. Grapes spread to the Far East by way of Persia and India. Many years later, when Europeans colonized new lands, the grape was always among the plants taken along."

From the middle ages to the 14th century, viticulture reached its major expression in France, Italy, and central Europe, a flourishing period that terminated with the phylloxera disaster that devastated primarily the vineyards of France. Grape phylloxera, *Daktulosphaira vitifoliae* (Fitch), was introduced into France from North America circa 1886 and spread to the Black Sea area of the former Soviet Union within 10 years. The renewal of devastated vineyards marked an important stage in improvement of cultural practices, especially with the production of hybrids and resistant rootstocks (Gonzalez, 1983).

About 10 million hectares of grapes are cultivated worldwide. Hundreds of clonal selections and hybrids have been produced for adapting the species to local climatic and

soil conditions in all the continents (Gonzalez, 1983). Diverse viticulture practices modify vine microclimates, which favor or inhibit pests indigenous to the ecosystem into which the grape has been introduced. Hence the ecological niches offered by the biocenosis of the vine have been occupied in each grape-growing region by different pest species (Bournier, 1976). Less commonly, pests from other *Vitis* species have been transported on plant samples and infested *V. vinifera* as was the case for phylloxera (Bournier, 1976). We would add that the diversity of growing conditions and pest species have likely given rise to a wealth of natural enemy species and biotypes which can be used to enhance biological control and pest management programs.

Interest in biological control has increased with the development of integrated pest management (IPM) systems for vineyards. This move was generally brought about by pest resistance to synthetic organic pesticides and the accompanying serious biological outbreaks (Flaherty *et al.*, 1985). In this chapter we present a general review of biological control of insects and mites in various grape-growing regions of the world, with emphasis on California.

BIOLOGICAL CONTROL OF INSECT AND MITE PESTS

Homoptera

Cicadellidae

Bournier (1976) lists nine species of cicadellids (leafhoppers) which are recorded on grapes worldwide. In California *Erythroneura elegantula* Osborn, grape leafhopper, is the most common pest of grapes. Economic damage is caused by leaf chlorophyll reduction, vine defoliation, marring of the surface of table grapes with excrement, and leafhoppers annoying pickers at harvest (Wilson *et al.,* 1992b).

Doutt and Nakata (1965, 1973) reported that large acreage of grapes planted near streams and rivers, where wild blackberries (*Rubus ursinus* Chamisso and Schlecht and *R. procerus* Mueller) flourish, seldom require control for grape leafhopper. This is because of the activity of a minute mymarid wasp, *Anagrus epos* Girault, which parasitizes the eggs of grape leafhopper and blackberry leafhopper, *Dikrella californica* (Lawson), a noneconomic species whose eggs are present throughout the year on wild blackberries. Williams (1984) verified the synchrony of blackberry leafhopper, grape leafhopper, and *A. epos* phenology reported by Doutt and Nakata (1973).

Efforts to establish effective blackberry refuges near commercial vineyards have not been successful. Flaherty *et al.* (1985) indicated that overwintering numbers of *A. epos* were few and insufficient due to inadequate *D. californica* egg production during a critical winter period. It was previously thought that effective overwintering of *A. epos* in blackberries necessitated continuous production of eggs by *D. californica.* Williams (1984) reported on a period of reproductive diapause of *D. californica* females during the winter. Species of leafhoppers which overwinter as diapausing eggs on other host plants in natural systems also need consideration as important hosts of *A. epos* (Flaherty *et al.,* 1985). Studies by McKenzie and Beirne (1972) in Canada and by Kido *et al.* (1984) in the San Joaquin Valley (California), who reported on the overwintering of *A. epos* on leafhopper eggs on wild rose, apple, and French prune, lend credence to the importance of other winter hosts of *A. epos.* For development of productive commercial refuges, studies in the San Joaquin Valley now place emphasis on a prune leafhopper species, *Edwardsiana prunicola* (Edwards), as a major overwintering host of *A. epos* (Flaherty *et al.,* 1985). Wilson *et al.* (1989) reported that French prune planted adjacent to vineyards resulted in *A. epos* invading the vineyards 4 weeks earlier than for vineyards without French prunes. The authors also showed that the establishment and buildup of *A. epos* on the prune trees were greatly affected by whether a tree was directly exposed to prevailing winds (windbreak trees) or sheltered downwind by the windbreak trees. Murphy *et al.* (1996) reported that a significant amount of variation in *A. epos* trap captures in vineyards was explained by *A. epos* densities in prune trees surrounding the vineyards. Corbett and Rosenheim (1996a, 1996b), using rubidium marking, concluded that vineyard colonization by *A. epos* was related to the distance of overwintering habitats to the vineyards. Pickett *et al.* (1990) provide a thorough review of the use of refuges in agroecosystem

management and how they affect the efficacy of biological control agents. *Typhlocyba pomaria* McAtee, a leafhopper on apples, is also being considered as an overwintering host.

Dependence on *A. epos* activity by grape growers is influenced by the grape cultivar and its intended use (Wilson *et al.,* 1992b). For example, high populations of grape leafhopper can be tolerated on Thompson Seedless vines destined for raisin or wine production and frequently *A. epos* provides sufficient control of the grape leafhopper. However, only low populations of leafhoppers are tolerated on Thompson Seedless vines and other grapes grown for fresh market consumption, making it more difficult to depend upon *A. epos* instead of applying insecticides for control of the leafhoppers. Excessive spotting (leafhopper droppings) determines the economic damage level on table grapes and this occurs long before any reduction in sugar or yield can be measured. Nonetheless, *A. epos* is highly effective on Thompson Seedless and its activity increases the possibility that grape leafhopper populations remain within tolerance limits, even in table grape vineyards (Wilson *et al.,* 1992b).

Anagrus epos is considerably less effective on Emperor and Ribier table grapes, particularly the latter. By late season, grape leafhopper populations increase to enormous numbers on these and some wine grape cultivars despite the presence of *A. epos.* Reasons for its effectiveness on Thompson Seedless, a cultivar harvested midseason, are not clear, but observations indicate that the smooth leaf surface of Thompson Seedless vines does not interfere with searching and oviposition by the parasitoid. Early maturing cultivars also produce fewer newly mature leaves, structures that are favored by grape leafhopper for egg deposition. In comparison, late cultivars such as Emperor and Ribier have hairy (tomentose) leaves that may interfere with the efficiency of the parasitoid. Development of newly matured leaves late into the season on these cultivars also favors grape leafhopper oviposition. Later season irrigation in such vineyards also favors grape leafhopper populations (Trichilo *et al.,* 1990).

A dryinid parasitoid, *Aphelopus albopictus* Ashmead (=*A. comesi* Fenton), attacks and oviposits in all instars of the grape leafhopper (Cate, 1975). This parasitoid places its egg between the nymph's second and third abdominal segments, where it remains undeveloped until the leafhopper nymph molts to the adult stage. During the parasitoid's larval development, the host appears normal, except for an elongating larval sack (thylacium) that increasingly protrudes from the abdomen with each parasitoid molt. By its fifth instar the parasitoid has developed mandibles and eviscerates the adult leafhopper. Prior to evisceration, the adult leafhopper is functionally nonreproductive since gonads fail to develop in parasitized adults. Usually *A. comesi* parasitism is between 10 and 40% (Cate, 1975; Wilson *et al.,* 1991a). Cate (1975) reported in one instance 77% parasitism. He speculated that vineyard cultural practice can be altered to favor parasitism and survival of the parasitoids. In a recent survey of leafhoppers and sharpshooters inhabiting grape vines and associated vegetation across most of western Texas, Wilson *et al.* (unpublished data) found dryinid parasitizing leafhoppers in each of the 13 vineyards that were studied.

Erythroneura variabilis Beamer, variegated grape leafhopper, is the principal vineyard pest in the desert areas of southern California, Arizona, and Mexico. Infestations are especially severe in the Coachella Valley in California, where high temperatures allow rapid development and up to six generations per year (Wilson *et al.,* 1992a). The low activity of *A. epos* in the Coachella Valley reportedly aggravates the problem. On the other hand, *A. epos* is active in the San Bernardino area of southern California where wild grapes and blackberries grow near commercial vineyards. Wilson *et al.* (1992a) reported parasitism as high as 90% in late summer.

Until recently, grape growers in the San Joaquin Valley had to contend only with grape leafhopper. Now, however, they also deal with the variegated grape leafhopper, which was first detected in the Valley about 1980, and populations have since increased and rapidly spread. Defoliation now occurs in many Thompson Seedless vineyards where grape leafhopper historically had been under good control by *A. epos.* It is evident that *A. epos* is not an effective parasitoid of variegated grape leafhopper. Adding to the problem is the fact that the dryinid parasitoid *A. comesi* does not attack variegated grape leafhopper (Wilson *et al.,* 1991a). Necessary chemical control causes serious outbreaks of secondary pests, such as spider mites and mealybugs. Annual costs for control of variegated grape leafhopper alone are projected to exceed $5–10 million, with yield and quality losses potentially reaching twice this amount (Wilson *et al.,* 1986, 1987). Grape IPM in the San Joaquin Valley has been dealt a serious setback and adjustments must be made (Flaherty *et al.,* 1985).

Reasons for the ineffectiveness of *A. epos* in controlling variegated grape leafhopper on Thompson Seedless in the San Joaquin Valley are not entirely clear, but Settle *et al.* (1986) stated "Variegated leafhopper is a more serious pest than the grape leafhopper, in part because of differences in egg-laying behavior. Variegated leafhopper eggs are deeply buried within the leaf tissue, and are less likely to be detected by *A. epos* than grape leafhopper eggs, which stand out as blisters on the leaf surface." Settle *et al.* (personal communication) also recorded over 90% parasitism of grape leafhopper eggs and less than 20% parasitism of variegated grape leafhopper eggs in San Joaquin Valley Thompson Seedless vineyards. Settle and Wilson (1990a,

1990b) developed a model of host preference by *A. epos* which verifies that the eggs of *E. variabilis* are probably more difficult to detect by *A. epos* than those of *E. elegantula.*

In 1985, a search was initiated for parasitoids of variegated grape leafhopper in southern California, Mexico, Arizona, Colorado, New Mexico, and Texas, with the object of collecting and introducing into the San Joaquin Valley many species and biotypes of leafhopper parasitoids collected from diverse climatic zones and different times of the season (Gonzalez *et al.,* 1988). Preliminary studies (Wilson *et al.,* unpublished) indicate that *A. epos* has evolved a wide range of biotypes, differing in relative preference for the grape leafhopper and variegated grape leafhopper. For example, studies by Pickett *et al.* (1987, 1989) indicated that the native, or San Joaquin Valley biotype, has a 6.9-fold preference for grape leafhopper than for variegated grape leafhopper. Moreover, on the basis of preference, biotypes from Coachella Valley in southern California and in Colorado would respectively choose, provided equal numbers of eggs of both leafhopper species, 2.0 and 4.3 times more variegated grape leafhopper eggs. The data suggest that prospects are good for introducing a more effective parasitoid of variegated grape leafhopper into the San Joaquin Valley. However, data are presently not available on the preference of these biotypes for overwintering hosts such as prune leafhopper and to determine their effectiveness if established in the San Joaquin Valley. This issue is made even more complex by studies which suggest that the collections from diverse climatic zones by Gonzalez *et al.* (1988) were actually a mixture of different species of *Anagrus* and not *A. epos* biotypes (Trjapitzin & Chiappini, 1994). Further studies are showing that different species of *Anagrus* exist in various grape-growing areas of California (Trjapitzin, 1995). A pressing need of biological control specialists is a quantitative tool that can be used to easily differentiate between closely related parasitoids, but which differ greatly in their ability to control pest species.

In the Okanagan Valley, British Columbia, Canada, *Erythroneura ziczac* Walsh is the most important insect on grape (McKenzie & Beirne, 1972). In their study to develop ways for preventing damaging leafhopper populations, they reported on two cultural procedures which would tend to keep *E. ziczac* populations below damaging levels: (1) destroy overwintering sites in and near the vineyards and (2) provide *A. epos* with overwintering leafhopper host eggs (*Edwardsiana rosae* (Linnaeus) and *T. pomaria*) by growing wild rose and apple in or near vineyards. Wells *et al.* (1988) found that *A. epos* is the main source of egg mortality for *E. elegantula* and *E. ziczac* in southcentral Washington and suggested that plum trees could be infested with nonpestiferous host species to provide a winter refuge for *A. epos.* Yigit and Erkilic (1987) reported that in southern

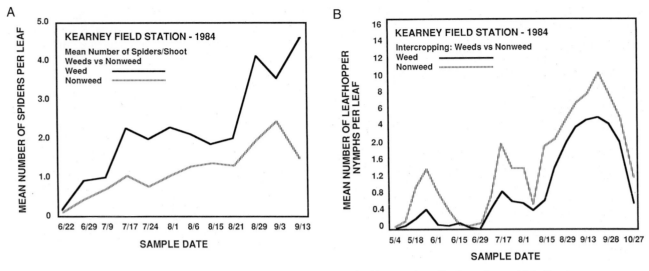

FIGURE 1   Impact of weedcover crop on (a) mean number of spiders per shoot (Settle *et al.,* unpublished) and (b) mean number of leafhopper nymphs per leaf (Settle *et al.,* 1986).

Turkey the egg parasites *Anagrus atomus* (Linnaeus) and the trichogrammatid *Oligosita pallida* Kryger appeared to control the grape leafhopper, *Arboridia adanae* (Dlabola). Blackberry (*Rubus* sp.) and dog rose (*Rosa* sp.), infested by the cicadellid *E. rosae,* were found to be winter refuges for the parasitoids. Pavan and Picotti (1994) reported higher levels of parasitism of *Empoasca vitis* (Gothe) in grape vineyards by *A. atomus* than in adjoining kiwi vineyards. Based on host oviposition studies and results from the parasitism studies, the authors concluded that *A. atomus* was responsible for suppressing *E. vitis* populations in the grape vineyards. Pinto and Viggiani (1987) reported two new Trichogrammatidae species emerged from eggs of grapevine cicadellids in Mexico and Arizona. Small releases

were made in the San Joaquin Valley without recovery (D. Gonzalez, personal communication).

Observations by Settle *et al.* (1986) indicated that generalist predators (primarily the theridiid spider *Theridion* sp.) might play a significant role in suppressing leafhoppers in vineyards having cover crops. In studies conducted in the San Joaquin Valley, approximately twice as many spiders were recorded in a vineyard with a weed cover crop. Correspondingly, only half as many leafhoppers were found compared to the numbers in the plots without cover crops (Fig. 1). Costello and Daane (1995) in a two-year study reported 27 species of spiders, representing 14 families inhabiting San Joaquin Valley vineyards. Polesny (1987) also discussed the importance of spiders as predators on insect pests in vineyards in Austria. In San Joaquin Valley vineyards, augmentative releases of the generalist predator *Chrysoperla* species suppressed variegated grape leafhopper, however, effectiveness varied greatly (Daane *et al.,* 1993, 1996).

Settle *et al.* (1986), when examining the effect of grape rootstock on leafhopper abundance, found that Thompson Seedless grapes grown on their own roots showed greatly reduced attractiveness of vines. This resulted in a nearly 8-fold reduction in leafhopper numbers compared with the more vigorous vines on Saltcreek rootstock (Fig. 2). Trichilo *et al.* (1990) reported an inverse linear relationship between the amount of water stress and leafhopper population growth, with longer season irrigation programs increasing late-season leafhopper densities in California. They suggested that the late-season leafhopper densities could be controlled by timing the last irrigation for mid- to late-July. The reduced leafhopper pressures afforded by cultural practices suggest considerable potential for enabling biological

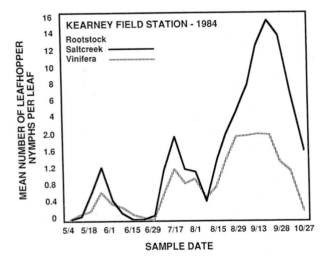

FIGURE 2   Impact of rootstock on leafhopper nymphal abundance (Settle *et al.,* 1986).

control programs to be incorporated more readily in grape management systems. However, cultural controls aimed at increasing the control of one pest species may have no effect or potentially a negative effect on the control of a second species. Buccholz and Schruft (1994) in a four-year study of generalist predators in two vineyards concluded that increased understory vegetation generally did not promote the abundance and effectiveness of generalist predators which control *Eupoecilia ambiguella* populations in vineyards in Germany.

Pseudococcidae

In California two mealybug species are known to infest vineyards. *Pseudococcus maritimus* (Ehrhorn), the grape mealybug, is primarily a pest of table grape cultivars whose bunches make contact with vine bark and become infested (Flaherty *et al.,* 1992a). Flaherty *et al.* (1992a) report that before the late 1940s occasional losses occurred in table grapes, and even though bothersome, infestations were mostly spotty and frequently disappeared the following year. Increasing and more persistent populations were noted in the late 1940s. Extensive use of DDT and other synthetic pesticides to control grape leafhoppers apparently had disrupted natural enemies of *P. maritimus.* The second species, *Pseudococcus affinis* (Mackell) (=*P. obscurus* Essig), the obscure mealybug, which has been recorded on a wide number of hosts, was recently found causing severe damage in unsprayed vineyards in San Luis Obispo County, on California's central south coast. The presence of a second species was suspected because native parasitoids would likely control the grape mealybug in unsprayed vineyards. Evidently, obscure mealybug has adapted to vineyards in this area. The absence of effective parasitoids attacking obscure mealybug has also been observed on other host plants, suggesting that it is an introduced species that will require importation of natural enemies for its control. Releases of the parasitoid *Pseudophycus flavidulus* (Brethes) are considered an important ingredient for the successful management of *P. affinis* in Chile (Ripa & Rojas, 1990). In 1993 *P. flavidulus* was introduced and recovered in coastal vineyards in California (K. M. Daane, personal communication).

Little work has been done to document the effectiveness of natural enemies of grape mealybug. Clausen (1924) reported rearing five primary endoparasitic encyrtid wasps from grape mealybugs in the Fresno area of the San Joaquin Valley. He also reported that in the late summer and fall of 1919 parasitism exceeded 90%. It is not clear, however, whether he considered mealybug populations to be effectively controlled by parasitoids. From 1084 grape mealybugs, Clausen (1924) reared nine species of parasitoids, five primary (one of which was gregarious) and four secondary (hyperparasites). In numbers of individuals, there were

TABLE 1 Parasitoids of Grape Mealybug, *Pseudococcus maritimus* (Ehrhorn)[a]

Parasitoids	Total reared	Number of hosts	Average per host
Acerophagus notativentris (Girault)	614	78	7.9
Zarhopalus corvinus (Girault)	350	350	1.0
Anagyrus subalbicornis (Girault)	280	280	1.0
Pseudoleptomastix squammulata Girault	40	40	1.0
Anagyrus clauseni Timberlake	1	1	1.0
Total	1285	749	

[a]Summarized from Clausen (1924)

1285 of the former and 515 of the latter. *Zarhopalus corvinus* (Girault) parasitized the most hosts in his samples (Table 1). Flaherty *et al.* (1976) reported that in the late 1960s, ranking of host parasitism from samples of Doutt and Nakata (unpublished data) differed from those given in Table 1. They showed that host parasitism by *Acerophagus notativentris* (Girault) clearly dominated; *Anagyrus subalbicornis* (Girault) was found occasionally, *Z. corvinus* was uncommon, and other species were rare. Studies in 1988 revealed *Pseudophycus angelicus* (Howard) commonly parasitizing grape mealybug in San Joaquin Valley vineyards (E. F. Legner, personal communication). This encrytid species was previously reared from *P. maritimus* in southern California (Clausen, 1924).

Flaherty *et al.* (1982) reported that parasitoids were not reared from mealybug samples collected from an Emperor vineyard with a history of heavy treatments in which over 25% of the bunches were badly infested, whereas in another Emperor vineyard that had a minimal insecticide program, parasitism was 46%, with mealybug activity below the acceptable economic level of 2% infested bunches. Parasitism by *A. notativentris* clearly dominated.

Serious outbreaks of the mealybug *Maconellicoccus hirsutus* (Green) have recently occurred on grapevines in India (Manjunath, 1985). It attacks different cultivars, including Thompson Seedless, Anab-e-Shahi, and Bangalore Blue. In severe attacks, up to 90% of the clusters are destroyed. Chemical control has not been effective. An encyrtid wasp, *Anagyrus dactylopii* Howard, reared from *M. hirsutus,* a new record from India, appears to offer promise as a biological control measure. In late-season (mid-March/April) samples at Bangalore, parasitization ranged from 60 to 70%, although these fields had been regularly sprayed with insecticides. Manjunath (1985) also recommended introduction of two encyrtid wasps, *Anagyrus kamali* Moursi and *Prochiloneurus* sp., which, according to Kamal (1951), effected complete control of *M. hirsutus* after being introduced into Egypt from Java.

Grasswitz and Burts (1995), in studies conducted in apple and pear orchards in the United Kingdom, reported that *Pseudococcus maritimus* was controlled by a complex of natural enemies including encrytid parasitoids (*Pseudophycus websteri* Timberlake and *Mayridia* spp.), a coccinellid beetle (*Hyperaspis lateralis* Mulsant), and a chamae myiid fly (*Leucopis verticalis* Malloch) in orchards that had not received pesticide controls during the previous 1–2 years. In the former Soviet Union, *Planococcus citri* (Risso) is injurious to over 20 species of plants, including grapes (Niyazov, 1969). The main parasitoid of *P. citri* is *Anagyrus pseudococci* (Girault), which occurs in the south of European Russia and Soviet Central Asia. It destroys up to 75% of the host population in areas not treated with insecticides. The platygasterid *Allotropa mecrida* (Walker) was responsible for 20% parasitism of *P. citri* in Turkmenia and Georgia. In 1960, two encyrtid wasps, *Leptomastidea abnormis* (Girault) and *Leptomastix dactylopii* Howard, were introduced into Georgia and Turkmenia from the United States, but native hyperparasites were reported to greatly reduce their effectiveness. In Transcaucasia and Soviet Central Asia, the signiphorid *Thysanus (Chartocerus) subaeneus* Förster is responsible for up to 20% hyperparasitism of *A. mecrida*. Rzaeva (1985) reported that *L. abnormis* had successfully established on *Planococcus ficus* Signoret, a mealybug pest of grapes in eastern Transcaucasus, USSR. Niyazov (1969) advocated the introduction of *Clausenia josefi* Rosen, an encyrtid, from the Mediterranean area. Russian entomologists also advocated the introduction of *A. notativentris* from California (K. S. Hagen, personal communication).

In California, little is known about the effectiveness of predators of mealybugs in vineyards. Cecidomyiid fly larvae were observed attacking mealybug egg masses in the above-mentioned lightly treated Emperor vineyard but not in the heavily treated vineyard (Flaherty *et al.,* 1982). Charles (1985) reported that the cecidomyiid *Diadiplosis koebelei* Koebele was recorded for the first time in New Zealand, with larvae feeding on adults of *Pseudococcus longispinus* (Targioni-Tozzetti) on grapevines. About 30% of adult female mealybugs in unsprayed fruit bunches at harvest were killed by the cecidomyiid. Adults of *Chrysoperla* (=*Chrysopa*) spp. frequently are abundant on grapevines harboring mealybugs in California. Chrysopid adults are attracted to the mealybug honeydew, but it is not known to what extent their egg laying and subsequent control of mealybugs is influenced by the presence of honeydew. Laboratory studies in India showed that the green lacewing *Mallada boninensis* (Okamoto) consumed 237.9 *M. hirsutus* nymphs during its larval development (Mani & Krishnamoorthy, 1989).

The coccinellid *Cryptolaemus montrouzieri* Mulsant, the mealybug destroyer, is rarely observed in California vineyards and the feasibility of mass releases for control of the grape mealybug has not been studied. The use of *C. montrouzieri* for the control of mealybugs on a range of crops, including grapes, in Georgia, USSR, is reviewed by Yasnosh and Mjavanadze (1983). Best control was achieved by releasing adult beetles; immature beetles apparently suffer high mortality during transportation. Niyazov (1969) reports that one of the most effective predators of mealybugs on grapevines in the former Soviet Union is *C. montrouzieri,* which was introduced into the Black Sea coastal area from Egypt in 1932. Other coccinellids noted in Turkmenia were *Coccinella septempunctata* Linnaeus, *Hyperaspis polita* Weise, *Scymnus apetzi* Mulsant, *S. subvileosus* (Goeze), *Nephus bipunctatus* Kugelann, and *S. biguttatus* Mulsant. The larvae of *Leucopis (Leucopomya) alticeps* Czerny (Diptera: Ochthiphilidae) and *Chrysoperla carnea* (Stephens) were observed to destroy all stages of *P. citri.* The coccinellids were parasitized by the encyrtid *Homalotylus* sp. and the chrysopids by the scelionid *Telenomus acrobates* Girard. Mani (1988) reported that the coccinellid *S. coccivora* Rumakrishna and the parasitoid *A. dactylopii* are of considerable importance in the control of *M. hirsutus* in India. It was also reported that *C. montrouzieri* was observed 6 weeks after the initial release, with 64.3% control of *M. hirsutus* when 10 predators were released per vine (Srinivasan & Babu, 1989). Orlinskii *et al.* (1989) suggested that the coccinellid *N. reunioni* Fürsch, introduced into the former Soviet Union in 1978, is a promising entomophage of *P. ficus* and *P. citri* and possesses certain advantages over *C. montrouzieri.* Initial controlled tests on effectiveness have given good results. The voracious appetite of mealybug-consuming coccinellids is illustrated by results from Padmaja *et al.* (1995), who showed that adult and immature *Scymnus coccivora* Aiyar consumed an average of 864 and 314 grape mealybug eggs, respectively, during their lifetime.

Because the grape mealybug problem in California is probably caused by the application of chemical insecticides to control grape leafhopper and variegated grape leafhopper in table grape vineyards, Flaherty *et al.* (1985) suggested developing treatment programs which leave untreated or minimally treated areas to encourage the activity and spread of natural enemies.

Coccidae

In Chile, Gonzalez (1983) reports that grape quality can be affected by copious amounts of honeydew produced by *Parthenolecanium persicae* (Fabricius). However, natural enemies play an important role in maintaining populations below economic levels. The principal enemies are two encrytid wasps, *Coccophagus caridei* (Brethes) and *Metaphycus flavus* (Howard). These and other parasitoids are common natural enemies of various lecanium coccids on other plant hosts, including *P. corni* (Bouché), which is also a pest of grapes in Chile. Therefore, it was important to maintain untreated reservoirs of those scales on ornamentals or natural vegetation to ensure a source from which natural enemies may move to nearby vineyards, particularly in the event that treatments for other pests severely reduced parasitoid activity.

Phylloxeridae

Wheeler and Jubb (1979) summarized their observations on the coccinellid *Scymnus cervicalis* Mulsant as a predator of the leaf-feeding form of grape phylloxera, *D. vitifoliae,* on wild grapes in Erie Co., Pennsylvania. Wheeler and Jubb (1979) believe that *S. cervicalis* is a potentially useful predator of a pest that is becoming increasingly important on wine grapes in Pennsylvania.

Lepidoptera

Tortricidae

Those species of tortricids that infest grape clusters are commonly severe pests in most grape-growing areas of the world. *Argyrotaenia citrana* (Fernald), orange tortrix, is a major pest in the cool, coastal regions of California. Larvae cause the primary damage by feeding in grape clusters and allowing rot-causing organisms to invade (Bettiga & Phillips, 1992). Kido *et al.* (1981) assessed the potential of its biological control. Orange tortrix populations were sampled on Gamay Beaujolais vines in two vineyards in the Salinas Valley. One, the Soledad vineyard, had a history of injurious infestations and required treatments. The other, the Greenfield vineyard, had very light infestations and no treatments. Samples of clusters from the Greenfield vineyard contained very few orange tortrix larvae and pupae, with 53.5% parasitism for the sampling period; the Soledad vineyard, with an extremely high orange tortrix population, had 16% parasitism for the sampling period. The ichneumonid *Exochus nigripalpus subobscurus* Townes was the dominant parasitoid in both vineyards, but the braconid *Apanteles aristoteliae* Viereck was also recovered, but much less frequently.

Orange tortrix infestations also occurred on coyote brush, *Baccharis pilularis* De Candolle, a native composite shrub commonly found in the Salinas Valley and growing in large numbers near the Greenfield vineyard. Large numbers of another tortricid species, *Aristoteliae argentifera* Busck, were found on coyote brush near the Greenfield vineyard, and several parasitic species of wasps and flies,

including *Exochus* sp. and *Apanteles* sp., were recovered from larvae and pupae. Coyote brush was much less abundant near the Soledad vineyard and consisted mostly of young plants and no infestations of *A. argentifera* were found. Studies are needed to determine whether the relative difference of orange tortrix infestations in the two vineyards was related to differences in pesticide history and the influence of the coyote brush ecosystem.

Since the 1960s the tortricid *Platynota stultana* Walsingham, the omnivorous leafroller, has become a major pest in the warmer inland valleys of California (Coviello *et al.,* 1992). Its rot-causing damage is similar to that of orange tortrix. A number of insect parasitoids have been recorded attacking omnivorous leafroller larvae in vineyards. However, parasitoids seldom accounted for more than 10% mortality, even in the presence of very high larval populations. Fortunately, omnivorous leafroller is controlled with the use of selective pesticides. Flaherty *et al.* (1985) recommended the importation and augmentation of natural enemies of omnivorous leafroller for the improvement of grape IPM in the San Joaquin Valley.

Mass release of *Trichogramma* spp. to augment biological control of omnivorous leafroller or orange tortrix has not been studied in California vineyards. Makhmudov *et al.* (1977) reported that releases of *Trichogramma* sp. gave good control of *Lobesia botrana* (Denis and Schiffermuller), a tortricid pest of grapes in the former Soviet Union. Marcelin (1985) reports on significant reductions of *L. botrana* and *Eupoecilia ambiguella* (Hübner) populations with *Trichogramma* sp. releases, but its use in vineyards was not suggested. No reasons were given. Sengonca and Leisse (1989) reported that when the native egg parasite *Trichogramma semblidis* (Aurivillius) was augmented in vineyards in Germany, infestations of *E. ambiguella* and *L. botrana* larvae were reduced about 25% in both generations. Recent studies in Europe have emphasized the selective control of tortricid pests of grapes with *Bacillus thuringiensis* Berliner treatment and mating disruption (G. A. Schruft, personal communication). Nasr *et al.* (1995) reported that endemic populations of *Trichogramma evanescens* Westwood in Egypt parasitized 22–64% on the eggs of *L. botrana,* and this species was considered to be an important biological control agent. Castaneda-Samayoa *et al.* (1993) in a four-year study in Guatemala on the effectiveness of four *Trichogramma* spp. reported that *Trichogramma* releases reduced damage by *E. ambiguella* and *L. botrana* from 23 to 83%, with *T. cacoeciae* the most efficient species.

In Chile, Gonzalez (1983) reports that damage by *Proeulia auraria* (Clarke), a leaf- and fruit-damaging tortricid, increases when natural enemies are destroyed by pesticides. A complex of five species of egg and larval parasitoids was identified. An aphelinid wasp, *Encarsia* sp., was the egg parasitoid. The larval parasitoids included a eulophid wasp of the genus *Elachertus* or *Bryopezus,* a braconid wasp of the genus *Apanteles,* an unidentified ichneumonid, and a tachinid fly, *Ollacheryphe aenea* (Aldrich), which was the most abundant enemy of *P. auraria* larvae.

Zygaenidae

Flaherty *et al.* (1985) report that western grapeleaf skeletonizer, *Harrisina brillians* Barnes and McDunnough, was originally distributed throughout the southwestern United States (excluding California) and Mexico's Sonora, Chihuahua, Coahuila, and Aguascalientes states. It was first found in California near San Diego in 1941, severely defoliating wild grapes, *Vitis girdiana* Munson, in canyon areas. In a short time it became a serious pest in commercial vineyards. The larvae are voracious feeders and may cause complete loss of crop by defoliating entire vineyards.

In 1950, the Division of Biological Control at the University of California, Riverside, initiated a program to search for natural enemies of grapeleaf skeletonizer. A number of parasitoids were introduced but only two, the braconid *Apanteles harrisinae* Muesebeck and the tachinid *Ametadoria miscella* (Wulp) (=*Sturmia harrisinae* Coquillett), became established. An extremely virulent granulosis virus, which became an important component in the successful biological control of grapeleaf skeletonizer, was also accidentally introduced (Clausen, 1961).

Flaherty *et al.* (1985) reported that surveys in San Diego County in 1982 and 1983 revealed that it was necessary to spray grapeleaf skeletonizer in commercial vineyards. Abandoned untreated vineyards and backyard vines were severely defoliated despite the activity of the introduced parasitoids. Symptoms of virus infection were not observed in the survey. Curiously, the survey revealed a total absence of grapeleaf skeletonizer in wild grapes, *V. girdiana,* except where they were in close proximity to heavily infested commercial *V. vinifera* vineyards.

The first grapeleaf skeletonizer infestation in the San Joaquin Valley was found in June 1961 (Clausen, 1961). New infestations appeared thereafter throughout the Valley despite eradication attempts. It became a widespread and serious pest of commercial vineyards and backyard vines, as well as wild grapes, *Vitis californica* Bentham.

Flaherty *et al.* (1985) reported that establishment of *A. harrisinae* and *A. miscella* was unsuccessful in the San Joaquin Valley. Only a few parasitoid recoveries were made at the release sites. Moreover, samples of larvae taken from three heavily infested and abandoned vineyards in San Diego County produced only 13% parasitism, which is far less than the percentage parasitism (42% in 1953 and 62% in 1954) reported by Clausen (1961). No evidence of the virus was present in these vineyards. Clausen (1961) com-

mented that the virus must be credited with the major role in reducing grapeleaf skeletonizer populations to low levels and exterminating many small infestations. Flaherty *et al.* (1985) commented that perhaps at that time the virus was more widespread and had reduced grapeleaf skeletonizer populations to levels that made it more manageable by the parasitoids. This may account for the greater parasitism reported by Clausen (1961) than that found by Flaherty *et al.* (1985). However, the absence of virus in the abandoned vineyards in San Diego County requires an explanation, as does the absence of observable grapeleaf skeletonizer in wild grapes that were once heavily infested. Studies in the San Joaquin Valley confirmed that the virus of grapeleaf skeletonizer is extremely virulent and has the potential to be incorporated into an areawide biological control program, including wild grapes, backyard vines, and commercial vineyards (Flaherty *et al.,* 1985). Recent surveys by California Department of Agriculture entomologists in the San Joaquin Valley revealed significantly lower skeletonizer populations and damage to wild grapes. The virus and parasitism by *M. miscella* are thought to be responsible (B. Villegas, personal communication). Apparently similar, but less dramatic, action has taken place in commercial vinifera vineyards (W. W. Barnett, personal communication). In Italy, Pucci and Dominici (1986) reported that the percentage parasitism of the zygaenid *Theresimima ampelophaga* Bayle-Barelle by six species of parasitoids was not high enough to control the pest. Perhaps the missing ingredient for control is a pathogen.

Pyralidae

Desmia funeralis (Hübner), the grape leaffolder, is a pest of grapes in the central and southern San Joaquin Valley. Larvae cause injury by rolling and feeding on leaves, reducing photosynthesis. Under extreme population densities, it may feed on fruit, but economic damage usually occurs only with massive, late-season infestations (Jensen *et al.,* 1992b). The most commonly observed natural enemy of grape leaffolder is a larval braconid parasitoid, *Bracon cushmani* (Muesebeck). Parasitism ordinarily is in the range of 30–40% but often is higher. *Bracon cushmani* usually increases in the summer and frequently reduces the size of the second and third brood to such small numbers that little increase in grape leaffolder populations is detectable. When necessary, grape leaffolder is effectively controlled with selective insecticides (Jensen *et al.,* 1992b).

Sesiidae

Vitacea polistiformis (Harris), grape root borer, is a pest of grapes east of the Rocky Mountains in North America. Larvae severely prune and girdle grape roots by excavating irregular burrows (Jubb, 1982). Saunders and All (1985) showed an inverse correlation between the severity of *V. polistiformis* and the activity of entomophilic rhabditoid nematode fauna in vineyard soils. Laboratory and field bioassays determined the susceptibility of first-instar larvae to the nematode *Steinernema feltiae* Filipjev, and the insect–nematode interaction was posited as a mechanism of natural control of the larval populations. Augmentation of entomopathic rhabditoid nematode populations during the critical period of oviposition and eclosion (egg hatch) is suggested as a technique for the control of *V. polistiformis* (Saunders & All, 1985).

Heliozelidae

Although leafminers are not a problem in California vineyards, recent research by Alma (1995) has shown the widespread subeconomic presence of *Holocascista rivillei* Stanton in northwestern vineyards in Italy. Alma reported the presence of eight eulophid parasitoids attacking this species, three of which had previously not been reported associated with this species. The author also reported the presence of a predaceous mirid and an anthocorid feeding on the leafminer. Mortality caused by the parasitoids was 37.5%, with pathogens responsible for an additional 16.4%.

Coleoptera

Curculionidae

In Chile, adults of *Naupactus xanthographus* (Germar) voraciously consume grape buds and leaves (Gonzalez, 1983). Additional damage is caused by feces which adhere to foliage and fruit clusters. Combined damage by adult and larval feeding on roots may weaken vines. A complex of pathogens (bacteria and fungi), nematodes, and insects attack larvae and pupae in the soil. An undescribed nematode of the family Rhabditidae parasitizes fourth and fifth instars; however, it is unknown whether death is caused by the nematode or by septicemia. The same nematode attacks other coleopteran species and may be reared on larvae of the pyralid moth, *Galleria melonella* (Linnaeus). Gonzalez (1983) also reports that larvae are frequently attacked by nematodes transported in soils carried in irrigation water from the Maipo River. However, the degree of control produced by nematodes has not been evaluated. Of particular importance as a natural enemy of *N. xanthographus* is a platygastrid wasp *Platystasius* (=*Fidiobia*) sp. Up to 60% of the egg masses under the bark have been found attacked by this parasitoid. Gonzalez (1983) reports that its action in conjunction with the complex of other natural enemies is sufficient to maintain *N. xanthographus* below economically damaging levels.

The black vine weevil, *Otiorhynchus sulcatus* (Fabricius), is an important pest of a wide range of horticultural crops in Europe, the United States, Canada, Australia, and New Zealand. Serious damage to grapes by adults feeding on berry pedicels and cluster stems and larvae feeding on roots has been reported in Europe and central Washington (Bedding & Miller, 1981). Their studies showed that application of aqueous suspensions of infective juvenile *Heterorhabditis heliothidis* (Khan, Brooks, and Hirschmann) to the soil resulted in up to 100% parasitism of larvae of *O. sulcatus* in potted grapes in nurseries. Pupae and newly emerged adults were also parasitized. *Steinernema bibionis* (Bovien) was found to be less effective. The use of nematodes provides an effective method for controlling *O. sulcatus* on potted plants in glasshouses and nurseries.

Bostrichidae

In California *Melalgus confertus* (LeConte), a branch and twig borer, is found associated with many species of cultivated and native trees and shrubs. In grapes both adult and larval stages cause injury to grapevines (Zalom *et al.*, 1992). Little is known about the natural enemies of *M. confertus;* however, a neuropteran predator in the family Rhaphidiidae and two coleopterans in the families Carabidae and Ostomidae may assist in maintaining low population levels. Recent studies have shown that the entomophagous nematode *Steinernema feltiae* Filipjev can move through frass tubes to infect larvae. Field studies are being conducted to determine the potential of the nematode as an agent for control (F. G. Zalom, personal communication).

Thysanoptera

Thripidae

Several thrips (including *Frankliniella* spp. and *Drepanothrips reuteri* Uzel) are serious pests on grapes worldwide (Flaherty & Wilson, 1988). Little is known about their biological control in California vineyards (Jensen *et al.*, 1992a) and, based on the lack of published information, elsewhere as well. However, studies in citrus in California suggest that the phytoseiid mite *Euseius tularensis* Congdon can reduce citrus thrips *Scirtothrips citri* (Moulton) populations (Grafton-Cardwell & Ouyang, 1995). Moreover, Schwartz (1987) reported that *Amblyseius citri* Van der Merwe and Ryke preys upon *Scirtothrips aurantii* Faure in South Africa. Schwartz (1987) also reported that *Amblyseius addoensis* Van der Merwe and Ryke probably preys upon *Thrips tabaci* Lindeman in table grapes in South Africa, although it did not show an ability to reduce midsummer population increases of *T. tabaci*.

Acari

Tetranychidae

Tetranychids (spider mites) became serious pests of grapes as well as other crops worldwide after World War II. Reasons for this development have not been fully documented, but it is believed that the use of synthetic organic insecticides has upset natural controls (Flaherty *et al.,* 1985).

In California, two spider mite species are commonly abundant on grapes, Pacific spider mite, *Tetranychus pacificus* McGregor, which deserves serious consideration, and Willamette spider mite, *Eotetranychus willamettei* (McGregor), whose populations only occasionally become large enough in San Joaquin Valley vineyards to cause concern (Flaherty *et al.,* 1992b). However, recent studies have shown that Willamette spider mite can cause serious yield reductions in wine grapes grown in coastal valleys and in the foothills of northern California (Welter *et al.,* 1991). Two-spotted spider mite, *T. urticae* Koch, is rarely found on grapes in California; however, it is reported as serious on grapes in other countries such as the former Soviet Union, South Africa, and Australia.

In the eastern United States, European red mite, *Panonychus ulmi* (Koch), is the principal spider mite pest (Jubb *et al.,* 1985). It is also a pest of grapes in Europe. Schruft (1986) lists *E. carpini vitis* Boisduval, *T. urticae, T. mcdanieli* McGregor, and *T. turkestani* Ugarov and Nikolski as other pest species in Europe. *Oligonychus vitis* Zaher & Shehata is reported as a serious pest of grapes in Egypt and Chile (Rizk *et al.,* 1978; Gonzalez, 1983). A number of spider mites, including *P. ulmi, E. pruni* (Oudemans), *T. turkestani,* and *Bryobia praetiosa* Koch, are reported as infesting vineyards in the former Soviet Union (S. R. Tuganow, personal communication). *Eotetranychus pruni* is also reported as a pest in Bulgaria (Schruft, 1986). In India, four species of tetranychid mites, namely *O. mangiferus* (Rahman and Sapra), *O. punicae* (Hirst), *T. urticae,* and *E. truncatus* Estebanes and Baker, are known to attack grapevines (Schruft, 1986). Schruft (1986) also reports grapes as hosts for *B. praetiosa, E. smithi* Pritchard and Baker, *T. kanzawai* Kishida, and *T. urticae* in Japan, but we are uncertain if they are serious pests.

The most important natural enemy of spider mites in San Joaquin Valley vineyards is the phytoseiid *Metaseiulus occidentalis* (Nesbitt) (Flaherty *et al.,* 1992b). Other phytoseiid species observed in California vineyards include *Amblyseius californicus* (McGregor) in the Salinas Valley and *M. mcgregori* (Chant) in the San Joaquin and Sacramento valleys. Both species have been observed to prey upon Willamette spider mite but little is known about their effectiveness. In the eastern United States the most common predatory mites found on Concord grapes (*Vitis labrusca* Linnaeus) are the two phytoseiids *Neoseiulus* (=*Amblyseius) fallacis* (Garman) and *Amblyseius andersoni* (Chant) and the stigmaeid *Zetzellia mali* (Ewing). *Neoseiulus fallacis* and *Z. mali* may be potentially important in the natural control of *P. ulmi* (Jubb *et al.,* 1985). In Europe, the most frequently encountered phytoseiids are *Typhlodromus pyri* Scheuten, *Euseius* (=*Amblyseius) finlandicus* (Oudemans), *Amblyseius aberrans* (Oudemans), and *A. andersoni,* but Schruft (1986) reports only *T. pyri* is of importance for biological control of phytophagous mites. Gambaro (1972) reported that *A. aberrans* in Italy is able to check and keep spider mite populations at low densities if not disrupted by pesticides.

Duso (1992) reported that *A. aberrans* and *T. pyri* populations were more abundant on cultivars with hairy leaf undersurfaces. The author concluded that under conditions of prey scarcity, the preference for cultivars was largely independent of prey availability and appeared to be associated with favorable microclimate, improved protection from rain or predation, availability of oviposition sites and shelters, and "an increased capacity for retaining wind pollen." Camporese and Duso (1996) in an extensive five-year study evaluated the competitiveness of *T. pyri, A. aberrans,* and *A. andersoni* on three grape cultivars. The authors found that *A. aberrans* was able to displace the other phytoseiids on all cultivars and reached higher densities on cultivars with pubescent leaf undersurfaces. However, *T. pyri* and *A. andersoni* were able to persist in vineyards having low spider mite densities. James and Whitney (1993) in studies of mite populations in southeastern Australia recorded nine species of phytoseiids with *Typhlodromus doreenae* Schicha and *Amblyseius victoriensis* (Womersley) in vineyards on which insecticide was not used and on which sulfur and copper were used to control vine diseases. In a second study, in the Canberra district of Australia, James *et al.* (1995) recorded six phytoseiid species (*Typhlodromus doreenae* Schicha, *T. dossei* Schicha, *T. australicus* (Womersley), *Amblyseius wattersi* Schicha, *A. elinae* Schicha, and *Phytoseius foiheringhamiae* Denmark and Schicha). *Typhlodromus doreenae* was the dominant species in all monitored vineyards. The results of these authors have considerable implications for choosing which phytoseiids to use in inoculative release programs for different geographical regions and/or climate types.

Karban *et al.* (1995) in a study of 20 North American species of grapes plus *Vitis vinifera* explained 25% of the variability in abundance of *Typhlodromus caudiglans* by leaf characteristics, including density of leaf hairs and the presence of leaf domatia. They also failed to find a significant correlation between spider mite abundance and phytoseiid abundance. However, the spider mite densities recorded in their studies were extremely low and probably not sufficient for the effect of prey abundance on phytoseiid abundance to be accurately measured. In contrast, Hanna and Wilson (1991) and Hanna *et al.* (1997a) working with

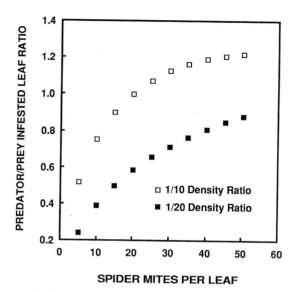

FIGURE 3   Estimated predator–prey binomial ratio required for control of *Tetranychus pacificus* on grapes (Wilson *et al.*, unpublished).

much higher populations of spider mites provide evidence that phytoseiid abundance is clearly associated with spider mite abundance and that increased phytoseiid abundance inevitably results in a rapid decrease in spider mite population size.

Rizk *et al.* (1978) report that the stigmaeids *Agistemus exsertus* Gonzalez and *Amblyseius gossipi* El-Badry and the tydeid *Tydeus californicus* (Banks) are the most abundant predatory mites found on grapevines in middle Egypt. In Chile, *Amblyseius chilenensis* (Dosse) is the most important predatory mite (Gonzalez, 1983). Gaponyuk and Asriev (1986) reported that since its introduction into the Crimea, USSR, in 1976, *M. occidentalis* has acclimatized and spread widely, exerting considerable influence on *E. pruni* populations in vineyards.

Predicting Effectiveness of Predaceous Mites

The potential for spider mite control can be determined in part by the ratio of the number of predators to spider mites. For any system there exists a particular predator–prey ratio at which the pest population will be controlled. For example, Tanigoshi *et al.* (1983) reported that a 1:10 ratio of *Neoseiulus fallacis* to *P. ulmi* provided control in orchards of Red Delicious apples. In an orchard of another apple cultivar, they found that a 1:20 ratio was sufficient and indicated that relatively fewer predaceous mites were required to provide control as a result of reduced spider mite fecundity on this apple cultivar. Wilson *et al.* (1984) similarly reported that a 1:11 ratio of *Metaseiulus occidentalis* to *Tetranychus* spp. provided control in almonds within 2 weeks when the spider mites were at densities greater than 5 mites per leaf. At lower densities, an increas-

ingly higher predator–prey ratio was required. The utility of the predator–prey ratio concept obviously requires that the spider mite density be below economically damaging levels. An equally important constraint is that the monitoring procedure utilizing this information must be easy to use.

Estimating spider mite abundance and predator effectiveness using standard monitoring procedures is particularly tedious, since mites are extremely small, often numerous, and therefore difficult and time-consuming to count. Schruft (1986) reports that it is possible to estimate the risk of damage by *P. ulmi* or *E. carpini vitis* using a method developed by Baillod *et at.* (1979). Its principle is based on the relationship between the number of spider mites per leaf and the proportion of leaves in a sample occupied by mites; instead of counting the mites, the number of leaves with one or more mites is recorded. The same sampling procedure has been used to evaluate predaceous mites (Baillod & Venturi, 1980). Flaherty *et al.* (1992b) similarly addressed the problem of counting in determining the need to initiate control actions by using an infested leaf (binomial) predator–prey ratio, in conjunction with information on the relative level of spider mites in the vineyard. Based on observations over several years, they determined that a 1:2 (0.5) binomial ratio of predator to spider mite infested leaves was sufficient for control. This ratio is less conservative than the ratios derived using Eq. (1) of Wilson *et al.* (1984) for almonds. Figure 3 indicates that a binomial ratio near 1:1 is equivalent to a 1:10 count ratio at densities between 15 and 50 spider mites per leaf. A binomial ratio closer to that reported by Flaherty *et al.* (1992b) was calculated with a 1:20 count ratio. The lower ratio for grapes may indicate a lowered reproductive capacity for spider mites on grapes compared to that found on almonds or possibly for that found by Tanigoshi *et al.* (1983) on apples.

Effect of Alternate Foods of Predaceous Mites on Predator–Prey Relations

Flaherty (1969) reported that *M. occidentalis* is better able to regulate low densities of Willamette mites in the presence of small numbers of *T. urticae* that moved from weeds onto grape leaves. Flaherty *et al.* (1992b) recommend that Willamette spider mite be considered as an important alternate prey since it is a much less serious pest of grapes than Pacific spider mite. Flaherty and Hoy (1971) found that tydeid mites were particularly important as a food source for *M. occidentalis* when spider mie prey were scarce. Karban *et al.* (1994) and Hanna *et al.* (1997a) showed that predatory mites were more effective at controlling Pacific spider mite when released in conjunction with low levels of Willamette spider mite. *Metaseiulus occidentalis* also preys upon other mites, such as eriophyids, tydeids, and possibly tarsonemids. In Europe, additional food for *T. pyri* includes eriophyids, tydeids, pollen, and pearls of the grape (Schruft, 1972). Flaherty and Hoy (1971) and Calvert and Huffaker (1974) speculated on possible benefits of augmenting pollen-feeding tydeids by pollen applications or planting cover crops that produce wind-borne pollen. Gambaro (1972) reported that the ability of *A. aberrans* to live and reproduce in the absence of prey enables it to keep spider mite populations at low densities. However, the alternate source of food was not clearly defined. Wiedner and Boller (1990) emphasized the need for supplying pollen for the survival of *T. pyri* in vineyards in eastern Switzerland. Corbett *et al.* (1991) showed that strip-planting of alfalfa within a cotton field, followed by releases of *Tetranychus urticae* as a food source and *Metaseiulus occidentalis* as a predator, resulted in spread of both mite species into the cotton and maintenance of spider mite populations at subeconomic levels.

Influence of Viticulture Practices on Predator–Prey Relations

Flaherty *et al.* (1992b) point out that viticultural practices influence spider mite outbreaks. They place considerable emphasis on correcting problems concerned with low vine vigor, dusty conditions, and water stress, all of which greatly increase the chance of Pacific spider mite outbreaks if not corrected. They also discuss how insecticides applied for control of other vineyard pests may disturb natural control of spider mites.

Flaherty and Huffaker (1970) collected 10 species of phytoseiid mites from commercial vineyards in the San Joaquin Valley but only *M. occidentalis* apparently played a significant role in the control of spider mites. However, observations in 1983 (Flaherty, unpublished data) suggested that the predaceous mite *Amblyseius* near *hibisci,* previously rare in commercial vineyards, was abundant in some vineyards where triadimefon replaced sulfur for control of powdery mildew, *Uncinula necator* Burrill. *Amblyseius* nr. *hibisci* is commonly found in wild grapes where sulfur is not applied (Flaherty *et al.*, 1985).

English-Loeb *et al.* (1986) showed that *A.* near *hibisci* was the dominant phytoseiid species where sulfur was not applied and maintained lower numbers of Willamette spider mites than *M. occidentalis* where sulfur was used. However, in the following two years, similar studies (Hanna *et al.*, 1997b) did not detect *A.* nr. *hibisci* in significant numbers, underscoring the unpredictability and unreliability of *A.* nr. *hibisci* as a major predatory mite of spider mites in grapes. Another phytoseiid, *Typhloseiopsis smithi* (Schuster), was also recorded on the non-sulfur-treated vine foliage, but not in abundance (English-Loeb *et al.*, 1986).

Effectiveness of Insect Predators of Spider Mites

Flaherty *et al.* (1992b) reported that predaceous insects and spiders are generally considered as ineffective natural enemies of spider mites; it is assumed they appear too late in the season or increase too slowly. However, their contri-bution to natural control in vineyards should not be discounted. Observers are often impressed with how quickly six-spotted thrips, *Scolothrips sexmaculatus* (Pergande), destroys Pacific spider mite populations. However, the appearance of this predator in vineyards is unpredictable. Its inconsistency is possibly related to periodic low prey densities. In Italy, anthocorids and coccinellids are considered more effective at high prey densities (Duso & Girolami, 1985). Schruft (1986) reports that various predaceous insects such as the coccinellid *Scymnus* sp., the staphylinid *Oligota* sp., *Scolothrips longicornis* Priesner, and the anthocorids *Anthocoris nemorum* (Linnaeus) and *Orius minutus* (Linnaeus) are found on grapevines infested with *P. ulmi,* but their importance for biological control of red spider mite populations is not evident. On the other hand, it is certain that some chrysopids, especially *Chrysoperla carnea,* are effective predators of red spider mites during summer and late autumn (Schruft, 1986).

The amount of spider mite control by insect predators may be subtle as well as important. For example, observations in vineyards and studies in the laboratory revealed that western flower thrips, *Frankliniella occidentalis* (Pergande), which is considered a pest of grapes, feeds on Pacific spider mite eggs and may actually affect the pest's population in vineyards (Flaherty *et al.*, 1992b). The importance of predation by *F. occidentalis* of spider mites has also been observed and partially quantified by researchers in cotton (Gonzalez *et al.*, 1982; Gonzalez & Wilson, 1982; Trichilo 1986; Trichilo & Wilson, 1993; Wilson *et al.*, 1991a, 1991b).

Augmentation of Predaceous Mites

Releases of predatory mites to control spider mites have been tried in several crops, with promising results. However, these releases were often designed to overwhelm or inundate the spider mite population to provide near-term control. An alternative to the use of inundative releases is augmentative releases where lower release rates are used, with the progeny of the released predators and not the released predators themselves being responsible for enhancing control to a level sufficient to prevent economic loss. Results from several horticultural and field crops strongly suggest that augmentative release of predatory mites will become economically viable once rates and timing of releases are perfected (Corbett *et al.*, 1991; Flaherty *et al.*, 1992b; Oatman 1970; Oatman *et al.*, 1976, 1977a, 1977b; Schruft, 1972; Tanigoshi *et al.*, 1983; Tijerina-Chavez, 1991; Wilson *et al.*, 1984). Results with different predatory mite species indicate control of spider mites with phytoseiid: spider mite ratios ranging from 1:20 to 1:10 (Flaherty *et al.*, 1992b; Tanigoshi *et al.*, 1983; Wilson *et al.*, 1984). Tijerina-Chavez (1991) reported that the spider mite population on cotton began to decrease when a treatment's phytoseiid: spider mite ratio reached a range of 1:20 to 1:10 (see Wilson *et al.*, 1991c). Analysis of data presented in

Tijerina-Chavez (1991) showed that the monetary value of increased cotton yield in plots receiving predatory mite releases greatly exceeded the cost of the predatory mites (Wilson et al., 1991c). Predatory mite releases in field crops will become cost-effective if targeted to augment naturally occurring predation, with the number of predators released dependent on the abundance of the naturally occurring predators rather than dependent on prohibitively costly inundative releases or disruptive acaricides.

Flaherty et al. (1985) concluded that there are too many problems to overcome in the development of timely mass releases of M. occidentalis for the control of increasing Pacific spider mites in early summer. They suggest that a fall release program may be more fruitful than early summer releases in correcting a Pacific spider mite imbalance. Flaherty and Huffaker (1970) showed that late-season predator activity in vineyards is an essential ingredient in spider mite balance. In their studies, a fall release of M. occidentalis resulted in excellent biological control of Willamette spider mite the following spring and summer. Hoy and Flaherty (1970, 1975) established the importance of late-season diapause induction for successful overwintering of M. occidentalis populations. Flaherty et al. (1985) reason that a fall release of predators reared under diapausing conditions would minimize the timing and survivorship problems encountered in early summer releases because immediate control of Pacific spider mite in the fall is not a factor and diapausing predators require little food. However, Hanna et al. (unpublished) found that the success of a fall release requires not only that the predatory mites are ready to diapause, but also that alternative prey (i.e., tydeid or Willamette mites) are present for successful overwintering of M. occidentalis. In Italy, Girolami and Duso (1985) report on the establishment of predator–prey equilibrium in pesticide-disturbed vineyards with reintroductions of A. aberrans. Baillod et al. (1982) describe methods for reintroduction of T. pyri into vineyards in Switzerland and recommend their use for biological control of phytophagous mites. Schruft (1986) also reports that artificial release of T. pyri by the introduction of infested canes or foliage has been successful. Duso et al. (1991) reported that releases of A. aberrans controlled P. ulmi at low densities and T. pyri was more efficient in controlling populations of P. ulmi than populations of E. carpini. Boller et al. (1988) concluded that hedges can provide ecological refuges and sources of T. pyri in the vineyards of northern Switzerland. Amblyseius nr. hibisci is frequently found in high numbers in vineyards bordering natural vegetation of riparian systems in the San Joaquin Valley. Hoy (1982) reported on the aerial dispersal of M. occidentalis.

Tenuipalpidae

Brevipalpus chilensis Baker is a serious pest of grapes in Chile. Gonzalez (1983) reports that the problem developed as a consequence of the unnecessary use of pesticides. Selective pesticide programs which take advantage of natural and cultural control are recommended to avoid strict dependence on chemical control. An important factor of natural control and perhaps the only natural enemy of B. chilensis is the predaceous mite Amblyseius chilenensis (Dosse). In the Mildura district of Victoria, Australia, B. lewisi McGregor causes a superficial scarring of bunch and berry stems (Buchanan et al., 1980). A close numerical relation between B. lewisi and its most common phytoseiid predator, Amblyseius reticulatus (Oudemans), did not reveal regulation of B. lewisi numbers by A. reticulatus during the growing season (September to April). However, large numbers of A. reticulatus during April may reduce the number of B. lewisi that overwinter. The false spider mite Tenuipalpus granati Sayed has been recognized as a serious pest of grapes in Egypt (Rizk et al., 1978). It was suggested that certain pesticides were in part responsible for the problem. Agistemum exsertus, Amblyseius gossipi, and Tydeus californicus were observed in association with T. granati, but the data did not reveal that pesticides disrupted regulatory action by these predatory mites.

Eriophyidae

Colomerus vitis (Pagenstecher) is an eriophyid mite that attacks various species and hybrids of grapes and is probably widely distributed. Different biotypes of C. vitis have been recognized in California by their characteristic injury (Barnes, 1992). The phytoseiid mite M. occidentalis was reported to be effective in reducing populations of C. vitis. Schruft (1972) reported that C. vitis and Calepitrimerus vitis (Nalepa), also eriophyid pests on grapes in Europe, are fed on by the tydeids Tydeus götzi Schruft and Pronematus stärki Schruft.

CONCLUSION

It is apparent to us that this review documents a rather weak worldwide biological control research program for a crop of such importance as grape. Instead of a prior and continuing commitment to biological control, the review reveals an emphasis on studies concerned with disruptions of secondary pests by pesticides and a paucity of studies dealing with biological control of primary pests. The relatively large number of studies dealing with spider mites as secondary pests and little in the way of studies on the importation, augmentation, and evaluation of natural enemies of primary pests lends credence to a lack of commitment by grape entomologists and the viticulture industry to biological control research.

Flaherty et al. (1985) recommended that grape IPM in the San Joaquin Valley of California would be improved by augmentation of biological control, including importation of natural enemies of some primary pests. The initiation of studies on the importation and augmentation of natural enemies of cicadellid pests of grapes discussed earlier has been an important step in California. The increasing emphasis on the use of augmentative releases to control spider mites on a range of crops shows particular promise for management of spider mites in grapes as well. However, we further recommend a greatly expanded cooperative effort in biological control studies by grape entomologists throughout the world. The efforts expended by citrus entomologists worldwide would serve as a working model for accelerating the importation and augmentation of natural enemies of grape pests.

Acknowledgment

The authors thank G. Gordh, K. Hagen, R. Hanna, H. Kaya, J. McMurtry, C. Pickett, and J. Smilanick for their generous assistance.

References

Alma, A. (1995). Bio-ethologic and epidemiologic researches on *Holocacista rivillei* Stanton (Lepidoptera Heliozelidae). Redia, 78, 373–378.

Baillod, M., & Venturi, I. (1980). Lutte biologique contre l'acarien rouge en viticulture. I. Répartition, distribution et méthode de contrôle des populations de prédateurs typhlodromes. Revue Suisse de Viticulture, Arboriculture, Horticulture, 12, 231–238.

Baillod, M., Bassino, J. P., & Piganeau, P. (1979). L'estimation du risque provoqué par l'acarien rouge (*Panonychus ulmi* Koch) et l'acarien des charmilles (*Eotetranychus carpini* Oud.) en viticulture. Revue Suisse de Viticulture, Arboriculture, Horticulture, 11, 123–130.

Baillod, M., Schmid, A., Guignard, E., Antonin, P., & Caccia, R. (1982). Lutte biologique contre l'acarien rouge en viticulture. II. Equilibres naturels, dynamique des populations et expériences de lâchers de typhlodromes. Revue Suisse de Viticulture, Arboriculture, Horticulture, 14, 345–352.

Barnes, M. M. (1992). Grape erineum mite. In "Grape Pest Management" (D. L. Flaherty, L. P. Christensen, W. T. Lanini, J. J. Marois, P. A. Phillips, & L. T. Wilson, Eds.), 2nd ed., pp. 262–264. Univ. of California, Berkeley, Agricultural Science Publication 3343.

Bedding, R. A., & Miller, L. A. (1981). Use of a nematode, *Heterorhabditis heliothidis,* to control black vine weevil, *Otiorhynchus sulcatus,* in potted plants. Annals of Applied Biology, 99, 211–216.

Bettiga, L. J., & Phillips, P. A. (1992). Orange tortrix. In "Grape Pest Management" (D. L. Flaherty, L. P. Christensen, W. T. Lanini, J. J. Marois, P. A. Phillips, & L. T. Wilson, Eds.), 2nd ed., pp. 174–179. Univ. of California, Berkeley, Agricultural Science Publication 3343.

Boller, E. F., Remund, U., & Candolfi, M. P. (1988). Hedges as potential sources of *Typhlodromus pyri,* the most important predatory mite in vineyards of northern Switzerland. Entomophaga, 33, 240–255.

Bournier, A. (1976). Grape insects. Annual Review of Entomology, 22, 355–376.

Buccholz, U., & Schruft, G. (1994). Predatory arthropods on flowers and fruits of the grapevine as antagonists of the grape moth, *Eupoecilia ambiguella* Hbn.) (Lep., Cochylidae). Journal of Applied Entomology, 118, 31–37.

Buchanan, G. A., Bengston, M., & Exley, E. M. (1980). Population growth of *Brevipalpus lewisi* McGregor (Acarina: Tenuipalpidae) on grapevines. Australian Journal of Agricultural Research, 31, 957–965.

Calvert, D. J., & Huffaker, C. B. (1974). Predator (*Metaseiulus occiden-*

talis)-prey (*Pronematus* spp.) interactions under sulfur and cattail pollen applications in a non-commercial vineyard. Entomophaga, 19, 361–369.

Camporese, P., & Duso, C. (1996). Different colonization patterns of phytophagous and predatory mites (Acari: Tetranychidae, Phytoseiidae) on three grape varieties: A case study. Experimental & Applied Acarology, 20, 1–22.

Castaneda-Samayoa, O., Holst, H., & Ohnesorge, B. (1993). Evaluation of some *Trichogramma* species with respect to biological control of *Eupoecilia ambiguella* Hb. and *Lobesia botrana* Schiff. (Lep., Tortricidae).Zeitschrift fuer Pflanzenkrankheiten und Pflanzenschutz, 100, 599–610.

Cate, J. R. (1975). Ecology of *Erythroneura elegantula* Osborn (Homoptera: Cicadellidae) in grape agroecosystems in California, Ph.D. Thesis, Univ. of California, Berkeley.

Charles, J. G. (1985). *Diadiplosis koebelei* Koebele (Diptera: Cecidomyiidae), a predator of *Pseudococcus longispinus* T.-T. (Homoptera: Pseudococcidae), newly recorded from New Zealand. New Zealand Journal of Zoology, 12, 331–333.

Clausen, C. P. (1924). The parasites of *Pseudococcus maritimus* (Ehrhorn) in California (Hymenoptera, Chalciodoidea). Part II. Biological studies and life histories. University of California Publications in Entomology, 3, 253–288.

Clausen, C. P. (1961). Biological control of western grape leaf skeletonizer (*Harrisina brillians* B. & McD.) in California. Hilgardia, 31, 613–637.

Corbett, A., & Rosenheim, J. A. (1996a). Impact of a natural enemy overwintering refuge and its interaction with the surrounding landscape. Ecological Entomology, 21, 155–164.

Corbett, A., & Rosenheim, J. A. (1996b). Quantifying movement of a minute parasitoid, *Anagrus epos* (Hymenoptera: Mymaridae), using fluorescent dust marking and recapture. Biological Control, 6, 35–44.

Corbett, A., Leigh, T. F., & Wilson, L. T. (1991). Evaluation of alfalfa interplantings as a source of *Metaseiulus occidentalis* (Acari: Phytoseiidae) for management of spider mites in cotton. Biological Control, 1, 188–196.

Costello, M. J., & Daane, K. M. (1995). Spider (Araneae) species composition and seasonal abundance in San Joaquin Valley Grape vineyards. Environmental Entomology, 24, 823–831.

Coviello, R., Hirschfelt, D. J., & Barnett, W. W. (1992). Omnivorous leafroller. In "Grape Pest Management" (D. L. Flaherty, L. P. Christensen, W. T. Lanini, J. J. Marois, P. A. Phillips, & L. T. Wilson, Eds.), 2nd ed., pp. 166–173. Univ. of California, Berkeley, Agricultural Science Publication 3343.

Daane, K. M., Yokota, G. Y., Zheng, Y., & Hagen, K. S. (1996). Inundative release of common green lacewings (Neuroptera: Chrysopidae) to suppress *Erythroneura variabilis* and *E. elegantula* (Homoptera: Cicadellidae) in vineyards. Environmental Entomology, 25, 1224–1234.

Daane, K. M., Yokota, G. Y., Rasmussen, Y. D., Zheng, Y., & Hagen, K. S. (1993). Effectiveness of leafhopper control varies with lacewing release methods. California Agriculture, 47, 19–23.

Doutt, R. L., & Nakata, J. (1965). Overwintering refuge of *Anagrus epos* (Hymenoptera: Mymaridae). Journal of Economic Entomology, 58, 586.

Doutt, R. L., & Nakata, J. (1973). The *Rubus* leafhopper and its egg parasitoid: An endemic biotic system useful in grape-pest management. Environmental Entomology, 2, 381–386.

Duso, C. (1992). Role of *Amblyseius aberrans* Oud., *Typhlodromus pyri* Scheuten and *Amblyseius andersoni* Chant (Acari: Phytoseiidae) in vineyards. III. Influence of variety characteristics on the success of *Amblyseius aberrans* and *Typhlodromus pyri* releases. Journal of Applied Entomology, 114, 455–462.

Duso, C., & Girolami, V. (1985). Strategie di controllo biologico degli acari tetranichidi su vite. Atti XIV Cong. Naz. Ital. Entomol., Palermo, Erice, Bagheria, pp. 719–728.

Duso, C., Pasqualetto, C., & Camporeses, P. (1991). Role of the predatory mites *Amblyseius aberrans* (Oud.), *Typhlodromus pyri* Scheuten and *Amblyseius andersoni* (Chant) (Acari, Phytoseiidae) in vineyards. Journal of Applied Entomology, 112, 298–308.

English-Loeb, G. M., Flaherty, D. L., Wilson, L. T., Barnett, W. W., Leavitt, G. M., & Settle, W. H. (1986). Pest management changes affect spider mites in vineyards. California Agriculture, 40, 28–30.

Flaherty, D. L. (1969). Ecosystem trophic complexity and Willamette mite, *Eotetranychus willamettei* Ewing (Acarina: Tetranychidae), densities. Ecology, 50, 911–916.

Flaherty, D. L., & Hoy, M. A. (1971). Biological control of Pacific mites and Willamette mites in San Joaquin Valley vineyards. III. Role of tydeid mites. Res. Popul. Ecol., 8, 80–96.

Flaherty, D. L., & Huffaker, C. B. (1970). Biological control of Pacific mites and Willamette mites in San Joaquin Valley vineyards. I. Role of *Metaseiulus occidentalis*. II. Influence of dispersion patterns of *Metaseiulus occidentalis*. Hilgardia, 40, 267–330.

Flaherty, D. L., & Wilson, L. T. (1988). Part II. Mites and insects that cause diseaselike symptoms in grapes. *In* "Compendium of Grape Diseases" (R. C. Pearson & A. C. Goheen, Eds.), pp. 60–62. American Phytopathological Society, St. Paul, MN.

Flaherty, D., Jensen, F., & Nakata, J. (1976). "Grape mealybug. Grape pest management in the southern San Joaquin Valley." San Joaquin Valley Agriculture Research Extension Center Publication.

Flaherty, D. L., Peacock, W. L., Bettiga, L., & Leavitt, G. M. (1982). Chemicals losing effect against grape mealybug. California Agriculture, 36, 15–16.

Flaherty, D. L., Wilson, L. T., Stern, V. M., & Kido, H. (1985). Biological control in San Joaquin Valley vineyards. *In* "Biological Control in Agricultural IPM Systems" (M. A. Hoy & D. C. Herzog, Eds.), pp. 501–520. Academic Press, San Diego.

Flaherty, D. L., Phillips, P. A., Legner, E. F., Peacock, W. L., & Bentley, W. J. (1992a). Mealybugs. *In* "Grape Pest Management" (D. L. Flaherty, L. P. Christensen, W. T. Lanini, J. J. Marois, P. A. Phillips, & L. T. Wilson, Eds.), 2nd ed., pp. 159–165. Univ. of California, Berkeley, Agricultural Science Publication 3343.

Flaherty, D. L., Wilson, L. T., Welter, S. C., Lynn, C. D., & Hanna, R. (1992b). Spider mites. *In* "Grape Pest Management" (D. L. Flaherty, L. P. Christensen, W. T. Lanini, J. J. Marois, P. A. Phillips, & L. T. Wilson, Eds.), 2nd ed., pp. 180–192. Univ. of California, Berkeley, Agricultural Science Publication 3343.

Gambaro, P. I. (1972). Il ruolo del *Typhlodromus aberrans* Oudm. (Acarina Phytoseiidae) nel controllo biologico degli Acari fitofagi dei vigneti del Veronese. Bollettino di Zoologia Agraria e di Bachicoltura, 11, 151–165.

Gaponyuk, I. L., & Asriev, E. A. (1986). *Metaseiulus occidentalis* in vineyards. Zashchita Rastenii, 8, 22–23 (in Russian).

Girolami, V., & Duso, C. (1985). Controllo biologico degli acari nei vigneti. L'Informatore Agrario, 41, 83–89.

Gonzalez, D., & Wilson, L. T. (1982). A food-web approach to economic thresholds: A sequence of pests/predaceous arthropods on California cotton. Entomophaga, 27, 31–43.

Gonzalez, D., Patterson, B. R., Leigh, T. F., & Wilson, L. T. (1982). Mites: A primary food source for two predators in San Joaquin Valley cotton. California Agriculture, 36, 18–20.

Gonzalez, D., White, W., Pickett, C., Cervenka, V., Moratorio, M., & Wilson, L. T. (1988). Biological control of variegated leafhopper in grape IPM program. California Agriculture, 42, 23–25.

Gonzalez, R. H. (1983). Manejo de plagas de la vid. Ciencies Agri. 13, Univ. de Chile.

Grafton-Cardwell, E. E., & Ouyang, Y. (1995). Augmentation of *Euseius tularensis* (Acari: Phytoseiidae) in citrus. Environmental Entomology, 24, 738–747.

Grasswitz, T. R., & Burts, E. C. (1995). Effect of native natural enemies on the population dynamics of the grape mealybug, *Pseudococcus maritimus* (Hom.: Pseudococcidae), in the apple and pear orchards. Entomophaga, 40, 105–117.

Hanna, R., & Wilson, L. T. (1991). Prey preference by *Metaseiulus occidentalis* (Acari: Phytoseiidae) and the role of prey aggregation. Biological Control, 1, 51–58.

Hanna, R., Wilson, L. T., Zalom, F. G., & Flaherty, D. L. (1997a). Effects of predation and competition on population dynamics of *Tetranychus pacificus* on grapevines. Journal of Applied Ecology, 34, 878–888.

Hanna, R., Zalom, F. G., Wilson, L. T., & Leavitt, G. M. (1997b). Sulfur can suppress mite predators in vineyards. California Agriculture, 51, 19–21.

Hoy, M. A. (1982). Aerial dispersal and field efficacy of a genetically improved strain of the spider mite predator, *M. occidentalis*. Entomologica Experimentalis Applicata, 32, 205–212.

Hoy, M. A., & Flaherty, D. L. (1970). Photoperiodic induction of diapause in a predaceous mite, *Metaseiulus occidentalis*. Annals of the Entomological Society of America, 63, 960–963.

Hoy, M. A., & Flaherty, D. L. (1975). Diapause induction and duration in vineyard-collected *Metaseiulus occidentalis*. Environmental Entomology, 4, 262–264.

James, D. G., & Whitney, J. (1993). Mite populations on grapevines in south-eastern Australia: Implications for biological control of grapevine mites acarina tenuipalpidae eriophyidae. Experimental & Applied Acarology, 17, 259–270.

James, D. G., Whitney, J., & Rayner, M. (1995). Phytoseiids (Acari: Phytoseiidae) dominate the mite fauna on grapevines in Canberra district vineyards. Journal of the Australian Entomological Society, 34, 79–82.

Jensen, F. L., Flaherty, D. L., & Luvisi, D. A. (1992a). Thrips. *In* "Grape Pest Management" (D. L. Flaherty, L. P. Christensen, W. T. Lanini, J. J. Marois, P. A. Phillips, & L. T. Wilson, Eds.), 2nd ed., pp. 193–201. Univ. of California, Berkeley, Agricultural Science Publication 3343.

Jensen, F. L., Hirschfelt, D. J., & Flaherty, D. L. (1992b). Grape leaffolder. *In* "Grape Pest Management" (D. L. Flaherty, L. P. Christensen, W. T. Lanini, J. J. Marois, P. A. Phillips, & L. T. Wilson, Eds.), 2nd ed., pp. 133–139. Univ. of California, Berkeley, Agricultural Science Publication 3343.

Jubb, G. L., Jr. (1982). Occurrence of the grape root borer *Vitacea polistiformis,* in Pennsylvania. Melsheimer Entomological Series, 32, 20–24.

Jubb, G. L., Jr., Masteller, E. C., & Lehman, R. D. (1985). Survey of arthropods in vineyards of Erie County, Pennsylvania. Acari. Int. J. Acarol., 11, 201–208.

Kamal, M. (1951). Biological control projects in Egypt, with a list of introduced parasites and predators. Bulletin de la Societe Fouad I d'Entomologie, 35, 205–220.

Karban, R., English-Loeb, G., Walker, M. A., & Thaler, J. (1995). Abundance of phytoseiid mites on Vitis species: Effects of leaf hairs, domatia, prey abundance and plant phylogeny. Experimental & Applied Acarology, 19, 189–197.

Karban, R., Hougen-Eitzmann, D., & English-Loeb, G. (1994). Predator-mediated apparent competition between two herbivores that feed on grapevines. Oecologia, 97, 508–511.

Kido, H., Flaherty, D. L., Bosch, D. F., & Valero, K. A. (1984). French prune trees as overwintering sites for grape leafhopper egg parasite. American Journal of Enology and Viticulture, 35, 156–160.

Kido, H., Flaherty, D. L., Kennett, C. E., McCalley, N. F., & Bosch, D. F. (1981). Seeking the reasons for differences in orange tortrix infestations. California Agriculture, 35, 27–28.

Mani, M. (1988). "Bioecology and Management of Grapevine Mealybug." Indian Institute of Horticultural Research, Technical Bulletin 5.

Mani, M., & Krishnamoorthy, A. (1989). Feeding potential and development of green lacewing *Mallada boninensis* (Okamoto) on the grape mealybug, *Maconellicoccus hirsutus* (Green). Entomon, 14, 19–20

Makhmudov, D., Azimov, A., Abdulagatov, A. Z., & Ataev, K. G. (1977). Control of the grape moth. Zashchita Rastenii, 7, 24–25 (in Russian).

Manjunath, T. M. (1985). India—*Maconellicoccus hirsutus* on grapevine. FAO Plant Protection Bulletin, 33, 74.

Marcelin, H. (1985). La lutte contre les tordeuses de la grape. Phytoma, 370, 29–32.

McKenzie, L. M., & Beirne, B. P. (1972). A grape leafhopper, *Erythroneura ziczac* (Homoptera: Cicadellidae) and its mymarid (Hymenoptera) egg-parasite in the Okanagan Valley, British Columbia. Canadian Entomologist, 104, 1229–1233.

Murphy, B. C., Rosenheim, J. A., & Granett, J. (1996). Habitat diversification for improving biological control: Abundance of *Anagrus epos* (Hymenoptera: Mymaridae) in grape vineyards. Environmental Entomology, 25, 495–504.

Nasr, F. N., Korashy, M. A., & Rashed, F. F. M. (1995). *Trichogramma evanescens* West. (Hym., Trichogrammatidae) as an egg parasitoid of grape moth *Lobesia botrana* (Den. & Schiff.) (Lep., Tortricidae). Anzeiger fuer Schaedlingskunde Pflanzenschutz Umweltschutz, 68, 44–45.

Niyazov, O. D. (1969). The parasites and predators of grape mealybug. Zashchita Rastenii, 14, 38–40 (in Russian).

Oatman, E. R. (1970). Integration of *Phytoseiulus persimilis* with native predators for control of the two-spotted spider mite on rhubarb. Journal of Economic Entomology, 63, 1177–1180.

Oatman, E. R., Gilstrap, F. E., & Voth, V. (1976). Effect of different release rates of *Phytoseiulus persimilis* (Acari: Phytoseiidae) on the two-spotted spider mite on strawberry in southern California. Entomophaga, 21, 269–273.

Oatman, E. R., McMurtry, J. A., Gilstrap, F. E., & Voth, V. (1977a). Effect of releases of *Amblyseius californicus* on the two-spotted spider mite on strawberry in southern California. Journal of Economic Entomology, 70, 638–640.

Oatman, E. R., McMurtry, J. A., Gilstrap, F. E., & Voth, V. (1977b). Effect of releases of *Ambylseius californicus, Phytoseiulus persimilis,* and *Typhlodromus occidentalis* on the two-spotted spider mite on strawberry in southern California. Journal of Economic Entomology, 70, 45–47.

Orlinskii, A. D., Rzaeva, L. M., & Shakhramanov, I. K. (1989). A promising entomophage. Zashchita Rastenii, 11, 25–26 (in Russian).

Padmaja, C., Babu, T. R., Reddy, D. D. R., & Sriramulu, M. (1995). Biology and predation potential of *Scymnus coccivora* Aiyar (Coleoptera: Coccinellidae) on mealybugs. Journal of Entomological Research, 19, 79–81.

Pavan, F., & Picotti, P. (1994). Population dynamics of *Empoasca vitis* (Goethe) (Homoptera Cicadellidae) and *Anagrus atomus* (Linnaeus) (Hymenoptera Mymaridae) in vineyards and kiwifruit orchards. Memorie della Societa Entomologica Italiana, 72, 163–173.

Pickett, C. H., Wilson, L. T., Gonzalez, D., & Flaherty, D. L. (1987). Biological control of variegated grape leafhopper. California Agriculture, 41, 14–16.

Pickett, C. H., Wilson, L. W., Flaherty, D. L., & Gonzalez, D. (1989). Measuring the host preference of parasites: An aid in evaluating biotypes of *Anagrus epos* (Hym.: Mymaridae). Entomophaga, 34, 551–558.

Pickett, C. H., Wilson, L. T., & Flaherty, D. L. (1990). Role of refuges in crop protection, with reference to plantings of French prune trees in a grape agroecosystem. *In* "Monitoring and Integrated Management of Arthropod Pests of Small Fruit Crops" (N. J. Bostanian, L. T. Wilson, & T. J. Dennehy, Eds.), pp. 161–155. Intercept, Dorset, England.

Pinto, J. D., & Viggiani, G. (1987). Two new Trichogrammatidae (Hymenoptera) from North America: *Ittysella lagunera* Pinto & Viggiani (n. gen., n. sp.) and *Epoligosita mexicana* Viggiani (n. sp.). Pan-Pacific Entomologist, 63, 371–376.

Polesny, F. (1987). Die Bedeutung von Spinnen im Weinbau. Pflanzenschutz, 3, 9–10.

Pucci, C., & Dominici, M. (1986). Biological notes and cyclical outbreaks of *Theresimima ampelophaga* Bayle-Barelle (Lep., Zygaenidae). Journal of Applied Entomology, 101, 479–491.

Ripa, S. R., & Rojas, P. S. (1990). Manejo y control biológico del chanchito blanco de la vid. Revista Frutícola, 11, 82–87.

Rizk, G. A., Sheta, I. B., & Ali, M. A. (1978). Chemical control of mites infesting grape-vine in middle Egypt. Bulletin of the Entomological Society of Egypt, Economic Series, 11, 105–111.

Rzaeva, L. M. (1985). Parasites and predators of the grape mealybug (*Planococcus ficus* Signoret) and introduction of new natural enemies into the eastern Transcacaus. Biologicheskikh Nauk, 4, 34–39 (in Russian).

Saunders, M. C., & All, J. N. (1985). Association of entomophilic rhabditoid nematode populations with natural control of first instar larvae of the grape root borer, *Vitacea polistiformis,* in concord grape vineyards. Journal of Invertebrate Pathology, 45, 147–151.

Schruft, G. (1972). Les tydéidés (Acari) sur vigne. OEPP/EPPO Bulletin, 3, 51–55.

Schruft, G. (1986). Grape. *In* "Spider Mites: Their Biology, Natural Enemies and Control" (W. Helle & M. W. Sabelis, Eds.), Vol. 1B, pp. 359–366. Elsevier, Amsterdam.

Schwartz, A. (1987). Seasonal occurrence of a predaceous mite *Amblyseius addoensis* Van der Merwe & Ryke (Acari: Phytoseiidae) on table grapes. South African Journal for Enology and Viticulture, 8, 78–79.

Sengonca, C., & Leisse, N. (1989). Enhancement of the egg parasite *Trichogramma semblidis* (Auriv.) (Hym., Trichogrammatidae) for control of both grape vine moth species in the Ahr Valley. Journal of Applied Entomology, 107, 41–45.

Settle, W. H., & Wilson, L. T. (1990a). Bahavioural factors affecting differential parasitism by *Anagrus epos* (Hymenoptera: Mymaridae), of two species of erythroneuran leafhoppers (Homoptera: Cicadellidae). Journal of Animal Ecology, 59, 877–891.

Settle, W. H., & Wilson, L. T. (1990b). Variation by the variegated leafhopper and biotic interactions: Parasitism, competition, and apparent competition. Ecology, 71, 1461–1470.

Settle, W. H., Wilson, L. T., Flaherty, D. L., & English-Loeb, G. M. (1986). The variegated leafhopper, an increasing pest of grapes. California Agriculture, 40, 30–32.

Srinivasan, T. R., & Sundara Babu, P. C. (1989). Field evaluation of *Cryptolaemus montrouzieri* Mulsant, the coccinellid predator against grapevine mealybug, *Maconellicoccus hirsutus* (Green). So. Indian Hort., 37, 50–51.

Tanigoshi, L. K., Hoyt, S. C., & Croft, B. A. (1983). Basic biological and management components for mite pests and their natural enemies. *In* "Integrated Management of Insect Pests of Pome and Stone Fruits" (B. A. Croft & S. A. Hoyt, Eds.), pp. 153–202. Wiley-Interscience, New York.

Tijerina-Chavez. (1991). Biological control of spider mites (Acari: Tetranychidae) on cotton through inoculative release of predatory mites *Metaseiulus occidentalis* and *Amblyseius californicus* (Acari: Phytoseiidae) in the San Joaquin Valley of California, Ph.D. Thesis, Univ. of California, Davis.

Trichilo, P. (1986). Influence of the host plant on the interaction of spider mites with their natural enemies in a cotton agroecosystem, Ph.D. Dissertation, Univ. of California, Davis.

Trichilo, P. J., & Wilson, L. T. (1993). An ecosystem analysis of spider mite outbreaks: Nutritional stimulation or natural enemy suppression? Experimental & Applied Acarology, 17, 291–314.

Trichilo, P. J., Wilson, L. T., & Grimes, D. W. (1990). Influence of grapevine water deficit on distribution of leafhoppers (Homoptera: Cicadellidae) in San Joaquin Valley Vineyards. Environmental Entomology, 19, 1803–1809.

Trjapitzin, S. V. (1995). The identities of *Anagrus* (Hymenoptera: Mymaridae) egg parasitoids of the grape and blackberry leafhoppers (Homoptera: Cicadellidae) in California. Pan-Pacific Entomologist, 71, 250–251.

Trjapitzin, S. V., & Chiappini, E. (1994). A new *Anagrus* (Hymenoptera: Mymaridae) egg parasitoid of *Erythroneura* spp. (Homoptera: Cicadellidae). Entomological News, 105, 137–140.

Wells, J. D., Cone, W. W., & Conant, M. M. (1988). Chemical and biological control of *Erythroneura* leafhoppers on *Vitis vinifera* in southcentral Washington. Journal of the Entomological Society of British Columbia, 85, 45–52.

Welter, S. C., Freeman, R., & Farnham, D. S. (1991). Recovery of 'Zinfandel' grapevines from feeding damage by Willamette mite (Acari: Tetranychidae): Implications for economic injury level studies in perennial crops. Environmental Entomology, 20, 104–109.

Wheeler, A. G., Jr., & Jubb, G. L., Jr. (1979). *Scymnus cervicalis* Mulsant, a predator of grape phylloxera, with notes on *S. brullei* Mulsant as a predator of woolly aphids on Elm (Coleoptera: Coccinellidae). Coleop. Bull., 33, 199–204.

Wiedner, U., & Boller, E. (1990). Blühende Rebberge in der Ostschweiz. 2. Zum Pollenangebot auf den Rebblättern. Schweizerische Zeitschrift für Obst- und Weinbau, 126, 426–431.

Williams, D. W. (1984). Ecology of a blackberry-leafhopper-parasite system and its relevance to California grape agroecosystems. Hilgardia, 51, 1–32.

Wilson, L. T., Carmen, I., & Flaherty, D. L. (1991a). *Aphelopus albopictus* Ashmead (Hymenoptera: Dryinidae): Abundance, parasitism, and distribution in relation to leafhopper hosts in grapes. Hilgardia, 59.

Wilson, L. T., Flaherty, D. L., & Peacock, W. L. (1992b). Grape leafhopper. *In* "Grape Pest Management" (D. L. Flaherty, L. P. Christensen, W. T. Lanini, J. J. Marois, P. A. Phillips, & L. T. Wilson, Eds.), 2nd ed., pp. 140–152. Univ. of California, Berkeley, Agricultural Science Publication 3343.

Wilson, L. T., Trichilo, P. J., & Gonzalez, D. (1991b). Natural enemies of spider mites (Acari: Tetranychidae) on cotton: Density regulation or casual association? Environmental Eology, 20, 849–856.

Wilson, L. T., Hoy, M. A., Zalom, F. G., & Smilanick, J. M. (1984).

Sampling mites in almonds: I. Within-tree distribution and clumping pattern of mites with comments on predator-prey interactions. Hilgardia, 51, 1–13.

Wilson, L. T., Pickett, C. H., Flaherty, D. L., & Bates, T. A. (1989). French prune trees: Refuge for grape leafhopper parasite. California Agriculture, 43, 7–8.

Wilson, L. T., Flaherty, D. L., Settle, W., Pickett, C., & Gonzalez, D. (1987). Managing the variegated leafhopper and grape leafhopper. 1987 Proceedings of the Table Grape Symposium, pp. 5–11.

Wilson, L. T., Trichilo, P. J., Flaherty, D. L., Hanna, R., & Corbett, A. (1991c). Natural enemy-spider mite interactions: Comments on implications for population assessment. Modern Acarology, 1, 167–173.

Wilson, L. T., Barnes, M. M., Flaherty, D. L., Andris, H. L., & Leavitt, G. M. (1992a). Variegated grape leafhopper. *In* "Grape Pest Management" (D. L. Flaherty, L. P. Christensen, W. T. Lanini, J. J. Marois, P. A. Phillips, & L. T. Wilson, Eds.), 2nd ed., pp. 202–213. Univ. of California, Berkeley, Agricultural Science Publication 3343.

Wilson, L. T., Flaherty, D. L., Andris, H., Barnett, W. W., Gonzalez, D., & Settle, W. (1986). The variegated leafhopper: An increasing pest of grapes in the San Joaquin Valley. 1986 Proceedings of the San Joaquin Valley Grape Symposium, pp. 27–29.

Winkler, A. J., Cook, J. A., Kliewer, W. M., & Lider, L. A. (1974). "General Viticulture." Univ. of Calif. Press, Berkeley.

Yasnosh, V. A., & Mjavanadze, V. I. (1983). On the efficiency and rational use of *Cryptolaemus montrouzieri* against plant pests in the Georgia SSR. *In* "Proceedings, 10th International Congress of Plant Protection, 2."

Yigit, A., & Erkilic, L. (1987). Studies on egg parasitoids of grape leafhopper, *Arboridia adanae* Dlab. (Hom., Cicadellidae) and their effects in the region of South Anatolia. Türkiye I. Entomoloji Kongresi Bildirileri, 13–16 (In Turkish).

Zalom, F. G., Bettiga, L. J., & Donaldson, D. R. (1992). Branch and twig borer. *In* "Grape Pest Management" (D. L. Flaherty, L. P. Christensen, W. T. Lanini, J. J. Marois, P. A. Phillips, & L. T. Wilson, Eds.), 2nd ed., pp. 121–123. Univ. of California, Berkeley, Agricultural Science Publication 3343.

34

Biological Control of Weeds in Terrestrial and Aquatic Environments

R. D. GOEDEN

Department of Entomology
University of California
Riverside, California

L. A. ANDRÉS

USDA–ARS, Western Research Center
800 Buchanan Street
Albany, California

INTRODUCTION

In the United States, 500 major species of weeds cause estimated annual losses of about $8 billion (Chandler, 1980). These weeds infest cropland, rangeland, recreational, and aquatic areas and vary in their economic importance and need for control. Cultural and chemical controls for weeds are effective, but relief is temporary at best and often uneconomical. In fact, the more effective the chemical control, the sooner other more herbicide-resistant or herbicide-tolerant weeds may fill the vacant niches. Thus, improved weed control is no longer a matter of improved plant kill. Planning and ingenuity are required to minimize immediate losses without inviting incursions by replacement weed species. As Aldrich (1984) noted, weeds are part of dynamic ecosystems continually evolving in response to natural and cultural control pressures. Biological control is a proven method of weed control. An increasing volume of literature is devoted to this approach, including a quadrennial international symposium and proceedings which update and highlight current activities and thought. Biological control continues to offer promise and expanded application in reducing weed losses.

WHAT IS BIOLOGICAL CONTROL OF WEEDS?

Definition

Biological control is the study of relationships among weeds, their associated organisms, and the environment, followed by the manipulation of selected species of these organisms (natural enemies) to the detriment of a target weed species. Attention is focused on those weed–natural enemy relations that have coevolved to the degree that the

weed's natural enemies cannot exist or would have little environmental impact in the absence of their host. In other words, coevolved natural enemies that have developed a high degree of host specificity have proven the safest to use, are least likely to damage nontarget plant species, and are most suitable for regulating weed abundance. Biological control scientists go to considerable effort to match natural enemies to their weedy host plants in problem environments, seeking combinations and devising manipulations most detrimental to the weeds.

The natural enemies used in biological control are self-perpetuating only in the presence of their weed hosts and then only within the limits set by the environment. According to definitions for biological control offered by Smith (1919, 1948) and DeBach (1964), the ability of natural enemies to regulate weed or pest arthropod populations in a self-sustaining, density-dependent manner sets biological control apart from other methods of control.

Polyphagous agents (plant-feeding fish, sheep, cattle, geese, and other grazing animals) are useful in removing weeds in some situations (see later), but their numbers and actions must be carefully regulated to avoid damage to nontarget plants. Although these vertebrate grazers remove unwanted foliage, their inability to selectively regulate weed numbers limits their use in biological control.

Methodology

Developing a biological control program begins with a systematic assessment of the weed problem: (1) assuring proper identification of the target weed, (2) charting the weed's geographic range, (3) characterizing the habitats it infests, (4) ascertaining the losses caused by the weed, (5) determining the degree of control required, and (6) compiling a list of natural enemies already present or reported elsewhere. Answers are sought to certain questions; e.g., why has the plant become a problem? If natural enemies are lacking, can they be introduced? Or, if natural enemies are already present, can their abundance be enhanced by conservation or augmentation?

Introducing New Natural Enemies

Naturalized weeds often have few host-specific natural enemies capable of effectively regulating their abundance. To correct this, new natural enemies can be sought in the weed's native range and introduced into the problem areas. This approach has been followed in many countries, which has led to the introduction of numerous weed-feeding insects and mites and, more recently, plant pathogens and nematodes (Julien, 1982, 1987, 1992; see later).

Although the concept of finding and introducing exotic phytophagous organisms is simple, it requires detailed prerelease studies to ensure that control is achieved and that economically and ecologically important plants will not be harmed. Guidelines for introducing natural enemies were described in detail by Zwölfer and Harris (1971), Frick (1974), Andres *et al.* (1976), Klingman and Coulson (1983), Schroeder and Goeden (1986), and Harris (1991) and include the following steps: *(1) Project selection.* Once released, introduced natural enemies cannot be restricted to parts of the weed's geographic range. Before studies that may lead to natural enemy introduction are undertaken, there must be assurance that the weed has few, if any, redeeming virtues and that there is little or no public opposition to the project (Turner, 1984). *(2) Search for natural enemies.* A list of organisms recorded from the target weed is compiled from literature and museum records. This is followed by field surveys and studies of associated organisms in selected parts of the weed's native range. Organisms are collected and identified, and their names are checked against literature and museum records. Promising candidate species are selected for further study. *(3) Tests and biological studies of host range.* Behavioral and biological studies and choice–no-choice feeding and oviposition tests are conducted in the laboratory to determine the host range of each candidate. Plants tested include cultivated and ecologically important species, with special focus on close taxonomic relatives of the weed (Wapshere, 1974a). *(4) Evaluation of host-range studies.* A report summarizing the candidate biological control agent's taxonomy, behavior, biology, and host-plant relations is submitted to regulatory agencies for review. In the United States, the U.S. Department of Agriculture, Animal and Plant Health Inspection Service, Technical Advisory Group (USDA, APHIS-TAG), Hyattsville, Maryland, plus the relevant state Departments of Agriculture, and, sometimes, state universities, e.g., University of California, perform these reviews

and recommend whether to provide the release clearances. *(5) Importation and release.* Organisms approved for release are collected from the same field populations from which the test material was obtained and are imported into a domestic quarantine facility. Species that are univoltine or difficult to culture are identified and examined to ensure that they are free from parasites and entomogenous pathogens and may then be released directly into the field. Species amenable to culture are reared for one generation before their release. The quarantine handling of individual candidate specimens is labor intensive and often limits the number of biological control agents released. Release sites are selected on the basis of climate, habitat, freedom from disturbance, and other factors that optimize establishment. *(6) Evaluation.* For biological control to be effective, the imported agents must establish, increase in number, and inflict meaningful damage at a time critical to the weed's development. The released agents are evaluated against each of these points. Detailed evaluations are important to refine and improve methods and guidelines, test ecological theory, and document results of projects on the biological control of weeds.

Weeds may have dozens of associated natural enemies. Some species attack only the flowers and fruit, whereas others may attack the leaves, stems, branches, crowns, or roots. The guild of agents attacking the flower heads and achenes of an asteraceous thistle, for example, may include monophagous to oligophagous species which vary in their impact on the host (Zwölfer, 1973, 1988). Prerelease studies help determine which species are sufficiently host specific for biological control purposes and suggest the best sequence for release. An agent's host-finding capability and its competitiveness with other flower head-infesting species can be important in this regard. The selection of natural enemies and determination of their order of release have received some attention but need further study (Zwölfer & Harris, 1971; Harris, 1973, 1991; Goeden, 1983).

Conservation Methods in Biological Control

Potentially effective indigenous or exotic natural enemies may feed and reproduce on a weed host, yet fail to provide effective biological control. In such instances, the abundance of an associated agent may be effectively enhanced to provide localized reduction of the weed. For example, a low-dosage application of DDT was used selectively to eliminate *Exochomus* sp., a coccinellid beetle predator restricting the potential of an introduced cochineal insect, *Dactylopius opuntiae* (Cockerell) (Hemiptera-Homoptera: Dactylopiidae), to control the prickly pear cacti *Opuntia ficus-indica* (Linnaeus) Miller and *O. tardispina* Griffiths (Cactaceae) in South Africa (Annecke *et al.,* 1969; Moran & Zimmerman, 1984). Once the coccinellid predators were reduced, the cochineal insects increased in numbers and the cacti were controlled. This method of natural enemy enhancement is called biological control through conservation.

In contrast to naturalized weeds, many indigenous species have complements of natural enemies which themselves are held in check by parasites, predators, and other environmental factors. However, identifying the factors prompting these natural and often wide-ranging outbreaks, knowing how to manipulate them, and determining whether this manipulation is ecologically desirable, let alone economically feasible, so far remain unresolved, hypothetical, and untested (Wapshere *et al.,* 1989).

Implementing biological control through conservation of natural enemies can be complicated. It requires a thorough understanding of the weed–natural enemy–environment relationship, plus the technical ability to manipulate aspects of this relationship to increase agent impact. Conservation methods also can be labor intensive and costly, which limits their usefulness to weed problems that promise high economic return.

Augmentation Methods in Biological Control

The abundance of a natural enemy and the timing of its impact may be altered by supplemental releases. Biological control through the conservation of natural enemies relies on enhancing the buildup of populations already present; augmentation utilizes large-scale releases or applications of natural enemies collected or reared elsewhere. The augmentation approach to biological control of weeds was not widely practiced until the discovery by Daniel *et al.* (1973) that the endemic fungal pathogen *Colletotrichum gloeosporioides* (Penz.) Sacc. (Melanconiales) could be cultured and sprayed on northern jointvetch, *Aeschynomene virginica* (Linnaeus) B. S. P. (Leguminosae), a weed of rice in Arkansas and nearby states. The sprays augmented the early seasonal, naturally occurring, low levels of the fungus and provided excellent biological control. This study sparked the search for indigenous pathogens associated with other weeds and assessments of their production and implementation potential (Templeton *et al.,* 1978; Charudattan & Walker, 1982; Wapshere *et al.,* 1989; also see later).

Frick and Chandler (1978) augmented the impact of the moth *Bactra verutana* Zeller (Lepidoptera: Tortricidae) by supplementing existing field populations with large numbers of this insect against purple nutsedge, *Cyperus rotundus* Linnaeus (Cyperaceae), infesting cotton. They demonstrated experimental control of the weed, but the method proved too costly for practical implementation.

Advantages and Limitations of Biological Control

Any person who has spent time hoeing or pulling weeds can appreciate the advantages and limitations of these time-tested methods. Unfortunately, shortcomings are inherent in all methods of weed control, making each method appropriate under one or more set of circumstances but inappropriate under others. Advantages of the biological method of weed control include the following: (1) the introduced agents can perpetuate and distribute themselves throughout the weed's range; (2) the impact of host-specific agents is focused on a single weed species without harm to other plants; (3) the cost of developing biological control is relatively inexpensive ($1.5 million) compared to much higher costs for other approaches (Andres, 1977; Harris, 1979); (4) the agents are nonpollutive, energy efficient, and biodegradable; and (5) the knowledge generated during prerelease and evaluation studies contributes to improved understanding of weed ecosystems and environmental factors regulating natural communities.

Some of the shortcomings of biological control are that: (1) once established in an area, an introduced agent cannot be recalled or limited to parts of the target weed's range; (2) a host-specific agent will control only one species in a weed species complex; (3) agent impact is often slow and may require 3–4 years before local control is attained; (4) an agent may feed and reproduce on closely related nontarget plants; and (5) the establishment, buildup, and impact of a biological control agent is determined by the quality of the environment and the host and cannot be predicted. Fortunately, by combining the several methods of biological control with appropriate timing, many of these shortcomings can be minimized (Wapshere *et al.,* 1989).

HISTORY OF BIOLOGICAL CONTROL OF WEEDS

Much can be learned from analyses of case histories of projects on the biological control of weeds such as those reviewed in the section starting on page 880. Not apparent from these abbreviated case histories are the pulse and temper of the historical periods during which projects were initiated, the ebb and flow of individual and collective research efforts, the diverse personalities and research approaches involved, and the trumpeted successes and reluctant failures, all of which constitute a rich history for the biological control of weeds. In its rite of passage from an art to an accepted subdiscipline of applied ecology, biological control of weeds has grown at different rates and in different directions in various areas of the world.

We provide a brief overview of biological control of weeds, one of three subdisciplines that, until recent times, comprised the field of biological control (DeBach, 1964). This account is largely taken from Goeden (1978, 1988, 1993). For most of its history, biological control of weeds was the domain of a rather small, dedicated group of broadly versed entomologists (DeBach, 1964), and, like certain other aspects of science, it began by accident. The earliest record of the biological control of a weed involved the intentional introduction of the cochineal insect *Dactylopius ceylonicus* (Green) (Hemiptera-Homoptera: Dactylopiidae) into northern India from Brazil in 1795 in the mistaken belief that it was *D. coccus* Costa, a species cultured commercially as a source of carmine dye. Instead of reproducing well on the cultivated, spineless prickly pear cactus, *Opuntia ficusindica* (Linnaeus) Miller, *D. ceylonicus* readily transferred to its natural host plant, *O. vulgaris* Miller, which had become a widespread, introduced weed in India when it escaped from cultivation in the absence of its South American natural enemies. Once the value of *D. ceylonicus* as a biological control agent was recognized, it was introduced from 1836 to 1838 into southern India, where it brought about the first successful, intentional use of an insect to control a weed. Shortly before 1865, *D. ceylonicus* also was transferred from India to Sri Lanka (then Ceylon), which again resulted in the successful control of *O. vulgaris* throughout that island and, accordingly, registered the first intentional transfer of a natural enemy between countries for the biological control of a weed (Goeden, 1978; Moran & Zimmerman, 1984).

The historical scene next shifted to Hawaii, where Albert Koebele was hired as foreign explorer after helping to achieve the spectacularly successful biological control of the cottony cushion scale on citrus in California during the 1890s (Doutt, 1958, 1964). He explored the jungles of southcentral Mexico during 1902 for insects feeding on lantana (*Lantana camara* Linnaeus, Verbenaceae). Lantana was an ornamental plant of Central and South American origins that had escaped from cultivation to become a serious weed in Hawaii. Koebele shipped to Hawaii 23 species of insects that he considered important natural enemies of this plant in the parts of Mexico he had explored (Goeden, 1978). The published letters of Koebele describing his labors and hardships during this foreign exploration make adventurous reading (Perkins & Swezey, 1924). Extreme temperatures, unscheduled shipping delays, pathogens, and other contaminants took their toll of shipments. Upon their arrival by ship in 1903, the insects were liberated directly on lantana plants in the field without host-specificity testing. Eight species, including some of the most effective natural enemies of lantana, reportedly were established on this weed throughout the island group by 1905 (Andres & Goeden, 1971; Goeden, 1978; Julien, 1982).

The procedure of foreign exploration for natural enemies of an alien weed in its country or countries of origin was pioneered in this early project on lantana. Also, the lantana seed fly, *Ophiomyia lantanae* (Froggatt) (Diptera: Agromyzidae), was transferred from Hawaii to New Caledonia in 1908–1909 and to Fiji in 1911 (Rao *et al.,* 1971). These shipments marked the beginnings of a tradition of "transfer projects" (DeBach, 1964), involving transfers of biological control agents of proven worth to other countries with the same weeds. For example, three more species of lantana insects were transferred from Hawaii to Fiji during 1922–1928 (Rao *et al.,* 1971). This tradition of international cooperation in natural enemy transfers continues today.

The introduction of *D. ceylonicus* to Australia from Ceylon and India was attempted again in 1903, but without establishment (Goeden, 1978; Moran & Zimmerman, 1984). Australian work on the biological control of prickly pear cacti (*Opuntia* spp.) began in earnest during the period 1913–1914, when the two-membered Prickly Pear Travelling Commission surveyed the insects and pathogens associated with these plants in Java, Ceylon, India, East Africa, South Africa, the Canary Islands, the Mediterranean littoral, the United States, Mexico and other parts of Central America, the West Indies, South America, and Hawaii (Johnston & Tryon, 1914). This feat of worldwide exploration for natural enemies of a group of weeds remains unequaled in scope of geographic coverage, though often since surpassed regionally in depth.

The control of the prickly pear cacti *Opuntia inermis* deCandolle and *O. stricta* Haworth in Australia ranks as one of the most successful projects in biological control of weeds. This project followed the initial efforts of the Prickly Pear Travelling Commission, which first recognized the potential value of what was later to become the principal natural enemy, the cladode-mining moth, *Cactoblastis cactorum* (Berg) (Lepidoptera: Pyralidae). The major entomological efforts in this biological control program occurred during the 1920s, when North and South America, particularly the southern United States, Mexico, and Argentina, were thoroughly explored for potentially useful, cactus-feeding insects. More than 150 species of cactus insects eventually were collected and studied, many of which were new to science. From 1921 to 1925, 48 species were imported into Australia, of which 19 were liberated and 11 became established.

A single consignment of *C. cactorum* was imported from Argentina in 1925. Large-scale mass rearings and host-plant specificity testing with useful and weedy plant species were undertaken for the first time in a biological control program. *Cactoblastis cactorum* became widely established following the distribution of more than 2.7 billion mass-cultured and field-collected eggs between 1925 and 1933. About 90% of the original stands of *O. inermis* and *O. stricta* were destroyed by 1934 through larval feeding by this moth, supplemented by airborne, soft-rot bacteria for which the borers provided portals of entry into infested plants. Virtually complete control of these weeds was achieved in Queensland and northern New South Wales, involving 24 million hectares of formerly infested land that was restored to agricultural use (Dodd, 1940; Goeden, 1978; Moran & Zimmerman, 1984).

The spectacular success of *Cactoblastis cactorum* tended

to eclipse the benefits derived from other cactus insects used in biological control, notably several species of cochineal insects (Moran & Zimmerman, 1984). For example, *Dactylopius ceylonicus* was successfully reintroduced into Australia during 1913–1915 and it virtually eliminated *O. vulgaris* as a rangeland weed in Queensland. "Nothing succeeds like success," and so *C. cactorum* and *Dactylopius* spp. were transferred during the mid-1920s and 1930s to countries where prickly pear cacti also were introduced weeds: Indonesia, Mauritius, New Caledonia, Reunion, and South Africa (Greathead, 1971; Rao *et al.*, 1971; Goeden, 1978). In South Africa and Mauritius, these early, successful transfer projects led to the independent development of other successful research projects in the biological control of weeds (Greathead *et al.*, 1971; Goeden, 1978; Julien, 1982, 1987, 1992).

The biological control of Koster's curse, *Clidemia hirta* (Linnaeus) D. Don (Melastomataceae), was the next noteworthy success. This is a shrub native to the West Indies and tropical South and Central America that became a serious introduced weed on Fiji. After a preliminary survey of insects attacking *C. hirta* and allied Melastomataceae, *Liothrips urichi* Karny (Thysanoptera: Thripiidae) was selected as a promising biological control agent and its life history and host-plant specificities were intensively studied in Trinidad during 1927 and 1928. Potted *C. hirta* infested with the thrips were shipped under cold storage to Fiji in 1930. Upon arrival, the thrips were transferred directly to plants in the field. Field releases continued throughout 1930. By 1932–1933, several hundred hectares of thrips-stunted *C. hirta* had been overgrown by plant competitors of greater forage value. Shaded and greatly weakened by thrips attack, these weeds were soon defoliated and killed. Regrowth was readily located and attacked by *L. urichi*. By 1937, the competitive ability of the weeds was permanently impaired by continued thrips attack except in a few shaded and wet areas, and successful biological control was achieved (Simmonds, 1937; Rao *et al.*, 1971; Goeden, 1978; Julien, 1982).

During the 1920s, New Zealand joined the list of Commonwealth countries sponsoring original research on biological control of weeds. Studies of insects attacking gorse, *Ulex europaeus* Linnaeus (Leguminosae), were initiated in England in 1926 and the introduction and successful colonization of the seed weevil, *Apion ulicis* Forster (Coleoptera: Apionidae), was accomplished during 1929–1931. Surveys of insects attacking blackberries (*Rubus* spp., Rosaceae) in Europe and North America were conducted during the mid-1920s, but no species was thought safe enough for introduction and the project was abandoned. Beginning in 1927 and continuing into 1930, diapausing pupae of the cinnabar moth, *Tyria jacobaeae* (Linnaeus) (Lepidoptera: Arctiidae), a defoliator of tansy ragwort, *Senecio jacobaea* Linnaeus (Asteraceae), were introduced into New Zealand from England, but establishment was not achieved. Australia also received *T. jacobaeae* from New Zealand during 1929–1932, beginning a series of colonizations continued in the 1950s and 1960s. Efforts to establish this moth were thwarted by strong predation by native insects, mainly scorpion flies. All attempts to establish this moth were unsuccessful (Goeden, 1978; Julien, 1982, 1992). It was not until the USDA study and introduction in 1969 of the chrysomelid beetle, *Longitarsus jacobaeae* (Waterhouse), for the control of tansy ragwort in northern California and Oregon that efforts against this weed became successful (Frick, 1970; Hawkes & Johnson, 1978; McEvoy *et al.*, 1991). In fact, the high degree of biological control attained rates it among the major successes (Julien, 1992).

The success of the prickly pear program led Australia to undertake biological control of other widespread, introduced, rangeland weeds, including cocklebur (*Xanthium strumarium* Linnaeus, Asteraceae). Foreign exploration for Australia began in the United States during 1929. St. Johnswort (*Hypericum perforatum* Linnaeus, Hypericaceae) was another pest for which foreign exploration for Australia began in England during 1926. The former project yielded only a limited measure of biological control. The latter project produced varying results with different introduced insects in different situations, analyzed in part by careful, quantitative, ecological research (N. Clark, 1953; L. R. Clark, 1953). This type of research has come to characterize contemporary Australian projects in the biological control of weeds. The insect agents and technology transferred from the *Hypericum* project contributed greatly to the development of biological control of weeds in North America. Before World War II, groundwork was laid for the rapid expansion of biological control of weeds that occurred during the 1950s, 1960s, and 1970s. Professor H. S. Smith [who first coined the term "biological control" (DeBach, 1964)] initiated the first project in the biological control of weeds in the continental United States. This involved the introduction of insects to control the native prickly pear cacti, *Opuntia littoralis* (Engelmann) Cockerell and *O. oricola* Philbrick, and their hybrids on rangeland on Santa Cruz Island off the coast of southern California. Several insect species from the California mainland and Texas were introduced to the island beginning in 1940, but successful biological control was attained only after *Dactylopius opuntiae*, native to Mexico and the southern California coast, was introduced to the island in 1951 from Hawaii ex Australia ex Mexico (Goeden *et al.*, 1967; Goeden, 1978; Goeden & Ricker, 1981)!

During World War II, activities on the biological control of weeds were reduced to a few transfer projects. For example, the leaf beetle *Chrysolina hyperici* (Förster) (Coleoptera: Chrysomelidae) was transferred from Australia to New Zealand in 1943 for the biological control of St. Johnswort. In 1944, J. K. Holloway of the USDA and

H. S. Smith arranged the introduction of several insect species from Australia for specificity testing in California and release during 1945–1946 for the biological control of St. Johnswort, also known as Klamath weed. The successful biological control of Klamath weed, primarily caused by the defoliating leaf beetle *Chrysolina quadrigemina* (Suffrian) (Coleoptera: Chrysomelidae), rivaled the Australian success with prickly pear cacti. This success primarily was responsible for fostering the establishment and expansion of the subdiscipline of biological control of weeds in North America (Huffaker, 1957). Within a decade after the liberation of *C. hyperici* and *C. quadrigemina,* the Klamath weed had been reduced in status from an extremely important rangeland weed to that of an occasional roadside plant. It now occurs at less than 1% of its former density and has been removed from the list of noxious weeds in California (Holloway & Huffaker, 1949, 1951). In 1950, Canada made its first intentional introduction of an insect for weed control when it imported *Chrysolina quadrigemina* and *C. hyperici* from California for control of St. Johnswort (Smith, 1951).

In 1945, Hawaii resumed an earlier interest in the biological control of *Eupatorium adenophorum* Sprengel (Asteraceae) and introduced the stem-gall-forming fly, *Procecidochares utilis* Stone (Diptera: Tephritidae), presumably one of those natural enemies recommended for introduction to Hawaii by Koebele 20 years earlier. This successful introduction was followed by a series of projects undertaken by the Entomology Division of the Hawaii Department of Agriculture, which made Hawaii a center of activity in biological control of weeds during the 1950s and 1960s. Among the weeds on which foreign exploration was initiated and insects were introduced and established for biological control during the 1950s were the Christmas berry (*Schinus terebinthifolius* Raddi; Anacardiaceae), elephant's foot (*Elephantopus mollis* Humboldt, Bonplaud & Kuth; Anacardiaceae), sourbush (*Pluchea odorata* (Linnaeus) Cassini; Asteraceae), melastoma (*Melastoma malabathricum* Linnaeus; Melastomataceae), firebush (*Myrica faya* Aiton; Myricaceae), and emex (*Emex australis* Steinheil and *E. spinosa* Campdera; Polygonaceae). Substantial to complete biological control of emex reportedly was achieved at 600- to 1200-m elevations with the weevil *Apion antiquum* (Gyllenhal) introduced from South Africa in 1957 (Davis, 1966). Otherwise, these six projects and several other contemporary transfer projects on biological control of weeds in Hawaii unfortunately are characterized by a paucity of published results (Goeden, 1978; Julien, 1982, 1987, 1992).

Another successful project was initiated at this time on black sage, *Cordia macrostachya* (Jacquin) Roemer & Schultes (Boraginaceae), an introduced weed in sugarcane fields on the island of Mauritius. A preliminary survey of the insect fauna of black sage and related plant species was

conducted in the West Indies during 1944–1946. Following detailed life history studies and host-specificity tests, the leaf beetle *Metrogaleruca obscura* DeGeer (Coleoptera: Chrysomelidae) was introduced to Mauritius from Trinidad in 1947 (Simmonds, 1950). The beetle populations multiplied explosively and by 1950 had spread islandwide, causing heavy defoliation which killed plants or weakened them so they were replaced by competing plant species. A seed-feeding wasp, *Eurytoma attiva* Burks (Hymenoptera: Eurytomidae), also was introduced in 1949 and 1950 (Williams, 1960). Defoliation and seed destruction by these introduced insects have continued to prevent the regeneration of black sage (Simmonds, 1967; Goeden, 1978; Julien, 1992). This also was one of the first projects that involved entomologists affiliated with what eventually became the CAB International Institute of Biological Control (IIBC), a major worldwide, contract research organization now headquartered at Silwood Park, U.K.

Most weeds targeted for biological control up to this time were introduced, perennial weeds of relatively undisturbed rangeland. However, beginning in Australia, Canada, and the United States in the late 1950s and early 1960s, projects were initiated on aquatic and semiaquatic weeds, annuals and biennials, and cropland and ruderal weeds. For example, the first aquatic weed targeted for biological control with insects was alligatorweed, *Alternanthera phylloxeroides* (Martius) Grisebach (Amaranthaceae), a pest in the southeastern United States. The first annual weed targeted for biological control with insects in North America was puncturevine, *Tribulus terrestris* Linnaeus (Zygophyllaceae). Both projects are reviewed in Section IV.

The classic CIBC study by Zwölfer (1965) of the insect fauna of Canada thistle (*Cirsium arvense* [Linnaeus] Scopoli) and other wild Cynareae (Asteraceae) in Europe sponsored by the (then) Canada Department of Agriculture during the early 1960s formed the foundation for a number of biological control projects targeted on introduced thistles in Canada and the United States (Schröder, 1980). These included annual, biennial, and perennial species of *Carduus, Centaurea, Cirsium,* and *Silybum* (Goeden *et al.,* 1974a). These projects continue with mixed results. Especially noteworthy is the successful biological control of musk thistle, *Carduus nutans* Linnaeus, attained with the introduced seedhead weevil, *Rhinocyllus conicus* Linnaeus (Coleoptera: Curculionidae), in Canada (Harris, 1984a) and Virginia (Kok & Surles, 1975) and the crown-mining weevil, *Trichosirocalus horridus* (Panzer) (Coleoptera: Curculionidae) (Kok, 1986).

A biological control program on introduced spurges (*Euphorbia* ssp., Euphorbiaceae) of Eurasian origins was initiated by Canada in the early 1960s, with the first introduction of the hawk moth, *Hyles euphorbiae* Linnaeus (Lepidoptera: Sphingidae), from Germany in 1965 (Harris, 1984b). One lesson from biological control of spurges (also demonstrated with salvinia) is the value of correctly identifying and matching the targeted weed with related populations in its native geographical range to facilitate the collection and establishment of coevolved natural enemies. This demonstrates the value of plant taxonomy and insect taxonomy to the biological control of weeds.

The biological control of rush skeletonweed, *Chondrilla juncea* Linnaeus (Asteraceae), in Australia is another noteworthy milestone. This project, launched in 1971 with cooperation between Italy and Australia, was the first to involve the intentional international transfer of a phytopathogen, i.e., the rust fungus (Uredinales) *Puccinia chondrillina* Bubak & Sydenham, for the successful biological control of a weed. This project also was one of the first to target a weed of cropland (nonirrigated wheatland). It established procedures for testing phytopathogens for host specificity under quarantine conditions and involved the first intentional importation and release in 1971 of a phytophagous mite [*Eriophyes chondrillae* (Canestrini) (Acarina: Eriophyidae)] for biological control (Cullen, 1974, 1978).

Other projects on biological control of weeds have been initiated, including more than 80 in the United States and Canada (Goeden *et al.*, 1974a; Julien, 1987, 1992). Several could be singled out as representing additional "firsts." Programs in classical biological control of weeds, i.e., those involving the importation of exotic natural enemies, have become scientifically exacting and quantitatively precise, but more and more enmeshed in governmental regulation (Harris, 1991; Goeden, 1993). In part this is due to the increasing number of workers in this subdiscipline (19 attendees at the first biological control of weeds symposium in Switzerland in 1968 versus 134 attendees at the sixth symposium in Canada in 1984) and in part to records of several early-introduced natural enemies from nontargeted, indigenous plant species (Turner, 1984). Conflicts of interest between ranchers and beekeepers in Australia over the introduction of biological control agents for the control of *Echium plantagineum* Linnaeus (Boraginaceae) (called Paterson's curse or salvation Jane, respectively, by these contending parties) caused the enactment of the Biological Control Act of 1984. This Act details how weeds targeted for biological control are to be open to public review before the release of biological control agents, among other comprehensive regulations (Cullen & Delfosse, 1985).

SCOPE OF BIOLOGICAL CONTROL OF WEEDS

Target Weeds

Julien (1982) and Julien *et al.* (1984), list 101 species of weeds that have been targeted for biological control. Thirty-three plant families are represented among these 101 weed species, 25 of which belong to the Asteraceae and 19 of

which are Cactaceae; the other 31 families are represented by five or fewer species (Julien *et al.*, 1984). Historically, biological control with introduced natural enemies has primarily been directed against perennial, terrestrial weeds of pastures and rangeland.

Pemberton (1981) noted that only four (22%) of what Holm *et al.* (1977) considered to be 18 of the world's worst weed species have been targeted for biological control, i.e., *Convolvulus arvensis* Linnaeus (Convolvulaceae), *Cyperus esculentus* Linnaeus and *C. rotundus* (Cyperaceae), and *Eichhornia crassipes* (Martius) Solms-Laubach (Pontederiaceae). The successful biological control of the aquatic weed water hyacinth (*E. crassipes*) is reviewed here; projects on the other three species have been unsuccessful (Julien, 1982). Pemberton (1981) also noted "that biological control of weeds has not yet had a single project against a weedy grass . . . ," this despite grasses (Graminaceae) comprising 10 of the world's 18 worst weeds according to Holm *et al.* (1977). Two grasses, *Digitaria sanguinalis* (Linnaeus) Scopoli and *Panicum dichotomiflorum* Michaux, were included among the ca. 80 species or species groups of weeds listed by Goeden *et al.* (1974a) as targeted for biological control in the United States and Canada. Nothing resulted from these two projects, however, because as Schroeder and Goeden (1986) remarked, one of the first and easiest steps in any project on the biological control of weeds is to designate a weed as a target species.

Weedy grasses traditionally have not been considered suitable for biological control because many are close relatives of important cultivars. The chances of finding arthropod natural enemies able to discriminate among such closely related, potential host plants were remote. However, phytopathogens offer promise for control of graminaceous weeds, as some are extremely host specific, e.g., rust fungi (Section 3). Weeds not amenable to biological control include (1) weeds of highly disturbed habitats, e.g., cropland, subject to host-plant removal by tillage and to pesticide applications, (2) submersed aquatic weeds, which apparently have few host-specific natural enemies, (3) highly toxic weeds for which tolerable densities are too low to be obtained by use of natural enemies, (4) minor weeds of limited distribution that do not threaten to invade other areas, and (5) weeds whose eradication is sought, because the goal of biological control is only the reduction of a pest's densities to tolerable levels, never its eradication (Harris, 1971; Frick, 1974; Goeden, 1977).

Most weeds successfully controlled with introduced natural enemies were introduced plant species (Julien, 1982; Julien *et al.*, 1984). Only four species of native weeds have been successfully controlled with intentionally introduced organisms: *Opuntia dillenii* (Ker-Gawler) Haworth (Cactaceae) on the island of Nevis in the West Indies (Simmonds & Bennett, 1966); *O. littoralis* and *O. oricola* on Santa Cruz Island off southern California (Goeden *et al.*, 1967;

Goeden & Ricker, 1981); and *O. triacantha* (Willdenow) Sweet on the islands of Antigua, Monserrat, and, again, Nevis in the West Indies (Simmonds & Bennett, 1966; Bennett, 1971). All four of these native weeds are prickly pear cacti (subgenus *Platyopuntia*), which along with other Cactaceae, as Moran and Zimmerman (1984) noted, are unusual among terrestrial weeds, e.g., as regards their insect relations. Before their biological control began, cacti were among the world's most economically important and troublesome pasture and rangeland weeds.

Insect Agents Used

Unless otherwise indicated, the following characterizations were taken from Julien *et al.* (1984) (see Chapter 17). Julien (1982) lists 174 biological control projects directed against the 101 weed species discussed earlier. Of these 174 projects, 151 (87%) used exotic organisms introduced against 82 weed species, and 23 (13%) used native organisms against 26 weed species.

One hundred seventy-one species of insects from 7 orders and 38 families comprised 98% of all releases of natural enemies and 96% of all species of natural enemies released for biological control of these 101 weeds. Most released species were Coleoptera, Lepidoptera, Diptera, and Hemiptera (Homoptera and Heteroptera) in this order of decreasing use. Few species of Orthoptera, Thysanoptera, and Hymenoptera have found use as agents for biological control of weeds.

Among the 69 colonized species of Coleoptera, 60 colonized species of Lepidoptera, 20 colonized species of Diptera, and 16 colonized species of Hemiptera, 65, 55, 70, and 66% became established and 29, 20, 19, and 44% proved effective as agents for biological control of weeds, respectively. The 12 families of insects that contained the most species released for biological control of weeds, in decreasing order of magnitude, were Chrysomelidae (Coleoptera), Curculionidae (Coleoptera), Pyralidae (Lepidoptera), Dactylopiidae (Hemiptera-Homoptera), Tingidae (Hemiptera-Heteroptera), Tephritidae (Diptera), Cerambycidae (Coleoptera), Noctuidae (Lepidoptera), Apionidae (Coleoptera), Agromyzidae (Diptera), Gelechiidae (Lepidoptera), and Tortricidae (Lepidoptera).

Moran and Zimmerman (1984) reported that 63 species of cactophagous insects were introduced worldwide for biological control of 22 species of Cactaceae. Nineteen (30%) of those species were successfully established. Australian entomologists introduced 54 species of insects for cactus control; South African entomologists introduced 24 species of insects for cactus control.

Phytopathogenic Agents Used

When the status of projects on the biological control of weeds in the United States and Canada was assessed for the

Weed Science Society of America by Goeden *et al.* (1974a), very few weed pathogens were in use as biological control agents. However, the decade of the 1970s saw a flurry of research efforts to use phytopathogens, especially fungi, for biological control of aquatic and terrestrial weeds (Charudattan, 1978; Freeman & Charudattan, 1981; Charudattan & Walker, 1982).

Julien (1982) listed four phytopathogens as imported for biological control of weeds worldwide. Two of these pathogens were accidental introductions; however, the other two, both rust fungi, provide excellent examples of successful control of alien terrestrial weeds with intentionally introduced natural enemies. The historically significant introduction of *Puccinia chondrillina* into Australia in 1971 for the biological control of rush skeletonweed has already been noted. The high degree of host specificity exhibited by *P. chondrillina* prevented the direct transfer of the Australian material to control the two forms of rush skeletonweed found in the western United States. Surveys in Europe located a strain of *P. chondrillina* that attacked the predominant form of the weed in the United States (Emge *et al.*, 1981).

The second example was the successful biological control of the weedy blackberries *Rubus constrictus* Lefevre & Mueller and *R. ulmifolius* Schott (Rosaceae) with *Phragmidium violaceum* (Schultz) Winter introduced from Germany to Chile in 1973 (Oehrens, 1977). A more recently reported example of the successful use of an intentionally introduced pathogen in the biological control of Hamakua pamakani, *Aegeratina riparia* (Regel) King & Robinson (Asteraceae), was achieved by the *Cercosporella* sp. (Uredinales) imported from Mexico to Hawaii and released in 1975 (Trujillo, 1985).

Charudattan (1984) discussed plant pathogens employed as "microbial herbicides." This strategy employs alien or native pathogens mass-cultured and applied as inundative inocula on target weeds. Successful examples involve (1) *Colletotrichum gloeosporioides* f. sp. *aeschynomene*, registered and sold as Collego® for control of northern jointvetch, *Aeschynomene virginica*, on rice and soybeans in Arkansas (Templeton *et al.*, 1978); and (2) *Phytophthora citrophthora* (R. E. & E. H. Smith) Leonian (Peronosporales), registered and sold as DeVine® for control of milkweed vine, *Morrenia odorata* (Hook. and Arn.) Lindle, in citrus in Florida (Ridings *et al.*, 1978).

Other Agents Used

Two species of mites (Acarina) have been successful as agents for biological control of weeds. *Tetranychus opuntiae* Banks (Tetranychidae) was an accidental introduction on prickly pear cacti in Australia. The association resembles many examples of accidentally introduced phytophagous insects reported to feed on alien and native plants. *Eriophyes chondrillae* (see p. 877) was the first mite species intentionally transferred between continents for biological control of weeds (Cullen, 1974, 1978). As with the rust fungus, workers returned to Europe to obtain a biotype of *E. chondrillae* that would attack rush skeletonweed in the western United States; one was found in Italy by Sobhian and Andres (1978). Another eriophyid mite, *E. boycei* Keifer (Eriophyidae), was exported from California to the USSR in 1971 and 1972 for the biological control of ragweeds (*Ambrosia* spp., Asteraceae), but it was not released (Goeden *et al.,* 1974b). Cromroy (1982) cited additional examples of native and introduced mites attacking weeds in an attempt to demonstrate their efficacy as biological control agents. Again, few substantive examples of accidental biological control are known.

Nematodes are well known as plant pests, but few species are used as agents for biological control of weeds worldwide. Only *Paranguina picridis* Kirjanova & Ivanova (Nematoda: Tylenchidae) was cited as an introduced agent by Julien (1982) and Julien *et al.* (1984). In 1976, it was released in restricted field trials in Quebec and Saskatchewan, following its introduction from the USSR for the biological control of Russian knapweed, *Centaurea repens* Linnaeus (Asteraceae). *Paranguina picridis* was successfully moved from central Asia to the Crimea and reportedly yielded good control of Russian knapweed (Kovalev, 1973). Experimental field use was made of *Nothanguina phyllobia* Thorne (Nematoda: Neotylenchidae) by augmenting its naturally occurring populations with large numbers of infectious larvae to control silverleaf nightshade, *Solanum elaeagnifolium* Cavanilles (Solanaceae), its native weedy host plant in Texas (Orr, 1980). The introduction of *N. phyllobia* into Australia and South Africa, where silverleaf nightshade is an alien weed, has been considered.

Other invertebrate natural enemies have limited use as nonselective grazers in biological control of aquatic weeds. These include crayfish, snails, and tadpole shrimp (Andres & Bennett, 1975; Takahashi, 1977). A vertebrate herbivore, the grass carp, *Ctenopharyngodon idella* (Curvier & Valenciennes) (Pisces: Cyprinidae), has yielded mixed results in different countries when introduced against mixes of aquatic weeds (Julien, 1982, 1992). Seven other fish species and the manatee complete the list of vertebrates that have limited use in aquatic weed control (Andres & Bennett, 1975; Julien, 1982). Geese, sheep, and goats have long been used as managed grazers of terrestrial weeds (King, 1966).

Countries

More than 70 countries were involved in 499 releases of introduced natural enemies for the biological control of weeds (Julien, 1982; Julien *et al.,* 1984). The principal countries involved through 1970 were mentioned earlier in this chapter. Julien *et al.* (1984) listed world regions by the number of direct and indirect (transfer projects) releases of natural enemies in the following descending order of activity: Australia, with 108 releases; North America, with 80; Hawaii, with 71; all other Pacific islands, with 69; Africa exclusive of the Republic of South Africa, with 48; South Africa, with 34; the Caribbean islands, with 32; the Indian Ocean islands, with 23; Asia, with 20; South America, with 7; the South China Sea islands, with 6; and Europe, with 1 release. Europe has long served as a source of natural enemies for biological control of weeds as well as a principal source of the target weeds themselves (Julien, 1982), reflecting patterns of world commerce and human immigration (Schroeder & Goeden, 1986).

Pemberton (1981) developed a separate system of ranking world regions by the number of projects that resulted in releases of natural enemies. From greatest to least active, these regions were North America, the Pacific islands (including Hawaii), Africa, Asia, Australia, Europe, the Caribbean islands, South America, Central America, and the Middle East. In terms of countries targeting the most weeds against which natural enemies were released, Africa and Australia exchanged rankings, while the other countries retained their order. Perhaps it is only a coincidence that the first set of rankings was compiled by an Australian, and the second system by a North American. Or, indeed, it may reflect how projects are composed and priorities assigned by research administrators in the several countries. The focusing of research personnel and resources on fewer weeds could result in the importation of more agents per weed.

Agencies

Many of the agencies involved in the biological control of weeds were mentioned previously. Julien *et al.* (1984) ranked research organizations involved in biological control of weeds by the total numbers of releases of natural enemies (repeated releases excluded). Ranked accordingly, in descending order, along with percentages judged effective (established and successful) by Julien *et al.* (1984) in parentheses, these mainly nonprofit governmental or international agencies were as follows: the Commonwealth Institute of Biological Control (CIBC [now IIBC]), with 79 releases (11% effective); the Hawaii Department of Agriculture, with 69 (43%); the U.S. Department of Agriculture (USDA), with 41 (30%); the Commonwealth Scientific and Industrial Research Organization (CSIRO), Australia, with 40 (33%); the Commonwealth Prickly Pear Board, Australia, with 37 (24%); Agriculture Canada, with 35 (17%); the Department of Agriculture and Fisheries, South Africa, with 33 (35%); the Queensland Department of Lands, Australia, with 28 (15%); the Ministry of Agriculture and Fisheries, Fiji, with 16 (7%); the Department of Scientific and Industrial Research, New Zealand, with 12 (20%); plus unnamed miscellaneous agencies, with 80 (40%), and unknown agencies, with 34 (39%).

Most domestic research by Agriculture Canada on biological control of weeds with insects is conducted at the Regina, Saskatchewan, Research Station. Canadian universities and provincial agencies also actively collaborate with Agriculture Canada scientists on domestic phases of projects. Most overseas research for Agriculture Canada on candidate agents is conducted by IIBC entomologists in Europe. IIBC laboratories where research on biological control of weeds is conducted on a contract basis currently include those located at Rawalpindi, Pakistan, at Delémont, Switzerland, and on Trinidad in the West Indies. The CSIRO maintains a European laboratory at Montpellier, France (Schroeder & Goeden, 1986). The Queensland, Australia, Department of Primary Lands has maintained a North American Field Station at Temple, Texas, in recent years.

The United States effort on biological control of weeds through the introduction of new natural enemies centered around the USDA laboratory at Albany, California, from the late 1950s to 1988. With the release of control agents from the Albany quarantine throughout the United States, other USDA personnel or laboratories were assigned biological control responsibilities at Fort Lauderdale and Gainesville, Florida; Beltsville and Frederick, Maryland; Stoneville, Mississippi; Columbia, Missouri; Bozeman, Montana; Lincoln, Nebraska; and Lubbock and Temple, Texas. A European laboratory for research on biological control of weeds was located at Rome, Italy, in 1959, which was later moved to Montpellier, France, in the mid-1980s, with substations at Rome and Thessaloniki, Greece. Other USDA laboratories for weed research are located at Hurlingham, Argentina (Coulson, 1985), and Townsville, Queensland, Australia.

State institutions in the United States that figured in the miscellaneous category of Julien et al. (1984) noted earlier and where at least one project on biological control of weeds was active during 1967–1987 were the University of Arkansas; the University of California, Berkeley, Davis, and Riverside; the California Department of Food and Agriculture, Sacramento; the University of Florida, Gainesville; the University of Idaho; Kansas State University, Manhattan; Montana State University; North Dakota State University, Fargo; Oregon State University; the Oregon State Department of Agriculture, Salem; Virginia Polytechnic Institute and State University; Washington State University; and the University of Wyoming, Laramie.

CASE HISTORIES OF RECENT SUCCESSES IN NORTH AMERICA AND OTHER PARTS OF THE WORLD

Alligatorweed, *Alternanthera phylloxeroides,* in the Southeastern United States and Australia

Alligatorweed is an emersed, perennial, aquatic plant from South America whose hollow, segmented stems allow it to form dense floating mats on the surface of rivers and other bodies of water. The floating mats block navigation, inhibit water use, and limit water flow. The rooted, segmented stems often break and allow the mats to float freely, spread, and root at new sites. The ability of the stems to root freely at the nodes renders mechanical removal of the mats ineffective, as the remaining fragments merely compound the problem. Stem segmentation also limits herbicide translocation and effectiveness. In fact, attempts to control alligatorweed with herbicides have often worsened the problem by killing neighboring plants and allowing the alligatorweed to flourish (Maddox et al., 1971; Coulson, 1977).

In view of the difficulties encountered in controlling alligatorweed, and as part of their expanded aquatic weed control program, the U.S. Army Corps of Engineers sought the assistance of the USDA, Agricultural Research Service (ARS), in assessing the potential for controlling this pest by biological means. In 1960, G. B. Vogt traveled to Argentina and adjacent countries to the north in search of phytophagous arthropods and plant pathogens that could be used to control this weed. He recorded over 40 species of natural enemies attacking alligatorweed, 3 of which he singled out as particularly important: *Agasicles hygrophila* Selman & Vogt (Coleoptera: Chrysomelidae), *Amynothrips andersoni* O'Neill (Thysanoptera: Phlaeothripidae), and *Vogtia malloi* Pastrana (Lepidoptera: Phyctinae). In 1962, the USDA opened a laboratory near Buenos Aires, Argentina, to study the biologies and host-plant specificities of these candidate biological control agents (Coulson, 1977).

Adults of the alligatorweed fleabeetle, *Agasicles hygrophila,* feed on the emersed leaves and stems of alligatorweed. The eggs are oviposited in clusters on the undersides of young leaves. The developing larvae feed on the leaves and stems; the third or last instar tunnels into the hollow stems to pupate. The imago later chews its way out through the stem wall and the life cycle begins again. Maddox (1968) reported as many as five generations per year in Argentina. Beetle feeding destroys both leaves and stems, and the latter become waterlogged after repeated perforations with adult emergence holes, causing the mats to sink.

The small (2.2 mm), black, short-winged or macropterous adults of *Amynothrips andersoni* feed among the bracts of the young buds or in the leaf axils. The larvae are reddish and complete their development in about 30 days. There are three to five generations annually. The thrips overwinters mainly as adults. Thrips feeding scars the leaf surface and stunts and distorts the stem growth (Maddox et al., 1971).

Vogtia malloi is nocturnal. Oviposition occurs on the terminal leaves. The neonate larvae tunnel into the stems. Larvae may exit periodically, reenter, and consequently damage a number of stems as they pass through five instars. Pupation occurs inside the hollow stem. There are three to five generations per year. *Vogtia* feeding causes extensive stem collapse and it develops well on both rooted and free-floating plants (O'Neill, 1968; Maddox *et al.,* 1971).

Alligatorweed fleabeetles were introduced into the United States from 1964 through 1970 from waterways near Buenos Aires (Coulson, 1977). The beetle is now established in Alabama, Florida, Georgia, Louisiana, Mississippi, South Carolina, and Texas but failed to establish in Arkansas, California, North Carolina, and Tennessee. Attempts were made to establish fleabeetles from a site south of Buenos Aires in hopes they would be better cold-adapted to the more northern alligatorweed infestations. This proved unsuccessful.

A laboratory colony of *V. malloi* established at Albany, California, from collections near Buenos Aires served as the source of North American colonies during 1971. A second laboratory colony originating from Neocochea, 156 km south of Buenos Aires, provided stock for releases through 1972 (Coulson, 1977). The moth is established in Arkansas, Florida, Louisiana, Mississippi, South Carolina, and Texas (Julien, 1987).

Agasicles gave good initial control in many coastal areas of the southeastern United States, but it has recently exhibited sensitivity to hot, dry summers and cold winters. Early-season supplemental releases of adult fleabeetles have enhanced its impact in inland areas. No other types of alligatorweed control are now needed in Florida, Louisiana, and Texas. Although sizable populations of *Amynothrips* are active early in the season, their impact is uncertain. *Vogtia malloi* has reduced mats by 70–80% in coastal areas of Mississippi, but control is uneven and appears cyclical over a period of several years (Julien, 1987).

Although insects had been used to control terrestrial weeds on numerous occasions, the introduction of *Agasicles* into the southeastern United States in 1964 marked the first use of an insect as an aquatic weed control agent. The relative success of this project reduced skepticism over whether insects associated with aquatic weeds were sufficiently monophagous to permit their use for biological control and focused increased attention on other species of emersed and floating weeds (Andres & Bennett, 1975). The initial buildup and control of alligatorweed on the St. Johns River, Florida, occurred within 15 months of initial release, while at other sites it took several years. The different rates of control may relate to the carbohydrate reserves in the stems composing the mats, the growth rate of the plant, the length of the growing season, and the percent of the mat renewed each season (Andres & Bennett, 1975; Coulson, 1977).

Fleabeetles from U.S. colonies also were transferred to Australia in 1977, New Zealand in 1982, and Thailand in 1982. The beetle is established in all three countries. A fourth species, *Disonycha argentinensis* Jacoby (Coleoptera: Chrysomelidae), was introduced into Australia in 1980 and New Zealand in 1982 directly from Brazil, but it failed to establish. *Agasicles* spread quickly through the infesta-

tions of alligatorweed in Australia and provided substantial control of the weed within 14 months. *Vogtia* impact on alligatorweed there is masked by *Agasicles*-caused plant injury. The moth is ineffective in terrestrial areas (Julien, 1987).

Puncturevine, *Tribulus terrestris,* in the Southwestern United States, Hawaii, and St. Kitts

This cosmopolitan weed of Old World origins is a prostrate, annual herb that bears an abundance of small yellow flowers and troublesome spiny fruit. These fruit separate at maturity into five, boney, one- to four-seeded segments, each of which is studded with two or four, sharp, rigid, divergent spines. The seeds apparently survive burial in soils for 20 years. The spines penetrate and lodge in tires, shoes, and the feet and fur of animals, which aids seed dissemination (Johnson, 1932; Goeden & Ricker, 1973).

The natural range of puncturevine includes the Mediterranean regions of Europe and Africa and drier parts of Asia (Andres & Angelet, 1963). It was accidentally introduced into the midwestern United States with livestock, especially sheep, imported from the Mediterranean area. Puncturevine now occurs from coast to coast, but is most common in the Southwest. It arrived in California about 1900, apparently as a railroad ballast contaminant, and spread rapidly along railroad and highway rights-of-way. As an agricultural weed, its spiny fruit interfere with hand harvesting, injure livestock, and contaminate seed, feed, and wool (Johnson, 1932). It also is a weed of disturbed residential and industrial land and, like crabgrass, is a weed that many city dwellers recognize. As an annual, nonwoody, nonrangeland weed, puncturevine represents a departure from the traditional perennial range and pasture weed targeted for biological control.

The natural enemies of puncturevine were surveyed in India, southern France, and Italy during 1957–1959. The seed-feeding weevil *Microlarinus lareynii* (Jacquelin duVal) (Coleoptera: Curculionidae) and the stem- and crown-mining weevil *M. lypriformis* (Wollaston) were selected as the most promising candidates for use as biological control agents. Field and laboratory studies conducted in France, Italy, and California during 1959–1961 demonstrated that the adults fed on a wide range of plant species, but reproduction was successful only on puncturevine, other species of *Tribulus,* and a few herbaceous annual Zygophyllaceae native to the southwestern United States (*Kallstroemia* sp.; Andres & Angelet, 1963). Although minor concern was expressed over its potential detrimental effect on the native plants, this conflict of interest was resolved by weighing the potential benefits of biological control of puncturevine against these potential losses. With the need for action recognized, both weevils were approved for release in compliance with less complicated federal regulatory procedures then in use. Since their release, the weevils have been recorded feeding on some nonhost plants but reproduce only on *Tribulus* or closely related Zygophyllaceae (Andres, 1978).

The immature stages of both weevils were described by Kirkland and Goeden (1977). The biology of *M. lareynii* was described by Andres and Angelet (1963) and Kirkland and Goeden (1978a); that of *M. lypriformis* by Andres and Angelet (1963) and Kirkland and Goeden (1978b). The egg of *M. lareynii* is deposited in a pit chewed in the pericarp of an immature fruit or occasionally in a floral bud or flower and capped with an anal secretion, often stained dark with feces. The larva feeds on the seeds and surrounding tissues, destroying seeds directly by mastication or indirectly by inducing abortion. Pupation occurs in an open cell in the fruit. The adult chews an emergence hole between adjacent carpels. The eggs hatch in 2–3 days, and larval development lasts 13–16 days; the pupal stadium lasts 4–5 days in southern California. The biology of *M. lypriformis* is similar, only most oviposition occurs in the undersides of the central, older parts of the prostrate, spreading, mat-like plants (i.e., root crowns, primary branches, and stem bases). The young larvae tunnel into the pith, where they largely confine their feeding, eventually pupating in open cells in the larval mine. The adults emerge from circular holes chewed mainly in the upper surfaces of stems, branches, and crowns. Both weevil species are multivoltine and produce a generation each month in the summer by reinfesting plants and attacking new plants as dispersed adults. Both species overwinter as adults in reproductive diapause among surface debris and plant litter and on or around associated nonhost plant species.

The weevil adults were initially imported from Italy and released directly in the field in Arizona, California, Colorado, Nevada, Utah, and Washington in July and August 1961 (Huffaker *et al.*, 1961; Andres & Angelet, 1963); establishment occurred in Arizona, California, and Nevada (Maddox, 1976). The weevils established readily in California and spread rapidly and widely, aided by extensive transfers of field-collected adults (Goeden & Ricker, 1967). Maddox (1976) reported the subsequent spread and establishment of both weevils in Kansas, New Mexico, Oklahoma, Texas, and Utah and of the stem weevil in Florida as well as the spread of both weevils into Mexico.

Soon after both weevils were established in southern California, Goeden and Ricker (1967, 1970) reported substantial egg predation by native Heteroptera and larval and pupal parasitism by indigenous chalcidoid Hymenoptera. Goeden and Kirkland (1981) assessed this predation in irrigated and nonirrigated field plots and determined that about half the seed weevil eggs infesting puncturevine fruit were killed by egg predation that reduced fruit infestation rates from 50 to 25%. Maddox (1981) determined that seed

germination in infested fruit was drastically reduced. Kirkland and Goeden (1978c) used the insecticide check method to assess the effects of both weevils acting in concert on irrigated and nonirrigated plants in field plots. Their results showed that water stress was the principal cause of early-season plant mortality; however, weevil attack caused a 60% reduction of flower production on surviving plants in nonirrigated plots. Moreover, only half of these flowers on nonirrigated plants produced fruit late in the growing season. Maddox (1981) also used insecticidal check plots to demonstrate that the stem weevils had a greater impact than the seed weevils on puncturevine plants per se, as measured by stem growth rates, metered water stress, and whole-plant biomass. Maddox (1981) and Huffaker *et al.* (1983) reported that seed weevils, largely acting alone in experimental field plots, increased flower production by puncturevine, which they attributed to some vaguely described "survival strategy" of the weevil. Huffaker *et al.* (1983) reported that 15 years after introduction of the weevils, puncturevine coverage and seed production declined in more than 80% of 1200 field plots monitored in California. They attributed this decline to the actions of both species of weevils. The biological control of puncturevine in California generally is considered a partial success or a substantial success under field conditions where weevil attacks intensify moisture stress on nonirrigated plants (Kirkland & Goeden, 1978c; Maddox & Andres, 1979; Julien, 1982).

Puncturevine also provides examples of successful transfer projects. Both species of weevils were transferred as field-collected adults in 1962 from California to Hawaii, where puncturevine and the perennial *Tribulus cistoides* Linnaeus were brought under complete biological control within a few years on all islands (Julien, 1982). Stem weevils subsequently were transferred to the island of St. Kitts in the West Indies from Hawaii in 1966, and seed weevils were transferred from southern California to St. Kitts in 1969. The latter species failed to establish, but the former species alone provided complete control of *T. cistoides* (Julien, 1982).

Water Hyacinth, *Eichhornia crassipes,* in the Southeastern United States

Eichhornia crassipes, a floating plant native to the American tropics, has become a naturalized weed throughout tropical, subtropical, and temperate climatic areas of the world (Bock, 1969). The plant favors vegetative reproduction, daughter plants forming on stolons originating from the central rhizomes. Enormous mats can form quickly, the plant doubling in volume every 10–15 days under favorable conditions (Penfound & Earle, 1948). Water hyacinth's ability to completely cover lakes and slow-moving streams can cause major navigational, agricultural, and health-related problems and otherwise disrupt the normal chains of events that occur in the aquatic habitat. Although herbicides can hold the plant in check, chemical control can be costly and only temporary at best. Interest in biological control of water hyacinth developed simultaneously between the U.S. Army Corps of Engineers and the British Ministry of Overseas Development (Bennett & Zwölfer, 1968).

Research developed along two lines. Surveys for natural enemies were carried out in Guyana, Surinam, the Amazon region of Brazil, Uruguay, the West Indies, and British Honduras (Bennett & Zwölfer, 1968). Of several dozen phytophagous arthropods found on this aquatic weed, seven were thought to have some biological control potential. Of these, three were tested and cleared for importation to North America: *Neochetina bruchi* Hustache (Coleoptera: Curculionidae), *N. eichhorniae* Warner, and *Sameodes albiguttalis,* Warren (Lepidoptera: Pyralidae). A fourth insect, *Acigona infusella* (Walker) (Pyralidae), was eventually released in Australia (Julien, 1987).

In the United States, research also involved the study and use of indigenous organisms attacking water hyacinth, i.e., *Bellura densa* (Walker) (Lepidoptera: Noctuidae), *Cercospora rodmani* Conway (Hyphomycetes), and *Acremonium zonatum* (Sawada) Gams (Hyphomycetes). Although the native host of *B. densa* is pickerelweed, *Pontederia cordata* Linnaeus (Pontederiaceae), a close relative of water hyacinth, the larvae can severely damage the latter weed at times. Biological studies were conducted and a diet on which to mass rear *B. densa* was developed, followed by the experimental release of many eggs and first instars to augment native populations. The moth had little impact in the United States (Julien, 1987).

Cercospora rodmani is a fungus indigenous to Florida that causes leaf spot, leaf necrosis, and secondary root rot of hyacinth plants. It also causes plant death when sprayed on mats of water hyacinth. The fungus is being registered as a commercial mycoherbicide (Charudattan, 1986). Another fungus, *Acremonium zonatum,* damages water hyacinth in southern Florida and Louisiana but seems to have its greatest impact when associated with the mite *Orthogalumna terebrantis* Wallwork (Acarina: Galumnidae). This mite is native to South America and accidentally entered the United States after its host plant was intentionally introduced as an ornamental.

The females of *Neochetina eichhorniae* and *N. bruchi* chew holes in the leaf petioles into which they insert one or several eggs, respectively (Center, 1982). The larvae tunnel beneath the epidermis and work their way down to the base of the petiole or the rhizome to which the leaf is attached, by which time they are in their third and final instar. The fully grown larvae chew their way out of the stems and move toward the surface of the water. They cut several lateral rootlets which are incorporated into an underwater pupal cocoon attached to the hyacinth roots. The

emerging adults leave the water from emergent plant parts, where they feed, mate, and oviposit. The weevil overwinters as larvae, pupae, or adult. There is one generation per year (DeLoach & Cordo, 1976).

Sameodes albiguttalis oviposits into the spongy leaf petioles, favoring areas with cuts in the epidermis or injuries made by other organisms. The young larvae feed under the epidermis, periodically exiting onto the petiole surface, crawling downward, and then reentering the globose area of the petiole to continue feeding. The fifth and final instar excavates a pupal cell in the petiole. DeLoach and Cordo (1978) estimated that there are five generations per year in Argentina. Larval feeding causes the petioles to break and die and results in heavy dieback the following winter.

Neochetina eichhorniae was introduced into the United States in 1972 from the vicinity of Buenos Aires (Center, 1982). It has established throughout the range of water hyacinth in North America. This weevil also was subsequently transferred from the United States to several other countries, where it is now established: Australia, Fiji, India, Indonesia, New Guinea, South Africa, Sudan, and Thailand (Julien, 1987). *Neochetina bruchi* was introduced into the United States from Argentina in 1974 and is now established in California, Florida, Louisiana, and Texas. Establishment has also been confirmed in India and Sudan (Julien, 1987).

Sameodes albiguttalis was released in the United States in 1977 (Center & Durden, 1981) and is established in California, Florida, Louisiana, and Mississippi. Establishment has also been confirmed in India and Sudan (Julien, 1987).

Acigona infusella was introduced into Australia from Brazil but failed to establish. *Orthogalumna terebrantis* was introduced into Egypt and Zambia, but again establishment has not been confirmed (Julien, 1987).

Neochetina eichhorniae has probably been the major contributor to the control of water hyacinth in the United States, Australia, and Sudan. Cofrancesco *et al.* (1985) documented the reduction of water hyacinth in Louisiana from about 445,000 ha in 1974 to 122,000 ha by 1980. *Sameodes albiguttalis* retards growth in the early stages of mat development, although its action may be sporadic and patchy (Center, 1985; Julien, 1987). When introduced from Buenos Aires, *N. bruchi* was observed successfully controlling water hyacinth in an isolated reservoir in La Rioja Province, Argentina, suggesting that its control potential should not be discounted (DeLoach & Cordo, 1983).

The water hyacinth project was the second attempt at biological control of an aquatic weed with introduced arthropods. Although focus remained on the search and importation of natural enemies from the native South American range of the weed, the discovery of the indigenous *Bellura densa* and *Cercospora rodmani* added another dimension to the research on biological control of aquatic weeds in the United States. Despite the limited success of attempts to augment natural *B. densa* population levels in Florida, the moth may eventually prove useful as an introduced agent in other countries.

Biological control of water hyacinth also differed from earlier projects on the biological control of weeds because the benefits realized were not confined to areas of low economic worth, e.g., rangeland, pastures, and waste areas, but rather involved the management of valuable water resources. As such, it offered one of the first real opportunities to develop a weed management program that could integrate imported natural enemies with other, more traditional and costly, weed control methods.

On the other hand, the reduction of water hyacinth to low levels has also tempted weed control workers to "clean up" remaining pockets of plants with herbicides. Without proper integration, this action can upset the balance between the natural enemies and their weed host and has resulted in further hyacinth outbreaks (Center, 1982). Buckingham and Passoa (1985) studied flight muscle development in the water hyacinth weevils. They suggested that in an integrated weed control program when conservation of the weevils is desired, summer herbicide applications should be delayed until the greatest number of newly emerged weevils with well-developed wing muscles are present. These agents would then be better able to migrate to unsprayed areas. They also suggested that herbicides should not be applied in the spring until water temperatures have risen above 18°C, the threshold for wing development. Unfortunately, herbicidal control is more efficient if begun early in the season.

Three Recent Successes Outside of North America

Cuscuta spp. in the People's Republic of China

The following account is largely taken from a translation of Wang (1986) provided to us by the author. Two species of parasitic weeds called dodders, *Cuscuta australis* R. Brown and *C. chinensis* Lamarck (Convolvulaceae), are serious weeds of oilseed, peanut, potato, and soybean crops in Shandong, Anhui, Hubei, Jilin, Liaoning, and Helongjiang Provinces and the Xinjiang autonomous Region, PRC. The fungal pathogen *Colleotrichum gloeosporioides* was isolated from dodder in soybeans in Jinan, Shandong Province, in 1963. The biology and mass culture of the fungus were studied from 1963 to 1966, which led to the development of a mycoherbicide named Lubao No.1. From its initial use in soybean fields in five provinces in 1966, production and application of this mycoherbicide grew until by the late 1970s, ca. 670,000 ha of soybeans were treated in 30 provinces. Reportedly, 85% control of these weeds

was realized in all fields and soybean losses were reduced by 30–50%.

During the 11 years of the "Cultural Revolution," the production of Lubao No. 1 was severely curtailed and problems were experienced with the mycoherbicide; e.g., it had a short shelf life and required high humidity for infection. An improved strain called Lubao No. 1 S22 is currently in mass production and is widely used. Microbial herbicides are expected to receive considerable attention in the future in the PRC.

Ragweeds, *Ambrosia* spp. (in the USSR)

Three species of ragweeds, *Ambrosia artemisiifolia* Linnaeus, *A. psilostachya* deCandolle, and *A. trifida* Linnaeus (Asteraceae), were accidentally introduced into the USSR from their native North America sometime during the 1920s to 1940s. They fast became noxious weeds and by 1980 had spread into the southern part of European USSR, Transcaucasia, Kazakhstan, and the Maritime Territory. *Ambrosia artemisiifolia* apparently is the most important of the three ragweeds, as it infests most crops and is costly and difficult to control with chemicals or by cultural and mechanical means. These ragweeds also have become medically important hay fever plants (Kovalev, 1974, 1980).

Thanks to an informal cooperative program involving Canadian, United States, and USSR scientists and begun in 1965, about 450 species of insects, mites, and fungi have been identified as natural enemies of 17 species of *Ambrosia* in North and South America (Harris & Piper, 1970; Kovalev, 1971, 1980; Goeden & Ricker, 1976). More than 30 species of natural enemies were introduced into the Soviet Union from 1967 to 1979 (Kovalev, 1971, 1980; Goeden *et al.*, 1974b). The host-plant specificities of these candidate agents were tested in quarantine in the USSR, using eight varieties of sunflower, *Helianthus anuus* Linnaeus (Asteraceae), as the "critical test plant," 18 species of *Helianthus,* and 80 species of cultivars representing 46 genera and 18 families of plants (Kovalev, 1970), in accordance with the so-called "centrifugal testing method" (Wapshere, 1974a) in general use today (see earlier).

By 1980, four species of insects obtained from the same species of weeds in their native North America had been released to establish a complex of natural enemies in the Soviet Union. These introduced ragweed insects included *Tarachidia candefacta* Hübner (Lepidoptera: Noctuidae), a defoliator obtained from Canada and California and released on *A. artemisiifolia* in 1969 and on *A. psilostachya* in 1972, respectively (Kovalev & Runeva, 1970; Kovalev & Samus, 1972; Gilstrap & Goeden, 1974; Goeden *et al.,* 1974b); *Brachytarsus tomentosus* (Say) (Coleoptera: Anthribidae), a pollen-feeding beetle first released in the northern Caucasus in 1978; *Euaresta bella* Loew (Diptera: Tephritidae), a monophagous seed feeder apparently obtained from Canada and released in the northern Caucasus in the 1970s with unknown results; and *Zygogramma suturalis* (Fabricius) (Coleoptera: Chrysomelidae), another defoliator obtained from Canada and the United States and first released in 1978 (Kovalev, 1980; Julien, 1982).

Tarachidia candefacta was the first natural enemy released for ragweed control in the USSR and the first natural enemy intentionally introduced into Europe from North America for biological control of a weed. In mass rearing *T. candefacta,* specially designed oviposition units were used to harvest large numbers of eggs deposited on hanging threads. Larvae were fed a prepared diet (Kovalev & Runeva, 1970; Kovalev & Nayanov, 1971). Although successfully established on both *A. artemisiifolia* and *A. psilostachya,* it apparently has been unsuccessful as a biological control agent, largely because of predation of the exposed larvae (O. V. Kovalev *in lit.*)

The results obtained with *Zygogramma suturalis* have been spectacularly successful under certain conditions (Reznik, 1991; Julien, 1992). This beetle was mass reared at several laboratories beginning in 1979 following its initial winter's survival. It was released south of the European part of the USSR and Transcaucasia to the Far East (Kovalev & Medvedev, 1983; Kovalev & Vechernin, 1986). Larvae and adults feed on leaves and inflorescence throughout the growing season and range of *A. artemisiifolia* from April to mid-September. The beetle produces two complete and a partial third generation annually (Kovalev, 1980). Kovalev and Vechernin (1986) reported that seventh-generation *Z. suturalis* formed feeding fronts consisting of a stable nondeclining wave of beetles which moved at a constant rate and which they termed an "isolated population wave (IPW)." On one state farm at Stravropol, the insects were concentrated in a narrow band at 5000 individuals per m^3, which completely destroyed all the ragweeds as it moved across an infested field at a rate of 3 m per day. About 10 million beetles were concentrated in a circular feeding front with a length of 1.5 km and a breadth of ca. 10 m. The weed-controlling action of these feeding fronts sharply increased yields of lucerne, sainfoin, and maize by two- to threefold.

Salvinia, *Salvinia molesta* D. S. Mitchell, in Australia and New Guinea

Unless otherwise indicated, the following account was summarized from the review of Thomas and Room (1986). The free-floating aquatic fern *Salvinia molesta* (Salviniaceae) is native to Brazil. It was spread as an aquarium plant and botanical curiosity to tropical and subtropical Africa, Asia, Australia, Fiji, and New Guinea, where it rapidly propagated vegetatively to become one of the world's most important aquatic weeds.

In 1955, it first appeared in Lake Kariba, a huge reservoir on the Zambia–Zimbabwe border, and quickly spread and soon covered 1000 km^2 or 21% of the lake's surface. CIBC entomologists conducted foreign exploration in South America for *Salvinia auricularia* Aublet, also called Kariba weed. Three insect species, the moth *Samea multiplicatus* Guenee (Lepidoptera: Pyralidae), the grasshopper *Paulinia acuminata* (DeGeer) (Orthoptera: Acrididae), and the weevil *Cyrtobagous singularis* Hustache (Coleoptera: Curculionidae), were recommended as the most promising agents for biological control of this fern. During the 1960s and 1970s, all three species of insects were released on Lake Kariba, elsewhere in Africa, and in Fiji. Only *P. acuminata* became established on Lake Kariba and provided limited control of salvinia.

By 1970, Kariba weed was cited as one of the best documented contemporary examples of the importance of taxonomy to biological control of weeds. However, a pressed specimen of Kariba weed collected in 1941 and preserved in the Rio de Janeiro Botanic Garden was not *Salvinia auricularia* but rather a new species described as *S. molesta* in 1972. By 1978, this weed was determined to be a sterile hybrid species that originated in southeastern Brazil. Exploration by CSIRO entomologists resulted in the detection, study, and recommendation of *Samea multiplicatus, P. acuminata,* and what then was thought to be a separate biotype of *C. singularis* as the most promising candidate biological control agents. The weevil was introduced into Australia in 1980, where it established readily, and in less than 2 years the weevil destroyed a 200-ha infestation on Lake Moondarra in northern Australia. After additional study, this weevil was described as *Cyrtobagous salviniae* Calder & Sands. These results support the concept that the best natural enemies are obtained from the target weed species origin. This concept was challenged by Hokkanen and Pimentel (1984), but supported by Goeden and Kok (1986) and Moran *et al.* (1986), and most recently discussed by Dennill and Hokkanen (1990).

Cyrtobagous salviniae was transferred to Papua, New Guinea, from Australia in 1982 and established only after the novel manipulation of the weevil's environment in field-release cages. This involved regular applications of urea fertilizer to the caged weed, which increased its nitrogen content and enhanced its nutritional suitability such that the weevils multiplied explosively. The weevils continued to multiply in phenomenal numbers once the cages were opened and this critical mass of weevils was given access to the large volume of unfertilized weed that covered Sepik Lake to a depth of up to 1 m. Following a large-scale redistribution program in August 1985, salvinia had been reduced from covering 250 km^2 of water surface on the lower floodplain of the Sepik River to 2 km². This represented the destruction of 2 million metric tons of salvinia in 2 years!

FUTURE DIRECTION OF BIOLOGICAL CONTROL OF WEEDS

The direction of biological control depends upon how the problems, tasks, and priorities of this subdiscipline are perceived (Harris, 1991, Goeden, 1993). The pressures of a rapidly expanding human population, increased environmental pollution, scarcity of natural resources, and increasingly stringent regulations governing pesticide use will speed the movement to biological control and other safe alternative control methods. The search for new, coevolved foreign and domestic natural enemies will continue, as will the development and registration of new mycoherbicides. Current techniques for finding, testing, introducing, colonizing, and manipulating these new natural enemies will continue (Zwölfer & Harris, 1971; Frick, 1974; Andres *et al.,* 1976; Klingman & Coulson, 1983; Schroeder & Goeden, 1986; Harris, 1991) and be refined. However, even these techniques are not without problems. Increasing human presence in many of the areas (such as western Europe) where weeds originated adds to the difficulty of locating natural populations of the weeds and their natural enemies (Schroeder & Goeden, 1986). The inability to find suitable natural enemies will promote attempts to artificially alter the host selectivity and adaptability of existing natural enemies by selective breeding or genetic engineering. Present efforts at genetic improvement of natural enemies are costly, are time-consuming, and may be of limited practical value (Beckendorf & Hoy, 1985).

Trial feeding in caged host-specificity studies, though limited in extent, may provide sufficient grounds for the rejection of a candidate natural enemy, even if there was no previous record of nontarget plant attack (Dunn, 1978). In the future, the interpretation of such laboratory studies will be balanced with the increased use of multiple-choice, open-field experiments in which potted test plants will be intermingled with populations of the target weed and its associated natural enemies. Despite the potential of new biochemical and physiological techniques for unraveling the myriad factors involved in host plant–natural enemy interactions, it is doubtful that these will ever replace field and laboratory organism–population studies.

Perhaps, biological control of weeds, like the so-called "dismal science" of economics (Stein, 1987), is destined to remain an art for the foreseeable future (Harris, 1976; Goeden, 1993). However, standardizing data collection techniques among countries will strengthen the significance of future project comparisons and generalizations (Harris, 1991; Julien, 1992).

Biological control will expand to new countries and include the transfer of proven biological control agents from other areas. As in the past, the transfer and establishment of successful agents is followed by the development of new facilities, expertise, and the undertaking of new projects. In developed countries, the main challenge will be to provide new biological control successes that will engage the imagination and support of research administrators and a public increasingly oriented toward high-technology solutions. Attempts to develop weed project cost–benefit analyses will improve, although efforts to include ecological and aesthetic benefits will continue to pose difficulties. To optimize the benefits of biological control, changes in agricultural production, resource use, and perceptions of what constitutes damaging weed populations will be required.

References

Aldrich, R. J., Ed. (1984). "Weed Crop Ecology." Breton, North Scituate, MA.

Andres, L. A. (1977). The economics of biological control of weeds. Aquatic Botany, 3, 111–123.

Andres, L. A. (1978). Biological control of puncturevine, *Tribulus terrestris* (Zygophyllaceae): Post introduction collection records of *Microlarinus* spp. (Coleoptera: Curculionidae). *In* "Proceedings of the IV International Symposium on Biological Control of Weeds, 1976, Gainesville, FL" (T. E. Freeman, Ed.), pp. 132–136.

Andres, L. A., & Angelet, G. W. (1963). Notes on the ecology and host specificity of *Microlarinus lareynii* and *M. lypriformis* (Coleoptera: Curculionidae) and the biological control of puncture vine, *Tribulus terrestris.* Journal of Economic Entomology, 56, 333–340.

Andres, L. A., & Bennett, F. D. (1975). Biological control of aquatic weeds. Annual Review of Entomology, 20, 31–46.

Andres, L. A., & Goeden, R. D. (1971). The biological control of weeds by introduced natural enemies. *In* "Biological Control" (C. B. Huffaker, Ed.), pp. 143–164. Plenum, New York.

Andres, L. A., Davis, C. J., Harris, P., & Wapshere, A. J. (1976). Biological control of weeds. *In* "Theory and Practice of Biological Control" (C. B. Huffaker & P. S. Messenger, Eds.), pp. 481–493. Academic Press, New York.

Annecke, D. P., Karny, M., & Burger, W. A. (1969). Improved biological control of the prickly pear, *Opuntia megacantha* Salm-Dyck, in South Africa through use of an insecticide. Phytophylactica, 1, 9–13.

Beckendorf, S. K., & Hoy, M. A. (1985). Genetic improvement of arthropod natural enemies through selection, hybridization or genetic engineering techniques. *In* "Biological Control in Agricultural IPM Systems" (M. A. Hoy & D. C. Herzog, Eds.), pp. 167–187. Academic Press, Orlando, FL.

Bennett, F. D. (1971). Some recent successes in the field of biological control in the West Indies. Entomologia, 14, 369–373.

Bennett, F. D., & Zwölfer, H. (1968). Exploration for natural enemies of water hyacinth in northern South America and Trinidad. Water Hyacinth Control Journal, 7, 44–52.

Bock, H. H. (1969). Productivity of the waterhyacinth, *Eichhornia crassipes* (Mart.) Solms. Ecology, 50, 460–464.

Buckingham, G., & Passoa, S. (1985). Flight muscle and egg development in waterhyacinth weevils. *In* "Proceedings of the VI International Symposium on Biological Control of Weeds, 1975, Vancouver, BC" (E. S. Delfosse, Ed.), pp. 497–510.

Center, T. D. (1982). The waterhyacinth weevils, *Neochetina eichhorniae* and *N. bruchi.* Aquatics, 4, 8, 16–19.

Center, T. D. (1985). Leaf life tables: A viable method for assessing sublethal effects of herbivory on waterhyacinth shoots. *In* "Proceedings of the VI International Symposium on Biological Control of Weeds, 1975, Vancouver, BC" (E. S. Delfosse, Ed.), pp. 511–524.

Center, T. D., & Durden, W. C. (1981). Release and establishment of *Sameodes albiguttalis* for the biological control of waterhyacinth. Environmental Entomology, 10, 75–80.

Chandler, J. M. (1980). Assessing losses caused by weeds. *In* "Proceedings of the E. C. Stakman Communication Symposium" (R. J. Aldrich, Ed.), pp. 234–240. Univ. of Minnesota, St. Paul, Agricultural Experiment Station Miscellaneous Publication No. 7.

Charudattan, R. (1978). "Biological Control Projects in Plant Pathology: A Directory." Miscellaneous Publications of the Plant Pathology Department, Univ. of Florida, Gainesville.

Charudattan, R. (1984). Microbial control of plant pathogens and weeds. Journal of the Georgia Entomological Society, 19 (Suppl.), 28–40.

Charudattan, R. (1986). Integrated control of waterhyacinth (*Eichhornia crassipes*) with a pathogen, insects and herbicides. Weed Science, 34 (Suppl. 1), 26–30.

Charudattan, R., & Walker, H. L., Eds. (1982). "Biological Control of Weeds with Plant Pathogens." Wiley, New York.

Clark, L. R. (1953). The ecology of *Chrysomela gemellata* Rossi and *C. hyperici* Forst., and their effect on St. John's wort in the Bright District, Victoria. Australian Journal of Zoology, 1, 1–69.

Clark, N. (1953). The biology of *Hypericum perforatum* Linnaeus var. *angustifolium* DC. (St. John's wort) in the Ovens Valley, Victoria, with particular reference to entomological control. Australian Journal of Botany, 1, 95–120.

Cofrancisco, A. F., Jr., Stewart, R. M., & Sanders, D. R., Sr. (1985). The impact of *Neochetina eichhorniae* (Coleoptera: Curculionidae) on waterhyacinth in Louisiana. *In* "Proceedings of the VI International Symposium on Biological Control of Weeds, 1984, Vancouver, BC" (E. S. Delfosse, Ed.), pp. 525–535.

Coulson, J. R. (1977). "Biological Control of Alligatorweed, 1959–1972. A Review and Evaluation." U.S. Department of Agriculture Technical Bulletin No. 1547.

Coulson, J. R. (1985). "Information on U.S. and Canadian Biological Control Programs Using Natural Enemies and Other Beneficial Organisms." U.S. Department of Agriculture, Agricultural Research Service, Biological Control Documentation Center, Information Document.

Cromroy, H. L. (1982). Potential use of mites in biological control of terrestrial and aquatic weeds. *In* "Biological Control of Pests by Mites" (M. A. Hoy, G. L. Cunningham, & L. Knutson, Eds.), pp. 61–66. Univ. of California Press, Berkeley.

Cullen, J. M. (1974). Seasonal and regional variation in the success of organisms imported to combat skeleton weed *Chondrilla juncea* L. in Australia. *In* "Proceedings of the III International Symposium on Biological Control of Weeds, 1973, Montpellier, France" (A. J. Wapshere, Ed.), pp. 111–117.

Cullen, J. M. (1978). Evaluating the success of the programme for the biological control of *Chondrilla juncea* in Australia. *In* "Proceedings of the IV International Symposium on Biological Control of Weeds, 1976, Gainesville, FL" (T. E. Freeman, Ed.), pp. 233–239.

Cullen, J. M., & Delfosse, E. S. (1985). *Echium plantagineum:* Catalyst for conflict and change in Australia. *In* "Proceedings of the VI International Symposium on Biological Control of Weeds, 1984, Vancouver, BC" (E. S. Delfosse, Ed.), pp. 19–25.

Daniel, J. T., Templeton, G. E., Smith, R. J., Jr., & Fox, W. T. (1973). Biological control of northern jointvetch in rice with endemic fungal disease. Weed Science, 21, 303–307.

Davis, C. J. (1966). "Progress Report: Biological Control Status of Noxious Weed Pests in Hawaii—1965–1966." Hawaiian Department of Agriculture Mimeographed Report.

DeBach, P. (1964). The scope of biological control. *In* "Biological Control

of Insect Pests and Weeds" (P. DeBach, Ed.), pp. 1–20. Reinhold, New York.

DeLoach, C. J., & Cordo, H. A. (1976). Life cycle and biology of *Neochetina bruchi,* a weevil attacking waterhyacinth in Argentina, with notes on *N. eichhorniae.* Annals of the Entomological Society of America, 69, 643–652.

DeLoach, C. J., & Cordo, H. A. (1978). Life history and ecology of the moth *Sameodes albiguttalis,* a candidate for biological control of waterhyacinth. Environmental Entomology, 7, 309–321.

DeLoach, C. J., & Cordo, H. A. (1983). Control of waterhyacinth by *Neochetina bruchi* (Coleoptera: Curculionidae: Bagoini) in Argentina. Environmental Entomology, 12, 19–23.

Dennill, G. B., & Hokkanen, H. M. T. (1990). Homeostasis and success in biological control of weeds—A question of balance. Agriculture, Ecosystems, & Environment, 33, 1–10.

Dodd, A. P. (1940). "The Biological Campaign against Prickly Pear." Commonwealth Prickly Pear Board, Brisbane, Australia.

Doutt, R. L. (1958). Vice, virtue, and the Vedalia. Bulletin of the Entomological Society of America, 4, 119–123.

Doutt, R. L. (1964). The historical development of biological control. *In* "Biological Control of Insect Pests and Weeds" (P. DeBach, Ed.), pp. 21–42. Reinhold, New York.

Dunn, P. H. (1978). Shortcomings in the classic tests of candidate insects for the biocontrol of weeds. *In* "Proceedings of the IV International Symposium on Biological Control of Weeds, 1976, Gainesville, FL" (T. E. Freeman, Ed.), pp. 51–56.

Emge, R. G., Melching, J. S., & Kingsolver, C. H. (1981). Epidemiology of *Puccina chondrillina,* a rust pathogen for the biological control of rush skeleton weed in the United States. Phytopathology, 71, 839–843.

Freeman, T. E., & Charudattan, R. (1981). Biological control of weeds with plant pathogens. Prospectus. *In* "Proceedings of the V International Symposium on Biological Control of Weeds, 1980, Brisbane, Australia" (E. S. Delfosse, Ed.), pp. 293–299.

Frick, K. E. (1970). *Longitarsus juacobaeae* (Coleoptera: Chrysomelidae), a flea beetle for the biological control of tansy ragwort: Host plant specificity studies. Annals of the Entomological Society of America, 63, 284–296.

Frick, K. E. (1974). Biological control of weeds: Introduction, history, theoretical and practical applications. *In* "Proceedings of the Summer Institute of Biological Control of Plant Insects and Diseases" (F. G. Maxwell & F. A. Harris, Eds.), pp. 204–223. Univ. of Mississippi Press, Jackson.

Frick, K. E., & Chandler, J. M. (1978). Augmenting the moth (*Bactra verutana*) in field plots for early-season suppression of purple nutsedge (*Cyperus rotundus*). Weed Science, 26, 703–710.

Gilstrap, F. E., & Goeden, R. D. (1974). Biology of *Tarachidia candefacta,* a Nearctic noctuid introduced into the U.S.S.R. for ragweed control. Annals of the Entomological Society of America, 67, 265–270.

Goeden, R. D. (1977). Biological control of weeds. *In* "Research Methods in Weed Science" (B. Truelove, Ed.), Chap. 4, pp. 43–47. Southern Weed Science Society, Auburn Printing, Auburn, GA.

Goeden, R. D. (1978). Biological control of weeds. *In* "Introduced Parasites and Predators of Arthropod Pests and Weeds" (C. P. Clausen, Ed.), Part II, pp. 357–545. U.S. Department of Agriculture Handbook No. 480.

Goeden, R. D. (1983). Critique and revision of Harris' scoring system for selection of insect agents in biological, control of weeds. Protection Ecology, 5, 287–301.

Goeden, R. D. (1988). History of biological weed control. Biocontrol News & Information, 9, 55–61.

Goeden, R. D. (1993). Arthropods for suppression of terrestrial weeds. *In* "Proceedings of Beltsville Symposium XVIII, Agricultural Research

Service, U.S. Department of Agriculture, Beltsville, MD" (R. D. Lumsden & J. L. Vaughn, Eds.), pp. 231–237. American Chemical Society, Washington, DC.

Goeden, R. D., & Kirkland, R. L. (1981). Interactions of field populations of indigenous egg predators, imported *Microlarinus* weevils, and puncturevine in southern California. *In* "Proceedings of the V International Symposium on Biological Control of Weeds, 1980, Brisbane, Australia" (E. S. Delfosse, Ed.), pp. 515–527.

Goeden, R. D., & Kok, L. T. (1986). Comments on a proposed "new" approach for selecting agents for the biological control of weeds. Canadian Entomologist, 118, 51–58.

Goeden, R. D., & Ricker, D. W. (1967). *Geocoris pallens* found to be predaceous on *Microlarinus* spp. introduced to California for the biological control of puncturevine, *Tribulus terrestris*. Journal of Economic Entomology, 60, 725–729.

Goeden, R. D., & Ricker, D. W. (1970). Parasitization of introduced puncturevine weevils by indigenous Chalcidoidea in southern California. Journal of Economic Entomology, 63, 827–831.

Goeden, R. D., & Ricker, D. W. (1973). A soil profile analysis for puncturevine fruit and seed. Weed Science, 21, 504–507.

Goeden, R. D., & Ricker, D. W. (1976). The phytophagous insect fauna of the ragweed, *Ambrosia psilostachya*, in southern California. Environmental Entomology, 5, 1169–1177.

Goeden, R. D., & Ricker, D. W. (1981). Santa Cruz Island—Revisited. Sequential photography records the causation, rates of progress, and lasting benefits of successful biological weed control. *In* "Proceedings of the V International Symposium on Biological Control of Weeds, 1980, Brisbane, Australia" (E. S. DelFosse, Ed.), pp. 355–365.

Goeden, R. D., Fleschner, C. A., & Ricker, D. W. (1967). Biological control of prickly pear cacti on Santa Cruz Island, California. Hilgardia, 38, 579–606.

Goeden, R. D., Andres, L. A., Freeman, T. E., Harris, P., Pienkowski, R. L., & Walker, C. R. (1974a). Present status of projects on the biological control of weeds with insects and plant pathogens in the United States and Canada. Weed Science, 22, 490–495.

Goeden, R. D., Kovalev, O. V., & Ricker, D. W. (1974b). Arthropods exported from California to the USSR for ragweed control. Weed Science, 22, 156–158.

Greathead, D. J. (1971). "A Review of Biological Control in the Ethiopian Region." Commonwealth Institute of Biological Control Technical Communication No. 5, pp. 1–172.

Harris, P. (1971). Current approaches to biological control of weeds. *In* "Biological Control Programmes against Insects and Weeds in Canada 1959–1968." Commonwealth Institute of Biological Control Technical Communication No. 4, pp. 79–83.

Harris, P. (1973). The selection of effective agents for the biological control of weeds. Canadian Entomologist, 105, 1495–1503.

Harris, P. (1976). Biological control of weeds: From art to science. *In* "Proceedings of the 15th International Congress of Entomology, Washington, DC," pp. 478–483.

Harris, P. (1979). Cost of biological control of weeds by insects in Canada. Weed Science, 27, 242–250.

Harris, P. (1984a). *Carduus nutans* L., nodding thistle, and *C. acanthoides* L., plumeless thistle (Compositae). *In* "Biological Control Programmes against Insects and Weeds in Canada 1969–1980" (J. S. Kelleher & M. A. Hume, Eds.), pp. 115–126. Commonwealth Agricultural Bureau, Slough, UK.

Harris, P. (1984b). *Euphorbia esula-virgata* complex, leafy spurge and *E. cuparissias* L., cypress spurge (Euphorbiacea). *In* "Biological Control Programmes against Insects and Weeds in Canada 1969–1980" (J. S. Kelleher & M. A. Hume, Eds.), pp. 159–169. Commonwealth Agricultural Bureau, Slough, UK.

Harris, P. (1991). Invitation paper (C. P. Alexander Fund): Classical biocontrol of weeds: Its definition, selection of effective agents, and administrative-political problems. Canadian Entomologist, 123, 827–849.

Harris, P., & Piper, G. L. (1970). "Ragweed (*Ambrosia* spp.: Compositae): Its North American Insects and Possibilities for Its Biological Control." Commonwealth Institute of Biological Control Technical Communication No. 13, pp. 117–140.

Hawkes, R. B., & Johnson, G. R. (1978). *Longitarsus jacoaeae* aids moth in biological control of tansy ragwort. *In* "Proceedings of the IV International Symposium on Biological Control of Weeds, 1976, Gainesville, FL" (T. E. Freeman, Ed.), pp. 193–196.

Hokkanen, H., & Pimentel, D. (1984). New approach for selecting biological control agents. Canadian Entomologist, 116, 1109–1121.

Holloway, J. K., & Huffaker, C. B. (1949). Klamath weed beetles. California Agriculture, 3, 3–10.

Holloway, J. K., & Huffaker, C. B. (1951). The role of *Chrysolina gemellata* in the biological control of Klamath weed. Journal of Economic Entomology, 44, 244–247.

Holm, L. G., Plucknett, D. L., Pancho, J. V., & Herberger, J. P. (1977). "The World's Worst Weeds, Distribution, and Biology." Univ. of Hawaii Press, Honolulu.

Huffaker, C. B. (1957). Fundamentals of biological control of weeds. Hilgardia, 27, 101–157.

Huffaker, C. B., Hamai, J., & Nowierski, R. M. (1983). Biological control of puncturevine, *Tribulus terrestris* in California after twenty years of activity of introduced weevils. Entomophaga, 28, 387–400.

Huffaker, C. B., Ricker, D., & Kennett, C. (1961). Biological control of puncture vine with imported weevils. California Agriculture, 15, 11–12.

Johnson, E. (1932). "The Puncture Vine in California." University of California, College of Agriculture, Agricultural Experiment Station, Bulletin 528.

Johnston, T. H., & Tryon, H. (1914). "Report on the Prickly-Pear Travelling Commission, 1st November, 1912-30th April, 1914. Parliamentary Paper." Cumming, Brisbane, Australia.

Julien, M. H., Ed. (1982). "Biological Control of Weeds: A World Catalogue of Agents and Their Target Weeds," 1st ed. Commonwealth Agricultural Bureau, Slough, UK.

Julien, M. H., Ed. (1987). "Biological Control of Weeds: A World Catalogue of Agents and Their Target Weeds," 2nd ed. Commonwealth Agricultural Bureau International, Wallingford, UK.

Julien, M. H., Ed. (1992). "Biological Control of Weeds: A World Catalogue of Agents and Their Target Weeds," 3rd ed. Commonwealth Agricultural Bureau International, Wallingford, UK.

Julien, M. H., Kerr, J. D., & Chan, R. R. (1984). Biological control of weeds: An evaluation. Protection Ecology, 7, 3–25.

King, L. J. (1966). "Weeds of the World, Biology and Control." Interscience, New York.

Kirkland, R. L., & Goeden, R. D. (1977). Descriptions of the immature stages of imported puncturevine weevils, *Microlarinus lareynii* and *M. lypriformis*. Annals of the Entomological Society of America, 70, 583–587.

Kirkland, R. L., & Goeden, R. D. (1978a). Biology of *Microlarinus lareynii* (Col.: Curculionidae) on puncturevine in southern California. Annals of the Entomological Society of America, 70, 13–18.

Kirkland, R. L., & Goeden, R. D. (1978b). Biology of *Microlarinus lypriformis* (Col.: Curculionidae) on puncturevine in southern California. Annals of the Entomological Society of America, 70, 65–69.

Kirkland, R. L., & Goeden, R. D. (1978c). An insecticidal-check study of the biological control of puncturevine (*Tribulus terristris*) by imported weevils, *Microlarinus lareynii* and *M. lypriformis* (Col.: Curculionidae). Environmental Entomology, 7, 349–354.

Klingman, D. L., & Coulson, J. R. (1983). Guidelines for introducing foreign organisms into the United States for biological control of weeds. Bulletin of the Entomological Society of America, 29, 55–61.

Kok, L. T. (1986). Impact of *Trichosirocalus horridus* (Coleoptera: Curculionidae) on *Carduus* thistles in pastures. Crop Protection, 5, 214–217.

Kok, L. T., & Surles, W. W. (1975). Successful biological control of musk thistle by an introduced weevil, *Rhinocyllus conicus*. Environmental Entomology, 4, 1025–1027.

Kovalev, O. V. (1970). Biological control of *Ambrosia* weeds. *In* "Proceedings of the 7th International Congress of Plant Protection, 1969, Paris, France," pp. 354–355.

Kovalev, O. V. (1971). Phytophages of ragweeds (*Ambrosia* L.) in North America and their application in biological control in the U.S.S.R.. Zoologicheskii Zhurnal, 50, 199–209 (in Russian).

Kovalev, O. V. (1973). Modern outlooks of biological control of weed plants in the U.S.S.R. and the international phytophagous exchange. *In* "Proceedings of the III International Symposium on Biological Control of Weeds, 1971, Rome, Italy" (P. H. Dunn, Ed.), pp. 166–172.

Kovalev, O. V. (1974). Development of a biological method of controlling weeds in the USSR and the countries of Europe. *In*, "Biological Agents for Plant Protection" (E. M. Shumakov, G. V. Gusev, & N. S. Fedorinchik, Eds.), pp. 302–309. Publ. House "Kolos," Moscow (in Russian).

Kovalev, O. V. (1980). Biological control of weeds: Accomplishments, problems and prospects. Zashchita rasteniy, 5, 18–21 (in Russian).

Kovalev, O. V., & Medvedev, L. N. (1983). Theoretical basis of introduction of ragweed leaf beetles of the genus *Zygogramma* Chevr. (Coleoptera: Chrysomelidae) in the USSR for biological control of ragweed. Entomologicheskoe Obozrenie, 62, 17–32 (in Russian).

Kovalev, O. V., & Nayanov, N. I. (1971). A moth against ragweed. Zemledelie, 6, 36.

Kovalev, O. V., & Runeva, T. D. (1970). *Tarachidia candefacta* Hubn. (Lepidoptera: Noctuidae), an efficient phytophagous insect for biological control of weeds of the genus *Ambrosia* L. Academy of Sciences U.S.S.R. Entomological Review, 49, 23–36 (in Russian).

Kovalev, O. V., & Samus, V. I. (1972). Biology of *Tarachidia candefacta* Hubn. and prospects of its use to control common ragweed. U.S.S.R. Agricultural Biology, 7, 281–284 (in Russian).

Kovalev, O. V., & Vechernin, V. V. (1986). Description of a new wave process in populations with reference to introduction and spread of the leaf beetle *Zygogramma suturalis* F. (Coleoptera: Chrysomelidae). Entomologicheskoe Obozrenie, 65, 21–38 (in Russian).

Maddox, D. M. (1968). Bionomics of an alligatorweed fleabeetle, *Agasicles* sp., in Argentina. Annals of the Entomological Society of America, 61, 1299–1305.

Maddox, D. M. (1976). History of weevils on puncturevine in and near the United States. Weed Science, 24, 414–416.

Maddox, D. M. (1981). Seed and stem weevils of puncturevine: A comparative study of impact, interaction, and insect strategy. *In* "Proceedings of the V International Symposium on Biological Control of Weeds, 1980, Brisbane, Australia" (E. S. Delfosse, Ed.), pp. 447–467.

Maddox, D. M., & Andres, L. A. (1979). Status of puncturevine weevils and their host plant in California. California Agriculture, 33, 7–8.

Maddox, D. M., Andres, L. A., Hennessey, R. D., Blackburn, R. D., & Spencer, N. R. (1971). Insects to control alligatorweed, an invader of aquatic ecosystems in the United States. BioScience, 21, 985–991.

McEvoy, P., Cox, C., & Coombs, E. (1991). Successful biological control of ragwort, *Senecio jacobaea*, by introduced insects in Oregon. Ecological Applications, 1, 430–432.

Moran, V. C., & Zimmerman, H. G. (1984). The biological control of cactus weeds: Achievements and prospects. Biocontrol News & Information, 5, 297–320.

Moran, V. C., Neser, S., & Hoffmann, J. H. (1986). The potential of insect herbivores for the biological control of invasive plants in South Africa. *In* "The Ecology and Management of Biological Invasions in Southern Africa" (I. A. W. Macdonald, F. J. Kruger, & A. A. Ferrar, Eds.), pp. 261–268. Oxford Univ. Press, Capetown, South Africa.

Oehrens, E. (1977). Biological control of the blackberry through the introduction of rust, *Phragmidium violaceum*, in Chile. FAO Plant Protection Bulletin, 25, 26–28.

O'Neill, K. (1968). *Amynothrips andersoni*, a new genus and species injurious to alligatorweed. Proceedings of the Entomological Society of Washington, 70, 175–183.

Orr, C. C. (1980). *Nothanguina phyllobia*, a nematode biocontrol of silverleaf nightshade. *In* "Proceedings of the V International Symposium on Biological Control of Weeds, 1980, Brisbane, Australia" (E. S. Delfosse, Ed.), pp. 389–391.

Pemberton, R. W. (1981). International activity in biological control of weeds: Patterns, limitations and needs. *In* "Proceedings of the V International Symposium on Biological Control of Weeds, 1980, Brisbane, Australia" (E. S. Delfosse, Eds.), pp. 57–71.

Penfound, W. T., & Earle, T. T. (1948). The biology of the waterhyacinth. Ecological Monographs, 18, 447–472.

Perkins, R. C. L., & Swezey, O. H. (1924). "The Introduction into Hawaii of Insects That Attack Lantana." Bulletin of the Hawaiian Sugar Plantation Association Experimental Station No. 16.

Rao, V. P., Ghani, M. A., Sankaran, T., & Mathur, K. C. (1971). "A Review of the Biological Control of Insects and Other Pests in Southeast Asia and the Pacific Region." Commonwealth Institute of Biological Control Technical Communication No. 6, pp. 1–149.

Reznick, S. Ya. (1991). The effects of feeding damage in ragweed *Ambrosia artemisiifolia* (Asteraceae) on populations of *Zygogramma suturalis* (Coleoptera, Chyrsomelidae). Oecologia, 88, 204–210.

Ridings, W. H., Mitchell, D. J., Schoulties, C. L., & El-Gholl, N. E. (1978). Biological control of milkweed vine in Florida citrus groves with a pathotype of *Phytophthora citrophthora*. *In* "Proceedings of the IV International Symposium on Biological Control of Weeds, 1976, Gainesville, FL" (T. E. Freeman, Ed.), pp. 224–240.

Schröder, D. (1980). The biological control of thistles. Biocontrol News & Information, 5, 297–320.

Schroeder, D., & Goeden, R. D. (1986). The search for arthropod natural enemies of introduced weeds for biological control—In theory and practice. Biocontrol News & Information, 7, 147–155.

Simmonds, F. J. (1950). Insects attacking *Cordia macrostachya* (Jacq.) Roem. and Schult. in the West Indies. II. *Schematiza cordiae* Barber (Coleoptera: Galerucidae). Canadian Entomologist, 81, 275–282.

Simmonds, F. J. (1967). Biological control of pests of veterinary importance. Veterinary Bulletin, 37, 71–85.

Simmonds, F. J., & Bennett, F. D. (1966). Biological control of *Opuntia* spp. by *Cactoblastis cactorum* in the Leeward Islands (West Indies). Entomophaga, 11, 183–189.

Simmonds, H. W. (1937). The biological control of the weed, *Clidemia hirta*, commonly known in Fiji as "the curse." Agricultural Journal (Suva, Fiji), 8, 37–39.

Smith, H. S. (1919). On some phases of insect control by the biological method. Journal of Economic Entomology, 12, 288–292.

Smith, H. S. (1948). Biological control of insect pests. *In* "Biological Control of Citrus Insects. The Citrus Industry" (W. E. Reuther, E. C. Calaran, & G. E. Carman, Eds.), Vol. IV, pp. 276–320. Univ. of California, Division of Agricultural Science, Berkeley.

Smith, J. M. (1951). Biological control of weeds in Canada. Canadian National Weed Commission, East Sector Proceedings, 5, 95–97.

Sobhian, R., & Andres, L. A. (1978). The response of the skeletonweed gall midge, *Cystiphora schmidti* (Diptera: Cecidomyiidae), and gall mite, *Aceria chondrillae* (Eriophyidae) to North American strains of rush skeletonweed (*Chondrilla juncea*). Environmental Entomology, 7, 506–508.

Stein, H. (1987). Is the dismal science really a science? *Discover* (Nov.), pp. 96–99.

Takahashi, F. (1977). *Triops* spp. (Notostraca: Triopsidae) for the biological agents of weeds in rice paddies in Japan. Entomophaga, 22, 351–357.

Templeton, G. E., TeBeest, D. O., & Smith, R. J., Jr. (1978). Development of an endemic fungal pathogen as a mycoherbicide for control of northern jointvetch in rice. *In* "Proceedings of the IV International Symposium on Biological Control of Weeds, 1976, Gainesville, FL" (T. E. Freeman, Ed.), pp. 214–216.

Thomas, P. A., & Room, P. M. (1986). Taxonomy and control of *Salvinia molesta*. Nature, 320, 581–584.

Trujillo, E. E. (1985). Biological control of Hamakua Pa-Makani with *Cercosporella* sp. in Hawaii. *In* "Proceedings of the VI International Symposium on Biological Control of Weeds, 1984, Vancouver, BC" (E. S. Delfosse, Ed.), pp. 661–671.

Turner, C. E. (1984). Conflicting interests and biological control of weeds. *In* "Proceedings of the VI International Symposium on Biological Control of Weeds, 1984, Vancouver, BC" (E. S. Delfosse, Ed.), pp. 203–225.

Wang, R. (1986). Current status and perspectives of biological weed control in China. Chinese Journal of Biological Control, 1, 173–177.

Wapshere, A. J. (1974a). A strategy for evaluating the safety of organisms for biological weed control. Annals of Applied Biology, 77, 201–211.

Wapshere, A. J., Delfosse, E. S., & Cullen, J. M. (1974b). Host specificity of phytophagous organisms and the evolutionary centers of plant genera and subgenera. Entomophaga, 19, 301–309.

Wapshere, A. J., Delfosse, E. S., & Cullen, J. M. (1989). Recent developments in biological control of weeds. Crop Protection, 8, 227–250.

Williams, J. R. (1960). The control of black sage (*Cordia macrostachya*) in Mauritius: The introduction, biology and bionomics of a species of *Eurytoma* (Hymenoptera, Chalcidoidea). Bulletin of Entomological Research, 51, 123–132.

Zwölfer, H. (1965). Preliminary list of phytophagous insects attacking wild Cynareae (Compositae) in Europe. Commonwealth Institute of Biological Control Technical Communication No. 1, pp. 81–154.

Zwölfer, H. (1973). Competitive coexistence of phytophagous insects in the flower heads of *Carduus nutans*. *In* "Proceedings of the II International Symposium on Biological Control of Weeds, 1971, Rome, Italy" (P. H. Dunn, Ed.), pp. 74–81.

Zwölfer, H. (1988). Evolutionary and ecological relationships of the insect fauna of thistles. Annual Review of Entomology, 33, 103–122.

Zwölfer, H., & Harris, P. (1971). Host specificity determination of insects for biological control of weeds. Anual Review of Entomology, 16, 159–178.

35

Use of Plant Pathogens in Weed Control

E. N. ROSSKOPF
USDA–ARS
2199 South Rock Road
Fort Pierce, Florida

R. CHARUDATTAN
Plant Pathology Department
University of Florida
Gainesville, Florida

J. B. KADIR
Department of Plant Protection
Faculty of Agriculture
Universiti Putra Malaysia
Selangor, Malaysia

INTRODUCTION

The significant advances made in agricultural technology during the past 40 years have made it possible for U.S. farmers to produce surplus food for millions of people worldwide, despite only 2% of the American population being engaged in agriculture (Peeples, 1994). Weeds represent one of the most costly and limiting factors in crop production. Weeds can cause substantial losses in the form of yield reductions through competition with crop plants for light, water, nutrients, heat energy, carbon dioxide, and space. Reductions may also be caused by the production of growth-inhibiting compounds, a phenomenon referred to as allelopathy. Weeds can also cause losses by reducing food, feed, and fiber quality as well as increasing the cost of land preparation or harvesting. Over the past four decades, chemical herbicides have dominated weed management strategies in developing countries (Abernathy & Bridges, 1994; Wyse, 1992). In the United States, chemical herbicides have accounted for nearly 60% of the total amount of all agricultural pesticides applied (Aspelin, 1994). Although herbicides have proven to be an extremely effective means of vegetation control, their use has come with a number of direct and indirect costs that may now be considered as outweighing the benefits in many cases. Public concern has increased over the contamination of water sources as well as agronomic problems caused by the overapplication and soil persistence of many herbicides. Herbicide resistance, which has been noted in more than 100 weed species, continues to be of increasing concern (Holt, 1992). These factors, coupled with the banning of many chemical pesticides, more stringent and costly registration and regulation, and the necessity for nonchemical alternatives in environmentally sensitive areas, have further opened avenues for the use of plant pathogens as biological control agents for weeds.

BIOLOGICAL CONTROL STRATEGIES

Microbial biocontrol agents may be introduced and used in one of three ways, based on the types and numbers of applications required. Strategies are grouped into the classical approach, which consists of a single, inoculative introduction; the augmentative approach, which consists of periodic releases; and the inundative, or biopesticide, tactic. Epidemiological, density-dependent relationships between the biological control target and the biological control agent can be used to describe and distinguish these strategies (Charudattan, 1989). The classical approach is primarily directed toward plants that are exotic introductions which have become weed problems as a result of their growing in the absence of natural enemies. The classical approach may also be appropriate where cost of weed control expense is a limiting factor, such as in pasture or rangelands. Organisms used in the classical approach are assumed to require little or no additional manipulation. Many of the most important and difficult-to-control weeds have been intentionally or accidentally introduced into the area in which they are now a major problem. Approximately half of the weeds in the United States and 13 of the top 15 weeds are introduced species (Watson, 1991). Exotic weed species are able to thrive in the new area, due, in part, to the absence of natural enemies. The search for biocontrol agents to be used in the classical approach is usually performed in the original geographical range of the weed. Potential agents are then screened for virulence and those with potential are introduced into the new region with the expectation that the organisms' population will establish itself to provide weed suppression without complete eradication. In this manner it is expected that the pathogen and weed populations will equilibrate (Watson, 1991). These agents must have adequate host specificity that is limited to the target weed or to a small, related group of plants which contains no economically important members. The host range can be established by using a variety of protocols. Wapshere (1974) outlined a strategy that utilizes a centrifugal phylogenetic scheme. This begins with the testing of the organism, under optimal conditions for disease development, on a group of closely related plant species. The group should include not only important, closely related crop plants but also additional weed and native plant species. The testing of organisms under optimal conditions may artificially expand the host range of the organism, thus eliminating some potential biological control agents due to the risk assumed to be associated with their release (Adams, 1988).

SUCCESSFUL EXAMPLES USING THE CLASSICAL APPROACH

There are several outstanding examples of successful control achieved with this approach. One of the most commonly cited successes involves the use of *Puccinia chondrillina* Bubák & Syd. for the control of skeletonweed, *Chondrilla juncea* Linnaeus. This weed was a major problem in wheat-growing areas of Australia (Hasan, 1988). Strains of the rust, found in the Mediterranean, were tested for their specificity by inoculation of important Australian crop plants as well as members of the Asteraceae that are closely related to skeletonweed. Tests were performed under a variety of environmental conditions and none of the plants were found to be infected with the rust. Thus, the rust fungus was considered adequately specific to skeletonweed (Hasan, 1988). It was subsequently learned that the single strain of the rust released into Australia was only capable of controlling the predominant narrow-leaf phenotype of the weed; thus, multiple strains were required for successful control of the broad- and intermediate-leaf phenotypes (Hasan, 1985b). The first strain of this fungus was released in Australia in 1971. Within several months, the density of narrow-leaved skeletonweed started to decline substantially and there were inoculated areas in which the weed was no longer considered a problem (Hasan, 1988). The decrease in the narrow-leaved population allowed for the increase of the intermediate- and broad-leaved types which were resistant to the introduced rust strain. Addi-

tional strains were subsequently released. The search for strains of the rust that are virulent to the broad-leaved form continues. *Puccinia chondrillina* has also been released in the western United States, where it has proved to be the most effective of the biocontrol agents released against this weed (Supkoff *et al.,* 1988).

The second successful example is the control of European blackberries that had become weed problems in Australia and Chile. The rust fungus *Phragmidium violaceum* (Schultz) Winter was successfully established in Chile in 1973 and has controlled *Rubus constrictus* Lefevre et R. J. Mueller effectively and *R. ulmifolius* Schott. to a lesser extent (Hasan, 1988). This pathogen is effective by causing the plants to be more sensitive to frost damage and allowing for the invasion of secondary pathogens. This fungus was also recently released into Australia (Bruzzese & Field, 1985; Bruzzese & Hasan, 1986a, 1986b, 1986c).

Ageratina riparia (Regel) K. & R., also known as hamakua pamakani or mistflower, is a considerable problem on several Hawaiian Islands. This weed has been successfully controlled using *Entyloma ageratinae* Barreto and Evans, which was introduced to Hawaii in 1975. Greater than 95% control of the weed was achieved within the first year of release (Trujillo, 1985).

An isolate *Puccinia carduorum* Jacky collected in Turkey was released in the United States to control *Carduus theormeri* Weinman, a problematic weed in pastures and rangelands (Bruckart and Dowler, 1986). Although the host range of this fungus was initially determined to include several important species in *Carduus* and related genera, these were found to be resistant when inoculated in restricted field trials (Bruckart *et al.,* 1985). The fungus was tested for pathogenicity to 16 accessions of *C. thoermeri,* 10 accessions of related weedy *Carduus* spp., 22 native and 2 weedy *Cirsium* species, and *Cynara scolymus* (artichoke). It was determined that species outside the genus *Carduus* are unsuitable hosts for this rust fungus. This group of unsuitable hosts includes 25% of the *Cirsium* species from North America tested, artichoke, and representatives from two other genera, *Saussurea* and *Silybum*. Considering the host-range data and the very susceptible nature of *C. thoermeri,* a judgment was made that species other than *C. thoermeri* would not be threatened by the introduction of *P. carduorum* into North America to control musk thistle.

Musk thistle plots were inoculated successfully in the fall and spring of each year in 1987–1989. Artichoke and selected *Cirsium* spp. (nontarget plants) were planted between plots of inoculated musk thistle in two of three years. Only one rust pustule was found on an artichoke plant in 1989; all other plants remained rust-free despite severe disease on surrounding musk thistle plants. This indicated that *P. carduorum* poses no threat to these to these nontarget plants. The rust overwintered on musk thistle. Its spread

TABLE 1   Weed Targets and Exotic Pathogens Intentionally Released for Biological Weed Control
Projects (Watson, 1991)

Target weed	Pathogen	Country of introduction	References
Asteraceae			
Acroptilon repens (L.) DC (Russian knapweed)	*Subanguina picridis* (Kirj.) Brezeski	Canada and U.S. (continental)	Watson, 1986a, 1986b
Ageratina riparia (Regel) K. & R. (hamakua pamakani)	*Entyloma ageratinae* Barr. & Evans	U.S. (Hawaii)	Barreto & Evans, 1988; Trujillo, 1985; Trujillo *et al.,* 1988
Carduus nutans L. (musk thistle, nodding thistle)	*Puccinia carduorum* Jacky	U.S. (continental)	Bruckart & Dowler, 1986; Bruckart *et al.,* 1985; Politis & Bruckart, 1986; Politis *et al.,* 1984
Chondrilla juncea L. (skeletonweed)	*Puccinia chondrillina* Buback & Syd.	Australia and U.S. (continental)	Adams, 1988; Adams & Line, 1984a, 1984b; Blanchette & Lee, 1981; Burdon *et al.,* 1984; Cullen, 1978; Emge *et al.,* 1981; Hasan, 1972, 1981, 1985b; Hasan & Wapshere, 1973; Supkoff *et al.* 1988
Eupatorium adenophorum Spreng. (crofton weed)	*Phaeoramularia* sp.	South Africa	Morris, 1991
Fabaceae (Leguminosae)			
Acacia saligna (Labill.) Wendl. (Acacia)	*Uromycladium tepperianum* (Sacc.) McAlp.	South Africa	Morris, 1991, 1997
Galega officinalis L. (galega, goat's rue)	*Uromyces galegae* (Opic) Sacc.	Chile	Oehrens & Gonzáles, 1975
Rosaceae			
Rubus spp. (blackberry)	*Phragmidium violaceum* (Schultz) Winter	Chile	Oehrens, 1977; Oehrens & Gonzáles, 1974

was limited during the rosette stage, and the disease became severe only during the bolting stage (i.e., stem elongation). Senescence of rust-infected musk thistles was accelerated and seed production was reduced by 20–57%. Thus, the results confirmed that *P. carduorum* can contribute to the control of musk thistle (Baudoin *et al.,* 1993). Accordingly, permission was granted in 1987 for a limited field release of the rust in Virginia.

Acacia saligna (Labill.) Wendl. poses a significant weed problem in South Africa. A gall-forming rust fungus, *Uromycladium tepperianum* (Sacc.) McAlp. was found in Australia and introduced into South Africa in 1987. Since then, the fungus has spread to nearly all areas in which the weed occurs and has decreased the density of this invasive tree by at least 80%. The rust fungus is reported to be successfully spreading and suppressing the weed populations (Morris, 1997).

Projects leading to intentional release of exotic phytopathogens for weed control are listed in Table 1. In addition to the foregoing examples, there are a substantial number of projects aimed at developing classical biological control agents (Table 2) and this approach is becoming increasingly important as more and more exotic weeds are becoming

invasive pests and an economical alternative is sought to the costly chemical controls.

COMMERCIALIZATION AND THE BIOHERBICIDE APPROACH

The concept of using inundative inoculation with a pathogen for weed control has gained momentum and attention in recent years. Research in the late 1960s through the 1980s culminated in the successful registration of two plant pathogens as bioherbicides and their subsequent use in commercial agriculture for nearly two decades (Charudattan, 1991). Between 1980 and 1997, five bioherbicides were registered in the United States, Canada, and Japan. DeVine®, composed of the pathogen *Phytophthora palmivora* [Butler] Butler to control *Morrenia odorata* (Hook. & Arn.) Lindl. in Florida citrus, has been extremely successful in controlling that weed. This fungus was the first commercially available mycoherbicide (Charudattan, 1991; Kenney, 1986). It is applied as a liquid containing chlamydospores and can provide as high as 96% weed control within

TABLE 2   Weed Targets and Exotic Pathogens Considered or under Investigation in Classical Biological
Weed Control Programs (Watson, 1991)

Target weed	Pathogen	Intended country of introduction	References
Asclepiadaceae			
Cryptostegia grandiflora R. Br.	*Maravalia cryptostegiae* (Cumm.) Ono	Australia	Tomley, 1989
Morrenia odorata (H. & A.) Lindle. (stranglervine, milkweed vine)	*Aecidium asclepiadinum* Speg.	U.S. (continental)	Charudattan *et al.*, 1978b
	Araujia mosaic virus	U.S. (continental)	Charudattan *et al.*, 1980
	Puccinia araujae Lév.	U.S. (continental)	Charudattan *et al.*, 1978a
Asteraceae			
Centaurea diffusa Lam. (diffuse knapweed)	*Puccinia jaceae* Otth	Canada	Hasan *et al.*, 1989; Watson & Alkhoury, 1981; Watson & Clément, 1986; Watson *et al.*, 1981
Centaurea maculosa Lam. (spotted knapweed)	*Puccinia centaureae* DC	Canada	Watson & Clément, 1986
Centaurea solstitialis L. (yellow starthistle)	*Puccinia jaceae* Otth	U.S. (continental)	Bruckart, 1989; Bruckart & Dowler, 1986
Chondrilla juncea L. (skeletonweed)	*Erysiphe cichoracearum* DC ex Marat	Australia	Hasan, 1980, 1983
	Leveillula taurica (Lév.) Arnaud f. sp. *chondrillae*	Australia	Hasan, 1980, 1983
Parthenium hysterophorus L. (parthenium)	*Puccinia abrupta* Diet. & Holw. var. *partheniicola* (Jackson) Parmelee	Australia	Evans, 1986; McClay, 1985; Parker, 1989
Senecio jacobaea L. (tansy ragwort)	*Puccinia expansa* Link. (=*P. glomerata* Grev.)	Australia	Alber *et al.*, 1986
Xanthium spp. (cocklebur)	*Puccinia xanthii* Schw.	Australia	Hasan, 1980, 1983
Boraginaceae			
Heliotropium europaeum L. (common heliotrope)	*Cercospora* sp.	Australia	Hasan, 1985a
	Uromyces heliotropii Sredinski	Australia	Hansan, 1985a
Euphorbiaceae			
Euphorbia esula L. complex (leafy spurge)	*Melampsora euphorbiae* (Schub.) Cast.	U.S. (continental)	Bruckart *et al.*, 1986
	Uromyces scutellatus (Pers.) Lév.	U.S. (continental)	Défago *et al.*, 1985
Hydrocharitaceae			
Hydrilla verticillata (L.f.) Royle (hydrilla)	*Fusarium roseum* "Culmorum" (Link ex Fr.) Snyder & Hans.	U.S. (continental)	Charudattan *et al.*, 1981; Smither-Kopperl *et al.*, 1998
Liliaceae			
Asphodelus fistulosus L. (onion weed)	*Puccinia barbeyi* (Roum.) P. Magn.	Australia	Anonymous, 1988
Polygonaceae			
Emex spp.	*Cercospora tripolitana* Sacc. et Trott.	Australia	Anonymous, 1988
	Peronospora rumicis Cda.	Australia	Anonymous, 1988
Rumex crispus L. (curly dock)	*Uromyces rumicis* (Schum.) Winter	U.S. (continental)	Inman, 1971; Schubiger *et al.*, 1985
Pontederiaceae			
Eichhornia crassipes (Mart.) Solms-Laub. (waterhyacinth)	*Uredo eichhorniae* Gonz-Frag. & Cif.	U.S. (continental)	Charudattan & Conway, 1975; Charudattan *et al.*, 1978b
Rosaceae			
Rubus fruticosus L. (European blackberry)	*Phragmidium violaceum* (Schultz) Winter	Australia	Bruzzese & Field, 1985; Bruzzese & Hasan, 1986a, 1986b, 1986c, 1987

10 weeks (Ridings, 1986). DeVine® is available by special order through Abbott Laboratories (Abbott Park, IL).

Collego®, composed of the fungus *Colletotrichum gloeosporioides* (Penz.) Penz. & Sacc. f. sp. *aeschynomene,* is registered for the control of northern jointvetch (*Aeschynomene virginica* [L.] B. S. P.) in rice and soybeans (TeBeest & Templeton, 1985). Collego® was developed by scientists at the University of Arkansas, the U.S. Department of Agriculture–Agricultural Research Service, and the Upjohn Company (Bowers, 1986). This fungus is formulated as a wettable powder and results generally in greater than 85% control (Smith, 1986). Collego® is now available from Encore Technologies (Minnetonka, MN). The host range of this fungus is more broad than was originally determined and risks posed to nontarget crop plants have been addressed through label restrictions (Te-Beest & Templeton, 1985).

Colletotrichum gloeosporioides f. sp. *malvae* was developed for control of round-leaved mallow (*Malva pusilla* Sm.) and was registered in Canada in 1992 for use in field crops (Makowski & Mortensen, 1992; Mortensen & Makowski, 1997). This product is not currently available, as the industrial partner, PhilomBios (Saskatoon, SK, Canada), has been unable to establish a viable market for this product (Boyetchko, 1998).

Puccinia canaliculata (Schw.) Lagerh., a rust fungus, has been registered as the bioherbicide Dr. BioSedge® for the control of yellow nutsedge (*Cyperus esculentus* L.) (Greaves & MacQueen, 1992; S. C. Phatak, personal communication). Work on this system began in 1979 (Phatak *et. al.,* 1981, 1983) and the bioherbicide was registered in 1994. This rust pathogen is reported to spread rapidly and can cause as high as 90% reduction in fresh weight of the weed when applied at the rate of 5 mg of uredospores per hectare. However, the strain of the rust fungus from the state of Georgia registered as Dr. BioSedge® does not infect purple nutsedge (Scheepens & Hoogerbrugge, 1991) or several yellow nutsedge biotypes that are naturally resistant to specialized strains of the fungus (Bruckhart *et al.,* 1988). When applied in early spring, *P. canaliculata* inhibits flowering and reduces yellow nutsedge stand by as much as 46%. It is also reported to inhibit new tuber formation by as much as 66% (Phatak *et al.,* 1983). The obligately parasitic nature of this fungus makes production of inoculum a problem and therefore Dr. BioSedge® is currently unavailable for commercial use.

Another isolate of *Colletotrichum gloeosporioides* has been developed for use in China for the control of *Cuscuta* L. spp. (dodder). It is currently being marketed under the name Luboa No. 1 and provides control of 80% or more of the weed (Mortensen, 1998).

Two different products based on the fungus *Chondrostereum purpureum* (Pers. ex Fr.) Pouzar, a wound-invading pathogen of broad-leaved trees, are currently in use or under development as bioherbicides. Biochon™ is used (not registered) for the control of *Prunus serotina* Ehrh. in Dutch forests and ECO-clear™ is under development for North American registration and use as a stump application in Canadian forests (Dumas *et al.,* 1997; Prasad, 1994; Shamoun & Hintz, 1998).

Another tree pathogen, *Cylindrobasidium laeve* (Pers: Fr.) Chamuris, has been registered in South Africa and is commercially available for use for stump treatment to prevent regrowth of wattles, *Acacia mearnsii* de Wildman and *A. pycnantha* Benth., in the Cape Province (Morris *et al.,* 1998). Methods for mass production of inoculum, based on

a laboratory-scale pilot plant, have been developed. Basidiospores of the pathogen are produced on solid substrate and a concentrated spore suspension in oil is packed in plastic sachets for marketing and use (Morris *et al.,* 1998). This bioherbicide, named Stumpout®, is now available for commercial use in wattle-clearing programs.

The bacterial pathogen *Xanthomonas campestris* Migula pv. *poae* has recently been registered as a bioherbicide. Japan Tobacco, Inc., has registered this bacterial agent under the name Camperico® to control annual bluegrass (*Poa annua* L.) in bermudagrass (*Cynodon dactylon* [L.] Pers.), zoysia grass (*Zoysia matrella* Merr.), and Kentucky bluegrass (*Poa pratensis* L.) turfs and golf greens (Imaizumi *et al.,* 1997). It is necessary to mow the turf prior to application to provide a means of host entry for the pathogen (Johnson, 1994).

A second bacterium, *Pseudomonas syringae* van Hall pv. *tagetis,* has potential as a bioherbicide for the control of Canada thistle in soybean (Johnson *et al.,* 1995). This pathogen causes apical chlorosis and is currently being tested for control of several members of the Asteraceae (Johnson *et al.,* 1996). Encore Technologies has licensed *P. syringae* pv. *tagetis* from the University of Minnesota and is pursuing its commercial development (D. R. Johnson, Encore Technologies, personal communication).

Among other agents under development for possible commercialization is *Alternaria cassiae* Jurair & Khan, which was researched extensively for possible use as a bioherbicide for Acacia Mearnsii sicklepod [*Senna obtusifolia* (L.) H. S. Irwin & Bar.], coffee senna (*Cassia occidentalis* L.), and showy crotalaria (*Crotalaria spectabilis* Roth) (Walker & Boyette, 1985). It was initially developed under the trade name CASST by Mycogen Corp. (San Diego, CA) (Bannon, 1988). In regional field trials, conducted under the auspices of the USDA–CSREES Cooperative Regional Research Project S-136, this fungus provided control that was comparable to that of chemical herbicides (Charudattan *et al.,* 1986). Mycogen has since relinquished the license to develop this fungus (Boyetchko, 1999). It is currently under investigation for use in Brazil (R. Charudattan, personal communication), possibly as a preemergent, soil-incorporated bioherbicide (Pitelli *et al.,* 1998).

Cercospora rodmanii Conway, a host-specific pathogen of waterhyacinth [*Eichhornia crassipes* (Mart.) Solms], was field tested between 1974 and 1984. The University of Florida patented the use of this pathogen and Abbott Laboratories was licensed to develop it as a bioherbicide. The pathogen was formulated as a wettable powder and was capable of reducing waterhyacinth growth by as much as 90% when used in combination with insect biological control agents (Charudattan *et al.,* 1985). However, when used alone, the efficacy of this pathogen was inadequate when compared to the available chemical herbicides. This inadequate efficacy, coupled with economic consider-ations, halted attempts at commercialization (Charudattan, 1990).

Two fungi have recently been patented by the University of Florida for use as biological control agents of weeds. *Phomopsis amaranthicola* sp. nov. [U.S. Patents 5,393,728 and 5,510,316 (Charudattan *et al.,* 1996b)] is a pathogen of pigweeds (*Amaranthus* spp.) that is effective against a num-ber of species in the genus. This fungus has been tested in greenhouse and field trials and found to be highly promis-ing (Rosskopf, 1997; Rosskopf *et al.,* 1996). Symptoms produced by this fungus begin as leaf lesions, which co-alesce (Fig. 1a), causing premature leaf abscission. Lesions then develop on the stem (Fig. 1b), which girdle the stem and cause the plant to topple. The second fungus, *Dacty-laria higginsii* (Luttrell) M. B. Ellis [U.S. Patent 5,698,491 (Kadir & Charudattan, 1996)], is effective in controlling several *Cyperus* spp., including two of the worst agricul-tural weeds, yellow nutsedge and purple nutsedge (*Cyperus esculentus* L. and *C. rotundus* L.). The fungus not only kills the above-ground shoots but also limits tuber production and the ability of the weed to compete with crops, such as tomato and pepper (Kadir, 1997). The pathogen produces a leaf blight (Fig. 2a), which has been successful in field trials (Fig. 2b.)

Colletotrichum truncatum (Schwein) Andrus & W. D. Moore is under development for control of hemp sesbania (*Sesbania exaltata* (Raf.) Rydb. ex A. W. Hill) in cotton, rice, and soybean (Boyette, 1994; Boyette *et al.,* 1993b; Egley *et al.,* 1993; Jackson *et al.,* 1996; Jackson & Slinin-ger, 1993; Schisler *et al.,* 1991, 1995; Schisler & Jackson, 1993). In small-scale plots hemp sesbania was effectively controlled in soybean and rice with a 1:3 (v:v) unrefined corn oil/*Colletotrichum truncatum* formulation containing 0.2% Silwet L-77. A dried formulation consisting of dex-trose and hydrated silica gel suspended in experimental oil formulations was also effective in controlling hemp ses-bania. Pesta formulations (explained later) containing co-nidia and microsclerotia provided some control when ap-plied preemergence or when preplant incorporated but were

A   B

FIGURE 1   Leaf spots (a) and stem lesions (b) caused on *Amaranthus hybridus* by *Phomopsis amaranthicola,* a biological control agent for pigweeds and amaranths.

FIGURE 2 (a) Leaf blight on *Cyperus esculentus* (purple nutsedge) resulting from application of the potential biological control agent *Dactylaria higginsii.* (b) Control of purple nutsedge in miniplot field trials with application of *Dactylaria higginsii.* Two applications of the fungus resulted in complete control of the weed.

significantly less effective than postemergence application of fungus–oil formulations. Control of hemp sesbania in cotton was significantly reduced by fungicide seed treatments and in-furrow fungicide and insecticide treatments. Rice formulations of the fungus were tested for viability and virulence to hemp sesbania. The fungus was highly virulent to weed seedlings even after 8 years of storage at −20°C and only slightly less virulent at 4°C. Survival of *C. truncatum* conidia in formulations stored at room temperature was significantly less than at the lower temperatures (C.D. Boyette, personal communication).

An *Alternaria* sp. is undergoing evaluation as a bioherbicide for dodder (*Cuscuta* spp.) in cranberries. Results from field trials and greenhouse studies are promising. Plots that had been treated the previous year with Pesta formulation of the pathogen experienced no infestation of dodder the following year, suggesting a possible residual effect of the pathogen (Bewick *et al.,* 1986; T. A. Bewick, Cranberry

Research Station, University of Massachusetts, personal communication).

Exserohilum monoceras is undergoing evaluations for development as a bioherbicide to control barnyard grass (*Echinochloa crus-galli* (L.) Beauv. var. *crus-galli*) and other *Echinochloa* species (*Echinochloa crus-galli* var. *formosensis, E. colona* (L.) Link, and *E. oryzicola* (Ard.) Fritsch) in Japan (Tsukamoto *et al.,* 1997). The same pathogen is being developed cooperatively by scientists from Australia, Canada, the Philippines, and Vietnam (Zhang *et al.,* 1996; Tuat *et al.,* 1998).

The foliar fungal pathogen *Ascochyta caulina* appears to have good prospects of being developed as a bioherbicide to control common lambsquarters (*Chenopodium album* L.). The ability of this fungus to control common lambsquarters in corn and sugar beet was studied in Europe. Both preemergence soil incorporation and postemergence spray application of the fungus were studied. Application of *A. caulina* spores to soils resulted in diseased, stunted, and dead seedlings of the weed. Disease incidence and weed mortality were influenced by spore density, soil water content, and soil type but not by sowing depth. Spore densities of 10^9-10^{10} m^{-2} were required for 50% mortality of emerged common lambsquarters plants (Kempenaar, 1995; Kempenaar *et al.,* 1996b). The postemergence application of the fungus caused necrosis and mortality of the weed seedlings. Up to 65% mortality of treated seedlings was obtained and sublethally diseased plants were reduced in growth. Maximum dry matter, number of fruits per plant, and seed weight of the weed were reduced as a result of fungal infection. Competitiveness of the weed was also reduced. Yield reduction was prevented by application of *A. caulina* in corn but not in sugar beet (Kempenaar *et al.,* 1996a). Attempts are underway to integrate the use of *A. caulina* in weed management systems by making use of pre- and postemergent applications of the fungus to reduce the number of emerging seedlings of common lambsquarters, the number of surviving plants, and the number of seeds produced (Scheepens *et al.,* 1998).

A list of bioherbicides, including the presently registered agents, those that are said to be used as unregistered, local-use agents, and experimental candidates, is provided in Table 3.

DELETERIOUS RHIZOBACTERIA AND WEED CONTROL

This group of microorganisms has been characterized as nonparasitic bacteria that colonize plant roots and cause decreases in plant growth. Many of these bacteria are specific in their growth-suppressive activity. The initial discovery of these bacteria was on crop plants. A weed association was first discovered on downy brome (*Bromus tectorum* L.)

TABLE 3   List of Registered and Unregistered Bioherbicides or Agents under Development

Weed	Pathogen (registered or trade mark name)	Target crop(s)	Country[a]	Status[b]
Abutilon theophrasti	*Colletotrichum coccodes*	Soybean	Canada	4*
Acacia mearnsii	*Cylindrobasidium laeve* (Stumpout®)	Tree plantations	South Africa	5
Aeschynomene virginica	*Colletotrichum gloeosporioides* f. sp. *aeschynomene* (Collego®)	Rice and soybean	U.S., Arkansas	5
Amaranthus spp.	*Phomopsis amaranthicola*	Vegetables	U.S., Florida	3
Chenopodium album	*Ascochyta caulina*	Various	Holland	4
Cuscuta spp.	*Alternaria* sp.	Cranberries	U.S., Florida	4
Cyperus spp.	*Dactylaria higginsii*	Various	U.S., Florida	3
Cyperus esculentus	*Puccinia canaliculata* (Dr. BioSedge®)	Various	U.S., Georgia	5*
Cyperus rotundus	*Cercospora caricus*	Various	Brazil; Israel	3
Cytisus scoparius	*Fusarium tumidum*	Tree plantations	New Zealand	3
			Canada	3
Echinochloa spp.	*Exserohilum monoceras*	Rice	Australia; Japan	3
			Philippines–Canada[c]	3
Eichhornia crassipes	*Cercospora rodmanii*	Aquatic habitats	U.S., Florida;	4*
			South Africa;	3
			Egypt; India; SE Asia	3
Euphorbia heterophylla	*Helminthosporium* sp.	Various	Brazil	3
Grass weeds	*Drechslera* spp. and *Exserohilum* spp.	Cereals	Vietnam–Australia	2
		Citrus	U.S., Florida	3
Hakea sericea	*Colletotrichum gloeosporioides*	Tree plantations	South Africa	4
Imperata cylindrica	*Colletotrichum caudatum*	Various	Malaysia	1
Malva pusilla	*Colletotrichum gloeosporioides* f. sp. *malvae* (BioMal®)	Various	Canada	5*
Mikania micrantha	*Cercospora mikaniicola*	Plantation crops	Malaysia	2
Morrenia odorata	*Phytophthora palmivora* (De Vine®)	Citrus	U.S., Florida	5
Poa annua	*Xanthomonas campestris* pv. *poae* (Comperico®)	Turf grass	Japan	4
Pteridium aquilinum	*Ascochyta pteridis*	Pastures	Scotland, U.K.	3
Rottboellia chochinchinensis	*Sporisorium ophiuri*	Cereals	Thailand–U.K.	3
	Colletotrichum sp. nov. nr. *graminicola*	Cereals	Thailand–U.K.	3
Sagittaria spp.	*Rhynchosporium alismatis*	Rice	Australia	2
Senecio vulgaris	*Puccinia lagenophorae*	Various	U.K.–Switzerland	4
Senna obtusifolia	*Alternaria cassiae*	Soybean	Brazil	3
Sesbania exaltata	*Colletotrichum truncatum*	Soybean and rice	U.S., Mississippi	4
Solanum viarum	*Pseudomonas solanacearum*	Citrus and sod	U.S., Florida	3
	Colletotrichum spp.		Brazil–U.S., Florida	2
Sphenoclea zeylanica	*Alternaria* sp.	Rice	Philippines	2
	Colletotrichum gloeosporioides	Rice	Malaysia	2
Striga hermonthica	*Fusarium nygamai*	Various	Sudan–Germany	3
	Fusarium oxysporum	Cereals	West Africa–Canada	3
	Fusarium semitectum var. *majus*	Sorghum	Sudan–Germany	3
Taraxacum officinale	*Sclerotinia sclerotiorum*	Lawn and garden	Canada	4
Ulex europaeus	*Fusarium tumidum*	Plantation crops	New Zealand	3
Various annual weeds	*Myrothecium verrucaria*	Various	U.S., Maryland	3
Various broad-leaved trees	*Chondrostereum purpureum* (Biochon™)	Tree plantations	Holland	5
	(ECO-clear™)		Canada	4
Various composite weeds	*Pseudomonas syringae* pv. *tagetis*	Various	U.S.; Canada	4
Xanthium spp.	*Colletotrichum orbiculare*	Various	Australia	4*

Note: Compiled from unpublished and published reports.

[a]"–", cooperative work is in progress between the indicated countries; ";", the indicated countries are engaged in independent work.

[b]Status: 1, in exploratory phase; 2, laboratory and/or greenhouse testing underway; 3, field trials in progress; 4, under early commercial or practical development; 4*, commercial development tried but registration uncertain; 5, available for commercial or practical use; 5*, available registered as a microbial herbicide but currently unavailable for use due to economic reasons.

[c]Countries joined by a hyphen are engaged in a cooperative project to develop the bioherbicide agent.

TABLE 4   Biological Weed Control Projects Based on Rhizobacteria (Kremer & Kennedy, 1994)

Investigator/location	Weeds[a]	Ecosystem[b]	Rhizobacteria	Status[c]	References
A. Caesar USDA–ARS Bozeman, MT	EPHES CENRE	Rangelands	*Agrobacterium* *Pseudomonas* *Xanthomonas*	A, B, C	Caesar, 1994 Caesar *et al.,* 1992 Pers. comm.[d]
J. Clapperton & R. Blackshaw Agriculture Canada Lethbridge, AB	SETVI AVEFA	Cereal crops Rangelands	*Pseudomonas*	A, B,	Pers. comm.
C. Dorworth, D. Macy, & R. Winder Canadian Forest Service Victoria, BC	CLMCD	Forest nursery	*Pseudomonas*	A	Pers. comm.
M. Dumas Canadian Forest Service Sault Ste. Marie, ON	*Rubus* spp.	Forest	Unidentified	A	Pers. comm.
L. Elliott USDA–ARS Corvallis, OR	POANN	Seed crops Turf grass	*Pseudomonas*	A, B, C	Elliott *et al.,* 1994
J. Frey Mankato State University Mankato, MN	XANST	Row crops	*Pseudomonas*	A, B,	Pers. comm.
P. Harris & P. Stahlman Kansas State University Hays, KS	BROTE, BROJA BROSE, AEGCY	Cereal crops	*Enterobacter* *Pseudomonas* *Xanthomonas*	A, B, C, D	Harris & Stahlman, 1991, 1992, 1993a, 1993b
A. Kennedy USDA–ARS Pullman, WA	BROTE, BROJA AEGCY, SSYAL AMARE	Cereal crops	*Pseudomonas flu- orescens*	A, B, C, D	Johnson *et al.,* 1993 Kennedy *et al.,* 1991 Kennedy *et al.,* 1989
R. Kremer USDA–ARS Columbia, MO	ABUTH, AMARE *Ipomoea* spp. *Setaria* spp. EPHES	Row crops Rangelands	*Pseudomonas* *Enterobacter* *Erwinia herbicola* *Flavobacterium*	A, B, C	Kremer, 1993 Kremer *et al.,* 1990 Souissi & Kremer, 1994
K. Mortensen & S. Boyetchko Agriculture Canada Saskatoon, SK	BROTE SETVI	Cereal crops Pastures	*Pseudomonas*	A, B	Boyetchko & Morten- sen, 1993
H. D. Skipper Clemson University Clemson, SC	BROTE	Cereal crops	*Pseudomonas*	A	Skipper *et al.,* 1996 Pers. comm.

[a] ABUTH, *Abutilon theophrasti* Medicus, velvetleaf; AEGCY, *Aegilops cylindrica* Host., joined gaotgrass; AMARE, *Amaranthus retroflexus* L., redroot pigweed; AVEFA, *Avena fatua* L., wild oat; BROJA, *Bromus japonicus* Thunb. ex Murr., Japanese brome; BROSE, *Bromus secalinus* L., cheat; BROTE, *Bromus tectorum* L., downy brome; CENRE, *Acroptilon repens* L., Russian knapweed; CLMCD, *Calamagrostis canadensis* var. *canadensis* Michaux, Canada reedgrass; EPHES, *Euphorbia esula* L., leafy spurge; POANN, *Poa annua* L., annual bluegrass; SETVI, *Setaria viridis* (L.) Beauv., green foxtail; SSYAL, *Sisymbrium altissimum* L., tumble mustard; XANST, *Xanthium strumarium* L., common cocklebur.

[b] Situation in which target weed occurs.

[c] Status: A, laboratory screening; B, greenhouse/growth chamber screening; C, small-scale field testing; D, large-scale field testing.

[d] Personal communication by authors with original investigators, 1993–1994.

in winter wheat (Cherrington & Elliott, 1987). These bacteria are used similarly to the fungal bioherbicides in that they are applied inundatively, although the application is made to the soil prior to weed emergence. This method does not result in eradication of the target weed species but significantly depresses its competitive ability (Kremer & Kennedy, 1996).

Table 4 provides a partial listing of programs using rhizobacteria. Accomplishments in this area, as listed by Kremer and Kennedy (1996) include control of downy brome using *Pseudomonas* spp. This system has been successful under field conditions when bacteria were applied to the soil. Grain yields of the winter wheat in plots treated with the bacteria were significantly higher than in plots where the downy brome was not suppressed. Isolates Pp001 and Pf239 of another species of *Pseudomonas* were tested for their effects on velvetleaf seedling emergence and found to reduce seedling emergence to less than 55% in treated soil. In both of these examples, the bacterial isolates also reduced the numbers of seeds produced, thus potentially

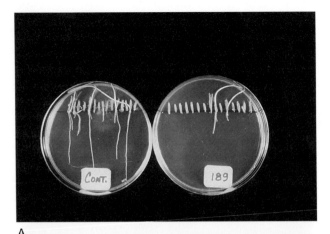

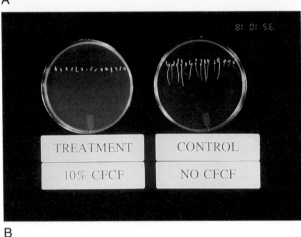

FIGURE 3 Bioassay of deleterious rhizobacteria effects on (a) downy brome (*Bromus tectorum* L.) and (b) green foxtail [*Setaria viridis* (L.) Beauv.] seed germination. Photographs provided by S. Boyetchko.

contributing to long-term weed control (Kremer & Kennedy, 1996).

The use of these agents is also being pursued in Canada, where more than 2000 isolates of rhizobacteria obtained from prairie soils are under investigation (Boyetchko, 1999). Some isolates, tested in a variety of bioassays (Figs. 3a and 3b), have proven to be extremely successful in minimizing weed seed germination. Host-range tests of many of these isolates have shown little or no detrimental effects on nontarget hosts, including spring and winter cereal crops. In some cases, the bacteria had growth-promoting effects on the tested crops (Boyetchko, 1999).

DEVELOPMENTAL TESTING OF MYCOHERBICIDES

The most commonly cited explanation for the failure of fungal plant pathogens used for biological control agents of weeds to cause severe disease is the requirement of an extended dew period for adequate infection. Some pathogens, such as *Sphacelotheca holci* Jack (= *S. cruenta* [Kuhn.] Potter), long considered a potential biological control agent for johnsongrass (*Sorghum halepense* [L.] Pers.), do not have this as a major constraint. Researchers working with this pathogen have reported that, although initial greenhouse studies indicated a need for free moisture for infection, a 55% infection of plants occurred regardless of the availability of free moisture. Although this is a relatively low infection rate, the authors found it to be adequate to reduce the competitive ability of the weed (Massion & Lindow, 1986).

Daniel and coworkers (1973), studying the use of a strain of *Colletotrichum gloeosporioides* (the Collego® pathogen) to control northern jointvetch (*Aeschynomene virginica*), found that infection was not limited by the absence of a dew period but that fastest onset of disease was achieved if the inoculated plants were exposed to 80% relative humidity overnight. This fungus was reported to be capable of causing infection at greenhouse temperatures ranging from 23 to 32°C. The same organism was capable of infecting plants ranging from 5.0 to 30.5 cm tall, although the greatest percentage of mortality was achieved with the smallest-size class of plants.

A species of *Phomopsis* (Sacc.) Bubák, *P. convolvulus* Ormeno, was evaluated for its potential for the control of field bindweed (*Convolvulus arvensis* L.) (Morin *et al.,* 1990a; Ormeno-Nunez *et al.,* 1988; Sparace *et al.,* 1991; Vogelsang *et al.,* 1994). This fungus caused the highest levels of plant mortality (55%) when the inoculated plants were exposed to a minimum of 18 h of dew and an inoculum concentration of 10^9 conidia/m². This treatment resulted in 55% plant mortality (Morin *et al.,* 1990b). Isolates of *Bipolaris setariae* (Saw.) Shoemaker examined for the control of goosegrass (*Eleusine indica* [L.] Gartner) required a 48-h dew period for 100% infectivity. The optimal temperature for disease development in this system was 24°C. The isolate was able to control goosegrass plants of the 2- and 4-week growth stages (Figliola *et al.,* 1988).

Similar results were obtained in studies involving the development of anthracnose of spiny cockleburr (*Xanthium spinosum* L.), caused by *Colletotrichum orbiculare* (Berk. et Mont.) v. Arx (Auld *et al.,* 1988, 1996). The optimal temperatures for disease development by this organism were between 20 and 25°C, with a dew period of 48 h. As with several other potential biological control organisms, if this dew period was split into 12-h exposures with 12 h separating dew exposure, the disease severity was decreased. A significant level of mortality of cockleburr was not achieved with any of the treatments, with disease ratings approaching five, which did not constitute plant death (McRae & Auld, 1988).

Colletotrichum truncatum (Schw.) Andrus and Moore,

used for the control of Florida beggarweed (*Desmodium tortuosum* [Sw.] DC), had optimal disease development with 14–16 h of dew at temperatures between 24 and 29°C. Conidial suspensions with concentration of 10^5–10^7 conidia/ml were most effective in controlling plants in the cotyledon stage of growth. The efficacy of the pathogen diminished as the plants matured (Cardina et al., 1988). The percentage of control obtained at 18°C was significantly lower.

Microsphaeropsis amaranthi (Ell. and Barth.) Heiny and Mintz (= *Aposphaeria amaranthi* Ell. and Barth.), a pathogen of tumble pigweed (*Amaranthus albus* L.), was found to cause the most significant levels of pigweed mortality in a series of growth-chamber experiments when applied to plants of the four-leaf stage and exposed to a minimum of 8 h of dew. There were no significant differences in the mortality of inoculated plants exposed to 8, 12, and 24 h of dew. This dew period requirement is substantially lower than those for many of the potential biological control agents that have been evaluated. There were also no significant differences in mortality when conidial suspensions ranging from 10^4 to 10^7 were used for inoculation. In this case, there was no detrimental effect produced by delaying the onset of dew and therefore the disease resulted in 100% plant mortality (Mintz et al., 1992).

Thus, it is clear that an understanding of the epidemiological parameters is essential for determining the real potential of biological control agents for weeds. Although many organisms that are effective at initial, small-scale testing do not prove to be efficacious in larger fields, determination of the basic environmental conditions necessary for disease development will assure some measure of success in field use.

FIELD ASSESSMENT

Field evaluation is one of the most important components of a complete evaluation of the efficacies of pathogens as biological control agents for weeds. Although an organism may perform effectively in greenhouse or growth-chamber experiments, the use of the fungus in the field is often characterized by variable activity against the target weed. *Puccinia carduorum* was evaluated in the field for the control of musk thistle (*Carduus theormeri*). Although this pathogen had performed effectively in the greenhouse, providing significant reductions in plant-matter accumulation after both single and multiple applications of the pathogen, field performance was limited to accelerated senescence and reductions in seed production (Baudoin et al., 1993).

Colletotrichum orbiculare was tested in small-scale plots for the control of *Xanthium spinosum*, with promising results. It was further tested at four sites with variable environmental conditions. At two sites with artificial dew periods of 18 h, the pathogen produced 100% plant mortality. In sites that were not treated with an artificial dew period, a maximum plant mortality of 50% was reached (Auld et al., 1990). The pathogen applied in oil emulsions proved to be more effective in the field (Klein et al., 1995a, 1995b, 1995c). This pathogen has also been tested in combination with the rust fungus *Puccinia xanthi* (Morin et al., 1993a, 1993b), with synergistic effects resulting.

Field trials testing the effect of *Colletotrichum truncatum* (Schw.) Andrus and Moore on the growth of hemp sesbania (*Sesbania exaltata* [Rydb.] ex Hill) included the application of the pathogen to plots at two rates in either an aqueous suspension or an invert emulsion. The maximum level of weed mortality in plots treated with the aqueous suspension was 42%, whereas in invert emulsion suspensions of the pathogen there was 97% mortality in the plots treated with the higher application rate. This was comparable to the control with the herbicide acifluorfen. The effect of the invert emulsion applied without the fungus was statistically similar to the effect of the pathogen in the aqueous suspension (Boyette et al., 1993b).

Alternaria crassa (Sacc.) Rands was evaluated for the control of jimsonweed (*Datura stramonium* L.). It was effective in one field-trial location using a mycelial preparation, whereas at the second site plant mortality reached a maximum of only 29% (Boyette et al., 1986). This system demonstrates a common phenomenon that of the reduced efficacy of many pathogens in the field, as well as the variability associated with site differences.

The control of *Amaranthus albus* L. by the agent *Microsphaeropsis amaranthi* is an example of a successful transition from greenhouse to field use. This pathogen, which was extremely efficacious on the target weed in controlled growth-chamber studies, also performed well in the field, with 96–99% mortality after applications of 1×10^6 and 6×10^6 conidia/ml (Mintz et al., 1992).

FORMULATION OF BIOLOGICAL CONTROL AGENTS

The use of formulations for biological control agents is a relatively new application of technology that is well known to the chemical herbicide industry. Formulations or formulating agents are considered to be the crucial difference between the product being able to reach the general public and remaining in the laboratory. Most chemical herbicides could not be distributed without being combined with some other material. Thus, the active constituent is combined with a solvent, carrier, or surfactant in order to make its delivery and dispersal convenient, which is the broad definition of formulation (Anderson, 1983). Although this allows the user to spread a very small amount of active ingredient over a wide area, this may not be the only advantage to a formulated chemical. Chemical herbicides may gain their activity or have enhanced phytotoxicity due to the agent with which they are mixed. The formulation may also allow the agent to have a prolonged shelf life or to be transported easily and without altering the effectiveness of the product. In some cases, the carrier may serve as the formulation. This is usually the case with dry pellet or powder formulations. In addition to the basic formulation, materials that act as surfactants may be added to the formulated product to further enhance the activity or usability of the product.

The same basic principles underlie the use of formulations for the application of biological control agents. Formulations are used most frequently to enhance the activity of the organism (Boyette & Abbas, 1994; Green *et al.*, 1998). The need for high levels of inoculum in the field and the normal requirement of an extended dew period for efficacy of fungal pathogens are two aspects of the development of a biological control agent that may be alleviated by the addition of a formulating agent. The formulation of biologicals is complicated by the need for the organism to remain alive and virulent through the processing. The convenience of application, which is so important for chemical formulation, is almost a secondary consideration when a biological control organism is the active ingredient. The more important focus in the formulation of biological control agent is improving the longevity and efficacy. In addition, the formulation of a biological may address the need for using combinations of biologicals and chemical pesticides.

The most commonly used addition to a suspension containing a biological control agent, which for convenience and likelihood is assumed to be a conidial suspension in water, is that of surfactants. Since water is the most common carrier of biological control agents during the preliminary stages of development, the addition of a water-compatible agent is the most convenient next step. Surfactants may be of various forms that result in a variety of effects, but focus is on modifying the surface properties of a liquid (Anderson, 1983). This may facilitate the spreading, emulsifying, or sticking properties of an agent in a water solution. Surfactants may themselves have phytotoxic effects or may predispose the plant to infection as well as providing increased coverage. In some cases, products such as Tween 80® or Tween 20® may have detrimental effects on the biological control agent (Prasad, 1993). These are important characteristics to evaluate during the experimental phase of surfactant use, as they may enhance or inhibit the effect of the biocontrol agent.

Surfactants can be classified into groups based on their ionization in water and this may give some clue as to how a new or untried surfactant will behave in a biological control setting. The commonly used surfactants, Tween

20®, Tween 80®, Triton X-100®, and Tergitol®, are all nonionic surfactants. These have been the most commonly used surfactants for initial experimentation. Silwet L-77® is an example of a new generation of surfactants which are organosilicones. These surfactants reduce the surface tension far more than other types of surfactants and may prove to be an important advancement for biological control agent applications, as they may allow for penetration of the cuticular area (Stevens, 1994; Zidack & Backman, 1996). One drawback with this type of nonionic surfactant is that in helping to breach the cuticle, it may cause the plant to produce resistance-inducing compounds or promote entry of microorganisms capable of inducing cross-protection, both of which can retard the growth of the pathogen (Zidack *et al.*, 1992; Womack & Burge, 1993). Other surfactants that act as emulsifying agents, which are used with more complex formulations such as those that involve oil-based spore suspensions, are now more commonly used.

In addition to the use of surfactants in water-carried systems, a number of other formulations are being explored. Material capable of forming aqueous gels when added to spore suspensions serves to stick the spores to the leaf surface, while providing a hydrophilic matrix in which the spores may germinate more readily in the absence of an adequate dew period (Shabana *et al.*, 1997). This type of formulation appears to have lost much of its attraction, as the solution is viscous and difficult to apply as well as being relatively costly. Other simple adjuvants could include the addition of a water-soluble nutrient base that can be mixed with the water suspension prior to application. This is the case with Collego® (Encore Technologies), which consists of a dried spore preparation that is resuspended in water and a prepackaged amount of sugar as an osmoticum and nutrient source to allow for the gradual rehydration and germination of spores of *Colletotrichum gloeosporioides* f. sp. *aeschynomene*. Use of sugar also has been reported for control of spurred anoda (*Anoda cristata* [L.] Schlect.) with *Alternaria macrospora* Zimm. Control of *Desmodium tortuosum* (Florida beggarweed) was enhanced when sucrose was added to the *Colletotrichum truncatum* (Pers.) Grove spore suspension prior to application. Surfactants may increase the even distribution of fungal spores within an aqueous solution by overcoming their hydrophobic nature (Bannon *et al.*, 1990). These are essentially cases where adjuvants or surfactants have been used to increase the efficacy of a pathogen, attributable to either the increase in germination or the more efficient application and adhesion of the biological control agent on the host plant.

Within the past 5 or 10 years, there has been substantial progress in developing effective formulations for biological control agents (Boyette *et al.*, 1996; Connick *et al.*, 1991; Daigle & Connick, 1990; Daigle & Cotty, 1992; Saad *et al.*, 1993; Smith, 1991). The first attempts were complicated by the inability of many fungal propagules to endure a great deal of processing, but to increase the long-term viability and to facilitate long-distance transport, this aspect of the development of a biological control agent has had to be addressed. Two general approaches have been taken that are directly related to the way in which the biological control agent is envisioned to act, either in the biological control of weeds or in the biological control of diseases. If the target site is the foliage portion of the plant, formulation tactics have been directed at a fluid formulation or a formulation that is readily suspended in an aqueous solution. For organisms that act at the soil level or are incorporated into the soil, solid formulations have been developed (Connick *et al.,* 1991). In some cases, the formulated product may be a solid, but the organism may multiply on this substrate and be disseminated into the foliage in this manner.

One type of formulation used for biological control consists of encapsulation of the propagule in starch granules. This allows the biocontrol agent to be delivered in a conveniently applied form; however, adherence to the leaf surface requires that there be preapplication wetting, which is not realistic for large-scale commercial use (McGuire & Shasha, 1992). A second type of formulation that has been the focus of most workers in the field of biological control of weeds utilizes invert emulsion. This method involves the use of a variety of mixtures composed of combinations of paraffin wax, mineral oil, soybean oil, and lecithin. These combinations are mixed with a spore suspension and an emulsifying agent. The first application of this type of mixture was with spores of *Alternaria cassiae* to control sicklepod (*Senna obtusifolia*) (Daigle *et al.,* 1990). The ability of the oils to significantly depress the rate of water evaporation facilitates germination of the applied spores when there is an inadequate dew period. The amendment of the invert emulsion with nutrients increased the germinability, as did establishing the optimal pH of the solution (Daigle & Cotty, 1991, 1992).

Since this first development of invert emulsion based formulations, numerous potential biological control agents have been applied using this type of formulation with different combinations of components (Womack *et al.,* 1996; Yang & Schaad, 1996). The knowledge gained from past experimentation in invert emulsion formulation has led to the extensive use of vegetable oils, and this treatment appears to cause no alteration to the cuticular layer of the leaf surfaces of the plants that have been examined (Auld, 1993; Egley & Boyette, 1995). Additional work with *Alternaria* Nees spp., *A. crassa* (Sacc.) Rands, and *A. cassiae* indicated that the formulation of these mycoherbicides as invert emulsions reduced the necessary inoculum threshold (Amsellam *et al.,* 1990). The work with these two pathogens indicated that conidia were receiving additional benefits from the formulation ingredients beyond that provided by

the prolonged availability of water. However, few experiments included adequate controls with the emulsion alone to determine what, if any, effects these oil and wax mixtures have on the plant. In the case of *Alternaria* spp., the assumption was that the fungus was enhanced in some way, rather than the plant being predisposed (Amsellam *et al.,* 1990).

The potential for phytotoxicity has been assessed in other studies and this has affected the choices of the components for the formulations. A great deal of work has been done in this area with *Colletotrichum truncatum* for the control of hemp sesbania (*Sesbania exaltata* [Raf.] Rydb. ex Hill) and *Colletotrichum orbiculare* for the control of *Xanthium spinosum.* Control of hemp sesbania was dramatically improved by the use of the invert emulsion in the absence of dew, again supporting the hypothesis that greater control is achieved through providing a prolonged period of water availability (Boyette, 1995). Additional work with the *Colletotrichum* spp. indicates that there is a direct effect on the plant from some types of oils (Auld, 1993; Klein *et al.,* 1995c). Reports of host-range expansion through the use of the invert emulsion formulation further support the idea of the effect of oils on the plant contributing to disease progress. Although the use of oils to increase the host range of biological control agents that have extremely limited host ranges is a positive prospect, host ranges might also be altered to include crops in which the agent is applied. In the case of *Alternaria crassa,* a biological control agent for jimsonweed (*Datura stramonium* L.), the host-range expansion resulted in a number of crops being affected, including tomato (*Lycopersicon esculentum* Mill.), eggplant (*Solanum melongena* L.), potato (*S. tuberosum* L.), and tobacco (*Nicotiana tabacum* L.) (Yang *et al.,* 1993).

Two additional formulations that have become increasingly popular for biological control agents of weeds and diseases are alginate pellets and one referred to as Pesta. These solid formulations are easily adapted to a number of different fungi. Alginate pellets have been used to deliver a substantial number of biological control agents, including *Alternaria macrospora, A. cassiae, Fusarium lateritium* Nees:Fr, *Colletotrichum malvarum* (Braun. and Casp.) South., *Fusarium solani* (Mart.) Sacc. f. sp. *cucurbitae* Snyd. and Hans., *Talaromyces flavus* (Klöcker) Stolk and Samson, *Gliocladium virens* Mill., Gold., and Foster (DeLucca *et al.,* 1990), *Penicillium oxalicum* Currie and Thom., and *Trichoderma viride* Pers:Fr. (Daigle & Cotty, 1992). This formulation can be manipulated in a number of ways to optimize the development of the agent in the field, as well as increasing the ease of application. Organisms that have been formulated in this manner have excellent shelf life compared to the unformulated organisms, and the composition actually allows the formula to act as a spore-production system. This type of system optimizes the potential of the agent by providing high concentrations of freshest inoculum possible, and it can be manipulated to provide optimal conditions for sporulation. Rather than quantifying the inoculum at the time of application and introducing it into the environment, depending on that source to provide adequate infection, this system allows a new crop of fresh spores to be produced from the mycelium or spores that have been incorporated into the pellet. Nutritional amendments, such as sugars or corn, soy, or rice meal as a carbon source or amino acids, just as in a growing medium, may be incorporated into the formulation to provide a base for improved germination as well. This method of formulation appears to be relatively inexpensive and versatile and can produce inoculum to virtually any size or weight standard (Gemeiner *et al.*, 1991). This also has the advantage that additional stickers may be added with an encapsulated biocontrol agent and allow the alginate to adhere to the surfaces of seeds to protect them against a pathogen (Andrews, 1992).

The Pesta formulation is also a means of producing inoculum within the formulated material. This material consists of a wheat gluten mixed with the fungus to prepare sheets of material that are then dried and crumbled. This material, used with *Fusarium oxysporum* Schlecht.:Fr. to control sicklepod, was found to retain viability for 1 year, although at a reduced level (Boyette *et al.*, 1993a). This formulation allows for a convenient means of delivery and results in the production of secondary cycles of conidia in the field. When agricultural byproducts, such as corn cobs or rice husks, are used to produce inoculum, the pathogen may be stored on this ground material and the material then applied directly to the field (Batson & Trevethan, 1988). This is a case of the medium of spore production acting as a carrier rather than a true formulation. This could also be said of the use of a liquid medium that is incorporated with Vermiculite or Perlite for field application. Although these methods may provide inexpensive means of producing inoculum for field use, prolonged storage of these types of materials does not seem feasible. Studies on the viability of the encapsulated pathogen after prolonged storage have not addressed the potential for contamination of this material or the degradation of nutrients in the mixture. These factors could result in not only a loss of viability but also a loss of infectivity.

One interesting aspect that has been explored recently is the use of sunscreens to protect spores that are not melanized (Prasad, 1993). For instance, *Chondrostereum purpureum* formulated with an ultraviolet protectant was able to withstand a wider range of environmental conditions.

Finally, there are a number of cases in which fungi tested as biocontrol agents were successful in greenhouse trials but failed to perform equally efficaciously in the field. Although the lack of an adequate dew period is most commonly cited as the explanation for this phenomenon, it is possible that the propellant used to pressurize the sprayer may play a role. The most commonly used sprayer systems rely on CO_2 as the spray propellant and the generation of carbonic acid in CO_2-pressurized suspensions can lower the pH of the suspension. Rosskopf *et al.* (1997) studied the effect of CO_2-induced changes in pH on spore germination or disease development with three bioherbicidal fungi. Germination of conidia of *Phomopsis amaranthicola*, a pathogen of pigweeds (*Amaranthus* spp.), and *Alternaria cassiae*, a pathogen of sicklepod (*Senna obtusifolia*), was significantly reduced when the suspensions were of pH 4.0–5.0 compared to pH 7.0. In the case of *P. amaranthicola*, the germination dropped from 93% at pH 7.0 to 73% at pH 4.0. The CO_2 propellant had a significant effect on *P. amaranthicola*, causing as much as a 50% reduction in germination of conidia, but did not similarly affect conidia of *A. cassiae*. Disease severity, rated on a scale of 0–12, was significantly reduced when *A. cassiae* was applied to plants with CO_2 rather than compressed air. In the case of *P. amaranthicola*, disease severity was reduced from 10.6 when the fungus was pressurized with air for 30 min to 5.5 when pressurized with CO_2 for the same duration. The third fungus, *Microsphaeropsis amaranthi*, also a pathogen of pigweed, was not affected by changes in pH or CO_2 pressurization. Six commonly used adjuvants and surfactants lowered the pH of pressurized solutions even more than CO_2 alone, which may additionally compromise field efficacy of bioherbicide agents.

BIOHERBICIDES AND INTEGRATED WEED MANAGEMENT

An important area of research that is beginning to receive more attention is the compatibility of biological control agents with multiple crop protection strategies in an integrated pest control program (Smith, 1982; Watson & Wymore, 1989). This may include the use of multiple biological control agents (Chandramohan & Charudattan, 1996) or the timing of applications of biological control agents with cultural management practices for diseases and weeds (TeBeest & Templeton, 1985). For an agent to be successfully incorporated into agricultural practices that are already established, such as the use of chemical fungicides or herbicides, it is important to determine what the effects of these practices might be on the biological agent. It has been found that chemical pesticides may serve as a nutritional base for some microorganisms or otherwise enhance the disease effects that plants incur from fungal plant pathogens (Altman & Campbell, 1977; Altman & Rovira, 1989; Cerkauska, 1988; Johal & Rahe, 1990).

The primary goal of testing the effect of chemicals on potential biological control agents is to determine the incompatible reactions that may limit the use of the organism.

In the case of *C. gloeosporioides* f. sp. *aeschynomene,* or Collego®, it was found that the fungus was compatible with tank-mixed applications of herbicides, such as bentazon (3-(1-methylethyl)-1*H*-2,1,3-benzothiadiazin-4(3*H*)-one 2,2-dioxide) or acifluorfen (5-(2-chloro-4-(trifluoromethyl) phenoxy)-2-nitrobenzoic acid). This organism was also found to be compatible with a number of insecticides and fungicides (Smith, 1991). The agent DeVine®, on the other hand, was found to be detrimentally affected by the addition of any wetting agents, fertilizers, or chemical pesticides (Kannwischer *et al.,* 1980). Charudattan (1993) and Hoagland (1996) provide reviews of the materials that have been tested with several biological control agents. Although fungal plant pathogens that are developed as biological control agents appear to be most commonly compatible with insecticides, there are few other generalizations that can be drawn. Thus, it is imperative that each individual agent be tested before it is tank-mixed or other combined use is recommended.

It is possible that the efficacy of microbial weed control agents could be enhanced when used in combination with chemical herbicides (Smith, 1991) as well as in conjunction with other weed management practices. For example, soil solarization has been shown to provide good control of several weeds, including winter annual grasses and field bindweed, but only marginal control of nutsedge (Chellemi *et al.,* 1997). Several weeds are not completely controlled by the heat produced, including bermudagrass (*Cynodon dactylon* [L.] Pers.), johnsongrass (*Sorghum halepense* [L.] Pers.), and common purslane (*Portulaca oleracea* L.) (Pullman *et al.,* 1984). Biological control agents could be applied for control of the weeds that are less affected by soil solarization and tend to come through the plastic.

Rhizobacteria also have the potential for use in combination with other control measures (Fernando *et al.,* 1994) and may be enhanced by management practices. Boyetchko (1999) has suggested that reduced tillage practices could allow for increased populations of rhizobacteria, with crop residues providing excellent habitats and substrate for overwintering of these bacteria. The weed-suppressive activity of bacteria applied to winter annual grass weeds was enhanced by the application of reduced rates of herbicides (Harris & Stahlman, 1992). This was also true when deleterious bacteria were applied to downy brome in combination with diclofop and metribuzin (Stubbs & Kennedy, 1993). There is also the potential for combining different strains of rhizobacteria and rhizobacteria with other biocontrol agents. A synergistic effect was observed when *Pseudomonas fluorescens* Migula was combined with *Flavobacterium balustinum* Harrison and applied to soil containing seedlings of leafy spurge (Kremer & Kennedy, 1996).

In situations where complete eradication of the target weed is not accomplished, biological agents may adequately affect the growth components of the weed, so that it is no longer a significant competitor (Kempenaar *et al.,* 1996a). Kennedy *et al.* (1991) conducted laboratory and field studies to examine the effects of *P. fluorescens* strain D7 on downy brome (*Bromus tectorum*), which is a major pest in winter wheat in the northwestern United States. The application of this rhizosphere bacterium reduced germination, biomass production, and seed production of the target weed but had no effect on the growth of wheat.

Paul and Ayres (1990) conducted field experimentation on the effect of common groundsel (*Senecio vulgaris* L.) competition on the growth and yield of lettuce (*Lactuca sativa* L.). The rust pathogen *Puccinia lagenophorae* Cooke, which infects groundsel, had a significant impact on the competitive ability of the weed when compared to noninoculated plants. Lettuce yields in plots containing diseased groundsel were between two and three times greater than in plots with noninfected plants. This rust fungus may have potential as a classical biocontrol agent for common groundsel in the United States (Wyss & Müller-Schärer, 1998).

Sclerotinia sclerotiorum (Lib.) de Bary was able to effect a decrease in competition between spotted knapweed (*Centaurea maculosa* Lam.) and bluebunch wheatgrass [*Agropyron spicatum* (Pursh.) Scribn. and Smith]. *Sclerotinia sclerotiorum* reduced the spotted knapweed density by 68–80% without reducing the density of the wheatgrass (Jacobs *et al.,* 1996).

Colletotrichum coccodes (Wallr.) S. J. Hughes, which has limited efficacy on velvetleaf (*Abutilon theophrasti* Medik.) monocultures (Anderson & Walker, 1985), was found to significantly reduce the plant growth and competitive ability when used in soybean production fields (DiTommaso *et al.,* 1996). Yield losses in soybean were significantly lower in plots in which the velvetleaf plants were treated with the pathogen.

BIOTECHNOLOGY AND BIOLOGICAL CONTROL

Successful biological control using microbial agents requires several complex and often specific interactions between the biocontrol target and the agent. This complexity of interactions is one reason for the inconsistency and unpredictability of biological control systems. Understanding these interactions at the molecular and biochemical levels will render the biological control approach more predictable and less arbitrary. It will also help to facilitate the development of agents through characterization of the mechanisms of plant pathogenesis. It is hoped that these types of understandings will lead to genetic engineering of highly effective biological control agents, enable the selection of agents with desired traits, and help to determine environmental fate of biological control agents.

Plant pathogens produce a variety of secondary metabolites capable of inflicting characteristic phytotoxic damage in their plant hosts. Two kinds of pathogen-produced phytotoxins are recognized: host-selective or host-specific toxins and host-nonselective or host-nonspecific toxins (Turgeon & Yoder, 1985; Yoder, 1983). Host-selective toxins are toxic only to the hosts that are normally attacked by the pathogen, and nonselective toxins are phytotoxins that affect a broader range of hosts than the toxin-producing pathogen does. Approximately 20 host-selective toxins have been identified and are often determinants of host range (Walton, 1996). A phytotoxin can be further classified on the basis of its biological role as a pathogenicity factor, a toxin that is essential for a pathogen to cause disease or a virulence factor, or a toxin that increases the level of disease once disease develops (Yoder, 1983). Finally, many pathogenic microorganisms are known to synthesize plant growth hormones. However, a growth hormone is different from a toxin; whereas a toxin is a pathogen-produced compound that the plant is unable to synthesize, many compounds with growth-hormonal activity are products of a plant's normal metabolism. Although such compounds are generally not toxic when the plant produces them, the physiological imbalance created by their synthesis during the disease process serves as a virulence factor of the pathogen.

It should be possible to genetically engineer two important factors affecting bioherbicides, level of virulence and alteration in host range. Theoretically, it should be possible to engineer for increased virulence or host range or both, provided the following conditions can be met: (1) availability of a suitable pathogen for genetic engineering; (2) availability of suitable vectors and selectable markers; (3) availability of cloned toxin genes; (4) availability of a suitable gene-expression system for the toxin genes and for any genes necessary to confer resistance to the toxin to the transformed agent; (5) in the case of fungal pathogens, ability to produce and regenerate protoplasts or feasibility of direct injection of genes; and (6) availability of suitable containment facilities to conduct tests under safe conditions.

Attempts to engineer bioherbicides have just begun, but several projects can serve as examples of the possibilities. In the first example, Charudattan *et al.* (1996a) attempted to alter the virulence and host range of the ubiquitous bacterial plant pathogen *Xanthomonas campestris* pv. *campestris* (Pammel) Dowson with genes encoding the production of bialaphos, which is a secondary metabolite and a host-nonselective phytotoxin produced by the soil saprophyte *Streptomyces hygroscopicus* (Jensen) Waksman & Henrici. Bialaphos and the active moiety phosphinothricin are registered as a new class of biologically derived herbicides. These biochemicals also have antimicrobial effect. A 33.5-kb *Hind* III-treated segment of a reconstructed plasmid (41 kb) containing bialaphos production genes was introduced into the genome of the bacterium, and the altered

and the parent bacteria were tested for alterations to their host range. The pathogenic host reaction was characterized by a typical black-rot disease caused by the parent and the transformed bacterium on its natural hosts, cabbage and broccoli, whereas the nonhost reaction caused by the parent was a necrotic hypersensitive reaction and a spreading, chlorotic, compatible host reaction caused by the transformed bacterium (Charudattan *et al.*, 1996).

The bioherbicide pathogen *Colletotrichum gloeosporioides* f. sp. *aeschynomene* (CGA) has been transformed with the *bar* gene, which confers resistance to bialaphos and phosphinothricin. This gene was also cloned from *S. hygroscopicus,* in which it serves to protect against the autocidal effects of bialaphos. In this case, the *bar*-gene-transformed fungus could be exposed to bialaphos without being affected and the *bar*-transformed biocontrol agent can then be coapplied with the chemical herbicide. This allows for the control of both northern jointvetch (*Aeschynomene virginica*), which is susceptible to CGA, and Indian jointvetch (*A. indica*), which is controlled by bialaphos, but not CGA (Brooker *et al.,* 1996).

In addition to the possibilities associated with the transformation of biological control agents for the purposes of increased virulence or host-range alteration, there is the application for monitoring introduced strains and for risk assessment. Characterization of isolates of *Chondrostereum purpureum,* used as a mycoherbicide for control of hardwood weed species, through the use of DNA fingerprinting has made it possible to differentiate between the isolate released as mycoherbicide and those that are naturally occurring (Gosselin *et al.,* 1996; Ramsfield *et al.,* 1996). This type of work has also been applied to the biological control agent *Phomopsis subordinaria* (Desmaz) Traverso, evaluated for the control of plantain (*Plantago lanceolata* L.). In addition to isozyme characterization of isolates to be used for biocontrol, randomly amplified polymorphic deoxyribonucleic acid (RAPD) analysis was used (de Nooij & van Damne, 1988; Meijer *et al.,* 1994).

In a recent study by Berthier *et al.* (1996), the internal transcribed spacer (ITS) polymorphic restriction patterns of the DNA of the biological control agent *Puccinia carduorum* Jacky were found to be distinct for *P. carduorum* strains from *Carduus acanthoides* L. and *C. thoermeri* compared to those of the same fungal species isolated from *C. tenuiflorus* Curtis and *C. pycnocephalus* L.

DNA fingerprinting techniques and the development of reliable markers and tracking systems can greatly contribute to further success of bioherbicides as well as the general knowledge base related to the biology of these organisms. Molecular markers may be used to monitor specific strains of microbes using polymerase chain reaction (PCR) amplification to detect specific DNA sequences. Tools such as bioluminescence markers and green fluorescent protein can contribute to this process and have been applied successfully in several cases (Kennedy, 1996; Poppenborg *et al.*, 1997). Although this technology has not been widely applied as a regular component in the selection and development of biological control agents, it is anticipated that utilization of this technology will facilitate the registration of plant pathogens as bioherbicides in the future.

LAWS GOVERNING BIOLOGICAL CONTROL AGENTS

The Federal Plant Pest Act (FPPA) of 1957, as amended, and the Federal Plant Quarantine Act (FPQA) of 1912, as amended, grant the Secretary of Agriculture broad authority to carry out operations or measures to detect, eradicate, suppress, control, prevent, or retard the spread of plant pests (Charudattan & Browning, 1992). This authority gives the United States Department of Agriculture (USDA) the flexibility to respond appropriately to protect American agriculture against foreign pests. According to this Act, a plant pest is any organism or infectious agent which directly or indirectly injures plants or causes disease or damage to any plants or plant parts. The authority to administer the regulations and activities of controlling and eradicating plant pests rests with the Animal and Plant Health Inspection Service (APHIS) of the USDA. The Plant Protection and Quarantine (PPQ) branch under USDA-APHIS bears the primary responsibility for the control and eradication activities. Plant Protection and Quarantine also regulates the legitimate importation, interstate movement, and use of foreign plants and plant products as well as plant pests and potential plant pests used for research or commercial purposes. This is done under a permitting process with which APHIS grants what is commonly known as a quarantine permit.

Microbial biological control agents used as pesticides are regulated under the Federal Insecticide, Fungicide, and Rodenticide Act (FIFRA). FIFRA, in its revised form, was enacted in October of 1988. Under this Act, the Environmental Protection Agency (EPA) has been given the responsibility for ensuring that the distribution and use of pesticides are done in such a manner as to prevent unacceptable risks to human health and the environment. The EPA regulates the use of pesticides primarily through registration of pesticides, which it does after a thorough review of the health and safety data, and through prohibition of distribution and sale of unregistered pesticides, or pesticides that are improperly labeled, or have improper composition. In addition, it is illegal to use a pesticide in a manner inconsistent with its labeling.

A coordinated framework for regulating products arising from biotechnology, including biocontrol agents, produced through genetic engineering was developed in 1986 under the authority of the executive Office of the President, Office

of Science and Technology Policy. Under this framework, the authority to regulate recombinant technology products and recombinant biocontrol agents is given to the USDA-APHIS' Biotechnology, Biologics, and Environmental Protection (BBEP) division. The USDA policy on the regulation of biotechnology does not view genetically engineered biocontrol organisms as fundamentally different from those that have been isolated from nature and introduced into a new environment. The organisms and products produced through the new techniques of biotechnology are regulated under existing laws that have applied to naturally occurring organisms and products of traditional technologies. To address the need for specific information necessary for the assessment of the products of the new technologies, a few new regulations have been promulgated and some old ones have been updated. The USDA regulations consider the risks posed by an organism or a product under a specific use rather than on the process (i.e., genetic engineering) used in production.

Permits are required for any introduction of organisms modified through genetic engineering. Under these regulations, "introduction" means to move a regulated article into or through the United States, to release into the environment, or to move between states. "Release into the environment" means the use of the regulated article outside the constraints of physical confinement of a laboratory, contained greenhouse, a fermentor, or other contained structure. A "regulated article" is any product which contains an unclassified organism and may be a plant pest or any organism which has been altered or produced through genetic engineering, if the donor organism, recipient organism, or vector or vector agent belongs to any genera or taxa designated by the USDA as meeting the definition of plant pest or if it is an organism whose classification is unknown.

The regulations have been amended to include a specific exemption from the need for a permit for organisms that meet the definition of a regulated article. A limited permit is not required for interstate movement for genetic material from any plant pest contained in *Escherichia coli* (Migula) Castellani and Chalmers genotype K-12, sterile strains of *Saccharomyces cerevisiae* Hansen, or asporogenic strains of *Bacillus subtilis* (Ehrenberg) Cohn, provided all of the specific conditions recognized by the USDA list are met.

As scientific knowledge with the products of biotechnology increases and we achieve familiarity with the interaction between modified organisms and their environment, additional amendments to the regulations are expected. These amendments will be added either by APHIS at the Agency's initiative or through a public petition process, which is an innovative feature of these regulations. This process is expected to be used by applicants and other members of the interested public to seek changes to the list of organisms on the USDA's list or to obtain specific exclusions of genetically engineered organisms from regulation.

The petition to amend must contain a statement of grounds and supporting literature, data, or unpublished studies as well as opposing views or contradictory data. A petition that meets these requirements will be published in the Federal Register for comment. When a petition is approved, the changes to the regulation are published with a complete response to public comments.

The National Environmental Policy Act (NEPA) was enacted in 1969 to protect against adverse environmental effects resulting from major actions implemented by federal agencies. Under this Act, all federal and federally supported agencies must consider the environmental impact of major actions, such as release of a foreign pathogen for weed control or a genetically engineered microbe into the environment, that may significantly affect the quality of the human environment in the United States. With certain exceptions, an Environmental Assessment (EA) is required for the initial field release of a foreign or engineered biocontrol agent into the United States or into the field. The necessary permits for importation or field release of the biocontrol agent will be granted only after the EA is approved.

The reviewing scientist in APHIS prepares the EA, and the environmental assessment is subject to peer review by the APHIS scientific staff. When the environmental assessment results in a Finding of No Significant Impact (FONSI) on the environment, a notice of availability of the environmental assessment and FONSI is published in the Federal Register, and the permit to go ahead with the planned introduction or field release is issued. The permit may contain special conditions which must be met in addition to those specified in the protocol submitted as part of the application. As a condition of every permit for release to the environment, scientific data must be collected and submitted to APHIS.

The EA document will contain information describing the purpose of the document, Departmental regulations, the conditions under which the permit is issued or denied, precautions against environmental risk, the background biology of the organisms, and the possible environmental consequences of the field test. The environment that could be affected by the field test is described and the precautions developed for protecting that environment, including field plot design, field inspection and monitoring, test plot security, and disposal plans, are analyzed. The environmental consequences of the test are examined from all possible perspectives. Consideration is given to the biology of the recipient, donor, and vector and to the potential for biological containment based on knowledge of this biology. Any possibility of risk to native flora or fauna is evaluated, with special consideration of organisms which are threatened or endangered. Any potential impact on human health is examined. A detailed report describing the effects of the proposed field release on the environment is called the Environmental Impact Statement (EIS). Requirement of an EIS is initiated if any adverse risk is perceived.

The Endangered Species Act (ESA) of 1973 was enacted to protect the many threatened and endangered (T&E) species, both plants and animals, in the United States. The Fish and Wildlife Service (FWS) of the U.S. Department of the Interior (USDI) administers this Act. Under the authority of this act, the impact of federal actions on native and endangered plant and animal species must be determined by using a safety evaluation protocol developed under this Act. Comments on concerns and any test results developed to assess risk to T&E species should be included in the EA or EIS documents.

The federal government has a legal means of designating weeds considered to be extremely harmful to the nation or a region. The Federal Noxious Weed List allows for such listing. Once a plant is designated as a noxious weed, certain regulatory actions can be enforced, such as banning the collection, use, and transport of such plants and the requirement of permits to collect and transport such plants for legitimate purposes. The federal Noxious Weed Act provides the authority for these regulations.

In addition to federal laws, several states have laws and regulations regarding environmental policy and/or endangered species within their boundaries, similar to NEPA and ESA. Certain states have regulations requiring permits or approval prior to shipment or release of arthropod and nematode biological control agents within their borders or otherwise have formal requirement for notification prior to importations or releases. Some states have their own Noxious Weed Laws that enable them to designate and regulate weeds of local importance.

In 1959, the USDA and the USDI established a joint committee that developed guidelines about kinds of data that would be necessary to assess insect species introduced into the United States for biological control of weeds. Later, plant pathogens considered for use as biological weed control agents were also evaluated by this group. The group, a voluntary assembly of scientists, formed a relationship with Canada and Mexico so decisions to release new organisms could be made on a continental basis. In 1966, this working group was put under APHIS-PPQ and it was reorganized in 1987 as the Technical Advisory Group on the Introduction of Biological Control Agents of Weeds (TAG) to make recommendations to scientists and to give assessments to regulating agencies. TAG membership includes representatives from 10 designated federal agencies and additional representatives invited to review applications to import and/ or release biocontrol agents (Coulson, 1992).

TAG deals only with naturally occurring organisms, not those which have been genetically engineered. Its responsibilities are to (1) review research proposals on biological control of weeds involving introduction of exotic organisms into the United States to ensure that all aspects of safety of the potential introductions are considered in the research plan; (2) provide guidelines for researchers that must be met before an organism is considered for entry into or approved for release from quarantine in the United States; (3) review the documentation supporting proposals to release exotic organisms for biological control of weeds in the United States and to evaluate the adequacy of the data showing the safety of the proposed release; and (4) recommend actions to be considered by PPQ to release or not to release biological control organisms from quarantine.

There are essentially three types of petitions to the TAG: (1) Proposals to conduct research on the classical biological control of a plant species not previously targeted for such research in the United states. It is at this point that conflicts of interest can be identified that may indicate to the researcher that approval for the introduction of exotic natural enemies of the plant would be unlikely or at most difficult to obtain, and consideration can thus be given to aborting the project before needless expenditure of research funds. (2) Proposals for the introduction of exotic organisms into quarantine in the United States for research purposes, review of which may be requested by PPQ. (3) Proposals for the field release of an exotic organism in the United States (or Canada or Mexico). In connection with the first or second type of petition, a list of plants to be included in host specificity testing is provided for review by the TAG.

The major issues evaluated by the TAG reviewers are (1) the adequacy of taxonomic information about both the biological control agent and its target weed, (2) the adequacy of the range of plants included in the host specificity tests, (3) the adequacy of the results of research on the host range of the agent proposed for field release, from both field and laboratory studies and literature search, and (4) the potential impact of the agent on nontarget plants.

The TAG Chair evaluates the comments of TAG and other outside reviewers of the petition and, if a clear consensus is evident, prepares a recommendation on the petition. Recommendations concerning a new plant for biological control or test plant lists are made to the petitioner. Recommendations concerning quarantine importation or field release of organisms are made to the Deputy Administrator of APHIS-PPQ. Often the review comments indicate the need for additional information or additional testing, in which case this need is expressed to the petitioner, with copy to the PPQ Deputy Administrator.

As stated previously, microbial biocontrol agents used as pesticides are regulated under the FIFRA by the EPA. According to FIFRA, biological control agents fall under the definition of a pesticide if they are intended for "preventing, destroying, repelling, or mitigating any pest" or are intended for "use as a plant regulator, defoliant, or desiccant." However, FIFRA provides authority to exempt from regulation any pesticide which is adequately regulated by another federal agency (such as USDA-APHIS) or which is

otherwise of a nature not requiring oversight under FIFRA. Under this authority, EPA exempted all biological control agents from FIFRA oversight, with the exception of eukaryotic microorganisms, including protozoa, algae, and fungi; prokaryotic microorganisms, including bacteria; and viruses. However, EPA can revoke this exemption if it determines that a biocontrol agent or class of biocontrol agents is not adequately regulated by another federal agency and the agent(s) should not otherwise be exempt from FIFRA.

EPA has recognized that regulatory oversight of microbial pesticides must consider ways in which these pesticides differ from chemical pesticides. Among microbial pesticides, those involving pathogenic, nonindigenous, or genetically modified microorganisms are considered to pose greater potential risks than indigenous, nonpathogenic, unmodified organisms.

The use of biocontrol organisms that will result in the appearance of the organism (or any parts of the organism) in food or feed is regulated by the EPA under the Federal Food, Drug and Cosmetic Act (FFDCA). Any biocontrol agent that poses toxic hazard will fall under the Toxic Substances Control Act (TSCA). TSCA becomes involved when a living recombinant is being produced; however, this will not affect most researchers, as materials produced for research and development are exempted from this regulation. Finally, the Endangered Species Act may require that labeling warn users of potential adverse effects on endangered species. The DeVine® label, for example, specifically instructs that this bioherbicide is not to be used in areas where the endangered native azalea, *Rhododendron chapmanii* Gray, naturally occurs.

In general, there are four kinds of applications which are processed by EPA under FIFRA. These are

Registrations: Registration allows the unrestricted use and marketing of a pesticide product as long as the accepted labeling is followed. Registrations are handled under Section 3 of FIFRA.

Experimental Use Permits: An experimental use permit (EUP) may be issued to a person, company, or institution for the purpose of obtaining the required data for registration. EUPs generally have a one- or two-year time limit, and the organism can only be used by cooperators of the experimental program. The organism can only be shipped to those states as stipulated in the program and the conditions of use are tightly controlled. EUPs are issued under Section 5 of FIFRA. Often researchers are confused as to whether the proposed work will require an EUP or a Notification. As a rule of thumb, applications to greater than 10 acres of land or 1 acre of water require an EUP. Also, if the use will result in residues of the pesticide in food or feed, an EUP is usually required. Smaller size experiments which are crop-destruct or designed for noncrop uses (such as forestry or ornamental plants) usually can be handled by a Notification.

Emergency Exemptions: Emergency exemptions are issued in situations where there is insufficient time to process a registration or the data base for registration is incomplete and significant adverse effects (such as public health impacts) may occur if the use of the pesticide is prohibited. Emergency exemptions are regulated under Section 18 of FIFRA and are only issued to Federal and State Agencies.

Special Local Needs: Section 24(c) provides that a State may register additional uses of registered pesticides for use and distribution within that State, provided that such use has not been previously denied, disapproved, or canceled.

In addition to formal applications, the following are regulated by the EPA:

Notifications: A notification is essentially a request that EPA make a determination as to whether the proposed small-scale testing of a nonindigenous, genetically altered or genetically engineered organism is of such a nature that an EUP should be required prior to the initiation of the study. Notifications are handled by EPA under the Coordinated Framework for Biotechnology.

Tolerances: EPA is responsible for regulating the residues of pesticides which may appear in food. This is done by establishing a tolerance level, which is the maximum permissible residue allowed. Because of their nature, microbial pesticides usually are regulated through an Exemption from the Requirement of a Tolerance. To qualify for the exemption, the applicant must provide data or information that indicates that the pesticide in any amount will be safe when used on food or feed crops. For EUPs where the microbial pesticide will be used on food or feed, a Temporary Exemption from the Requirement of a Tolerance may be issued, with somewhat less data. In general, microbial pesticides have been determined by EPA to be in the public interest (due to host specificity, low toxicity, and unique modes of action) and the tolerance fees for these submissions are often refunded.

Subdivision M of the FIFRA Guidelines: Subdivision M of the FIFRA guidelines describes in detail the studies and data required for registration of a microbial pesticide. If a particular requirement is inappropriate, the applicant may request that the requirement be waived. Such waiver requests should be accompanied by the pertinent information that provides a basis for the waiver request. Subdivision M guidelines have been recently updated and replaced. For the latest information, see Environmental subset at Gopher.epa.gov.

Some simple rules to follow are

1. Always strictly follow the rules and regulations that are in force.

2. Whenever applicable, obtain proper permits and institutional clearances *before* doing "controlled" or "regulated" type of activity.

3. Keep institutional, state, and federal agencies fully informed of progress and promptly report back to the agencies when the work is completed.

4. Always keep the public fully informed if your research might involve public sentiments.

PRODUCTION OF BIOHERBICIDE AGENTS

One of the important challenges facing the developers of biological control agents is to establish protocols by which the agents can be produced and utilized in a manner that is competitive or compatible with the tools that are currently available. The cost of producing large quantities of biomass can be inhibitory if there does not exist a rather substantial market. It was once thought that if an organism could not be produced using submerged fermentation, it would not have promise as a viable biocontrol agent. Through improvements in existing technologies, this is no longer the case, although the information base is more fully developed for liquid fermentation (Churchill, 1982). The majority of the information available on fungal nutrition and the conditions necessary for sporulation may apply to any production system, but most have been evaluated using liquid fermentation. Some examples of fungi that have been cultured on a large scale using submerged fermentation include *C. gloeosporioides* f. sp. *aeschynomene* (Collego®) for the control of northern jointvetch and *Phytophthora palmivora* (DeVine®) for the control of stranglervine (Churchill, 1982).

In the submerged fermentation system, a high yield of biomass is produced and this biomass is easily separated from the culture medium. At the laboratory experimentation scale, shake flasks may be used to culture ample amounts of inoculum but may not allow for predicting how the organism will grow on a larger scale. A number of factors can contribute to the proper growth of the cultured organism and these may be difficult to monitor in a small, shake flask system. Laboratory fermentors are available that monitor parameters such as pH, temperature, dissolved oxygen, and the amount of agitation (Bannon *et al.,* 1990). Some fungi require manipulation of one or more of these factors to induce sporulation in liquid culture. Spores are favored over mycelium for long-term storage (Churchill, 1982).

Depending on the conditions, some fungi will produce a number of different types of propagules in a single liquid culture. This is the case with *Colletotrichum gloeosporioi-*

des f. sp. *aeschynomene,* in which the majority of spores produced in liquid fermentation are yeast like spores, produced by budding (Churchill, 1982). Conidia and blastospores and arthrospores are also produced by this fungus, a phenomenon referred to as fungal dimorphism. Although in the case of Collego® all spore types are infective, it is possible in other cases, when dimorphism is induced, that the resulting propagules may not have the same efficacy. In some studies, the nutritional components of the culture media have been manipulated to induce sporulation in fungi that do not normally produce spores in liquid culture (Stowell, 1991). With some *Fusarium* species, it was found that potassium phosphate, magnesium sulfate, and potassium chloride were particularly important in inducing sporulation (Diarra *et al.,* 1996; Hallsworth & Magan, 1994). In general, among the fungi tested, sporulation often occurs in the presence of various types of stress, most commonly nutritional (Mckoy & Trinci, 1987). This means of inducing sporulation may not be applicable for production of biological control agents because the postprocessing viability and efficacy can be detrimentally affected.

To address this complication, many types of media may need to be tested to determine the optimal conditions for a large quantity of biomass that must also maintain viability and activity. The first aspect may be addressed in terms of the specific growth rate, which is a measure of the biomass increase over time. This can then be used to compare different compositions of media for optimal growth. The type of inoculum used to seed the fermentors may also affect the end product. In some cases, as much as a twofold increase in fermentation time is required when conidia are used for fermentor inoculation rather than mycelium (Churchill, 1982). The quantity of conidia produced, as well as their tolerance to the drying process, is affected by the carbon-to-nitrogen ratio as well as the sources of those nutrients (Jackson & Bothast, 1990). Although a large portion of the work done in this area has been done with *Colletotrichum* species used for biological control of weeds, the foundation lies in the development of yeasts used in food preparation (D'Amore *et al.,* 1991). It has been found that if a high carbon-to-nitrogen ratio is used, excessive availability of carbohydrates results in lipid accumulation. A high lipid concentration is associated with poor germination in many fungi (Griffin, 1994). The physiological role of the carbon source may be even more important than the ratio. For example, glucose is often used as a component of many types of media (Dhingra & Sinclair, 1995). Some fungi may cease to produce a number of inducible enzymes when they are in the presence of glucose (Griffin, 1994). These enzymes may be crucial to pathogenesis. The levels of specific minerals may be as important as the carbon source. Zinc and magnesium, for example, play an important role in the production of pathogenesis-related toxins in species of *Alternaria* (Griffin, 1994).

After a system is established that results in an adequate amount of the most desirable type of inoculum, it must be determined if the propagules will be tolerant to the drying process. This may be highly dependent upon the culturing conditions and the absence or presence of an osmoticum in the medium (Abbas *et al.,* 1995b; Dillard, 1988; Harman *et al.,* 1991). Several studies have been conducted to determine the medium components that might contribute to desiccation tolerance (Burleigh & Dawson, 1994). Some substances that have been used to influence the water potential and influence spore viability include polyethylene glycol, glycerol, sucrose, glucose, and maltose, as well as a variety of carbohydrate sources. An increase in tolerance to the drying process, as represented by increased viability and maintenance of efficacy, has been associated with an increase in the cellular concentration of trehalose (Burleigh & Dawson, 1994; D'Amore *et al.,* 1991; Hallsworth & Magan, 1994).

In addition to submerged fermentation, other methods have been used for large-scale production of inoculum, although the technology has not progressed as rapidly. Solid substrate fermentation utilizes materials such as cereal grains, straw, or other inexpensive agricultural wastes, such as corn cobs or rice sheaths. This material is used for growing the fungus and then is applied directly to the field with little or no additional processing. Although this may be the only alternative for fungi that will not produce easily dispersed propagules in any other system, this method has a number of drawbacks. Inoculum produced by this method is difficult to quantify because the fungus is not separated from the culturing material. It is often difficult to maintain sterility in these systems and storage of the resulting bulky material may be limiting.

For some fungi, a combination of solid substrate and submerged fermentation is necessary to produce the desired inoculum. The potential mycoherbicidal agent *Alternaria macrospora* has been produced by first using liquid medium to generate mycelium that is then blended and combined with vermiculite (Walker, 1980). This mixture is then spread in a thin layer and exposed to cycling light. This method produces an ample amount of conidia but is considerably more labor intensive and expensive than liquid fermentation alone. The additional handling of the fungus also provides more opportunities for contamination. Regardless of the drawbacks of this system, it has been used for midscale production of several biological control agents (Churchill, 1982).

References

Abbas, H. K., Egley, G. H., & Paul, R. N. (1995b). Effect of conidia production temperature on germination and infectivity of *Alternaria helianthi.* Phytopathology, 85, 677–682.
Abernathy, J. R., & Bridges, D. C. (1994). Research priority dynamics in weed science. Weed Technology, 8, 396–399.

Adams, E. B. (1988). Fungi in classical biocontrol of weeds. *In* "Fungi in Biological Control Systems" (M. N. Burge, Ed.), pp. 11–124. Manchester Univ. Press, Manchester, England.

Adams, E. B., & Line, R. F. (1984a). Epidemiology and host morphology in the parasitism of rusk skeleton weed by *Puccinia chondrillina*. Phytopathology, 74, 745–748.

Adams, E. B., & Line, R. F. (1984b). Biology of *Puccinia chondrillina* in Washington. Phytopathology, 74, 742–745.

Alber, G., Defago, C., Kern, H., & Sedlar, L. (1986). Host range of *Puccinia expansa* Link (=*P. glomerata* GREv), a possible fungal biocontrol against *Senecio* weeds. Weed Research, 26, 69–74.

Altman, J., & Campbell, C. L. (1977). Effect of herbicides on plant disease. Annual Review of Phytopathology, 15, 361–385.

Altman, J., & Rovira, A. D. (1989). Herbicide-pathogen interactions in soil-borne root diseases. Canadian Journal of Plant Pathology, 11, 166–172.

Amsellam, Z., Sharon, A., Gressel, J., & Quimby, P. C. (1990). Complete abolition of high inoculum threshold of two mycoherbicides (*Alternaria cassiae* and *A. crassa*) when applied in invert emulsion. Phytopathology, 80, 925–929.

Anderson, R. N., & Walker, H. L. (1985). *Colletotrichum coccodes*: A pathogen of eastern nightshade (*Solanum ptycanthum*). Weed Science, 33, 902–905.

Anderson, W. P. (1983). "Weed Science: Principle." West Publishing Co., St. Paul, MN.

Andrews, J. H. (1992). Biological control in the phyllosphere. Annual Review of Phytopathology, 30, 603–635.

Anonymous. (1988). Biennial Report 1985–97, Division of Entomolgy, CSIRO, Canberra. Aspelin, A. L. (1994). "Pesticide Industry Sales and Usage—1992 and 1993 Market Estimates." U.S. Environmental Protection Agency, Washington, DC.

Aspelin, A. L. (1994). Pesticide industry sales and usage—1992 and 1993 market estimates. U.S. EPA, Washington, D.C.

Auld, B. (1993). Vegetable oil suspension emulsions reduce dew dependence of a mycoherbicide. Crop Protection, 12, 477–479.

Auld, B. A., McRae, C. F., & Say, M. M. (1988). Possible control of *Xanthium spinosus* by a fungus. Agriculture, Ecosystems, & Environment, 21, 513–519.

Auld, B. A., Say, M. M., Ridings, H. I., & Andrews, J. (1990). Field application of *Colletotrichum orbiculare* to control *Xanthium spinosum*. Agriculture, Ecosystems, & Environment, 32, 315–323.

Auld, B. A., Womack, J. G., Burge, M. N., Eccleston, G. M., Smith, H. E., & Sheng, Q. (1996). Success of an invert emulsion formulation in two bioherbicide systems. "Biological Control of Weeds, IX International Symposium, 19–26 January 1996, Stellenbosch, South Africa" (V. C. Moran & J. H. Hoffmann, Eds.), p. 541 (Abstract). Univ. of Cape Town, Cape Town, South Africa.

Bannon, J. S. (1988). CASST™ herbicide (*Alternaria cassiae*): A case history of a mycoherbicide. Journal of Alternative Agriculture, 3, 73–76.

Bannon, J. S., White, J. C., Long, D., Riley, J. A., Baragona, J., Atkins, M., & Crowley, R. C. (1990). Bioherbicide technology: An industrial perspective. *In* "Microbes and Microbial Products as Herbicides" [ACS Symposium Series 439] (R. E. Hoagland, Ed.), pp. 305–319. American Chemical Society, Washington, DC.

Barreto, R. W., & Evans, H. C. (1988). Taxonomy of a fungus introduced into Hawaii for biological control of *Ageratina riparia* (Eupatorieae: Compositae), with observations on related weed pathogens. Transactions of the British Mycological Society, 91, 81–97.

Batson, W. E., & Trevathan, L. E. (1988). Suitability and efficacy of ground corncobs as a carrier of *Fusarium solani* spores. Plant Disease, 72, 222–225.

Baudoin, A. B. A. M., Abad, R. G., Kok, L. T., & Bruckart, W. L. (1993). Field evaluation of *Puccinia carduorum* for biological control of musk thistle. Biological Control, 3, 53–60.

Berthier, Y. T., Bruckart, W. L., Chaboudez, P., & Luster, D. G. (1996). Polymorphic restriction patterns of ribosomal transcribed spacers in the biocontrol fungus *Puccinia carduorum* correlates with weed host origin. Applied and Environmental Microbiology, 62, 3037–3041.

Bewick, T. A., Binning, L. K., Stevenson, W. R., & Stewart, J. (1986). Development of biological control for swamp dodder. Proceedings of the North Central Weed Control Conference, 41, 24.

Blanchette, B. L., & Lee, G. A. (1981). The influence of environmental factors on infection of rush skeleton weed (*Chondrilla juncea*) by *Puccinia chondrillina*. Weed Science, 29, 364–367.

Bowers, R. C. (1986). Commercialization of Collego—An industrialist's view. Weed Science, 34 (Suppl. 1), 24–25.

Boyetchko, S. M. (1999). Innovative applications of microbial agents for biological weed control. *In* "Biotechnological Approaches in Biocontrol of Plant Pathogens" (K. G. Mukerji, B. P. Chamola, & R. K. Upadhyay, Eds.), pp. 73–97. Plenum, London.

Boyetchko, S. M., & Mortensen, K. (1993). Use of rhizobacteria as biological control agents of downy brome. *In* "Proceedings of the Soils and Crop Workshop, Saskatoon, Saskatchewan, Canada," pp. 443–448.

Boyette, C. D. (1994). Unrefined corn oil improves the mycoherbicidal activity of *Colletotrichum truncatum* for hemp sesbania (*Sesbania exaltata*) control. Weed Technology, 8, 526–529.

Boyette, C. D. (1995). Improvement of mycoherbicidal activity of anthracnose pathogens with adjuvants and surfactants. Proceedings of the Southern Weed Science Society, 48, 167 (Abstract).

Boyette, C. D., & Abbas, H. K. (1994). Host range alteration of the bioherbicidal fungus *Alternaria crassa* with fruit pectin and plant filtrates. Weed Science, 42, 487–491.

Boyette, C. D., Abbas, H. K., & Connick, W. J. (1993a). Evaluation of *Fusarium oxysporum* as a potential bioherbicide for sicklepod (*Cassia obtusifolia*), coffee senna (*C. occidentalis*), and hemp sesbania (*Sesbania exaltata*). Weed Science, 41, 678–681.

Boyette, C. D., Quimby, P. C., Bryson, C. T., Egley, G. H., & Fulgham, F. E. (1993b). Biological control of hemp sesbania (*Sesbania exaltata*) under field conditions with *Colletotrichum truncatum* formulated in an invert emulsion. Weed Science, 41, 497–500.

Boyette, C. D., Quimby, P. C., Caesar, A. J., Birdsall, J. L., Connick, W. J., Daigle, D. J., Jackson, M. A., Egley, G. H., & Abbas, H. K. (1996). Adjuvants, formulations, and spraying systems for improvement of mycoherbicides. Weed Technology, 10, 637–644.

Boyette, C. D., Weidemann, G. J., TeBeest, D. O., & Turfitt, L. B. (1986). Biocontrol of jimsonweed in the field with *Alternaria crassa*. Proceedings of the Southern Weed Science Society, 39, 388.

Brooker, N. L., Mischke, C. F., Patterson, C. L., Mischke, S., Bruckart, W. L., & Lydon, J. (1996). Pathogenicity of *bar*-transformed *Colletotrichum gloeosporioides* f. sp. *aeschynomene*. Biological Control, 7, 159–166.

Bruckart, W. L. (1989). Host range determination of *Puccinia jaceae* from yellow starthistle. Plant Disease, 73, 155–160.

Bruckart, W. L., & Dowler, W. M. (1986). Evaluation of exotic rust fungi in the United States for classical biological control of weeds. Weed Science, 34 (Suppl. 1), 11–14.

Bruckart, W. L., Johnson, D. R., & Frank, R. J. (1988). Bentazon reduces rust-induced disease in yellow nutsedge (*Cyperus esculentus*). Weed Technology, 2, 299–303.

Bruckart, W. L., Politis, D. J., & Sutker, E. M. (1985). Susceptibility of *Cynara scolymus* (artichoke) to *Puccinia carduorum* observed under greenhouse conditions. *In* "Proceedings of the VI International Symposium on Biological Control of Weeds, Vancouver, Aug. 19–25, 1984" (E. S. Delfosse, Ed.), pp. 603–607. Agriculture Canada, Ottawa.

Bruckart, W. L., Turner, S. K., Sutker, E. M., Vonmoos, R., Sedlar, L., & Défago, G. (1986). Relative virulence of *Melampsora euphorbiae* from Central Europe toward North America and European spurges. Plant Disease, 70, 847–850.

Bruzzese, E., & Field, R. P. (1985). Occurrence and spread of *Phragmidium violaceum* on blackberry (*Rubus fruticosus*) in Victoria, Australia. *In* "Proceedings of the VI International Symposium on Biological Control of Weeds, Vancouver, Aug. 19–25, 1984" (E. S. Delfosse, Ed.), pp. 609–612. Agriculture Canada, Ottawa.

Bruzzese, E., & Hasan, S. (1986a). The collection and selection in Europe of the isolates *Phragmidium violaceum* (Uredinales) pathogenic to species of European blackberry naturalized in Australia. Annals of Applied Biology, 108, 527–533.

Bruzzese, E., & Hasan, S. (1986b). Host specificity of the rust *Phragmidium violaceum,* a potential biological control agent of European blackberry. Annals of Applied Biology, 108, 585–596.

Bruzzese, E., & Hasan, S. (1986c). Infection of Australian and New Zealand *Rubus subgenera Dalebarda* and *Lampobatus* by the European blackberry rust fungus *Phragmidium violaceum.* Plant Pathology, 35, 413–416.

Bruzzese, E., & Hasan, S. (1987). Infection of blackberry cultivars by the European blackberry rust fungus, *Phragmidium violaceum.* Journal of Horticultural Science, 64, 475–479.

Burdon, J. J., Groves, R. H., Kaye, R. E., & Speer, S. S. (1984). Competition in mixtures of susceptible and resistant genotypes of *Chondrilla juncea* differentially infected with rust. Oecologia, 64, 199–203.

Burleigh, S. H., & Dawson, J. O. (1994). Desiccation tolerance and trehalose production in *Frankia* hyphae. Soil Biology & Biochemistry, 26, 539–598.

Caesar, A. J. (1994). Pathogenicity of *Agrobacterium* species from the noxious rangeland weeds *Euphorbia esula* and *Centaurea repens.* Plant Disease, 78, 796–800.

Caesar, A. J., Quimby, P. C., Rees, N. E., & Spencer, N. R. (1992). Diseases of leafy spurge in the northern Great Plains. *In* Great Plains Agriculture Council Publication No. 44, pp. 37–40.

Cardina, J., Littrel, R. H., & Hanlin, R. T. (1988). Anthracnose of Florida beggarweed (*Desmodium tortuosum*) caused by *Colletotrichum truncatum.* Weed Science, 36, 329–334.

Cerkauskas, R. F. (1988). Latent colonization by *Colletotrichum* spp.: Epidemiological considerations and implications for mycoherbicides. Canadian Journal of Plant Pathology, 10, 297–310.

Chandramohan, S., & Charudattan, R. (1996). Multiple-pathogen strategy for bioherbicidal control of several weeds. WSSA Abstracts, 36, 49.

Charudattan, R. (1986). Integrated control of waterhyacinth (*Eichhornia crassipes*) with a pathogen, insects and herbicides. Weed Science, 34 (Suppl. 1), 26.

Charudattan, R. (1989). Inundative control of weeds with indigenous fungal pathogens. *In* "Fungi in Biological Control Systems" (M. N. Burge, Ed.), pp. 86–110. Manchester Univ. Press, Manchester, England.

Charudattan, R. (1990). Biological control by means of fungi. *In* "Aquatic Weeds" (K. J. Murphy & A. Pieterse, Eds.), pp. 186–201. Oxford Univ. Press, Oxford, UK.

Charudattan, R. (1991). The mycoherbicide approach with plant pathogens. *In* "Microbial Control of Weeds" (D. O. TeBeest, Ed.), pp. 24–57. Chapman & Hall, New York.

Charudattan, R. (1993). The role of pesticides in altering biocontrol efficacy. *In* "Pesticide Interactions in Crop Production: Beneficial and Deleterious Effects" (J. Altman, Ed.), pp. 422–430. CRC Press, Boca Raton, FL.

Charudattan, R., & Browning, W. H. (1992). "Regulations and Guidelines: Critical Issues in Biological Control. Proceedings of a USDA-CSRS National Workshop, June 10–12, 1991, Vienna, VA." Institute of Food and Agricultural Sciences, Univ. of Florida, Gainesville, FL.

Charudattan, R., & Conway, K. E. (1975). Comparison of *Uredo eichhor-niae,* the waterhyacinth rust, and *Uromyces pontederiae.* Mycologia, 67, 653–657.

Charudattan, R., Cordo, H. A., Silveira-Guido, A., & Zettler, F. W. (1978a). Obligate pathogens of the milkweed vine *Morrenia odorata* as biological control agents (Abstract). *In* "Proceedings of the IV International Symposium on Biological Control of Weeds, Gainesville, Aug. 30–Sept. 2, 1976" (T. E. Freeman, Ed.), p. 241. Univ. of Florida, Gainesville, FL.

Charudattan, R., Freeman, T. E., Cullen, R. E., & Hofmeister, F. M. (1981). Evaluation of *Fusarium roseum* "Culmorum" as a biological control for *Hydrilla verticillata*: Safety. *In* "Proceedings of the V International Symposium on Biological Control of Weeds, Brisbane, July 22–27, 1980" (E. S. Delfosse, Ed.), pp. 307–323. CSIRO, Melbourne.

Charudattan, R., Linda, S. B., Kluepfel, M., & Osman, Y. A. (1985). Biocontrol efficacy of *Cercospora rodmanii* on waterhyacinth. Phytopathology, 75, 1263–1269.

Charudattan, R., McKenney, D. E., Cordo, H. A., & Silveria-Guido, A. (1978b). *Uredo eichhorniae,* a potential biocontrol agent for waterhyacinth. *In* "Proceedings of the IV International Symposium on Biological Control of Weeds, Gainesville, Aug. 30–Sept. 2, 1976" (T. E. Freeman, Ed.), pp. 210–213. Univ. of Florida, Gainesville, FL.

Charudattan, R., Prange, V. J., & DeValerio, J. T. (1996a). Exploration of the "bialophos genes" for improving bioherbicide efficacy. Weed Technology, 10, 625–636.

Charudattan, R., Shabana, Y. M., DeValerio, J. T., & Rosskopf, E. N. (1996b). *Phomopsis* species fungus useful as a broad-spectrum bioherbicide to control several species of pigweeds. U.S. Patent No. 5,510,316.

Charudattan, R., Walker, H. L., Boyette, C. D., Ridings, W. H., TeBeest, D. O., VanDyke, C. G., & Worsham, A. D. (1986). Evaluation of *Alternaria cassiae* as a mycoherbicide for sicklepod (*Cassia obtusifolia*) in regional field tests. Southern Cooperative Series Bulletin 317. Department of Information, Alabama Agricultural Station, Auburn Univ., Auburn, AL.

Charudattan, R., Zettler, F. W., Cordo, H. A., & Christie, R. G. (1980). Partial characterization of a potyvirus infecting milkweed vine, *Morrenia odorata.* Phytopathology, 70, 909–913.

Chellemi, D. O., Olson, S. M., Mitchell, D. J., Secker, I., & McSorley, R. (1997). Adaptation of soil solarization to the integrated management of soilborne pests of tomato under humid conditions. Phytopathology, 87, 250–258.

Cherrington, C. A., & Elliott, L. F. (1987). Incidence of inhibitory pseudomonads in the Pacific Northwest. Plant and Soil, 101, 159–165.

Churchhill, B. W. (1982). Mass production of microorganisms for biological control. *In* "Biological Control of Weeds with Plant Pathogens" (R. Charudattan and H. L. Walker, Eds.), pp. 139–156. Wiley, New York.

Connick, W. J., Jr., Boyette, C. D., & McAlpine, J. R. (1991). Formulation of mycoherbicides using a pasta-like process. Biological Control, 1, 281–287.

Coulson, J. R. (1992). The TAG: Development, functions, procedures, and problems. *In* "Regulations and Guidelines: Critical Issues in Biological Control. Proceedings of a USDA–CSRS National Workshop, June 10–12, 1991, Vienna, VA" (R. Charudattan & H. W. Browning, Eds.), pp. 54–60. Institute of Food and Agricultural Sciences, Univ. of Florida, Gainesville, FL.

Cullen, J. M. (1978). Evaluating the success of the programme for the biological control of *Chondrilla juncea* L. *In* "Proceedings of the IV International Symposium on Biological Control of Weeds, Gainesville, Aug. 30–Sept. 2, 1976" (T. E. Freeman, Ed.), pp. 117–121. Univ. of Florida, Gainesville, FL.

D'Amore, T., Crumplen, R., & Stewart, G. G. (1991). The involvement of trehalose in yeast stress tolerance. Journal of Industrial Microbiology, 7, 191–196.

Daigle, D. J., & Connick, W. J., Jr. (1990). Formulation and application technology for microbial weed control. *In* "Microbes and Microbial Products as Herbicides" (Hoagland, R. E., Ed.). American Chemical Society, Washington, DC.

Daigle, D. J., Connick, W. J., Jr., Quimby, P. C., Jr., Evans, J., Trask-Morrell, B., & Fulgham, F. E. (1990). Invert emulsions: Carrier and water source for the mycoherbicide, *Alternaria cassiae*. Weed Technology, 4, 327–331.

Daigle, D. J., & Cotty, P. J. (1991). Factors that influence germination and mycoherbicidal activity of *Alternaria cassiae*. Weed Technology, 5, 82–86.

Daigle, D. J., & Cotty, P. J. (1992). Production of conidia of *Alternaria cassiae* with alginate pellets. Biological Control, 2, 278–281.

Daniel, J. T., Templeton, G. E., Smith, R. J., Jr., & Fox, W. T. (1973). Biological control of northern jointvetch in rice with an endemic fungal disease. Weed Science, 21, 303–307.

Défago, G., Kern, H., & Sedlar, L. (1985). Potential control of weedy spurges by the rust *Uromyces scutellatus* s.l. Weed Science, 33, 857–860.

DeLucca, A. J., II, Connick, W. J., Jr., Fravel, D. R., Lewis, J. A., & Bland, J. M. (1990). The use of bacterial alginates to prepare biocontrol formulations. Journal of Industrial Microbiology, 6, 129–134.

de Nooij, M. P., & van Damme, J. M. M. (1988). Variation in pathogenicity among and within populations of the fungus *Phomopsis subordinaria* infecting *Plantago lanceolata*. Evolution, 42, 1166–1171.

Diarra, C., Ciotola, M., Hallett, S. G., Hess, D. E., & Watson, A. K. (1996). Mass production of *Fusarium oxysporum* (M12-4a), a biocontrol agent for *Striga hermonthica*. *In* "Biological Control of Weeds, IX International Symposium, 19–26 January 1996, Stellenbosch, South Africa" (V. C. Moran & J. H. Hoffmann, Eds.), pp. 149–152. Univ. of Cape Town, Cape Town, South Africa.

Dillard, H. M. (1988). Influence of temperature, pH, osmotic potential and fungicide sensitivity on germination of conidia and growth from sclerotia of *Colletotrichum coccodes* in vitro. Phytopathology, 78, 1357–1361.

Dhingra, O. D., & Sinclair, J. B. (1995). "Basic Plant Pathology Methods," 2nd ed. Lewis Publishers, Boca Raton, FL.

DiTommaso, A., Watson, A. K., & Hallett, S. G. (1996). Infection by the fungal pathogen *Colletotrichum coccodes* affects velvetleaf (*Abutilon theophrasti*)-soybean competition in the field. Weed Science, 44, 924–933.

Dumas, M. T., Wood, J. E., Mitchell, E. G., & Boyonoski, N. W. (1997). Control of stump sprouting of *Populus tremuloides* and *P. grandidentata* by inoculation with *Chondrostereum purpureum*. Biological Control, 10, 37–41.

Egley, G. H., & Boyette, C. D. (1995). Water-corn oil emulsion enhances conidia germination and mycoherbicidal activity of *Colletotrichum truncatum*. Weed Science, 43, 312–317.

Egley, G. H., Hanks, J. E., & Boyette, C. D. (1993). Invert emulsion droplet size and mycoherbicidal activity of *Colletotrichum truncatum*. Weed Technology, 7, 417–424.

Elliot, L. F., Horwath, W. R., & Muller-Warrant, G. W. (1994). Biocontrol of annual bluegrass with deleterious rhizobacteria. Agronomy Abstracts, 58, 280.

Emge, R. G., Melching, J. S., & Kingsolver, C. H. (1981). Epidemiology of *Puccinia chondrillina*, a rust pathogen for the biological control of rush skeleton weed in the United States. Phytopathology, 71, 839–843.

Evans, H. C. (1986). The life cycle of *Puccinia abrupta* var. *partheniicola*, a potential biological control agent of *Parthenium hysterophorus*. Transactions of the British Mycological Society, 88, 105–111.

Fernando, W. G. D., Watson, A. K., & Paulitz, T. C. (1994). Phylloplane *Pseudomomas* spp. enhance disease caused by *Colletotrichum coccodes* in velvetleaf. Biological Control, 4, 125–131.

Figliola, S. S., Camper, N. D., & Ridings, W. H. (1988). Potential biological control agents for goosegrass (*Eleusine indica*). Weed Science, 36, 830–835.

Gemeiner, P., Kurillova, L., Markovic, O., Malovikova, A., Uhrin, D., Ilavsky, M., Stefuca, V., Polakovic, M., & Bales, V. (1991). Calcium pectate gel beads for cell entrapment. 3. Physical properties of calcium pectate and calcium alginate gel beads. Biotechnology and Applied Biochemistry, 13, 335–345.

Gosselin, L., Jobidon, R., & Bernier, L. (1996). Assessment of genetic variation within *Chondrostereum purpureum* from Quebec by random amplified polymorphic DNA analysis. Mycological Research, 100, 151–158.

Greaves, M. P., & MacQueen, M. D. (1992). Bioherbicides: Their role in tomorrow's agriculture. *In* "Resistance: Acheivements and Developments in Combating Pesticide Resistance Science Symposium," pp. 295–306. Harpenden, England, UK.

Green, S., Stewart-Wade, S. M., Boland, G., Teshler, M. P., & Liu, S. H. (1998). Formulating microorganisms for biological control of weeds. *In* "Plant-Microbe Interactions and Biological Control" (G. J. Boland and D. Kuykendall, Eds.), pp. 249–281. Dekker, New York.

Griffin, D. (1994). "Fungal Physiology," 2nd ed. Wiley-Liss, New York.

Hallsworth, J. E., & Magan, N. (1994). Effect of carbohydrate type and concentration on polyhydroxy alcohol and trehalose content of conidia of three entomopathogenic fungi. Microbiology, 140, 2705–2713.

Harman, G. E., Jin, X., Stasz, E., Peruzzotti, G., Leopold, A. C., & Tayler, A. G. (1991). Production of conidial biomass of *Trichoderma harzianum* for biological control. Biological Control, 1, 23–28.

Harris, P. A., & Stahlman, P. W. (1991). Biocontrol of *Bromus* and *Aegilops* spp. in winter wheat using deleterious rhizobacteria. Agronomy Abstracts, 55, 266.

Harris, P. A., & Stahlman, P. W. (1992). Biological weed control using deleterious rhizobacteria. Abstracts, Weed Science Society of America, 32, 50.

Harris, P. A., & Stahlman, P. W. (1993a). Selective biocontrol of winter annual grass weeds in winter wheat. North Central Weed Science Society Proceedings, 58, 6.

Harris, P. A., & Stahlman, P. W. (1993b). Soil bacteria selectively inhibit winter annual grass weeds in winter wheat. Agronomy Abstracts, 57, 250.

Hasan, S. (1972). Specificity and host specialization of *Puccinia chondrillina*. Annals of Applied Biology, 72, 257–263.

Hasan, S. (1980). Plant pathogens and biological control of weeds. Review of Plant Pathology, 59, 349–356.

Hasan, S. (1981). A new strain of the rust fungus *Puccinia chondrillina* for biological control of skeleton weed in Australia. Annals of Applied Biology, 99, 119–124.

Hasan, S. (1983). Biological control of weeds with plant pathogens—Status and prospects. Proceedings, 10th International Congress on Plant Protection, 2, 759–776.

Hasan, S. (1985a). Prospects for biological control of *Heliotropium europaeum* by fungal pathogens. *In* "Proceedings of the VI International Symposium on Biological Control of Weeds, Vancouver, Aug. 19–25, 1984" (E. S. Delfosse, Ed.), pp. 617–623. Agriculture Canada, Ottawa.

Hasan, S. (1985b). Search in Greece and Turkey for *Puccinia chondrillina* strains suitable to Australian forms of skeleton weed. *In* "Proceedings of the VI International Symposium on Biological Control of Weeds, Vancouver, Aug. 19–25, 1984" (E. S. Delfosse, Ed.), pp. 625–632. Agriculture Canada, Ottawa.

Hasan, S. (1988). Biocontrol of weeds with microbes. *In* "Biocontrol of Plant Diseases" (K. G. Mukerji & K. L. Garg, Eds.), Vol. 1, pp. 129–151. CRC Press, Boca Raton, FL.

Hasan, S., & Wapshere, A. J. (1973). The biology of *Puccinia chondrillina*, a potential biological control agent of skeletonweed. Annals of Applied Biology, 74, 325–332.

Hasan, S., Chaboudez, P., & Mortensen, K. (1989). Field experiment with the European knapweed rust (*Puccinia jaceae*) on safflower, sweet sultan and bachelor's button. *In* "Proceedings of the VII International Symposium on Biological Control of Weeds" (E. S. Delfosse, Ed.), pp. 499–511. Ist. Sper. Patol. Veg. (MAF).

Hoagland, R. E. (1996). Chemical interactions with bioherbicides to improve efficacy. Weed Technology, 10, 651–674.

Holt, J. S. (1992). History of identification of herbicide resistant weeds. Weed Technology, 6, 615–620.

Imaizumi, S., Nishino, T., Miyabe, K., Fujimori, T., & Yamada, M. (1997). Biological control of annual bluegrass (*Poa annua* L.) with a Japanese isolate of *Xanthomonas campestris* pv. *poae* (JT-P482). Biological Control, 8, 7–14.

Inman, R. E. (1971). A preliminary evaluation of *Rumex* rust as a biological control agent for curly dock. Phytopathology, 61, 102–107.

Jackson, M. A., & Bothast, R. J. (1990). Carbon concentration and carbon-to-nitrogen ratio influence submerged-culture conidiation by the potential bioherbicide *Colletotrichum truncatum*. *In* "Microbial Control of Weeds" (D. O. TeBeest, Ed.), pp. 225–261. Chapman & Hall, New York.

Jackson, M. A., & Slininger, P. J. (1993). Submerged culture conidial germination and conidiation of the bioherbicide *Colletotrichum truncatum* are influenced by the amino acid composition of the medium. Journal of Industrial Microbiology, 12, 417–422.

Jackson, M. A., Shasha, B. S., & Schisler, D. A. (1996). Formulation of *Colletotrichum truncatum* microsclerotia for improved biocontrol of the weed hemp sesbania (*Sesbania exaltata*). Biological Control, 7, 107–113.

Jacobs, J. S., Sheley, R. L., & Maxwell, B. D. (1996). Effect of *Sclerotinia sclerotiorum* on the interference between bluebunch wheatgrass (*Agropyron spicatum*) and spotted knapweed (*Centaurea maculosa*). Weed Technology, 10, 13–21.

Johal, G. S., & Rahe, J. E. (1990). Role of phytoalexins in the suppression of resistance of *Phaseolus vulgaris* to *Colletotrichum lindemuthianum* by glyphosate. Canadian Journal of Plant Pathology, 12, 225–235.

Johnson, B. J. (1994). Biological control of annual bluegrass with *Xanthomonas campestris* pv. *poannua* in bermudagrass. HortScience, 29, 659–662.

Johnson, B. N., Doty, J. A., Stubbs, T. L., & Kennedy, A. C. (1993). Efficacy and specificity of rhizobacteria for the control of weeds. Agronomy Abstracts, 57, 252.

Johnson, D. R., Wyse, D. L., & Jones, K. J. (1995). Efficacy of spring and fall applications of *Pseudomonas syringae* pv. *tagetis* for Canada thistle (*Cirsium arvense* L.) control in soybean. WSSA Abstracts, 35, 61.

Johnson, D. R., Wyse, D. L., & Jones, K. J. (1996). Controlling weeds with phytopathogenic bacteria. Weed Technology, 10, 621–624.

Kadir, J. (1997). Development of a bioherbicide for the control of purple nutsedge. Ph.D. Dissertation, Univ. of Florida, Gainesville.

Kadir, J., & Charudattan, R. (1996). *Dactylaria higginsii* (Lutrell) M. B. Ellis: A potential bioherbicide for nutsedge (*Cyperus* spp.). WSSA Abstracts, 36, 49.

Kannwischer, M. E., Ridings, W. H., & Mitchell, D. J. (1980). Evaluation of the effects of various herbicides on the survival of *Phytophthora citrophthora*. Phytophthora Newsletter, 8, 31, Feb. 1980.

Kempenaar, C. (1995). "Studies on Biological Control of *Chenopodium album* by *Ascochyta caulina*." Cip-Data Koniklijke Bibliotheek, Den Haag, Wageningen, The Netherlands.

Kempenaar, C., Horsten, P. J. F. M., & Scheepens, P. C. (1996a). Growth and competitiveness of common lambsquarters (*Chenopodium album*) after foliar application of *Ascochyta caulina* as a mycoherbicide. Weed Science, 44, 609–614.

Kempenaar, C., Wanningen, R., & Scheepens, P. C. (1996b). Control of *Chenopodium album* by soil application of *Ascochyta caulina* under greenhouse conditions. Annals of Applied Biology, 129, 343–354.

Kennedy, A. C. (1996). Molecular biology of bacteria and fungi for biological control of weeds. *In* "Molecular Biology of the Biological Control of Pests and Diseases of Plants" (M. Gunasekaran & D. J. Weber, Eds.), pp. 155–172. CRC Press, Boca Raton, FL.

Kennedy, A. C., Elliott, L. F., Young, F. L., & Douglas, C. L. (1991). Rhizobacteria suppressive to the weed downy brome. Soil Science Society of America Journal, 55, 722–727.

Kennedy, A. C., Stubbs, T. L., & Young, F. L. (1989). Rhizobacterial colonization of winter wheat and grass weeds. Agronomy Abstracts, 53, 220.

Kenney, D. S. (1986). DeVine: The way it was developed—An industrialist's view. Weed Science, 34 (Suppl. 1), 15–16.

Klein, T. A., & Auld, B. A. (1995a). Influence of spore dose and water volume on a mycoherbicide's efficacy in field trials. Biological Control, 5, 173–178.

Klein, T. A., & Auld, B. A. (1995b). Evaluation of Tween 20 and glycerol as additives to mycoherbicide suspensions applied to bathurst burr. Plant Protection Quarterly, 10, 14–16.

Klein, T. A., Auld, B. A., & Fang, W. (1995). Evaluation of oil suspension emulsions of *Colletotrichum orbiculare* as a mycoherbicide in field trials. Crop Protection, 14, 193–197.

Kremer, R. J. (1993). Management of weed seed banks with microorganisms. Ecological Applications, 3, 42–52.

Kremer, R. J., & Kennedy, A. C. (1996). Rhizobacteria as biocontrol agents of weeds. Weed Technology, 10, 601–609.

Kremer, R. J., Begonia, M. F. T., Stanley, L., & Lanham, E. T. (1990). Characterization of rhizobacteria associated with weed seedlings. Applied and Environmental Microbiology, 56, 1649–1655.

Makowski, R. M. D., & Mortensen, K. (1992). The first mycoherbicide in Canada: *Colletotrichum gloeosporioides* f. sp. *malvae* for round-leaved mallow control. *In* "Proceedings of the First International Weed Control Congress, Monash University, Melbourne, Australia, February 17–21, 1992," pp. 298–300. International Weed Science Society, Corvallis, Oregon.

Massion, C. L., & Lindow, S. E. (1986). Effects of *Sphacelotheca holci* infection on morphology and competitiveness of johnsongrass (*Sorghum halepense*). Weed Science, 34, 883–888.

McClay, A. S. (1985). Biocontrol agents for *Parthenium hysterophorus* from Mexico. *In* "Proceedings of the VI International Symposium on Biological Control of Weeds, Vancouver, Aug. 19–25, 1984" (E. S. Delfosse, Ed.), pp. 771–778. Agriculture Canada, Ottawa.

McGuire, M. R., & Shasha, B. S. (1992). Adherent starch granules for encapsulation of insect control agents. Journal of Economic Entomology, 85, 1425–1433.

McKoy, J. F., & Trinci, A. P. J. (1987). Sporulation of *Verticillium agaricinum* and *Schizosaccharomyces pombe* in batch and chemostat culture. Transactions of the British Mycological Society, 88, 299–307.

McRae, C. F., & Auld, B. A. (1988). The influence of environmental factors on anthracnose of *Xanthium spinosum*. Phytopathology, 78, 1182–1186.

McRae, C. F., & Stevens, G. R. (1990). Role of conidial matrix of *Colletotrichum orbiculare* in pathogenesis of *Xanthium spinosum*. Mycological Research, 94, 890–896.

Meijer, G., Megnegneau, B., & Linders, E. (1994). Variability for isozyme: Vegetative compatibility and RAPD markers in natural populations of *Phomopsis subordinaria*. Mycological Research, 98, 267–276.

Mintz, A. S., Heiny, D. K., & Weidemann, G. J. (1992). Factors influencing the biocontrol of tumble pigweed (*Amaranthus albus*) with *Aposphaeria amaranthi*. Plant Disease, 76, 267–269.

Morin, L., Auld, B. A., & Brown, J. F. (1993a). Interaction between *Puccinia xanthii* and facultative parasitic fungi on *Xanthium occidentale*. Biological Control, 3, 288–295.

Morin, L., Auld, B. A., & Brown, J. F. (1993b). Synergy between *Pucci-*

nia xanthii and *Colletotrichum orbiculare* on *Xanthium occidentale.* Biological Control, 3, 296–310.

Morin, L., Watson, A. K., & Reeleder, R. D. (1990a). Effect of dew, inoculum density, and spray additives on infection of field bindweed by *Phomopsis convolvulus.* Canadian Journal of Plant Pathology, 12, 48–52.

Morin, L., Watson, A. K., & Reeleder, R. D. (1990b). Production of conidia by *Phomopsis convolvulus.* Canadian Journal of Microbiology, 36, 86–91.

Morris, M. J. (1991). The use of plant pathogens for biological weed control in South African Agriculture, Ecosystems and Environment, 37, 234–255.

Morris, M. J. (1997). Impact of the gall-forming rust fungus *Uromycladium tepperianum* on the invasive tree *Acacia saligna* in South Africa. Biological Control, 10, 75–82.

Morris, M. J., Wood, A. R., & Den Breëyen, A. (1998). Development and registration of a fungal inoculant to prevent regrowth of cut wattle tree stump in South Africa, and a brief overview of other bioherbicide projects currently in progress. "IV International Bioherbicide Workshop Programme and Abstracts, 6–7 August 1998," p. 15. Univ. of Strathclyde, Glasgow, Scotland.

Mortensen, K. (1998). Biological control of weeds using microorganisms. *In* "Plant-Microbe Interaction and Biological Control" (G. J. Boland & L. D. Kuykendall, Eds.), pp. 223–248. Dekker, New York.

Mortensen, K., & Makowski, R. M. D. (1997). Effects of *Colletotrichum gloeosporioides* f. sp. *malvae* on plant development and biomass of non-target field crops under controlled and field conditions. Weed Research, 37, 351–360.

Oehrens, E. (1977). Biological control of the blackberry through the introduction of rust, *Phragmidium violaceum* in Chile. FAO Plant Protection Bulletin, 25, 26–28.

Oehrens, E. B., & Gonzales, S. M. (1974). Introducción de *Phragmidium violaceum* (Schulz) Winter como factor de biologico de zarzamora (*Rubus constrictus* Lef. et M. y R. *ulmifolius* Schott.). Agro Sur, 2, 30–33.

Oehrens, E. B., & Gonzales, S. M. (1975). Introducción de *Uromyces galegae* (Opiz) Saccardo como factor de control biologico de galega (*Galega officinalis* L.). Agro Sur, 3, 87–91.

Ormeno-Nunez, J., Reeleder, R. D., & Watson, A. K. (1988). A foliar disease of field bindweed (*Convolvulus arvensis*) caused by *Phomopsis convolvulus.* Plant Disease, 72, 338–342.

Parker, A. (1989). Biological control of parthenium weed using two rust fungi. *In* "Proceedings of the VII International Symposium on Biological Control of Weeds" (E. S. Delfosse, Ed.), pp. 531–538. Ist. Sper. Patol. Veg. (MAF).

Paul, N. D., & Ayres, P. G. (1990). Effects of interactions between nutrient supply and rust infection of *Senecio vulgaris* L. on competition with *Capsella bursa-pastoris* (L.) Medic. New Phytologist, 114, 667–674.

Peeples, K. A. (1994). Agriculture's challenge to develop a vision for the future. Weed Technology, 8, 372–375.

Phatak, S. C., Wells, H. D., Sumner, D. R., Bell, D. K., & Glaze, N. C. (1981). Observation of rust on yellow nutsedge (*Cyperus esculentus* L.). Phytopathology, 71: 899.

Phatak, S. C., Wells, H. D., Sumner, D. R., Bell, D. K., & Glaze, N. C. (1983). Biological control of yellow nutsedge with the indigenous rust fungus *Puccinia caniliculata.* Science, 219, 1446–1447.

Pitelli, R., Charudattan, R., & DeValerio, J. T. (1998). Effect of *Alternaria cassiae, Pseudocercospora nigricans,* and soybean (*Glycine max*) planting density on the biological control of sicklepod (*Senna obtusifolia*). Weed Technology, 12, 37–40.

Politis, D. J., & Bruckart, W. L. (1986). Infection of musk thistle by *Puccinia carduorum* influenced by conditions of dew and plant age. Plant Disease, 70, 288–290.

Politis, D. J., Watson, A. K., & Bruckart, W. L. (1984). Susceptibility of musk thistle and related composites to *Puccinia carduorum.* Phytopathology, 74, 687–691.

Poppenborg, L., Friehs, K., & Flaschel, E. (1997). The green fluorescent protein is a versatile reporter for bioprocess monitoring. Journal of Biotechnology, 58, 79–88.

Prasad, R. (1993). Role of adjuvants in modifying the efficacy of a bioherbicide on forest species: Compatibility studies under laboratory conditions. Pesticide Science, 37, 427–433.

Prasad, R. (1994). Influence of several pesticides and adjuvants on *Chondrostereum purpureum*—A bioherbicide agent for control of forest weeds. Weed Technology, 8, 445–449.

Pullman, G. S., DeVay, J. E., Elmore, C. L., & Hart, W. H. (1984). "Soil Solarization: A Nonchemical Method for Controlling Diseases and Pests." Univ. of California Cooperative Extension Leaflet 21377.

Ramsfield, T. D., Becker, E. M., Rathlef, S. M., Tang, Y., Vrain, T. C., Shamoun, S. F., & Hintz, W. E. (1996). Geographic variation of *Chondrostereum purpureum* detected by polymorphisms in the ribosomal DNA. Canadian Journal of Botany, 74, 1919–1929.

Ridings, W. H. (1986). Biological control of stranglervine in citrus—A researcher's view. Weed Science, 34 (Suppl. 1), 31–32.

Rosskopf, E. N. (1997). Evaluation of *Phomopsis amaranthicola* sp. nov. as a biological control agent for *Amaranthus* spp. Ph.D. Thesis, Univ. of Florida, Gainesville, FL.

Rosskopf, E. N., Charudattan, R., DeValerio, J. T., & Stall, W. B. (1996). Control of pigweeds and amaranths (*Amaranthus* spp.) with a fungus: Three years of field experimentation. WSSA Abstracts, 36, 49.

Rosskopf, E. N., Gaffney, J. F., & Charudattan, R. (1997). The effect of spray propellant on the efficacy of bioherbicide candidates. WSSA Abstracts, 37, 62.

Saad, F., Watson, A. K., Smith, J. P., & Sparace, S. (1993). Formulation of *Colletotrichum coccodes* as a bioherbicide. Abstracts, II International Bioherbicide Workshop: "Bioherbicides—Applying the Temperate Experience to the Tropics," July 31–Aug. 1, 1993, MacDonald College of McGill University, Quebec.

Scheepens, P. C., & Hoogerbrugge, A. (1991). Host specificity of *Puccinia caniliculata,* a potential biocontrol agent for *Cyperus esculentus.* Netherlands Journal of Plant Pathology, 97, 245–250.

Scheepens, P. C., Kempenaar, C., & Van der Zwerde, W. (1998). Integrating biological control of *Chenopodium album* in weed management systems. Abstracts, IV International Bioherbicide Workshop, Univ. of Strathclyde, Glasgow, Scotland.

Schisler, D. A., & Jackson, M. A. (1993). Storage and efficacy of liquid culture produces microsclerotia of the bioherbicidal fungus, *Colletotrichum truncatum.* Abstracts, II International Bioherbicide Workshop: "Bioherbicides—Applying the Temperate Experience to the Tropics," July 31–Aug. 1, 1993, MacDonald College of McGill University, Quebec.

Schisler, D. A., Jackson, M. A., & Bothast, R. J. (1991). Influence of nutrition during conidiation of *Colletotrichum truncatum* on conidial germination and efficacy in inciting disease in *Sesbania exaltata.* Phytopathology, 81, 458–461.

Schisler, D. A., Jackson, M. A., McGuire, M. R., & Bothast, R. J. (1995). Use of pregelatinized starch and casamino acids to improve the efficacy of *Colletotrichum truncatum* conidia produced in differing nutritional environments. *In* "Proceedings of the VIII International Symposium on Biological Control of Weeds, Feb. 2–7, 1992, Canterbury, New Zealand" (E. S. Delfosse and R. R. Scott, Eds.), pp. 659–664. DSIR/CSIRO, Melbourne.

Schubiger, F. X., Défago, G., Sedlar, L., & Kern, H. (1985). Host range of the haplontic phase of *Uromyces rumicis.* In "Proceedings of the VI International Symposium on Biological Control of Weeds, Vancouver, Aug. 19–25, 1984" (E. S. Delfosse, Ed.), pp. 653–659. Agriculture Canada, Ottawa.

Shabana, Y. M., Charudattan, R., DeValerio, J. T., & Elwakil, M. A. (1997). An evaluation of hydrophilic polymers for formulating the bioherbicide agents *Alternaria cassiae* and *A. eichhorniae*. Weed Technology, 11, 212–220.

Shamoun, S. F., & Hintz, W. E. (1998). Development and registration of *Chondrostereum purpureum* as a mycoherbicide for hardwood weeds in conifer reforestation sites and utilities rights-of-way. "IV International Bioherbicide Workshop Programme and Abstracts, 6–7 Aug., 1998," p. 14. Univ. of Strathclyde, Glasgow, Scotland.

Skipper, H. D., Ogg, A. G., Jr., & Kennedy, A. C. (1996). Root biology of grasses and ecology of rhizobacteria for biological control. Weed Technology, 10, 610–620.

Smith, R. J. (1982). Integration of microbial herbicides with existing pest management programs. *In* "Biological Control of Weeds with Plant Pathogens" (R. Charudattan & H. L. Walker, Eds.), p. 189. Wiley, New York.

Smith, R. J., Jr. (1986). Biological control of northern jointvetch in rice and soybeans—A researcher's view. Weed Science, 34 (Suppl. 1), 17–23.

Smith, R. J., Jr. (1991) Integration of biological control agents with chemical pesticides. *In* "Microbial Control of Weeds" (D. O. TeBeest, Ed.), pp. 189–208. Chapman & Hall, New York.

Smither-Kopperl, M. L., Charudattan, R., & Berger, R. D. (1998). Dispersal of spores of *Fusarium culmorum* in aquatic system. Phytopathology, 88, 382–388.

Souissi, T., & Kremer, R. J. (1994). Leafy spurge (*Euphorbia escula*) cell cultures for screening deleterious rhizobacteria. Weed Science, 42, 310–315.

Sparace, S. A., Wymore, L. A., Menassa, R., & Watson, A. K. (1991). Effects of the *Phomopsis convolvulus* conidial matrix on conidia germination and the leaf anthracnose disease of field bindweed (*Convolvulus arvensis*). Plant Disease, 75, 1175–1179.

Stevens, P. J. G. (1994). Silwet L-77 organosilicone surfactant to maximize infection of weeds with pathogens. Regional Research Project S234 Annual Meeting.

Stowell, L. J. (1991). Submerged fermentation of biological herbicides. *In* "Microbial Control of Weeds" (D. O. TeBeest, Ed.), pp. 225–261. Chapman & Hall, New York.

Stubbs, T. L., & Kennedy, A. C. (1993). Effect of bacterial and chemical stresses in biological weed control systems. Agronomy Abstracts, 57, 261.

Supkoff, D. M., Joley, D. B., & Marois, J. J. (1988). Effect of introduced biological control organisms on the density of *Chondrilla juncea* in California. Journal of Applied Ecology, 25, 1089–1095.

TeBeest, D. O., & Templeton, G. E. (1985). Mycoherbicides: Progress in the biological control of weeds. Plant Disease, 69, 6–10.

Tomley, A. J. (1989). The biological control program for *Cryptostegia grandiflora* in Australia. *In* "Proceedings of the VII International Symposium on Biological Control of Weeds" (E. S. Delfosse, Ed.), pp. 685–688. Ist. Sper. Patol. Veg. (MAF).

Trujillo, E. E. (1985). Biological control of hamakua pa-makani with *Cercosporella* sp. in Hawaii. *In* "Proceedings of the VI International Symposium on Biological Control of Weeds, Vancouver, Aug. 19–25, 1984" (E. S. Delfosse, Ed.), pp. 661–671. Agriculture Canada, Ottawa.

Trujillo, E. E., Aragaki, M., & Shoemaker, R. A. (1988). Infection, disease development, and axenic cultures of *Entyloma compositarum,* the cause of hamakua pamakani blight in Hawaii. Plant Disease, 72, 355–357.

Tsukamoto, H. Gohbara, M., Tsuda, M., & Fujimori, T. (1997). Evaluation of fungal pathogens as biological control agents for the paddy weed, *Echinochloa* species by drop inoculation. Annals of the Phytopathological Society of Japan, 63, 366–372.

Tuat, N. A., Trung, H. A. M., Hetherington, S. D., & Auld, B. A. (1998). Potential bioherbicide for *Echinochloa* in Vietnam. Abstracts, IV International Bioherbicide Workshop, Univ. of Strathclyde, Glasgow, Scotland.

Turgeon, G., & Yoder, O. C. (1985). Genetically engineered fungi for weed control. *In* "Biotechnology: Applications and Research" (P. N. Cheremisinoff & R. P. Ouellette, Eds.), pp. 221–230. Technomic, Lancaster, PA.

Vogelgsang, S., Watson, A. K., & Hurle, K. (1994). The efficacy of *Phomopsis convolvulus* against field bindweed (*Convolvulus arvensis*) applied as a preemergence bioherbicide. Zeitschrift fuer Pflanzenkrankheiten Pflanzenschutz, 14, 253–260.

Walker, H. L. (1980). *Alternaria macrospora* as a potential biocontrol agent for spurred anoda: Production of spores for field studies. Advances in Agricultural Technology, AAT-S-12. USDA–SEA–AR, New Orleans, LA.

Walker, H. L., & Boyette, C. D. (1985). Biocontrol of sicklepod (*Cassia obtusifolia*) in soybeans (*Glycine max*) with *Alternaria cassiae*. Weed Science, 33, 212–215.

Walton, D. (1996). Host-selective toxins: Agents of compatibility. Plant Cell, 8, 1723–1733.

Wapshere, A. J. (1974). A strategy for evaluating the safety of organisms for biological weed control. Annals of Applied Biology, 77, 201–211.

Watson, A. K. (1986a). Biology of *Subanguina picridis,* a potential biological control agent of Russian knapweed. Journal of Nematology, 18, 149–154.

Watson, A. K. (1986b). Host range of, and plant reaction to, *Subanguina picridis*. Journal of Nematology, 18, 112–120.

Watson, A. K. (1991). The classical approach with plant pathogens. *In* "Microbial Control of Weeds" (D. O. TeBeest, Ed.), pp. 3–23. Chapman & Hall, New York.

Watson, A. K., & Alkhoury, I. (1981). Response of safflower cultivars to *Puccinia jaceae* collected from diffuse knapweed in eastern Europe. *In* "Proceedings of the V International Symposium on Biological Control of Weeds, Brisbane, July 22–27, 1980" (E. S. Delfosse, Ed.), pp. 301–305. CSIRO, Melbourne.

Watson, A. K., & Clément, M. (1986). Evaluation of rust fungi as biological control agents of weedy *Centaurea* in North America. Weed Science, 34 (Suppl. 1), 7–10.

Watson, A. K., & Wymore, L. A. (1989). Biological control, a component of integrated weed management. *In* "Proceedings of the VII International Symposium on Biological Control of Weeds" (E. S. Delfosse, Ed.), pp. 101–106. Ist. Sper. Patol. Veg. (MAF).

Watson, A. K., Schroeder, D., & Alkhoury, I. (1981). Collection of *Puccinia* species from diffuse knapweed in eastern Europe. Canadian Journal of Plant Pathology, 3, 6–8.

Womack, J. G., & Burge, M. N. (1993). Mycoherbicide formulation and the potential for bracken control. Pesticide Science, 37, 337–341.

Womack, J. G., Eccleston, G. M., & Burge, M. N. (1996). A vegetable oil-based invert emulsion for mycoherbicide delivery. Biological Control, 6, 23–28.

Wyse, D. L. (1992). Future of weed science research. Weed Technology, 6, 162–165.

Wyss, G. S., & Müller-Schärer, H. (1998). *Puccinia lagenophorae* as a classical biocontrol agent for common groundsel (*Senecio vulgaris*) in the United States? Phytopathology, 88 (Suppl.), S99.

Yang, S., & Schaad, N. W. (1996). Combined non- or low-virulent pathogens and special formulated carriers as broad-spectrum bioherbicides. "Biological Control of Weeds, IX International Symposium, 19–26 January 1996, Stellenbosch, South Africa" (V. C. Moran & J. H. Hoffmann, Eds.), p. 482 (Abstract). Univ. of Cape Town, Cape Town, South Africa.

Yang, S., Johnson, D. R., Dowler, W. M., & Connick, W. J., Jr. (1993). Infection of leafy spurge by *Alternaria alternata* and *A. angustiovoidea* in the absence of dew. Phytopathology, 83, 953–958.

Yoder, O. C. (1983). Use of pathogen-produced toxins in genetic engineering of plants and pathogens. *In* "Genetic Engineering of Plants" (T. Kosuge, C. P. Meredith, & A. Hollaender. Eds.), pp. 335–353. Plenum Press, New York.

Zhang, W. M., Moody, K., & Watson, A. K. (1996). Responses of *Echinochloa* species in rice (*Oryza satira*) to indigenous pathogenic fungi. Plant Disease, 80, 1053–1058.

Zidack, N. K., & Backman, P. A. (1996). Biological control of kudzu (*Pueraria lobata*) with the plant pathogen *Pseudomonas syringae* pv. *phaseolicola.* Weed Science, 44, 645–649.

Zidack, N. K., Backman, P. A., & Shaw, J. J. (1992). Promotion of bacterial infection of leaves by an organosilicone surfactant: Implication for biological weed control. Biological Control, 2, 111–117.

36

Biological Control of Insects in
Urban Environments

D. L. DAHLSTEN

Center for Biological Control
University of California
Berkeley, California

R. W. HALL

Department of Entomology
Ohio State University
Columbus, Ohio

INTRODUCTION

The urban environment is perhaps the most complex with which the pest manager must deal. Although the pest categories—vertebrate, weed, pathogen, and arthropod—are the same as in agriculture, the habitat is extremely diverse and discontinuous. The urban environment provides special opportunities for pests and for pest control and specifically biological control. The management goals in the urban environment are almost impossible to define because of the subjectivity involved in determining economic value of aesthetic or nuisance pests. Although the principles of biology and ecology pertain to the diverse, human-structured environment, the diffuse management and economic goals have made it difficult for pest managers to apply ecological concepts to control problems in urban environments. Therefore, it is difficult to apply modern pest management philosophy to the urban environment.

Urban pest management has only recently adopted (in the past 15 years) integrated pest management approaches. This is due in large part to the public's increased concern for pesticides in the environment. Because economic values are not easily defined, individuals either did nothing or contracted with pest control operators to apply prophylactic treatments on a yearly or quarterly basis. Since the 1940s, most pest control in urban environments has been essentially chemical. From 5.3 to 10.6 pounds of pesticide per acre are applied to urban gardens and lawns (National Research Council, 1980). These rates are higher than the average for agriculture or other managed lands.

Because the fields of urban entomology and urban pest management are just beginning to develop, entomology departments have begun to hire specialists on insects in urban environments. Ebeling's (1975) book on urban entomology excluded pests of ornamental plants except for a treatment of house plant pests. The main emphasis was on structural, stored product, food, fabric and paper, veterinary, and vertebrate (rats and mice) pests. There is also a chapter on entomophobia, the cause for many calls to the pest control operator or urban pest manager.

Emphasis on developing ecological approaches for urban insect pests has been with urban plant pests. Frankie and Ehler (1978) cited a dearth of ecological studies in urban environments. Olkowski *et al.* (1976, 1978) attempted to establish an ecological framework for control of urban tree pests. Three books (Frankie & Koehler, 1978, 1983; Bennett & Owens, 1986) have been published since Ebeling's 1975 work dealing with various aspects of urban entomology and urban pest management.

The field of urban pest management can be divided into seven areas: medical, psychological, architectural, agricultural, floricultural, silvicultural, and horticultural (Olkowski *et al.*, 1978). Several of these areas are treated in other sections of this volume. Medical and veterinary problems are traditionally dealt with by medical entomologists. We discuss biological control of cockroaches as an example of the diversity of insect pest problems in the urban environment. Also, the Mackinac Island outbreak of scale insects because of spraying for flies is a special example of a secondary outbreak due to disruption of natural enemies in an urban environment. Architectural pests include structural, fabric, paper, and other household pests. Little has been done on the biological control of these organisms.

Agriculture and forest pests occur in urban environments and most of these are included in other chapters in this volume. Urban forestry is developing rapidly and perhaps trees will be harvested for commercial timber in urban environments in the future. This would drastically change the management goals for urban trees and possibly pest management practices.

Agricultural pests can become pests on backyard ornamental and fruit trees. For example, cottony cushion scale, *Icerya purchasi* Maskell, infests *Citrus* spp., *Pittosporum* spp., and *Acacia* spp. Biological control successes on ornamentals are often offshoots of biological control successes in nearby agriculture.

Floriculture includes house plants and plants used for cut flowers. Biological control is gaining credibility as a highly useful tool in integrated control programs directed against certain pests in commercial glasshouses (see Chapter 31). However, there are few examples of biological control of pests of flowers in urban plantings. Most examples discussed herein fall into the ornamental horticulture area, which includes trees and other plants used in landscaping in yards, parks, cemeteries, botanical gardens, turf areas, streets, and highway right-of-ways.

Finally, urban pest management also involves psychology. Entomophobia is a human problem rather than a pest problem. Education, rather than biological control, is an important step in eliminating entomophobia.

ECOLOGICAL ATTRIBUTES OF URBAN OUTDOOR ENVIRONMENTS

The urban arthropod fauna is rich and diverse (Frankie & Ehler, 1978). One striking example comes from a small suburban yard near New York City where 1402 insect species were collected over several years (Lutz, 1941). Urbanization has provided unique opportunities for insects to exploit, which include construction of buildings, maintenance of environmental conditions favorable to year-round insect growth and development, storage of food products, generation of solid and liquid wastes, and planting of host plants. Since most of the biological control successes have been with outdoor plants, we focus on what humans have done to alter urban habitats with plants and how this influences the potential for biological control in urban environments.

Frankie and Ehler (1978) conclude that arthropod communities in urban environments are similar to those in natural environments. However, there are a number of human influences that affect the distribution and abundance of arthropods in urban environments. One factor that complicates urban pest management is the size of the area managed, which may range from a backyard garden with one or two fruit trees to miles of highway right-of-way planted with ground cover. The pest manager is confronted with a diverse habitat and clientele with multiple goals. It is difficult to organize diverse programs without involving public

agencies that take an interest in specific problems such as a department of transportation with roadside plants, a local tree commission, or a user group concerned with a particular habitat type such as turf grass. This may also explain why a number of the biological control examples were first tried or were a success in agricultural or forest environments before being used in urban areas.

The native and introduced flora in urban environments is dramatically diverse. For example, there are 132 tree, 147 shrub, and 53 ground cover or vine species in Austin, Texas (Frankie & Ehler, 1978). In Berkeley, California, 10 tree species constituted the native tree flora; now 123 tree species exist in Berkeley (Olkowski, 1974; Frankie & Ehler, 1978) and there are nearly 300 species on the Berkeley campus of the University of California (Cockrell & Warnke, 1976). There is tremendous variability in the number of species, depending on where cities or towns are located. One example is the Siberian and English elms that were planted in many small towns and ranches in northeastern California, creating small islands of elms separated by 15–40 km. Two other examples are given by McBride and Jacobs (1986), who contrasted two California cities, Menlo Park, with a presettlement oak forest, and South Lake Tahoe, with a presettlement Jeffrey pine forest. Urbanization resulted in a change of the number of tree species from 5 to 145 at Menlo Park and from 1 to 6 at South Lake Tahoe. These examples represent three very different ecological situations: an open grass–oak woodland in northeastern California where elms are planted, an oak savanna woodland in Menlo Park, and an urban center in a large contiguous forest in South Lake Tahoe. The areas surrounding cities, either as natural ecosystems or as large areas of agricultural development, are extremely important to the type of insect problems that might be anticipated on the cities' trees.

The physical environment within cities influences development, distribution, and abundance of insects and their natural enemies. Buildings and other structures such as fences can restrict movement or create wind fields and may also provide insect refuges. Buildings and paved areas such as streets, sidewalks, and parking lots can influence the ambient temperature. Runoff from rain is also increased in paved areas. Thus, in addition to modifying existing habitats, urbanization has also formed new ones.

POTENTIAL FOR URBAN BIOLOGICAL CONTROL

Until recently, there has been little incentive for biological control projects in urban environments. However, public concerns about chemical pollutants in air and water may limit urban pesticide applications and provide incentives for initiation of urban biological control programs (Dahlsten *et al.*, 1985).

Most exotic pest introductions occur in the discontinuous, diverse urban environment, with many species of exotic plants (Dahlsten, 1986). Thus, urban environments may be well suited to the classical biological control importation approach. Augmentation through inoculative releases and relocation of natural enemies would also be a useful tactic for urban environments. Because of the potential for pesticide-induced disruption, plant diversity, multiple ownerships, and diverse objectives, conservation of natural enemies is as important or more so than in other environments.

Many examples of successful biological control in urban environments are not well documented, but this lack of critical assessment is characteristic of most pest control tactics. Documentation, sound sampling and monitoring procedures, and quantification are as important in urban environments as they are in forestry and agriculture. Because of the diversity and the interrupted distribution of habitat type, the urban environment offers a unique opportunity for experimentation to address such questions in biological control as single vs multiple species introductions and predators vs parasites. A good example of this is provided in a series of studies evaluating three different strains of an egg parasitoid, *Oomyzus gallerucae* (Fonscolmbe) (Eulophidae), of the elm leaf beetle, *Xantherogaleruca luteola* (Müller), in northeastern California (Clair *et al.*, 1987, 1988).

Generally, the urban environment has received little attention with respect to classical biological control. The National Research Council (1980) reports 70 species of natural enemies released against 15 urban pests (excluding agricultural, medical/veterinary, forest, and greenhouse pests that also occur in urban environments), with an establishment rate of 34.3%. This is compared to 915 species of parasitoids and predators released against 106 agricultural, medical/veterinary, forest, and greenhouse pests, with an establishment rate of 29.4%. The determination of "success" is difficult (see Chapter 40), but most classical biological control successes in the urban environment have been against scale insects on perennial plants. This is true in agriculture as well (Flanders, 1986). Some degree of success has been reported for classical biological control with natural enemies against 43 urban and ornamental pests (Table 1).

The importance of naturally occurring biological control should not be underestimated (Hagen *et al.*, 1971). Often the importance of biological control agents is not realized until they are disrupted. Such was the case with the pine needle scale (*Chionaspis pinifoliae* [Fitch]) at South Lake Tahoe (discussed later). In other cases, the documentation of the importance of natural enemies was fortuitous. Such was the case with the elm spanworm, *Ennomos subsignarius* (Hübner), in Connecticut. The elm spanworm is occasionally a serious defoliator in the eastern hardwood forests. Like the gypsy moth, it is also a pest of shade trees in towns and cities and therefore an urban pest. Kaya and Anderson (1974) documented the importance of an undes-

cribed encrytid egg parasitoid, *Ooencyrtus* sp., in the collapse of elm spanworm populations in the woodlands and towns of southwestern Connecticut.

Shade tree pests may be the most fruitful area for biological control in the future. With the rapidly expanding concept of the urban forest, it may be possible to organize larger programs and generate public support for biological control attempts. The urban forest is loosely defined as trees as well as other vegetation growing in close association with people (Kielbaso & Kennedy, 1983).

The amount of money spent on tree care programs in the United States is substantial, and in 1980 it amounted to $2.19 per capita or $10.78 per tree (Kielbaso & Kennedy, 1983). The value of street trees in 1974 in the United States was conservatively valued at $15 billion. The economic incentive appears present for the development of strong programs in biological control. Certain problems could be focused on regionally or nationally. Although many trees are planted in the urban environment, there has been a tendency to develop monocultures and there is concern over the low diversity of urban forest trees (Kielbaso & Kennedy, 1983). Six of the ten most important urban forest pests nationally are defoliators and the top two pests are aphids and scale insects (Kielbaso & Kennedy, 1983). The focus of developing biological control programs in the urban forests has therefore been on defoliators and homopterans (Olkowski *et al.*, 1978; Dahlsten *et al.*, 1985).

Many of the examples given herein are for problems on woody plants or trees, either native or introduced. The trees may be part of the forest into which the towns have encroached, such as with the elm spanworm in the eastern hardwood forest or the pine needle scale in the western coniferous forest. In other cases, fruit or forest trees are planted and pests of agriculture or forestry become urban ornamental pests, such as the woolly whitefly, *Aleurothrixus floccosus* (Maskell), on citrus and the Nantucket pine tip moth, *Rhyacionia frustana* (Comstock), on pine in southern California. The brief case histories are examples of classical biological control, augmentation, and naturally occurring biological control. Although precise documentation is lacking in many cases, the available evidence strongly suggests that persistent, well-funded research on the use of natural enemies to control plant pests in the outdoor urban environment may lead to more successes.

BIOLOGICAL CONTROL CASE HISTORIES

Blattaria: Blattidae and Blattellidae (Cockroaches)

Cockroaches are among the most important pests in urban environments (Ebeling, 1975). Most of the cockroaches that are household and structural pests in temperate regions are tropical or subtropical in origin and because they are

TABLE 1   Pests Found in Urban Areas for Which Some Success Has Been Reported for Introduced Natural
Enemies in Urban or Unspecified (*) Environments

Pest	Common name	Location	Some urban or ornamental success reported (reference)
Blattaria			
Periplaneta americana	American cockroach	Hawaii	DeBach, 1964; Luck, 1981
Periplaneta australasiae	Australian cockroach	Hawaii	DeBach, 1964; Luck, 1981
Supella longipalpa	Brownbanded cockroach	U.S.	Slater, 1984
Dermaptera			
Forficula auricularia	European earwig	Canada	Turnbull & Chant, 1961
Thysanoptera			
Gynaikothrips ficorum	Cuban laurel thrips	Hawaii	Rao *et al.,* 1971
Hemiptera			
Nezara viridula	Southern green stink bug	Australia, Hawaii	*Luck, 1981
Homoptera			
Acizzia uncatoides	Acacia psyllid	U.S.	Pinnock *et al.,* 1978; Dreistadt & Hagen, 1994
Trioza eugeniae	Eugenia psyllid	U.S.	Dahlsten *et al.,* 1995
Ctenarytaina eucalypti	Blue gum psyllid	U.S.	Dahlsten *et al.,* 1998a, 1998b
Aleurothrixus floccosus	Woolly white fly	U.S.	DeBach & Rose, 1976, 1977
Antonina graminis	Rhodesgrass mealybug	U.S.	Dean *et al.,* 1979
Aonidiella aurantii	California red scale	U.S.	*Luck, 1981
Aonidiella citrina	Yellow scale	U.S.	*Luck, 1981
Asterodiaspis variolosum	Golden oak scale	Tasmania, New Zealand	Wilson, 1960; Ferguson, 1989
Asterolecanium pustulans	Bamboo (pustule) scale	Puerto Rico	DeBach, 1964
Chloropulvinaria psidii	Green guava mealybug	Puerto Rico, Bermuda	DeBach, 1964; Simmonds, 1969
Chromaphis juglandicola	Walnut aphid	U.S.	van den Bosch *et al.,* 1970
Coccus hesperidum	Soft brown scale	New Zealand, USSR	Saakian-Baranova, 1966
Cornuaspis beckii	Purple scale	U.S.	DeBach, 1964; DeBach & Rose, 1977
Eucallipterus tiliae	Linden aphid	U.S.	Olkowski *et al.,* 1982a; Zuparko, 1983
Icerya purchasi	Cottony cushion scale	U.S.	Quezada & DeBach, 1973
Lecanium corni	European fruit lecanium	British Columbia	DeBach, 1964
Maconellicoccus hirsutus	Hisbiscus mealybug	Egypt	Clausen, 1978
Myzocallis annulata	Oak aphid	Tasmania, New Zealand	*Luck, 1981; Walker, 1989
Nipaecoccus nipae	Coconut mealybug	Hawaii	*Luck, 1981
Nipaecoccus viridis	Lebbeck mealybug	Hawaii, Egypt	DeBach, 1964; Laing & Hamai, 1976
Orthezia insignis	Greenhouse orthezia	Kenya	DeBach 1964
Parasaisetia nigra	Nigra scale	U.S.	Flanders, 1959
Parlatoria oleae	Olive scale	U.S.	DeBach, 1964
Planococcus citri	Citrus mealybug	U.S., Chile	*Luck, 1981
Planococcus kenyae	Coffee mealybug	Kenya	*Luck, 1981
Pseudaulacaspis pentagona	White peach scale	Bermuda	Bennett & Hughes, 1959
Pseudococcus comstocki	Comstock mealybug	U.S.	Meyerdirk *et al.,* 1981
Pseudococcus fragilis	Citrophilis mealybug	U.S.	DeBach, 1964
Pseudococcus longispinus	Long-tailed mealybug	U.S., Bermuda, Israel	*Luck, 1981
Pulvinaria delottoi	Ice plant scale	U.S.	Frankie & Hagen, 1986
Pulvinariella mesembryanthemi	Ice plant scale	U.S.	Frankie & Hagen, 1986
Quadraspidiotus perniciosus	San Jose scale	U.S., Europe	Sailer, 1972
Saisettia coffeae	Hemispherical scale	Chile	Beingolea, 1969
Saisettia oleae	Black scale	U.S., Australia	*Luck, 1981
Siphoninus phillyreae	Ash whitefly	U.S.	Bellows *et al.,* 1992
Tinocalis plantani	Elm aphid	U.S.	Olkowski *et al.,* 1982b
Trialeurodes vaporariorum	Greenhouse whitefly	Australia, Tasmania	*Luck, 1981
Coleoptera			
Adoretus sinicus	Chinese rose beetle	Hawaii	Pemberton 1954
Popillia japonica	Japanese beetle	U.S.	Fleming 1968
Xanthogaleruca luteola	Elm leaf beetle	U.S.	Luck & Scriven, 1976; Cranshaw *et al.,* 1989

(continues)

TABLE 1 *(continued)*

Pest	Common name	Location	Some urban or ornamental success reported (reference)
Lepidoptera			
Cnidocampa flavescens	Oriental moth	U.S.	DeBach, 1964; Laing & Hamai, 1976
Lithocolletis saniella	Oak leaf miner	New Zealand	Swan, 1973
Lymantria dispar	Gypsy moth	U.S.	DeBach, 1964
Mygmia phaerorrhoea	Brown-tail moth	U.S.	DeBach, 1964
Operophtera brumata	Winter moth	Canada	Embree, 1971
Rhyacionia frustrana	European pine shoot moth	U.S.	Scriven & Luck, 1978
Spodoptera mauritia acronyctoides	Lawn armyworm	Hawaii	Clausen, 1978
Stilpnotia salicis	Stain moth	U.S., Canada	Clausen, 1978
Diptera			
Phytomyza ilicis	Holly leaf miner	Canada	Turnbull & Chant, 1961
Hymenoptera			
Diprion similis	Introduced pine sawfly	Canada	McGugan & Coppel, 1962

introduced, they seem to be ideal candidates for classical biological control. Cockroaches contaminate food, impart an unpleasant odor to residences, and may transmit disease. Also, some people are allergic to cockroaches and many exhibit reactions to whole-body extracts of cockroaches or cockroach feces (Ebeling, 1975). The major urban pest species in the United States include the German cockroach, *Blatella germanica* (Linnaeus), the American cockroach, *Periplaneta americana* (Linnaeus), the brownbanded cockroach, *Supella longipalpa* (Serville), the Oriental cockroach, *Blatta orientalis* Linnaeus, and the smokybrown cockroach, *Periplaneta fuliginosa* (Serville).

Several authors have noted predation or parasitism of cockroaches (Ebeling, 1975), but little evidence exists to indicate that natural enemies can reduce or maintain cockroach densities beneath economic or aesthetic injury levels. However, the potential exists and there is need for serious consideration of the use of natural enemies for control of cockroach populations (Piper & Frankie, 1978). Hymenopteran egg parasitoids appear to offer the greatest potential for short- and long-term population regulation of cockroaches (Piper *et al.,* 1978). Inundative releases have enormous potential for control of indoor and outdoor cockroach populations (Piper & Frankie, 1978). However, use of such techniques requires a knowledge of the patterns of distribution and abundance of the pest species involved.

A complicating factor in use of natural enemies for cockroach control is that not all urbanites will be amenable to indoor parasitoid releases (Piper & Frankie, 1978). In fact, Edmunds (1957) noted that one family was more disturbed by the cockroach parasitoids in their home than by the infestation of Oriental cockroaches.

Pemberton (1948) noted that two species of exotic ampulicid wasps, *Ampulex compressa* (Fabricius) and *Dolicurus stantoni* (Ashmead), played a significant role in controlling some species of cockroaches in Hawaii. The cockroach egg parasitoid *Comperia merceti* (Compere) has potential for use in urban cockroach biological control. *Comperia merceti* appears to have a substantial impact on brownbanded cockroach when densities of oothecae are high (Coler *et al.,* 1984). At lower densities, parasitism rates are low. Thus, at such densities, inoculative or inundative releases may be necessary to achieve satisfactory levels of control. The use of *C. merceti* for brownbanded cockroach control has enjoyed great success at the University of California, Berkeley (Slater, 1984). Slater *et al.* (1980) note that in an institutional setting, workers were willing to tolerate *C. merceti*.

Another species of egg parasitoid, *Aprostocetus hagenowii* (Ratzeburg), is thought to be an important natural enemy of cockroaches (Cameron, 1955). When released into a room with high densities of American cockroach oothecae, *A. hagenowii* parasitized 83% of the oothecae (Roth & Willis, 1954). Fleet and Frankie (1975) and Piper *et al.* (1978) found significant mortality of oothecae of American and smokybrown cockroaches due to parasitism by *A. hagenowii*.

Dermaptera: European Earwig (*Forficula auricularia* Linnaeus)

The European earwig is a nuisance in close proximity to homes and gardens. It feeds on fruits, vegetables, and flowering plants and may enter houses. Introductions of *Bigonicheta spinipennis* (Meigen) (Diptera: Tachinidae) were made in British Columbia from the United States (this material was originally from Europe). *Bigonicheta spinipennis* became established and contributed to mortality of the earwig. Turnbull and Chant (1961) rated this program as a success. However, complete biological control of European earwig may be undesirable as it has value as a predator, consuming large numbers of aphids and scale insects (Clausen, 1978).

Thysanoptera: Cuban Laurel Thrips (*Gynaikothrips ficorum* Marchal)

Cuban laurel thrips infested *Ficus retusa* Linnaeus and *Ficus benjamina* Linnaeus in Hawaii and feeding by the thrips caused a heavy drop of young foliage. In addition, biting and swarming of the thrips made it a public nuisance (Rao *et al.*, 1971). The anthocorid predator *Montandoniola moraguesi* Puton was introduced into Hawaii from the Philippines in 1964. The predator became established on Oahu and to date has been important in suppressing the thrips (Rao *et al.*, 1971).

Homoptera

Psyllidae: The Acacia Psyllid (*Acizzia uncatoides* (Ferris & Klyver))

The acacia psyllid project is an example of a successful classical biological pest control project in roadside plantings. The psyllid was first found in California in 1955 and by 1971 became so abundant on the acacias along some highways that the California Department of Transportation contracted with the Division of Biological Control, University of California, Berkeley, for assistance (Pinnock *et al.*, 1978). This led to exploration for natural enemies in Victoria, Australia, the native home of many acacias and apparently the acacia psyllid. Five predacious and two parasitoid species were initially collected and imported from 1972 to 1975, but only the predators were released and of these, only *Diomus pumilio* Weise was recovered (Pinnock *et al.*, 1978). A subsequent study confirmed that *D. pumilio* was the only purposely introduced species established, but the most numerous predator attacking the psyllid was an unintentionally introduced species, *Anthocoris nemoralis* (Fabricius) (Dreistadt & Hagen, 1994). These researchers found that the combination of both introduced species provided acceptable biological control of the psyllid in the San Francisco Bay area.

One interesting aspect of this project is that an exotic coccinellid species, *Harmonia conformis* (Boisduval), which failed to become established in California, successfully brought the acacia psyllid under control in koa tree forests when sent from California to Hawaii (Pinnock *et al.*, 1978). *Diomus pumilio*, which did so well in California, did not control the psyllid when introduced into Hawaii. This is a good example of matching a natural enemy to its hosts in different habitats.

The Eugenia Psyllid (*Trioza eugeniae* Froggat)

The eugenia psyllid was introduced into California from Australia and was first observed in May of 1988 in Inglewood, which is near Los Angeles. It is a pest of eugenia (also called Australian brush cherry or lillypilly), *Syzygium paniculatum,* a widely planted ornamental shrub or small tree, oftentimes used as a hedge in the coastal counties of California. The psyllid spread to neighboring counties by the end of 1988 and was a pest in northern California by 1989. Monitoring procedures were developed that were effective for both the psyllid and its primary parasitoid, a eulophid, *Tamarixia* n. sp., that was introduced from Australia in 1992 (Dahlsten *et al.*, 1995). Initial results were spectacular but psyllid populations became fairly abundant again in 1996 and 1997, particularly in cool regions such as San Francisco. Although the parasitoid is very effective in most regions of the state, it appears that the psyllid populations become very abundant in some cool areas, but even in these areas populations are much lower than they were prior to the release of *Tamarixia.*

The Blue Gum Psyllid (*Ctenarytaina eucalypti* (Maskell))

The blue gum psyllid, native to Australia, was first discovered in North America in January of 1991 in Monterey County, California (Dahlsten *et al.*, 1998a). The psyllid feeds on several species of *Eucalyptus* and in particular those with waxy blue foliage such as *E. cinerea, E. glaucescens, E. viminalis, E. pulverulenta,* and *E. globulus.* It was found on *E. pulverulenta,* which has waxy blue foliage on the mature as well as the juvenile leaves, and this species is known as silver-leaved mountain gum or baby blue gum in the foliage industry in California. Tasmanian blue gum, *E. globulus,* is planted as an ornamental throughout California and is a prominent feature of the landscape in the state, having been planted and distributed widely in California over 100 years ago (Dahlsten, 1996). *Eucalyptus globulus* has waxy blue foliage only on the juvenile leaves and new shoots and foliage, which, when heavily infested by the psyllid, may cause homeowners concern because of growth loss, honeydew, sooty mold, and waxy secretions from nymphs. The main concern was with growers producing *E. pulverulenta,* which accounted for $2.3 billion in sales in 1996 in California (Dahlsten *et al.*, 1998b).

Blue gum psyllid populations exploded and by the end of 1992 could be found everywhere in the state on *E. globulus* and *E. pulverulenta.* The Eucalyptus Growers Association raised money to finance a biological control program with the University of California Center for Biological Control in Berkeley. A single species of primary parasitoid, *Psyllaephagous pilosus* Noyes (Hymenoptera:

Eucrytidae), was found in late 1991 in Southern Australia and in New Zealand and sent to the University of California quarantine facility in Albany, California. Six species of secondary or hyperparasitoids were reared in quarantine in low numbers.

In the spring and summer of 1993, approximately 7000 parasitoids were released at eight locations in California from Alameda County in the north (San Francisco Bay Area) to San Diego County in the south. The parasitoids spread as rapidly as the psyllids had done earlier. Psyllid populations declined at most sites in 1993, and in 1994 the blue gum psyllid was no longer a problem (Dahlsten *et al.*, 1998a). This highly successful biological control program has eliminated the need for chemical sprays for the psyllid in the *E. pulverulenta* plantations in California and has resulted in a benefit–cost ratio ranging from at least 9:1 to 24:1 based solely on the reduction of insecticide treatments (Dahlsten *et al.*, 1998b).

Aleyrodidae: Woolly Whitefly (*Aleurothrixus floccosus* (Maskell))

The origin of the woolly whitefly is uncertain but it is widely distributed in the West Indies and in Central and South America (Clausen, 1978). It is a pest primarily of citrus. A platygasterid, *Amitus spiniferus* (Brethes), and two aphelinids, *Eretmocerus paulistis* Heimpel and *Cales noacki* De Santis, have been used successfully as biological control agents in Chile, Mexico, and France (Luck, 1981).

In San Diego, California, citrus is planted as a garden and ornamental tree. The woolly whitefly was first discovered there in 1966 (DeBach & Rose, 1976). DeBach had observed the whitefly to be under good control in mainland Mexico and *A. spiniferus*, *E. paulistis,* and an *Encarsia* sp. were introduced in small numbers in San Diego in 1967 (DeBach & Rose, 1976). The first two parasitoids became well established and showed promise as biological control agents. In 1969, the California Department of Food and Agriculture decided to eradicate the whitefly with chemical insecticides. The biological control program was therefore halted in San Diego and shifted to Baja California Norte in Mexico, where the whitefly was also found shortly after its discovery in San Diego (DeBach & Rose, 1976). Thirty *A. spiniferus* and 169 *E. paulistis* were colonized in Tijuana in 1969. Two of the original colonized trees were used as nursery trees. In one year, by transferring twigs with parasitized whitefly, more than 27,000 of the two parasitoids were distributed around Tijuana (DeBach & Rose, 1976). A fourth parasitoid, *C. noacki,* from Chile was also released. By late 1970, these three parasitoids were well established and it appeared that the biological control program was a success.

In San Diego, the biological control program was re-

sumed in 1971 when the failed eradication project was abandoned (DeBach & Rose, 1976). The area infested with woolly whitefly in California was about 200 square miles. The University of California at Riverside and the California Department of Food and Agriculture cooperated in releasing the parasitoids that were collected in Tijuana or imported from mainland Mexico, Chile, and Brazil. Parasitoids were colonized at all release sites in San Diego in 1971 and these sites provided nursery trees for further parasitoid releases in 1972. The woolly whitefly continued to increase in California but by 1973 the biological control program was a success in San Diego County and Tijuana (DeBach & Rose, 1976).

In 1973, the Japanese beetle (*Popillia japonica* Newman) was found unexpectedly in San Diego and an eradication zone of about 100 city blocks was established. A study was set up to investigate the effects on the woolly whitefly natural enemies of the three chemicals (carbaryl, chlordane, and dicofol) that were to be used for the beetle eradication program (DeBach & Rose, 1977). Dicofol was used for mite control because of disruption caused by eradication treatments. In this study it was shown that the eradication efforts against the Japanese beetle disrupted the biological control of the woolly whitefly. Before the chemical treatments began the two whitefly parasitoids *A. spiniferus* and *C. noacki* were generally distributed and other citrus pests were also under good biological control. After treatment woolly whitefly populations increased dramatically and were up to 1200 times higher in treated areas than in untreated areas. Populations of citrus red mite, *Panonychus citri* (McGregor), and purple scale, *Cornuapis* (=*Lepidosaphes*) *beckii* (Newman), also increased (DeBach & Rose, 1977). After the eradication treatments ended, *A. spiniferus* and *C. noacki* began to return to the previously treated areas.

The side effects of eradication programs have not been well documented. This is partly because most of these programs are in urban environments. Eradication projects employing broad-spectrum insecticides provide a unique opportunity to study biological control in urban environments. The 1980–1982 Mediterranean fruit fly, *Ceratitis capitata* (Wiedemann), eradication program employing malathion bait spray in northern California resulted in disruption of biological control of various insects in an urban area (Dreistadt & Dahlsten, 1986). Gardens in treated areas had more mites, aphids, and whiteflies than gardens in unsprayed areas (Troetschler, 1983). Decreased control of other insects in urban areas such as the walnut aphid, *Chromaphis juglandicola* (Kaltenbach), and black scale, *Saissetia oleae* (Bernard), was also attributed to the effects of the bait spray on their natural enemies (Ehler & Endicott, 1984). In another study, parasitoids of iris whitefly, *Aleyrodes spiraeoides* Quaintance, were significantly more susceptible to malathion bait spray than were the whiteflies and parasitism was significantly lower on plants in the sprayed areas (Hoelmer & Dahlsten, 1993).

Ash Whitefly (*Siphoninus phillyreae* (Haliday))

The discovery of the Old World ash whitefly in California in 1988 led to the prompt inception of a biological control program. Foreign exploration was carried out in 1989 and 1990, leading to the successful importation and release of two species, *Encarsia inaron* (Walker) (= *partenopea* Masi) (Hymenoptera: Aphelinidae) and *Clitostethus arcuatus* Rossi (Coleoptera: Coccinellidae) (Bellows *et al.*, 1992). Although both species became established, the aphelinid colonized a greater area and thus received greater attention (Bellows *et al.*, 1992). Subsequent life-table analyses of field sites documented that *E. inaron* was playing a key role in the reduction of *S. phillyreae* populations (Gould *et al.*, 1992a, 1992b).

By 1992, whitefly populations had been reduced by the wasp to undetectable levels. The loss of aesthetic benefits due to ash whitefly damage was analyzed, and the total benefits of the biological control program ranged from $324 million at wholesale values to $412 million at retail. The direct cost of the program was around $1.2 million. The benefit to cost ratios were 270:1 and 344:1, respectively (Jetter *et al.*, 1997).

Aphididae and Drepanosiphidae: Oak Aphid (*Myzocallis annulata* (Hartig))

In Australia, the oak aphid causes damage to oaks (*Quercus* spp.) in Victoria and Tasmania. *Aphelinus subflavescens* (Westwood) (= *flavus* Nees) was imported from England and released in 1939. Shortly thereafter, *A. subflavescens* from an unknown source was recovered in Australia and Tasmania. Parasitism was heavy in Tasmania and the health of the oaks gradually improved (Wilson, 1960).

The parasitoid was also introduced into New Zealand in 1939, although there was evidence that it had already been unintentionally imported (Walker, 1989). Here the aphid is considered strictly an aesthetic pest and has not been of any apparent importance since the establishment of *A. subflavescens* was documented, although no detailed studies have been published.

Linden and Elm Aphids (*Eucallipterus tiliae* Linnaeus and *Tinocallis platani* (Kaltenbach))

Both lindens and elms are planted as street trees, and the linden aphid, *E. tiliae,* and elm aphid, *T. platani,* have been the targets of urban biological control projects in northern California.

The linden aphid is common in Europe and has spread across the United States and Canada since its discovery in Washington, D.C., in 1886 (Olkowski *et al.*, 1982a). In 1970, several species were introduced in Berkeley, but only *Trioxys curvicaudus* Mackauer became established (Olkowski *et al.*, 1982a); it was subsequently established in San Jose, California (Zuparko, 1983). Since aphid populations, honeydew, and citizen complaints were reduced, this project was considered a biological control success (Olkowski *et al.*, 1982a). However, this judgment may not be valid. Populations of the aphid were low in 1971, with 40% parasitization, but in 1972, aphid populations were higher, with parasitism fluctuating between 20 and 50%; populations were not sampled again until 1978, when parasitism did not exceed 30%, and about half of the aphid mummies yielded hyperparasitoids (Olkowski *et al.*, 1982a). Additional research from 1991 to 1994 found that aphid populations and parasitism rates varied tremendously between different sites in the San Francisco Bay area (Zuparko & Dahlsten, 1995). Analysis of these results reveals the importance of host plant resistance against the linden aphid, with lower aphid populations correlated with linden species with relatively denser leaf pubescence that have *T. curvicaudus* present (Zuparko & Dahlsten, 1994). Although host plant resistance is considered to be the more important factor in limiting aphid populations, the parasitoid contributes a significant mortality factor and thus may be considered a partial biological control success. The controversy over the success of this program illustrates the importance of careful quantitative documentation of the release and efficacy of natural enemies.

The elm aphid is monophagous on elms. It is known from Europe and Russia and has been introduced into western North America (Richards, 1967). In Berkeley the aphid reaches high densities and its honeydew creates a nuisance. Chemical insecticides were commonly used from 1945 to 1971, although with decreasing effectiveness (Olkowski *et al.*, 1982b). In 1972, two parasitoids, *Trioxys hortorum* Stary and *T. tenuicaudus* Stary, were introduced from Czechoslovakia. *Trioxys tenuicaudus* became established and spread very slowly. Resident complaints of honeydew beneath elms diminished and this project was reported to be a successful classical biological control project (Olkowski *et al.*, 1982b).

Diaspididae

Pine Needle Scale (Chionaspis pinifoliae *(Fitch)*)

The pine needle scale is a native insect with many coniferous hosts. The outbreak of this scale in 1968 in South Lake Tahoe, California, provided an opportunity to document the importance of naturally occurring biological control in an urban environment. The city of South Lake Tahoe has encroached into the natural westside Sierra forests and both lodgepole pine, *Pinus contorta* Douglas, and Jeffrey pine, *P. jeffreyi* Greville & Balfour, are common along the streets and in the yards of homes. The scale normally occurs in low numbers on both pine species but in 1968, populations erupted in areas that had been fogged with malathion for control of a mosquito, *Culex tarsalis* Coquillett. The fogging program was halted and the subsequent decline of the scale population was documented (Luck & Dahlsten, 1975).

The scale was found to have a uniparental population on lodgepole pine and a biparental population on Jeffrey pine (Luck & Dahlsten, 1974). Not only was the biology of the scale different but the parasitoids varied depending on the host tree species of the scale. The two scale parasitoids on lodgepole pine were the aphelinids *Physcus howardi* Compere and *Prospaltella bella* Gahan. The parasitoid on scale on Jeffrey pine was *Achrysocharis phenacapsia* Yoshimoto (Eulophidae) (Luck & Dahlsten, 1974). Two coccinellid predators, *Chilocorus orbus* Casey and *Cryptoweisia atronitens* (Casey), were the dominant biotic mortality factors on both pine species along with parasitism by *P. bella* on lodgepole pine (Luck & Dahlsten, 1975). Other scales, *Nuculaspis californica* (Coleman), *Pineus* sp., *Physokermes* sp., and *Matsucoccus* sp., two species of mites, *Oligonychus* sp. and *Brevipalpus* sp., and an aphid, *Schizolachnus* sp., were found to be very abundant in the fogged area as well (Dahlsten *et al.*, 1969).

White Peach Scale (Pseudaulacaspis pentagona *(Targ.)*)

The white peach scale was introduced into Bermuda in 1917. By 1920, the scale threatened oleander (*Nerium oleander* Linnaeus) with extinction (Bennett & Hughes, 1959). Several species of natural enemies were imported from Italy and the United States. *Aphytis diaspidis* (Howard) from Italy was an immediate success and brought *P. pentagona* under substantial control (Bennett & Hughes, 1959).

Asterolecaniidae

Golden Oak Scale (Asterodiaspis variolosa *(Ratzeburg)*)

The golden oak scale is a pit-making species that became an important pest of oaks in New Zealand and Tasmania in the 1920s and in Australia in the 1930s. The principal problems were on the golden oak, *Quercus pedunculata* Ehr. Severely infested trees grow poorly and suffer dieback; young trees are sometimes killed (Johnson & Lyon, 1976). The parasitoid *Habrolepis dalmanni* (Westwood) (Encyrtidae) was introduced from the eastern United States into New Zealand in the 1920s. Later, *H. dalmanni* was introduced into Tasmania and Australia from New Zealand and

into Chile from the United States. Biological control was partially to substantially successful in Australia, Tasmania, and Chile (Clausen, 1978) and almost completely successful in New Zealand (Ferguson, 1989).

Coccidae

European Fruit Lecanium (Lecanium corni *(Bouché)*)

In the 1970s the *Lecanium corni* (Bouché) complex was causing serious injury to ornamental and fruit trees on Mackinac Island in Michigan. Trees with high *L. corni* population densities exhibited dieback and a decline in vigor. In addition, the trees, sidewalks, and park benches were aesthetically injured by honeydew deposition and growth of sooty mold. Merritt *et al.* (1983) attributed the scale infestation to destruction of natural enemies by weekly applications of dimethoate for fly control. When dimethoate treatments were reduced after initiation of an IPM program for flies, there was an immediate decline in *L. corni* densities. In the two years following the reduction in pesticide use, a dramatic decline in honeydew deposition was also noted (Merritt *et al.,* 1983).

Green Guava Mealybug (Chloropulvinaria (=Pulvinaria) psidii *(Maskell)*)

The green guava mealybug, also called the green shield scale, is a tropical species that is common in California, Southeast Asia, Hawaii, the West Indies, and Florida. It has a wide host list of tropical and subtropical plants; *Ficus* spp., guava, coffee, and oleander are favored hosts (Clausen, 1978). Following an increase of *C. psidii* on ornamentals in Bermuda, several species of natural enemies were introduced from Hawaii and California and four species became established (Bennett & Hughes, 1959). *Microterys kotinskyi* Fullaway (Encyrtidae) is thought to be important in control of the scale. Simmonds (1969) cited biological control of *C. psidii* as complete.

Ice Plant Scales (Pulvinariella mesembryanthemi *(Vallot)* and Pulvinaria delottoi *Gill*)

Both of these ice plant scales originated in South Africa and attack plants in the families Aizoaceae and Crassulaceae. A scale was first detected in California in 1949 at the University of California Botanical Garden on ice plants, *Carpobrotus* spp., and several more reports of the scales were made in the Bay Area between 1949 and 1970 (Tassan *et al.*, 1982). Ice plants are used widely in California as ornamental ground covers and the California Department of Transportation (Caltrans) alone maintains approximately 6000 acres of ice plants along highways (Frankie & Hagen, 1986).

Pulvinariella mesembryanthemi has a wide distribution and has moved from its origin in South Africa to southern

Europe along the Mediterranean, the Canary Islands, Germany, and several countries in South America. At present, *P. delottoi* has moved only into California (Frankie & Hagen, 1986). Under some circumstances the ice plants can become pests by encroaching on native vegetation and the scales could then be considered biological control agents.

It was not until 1971 that dense populations of *P. mesembryanthemi* were discovered in Napa, California, damaging ice plants at private residences. In 1973, a survey of ice plants along highways in Alameda County was conducted by Caltrans with entomologists from the county, state, and federal government and the University of California, Berkeley (Tassan *et al.*, 1982). It was discovered at that time that there were actually two scale species present. Biological studies showed that *P. delottoi* had one generation per year and colonized the mature lower portions of the plant, whereas *P. mesembryanthemi* had two generations per year and usually attacked the new terminal growth (Donaldson *et al.*, 1978).

By 1976, the scales were causing considerable damage to ice plants along highways. In 1978, the Division of Biological Control at the University of California, Berkeley, embarked on a three-year biological control program funded by Caltrans. As a result of this program, seven natural enemies of the scales were collected in South Africa and released in California. Two coccinellid species, *Hyperaspis senegalensis hottentotta* Mulsant and *Exochomus flavipes* (Thunberg), failed to establish (Frankie & Hagen, 1986). Five wasps were released: the encyrtids *Metaphycus funicularis* Annecke, *M. stramineus* Compere, *Metaphycus* spp., and *Encyrtus saliens* Prinsloo & Annecke and the aphelinid *Coccophagus cowperi* Girault. The first two have established throughout California and *E. saliens* is probably established in northern California (Frankie & Hagen, 1986). These three introduced natural enemies have largely eliminated the need for chemical sprays for ice plant scales along freeways in California (Frankie & Hagen, 1986).

Native aphelinids (*Coccophagus* spp.), an exotic encyrtid (*Metaphycus helvolus* (Compere), introduced from South Africa in the 1930s for control of black scale on citrus), and an exotic coccinellid (*Cryptolaemus montrouzieri* Mulsant, released for control of mealybugs on citrus) also attack the ice plant scales in California.

The Mediterranean fruit fly eradication program in California in 1980–1982 provided an opportunity to study the effect of malathion bait spray on the ice plant scales and their natural enemies in 1981–1982. A single insecticide application reduced the number of parasitoids, *Coccophagus lycimnia* (Walker), by 79% and *M. stramineus* by 90% (Washburn *et al.*, 1983). The timing and number of applications of bait spray were critical as a number of applications would be detrimental to the parasites. Also, immature scale (crawlers) was affected by the bait spray (Washburn

et al., 1983). In the laboratory, malathion bait spray at very low concentrations produced sublethal effects on one of the ice plant scale parasitoids, *E. saliens* (Hoy & Dahlsten, 1984).

The ice plant scale project demonstrates that classical biological control can be used for pests of roadside plantings of ground cover. It was also shown that secondary outbreaks of ice plant scale could occur due to the aerial application of bait sprays. Finally, the ice plant scales could be considered as agents of biological control in areas where ice plant is a pest.

Miscellaneous Scales

Several homopteran pests of ornamentals have been controlled by natural enemies introduced for biological control on agricultural crops. Natural enemies imported for control of cottony cushion scale, *Icerya purchasi* Maskell (Margarodidae), on citrus controlled it on ornamentals as well (Clausen, 1978). Another example is the nigra scale, *Parasaissetia nigra* (Nietner), which was controlled by *Metaphycus helvolus* (Compere), a parasitoid introduced against black scale, *Saissetia oleae* (Bernard), on citrus (Clausen, 1978).

Pseudococcidae

Rhodesgrass Mealybug (Antonina graminis *(Maskell)*)

In the 1940s the rhodesgrass mealybug became a major pest of forage and lawn grasses in Texas. The mealybug is of Asiatic origin and infests at least 69 species of lawn and turf grasses (Dean *et al.*, 1979). Several parasitoids were introduced as a part of a biological control program (Schuster *et al.*, 1971). Complete control resulted from the action of *Neodusmetia sangwani* (Rao), an encyrtid from India. Dean *et al.* (1979) estimated that in 1976, the parasitoid saved nearly $17 million per year in turf grass management costs in Texas.

Comstock Mealybug (Pseudococcus comstocki *(Kuwana)*)

Comstock mealybug was introduced into a residential area of the San Joaquin Valley of California and caused considerable damage to ornamentals. Three species of parasitoids were introduced from Japan and became established (Meyerdirk *et al.*, 1981). These species, *Pseudaphycus malinus* Gahan (Encyrtidae), *Allotropa burrelli* Muesebeck (Platygasteridae), and *Allotropa convexifrons* Muesebeck (Platygasteridae), combined with *Zarhopalus corvinus* (Girault) (Encyrtidae) and the native predators *Leucopsis ocellaris* Malloch (Leucospididae) and *Chrysopa* spp. (Chrysopidae) to provide biological control. Population densities of the Comstock mealybug were reduced 68–73% by the natural enemy complex. After establishment of the natural enemies, mealybug population densities were below damage thresholds for ornamentals in residential areas (Meyerdirk *et al.*, 1981). The program is considered a success (Ervin *et al.*, 1983)

Lebbeck Mealybug (Nipaecoccus virdis *(Maskell)*)

The lebbeck mealybug infests several species of ornamental plants but was especially destructive to the lebbeck tree, *Albizia lebbeck,* in Cairo and other cities in Egypt. From 1933 to 1939, natural enemies were introduced from Java. Two encyrtid parasitoids, *Anagyrus aegyptiacus* Moursi and *Leptomastix phenacocci* Compere, were established and provided complete control of the mealybug. In some areas, parasitism reached 98% and in many localities the mealybug almost disappeared (Clausen, 1978).

Hibiscus Mealybug (Maconellicoccus hirsutus *(Green)*)

The hibiscus mealybug is thought to be native to the Far East. It was introduced into Egypt and was considered to be the most injurious mealybug in Egypt. The mealybug severely infests *Hibiscus* spp., mulberry (*Morus* sp.), *Albizia lebbeck* (Linnaeus), *Acacia arabica* (Willdenow), *Bauhinia* spp., *Grevillia* spp., guava, *Parkinsonia aculeatus* Linnaeus, *Robinia pseudacacia* Linnaeus, and *Cajanus cajan* (Linnaeus) Millsp. After a severe outbreak in and around Cairo in 1920, several species of natural enemies of other mealybugs were introduced. The encyrtid *Anagyrus kamali* Moursi, imported from Java, became established and provided control (Clausen, 1978).

Coleoptera

Chrysomelidae

Elm Leaf Beetle (Xanthogaleruca luteola *(Müller)*)

The elm leaf beetle was introduced into the United States from Europe in the 1830s (Howard, 1908). Since its introduction, elm leaf beetle has spread across the United States and now defoliates elms in urban and suburban plantings nearly everywhere they are grown.

The eulophid egg parasitoid *Oomyzus gallerucae* (Fonscolombe) (=*Tetrastichus xanthomelaenae* Rondani) has been repeatedly introduced from Europe and the Middle East (Berry, 1938a; Clair *et al.*, 1988). *Oomyzus gallerucae* is established in central Ohio, apparently as a result of an introduction made in 1932 (Hall & Johnson, 1983). Unfortunately, *O. gallerucae* does not prevent the elm leaf beetle from causing severe aesthetic injury to elms in central Ohio (Hamerski *et al.*, 1990).

Oomyzus gallerucae has also been introduced and recovered from northern and southern California (Luck & Scriven, 1976; Clair *et al.*, 1988; Dahlsten *et al.*, 1990). However, in California, *O. gallerucae* apparently does not

overwinter well and parasitism rates early in the season are extremely low. In many areas it has not been recovered even late in the season. *Aprostocetus brevistigma* (Gahan), which is apparently native to the northeastern United States, has been reported as parasitizing 50–80% of elm leaf beetle pupae in the northeastern United States (Berry, 1938b). *Aprostocetus brevistigma* has been established in California but has little impact on elm leaf beetle populations (Luck & Scriven, 1976; Dreistadt & Dahlsten, 1990). Recent surveys in Ohio suggest that *A. brevistigma* is unimportant there as well (R. W. Hall, unpublished).

Erynniopsis antennata Rondani (Diptera: Tachinidae) was imported from several European countries and established in central California in 1939. It is now well distributed throughout northern California, but parasitism rates do not exceed 40% and average under 5% (Dreistadt & Dahlsten, 1990). Luck and Scriven (1976) found that *E. antennata* parasitized a significant portion of the elm leaf beetle larval population late in the season. However, early season parasitism was insufficient to prevent aesthetically damaging levels of defoliation.

A new strain ("San Diego") of *Bacillus thuringiensis* Berliner has been developed for use against Chrysomelidae; Cranshaw *et al.* (1989) report it appears to be highly effective against both larval and adult elm leaf beetles, and Dahlsten *et al.* (1998c) have found it to be effective with two applications per generation, timed 1 week apart, at first instar larval peak, at one location in California.

Scarabaeidae

Japanese Beetle (Popillia japonica *Newman*)

The Japanese beetle is a major pest of turf grass on golf courses, recreation areas, and home lawns in the northeastern and midwestern United States (Tashiro, 1987). Prior to introduction into the United States, the Japanese beetle was known to occur only on the four main islands of Japan, where it was thought to be of little economic importance (Fleming, 1972). Damage to turf is caused by the larva (grub), which feeds on the roots of grasses. Turf injury depends on grub density and degree of maintenance of the lawn. Poorly maintained lawns may be damaged at densities of 4–5 grubs/ft^2 whereas on well-maintained turf, damage is usually not apparent until densities exceed 10 grubs/ft^2.

The most effective biological control agent of the Japanese beetle is *Bacillus popilliae* Dutky, which causes milky spore disease in the larva (Tashiro, 1987). *Bacillus popilliae* was first identified in central New Jersey in 1933 (White, 1941). The name milky spore disease refers to the milky color assumed by the hemolymph of the grub as sporulation of the bacteria occurs. When sufficient quantities of *B. popilliae* spores are ingested, grubs become infected. Although infected grubs may live and continue to feed for weeks or months after infection, eventually they become weakened and die. Upon death and decay of the grub, a high concentration of spores is released into the soil and other grubs ingesting the spores are subject to infection. Commercially available formulations of spore powder (marketed as "Milky Spore" and "Japanese Beetle Attack") are applied by depositing approximately 2 g of spore powder at intervals of 1.5–3 m on turf grass areas with high population densities of grubs (Tashiro, 1987).

Strains NJ-43 and SI-12 of the nematode *Steinernema glaseri* Steiner are capable of reducing the Japanese beetle in turf grass by two-thirds if the nematodes are rinsed from the grass surface after application (Selvan *et al.*, 1994).

Lepidoptera

Tortricidae

Nantucket Pine Tip Moth (Rhyacionia frustrana *(Comstock)*)

The Nantucket pine tip moth is native to the eastern and southern United States. It is widely distributed and infests most native and exotic pines in that region. Its primary importance is in the south, where it is a pest in loblolly or shortleaf pine plantations. However, this moth is also a common pest of ornamentals and nursery stock. Usually infestations occur on seedlings and saplings less than 3 m in height. Although trees are rarely killed by this insect, the feeding on the terminal and lateral buds and stems causes discoloration and deformation of the branches and stem (Coulson & Witter, 1984).

In 1971, the Nantucket pine tip moth was discovered on Monterey pine, *Pinus radiata,* on a golf course in San Diego County (Scriven & Luck, 1978). Eventually a major portion of southwestern San Diego County was infested. Other species of ornamental pine were attacked but Monterey pine was the most severely attacked.

A native tachinid, *Erynnia tortricis* (Coquillett), and an ichneumonid, *Scambus aplopappi* (Ashmead), were found to parasitize the tip moth, but only at a rate of up to 10% (Scriven & Luck, 1978). Natural enemies were introduced from the eastern and southern United States and in 1979, an ichneumonid, *Campoplex frustranae* Cushman, and a tachinid, *Lixophaga mediocris* Aldrich, were introduced from the southeastern United States. The tachinid did not become established, but *C. frustranae* did and increased rapidly throughout 1976 and 1977 (Scriven & Luck, 1978). By the end of 1979, parasitism of overwintering tip moth had reached 50%. Many Monterey pines at the original parasite establishment site improved in appearance and vigor, indicating that the biological control program with *C. frustranae* was promising (Scriven & Luck, 1978). Noth-

ing has been published recently, but *C. frustranae* has continued to spread in southern California with the tip moth.

Miscellaneous Lepidoptera

Applications of *Bacillus thuringiensis* have been shown to be an effective control for several species of lepidopteran defoliators in urban environments, including gypsy moth, *Lymantria dispar* (Linnaeus), fall cankerworm, *Alsophila pometaria* (Harris), spring cankerworm, *Paleacrita vernata* (Peck), fall webworm, *Hyphantria cunea* (Drury), bagworm, *Thyridopteryx ephemeraeformis* (Haworth), California oakworm, *Phryganidia californica* Packard, and red humped caterpillar, *Schizura coccina* (Smith).

Diptera

Agromyzidae

Holly Leafminer (Phytomyza ilicis *Curtis*)

The holly leafminer was introduced into British Columbia and damaged leaves of English holly in commercial and urban plantings. Five species of parasitoids were released on Vancouver Island from 1936 to 1938 and on the mainland in 1939 (Clausen, 1978). Although four species became established, parasitism was primarily due to *Chrysocharis gemma* (Walker) (Eulophidae) and *Opius ilicis* Nixon (Braconidae). *Chrysocharis gemma* caused approximately 90% of the parasitism on Vancouver Island whereas *O. ilicis* was responsible for approximately 90% of the parasitism on the mainland.

On English holly in ornamental plantings, the parasitoids usually prevent serious disfigurement of the plants and chemical controls are rarely needed. Therefore, Turnbull and Chant (1961) deemed biological control to be a complete success in these situations. However, on commercially produced holly, parasitoids did not ensure a blemish-free plant and pesticides are still used. Turnbull and Chant (1961) and Munroe (1971) consider biological control as partially to substantially successful in commercial plantings.

References

Beingolea, G. O. (1969). Biological control of citrus pests in Peru. Proceedings of the 1st International Citrus Symposium, Riverside, California, 2, 827–838.

Bellows, T. S., Paine, T. D., Gould, J. R., Bezark, L. G., & Ball, J. C. (1992). Biological control of ash whitefly: A success in progress. California Agriculture, 46(1), 24, 27–28.

Bennett, F. D., & Hughes, I. W. (1959). Biological control of insect pests in Bermuda. Bulletin of Entomological Research, 50, 423–436.

Bennett, G. W., & Owens, J. M., Eds. (1986). "Advances in Urban Pest Management." Van Nostrand-Reinhold, New York.

Berry, P. A. (1938a). Laboratory studies on *Tetrastichus xanthomelaenae*

Rond. and *Tetrastichus* sp., two hymenopterous egg parasites of the elm leaf beetle. Journal of Agricultural Research, 57, 859–863.

Berry, P. A. (1938b). "*Tetrastichus brevistigma* Gahan, a Pupal Parasite of the Elm Leaf Beetle." USDA Circular 485.

Cameron, E. (1955). On the parasites and predators of the cockroach. I. *Tetrastichus hagenowii* (Ratz.). Bulletin of Entomological Research, 46, 137–147.

Clair, D. J., Dahlsten, D. L., & Dreistadt, S. H. (1988). Biological control of the elm leaf beetle, *Xanthogaleruca luteola*, in California—A case study. *In* "Global Perspectives on Agroecology and Sustainable Agricultural Systems" (P. Allen & D. van Dusen, Eds.), Vol. 2, pp. 497–504. Agroecology Program, Univ. of California, Santa Cruz.

Clair, D. J., Dahlsten, D. L., & Hart, E. R. (1987). Rearing *Tetrastichus gallerucae* (Hymenoptera: Eulophidae) for biological control of the elm leaf beetle, *Xanthogaleruca luteola*. Entomophaga, 32, 457–461.

Clausen, C. P., Ed. (1978). "Introduced Parasites and Predators of Arthropod Pests and Weeds: A World Review." USDA–ARS Agricultural Handbook No. 480.

Cockrell, R. A., & Warnke, F. F. (1976). "Trees of the Berkeley Campus." Division of Agricultural Sciences, Univ. of California, Berkeley.

Coler, R. R., Van Driesche, R. G., & Elkinton, J. S. (1984). Effect of an oothecal parasitoid *Comperia merceti* (Compere) (Hymenoptera: Encyrtidae) on a population of the brownbanded cockroach (Orthoptera: Blattellidae). Environmental Entomology, 13, 603–606.

Coulson, R. N., & Witter, J. A. (1984). "Forest Entomology." Wiley, New York.

Cranshaw, W. S., Day, S. J., Gritzmacher, T. J., & Zimmerman, R. J. (1989). Field and laboratory evaluation of *Bacillus thuringiensis* strains for control of elm leaf beetle. Journal of Aboriculture, 15, 31–34.

Dahlsten, D. L. (1986). Control of invaders. *In* "Ecological Studies: Ecology of Biological Invasions of North America and Hawaii" (H. A. Mooney & J. A. Drake, Eds.), Vol. 58, pp. 275–302. Springer-Verlag, New York.

Dahlsten, D. L. (1996). Introduced *Eucalyptus* foliage feeders in California. *In* "Proceedings of the California Forest Pest Council 44th Annual Meeting, Nov. 15–16, 1995," pp. 32–35.

Dahlsten, D. L., Dreistadt, S. H., Geiger, J. R., Tait, S. M., Rowney, D. L., Yokota, G. Y., & Copper, W. A. (1990). "Elm Leaf Beetle Biological Control and Management in Northern California." Final Report to the California Department of Forestry and Fire Protection.

Dahlsten, D. L., Garcia, R., Prine, J. E., & Hunt, R. (1969). Insect problems in forest recreation areas. California Agriculture, 23(7), 4–6.

Dahlsten, D. L., Hajek, A. E., Clair, D. J., Dreistadt, S. H., Rowney, D. L., & Lewis, V. R. (1985). Pest management in the urban forest. California Agriculture, 39(1–2), 21–22.

Dahlsten, D. L., Hansen, E. P., Zuparko, R. L., & Norgaard, R. B. (1998b). Biological control of the blue gum psyllid proves economically beneficial. California Agriculture, 52(1), 35–40.

Dahlsten, D. L., Kent, D. M., Rowney, D. L., Copper, W. A., Young, T. E., & Tassan, R. L. (1995). Parasitoid shows potential for biocontrol of eugenia psyllid. California Agriculture, 49(4), 36–40.

Dahlsten, D. L., Rowney, D. L., Copper, W. A., Tassan, R. L., Chaney, W. E., Robb, K. L., Tjosvold, S., Bianchi, M., & Lane, P. (1998a). Parasitoid wasp controls blue gum psyllid. California Agriculture, 52(1), 31–34.

Dahlsten, D. L., Rowney, D. L., & Lawson, A. B. (1998c). IPM helps control elm leaf beetle. California Agriculture, 52(2), 18–23.

Dean, H. A., Schuster, M. F., Boling, J. C., & Riherd, P. T. (1979). Complete biological control of *Antonina graminis* in Texas with *Neodusmetia sangwani* (a classic example). Bulletin of the Entomological Society of America, 25, 262–267.

DeBach, P. (1964). Successes, trends, and future possibilities. *In* "Biological Control of Insect Pests and Weeds" (P. DeBach, Ed.), pp. 673–713. Reinhold, New York.

DeBach, P., & Rose, M. (1976). Biological control of woolly whitefly. California Agriculture, 30(5), 4–7.

DeBach, P., & Rose, M. (1977). Environmental upsets caused by chemical eradication. California Agriculture, 31(7), 8–10.

Donaldson, D. R., Moore, W. S., Koehler, C. S., & Joos, J. L. (1978). Scales threaten ice plant in Bay Area. California Agriculture, 32(10), 4, 7.

Dreistadt, S. H., & Dahlsten, D. L. (1986). Medfly eradication in California—Lessons from the field. Environment, 28(6), 18–20, 41–44.

Dreistadt, S. H., & Dahlsten, D. L. (1990). Distribution and abundance of *Erynniopsis antennata* (Dipt.: Tachinidae) and *Tetrastichus brevistigma* (Hym.: Eulophidae), two introduced elm leaf beetle parasites in northern California. Entomophaga, 35, 527–536.

Dreistadt, S. H., & Hagen, K. S. (1994). Classical biological control of the acacia psyllid, *Acizzia uncatoides* (Homoptera: Psyllidae), and predator–prey–plant interactions in the San Francisco Bay area. Biological Control, 4, 319–327.

Ebeling, W. (1975). "Urban Entomology." Univ. of California, Division of Agricultural Sciences, Richmond, CA.

Edmunds, L. R. (1957). Observations on the biology and life history of the brown cockroach *Periplaneta brunnea* Burmeister. Proceedings of the Entomological Society of Washington, 59, 283–286.

Ehler, L. E., & Endicott, P. C. (1984). Effect of malathion-bait sprays on biological control of insect pests of olive, citrus, and walnut. Hilgardia, 52(5), 1–47.

Embree, D. G. (1971). *Operophtera brumata* (L.), winter moth (Lepidoptera: Geometridae). *In* "Biological Control Programmes against Insects and Weeds in Canada, 1959–1968," pp. 167–175. Commonwealth Institute of Biological Control Technical Communication No. 4.

Ervin, R. T., Moffitt, L. J., & Meyerdirk, D. E. (1983). Comstock mealybug (Homoptera: Pseudococcidae): Cost analysis of a biological control program in California. Journal of Economic Entomology, 76, 605–609.

Ferguson, A. M. (1989). *Asterodiospis* spp., pit scales (Homoptera: Asterolecaniidae). *In* "A Review of Biological Control of Invertebrate Pests and Weeds in New Zealand, 1874–1987" (P. J. Cameron, R. L. Hill, J. Bain, & W. P. Thomas, Eds.), pp. 257–262. Commonwealth Institute of Biological Control Technical Communication No. 10.

Flanders, R. V. (1986). Potential for biological control in urban environments. *In* "Advances in Urban Pest Management" (G. W. Bennett & J. M. Owens, Eds.), pp. 95–127. Van Nostrand-Reinhold, New York.

Flanders, S. E. (1959). Biological control of *Saissetia nigra* (Nietn.) in California. Journal of Economic Entomology, 52, 596–600.

Fleet, R. R., & Frankie, G. W. (1975). Behavioral and ecological characteristics of a eulophid egg parasite of two species of domiciliary cockroaches. Environmental Entomology, 4, 282–284.

Fleming, W. E. (1968). "Biological Control of Japanese Beetle." USDA Technical Bulletin No. 1383.

Fleming, W. E. (1972). "Biology of Japanese Beetle." USDA Technical Bulletin No. 1449.

Frankie, G. W., & Ehler, L. E. (1978). Ecology of insects in urban environments. Annual Review of Entomology, 23, 367–387.

Frankie, G. W., & Hagen, K. S. (1986). "Ecology and Biology of Iceplant Scales, *Pulvinaria* and *Pulvinariella* in California." California Department of Transportation, Division of Highway Maintenance, Sacramento, CA (Report No. FHWA-CA-HM-OZ).

Frankie, G. W., & Koehler, C. S., Eds. (1978). "Perspectives in Urban Entomology." Academic Press, New York.

Frankie, G. W., & Koehler, C. S., Eds. (1983). "Urban Entomology: Interdisciplinary Perspectives." Praeger, New York.

Gould, J. R., Bellows, T. S., & Paine, T. D. (1992a). Population dynamics of *Siphoninus phillyreae* in California in the presence and absence of a parasitoid, *Encarsia partenopea*. Ecological Entomology, 17, 127–134.

Gould, J. R., Bellows, T. S., & Paine, T. D. (1992b). Evaluation of biological control of *Siphoninus phillyreae* (Haliday) by the parasitoid *Encarsia partenopea* (Walker), using life-table analysis. Biological Control, 2, 257–265.

Hagen, K. S., van den Bosch, R., & Dahlsten, D. L. (1971). The importance of naturally occurring biological control in the western United States. *In* "Biological Control" (C. B. Huffaker, Ed.), pp. 253–293. Plenum, New York.

Hall, R. W., & Johnson, N. F. (1983). Recovery of *Tetrastichus gallerucae* (Hymenoptera: Eulophidae), an introduced egg parasitoid of the elm leaf beetle (*Pyrrhalta luteola*) (Coleoptera: Chrysomelidae). Journal of the Kansas Entomological Society, 56, 297–298.

Hamerski, M. R., Hall, R. W., & Keeney, G. D. (1990). Laboratory biology and rearing of *Tetrastichus brevistigma* (Hymenoptera: Eulophidae), a larval-pupal parasitoid of the elm leaf beetle (Coleoptera: Chrysomelidae). Journal of Economic Entomology, 83, 2196–2199.

Hoelmer, K. A., & Dahlsten, D. L. (1993). Effects of malathion bait spray on *Aleyrodes spiraeoides* (Homoptera: Aleyrodidae) and its parasitoids in northern California. Environmental Entomology, 22, 49–56.

Howard, L. O. (1908). The importation of *Tetrastichus xanthomelaenae* (Rond.). Journal of Economic Entomology, 1, 281–289.

Hoy, J. B., & Dahlsten, D. L. (1984). Effects of malathion and Staley's bait on the behavior and survival of parasitic Hymenoptera. Environmental Entomology, 13, 1483–1486.

Jetter, K., Klonsky, K., & Pickett, C. H. (1997). A cost-benefit analysis of the ash whitefly biological control program in California. Journal of Arboriculture, 23, 65–72.

Johnson, W. T., & Lyon, H. H. (1976). "Insects that Feed on Trees and Shrubs." Comstock Publ. Assoc., Cornell Univ. Press, Ithaca, NY.

Kaya, H. K., & Anderson, J. F. (1974). Collapse of the elm spanworm outbreak in Connecticut: Role of *Ooencyrtus*. Environmental Entomology, 3, 659–663.

Kielbaso, J. J., & Kennedy, M. K. (1983). Urban forestry and entomology: A current appraisal. *In* "Urban Entomology: Interdisciplinary Perspectives" (G. W. Frankie & C. S. Koehler, Eds.), pp. 423–440. Praeger, New York.

Laing, J. E., & Hamai, J. (1976). Biological control of insect pests and weeds by imported parasites, predators and pathogens. *In* "Theory and Practice of Biological Control" (C. B. Huffaker & P. S. Messenger, Eds.), pp. 685–743. Academic Press, New York.

Luck, R. F. (1981). Parasitic insects introduced as biological control agents for arthropod pests. *In* "CRC Handbook of Pest Management in Agriculture" (D. Pimentel, Ed.), Vol. II, pp. 125–284. CRC Press, Boca Raton, FL.

Luck, R. F., & Dahlsten, D. L. (1974). Bionomics of the pine needle scale, *Chionaspis pinifoliae* (Fitch), and its natural enemies at South Lake Tahoe, California. Annals of the Entomological Society of America, 66, 309–316.

Luck, R. F., & Dahlsten, D. L. (1975). Natural decline of a pine needle scale (*Chionaspas pinifoliae* [Fitch]) outbreed at South Lake Tahoe, California, following cessation of adult mosquito control with malathion. Ecology, 56, 893–904.

Luck, R. F., & Scriven, G. T. (1976). The elm leaf beetle, *Pyrrhalta luteola* in southern California: Its pattern of increase and its control by introduced parasites. Environmental Entomology, 5, 409–416.

Lutz, F. F. (1941). "A Lot of Insects (Entomology in a Suburban Garden)." Putnam, New York.

McBride, J. R., & Jacobs, D. F. (1986). Presettlement forest structure as a factor in urban forest development. Urban Ecology, 9, 245–266.

McGugan, B. M., & Coppel, H. C. (1962). A review of the biological control attempts against insects and weeds in Canada. *In* "Biological Control of Forest Insects, 1910–1958," pp. 35–216. CAB, Commonwealth Institute of Biological Control Technical Communication No. 2.

Merritt, R. W., Kennedy, M. K., & Gersabeck, E. F. (1983). Integrated pest management of nuisance and biting flies in a Michigan resort: Dealing with secondary pest outbreaks. *In* "Urban Entomology: Interdisciplinary Perspectives" (G. W. Frankie & C. S. Koehler, Eds.), pp. 277–299. Praeger, New York.

Meyerdirk, D. E., Newell, I. M., & Warkentin, R. W. (1981). Biological control of Comstock mealybug. Journal of Economic Entomology, 74, 79–84.

Munroe, E. G. (1971). Status and potential of biological control in Canada. *In* "Biological Control Programmes against Insects and Weeds in Canada," Chap. 48, pp. 213–255. Commonwealth Institute of Biological Control Technical Communication No. 4.

National Research Council. (1980). "Urban Pest Management." Report by Committee on Urban Pest Management, Environmental Studies Board, Commission on National Resources. National Academy Press, Washington, DC.

Olkowski, W. (1974). A model ecosystem management program. Proceedings of the Tall Timbers Conference, Ecology, Animal Control Habitat Management, 5, 103–117.

Olkowski, W., Olkowski, H., Kaplan, A. I., & van den Bosch, R. (1978). The potential for biological control in urban areas: Shade tree insect pests. *In* "Perspectives in Urban Entomology" (G. W. Frankie & C. S. Koehler, Eds.), pp. 311–349. Academic Press, New York.

Olkowski, W., Olkowski, H., & van den Bosch, R. (1982a). Linden aphid parasite establishment. Environmental Entomology, 11, 1023–1025.

Olkowski, W., Olkowski, H., van den Bosch, R., & Hom, R. (1976). Ecosystem management: A framework for urban pest control. Bioscience, 26, 384–389.

Olkowski, W., Olkowski, H., van den Bosch, R., Hom, R., Zuparko, R., & Klitz, W. (1982b). The parasitoid *Trioxys tenuicaudus* Stary (Hymenoptera: Aphidiidae) established on the elm aphid *Tinocallis platani* Kaltenbach (Homoptera: aphididae) in Berkeley, California. Pan-Pacific Entomologist, 58, 59–63.

Pemberton, C. E. (1948). History of the entomology department, experiment station, H.S.P.A., 1904–1945. Hawaii Planters' Record, 52, 53–90.

Pemberton, C. E. (1954). Invertebrate Consultants Committee for the Pacific. Pacific Science Board, National Academy of Sciences, National Research Council.

Pinnock, D. E., Hagen, K. S., Cassidy, D. V., Brand, R. J., Milstead, J. E., & Tassan, R. L. (1978). Integrated pest management in highway landscapes. California Agriculture, 32(2), 33–34.

Piper, G. L., & Frankie, G. W. (1978). Integrated management of urban cockroach populations. *In* "Perspectives in Urban Entomology" (G. W. Frankie & C. S. Koehler, Eds.), pp. 249–266. Academic Press, New York.

Piper, G. L., Frankie, G. W., & Loehr, J. (1978). Incidence of cockroach egg parasitoids in urban environments in Texas and Louisiana. Environmental Entomology, 7, 289–293.

Quezada, J. R., & DeBach, P. (1973). Bioecological and population studies of the cottony cushion scale, *Icerya purchasi* Mask., and its natural enemies, *Rodalia cardinalis* Muls. and *Cryptochaetum iceryae* Will., in southern California. Hilgardia, 41, 631–688.

Rao, V. P., Ghani, M. A., Sankaran, T., & Mathur, K. C. (1971). "A Review of the Biological Control of Insects and Other Pests in the Southeast Asia and the Pacific Region." Commonwealth Institute of Biological Control Technical Communication No. 6.

Richards, W. R. (1967). A review of the *Tinocallis* of the world (Homoptera: Aphididae). Canadian Entomologist, 99, 536–553.

Roth, L. M., & Willis, E. R. (1954). The biology of the cockroach parasite *Tetrastichus hagenowii* (Ratzeburg), a chalcidoid egg parasite (Hymenoptera: Eulophidae). Transactions of the American Entomological Society, 80, 53–72.

Saakian-Baranova, A. A. (1966). The life cycle of *Metaphycus luteolus* Timb. (Hymenoptera: Encyrtidae), parasite of *Coccus hesperidum* (Homoptera: Coccidae), and the attempt of its introduction into the USSR. Entomological Review, 45, 733–751 (in Russian).

Sailer, R. I. (1972). Concepts, principles and potentials of biological control: Parasites and predators. Proceedings of the North Central Entomological Society of America, 27, 35–39.

Schuster, M. F., Boling, J. C., & Marony, J. J., Jr. (1971). Biological control of rhodesgrass scale by airplane releases of an introduced parasite of limited dispersing activity. In "Biological Control" (C. B. Huffaker, Ed.), pp. 227–250. Plenum, New York.

Scriven, G. T., & Luck, R. F. (1978). Natural enemy promises control of Nantucket pine tip moth. California Agriculture, 32(10), 19–20.

Selvan, S., Grewal, P. S., Gaugler, R., & Tomalak, M. (1994). Evaluation of Steinernematid nematodes against *Popilla japonica* (Coleoptera: Scarabaeidae) larvae: Species, strains and rinse after application. Journal of Economic Entomology, 87, 605–609.

Simmonds, F. J. (1969). "Brief Resume of Activities and Recent Successes Achieved." Commonwealth Institute of Biological Control. Commonwealth Agricultural Bureaux Publications. 16 pp.

Slater, A. J. (1984). Biological control of the brownbanded cockroach, *Supella longipalpa* (Serville) with an encyrtid wasp, *Comperia merceti* (Compere). Pest Management, 3(4), 14–17.

Slater, A. J., Hurlbert, M. J., & Lewis, V. R. (1980). Biological control of the brownbanded cockroaches. California Agriculture, 34(8), 16–18.

Swan, D. I. (1973). Evaluation of biological control of the oak leaf miner, *Phyllonorycter messeniella* (Zell.) (Lepidoptera: Gracillariidae), in New England. Bulletin of Entomological Research, 63, 49–55.

Tashiro, H. (1987). "Turfgrass Insects of the United States and Canada." Comstock Publ. Assoc., Cornell Univ. Press, Ithaca, NY.

Tassan, R. L., Hagen, K. S., & Cassidy, D. V. (1982). Imported natural enemies established against ice plant scales in California. California Agriculture, 36(9–10), 16–17.

Troetschler, R. G. (1983). Effects on nontarget arthropods of malathion bait sprays used in California to eradicate the Mediterranean fruit fly, *Ceratis capitata* (Wiedemann) (Diptera: Tephritidae). Environmental Entomology, 12, 1816–1822.

Turnbull, A. L., & Chant, D. A. (1961). The practice and theory of biological control in Canada. Canadian Journal of Zoology, 39, 697–753.

van den Bosch, R., Frazer, B. D., Davis, C. S., Messenger, P. S., & Hom, R. (1970). *Trioxys pallidus,* an effective new walnut aphid parasite from Iran. California Agriculture, 24(11), 8–10.

Walker, J. T. S. (1989). *Tuberculatus annulatus* (Hartig), oak aphid (Homoptera: Callaphididae). In "A Review of Biological Control of Invertebrate Pests and Weeds in New Zealand, 1874–1987" (P. J. Cameron, R. L. Hill, J. Bain, & W. P. Thomas, Eds.), pp. 307–308. Commonwealth Institute of Biological Control Technical Communication No. 10.

Washburn, J. A., Tassan, R. L., Grace, K., Bellis, E., Hagen, K. S., & Frankie, G. W. (1983). Effects of malathion sprays on the ice plant insect system. California Agriculture, 37(1–2), 30–32.

White, R. T. (1941). Development of milky disease on Japanese beetle larvae under field conditions. Journal of Economic Entomology, 34, 213–215.

Wilson, F. (1960). "A Review of the Biological Control of Insects and Weeds in Australia and Australian New Guinea." Commonwealth Institute of Biological Control Technical Communication No. 1.

Zuparko, R. L. (1983). Biological control of *Eucallipterus tiliae* [Hom.: Aphididae] in San Jose, Calif., through establishment of *Trioxys curvicaudus* [Hym.: Aphidiidae]. Entomophaga, 28, 325–330.

Zuparko, R. L., & Dahlsten, D. L. (1994). Host plant resistance and biological control for linden aphids. Journal of Arboriculture, 20, 278–281.

Zuparko, R. L., & Dahlsten, D. L. (1995). Parasitoid complex of *Eucallipterus tiliae* (Homoptera: Drepanosiphidae) in northern California. Environmental Entomology, 24, 730–737.

Biological Control of Medical and Veterinary Pests

R. GARCIA

Division of Biological Control
University of California
Albany, California

E. F. LEGNER

Department of Entomology
University of California
Riverside, California

INTRODUCTION

The manipulative use of natural enemies for the control of medical and veterinary invertebrate pests has been restricted largely to various species of Diptera. Some work has been conducted on ants, cockroaches, wasps, ticks, and snails, but work on these animals has been limited. Here we review the biological control agents that can be manipulated, that have been used successfully, that are being researched, and that show at least some promise for successful application.

Bay *et al.* (1976) indicate that medically important pests differ from agricultural pests in fundamental ways. First, pests that affect humans are usually in the adult stage while those that attack crops are usually in the immature stage. This is of some advantage for control of medically important pests because it allows the control action to be taken against the immatures, thus eliminating the adult before it can cause problems. A second difference, however, is not favorable as it relates to setting tolerance levels. Whereas an allowable number of pests (tolerance level) can be established for the biological control of a crop pest, such levels are far more difficult to establish for pests attacking humans. For example, an individual mosquito can be of great annoyance and can precipitate a reaction for control. In addition, low population levels of a vector may still transmit a disease and, therefore, cannot be tolerated (Service, 1983). However, setting tolerance levels for veterinary pests would be more in line with those for agricultural pests. A third difference, usually a distinct disadvantage for biological control, is that the habitat utilized by medically important pests is frequently temporary, as opposed to that of an agricultural crop, which is more permanent. In the

agricultural situation, natural enemies can coexist with pests and thus may regulate the pest populations. Additionally, in many situations the habitat exploited by the medically important pests is only an undesirable extension of human activity. An example would be the cultivation of rice, where the production of pests such as mosquitoes is usually of little concern to the grower.

Interest in biological control of medical pests and vectors had its modest beginning nearly 100 years ago (Lamborn, 1890). At that time, the possible use of dragonflies as natural enemies for the control of mosquitoes was clearly recognized. However, as is true even today, the enormous difficulties associated with the colonization and management of these insects quickly extinguished any idea for the practical use of these predators for mosquito control. Shortly after the turn of the century, the mosquitofish, *Gambusia affinis* (Baird & Girard), came to the forefront of biological control. This small fish, being much easier to deal with than dragonflies, was quickly utilized and transported throughout the world during the early decades of this century in attempts to control mosquitoes.

The mosquitofish, *G. affinis,* and other naturalistic methods of control were employed with some vigor for about the first 40 years of the century. All these control measures were curtailed sharply with the introduction of synthetic organic insecticides after World War II. The convenience and quick killing power of these chemicals was so dramatic for mosquitoes, flies, and lice that other control tactics were quickly reduced to a minor role. Interest in biological control arose again when the succession of chemicals developed during the 1940s and 1950s began to fail due to the development of genetic resistance in vector and pest populations. The biological control of medically important pests and vectors has made slow progress since its revival and is still behind that which has occurred in agricultural systems (Service, 1983). This disparity is due to the problems of establishing pest tolerance levels and the temporary, unstable habitats exploited by medically important pests (Legner & Sjogren, 1984).

While progress in the development of biological control agents has been substantial and work in progress appears promising, our overall evaluation at this point is that biological control will rarely be a panacea for medically important pests. However, with continued effort it can be a major component in the overall strategy for the control of some of these important pests (Legner & Sjogren, 1984). The literature reviewed in this chapter is organized by taxonomic groups because of the wide assemblage of different habitats occupied by medically important pests. The major groups where some success has been achieved or where work is currently being conducted are the mosquitoes, blackflies, synanthropic flies, intermediate host snails, and cockroaches. Most effort has been directed against mosquitoes because of the human disease agents they transmit. Consequently, much of this chapter is devoted to mosquitoes. The synanthropic flies and snails are also covered in this chapter, but the biological control of cockroaches is reviewed in the chapter on urban pests (Chapter 36).

MOSQUITOES

The successful widespread use of biological control agents against mosquitoes requires a much better understanding of the ecology of predator–prey and pathogen–host relationships (Service, 1983). The opportunistic characteristics of many species (i.e., their ability to exploit temporary habitats, coupled with their short generation time, high natural mortality, great dispersal potential, and other R-strategist characteristics) pose difficult problems for any biotic regulatory mechanism. Mosquitoes typically exploit many aquatic habitats. Often a biological control agent has a much narrower range of environmental activity than the target mosquito has. Thus, in many situations, a number of different biological control agents and/or appropriate methods will be necessary to control even one species of mosquito across its range of exploitable breeding sources.

Fish

Several species of fish are used for the biological control of mosquitoes, and these species together form the major successes in the field. Unfortunately, their usefulness is limited to more permanent bodies of water, and even under these situations, their impact on the target species has been only partially successful. Bay et al. (1976) point out that many species of fish consume mosquito larvae, but only a few fish species have been manipulated to manage mosquito populations.

The mosquitofish, G. affinis, is the best-known agent for mosquito control. This fish, which is native to the southern United States, Mexico, and the Caribbean area, was first used as an introduced agent for mosquito control when it was transported from North Carolina to New Jersey in 1905 (Lloyd, 1987). Shortly thereafter, it was introduced to the Hawaiian Islands to control mosquitoes that had been introduced during the nineteenth century. During the next 70 years, the mosquitofish was transported to over 50 countries and today stands as the most widely disseminated biological control agent (Bay, 1969; Lloyd, 1987). Many of these introductions were aimed at Anopheles species that were transmitting malaria. Hackett (1937) described its usefulness in malaria control programs in Europe. He commented that its effects were cumulative and not sufficient by themselves, but that the fish had a definite impact on the suppression of the disease. Tabibzadeh et al. (1970) reported a rather extensive release program in Iran and concluded that the fish was an important component in the malaria eradication program. Sasa and Kurihara (1981) and Service (1983) believed that the fish had little impact on the disease and that most evidence supporting these claims is circumstantial. However, Inci et al. (1992) reported a 50% reduction in malaria cases in southeast Turkey after releases of G. affinis. Gambusia is not recommended by the World Health Organization (WHO) for malaria control programs, primarily because of its impact on indigenous species of fish (Services, 1983; Lloyd, 1987).

The biological attributes of G. affinis, namely, a high reproductive capability; high survivorship; small size; omnivorous foraging in shallow water; and relatively high tolerance to variations in temperature, salinity, and organic waste, would seemingly make this species an excellent biological control agent (Bay et al., 1976; Moyle, 1976). However, whether this fish leads to effective mosquito control at practical costs in many situations is still debated. Kligler's (1930) statement that . . . their usefulness as larvae destroyers under local conditions where vegetation is abundant and micro fauna rich enough to supply their needs without great trouble, is limited. In moderately clear canals, on the other hand, or in pools having a limited food supply, they yielded excellent results. . . . is probably one of the most accurate.

In California, this fish has been used rather extensively over the years for control of mosquitoes in various habitats (Bay et al., 1976). Many mosquito abatement districts in this state have developed systems for culturing, harvesting, and winter storage of the mosquitofish to have enough available for planting early in the spring (Coykendall, 1980). This is particularly important in the rice growing areas of California where early stocking appears to be of critical importance for buildup of fish populations to control mosquitoes during late summer. The results of the use of *G. affinis* in California rice fields will be summarized later as an illustrative example of the mixed successes achieved in the field.

Rice cultivation continuously poses one of the most difficult control problems for *Anopheles* and *Culex* species. Hoy and Reed (1970) showed that good to very good control of *C. tarsalis* Coquillett could be achieved at stocking rates of about 480 or more females per hectare and Stewart *et al.* (1983) reported excellent control with a similar stocking rate against this species in the San Joaquin Valley.

Although *C. tarsalis* appears to be controlled effectively by *G. affinis,* the control of its frequent companion in northern California rice fields, *A. freeborni* Aitken, is less apparent. Hoy *et al.* (1971) showed a reduction of *A. freeborni* populations at various stocking rates of about 120 to 720 fish per hectare, but the reduction was not nearly as striking as that for *C. tarsalis.* These workers surmised that improvement in control could be achieved by earlier season stocking, possibly multiple release points in fields and a reliable source of healthy fish for stocking. Despite an extensive research effort in mass culture, management, and storage for *G. affinis* by the state of California (Hoy & Reed, 1971), a mass production method has not been satisfactorily achieved (Downs *et al.,* 1986; Cech & Linden, 1987).

Studies of *G. affinis* for control of mosquitoes in wild rice show that relatively high stocking rates can effectively reduce *A. freeborni* and *C. tarsalis* populations within a 3-month period (Kramer *et al.* 1987). The commercial production of wild rice, which is a more robust and taller plant than white rice and requires only 90 instead of 150 days to mature, has been increasing over the last few years in California (Kramer *et al.,* 1987). In the preceding study, stocking rates of 1.7 kg/ha (ca. 2400 fish per kilogram) released in 1/10 ha wild rice plots failed to show a significant difference in reduction of mosquitoes from plots with no fish. A decrease in numbers of larvae was noted just prior to harvest, which suggested that the fish were beginning to have an impact on mosquito numbers (Kramer *et al.,* 1987). Fish in these plots, based on recovery after drainage, was about 100,000 individuals per hectare (ca. 32 kg/ha) or a density of about 10 fish per square meter. However, significant control was not achieved.

During 1987, this study was repeated at the rates of 1.7 and 3.4 kg/ha of fish. Results showed an average suppression of larvae (primarily *A. freeborni*) of < 1 and 0.5 per dip for the low and high rate, respectively, compared with control plots that averaged > 4.5 per dip. Fish densities in the 1987 study surpassed those of 1986 by about twofold at the 1.7 kg/ha rate and threefold at the 3.4 kg/ha rate. It is believed that these greater fish numbers accounted for the control differences observed in the second year, although mosquitoes were not eliminated. Differences between test and control plots were first observed 8 weeks after the fish had been planted and mosquitoes remained under control until drainage of the fields (Kramer *et al.,* 1988).

Davey and Meisch (1977a, 1977b) showed that the mosquitofish, at inundative release rates of 4800 fish per hectare, was effective for control of *Psorophora columbiae* (Dyar & Knab) in Arkansas rice fields. Fish released at the water flow inlets dispersed quickly throughout the fields. This is an important attribute for controlling species of *Psorophora* and *Aedes,* whose hatch and larval development are completed within a few days. A combination of 1200 *G. affinis* and about 300 sunfish (*Lepomis cyanellus* Rafinesque) gave better control than either four times the amount of *G. affinis* or *L. cyanellus* used separately. This synergistic effect reduces logistic problems associated with having enough fish available at the times fields are inundated. Blaustein (1986) found enhanced control of *A. freeborni* by mosquitofish in California rice fields after the addition of green sunfish. He speculated that the increased control was the result of the mosquitofish spending more time in protected areas where mosquitoes were more abundant and the green sunfish was avoided. The availability of fish for stocking fields either inundatively, such as in Arkansas, or for control later in the season, as practiced in California, has been a fundamental reason why fish have not been used more extensively in rice fields.

A unique use of the mosquitofish by inundative release was reported by Farley and Caton (1982). These workers released fish in subterranean urban storm drains to control *C. quinquefasciatus* Say breeding in entrapped water at low points in the system. Fish releases were made following the last major rains to avoid having them flushed out of the system. Fish survived for more than 3 months during the summer and were found throughout the system. Gravid females produced progeny. However, no mating occurred, and after the initial increase in numbers, populations of fish diminished as summer progressed. Reduction of mosquitoes from 75 to 94% were observed for 3 months compared with untreated areas (Mulligan *et al.,* 1983). This control practice is now conducted on a routine basis by the Fresno Mosquito Abatement District (J. R. Caton, personal communication 1987).

Although *G. affinis* has been useful for control of mosquitoes in a number of situations, clearly there are drawbacks to its use. In fact, if today's environmental awareness existed at the turn of the century, this fish probably never would have been intentionally introduced into exotic areas (Pelzman, 1975; Lloyd, 1987). The major objection to this fish has been its direct impact on native fish through predation, or its indirect impact through competition (Bay *et al.,* 1976; Schoenherr, 1981; Lloyd, 1987). More than 30 species of native fish have been adversely affected by the introduction of *Gambusia* (Schoenherr, 1981; Lloyd, 1987).

Gambusia, a general predator, can also substantially reduce zooplankton and thus lead to algal blooms in certain situations (Hurlbert *et al.,* 1972). Introductions of *Gambusia* have also reduced numbers of other aquatic invertebrates coinhabiting the same waters (Hoy *et al.,* 1972; Farley & Younce, 1977; Rees, 1979; Walters & Legner, 1980; Hurlbert & Mulla, 1981).

The next most widely used fish for mosquito control is the common guppy, *Poecilia reticulata* (Peters). It has been deployed successfully in Asia for the control of wastewater mosquitoes, especially *C. quinquefasciatus.* Like its poeciliid relative, *Gambusia,* it is native to the Americas (tropical South America). However, instead of being intentionally introduced to control mosquitoes, it was taken to other parts of the world by tropical fish fanciers. Sasa *et al.* (1965) observed wild populations of this fish breeding in drains in Bangkok and concluded that it was controlling mosquitoes common to that habitat. The practical use of guppies is primarily restricted to subtropical climates because of an inability to tolerate temperate-zone water temperatures (Sasa & Kurihara, 1981). However, their most important attribute is a tolerance to relatively high levels of organic pollutants, which makes them ideal for urban water sources that are rich in organic wastes. In Sri Lanka, wild populations have been harvested and used for the control of mosquitoes in abandoned wells, coconut husk pits, and other sources rich in organics (Sasa & Kurihara, 1981). The fish occurs widely in India, Indonesia, and China and has been intentionally introduced for filariasis control into Rangoon and Burma (Sasa & Kurihara, 1981). On the Comoro Islands, it has shown good potential for the control of *A. gambiae* Giles in cisterns (Sabatinelli *et al.,* 1990). Mian *et al.* (1985) evaluated its use for control of mosquitoes in sewage treatment facilities in southern California and concluded that guppies showed great potential for mosquito control in these situations. In Cuba, investigators have successfully controlled mosquitoes in polluted ditches and oxidation ponds with guppies (Koldenkova *et al.,* 1988; García *et al.,* 1991).

Exotic fish have also been used for clearing aquatic vegetation from waterways, which has resulted in excellent mosquito control. In the irrigation systems of southeastern California, three species of subtropical cichlids, *Tilapia zillii* (Gervais), *Oreochromis mossambica* (Peters), and *O. hornorum* (Trewazas) were introduced and have become established over some 2000 ha of *C. tarsalis* breeding habitat (Legner & Sjogren, 1984). In this situation, mosquito populations are under control by a combination of direct predation and consumption of aquatic plants by these omnivorous fish (Legner & Medved, 1973; Legner 1978a, 1983; Legner & Fisher, 1980; Legner & Murray, 1981; Legner & Pelsue, 1980, 1983). As Legner and Sjogren (1984) indicate, this is a unique example of persistent biological control and is probably only applicable for relatively

sophisticated irrigation systems where a permanent water supply is assured, and water conditions are suitable to support the fish (Legner *et al.,* 1980b). There is a threefold advantage in the use of these fish: (1) they clear vegetation to keep waterways open, (2) they control mosquitoes, and (3) they grow large enough to be caught for human consumption. Some sophistication is necessary when stocking these cichlids for aquatic weed control, which is often not understood by irrigation district personnel (Hauser *et al.,* 1976, 1977; Legner, 1978b). Otherwise competitive displacement may eliminate *T. zillii,* the most efficient weedeating species (Legner, 1986).

Household storage of water in open containers has frequently been the cause for outbreaks of human disease transmitted by *Aedes aegypti* (Linnaeus) in less developed parts of the world. While conducting *A. aegypti* surveys in Malaysia during the mid-1960s, one of us (Garcia) observed what were apparently *P. reticulata* guppies being utilized by town residents for the control of mosquitoes in bath and drinking water storage containers. The origin of this control technique was not clear, but it appeared to be a custom brought to the area by Chinese immigrants. Not all residents used fish, but those that did had no breeding populations of *A. aegypti.*

Neng (1986) reported on the use of catfish of the *Claris* sp. for the control of *A. aegypti* in water storage tanks in coastal villages of southern China. This fish was considered appropriate because it was indigenous, edible, consumed large numbers of mosquito larvae; had a high tolerance for "adverse conditions"; and could be obtained from the local market. One fish was placed in each water source and later checked for its presence by larval survey teams about every 10 to 15 days. If fish were not found on inspection, the occupant was told to replace the fish or be fined. The investigation was conducted from 1981 to 1985, and surveys over this period showed a sharp initial reduction in *A. aegypti* followed by a low occurrence of the mosquito over the 4-year study period. Outbreaks of dengue were observed in neighboring provinces during this period, but not in the fishing villages under observation. The cost of the program was estimated to be about 1/15th that of indoor house spraying (Neng, 1986).

Alio *et al.* (1985) described another use of a local species of fish for the control of a malaria vector similar to the method reported by Kligler (1930). *Oreochromis* sp., a tilapine, was introduced into man-made water catchment basins called "barkits" in this semiarid region of northern Somalia. These small scattered impoundments serve as the only sources of water during the dry season for the large pastoral population of the area. *Anopheles arabiensis* Patton, the vector of malaria in that area, is essentially restricted to these sites. Release of fish into the "barkits" dramatically reduced both the vector and nonvector populations of mosquitoes rather quickly. Treatment of the human population with antimalarial drugs during the initial phase of this 2-year study, combined with the lower vector population, reduced the transmission rate of malaria to insignificance over a 21-month period whereas the control villages remained above 10%. Alio *et al.* (1985) commented that the added benefits of reduced vegetation and insects in the water sources were also recognized by the local population. This resulted in community cooperation and was expected to further benefit the control strategy by providing assistance in fish distribution and maintenance as the program expanded to other areas.

The last two examples involve the use of indigenous over exotic fish, where feasible, in vector control programs. There are other examples where native fish have been used in specialized circumstances or are being evaluated for such use (Kligler, 1930; Legner *et al.,* 1974, 1975a; Menon & Rajagopalan, 1978; Walters & Legner, 1980; Ataur-Rahim, 1981; Luh, 1981; Koldenkova *et al.,* 1988; García *et al.,* 1991; Lounibos *et al.,* 1992; Nelson & Keenan, 1992; Yadav *et al.,* 1992; Fletcher *et al.,* 1993). Lloyd (1987) argued that only indigenous fish should be employed for mosquito control because of the environmental disruption induced by exotics such as *G. affinis.* However, he suggested that native fish should be analyzed carefully for prey selectivity, reproductive potential, and effectiveness in suppressing pest populations before attempting their use. Lloyd (1987) also pointed out that a multidisciplinary approach involving fishery biologists and entomologists should be employed when developing indigenous fish for mosquito control. However, in California where native pup fish in the genus *Cyprinodon* may afford a greater potential for mosquito control under a wider range of environmental stresses than afforded by *Gambusia* (Walters & Legner, 1980), the California Department of Fish and Game discourages their use on the basis that unknown harmful effects might result to other indigenous fish. There is also the concern that certain rare species of *Cyprinodon* might be lost through hybridization.

Perhaps China's example of a multipurpose use of native fish for mosquito control and human protein source is the most resourceful strategy. This application for mosquito control is not new. Kligler (1930) used a tilapine fish to control *Anopheles* sp. in citrus irrigation systems in old Palestine, where farmers cared for the fish, consuming the larger ones. According to Luh (1981), the culture of edible fish for the purpose of mosquito control and human food is now widely encouraged in China. The old Chinese peasant custom of raising edible fish in rice fields has received greater attention in recent times because of the benefits made possible through this practice. The common carp, *Cyprinus carpio* Linnaeus, and the grass carp, *Ctenopharygodon idella* Valenciennes, are most commonly used. Fish are released as fry at the time rice seedlings are planted. Fields are specially prepared with a central "fish pit" and radiating ditches for refuge when water levels are low.

Pisciculture in rice fields, as noted by Luh (1981), has three major benefits: (1) there is a significant reduction in culicine and to a lesser extent anopheline larvae, (2) fish are harvested as food, and (3) rice yields are increased apparently by a reduction in competitors and possibly by fertilization of the plants by fish excreta.

Another group of fish, the so-called "instant" or annual fish (Cyprinodontidae), which are native to South America and Africa, have been considered as possible biological control agents for mosquitoes (Vanderplank, 1941, 1967; Hildemann & Wolford, 1963; Bay, 1965, 1972; Markofsky & Matias, 1979). The relatively drought-resistant eggs of these cyprinodontids, which allow them to utilize temporary water sources as habitat, would seem to make them ideal candidates for mosquito control. There is also some evidence that they do impact mosquito populations in native areas (Vanderplank, 1941; Hildemann & Wolford, 1963; Markofsky & Matias, 1979). Research on the biology and ecology of several species has been conducted; however, there are no published accounts, to our knowledge, on the successful use of these fish in field situations. In California, the South American species *Cynolebias nigripinnis* Regan and *C. bellottii* (Steindachner) survived the summer in rice fields, but no reproduction was observed over a 3-year period (Coykendall, 1980). It was speculated that they may play a future role in California's mosquito control program in temporary pools and possibly in rice fields. *Cynolebias bellottii* was observed to reproduce repeatedly and to persist in small, intermittently dried ponds in Riverside, California, for 11 consecutive years, 1968 to 1979 (E. F. Legner, personal communication, 1999). Four drying and flooding operations over 2 months were required to eliminate this species from ponds that were to be used for native fish studies (Walters & Legner, 1980). It seems logical, given the biological capability of surviving an annual dry period, that these fish could be successfully integrated into mosquito control programs, especially in newly created water sources in geographic areas where they naturally occur (Vaz-Ferreira *et al.,* 1963; Anonymous, 1981; Gerberich, 1985).

Arthropods

Numerous species of predatory arthropods have been observed preying on mosquitoes, and in some cases, are believed to be important in controlling mosquito populations (James, 1964; Service, 1977; Collins & Washino, 1979; McDonald & Buchanan, 1981). However, among the several hundred predatory species observed, only a few have been used in a manipulative way to control mosquitoes. Dragonflies, sometimes referred to as mosquito hawks, were one of the first arthropods to be examined before the turn of the century. Difficulties in colonization, production, and handling have restricted their use to experimental observation. It is unlikely that they will be used extensively (Lamborn, 1890; Beesley, 1974; El Rayah, 1975; Riviere *et al.,* 1987a).

There are a few cases where the difficulties associated with the manipulative use of arthropods have been at least partly overcome. More than 50 years ago, in a classic use of biological control, the mosquito *Toxorhynchites,* whose larvae are predators of other mosquitoes, was released on several Pacific Islands in an effort to control natural and artificial container breeding mosquitoes such as *A. aegypti* and *A. albopictus* (Skuse) (Paine, 1934; Bonnet & Hu, 1951; Peterson, 1956). The releases were not considered successful; however, the mosquitoes did establish in some areas (Steffan, 1975). Several reasons given to explain why these releases failed were low egg production, lack of synchrony between predator and prey life cycles, and selection of only a relatively small number of prey breeding sites (Muspratt, 1951; Nakagawa, 1963; Trpis, 1973; Bay, 1974; Riviere, 1985).

Although not apparently a suitable predator in the classical sense, there is still interest in the use of various *Toxorhynchites* spp. for inundative release (Gerberg & Visser, 1978). Trpis (1981), working with *T. brevipalpis* (Theobald), showed that the high daily consumption rate and long survival of the larvae without prey made it a prime candidate for biological control use. Observations on adult females indicated a 50% survivorship over a 10-week period with a relatively high oviposition rate per female. All these preceding attributes suggest that this species would be useful for inundative release programs against container breeding mosquitoes. Studies by Focks *et al.* (1979) in Florida, working with *T. r. rutilis* Coquillett, showed that this species had a high success rate in artificial breeding containers. In a 12.6-ha residential area, about 70% of the available oviposition sites were located over a 14-day period by two releases of 175 females. Mass-culturing techniques have been developed for this species and *T. amboinensis* (Doleschall) (Focks & Boston, 1979; Riviere *et al.,* 1987b).

Focks *et al.* (1986), working with *T. amboinensis,* reported that release of 100 females per block for several weeks, combined with ultra low volume (ULV) application of malathion, reduced *A. aegypti* populations by about 96% in a residential area of New Orleans. The *Toxorhynchites* releases and not the insecticide treatment apparently accounted for most of the reduction. These workers noted that the procedure could be further refined by reducing both the number of predators and malathion applications without lowering efficacy. Mosquitoes such as *A. aegypti* and *A. albopictus,* which breed in and whose eggs are dispersed via artificial containers, pose major health hazards as vectors of human diseases throughout much of the warmer climates of the world. The massive quantities of containerized products and rubber tires that are discarded without care or are stockpiled, have given these mosquito species a

tremendous ecological advantage. The establishment and extensive spread of *A. albopictus* in the United States underline this point (Sprenger & Wuithironyagool, 1986). The apparent inability of governments to appropriately control disposal of these containers and difficulties in location once they are discarded makes inundative releases of *Toxorhynchites,* either alone or in combination with other control tactics, a much more plausible approach (Focks *et al.,* 1986; Riviere *et al.,* 1987a; Miyagi *et al.,* 1992; Tikasingh, 1992; Tikasingh & Eustace, 1992; Toma & Miyagi, 1992).

Notonectids are voracious predators of mosquito larvae under experimental conditions (Ellis & Borden, 1970; Garcia *et al.,* 1974; Hazelrig, 1974). *Notonecta undulata* Say and *N. unifasciata* Guerin have been colonized in the laboratory. In addition, the collection of large numbers of eggs, nymphs, and adults is feasible from such breeding sites as sewage oxidation ponds (Ellis & Borden, 1969; Garcia, 1973; Sjogren & Legner, 1974; Hazelrig, 1975; Miura, 1986). Some studies have been conducted on storage of eggs at low temperatures; however, viability decreased rapidly with time (Sjogren & Legner, 1989). At present, the most feasible use of these predators appears to lie in the recovery of eggs from wild populations on artificial oviposition materials and their redistribution to mosquito breeding sites. Such investigations were conducted in central California rice fields by Miura (1986). Floating vegetation such as algal mats and sometimes duckweed (*Lemna* spp.) form protective refugia for mosquito larvae; and consequently, populations of mosquitoes can be high in the presence of notonectids (Garcia *et al.,* 1974). It appears that colonization and mass production costs, coupled with the logistics of distribution, handling, and timing of release at the appropriate breeding site, are almost insurmountable problems for routine use of notonectids in mosquito control.

In addition to insect predators, several crustaceans feed on mosquito larvae. Among these are the tadpole shrimp, *Triops longicaudatus* (LeConte), and several copepod species. Mulla *et al.* (1986) and Tietze and Mulla (1987, 1991), investigating the tadpole shrimp, showed that it was an effective predator under laboratory conditions and speculated that it may play an important role in the field against floodwater *Aedes* and *Psorophora* species in southern California. Drought resistance in predator eggs is an appealing attribute for egg production, storage, and manipulation in field situations against these mosquitoes. However, synchrony in hatch and development between the predator and the prey is crucial if this is to be a successful biological control agent for the rapidly developing *Aedes* and *Psorophora* spp. In addition, the tadpole shrimp is considered an important pest in commercial rice fields.

Miura and Takahashi (1985) reported that *Cyclops vernalis* Fisher was an effective predator on early-instar *Culex tarsalis* larvae in the laboratory. These workers speculated that copepods could have an important role in suppressing mosquito populations in rice fields because of their feeding behavior and abundance.

Another crustacean with promise for more extensive application is cyclopoid predator *Mesocyclops aspericornis* Daday (Riviere *et al.,* 1987b; Lardeux *et al.,* 1992). Work by Riviere and colleagues has indicated reductions of *A. aegypti* and *A. polynesiensis* Marks by more than 90% after inoculative release of this organism into artificial containers, wells, tree holes, and land crab burrows. Although not able to withstand desiccation, this small predator has persisted almost 2 1/2 years in crab holes and up to 5 years in wells, tires, and tree holes under subtropical conditions. This species can be mass-produced, but its occurrence in large numbers in local water sources allows for the inexpensive and widespread application to mosquito breeding sites in Polynesia (Riviere *et al.,* 1987a,b). The species is also very tolerant of salinities greater than 50 parts per thousand. The benthic feeding behavior of *Mesocyclops* makes it an effective predator of the bottom-foraging *Aedes,* but limits effectiveness against surface-foraging mosquitoes. Riviere *et al.* (1987a, 1987b) believed that the effectiveness against *Aedes* is due to a combination of predation and competition for food. Perhaps the greatest utility of this *Mesocyclops* will lie in the control of crab hole breeding species, such as *A. polynesiensis* in the South Pacific. Further investigations may uncover additional cyclopods that can impact other mosquito species (Kay *et al.,* 1992).

The most important nonarthropod invertebrate predators to draw attention for mosquito control are the turbellarian flatworms and a coelenterate. Several flatworm species have been reported to be excellent predators of mosquito larvae in a variety of aquatic habitats (Legner & Medved, 1974; Yu & Legner, 1976; Legner, 1977, 1979; Collins & Washino, 1978; Case & Washino, 1979; Ali & Mulla, 1983; George *et al.,* 1983). Several biological and ecological attributes of flatworms would seem to make them ideal candidates for manipulative use: these flatworms have ease of mass production, overwintering embryos, effective predatory behavior in shallow waters with emergent vegetation, on-site exponential reproduction following inoculation (Medved & Legner, 1974; Tsai & Legner, 1977; Legner & Tsai, 1978; Legner, 1979), and tolerance to environmental contaminants (Levy & Miller, 1978; Nelson, 1979).

Collins and Washino (1978) and Case and Washino (1979) hypothesized that flatworms, particularly *Mesostoma,* may play an important role in the natural regulation of mosquitoes in some California rice fields because of their densities and their predatory attack on mosquito larvae in sentinel cages. Preliminary analysis showed a significant negative correlation between the presence of flatworms and population levels of *C. tarsalis* and *Anopheles freeborni*

(Case & Washino, 1979). However, these workers cautioned that an alternative hypothesis related to the ecology of these species may have accounted for the correlations. Later investigations by Palchick and Washino (1984) were not able to confirm the correlations between flatworms and mosquito populations. However, the enormity of the problem associated with sampling in California rice fields, coupled with the complexity of the prey and predator interactions, makes further studies necessary before the role of flatworms in rice fields can be clearly established.

The important attributes for manipulative use of flatworms mentioned earlier raise the question of why they have not been developed further for use in mosquito control. Perhaps the contemporary development of *Bacillus thuringiensis* var. *israelensis* (B.t.i.) DeBarjac (H-14)—a highly selective, easily applied, microbial insecticide—may have been at least partially responsible for slowing further work and development of these predators. Their mass culture must be continuous and demands skilled technical assistants (Legner & Tsai, 1978). Their persistence in field habitats may also depend on the presence of other organisms such as ostracods that can be utilized for food during low mosquito abundance (Legner *et al.,* 1976).

The coelenterates, like the flatworms, showed great promise for further development and use in selected breeding habitats. *Chlorohydra viridissima* (Pallas) is efficient in suppressing culicine larvae in ponds with dense vegetation, and this species also can be mass-produced (Lenhoff & Brown, 1970; Yu *et al.,* 1974a, 1974b, 1975). However, like the flatworms, work on these predators has waned, perhaps for reasons similar to those speculated for the flatworms. Microbial pesticides can be employed over an extensive range of different mosquito-breeding habitats.

Fungi

The most promising fungal pathogen is a highly selective and environmentally safe oomycete, *Lagenidium giganteum* Couch. First tested for its pathogenicity to mosquitoes in the field by McCray *et al.* (1973), it has been applied by aircraft to rice fields (Kerwin & Washino, 1987). *Lagenidium* develops asexually and sexually in mosquito larvae and is capable of recycling in standing bodies of water. This creates the potential for prolonged infection in overlapping generations of mosquitoes. *Lagenidium* may also remain dormant after the water source has dried up and then may become active again when water returns. The sexually produced oospore offers the most promising stage for commercial production because of its resistance to desiccation and long-term stability. However, problems in production and activation of the oospores still remain (Axtell *et al.,* 1982; Merriam & Axtell, 1982a, 1982b, 1983; Jaronski & Axtell, 1983a, 1983b; Jaronski *et al.,* 1983; Kerwin *et al.,* 1986; Kerwin & Washino, 1987). Field trials with the asexual oospore and the sexual zoospore indicate that this mosquito pathogen is near the goal of practical utilization. Kerwin *et al.* (1986) report that the asynchronous germination of the oospore is of particular advantage in breeding sources where larval populations of mosquitoes are relatively low, but recruitment of mosquitoes is continuous due to successive and overlapping generations, as in California rice fields. The germination of oospores over several months provides long-term control for these continuous low-level populations. In addition, the asexual zoospores arising from the oospore-infected mosquito are available every 2 to 3 days to respond in a density-dependent manner to suppress any resurging mosquito population. This stage survives about 48 hours after emerging from the infected host.

Kerwin *et al.* (1986) indicate that laboratory fermentation production of the asexual stage of *Lagenidium* for controlling mosquitoes in the field is approaching the development requirements and costs for the production of B.t.i. A distinct advantage of this pathogen over B.t.i. is its potential to recycle through successive host generations. The disadvantage of the asexual stage is that it is relatively fragile, cannot be dried, and has a maximum storage life of only 8 weeks (Kerwin & Washino, 1987). Thus, the focus of attention for commercial production by Kerwin and colleagues is on the oospore, which is resistant to desiccation and can be easily stored. Axtell and Guzman (1987) have encapsulated both the sexual and asexual stages in calcium alginate and have reported activity against mosquito larvae after storage for up to 35 and 75 days, respectively. Further refinement in techniques of production and encapsulation might make this approach a viable option for future commercial production and application.

Limitations on the use of this pathogen include intolerance to polluted water, salinity, and other environmental factors (Jaronski & Axtell, 1982; Lord & Roberts, 1985; Kerwin & Washino, 1987). However, there are numerous mosquito breeding sources where these limitations do not exist; therefore, we should expect to see this selective and persistent pathogen available for routine mosquito control in the future.

The fungus *Culicinomyces clavosporus* Couch, Romney & Rao, first isolated from laboratory mosquito colonies and later from field habitats, has been under research and development for more than a decade (Sweeney *et al.,* 1973; Couch *et al.,* 1974; Russel *et al.,* 1979; Frances *et al.,* 1985). The fungus is active against a wide range of mosquito species and also causes infections in other aquatic Diptera (Knight, 1980; Sweeney, 1981). The ease of production with relatively inexpensive media in fermentation tanks is an extremely desirable trait. However, problems in storage must be overcome if this fungus is to be widely used. Perhaps a drying process, now being investigated,

will solve storage requirements (Sweeney, 1987). Although the fungus has high infection rates in field trials, dosage rates have been high and appreciable persistence at the site has not been demonstrated (Lacey & Undeen, 1986; Sweeney, 1987). For a more thorough review of this pathogen, see Sweeney *et al.* (1973), Lacey and Undeen (1986), and Sweeney (1983, 1987).

Various species of *Coelomomyces* have been studied over the last two decades for use in mosquito control. Natural epizootics with infection rates in excess of 90% have been recorded. These fungi persist in certain habitats for long periods; however, factors triggering outbreaks in these situations are not well understood (Chapman, 1974). Some field testing has been done, but results have been highly variable (Federici, 1981). In general, difficulties associated with the complex life cycle of these fungi have encumbered research on them. Federici (1981) and Lacey and Undeen (1986) have reviewed the potential of these fungi for mosquito control.

Nematodes

Among the various nematodes pathogenic for mosquitoes, *Romanomermis culicivorax* Ross & Smith has received the most attention (Petersen & Willis, 1970, 1972a, 1972b, 1975; Brown & Platzer, 1977; Brown *et al.,* 1977; Poinar, 1979; Petersen, 1980a, 1980b; Brown-Westerdahl *et al.,* 1982; Kerwin & Washino, 1984). This mermithid, which is active against a wide range of mosquito species, has been mass-produced (Petersen & Willis, 1972a) and has been utilized in a number of field trials. The nematode was commercially produced and sold under the name Skeeter Doom®; however, according to Service (1983), eggs showed reduced viability in transport and the product currently is no longer sold. However, the nematode's ability to recycle through multigenerations of mosquitoes and overwinter in various habitats, including drained, harvested, stubble-burned, cultivated, and replanted rice fields, are strong attributes favoring its further research and development for biological control (Petersen & Willis, 1975; Brown-Westerdahl *et al.,* 1982). Several field applications have produced good results and have included both the preparasitic stage and post-parasitic stages, with the former being more applicable for the "quick kill" and the latter being more long-term for continuous control such as in California rice fields (Petersen *et al.,* 1978a, 1978b; Levy *et al.,* 1979; Brown-Westerdahl *et al.,* 1982). Some drawbacks to its widespread use include intolerance to low levels of salinity, polluted water, and low oxygen levels; predation by aquatic organisms; and potential for development of resistance by the host (Petersen & Willis, 1970; Brown & Platzer, 1977; Brown *et al.,* 1977; Petersen, 1978; Brown-Westerdahl, 1982). However, these environmental problems are not generally an issue for anopheline control.

For control of these species the cost of *in vivo* mass production clearly stands as the major drawback for this pathogen. Perhaps its most plausible use will be in specialized habitats integrated with other control strategies (Brown-Westerdahl *et al.,* 1982).

Bacteria

The spore-forming bacterial pathogen B.t.i. (H-14) was isolated by Goldberg and Margalit (1977) and the produced toxin has been demonstrated by numerous studies to be an effective and environmentally sound microbial insecticide against mosquitoes and blackflies. Its high degree of specificity and toxicity, coupled with its relative ease of production, has made it the most widely used commercial product to date for mosquito and blackfly control. Several formulations are currently available from commercial firms throughout the world. Its efficacy under different environmental conditions and problems associated with its use have been reviewed by Garcia and Sweeney (1986), Lacey and Undeen (1986), Garcia (1987), de Barjac and Sutherland (1990), Priest (1992), and Becker and Margalit (1993).

Another spore-forming bacterium, *Bacillus sphaericus* Neide, has also shown great promise as a larvicide against certain mosquito species (Mulla *et al.,* 1984). In general, several strains of this pathogen show a much higher degree of toxic variability among species of mosquitoes. *Culex* spp. appear to be highly susceptible, whereas other species such as *Aedes aegypti* are highly refractory. Unlike the ephemeral larvacidal activity of B.t.i. toxin, some strains of *B. sphaericus* have shown persistence in certain habitats (Des Rochers & Garcia, 1984; Matanmi *et al.,* 1990). For further detail, see reviews by Lacey and Undeen (1986), de Barjac and Sutherland (1990), and Priest (1992).

Protozoa

A large number of protozoa have been isolated from mosquitoes and other medically important arthropods (Roberts *et al.,* 1983; Lacey & Undeen, 1986). Of this assemblage, the microsporidians have been studied rather intensively. Because of their complex life cycle and the *in vivo* production methods necessary for maintaining them, research on their practical utility has been limited. However, as Lacey and Undeen (1986) point out, if more information is developed on their life cycle, it may be found that they could play a role in suppressing mosquitoes through inoculative and augmentative releases in certain habitats.

Among other protozoa that show promise is endoparasitic ciliate *Lambornella clarki* Corliss & Coats, a natural pathogen of the tree hole mosquito, *Aedes sierrensis* (Ludlow). This pathogen has received considerable attention over the last few years as a potential biological control agent for container-breeding mosquitoes (Egeter *et al.,* 1986; Washburn & Anderson, 1986). Desiccation-resistant cysts allow persistence of the ciliate from one year to the next. Currently, *in vitro* production methods are being developed and small field trials are being initiated to determine its efficacy and practicability for field use (Anderson *et al.,* 1986a, 1986b).

Viruses

Numerous pathogenic viruses have been isolated from mosquitoes and blackflies. However, to date none look promising for practical use in control (Lacey & Undeen, 1986).

SYNANTHROPIC DIPTERA

These flies, the most important of which are muscoid species, can be defined broadly as those most closely associated with human activities. Breeding habitats vary from the organic wastes of urban and rural settlements to those provided by various agricultural practices, particularly ones related to the management and care of domestic and range animals. Their degree of relationship to humans varies considerably with the ecology and behavior of the fly involved. Some are more often found inside dwellings (endophilic) while others remain mostly outdoors (exophilic). The discussion that follows separates these flies by their general endophilic and exophilic habits. This discussion is restricted to brief comments because the potential for biological control of these flies has been reviewed by Legner *et al.* (1974), Bay *et al.* (1976), and Legner (1986).

Endophilic Flies

Povolny (1971) describes these flies as primarily dependent on human and domestic animal wastes. *Musca domestica* Linnaeus is by far the best known example; however, some *Drosophila* and *Psychoda* spp. also fall into this category. Certain *Fannia* spp. are more on the periphery but also are included here.

The common housefly, *M. domestica,* has been a constant associate of humans over much of our modern history. Attempts to control its populations by biological means have been extensive and on occasion successful in special situations. More frequently, they have failed to reduce numbers to acceptable levels. It should be emphasized that control of *M. domestica* populations, as well as most other endophilic flies pestiferous to humans, would be largely unnecessary if waste products produced by human activities could be appropriately managed. Because this is not the case, efforts toward the biological control of these species have continued.

Starting around the turn of this century to approximately 1968, biological control of these flies was attempted by the introduction of a broad range of different natural enemies into areas where the flies presented problems. The Pacific Islands were a focus of much attention with the introduction of dung beetles, several parasitoids, and predators during this period. It was believed that the accidental introduction of the ant *Pheidole megacephala* Fabricius, combined with the introduction of the coprophagous dung beetle, *Hister chinensis* Quensel, caused significant fly reductions on the islands of Fiji and Samoa (Simmonds, 1958). The Hawaiian islands had 16 introductions from 1909 to 1967, of which 12 became established. However, the exact role of these natural enemies in overall regulation of flies on the islands is still not well understood (Legner *et al.,* 1974; Legner, 1978c).

Rodriguez and Riehl (1962), in California, used the novel and successful approach of chicken cockerels as direct predators of fly larvae in chicken and rabbit manure. However, this technique is utilized very little today because of the threat that roving birds pose to the spread of avian pathogens.

Research over the last two decades has been centered on the more highly destructive parasitoid and predatory species. Examples, such as the encyrtid *Tachinaephagus zealandicus* Ashmead, five species of the pteromalid genus *Muscidifurax,* and *Spalangia* sp., were evaluated for their capabilities of attacking dipterous larvae and pupae in various breeding sources. They are believed to be capable of successful fly suppression if the right species and strains are applied in the right locality (Legner & Brydon, 1966; Legner & Dietrick, 1972, 1974; Morgan *et al.,* 1975, 1977; Olton & Legner, 1975; Pickens *et al.,* 1975; Morgan & Patterson, 1977; Rutz & Axtell, 1979; Gold & Dahlsten, 1981; Propp & Morgan, 1985; Axtell & Rutz, 1986; Legner, 1988b; Mandeville *et al.,* 1988; Pawson & Petersen, 1988). Other approaches have included the use of pathogens and predatory mites, and inundative releases of parasitoids and predators (Ripa, 1986). Although partially successful, none of these strategies has become the sole method for fly control, and the wrong choice of a parasitoid strain may have detrimental results (Legner, 1988b). Instead, the focus is on integrated controls including other methods such as cultural, adult baiting, and aerosol treatments with short residual insecticides. However, it is generally agreed that existing predatory complexes exert great influences on fly densities (Legner *et al.,* 1975b, 1980a; Geden, 1984; Geden *et al.,* 1987, 1988; Geden & Axtell, 1988), that many biological control agents of endophilous flies have not been thoroughly surveyed, and that their potential has not been adequately assessed (Mullens, 1986; Mullens *et al.,* 1986).

Exophilic Flies

These species include flies that persist in nature in the absence of humans, but whose populations can increase dramatically as a result of certain human activities such as providing more breeding habitat. They include several species in the genera *Calliphora, Hippelates, Musca, Muscina, Phaenicia, Stomoxys,* and others.

Some success has been recorded with the use of natural enemies against the calliphorid species in California and Hawaii, but attempts elsewhere in the world have not been effective (Bay *et al.,* 1976). The braconid parasitoid *Alysia ridibunda* Say, indigenous to parts of the United States, was released into an area of Texas new to its range and did parasitize the blowflies, *Phaenicia sericata* (Meigen), and a *Sarcophaga* species. However, the parasitoid did not maintain control and became rare within a couple of years (Lindquist, 1940).

The gregarious parasitoid. *T. zealandicus,* may have considerable potential for biological control of exophilic flies (Olton & Legner, 1975). The range of habitats utilized by this natural enemy is considered unparalleled by any other fly parasitoid. However, extensive field work with this genus has not been given much attention. However, one species, *T. stomoxcida* Subba-Rao, provides overall permanent reductions of *Stomoxys* in Mauritius (Greathead & Monty, 1982).

The complex of problems that confronts field programs in biological control of exophilic flies has clearly had a dampening effect on research in this area. The unforeseen problems associated with attempts to biologically control the eye gnat, *Hippelates collusor* (Townsend), in California exemplify those problems. In the early 1960s, a concerted effort was mounted to control this gnat with the use of both indigenous and exotic parasitoids in orchards in southern California. About a dozen species and strains were evaluated for several years. Some of the exotics became established, but eye gnat reductions were obvious only where cultivation practices were curtailed (Legner *et al.,* 1966; Legner, 1970). Cultivation of the orchards buried the larvae and pupae of the eye gnat below the search zone of the parasitoids and cultivation also removed vegetation that offered the parasitoids protection and possibly nutrients (Legner, 1968; Legner & Olton, 1969; Legner & Bay, 1970). Buried eye gnats emerged from several centimeters below the soil surface and thus continued to pose a serious problem (Bay *et al.,* 1976).

The discovery of a group of genes, tentatively called "wary genes," in parasitoids of synanthropic Diptera affords greater opportunities for biological control (Legner, 1987, 1988a, 1989, 1993). Inheritance of quantitative behavior associated with gregarious oviposition and fecundity in the South American parasitoid *Muscidifurax raptorellus* Kogan & Legner is accompanied by unique extranuclear influences that cause changes in the oviposition and larval survival phenotypes of females prior to the production of their progeny (Legner, 1989). Males can change a female's fecundity phenotype when mating, by transferring an unknown substance. Some genes in the female apparently have the phenotypic plasticity to change expression under the influence of substances in the male seminal fluid. The intensity of this response depends on the genetic composition of the male and female. Full expression occurs in the F_1 virgin female (Legner, 1989). The mated female receives a message from the male after mating that expresses *his* genome for the presence or absence of polygenes governing quantitative behavior, such as fecundity. The discovery of this behavior in *M. raptorellus* has opened questions into the stability of polygenic loci (Legner, 1993). The ability of the male substance to apparently switch loci on or off in the female suggests active and inactive states for such loci. Polygenic loci generally are occupied by genes coded for a fixed kind of expression.

Greater importance may be placed on liberated males during mass release strategies that seek seasonally to accelerate and increase the magnitude of parasitism, because it may be possible to convey directly to unmated females already resident in the environment certain desirable strain characteristics. In the process of hybridization, wary genes may serve to quicken the pace of evolution by allowing natural selection to begin to act in the parental generation (Legner, 1987, 1988a).

Tabanidae, or horseflies, although widespread and on occasion serious pests and vectors of disease to livestock, have not received much attention. Only one successful inundative release of egg parasitoid *Phanurus emersoni* Girault has been recorded (Parman, 1928). Apparently, this effort was precipitated by a severe outbreak of anthrax at the time and because this disease diminished and other control tactics are available, interest in their biological control has not been fostered.

Of the flies associated with cattle droppings, symbovine flies (Povolny, 1971) have received the most attention for biological control over the last two decades. The primary targets for control have been the bush fly, *Musca vetustissima* Walker; the horn fly, *Haematobia irritans* (Linnaeus); and the face fly, *Musca autumnalis* DeGeer (Wallace & Tyndale-Biscoe, 1983; Ridsdill-Smith *et al.*, 1986; Ridsdill-Smith & Hayles, 1987). The primary emphasis of control has been on habitat destruction through the use of introduced dung-burying scarab beetles. Biological control through dung destruction has been reviewed by Legner (1986). Although the importation and introduction of dung beetles have clearly aided agriculture by reducing operating costs and increasing grazing areas through dung removal, they have not had a great impact on the densities of flies in any area. Because there are no practical nonbiological control methods to reduce fly numbers and because the addition

of more scarabs may actually exacerbate the problem, we believe that the most logical direction for research is to intensify worldwide search for more effective natural enemies, especially predators and pathogens.

A number of pathogens have been isolated from various species of muscids and some studies have been conducted evaluating their role as control agents. For example, the exotoxin of B.t. Berliner has been reported to reduce fly production under certain conditions. However, only a few of these agents appear to show great promise for manipulative use (Daoust, 1983a, 1983b; Mullens, 1986; Mullens *et al.*, 1987a, 1987b, 1987c.)

SNAILS

Berg (1973), Bay *et al.* (1976), Garcia and Huffaker (1979), and McCullough (1981) have reviewed developments in biological control of mainly freshwater snails, especially as they relate to the transmission of trematode parasites of humans and their domestic animals. Discussion here is restricted to some pertinent points of those reviews and to some developments that have occurred since their completion.

Predators

Many general predators, including species of fish, frogs, birds, and certain aquatic insects, consume freshwater snails. Domestic ducks have been used with some success in China by herding them through rice fields to forage for food. However, of all these general predators, only certain tilapine fish have been given research consideration as possible biological control agents. Fish in the genera *Oreochromis, Sarotherodon,* and *Tilapia* feed directly on snails during various stages of their life cycle. This occurs primarily because the feeding behavior of these fish is frequently in the vegetation/detrital zone that is also utilized for feeding by snails. Larger adult species of *Oreochromis* and *Sarotherodon* feed directly on adult snails but this predation has not been observed for *Tilapia* adults. *Tilapia* adults only consume snails incidentally during their normal foraging on plant materials (Roberts & Sampson, 1987).

Possibly the greatest impact of these fish on snail populations is through competition for resources. Roberts and Sampson (1987) state that, in general, *Tilapia* spp. compete directly with the snails that feed on higher plants while *Oreochromis* competes with snails that feed on algae. In addition to competition for food, these fish alter the habitat, and therefore have a disruptive effect on the snails' life cycle (habitat modification); see Roberts & Sampson (1987) for details of culturing techniques and other biological and ecological attributes of the various tilapiine species.

Certain species of sciomyzid flies are probably the most host-specific predators of snails. Several hundred species have been described, of which the larvae depend on mollusks for food. Of six species released in Hawaii primarily against *Lymnaea ollula* Gould, the intermediate host of cattle liver flukes (*Fasciola* spp.), two—*Sepedon sauteri* Walker and *S. macropus* Hendel—were partially successful as indicated by a reduction in liver infections at slaughterhouses (Bay *et al.*, 1976; Garcia & Huffaker, 1979). Berg (1973) emphasized that because there are several hundred species in this family with a wide range of biological attributes, they offer great opportunity for matching a certain sciomyzid with the appropriate ecotype snail. Unfortunately, the scope of opportunities for use of these flies for snail control has not been given the attention it deserves.

Antagonists

Another approach for control of snails has been through interspecific competition. The large predatory snail, *Marisa cornuarietis* Linnaeus, has been evaluated rather extensively in Puerto Rico and has demonstrated effectiveness for control of *Biomphalaria glabrata* Say, the intermediate host of human schistosomiasis, in certain habitats, especially ponds. Suppression of *B. glabrata* by *Marisa* is primarily due to competitive feeding and to incidental predation on the immature stages of this snail (McCullough, 1981; Madsen, 1990). Elsewhere in the Caribbean, the thiarid snails *Melanoides tuberculata* (Müller) and *Thiara granifera* (Lamarck) have spread after introduction and have reduced *B. glabrata* populations in certain areas (Madsen, 1990).

In Africa, *M. cornuarietis* eliminates three species of pulmonate snails (*Biomphalaria* sp., *Bulinis* sp., and *Lymnaea* sp.) in a water impoundment in northern Tanzania. Prior to release of *M. cornuarietis*, the three pulmonate species, in addition to a melaniid snail, *Melanoides* sp., existed in large, thriving populations. Two years after the introduction, only *M. cornuarietis* and the melaniid snail remained, the latter in population densities similar to preintroduction levels (Nguma *et al.*, 1982; Madsen, 1990). No adverse environmental effect was recorded in this situation; however, the authors stressed that a careful examination of potential environmental risks should be made before introduction to a new area.

Another competitor snail, *Helisoma duryi* (Wetherby), has promise for the control of *B. glabrata*. Christie *et al.* (1981) working with the ram's horn snail, *H. duryi*, showed that it controlled *B. glabrata* in artificial outdoor drains on the Caribbean Island of St. Lucia. The elimination of *B. glabrata* may have been due to inhibition of reproduction by adults and possibly to increased mortality of immature snails. The time required for elimination was related to environmental temperature and the number of *H. duryi* ini-

tially released. In Africa, Madsen (1983) surveyed *H. duryi* as an introduced species in an irrigation scheme in northern Tanzania and found it restricted to just a few drains 10 years after it had been established in the area. He noted that its failure to spread may have been related to the routine molluscacide applications to the irrigation canal system.

Moens (1980, 1982) achieved successful biological control of *L. truncatula* Müller, an intermediate host of the trematode *Fasciola hepatica* Linneaus in watercress in Belgium, with the predatory snail *Zonitoides nitidus* Müller. Predation was related to temperature, soil moisture, and cover.

It is obvious from this review that the role of biological control of snails as intermediate hosts of human diseases is limited. McCullough (1981) believed it to be restricted to specific situations and not to have widespread applicability. We conclude that the natural-enemy component is to play a supportive role in almost all geographic areas where schistosomiasis and other snail-transmitted diseases exist, because any method that reduces transmission of a disease in a self-sustaining fashion is of major benefit to a community.

References

Ali, A., & Mulla, M. S. (1983). Evaluation of the planarian, *Dugesia dorotocephala,* as a predator of chironomid midges and mosquitoes in experimental ponds. Mosquito News, 43, 46–49.

Alio, A. Y., Isaq, A., & Delfini, L. F. (1985). Field trial on the impact of Oreochromis spilurus spilurus on malaria transmission in northern Somalia (WHO/VBC/ 85.910, pp. 18). Geneva: World Health Organization.

Anderson, J. R., Washburn, J. O., & Egerter, D. E. (1986a). Life cycle of the pathogen *Lambornella clarki* and its impact on *Aedes sierrensis*. Mosquito Control Research Annual Report, 1986 (pp. 20–21). University of California, Davis.

Anderson, J. R., Washburn, J. O., & Gross, M. E. (1986b). Mass production, storage and field release of *Lambornella clarki*, a pathogen of *Aedes sierrensis*. Mosquito Control Research Annual Report, 1986 (pp. 21–22). University of California, Davis.

Anonymous. (1981). Data sheet on Nothobranchius spp., N. guentheri and N. rachovi [as a predator of mosquito larvae] distribution, life cycle, biology, growth, reproduction, aging and behaviour (WHO/VBC/ 81.829; VBC/BCDS/81.16). Geneva: World Health Organization.

Ataur-Rahim, M. (1981). Observations on *Aphanius dispar* (Ruppell 1828), a mosquito larvivorous fish in Riyadh, Saudi Arabia. Annals of Tropical Medicine Parasitology, 75, 359–362.

Axtell, R. C., & Guzman, D. R. (1987). Encapsulation of the mosquito fungal pathogen *Lagenidium giganteum* (Oomycetes: Lagenidiales) in calcium alginate. Journal of American Mosquito Control Association, 3, 450–459.

Axtell, R. C., & Rutz, D. A. (1986). Role of parasites and predators as biological control agents in poultry production facilities. Miscellaneous Publications of the Entomological Society of America, 61, 88–100.

Axtell, R. C., Jaronski, S. T. & Merriam, T. L. (1982). Efficacy of the mosquito fungal pathogen, *Lagenidium giganteum* (Oomycetes: Lagenidiales). Proceedings of the California Mosquito Vector Control Association, 50, 41–42.

Bay, E. C. (1965). Preliminary findings concerning the adaptability of annual fishes to California mosquito habitats. Proceedings of the California Mosquito Control Association, 33, 29–30.

Bay, E. C. (1969). Fish predators. Proceedings of the California Mosquito Control Association, 37, 15–16.

Bay, E. C. (1972). Mosquitofish a controversial friend to mosquito control. Pest Control, 40, 32–33.

Bay, E. C. (1974). Predator-prey relationships among aquatic insects. Annual Review of Entomology, 19, 441–453.

Bay, E. C., Berg, C. O., Chapman, H. C., & Legner, E. F. (1976). Biological control of medical and veterinary pests. In C. B. Huffaker & P. S. Messenger (Eds.), Theory and practice of biological control (pp. 457–479). New York: Academic Press.

Becker, N., & Margalit, J. (1993). Use of *Bacillus thuringiensis* against mosquitoes and blackflies. In P. F. Entwisle, J. S. Cory, & M. J. Bailey (Eds.), Bacillus thuringiensis, an environmental biopesticide: Theory and practice (pp. 147–170). Chichester: John Wiley & Sons.

Beesley, C. (1974). Simulated predation of single-prey (*Culex peus*) and alternative prey (*Culex peus; Chironomus* sp. 51) by *Anax junius* Drury (Odonata: Aeschnidae). Proceedings of the California Mosquito Control Association, 42, 73–76.

Berg, C. O. (1973). Biological control of snail borne diseases: A review. Experimental Parasitology, 33, 318–330.

Blaustein, L. (1986). Green sunfish: Friend or foe of rice field mosquitoes? Proceedings of the California Mosquito Vector Control Association, 54, 90.

Bonnet, D. D., & Hu, A. M. K. (1951). The introduction of *Toxorhynchites brevipalpis* (Theobald) into the Territory of Hawaii. Proceedings of the Hawaiian Entomological Society, 14, 237–242.

Brown, B. J., & Platzer, E. G. (1977). The effects of temperature on the infectivity of *Romanomermis culicivorax*. Journal of Nematology, 9, 166–172.

Brown, B. J., Platzer, E. G., & Hughes, D. S. (1977). Field trials with the mermithid nematode *Romanomermis culicivorax* in California. Mosquito News, 37, 603–608.

Brown-Westerdahl, B., Washino, R. K., & Platzer, E. G. (1982). Successful establishment and subsequent recycling of *Romanomermis culicivorax* (Mermithidae: Nematoda) in a California rice field following post parasite applications. Journal of Medical Entomology, 19, 34–41.

Case, T. J., & Washino, R. K. (1979). Flatworm control of mosquito larvae in rice fields. Science, 206, 1412–1414.

Cech, J. J., Jr., & Linden, A. L. (1987). Comparative larvivorous performances of mosquitofish, *Gambusia affinis,* and juvenile Sacramento blackfish, *Orthodon microlepidotus,* in experimental paddies. Journal of American Mosquito Control Association, 3, 35–41.

Chapman, H. C. (1974). Biological control of mosquito larvae. Annual Review of Entomology, 19, 33–59.

Christie, J. D., Edward, J., Goolaman, K., James, B. O., Simm, J., & Dugat, P. S. (1981). Interactions between St. Lucian *Biomphalaria glabrata* and *Helisoma duryi,* a possible competitor snail in a semi-natural habitat. Acta Tropica, 38, 395–417.

Collins, F. H., & Washino, R. K. (1978). Microturbellarians as natural predators of mosquito larvae in northern California rice fields. Proceedings of the California Mosquito Vector Control Association, 46, 91.

Collins, F. H., & Washino, R. K. (1979). Factors affecting the density of *Culex tarsalis* and *Anopheles freeborni* in northern California rice fields, Proceedings of the California Mosquito Control Association, 47, 97–98.

Couch, J. N., Romney, S. V., & Rao, B.(1974). A new fungus which attacks mosquitoes and related Diptera. Mycologia, 66, 374–379.

Coykendall, R. L. (Ed.). (1980). Fishes in California mosquito control, (pp. 1–90). Sacramento: California Mosquito Control Association Press.

Daoust, R. A. (1983a). Pathogens of Tabanidae (horse flies). In D. W.

Roberts, R. A. Daoust, & S. P. Wraight (Eds.), Bibliography on pathogens of medically important arthropods (1981) (WHO, VBC/83.1, pp. 223–229). Geneva: World Health Organization.

Daoust, R. A. (1983b). Pathogens of Muscidae (muscid flies). In D. W. Roberts, R. A. Daoust, & S. P. Wraight (Eds.); Bibliography on pathogens of medically important arthropods (1981) (WHO, VBC/83.1, pp. 230–239). Geneva: World Health Organization.

Davey, R. B., & Meisch, M. V. (1977a). Dispersal of mosquitofish, *Gambusia affinis,* in Arkansas rice fields. Mosquito News, 37, 777–778.

Davey, R. B., & Meisch, M. V. (1977b). Control of dark rice-field mosquito larvae, *Psorophora columbiae* by mosquitofish, *Gambusia affinis* and green sunfish, *Lepomis cyanellus,* in Arkansas rice fields. Mosquito News, 37, 258–262.

de Barjac, H., & Sutherland, D. J. (1990). Bacterial control of mosquitos and blackflies: Biochemistry, genetics and application of *Bacillus thuringiensis* var. *israelensis* and *Bacillus sphaericus.* New Brunswick, NJ: Rutgers University Press.

Des Rochers, B., & Garcia, R. (1984). Evidence for persistence and recycling of *Bacillus sphaericus.* Mosquito News, 44, 160–173.

Downs, C. W., Beesley, C., Fontaine, R. E., & Cech, J. J., Jr. (1986). Operational mosquito production: Brood stock management. Proceedings of the California Mosquito Control Association, 54, 86–88.

Egeter, D. E., Anderson, J. R., & Washburn, J. O. (1986). Dispersal of the parasitic ciliate *Lambornella clarki:* Implications for ciliates in the biological control of mosquitoes. Proceedings of the National Academy of Sciences, USA, 83, 7335–7339.

Ellis, R. A., & Borden, J. H. (1969). Laboratory rearing of *Notonecta undulata* Say (Hemiptera: Notonectidae). Journal of the Entomological Society of British Columbia, 66, 51–53.

Ellis, R. A., & Borden, J. H. (1970). Predation by *Notonecta undulata* (Heteroptera: Notonectidae) on larvae of the yellow fever mosquito. Annals of the Entomological Society of America, 63, 963–973.

El Rayah, E. A. (1975). Dragonfly nymphs as active predators of mosquito larvae. Mosquito News, 35, 229–230.

Farley, D. G., & Caton, J. R. (1982). A preliminary report on the use of mosquito fish to control mosquitoes in an urban storm drain system. Proceedings of the California Mosquito Control Association, 50, 57–81.

Farley, D. G., & Younce, L. C. (1977). Effects of *Gambusia affinis* (Baird & Girard) on selected non-target organisms in Fresno county rice fields. Proceedings of the California Mosquito Control Association, 45, 87–94.

Federici, B. A. (1981). Mosquito control by the fungi *Culicinomyces, Lagenidium* and *Coelomomyces* In H. D. Burges (Ed.), Microbial control of pests and plant diseases 1970–1980 (pp. 555–572). London: Academic Press.

Fletcher, M., Teklehaimanot, A., Yemane, G., Kasahun, A., Kiane, G., & Beyene, Y. (1993). Prospects for the use of larvivorous fish for malaria control in Ethiopia: Search for indigenous species and evaluation of their capacity for mosquito larvae. Journal of Tropical Medicine and Hygiene, 96, 12–21.

Focks, D. A., & Boston, M. D. (1979). A quantified mass-rearing technique for *Toxorhynchites rutilus rutilus* Coquillett. Mosquito News, 39, 616–619.

Focks, D. A., Seawright, J. A., & Hall, D. W. (1979). Field survival, migration and ovipositional characteristics of laboratory-reared *Toxorhynchites rutilus rutilus* (Diptera: Culicidae). Journal of Medical Entomology, 16, 121–127.

Focks, D. A., Sacket, S. R., Klotter, K. O., Dame, D. A., & Carmichael, G. T. (1986). The integrated use of *Toxorhynchites amboinensis* and ground level ULV insecticide application to suppress *Aedes aegypti* (Diptera: Culicidae). Journal of Medical Entomology, 23, 513–519.

Frances, S. P., Leel, D. J., Russell, R. C., & Panter, C. (1985). Seasonal occurrence of the mosquito pathogenic fungus *Culicinomyces clavis-*

porus in a natural habitat. Journal of Australian Entomological Society, 24, 241–246.

García, A. I., Koldenkova, L., Santamarina, M. A., & Gonzalez, B. R. (1991). Introduction of the larvivorous fish *Poeciliareticulata* (Peters 1895) (Cyprinodontiformes: Poecillidae), a biological control agent of mosquitoes, to oxidation ponds and polluted ditches on the isla de la Juventud. Revista Cubana de Medicina Tropical, 43, 45–49.

Garcia, R. (1973). Strategy of mosquito control. In R. H. McCabe (Ed.), Integrated pest management—a curriculum report (pp. 43–47). Miami, FL: Associate Research Corporation.

Garcia, R. (1987). Strategies for the management of mosquito populations with *Bacillus thuringiensis* var. *israelensis* (H-14). In T. D. St. George, B. H. Kay, & J. Blok (Eds.), Proceedings, Fourth Symposium on Arbovirus Research in Australia (1986) (pp. 145–50). Brisbane: Q.I.M.R.

Garcia, R., & Huffaker, C. B. (1979). Ecosystem management for suppression of vectors of human malaria and schistosomiasis. Agro-Ecosystems, 5, 295–315.

Garcia, R., & Sweeney, A. W. (1986). The use of microbial pathogens for the control of mosquitoes. Agriculture Ecosystems and Environment, 15, 201–208.

Garcia, R., Voigt, W. G., & DesRochers, B. S. (1974). Studies of the predatory behavior of notonectids on mosquito larvae. Proceedings of the California Mosquito Control Association, 42, 67–69.

Geden, C. J. (1984). Population dynamics, spatial distribution, dispersal behavior and life history of the predaceous histerid, Carcinops pumilio (Erichson), with observations of other members of the poultry manure arthropod community. Unpublished doctoral dissertation, University of Massachusetts, Amherst.

Geden, C. J., & Axtell, R. C. (1988). Predation by *Carcinops pumilio* (Coleoptera: Histeridae) and *Macrocheles muscaedomesticae* (Acarina: Macrochelidae) on the housefly (Diptera: Muscidae): Functional response, effects of temperature, and availability of alternative prey. Environmental Entomology, 17, 739–744.

Geden, C. J., Stinner, R. F., & Axtell, R. C. (1988). Predation by predators of the house fly in poultry manure: Effects of predator density, feeding history, interspecific interference and field conditions. Environmental Entomology, 17(2), 320–329.

Geden, C. J., Stoffolano, J. G., Jr., & Elkinton, J. S. (1987). Prey-mediated dispersal behavior of Carcinops pumilio (Coleoptera: Histeridae). Environmental Entomology, 16, 415–419.

George, J. A., Nagy, B. A. L., & Stewart, J. W. (1983). Efficacy of *Dugesia tigrina* (Tricladida: Turbellaria) in reducing *Culex* numbers in both field and laboratory. Mosquito News, 43, 281–287.

Gerberg, E. J., & Visser, W. M. (1978). Preliminary field trial for the biological control of *Aedes aegypti* by means of *Toxorhynchites brevipalpis*, a predatory mosquito larva. Mosquito News, 38, 197–200.

Gerberich, J. B. (1985). Update of annotated bibliography of papers relating to the control of mosquitoes by the use of fish for the years 1965–1981 (WHO/VBC/917, pp. 1–33). Geneva: World Health Organization.

Gold, C. S., & Dalhsten, D. L. (1981). A new host record for *Tachinaephagus zealandicus* (Hymenoptera: Encyrtidae). Entomophaga, 26, 459–460.

Goldberg, L. H., & Margalit, J. (1977). A bacterial spore demonstrating rapid larvicidal activity against *Anopheles sergentii, Uranotaenia unguiculata, Culex univittatus, Aedes aegypti* and *Culex pipiens*. Mosquito News, 37, 355–358.

Greathead, D. J., & Monty, J. (1982). Biological control of stableflies (*Stomoxys* spp.): Results from Mauritius in relation to fly control in dispersed breeding sites. C.I.B.C. Biocontrol News and Information, 3(2), 105–109.

Hackett, L. W. (1937). Malaria in Europe. New York: Oxford University Press.

Hauser, W. J., Legner, E. F., & Robinson, F. E. (1977). Biological control of aquatic weeds by fish in irrigation channels (pp. 139–145). Proceedings, Water Management for Irrigation and Drainage. ASC/Reno, Nevada, July 20–22.

Hauser, W. J., Legner, E. F., Medved, R. A., & Platt, S. (1976). Tilapia—a management tool for biological control of aquatic weeds and insects. Bulletin of American Fisheries Society, 1, 15–16.

Hazelrig, J. E. (1974). Notonecta unifasciata as predators of mosquito larvae in simulated field habitats. Proceedings of the California Mosquito Control Association, 42, 60–65.

Hazelrig, J. E. (1975). Laboratory colonization and sexing of *Notonecta unifasciata* Guerin reared on *Culex peus* Speiser. Proceedings of the California Mosquito Control Association, 43, 142–144.

Hildemann, W. H., & Wolford, R. L. (1963). Annual fishes—promising species as biological control agents. Journal of Tropical Medicine, 66, 163–166.

Hoy, J. B., & Reed, D. E. (1970). Biological control of *Culex tarsalis* in a California rice field. Mosquito News, 30, 222–230.

Hoy, J. B., & Reed, D. E. (1971). The efficacy of mosquitofish for control of *Culex tarsalis* in California rice fields. Mosquito News, 31, 567–572.

Hoy, J. B., Kauffman, E. E., & O'Berg, A. G. (1972). A large-scale test of *Gambusia affinis* and chlorpyrifos for mosquito control. Mosquito News, 32, 161–171.

Hoy, J. B., O'Berg, A. G., & Kauffman, E. E. (1971). The mosquito fish as a biological control agent against *Culex tarsalis* and *Anopheles freeborni*. Mosquito News, 31, 146–152.

Hurlbert, S. H., & Mulla, M. S. (1981). Impacts of mosquitofish, *Gambusia affinis*, predation of plankton communities. Hydrobiologia, 83, 125–151.

Hurlbert, S. H., Zedler, J., & Fairbanks, D. (1972). Ecosystem alteration by mosquitofish (*Gambusia affinis*) predation. Science, 175, 639–641.

Inci, R., Yildirim, M., Bagci, N., & Inci, S. (1992). Biological control of mosquito larvae by mosquitofish (*Gambusia affinis*) in the Batman-Siirt Arva, Turkiye. Parazitoloji Dergisi, 16, 60–66.

James, H. G. (1964). Insect fauna associated with the rock pool mosquito *Aedes artropalpus* (Coq.). Mosquito News, 24, 325–329.

Jaronski, S. T., & Axtell, R. C. (1982). Effects of organic water pollution on the infectivity of the fungus *Lagenidium giganteum* (Oomycetes: Lagenidiales) for larvae of *Culex quinquefasciatus* (Diptera: Culicidae): Field and laboratory evaluation. Journal of Medical Entomology, 19, 255–262.

Jaronski, S. T., & Axtell, R. C. (1983a). Persistence of the mosquito fungal pathogen *Lagenidium giganteum* (Oomycetes: Lagenidiales) after introduction into natural habitats. Mosquito News, 43, 332–337.

Jaronski, S. T., & Axtell, R. C. (1983b). Effects of temperature on infection, growth and zoosporogenesis of *Lagenidium giganteum*, a fungal pathogen of mosquito larvae. Mosquito News, 43, 42–45.

Jaronski, S., Axtell, R. C., Fagan, S. M., & Domnas, A. J. (1983). *In vitro* production of zoospores by the mosquito pathogen *Lagenidium giganteum* (Oomycetes: Lagenidiales) on solid media. Journal of Invertebrate Pathology, 41, 305–309.

Kay, B. H., Cabral, C. P., Sleigh, A. C., Brown, M. D., Ribeiro, Z. M., & Vasconcelos, A. W. (1992). Laboratory evaluation of Brazilian *Mesocyclops* (Copepoda: Cyclopidae) for mosquito control. Journal of Medical Entomology, 29, 599–602.

Kerwin, J. L., & Washino, R. K. (1984). Efficacy of *Romanomermis culicivorax* and *Lagenidium giganteum* for mosquito control: Strategies for use of biological control agents in rice fields of the Central Valley of California. Proceedings of the California Mosquito Control Association, 52, 86–92.

Kerwin, J. L., & Washino, R. K. (1987). Ground and aerial application of the asexual stage of *Lagenidium giganteum* for control of mosquitoes associated with rice culture in the Central Valley of California. Journal of the American Mosquito Control Association, 3, 59–64.

Kerwin, J. L., Simmons, C. A., & Washino, R. K. (1986). Oosporogenesis by *Lagenidium giganteum* in liquied culture. Journal of Invertebrate Pathology 47, 258–270.

Kligler, I. J. (1930). The epidemiology and control of malaria in Palestine. Chicago: University of Chicago Press.

Knight, A. L. (1980). Host range and temperature requirements of *Culicinomyces clavisporus*. Journal of Invertebrate Pathology, 36, 423–425.

Koldenkova, L., García, A. I., Garces, F. J., & Gonzalez, B. R. (1988). Predatory capacity of the fish *Poecilia reticulata* Peters 1895) (Cyprinodontiformes: Poecilidae) in a natural habitat of the mosquito *Culex quinquefasciatus* Say 1823. Revista Cubana de Medicina Tropical, 40, 21–26.

Kramer, V. L., Garcia, R., & Colwell, A. E. (1987). An evaluation of the mosquitofish, *Gambusia affinis,* and the inland silverside, *Menidia beryllina,* as mosquito control agents in California wild rice fields. Journal of American Mosquito Control Association, 3, 626–632.

Kramer, V. L., Garcia, R., & Colwell, A. E. (1988). An evaluation of *Gambusia affinis* and *Bacillus thuringiensis* var. *israelensis* as mosquito control agents in California wild rice fields. Journal of the American Mosquito Control Association, 4, 470–478.

Lacey, L. A., & Undeen, A. H. (1986). Microbial control of blackflies and mosquitoes. Annual Review of Entomology, 31, 265–296.

Lamborn, R. H. (1890). Dragon flies vs. mosquitoes. New York: D. Appleton.

Lardeux, F., Riviere, F., Séchan, Y., & Kay, B. H. (1992). Release of *Mesocyclops aspercornis* (Copepoda) for control of larval *Aedes polynesiensis* (Diptera: Culicidae) in land crab burrows on an atoll of French Polynesia. Journal of Medical Entomology, 29, 571–576.

Legner, E. F. (1968). Parasite activity related to ovipositional responses in *Hippelates collusor*. Journal of Economic Entomology, 61, 1160–1163.

Legner, E. F. (1970). Advances in the ecology of *Hippelates* eye gnats in California indicate means for effective integrated control. Proceedings of the California Mosquito Control Association, 38, 89–90.

Legner, E. F. (1977). Response of *Culex* spp. larvae and their natural insect predators to two inoculation rates with *Dugesia dorotocephala* (Woodworth) in shallow ponds. Mosquito News, 37, 435–440.

Legner, E. F. (1978a). Efforts to control *Hydrilla verticillata* Royle with herbivorous *Tilapia zillii* (Gervais) in Imperial County irrigation canals. Proceedings of the California Mosquito Vector Control Association, 46, 103–104.

Legner, E. F. (1978b). Mass culture of *Tilapia zillii* (Cichlidae) in pond ecosystems. Entomophaga, 23, 51–56.

Legner, E. F. (1978c). Part I. Parasites and predators introduced against arthropod pests. Diptera, In C. P. Clausen (Ed.), Introduced parasites, and predators of arthropod pests and weeds: A world review (pp. 335–339; 346–355). Agriculture Handbook No. 480, Washington, DC: U.S. Department of Agriculture.

Legner, E. F. (1979). Advancements in the use of flatworms for biological mosquito control. Proceedings of the California Mosquito Control Association, 47, 42–43.

Legner, E. F. (1983). Imported cichlid behavior in California (pp. 8–13). International Symposium on *Tilapia* in Aquaculture, Nazareth, Israel.

Legner, E. F. (1986). Importation of exotic natural enemies. In J. M. Frane (Ed.), Biological control of plant pests and of vectors of human and animal diseases. Fortschritte der Zoologisch Bd, 32 (341), 19–30.

Legner, E. F. (1987). Inheritance of gregarious and solitary oviposition in *Muscidifurax rapturellus* Kogan & Legner (Hymenoptera: Pteromalidae). Canadian Entomologist, 119, 791–808.

Legner, E. F. (1988a). *Muscidifurax rapturellus* (Hymenoptera: Pteromalidae) females exhibit postmating oviposition behavior typical of the male genome. Annals of the Entomological Society of America, 81, 522–527.

Legner, E. F. (1988b). Hybridization in principal parasitoids of synanthopic Diptera: The genus *Muscidifurax* (Hymenoptera: Pteromalidae). Hilgardia, 56(4), 1–36.

Legner, E. F. (1989). Wary genes and accretive inheritance in Hymenoptera. Annals of the Entomological Society of America, 82, 245–249.

Legner, E. F. (1993). Theory for quantitative inheritance of behavior in a protelean parasitoid, *Muscidifurax raptorellus* (Hymenoptera: Pteromalidae). European Journal of Entomology, 90, 11–21.

Legner, E. F., & Bay, E. C. (1970). Dynamics of *Hippelates* eye gnat breeding in the southwest. California Agriculture, 24(5), l, 4–6.

Legner, E. F., & Brydon, H. W. (1966). Suppression of dung-inhabiting fly populations by pupal parasites. Annals of the Entomological Society of America, 59, 638–651.

Legner, E. F., & Dietrick, E. J. (1972). Inundation with parasitic insects to control filth breeding flies in California. Proceedings of the California Mosquito Control Association, 40, 129–130.

Legner, E. F., & Dietrick, E. J. (1974). Effectiveness of supervised control practices in lowering population densities of synanthropic flies on poultry ranches. Entomophaga, 19, 467–478.

Legner, E. F., & Fisher, T. W. (1980). Impact of *Tilapia zillii* (Gervais) on *Potamogeton pectinatus* L., *Myriophyllum spicatum* var. *exalbescens* Jepson, and mosquito reproduction in lower Colorado desert irrigation canals. Acta Oecologica, Oecologia Appl. 1, 3–14.

Legner, E. F., & Medved, R. A. (1973). Influence of *Tilapia mossambica* (Peters), *T. zillii* (Gervais) (Cichlidae) and *Mollienesia latipinna* LeSueur (Poeciliidae) on pond populations of *Culex* mosquitoes and chironomid midges. Mosquito News, 33, 354–364.

Legner, E. F., & Medved, R. A. (1974). Laboratory and small-scale field experiments with planaria (Tricladida: Turbellaria) as biological mosquito control agents. Proceedings of the California Mosquito Control Association, 42, 79–80.

Legner, E. F., & Murray, C. A. (1981). Feeding rates and growth of the fish *Tilapia zillii* (Cichlidae) on *Hydrilla verticillata, Potamogeton pectinatus* and *Myriophyllum spicatum* var *exalbescens,* and interactions in irrigation canals of southeastern California. Mosquito News, 41, 241–250.

Legner, E. F., & Olton, G. S. (1969). Migrations of *Hippelates collosor* larvae from moisture and trophic stimuli and their encounter by *Trybliographa* parasites. Annals of the Entomological Society of America, 62, 136–141.

Legner, E. F., & Pelsue, F. W., Jr. (1980). Bioconversion: *Tilapia* fish turn insects and weeds into edible protein. California Agriculture, 34(11–12), 13–14.

Legner, E. F., & Pelsue, F. W., Jr. (1983). Contemporary appraisal of the population dynamics of introduced cichlid fish in south California. Proceedings of the California Mosquito Control Association, 51, 38–39.

Legner, E. F., & Sjogren, R. D. (1984). Biological mosquito control furthered by advances in technology and research. Mosquito News, 44, 449–456.

Legner, E. F., & Tsai, S.-C. (1978). Increasing fission rate of the planarian mosquito predator, *Dugesia dorotocephala,* through biological filtration. Entomophaga, 23, 293–298.

Legner, E. F., Olton, G. A., & Eskafi, F. M. (1966). Influence of physical factors on the developmental stages of *Hippelates collusor* in relation to the activities of its natural parasites. Annals of the Entomological Society of America, 59, 851–861.

Legner, E. F., Sjogren, R. D., & Hall, I. M. (1974). The biological control of medically important arthropods. Critical Reviews in Environmental Control, 4, 85–113.

Legner, E. F., Medved, R. A., & Hauser, W. J. (1975a). Predation by the desert pupfish, *Cyprinodon macularius* on *Culex* mosquitoes and benthic chironomid midges. Entomophaga, 20, 23–30.

Legner, E. F., Olton, G. S., Eastwood, R. E., & Dietrick, E. J. (1975b). Seasonal density, distribution and interactions of predatory and scavenger arthropods in accumulating poultry wastes in coastal and interior southern California. Entomophaga, 20, 269–283.

Legner, E. F., Tsai, T.-C., & Medved, R. A. (1976). Environmental stimulants to asexual reproduction in the planarian, *Dugesia dorotoce-phala.* Entomophaga, 21, 415–423.

Legner, E. F., Greathead, D. J., & Moore, I. (1980a). Population density fluctuations of predatory and scavenger arthropods in accumulating bovine excrement of three age classes in equatorial East Africa. Bulletin of the Society of Vector Ecology, 5, 23–44.

Legner, E. F., Medved, R. A., & Pelsue, F. (1980b). Changes in chironomid breeding patterns in a paved river channel following adaptation of cichlids of the *Tilapia mossambica-hornorum* complex. Annals of the Entomological Society of America, 73, 293–299.

Lenhoff, M. H., & Brown, R. D. (1970). Mass culture of hydra: Improved method and application to other invertebrates. Laboratory Animal, 4, 139–154.

Levy, R., & Miller, T. W., Jr. (1978). Tolerance of the planarian *Dugesia dorotocephala* to high concentrations of pesticides and growth regulators. Entomophaga 23, 31–34.

Levy, R., Hertlein, B. C., Peterson, J. J., Doggett, D. W., & Miller, T. W., Jr. (1979). Aerial application of *Romanomermis culicivorax* (Mermithidae: Nematoda) to control *Anopheles* and *Culex* mosquitoes in southwest Florida. Mosquito News, 39, 20–25.

Lindquist, A. W. (1940). The introduction of an indigenous blowfly parasite, *Alysia ridibunda* Say, into Uvalde County, Texas. Annals of the Entomological Society of America, 33, 103–112.

Lloyd, L. (1987). An alternative to insect control by "mosquitofish" *Gambusia affinis,* In T. D. St. George, B. H. Kay, & J. Blok (Eds.), Proceedings, Fourth Symposium on Arbovirus Research in Australia (1986) (pp. 156–163). Brisbane, Australia: Q.I.M.R.

Lord, J. C., & Roberts, D. W. (1985). Effects of salinity, pH, organic solutes, anaerobic conditions and the presence of other microbes on production and survival of *Lagenidium giganteum* (Oomycetes: Lagenidiales) zoospores. Journal of Invertebrate Pathology, 45, 331–338.

Lounibos, L. P., Nishimura, N., & Dewald, L. B. (1992). Predation of *Mansonia* (Diptera: Culicidae) by native mosquito fish in southern Florida. Journal of Medical Entomology, 29, 236–241.

Luh, P. L. (1981). The recent status of biocontrol of mosquitoes in China. In M. Laird (Ed.), Biocontrol of medical and veterinary pests (pp. 54–77). New York: Praeger.

Madsen, H. (1983). Distribution of *Helisoma duryi,* an introduced competitor of intermediate hosts of schistosomiasis, in an irrigation scheme in northern Tanzania. Acta Tropica, 40, 297–306.

Madsen, H. (1990). Biological methods for the control of freshwater snails. Parasitology Today, 6, 237–240.

Mandeville, J. D., Mullens, B. A., & Meyer, J. A. (1988). Rearing and host age suitability of *Fannia canicularis* (L.) for parasitization by *M uscidifurax zaraptor* Kogan & Legner. Canadian Entomologist, 120, 153–159.

Markofsky, J., & Matias, J. R. (1979). Waterborne vectors of disease in tropical and subtropical areas; and a novel approach to mosquito control using annual fish, In Halasi-Kun (Ed.), Proceedings, University Seminar on Pollution and Water Resources (Vol. 21, pp. H-1, H-l7). New York: Columbia University.

Matanmi, B. A., Federici, B. A., & Mulla, M. S. (1990). Fate and persistence of *Bacillus sphaericus* used as a mosquito larvacide in dairy waste lagoons. Journal of the American Mosquito Control Association, 6, 384–389.

McCray, E. M., Jr., Womeldorf, D. J., Husbands, R. C., & Eliason, D. A. (1973). Laboratory observations and field tests with *Lagenidium* against California mosquitoes. Proceedings of the California Mosquito Control Association, 41, 123–128.

McCullough, F. S. (1981). Biological control of the snail intermediate host of human *Schistosoma* spp.: A review of its present stature and future proposals. Acta Tropica, 38, 5–13.

McDonald, G., & Buchanan, G. A. (1981). The mosquito and predatory insect fauna inhabiting fresh-water ponds, with particular reference to

Culex annulirostris Skuse (Diptera: Culicidae). Australian Journal of Ecology, 6, 21–27.

Medved, R. A., & Legner, E. F. (1974). Feeding and reproduction of the planarian, *Dugesia dorotocephala* (Woodworth), in the presence of *Culex peus* Speiser. Environmental Entomology, 3, 637–641.

Menon, P. K. B., & Rajagopalan, P. K. (1978). Control of mosquito breeding in wells by using *Gambusia affinis* and *Aplocheilus blockii* in Pondicherry town. Indian Journal of Medical Research, 68, 927–933.

Merriam, T. L., & Axtell, R. C. (1982a). Salinity tolerance of two isolates of *Lagenidium giganteum* (Oomycetes: Lagenidiales), a fungal pathogen of mosquito larvae. Journal of Medical Entomology, 19, 388–393.

Merriam, T. L., & Axtell, R. C. (1982b). Evaluation of the entomogenous fungi *Culicinomyces clavosporus* and *Lagenidium giganteum* for control of the salt marsh mosquito, *Aedes taeniorhynchus.* Mosquito News, 42, 594–602.

Merriam, T. L., & Axtell, R. C. (1983). Relative toxicity of certain pesticides to *Lagenidium giganteum* (Oomycetes: Lagenidiales), a fungal pathogen of mosquito larvae. Environmental Entomology, 12, 515–521.

Mian, L. S., Mulla, M. S., & Chavey, J. D. (1985). Biological strategies for control of mosquitoes associated with aquaphyte treatment of waste water. Mosquito Control Research Annual Report (pp. 91–92). University of California, Davis.

Miura, T. (1986). Rice field mosquito studies. Mosquito Control Research Annual Report (pp. 50–52). University of California, Davis.

Miura, T., & Takahashi, R. M. (1985). A laboratory study of crustacean predation on mosquito larvae. Proceedings of the California Mosquito Control Association, 52, 94–97.

Miura, T., Takahashi, R. M., & Mulligan, F. S., III. (1980). Effects of the bacterial mosquito larvicide *Bacillus thuringiensis* serotype H-14 on selected aquatic organisms. Mosquito News, 40, 619–622.

Miyagi, I., Toma, T., & Mogi, M. (1992). Biological control of container-breeding mosquitoes. *Aedes albopictus* and *Culex quinquefasciatus,* in a Japanese island by release of *Toxorhynchites splendens* adults. Medical Veterinary Entomology, 6, 290–300.

Moens, R. (1980). Le probleme des limaces dans la protection des vegetaux. Revue de L'Agriculture, 33, 117–133.

Moers, R. (1982). Mecanisme de reinfestation par *Lymnaea truncatula* Muller (Mollusca, Pulmonata) des terrains propices a la fasciolose. Malacologia, 22, 29–34.

Morgan, P. B., & Patterson, R. S. (1977). Sustained release of *Spalangia endius* to parasitize field populations of three species of filth breeding flies. Journal of Economic Entomology, 70, 450–452.

Morgan, P. B., Patterson, R. S., & LaBrecque, G. C. (1977). Controlling houseflies at a dairy installation by releasing a protelean parasitoid. *Spalangia endius* (Hymenoptera: Pteromalidae). Journal of Georgia Entomological Society, 11, 39–43.

Morgan, P. B., Patterson, R. S., LaBrecque, G. C., Weidhaas, D. E., & Benton, A. (1975). Suppression of a field population of houseflies with *Spalangia endius.* Science, 189, 388–389.

Moyle, P. B. (1976). Inland fishes of California. Berkeley: University of Calif. Press.

Mulla, M. S., Tietze, N. S., & Wargo, M. J. (1986). Tadpole shrimp (*Triops longicaudatus*), a potential biological control agent for mosquitoes. Mosquito Control Research Annual Report University of California, Davis (pp. 44–45).

Mulla, M. S., Darwazeh, H. A., Davidson, E. W., & Dulmage, H. T. (1984). Efficacy and persistence of the microbial agent *Bacillus sphaericus* against mosquito larvae in organically enriched habitats. Mosquito News, 44, 166–173.

Mullens, B. A. (1986). A method for infecting large numbers of *Musca domestica* (Diptera: Muscidae) with *Entomophthora muscae* (Entomophthorales: Entomophthoraceae). Journal of Medical Entomology, 23, 457–458.

Mullens, B. A., Meyer, J. A., & Mandeville, J. D. (1986). Seasonal and diel activity of filth fly parasites (Hymenoptera: Pteromalidae) in caged-layer poultry manure in southern California. Environmental Entomology, 15(1), 56–60.

Mullens, B. A., Rodriguez, J. L., & Meyer, J. A. (1987a). An epizootiological study of *Entomophthora muscae* in muscoid fly populations on southern California poultry facilities, with emphasis on *Musca domestica*. Hilgardia, 55(3), 1–41.

Mullens, B. A., Meyer, J. A., & Cyr, T. L. (1987b). Infectivity of insect parasitic nematodes (Rhabditida: Steinernematidae, Heterorhbditidae) for larvae of some manure-breeding flies. Environmental Entomology, 16, 769–773.

Mullens, B. A., Meyer, J. A., & Georgis, R. (1987c). Field tests of insect parasitic nematodes (Rhabditida: Steinernematidae, Heterorhabditidae) against larvae of manure-breeding flies (Diptera: Muscidae) on caged layer poultry facilities. Journal Economic Entomology, 80, 438–442.

Mulligan, F. S., III, Farley, D. G., Caton, J. R., & Schaefer, C. H. (1983). Survival and predatory efficiency of *Gambusia affinis* for control of mosquitoes in underground drains. Mosquito News, 43, 318–321.

Muspratt, J. (1951). The bionomics of an African *Megarhinus* (Dipt., Culicidae) and its possible use in biological control. Bulletin of Entomology Research, 42, 355–370.

Nakagawa, P. Y. (1963). Status of *Toxorhynchites* in Hawaii. Proceedings of the Hawaiian Entomological Society, 18, 291–293.

Nelson, F. R. S. (1979). Comparative predatory potential and asexual reproduction of sectioned *Dugesia dorotocephala* as they relate to biological control of mosquito vectors. Environmental Entomology, 8, 679–681.

Nelson, S. M., & Keenan, L. C. (1992). Use of an indigenous fish species, *Fundulus zebrinus*, in a mosquito abatement program: Field comparison with the mosquitofish *Gambusia affinis*. Journal of the American Mosquito Control Association, 8, 301–304.

Neng, W. (1986). Vector control by fish for the prevention of dengue fever in Guangxi, Peoples' Republic of China. In T. D. St. George, B. H. Kay, & J. Blok (Eds.), Proceedings, Fourth Symposium on Arbovirus Research in Australia (1986) (pp. 155–156). Brisbane, Australia: Q.I.M.R.

Nguma, J. F. M., McCullough, F. S., & Masha, E. (1982). Elimination of *Biomphalaria pfeifferi*, *Bulinus tropicus*, *Lymnaea natalensis* by the ampullorid, *Marisa cornuarietis* in a man-made dam in northern Tanzania. Acta Tropica, 39, 85–90.

Olton, G. S., & Legner, E. F. (1975). Winter inoculative releases of parasitoids to reduce houseflies in poultry manure. Journal of Economic Entomology, 68, 35–38.

Paine, R. W. (1934). The introduction of *Megarhinus* mosquitoes into Fiji. Bulletin of Entomological Research, 25, 1–32.

Palchick, S., & Washino, R. K. (1984). Factors affecting mosquito larval abundance in northern California rice fields. Proceedings of the California Mosquito Control Association, 52, 144–147.

Parman, D.C. (1928). Experimental dissemination of the tabanid egg parasite Phanurus emersoni Girault and biological notes on the species (Circular No. 18). Washington, DC: U.S. Department of Agriculture.

Pawson, B. M., & Petersen, J. J. (1988). Dispersal of *Muscidifurax zaraptor* (Hymenoptera: Pteromalidae), a filth fly parasitoid, at dairies in eastern Nebraska. Environmental Entomology, 17, 398–402.

Pelzman, R. J. (1975). California State Department of Fish and Game policy on the use of native and exotic fish as biological control agents. Proceedings of the California Mosquito Control Association, 43, 49–50.

Petersen, J. J. (1978). Development of resistance by the southern house mosquito to the parasitic nematode *Romanomermis culicivorax*. Environmental Entomology, 7, 518–520.

Petersen, J. J. (1980a). Mass production of the mosquito parasite *Romanomermis culicivorax*: Effect of density. Journal of Nematology, 12, 45–48.

Petersen, J. J. (1980b). The effect of culture age on the infectivity of preparasites of the mosquito parasite *Romanomermis culicivorax*. Mosquito News, 40, 640–641.

Petersen, J. J., & Willis, O. R. (1970). Some factors affecting parasitism by mermithid nematodes in southern house mosquitoes. Journal of Economic Entomology, 63, 175–178.

Petersen, J. J., & Willis, O. R. (1972a). Procedures for the mass rearing of a mermithid parasite of mosquitoes. Mosquito News, 32, 226–230.

Petersen, J. J., & Willis, O. R. (1972b). Results of preliminary field applications of *Reesimermis nielseni* (Mermithidae: Nematoda) to control mosquito larvae. Mosquito News, 32, 312–313.

Petersen, J. J., & Willis, O. R. (1975). Establishment and recycling of a mermithid nematode for control of larval mosquitoes. Mosquito News, 35, 526–532.

Petersen, J. J., Willis, O. R. & Chapman, H.C. (1978a). Release of *Romanomermis culicivorax* for the control of *Anopheles albimanus* in El Salvador. I. American Journal of Tropical Medicine and Hygiene, 27, 1265–1266.

Petersen, J. J., Chapman, H. C., Willis, O. R., & Fukada, T. (1978b). Release of *Romanomermis culicivorax* for the control of *Anopheles albimanus* in El Salvador. II. American Journal of Tropical Medicine and Hygiene, 27, 1268–1273.

Peterson, G. D. (1956). The introduction of mosquitoes of the genus *Toxorhynchites* into American Samoa. Journal of Economic Entomology, 49, 786–789.

Pickens, L. B., Miller, R. W., & Centala, M. M. (1975). Biology, population dynamics, and host finding efficiency of *Pachycrepoideus vindemiae* in a box stall and a poultry house. Environmental Entomology, 4, 975–979.

Poinar, G. O., Jr. (1979). Nematodes for biological control of insects. Boca Raton, FL: CRC Press.

Povolny, D. (1971). Synanthropy: Definition, evolution and classification. In B. Greenberg (Ed.), Flies and disease: Vol. 1. Ecology, classification and biotic associations (pp. 17–54). Princeton, NJ: Princeton University Press.

Priest, F. G. (1992). Biological control of mosquitoes and other biting flies by *Bacillus sphaericus* and *Bacillus thuringiensis*. Journal of Applied Bacteriology, 72, 357–369.

Propp, G., & Morgan, P. B. (1985). Effect of host distribution on parasitoidism of housefly (Diptera: Muscidae) pupae by *Spalangia* spp. and *Muscidifurax raptor* (Hymenoptera: Pteromalidae). Canadian Entomologist, 117, 515–524.

Rees, T. J. (1979). Community development in freshwater microcosms. Hydrobiologia, 63, 113–128.

Ridsdill-Smith, T. J., & Hayles, L. (1987). Mortality of eggs and larvae of the bush fly, *Musca vetustissima* Walker (Diptera: Muscidae), caused by scarabaeine dung beetles (Coleoptera: Scarabaeidae) in favourable cattle dung. Bulletin of Entomological Research, 77, 731–736.

Ridsdill-Smith, T. J., Hayles, L., & Palmer, M.J. (1986). Competition between the bush fly and a dung beetle in dung of differing characteristics. Entomologia Experimentalis et Applicata, 41, 83–90.

Ripa, R. (1986). Survey and use of biological control agents on Easter Island and in Chile. Miscellaneous Publications of the Entomological Society of America, 61, 39–44.

Riviere, F. (1985). Effects of two predators on community composition and biological control of *Aedes aegypti* and *Aedes polynesiensis*. In L. P. Lounibos, J. R. Rey, & J. H. Frank (Eds.), Ecology of mosquitoes: Proceedings of a workshop (pp. 121–135). Florida Medical Entomology Laboratory. Vero Beach, Florida.

Riviere, F., Kay, B. H., Klein, J.-M., & Sechan, Y. (1987a). *Mesocyclops aspericornis* (Copepoda) and *Bacillus thuringiensis* var. *israelensis* for the biological control of *Aedes* and *Culex* vectors (Diptera: Culicidae) breeding in crab holes, tree holes, and artificial containers. Journal of Medical Entomology, 24, 425–430.

Riviere, F., Sechan, Y., & Kay, B. H. (1987b). The evaluation of predators for mosquito control in French Polynesia. In T. D. St. George, B. H. Kay, & J. Blok (Eds.), Proceedings of the Fourth Symposium on Arbovirus Research in Australia (1986) (pp. 150–154). Brisbane, Australia: Q.I.M.R.

Roberts, D. W., Daoust, R. A., & Wraight, S. P. (1983). Bibliography on pathogens of medically important arthropods (1981) (WHOVBC/83.1). Geneva: World Health Organization.

Roberts, J., & Sampson, D. R. T. (1987). Data sheet on biological control agents: Tilapiine fish (WHO/VBC/945). Geneva: World Health Organization.

Rodriguez, J. L., & Riehl, L. A. (1962). Control of flies in manure of chickens and rabbits by cockerels in southern California. Journal of Economic Entomology, 55, 473–477.

Russell, R. C., Debenham, M. L., & Lee, D. J. (1979). A natural habitat of the insect pathogenic fungus *Culicimomyces* in the Sidney area. Proceedings of the Linnean Society of New South Wales, 103, 71–73.

Rutz, D. A., & Axtell, R. C. (1979). Sustained releases of *Muscidifurax raptor* (Hymenoptera: Pteromalidae) for house fly (*Musca domestica*) control in two types of caged layer poultry houses. Environmental Entomology, 8, 1105–1110.

Sabatinelli, G., Majori, G., Blanchy, S., Fayaerts, P., & Papakay, M. (1990). Testing of the larvivorous fish, *Poecillia reticulata* in the control of malaria in the Islamic Federal Republic of the Comoros. World Health Organization WHO/MAL, 90, 1–10.

Sasa, M., & Kurihara, T. (1981). The use of poeciliid fish in the control of mosquitoes. In M. Laird (Ed.), Biocontrol of medical and veterinary pests (pp. 36–53). New York: Praeger.

Sasa, M., Harinasuta, C., Purived, Y., & Kurihara, T. (1965). Studies on mosquitoes and their natural enemies in Bangkok. III. Observations on a mosquito-eating fish "guppy," *Lebistes reticulatus,* breeding in polluted waters. Japanese Journal of Sanitation and Zoology, 23, 113–127 (in Japanese).

Schoenherr, A. A. (1981). The role of competition in the displacement of native fishes by introduced species. In R. J. Naiman & D. L. Stoltz (Eds.), Fishes in North American deserts (pp. 173–203). New York: Wiley Interscience.

Service, M. W. (1977). Mortalities of the immature stages of species B of the *Anopheles gambiae* complex in Kenya: Comparison between rice fields and temporary pools, identification of predators and effects of insecticidal spraying. Journal of Medical Entomology, 13, 535–545.

Service, M. W. (1983). Biological control of mosquitoes—has it a future? Mosquito News, 43, 113–120.

Simmonds, H. W. (1958). The housefly problem in Fiji and Samoa. South Pacific Communications Quarterly Bulletin, 8, 29–30, 47.

Sjogren, R. D., & Legner, E. F. (1974). Studies of insect predators as agents to control mosquito larvae, with emphasis on storage of *Notonecta* eggs. Proceedings of the California Mosquito Control Association, 42, 71–72.

Sjogren, R. D., & Legner, E. F. (1989). Survival of the mosquito predator, *Notonecta unifasciata* (Hemiptera: Notonectidae) embryos at low thermal gradients. Entomophaga, 34(2), 201–208.

Sprenger, D., & Wuithironyagool, T. (1986). The discovery and distribution of *Aedes albopictus* in Harris County, Texas. Journal of American Mosquito Control Association, 2, 217–219.

Steffan, W. A. (1975). Systematics and biological control potential of Toxorhynchites. Mosquito Systematics, 7, 59–67.

Stewart, R. J., Schaefer, C. H., & Miura, T. (1983). Sampling *Culex tarsalis* immatures on rice fields treated with combinations of mosquitofish and *Bacillus thuringiensis* H-14 toxin. Journal of Economic Entomology, 76, 91–95.

Sweeney, A. W. (1981). Prospects for the use of *Culicinomyces* fungi for biocontrol of mosquitoes. In M. Laird (Ed.), Biocontrol of medical and veterinary pests (pp. 105–121). New York: Praeger.

Sweeney, A. W. (1983). The time-mortality response of mosquito larvae infected with the fungus *Culicinomyces.* Journal of Invertebrate Pathology, 42, 162–166.

Sweeney, A. W. (1987). Prospects of pathogens and parasites for vector control. In T. D. St. George, B. H. Kay, & J. Blok (Eds.), Proceedings, Fourth Symposium on Arbovirus Research in Australia (1986) (pp. 163–165). Brisbane, Australia: Q.I.M.R.

Sweeney, A. W., Lee, K. J., Panter, C., & Burgess, L. W. (1973). A fungal pathogen for mosquito larvae with potential as a microbial insecticide. Search, 4, 344–345.

Tabibzadeh, I., Behbehani, C., & Nakhai, R. (1970). Use of *Gambusia* fish in the malaria eradication programme of Iran. Bulletin of the World Health Organization, 43, 623–626.

Tietze, N. S., & Mulla M. S. (1987). Tadpole shrimp (*Triops longicaudatus*), new candidates as biological control agents for mosquitoes. California Mosquito Vector Control Association, Biological Briefs, 13(2), 1.

Tietze, N. S. & Mulla, M. S. (1991). Biological control of *Culex* mosquitoes (Diptera: Culicidae) by the tadpole shrimp, *Triops longicaudatus* (Nostostraca: Tiopsidae). Journal of Medical Entomology, 28, 24–31.

Tikasingh, E. S. (1992). Effects of *Toxorhynchites moctezuma* larval predation on *Aedes aegypti* populations: Experimental evaluation. Medical Veterinary Entomology 6, 266–271.

Tikasingh, E. S., & Eustace, A. (1992). Suppression of *Aedes aegypti* by predatory *Toxorhynchites moctezuma* in an island habitat. Medical Veterinary Entomology 6, 272–280.

Toma, T., & Miyagi, I. (1992). Laboratory evaluation of *Toxorhynchites splendens* (Diptera: Culicidae) for predation of *Aedes albopictus* mosquito larvae. Medical Veterinary Entomology 6, 281–289.

Trpis, M. (1973). Interactions between the predator *Toxorhynchites brevipalpus* and its prey *Aedes aegypti.* Bulletin of the World Health Organization, 49, 359–365.

Trpis, M. (1981). Survivorship and age specific fertility of *Toxorhynchites brevipalpis* females (Diptera: Culicidae). Journal of Medical Entomology, 18, 481–486.

Tsai, S. C., & Legner, E. F. (1977). Exponential growth in culture of the planarian mosquito predator *Dugesia dorotocephala* (Woodworth). Mosquito News, 37, 474–478.

Vanderplank, F. L. (1941). *Nothobranchius* and *Barbus* species: Indigenous antimalarial fish in East Africa. East African Medical Journal, 17, 431–436.

Vanderplank, F. L. (1967). Why not "instant fish" farms? New Science, 33, 42.

Vaz-Ferreira, R., Sierra, B., & Scaglia, S. (1963). Eco-etologia de la reproduccion en los peces del genera *Cynolebias* Steindachner. Aptdo. de los Archivos de la Sociedad de Biologia de Montevideo, 26, 44–49.

Wallace, M. M. H., & Tyndale-Biscoe, M. (1983). Attempts to measure the influence of dung beetles (Coleoptera: Scarabaeidae) on the field mortality of the bush fly *Musca vetustissima* Walker (Diptera: Muscidae) in south-eastern Australia. Bullettin of Entomolgical Research, 73, 33–44.

Walters, L. L., & Legner, E. F. (1980). Impact of the desert pupfish, *Cyprinodon macularius,* and *Gambusia affinis affinis* on fauna in pond ecosystems. Hilgardia, 48(3), 1–8.

Washburn, J. O., & Anderson, J. R. (1986). Distribution of *Lambornella clarki* (Ciliophora: Tetrahymenidae) and other mosquito parasites in California treeholes. Journal of Invertebrate Pathology, 48, 296–309.

Yadav, R. S., Padhan, K., & Sharma, V. P. (1992). Fishes of the District Sundargarh, Orissa, with special reference to their potential in mosquito control. Indian Journal of Malariology, 29, 225–233.

Yu, H.-S., & Legner, E. F. (1976). Regulation of aquatic Diptera by planaria. Entomophaga, 21, 3–12.

Yu, H.-S., Legner, E. F., & Pelsue, F. (1975). Control of Culex mosquitoes in weedy lake habitats in Los Angeles with Chlorohydra viridissima. Proceedings of the California Mosquito Control Association, 43, 123–126.

Yu, H.-S., Legner E. F., & Sjogren, R. D. (1974a). Mosquito control with European green hydra in irrigated pastures, river seepage and duck club ponds in Kern County. Proceedings of the Mosquito Control Association, 42, 77–78.

Yu, H.-S., Legner, E. F., & Sjogren, R. D. (1974b). Mass release effects of Chlorohydra viridissima (Coelenterata) on field populations of Aedes nigromaculis and Culex tarsalis in Kern County, California. Entomophaga, 19, 409–420.

38

Biological Control of Vertebrate Pests

MARK S. HODDLE

Department of Entomology
University of California
Riverside, California

INTRODUCTION

Several species of vertebrates, especially mammals, have been successful invaders and colonizers of new territories, in particular, insular island ecosystems. Others have exhibited high environmental tolerance and adaptability after careful and repeated introductions to new locales by humans. Following establishment, several vertebrate species have become important pests. These pests harm agricultural systems by damaging agricultural lands [e.g., rabbits (*Oryctolagus cuniculus* Linnaeus) in Australia and New Zealand], by attacking crops [e.g., European starlings (*Sturnus vulgaris* Linnaeus) in the United States], and by acting as sources for communicable diseases [e.g., brushtail possums, (*Trichosurus vulpecula* Kerr), are reservoirs for bovine tuberculosis (*Mycobacterium bovis* Karlson and Lessel) in New Zealand]. Other pest vertebrates damage natural systems by threatening the continued existence of endangered flora [e.g., goats (*Capra hircus* Linnaeus) on the Galapagos Islands] and fauna [e.g., brown tree snake (*Boiga irregularis* {Merrem}) on Guam], and by adversely affecting wilderness areas by changing ecosystem functions and diversity (Vitousek *et al.,* 1996). The continued relocation of vertebrates exacerbates the ongoing problem of global homogenization of biota (Lodge, 1993).

Movement of particular vertebrates into areas where they had not previously existed has, in some instances, occurred naturally without human intervention [e.g., the passerine bird (*Zosterops lateralis* {Latham}) arrived in New Zealand unassisted from Australia]. The vast majority of vertebrate translocations have been human assisted. Accidental introduction has occurred as a consequence of human transportation [e.g., brown tree snakes, mice (*Mus musculus* Lin-

naeus), and rats such as *Rattus rattus* Linnaeus and *R. norvegicus* Berkenhout)]. Some releases have been intentional (but illegal) to serve self-centered private interests [e.g., monk parakeets (*Myiopsitta monachus* {Boddaert}) in New York, Florida, and Texas]. Other species have been legitimately introduced to procure public benefit by providing: (1) new agricultural products [e.g., European wild boars (*Sus scrofa* Linnaeus); sheep (*Ovis aries* Linnaeus); cows (*Bos taurus* Linnaeus), goats and rabbits for meat, and brushtail possums for fur in New Zealand], (2) recreation [e.g., red deer (*Cervus elaphus* Linnaeus), fallow deer (*Dama dama* Linnaeus), and trout (*Salmo trutta* Linnaeus and *S. gairdneri* Richardson) in New Zealand]; (3) companionship [e.g., cats (*Felis domesticus* Linnaeus) and dogs (*Canis familiaris* Linnaeus)]; or (4) biological control agents [e.g., the European fox (*Vulpes vulpes* Linnaeus), stoats (*Mustela erminea* Linnaeus), weasels (*M. nivalis* Linnaeus), and ferrets (*M. putorius furo* Linnaeus) for the control of rabbits in Australia or New Zealand; cane toads (*Bufo marinus* Linnaeus) for control of gray-backed cane beetles (*Dermolepida albohirtum* {Waterhouse}) in sugarcane plantations in Australia; and the small Indian mongoose (*Herpestes javanicus* {Saint-Hilaire}, {= *auropunctatus* (Hodgson)}) for control of rats in Hawaii].

Chemical and cultural control of vertebrate pests is expensive and nonsustainable, and at best provides a temporary local solution to problems (Hone, 1994; Williams & Moore, 1995). Biological control of vertebrates, a potentially less expensive and self-sustaining method of population suppression, has focused primarily on mammalian pests. Predators, parasites, and pathogens specific to mammals with two notable exceptions (the myxoma and calici viruses that infect rabbits) have failed to provide satisfactory control (Shelford, 1942; Howard, 1967; Davis *et al.,* 1976; Wood, 1985; Smith & Remington, 1996; Whisson, 1998). Historical records indicate that the majority of attempts at vertebrate biological control have been *ad hoc* efforts and not the product of careful studies designed to elucidate factors and conditions likely to affect the impact of natural enemy introductions on pest populations. Furthermore, failure of biological control of vertebrates by predatory vertebrates has compounded problems associated with exotic vertebrates because control attempts result in addition of new species that cause biological and conservation problems.

The level of control achieved by natural enemies is dependent on ratios of natality to mortality of control agents and their host species (Davis *et al.,* 1976). For vertebrates these ratios are affected by many factors: advanced learning; social, territorial, and breeding behaviors; chemical, physical, and immunological defenses; temporal and spatial escape strategies; and genetic selection in both natural enemy and host populations for persistent coexistence. These complex interrelated factors, coupled with opportunistic feeding habits, have made vertebrate pests difficult targets for biological control with natural enemies.

Advances in understanding of mammalian fertilization biology have provided molecular biologists with necessary information to develop and investigate the concept of immunocontraception for vertebrate pest control. Immunocontraception utilizes genetically modified pathogens that express surface proteins from the target pest's egg or sperm to induce an immune response in the host. Antibodies then attack gametes in the host's reproductive tract causing sterilization (Tyndale-Biscoe, 1994b, 1995). Computer models indicate that immunocontraception may provide long-term control of vertebrate pests because genetically modified pathogens reduce net reproductive rates without killing hosts (Barlow, 1994, 1997).

In this chapter, I discuss attributes that have aided vertebrate establishment; damage resulting from colonization and uncontrolled population growth; biological control of mammalian pest species with predators, parasites, and pathogens; and future directions that biological control research for vertebrates is taking with genetically engineered microorganisms.

FACTORS THAT FACILITATE VERTEBRATE ESTABLISHMENT

Community Characteristics

Communities vary in their ability to accommodate the establishment and proliferation of new species (Primack, 1995). Elton (1958) suggested that species-poor communities (e.g., islands) or highly disturbed habitats are more permissive to successful introductions of new species. Elton's (1958) predictions have been substantiated in part by paleobiological reconstructions of invasions between newly joined communities (Vermeiji, 1991) and by mathematical modeling describing multi-species interactions in communities (MacArthur, 1955; Case, 1991). Isolated oceanic islands (e.g., New Zealand), and insular continents and habitats (e.g., Australia and lakes) often have a low diversity of native species. Such environments have typically experienced little immigration and are susceptible to invasion by vertebrates (Brown, 1989). Stable, speciose communities with high levels of interspecific competition appear to resist invasion by new species and are sources of successful colonists into less speciose or disturbed communities (MacArthur, 1955; Brown, 1989; Lodge 1993). This phenomenon has produced asymmetrical patterns of colonization, with successful vertebrate invaders usually being native to continents or extensive nonisolated habitats within continents with more diverse biotas (Brown, 1989; Vermeiji, 1991).

Continental herbivores and predators have been very successful in establishing self-sustaining populations in insular habitats, in part because such habitats often lack large generalist vertebrates and essentially have just two trophic levels, producers and decomposers (if specialist herbivores and their predators are excluded) (Vitousek, 1990). Niches equivalent to those on the mainland are largely unoccupied (Brown, 1989). Insular ecosystems therefore often appear to readily accommodate generalist herbivores and predators, perhaps because of low levels of competition for resources that are often inadequately defended chemically or physically (Vitousek, 1990; Vermeij, 1991; Bowen & van Vuren, 1997). Organisms from insular habitats that have not coevolved closely with predators or herbivores lack life history features that deter attack or permit survival despite high mortality from predation or herbivory (Bowen & van Vuren, 1997). Lack of such biological attributes may increase the competitive advantage of exotics (Case, 1996; Coblentz, 1978; Vitousek, 1990).

Introduced vertebrates can also be extremely disruptive in continental regions when habitat disturbance by urbanization or agriculture occurs. European wild boars have detrimental effects on gray beech forests in the Great Smoky Mountains in the southeast United States. Habitat disturbance—through pig rooting, trampling, and browsing—and human removal of predators (pumas and wolves) aided pig establishment and spread in this area (Bratton, 1975; Singer *et al.,* 1984; Vitousek, 1990).

Species Characteristics

Although there are well-recognized exceptions to general rules that characterize successful vertebrate colonists (Ehrlick, 1989; Williamson & Fitter, 1996b), species that establish self-sustaining populations outside their native range typically exhibit some of the following general characters: (1) short generation times, (2) high dispersal rates, (3) high tolerance for varying geographic and climatic conditions, (4) polyphagy, (5) low attack rates from upper trophic level organisms, and (6) human commensalism (Ehrlich, 1986, 1989; Lodge, 1993; Williamson & Fitter, 1996a). The assumption of a high intrinsic rate of increase is generally unnecessary for establishment (Ehrlich, 1986; Lodge, 1993), although it is important for the establishment of some exotic bird species (Veltman *et al.*, 1996). The spread of the European rabbit in Australia and New Zealand, countries that both historically lacked significant eutherian mammal fauna, illustrates some of the preceding points.

The European rabbit originated in Spain and Portugal (Corbet, 1994) and spread through most of Europe over 3000 years ago following deforestation by humans for agriculture and overgrazing by livestock. Animals reared in captivity were carried through Europe in advance of naturally spreading populations (Flux, 1994). The spread of the European rabbit throughout Australia following its introduction in 1788 varied from 15 km/year to approximately 300 km/year. Rate of spread was fastest across dry savannas where conditions were most similar to Mediterranean climates and slowest through woodlands (Myers *et al.*, 1994).

In Australia, habitat alteration by humans such as conversion of land to pasture, overgrazing of rangelands, and predator eradication [e.g., dingos (*Canis familiaris dingo* Meyer)] aided rabbit survivorship and spread. Rabbits are polyphagous and feed on grasses and browse shrubs. During severe food and water shortages, bark, fallen leaves, seed pods, tree roots, and termites are consumed (Myers *et al.*, 1994).

Rabbits are highly fecund and exhibit rapid population growth under good conditions. Individual female rabbits 10 to 12 months of age can produce litters averaging 5 to 6 young, and with abundant food 23 to 48 offspring per year are produced (Gibb & Williams, 1994). Behavioral adaptability, sociality, and territoriality, in addition to use of elaborate underground warrens, have also aided rabbit proliferation in marginal habitats (Myers *et al.*, 1994).

Relocation of Vertebrates and Human Assistance in Establishment

Some of the first introductions of exotic vertebrates were those commensal with humans as they colonized new areas. Among well-documented early human introductions are dingoes in Australia (Brown, 1989) and Polynesian rats [*Rattus exulans* (Peale)] on Pacific islands (Roberts *et al.*, 1992). As early Europeans explored the planet, other commensal species such as *R. rattus, R. norvegicus,* and *Mus*

musculus expanded their geographic range without deliberate human assistance and are now cosmopolitan in distribution (Brown, 1989). Not all vertebrate pests cohabitat with or are associated with human disturbance of the environment. Red deer and feral cats, for example, inhabit much of New Zealand's pristine habitats with no human management.

Most vertebrate translocations fail even with human assistance. Failure of exotics to establish may depend on life history parameters, responses to abiotic factors, inability to outcompete native species for resources or enemy free space, or chance (Cornell & Hawkins, 1993). Deliberate releases of exotic birds have establishment rates of around 20 to 35% (Veltman *et al.*, 1996). Establishment estimates for intentional vertebrate releases are biased because successes are more often recorded than failures (Ehrlich, 1989; Veltman *et al.*, 1996).

The amount of effort directed toward introduction is an important variable affecting successful colonization by vertebrates, and establishment rates increase with high levels of management and numbers of individuals released (Ehrlick, 1989; Griffith *et al.*, 1989; Williamson & Fitter, 1996b; Veltman *et al.*, 1996). In contrast, organisms that are casually introduced into new areas have much lower probabilities of establishing and proliferating. This phenomenon is expressed in the tens rule, a statistical characterization of probability outcomes for different levels of invasion success (Williamson & Fitter, 1996a, 1996b). For a variety of plants and animals, a general rule holds that 1 in 10 species imported (i.e., brought into new areas intentionally or accidentally) appear in the wild, 1 in 10 of those now found in the wild become established, and that 1 in 10 of those established with self-sustaining populations become pests (Williamson & Fitter, 1996b).

VERTEBRATE PEST MANAGEMENT

Following establishment, proliferation, and rise to pest status, control of exotic vertebrates is often prompted by economic, environmental, or conservation concerns. Several control strategies may be pursued, the most common being chemical control (e.g., poisoning) and cultural control (e.g., trapping, fencing, and shooting). The least used option has been biological control. Chemical and cultural control of vertebrate pests has been covered by Hone (1994) and Williams and Moore (1995).

Biological control is the intentional use of populations of upper trophic level organisms (e.g., predators, parasites, and pathogens) commonly referred to as natural enemies to suppress populations of pests to lower densities than would occur in the absence of natural enemies (DeBach, 1964; Van Driesche & Bellows, 1996). Biological control programs for vertebrates have employed all three classes of natural enemies: predators, parasites, and pathogens. In contrast to weeds and pestiferous arthropods, however, biological control as a population suppression tactic for management of vertebrate pests has historically received much less attention. Predators of vertebrates in the few instances they have been used have not been particularly successful. Some vertebrate predator introductions also had severe impacts on nontarget organisms. Consequently, early unpromising results discouraged intensive development of this technology (Howard, 1967; Davis *et al.,* 1976; Wood, 1985).

There is a need for increased effort using biological control agents against vertebrates, especially where resistance to toxins has developed, or behavior and terrain makes chemical and cultural control difficult and expensive (Wood, 1985; Bloomer & Bester, 1992). Biological control should be fostered internationally because many countries experience similar problems (e.g., rabbits are agricultural pests in Argentina, Australia, Chile, Europe, and New Zealand; rats, cats, and dogs attack endangered faunas on many oceanic islands; feral pigs and goats in New Zealand, Australia, and the United States degrade habitat and threaten endangered flora).

Biological control can be aided by the establishment of institutions to help coordinate regional and international research activities. For example, in 1992, the Australian federal government supported the creation of the Cooperative Research Center for Biological Control of Vertebrate Pest Populations (also known as the Vertebrate Biocontrol Center), an unincorporated collaborative venture between state and federal organizations with international cooperators (Anonymous, 1997a). The principle research goal of this institution is population suppression of noxious vertebrate species by regulating reproductive rates (Tyndale-Biscoe, 1994a).

Predators as Biological Control Agents

A fundamental issue that has important implications for biological control is understanding the regulatory effects predators have on prey populations. Determining impact of introduced predators on pest and nontarget populations is becoming increasingly more important as public awareness of potential nontarget impacts increases and the impact of past introductions on nontarget organisms has become clearer (Howarth, 1983; Van Driesche & Hoddle, 1997).

Introductions of vertebrate predators as biological control agents against vertebrates have in some instances had disastrous impacts on nontarget wildlife, especially insular communities that have lacked an evolutionary history with generalist predators (Case, 1996). An example is the impact the small Indian mongoose on native rail populations in Hawaii following its release in the 1880s for rat control

(Loope *et al.,* 1988). The mongoose has little demonstrable effect on rats (Cagne, 1988) and mongoose populations are now poisoned to protect native birds (Loope *et al.,* 1988). Exotic predators may enhance the success of introduced pest species by moderating the competitive impact of natives on introduced pests should predators reduce densities of native species (Case, 1996).

On the other hand, there is very good evidence that under certain circumstances introduced exotic predators can regulate target vertebrate pests. In such cases, predator efficacy may be affected by ecosystem complexity; by influence of such extrinsic factors as weather, disease, or human intervention on pest population growth; and by the availability of alternative food sources to sustain predators when pest populations are low.

In simple ecosystems such as islands, establishment of reproducing predator populations can result in the extinction of target pests. In the absence of alternative prey, releases of cats for rabbit control on Berlinger Island near Portugal by a lighthouse keeper resulted in the eradication of the rabbits, and subsequently cats died from starvation (Elton, 1927). In complex communities, alternate prey may be taken when primary prey populations are low for prolonged periods, as is the case for some microtine rodent species in Europe. European rodent populations exhibit fluctuations in size as a response to variation in food availability. These fluctuations are also influenced by predation. Population cycles are not observed in Southern Hemisphere countries where European pests and predators have been transferred (Korpimäki & Krebs, 1996; Sinclair, 1996).

Rabbit populations in Australia and New Zealand are maintained at low levels by introduced predators, but regulation only occurs after pest numbers have been reduced by other means. Poisoning programs in New Zealand in the 1950s and 1960s substantially reduced rabbit densities and populations were maintained at low levels by introduced predators, in particular, ferrets and cats (Newsome, 1990). Similarly in Australia, European foxes and cats maintain rabbit populations at low densities following population crashes caused by prolonged hot summers that reduce forage and browse (Newsome *et al.,* 1989; Newsome, 1990). Mouse populations are regulated in a similar fashion by predators (raptors and foxes) in Australia (Sinclair *et al.,* 1990). Introduction of two foxes *Dusicyon culpaeus* (Molina) and *D. griseus* (Gray) native to mainland Chile onto the Chilean side of Tierra del Fuego Island regulated rabbit populations after rabbit densities were substantially reduced by the myxoma virus (Jaksic & Yànez, 1983).

The suppressive action of predators on rabbits in Australia has been demonstrated through predator removal experiments [referred to as perturbation experiments by Sinclair, (1996)] in which European foxes and cats were shot from four-wheel drive vehicles at night. Removal of predators resulted in rapid rabbit population growth compared with rabbit densities in control plots in which predators were not removed (Newsome *et al.,* 1989).

The "predator pit" first conceptualized by May (1977) describes rabbit regulation in Australia and New Zealand by generalist predators. The model suggests that once prey populations fall below certain densities (i.e., because of culling or disease) predators can prevent recovery to higher levels. Generalist predators achieve this by maintaining relatively high numbers by attacking alternate prey species, and low but persistent levels of predation on species in the pit prevents pest populations from outbreaking. For the rabbit–fox system in Australia, a predator pit operates at densities of 8 to 15 rabbits per kilometer of linear transect. Below these densities foxes utilize alternate food sources (e.g., native animals) and above this critical density rabbit populations escape regulation by predators (Newsome, 1990). In New Zealand, rabbits are contained in the pit by cats and ferrets when densities are 0.4 rabbits per hectare (Barlow & Wratten, 1996). The effect of predators on long-term population dynamics of alternate prey species is generally difficult to disentangle from confounding effects of habitat degradation and competition for food and breeding sites from other introduced species (Pech *et al.,* 1995).

Foxes regulate mice and rabbits through positive density dependence at low prey densities. Increasing pest densities during outbreaks results in inverse density-dependent predation and type III functional responses (Sinclair *et al.,* 1990; Pech *et al.,* 1992). Rabbit and mouse populations escape predator regulation when favorable weather provides good breeding conditions, or when predators are controlled by shooting or poisoning (Newsome, 1990; Sinclair *et al.,* 1990; Pech *et al.,* 1992).

Predator control may be necessary for livestock protection or for conservation of endangered wildlife and when implemented resurgence of pest populations occurs (Newsome *et al.,* 1989; Newsome, 1990). Predator numbers may increase when primary prey (e.g., rabbits) are abundant. Predation by abundant predators on secondary prey (e.g., native animals) result, leading to declines in secondary prey density. Under such circumstances native prey species may only persist in refugia or in areas with artificially reduced predator pressure (Pech *et al.,* 1995). Alternatively, declining densities of a primary pest prey species (either through management or disease) can intensify predator attacks on nontarget organisms. Therefore, when conservation of endangered natives is a concern, culling of predators may be undertaken either concurrently with the decline in prey density or in anticipation of such a decline (Grant Norbury, personal communication, 1997). In such situations an integrated approach to managing vertebrate pests and their predators is necessary (Newsome, 1990).

Predator efficacy can be enhanced either through habitat modification or resource provisioning. Cats can maintain rat and mouse populations around farm buildings below environmental carrying capacity as long as they are provisioned with additional food (e.g., milk). Dietary supplementation prevents rodent extermination and the subsequent extinction of cats. Sustaining a cat population prevents uncontrolled invasions by rodents and low pest densities are maintained (Elton, 1953). Provision of nesting boxes for barn owls (*Tyto alba javanica*) reduces crop damage by rats in Malaysian oil palm plantations; rodenticide use declined, and in some instances use was eliminated (Wahid *et al.,* 1996).

Changes in management practices can improve predator efficacy. Rodent control by *Tyto alba* (Scopoli) in *Pinus radiata* Don plantations in Chile was enhanced by clearing 4-m wide strips between trees (for owls to maneuver in while in flight) and construction of perches in forests (for resting and surveillance). Barn owl numbers and predation rates on rodents increased following habitat modification (Muñoz & Murúa, 1990). Under increased predation pressure, rodents will modify foraging behaviors by reducing activity when owls are flying or making hunger calls (Abramsky *et al.,* 1996). Learning in this manner produces behavioral adaptations because of strong selection pressures to minimize predation risks on pest populations (Davis *et al.,* 1976).

Parasites as Biological Control Agents

Parasites or macroparasites (e.g., helminths, lice, ticks, fleas, and other metazoans) do not typically kill their hosts as a prerequisite for successful development as insect parasitoids do. They tend to be enzootic (i.e., remain at fairly constant levels through time) and usually must pass through a free-living stage to complete an entire life cycle (Anderson, 1979; McCallum, 1994). The potential of parasites to regulate vertebrate host populations was first proposed as early as 1911 (Lack, 1954) and later was demonstrated theoretically with Lotka–Voltera models in which parasites increased host mortality rates (Anderson & May, 1978; May & Anderson, 1978; May, 1980).

Parasites act in a positive density-dependent manner by adversely affecting host survival or reproduction (Anderson & May, 1978; Dobson & Hudson, 1986; Scott & Dobson, 1989). Host parasite load also affects the ability of individual parasites to grow, reproduce, and survive in definitive hosts; and the severity of density dependence on host and parasite fitness is affected by patterns of parasite distribution in host populations (Scott & Lewis, 1987). Helminths, for example, tend to be aggregated within host populations so that few hosts are heavily burdened while most are lightly infected (Scott & Lewis, 1987). Density-dependent constraints on parasite survival and reproduction occur in the few heavily infected hosts, and under such conditions, helminth population stability is enhanced (Anderson & May, 1978; May & Anderson, 1978). Furthermore, parasites with low-to-moderate pathogenicity exert stronger regulatory actions on populations than highly pathogenic species that cause their own extinction by killing hosts before transmission (Anderson, 1979).

Parasite regulation of vertebrate populations has been observed under field conditions. Botfly (*Cuterebra grisea* Coquillett) parasitism of voles [*Microtus townsendii* (Bachman)] in Vancouver Canada, is inversely density dependent; and botfly infestation significantly reduces vole survival, reproduction, and development (Boonstra et al., 1980). The parasitic helminth *Trichostrongylus tenuis* (Cobbold) is the primary agent responsible for long-term population cycles in red grouse [*Lagopus lagopus scoticus* (Latham)] inhabiting Scottish heathlands (Dobson & Hudson, 1994). The regulatory effect of *T. tenuis* has been demonstrated by reducing parasite infestations with helminthicides in experimental birds. Treated grouse showed increased overwintering survival, clutch sizes, and hatching rates when compared with untreated birds (Dobson & Hudson, 1994). In the laboratory, introduction of the nematode *Heligmosomoides polygyrus* Dujardin reduced mice densities by 94% in comparison with control populations. Reduction of nematode transmission rates and elimination of parasites with helminthicides allowed infested mouse populations to increase (Scott, 1987). Although host and parasite densities in this study were higher than those found in nature, the data showed that introduction of a parasite regulated host population abundance. The potential effectiveness of nematodes as biological control agents in field situations has been evaluated for control of the house mouse, an introduced pest in Australia (Singleton & McCallum, 1990; Spratt, 1990).

Mouse (*Mus domesticus*) populations erupt every 7 to 9 years in cereal-growing regions of southeastern Australia (Singleton & McCallum, 1990; McCallum, 1993) and economic losses to mouse plagues exceed $50 million (Australian) (Beckman, 1988; Singleton, 1989). Outbreaks are associated with high autumn rainfalls following prolonged periods of drought that extend the growing season for grasses that set seeds. This high-quality food source increases high mouse survivorship and breeding throughout winter. Population crashes occur when food supplies are exhausted (Singleton, 1989). Saunders and Giles (1977) suggested that droughts are necessary to remove the regulating effects of natural enemies, and this removal combined with favorable weather conditions permitted mouse numbers to increase rapidly.

Capillaria hepatica (Bancroft), a parasitic nematode that infests mice, is naturally occurring and widely distributed in pestiferous rodents in coastal areas of Australia. It is,

however, absent in mouse populations in cereal-growing areas (Singleton et al., 1991). This nematode is unique because it is the only known helminth with a direct life cycle that requires host death for transmission. Female nematodes deposit eggs in the host's liver; and these eggs are liberated by predation, cannibalism, or necrophagy with subsequent digestion of infected liver. Unembryonated nematode eggs voided after ingestion undergo embryonation to become infective and are probably consumed when mice preen their fur and feet (Singleton et al., 1991) (Fig. 1). Ground beetles (Carabidae) may vector *C. hepatica* eggs after they have been eaten (Mobedi & Arfaa, 1971). First-stage larvae emerge from ingested embryonated eggs and move into the liver through the hepatic portal system (Wright, 1961). Nematode infestation significantly reduces natality and numbers of young mice weaned by infected females (McCallum & Singleton, 1989; Singleton & McCallum, 1990; Singleton & Spratt, 1986; Spratt & Singleton, 1986). *Capillaria hepatica* is associated with introduced rat and mouse species in urban areas, and naturally occurring infections in native Australian mammals are rare probably because of the susceptibility of nematode eggs to ultraviolet radiation and desiccation (Spratt & Singleton, 1986; Singleton et al., 1991). Native Australian mice and marsupials are susceptible to experimental infection in laboratories (Spratt & Singleton, 1986).

Rats (*R. norvegicus* and *R. rattus*) are major reservoirs for *C. hepatica* in urban areas and infestation rates range from 40 to 80% (Childs et al., 1988; Singleton et al., 1991). Infestation levels are lower in sympatric mice populations (0 to 40%, Singleton et al., 1991). Low rat numbers in cereal-growing regions of Australia may be a factor contributing to the nonpersistence of *C. hepatica* in these areas (Singleton et al., 1991). Infestation of nonrodent mammals by *C. hepatica* is rare (Singleton et al., 1991) but has occurred in rabbits (Gevrey & Chirol, 1978), dogs (LeBlanc & Fagin, 1983), horses (Munroe, 1984), and humans (Pannenbecker et al., 1990). Human infections can be treated successfully (Pereira & Franca, 1983).

Exploratory models investigating the impact of *C. hepatica* on mouse populations indicated that the requirement of host death for parasite transmission is strongly destabilizing. In the absence of resource limitation mouse densities increase similarly to disease-free populations before parasites have an impact and infected populations decline in density (McCallum & Singleton, 1989). Slow regulation of mouse populations occurs because of the need for host death for transmission. Consequently, the nematode's life cycle operates on the same time scale as that of its host instead of being orders of magnitude faster, as is the case with other parasites that do not require host death for transmission (McCallum & Singleton, 1989; Singleton & McCallum, 1990).

The destabilizing influence of *C. hepatica* on mouse

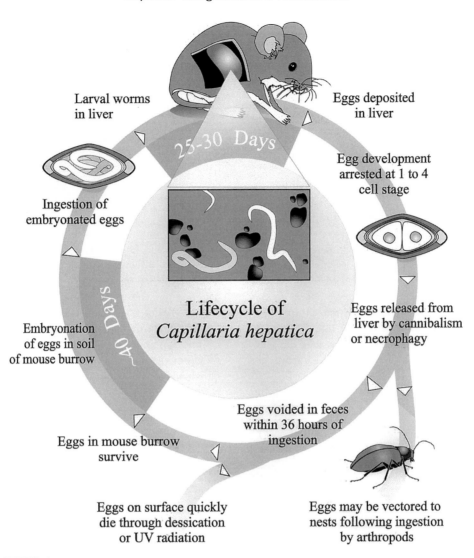

FIGURE 1   Life cycle of *Capillaria hepatica* (Nematoda) in the house mouse [*Mus musculus* (Rodentia: Muridae)] (artwork prepared by Vincent D'Amico, III). [After Singleton, G. R., & McCallum, H. I. (1990). *Parasitology Today, 6,* 190–193.]

populations may contribute to localized host and parasite extinctions. These extinctions, coupled with very low mouse densities in nonoutbreak years, result in population bottlenecks and may explain why nematodes do not persist in regions where mouse outbreaks occur. Soil type, temperature, and moisture content do not affect nematode egg survival and embryonation under favorable conditions in outbreak regions (Spratt & Singleton, 1987). Outbreak intensity can theoretically be reduced by *C. hepatica* if populations are inoculated early, preferably 1 year before an outbreak is expected (McCallum, 1993). Releases of high doses of nematode eggs in the summer or autumn when mouse densities are sufficient to enable high levels of transmission may offer the best chance for successful control (McCallum & Singleton, 1989; McCallum, 1993).

Field experiments in enclosures and with increasing populations of free-ranging mice have failed to demonstrate long-term regulation on mouse population growth with periodic inoculative releases of *C. hepatica* eggs. Unexpected declines in control populations (i.e., populations not treated with nematodes) have to some degree masked the effect of *C. hepatica* on mice populations (Barker *et al.,* 1991; Singleton *et al.,* 1995; Singleton & Chambers, 1996). Transmission of *C. hepatica* in treated populations is not density dependent and can occur at low levels for 12 to 18 months. Transmission rates show seasonal trends influenced by soil temperatures and increasing aridity (Barker *et al.,* 1991; Singleton & Chambers, 1996). Improved understanding of the influence of factors (such as temperature and rainfall on nematode persistence, survival, and transmission in field situations) and timing of releases of parasite eggs may improve releases of *C. hepatica* for control of mouse outbreaks (Singleton *et al.,* 1995; Singleton & Chambers, 1996).

Vertebrate species that successfully colonize new habitats have reduced parasite loads in comparison with mother populations from which they originated (Dobson & May, 1986). Lower infestation levels probably occur because individuals that make up small founding populations either were uninfected or had only a limited subset of the total potential parasite species found in the area of origin, or intermediate hosts required for parasite persistence were absent in the new range. Sparrows and starlings, both successful colonizing species from Europe, have two to three times fewer parasites in North America compared with populations from which they originated. Populations established outside of Europe may have benefited from reduced parasite burdens, although there are no quantitative data to indicate that this aided establishment and proliferation (Dobson & May, 1986). Investigating the role of parasites on population dynamics of rabbits in Europe with the view for possible introduction into countries where rabbits are pests is also warranted (Boag, 1989).

Introduced mammals such as rats, goats, and cats on oceanic islands exhibit depauperate parasite faunas (Dobson, 1988). Fewer parasites coupled with presumed low genetic diversity of small founding populations, and reduced selection pressures for parasite resistance may make these pest vertebrates vulnerable to introduced host-specific parasites. The ideal parasite introduced into a high-density pest population that originated from a small founding population should have low-to-intermediate pathogenicity, because such parasites establish and maintain themselves in populations at lower densities than more pathogenic species do (Anderson & May, 1978; May & Anderson, 1978). Macroparasites that reduce both host longevity and fecundity may have the potential to cause sustained reductions of host population densities (Dobson, 1988). Low genetic variability among target populations should theoretically enable introduced parasites to become more evenly distributed among hosts, and reduction in parasite aggregation would increase natural-enemy efficacy (Dobson & Hudson, 1986).

The possibility of reassociating parasites with vertebrate pests is not limited to mammals and birds. Host-specific parasites may have the potential to reduce reproduction and longevity of pest reptile (Dobson, 1988) and amphibian species (Freeland, 1985). The brown tree snake is the proximate cause of 12 native bird extinctions on Guam following its accidental introduction after World War II on military equipment (Pimm, 1987; Savidge, 1987; Jaffe, 1994; Rodda *et al.,* 1997). The snake also has caused declines of native reptile and small mammal populations, and enters houses and attacks sleeping human infants (Rodda *et al.,* 1997). Additionally, the brown tree snake has caused eco-nomic losses by adversely affecting domestic animals (e.g., chickens and pets), and high densities of snakes on power lines regularly cause short circuits that interrupt electrical supplies and necessitate repairs. Control of the brown tree snake has been attempted through trapping, but the snake's extreme preference for live bait over artificial lures has made this approach impractical (Rodda *et al.,* 1997).

The brown tree snake—native to eastern Indonesia, the Solomon islands, New Guinea, and northeastern Australia—belongs to the family Colubridae. It is the only member of this family on Guam. There is one native species of snake on Guam, the blind snake, *Rhamphotyphlopys braminus* (Daudin), which belongs to the family Typhlopidae and is the only snake occurring on many islands in the central Pacific region (T. Fritts, personal communication, 1998). The brown tree snake has extended its range and is now established on the previously snake-free island of Saipan, and this snake has been intercepted in Hawaii; Corpus Christi, Texas, and Spain (Rodda *et al.,* 1997). Given the propensity for the brown tree snake to be dispersed to new habitats within cargo loads on planes and ships, the major social, economic, and ecological problems that are caused on islands after colonization, in addition to its distant taxonomic relationship to snakes common to Pacific islands, make the brown tree snake an excellent target for biological control.

The taxonomic relationship between colubrids and typhlopids may simplify the task and reduce the cost of finding natural enemies unique to the brown tree snake. Parasites or pathogens that are host specific just to the family (i.e., Colubridae) or genus (i.e., *Boiga*) level may be safe to nontarget snakes (e.g., typhlopids) because these organisms have not evolved the ability to cause disease in distantly related hosts.

Extreme caution should be exercised when implementing a biological control program against vertebrate pests with parasites. Parasites and pathogens can pose major threats to populations of endangered animals (McCallum, 1994; McCallum & Dobson, 1995). The susceptibility of nontarget organisms, especially endemic species, to infection by candidate biological control agents should be investigated thoroughly prior to parasite releases. Reassociating parasites that preferentially infect a competitively dominant pest species may increase species diversity of invaded communities by reducing the pest's prevalence. In this instance, the natural enemy would assume the position of a keystone parasite (Marcogliese & Cone, 1997).

Pathogens as Biological Control Agents

Pathogens or microparasites include viruses, bacteria, and protozoans. Pathogens tend to be unicellular and exhibit epizootic (i.e., boom or bust) life cycles due to rapid proliferation in hosts (Anderson, 1979; McCallum, 1994).

The potential of pathogens to regulate vertebrate population densities by reducing the longevity and fecundity of infected hosts has been demonstrated theoretically with mathematical models and by perturbation experiments using vaccines (Smith, 1994). As with macroparasites, models indicate that microparasites of intermediate pathogenicity are more effective biological control agents (Anderson, 1982). Highly virulent pathogens kill themselves by destroying hosts before they can be transmitted and avirulent strains are not transmitted because they are removed by the immune system. The immune system is theorized as being responsible for maintaining the intermediate virulence of vertebrate microparasites (Anita *et al.*, 1994). Pathogens that are readily transmitted (i.e., microparasites spread by water, air, and vectors) or have high-density host populations are more contagious than those with low transmission rates (i.e., spread is by host-to-host contact) or low host densities (Ebert & Herre, 1996).

New associations between pathogens and novel hosts are generally not more harmful than those that have evolved closely with the host. Experimental evidence indicates that novel disease-causing organisms are on average less harmful, less infectious, and less fit than the same parasite strain infecting the host it is adapted to (Ebert & Herre, 1996). Also, a microparasite's ability to infect and exploit novel hosts decreases with increasing geographic and presumably genetic distance from the host to which the pathogen is adapted (Ebert, 1994). Exceptions do occur, however, and pathogens can have devastating impacts on hosts that have no evolutionary history with the disease organism. An example is the myxoma virus, the causative agent of myxomatosis in European rabbits. The use of this natural enemy against rabbits in Australia and Europe has been the most thorough biological control program against a vertebrate pest.

The myxoma virus is a member of the genus *Leporipoxvirus* (Poxviridae) and originated from South America where it was first recognized as an emerging disease of European rabbits in laboratories in Montevideo, Uruguay, in 1896. Infected laboratory rabbits died of a fatal febrile disease that caused tumors on the head and ears. The tumors resembled myxomas (a benign tumor composed of connective tissue and mucous elements) and the disease was subsequently named infectious myxomatosis of rabbits (Fenner & Marshall, 1957; Fenner & Ratcliffe, 1965; Fenner, 1994).

The indigenous host for myxoma virus in South America is the forest rabbit [*Sylvilagus brasiliensis* (Linnaeus)]. Unlike its effect on European rabbits, myxoma inoculum injected into forest rabbits caused benign fibromas at the site of inoculation that persisted for many months, although death did not occur. Mosquitoes were implicated in vectoring the disease from forest rabbits to European rabbits being bred in South American rabbitries. Another leporipox-

virus has been isolated in California from the brush rabbit, *Sylvilagus bachmani* (Waterhouse), and is closely related to the myxoma virus (Fenner & Marshall, 1957; Fenner & Ratcliffe, 1965; Fenner, 1994; Fenner & Ross, 1994; Ross & Tittensor, 1986).

Myxoma virus has been used in Australia, Europe, Chile, and Argentina for biological control of European rabbits. The virus was first imported into Australia from Brazil in 1919 and 1926 but was not released (Fenner & Ratcliffe, 1965). Work by Australians with the virus began again in the United Kingdom in 1934 and continued with caged rabbits on Wardang Island off the south coast of Australia. The virus was successfully established on mainland Australia in 1950 (Fenner, 1994) and within 2 years it had established itself over most of the rabbit's range (Fenner & Ratcliffe, 1965). The virus initially had a major impact on the estimated 600 million rabbits and on the damage they caused, reducing population density by 75 to 95%. Efficacy was dependent on climate and rabbit population susceptibility. Populations have subsequently increased and stabilized at around 300 million because of myxomatosis.

Damage attributable to rabbits still amounts to $600 million (Australian) annually, including both lost agricultural production and cost of control applications (Robinson *et al.*, 1997). In addition to agricultural losses, rabbits severely affect native flora by eating foliage and inducing wind and water erosion of soils by overgrazing. Native fauna are also affected as rabbits out-compete indigenous herbivores and dense rabbit populations sustain exotic predator populations that feed on native animals (Gibb & Williams, 1994; Myers *et al.*, 1994; Robinson *et al.*, 1997).

Within a few years of the initial panzootic, field isolates of the virus showed less virulence when compared with the original strain that had been released. The original strain killed >99% of laboratory rabbits on average 10.8 days after infection, while circulating strains caused 90% mortality after 21.5 days. Genetic resistance in rabbits was also detected (Fenner & Marshall, 1957; Fenner & Ratcliffe, 1965). Dual natural selection had occurred, the virus had attenuated, and rabbits had increased in resistance to the disease.

Mosquitoes have been responsible for vectoring myxoma virus in Australia. The European rabbit flea, *Spilopsylus cuniculi* (Dale), an important vector in Europe, was introduced into Australia in 1968 and increased the geographic distribution of the disease. This flea did not persist in areas with rainfall <200 mm. The xeric adapted Spanish rabbit flea, *Xenopsylla cunicularis* Smit, was introduced in 1993 and active redistribution is still ongoing (Fenner & Ross, 1994).

New Zealand also has inordinate numbers of rabbits, and attempts to establish the myxoma virus from 1951 to 1953 failed because of inclement weather and a paucity of suitable arthropod vectors. Further attempts at establishment were not undertaken because poisoning programs had reduced rabbits to very low numbers, additional control expenditure was unjustifiable, and the New Zealand public was not in favor of using lethal myxoma virus for rabbit control on humanitarian grounds (Gibb & Williams, 1994).

Until the 1980s, myxomatosis was the only disease known to severely affect rabbit numbers. A second highly contagious viral disease emerged in the mid-1980s and was accidentally introduced onto mainland Australia (O'Brien, 1991). It is the first pathogenic natural enemy to have established in New Zealand for biological control of rabbits. Rabbit calicivirus disease (RCD) [also known as rabbit hemorrhagic disease virus (RHDV)] emerged as a fatal disease in 1984 in Angora rabbits exported from East Germany to Jiangsu Province of China (Liu *et al.*, 1984). In 1986, the disease appeared in Italy where 38 million rabbits were estimated to have died. The disease spread rapidly through rabbit populations in Europe reaching the United Kingdom in 1992 (Chasey, 1994). The probable mechanism for dispersal in continental Europe was the movement of live rabbits and rabbit products. Transmission of RCD from France into coastal areas of southeast England is thought to have occurred by wind-borne aerosols containing virus, birds, and transchannel ferry traffic (Chasey, 1994). Outbreaks of RCD occurred in Mexico in 1988 and 1989 (Gregg *et al.*, 1991) and in Réunion Island in the Indian Ocean in 1989. Movement of RCD to these areas probably occurred with imports of frozen rabbit carcasses from China because the virus can survive freezing to temperatures of −20°C (Chasey, 1994).

The RCD virus belongs to the Caliciviridae and consists of a positive sense, single-stranded RNA genome, enclosed by a sculptured capsid composed of multiple copies of a single major protein of 60 kDa, and is 30 to 40 nm in diameter (Ohlinger *et al.*, 1990; Parra & Prieto, 1990). Disease symptoms are characterized by high morbidity and mortality in rabbits over 8 weeks of age. Younger rabbits often survive infection and may develop antibodies to RCD virus (Nagesha *et al.*, 1995). Clinically, RCD symptoms are expressed after an incubation period of 24 to 48 h in which a febrile response and increasing lethargy are observed. Infected rabbits typically die within 12 to 72 h postinfection and 90% mortality is observed after 5 days. Necropsies show a pale swollen friable liver, enlarged spleen, and clots in blood vessels. Death is ascribed to acute necrotizing hepatitis and possible hemorrhaging (Fuchs & Weissenböck, 1992; Studdert, 1994). However, necropsies close to the time of death show an absence of hemorrhaging and inclusion of hemorrhagic in the name of this rabbit disease that indicates the cause of death is misleading (Studdert, 1994).

A different viral disease is responsible for European brown hare syndrome (EBHS) which causes severe hepatic necrosis in hares (*Lepus europaeus* Pallas and *L. timidus* Linnaeus). The disease was first recorded in Sweden in 1980, and spread through continental Europe and reached the United Kingdom in 1990 (Fuller *et al.*, 1993). In Sweden, losses of hares to EBHS occurred 10 years prior to sympatric rabbit populations developing RCD. Similar observations were made in the United Kingdom where hares began dying from EBHS 2 years before RCD was observed in rabbit populations (Fuller *et al.*, 1993). Electron microscopy, nucleotide sequencing, and experimental cross-transmission studies have indicated that RCD virus and EBHS virus are closely related (Le Gall *et al.*, 1996) but distinct members of the Caliciviridae (Chasey *et al.*, 1992; Nowotny *et al.*, 1997). Disease symptoms are generally similar for rabbits and hares but show distinguishing characteristics in necrosis of liver lobules and clotting of blood vessels (Fuchs & Weissenböck, 1992). Serological studies on rabbit sera collected in 1961 from Czechoslovakia and Austria indicate that RCD virus probably evolved from an apathogenic strain endemic to Europe from at least this time (Nowotny *et al.*, 1997).

Studdert (1994) speculates that the causative agent of RCD probably existed in Europe as a quasi species, a collection of indifferent mutants with a variety of accumulated nucleotide changes. In this scenario, mutants occupied a specific ecological niche until one strain better adapted to prevailing conditions became the dominant member of the population. Adaptation may have occurred because mutations caused increased virulence in an avirulent rabbit virus or increased the host range of hare-infecting viruses by allowing mutant strains to bind more efficiently to surface receptors on rabbit hepatocytes. Given Studdert's (1994) speculative scenario, RCD virus may be a highly evolvable organism.

European rabbits appear to be the only animals susceptible to infection by RCD virus, and vaccines have been developed to protect domestic animals (Boga *et al.*, 1997). Other rabbits including cottontail rabbits (*Sylvilagus* spp.), black-tailed jack rabbits (*Lepus californicus* Gray), volcano rabbits [*Romerolagus diazi* (Ferrari-Pérez)] (Gregg *et al.*, 1991), and hares (Gould *et al.*, 1997) are not affected. The limited host range of RCD virus makes it an obvious candidate for use in a biological control program against European rabbits in New Zealand and Australia. A joint biological control program between these two countries using RCD virus was initiated in 1989 and a strain of virus from the Czechoslovakia Republic was imported into Australian quarantine facilities in 1991 to test effects on nontarget species (Robinson & Westbury, 1996).

Host-specificity testing of 28 nontarget species in Australia for susceptibility to RCD virus further verified the limited host range of this natural enemy. Test subjects included domestic lifestock (horses, cattle, sheep, deer, goats, pigs, cats, dogs, and fowls), noxious exotic vertebrates (foxes, hares, ferrets, rats, and mice), native mammals (eight species), birds (five species), and reptiles (one species). There was no evidence for viral replication, clinical signs, or lessions in any organisms tested (Gould *et al.*, 1997). Artificial inoculation of RCD virus in North Island brown kiwis (*Apteryx australis mantelli* Bartlett) and lesser short-tailed bats (*Mystacina tuberculata* Gray), both native to New Zealand, also failed to produce disease symptoms (Buddle *et al.* 1997).

The apparent host specificity of RCD virus to rabbits, rapidity of action, and the capacity to infect rabbits from other rabbits, [through feed and feces, or from a contaminated environment (O'Brien, 1991)] prompted further evaluation of this biological control agent under field quarantine conditions in Australia. Studies monitoring the effects of RCD virus on rabbit populations were initiated on Wardang Island near Adelaide off the south coast of Australia in 1995 (Rudzki, 1995; Robinson & Westbury, 1996). In September 1995, RCD breached quarantine and appeared on mainland Australia, possibly carried there by calliphorid flies and onshore winds (Cooke, 1996; Lawson, 1995). Attempts at containment failed (Seife, 1996). Within 2 months of the initial discovery of RCD virus on the mainland, an estimated 5 million rabbits were killed in South Australia. In dry areas, 80 to 95% of infected populations died (Anderson, 1995) with dead rabbits averaging 15 per hectare. Elsewhere, fatality rates were closer to 65% (Anonymous, 1997b). In the period from October to November 1995, an estimated total of 30 million rabbits died from RCD in South Australia and the majority of surviving rabbits were less than 6 weeks of age (Cooke, 1996). The development of resistance in young rabbits may have profound effects on the long-term population dynamics on the rabbit–RCD virus system. Ten arthropod vectors of RCD virus have been identified and include flies, mosquitoes, and rabbit fleas (Anonymous, 1997b).

Rates of spread of RCD are greatest in spring and autumn at 10 to 18 km a day and are correlated with peaks of insect activity. Dispersal of the disease probably has been assisted by humans moving contaminated material to new areas (Cooke, 1996). Increased attacks on native fauna by exotic predators such as foxes because of declines in rabbit numbers do not appear to have occurred because predator populations have declined with rabbit numbers (Anonymous, 1997b). The virus is now endemic in Australia and will probably be officially declared as a biological control agent under the Biological Control Acts of the Commonwealth and States (Robinson & Westbury, 1996).

RCD virus was smuggled into the South Island of New Zealand by high country farmers in August 1997 and illegally disseminated by feeding rabbits carrots and oats satu-

rated with contaminated liquefied rabbit livers. A network of cooperators spread the virus over large areas of the South Island and its subsequent spread (human assisted through the movement of carcasses, baiting, and insect vectors) made containment and eradication of the disease impossible. Such actions by farmers clearly violated New Zealand's Biosecurity Act, which was enacted in part to protect agriculture from unwanted introductions of pests. The New Zealand government has sanctioned controlled virus releases into new areas. The short-term impact of RCD on New Zealand rabbit populations has resulted in 47 to 66% mortality in central Otago and large-scale field studies are planned (G. Norbury, personal communication, 1997).

Cats on oceanic islands have been subjected to biological control with pathogens. Six cats were introduced onto Marion Island in the Indian Ocean in 1949 (Howell, 1984); by 1977, numbers were in excess of 3000 and were increasing an average of 23% per year (van Rensburg *et al.*, 1987). Populations were sustained by consuming approximately 450,000 seabirds yearly and cats were probably responsible for the local extinction of the common diving petrel *Pelecanoides urinatrix* (Gmelin) (Bloomer & Bester, 1992). Surveys of cats on Marion Island revealed the presence of feline herpes virus and feline corona virus, but the highly contagious feline parvo virus was absent in the population (Howell, 1984).

Initiation of a biological control program with feline parvo virus, the causative agent of feline panleucopenia, began in 1977 with the release of 93 artificially inoculated feral cats collected from the island (Howell, 1984). The disease reduced cat numbers by 82% after 5 years by reducing fecundity and increasing mortality of juvenile cats (van Rensburg *et al.*, 1987). Virions found in high concentrations in feces, urine, saliva, and vomit were transmitted through direct contact between cats or contact with contaminated objects (Howell, 1984). Annual declines of cat numbers were 29% from 1977 to 1982. This rate decreased to 8% per year from 1981 to 1983 and was accompanied by lower titers of virus in serum samples collected from feral cats, indicating that viral efficacy was decreasing (van Rensburg *et al.*, 1987). At reduced densities, hunting and trapping became viable and have been incorporated into an ongoing eradication program that may be assisted by the use of trained dogs (Bloomer & Bester, 1992).

Sexually transmitted diseases have adverse effects on domestic and wild vertebrates by reducing survival, conception rates, and numbers of offspring born and successfully weaned (Smith & Dobson, 1992). Rabbits are susceptible to infections of venereal spirochetosis (*Treponema cuniculi*), which causes sterility (Smith & Dobson, 1992). Goats can develop trichomoniasis, a sexually transmitted disease caused by the flagellated protozoan *Trichomonas foetus* (Reidmuller). This pathogen has been suggested as a biological control agent for goat populations on oceanic islands that lack this microparasite (Dobson, 1988).

NEW AVENUES FOR BIOLOGICAL CONTROL OF VERTEBRATES

Sexual transmission of diseases may further guarantee host specificity in biological control programs. It also enhances the ability of parasites and pathogens to persist in low-density populations or solitary species (e.g., predators). The rate of spread of sexually transmitted organisms is tightly correlated with mean and variance of the numbers of sexual partners per host because of the need for host-to-host contact (horizontal transmission) for transmission. Host population density is not important with respect to persistence or rate of spread of sexually transmitted diseases. This property, coupled with asymptomatic carrier states, long infectious periods, or vertical transmission (infective propagules are passed from mother to offspring), greatly enhances the ability of pathogens to persist in low-density host populations (Smith & Dobson, 1992).

Because sexually transmitted organisms can persist in low-density populations or populations of declining density, the potential of genetically engineering sexually transmitted viruses to sterilize infected hosts is being investigated (Barlow, 1994). Viruses that have antigens from the host sperm, or the zona pellucida around host eggs engineered into the genome provoke an immune system response that renders the recipient sterile. Immunocontraception (also referred to as immunosterilization) as a means to control noxious vertebrates is being actively pursued by Australia and New Zealand (McCallum, 1996). An alternative approach to immunocontraception is to use genetically modified microparasites to prevent lactation in females so that juveniles are not successfully weaned or to interfere with hormonal control of reproduction (Cowan, 1996; Jolly, 1993; Rodger, 1997).

Immunocontraception for Control of Vertebrate Pests

Many species that become pests are distinguished from nonpestiferous species by their higher intrinsic rates of increase (r_m). Pest vertebrates have high r_m values characterized by large litters, and by maturing sexually at young ages. Agents that reduce reproductive rates may be more effective for control than mortality-inducing biological control agents are because resistance development should take longer to occur and population recovery would be slower (Tyndale-Biscoe, 1994b). Resistance development may be further delayed by combined use of multiple agents that affect fertility in different ways [e.g., using agents that cause sterilization, alter levels of reproductive hormones, or affect lactation (Cowan & Tyndale-Biscoe, 1997)]. In sexually reproducing vertebrates, proteins associated with male and female gametes are potentially foreign antigens in the opposite sex. Exposure to reproductive antigens occurs when females receive sperm and accessory fluids from males during copulation. As a general rule, females do not develop antibodies to these antigens because physiological and immunological mechanisms have evolved to prevent this (Robinson & Holland, 1995). Inoculation of sperm into females of the same species either subcutaneously or intramuscularly produces high sperm antigen antibody titers in recipients. In most cases, this causes either permanent or temporary infertility in females. Such results indicate that sperm antigens in the reproductive tract are tolerated and that exposure to these antigens by different routes overcomes protective mechanisms, with infertility resulting (Robinson & Holland, 1995).

Sperm antibodies in females that can arise from either systemic or local immune responses are found in cervical mucus, genital fluids (e.g., endometrial, tubal, and follicular fluids), and blood. Antibodies bind to sperm, often in specific locations such as the head, midpiece, tail shaft, or tail tip. Once bound to sperm, antibodies cause agglutination (e.g., irreversible binding to cervical mucus that normally aids sperm transport) or immobilization of sperm. Antibodies may also interfere with acrosome reactions preventing ovum penetration and fertilization, or they block the binding of sperm to the zona pellucida (Shulman, 1996).

The zona pellucida (zona) that surrounds growing oocytes and ovulated eggs is antigenic and available to circulating antibodies during oocyte growth and ovulation. Nonreproductive tract inoculation of females with zona preparations leads to infertility (Millar et al., 1989). Antibodies produced in response to administered zona antigens bind to the zona and prevent sperm penetration (Millar et al., 1989). Zona glycoproteins are highly conserved among mammals, for example, nonspecific pig zona preparations cause infertility in humans, primates, dogs, rabbits, horses, and deer (Robinson & Holland, 1995). A major objective in immunocontraception research is isolation of species-specific zona glycoproteins that do not cause sterility induced by immune response in species from which zona preparations were not derived. Low variability among zona glycoproteins may limit the number of species-specific zona preparations for immunocontraception (Millar et al., 1989).

Immunocontraception for wildlife population control has been successfully implemented for horses (Equus caballus Linnaeus) (Kirkpatrick et al., 1992, 1997). Free-ranging feral mares inoculated by dart gun with porcine zona pellucida showed depressed urinary estrogen concentrations and failure to ovulate. Zona booster inoculations given 2 years after initial inoculations prevented conception in treated horses for a third year compared with control populations that were not vaccinated. Contraceptive effects were reversible after 4 years of consecutive treatment, but prolonged treatment (5 to 7 years) with zona preparations caused irreversible ovarian dysfunction and fertility loss (Kirkpatrick *et al.,* 1992, 1997). Similar results have been achieved with porcine zona pellucida inoculations in whitetail deer [*Odocoileus virginianus* (Zimmerman)] (Kirkpatrick *et al.,* 1997).

Gametic antigens that induce immune response can be administered by baits that are ingested by target organisms or can be inoculated directly into hosts with darts or bullets (Tyndale-Biscoe, 1994b). Injection of foxes with sperm antigens reduces fertility from 75 to 35%. Baits are considered the favorable method for delivering antigens to foxes in Australia. Potential baits include dried meats that contain microencapsulated antigens. Use of recombinant bacterial vectors (e.g., *Salmonella typhimurium*) also are being considered. An orally administered agent needs to reach the lower gastrointestinal tract to stimulate a response in the common mucosal immune system in the gut-associated lymphoid tissue. This in turn induces mucosal immunity in the reproductive tract of female foxes and causes sterilization (Bradley *et al.,* 1997).

At present, an effective bait specific to foxes that is environmentally stable and easy to manufacture has not been developed. Nontarget impact is a concern because most antigens exhibit some specificity to the family level only. Effective vaccines for rabies have been delivered as oral baits to foxes in Europe, demonstrating the baiting technique is an effective dissemination method (Bradley *et al.,* 1997). Models indicate density-independent factors such as drought and rain (which affect pasture growth and rabbit numbers) strongly influence the effectiveness of bait-delivered fertility control in reducing fox abundance (Pech *et al.,* 1997).

An alternative proposal to deliver antigens orally is to develop transgenic plants to produce and deliver gametic antigens in palatable form to herbivorous pests. Plants could be sown over target areas and allowed to become self-propagating vaccines. Transgenic seeds, fruits, or leaves (e.g., transgenic carrots or maize) could be harvested and used as oral baits delivered to specific sites such as fenced watering points that allow pest animals access while excluding lifestock (Smith *et al.,* 1997).

Baiting is an expensive form of control that requires monitoring of dosage and uptake rates and multiple area-wide applications. Problems of hormonal modification of behavior and delayed population control are additional drawbacks. One advantage is that baits can be used to treat localized populations that are problematic. Similar shortcomings exist with antigen inoculations by projectiles where cost estimates are significant. To control the estimated 300,000 wild horses in Australia with dart-delivered

porcine zona pellucida would cost $20 (Australian) per horse per year compared with 50 cents for permanent control with a bullet (Tyndale-Biscoe, 1991). Lethal methods of control provide immediate impact on pest populations and reduce pest status rapidly, with control being quickly observable. In contrast, fertility impairment is not immediate, population responses are delayed, and large proportions of populations need to be sterilized for this technique to be effective. Large-scale distribution of gametic antigens might be possible through releases of host-specific microparasites expressing species-specific antigen genes (Tyndale-Biscoe, 1994a, 1994b). Self-spreading and replicating parasitic vectors that have been genetically engineered and that may require periodic reinoculation into populations are analogous to augmentative biological control programs with traditional natural enemies (e.g., parasitoids, predators, or pathogens) released periodically for the control of pest arthropods.

Host-specific viruses carrying foreign DNA could be cheap and effective biological control agents that have the potential to disseminate widely by sexual transmission, contagion, or arthropod vectors. The selected micro- or macroparasite must be able to carry foreign DNA coding for gametic antigens, as well as promoters to express foreign genes and cytokines to enhance effectiveness (Tyndale-Biscoe, 1994b). Such agents must be able to reduce growth rates of infected populations and to maintain reproductive rates at lower levels (Caughley *et al.,* 1992), and should not interfere with sexual behavior or social organization (Caughley *et al.,* 1992; Robinson & Holland, 1995; Tyndale-Bisoce, 1994b). With some pests such as rabbits and foxes, dominant members of populations make the main contribution to reproduction and inhibit breeding by subordinate members by occupying prime territories.

Ideally, a sterilizing agent should not change social hierarchies by allowing individuals with lower social status to successfully rear more offspring because this will cause pest populations to increase substantially (Caughley *et al.,* 1992). Genetically engineered agents should sterilize females because models predict greater population suppression with infertile females than with sterilized males (Barlow, 1994; Caughley *et al.,* 1992). In the absence of arthropod vectors, sexually transmitted diseases engineered to cause sterilization are superior to nonsexually transmitted ones because multiple matings with sterilized females increases contact rates and the competitive ability of the engineered agent with nonsterilizing strains. The impact of immunocontraception is further enhanced if the sterilizing agent causes limited host mortality and there is low naturally occurring immunity to sexually transmitted diseases (Barlow, 1997).

Sexually transmitted herpes-type viruses are being proposed as vectoring agents to induce sterilization in brushtail possums in New Zealand (Barlow, 1994; Barlow, 1997).

The recently identified borna disease virus that causes wobbly possum disease in New Zealand may be a suitable alternative to a herpes virus (Atkinson, 1997).

The myxoma virus and murine cytomegalovirus are being investigated as gamete antigen delivery agents for rabbits and mice, respectively, in Australia (McCallum, 1996; Tyndale-Bisoce, 1994b; Shellam, 1994). Four potential insertion sites for genes coding for gametic antigens have been identified in myxoma virus and recombinants have been constructed to express two *Esherichia coli* (Escherich) enzymes and influenza virus hemagglutinin genes. The ability of a novel myxoma virus to compete and spread among existing myxoma strains in field situations has been demonstrated by monitoring the spread of virus containing identifiable gene deletions (Robinson *et al.,* 1997). The myxoma virus that can express foreign genes may operate as a vector for gametic proteins (Robinson *et al.,* 1997). Work is continuing on isolating and inserting rabbit gamete antigen genes into the myxoma virus genome (Robinson *et al.,* 1997).

The responses of experimental rabbit and fox populations in Australia to imposed sterility by surgical ligation of fallopian tubes in females have been studied in an attempt to simulate the effects of virally mediated immunocontraception after recombinant virus establishment in wild populations. This technique prevents conception among predetermined proportions of females in populations without interfering with hormones or reproductive behavior (Williams & Twigg, 1996). The dynamics of 12 rabbit populations enclosed by rabbit-proof fencing that exhibited 0, 40, 60, or 80% sterilization of females were studied in each of two locations in western and eastern Australia where climate patterns differed. Females born into treatment populations were trapped and sterilized to maintain the same overall sterility levels (Williams & Twigg, 1996). Juvenile rabbits born into populations with sterilized females exhibited greater survivorship because of lowered competition for resources. This greater survival compensated for decreased fertility, but recruitment rates were ultimately constrained by environmental factors (e.g., depletion of vegetation). In populations with 80% sterility, reduced juvenile mortality did not compensate fully for lowered reproduction, smaller numbers of rabbits were recruited into these populations, and numbers subsequently declined. These results indicate that levels of sterilization with a genetically altered micorparasite have to reach at least 80% to achieve reductions in population density (Williams & Twigg, 1996).

Surgical sterilization does not affect reproductive behavior in treated populations. Sterile dominant female rabbits maintain hierarchical dominance, increased body weight over control females, continued to defend prime territory, and engaged in normal reproductive behavior including breeding burrow construction (Tyndale-Bisoce, 1994b).

Birth rates of sexually mature females were in direct proportion to the level of fertility in experimental populations, indicating that fertile females did not respond to female infertility or decreased densities of young by producing larger litters (Williams & Twigg, 1996).

Sterilized females tended to live longer than unsterilized females. This increased longevity suggests that sterile females may proportionately increase as treated populations reach an equilibrium density. Obviously, larger proportions of sterile females reduce population productivity and the numbers of fertile females that a sterilizing microparasite would have to infect and sterilize. Higher proportions of sterile females may reduce numbers of infective individuals harboring sterilizing microparasites and numbers of vectors (e.g., fleas that would spread an engineered myxoma virus), and may contribute to decline of transmission rates. These interactions need to be clarified and mathematical models may be of use here (Williams & Twigg, 1996).

Engineered microparasites that sterilize pest animals offer the possibility of humane control without killing or causing animals to suffer the effects of debilitating disease. As a form of biological control, immunocontraception may also reduce the need for broadcast distribution of toxins for pest suppression, thereby reducing environmental contamination and nontarget mortality. This is of special concern when pests inhabit suburbs, urban parks, government and state campuses, nature reserves, military bases, or other areas where lethal controls may no longer be legal or safe (Kirkpatrick *et al.,* 1997; Williams, 1997). The concept of virally mediated immunocontraception has generated considerable debate on legal and ethical issues concerning releases of engineered microorgansims into the environment. Once contagious recombinant agents that cause permanent sterilization in animals are released into the environment they cannot be recalled (Tyndale-Bisoce, 1995).

Several potential risks are recognized. First, engineered viruses that are host specific and contain species-specific antigens could mutate and infect and sterilize nontarget species after release (Anderson, 1997). Under such conditions it may be difficult if not impossible to contain and eradicate a mutant virus from an infected animal population that is abundant, secretive, and free ranging. Second, sterilizing viruses either might cross international boundaries accidentally or be maliciously moved to sterilize desirable organisms in new areas (Tyndale-Bisoce, 1994b). For example, viruses engineered with little host specificity to sterilize widely dispersed marsupial pests in New Zealand may enter Australia and infect endangered wildlife (Rodger, 1997; McCallum, 1996); engineered myxoma viruses may spread from Australia into the Americas and sterilize native rabbit species (Tyndale-Bisoce, 1995).

Third, dart-delivered contraceptives used for wildlife control in the past have had adverse effects on individuals within target populations. Changes in morphology of reproductive organs, secondary sexual characteristics, and behavior have been observed. Viruses that induce sterility could alter genetic profiles of target populations because infectious agents may act as a new reproductive disease and individuals may exhibit differential susceptibility (Nettles, 1997). Fourth, public concerns over the use of viruses and genetic engineering indicate substantial apprehension about the use of sterilizing viruses for pest management, these fears that need to be fully alleviated may delay or prevent field trials and widespread application (Lovett, 1997).

Despite potential drawbacks, immunocontraception is a potentially cost-effective method for reducing pest impact on endangered native species (Sinclair, 1997) and on agricultural yields, and is an additional tool for sustainable pest management (Williams, 1997). A sterilizing agent that does not cause painful disease symptoms is an ethically acceptable form of pest control that is justifiable from animal rights perspectives, because it does not cause the suffering typical of current lethal methods (e.g., trapping, shooting, poisoning, and introduced disease) (Oogjes, 1997; Singer, 1997). Under certain circumstances, the use of vectors to disseminate genetically engineered viruses is warranted (McCallum, 1996). Experience with the myxoma virus in Australia indicates that it has not been deliberately or accidentally spread to any other country since its introduction in the 1950s because of either the lack of suitable arthropod vectors or the inability of the virus to establish where different strains are already present. This history may indicate possible difficulty for unintentional establishment of genetically engineered microparasites in new areas, and establishment of engineered myxoma viruses may be possible only with carefully timed and repeated releases into rabbit populations (Tyndale-Biscoe, 1995).

However, such safeguards may be moot if a highly competitive sterilizing strain is engineered and released. Quarantine legislation designed to prevent accidental or intentional but illegal importation of unwanted organisms would be exercised by countries under current international obligations and should impede establishment in new countries if rigorously enforced. However, current legal safeguards may be insufficient. New Zealand's experience with RCD indicates it is possible for lay people to illegally import and establish reproducing populations of exotic pathogens. In Australia, RCD breached a carefully planned quarantine on an offshore island. Unintended establishment and proliferation of engineered viruses may be contained if outbreaks are recognized early, and if proportions of susceptible individuals are removed rapidly from the population either by culling or by immunizing against the pathogen (Tyndale-Biscoe, 1995). This has never been tried with wild animal populations. The containment of contagious pathogens, such as foot and mouth disease in livestock, indicates such an approach may be possible. Highly attenuated forms of myxoma virus are used to protect wild and domestic rabbits

in France and the United States, indicating the availability of such technology for this virus at least (Fenner & Ross, 1994; Tyndale-Biscoe, 1995). Limited field trails with sterilizing microorgansisms are unlikely before 2005 (Anderson, 1997).

REGULATING VERTEBRATE INTRODUCTIONS

There is abundant evidence that introduced exotic vertebrates that establish feral reproducing populations have disastrous consequences for agriculture and preservation programs for native plants and animals. Sources of current vertebrate introductions include sellers and buyers of exotic pets; acclimatization societies that import, establish, and relocate game animals and whose constituents include hunters and fishermen; and farmers and ranchers who import and experiment with novel lifestock (e.g., fitch farming). Exotic vertebrates have in some instances great economic importance (as with lifestock and game animals), they also enjoy public popularity because of interest in hunting, fishing, eating, or viewing large and unusual animals in familiar environments. The negative ecological aspects of introduced vertebrates may be poorly understood by the public at large. Such limited understanding may hinder control efforts and prevention of importation (Bland & Temple, 1993).

Legislation has been passed in the United States to minimize risks of importing new and relocating existing vertebrate species. The Lacey Act passed in 1900 and ammended in 1981 was enacted to protect certain animals and endangered habitats, and to prevent introduction of noxious pests. Under the act, violation of the law can result in fines and imprisonment [see 18 USC §42; Importation or shipment of injurious mammals, birds, fish (including mollusks and crustacea), amphibia, and reptiles; permits, specimens for museums; regulations—for more details]. Similar legislation has been developed in New Zealand. The Biosecurity (1993) and Hazardous Substances and New Organisms (HSNO) (1996) Acts were devised to protect the environment by preventing or managing the adverse effects of hazardous substances and exotic organisms.

Campbell (1993) points out that existing laws have many loopholes and are not effective when applied, indicating a need to improve existing regulations and to develop new laws to curtail unwanted entry by alien vertebrates. One proposal is to require importers of exotic organisms to develop "clean lists" and to prove that organisms are not potentially invasive and disrupting to native ecosystems (Campbell, 1993). Legislative approaches limiting imports and exports of organisms may encounter complaints under the General Agreement of Tariffs and Trade (GATT) that stricter quarantine measures are an unacceptable imposition of one country's environmental standards on others (Campbell, 1993). There is an obvious need for greater cooperation among interest groups, scientists, and legislators to devise solutions to problems associated with continuing introductions of exotic species and to provide direction for future action.

BIOLOGICAL CONTROL OF EXOTIC PESTS AS AN EVOLUTIONARILY STABLE CONTROL STRATEGY

Development of resistance (behavioral or physiological) to pesticides (e.g., rodenticides) by vertebrates, and the need for repeated or multiple simultaneous control strategies (e.g., poisoning combined with trapping and hunting) indicate that control of vertebrates is an ongoing endeavor that attempts to reduce agricultural damage and losses (Greaves, 1994) or to protect wilderness areas (Cowan, 1992; Morgan *et al.,* 1986; Payton *et al.,* 1997) from pest damage. Biological control has several advantages over chemical and cultural control practices (Van Driesche & Bellows, 1996): (1) it is relatively cheap and biological control programs are often quicker to implement than to develop and to register new pesticides; (2) use of carefully screened natural enemies increases selectivity of attack toward target pests; (3) natural enemies in many instances are self-perpetuating and self-distributing; and (4) development of resistance to natural enemies is extremely rare.

One documented case of pests developing resistance to natural enemies is the development of resistance to myxomatosis by rabbits and corresponding attenuation of highly virulent strains of the myxoma virus to strains of intermediate virulence (Fenner & Ross, 1994). The myxoma virus–rabbit system in Australia and Europe is dynamic with increasing rabbit resistance selecting for more virulent strains of virus. This suggests that for the short-term, at least, the system is coupled in an antagonistic coevolutionary arms race (Dwyer *et al.,* 1990).

Flexible natural-enemy behavior patterns and physiology have the potential to weaken evolutionary responses that can cause pest resistance to introduced control agents (Holt & Hochberg, 1997; Jervis, 1997). In comparison, pesticides and cultural controls tend to target a fixed physiological or behavioral function or pattern, and the resulting selection regime is constant allowing pests either to increase tolerance to poisons or to learn and develop avoidance behaviors (e.g., bait and trap shyness).

Spatial heterogeneity of natural-enemy attack limits selection pressure on hosts by natural enemies, thus reducing the rate of resistance development by pests compared with uniformly applied selection pressures such as pesticides. Pests that escape attack move into enemy-free areas and continue breeding; thus, the rate of coevolution is reduced by susceptible pests in transient refuges (Jervis, 1997). At the metapopulation level, natural enemies may be ineffective selection agents because of widespread extinction and establishment of pest subpopulations that maintain pest susceptibility. Additionally, resistance development may involve costs leading to a corresponding decrease in fitness. For example, increased tolerance to attack may reduce the pest's reproductive capacity and ability to compete for resources, or may increase susceptibility to other mortality agents (Holt & Hochberg, 1997).

There are opportunities to enhance biological control programs against vertebrate pests that cause social, agricultural, and conservation problems. In many instances, biological control offers the best chances for long-term control, particularly in isolated areas with rugged terrain, in suburban areas with high-density human populations, or in places where pests are nocturnal or secretive. Biological control will not totally alleviate vertebrate pest problems. It may, however, reduce the vigor of pest populations, thereby reducing damage, minimizing nuisance value, or allowing native species to compete more effectively for food and breeding sites. Programs could be initiated to simply reassociate host-specific micro- and macroparasites with pest populations that have depauperate natural-enemy faunas (Dobson & May, 1986), and there is no shortage of targets as small founding populations of vertebrates continue to invade and proliferate in new habitats. Genetically engineered natural enemies are additional tools to aid biological control efforts. Research with agents that cause immunocontraception will likely diversify as advances in molecular biology continue, and routes alternative to sterilization may be taken. This area of vertebrate biological control will be tested more thoroughly once small-scale and long-term field trials begin with sterilizing microorganisms.

Acknowledgments

I thank Vincent D'Amico, III of Bean's Art Ink for preparing the *Capillaria hepatica* life cycle schematic. Thomas Fritts of the United States Geological Survey provided information on the brown tree snake.

References

Abramsky, Z., Strauss, E., Subach, A. Kotler, B. P., & Reichman, A. (1996). The effect of barn owls (*Tyto alba*) on the activity and microhabitat selection of *Gerbillus allenbyi* and *G. pyramidum.* Oecologia (Berlin), 105, 313–319.

Anderson, I. (1995). Runaway rabbit virus kills millions. New Scientist, 152, 4.

Anderson, I. (1997). Alarm greets contraceptive virus. New Scientist, 154, 4.

Anderson, R. M., & May, R. M. (1978). Regulation and stability of host-parasite population interactions I. Regulatory processes. Journal of Animal Ecology, 47, 219–247.

Anderson, R. M. (1979). Parasite pathogenicity and the depression of host equilibria. Nature (London). 279, 150–152.

Anderson, R. M. (1982). Theoretical basis for the use of pathogens as biological control agents. Parasitology, 84, 3–33.

Anita, R., Levin, B. R., & May, R. M. (1994). Within-host population dynamics and the evolution and maintenance of microparasite virulence. American Naturalist, 144, 457–472.

Anonymous. (1997a). Vertebrate Biocontrol Center Annual Report 1995–1996. Lyneham, Australia: The Cooperative Research Center for Biological Control of Vertebrate Pest Populations.

Anonymous. (1997b). Rabbit virus vectors named. Science, 278, 229.

Atkinson, K. (1997). New Zealand grapples with possum virus. Search, 28, 260.

Barker, S. C., Singleton, G. R., & Spratt, D. M. (1991). Can the nematode Capillaria hepatica regulate abundance in wild house mice? Results of enclosure experiments in southeastern Australia. Parasitology, 103, 439–449.

Barlow, N. D. (1994). Predicting the effect of a novel vertebrate biocontrol agent: A model for viral-vectored immunocontraception of New Zealand possums. Journal of Applied Ecology, 31, 454–462.

Barlow, N. D., & Wratten, S. D. (1996). Ecology of predator-prey and parasitoid-host systems: Progress since Nicholson. In R. B. Floyd, A. W. Sheppard, & P. J. De Barro (Eds.), Frontiers of Population Ecology. Collingwood, Australia: CSIRO Publishing.

Barlow, N. D. (1997). Modeling immunocontraception in disseminating systems. Reproduction, Fertility and Development, 9, 51–60.

Beckman, R. (1988). Mice on the farm. Rural Research, 138, 23–27.

Bland, J. D., & Temple, S. A. (1993). The Himalayan snowcock: North America's newest exotic bird. In B. N. McKnight (Ed.), Biological pollution: The control and impact of invasive exotic species. Indianapolis: Indiana Academy of Science.

Bloomer, J. P., & Bester, M. N. (1992). Control of feral cats on sub-Antarctic Marion Island, Indian Ocean. Biological Conservation, 60, 211–219.

Boag, B. (1989). Population dynamics of parasites of the wild rabbit. In P. J. Putman (Ed.), Mammals as pests. London: Chapman & Hall.

Boga, J. A., Alonso, J. M. M., Casais, R., & Parra, F. (1997). A single dose immunization with rabbit haemorrhagic disease virus major capsid protein produced in Saccharomyces cerevisiae induces protection. Journal of General Virology, 78, 2315–2318.

Boonstra, R., Krebs, C. J., & Beacham, T. D. (1980). Impact of botfly parasitism on Microtus townsendii populations. Canadian Journal of Zoology, 58, 1683–1692.

Bowen, L., & van Vuren, D. (1997). Insular endemic plants lack defenses against herbivores. Conservation Biology, 11, 1249–1254.

Bradley, M. P., Hinds, L. A., & Bird, P. H. (1997). A bait delivered immunocontracpetive vaccine for the European fox (Vulpes vulpes) by the year 2002? Reproduction, Fertility and Development, 9, 111–116.

Bratton, S. P. (1975). The effect of the European wild boar Sus scrofa, on gray beech forest in the Great Smoky Mountains. Ecology, 56, 1356–1366.

Brown, J. H. (1989). Patterns, modes and extents of invasions by vertebrates. In J. A. Drake, H. A. Mooney, F. di Castri, R. H. Groves, F. J., Kruger, & M. Rejmanek (Eds.), Biological invasions: a global perspective. Chichester: John Wiley & Sons.

Buddle, B. M., de Lisle, G. W., McColl, K., Collins, B. J., Morrissy, C., & Westbury, H. A. (1997). Response of the North Island brown kiwi, Apteryx australis mantelli and the lesser short-tailed bat, Mystacina tuberculata to a measured dose of rabbit haemorrhagic disease virus. New Zealand Veterinary Journal, 45, 109–113.

Cagne, W. C. (1988). Conservation priorities in Hawaiian natural systems. BioScience, 38, 264–271.

Campbell, F. T. (1993). Legal avenues for controlling exotics. In B. N. McKnight (Ed.), Biological pollution: The control and impcat of invasive exotic species. Indianapolis: Indiana Academy of Science.

Case, T. J. (1991). Invasion resistance, species build up and community collapse in metapopulation models with interspecies competition. Biological Journal of the Linnaean Society, 42, 239–266.

Case, T. J. (1996). Global patterns in the establishment and distribution of exotic birds. Biological Conservation, 78, 69–96.

Caughley, G., Pech, R., & Grice, D. (1992). Effect of fertility control on a population's productivity. Wildlife Research, 19, 623–627.

Chasey, D., Lucas, M., Westcott, D., & Williams, M. (1992). European brown hare syndrome in the UK: A calicivirus related to but distinct from that of viral haemorrhagic disease in rabbits. Archives of Virology, 124, 363–370.

Chasey, D. (1994). Possible origin of rabbit haemorrhagic disease in the United Kingdom. Veterinary Record, 135, 496–499.

Childs, J. E., Glass, G. E., & Korch, G. W., Jr. (1988). The comparative epizootiology of Capillaria hepatica (Nematoda) in urban rodents from different habitats of Baltimore, Maryland. Canadian Journal of Zoology, 66, 2769–2775.

Coblentz, B. E. (1978). Effects of feral goats on island ecosystems. Biological Conservation, 13, 279–286.

Cooke, B. D. (1996). Field epidemiology of rabbit calicivirus disease in Australia. In ESVV Symposium on caliciviruses, abstracts of oral and poster presentations. European Society for Veterinary Virology.

Corbet, G. B. (1994). Taxonomy and origins. In H. V. Thompson & C. M. King (Eds.), The European rabbit, the history and biology of a successful colonizer. Oxford: Oxford University Press.

Cornell, H. V. & Hawkins, B. A. (1993). Accumulation of native parasitoid species on introduced herbivores: A comparison of hosts as natives and hosts as invaders. American Naturalist, 141, 847–865.

Cowan, P. E. (1992). The eradication of introduced Australian brushtail possums, Trichosurus vulpecula, from Kapiti Island, a New Zealand Nature Reserve. Biological Conservation, 61, 217–226.

Cowan, P. E. (1996). Possum biocontrol: Prospects for fertility control. Reproduction, Fertility and Development, 8, 655–660.

Cowan, P. E., & Tyndale-Biscoe, C. H. (1997). Australian and New Zealand mammal species considered to be pests or problems. Reproduction, Fertility and Development, 9, 27–36.

Davis, D. E., Myers, K., and Hoy J. B. (1976). Biological control among vertebrates. In C. B. Huffaker & P. S. Messenger (Eds.), Theory and practice of biological control. New York: Academic Press.

DeBach, P. (1964). The scope of biological control. In Biological control of insect pests and weeds. New York: Reinhold.

Dobson, A. P. & May, R. M. (1986). Patterns of invasion by pathogens and parasites. In H. A. Mooney & J. A. Drake (Eds.), Ecology of biological invasions of North America and Hawaii. New York: Springer-Verlag.

Dobson, A. P. & Hudson, P. J. (1986). Parasites, disease and the structure of ecological communities. Trends in Ecology and Evolution, 1, 11–15.

Dobson, A. P. (1988). Restoring island ecosystems: The potential of parasites to control introduced mammals. Conservation Biology, 2, 31–39.

Dobson, A. P. & Hudson, P. J. (1994). Population biology of Trichostrongylus tenuis in the red grouse Lagopus lagopus scoticus. In M. E. Scott & G. Smith (Eds.), Parasitic and infectious diseases, epidemiology and ecology. San Diego: Academic Press.

Dwyer, G., Levin, S. A., & Buttel, L. (1990). A simulation model of the population dynamics and evolution of myxomatosis. Ecological Monographs, 60, 423–447.

Ebert, D. (1994). Virulence and local adaption of a horizontally transmitted parasite. Science, 265, 1084–1086.

Ebert, D., & Herre, E. A. (1996). The evolution of parasitic diseases. Parasitology Today, 12, 96–101.

Elton, C. S. (1927). Animal ecology. Sidgwick & Jackson. London.

Elton, C. S. (1953). The use of cats in farm rat control. British Journal of Animal Behavior, 1, 151–154.

Elton, C. S. (1958). The ecology of invasions by animals and plants. London: Methuen.

Ehrlich, P. R. (1986). Which animal will invade? In H. A. Mooney & J. A. Drake (Eds.), Ecology of biological invasions of North America and Hawaii. New York: Springer-Verlag.

Ehrlich, P. R. (1989). Attributes of invaders and the invading process: Vertebrates. In J. A. Drake, H. A. Mooney, F. di Castri, R. H. Groves, F. J. Kruger, & M. Rejmanek (Eds.), Biological invasions: A global perspective. Chichester: John Wiley & Sons.

Fenner, F. (1994). Myxomatosis. In M. E. Scott & G. Smith (Eds.), Parasitic and infectious diseases, epidemiology and ecology. San Diego: Academic Press.

Fenner, F., & Marshall, I. D. (1957). A comparison of the virulence for European rabbits (Oryctolagus cuniculus) of strains of myxoma virus recovered in the field in Australia, Europe, and America. Hygiene, 55, 149–191.

Fenner, F., & Ratcliffe, F. N. (1965). Myxomatosis. Cambridge: Cambridge University Press.

Fenner, F., & Ross, J. (1994). Myxomatosis. In H. V. Thompson & C. M. King (Eds.), The European rabbit, the history and biology of a successful colonizer. Oxford: Oxford University Press.

Flux, J. E. C. (1994). World distribution. In H. V. Thompson & C. M. King (Eds.), The European rabbit, the history and biology of a successful colonizer. Oxford: Oxford University Press.

Freeland, W. J. (1985). The need to control cane toads. Search, 16, 211–215.

Fuchs, A., & Weissenböck, H. (1992). Comparative histopathological study of rabbit haemorrhagic disease (RHD) and European brown hare syndrome (EBHS). Journal of Comparative Pathology, 107, 103–113.

Fuller, H. E., Chasey, D. M., Lucas, M. H., & Gibbens, J. C. (1993). Rabbit haemorrhagic disease in the United Kingdom. Veterinary Record, 133, 611–613.

Gevry, J., & Chirol, C. (1978). A propos d'un cas de capillariose a Capillaria hepatica observe dans un elevage de lapins croises garenne. Revue de Medecine Veterinaire, 129, 1019–1026.

Gibb, J. A., & Williams, J. M. (1994). The rabbit in New Zealand. In H. V. Thompson & C. M. King (Eds.), The European rabbit, the history and biology of a successful colonizer. Oxford: Oxford University Press.

Gould, A. R., Kattenbelt, J. A., Lenghaus, C., Morrissy, C., Chamberlain, T., & Collins, B. J. (1997). The complete nucleotide sequence of rabbit haemorrhagic disease virus (Czech stain C V351): Use of the polymerase chain reaction to detect replication in Australian vertebrates and analysis of viral population sequence variation. Virus Research, 47, 7–17.

Greaves, J. H. (1994). Resistance to anticoagulant rodenticides. In A. P. Buckle & R. H. Smith (Eds.), Rodent pests and their control. Wallingford: CAB International.

Gregg, D. A., House, C., Myer, R., & Berninger, M. (1991). Viral haemorrhagic disease or rabbits in Mexico: Epidemiology and viral characterization. Revue Scientifique et Technique Office International des Epizooties, 10, 435–451.

Griffith, B., Scott, J. M., Carpenter, J. W., & Reed, C. (1989). Translocation and a species conservation tool: Status and strategy. Science, 245, 477–480.

Holt, R. D., & Hochberg, M. E. (1997). When is biological control evolutionary stable (or is it)? Ecology, 78, 1673–1683.

Hone, J. (1994). Analysis of vertebrate pest control. Cambridge: Cambridge University Press.

Howard, W. E. (1967). Biological control of vertebrate pests, (pp. 137–157). In Proceedings of third vertebrate pest conference, San Francisco.

Howarth, F. G. (1983). Classical biocontrol: Panacea or Pandora's box? Proceedings of the Hawaiian Entomological Society, 24, 239–244.

Howell, P. G. (1984). An evaluation of the biological control of the feral cat Felis catus (Linnaeus, 1758). Acta Zoologica Fennica, 172, 111–113.

Jaffe, M. (1994). And no birds sing: the story of ecological disaster in a tropical paradise. New York: Simon & Schuster.

Jaksic, F. M. & Yànez, J. L. (1983). Rabbit and fox introductions in Tierra del Fuego: History and assessment of the attempts at biological control of the rabbit infestation. Biological Conversation, 26, 367–374.

Jervis, M. (1997). Parasitoids as limiting and selective factors: Can biological control be evolutionary stable? Trends in Ecology and Evolution, 12, 378–380.

Jolly, S. E. (1993). Biological control of possums. New Zealand Journal of Zoology, 20, 335–339.

Kirkpatrick, J. F., Liu, I. M. K., Turner, J. W., Jr., Naugle, R., & Keiper, R. (1992). Long-term effects of porcine zonae pellucidae immunocontraception on ovarian function in feral horses (Equus caballus). Journal of Reproduction and Fertility, 94, 437–444.

Kirkpatrick, J. F., Turner, J. W. Jr., Liu I. K. M., Fyrer-Hosken, R., & Rutberg, A. T. (1997). Case studies in wildlife immunocontraception: Wild and feral equids and white-tailed deer. Reproduction, Fertility and Development, 9, 105–110.

Korpimäki, E., & Krebs, C. J. (1996). Predation and population cycles of small mammals. BioScience, 46, 745–764.

Lack, D. (1954). The natural regulation of animal numbers. Oxford: Clarendon Press.

Lawson, M. (1995). Rabbit virus threatens ecology after leaping the fence. Nature (London), 378, 531.

LeBlanc, P., & Fagin, B. (1983). Capillaria hepatica infection: Incidental finding in a dog with renal insufficiency. Canine Practice, 10, 12–14.

Le Gall, G., Huguet, S., Vende, P., Vautherot, J. F., & Rasschaert, D. (1996). European brown hare syndrome virus: Molecular cloning and sequencing of the gonome. Journal of General Virology, 77, 1693–1697.

Liu, S. J., Xue, H. P., Pu, B. Q., & Qian, N. H. (1984). A new viral disease in rabbits. Animal Husbandry and Veterinary Medicine, 16, 253–255.

Lodge, D. M. (1993). Biological invasions: Lessons for ecology. Trends in Ecology and Evolution, 8, 133–137.

Loope, L. L., Hamann, O., & Stone, C. P. (1988). Comparative conservation biology of oceanic archipelagoes. BioScience, 38, 272–282.

Lovett, J. (1997). Birth control for feral pests. Search, 28, 209–211.

MacArthur, R. (1955). Fluctuations of animal populations and a measure of community stability. Ecology, 36, 533–536.

Marcogliese, D. J., & Cone, D. K. (1997). Food webs: A plea for parasites. Trends in Ecology and Evolution, 12, 320–325.

May, R. M. (1977). Thresholds and breakpoints in ecosystems with a multiplicity of stable states. Nature (London), 269, 471–477.

May, R. M., & Anderson, R. M. (1978). Regulation and stability of host-parasite population interactions II. Destabilizing processes. Journal of Animal Ecology, 47, 249–267.

May, R. M. (1980). Depression of host population abundance by direct life cycle macroparasites. Journal of Theoretical Biology, 82, 283–311.

McCallum, H. I., & Singleton, G. R. (1989). Models to assess the potential of Capillaria hepatica to control population outbreaks of house mice. Parasitology, 98, 425–437.

McCallum, H. I. (1993). Evaluation of a nematode (Capillaria hepatica Bancroft, 1893) as a control agent for populations of house mice (Mus musculus domesticus Schwartz and Schwartz, 1943). Revue Scientifique et Technique Office International des Epizooties, 12, 83–93.

McCallum, H. (1994). Quantifying the impact of disease on threatened species. Pacific Conservation Biology, 1, 107–117.

McCallum, H. I. (1996). Immunocontraception for wildlife population control. Trends in Ecology and Evolution, 11, 491–493.

McCallum, H., & Dobson, A. (1995). Detecting disease and parasite threats to endangered species and ecosystems. Trends in Ecology and Evolution, 10, 190–194.

Millar, S. E., Chamow, S. M., Baur, A. W., Oliver, C., Robey, F., & Dean, J. (1989). Vaccination with a synthetic zona pellucida peptide produces long-term contraception in female mice. Science, 246, 935–938.

Mobedi, I., & Arfaa, F. (1971). Probable role of ground beetles in the transmission of Capillaria hepatica. Journal of Parasitology, 57, 1144–1145.

Morgan, D. R., Batcheler, C. L., & Peters, J. A. (1986). Why do possums survive aerial poisoning operations? In T. P. Salmon, R. E. Marsh, & D. E. Beadle (Eds.), Proceedings of the Twelfth Vertebrate Pest Control Conference, San Diego CA, (pp. 210–214). Davis: University of California.

Muñoz, A., & Murúa, R. (1990). Control of small mammals in a pine plantation (Central Chile) by modification of the habitat of predators (Tyto alba, Stringiforme and Pseudalopex sp. Canidae). Acta Oecologia, 11, 251–261.

Munroe, G. A. (1984). Pyloric stenosis in a yearling with an incidental finding of Capillaria hepatica in the liver. Equine Veterinary Journal, 16, 221–222.

Myers, K., Parer, I., Wood, D., & Cooke, B. D. (1994). The rabbit in Australia. In H. V. Thompson & C. M. King (Eds.), The European rabbit, the history and biology of a successful colonizer. Oxford: Oxford University Press.

Nagesha, H. S., Wang, L. F., Hyatt, A. D., Morrissy, C. J., Lenghaus, C. & Westbury, H. A. (1995). Self assembly, antigenicity, and immunogenicity of the rabbit haemorrhagic disease virus (Cezchoslovakia strain V-351) capsid protein expressed in baculovirus. Archives of Virology, 140, 1095–1108.

Nettles, V. F. (1997). Potential consequences and problems with wildlife contraceptives. Reproduction, Fertility and Development, 9, 137–143.

Newsome, A. E., Parer, I., & Catling, P. C. (1989). Prolonged prey suppression by carnivores: predator-removal experiments. Oecologia, (Berlin) 78, 458–467.

Newsome, A. (1990). The control of vertebrate pests by vertebrate predators. Trends in Ecology and Evolution, 5, 187–191.

Nowotny, N. B., Ballagi-Pordany, A., Gavier-Widen, D., Uhlen, M., & Belak, S. (1997). Phylogenetic analysis of rabbit haemorrhagic disease and European brown hare syndrome viruses by comparison of sequences from the capsid protein gene. Archives of Virology, 142, 657–673.

O'Brien, P. (1991). The social and economic implications of RHD introduction. Search, 22, 191–193.

Ohlinger, V. F., Haas, B., Meyers, G., & Thiel, H. J. (1990). Identification and characterization of the virus causing rabbit haemorrhagic disease. Journal of Virology, 64, 3331–3336.

Oogjes, G. (1997). Ethical aspects and dilemmas of fertility control of unwanted wildlife: An animal welfarist's perspective. Reproduction, Fertility and Development, 9, 163–167.

Pannenbecker, J., Miller, T. C., Muller, J., & Jeschke, R. (1990). Schwerer leberbefall durch Capillaria hepatica. Monatsschrift fur Kinderheilkunde, 138, 767–771.

Parra, F., & Prieto, M. (1990). Purification and characterization of a calicivirus as the causative agent of lethal haemorrhagic disease in rabbits. Journal of Virology, 68, 4013–4015.

Payton, I. J., Forester, L., Frampton, C. M., & Thomas, M. D. (1997). Response of selected tree species to culling of introduced Australian brushtail possums Trichosurus vulpecula at Waipoua Forest, Northland, New Zealand. Biological Conservation, 81, 247–255.

Pech, R. P., Sinclair, A. R. E., Newsome, A. E., & Catling, P. C. (1992). Limits to predator regulation of rabbits in Australia: Evidence from predator removal experiments. Oecologia, (Berlin), 89, 102–112.

Pech, R. P., Sinclair, A. R. E., & Newsome, A. E. (1995). Predation models for primary and secondary prey species. Wildlife Research, 22, 55–64.

Pech, R. P., Hood, G. M., McIlroy, J., & Saunders, G. (1997). Can foxes be controlled by reducing their fertility? Reproduction, Fertility and Development, 9, 41–50.

Pereira, V. G., & Franca, L. C. M. (1983). Successful treatment of Capillaria hepatica infection in an acutely ill adult. American Journal of Tropical Medicine and Hygiene, 32, 1272–1274.

Pimm, S. L. (1987). The snake that ate Guam. Trends in Ecology and Evolution, 2, 293–295.

Primack, R. B. (1993). Essentials of conservation biology. Sunderland, MA: Sinauer Associates.

Primack, R. B. (1995). A primer of conservation biology. Sunderland, MA: Sinauer Associates.

Roberts, M., Rodrigo, A., McArdle, B., & Charleston, W. A. G. (1992). The effect of habitat on the helminth parasites of an island population of the Polynesian rat (Rattus exulans). Journal of Zoology London, 227, 109–125.

Robinson, A. J., & Holland, M. K. (1995). Testing the concept of virally vectored immunosterilization for the control of wild rabbit and fox populations in Australia. Australian Veterinary Journal, 72, 65–68.

Robinson, A. J., Jackson, R., Kerr, P., Mercahnt, J., Parer, I., & Pech, R. (1997). Progress towards using a recombinant myxoma virus as a vector for fertility control in rabbits. Reproduction, Fertility and Development, 9, 77–83.

Robinson, T., & Westbury, H. (1996). The Australian and New Zealand calicivirus disease program. In ESVV Symposium on caliciviruses, abstracts of oral and poster presentations. European Society for Veterinary Virology.

Rodda, G. H., Fritts, T. H., & Chiszar, D. (1997). The disappearance of Guam's wildlife. BioScience, 47, 565–574.

Rodger, J. C. (1997). Likely targets for immunocontraception in marsupials. Reproduction, Fertility and Development, 9, 131–136.

Ross, J., & Tittensor, A. M. (1986). The establishment and spread of myxomatosis and its effect on rabbit populations. Philosophical Transactions of the Royal Society of London B, 314, 599–606.

Rudzki, S. (1995). Escaped rabbit calicivirus highlights Australia's chequered history of biological control. Search, 26, 287.

Saunders, G. R., & Giles, J. R. (1977). A relationship between plagues of the house mouse Mus musculus (Rodentia: Muridae) and prolonged periods of dry weather in south-eastern Australia. Australian Journal of Wildlife Research, 4, 241–247.

Savidge, J. A. (1987). Extinction of an island forest avifauna by an introduced snake. Ecology, 68, 660–668.

Scott, M. E. (1987). Regulation of mouse colony abundance by Heligmosomoides polygyrus. Parasitology, 95, 111–124.

Scott, M. E., & Lewis, J. W. (1987). Population dynamics in wild and laboratory rodents. Mammal Review, 17, 95–103.

Scott, M. E., & Dobson, A. 1989. The role of parasites in regulating host abundance. Parasitology Today, 5, 176–183.

Seife, C. (1996). A harebrained scheme. Scientific American, 274, 24–26.

Shelford, V. E. (1942). Biological control of rodents and predators. Scientific Monthly, 55, 331–341.

Shellam, G. R. (1994). The potential of murine cytomegalovirus as a vital vector for immunocontraception. Reproduction, Fertility and Reproduction, 6, 401–409.

Shulman, S. (1996). Immunological reactions and infertility. In M. Kurpisz & N. Fernandez (Eds.), Immunology of human reproduction. Oxford: BIOs Scientific.

Sinclair, A. R. E., Olsen, P. D., & Redhead, T. D. (1990). Can predators regulate small mammal populations? Evidence from house mouse outbreaks in Australia. Oikos, 59, 382–392.

Sinclair, A. R. E. (1996). Mammal populations: Fluctuation, regulation, life history theory and their implications for conservation. In R. B. Floyd, A. W. Sheppard, & P. J. De Barro (Eds.), Frontiers of Population Ecology. Collingwood Australia: CSIRO Publishing.

Sinclair, A. R. E. (1997). Fertility control of mammal pests and the conservation of endangered marsupials. Reproduction, Fertility and Development, 9, 1–16.

Singer, F. J., Swank, W. T., & Clebsch, E. E. C. (1984). Effects of wild pig rooting in a deciduous forest. Journal of Wildlife Management, 48, 463–473.

Singleton, G. R., & Spratt, D. M. (1986). The effects of Capillaria hepatica (Nematoda) on natality and survival to weaning in BALB/c mice. Australian Journal of Zoology, 34, 677–681.

Singleton, G. R. (1989). Population dynamics of an outbreak of house mice (Mus domesticus) in the mallee wheatlands of Australia: hypothesis of plague formation. Journal of Zoology London, 219, 495–515.

Singleton, G. R., & McCallum, H. I. (1990). The potential of Capillaria hepatica to control mouse plagues. Parasitology Today, 6, 190–193.

Singleton, G. R., Spratt, D. M., Barker, S. C., & Hodgson, P. F. (1991). The geographic distribution and host range of Capillaria hepatica (Bancroft) (Nematoda) in Australia. International Journal for Parasitology, 21, 945–957.

Singleton, G. R., Chambers, L. K., & Spratt, D. M. (1995). An experimental field study to examine whether Capillaria hepatica (Nematoda) can limit house mouse populations in eastern Australia. Wildlife Research, 22, 31–53.

Singleton, G. R., & Chambers, L. K. (1996). A manipulative field experiment to examine the effect of Capillaria hepatica (Nematoda) on wild mouse populations in southern Australia. International Journal for Parasitology, 26, 383–398.

Singer, P. (1997). Neither human nor natural: Ethics and feral animals. Reproduction, Fertility and Development, 9, 157–162.

Smith, G. (1994). Parasite population density is regulated. In M. E. Scott & G. Smith (Eds.), Parasitic and infectious diseases, epidemiology and ecology. San Diego: Academic Press.

Smith, G., & Dobson, A. P. (1992). Sexually transmitted diseases in animals. Parasitology Today, 8, 159–166.

Smith, G., Walmsley, A., & Polkinghorne, I. (1997). Plant-derived immunocontraceptive vaccines. Reproduction, Fertility and Development, 9, 85–89.

Smith, H. R., & Remington, C. L. (1996). Food specificity in interspecies competition. BioScience, 46, 436–447.

Spratt, D. M., & Singleton, G. R. (1986). Studies of the life cycle infectivity and clinical effects of Capillaria hepatica (Bancroft) (Nematoda) in mice, Mus musculus. Australian Journal of Zoology, 34, 663–675.

Spratt, D. M., & Singleton, G. R. (1987). Experimental embryonation and survival of eggs of Capillaria hepatica (Nematoda) under mouse burrow conditions in cereal growing soils. Australian Journal of Zoology, 35, 337–341.

Spratt, D. M. (1990). The role of helminths in the biological control of mammals. International Journal for Parasitology, 20, 543–550.

Studdert, M. J. (1994). Rabbit haemorrhagic disease virus: A calicivirus with differences. Australian Veterinary Journal, 71, 264–266.

Tyndale-Biscoe, C. H. (1991). Fertility control in wildlife. Reproduction, Fertility and Development, 3, 339–343.

Tyndale-Biscoe, C. H. (1994a). The CRC for biological control of vertebrate pest populations: Fertility control of wildlife for conservation. Pacific Conservation Biology, 1, 160–162.

Tyndale-Bisoce, C. H. (1994b). Virus-vectored immunocontraception of feral mammals. Reproduction, Fertility and Development, 6, 281–287.

Tyndale-Biscoe, C. H. (1995). Vermin and viruses: Risks and benefits of viral-vectored immunosterilization. Search, 26, 239–244.

Van Driesche, R. G., & Bellows, T. S., Jr. (1996). Biological control. New York: Chapman & Hall.

Van Driesche, R. G., & Hoddle, M. S. (1997). Changing social views of the desired degree of host range specificity of natural enemies of arthropods. Agriculture and Human Values, 14, 211–226.

van Rensburg, P. J. J., Skinner, J. D., & van Aarde, R. J. (1987). Effects of feline panleucopaenia on the population characteristics of feral cats on Marion Island. Journal of Applied Ecology, 24, 63–73.

Veltman, C. J., Nee, S. & Crawley, M. J. (1996). Correlates of introduction in exotic New Zealand birds. American Naturalist, 147, 542–557.

Vermeiji, G. J. (1991). When biotas meet: Understanding biotic interchange Science, 253, 1099–1104.

Vitousek, P. M. (1990). Biological invasions and ecosystem processes: Towards an integration of population and ecosystem studies. Oikos, 57, 7–13.

Vitousek, P. M., D'Antonio, C. M., Loope, L. L., & Westbrooks, R. (1996). Biological invasions as global environmental change. American Scientist, 84, 468–478.

Wahid, M. B., Ismail, S., & Kamarudin, N. (1996). The extent of biological control of rats with barn owls, Tyto alba javanica in Malaysian oil palm plantations. The Planter, 72, 5–18.

Whisson, D. (1998). Biological control of vertebrate pest. In M. S. Hoddle (Ed.), California Conference on Biological Control. University of California, Berkeley, June 10–11, 1998 (pp. 28–32). University of California at Davis.

Williams, C. K. & Moore, R. J. (1995). Effectiveness and cost-efficiency of control of the wild rabbit, Oryctolagus cuniculus (L.), by combinations of poisoning, ripping, fumigation, and maintenance fumigation. Wildlife Research, 22, 253–269.

Williams, C. K., & Twigg, L. E. (1996). Response of wild rabbit populations to imposed sterility. In R. B. Floyd, A. W. Sheppard, & P. J. De Barro (Eds.), Frontiers of Population Ecology. Collingwood, Australia: CSIRO Publishing.

Williams, C. K. (1997). Development and use of virus-vectored immunocontraception. Reproduction, Fertility and Development, 9, 169–78.

Williamson, M. H., & Fitter, A. (1996a). The characters of successful invaders. Biological Conservation, 78, 163–170.

Williamson, M., & Fitter, A. (1996b). The varying success of invaders. Ecology, 77, 1661–1666.

Wood, B. J. (1985). Biological control of vertebrates—a review, and an assessment of prospects for Malaysia. Journal of Plant Protection in the Tropics, 2, 67–79.

Wright, K. A. (1961). Observations of the lifecycle of Capillaria hepatica (Bancroft, 1893) with a description of the adult. Canadian Journal of Zoology, 38, 167–182.

39

Classical Biological Control in Latin America
Past, Present, and Future

MIGUEL A. ALTIERI

Division of Insect Biology
University of California
Berkeley, California

CLARA I. NICHOLLS

Cooperative Extension
University of California
Alameda, California

INTRODUCTION

Pesticide expenses in Latin America, which reached about US$300 million between 1980 and 1984, are expected to nearly triple by the end of the millennium. Prevailing economic policies in Lain America encourage the production of export and/or commercial crops primarily in large-scale monocultures, with production systems of sugarcane, cotton, maize, soybeans, rice, citrus, and tomatoes as the major recipients of insecticide applications, especially in Brazil, Colombia, Argentina, and Mexico. The growing nontraditional sector in Central America has kept pesticide inputs relatively high and thus has become a dynamic contributor to the regional pesticide market (Murray, 1994). Predictably, the emphasis of the chemical-intensive agricultural export model has intensified ecologically based crisis conditions with serious environmental and health consequences (Belloti *et al.,* 1990).

Despite these trends, there are several documented cases scattered throughout the region where withdrawal from the pesticide treadmill through the implementation of biological control programs has resulted in viable crop production. These approaches have been developed by national research centers, leading in some cases to substantially reducing the need for pesticide imports. Although still restricted to a few countries and actually given low research priority, we believe that biological control represents one of the most economically viable, environmentally sound, and self-sustained methods of insect pest control in Latin America.

After providing a brief historical overview of biological control in Latin America, we describe some current programs, as well as suggestions on how to scale up classical biological control in the region.

HISTORICAL OVERVIEW

The earliest recorded efforts of classical biological control in the region date to the beginning of the twentieth century. For example, in 1903, *Hippodamia convergens* Guérin-Méneville and *Rhizobius ventralis* (Erichson) were introduced into Chile from California for the control of scale insects (Gonzalez & Rojas, 1966). In 1904, natural enemies were introduced into Peru for the control of white scale, *Pinnaspis strachani* (Cooley), in cotton; in 1908, *Encarsia (Prospaltella) berlesi* (Howard) was introduced into Argentina to combat the white peach scale, *Pseudaulacaspis pentagona* (Targioni-Tozzetti) (Hagen & Franz, 1973). These efforts were supplemented by the establishment of specialized insectaries in Mexico in 1928; in Chile (La Cruz) in 1929; and later in Peru, Argentina, Brazil, Colombia, and Nicaragua.

Most of the early work on biological control was concentrated on homopteran citrus pests, mainly because citrus marked the beginning of biological control history in 1888. Cooperative work with Latin American entomologists was initiated later and supported by the Divisions of Biological Control, University of California, Riverside, which for decades was a world center for mass-rearing and distributing natural enemies of citrus pests. Other efforts were later initiated, by decreasing order of magnitude, in sugarcane with the U.S. Department of Agriculture (USDA); and in apple, peach, olive, alfalfa, cotton, and other field crops with the University of California.

The main successes in classical biological control programs in the region include the citrus blackfly, *Aleurocanthus woglumi* Ashby, in Mexico; the sugarcane borer, *Diatraea saccharalis* (Fabricius), in Cuba, Peru, Brazil, and the

---

TABLE 1   Insecticide Use (1978) and Forecast (1988) for
Latin America

	Metric tons active ingredient		Change (%)
	1978	1988	
Organochlorines			
DDT	15,988	11,605	−27.4
BHC	4,920	4,698	−4.5
Aldrin	1,141	925	−18.9
Toxaphene	17,338	11,000	−36.6
Endosulfan	1,286	1,682	+23.5
Endrin	1,716	1,377	−19.8
Heptachlor	755	574	−24.0
Subtotal	43,144	31,861	−26.2
Organophosphates			
Methyl parathion	14,274	17,948	+25.0
Parathion	7,388	9,029	+22.2
Malathion	2,856	3,017	+5.6
Dimethoate	1,907	2,247	+17.8
Disulfoton	276	283	+2.5
Fenitrothion	772	824	+6.7
Monocrotophos	3,284	4,467	+36.0
Phosfamidon	289	285	−1.4
Chlorpyrifos	814	781	−4.1
Trichlofon	1,201	1,550	+29.1
DDVP	318	404	+27.0
Azinphos	615	610	−0.1
Methamidophos	1,488	1,874	+25.9
Profenphos	500	600	+20.0
Diazinon	255	220	−13.7
Subtotal	36,237	44,038	+21.5
Carbamates			
Carbaryl	3,463	4,001	+15.5
Methomyl	863	1,007	+16.7
Carbofuran	680	764	+12.4
Aldicarb	147	150	+12.0
Subtotal	5,153	5,922	+14.9
Pyrethroids	39	150	+284.6
Others	4,567	3,996	−12.5
Total	89,140	85,967	−3.8

After Maltby, C. (1980). *Report on the use of pesticides in Latin America* (UNIDO/IOD 353). Geneva: United Nations Industrial Development Organization (UNIDO).

Caribbean; the cottony-cushion scale, *Icerya purchasi* Maskell, in Chile; the woolly apple aphid, *Eriosoma lanigerum* (Hausmann), in Uruguay, Chile, and Argentina; the black scale, *Saissetia oleae* (Olivier), in Chile and Peru; and several species of mealybugs and scale insects in various countries (Gonzalez, 1976; MacPhee *et al.*, 1976).

With the advent of chemical insecticides after World War II, interest in biological control noticeably declined for about two decades in Latin America. However, the environ-mental costs associated with many organochlorine insecticides, in addition to the restrictions placed on residue levels of export meats, vegetables, and fruits by Europe and the United States (in 1981 more than 1,105,000 kg of meat from Central America were rejected by the United States), awakened a renewed interest in biological control, mostly as a component strategy of integrated pest management (IPM) (Gomero & Lizárraga, 1995).

PESTICIDE USE IN LATIN AMERICA AND IMPLIED ENVIRONMENTAL AND HEALTH IMPACTS

Between 1980 and 1984, Latin America imported about $430 million worth of pesticides; these expenditures are expected to triple over the next 10 years, especially in Brazil, Mexico, Argentina, and Colombia (Maltby, 1980). Use of organochlorine insecticides are, with the exception of endosulfan, expected to decline while use of organophosphates, carbamates, and especially pyrethroids will increase considerably (Table 1). While some pesticide formulation and production does occur at the local level, Latin America must rely on importing most of its chemically based pest control products from industrialized countries (Murray, 1994).

Traditionally, cotton accounted for most of the insecticide use in Latin America at a level of about 6 kg of pesticide per hectare. A few years ago, 75% of total pesticide consumption in El Salvador and Guatemala was devoted to cotton, which received up to 35 applications per season (Burton & Philogene, 1984). Such excessive numbers of pesticide treatments resulted in serious public health problems as well as ecological disturbances (Leonard, 1986). Apple and pear orchards still receive from 8 to 16 treatments per season in the southern countries (Chile, Argentina, Uruguay, and southern Brazil) (González & Rojas, 1976), and most fruit trees in the subtropical and tropical countries are routinely sprayed for protection against fruit flies (Maltby, 1980). Among the vegetable crops, tomatoes and potatoes by far account for the greatest pesticide use. Soybeans received more insecticide inputs than any other crop, primarily due to extensive cultivation in Brazil, Argentina, and Bolivia (Fig. 1). The growing nontraditional export sector in Central America generated increasing demand and kept pesticide inputs relatively high throughout the late 1980s and early 1990s (Conroy *et al.*, 1996)

Although there has been general concern about the environmental and public health impacts of pesticides and their toxic residues in the region, comparatively little information is available on the dimensions of environmental contamination (Burton & Philogene, 1984). This lack of data has led some policymakers to believe that pesticides

A

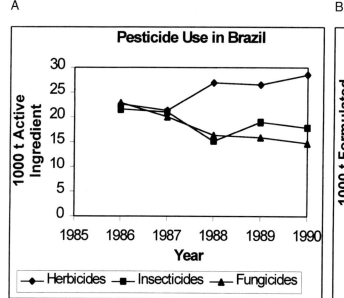

B

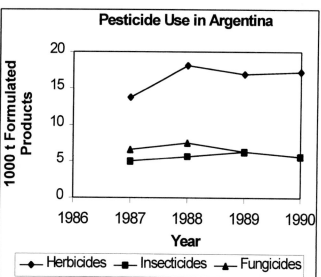

FIGURE 1   (a) Pesticide use in Brazil. Use of main classes of pesticides, 1986–1990 (after Murray, 1994); (b) Pesticide use (formulated products) in Argentina, 1986–1990 (after Murray, 1994).

are not likely to cause significant environmental disruption or to seriously affect the continued growth of agriculture in Latin America (Murdoch, 1980). The few available data, however, contradict this viewpoint (Leonard, 1986). The data suggest that indiscriminate use of pesticides has taken a heavy toll of biotic resources, wildlife, and the human population. The impacts of pesticide use include human poisoning, pollution of water tables, and resurgence of illness (such as malaria) resulting from insect resistance (Leonard, 1986) Pesticides have also increased the economic costs of farming because of the need for more intensive doses and/or the reduction in yields resulting from increased insect resistance in monocultures. Between 1971 and 1976, more than 19,000 pesticide poisonings were reported in Central America by the Instituto Centro Americano de Investigacíon y Tecnologia Industrial (ICAITI), mostly occurring in Guatemala and El Salvador (ICAITI, 1977). In Nicaragua, more than 3000 cases of poisoning and over 400 deaths occurred yearly from 1962 to 1972 (ICAITI, 1977). In Costa Rica, pesticide poisonings average about 550 per year (Leonard, 1986). Parathion has been largely responsible for poisonings in many countries (Almeida & Pereira, 1963). Self-reported rates of pesticide intoxication from surveys in Latin America run at about 13% of agricultural workers per year. Pesticide poisoning among children under 18 years of age accounts for roughly 10 to 20% of all poisonings. Several studies conducted throughout the region alarmingly confirm the widespread risks that pesticide exposure entails for farmworkers and

also for their families and children (McConnell et al., 1993). Organochlorine concentrations in human blood, fat tissue, and mothers' milk also have reached alarming levels in many countries (ICAITI, 1977).

The plummeting of cotton production in Central America was a direct result of an ecological crisis prompted by pest resistance to insecticides; this accelerating process ended in the well-known "pesticide treadmill," which involves (in addition to pesticide resistance) secondary pest outbreaks, pest resurgence, and elimination of natural enemies. Other ecological disturbances and serious public health problems also became prominent (Leonard, 1986). In cotton-growing areas of Central America, resurgence of malaria has occurred mainly because mosquitoes have developed pesticide resistance (Leonard, 1986). By 1970, over 35 cases of insecticide resistance had been detected, including important pests of cotton, banana, and stored grain (Gonzalez, 1976). Residues of organochlorine insecticides have been detected in fish and several invertebrate species, especially in estuaries and in areas near cotton fields (Giam, 1971).

Chemical pesticide technology has spread more quickly in Latin America than has the capability to ensure its effective and safe use. Pesticide use is rapidly increasing, and pesticide exporters from the industrialized nations are increasing their sales to Latin America (Nicholls & Altieri, 1997). Many pesticides considered too dangerous for unrestricted use in the western nations are still being exported from these nations to Latin America (Norris, 1982); examples include dibromo-chloro-propene (DBCP), leptophos, and BHC. Under current law, it is legal for companies to export them. Some countries have not enacted legislation to govern the importation, domestic use, and disposal of these pesticide materials. Even with laws, the governments frequently lack the infrastructures required to enforce them (Bottrell, 1984). Furthermore, many countries seldom have the medical personnel and facilities required for diagnosing and treating cases of pesticide poisoning (Bottrell, 1984). Programs to train farmers on the correct use of pesticides and alternative pest control methods are often inadequate (Bottrell, 1984). Norris (1982) dramatically documents the practice and consequences of pesticide dumping in Latin American and in other developing countries.

The growing use of pesticides is influenced by government subsidies that lower the costs of supplying pesticides to farmers. In Honduras, Colombia, and Ecuador, the rate of subsidies can be as high as 45% of the full retail costs (Repetto, 1985). These subsidies are large enough to affect a farmer's decision about pesticide use. By lowering the financial cost, subsidies raise the expected net returns from heavier and more frequent pesticide applications, and induce farmers to substitute chemicals for nonchemical methods of pest management. At the same time, subsidies deprive governments of funds that could be used to do a better job of monitoring, training, regulation, research, and extension for safe and effective alternative pest control methods.

Export promotion and trade liberalization policies are reinforcing the trend toward larger plantations, increased mechanization, and utilization of agricultural chemicals. With the reduction of farm subsidies in Europe and in the United States, and therefore higher competition, industry expects farmers in Latin America to increase pesticide consumption in the hope of boosting yields (Belloti *et al.,* 1990). Also, some countries in Latin America, in response to structural adjustment programs, have reduced pesticide subsidies. By depressing pesticide costs such policies can lead to overuse, such as the practice of frequent spraying as "insurance," although the benefits in terms of increased crop yields may be marginal (Conway & Pretty, 1991). Given the preceding conditions, it is not surprising to observe that technological change has mainly benefited the production for export and/or commercial crops produced primarily in the large farm sector, marginally impacting productivity of food crops, and thus bypassing the peasant sector (Convoy *et al.,* 1996).

BRIEF COUNTRY PROFILE ON CLASSICAL BIOLOGICAL CONTROL

With the foregoing account in mind, it is amazing that introduced biological control agents could become established at all, much less exert meaningful control of their hosts despite a heavy pesticide load in the agricultural environment. The following selected profiles [also see FAO (1990)] give clear indication that biological control could well play a much larger role in the control of agricultural pests in Latin America.

Argentina

From 1900 to 1979, 46 natural-enemy species were imported to control 21 pest species. Of these 46 imported natural enemies, 18 became established, of which 14 achieved partial control and 4 achieved complete control. Of the 21 main pests, 7 are targeted under permanent biological control. Among the successful introductions were *Encarsia* (= *Prospaltella*) *berlesi* against the white peach scale, *Aphelinus mali* (Haldeman), against the woolly apple aphid, and *Rodolia cardinalis* (Mulsant) against the cottony-cushion scale (Crouzel, 1984).

Brazil

There have been relatively few natural enemies imported into Brazil. *Aphelinus mali,* introduced in 1923, gave substantial control of the woolly apple aphid, and *P. berlesi* achieved complete control of the white peach scale in 1921. Poor results were obtained with the introduction of *Prorops nasuta* Waterston and *Tetrastichus giffardianus* Silvestri against the coffee berry borer, *Hypothenemus hampei* (Ferris), and the Mediterranean fruit fly, *Ceratitis capitata* (Wiedemann), respectively (Clausen, 1978). Three species of tachinids, *Lixophaga diatraeae* (Townsend), *Metagonistylum minense* Townsend, and *Paratheresia claripalpis* (Wulp), have been released against the sugarcane borer and are still used on plantations. *Apanteles flavipes* (Cameron) was introduced and achieved up to 62% parasitization of sugarcane borer in south central Brazil (Macedo, 1983). In 1978, a large biological control program against cereal aphids was initiated in southern Brazil. Fourteen species of hymenopterous parasites and two coccinellids (*Hippodamia quinquesignata* Kirby and *Coccinella septempunctata* Linnaeus) were introduced from Europe and Chile. Good adaptation and significant impact were observed on *Sitobium avenae* (Linnaeus) by *Aphidius uzbekistanicus* Luzhetski and *A. rhopalosiphi* de Stephani, and on *S. avenae* and *Metopolophium dirhodum* (Walker) by *Praon volucre* (Haliday) (Gassen, 1983). Research programs have been initiated in cassava, soybean, coffee, and cotton (Campanhola *et al.,* 1995).

Chile

Like the rest of South America in the early 1900s, Chile imported *Aphelinus mali* against *Eriosoma lanigerum* and *R. cardinalis* against *Icerya purchasi,* which also controlled

I. palmeri Riley & Howard (Hagen & Franz, 1973). In Chile between 1903 and 1984, 66 species of beneficial insects were introduced against several pest species of such crops as citrus, grape, peach, apple, and potato. Of these species, 42 became established. Of the targeted pests, 60% are under complete or substantial control, and 38% of the introduced predators and 24% of the parasitoids are responsible for maintaining this successful degree of control. Efforts at La Cruz have resulted in substantial control of whiteflies by *Amitus spiniferus* (Bréthes); of various Lepidoptera pests by *Trichogramma* spp.; of alfalfa aphid, *Acyrthosiphon pisum* (Harris), by several species of *Aphidius* and of *Pieris brassicae* (Linnaeus) by *Cotesia (Apanteles) glomeratus* (Linnaeus). Biological control of several pests [*Aonidiella* sp., the purple scale, *Lepidosaphes beckii* (Newman), and several species of the family Aphididae] has saved the Chilean citrus industry approximately $900,000 per year in pesticide costs (Gonzalez & Rojas, 1966; Zuñiga 1985, 1986).

Colombia

In 1933, *Aphelinus mali* was introduced from the United States and complete control of *E. lanigerum* was obtained. In neighboring Venezuela, *A. mali* also was introduced against *E. lanigerum,* and *R. cardinalis* was introduced in 1941 for control of *I. purchasi.* Attempts to control *Diatraea saccharalis* have involved the introduction and mass release of the Peruvian biotype of *Paratheresia claripalpis* Wulp., which has a shorter life cycle than the native biotype has (Hagen & Franz, 1973). *Metagonistylum minense* and *C. flavipes* also are produced by the sugar mills for use on their own plantations. Periodical releases of the parasitoid wasp *Diglyphus begini* also has proved key in the control of leaf miners in greenhouse chrysanthemums (Parrella & Nicholls, 1997).

Cuba

A most outstanding biological control success was the 1930 introduction of parasite *Eretmocerus serius* Silvestri against the citrus blackfly, which had become a serious pest of citrus and other trees. Full economic control was rapidly accomplished by the parasite (Hagen & Franz, 1973).

Since Cuba's trade relations with the socialist bloc collapsed in 1990, pesticide imports have dropped by more than 60%. Since then, the Cuban government has created about 220 Centers for the Production of Entomophages and Entomopathogens (CREEs) where decentralized, "artesanal" production of biocontrol agents takes place. These centers produce a number of entomopathogens (*Bacillus thuringiensis, Beauvaria bassiana, Metarhizium anisopliae,* and *Verticillium lecanii*), as well as one or more species of *Trichogramma,* depending on the crops grown in each area. *Trichogramma* is used for the control of Lepidopteran pests, principally *Mocis latipes* in improved pastures, *Heliothis* sp. in tobacco, and *Erinnys ello* in cassava. In 1994, levels of production of *Bacillus* and *Beauvaria* reached 1321 and 781 tons, respectively, and production of *Trichoderma* for the control of various plant diseases reached 2842 metric tons (Rosset & Benjamin, 1994).

Mexico

In 1948, *Eretmocerus serius* was introduced against the citrus blackfly. It became established and controlled the pest mostly in humid areas. A search for parasites was made in semiarid regions of Asia and four additional parasites were found and established, three of which became dominant in humid and dry areas of Mexico. *Amitus hesperidium* Silvestri became by far the most effective parasite, which then was released extensively in the 1950s by the newly organized Departmento de Control Biologico de Defensa Agricola (Hagen & Franz, 1973).

During 1954 and 1955, several parasites were introduced from Hawaii for the control of the Mexican fruit fly, *Anastrepha ludens* (Loew), native to Mexico. A large-scale production program was initiated, and in 5 years more than 7 million of the parasite *Aceratoneuromyia indicum* (Silvestri) were released in the field. *Aceratoneuromyia indicum* became quickly established, accounting at times for parasitism up to 80%; this lowered fruit damage to about 30% in Morelos, Oaxaca, Veracruz, Michoacan, and other states (Clausen, 1978).

Aphytis holoxanthus DeBach was released against the Florida red scale, *Chrysomphalus aonidium* (Linnaeus), in 1957 in Morelos, drastically reducing infestations in citrus groves within 1 year. Releases in 1954 of *A. lepidosaphes* Compere against the purple scale also resulted in effective biological control (Clausen, 1978).

Peru

As in many other countries, in Peru the woolly apple aphid was controlled by *Aphelinus mali,* and *I. purchasi* was controlled by *R. cardinalis.* The black scale was controlled by three imported parasites from the United States. Biological control of the cotton white scale by several parasites became successful in the Piura district following changes in cultivation practices. In Peru, there have been 12 cases of successful classical biological control against one pest of cotton, five pests of citrus, two pests of olive, one pest of alfalfa, and one pest of sugarcane (Aguilar, 1980). The most recent case studies occurred in the 1970s with the introduction of *Aphytis roseni* DeBach and *Cales noacki* Howard against *Selenaspidus articulatus* Morgan and *Aleurothrixus floccosus* (Maskell), respectively, in citrus. In that decade, *Aphidius smithi* Sharma & Subba Rao was also introduced against *Acrythosiphon pisum* (Harris)

TABLE 2 Cases of Partial, Substantial, or Completely Successful Biological Control Projects in Latin America

Crop	Pest	Order: family	Natural enemy	Order: family	Origin	Degree of success[a]	References
Argentina							
Sugarcane, corn, sorghum	Diatraea saccharalis	Lepidoptera: Pyralidae	Paratheresia claripalpis	Diptera: Tachinidae	Mexico via Peru	E, P	Crouzel, 1982
Field crops	D. elongatus Giglio-Tos	Orthoptera: Acarididae	Microsporida: Nosema locusta	Nosematidae	United States	E	Crouzel, 1982
	D. macalipenis Blanchard	Orthoptera: Acarididae	Nosema locusta				
	Schistocerca cancellata Serville						
Alfalfa	Acyrthosiphon pisum	Homoptera: Aphididae	Aphidius smithi / A. ervi	Hymenoptera: Aphididae	India via United States, Libano via United States	E, S	Crouzel, 1982
Apple	Eriosoma lanigerum	Homoptera: Aphididae	Aphelinus mali	Hymenoptera: Aphelinidae	United States via Uruguay	E, C	Crouzel, 1982
Peach	Pseudaulacaspis pentagona	Homoptera: Diaspididae	Encarsia (= Prospaltella) berlesei	Hymenoptera: Aphelinidae	United States	E, C	Crouzel, 1982
	Grapholita molesta (Busck)	Lepidoptera: Olethreutidae	Macrocentrus ancylivorus Rohwer	Hymenoptera: Braconidae	United States	E	Clausen, 1982
Fruit trees	Ceratitis capitata	Diptera: Tephritidae	Aceratoneuromyia indicua (Silv.)	Hymenoptera: Eulophidae	India (Mexico, Costa Rica)	E, P	Crouzel, 1982
			Biosteres (= Opius) longicaudatus Ash.	Hymenoptera: Braconidae	Mexico	E, P	Crouzel, 1982
	Anastrepha fraterculus (Wied.)	Hymenoptera: Braconidae	A. indicua Opius longicaudatus	Hymenoptera: Braconidae	Mexico (Mexico)	E / E	Clausen, 1978 / Clausen, 1978
Citrus	Icerya purchasi	Homoptera: Margarodidae	Rodolia cardinalis		Australia (Uruguay)	E, C	Clausen, 1978
	Aonidiella aurantii (Maskell)	Homoptera: Diaspididae	Sphaerotilbe coccophila	Hypocreaceae	United States	E, P	Crouzel, 1982
			Aphytis lingnanensis Comp.	Hymenoptera: Aphelinidae	China (United States, Chile)	E, P	Crouzel, 1982
			Aphytis melinus	Hymenoptera: Aphelinidae	India, Pakistan (United States)	E, S	Crouzel, 1982
			Comperiella bifasciata How.	Hymenoptera: Encyrtidae	China (United States)	E, P	Crouzel, 1982
	Unaspis lingnanensis (Kuwana)	Homoptera: Diaspididae	Aphytis lingnanensis		Hong Kong (United States)	E	Crouzel, 1982
	Lepidosaphes beckii	Homoptera: Diaspididae	Aphytis lepidosaphes		China (United States)	E	Crouzel, 1982
	Chrysomphalus ficus Ashmead	Homoptera: Diaspididae	Aphytis holoxanthus	Hymenoptera: Aphelinidae	United States	E	Crouzel, 1982
Bolivia							
Corn	Heliothis sp.	Lepidoptera: Noctuidae	Apanteles flavipes	Hymenoptera: Braconidae	not reported	E	Squire, 1972
Sugarcane	Diatraea saccharalis		Metagonistylum minense		not reported	E	Squire, 1972
Fruit trees	Anastrepha spp.		Opius concolor var. siculus (Szepl.)	Hymenoptera: Braconidae	not reported	E	Squire, 1972
			Aceratoneuromyia indicum	Hymenoptera: Eulophidae	not reported	E	Squire, 1972
			Pachycrepoideus vindemiae (Rond.)	Hymenoptera: Pteromalidae	not reported	E	Squire, 1972

(continues)

Brazil

Crop	Pest	Pest Order: Family	Natural enemy	Enemy Order: Family	Origin	Code	Reference
Sugarcane	*D. saccharalis*		*Apanteles flavipes*		West Indies	S	Bennett & Street, 1984, Planalsucar, 1980
			M. minense	Diptera: Tachinidae	Native	S	Gallo, 1980
			Lixophaga diatraeae	Diptera: Tachinidae	Cuba	S	Gallo, 1980
			Paratheresia claripalpis	Diptera: Tachinidae	Native	S	Gallo, 1980
			Myobiopsis diadema (Wiedemann)	Diptera: Tachinidae	Native	S	Gallo, 1980
	Mahanarva posticata Stål	Hemiptera: Cercopidae	*Metarrhizium anisopliae*	Hyphomycetes	not reported	S	Planalsucar, 1980
			Acmopolynema spp.	Hymenoptera: Mymaridae	not reported	S	Planalsucar, 1980
Apple	*Eriosoma lanigerum*		*Aphelinus mali*		Uruguay	S	DeBach, 1964
Peach	*Pseudaulacaspis pentagona*		*Encarsia berlesei*		United States	C	DeBach, 1964
Citrus	*Lepidosaphes beckii*		*Aphytis lepidosaphes*		China	S–C	Laing & Hamai, 1976

Central America

Chile

Crop	Pest	Pest Order: Family	Natural enemy	Enemy Order: Family	Origin	Code	Reference
Many fruits	*Ceratitis capitata*	Diptera Tephritidae	*Opius* spp.			P	Laing & Hamai, 1976
Potato	*Phthorimaea operculella* (Zeller)	Lepidoptera: Gelechiidae	*Apanteles subandinus* Blanch.	Hymenoptera: Braconidae	Argentina	S	Bennett & Street, 1984
Several	Aphids	Aphididae	*Adalia bipunctata* (Linnaeus)	Coleoptera: Coccinellidae	not reported	E, P	Gonzalez & Rojas, 1966
	Tetranychus urticae Koch	Acari: Tetranychidae	*Adalia bipunctata*		not reported	P	Gonzalez & Rojas, 1966
	Oligonychus vothersi (McGregor)	Acari: Tetranychidae					
	Several caterpillars		*Trichogramma achaeae* Nagajara & Nagarkatti	Hymenoptera: Trichogrammatidae	not reported	S	Bennett & Street, 1984
Citrus and deciduous trees	*Quadraspidiotus perniciosus* Comstock	Homoptera: Diaspididae	*Encarsia* (= *Prospaltella*) *perniciosi* Tower	Hymenoptera: Aphelinidae	United States	E	Clausen, 1978
			Aphytis mytilaspidis (LeBaron)	Hymenoptera: Aphelinidae	Caribbean Islands	P	Gonzalez & Rojas, 1966
			Aphytis procila (Walker)	Hymenoptera: Aphelinidae	Caribbean Islands	P	Gonzalez & Rojas, 1966
			Coccidophilus spp.	Coleoptera: Coccinellidae	Caribbean Islands	P	Gonzalez & Rojas, 1966
			Microwesia spp.	Coleoptera: Coccinellidae	Caribbean Islands	P	Gonzalez & Rojas, 1966
Citrus	*Icerya purchasi*		*Cryptocheta iceryae* (Williston)	Diptera: Cryptochetidae	Australia	C	Gonzalez & Rojas, 1966
	I. palmeri		*R. cardinalis*		Australia	C	DeBach, 1964
	Pseudococcus gahani Green	Homoptera: Pseudococcidae	*Coccophagus gurneyi* Comp.	Hymenoptera: Aphelinidae	Australia	S	Gonzalez & Rojas, 1966
			Hungariella (=*Tetracnemus*) *pretiosa* (Timb.)	Hymenoptera: Encyrtidae	Australia	S	Gonzalez & Rojas, 1966
	Lepisosaphes beckii		*A. lepidosaphes*		China	S, P	Gonzalez & Rojas, 1966
			Lindorus lophanthae (Blaisd.)	Coleoptera: Coccinellidae	Australia	P	Gonzalez & Rojas, 1966
	Aonidiella aurantii		*A. lingnanensis*		China	P	Gonzalez & Rojas, 1966
			Coccidophilus citricola Brethes	Coleoptera: Coccinellidae	Caribbean Islands	P	Gonzalez & Rojas, 1966

TABLE 2 (continued)

Crop	Pest	Order: family	Natural enemy	Order: family	Origin	Degree of success[a]	References
	Planococcus citri (Risso)	Homoptera: Pseudococcidae	Leptomastix dactylopii How.	Hymenoptera: Encyrtidae	not reported	P	Gonzalez & Rojas, 1966
	Pseudococcus adonidum Linnaeus	Homoptera: Pseudococcidae	Leptomastidea abnormis (Grlt.)	Hymenoptera: Encyrtidae	Italy	P	Gonzalez & Rojas, 1966
			Coccophagous gurneyi		Australia	P	Gonzalez & Rojas, 1966
			Cryptolaemus montrouzieri Muls.	Coleoptera: Coccinellidae	Australia	S	Gonzalez & Rojas, 1966
			C. montrouzieri		Australia	S	Gonzalez & Rojas, 1966
	Pseudococcus sp.	Homoptera: Pseudococcidae	Chrysoperla spp.	Neuroptera: Chrysopidae	Peru	S	Gonzalez & Rojas, 1966
			Sympherobius spp.	Neuroptera: Hemerobiidae	Caribbean Islands	S	Gonzalez & Rojas, 1966
Apple	Eriosoma lanigerum		Aphelinus mali		United States	C	DeBach, 1964
Peach	Scolytus rugulosus (Ratzeburg)	Coleoptera: Scolytidae	Cheiropachus colon L.	Hymenoptera: Cleonymidae	United States	P	Gonzalez & Rojas, 1966
Olive and citrus	Saissetia oleae		Metaphycus lounsbury	Hymenoptera: Pteromalidae	Australia	P	Gonzalez & Rojas, 1966
			Scutellista cyanea Mots.		South Africa	P	Gonzalez & Rojas, 1966
	S. coffeae (Walker)	Homoptera: Coccidae	M. helvolus (Comp.)	Hymenoptera: Encyrtidae	South Africa	S, P	Gonzalez & Rojas, 1966
			M. helvolus	Hymenoptera: Eupelmidae	South Africa	S	Gonzalez & Rojas, 1966
			Lecaniobius utilis Compere		United States via Peru	P	Gonzalez & Rojas, 1966
Colombia							
Corn, sorghum	Spodoptera spp.	Lepidoptera: Noctuidae	Telenomus remus Nixon	Hymenoptera: Scelionidae	not reported	S	Bennett & Street, 1984
Corn, sorghum, sugarcane	Diatraea saccharalis		Trichogramma perkinsi Grlt.	Hymenoptera: Trichogrammatidae	not reported	S	Amaya, 1982
Bean, soybean	Heliothis spp.		T. pretiosum Riley	Hymenoptera: Trichogrammatidae	not reported	S	Amaya, 1982
			T. australicum Grlt.	Hymenoptera: Trichogrammatidae	not reported	S	Amaya, 1982
	Anticarsia gemmatalis Hübner	Lepidoptera: Noctuidae	T. pretiosum			S	Amaya, 1982
			T. australicum			S	Amaya, 1982
Cassava	Erinnyia ello Linnaeus	Lepidoptera: Sphingidae	T. perkinsi			S	Amaya, 1982
			T. australicum			S	Amaya, 1982
Sugarcane	D. saccharalis		Lixophaga diatraeae		Cuba	S	Bennett & Street, 1982
Tomato	Scrobipalpula sp.	Lepidoptera: Gelechiidae	T. australicum		Cuba	S	Amaya, 1982
Cotton	Alabama argillacea Hübner	Lepidoptera: Noctuidae	T. australicum		Cuba	E, S	Amaya, 1982
	Heliothis sp.		T. pretiosum			S	Amaya, 1982
			T. australicum			E, S	Amaya, 1982
Apple	Eriosoma lanigerum		Aphelinus mali		United States	C	DeBach, 1964
Costa Rica							
Citrus	Aleurocanthus woglumi		Eretmocerus serius Silv.	Hymenoptera: Aphelinidae	Cuba	C	DeBach, 1964
Apple	E. lanigerum		A. mali		United States	C	DeBach, 1964

Cuba

Crop	Pest	Pest (Order: Family)	Natural enemy	Natural enemy (Order: Family)	Origin	Status	Reference
Potato	Cylas formicarius elegantulus (Summers)	Coleoptera: Curculionidae	Pheidole megacephala (Fabricius)	Hymenoptera: Formicidae		S	Castineiras et al., 1982
Banana	Cosmopolites sordidus (Germar)	Coleoptera: Curculionidae	Tetramorium guineese Fabricius	Hymenoptera: Formicidae		S	Roche & Abreu, 1983
Sugarcane	Diatraea spp.		Lixophaga diatraeae		Native	S	Bennett, 1971
Citrus	Aleurocanthus woglumi		E. serius		Asia	C	Laing & Hamai, 1976

Ecuador

Crop	Pest	Pest (Order: Family)	Natural enemy	Natural enemy (Order: Family)	Origin	Status	Reference
Citrus	Lepidosaphes beckii		Aphytis lepidosaphes			S	Jimenez, 1961
			Encarsia sp.			S	
Sugarcane	D. saccharalis		Paratheresia claripalpis		Peru	S	Bennett, 1971
	I. montserratensis (Riley & Howard)	Homoptera: Margarodidae	Rodolia cardinalis		(United States)	S	DeBach, 1964

El Salvador

Crop	Pest	Pest (Order: Family)	Natural enemy	Natural enemy (Order: Family)	Origin	Status	Reference
Citrus	A. woglumi	Homoptera: Aleyrodidae	Encarsia spp.			C	Quezada, 1974
	Lepidosaphes beckii		A. lepidosaphes		China	S–C	Laing & Hamai, 1976

British Guyana

Crop	Pest	Pest (Order: Family)	Natural enemy	Natural enemy (Order: Family)	Origin	Status	Reference
Sugarcane	D. saccharalis		Metagonistylum spp.	Diptera: Tachinidae	Brazil	C	Bennett & Street, 1984

Haiti

Crop	Pest	Pest (Order: Family)	Natural enemy	Natural enemy (Order: Family)	Origin	Status	Reference
Citrus	A. woglumi		Eretmocerus serius		Cuba	C	DeBach, 1964

Jamaica

Crop	Pest	Pest (Order: Family)	Natural enemy	Natural enemy (Order: Family)	Origin	Status	Reference
Banana	C. sordidus		Plaesius javanus Erich.	Coleoptera: Histeridae	Malaya	E	Clausen, 1978
Citrus	A. woglumi		E. serius		Cuba	C	DeBach, 1964
			Encarsia opulenta (Silvestri)	Hymenoptera: Aphelinidae	Cuba	C	Laing & Hamai, 1976

Mexico

Crop	Pest	Pest (Order: Family)	Natural enemy	Natural enemy (Order: Family)	Origin	Status	Reference
Several	Aphids	Homoptera: Aphididae	Hippodamia convergens		not reported	E	Jimenez, 1961
Wheat	Schizaphis (= Toxoptera) graminum (Rondani)	Homoptera: Aphididae	H. convergens		not reported	C	Oliva, 1961
Alfalfa	Therioaphis maculata (Buckton)	Homoptera: Aphididae	H. convergens		not reported	E	Oliva, 1961
			Praon palitans Muesebeck	Hymenoptera: Aphididae	not reported	P	Jimenez, 1961
			Trioxys utilis Muesebeck	Hymenoptera: Aphididae	not reported	P	Jimenez, 1961
			Aphelinus semiflavus How.	Hymenoptera: Aphididae	not reported	P	Jimenez, 1961
Sugarcane	Diatraea spp.		Trichogramma pretiosum		not reported	E	Jimenez, 1961
			Chelonus sp.	Hymenoptera: Braconidae	not reported	E	Jimenez, 1961
			Trichogramma spp.		not reported	S	Sanchez et al., 1979
Cotton	Trichoplusia sp.	Lepidoptera: Noctuidae	T. pretiosum		not reported	E	Jimenez, 1961
	Pectinophora gossypiella (Saunders)	Lepidoptera: Gelechiidae	Chelonus sp.		not reported	E	Jimenez, 1961
	Aphids		Hippodamia convergens		not reported	E	Oliva, 1961

(continues)

TABLE 2 *(continued)*

Country and crop	Pest	Order: family	Natural enemy	Order: family	Origin	Degree of success[a]	References
Cotton, maize, sorghum, wheat, rice, lucerne, bean, soybean, sugarcane	Noctuids		*Trichogramma* spp.			S	Anonymous, 1978
Fruit trees	*Anastrepha ludens*		*Opius* spp.		not reported	S	Jimenez, 1961
			Aceratoneuromyia indicua		not reported	S	Jimenez, 1961
Citrus	*A. woglumi*		*Encarsia opulenta*		India, Pakistan	C	Jimenez, 1961
			E. clypealis (Silvestri)		India, Pakistan	S	Jimenez, 1961
			Amitus hesperidum			S	Jimenez, 1961
			Eretmocerus serius			P	Jimenez, 1961
	Aleurothrixus floccosus		*A. spiniferus*		United States, Mexico	S	Laing & Hamai, 1976
	Aomidiella aurantii		*Encarsia* spp.			S	Jimenez, 1961
			Aphytis chrysomphali (Mercet)		United States, Mexico	S	Jimenez, 1961
			A. lingnanensis			S	Jimenez, 1961
			E. perniciosi			S	Jimenez, 1961
			Comperiella bifasciata		not reported	S	Jimenez, 1961
	Chrysomphalus aonidum (Linnaeus)	Homoptera: Diaspididae	*A. lingnanensis*			S	Jimenez, 1961
			A. holoxanthus			S	Jimenez, 1961
			Pteroptrix smithi (Compere)	Hymenoptera: Aphelinidae	not reported	S	Jimenez, 1961
			E. (=Prospaltella) aurantii (How.)		Native	S	Jimenez, 1961
			Pseudohomolopoda prima (Girault)	Hymenoptera: Encyrtidae	not reported	S	Jimenez, 1961
	Lepisosaphes beckii		*A. lepidosaphes*		Native	S	Jimenez, 1961
			Encarsia sp.	Hymenoptera: Encyrtidae		S	Jimenez, 1961
Apple	*Eriosoma lanigerum*		*Aphelinus mali*		United States	S	Jimenez, 1961
Panama							
Sugarcane	*D. saccharalis*		*Apanteles flavipes*			S	Bennett & Street, 1982
Citrus	*Aleurocanthus woglumi*		*Eretmocerus serius*		Cuba	C	DeBach, 1964
Peru							
Alfalfa	*Acyrthosiphon pisum*		*Aphidius smithi*		not reported	S	Aguilar, 1980
Sugarcane	*D. saccharalis*		*T. fasciatum* (Perk.)	Hymenoptera: Trichogrammatidae		S	Cueva, 1979
			Apanteles flavipes		not reported	S	Cueva et al., 1981
			T. fasciatum		not reported		
			T. brasiliensis (Ashmead)			S	Cueva, 1979
			Telemos alecto Crawford	Hymenoptera: Scelionidae	not reported		
Cotton	*Aphis gossypii* Glover	Hymenoptera: Aphididae	*Hippodamia convergens*	Coccinellidae	not reported	S	Cueva, 1979
	Pinnaspis minor Maskell	Homoptera: Diaspididae	*Aspidiotiphagus citrinus* (Craw.)	Hymenoptera: Aphelinidae	Barbados, Italy, Japan, United States	P	Laing & Hamai, 1976
			Arrhenophagous chionaspidis Aurivillius	Hymenoptera: Aphelinidae		P	Laing & Hamai, 1976
			Microweisia spp.	Coleoptera: Coccinellidae	not reported	P	Laing & Hamai, 1976

Crop	Pest	Order: Family	Natural enemy	Origin	Result[a]	Reference
Citrus	Chrysomphalus aonidum		Aphytis holoxanthus	not reported	S	Cueva, 1979
	Aleurothrixus floccosus		Cales noacki	not reported	S	Cueva, 1979
	Selenaspidus articulatus		A. roseni	Kenya	S	Aguilar, 1980; Bennett & Street, 1984
	Cornuaspis (Lepidosaphes) beckii		A. lepidosaphes	China	S	Laing & Hamai, 1976
	I. purchasi		Rodolia cardinalis	Australia	S	Aguilar, 1980
	Saccharicoccus sacchari (Cockerell)	Homoptera: Pseudococcidae	Anagyrus saccharicola (Timb.)	not reported	S	Aguilar, 1980
Olive and citrus	Saissetia oleae	Hymenoptera: Encyrtidae	Metaphycus helvolus	not reported	S	Aguilar, 1980
			M. lounsburyi	not reported	S	Aguilar, 1980
			Scutellista cyanea	not reported	S	DeBach, 1964
			Lecaniobius utilis	not reported	S	Aguilar, 1980
	S. coffeae	Hymenoptera: Aphelinidae	Coccophagus rusti	not reported	S	Aguilar, 1980
			M. helvolus	not reported	S	DeBach, 1964
Apple	E. lanigerum		Aphelinus mali	United States	C	DeBach, 1964
Peach	Pseudaulacaspis pentagona		Encarsia berlesei	United States, Italy	C	DeBach, 1964
Fruit trees	Anastrepha striata Schiner	Hymenoptera: Cynipidae	Ganaspis pellaranoi (Bréthes)	(Argentina)	S	Clausen, 1978
Puerto Rico						
Sugarcane	Phyllophaga portoricencis	Procoela: Bufidae (toad)	Bufo marinus	not reported	S–C	Laing & Hamai, 1976
Citrus	I. purchasi		R. cardinalis	Australia (United States	C	DeBach, 1964
	Lepidosaphes beckii		Aphytis lepidosaphes	China	S–C	Laing & Hamai, 1976
Papaya and mulberry	Pseudaulacaspis pentagona	Coleoptera: Coccinellidae	Chilocorus cacti Linnaeus	Cuba	S	DeBach, 1964
Uruguay						
Citrus	I. purchasi		R. cardinalis	Australia via Portugal	C	DeBach, 1964
Apple	E. lanigerum		Aphelinus mali			DeBach, 1964
Peach	P. pentagona		E. berlesei	United States, Italy	C	DeBach, 1964
Venezuela						
Sugarcane	D. saccharalis		Metagonistylum minense	Brazil	S	Bennett & Street, 1984
Citrus	I. purchasi		R. cardinalis	Australia (United States) (Mexico)	C	DeBach, 1964
	Aleurocanthus woglumi		Encarsia sp.		C	Geraud et al., 1977
Apple	E. lanigerum		A. mali	United States	P	DeBach, 1964

[a] Abbreviations: C = complete control; E = established natural enemy; P = partial control; S = substantial control.

TABLE 3   Estimated Costs of Chemical Control of Seven Important South American Pests in the Absence of Imported Natural Enemies That Keep Them under Complete or Substantial Control

Pest	Area of host plants (ha)	Sprays per year	US$ per spray (per ha)	US$
Icerya purchasi	12,928	2	44.67	1,155,179
Saissetia oleae	40,783	1	44.67	1,822,079
Eriosoma lanigerum	18,766	2	12.60	472,177
Pseudococcus and *Planococcus*				
Citrus	12,492	2	33.87	463,423
Other fruits	6,841	2	33.87	846,232
Aleurothrixus floccosus	12,492	3	33.87	1,269,348
Sitobium avenae and *Metopolophium dirhodum*	463,594	2	12.60	11,664,623
Total	567,896			17,693,061

After Campanhola, C., *et al.* (1995). In Mengech, A. N., *et al.* (Eds.), *Integrated pest management in the Tropics: Current status and future prospects.* New York: John Wiley & Sons.

in alfalfa. At the national insectary Centro de Introduccíon y Cría de Insectos Utiles (CICIU) (see later text), yearly production of *Trichogramma* reached 131 million wasps in 1976, which were distributed to about 1300 ha at a rate of 100,000 wasps per hectare (Klein-Koch, 1977). *Metaphycus helvolus* Compere and *Coccophagus rusti* Compere have been introduced against *Saissetia oleae* Bern in olives, causing high levels of mortality (Gomero & Lizárraga, 1995).

Other Programs

In all the preceding countries, many more introductions have been done (against fruit flies, coffee berry borer, oriental fruit moth, etc.), but no documentation on the scale of release or on the impact of released enemies on pest damage is available (Clausen, 1978; Andrews & Quezada, 1989; Campanhola *et al.,* 1995).

In the countries not included earlier, the efforts have been very limited. For example, in Uruguay introductions of enemies well adapted in neighboring countries resulted in complete control of *I. purchasi, Eriosoma lanigerum,* and *Pseudaulacaspis pentagona.* In Venezuela, sugarcane borers have been successfuly controlled with *Metagonistylum minense* and *Cotesia flavipes.* In Central America, the most outstanding case is the complete biological control of the citrus blackfly by *Eretmocerus serius* in Costa Rica and Panama in 1932 (Hagen & Franz, 1973). In 1985 in Honduras, *C. flavipes* was imported and released against corn and sugarcane borers. Table 2 lists all the examples of classical biological control with varying degrees of success in Latin America during the last 60 years. Table 3 shows estimated costs of chemical control for seven key pests had biological control not been introduced (Campanhola *et al.,* 1995).

THREE CONTEMPORARY CASE STUDIES

Biological Control of Cereal Aphids in Chile and in Brazil

In 1972, populations of two aphid species (*Sitobium avenae* and *Metopolophium dirhodum*) were detected in cereal fields in Chile. Despite the presence of resident natural enemies, these aphids reached outbreak proportions that led to aerial application of insecticides over 120,000 ha of wheat. In 1975, the aphids and the barley yellow dwarf virus (BYDV) they transmit caused a loss of about 20% of the national wheat production (Zuñiga, 1986). In 1976, the Chilean government's agricultural research center, Instituto de Investigaciones Agropecuarias (INIA), in conjunction with the Food and Agriculture Organization (FAO) of the United Nations, initiated an IPM program. As part of this program, several aphidophagous insects were introduced against *M. dirhodum* and *S. avenae.* Five species of predators were introduced from South Africa, Canada, and Israel; and nine species of parasitoids of the families Aphidiidae and Aphelinidae were brought from Europe, California, Israel, and Iran (Zuñiga, 1986). In 1975, more than 300,000 Coccinellidae were mass-reared and released, and from 1976 to 1981 more than 4 million parasitoids were distributed throughout the cereal areas of the country. Today aphid populations are maintained below the economic threshold level by the action of biological control agents (Zuñiga, 1986).

The success in Chile prompted Brazilian researchers in 1978 to introduce 14 species of hymenopteran parasites and two of the Coccinellidae family. About 3.8 million parasites were released throughout the wheat-producing regions of Rio Grande do Sul, Paraná, and Santa Catarina. *Aphidius uzbekistanicus* became established and efficiently adapted against *S. avenae* and *Praon volucre* attacked *S. avenae* and *M. dirhodum* (Gassen, 1983). This program prevented the use of millions of liters of insecticides representing savings of more than US$11 million (Campanhola *et al.,* 1995). In 1981, Brazilian researchers shipped parasite colonies of *Ephedrus plagiator* (Nees) and *P. gallicum* Stary to Argentina where personnel of Instituto Nacional de Tecnología Agropecuaria (INTA) initiated mass rearing and release activities at their insectary in Castelar.

Biological Control of Sugarcane Borer in Brazil

In Latin American sugarcane-producing countries (Brazil, Colombia, and Venezuela), populations of *Diatraea saccharalis* and *D. flavipenella* Box traditionally have been subjected to biological control through liberations of tachinid parasites mass-reared in private or in government insectaries. In Brazil, an additional enemy was added, the hymenopteran parasitoid *Apanteles flavipes,* which was introduced in 1974 from Trinidad and Tobago. From 1975 to 1982, more than 1 billion *A. flavipes* females were mass-reared and then released in the eastern areas of Brazil located between 0 and 25 degrees south latitudes (Macedo, 1983). Parasitism rates were increased from 12.8%, which was the level attained by resident parasitoids, to 26.0% after the introduction of *A. flavipes.* The key to the success of *A. flavipes* is that it attacks young larvae, thus preventing them from becoming adults. Today the level of damage of *D. saccharalis* remains around 5% below what is considered economically damaging. The sugarcane plant hopper, *Mahanarva posticata,* a serious pest in northeastern Brazil, has been brought under control for almost 20 years by applications of the fungus *Metarhizium anisopliae* (Campanhola *et al.,* 1995).

Biological Control of Alfalfa Aphids in Argentina

In 1972, two parasitoids were introduced from California: *Aphidius smithi* Sharma & Subba Rao and *A. ervi* Haliday against *Acyrthosiphon pisum.* Large numbers of these species were reared and released and despite the presence of native hyperparasites, substantial control has been achieved in many areas. Today *A. pisum* is not a significant problem in alfalfa.

In 1976, another aphid pest, the blue alfalfa aphid (*A. kondoi* Shinji), was detected in alfalfa throughout Argentina exerting more damage than *A. pisum.* In 1978, five biotypes of the aphidiids *E. plagiator* and *Aphidius ervi* were introduced into Argentina, mass-reared in Anguil, and distributed throughout the alfalfa regions (Crouzel, 1982). Such efforts resulted in partial control of the blue alfalfa aphid.

CENTERS OF BIOLOGICAL CONTROL IN LATIN AMERICA

In Latin America, there are few research centers devoted solely to biological control. In the early part of the century there were only three centers: the INTA insectary at Castelar in Argentina; the INIA experimental substation in La Cruz, Chile; and the CICIU in Lima, Peru. In Argentina, the USDA opened a subsidiary laboratory for biological control of weeds in Hurlingham near Buenos Aires, which (although devoted mostly to quarantine and selectivity studies of weed herbivores to be introduced into the United States) sponsored activities that led to the successful biological control of the water hyacinth, *Eichornia crassipes* Solms-Laubach by *Neochetina bruchi* Hust. in the La Rioja region. In 1970, the Universidad de Tucuman established in San Miguel de Tucuman the Centro de Investigaciones sobre Regulaciones de Poblaciones de Organismos Nocivos (CIRPON), a center devoted to the integrated and biological control of citrus and soybean pests. CIRPON also conducts regular training courses in IPM, biological control, and agroecology with participation of graduates from throughout Argentina.

In Brazil, 4 laboratories and 23 multiplication units were established by the Programa Nacional de Melhoramiento de Cana de Azucar, for the mass rearing and release of *Apanteles flavipes* and tachinid parasitoids of sugarcane borers. Brazil's agricultural research center [Empresa Brasileira do Pesquisa Agropacuaria (EMBRAPA)], has also set up insectaries and laboratory facilities in southern Brazil to support the cereal aphid biological control program initiated in the late 1970s. The Sugar and Alcohol Institute (IAA/Planalsucar), State of Pernanbuco Agricultural Research Enterprise (IPA), and several private laboratories have engaged in production of *Metarhizium* for sugarcane plant hopper control.

In Colombia, private sugarcane plantations have small insectaries for the mass rearing of *Diatraea* spp. parasites and *Trichogramma* spp. In Mexico, the government established a national system for the mass production of *Trichogramma* spp. and other beneficial organisms. In the rest of the countries, private or government groups have initiated small efforts to deal with specific pest problems. Examples are the projects in Venezuela against *Diatraea* spp. in sugarcane that resulted in 50% damage reduction following the introduction of *Metagonistylum minense* (Clausen, 1978), and the releases of two predators in the Dominican Republic against the coconut scale in 1937 (Gomez-Menor, 1937) and in Colombia against the cypress sawfly (Drooz *et al.,* 1977). In Paraguay, various cooperatives have initiated production of baculovirus for control of velvet bean caterpillar in soybeans. The most dramatic example of massive production of beneficial insects and microbial insecticides are the current efforts in Cuba.

TABLE 4   Some Insect Pests Attacking Crops in Latin America That Potentially Can Be Controlled with Natural Enemies

Insect pest	Potential natural enemies	Region
Aspidiotus destructor Signoret (Homoptera: Diaspididae)[a]		Venezuela, Colombia
Parlatoria oleae (Colvé) (Homoptera: Diaspididae)	*Aphytis maculicornis* (Masi) (Hymenoptera: Aphelinidae) and *Coccophagoides utilis* Doutt (Hymenoptera: Aphelinidae)[b]	Argentina, Brazil
Quadraspidiotus perniciosus (Comstock) (Homoptera: Diaspididae)	*Encarsia perniciosi* (Tower) (Hymenoptera: Aphelinidae)	Southern cone (Argentina, Chile, Uruguay, southern Brazil)
Unaspis citri (Comstock) (Homoptera: Diaspididae)	*A. lingnanensis* Compere	Entire region
Aphis citricola v.d. Goot (Homoptera: Aphididae)[a]		Neotropical areas
Pseudococcus maritimus (Ehrhorn) (Homoptera: Pseudococcidae)	*Acerophagas notativentris* (Girault) (Hymenoptera: Encyrtidae)[b]	Grape growing areas
Nezara viridula (Linnaeus) (Heteroptera: Pentatomidae)	*Trissolcus basalis* (Wollaston) (Hymenoptera: Scelionidae)	Tropical areas
Phoracantha semipunctata Fabricius (Coleoptera: Cerambycidae)[a]		
Ceratitis capitata[a]		
Bemisia tabaci (Gennadius) (Homoptera: Aleyrodidae)[a]		
Grapholitha molesta[a]		
Scrobipalpula absoluta Meyr (Lepidoptera: Gelechiidae)[a]		Potato and tomato growing areas
Epinotia aporema Wals. (Lepidoptera: Olethreutidae)[a]		Bean, soybean, and alfalfa growing areas

[a] Foreign exploration needed to search for more natural enemies.
[b] Enemies already established in California.

POSSIBILITIES FOR FURTHER WORK AND CONCLUSIONS

There is a long and rich tradition of biological control in Latin America, especially in Chile, Argentina, Peru, Brazil, Colombia, and Mexico. The early success of biological control of citrus pests obtained in California triggered a number of introductions into the citrus-growing areas of the continent, thus promoting wide interest in biological control. Other projects followed, such as those involving sugarcane, cotton, peach, olive, and wheat crops.

Given the economic crises and the degree of environmental degradation in the region, governments should expand the efforts in biological control, and tailor specific programs to fit the needs of the large peasant–farmer sector that lacks the necessary capital to purchase pesticides or to set up insectaries. In this regard, the efforts in Cuba and also of several nongovernment organizations currently assisting the peasant–farmer in alternative agricultural technologies (Altieri & Anderson, 1986) should promote bio-

logical control at the village level utilizing simple mass rearing and release techniques against specific pests (Rosset & Benjamin, 1994).

In Table 4, a number of prevalent insect pests in Latin America that can be potentially controlled with classical biological control have been identified. Some of these pests are already under successful biological control in other areas of the world, and some require further foreign exploration to search for new or more effective natural enemies. There is also tremendous potential to engage in exchanges of beneficial organisms between areas that share ecological commonalities. For example, Chile and California share similar Mediterranean climates, cropping systems, and a number of pest organisms. An exchange of effective natural enemies between both regions could be initiated. A list of potential organisms is presented in Table 5. Similar projects could be proposed between the tropical areas of Latin America and Florida, and the Caribbean and tropical regions of southeast Asia.

Despite trends toward economic globalization, the cur-

TABLE 5   List of Natural Enemies of Insect Pests of Mutual Interest to Chile and California

Natural enemy	Pest to be regulated	For possible introduction to	
		Chile	California
Macrocentrus delicatus Cress. (Hymenoptera: Braconidae)	*Grapholitha molesta*	X	X
Lysiphlebus testaceipes (Cress.) (Hymenoptera: Aphididae)	Several aphid species including *Aphis citricola* Van der Goot (Homoptera: Aphididae)	X	
Metaphycus (=*Aphycus*) *zebratus* Mercet (Hymenoptera: Encyrtidae)	*Saissetia oleae*	X	
M. bartletti Annecke & Mynhardt			
Cryptolaemus montrouzieri	*Pseudococcus* sp.		
Praon volucre	Several aphid species, *Macrosiphum rosae* (Linnaeus) (Homoptera: Aphididae) *A. pisum, A. kondoi,* etc.		X
Several species of Phytoseiidae	Several species of mites and thrips	X	
Acerophagus notativentris	*P. maritimus* (Ehrhorn) (Homoptera: Pseudococcidae) and *P. obscurus* Essig	X	
Granulosis virus	*Rhyacionia buoliana* (Schiff.) and *Cydia pomonella* (Linnaeus) (Lepidoptera: Oletreutidae)	X	
Parasitoids of synanthropic flies	*Musca domestica* (Linnaeus) (Diptera: Muscidae) *Stomoxys calcitrans* (Linnaeus)	X	X
Several predators and parasites including *Tetrastichus* sp.	Pepper tree (*Schinus molle*) (Sapinales: Anacardiaceae) psyllid (Homoptera: Psyllrdae)	X	
Copidosoma truncatella (Dalman) (Hymenoptera: Encyrtidae)	*Rachiplusia ou* (Guené) and *Tricoplusia ni* (Hübner) (Lepidoptera: Noctuidae)	X	
Stethorus punctum (LeConte) (Coleoptera: Coccinellidae)	Spider mites	X	
Several predators	Spider mites	X	X
Oligota (=*Pygmaea*) *parva* Kraatz (Coleoptera: Staphylinidae) and *S. histrio* Fall (Coleoptera: Coccinellidae)	Spider mites	X	
Several predators	Thrips	X	
Several parasitoids	*Therioaphis maculata* (Homoptera: Aphididae)	X	
Several predators	*Icelococcus* sp. (Homoptera: Eriococcidae)	X	

rent economic and social juncture in the region calls for more low-input approaches to agriculture. Classical biological control should be at the forefront of any sustainable agricultural development effort, complemented by agroecosystem management schemes (intercropping, crop rotations, cover crops, etc.) not only that aid biological control agents to perform more efficiently, but also that conserve the soil and make the agroecosystems less dependent on fertilizers, herbicides, and other chemical inputs.

Acknowledgments

The authors and the editors gratefully acknowledge the following entomologists for their contributions to the information presented in this chapter: Javier Trujillo (Mexico), Luciano Campos S. (deceased, Chile), Carlos Klein-Koch (Chile), Clifford S. Gold (United States), and Jose R. Quezada (El Salvador).

References

Aguilar, P. G. (1980). Apuntes sobre el control biologico y el control integrado de las plagas agricolas en el Peru. Revista Peruviana de Entomologia, 23, 85–110.

Almeida, W. F., & Pereira, A. P. (1963). Paration como principais responsaiveis pelos casos accidentais de intoxicacao por insecticidas de uso agricola. O Biologico, 29, 249–257.

Altieri, M. A., & Anderson, M. K. (1986). An ecological basis for the development of alternative agricultural systems for small farmers in the Third World. American Journal of Alternative Agriculture, 1, 30–38.

Amaya, N. M. (1982). Investigacion, utilizacion y resultados obtenidos en diferentes cultivos con el uso de *Trichogramma* en Colombia Sur

America. In Les Trichogrammes. Ier Symposium International. Antibes, France: Institut National de la Recherche Agronomique.

Andrews, K. L., & Quezada, J. R. (1989). Manejo integrado de plagas insectiles en la agricultura. Estado actual y futuro. El Zamorano, Honduras: Departamento de Proteccion Vegetal. Escuela Agricola Panamericana.

Anonymous, (1978). VII Reunion Nacional de Control Biologico (Memorias), Veracruz, Mexico.

Belloti, A. C., Cardona, C., & Lapointe, S. L. (1990). Trends of Pesticide use in Colombia and Brazil. Journal of Agricultural Entomology, 7, 191–201.

Bennett, F. D. (1971). Current status of biological conrol of the small moth borers of sugar cane Diatraea spp. (Lep.: Pyralidae). Entomophaga, 16, 111–124.

Bennett, D., & Street, G. (1984). The commonwealth institute of biological control in integrated pest management programs in Latin America. In G. Allen & A. Rada (Eds.), The role of biological control in pest management. Proceedings, International Symposium of IOBC/WHRS, Santiago, Chile. Ottawa: Ottawa University Press.

Bottrell, D. G. (1984). Government influence on pesticide use in developing countries. Insect Science Applications, 5, 151–155.

Burton, D. K., & Philogene, B. J. R. (1984). An overview of pesticide usage in Latin America. A report to the Canadian Wildlife Service Latin American Program. Ottawa, Canada: Canadian Wildlife Service.

Campanhola, C., Jose De Moraes, G., & DeSa, L. A. N. (1995). Review of IPM in South America. In A. N. Mengech, et al. (Eds.). Integrated pest management in the tropics: Current status and future prospects. New York: John Wiley & Sons.

Castineiras, A., Caballero, S., Rego, G., & Gonzalez, M. (1982). Efectividad tecnico economico del empleo de la hormiga leona Pheidole megacephala en el control del tetuan del boniato Cylasformicarius elegantulus. Habana, Cuba: Ciencia y Tecnica en la Agricultura. Proteccion de Plantas.

Clausen, C. P. (1978). Introduced parasites and predators of arthropod pests and weeds: A world review. Agriculture Handbook 480, Washington, DC: U.S. Department of Agriculture.

Conroy, M. E., Murray, D. L., & Rosset, P. M. (1996). A cautionary tale: Failed US development policy in Central America. Boulder, CO: Lynne Rienner Publisher.

Conway, G., & Pretty, J. (1991). Unwelcome harvest: Agriculture and pollution. London: Earthscan.

Crouzel, I. S. (1984). Biological control in Argentina. In G. Allen, & A. Rada (Eds.), The role of biological control in pest management. Proceedings, International Symposium of IOBC/WHRS, Santiago, Chile: Ottawa: Ottawa University Press.

Crouzel, I. S. (1982). El control biologico en la Argentina. Informe Final IX CLAZ, Peru (pp. 169–174). Arequipa, Peru.

Cueva, C. M. A. (1979). Estudio preliminar de las poblaciones de huevos de Diatraea sacharalis (F.) y sus parasitos naturales en la cana de azucar. Revista Peruana de Entomologia, 22, 25–28.

Cueva, C. M., Ayquipa, A., & Mescua, B. V. (1981). Estudios sobre Apanteles flavipes (Cameron) introducido para controlar Diatraea sacharalis (F.) en el Peru. Revista Peruana de Entomologia, 23, 73–76.

DeBach, P. (Ed.). (1964). Biological control of insect pests and weeds. New York: Reinhold.

Drooz, A. T., Bustillo, A. E., Fedde, G. F., & Fedde, V. H. (1977). North American egg parasite successfully controls a different host genus in South America. Science, 197, 340–41.

Food and Agriculture Organization (FAO). (1990). Second roundtable on biological control in the neotropics. RFLAC/90/18-PROVEG-26. San Miguel de Tucumán. Argentina, September 4–8, 1989.

Gallo, D. (1980). Situacao do controle biologico da broca da cana-de-acucar no Brasil. Anais da Sociedade Entomologia do Brasil, 9, 303–308.

Gassen, D. N. (1983). Controle biologico des pulgoes do trigo no Brasil. Informe Agropecuario, 104, 44–51.

Geraud, F., Perez, G., Boscan de Martinez, N., & Teran, J. (1977). La mosca prieta de los citricos en Venezuela y su control biologico. In Memorias de la V reunion Nacional de Control Biologico y Sector Agropecuario Organizado. SAG, Mexico.

Giam, C. S. (1971). DDT, DDE and polychlorinated biphenyls in biota from the Gulf of Mexico and Caribbean Sea. Pesticides Monitoring Journal, 6, 139–43.

Gomero, L. O., & Lizárraga, T. A. (1995). Aporte del control biológico a la agricultura sostenible. Red de Acción en Alternativas al Uso de Plaguicidas. Lima, Perú.

Gomez-Menor, J. (1937). Actividades de control biologico en la Republica Dominicana. Revista Agricola, 28, 372–374.

Gonzalez, R. H. (1976). Plant protection in Latin America. PANS, 22, 26–34.

Gonzalez, R. H., & Rojas, S. (1966). Estudio analitico del control biologico de plagas agricolas en Chile. Agricultura Tecnica, 26, 133–147.

Hagen, K. S., & Franz, J. M. (1973). A history of biological control. In R. F. Smith, T. E. Mittler, & C. N. Smith (Eds.), History of entomology (pp. 443–476). Palo Alto, CA: Annual Reviews.

Instituto Centro Americano de Investigacíon y Tecnologia Industrial (ICAITI). (1977). An environmental and economic study of the consequences of pesticide use in Central American cotton production (ICAITI Report 1412, pp. 1–295). ICAITI.

Jimenez, J. E. (1961). Resumen de los trabajos de control biologico que se efectuan en Mexico para el combate de plagas agricolas. Fitofilo, 32, 9–15.

Klein-Koch, C. (1977). Consideraciones sobre la riá de entomofagos en Chile y Peru, especialmente contra la mosca blanca de los citricos Aleurothrixus floccosus Mask. (Boletin Vol. 3, pp. 101–109). Spain: Servicio de Defensa Contra Plagas e Inspeccion Fitopatologica.

Laing, J. E., & Hamai, J. (1976). Biological control of insect pests and weeds by imported parasites, predators, and pathogens. In C. B. Huffaker & P. S. Messenger (Eds.), Theory and practice of biological control (pp. 685–743). New York: Academic Press.

Leonard, H. J. (1986). Recursos naturales y desarrollo economico en America Central. Washington, DC: IIED.

Macedo, N. (1983). Controle biologico de pragas da cana de azucar. Informe Agropecuario, 104, 20–23.

MacPhee, A. W., Caltagirone, L. E., van de Vrie, M., & Collyer, E. (1976). Biological control of pests of temperate fruits and nuts. In C. B. Huffaker & P. S. Messenger (Eds.), Theory and practice of biological control (pp. 337–358). New York: Academic Press.

Maltby, C. (1980). Report on the use of pesticides in Latin America (UNIDO/IOD 353). Geneva: United Nations Industrial Development Organization (UNIDO).

McConnell, R., Corey, G., Henao, S., Zapata, O. N., Rosenstock, L., & Trape, A. (1993). Pesticides. In Finkelman, J. (Ed.), Environmental epidemiology: A project for Latin America and the Caribbean (pp. 147–200). PAHO, Mexico City: Pan American Center for Human Ecology and Health.

Murdoch, W. W. (1980). The poverty of nations. Baltimore: The Johns Hopkins University Press.

Murray, D. (1994). Cultivating crisis: The human cost of pesticides in Latin America. Austin: University of Texas Press.

Nicholls, C. I., & Altieri, M. A. (1997). Conventional agricultural development models and the persistence of the pesticide treadmill in Latin America. International Journal of Sustainable Development and World Ecology, 4, 93–111.

Norris, R. (Ed.). (1982). Pills, pesticides, and profits. New York: North River Press.

Oliva, A. J. (1961). Posibilidades de combate biologico de las plagas del algodonero en la region agricola del Valle de Mexicali. Fitofilo, 32, 25–28.

Ortega, E. (1986). Peasant agriculture in Latin America. Santiago, Chile: Joint ECLAC/FAO Agriculture Division.

Parrella, M. P., & Nicholls, C. I. (1997). El control biologico de las plagas de invernadero en Colombia. In Marta Pizano (Ed.), Floricultura y Medio Ambiente: La experiencia Colombiana (pp. 221–254). Bogotá; Ed. Hortitecnia.

Planalsucar. (1980). Relatorioanual, 1979. Brasil: Programa Nacional del Melhoramentoda cana de Acucar.

Quezada, J. R. (1974). Biological control of Aleurocanthus woglumi (Homoptera: Aleyrodidae) in El Salvador. Entomophaga, 19, 243–254.

Repetto, R. (1985). Paying the price: Pesticide subsidies in developing countries (Report No. 2). Washington, DC: World Resources Institute Research.

Risco, B. S. H. (1980). Biological control of the leaf froghoppers. Mahanarva postica Stal, with the fungus Metarhizium anisopliae in the state of Alagoas. Entomology Newsletter (International Society of Sugarcane Technologists), No. 9, 10.

Roche, R., & Abreau, S. (1983). Control del picudo negro del platano Cosmopolites sordidus por la hormiga Tetramorium guineense (Ciencias de la agricultura. Instituto de Investigaciones Fundamentales en Agricultura Tropical "Alejandro Humboldt," Academia de Ciencias de Cuba, Havana, Cuba.

Rosset, P., & Benjamin, M. (1994). The greening of Cuba: A national experiment in organic agriculture. San Francisco: Ocean Press.

Sanchez, N. F., Salazar, M. A., Jimenez, H., & Rodriguez, R. (1979). Combate biologico del barrenador de la cana de azucar. Seventh Reunion Nacional de Control Biologico. Veracruz, SARH, Mexico.

Squire, F. A. (1972). Entomological problems in Bolivia. PANS, 18(3), 249–268.

Zuñiga, E. (1985). Ochenta años de control biologico en Chile. Revisión histórica y evaluación de los proyectos desarrollados (1903–1983). Agricultura Tecnica, 45, 175–183.

Zuñiga, E. (1986). Control biologico de los afidos de los cereales en Chile. I. Revision historica y lineas de trabajo. Agricultura Tecnica, 46, 475–477.

Social and Economic Factors Affecting Research and Implementation of Biological Control

JOHN H. PERKINS

The Evergreen State College
Olympia, Washington

RICHARD GARCIA

University of Californa
Division of Biological Control
Berkeley, California

INTRODUCTION

The ability to suppress populations of pest organisms in a socially and economically desirable way is central to successful biological control programs. Biological control workers have two sets of considerations in their work. First, is a proposed biological control scheme manageable in a biological sense (do the organisms behave in predictable, reliable ways)? This question is treated in other chapters in this book. Second, can the organisms be manipulated in ways that are socially acceptable and economically feasible? This question that raises issues in social sciences, politics, and philosophy constitutes the focus of this chapter.

Biological control researchers have a history of successful practice, but advocates of this approach to pest control believe their knowledge has not been fully utilized. Since the discovery of DDT's insecticidal properties in 1939, researchers in biological control have been sensitive to competition from chemicals. They have believed that limited adoption of biological control stemmed from social and economic issues instead from a lack of biological knowledge.

To explore how social and economic factors affect biological control, we must define their meaning and scope. We outline the economic framework of biological control science. The definition of biological control itself has been contested, and it is important to state clearly the definition used in our analysis. Some conceptual and practical problems attend analyses of social and economic factors, and we identify these issues. Finally, we synthesize our findings into some pratical, forward-looking recommendations on how biological control will best fit into the world's complex social and economic systems.

THE POLITICAL ECONOMIC FRAMEWORK FOR BIOLOGICAL CONTROL

Political economy examines the interactions between how resources are created, distributed, and used and how power and control are exercised. It is easy to see the links between economic and political power that derive from the ownership of factories and machines. The owners, individuals or corporations, decide what will be made, how products will be distributed, and how proceeds from sales will be allocated. The power of ownership is not absolute but, compared with the work force of the company, the owners have more authority within the boundaries of the manufacturing plant. This power and wealth can be used to influence the general political process of a country and may be more influential than that exercised by the nonowning class. Similarly, ownership of land creates power to make economic decisions that affect the welfare of the work force and of consumers of the land's products. Owners of land thus tend to be wealthier than nonowners, and they may exercise influence in the political process that is not available to nonowners.

The creation and use of scientific and technological knowledge have attributes similar to the creation of other forms of wealth. Research and development occurs in laboratories and field stations that are owned and controlled by corporations, government agencies, or universities. The working scientist has more autonomy than a factory worker, but this should not obscure the employer–employee relationship that exists between the working scientist and the laboratory administration. The ability of a scientist to work depends critically on convincing the administration that proposed research would yield a useful product or knowledge that the administration wants created. Once developed, the scientific or technical knowledge may be owned and controlled by the administration. Alternatively, the knowledge may become part of the public domain and may pass into the hands of economic decision makers, who have interest in and influence with the laboratory administration.

Political neutrality is not an inherent attribute of applied science and technology. Most scientific work and products are subject to the political and economic considerations that have little to do with the scientific subject matter.

Pest control has been developed primarily in agricultural research stations, public health laboratories, and the private chemical industry. Historically biological control had its origins almost exclusively within agricultural research stations supported by government and university funds. Sawyer established in detail the important links between the California citrus industry and the development of biological control science at what was then the Citrus Experiment Station and is now the University of California at Riverside. Since 1980, some aspects of biological control knowledge have been developed by private, profit-seeking firms, but the contributions of these companies have been small.

Despite the "free" appearance of biological control knowledge, it would be an error to assume that: (1) issues of power and control were not involved in the creation of this expertise or (b) future developments in biological control will be remote from questions about the exercise of political power. The allocations of budgets for agricultural research are highly politicized events (Guttman, 1978; Rose-Ackerman & Evenson, 1985). Some lines of research are favored over others. Political leaders in legislatures, executive branches, and university administrations may be sensitive to the demands of powerful constituents.

Commercial agriculture is becoming increasingly competitive, because farmers, particularly in North America, have had productive capacities in excess of markets. As a result, farmers have been in an economic race to use the best technology to lower production costs and increase profits (Buttel, 1980; deJanvry & Vandeman, 1982).

Biological control expertise must be applied to this highly competitive farm industry. Some research has addressed problems of urban, forest, and public health issues. These research lines will expand in the future. Nevertheless, much of the political fortune of biological control continues to be based on its ability to serve the farming industry. Farm firms may be controlled by individuals, partnerships, corporations, cooperatives, or the state, but in each case they must behave as profit centers and entrepreneurs competing against other farm firms. Other forms of pest control technology compete with biological control in the sense that farmers usually have options among several technical practices. The exercise of political power around biological

control research revolves about the abilities of the expertise to function within the economic framework of agricultural enterprise that produces for a competitive, global market.

A political economic analysis of the creation of biological control technologies must examine several issues that follow:

1. Resources for scientific investigation must be allocated before scientific knowledge can be developed. Part of understanding how social and economic factors affect biological control involves understanding the resource allocation process for biological control research. The allocation process is political and influential parties try to direct research resources in ways that will protect and enhance their interests.

2. Once knowledge is articulated, questions arise about its usefulness. These questions center on the "goodness of fit" of the new technical knowledge to the complex of operations involved in agriculture. Is the technology cost-effective? Can the user receive training and advice on how to use it? Is the new technology compatible with the user's other production practices? Does the new practice fit within the user's traditional activities? Does the new practice fit the habits of how the user relates to government authority and traditions? Does the new user have to adopt new assumptions about nature or the state to feel positive about trying the new knowledge? Answers to these questions about "fit" are essential to understanding how social and economic factors influence the fortunes of biological control in practical operations.

DEFINITION OF BIOLOGICAL CONTROL

Contemporary U.S. researchers in biological control trace its definition to Smith (1919):

> The biological method of insect pest control . . . embraces the use of all natural organic checks, bacterial and fungous diseases as well as parasitic and predaceous insects . . . . From a practical stand point, the biological method may be arbitrarily divided into two sections: *First,* is the introduction of new entomophagous insects which do not occur in the infested region; and *second,* the increasing by artificial manipulation, of the individuals of a species already present in the infested region, in such a way as to bring about a higher mortality in their host than would have occurred if left to act under normal conditions.

Increases in knowledge since 1919 have required researchers to expand and refine the definition of biological control. The scope and content of the definition of biological control have become important public policy. In 1987, the Committee on Science, Engineering and Public Policy (COSEPUP) of the National Academy of Sciences, the National Academy of Engineering, and the Institute of Medicine advocated an expanded definition of biological control:

". . . the use of natural or modified organisms, genes, or gene products to reduce the effects of undesirable organisms (pests), and to favor desirable organisms such as crops, trees, animals, and beneficial insects and microorganisms." (Anonymous, 1987).

Researchers at the Division of Biological Control (University of California, Berkeley) and the Department of Entomology (University of California, Riverside) have argued strongly against the conceptual integrity and the utility of the proposed COSEPUP definition. This group, with historical ties to biological control as defined by Smith in 1919, believes the COSEPUP definition fails to provide essential and clear distinctions between different pest control technologies (Garcia *et al.,* 1988):

1. Self-sustaining control compared with control requiring continual input
2. Density-dependent action characteristic of true biological control compared with the density-independent action of other suppression technologies

The University of California group proposed that the essence of biological control was still best captured in a definition provided by DeBach (1964): ". . . the action of parasites, predators, or pathogens in maintaining another organism's population density at a lower average than would occur in their absence." Garcia *et al.* (1988) provided a schematic (Fig. 1) that illustrates the distinctions they consider necessary for an intellectually adequate definition of biological control. In addition, the scheme by Garcia *et al.* (1988) shows how true biological control is related to other modes of insect suppression, including the methods that COSEPUP would prefer to call biological control.

Cook (1993) has urged that biological control should encompass bioengineering and other biotechnologies. He believes such a broad definition could make biological control the most important pest control technology in the twenty-first century.

In this discussion, we use the definition of biological control maintained over 80 years by University of California researchers. Our choice does not cast aspersions on the usefulness of research lines advocated by COSEPUP and Cook, but it is intended to keep the concept of biological control focused on the notion of suppression provided by the interactions between the populations of different species. Within this definition of biological control, three modes of activity must be distinguished:

1. Classical biological control: The identification of indigenous and exotic natural enemies, the importation and release of exotic natural enemies, and the evaluation of the abilities of natural enemies to suppress a pest must be performed.

2. Augmentation of natural enemies: The culture and release of natural enemies to suppress a pest when a natural enemy is present but in numbers insufficient to provide adequate suppression must be accomplished. Waage and Greathead (1987) distinguish three subsets of this mode: inoculation, augmentation, and inundation.

3. Conservation of natural enemies: Action must be taken to conserve existing natural enemies by preventing their destruction from other practices.

PROBLEMS IN MEASURING BIOLOGICAL CONTROL

It is impossible to know how social and economic factors affect research and implementation in biological control without knowing how these activities have fared in the past. Unfortunately, our abilities to trace research and implementation in biological control are limited, especially when attempting to quantify the trends.

It is possible to make quantitative estimates of research output and personnel levels in biological control, at least for some periods and countries. Quantitative estimates of research output, levels of research support, and number of scientifically trained personnel engaged give only partial insights into the success of a scientific enterprise.

Qualitative considerations are important in assessing the progress of a research area. Prominent governing factors are the goals and methods involved, the quality of personnel training, morale, the location of the institutional base within the framework of power, and the relationships between the scientific personnel and their clients (pest control decision makers) who must ultimately use the knowledge generated. Numbers of papers published, personnel, and amounts of money spent on biological control research do not necessarily indicate the quality of a research operation. Conversely, low levels of publication, limited scientific personnel, and low financial support levels may have results in the field that surpass expectations from a program that appears insignificant.

Similar but more complex considerations surround our ability to understand the fate of biological control at the implementation stage. Biological control scientists have periodically issued compilations of "successes," sometimes as part of an effort to generate social and political support for their research programs (DeBach, 1974; Huffaker & Messenger, 1976; Commonwealth Agricultural Bureaus, 1980). Unfortunately, information gathered in listing successful biological control events often is limited to the amount of damage done by the pest before and after the introduction of a biological agent. The difference between the before and after damages are then imputed to be the value of the

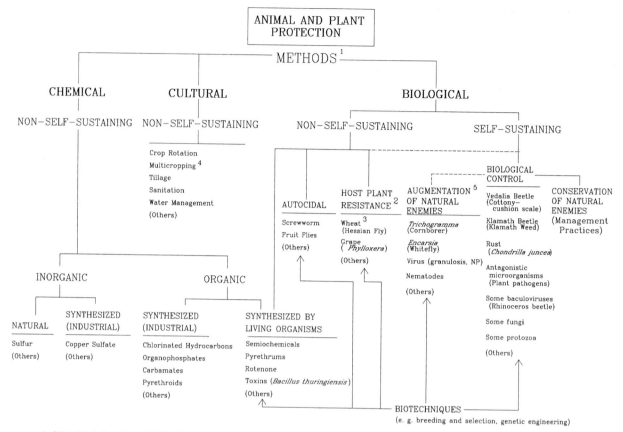

FIGURE 1   Schematic illustrating the distinctions considered necessary for adequate definition of biological control. [From Garcia, R., *et al.* (1988). *Bioscience, 38,* 692–694].

biological control agent. Such figures are often impressive because some examples of biological control show enormous returns for small amounts of money invested in research.

Insights into the factors affecting the use of biological control are difficult to draw from such studies. The behavior of all the organisms involved is not established and the interests of the pest control decision makers are often confounded with those of the biological control researcher. Such confusion is understandable, because in classical biological control the researcher and the implementer are usually the same person; release of an exotic natural enemy constitutes implementation of control, if successful.

If classical biological control were the only valuable mode of biological control, then we would not be concerned with factors affecting decision makers such as farmers. Only the forces governing the amount of research in biological control would matter, because farmers would be

the recipients of a new technology without having to take positive action.

Implementation of biological control through augmentation and conservation of natural enemies is virtually certain to require changed behaviors on the part of a pest control decision maker, who is different from the researcher. In such cases, we must distinguish the behavior and interests of the implementer from the scientist, or we are unable to analyze the factors affecting implementation. We must know how the pest control decision maker formulates long-term goals. What sorts of knowledge are likely to appeal to the aspirations, experience, and constraints within which the decision maker works? To what extent do economic factors interact with more subtle social, political, and philosophical considerations? Failure to understand the actions of decision makers leads to frustration for researchers and policy makers who believe that biological control offers substantial benefits.

ALLOCATION OF RESOURCES TO BIOLOGICAL CONTROL RESEARCH

Research requires trained personnel, supportive institutions, and money. Sources of public and private money are primary social and economic factors affecting the research enterprise in biological control.

Several indicators suggest that a biological control research community should be a vigorous and vital group generating new results, conceptual and methodological tools, and successful control schemes. These indicators include (1) the output of literature in biological control, (2) the staffing levels in research organizations, (3) the signs of intellectual vigor in institutions essential to biological control research, and (4) the introduction of exotic species in programs of classical biological control.

Magnitude of the Research Effort

Indications of vigor notwithstanding, it is difficult to estimate the size of the biological control research community and its productivity. No agency currently tracks the number of international scientists involved, their levels of productivity, the levels of funding provided, and the number of projects completed. Other sources, however, permit some educated guesses about these parameters.

Abstracts of papers, reports and books in biological control are published in *Biocontrol News and Information* (BNI), a publication of CAB International Institute of Biological Control (CIBC, formerly the Commonwealth Institute for Biological Control). BNI has been published regularly since 1980, and the number of abstracts published per year is the only global estimate available for the size of the world's biological control literature. The number of abstracts may be constrained more by budget limitations of CIBC than by the number of literature entries available. The BNI database provides a minimal estimate of scientific activity in biological control.

Between 1980 and 1986, the average number of abstracts per year in BNI was 2421. Table 1 provides an annual listing of number of abstracts published (Anonymous, 1985b; D. J. Girling, personal communication, D. J. Greathead, personal communication).

That about 2400 "literature messages" are produced in biological control per year is of interest because it allows a crude estimate of the number of scientist-years involved in the biological control research enterprise. If we assume that one full-time efficient scientist can produce one to four messages per year, then a production of 2400 messages per year implies that the world has a minimum of about 600 to 2400 scientist-years working in biological control.

Many personnel involved are part-time in their research activities, so more individuals are involved than scientist-

TABLE 1 Estimated World Production of Literature in Biological Control[a]

Year	Number
1980	2403
1981	2200
1982	2076
1983	2751
1984	2741
1985	2705
1986	2070

[a] Number of articles, reports, and books abstracted per year in *Biocontrol News and Information,* a publication of CAB International Institute of Biological Control.

years indicate. In addition, the estimate of one to four messages per year for the average scientist cannot be verified. Furthermore, some work in biological control does not result in publication. Judd *et al.* (1987) estimated the global resources for agricultural research to be 148000 scientist-years in 1980. Research in biological control, based on our estimates, is about 0.4 to 1.6% of the total research in agriculture in terms of manpower allocations.

Agricultural science resources are not evenly distributed over the globe (Judd *et al.,* 1987). Historically, agricultural research was conducted primarily in the industrialized countries. In 1959, 69% of the manpower and 76% of the funds for agricultural research were spent in Europe, the (former) USSR, North America, and Oceania (Judd *et al.,* 1987). By 1980, increases in third world agricultural research caused the proportion of resources expended by the industrialized world to drop to 57% of the manpower and 69% of the funds. Agricultural research is still an activity dominated by developed countries. (All figures are calculated from Judd *et al.,* 1987.)

Biological control research scientists also are concentrated in certain areas. A report from the U.S. Department of Agriculture (USDA) estimated that approximately 190 scientist-years were devoted to biological control work in the USDA laboratories and agencies (calculated from USDA, 1985).

Biological Control within the Pest Control Sciences

Research activity alone does not indicate the health and respectability of biological control research. Instead, only the role of biological control research within the general framework of research in the pest control sciences can indicate how the field rates within the professional peer group. By this criterion, too, biological control appears to

TABLE 2 Number and Percentage of Entomological Literature Devoted to Biological Control, 1915–1986[a]

Year	Number entomological literature	Number biological control literature	Percentage biological control
1915	1,482	178	12
1920	1,632	82	5
1925	1,120	123	11
1930	1,410	296	21
1935	1,692	288	17
1940	972	233	24
1945	960	269	28
1950	876	105	12
1955	833	33	4
1960	1,068	176	16
1965	2,370	332	14
1970	3,760	714	19
1975	5,441	1,110	20
1980	9,990	2,403	24
1981	10,942	2,200	20
1982	10,865	2,076	19
1983	11,708	2,751	24
1984	11,639	2,741	24
1985	11,819	2,705	23
1986	8,393	2,070	25

[a] For 1915–1975, the numbers of entomological literature and biological control literature are based on samples of 10% of the abstracts in the *Review of Applied Entomology*. For 1980–1986, the numbers are the number of abstracts in *Biocontrol News and Information* and in the *Review of Applied Entomology*, Series A and B.

Source: *Biocontrol News and Information* 2(1981): David Girling, CIBC, Personal Communication, 1987.

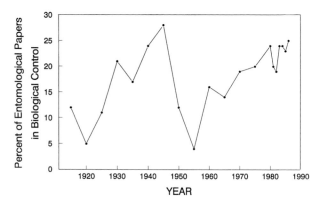

FIGURE 2 The proportion of entomological literature devoted to biological control investigations from 1915 to 1986.

1955 had fewer than 100 papers. Now, each year has over 2000 (Anonymous, 1985b).

Confirmation that enthusiasm for research on insecticides eclipsed biological control work also was noted by Price-Jones (1973), who sampled articles from the *Journal of Economic Entomology*. Similar conclusions were reached by Perkins (1978) in a study on how the introduction of DDT to the United States affected research by American economic entomologists. Perkins (1982) analyzed the changes in direction of one American research entomologist in the 1940s and 1950s and concluded that the technical capabilities of insecticides were responsible for a strong shift in research interests away from biologically based means of control toward chemically oriented technologies.

Organization of Biological Control Research

The rising number of papers and a number of developments in organizations and research also indicate that biological control has been reinvigorated. CIBC is the largest multinational network of scientists engaged in biological control research. It was reorganized in 1985 to make it more useful to a wider range of clients (Anonymous, 1985a, CAB International Institute of Biological Control, 1986). This institute currently operates on about £1 million per year (US$1.7 million), up 240% from its 1979 levels (Commonwealth Agricultural Bureaux, 1985). Many of the funds are expended for projects in developing countries and in Canada (J. Waage, personal communication). Table 3 summarizes the activities of CIBC in 1985 and 1986 in terms of projects conducted and of shipments of natural enemies to other locations. The pattern of work varies from year to year, so the data in Table 3 are merely indicative of the types of work conducted (CAB International Institute of Biological Control, 1986; D. J. Greathead, personal communication).

USDA, the world's largest agricultural research organization, has introduced substantial changes in its biological

be in high repute compared with various periods within the last 50 years.

Biological control research in insects has enjoyed a recent renaissance. Table 2 and Fig. 2 list at 5-year intervals the numbers and proportions of entomological literature devoted to biological control investigations, from 1915 to 1980, and 1-year intervals from 1980 to 1986. These suggest that biological control enjoyed a wave of rising popularity among researchers from 1920 to 1945 and then went into a decline, probably as a result of enthusiasm for research on the newly introduced synthetic organic insecticides. After a low point in 1955, the fashion of doing research in biological control again began to climb, and the proportion of entomological papers now devoted to biological control is about 25%, approximately equal to the previous high of about 28% in 1940 (Fig. 2) (Anonymous, 1981; D. J. Girling, personal communication).

The proportion of entomological papers devoted to biological control increased during the last three decades and the numbers of papers increased dramatically. Before 1955, no year had more than 400 papers in biological control;

TABLE 3 Activities of Commonwealth Institute of Biological Control, 1985–1986

Area	Number projects	Number of shipments of natural enemies sent	Number of countries to which shipments were sent	Number of species	Number of individuals
Europe	11	34	5	24	35,000
India	6	57	4	18	>14,500
Kenya	5	8	3	6	2,750
Pakistan	7	27	7	12	26,046
United Kingdom	12	10	5	15	7,000
West Indies	3	68	7	20	12,000

control work during the past 45 to 50 years. It had an active program of foreign exploration that was reduced during World War II. For 15 years, no effort was made to revive the former program, but plans were developed in 1955 to expand work, primarily in augmentative biological control. A major laboratory began operations in Columbia, Missouri, in 1963 (Perkins, 1982). The USDA began a comprehensive effort during the 1980s to rationalize and coordinate its biological control work (USDA, 1984, 1985), and selected elements were reviewed in the mid-1990s (USDA, 1994; 1997, pp. 159–166).

A final example of the continuing vitality of biological control comes from publications. *Entomophaga* has been published in France by the International Organization for Biological Control since 1956. This journal is supplemented by publications such as the *Chinese Journal of Biological Control* (begun in 1985), BNI (begun 1980 as mentioned earlier), and *Biological Control* (begun in 1990).

New Areas of Study

More important than new organizations and publications, however, has been the opening of completely new industries and new areas of study during the last 15 years. The emergence of new companies for supplying biological control agents is promising. Some of these enterprises are oriented toward the production and sale of long-recognized biological control agents, such as *Bacillus thuringiensis* Berliner and *Trichogramma spp.* Other companies search for new agents, and modify existing agents by genetic engineering (Anonymous, 1985a, 1985c, 1985d, 1986, 1987; Hussey, 1985; Roberts, 1989; Lambert & Peferoen, 1992). Microbial agents now take less than 1% of the world's pesticide market (Anonymous, 1985c), but interest showed by new companies suggests that at least some sales of chemicals may be shifted to products derived from organisms (Moffat, 1991). Recombinant DNA technologies have opened the possibility of creating transgenic predators that could be "customized" to fit the requirements of particular agricultural industries (Goodman, 1993).

Biological control research 30 years ago was almost completely confined to the use of insects to control insect pests and weeds. The methods used were largely those of classical biological control: foreign exploration for exotic natural enemies, importation of natural enemies, and release in the field followed by evaluation to see if control of the pest was affected.

In the fields of pathology, a few useful cases were known of the use of pathogens against plant (Andres *et al.,* 1976) and animal species (Weiser *et al.,* 1976). In addition, work before World War II had demonstrated the utility of indigenous natural enemies. Rudimentary ideas began to emerge during the 1930s and the 1940s concerning the need to use insecticides in ways that would not interfere with the suppressive power of insect natural enemies (Michelbacher & Smith, 1943). Nevertheless, the field of biological control was largely "classical," and research was oriented toward finding new natural enemies that would provide dramatic suppression of a pest comparable to that provided by the vedalia beetle against the cottony-cushion scale.

Three new areas of research have developed since the 1950s: use of pathogens for the suppression of weeds, biological control of plant pathogens, and integrated pest management (IPM).

The use of plant pathogens to control weeds is an active area of research that unites the study of plant pathology, weed science, and plant physiology. Charudattan and Walher (1982) produced a landmark monograph on the subject. They cited 55 projects involving the use of pathogens, including bacteria, fungi, nematodes, and viruses. Five of these projects were considered "operational." Control of skeletonweed in Australia by the rust *Puccinia chondrillina* Bubak & Sydershan from the Mediterranean region returned an estimated annual savings of $25.96 million. Water hyacinth control by *Cercospora rodmanii* Conway reached the stage of pilot tests by the U.S. Corps of Engineers in 1982.

A second new field is the use of biological control for the control of plant pathogens. A major work by Cook and Baker (1983) notes that 20 years earlier only three examples of the use of antagonistic organisms to control plant pathogens could be cited; 10 years earlier only six examples could be cited, and only two were used commercially. The monograph in 1983 had 1081 references, 60% of which were post-1974. At the time of publication, Cook and Baker had 15 examples of successful biological control of plant pathogens that could be illustrated in detail. Further work in this aspect of biological control continued to encourage researchers (Tjamos *et al.*, 1992).

IPM, which heavily involves biological control, is a promising new frontier. IPM as a pest control strategy was profoundly influenced by classical biological control (Perkins, 1982), but it is doubtful that the roots of IPM helped encourage research in biological control between 1960 and 1980. The U.S. National Science Foundation removed classical biological control from the large research project, Principles, Strategies, and Tactics of Pest Population Regulation and Control in Major Crop Ecosystems, in favor of research on the ecological theory of why and how biological control works (Huffaker, 1985). Thus, the first major research effort in IPM was handicapped by not being allowed to build one of the component techniques for pest suppression into the basic design of the new research. Systems analysis and computer modeling instead were favored.

Combining biological control with the use of pesticides was the cornerstone on which the concept of "integrated control" was founded (Perkins, 1982), but later definitions of IPM obscured the importance of biological control. The current definition of IPM that still has the greatest currency does not even mention biological control, or any other specific control technology explicitly:

> Integrated pest control is a pest population management system that utilizes all suitable techniques in a compatible manner to reduce pest populations and maintain them at levels below those causing economic injury. Integrated control achieves this ideal by harmonizing techniques in an organized way, by making control practices compatible, and by blending them in a multi-faceted, flexible, evolving system. (Smith & Reynolds, 1967, quoted in Frisbie & Adkisson 1985, p. 41).

Researchers, however, have begun to ask whether biological control ought to be seen as fundamental to IPM, and ought to receive the funding levels appropriate to such a critically important technology. Some of these researchers believe biological control is fundamental to IPM, but funding for biological control research is less than 20% of the total given to IPM. Most funds support pesticide timing, modeling of plant–pest interactions, defining economic thresholds, and predicting the size of pest populations (Hoy & Herzog, 1985). Work on biological control must be built into IPM research from the beginning if biological control practice is to be successful:

> In many, if not most cases, biological control by itself does not provide economically acceptable pest suppression in agricultural cropping systems. Therefore, biological control must be developed and implemented as a component of IPM. However, if it is to be an integral part of IPM (along with plant resistance, cultural methods, and pesticidal controls) biological control must be nurtured to become a strong vital entity. (Tauber *et al.*, 1985).

Factors Affecting Biological Control Research

Many indicators suggest that the biological control research effort of the mid- to late-1980s is healthy and vibrant: substantial research productivity, new and revitalized institutions and organizations, continuing strengths of longer established organizations, new industrial concerns seeking part of the pesticide market for biological control agents, and dynamic new frontiers of research. These indicators suggest that governing factors are moving in favor of biological control. What factors are acting favorably? Complex social phenomena are impossible to attribute precisely to clear-cut causes, but several factors seem particularly relevant during the past 10 to 20 years. Some arise from events removed from the activities of the biological control workers, but others are due to activities of the research community.

Scientific research requires resources, so it is not surprising that the amount of research in biological control is highly correlated with the gross domestic product (GDP) of a country. Table 4 presents the GDP (in billions of US$, 1979) of 58 countries (not a random sample), each of which produced at least one paper in biological control, of which an abstract was published during the period of 1984 to 1986. The distribution of the number of papers per country and the amounts of GDP are highly skewed, so they were transformed to $\log_{10}$ of the number of papers produced in 3 years and the annual GDP in billions, respectively. Figure 3 displays the relationship between publication numbers and GDP. The best straight line (least-squares method) is indicated and has the equation: $\log_{10}$ (publications) = $0.79 \log_{10}$ (GDP) + 0.19 ($R = 0.83$, $p < 0.01$). This equation between production of research papers in biological control and GDP suggests that in these 58 countries, an annual GDP of $10 billion will result in about 9.5 papers in biological control every 3 years, or 3.2 papers per year. Alternatively, for $2.3 billion of annual GDP, one would expect to see one paper in biological control published each year.

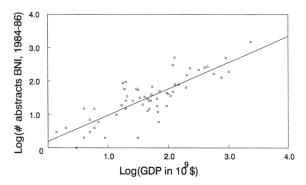

FIGURE 3 Relationship between publications in biological control and Gross Domestic Product, early 1980s.

TABLE 4   Gross Domestic Products and Production of Publications in Biological Control, early 1980s

Country	Gross domestic product[a] (US $ × 10^9)	Publications in biological control[b]	Country	Gross domestic product[a] (US $ × 10^9)	Publications in biological control[b]
Algeria	31.612	2	Kenya	6.041	15
Argentina	53.000	40	Lebanon	3.700	2
Australia	129.805	303	Malaysia	20.267	26
Austria	68.384	12	Mexico	119.850	48
Belgium	110.910	36	Mozambique	1.400	3
Brazil	204.520	152	Netherlands	149.060	79
Bulgaria	32.000	55	New Zealand	20.100	95
Burma	5.050	3	Nicaragua	2.500	1
Canada	228.610	245	Nigeria	49.000	26
Chile	17.500	19	Norway	46.266	13
China	450.000	276	Pakistan	21.360	35
Colombia	25.800	26	Peru	13.620	10
Czechoslovakia	72.000	51	Philippines	30.080	59
Denmark	66.229	18	Poland	130.000	80
Egypt	18.000	88	Portugal	20.098	15
Finland	41.417	34	Romania	46.000	29
France	571.370	164	South Africa	55.740	68
German Democratic Republic	95.000	18	Spain	197.065	65
German Federal Republic	763.880	129	Sweden	101.490	42
			Switzerland	94.991	45
Greece	38.386	20	Syria	8.896	2
Guatemala	6.886	6	Tanzania	4.564	7
Hungary	34.000	32	Turkey	66.131	30
India	128.000	514	USSR	900.000	498
Indonesia	49.215	15	United Kingdom	402.472	258
Iran	70.000	6	United States	2343.500	1524
Israel	18.785	58	Yugoslavia	52.000	26
Italy	323.586	217	Zaire	6.162	4
Japan	1,011.333	193	Zimbabwe	3.863	15
Jordan	2.279	4	Total	9574.773	5856
			World	10,223.000	7515

[a]Figures are for 1979. Taken from The Economist, *The World in Figures* (London: The Economist Newspaper Ltd., 1981), page 9.

[b]Figures represent 3 years of publication, approximately from 1982 and 1983 through 1984 and 1985. Numbers reflect the number of abstracted literature entries in *Biocontrol News and Information* for which a geographic location (country) of the work was indicated, for each of volumes 5 (1984), 6 (1985), and 7 (1986). Some abstracts have no geographic location entered, because the work was laboratory based; such abstracts are not counted in this method of estimating national productivity in biological control research. It is also assumed that a 1- to 2-year lag separates publication from abstract entry in the CAB International database.

Correlation between productivity of research in biological control and GDP indicates that this form of research is similar to others in the sense that wealthy countries do more of it. Correlation between a country's wealth and its research productivity, does not, however, tell us everything about the ways in which each country may decide how much and what kind of biological control research to perform. Moreover, the data in Table 4 and Figure 3 suggest that some countries are particularly high in their productivity of biological control research given their GDPs (e.g., India, Canada, and Australia), while others may be low in output compared with their GDPs [e.g., France, Japan, and the former Federal Republic of Germany and the German Democratic Republic (now Germany)].

Explanations for why some countries are high producers compared with others are not obvious, but one possibility

is that membership in an international network such as CIBC is conducive to productivity in biological control research. Hence, countries such as Australia, Canada, and India—all long-term members of CIBC—are comparatively high producers. Conversely, countries that are not in coordinated networks may have research productivities considerably below what the sizes of their economies might suggest. France and Japan have, for example, GDPs 2.5 and 4.4 times the size of Canada's GDP, respectively, but these two countries have research outputs of 0.67 and 0.79 the size of Canada's, respectively.

A second possibility to explain high interest in biological control in such countries as Canada and Australia is that both areas were subject to European invasion and conquest starting in the seventeenth century. European people brought insect and weed pests (Crosby, 1986). Much biological control work in these areas has been an effort to reassociate natural enemies with imported pests. Europe, in contrast, has had fewer invasive pests and hence may be a place in which classical biological control has a smaller role (D. J. Whitehead, personal communication).

Environmental concerns about potential pollution from pesticides or from failure of chemical control through resistance and destruction of natural enemies may also affect allocations for biological control research. For example, Malaysia has recently indicated interest in biological control for conservation purposes, despite some anxiety about introducing exotic pests (D. J. Whitehead, personal communication). Similarly, Indonesia has an official government policy to encourage implementation of IPM and conservatory biological control, in place of insecticides, because of concerns about insecticide-induced outbreaks of the brown plant hopper on rice (P. Kenmore, personal communication; D. J. Whitehead, personal communication). The problem with shortage of foreign exchange to import chemicals has also been a factor in Indonesia and elsewhere (Repetto, 1985; England, 1987).

Interest in environmental protection created barriers to research in biological control. Ecologists and the public realize that the introduction of any new agent, even entirely beneficial, can have undesirable features (Gillis, 1992; Simberlof & Stiling, 1996). Capabilities of producing genetically engineered agents have complicated this issue further. As a result, biological control researchers must now contend with regulations from which they were previously exempt, such as the Endangered Species Act; the National Environmental Protection Act; and the Federal Food, Drug and Cosmetic Act (Waage & Greathead, 1987; Coulson & Soper, 1989).

Research Needs on the Political Economy of Biological Control Research

Work is needed to determine how social and economic factors affect the allocation of resources to biological control. Factors likely to be of importance are the philosophical world views prevalent in the research community, the political strength and organization of agriculturalists, the nature and size of the agricultural economy (e.g., the importance of exports, the type of produce and its cosmetic appeal, and the importance of agriculture in the nation's economy), the research activities of neighboring countries, and the political currents favoring environmental protection (e.g., strict regulation of pesticides might be conducive to research in biological control). Judd *et al.* (1987) analyzed socioeconomic factors determining research investments in agricultural science, but it is not possible to make predictions about the factors affecting research in biological control.

Factors affecting research, both positive and negative, need analysis on a country-by-country basis. Recent events

offer challenges to increasing research in biological control, particularly on weeds. The basic conflict is that one person's weed may be another's valuable crop plant. For example, Australian court injunctions recently blocked the importation of insect control agents for the plant *Echium plantegineum* (Linnaeus). Livestock interests know *E. plantagineum* as "Patterson's curse," but beekeepers call it "salvation Jane." The popular names imply strongly differing attitudes toward the plant (Waage & Greathead, 1987).

A commercial biological control industry based on release of mass-produced pathogens, parasites, and predators indicates other factors may be important in the future: patent, tax, commercial policies, and laws. Countries encouraging the entrepreneurship of biological control manufacturers, through policies favorable to their enterprises or through direct subsidies, may find that the manufacturers make significant additions to research in biological control. At the moment, however, the contributions of the private sector are relatively small.

SOCIOECONOMICS OF BIOLOGICAL CONTROL

Biological Control as an Investment

Understanding how social and economic factors affect the use of biological control is more difficult than analyzing their effects on research. Users number in the hundreds of millions of individuals, including all the farmers of the world and other decision makers who have responsibilities for limiting the impact of pests. The research community, in contrast, is minute, with numbers limited to a few thousand individuals. Moreover, researchers deliberately leave a trace of their activities (i.e., publications), but farmers and other pest control decision makers generally leave no written record of their actions, certainly not a public record that is easily accessible for analysis.

Pest control decisions are also complex. Probably most are made for the private benefit of the decision makers. Some, however, are made by officials in public agencies for the benefit of the public. Thus, the social and economic factors affecting the choices made by the decision maker are complicated by the nature of the institution in which he or she works. A final complication in assessing the degree of implementation in biological control emerges from the complexity of the technology. Each of the three categories of biological control (classical, augmentative, and conservatory) is likely to have different factors affecting its respective implementations.

We can construct a theoretical model of the relevant social and economic factors. The most useful perspective on the subject begins with the recognition that if someone decides to use biological control, then he or she has chosen a particular technology to achieve specific material ends. The choice is made in the context of alternatives ranging from doing nothing to selecting another way of mitigating the damage from the pest. A decision to adopt biological control is an investment decision (i.e., a choice from which the decision maker expects a better return than could be expected from other choices). The return could be monetary or could have a high monetary component. Nonmaterial returns also could be important to the decision maker.

When use of biological control is perceived as an investment decision, the factors affecting the use of the three categories of biological control differ in terms of the risks accepted by the decision maker. Classical biological control is the least problematic for the researcher and the farmer (or other affected person). If an introduced exotic control agent has been screened for adverse properties, the only monetary risks involved are the costs for exploration, shipment, quarantine, and release. If the organism succeeds, then most likely it will return substantial rewards over a period of years for a relatively small investment. Moreover, most classical biological control is conducted by the state, so that the individual pest control decision maker bears no personal financial risk. In many cases the individual grower may not be consulted about the decision to release an exotic agent because the research community bears full responsibility and power to identify pests, seek solutions, and implement them as part of a research implementation program. DeBach (1974) and Coulson and Soper (1989) argue that the returns from this sort of biological control research and development have been extraordinarily high compared with the amounts spent on biological control research (30: 1), including expenses for those introductions that resulted in no significant suppression of the pest.

Classical Biological Control

The close relationship of classical biological control to the research enterprise makes the factors affecting implementation likely identical to those affecting research. The larger the economy the more likelihood there is that classical biological control research and implementation will be done. Size of economy, however, cannot be the sole determinant of the efforts invested in classical biological control. Decisions within the research community could lead to resources going into augmentative or conservatory biological control research. Furthermore, classical biological control requires either foreign exploration or collaboration with foreign scientists who agree to ship exotic organisms. Political relationships between countries therefore can influence the development of a program.

Augmentative Biological Control

Augmentative biological control involves an investment decision that is like the investment decisions needed for other types of pest control technology. Practical schemes for artificially releasing organisms to attack a pest are de-

pendent on the industrial-scale culture of the controlling organism (insect, fungus, bacterium, or virus). The control agents are packaged, sold, and dispersed much as a pesticide product. Factors affecting use of such a product are not different from those affecting the use of a pesticide: cost, ease of use, effectiveness, safety, and ability of the agent to mesh easily with other parts of the business operation.

Conservatory Biological Control

Conservatory biological control, the preservation of indigenous natural enemies that tend to be destroyed by existing modes of operation, is the most problematic investment from the viewpoint of the individual pest control decision maker. Perhaps the difficulties can best be explained by reference to a typical IPM situation. A natural enemy present in the farmer's fields provides less suppression of a pest because pesticides or cultural practice destroys the natural enemies. If the grower would alter existing practices, the natural-enemy population might provide a higher level of suppression of the pest, perhaps enough to obviate need for other pest control measures.

Two events must occur to make use of the existing natural-enemy populations. First, the grower must stop destroying the effectiveness of the existing natural enemies. Unfortunately, altering existing practices may open the grower's entire mode of operation to risk of loss or severe disruption. Ending use of a pesticide exposes the grower to catastrophic losses from a pest, a loss that might be prevented by the preserved natural enemies. The grower, therefore, must have confidence that the foregone practice will not result in devastating losses. Second, the grower must monitor the pest and natural-enemy populations to ensure the pests are suppressed.

Changes in practices are important to biological control based on conservation of natural enemies. Pest control decision makers have adopted new technology, but any change of established practices is accepted only if the decision maker thinks the change is for the better. Assessments of risk, faith in new technology, income and education, age, size of farm operation, impacts on personal time, and implications for other parts of the operation are major factors governing change.

Studies have examined the inclinations and abilities of farmers to adopt IPM schemes, most of which have a component of conserving natural enemies (Wearing, 1988). Fernandez-Cornejo (1996) found that tomato growers in eight states had about the same yields and profits from using IPM compared with conventional insecticides and fungicides. Their use of the chemicals declined. Peanut farmers in Georgia evidenced reluctance to shift to IPM despite objective data indicating the new technology was more efficient than conventional pest control (Musser *et al.,* 1986). Georgia cotton growers received no increase in net returns from

IPM (Hatcher *et al.,* 1984), so the attractiveness of this technology for them is probably slight. Other studies, both empirical and theoretical, indicate that IPM in cotton increases net returns but that many growers still do not use it (White & Wetzstein, 1995). Adoption of IPM by Iowa growers was related to opinion leadership, adoption orientation, and prior knowledge of IPM. In turn, these variables were positively related to income, education, farm size, and attitudes toward chemicals (Salama, 1983). Hardison (1986), a citrus grove manager in California, reports savings of $500 to 600 per hectare by using IPM instead of chemicals alone. He reduced pest control costs from 35 to 40% of cash production costs to 8%. Characteristics of vegetable growers who adopted IPM included less risk aversion, more managerial time on farm activities, larger size, and use of irrigation and more family labor (Fernandez-Cornejo *et al.,* 1994).

Risks for adopting conservatory biological control may be particularly high when the crop price depends on cosmetic quality. Fenmore and Norton (1985) analyzed the economics of the production of English dessert apples and the potentials for adopting conservatory biological control instead of calendar-based spray schedules. They conclude that the recent price history of these apples is such that a farmer would be taking a grave risk to switch from automatic sprays for insects to IPM: as little as a 1 to 2% shift of the crop from cosmetically perfect to damaged but usable fruit would be sufficient to eliminate all savings of the costs of insecticides obtained from IPM practices. This case study exemplifies the concept that pesticides provide cheap insurance against catastrophic economic losses, even when an argument can be made that the chemicals are not needed on biological grounds.

Despite the risks faced by producers of fresh-market fruit, Ridgley and Brush (1992) found that California pear growers selectively adopted IPM practices. For example, many growers adopted monitoring methods for codling moth (*Cydia pomonella*), the major pest of pear, and use of economic thresholds to initiate treatments. Farmers with the highest adoption of IPM practices tended to have higher levels of education and stronger collaboration with Cooperative Extension Services than growers who did not adopt IPM.

Crops not heavily dependent on cosmetic quality may be better candidates for conservatory biological control. Masud and Lacewell (1985), for example, analyzed adoptions of IPM in southern and southeastern United States cotton production. They concluded that various changes in production practices, including IPM, could reduce insecticide use and produce significant savings at an acceptable level of risk. Burrows (1983) also concluded IPM could reduce pesticide use in California cotton production by 31%. Thomas *et al.* (1990) found that Texas cotton growers adopting IPM tended to combine the use of IPM with de-

cisions about use of irrigation and cotton variety. Adoption of IPM increased yields.

The previously cited examples do not necessarily mean that biological control will have little utility on crops in which cosmetic quality is important. Examples from many countries demonstrate the success of biological control, usually classical, in fruits such as citrus, apples, and olives (Huffaker & Messenger, 1976; Hardison, 1986). The critical factor may lie in the notion that if biological control is extremely effective (i.e., suppresses the pest with no further planning or management required by the grower), then it is easily incorporated into accepted pest control practices. Cases in which the natural enemy is not as effective (i.e., ongoing management is required by the grower) may have a considerably more difficult time in adoption. Entomological researchers may correctly argue that suppressive power is available from the natural enemy, but a grower less skilled in entomology and dependent on each year's crop to stay in business will see the situation differently. Alternatively, biological control may be useful on crops in which cosmetic quality is important, provided the pest does not attack the marketed portion of the plant directly (G.A. Norton, personal communication).

Our discussion has focused on the "microdecision-making process" of the pest control decision maker. Under some circumstances, factors transcending the individual decision maker may become of paramount importance. Attitudes and knowledge of the individual pest control decision maker may be immaterial to the ability of a particular biological control technology to function. Examples of these transcendent considerations include (1) how the pest control decision makers are organized and relate to each other and to their supporting scientists, (2) the relationships between the decision makers and the authority of the state, and (3) the nature of the market for the decision maker's product.

Several countries provide examples of how political factors can influence the fortunes of biological control. Swezey and Daxl (1983) argued that little headway could be made on the development of IPM practices for Nicaraguan cotton until after the Samoza government was overthrown in 1979. Earlier, the political and economic structures of Nicaragua were conducive to overuse of pesticides. Later, the collapse of the USSR sent researchers and administrators in Cuba into a strong program of biological control; once pesticides could no longer be imported, the Cubans had to turn to biological control in their efforts to keep their agricultural yields high (Rosset & Benjamin, 1993).

Effectiveness of an IPM scheme may depend on cooperation among farmers. When some farmers spray, pesticide drift could destroy the natural enemies being conserved by other farmers. Under such circumstances, an individual grower would be powerless to adopt biological control without convincing all growers in the neighborhood that they, too, should conserve their natural enemies by not spraying.

Collaboration between growers may be more difficult to achieve than to convince individual growers to conserve their natural enemies. An apparently simple question of technological choice by a farmer may be a more complicated matter involving the question of social and political relationships among growers in a particular region.

Theoretical considerations suggest that a complex of social and economic factors can influence the implementation of biological control efforts, positively or negatively. Making sense of why a particular scheme of biological control was adopted, however, generally requires an analysis of the details of the individual cases. Relatively little research of this sort has been done, and our understanding of the adoption process for this technology is rudimentary.

SUMMARY OF FACTORS AFFECTING BIOLOGICAL CONTROL

Factors affecting biological control can be recognized as social, economic, political, and philosophical.

Social

- Some means of biological control, particularly methods based on conservation of natural enemies, may depend on cooperation among decision makers that is not required for other modes of pest control.
- Change of practices may subject a decision maker to criticism by peers. This is not unique to change of biological control, but it works against research and implementation of biological control.
- Research in classical biological control requires ability and willingness to spend time abroad, often in nonindustrialized cultures. University practices and the researcher's personal life may be unsuited to such demands.
- Use of biological control, particularly such conservation schemes as IPM, may require decision makers to involve new forms of labor, such as insect pest surveyors, in the farm operation. Most decision makers are reluctant to involve themselves in new dependencies except when absolutely necessary.

Economic

- No evidence suggests that research in biological control is more expensive than alternative forms of research in pest control science. Moreover, many implementations of biological control, especially classical biological con-

trol, are spectacularly cost-effective. Thus, no intrinsic economic barriers stand in the way of biological control. However, the form of economic risk taking may change in subtle but significant ways. For example, with chemical control a farmer may see a pest hazard and then act to avoid it by "insurance" spraying. With biological control, the farmer may see the hazard but then "act" by doing nothing to see whether natural enemies will suppress the pest. Most economic behavior involves action, not waiting for something to work. Biological control may require psychological habits that are unfamiliar to commercially aggressive farmers. Anxiety from waiting may be aggravated by intensive advertising and advice from chemical companies.

- Historically, the transformation of agriculture from heavy dependency on labor inputs to more dependency on capital inputs was facilitated by the use of insecticides (Perkins, 1982). Any technology that is to replace or supplement insecticides must preserve the abilities of capital-intensive farmers to maintain their operations. Otherwise, the farmers are reluctant to change their practices.

Political

- Classical biological control, unless state operated, requires no governmental involvement other than to support research, foreign exploration, quarantine, and release activities. If the release is successful, then the state's involvement in the operation is complete. Therefore, classical biological control has no significant political barriers other than convincing the state that biological control is worthwhile. Effective work in classical biological control is aided by friendly foreign relations, so belligerent state policy may hinder classical biological control work.
- Augmentative biological control involves the state in research on the production and use of suitable agents; and in the regulation of production, marketing, and use of the products. These activities are similar to those played by most industrialized states in the promotion of chemical control technologies and therefore require little change in state behavior. Appropriate modes of regulating commercial augmentation products must be implemented. For example, the protocols for safety and efficacy testing appropriate to a new chemical product may be different from the modes of regulating commercial augmentation products (Rispen, 1988); while appropriate to a new chemical product, these protocols may be quite inappropriate to a biological agent. New quarantine procedures may be needed, particularly for augmentation products that have been genetically engineered (Coulson & Soper, 1989).

Some augmentation projects require the state to rear

and release the control agent, a practice unfamiliar to many governments. The use of autocidal techniques requires state production and release to be effective. In addition, some augmentation schemes require 100% participation by pest control decision makers to be effective, a condition that may mandate heavy state involvement. Autocidal techniques are not within the definition of biological control used in this chapter, but their dependence on state involvement in ongoing pest control operations exemplifies a barrier to implementation. This hindrance may be stronger in market economies than in centrally planned economies, because socialist governments are inclined toward substantial participation in economic activities.

• Biological control through conservation of natural enemies requires state support for research but may have no necessary involvement of the state after appropriate research and education activities are concluded. Efforts to foster the use of IPM in the United States, however, have used state-supported insect pest surveyors as a subsidy to growers to persuade them of the technology's utility. Grower skepticism may indicate a need for state involvement in the pest control operation, even in economies in which pest control is considered the private responsibility of the pest control decision maker. Such involvement may be a barrier to use of conservatory biological control.

Philosophical

• Industrialized cultures have an ambivalent relationship with the natural world. Modern technology is regarded as the prerequisite for moving from a "wilderness" and "primitiveness" to "civilization." Most technical knowledge pictures nature as atomistic matter that works as a machine: moved by energy but devoid of soul and vitality. However, a rich literature treats nature and other species as sacred and with intrinsic rights; nature is not a dead machine but a vitally alive being (Merchant, 1980).

• Biological control, based on modern biological sciences, draws from the mechanistic traditions of modern technological societies but it places the fate of economic activities at the mercy of another species, which is not entirely within human control. This latter quality, reliance on another species not entirely within our control, may be philosophically alien to researchers, commercially aggressive farmers, and regulatory officials. A different world view that allows comfort with reliance on a species not within our complete control may be necessary for enthusiasm about biological control. Absence of enthusiasm poses a barrier to research and implementation of biological control technology.

RECOMMENDATION

Biological control has a place in the most modern and sophisticated of pest control technologies, but researchers and advocates of the technology must be sensitive to the factors working against an easy transition to more reliance on biological control. Failure to be realistic about the social and economic factors weighing against use of biological control will, in the long run, be detrimental to the research enterprise. Despite the problems, several biological and cultural trends could improve biological control technology. An understanding of these trends could promote research and use of biological control in many areas.

Biological considerations of importance to biological control include problems associated with the use of pesticides: resistance, destruction of natural enemies accompanied by pest outbreaks, and damage to the health of nontarget organisms. Carson (1962) presented the first analysis of these problems, and many policy studies since her landmark work have confirmed the correctness of her thinking. Pesticide resistance and destruction of natural enemies often make chemicals technologically ineffective for the pest controller, thus providing an incentive to look elsewhere for relief from pest damage. Harm to nontarget organisms leads to more stringent regulations, which places the pest controller under severe political pressure to find an alternative. In either case, the biological control researcher can find an opportunity to provide a less dangerous mode of pest control, and the client audience of pest controllers will be a willing audience.

Cultural factors affecting the fortunes of biological control are more complex, than the biological considerations. The most important trends can be clustered into two main categories: regulatory and pricing. Many countries have increased regulation of activities related to pest control since the 1960s, mostly as a response to increased awareness of environmental and occupational hazards associated with pesticides. Regulations affecting the manufacture, sale, and use of pesticides are now common. In the United States, for example, the sale and use of pesticides were changed considerably in 1972 when the Federal Environmental Pesticide Control Act completely amended the 1947 Federal Insecticide, Fungicide, and Rodenticide Act. Explicit rules were authorized to protect nontarget organisms, and government regulators were ordered to consider the costs and benefits of the use of a prospective chemical before granting a license to sell and use it. The law was aimed at eliminating from use those chemicals for which environmental costs were out of proportion to benefits. Other legislation aimed to increase the safety of workers engaged in the manufacture of hazardous materials, including pesticides.

The new laws affecting the manufacture, sale, and use

of pesticides were an addition to older laws on the quality and purity of food products. Pesticide residues were regulated through the amendments to the Federal Food, Drug, and Cosmetic Act in 1953. Older parts of that act provided the authority to regulate marketed food for damages or remains of pests. The combination of regulations on pesticide residues and on the levels of pest damage permitted in food reflected a trade-off between the dangers posed by pesticide residues and those posed by the consumption of pest-damaged or contaminated food. Use of pesticides might provide damage-free, cosmetically perfect food products, but only at the expense of increased residues of chemicals on the food. No easy way exists to reconcile the relative danger of residues compared with insect contamination other than a case-by-case study of the particulars. Unfortunately, the ability of pesticides to produce cosmetically perfect food products created a standard in the marketplace for such produce and engendered habits within consumers for always demanding pest-free produce, which lacked visible damage but also contained the invisible residues of chemicals.

In 1996, Congress passed the Food Quality Protection Act, which amended the laws governing pesticide use and sales as well as residues of pesticides on food. Some of the major objectives of Congress were to provide better protection of children's health from pesticides and to protect against growing concern that pesticides might disrupt endocrine systems of people and other animals (Mintzer & Osteen, 1997). No evidence suggests that such safety regulations will abate in the future. Increasing research in toxicology continually finds damages from ever smaller exposures to hazardous substances, and an increasingly informed and sensitive public continues to demand protection through legislation.

Can this trend be utilized to promote research and implementation of biological control? Almost certainly yes, provided researchers and administrators seek the cases of toxicological hazard that are of most concern and design research programs that could "solve" the problem by creating an alternative technology less reliant on the chemicals. Judicious public relations with environmental and labor advocates could forge alliances to aid the biological control enterprise in the legislative arena.

Farming is very sensitive to price changes. Oil prices went up markedly in 1973 due to the actions of the Organization of Petroleum Exporting Countries (OPEC). Farmers' expenses thus tended to rise for such essentials as fuel, fertilizers, and pesticides. Financial stress, sometimes considerable, was the result for farmers (USDA, 1974). The shocks of OPEC price rises were sometimes dampened by government support subsidies to farmers, particularly in North America, Europe, and Japan. Instability in the OPEC cartel and energy conservation also produced some downward pressure on oil prices that provided limited relief to farmers (Brown & Wolf, 1987). Nevertheless, the general trend in agriculture since 1973 has been for a squeeze between rising input and output prices. Increased grain production stemming from the modernization of agriculture in previously low-yield countries (e.g., India rapidly increased wheat output in the 1970s and 1980s) further exacerbated the pressure by forcing prices of major food commodities down in world markets (International Institute for Environment and Development, 1987).

No evidence suggests that the pressure on profits of market-oriented agriculture will ease in the immediate future. Biological control, however, may play an increased role because of its cost-effectiveness. Classical and conservatory biological control are open to this line of argument, because purchased inputs are substituted by essentially "free" use of naturally reproducing biological control agents. Successful schemes allow the farmer to substitute a free input for an increasingly costly input. Provided the technology is reliable, the farmer has an economic incentive to seek and to adopt a new mode of production.

It is also possible that a successful political argument can increase state support of biological control. The tendency for increased regulation and for increased downward pressure on farm profits can be used to justify such subsidies. The state can aid biological control research through increased funds for research and for education. The research community will be well attuned to the needs for research, but the educational dimension should be broadly conceived. Programs to train pest control scientists and practitioners should be enhanced; and a broader audience in environmental studies, economics, and public affairs should be addressed. Moreover, educational needs include nonformal situations such as extension education and the media as well as the traditional classroom activities.

Subsidies for biological control technology should also be considered. Research and education to farmers are the easiest subsidies to provide, but it is more difficult to subsidize operating expenses. Heterogeneity in farm operations due to geographic and social factors makes direct operating subsidies extremely difficult to conceive and implement in a just way. Subsidies could be directed somewhat more easily toward industries producing commercial biological control agents, which would be of assistance to schemes based on augmentative biological control. Such subsidies might take the form of tax credits or other incentives to alleviate the start-up and operating expenses of such ventures.

Pricing trends and regulatory policies thus provide opportunities for research in biological control. For example, theoretical studies suggest that incentive payments to farmers or taxes on pesticides both could promote the use of IPM or other non-chemical pest control (Cooper & Keim,

1996; Underwood & Caputo, 1996). Large farms, however, may be more sensitive than small farms to loss of profits from restrictions on pesticide use (Whittaker *et al.,* 1995). Multiple factors, however, make it hard to dislodge chemical control from its role as the dominant technology for pest control (Cowan & Gunby, 1996). Organic, sustainable plant and animal crop production in the U.S. is now a $3 billion industry and growing.

Acknowledgments

We are grateful to the following people for reading a draft of this chapter in manuscript form: Paul R. Butler, Gordon R. Conway, Jack R. Coulson, David J. Girling, David J. Greathead, Carl B. Huffaker, John Mumford, Ralph Murphy, G. A. Norton, Robert Sluss, Edward Smith, and Jeffery K. Waage. We did not accept all their suggestions, but we are deeply indebted to them for stimulating further discussion. The manuscript was much improved with their help, but we remain responsible for any shortcomings it may have.

Special thanks go to the library and publications staff of the CAB International Institute of Entomology; the library staff of the Evergreen State College, particularly to C. J. Hamilton, Kate Howard, Theresa Kewell, Ernestine Kimbro, Sara Rideout, and Andrea Winship; the Division of Biological Control, University of California, Berkeley; W. G. Voigt for graphic illustrations; and D. DeMars for manuscript preparation. Ann Vandeman, Assistant Director for Management, Economic Research Service, USDA, generously helped us locate several recent studies on biological control and IPM.

Financial support from The Evergreen State College and the National Science Foundation (NSF SES-8608372) made part of the work for this chapter possible and is gratefully acknowledged.

References

Andrés, L. A., Davis, C. J., Harris, P., & Wapshere, A. J. (1976). Biological control of weeds. In C. B. Huffaker & P. S. Messenger (Eds.), Theory and practice of biological control (pp. 481–499). New York: Academic Press.

Anonymous. (1981). Editorial. Biocontrol News and Information, 2(4), 273.

Anonymous. (1985a). Editorial. Biocontrol News and Information, 6, 297.

Anonymous. (1985b). Editorial. Biocontrol News and Information, 6(1).

Anonymous. (1985c). Search for microbial insecticide at Microbial Resources Ltd. in UK. Biocontrol News and Information, 6(4), 298–299.

Anonymous. (1985d). Editorial. Biocontrol News and Information, 6(2), 87.

Anonymous. (1986). Editorial. Biocontrol News and Information, 7(4), 219.

Anonymous. (1987). Editorial. Biocontrol News and Information, 8(1), 5.

Anonymous. (1987). Research briefings 1987: Report of the research briefing panel on biological control in managed ecosystems. Washington, DC: COSEPUP, National Academy Press.

Brown, L. R., & Wolf, E. C. (1987). Charting a sustainable course. In L. R. Brown (Ed.), State of the world 1987 (pp. 196–213). New York: W. W. Norton.

Burrows, T. M. (1983). Pesticide demand and integrated pest management: A limited dependent variable analysis. American Journal of Agricultural Economics, 65, 806–810.

Buttel, F. H. (1980). Agriculture, environment and social change: Some emergent issues. In F. H. Buttel & H. Newby (Eds.), The rural sociology of the advanced societies. Montclair, NJ: Allanheld, Osmun.

CAB International Institute of Biological Control. (1986). CIBC annual report 1985–86. Slough, United Kingdom: CAB International.

Carson, R. (1962). Silent spring. Boston: Houghton Mifflin.

Charudattan, R., & Walker, H. L. (Eds.). (1982). Biological control of weeds with plant pathogens. New York: John Wiley & Sons.

Commonwealth Agricultural Bureaux. (1980). Biological control service, 25 years of achievement. Slough, United Kingdom: Commonwealth Agricultural Bureaux.

Commonwealth Agricultural Bureaux. (1985). CAB's institutes and scientific services, the future—a consultative paper. Slough, United Kingdom: Commonwealth Agricultural Bureaux.

Conway, G. R. (1972). Ecological aspects of pest control in Malaysia. In M. T. Farvar & J. P. Milton (Eds.), The careless technology (pp. 467–488). New York: The Natural History Press.

Cook, R. J. (1993). Diversity is key to biocontrol success. Agricultural Research, 41, 2.

Cook, R. J., & Baker, K. F. (1983). The nature and practice of biological control of plant pathogens. St. Paul, MN: American Phytopathological Society.

Cooper, J. C., & Keim, R. W. (1996). Incentive payments to encourage farmer adoption of water quality protection practices, American Journal of Agricultural Economics, 78(1), 54–64.

Coulson, J. R., & Soper, R. S. (1989). Protocols for the introduction of biological control agents in the United States. In R. P. Kahn (Ed.), Plant quarantine (Vol. 3). Boca Raton, FL: CRC Press.

Cowan, R., & Gunby, P. (1996). Sprayed to death: Path dependence, lock-in and pest control strategies, Economic Journal, 106(436), 521–542.

Crosby, A. W. (1986). Ecological imperialism, the biological expansion of Europe, 900–1900. Cambridge: Cambridge University Press.

DeBach, P. H. (1964). The scope of biological control. In P. H. DeBach (Ed.), Biological control of insect pests and weeds (p. 6). New York: Reinhold.

DeBach, P. H. (1974). Biological control by natural enemies. New York: Cambridge University Press.

deJanvry, A., & Vandeman, A. (1982). The macrocontext of rural development: Lessons from the United States experience. Conference of U.S.—Mexico Agriculture and Rural Development, Mexico City, Mexico.

England, V. (1987). Bugs in the system. Far Eastern Economic Review, 135(12) (March 19, 1987), 116–117.

Fenmore, P. G., & Norton, G. A. (1985). Problems of implementing improvements in pest control: A case study of apples in the UK. Crop Protection, 4, 50–70.

Fernandez-Cornejo, J. (1996). The microeconomic impact of IPM adoption: theory and application. Agricultural and Resource Economics Review, 25(2), 149–160.

Fernandez-Cornejo, J., Jorge, Beach, E. D., & Huang, W. H. (1994). The adoption of IPM techniques by vegetable growers in Florida, Michigan, and Texas, Journal of Agricultural and Applied Economics, 26(1), 158–172.

Frisbie, R. E., & Adkisson, P. L. (1985). IPM: Definitions and current status in U.S. Agriculture. In M. A. Hoy & D. C. Herzog (Eds.), Biological control in agricultural IPM systems (pp. 41–51). Orlando, FL: Academic Press.

Garcia, R., Caltagirone, L. E., & Gutierrez, A. P. (1988). Comments on a redefinition of biological control. BioScience, 38, 692–694.

Gillis, A. M. (1992). Keeping aliens out of paradise. BioScience, 42, 482–485.

Goodman, B. (1993). Debating the use of transgenic predators. Science, 262, 1507.

Guttman, J. M. (1978). Interest groups and the demand for agricultural research. Journal of Political Economics, 86, 467–484.

Hardison, A. C. (1986). A different viewpoint on integrated pest management. Citrograph, January, 48–49.

Hatcher, J. E., Wetzstein, M. E., & Douce, G. K. (1984). An economic evaluation of integrated pest management for cotton, peanuts, and soybeans in Georgia (Research Bulletin No. 318). University of Georgia Agriculture Experiment Station. Athens, GA.

Hoy, M. A., & Herzog, D. C. (Eds.). (1985). Biological control in agricultural IPM systems. Orlando, FL: Academic Press.

Huffaker, C. B. (1985). Biological control in integrated pest management: An entomological perspective. In M. A. Hoy & D. C. Herzog (Eds.), Biological control in agricultural IPM systems (pp. 13–23). Orlando, FL: Academic Press.

Huffaker, C. B., & P. S. Messenger. (Eds.). (1976). Theory and practice of biological control. New York: Academic Press.

Hussey, N. W. (1985). Biological control—a commercial evaluation. Biocontrol News and Information, 6(2), 93–99.

International Institute for Environment and Development. (1987). World resources 1987. New York: Basic Books.

Judd, M. A., Boyce, J. K., & Evenson, R. E. (1987): Investment in agricultural research and extension, In V. W. Ruttan & C. E. Pray (Eds.); Policy for agricultural research (pp. 7–38). Boulder, Co: Westview Press.

Lambert, B., & Peferoen, M. (1992). Insecticidal promise of Bacillus thuringiensis. BioScience, 42, 112.

Masud, S. M., & Lacewell, R. D. (1985). Economic implications of alternative cotton IPM strategies in the United States (Report DIR 85–5) Texas Agricultural Experiment Station. College Station, TX.

Merchant, C. (1980). The death of nature. San Francisco: Harper & Row.

Michelbacher, A. E., & Smith, R. F. (1943). Some factors limiting the abundance of the alfalfa butterfly. Hilgardia, 15, 369–397.

Mintzer, E. S., & Osteen, C. (1997). New Uniform standards for pesticide residues in food. Food Review, 20(1)(January–April), 18–26.

Moffat, A. S. (1991). Research on biological pest control moves ahead. Science, 252, 211–212.

Musser, W. N., Wetzstein, M. E., Reece, S. Y., Varca, P. E., Edwards, D. M., & Douce, G. K. (1986). Beliefs of farmers and adoption of integrated pest management. Agricultural Economic Research, 38(1), 34–44.

Perkins, J. H. (1982). Insects, experts, and the insecticide crisis: The quest for new pest management strategies. New York: Plenum Press.

Perkins, J. H. (1978). Reshaping technology in wartime: The effect of military goals on entomological research and insect-control practices. Technology & Culture, 19(2), 169–186.

Price-Jones, D. (1973). Agricultural entomology. In R. F. Smith, T. E. Mittler, & C. N. Smith (Eds.), History of entomology (pp. 307–332). Palo Alto, CA: Annual Reviews.

Repetto, R. (1985). Paying the price: Pesticide subsidies in developing countries (Research Report No. 2). Washington, DC: World Research Institute.

Ridgley, A.-M., & Brush, S. B. (1992): Social factors and selective technology adoption: The case of integrated pest management. Human Organization, 51, 367–378.

Roberts, D. W. (1989). Proceedings, conference on biotechnology, biological pesticides, and novel plant-pest resistance for insect pest management. Ithaca, NY: Boyce Thompson Institute.

Rose-Ackerman, S., & Evenson, R. (1985). The political economy of agricultural research and extension: Grants, votes, and reapportionment. American Journal of Agricultural Economics, 67, 1–14.

Rosset, P., & Benjamin, M. (Eds.). (1993). Two steps backward, one step forward: Cuba's national experiment with organic agriculture. San Francisco: Global Exchange.

Salama, F. A. L. (1983). A causal model of integrated pest management adoption among Iowa farmers. Unpublished doctoral thesis, Iowa State University, Ames, IA.

Sawyer, R. C. (1996). To make a spotless orange: Biological control in California. Ames, Iowa: Iowa State University Press.

Simberlof, D., & Stiling, P. (1996). How risky is biological control. Ecology, 77(7), 1965–1974.

Smith, H. S. (1919). On some phases of insect control by the biological method. Journal of Economic Entomology, 12, 288–292.

Swezey, S. L., & Daxl, R. (1983). Breaking the circle of poison: The integrated pest management revolution in Nicaragua. San Francisco: Institute for Food and Development Policy.

Tauber, M. J., Hoy, M. A., & Herzog, D. C. (1985). Biological control in agricultural IPM systems: A brief overview of the current status and future prospect. In M. A. Hoy & D. C. Herzog (Eds.), Biological control in agricultural IPM systems (pp. 3–9): Orlando, FL: Academic Press.

Thomas, J. K., Ladewig, H., & McIntosh, W. A. (1990). The adoption of integrated pest management practices among Texas cotton growers. Rural Society, 55, 395–410.

Tjamos, E. C., Papavizas, G. C., & Cook, R. J. (1992). Biological control of plant diseases: Progress and challenges for the future, New York: Plenum Press.

Underwood, N. A., & Caputo, M. R. (1996). Environmental and agricultural policy effects on information acquisition and input choice, Journal of Environmental Economics and Management, 31(2), 198–218.

U.S. Department of Agriculture (USDA). (1974). The world food situation and prospects to 1985 (Foreign Agriculture Economic Report No. 98). Washington, DC: USDA, Economic Research Service.

U.S. Department of Agriculture (USDA). (1984). Biological control documentation activities of the insect identification and beneficial insect introduction institute. [Document 008F (mimeo)]. Washington, DC: USDA, Agricultural Research Service.

U.S. Department of Agriculture (USDA). (1985). Biological control information document (Document 00061). Beltsville, MD: USDA, ARS Biological Control Documentation Center.

U.S. Department of Agriculture (USDA). (1994). Adoption of integrated pest management in U.S. agriculture (Agriculture Information Bulletin No. 707). Washington, DC: U.S. Department of Agriculture.

U.S. Department of Agriculture (USDA). (1997). Proceedings of the Third National IPM Symposium/Workshop, Broadening Support for 21st Century IPM (Miscellaneous Publication No. 1542). (Washington, DC: U.S. Department of Agriculture.

Wearing, C. H. (1988). Evaluating the IPM implementation process. Annual Review of Entomology, 33, 17–38.

Waage, J. K., & Greathead, D. J. (1987). Biological control challenges and opportunities, delivered to the discussion meeting, biological control of pests, pathogens and weeds: Development and prospects, February, 18–19, 1987. London: The Royal Society.

Weiser, J., Bucher, G. E., & Poinar, G. O., Jr. (1976). Host relationships and utility of pathogens. In C. B. Huffaker & P. S. Messenger (Eds.), Theory and practice of biological control (pp. 169–185). New York: Academic Press.

White, F. C., & Wetzstein, M. E. (1995). Market effects of cotton integrated pest management, American Journal of Agricultural Economics, 77(3), 602–612.

Whittaker, G., Lin, B. H., & Vasavada, U. (1995). Restricting pesticide use: The impact on profitability by farm size, Journal of Agricultural and Applied Economics, 27(2), 352–362.

41

Whither Hence, Prometheus?

The Future of Biological Control

T. S. BELLOWS

Department of Entomology
University of California
Riverside, California

A GIFT OF FIRE

An interesting character was Prometheus. He was one of the first individuals in classical mythology to express compassion for humanity, in its huddled, cold, and unenlightened state. He saw the solution to humanity's need for growth in the gift of fire. Being an impetuous fellow, he stole some fire from the powers that were and gave it to humankind, along with civilization and the arts. So humanity began to develop. The powers that were decided some sort of balancing response was necessary and visited upon humanity, in the guise of curiosity, all manner of pestilence and plague. Thus began our ecological struggles in producing flowers, food, and fiber.

Biological control has, like civilization, grown from stages of initial garnering of knowledge to something approaching civilized application of that knowledge. Predatory arthropods were used in China and Yemen long before the development of the natural sciences in Europe during the Renaissance (Van Driesche & Bellows, 1996). Aldrovani in 1602 (Bodenheimer, 1931) and Goedaert (1662) were among the first to chronicle the appearance of parasitoids emerging from a lepidopteran host. Carl Linnaeus provided one of the first written proposals recommending the use of predaceous insects for pest control when he remarked in 1752 that "Every insect has its predator which follows and destroys it. Such predatory insects should be caught and used for disinfesting crop-plants." (Hörstadius, 1974). In the nineteenth century, important investigations into insect diseases (Kirby & Spence, 1815) were undertaken, and recommendations were made for their use [L. Pasteur, J. L. LeConte, A. Bassi, and E. Metchnikoff all urged the use of pathogens against insects (Bellows 1993)]. During this period there was a phenomenal increase in the number of scientific works on the taxonomy, biology, and ecology of insect parasitoids and predators by such scientists as M. M. Spinola, J. W. Dalman, J. L. C. Gravenhorst, J. O. Westwood, Francis Walker, C. Rondani, A. Förster, J. T. C. Ratzeburg, and many others (DeBach & Rosen, 1991; Van Driesche & Bellows, 1996). In addition, recommendations for the use of parasitoids and predators from native lands to control invasive pests began to appear, as in 1855 when Asa Fitch recommended the need for natural enemies from Europe to control the wheat midge *Sitodiplosis mosellana* (Géhin) in New York (Bellows, 1993). Riley (1885) reported the first successful movement of a parasitoid species, *Cotesia glomerata* (Linnaeus), between continents, from England to the United States.

Thus, the ideas surrounding the natural control of pestilence gathered momentum, waiting for a propitious moment when knowledge, need, expertise, and providence would come together to provide an earth-shattering demonstration that here, in the shape of tiny, dependable organisms, existed a mechanism that could permanently liberate humanity from at least some of the pestilence it was battling. Enter the Vedalia beetle. In the late 1880s, the beetle was brought to North America, together with its colleague in arms, a tiny parasitic fly called Cryptochaetum, to combat the ravages of a devastating homopteran pest. In little more than a few months this pair rescued an agricultural industry on the verge of collapse. The humans involved in the program (Albert Koebelle who worked for C. V. Riley) were hailed and thanked at testimonial dinners and, after their 15 minutes in the spotlight, went back to work on applying this idea to other systems. The vedalia beetle and Cryptochaetum have continued their labors ever since, without the benefit of testimonial dinners. Their control of the pest has lasted more than a century.

It is impossible to overstate the impact this simple introduction had on biological control and pest management science. The results of this program, in part because of its high visibility and in part because the principals involved were employed by a federal agency (and thus were able to influence programs across much of an entire continent), thundered like a lightning bolt through entomology and launched a century of natural-enemy warfare on both exotic and endemic pests throughout the world. Awareness of the importance of natural enemies expanded with work on conservation of natural enemies, and further developments led to technologies for mass-rearing and delivering natural enemies for augmentation of their numbers.

The first practical attempts at biological control of weeds predate the Vedalia story. Fitch in 1855 proposed using herbivorous insects from Europe to combat European weeds, such as toadflax, in North America. The prickly pear cactus, *Opuntia vulgaris* Miller, was controlled in southern India by introducing to that region in 1863 the cochineal insect *Dactylopius ceylonicus* (Green), after this insect had unexpectedly decimated cultivated plantings of the same cactus in northern India (Goeden, 1978). Other programs followed. A widely known and very successful program was conducted against two prickly pear cactus species in Australia. In 1925 and 1926, herbivores were collected in South America and shipped to Australia. One of these, the moth *Cactoblastis cactorum* (Bergroth), became established and caused widespread death to the cactus plants, recovering vast areas of natural grassland and forest from the cacti.

Plant pathology was not to be left behind. Sanford (1926) was among the first to recognize that competition between saprotrophic and pathogenic organisms could lead to biological suppression of pathogens and disease. Investigations into the mechanisms and applications of microbes for control of plant pathogens expanded from the early work on potato scab to encompass research and application against diseases of foliage, flowers, fruit, stems, roots, and diseases and infections in timber (Cook & Baker, 1983; Baker, 1985; Campbell, 1989). As the worldwide awareness of the impact of introduced or adventive organisms grew, pest vertebrates (such as rats, snakes, goats, and feral cats) also became the target of biological control programs.

WHITHER HENCE?

With a foundation of experience spanning a century, and a depth of knowledge going back even farther, where will this field go now? Some things are certain: we will see continued application of biological control to adventive pests throughout the world, expanded use of natural-enemy conservation in pest management programs, continued development of microbial agents as biological pesticides, and broadening of the field as it grows to encompass more and

more of life sciences and environmental protection. Other matters are less easily divined: the possibility of expanding the impact of biological control into urban or other nontraditional settings, the extent to which we may be able to modify or alternatively employ the genomes of natural enemies, the future of agencies devoted to these endeavors, and the training of future scientists and practitioners in this field.

Combating Adventive and Native Pests

Biological control clearly holds great promise for solutions to pest problems affecting agricultural. The ecological principles that underlie biological control in both managed and natural ecosystems do not change with the passage of time; they are basic to interactions among species. Consequently, biological control continues to provide productive, efficient, and economical solutions to pest problems. The record to date indicates that biological control is of significant value and widely successful, providing either partial or complete control in 60% of the cases were it has been used (Greathead & Greathead, 1992). Adventive insects make up a large portion of the world's most serious pests (Van Driesche & Bellows, 1996), yet biological control has been applied against only 5% of the world's pest species (Van Driesche & Ferro, 1987). With a 60% success rate and 95% of pest species still waiting to be addressed, this is clearly a field where application will bear a marvelous return on investment.

New biological control efforts are currently needed for many existing pest problems, both for programs targeted against introduced pests and for additional work toward natural-enemy conservation in pest management systems. International man-assisted movement of plant material and insect pests will likely continue, and cause the unintentional eruption of new pests by shipping pest species to new locations while separating them from their natural enemies. Such pests eruptions will require research and action to locate and introduce natural enemies suitable for limiting the pests in their new environments. The dependence of civilization on agriculture, both in developed and developing countries, continues to press the need for integrated management systems for efficient and economical production. Such integrated systems increasingly rely on natural enemies to reduce both costs and pesticide use.

Microbial Agents as Biological Pesticides

The use of microbial agents as biological pesticides continues to grow. This field, which was focused largely on *Bacillus thuringiensis* Berliner for many years, has expanded remarkably. There is a renewed interest in the use of fungi, for example, and in developing delivery technologies that allow fungal spores to germinate even in arid environments for use against insects. The possibilities of using plant pathogens as microbial herbicides is similarly undergoing development. Some of the more significant hurdles that these areas must overcome are a direct result of the biological features that make microbes so attractive. For example, many plant pathogens are quite host specific. This is often a desirable biological characteristic, but has the added effect of limiting the application of a particular product to very narrow uses. Economies of scale and broad markets are thus unavailable, and these restrictions may limit the application or development of some of these technologies. Partnering with government agencies for development or production has been done in the past to overcome these difficulties, and may play a role in the future.

Protection of Natural Ecosystems

Natural ecosystems are becoming increasingly affected by nonnative plants and animals. Examples span a wide range of habitats, including herbaceous weed invasion of natural grasslands, woody weed species in forests and shrub lands, invasion of seabird breeding colonies by adventive predators (cats, rats, etc.), and introduction of predaceous reptiles into ecosystems previously lacking such species. Such harmful species find their way into ecosystems through many channels: some are accidentally introduced as seeds in produce; other weedy plants are sometimes introduced as ornamentals and then become established in native habitats.

Controlling the impact of such species once they are established in an environment is a very challenging proposition. Biological control has proved to be a highly selective and successful strategy. By using suitable natural enemies, biological control programs can target the offending introduced species and limit harm to nontarget species. Records of remarkable successes include recoveries of grasslands after invasion by herbaceous weeds, and recovery of aquatic ecosystems after invasion by nonnative aquatic plants. Work on vertebrates is more recent, and needs more thorough evaluation, but new technologies in this area are promising. A variety of ecosystems the world over have been affected by introduced species, and the opportunities to provide ecologically and environmentally acceptable solutions to protect these natural environments through biological control abound.

Nontraditional Settings

There are numerous opportunities for use of biological control in areas that historically have received little attention. These include, for example, pest problems in urban environments. Ornamental landscape plants and indoor plants in many countries, for example, are introduced from other parts of the world, and are often attacked by arthropods native to the plants' original homes. These are perfect opportunities for the introduction of natural enemies to control these adventive pests. Indeed, introduction of natural enemies is often the only viable pest management solution for these landscape pests, in part because of the cost of other solutions being larger than the value of the plants, and in part also because of a growing aversion to the use of insecticides in or near homes and commercial buildings. In landscape settings, particularly of landscape trees, biological control has been applied to a few pests with great success, but many more opportunities await appropriate attention.

Another essentially untouched arena is the area of adventive structural or domicile pests. The vast majority of household insect pests are introduced in most locations in the world. One illustrative example is the list of pest cockroaches in North America: all pest cockroaches in North America are adventive, while none of the cockroaches native to North America ever reach pest status. This is an obvious and classical setting for biological control research. Only limited work has been done on biological control of structural pests relative to the attention received by agricultural pests. This is in part due to the lack of organized funding directed at such solutions for homeowners: individual homeowners are generally responsible for their own pest management, instead of belonging to a large cooperative research organization, as is the case for many agricultural commodities. This lack of biological control work is also due in part to a misconception that biological control of household pests is fundamentally flawed because homeowners would not tolerate the presence of natural enemies any more than they tolerate pests. This misconception fails to recognize that pest and natural-enemy populations together are usually present in only one hundredth or one thousandth of the numbers of pests alone when the pest is not controlled by natural enemies. Hence, insect natural enemies and household pests together would likely be much, much rarer than the pests the natural enemies are introduced to control. In addition, education concerning "good" and "bad" insects is already affecting public perception of arthropods (in the garden, ladybird beetles are tolerated, while pest insects are not), and additional education concerning insects beneficial to households would naturally become a part of any biological control program aimed at such pests.

Gene Use and Modification

Genes are useful entities. They are responsible for defining biological and environmental limits that natural enemies and pests can inhabit, for coding receptors that mediate interactions between organisms, for allowing the presence of pesticide resistance in populations, and for coding various reproductive ontogenies. In some species, such as *B.*

thuringiensis, they also code for products that are useful in themselves, such as proteins toxic to insects.

Our ability to recognize, isolate, translate, and utilize genes has exploded indescribably in the last 25 years. Various genes coding for insect-toxic proteins have been moved out of their original species and placed in other microbes for mass culture, and have been moved into various plants to create plants that contain natural titers of insecticidal proteins within their own tissues. Genes for pesticide resistance have been selected or discovered in several natural-enemy species, and strains with these genotypes have become vital in integrated pest management (IPM) programs where some limited pesticide use is necessary but reliance on natural enemies is also critical. The direction this field will take biological control in the near future is difficult to even imagine. Our basic knowledge of the genomes of our natural enemies is extremely limited, so considerable basic research will probably take place before the structured use of genes or their modifications begins. As interrelated as natural-enemy and pest interactions are, however, the possibilities seem nearly endless. As just one thought, perhaps the ability to provide a natural enemy with extra copies of genes coding for receptors to pest kairomones will develop, and thus lead to a strengthened searching response. Whether or not this would increase a natural enemy's effectiveness is not knowable in advance, but could be testable in a field trial. Pest and natural-enemy interactions are too intricate to permit extrapolation about the effects of a single or even multiple gene changes, and because natural enemies have been under selection for ages, in many ways they may be already optimized for interactions with their hosts. However, in our modified agricultural systems, we may discover areas where simple changes (such as added pesticide resistance) vastly improve the effectiveness of a natural enemy.

Experts and Institutions

Amid a growing worldwide awareness of the significance of biological control in limiting damage from pests, some important challenges to its continued development arise. One of these challenges is the widespread nature of its potential application, in that biological control can and should be applied in every setting where population management is a crucial part of a natural or an agricultural ecosystem. This ubiquitous potential for application can attract the attention, however fleeting, of scientists, administrators, and others. Such attention can give the impression that biological control research is indeed already pervasive and widely applied. Nevertheless, fleeting or partial attention to a problem can be deceptive. Ad hoc attempts at biological control solutions have rarely proved to be successful (Goeden, 1993), perhaps in part because biological control solutions may (but not always) require many years to be identified, developed, and put into place.

Successful development of biological control programs generally has required the extended commitment of full-time career biological control scientists with adequate support facilities and institutions, and it seems likely that this will continue to be the case. This is largely because of the long-term nature of biological control programs, which require an understanding of population-level processes and ecological functioning of each particular system. This is clear in the case of conservation and augmentation research, where a sound knowledge of structure and function of the natural enemy and pest fauna is critical to making informed decisions concerning modifications to system management.

Similarly, but for different reasons, long-term commitment is generally necessary in the effective functioning of a program designed at introduction of natural enemies against an adventive pest. This commitment is in part a matter of professional focus, where the time given to the study of natural enemy groups over the years of professional development provides the exploration entomologist or pathologist a worldwide perspective on a particular fauna and flora. In part it is also a matter of philosophical commitment (Goeden, 1993); travel necessary to secure natural enemies reduces time for research on other topics by the explorer, and may place the explorer at increased risk of personal injury or harm. Quantifying the effects of any changes to a natural or agricultural system, such as the addition of a natural enemy, requires detailed and long-term study both before and after the addition. A long-term commitment and a stable institution are vital to the eventual success of most programs.

Training Future Scientists and Practitioners

Training future scientists and development of research positions specifically for biological control are necessary for fulfilling its promise. Relatively few educational institutions offer formal programs in this field, but where available such programs provide unparalleled training for future professionals. Where such programs are not available, sound academic background in its component fields (which might include pathology, ecology, entomology, and agriculture), together with on-the-job training will offer opportunities for progress in the field. As increasing numbers of scientists obtain training and experience in biological control in all its facets, there will be an increase worldwide in biological control programs and in public awareness of its value. It is likely that public funding will, in the future as in the past, be a vital part of the support for such programs.

TOWARD THE LIGHT

In Prometheus's tale, humanity is provided with a counter to pestilence and plague. After dispersal of the terrors, humanity discovers hope. From this hope for bettering ourselves have sprung through the ages homes, medicine, the arts, cultivation, civilization, and industry. Buried among these endeavors lies biological control, a blend of science and art that serves to better our world, from food and fiber production to environmental protection and conservation. Biological control has in its deepest roots a theme of serving humanity, by finding permanent solutions to problems in ways that will benefit everyone. This is the most certain part of the future of biological control: it will continue in its theme of service to find solutions to pestilence in ways harmonious and consistent with the civilization at hand. After all, bettering humanity was really all Prometheus had in mind.

References

Baker, R. (1985). Biological control of plant pathogens: Definitions. In M. A. Hoy & D. C. Herzog (Eds.), Biological control in agricultural IPM systems (pp. 25–39). London: Academic Press.

Bellows, T. S., Jr. (1993). Introduction of natural enemies for suppression of arthropod pests. In R. D. Lumsden & J. L. Vaughn (Eds.). Pest management: Biologically based technologies (pp. 82–89). Washington, DC: American Chemical Society.

Bodenheimer, F. S. (1931). Zur Fruhgeschichte der Enforschung des Insektenparasitismus. (To the early history of the study of insect parasitism). Archiv fuer Geschichte der Mathematik, der Naturwissenschaften und der Technik, 13, 402–416.

Cook, R. J., & Baker, K. F. (1983). The Nature and Practice of Biological Control of Plant Pathogens. St. Paul, MN: American Phytopathological Society.

DeBach, P., & Rosen, D. (1991). Biological control by natural enemies (2nd ed.). Cambridge: Cambridge University Press.

Goedaert, J. (1662). Metamorphosis et historia naturales insectorum I (pp. 175–178). Medisburg: Jacobum Fierensium.

Goeden, R. D. (1978). Part II. Biological control of weeds. In C. P. Clausen (Ed.). Introduced parasites and predators of arthropod pests and weeds: A world review (pp. 357–414). Agricultural Handbook No. 480. Washington, DC: U.S. Department of Agriculture.

Goeden, R. D. (1993). Arthropods for suppression of terrestrial weeds. In R. D. Lumsden & J. L. Vaughn (Eds.), Pest management: Biologically based technologies (pp. 231–237). Washington, DC: American Chemical Society.

Greathead, D. J., & Greathead, A. H. (1992). Biological control of insect pests by parasitoids and predators: The BIOCAT database. Biocontrol News and Information, 13(4), 61N–68N.

Hörstadius, S. (1974). Linnaeus, animals and man. Biological Journal of the Linnaean Society, 6, 269–275.

Kirby, W., & Spence, W. (1815). An introduction to entomology. London: Longman, Brown, Green & Longmans.

Riley, C. V. (1885). Fourth report of the U.S. Entomological Commission. In S. H. Scudder (1889). Butterflies of Eastern United States and Canada (p. 323). Cambridge University Press.

Sanford, G. B. (1926). Some factors affecting the pathogenicity of Actinomyces scabies. Phytopathology, 16, 525–547

Van Driesche, R. G., & Bellows, T. S. (1996). Biological control. New York: Chapman & Hall.

Van Driesche, R. G., & Ferro, D. (1987). Will the benefits of classical biological control be lost in the 'biotechnology stampede.' American Journal of Alternative Agriculture, 2, 50, 96.

Subject Index

Entries followed by f or t denote figures and tables, respectively.

A

Acacia psyllid, 924
Acari, 386–391
  biological control on grapes
    Eriophyidae, 865
    Tenuipalpidae, 865
    Tetranychidae, 862–865
Acariformes, 390–391
Accessory gland secretions, in Hymenoptera, 357
Acephate, 309
Acizzia uncatoides, control in urban environments, 924
Actinedida, 390–391
Aculeata
  Chrysidoidea, 360, 457
  Formicoidea, 457–459
  Pompiloidea, 459
  Scolioidea, 459
  Sphecoidea, 459
  Vespoidea, 459–460
Adephaga, 425–429
Aeolothripidae, 395
Agistemus, 391
AgNPV, *see Anticarsia gemmatalis* nuclear polyhedrosis virus
Agroecosystems, *see also* Crops
  classification, 320, 323
  components of biodiversity in, 322
  mechanical crop management, 320–332
Agromyzidae, in weed control, 511
Alcohol, as preservative, 53
Aldicarb, 310
Aleurothrixus floccosus, control in urban environments, 925
Aleyrodidae, biological control in citrus, 722–724
Alfalfa
  aphid control in Latin America, 987
  entomophage diet supplementation, 634
  strip cropping, 320
Alfalfa mosaic virus, CP-mediated protection, 559

Alfalfa weevil
  classical biological control, 796–797
  economic cost of, 248
Alligatorweed, biological control, 880–881
Allozymes, as measure of specific status, 59, 65, 68, 71–72
Alternanthera phylloxeroides, biological control, 880–881
Ambrosia spp., biological control, 884–885
Amitraz, 309
Animal and Plant Health Inspection Service, 568, 907, 908
  certification of quarantine facilities, 104, 106
  importation permits, 111
Anthocoridae, 397–401
Anticarsia gemmatalis nuclear polyhedrosis virus, 525–526, 809
Ants
  fire, 542
    baiting of, 310
  red wood, in forest ecosystems, 763–764
  and tillage practices, 331
Anystidae, 390
Aphelenchidae, 701
Aphid
  alfalfa, biological control in Argentina, 986–987
  balsam woolly, 763
  cereal, biological control in Brazil and Chile, 986–987
  control in glasshouses, 826–827
    on chrysanthemums, 830–831
  in deciduous fruit crops, 746, 748–749
  elm, control in urban environments, 926
  linden, control in urban environments, 926
  oak, control in urban environments, 926
  walnut, 749
Aphidius spp., size and fitness, host effects, 594–595
Apocrita, 456–457
Apple, 461
  integrated mite control programs, 745t
Arachnida, 384–386

Araneae, 384–386
Araneidae, 385
Arctiidae, in weed control, 510
Argentina, classical biological control in, 247, 978
Arrhenotoky, 653
Arthropods, *see also* Entomophages
  design criteria for quarantine facilities, 106
  parasitic/predaceous, exploration for, 93–94
  predators, 383
Artichoke mottled crinkle tombusvirus, 565
Ascidae, 388–389
Augmentation/augmentive release, 226–227
  in biological control of weeds, 873
  in deciduous fruit crops, 750
  economic analysis of projects, 246t
  evaluation of natural enemies as candidates for, 257–259
  field evaluation/assessment, 259–260
  history of, 461
  microbial agents, 705–706
    plant-parasitic nematodes, 707–708
    root diseases, 706–707
  natural enemies in forest ecosystems, 766–768
  parasitoid/predator species showing feasibility in U.S., 254t
  versus pesticide use, 249
  predaceous mites, in grape biological control, 864–865
  socioeconomics, 1003
  steps for effecting, 255f
  Trichogramma spp., 253, 256–257
Australia
  biological control organizations in forestry, 770
  Echium plantagineum control, 278, 279, 877
  rabbit control in, 958, 959, 963
Autographa californica nuclear polyhedrosis virus, 522
Autoparasitism
  in entomophage culture, and sex-ratio determination, 158

Taxonomic Index

ISBN 0-12-257305-6